Prealgebra and Introductory Algebra

Second Edition

Margaret L. Lial
American River College

Diana L. Hestwood
Minneapolis Community and Technical College

John Hornsby
University of New Orleans

Terry McGinnis

PEARSON
Addison
Wesley

Boston   San Francisco   New York
London   Toronto   Sydney   Tokyo   Singapore   Madrid
Mexico City   Munich   Paris   Cape Town   Hong Kong   Montreal

Publisher	Greg Tobin
Editor in Chief	Maureen O'Connor
Senior Project Editor	Lauren Morse
Editorial Assistant	Caroline Case
Managing Editor	Ron Hampton
Senior Production Supervisor	Kathleen A. Manley
Senior Designer (text and cover)	Dennis Schaefer
Photo Researcher	Beth Anderson
Digital Assets Manager	Marianne Groth
Media Producer	Sharon Smith
Software Development	Janet Szykowny and Ted Hartman
Marketing Manager	Michelle Renda
Marketing Coordinator	Alexandra Waibel
Senior Prepress Supervisor	Caroline Fell
Senior Manufacturing Buyer	Evelyn Beaton
Production Coordination, Composition, and Illustrations	Pre-Press Company, Inc.
Cover Photo	© Daryl Benson/Masterfile
Cover Image	Boardwalk and Lake at Sunset, Kouchibouguac National Park, New Brunswick, Canada.

Photo Credits All photos from PhotoDisc, except the following: Thinkstock RF, pp. 1 left, 308, 380, 493, 527 right, 548 left; PictureQuest/Image Farm RF, pp. 1 right, 28, 420; Corbis RF, pp. 13, 165 left, 166 right, 176 right, 255 right, 265, 316 top and bottom, 355 bottom left, 392, 504 left and right, 589 top left, 591 bottom left, 592 bottom right, 595 left, 595 right, 636, 726, 835, 843, 844; GE Healthcare, p. 14; Troy Wayrynen/NewSport/Corbis, p. 24; Brand X Pictures, pp. 38 left, 83 bottom right, 91, 117, 348 right, 572 bottom left, 604 right; Beth Anderson, pp. 38 right, 47 right, 79, 92 left, 102 left, 147, 315, 390 left, 535 right, 536 left, 589 top right, 589 bottom right, 590 left, 590 right, 591 top left, 592 bottom left, 596, 612 top right, 612 bottom right, 717, 726, 858, 994, 1067, 1071 left, 1188; Ralph White/Corbis, p. 48 right; Comstock RF, pp. 83, 91 left, 94, 156, 346 left, 1128; Hattie Young/Photo Researchers (PAL), p. 102 top right; Digital Vision, pp. 147, 354 right, 355 bottom right, 411, 430, 440 right, 508 right, 548 right, 573 right, 632, 757, 807, 824, 922 right, 933, 985 top, 998; Bohemian Nomad Picturemakers/Corbis, p. 165 right; Wally McNamee/Corbis, p. 182 left; Laurel Beager, p. 182 right; Getty Images/Taxi, pp. 193, 276; Reuters/Corbis, pp. 201 right, 345, 440 left, 443, 1056, 1064 right; Bettman/Corbis, p. 229; Phillip Gould/Corbis, p. 230 right; Diana Hestwood, p. 242; Public domain, pp. 258, 537; NationBill/Corbis Sygma, p. 346 right; Dorling Kindersley, pp. 348 center; NASA, pp. 355, 390 right, 538 right, 563 chapter opener, 574, 921 right, 1118, 1176; Bob Daemmrich/Image Works, p. 404 right; Robert Brenner/PhotoEdit, p. 535 left; Martin Harvey/Corbis, p. 571 left; Dwi Oblo/Stringer/Reuters/Corbis, p. 571 right; Richard Carson/Reuters/Corbis, p. 572 right; Custom Medical Stock Photography, p. 591 bottom right; Thierry Orban/Corbis Sygma, p. 601; Michael Newman/PhotoEdit, p. 612 top left; Reprinted courtesy of Caterpillar, Inc., p 614; Stone/Getty Images, p. 707; AP/Wideworld, p. 842; PunchStock/BrandX Pictures, p. 832; Corbis RF, p. 835, 843, 844; 20th Century Fox/Paramount/The Kobal Collection, p. 836; Steve Starr/Corbis, p. 862; James Leynse/Corbis, p. 863; U.S. Department of Commerce, p. 985 bottom; Tom Wagner/Corbis Saga, p. 996; The Kobal Collection, pp. 999, 1082; David Bergman/Corbis, p. 1064 left; Bettmann/Corbis, p. 1069; Mitchell Layton/Corbis, p. 1071 right; John Hornsby, p. 1096; Ben Woods/Corbis, p. 1085; Nik Wheeler/Corbis, p. 1124 top and bottom left; Associated Press/Goldsboro News-Argus, p. 1150; Roger Ressmeyer/Corbis, pp. 1153, 1183; Guy Motil/Corbis, p. 1204.

Many of the designations used by manufacturers and sellers to distinguish their products are claimed as trademarks. Where those designations appear in this book, and Addison-Wesley was aware of a trademark claim, the designations have been printed in initial caps or all caps.

Library of Congress Cataloging-in-Publication Data
Prealgebra and introductory algebra / Margaret L. Lial ... [et al.]. --2nd ed.
   p.cm.
   Includes index
   Rev. ed. of: Prealgebra and introductory algebra / Margaret L. Lial, Diana L. Hestwood, John Hornsby. c2001.
   ISBN 0-321-43346-7
   1. Algebra--Textbooks. I. Lial, Margaret L. II. Lial, Margaret L. Prealgebra and introductory algebra.
QA152.3.P742006
512.9--dc 22
                                                     2006044628

Copyright © 2007 Pearson Education, Inc. All rights reserved. No part of this publication may be reproduced, stored in a retrieval system, or transmitted, in any form or by any means, electronic, mechanical, photocopying, recording, or otherwise, without the prior written permission of the publisher. Printed in the United States of America. For information on obtaining permission for use of material in this work, please submit a written request to Pearson Education, Inc., Rights and Contracts Department, 75 Arlington Street, Suite 300, Boston, MA 02116, fax your request to 617-848-7047, or e-mail at http://www.pearsoned.com/legal/permissions.htm.

4 5 6 7 8 9 10—DOW—10 09 08

Contents

Index of Focus on Real-Data Applications		vii
Preface		ix
To the Student		xvii
Diagnostic Pretest		xix

CHAPTER 1 Introduction to Algebra: Integers — 1

1.1	Place Value	2
1.2	Introduction to Signed Numbers	9
1.3	Adding Integers	15
1.4	Subtracting Integers	25
1.5	Problem Solving: Rounding and Estimating	29
1.6	Multiplying Integers	39
1.7	Dividing Integers	49
1.8	Exponents and Order of Operations	59
	Summary Exercises on Operations with Integers	69
	Summary	71
	Review Exercises	77
	Test	81

CHAPTER 2 Understanding Variables and Solving Equations — 83

2.1	Introduction to Variables	84
2.2	Simplifying Expressions	95
2.3	Solving Equations Using Addition	107
2.4	Solving Equations Using Division	119
2.5	Solving Equations with Several Steps	127
	Summary	137
	Review Exercises	143
	Test	145

CHAPTER 3 Solving Application Problems — 147

3.1	Problem Solving: Perimeter	148
3.2	Problem Solving: Area	157
3.3	Solving Application Problems with One Unknown Quantity	167
3.4	Solving Application Problems with Two Unknown Quantities	177
	Summary	183
	Review Exercises	189
	Test	191

CHAPTER 4 Rational Numbers: Positive and Negative Fractions — 193

4.1	Introduction to Signed Fractions	194
4.2	Writing Fractions in Lowest Terms	207
4.3	Multiplying and Dividing Signed Fractions	219
4.4	Adding and Subtracting Signed Fractions	231
4.5	Problem Solving: Mixed Numbers and Estimating	243
	Summary Exercises on Fractions	257
4.6	Exponents, Order of Operations, and Complex Fractions	259
4.7	Problem Solving: Equations Containing Fractions	267
4.8	Geometry Applications: Area and Volume	277
	Summary	287
	Review Exercises	295
	Test	297

CHAPTER 5 Rational Numbers: Positive and Negative Decimals — 299

5.1	Reading and Writing Decimal Numbers	300
5.2	Rounding Decimal Numbers	309
5.3	Adding and Subtracting Signed Decimal Numbers	317
5.4	Multiplying Signed Decimal Numbers	327
5.5	Dividing Signed Decimal Numbers	335
	Summary Exercises on Decimals	347
5.6	Fractions and Decimals	349
5.7	Problem Solving with Statistics: Mean, Median, Mode, and Variability	357
5.8	Geometry Applications: Pythagorean Theorem and Square Roots	367

5.9	Problem Solving: Equations Containing Decimals	375
5.10	Geometry Applications: Circles, Cylinders, and Surface Area	381
	Summary	395
	Review Exercises	403
	Test	409

CHAPTER 6 Ratio, Proportion, and Line/Angle/Triangle Relationships 411

6.1	Ratios	412
6.2	Rates	423
6.3	Proportions	431
Summary Exercises on Ratios, Rates, and Proportions		443
6.4	Problem Solving with Proportions	445
6.5	Geometry: Lines and Angles	453
6.6	Geometry Applications: Congruent and Similar Triangles	467
	Summary	477
	Review Exercises	485
	Test	491

CHAPTER 7 Percent 493

7.1	The Basics of Percent	494
7.2	The Percent Proportion	509
7.3	The Percent Equation	517
Summary Exercises on Percent		527
7.4	Problem Solving with Percent	529
7.5	Consumer Applications: Sales Tax, Tips, Discounts, and Simple Interest	539
	Summary	551
	Review Exercises	557
	Test	561

CHAPTER 8 Measurement 563

8.1	Problem Solving with English Measurement	564
8.2	The Metric System—Length	575
8.3	The Metric System—Capacity and Weight (Mass)	583
8.4	Problem Solving with Metric Measurements	593
8.5	Metric−English Conversions and Temperature	597
	Summary	605
	Review Exercises	611
	Test	615

CHAPTER 9 Graphs 617

9.1	Problem Solving with Tables and Pictographs	618
9.2	Reading and Constructing Circle Graphs	627
9.3	Bar Graphs and Line Graphs	637
9.4	The Rectangular Coordinate System	645
9.5	Introduction to Graphing Linear Equations	651
	Summary	665
	Review Exercises	673
	Test	679

CHAPTER 10 Real Numbers, Equations, and Inequalities 683

10.1	Real Numbers and Expressions	684
10.2	More on Solving Linear Equations	694
10.3	Formulas and Solving for a Specified Variable	703
10.4	Solving Linear Inequalities	709
	Summary	719
	Review Exercises	723
	Test	727

CHAPTER 11 Graphs of Linear Equations and Inequalities in Two Variables 729

11.1	Reading Graphs; Linear Equations in Two Variables	730
11.2	Graphing Linear Equations in Two Variables	745
11.3	Slope of a Line	759
11.4	Equations of Lines	771
Summary Exercises on Graphing Linear Equations		783
11.5	Graphing Linear Inequalities in Two Variables	785
	Summary	793
	Review Exercises	797
	Test	803

CHAPTER 12 Systems of Linear Equations and Inequalities 807

12.1	Solving Systems of Linear Equations by Graphing	808
12.2	Solving Systems of Linear Equations by Substitution	817
12.3	Solving Systems of Linear Equations by Elimination	825

Summary Exercises on Solving Systems of Linear Equations	833
12.4 Applications of Linear Systems	835
12.5 Solving Systems of Linear Inequalities	847
Summary	853
Review Exercises	857
Test	861

CHAPTER 13 Exponents and Polynomials — 863

13.1 Adding and Subtracting Polynomials	864
13.2 The Product Rule and Power Rules for Exponents	873
13.3 Multiplying Polynomials	881
13.4 Special Products	889
13.5 Integer Exponents and the Quotient Rule	895
Summary Exercises on the Rules for Exponents	905
13.6 Dividing a Polynomial by a Monomial	907
13.7 Dividing a Polynomial by a Polynomial	911
13.8 An Application of Exponents: Scientific Notation	917
Summary	923
Review Exercises	927
Test	931

CHAPTER 14 Factoring and Applications — 933

14.1 Factors; The Greatest Common Factor	934
14.2 Factoring Trinomials	943
14.3 Factoring Trinomials by Grouping	949
14.4 Factoring Trinomials Using FOIL	953
14.5 Special Factoring Techniques	959
Summary Exercises on Factoring	965
14.6 Solving Quadratic Equations by Factoring	967
14.7 Applications of Quadratic Equations	975
Summary	987
Review Exercises	991
Test	997

CHAPTER 15 Rational Expressions and Applications — 999

15.1 The Fundamental Property of Rational Expressions	1000
15.2 Multiplying and Dividing Rational Expressions	1009
15.3 Least Common Denominators	1017
15.4 Adding and Subtracting Rational Expressions	1023
15.5 Complex Fractions	1033
15.6 Solving Equations with Rational Expressions	1041
Summary Exercises on Rational Expressions and Equations	1053
15.7 Applications of Rational Expressions	1055
15.8 Variation	1067
Summary	1073
Review Exercises	1079
Test	1083

CHAPTER 16 Roots and Radicals — 1085

16.1 Evaluating Roots	1086
16.2 Multiplying, Dividing, and Simplifying Radicals	1097
16.3 Adding and Subtracting Radicals	1107
16.4 Rationalizing the Denominator	1111
16.5 More Simplifying and Operations with Radicals	1119
Summary Exercises on Operations with Radicals	1129
16.6 Solving Equations with Radicals	1131
Summary	1141
Review Exercises	1145
Test	1151

CHAPTER 17 Quadratic Equations — 1153

17.1 Solving Quadratic Equations by the Square Root Property	1154
17.2 Solving Quadratic Equations by Completing the Square	1161
17.3 Solving Quadratic Equations by the Quadratic Formula	1169
Summary Exercises on Quadratic Equations	1177
17.4 Graphing Quadratic Equations	1179
17.5 Introduction to Functions	1187
Summary	1197
Review Exercises	1201
Test	1205

Contents

Whole Numbers Computation: Pretest   **1207**

Chapter R — Whole Numbers Review   **1209**

R.1 Adding Whole Numbers   1209
R.2 Subtracting Whole Numbers   1217
R.3 Multiplying Whole Numbers   1225
R.4 Dividing Whole Numbers   1233
R.5 Long Division   1243
    Summary   1249
    Review Exercises   1251
    Test   1253

Appendix A: Inductive and Deductive Reasoning   A-1

Appendix B: Graphing Direct and Inverse Variation   A-7

Appendix C: Graphing Quadratic Inequalities   A-15

Answers to Selected Exercises   A-19

Index   I-1

THEA Index   I-17

Index of Focus on Real-Data Applications

SECTION	PAGE	TITLE	APPLICATION TOPIC	OBJECTIVE
Chapter 1: Introduction to Algebra: Integers				
1.3	20	Auto Aide	Tracking auto assistance services on an Interstate highway	Use integers to record movement on a number line; absolute values
1.7	54	'Til Debt Do You Part!	Wedding expenses	Add, subtract, divide; interpret remainders
Chapter 2: Understanding Variables and Solving Equations				
2.2	102	Expressions	Tickets for graduation; length of yellow traffic light; words in a child's vocabulary	Use and interpret expressions that model real-world events
Summary	142	Algebraic Expressions and Tuition Costs	Calculating college tuition and fees	Add and multiply; write expressions representing calculation steps
Chapter 3: Solving Application Problems				
3.3	172	Formulas	Perimeter of triangles; child's vocabulary	Solve equations and interpret the solutions
3.4	180	Connections: Arithmetic to Algebra	Renting a chainsaw; monthly car payments	Compare arithmetic and algebraic solutions
Chapter 4: Rational Numbers: Positive and Negative Fractions				
4.4	238	Music	Time signatures	Add unlike fractions
4.5	252	Recipes	Increasing/decreasing recipe amounts	Multiply and divide fractions, mixed numbers
4.6	262	Heart-Rate Training Zone	Appropriate heart rate for exercise, based on age	Multiply fraction times whole or mixed number
4.7	272	Hotel Expenses	Sharing hotel room costs between 2, 3, or 4 people	Find a fraction of a whole number; add, divide
4.8	282	Quilt Patterns	Authentic quilt designs	Identify fractional parts of a whole (geometric shapes)
Chapter 5: Rational Numbers: Positive and Negative Decimals				
5.2	314	Lawn Fertilizer	Amounts of fertilizer used by Americans	Divide with decimal quotients; round decimals
5.4	330	Life Insurance Benefits	Calculating employees' insurance premiums	Multiply and divide decimals
5.5	342	Dollar-Cost Averaging	Maximizing stock market investments	Add and divide decimals; find averages
Chapter 6: Ratio, Proportion, and Line/Angle/Triangle Relationships				
6.1	418	Historical Ratios	Compare costs and earnings from 1960 to the present	Write and compare ratios; change fractions to decimals
6.3	438	Tour of the West	Travel distances and times	Convert hour and minute times to decimals; divide decimals to find rates
6.4	448	Feeding Hummingbirds	Increasing/decreasing recipe amounts	Apply ratios; solve proportions
Chapter 7: Percent				
7.1	502	Decimals, Percents, and Quilt Patterns	Authentic quilt designs	Convert among fractions, decimals, and percents
7.5	550	Make Your Investments Grow—Compound Interest	Calculating compound interest on investments	Use a compound interest table; decimal calculation
Summary	556	Educational Tax Incentives	Education costs and eligibility for tax credits and scholarships	Add and multiply; find a percent of a number
Chapter 8: Measurement				
8.1	570	Growing Sunflowers	Use information on a packet of seeds (English units)	Convert among inches, feet, yards; use percents and decimals
8.2	580	Measuring Up	Hair and nail growth; new measuring device (metric units)	Convert among metric length units; use unit fractions

Chapter 9: Graphs

9.1	622	Currency Exchange	Using currency exchange rates	Set up and solve proportions
9.3	640	Grocery Shopping	Saving strategies for grocery shopping	Interpret a double-bar graph; collect data and make a graph
Summary	672	Surfing the Net	Online usage	Interpret a bar graph; use percents; collect data and make a graph

Chapter 11: Graphs of Linear Equations and Inequalities in Two Variables

11.3	766	Linear or Nonlinear? That Is the Question about Windchill	Find windchill from air temperature and wind speed	Read and graph data from a table

Chapter 12: Systems of Linear Equations and Inequalities

12.1	812	Estimating Fahrenheit Temperature	Convert Celsius to Fahrenheit temperatures	Use formulas
12.5	850	Sales of Compact Discs versus Cassettes	Compare sales	Solve inequalities and interpret results
Summary	856	Systems of Linear Equations and Modeling	Relate Kelvin and Fahrenheit temperatures	Use a system of equations as a model

Chapter 13: Exponents and Polynomials

13.3	886	Algebra as Generalized Arithmetic	Compare rules in algebra and arithmetic	Use the FOIL method
13.4	892	Using a Rule of Thumb	Determine dimensions of a patio	Evaluate algebraic expressions
13.5	902	Numbers BIG and SMALL	Apply powers of numbers	Write and interpret numbers using exponents
Summary	926	Earthquake Intensities Measured by the Richter Scale	Find earthquake intensity	Use scientific notation

Chapter 14: Factoring and Applications

14.1	940	Idle Prime Time	Explore prime numbers	Determine prime and composite numbers
14.6	972	Factoring Trinomials Made Easy	Use FOIL to factor	Factor by grouping
Summary	990	Stopping Distance	Calculate distance to stop a vehicle	Write and interpret equations

Chapter 15: Rational Expressions and Applications

15.2	1014	Is 5 or 10 Minutes Really Worth the Risk?	Find time saved by speeding	Use the formula $t = \dfrac{d}{r}$
15.7	1062	Upward Mobility	Calculate the cost to own a new vehicle	Perform operations with rational expressions

Chapter 16: Roots and Radicals

16.4	1116	The Golden Ratio—A Star Number	Investigate the golden ratio	Use radical quotients
16.5	1124	Spaceship Earth—A Geodesic Sphere	Investigate Spaceship Earth at Walt Disney World	Evaluate a radical expression
16.6	1136	On a Clear Day	Find the visible distance from a tall building	Use and interpret a radical expression

Chapter 17: Quadratic Equations

17.3	1174	Estimating Interest Rates	Find an average interest rate	Solve a quadratic equation with a variable expression

Preface

The second edition of *Prealgebra and Introductory Algebra* continues our ongoing commitment to provide the best possible text and supplements package that will help instructors teach and students succeed. To that end, we have addressed the diverse needs of today's students through an attractive design, updated applications and graphs, helpful features, careful explanation of concepts, and an expanded package of supplements and study aids. We have also responded to the suggestions of users and reviewers and have added new examples and exercises based on their feedback.

The text is designed for mathematics students who are new to algebra, are relearning the algebra they studied in the past, or are anxious about their ability to learn algebra. The text interweaves arithmetic review and geometry topics, as appropriate, into the algebraic themes of integers, variables and expressions, linear and quadratic equations, inequalities, solving application problems, positive and negative fractions and decimals, proportions, percents, measurements, and graphing. The emphasis is on building a solid understanding of the foundations of algebra. This is accomplished by tying the content to students' experiences and previous knowledge, explaining important terminology in everyday English, showing *why* things work the way they do, and providing carefully sequenced exercises. All objectives from the Texas THEA Test are covered (see THEA Index).

This text is part of a series that also includes the following books:

- *Essential Mathematics,* Second Edition, by Lial and Salzman
- *Basic College Mathematics,* Seventh Edition, by Lial, Salzman, and Hestwood
- *Prealgebra,* Third Edition, by Lial and Hestwood
- *Introductory Algebra,* Eighth Edition, by Lial, Hornsby, and McGinnis
- *Intermediate Algebra,* Eighth Edition, by Lial, Hornsby, and McGinnis
- *Introductory and Intermediate Algebra,* Third Edition, by Lial, Hornsby, and McGinnis.

HALLMARK FEATURES

We have retained the popular features of the previous edition of the text, including the following:

Meaningful Real-Life Applications We are always on the lookout for interesting data to use in real-life applications. As a result, we have included many new or updated examples and exercises throughout the text that focus on real-life applications of mathematics. Students are often asked to find data in a table, chart, graph, or advertisement. (See pp. 692, 730, and 732.) These applied problems provide an up-to-date flavor that will appeal to and motivate students.

Helpful Figures and Attention-getting Photos Today's students are more visually oriented than ever. Thus, we have made a concerted effort to add mathematical figures, diagrams, tables, and graphs whenever possible. Many of the graphs use a style similar to that seen by students in today's print and electronic media. Photos have been incorporated to enhance applications in examples and exercises. (See pp. 731, 739, and 740.)

▶ ***Emphasis on Problem Solving*** Chapter 3 introduces students to our six-step process for solving application problems algebraically: *Read, Assign a Variable, Write an Equation, Solve, State the Answer,* and *Check.* By devoting an entire chapter to this process, students build a strong foundation for problem solving, which is then reinforced through specific problem-solving sections in Chapters 4, 5, 6, 7, 8, 12, 14, and 15. (See pp. 169, 177–178 and 530.) The same six steps are also used throughout the other algebra titles in this textbook series.

▶ ***Learning Objectives*** Each section begins with clearly stated numbered objectives, and the material within sections is keyed to these objectives so that students know exactly what concepts are covered. (See pp. 684, 745, and 771.)

▶ ***Cautions and Notes*** These color-coded, boxed comments, one of the most popular features of the previous edition, warn students about common errors and emphasize important ideas throughout the exposition. (See pp. 771, 773, and 810.) There are more of these in the second edition, and the text design makes them easier to spot; Cautions are highlighted in yellow and Notes are highlighted in purple.

▶ ***Calculator Tips*** These optional tips, marked with a red calculator icon, offer basic information and instruction for students using scientific calculators in the course. The calculator is used to emphasize such concepts as order of operations, commutativity of addition and multiplication, rounding of decimal numbers, division by zero as undefined, and so on. (See pp. 53, 212, and 331.)

▶ ***Margin Problems*** Margin problems, with answers immediately available on the bottom of the page, are found in every section of the text. (See pp. 817, 821, and 826.) This key feature allows students to immediately practice the material covered in the examples in preparation for the exercise sets.

▶ ***Ample and Varied Exercise Sets*** The text contains a wealth of exercises to provide students with opportunities to practice, apply, connect, and extend the skills they are learning. Numerous illustrations, tables, graphs, and photos have been added to the exercise sets to help students visualize the problems they are solving. Problem types include skill building, writing, estimation, and calculator exercises, as well as applications and correct-the-error problems. In the Annotated Instructor's Edition of the text, the writing exercises are marked with an icon for writing so that instructors may assign these problems at their discretion. Exercises suitable for calculator work are marked in both the student and instructor editions with a red calculator icon. (See pp. 829, 841, and 851.)

▶ ***Relating Concepts Exercises*** These sets of exercises help students tie concepts together and develop higher level problem-solving skills as they compare and contrast ideas, identify and describe patterns, and extend concepts to new situations. (See pp. 166, 240, and 353.) These exercises make great collaborative activities for pairs or small groups of students.

▶ ***Ample Opportunity for Review*** Each chapter ends with a Chapter Summary featuring: Key Terms with definitions and helpful graphics, New Formulas, New Symbols, Test Your Word Power, and a Quick Review of each section's content with additional examples. Also included is a comprehensive set of Chapter Review Exercises keyed to individual sections, a set of Mixed Review Exercises, and a Chapter Test. (See pp. 719, 793, and 853.)

▶ ***Use of Raised Red Negative Signs in Chapters 1–3*** A red dash in the raised position is used to designate negative numbers in Chapters 1–3; this makes a negative number visually distinct from the dash used to indicate the operation of subtraction. Traditional notation, using parentheses around negative numbers when necessary, is introduced in Chapter 4 and used throughout the rest of the text. The raised red negative sign helps students develop a clear understanding of the difference between a negative number and the subtraction sign. It also helps students avoid errors related to "forgetting" the negative sign. The authors and reviewers have found that students make the transition to traditional notation without difficulty.

	Chapters 1–3	Chapters 4–17
	$^{-}7 - {}^{-}3$	$-7 - (-3)$

▶ ***Interweaving of Arithmetic Review and Geometry Topics throughout the Text*** Developmental math students at the college level are tired of rehashing arithmetic. The backbone of this text is algebra, with necessary arithmetic review in fractions, decimals, ratios, and percents included at points where it is needed. Skills in whole number computation can be assessed at the beginning of the course with the Whole Numbers Computation Pretest (following Chapter 17). The student then works only the needed sections in Chapter R: Whole Numbers Review before starting Chapter 1. Geometry topics introduced at appropriate points include perimeter, area, volume, surface area, circles, Pythagorean theorem, line and angle relationships, and congruent and similar triangles.

WHAT'S NEW IN THIS EDITION

▶ ***Diagnostic Pretest*** A diagnostic pretest starts on p. xix of the text and covers all the material in the book, much like a sample final exam. This pretest can be used to facilitate student placement in the correct chapter according to skill level. The pretest also exposes students to the scope of the course content.

▶ ***Chapter Openers*** The chapter openers feature real-world applications of mathematics that are relevant to students and tied to specific material within the chapters. Examples of topics include finding the best buy on cell phone service, weather, home improvements, driving, credit card debt, discounts and sales tax, fishing, popular movies, and work/career applications. (See pp. 683, 729, and 807.)

▶ ***More Summary Exercises*** Nine sets of in-chapter summary exercises have been added, for a total of twelve sets, to provide students with the all-important *mixed* practice they need at critical points in their skill development. (See pp. 69, 257, and 347.)

▶ ***Test Your Word Power*** This feature, now incorporated into each chapter summary, helps students understand and master mathematical vocabulary. Key terms from the chapter are presented along with four possible definitions in a multiple-choice format. Answers and examples illustrating each term are provided. (See pp. 288, 478, and 551.)

▶ ***Focus on Real-Data Applications*** Each one-page activity presents a relevant and in-depth look at how mathematics is used in the real world. Designed to help instructors answer the often-asked question, "When will I ever use this stuff?," these activities ask students to read and interpret data from newspaper articles, the Internet, and other familiar, real-world sources. (See pp. 102, 238, 580, and 1136.) The activities are well-suited for collaborative work, or they can be completed by individuals or used for open-ended class discussions. A helpful index of Focus on Real-Data Applications appears on page vii.

▶ ***Study Skills Component*** Poor study skills are a major reason why students do not succeed in math. A few generic tips sprinkled here and there are not enough to help students change their behavior. So, in this text, a desk-light icon at key points in the text directs students to one of twelve activities in a separate *Study Skills Workbook*. These carefully designed activities, correlated directly to the text material, cover such essential skills as using the textbook, note taking, completing homework, making study cards, test preparation, test taking, and preparing for a final exam. (See pp. 84, 145, and 191.) This unique workbook explains *how* the brain actually learns and remembers so students understand *why* the study skills activities will help them succeed in the course. Students are introduced to the workbook in a To the Student section at the beginning of the text.

▶ *Sequence of Topics* The scope of topics in *Prealgebra and Introductory Algebra* is highly rated by our reviewers and has been retained, but you will notice several changes in the sequence. Based on reviewer feedback, graphs of linear equations and inequalities are now presented in Chapter 11, and systems of linear equations is now Chapter 12 (instead of Chapters 14 and 15 in the previous edition). Two topics that are not used by all instructors at this level have been moved to the Appendices. Graphing direct and inverse variation is now Appendix B, and graphing quadratic inequalities is now Appendix C. You will also find many places in the text where we have polished individual presentations, added help for common trouble spots, and updated examples, exercises, and applications.

WHAT SUPPLEMENTS ARE AVAILABLE

For a comprehensive list of the supplements and study aids that accompany *Prealgebra and Introductory Algebra,* Second Edition, see pages xiii and xiv.

Student Supplements

Student's Solutions Manual
- Provides detailed solutions to the odd-numbered section-level exercises and to all margin, Relating Concepts, Summary, Chapter Review, Chapter Test, and Cumulative Review Exercises
ISBN: 0-321-44820-0

Study Skills Workbook
- The twelve activities in the workbook teach students how to use the textbook effectively, plan their homework, take notes, make mind maps and study cards, manage study time, and prepare for and take tests
ISBN: 0-321-45921-0

Digital Video Tutor
- Complete set of digitized videos on CD-ROM for students to use at home or on campus
- Ideal for distance learning or supplemental instruction
ISBN: 0-321-44819-7

MathXL® Tutorials on CD
- This interactive tutorial CD-ROM provides algorithmically generated practice exercises that are correlated at the objective level to the exercises in the textbook
- Every practice exercise is accompanied by an example and a guided solution designed to involve students in the solutions process
- Selected exercises may also include a video clip to help students visualize concepts
- The software provides helpful feedback for incorrect answers and can generate printed summaries of students' progress
ISBN 0-321-44285-7

Addison-Wesley Math Tutor Center
- Staffed by qualified mathematics instructors
- Provides tutoring on examples and odd-numbered exercises from the textbook through a registration number with a new textbook or purchased separately
- Accessible via toll-free telephone, toll-free fax, e-mail, or the Internet
www.aw-bc.com/tutorcenter

InterAct Math Tutorial Website
- Get practice and tutorial help online!
- This interactive tutorial website provides algorithmically generated practice exercises that correlate directly to the exercises in the textbook.
- Students can retry an exercise as many times as they like with new values each time for unlimited practice and mastery
- Every exercise is accompanied by an interactive guided solution that provides helpful feedback for incorrect answers, and students can also view a worked-out sample problem that steps them through an exercise similar to the one they're working on
www.interactmath.com

Instructor Supplements

Annotated Instructor's Edition
- Provides answers to all text exercises in color next to the corresponding problems
- Icons identify writing and calculator exercises
ISBN: 0-321-44360-8

Instructor's Solutions Manual
- Provides complete solutions to all even-numbered section-level exercises
ISBN: 0-321-44817-0

Printed Test Bank and Instructor's Resource Guide
- The test bank contains two diagnostic pretests, six free-response and two multiple-choice test forms per chapter, and two final exams
- The resource guide contains teaching suggestions for each chapter, additional practice exercises for most objectives of every section, a correlation guide from the first to the second edition, phonetic spellings for all key terms in the text, and teaching notes and extensions for the Focus on Real-Data Applications in the text
ISBN: 0-321-44818-9

Videotape Series
- Features an engaging team of lecturers
- Provides comprehensive coverage of each section and topic in the text
ISBN: 0-321-44978-9

TestGen
- Enables instructors to build, edit, print, and administer tests using a computerized bank of questions developed to cover all the objectives of the text
- Algorithmically based, allowing instructors to create multiple but equivalent versions of the same question or test with the click of button
- Instructors can also modify test bank questions or add new questions
- Tests can be printed or administered online.
- Available on a dual-platform Windows/Macintosh CD-ROM
ISBN: 0-321-44573-2

ADDITIONAL INSTRUCTOR SUPPLEMENTS

MathXL®: www.mathxl.com MathXL® is a powerful online homework, tutorial, and assessment system that accompanies your Addison-Wesley textbook in mathematics or statistics. With MathXL, instructors can create, edit, and assign online homework and tests using algorithmically generated exercises correlated at the objective level to the textbook. They can also create and assign their own online exercises and import TestGen test for added flexibility. All student work is tracked in MathXL's online gradebook. Students can take chapter tests in MathXL and receive personalized study plans based on their test results. The study plan diagnoses weaknesses and links students directly to tutorial exercises for the objectives they need to study and retest. Students can also access supplemental video clips directly from selected exercises. MathXL is available to qualified adopters. For more information, visit our Web site at www.mathxl.com, or contact your Addison-Wesley sales representative.

MyMathLab MyMathLab is a series of text-specific, easily customizable online courses for Addison-Wesley textbooks in mathematics and statistics. MyMathLab is powered by CourseCompass™—Pearson Education's online teaching and learning environment—and by MathXL®—our online homework, tutorial, and assessment system. MyMathLab gives instructors the tools they need to deliver all or a portion of their course online, whether students are in a lab setting or working from home. MyMathLab provides a rich and flexible set of course materials, featuring free-response exercises that are algorithmically generated for unlimited practice and mastery. Students can also use online tools, such as video lectures, animations, and a multimedia textbook, to independently improve their understanding and performance. Instructors can use MyMathLab's homework and test managers to select and assign online exercises correlated directly to the textbook, and they can also create and assign their own online exercises and import TestGen tests for added flexibility. MyMathLab's online gradebook—designed specifically for mathematics and statistics—automatically tracks students' homework and test results and gives the instructor control over how to calculate final grades. Instructors can also add offline (paper-and-pencil) grades to the MathXL gradebook. MyMathLab is available to qualified adopters. For more information, visit our Web site at www.mymathlab.com or contact your Addison-Wesley sales representative.

Adjunct Support Center The Addison-Wesley Math Adjunct Support Center is staffed by qualified mathematics instructors with over 50 years combined experience at both the community college and university level. Assistance is provided for faculty in the following areas:

- Suggested syllabus consultation
- Tips on using materials packaged with your book
- Book-specific content assistance
- Teaching suggestions including advice on classroom strategies

For more information, visit www.aw-bc.com/tutorcenter/math-adjunct.html

ACKNOWLEDGMENTS

The comments, criticisms, and suggestions of users, nonusers, instructors and students have positively shaped this textbook. We especially wish to thank these individuals who provided invaluable suggestions for this and previous editions:

Carla Ainsworth, *Salt Lake Community College*
Randall Allbritton, *Daytona Beach Community College*
Jannette Avery, *Monroe Community College*
Pam Baenziger, *Kirkwood Community College*
Linda Beattie, *Western New Mexico University*
Jean Bolyard, *Fairmont State College*

Barbara Brown, *Anoka-Ramsey Community College*
Kim Brown, *Tarrant County College—Northeast Campus*
Hien Bui, *Hillsborough Community College*
Tim C. Caldwell, *Meridian Community College*
Russell Campbell, *Fairmont State College*
John Close, *Salt Lake Community College*
Jane Cuellar, *Taft College*
Ky Davis, *Muskingum Area Technical College*
Bill Dunn, *Las Positas College*
Lucy Edwards, *Las Positas College*
Randy Gallaher, *Lewis and Clark Community College*
Veronica Gold, *Assumption College*
J. Lloyd Harris, *Gulf Coast Community College*
Terry Haynes, *Eastern Oklahoma State College*
Edith Hays, *Texas Woman's University*
Karen Heavin, *Morehead State University*
Elizabeth Heston, *Monroe Community College*
Scott Higinbotham, *Middlesex Community College*
Lori Holdren, *Manatee Community College*
Rosemary Karr, *Collin County Community College*
Harriet Kiser, *Floyd College*
Valerie Lazzara, *Palm Beach Community College*
Christine Heinecke Lehmann, *Purdue University—North Central*
Lou Ann Mahaney, *Tarrant County College—Northeast Campus*
Valerie H. Maley, *Cape Fear Community College*
Linda Marable, *Nashville State Technical Institute*
Susan McClory, *San Jose State University*
Pam Miller, *Phoenix College*
Jeffrey Mills, *Ohio State University*
Michael Montano, *Riverside Community College*
Elizabeth Morrison, *Valencia Community College—West Campus*
Linda J. Murphy, *Northern Essex Community College*
Celia Nippert, *Western Oklahoma State College*
Elizabeth Olgilvie, *Horry-Georgetown Technical College*
Faith Peters, *Broward Community College*
Larry Pontaski, *Pueblo Community College*
Manoj Raghunandanan, *Temple University*
Janalyn Richards, *Idaho State University*
Diann Robinson, *Ivy Tech State College—Lafayette*
Rachael Schettenhelm, *Southern Connecticut State University*
Julia Simms, *Southern Illinois University—Edwardsville*
Sounny Slitine, *Palo Alto College*
Lee Ann Spahr, *Durham Technical Community College*
Carol Stewart, *Fairmont State College*
Sharon Testone, *Onondaga Community College*
Sam Tinsley, *Richland College*
Shae Thompson, *Montana State University*
Cora S. West, *Florida Community College at Jacksonville*
Cheryl Wilcox, *Diablo Valley College*
Johanna Windmueller, *Seminole Community College*
Gabriel Yimesghen, *Community College of Philadelphia*
Kevin Yokoyama, *College of the Redwoods*
Karl Zilm, *Lewis and Clark Community College*

Our sincere thanks go to these dedicated individuals at Addison-Wesley who have worked to make this revision a success: Maureen O'Connor, Sharon Smith, Ron Hampton, Dennis Schaefer, Kathy Manley, Michelle Renda, Alexandra Waibel, Caroline Case, Lauren Morse, and Beth Anderson. Sam Blake of Pre-Press Company, Inc. provided excellent production work. We are most grateful to Peg Crider for researching and writing many of the Focus on Real-Data Applications; Kay Schlembach for her accurate and useful index; Abby Tanenbaum for writing the Diagnostic Pretest; and Perian Herring, Cheryl Davids, and Cheryl Cantwell for accuracy checking the page proofs.

Special thanks go to Barbara Brown, *Anoka-Ramsey Community College*, who reviewed all the new material for this edition in great detail. Her expertise in teaching developmental algebra students resulted in greater clarity, consistency, and creativity. Finally, it is the synergistic collaboration with Linda Russell, reading and study skills specialist at *Minneapolis Community and Technical College*, which resulted in the Study Skills Workbooks to accompany this text and the others in the series.

The ultimate measure of this textbook's success is whether it helps students master algebra skills, develop problem-solving techniques, and increase their confidence in learning and using mathematics. Please let us know how we are doing by sending an e-mail to math@aw.com.

This book is dedicated to my husband, Earl Orf, who is the wind beneath my wings, and to my students at Minneapolis Community and Technical College, from whom I have learned so much.

Diana L. Hestwood

To the Student: Success in Mathematics

There are two main reasons why students have difficulty with mathematics:

- Students start in a course for which they do not have the necessary background knowledge.
- Students don't know how to study mathematics effectively.

Your instructor can help you decide whether this is the right course for you. We can give you some study tips.

Studying mathematics *is* different from studying subjects like English and history. The key to success is regular practice. This should not be surprising. After all, can you learn to play the piano or ski well without a lot of regular practice? The same is true for learning mathematics. Working problems nearly every day is the key to becoming successful. Here is a list of things that will help you succeed in studying mathematics.

1. *Attend class regularly.* Pay attention to what your instructor says and does in class, and take careful notes. In particular, note the problems the instructor works on the board and copy the complete solutions. Keep these notes separate from your homework to avoid confusion when you review them later.

2. Don't hesitate to *ask questions in class.* It is not a sign of weakness but of strength. There are always other students with the same question who are too shy to ask.

3. *Read your text carefully.* Many students read only enough to get by, usually only the examples. Reading the complete section will help you solve the homework problems. Most exercises are keyed to specific examples or objectives that will explain the procedures for working them.

4. Before you start on your homework assignment, *rework the problems the teacher worked in class.* This will reinforce what you have learned. Many students say, "I understand it perfectly when you do it in class, but I get stuck when I try to do the homework problems by myself."

5. Do your homework assignment only *after reading the text* and reviewing your notes from class. Check your work against the answers in the back of the book. If you get a problem wrong and are unable to understand why, mark that problem and ask your instructor about it. Then practice working additional problems of the same type to reinforce what you have learned.

6. *Work as neatly as you can.* Write your symbols clearly, and make sure the problems are clearly separated from each other. Working neatly will help you to think clearly and also make it easier to review the homework before a test.

7. After you complete a homework assignment, *look over the text again.* Try to identify the main ideas that are in the lesson. Often they are clearly highlighted or boxed in the text.

8. *Use the chapter test at the end of each chapter as a practice test.* Work through the problems under test conditions, without referring to the text or the answers until you are finished. You may want to time yourself to see how long it takes you. When you finish, check your answers against those in the back of the book, and rework the problems you missed.

9. *Keep all quizzes and tests that are returned to you,* and use them when you study for future tests and the final exam. These quizzes and tests indicate those concepts your instructor considers to be most important. Be sure to correct any problems that you missed on these tests, so you will have the corrected work to study.

10. *Don't worry if you do not understand a new topic right away.* As you read more about it and work through the problems, you will gain understanding. Each time you review a topic, you will understand it a little better. Few people understand each topic completely right from the start.

> **NOTE**
> Reading a list of study tips is a good start, but you may need some help actually *applying* the tips to your work in this math course.
>
> 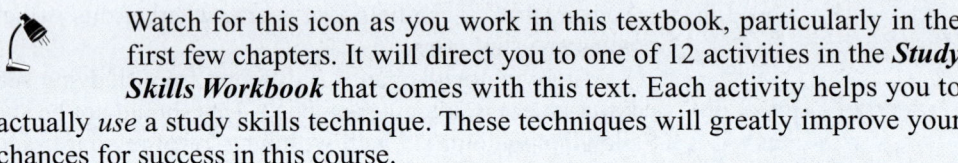 Watch for this icon as you work in this textbook, particularly in the first few chapters. It will direct you to one of 12 activities in the ***Study Skills Workbook*** that comes with this text. Each activity helps you to actually *use* a study skills technique. These techniques will greatly improve your chances for success in this course.
>
> - Find out *how your brain learns new material.* Then use that information to set up effective ways to learn math.
> - Find out *why short-term memory is so short* and what you can do to help your brain remember new material weeks and months later.
> - Find out *what happens when you "blank out" on a test* and simple ways to prevent it from happening.
>
> All the activities in the *Study Skills Workbook* are brain-friendly ways to enjoy and succeed at math. Whether you need help with note taking, managing homework, taking tests, or preparing for a final exam, you'll find specific, clearly explained ideas that really work because they're based on research about how the brain learns and remembers.

> **CAUTION**
> Be sure to complete **Activity 1: Your Brain** in the *Study Skills Workbook* before starting Chapter 1. You'll learn the basics about how your brain works and be ready to understand and use the techniques explained in the workbook.

Diagnostic Pretest

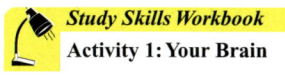
Study Skills Workbook
Activity 1: Your Brain

[Chapter 1]

1. Write this number using digits: twenty-five million, three thousand, seven hundred one

2. Simplify $-3(-4)^2 \div 12 - (-10)$.

3. Round 4,396,028 to the nearest ten-thousand.

4. Max had $185 in his checking account. He deposited his $372 paycheck, and then wrote a $575 check for rent. What is the new balance in his account?

[Chapter 2]

5. Evaluate $-5a^3b^2$ when a is 2 and b is -1.

6. Simplify. $-5 + 4(r - 7)$

7. Solve. $-18 = -2(t + 4)$

8. Solve. $7n - 6 = -2n + 3$

[Chapter 3]

9. Find the perimeter.

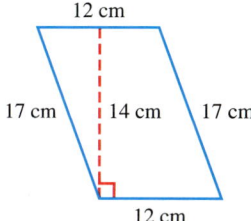

10. Find the area.

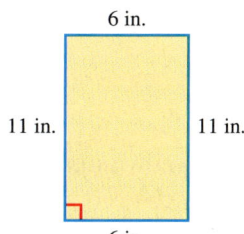

11. Translate this sentence into an equation and solve it. If 13 is added to four times a number, the result is 8 less than the number. What is the number?

12. Write an equation and solve it to answer this problem. A rope is 238 ft long. Marc cut it into two pieces, with one piece 12 ft longer than the other. Find the lengths of both pieces.

1. _____
2. _____
3. _____
4. _____

5. _____
6. _____
7. _____
8. _____

9. _____
10. _____

11. _____

12. _____

xix

[Chapter 4]

13. Write $\dfrac{39}{52}$ in lowest terms.

14. Add: $-\dfrac{5}{6} + \dfrac{7}{12}$

15. Solve. $\dfrac{4}{5}x + 9 = -7$

16. Find the area.

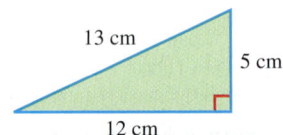

[Chapter 5]

17. Anita bought 2.6 pounds of salmon at $6.99 per pound. How much did she pay for the salmon, to the nearest cent?

18. Solve. $-3r + 5.8 = -1.1$

19. The hourly wages for six employees of a small company are $8.50, $9.40, $7.30, $13.75, $11, and $12. Find the median hourly wage for these employees.

20. Find the circumference of this circle. Use 3.14 for π and round your answer to the nearest tenth.

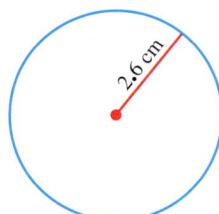

[Chapter 6]

21. Find the unknown number.

$$\dfrac{9}{8} = \dfrac{x}{96}$$

22. On a road map, 1 inch represents 32 miles. If two towns are 3.8 inches apart on the map, what is the actual distance between them?

23. Find the supplement of a 63° angle.

24. Find the unknown lengths in these similar triangles.

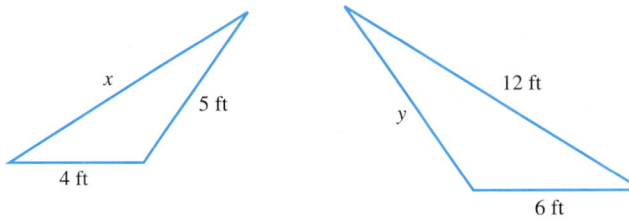

[Chapter 7]

25. (a) Write 250% as a decimal.

    (b) Write 0.3 as a percent.

25. (a) _____

    (b) _____

26. (a) Write 38% as a fraction in lowest terms.

    (b) Write $\frac{3}{40}$ as a percent.

26. (a) _____

    (b) _____

27. The price of a digital camera is $235 plus sales tax of 6%. Find the total cost of the camera including sales tax.

27. _____

28. Suppose that you go out to lunch with three friends and the bill comes to $48. You decide to add a 15% tip, and then, split the bill evenly. How much should each of you pay?

28. _____

[Chapter 8]

29. Convert each measurement.

    (a) 432 cm to meters

    (b) 0.08 kg to grams

29. (a) _____

    (b) _____

30. The average annual precipitation over the period 1971–2000 is 64.16 inches in New Orleans, Louisiana, and 4.16 inches in Barrow, Alaska. What was the difference in precipitation between the two cities in feet? (*Source: World Almanac and Book of Facts.*)

30. _____

31. Each serving of punch at a graduation party is to be 160 mL. How many liters of punch are needed for 65 servings?

31. _____

32. Write the most reasonable metric unit in each blank. Choose from km, m, cm, mm, L, mL, kg, g, and mg.

    (a) The length of Rosie's bedroom is about 4 _____.

    (b) Satish takes a pill containing 500 _____ of vitamin C every morning.

32. (a) _____

    (b) _____

Diagnostic Pretest   **xxi**

[Chapter 9]

33. This circle graph shows the budget for the Patel family. The total budget for one month is $3600. Find the monthly amount budgeted for savings.

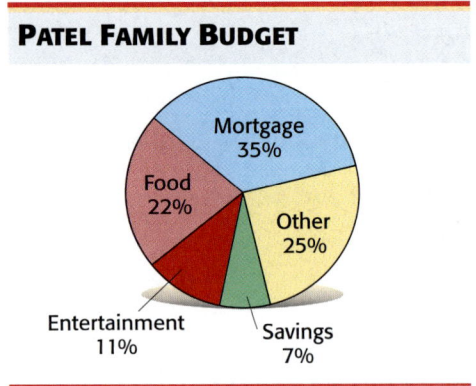

PATEL FAMILY BUDGET
- Mortgage 35%
- Food 22%
- Other 25%
- Entertainment 11%
- Savings 7%

34. This graph shows the number of sections of Prealgebra and Introductory Algebra at River Bend Community College over a four-year period.

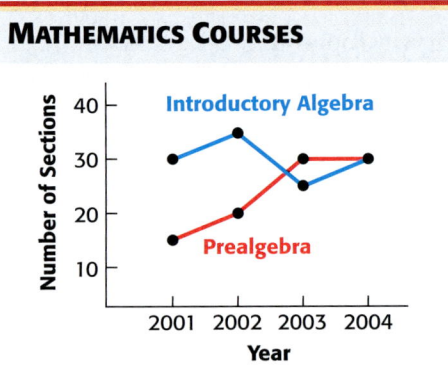

MATHEMATICS COURSES

(a) In what year(s) were there more sections of Introductory Algebra than of Prealgebra?
(b) In what year(s) were there the same number of sections of both courses? How many sections of each course were there in that year?

35. Name the quadrant (if any) in which each point is located.
   (a) $(-3, 5)$
   (b) $(6, 0)$
   (c) $(-2, -6)$
   (d) $(10, 10)$

36. Graph $y = 2x - 1$ on the grid at the left. Complete the table at the right using 0, 1, and 2 as the values of x.

x	y	(x, y)

[Chapter 10]

37. Simplify, then label the simplified statement as true or false.
$$-(3^2)[-1 + 2(-8 \div 2)] \geq (-9)^2$$

38. Solve $-2(1 - 3y) - y = -y - (-6y - 1)$

39. Solve $A = p + prt$ for t.

40. Solve $-4x + 8 \geq -12$, and graph the solutions.

[Chapter 11]

41. Graph $2x - 5y = 10$. Give the x- and y-intercepts.

41. x-intercept: _____

    y-intercept: _____

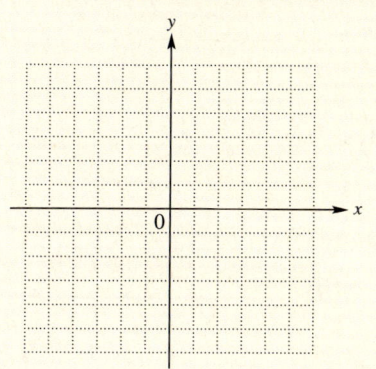

42. Find the slope of the line through $(-2, 5)$ and $(1, -7)$.

42. _____

43. Write an equation in slope-intercept form for the line through $(-5, 6)$ and $(1, 0)$.

43. _____

44. Graph $2x + y > 4$.

44.

[Chapter 12]

Solve each system of equations.

45. $3x + y = 9$
    $x - y = -1$

45. _____

46. $-4x + 7y = 3$
    $12x - 21y = 9$

46. _____

47. Write a system of equations and use it to solve the problem.

Marla and Rick left from the same place at the same time and traveled in opposite directions. Marla drove 8 mph faster than Rick. After 2 hr, they were 228 miles apart. Find Marla's and Rick's speeds.

47. _____

48. Graph the solution of the system of inequalities.
$$3x - 5y < 15$$
$$y \geq -2x$$

48.

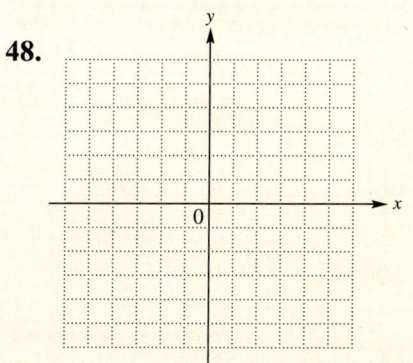

[Chapter 13]

49. Subtract $(4m^3 - 5m^2 + m - 8) - (6m^3 - 5m^2 + 10m - 3)$.

50. Multiply $(7z + 3w)^2$.

51. Evaluate the expression $5^{-1} + 2^{-1} - 3^0$.

52. (a) Write 445,000,000 in scientific notation.
    (b) Write 2.34×10^{-4} without exponents.

[Chapter 14]

53. Factor $3x^2 + 2x - 8$.

54. Factor $16n^2 - 49$.

55. Solve $t^2 - 2t = 15$.

56. The length of the cover of a road atlas is 4 in. more than the width. The area is 165 in.^{2}. Find the dimensions of the cover.

[Chapter 15]

57. Write $\dfrac{x^2 + x - 20}{x^2 - 16}$ in lowest terms.

58. Divide. Write your answer in lowest terms.
$$\frac{2r + 1}{r - 4} \div \frac{6r^2 + 3r}{4 - r}$$

59. Subtract. Write your answer in lowest terms.
$$\frac{z^2}{z - 3} - \frac{z}{z + 3}$$

60. Simplify.
$$\frac{\dfrac{1}{y} + \dfrac{1}{y + 2}}{\dfrac{1}{y} - \dfrac{1}{y + 2}}$$

[Chapter 16]

Simplify when possible.

61. $4\sqrt{300} - 8\sqrt{75}$

61. _____

62. $(\sqrt{5} - \sqrt{7})^2$

62. _____

63. Rationalize the denominator.

$$\frac{4\sqrt{5}}{\sqrt{2}}$$

63. _____

64. Solve $\sqrt{3r} + 6 = 15$.

64. _____

[Chapter 17]

65. Solve $(x - 3)^2 = 20$.

65. _____

66. Solve $2r^2 + 3r - 1 = 0$.

66. _____

67. Sketch the graph of $y = -x^2 + 5$. Identify the vertex.

67. vertex: _____

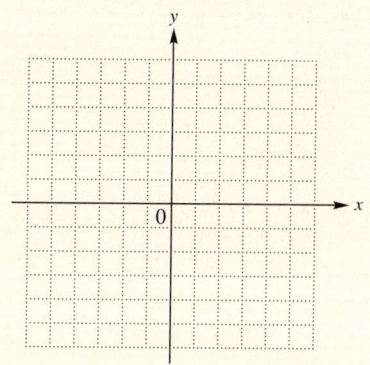

68. Decide whether the relation

$$\{(-2, 4), (-1, 1), (0, 0), (1, 1), (2, 4)\}$$

represents a function. Give the domain and range.

68. _____

domain: _____

range: _____

Introduction to Algebra: Integers

1.1 Place Value

1.2 Introduction to Signed Numbers

1.3 Adding Integers

1.4 Subtracting Integers

1.5 Problem Solving: Rounding and Estimating

1.6 Multiplying Integers

1.7 Dividing Integers

1.8 Exponents and Order of Operations

Summary Exercises on Operations with Integers

The weather—we all talk about it and often complain about it. When reporting temperatures, we need both *positive* and *negative* numbers. If you live in Chicago, the highest temperature ever recorded is 104 °F and the lowest is −27 °F, a difference of 131 degrees! In Atlanta, the record high and low are 105 °F and −8 °F. And in Barrow, Alaska, the range of extreme temperatures is 79 °F to −56 °F, a difference of 135 degrees. To make things even more uncomfortable, humidity can make hot temperatures feel hotter and wind can make cold temperatures feel colder. Learn more about "windchill" as you work Exercises 47–48 in **Section 1.4**. (*Source:* National Climatic Data Center.)

1.1 Place Value

OBJECTIVES

1. Identify whole numbers.
2. Identify the place value of a digit through hundred-trillions.
3. Write a whole number in words or digits.

It would be nice to earn millions of dollars like some of our favorite entertainers or sports stars. But how much is a million? If you received $1 every second, 24 hours a day, day after day, how many days would it take for you to receive a million dollars? How long to receive a billion dollars? Or a trillion dollars? Make some guesses and write them here.

It would take _____ to receive a million dollars.

It would take _____ to receive a billion dollars.

It would take _____ to receive a trillion dollars.

The answers are at the bottom left of the page. Later, in the exercises for **Section 1.7,** you'll find out how to calculate the answers.

OBJECTIVE 1 Identify whole numbers. First we have to be able to write the number that represents *one million*. We can write *one* as 1. How do we make it 1 *million*? Our number system is a **place value system.** That means that the location, or place, in which a number is written gives it a different value. Using money as an example, you can see that

$1 is one dollar.

$10 is ten dollars.

$100 is one hundred dollars.

$1000 is one thousand dollars.

Each time the 1 moved to the left one place, it was worth *ten times* as much. Can you keep moving it to the left? Yes, as many times as you like.

The chart below shows the *value* of each *place*. In other words, you write the 1 in the correct place to represent the number you want to express. It is important to memorize the place value names shown on the chart.

Whole Number Place Value Chart

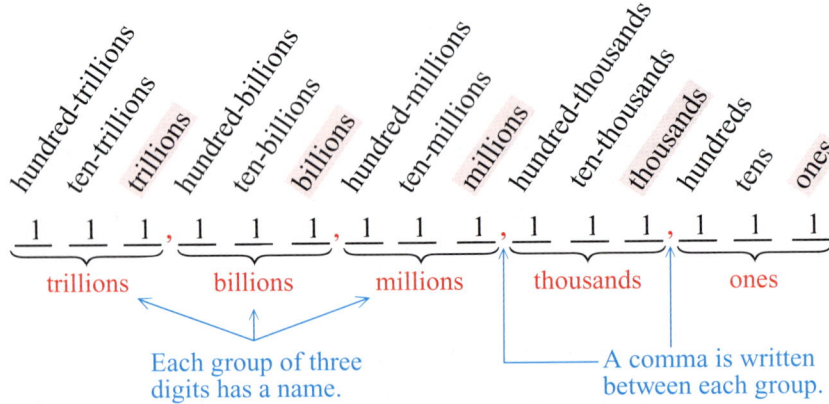

If there had been more room on this page, we could have continued to the left with quadrillions, quintillions, sextillions, septillions, octillions, and more.

Of course, we can use other *digits* besides 1. In our decimal system of writing numbers, we can use these ten **digits:** 0, 1, 2, 3, 4, 5, 6, 7, 8, and 9. In this section we will use the digits to write **whole numbers.**

ANSWERS

About $11\frac{1}{2}$ days to receive a million dollars; nearly 32 *years* to receive a billion dollars; about 31,710 *years* to receive a trillion dollars.

Section 1.1 Place Value **3**

These are whole numbers.	These are *not* whole numbers.
0 8 37 100 24,014	-6 $\frac{3}{4}$ 7.528 0.3 $5\frac{2}{3}$

① Circle the whole numbers.

0.8   −14   502
$\frac{7}{9}$ 3 $\frac{3}{2}$
14   0 $6\frac{4}{5}$
9.082 $-\frac{8}{3}$ 60,005

EXAMPLE 1 Identifying Whole Numbers

Circle the whole numbers in this list.

75   −4   0   1.5 $\frac{5}{8}$ 300   0.666 $7\frac{1}{2}$ 2

Whole numbers *do* include zero. If we started a list of *all* the whole numbers, it would look like this: 0, 1, 2, 3, 4, 5, . . . with the three dots indicating that the list goes on and on. So the whole numbers in this example are: 75, 0, 300, and 2.

Work Problem 1 at the Side.

OBJECTIVE 2 Identify the place value of a digit through hundred-trillions. The estimated world population in 2004 was 6,323,979,176 people. (*Source:* U.S. Census Bureau.) There are two 6s in this number but the value of each 6 is very different. The 6 on the right is in the ones place, so its value is simply 6. But the 6 on the left is worth a great deal more because of the *place* where it is written. Looking back at the place value chart, we see that this 6 is in the billions place, so its value is *6 billion*.

6 , 3 2 3 , 9 7 9 , 1 7 6 people in the world
↑   ↑
Value of   Value of
6 *billion*   6 *ones*

② Identify the place value of the digit 8 in each number.

(a) 45,628,665

(b) 800,503,622

EXAMPLE 2 Identifying Place Value

Identify the place value of each 7 in the number of people in the world.

6 , 3 2 3 , 9 7 9 , 1 7 6
Ten-thousands place ↑   ↑ Tens place

Work Problem 2 at the Side.

(c) 428,000,000,000

OBJECTIVE 3 Write a whole number in words or digits. To write a whole number in words, or to say it aloud, begin at the left. Write or say the number in each group of three, followed by the name for that group. When you get to the ones group, do *not* include the group name. Hyphens (dashes) are used whenever you write a number from 21 to 99, like twenty–one thousand, or thirty–seven million, or ninety–four billion.

(d) 2,385,071

EXAMPLE 3 Writing Numbers in Words

(a) Write 6,058,120 in words.
Start at the left.

6 , 0 5 8 , 1 2 0

six million, fifty-eight thousand, one hundred twenty
   Group name   Group name   ↑ Do not use group name "ones."

Continued on Next Page

ANSWERS
1. 502; 3; 14; 0; 60,005
2. (a) thousands (b) hundred-millions (c) billions (d) ten-thousands

4 Chapter 1 Introduction to Algebra: Integers

3 Write these numbers in words.

(a) 23,605

(b) 400,033,007

(c) 193,080,102,000,000

4 Write each number using digits.

(a) Eighteen million, two thousand, three hundred five

(b) Two hundred billion, fifty million, six hundred sixteen

(c) Five trillion, forty-two billion, nine million

(d) Three hundred six million, seven hundred thousand, nine hundred fifty-nine

ANSWERS
3. (a) twenty-three *thousand*, six hundred five
   (b) four hundred *million*, thirty-three *thousand*, seven
   (c) one hundred ninety-three *trillion*, eighty *billion*, one hundred two *million*
4. (a) 18,002,305 (b) 200,050,000,616
   (c) 5,042,009,000,000 (d) 306,700,959

(b) Write 50,588,000,040,000 in words.
Start at the left.

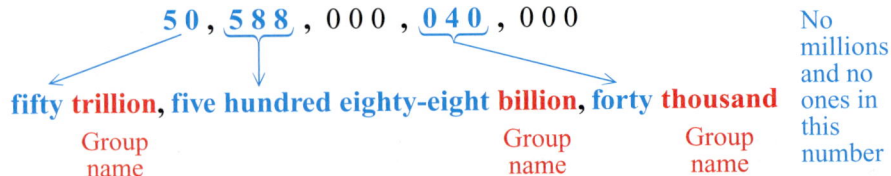

CAUTION
You often hear people say "and" when reading a group of three digits. For example, you may hear 120 as "one hundred *and* twenty," but this is *not* correct. The word *and* is used only when reading a **decimal point**, which we do not have here. The correct wording for 120 is "one hundred twenty."

▶◀ **Work Problem 3 at the Side.**

When you read or hear a number and want to write it in digits, look for the group names: **trillion**, **billion**, **million**, and **thousand**. Write the number in each group, followed by a comma. Do *not* put a comma at the end of the ones group.

EXAMPLE 4 Writing Numbers in Digits

Write each number using digits.

(a) Five hundred sixteen **thousand**, nine

The first group name is *thousand*, so you need to fill *two groups* of three digits: thousands and ones.

$$\underbrace{5\ 1\ 6}_{\text{thousands}} , \underbrace{0\ 0\ 9}_{\text{ones}}$$
comma between groups

The number is 516,009.

(b) Seventy-seven **billion**, thirty **thousand**, five hundred

The first group name is *billion*, so you need to fill *four groups* of three digits: billions, millions, thousands, and ones.

$$\underbrace{0\ 7\ 7}_{\text{billions}} , \underbrace{0\ 0\ 0}_{\text{millions}} , \underbrace{0\ 3\ 0}_{\text{thousands}} , \underbrace{5\ 0\ 0}_{\text{ones}}$$

There are no millions, so fill the millions group with zeros.
When writing the number, you can omit the leading **0** in the billions group. The number is 77,000,030,500.

▶◀ **Work Problem 4 at the Side.**

1.1 Exercises

Circle the whole numbers. See Example 1.

1. 15 $8\frac{3}{4}$ 0   3.781   2. 33.7   −5   457 $\frac{8}{5}$

   83,001   −8 $\frac{7}{16}$ $\frac{9}{5}$ 0   6 $1\frac{3}{4}$ −14.1

3. 5.8   −6   7 $\frac{5}{4}$ 4. 75,039 $\frac{1}{3}$ −87   6.49

   $\frac{1}{10}$ 362,049   0.1 $7\frac{7}{8}$ −0.5 $2\frac{7}{10}$ $\frac{15}{8}$ 4

Identify the place value of the digit 2 in each number. See Example 2.

5. 61,284

6. 82,110

7. 284,100

8. 823,415

9. 725,837,166

10. 442,653,199

11. 253,045,701,000

12. 823,000,419,567

13. From left to right, name the place value for each 0 in this number: 302,016,450,098,570.

14. From left to right, name the place value for each 0 in this number: 810,704,069,809,035.

In Exercises 15–26, write each number in words. See Example 3.

15. 8421

16. 1936

17. 46,205

18. 75,089

19. 3,064,801

20. 7,900,408

21. 840,111,003

22. 304,008,401

23. 51,006,888,321

24. 99,046,733,214

25. 3,000,712,000,000

26. 50,918,000,000,600

In Exercises 27–36, write each number using digits. See Example 4.

27. Forty-six thousand, eight hundred five

28. Seventy-nine thousand, forty-six

29. Five million, six hundred thousand, eighty-two

30. One million, thirty thousand, five

31. Two hundred seventy-one million, nine hundred thousand

32. Three hundred eleven million, four hundred

33. Twelve billion, four hundred seventeen million, six hundred twenty-five thousand, three hundred ten

34. Seventy-five billion, eight hundred sixty-nine million, four hundred eighty-eight thousand, five hundred six

35. Six hundred trillion, seventy-one million, four hundred

36. Four hundred forty trillion, thirty-six thousand, one hundred two

In Exercises 37–46, if the number is given in digits, write it in words. If the number is given in words, write it in digits. See Examples 3 and 4. (Source: The World Almanac, Statistical Abstract of the United States, Scholastic Book of World Records.)

37. In the United States, 6375 couples get married every day.

38. There are 3582 pairs of bowling shoes purchased each day.

39. One hundred one million, two hundred eighty thousand adults are on diets at any one time in the United States.

40. Two million, five thousand Americans suffer from heartburn on any given day.

41. The average amount spent each day by individuals in the United States on computers and software is $93,972,602.

42. Americans spend an average of $176,986,300 daily on toys.

43. People eat fifty-five million, eight hundred hot dogs each day.

44. Five hundred twenty-four million servings of cola drinks are consumed every day.

45. Dunkin' Donuts sells nearly 6,400,000 doughnuts every day. That's 2,336,000,000 doughnuts in one year!

46. The Hostess company bakeries are able to make 60,000 Twinkies every hour. That adds up to 525,600,000 Twinkies per year.

Chapter 1 Introduction to Algebra: Integers

RELATING CONCEPTS (EXERCISES 47–50) For Individual or Group Work

Use your knowledge of place value to work Exercises 47–50 in order.

47. Here is a group of digits: 6, 0, 9, 1, 5, 0, 7, 1. Using each digit exactly once, arrange them to make the largest possible whole number and the smallest possible whole number. Then write each number in words.

48. Write these numbers in digits and in words.
   (a) Your house number or apartment building number
   (b) Your phone number, including the area code
   (c) The approximate cost of your tuition and books for this quarter or semester
   (d) Your zip code

   Now tell how you *usually* say each of the numbers in parts (a)–(d) above. Why do you think we ignore the rules when saying these numbers in everyday situations?

49. Look again at the Whole Number Place Value Chart at the beginning of this section. As you move to the left, each place is worth *ten* times as much as the previous place. Computers work on a *binary* system where each place is worth *two* times as much as the previous place. Complete this place value chart based on 2s.

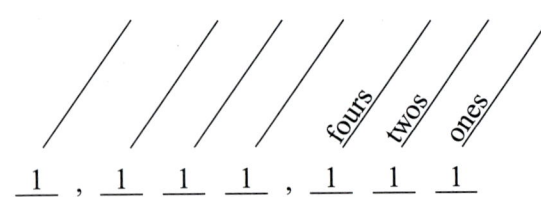

   $\underline{1}\,,\,\underline{1}\ \underline{1}\ \underline{1}\,,\,\underline{1}\ \underline{1}\ \underline{1}$

   Try writing some numbers using the 2s (binary) place value chart.

   (a) 5 = _____
   (b) 10 = _____
   (c) 15 = _____

   The *only* digits you may use in the binary system are 0 and 1.
   Here is an example.
   $$6 = 110$$
   one 4 + one 2 + zero 1s = 4 + 2 + 0 = 6

50. (a) Explain in your own words why our number system is called a *place value system*. Include an example as part of your explanation.

   (b) Find information on the Roman numeral system. Write these numbers using Roman numerals.

   8 = _____   38 = _____
   275 = _____   3322 = _____

   (c) Explain why the Roman system is *not* a place value system. What are the disadvantages of the Roman system?

1.2 Introduction to Signed Numbers

OBJECTIVE 1 Write positive and negative numbers used in everyday situations. The whole numbers in **Section 1.1** were either 0 or greater than 0. Numbers *greater* than 0 are called *positive numbers*. But many everyday situations involve numbers that are *less* than 0, called *negative numbers*. Here are a few examples.

OBJECTIVES

1. Write positive and negative numbers used in everyday situations.
2. Graph signed numbers on a number line.
3. Use the < and > symbols to compare integers.
4. Find the absolute value of integers.

Study Skills Workbook
Activity 2: Your Textbook

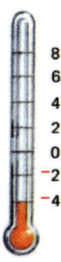

At midnight the temperature dropped to 4 degrees *below* zero.

-4 degrees

Jean had $30 in her checking account. She wrote a check for $40.75. We say she is now $10.75 "in the hole," or $10.75 "in the red," or *overdrawn* by $10.75.

$-\$10.75$

The Packers *gained* 6 yards on the first play. On the second play they *lost* 9 yards. We can write the results using a positive and a negative number.

$+6$ yards   and -9 yards

A plane took off from the airport and climbed to 20,000 feet *above* sea level. We can write this using a positive number.

$+20,000$ feet

A scuba diver swam down to $25\frac{1}{2}$ feet *below* the surface. We can write this using a negative number.

$-25\frac{1}{2}$ feet

NOTE
To write a negative number, put a negative sign (a dash) in front of it: -10. Notice that the negative sign looks exactly like the subtraction sign, as in $5 - 3 = 2$. The negative sign and subtraction sign do *not* mean the same thing (more on that in the next section). To avoid confusion for now, we will write negative signs in **red** and put them up higher than subtraction signs.

$^-10$ means **negative** 10 $14 - 10$ means 14 **minus** 10

— Raised dash

Starting in **Chapter 4**, we will write negative signs in the traditional way. However, if you use a graphing calculator, it may show negative signs in the raised position.

Positive numbers can be written two ways:

1. Write a positive sign in front of the number: $^+2$ is *positive* 2. We will write the sign in the raised position to avoid confusion with the sign for addition, as in $6 + 3 = 9$.

2. Do not write any sign. For example, 16 is assumed to be *positive* 16.

10 Chapter 1 Introduction to Algebra: Integers

❶ Write each negative number with a raised negative sign. Write each positive number in two ways.

(a) The temperature is $5\frac{1}{2}$ degrees below zero.

(b) Cameron lost 12 pounds on a diet.

(c) I deposited $210.35 in my checking account.

(d) I wrote too many checks, so my account is overdrawn by $65.

(e) The submarine dived to 100 feet below the surface of the sea.

(f) In this round of the card game, I won 50 points.

❷ Graph each set of numbers.

(a) $^-2$ (b) 2 (c) 0

(d) $^-4$ (e) 4

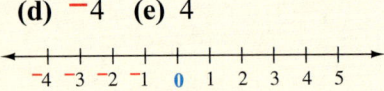

(f) $^-3\frac{1}{2}$ (g) $\frac{1}{2}$

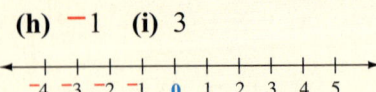

(h) $^-1$ (i) 3

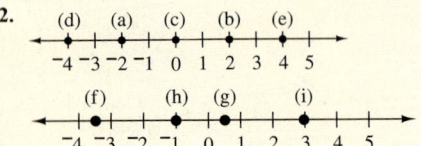

ANSWERS

1. (a) $^-5\frac{1}{2}$ degrees (b) $^-12$ pounds
   (c) $210.35 or $^+210.35 (d) $^-$65
   (e) $^-100$ feet (f) 50 points or $^+50$ points

2.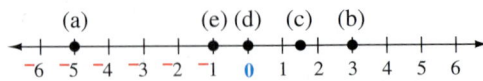

EXAMPLE 1 Writing Positive and Negative Numbers

Write each negative number with a raised negative sign. Write each positive number in two ways.

(a) The river rose to 8 feet above flood stage.

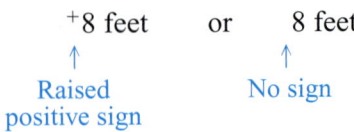

(b) Michael lost $500 in the stock market.

$$^-\$500$$
↑
Raised negative sign

◀◀◀ **Work Problem 1 at the Side.**

OBJECTIVE ❷ Graph signed numbers on a number line. Mathematicians often use a **number line** to show how numbers relate to each other. A number line is like a thermometer turned sideways. *Zero is the dividing point between the positive and negative numbers.*

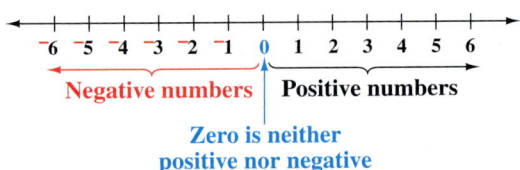

The number line could be shown with positive numbers on the left side of 0 instead of the right side. But it helps if everyone draws it the same way, as shown above. This method will also match what you do when graphing points and lines in **Chapter 9**.

EXAMPLE 2 Graphing Numbers on a Number Line

Graph each number on the number line.

(a) $^-5$ (b) 3 (c) $1\frac{1}{2}$ (d) 0 (e) $^-1$

Draw a dot at the correct location for each number.

```
       (a)           (e)(d) (c)  (b)
   ←——●——+——+——+——●—●——+——●——+——●——+——→
      ⁻6 ⁻5 ⁻4 ⁻3 ⁻2 ⁻1  0  1  2  3  4  5  6
```

◀◀◀ **Work Problem 2 at the Side.**

OBJECTIVE ❸ Use the < and > symbols to compare integers. In **Chapters 4 and 5** you will work with fractions and decimals. For the rest of this chapter, you will work only with *integers*. A list of **integers** can be written like this:

$$\ldots, {}^-6, {}^-5, {}^-4, {}^-3, {}^-2, {}^-1, 0, 1, 2, 3, 4, 5, 6, \ldots$$

The dots show that the list goes on forever in both directions.

We can use the number line to compare two integers.

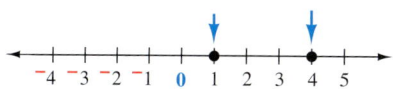

1 is to the *left* of 4.
1 is *less than* 4.
Use < to mean "is less than."

$$1 < 4$$

1 **is less than** 4

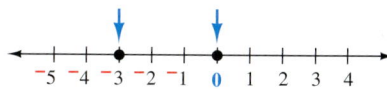

0 is to the *right* of ⁻3.
0 is *greater than* ⁻3.
Use > to mean "is greater than."

$$0 > {}^-3$$

0 **is greater than** ⁻3

NOTE
One way to remember which symbol to use is that the "smaller end of the symbol" points to the "smaller number" (the number that is less).

$1 < 4$ $0 > {}^-3$
Smaller number — Smaller end of symbol   Smaller end of symbol — Smaller number

EXAMPLE 3   Comparing Integers, Using the < and > Symbols

Write < or > between each pair of numbers to make a true statement.

(a) 0 ___ 2

0 is to the *left* of 2 on the number line, so 0 is *less than* 2. Write $0 < 2$.

(b) 1 ___ ⁻4

1 is to the *right* of ⁻4, so 1 is *greater than* ⁻4. Write $1 > {}^-4$.

(c) ⁻4 ___ ⁻2

⁻4 is to the *left* of ⁻2, so ⁻4 is *less than* ⁻2. Write ${}^-4 < {}^-2$.

▶▶ **Work Problem 3 at the Side.** ▶▶

OBJECTIVE 4   Find the absolute value of integers.

In order to graph a number on the number line, you need to know two things:

1. Which *direction* it is from 0. It can be in a positive direction or a negative direction. You can tell the direction by looking for a positive sign (or no sign, which means positive), or a negative sign.

2. How *far* it is from 0. The distance from 0 is the *absolute value* of a number.

Absolute Value
The **absolute value** of a number is its distance from 0 on the number line. *Absolute value* is indicated by two vertical bars. For example,

$|6|$ is read   "the **absolute value** of 6."

---

3 Write < or > between each pair of numbers to make a true statement.

(a) 5 ___ 4

(b) 0 ___ 2

(c) ⁻3 ___ ⁻2

(d) ⁻1 ___ ⁻4

(e) 2 ___ ⁻2

(f) ⁻5 ___ 1

ANSWERS
3. (a) >   (b) <   (c) <   (d) >
    (e) >   (f) <

4 Find each absolute value.

(a) $|13|$

(b) $|{-7}|$

(c) $|0|$

(d) $|{-350}|$

(e) $|6000|$

The absolute value of a number will *always* be positive (or 0), because it is the *distance* from 0. A distance is never negative. (You wouldn't say that your living room is -16 feet long.) So absolute value concerns only *how far away* the number is from 0; we don't care which direction it is from 0.

EXAMPLE 4 Finding Absolute Values

Find each absolute value.

(a) $|4|$ The distance from 0 to 4 on the number line is 4 spaces. So, $|4| = 4$.

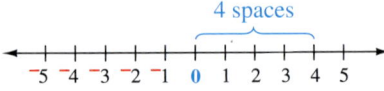

(b) $|{-4}|$ The distance from 0 to -4 on the number line is also 4 spaces. So, $|{-4}| = 4$.

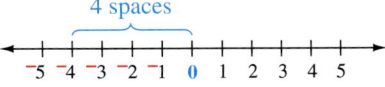

(c) $|0|$ $|0| = 0$ because the distance from 0 to 0 on the number line is 0 spaces.

Work Problem 4 at the Side.

ANSWERS
4. (a) 13   (b) 7   (c) 0   (d) 350   (e) 6000

1.2 Exercises

Write each negative number with a raised negative sign. Write each positive number in two ways. See Example 1.

1. Mount Everest, the tallest mountain in the world, rises 29,035 feet above sea level. (*Source: World Almanac.*)

2. The bottom of Lake Baikal in Siberia, Russia, is 5371 feet below the surface of the water. (*Source: Guinness Book of World Records.*)

3. The coldest temperature ever recorded on Earth is 128.6 degrees below zero in Antartica. (*Source: Fact Finder.*)

4. Normal body temperature is 98.6 degrees Fahrenheit, although it varies slightly for some people. (*Source:* Mayo Clinic *Health Letter.*)

5. During the first three plays of the football game, the Trojans lost a total of 18 yards.

6. The Jets gained 25 yards on a pass play.

7. Angelique won $100 in a prize drawing at the shopping mall.

8. Derice overdrew his checking account by $37.

9. Keith lost $6\frac{1}{2}$ pounds while he was sick with the flu.

10. The price of Mathtronic stock went up $2\frac{1}{2}$ dollars yesterday.

Graph each set of numbers. See Example 2.

11. $^-3, 3, 0, ^-5$

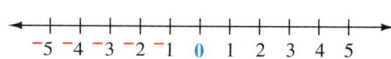

12. $^-2, 2, 0, 5$

13. $^-1, 4, ^-2, 5$

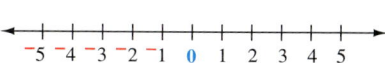

14. $3, ^-4, 1, ^-5$

15. $^-4\frac{1}{2}, \frac{1}{2}, 0, ^-8$

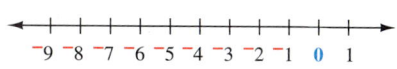

16. $^-7, 1\frac{1}{2}, -\frac{1}{2}, ^-9$

Write < or > between each pair of numbers to make a true statement. See Example 3.

17. 10 ___ 2
18. 6 ___ 0
19. ⁻1 ___ 0
20. ⁻3 ___ ⁻1

21. ⁻10 ___ 2
22. ⁻9 ___ 7
23. ⁻3 ___ ⁻6
24. 0 ___ ⁻1

25. ⁻10 ___ ⁻2
26. ⁻1 ___ ⁻5
27. 0 ___ ⁻8
28. 6 ___ ⁻4

29. 10 ___ ⁻2
30. ⁻2 ___ 1
31. ⁻4 ___ 4
32. 9 ___ ⁻9

Find each absolute value. See Example 4.

33. |15|
34. |10|
35. |⁻3|
36. |⁻8|

37. |0|
38. |100|
39. |200|
40. |⁻99|

41. |⁻75|
42. |⁻6320|
43. |⁻8042|
44. |0|

RELATING CONCEPTS (EXERCISES 45–48) For Individual or Group Work

*An ultrasound of a person's heel bone is a quick way to screen patients for osteoporosis (brittle bone disease). Use the information on the Patient Report Form to **work Exercises 45–48 in order.***

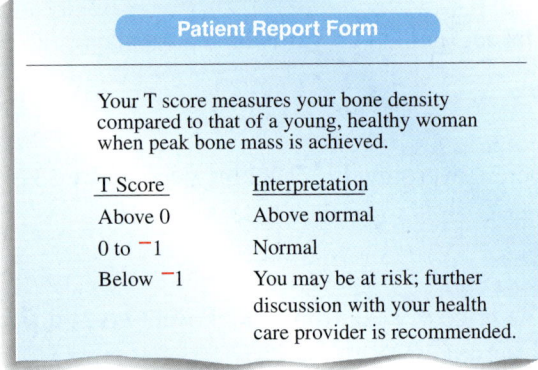

Patient Report Form

Your T score measures your bone density compared to that of a young, healthy woman when peak bone mass is achieved.

T Score	Interpretation
Above 0	Above normal
0 to ⁻1	Normal
Below ⁻1	You may be at risk; further discussion with your health care provider is recommended.

Source: Health Partners, Inc.

45. Here are the T scores for four patients. Draw a number line and graph the four scores.

    Patient A: ⁻1.5   Patient C: ⁻1
    Patient B: 0.5   Patient D: 0

46. List the patients' scores in order from lowest to highest.

47. What is the interpretation of each patient's score?

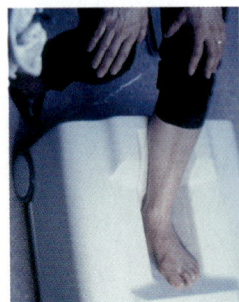

Taking an ultrasound of a patient's heel bone.

48. (a) What could happen if patient A did not understand the importance of a negative sign?

   (b) For which patient does the sign of the score make no difference? Explain your answer.

1.3 Adding Integers

OBJECTIVE 1 Add integers. Numbers that you are adding are called **addends,** and the result is called the **sum.** You can add integers while watching a football game. On each play, you can use a *positive* integer to stand for the yards *gained* by your team, and a *negative* integer for yards *lost.* Zero indicates no gain or loss. For example,

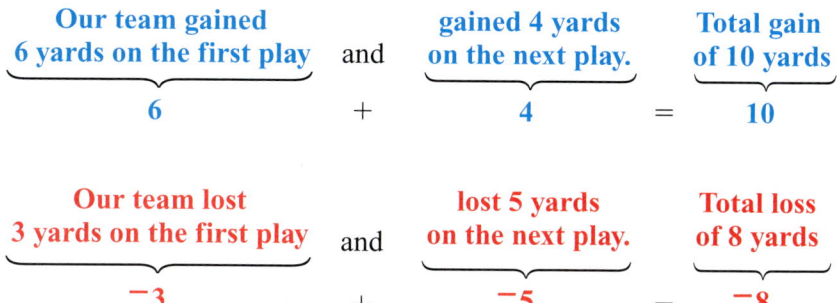

A drawing of a football field can help you add integers. Notice how similar the drawing is to a number line. Zero marks your team's starting point.

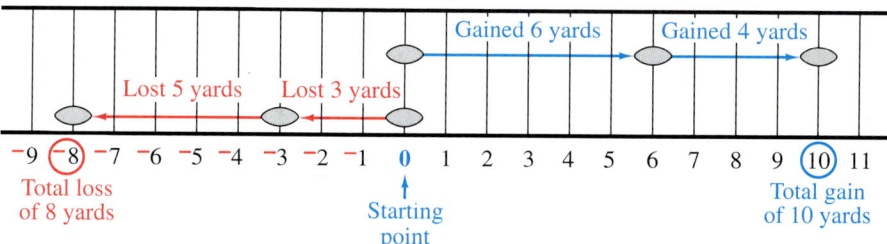

EXAMPLE 1 Using a Number Line to Add Integers

Use a number line to find $^-5 + {}^-4$.

Think of the number line as a football field. Your team starts at 0. On the first play it lost 5 yards. On the next play it lost 4 yards. The total loss is 9 yards.

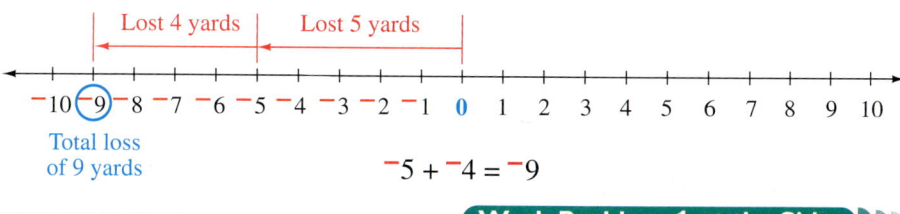

$$^-5 + {}^-4 = {}^-9$$

▶▶▶ **Work Problem 1 at the Side.** ▶▶▶

Do you see a pattern in the margin problems you just did? The answers to the first two problems are the same, *except for the sign.* The same is true for the next two problems and for the last two problems. This pattern leads to a rule for adding two integers when the signs are the same.

Adding Two Integers with the Same Sign

Step 1 *Add* the absolute values of the numbers.

Step 2 Use the *common sign* as the sign of the sum. If both numbers are positive, the sum is positive. If both numbers are negative, the sum is negative.

OBJECTIVES

1. Add integers.
2. Identify properties of addition.

1 Find each sum. Use the number line to help you.

(a) $^-2 + {}^-2$

(b) $2 + 2$

(c) $^-10 + {}^-1$

(d) $10 + 1$

(e) $^-3 + {}^-7$

(f) $3 + 7$

ANSWERS
1. (a) $^-4$ (b) 4 (c) $^-11$
   (d) 11 (e) $^-10$ (f) 10

2 Find each sum.

(a) $^-6 + {}^-6$

(b) $9 + 7$

(c) $^-5 + {}^-10$

(d) $^-12 + {}^-4$

(e) $13 + 2$

ANSWERS
2. (a) $^-12$ (b) 16 (c) $^-15$
   (d) $^-16$ (e) 15

EXAMPLE 2 Adding Two Integers with the Same Sign

Add.

(a) $^-8 + {}^-7$

  Step 1 Add the absolute values.

  $$|{}^-8| = 8 \quad \text{and} \quad |{}^-7| = 7$$

  Add $8 + 7$ to get 15.

  Step 2 Use the *common sign* as the sign of the sum. Both numbers are negative, so the sum is negative.

  $$^-8 + {}^-7 = {}^-15$$

  Both negative — Sum is negative.

(b) $3 + 6 = 9$

  Both positive — Sum is positive.

  In *Step 1*, when both numbers are positive, their absolute values are also positive, so we only need to show *Step 2*.

Work Problem 2 at the Side.

You can also use a number line (or drawing of a football field) to add integers with *different* signs. For example, suppose that your team gained 2 yards on the first play and then lost 7 yards on the next play. We can represent this as $2 + {}^-7$.

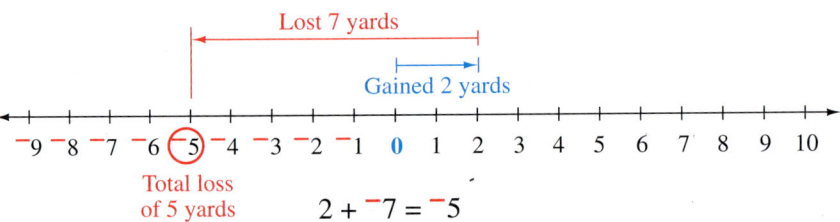

$2 + {}^-7 = {}^-5$

Or, try this one. On the first play your team gained 10 yards, but then it lost 4 yards on the next play. We can represent this as $10 + {}^-4$.

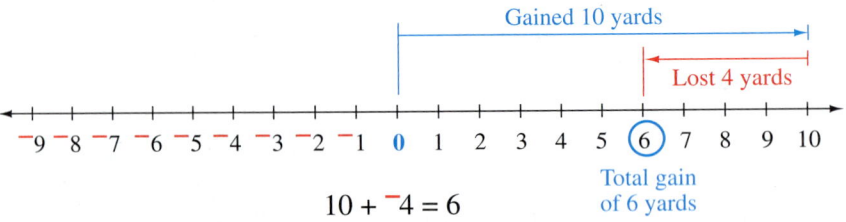

$10 + {}^-4 = 6$

These examples illustrate the rule for adding two integers with unlike, or different, signs.

Adding Two Integers with Unlike Signs

Step 1 *Subtract* the smaller absolute value from the larger absolute value.

Step 2 Use the sign of the number with the *larger absolute value* as the sign of the sum.

EXAMPLE 3 Adding Two Integers with Unlike Signs

Add.

(a) $^-8 + 3$

  Step 1 $|{^-8}| = 8$ and $|3| = 3$

  Subtract $8 - 3$ to get 5.

  Step 2 $^-8$ has the *larger absolute value* and is negative, so the sum is also negative.

  $$^-8 + 3 = {^-5}$$

(b) $^-5 + 11$

  Step 1 $|{^-5}| = 5$ and $|11| = 11$

  Subtract $11 - 5$ to get 6.

  Step 2 11 has the *larger absolute value* and is positive, so the sum is also positive.

  $$^-5 + 11 = {^+6} \text{ or } 6$$

Work Problem 3 at the Side.

EXAMPLE 4 Adding Several Integers

A football team has to gain at least 10 yards during four plays in order to keep the ball. Suppose that your college team lost 6 yards on the first play, gained 8 yards on the second play, lost 2 yards on the third play, and gained 7 yards on the fourth play. Did the team gain enough to keep the ball?

When you're adding several integers, work from left to right.

```
  1st play   2nd play        3rd play   4th play
     ↓          ↓                ↓         ↓
  ⎯⎯⎯⎯⎯⎯⎯⎯⎯⎯⎯⎯⎯⎯⎯
  ⁻6 yards + 8 yards   +   ⁻2 yards + 7 yards    First add ⁻6 + 8.
       2 yards         +   ⁻2 yards + 7 yards    Add 2 + ⁻2.
              0 yards              + 7 yards    Add 0 + 7.
                          7 yards
```

No, the team didn't gain enough yards to keep the ball.

Work Problem 4 at the Side.

OBJECTIVE 2 Identify properties of addition. In our football model for adding integers, we said that 0 indicated no gain or loss, that is, no change in the position of the ball. This example illustrates one of the *properties of addition*. A property of addition is something that applies to all addition problems, regardless of the specific numbers you use.

Addition Property of 0

Adding 0 to any number leaves the number unchanged.

Some examples are shown below.

$$0 + 6 = 6 \qquad ^-25 + 0 = {^-25} \qquad 72{,}399 + 0 = 72{,}399$$
$$0 + {^-100} = {^-100}$$

---

❸ Find each sum.

(a) $^-3 + 7$

(b) $6 + {^-12}$

(c) $12 + {^-7}$

(d) $^-10 + 2$

(e) $5 + {^-9}$

(f) $^-8 + 9$

❹ Write an addition problem and solve it for each situation.

(a) The temperature was $^-15$ degrees this morning. It rose 21 degrees during the day, then dropped 10 degrees. What is the new temperature?

(b) Andrew had $60 in his checking account. He wrote a $20 check for gas and a $75 check for groceries. Later in the day he deposited an $85 tax refund in his account. What is the balance in his account?

ANSWERS

3. (a) 4 (b) $^-6$ (c) 5
   (d) $^-8$ (e) $^-4$ (f) 1
4. (a) $^-15 + 21 + {^-10} = {^-4}$ degrees
   (b) $60 + {^-20} + {^-75} + 85 = \$50$

Chapter 1 Introduction to Algebra: Integers

5 Rewrite each sum using the commutative property of addition. Check that the sum is unchanged.

(a) $175 + 25 =$ __ $+$ __

Both sums are ____.

(b) $7 + {}^-37 =$ __ $+$ __

Both sums are ____.

(c) ${}^-16 + 16 =$ __ $+$ __

Both sums are ____.

(d) ${}^-9 + {}^-41 =$ __ $+$ __

Both sums are ____.

Another property of addition is that you can change the *order* of the addends and still get the same sum. For example,

Gaining 2 yards, then losing 7 yards, gives a result of ${}^-5$ yards.

Losing 7 yards, then gaining 2 yards, also gives a result of ${}^-5$ yards.

Commutative Property of Addition

Changing the *order* of two addends does *not* change the sum.

Here are some examples.

$$84 + 2 = 2 + 84 \quad \text{Both sums are 86.}$$

$${}^-10 + 6 = 6 + {}^-10 \quad \text{Both sums are } {}^-4.$$

EXAMPLE 5 Using the Commutative Property of Addition

Rewrite each sum, using the **commutative property of addition.** Check that the sum is unchanged.

(a) $65 + 35$

$$65 + 35 = 35 + 65$$
$$100 = 100$$

Both sums are 100, so the sum is unchanged.

(b) ${}^-20 + {}^-30$

$${}^-20 + {}^-30 = {}^-30 + {}^-20$$
$${}^-50 = {}^-50$$

Both sums are ${}^-50$, so the sum is unchanged.

◀◀ Work Problem 5 at the Side.

When there are three addends, parentheses may be used to tell you which pair of numbers to add first, as shown below.

$(3 + 4) + 2$ First add $3 + 4$. $3 + (4 + 2)$ First add $4 + 2$.
$7 + 2$ Then add $7 + 2$. $3 + 6$ Then add $3 + 6$.
9 9

Both sums are 9. This example illustrates another property of addition.

Associative Property of Addition

Changing the *grouping* of addends does *not* change the sum.

Some examples are shown below.

$({}^-5 + 5) + 8 = {}^-5 + (5 + 8)$ | $3 + ({}^-4 + {}^-6) = (3 + {}^-4) + {}^-6$
$0 + 8 = {}^-5 + 13$ | $3 + {}^-10 = {}^-1 + {}^-6$
$8 = \phantom{{}^-5 + 1}8$ | ${}^-7 = {}^-7$

Both sums are 8. | Both sums are ${}^-7$.

ANSWERS
5. (a) $175 + 25 = 25 + 175$; 200
   (b) $7 + {}^-37 = {}^-37 + 7$; ${}^-30$
   (c) ${}^-16 + 16 = 16 + {}^-16$; 0
   (d) ${}^-9 + {}^-41 = {}^-41 + {}^-9$; ${}^-50$

We can use the associative property to make addition problems easier. Notice in the first example in the box on the previous page that it is easier to group $^{-}5 + 5$ (which is 0) and then add 8. In the second example in the box, it is helpful to group $^{-}4 + {}^{-}6$ because the sum is $^{-}10$, and it is easy to work with multiples of 10. (See **Section R.3** for more on multiples of 10.)

EXAMPLE 6 Using the Associative Property of Addition

In each addition problem, pick out the two addends that would be easiest to add. Write parentheses around those addends. Then find the sum.

(a) $6 + 9 + {}^{-}9$

Group $9 + {}^{-}9$ because the sum is 0.

$$6 + (9 + {}^{-}9)$$
$$6 + 0$$
$$6$$

(b) $17 + 3 + {}^{-}25$

Group $17 + 3$ because the sum is 20, which is a multiple of 10.

$$(17 + 3) + {}^{-}25$$
$$20 + {}^{-}25$$
$$^{-}5$$

> **NOTE**
> When using the associative property to make the addition of a group of numbers easier:
>
> 1. Look for two numbers whose sum is 0.
> 2. Look for two numbers whose sum is a multiple of 10 (the sum ends in 0, such as 10, 20, 30, or $^{-}100$, $^{-}200$, etc.).
>
> If neither of these occurs, look for two numbers that are easier for you to add. For example, in $98 + 42 + 5$, you may find that adding $42 + 5$ is easier than adding $98 + 5$.

Work Problem 6 at the Side.

---

6 In each problem, write parentheses around the two addends that would be easiest to add. Then find the sum.

(a) $^{-}12 + 12 + {}^{-}19$

(b) $31 + 75 + {}^{-}75$

(c) $16 + {}^{-}1 + {}^{-}9$

(d) $^{-}8 + 5 + {}^{-}25$

ANSWERS

6. (a) $(^{-}12 + 12) + {}^{-}19$
$\qquad 0 + {}^{-}19$
$\qquad {}^{-}19$

(b) $31 + (75 + {}^{-}75)$
$\qquad 31 + 0$
$\qquad 31$

(c) $16 + ({}^{-}1 + {}^{-}9)$
$\qquad 16 + {}^{-}10$
$\qquad 6$

(d) $^{-}8 + (5 + {}^{-}25)$
$\qquad {}^{-}8 + {}^{-}20$
$\qquad {}^{-}28$

Focus on Real-Data Applications

Auto Aide

The *Auto Aide Service* is a hypothetical business located in downtown Houston, Texas. It specializes in assisting motorists who have minor mechanical problems, such as a flat tire or a dead battery. The I-10 corridor in the greater Houston area is a major highway with constant heavy traffic, so the company assigns five vans to concentrate on I-10 assistance needs. The owner is a retired mathematics teacher who believes in using integers to track the actions of the auto aides along their service routes.

A schematic map of the I-10 corridor is shown below. Each tick mark represents 1 mile. The home office is located at 0. Locations to the west are represented by negative integers, and locations to the east are represented by positive integers.

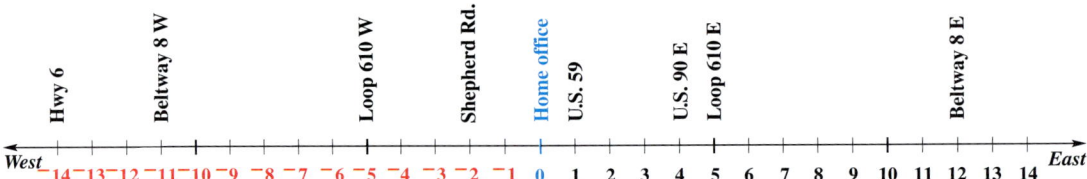

A radio dispatcher notifies the aide to assist a motorist at a given location. (Only a few locations are identified in this hypothetical problem, for simplicity.) Each aide must include four calculations on the daily report. First, the aide must translate the route instructions into an integer sentence that gives the direction and miles between locations. Second, the aide must record the displacement at the end of the route (location relative to home office). Third, the aide must record the distance from the home office at the end of the route. Fourth, the aide must record the total distance traveled on the route.

Aide Anne is shown as an example. Complete the table for the remaining aides.

Aide	Route Instructions	Integer Sentence	Displacement (final location)	Distance from Home Office	Total Distance
Anne	Home office to U.S. 90 E; to Loop 610 W; to Hwy 6; to Shepherd Rd.	4 + ⁻9 + ⁻9 + 12	⁻2 (Shepherd) (*Hint:* Find the integer sum.)	2 miles (*Hint:* Find the absolute value of displacement.)	34 miles (*Hint:* Find sum of absolute values of trip segments.)
Bill	Home office to Hwy 6; to U.S. 59; to Shepherd Rd.; to Loop 610 W				
Carlos	Home office to Shepherd Rd.; to Loop 610 W; to Beltway 8 W; to Hwy 6				
Dylan	Home office to Beltway 8 E; to U.S. 90 E; to Home office				
Ellen	Home office to U.S. 59; to Shepherd Rd.; to U.S. 59; to Loop 610 W; to U.S. 59				

1.3 Exercises

Add by using the number line. See Example 1.

1. $^-2 + 5$

2. $^-3 + 4$

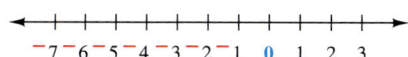

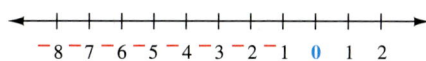

3. $^-5 + {}^-2$

4. $^-2 + {}^-2$

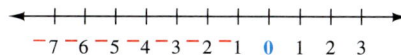

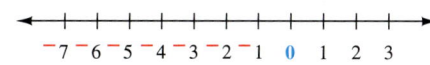

5. $3 + {}^-4$

6. $5 + {}^-1$

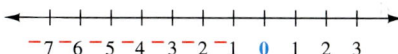

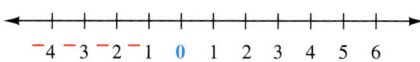

Add. See Example 2.

7. (a) $^-5 + {}^-5$
    (b) $5 + 5$

8. (a) $^-9 + {}^-9$
    (b) $9 + 9$

9. (a) $7 + 5$
    (b) $^-7 + {}^-5$

10. (a) $3 + 6$
    (b) $^-3 + {}^-6$

11. (a) $^-25 + {}^-25$
    (b) $25 + 25$

12. (a) $^-30 + {}^-30$
    (b) $30 + 30$

13. (a) $48 + 110$
    (b) $^-48 + {}^-110$

14. (a) $235 + 21$
    (b) $^-235 + {}^-21$

15. What pattern do you see in your answers to Exercises 7–14? Explain why this pattern occurs.

16. In your own words, explain how to add two integers that have the same sign.

Add. See Example 3.

17. (a) ⁻6 + 8
    (b) 6 + ⁻8

18. (a) ⁻3 + 7
    (b) 3 + ⁻7

19. (a) ⁻9 + 2
    (b) 9 + ⁻2

20. (a) ⁻8 + 7
    (b) 8 + ⁻7

21. (a) 20 + ⁻25
    (b) ⁻20 + 25

22. (a) 30 + ⁻40
    (b) ⁻30 + 40

23. (a) 200 + ⁻50
    (b) ⁻200 + 50

24. (a) 150 + ⁻100
    (b) ⁻150 + 100

25. What pattern do you see in your answers to Exercises 17–24? Explain why this pattern occurs.

26. In your own words, explain how to add two integers that have different signs.

Add. See Examples 2–4.

27. ⁻8 + 5

28. ⁻3 + 2

29. ⁻1 + 8

30. ⁻4 + 10

31. ⁻2 + ⁻5

32. ⁻7 + ⁻3

33. 6 + ⁻5

34. 11 + ⁻3

35. 4 + ⁻12

36. 9 + ⁻10

37. ⁻10 + ⁻10

38. ⁻5 + ⁻20

39. ⁻17 + 0

40. 0 + ⁻11

41. 1 + ⁻23

42. 13 + ⁻1

43. ⁻2 + ⁻12 + ⁻5

44. ⁻16 + ⁻1 + ⁻3

45. 8 + 6 + ⁻8

46. ⁻5 + 2 + 5   **47.** ⁻7 + 6 + ⁻4   **48.** ⁻9 + 8 + ⁻2

49. ⁻3 + ⁻11 + 14   **50.** 15 + ⁻7 + ⁻8   **51.** 10 + ⁻6 + ⁻3 + 4

52. 2 + ⁻1 + ⁻9 + 12   **53.** ⁻7 + 28 + ⁻56 + 3   **54.** 4 + ⁻37 + 29 + ⁻5

Write an addition problem for each situation and find the sum.

55. The football team gained 13 yards on the first play and lost 17 yards on the second play. How many yards did the team gain or lose in all?

56. At penguin breeding grounds on Antarctic islands, temperatures routinely drop to ⁻15 °C. Temperatures in the interior of the continent may drop another 60 °C below that. What is the temperature in the interior?

57. Nick's checking account was overdrawn by $62. He deposited $50 in his account. What is the balance in his account?

58. Cynthia had $100 in her checking account. She wrote a check for $83 and was charged $17 for overdrawing her account last month. What is her account balance?

59. $88 was stolen from Jay's car. He got $35 of it back. What was his net loss?

60. Marion lost 4 pounds in April, gained 2 pounds in May, and gained 3 pounds in June. How many pounds did she gain or lose in all?

61. Use the score sheet to find each player's point total after three rounds in a card game.

	Jeff	Terry
Round 1	Lost 20 pts	Won 42 pts
Round 2	Won 75 pts	Lost 15 pts
Round 3	Lost 55 pts	Won 20 pts

62. Use the information in the table on flood water depths to find the new flood level for each river.

	Red River	Mississippi
Monday	Rose 8 ft	Rose 4 ft
Tuesday	Fell 3 ft	Rose 7 ft
Wednesday	Fell 5 ft	Fell 13 ft

At the 2003 U.S. Women's Open golf tournament in North Plains, Oregon, 71 strokes was the "par" score for each round of play. A negative score indicates that the player had fewer than 71 strokes for that round, and a positive score indicates more than 71 strokes. Find the total score for each of these players in the tournament.

	Player	Round 1	Round 2	Round 3	Round 4	Total
63.	Hilary Lunke	0	⁻2	⁻3	4	
64.	Kelly Robbins	3	⁻2	0	⁻2	
65.	Angela Stanford	⁻1	⁻1	⁻2	3	
66.	Annika Sorenstam	1	1	⁻4	2	

Source: Associated Press.

Hilary Lunke won the 2003 U.S. Women's Open in a playoff round.

Rewrite each sum, using the commutative property of addition. Show that the sum is unchanged. See Example 5.

67. ⁻18 + ⁻5 = ____ + ____

    Both sums are ____.

68. ⁻12 + 20 = ____ + ____

    Both sums are ____.

69. ⁻4 + 15 = ____ + ____

    Both sums are ____.

70. 17 + 1 = ____ + ____

    Both sums are ____.

In each addition problem, write parentheses around the two addends that would be easiest to add. Then find the sum. See Example 6.

71. 6 + ⁻14 + 14

72. 9 + ⁻9 + ⁻8

73. ⁻14 + ⁻6 + ⁻7

74. ⁻18 + 3 + 7

75. Make up three of your own examples that illustrate the addition property of 0.

76. Make up three of your own examples that illustrate the associative property of addition. Show that the sum is unchanged.

Find each sum.

77. ⁻7081 + 2965

78. ⁻1398 + 3802

79. ⁻179 + ⁻61 + 8926

80. 36 + ⁻6215 + 428

81. 86 + ⁻99,000 + 0 + 2837

82. ⁻16,719 + 0 + 8878 + ⁻14

1.4 Subtracting Integers

OBJECTIVE 1 Find the opposite of a signed number. Look at how the integers match up on this number line.

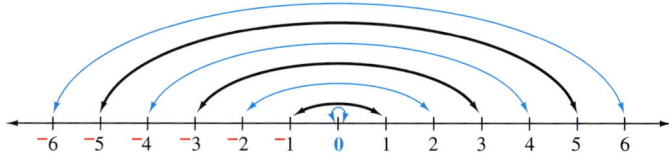

Each integer is matched with its *opposite*. **Opposites** are the same *distance* from 0 on the number line but are on *opposite sides* of 0.

$^+2$ is the opposite of $^-2$ and $^-2$ is the opposite of $^+2$

When you add opposites, the sum is always 0. The opposite of a number is also called its *additive inverse*.

$$2 + {}^-2 = 0 \quad \text{and} \quad {}^-2 + 2 = 0$$

EXAMPLE 1 Finding the Opposites of Signed Numbers

Find the opposite (additive inverse) of each number. Show that the sum of the number and its opposite is 0.

(a) 6   The opposite of 6 is $^-6$ and $6 + {}^-6 = 0$
(b) $^-10$ The opposite of $^-10$ is 10 and ${}^-10 + 10 = 0$
(c) 0   The opposite of 0 is 0   and $0 + 0 = 0$

▶▶▶ Work Problem 1 at the Side. ▶▶▶

OBJECTIVE 2 Subtract integers. Now that you know how to add integers and how to find opposites, you can subtract integers. Every subtraction problem has the same answer as a related addition problem. The problems below illustrate how to change subtraction problems into addition problems.

$$6 - 2 = 4 \quad\quad 8 - 3 = 5$$
$$\downarrow\ \downarrow \quad\quad\quad\quad \downarrow\ \downarrow$$
$$6 + {}^-2 = 4 \quad\quad 8 + {}^-3 = 5$$

Same answer   Same answer

Subtracting Two Integers

To subtract two numbers, *add* the first number to the *opposite* of the second number. Remember to change *two* things:

Step 1 Make one pencil stroke to change the subtraction symbol to an addition symbol.

Step 2 Make a second pencil stroke to change the *second* number to its *opposite*. If the second number is positive, change it to negative. If the second number is negative, change it to positive.

CAUTION
When changing a subtraction problem to an addition problem, do *not* make any change in the *first* number. The pattern is

1st number − 2nd number = 1st number + opposite of 2nd number.

OBJECTIVES
1. Find the opposite of a signed number.
2. Subtract integers.
3. Combine adding and subtracting of integers.

1 Find the additive inverse (opposite) of each number. Show that the sum of the number and its additive inverse is 0.

(a) 5

(b) 48

(c) 0

(d) $^-1$

(e) $^-24$

ANSWERS
1. (a) $^-5$; $5 + {}^-5 = 0$
   (b) $^-48$; $48 + {}^-48 = 0$
   (c) 0; $0 + 0 = 0$
   (d) 1; $^-1 + 1 = 0$
   (e) 24; $^-24 + 24 = 0$

2 Subtract by changing subtraction to adding the opposite. (Make *two* pencil strokes.)

(a) $^-6 - 5$

(b) $3 - {^-10}$

(c) $^-8 - {^-2}$

(d) $0 - 10$

(e) $^-4 - {^-12}$

(f) $9 - 7$

3 Simplify.

(a) $6 - 7 + {^-3}$

(b) $^-2 + {^-3} - {^-5}$

(c) $7 - 7 - 7$

(d) $^-3 - 9 + 4 - {^-20}$

ANSWERS
2. (a) $^-11$ (b) 13 (c) $^-6$ (d) $^-10$
 (e) 8 (f) 2
3. (a) $^-4$ (b) 0 (c) $^-7$ (d) 12

EXAMPLE 2 Subtracting Two Integers

Make *two* pencil strokes to change each subtraction problem into an addition problem. Then find the sum.

(a) Change 10 to $^-10$.
$$4 - 10 = 4 + {^-10} = {^-6}$$
Change subtraction to addition.

(b) Change $^-6$ to $^+6$.
$$^-9 - {^-6} = {^-9} + {^+6} = {^-3}$$
Change subtraction to addition.

(c) Change $^-5$ to $^+5$.
$$3 - {^-5} = 3 + {^+5} = 8 \quad \text{Make } two \text{ pencil strokes.}$$
Change subtraction to addition.

(d) Change 9 to $^-9$.
$$^-2 - 9 = {^-2} + {^-9} = {^-11} \quad \text{Make } two \text{ pencil strokes.}$$
Change subtraction to addition.

Work Problem 2 at the Side.

OBJECTIVE 3 Combine adding and subtracting of integers. When adding and subtracting more than two signed numbers, first change all subtractions to additions. Then add from left to right.

EXAMPLE 3 Combining Addition and Subtraction

Simplify by completing all the calculations.

$$^-5 - 10 - 12 + 1$$
Change all subtractions to addition. Change 10 to $^-10$. Change 12 to $^-12$.

$$^-5 + {^-10} + {^-12} + 1$$
Add from left to right. First add $^-5 + {^-10}$.

$$\underbrace{^-15} + {^-12} + 1$$
Then add $^-15 + {^-12}$.

$$\underbrace{^-27} + 1$$
Finally, add $^-27 + 1$.

$$^-26$$

Work Problem 3 at the Side.

🖩 **Calculator Tip** You can use the *change of sign* key (+/−) or (+/−) on your *scientific* calculator to enter negative numbers. To enter $^-5$, press ⑤ (+/−). To enter $^+5$, just press ⑤. To enter Example 3 above, press the following keys.

5 (+/−) (−) 10 (−) 12 (+) 1 (=)   The answer is $^-26$.

$^-5$ Subtract.

When using a calculator, you do *not* need to change subtraction to addition.

1.4 Exercises

Find the opposite (additive inverse) of each number. Show that the sum of the number and its opposite is 0. See Example 1.

1. 6
2. 10
3. ⁻13

4. ⁻3
5. 0
6. 1

Subtract by changing subtraction to addition. See Example 2.

7. 19 − 5
8. 24 − 11
9. 10 − 12
10. 1 − 8

11. 7 − 19
12. 2 − 17
13. ⁻15 − 10
14. ⁻10 − 4

15. ⁻9 − 14
16. ⁻3 − 11
17. ⁻3 − ⁻8
18. ⁻1 − ⁻4

19. 6 − ⁻14
20. 8 − ⁻1
21. 1 − ⁻10
22. 6 − ⁻1

23. ⁻30 − 30
24. ⁻25 − 25
25. ⁻16 − ⁻16
26. ⁻20 − ⁻20

27. 13 − 13
28. 19 − 19
29. 0 − 6
30. 0 − 12

31. (a) 3 − ⁻5
    (b) 3 − 5
    (c) ⁻3 − ⁻5
    (d) ⁻3 − 5

32. (a) 9 − 6
    (b) ⁻9 − 6
    (c) 9 − ⁻6
    (d) ⁻9 − ⁻6

33. (a) 4 − 7
    (b) 4 − ⁻7
    (c) ⁻4 − 7
    (d) ⁻4 − ⁻7

34. (a) 8 − ⁻2
    (b) ⁻8 − ⁻2
    (c) 8 − 2
    (d) ⁻8 − 2

Simplify. See Example 3.

35. ⁻2 − 2 − 2
36. ⁻8 − 4 − 8
37. 9 − 6 − 3 − 5

38. 12 − 7 − 5 − 4
39. 3 − ⁻3 − 10 − ⁻7
40. 1 − 9 − ⁻2 − ⁻6

41. ⁻2 + ⁻11 − ⁻3
42. ⁻5 − ⁻2 + ⁻6
43. 4 − ⁻13 + ⁻5

44. 6 − ⁻1 + ⁻10
45. 6 + 0 − 12 + 1
46. ⁻10 − 4 + 0 + 18

WINDCHILL
Temperature (degrees Fahrenheit)

Wind Speed (miles per hour) \ Calm	40	35	30	25	20	15	10	5	0	−5	−10	−15	−20	−25	−30
5	36	31	25	19	13	7	1	−6	−11	−16	−22	−28	−34	−40	−46
10	34	27	21	15	9	3	−4	−10	−16	−22	−28	−35	−41	−47	−53
15	32	25	19	13	6	0	−7	−13	−19	−26	−32	−39	−45	−51	−58
20	30	24	17	11	4	−2	−9	−15	−22	−29	−35	−42	−48	−55	−61
25	29	23	16	9	3	−4	−11	−17	−24	−31	−37	−44	−51	−58	−64
30	28	22	15	8	1	−5	−12	−19	−26	−33	−39	−46	−53	−60	−67
35	28	21	14	7	0	−7	−14	−21	−27	−34	−41	−48	−55	−62	−69
40	27	20	13	6	−1	−8	−15	−22	−29	−36	−43	−50	−57	−64	−71

Shaded area: Frostbite occurs in 15 minutes or less.

Source: National Weather Service

This windchill table shows how wind increases a person's heat loss. For example, find the temperature of 15 °F along the top of the table. Then find a wind speed of 20 mph along the left side of the table. This column and row intersect at ⁻2 °F, the "windchill temperature." The actual temperature is 15 °F but the wind makes it feel like ⁻2 °F. The difference between the actual temperature and the windchill temperature is $15 - {}^-2 = 15 + {}^+2 = 17$ degrees difference.

Use the table to find the windchill temperature under each set of conditions in Exercises 47–48. Then write and solve a subtraction problem to calculate actual temperature minus windchill temperature.

47. (a) 30 °F; 10 mph wind

(b) 15 °F; 15 mph wind

(c) 5 °F; 25 mph wind

(d) ⁻10 °F; 35 mph wind

48. (a) 40 °F; 20 mph wind

(b) 20 °F; 35 mph wind

(c) 10 °F; 15 mph wind

(d) ⁻5 °F; 30 mph wind

49. Find, correct, and explain the mistake made in this subtraction.

$$\begin{array}{c} {}^-6 - 6 \\ \downarrow \ \downarrow \\ {}^-6 + 6 = 0 \end{array}$$

50. Find, correct, and explain the mistake made in this subtraction.

$$\begin{array}{c} {}^-7 - \ \ 5 \\ \downarrow \ \ \ \downarrow \\ {}^+7 + {}^-5 = 2 \end{array}$$

Simplify. Begin each exercise by working inside the absolute value bars or the parentheses.

51. ⁻2 + ⁻11 + |⁻2|

52. 5 − |⁻3| + 3

53. 0 − |⁻7 + 2|

54. |1 − 8| − |0|

55. ⁻3 − (⁻2 + 4) + ⁻5

56. 5 − 8 − (6 − 7) + 1

RELATING CONCEPTS (EXERCISES 57–58) For Individual or Group Work

*Use your knowledge of the properties of addition to **work Exercises 57 and 58 in order**.*

57. Look for a pattern in these pairs of subtractions.

⁻3 − 5 = _____   ⁻4 − ⁻3 = _____

5 − ⁻3 = _____   ⁻3 − ⁻4 = _____

Explain what happens when you try to apply the commutative property to subtraction.

58. Recall the addition property of 0. Can 0 be used in a subtraction problem without changing the other number? Explain what happens and give several examples. (*Hint:* Think about *order* in a subtraction problem.)

1.5 Problem Solving: Rounding and Estimating

One way to get a rough check on an answer is to *round* the numbers in the problem. **Rounding** a number means finding a number that is close to the original number, but easier to work with.

For example, a superintendent of schools in a large city might be discussing the need to build new schools. In making her point, it probably would not be necessary to say that the school district has 152,807 students—it probably would be sufficient to say that there are 153,000 students, or even 150,000 students.

OBJECTIVE 1 Locate the place to which a number is to be rounded. The first step in rounding a number is to locate the *place to which the number is to be rounded*.

EXAMPLE 1 Finding the Place to Which a Number Is to Be Rounded

Locate and draw a line under the place to which each number is to be rounded. Then answer the question.

(a) Round −23 to the nearest ten. Is −23 closer to −20 or −30?

−23 is closer to −20

Tens place

(b) Round $381 to the nearest hundred. Is it closer to $300 or $400?

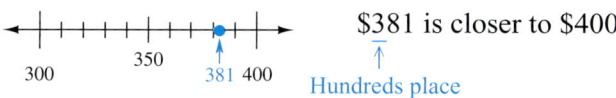

$381 is closer to $400

Hundreds place

(c) Round −54,702 to the nearest thousand. Is it closer to −54,000 or −55,000?

−54,702 is closer to −55,000

Thousands place

▶▶▶ **Work Problem 1 at the Side.**

OBJECTIVE 2 Round integers. Use the following steps for rounding integers.

Rounding an Integer
Step 1 Locate the *place* to which the number is to be rounded. Draw a line under that place.

Step 2 Look only at the next digit to the right of the one you underlined. If the next digit is *5 or more*, increase the underlined digit by 1. If the next digit is *4 or less*, do *not* change the digit in the underlined place.

Step 3 Change all digits to the right of the underlined place to zeros.

CAUTION
If you are rounding a negative number, be careful to write the negative sign in front of the rounded number. For example, −79 rounds to −80.

OBJECTIVES
1. Locate the place to which a number is to be rounded.
2. Round integers.
3. Use front end rounding to estimate answers in addition and subtraction.

1 Locate and draw a line under the place to which the number is to be rounded. Then answer the question.

(a) −746 (nearest ten)

Is it closer to −740 or −750? _____

(b) 2412 (nearest thousand)

Is it closer to 2000 or 3000? _____

(c) −89,512 (nearest hundred)

Is it closer to −89,500 or −89,600? _____

(d) 546,325 (nearest ten-thousand)

Is it closer to 540,000 or 550,000? _____

ANSWERS
1. (a) −7<u>4</u>6 is closer to −750
   (b) <u>2</u>412 is closer to 2000
   (c) −89,<u>5</u>12 is closer to −89,500
   (d) 54<u>6</u>,325 is closer to 550,000

2 Round to the nearest ten.

(a) 34

(b) ⁻61

(c) ⁻683

(d) 1792

3 Round to the nearest thousand.

(a) 1725

(b) ⁻6511

(c) 58,829

(d) ⁻83,904

ANSWERS
2. (a) 30 (b) ⁻60 (c) ⁻680 (d) 1790
3. (a) 2000 (b) ⁻7000 (c) 59,000
   (d) ⁻84,000

EXAMPLE 2 Using the Rounding Rule for 4 or Less

Round 349 to the nearest hundred.

Step 1 Locate the place to which the number is being rounded. Draw a line under that place.

$$349$$
↑
└──── Hundreds place

Step 2 Because the next digit to the right of the underlined place is 4, which is *4 or less*, do *not* change the digit in the underlined place.

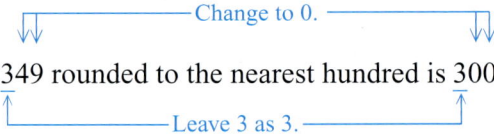

Step 3 Change all digits to the right of the underlined place to zeros.

┌──── Change to 0. ────┐
⇓⇓
349 rounded to the nearest hundred is 300
↑   ↑
└──────── Leave 3 as 3. ──────────┘

In other words, 349 is closer to 300 than to 400.

◀◀◀ **Work Problem 2 at the Side.**

EXAMPLE 3 Using the Rounding Rule for 5 or More

Round 36,833 to the nearest thousand.

Step 1 Find the place to which the number is to be rounded. Draw a line under that place.

Step 2 Because the next digit to the right of the underlined place is 8, which is *5 or more*, add 1 to the underlined place.

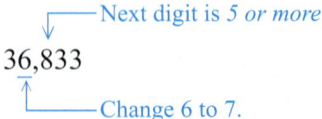

Step 3 Change all digits to the right of the underlined place to zeros.

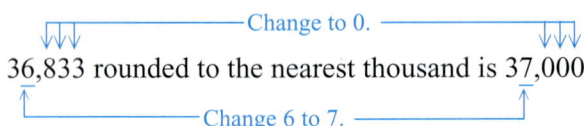

In other words, 36,833 is closer to 37,000 than to 36,000.

◀◀◀ **Work Problem 3 at the Side.**

EXAMPLE 4 Using the Rules for Rounding

(a) Round ⁻2382 to the nearest ten.

Step 1 ⁻2382
         ↑
         └─ Tens place

Step 2 The next digit to the right is 2, which is *4 or less*.

┌─ Next digit is *4 or less*.
        ↓
⁻2382
    ↑
    └─ Leave 8 as 8.

Step 3   ┌─ Change to 0. ─┐
           ↓   ↓
   ⁻2382 rounds to ⁻2380
    ↑   ↑
    └─ Leave 8 as 8. ─┘

⁻2382 rounded to the nearest ten is ⁻2380.

(b) Round 13,961 to the nearest hundred.

Step 1 13,961
           ↑
           └─ Hundreds place

Step 2 The next digit to the right is 6, which is *5 or more*.

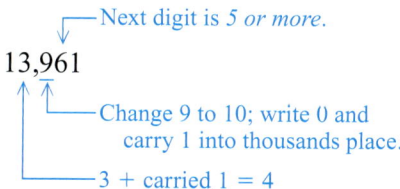

Step 3   ┌─ Change to 0. ─┐
            ↓   ↓
    13,961 rounds to 14,000
       ↑   ↑
       └─ Change 39 to 40. ─┘

13,961 rounded to the nearest hundred is 14,000.

> **NOTE**
> In Step 2 above, when you added 1 to the hundreds place, notice that the first three digits increased from 139 to 140.
>
> 13,**9**61   rounded to   14,**0**00

Work Problem 4 at the Side.

4 Round as indicated.

(a) ⁻6036 to the nearest ten

(b) 31,968 to the nearest hundred

(c) ⁻73,077 to the nearest thousand

(d) 4952 to the nearest thousand

(e) 85,949 to the nearest hundred

(f) 40,387 to the nearest thousand

ANSWERS
4. (a) ⁻6040 (b) 32,000 (c) ⁻73,000
   (d) 5000 (e) 85,900 (f) 40,000

32 Chapter 1 Introduction to Algebra: Integers

5 Round as indicated.

(a) ⁻14,679 to the nearest ten-thousand

(b) 724,518,715 to the nearest million

(c) ⁻49,900,700 to the nearest million

(d) 306,779,000 to the nearest hundred-million

ANSWERS
5. (a) ⁻10,000 (b) 725,000,000
   (c) ⁻50,000,000 (d) 300,000,000

EXAMPLE 5 Rounding Large Numbers

(a) Round ⁻37,892 to the nearest ten-thousand.

Step 1   ⁻37,892
              ↑
              └── Ten-thousands place

Step 2 The next digit to the right is 7, which is *5 or more.*

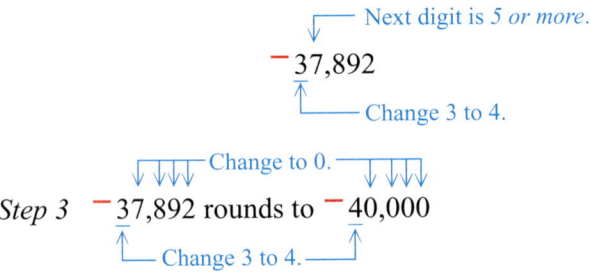

Step 3 ⁻37,892 rounds to ⁻40,000.

⁻37,892 rounded to the nearest ten-thousand is ⁻40,000.

(b) Round 528,498,675 to the nearest million.

Step 1 528,498,675
           ↑
           └── Millions place

                        ┌── Next digit is *4 or less.*
                        ↓
Step 2 528,498,675
           ↑
           └── Leave 8 as 8.

                           ┌── Change to 0. ──┐
                           ↓↓↓ ↓↓↓
Step 3 528,498,675 rounds to 528,000,000
           ↑
           └── Leave 8 as 8. ──┘

528,498,675 rounded to the nearest million is 528,000,000.

◀◀◀ **Work Problem 5 at the Side.**

OBJECTIVE 3 Use front end rounding to estimate answers in addition and subtraction. In many everyday situations, we can round numbers and **estimate** the answer to a problem. For example, suppose that you're thinking about buying a sofa for $988 and a chair for $209. You can round the prices and estimate the total cost as $1000 + $200 ≈ $1200. The ≈ symbol means "approximately equal to." The estimated total of $1200 is close enough to help you decide whether you can afford both items. Of course, when it comes time to pay the bill, you'll want the *exact* total of $988 + $209 = $1197.

Sofa		Chair			
$988	+	$209	=	$1197	← Exact cost
Rounds to		Rounds to			
↓		↓			
$1000	+	$200	=	$1200	← Estimated cost

Front end rounding is often used to estimate answers. Each number is rounded to the highest possible place, so all the digits become 0 except the first digit. Once the numbers have lots of zeros, working with them is easy.

EXAMPLE 6 Using Front End Rounding

Use front end rounding to round each number.

(a) $^-216$

Round to the highest possible place, that is, the leftmost digit. In this case, the leftmost digit, 2, is in the hundreds place, so round to the nearest hundred.

$$^-216 \quad\quad ^-216 \text{ rounds to } ^-200$$

- Next digit is *4 or less*. Leave 2 as 2.
- Change to 0. Leave 2 as 2.

The rounded number is $^-200$. Notice that all the digits in the rounded number are 0, except the first digit. Also, remember to write the negative sign.

(b) 97,203

The leftmost digit, 9, is in the ten-thousands place, so round to the nearest ten-thousand.

$$97{,}203 \quad\quad 97{,}203 \text{ rounds to } 100{,}000$$

- Next digit is *5 or more*. Change 9 to 10.
- Change to 0. Change 9 to 10. Carry 1 into the hundred-thousands place.

The rounded number is 100,000. Notice that all the digits in the rounded number are 0, except the first digit.

Work Problem 6 at the Side.

EXAMPLE 7 Using Front End Rounding to Estimate an Answer

Use front end rounding to estimate an answer. Then find the exact answer.

Meisha's paycheck showed gross pay of $823. It also listed deductions of $291. What is her net pay after deductions?

Estimate: Use front end rounding to round $823 and $291.

$823 rounds to $800 $291 rounds to $300
- Next digit is *4 or less*. Leave 8 as 8.
- Next digit is *5 or more*. Change 2 to 3.

Use the rounded numbers and subtract to *estimate* Meisha's net pay.

$$\$800 - \$300 = \$500 \leftarrow \text{Estimate}$$

Exact: Use the original numbers and subtract to find the *exact* amount.

$$\$823 - \$291 = \$532 \leftarrow \text{Exact}$$

Meisha's paycheck will show the *exact* amount of $532. Because $532 is fairly close to the *estimate* of $500, Meisha can quickly see that the amount shown on her paycheck probably is correct. She might also use the estimate when talking to a friend, saying, "My net pay is about $500."

Continued on Next Page

---

6 Use front end rounding to round each number.

(a) $^-94$

(b) 508

(c) $^-2522$

(d) 9700

(e) 61,888

(f) $^-963{,}369$

ANSWERS
6. (a) $^-90$ (b) 500 (c) $^-3000$
(d) 10,000 (e) 60,000 (f) $^-1{,}000{,}000$

7 Use front end rounding to estimate an answer. Then find the exact answer.

Pao Xiong is a bookkeeper for a small business. The company checking account is overdrawn by $3881. He deposits a check for $2090. What is the balance in the account?

Estimate:

Exact:

> **CAUTION**
> Always *estimate* the answer first. Then, when you find the *exact* answer, check that it is close to the estimate. If your exact answer is very far off, rework the problem because you probably made an error.

Calculator Tip It's easy to press the wrong key when using a calculator. If you use front end rounding and estimate the answer *before* entering the numbers, you can catch many such mistakes. For example, a student thought that he entered this problem correctly.

$$7836 \; \boxed{+} \; 5060 \; \boxed{=} \quad \boxed{2776}$$

Front end rounding gives an estimated answer of $8000 + 5000 = 13{,}000$, which is very different from 2776. Can you figure out which key the student pressed incorrectly? (Answer: The student pressed $\boxed{-}$ instead of $\boxed{+}$.)

◀◀ **Work Problem 7 at the Side.**

ANSWERS
7. *Estimate:* $^-\$4000 + \$2000 = {}^-\$2000$
   Exact: $^-\$3881 + \$2090 = {}^-\$1791$
   The account is overdrawn by $1791, which is fairly close to the estimate of $^-\$2000$.

1.5 Exercises

FOR EXTRA HELP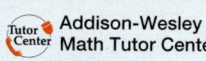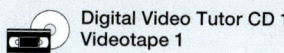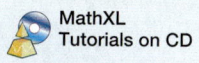

Round each number to the indicated place. See Examples 1–5.

1. 625 to the nearest ten
2. 206 to the nearest ten
3. ⁻1083 to the nearest ten
4. ⁻2439 to the nearest ten
5. 7862 to the nearest hundred
6. 6746 to the nearest hundred
7. ⁻86,813 to the nearest hundred
8. ⁻17,211 to the nearest hundred
9. 42,495 to the nearest hundred
10. 18,273 to the nearest hundred
11. ⁻5996 to the nearest hundred
12. ⁻8451 to the nearest hundred
13. 15,758 to the nearest hundred
14. 28,065 to the nearest hundred
15. ⁻78,499 to the nearest thousand
16. ⁻14,314 to the nearest thousand
17. 5847 to the nearest thousand
18. 49,706 to the nearest thousand
19. 53,182 to the nearest thousand
20. 13,124 to the nearest thousand
21. 595,008 to the nearest ten-thousand
22. 725,182 to the nearest ten-thousand
23. ⁻8,906,422 to the nearest million
24. ⁻13,713,409 to the nearest million
25. 139,610,000 to the nearest million
26. 609,845,500 to the nearest million

Use front end rounding to round each number. See Example 6.

27. Tyrone's truck shows this number on the odometer.

28. Ezra bought a used car with this odometer reading.

29. From summer to winter the average temperature drops 56 degrees.

30. The flood waters fell 42 inches yesterday.

31. Jan earned $9942 working part time.

32. Carol deposited $285 in her checking account.

33. 60,950,000 Americans go to a video store each week. (*Source:* Video Software Dealer's Assoc.)

34. 99,375,000 U.S. households have at least one TV. (*Source:* Nielsen Media Research.)

35. The submarine will dive to 255 feet below the surface of the ocean.

36. DeAnne lost $1352 in the stock market.

37. The population of Alaska is 635,478 people, and the population of California is 34,501,994 people. (*Source:* U.S. Bureau of the Census.)

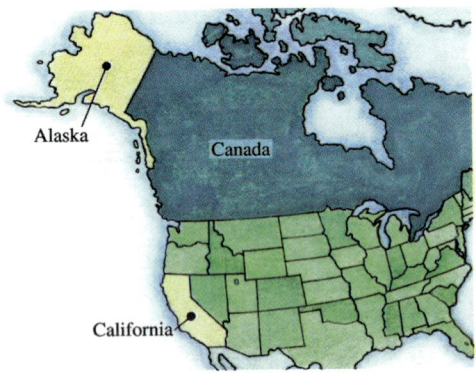

38. By 2010, it is estimated that the U.S. population will be 300,118,000 people and Canada's population will be 34,253,000 people. (*Source:* U.S. Bureau of the Census.)

39. Explain in your own words how to do front end rounding. Also show two examples of numbers and how you round them.

40. Describe two situations in your own life when you might use rounded numbers. Describe two situations in which exact numbers are important.

Section 1.5 Problem Solving: Rounding and Estimating   37

First, use front end rounding to estimate each answer. Then find the exact answer.
In Exercises 49–54, change subtraction to addition. See Example 7.

41. ⁻42 + 89

Estimate: _____ + _____ = _____

Exact:

42. ⁻66 + 25

Estimate: _____ + _____ = _____

Exact:

43. 16 + ⁻97

Estimate: _____ + _____ = _____

Exact:

44. 58 + ⁻19

Estimate: _____ + _____ = _____

Exact:

45. ⁻273 + ⁻399

Estimate:

Exact:

46. ⁻311 + ⁻582

Estimate:

Exact:

47. 3081 + 6826

Estimate:

Exact:

48. 4904 + 1181

Estimate:

Exact:

49. 23 − 81

Estimate:

Exact:

50. 72 − 84

Estimate:

Exact:

51. ⁻39 − 39

Estimate:

Exact:

52. ⁻91 − 91

Estimate:

Exact:

53. ⁻106 + 34 − ⁻72

Estimate:

Exact:

54. 52 − ⁻87 − 139

Estimate:

Exact:

First use front end rounding to estimate the answer to each application problem. Then find the exact answer. See Example 7.

55. The community has raised $52,882 for the homeless shelter. If the amount needed for the shelter is $78,650, how much more needs to be collected?

Estimate:

Exact:

56. A truck weighs 9250 pounds when empty. After being loaded with firewood, it weighs 21,375 pounds. What is the weight of the firewood?

Estimate:

Exact:

57. Dorene Cox decided to establish a monthly budget. She will spend $685 for rent, $325 for food, $320 for child care, $182 for transportation, and $150 for other expenses, and she will put the remainder in savings. If her monthly take-home pay is $1920, find her monthly savings.

Estimate:

Exact:

58. Jared Ueda had $2874 in his checking account. He wrote checks for $308 for auto repairs, $580 for child support, and $778 for tuition. Find the amount remaining in his account.

Estimate:

Exact:

59. In a laboratory experiment, a mixture started at a temperature of $^-102$ degrees. First the temperature was raised 37 degrees and then raised 52 degrees. What was the final temperature?

Estimate:

Exact:

60. A scuba diver was photographing fish at 65 feet below the surface of the lagoon. She swam up 24 feet and then swam down 49 feet. What was her final depth?

Estimate:

Exact:

61. The White House in Washington, D.C., has 132 rooms, 412 doors, and 147 windows. What is the total number of doors and windows? (*Source: Scholastic Book of World Records.*)

Estimate:

Exact:

62. There are 20,802 McDonald's restaurants throughout the world in 119 different countries. Burger King has 10,529 restaurants. How many restaurants do the two companies have together? (*Source: Scholastic Book of World Records.*)

Estimate:

Exact:

1.6 Multiplying Integers

OBJECTIVES
1. Use a raised dot or parentheses to express multiplication.
2. Multiply integers.
3. Identify properties of multiplication.
4. Estimate answers to application problems involving multiplication.

OBJECTIVE 1 Use a raised dot or parentheses to express multiplication. In arithmetic we usually use "×" when writing multiplication problems. But in algebra, we use a raised dot or parentheses to show multiplication. The numbers being multiplied are called **factors** and the answer is called the **product**.

Arithmetic
$3 \times 5 = 15$
Factors Product

Algebra
$3 \cdot 5 = 15$ or $3(5) = 15$ or $(3)(5) = 15$
Factors Product   Factors Product   Factors Product

EXAMPLE 1 Expressing Multiplication in Algebra

Rewrite each multiplication in three different ways, using a dot or parentheses. Also identify the factors and the product.

(a) 10×7

Rewrite it as $10 \overset{\text{Raised dot}}{\cdot} 7$ or $10(7)$ or $(10)(7)$

The factors are 10 and 7. The product is 70.

(b) 4×80

Rewrite it as $4 \cdot 80$ or $4(80)$ or $(4)(80)$

The factors are 4 and 80. The product is 320.

Work Problem 1 at the Side.

① Rewrite each multiplication in three different ways using a dot or parentheses. Also identify the factors and the product.

(a) 100×6

(b) 7×12

NOTE
Parentheses are used to show several different things in algebra. When we discussed the associative property of addition earlier in this chapter, we used parentheses as shown below.

$6 + \underbrace{(9 + {}^-9)}$ ← Parentheses show which numbers to add first.
$\underbrace{6 + \quad 0}$
$\quad\;\; 6$

Now we are using parentheses to indicate multiplication, as in $3(5)$ or $(3)(5)$.

OBJECTIVE 2 Multiply integers. Suppose that our football team gained 5 yards on the first play, gained 5 yards again on the second play, and gained 5 yards again on the third play. We can add to find the result.

$$5 \text{ yards} + 5 \text{ yards} + 5 \text{ yards} = 15 \text{ yards}$$

A quick way to add the same number several times is to multiply.

$\underbrace{\text{Our team made 3 plays}}_{3}$ and $\underbrace{\text{gained 5 yards each time.}}_{5}$ $=$ $\underbrace{\text{Our team gained a total of 15 yards.}}_{15}$

ANSWERS
1. (a) $100 \cdot 6$ or $100(6)$ or $(100)(6)$
   The factors are 100 and 6; the product is 600.
   (b) $7 \cdot 12$ or $7(12)$ or $(7)(12)$
   The factors are 7 and 12; the product is 84.

Here are the rules for multiplying two integers.

> **Multiplying Two Integers**
>
> If two factors have *different signs*, the product is *negative*.
> For example,
> $$^-2 \cdot 6 = {}^-12 \quad \text{and} \quad 4 \cdot {}^-5 = {}^-20$$
> If two factors have the *same sign*, the product is *positive*.
> For example,
> $$7 \cdot 3 = 21 \quad \text{and} \quad ^-3 \cdot {}^-10 = 30$$

There are several ways to illustrate these rules. First we'll continue with football. Remember, you are interested in the results for *our* team. We will designate **our team** with a **positive sign** and **their team** with a **negative sign**.

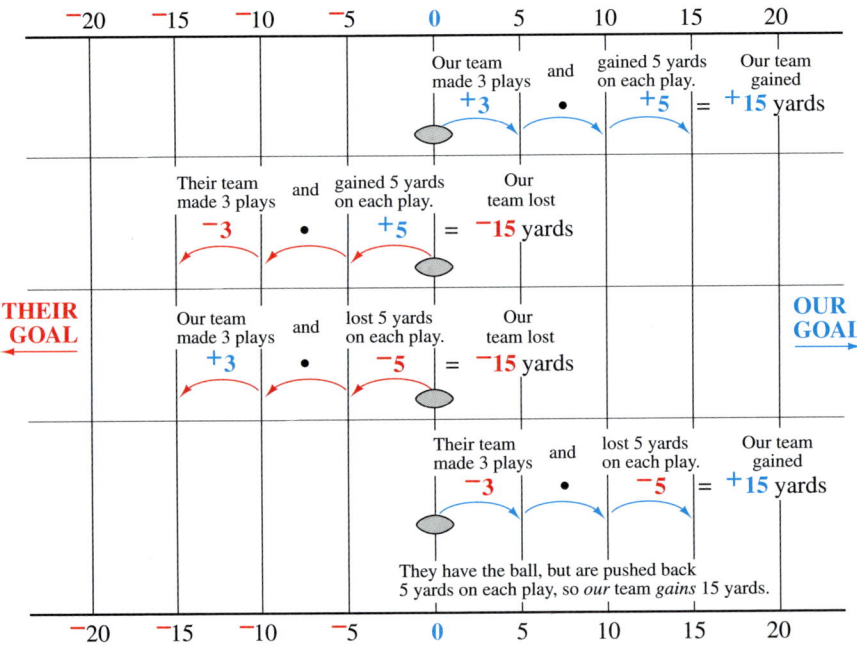

Here is a summary of the football examples.

When two factors have the *same* sign, the product is *positive*.

Both positive
$$3 \cdot 5 = 15 \quad \leftarrow \text{Product is positive.}$$
$$^-3 \cdot {}^-5 = 15 \quad \leftarrow$$
Both negative

When two factors have *different* signs, the product is *negative*.

$$^-3 \cdot 5 = {}^-15 \quad \leftarrow \text{Product is negative.}$$
$$3 \cdot {}^-5 = {}^-15 \quad \leftarrow$$

There is another way to look at these multiplication rules. In mathematics, the rules or patterns must always be consistent.

Look for a pattern in this list of products.	Look for a pattern in this list of products.
$4 \cdot 2 = 8$	$4 \cdot {}^-2 = {}^-8$
$3 \cdot 2 = 6$	$3 \cdot {}^-2 = {}^-6$
$2 \cdot 2 = 4$	$2 \cdot {}^-2 = {}^-4$
$1 \cdot 2 = 2$	$1 \cdot {}^-2 = {}^-2$
$0 \cdot 2 = 0$	$0 \cdot {}^-2 = 0$
${}^-1 \cdot 2 = ?$	${}^-1 \cdot {}^-2 = ?$

Blue numbers decrease by 1.   Red numbers *decrease* by 2.

Blue numbers decrease by 1.   Red numbers *increase* by 2.

To keep the red pattern going, replace the **?** with a number that is 2 *less than* 0, which is $^-2$.

So, $^-1 \cdot 2 = {}^-2$. This pattern illustrates that the product of two numbers with *different* signs is *negative*.

To keep the red pattern going, replace the **?** with a number that is 2 *more than* 0, which is $^+2$.

So, $^-1 \cdot {}^-2 = {}^+2$. This pattern illustrates that the product of two numbers with the *same* sign is *positive*.

EXAMPLE 2 Multiplying Two Integers

(a) $^-2 \cdot 8 = {}^-16$ The factors have *different signs,* so the product is *negative*.
  ↑ Negative ↑ Positive

(b) $^-10\,({}^-6) = 60$ The factors have the *same sign,* so the product is *positive*.
  Both negative

(c) $(9)\,({}^-11) = {}^-99$ The factors have *different signs,* so the product is *negative*.
  ↑ Positive ↑ Negative

Work Problem 2 at the Side.

Sometimes there are more than two factors in a multiplication problem. If there are parentheses around two of the factors, multiply them first. If there aren't any parentheses, start at the left and work with two factors at a time.

EXAMPLE 3 Multiplying Several Factors

Multiply.

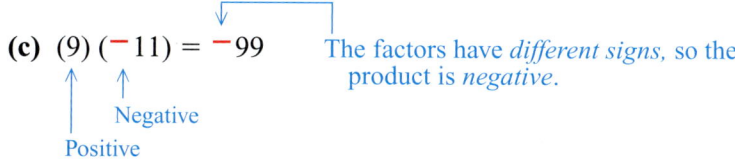

(a) $^-3 \cdot (4 \cdot {}^-5)$
   $\;\;{}^-3 \cdot \;\;{}^-20$
   $\;\;\;\;\;\;60$

Parentheses tell you to multiply $4 \cdot {}^-5$ first. The factors have *different* signs, so the product is *negative*. Then multiply $^-3 \cdot {}^-20$. Both factors have the *same* sign, so the product is *positive*.

(b) $^-2 \cdot {}^-2 \cdot {}^-2$
   $\;\;\;\;4\;\;\; \cdot \;{}^-2$
   $\;\;\;\;\;\;\;{}^-8$

There are no parentheses, so multiply $^-2 \cdot {}^-2$ first. The factors have the *same* sign, so the product is *positive*. Then multiply $4 \cdot {}^-2$. The factors have *different* signs, so the product is *negative*.

---

2 Multiply.

(a) $7({}^-2)$

(b) $^-5 \cdot {}^-5$

(c) $^-1(14)$

(d) $10 \cdot 6$

(e) $({}^-4)({}^-9)$

ANSWERS
2. (a) $^-14$ (b) 25 (c) $^-14$
   (d) 60 (e) 36

3 Multiply.

(a) $5 \cdot (^-10 \cdot 2)$

(b) $^-1 \cdot 8 \cdot ^-5$

(c) $^-3 \cdot ^-2 \cdot ^-4$

(d) $^-2 \cdot (7 \cdot ^-3)$

(e) $(^-1)(^-1)(^-1)$

4 Multiply. Then name the property illustrated by each example.

(a) $819 \cdot 0$

(b) $1(^-90)$

(c) $25 \cdot 1$

(d) $(0)(^-75)$

ANSWERS
3. (a) $^-100$ (b) 40 (c) $^-24$
   (d) 42 (e) $^-1$
4. (a) 0; multiplication property of 0
   (b) $^-90$; multiplication property of 1
   (c) 25; multiplication property of 1
   (d) 0; multiplication property of 0

CAUTION
In Example 3(b) you may be tempted to think that the final product will be *positive* because all the factors have the *same* sign. Be careful to work with just two factors at a time and keep track of the sign at each step.

Calculator Tip You can use the *change of sign* key for multiplication and division, just as you did for adding and subtracting. To enter Example 3(b) on your scientific calculator, press the following keys.

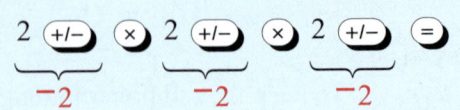

The answer is $^-8$.

◀◀◀ **Work Problem 3 at the Side.**

OBJECTIVE 3 Identify properties of multiplication. Addition involving 0 is unusual because adding 0 does *not* change the number. For example, $7 + 0$ is still 7. (See **Section 1.3**.) But what happens in multiplication? Let's use our football team as an example.

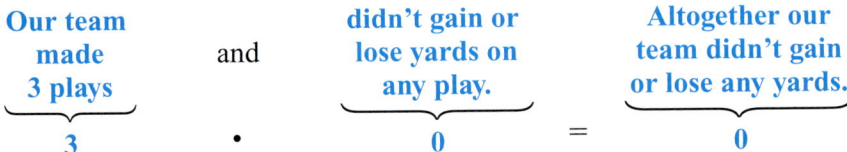

This example illustrates one of the properties of multiplication.

Multiplication Property of 0
Multiplying any number by 0 gives a product of 0.
Some examples are shown below.
$$^-16 \cdot 0 = 0 \qquad (0)(5) = 0 \qquad 32{,}977(0) = 0$$

So, can you multiply a number by something that will *not* change the number?

$$6 \cdot ? = 6 \qquad ^-12(?) = ^-12 \qquad (?)(5876) = 5876$$

The number 1 can replace the ? in each example. This illustrates another property of multiplication.

Multiplication Property of 1
Multiplying a number by 1 leaves the number unchanged.
Some examples are shown below.
$$6 \cdot 1 = 6 \qquad ^-12(1) = ^-12 \qquad (1)(5876) = 5876$$

EXAMPLE 4 Using Properties of Multiplication

Multiply. Then name the property illustrated by each example.

(a) $(0)(^-48) = 0$ Illustrates the **multiplication property of 0**.

(b) $615(1) = 615$ Illustrates the **multiplication property of 1**.

◀◀◀ **Work Problem 4 at the Side.**

When adding, we said that changing the order of the addends did not change the sum (commutative property of addition). We also found that changing the *grouping* of addends did not change the sum (associative property of addition). These same ideas apply to multiplication.

> **Commutative Property of Multiplication**
> Changing the *order* of two factors does not change the product.
> For example,
> $$2 \cdot 5 = 5 \cdot 2 \quad \text{and} \quad {}^-4 \cdot 6 = 6 \cdot {}^-4$$

> **Associative Property of Multiplication**
> Changing the *grouping* of factors does not change the product.
> For example,
> $$9 \cdot (1 \cdot 2) = (9 \cdot 1) \cdot 2$$

EXAMPLE 5 Using the Commutative and Associative Properties

Show that the product is unchanged and name the property that is illustrated in each case.

(a) $\underbrace{{}^-7 \cdot {}^-4}_{28} = \underbrace{{}^-4 \cdot {}^-7}_{28}$ Both products are 28.

This example illustrates the **commutative property of multiplication.**

(b) $5 \cdot (10 \cdot 2) = (5 \cdot 10) \cdot 2$

$\underbrace{5 \cdot \underbrace{20}}_{100} = \underbrace{\underbrace{50} \cdot 2}_{100}$ Both products are 100.

This example illustrates the **associative property of multiplication.**

▶▶▶ **Work Problem 5 at the Side.** ▶▶▶

Now that you are familiar with multiplication and addition, we can look at a property that involves both operations.

> **Distributive Property**
> Multiplication *distributes* over addition. An example is shown below.
> $$3(6 + 2) = 3 \cdot 6 + 3 \cdot 2$$

What is the **distributive property** really saying? Notice that there is an understood multiplication symbol between the 3 and the parentheses. To "distribute" the 3 means to multiply 3 times each number inside the parentheses.

Understood to be
multiplying by 3

$3(6 + 2)$
$\downarrow$
$3 \cdot (6 + 2)$

Using the distributive property,

$3 \cdot (6 + 2)$ can be rewritten as $3 \cdot 6 + 3 \cdot 2$

---

5 Show that the product is unchanged and name the property that is illustrated in each case.

(a) $(3 \cdot 3) \cdot 2 = 3 \cdot (3 \cdot 2)$

(b) $11 \cdot 8 = 8 \cdot 11$

(c) $0 \cdot {}^-15 = {}^-15 \cdot 0$

(d) ${}^-4 \cdot (1 \cdot 5) = ({}^-4 \cdot 1) \cdot 5$

ANSWERS
5. (a) $18 = 18$; associative property of multiplication
   (b) $88 = 88$; commutative property of multiplication
   (c) $0 = 0$; commutative property of multiplication
   (d) ${}^-20 = {}^-20$; associative property of multiplication

6 Rewrite each product, using the distributive property. Show that the result is unchanged.

(a) $3(8 + 7)$

(b) $10(6 + {}^-9)$

(c) ${}^-6(4 + 4)$

EXAMPLE 6 Using the Distributive Property

Rewrite each product, using the distributive property. Show that the result is unchanged.

(a) $4(3 + 7)$

$$4(3 + 7) = 4 \cdot 3 + 4 \cdot 7$$
$$4(10) = 12 + 28$$
$$40 = 40 \quad \text{Both results are 40.}$$

(b) ${}^-2({}^-5 + 1)$

$$^-2({}^-5 + 1) = {}^-2 \cdot {}^-5 + {}^-2 \cdot 1$$
$$^-2({}^-4) = 10 + {}^-2$$
$$8 = 8 \quad \text{Both results are 8.}$$

Work Problem 6 at the Side.

OBJECTIVE 4 Estimate answers to application problems involving multiplication. Front end rounding can be used to estimate answers in multiplication, just as we did when adding and subtracting (see **Section 1.5**). Once the numbers have been rounded so that there are lots of zeros, we can use the multiplication shortcut described in the Review chapter (see **Section R.3**). As a brief review, look at the pattern in these examples.

$${}^-3 \cdot 2 \text{ is } {}^-6$$
$${}^-30 \cdot 200 = {}^-6000$$
Total of three zeros   Write three zeros after the ${}^-6$.

$${}^-2 \cdot {}^-5 \text{ is } 10$$
$${}^-2000 \cdot {}^-5000 = 10{,}000{,}000$$
Total of six zeros   Write six zeros after the 10.

7 Use front end rounding to estimate an answer. Then find the exact answer.

An average of 27,095 baseball fans attended each of the 81 home games during the season. What was the total home game attendance for the season?

Estimate:

Exact:

EXAMPLE 7 Using Front End Rounding to Estimate an Answer

Use front end rounding to estimate an answer. Then find the exact answer.

Last year the Video Land store had to replace 392 defective videos at a cost of $19 each. How much money did the store lose on defective videos? (*Hint:* Because it's a loss, use a negative number for the cost.)

Estimate: Use front end rounding: 392 rounds to 400 and ${}^-\$19$ rounds to ${}^-\$20$. Use the rounded numbers and multiply to estimate the total amount of money lost.

$$4 \cdot {}^-2 \text{ is } {}^-8$$
$$400 \cdot {}^-\$20 = {}^-\$8000 \quad \text{Estimate}$$
Total of three zeros   Write three zeros after the ${}^-8$.

Exact: $392 \cdot {}^-\$19 = {}^-\$7448$

Because the exact answer of ${}^-\$7448$ is fairly close to the estimate of ${}^-\$8000$, you can see that ${}^-\$7448$ probably is correct. The store manager could also use the estimate to say, "We lost about $8000 on defective videos last year."

Work Problem 7 at the Side.

ANSWERS

6. (a) $3 \cdot 8 + 3 \cdot 7$; both results are 45.
   (b) $10 \cdot 6 + 10 \cdot {}^-9$; both results are ${}^-30$.
   (c) ${}^-6 \cdot 4 + {}^-6 \cdot 4$; both results are ${}^-48$.
7. *Estimate:* $30{,}000 \cdot 80 = 2{,}400{,}000$ fans
   Exact: $27{,}095 \cdot 81 = 2{,}194{,}695$ fans

1.6 Exercises

FOR EXTRA HELP 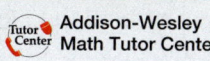 Addison-Wesley Math Tutor Center 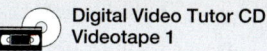 MathXL  Digital Video Tutor CD 1 Videotape 1 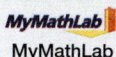 Student's Solutions Manual MyMathLab 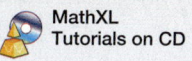 MathXL Tutorials on CD

Multiply. See Examples 1–4.

1. (a) $9 \cdot 7$
   (b) $^-9 \cdot {}^-7$
   (c) $^-9 \cdot 7$
   (d) $9 \cdot {}^-7$

2. (a) $^-6 \cdot 9$
   (b) $6 \cdot {}^-9$
   (c) $^-6 \cdot {}^-9$
   (d) $6 \cdot 9$

3. (a) $7(^-8)$
   (b) $^-7(8)$
   (c) $7(8)$
   (d) $^-7(^-8)$

4. (a) $8(6)$
   (b) $^-8(^-6)$
   (c) $^-8(6)$
   (d) $8(^-6)$

5. $^-5 \cdot 7$
6. $^-10 \cdot 2$
7. $(^-5)(9)$
8. $(^-9)(4)$

9. $3(^-6)$
10. $8(^-9)$
11. $10(^-5)$
12. $5(^-11)$

13. $(^-1)(40)$
14. $(75)(^-1)$
15. $^-56 \cdot 1$
16. $1 \cdot {}^-87$

17. $^-8(^-4)$
18. $^-3(^-9)$
19. $11 \cdot 7$
20. $4 \cdot 25$

21. $25 \cdot 0$
22. $0 \cdot 30$
23. $^-19(^-7)$
24. $^-21(^-3)$

25. $^-13(^-1)$
26. $^-1(^-31)$
27. $(0)(^-25)$
28. $(^-50)(0)$

29. $^-4 \cdot {}^-6 \cdot 2$
30. $^-9 \cdot 3 \cdot {}^-3$
31. $(^-4)(^-2)(^-7)$

32. $(^-6)(^-2)(^-3)$
33. $5(^-8)(4)$
34. $5(4)(^-6)$

Write an integer in each blank to make a true statement.

35. $(^-3)(\_\_\_) = {}^-15$
36. $6 \cdot \_\_\_ = {}^-24$
37. $\_\_\_ \cdot 10 = {}^-30$

38. $(\_\_\_)(^-4) = 16$
39. $^-17 = 17(\_\_\_)$
40. $29 = {}^-29(\_\_\_)$

41. $(\_\_\_)(^-350) = 0$
42. $\_\_\_ \cdot 99 = 99$
43. $5 \cdot {}^-4 \cdot \_\_\_ = {}^-100$

44. $\_\_\_ \cdot 2 \cdot {}^-2 = {}^-24$
45. $(\_\_\_)(^-5)(^-2) = {}^-40$
46. $(^-3)(\_\_\_)(^-3) = {}^-27$

47. In your own words, explain the difference between the commutative and associative properties of multiplication. Show an example of each.

48. A student did this multiplication.
$$^-3 \cdot {}^-3 \cdot {}^-3 = 27$$
He knew that $3 \cdot 3 \cdot 3$ is 27. Since all the factors have the same sign, he made the product positive. Do you agree with his reasoning? Explain.

RELATING CONCEPTS (EXERCISES 49–50) For Individual or Group Work

*Look for patterns as you **work Exercises 49 and 50 in order**.*

49. Write three numerical examples for each of these situations:

(a) A positive number multiplied by $^-1$

(b) A negative number multiplied by $^-1$

Now write a rule that explains what happens when you multiply a signed number by $^-1$.

50. Do these multiplications.
$$^-2 \cdot {}^-2 = \underline{}$$
$$^-2 \cdot {}^-2 \cdot {}^-2 = \underline{}$$
$$^-2 \cdot {}^-2 \cdot {}^-2 \cdot {}^-2 = \underline{}$$
$$^-2 \cdot {}^-2 \cdot {}^-2 \cdot {}^-2 \cdot {}^-2 = \underline{}$$

Describe the pattern in the products. Then find the next three products without multiplying all the $^-2$s.

Rewrite each multiplication, using the stated property. Show that the result is unchanged. See Examples 5 and 6.

51. Distributive property
$9(^-3 + 5)$

52. Distributive property
$^-6(4 + {}^-5)$

53. Commutative property
$25 \cdot 8$

54. Commutative property
$^-7 \cdot {}^-11$

55. Associative property
$^-3 \cdot (2 \cdot 5)$

56. Associative property
$(5 \cdot 5) \cdot 10$

First use front end rounding to estimate the answer to each application problem. Then find the exact answer. See Example 7.

57. Alliette receives $324 per week for doing child care in her home. How much income will she have for an entire year? There are 52 weeks in a year.

Estimate:

Exact:

58. Enrollment at our community college has increased by 875 students each of the last four semesters. What is the total increase?

Estimate:

Exact:

59. A new computer software store had losses of $9950 during each month of its first year. What was the total loss for the year?

Estimate:

Exact:

60. A long-distance phone company estimates that it is losing 95 customers each week. How many customers will it lose in a year?

Estimate:

Exact:

61. Tuition at the state university is $182 per credit for undergraduates. How much tuition will Wei Chen pay for 13 credits?

Estimate:

Exact:

62. Pat ate a dozen crackers as a snack. Each cracker had 17 calories. How many calories did Pat eat?

Estimate:

Exact:

63. There are 24 hours in one day. How many hours are in one year (365 days)?

Estimate:

Exact:

64. There are 5280 feet in one mile. How many feet are in 17 miles?

Estimate:

Exact:

Simplify.

65. $^-8 \cdot |^-8 \cdot 8|$

66. $^-7 \cdot |7| \cdot |^-7|$

67. $(^-37)(^-1)(85)(0)$

68. $^-1(9732)(^-1)(^-1)$

69. $|6-7| \cdot {}^-355{,}299$

70. $987 \cdot {}^-65{,}432 \cdot |9-9|$

Each of these application problems requires several steps and may involve addition and subtraction as well as multiplication.

71. Each of Maurice's four cats needed a $24 rabies shot and a $29 shot to prevent respiratory infections. There was also one $35 office visit charge. What was the total amount of Maurice's bill?

72. Chantele has three children. Her older daughter had a throat culture taken at the clinic today. Her baby received three immunization shots and her son received two shots. The co-pay amounts were $8 for each shot, an $18 office charge for each child, and $12 for the throat culture. How much did Chantele pay?

73. There is a 3-degree drop in temperature for every thousand feet that an airplane climbs into the sky. If the temperature on the ground is 50 degrees, what will be the temperature when the plane reaches an altitude of 24,000 feet? (*Source:* Lands' End.)

74. An unmanned research submarine descends to 150 feet below the surface of the ocean. Then it continues to go deeper, taking a water sample every 25 feet. What is its depth when it takes the 15th sample?

75. In Ms. Zubero's algebra class, there are six tests of 100 points each, eight quizzes of six points each, and 20 homework assignments of five points each. There are also four "bonus points" on each test. What is the total number of possible points?

76. In Mr. Jackson's prealgebra class, there are three group projects worth 25 points each, five 100-point tests, a 150-point final exam, and seven quizzes worth 12 points each. Find the total number of possible points.

1.7 Dividing Integers

OBJECTIVE 1 Divide integers. In arithmetic, we usually use $\overline{)}$ to write division problems so that we can do them by hand. Calculator keys use the ÷ symbol for division. In algebra, we usually show division by using a fraction bar, a slash mark, or the ÷ symbol. The answer to a division problem is called the **quotient**.

OBJECTIVES

1. Divide integers.
2. Identify properties of division.
3. Combine multiplying and dividing of integers.
4. Estimate answers to application problems involving division.
5. Interpret remainders in division application problems.

Arithmetic

Divisor → $2\overline{)16}$ ← Dividend, Quotient 8

Calculator and Algebra

Dividend ↓, Divisor ↓
$16 ÷ 2 = 8$ ↑ Quotient

Dividend → $\dfrac{16}{2} = 8$ ← Quotient
Divisor →

For every division problem, we can write a related multiplication problem. (See **Section R.4**.) Because of this relationship, the rules for dividing integers are the same as the rules for multiplying integers. For example,

$\dfrac{16}{8} = 2$ because $(2)(8) = 16$

$\dfrac{^-16}{^-8} = 2$ because $(2)(^-8) = {^-16}$

$\dfrac{^-16}{8} = {^-2}$ because $(^-2)(8) = {^-16}$

$\dfrac{16}{^-8} = {^-2}$ because $(^-2)(^-8) = 16$

Dividing Two Integers

If two numbers have *different signs,* the quotient is *negative.* Some examples are shown below.

$\dfrac{^-18}{3} = {^-6}$ $\dfrac{40}{^-5} = {^-8}$

If two numbers have the *same sign,* the quotient is *positive.* Some examples are shown below.

$\dfrac{^-30}{^-6} = 5$ $\dfrac{48}{8} = 6$

EXAMPLE 1 Dividing Two Integers

Divide.

(a) $\dfrac{^-20}{5}$ ⎱ Numbers have *different* signs, so the quotient is *negative.* $\dfrac{^-20}{5} = {^-4}$

(b) $\dfrac{^-24}{^-4}$ ⎱ Numbers have the *same* sign, so the quotient is *positive.* $\dfrac{^-24}{^-4} = 6$

(c) $60 ÷ {^-2}$ Numbers have *different* signs, so the quotient is *negative.* $60 ÷ {^-2} = {^-30}$

Work Problem 1 at the Side.

1 Divide.

(a) $\dfrac{40}{^-8}$

(b) $\dfrac{49}{7}$

(c) $\dfrac{^-32}{4}$

(d) $\dfrac{^-10}{^-10}$

(e) $^-81 ÷ 9$

(f) $^-100 ÷ {^-50}$

ANSWERS
1. (a) $^-5$ (b) 7 (c) $^-8$ (d) 1
   (e) $^-9$ (f) 2

② Divide. Then state the property illustrated by each division.

(a) $\dfrac{-12}{0}$ undefined

(b) $\dfrac{0}{39}$ 0

(c) $\dfrac{-9}{1}$

(d) $\dfrac{21}{21}$

ANSWERS
2. (a) undefined; division by 0 is undefined.
   (b) 0; 0 divided by any nonzero number is 0.
   (c) −9; any number divided by 1 is the number.
   (d) 1; any nonzero number divided by itself is 1.

OBJECTIVE 2 Identify properties of division. You have seen that 0 and 1 are used in special ways in addition and multiplication. This is also true in division.

Examples			Pattern (Division Property)
$\dfrac{5}{5}=1$	$\dfrac{-18}{-18}=1$	$\dfrac{-793}{-793}=1$	When a nonzero number is divided by itself, the quotient is 1.
$\dfrac{5}{1}=5$	$\dfrac{-18}{1}=-18$	$\dfrac{-793}{1}=-793$	When a number is divided by 1, the quotient is the number.
$\dfrac{0}{5}=0$	$\dfrac{0}{-18}=0$	$\dfrac{0}{-793}=0$	When 0 is divided by any other number (except 0), the quotient is 0.
$\dfrac{5}{0}$ is undefined.	$\dfrac{-18}{0}$ is undefined.		Division by 0 is *undefined*. There is no answer.

The most surprising property is that division by 0 *cannot be done*. Let's review the reason for that by rewriting this division problem as a related multiplication problem.

$$\dfrac{-18}{0} = ? \quad \text{can be written as the multiplication} \quad ? \cdot 0 = {}^-18$$

If you thought the answer to $\dfrac{-18}{0}$ should be 0, try replacing **?** with 0. It doesn't work in the related multiplication problem! Try replacing **?** with any number you like. The result in the related multiplication problem is always 0 instead of ⁻18. That is how we know that dividing by 0 cannot be done. Mathematicians say that it is *undefined* and have agreed never to divide by 0.

EXAMPLE 2 Using the Properties of Division

Divide. Then state the property illustrated by each example.

(a) $\dfrac{-312}{-312} = 1$ Any nonzero number divided by itself is 1

(b) $\dfrac{75}{1} = 75$ Any number divided by 1 is the number.

(c) $\dfrac{0}{-19} = 0$ Zero divided by any nonzero number is 0

(d) $\dfrac{48}{0}$ is undefined.   Division by 0 is undefined.

🖩 **Calculator Tip** Try Examples 2(c) and 2(d) above on your calculator. Use the change of sign key to enter ⁻19 on a *scientific* calculator.

0 ÷ 19 +/− =   Answer is 0

⁻19

48 ÷ 0 =   Calculator shows "Error" or "ERR" or "E" for error because it cannot divide by 0.

◀◀◀ **Work Problem 2 at the Side.**

OBJECTIVE 3 Combine multiplying and dividing of integers. When a problem involves both multiplying and dividing, first check to see if there are any parentheses. Do what is inside parentheses first. Then start at the left and work toward the right, using two numbers at a time.

EXAMPLE 3   Combining Multiplication and Division of Integers

Simplify.

(a) 6(⁻10) ÷ (⁻3 • 2)   Work inside parentheses first: ⁻3 • 2 is ⁻6.
     6(⁻10) ÷ ⁻6   Start at the left: 6(⁻10) is ⁻60.
     ⁻60 ÷ ⁻6   Finally, ⁻60 ÷ ⁻6 is 10.
     10

(b) 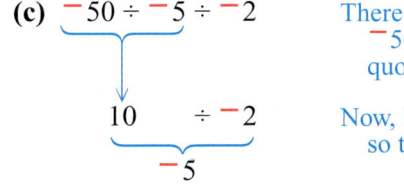   There are no parentheses, so start at the left: ⁻24 ÷ ⁻2 is 12.
    12(4) ÷ ⁻6   Next, 12(4) is 48.
    48 ÷ ⁻6   Finally, 48 ÷ ⁻6 is ⁻8.
    ⁻8

(c) ⁻50 ÷ ⁻5 ÷ ⁻2   There are no parentheses, so start at the left: ⁻50 ÷ ⁻5 is 10. The signs are the *same*, so the quotient is *positive*.
    10 ÷ ⁻2   Now, 10 ÷ ⁻2 is ⁻5. The signs are *different*, so the quotient is *negative*.
    ⁻5

Work Problem 3 at the Side.

OBJECTIVE 4 Estimate answers to application problems involving division. Front end rounding can be used to estimate answers in division just as you did when multiplying (see **Section 1.6**). Once the numbers have been rounded so that there are lots of zeros, you can use the division shortcut described in the Review chapter (see **Section R.5**). As a brief review, look at the pattern in these examples.

$$4000 \div {}^-50 = 400 \div {}^-5 = {}^-80$$

Drop one 0 from both dividend and divisor.

$$^-6000 \div {}^-3000 = {}^-6 \div {}^-3 = 2$$

Drop three zeros from both dividend and divisor.

---

3 Simplify.

(a) 60 ÷ ⁻3(⁻5)

(b) ⁻6(⁻16 ÷ ⁻8) • 2

(c) ⁻8(10) ÷ 4(⁻3) ÷ ⁻6

(d) 56 ÷ ⁻8 ÷ ⁻1

ANSWERS
3. (a) 100   (b) ⁻24   (c) ⁻10   (d) 7

4 First use front end rounding to estimate an answer. Then find the exact answer.

Laurie and Chuck Struthers lost $2724 on their stock investments last year. What was their average loss each month?

Estimate:

Exact:

EXAMPLE 4 Using Front End Rounding to Estimate an Answer in Division

First use front end rounding to estimate an answer. Then find the exact answer.

During a 24-hour laboratory experiment, the temperature of a solution dropped 96 degrees. What was the average drop in temperature each hour?

Estimate: Use front end rounding: ⁻96 degrees rounds to ⁻100 degrees and 24 hours rounds to 20 hours. To estimate the average, divide the rounded number of degrees by the rounded number of hours.

⁻100 degrees ÷ 20 hours = ⁻5 degrees each hour   ← Estimate

Exact: ⁻96 degrees ÷ 24 hours = ⁻4 degrees each hour   ← Exact

Because the exact answer of ⁻4 degrees is close to the estimate of ⁻5 degrees, you can see that ⁻4 degrees probably is correct.

> **Calculator Tip**   The answer in Example 4 above "came out even." In other words, the quotient was an integer. Suppose that the drop in temperature had been 97 degrees. Do the division on your calculator.
>
> 97 +/– ÷ 24 =   Calculator shows −4.041666667
> ⎵
> ⁻97
>
> The quotient is *not* an integer. We will work with numbers like these in **Chapter 5**, Positive and Negative Decimals.

Work Problem 4 at the Side.

OBJECTIVE 5 Interpret remainders in division application problems.

In arithmetic, division problems often have a remainder, as shown below.

$$
\begin{array}{r}
14\ \mathbf{R10} \\
25\overline{\smash{)}360} \\
\underline{25} \\
110 \\
\underline{100} \\
10 \leftarrow \text{Remainder}
\end{array}
$$

But what does **R10** really mean? Let's look at this same problem by using money amounts.

EXAMPLE 5 Interpreting Remainders in Division Applications

Divide; then interpret the remainder in each application.

(a) The math department at Lake Community College has $360 in its budget to buy scientific calculators for the math lab. If the calculators cost $25 each, how many can be purchased? How much money will be left over?

We can solve this problem by using the same division as shown above. But this time we can decide what the remainder really means.

$$
\begin{array}{r}
14 \leftarrow \text{Number of calculators purchased} \\
\text{Cost of one calculator} \rightarrow \$25\overline{\smash{)}\$360} \leftarrow \text{Budget} \\
\underline{25} \\
110 \\
\underline{100} \\
\$10 \leftarrow \text{Money left over}
\end{array}
$$

The department can buy 14 calculators. There will be $10 left over.

Continued on Next Page

ANSWERS

4. *Estimate:* ⁻$3000 ÷ 10 months = ⁻$300 each month
   Exact: ⁻$2724 ÷ 12 months = ⁻$227 each month

Calculator Tip You can use your calculator to solve Example 5(a) on the previous page. Recall that digits on the *right* side of the decimal point show *part* of one whole. You cannot order *part* of one calculator, so ignore those digits and use only the *whole number part* of the quotient.

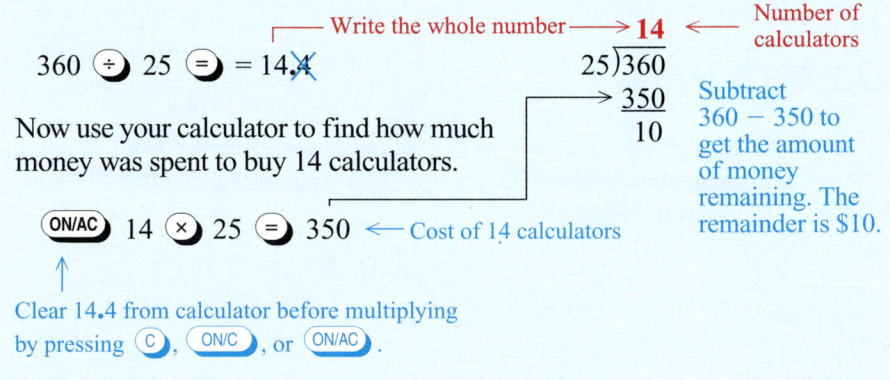

Now use your calculator to find how much money was spent to buy 14 calculators.

ON/AC 14 × 25 = 350 ← Cost of 14 calculators

Clear 14.4 from calculator before multiplying by pressing C, ON/C, or ON/AC.

5 Divide; then interpret the remainder in each of these applications.

(a) Chad and Martha are baking cookies for a fund-raiser. They baked 116 cookies and are putting them into packages of a dozen each. How many packages will they have for the fund-raiser? How many cookies will be left over for them to eat?

(b) Luke's son is going on a Scout camping trip. There are 135 Scouts. Luke is renting tents that sleep 6 people each. How many tents should he rent?

We again use division to solve the problem. There is a remainder, but this time it must be interpreted differently than in the calculator example.

```
                        22  ← Number of tents with 6 Scouts each
   Each tent holds → 6)135  ← Total number of Scouts
                        12
                        15
                        12
                         3  ← Scouts left over
```

If Luke rents 22 tents, 3 Scouts will have to sleep out in the rain. He must rent 23 tents to accommodate all the Scouts. (One tent will have only 3 Scouts in it.)

Work Problem 5 at the Side.

(b) Coreen is a dispatcher for a bus company. A group of 249 senior citizens is going to a baseball game. If the buses will each hold 44 people, how many buses should she send to pick up the seniors?

ANSWERS
5. (a) 9 packages, with 8 cookies left over to eat
   (b) 6 buses, because 5 buses would leave 29 seniors standing on the curb

Focus on Real-Data Applications

'Til Debt Do You Part!

Bride's magazine reported that average wedding costs in the United States increased from $15,208 in 1990 to the grand total you will calculate in the table below (in 2002). The average number of wedding guests also increased, from 170 guests to 200 guests.

Category	Average Cost in 2002
Miscellaneous expenses (stationery, clergy, gifts, limousine)	$ 1260
Bouquets and other flowers	$ 855
Photography and videography	$ 3000
Music	$ 830
Engagement and wedding rings (bride and groom)	$ 4060
Rehearsal dinner	$ 698
Bride's wedding dress and headpiece	$ 989
Bridal attendants' dresses (average of 5 bridesmaids)	$ 790
Mother of the bride's apparel	$ 231
Men's formalwear (ushers, best man, groom)	$ 544
Wedding reception	$10,320
Grand Total	

Source: *Bride's* magazine.

1. What is the grand total of expenses shown in the chart?
2. How much more expensive was a wedding in 2002 compared to 1990?
3. If the groom pays for the bouquets and flowers, the rehearsal dinner, the bride's engagement and wedding rings ($3500), the clergy ($232), and his formalwear ($95), what is the total amount spent by him?
4. If you budgeted $48 per person for the wedding reception and you invited 200 guests to a wedding in 2005, how much money would you have spent compared to the 2002 wedding reception costs?
5. If you budget $5000 for the reception and the caterer charges $48 per person, how many guests can you invite? How much of your budget is left over?
6. If you budget $8500 for the reception and the caterer charges $48 per person, how many guests can you invite? How much of your budget is left over?
7. What type of arithmetic problem did you work to get the answers to Problems 5 and 6? What is the mathematical term for the "left over" budget?
8. What is the average cost of each bridesmaid's dress? When you solved this problem, which number was the dividend? the divisor? the quotient? Rewrite the division as a related multiplication.

Section 1.7 Dividing Integers 55

1.7 Exercises

FOR EXTRA HELP: Addison-Wesley Math Tutor Center, 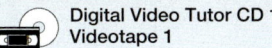 MathXL, Digital Video Tutor CD 1 Videotape 1, Student's Solutions Manual,  MyMathLab, 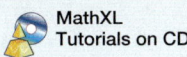 MathXL Tutorials on CD

Divide. See Examples 1 and 2.

1. (a) $14 \div 2$
   (b) $-14 \div -2$
   (c) $14 \div -2$
   (d) $-14 \div 2$

2. (a) $-18 \div -3$
   (b) $18 \div 3$
   (c) $-18 \div 3$
   (d) $18 \div -3$

3. (a) $-42 \div 6$
   (b) $-42 \div -6$
   (c) $42 \div -6$
   (d) $42 \div 6$

4. (a) $45 \div 5$
   (b) $45 \div -5$
   (c) $-45 \div -5$
   (d) $-45 \div 5$

5. (a) $\dfrac{35}{35}$
   (b) $\dfrac{35}{1}$
   (c) $\dfrac{-13}{1}$
   (d) $\dfrac{-13}{-13}$

6. (a) $\dfrac{-23}{1}$
   (b) $\dfrac{-23}{-23}$
   (c) $\dfrac{17}{1}$
   (d) $\dfrac{17}{17}$

7. (a) $\dfrac{0}{50}$
   (b) $\dfrac{50}{0}$
   (c) $\dfrac{-11}{0}$
   (d) $\dfrac{0}{-11}$

8. (a) $\dfrac{-85}{0}$
   (b) $\dfrac{0}{-85}$
   (c) $\dfrac{6}{0}$
   (d) $\dfrac{0}{6}$

9. $\dfrac{-8}{2}$

10. $\dfrac{-14}{7}$

11. $\dfrac{21}{-7}$

12. $\dfrac{30}{-6}$

13. $\dfrac{-54}{-9}$

14. $\dfrac{-48}{-6}$

15. $\dfrac{55}{-5}$

16. $\dfrac{70}{-7}$

17. $\dfrac{-28}{0}$

18. $\dfrac{-40}{0}$

19. $\dfrac{14}{-1}$

20. $\dfrac{25}{-1}$

21. $\dfrac{-20}{-2}$

22. $\dfrac{-80}{-4}$

23. $\dfrac{-48}{-12}$

24. $\dfrac{-30}{-15}$

25. $\dfrac{-18}{18}$

26. $\dfrac{50}{-50}$

27. $\dfrac{0}{-9}$

28. $\dfrac{0}{-4}$

29. $\dfrac{-573}{-3}$

30. $\dfrac{-580}{-5}$

31. $\dfrac{163,672}{-328}$

32. $\dfrac{-69,496}{1022}$

Simplify. See Example 3.

33. $^-60 \div 10 \div {}^-3$

34. $36 \div {}^-4 \div 3$

35. $^-64 \div {}^-8 \div {}^-2$

36. $^-72 \div {}^-9 \div {}^-4$

37. $100 \div {}^-5({}^-2)$

38. $^-80 \div 4({}^-5)$

39. $48 \div 3 \cdot (12 \div {}^-4)$

40. $^-2 \cdot ({}^-3 \cdot {}^-7) \div 7$

41. $^-5 \div {}^-5({}^-10) \div {}^-2$

42. $^-9(4) \div {}^-36(50)$

43. $64 \cdot 0 \div {}^-8(10)$

44. $^-88 \div {}^-8 \div {}^-11(0)$

RELATING CONCEPTS (EXERCISES 45–50) For Individual or Group Work

*Use your knowledge of the properties of multiplication as you **work Exercises 45–50 in order.***

45. Explain whether or not division is commutative like multiplication. Start by doing these two divisions on your calculator: $2 \div 1$ and $1 \div 2$.

46. Explain whether or not division is associative like multiplication. Start by doing these two divisions: $(12 \div 6) \div 2$ and $12 \div (6 \div 2)$.

47. Explain what is different and what is similar about multiplying and dividing two signed numbers.

48. In your own words, describe at least three division properties. Include examples to illustrate each property.

49. Write three numerical examples for each situation.
    (a) A negative number divided by $^-1$

    (b) A positive number divided by $^-1$

    Now write a rule that explains what happens when you divide a signed number by $^-1$.

50. Explain why $\frac{0}{-3}$ and $\frac{-3}{0}$ do not give the same result.

Solve these application problems by using addition, subtraction, multiplication, or division. First use front end rounding to estimate the answer. Then find the exact answer. See Example 4.

51. The greatest ocean depth is 35,836 feet below sea level. If an unmanned research sub dives to that depth in 17 equal steps, how far does it dive in each step? (*Source: The Top 10 of Everything.*)

Estimate:

Exact:

52. Our college enrollment dropped by 3245 students over the last 11 years. What was the average drop in enrollment each year?

Estimate:

Exact:

53. When Ashwini discovered that her checking account was overdrawn by $238, she quickly transferred $450 from her savings to her checking account. What is the new balance in her checking account?

Estimate:

Exact:

54. The Tigers offensive team lost a total of 48 yards during the first half of the football game. During the second half they gained 191 yards. How many yards did they gain or lose during the entire game?

Estimate:

Exact:

55. The foggiest place in the United States is Cape Disappointment, Washington. It is foggy there an average of 106 days each year. How many days is it not foggy each year? (*Source:* National Weather Service.)

Estimate:

Exact:

56. The number of cellular phone users worldwide in 1993 was 34 million. The number of users reached 946 million in 2001. What was the increase in the number of users during this 8-year period? (*Source: The World Almanac.*)

Estimate:

Exact:

57. A plane descended an average of 730 feet each minute during a 37-minute landing. How far did the plane descend during the landing?

Estimate:

Exact:

58. A discount store found that 174 items were lost to shoplifting last month. The average value of each item was $24. What was the total loss due to shoplifting?

Estimate:

Exact:

59. Mr. and Mrs. Martinez drove on the Interstate for five hours and traveled 315 miles. What was the average number of miles they drove each hour?

Estimate:

Exact:

60. Rochelle has a 48-month car loan for $12,048. How much is her monthly payment?

Estimate:

Exact:

Find the exact answer in Exercises 61–66. Solving these problems requires more than one step.

61. Clarence bowled four games and had scores of 143, 190, 162, and 177. What was his average score? (*Hint:* To find the average, add all the scores and divide by the number of scores.)

62. Sheila kept track of her grocery expenses for six weeks. The amounts she spent were $84, $111, $82, $110, $98, and $79. What was the average weekly cost of her groceries?

63. On the back of an oatmeal box, it says that one serving weighs 40 grams and that there are 13 servings in the box. On the front of the box, it says that the weight of the contents is 510 grams. What is the difference in the total weight on the front and the back of the box? (*Source:* Quaker Oats.)

64. A 2000-calorie-per-day diet recommends that you eat no more than 65 grams of fat. If each gram of fat is 9 calories, how many calories can you consume in other types of food? (*Source:* U.S. Department of Agriculture.)

65. Stephanie had $302 in her checking account. She wrote a $116 check for day care and a $548 check for rent. She also deposited her $347 paycheck. What is the balance in her account?

66. Gary started a new checking account with a $500 deposit. The bank charged him $18 to print his checks. He also wrote a $193 check for car repairs and a $289 check to his credit card company. What is the balance in his account?

Divide; then interpret the remainder in each application. See Example 5.

67. A cellular phone company is offering 1000 free minutes of air time to new subscribers. How many hours of free time will a new subscriber receive?

68. Nikki is catering a large party. If one pie will serve eight guests, how many pies should she make for 100 guests?

69. Hurricane victims are being given temporary shelter in a hotel. Each room can hold five people. How many rooms are needed for 163 people?

70. A college has received a $250,000 donation to be used for scholarships. How many $3500 scholarships can be given to students?

Simplify:

71. $|{-8}| \div {-4} \cdot |{-5}| \cdot |1|$

72. $-6 \cdot |{-3}| \div |9| \cdot {-2}$

73. $-6(-8) \div (-5 - {-5})$

74. $-9 \div -9(-9 \div 9) \div (12 - 13)$

75. Look back at page 2 of this chapter. You guessed how many days it would take to receive a million dollars if you got $1 each second. Here's how to use your calculator to get the answer. If you get $1 per second, it would take 1,000,000 seconds to receive $1,000,000. Press these keys.

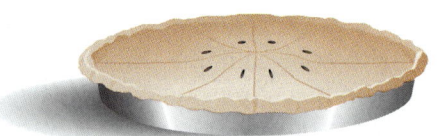

Notice that you do *not* have to re-enter the intermediate answers. When the answer 16666.66667 appears on your calculator display, just go ahead and enter ÷ 60.

Now use a *scientific* calculator to find how long it takes to receive a *billion* dollars. Start by entering 1000000000. Then follow the pattern shown above. You will need to do one more division step to get the number of years. (Assume that there are 365 days in one year.)

1.8 Exponents and Order of Operations

OBJECTIVE 1 Use exponents to write repeated factors. An **exponent** is a quick way to write repeated multiplication. Here is an example.

$2 \cdot 2 \cdot 2 \cdot 2 \cdot 2$ can be written 2^5 ← Exponent
↑
Base

The *base* is the number being multiplied over and over, and the exponent tells how many times to use the number as a factor. This is called *exponential notation* or *exponential form*.

To simplify 2^5, actually do the multiplication.

$$2^5 = 2 \cdot 2 \cdot 2 \cdot 2 \cdot 2 = 32$$

Here are some more examples, using 2 as the base.

$2 = 2^1$ is read "2 to the **first power**."
$2 \cdot 2 = 2^2$ is read "2 to the **second power**" or, more commonly, "2 **squared**."
$2 \cdot 2 \cdot 2 = 2^3$ is read "2 to the **third power**" or, more commonly, "2 **cubed**."
$2 \cdot 2 \cdot 2 \cdot 2 = 2^4$ is read "2 to the **fourth power**."
$2 \cdot 2 \cdot 2 \cdot 2 \cdot 2 = 2^5$ is read "2 to the **fifth power**."

and so on.

We usually don't write an exponent of 1, so if no exponent is shown, you can assume that it is 1. For example, 6 is actually 6^1, and 4 is actually 4^1.

> **NOTE**
> Exponents can also be negative numbers or 0, for example, 2^{-3} and 2^0. You will learn more about these exponents in **Chapter 10**.

EXAMPLE 1 Using Exponents

Rewrite each multiplication using exponents. Also indicate how to read the exponential form.

(a) $5 \cdot 5 \cdot 5$ can be written as 5^3, which is read "5 cubed" or "5 to the third power."

(b) $(4)(4)$ can be written as 4^2, which is read "4 squared" or "4 to the second power."

(c) 7 can be written as 7^1, which is read "7 to the first power."

▶▶▶ **Work Problem 1 at the Side.** ▶▶▶

OBJECTIVE 2 Simplify expressions containing exponents. Exponents are also used with signed numbers, as shown below.

$(^-3)^2 = (^-3)(^-3) = 9$ The factors have the same sign, so the product is positive.

$(^-4)^3 = \underbrace{(^-4)(^-4)}_{16}(^-4)$ Multiply two numbers at a time.

$ 16 \quad (^-4)$ First, $(^-4)(^-4)$ is positive 16.

$ ^-64$ Then, $16(^-4)$ is $^-64$.

OBJECTIVES

1. Use exponents to write repeated factors.
2. Simplify expressions containing exponents.
3. Use the order of operations.
4. Simplify expressions with fraction bars.

Study Skills Workbook
Activity 4: Note Taking

1 Write each multiplication using exponents. Indicate how to read the exponential form.

(a) $3 \cdot 3 \cdot 3 \cdot 3$

(b) $6 \cdot 6$

(c) 9

(d) $(2)(2)(2)(2)(2)(2)$

ANSWERS
1. (a) 3^4 is read "3 to the fourth power."
   (b) 6^2 is read "6 squared" or "6 to the second power."
   (c) 9^1 is read "9 to the first power."
   (d) 2^6 is read "2 to the sixth power."

2 Simplify.

(a) $(^-2)^3$

(b) $(^-6)^2$

(c) $2^4(^-3)^2$

(d) $3^3(^-4)^2$

Simplify exponents before you do other multiplications, as shown below. Notice that the exponent applies only to the *first* thing to its *left*.

Exponent applies only to the 2. $2^3(5)(4^2)$ Exponent applies only to the 4.

$(2)(2)(2)$ is 8. → $8\,(5)(16)$ ← $(4)(4)$ is 16.

$40\,(16)$

640

EXAMPLE 2 Using Exponents with Negative Numbers

Simplify.

(a) $(^-5)^2 = (^-5)(^-5) = 25$

(b) $(^-5)^3 = (^-5)(^-5)(^-5)$
$= 25(^-5)$
$= {}^-125$

(c) $(^-2)^4 = (^-2)(^-2)(^-2)(^-2) = 16$

(d) $2^3(^-3)^2 = (2)(2)(2)(^-3)(^-3)$
$= (8)(9)$
$= 72$

Calculator Tip On a *scientific* calculator, use the exponent key y^x to enter exponents. To enter 5^8, press the following keys.

5 (Base) y^x 8 (Exponent)   =   Answer is 390,625.

Be careful when using your calculator's exponent key with a negative number, such as $(^-5)^3$. Different calculators use different keystrokes, so check the instruction manual, or experiment to see how your calculator works.

◀◀◀ **Work Problem 2 at the Side.**

OBJECTIVE 3 Use the order of operations. In Sections 1.4 and 1.7, you worked examples that mixed either addition and subtraction or multiplication and division. In those situations, you worked from left to right. Example 3 below is a review.

EXAMPLE 3 Working from Left to Right

Simplify.

(a) $^-8 - {}^-6 + {}^-11$ Do additions and subtractions from left to right.
   ${}^-2 + {}^-11$
   ${}^-13$

Continued on Next Page

ANSWERS
2. (a) $^-2 \cdot {}^-2 \cdot {}^-2 = {}^-8$
   (b) $^-6 \cdot {}^-6 = 36$
   (c) $16 \cdot 9 = 144$
   (d) $27 \cdot 16 = 432$

(b) $\overbrace{{}^-15 \div {}^-3}(6)$ Do multiplications and divisions from left to right.
    $\underbrace{5(6)}$
    30

Work Problem 3 at the Side.

Now we're ready to do problems that use a mix of the four operations, parentheses, and exponents. Let's start with a simple example: $4 + 2 \cdot 3$.

If we work from left to right	If we multiply first
$4 + 2 \cdot 3$	$4 + 2 \cdot 3$
$\ \ 6\ \ \cdot 3$	$4 +\ \ 6$
$\ \ \ \ 18$	$\ \ \ \ 10$

To be sure that everyone gets the same answer to a problem like this, mathematicians have agreed to do things in a certain order. The following order of operations shows that multiplying is done ahead of adding, so the correct answer is 10.

Order of Operations
Step 1 Work inside **parentheses** or **other grouping symbols**.
Step 2 Simplify expressions with **exponents**.
Step 3 Do the remaining **multiplications and divisions** as they occur from left to right.
Step 4 Do the remaining **additions and subtractions** as they occur from left to right.

Calculator Tip Enter the example above in your calculator.

$4\ \oplus\ 2\ \otimes\ 3\ \circledcirc$

Which answer do you get? If you have a scientific calculator, it automatically uses the order of operations and multiplies first to get the correct answer of 10. Some standard, four-function calculators may *not* have the order of operations built into them and will give the *incorrect* answer of 18.

3 Simplify.

(a) ${}^-9 + {}^-15 - 3$

(b) ${}^-4 - 2 + {}^-6$

(c) $3({}^-4) \div {}^-6$

(d) ${}^-18 \div 9({}^-4)$

ANSWERS
3. (a) ${}^-27$ (b) ${}^-12$ (c) 2 (d) 8

Chapter 1 Introduction to Algebra: Integers

4 Simplify.

(a) $8 + 6(14 \div 2)$

(b) $4(1) + 8(9 - 2)$

(c) $3(5 + 1) + 20 \div 4$

EXAMPLE 4 Using the Order of Operations with Whole Numbers

Simplify.

$9 +$	$3(20 - 4)$	$\div 8$	Work inside parentheses first: $20 - 4$ is 16. Bring down the other numbers and signs that you haven't used.
$9 +$	$3(16)$	$\div 8$	Look for exponents: none. Move from left to right, looking for multiplying and dividing.
$9 +$	$3(16)$	$\div 8$	Yes, here is multiplying: $3(16)$ is 48.
$9 +$	48	$\div 8$	Here is dividing: $48 \div 8$ is 6. There is no other multiplying or dividing, so look for adding and subtracting.
$9 +$	6		Add last: $9 + 6$ is 15.
15			

Work Problem 4 at the Side.

5 Simplify.

(a) $2 + 40 \div (^-5 + 3)$

(b) $^-5(5) - (15 + 5)$

(c) $(^-24 \div 2) + (15 - 3)$

(d) $^-3(2 - 8) - 5(4 - 3)$

(e) $3(3) - (10 \cdot 3) \div 5$

(f) $6 - (2 + 7) \div (^-4 + 1)$

EXAMPLE 5 Using the Order of Operations with Integers

Simplify.

(a) $^-8 \div (7 - 5) - 9$ Work inside parentheses first: $7 - 5$ is 2. Bring down the other numbers and signs you haven't used.

$^-8 \div (2) - 9$ Look for exponents: none. Move from left to right, looking for multiplying and dividing.

$^-8 \div (2) - 9$ Here is dividing: $^-8 \div 2$ is $^-4$. No other multiplying or dividing, so look for adding and subtracting.

$^-4 - 9$ Change subtracting to adding. Change 9 to its opposite.

$^-4 + ^-9$ Add $^-4 + ^-9$.

$^-13$

(b) $3 + 2(6 - 8) \cdot (15 \div 3)$ Work inside first set of parentheses. Change $6 - 8$ to $6 + ^-8$ to get $^-2$.

$3 + 2(^-2) \cdot (15 \div 3)$ Work inside second set of parentheses: $15 \div 3$ is 5.

$3 + 2(^-2) \cdot 5$ Multiply and divide from left to right. First multiply $2(^-2)$ to get $^-4$.

$3 + ^-4 \cdot 5$ Then multiply $^-4 \cdot 5$ to get $^-20$.

$3 + ^-20$ Add last: $3 + ^-20$ is $^-17$.

$^-17$

Work Problem 5 at the Side.

ANSWERS
4. (a) 50 (b) 60 (c) 23
5. (a) $^-18$ (b) $^-45$ (c) 0
   (d) 13 (e) 3 (f) 9

Section 1.8 Exponents and Order of Operations

EXAMPLE 6 Using the Order of Operations with Exponents

Simplify.

(a) $4^2 - (^-3)^2$ The only parentheses are around $^-3$, but there is no work to do inside these parentheses.

$\underbrace{4^2}_{16} - \underbrace{(^-3)^2}_{9}$ Simplify the exponents: $4^2 = (4)(4) = 16$, and $(^-3)^2 = (^-3)(^-3) = 9$.

7 There is no multiplying or dividing, so add and subtract: $16 - 9$ is 7.

(b) $(^-4)^3 - (\mathbf{4 - 6})^2(^-3)$ Work inside parentheses: $4 - 6$ becomes $4 + {^-6}$, which is $^-2$.

$\underbrace{(^-4)^3}_{^-64} - \underbrace{(^-\mathbf{2})^2}_{4}(^-3)$ Simplify the exponents next: $(^-4)^3$ is $(^-4)(^-4)(^-4) = {^-64}$, and $(^-2)^2$ is $(^-2)(^-2) = 4$.

$^-64 - \;\;\; 4(^-3)$ Look for multiplying and dividing. Multiply $4(^-3)$ to get $^-12$.

$^-64 \;\; - \;\;\; ^-12$ Change subtraction to addition. Change $^-12$ to its opposite.
$\;\;\downarrow \;\;\;\;\;\;\;\;\;\;\; \downarrow$
$^-64 \;\; + \;\;\; ^+12$ Add: $^-64 + 12$ is $^-52$.

$^-52$

> **CAUTION**
> To help in remembering the order of operations, you may have memorized the letters **PEMDAS**, or the phrase "Please Excuse My Dear Aunt Sally."
>
>   **P**lease   **E**xcuse   **M**y   **D**ear   **A**unt   **S**ally
>
>   **P**arentheses; **E**xponents; **M**ultiply and **D**ivide; **A**dd and **S**ubtract
>
> Be careful! Do *not* automatically do all multiplication before division. Multiplying and dividing are done *from left to right* (after parentheses and exponents).

Work Problem 6 at the Side. ▶▶▶

6 Simplify.

(a) $2^3 - 3^2$

(b) $6^2 \div (^-4)(^-3)$

(c) $(^-4)^2 - 3^2(5 - 2)$

(d) $(^-3)^3 + (3 - 9)^2$

ANSWERS
6. (a) $^-1$ (b) 27 (c) $^-11$ (d) 9

7 Simplify.

(a) $\dfrac{-3\,(2^3)}{-10 - 6 + 8}$

(b) $\dfrac{(-10)(-5)}{-6 \div 3(5)}$

(c) $\dfrac{6 + 18 \div (-2)}{(1 - 10) \div 3}$

(d) $\dfrac{6^2 - 3^2(4)}{5 + (3 - 7)^2}$

OBJECTIVE 4 **Simplify expressions with fraction bars.** A fraction bar indicates division, as in $\dfrac{-6}{2}$, which means $-6 \div 2$. In an expression such as

$$\dfrac{-5 + 3^2}{16 - 7(2)}$$

the fraction bar also acts as a grouping symbol, like parentheses. It tells us to do the work in the numerator (above the bar) and then the work in the denominator (below the bar). The last step is to divide the results.

$$\dfrac{-5 + 3^2}{16 - 7(2)} \longrightarrow \dfrac{-5 + 9}{16 - 14} \longrightarrow \dfrac{4}{2} \longrightarrow 4 \div 2 = 2$$

The final result is 2.

EXAMPLE 7 Using the Order of Operations with Fraction Bars

Simplify $\dfrac{-8 + 5(4 - 6)}{4 - 4^2 \div 8}$.

First do the work in the numerator.

$-8 + 5(4 - 6)$ Work inside the parentheses.

$-8 + 5(-2)$ Multiply.

$-8 + (-10)$ Add.

Numerator $\longrightarrow -18$

Now do the work in the denominator.

$4 - 4^2 \div 8$ There are no parentheses; simplify the exponent.

$4 - 16 \div 8$ Divide.

$4 - 2$ Subtract.

Denominator $\longrightarrow 2$

The last step is the division.

$\dfrac{\text{Numerator} \longrightarrow -18}{\text{Denominator} \longrightarrow 2} = -9$

Work Problem 7 at the Side.

ANSWERS

7. (a) $\dfrac{-24}{-8} = 3$ (b) $\dfrac{50}{-10} = -5$

   (c) $\dfrac{-3}{-3} = 1$ (d) $\dfrac{0}{21} = 0$

1.8 Exercises

FOR EXTRA HELP Addison-Wesley Math Tutor Center 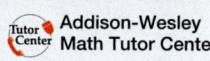 MathXL  Digital Video Tutor CD 1 Videotape 1  Student's Solutions Manual 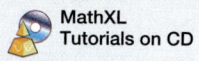 MyMathLab MathXL Tutorials on CD

Complete this table. See Example 1.

	Exponential Form	Factored Form	Simplified	Read as
1.	4^3		64	
2.	10^2		100	
3.		$2 \cdot 2 \cdot 2 \cdot 2 \cdot 2 \cdot 2 \cdot 2$		
4.		$3 \cdot 3 \cdot 3 \cdot 3 \cdot 3$		
5.		$5 \cdot 5 \cdot 5 \cdot 5$		
6.		$2 \cdot 2 \cdot 2 \cdot 2 \cdot 2 \cdot 2$		
7.				7 squared
8.				6 cubed
9.				10 to the first power
10.				4 to the fourth power

Simplify. See Examples 1 and 2.

11. (a) 10^1 (b) 10^2 (c) 10^3 (d) 10^4

12. (a) 5^1 (b) 5^2 (c) 5^3 (d) 5^4

13. (a) 4^1 (b) 4^2 (c) 4^3 (d) 4^4

14. (a) 3^1 (b) 3^2 (c) 3^3 (d) 3^4

15. 5^{10}

16. 4^9

17. 2^{12}

18. 3^{10}

19. $(^-2)^2$

20. $(^-4)^2$

21. $(^-5)^2$

22. $(^-10)^2$

23. $(^-4)^3$

24. $(^-2)^3$

25. $(^-3)^4$

26. $(^-2)^4$

27. $(^-10)^3$

28. $(^-5)^3$

29. 1^4

30. 1^5

31. $3^3 \cdot 2^2$ **32.** $4^2 \cdot 5^2$ **33.** $2^3(^-5)^2$ **34.** $3^2(^-2)^2$

35. $6^1(^-5)^3$ **36.** $7^1(^-4)^3$ **37.** $(^-2)(^-2)^4$ **38.** $^-6(^-6)^2$

39. Simplify.

$(^-2)^2 = \underline{\quad}$ $(^-2)^6 = \underline{\quad}$
$(^-2)^3 = \underline{\quad}$ $(^-2)^7 = \underline{\quad}$
$(^-2)^4 = \underline{\quad}$ $(^-2)^8 = \underline{\quad}$
$(^-2)^5 = \underline{\quad}$ $(^-2)^9 = \underline{\quad}$

(a) Describe the pattern you see in the signs of the answers.

(b) What would be the sign of $(^-2)^{15}$ and the sign of $(^-2)^{24}$?

40. Explain why it is important to have rules for the order of operations. Why do you think our "natural instinct" is to just work from left to right?

Simplify. See Examples 3–7.

41. $6 + 3(^-4)$ **42.** $10 - 30 \div 2$ **43.** $^-1 + 15 + ^-7(2)$

44. $9 + ^-5 + 2(^-2)$ **45.** $10 - 7^2$ **46.** $5 - 5^2$

47. $2 - ^-5 + 3^2$ **48.** $6 - ^-9 + 2^3$ **49.** $3 + 5(6 - 2)$

50. $4 + 3(8 - 3)$ **51.** $^-7 + 6(8 - 14)$ **52.** $^-3 + 5(9 - 12)$

53. $2(^-3 + 5) - (9 - 12)$ **54.** $3(2 - 7) - (^-5 + 1)$ **55.** $^-5(7 - 13) \div ^-10$

56. $^-4(9-17) \div {^-8}$

57. $9 \div (^-3)^2 + {^-1}$

58. $^-48 \div (^-4)^2 + 3$

59. $2 - {^-5}(^-2)^3$

60. $1 - {^-10}(^-3)^3$

61. $^-2(^-7) + 3(9)$

62. $4(^-2) + {^-3}(^-5)$

63. $30 \div {^-5} - 36 \div {^-9}$

64. $8 \div {^-4} - 42 \div {^-7}$

65. $2(5) - 3(4) + 5(3)$

66. $9(3) - 6(4) + 3(7)$

67. $4(3^2) + 7(3+9) - {^-6}$

68. $5(4^2) - 6(1+4) - {^-3}$

69. $(^-4)^2 \cdot (7-9)^2 \div 2^3$

70. $(^-5)^2 \cdot (9-17)^2 \div (^-10)^2$

71. $\dfrac{^-1 + 5^2 - {^-3}}{^-6 - 9 + 12}$

72. $\dfrac{^-6 + 3^2 - {^-7}}{7 - 9 - 3}$

73. $\dfrac{^{-}2(4^2) - 4(6 - 2)}{^{-}4(8 - 13) \div {^{-}5}}$

74. $\dfrac{3(3^2) - 5(9 - 2)}{8(6 - 9) \div {^{-}3}}$

75. $\dfrac{2^3 \cdot (^{-}2 - 5) + 4(^{-}1)}{4 + 5(^{-}6 \cdot 2) + (5 \cdot 11)}$

76. $\dfrac{3^3 + 4(^{-}1 - 2) - 25}{^{-}4 + 4(3 \cdot 5) + (^{-}6 \cdot 9)}$

77. $5^2(9 - 11)(^{-}3)(^{-}3)^3$

78. $4^2(13 - 17)(^{-}2)(^{-}2)^3$

79. $|^{-}12| \div 4 + 2 \cdot |(^{-}2)^3| \div 4$

80. $6 - |2 - 3 \cdot 4| + (^{-}5)^2 \div 5^2$

81. $\dfrac{^{-}9 + 18 \div {^{-}3}(^{-}6)}{32 - 4(12) \div 3(2)}$

82. $\dfrac{^{-}20 - 15(^{-}4) - {^{-}40}}{14 + 27 \div 3(^{-}2) - {^{-}4}}$

Summary Exercises on Operations with Integers

Simplify each expression.

1. $2 - 8$

2. $(^-16)(0)$

3. $^-14 - {}^-7$

4. $\dfrac{^-42}{6}$

5. $^-9(^-7)$

6. $\dfrac{^-12}{12}$

7. $(1)(^-56)$

8. $1 + {}^-23$

9. $5 - {}^-7$

10. $^-88 \div {}^-11$

11. $^-18 + 5$

12. $\dfrac{0}{^-10}$

13. $^-40 - {}^-40$

14. $^-17 + 0$

15. $8(^-6)$

16. $^-1 - 9$

17. $^-5(10)$

18. $\dfrac{30}{0}$

19. $0 - 14$

20. $\dfrac{18}{^-3}$

21. $^-13 + 13$

22. $\dfrac{^-16}{^-1}$

23. $20 - 50$

24. $\dfrac{^-7}{0}$

25. $(^-4)(^-6)(2)$

26. $^-2 + {}^-12 + {}^-5$

27. $^-60 \div 10 \div {}^-3$

28. $^-8 - 4 - 8$

29. $64(0) \div {}^-8$

30. $2 - {}^-5 + 3^2$

31. $^-9 + 8 + {}^-2$

32. $(^-6)(^-2)(^-3)$

33. $8 + 6 + {}^-8$

34. $3^2 - 2^4$

35. $(^-5)^2 \div {}^-5$

36. $(^-2)^5 + 1^3$

37. $^-72 \div {}^-9 \div {}^-4$

38. $^-7 + 28 + {}^-56 + 3$

39. $9 - 6 - 3 - 5$

40. $^-6(^-8) \div (^-5 - 7)$

41. $^-1(9732)(^-1)(^-1)$

42. $^-80 \div 4(^-5)$

43. $^-10 - 4 + 0 + 18$

44. $^-7 \cdot |7| \cdot |^-7|$

45. $5 - |^-3| + 3$

46. $^-2(^-3)(7) \div {}^-7$

47. $^-3 - (^-2 + 4) - 5$

48. $0 - |^-7 + 2|$

49. $(^-4)^2(7-9)^2 \div 2^3$

50. $12 \div 4 + 2(^-2)^2 \div {}^-4$

51. $\dfrac{^-1(5^2) - {}^-3}{(^-3)^3 + 4^2}$

52. $\dfrac{^-6 \div 3(^-2) + 4}{2 + 8 \div 4 - 6}$

53. $\dfrac{^-9 + 24 \div (^-4)(^-6)}{32 - 4(12) \div 3(2)}$

54. $\dfrac{5 - |2 - 4(4)| + (^-5)^2 \div 5^2}{^-9 \div 3(2-2) - {}^-8}$

Chapter 1
SUMMARY

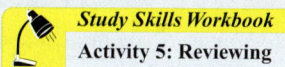
Study Skills Workbook
Activity 5: Reviewing

KEY TERMS

1.1	place value system	A place value system is a number system in which the location, or place, where a digit is written gives it a different value.
	digits	The 10 digits in our number system are 0, 1, 2, 3, 4, 5, 6, 7, 8, and 9.
	whole numbers	The whole numbers are 0, 1, 2, 3, and so on.
1.2	number line	A number line is like a thermometer turned sideways. It is used to show how numbers relate to each other.
	integers	Integers are the whole numbers and their opposites.
	absolute value	The absolute value of a number is its distance from 0. Absolute value is indicated by two vertical bars and is always positive (or 0) but never negative.
1.3	addends	In an addition problem, the numbers being added are called addends.
	sum	The answer to an addition problem is called the sum.
	addition property of 0	Adding 0 to any number leaves the number unchanged.
	commutative property of addition	Changing the *order* of two addends does not change the sum.
	associative property of addition	Changing the *grouping* of addends does not change the sum.
1.4	opposite	The opposite of a number is the same distance from 0 on the number line but on the opposite side of 0. It is also called the *additive inverse* because a number plus its opposite equals 0.
1.5	rounding	Rounding a number means finding a number that is close to the original number but easier to work with.
	estimate	Use rounded numbers to get an approximate answer, or estimate.
	front end rounding	Front end rounding is rounding numbers to the highest possible place, so all the digits become 0 except the first digit.
1.6	factors	In a multiplication problem, the numbers being multiplied are called factors.
	product	The answer to a multiplication problem is called the product.
	multiplication property of 0	Multiplying any number by 0 gives a product of 0.
	multiplication property of 1	Multiplying a number by 1 leaves the number unchanged.
	commutative property of multiplication	Changing the *order* of two factors does not change the product.
	associative property of multiplication	Changing the *grouping* of factors does not change the product.
	distributive property	Multiplication distributes over addition. For example, $3(6 + 2) = 3 \cdot 6 + 3 \cdot 2$.
1.7	quotient	The answer to a division problem is called the quotient.
1.8	exponent	An exponent tells how many times a number is used as a factor in repeated multiplication.

NEW SYMBOLS

<	is less than	$\|6\|$; $\|{-2}\|$	absolute value of 6; absolute value of -2
>	is greater than		

Chapter 1 Introduction to Algebra: Integers

TEST YOUR WORD POWER

See how well you have learned the vocabulary in this chapter. Answers follow the Quick Review.

1. In $(4)(^-6) = ^-24$, the 4 and $^-6$ are called
   A. products
   B. factors
   C. addends
   D. opposites.

2. A list of all the **whole numbers** is
   A. 1, 2, 3, 4, . . .
   B. . . . , $^-4$, $^-3$, $^-2$, $^-1$, 0, 1, 2, 3, . . .
   C. 0, 1, 2, 3, 4, 5, 6, 7, 8, 9
   D. 0, 1, 2, 3, 4, . . .

3. The **absolute value** of a number is
   A. its distance from 0
   B. never positive
   C. used when multiplying
   D. less than 0.

4. An **exponent**
   A. tells how many times a number is added
   B. is the number being multiplied
   C. tells how many times a number is multiplied
   D. applies only to the first thing to its right.

5. The **opposite** of a number is
   A. never negative
   B. called the additive inverse
   C. called the absolute value
   D. at the same point on the number line.

6. **Front end rounding** is rounding numbers
   A. to the nearest thousand
   B. so there are no zeros
   C. to the highest possible place
   D. so they become integers.

7. A list of all the **integers** is
   A. 1, 2, 3, 4, . . .
   B. . . . , $^-4$, $^-3$, $^-2$, $^-1$, 0, 1, 2, 3, . . .
   C. 0, 1, 2, 3, 4, 5, 6, 7, 8, 9
   D. 0, 1, 2, 3, 4, . . .

8. The **associative property of multiplication** says that
   A. changing the order of two factors does not change the product
   B. multiplication distributes over addition
   C. multiplying a number by 0 is undefined
   D. changing the grouping of factors does not change the product.

QUICK REVIEW

Concepts	Examples
1.1 Reading and Writing Whole Numbers Do not use the word *and* when reading whole numbers. Commas separate groups of three digits. The first few group names are ones, thousands, millions, billions, trillions.	Write 3,008,160 in words. **three million, eight thousand, one hundred sixty** Write this number using digits: twenty **billion**, sixty-five **thousand**, eighteen. $\underbrace{2\ 0}_{\text{billions}}, \underbrace{0\ 0\ 0}_{\text{millions}}, \underbrace{0\ 6\ 5}_{\text{thousands}}, \underbrace{0\ 1\ 8}_{\text{ones}}$
1.2 Graphing Signed Numbers Place a dot at the correct location on the number line.	Graph: (a) $^-4$ (b) 0 (c) $^-1$ (d) $\frac{1}{2}$ (e) 2 [number line with points plotted at $^-4$, $^-1$, 0, $\frac{1}{2}$, 2]
1.2 Comparing Integers When comparing two integers, the one that is farther to the left on the number line is less than the other. Use the < symbol for "is less than" and the > symbol for "is greater than."	Write < or > between each pair of numbers to make a true statement. $^-3 < ^-2$ $\qquad$ $0 > ^-4$ $^-3$ **is less than** $^-2$ $\qquad$ 0 **is greater than** $^-4$ because $^-3$ is to the *left* of $^-2$ on the number line. $\qquad$ because 0 is to the *right* of $^-4$ on the number line.

Chapter 1 Summary 73

Concepts	Examples
1.2 Finding the Absolute Value of a Number Find the distance on the number line from 0 to the number. The absolute value is always positive (or 0) but never negative.	Find each absolute value. $\lvert {}^-5 \rvert = 5$ because $^-5$ is 5 steps away from 0 on the number line. $\lvert 3 \rvert = 3$ because 3 is 3 steps away from 0 on the number line.
1.3 Adding Two Integers When both integers have the *same sign*, add the absolute values and use the common sign as the sign of the sum. When the integers have *different signs*, subtract the smaller absolute value from the larger absolute value. Use the sign of the number with the larger absolute value as the sign of the sum.	Add. (a) $^-6 + {}^-7$ Add the absolute values. $\lvert {}^-6 \rvert = 6$ and $\lvert {}^-7 \rvert = 7$ Add $6 + 7 = 13$ and use the common sign as the sign of the sum: $^-6 + {}^-7 = {}^-13$. (b) $^-10 + 4$ Subtract the smaller absolute value from the larger. $\lvert {}^-10 \rvert = 10$ and $\lvert 4 \rvert = 4$ Subtract $10 - 4 = 6$; the number with the larger absolute value is negative, so the sum is negative: $^-10 + 4 = {}^-6$.
1.3 Using Properties of Addition *Addition Property of 0:* Adding 0 to any number leaves the number unchanged. *Commutative Property of Addition:* Changing the *order* of two addends does not change the sum. *Associative Property of Addition:* Changing the *grouping* of addends does not change the sum.	Name the property illustrated by each case. (a) $^-16 + 0 = {}^-16$ (b) $4 + 10 = 10 + 4$ Both sums are 14. (c) $2 + ({}^-6 + 1) = (2 + {}^-6) + 1$ Both sums are $^-3$. (a) Addition property of 0 (b) Commutative property of addition (c) Associative property of addition

1.4 Subtracting Two Integers
To subtract two numbers, add the first number to the opposite of the second number.
Step 1 Make one pencil stroke to change the subtraction symbol to an addition symbol.
Step 2 Make a second pencil stroke to change the *second* number to its *opposite*.
	Subtract.
	(a) $7 - {}^-2$ Change subtraction to addition.
	     $\downarrow \downarrow \downarrow$ Change $^-2$ to its opposite, $^+2$.
	     $7 + {}^+2$
	     9
	(b) $^-9 - 12$ Change subtraction to addition.
	     $\downarrow \downarrow \downarrow$ Change 12 to its opposite, $^-12$.
	     $^-9 + {}^-12$
	     $^-21$

Concepts	Examples
1.5 Rounding Integers *Step 1* Draw a line under the place to which the number is to be rounded. *Step 2* Look only at the next digit to the right of the underlined place. If the next digit is 5 or more, increase the underlined digit by 1. If the next digit is 4 or less, do not change the digit in the underlined place. *Step 3* Change all digits to the right of the underlined place to zeros.	Round 36,833 to the nearest thousand. Next digit is *5 or more*. ↓ Changed to 0 ↓↓↓ 36,<u>8</u>33 rounds to 37,000 ↑ Thousands place ↑ Change 6 to 7. Round $^-3582$ to the nearest ten. Next digit is *4 or less*. ↓ Changed to 0 ↓ $^-358\underline{2}$ rounds to $^-3580$ ↑ Tens place ↑ Leave 8 as 8.
1.5 Front End Rounding Round to the highest possible place so that all the digits become 0 except the first digit.	Use front end rounding. Next digit is *5 or more*. ↓ Changed to 0 ↓↓↓ <u>9</u>7,203 rounds to 100,000 ↑— Change 9 to 10. Carry 1 —↑ into hundred-thousands place.
1.5 Estimating Answers in Addition and Subtraction Use front end rounding to round the numbers in a problem. Then add or subtract the rounded numbers to estimate the answer.	First use front end rounding to estimate the answer. Then find the exact answer. The temperature was 48 degrees below zero. During the morning it rose 21 degrees. What was the new temperature? *Estimate:* $^-50 + 20 = {}^-30$ degrees *Exact:* $^-48 + 21 = {}^-27$ degrees
1.6 Multiplying Two Integers If two factors have *different signs*, the product is *negative*. If two factors have the *same sign*, the product is *positive*.	Multiply. (a) $^-5(6) = {}^-30$ The factors have *different* signs, so the product is *negative*. (b) $(^-10)(^-2) = 20$ The factors have the *same* sign, so the product is *positive*.
1.6 Using Properties of Multiplication *Multiplication property of 0:* Multiplying any number by 0 gives a product of 0. *Multiplication property of 1:* Multiplying any number by 1 leaves the number unchanged. *Commutative property of multiplication:* Changing the *order* of two factors does not change the product. *Associative property of multiplication:* Changing the *grouping* of factors does not change the product. *Distributive property:* Multiplication distributes over addition.	Name the property illustrated by each case. (a) $^-49 \cdot 0 = 0$ (b) $1(675) = 675$ (c) $(^-8)(2) = (2)(^-8)$ Both products are $^-16$. (d) $(^-3 \cdot {}^-2) \cdot 4 = {}^-3 \cdot (^-2 \cdot 4)$ Both products are 24. (e) $5(2 + 4) = 5 \cdot 2 + 5 \cdot 4$ Both results are 30. (a) Multiplication property of 0 (b) Multiplication property of 1 (c) Commutative property of multiplication (d) Associative property of multiplication (e) Distributive property

Concepts	Examples
1.6 Estimating Answers in Multiplication First use front end rounding. Then multiply the rounded numbers using a shortcut: Multiply the nonzero digits in each factor; count the total number of zeros in the two factors and write that number of zeros in the product.	First use front end rounding to estimate the answer. Then find the exact answer. At a PTA fund-raiser, Lionel sold 96 photo albums at $22 each. How much money did he take in? *Estimate:* 100 • $20 = $2000 *Exact:* 96 • $22 = $2112
1.7 Dividing Two Integers Use the same rules for dividing two integers as for multiplying two integers. If two numbers have *different signs*, the quotient is *negative*. If two numbers have the *same sign*, the quotient is *positive*.	Divide. $\dfrac{-24}{6} = -4$ Numbers have *different* signs, so the quotient is *negative*. $-72 \div -8 = 9$ Numbers have the *same* sign, so the quotient is *positive*. $\dfrac{50}{-5} = -10$ Numbers have *different* signs, so the quotient is *negative*.
1.7 Using Properties of Division (a) When a nonzero number is divided by itself, the quotient is 1. (b) When a number is divided by 1, the quotient is the number. (c) When 0 is divided by any other number (except 0), the quotient is 0. (d) Division by 0 is *undefined*. There is no answer.	State the property illustrated by each case. (a) $\dfrac{-4}{-4} = 1$ (b) $\dfrac{65}{1} = 65$ (c) $\dfrac{0}{9} = 0$ (d) $\dfrac{-10}{0}$ is undefined. The examples are in the same order as the properties listed at the left. Note that division is *not* commutative nor associative.
1.7 Estimating Answers in Division First use front end rounding. Then divide the rounded numbers, using a shortcut: Drop the same number of zeros in both the divisor and the dividend.	First use front end rounding to estimate the answer. Then find the exact answer. Joan has one year to pay off a $1020 loan. What is her monthly payment? *Estimate:* $1000 ÷ 10 = $100 *Exact:* $1020 ÷ 12 = $85
1.7 Interpreting Remainders in Division In some situations the remainder tells you how much is left over. In other situations, you must increase the quotient by 1 in order to accommodate the "left over."	Divide; then interpret the remainder. Each chemistry student needs 35 milliliters of acid for an experiment. How many students can be served from a bottle holding 500 milliliters of acid? $$\begin{array}{r} 14 \\ 35\overline{)500} \\ \underline{35} \\ 150 \\ \underline{140} \\ 10 \end{array}$$ 14 → 14 students served 10 → 10 milliliters of acid left over

Concepts	Examples
1.8 Using Exponents An exponent tells how many times a number is used as a factor in repeated multiplication. An exponent applies only to its base (the first thing to the left of the exponent).	Simplify. Exponent ↓ (a) $2^5 = 2 \cdot 2 \cdot 2 \cdot 2 \cdot 2 = 32$ (b) $(^-3)^2 = (^-3)(^-3) = 9$
1.8 Order of Operations Mathematicians have agreed to follow this order. *Step 1* Work inside parentheses or other grouping symbols. *Step 2* Simplify expressions with exponents. *Step 3* Do the remaining multiplications and divisions as they occur from left to right. *Step 4* Do the remaining additions and subtractions as they occur from left to right.	Simplify. $(^-2)^4 + 3(^-4 - ^-2)$ Work inside parentheses. $(^-2)^4 + \;\;\;\;\; 3(^-2)$ Simplify exponents. $16 \;\;+\;\;\;\;\;\; 3(^-2)$ Multiply. $16 \;\;+\;\;\;\;\;\; ^-6$ Add. $\;\;\;\;\;\;\;\;\; 10$
1.8 Using the Order of Operations with Fraction Bars When there is a fraction bar, do all the work in the numerator. Then do all the work in the denominator. Finally, divide numerator by denominator.	Simplify. $$\frac{^-10 + 4^2 - 6}{2 + 3(1 - 4)} = \frac{^-10 + 16 - 6}{2 + 3(^-3)} = \frac{0}{^-7} = 0$$

ANSWERS TO TEST YOUR WORD POWER

1. B. *Example:* In $7(10) = 70$, the factors are 7 and 10.
2. D. *Example:* 12, 0, 710, and 89,475 are all whole numbers.
3. A. *Example:* $|^-3| = 3$ and $|3| = 3$ because both $^-3$ and 3 are 3 steps away from 0.
4. C. *Example:* In 2^5, the exponent is 5, which indicates that 2 is multiplied by itself 5 times, so $2^5 = 2 \cdot 2 \cdot 2 \cdot 2 \cdot 2 = 32$.
5. B. *Example:* The opposite of 5 is $^-5$; it is the additive inverse because $5 + (^-5) = 0$.
6. C. *Example:* Using front end rounding, round 48,299 to the highest possible place, which is ten-thousands. So, 48,299 rounds to 50,000. All digits are 0 except the first digit.
7. B. *Example:* $^-10$, 0, and 6 are all integers.
8. D. *Example:* $4 \cdot (^-2 \cdot 3)$ can be rewritten as $(4 \cdot ^-2) \cdot 3$; both products are $^-24$.

Chapter 1
REVIEW EXERCISES

If you need help with any of these Review Exercises, look in the section indicated in the red brackets.

[1.1] 1. Circle the whole numbers: 86   2.831   ⁻4   0 $\frac{2}{3}$ 35,600

Write these numbers in words.

2. 806

3. 319,012

4. 60,003,200

5. 15,749,000,000,006

Write these numbers using digits.

6. Five hundred four thousand, one hundred

7. Six hundred twenty million, eighty thousand

8. Ninety-nine billion, seven million, three hundred fifty-six

[1.2] 9. Graph these numbers: $^-3\frac{1}{2}$, 2, ⁻5, 0.

Write < or > between each pair of numbers to make a true statement.

10. 0 ___ ⁻4
11. ⁻3 ___ ⁻1
12. 2 ___ ⁻2
13. ⁻2 ___ 1

Find each absolute value.

14. |⁻5|
15. |9|
16. |0|
17. |⁻125|

[1.3] *Add.*

18. ⁻9 + 8
19. ⁻8 + ⁻5
20. 16 + ⁻19
21. ⁻4 + 4

22. 6 + ⁻5
23. ⁻12 + ⁻12
24. 0 + ⁻7
25. ⁻16 + 19

26. 9 + ⁻4 + ⁻8 + 3

27. ⁻11 + ⁻7 + 5 + ⁻4

78 Chapter 1 Introduction to Algebra: Integers

[1.4] *Find the opposite (additive inverse) of each number. Show that the sum of the number and its opposite is 0.*

28. ⁻5

29. 18

Subtract by changing subtraction to addition.

30. 5 − 12

31. 24 − 7

32. ⁻12 − 4

33. 4 − ⁻9

34. ⁻12 − ⁻30

35. ⁻8 − 14

36. ⁻6 − ⁻6

37. ⁻10 − 10

38. ⁻8 − ⁻7

39. 0 − 3

40. 1 − ⁻13

41. 15 − 0

Simplify.

42. 3 − 12 − 7

43. ⁻7 − ⁻3 + 7

44. 4 + ⁻2 − 0 − 10

45. ⁻12 − 12 + 20 − ⁻4

[1.5] *Round each number as indicated.*

46. 205 to the nearest ten

47. 59,499 to the nearest thousand

48. 85,066,000 to the nearest million

49. ⁻2963 to the nearest hundred

50. ⁻7,063,885 to the nearest ten-thousand

51. 399,712 to the nearest thousand

Use front end rounding to round each number.

52. The combined weight loss of 10 dieters was 197 pounds.

53. The land on the shore of the Dead Sea in the Middle East is 1312 feet below sea level. (*Source: Goode's World Atlas.*)

54. There are 362,000,000 Yellow Pages directories published in the United States each year. (*Source:* Yellow Pages Publishers Association.)

55. By 2050, the Census Bureau predicts a world population of 9,104,000,000 people. (*Source:* U.S. Bureau of the Census.)

Chapter 1 Review Exercises 79

[1.6] *Multiply.*

56. $^-6(9)$ **57.** $(^-7)(^-8)$ **58.** $10(^-10)$ **59.** $^-45 \cdot 0$

60. $^-1(^-24)$ **61.** $17 \cdot 1$ **62.** $4(^-12)$ **63.** $(^-5)(^-25)$

64. $^-3(^-4)(^-3)$ **65.** $^-5(2)(^-5)$ **66.** $(^-8)(^-1)(^-9)$

[1.7] *Simplify.*

67. $\dfrac{^-63}{^-7}$ **68.** $\dfrac{70}{^-10}$ **69.** $\dfrac{^-15}{0}$ **70.** $^-100 \div ^-20$

71. $18 \div ^-1$ **72.** $\dfrac{0}{12}$ **73.** $\dfrac{^-30}{^-2}$ **74.** $\dfrac{^-35}{35}$

75. $^-40 \div ^-4 \div ^-2$ **76.** $^-18 \div 3(^-3)$ **77.** $0 \div ^-10(5) \div 5$

78. Divide; then interpret the remainder. It took 1250 hours to build the set for a new play. How many working days of eight hours each did it take to build the set?

[1.8] *Simplify.*

79. 10^4 **80.** 2^5 **81.** 3^3 **82.** $(^-4)^2$

83. $(^-5)^3$ **84.** 8^1 **85.** $6^2 \cdot 3^2$ **86.** $5^2(^-2)^3$

87. $^-30 \div 6 - 4(5)$ **88.** $6 + 8(2 - 3)$ **89.** $16 \div 4^2 + (^-6 + 9)^2$

90. $^-3(4) - 2(5) + 3(^-2)$ **91.** $\dfrac{^-10 + 3^2 - ^-9}{3 - 10 - 1}$

92. $\dfrac{^-1(1 - 3)^3 + 12 \div 4}{^-5 + 24 \div 8 \cdot 2(6 - 6) + 5}$

MIXED REVIEW EXERCISES

Name the property illustrated by each case.

93. $^-3 + (5 + 1) = (^-3 + 5) + 1$

94. $^-7(2) = 2(^-7)$

95. $0 + 19 = 19$

96. $^-42 \cdot 0 = 0$

97. $2(^-6 + 4) = 2 \cdot {}^-6 + 2 \cdot 4$

98. $(^-6 \cdot 3) \cdot {}^-1 = {}^-6 \cdot (3 \cdot {}^-1)$

First use front end rounding to estimate each answer. Then find the exact answer.

99. Last year, 192 Elvis jukeboxes were sold at a price of $11,900 each. What was the total value of the jukeboxes?

Estimate:

Exact:

100. Chad had $185 in his checking account. He deposited his $428 paycheck and then wrote a $706 check for car repairs. What is the balance in his account?

Estimate:

Exact:

101. Georgia's car used 24 gallons of gas on her 840-mile vacation trip. What was the average number of miles she drove on each gallon of gas?

Estimate:

Exact:

102. When inventory was taken at Mathtronic Company, 19 calculators and 12 computer modems were missing. Each calculator is worth $39 and each modem is worth $85. What is the total value of the missing items?

Estimate:

Exact:

Elena Sanchez opened a shop that does alterations and designs custom clothing. Use the table of her income and expenses to answer Exercises 103–106.

Month	Income	Expenses	Profit or Loss
Jan.	$2400	$3100	
Feb.	$1900	$2000	
Mar.	$2500	$1800	
Apr.	$2300	$1400	
May	$1600	$1600	
June	$1900	$1200	

103. Complete the table by finding Elena's profit or loss for each month.

104. Which month had the greatest loss? Which month had the greatest profit?

105. What was Elena's average monthly income?

106. What was the average monthly amount of expenses?

Chapter 1
TEST

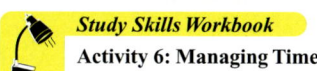
Study Skills Workbook
Activity 6: Managing Time

1. Write this number in words: 20,008,307

2. Write this number using digits:
   thirty billion, seven hundred thousand, five

3. Graph the numbers 3, ⁻2, 0, $-\frac{1}{2}$ on the number line at the right.

4. Write < or > between each pair of numbers to make a true statement.
   0 ___ ⁻3   ⁻2 ___ ⁻1

5. Find each absolute value: |10| and |⁻14|.

Add, subtract, multiply, or divide.

6. 3 − 9

7. ⁻12 + 7

8. $\frac{-28}{-4}$

9. ⁻1(40)

10. ⁻5 − ⁻15

11. (⁻8)(⁻8)

12. ⁻25 + ⁻25

13. $\frac{17}{0}$

14. ⁻30 − 30

15. $\frac{50}{-10}$

16. 5(⁻9)

17. 0 − ⁻6

Simplify.

18. ⁻35 ÷ 7(⁻5)

19. ⁻15 − ⁻8 + 7

20. 3 − 7(⁻2) − 8

21. (⁻4)² • 2³

22. $\frac{5^2 - 3^2}{(4)(-2)}$

23. ⁻2(⁻4 + 10) + 5(4)

24. ⁻3 + (⁻7 − ⁻10) + 4(6 − 10)

25. Explain how an exponent is used. Include two examples.

1. _____
2. _____
3. [number line from ⁻3 to 3]
4. _____
5. _____
6. _____
7. _____
8. _____
9. _____
10. _____
11. _____
12. _____
13. _____
14. _____
15. _____
16. _____
17. _____
18. _____
19. _____
20. _____
21. _____
22. _____
23. _____
24. _____
25. _____

26. Explain the commutative and associative properties of addition. Also give an example to illustrate each property.

Round each number as indicated.

27. 851 to the nearest hundred

28. 36,420,498,725 to the nearest million

29. 349,812 to the nearest thousand

First use front end rounding to estimate the answer to each application problem. Then find the exact answer.

30. Lorene had $184 in her checking account. She deposited her $293 paycheck and then wrote a $506 check for tuition. What is the balance in her account?

31. The Cardinals football team had a bad year. It lost a total of 1140 yards in 12 games. What was the average loss in each game?

32. One kind of cereal has 220 calories in each serving. Another kind has 110 calories in each serving. During a month with 31 days, how many calories would you save by eating the second kind of cereal each morning for breakfast? (*Source:* General Mills and Post Cereals.)

Write and solve a subtraction problem to answer this question.

33. On the planet Mars the average high temperature is ⁻10 °F and the average low is ⁻100 °F. What is the difference between the high and low temperatures? (*Source:* NASA.)

Divide; then interpret the remainder.

34. Anthony has a part-time job as a shipping clerk. He is sending 1276 pounds of books to a bookstore. Each shipping carton can safely hold 48 pounds. What is the minimum number of cartons he will need?

Understanding Variables and Solving Equations

2

2.1 **Introduction to Variables**
2.2 **Simplifying Expressions**
2.3 **Solving Equations Using Addition**
2.4 **Solving Equations Using Division**
2.5 **Solving Equations with Several Steps**

As many as 50,000 thunderstorms occur worldwide in a single day. Thunderstorms always include lightning, and in the United States an average of 300 people are struck by lightning each year. (*Source:* AccuWeather.com) But you don't need a weather forecaster to tell you how far away a thunderstorm is. You can use the algebraic expression in **Section 2.1,** Exercises 59–60, to estimate your distance from the storm and, if it is close, move to shelter.

There are other algebraic expressions related to weather. For example, in summertime, you can find the temperature by counting the chirps of a cricket and using the expression in **Section 2.1,** Exercise 14.

2.1 Introduction to Variables

OBJECTIVES

1. Identify variables, constants, and expressions.
2. Evaluate variable expressions for given replacement values.
3. Write properties of operations using variables.
4. Use exponents with variables.

Study Skills Workbook
Activity 7: Study Cards

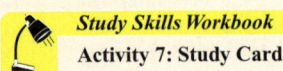

OBJECTIVE 1 Identify variables, constants, and expressions. You probably know that algebra uses letters, especially the letter x. But why use letters when numbers are easier to understand? Here is an example.

Suppose that you run your college bookstore. When deciding how many books to order for a certain class, you first find out the class limit, that is, the maximum number of students allowed in the class. You will need at least that many books. But you decide to order 5 extra copies for emergencies.

Rule for ordering books: Order the class limit + 5 extra

How many books would you order for a prealgebra class with a limit of 25 students?

Class limit ⟶   ⟵ Extra
$$25 + 5$$
You would order 30 prealgebra books.

How many books would you order for a geometry class that allows 40 students to register?

Class limit ⟶   ⟵ Extra
$$40 + 5$$
You would order 45 geometry books.

You could set up a table to keep track of the number of books to order for various classes.

Class	Rule for Ordering Books: Class Limit + 5 Extra	Number of Books to Order
Prealgebra	25 + 5	30
Geometry	40 + 5	45
College algebra	35 + 5	40
Calculus 1	50 + 5	55

A shorthand way to write your rule is shown below.

$$c + 5$$
↑
The c stands for class limit.

You can't write your rule by using just numbers because the class limit varies, or changes, depending on which class you're talking about. So you use a letter, called a **variable**, to represent the part of the rule that varies. Notice the similarity in the words *varies* and *variable*. When part of a rule does *not* change, it is called a **constant.**

The variable, or the part of
the rule that varies or changes
↓
$$c + 5$$
↑
The constant, or the part of
the rule that does *not* change

$c + 5$ is called an **expression**. It expresses (tells) the rule for ordering books. You could use any letter you like for the variable part of the expression, such as $x + 5$, or $n + 5$, and so on. But one suggestion is to use a letter that reminds you of what it stands for. In this situation, the letter c reminds us of "class limit."

EXAMPLE 1 Writing an Expression and Identifying the Variable and Constant

Write an expression for this rule. Identify the variable and the constant.
Order the class limit minus 10 books because some students will buy used books.

$$\underset{\uparrow}{c} - \underset{\uparrow}{10}$$
Variable — Constant

Work Problem 1 at the Side.

OBJECTIVE 2 Evaluate variable expressions for given replacement values. When you need to figure out how many books to order for a particular class, you use a specific value for the class limit, like 25 students in prealgebra. Then you **evaluate the expression,** that is, you follow the rule.

Ordering Books for a Prealgebra Class

$c + 5$ Expression (rule for ordering books) is $c + 5$.
↓   Replace c with 25, the class limit for prealgebra.
$\underbrace{25 + 5}$ Follow the rule. Add $25 + 5$.
30   Order 30 prealgebra books.

EXAMPLE 2 Evaluating an Expression

Use this rule for ordering books: Order the class limit minus 10. The expression is $c - 10$.

(a) Evaluate the expression when the class limit is 32.

$c - 10$ Replace c with 32.
↓
$\underbrace{32 - 10}$ Follow the rule. Subtract to find $32 - 10$.
22   Order 22 books.

(b) Evaluate the expression when the class limit is 48.

$c - 10$ Replace c with 48.
↓
$\underbrace{48 - 10}$ Follow the rule. Subtract to find $48 - 10$.
38   Order 38 books.

Work Problem 2 at the Side.

In any career you choose, there will be many useful "rules" that need to be written using variables (letters) because part of the rule changes depending on the situation. This is one reason algebra is such a powerful tool. Here is another example.

Suppose that you work in a landscaping business. You are putting a fence around a square-shaped garden. Each side of the garden is 6 feet long. How much fencing material should you bring to finish the job? You could add the lengths of the four sides.

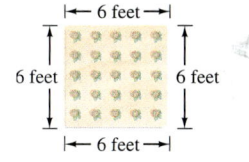

6 feet + 6 feet + 6 feet + 6 feet = 24 feet of fencing

---

1 Write an expression for this rule. Identify the variable and the constant.

Order the class limit plus 15 extra books because it is a very large class.

2 Use this expression for ordering books: $c + 3$.

(a) Evaluate the expression when the class limit is 25.

(b) Evaluate the expression when the class limit is 60.

ANSWERS
1. The expression is $c + 15$. The variable is c and the constant is 15.
2. **(a)** $25 + 3$ is 28; order 28 books
   (b) $60 + 3$ is 63; order 63 books

3 (a) Evaluate the expression $4s$ when the length of one side of a square table is 3 feet.

Or, recall that multiplication is a quick way to do repeated addition. The square garden has 4 sides, so multiply by 4.

$$4 \cdot 6 \text{ feet} = 24 \text{ feet of fencing}$$

So the rule for calculating the amount of fencing for a square garden is

$$4 \cdot \text{length of one side.}$$

Other jobs may require fencing for larger or smaller square shapes. The following table shows how much fencing you will need.

Length of One Side of Square Shape	Expression (Rule) to Find Total Amount of Fencing: 4 • Length of One Side	Total Amount of Fencing Needed
6 feet	4 • 6 feet	24 feet
9 feet	4 • 9 feet	36 feet
10 feet	4 • 10 feet	40 feet
3 feet	4 • 3 feet	12 feet

The expression (rule) can be written in shorthand form as shown below.

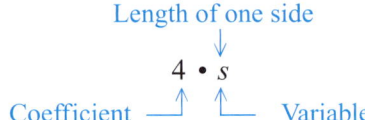

The number part in a *multiplication* expression is called the numerical coefficient, or just the **coefficient.** We usually don't write multiplication dots in expressions, so we do the following.

$$4 \cdot s \quad \text{is written as} \quad 4s$$

(b) Evaluate the expression $4s$ when the length of one side of a square park is 7 miles.

You can use the expression $4s$ any time you need to know the *perimeter* of a square shape, that is, the total distance around all four sides of the square.

> **CAUTION**
> If an expression involves adding, subtracting, or dividing, then you *do* have to write $+$, $-$, or $\div$. It is *only* multiplication that is understood without writing an operation symbol.
>
> $\underset{\text{Add } s.}{4 + s} \qquad \underset{\text{Subtract } s.}{4 - s} \qquad \underset{\text{Divide by } s.}{4 \div s} \qquad \underset{\text{Multiply by } s.}{4s}$

EXAMPLE 3 Evaluating an Expression with Multiplication

The expression (rule) for finding the perimeter of a square shape is $4s$. Evaluate the expression when the length of one side of a square parking lot is 30 yards.

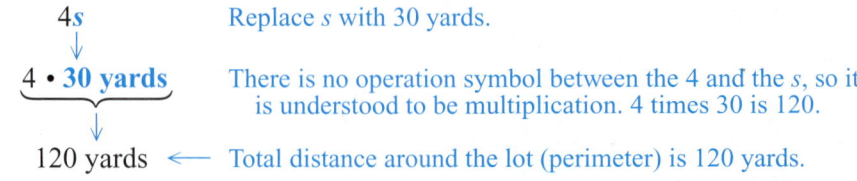

ANSWERS
3. (a) 4 • 3 feet; 12 feet
   (b) 4 • 7 miles; 28 miles

Some expressions (rules) involve several different steps. An expression for finding the approximate systolic blood pressure of a person of a certain age is shown below.

$$100 + \frac{a}{2} \leftarrow \text{Age of person (the variable)}$$

Remember that a fraction bar means division, so $\frac{a}{2}$ is the person's age divided by 2. You also need to remember the order of operations, which means doing division before addition.

EXAMPLE 4 Evaluating an Expression with Several Steps

Evaluate the expression $100 + \frac{a}{2}$ when the age of the person is 24.

$100 + \frac{a}{2}$ Replace *a* with 24, the age of the person.

$100 + \frac{24}{2}$ Follow the rule using the order of operations. First divide: 24 ÷ 2 is 12.

$100 + 12$ Now add: 100 + 12 is 112.

112 ←— The approximate systolic blood pressure is 112.

Work Problem 4 at the Side.

Calculator Tip If you like to fish, you can use an expression (rule) like the one below to find the approximate weight (in pounds) of a fish you catch. Measure the length of the fish (in inches) and then use the correct expression for that type of fish. For a northern pike, the weight expression is shown below.

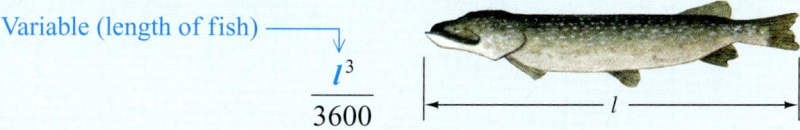

Variable (length of fish) — $\frac{l^3}{3600}$

where *l* is the length of the fish in inches. (*Source: InFisherman.*)

To evaluate this expression for a fish that is 43 inches long, follow the rule by calculating as follows.

$\frac{43^3}{3600}$ Replace *l* with 43, the length of the fish in inches.

In the numerator, you can multiply 43 • 43 • 43 or use the y^x key on your calculator. Then divide by 3600.

Enter 43 y^x 3 ÷ 3600 =   Calculator shows 22.08527778
↑   ↑
Base   Exponent

The fish weighs about 22 pounds.

Now use the expression to find the approximate weight of a northern pike that is 37 inches long. (Answer: about 14 pounds.)

(*continued*)

---

4 Evaluate the expression $100 + \frac{a}{2}$ when the age of the person is 40.

ANSWERS

4. $100 + \frac{40}{2}$ ←— Replace *a* with 40.

     $100 + 20$

     120   ←— Approximate systolic blood pressure is 120.

5 **(a)** Use the expression for finding your average bowling score. Evaluate the expression if your total score for 4 games is 532.

(b) Complete this table.

Value of x	Value of y	Expression x − y
16	10	16 − 10 is 6
100	5	
3	7	
8	0	

Notice that variables are used on your calculator keys. On the y^x key, y represents the base and x represents the exponent. You first evaluated y^x by entering 43 as the base and 3 as the exponent for the first fish. Then you evaluated y^x again by entering 37 as the base and 3 as the exponent for the second fish.

Some expressions (rules) involve several variables. For example, if you bowl three games and want to know your average score, you can use this expression.

$\dfrac{t}{g}$ ← Total score for all games (variable)
← Number of games (variable)

EXAMPLE 5 **Evaluating Expressions with Two Variables**

(a) Find your average score if you bowl three games and your total score for all three games is 378.

Use the expression (rule) for finding your average score.

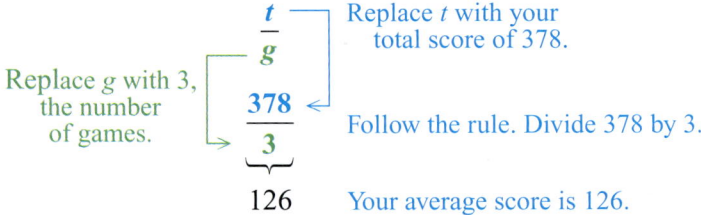

Your average score is 126.

(b) Complete these tables to show how to evaluate each expression.

Value of x	Value of y	Expression (Rule) x + y
2	5	2 + 5 is 7
−6	4	__ + __ is __
0	16	__ + __ is __

Value of x	Value of y	Expression (Rule) xy
2	5	2 • 5 is 10
−6	4	__ • __ is __
0	16	__ • __ is __

The expression (rule) is to *add* the two variables. So the completed table is:

Value of x	Value of y	Expression (Rule) x + y
2	5	2 + 5 is 7
−6	4	−6 + 4 is −2
0	16	0 + 16 is 16

The expression (rule) is to *multiply* the two variables. We know that it's multiplication because there is no operation symbol between the *x* and *y*. So the completed table is:

Value of x	Value of y	Expression (Rule) xy
2	5	2 • 5 is 10
−6	4	−6 • 4 is −24
0	16	0 • 16 is 0

◀◀◀ **Work Problem 5 at the Side.**

ANSWERS

5. **(a)** $\dfrac{532}{4}$; average score is 133.

(b) 100 − 5 is 95
3 − 7 is −4
8 − 0 is 8

OBJECTIVE 3 Write properties of operations using variables. Now you can use variables as a shorthand way to express the properties you learned about in **Sections 1.3, 1.6,** and **1.7.** We'll use the letters a and b to represent any two numbers.

Commutative Property of Addition

$a + b = b + a$

Commutative Property of Multiplication

$a \cdot b = b \cdot a$

To get specific examples, you can pick values for a and b. For example, if a is $^-3$, replace every a with $^-3$. If b is 5, replace every b with 5.

$$a + b = b + a$$
$$^-3 + 5 = 5 + {^-3}$$
$$2 = 2$$

Both sums are 2

$$a \cdot b = b \cdot a$$
$$^-3 \cdot 5 = 5 \cdot {^-3}$$
$$^-15 = {^-15}$$

Both products are $^-15$

Of course, you could pick many different values for a and b, because the commutative "rule" will always work for adding *any* two numbers or multiplying *any* two numbers.

EXAMPLE 6   Writing Properties of Operations Using Variables

Use the variable b to state this property: When any number is divided by 1, the quotient is the number.

Use the letter b to represent any number.

$$\frac{b}{1} = b$$

▶▶▶ **Work Problem 6 at the Side.**

OBJECTIVE 4 Use exponents with variables. In **Section 1.8** we used an exponent as a quick way to write repeated multiplication. For example,

$\underbrace{3 \cdot 3 \cdot 3 \cdot 3 \cdot 3}_{\text{3 is used as a factor 5 times.}}$ can be written $3^5 \leftarrow$ Exponent
$\phantom{3 \cdot 3 \cdot 3 \cdot 3 \cdot 3 \text{ can be written } 3}\uparrow$ Base

The meaning of an exponent remains the same when a variable (a letter) is the base.

$\underbrace{c \cdot c \cdot c \cdot c \cdot c}_{c \text{ is used as a factor 5 times.}}$ can be written $c^5 \leftarrow$ Exponent
$\phantom{c \cdot c \cdot c \cdot c \cdot c \text{ can be written } c}\uparrow$ Base

m^2 means $m \cdot m$ Here m is used as a factor 2 times.

$x^4 y^3$ means $x^4 \cdot y^3$ or $\underbrace{x \cdot x \cdot x \cdot x}_{x^4} \cdot \underbrace{y \cdot y \cdot y}_{y^3}$

$7b^2$ means $7 \cdot b \cdot b$ The exponent applies *only* to b.

$^-4xy^2z$ means $^-4 \cdot x \cdot y \cdot y \cdot z$ The exponent applies *only* to y.

---

6 (a) Use the variable a to state this property: Multiplying any number by 0 gives a product of 0.

(b) Use the variables a, b, and c to state the associative property of addition: Changing the grouping of addends does not change the sum.

ANSWERS
6. (a) $0 \cdot a = 0$ or $a \cdot 0 = 0$
   (b) $(a + b) + c = a + (b + c)$

Chapter 2 Understanding Variables and Solving Equations

7 Rewrite each expression without exponents.

(a) x^5

(b) $4a^2b^2$

(c) $-10xy^3$

(d) s^4tu^2

EXAMPLE 7 Understanding Exponents Used with Variables

Rewrite each expression without exponents.

(a) y^6 can be written as $y \cdot y \cdot y \cdot y \cdot y \cdot y$

y is used as a factor 6 times.

(b) $12bc^3$ can be written as $12 \cdot b \cdot \underbrace{c \cdot c \cdot c}_{c^3}$

Coefficient is 12. The exponent applies *only* to c.

(c) $-2m^2n^4$ can be written as $-2 \cdot \underbrace{m \cdot m}_{m^2} \cdot \underbrace{n \cdot n \cdot n \cdot n}_{n^4}$

Coefficient is -2.

◀◀◀ **Work Problem 7 at the Side.**

To evaluate an expression with exponents, multiply all the factors.

EXAMPLE 8 Evaluating Expressions with Exponents

Evaluate each expression.

(a) x^2 when x is -3

x^2 means $x \cdot x$ — Replace each x with -3.

$\underbrace{-3 \cdot -3}_{9}$ Multiply -3 times -3.

So x^2 becomes $(-3)^2$, which is $(-3)(-3)$, or 9.

(b) x^3y when x is -4 and y is -10

x^3y means $x \cdot x \cdot x \cdot y$ — Replace x with -4, and replace y with -10.

$\underbrace{-4 \cdot -4}_{16} \cdot -4 \cdot -10$ Multiply two factors at a time.

$\underbrace{16 \cdot -4}_{-64} \cdot -10$

$\underbrace{-64 \cdot -10}_{640}$

So x^3y becomes $(-4)^3(-10)$, which is $(-4)(-4)(-4)(-10)$, or 640.

8 Evaluate each expression.

(a) y^3 when y is -5

(b) r^2s^2 when r is 6 and s is 3

(c) $10xy^2$ when x is 4 and y is -3

(d) $-3c^4$ when c is 2

(c) $-5ab^2$ when a is 5 and b is 3

$-5ab^2$ means $-5 \cdot a \cdot b \cdot b$ — Replace a with 5, and replace b with 3.

$\underbrace{-5 \cdot 5}_{-25} \cdot 3 \cdot 3$ Multiply two factors at a time.

$\underbrace{-25 \cdot 3}_{-75} \cdot 3$

-225

So $-5ab^2$ becomes $-5(5)(3)^2$, which is $-5(5)(3)(3)$, or -225.

Coefficient is -5.

ANSWERS
7. (a) $x \cdot x \cdot x \cdot x \cdot x$
   (b) $4 \cdot a \cdot a \cdot b \cdot b$
   (c) $-10 \cdot x \cdot y \cdot y \cdot y$
   (d) $s \cdot s \cdot s \cdot s \cdot t \cdot u \cdot u$
8. (a) $(-5)^3$ is -125. (b) $(6)^2(3)^2$ is 324.
   (c) $10(4)(-3)^2$ is 360. (d) $-3(2)^4$ is -48.

◀◀◀ **Work Problem 8 at the Side.**

2.1 Exercises

Identify the parts of each expression. Choose from these labels: variable, constant, and coefficient. See Example 1.

1. $c + 4$ **2.** $d + 6$ **3.** $5h$ **4.** $3s$

5. $^-3 + m$ **6.** $^-4 + n$ **7.** $2c - 10$ **8.** $6b - 1$

9. $x - y$ **10.** xy **11.** $^-6g + 9$ **12.** $^-10k + 15$

Evaluate each expression. See Examples 2–5.

13. The expression (rule) for ordering robes for the graduation ceremony at West Community College is $g + 10$, where g is the number of graduates. Evaluate the expression when

   (a) there are 654 graduates.

   (b) there are 208 graduates.

   (c) there are 95 graduates.

14. Crickets chirp faster as the weather gets hotter. The expression (rule) for finding the approximate temperature (in degrees Fahrenheit) is $c + 37$ where c is the number of chirps made in 15 seconds by a field cricket. (*Source:* AccuWeather.com) Evaluate the expression when the cricket

   (a) chirps 45 times.

   (b) chirps 33 times.

   (c) chirps 58 times.

15. The expression for finding the perimeter of a triangle with sides of equal length is $3s$, where s is the length of one side. Evaluate the expression when

   (a) the length of one side is 11 inches.

   (b) the length of one side is 3 feet.

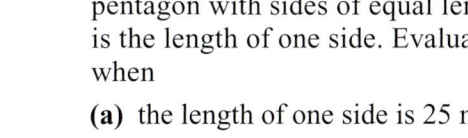

16. The expression for finding the perimeter of a pentagon with sides of equal length is $5s$, where s is the length of one side. Evaluate the expression when

   (a) the length of one side is 25 meters.

   (b) the length of one side is 8 inches.

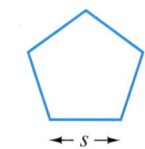

17. The expression for ordering brushes for an art class is $3c - 5$, where c is the class limit. Evaluate the expression when

   (a) the class limit is 12.

   (b) the class limit is 16.

18. The expression for ordering doughnuts for the office staff is $2n - 4$, where n is the number of people at work. Evaluate the expression when

   (a) there are 13 people at work.

   (b) there are 18 people at work.

19. The expression for figuring a student's average test score is $\frac{p}{t}$, where p is the total points earned on all the tests and t is the number of tests. Evaluate the expression when

   (a) 332 points were earned on 4 tests.

   (b) there were 7 tests and 637 points were earned.

20. The expression for deciding how many buses are needed for a group trip is $\frac{p}{b}$, where p is the total number of people and b is the number of people that one bus will hold. Evaluate the expression when

   (a) 176 people are going on a trip and one bus holds 44 people.

   (b) A bus holds 36 people and 72 people are going on a trip.

Complete each table by evaluating the expressions. See Example 5.

21.

Value of x	Expression $x + x + x + x$	Expression $4x$
$^-2$	$^-2 + {}^-2 + {}^-2 + {}^-2$ is $^-8$	$4 \cdot {}^-2$ is $^-8$
12		
0		
$^-5$		

22.

Value of y	Expression $3y$	Expression $y + 2y$
$^-6$	$3({}^-6)$ is $^-18$	$^-6 + 2({}^-6)$ is $^-6 + {}^-12$, or $^-18$
10		
$^-3$		
0		

23.

Value of x	Value of y	Expression $^-2x + y$
3	7	$^-2(3) + 7$ is $^-6 + 7$, or 1
$^-4$	5	
$^-6$	$^-2$	
0	$^-8$	

24.

Value of x	Value of y	Expression ^-2xy
3	7	$^-2 \cdot 3 \cdot 7$ is $^-42$
$^-4$	5	
$^-6$	$^-2$	
0	$^-8$	

25. Explain the words *variable* and *expression*.

26. Explain the words *coefficient* and *constant*.

Use the variable b to express each of these properties. See Example 6.

27. Multiplying a number by 1 leaves the number unchanged.

28. Adding 0 to any number leaves the number unchanged.

29. Any number divided by 0 is undefined.

30. Multiplication distributes over addition. (Use *a*, *b*, and *c* as the variables.)

Rewrite each expression without exponents. See Example 7.

31. c^6

32. d^7

33. $x^4 y^3$

34. $c^2 d^5$

35. $^-3a^3 b$

36. $^-8m^2 n$

37. $9xy^2$

38. $5ab^4$

39. $^-2c^5 d$

40. $^-4x^3 y$

41. $a^3 bc^2$

42. $x^2 y z^6$

Evaluate each expression when r is $^-3$, s is 2, and t is $^-4$. See Example 8.

43. t^2

44. r^2

45. rs^3

46. $s^4 t$

47. $3rs$

48. $6st$

49. $-2s^2t^2$

50. $-4rs^4$

51. $r^2s^5t^3$

52. $r^3s^4t^2$

53. $-10r^5s^7$

54. $-5s^6t^5$

Evaluate each expression when x is 4, y is -2, and z is -6.

55. $|xy| + |xyz|$

56. $x + |y^2| + |xz|$

57. $\dfrac{z^2}{-3y + z}$

58. $\dfrac{y^2}{x + 2y}$

RELATING CONCEPTS (EXERCISES 59–60) For Individual or Group Work

At the beginning of this chapter, you read that there is an expression that tells your approximate distance from a thunderstorm. First, count the number of seconds from the time you see a lightning flash until you start to hear the thunder. Then, to estimate the distance (in miles), use the expression $\dfrac{s}{5}$, where s is the number of seconds. Use this expression as you **work Exercises 59 and 60 in order.** *(This expression is based on the fact that the light from the lightning travels faster than the sound from the thunder.)*

59. Evaluate the thunderstorm expression for each number of seconds. How far away is the storm?

(a) 15 seconds

(b) 10 seconds

(c) 5 seconds

60. Explain how you can use your answers from Exercise 59 to:

(a) Estimate the distance when the time is $2\tfrac{1}{2}$ seconds.

(b) Find the number of seconds when the distance is $1\tfrac{1}{2}$ miles.

(c) Find the number of seconds when the distance is $2\tfrac{1}{2}$ miles.

2.2 Simplifying Expressions

OBJECTIVE 1 **Combine like terms, using the distributive property.**
In **Section 2.1**, the expression for ordering math textbooks was $c + 5$. This expression was simple and easy to use. Sometimes expressions are *not* written in the simplest possible way. For example:

Evaluate this expression when c is 20.

$c + 5$ Replace c with 20.
$20 + 5$ Add $20 + 5$.
25

Evaluate this expression when c is 20.

$2c - 10 - c + 15$ Replace c with 20.
$2 \cdot 20 - 10 - 20 + 15$ Multiply $2 \cdot 20$.
$40 - 10 - 20 + 15$ Change subtraction to adding the opposite.
$40 + {}^-10 + {}^-20 + 15$ Add from left to right.
$30 + {}^-20 + 15$
$10 + 15$
25

OBJECTIVES

1. Combine like terms, using the distributive property.
2. Simplify expressions.
3. Use the distributive property to multiply.

These two expressions are actually equivalent. When you evaluate them, the final result is the same, but it takes a lot more work when you use the right-hand expression. To save a lot of work, you need to learn how to *simplify expressions*. Then you can rewrite $2c - 10 - c + 15$ in the simplest way possible, which is $c + 5$.

The basic idea in **simplifying expressions** is to *combine*, or *add*, like terms. Each addend in an expression is a **term**. Here are two examples.

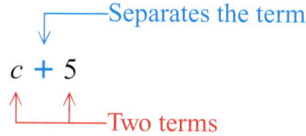

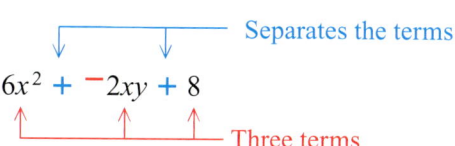

In $6x^2 + {}^-2xy + 8$, the 8 is the *constant term*. There are also two *variable terms* in the expression: $6x^2$ is a variable term, and ${}^-2xy$ is a variable term. A **variable term** has a number part (coefficient) and a letter part (variable).

If no coefficient is shown, it is assumed to be 1. Remember from **Section 1.6** that multiplying any number by 1 does *not* change the number.

c can be written $1 \cdot c$ or just $1c$

Coefficient is understood to be 1.   Variable part is c.

Also, ${}^-c$ can be written ${}^-1 \cdot c$. The coefficient of ${}^-c$ is understood to be ${}^-1$.

Like Terms

Like terms are terms with exactly the same variable parts (the same letters and exponents). The coefficients do *not* have to match.

96 Chapter 2 Understanding Variables and Solving Equations

1 List the like terms in each expression. Then identify the coefficients of the like terms.

(a) $3b^2 + {}^-3b + 3 + b^3 + b$

(b) $^-4xy + 4x^2y + {}^-4xy^2 + {}^-4 + 4$

(c) $5r^2 + 2r + {}^-2r^2 + 5 + 5r^3$

(d) $^-10 + {}^-x + {}^-10x + {}^-x^2 + {}^-10y$

	Like Terms		Unlike Terms	
$5x$ and $3x$	Variable parts match; both are x.	$3x$ and $3x^2$	Variable parts do *not* match; exponents are different.
$^-6y^3$ and y^3	Variable parts match; both are y^3.	^-2x and ^-2y	Variable parts do *not* match; letters are different.
$4a^2b$ and $5a^2b$	Variable parts match; both are a^2b.	a^3b and a^2b	Variable parts do *not* match; exponents are different.
$^-8$ and 4	There are no variable parts; numbers are like terms.	^-8c and 4	Variable parts do *not* match; one term has a variable part, but the other term does not.

EXAMPLE 1   Identifying Like Terms and Their Coefficients

List the like terms in each expression. Then identify the coefficients of the like terms.

(a) $^-5x + {}^-5x^2 + 3xy + x + {}^-5$
The like terms are ^-5x and x.
The coefficient of ^-5x is $^-5$, and the coefficient of x is understood to be 1.

(b) $2yz^2 + 2y^2z + {}^-3y^2z + 2 + {}^-6yz$
The like terms are $2y^2z$ and $^-3y^2z$.
The coefficients are 2 and $^-3$.

(c) $10ab + 12 + {}^-10a + 12b + {}^-6$
The like terms are 12 and $^-6$.
The like terms are constants (there are no variable parts).

Work Problem 1 at the Side.

The distributive property (see **Section 1.6**) can be used "in reverse" to combine like terms. Here is an example.

$$\underbrace{3 \cdot x + 4 \cdot x}_{3x \;+\; 4x} \quad \text{can be written as} \quad \underbrace{(3+4) \cdot x}_{\underbrace{7 \cdot x}_{7x}}$$

Thus, $3x + 4x$ can be written in *simplified form* as $7x$. To check that $3x + 4x$ is the same as $7x$, evaluate each expression when x is 2.

$$\begin{array}{cc} 3x \;+\; 4x & 7x \\ \downarrow \quad\;\; \downarrow & \downarrow \\ 3 \cdot 2 + 4 \cdot 2 & 7 \cdot 2 \\ \underbrace{6 \;+\; 8}_{14} & 14 \end{array}$$

Both results are 14, so the expressions are equivalent. But you can see how much easier it is to work with $7x$, the simplified expression.

ANSWERS
1. (a) ^-3b and b; the coefficients are $^-3$ and 1.
   (b) $^-4$ and 4; they are constants.
   (c) $5r^2$ and $^-2r^2$; the coefficients are 5 and $^-2$.
   (d) ^-x and ^-10x; the coefficients are $^-1$ and $^-10$.

> **CAUTION**
> Notice that $3x + 4x$ is simplified to $7x$, **not** to $7x^2$.
>
> ↑ ↑ ↑   ↑
> Variable part is unchanged.   Do *not* change x to x^2.

Combining Like Terms

Step 1   If there are any variable terms with no coefficient, write in the understood 1.

Step 2   If there are any subtractions, change each one to adding the opposite.

Step 3   Find *like* terms (the variable parts match).

Step 4   Add the coefficients (number parts) of like terms. *The variable part stays the same.*

EXAMPLE 2   Combining Like Terms

Combine like terms.

(a) $2x + 4x + x$

$2x + 4x + x$ No coefficient; write understood 1.
                There are no subtractions to change.

$2x + 4x + 1x$ Find like terms: $2x$, $4x$, and $1x$ are like terms,
                 so add the coefficients, $2 + 4 + 1$.

$(2 + 4 + 1)x$ The variable part, x, stays the same.

$7x$

Therefore, $2x + 4x + x$ can be written as $7x$.

(b) $^-3y^2 - 8y^2$

$^-3y^2 - 8y^2$ Both coefficients are shown.
             Change subtraction to adding the opposite.

$^-3y^2 + {}^-8y^2$ Find like terms: $^-3y^2$ and $^-8y^2$ are like terms,
              so add the coefficients, $^-3 + {}^-8$.

$(^-3 + {}^-8)y^2$ The variable part, y^2, stays the same.

$^-11y^2$

Therefore, $^-3y^2 - 8y^2$ can be written as $^-11y^2$.

Work Problem 2 at the Side.

OBJECTIVE 2   **Simplify expressions.** When simplifying expressions, be careful to combine only *like* terms—those having variable parts that match. You *cannot* combine terms if the variable parts are different.

---

2 Combine like terms.

(a) $10b + 4b + 10b$

(b) $y^3 + 8y^3$

(c) $^-7n - n$

(d) $3c - 5c - 4c$

(e) $^-9xy + xy$

(f) $^-4p^2 - 3p^2 + 8p^2$

(g) $ab - ab$

ANSWERS
2. **(a)** $24b$ **(b)** $9y^3$ **(c)** ^-8n **(d)** ^-6c
    (e) ^-8xy **(f)** $1p^2$, or just p^2
    (g) 0, because $1ab + {}^-1ab$ is $(1 + {}^-1)\,ab$, or $0ab$, and 0 times anything is 0.

3 Simplify each expression by combining like terms.

(a) $3b^2 + 4d^2 + 7b^2$

(b) $4a + b - 6a + b$

(c) $^-6x + 5 + 6x + 2$

(d) $2y - 7 - y + 7$

(e) $^-3x - 5 + 12 + 10x$

EXAMPLE 3 Simplifying Expressions

Simplify each expression by combining like terms.

(a) $6xy + 2y + 3xy$

The *like* terms are $6xy$ and $3xy$. We can use the commutative property to rewrite the expression so that the like terms are next to each other. This helps to organize our work.

$6xy + 3xy + 2y$	Combine *like* terms only.
$(6 + 3)xy + 2y$	Add the coefficients, $6 + 3$. The variable part, xy, stays the same.
$9xy + 2y$	Keep writing $2y$, the term that was *not* combined; it is still part of the expression.

The simplified expression is $9xy + 2y$.

(b) Here is the expression from the first page in this section.

$2c - 10 - c + 15$	Write the understood 1 as the coefficient of c.
$2c - 10 - 1c + 15$	Change subtractions to adding the opposite.
$2c + {}^-10 + {}^-1c + 15$	Rewrite the expression so that like terms are next to each other.
$2c + {}^-1c + {}^-10 + 15$	Combine $2c + {}^-1c$. Also combine $^-10 + 15$.
$(2 + {}^-1)c + 5$	
$1c + 5$	

The simplified expression is $1c + 5$ or just $c + 5$.

$1c$ is the same as c.

> **NOTE**
> In this book, when combining like terms we will usually write the variable terms in alphabetical order. A constant term (number only) will be written last. So, in Examples 3(a) and 3(b) above, the preferred and alternative ways of writing the expressions are as follows.
>
> The simplified expression is $9xy + 2y$ (alphabetical order). However, by the commutative property of addition, $2y + 9xy$ is also correct.
>
> The simplified expression is $c + 5$ (constant written last). However, by the commutative property of addition, $5 + c$ is also correct.

◀◀◀ **Work Problem 3 at the Side.**

We can use the associative property of multiplication to simplify an expression such as $4(3x)$.

$4(3x)$ can be written as $4 \cdot (3 \cdot x)$

Understood multiplications

ANSWERS
3. (a) $10b^2 + 4d^2$ (b) $^-2a + 2b$
(c) 7 (d) y (e) $7x + 7$

Using the associative property, we can regroup the factors.

4 • (3 • x)   can be written as   (4 • 3) • x   To simplify, multiply 4 • 3.
                                  12 • x   Write 12 • x without the multiplication dot.
                                  12x

The simplified expression is 12x.

EXAMPLE 4 Simplifying Multiplication Expressions

Simplify.

(a) $5(10y)$

Use the associative property.

5 • (10 • y)   can be written as   (5 • 10) • y   Multiply 5 • 10.
                                    50 • y   Write 50 • y without the multiplication dot.
                                    50y

So, $5(10y)$ simplifies to $50y$.

(b) $^-6(3b)$

Use the associative property.

$^-6(3b)$ can be written as $(^-6 \cdot 3)b$
                                ^-18b

So, $^-6(3b)$ simplifies to ^-18b.

(c) $^-4(^-2x^2)$

Use the associative property.

$^-4(^-2x^2)$ can be written as $(^-4 \cdot {}^-2)x^2$
                                    $8x^2$

So, $^-4(^-2x^2)$ simplifies to $8x^2$.

▶▶▶ **Work Problem 4 at the Side.**

OBJECTIVE 3 Use the distributive property to multiply. The distributive property can also be used to simplify expressions such as $3(x + 5)$. You *cannot* add the terms inside the parentheses because x and 5 are *not* like terms. But notice the understood multiplication dot between the 3 and the parentheses.

$$3(x + 5)$$
$$3 \cdot (x + 5)$$

Thus you can *distribute* multiplication over addition, as you did in **Section 1.6.** That is, multiply 3 times each term inside the parentheses.

$3 \cdot (x + 5)$ can be written as $3 \cdot x + 3 \cdot 5$
                                        $3x + 15$

So, $3(x + 5)$ simplifies to $3x + 15$.
    ↑————— Stays as addition —————↑

4 Simplify.

(a) $7(4c)$

(b) $^-3(5y^3)$

(c) $20(^-2a)$

(d) $^-10(^-x)$

ANSWERS
4. (a) $28c$ (b) $^-15y^3$ (c) ^-40a (d) $10x$

5 Simplify.

(a) $7(a + 10)$

(b) $3(x - 3)$

(c) $4(2y + 6)$

(d) $^-5(3b + 2)$

(e) $^-8(c + 4)$

Multiplication also distributes over subtraction.

$$4(y - 2) \quad \text{means} \quad 4 \cdot (y - 2) \quad \text{and can be written} \quad 4 \cdot y - 4 \cdot 2$$
$$4y - 8$$

So, $4(y - 2)$ simplifies to $4y - 8$.   Notice that we did *not* need to change
— Stays as subtraction —   subtraction to adding the opposite.

EXAMPLE 5 Using the Distributive Property

Simplify.

(a) $6(y - 4)$ can be written as $6 \cdot y - 6 \cdot 4$
$$6y - 24$$
— Stays as subtraction —

So, $6(y - 4)$ simplifies to $6y - 24$

(b) $5(3x + 2)$ can be written as $5 \cdot 3x + 5 \cdot 2$
$$5 \cdot 3 \cdot x + 10$$
$$15 \cdot x + 10$$
$$15x + 10$$

So, $5(3x + 2)$ simplifies to $15x + 10$

(c) $^-2(4a + 3)$ can be written as $^-2 \cdot 4a + {}^-2 \cdot 3$
$$^-2 \cdot 4 \cdot a + {}^-6$$
$$^-8 \cdot a + {}^-6$$
$$^-8a + {}^-6$$

Now we will use the definition of subtraction "in reverse" to rewrite $^-8a + {}^-6$.

Change $^-6$ to its
opposite, $^+6$.

Write $^-8a + {}^-6$ as $^-8a - 6$

Change addition
to subtraction.

Think back to the way we changed subtraction to adding the opposite. Here we are "working backward." From now on, whenever addition is followed by a negative number, we will change it to subtracting a positive number.

$^-8a + {}^-6$
$\updownarrow \quad \updownarrow$ Equivalent expressions
$^-8a - 6$

So, $^-2(4a + 3)$ simplifies to $^-8a - 6$.

◀◀◀ **Work Problem 5 at the Side.**

ANSWERS
5. (a) $7a + 70$ (b) $3x - 9$ (c) $8y + 24$
(d) $^-15b - 10$ (e) $^-8c - 32$

Sometimes you need to do several steps to simplify an expression.

EXAMPLE 6 Simplifying a More Complex Expression

Simplify: $8 + 3(x - 2)$

$8 + 3(x - 2)$	Do *not* add $8 + 3$. Use the distributive property first because multiplying is done *before* adding.
$8 + 3 \cdot x - 3 \cdot 2$	Do the multiplications.
$8 + 3x - 6$	Rewrite so that like terms are next to each other.
$3x + 8 - 6$	Subtract to find $8 - 6$ or change to adding $8 + {}^-6$.
$3x + 2$	

The simplified expression is $3x + 2$.

CAUTION

Do *not* add $8 + 3$ as the first step in Example 6 above. Remember that the order of operations tells you to do addition *last*.

▶▶▶ **Work Problem 6 at the Side.**

6 Simplify.

(a) $^-4 + 5(y + 1)$

(b) $2(3w + 4) - 5$

(c) $5(6x - 2) + 3x$

(d) $21 + 7(a^2 - 3)$

(e) $^-y + 3(2y + 5) - 18$

ANSWERS
6. (a) $5y + 1$ (b) $6w + 3$ (c) $33x - 10$
   (d) $7a^2$ (e) $5y - 3$

Focus on Real-Data Applications

Expressions

A college with four campuses uses an auditorium for graduation that has 1500 seats. Each campus hosts its own graduation ceremony. Students are allocated a whole number of tickets. Round each result **down** to the nearest whole number.

1. How many tickets are allocated to each of 458 graduates at the **North** Campus graduation?
2. How many tickets are allocated to each of 297 graduates at the **East** Campus graduation?
3. How many tickets are allocated to each of 315 graduates at the **South** Campus graduation?
4. How many tickets are allocated to each of 186 graduates at the **West** Campus graduation?
5. Write an *expression* that represents the number of tickets that are allocated to each of g graduates.
6. What recommendations do you have for unallocated tickets?

Traffic engineers have to decide how long to have the red, yellow, and green lights showing on a traffic signal. To decide the number of seconds that a yellow light should be on, the engineers use the expression

$$\frac{5v}{100} + 1$$

where v is the speed limit in miles per hour (mph).

7. How many seconds should the yellow light be on if the speed limit is 20 mph? 40 mph? 60 mph?
8. Based on the answers you just calculated, how could you estimate the time for a yellow light if the speed limit is 30 mph? 50 mph?
9. Use the given expression to find the number of seconds that the yellow light should be on if the speed limit is 30 mph and 50 mph. Did you get the same result as in Problem 8?

To estimate the number of words in a child's vocabulary, a speech therapist uses the expression

$$60A - 900$$

where A is the child's age in months.

10. Estimate the number of words that a child aged 20 months knows.
11. Estimate the number of words that a child aged 2 years knows. (*Hint:* How many months are in two years?)
12. How many words does a child learn between the ages of 20 months and 2 years?
13. Estimate the number of words in the vocabulary of a 3-year-old child.
14. How many words does a child learn between the ages of 2 years and 3 years?
15. Evaluate the expression for a child who is 15 months old. Do you think that the answer is reasonable? Explain why or why not.
16. Evaluate the expression for a child who is 12 months old. Do you think that the answer makes sense? Explain why or why not.

Section 2.2 Simplifying Expressions   103

2.2 Exercises

Circle the like terms in each expression. Then identify the coefficients of the like terms. See Example 1.

1. ⓩb^2 + 2b + 2b^3 + ⓑ^{2} + 6

2. 3x + x^3 + 3x^2 + 3 + 2x^3

3. $^-x^2y$ + ⓧy + ⓧy + $^-$2xy^2

4. ab^2 + $^-a^2b$ + 2ab + $^-3a^2b$

5. ⑦ + 7c + ③ + 7c^3 + $^-$4

6. 4d + $^-$5 + 1 + $^-5d^2$ + 4

Simplify each expression. See Example 2.

7. 6r + 6r

8. 4t + 10t

9. x^2 + 5x^2

10. 9y^3 + y^3

11. p − 5p

12. n − 3n

13. $^-2a^3$ − a^3

14. $^-10x^2$ − x^2

15. c − c

16. b^2 − b^2

17. 9xy + xy − 9xy

18. r^2s − 7r^2s + 7r^2s

19. 5t^4 + 7t^4 − 6t^4

20. 10mn − 9mn + 3mn

21. y^2 + y^2 + y^2 + y^2

22. a + a + a

23. ^-x − 6x − x

24. ^-y − y − 3y

Chapter 2 Understanding Variables and Solving Equations

Simplify by combining like terms. Write each answer with the variables in alphabetical order and any constant term last. See Example 3.

25. $8a + 4b + 4a$

26. $6x + 5y + 4y$

27. $6 + 8 + 7rs$
    $14 + 7rs$

28. $10 + 2c^2 + 15$

29. $a + ab^2 + ab^2$

30. $n + mn + n$

31. $6x + y - 8x + y$

32. $d + 3c - 7c + 3d$

33. $8b^2 - a^2 - b^2 + a^2$

34. $5ab - ab + 3a^2b - 4ab$

35. $^-x^3 + 3x - 3x^2 + 2$

36. $a^2b - 2ab - ab^3 + 3a^3b$

37. $^-9r + 6t - s - 5r + s + t - 6t + 5s - r$

38. $^-x - 3y + 4z + x - z + 5y - 8x - y$

Simplify by using the associative property of multiplication. See Example 4.

39. $3(10a)$

40. $8(4b)$

41. $^-4(2x^2)$

42. $^-7(3b^3)$

43. $5(^-4y^3)$

44. $2(^-6x)$

45. $^-9(^-2cd)$

46. $^-6(^-4rs)$

47. $7(3a^2bc)$

48. $4(2xy^2z^2)$

49. $^-12(^-w)$

50. $^-10(^-k)$

Use the distributive property to simplify each expression. See Example 5.

51. $6(b + 6)$

52. $5(a + 3)$

53. $7(x - 1)$

54. $4(y - 4)$

55. $3(7t + 1)$

56. $8(2c + 5)$

57. $-2(5r + 3)$

58. $-5(6z + 2)$

59. $-9(k + 4)$

60. $-3(p + 7)$

61. $50(m - 6)$

62. $25(n - 1)$

Simplify each expression. See Example 6.

63. $10 + 2(4y + 3)$

64. $4 + 7(x^2 + 3)$

65. $6(a^2 - 2) + 15$

66. $5(b - 4) + 25$

67. $2 + 9(m - 4)$

68. $6 + 3(n - 8)$

69. $-5(k + 5) + 5k$

70. $-7(p + 2) + 7p$

71. $4(6x - 3) + 12$

72. $6(3y - 3) + 18$

73. $5 + 2(3n + 4) - n$

74. $8 + 8(4z + 5) - z$

75. $-p + 6(2p - 1) + 5$

76. $-k + 3(4k - 1) + 2$

77. Explain the difference between *simplifying* an expression and *evaluating* an expression.

78. Simplify each expression. Are the answers equivalent? Explain why or why not.
$$5(3x + 2) \qquad 5(2 + 3x)$$

79. Explain what makes two terms *like* terms. Include several examples in your explanation.

80. Explain how to combine like terms. Include an example in your explanation.

81. Explain and correct the error made by a student who simplified this expression.
$$\underbrace{{}^-2x + 7x}_{5x^2} + 8$$
$$+ 8$$

82. Explain and correct the error made by a student who simplified this expression.
$$\underbrace{{}^-10a + 6a}_{-4a} - \underbrace{7 + 2}_{{}^-7 + 2}$$
$$-4a \quad + \quad {}^-7 + 2$$
$$-4a \quad + \quad {}^-5$$
$$4a - 5$$

Simplify.

83. $^-4(3y) - 5 + 2(5y + 7)$

84. $6(^-3x) - 9 + 3(^-2x + 6)$

85. $^-10 + 4(^-3b + 3) + 2(6b - 1)$

86. $12 + 2(4a - 4) + 4(^-2a - 1)$

87. $^-5(^-x + 2) + 8(^-x) + 3(^-2x - 2) + 16$

88. $^-7(^-y) + 6(y - 1) + 3(^-2y) + 6 - y$

2.3 Solving Equations Using Addition

OBJECTIVES

1. Determine whether a given number is a solution of an equation.
2. Solve equations, using the addition property of equality.
3. Simplify equations before using the addition property of equality.

Now you are ready for a look at the "heart" of algebra, writing and solving **equations.** *Writing* an equation is a way to show the relationship between what you *know* about a problem and what you *don't* know. Then, *solving* the equation is a way to figure out the part that you didn't know and answer your question.

The questions you can answer by writing and solving equations are as varied as the careers people choose. A zookeeper can solve an equation that answers the question of how long to incubate the egg of a particular tropical bird. An aerobics instructor can solve an equation that answers the question of how hard a certain person should exercise for maximum benefit.

OBJECTIVE 1 Determine whether a given number is a solution of an equation. Let's start with the example from the beginning of this chapter: ordering textbooks for math classes. The expression we used to order books was $c + 5$, where c was the class limit (the maximum number of students allowed in the class). Suppose that 30 prealgebra books were ordered. What is the class limit for prealgebra? To answer this question, write an equation showing the relationship between what you know and what you don't know.

You don't know the class limit. ↓   ↓ You do know the total number of books ordered.

$$c + 5 = 30$$

↑ You do know that 5 extra books were ordered.

> **NOTE**
> An equation has an equal sign. Notice the similarity in the words **equa**tion and **equa**l. An expression does *not* have an equal sign.

The equal sign in an equation is like the balance point on a playground teeter-totter, or seesaw. To have a true equation, the two sides must balance.

These equations balance, so we can use the = sign.

$6 + 8 = 14$

$10 = 5 \cdot 2$

$3 \cdot 2 = 5 + 1$

These equations do *not* balance, so we write ≠ to mean "not equal to."

$6 + 8 \neq 15$

$10 \neq 4 \cdot 2$

$4 + 5 \neq 5 \cdot 4$

108 Chapter 2 Understanding Variables and Solving Equations

❶ **(a)** Which of these numbers, 95, 65, or 70, is the solution of the equation $c + 15 = 80$?

When an equation has a variable, we **solve the equation** by finding a number that can replace the variable and make the equation balance. For the example about ordering prealgebra textbooks:

$$c + 5 = 30$$

What number can replace c so that the equation balances?

Try replacing c with **15**. $15 + 5 \neq 30$ Does *not* balance: $15 + 5$ is only 20.

Try replacing c with **40**. $40 + 5 \neq 30$ Does *not* balance: $40 + 5$ is more than 30.

Try replacing c with **25**. $25 + 5 = 30$ Balances: $25 + 5$ is 30.

The **solution** is **25** because **25** is the *only* number that makes the equation balance. By solving the equation, you have answered the question about the class limit for prealgebra. The class limit is 25.

> **NOTE**
> The equations that you will solve in Chapters 2–8 of this book have only one solution, that is, one number that makes the equation balance. In later chapters, and in other algebra courses, you will solve equations that have two or more solutions.

(b) Which of these numbers, 20, 24, or 32, is the solution of the equation $28 = c - 4$?

EXAMPLE 1 **Identifying the Solution of an Equation**

Which of these numbers, 70, 40, or 60, is the solution of the equation $c - 10 = 50$?

Replace c with each of the numbers. The one that makes the equation balance is the solution.

$70 - 10 \neq 50$ $40 - 10 \neq 50$ $60 - 10 = 50$

Does *not* balance: $70 - 10$ is more than 50.

Does *not* balance: $40 - 10$ is only 30.

Balances: $60 - 10$ is 50.

The solution is 60 because, when c is 60, the equation balances.

◀◀◀ **Work Problem 1 at the Side.**

ANSWERS
1. **(a)** The solution is 65.
   (b) The solution is 32.

OBJECTIVE 2 **Solve equations, using the addition property of equality.** When solving the book ordering equation, $c + 5 = 30$, you could just look at the equation and think, "What number, plus 5, would balance with 30?" You could easily see that c had to be 25. Not all equations can be solved this easily, so you'll need some tools for the harder ones. The first tool, called the **addition property of equality,** allows us to add the *same* number to *both* sides of an equation.

> **Addition Property of Equality**
> If $a = b$, then $a + c = b + c$.
>
> In other words, you may add the same number to both sides of an equation and still keep it balanced.

Think of the teeter-totter. If there are 3 children of the same size on each side, it will balance. If 2 more children climb onto the left side, the only way to keep the balance is to have 2 more children of the same size climb onto the right side as well.

$$3 = 3 \qquad\qquad 3 + 2 = 3 + 2$$

All the tools you will learn to use with equations have one goal.

> **Goal in Solving an Equation**
> The goal is to end up with the variable (letter) on one side of the equal sign balancing a number on the other side.
>
> We work on the original equation until we get:
>
> variable = number   or   number = variable
>
> Once we have arrived at that point, the number balancing the variable is the solution to the original equation.

EXAMPLE 2   Using the Addition Property of Equality

Solve each equation and check the solution.

(a) $c + 5 = 30$

We want to get the variable, c, by itself on the left side of the equal sign. To do that, we add the *opposite* of 5, which is $^-5$. Then $5 + {}^-5$ will be 0.

$$c + 5 = 30$$

Add $^-5$ to the left side. → $\underline{{}^-5 \quad {}^-5}$ ← To keep the balance, add $^-5$ to the right side also.

$$c + 0 = 25 \quad \leftarrow 30 + {}^-5 \text{ is } 25.$$

$5 + {}^-5$ is 0.

Recall that adding 0 to any number leaves the number unchanged, so $c + 0$ is c.

$$c + 0 = 25$$
$$c = 25$$

Because c *balances* with 25, the *solution* is 25.

Check the solution by replacing c with 25 in the *original equation*.

$$c + 5 = 30 \quad \text{Original equation}$$
$$25 + 5 = 30 \quad \text{Replace } c \text{ with } 25.$$
$$30 = 30 \quad \text{Balances}$$

Because the equation balances when we use 25 to replace the variable, we know that **25 is the correct solution**. If it had *not* balanced, we would need to rework the problem, find our error, and correct it.

(b) $^-5 = x - 3$

We want the variable, x, by itself on the right side of the equal sign. (Remember, it doesn't matter which side of the equal sign the variable is on, just so it ends up by itself.) To see what number to add, we change the subtraction to adding the opposite.

$$^-5 = x - 3 \quad \text{Change subtraction to adding the opposite.}$$

$$^-5 = x + {}^-3 \quad \text{To get } x \text{ by itself on the right side, add the opposite of } {}^-3, \text{ which is } 3. \text{ Then } {}^-3 + 3 \text{ is } 0.$$

To keep the balance, add 3 to the left side also. $\underline{3 \qquad 3}$

$^-5 + 3$ is $^-2$. $\quad ^-2 = x + 0 \quad$ Adding 0 to x leaves x unchanged.

$$^-2 = x \quad x \text{ balances with } {}^-2, \text{ so } {}^-2 \text{ is the solution.}$$

We check the solution by replacing x with $^-2$ in the *original equation*. If the equation balances when we use $^-2$, we know that it is the correct solution. If the equation does *not* balance when we use $^-2$, we made an error and need to try solving the equation again.

Continued on Next Page

Section 2.3 Solving Equations Using Addition 111

Check
$$-5 = x - 3 \quad \text{Original equation}$$
$$-5 = -2 - 3 \quad \text{Replace } x \text{ with } -2.$$
$$-5 = \underbrace{-2 + -3} \quad \text{Change subtraction to adding the opposite.}$$
$$-5 = -5 \quad \text{Balances; this shows } -2 \text{ is the correct solution.}$$

When x is replaced with -2, the equation balances, so **-2 is the correct solution**.

> **CAUTION**
> When checking the solution to Example 2(b) above, we ended up with $-5 = -5$. Notice that -5 is *not* the solution. The solution is -2, the number used to replace x in the original equation.

Work Problem 2 at the Side.

OBJECTIVE 3 Simplify equations before using the addition property of equality. Sometimes you can simplify the expression on one or both sides of the equal sign. Doing so will make it easier to solve the equation.

EXAMPLE 3 Simplifying before Solving Equations

Solve each equation and check each solution.

(a) $y + 8 = 3 - 7$

You cannot simplify the left side because y and 8 are *not* like terms.

$$y + 8 = 3 - 7 \quad \text{Simplify the right side by changing subtraction to adding the opposite.}$$
$$y + 8 = 3 + -7 \quad \text{Add } 3 + -7.$$
$$y + 8 = -4$$

To get y by itself on the left side, add the opposite of 8, which is -8.

$$y + 8 = -4$$
$$ -8 -8 \quad \longleftarrow \text{To keep the balance, add } -8 \text{ to the right side also.}$$
$$\underbrace{y + 0} = -12 \quad -4 + -8 \text{ is } -12.$$
$$y = -12$$

$8 + -8$ is 0.

The solution is -12. Now check the solution.

Check
Add $-12 + 8$.
$$y + 8 = 3 - 7 \quad \text{Go back to the } original \text{ equation and replace } y \text{ with } -12.$$
$$-12 + 8 = 3 - 7 \quad \text{Change } 3 - 7 \text{ to } 3 + -7.$$
$$-4 = 3 + -7 \quad \text{Add } 3 + -7.$$
$$-4 = -4 \quad \text{Balances; so } -12 \text{ is the correct solution.}$$

When y is replaced with -12, the equation balances, so **-12 is the correct solution** (not -4).

—— **Continued on Next Page**

---

2 Solve each equation and check each solution.

(a) $12 = y + 5$

Check

(b) $b - 2 = -6$

Check

ANSWERS
2. (a) $y = 7$
   Check
   $12 = y + 5$
   $12 = 7 + 5$
   Balances $12 = 12$

   (b) $b = -4$
   Check
   $b - 2 = -6$
   $-4 + -2 = -6$
   Balances $-6 = -6$

Chapter 2 Understanding Variables and Solving Equations

❸ Simplify each side of the equation when possible. Then solve the equation and check the solution.

(a) $2 - 8 = k - 2$

Check

(b) $4r + 1 - 3r = {}^-8 + 11$

Check

(b) ${}^-2 + 2 = {}^-4b - 6 + 5b$

Simplify the left side by adding ${}^-2 + 2$.	${}^-2 + 2 = {}^-4b - 6 + 5b$	Simplify the right side by changing subtraction to adding the opposite.
	$0 = {}^-4b + {}^-6 + 5b$	Find like terms.
	$0 = {}^-4b + 5b + {}^-6$	Combine ${}^-4b + 5b$.
To keep the balance, add 6 to the left side also. → 6	$0 = 1b + {}^-6$	To get $1b$ by itself, add the opposite of ${}^-6$, which is 6. 6 ←
	$6 = 1b + 0$	
	$6 = 1b$	$1b$ is equivalent to b.
	$6 = b$	

The solution is 6.

Check	${}^-2 + 2 = {}^-4b - 6 + 5b$	Go back to the *original* equation and replace each b with 6.
Add ${}^-2 + 2$.	${}^-2 + 2 = {}^-4 \cdot 6 - 6 + 5 \cdot 6$	On the right side, do multiplications first.
	$0 = {}^-24 + {}^-6 + 30$	Change subtraction to adding the opposite.
	$0 = {}^-30 + 30$	Add from left to right.
	$0 = 0$	Balances

When b is replaced with 6, the equation balances, so **6 is the correct solution (not 0)**.

> **CAUTION**
> When checking a solution, always go back to the *original* equation. That way, you will catch any errors you made when simplifying each side of the equation.

◀◀◀ **Work Problem ❸ at the Side.**

ANSWERS
3. (a) $k = {}^-4$
Check $2 - 8 = k - 2$
$2 + {}^-8 = {}^-4 + {}^-2$
Balances ${}^-6 = {}^-6$

(b) $r = 2$
Check $4r + 1 - 3r = {}^-8 + 11$
$4 \cdot 2 + 1 - 3 \cdot 2 = {}^-8 + 11$
$8 + 1 - 6 = 3$
$9 + {}^-6 = 3$
Balances $3 = 3$

2.3 Exercises

In each list of numbers, find the one that is a solution of the given equation. See Example 1.

1. $n + 50 = 8$

    58, 42, 60

2. $r - 20 = 5$

    15, 30, 25

3. $-6 = y + 10$

    $-4, -16, 16$

4. $-4 = x + 13$

    $17, -17, -9$

5. $t + 12 = 0$

    $0, -12, -24$

6. $b - 8 = 0$

    $8, 0, -8$

Solve each equation and check each solution. See Example 2.

7. $p + 5 = 9$ Check $p + 5 = 9$

8. $a + 3 = 12$ Check $a + 3 = 12$

9. $8 = r - 2$ Check $8 = r - 2$

10. $3 = b - 5$ Check $3 = b - 5$

11. $-5 = n + 3$ Check

12. $-1 = a + 8$ Check

13. $-4 + k = 14$ Check

14. $-9 + y = 7$ Check

15. $y - 6 = 0$ Check

16. $k - 15 = 0$ Check

17. $7 = r + 13$ Check

18. $12 = z + 19$ Check

19. $x - 12 = {}^-1$ Check

20. $m - 3 = {}^-9$ Check

21. ${}^-5 = {}^-2 + t$ Check

22. ${}^-1 = {}^-10 + w$ Check

A solution is given for each equation. Show how to check the solution. If the solution is correct, leave it. If the solution is not correct, solve the equation and check your new solution. See Example 2.

23. $z - 5 = 3$ Check $z - 5 = 3$
The solution is ${}^-2$.   ↓

24. $x - 9 = 4$ Check $x - 9 = 4$
The solution is 13.   ↓

25. $7 + x = {}^-11$ Check
The solution is ${}^-18$.

26. $2 + k = {}^-7$ Check
The solution is ${}^-5$.

27. ${}^-10 = {}^-10 + b$ Check
The solution is 10.

28. $0 = {}^-14 + a$ Check
The solution is 0.

Section 2.3 Solving Equations Using Addition

Simplify each side of the equation when possible. Then solve the equation and check the solution. Show your work. See Example 3.

29. $c - 4 = {}^-8 + 10$ Check

30. $b - 8 = 10 - 6$ Check

31. ${}^-1 + 4 = y - 2$ Check

32. $2 + 3 = k - 4$ Check

33. $10 + b = {}^-14 - 6$ Check

34. $1 + w = {}^-8 - 8$ Check

35. $t - 2 = 3 - 5$ Check

36. $p - 8 = {}^-10 + 2$ Check

37. $10z - 9z = {}^-15 + 8$ Check

38. $2r - r = 5 - 10$ Check

39. ${}^-5w + 2 + 6w = {}^-4 + 9$ Check

40. ${}^-2t + 4 + 3t = 6 - 7$ Check

Solve each equation. Show your work. See Examples 2 and 3.

41. $^-3 - 3 = 4 - 3x + 4x$

42. $^-5 - 5 = {}^-2 - 6b + 7b$

43. $^-3 + 7 - 4 = {}^-2a + 3a$

44. $6 - 11 + 5 = {}^-8c + 9c$

45. $y - 75 = {}^-100$

46. $a - 200 = {}^-100$

47. $^-x + 3 + 2x = 18$

48. $^-s + 2s - 4 = 13$

49. $82 = {}^-31 + k$

50. $^-5 = 72 + w$

51. $^-2 + 11 = 2b - 9 - b$

52. $^-6 + 7 = 2h - 1 - h$

53. $r - 6 = 7 - 10 - 8$

54. $m - 5 = 2 - 9 + 1$

55. $^-14 = n + 91$

56. $66 = x - 28$

57. $^-9 + 9 = 5 + h$

58. $18 - 18 = 6 + p$

59. A student did this work when solving an equation. Do you agree that the solution is $^-7$? Explain why or why not.

$$\underbrace{^-8 + 1} = x + 7$$
$$^-7 = x + 7$$
$$\underline{^-7} = \underline{^-7}$$
$$^-14 = \underbrace{x + 0}$$
$$^-14 = x$$

Check

$^-8 + 1 = \quad x + 7$

$\underbrace{^-8 + 1} = \underbrace{^-14 + 7}$

$^-7 = \quad ^-7$

Balances, so $^-7$ is the solution.

60. A student did this work when solving an equation. Show how to check the solution. If the solution does not check, find and correct the errors.

$$^-3 - 6 = n - 5$$
$$\underbrace{^-3 + 6} = n - 5$$
$$3 = n - 5$$
$$\underline{^-5} \quad \underline{^-5}$$
$$^-2 = \underbrace{n + 0}$$
$$^-2 = n$$

61. West Community College always orders 10 extra robes for the graduation ceremony. The college ordered 305 robes this year. Solving the equation $g + 10 = 305$ will give you the number of graduates (g) this year. Solve the equation.

62. Refer to Exercise 61. The college ordered 278 robes last year. Solve the equation $g + 10 = 278$ to find the number of graduates last year.

63. The warmer the temperature, the faster a field cricket chirps. Solving the equation $92 = c + 37$ will give you the number of chirps (in 15 seconds) when the temperature is 92 degrees. Solve the equation.

64. Refer to Exercise 63. Solve the equation $77 = c + 37$ to find the number of times a field cricket chirps (in 15 seconds) when the temperature is 77 degrees.

65. During the summer months, Ernesto spends an average of only $45 per month on parking fees by riding his bike to work on nice days. This is $65 less per month than what he spends for parking in the winter. Solving the equation $p - 65 = 45$ will give you his monthly parking fees in the winter. Solve the equation.

66. By walking to work several times a week in the summer, Aimee spends an average of $56 less per month on parking fees. If she spends $98 per month on parking in the summer, solve the equation $p - 56 = 98$ to find her monthly parking fees in the winter.

Solve each equation. Show your work.

67. $^{-}17 - 1 + 26 - 38 = {}^{-}3 - m - 8 + 2m$

68. $19 - 38 - 9 + 11 = {}^{-}t - 6 + 2t - 6$

69. $^{-}6x + 2x + 6 + 5x = |0 - 9| - |{}^{-}6 + 5|$

70. $^{-}h - |{}^{-}9 - 9| + 8h - 6h = {}^{-}12 - |{}^{-}5 + 0|$

RELATING CONCEPTS (EXERCISES 71–72) For Individual or Group Work

Use what you have learned about solving equations to work Exercises 71 and 72 in order.

71. (a) Write two *different* equations that have $^{-}2$ as the solution. Be sure that you have to use the *addition property of equality* to solve the equations. Show how to solve each equation. Use Exercises 7 to 22 as models.

(b) Follow the directions in part (a), but this time write two equations that have 0 as the solution.

72. Not all equations have solutions that are integers. Try solving these equations.

(a) $x + 1 = 1\frac{1}{2}$

(b) $\frac{1}{4} = y - 1$

(c) $\$2.50 + n = \$3.35$

(d) Write two more equations that have fraction or decimal solutions.

2.4 Solving Equations Using Division

OBJECTIVE 1 Solve equations, using the division property of equality. In **Section 2.1** you worked with the expression for finding the perimeter of a square-shaped garden, that is, finding the total distance around all four sides of the garden:

$4s$, where s is the length of one side of the square.

Suppose you know that 24 feet of fencing was used around a square-shaped garden. What was the length of one side of the garden? To answer this question, write an equation showing the relationship between what you know and what you don't know.

You don't know the length of one side.
You do know that there are 4 sides. — You do know the perimeter.
$4s = 24$

To solve the equation, what number can replace s so that the equation balances? You can see that s is 6 feet.

$4 \cdot 6 = 24$ Balances: $4 \cdot 6$ is exactly 24.

The **solution** is **6 feet** because 6 is the *only* number that makes the equation balance. You have answered the question about the length of one side: The length is 6 feet.

There is a tool that you can use to solve equations such as $4s = 24$. Called the **division property of equality,** it allows you to *divide* both sides of an equation by the *same* number. (The only exception is that you cannot divide by 0.)

OBJECTIVES

1. Solve equations, using the division property of equality.
2. Simplify equations before using the division property of equality.
3. Solve equations such as $^-x = 5$.

Division Property of Equality

If $a = b$, then $\dfrac{a}{c} = \dfrac{b}{c}$ as long as c is not 0.

In other words, you may divide both sides of an equation by the same nonzero number and still keep it balanced.

In **Section 2.3,** you saw that *adding* the same number to both sides of an equation kept it balanced. We could also have *subtracted* the same number from both sides because subtraction is defined as adding the opposite. Now we're saying that you can *divide* both sides by the same number. In **Chapter 4** we'll *multiply* both sides by the same number.

Equality Principle for Solving an Equation

As long as you do the *same* thing to *both* sides of an equation, the balance is maintained and you still have a true equation. (The only exception is that you cannot divide by 0.)

1 Solve each equation and check each solution.

(a) $4s = 44$

Check

(b) $27 = {}^{-}9p$

Check

(c) ${}^{-}40 = {}^{-}5x$

Check

(d) $7t = {}^{-}70$

Check

ANSWERS
1. (a) $s = 11$
   Check $4s = 44$
   $4 \cdot 11 = 44$
   Balances $44 = 44$

   (b) $p = {}^{-}3$
   Check $27 = {}^{-}9p$
   $27 = {}^{-}9 \cdot {}^{-}3$
   Balances $27 = 27$

   (c) $x = 8$
   Check ${}^{-}40 = {}^{-}5x$
   ${}^{-}40 = {}^{-}5 \cdot 8$
   Balances ${}^{-}40 = {}^{-}40$

   (d) $t = {}^{-}10$
   Check $7t = {}^{-}70$
   $7 \cdot {}^{-}10 = {}^{-}70$
   Balances ${}^{-}70 = {}^{-}70$

EXAMPLE 1 Using the Division Property of Equality

Solve each equation and check each solution.

(a) $4s = 24$

As with any equation, the goal is to get the variable by itself on one side of the equal sign. On the left side we have $4s$, which means $4 \cdot s$. The variable is multiplied by 4. Division is the opposite of multiplication, so dividing by 4 can be used to "undo" multiplying by 4.

Divide $4s$ by 4. The fraction bar indicates division: $4s \div 4$ is s.
$$\frac{4s}{4} = \frac{24}{4}$$
To keep the balance, divide the right side by 4 also: $24 \div 4$ is 6:
$$s = 6$$

So, as we already knew, 6 is the solution. We check the solution by replacing s with 6 in the original equation.

Check $4s = 24$ Original equation
$4 \cdot 6 = 24$ Replace s with 6.
$24 = 24$ Balances

When s is replaced with 6, the equation balances, so **6 is the correct solution** (**not** 24).

(b) $42 = {}^{-}6w$

On the right side of the equation, the variable is *multiplied* by ${}^{-}6$. To undo the multiplication, *divide* by ${}^{-}6$.

To keep the balance, divide by ${}^{-}6$ on the left side also.
$$\frac{42}{{}^{-}6} = \frac{{}^{-}6w}{{}^{-}6}$$
Use division to undo multiplication: ${}^{-}6w \div {}^{-}6$ is w.
$${}^{-}7 = w$$

The solution is ${}^{-}7$.

Check $42 = {}^{-}6w$ Original equation
$42 = {}^{-}6 \cdot {}^{-}7$ Replace w with ${}^{-}7$.
$42 = 42$ Balances

When w is replaced with ${}^{-}7$, the equation balances, so **${}^{-}7$ is the correct solution** (**not** 42).

> **CAUTION**
> Be careful to divide both sides by the *same* number as the coefficient of the variable term. In Example 1(b) above, the coefficient of ${}^{-}6w$ is ${}^{-}6$, so divide both sides by ${}^{-}6$. (Do *not* divide by the *opposite* of ${}^{-}6$, which is 6. Use the opposite only when you're *adding* the same number to both sides.)

◀◀◀ **Work Problem 1 at the Side.**

OBJECTIVE 2 Simplify equations before using the division property of equality. You can sometimes simplify the expression on one or both sides of the equal sign, as you did in **Section 2.3**.

EXAMPLE 2 Simplifying before Solving Equations

Solve each equation and check each solution.

(a) $4y - 7y = {}^-12$

Simplify the left side by combining like terms.	$4y - 7y = {}^-12$	The right side cannot be simplified.
Change subtraction to adding the opposite.	$4y + {}^-7y = {}^-12$	
Divide by the coefficient, which is $^-3$.	$\dfrac{{}^-3y}{{}^-3} = \dfrac{{}^-12}{{}^-3}$	To keep the balance, divide by $^-3$ on the right side.
	$y = 4$	${}^-12 \div {}^-3$ is 4.

The solution is 4.

Check

	$4y - 7y = {}^-12$	Go back to the *original* equation and replace each y with 4.
Do multiplications first.	$4 \cdot 4 - 7 \cdot 4 = {}^-12$	
Change subtraction to adding the opposite.	$16 - 28 = {}^-12$	
	$16 + {}^-28 = {}^-12$	
	${}^-12 = {}^-12$	Balances

When y is replaced with 4, the equation balances, so **4 is the correct solution**.

(b) $3 - 10 + 7 = h + 7h$

Change subtraction to adding the opposite.	$3 - 10 + 7 = h + 7h$	Write the understood 1 as the coefficient of h.
Add from left to right.	$3 + {}^-10 + 7 = 1h + 7h$	Combine like terms.
	${}^-7 + 7 = 8h$	
To keep the balance, divide by 8 on the left side also.	$\dfrac{0}{8} = \dfrac{8h}{8}$	Divide by the coefficient, which is 8.
$0 \div 8$ is 0.	$0 = h$	

The solution is 0.

Check

	$3 - 10 + 7 = h + 7h$	Go back to the *original* equation and replace each h with 0.
	$3 + {}^-10 + 7 = 0 + 7 \cdot 0$	
	${}^-7 + 7 = 0 + 0$	
	$0 = 0$	Balances

When h is replaced with 0, the equation balances, so **0 is the correct solution**.

▶▶▶ **Work Problem 2 at the Side.**

---

2 Simplify each side of the equation when possible. Then solve the equation and check the solution.

(a) ${}^-28 = {}^-6n + 10n$

Check

(b) $p - 14p = {}^-2 + 18 - 3$

Check

ANSWERS

2. (a) $n = {}^-7$
Check
${}^-28 = {}^-6n + 10n$
${}^-28 = {}^-6 \cdot {}^-7 + 10 \cdot {}^-7$
${}^-28 = 42 + {}^-70$
Balances ${}^-28 = {}^-28$

(b) $p = {}^-1$
Check $p - 14p = {}^-2 + 18 - 3$
${}^-1 - 14({}^-1) = 16 - 3$
${}^-1 - {}^-14 = 13$
${}^-1 + {}^+14 = 13$
Balances $13 = 13$

3 Solve each equation and check each solution.

(a) $^-k = {^-12}$

Check

(b) $7 = {^-t}$

Check

(c) $^-m = {^-20}$

Check

ANSWERS

3. (a) $k = 12$
   Check $\quad ^-1k = {^-12}$
   $\underbrace{^-1 \cdot 12}_{} = {^-12}$
   Balances $^-12 = {^-12}$

(b) $t = {^-7}$
   Check $\quad 7 = {^-1t}$
   $7 = \underbrace{^-1 \cdot {^-7}}_{}$
   Balances $7 = 7$

(c) $m = 20$
   Check $\quad ^-1m = {^-20}$
   $\underbrace{^-1 \cdot 20}_{} = {^-20}$
   Balances $^-20 = {^-20}$

OBJECTIVE 3 Solve equations such as $^-x = 5$. When solving equations, do *not* leave a negative sign in front of the variable.

EXAMPLE 3 Solving an Equation of the Type $^-x = 5$

Solve $^-x = 5$ and check the solution.

It may look as if there is nothing more we can do to the equation $^-x = 5$, but ^-x is *not* the same as x. To see this, we write in the understood $^-1$ as the coefficient of ^-x.

Coefficient is understood to be $^-1$.

$^-x = 5 \quad$ can be written $\quad ^-1x = 5$

We want the coefficient of x to be $^+1$, not $^-1$. To accomplish that, we can divide both sides by the coefficient of x, which is $^-1$.

$$\frac{^-1x}{^-1} = \frac{5}{^-1} \quad \text{Divide both sides by } ^-1.$$

On the left side, $\quad 1x = {^-5} \quad$ On the right side, $5 \div {^-1}$ is $^-5$.
$^-1 \div {^-1}$ is 1.

Now x is by itself on one side of the equal sign and has a coefficient of $^+1$. The solution is $^-5$.

Check $\quad ^-x = 5 \quad$ Go back to the *original* equation.

$\quad ^-1 x = 5 \quad$ Write in the understood $^-1$ as the coefficient of ^-x.

$\underbrace{^-1 \cdot {^-5}}_{5} = 5 \quad$ Replace x with $^-5$.

$\quad\quad 5 = 5 \quad$ Balances

When x is replaced with $^-5$, the equation balances, so **$^-5$ is the correct solution**.

CAUTION
As the last step in solving an equation, do *not* leave a negative sign in front of a variable. For example, do *not* leave $^-y = {^-8}$. Write in the understood $^-1$ as the coefficient, so that

$^-y = {^-8} \quad$ is written as $\quad ^-1y = {^-8}$

Then divide both sides by $^-1$ to get $y = 8$. The solution is 8.

Work Problem 3 at the Side.

2.4 Exercises

Solve each equation and check each solution. See Example 1.

1. $6z = 12$ Check $6z = 12$
2. $8k = 24$ Check $8k = 24$

3. $48 = 12r$ Check
4. $99 = 11m$ Check

5. $3y = 0$ Check
6. $5a = 0$ Check

7. $-7k = 70$ Check
8. $-6y = 36$ Check

9. $-54 = -9r$ Check
10. $-36 = -4p$ Check

11. $-25 = 5b$ Check
12. $-70 = 10x$ Check

Simplify where possible. Then solve each equation and check each solution. See Example 2.

13. $2r = -7 + 13$ Check $2r = \underbrace{-7 + 13}$
14. $6y = 28 - 4$ Check $6y = \underbrace{28 - 4}$

15. $-12 = 5p - p$ Check
16. $20 = z - 11z$ Check

Chapter 2 Understanding Variables and Solving Equations

Solve each equation. Show your work. See Examples 1 and 2.

17. $3 - 28 = 5a$

18. $^-55 + 7 = 8n$

19. $x - 9x = 80$

20. $4c - c = {^-27}$

21. $13 - 13 = 2w - w$

22. $^-11 + 11 = 8t - 7t$

23. $3t + 9t = 20 - 10 + 26$

24. $6m + 6m = 40 + 20 - 12$

25. $0 = {^-9t}$

26. $^-10 = 10b$

27. $^-14m + 8m = 6 - 60$

28. $7w - 14w = 1 - 50$

29. $100 - 96 = 31y - 35y$

30. $150 - 139 = 20x - 9x$

Use multiplication to simplify the side of the equation with the variable. Then solve each equation.

31. $3(2z) = {}^-30$

32. $2(4k) = 16$

33. $50 = {}^-5(5p)$

34. $60 = 4({}^-3a)$

35. ${}^-2({}^-4k) = 56$

36. ${}^-5(4r) = {}^-80$

37. ${}^-90 = {}^-10({}^-3b)$

38. ${}^-90 = {}^-5({}^-2y)$

Solve each equation. See Example 3.

39. ${}^-x = 32$

40. ${}^-c = 23$

41. ${}^-2 = {}^-w$

42. ${}^-75 = {}^-t$

43. ${}^-n = {}^-50$

44. ${}^-x = {}^-1$

45. $10 = {}^-p$

46. $100 = {}^-k$

47. Look again at the solutions to Exercises 39–46. Describe the pattern you see. Then write a rule for solving equations with a negative sign in front of the variable, such as $^-x = 5$.

48. Explain the division property of equality in your own words.

49. Explain and correct the error made by a student who solved this equation.

$$3x = \underbrace{16 - 1}$$
$$\frac{3x}{-3} = \frac{15}{-3}$$
$$x = {}^-5$$

50. Write two *different* equations that have $^-4$ as the solution. Be sure that you have to use the division property of equality to solve the equations. Show how to solve each equation. Use Exercises 1–14 as models.

51. The perimeter of a triangle with sides of equal length is 3 times the length of one side (s). If the perimeter is 45 ft, solving the equation $3s = 45$ will give the length of one side. Solve the equation.

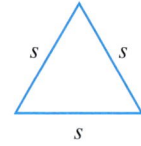

52. Refer to Exercise 51. If the perimeter of the triangle is 63 inches, solve the equation $3s = 63$ to find the length of one side.

53. The perimeter of a pentagon with sides of equal length is 5 times the length of one side (s). If the perimeter is 120 meters, solve the equation $120 = 5s$ to find the length of one side.

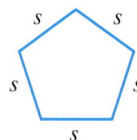

54. Refer to Exercise 53. If the pentagon has a perimeter of 335 yards, solving the equation $335 = 5s$ will give the length of one side. Solve the equation.

Solve each equation. Show your work.

55. $89 - 116 = {}^-4({}^-4y) - 9(2y) + y$

56. $58 - 208 = {}^-b + 8({}^-3b) + 5({}^-5b)$

57. $^-37(14x) + 28(21x) = |72 - 72| + |{}^-166 + 96|$

58. $6a - 10a - 3(2a) = |{}^-25 - 25| - 5(8)$

2.5 Solving Equations with Several Steps

OBJECTIVE 1 Solve equations, using the addition and division properties of equality. To solve some equations, you need to use both the addition property of equality (see **Section 2.3**) and the division property of equality (see **Section 2.4**). Here are the steps.

Solving an Equation Using the Addition and Division Properties

Step 1 *Add* the same amount to both sides of the equation so that the variable term (the variable and its coefficient) ends up by itself on one side of the equal sign.

Step 2 *Divide* both sides by the coefficient of the variable term to find the solution.

Step 3 *Check* the solution by going back to the *original* equation.

EXAMPLE 1 Solving an Equation with Several Steps

Solve this equation and check the solution: $5m + 1 = 16$.

Step 1 Get the variable term by itself on one side of the equal sign. The variable term is $5m$. Adding $^-1$ to the left side of the equation will leave $5m$ by itself. To keep the balance, add $^-1$ to the right side also.

$$5m + 1 = 16$$
$$\underline{^-1 \quad ^-1}$$
$$5m + 0 = 15$$
$$5m = 15$$

Step 2 Divide both sides by the coefficient of the variable term. In $5m$, the coefficient is 5, so divide both sides by 5.

$$\frac{5m}{5} = \frac{15}{5}$$
$$m = 3$$

Step 3 Check the solution by going back to the *original* equation.

$$5m + 1 = 16 \quad \text{Use the original equation}$$
$$\text{and replace } m \text{ with 3.}$$
$$5(3) + 1 = 16$$
$$15 + 1 = 16$$
$$16 = 16 \quad \text{Balances}$$

When m is replaced with 3, the equation balances, so **3 is the correct solution** (not 16).

▶ **Work Problem 1 at the Side.**

So far, variable terms have appeared on just one side of the equal sign. But some equations start with variable terms on both sides. In that case, you can use the addition property of equality to add the same *variable term* to both sides of the equation, just as you have added the same *number* to both sides.

OBJECTIVES

1. Solve equations, using the addition and division properties of equality.
2. Solve equations, using the distributive, addition, and division properties.

1 Solve each equation and check each solution.

(a) $2r + 7 = 13$

Check

(b) $20 = 6y - 4$

Check

(c) $^-10z - 9 = 11$

Check

ANSWERS

1. **(a)** $r = 3$
   Check $\quad 2(3) + 7 = 13$
   $ 6 + 7 = 13$
   Balances $\quad 13 = 13$

   (b) $y = 4$
   Check $\quad 20 = 6(4) - 4$
   $ 20 = 24 - 4$
   Balances $\quad 20 = 20$

   (c) $z = ^-2$
   Check $\quad ^-10(^-2) - 9 = 11$
   $ 20 - 9 = 11$
   Balances $\quad 11 = 11$

② Solve each equation *two ways*. First keep the variable term on the *left* side when you solve. Then solve again, keeping the variable term on the *right* side. Compare the solutions.

(a) $3y - 1 = 2y + 7$

$3y - 1 = 2y + 7$

(b) $3p - 2 = p - 6$

$3p - 2 = p - 6$

ANSWERS
2. (a) $y = 8$ and $8 = y$
 (b) $p = {}^-2$ and ${}^-2 = p$

First decide whether to keep the variable term on the left side, or to keep the variable term on the right side. It doesn't matter which one you keep; just pick one side or the other. Then use the addition property to "get rid of" the variable term on the *other* side by adding its opposite.

EXAMPLE 2 Solving an Equation with Variable Terms on Both Sides

Solve this equation and check the solution: $2x - 2 = 5x - 11$.

First let's keep $2x$, the variable term on the *left* side. That means we need to "get rid of" $5x$ on the *right* side. We can do that by adding the opposite of $5x$, which is ${}^-5x$.

To keep the balance, add ${}^-5x$ to the left side also. Write ${}^-5x$ under $2x$, *not* under 2.

$$2x - 2 = 5x - 11$$
$$\underline{{}^-5x \qquad\qquad {}^-5x}$$
$${}^-3x - 2 = 0 - 11$$
$$\downarrow \downarrow \downarrow \downarrow$$
$${}^-3x + {}^-2 = 0 + {}^-11$$

Write ${}^-5x$ under $5x$, *not* under 11. $5x + {}^-5x$ is $0x$, or 0.

Change subtractions to adding the opposite.

To get ${}^-3x$ by itself, add 2 to both sides.

$$\underline{2 2}$$
$${}^-3x + 0 = 0 + {}^-9$$

Divide both sides by ${}^-3$, the coefficient of the variable term.

$$\frac{{}^-3x}{{}^-3} = \frac{{}^-9}{{}^-3}$$
$$x = 3$$

Suppose that, in the first step, we decided to keep $5x$ on the *right* side and "get rid of" $2x$ on the *left* side. Let's see what happens.

$$2x - 2 = 5x - 11$$

Add ${}^-2x$ to both sides.

$$\underline{{}^-2x {}^-2x }$$
$$0 - 2 = 3x - 11$$
$$\downarrow \downarrow \downarrow \downarrow$$
$$0 + {}^-2 = 3x + {}^-11$$

Change subtractions to adding the opposite. To get $3x$ by itself, add 11 to both sides.

$$\underline{11 11}$$
$$0 + 9 = 3x + 0$$

$$\frac{9}{3} = \frac{3x}{3}$$

Divide both sides by 3.

$$3 = x$$

The two solutions are the same. In both cases, x balances with 3.

Notice that we used the addition principle *twice*: once to "get rid of" the variable term $2x$ and once to "get rid of" the number ${}^-11$. We could have done those steps in the reverse order without changing the result.

> **NOTE**
> More than one sequence of steps will work to solve complicated equations. The basic approach is the following:
> - Simplify each side of the equation, if possible.
> - Get the variable term by itself on one side of the equal sign and a number by itself on the other side.
> - Divide both sides by the coefficient of the variable term.

◀◀◀ **Work Problem 2 at the Side.**

OBJECTIVE 2 Solve equations, using the distributive, addition, and division properties. If an equation contains parentheses, check to see whether you can use the distributive property to remove them.

EXAMPLE 3 Solving an Equation Using the Distributive Property

Solve this equation and check the solution: $^-6 = 3(y - 2)$.

We can use the distributive property to simplify the right side of the equation. Recall from **Section 2.2** that

$$3(y - 2) \text{ can be written as } 3 \cdot y - 3 \cdot 2 = 3y - 6$$

So the original equation $^-6 = 3(y - 2)$ becomes $^-6 = 3y - 6$.

$$^-6 = 3y - 6 \quad \text{Change subtraction to adding the opposite.}$$
$$^-6 = 3y + {}^-6 \quad \text{To get } 3y \text{ by itself, add 6 to both sides.}$$
$$\underline{6 6}$$
$$\frac{0}{3} = \frac{3y}{3} \quad \text{Divide both sides by 3, the coefficient of } 3y.$$
$$0 = y$$

The solution is 0.

Check
$$^-6 = 3(y - 2) \quad \text{Go back to the } \textit{original} \text{ equation and replace } y \text{ with 0.}$$
$$^-6 = 3(0 - 2) \quad \text{Follow the order of operations: work inside parentheses first.}$$
$$^-6 = 3(0 + {}^-2) \quad \text{Change subtraction to addition.}$$
$$^-6 = 3({}^-2)$$
$$^-6 = {}^-6 \quad \text{Balances}$$

When y is replaced with 0, the equation balances, so **0 is the correct solution**.

▶▶▶ **Work Problem 3 at the Side.**

Here is a summary of all the steps you can use to solve an equation. Sometimes you will use only two or three steps, and sometimes you will need all five steps.

Solving an Equation

Step 1 If possible, use the **distributive property** to remove parentheses.

Step 2 **Combine** any like terms on the left side of the equation. Combine any like terms on the right side of the equation.

Step 3 **Add** the same amount to both sides of the equation so that the variable term ends up by itself on one side of the equal sign and a number is by itself on the other side. You may have to do this step more than once.

Step 4 **Divide** both sides by the coefficient of the variable term to find the solution.

Step 5 **Check** your solution by going back to the *original* equation. Replace the variable with your solution. Follow the order of operations to complete the calculations. If the two sides of the equation balance, your solution is correct.

❸ Solve each equation and check each solution.

(a) $^-12 = 4(y - 1)$

(b) $5(m + 4) = 20$

(c) $6(t - 2) = 18$

ANSWERS
3. (a) $y = {}^-2$ (b) $m = 0$ (c) $t = 5$

4 Solve each equation and check each solution.

(a) $3(b + 7) = 2b - 1$

(b) $6 - 2n = 14 + 4(n - 5)$

> **EXAMPLE 4** Solving an Equation
>
> Solve this equation and check the solution: $8 + 5(m + 2) = 6 + 2m$.
>
> *Step 1* Use the distributive property on the left side.
>
> $8 + 5(m + 2) = 6 + 2m$
>
> *Step 2* Combine like terms on the left side.
>
> $8 + 5m + 10 = 6 + 2m$ No like terms on the right side.
>
> $5m + 18 = 6 + 2m$
>
> *Step 3* Add ^-2m to both sides.
>
> $\ ^-2m \ ^-2m$
>
> $3m + 18 = 6 + 0$
>
> $3m + 18 = 6$
>
> *Step 3* To get $3m$ by itself, add $^-18$ to both sides.
>
> $\ ^-18 \ ^-18$
>
> $3m + 0 = {^-12}$
>
> *Step 4* Divide both sides by 3, the coefficient of the variable term $3m$.
>
> $\dfrac{3m}{3} = \dfrac{^-12}{3}$
>
> $m = {^-4}$
>
> The solution is $^-4$.
>
> *Step 5* Check
>
> $8 + 5(m + 2) = 6 + 2m$ Replace each m with $^-4$.
>
> $8 + 5(^-4 + 2) = 6 + 2(^-4)$
>
> $8 + 5(^-2) = 6 + {^-8}$
>
> $8 + ^-10 = {^-2}$
>
> $^-2 = {^-2}$ Balances
>
> When m is replaced with $^-4$, the equation balances, so **$^-4$ is the correct solution**.

Work Problem 4 at the Side.

ANSWERS

4. (a) $b = {^-22}$ (b) $n = 2$

Section 2.5 Solving Equations with Several Steps   131

2.5 Exercises

Solve each equation and check each solution. See Example 1.

1. $7p + 5 = 12$ Check $7p + 5 = 12$
2. $6k + 3 = 15$ Check $6k + 3 = 15$

3. $2 = 8y - 6$ Check
4. $10 = 11p - 12$ Check

5. $^-3m + 1 = 1$ Check
6. $^-4k + 5 = 5$ Check

7. $28 = ^-9a + 10$ Check
8. $75 = ^-10w + 25$ Check

9. $^-5x - 4 = 16$ Check
10. $^-12b - 3 = 21$ Check

Chapter 2 Understanding Variables and Solving Equations

*In Exercises 11–16, solve each equation **two** ways. First keep the variable term on the left side when you solve it. Then solve it again, keeping the variable term on the right side. Finally, check your solution. See Example 2.*

11. $6p - 2 = 4p + 6$ $\qquad$ $6p - 2 = 4p + 6$ $\qquad$ Check $\quad 6p - 2 = 4p + 6$

$(6p - 4p) = 2p$

12. $5y - 5 = 2y + 10$ $\qquad$ $5y - 5 = 2y + 10$ $\qquad$ Check $\quad 5y - 5 = 2y + 10$

13. $^-2k - 6 = 6k + 10$ $\qquad$ $^-2k - 6 = 6k + 10$ $\qquad$ Check

14. $5x + 4 = {}^-3x - 4$ $\qquad$ $5x + 4 = {}^-3x - 4$ $\qquad$ Check

15. $^-18 + 7a = 2a + 7$ $\qquad$ $^-18 + 7a = 2a + 7$ $\qquad$ Check

16. $^-9 + 2z = 9z + 12$ $^-9 + 2z = 9z + 12$ *Check*

Use the distributive property to help you solve each equation. Show your work. See Example 3.

17. $8(w - 2) = 32$ **18.** $9(b - 4) = 27$ **19.** $^-10 = 2(y + 4)$

20. $^-3 = 3(x + 6)$ **21.** $^-4(t + 2) = 12$ **22.** $^-5(k + 3) = 25$

23. $6(x - 5) = ^-30$ **24.** $7(r - 7) = ^-49$ **25.** $^-12 = 12(h - 2)$

26. $^-11 = 11(c - 3)$ **27.** $0 = ^-2(y + 2)$ **28.** $0 = ^-9(b + 1)$

Solve each equation. Show your work. See Example 4.

29. $6m + 18 = 0$

30. $8p - 40 = 0$

31. $6 = 9w - 12$

32. $8 = 8h + 24$

33. $5x = 3x + 10$

34. $7n = {}^-2n - 36$

35. $2a + 11 = 8a - 7$

36. $r - 10 = 10r + 8$

37. $7 - 5b = 28 + 2b$

38. $1 - 8t = {}^-9 - 3t$

39. ${}^-20 + 2k = k - 4k$

40. $6y - y = {}^-16 + y$

41. $10(c - 6) + 4 = 2 + c - 58$

42. $8(z + 7) - 6 = z + 60 - 10$

43. ${}^-18 + 13y + 3 = 3(5y - 1) - 2$

44. $3 + 5h - 9 = 4(3h + 4) - 1$

45. $6 - 4n + 3n = 20 - 35$

46. $^-19 + 8 = 6p - 7p - 5$

47. $6(c - 2) = 7(c - 6)$

48. $^-3(5 + x) = 4(x - 2)$

49. $^-5(2p + 2) - 7 = 3(2p + 5)$

50. $4(3m - 6) = 72 + 3(m - 8)$

51. $^-6b - 4b + 7b = 10 - b + 3b$

52. $w + 8 - 5w = \,^-w - 15w + 11w$

53. Solve $^-2t - 10 = 3t + 5$. Show each step you take while solving it. Next to each step, write a sentence that explains what you did in that step. Be sure to tell when you used the addition property of equality and when you used the division property of equality.

54. Explain the distributive property in your own words. Show two examples of using the distributive property to remove parentheses in an expression.

55. Here is one student's solution to an equation. Show how to check the solution. If the solution doesn't check, explain the error and correct it.

$$
\begin{aligned}
-8 + 4a &= 2a + 2 \\
-2a & -2a \\
\hline
-10 + 4a &= 0 + 2 \\
-10 + 4a &= 2 \\
10 & 10 \\
\hline
0 + 4a &= 12 \\
\frac{4a}{4} &= \frac{12}{4} \\
a &= 3
\end{aligned}
$$

56. Here is one student's solution to an equation. Show how to check the solution. If the solution doesn't check, explain the error and correct it.

$$
\begin{aligned}
2(x + 4) &= -16 \\
2x + 4 &= -16 \\
-4 & -4 \\
\hline
2x + 0 &= -20 \\
\frac{2x}{2} &= \frac{-20}{2} \\
x &= -10
\end{aligned}
$$

RELATING CONCEPTS (EXERCISES 57–60) For Individual or Group Work

Work Exercises 57–60 in order.

57. (a) Suppose that the sum of two numbers is negative and you know that one of the numbers is positive. What can you conclude about the other number?

(b) How can you tell, just by looking, that the solution to $x + 5 = -7$ must be a negative number? Recall your answer from part (a).

58. (a) Suppose that the sum of two numbers is positive, and you know that one of the numbers is negative. What can you conclude about the other number?

(b) How can you tell, just by looking, that the solution to $-8 + d = 2$ must be a positive number? Recall your answer from part (a).

59. (a) Suppose the product of two numbers is negative and you know that one of the numbers is negative. What can you conclude about the other number?

(b) How can you tell, just by looking, that the solution to $-15n = -255$ must be positive?

60. (a) Suppose the product of two numbers is positive and you know that one of the numbers is negative. What can you conclude about the other number?

(b) How can you tell, just by looking, that the solution to $437 = -23y$ must be negative?

Chapter 2
SUMMARY

KEY TERMS

2.1 **variable** — A variable is a letter that represents a number that varies or changes, depending on the situation.

constant — A constant is a number that is added or subtracted in an expression. It does not vary. For example, 5 is the constant in the expression $c + 5$.

expression — An expression expresses, or tells, the rule for doing something. It is a combination of operations on variables and numbers. An expression does *not* have an equal sign.

evaluate the expression — To evaluate an expression, replace each variable with specific values (numbers) and follow the order of operations.

coefficient — The number part in a multiplication expression is the coefficient. For example, 4 is the coefficient in the expression $4s$.

2.2 **simplifying expressions** — To simplify an expression, write it in a simpler way by combining all the like terms.

term — Each addend in an expression is a term.

variable term — A variable term has a number part (called the coefficient) multiplied by a variable part (a letter). An example is $4s$.

like terms — Like terms are terms with exactly the same variable parts (the same letters and exponents). The coefficients may be different.

2.3 **equations** — An equation has an equal sign. It shows the relationship between what is known about a problem and what isn't known.

solve the equation — To solve an equation, find a number that can replace the variable and make the equation balance.

solution — A solution of an equation is a number that can replace the variable and make the equation balance.

addition property of equality — The addition property of equality states that adding the same quantity to both sides of an equation will keep it balanced.

check the solution — To check the solution of an equation, go back to the *original* equation and replace the variable with the solution. If the equation balances, the solution is correct.

2.4 **division property of equality** — The division property of equality states that dividing both sides of an equation by the same nonzero number will keep it balanced.

TEST YOUR WORD POWER

See how well you have learned the vocabulary in this chapter. Answers follow the Quick Review.

1. A **variable**
   A. can only be the letter x
   B. is never an addend in an expression
   C. is the solution of an equation
   D. represents a number that varies.

2. Which expression has 2 as a **coefficient**?
   A. x^2
   B. $2x$
   C. $x + 2$
   D. $2 - x$

3. Which expression has 4 as a **constant** term?
   A. $4y$
   B. y^4
   C. $4 + y$
   D. $\dfrac{y}{4}$

4. Which expression has four **terms**?
   A. $2 + 3x = {}^-6 + x$
   B. $2 + 3x - 6 + x$
   C. $(2)(3x)({}^-6)(x)$
   D. $2(3x) = {}^-6(x)$

5. **Like terms**
   A. can be multiplied but not added
   B. have the same coefficients
   C. have the same solutions
   D. have the same variable parts.

6. To **simplify an expression,**
   A. combine all the like terms
   B. multiply the exponents
   C. add all the numbers in the expression
   D. add the same quantity to both sides.

137

QUICK REVIEW

Concepts

2.1 Evaluating Expressions
Replace each variable with the specified value. Then follow the order of operations to simplify the expression.

2.1 Using Exponents with Variables
An exponent next to a variable tells how many times to use the variable as a factor in multiplication.

2.1 Evaluating Expressions with Exponents
Rewrite the expression without exponents, replace each variable with the specified value, and multiply all the factors.

2.2 Identifying Like Terms
Like terms have *exactly* the same letters and exponents. The coefficients may be different.

2.2 Combining Like Terms
Step 1 If there are any variable terms with no coefficient, write in the understood 1.

Step 2 Change any subtractions to adding the opposite.

Step 3 Find like terms.

Step 4 Add the coefficients of like terms, keeping the variable part the same.

Examples

The expression for ordering textbooks for two prealgebra classes is $2c + 10$, where c is the class limit. Evaluate the expression when the class limit is 24.

$$2c + 10 \qquad \text{Replace } c \text{ with } 24.$$
$$2 \cdot 24 + 10 \qquad \text{Multiply first.}$$
$$48 + 10 \qquad \text{Add last.}$$
$$58 \qquad \text{Order 58 books.}$$

Rewrite $^-6x^4$ without exponents.

$^-6x^4$ can be written as $^-6 \cdot x \cdot x \cdot x \cdot x$

Coefficient is $^-6$; x is used as a factor 4 times.

Evaluate x^3y when x is $^-4$ and y is 5.

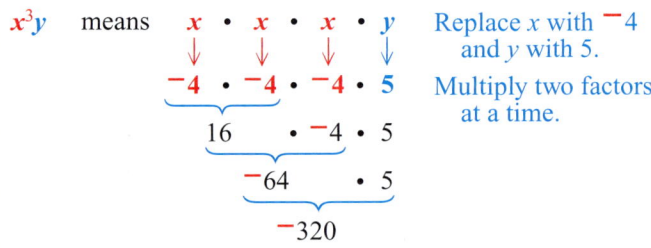

x^3y means $x \cdot x \cdot x \cdot y$ — Replace x with $^-4$ and y with 5.

$^-4 \cdot {^-4} \cdot {^-4} \cdot 5$ — Multiply two factors at a time.

$16 \cdot {^-4} \cdot 5$

$^-64 \cdot 5$

$^-320$

List the like terms in this expression. Then identify the coefficients of the like terms.

$$^-3b + {^-3b^2} + 3ab + b + 3$$

The like terms are ^-3b and b. The coefficient of ^-3b is $^-3$, and the coefficient of b is understood to be 1.

Simplify $4x^2 - 10 + x^2 + 15$.

$4x^2 - 10 + \mathbf{1}x^2 + 15$ — Write understood 1.

$4x^2 + {^-10} + 1x^2 + 15$ — Change subtraction to adding the opposite.

$4x^2 + 1x^2 + {^-10} + 15$ — Combine $4x^2 + 1x^2$. The variable part stays the same.

$(4 + 1)x^2 + 5$ — Also combine $^-10 + 15$.

$5x^2 + 5$

The simplified expression is $5x^2 + 5$.

Concepts

2.2 Simplifying Multiplication Expressions
Use the associative property to rewrite the expression so that the two number parts can be multiplied. The variable part stays the same.

2.2 Using the Distributive Property
Multiplication distributes over addition and over subtraction. Be careful to multiply *every* term inside the parentheses by the number outside the parentheses.

2.3 Solving and Checking Equations Using the Addition Property of Equality
If possible, *simplify* the expression on one or both sides of the equal sign.

Next, to get the variable by itself on one side of the equal sign, *add* the same number to both sides.

Finally, *check* the solution by going back to the original equation and replacing the variable with the solution. If the equation balances, the solution is correct.

Examples

Simplify: $^-7(5k)$.
Use the associative property of multiplication.

$$^-7 \cdot (5 \cdot k) \text{ can be written as } (^-7 \cdot 5) \cdot k$$
$$\underbrace{^-35} \cdot k$$
$$^-35k$$

The simplified expression is ^-35k.

Simplify.

(a) $6(w - 4)$ can be written as $\underbrace{6 \cdot w}_{6w} - \underbrace{6 \cdot 4}_{24}$

The simplified expression is $6w - 24$.

(b) $^-3(2b + 5)$ can be written as $\underbrace{^-3 \cdot 2b}_{^-6b} + \underbrace{^-3 \cdot 5}_{^-15}$

Use the definition of subtraction "in reverse" to write $^-6b + {^-15}$ as $^-6b - 15$.

The simplified expression is $^-6b - 15$.

Solve this equation and check the solution.

$$\underbrace{^-5 + 8}_{3} = 9 + r \quad \text{Simplify the left side by adding } ^-5 + 8.$$
$$3 = 9 + r \quad \text{To get } r \text{ by itself, add}$$
$$\frac{^-9}{^-6} = \frac{^-9}{\underbrace{0 + r}} \quad \text{the opposite of 9, which is } ^-9, \text{ to both sides.}$$
$$^-6 = r$$

The solution is $^-6$.

Check $^-5 + 8 = 9 + r$ Use the original equation and replace r with $^-6$.
$$\underbrace{^-5 + 8}_{3} = \underbrace{9 + {^-6}}_{3}$$
$$3 = 3 \quad \text{Balances}$$

When r is replaced with $^-6$, the equation balances, so $^-6$ is the correct solution.

Concepts	Examples

2.4 Solving and Checking Equations Using the Division Property of Equality

If possible, *simplify* the expression on one or both sides of the equal sign.

Next, to get the variable by itself on one side of the equal sign, *divide* both sides by the coefficient of the variable term.

Finally, *check* the solution by going back to the original equation and replacing the variable with the solution. If the equation balances, the solution is correct.

Solve this equation and check the solution.

Simplify the left side. Change subtraction to adding the opposite.
$$2h - 6h = 18 + 22$$ Simplify the right side; add $18 + 22$.
$$2h + {}^-6h = 40$$

Divide by $^-4$, the coefficient of ^-4h.
$$\frac{^-4h}{^-4} = \frac{40}{^-4}$$ Also divide 40 by $^-4$ to keep the balance.
$$h = {}^-10$$

The solution is $^-10$.

Check
$$2h - 6h = 18 + 22$$ Original equation; replace h with $^-10$.
$$2(^-10) - 6(^-10) = 40$$
$$^-20 - {}^-60 = 40$$
$$^-20 + {}^+60$$
$$40 = 40$$ Balances

When h is replaced with $^-10$, the equation balances, so $^-10$ is the correct solution.

2.4 Solving Equations Such as $^-x = 5$

As the last step in solving an equation, do **not** leave a negative sign in front of the variable, such as $^-x = 5$, because ^-x is **not** the same as x. Divide both sides by $^-1$, the understood coefficient of ^-x.

Solve this equation and check the solution.
$$9 = {}^-n$$

Write the understood $^-1$ as the coefficient of n.

$$9 = {}^-n \text{ can be written as } 9 = {}^-1n$$

Now divide both sides by $^-1$.

$$\frac{9}{^-1} = \frac{^-1n}{^-1}$$

$$^-9 = n$$

The solution is $^-9$.

Check
$$9 = {}^-n$$ Original equation
$$9 = {}^-1n$$ Write understood $^-1$.
$$9 = {}^-1(^-9)$$ Replace n with $^-9$.
$$9 = 9$$ Balances

When n is replaced with $^-9$, the equation balances, so $^-9$ is the correct solution.

Concepts	Examples
2.5 Solving Equations with Several Steps	Solve this equation and check the solution.

Step 1 If possible, use the distributive property to remove parentheses.

Step 2 Combine any like terms on the left side of the equal sign. Combine any like terms on the right side of the equal sign.

Step 3 Add the same amount to both sides of the equation so that the variable term ends up by itself on one side of the equal sign, and a number is by itself on the other side. You may have to do this step more than once.

Step 4 Divide both sides by the coefficient of the variable term to find the solution.

Step 5 Check the solution by going back to the original equation. Replace the variable with the solution. If the equation balances, the solution is correct.

Step 1: $3 + 2(y + 8) = 5y + 4$

Step 2: $3 + 2y + 16 = 5y + 4$

$2y + 19 = 5y + 4$

Step 3:
$$\,\,\underline{-2y}\,\,\underline{-2y}$$
$$0 + 19 = 3y + 4$$

Step 3:
$$\,\,\underline{-4}\,\,\underline{-4}$$
$$0 + 15 = 3y + 0$$

Step 4: $\dfrac{15}{3} = \dfrac{3y}{3}$

$5 = y$

The solution is 5.

Check Step 5
$3 + 2(y + 8) = 5y + 4$ Original equation
$3 + 2(5 + 8) = 5(5) + 4$
$3 + 2(13) = 25 + 4$
$3 + 26 = 29$
$29 = 29$ Balances

When y is replaced with 5, the equation balances, so 5 is the correct solution.

Answers to Test Your Word Power

1. D; *Example:* In $c + 5$, the variable is c.
2. B, *Example:* $2y^3$ and $2n$ also have 2 as a coefficient.
3. C; *Example:* $-5a^2 + 4$ also has 4 as a constant term.
4. B; *Example:* $3y^2 - 6y + 2y - 5$ also has four terms. Choices (A) and (D) are equations, not expressions; choice (C) has four *factors*.
5. D; *Example:* $7n^2$ and $-3n^2$ are like terms.
6. A; *Example:* To simplify $4a - 9 + 6a$, combine $4a$ and $6a$ by adding the coefficients. The simplified expression is $10a - 9$.

Focus on Real-Data Applications

Algebraic Expressions and Tuition Costs

Algebraic expressions are useful in real-life scenarios in which the same set of instructions are repeated for different choices of numbers. Below is the description of how tuition and fees are calculated for "Resident of District" students at North Harris Montgomery Community College District (NHMCCD) in Texas for spring semester 2004. The information is given in the college's schedule and can be found at the Web site www.nhmccd.edu.*

Fees Required at NHMCCD
[Residents of the district pay] tuition at the rate of $32 per credit hour, a $6 per credit hour technology fee, a $2 per credit hour student activity fee, and a registration fee of $12.

For Group Discussion

1. Calculate the tuition and fees for a student who is a resident of the district and who enrolls at NHMCCD for the number of credit hours listed in parts **(a), (b),** and **(c)** below. Then, for part **(d)** let x represent the number of credit hours. Pay attention to the *process* you used in your calculations so that you can write the algebraic expression for x credit hours.

   (a) 3 credit hours: _____   (b) 9 credit hours: _____
   (c) 12 credit hours: _____   (d) x credit hours: _____ dollars

Write the algebraic expression that represents the tuition and fees for each institution for one semester. Let x represent the number of credit hours. If you have difficulty, first calculate the costs for 3 or 9 credit hours and focus on the process that you used to get the answer.

2. American River College, California (nonresident student) www.arc.losrios.edu*
   Enrollment: $18 per credit hour; parking: $30 per semester; an additional nonresident enrollment: $149 per credit hour; other fees: $6

3. Austin Community College, Texas (out-of-district student) www.austin.edu*
   Tuition: $34 per credit hour; building fee: $13 per credit hour; parking: $10 per semester; an additional out-of-district tuition: $55 per credit hour; student service fee: $3

4. Valdosta State University, Georgia (in-state student) www.valdosta.edu*
   Tuition: $93 per credit hour; health fee: $73; student services fee: $99; athletics fees: $109; technology fee: $38; parking fee: $50.

5. Your college tuition and fees

---

*__Note__ that URLs sometimes change, although that is unlikely for academic institutions. If the Web address given does not work, use a search engine, such as www.yahoo.com, to find the new URL.

Chapter 2
REVIEW EXERCISES

[2.1]

1. (a) Identify the variable, the coefficient, and the constant in this expression.
   $$-3 + 4k$$

   (b) Circle the expression that has 20 as the constant term and -9 as the coefficient.

   $20m - 9$ $\quad -9 + 20x$

   $-9y + 20$ $\quad 20 + 9n$

2. The expression for ordering test tubes for a chemistry lab is $4c + 10$, where c is the class limit. Evaluate the expression when

   (a) the class limit is 15.

   (b) the class limit is 24.

3. Rewrite each expression without exponents.
   (a) $x^2 y^4$
   (b) $5ab^3$

4. Evaluate each expression when m is 2, n is -3, and p is 4.
   (a) n^2
   (b) n^3
   (c) $-4mp^2$
   (d) $5m^4 n^2$

[2.2] *Simplify.*

5. $ab + ab^2 + 2ab$

6. $-3x + 2y - x - 7$

7. $-8(-2g^3)$

8. $4(3r^2 t)$

9. $5(k + 2)$

10. $-2(3b + 4)$

11. $3(2y - 4) + 12$

12. $-4 + 6(4x + 1) - 4x$

13. Write an expression with four terms that *cannot* be simplified.

[2.3] *Solve each equation and check each solution.*

14. $16 + n = 5$ \quad Check

15. $-4 + 2 = 2a - 6 - a$ \quad Check

[2.4] *Solve each equation. Show your work.*

16. $48 = -6m$

17. $k - 5k = -40$

18. $-17 + 11 + 6 = 7t$

19. $^-2p + 5p = 3 - 21$ **20.** $^-30 = 3(^-5r)$ **21.** $12 = ^-h$

[2.5] *Solve each equation. Show your work.*

22. $12w - 4 = 8w + 12$ **23.** $0 = ^-4(c + 2)$

24. Every Wednesday is "treat day" at the office. The person who brings the treats buys two treat items for each employee plus four extras. If Roger brings 34 doughnuts, solve the equation $34 = 2n + 4$ to find the number of employees.

MIXED REVIEW EXERCISES

Solve each equation. Show your work.

25. $12 + 7a = 4a - 3$ **26.** $^-2(p - 3) = ^-14$ **27.** $10y = 6y + 20$

28. $2m - 7m = 5 - 20$ **29.** $20 = 3x - 7$ **30.** $b + 6 = 3b - 8$

31. $z + 3 = 0$ **32.** $3(2n - 1) = 3(n + 3)$ **33.** $^-4 + 46 = 7(^-3t + 6)$

34. $6 + 10d - 19 = 2(3d + 4) - 1$ **35.** $^-4(3b + 9) = 24 + 3(2b - 8)$

Chapter 2
TEST

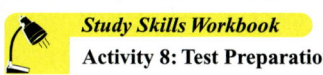
Study Skills Workbook
Activity 8: Test Preparation

1. Identify the parts of this expression: $^-7w + 6$
   Choose from these labels: variable, constant, coefficient.

   1. _____

2. The expression for buying hot dogs for the company picnic is $3a + 2c$, where a is the number of adults and c is the number of children. Evaluate the expression when there are 45 adults and 21 children.

   2. _____

Rewrite each expression without exponents.

3. x^5y^3

4. $4ab^4$

   3. _____

   4. _____

5. Evaluate $^-2s^2t$ when s is $^-5$ and t is 4.

   5. _____

Simplify each expression.

6. $3w^3 - 8w^3 + w^3$

7. $xy - xy$

   6. _____

   7. _____

8. $^-6c - 5 + 7c + 5$

9. $3m^2 - 3m + 3mn$

   8. _____

   9. _____

10. $^-10(4b^2)$

11. $^-5(^-3k)$

    10. _____

    11. _____

12. $7(3t + 4)$

13. $^-4(a + 6)$

    12. _____

    13. _____

14. $^-8 + 6(x - 2) + 5$

15. $^-9b - c - 3 + 9 + 2c$

    14. _____

    15. _____

Chapter 2 Test 145

Solve each equation and check each solution.

16. $^-4 = x - 9$ Check

17. $^-7w = 77$ Check

18. $^-p = 14$ Check

19. $^-15 = ^-3(a + 2)$ Check

Solve each equation. Show your work.

20. $6n + 8 - 5n = ^-4 + 4$

21. $5 - 20 = 2m - 3m$

22. $^-2x + 2 = 5x + 9$

23. $3m - 5 = 7m - 13$

24. $2 + 7b - 44 = ^-3b + 12 + 9b$

25. $3c - 24 = 6(c - 4)$

26. Write an equation that requires the *addition* property of equality to solve it and has $^-4$ as its solution. Then write a different equation that requires the *division* property of equality to solve it and has $^-4$ as its solution. Show how to solve each equation.

Solving Application Problems

3.1 **Problem Solving: Perimeter**

3.2 **Problem Solving: Area**

3.3 **Solving Application Problems with One Unknown Quantity**

3.4 **Solving Application Problems with Two Unknown Quantities**

A century ago there were only 8000 cars and 144 miles of paved roads in the United States. But now there are 132,000,000 registered cars and 4,000,000 miles of paved roads. One 12-lane segment of Interstate Highway 5 in California carries over 365,000 vehicles every day! (*Source:* Federal Highway Administration.)

A handy formula for drivers to use is the distance formula, $d = rt$. In **Section 3.1,** Exercises 51–54, you'll see how to use the formula to find your driving time, rate (speed), or distance traveled on long trips.

3.1 Problem Solving: Perimeter

OBJECTIVES

1. Use the formula for perimeter of a square to find the perimeter or the length of one side.
2. Use the formula for perimeter of a rectangle to find the perimeter, the length, or the width.
3. Find the perimeter of parallelograms, triangles, and irregular shapes.

OBJECTIVE 1 Use the formula for perimeter of a square to find the perimeter or the length of one side. If you have ever studied geometry, you probably used several different formulas such as $P = 2l + 2w$ and $A = lw$. A **formula** is just a shorthand way of writing a rule for solving a particular type of problem. A formula uses variables (letters) and it has an equal sign, so it is an equation. That means you can use the equation-solving techniques you learned in **Chapter 2** to work with formulas.

But let's start at the beginning. Geometry was developed centuries ago when people needed a way to measure land. The name *geometry* comes from the Greek words *ge,* meaning earth, and *metron,* meaning measure. Today we still use geometry to measure land. It is also important in architecture, construction, navigation, art and design, physics, chemistry, and astronomy. You can use geometry at home when you buy carpet or wallpaper, hang a picture, or do home repairs. In this chapter you'll learn about two basic ideas, perimeter and area. Other geometry concepts will appear in later chapters.

In **Section 2.1,** you found the *perimeter* of a square garden.

> **Perimeter**
> The distance around the outside edges of any flat shape is called the **perimeter** of the shape.

To review, a **square** has four sides that are all the same length. Also, the sides meet to form *right angles,* which measure 90° (90 degrees). This means that the sides form "square corners." (For more information on angles, see **Section 6.5.**) Two examples of squares are shown below.

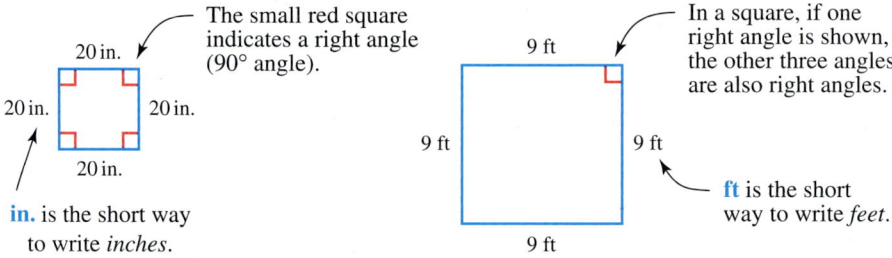

To find the *perimeter* of a square, we can "unfold" the shape so the four sides lie end-to-end, as shown below.

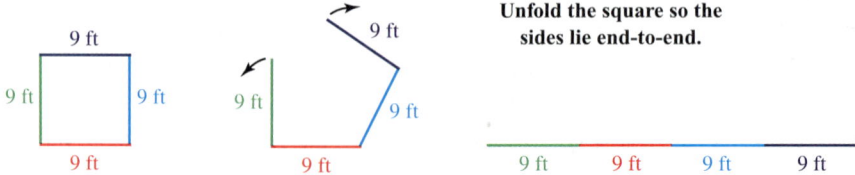

Now we can see the total length of the four sides. The total length is the *perimeter* of the square. We can find the perimeter by adding.

$$\text{Perimeter} = 9 \text{ ft} + 9 \text{ ft} + 9 \text{ ft} + 9 \text{ ft} = 36 \text{ ft}$$

A shorter way is to multiply the length of one side times 4, because all 4 sides are the same length.

Finding the Perimeter of a Square

Perimeter of a square = side + side + side + side

or   P = 4 • side

P = 4s

EXAMPLE 1 Finding the Perimeter of a Square

Find the perimeter of the square on the previous page that measures 9 ft on each side.

Use the formula for perimeter of a square, $P = 4s$. You know that for this particular square, the value of s is 9 ft.

$P = 4s$ Formula for perimeter of a square

$P = 4 \cdot 9 \text{ ft}$ Replace s with 9 ft. Multiply 4 times 9 ft.

$P = 36 \text{ ft}$ Write 36 ft; ft is the unit of measure.

The perimeter of the square is 36 ft. Notice that this answer matches the result obtained from adding the four sides.

Work Problem 1 at the Side.

EXAMPLE 2 Finding the Length of One Side of a Square

If the perimeter of a square is 40 cm, find the length of one side. (Note: **cm** is the short way to write *centimeters*.)

Use the formula for perimeter of a square, $P = 4s$. This time you know that the value of P (the perimeter) is 40 cm.

$P = 4s$ Formula for perimeter of a square

$40 \text{ cm} = 4s$ Replace P with 40 cm

$\dfrac{40 \text{ cm}}{4} = \dfrac{4s}{4}$ To get the variable by itself on the right side, divide both sides by 4.

$10 \text{ cm} = s$

The length of one side of the square is 10 cm.

Check Check the solution by drawing a square and labeling the length of each side as 10 cm. The perimeter is 10 cm + 10 cm + 10 cm + 10 cm = 40 cm. This result matches the perimeter given in the problem.

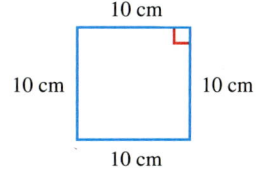

Work Problem 2 at the Side.

---

1 Find the perimeter of each square, using the appropriate formula.

(a) The 20 in. square shown on the previous page

(b) A square measuring 14 miles (14 mi) on each side (*Hint*: Draw a sketch of the square and label each side with its length.)

2 Use the perimeter of each square and the appropriate formula to find the length of one side. Then check your solution by drawing a square, labeling each side, and finding the perimeter.

(a) Perimeter is 28 in.

(b) Perimeter is 100 ft.

(c) Perimeter is 64 cm.

ANSWERS

1. (a) $P = 80$ in.
   (b) $P = 56$ miles

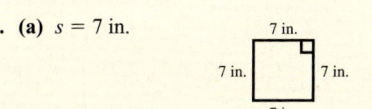

2. (a) $s = 7$ in.

Check
$P = 7 \text{ in.} + 7 \text{ in.} + 7 \text{ in.} + 7 \text{ in.} = 28 \text{ in.}$

(b) $s = 25$ ft

Check
$P = 25 \text{ ft} + 25 \text{ ft} + 25 \text{ ft} + 25 \text{ ft} = 100 \text{ ft}$

(c) $s = 16$ cm

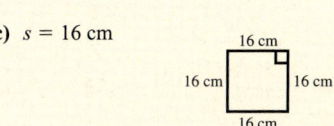

Check
$P = 16 \text{ cm} + 16 \text{ cm} + 16 \text{ cm} + 16 \text{ cm} = 64 \text{ cm}$

3 Find the perimeter of each rectangle using the appropriate formula. Check your solutions by adding the lengths of the four sides.

(a)

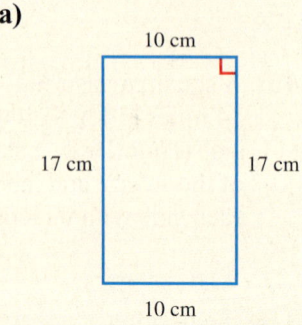

(b) A rectangle 6 m wide and 11 m long (*Hint:* First draw a sketch of the rectangle and label the length of each side.)

(c)

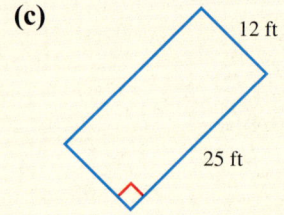

ANSWERS

3. (a) $P = 54$ cm
   Check 17 cm + 17 cm + 10 cm + 10 cm = 54 cm

(b)
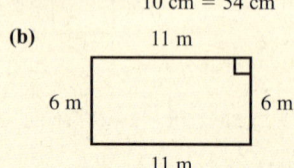

$P = 34$ m
Check 11 m + 11 m + 6 m + 6 m = 34 m

(c) $P = 74$ ft
Check 25 ft + 25 ft + 12 ft + 12 ft = 74 ft

OBJECTIVE 2 **Use the formula for perimeter of a rectangle to find the perimeter, the length, or the width.** A **rectangle** is a figure with four sides that meet to form right angles, which measure 90°. Each set of opposite sides is parallel and congruent (has the same length). Three examples of rectangles are shown below.

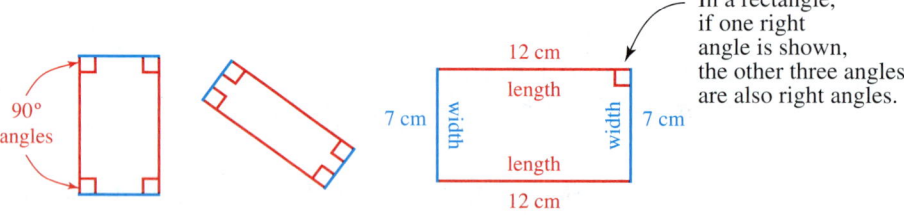

In a rectangle, if one right angle is shown, the other three angles are also right angles.

Each longer side of a rectangle is called the **length** (*l*) and each shorter side is called the **width** (*w*).

Look at the rectangle above with the lengths of the sides labeled. To find the perimeter (distance around), you could unfold the shape so the sides lie end-to-end. Then add the lengths of the sides.

$P = $ **12 cm** + **7 cm** + **12 cm** + **7 cm** $= 38$ cm

or $P = $ **12 cm** + **12 cm** + **7 cm** + **7 cm** $= 38$ cm Commutative property

Because the two long sides are both 12 cm, and the two short sides are both 7 cm, you can also use the formula below.

> **Finding the Perimeter of a Rectangle**
>
> Perimeter of a rectangle = length + length + width + width
>
> $P = $ (2 · length) + (2 · width)
> $P = $ $2l$ + $2w$

EXAMPLE 3 Finding the Perimeter of a Rectangle

Find the perimeter of this rectangle.

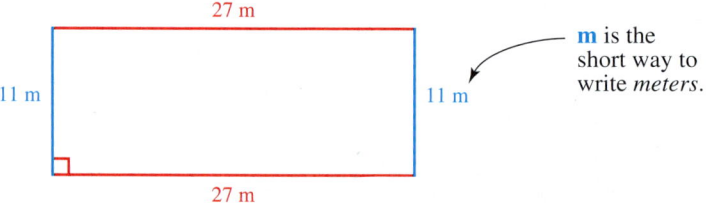

m is the short way to write *meters*.

The length is **27 m**, and the width is **11 m**.

$P = \quad 2l \quad + \quad 2w$ Replace *l* with 27 m and *w* with 11 m.

$P = 2 \cdot \text{27 m} + 2 \cdot \text{11 m}$ Do the multiplications first.

$P = \quad 54 \text{ m} \quad + \quad 22 \text{ m}$ Add last.

$P = 76$ m

The perimeter of the rectangle (the distance you would walk around the outside edges of the rectangle) is 76 m.

Check To check the solution, add the lengths of the four sides.

$P = $ **27 m** + **27 m** + **11 m** + **11 m**

$P = 76$ m Matches the solution above

◀◀◀ **Work Problem** 3 **at the Side.**

Section 3.1 Problem Solving: Perimeter 151

EXAMPLE 4 Finding the Length or Width of a Rectangle

If the perimeter of a rectangle is 20 ft and the width is 3 ft, find the length.

First draw a sketch of the rectangle and label the widths as 3 ft.

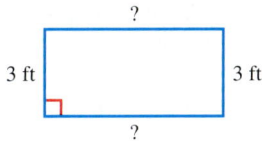

Then use the formula for perimeter of a rectangle, $P = 2l + 2w$. The value of P is 20 ft and the value of w is 3 ft.

$P = 2l + 2w$ Formula for perimeter of a rectangle

$20 \text{ ft} = 2l + 2 \cdot 3 \text{ ft}$ Replace P with 20 ft and w with 3 ft. Simplify the right side by multiplying $2 \cdot 3$ ft.

$20 \text{ ft} = 2l + 6 \text{ ft}$
$\underline{-6 \text{ ft} -6 \text{ ft}}$ To get $2l$ by itself, add $^-6$ ft to both sides: $6 + {^-6}$ is 0.
$14 \text{ ft} = 2l + 0$

$\dfrac{14 \text{ ft}}{2} = \dfrac{2l}{2}$ To get l by itself, divide both sides by 2.

$7 \text{ ft} = l$

The length is 7 ft.

Check   To check the solution, put the length measurements on your sketch. Then add the four measurements.

$P = 7 \text{ ft} + 7 \text{ ft} + 3 \text{ ft} + 3 \text{ ft}$

$P = 20 \text{ ft}$

A perimeter of 20 ft matches the information in the original problem, so 7 ft is the correct length of the rectangle.

Work Problem 4 at the Side.

OBJECTIVE 3 **Find the perimeter of parallelograms, triangles, and irregular shapes.** A **parallelogram** is a four-sided figure in which opposite sides are both parallel and equal in length. Some examples are shown below. Notice that opposite sides have the same length.

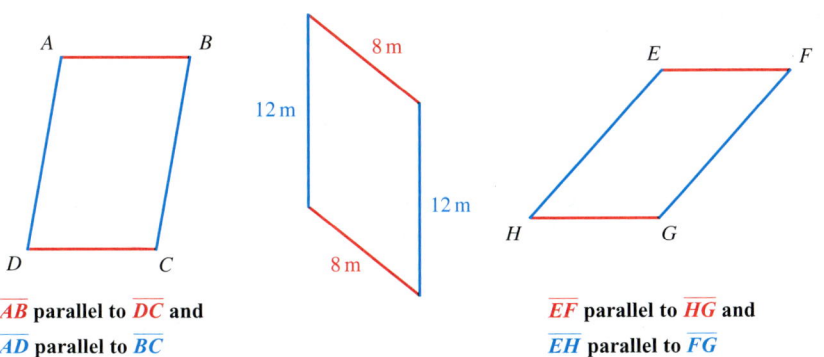

$\overline{AB}$ parallel to $\overline{DC}$ and
$\overline{AD}$ parallel to $\overline{BC}$

$\overline{EF}$ parallel to $\overline{HG}$ and
$\overline{EH}$ parallel to $\overline{FG}$

Perimeter is the distance around a flat shape, so the easiest way to find the perimeter of a parallelogram is to add the lengths of the four sides.

4 Use the perimeter of each rectangle and the appropriate formula to find the length or width. Draw a sketch of each rectangle and use it to check your solution.

(a) The perimeter of a rectangle is 36 in. and the width is 8 in. Find the length.

(b) A rectangle has a width of 4 cm. The perimeter is 32 cm. Find the length.

(c) A rectangle with a perimeter of 14 ft has a length of 6 ft. Find the width.

ANSWERS
4. (a) $l = 10$ in.

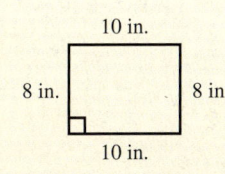

Check   10 in. + 10 in. + 8 in. + 8 in. = 36 in.

(b) $l = 12$ cm

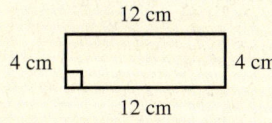

Check   12 cm + 12 cm + 4 cm + 4 cm = 32 cm

(c) $w = 1$ ft

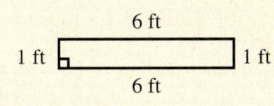

Check   6 ft + 6 ft + 1 ft + 1 ft = 14 ft

5 Find the perimeter of each parallelogram.

(a)
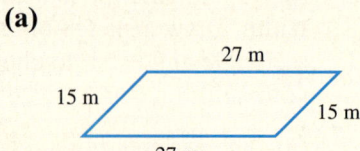

(b)
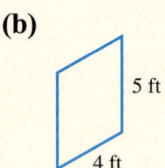

6 Find the perimeter of each triangle.

(a)
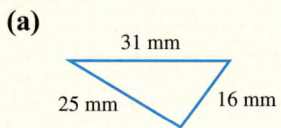

(b) A triangle with sides that each measure 5 in. Draw a sketch of the triangle and label the length of each side.

7 How much fencing will be needed to go around a flower bed with the measurements shown below?

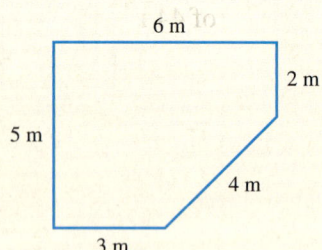

ANSWERS
5. (a) $P = 84$ m (b) $P = 18$ ft
6. (a) $P = 72$ mm
   (b) $P = 15$ in.
   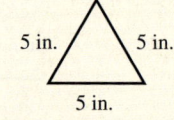
7. 20 m of fencing are needed.

EXAMPLE 5 Finding the Perimeter of a Parallelogram

Find the perimeter of the middle parallelogram on the previous page.

$P = 12\text{ m} + 12\text{ m} + 8\text{ m} + 8\text{ m}$
$P = 40\text{ m}$

◀◀◀ Work Problem 5 at the Side.

A **triangle** is a figure with exactly three sides. Some examples are shown below.

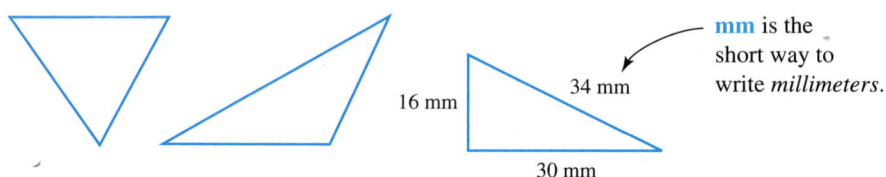

mm is the short way to write *millimeters*.

To find the perimeter of a triangle (the distance around the edges), add the lengths of the three sides.

EXAMPLE 6 Finding the Perimeter of a Triangle

Find the perimeter of the triangle above on the right.

To find the perimeter, add the lengths of the sides.

$P = 16\text{ mm} + 30\text{ mm} + 34\text{ mm}$
$P = 80\text{ mm}$

◀◀◀ Work Problem 6 at the Side.

As with any other shape, you can find the perimeter of (distance around) an irregular shape by adding the lengths of the sides.

EXAMPLE 7 Finding the Perimeter of an Irregular Shape

The floor of a room has the shape shown below.

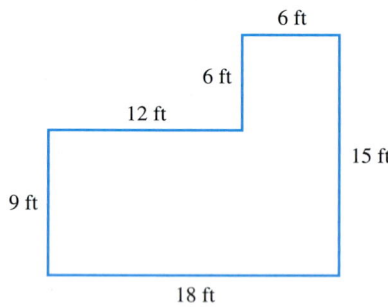

Suppose you want to put a new wallpaper border along the top of all the walls. How much material do you need?

Find the perimeter of the room by adding the lengths of the sides.

$P = 9\text{ ft} + 12\text{ ft} + 6\text{ ft} + 6\text{ ft} + 15\text{ ft} + 18\text{ ft}$
$P = 66\text{ ft}$

You need 66 ft of wallpaper border.

◀◀◀ Work Problem 7 at the Side.

Section 3.1 Problem Solving: Perimeter 153

3.1 Exercises

FOR EXTRA HELP 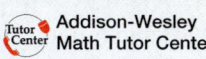 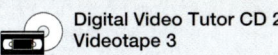  Student's Solutions Manual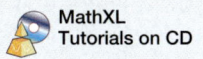

Find the perimeter of each square, using the appropriate formula. See Example 1.

1.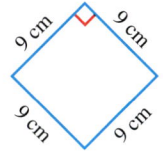
2.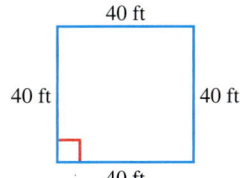
3.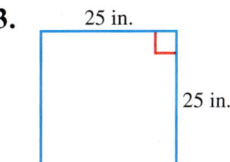
4.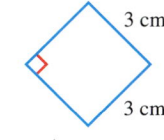

Draw a sketch of each square and label the lengths of the sides. Then find the perimeter. (Sketches may vary; show your sketches to your instructor.)

5. A square park measuring 1 mile on each side
6. A square garden measuring 4 meters on each side
7. A 22 mm square postage stamp
8. A 10 in. square piece of cardboard

For the given perimeter of each square, find the length of one side using the appropriate formula. See Example 2.

9. The perimeter is 120 ft.
10. The perimeter is 52 cm.
11. The perimeter is 4 mm.
12. The perimeter is 20 miles.

13. A square parking lot with a perimeter of 92 yards
14. A square building with a perimeter of 144 meters
15. A square closet with a perimeter of 8 ft
16. A square bedroom with a perimeter of 44 ft

Find the perimeter of each rectangle, using the appropriate formula. Check your solutions by adding the lengths of the four sides. See Example 3.

17.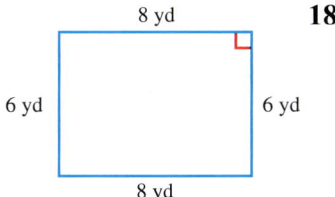
18.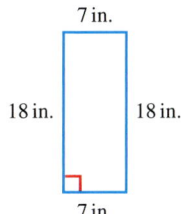
19.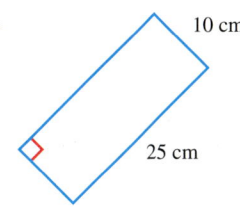
20.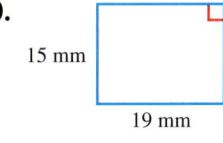

Draw a sketch of each rectangle and label the lengths of the sides. Then find the perimeter by using the appropriate formula. (Sketches may vary; show your sketches to your instructor.)

21. A rectangular living room 20 ft long by 16 ft wide

22. A rectangular placemat 45 cm long by 30 cm wide

23. An 8 in. by 5 in. rectangular piece of paper

24. A 2 ft by 3 ft rectangular window

For each rectangle, you are given the perimeter and either the length or width. Find the unknown measurement using the appropriate formula. Draw a sketch of each rectangle and use it to check your solution. See Example 4. (Show your sketches to your instructor.)

25. The perimeter is 30 cm and the width is 6 cm.

26. The perimeter is 48 yards and the length is 14 yards.

27. The length is 4 miles and the perimeter is 10 miles.

28. The width is 8 meters and the perimeter is 34 meters.

29. A 6 ft long rectangular table has a perimeter of 16 ft.

30. A 13 in. wide rectangular picture frame has a perimeter of 56 in.

31. A rectangular door 1 meter wide has a perimeter of 6 meters.

|←1 m→|

32. A rectangular house 33 ft long has a perimeter of 118 ft.

In Exercises 33–44, find the perimeter of each shape. The figures in Exercises 33–36 are parallelograms. See Examples 5–7.

33.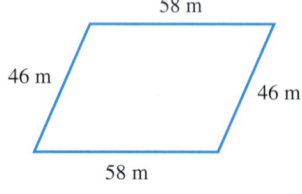

34.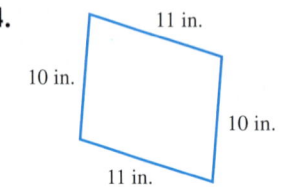

35.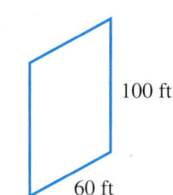

36.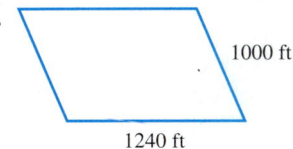
37.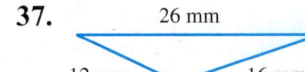
38.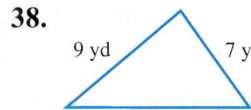

39.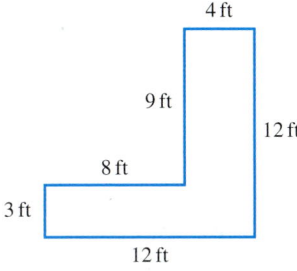
40.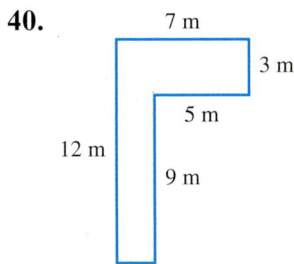
41.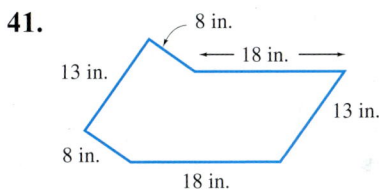

42.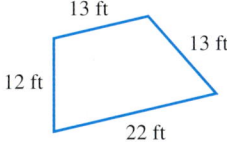
43.
44.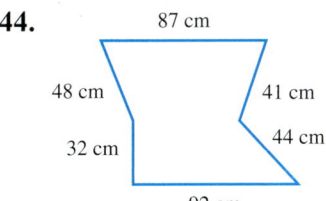

For each shape you are given the perimeter and the lengths of all sides except one. Find the length of the unlabeled side.

45. The perimeter is 115 cm.

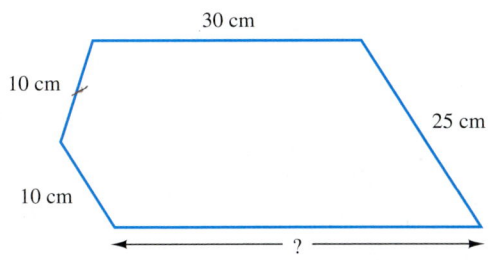

46. The perimeter is 63 in.

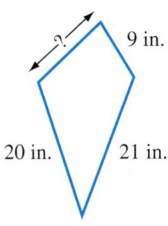

47. The perimeter is 78 in.

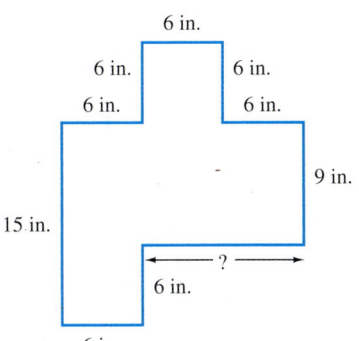

48. The perimeter is 196 ft.

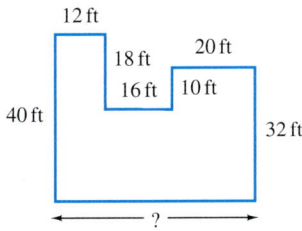

49. In an *equilateral* triangle, all sides have the same length.

(a) Draw sketches of four different equilateral triangles, label the lengths of the sides, and find the perimeters.

(b) Write a "shortcut" rule (a formula) for finding the perimeter of an equilateral triangle.

(c) Will your formula work for other kinds of triangles that are not equilateral? Explain why or why not.

50. Be sure that you have done Exercise 49 first.

(a) Draw a sketch of a figure with five sides of equal length. Write a "shortcut" rule (a formula) for finding the perimeter of this shape.

(b) Draw a sketch of a figure with six sides of equal length. Write a formula for finding the perimeter of the shape.

(c) Write a formula for finding the perimeter of a shape with 10 sides of equal length.

(d) Write a formula for finding the perimeter of a shape with n sides of equal length.

RELATING CONCEPTS (EXERCISES 51–54) For Individual or Group Work

A formula that has many uses for drivers is $d = rt$, called the distance formula. If you are driving a car, then

> d is the *distance* you travel (how many miles)
> r is the *rate* (how fast you are driving in miles per hour)
> t is the *time* (how many hours you drive).

Use the distance formula as you **work Exercises 51–54 in order.**

51. Suppose you are driving on Interstate highways at an average rate of 70 miles per hour. Use the distance formula to find out how far you will travel in (a) 2 hours; (b) 5 hours; (c) 8 hours.

52. If an ice storm slows your driving rate to 35 miles per hour, how far will you travel in (a) 2 hours, (b) 5 hours, (c) 8 hours? Show how to find each answer using the formula. (d) Explain how to find each answer using the results from Exercise 51 instead of the formula.

53. Use the distance formula to find out how many hours you would have to drive to travel the 3000 miles from Boston to San Francisco if your average rate is (a) 60 miles per hour; (b) 50 miles per hour; (c) 20 miles per hour (which was the speed limit 100 years ago).

54. Use the distance formula to find the average driving rate (speed) on each of these trips. (Distances are from *World Almanac.*)

(a) It took 11 hours to drive 671 miles from Atlanta to Chicago.

(b) Sam drove 1539 miles from New York City to Dallas in 27 hours.

(c) Carlita drove 16 hours to travel 1040 miles from Memphis to Denver.

3.2 Problem Solving: Area

OBJECTIVE 1 Use the formula for area of a rectangle to find the area, the length, or the width.

OBJECTIVES

1. Use the formula for area of a rectangle to find the area, the length, or the width.
2. Use the formula for area of a square to find the area or the length of one side.
3. Use the formula for area of a parallelogram to find the area, the base, or the height.
4. Solve application problems involving perimeter and area of rectangles, squares, or parallelograms.

> **Knowing the Difference between Perimeter and Area**
> **Perimeter** is the *distance around the outside edges* of a flat shape.
> **Area** is the amount of *surface inside* a flat shape.

The *perimeter* of a rectangle is the distance around the *outside edges*. Recall that we unfolded a shape and laid the sides end-to-end so we could see the total distance. The *area* of a rectangle is the amount of surface *inside* the rectangle. We measure area by finding the number of squares of a certain size needed to cover the surface inside the rectangle. Think of covering the floor of a living room with carpet. Carpet is measured in square yards, that is, square pieces that measure 1 yard along each side. Here is a drawing of a rectangular living room floor.

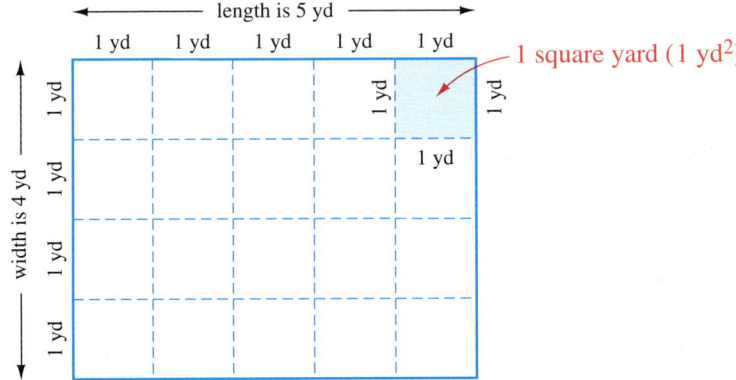

You can see from the drawing that it takes 20 squares to cover the floor. We say that the area of the floor is 20 *square yards*. A short way to write square yards is yd^2.

    20 **square yards**   can be written as   20 **yd^2**

To find the number of squares, you can count them, or you can multiply the number of squares in the length (5) times the number of squares in the width (4) to get 20. The formula is given below.

> **Finding the Area of a Rectangle**
>   Area of a rectangle = length • width
> $A = lw$
> Remember to use *square units* when measuring area.

Squares of many sizes can be used to measure area. For smaller areas, you might use the ones shown at the right.

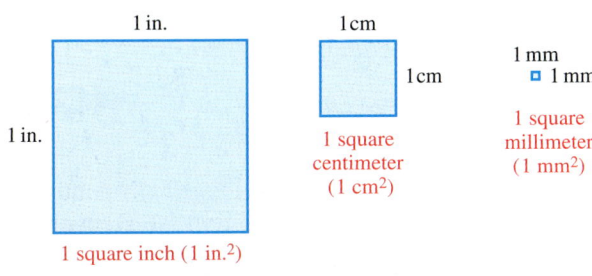

Actual-size drawings

1 Find the area of each rectangle, using the appropriate formula.

(a)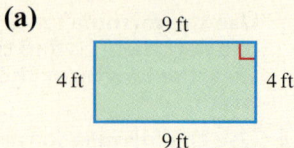

(b) A rectangle is 35 yd long and 20 yd wide. (First make a sketch of the rectangle and label the lengths of the sides.)

(c) A rectangular patio measures 3 m by 2 m. (First make a sketch of the patio and label the lengths of the sides.)

ANSWERS
1. (a) $A = 36$ ft^2
   (b) 20 yd, 35 yd, $A = 700$ yd^2
   (c) 2 m, 3 m, $A = 6$ m^2

Other sizes of squares that are often used to measure area are listed here, but they are too large to draw on this page.

1 square meter (1 m^2)   1 square foot (1 ft^2)
1 square kilometer (1 km^2) 1 square yard (1 yd^2)
                              1 square mile (1 mi^2)

CAUTION
The raised 2 in 4^2 means that you multiply 4 • 4 to get 16. The raised 2 in cm^2 or yd^2 is a short way to write the word "square." It means that you multiplied cm times cm to get cm^2, or yd times yd to get yd^2. Recall that a short way to write $x \cdot x$ is x^2. Similarly, cm • cm is cm^2. When you see 5 cm^2, say "five square centimeters." Do *not* multiply 5 • 5, because the exponent applies only to the *first* thing to its left. The exponent applies to cm, *not* to the number.

EXAMPLE 1 Finding the Area of Rectangles

Find the area of each rectangle.

(a)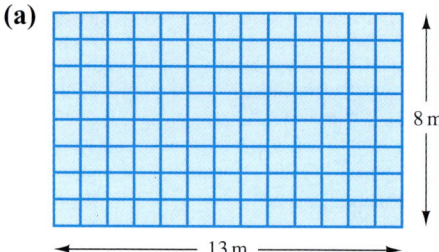

The length of this rectangle is 13 m and the width is 8 m. Use the formula $A = lw$, which means $A = l \cdot w$.

$A = l \cdot w$ Replace l with 13 m and w with 8 m.
$A = $ **13 m • 8 m**   Multiply 13 times 8 to get 104.
$A = 104$ m^2 Multiply m times m to get m^2.

The area of the rectangle is 104 m^2. If you count the number of squares in the sketch, you will also get 104 m^2. Each square in the sketch represents 1 m by 1 m, which is 1 square meter (1 m^2).

(b) A rectangle measuring 7 cm by 21 cm
First make a sketch of the rectangle. The length is 21 cm (the longer measurement) and the width is 7 cm. Then use the formula for area of a rectangle, $A = lw$.

$A = l \cdot w$ Replace l with 21 cm and w with 7 cm.
$A = $ **21 cm • 7 cm**   Multiply 21 • 7 to get 147.
$A = 147$ cm^2 Multiply cm • cm to get cm^2.

The area of the rectangle is 147 cm^2.

Work Problem 1 at the Side.

CAUTION
The units for *area* will always be *square* units (cm^2, m^2, yd^2, mi^2, and so on). The units for *perimeter* will always be *linear* units (cm, m, yd, mi, and so on), *not* square units.

EXAMPLE 2 Finding the Length or Width of a Rectangle

If the area of a rectangular rug is 12 yd² and the length is 4 yd, find the width.

First draw a sketch of the rug and label the length as 4 yd.

Use the formula for area of a rectangle, $A = lw$.
The value of A is 12 yd^2 and the value of l is 4 yd.

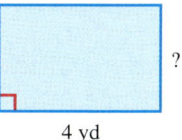
4 yd

$A = l \cdot w$ Replace A with 12 yd² and replace l with 4 yd.

$12 \text{ yd}^2 = 4 \text{ yd} \cdot w$ To get w by itself, divide both sides by 4 yd.

$\dfrac{12 \text{ yd} \cdot \text{yd}}{4 \text{ yd}} = \dfrac{4 \text{ yd} \cdot w}{4 \text{ yd}}$ On the left side, rewrite yd² as yd · yd. Then $\dfrac{\text{yd}}{\text{yd}}$ is 1, so they "cancel out."

$3 \text{ yd} = w$ On the left side, 12 yd ÷ 4 is 3 yd.

The width of the rug is 3 yd.

Check To check the solution, put the width measurement on your sketch. Then use the area formula.

$A = l \cdot w$
$A = 4 \text{ yd} \cdot 3 \text{ yd}$
$A = 12 \text{ yd}^2$

An area of 12 yd² matches the information in the original problem. So 3 yd is the correct width of the rug.

Work Problem 2 at the Side.

OBJECTIVE 2 Use the formula for area of a square to find the area or the length of one side. As with a rectangle, you can multiply length times width to find the area (surface inside) of a square. Because the length and the width are the same in a square, the formula is written as shown below.

Finding the Area of a Square

Area of a square = side · side

$A = s \cdot s$

$A = s^2$

Remember to use *square units* when measuring area.

EXAMPLE 3 Finding the Area of a Square

Find the area of a square highway sign that is 4 ft on each side.
Use the formula for area of a square, $A = s^2$.

$A = s^2$ Remember that s^2 means $s \cdot s$.

$A = s \cdot s$ Replace s with 4 ft.

$A = 4 \text{ ft} \cdot 4 \text{ ft}$ Multiply 4 · 4 to get 16.

$A = 16 \text{ ft}^2$ Multiply ft · ft to get ft².

The area of the sign is 16 ft².

Continued on Next Page

---

2 Use the area of each rectangle and the appropriate formula to find the length or width. Draw a sketch of each rectangle and use it to check your solution.

(a) The area of a microscope slide is 12 cm², and the length is 6 cm. Find the width.

(b) A child's play lot is 10 ft wide and has an area of 160 ft². Find the length.

(c) A hallway is 31 m long and has an area of 93 m². Find the width of the hall.

ANSWERS

2. (a) $w = 2 \text{ cm}$

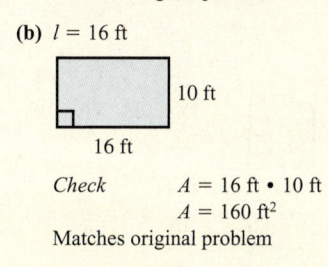

Check $A = 6 \text{ cm} \cdot 2 \text{ cm}$
           $A = 12 \text{ cm}^2$
Matches original problem

(b) $l = 16 \text{ ft}$

Check $A = 16 \text{ ft} \cdot 10 \text{ ft}$
           $A = 160 \text{ ft}^2$
Matches original problem

(c) $w = 3 \text{ m}$

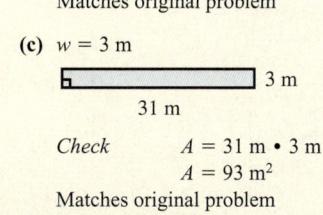

Check $A = 31 \text{ m} \cdot 3 \text{ m}$
           $A = 93 \text{ m}^2$
Matches original problem

160 Chapter 3 Solving Application Problems

3 Find the area of each square, using the appropriate formula. Make a sketch of each square.

(a) A 12 in. square piece of fabric

(b) A square township 7 miles on a side

(c) A square earring measuring 20 mm on each side

4 Given the area of each square, find the length of one side by inspection.

(a) The area of a square-shaped nature center is 16 mi².

(b) A square floor has an area of 100 m².

(c) A square clock face has an area of 81 in.²

ANSWERS
3. (a) $A = 144$ in²
   12 in. × 12 in.
   (b) $A = 49$ mi²
   7 mi × 7 mi
   (c) $A = 400$ mm²
   20 mm × 20 mm
4. (a) $s = 4$ mi   (b) $s = 10$ m   (c) $s = 9$ in.

> **CAUTION**
> Be careful! s^2 means $s \cdot s$. It does *not* mean $2 \cdot s$. In this example s is 4 ft, so $(4 \text{ ft})^2$ is 4 ft · 4 ft = 16 ft². It is *not* 2 · 4 ft = 8 ft.

Check Check the solution by drawing a square and labeling each side as 4 ft. You can multiply length (4 ft) times width (4 ft), as for a rectangle. So the area is 4 ft · 4 ft, or 16 ft². This result matches the solution you got by using the formula $A = s^2$.

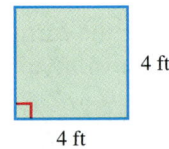

▸▸▸ **Work Problem 3 at the Side.**

EXAMPLE 4 Finding the Length of One Side of a Square

If the area of a square township is 49 mi², what is the length of one side of the township?

Use the formula for area of a square, $A = s^2$. The value of A is 49 mi².

$A = s^2$ Replace A with 49 mi².

$49 \text{ mi}^2 = s^2$ To get s by itself, we have to "undo" the squaring of s. This is called *finding the square root* (more on square roots in **Chapter 5**).

$49 \text{ mi}^2 = s \cdot s$ For now, solve by inspection. Ask, what number times itself gives 49?

$49 \text{ mi}^2 = 7 \text{ mi} \cdot 7 \text{ mi}$ 7 · 7 is 49, so 7 mi · 7 mi is 49 mi².

The value of s is 7 mi, so the length of one side of the township is 7 mi. Notice how this result matches the information about the township in Margin Problem 3(b) at the left.

▸▸▸ **Work Problem 4 at the Side.**

OBJECTIVE 3 Use the formula for area of a parallelogram to find the area, the base, or the height. To find the area of a parallelogram, first draw a dashed line inside the figure, as shown here.

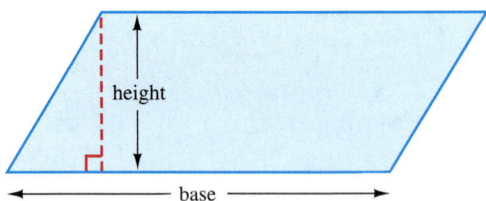

As an experiment, try this yourself by tracing this parallelogram onto a piece of paper.

The length of the dashed line is the *height* of the parallelogram. It forms a 90° angle (a right angle) with the base. The height is the shortest distance between the base and the opposite side.

Now cut off the triangle created on the left side of the parallelogram and move it to the right side, as shown below.

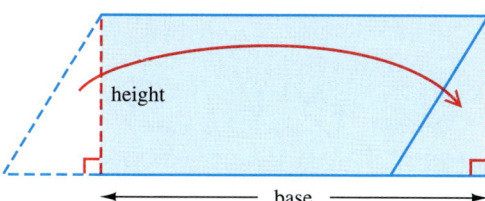

You have made the parallelogram into a rectangle. You can see that the area of the parallelogram and the rectangle are the same.

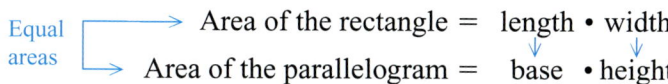

Finding the Area of a Parallelogram

Area of a parallelogram = base • height

$A = bh$

Remember to use *square units* when measuring area.

EXAMPLE 5 Finding the Area of Parallelograms

Find the area of each parallelogram.

(a) 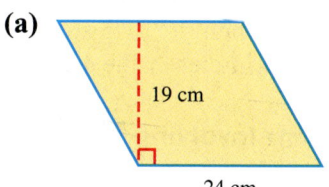   (b)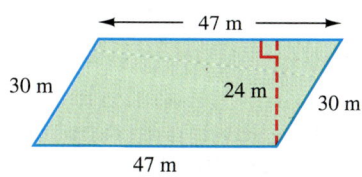

(a) The base is 24 cm and the height is 19 cm. The formula for the area of a parallelogram is $A = bh$.

$A = \ b \ \cdot \ h$ Replace *b* with 24 cm and *h* with 19 cm.

$A = 24 \text{ cm} \cdot 19 \text{ cm}$ Multiply 24 • 19 to get 456.

$A = 456 \text{ cm}^2$ Multiply cm • cm to get cm².

The area of the parallelogram is 456 cm².

(b) Use the formula for area of a parallelogram, $A = bh$.

$A = \ b \ \cdot \ h$ Replace *b* with 47 m and *h* with 24 m.

$A = 47 \text{ m} \cdot 24 \text{ m}$ Multiply 47 • 24 to get 1128.

$A = 1128 \text{ m}^2$ Multiply m • m to get m².

Notice that the 30 m sides are *not* used in finding the area. But you would use them when finding the *perimeter* of the parallelogram.

Work Problem 5 at the Side.

EXAMPLE 6 Finding the Base or Height of a Parallelogram

The area of a parallelogram is 24 ft² and the base is 6 ft. Find the height.

First draw a sketch of the parallelogram and label the base as 6 ft.

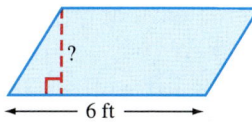

Continued on Next Page

Section 3.2 Problem Solving: Area   **161**

5 Find the area of each parallelogram.

(a)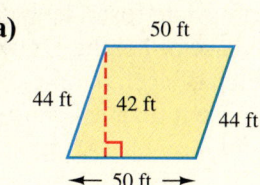

(b)
```
         11 in.
      ┌────┐
18 in.│10 in.│18 in.
      │      │
      └──────┘
         11 in.
```

(c) A parallelogram with base 8 cm and height 1 cm.

ANSWERS

5. (a) $A = 2100 \text{ ft}^2$ (b) $A = 180 \text{ in.}^2$
   (c) $A = 8 \text{ cm}^2$

162 Chapter 3 Solving Application Problems

6 Use the area of each parallelogram and the appropriate formula to find the base or height. Draw a sketch of each parallelogram and use it to check your solution.

(a) The area of a parallelogram is 140 in.² and the base is 14 in. Find the height.

(b) A parallelogram has an area of 4 yd². The height is 1 yd. Find the base.

7 If sod costs $3 per square yard, how much will the neighbors in Example 7 spend to cover the playground with grass?

ANSWERS
6. (a) $h = 10$ in.

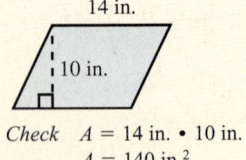

Check $A = 14$ in. • 10 in.
$A = 140$ in.²
Matches original problem

(b) $b = 4$ yd

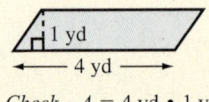

Check $A = 4$ yd • 1 yd
$A = 4$ yd²
Matches original problem

7. $1056 (Find the area by multiplying 22 yd • 16 yd to get 352 yd². Then multiply $3 • 352 to get $1056.)

Use the formula for the area of a parallelogram, $A = bh$. The value of A is 24 ft², and the value of b is 6 ft.

$A = b \cdot h$ Replace A with 24 ft² and b with 6 ft.

24 ft² = 6 ft • h To get h by itself, divide both sides by 6 ft.

$\dfrac{24 \text{ ft} \cdot \text{ft}}{6 \text{ ft}} = \dfrac{6 \text{ ft} \cdot h}{6 \text{ ft}}$ On the left side, rewrite ft² as ft • ft. Then $\dfrac{\text{ft}}{\text{ft}}$ is 1, so they "cancel out."

4 ft = h On the left side, 24 ft ÷ 6 is 4 ft.

The height of the parallelogram is 4 ft.

Check To check the solution, put the height measurement on your sketch. Then use the area formula.

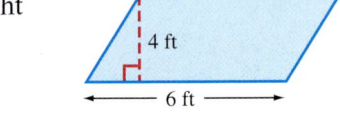

$A = b \cdot h$
$A = 6 \text{ ft} \cdot 4 \text{ ft}$
$A = 24 \text{ ft}^2$

An area of 24 ft² matches the information in the original problem. So 4 ft is the correct height of the parallelogram.

▶◀ **Work Problem 6 at the Side.**

OBJECTIVE 4 Solve application problems involving perimeter and area of rectangles, squares, or parallelograms. When you are solving problems, first decide whether you need to find the perimeter or the area.

EXAMPLE 7 Solving an Application Problem Involving Perimeter or Area

A group of neighbors is fixing up a playground for their children. The rectangular lot is 22 yd by 16 yd. If chain-link fencing costs $6 per yard, how much will they spend to put a fence around the lot?

First draw a sketch of the rectangular lot and label the lengths of the sides. The fence will go around the edges of the lot, so you need to find the *perimeter* of the lot.

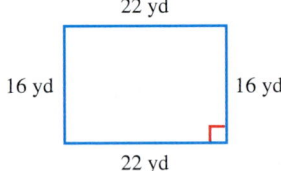

$P = 2l + 2w$ Formula for perimeter of a rectangle
$P = 2 \cdot 22 \text{ yd} + 2 \cdot 16 \text{ yd}$ Replace l with 22 yd and w with 16 yd.
$P = 44 \text{ yd} + 32 \text{ yd}$
$P = 76 \text{ yd}$

The perimeter of the lot is 76 yd, so the neighbors need to buy 76 yd of fencing. The cost of the fencing is $6 *per yard*, which means $6 *for 1 yard*. To find the cost for 76 yd, multiply $6 • 76. The neighbors will spend $456 on the fence.

An application involving the *area* of the playground is found in Margin Problem 7 at the left.

▶◀ **Work Problem 7 at the Side.**

3.2 Exercises

FOR EXTRA HELP 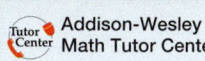 MathXL  Digital Video Tutor CD 2 Videotape 3 Student's Solutions Manual  MyMathLab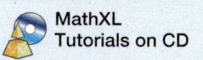

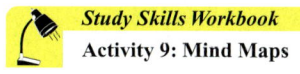

Find the area of each rectangle, square, or parallelogram using the appropriate formula. See Examples 1, 3, and 5.

1.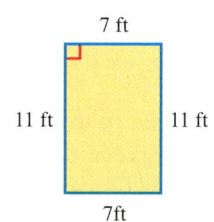
2.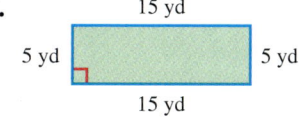
3.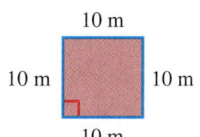
4.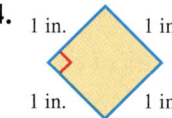

5.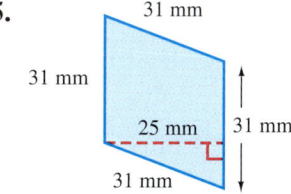
6.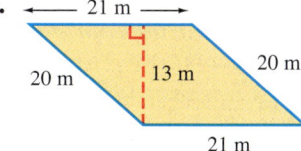
7.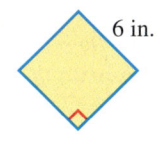
8.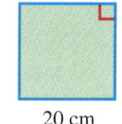

In Exercises 9–16, first draw a sketch of the shape and label the lengths of the sides or base and height. Then find the area. (Sketches may vary; show your sketches to your instructor.)

9. A rectangular calculator that measures 15 cm by 7 cm

10. A rectangular piece of plywood that is 8 ft long and 2 ft wide

11. A parallelogram with height of 9 ft and base of 8 ft

12. A parallelogram measuring 18 mm on the base and 3 mm on the height

13. A fire burned a square-shaped forest 25 mi on a side.

14. An 11 in. square pillow

15. A piece of window glass 1 m on each side

16. A table 12 ft long by 3 ft wide

Use the area of each rectangle and either its length or width, and the appropriate formula, to find the other measurement. Draw a sketch of each rectangle and use it to check your solution. See Example 2. (Sketches may vary; show your sketches to your instructor.)

17. The area of a desk is 18 ft^2, and the width is 3 ft. Find its length.

18. The area of a classroom is 630 ft^2, and the length is 30 ft. Find its width.

19. A parking lot is 90 yd long and has an area of 7200 yd^2. Find its width.

20. A playground is 60 yd wide and has an area of 6000 yd². Find its length.

21. A 154 in.² photo has a width of 11 in. Find its length.

22. A 15 in.² note card has a width of 3 in. Find its length.

Given the area of each square, find the length of one side by inspection. See Example 4.

23. A square floor has an area of 36 m².

24. A square stamp has an area of 9 cm².

25. The area of a square sign is 4 ft².

26. The area of a square piece of metal is 64 in.².

Use the area of each parallelogram and either its base or height, and the appropriate formula, to find the other measurement. Draw a sketch of each parallelogram and use it to check your solution. See Example 6. (Sketches may vary; show your sketches to your instructor.)

27. The area is 500 cm², and the base is 25 cm. Find the height.

28. The area is 1500 m², and the height is 30 m. Find the base.

29. The height is 13 in. and the area is 221 in.². Find the base.

30. The base is 19 cm, and the area is 114 cm². Find the height.

31. The base is 9 m, and the area is 9 m². Find the height.

32. The area is 25 mm², and the height is 5 mm. Find the base.

*Explain and correct the **two** errors made by students in Exercises 33 and 34.*

33.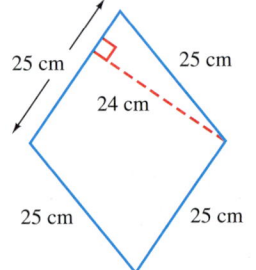

34.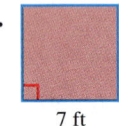

7 ft

$A = s^2$

$A = 2 \cdot 7$ ft

$A = 14$ ft

$P = 25 \text{ cm} + 24 \text{ cm} + 25 \text{ cm} + 25 \text{ cm} + 25 \text{ cm}$

$P = 124 \text{ cm}^2$

Name each figure and find its perimeter and area.

35.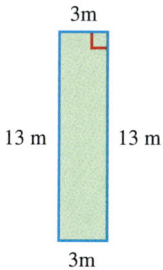

36.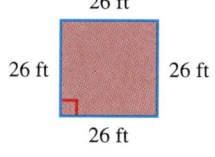

37.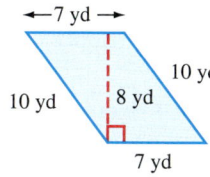

38.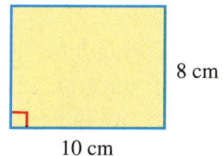

39.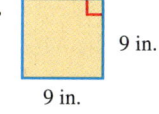

40.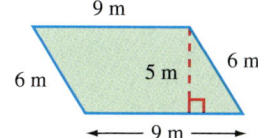

Solve each application problem. You may need to find the perimeter, the area, or one of the side measurements. In Exercises 41–46 draw a sketch for each problem and label the sketch with the appropriate measurements. See Example 7. (Sketches may vary; show your sketches to your instructor.)

41. Gymnastic floor exercises are performed on a square mat that is 12 meters on a side. Find the perimeter and area of a mat. (*Source:* www.nist.gov)

42. A regulation volleyball court is 18 meters by 9 meters. Find the perimeter and area of a regulation court. (*Source:* www.nist.gov)

43. Tyra's kitchen is 4 m wide and 5 m long. She is pasting a decorative strip that costs $6 per meter around the top edge of all the walls. How much will she spend?

44. The Wang's family room measures 20 ft by 25 ft. They are covering the floor with square tiles that measure 1 ft on a side and cost $1 each. How much will they spend on tile?

45. Mr. and Mrs. Gomez are buying carpet for their square-shaped bedroom, which measures 5 yd along each wall. The carpet is $23 per square yard and padding and installation is another $6 per square yard. How much will they spend in all?

46. A page in this book measures about 27 cm from top to bottom and 21 cm from side to side. Find the perimeter and the area of the page.

47. A regulation football field is 100 yd long (excluding end zones) and has an area of 5300 yd². Find the width of the field. (*Source:* NFL.)

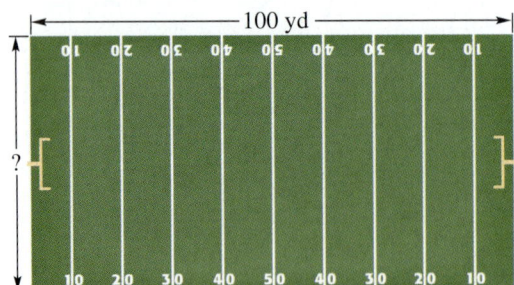

48. There are 14,790 ft² of ice in the rectangular playing area for a major league hockey game (excluding the area behind the goal lines). If the playing area is 85 ft wide, how long is it? (*Source:* NHL.)

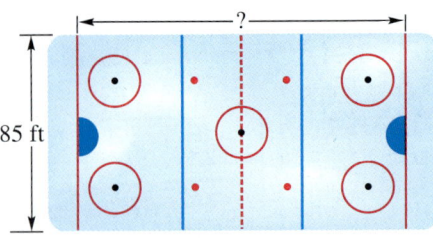

49. Advanced Photo System (APS) cameras allow you to choose from three different print sizes each time you snap a photo. The choices are shown below. Find the perimeter and area of each size print. (*Source:* Kodak.)

Panoramic 4 in. × 14 in.

4 in. × 7 in.

4 in. × 6 in.

50. The Monterey Bay Aquarium in California lets visitors look into a million-gallon tank through an acrylic panel that is 13 in. thick. The panel is 54 ft long and 15 ft high. What is the perimeter and the area of the panel? (*Source:* AAA California Tour Book.)

RELATING CONCEPTS (EXERCISES 51–54) For Individual or Group Work

Use your knowledge of perimeter and area to **work Exercises 51–54 in order.**

51. Suppose you have 12 ft of fencing to make a square or rectangular garden plot. Draw sketches of *all* the possible plots that use exactly 12 ft of fencing and label the lengths of the sides. Use only *whole number* lengths. (*Hint:* There are three possibilities.)

52. (a) Find the area of each plot in Exercise 51.

(b) Which plot has the greatest area?

53. Repeat Exercise 51 using 16 ft of fencing. Be sure to draw *all* possible square or rectangular plots that have whole number lengths for the sides.

54. (a) Find the area of each plot in Exercise 53.

(b) Compare your results to those from Exercise 52. What do you notice about the plots with the greatest area?

3.3 Solving Application Problems with One Unknown Quantity

OBJECTIVE 1 Translate word phrases into algebraic expressions. In Sections **3.1** and **3.2** you worked with applications involving perimeter and area. You were able to use well-known rules (formulas) to set up equations that could be solved. However, you will encounter many problems for which no formula is available. Then you need to analyze the problem and translate the words into an equation that fits the particular situation. We'll start by translating word phrases into algebraic expressions.

OBJECTIVES

1. Translate word phrases into algebraic expressions.
2. Translate sentences into equations.
3. Solve application problems with one unknown quantity.

EXAMPLE 1 Translating Word Phrases into Algebraic Expressions

Write each phrase as an algebraic expression. Use x as the variable.

Words	Algebraic Expression	
A number **plus** 2	$x + 2$ or $2 + x$	⎫
The **sum** of 8 and a number	$8 + x$ or $x + 8$	⎬ Two correct ways to write each addition expression
5 **more than** a number	$x + 5$ or $5 + x$	
-35 **added to** a number	$-35 + x$ or $x + -35$	
A number **increased by** 6	$x + 6$ or $6 + x$	⎭
9 **less than** a number	$x - 9$	⎫
A number **subtracted from** 3	$3 - x$	⎬ Only one correct way to write each subtraction expression
3 **subtracted from** a number	$x - 3$	
A number **decreased by** 4	$x - 4$	
10 **minus** a number	$10 - x$	⎭

> **CAUTION**
> Recall that addition can be done in any order, so $x + 2$ gives the same result as $2 + x$. This is *not* true in subtraction, so be careful. $10 - x$ does *not* give the same result as $x - 10$.

Work Problem 1 at the Side.

EXAMPLE 2 Translating Word Phrases into Algebraic Expressions

Write each phrase as an algebraic expression. Use x as the variable.

Words	Algebraic Expression
8 **times** a number	$8x$
The **product** of 12 and a number	$12x$
Double a number (meaning "2 times")	$2x$
The **quotient** of -6 and a number	$\dfrac{-6}{x}$
A number **divided by** 10	$\dfrac{x}{10}$
15 **subtracted from** 4 **times** a number	$4x - 15$
The result **is**	$=$

Work Problem 2 at the Side.

---

1 Write each phrase as an algebraic expression. Use x as the variable.

(a) 15 less than a number

(b) 12 more than a number

(c) A number increased by 13

(d) A number minus 8

(e) 10 plus a number

(f) A number subtracted from 6

(g) 6 subtracted from a number

2 Write each phrase as an algebraic expression. Use x as the variable.

(a) Double a number

(b) The product of -8 and a number

(c) The quotient of 15 and a number

(d) 5 times a number subtracted from 30

ANSWERS
1. (a) $x - 15$ (b) $x + 12$ or $12 + x$
   (c) $x + 13$ or $13 + x$ (d) $x - 8$
   (e) $10 + x$ or $x + 10$ (f) $6 - x$
   (g) $x - 6$
2. (a) $2x$ (b) $-8x$ (c) $\dfrac{15}{x}$
   (d) $30 - 5x$ (*not* $5x - 30$)

3 Translate each sentence into an equation and solve it. Check your solution by going back to the words in the original problem.

(a) If 3 times a number is added to 4, the result is 19. Find the number.

(b) If 7 is subtracted from 6 times a number, the result is $^-25$. Find the number.

OBJECTIVE 2 Translate sentences into equations. The next example shows you how to translate a sentence into an equation that you can solve.

EXAMPLE 3 Translating a Sentence into an Equation

If 5 times a number is added to 11, the result is 26. Find the number.

Let x represent the unknown number. Use the information in the problem to write an equation.

$$\underbrace{5 \text{ times a number}}_{5x} \underbrace{\text{added to } 11}_{+\ 11} \underbrace{\text{is } 26}_{=\ 26}$$

Next, solve the equation.

$$5x + 11 = 26$$
$$\underline{\ -11\ \ \ -11\ }$$ To get $5x$ by itself, add $^-11$ to both sides.
$$5x + 0 = 15$$
$$\frac{5x}{5} = \frac{15}{5}$$ To get x by itself, divide both sides by 5.
$$x = 3$$

The number is 3.

Check Go back to the words of the original problem.

If 5 times a number is added to 11, the result is 26.
$$5 \cdot 3 + 11 = 26$$

Does $5 \cdot 3 + 11$ really equal 26? Yes, $5 \cdot 3 + 11 = 15 + 11 = 26$.

So 3 is the correct solution because it "works" when you put it back into the original problem.

Work Problem 3 at the Side.

OBJECTIVE 3 Solve application problems with one unknown quantity. Now you are ready to tackle application problems. The steps we will use are summarized below.

Solving an Application Problem

Step 1 **Read** the problem once to see what it is about. Read it carefully a second time. As you read, make a sketch or write word phrases that identify the known and the unknown parts of the problem.

Step 2(a) If there is one unknown quantity, **assign a variable** to represent it. Write down what your variable represents.

Step 2(b) If there is more than one unknown quantity, **assign a variable** to represent "the thing you know the least about." Then write variable expression(s), using the same variable, to show the relationship of the other unknown quantities to the first one.

Step 3 **Write an equation,** using your sketch or word phrases as the guide.

Step 4 **Solve** the equation.

Step 5 **State the answer** to the question in the problem and label your answer.

Step 6 **Check** whether your answer fits all the facts given in the *original* statement of the problem. If it does, you are done. If it doesn't, start again at Step 1.

ANSWERS

3. (a) $3x + 4 = 19$
$x = 5$
Check $3 \cdot 5 + 4$ does equal 19
$\underline{15\ +\ 4}$
19

(b) $6x - 7 = {}^-25$
$x = {}^-3$
Check $6 \cdot {}^-3 - 7$ does equal $^-25$
$\underline{{}^-18\ +\ {}^-7}$
$^-25$

EXAMPLE 4 Solving an Application Problem with One Unknown Quantity

Heather had put some money aside in an envelope for household expenses. Yesterday she took out $20 for groceries. Today a friend paid back a loan and Heather put the $34 in the envelope. Now she has $43 in the envelope. How much was in the envelope at the start?

Step 1 **Read** the problem once. It is about money in an envelope. Read it a second time and write word phrases.

Unknown: amount of money in the envelope at the start
Known: took out $20; put in $34; ended up with $43

Step 2(a) There is only one unknown quantity, so **assign a variable** to represent it. Let m represent the money at the start.

Step 3 **Write an equation,** using the phrases you wrote as a guide.

Money at the start	Took out $20	Put in $34	Ended up with $43
m	$- \$20$	$+ \$34$	$= \$43$

Step 4 **Solve** the equation.

$$m - 20 + 34 = 43 \quad \text{Change subtraction to adding the opposite.}$$
$$m + {}^-20 + 34 = 43 \quad \text{Simplify the left side.}$$
$$m + 14 = 43$$
$$ {}^-14 \quad {}^-14 \quad \text{To get } m \text{ by itself, add } {}^-14 \text{ to both sides.}$$
$$m + 0 = 29$$
$$m = 29$$

Step 5 **State the answer** to the question, "How much was in the envelope at the start?" There was $29 in the envelope.

Step 6 **Check** the solution by going back to the *original* problem and inserting the solution.

Started with $29 in the envelope
Took out $20, so $29 − $20 = $9 in the envelope
Put in $34, so $9 + $34 = $43
Now has $43 ← Matches ↑

Because $29 "works" when you put it back into the original problem, you know it is the correct solution.

Work Problem 4 at the Side.

---

4 Some people got on an empty bus at its first stop. At the second stop, 3 people got on. At the third stop, 5 more people got on. At the fourth stop, 10 people got off, but 4 people were still on the bus. How many people got on at the first stop? Show your work for each of the six problem-solving steps.

ANSWERS

4. *Step 1* **Read.**
Unknown: number of people who got on at first stop
Known: 3 got on; 5 got on; 10 got off; 4 people still on bus

Step 2(a) **Assign a variable.**
Let p be people who got on at first stop. (You may use any letter you like as the variable.)

Step 3 **Write an equation.**
$p + 3 + 5 - 10 = 4$

Step 4 **Solve.**
$$p + 3 + 5 + {}^-10 = 4$$
$$p + {}^-2 = 4$$
$$ +2 \quad +2$$
$$p + 0 = 6$$
$$p = 6$$

Step 5 **State the answer.**
6 people got on at the first stop.

Step 6 **Check.**
6 got on at first stop
3 got on at 2nd stop: 6 + 3 = 9
5 got on at 3rd stop: 9 + 5 = 14
10 got off at 4th stop: 14 − 10 = 4
4 people are left. ← Matches ↑

Chapter 3 Solving Application Problems

5 Five donors each gave the same amount of money to a college to use for scholarships. From the money, scholarships of $1250, $900, and $850 were given to students; $250 was left. How much money did each donor give to the college? Show your work for each of the six problem-solving steps.

EXAMPLE 5 Solving an Application Problem with One Unknown Quantity

Three friends each put in the same amount of money to buy a gift. After they spent $2 for a card and $31 for the gift, they had $6 left. How much money had each friend put in originally?

Step 1 **Read** the problem. It is about 3 friends buying a gift.

Unknown: amount of money each friend contributed
Known: 3 friends put in money; spent $2 and $31; had $6 left

Step 2(a) There is only one unknown quantity. **Assign a variable**, m, to represent the amount of money each friend contributed.

Step 3 **Write an equation.**

Number of friends	Amount each friend put in	Spent on card	Spent on gift	Left over
3 •	m	$-$ $2	$-$ $31 =	$6

To see why this is multiplication, think of an example. If each friend put in $10, how much money would there be? 3 • $10, or $30

Step 4 **Solve.**

$3m - 2 - 31 = 6$ Changes subtractions to adding the opposite.
$3m + {}^-2 + {}^-31 = 6$ Simplify the left side.
$3m + {}^-33 = 6$
$\ {+}33 \ {+}33$ To get $3m$ by itself, add 33 to both sides.
$3m + 0 = 39$

$\dfrac{3m}{3} = \dfrac{39}{3}$ To get m by itself, divide both sides by 3.

$m = 13$

Step 5 **State the answer.** Each friend put in $13.

Step 6 **Check** the solution by putting it back into the *original* problem.

3 friends each put in $13, so 3 • $13 = $39.
Spent $2, spent $31, so $39 − $2 − $31 = $6.
Had $6 left ⟵ Matches

$13 is the correct solution because it "works."

Work Problem 5 at the Side.

ANSWERS

5. *Step 1* **Read.**
Unknown: money given by each donor
Known: 5 donors; gave out $1250, $900, $850; $250 left

Step 2(a) **Assign a variable.**
Let m be each donor's money.

Step 3 **Write an equation.**
5 • m − $1250 − $900 − $850 = $250

Step 4 **Solve.**
$5m + {}^-1250 + {}^-900 + {}^-850 = 250$
$5m + {}^-3000 = 250$
$\ {+}3000 \ {+}3000$
$5m + 0 = 3250$

$\dfrac{5m}{5} = \dfrac{3250}{5}$

$m = 650$

Step 5 **State the answer.**
Each donor gave $650.

Step 6 **Check.**
5 donors each gave $650, so
5 • $650 = $3250
Gave out $1250, $900, $850, so
$3250 − 1250 − 900 − 850 = $250
Had $250 left ⟵ Matches

EXAMPLE 6 Solving a More Complex Application Problem with One Unknown Quantity

Michael has completed 5 less than three times as many lab experiments as David. If Michael has completed 13 experiments, how many experiments has David completed?

Step 1 **Read** the problem. It is about the number of experiments done by two students.

Unknown: number of experiments David did
Known: Michael did 5 less than 3 times the number David did; Michael did 13.

Step 2(a) **Assign a variable.** Let n represent the number of experiments David did.

Step 3 **Write an equation.**

The number Michael did	is	5 less than 3 times David's number.
13	=	$3n - 5$

Step 4 **Solve.**

$13 = 3n - 5$ Change subtraction to adding the opposite.

$13 = 3n + {}^-5$

$\underline{{}^+5 \qquad\qquad {}^+5}$ To get $3n$ by itself, add 5 to both sides.

$18 = 3n + 0$

$\dfrac{18}{3} = \dfrac{3n}{3}$ To get n by itself, divide both sides by 3.

$6 = n$

Step 5 **State the answer.** David did 6 experiments.

Step 6 **Check** the solution by putting it back into the *original* problem.

3 times David's number	$3 \cdot 6 = 18$
Less 5	$18 - 5 = 13$
Michael did 13. ← Matches ↑	

The correct solution is: David did 6 experiments.

Work Problem 6 at the Side.

---

6 Susan donated $10 more than twice what LuAnn donated. If Susan donated $22, how much did LuAnn donate?

Show your work for each of the six problem-solving steps.

ANSWERS

6. *Step 1* **Read.**
Unknown: LuAnn's donation
Known: Susan donated $10 more than twice what LuAnn donated; Susan donated $22.

Step 2(a) **Assign a variable.**
Let d be LuAnn's donation.

Step 3 **Write an equation.**
$2d + \$10 = \$22$
(or $\$10 + 2d = \$22$)

Step 4 **Solve.**

$2d + 10 = 22$
$\underline{ {}^-10 = {}^-10}$
$2d + 0 = 12$

$\dfrac{2d}{2} = \dfrac{12}{2}$

$d = 6$

Step 5 **State the answer.**
LuAnn donated $6.

Step 6 **Check.**
$10 more than twice $6 is

$\$10 + 2 \cdot \$6 = \$10 + \$12 = \$22.$

Susan donated $22. ← Matches ↑

Focus on Real-Data Applications

Formulas

In **Section 3.1**, you found the perimeter of a triangle by adding the lengths of the three sides. A formula for this is shown here.

$$P = a + b + c$$

where P is the perimeter and a, b, and c are the lengths of the sides.

1. Use the formula to find the length of the third side in this triangle. Replace P with 33 ft, replace a with 12 ft, and replace b with 7 ft. Then solve the equation.

   12 ft, 7 ft, ?
   Perimeter is 33 ft.

Use the formula to find the length of the third side in each of these triangles.

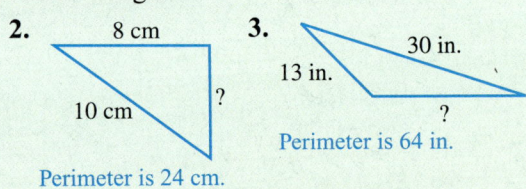

2. 8 cm, 10 cm, ?
   Perimeter is 24 cm.

3. 30 in., 13 in., ?
   Perimeter is 64 in.

4. Describe *in words* what you were doing as you found the length of the third side. What things were being added? What were you doing with the perimeter?

5. How can you simplify the perimeter formula when *two* sides of the triangle are the *same* length?

Use the simplified formula to find the unknown lengths in these triangles.

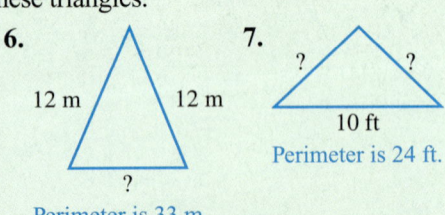

6. 12 m, 12 m, ?
   Perimeter is 33 m.

7. ?, ?, 10 ft
   Perimeter is 24 ft.

In **Section 2.2** you worked with the expression $60A - 900$ for the vocabulary of a child. Now we can use this formula.

$$V = 60A - 900$$

where V is the number of vocabulary words and A is the age of the child in months.

Use the formula to find the *age* of each child.

8. The child's vocabulary is 180 words.

9. The child's vocabulary is 1440 words.

10. The child's vocabulary is 60 words.

11. Describe *in words* what you were doing as you found each child's age.

12. Find the child's age if the child's vocabulary is 0 words.

13. Refer to Problem 12 above. Without solving an equation, are there any other ages at which children would have a vocabulary of 0 words?

14. Is the formula useful for all ages of children? Explain your answer.

3.3 Exercises

Write an algebraic expression, using x as the variable. See Examples 1 and 2.

1. 14 plus a number

2. The sum of a number and −8

3. −5 added to a number

4. 16 more than a number

5. 20 minus a number

6. A number decreased by 25

7. 9 less than a number

8. A number subtracted from −7

9. Subtract 4 from a number.

10. 3 fewer than a number

11. −6 times a number

12. The product of −3 and a number

13. Double a number

14. A number times 10

15. A number divided by 2

16. 4 divided by a number

17. Twice a number added to 8

18. Five times a number plus 5

19. 10 fewer than seven times a number

20. 12 less than six times a number

21. The sum of twice a number and the number

22. Triple a number subtracted from the number

Translate each sentence into an equation and solve it. Check your solution by going back to the words in the original problem. See Example 3.

23. If four times a number is decreased by 2, the result is 26. Find the number.

24. The sum of 8 and five times a number is 53. Find the number.

25. If a number is added to twice the number, the result is −15. What is the number?

26. If a number is subtracted from three times the number, the result is −8. What is the number?

27. If the product of some number and 5 is increased by 12, the result is seven times the number. Find the number.

28. If eight times a number is subtracted from eleven times the number, the result is −9. Find the number.

29. When three times a number is subtracted from 30, the result is 2 plus the number. What is the number?

30. When twice a number is decreased by 8, the result is the number increased by 7. Find the number.

Solve each application problem. Use the six problem-solving steps you learned in this section. See Examples 4–6.

31. Ricardo gained 15 pounds over the winter. He went on a diet and lost 28 pounds. Then he regained 5 pounds and weighed 177 pounds. How much did he weigh originally?

32. Mr. Chee deposited $80 into his checking account. Then, after writing a $23 check for gas and a $90 check for his child's day care, the balance in his account was $67. How much was in his account before he made the deposit?

33. There were 18 cookies in Magan's cookie jar. While she was busy in another room, her children ate some of the cookies. Magan bought three dozen cookies and added them to the jar. At that point she had 49 cookies in the jar. How many cookies did her children eat?

34. The Greens had a 20-pound bag of bird seed in their garage. Mice got into the bag and ate some of it. The Greens then bought an 8-pound bag of seed and put all the seed in a metal container. They now have 24 pounds of seed. How much did the mice eat?

35. A college bookstore ordered six boxes of red pens. The store sold 32 red pens last week and 35 red pens this week. Five pens were left on the shelf. How many pens were in each box?

36. A local charity received a donation of eight cartons filled with cans of soup. The charity gave out 100 cans of soup yesterday and 92 cans today before running out. How many cans were in each carton?

37. The 14 music club members each paid the same amount for dues. The club also earned $340 selling magazine subscriptions. They spent $575 to organize a jazz festival. Now their bank account is overdrawn by $25. How much did each member pay in dues?

38. The manager of an apartment complex had 11 packages of lightbulbs on hand. He replaced 29 burned out bulbs in hallway lights and 7 bulbs in the party room. Eight bulbs were left. How many bulbs were in each package?

39. When 75 is subtracted from four times Tamu's age, the result is Tamu's age. How old is Tamu?

40. If three times Linda's age is decreased by 36, the result is twice Linda's age. How old is Linda?

41. While shopping for clothes, Consuelo spent $3 less than twice what Brenda spent. Consuelo spent $81. How much did Brenda spend?

42. Dennis weighs 184 pounds. His weight is 2 pounds less than six times his child's weight. How much does his child weigh?

43. Paige bought five bags of candy for Halloween. Forty-eight children visited her home and she gave each child three pieces of candy. At the end of the night she still had one bag of candy. How many pieces of candy were in each bag?

44. A restaurant ordered four packages of paper napkins. Yesterday they used up one package, and today they used up 140 napkins. Two packages plus 60 napkins remain. How many napkins are in each package?

45. The recommended daily intake of iron for an adult female is 3 mg more than twice the recommended amount for a newborn infant. The amount for an adult female is 15 mg. How much should the infant receive? (*Source:* Food and Nutrition Board.)

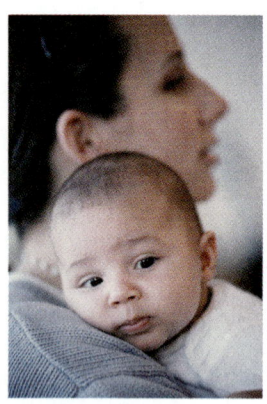

46. A cheetah's sprinting speed is 61 miles per hour less than three times a zebra's running speed. A cheetah can sprint 68 miles per hour. Find the zebra's running speed. (*Source: Grolier Multimedia Encyclopedia.*)

3.4 Solving Application Problems with Two Unknown Quantities

OBJECTIVE 1 **Solve application problems with two unknown quantities.** In the preceding section, the problems had only one unknown quantity. As a result, we used Step 2(a) rather than Step 2(b) in the problem-solving steps. For easy reference, we repeat the steps here.

OBJECTIVES
1. Solve application problems with two unknown quantities.

Solving an Application Problem

Step 1 **Read** the problem once to see what it is about. Read it carefully a second time. As you read, make a sketch or write word phrases that identify the known and the unknown parts of the problem.

Step 2(a) If there is one unknown quantity, **assign a variable** to represent it. Write down what your variable represents.

Step 2(b) If there is more than one unknown quantity, **assign a variable** to represent "the thing you know the least about." Then write variable expression(s), using the same variable, to show the relationship of the other unknown quantities to the first one.

Step 3 **Write an equation**, using your sketch or word phrases as the guide.

Step 4 **Solve** the equation.

Step 5 **State the answer** to the question in the problem and label your answer.

Step 6 **Check** whether your answer fits all the facts given in the *original* statement of the problem. If it does, you are done. If it doesn't, start again at Step 1.

Now you are ready to solve problems with two unknown quantities.

EXAMPLE 1 Solving an Application Problem with Two Unknown Quantities

Last month, Sheila worked 72 hours more than Russell. Together they worked a total of 232 hours. Find the number of hours each person worked last month.

Step 1 **Read** the problem. It is about the number of hours worked by Sheila and by Russell.

Unknowns: hours worked by Sheila;
hours worked by Russell
Known: Sheila worked 72 hours more than Russell;
232 hours total for Sheila and Russell

Step 2(b) There are *two* unknowns so **assign a variable** to represent "the thing you know the least about." You know the *least* about the hours worked by Russell, so let h represent Russell's hours.

Sheila worked 72 hours more than Russell, so her hours are $h + 72$, that is, Russell's hours (h) plus 72 more.

Step 3 **Write an equation.**

Hours worked by Russell		Hours worked by Sheila		Total hours worked
h	$+$	$h + 72$	$=$	232

Continued on Next Page

❶ In a day of work, Keonda made $12 more than her daughter. Together they made $182. Find the amount that each person made. (*Hint:* Which amount do you know the *least* about, Keonda's or her daughter's? Let m be that amount.) Use the six problem-solving steps.

Step 4 **Solve.**

$$\underbrace{h + h} + 72 = 232$$
$$2h + 72 = 232$$

Simplify the left side by combining like terms.

$$\begin{array}{r} 2h + 72 = 232 \\ -72 \quad -72 \end{array}$$

To get $2h$ by itself, add -72 to both sides.

$$\underbrace{2h + 0} = 160$$
$$\frac{2h}{2} = \frac{160}{2}$$

To get h by itself, divide both sides by 2.

$$h = 80$$

Step 5 **State the answer.**
Because h represents Russell's hours, and the solution of the equation is $h = 80$, Russell worked 80 hours.

$h + 72$ represents Sheila's hours. Replace h with 80.

$80 + 72 = 152$, so Sheila worked 152 hours.

The final answer is: Russell worked 80 hours and Sheila worked 152 hours.

Step 6 **Check** the solution by putting both numbers back into the *original* problem.

"**Sheila worked 72 hours more than Russell.**"
Sheila's 152 hours are 72 more than Russell's 80 hours, so the solution checks.

$$\begin{array}{r} 152 \\ -72 \\ \hline 80 \end{array}$$

"**Together they worked a total of 232 hours.**"
Sheila's 152 hours + Russell's 80 hours = 232 hours so the solution checks.

$$\begin{array}{r} 152 \\ +80 \\ \hline 232 \end{array}$$

You've answered the question correctly because 80 hours and 152 hours fit all the facts given in the problem.

> **CAUTION**
> Check the solution to an application problem by putting the numbers back in the *original* problem. If they do *not* work, recheck your work or try solving the problem in a different way.

◀◀◀ Work Problem 1 at the Side.

EXAMPLE 2 Solving a Geometry Application with Two Unknown Quantities

The length of a rectangle is 2 cm more than the width. The perimeter is 68 cm. Find the length and width.

Step 1 **Read** the problem. It is about a rectangle. Make a sketch of a rectangle.

Unknowns: length of the rectangle; width of the rectangle
Known: The length is 2 cm more than the width; the perimeter is 68 cm.

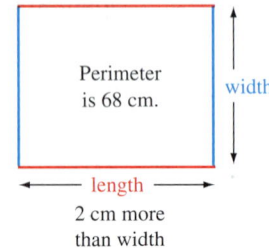
Perimeter is 68 cm.
width
length
2 cm more than width

Continued on Next Page

ANSWERS
1. Daughter made m.
Keonda made $m + 12$.
$m + m + 12 = 182$
Daughter made $85.
Keonda made $97.
Check $97 − $85 = $12
and $97 + $85 = $182

Section 3.4 Solving Application Problems with Two Unknown Quantities 179

Step 2(b)   There are *two* unknowns so **assign a variable** to represent "the thing you know the least about." You know the *least* about the width, so let *w* represent the **width**.

The length is 2 cm more than the width, so the **length** is *w* + 2.

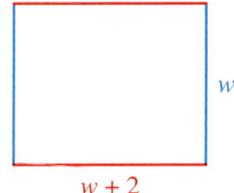

Step 3   **Write an equation.**
Use the formula for perimeter of a rectangle, $P = 2l + 2w$, to help you write the equation.

$$P = 2 \quad l \quad + 2 \quad w \qquad \text{Replace } P \text{ with 68.}$$
$$\downarrow \quad \downarrow \quad \downarrow \quad \quad \downarrow \quad \downarrow \qquad \text{Replace } l \text{ with } (w + 2).$$
$$68 = 2(w + 2) + 2 \cdot w$$

Step 4   **Solve.**

$$68 = 2(w + 2) + 2w \qquad \text{Use the distributive property.}$$
$$68 = 2w + 4 + 2w \qquad \text{Combine like terms.}$$
$$68 = 4w + 4$$
$$\underline{\,^{-}4 \qquad \quad ^{-}4} \qquad \text{To get } 4w \text{ by itself, add } ^{-}4 \text{ to both sides.}$$
$$64 = 4w + 0$$
$$\frac{64}{4} = \frac{4w}{4} \qquad \text{To get } w \text{ by itself, divide both sides by 4.}$$
$$16 = w$$

Step 5   **State the answer.**

w represents the width, and *w* = 16, so the width is 16 cm.

w + 2 represents the length. Replace *w* with 16.

16 + 2 = 18, so the length is 18 cm. ← The label for both answers is cm.

The final answer is: The width is 16 cm and the length is 18 cm.

Step 6   **Check** the solution by putting the measurements on your sketch and going back to the original problem.

"**The length of a rectangle is 2 cm more than the width.**"

18 cm is 2 cm more than 16 cm, so the solution checks.

"**The perimeter is 68 cm.**"

$$P = \underbrace{2 \cdot 18 \text{ cm}} + \underbrace{2 \cdot 16 \text{ cm}}$$
$$P = \quad 36 \text{ cm} \quad + \quad 32 \text{ cm}$$
$$P = 68 \text{ cm} \leftarrow \text{This matches the perimeter given in the original problem, so the solution checks.}$$

Work Problem 2 at the Side. ▶▶▶

② Make a sketch to help solve this problem.

The length of Ann's rectangular garden plot is 3 yd more than the width. She used 22 yd of fencing around the entire garden. Find the length and the width of the garden, using the six problem-solving steps.

ANSWERS

2.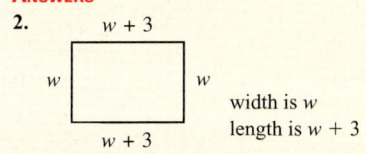

width is *w*
length is *w* + 3

$22 = 2(w + 3) + 2 \cdot w$
width is 4 yd
length is 7 yd

Check   7 yd is 3 yd more than 4 yd.
$P = 2 \cdot 7 \text{ yd} + 2 \cdot 4 \text{ yd} = 22 \text{ yd}$
Matches perimeter given in the original problem

Focus on Real-Data Applications

Connections: Arithmetic to Algebra

As you begin to learn algebra, you may be frustrated when asked to work problems algebraically that are simple enough to be solved using arithmetic. Why bother with algebra? This activity shows that the *process* you use when solving a problem using arithmetic becomes *automated* when you use an **algebraic equation.** As the applications become more complicated and you find it more and more difficult to solve a problem using just arithmetic, the algebraic process becomes a powerful tool. In the example, compare the steps of the arithmetic process to the steps involved in solving the algebraic equation. Note that the step-by-step *operations* are identical.

Problem: Kevin rented a chain saw for a one-time $15 sharpening fee plus $18-a-day rental. His total bill was $105. For how many days did Kevin rent the chain saw?

Arithmetic Solution	Algebraic Solution
To solve the problem, we have to first find the amount spent on rental fee alone. Then we have to find the number of days that gives that rental fee. The process is as follows.	First, let n be the number of rental days. Then, the daily rental fee times the number of days, plus the sharpening fee, equals the total charge: $18n + 15 = 105$. Solve the equation.
1. First subtract the sharpening fee from the total charge to find the rental fee: $105 - 15 = 90$.	1. Add $^-15$ to both sides. $$18n + 15 = 105$$ $$\underline{ -15 -15}$$ $$18n = 90$$
2. Divide the rental fee by the daily rental charge to find the number of days: $90 \div 18 = 5$.	2. Divide both sides by 18. $$\frac{18n}{18} = \frac{90}{18}$$ $$n = 5$$
3. *Answer:* Kevin rented the saw for 5 days.	3. *Answer:* Kevin rented the saw for 5 days.

In the following problems, the first one can be solved easily using both arithmetic and algebraic methods. Show that the processes are the same, in a manner similar to the above example. The second problem should be much easier to work using an algebraic equation. Show your work on separate paper, using a format similar to the one in the example.

1. Jose made a $300 down payment on a used car. His monthly payment was $150, and he paid $3900 in all. For how many months did he make payments?

Arithmetic Solution	Algebraic Solution
	Let m be the number of months of payments.

2. A 295 ft long rope is to be cut into three parts. Two parts are the same length, and the third part is 47 ft shorter than either of the other two parts. Find the length of each part.

Arithmetic Solution	Algebraic Solution
	Let x be the length of Part 1 and also of Part 2. _____ is the length of Part 3.

3.4 Exercises

Solve each application problem using the six problem-solving steps you learned in this section. See Example 1.

1. My sister is 9 years older than I am. The sum of our ages is 51. Find our ages.

2. Ed and Marge were candidates for city council. Marge won, with 93 more votes than Ed. The total number of votes cast in the election was 587. Find the number of votes received by each candidate.

3. Last year, Lien earned $1500 more than her husband. Together they earned $37,500. How much did each of them earn?

4. A $149,000 estate is to be divided between two charities so that one charity receives $18,000 less than the other. How much will each charity receive?

5. Jason paid five times as much for his computer as he did for his printer. He paid a total of $1320 for both items. What did each item cost?

6. The attendance at the Saturday night baseball game was three times the attendance at Sunday's game. In all, 56,000 fans attended the games. How many fans were at each game?

7. A board is 78 cm long. Rosa cut the board into two pieces, with one piece 10 cm longer than the other. Find the length of both pieces. (*Hint:* Make a sketch of the board. Which piece do you know the least about? Let x represent the length of that piece.)

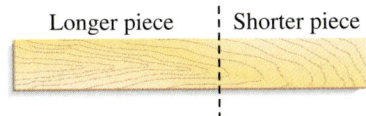
Longer piece | Shorter piece

8. A rope is 21 yd long. Marcos cut it into two pieces so that one piece is 3 yd longer than the other. Find the length of each piece.

Longer piece | Shorter piece

9. A wire is cut into two pieces, with one piece 7 ft shorter than the other. The wire was 31 ft long before it was cut. How long was each piece?

10. A 90 cm pipe is cut into two pieces so that one piece is 6 cm shorter than the other. Find the length of each piece.

11. In the U.S. Congress, the number of Representatives is 65 less than five times the number of Senators. There are a total of 535 members of Congress. Find the number of Senators and the number of Representatives. (*Source: World Almanac.*)

12. Florida's record low temperature is 68 degrees higher than Montana's record low. The sum of the two record lows is ⁻72 degrees. What is the record low for each state? (*Source*: National Climatic Data Center.)

13. A fence is 706 m long. It is to be cut into three parts. Two parts are the same length, and the third part is 25 m longer than either of the other two. Find the length of each part.

14. A wooden railing is 82 m long. It is to be divided into four pieces. Three pieces will be the same length, and the fourth piece will be 2 m longer than each of the other three. Find the length of each piece.

In Exercises 15–20, use the formula for the perimeter of a rectangle, $P = 2l + 2w$. Make a sketch to help you solve each problem. See Example 2. (Sketches may vary; show your sketches to your instructor.)

15. The perimeter of a rectangle is 48 yd. The width is 5 yd. Find the length.

16. The length of a rectangle is 27 cm, and the perimeter is 74 cm. Find the width of the rectangle.

17. A rectangular dog pen is twice as long as it is wide. The perimeter of the pen is 36 ft. Find the length and the width of the pen.

18. A new city park is a rectangular shape. The length is triple the width. It will take 240 meters of fencing to go around the park. Find the length and width of the park.

19. The length of a rectangular jewelry box is 3 in. more than twice the width. The perimeter is 36 in. Find the length and the width.

20. The perimeter of a rectangular house is 122 ft. The width is 5 ft less than the length. Find the length and the width.

21. A photograph measures 8 in. by 10 in. Earl put it in a frame that is 2 in. wide. Find the area of the frame and its outside perimeter.

22. Barb had a 16 in. by 20 in. photograph. She cropped 3 in. off each side of the photo. What are the perimeter and the area of the cropped photo?

Chapter 3
SUMMARY

KEY TERMS

3.1 **formula** Formulas are well-known rules for solving common types of problems. They are written in a shorthand form that uses variables.

perimeter Perimeter is the distance around the outside edges of a flat shape. It is measured in linear units such as in., ft, yd, mm, cm, m, and so on.

square A square is a figure with four sides that are all the same length and meet to form right angles, which measure 90°. See example at the right.

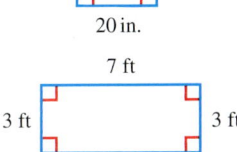

rectangle A rectangle is a four-sided figure in which all sides meet to form right angles, which measure 90°. The opposite sides are the same length. See example at the right.

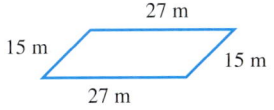

parallelogram A parallelogram is a four-sided figure in which opposite sides are both parallel and equal in length. See example at the right.

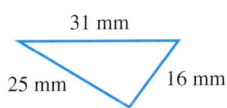

triangle A triangle is a figure with exactly three sides. See example at the right.

3.2 **area** Area is the surface inside a two-dimensional (flat) shape. It is measured by determining the number of squares of a certain size needed to cover the surface inside the shape. Some of the commonly used units for measuring area are square inches (in.²), square feet (ft²), square yards (yd²), square centimeters (cm²), and square meters (m²).

NEW FORMULAS

Perimeter of a square: $P = 4s$

Perimeter of a rectangle: $P = 2l + 2w$

Area of a square: $A = s^2$

Area of a rectangle: $A = lw$

Area of a parallelogram: $A = bh$

NEW SYMBOLS

Right angle:
(90° angle)

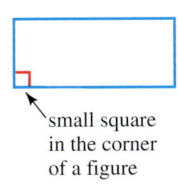

small square in the corner of a figure

Square units: in.² ft² yd² mi²
(for measuring area) mm² cm² m² km²

183

Chapter 3 Solving Application Problems

Test Your Word Power

See how well you have learned the vocabulary in this chapter. Answers follow the Quick Review.

1. The **perimeter** of a flat shape is
   A. measured in square units
   B. found by adding the lengths of the sides
   C. measured in cubic units
   D. found by multiplying length times width.

2. The **area** of a flat shape is
   A. found by adding length plus width
   B. measured in linear units
   C. found by squaring the height
   D. measured in square units.

3. When working with a **square** shape,
   A. the perimeter formula is $P = s^2$
   B. the area formula is $A = \frac{1}{2} bh$
   C. all sides have the same length
   D. all angles measure 180°.

4. When working with a **rectangular** shape,
   A. the area formula is $A = lw$
   B. opposite sides have different lengths
   C. the perimeter formula is $P = bh$
   D. the width is the longer measurement.

5. In *all* **triangles**
   A. the sides meet at 90° angles
   B. the sides have the same length
   C. there are exactly three sides
   D. the area formula is $A = s^2$.

6. In *all* **parallelograms**
   A. there are exactly six sides
   B. the height is parallel to the base
   C. all sides have the same length
   D. opposite sides are both parallel and equal in length.

Quick Review

Concepts

Examples

3.1 Finding Perimeter

To find the perimeter of *any* shape, add the lengths of the sides. Perimeter is measured in linear units (cm, m, ft, yd, and so on).

Or, for squares and rectangles, you can use the formulas shown below.

Find the perimeter of each figure.

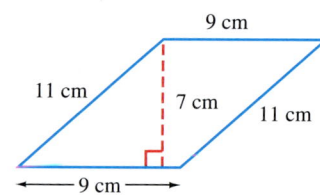

$P = 9 \text{ cm} + 11 \text{ cm} + 9 \text{ cm} + 11 \text{ cm}$
$P = 40 \text{ cm}$

Perimeter of a square: $P = 4s$

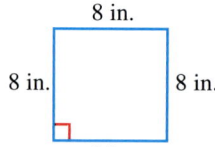

$P = 4s$
$P = 4 \cdot 8 \text{ in.}$
$P = 32 \text{ in.}$

Perimeter of a rectangle: $P = 2l + 2w$

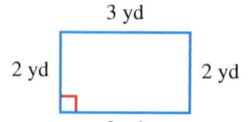

$P = 2 \cdot l + 2 \cdot w$
$P = 2 \cdot 3 \text{ yd} + 2 \cdot 2 \text{ yd}$
$P = 6 \text{ yd} + 4 \text{ yd}$
$P = 10 \text{ yd}$

Concepts	Examples
3.1 Finding the Length of One Side of a Square If you know the perimeter of a square, use the formula $P = 4s$. Replace P with the value for the perimeter and solve the equation for s.	If the perimeter of a square room is 44 ft, find the length of one side. $P = 4s$ Replace P with 44 ft. $44 \text{ ft} = 4s$ $\dfrac{44 \text{ ft}}{4} = \dfrac{4s}{4}$ Divide both sides by 4. $11 \text{ ft} = s$ The length of one side is 11 ft. *Check* Check the solution by drawing a sketch of the room. The perimeter is $11 \text{ ft} + 11 \text{ ft} + 11 \text{ ft} + 11 \text{ ft} = 44 \text{ ft}$. The result matches the perimeter given in the problem. 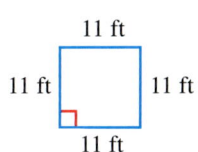
3.1 Finding the Length or Width of a Rectangle If you know the perimeter of a rectangle and either its width or length, use the formula $P = 2l + 2w$. Replace P and either l or w with the values that you know. Then solve the equation.	The width of a rectangular rug is 8 ft. The perimeter is 36 ft. Find the length. $P = 2l + 2w$ Replace P with 36 ft and w with 8 ft. $36 \text{ ft} = 2l + 2 \cdot 8 \text{ ft}$ $36 \text{ ft} = 2l + 16 \text{ ft}$ $\underline{^{-}16 \text{ ft}^{-}16 \text{ ft}}$ Add $^{-}16$ ft to both sides. $20 \text{ ft} = 2l + 0$ $\dfrac{20 \text{ ft}}{2} = \dfrac{2l}{2}$ Divide both sides by 2. $10 \text{ ft} = l$ The length is 10 ft. *Check* To check the solution, draw a sketch of the rectangle and label the lengths of the sides. Then add the four measurements. 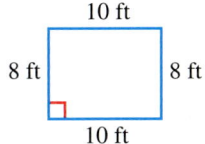 $10 \text{ ft} + 10 \text{ ft} + 8 \text{ ft} + 8 \text{ ft} = 36 \text{ ft}$ The result matches the perimeter given in the problem.

Concepts	Examples
3.2 Finding Area Use the appropriate formula. Remember to measure area in *square* units (cm², m², ft², yd², and so on). Area of a rectangle: $A = lw$, where l is the length and w is the width.	Find the area of each figure. 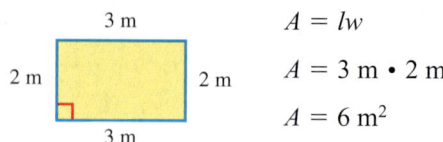 $A = lw$ $A = 3\text{ m} \cdot 2\text{ m}$ $A = 6\text{ m}^2$
Area of a square: $A = s^2$, which means $s \cdot s$, where s is the length of one side.	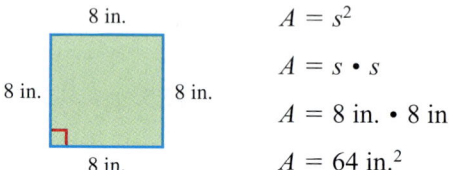 $A = s^2$ $A = s \cdot s$ $A = 8\text{ in.} \cdot 8\text{ in.}$ $A = 64\text{ in.}^2$
Area of a parallelogram: $A = bh$, where b is the base and h is the height.	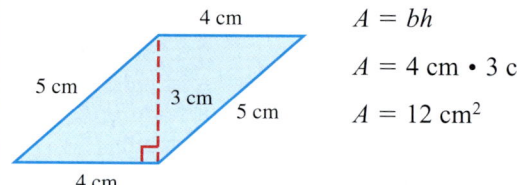 $A = bh$ $A = 4\text{ cm} \cdot 3\text{ cm}$ $A = 12\text{ cm}^2$
3.2 Finding the Unknown Length in a Rectangle or Parallelogram If you know the area of a rectangle or parallelogram and one of the other measurements, use the appropriate area formula (see above). Replace A and one of the other variables with the values that you know. Then solve the equation.	The area of a parallelogram is 72 yd², and its height is 9 yd. Find the base. $A = bh$ Replace A with 72 yd² and h with 9 yd. $72\text{ yd}^2 = b \cdot 9\text{ yd}$ $\dfrac{72\text{ yd} \cdot \text{yd}}{9\text{ yd}} = \dfrac{b \cdot 9\text{ yd}}{9\text{ yd}}$ Divide both sides by 9 yd. $8\text{ yd} = b$ The base of the parallelogram is 8 yd.
3.2 Finding the Length of One Side of a Square If you know the area of a square, use the formula $A = s^2$. Replace A with the value that you know. Then solve the equation by asking, "What number, times itself, gives the value of A?"	A square ceiling has an area of 100 ft². What is the length of each side of the ceiling? $A = s^2$ Replace A with 100 ft². $100\text{ ft}^2 = s^2$ Rewrite s^2 as $s \cdot s$. $100\text{ ft}^2 = s \cdot s$ Ask, "What number times itself gives 100?" $100\text{ ft}^2 = 10\text{ ft} \cdot 10\text{ ft}$ The value of s is 10 ft, so each side of the ceiling is 10 ft long.

Concepts

3.3 Translating Sentences into Equations
Translate word phrases into symbols using x (or any other letter) as the variable. Then solve the equation. Check the solution by putting it back in the original problem.

3.3 Solving Application Problems with One Unknown Quantity
Use the six problem-solving steps outlined in **Section 3.3**. They are listed below in abbreviated form.

Step 1 **Read** the problem and identify what is known and what is unknown.

Step 2 **Assign a variable** to represent the unknown quantity.

Step 3 **Write an equation.**

Step 4 **Solve** the equation.

Step 5 **State the answer.**

Step 6 **Check** whether your answer fits all the facts given in the *original* statement of the problem.

Examples

If 10 is subtracted from three times a number, the result is 14. Find the number.

Let x represent the unknown number.

$$3x - 10 = 14$$
$$3x + {}^-10 = 14$$
$$\underline{ {}^+10 \quad {}^+10} \quad \text{Add 10 to both sides.}$$
$$3x + 0 = 24$$
$$\frac{3x}{3} = \frac{24}{3} \quad \text{Divide both sides by 3.}$$
$$x = 8$$

The number is 8.

Check If 10 is subtracted from three times 8, do you get 14? Yes, $3 \cdot 8 - 10 = 24 - 10 = 14$.

Denise had some money in her purse this morning. She gave $15 to her daughter and paid $4 to park in the lot at work. At that point she still had $27. How much was in her purse this morning?

Step 1 **Unknown:** money in purse this morning
Known: took out $15; took out $4; still had $27.

Step 2 Let m represent the money in her purse this morning.

Step 3 $m - \$15 - \$4 = \$27$

Step 4 $m - 15 - 4 = 27$
$m + {}^-15 + {}^-4 = 27$ Combine like terms.
$m + {}^-19 = 27$
$\underline{ {}^+19 \quad {}^+19}$ Add 19 to both sides.
$m + 0 = 46$
$m = 46$

Step 5 She had $46 in her purse this morning.

Step 6 Started with $46.
Took out $15, so $46 − $15 = $31
Took out $4, so $31 − $4 = $27
Had $27 left ⟵ Matches

Concepts

3.4 Solving Application Problems with Two Unknown Quantities

Use the six problem-solving steps outlined in **Section 3.3**.

In *Step 2*, there are *two* unknown quantities, so assign a variable to represent "the thing you know the least about." Then write a variable expression, using the same variable, to show the relationship of the other unknown quantity to the first one.

Examples

Last week, Brian earned $50 more than twice what Dan earned. How much did each person earn if the total for both of them was $254?

Step 1 **Unknowns:** Brian's earnings; Dan's earnings

Known: Brian earned $50 more than twice what Dan earned; the sum of their earnings was $254.

Step 2 You know the least about Dan's earnings, so let m represent Dan's earnings.

Brian earned $50 more than twice what Dan earned, so Brian's earnings are $2m + \$50$.

Step 3 $\underbrace{m + 2m}_{} + 50 = 254$

Step 4 $\quad\quad 3m + 50 = 254$
$\quad\quad\quad\quad\quad -50 \quad -50 \quad$ Add $^-50$ to both sides.
$\quad\quad \underbrace{3m + 0}_{} = 204$

$\quad\quad \dfrac{3m}{3} = \dfrac{204}{3} \quad$ Divide both sides by 3.

$\quad\quad\quad m = 68$

Step 5 m represents Dan's earnings, so Dan earned $68.

$2m + \$50$ represents Brian's earnings, and $2 \cdot \$68 + \$50$ is $\$136 + \$50 = \$186$.

Dan earned $68; Brian earned $186.

Step 6 Is $186 actually $50 more than twice $68? Yes, the solution checks.

Does $68 + $186 = $254? Yes, the solution checks.

ANSWERS TO TEST YOUR WORD POWER

1. B; *Example:* If a triangle has sides measuring 4 ft, 10 ft, and 8 ft, then $P = 4$ ft $+ 10$ ft $+ 8$ ft $= 22$ ft.
2. D; *Example:* Area is measured in square units such as in.^{2}, ft^2, yd^2, cm^2, and km^2.
3. C; *Example:* If one side of a square is 5 in. long, all the other sides will also be 5 in. long.
4. A; *Example:* If the length of a rectangle is 12 cm and the width is 8 cm, then $A = 12$ cm $\cdot$ 8 cm $= 96$ cm^2.
5. C; *Examples:*

6. D; *Example:* In parallelogram *ABCD*, sides *AB* and *DC* are parallel and equal in length. Also, sides *AD* and *BC* are parallel and equal in length.

Chapter 3
REVIEW EXERCISES

[3.1] *In Exercises 1–4, find the perimeter of each figure. Also, name each figure in Exercises 1–3.*

1.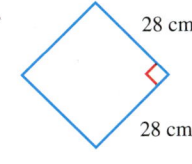
2.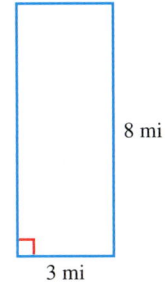
3.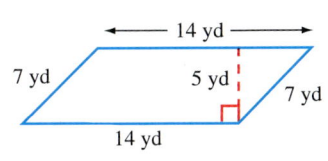
4.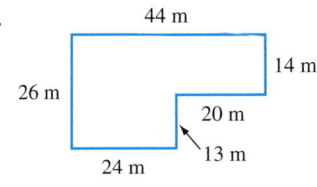

In Exercises 5–7, find the unknown measurement by using the appropriate formula.

5. A square card table has a perimeter of 12 ft. Find the length of one side of the table.

6. A rectangular playground has a perimeter of 128 yd. Find its length if it is 31 yd wide.

7. A rectangular watercolor painting that is 21 in. long has a perimeter of 72 in. What is the width of the painting?

[3.2] *In Exercises 8–10, draw a sketch of each shape and label the lengths of the sides or the base and height. Then find the area, using the appropriate formula. Sketches may vary; show your sketches to your instructor.*

8. A tablecloth that measures 5 ft by 8 ft

9. A 25 m square dance floor

10. A parallelogram-shaped lot with a base of 16 yd and a height of 13 yd

11. A rectangular patio that is 14 ft long has an area of 126 ft². Find its width.

12. A parallelogram has an area of 88 cm². If the base is 11 cm, what is the height?

13. The area of a square piece of land is 100 mi². What is the length of one side?

[3.3] *Write each phrase as an algebraic expression. Use x as the variable.*

14. A number subtracted from 57

15. The sum of 15 and twice a number

16. The product of ⁻9 and a number

Translate each sentence into an equation and solve it. Show your work.

17. The sum of four times a number and 6 is $^-30$. What is the number?

18. When twice a number is subtracted from 10, the result is 4 plus the number. Find the number.

[3.3–3.4] *Use the six problem-solving steps to solve each problem.*

19. Grace wrote a $600 check for her rent. Then she deposited her $750 paycheck and a $75 tax refund into her account. The new balance was $309. How much was in her account before she wrote the rent check?

20. Yoku ordered four boxes of candles for his restaurant. One candle was put on each of the 25 tables. There were 23 candles left. How many candles were originally in each box?

21. $1000 in prize money in an essay contest is being split between Reggie and Donald. Donald should get $300 more than Reggie. How much will each man receive?

22. A rectangular photograph is twice as long as it is wide. The perimeter of the photograph is 84 cm. Find the length and the width of the photograph.

MIXED REVIEW EXERCISES

Use the information in the advertisement and the appropriate formulas to work Exercises 23–26.

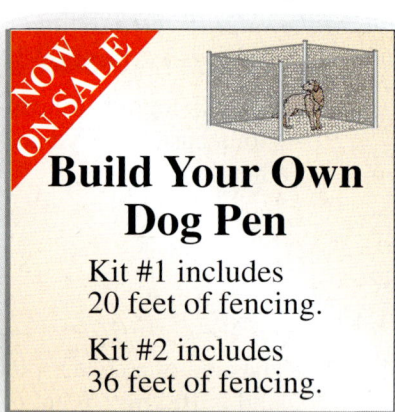

Build Your Own Dog Pen

Kit #1 includes 20 feet of fencing.

Kit #2 includes 36 feet of fencing.

NOW ON SALE

23. Anthony made a square dog pen using Kit #2.

(a) What was the length of each side of the pen?

(b) What was the area of the pen?

24. First draw sketches of two *different* rectangular dog pens that you could build using all the fencing in Kit #1. Label the lengths of the sides. Then find the area of each pen.

25. Timotha bought Kit #2. But she used some of the fencing around her garden, so she went back and bought Kit #1. Now she has 41 ft of fencing for a dog pen. How much fencing did she use around her garden? Use the six problem-solving steps.

26. Diana bought Kit #2. The pen she built had a length that was 2 ft more than the width. Find the length and width of the pen. Use the six problem-solving steps.

Chapter 3
TEST

Study Skills Workbook
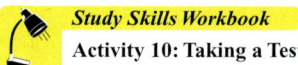
Activity 10: Taking a Test

Find the perimeter of each shape.

1.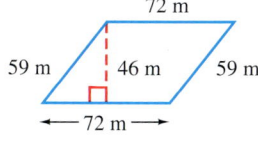

2.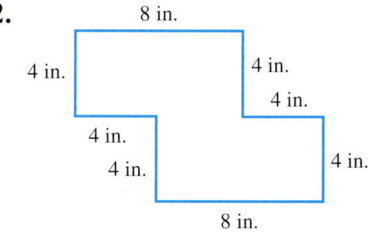

3. A square wetland 3 miles on a side

4. A rectangular mirror that measures 2 ft by 4 ft

5.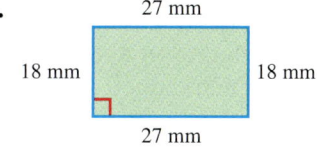

Find the area of each shape.

6.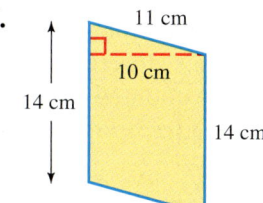

7. (figure: trapezoid with parallel sides 11 cm and 11 cm, height 10 cm, slant sides 14 cm, overall height 14 cm)

8. A rectangular animal preserve is 55 mi wide and 68 mi long.

9. A square room measures 6 m on each side.

Solve Exercises 10–14 using the appropriate formulas. Show your work.

10. A square table has a perimeter of 12 ft. Find the length of one side.

11. The Mercado family has 34 ft of fencing to put around a garden plot. The plot is rectangular in shape. If it is 6 ft wide, find the length of the plot.

12. The area of a parallelogram is 65 in.^{2}, and the base is 13 in. What is the height of the parallelogram?

1. _____

2. _____

3. _____

4. _____

5. _____

6. _____

7. _____

8. _____

9. _____

10. _____

11. _____

12. _____

13. A rectangular postage stamp has a length of 4 cm and an area of 12 cm². Find its width.

14. A square bulletin board has an area of 16 ft². How long is each side of the bulletin board?

15. Explain the difference between ft and ft². For which types of problems might you use each of these units?

Translate each sentence into an equation and solve it. Show your work.

16. If 40 is added to four times a number, the result is zero. Find the number.

17. When 7 times a number is decreased by 23, the result is the number plus 7. What is the number?

Solve each application problem, using the six problem-solving steps.

18. Josephine had $43 in her wallet. Her son used some of it when he bought groceries. Josephine found $16 in her desk drawer and put it in her wallet. She counted $44 in the wallet. How much money did her son spend on groceries?

19. Ray is 39 years old. His age is 4 years more than five times his daughter's age. How old is his daughter?

20. A board is 118 cm long. Karin cut it into two pieces, with one piece 4 cm longer than the other. Find the length of both pieces.

21. The perimeter of a rectangular building is 420 ft. The length is four times as long as the width. Find the length and the width. Draw a sketch to help you solve this problem.

22. Marcella and her husband, Tim, spent a total of 19 hours redecorating their living room. Tim spent 3 hours less time than Marcella. How long did each person work on the room?

Rational Numbers: Positive and Negative Fractions

4

- **4.1** Introduction to Signed Fractions
- **4.2** Writing Fractions in Lowest Terms
- **4.3** Multiplying and Dividing Signed Fractions
- **4.4** Adding and Subtracting Signed Fractions
- **4.5** Problem Solving: Mixed Numbers and Estimating

Summary Exercises on Fractions

- **4.6** Exponents, Order of Operations, and Complex Fractions
- **4.7** Problem Solving: Equations Containing Fractions
- **4.8** Geometry Applications: Area and Volume

Americans spend over $270 billion each year buying materials for their home repair, construction, or remodeling projects. Both men and women take on home improvement projects in about equal numbers. One of the small but essential items for many projects is nails. There are many kinds and sizes of nails, and you'll need to buy the right size nail for each project. Nails are sized according to the "penny" system, which is a number followed by the abbreviation "d." (The "d" is the traditional British abbreviation for "penny.") An algebraic expression, and the ability to solve equations with fractions, will help you find the relationship between a nail's "penny size" and its length in inches. See **Section 4.7**, Exercises 33–36. (*Source:* U.S. Bureau of the Census; Opinion Research Corp.; *Season by Season Home Maintenance.*)

4.1 Introduction to Signed Fractions

OBJECTIVES

1. Use a fraction to name the part of a whole that is shaded.
2. Identify numerators, denominators, proper fractions, and improper fractions.
3. Graph positive and negative fractions on a number line.
4. Find the absolute value of a fraction.
5. Write equivalent fractions.

OBJECTIVE 1 Use a fraction to name the part of a whole that is shaded. In **Chapters 1–3** you worked with integers. Recall that a list of the integers can be written as follows.

$$\ldots, {}^-6, {}^-5, {}^-4, {}^-3, {}^-2, {}^-1, 0, 1, 2, 3, 4, 5, 6, \ldots$$

The dots show that the list goes on forever in both directions.

Now we will work with *fractions*.

Fractions
A **fraction** is a number of the form $\dfrac{a}{b}$ where a and b are integers and b is not 0.

One use for fractions is situations in which we need a number that is between two integers. Here is an example.

A recipe uses $\frac{2}{3}$ cup of milk.
$\frac{2}{3}$ is between 0 and 1.
$\frac{2}{3}$ is a fraction because it is of the form $\dfrac{a}{b}$ and 2 and 3 are integers.

The number $\frac{2}{3}$ is a fraction that represents 2 of 3 equal parts. In this example, the cup is divided into 3 equal parts and we use enough milk to fill 2 of the parts.

We read $\frac{2}{3}$ as "two-thirds."

EXAMPLE 1 Using Fractions to Represent Part of One Whole

Use fractions to represent the shaded portion and the unshaded portion of each figure.

(a)

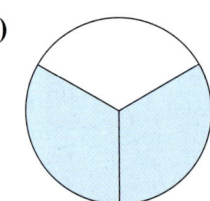

The figure has 3 equal parts. The 2 shaded parts are represented by the fraction $\frac{2}{3}$. The *un*shaded part is $\frac{1}{3}$ of the figure.

(b)

The figure has 7 equal parts. The 4 shaded parts are represented by the fraction $\frac{4}{7}$. The *un*shaded part is $\frac{3}{7}$ of the figure.

◀◀◀ Work Problem 1 at the Side.

---

1 Write fractions for the shaded portion and the unshaded portion of each figure.

(a)

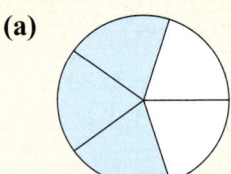

(b)

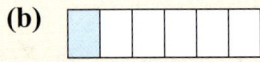

(c)

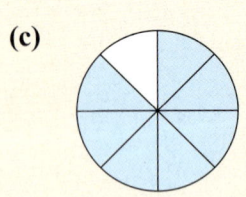

ANSWERS

1. (a) $\frac{3}{5}$; $\frac{2}{5}$ (b) $\frac{1}{6}$; $\frac{5}{6}$ (c) $\frac{7}{8}$; $\frac{1}{8}$

Fractions can also be used to represent more than one whole object.

EXAMPLE 2 Using Fractions to Represent More Than One Whole

Use a fraction to represent the shaded parts.

(a)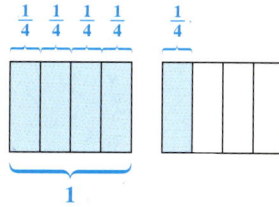

(b)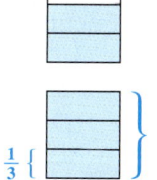

An area equal to 5 of the $\frac{1}{4}$ parts is shaded, so $\frac{5}{4}$ is shaded.

An area equal to 5 of the $\frac{1}{3}$ parts is shaded, so $\frac{5}{3}$ is shaded.

Work Problem 2 at the Side.

OBJECTIVE 2 Identify numerators, denominators, proper fractions, and improper fractions.
In the fraction $\frac{2}{3}$, the number 2 is the *numerator* and 3 is the *denominator*. The bar between the numerator and the denominator is the *fraction bar*.

$$\text{Fraction bar} \rightarrow \frac{2}{3} \begin{array}{l} \leftarrow \text{Numerator} \\ \leftarrow \text{Denominator} \end{array}$$

Numerator and Denominator

The **denominator** of a fraction shows the number of equal parts in the whole, and the **numerator** shows how many parts are being considered.

NOTE
Recall that a fraction bar, —, is a symbol for division and division by 0 is undefined. Therefore a fraction with a denominator of 0 is also undefined.

EXAMPLE 3 Identifying Numerators and Denominators

Identify the numerator and denominator in each fraction.

(a) $\frac{5}{9}$

(b) $\frac{11}{7}$

$\frac{5}{9} \begin{array}{l} \leftarrow \text{Numerator} \\ \leftarrow \text{Denominator} \end{array}$

$\frac{11}{7} \begin{array}{l} \leftarrow \text{Numerator} \\ \leftarrow \text{Denominator} \end{array}$

Work Problem 3 at the Side.

Fractions are sometimes called *proper* or *improper* fractions.

Proper and Improper Fractions

If the numerator of a fraction is *smaller* than the denominator, the fraction is a **proper fraction**. A proper fraction is less than 1.

If the numerator is *greater than or equal to* the denominator, the fraction is an **improper fraction**. An improper fraction is greater than or equal to 1.

---

2 Write fractions for the shaded portions.

(a)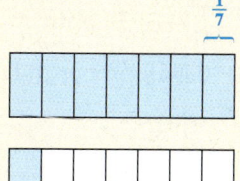

(b)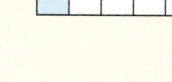

3 Identify the numerator and the denominator. Draw a picture with shaded parts to show each fraction. Your drawings may vary, but they should have the correct number of shaded parts.

(a) $\frac{2}{3}$

(b) $\frac{1}{4}$

(c) $\frac{8}{5}$

(d) $\frac{5}{2}$

ANSWERS
2. (a) $\frac{8}{7}$ (b) $\frac{7}{4}$
3. (a) N: 2; D: 3
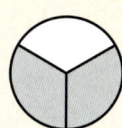

(b) N: 1; D: 4

(c) N: 8; D: 5
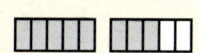

(d) N: 5; D: 2
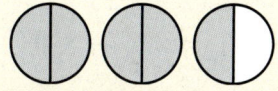

4 From this group of fractions:

$$\frac{3}{4} \quad \frac{8}{7} \quad \frac{5}{7} \quad \frac{6}{6} \quad \frac{1}{2} \quad \frac{2}{1}$$

(a) list all proper fractions.

(b) list all improper fractions.

ANSWERS

4. (a) $\frac{3}{4}, \frac{5}{7}, \frac{1}{2}$ (b) $\frac{8}{7}, \frac{6}{6}, \frac{2}{1}$

Proper Fractions			Improper Fractions		
$\frac{1}{2}$	$\frac{5}{11}$	$\frac{35}{36}$	$\frac{9}{7}$	$\frac{126}{125}$	$\frac{7}{7}$

EXAMPLE 4 Classifying Types of Fractions

(a) Identify all proper fractions in this list.

$$\frac{3}{4} \quad \frac{5}{9} \quad \frac{17}{5} \quad \frac{9}{7} \quad \frac{3}{3} \quad \frac{12}{25} \quad \frac{1}{9} \quad \frac{5}{3}$$

Proper fractions have a numerator that is *smaller* than the denominator. The proper fractions in the list are shown below.

$$\frac{3}{4} \leftarrow \text{3 is smaller than 4.} \qquad \frac{5}{9} \quad \frac{12}{25} \quad \frac{1}{9}$$

(b) Identify all improper fractions in the list in part (a).

Improper fractions have a numerator that is *equal to or greater* than the denominator. The improper fractions in the list are shown below.

$$\frac{17}{5} \leftarrow \text{17 is greater than 5.} \qquad \frac{9}{7} \quad \frac{3}{3} \quad \frac{5}{3}$$

◀◀◀ **Work Problem 4 at the Side.**

OBJECTIVE 3 Graph positive and negative fractions on a number line. Sometimes we need *negative* numbers that are between two integers. For example, $-\frac{3}{4}$ is between 0 and $^-1$. Graphing numbers on a number line helps us see the difference between $\frac{3}{4}$ and $-\frac{3}{4}$. Both represent 3 out of 4 equal parts, but they are in opposite directions from 0 on the number line. For $\frac{3}{4}$, divide the distance from 0 to 1 into 4 equal parts. Then start at 0, count over 3 parts, and make a dot. For $-\frac{3}{4}$, repeat the same process between 0 and $^-1$.

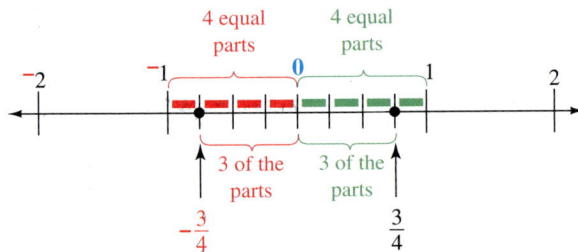

CAUTION

In **Chapters 1–3** we used a raised negative sign to help you avoid confusion between negative numbers and subtraction. Now you are ready to start writing the negative sign in the more traditional way. In this chapter, the negative sign will still be red, but it will be centered on the number instead of raised: for example, -2 instead of $^-2$. When the negative sign might be confused with the sign for subtraction, we will write parentheses around the negative number. Here is an example.

$$3 - (-2) \quad \text{means} \quad 3 \text{ minus (negative 2)}$$

For fractions, the negative sign will be written in front of the fraction bar: for example, $-\frac{3}{4}$. As with integers, the negative sign tells you that a fraction is *less than 0*; it is to the *left* of 0 on the number line. When there is *no* sign in front of a fraction, the fraction is assumed to be positive. For example, $\frac{3}{4}$ is assumed to be $+\frac{3}{4}$. It is to the *right* of 0 on the number line.

Section 4.1 Introduction to Signed Fractions   **197**

> **EXAMPLE 5** Graphing Positive and Negative Fractions
>
> Graph each fraction on the number line.
>
> (a) $\dfrac{2}{5}$
>
> There is *no* sign in front of $\tfrac{2}{5}$, so it is *positive*. Because $\tfrac{2}{5}$ is between 0 and 1, we divide that space into 5 equal parts. Then we start at 0 and count to the right 2 parts.
>
>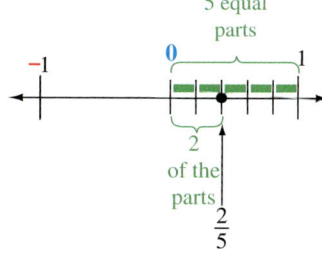
>
> (b) $-\dfrac{4}{5}$
>
> The fraction is *negative*, so it is between 0 and -1. We divide that space into 5 equal parts. Then we start at 0 and count to the left 4 parts.
>
>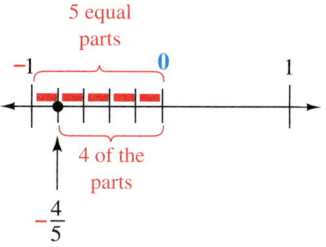
>
> **Work Problem 5 at the Side.**

OBJECTIVE 4 Find the absolute value of a fraction. In Section 1.2 we said that the *absolute value* of a number was its distance from 0 on the number line. Two vertical bars indicate absolute value, as shown below.

$$\left|-\dfrac{3}{4}\right| \text{ is read "the absolute value of negative three-fourths."}$$

As with integers, the absolute value of fractions will *always* be positive (or 0) because it is the *distance* from 0 on the number line.

> **EXAMPLE 6** Finding the Absolute Value of Fractions
>
> Find each absolute value: $\left|\dfrac{1}{2}\right|$ and $\left|-\dfrac{1}{2}\right|$.
>
> The distance from 0 to $\tfrac{1}{2}$ on the number line is $\tfrac{1}{2}$ space, so $\left|\tfrac{1}{2}\right| = \tfrac{1}{2}$.
>
> The distance from 0 to $-\tfrac{1}{2}$ is also $\tfrac{1}{2}$ space, so $\left|-\tfrac{1}{2}\right| = \tfrac{1}{2}$.
>
>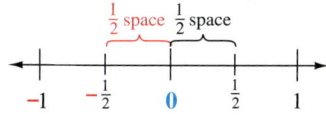
>
> **Work Problem 6 at the Side.**

⑤ Graph each fraction on the number line.

(a) $\dfrac{2}{4}$

(b) $\dfrac{1}{2}$

(c) $-\dfrac{2}{3}$

⑥ Find each absolute value.

(a) $\left|-\dfrac{3}{4}\right|$

(b) $\left|\dfrac{5}{8}\right|$

(c) $|0|$

ANSWERS

5. (a)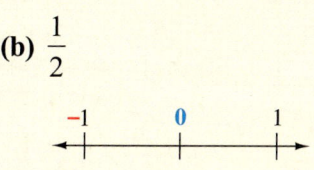

(b)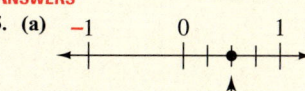

(c)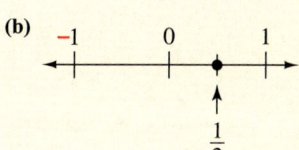

6. (a) $\dfrac{3}{4}$ (b) $\dfrac{5}{8}$ (c) 0

OBJECTIVE 5 **Write equivalent fractions.** You may have noticed in Margin Problems 5(a) and 5(b) on the previous page that $\frac{2}{4}$ and $\frac{1}{2}$ were at the same point on the number line. Both of them were halfway between 0 and 1. There are actually *many* different names for this point. We illustrate some of them below.

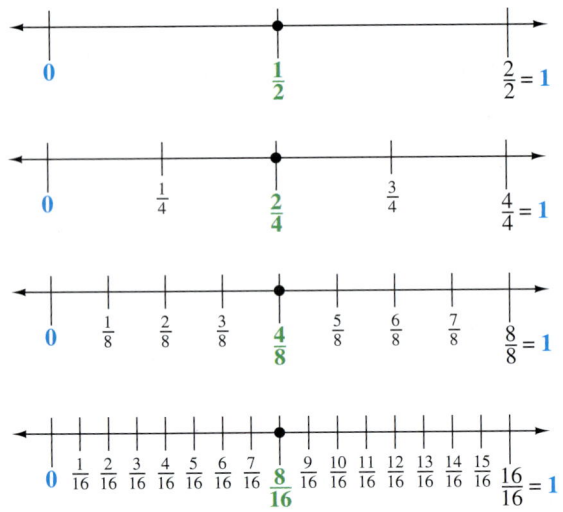

That is, $\frac{8}{16} = \frac{4}{8} = \frac{2}{4} = \frac{1}{2}$. If you have used a standard ruler with inches divided into sixteenths, you probably already noticed that these distances are the same. Although the fractions look different, they all name the same point that is halfway between 0 and 1. In other words, they have the same value. We say that they are *equivalent fractions*.

> **Equivalent Fractions**
>
> Fractions that represent the same number (the same point on a number line) are **equivalent fractions.**

Drawing number lines is tedious, so we usually find equivalent fractions by multiplying or dividing both the numerator and denominator by the same number. We can use some of the fractions that we just graphed to illustrate this method.

$$\frac{1}{2} = \frac{1 \cdot 2}{2 \cdot 2} = \frac{2}{4} \qquad\qquad \frac{8}{16} = \frac{8 \div 4}{16 \div 4} = \frac{2}{4}$$

Multiply both numerator and denominator by 2.   Divide both numerator and denominator by 4.

> **Writing Equivalent Fractions**
>
> If a, b, and c are numbers (and b and c are not 0), then
>
> $$\frac{a}{b} = \frac{a \cdot c}{b \cdot c} \quad \text{or} \quad \frac{a}{b} = \frac{a \div c}{b \div c}.$$

In other words, if the numerator and denominator of a fraction are multiplied or divided by the *same* nonzero number, the result is an *equivalent* fraction.

EXAMPLE 7 Writing Equivalent Fractions

(a) Write $-\dfrac{1}{2}$ as an equivalent fraction with a denominator of 16.

In other words, $-\dfrac{1}{2} = -\dfrac{?}{16}$.

The original denominator is 2. *Multiplying* 2 times 8 gives 16, the new denominator. To write an equivalent fraction, multiply *both* the numerator and denominator by 8.

$$-\dfrac{1}{2} = -\dfrac{1 \cdot 8}{2 \cdot 8} = -\dfrac{8}{16}$$

Keep the negative sign.

So, $-\dfrac{1}{2}$ is equivalent to $-\dfrac{8}{16}$.

(b) Write $\dfrac{12}{15}$ as an equivalent fraction with a denominator of 5.

In other words, $\dfrac{12}{15} = \dfrac{?}{5}$.

The original denominator is 15. *Dividing* 15 by 3 gives 5, the new denominator. To write an equivalent fraction, divide *both* the numerator and denominator by 3.

$$\dfrac{12}{15} = \dfrac{12 \div 3}{15 \div 3} = \dfrac{4}{5}$$

So, $\dfrac{12}{15}$ is equivalent to $\dfrac{4}{5}$.

Work Problem 7 at the Side.

7 (a) Write $\dfrac{2}{5}$ as an equivalent fraction with a denominator of 20.

(b) Write $-\dfrac{21}{28}$ as an equivalent fraction with a denominator of 4.

Look back at the set of four number lines on the previous page. Notice that there are many different names for 1.

$$\dfrac{2}{2} = 1 \qquad \dfrac{4}{4} = 1 \qquad \dfrac{8}{8} = 1 \qquad \dfrac{16}{16} = 1$$

Because a fraction bar is a symbol for division, you can think of $\dfrac{2}{2}$ as $2 \div 2$, which equals 1. Similarly, $\dfrac{4}{4}$ is $4 \div 4$, which also is 1, and so on. These examples illustrate one of the division properties from **Section 1.7**.

Division Properties

If a is any number (except 0), then $\dfrac{a}{a} = 1$. In other words, when a nonzero number is divided by itself, the result is 1.

For example, $\dfrac{6}{6} = 1$ and $\dfrac{-4}{-4} = 1$.

Also recall that when any number is divided by 1, the result is the number. That is, $\dfrac{a}{1} = a$.

For example, $\dfrac{6}{1} = 6$ and $-\dfrac{12}{1} = -12$.

ANSWERS

7. **(a)** $\dfrac{2}{5} = \dfrac{2 \cdot 4}{5 \cdot 4} = \dfrac{8}{20}$

   (b) $-\dfrac{21}{28} = -\dfrac{21 \div 7}{28 \div 7} = -\dfrac{3}{4}$

Chapter 4 Rational Numbers: Positive and Negative Fractions

⑧ Simplify each fraction by dividing the numerator by the denominator.

(a) $\dfrac{10}{10}$

(b) $-\dfrac{3}{1}$

(c) $\dfrac{8}{2}$

(d) $-\dfrac{25}{5}$

EXAMPLE 8 Using Division to Simplify Fractions

Simplify each fraction by dividing the numerator by the denominator.

(a) $\dfrac{5}{5}$ Think of $\dfrac{5}{5}$ as $5 \div 5$. The result is 1, so $\dfrac{5}{5} = 1$.

(b) $-\dfrac{12}{4}$ Think of $-\dfrac{12}{4}$ as $-12 \div 4$. The result is -3, so $-\dfrac{12}{4} = -3$.

Keep the negative sign.

(c) $\dfrac{6}{1}$ Think of $\dfrac{6}{1}$ as $6 \div 1$. The result is 6, so $\dfrac{6}{1} = 6$.

◀◀ Work Problem 8 at the Side.

NOTE
The title of this chapter is "Rational Numbers: Positive and Negative Fractions." *Rational numbers* are numbers that can be written in the form $\dfrac{a}{b}$, where a and b are integers and b is not 0. In Example 8(c) above, you saw that an integer can be written in the form $\dfrac{a}{b}$ (6 can be written as $\dfrac{6}{1}$). So rational numbers include all the integers and all the fractions. In **Chapter 5** you'll work with rational numbers that are in decimal form.

ANSWERS
8. (a) 1 (b) -3 (c) 4 (d) -5

4.1 Exercises

Write the fractions that represent the shaded and unshaded portions of each figure.
See Examples 1 and 2.

1.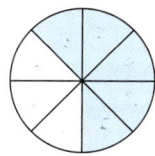

2.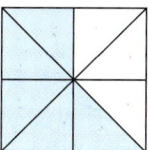

3.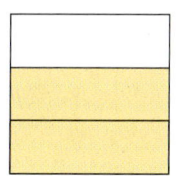

4.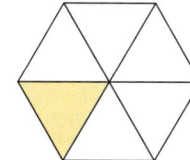

5.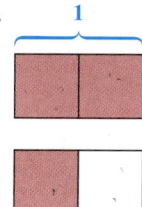

6.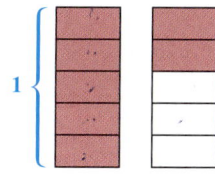

7.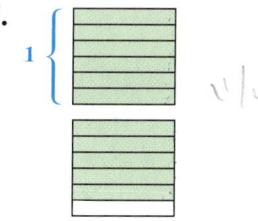

8.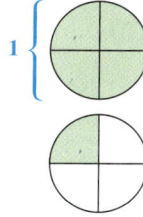

9. What fraction of these 11 coins are dimes? What fraction are pennies? What fraction are nickels?

10. What fraction of these six recording artists are men? What fraction are women? What fraction are wearing something orange?

11. In an American Sign Language (A.S.L.) class of 25 students, eight are deaf. What fraction of the students are deaf? What fraction are not deaf?

12. Of 35 motorcycles in the parking lot, 17 are Harley-Davidsons. What fraction of the motorcycles are *not* Harley-Davidsons? What fraction are Harley-Davidsons?

13. Of the 71 computers in the lab, 58 are laptops. What fraction of the computers are *not* laptops? What fraction are laptops?

14. A community college basketball team has 12 members. If five of the players are sophomores and the rest are freshmen, find the fraction of the members that are sophomores and the fraction that are freshmen.

The circle graph shows the results of a survey on where women would like to have flowers delivered on Valentine's Day. Use the graph to answer Exercises 15–18.

Out of every 20 women surveyed, the number who would like flowers delivered at home, at work, or elsewhere.

Source: FTD, Inc.

15. What fraction of the women would like flowers delivered at work?

16. What fraction of the women would like flowers delivered at home?

17. What fraction would like flowers delivered either at home or at work?

18. What fraction picked a location other than home or work?

Identify the numerator and denominator in each fraction. See Example 3.

19. $\dfrac{3}{4}$ **20.** $\dfrac{5}{8}$ **21.** $\dfrac{12}{7}$ **22.** $\dfrac{8}{3}$

List the proper and improper fractions in each group of numbers. See Example 4.

	Proper	**Improper**
23. $\dfrac{8}{5}, \dfrac{1}{3}, \dfrac{5}{8}, \dfrac{6}{6}, \dfrac{12}{2}, \dfrac{7}{16}$	_____	_____
24. $\dfrac{1}{6}, \dfrac{5}{8}, \dfrac{15}{14}, \dfrac{11}{9}, \dfrac{7}{7}, \dfrac{3}{4}$	_____	_____
25. $\dfrac{3}{4}, \dfrac{3}{2}, \dfrac{5}{5}, \dfrac{9}{11}, \dfrac{7}{15}, \dfrac{19}{18}$	_____	_____
26. $\dfrac{12}{12}, \dfrac{15}{11}, \dfrac{13}{12}, \dfrac{11}{8}, \dfrac{17}{17}, \dfrac{19}{12}$	_____	_____

27. Write a fraction of your own choice. Label the *three* parts of the fraction and write a sentence describing what each part represents. Draw a figure with shaded parts to illustrate your fraction.

28. Give one example of a proper fraction and one example of an improper fraction. What determines whether a fraction is proper or improper? Draw figures with shaded parts to illustrate your fractions.

Section 4.1 Introduction to Signed Fractions 203

Graph each pair of fractions on the number line. See Example 5.

29. $\dfrac{1}{4}, -\dfrac{1}{4}$

30. $-\dfrac{1}{3}, \dfrac{1}{3}$

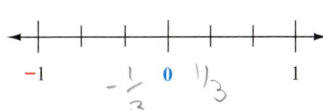

31. $-\dfrac{3}{5}, \dfrac{3}{5}$

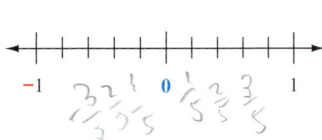

32. $\dfrac{5}{6}, -\dfrac{5}{6}$

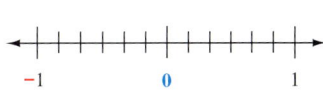

33. $\dfrac{7}{8}, -\dfrac{7}{8}$

34. $-\dfrac{3}{4}, \dfrac{3}{4}$

Write a positive or negative fraction to describe each situation.

35. The baby lost $\tfrac{3}{4}$ pound in weight while she was sick.

36. Greta needed $\tfrac{1}{3}$ cup of brown sugar for the cookie recipe.

37. The oil level in my car is $\tfrac{1}{2}$ quart below normal.

38. The mice who were on an experimental diet lost an average of $\tfrac{1}{4}$ ounce.

39. Barb Brown's driveway is $\tfrac{3}{10}$ mile long.

40. Marcel cut $\tfrac{5}{8}$ in. from the bottom of the door so that it wouldn't scrape the floor.

Find each absolute value. See Example 6.

41. $\left|-\dfrac{2}{5}\right|$

42. $\left|-\dfrac{2}{3}\right|$

43. $\left|\dfrac{9}{10}\right|$

44. $|0|$

45. $\left|-\dfrac{13}{6}\right|$

46. $\left|-\dfrac{4}{4}\right|$

47. Rewrite each fraction as an equivalent fraction with a denominator of 24. See Example 7.

(a) $\dfrac{1}{2} = \dfrac{}{24}$
(b) $\dfrac{1}{3} = \dfrac{}{}$
(c) $\dfrac{2}{3} = \dfrac{}{}$
(d) $\dfrac{1}{4} = \dfrac{}{}$
(e) $\dfrac{3}{4} = \dfrac{}{}$

(f) $\dfrac{1}{6} = \dfrac{}{}$
(g) $\dfrac{5}{6} = \dfrac{}{}$
(h) $\dfrac{1}{8} = \dfrac{}{}$
(i) $\dfrac{3}{8} = \dfrac{}{}$
(j) $\dfrac{5}{8} = \dfrac{}{}$

48. Rewrite each fraction as an equivalent fraction with a denominator of 36. See Example 7.

(a) $\dfrac{1}{2} = \dfrac{}{36}$
(b) $\dfrac{1}{3} = \dfrac{}{}$
(c) $\dfrac{2}{3} = \dfrac{}{}$
(d) $\dfrac{1}{4} = \dfrac{}{}$
(e) $\dfrac{3}{4} = \dfrac{}{}$

(f) $\dfrac{1}{6} = \dfrac{}{}$
(g) $\dfrac{5}{6} = \dfrac{}{}$
(h) $\dfrac{1}{9} = \dfrac{}{}$
(i) $\dfrac{4}{9} = \dfrac{}{}$
(j) $\dfrac{8}{9} = \dfrac{}{}$

49. Rewrite each fraction as an equivalent fraction with a denominator of 3. See Example 7.

(a) $-\dfrac{2}{6} = -\dfrac{1}{3}$
(b) $-\dfrac{4}{6} = \dfrac{}{}$
(c) $-\dfrac{12}{18} = \dfrac{}{}$
(d) $-\dfrac{6}{18} = \dfrac{}{}$
(e) $-\dfrac{200}{300} = \dfrac{}{}$

(f) Write two more fractions that are equivalent to $-\dfrac{1}{3}$ and two more fractions equivalent to $-\dfrac{2}{3}$.

50. Rewrite each fraction as an equivalent fraction with a denominator of 4. See Example 7.

(a) $-\dfrac{2}{8} = \dfrac{}{4}$
(b) $-\dfrac{6}{8} = \dfrac{}{}$
(c) $-\dfrac{15}{20} = \dfrac{}{}$
(d) $-\dfrac{50}{200} = \dfrac{}{}$
(e) $-\dfrac{150}{200} = \dfrac{}{}$

(f) Write two more fractions that are equivalent to $-\dfrac{1}{4}$ and two more fractions equivalent to $-\dfrac{3}{4}$.

RELATING CONCEPTS (EXERCISES 51–58) For Individual or Group Work

Use your calculator as you work Exercises 51–58 in order.

51. (a) Write $\dfrac{3}{8}$ as an equivalent fraction with a denominator of 3912.

(b) Explain how you solved part (a).

52. (a) Write $\dfrac{7}{9}$ as an equivalent fraction with a denominator of 5472.

(b) Explain how you solved part (a).

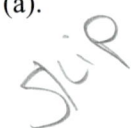

53. (a) Is $-\dfrac{697}{3485}$ equivalent to $-\dfrac{1}{2}, -\dfrac{1}{3},$ or $-\dfrac{1}{5}$?

(b) Explain how you solved part (a).

54. (a) Is $-\dfrac{817}{4902}$ equivalent to $-\dfrac{1}{4}, -\dfrac{1}{6},$ or $-\dfrac{1}{8}$?

(b) Explain how you solved part (a).

(continued)

Find a number to replace the ? that will make the two fractions equivalent.

55. $\dfrac{1183}{2028} = \dfrac{?}{12}$

56. $\dfrac{2775}{6105} = \dfrac{?}{11}$

57. $\dfrac{13}{?} = \dfrac{1157}{1335}$

58. $\dfrac{9}{?} = \dfrac{891}{1584}$

---

59. Explain how to write equivalent fractions. Show one example in which the new denominator is larger than the original denominator, and one example in which the new denominator is smaller.

60. Explain how to find the absolute value of a fraction. Draw a number line to illustrate your explanation and include a positive fraction and a negative fraction as examples.

61. Can you write $\tfrac{3}{5}$ as an equivalent fraction with a denominator of 18? Explain why or why not. If not, what denominators could you use instead of 18?

62. Can you write $\tfrac{3}{4}$ as an equivalent fraction with a denominator of 0? Explain why or why not.

Simplify each fraction by dividing the numerator by the denominator. See Example 8.

63. $\dfrac{10}{1}$

64. $\dfrac{9}{9}$

65. $-\dfrac{16}{16}$

66. $-\dfrac{7}{1}$

67. $-\dfrac{18}{3}$

68. $-\dfrac{40}{4}$

69. $\dfrac{24}{8}$

70. $\dfrac{42}{6}$

71. $\dfrac{12}{12}$

72. $-\dfrac{5}{5}$

73. $\dfrac{14}{7}$

74. $\dfrac{8}{2}$

75. $-\dfrac{5}{1}$

76. $-\dfrac{90}{10}$

77. $-\dfrac{45}{9}$

78. $\dfrac{16}{1}$

79. $\dfrac{150}{150}$

80. $\dfrac{55}{5}$

81. $-\dfrac{32}{4}$

82. $-\dfrac{200}{200}$

There are many correct ways to draw the answers for Exercises 83–100, so ask your instructor to check your work.

83. Shade $\frac{3}{5}$ of this figure. What fraction is unshaded?

84. Shade $\frac{5}{6}$ of this figure. What fraction is unshaded?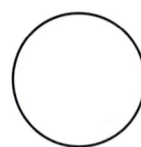

85. Shade $\frac{3}{8}$ of this figure. What fraction is unshaded?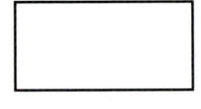

86. Shade $\frac{1}{3}$ of this figure. What fraction is unshaded?

87. Shade $\frac{7}{4}$ of this figure. What fraction is unshaded?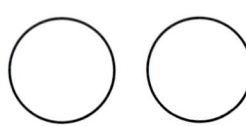

88. Shade $\frac{6}{5}$ of this figure. What fraction is unshaded?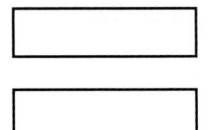

89. Shade $\frac{4}{3}$ of this figure. What fraction is unshaded?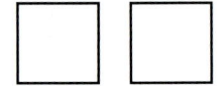

90. Shade $\frac{11}{8}$ of this figure. What fraction is unshaded?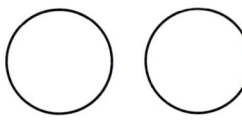

91. Shade $\frac{6}{6}$ of this figure. What fraction is unshaded?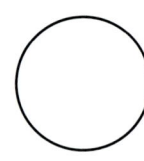

92. Shade $\frac{10}{10}$ of this figure. What fraction is unshaded?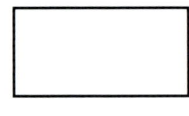

93. Shade $\frac{10}{5}$ of this figure.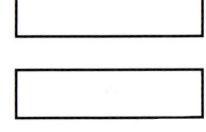

94. Shade $\frac{8}{4}$ of this figure.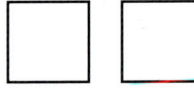

95. Draw a group of seven faces. Draw some of the faces smiling and some sad. What fraction of your faces are smiling? What fraction are sad?

96. Draw a group of eight apples. Draw some of the apples with a stem and some of them with a bite taken out. What fraction of your apples have stems? What fraction have a bite taken out?

97. Draw a group of figures. Make $\frac{1}{10}$ of the figures circles, $\frac{6}{10}$ of the figures squares, and $\frac{3}{10}$ of the figures triangles. Then shade $\frac{1}{6}$ of the squares and $\frac{2}{3}$ of the triangles.

98. Write a group of capital letters. Make $\frac{4}{9}$ of the letters A's, $\frac{2}{9}$ of the letters B's, and $\frac{3}{9}$ of the letters C's. Then draw a line under $\frac{3}{4}$ of the A's and $\frac{1}{2}$ of the B's.

99. Draw a group of punctuation marks. Make $\frac{5}{12}$ of them exclamation points, $\frac{1}{12}$ of them commas, $\frac{3}{12}$ of them periods, and add enough question marks to make a full $\frac{12}{12}$ in all. Then circle $\frac{2}{5}$ of the exclamation points.

100. Draw a group of symbols. Make $\frac{2}{15}$ of them addition signs, $\frac{4}{15}$ of them subtraction signs, $\frac{2}{15}$ of them division signs, and add enough equal signs to make a full $\frac{15}{15}$ in all. Then circle $\frac{3}{4}$ of the subtraction signs.

4.2 Writing Fractions in Lowest Terms

OBJECTIVE 1 Identify fractions written in lowest terms. You can see from these drawings that $\frac{1}{2}$ and $\frac{4}{8}$ are different names for the same amount of pizza.

$\frac{1}{2}$ of the pizza has pepperoni on it.

$\frac{4}{8}$ of the pizza has pepperoni on it.

You saw in the last section that $\frac{1}{2}$ and $\frac{4}{8}$ are equivalent fractions. But we say that the fraction $\frac{1}{2}$ is in *lowest terms* because the numerator and denominator have no *common factor* other than 1. That means that 1 is the only number that divides evenly into both 1 and 2. However, the fraction $\frac{4}{8}$ is *not* in lowest terms because its numerator and denominator have a common factor of 4. That means 4 will divide evenly into both 4 and 8.

> **NOTE**
> Recall that *factors* are numbers being multiplied to give a product. For example,
>
> 1 • 4 = 4, so 1 and 4 are factors of 4.
>
> 2 • 4 = 8, so 2 and 4 are factors of 8.
>
> 4 is a factor of both 4 and 8, so 4 is a *common factor* of those numbers.

Writing a Fraction in Lowest Terms

A fraction is written in **lowest terms** when the numerator and denominator have no common factor other than 1. Examples are $\frac{1}{3}, \frac{3}{4}, \frac{2}{5}$, and $\frac{7}{10}$.

When you work with fractions, always write the final answer in lowest terms.

EXAMPLE 1 Identifying Fractions Written in Lowest Terms

Are the following fractions in lowest terms?

(a) $\frac{3}{8}$

The numerator and denominator have no common factor other than 1, so the fraction is in lowest terms.

(b) $\frac{21}{36}$

The numerator and denominator have a common factor of 3, so the fraction is *not* in lowest terms.

Work Problem 1 at the Side.

OBJECTIVES

1. Identify fractions written in lowest terms.
2. Write a fraction in lowest terms using common factors.
3. Write a number as a product of prime factors.
4. Write a fraction in lowest terms using prime factorization.
5. Write a fraction with variables in lowest terms.

1 Are the following fractions in lowest terms? If not, find a common factor of the numerator and denominator (other than 1).

(a) $\frac{2}{3}$

(b) $-\frac{8}{10}$

(c) $-\frac{9}{11}$

(d) $\frac{15}{20}$

ANSWERS
1. (a) yes (b) No; 2 is a common factor.
   (c) yes (d) No; 5 is a common factor.

2 Write in lowest terms.

(a) $\dfrac{5}{10}$

(b) $\dfrac{9}{12}$

(c) $-\dfrac{24}{30}$

(d) $\dfrac{15}{40}$

(e) $-\dfrac{50}{90}$

OBJECTIVE 2 Write a fraction in lowest terms using common factors. We will show you two methods for writing a fraction in lowest terms. The first method, dividing by a common factor, works best when the numerator and denominator are small numbers.

EXAMPLE 2 Using Common Factors to Write Fractions in Lowest Terms

Write each fraction in lowest terms.

(a) $\dfrac{20}{24}$

The *largest* common factor of 20 and 24 is 4. Divide both numerator and denominator by 4.

$$\dfrac{20}{24} = \dfrac{20 \div 4}{24 \div 4} = \dfrac{5}{6}$$

(b) $\dfrac{30}{50} = \dfrac{30 \div 10}{50 \div 10} = \dfrac{3}{5}$ Divide both numerator and denominator by 10.

(c) $-\dfrac{24}{42} = -\dfrac{24 \div 6}{42 \div 6} = -\dfrac{4}{7}$ Divide both numerator and denominator by 6. Keep the negative sign.

(d) $\dfrac{60}{72}$

Suppose we made an error and thought that 4 was the largest common factor of 60 and 72. Dividing by 4 gives the following.

$$\dfrac{60}{72} = \dfrac{60 \div 4}{72 \div 4} = \dfrac{15}{18}$$

But $\dfrac{15}{18}$ is *not* in lowest terms because 15 and 18 have a common factor of 3. Therefore, divide the numerator and denominator by 3.

$$\dfrac{15}{18} = \dfrac{15 \div 3}{18 \div 3} = \dfrac{5}{6} \quad \leftarrow \text{Lowest terms}$$

The fraction $\dfrac{60}{72}$ could have been written in lowest terms in one step by dividing by 12, the *largest* common factor of 60 and 72.

$$\dfrac{60}{72} = \dfrac{60 \div 12}{72 \div 12} = \dfrac{5}{6} \quad \left\{ \begin{array}{l} \text{Same answer} \\ \text{as above} \end{array} \right.$$

Either way works. Just keep dividing until the fraction is in lowest terms.

This method of writing a fraction in lowest terms by dividing by a common factor is summarized below.

Dividing by a Common Factor to Write a Fraction in Lowest Terms

Step 1 Find the *largest* number that will divide evenly into both the numerator and denominator. This number is a **common factor**.

Step 2 **Divide** both numerator and denominator by the common factor.

Step 3 **Check** to see if the new numerator and denominator have any common factors (besides 1). If they do, repeat Steps 2 and 3. If the only common factor is 1, the fraction is in lowest terms.

ANSWERS

2. (a) $\dfrac{1}{2}$ (b) $\dfrac{3}{4}$ (c) $-\dfrac{4}{5}$ (d) $\dfrac{3}{8}$ (e) $-\dfrac{5}{9}$

Work Problem 2 at the Side.

OBJECTIVE 3 **Write a number as a product of prime factors.** In Example 2(d) on the previous page, the largest common factor of 60 and 72 was difficult to see quickly. You can handle a problem like that by writing the numerator and denominator as a product of *prime numbers*.

> **Prime Numbers**
>
> A **prime number** is a whole number that has exactly *two different* factors, itself and 1.

The number 3 is a prime number because it can be divided evenly only by itself and 1. The number 8 is *not* a prime number. The number 8 is a *composite number* because it can be divided evenly by 2 and 4, as well as by itself and 1.

> **Composite Numbers**
>
> A number with a factor other than itself or 1 is called a **composite number.**

> **CAUTION**
>
> A prime number has *only two* different factors, itself and 1. The number 1 is *not* a prime number because it does not have *two different* factors; the only factor of 1 is 1. Also, 0 is *not* a prime number. Therefore, 0 and 1 are *neither* prime nor composite numbers.

EXAMPLE 3 Finding Prime Numbers

Label each number as *prime* or *composite* or *neither*.

$$0 \quad 2 \quad 5 \quad 10 \quad 11 \quad 15$$

First, 0 is *neither* prime nor composite. Next, 2, 5, and 11, are *prime*. Each of these numbers is divisible only by itself and 1. The number 10 can be divided by 5 and 2, so it is *composite*. Also, 15 is a *composite* number because 15 can be divided by 5 and 3.

Work Problem 3 at the Side.

For reference, here are the prime numbers smaller than 100.

2	3	5	7	11
13	17	19	23	29
31	37	41	43	47
53	59	61	67	71
73	79	83	89	97

> **CAUTION**
>
> All prime numbers are odd numbers except the number 2. Be careful though, because *not all odd numbers are prime numbers*. For example, 9, 15, and 21 are odd numbers but they are *not* prime numbers.

The *prime factorization* of a number can be especially useful when working with fractions.

> **Prime Factorization**
>
> A **prime factorization** of a number is a factorization in which every factor is a prime number.

3 Label each number as *prime* or *composite* or *neither*.
1, 2, 3, 4, 7, 9, 13, 19, 25, 29

ANSWERS
3. 2, 3, 7, 13, 19, and 29 are *prime*.
4, 9, and 25 are *composite*.
1 is *neither* prime nor composite.

4 Find the prime factorization of each number.

(a) 8

(b) 42

(c) 90

(d) 100

(e) 81

ANSWERS
4. (a) 2 • 2 • 2 (b) 2 • 3 • 7
   (c) 2 • 3 • 3 • 5 (d) 2 • 2 • 5 • 5
   (e) 3 • 3 • 3 • 3

EXAMPLE 4 Factoring Using the Division Method

(a) Find the prime factorization of 48.

$2\overline{)48}$ ← Divide 48 by 2 (the first prime number); quotient is 24
$2\overline{)24}$ ← Divide 24 by 2; quotient is 12
$2\overline{)12}$ ← Divide 12 by 2; quotient is 6
$2\overline{)6}$ ← Divide 6 by 2; quotient is 3
$3\overline{)3}$ ← Divide 3 by 3; quotient is 1
1 ← Continue to divide until the quotient is 1

All the divisors are prime factors.

Because all the factors (divisors) are prime, the prime factorization of 48 is

2 • 2 • 2 • 2 • 3

Check by multiplying the factors to see if the product is 48.
Yes, 2 • 2 • 2 • 2 • 3 does equal 48.

NOTE
You may write the factors in any order because multiplication is commutative. So you could write the factorization of 48 as 3 • 2 • 2 • 2 • 2. We will show the factors from smallest to largest in our examples.

(b) Find the prime factorization of 225.

$3\overline{)225}$ ← 225 is not divisible by 2 (first prime) so use 3 (next prime)
$3\overline{)75}$ ← Divide 75 by 3
$5\overline{)25}$ ← 25 is not divisible by 3; use 5
$5\overline{)5}$ ← Divide 5 by 5
1 ← Quotient is 1

All the divisors are prime factors.

So, 225 = 3 • 3 • 5 • 5.

CAUTION
When you're using the division method of factoring, the last quotient is 1. Do *not* list 1 as a prime factor because 1 is not a prime number.

Work Problem 4 at the Side.

Another method of factoring uses what is called a *factor tree*.

EXAMPLE 5 Factoring Using a Factor Tree

Find the prime factorization of each number.

(a) 60

Try to divide 60 by the first prime number, 2. Write the factors under the 60. Circle the 2, because it is a prime.

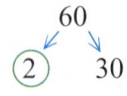

Continued on Next Page

Try dividing 30 by 2. Write the factors under the 30.

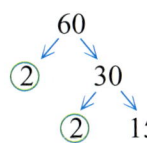

Because 15 cannot be divided evenly by 2, try dividing 15 by the next prime number, 3.

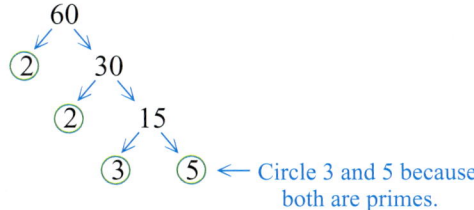

← Circle 3 and 5 because both are primes.

No uncircled factors remain, so you have found the prime factorization (the circled factors).

$$60 = 2 \cdot 2 \cdot 3 \cdot 5$$

(b) 72
Divide 72 by 2, the first prime number.

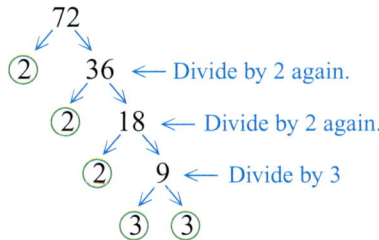

So, $72 = 2 \cdot 2 \cdot 2 \cdot 3 \cdot 3$.

(c) 45
Because 45 cannot be divided evenly by 2, try dividing by the next prime, 3.

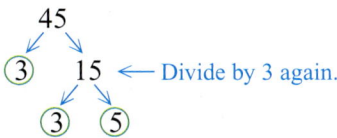

So, $45 = 3 \cdot 3 \cdot 5$.

> **NOTE**
> Here is a reminder about the quick way to see whether a number is *divisible* by 2, 3, or 5; in other words, there is no remainder when you do the division.
>
> A number is divisible by 2 if the ones digit is 0, 2, 4, 6, or 8.
> For example, 30, 512, 76, and 3018 are all divisible by 2.
>
> A number is divisible by 3 if the *sum* of the digits is divisible by 3. For example, 129 is divisible by 3 because $1 + 2 + 9 = 12$ and 12 is divisible by 3.
>
> A number is divisible by 5 if it has 0 or 5 in the ones place. For example, 85, 610, and 1725 are all divisible by 5.
>
> See **Section R.4** for more information.

Work Problem 5 at the Side.

---

5 Complete each factor tree and write the prime factorization.

(a) 28

(b) 35

(c) 90

ANSWERS

5.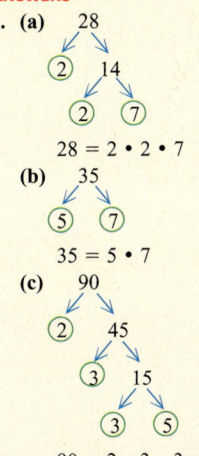

▦ **Calculator Tip** You can use your calculator to find the prime factorization of a number. Here is an example that uses 539.

Try dividing 539 by the first prime number, 2.

539 ÷ 2 = 269.5   Does not divide evenly

Try dividing 539 by the next prime number, 3.

539 ÷ 3 = 179.6666667   Does not divide evenly

Keep trying the next prime numbers until you find one that divides evenly.

539 ÷ 5 = 107.8   Does not divide evenly

539 ÷ 7 = 77   Divides evenly

Once you have found that 7 works, try using it again on the quotient 77.

77 ÷ 7 = 11   Divides evenly

Because 11 is prime, you're finished. The prime factorization of 539 is 7 • 7 • 11.

Now try factoring 2431 using your calculator. (The answer is at the bottom left of this page.)

OBJECTIVE 4 Write a fraction in lowest terms, using prime factorization. Now you can use the second method for writing fractions in lowest terms: prime factorization. This is a good method to use when the numerator and denominator are larger numbers.

EXAMPLE 6 Using Prime Factorization to Write Fractions in Lowest Terms

(a) Write $\frac{20}{35}$ in lowest terms.

20 can be written as 2 • 2 • 5 ← Prime factors
35 can be written as 5 • 7 ← Prime factors

$$\frac{20}{35} = \frac{2 \cdot 2 \cdot 5}{5 \cdot 7}$$

The numerator and denominator have 5 as a common factor. Dividing both numerator and denominator by 5 will give an equivalent fraction.

Any number divided by itself is 1

$$\frac{20}{35} = \frac{2 \cdot 2 \cdot 5}{5 \cdot 7} = \frac{2 \cdot 2 \cdot 5}{7 \cdot 5} = \frac{2 \cdot 2 \cdot \boxed{5 \div 5}}{7 \cdot \boxed{5 \div 5}} = \frac{2 \cdot 2 \cdot 1}{7 \cdot 1} = \frac{4}{7}$$

Multiplication is commutative.

$\frac{20}{35}$ is written in lowest terms as $\frac{4}{7}$.

To shorten the work, you may use slashes to indicate the divisions. For example, the work on $\frac{20}{35}$ can be shown as follows.

$$\frac{20}{35} = \frac{2 \cdot 2 \cdot \cancel{5}^{1}}{\cancel{5}_{1} \cdot 7}$$ Slashes indicate 5 ÷ 5, and the result is 1

Continued on Next Page

CALCULATOR TIP ANSWER
2431 = 11 • 13 • 17

(b) Write $\frac{60}{72}$ in lowest terms.

Use the prime factorizations of 60 and 72 from Examples 5(a) and 5(b) on pages 210–211.

$$\frac{60}{72} = \frac{2 \cdot 2 \cdot 3 \cdot 5}{2 \cdot 2 \cdot 2 \cdot 3 \cdot 3}$$

This time there are three common factors. Use slashes to show the three divisions.

$$\frac{60}{72} = \frac{\cancel{2} \cdot \cancel{2} \cdot \cancel{3} \cdot 5}{\cancel{2} \cdot \cancel{2} \cdot 2 \cdot \cancel{3} \cdot 3} = \frac{5}{6}$$

← Multiply 1 • 1 • 1 • 5 to get 5
← Multiply 1 • 1 • 2 • 1 • 3 to get 6

(2 ÷ 2 is 1; 2 ÷ 2 is 1; 3 ÷ 3 is 1)

(c) $\frac{18}{90}$

$$\frac{18}{90} = \frac{\cancel{2} \cdot \cancel{3} \cdot \cancel{3}}{\cancel{2} \cdot \cancel{3} \cdot \cancel{3} \cdot 5} = \frac{1}{5}$$

← Multiply 1 • 1 • 1 to get 1
← Multiply 1 • 1 • 1 • 5 to get 5

> **CAUTION**
> In Example 6(c) above, all factors of the numerator divided out. But 1 • 1 • 1 is still 1, so the final answer is $\frac{1}{5}$ (**not** 5).

This method of writing a fraction in lowest terms is summarized as follows.

Using Prime Factorization to Write a Fraction in Lowest Terms

Step 1 Write the *prime factorization* of both numerator and denominator.

Step 2 Use slashes to show where you are *dividing* the numerator and denominator by any common factors.

Step 3 **Multiply** the remaining factors in the numerator and in the denominator.

Work Problem 6 at the Side.

OBJECTIVE 5 Write a fraction with variables in lowest terms.
Fractions may have variables in the numerator or denominator. Examples are shown below.

$$\frac{6}{2x} \qquad \frac{3xy}{9xy} \qquad \frac{4b^3}{8ab} \qquad \frac{7ab^2}{n^2}$$

You can use prime factorization to write these fractions in lowest terms.

---

6 Use the method of prime factorization to write each fraction in lowest terms.

(a) $\frac{16}{48}$

(b) $\frac{28}{60}$

(c) $\frac{74}{111}$

(d) $\frac{124}{340}$

ANSWERS

6. (a) $\frac{\cancel{2} \cdot \cancel{2} \cdot \cancel{2} \cdot \cancel{2}}{\cancel{2} \cdot \cancel{2} \cdot \cancel{2} \cdot \cancel{2} \cdot 3} = \frac{1}{3}$

(b) $\frac{\cancel{2} \cdot \cancel{2} \cdot 7}{\cancel{2} \cdot \cancel{2} \cdot 3 \cdot 5} = \frac{7}{15}$

(c) $\frac{2 \cdot \cancel{37}}{3 \cdot \cancel{37}} = \frac{2}{3}$

(d) $\frac{\cancel{2} \cdot \cancel{2} \cdot 31}{\cancel{2} \cdot \cancel{2} \cdot 5 \cdot 17} = \frac{31}{85}$

214 Chapter 4 Rational Numbers: Positive and Negative Fractions

7 Write each fraction in lowest terms.

(a) $\dfrac{5c}{15}$

(b) $\dfrac{10x^2}{8x^2}$

(c) $\dfrac{9a^3}{11b^3}$

(d) $\dfrac{6m^2n}{9n^2}$

EXAMPLE 7 Writing Fractions with Variables in Lowest Terms

Write each fraction in lowest terms.

(a) $\dfrac{6}{2x}$ ← Prime factors of 6 are 2 · 3
   ← 2x means 2 · x.

$$\dfrac{6}{2x} = \dfrac{\overset{1}{\cancel{2}} \cdot 3}{\underset{1}{\cancel{2}} \cdot x} = \dfrac{3}{x} \quad \begin{array}{l} \leftarrow 1 \cdot 3 \text{ is } 3 \\ \leftarrow 1 \cdot x \text{ is } x \end{array}$$

(b) 3xy means 3 · x · y

$$\dfrac{3xy}{9xy} = \dfrac{3 \cdot x \cdot y}{\underbrace{3 \cdot 3}_{\text{The prime factors of 9 are 3 · 3}} \cdot x \cdot y} = \dfrac{\overset{1}{\cancel{3}} \cdot \overset{1}{\cancel{x}} \cdot \overset{1}{\cancel{y}}}{\underset{1}{\cancel{3}} \cdot 3 \cdot \underset{1}{\cancel{x}} \cdot \underset{1}{\cancel{y}}} = \dfrac{1}{3}$$

(c) b^3 means b · b · b

$$\dfrac{4b^3}{8ab} = \dfrac{2 \cdot 2 \cdot b \cdot b \cdot b}{\underbrace{2 \cdot 2 \cdot 2}_{\text{The prime factors of 8 are 2 · 2 · 2}} \cdot a \cdot b} = \dfrac{\overset{1}{\cancel{2}} \cdot \overset{1}{\cancel{2}} \cdot \overset{1}{\cancel{b}} \cdot b \cdot b}{\underset{1}{\cancel{2}} \cdot \underset{1}{\cancel{2}} \cdot 2 \cdot a \cdot \underset{1}{\cancel{b}}} = \dfrac{b^2}{2a} \quad \begin{array}{l} \leftarrow b \cdot b \text{ is } b^2 \\ \leftarrow 2 \cdot a \text{ is } 2a \end{array}$$

(d) $\dfrac{7ab^2}{n^2} = \dfrac{7 \cdot a \cdot b \cdot b}{n \cdot n}$ There are no common factors.

$\dfrac{7ab^2}{n^2}$ is already in lowest terms.

◀◀◀ **Work Problem 7 at the Side.**

ANSWERS

7. (a) $\dfrac{\overset{1}{\cancel{5}} \cdot c}{3 \cdot \underset{1}{\cancel{5}}} = \dfrac{1c}{3}$ or $\dfrac{c}{3}$

(b) $\dfrac{\overset{1}{\cancel{2}} \cdot 5 \cdot \overset{1}{\cancel{x}} \cdot \overset{1}{\cancel{x}}}{\underset{1}{\cancel{2}} \cdot 2 \cdot 2 \cdot \underset{1}{\cancel{x}} \cdot \underset{1}{\cancel{x}}} = \dfrac{5}{4}$

(c) already in lowest terms

(d) $\dfrac{2 \cdot \overset{1}{\cancel{3}} \cdot m \cdot m \cdot \overset{1}{\cancel{n}}}{\underset{1}{\cancel{3}} \cdot 3 \cdot \underset{1}{\cancel{n}} \cdot n} = \dfrac{2m^2}{3n}$

4.2 Exercises

Label each number as prime *or* composite *or* neither. *See Example 3.*

1. 9   2   8   1   5   11   10   21

2. 12   3   7   6   0   15   13   25

Find the prime factorization of each number. See Examples 4 and 5.

3. 6 **4.** 12 **5.** 20 **6.** 30

7. 25 **8.** 18 **9.** 36 **10.** 56

11. 44 **12.** 68 **13.** 88 **14.** 64

15. 75 **16.** 80

17. Write definitions of a composite number and a prime number. Give three examples of each. Which whole numbers are neither prime nor composite?

18. With the exception of the number 2, all prime numbers are odd numbers. Nevertheless, not all odd numbers are prime numbers. Explain why these statements are true.

Write each numerator and denominator as a product of prime factors. Then use the prime factorization to write the fraction in lowest terms. See Examples 1 and 6.

19. $\dfrac{8}{16}$ 20. $\dfrac{6}{8}$ 21. $\dfrac{32}{48}$

22. $\dfrac{9}{27}$ 23. $\dfrac{14}{21}$ 24. $\dfrac{20}{32}$

25. $\dfrac{36}{42}$ 26. $\dfrac{22}{33}$ 27. $\dfrac{63}{70}$

28. $\dfrac{72}{80}$ 29. $\dfrac{27}{45}$ 30. $\dfrac{36}{63}$

31. $\dfrac{12}{18}$ 32. $\dfrac{63}{90}$ 33. $\dfrac{35}{40}$

34. $\dfrac{36}{48}$ 35. $\dfrac{90}{180}$ 36. $\dfrac{16}{64}$

37. $\dfrac{210}{315}$ 38. $\dfrac{96}{192}$

39. $\dfrac{429}{495}$ 40. $\dfrac{135}{182}$

Write your answers to Exercises 41–46 in lowest terms.

41. There are 60 minutes in an hour. What fraction of an hour is

(a) 15 minutes? (b) 30 minutes?
(c) 6 minutes?   (d) 60 minutes?

42. There are 24 hours in a day. What fraction of a day is

(a) 8 hours?   (b) 18 hours?
(c) 12 hours? (d) 3 hours?

43. SueLynn's monthly income is $1500.

(a) She spends $500 on rent. What fraction of her income is spent on rent?
(b) She spends $300 on food. What fraction of her income is spent on food?
(c) What fraction of her income is left for other expenses?

44. There are 10,000 students at Minneapolis Community and Technical College.

(a) 7500 of the students receive some form of financial aid. What fraction of the students receive financial aid?
(b) 6000 of the students are women. What fraction are women?
(c) What fraction of the students are men?

45. What fraction of the time spent on household chores is done by (a) husbands, (b) wives, (c) children?

Source: Journal of Marriage and the Family.

46. A survey asked people whether certain types of advertising were believable. Out of every 100 people in the survey, here is the number who said the advertising was believable.

Type of Advertising	Number Who Said Advertising Was Believable
Computer software	35
Pharmaceutical companies	28
Auto manufacturers	18
Insurance companies	15

Source: Porter Novella.

What fraction of the people said each type of advertising was believable?

47. Explain the error in each of these problems and correct it.

(a) $\dfrac{9}{36} = \dfrac{\cancel{3} \cdot \cancel{3}}{2 \cdot 2 \cdot \cancel{3} \cdot \cancel{3}} = 4$

(b) $\dfrac{9}{16} = \dfrac{9 \div 3}{16 \div 4} = \dfrac{3}{4}$

48. (a) Explain how you could use your calculator to find the prime factorization of 437. Then find the prime factorization.

(b) The text lists all the prime numbers less than 100. Use the divisibility rules and your calculator to find at least five prime numbers between 100 and 150.

Write each fraction in lowest terms. See Example 7.

49. $\dfrac{16c}{40}$

50. $\dfrac{36}{54a}$

51. $\dfrac{20x}{35x}$

52. $\dfrac{21n}{28n}$

53. $\dfrac{18r^2}{15rs}$

54. $\dfrac{18ab}{48b^2}$

55. $\dfrac{6m}{42mn^2}$

56. $\dfrac{10g^2}{90g^2h}$

57. $\dfrac{9x^2}{16y^2}$

58. $\dfrac{5rst}{8st}$

59. $\dfrac{7xz}{9xyz}$

60. $\dfrac{6a^3}{23b^3}$

61. $\dfrac{21k^3}{6k^2}$

62. $\dfrac{16x^3}{12x^4}$

63. $\dfrac{13a^2bc^3}{39a^2bc^3}$

64. $\dfrac{22m^3n^4}{55m^3n^4}$

65. $\dfrac{14c^2d}{14cd^2}$

66. $\dfrac{19rs}{19s^3}$

67. $\dfrac{210ab^3c}{35b^2c^2}$

68. $\dfrac{81w^4xy^2}{300wy^4}$

69. $\dfrac{25m^3rt^2}{36n^2s^3w^2}$

70. $\dfrac{42a^5b^4c^3}{7a^4b^3c^2}$

71. $\dfrac{33e^2fg^3}{11efg}$

72. $\dfrac{21xy^2z^3}{17ab^2c^3}$

4.3 Multiplying and Dividing Signed Fractions

OBJECTIVES
1. Multiply signed fractions.
2. Multiply fractions that involve variables.
3. Divide signed fractions.
4. Divide fractions that involve variables.
5. Solve application problems involving multiplying and dividing fractions.

OBJECTIVE 1 Multiply signed fractions. Suppose that you give $\frac{1}{3}$ of your candy bar to your friend Ann. Then Ann gives $\frac{1}{2}$ of her share to Tim. How much of the bar does Tim get to eat?

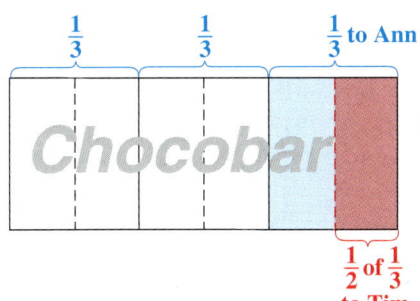

A sketch of the candy bar shows that Tim will get $\frac{1}{6}$ of the bar.

Tim's share is $\frac{1}{2}$ **of** $\frac{1}{3}$ candy bar. When used with fractions, the word **of** indicates multiplication.

$$\frac{1}{2} \text{ of } \frac{1}{3} \quad \text{means} \quad \frac{1}{2} \cdot \frac{1}{3}$$

Tim's share is $\frac{1}{6}$ bar, so $\frac{1}{2} \cdot \frac{1}{3} = \frac{1}{6}$.

This example illustrates the rule for multiplying fractions.

Multiplying Fractions
If a, b, c, and d are numbers (but b and d are not 0), then
$$\frac{a}{b} \cdot \frac{c}{d} = \frac{a \cdot c}{b \cdot d}$$
In other words, multiply the numerators and multiply the denominators.

When we apply this rule to find Tim's part of the candy bar, we get

$$\frac{1}{2} \cdot \frac{1}{3} = \frac{1 \cdot 1}{2 \cdot 3} = \frac{1}{6} \quad \leftarrow \text{Multiply numerators.}$$
$\leftarrow$ Multiply denominators.

EXAMPLE 1 Multiplying Signed Fractions

Multiply.

(a) $-\dfrac{5}{8} \cdot -\dfrac{3}{4}$ Recall that the product of two negative numbers is a positive number.

Multiply the numerators and multiply the denominators.

$$-\frac{5}{8} \cdot -\frac{3}{4} = \frac{5 \cdot 3}{8 \cdot 4} = \frac{15}{32} \quad \leftarrow \text{Lowest terms}$$

The product of two negative numbers is positive.

The answer is in lowest terms because 15 and 32 have no common factor other than 1.

Continued on Next Page

Chapter 4 Rational Numbers: Positive and Negative Fractions

1 Multiply

(a) $-\dfrac{3}{4} \cdot \dfrac{1}{2}$

(b) $\left(-\dfrac{2}{5}\right)\left(-\dfrac{2}{3}\right)$

(c) $\dfrac{3}{4}\left(\dfrac{3}{8}\right)$

(b) $\left(\dfrac{4}{7}\right)\left(-\dfrac{2}{5}\right) = -\dfrac{4 \cdot 2}{7 \cdot 5} = -\dfrac{8}{35}$

Recall that the product of a negative number and a positive number is negative.

Work Problem 1 at the Side.

Sometimes the result won't be in lowest terms. For example, find $\dfrac{3}{10}$ of $\dfrac{5}{6}$.

$$\dfrac{3}{10} \text{ of } \dfrac{5}{6} \text{ means } \dfrac{3}{10} \cdot \dfrac{5}{6} = \dfrac{3 \cdot 5}{10 \cdot 6} = \dfrac{15}{60} \quad \{\text{Not in lowest terms}$$

Now write $\dfrac{15}{60}$ in lowest terms.

$$\dfrac{15}{60} = \dfrac{\cancel{3} \cdot \cancel{5}}{2 \cdot 2 \cdot \cancel{3} \cdot \cancel{5}} = \dfrac{1}{4} \leftarrow \text{Lowest terms}$$

You used prime factorization in **Section 4.2** to write fractions in lowest terms. You can also use it when multiplying fractions. Writing the prime factors of the original fractions and dividing out common factors *before* multiplying usually saves time. If you divide out *all* the common factors, the result will automatically be in lowest terms. Let's see how that works when finding $\dfrac{3}{10}$ of $\dfrac{5}{6}$.

3 and 5 are already prime.

$$\dfrac{3}{10} \cdot \dfrac{5}{6} = \dfrac{3 \cdot 5}{2 \cdot 5 \cdot 2 \cdot 3} = \dfrac{\cancel{3} \cdot \cancel{5}}{2 \cdot \cancel{5} \cdot 2 \cdot \cancel{3}} = \dfrac{1}{4} \quad \{\text{Same result as above}$$

Write 10 as 2 · 5
Write 6 as 2 · 3
Divide out the common factors.

> **CAUTION**
> When you are working with fractions, always write the final result in lowest terms. Visualizing $\dfrac{15}{60}$ is hard to do. But when $\dfrac{15}{60}$ is written as $\dfrac{1}{4}$, working with it is much easier. A fraction is *simplified* when it is written in lowest terms.

EXAMPLE 2 Using Prime Factorization to Multiply Fractions

(a) $-\dfrac{8}{5}\left(\dfrac{5}{12}\right)$

Multiplying a negative number times a positive number gives a negative product.

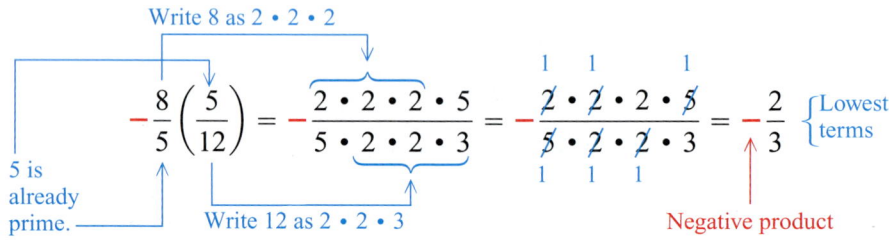

5 is already prime.
Write 8 as 2 · 2 · 2
Write 12 as 2 · 2 · 3
Negative product

Continued on Next Page

ANSWERS
1. (a) $-\dfrac{3}{8}$ (b) $\dfrac{4}{15}$ (c) $\dfrac{9}{32}$

(b) Find $\dfrac{2}{9}$ of $\dfrac{15}{16}$.

Recall that, when used with fractions, *of* indicates multiplication.

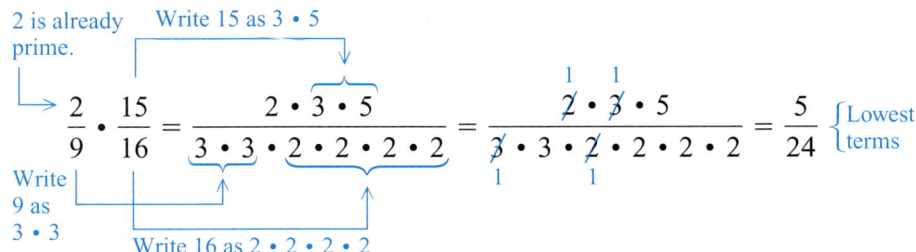

Work Problem 2 at the Side.

🖩 **Calculator Tip** If your *scientific* calculator has a fraction key [a^b/c], you can do calculations with fractions. You'll also need the change of sign key [+/−] to enter negative fractions, as you did to enter negative integers in **Section 1.4**.

Start by entering several different fractions. Clear your calculator after each one.

To enter $\frac{3}{4}$, press 3 [a^b/c] 4. The display will show $\boxed{3 \lrcorner 4}$

↑
Fraction bar

To enter $-\frac{9}{10}$, press 9 [a^b/c] 10 [+/−]. The display will show $\boxed{-9 \lrcorner 10}$.

Try entering a fraction that is *not* in lowest terms. As soon as you press an operation key, such as [×] or [÷], most calculators will automatically show the fraction in lowest terms. Suppose that you start to enter the multiplication problem $\frac{4}{16} \cdot \frac{2}{3}$. Press 4 [a^b/c] 16 [×]. The display shows $\boxed{1 \lrcorner 4}$, or $\frac{1}{4}$, which is $\frac{4}{16}$ in lowest terms. The calculator will always show fractions in lowest terms.

Let's check the result of Example 2(a) on the previous page: Multiply $-\frac{8}{5}\left(\frac{5}{12}\right)$ by pressing

8 [a^b/c] 5 [+/−] [×] 5 [a^b/c] 12 [=]. The display shows $\boxed{-2 \lrcorner 3}$.

$\underbrace{\qquad\qquad\qquad}_{-\frac{8}{5}}$ $\underbrace{\qquad\qquad}_{\frac{5}{12}}$ $\underbrace{\qquad}_{-\frac{2}{3}}$

Now try Example 2(b) above: Find $\frac{2}{9}$ of $\frac{15}{16}$. (Did you get $\frac{5}{24}$?)

There are some limitations to the calculations that you can do using the fraction key.

Try entering the fraction $\frac{9}{1000}$. What happens?

(You can't enter denominators >999.)

Try doing this multiplication: $\frac{7}{10} \cdot \frac{3}{100}$. The result should be $\frac{21}{1000}$. What happens?

(The answer is given in decimal form because the denominator is >999.)

---

2 Use prime factorization to multiply these fractions.

(a) $\dfrac{15}{28}\left(-\dfrac{6}{5}\right)$

(b) $\dfrac{12}{7} \cdot \dfrac{7}{24}$

(c) $\left(-\dfrac{11}{18}\right)\left(-\dfrac{9}{20}\right)$

Answers

2. (a) $-\dfrac{\overset{1}{3} \cdot \overset{1}{\cancel{5}} \cdot \cancel{2} \cdot 3}{2 \cdot \cancel{2} \cdot 7 \cdot \cancel{5}} = -\dfrac{9}{14}$

(b) $\dfrac{\overset{1}{\cancel{2}} \cdot \overset{1}{\cancel{2}} \cdot \overset{1}{\cancel{3}} \cdot \overset{1}{\cancel{7}}}{\cancel{7} \cdot \cancel{2} \cdot \cancel{2} \cdot 2 \cdot \cancel{3}} = \dfrac{1}{2}$

(c) $\dfrac{11 \cdot \overset{1}{\cancel{3}} \cdot \overset{1}{\cancel{3}}}{2 \cdot \cancel{3} \cdot \cancel{3} \cdot 2 \cdot 5} = \dfrac{11}{40}$

Chapter 4 Rational Numbers: Positive and Negative Fractions

3 Use prime factorization to find these products.

(a) $\dfrac{3}{4}$ of 36

(b) $-10 \cdot \dfrac{2}{5}$

(c) $\left(-\dfrac{7}{8}\right)(-24)$

4 Use prime factorization to find these products.

(a) $\dfrac{2c}{5} \cdot \dfrac{c}{4}$

(b) $\left(\dfrac{m}{6}\right)\left(\dfrac{9}{m^2}\right)$

(c) $\left(\dfrac{w^2}{y}\right)\left(\dfrac{x^2 y}{w}\right)$

ANSWERS

3. (a) $\dfrac{3 \cdot \cancel{2} \cdot \cancel{2} \cdot 3 \cdot 3}{\cancel{2} \cdot \cancel{2} \cdot 1} = \dfrac{27}{1} = 27$

(b) $-\dfrac{2 \cdot \cancel{5} \cdot 2}{1 \cdot \cancel{5}} = -\dfrac{4}{1} = -4$

(c) $\dfrac{7 \cdot \cancel{2} \cdot \cancel{2} \cdot \cancel{2} \cdot 3}{\cancel{2} \cdot \cancel{2} \cdot \cancel{2}} = \dfrac{21}{1} = 21$

4. (a) $\dfrac{\cancel{2} \cdot c \cdot c}{5 \cdot \cancel{2} \cdot 2} = \dfrac{c^2}{10}$

(b) $\dfrac{\cancel{m} \cdot \cancel{3} \cdot 3}{2 \cdot \cancel{3} \cdot \cancel{m} \cdot m} = \dfrac{3}{2m}$

(c) $\dfrac{\cancel{y} \cdot w \cdot x \cdot x \cdot \cancel{y}}{\cancel{y} \cdot \cancel{y}} = \dfrac{wx^2}{1} = wx^2$

CAUTION
The fraction key on a calculator is useful for *checking* your work. But knowing the rules for fraction computation is important because you'll need them when fractions involve variables. You *cannot* enter fractions such as $\dfrac{3x}{5}$ or $\dfrac{9}{m^2}$ on your calculator (see Example 4 below).

EXAMPLE 3 Multiplying a Fraction and an Integer

Find $\tfrac{2}{3}$ of 6.

We can write 6 in fraction form as $\tfrac{6}{1}$. Recall that $\tfrac{6}{1}$ means $6 \div 1$, which is 6. So we can write any integer a as $\tfrac{a}{1}$.

$$\dfrac{2}{3} \text{ of } 6 \text{ means } \dfrac{2}{3} \cdot \dfrac{6}{1} = \dfrac{2 \cdot 2 \cdot \cancel{3}}{\cancel{3} \cdot 1} = \dfrac{4}{1} = 4$$

Work Problem 3 at the Side.

OBJECTIVE 2 Multiply fractions that involve variables. The multiplication method that uses prime factors also works when there are variables in the numerators and/or denominators of the fractions.

EXAMPLE 4 Multiplying Fractions with Variables

Multiply.

(a) $\dfrac{3x}{5} \cdot \dfrac{2}{9x}$

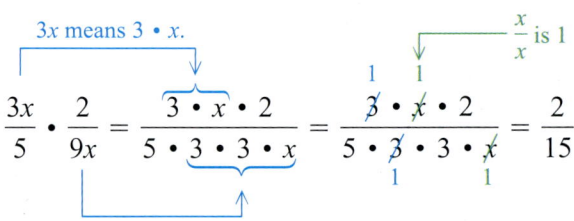

The prime factors of 9 are $3 \cdot 3$, so $9x$ is $3 \cdot 3 \cdot x$.

(b) $\left(\dfrac{3y}{4x}\right)\left(\dfrac{2x^2}{y}\right)$

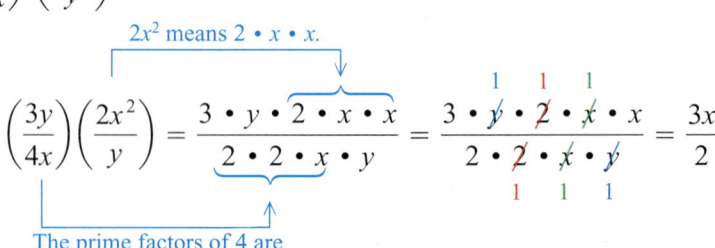

The prime factors of 4 are $2 \cdot 2$, so $4x$ is $2 \cdot 2 \cdot x$.

Work Problem 4 at the Side.

OBJECTIVE 3 Divide signed fractions. To divide fractions, you will rewrite division problems as multiplication problems. For division, you will leave the first number (the dividend) as it is, but change the second number (the divisor) to its *reciprocal*.

Reciprocal of a Fraction

Two numbers are **reciprocals** of each other if their product is 1.

The reciprocal of the fraction $\dfrac{a}{b}$ is $\dfrac{b}{a}$ because

$$\frac{a}{b} \cdot \frac{b}{a} = \frac{\cancel{a} \cdot \cancel{b}}{\cancel{b} \cdot \cancel{a}} = \frac{1}{1} = 1$$

Notice that you "flip" or "invert" a fraction to find its reciprocal. Here are some examples.

Number	Reciprocal	Reason
$\dfrac{1}{6}$	$\dfrac{6}{1}$	Because $\dfrac{1}{6} \cdot \dfrac{6}{1} = \dfrac{6}{6} = 1$
$-\dfrac{2}{5}$	$-\dfrac{5}{2}$	Because $\left(-\dfrac{2}{5}\right)\left(-\dfrac{5}{2}\right) = \dfrac{10}{10} = 1$
4	$\dfrac{1}{4}$	Because $4 \cdot \dfrac{1}{4} = \dfrac{4}{1} \cdot \dfrac{1}{4} = \dfrac{4}{4} = 1$

Think of 4 as $\dfrac{4}{1}$

> **NOTE**
> Every number has a reciprocal except 0. Why not 0? Recall that a number times its reciprocal equals 1. But that doesn't work for 0.
>
> $$0 \cdot (\text{reciprocal}) \neq 1$$
>
> Put any number here. When you multiply it by 0, you get 0, never 1

Dividing Fractions

If a, b, c, and d are numbers (but b, c, and d are not 0), then we have the following.

$$\frac{a}{b} \div \frac{c}{d} = \frac{a}{b} \cdot \frac{d}{c}$$

Reciprocals

In other words, change division to multiplying by the reciprocal of the divisor.

Use this method to find the quotient for $\dfrac{2}{3} \div \dfrac{1}{6}$. Rewrite it as a multiplication problem and then use the steps for multiplying fractions.

$$\frac{2}{3} \div \frac{1}{6} = \frac{2}{3} \cdot \frac{6}{1} = \frac{2 \cdot 2 \cdot \cancel{3}}{\cancel{3} \cdot 1} = \frac{4}{1} = 4$$

Change division to multiplication.

The reciprocal of $\dfrac{1}{6}$ is $\dfrac{6}{1}$

Does it make sense that $\frac{2}{3} \div \frac{1}{6} = 4$? Let's compare dividing fractions to dividing whole numbers.

$15 \div 3$ is asking, "How many 3s are in 15?"

$\frac{2}{3} \div \frac{1}{6}$ is asking, "How many $\frac{1}{6}$s are in $\frac{2}{3}$?"

The figure below illustrates $\frac{2}{3} \div \frac{1}{6}$.

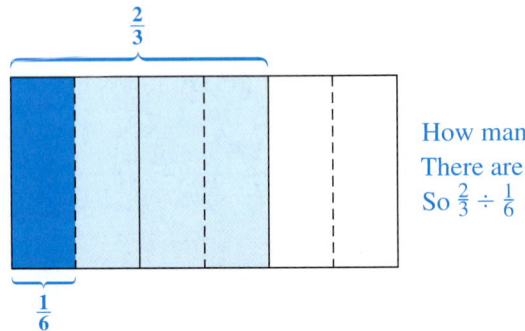

How many $\frac{1}{6}$s are in $\frac{2}{3}$?
There are 4 of the $\frac{1}{6}$ pieces in $\frac{2}{3}$.
So $\frac{2}{3} \div \frac{1}{6} = 4$

As a final check on this method of dividing, try changing $15 \div 3$ into a multiplication problem. You know that the quotient should be 5.

$$15 \div 3 = 15 \cdot \frac{1}{3} = \frac{15}{1} \cdot \frac{1}{3} = \frac{\cancel{3} \cdot 5 \cdot 1}{1 \cdot \cancel{3}} = \frac{5}{1} = 5 \quad \left\{\begin{array}{l}\text{The quotient}\\ \text{we expected}\end{array}\right.$$

Reciprocals

So, you can see that *dividing* by a fraction is the same as *multiplying* by the *reciprocal* of the fraction.

EXAMPLE 5 Dividing Signed Fractions

Rewrite each division problem as a multiplication problem. Then multiply.

(a) $\dfrac{3}{10} \div \dfrac{4}{5} = \dfrac{3}{10} \cdot \dfrac{5}{4} = \dfrac{3 \cdot \cancel{5}}{2 \cdot \cancel{5} \cdot 2 \cdot 2} = \dfrac{3}{8}$

Reciprocals

> **NOTE**
> When multiplying fractions, you don't always have to factor the numerator and denominator completely into prime numbers. In part (a) above, if you notice that 5 is a common factor of the numerator and denominator, you can write
>
> $$\dfrac{3}{10} \cdot \dfrac{5}{4} = \dfrac{3 \cdot \cancel{5}}{2 \cdot \cancel{5} \cdot 4} = \dfrac{3}{8}$$
>
> Factor 10 into $2 \cdot 5$
>
> Leave 4 as it is.
>
> If no common factors are obvious to you, then write out the complete prime factorization to help find the common factors.

Continued on Next Page

(b) $2 \div \left(-\dfrac{1}{3}\right)$

First notice that the numbers have different signs. In a division problem, different signs mean that the quotient is negative. Then write 2 in fraction form as $\frac{2}{1}$.

$$2 \div \left(-\dfrac{1}{3}\right) = \dfrac{2}{1} \cdot \left(-\dfrac{3}{1}\right) = -\dfrac{2 \cdot 3}{1 \cdot 1} = -\dfrac{6}{1} = -6$$

↑ Reciprocals ↑   ↑ Negative product

(c) $-\dfrac{3}{4} \div (-8) = -\dfrac{3}{4} \cdot \left(-\dfrac{1}{8}\right) = \dfrac{3 \cdot 1}{4 \cdot 8} = \dfrac{3}{32}$

↑ Reciprocals ↑   No common factor to divide out

The quotient is positive because both numbers in the problem were negative. When signs match, the quotient is positive.

(d) $\dfrac{9}{16} \div 0$

This is *undefined*, as dividing by 0 was undefined for integers (see **Section 1.7**). Recall that 0 does *not* have a reciprocal, so you can't change the division to multiplying by the reciprocal of the divisor.

(e) $0 \div \dfrac{9}{16} = 0 \cdot \dfrac{16}{9} = 0$ Recall that 0 divided by any nonzero number gives a result of 0.

↑ Reciprocals ↑

▶ **Work Problem 5 at the Side.**

OBJECTIVE 4 Divide fractions that involve variables. The method for dividing fractions also works when there are variables in the numerators and/or denominators of the fractions.

EXAMPLE 6 Dividing Fractions with Variables

Divide.

(a) $\dfrac{x^2}{y} \div \dfrac{x}{3y} = \dfrac{x^2}{y} \cdot \dfrac{3y}{x} = \dfrac{x \cdot x \cdot 3 \cdot y}{y \cdot x} = \dfrac{3x}{1} = 3x$

↑ Reciprocals ↑

(b) $\dfrac{8b}{5} \div b^2 = \dfrac{8b}{5} \cdot \dfrac{1}{b^2} = \dfrac{8 \cdot b \cdot 1}{5 \cdot b \cdot b} = \dfrac{8}{5b}$

↑ Reciprocals ↑

Think of b^2 as $\dfrac{b^2}{1}$ so the reciprocal is $\dfrac{1}{b^2}$.

▶ **Work Problem 6 at the Side.**

---

5 Rewrite each division problem as a multiplication problem. Then multiply.

(a) $-\dfrac{3}{4} \div \dfrac{5}{8}$

(b) $0 \div \left(-\dfrac{7}{12}\right)$

(c) $\dfrac{5}{6} \div 10$

(d) $-9 \div \left(-\dfrac{9}{16}\right)$

(e) $\dfrac{2}{5} \div 0$

6 Divide.

(a) $\dfrac{c^2 d^2}{4} \div \dfrac{c^2 d}{4}$

(b) $\dfrac{20}{7h} \div \dfrac{5h}{7}$

(c) $\dfrac{n}{8} \div mn$

ANSWERS

5. **(a)** $-\dfrac{3}{4} \cdot \dfrac{8}{5} = -\dfrac{3 \cdot 2 \cdot 4}{4 \cdot 5} = -\dfrac{6}{5}$

(b) $0 \cdot \left(-\dfrac{12}{7}\right) = 0$

(c) $\dfrac{5}{6} \cdot \dfrac{1}{10} = \dfrac{5 \cdot 1}{6 \cdot 2 \cdot 5} = \dfrac{1}{12}$

(d) $-\dfrac{9}{1} \cdot \left(-\dfrac{16}{9}\right) = \dfrac{9 \cdot 16}{1 \cdot 9} = \dfrac{16}{1} = 16$

(e) undefined; can't be written as multiplication because 0 doesn't have a reciprocal

6. **(a)** d **(b)** $\dfrac{4}{h^2}$ **(c)** $\dfrac{1}{8m}$

7 Look for indicator words or draw sketches to help you with these problems.

(a) How many times can a $\frac{2}{3}$ quart spray bottle be filled before 18 quarts of window cleaner are used up?

(b) A retiring police officer will receive $\frac{5}{8}$ of her highest annual salary as retirement income. If her highest annual salary is $48,000, how much will she receive as retirement income?

OBJECTIVE 5 Solve application problems involving multiplying and dividing fractions. When you're solving application problems, some indicator words are used to suggest multiplication and some are used to suggest division.

INDICATOR WORDS FOR MULTIPLICATION	INDICATOR WORDS FOR DIVISION
product	per
double	each
triple	goes into
times	divided by
twice	divided into
of (when *of* follows a fraction)	divided equally

Look for these indicator words in the following examples. However, *you won't always find an indicator word*. Then, you need to think through the problem to decide what to do. Sometimes, drawing a sketch of the situation described in the problem will help you decide which operation to use.

EXAMPLE 7 Using Indicator Words and Sketches to Solve Application Problems

(a) Lois gives $\frac{1}{10}$ of her income to her church. Last month she earned $1980. How much of that did she give to her church?

Notice the word **of**. Because the word **of** *follows the fraction* $\frac{1}{10}$, it indicates multiplication.

$$\frac{1}{10} \text{ of } 1980 = \frac{1}{10} \cdot \frac{1980}{1} = \frac{1 \cdot \cancel{10} \cdot 198}{\cancel{10} \cdot 1} = \frac{198}{1} = 198$$

Lois gave $198 to her church.

(b) The apparel design class is making infant snowsuits to give to a local shelter. A fabric store donated 12 yd of fabric for the project. If one snowsuit needs $\frac{2}{3}$ yd of fabric, how many suits can the class make?

The word **of** appears in the second sentence: "A fabric store donated 12 yd **of** fabric." But the word **of** does *not follow a fraction*, so it is *not* an indicator to multiply. Let's try a sketch. There is a 12 yd piece of fabric. One snowsuit will use $\frac{2}{3}$ yd. The question is, how many $\frac{2}{3}$ yd pieces can be cut from the 12 yards?

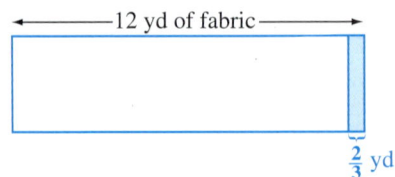

Cutting 12 yards into equal size pieces indicates division. How many $\frac{2}{3}$s are in 12?

$$12 \div \frac{2}{3} = \frac{12}{1} \cdot \frac{3}{2} = \frac{\cancel{2} \cdot 6 \cdot 3}{1 \cdot \cancel{2}} = \frac{18}{1} = 18$$

Reciprocals

The class can make 18 snowsuits.

Work Problem 7 at the Side.

ANSWERS

7. (a) $18 \div \frac{2}{3} = \frac{18}{1} \cdot \frac{3}{2} = \frac{\cancel{2} \cdot 9 \cdot 3}{1 \cdot \cancel{2}} = \frac{27}{1} = 27$

The bottle can be filled 27 times.

(b) $\frac{5}{8} \cdot \frac{48,000}{1} = \frac{5 \cdot \cancel{8} \cdot 6000}{\cancel{8} \cdot 1} = \frac{30,000}{1} = 30,000$

The officer will receive $30,000.

4.3 Exercises

Multiply. Write the products in lowest terms. See Examples 1–4.

1. $-\dfrac{3}{8} \cdot \dfrac{1}{2}$

2. $\left(\dfrac{2}{3}\right)\left(-\dfrac{5}{7}\right)$

3. $\left(-\dfrac{3}{8}\right)\left(-\dfrac{12}{5}\right)$

4. $\dfrac{4}{9} \cdot \dfrac{12}{7}$

5. $\dfrac{21}{30}\left(\dfrac{5}{7}\right)$

6. $\left(-\dfrac{6}{11}\right)\left(-\dfrac{22}{15}\right)$

7. $10\left(-\dfrac{3}{5}\right)$

8. $-20\left(\dfrac{3}{4}\right)$

9. $\dfrac{4}{9}$ of 81

10. $\dfrac{2}{3}$ of 48

11. $\left(\dfrac{3x}{4}\right)\left(\dfrac{5}{xy}\right)$

12. $\left(\dfrac{2}{5a^2}\right)\left(\dfrac{a}{8}\right)$

Divide. Write the quotients in lowest terms. See Examples 5 and 6.

13. $\dfrac{1}{6} \div \dfrac{1}{3}$

14. $-\dfrac{1}{2} \div \dfrac{2}{3}$

15. $-\dfrac{3}{4} \div \left(-\dfrac{5}{8}\right)$

16. $\dfrac{7}{10} \div \dfrac{2}{5}$

17. $6 \div \left(-\dfrac{2}{3}\right)$

18. $-7 \div \left(-\dfrac{1}{4}\right)$

19. $-\dfrac{2}{3} \div 4$

20. $\dfrac{5}{6} \div (-15)$

21. $\dfrac{11c}{5d} \div 3c$

22. $8x^2 \div \dfrac{4x}{7}$

23. $\dfrac{ab^2}{c} \div \dfrac{ab}{c}$

24. $\dfrac{mn}{6} \div \dfrac{n}{3m}$

25. Explain and correct the error in each of these calculations.

   (a) $\dfrac{3}{14} \cdot \dfrac{7}{9} = \dfrac{\cancel{3} \cdot \cancel{7}}{2 \cdot \cancel{7} \cdot \cancel{3} \cdot 3} = 6$

   (b) $8 \cdot \dfrac{2}{3} = \dfrac{8}{1} \cdot \dfrac{3}{2} = \dfrac{\overset{1}{\cancel{2}} \cdot 4 \cdot 3}{1 \cdot \underset{1}{\cancel{2}}} = \dfrac{12}{1} = 12$

26. Explain and correct the error in each of these calculations.

   (a) $\dfrac{3}{4} \cdot \dfrac{8}{9} = \dfrac{3 \cdot 8}{4 \cdot 9} = \dfrac{24}{36}$

   (b) $\dfrac{2}{5} \cdot \dfrac{3}{8} = \dfrac{\overset{1}{\cancel{2}}}{5} \cdot \dfrac{3}{\underset{2}{\cancel{8}}} = \dfrac{3}{10}$

27. Explain and correct the error in each of these calculations.

(a) $\dfrac{2}{3} \div 4 = \dfrac{2}{3} \cdot \dfrac{4}{1} = \dfrac{2 \cdot 4}{3 \cdot 1} = \dfrac{8}{3}$

(b) $\dfrac{5}{6} \div \dfrac{10}{9} = \dfrac{6}{5} \cdot \dfrac{10}{9} = \dfrac{2 \cdot \cancel{3} \cdot 2 \cdot \cancel{5}}{\cancel{5} \cdot \cancel{3} \cdot 3} = \dfrac{4}{3}$

28. Explain and correct the error in each of these calculations.

(a) $\dfrac{1}{2} \div 0 = 0$

(b) $\dfrac{\cancel{3}^{1}}{10} \div \dfrac{1}{\cancel{3}_{1}} = \dfrac{1}{10}$

29. Your friend missed class and is confused about how to divide fractions. Write a short explanation for your friend.

30. Mary spilled coffee on her math homework, and part of one problem is covered up.

$\dfrac{3}{\blacksquare} \div \dfrac{4}{5}$

She knows the answer given in the back of the book is $\tfrac{3}{4}$. Describe how to find the missing number.

Find each product or quotient. Write all answers in lowest terms. See Examples 1–6.

31. $\dfrac{4}{5} \div 3$

32. $\left(-\dfrac{20}{21}\right)\left(-\dfrac{14}{15}\right)$

33. $-\dfrac{3}{8}\left(\dfrac{3}{4}\right)$

34. $-\dfrac{8}{17} \div \dfrac{4}{5}$

35. $\dfrac{3}{5}$ of 35

36. $\dfrac{2}{3} \div (-6)$

37. $-9 \div \left(-\dfrac{3}{5}\right)$

38. $\dfrac{7}{8} \cdot \dfrac{25}{21}$

39. $\dfrac{12}{7} \div 0$

40. $\dfrac{5}{8}$ of (-48)

41. $\left(\dfrac{11}{2}\right)\left(-\dfrac{5}{6}\right)$

42. $\dfrac{3}{4} \div \dfrac{3}{16}$

43. $\dfrac{4}{7}$ of $14b$

44. $\dfrac{ab}{6} \div \dfrac{b}{9}$

45. $\dfrac{12}{5} \div 4d$

46. $\dfrac{18}{7} \div 2t$

47. $\dfrac{x^2}{y} \div \dfrac{w}{2y}$

48. $\dfrac{5}{6}$ of $18w$

Solve each application problem. See Example 7.

49. Al is helping Tim make a mahogany lamp table for Jill's birthday. Find the area of the rectangular top of the table if it is $\frac{4}{5}$ yd long by $\frac{3}{8}$ yd wide.

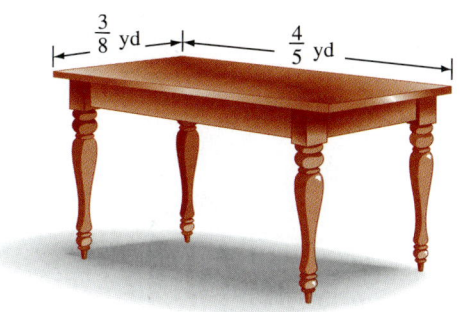

50. A rectangular dog bed is $\frac{7}{8}$ yd by $\frac{10}{9}$ yd. Find its area.

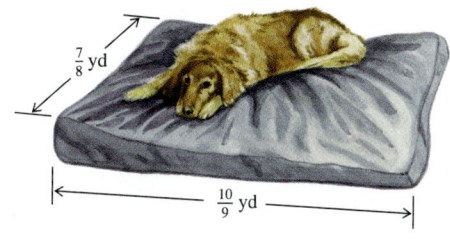

51. How many $\frac{1}{8}$-ounce eyedrop dispensers can be filled from a container that holds 10 ounces of eyedrops?

52. Ms. Shaffer has a piece of property with an area that is $\frac{9}{10}$ acre. She wishes to divide it into three equal parts for her children. How many acres of land will each child get?

53. Todd estimates that it will cost him $12,400 to attend a community college for one year. He thinks he can earn $\frac{3}{4}$ of the cost and borrow the balance. Find the amount he must earn and the amount he must borrow.

54. Joyce Chen wants to make vests to sell at a craft fair. Each vest requires $\frac{3}{4}$ yd of material. She has 36 yd of material. Find the number of vests she can make.

55. Pam Trizlia has a small pickup truck that can carry $\frac{2}{3}$ cord of firewood. Find the number of trips needed to deliver 6 cords of wood.

56. At the Garlic Festival Fun Run, $\frac{5}{12}$ of the runners are women. If there are 780 runners, how many are women? How many are men?

57. There are about 210 players in the Baseball Hall of Fame. One-third of these men played at infield positions (1st base, 2nd base, shortstop, or 3rd base). How many infield players are in the Hall of Fame? (*Source: National Baseball Hall of Fame.*)

58. Parking lot A is $\frac{1}{4}$ mile long and $\frac{3}{16}$ mile wide, and parking lot B is $\frac{3}{8}$ mile long and $\frac{1}{8}$ mile wide. Which parking lot has the larger area?

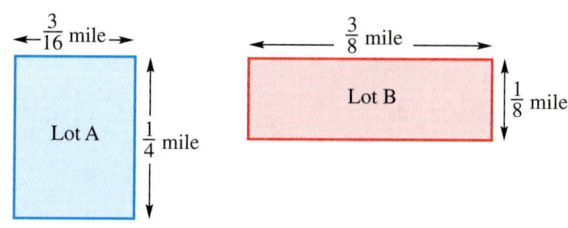

59. An adult male alligator may weigh 400 pounds. A newly hatched alligator weighs only $\frac{1}{8}$ pound. The adult weighs how many times the hatchling? (*Source:* St. Marks NWR.)

60. A female alligator lays about 35 eggs in a nest of marsh grass and mud. But $\frac{4}{5}$ of the eggs or hatchlings will fall prey to raccoons, wading birds, or larger alligators. How many of the 35 eggs will hatch and survive? (*Source:* St. Marks NWR.)

The table below shows the income for the Gomez family last year and the circle graph shows how they spent their income. Use this information to work Exercises 61–64.

Month	Income	Month	Income
January	$4575	July	$5540
February	$4312	August	$3732
March	$4988	September	$4170
April	$4530	October	$5512
May	$4320	November	$4965
June	$4898	December	$6458

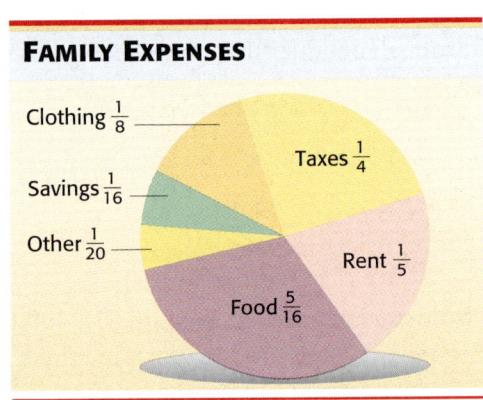

FAMILY EXPENSES

Clothing $\frac{1}{8}$, Taxes $\frac{1}{4}$, Savings $\frac{1}{16}$, Other $\frac{1}{20}$, Rent $\frac{1}{5}$, Food $\frac{5}{16}$

61. (a) What was the family's total income for the year?
(b) Find the amount of the family's rent for the year.

62. (a) How much did the family pay in taxes during the year?
(b) How much more did the family spend on taxes than on rent?

63. How much did the family spend for food and clothing last year?

64. Find the amount the family saved during the year.

There are a total of about 150 million companion pets in the United States. This table shows the fraction of pets that are dogs, cats, birds, and horses. Use the table to answer Exercises 65–68.

COMPANION PETS IN THE UNITED STATES

Type of Pet	Fraction of All Pets
Dog	$\frac{2}{5}$
Cat	$\frac{23}{50}$
Bird	$\frac{1}{10}$
Horse	$\frac{3}{100}$

Source: American Veterinary Medical Association.

65. How many U.S. pets are birds?

66. How many U.S. pets are dogs?

67. How many dogs and cats are pets?

68. How many birds and cats are pets?

4.4 Adding and Subtracting Signed Fractions

OBJECTIVE 1 Add and subtract like fractions. You probably remember learning something about "common denominators" in other math classes. When fractions have the *same* denominator, we say that they have a *common* denominator, which makes them **like fractions**. When fractions have different denominators, they are called **unlike fractions**. Here are some examples.

OBJECTIVES
1. Add and subtract like fractions.
2. Find the lowest common denominator for unlike fractions.
3. Add and subtract unlike fractions.
4. Add and subtract unlike fractions that contain variables.

Like Fractions

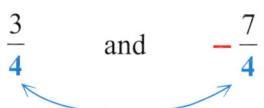
Common denominator

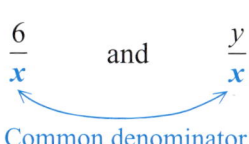
Common denominator

Unlike Fractions

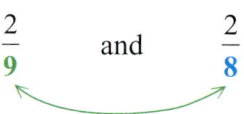
Different denominators

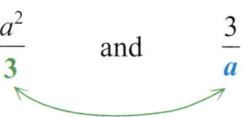
Different denominators

You can add or subtract fractions *only* when they have a common denominator. To see why, let's look at more pizzas.

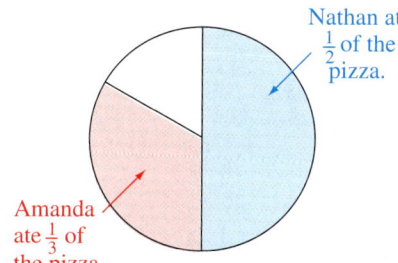

What fraction of the pizza has been eaten? We can't write a fraction until the pizza is cut into pieces of *equal* size. That's what the denominator of a fraction tell us: the number of *equal* size pieces in the pizza.

Now the pizza is cut into 6 *equal* pieces, and we can find out how much was eaten.

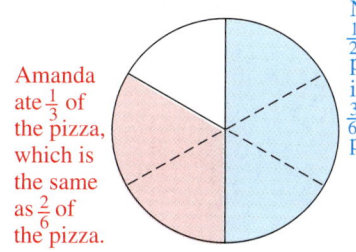

$$\frac{1}{2} + \frac{1}{3}$$
$$\downarrow \quad \downarrow$$
$$\frac{3}{6} + \frac{2}{6}$$
$$= \frac{5}{6} \leftarrow \text{Number of pieces eaten}$$
$$\phantom{= \frac{5}{6}} \leftarrow \text{Number of equal pieces in the pizza}$$

Adding and Subtracting Like Fractions

You can add or subtract fractions *only* when they have a common denominator. If a, b, and c are numbers (and b is not 0), then

$$\frac{a}{b} + \frac{c}{b} = \frac{a+c}{b} \quad \text{and} \quad \frac{a}{b} - \frac{c}{b} = \frac{a-c}{b}$$

In other words, add or subtract the numerators and write the result over the common denominator. Then check to be sure that the answer is in lowest terms.

1 Write each sum or difference in lowest terms.

(a) $\dfrac{1}{6} + \dfrac{5}{6}$

(b) $-\dfrac{11}{12} + \dfrac{5}{12}$

(c) $-\dfrac{2}{9} - \dfrac{3}{9}$

(d) $\dfrac{8}{ab} + \dfrac{3}{ab}$

ANSWERS

1. (a) $\dfrac{6}{6} = 1$ (b) $-\dfrac{1}{2}$
   (c) $-\dfrac{5}{9}$ (d) $\dfrac{11}{ab}$

EXAMPLE 1 Adding and Subtracting Like Fractions

Find each sum or difference.

(a) $\dfrac{1}{8} + \dfrac{3}{8}$

These are *like* fractions because they have a *common denominator of 8*. So they are ready to be added. Add the numerators and write the sum over the common denominator.

$$\underset{\text{Common denominator}}{\dfrac{1}{8} + \dfrac{3}{8} = \dfrac{1+3}{8} = \dfrac{4}{8}} \qquad \text{Now write } \tfrac{4}{8} \text{ in lowest terms.} \qquad \dfrac{4}{8} = \dfrac{\overset{1}{\cancel{2}} \cdot \overset{1}{\cancel{2}}}{\underset{1}{\cancel{2}} \cdot \underset{1}{\cancel{2}} \cdot 2} = \dfrac{1}{2}$$

The sum, in lowest terms, is $\tfrac{1}{2}$.

CAUTION

Add *only* the numerators. **Do not add the denominators.** In part (a) above we *kept the common denominator*.

$$\dfrac{1}{8} + \dfrac{3}{8} = \dfrac{1+3}{8} \qquad \textbf{not} \qquad \dfrac{1}{8} + \dfrac{3}{8} = \cancel{\dfrac{1+3}{8+8}} = \cancel{\dfrac{4}{16}}$$
Incorrect

To help you understand why we add *only* the numerators, think of $\tfrac{1}{8}$ as $1\left(\tfrac{1}{8}\right)$ and $\tfrac{3}{8}$ as $3\left(\tfrac{1}{8}\right)$. Then we use the distributive property.

$$\dfrac{1}{8} + \dfrac{3}{8} = 1\left(\dfrac{1}{8}\right) + 3\left(\dfrac{1}{8}\right) = \underbrace{(1+3)\left(\dfrac{1}{8}\right)}_{\text{Use the distributive property.}} = 4\left(\dfrac{1}{8}\right) = \dfrac{4}{8} = \dfrac{1}{2}$$

Or think about a pie cut into 8 equal pieces. If you eat 1 piece and your friend eats 3 pieces, together you've eaten 4 of the 8 pieces or $\tfrac{4}{8}$ of the pie (*not* $\tfrac{4}{16}$ of the pie, which would be 4 out of 16 pieces).

(b) $\underset{\text{Common denominator}}{-\dfrac{3}{5} + \dfrac{4}{5}} = \dfrac{-3+4}{5} = \dfrac{1}{5} \leftarrow \text{Lowest terms}$

(c) $\underset{\text{Common denominator}}{\dfrac{3}{10} - \dfrac{7}{10}} = \dfrac{3-7}{10} = \overset{\overset{\text{Rewrite subtraction as adding the opposite.}}{\downarrow}}{\dfrac{3 + (-7)}{10}} = \dfrac{-4}{10} \quad \text{or} \quad -\dfrac{4}{10}$

Now write $-\tfrac{4}{10}$ in lowest terms. $\qquad -\dfrac{4}{10} = -\dfrac{\overset{1}{\cancel{2}} \cdot 2}{\underset{1}{\cancel{2}} \cdot 5} = -\dfrac{2}{5} \leftarrow \text{Lowest terms}$

(d) $\underset{\text{Common denominator}}{\dfrac{5}{x^2} - \dfrac{2}{x^2}} = \dfrac{5-2}{x^2} = \dfrac{3}{x^2}$

Work Problem 1 at the Side.

OBJECTIVE 2 Find the lowest common denominator for unlike fractions. When we first tried to add the pizza eaten by Nathan and Amanda, we could *not* do so because the pizza was *not cut into pieces of the same size*. So we rewrote $\frac{1}{2}$ and $\frac{1}{3}$ as equivalent fractions that both had 6 as the common denominator.

$$\frac{1}{2} = \frac{1 \cdot 3}{2 \cdot 3} = \frac{3}{6}$$
$$\frac{1}{3} = \frac{1 \cdot 2}{3 \cdot 2} = \frac{2}{6}$$ Common denominator of 6

Then we could add the fractions, because the pizza was cut into pieces of *equal* size.

$$\frac{1}{2} + \frac{1}{3} = \frac{3}{6} + \frac{2}{6} = \frac{3 + 2}{6} = \frac{5}{6} \leftarrow \text{Lowest terms}$$

In other words, when you want to add or subtract unlike fractions, the first thing you must do is rewrite them so that they have a common denominator.

A Common Denominator for Unlike Fractions

To find a common denominator for two unlike fractions, find a number that is divisible by *both* of the original denominators.

For example, a common denominator for $\frac{1}{2}$ and $\frac{1}{3}$ is 6 because 2 goes into 6 evenly and 3 goes into 6 evenly.

Notice that 12 is also a common denominator for $\frac{1}{2}$ and $\frac{1}{3}$ because 2 and 3 both go into 12 evenly.

$$\frac{1}{2} = \frac{1 \cdot 6}{2 \cdot 6} = \frac{6}{12}$$
$$\frac{1}{3} = \frac{1 \cdot 4}{3 \cdot 4} = \frac{4}{12}$$ Common denominator of 12

Now that the fractions have a common denominator, we can add them.

$$\frac{1}{2} + \frac{1}{3} = \frac{6}{12} + \frac{4}{12} = \frac{6 + 4}{12} = \frac{10}{12} \left\{ \begin{array}{l} \text{Not in} \\ \text{lowest} \\ \text{terms} \end{array} \right. \quad \text{but} \quad \frac{10}{12} = \frac{\overset{1}{\cancel{2}} \cdot 5}{\underset{1}{\cancel{2}} \cdot 6} = \frac{5}{6} \left\{ \begin{array}{l} \text{Same} \\ \text{result as} \\ \text{above} \end{array} \right.$$

Both 6 and 12 worked as common denominators for adding $\frac{1}{2}$ and $\frac{1}{3}$, but using the smaller number saved some work. You should always try to find the smallest common denominator. If you don't for some reason, you can still work the problem—but it may take you longer. You'll have to divide out some common factors at the end in order to write the answer in lowest terms.

Least Common Denominator (LCD)

The **least common denominator** (LCD) for two fractions is the *smallest* positive number divisible by both denominators of the original fractions. For example, both 6 and 12 are common denominators for $\frac{1}{2}$ and $\frac{1}{3}$, but 6 is smaller, so it is the LCD.

There are several ways to find the LCD. When the original denominators are small numbers, you can often find the LCD by inspection. *Hint:* Always check to see if the larger denominator will work as the LCD.

Chapter 4 Rational Numbers: Positive and Negative Fractions

② Find the LCD for each pair of fractions.

(a) $\dfrac{3}{5}$ and $\dfrac{3}{10}$

(b) $\dfrac{1}{2}$ and $\dfrac{2}{5}$

(c) $\dfrac{3}{4}$ and $\dfrac{1}{6}$

(d) $\dfrac{5}{6}$ and $\dfrac{7}{18}$

③ Use prime factorization to find the LCD for each pair of fractions.

(a) $\dfrac{1}{10}$ and $\dfrac{13}{14}$

(b) $\dfrac{5}{12}$ and $\dfrac{17}{20}$

(c) $\dfrac{7}{15}$ and $\dfrac{7}{9}$

ANSWERS
2. (a) 10 (b) 10 (c) 12 (d) 18

3. (a) $10 = 2 \cdot 5$
$14 = 2 \cdot 7$
LCD $= 2 \cdot 5 \cdot 7 = 70$

(b) $12 = 2 \cdot 2 \cdot 3$
$20 = 2 \cdot 2 \cdot 5$
LCD $= 2 \cdot 2 \cdot 3 \cdot 5 = 60$

(c) $15 = 3 \cdot 5$
$9 = 3 \cdot 3$
LCD $= 3 \cdot 3 \cdot 5 = 45$

EXAMPLE 2 Finding the LCD by Inspection

(a) Find the LCD for $\dfrac{2}{3}$ and $\dfrac{1}{9}$.

Check to see if 9 (the larger denominator) will work as the LCD. Is 9 divisible by 3 (the other denominator)? Yes, so 9 is the LCD for $\dfrac{2}{3}$ and $\dfrac{1}{9}$.

(b) Find the LCD for $\dfrac{5}{8}$ and $\dfrac{5}{6}$.

Check to see if 8 (the larger denominator) will work. No, 8 is not divisible by 6. So start checking numbers that are multiples of 8, that is, 16, 24, and 32. Notice that 24 will work because it is divisible by 8 and by 6.

The LCD for $\dfrac{5}{8}$ and $\dfrac{5}{6}$ is 24.

◀◀ Work Problem 2 at the Side.

For larger denominators, you can use prime factorization to find the LCD. Factor each denominator completely into prime numbers. Then use the factors to build the LCD.

EXAMPLE 3 Using Prime Factors to Find the LCD

(a) What is the LCD for $\dfrac{7}{12}$ and $\dfrac{13}{18}$?

Write 12 and 18 as the product of prime factors. Then use prime factors in the LCD that "cover" both 12 and 18.

Factors of 12

$12 = 2 \cdot 2 \cdot 3$
$18 = 2 \cdot 3 \cdot 3$
LCD $= 2 \cdot 2 \cdot 3 \cdot 3 = 36$

Factors of 18

Check whether 36 is divisible by 12 (yes) and by 18 (yes). So 36 is the LCD for $\dfrac{7}{12}$ and $\dfrac{13}{18}$.

CAUTION
When finding the LCD, notice that we did *not* have to repeat the factors that 12 and 18 have in common. If we had used *all* the 2s and 3s, we would get a common denominator, but not the *smallest* one.

(b) What is the LCD for $\dfrac{11}{15}$ and $\dfrac{9}{70}$?

Factors of 15

$15 = 3 \cdot 5$
$70 = 2 \cdot 5 \cdot 7$
LCD $= 3 \cdot 5 \cdot 2 \cdot 7 = 210$

Factors of 70

Check whether 210 is divisible by 15 (yes) and divisible by 70 (yes). So 210 is the LCD for $\dfrac{11}{15}$ and $\dfrac{9}{70}$.

◀◀ Work Problem 3 at the Side.

OBJECTIVE 3 Add and subtract unlike fractions. Here are the steps for adding or subtracting unlike fractions. The key idea is that you must rewrite the fractions so that they have a common denominator before you can add or subtract them.

> **Adding and Subtracting Unlike Fractions**
> *Step 1* Find the LCD, the smallest number divisible by both denominators in the problem.
> *Step 2* Rewrite each original fraction as an equivalent fraction whose denominator is the LCD.
> *Step 3* Add or subtract the numerators of the like fractions. Keep the common denominator.
> *Step 4* Write the sum or difference in lowest terms.

EXAMPLE 4 Adding and Subtracting Unlike Fractions

Find each sum or difference.

(a) $\dfrac{1}{5} + \dfrac{3}{10}$

Step 1 The larger denominator (10) is the LCD.

Step 2 $\dfrac{1}{5} = \dfrac{1 \cdot 2}{5 \cdot 2} = \dfrac{2}{10}$ ←LCD and $\dfrac{3}{10}$ already has the LCD.

Step 3 Add the numerators. Write the sum over the common denominator.

$$\dfrac{1}{5} + \dfrac{3}{10} = \dfrac{2}{10} + \dfrac{3}{10} = \dfrac{2+3}{10} = \dfrac{5}{10}$$

Step 4 Write $\dfrac{5}{10}$ in lowest terms.

$$\dfrac{5}{10} = \dfrac{\cancel{5}}{2 \cdot \cancel{5}} = \dfrac{1}{2} \leftarrow \text{Lowest terms}$$

(b) $\dfrac{3}{4} - \dfrac{5}{6}$

Step 1 The LCD is 12.

Step 2 $\dfrac{3}{4} = \dfrac{3 \cdot 3}{4 \cdot 3} = \dfrac{9}{12}$ ←LCD and $\dfrac{5}{6} = \dfrac{5 \cdot 2}{6 \cdot 2} = \dfrac{10}{12}$ ←LCD

Step 3 Subtract the numerators. Write the difference over the common denominator.

$9 + (-10)$ is -1

$$\dfrac{3}{4} - \dfrac{5}{6} = \dfrac{9}{12} - \dfrac{10}{12} = \dfrac{9 - 10}{12} = \dfrac{-1}{12} \text{ or } -\dfrac{1}{12}$$

Step 4 $-\dfrac{1}{12}$ is in lowest terms.

Continued on Next Page

Chapter 4 Rational Numbers: Positive and Negative Fractions

4 Find each sum or difference. Write all answers in lowest terms.

(a) $-\dfrac{2}{3} + \dfrac{1}{6}$

(b) $\dfrac{1}{12} - \dfrac{5}{6}$

(c) $3 - \dfrac{4}{5}$

(d) $\dfrac{9}{16} + \left(-\dfrac{5}{12}\right)$

ANSWERS

4. (a) $-\dfrac{1}{2}$ (b) $-\dfrac{3}{4}$ (c) $\dfrac{11}{5}$ (d) $\dfrac{7}{48}$

(c) $-\dfrac{5}{12} + \dfrac{5}{9}$

Step 1 Use prime factorization to find the LCD.

$$12 = 2 \cdot 2 \cdot 3$$
$$9 = 3 \cdot 3$$

Factors of 12
$$\text{LCD} = 2 \cdot 2 \cdot 3 \cdot 3 = 36$$
Factors of 9

Step 2 $-\dfrac{5}{12} = -\dfrac{5 \cdot 3}{12 \cdot 3} = -\dfrac{15}{36}$ and $\dfrac{5}{9} = \dfrac{5 \cdot 4}{9 \cdot 4} = \dfrac{20}{36}$

Step 3 Add the numerators. Keep the common denominator.

$$-\dfrac{5}{12} + \dfrac{5}{9} = -\dfrac{15}{36} + \dfrac{20}{36} = \dfrac{-15 + 20}{36} = \dfrac{5}{36}$$

Step 4 $\dfrac{5}{36}$ is in lowest terms.

(d) $4 - \dfrac{2}{3}$

Step 1 Think of 4 as $\dfrac{4}{1}$. The LCD for $\dfrac{4}{1}$ and $\dfrac{2}{3}$ is 3, the larger denominator.

Step 2 $\dfrac{4}{1} = \dfrac{4 \cdot 3}{1 \cdot 3} = \dfrac{12}{3}$ and $\dfrac{2}{3}$ already has the LCD.

Step 3 Subtract the numerators. Keep the common denominator.

$$\dfrac{4}{1} - \dfrac{2}{3} = \dfrac{12}{3} - \dfrac{2}{3} = \dfrac{12 - 2}{3} = \dfrac{10}{3}$$

Step 4 $\dfrac{10}{3}$ is in lowest terms.

Work Problem 4 at the Side.

OBJECTIVE 4 Add and subtract unlike fractions that contain variables.
We use the same steps to add or subtract unlike fractions with variables in the numerators or denominators.

EXAMPLE 5 Adding and Subtracting Unlike Fractions with Variables

Find each sum or difference.

(a) $\dfrac{1}{4} + \dfrac{b}{5}$

Step 1 The LCD is 20.

Step 2 $\dfrac{1}{4} = \dfrac{1 \cdot 5}{4 \cdot 5} = \dfrac{5}{20}$ and $\dfrac{b}{5} = \dfrac{b \cdot 4}{5 \cdot 4} = \dfrac{4b}{20}$

Step 3 $\dfrac{1}{4} + \dfrac{b}{5} = \dfrac{5}{20} + \dfrac{4b}{20} = \dfrac{5 + 4b}{20}$ ← Add the numerators.
← Keep the common denominator.

Step 4 $\dfrac{5 + 4b}{20}$ is in lowest terms.

Continued on Next Page

> **CAUTION**
> In part (a), we could ***not*** add $5 + 4b$ in the numerator of the answer because 5 and $4b$ are ***not*** like terms. We *could* add $5b + 4b$ but ***not*** $5 + 4b$.
>
> Variable parts match.

(b) $\dfrac{2}{3} - \dfrac{6}{x}$

Step 1   The LCD is $3 \cdot x$, or $3x$.

Step 2 $\dfrac{2}{3} = \dfrac{2 \cdot x}{3 \cdot x} = \dfrac{2x}{3x}$ and $\dfrac{6}{x} = \dfrac{6 \cdot 3}{x \cdot 3} = \dfrac{18}{3x}$

Step 3 $\dfrac{2}{3} - \dfrac{6}{x} = \dfrac{2x}{3x} - \dfrac{18}{3x} = \dfrac{2x - 18}{3x}$ ← Subtract the numerators.
                                                      ← Keep the common denominator.

Step 4 $\dfrac{2x - 18}{3x}$ is in lowest terms.

> **NOTE**
> Notice in part (b) above that we found the LCD for $\dfrac{2}{3} - \dfrac{6}{x}$ by multiplying the two denominators. The LCD is $3 \cdot x$ or $3x$.
>
>   Multiplying the two denominators will *always* give you a common denominator, but it may not be the *smallest* common denominator. Here are more examples.
>
> $\dfrac{1}{3} - \dfrac{2}{5}$ If you multiply the denominators, $3 \cdot 5 = 15$ and 15 is the LCD.
>
> $\dfrac{5}{6} + \dfrac{3}{4}$ If you multiply the denominators, $6 \cdot 4 = 24$ and 24 will work. But you'll save some time by using the smallest common denominator, which is 12.
>
> $\dfrac{7}{y} + \dfrac{a}{4}$ If you multiply the denominators, $y \cdot 4 = 4y$ and $4y$ is the LCD.

Work Problem 5 at the Side.

---

5 Find each sum or difference.

(a) $\dfrac{5}{6} - \dfrac{h}{2}$

(b) $\dfrac{7}{t} + \dfrac{3}{5}$

(c) $\dfrac{4}{x} - \dfrac{8}{3}$

ANSWERS

5. (a) $\dfrac{5 - 3h}{6}$ (b) $\dfrac{35 + 3t}{5t}$ (c) $\dfrac{12 - 8x}{3x}$

Focus on Real-Data Applications

Music

The time signature at the beginning of a piece of music looks like a fraction. Commonly used time signatures are $\frac{2}{4}$, $\frac{3}{4}$, $\frac{4}{4}$, and $\frac{6}{8}$. Musicians use the time signature to tell how long to hold each note. The values of different notes can be written as fractions:

$\mathbf{o} = 1 \qquad \text{𝅗𝅥} = \frac{1}{2} \qquad \text{♩} = \frac{1}{4} \qquad \text{♪} = \frac{1}{8} \qquad \text{𝅘𝅥𝅯} = \frac{1}{16}$

Music is divided into measures. In $\frac{4}{4}$ time, each measure contains notes that add up to $\frac{4}{4}$ (or 1). In $\frac{2}{4}$ time the notes in each measure add up to $\frac{2}{4}$ (or $\frac{1}{2}$), and so on for $\frac{3}{4}$ time and $\frac{6}{8}$ time.

Write one or more notes in each measure to make it add up to its time signature. Use as many different kinds of notes as possible.

Below are excerpts from "Jingle Bells" and "The Star-Spangled Banner." Divide each line of music into measures based on the time signature.

4.4 Exercises

Find each sum or difference. Write all answers in lowest terms. See Examples 1–5.

1. $\dfrac{3}{4} + \dfrac{1}{8}$

2. $\dfrac{1}{3} + \dfrac{1}{2}$

3. $-\dfrac{1}{14} + \left(-\dfrac{3}{7}\right)$

4. $-\dfrac{2}{9} + \dfrac{2}{3}$

5. $\dfrac{2}{3} - \dfrac{1}{6}$

6. $\dfrac{5}{12} - \dfrac{1}{4}$

7. $\dfrac{3}{8} - \dfrac{3}{5}$

8. $\dfrac{1}{3} - \dfrac{3}{5}$

9. $-\dfrac{5}{8} + \dfrac{1}{12}$

10. $-\dfrac{13}{16} + \dfrac{13}{16}$

11. $-\dfrac{7}{20} - \dfrac{5}{20}$

12. $-\dfrac{7}{9} - \dfrac{5}{6}$

13. $0 - \dfrac{7}{18}$

14. $-\dfrac{7}{8} + 3$

15. $2 - \dfrac{6}{7}$

16. $5 - \dfrac{2}{5}$

17. $-\dfrac{1}{2} + \dfrac{3}{24}$

18. $\dfrac{7}{10} + \dfrac{7}{15}$

19. $\dfrac{1}{5} + \dfrac{c}{3}$

20. $\dfrac{x}{4} + \dfrac{2}{3}$

21. $\dfrac{5}{m} - \dfrac{1}{2}$

22. $\dfrac{2}{9} - \dfrac{4}{y}$

23. $\dfrac{3}{b^2} + \dfrac{5}{b^2}$

24. $\dfrac{10}{xy} - \dfrac{7}{xy}$

25. $\dfrac{c}{7} + \dfrac{3}{b}$

26. $\dfrac{2}{x} - \dfrac{y}{5}$

27. $-\dfrac{4}{c^2} - \dfrac{d}{c}$

28. $-\dfrac{1}{n} + \dfrac{m}{n^2}$

29. $-\dfrac{11}{42} - \dfrac{11}{70}$

30. $\dfrac{7}{45} - \dfrac{7}{20}$

31. A key step in adding or subtracting unlike fractions is to rewrite the fractions so that they have a common denominator. Explain why this step is necessary.

32. Explain how to write a fraction with an indicated denominator. As part of your explanation, show how to rewrite $\tfrac{3}{4}$ as an equivalent fraction with a denominator of 12.

33. Explain the error in each calculation and correct it.

(a) $\dfrac{3}{4} + \dfrac{2}{5} = \dfrac{3+2}{4+5} = \dfrac{5}{9}$

(b) $\dfrac{5}{6} - \dfrac{4}{9} = \dfrac{5}{18} - \dfrac{4}{18} = \dfrac{5-4}{18} = \dfrac{1}{18}$

34. Explain the error in each calculation and correct it.

(a) $-\dfrac{1}{4} + \dfrac{7}{12} = -\dfrac{3}{12} + \dfrac{7}{12} = \dfrac{-3+7}{12} = \dfrac{4}{12}$

(b) $\dfrac{3}{10} - \dfrac{1}{4} = \dfrac{3-1}{10-4} = \dfrac{2}{6} = \dfrac{1}{3}$

RELATING CONCEPTS (EXERCISES 35–36) For Individual or Group Work

As you work Exercises 35 and 36 in order, think about the properties you learned when working with integers. Then explain what each pair of problems illustrates. (For a review of the properties, see the **Chapter 1 Summary**.)

35. (a) $-\dfrac{2}{3} + \dfrac{3}{4} = $ _____ $\dfrac{3}{4} + \left(-\dfrac{2}{3}\right) = $ _____

(b) $\dfrac{5}{6} - \dfrac{1}{2} = $ _____ $\dfrac{1}{2} - \dfrac{5}{6} = $ _____

(c) $\left(-\dfrac{2}{3}\right)\left(\dfrac{9}{10}\right) = $ _____ $\left(\dfrac{9}{10}\right)\left(-\dfrac{2}{3}\right) = $ _____

(d) $\dfrac{2}{5} \div \dfrac{1}{15} = $ _____ $\dfrac{1}{15} \div \dfrac{2}{5} = $ _____

36. (a) $-\dfrac{7}{12} + \dfrac{7}{12} = $ _____ $\dfrac{3}{5} + \left(-\dfrac{3}{5}\right) = $ _____

(b) $-\dfrac{13}{16} \div \left(-\dfrac{13}{16}\right) = $ _____ $\dfrac{1}{8} \div \dfrac{1}{8} = $ _____

(c) $\dfrac{5}{6} \cdot 1 = $ _____ $1\left(-\dfrac{17}{20}\right) = $ _____

(d) $\left(-\dfrac{4}{5}\right)\left(-\dfrac{5}{4}\right) = $ _____ $7 \cdot \dfrac{1}{7} = $ _____

Solve each application problem. Write all answers in lowest terms.

37. When assembling shelving units, Pam Phelps must use the proper type and size of hardware. Find the total length of the bolt shown.

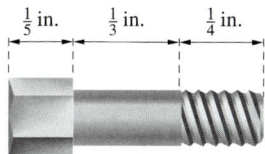

38. How much fencing will be needed to enclose this rectangular wildflower preserve?

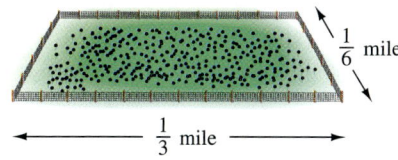

39. The owner of Racy's Feed Store ordered $\frac{1}{3}$ cubic yard of corn, $\frac{3}{8}$ cubic yard of oats, and $\frac{1}{4}$ cubic yard of washed medium mesh gravel. How many cubic yards of material were ordered?

40. A flower grower purchased $\frac{9}{10}$ acre of land one year and $\frac{3}{10}$ acre the next year. She then sold $\frac{7}{10}$ acre of land. How much land does she now have?

41. A forester planted $\frac{5}{12}$ acre of seedlings in the morning and $\frac{11}{12}$ acre in the afternoon. The next day, $\frac{7}{12}$ acre of seedlings were destroyed by a brush fire. How many acres of seedlings remained?

42. Adrian Ortega drives a tanker for the British Petroleum Company. He leaves the refinery with his tanker filled to $\frac{7}{8}$ of capacity. If he delivers $\frac{1}{4}$ of the tanker's capacity at the first stop and $\frac{1}{3}$ of the tanker's capacity at the second stop, find the fraction of the tanker's capacity remaining.

The circle graph shows the fraction of U.S. workers who used various methods to learn about computers. Use the graph to answer Exercises 43–46.

Source: John J. Heldrich Center for Workforce Development.

43. What fraction of workers are self-taught or learned from friends or family?

44. What fraction of workers learned at school or through work?

45. What is the difference in the fraction of workers who are self-taught and those who learned at work?

46. What is the difference in the fraction of workers who are self-taught and those who learned at school?

Refer to the circle graph to answer Exercises 47–50.

47. What fraction of the day was spent in class and study? How many hours is that?

THE DAY OF THE STUDENT

48. What fraction of the day was spent in work and travel and sleep? How many hours is that?

49. How much more of the day was spent sleeping than studying? Write your answer as a fraction of the day.

50. How much more of the day was spent working and traveling than in class? Write your answer as a fraction of the day.

Use the photo to answer Exercises 51–52. A nut driver is like a screwdriver but is used to tighten nuts instead of screws. The ends of the drivers are sized from $\frac{3}{16}$ inch to $\frac{1}{2}$ inch to fit smaller or larger nuts. (The symbol " is for inches.)

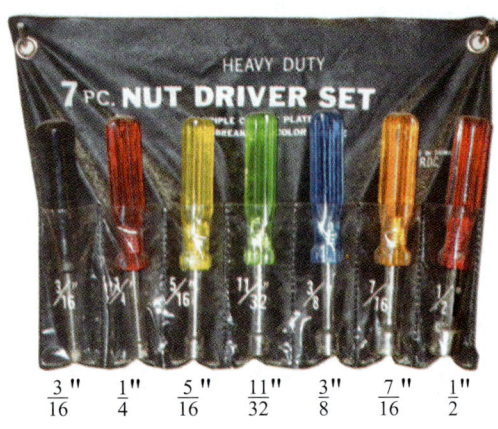

51. The rightmost driver fits a nut that is how much larger than the nut for the leftmost driver?

52. The nut size for the yellow-handled driver is how much less than the nut size for the blue-handled driver?

Source: Author's tool collection.

53. A hazardous waste dump site will require $\frac{7}{8}$ mile of security fencing. The site has four sides with three of the sides measuring $\frac{1}{4}$ mile, $\frac{1}{6}$ mile, and $\frac{3}{8}$ mile. Find the length of the fourth side.

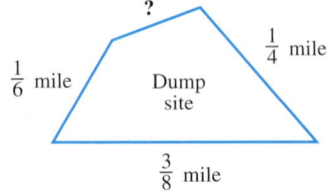

54. Chakotay is fitting a turquoise stone into a bear claw pendant. Find the diameter of the hole in the pendant. (The diameter is the distance across the center of the hole.)

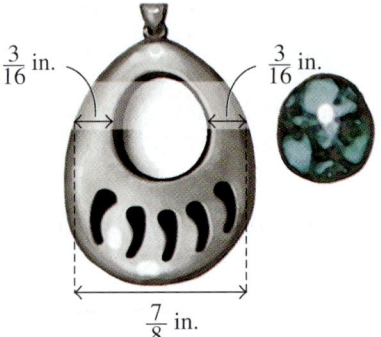

4.5 Problem Solving: Mixed Numbers and Estimating

OBJECTIVE 1 Identify mixed numbers and graph them on a number line. When a fraction and a whole number are written together, the result is a **mixed number.** For example, the mixed number

$$3\frac{1}{2} \quad \text{represents} \quad 3 + \frac{1}{2}$$

or 3 wholes and $\frac{1}{2}$ of a whole. Read $3\frac{1}{2}$ as "three and one half."

One common use of mixed numbers is to measure things. Examples are shown below.

Juan worked $5\frac{1}{2}$ hours. The box weighs $2\frac{3}{4}$ pounds.
The park is $1\frac{7}{10}$ miles long. Add $1\frac{2}{3}$ cups of flour.

EXAMPLE 1 Illustrating a Mixed Number with a Diagram and a Number Line

As this diagram shows, the mixed number $3\frac{1}{2}$ is equivalent to the improper fraction $\frac{7}{2}$.

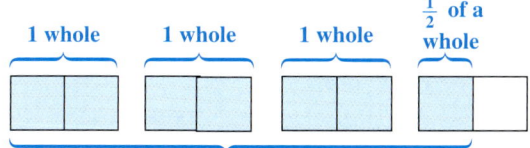

$\frac{7}{2}$ ← Number of shaded parts
← Number of equal parts in each whole

We can also use a number line to show mixed numbers, as in this graph of $3\frac{1}{2}$ and $-3\frac{1}{2}$.

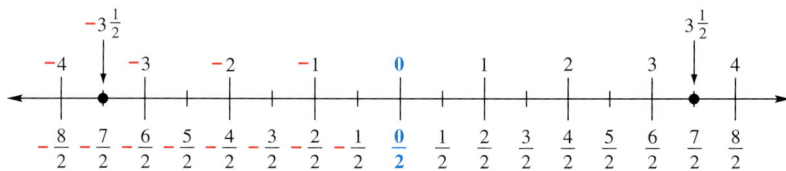

The number line shows the following.

$$3\frac{1}{2} \text{ is equivalent to } \frac{7}{2}$$

$$-3\frac{1}{2} \text{ is equivalent to } -\frac{7}{2}$$

> **NOTE**
> $3\frac{1}{2}$ represents $3 + \frac{1}{2}$.
>
> $-3\frac{1}{2}$ represents $-3 + (-\frac{1}{2})$, which can also be written as $-3 - \frac{1}{2}$.
>
> In algebra we usually work with the improper fraction form of mixed numbers, especially for negative mixed numbers. However, positive mixed numbers are frequently used in daily life, so it's important to know how to work with them. For example, we usually say $3\frac{1}{2}$ inches rather than $\frac{7}{2}$ inches.

Work Problem 1 at the Side. ▶▶▶

OBJECTIVES

1. Identify mixed numbers and graph them on a number line.
2. Rewrite mixed numbers as improper fractions, or the reverse.
3. Estimate the answer and multiply or divide mixed numbers.
4. Estimate the answer and add or subtract mixed numbers.
5. Solve application problems containing mixed numbers.

1 (a) Use these diagrams to write $1\frac{2}{3}$ as an improper fraction.

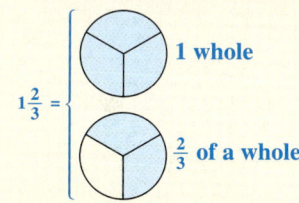

(b) Now graph $1\frac{2}{3}$ and $-1\frac{2}{3}$ on this number line.

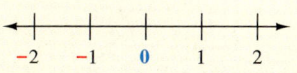

(c) Use these diagrams to write $2\frac{1}{4}$ as an improper fraction.

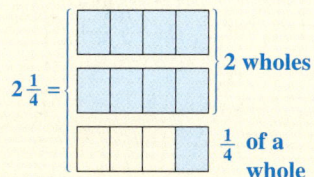

(d) Now graph $2\frac{1}{4}$ and $-2\frac{1}{4}$ on this number line.

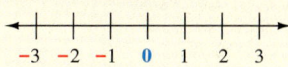

ANSWERS

1. (a) $\frac{5}{3}$ (b)

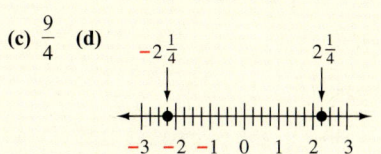

(c) $\frac{9}{4}$ (d)

2 Write each mixed number as an equivalent improper fraction.

(a) $3\frac{2}{3}$

(b) $4\frac{7}{10}$

(c) $5\frac{3}{4}$

(d) $8\frac{5}{6}$

OBJECTIVE 2 Rewrite mixed numbers as improper fractions, or the reverse. You can use the following steps to write $3\frac{1}{2}$ as an improper fraction without drawing a diagram or a number line.

Step 1 Multiply 2 times 3 and add 1 to the product.

$$3\frac{1}{2} \qquad 2 \cdot 3 = 6 \qquad \text{Then } 6 + 1 = 7$$

Step 2 Use 7 (from Step 1) as the numerator and 2 as the denominator.

$$3\frac{1}{2} = \frac{7}{2} \leftarrow (2 \cdot 3) + 1$$

Same denominator

To see why this method works, recall that $3\frac{1}{2}$ represents $3 + \frac{1}{2}$. Let's add $3 + \frac{1}{2}$.

$$3 + \frac{1}{2} = \frac{3}{1} + \frac{1}{2} = \frac{6}{2} + \frac{1}{2} = \frac{6 + 1}{2} = \frac{7}{2} \quad \left\{\begin{array}{l}\text{Same result}\\\text{as above}\end{array}\right.$$

Common denominator

In summary, use the following steps to *write a mixed number as an improper fraction*.

Writing a Mixed Number as an Improper Fraction

Step 1 **Multiply** the denominator of the fraction times the whole number and **add** the numerator of the fraction to the product.

Step 2 Write the result of Step 1 as the **numerator** and keep the original **denominator.**

EXAMPLE 2 Writing a Mixed Number as an Improper Fraction

Write $7\frac{2}{3}$ as an improper fraction (numerator greater than denominator).

Step 1 $7\frac{2}{3} \qquad 3 \cdot 7 = 21 \qquad \text{Then } 21 + 2 = 23$

Step 2 $7\frac{2}{3} = \frac{23}{3} \leftarrow (3 \cdot 7) + 2$

Same denominator

Work Problem 2 at the Side.

ANSWERS

2. (a) $\frac{11}{3}$ (b) $\frac{47}{10}$ (c) $\frac{23}{4}$ (d) $\frac{53}{6}$

We used *multiplication* for the first step in writing a mixed number as an improper fraction. To work in *reverse*, writing an improper fraction as a mixed number, we will use *division*. Recall that the fraction bar is a symbol for division.

> **Writing an Improper Fraction as a Mixed Number**
>
> To write an *improper fraction* as a mixed number, divide the numerator by the denominator. The quotient is the whole number part (of the mixed number), the remainder is the numerator of the fraction part, and the denominator remains the same.

Always check to be sure that the fraction part of the mixed number is in lowest terms. Then the mixed number is in *simplest form*.

EXAMPLE 3 **Writing Improper Fractions as Mixed Numbers**

Write each improper fraction as an equivalent mixed number in simplest form.

(a) $\dfrac{17}{5}$

Divide 17 by 5.

$$\begin{array}{r} 3 \\ 5\overline{)17} \\ \underline{15} \\ 2 \end{array} \begin{array}{l} \leftarrow \text{Whole number part} \\ \\ \leftarrow \text{Remainder} \end{array}$$

The quotient **3** is the whole number part of the mixed number. The remainder **2** is the numerator of the fraction, and the denominator stays as **5**.

$$\frac{17}{5} = 3\frac{2}{5} \leftarrow \text{Remainder}$$
$$\text{Same denominator}$$

Let's look at a drawing of $\frac{17}{5}$ to check our work.

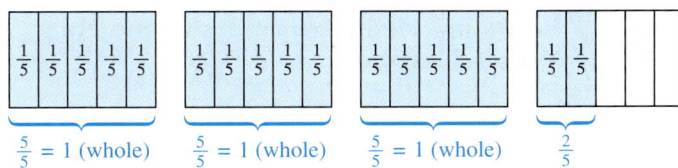

$\frac{5}{5} = 1$ (whole) $\frac{5}{5} = 1$ (whole) $\frac{5}{5} = 1$ (whole) $\frac{2}{5}$

Continued on Next Page

3 Write each improper fraction as an equivalent mixed number in simplest form.

(a) $\dfrac{5}{2}$

(b) $\dfrac{14}{4}$

(c) $\dfrac{33}{5}$

(d) $\dfrac{58}{10}$

4 Round each mixed number to the nearest whole number.

(a) $2\dfrac{3}{4}$

(b) $6\dfrac{3}{8}$

(c) $4\dfrac{2}{3}$

(d) $1\dfrac{7}{10}$

(e) $3\dfrac{1}{2}$

(f) $5\dfrac{4}{9}$

In other words,

$$\dfrac{17}{5} = \dfrac{5}{5} + \dfrac{5}{5} + \dfrac{5}{5} + \dfrac{2}{5}$$

$$= \underbrace{1 + 1 + 1}_{} + \dfrac{2}{5}$$

$$= 3 + \dfrac{2}{5}$$

$$= 3\dfrac{2}{5}$$

(b) $\dfrac{26}{4}$

Divide 26 by 4.

$$\begin{array}{r}6\\4\overline{)26}\\24\\\hline 2\end{array} \quad \text{so} \quad \dfrac{26}{4} = 6\dfrac{2}{4} = 6\dfrac{1}{2} \quad \{\text{Simplest form}\}$$

Write $\tfrac{2}{4}$ in lowest terms.

You could write $\tfrac{26}{4}$ in lowest terms first.

$$\dfrac{26}{4} = \dfrac{\overset{1}{\cancel{2}} \cdot 13}{\underset{1}{\cancel{2}} \cdot 2} = \dfrac{13}{2} \quad \text{Then} \quad \begin{array}{r}6\\2\overline{)13}\\12\\\hline 1\end{array} \quad \text{so} \quad \dfrac{13}{2} = 6\dfrac{1}{2} \quad \{\text{Same result as above}\}$$

Work Problem 3 at the Side.

OBJECTIVE 3 Estimate the answer and multiply or divide mixed numbers. Once you have rewritten mixed numbers as improper fractions, you can use the steps you learned in **Section 4.3** to multiply and divide. However, it's a good idea to estimate the answer before you start any other work.

EXAMPLE 4 Rounding Mixed Numbers to the Nearest Whole Number

To estimate answers, first round each mixed number to the *nearest whole number*. If the numerator is *half* of the denominator or *more*, round up the whole number part. If the numerator is *less* than half the denominator, leave the whole number as it is.

(a) Round $1\dfrac{5}{8}$ ← 5 is more than 4
← Half of 8 is 4 $1\dfrac{5}{8}$ rounds up to 2

(b) Round $3\dfrac{2}{5}$ ← 2 is less than $2\dfrac{1}{2}$
← Half of 5 is $2\dfrac{1}{2}$ $3\dfrac{2}{5}$ rounds to 3

Work Problem 4 at the Side.

ANSWERS

3. (a) $2\dfrac{1}{2}$ (b) $3\dfrac{1}{2}$ (c) $6\dfrac{3}{5}$ (d) $5\dfrac{4}{5}$

4. (a) 3   (b) 6   (c) 5   (d) 2   (e) 4   (f) 5

Section 4.5 Problem Solving: Mixed Numbers and Estimating   247

Multiplying and Dividing Mixed Numbers

Step 1 *Rewrite* each mixed number as an improper fraction.

Step 2 *Multiply* or *divide* the improper fractions.

Step 3 Write the answer in lowest terms and change it to a mixed number or whole number where possible. This step gives you an answer that is in *simplest form.*

EXAMPLE 5 Estimating the Answer and Multiplying Mixed Numbers

First, round the numbers and estimate each answer. Then find the exact answer. Write exact answers in simplest form.

(a) $2\frac{1}{2} \cdot 3\frac{1}{5}$

Estimate the answer by rounding the mixed numbers.

$2\frac{1}{2}$ rounds to 3   and $3\frac{1}{5}$ rounds to 3

$3 \cdot 3 = 9$ ← Estimated answer

To find the exact answer, first rewrite each mixed number as an improper fraction.

Step 1 $2\frac{1}{2} = \frac{5}{2}$ and $3\frac{1}{5} = \frac{16}{5}$

Next, multiply.

$$2\frac{1}{2} \cdot 3\frac{1}{5} = \frac{5}{2} \cdot \frac{16}{5} = \frac{\cancel{5} \cdot \cancel{2} \cdot 8}{\cancel{2} \cdot \cancel{5}} = \frac{8}{1} = 8 \quad \{\text{Simplest form}$$

The estimate was 9, so an exact answer of 8 is reasonable.

(b) $\left(3\frac{5}{8}\right)\left(4\frac{4}{5}\right)$

First, round each mixed number and estimate the answer.

$3\frac{5}{8}$ rounds to 4   and $4\frac{4}{5}$ rounds to 5

$4 \cdot 5 = 20$ ← Estimated answer

Now find the exact answer.

$$\left(3\frac{5}{8}\right)\left(4\frac{4}{5}\right) = \left(\frac{29}{8}\right)\left(\frac{24}{5}\right) = \frac{29 \cdot 3 \cdot \cancel{8}}{\cancel{8} \cdot 5} = \frac{87}{5} = 17\frac{2}{5} \quad \{\text{Simplest form}$$

The estimate was 20, so an exact answer of $17\frac{2}{5}$ is reasonable.

Work Problem 5 at the Side.

---

5 First, round the numbers and estimate each answer. Then find the exact answer. Write exact answers in simplest form.

(a) $2\frac{1}{4} \cdot 7\frac{1}{3}$

   ___ • ___ = ___ estimate

(b) $\left(4\frac{1}{2}\right)\left(1\frac{2}{3}\right)$

   (___)(___) = ___ estimate

(c) $3\frac{3}{5} \cdot 4\frac{4}{9}$

   ___ • ___ = ___ estimate

(d) $3\frac{1}{5} \cdot 5\frac{3}{8}$

   ___ • ___ = ___ estimate

ANSWERS

5. (a) *Estimate:* $2 \cdot 7 = 14$; *Exact:* $16\frac{1}{2}$

   (b) *Estimate:* $(5)(2) = 10$; *Exact:* $7\frac{1}{2}$

   (c) *Estimate:* $4 \cdot 4 = 16$; *Exact:* 16

   (d) *Estimate:* $3 \cdot 5 = 15$; *Exact:* $17\frac{1}{5}$

248 Chapter 4 Rational Numbers: Positive and Negative Fractions

6 First, round the numbers and estimate each answer. Then find the exact answer. Write exact answers in simplest form.

(a) $6\frac{1}{4} \div 3\frac{1}{3}$
$\downarrow \quad \downarrow$
____ ÷ ____ = ____ estimate

(b) $3\frac{3}{8} \div 2\frac{4}{7}$
$\downarrow \quad \downarrow$
____ ÷ ____ = ____ estimate

(c) $8 \div 5\frac{1}{3}$
$\downarrow \quad \downarrow$
____ ÷ ____ = ____ estimate

(d) $4\frac{1}{2} \div 6$
$\downarrow \quad \downarrow$
____ ÷ ____ = ____ estimate

ANSWERS

6. (a) Estimate: $6 \div 3 = 2$; Exact: $1\frac{7}{8}$

(b) Estimate: $3 \div 3 = 1$; Exact: $1\frac{5}{16}$

(c) Estimate: $8 \div 5 = 1\frac{3}{5}$; Exact: $1\frac{1}{2}$

(d) Estimate: $5 \div 6 = \frac{5}{6}$; Exact: $\frac{3}{4}$

EXAMPLE 6 Estimating the Answer and Dividing Mixed Numbers

First, round the numbers and estimate each answer. Then find the exact answer. Write exact answers in simplest form.

(a) $3\frac{3}{5} \div 1\frac{1}{2}$

To estimate the answer, round each mixed number to the nearest whole number.

$$3\frac{3}{5} \div 1\frac{1}{2}$$
Rounded
$$4 \div 2 = 2 \leftarrow \text{Estimate}$$

To find the exact answer, first rewrite each mixed number as an improper fraction.

$$3\frac{3}{5} \div 1\frac{1}{2} = \frac{18}{5} \div \frac{3}{2}$$

Now rewrite the problem as multiplying by the reciprocal of $\frac{3}{2}$.

$$\frac{18}{5} \div \frac{3}{2} = \frac{18}{5} \cdot \frac{2}{3} = \frac{\cancel{3} \cdot 6 \cdot 2}{5 \cdot \cancel{3}} = \frac{12}{5} = 2\frac{2}{5} \quad \begin{cases}\text{Simplest} \\ \text{form}\end{cases}$$
Reciprocals

The estimate was 2, so an exact answer of $2\frac{2}{5}$ is reasonable.

(b) $4\frac{3}{8} \div 5$

First, round the numbers and estimate the answer.

$$4\frac{3}{8} \div 5$$
Rounded
$$4 \div 5 \quad \text{Write } 4 \div 5 \text{ using a fraction bar.} \quad \frac{4}{5} \leftarrow \text{Estimate}$$

Now find the exact answer.

Write 5 as $\frac{5}{1}$

$$4\frac{3}{8} \div 5 = \frac{35}{8} \div \frac{5}{1} = \frac{35}{8} \cdot \frac{1}{5} = \frac{\cancel{5} \cdot 7 \cdot 1}{8 \cdot \cancel{5}} = \frac{7}{8} \quad \begin{cases}\text{Simplest} \\ \text{form}\end{cases}$$
Reciprocals

The estimate was $\frac{4}{5}$, so an exact answer of $\frac{7}{8}$ is reasonable. They are both less than 1.

Work Problem 6 at the Side.

Section 4.5 Problem Solving: Mixed Numbers and Estimating

OBJECTIVE 4 **Estimate the answer and add or subtract mixed numbers.** The steps you learned for adding and subtracting fractions in **Section 4.4** will also work for mixed numbers: Just rewrite the mixed numbers as equivalent improper fractions. Again, it is a good idea to estimate the answer before you start any other work.

EXAMPLE 7 Estimating the Answer and Adding or Subtracting Mixed Numbers

First, estimate each answer. Then add or subtract to find the exact answer. Write exact answers in simplest form.

(a) $2\frac{3}{8} + 3\frac{3}{4}$

To estimate the answer, round each mixed number to the nearest whole number.

$$2\frac{3}{8} + 3\frac{3}{4}$$
$$\downarrow \quad \quad \downarrow$$
$$2 + 4 = 6 \leftarrow \text{Estimate}$$

To find the exact answer, first rewrite each mixed number as an equivalent improper fraction.

$$2\frac{3}{8} + 3\frac{3}{4} = \frac{19}{8} + \frac{15}{4}$$

You can't add fractions until they have a common denominator. The LCD for $\frac{19}{8}$ and $\frac{15}{4}$ is 8. Rewrite $\frac{15}{4}$ as an equivalent fraction with a denominator of 8.

$$\frac{19}{8} + \frac{15}{4} = \frac{19}{8} + \frac{30}{8} = \frac{19 + 30}{8} = \frac{49}{8} = 6\frac{1}{8} \quad \left\{\begin{array}{l}\text{Simplest}\\ \text{form}\end{array}\right.$$

Common denominator

The estimate was 6, so an exact answer of $6\frac{1}{8}$ is reasonable.

(b) $4\frac{2}{3} - 2\frac{4}{5}$

Round each number and estimate the answer.

$$4\frac{2}{3} - 2\frac{4}{5}$$
$$\downarrow \quad \quad \downarrow$$
$$5 - 3 = 2 \leftarrow \text{Estimate}$$

To find the exact answer, rewrite the mixed numbers as improper fractions and subtract.

$$4\frac{2}{3} - 2\frac{4}{5} = \frac{14}{3} - \frac{14}{5} = \frac{70}{15} - \frac{42}{15} = \frac{70-42}{15} = \frac{28}{15} = 1\frac{13}{15} \quad \left\{\begin{array}{l}\text{Simplest}\\ \text{form}\end{array}\right.$$

LCD is 15

The estimate was 2, so an exact answer of $1\frac{13}{15}$ is reasonable.

Continued on Next Page

7 First, round the numbers and estimate each answer. Then add or subtract to find the exact answer.

(a) $5\frac{1}{3} - 2\frac{5}{6}$

$\underline{\quad} - \underline{\quad} = \underline{\quad}$ estimate

(b) $\frac{3}{4} + 3\frac{1}{8}$

$\underline{\quad} + \underline{\quad} = \underline{\quad}$ estimate

(c) $6 - 3\frac{4}{5}$

$\underline{\quad} - \underline{\quad} = \underline{\quad}$ estimate

ANSWERS
7. (a) *Estimate:* $5 - 3 = 2$; *Exact:* $2\frac{1}{2}$

(b) *Estimate:* $1 + 3 = 4$; *Exact:* $3\frac{7}{8}$

(c) *Estimate:* $6 - 4 = 2$; *Exact:* $2\frac{1}{5}$

(c) $5 - 1\frac{3}{8}$

$5 - 1\frac{3}{8}$
$\downarrow \quad \downarrow$
$5 - 1 = 4 \leftarrow$ Estimate

Write 5 as $\frac{5}{1}$

$5 - 1\frac{3}{8} = \frac{5}{1} - \frac{11}{8} = \frac{40}{8} - \frac{11}{8} = \frac{29}{8} = 3\frac{5}{8}$ {Simplest form}

LCD is 8

The estimate was 4, so an exact answer of $3\frac{5}{8}$ is reasonable.

▶ **Work Problem 7 at the Side.**

NOTE
In some situations the method of rewriting mixed numbers as improper fractions may result in very large numerators. Consider this example.

Last year Hue's child was $48\frac{3}{8}$ in. tall. This year the child is $51\frac{1}{4}$ in. tall. How much has the child grown?

First, estimate the answer by rounding each mixed number to the nearest whole number.

$51\frac{1}{4} - 48\frac{3}{8}$
$\downarrow \quad \downarrow$
$51 - 48 = 3$ in. $\leftarrow$ Estimate

To find the exact answer, rewrite the mixed numbers as improper fractions.

Rewrite $\frac{205}{4}$ as $\frac{410}{8}$

$51\frac{1}{4} - 48\frac{3}{8} = \frac{205}{4} - \frac{387}{8} = \frac{410}{8} - \frac{387}{8} = \frac{410 - 387}{8} = \frac{23}{8} = 2\frac{7}{8}$ in.

LCD is 8

You can also use the fraction key $\boxed{a^{b/c}}$ on your *scientific* calculator to solve this problem.

To enter $51\frac{1}{4}$, press

51 $\boxed{a^{b/c}}$ 1 $\boxed{a^{b/c}}$ 4 $\boxed{a^{b/c}}$. The display shows $\boxed{51\_1\_4}$.
↑   ↑   ↑
Whole   Numerator   Denominator
number

Then press $\boxed{-}$ 48 $\boxed{a^{b/c}}$ 3 $\boxed{a^{b/c}}$ 8 $\boxed{=}$. The display shows $\boxed{2\_7\_8}$.
↑
Subtract. $48\frac{3}{8}$ $2\frac{7}{8}$

Either way the exact answer is $2\frac{7}{8}$ in., which is close to the estimate of 3 in.

Another efficient method for handling large mixed numbers is to rewrite them in decimal form. You will learn how to do that in **Chapter 5**.

Section 4.5 Problem Solving: Mixed Numbers and Estimating **251**

OBJECTIVE 5 Solve application problems containing mixed numbers.
Rounding mixed numbers to the nearest whole number can also help you decide whether to solve an application problem by adding, subtracting, multiplying, or dividing.

EXAMPLE 8 Solving Application Problems with Mixed Numbers

First, estimate the answer to each application problem. Then find the exact answer.

(a) Gary needs to haul $15\frac{3}{4}$ **tons** of sand to a construction site. His truck can carry $2\frac{1}{4}$ **tons**. How many trips will he need to make?

First, round each mixed number to the nearest whole number.

$$15\frac{3}{4} \text{ rounds to } 16 \quad \text{and} \quad 2\frac{1}{4} \text{ rounds to } 2$$

Now read the problem again, *using the rounded numbers.*

Gary needs to haul **16 tons** of sand to a construction site. His truck can carry **2 tons**. How many trips will he need to make?

Using the rounded numbers in the problem makes it easier to see that you need to *divide*.

$$16 \div 2 = 8 \text{ trips} \leftarrow \text{Estimate}$$

To find the exact answer, use the original mixed numbers and divide.

$$15\frac{3}{4} \div 2\frac{1}{4} = \frac{63}{4} \div \frac{9}{4} = \frac{63}{4} \cdot \frac{4}{9} = \frac{7 \cdot \cancel{9} \cdot \cancel{4}}{\cancel{4} \cdot \cancel{9}} = \frac{7}{1} = 7 \quad \{\text{Simplest form}$$

Reciprocals

Gary needs to make 7 trips to haul all the sand. This result is close to the estimate of 8 trips.

(b) Zenitia worked $3\frac{5}{6}$ **hours** on Monday and $6\frac{1}{2}$ **hours** on Tuesday. How much longer did she work on Tuesday than on Monday?

First, round each mixed number to the nearest whole number.

$$3\frac{5}{6} \text{ rounds to } 4 \quad \text{and} \quad 6\frac{1}{2} \text{ rounds to } 7$$

Now read the problem again, *using the rounded numbers*.

Zenitia worked **4 hours** on Monday and **7 hours** on Tuesday. How much longer did she work on Tuesday than on Monday?

Using the rounded numbers in the problem makes it easier to see that you need to *subtract*.

$$7 - 4 = 3 \text{ hours} \leftarrow \text{Estimate}$$

To find the exact answer, use the original mixed numbers and subtract.

Write answer in simplest form.

$$6\frac{1}{2} - 3\frac{5}{6} = \frac{13}{2} - \frac{23}{6} = \frac{39}{6} - \frac{23}{6} = \frac{39 - 23}{6} = \frac{16}{6} = 2\frac{4}{6} = 2\frac{2}{3}$$

LCD is 6

Zenitia worked $2\frac{2}{3}$ hours longer on Tuesday. This result is close to the estimate of 3 hours.

Work Problem 8 at the Side. ▶▶▶

---

8 First, round the numbers and estimate the answer to each problem. Then find the exact answer.

(a) Richard's son grew $3\frac{5}{8}$ inches last year and $2\frac{1}{4}$ inches this year. How much has his height increased over the two years?

Estimate:

Exact:

(b) Ernestine used $2\frac{1}{2}$ packages of chocolate chips in her cookie recipe. Each package has $5\frac{1}{2}$ ounces of chips. How many ounces of chips did she use in the recipe?

Estimate:

Exact:

ANSWERS
8. (a) *Estimate:* $4 + 2 = 6$ inches;
    Exact: $5\frac{7}{8}$ inches

(b) *Estimate:* $3 \cdot 6 = 18$ ounces;
    Exact: $13\frac{3}{4}$ ounces

Focus on Real-Data Applications

Recipes

The side of the corn starch box shown below has useful recipes for Fun-Time Dough and for Great Gravy. Suppose you have made the recipes before and know from experience that 1 pound of Fun-Time Dough is enough for three children.

ARGO CORN STARCH FAVORITE RECIPES

Fun-Time Dough

$1\frac{1}{2}$ cups Argo Corn Starch
$\frac{1}{2}$ cup flour
2 cups water
2 tsp cream of tartar
1 cup salt
1 Tbsp vegetable oil

Mix all ingredients together in saucepan. Cook over medium heat, stirring constantly, until mixture gathers on the stirring spoon and forms dough. This will take about 6 minutes. Dump onto waxed paper until cool enough to handle and knead to form a pliable mass. Store in covered container or plastic bag. Food coloring may be added to make different colors.
Makes about 2 lbs. of Fun-Time Dough.

Great Gravy

3 Tbsp bacon fat or meat drippings
2 Tbsp Argo Corn Starch
$1\frac{1}{2}$ cups water
$\frac{3}{4}$ tsp. salt
$\frac{1}{8}$ tsp. pepper

Blend fat and Argo Corn Starch over low heat until it is a rich brown color, stirring constantly. Gradually add water, salt, and pepper. Heat to boiling over direct heat and then boil gently 2 minutes, stirring constantly.
Makes $1\frac{1}{2}$ cups.

— Satisfaction Guaranteed —

1. If you make the recipe as written, you will have enough Fun-Time Dough for how many children?

2. Suppose you work in a day care center. How many pounds of Fun-Time Dough will you need for nine children?

3. If you double the recipe, you would have 4 pounds of dough. By what number should you multiply each ingredient amount to make 3 pounds of dough?

4. Fill in the blanks with the ingredient amounts needed to make 3 pounds of Fun-Time Dough. Show your work.

   Corn starch: _____   Flour: _____

   Water: _____   Cream of tartar: _____

   Salt: _____   Vegetable oil: _____

You decide to make Great Gravy for Thanksgiving dinner.

5. How much gravy does the recipe make?

6. Suppose you decide to double the recipe. By what factor will you change the ingredient amounts?

7. You did not make enough! Suppose you decide to halve the recipe to make more gravy. Now, by what factor will you change the ingredient amounts?

8. If you need 4 cups of gravy, by what factor must you multiply each ingredient amount? Explain how you determined your answer.

4.5 Exercises

Graph the mixed numbers or improper fractions on the number line. See Example 1.

1. Graph $2\frac{1}{3}$ and $-2\frac{1}{3}$.

2. Graph $1\frac{3}{4}$ and $-1\frac{3}{4}$.

   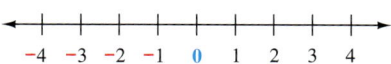

3. Graph $\frac{3}{2}$ and $-\frac{3}{2}$.

   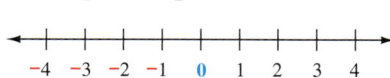

4. Graph $\frac{11}{3}$ and $-\frac{11}{3}$.

Write each mixed number as an improper fraction. See Example 2.

5. $4\frac{1}{2}$
6. $2\frac{1}{4}$
7. $-1\frac{3}{5}$
8. $-1\frac{5}{6}$
9. $2\frac{3}{8}$

10. $3\frac{4}{9}$
11. $-5\frac{7}{10}$
12. $-4\frac{5}{7}$
13. $10\frac{11}{15}$
14. $12\frac{9}{11}$

Write each improper fraction as a mixed number in simplest form. See Example 3.

15. $\frac{13}{3}$
16. $\frac{11}{2}$
17. $-\frac{10}{4}$
18. $-\frac{14}{5}$
19. $\frac{22}{6}$

20. $\frac{28}{8}$
21. $-\frac{51}{9}$
22. $-\frac{44}{10}$
23. $\frac{188}{16}$
24. $\frac{200}{15}$

First, round the mixed numbers to the nearest whole number and estimate each answer. Then find the exact answer. Write exact answers in simplest form. See Examples 4–7.

25. Exact:

    $2\frac{1}{4} \cdot 3\frac{1}{2}$

    Estimate:

    ___ • ___ = ___

26. Exact:

    $\left(1\frac{1}{2}\right)\left(3\frac{3}{4}\right)$

    Estimate:

    ___ • ___ = ___

27. Exact:

    $3\frac{1}{4} \div 2\frac{5}{8}$

    Estimate:

    ___ ÷ ___ = ___

28. Exact:

    $2\frac{1}{4} \div 1\frac{1}{8}$

    Estimate:

    ___ ÷ ___ = ___

29. Exact:

    $3\frac{2}{3} + 1\frac{5}{6}$

    Estimate:

    ___ + ___ = ___

30. Exact:

    $4\frac{4}{5} + 2\frac{1}{3}$

    Estimate:

    ___ + ___ = ___

31. *Exact:*

$$4\frac{1}{4} - \frac{7}{12}$$

Estimate:

____ − ____ = ____

32. *Exact:*

$$10\frac{1}{3} - 6\frac{5}{6}$$

Estimate:

____ − ____ = ____

33. *Exact:*

$$5\frac{2}{3} \div 6$$

Estimate:

____ ÷ ____ = ____

34. *Exact:*

$$1\frac{7}{8} \div 6\frac{1}{4}$$

Estimate:

____ ÷ ____ = ____

35. *Exact:*

$$8 - 1\frac{4}{5}$$

Estimate:

____ − ____ = ____

36. *Exact:*

$$7 - 3\frac{3}{10}$$

Estimate:

____ − ____ = ____

Find the perimeter and the area of each square or rectangle. Write all answers in simplest form.

37.

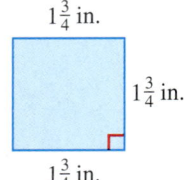

$1\frac{3}{4}$ in.

$1\frac{3}{4}$ in.

$1\frac{3}{4}$ in.

38.

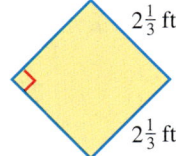

$2\frac{1}{3}$ ft

$2\frac{1}{3}$ ft

39.

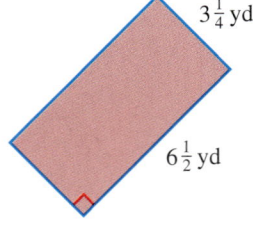

$3\frac{1}{4}$ yd

$6\frac{1}{2}$ yd

40.

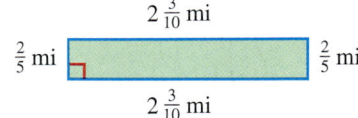

$2\frac{3}{10}$ mi

$\frac{2}{5}$ mi

$\frac{2}{5}$ mi

$2\frac{3}{10}$ mi

First, estimate the answer to each application problem. Then find the exact answer. Write all answers in simplest form. See Example 8.

41. A carpenter has two pieces of oak trim. One piece of trim is $12\frac{1}{2}$ ft long and the other is $8\frac{2}{3}$ ft long. How many feet of oak trim does he have in all?

Estimate:

Exact:

42. On Monday, $5\frac{3}{4}$ tons of cans were recycled, and, on Tuesday, $9\frac{3}{5}$ tons were recycled. How many tons were recycled on these two days?

Estimate:

Exact:

43. The directions for mixing an insect spray say to use $1\frac{3}{4}$ ounces of chemical in each gallon of water. How many ounces of chemical should be mixed with $5\frac{1}{2}$ gallons of water?

Estimate:

Exact:

44. Shirley Cicero wants to make 16 holiday wreaths to sell at the craft fair. Each wreath requires $2\frac{1}{4}$ yd of ribbon. How many yards does she need?

Estimate:

Exact:

45. The Boy Scout troop has volunteered to pick up trash along a 4-mile stretch of highway. So far they have done $1\frac{7}{10}$ miles. How much do they have left to do?

Estimate:

Exact:

46. The gas tank on a Jeep Cherokee has a capacity of $21\frac{3}{8}$ gallons. Scott started with a full tank and then used $8\frac{1}{2}$ gallons of gasoline. Find the number of gallons that remain.

Estimate:

Exact:

47. Suppose that a bridesmaid's floor-length dress requires $3\frac{3}{4}$ yd of material. How much material would be needed to make dresses for 5 bridesmaids?

Estimate:

Exact:

48. A cookie recipe uses $\frac{2}{3}$ cup brown sugar. How much brown sugar is needed to make $2\frac{1}{2}$ times the original recipe?

Estimate:

Exact:

49. A landscaper has $9\frac{5}{8}$ cubic yards of peat moss in a truck. If she unloads $1\frac{1}{2}$ cubic yards at the first stop and 3 cubic yards at the second stop, how much peat moss remains in the truck?

Estimate:

Exact:

50. Marv bought 10 yd of Italian silk fabric. He used $3\frac{7}{8}$ yd to make a jacket. How much fabric is left for other sewing projects?

Estimate:

Exact:

51. Melissa worked $18\frac{3}{4}$ hours over the last five days. If she worked the same amount each day, how long was she at work each day?

Estimate:

Exact:

52. Michael is cutting a $10\frac{1}{2}$ ft board into shelves for a bookcase. Each shelf will be $1\frac{3}{4}$ ft long. How many shelves can he cut?

Estimate:

Exact:

53. Find the length of the arrow shaft.

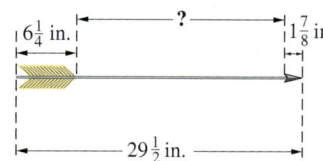

Estimate:

Exact:

54. Find the length of the indented section on this board.

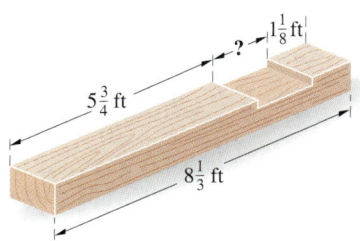

Estimate:

Exact:

First, estimate the answer to each application problem. Then use your calculator to find the exact answer.

55. A craftsperson must attach a lead strip around all four sides of a rectangular stained glass window before it is installed. Find the length of lead stripping needed for the window shown.

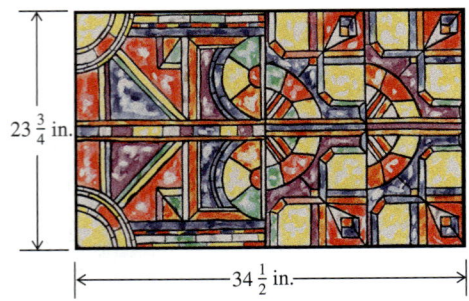

Estimate:

Exact:

56. To complete a custom order, Zak Morten of Home Depot must find the number of inches of brass trim needed to go around the four sides of the lamp base plate shown. Find the length of brass trim needed.

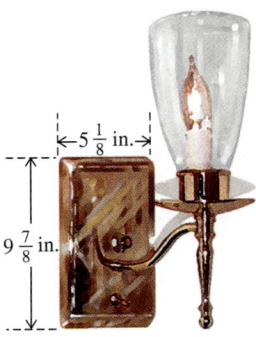

Estimate:

Exact:

57. A fishing boat anchor requires $10\frac{3}{8}$ pounds of steel. Find the number of anchors that can be manufactured with 25,730 pounds of steel.

Estimate:

Exact:

58. Each apartment requires $62\frac{1}{2}$ square yards of carpet. Find the number of apartments that can be carpeted with 6750 square yards of carpet.

Estimate:

Exact:

59. Claire and Deb create custom hat bands that people can put around the crowns of their hats. The finished bands are the lengths shown in the table. The strip of fabric for each band must include the finished length plus an extra $\frac{3}{4}$ in. for the seam.

Band Size	Finished Length
Small	$21\frac{7}{8}$ in.
Medium	$22\frac{5}{8}$ in.
Large	$23\frac{1}{2}$ in.

What length of fabric strip is needed to make 4 small bands, 5 medium bands, and 3 large bands, including the seam allowance?

Estimate:

Exact:

60. Three sides of a parking lot are $108\frac{1}{4}$ ft, $162\frac{3}{8}$ ft, and $143\frac{1}{2}$ ft. If the distance around the lot is $518\frac{3}{4}$ ft, find the length of the fourth side.

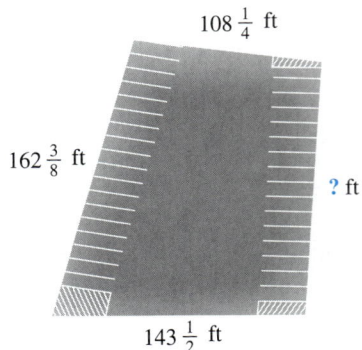

Estimate:

Exact:

Summary Exercises on Fractions

1. Write fractions that represent the shaded and unshaded portion of each figure.

   (a) 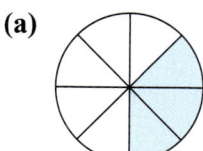 (b)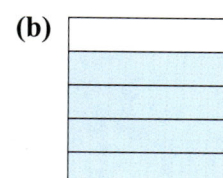

2. Graph $\dfrac{2}{3}$ and $-\dfrac{2}{3}$ on the number line.

   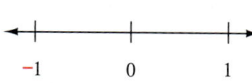

3. Rewrite each fraction with the indicated denominator.

   (a) $-\dfrac{4}{5} = -\dfrac{}{30}$ (b) $\dfrac{2}{7} = \dfrac{}{14}$

4. Simplify.

   (a) $\dfrac{15}{15}$ (b) $-\dfrac{24}{6}$ (c) $\dfrac{9}{1}$

5. Write the prime factorization of each number.

   (a) 72   (b) 105

6. Write each fraction in lowest terms.

   (a) $\dfrac{24}{30}$ (b) $\dfrac{175}{200}$

Simplify.

7. $\left(-\dfrac{3}{4}\right)\left(-\dfrac{2}{3}\right)$

8. $-\dfrac{7}{8} + \dfrac{2}{3}$

9. $\dfrac{7}{16} + \dfrac{5}{8}$

10. $\dfrac{5}{8} \div \dfrac{3}{4}$

11. $\dfrac{2}{3} - \dfrac{4}{5}$

12. $\dfrac{7}{12}\left(-\dfrac{9}{14}\right)$

13. $-21 \div \left(-\dfrac{3}{8}\right)$

14. $\dfrac{7}{8} - \dfrac{5}{12}$

15. $-\dfrac{35}{45} \div \dfrac{10}{15}$

16. $-\dfrac{5}{6} - \dfrac{3}{4}$

17. $\dfrac{7}{12} + \dfrac{5}{6} + \dfrac{2}{3}$

18. $\dfrac{5}{8}$ of 56

First round the numbers and estimate each answer. Then find the exact answer.

19. *Exact:*

    $4\dfrac{3}{4} + 2\dfrac{5}{6}$

    Estimate:

    ___ + ___ = ___

20. *Exact:*

    $2\dfrac{2}{9} \cdot 5\dfrac{1}{7}$

    Estimate:

    ___ • ___ = ___

21. *Exact:*

    $6 - 2\dfrac{7}{10}$

    Estimate:

    ___ − ___ = ___

22. *Exact:*

    $1\dfrac{3}{5} \div 3\dfrac{1}{2}$

    Estimate:

    ___ ÷ ___ = ___

23. *Exact:*

    $4\dfrac{2}{3} \div 1\dfrac{1}{6}$

    Estimate:

    ___ ÷ ___ = ___

24. *Exact:*

    $3\dfrac{5}{12} - \dfrac{3}{4}$

    Estimate:

    ___ − ___ = ___

Solve each application problem. Write all answers in simplest form.

25. When installing cabinets, Cecil Feathers must be certain that the proper type and size of mounting screw is used.

(a) Find the total length of the screw shown.

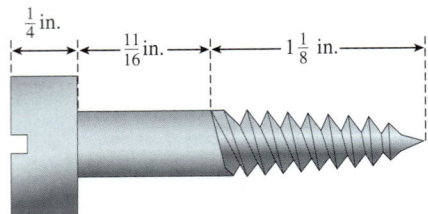

(b) If the screw is put into a board that is $1\frac{3}{4}$ in. thick, how much of the screw will stick out the back of the board?

26. Find the perimeter and the area of this postage stamp.

27. A batch of cookies requires $\frac{3}{4}$ pound of chocolate chips. If you have nine pounds of chocolate chips, how many batches of cookies can you make?

28. The Municipal Utility District says that the cost of operating a hair dryer is $\frac{1}{5}$¢ per minute. Find the cost of operating the hair dryer for a half hour.

29. A survey asked 1500 adults if UFOs (unidentified flying objects) are real. The circle graph shows the fraction of the adults who gave each answer. How many adults gave each answer?

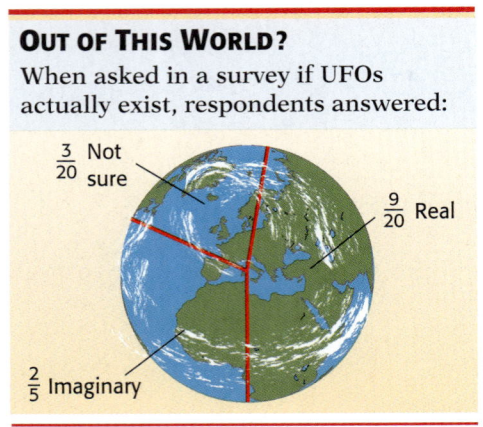

Source: Yankelovich Partners for *Life* magazine.

30. Find the diameter of the hole in the rectangular mounting bracket shown. (The diameter is the distance across the center of the hole.) Then find the perimeter of the bracket.

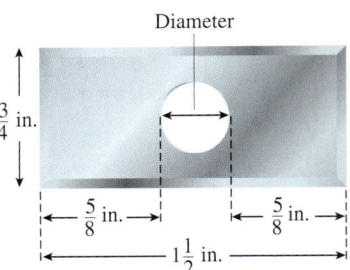

31. A bottle of contact lens daily cleaning solution holds $\frac{2}{3}$ fluid ounce. How many bottles can be filled with $15\frac{1}{3}$ fluid ounces? (*Source:* Alcon Laboratories, Inc.)

32. A home builder bought two parcels of land that were $5\frac{7}{8}$ acres and $10\frac{3}{4}$ acres. She is setting aside $2\frac{7}{8}$ acres for a park and using the rest for $1\frac{1}{4}$ acre home lots. How many lots will be in the development?

4.6 Exponents, Order of Operations, and Complex Fractions

OBJECTIVE 1 Simplify fractions with exponents. We have used exponents as a quick way to write repeated multiplication of integers and variables. Here are two examples as a review.

$$(-3)^2 = \underbrace{(-3)(-3)}_{\text{Two factors of }-3} = 9 \quad \text{and} \quad x^3 = \underbrace{x \cdot x \cdot x}_{\text{Three factors of }x}$$

Base — Exponent

The meaning of an exponent remains the same when a fraction is the base.

EXAMPLE 1 Simplifying Fractions with Exponents

Simplify.

(a) $\left(-\dfrac{1}{2}\right)^3$

The base is $-\dfrac{1}{2}$. The exponent indicates that there are three factors of $-\dfrac{1}{2}$.

$$\left(-\frac{1}{2}\right)^3 = \underbrace{\left(-\frac{1}{2}\right)\left(-\frac{1}{2}\right)\left(-\frac{1}{2}\right)}_{\text{Three factors of }-\frac{1}{2}}$$

Watch the signs carefully. Multiply $\left(-\frac{1}{2}\right)\left(-\frac{1}{2}\right)$ to get $\frac{1}{4}$

$$= \underbrace{\frac{1}{4} \left(-\frac{1}{2}\right)}$$

Now multiply $\frac{1}{4}\left(-\frac{1}{2}\right)$ to get $-\frac{1}{8}$

$$= -\frac{1}{8}$$

The product is negative.

(b) $\left(\dfrac{3}{4}\right)^2 \left(\dfrac{2}{3}\right)^3$

$$\left(\frac{3}{4}\right)^2 \left(\frac{2}{3}\right)^3 = \underbrace{\left(\frac{3}{4}\right)\left(\frac{3}{4}\right)}_{\text{Two factors of }\frac{3}{4}} \underbrace{\left(\frac{2}{3}\right)\left(\frac{2}{3}\right)\left(\frac{2}{3}\right)}_{\text{Three factors of }\frac{2}{3}}$$

$$= \frac{\cancel{3} \cdot \cancel{3} \cdot \cancel{2} \cdot \cancel{2} \cdot \cancel{2}}{\cancel{2} \cdot \cancel{2} \cdot \cancel{2} \cdot 2 \cdot \cancel{3} \cdot \cancel{3} \cdot 3}$$

Divide out all common factors.

$$= \frac{1}{6}$$

Work Problem 1 at the Side.

OBJECTIVE 2 Use the order of operations with fractions. The order of operations that you used in **Section 1.8** for integers also applies to fractions.

OBJECTIVES

1. Simplify fractions with exponents.
2. Use the order of operations with fractions.
3. Simplify complex fractions.

1 Simplify.

(a) $\left(-\dfrac{3}{5}\right)^2$

(b) $\left(\dfrac{1}{3}\right)^4$

(c) $\left(-\dfrac{2}{3}\right)^3 \left(\dfrac{1}{2}\right)^2$

(d) $\left(-\dfrac{1}{2}\right)^2 \left(\dfrac{1}{4}\right)^2$

ANSWERS

1. (a) $\dfrac{9}{25}$ (b) $\dfrac{1}{81}$ (c) $-\dfrac{2}{27}$ (d) $\dfrac{1}{64}$

Chapter 4 Rational Numbers: Positive and Negative Fractions

2 Simplify.

(a) $\dfrac{1}{3} - \dfrac{5}{9}\left(\dfrac{3}{4}\right)$

(b) $-\dfrac{3}{4} + \left(-\dfrac{1}{2}\right)^2 \div \dfrac{2}{3}$

(c) $\dfrac{12}{5} - \dfrac{1}{6}\left(3 - \dfrac{3}{5}\right)$

Order of Operations

Step 1 Work inside *parentheses* or *other grouping symbols*.

Step 2 Simplify expressions with *exponents*.

Step 3 Do the remaining *multiplications and divisions* as they occur from left to right.

Step 4 Do the remaining *additions and subtractions* as they occur from left to right.

EXAMPLE 2 Using the Order of Operations with Fractions

Simplify.

(a) $-\dfrac{1}{3} + \dfrac{1}{2}\left(\dfrac{4}{5}\right)$ There is no work to be done inside the parentheses. There are no exponents, so start with Step 3, multiplying and dividing.

$-\dfrac{1}{3} + \dfrac{1 \cdot 4}{2 \cdot 5}$ Multiply.

$-\dfrac{1}{3} + \dfrac{4}{10}$ Now add. The LCD is 30. $-\dfrac{1}{3} = -\dfrac{10}{30}$ and $\dfrac{4}{10} = \dfrac{12}{30}$

$\dfrac{-10 + 12}{30}$ Add the numerators; keep the common denominator.

$\dfrac{2}{30}$ Write $\dfrac{2}{30}$ in lowest terms: $\dfrac{\cancel{2}}{\cancel{2} \cdot 15} = \dfrac{1}{15}$

$\dfrac{1}{15}$ The answer is in lowest terms.

(b) $-2 + \left(\dfrac{1}{4} - \dfrac{3}{2}\right)^2$ Work inside parentheses. The LCD for $\dfrac{1}{4}$ and $\dfrac{3}{2}$ is 4. Rewrite $\dfrac{3}{2}$ as $\dfrac{6}{4}$ and subtract.

$-2 + \left(\dfrac{1 - 6}{4}\right)^2$

$-2 + \left(\dfrac{-5}{4}\right)^2$ Simplify the term with the exponent. Multiply $\left(-\dfrac{5}{4}\right)\left(-\dfrac{5}{4}\right)$. Signs match, so the product is positive.

$-2 + \left(\dfrac{25}{16}\right)$ Add last. Write -2 as $-\dfrac{2}{1}$

$-\dfrac{2}{1} + \dfrac{25}{16}$ The LCD is 16. Rewrite $-\dfrac{2}{1}$ as $-\dfrac{32}{16}$

$\dfrac{-32 + 25}{16}$ Add the numerators; keep the common denominator.

$-\dfrac{7}{16}$ The answer is in lowest terms.

◀◀◀ Work Problem 2 at the Side.

ANSWERS

2. (a) $-\dfrac{1}{12}$ (b) $-\dfrac{3}{8}$ (c) 2

Section 4.6 Exponents, Order of Operations, and Complex Fractions

OBJECTIVE 3 Simplify complex fractions. We have used both the symbol ÷ and a fraction bar to indicate division. For example,

Indicates division → $\dfrac{6}{2}$ can be written as $6 \div 2$ ← Indicates division

That means we could write $-\frac{4}{5} \div \left(-\frac{3}{10}\right)$ using a fraction bar instead of ÷.

$$-\frac{4}{5} \div \left(-\frac{3}{10}\right) \text{ can be written as } \dfrac{-\frac{4}{5}}{-\frac{3}{10}} \leftarrow \text{Indicates division}$$

The result looks a bit complicated, and its name reflects that fact. We call it a *complex fraction*.

Complex Fractions
A **complex fraction** is a fraction in which the numerator and/or denominator contain one or more fractions.

To simplify a complex fraction, rewrite it in horizontal format using the ÷ symbol for division.

EXAMPLE 3 Simplifying a Complex Fraction

Simplify: $\dfrac{-\frac{4}{5}}{-\frac{3}{10}}$

Rewrite the complex fraction using the ÷ symbol for division. Then follow the steps for dividing fractions.

$$\dfrac{-\frac{4}{5}}{-\frac{3}{10}} = -\frac{4}{5} \div -\frac{3}{10} = -\frac{4}{5} \cdot -\frac{10}{3} = \frac{4 \cdot 2 \cdot \cancel{5}}{\cancel{5} \cdot 3} = \frac{8}{3} \text{ or } 2\frac{2}{3}$$

↑ Reciprocals ↑

The quotient is positive because the numbers in the problem had matching signs (both were negative).

Work Problem 3 at the Side.

---

3 Simplify.

(a) $\dfrac{-\frac{3}{5}}{\frac{9}{10}}$

(b) $\dfrac{6}{\frac{3}{4}}$ ← *Hint:* Write 6 as $\frac{6}{1}$

(c) $\dfrac{-\frac{15}{16}}{-5}$

ANSWERS

3. (a) $-\frac{2}{3}$ (b) 8   (c) $\frac{3}{16}$

Focus on Real-Data Applications

Heart-Rate Training Zone

Performing aerobic exercise is beneficial both for improving aerobic fitness and for burning fat. For best results, you should keep your heart rate within the training zone for a minimum of 12 minutes. If you train at the higher end of the training zone, you will burn glycogen and improve aerobic fitness. Training for longer periods at the lower end of the training zone results in your body using fat reserves for energy.

Example: The training zone (TZ) is based on your heart rate (HR) for one minute. To see if you are in the training zone, measure your heart rate for 15 seconds. Compare it to the 15-second training zone. Find the exact answer, and then round to the nearest whole number.

Instruction	Calculation	Example (age 22)
Calculate maximum heart rate (MHR)	220 − your age	$220 - 22 = 198$
Calculate lower limit of training zone (TZ)	$\frac{3}{5} \times$ (MHR)	$\frac{3}{5} \times (198) = \frac{594}{5} = 118\frac{4}{5}$
Calculate upper limit of training zone (TZ)	$\frac{4}{5} \times$ (MHR)	$\frac{4}{5} \times (198) = \frac{792}{5} = 158\frac{2}{5}$
Calculate the exact 15-second training zone. Round the results to the nearest whole number.	$\left(\frac{1}{4} \times \text{lower TZ}, \frac{1}{4} \times \text{Upper TZ}\right)$	$\frac{1}{4} \times \frac{594}{5} = 29\frac{7}{10}; \frac{1}{4} \times \frac{792}{5} = 39\frac{3}{5}$ $29\frac{7}{10} < \text{HR} < 39\frac{3}{5}$ $30 < \text{HR} < 40$

Age	MHR	Lower Limit of TZ	Upper Limit of TZ	15-Second TZ (exact)	15-Second TZ (rounded)
18					
25					
30					
40					
50					
60					

1. Suppose you work in a physical fitness center and decide to design a poster to remind the clients of the training zone for their age. Compute the exact 15-second training zone for people of each age at the left. Write fractions in lowest terms. Then round the answers to the nearest whole number.

2. Explain why the lower and upper training zones (TZ) are multiplied by $\frac{1}{4}$.

3. Explain why the 15-second training zone is lower for a person aged 50 in comparison to a person aged 20.

4. Based on the chart, what would you tell a 45-year-old person about their 15-second training zone?

Section 4.6 Exponents, Order of Operations, and Complex Fractions

4.6 Exercises

Simplify. See Example 1.

1. $\left(-\dfrac{3}{4}\right)^2$
2. $\left(-\dfrac{4}{5}\right)^2$
3. $\left(\dfrac{2}{5}\right)^3$

4. $\left(\dfrac{1}{4}\right)^3$
5. $\left(-\dfrac{1}{3}\right)^3$
6. $\left(-\dfrac{3}{5}\right)^3$

7. $\left(\dfrac{1}{2}\right)^5$
8. $\left(\dfrac{1}{3}\right)^4$
9. $\left(\dfrac{7}{10}\right)^2$

10. $\left(\dfrac{8}{9}\right)^2$
11. $\left(-\dfrac{6}{5}\right)^2$
12. $\left(-\dfrac{8}{7}\right)^2$

13. $\dfrac{15}{16}\left(\dfrac{4}{5}\right)^3$
14. $-8\left(-\dfrac{3}{8}\right)^2$
15. $\left(\dfrac{1}{3}\right)^4\left(\dfrac{9}{10}\right)^2$

16. $\left(\dfrac{4}{5}\right)^2\left(\dfrac{1}{2}\right)^6$
17. $\left(-\dfrac{3}{2}\right)^3\left(-\dfrac{2}{3}\right)^2$
18. $\left(\dfrac{5}{6}\right)^2\left(-\dfrac{2}{5}\right)^3$

264 Chapter 4 Rational Numbers: Positive and Negative Fractions

RELATING CONCEPTS (EXERCISES 19–20) For Individual or Group Work

Use your knowledge of exponents as you **work Exercises 19 and 20 in order.**

19. (a) Evaluate this series of examples.

$\left(-\dfrac{1}{2}\right)^2 = $ _____ $\left(-\dfrac{1}{2}\right)^6 = $ _____

$\left(-\dfrac{1}{2}\right)^3 = $ _____ $\left(-\dfrac{1}{2}\right)^7 = $ _____

$\left(-\dfrac{1}{2}\right)^4 = $ _____ $\left(-\dfrac{1}{2}\right)^8 = $ _____

$\left(-\dfrac{1}{2}\right)^5 = $ _____ $\left(-\dfrac{1}{2}\right)^9 = $ _____

20. Several drops of ketchup fell on Ron's homework. Explain how he can figure out what real number is covered by each drop. Be careful. More than one number may work, or there may not be any real number that works.

(a) $(\blacksquare)^2 = \dfrac{4}{9}$ (b) $(\blacksquare)^3 = -\dfrac{1}{27}$

(c) $(\blacksquare)^4 = \dfrac{1}{16}$ (d) $(\blacksquare)^2 = -\dfrac{9}{16}$

(e) $(\blacksquare)^2 (\blacksquare)^2 = \dfrac{1}{36}$

(b) Explain the pattern in the sign of the answers.

Simplify. See Example 2.

21. $\dfrac{1}{5} - 6\left(\dfrac{7}{10}\right)$

22. $\dfrac{2}{9} - 4\left(\dfrac{5}{6}\right)$

23. $\left(\dfrac{4}{3} \div \dfrac{8}{3}\right) + \left(-\dfrac{3}{4} \cdot \dfrac{1}{4}\right)$

24. $\left(-\dfrac{1}{3} \cdot \dfrac{3}{5}\right) + \left(\dfrac{3}{4} \div \dfrac{1}{4}\right)$

25. $-\dfrac{3}{10} \div \dfrac{3}{5}\left(-\dfrac{2}{3}\right)$

26. $5 \div \left(-\dfrac{10}{3}\right)\left(-\dfrac{4}{9}\right)$

27. $\dfrac{8}{3}\left(\dfrac{1}{4} - \dfrac{1}{2}\right)^2$

28. $\dfrac{1}{3}\left(\dfrac{4}{5} - \dfrac{3}{10}\right)^3$

29. $-\dfrac{3}{8} + \dfrac{2}{3}\left(-\dfrac{2}{3} + \dfrac{1}{6}\right)$

30. $\dfrac{1}{6} + 4\left(\dfrac{2}{5} - \dfrac{7}{10}\right)$

31. $2\left(\dfrac{1}{3}\right)^3 - \dfrac{2}{9}$

32. $8\left(-\dfrac{3}{4}\right)^2 + \dfrac{3}{2}$

33. $\left(-\dfrac{2}{3}\right)^3\left(\dfrac{1}{8} - \dfrac{1}{2}\right) - \dfrac{2}{3}\left(\dfrac{1}{8}\right)$

34. $\left(\dfrac{3}{5}\right)^2\left(\dfrac{5}{9} - \dfrac{2}{3}\right) \div \left(-\dfrac{1}{5}\right)^2$

35. A square operation key on a calculator is $\frac{3}{8}$ inch on each side. What is the area of the key? Use the formula $A = s^2$. (*Source:* Texas Instruments.)

$\frac{3}{8}$ in.

36. A square lot for sale in the country is $\frac{3}{10}$ mile on a side. Find the area of the lot by using the formula $A = s^2$.

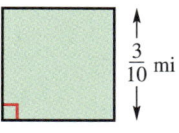

$\frac{3}{10}$ mi

37. A rectangular parking lot at the megamall is $\frac{7}{10}$ mile long and $\frac{1}{4}$ mile wide. How much fencing is needed to enclose the lot? Use the formula $P = 2l + 2w$.

38. A computer chip in a rectangular shape is $\frac{7}{8}$ inch long and $\frac{5}{16}$ inch wide. An insulating strip must be put around all sides of the chip. Find the length of the strip. Use the formula $P = 2l + 2w$.

Simplify. See Example 3.

39. $\dfrac{-\dfrac{7}{9}}{-\dfrac{7}{36}}$

40. $\dfrac{\dfrac{15}{32}}{-\dfrac{5}{64}}$

41. $\dfrac{-15}{\dfrac{6}{5}}$

42. $\dfrac{-6}{-\dfrac{5}{8}}$

43. $\dfrac{\dfrac{4}{7}}{8}$

44. $\dfrac{-\dfrac{11}{5}}{3}$

45. $\dfrac{-\dfrac{2}{3}}{-2\dfrac{2}{5}}$

46. $\dfrac{\dfrac{1}{2}}{3\dfrac{1}{3}}$

47. $\dfrac{-4\dfrac{1}{2}}{\dfrac{3}{4}}$

48. $\dfrac{-1\dfrac{2}{3}}{-\dfrac{4}{9}}$

49. $\dfrac{\left(\dfrac{2}{5}\right)^2}{\left(-\dfrac{4}{3}\right)^2}$

50. $\dfrac{\left(-\dfrac{5}{6}\right)^2}{\left(\dfrac{1}{2}\right)^3}$

4.7 Problem Solving: Equations Containing Fractions

OBJECTIVE 1 **Use the multiplication property of equality to solve equations containing fractions.** In **Section 2.4** you used the division property of equality to solve an equation such as $4s = 24$. The division property says that you may divide both sides of an equation by the same nonzero number and that the equation will still be balanced. Now that you have some experience with fractions, let's look again at how the division property works.

$$4s = 24$$

4s means $4 \cdot s$

$$\frac{4 \cdot s}{4} = \frac{24}{4} \quad \text{Divide both sides by 4}$$

Divide out the common factor of 4

$$\frac{\cancel{4} \cdot s}{\cancel{4}} = 6$$

$$s = 6$$

Because multiplication and division are related to each other, we can also *multiply* both sides of an equation by the same nonzero number and keep it balanced.

OBJECTIVES

1. Use the multiplication property of equality to solve equations containing fractions.
2. Use both the addition and multiplication properties of equality to solve equations containing fractions.
3. Solve application problems using equations containing fractions.

Multiplication and Division Properties of Equality

The **multiplication property of equality** says, if $a = b$, then $a \cdot c = b \cdot c$.

Also, if $a = b$, then $\dfrac{a}{c} = \dfrac{b}{c}$ as long as c is not 0.

This is the **division property of equality.**

In other words, you may multiply or divide both sides of an equation by the same nonzero number and it will still be balanced.

EXAMPLE 1 Using the Multiplication Property of Equality

Solve each equation and check each solution.

(a) $\dfrac{1}{2}b = 5$

As in **Chapter 2**, you want the variable by itself on one side of the equal sign. In this example, you want $1b$, not $\frac{1}{2}b$, on the left side. (Recall that $1b$ is equivalent to b.)

In **Section 4.3** you learned that the product of a number and its reciprocal is 1. Thus, multiplying $\frac{1}{2}$ by $\frac{2}{1}$ will give the desired coefficient of 1.

$$\frac{1}{2}b = 5$$

$$\frac{2}{1}\left(\frac{1}{2}b\right) = \frac{2}{1}(5) \quad \text{Multiply both sides by } \tfrac{2}{1} \text{ (the reciprocal of } \tfrac{1}{2}\text{).}$$

On the left side, use the associative property to regroup the factors.

$$\left(\frac{2}{1} \cdot \frac{1}{2}\right)b = \frac{2}{1}\left(\frac{5}{1}\right) \quad \text{On the right side, 5 is equivalent to } \tfrac{5}{1}$$

$$\left(\frac{\cancel{2}}{1} \cdot \frac{1}{\cancel{2}}\right)b = \frac{10}{1}$$

$1b$ is equivalent to b.

$$1b = 10$$

$$b = 10$$

Continued on Next Page

Once you understand the process, you don't have to show every step. Here is a shorthand solution of the same problem.

$$\frac{1}{2}b = 5$$

$$\frac{\cancel{2}}{1}\left(\frac{1}{\cancel{2}}b\right) = \frac{2}{1}(5)$$

$$b = 10$$

The solution is 10. Check the solution by going back to the *original* equation.

Check $\quad \frac{1}{2}b = 5 \quad$ Replace b with 10 in the original equation.

$\quad\quad\quad \frac{1}{2}(10) = 5 \quad$ Multiply on the left side: $\frac{1}{2}(10)$ is $\frac{1}{2} \cdot \frac{10}{1}$, or $\frac{1}{2} \cdot \frac{2 \cdot 5}{1}$

$$\frac{1 \cdot \cancel{2} \cdot 5}{\cancel{2} \cdot 1} = 5$$

$\quad\quad\quad 5 = 5 \quad$ Balances

When b is 10, the equation balances, so 10 is the correct solution (**not** 5).

(b) $12 = -\frac{3}{4}x$

$\quad\quad 12 = -\frac{3}{4}x \quad$ Multiply both sides by $-\frac{4}{3}$ (the reciprocal of $-\frac{3}{4}$). The reciprocal of a negative number is also negative.

$-\frac{4}{3}(12) = -\frac{\cancel{4}}{\cancel{3}}\left(-\frac{\cancel{3}}{\cancel{4}}x\right) \quad$ On the left side $-\frac{4}{3}(12)$ is $-\frac{4}{3} \cdot \frac{12}{1}$, or $-\frac{4 \cdot \cancel{3} \cdot 4}{\cancel{3} \cdot 1} = -16$

$\quad\quad -16 = x$

The solution is -16. Check the solution by going back to the *original* equation.

Check $\quad 12 = -\frac{3}{4}x \quad$ Replace x with -16 in the original equation.

$\quad\quad 12 = -\frac{3}{4}(-16) \quad$ The product of two negative numbers is positive.

$\quad\quad\quad\quad\quad\quad\quad\quad -\frac{3}{4}(-16)$ is $\frac{3}{4} \cdot \frac{16}{1}$, or $\frac{3}{4} \cdot \frac{4 \cdot 4}{1}$

$$12 = \frac{3 \cdot \cancel{4} \cdot 4}{\cancel{4} \cdot 1}$$

$\quad\quad 12 = 12 \quad$ Balances

When x is -16, the equation balances, so -16 is the correct solution (**not** 12).

Continued on Next Page

Section 4.7 Problem Solving: Equations Containing Fractions   **269**

(c) $-\dfrac{2}{5}n = -\dfrac{1}{3}$

$$-\dfrac{\cancel{5}}{\cancel{2}}\left(-\dfrac{\cancel{2}}{\cancel{5}}n\right) = \left(-\dfrac{5}{2}\right)\left(-\dfrac{1}{3}\right)$$ Multiply both sides by $-\dfrac{5}{2}$ (the reciprocal of $-\dfrac{2}{5}$).

$$n = \dfrac{5 \cdot 1}{2 \cdot 3}$$ The product of two negative numbers is positive.

$$n = \dfrac{5}{6}$$

Check $-\dfrac{2}{5}n = -\dfrac{1}{3}$ Original equation

$$\left(-\dfrac{2}{5}\right)\left(\dfrac{5}{6}\right) = -\dfrac{1}{3}$$ Replace n with $\dfrac{5}{6}$

$$-\dfrac{\cancel{2} \cdot \cancel{5}}{\cancel{5} \cdot \cancel{2} \cdot 3} = -\dfrac{1}{3}$$ Multiply on the left side.

$$-\dfrac{1}{3} = -\dfrac{1}{3}$$ Balances

When n is $\dfrac{5}{6}$, the equation balances, so $\dfrac{5}{6}$ is the correct solution (**not** $-\dfrac{1}{3}$).

▶ **Work Problem 1 at the Side.** ▶▶▶

OBJECTIVE 2 Use both the addition and multiplication properties of equality to solve equations containing fractions.   In **Section 2.5** you used both the addition and *division* properties of equality to solve equations. Now you can use both the addition and *multiplication* properties.

EXAMPLE 2 Using the Addition and Multiplication Properties of Equality

Solve each equation and check each solution.

(a) $\dfrac{1}{3}c + 5 = 7$

The first step is to get the variable term $\dfrac{1}{3}c$ by itself on the left side of the equal sign. Recall that to "get rid of" the 5 on the left side, add the opposite of 5, which is -5, to both sides.

$$\dfrac{1}{3}c + 5 = 7$$
$$\underline{\phantom{\dfrac{1}{3}c + 0 =\ } -5 = -5}\quad \text{Add } -5 \text{ to both sides.}$$
$$\dfrac{1}{3}c + 0 = 2$$
$$\dfrac{1}{3}c = 2$$

$$\dfrac{\cancel{3}}{1}\left(\dfrac{1}{\cancel{3}}c\right) = \dfrac{3}{1}\left(\dfrac{2}{1}\right)$$ Multiply both sides by $\dfrac{3}{1}$ (the reciprocal of $\dfrac{1}{3}$).

$$c = 6$$

The solution is 6. Check the solution by going back to the *original* equation.

Continued on Next Page

---

1 Solve each equation. Check each solution.

(a) $\dfrac{1}{6}m = 3$ Check

(b) $\dfrac{3}{2}a = -9$ Check

(c) $\dfrac{3}{14} = -\dfrac{2}{7}x$ Check

ANSWERS

1. (a) $m = 18$ Check $\dfrac{1}{6}m = 3$
  $\dfrac{1}{6}(18) = 3$
  Balances $3 = 3$

 (b) $a = -6$ Check $\dfrac{3}{2}a = -9$
  $\dfrac{3}{2}(-6) = -9$
  Balances $-9 = -9$

 (c) $x = -\dfrac{3}{4}$ Check $\dfrac{3}{14} = -\dfrac{2}{7}x$
  $\dfrac{3}{14} = -\dfrac{2}{7}\left(-\dfrac{3}{4}\right)$
  Balances $\dfrac{3}{14} = \dfrac{3}{14}$

Chapter 4 Rational Numbers: Positive and Negative Fractions

❷ Solve each equation. Check each solution.

(a) $18 = \frac{4}{5}x + 2$ Check

(b) $\frac{1}{4}h - 5 = 1$ Check

(c) $\frac{4}{3}r + 4 = -8$ Check

Check $\frac{1}{3}c + 5 = 7$ Original equation

$\frac{1}{3}(6) + 5 = 7$ Replace c with 6.

$2 + 5 = 7$

$7 = 7$ Balances

When c is 6, the equation balances, so 6 is the correct solution (**not** 7).

(b) $-3 = \frac{2}{3}y + 7$

To get the variable term $\frac{2}{3}y$ by itself on the right side, add -7 to both sides.

$$-3 = \frac{2}{3}y + 7$$

$$\underline{-7} \qquad \underline{-7}$$ Add -7 to both sides.

$$-10 = \frac{2}{3}y + 0$$

$$\frac{3}{2}(-10) = \frac{\cancel{3}}{\cancel{2}}\left(\frac{\cancel{2}}{\cancel{3}}y\right)$$ Multiply both sides by $\frac{3}{2}$ (the reciprocal of $\frac{2}{3}$).

$$-15 = y$$

Check $-3 = \frac{2}{3}y + 7$ Original equation

$-3 = \frac{2}{3}(-15) + 7$ Replace y with -15

$-3 = -10 + 7$

$-3 = -3$ Balances

When y is -15, the equation balances, so -15 is the correct solution (**not** -3).

◀◀◀ Work Problem ❷ at the Side.

OBJECTIVE ❸ Solve application problems using equations containing fractions. Use the six problem-solving steps from **Section 3.3** to solve application problems.

ANSWERS

2. (a) $x = 20$ Check $18 = \frac{4}{5}x + 2$
$18 = \frac{4}{5}(20) + 2$
$18 = 16 + 2$
Balances $18 = 18$

(b) $h = 24$ Check $\frac{1}{4}h - 5 = 1$
$\frac{1}{4}(24) - 5 = 1$
$6 - 5 = 1$
Balances $1 = 1$

(c) $r = -9$ Check $\frac{4}{3}r + 4 = -8$
$\frac{4}{3}(-9) + 4 = -8$
$-12 + 4 = -8$
Balances $-8 = -8$

EXAMPLE 3 Solving an Application Problem Using an Equation with Fractions

The expression for finding a person's approximate systolic blood pressure is $100 + \frac{age}{2}$. Suppose your friend's systolic blood pressure is 116 (and he has normal blood pressure). Find his age.

Step 1 **Read** the problem. It is about blood pressure and age.

Unknown: friend's age

Known: Blood pressure expression is $100 + \frac{age}{2}$; friend's pressure is 116.

Step 2 **Assign a variable.** Let a represent the friend's age.

Step 3 **Write an equation.**

$$100 + \frac{age}{2} \text{ is blood pressure}$$

$$100 + \frac{a}{2} = 116$$

Step 4 **Solve.**

$$100 + \frac{a}{2} = 116$$

$$\underline{-100 \qquad\qquad -100} \qquad \text{Add } -100 \text{ to both sides.}$$

$$0 + \frac{a}{2} = 16$$

$\frac{1}{2}a$ is equivalent to $\frac{a}{2}$ because $\frac{1}{2}a$ is $\frac{1}{2} \cdot \frac{a}{1}$

$$\frac{a}{2} = 16$$

$$\frac{1}{2}a = 16$$

$$\frac{\cancel{2}}{1}\left(\frac{1}{\cancel{2}}a\right) = \frac{2}{1}(16) \qquad \text{Multiply both sides by } \frac{2}{1} \text{ (the reciprocal of } \frac{1}{2}\text{).}$$

$$a = 32$$

Step 5 **State the answer.** Your friend is 32 years old.

Step 6 **Check** the solution by putting it back into the *original* problem.

Approximate systolic blood pressure is $100 + \frac{age}{2}$.

If the age is 32, then $100 + \frac{32}{2} = 100 + 16 = 116$.

Friend's blood pressure is 116. ← Matches

Age 32 is the correct solution because it "works" when you put it back into the original problem.

Work Problem 3 at the Side.

---

3 A woman's systolic blood pressure is 111. Find her age, using the expression for systolic blood pressure from Example 3 and the six problem-solving steps. (Assume that the woman has normal blood pressure.)

ANSWERS

3. Age is a.

$100 + \frac{a}{2} = 111$ The woman is 22 years old.

Check $100 + \frac{22}{2} = 100 + 11 = 111$

Focus on Real-Data Applications

Hotel Expenses

Mathematics teachers attending conferences in New Orleans, Louisiana, and Philadelphia, Pennsylvania, found the following information about hotel rates on their organization's Internet Web site.

Hotel	Single	Double	Triple	Quad	Suites
New Orleans (November 2004)					
Marriott	$116	$121	$121	$121	$603
Sheraton (Club)	$157	$169	$194	$219	$598
Philadelphia (April 2004)					
Windsor	$149	$169	$189	$209	—
Courtyard	$190	$210	$230	$250	—
Park Hyatt	$216	$241	$266	$291	—

Source: NCTM.

1. The double rate is for two people sharing a room. What fractional part does each person pay?

2. (a) Multiply the double rate at the Windsor in Philadelphia by $\frac{1}{2}$. What is the result?

   (b) Divide the double rate at the Windsor in Philadelphia by 2. What is the result?

   (c) Explain what happened. How much money would one person owe if he or she shared a double room at the Windsor in Philadelphia?

3. The triple rate is for three people sharing a room. What fractional part does each person pay?

4. How much money would one person owe if he or she shared a triple room at the Park Hyatt in Philadelphia? How much money would each person save if they could book a triple room at the Windsor instead of the Park Hyatt? Find your answer using two different methods, based on your observations in Problem 2.

5. The quad rate is for four people sharing a room. What fractional part does each person pay?

6. How much money would one person owe if he or she shared a quad room at the Sheraton in New Orleans? Find your answer using two different methods, based on your observations in Problem 2.

7. How many people would have to share a suite at the Sheraton in New Orleans for the cost per person to be less than sharing a quad room at the same hotel? Round the answer to the nearest whole number. (*Hint:* Estimate the cost per person for a quad room first. Then estimate the number of people needed to share the cost of the suite. Check your work using actual values.)

8. Suppose you have a travel allotment of $500 that can be spent on transportation, hotel, food, and registration fees. You and a colleague decide to attend the 3-day New Orleans conference and plan to share a room at the Marriott. Registration costs $150; the flight costs $129 round-trip; the taxi ride between the airport and the hotel costs $20 per person each way; and you budget $35 per day for meals. How much out-of-pocket expense will you have to pay? How much would you save if you could recruit a third person to share the room?

4.7 Exercises

Solve each equation and check each solution. See Examples 1 and 2.

1. $\frac{1}{3}a = 10$ Check

2. $7 = \frac{1}{5}y$ Check

3. $-20 = \frac{5}{6}b$ Check

4. $-\frac{4}{9}w = 16$ Check

5. $-\frac{7}{2}c = -21$ Check

6. $-25 = \frac{5}{3}x$ Check

7. $\frac{9}{16} = \frac{3}{4}m$ Check

8. $\frac{5}{12}k = \frac{15}{16}$ Check

9. $\frac{3}{10} = -\frac{1}{4}d$ Check

10. $-\frac{7}{8}h = -\frac{1}{6}$ Check

11. $\frac{1}{6}n + 7 = 9$ Check

12. $3 + \frac{1}{4}p = 5$ Check

13. $-10 = \dfrac{5}{3}r + 5$ Check

14. $0 = 6 + \dfrac{3}{2}t$ Check

15. $\dfrac{3}{8}x - 9 = 0$ Check

16. $\dfrac{1}{3}s - 10 = -5$ Check

Solve each equation. Show your work.

17. $7 - 2 = \dfrac{1}{5}y - 4$

18. $0 - 8 = \dfrac{1}{10}k - 3$

19. $4 + \dfrac{2}{3}n = -10 + 2$

20. $-\dfrac{2}{5}m - 3 = -9 + 0$

21. $3x + \dfrac{1}{2} = \dfrac{3}{4}$

22. $4y + \dfrac{1}{3} = \dfrac{7}{9}$

23. $\dfrac{3}{10} = -4b - \dfrac{1}{5}$

24. $\dfrac{5}{6} = -3c - \dfrac{2}{3}$

25. Check the solution given for each equation. If a solution doesn't check, show how to find the correct solution.

   (a) $\frac{1}{6}x + 1 = -2$

   $x = 18$

   (b) $-\frac{3}{2} = \frac{9}{4}k$

   $k = -\frac{2}{3}$

26. Check the solution given for each equation. If a solution doesn't check, show how to find the correct solution.

   (a) $-\frac{3}{4}y = -\frac{5}{8}$

   $y = \frac{5}{6}$

   (b) $16 = -\frac{7}{3}w + 2$

   $w = 6$

27. Write two different equations that have 8 as a solution. Write your equations with a fraction as the coefficient of the variable term. Use Exercises 1–10 as models.

28. Write two different equations that have -12 as the solution. Write your equations with a fraction as the coefficient of the variable term. Use Exercises 1–10 as models.

In Exercises 29–32, find each person's age using the six problem-solving steps and this expression for approximate systolic blood pressure: $100 + \frac{age}{2}$. Assume that all the people have normal blood pressure. See Example 3.

29. A man has systolic blood pressure of 109. How old is he?

30. A man has systolic blood pressure of 118. How old is he?

31. A woman has systolic blood pressure of 122. How old is she?

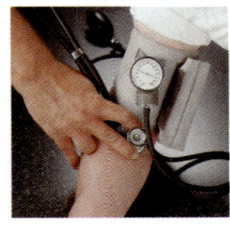

32. A woman has systolic blood pressure of 113. How old is she?

As you read at the start of this chapter, nails for home repair projects are classified by the "penny" system, which is a number that indicates the nail's length. The expression for finding the length of a nail, in inches, is $\frac{\text{penny size}}{4} + \frac{1}{2}$ inch. In Exercises 33–36, find the penny size for each nail using this expression and the six problem-solving steps. (Source: Season by Season Home Maintenance.)

33. The length of a common nail is 3 inches. What is its penny size?

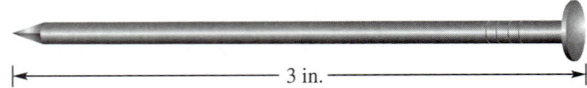

34. The length of a drywall nail is 2 inches. Find the nail's penny size.

35. The length of a box nail is $2\frac{1}{2}$ inches. What penny size would you ask for when buying these nails?

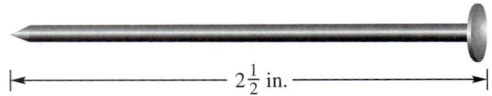

36. The length of a finishing nail is $1\frac{1}{2}$ inches. What is its penny size?

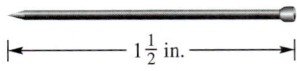

An expression for the recommended weight of an adult is $\frac{11}{2}$ (height in inches) − 220. In Exercises 37–40, find each person's height using this expression and the six problem-solving steps. Assume that all the people are at their recommended weight.

37. A man weighs 209 pounds. What is his height in inches?

38. A woman weighs 110 pounds. What is her height in inches?

39. A woman weighs 132 pounds. What is her height in inches?

40. A man weighs 176 pounds. What is his height in inches?

4.8 Geometry Applications: Area and Volume

OBJECTIVES

1. Find the area of a triangle.
2. Find the volume of a rectangular solid.
3. Find the volume of a pyramid.

OBJECTIVE 1 Find the area of a triangle. In Section 3.1 you worked with triangles, which are flat shapes that have exactly three sides. You found the perimeter of a triangle by adding the lengths of the three sides. Now you are ready to find the area of a triangle (the amount of surface inside the triangle).

You can find the *height* of a triangle by measuring the distance from one vertex of the triangle to the opposite side (the base). The height must be *perpendicular* to the base, that is, it must form a right angle with the base. Sometimes you have to extend the base before you can draw the height perpendicular to it, as shown on the right-hand figure below.

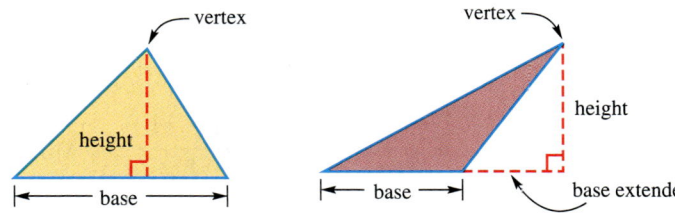

If you cut out two identical triangles and turn one upside down, you can fit them together to form a parallelogram.

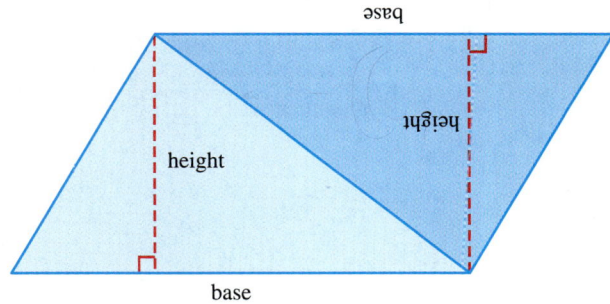

Recall from **Section 3.2** that the area of the parallelogram is *base* times *height*. Because each triangle is *half* of the parallelogram, the area of one triangle is

$\frac{1}{2}$ of base times height.

Use the following formula to find the area of a triangle.

Finding the Area of a Triangle

$$\text{Area of triangle} = \frac{1}{2} \cdot \text{base} \cdot \text{height}$$

$$A = \frac{1}{2}bh$$

Remember to use *square units* when measuring area.

1 Find the area of each triangle.

(a)
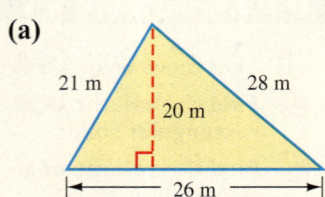

(b)
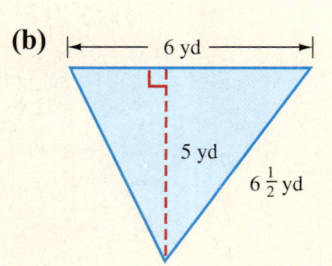

(c)
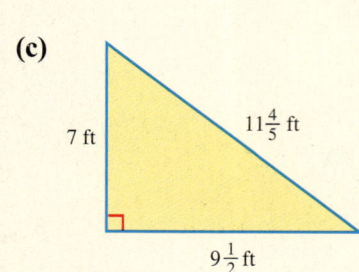

ANSWERS
1. (a) 260 m² (b) 15 yd²
(c) $\frac{133}{4}$ ft² or $33\frac{1}{4}$ ft²

EXAMPLE 1 Finding the Area of Triangles

Find the area of each triangle.

(a) The base is 47 ft and the height is 22 ft. You do *not* need the 26 ft or $39\frac{3}{4}$ ft sides to find the area.

$A = \frac{1}{2} \cdot b \cdot h$ Replace b with 47 ft and h with 22 ft.

$A = \frac{1}{2} \cdot 47 \text{ ft} \cdot 22 \text{ ft}$

$A = \frac{1}{2} \cdot \frac{47 \text{ ft}}{1} \cdot \frac{22 \text{ ft}}{1}$

$A = \frac{1 \cdot 47 \text{ ft} \cdot \cancel{2} \cdot 11 \text{ ft}}{\cancel{2} \cdot 1 \cdot 1}$ Divide out the common factor of 2. Multiply 47 · 11 to get 517

$A = 517 \text{ ft}^2$ Multiply ft · ft to get ft².

(b) Because two sides of the triangle are perpendicular to each other, use those sides as the base and the height. (Remember that the height must be perpendicular to the base.)

$A = \frac{1}{2} bh$ Formula for area of a triangle

$A = \frac{1}{2} \cdot 9 \text{ in.} \cdot 6\frac{1}{2} \text{ in.}$ Replace b with 9 in. and h with $6\frac{1}{2}$ in.

$A = \frac{1}{2} \cdot \frac{9 \text{ in.}}{1} \cdot \frac{13 \text{ in.}}{2}$ Write 9 in. and $6\frac{1}{2}$ in. as improper fractions.

$A = \frac{1 \cdot 9 \text{ in.} \cdot 13 \text{ in.}}{2 \cdot 1 \cdot 2}$ ← Multiply 9 · 13 to get 117 and in. · in. to get in.²

$A = \frac{117}{4} \text{ in.}^2$ or $29\frac{1}{4} \text{ in.}^2$

Work Problem 1 at the Side.

EXAMPLE 2 Using the Concept of Area

Find the area of the shaded part in this figure.

The *entire* figure is a rectangle.

$A = lw$

$A = 30 \text{ cm} \cdot 40 \text{ cm} = 1200 \text{ cm}^2$

Continued on Next Page

The *un*shaded part is a triangle.

$$A = \frac{1}{2} bh$$

$$A = \frac{1}{2} \cdot \frac{30 \text{ cm}}{1} \cdot \frac{32 \text{ cm}}{1}$$

$$A = \frac{1 \cdot \overset{1}{\cancel{2}} \cdot 15 \text{ cm} \cdot 32 \text{ cm}}{\underset{1}{\cancel{2}} \cdot 1 \cdot 1}$$

$$A = 480 \text{ cm}^2$$

Subtract to find the area of the shaded part.

$$A = \overbrace{1200 \text{ cm}^2}^{\text{Entire area}} - \overbrace{480 \text{ cm}^2}^{\text{Unshaded part}} = \overbrace{720 \text{ cm}^2}^{\text{Shaded part}}$$

Work Problem 2 at the Side. ▶▶▶

OBJECTIVE 2 Find the volume of a rectangular solid. A shoe box and a cereal box are examples of three-dimensional (or solid) figures. The three dimensions are length, width, and height. (A rectangle or square is a two-dimensional figure. The two dimensions are length and width.) If we want to know how much a shoe box will hold, we find its *volume*. We measure volume by seeing how many cubes of a certain size will fill the space inside the box. Three sizes of *cubic units* are shown here. Notice that all the edges of a cube have the same length and all sides meet at right angles.

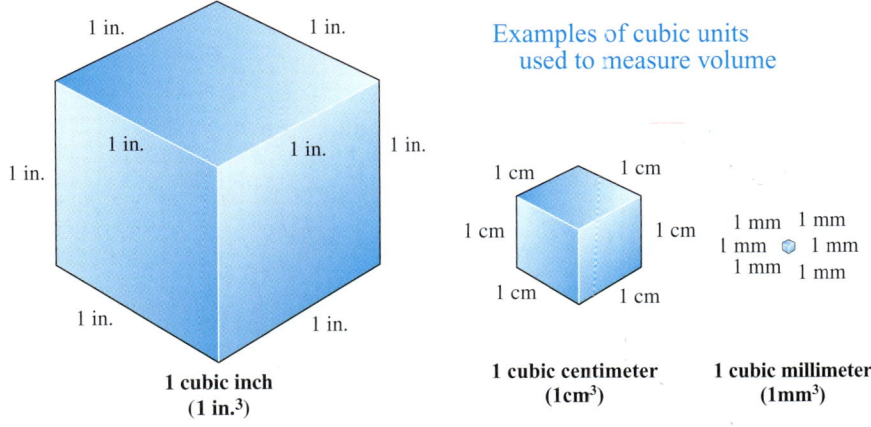

Examples of cubic units used to measure volume

Some other sizes of cubes that are used to measure volume are 1 cubic foot (1 ft^3), 1 cubic yard (1 yd^3), and 1 cubic meter (1 m^3).

> **CAUTION**
> The raised 3 in 4^3 means that you multiply 4 • 4 • 4 to get 64. The raised 3 in cm^3 or ft^3 is a short way to write the word "cubic." It means that you multiplied cm times cm times cm to get cm^3, or ft times ft times ft to get ft^3. Recall that a short way to write $x \cdot x \cdot x$ is x^3. Similarly, cm • cm • cm is cm^3. When you see 5 cm^3, say "five cubic centimeters." Do *not* multiply 5 • 5 • 5 because the exponent applies only to the *first* thing to its left. The exponent applies to cm, *not* to the 5.

Volume

Volume is a measure of the space inside a solid shape. The volume of a solid is the number of cubic units it takes to fill the solid.

---

② Find the area of the shaded part in this figure.

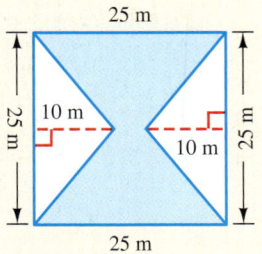

ANSWER
2. $A = 625 \text{ m}^2 - 125 \text{ m}^2 - 125 \text{ m}^2 = 375 \text{ m}^2$

3 Find the volume of each rectangular solid.

(a)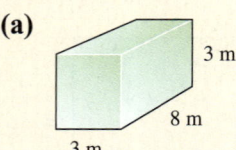

(b) Length $6\frac{1}{4}$ ft, width $3\frac{1}{2}$ ft, height 2 ft

Use the formula below to find the volume of *rectangular solids* (box-like shapes).

> **Finding the Volume of Rectangular Solids**
> Volume of a rectangular solid = length • width • height
> $$V = lwh$$
> Remember to use *cubic units* when measuring volume.

EXAMPLE 3 Finding the Volume of Rectangular Solids

Find the volume of each box.

(a)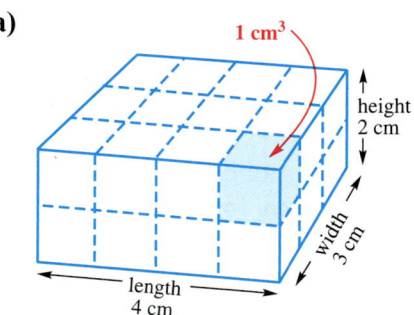

Each cube that fits in the box is 1 cubic centimeter (1 cm³). To find the volume, you can count the number of cubes.

Bottom layer has 12 cubes.
Top layer has 12 cubes.
} Total of 24 cubes (24 cm³)

Or you can use the formula for rectangular solids.

$V = l \cdot w \cdot h$
$V = 4 \text{ cm} \cdot 3 \text{ cm} \cdot 2 \text{ cm}$ Multiply 4 • 3 • 2 to get 24
$V = 24 \text{ cm}^3$ Multiply cm • cm • cm to get cm³.

(b)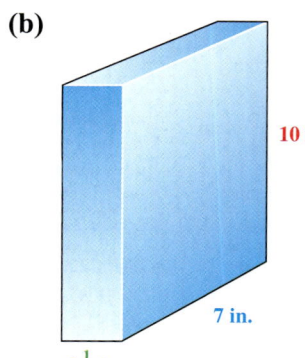

Use the formula $V = lwh$.

$V = 7 \text{ in.} \cdot 2\frac{1}{2} \text{ in.} \cdot 10 \text{ in.}$ Write each measurement as an improper fraction.

$V = \frac{7 \text{ in.}}{1} \cdot \frac{5 \text{ in.}}{2} \cdot \frac{10 \text{ in.}}{1}$

$V = \frac{7 \text{ in.} \cdot 5 \text{ in.} \cdot \overset{1}{\cancel{2}} \cdot 5 \text{ in.}}{1 \cdot \underset{1}{\cancel{2}} \cdot 1}$ Divide out the common factor of 2

$V = 175 \text{ in.}^3$ Cubic units for volume

Work Problem 3 at the Side.

OBJECTIVE 3 Find the volume of a pyramid. A pyramid is a solid shape like the one shown below. We will study pyramids with square or rectangular bases.

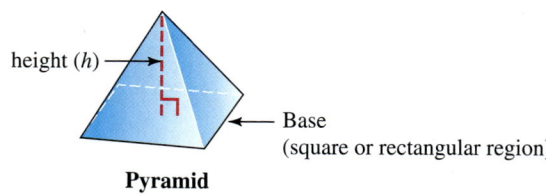

Pyramid

The height is the distance from the base to the highest point of the pyramid. The height must be perpendicular to the base.

ANSWERS

3. (a) 72 m³ (b) $\frac{175}{4}$ ft³ or $43\frac{3}{4}$ ft³

NOTE
In this book we will work only with pyramids that have a base with four sides that is square or rectangular. In later math courses you may work with pyramids that have a base with three sides (triangle), five sides (pentagon), six sides (hexagon), and so on.

Use this formula to find the volume of a pyramid.

Finding the Volume of a Pyramid

$$\text{Volume of a pyramid} = \frac{1}{3} \cdot B \cdot h$$

$$V = \frac{1}{3} Bh$$

where B is the area of the base of the pyramid and h is the height of the pyramid.

Remember to use *cubic units* when measuring volume.

EXAMPLE 4 Finding the Volume of a Pyramid

Find the volume of this pyramid with a rectangular base.

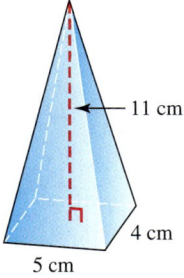

First find the value of B in the formula, which is the *area of the rectangular base*. Recall that the area of a rectangle is found by multiplying length times width.

$$B = 5 \text{ cm} \cdot 4 \text{ cm}$$
$$\mathbf{B = 20 \text{ cm}^2}$$

Next, find the volume.

$V = \dfrac{1}{3} Bh$ Formula for volume of pyramid

$V = \dfrac{1}{3} \cdot 20 \text{ cm}^2 \cdot 11 \text{ cm}$ Replace B with 20 cm² and h with 11 cm.

$V = \dfrac{1}{3} \cdot \dfrac{20 \text{ cm}^2}{1} \cdot \dfrac{11 \text{ cm}}{1}$

$V = \dfrac{1 \cdot 20 \text{ cm}^2 \cdot 11 \text{ cm}}{3 \cdot 1 \cdot 1}$ There are no common factors to divide out.

$V = \dfrac{220}{3} \text{ cm}^3 \quad \text{or} \quad 73\dfrac{1}{3} \text{ cm}^3$ Cubic units for volume

Work Problem 4 at the Side.

④ Find the volume of a pyramid with a square base measuring 10 ft by 10 ft. The height of the pyramid is 6 ft.

ANSWERS
4. $V = 200 \text{ ft}^3$

Focus on Real-Data Applications

Quilt Patterns

People who make quilts often base their designs on a block that is cut into a grid of 4, 9, 16, or 25 squares. The quilter chooses various colors for the pieces. Each quilt design shown below was selected from an Archive of American Quilt Designs.

1. Identify the makeup of the block as 4, 9, 16, etc. Each color is what fractional part of the block?

Aunt Eliza's Star

Block size: _____

Tan: _____ Purple: _____

Green: _____

Barbara Fritchie Star

Block size: _____

Blue: _____ White: _____

Cobwebs

Block size: _____

Mauve: _____ Purple: _____

Yellow: _____

Handy Andy

Block size: _____

Blue: _____ White: _____

Peace and Plenty

Block size: _____

Blue: _____ Yellow: _____

Whitish: _____

Prairie Queen

Block size: _____

Blue: _____ Gray: _____

Brown: _____ White: _____

2. Use the blocks to design and color your own quilt patterns. Tell the fractional part of the block that is represented by each color.

3. Find the next two numbers in this pattern: 4, 9, 16, 25, _____, _____.

4. Explain how the pattern works.

4.8 Exercises

Find the perimeter and area of each triangle. See Example 1.

1.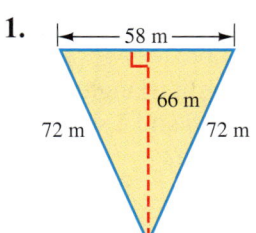
2.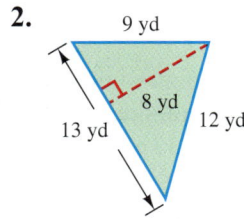
3.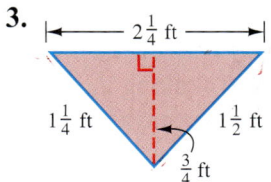
4.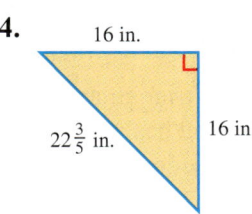

5.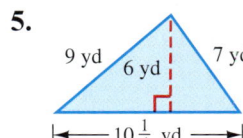
6.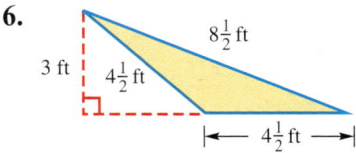
7.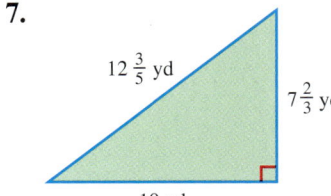
8.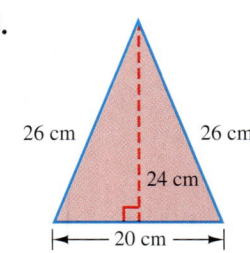

Find the shaded area in each figure. See Example 2.

9.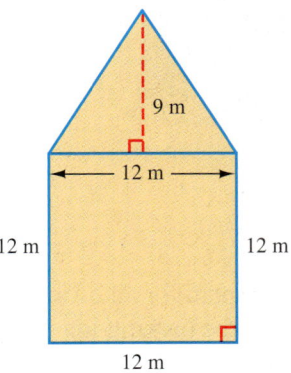

10.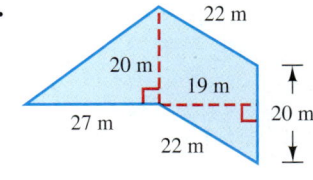

11.

12.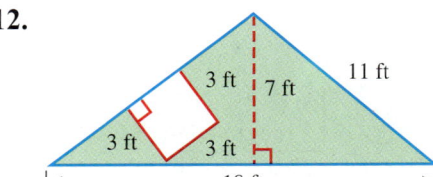

13. Explain the difference between perimeter, area, and volume.

14. Explain where the $\frac{1}{2}$ comes from in the formula for area of a triangle. Draw a figure to illustrate your explanation.

Solve each application problem.

15. A triangular tent flap measures $3\frac{1}{2}$ ft along the base and has a height of $4\frac{1}{2}$ ft. How much canvas is needed to make the flap?

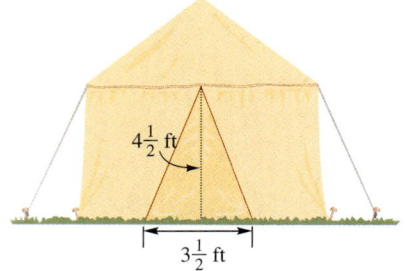

16. A wooden sign in the shape of a right triangle has perpendicular sides measuring $1\frac{1}{2}$ yd and 1 yd. How much surface area does the sign have?

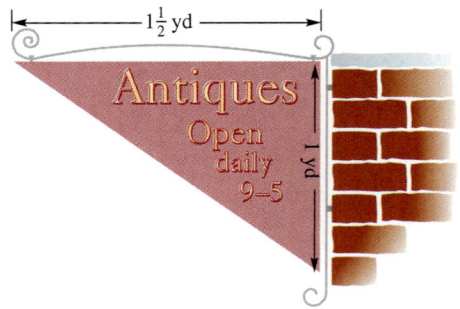

17. A triangular space between three streets has the measurements shown below. How much new curbing will be needed to go around the space? How much sod will be needed to cover the space?

18. Each gable end of a new house has a span of 36 ft and a rise of $9\frac{1}{2}$ ft. What is the total area of both gable ends of the house?

19. A city lot with an unusual shape is shown below.
    (a) How much frontage (distance along streets) does the lot have?
    (b) What is the area of the lot?

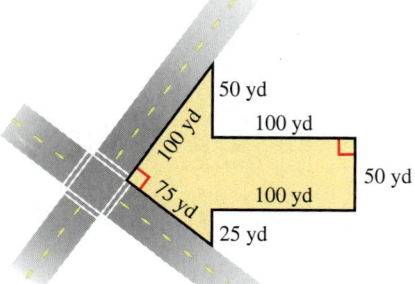

20. The sketch below shows the plan for an office building. The shaded part will be the parking lot. Find the area of the parking lot.

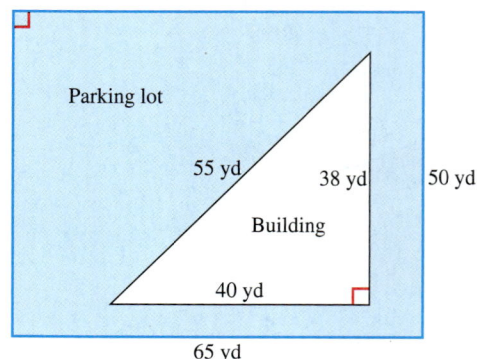

Section 4.8 Geometry Applications: Area and Volume 285

Name each solid and find its volume. See Examples 3 and 4.

21.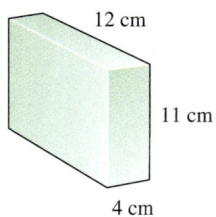

22.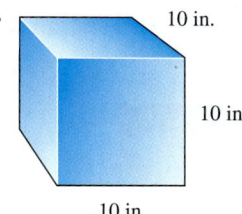

23.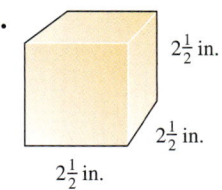

24.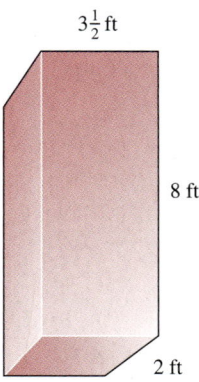

25.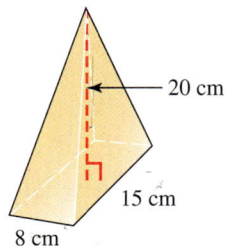

26.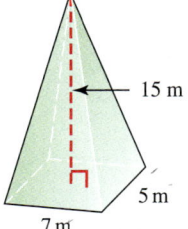

27. A box to hold pencils measures 3 in. by 8 in. by $\frac{3}{4}$ in. high. Find the volume of the box. (*Source:* Faber Castell.)

28. A train is being loaded with shipping crates. Each crate is 12 ft long, 8 ft wide, and $2\frac{1}{4}$ ft high. How much space will each crate take?

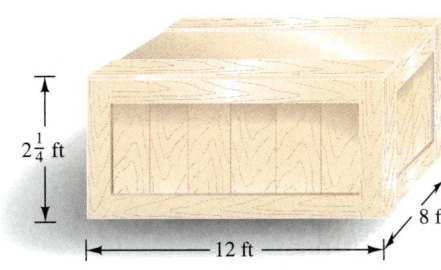

29. One of the ancient stone pyramids in Egypt has a square base that measures 145 m on each side. The height is 93 m. What is the volume of the pyramid? (*Source: Columbia Encyclopedia.*)

30. A cardboard model of an ancient stone pyramid has a square base that is $10\frac{3}{8}$ in. on each side. The height is $6\frac{1}{2}$ in. Find the volume of the model.

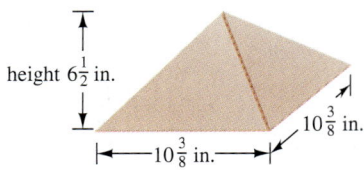

31. Find the volume.

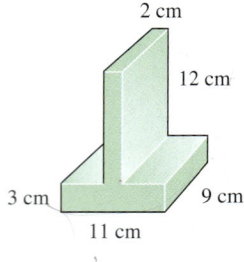

32. Find the volume. (*Hint:* Notice the square hole that goes through the center of the rectangular solid.)

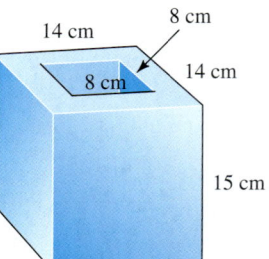

The following are some answers from a student's test. The number part of each answer is correct, but some of the units are not. Find the errors, explain what is wrong, and correct the errors.

33. There are *two* errors in this list of answers.

$$A = 135^2 \text{ ft}$$
$$V = 76 \text{ yd}^3$$
$$P = 5\frac{1}{2} \text{ in.}^2$$

34. There are *two* errors in this list of answers.

$$P = 215 \text{ cm}$$
$$A = 8\frac{1}{2} \text{ m}$$
$$V = 98^3 \text{ in.}$$

Chapter 4
SUMMARY

KEY TERMS

4.1 **fraction** — A fraction is a number of the form $\frac{a}{b}$, where a and b are integers and b is not 0.

numerator — The top number in a fraction is the numerator. It shows how many of the equal parts are being considered.

denominator — The bottom number in a fraction is the denominator. It shows how many equal parts are in the whole.

proper fraction — In a proper fraction, the numerator is smaller than the denominator. The fraction is less than 1.

improper fraction — In an improper fraction, the numerator is greater than or equal to the denominator. The fraction is greater than or equal to 1.

equivalent fractions — Equivalent fractions have the same value even though they look different. When graphed on a number line, they are names for the same point.

4.2 **lowest terms** — A fraction is written in lowest terms when its numerator and denominator have no common factor other than 1.

prime number — A prime number is a whole number that has exactly two different factors, itself and 1. The first few prime numbers are 2, 3, 5, 7, 11, 13, and 17.

composite number — A composite number has at least one factor other than itself and 1. Examples are 4, 6, 9, and 10. The numbers 0 and 1 are neither prime nor composite.

prime factorization — In a prime factorization, every factor is a prime number. For example, the prime factorization of 24 is 2 • 2 • 2 • 3.

4.3 **reciprocal** — Two numbers are reciprocals of each other if their product is 1. The reciprocal of $\frac{a}{b}$ is $\frac{b}{a}$ because $\frac{a}{b} \cdot \frac{b}{a} = 1$.

4.4 **like fractions** — Like fractions have the same denominator.

unlike fractions — Unlike fractions have different denominators.

least common denominator — The least common denominator (LCD) for two fractions is the smallest positive number that can be divided evenly by both denominators.

4.5 **mixed number** — A mixed number is a fraction and a whole number written together. It represents the sum of the whole number and the fraction. For example, $5\frac{1}{3}$ represents $5 + \frac{1}{3}$.

4.6 **complex fraction** — A complex fraction is a fraction in which the numerator and/or denominator contain one or more fractions.

4.7 **multiplication property of equality** — The multiplication property of equality states that you may multiply both sides of an equation by the same nonzero number and it will still be balanced.

division property of equality — The division property of equality states that you may divide both sides of an equation by the same nonzero number and it will still be balanced.

4.8 **volume** — Volume is a measure of the space inside a solid shape. Volume is measured in cubic units, such as in.^{3}, ft^3, yd^3, mm^3, cm^3, and so on.

NEW SYMBOLS

Cubic units (for measuring volume)   in.^{3} ft^3 yd^3 mm^3 cm^3 m^3

NEW FORMULAS

Area of a triangle: $A = \frac{1}{2}bh$

Volume of a rectangular solid: $V = lwh$

Volume of a pyramid: $V = \frac{1}{3}Bh$

Test Your Word Power

See how well you have learned the vocabulary in this chapter. Answers follow the Quick Review.

1. A **fraction** is in lowest terms if
   A. it has a value less than 1
   B. its numerator and denominator have no common factor other than 1
   C. its numerator and denominator are composite numbers
   D. it is rewritten as a mixed number.

2. The **denominator** of a fraction
   A. is written above the fraction bar
   B. is a prime number
   C. shows how many equal parts are in the whole
   D. is the smallest number divisible by the numerator.

3. The **LCD** of two fractions is the
   A. smallest number divisible by both denominators
   B. largest factor common to both denominators
   C. smallest number divisible by both numerators
   D. smallest prime number that divides evenly into both denominators.

4. Two numbers are **reciprocals** of each other if
   A. they have the same prime factorizations
   B. their sum is 0
   C. they are written in lowest terms
   D. their product is 1.

5. **Volume** is
   A. measured in square units
   B. the space inside a solid shape
   C. the sum of the lengths of the sides of a shape
   D. found by multiplying base times height.

6. A **mixed number**
   A. has a value equal to 1
   B. is the reciprocal of an improper fraction
   C. is the sum of a whole number and a fraction
   D. has a value less than 1.

7. A whole number is **prime** if
   A. it is divisible by itself and 1
   B. it has only composite factors
   C. it cannot be divided
   D. it has exactly two factors, itself and 1.

8. **Equivalent fractions**
   A. have the same denominators
   B. are written in lowest terms
   C. name the same point on a number line
   D. are reciprocals of each other.

Quick Review

Concepts	Examples

4.1 Understanding Fraction Terminology

The *numerator* is the top number. The *denominator* is the bottom number. In a *proper fraction*, the numerator is smaller than the denominator. In an *improper fraction*, the numerator is greater than or equal to the denominator.

Proper fractions $\quad \dfrac{2}{3}, \dfrac{3}{4}, \dfrac{15}{16}, \dfrac{1}{8} \quad \begin{array}{l}\leftarrow \text{Numerator} \\ \leftarrow \text{Denominator}\end{array}$

Improper fractions $\quad \dfrac{17}{8}, \dfrac{19}{12}, \dfrac{11}{2}, \dfrac{5}{3}, \dfrac{7}{7}$

4.1 Writing Equivalent Fractions

Multiply or divide the numerator and denominator by the same nonzero number. The result is an equivalent fraction.

$$\dfrac{1}{2} = \dfrac{1 \cdot 8}{2 \cdot 8} = \dfrac{8}{16} \leftarrow \text{Equivalent to } \tfrac{1}{2}$$

$$-\dfrac{12}{15} = -\dfrac{12 \div 3}{15 \div 3} = -\dfrac{4}{5} \leftarrow \text{Equivalent to } -\tfrac{12}{15}$$

Concepts	Examples
4.2 Finding Prime Factorizations A prime factorization of a number shows the number as the product of prime numbers. The first few prime numbers are 2, 3, 5, 7, 11, 13, and 17. You can use a division method or a factor tree to find the prime factorization.	Find the prime factorization of 24. **Division Method:** $2\overline{)24}$ ← Divide 24 by 2, the first prime; quotient is 12 $2\overline{)12}$ ← Divide 12 by 2; quotient is 6 $2\overline{)6}$ ← Divide 6 by 2; quotient is 3 $3\overline{)3}$ ← Divide 3 by 3; quotient is 1 1 ← Continue to divide until the quotient is 1 $24 = 2 \cdot 2 \cdot 2 \cdot 3$ **Factor Tree Method:** Circle each prime number. $24 = 2 \cdot 2 \cdot 2 \cdot 3$
4.2 Writing Fractions in Lowest Terms Write the prime factorization of both numerator and denominator. Divide out all common factors, using slashes to show the division. Multiply any remaining factors in the numerator and in the denominator.	$\dfrac{18}{90} = \dfrac{2 \cdot 3 \cdot 3}{2 \cdot 3 \cdot 3 \cdot 5} = \dfrac{1}{5}$ $\dfrac{2b^3}{8ab} = \dfrac{2 \cdot b \cdot b \cdot b}{2 \cdot 2 \cdot 2 \cdot a \cdot b} = \dfrac{b^2}{4a}$
4.3 Multiplying Fractions Multiply the numerators and multiply the denominators. The product must be written in lowest terms. One way to do this is to write each original numerator and each original denominator as the product of primes and divide out any common factors before multiplying.	$\left(-\dfrac{7}{10}\right)\left(\dfrac{5}{6}\right) = -\dfrac{7 \cdot 5}{2 \cdot 5 \cdot 2 \cdot 3} = -\dfrac{7}{12}$ $\dfrac{3x^2}{5} \cdot \dfrac{2}{9x} = \dfrac{3 \cdot x \cdot x \cdot 2}{5 \cdot 3 \cdot 3 \cdot x} = \dfrac{2x}{15}$
4.3 Dividing Fractions Rewrite the division problem as multiplying by the reciprocal of the divisor. In other words, the first number (dividend) stays the same and the second number (divisor) is changed to its reciprocal. Then use the steps for multiplying fractions. The quotient must be in lowest terms. Division by 0 is undefined.	$2 \div \left(-\dfrac{1}{3}\right) = \dfrac{2}{1} \cdot \left(-\dfrac{3}{1}\right) = -\dfrac{2 \cdot 3}{1 \cdot 1} = -6$ Reciprocals $\dfrac{x^2}{y^2} \div \dfrac{x}{3y} = \dfrac{x^2}{y^2} \cdot \dfrac{3y}{x} = \dfrac{x \cdot x \cdot 3 \cdot y}{y \cdot y \cdot x} = \dfrac{3x}{y}$ Reciprocals

Concepts	Examples
4.4 Adding and Subtracting Like Fractions You can add or subtract fractions *only* when they have the *same* denominator. Add or subtract the numerators and write the result over the common denominator. Be sure that the final result is in lowest terms.	$\dfrac{3}{10} - \dfrac{7}{10} = \dfrac{3-7}{10} = \dfrac{-4}{10}$ or $-\dfrac{4}{10}$ Write $-\dfrac{4}{10}$ in lowest terms. $\quad -\dfrac{4}{10} = -\dfrac{\cancel{2} \cdot 2}{\cancel{2} \cdot 5} = -\dfrac{2}{5}$ $\{$Lowest terms $\dfrac{5}{a} + \dfrac{7}{a} = \dfrac{5+7}{a} = \dfrac{12}{a}$ $\{$Lowest terms
4.4 Finding the Lowest Common Denominator (LCD) Write the prime factorization of each denominator. Then use enough prime factors in the LCD to "cover" both denominators.	What is the LCD for $\dfrac{5}{12}$ and $\dfrac{5}{18}$? $12 = 2 \cdot 2 \cdot 3$ $18 = 2 \cdot 3 \cdot 3$ $\text{LCD} = 2 \cdot 2 \cdot 3 \cdot 3 = 36$ The LCD for $\dfrac{5}{12}$ and $\dfrac{5}{18}$ is 36.
4.4 Adding and Subtracting Unlike Fractions Find the LCD. Rewrite each original fraction as an equivalent fraction whose denominator is the LCD. Then add or subtract the numerators and keep the common denominator. Be sure that the final result is in lowest terms.	$-\dfrac{5}{12} + \dfrac{7}{9}$ The LCD is 36. Rewrite: $\quad -\dfrac{5}{12} = -\dfrac{5 \cdot 3}{12 \cdot 3} = -\dfrac{15}{36}$ Rewrite: $\quad \dfrac{7}{9} = \dfrac{7 \cdot 4}{9 \cdot 4} = \dfrac{28}{36}$ Add: $\quad -\dfrac{15}{36} + \dfrac{28}{36} = \dfrac{-15 + 28}{36} = \dfrac{13}{36}$ $\{$Lowest terms $\dfrac{2}{3} - \dfrac{6}{x}$ The LCD is $3 \cdot x$ or $3x$. Rewrite: $\quad \dfrac{2}{3} = \dfrac{2 \cdot x}{3 \cdot x} = \dfrac{2x}{3x}$ Rewrite: $\quad \dfrac{6}{x} = \dfrac{6 \cdot 3}{x \cdot 3} = \dfrac{18}{3x}$ Subtract: $\quad \dfrac{2x}{3x} - \dfrac{18}{3x} = \dfrac{2x - 18}{3x}$ $\{$Lowest terms

Concepts	Examples
4.5 Mixed Numbers and Improper Fractions **Changing Mixed Numbers to Improper Fractions** Multiply the denominator by the whole number, add the numerator, and place the result over the original denominator. **Changing Improper Fractions to Mixed Numbers** Divide the numerator by the denominator and place the remainder over the original denominator.	Mixed to improper: $7\frac{2}{3} = \frac{23}{3} \leftarrow (3 \cdot 7) + 2$ Same denominator Improper to mixed: $\frac{17}{5} = 3\frac{2}{5}$ Same denominator
4.5 Multiplying Mixed Numbers First, round the numbers and estimate the answer. Then follow these steps to find the exact answer. *Step 1* Rewrite each mixed number as an improper fraction. *Step 2* Multiply. *Step 3* Write the answer in lowest terms and change the answer to a mixed number if desired. Then the answer is in simplest form.	Estimate: $1\frac{3}{5} \cdot 3\frac{1}{3}$ $\downarrow$ Rounded $\downarrow$ $2 \cdot 3 = 6$ Exact: $1\frac{3}{5} \cdot 3\frac{1}{3} = \frac{8}{5} \cdot \frac{10}{3}$ $= \frac{8 \cdot 2 \cdot \cancel{5}}{\cancel{5} \cdot 3}$ $= \frac{16}{3} = 5\frac{1}{3} \leftarrow$ Close to estimate
4.5 Dividing Mixed Numbers First, round the numbers and estimate the answer. Then follow these steps to find the exact answer. *Step 1* Rewrite each mixed number as an improper fraction. *Step 2* Divide. (Rewrite as multiplication using the reciprocal of the divisor.) *Step 3* Write the answer in lowest terms and change the answer to a mixed number if desired. Then the answer is in simplest form.	Estimate: $3\frac{3}{4} \div 2\frac{2}{5}$ $\downarrow$ Rounded $\downarrow$ $4 \div 2 = 2$ Exact: $3\frac{3}{4} \div 2\frac{2}{5} = \frac{15}{4} \div \frac{12}{5}$ Reciprocal of $\frac{12}{5}$ is $\frac{5}{12}$ $= \frac{15}{4} \cdot \frac{5}{12}$ $= \frac{\cancel{3} \cdot 5 \cdot 5}{4 \cdot \cancel{3} \cdot 4}$ $= \frac{25}{16} = 1\frac{9}{16} \leftarrow$ Close to estimate
4.5 Adding and Subtracting Mixed Numbers First round the numbers and estimate the answer. Then rewrite the mixed numbers as improper fractions and follow the steps for adding and subtracting fractions. Write the answer in simplest form.	Estimate: $2\frac{3}{8} + 3\frac{3}{4}$ $\downarrow \qquad \downarrow$ $2 + 4 = 6 \leftarrow$ Estimate Exact: $2\frac{3}{8} + 3\frac{3}{4} = \frac{19}{8} + \frac{15}{4} = \frac{19}{8} + \frac{30}{8} = \frac{19 + 30}{8}$ $= \frac{49}{8} = 6\frac{1}{8} \leftarrow$ Close to estimate

Concepts	Examples
4.6 Simplifying Fractions with Exponents The meaning of an exponent is the same for fractions as it is for integers. An exponent is a way to write repeated multiplication.	$\left(-\dfrac{2}{3}\right)^2$ means $\left(-\dfrac{2}{3}\right)\left(-\dfrac{2}{3}\right) = \dfrac{2 \cdot 2}{3 \cdot 3} = \dfrac{4}{9}$ The product of two negative numbers is positive.
4.6 Order of Operations The order of operations is the same for fractions as for integers. 1. Work inside *parentheses* or *other grouping symbols*. 2. Simplify expressions with *exponents*. 3. Do the remaining *multiplications and divisions* as they occur from left to right. 4. Do the remaining *additions and subtractions* as they occur from left to right.	Simplify. $-\dfrac{2}{3} + 3\left(\dfrac{1}{4}\right)^2$ Cannot work inside parentheses. Apply the exponent: $\frac{1}{4} \cdot \frac{1}{4}$ is $\frac{1}{16}$ $-\dfrac{2}{3} + 3\left(\dfrac{1}{16}\right)$ Multiply next: $3\left(\frac{1}{16}\right)$ is $\frac{3}{1} \cdot \frac{1}{16} = \frac{3}{16}$ $-\dfrac{2}{3} + \dfrac{3}{16}$ Add last. The LCD is 48 $-\dfrac{32}{48} + \dfrac{9}{48}$ Rewrite $-\frac{2}{3}$ as $-\frac{32}{48}$ Rewrite $\frac{3}{16}$ as $\frac{9}{48}$ $\dfrac{-32 + 9}{48}$ Add the numerators. Keep the common denominator. $-\dfrac{23}{48}$ The answer is in lowest terms.
4.6 Simplifying Complex Fractions Recall that the fraction bar indicates division. Rewrite the complex fraction using the ÷ symbol for division. Then follow the steps for dividing fractions.	Simplify. $\dfrac{-\frac{4}{5}}{10}$ Rewrite as $-\dfrac{4}{5} \div 10$. $-\dfrac{4}{5} \div 10 = -\dfrac{4}{5} \cdot \dfrac{1}{10} = -\dfrac{\overset{1}{\cancel{2}} \cdot 2 \cdot 1}{5 \cdot \underset{1}{\cancel{2}} \cdot 5} = -\dfrac{2}{25}$ Reciprocals

Chapter 4 Summary

Concepts

4.7 Solving Equations Containing Fractions

Step 1 If necessary, add the same number to both sides of the equation so that the variable term is by itself on one side of the equal sign.

Step 2 Multiply both sides by the reciprocal of the coefficient of the variable term.

Step 3 To check your solution, go back to the original equation and replace the variable with your solution. If the equation balances, your solution is correct. If it does not balance, rework the problem.

Examples

Solve the equation. Check the solution.

$$\frac{1}{3}b + 6 = 10$$
$$\phantom{\frac{1}{3}b + }-6 -6 \quad \text{Add } -6 \text{ to both sides.}$$
$$\frac{1}{3}b + 0 = 4$$
$$\frac{1}{3}b = 4$$

$$\frac{\cancel{3}}{1}\left(\frac{1}{\cancel{3}}b\right) = \frac{3}{1}(4) \quad \text{Multiply both sides by } \tfrac{3}{1} \text{ (the reciprocal of } \tfrac{1}{3}\text{).}$$

$$b = 12$$

The solution is 12.

Check $\frac{1}{3}\,b\, + 6 = 10 \quad$ Original equation

$\frac{1}{3}(12) + 6 = 10 \quad$ Replace b with 12

$\phantom{\frac{1}{3}}4 + 6 = 10$

$\phantom{\frac{1}{3}(12) + 6} 10 = 10 \quad$ Balances

When b is 12, the equation balances, so 12 is the correct solution (**not** 10).

4.8 Finding the Area of a Triangle

Use this formula to find the area of a triangle.

$$\text{Area} = \frac{1}{2} \cdot \text{base} \cdot \text{height}$$
$$A = \frac{1}{2}bh$$

Remember that area is measured in **square units**.

Find the area of this triangle.

$$A = \frac{1}{2}bh$$

$$A = \frac{1}{2} \cdot 20 \text{ ft} \cdot 5 \text{ ft}$$

$$A = \frac{1}{2} \cdot \frac{20 \text{ ft}}{1} \cdot \frac{5 \text{ ft}}{1}$$

$$A = \frac{1 \cdot \cancel{2} \cdot 10 \text{ ft} \cdot 5 \text{ ft}}{\cancel{2} \cdot 1 \cdot 1}$$

$$A = 50 \text{ ft}^2 \quad \text{Measure area in square units.}$$

4.8 Finding the Volume of a Rectangular Solid

Use this formula to find the volume of box-like solids.

$$\text{Volume} = \text{length} \cdot \text{width} \cdot \text{height}$$
$$V = lwh$$

Volume is measured in **cubic units**.

Find the volume of this box.

$V = l \cdot w \cdot h$

$V = 5 \text{ cm} \cdot 3 \text{ cm} \cdot 6 \text{ cm}$

$V = 90 \text{ cm}^3$

Measure volume in cubic units.

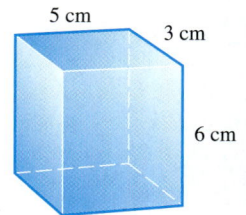

Concepts	Examples
4.8 Finding the Volume of a Pyramid Use this formula to find the volume of a pyramid. $$\text{Volume} = \frac{1}{3} \cdot B \cdot h$$ $$V = \frac{1}{3} Bh$$ where B is the area of the base and h is the height of the pyramid. Volume is measured in **cubic units**.	Find the volume of a pyramid with a square base 2 cm by 2 cm and a height of 6 cm. Area of square base = 2 cm • 2 cm $B = \mathbf{4\ cm^2}$ $V = \frac{1}{3} \cdot B \cdot h$ $V = \frac{1}{3} \cdot \mathbf{4\ cm^2} \cdot \mathbf{6\ cm}$ $V = \frac{1}{3} \cdot \frac{4\ cm^2}{1} \cdot \frac{6\ cm}{1}$ $V = \dfrac{1 \cdot 4\ cm^2 \cdot \overset{1}{\cancel{3}} \cdot 2\ cm}{\underset{1}{\cancel{3}} \cdot 1 \cdot 1}$ $V = 8\ \mathbf{cm^3}$ Measure volume in cubic units.

Answers to Test Your Word Power

1. B; *Example:* $\frac{2}{5}$ is in lowest terms, but $\frac{4}{10}$ is not.

2. C; *Example:* In $\frac{3}{4}$ the denominator, 4, shows that the whole is divided into 4 equal parts.

3. A; *Example:* The LCD of $\frac{1}{4}$ and $\frac{5}{6}$ is 12, because 12 is the smallest number that can be divided evenly by 4 and by 6.

4. D; *Example:* The reciprocal of $\frac{3}{8}$ is $\frac{8}{3}$ because $\frac{3}{8} \cdot \frac{8}{3} = \frac{24}{24} = 1$.

5. B; *Example:* The volume of a rectangular solid (box-like shape) is the space inside the box, measured in cubic units.

6. C; *Example:* The mixed number $2\frac{3}{4}$ represents $2 + \frac{3}{4}$.

7. D; *Example:* 7 is prime because it has exactly two factors, 7 and 1.

8. C; *Example:* $\frac{1}{2}$ and $\frac{2}{4}$ are equivalent fractions because they both name the point halfway between 0 and 1.

Chapter 4
REVIEW EXERCISES

[4.1] 1. What fraction of these figures are squares? What fraction are circles?

2. Write fractions to represent the shaded and unshaded portions of this figure.

3. Graph $-\frac{1}{2}$ and $1\frac{1}{2}$ on the number line.

4. Simplify each fraction.
   (a) $-\frac{20}{5}$ (b) $\frac{8}{1}$ (c) $-\frac{3}{3}$

[4.2] *Write each fraction in lowest terms.*

5. $\frac{28}{32}$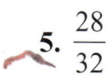

6. $\frac{54}{90}$

7. $\frac{16}{25}$

8. $\frac{15x^2}{40x}$

9. $\frac{7a^3}{35a^3b}$

10. $\frac{12mn^2}{21m^3n}$

[4.3] *Multiply or divide. Write all answers in lowest terms.*

11. $-\frac{3}{8} \div (-6)$

12. $\frac{2}{5}$ of (-30)

13. $\frac{4}{9}\left(\frac{2}{3}\right)$

14. $\left(\frac{7}{3x^3}\right)\left(\frac{x^2}{14}\right)$

15. $\frac{ab}{5} \div \frac{b}{10a}$

16. $\frac{18}{7} \div 3k$

[4.4] *Add or subtract. Write all answers in lowest terms.*

17. $-\frac{5}{12} + \frac{5}{8}$

18. $\frac{2}{3} - \frac{4}{5}$

19. $4 - \frac{5}{6}$

20. $\frac{7}{9} + \frac{13}{18}$

21. $\frac{n}{5} + \frac{3}{4}$

22. $\frac{3}{10} - \frac{7}{y}$

296 Chapter 4 Rational Numbers: Positive and Negative Fractions

[4.5] *First, round the mixed numbers to the nearest whole number and estimate each answer. Then find the exact answer.*

23. Exact:
$2\frac{1}{4} \div 1\frac{5}{8}$
Estimate:
___ ÷ ___ = ___

24. Exact:
$7\frac{1}{3} - 4\frac{5}{6}$
Estimate:
___ − ___ = ___

25. Exact:
$1\frac{3}{4} + 2\frac{3}{10}$
Estimate:
___ + ___ = ___

[4.6] *Simplify.*

26. $\left(-\frac{3}{4}\right)^3$

27. $\left(\frac{2}{3}\right)^2\left(-\frac{1}{2}\right)^4$

28. $\frac{2}{5} + \frac{3}{10}(-4)$

29. $-\frac{5}{8} \div \left(-\frac{1}{2}\right)\left(\frac{14}{15}\right)$

30. $\dfrac{\frac{5}{8}}{\frac{1}{16}}$

31. $\dfrac{\frac{8}{9}}{-6}$

[4.7] *Solve each equation. Show your work.*

32. $-12 = -\frac{3}{5}w$

33. $18 + \frac{6}{5}r = 0$

34. $3x - \frac{2}{3} = \frac{5}{6}$

[4.8] *Find the area of the triangle. Name each solid and find its volume.*

35.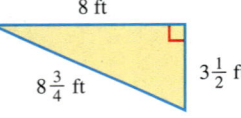
8 ft, $8\frac{3}{4}$ ft, $3\frac{1}{2}$ ft

36.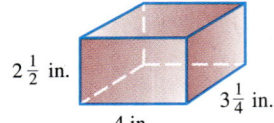
$2\frac{1}{2}$ in., 4 in., $3\frac{1}{4}$ in.

37.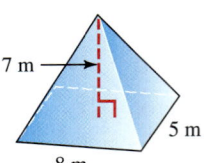
7 m, 5 m, 8 m

MIXED REVIEW EXERCISES

Solve each application problem.

38. A chili recipe that makes 10 servings uses $2\frac{1}{2}$ pounds of meat. How much meat will be in each serving? How much meat would be needed to make 30 servings?

39. Yanli worked as a math tutor for $4\frac{1}{2}$ hours on Monday, $2\frac{3}{4}$ hours on Tuesday, and $3\frac{2}{3}$ hours on Friday. How much longer did she work on Monday than on Friday? How many hours did she work in all?

40. There are 60 children in the day care center. If $\frac{1}{5}$ of the children are preschoolers, $\frac{2}{3}$ of the children are toddlers, and the rest are infants, find the number of children in each age group.

41. A rectangular city park is $\frac{3}{4}$ mile long and $\frac{3}{10}$ mile wide. Find the perimeter and area of the park.

Chapter 4
TEST

1. Write fractions to represent the shaded and unshaded portions of the figure.

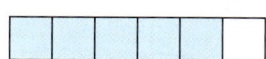

2. Graph $-\frac{2}{3}$ and $2\frac{1}{3}$ on the number line at the right.

Write each fraction in lowest terms.

3. $\dfrac{21}{84}$ **4.** $\dfrac{25}{54}$ **5.** $\dfrac{6a^2b}{9b^2}$

Add, subtract, multiply, or divide, as indicated. Write all answers in lowest terms.

6. $\dfrac{1}{6} + \dfrac{7}{10}$ **7.** $-\dfrac{3}{4} \div \dfrac{3}{8}$

8. $\dfrac{5}{8} - \dfrac{4}{5}$ **9.** $(-20)\left(-\dfrac{7}{10}\right)$

10. $\dfrac{\frac{4}{9}}{-6}$ **11.** $4 - \dfrac{7}{8}$

12. $-\dfrac{2}{9} + \dfrac{2}{3}$ **13.** $\dfrac{21}{24}\left(\dfrac{9}{14}\right)$

14. $\dfrac{12x}{7y} \div 3x$ **15.** $\dfrac{6}{n} - \dfrac{1}{4}$

16. $\dfrac{2}{3} + \dfrac{a}{5}$ **17.** $\left(\dfrac{5}{9b^2}\right)\left(\dfrac{b}{10}\right)$

18. Simplify.

$\left(-\dfrac{1}{2}\right)^3 \left(\dfrac{2}{3}\right)^2$

19. Simplify.

$\dfrac{1}{6} + 4\left(\dfrac{2}{5} - \dfrac{7}{10}\right)$

1. _____
2. (number line from −3 to 3)
3. _____
4. _____
5. _____
6. _____
7. _____
8. _____
9. _____
10. _____
11. _____
12. _____
13. _____
14. _____
15. _____
16. _____
17. _____
18. _____
19. _____

Study Skills Workbook
Activity 11: Using Test Results

First, round the numbers and estimate each answer. Then find the exact answer. Write exact answers in simplest form.

20. Estimate: _____ Exact: _____

20. $4\dfrac{4}{5} \div 1\dfrac{1}{8}$

21. Estimate: _____ Exact: _____

21. $3\dfrac{2}{5} - 1\dfrac{9}{10}$

Solve each equation. Show your work.

22. $7 = \dfrac{1}{5}d$

23. $-\dfrac{3}{10}t = \dfrac{9}{14}$

24. $0 = \dfrac{1}{4}b - 2$

25. $\dfrac{4}{3}x + 7 = -13$

Find the area of each triangle.

26.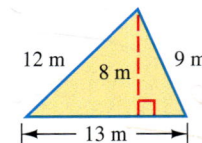

27.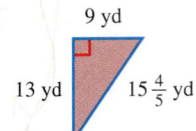

Name each solid and find its volume.

28.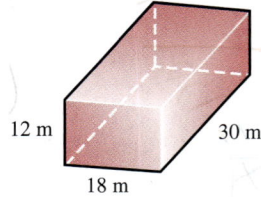

29.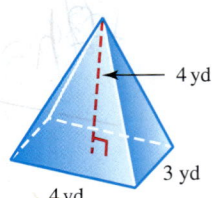

Solve each application problem.

30. Ann-Marie Sargent is training for an upcoming wheelchair race. She rides $4\dfrac{5}{6}$ hours on Monday, $6\dfrac{2}{3}$ hours on Tuesday, and $3\dfrac{1}{4}$ hours on Wednesday. How many hours did she spend in all? How many more hours did she train on Tuesday than on Monday?

31. A new vaccine is synthesized at the rate of $2\dfrac{1}{2}$ ounces per day. How long will it take to synthesize $8\dfrac{3}{4}$ ounces?

32. There are 8448 students at the Metro Community College campus. If $\dfrac{7}{8}$ of the students work either full time or part time, find the total number of students who work.

Rational Numbers: Positive and Negative Decimals

5

- **5.1** Reading and Writing Decimal Numbers
- **5.2** Rounding Decimal Numbers
- **5.3** Adding and Subtracting Signed Decimal Numbers
- **5.4** Multiplying Signed Decimal Numbers
- **5.5** Dividing Signed Decimal Numbers

Summary Exercises on Decimals

- **5.6** Fractions and Decimals
- **5.7** Problem Solving with Statistics: Mean, Median, Mode, and Variability
- **5.8** Geometry Applications: Pythagorean Theorem and Square Roots
- **5.9** Problem Solving: Equations Containing Decimals
- **5.10** Geometry Applications: Circles, Cylinders, and Surface Area

Nearly 49 million Americans go fishing at least once a year, making it America's fourth most popular recreational activity. (*Source:* National Sporting Goods Association.) In **Section 5.3,** Exercises 57–60, these friends will use decimal numbers when paying for new equipment. But will decimals help them catch their limit? (See **Section 5.1,** Exercises 59–62, and **Section 5.6,** Exercises 65–68.)

299

5.1 Reading and Writing Decimal Numbers

OBJECTIVES

1. Write parts of a whole using decimals.
2. Identify the place value of a digit.
3. Read decimal numbers.
4. Write decimals as fractions or mixed numbers.

In **Chapter 4**, you worked with rational numbers written in fraction form to represent parts of a whole. In this chapter, we will use rational numbers written as **decimals** to show parts of a whole. For example, our money system is based on decimals. One dollar is divided into 100 equivalent parts. One cent ($0.01) is one of the parts, and a dime ($0.10) is 10 of the parts. Metric measurement (see **Chapter 8**) is also based on decimals.

OBJECTIVE 1 Write parts of a whole using decimals. Decimals are used when a whole is divided into 10 equivalent parts or into 100 or 1000 or 10,000 equivalent parts. In other words, decimals are fractions with denominators that are a power of 10. For example, the square at the right is cut into 10 equivalent parts. Written as a fraction, each part is $\frac{1}{10}$ of the whole. Written as a decimal, each part is **0.1**. Both $\frac{1}{10}$ and 0.1 are read as "one tenth."

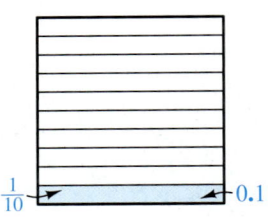

One tenth of the square is shaded.

1 There are 10 dimes in one dollar. Each dime is $\frac{1}{10}$ of a dollar. Write a fraction, a decimal, and the words that name the yellow shaded portion of each dollar.

(a)

(b)

(c)

The dot in 0.1 is called the **decimal point**.

0.1
↑
Decimal point

The square at the right has **7** of its 10 parts shaded.
Written as a *fraction*, $\frac{7}{10}$ of the square is shaded.
Written as a *decimal*, **0.7** of the square is shaded.
Both $\frac{7}{10}$ and 0.7 are read as "seven tenths."

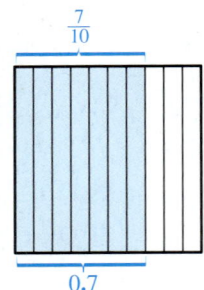

Seven tenths of the square is shaded.

◀◀◀ **Work Problem 1 at the Side.**

Each square below is cut into 100 equivalent parts.
Written as a fraction, each part is $\frac{1}{100}$ of the whole.
Written as a decimal, each part is **0.01** of the whole.
Both $\frac{1}{100}$ and 0.01 are read as "one hundredth."

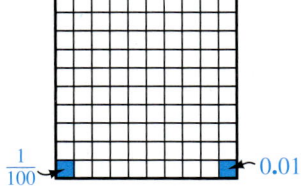 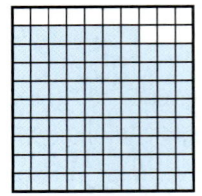

Eighty-seven hundredths of the square is shaded.

The square above on the right has 87 of its 100 parts shaded.
Written as a fraction, $\frac{87}{100}$ of the total area is shaded.
Written as a decimal, **0.87** of the total area is shaded.
Both $\frac{87}{100}$ and 0.87 are read as "eighty-seven hundredths."

ANSWERS

1. (a) $\frac{1}{10}$; 0.1; one tenth

   (b) $\frac{3}{10}$; 0.3; three tenths

   (c) $\frac{9}{10}$; 0.9; nine tenths

Section 5.1 Reading and Writing Decimal Numbers

Work Problem 2 at the Side.

The example below shows several numbers written as fractions, as decimals, and in words.

EXAMPLE 1 Using the Decimal Forms of Fractions

	Fraction	Decimal	Read As
(a)	$\dfrac{4}{10}$	0.4	four tenths
(b)	$-\dfrac{9}{100}$	−0.09	negative nine hundredths
(c)	$\dfrac{71}{100}$	0.71	seventy-one hundredths
(d)	$\dfrac{8}{1000}$	0.008	eight thousandths
(e)	$-\dfrac{45}{1000}$	−0.045	negative forty-five thousandths
(f)	$\dfrac{832}{1000}$	0.832	eight hundred thirty-two thousandths

Work Problem 3 at the Side.

OBJECTIVE 2 Identify the place value of a digit. The decimal point separates the *whole number part* from the *fractional part* in a decimal number. In the chart below, you see that the **place value** names for fractional parts are similar to those on the whole number side but end in "*ths*."

Decimal Place Value Chart

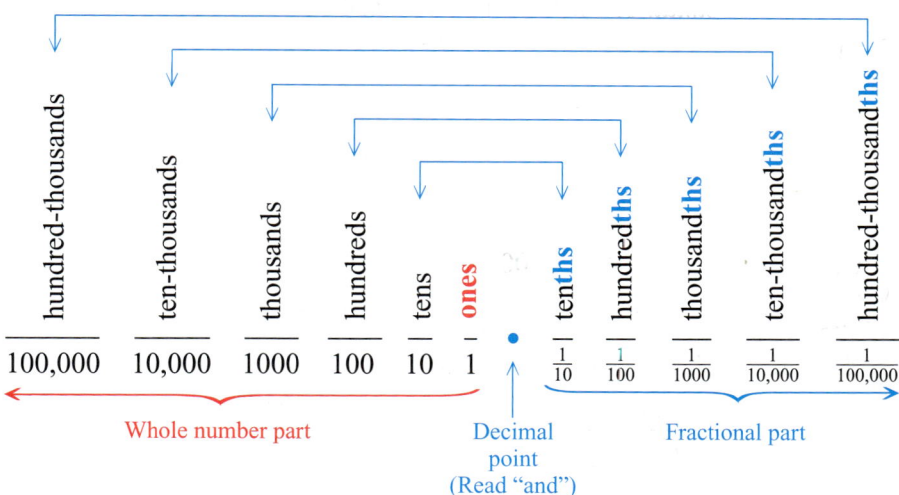

NOTE

Notice that the **ones** place is at the center of the place value chart above. There is no "oneths" place.

Also notice that each place is 10 times the value of the place to its right.

Finally, be sure to write a hyphen (dash) in ten-thousand**ths** and hundred-thousand**ths**.

---

2 Write the portion of each square that is shaded as a fraction, as a decimal, and in words.

(a)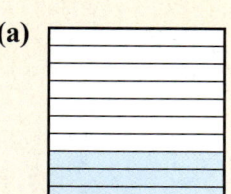

(b)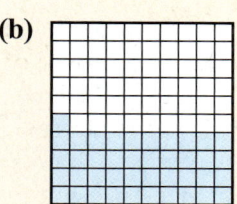

3 Write each decimal as a fraction.

(a) −0.7

(b) 0.2

(c) −0.03

(d) 0.69

(e) 0.047

(f) −0.351

ANSWERS

2. (a) $\dfrac{3}{10}$; 0.3; three tenths

   (b) $\dfrac{41}{100}$; 0.41; forty-one hundredths

3. (a) $-\dfrac{7}{10}$ (b) $\dfrac{2}{10}$ (c) $-\dfrac{3}{100}$

   (d) $\dfrac{69}{100}$ (e) $\dfrac{47}{1000}$ (f) $-\dfrac{351}{1000}$

302 Chapter 5 Rational Numbers: Positive and Negative Decimals

4 Identify the place value of each digit.

(a) 971.54

(b) 0.4

(c) 5.60

(d) 0.0835

5 Tell how to read each decimal in words.

(a) 0.6

(b) 0.46

(c) 0.05

(d) 0.409

(e) 0.0003

(f) 0.0703

(g) 0.088

ANSWERS

4. (a) 9 7 1 . 5 4 — hundreds, tens, ones, tenths, hundredths
   (b) 0 . 4 — ones, tenths
   (c) 5 . 6 0 — ones, tenths, hundredths
   (d) 0 . 0 8 3 5 — ones, tenths, hundredths, thousandths, ten-thousandths

5. (a) six tenths
   (b) forty-six hundredths
   (c) five hundredths
   (d) four hundred nine thousandths
   (e) three ten-thousandths
   (f) seven hundred three ten-thousandths
   (g) eighty-eight thousandths

> **CAUTION**
> If a number does *not* have a decimal point, it is an *integer*. An integer has no fractional part. If you want to show the decimal point in an integer, it is just to the **right** of the digit in the ones place. Here are three examples.
>
> 8 = 8.   306 = 306.   −42 = −42.
>   ↑   ↑   ↑
> Decimal point   Decimal point   Decimal point

EXAMPLE 2 Identifying the Place Value of a Digit

Identify the place value of each digit.

(a) 178.36

1 7 8 . 3 6 — hundreds, tens, ones, tenths, hundredths

(b) 0.00935

0 . 0 0 9 3 5 — ones, tenths, hundredths, thousandths, ten-thousandths, hundred-thousandths

Notice in Example 2(b) that we do *not* use commas on the right side of the decimal point.

Work Problem 4 at the Side.

OBJECTIVE 3 Read decimal numbers. A decimal number is read according to its form as a fraction.

0.9 — ones, tenths

We read 0.9 as "nine tenths" because 0.9 is the same as $\frac{9}{10}$. Notice that 0.9 ends in the tenths place.

0.02 — ones, tenths, hundredths

We read 0.02 as "two hundredths" because 0.02 is the same as $\frac{2}{100}$. Notice that 0.02 ends in the hundredths place.

EXAMPLE 3 Reading Decimal Numbers

Tell how to read each decimal in words.

(a) 0.3

Because $0.3 = \frac{3}{10}$, read the decimal as: three ten**ths**.

(b) 0.49   Read it as: forty-nine hundred**ths**.

(c) 0.08   Read it as: eight hundred**ths**.

(d) 0.918   Read it as: nine hundred eighteen thousand**ths**.

(e) 0.0106   Read it as: one hundred six ten-thousand**ths**.

Work Problem 5 at the Side.

Reading a Decimal Number

Step 1 Read any whole number part to the *left* of the decimal point as you normally would.

Step 2 Read the decimal point as "*and*."

Step 3 Read the part of the number to the *right* of the decimal point as if it were an ordinary whole number.

Step 4 Finish with the place value name of the rightmost digit; these names all end in "*ths*."

NOTE
If there is *no whole number part*, you will use only Steps 3 and 4.

EXAMPLE 4 Reading Decimal Numbers

Read each decimal.

(a)

9 is in tenths place.

16.9

sixteen **and** nine **tenths**

16.9 is read "sixteen and nine tenths."

(b)

5 is in hundredths place.

482.35

four hundred eighty-two **and** thirty-five **hundredths**

482.35 is read "four hundred eighty-two and thirty-five hundredths."

3 is in thousandths place.

(c) 0.063 is "sixty-three **thousandths**." (No whole number part)

(d) 11.1085 is "eleven **and** one thousand eighty-five **ten-thousandths**."

CAUTION
Use "and" *only* when reading a decimal point. A common mistake is to read the whole number 405 as "four hundred *and* five." But there is *no decimal point* shown in 405, so it is read "four hundred five."

▶ **Work Problem 6 at the Side.** ▶▶▶

OBJECTIVE 4 Write decimals as fractions or mixed numbers. Knowing how to read decimals will help you when writing decimals as fractions or mixed numbers.

Writing a Decimal as a Fraction or Mixed Number

Step 1 The digits to the right of the decimal point are the numerator of the fraction.

Step 2 The denominator is 10 for tenths, 100 for hundredths, 1000 for thousandths, 10,000 for ten-thousandths, and so on.

Step 3 If the decimal has a whole number part, the fraction will be a mixed number with the same whole number part.

6 Tell how to read each decimal in words.

(a) 3.8

(b) 15.001

(c) 0.0073

(d) 64.309

ANSWERS
6. **(a)** three and eight tenths
   (b) fifteen and one thousandth
   (c) seventy-three ten-thousandths
   (d) sixty-four and three hundred nine thousandths

304 Chapter 5 Rational Numbers: Positive and Negative Decimals

7 Write each decimal as a fraction or mixed number.

(a) 0.7

(b) 12.21

(c) 0.101

(d) 0.007

(e) 1.3717

8 Write each decimal as a fraction or mixed number in lowest terms.

(a) 0.5

(b) 12.6

(c) 0.85

(d) 3.05

(e) 0.225

(f) 420.0802

ANSWERS

7. (a) $\frac{7}{10}$ (b) $12\frac{21}{100}$ (c) $\frac{101}{1000}$
(d) $\frac{7}{1000}$ (e) $1\frac{3717}{10,000}$
8. (a) $\frac{1}{2}$ (b) $12\frac{3}{5}$ (c) $\frac{17}{20}$ (d) $3\frac{1}{20}$
(e) $\frac{9}{40}$ (f) $420\frac{401}{5000}$

EXAMPLE 5 Writing Decimals as Fractions or Mixed Numbers

Write each decimal as a fraction or mixed number.

(a) 0.19

The digits to the right of the decimal point, 19, are the numerator of the fraction. The denominator is 100 for hundredths because the rightmost digit is in the hundredths place.

$0.19 = \frac{19}{100}$ ← 100 for hundredths

↑ Hundredths place

(b) 0.863

$0.863 = \frac{863}{1000}$ ← 1000 for thousandths

↑ Thousandths place

(c) 4.0099

The whole number part stays the same.

$4.0099 = 4\frac{99}{10,000}$ ← 10,000 for ten-thousandths

↑ Ten-thousandths place

◀◀◀ **Work Problem 7 at the Side.**

EXAMPLE 6 Writing Decimals as Fractions or Mixed Numbers

Write each decimal as a fraction or mixed number in lowest terms.

(a) $0.4 = \frac{4}{10}$ ← 10 for tenths

Write $\frac{4}{10}$ in lowest terms. $\frac{4}{10} = \frac{4 \div 2}{10 \div 2} = \frac{2}{5}$ ← Lowest terms

(b) $0.75 = \frac{75}{100} = \frac{75 \div 25}{100 \div 25} = \frac{3}{4}$ ← Lowest terms

(c) $18.105 = 18\frac{105}{1000} = 18\frac{105 \div 5}{1000 \div 5} = 18\frac{21}{200}$ ← Lowest terms

(d) $42.8085 = 42\frac{8085}{10,000} = 42\frac{8085 \div 5}{10,000 \div 5} = 42\frac{1617}{2000}$ ← Lowest terms

CAUTION
Always check that your fraction answers are in lowest terms.

◀◀◀ **Work Problem 8 at the Side.**

Calculator Tip In this book, we will write 0.45 instead of just .45, to emphasize that there is no whole number. Your *scientific* calculator shows these zeros also. Enter 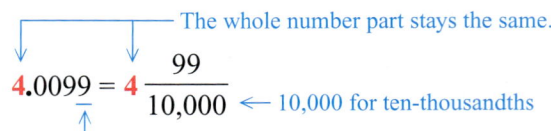 and notice that the display automatically shows 0.45 even though you did not press 0. For comparison, enter the whole number 45 by pressing ④ ⑤ ⊕ and notice that the decimal point automatically appears to the *right* of the 5. (*Graphing* calculators may not automatically show a 0 in the ones place.)

5.1 Exercises

Identify the digit that has the given place value. See Example 2.

1. 70.489
   tens
   ones
   tenths

2. 135.296
   ones
   tenths
   tens

3. 0.2518
   hundredths
   thousandths
   ten-thousandths

4. 0.9347
   hundredths
   thousandths
   ten-thousandths

5. 93.01472
   thousandths
   ten-thousandths
   tenths

6. 0.51968
   tenths
   ten-thousandths
   hundredths

7. 314.658
   tens
   tenths
   hundreds

8. 51.325
   tens
   tenths
   hundredths

9. 149.0832
   hundreds
   hundredths
   ones

10. 3458.712
    hundreds
    hundredths
    tenths

11. 6285.7125
    thousands
    thousandths
    hundredths

12. 5417.6832
    thousands
    thousandths
    ones

Write the decimal number that has the specified place values. See Example 2.

13. 0 ones, 5 hundredths, 1 ten, 4 hundreds, 2 tenths

14. 7 tens, 9 tenths, 3 ones, 6 hundredths, 8 hundreds

15. 3 thousandths, 4 hundredths, 6 ones, 2 ten-thousandths, 5 tenths

16. 8 ten-thousandths, 4 hundredths, 0 ones, 2 tenths, 6 thousandths

17. 4 hundredths, 4 hundreds, 0 tens, 0 tenths, 5 thousandths, 5 thousands, 6 ones

18. 7 tens, 7 tenths, 6 thousands, 6 thousandths, 3 hundreds, 3 hundredths, 2 ones

Write each decimal as a fraction or mixed number in lowest terms. See Examples 1, 5, and 6.

19. 0.7 **20.** 0.1 **21.** 13.4 **22.** 9.8 **23.** 0.35

24. 0.85 **25.** 0.66 **26.** 0.33 **27.** 10.17 **28.** 31.99

29. 0.06 **30.** 0.08 **31.** 0.205 **32.** 0.805

33. 5.002 **34.** 4.008 **35.** 0.686 **36.** 0.492

Tell how to read each decimal in words. See Examples 1, 3, and 4.

37. 0.5 **38.** 0.2

39. 0.78 **40.** 0.55

41. 0.105 **42.** 0.609

43. 12.04 **44.** 86.09

45. 1.075 **46.** 4.025

Write each decimal in numbers. See Examples 3 and 4.

47. Six and seven tenths

    6.7

48. Eight and twelve hundredths

    8.12

49. Thirty-two hundredths

    .032

50. One hundred eleven thousandths

    .0111

51. Four hundred twenty and eight thousandths

    420.008

52. Two hundred and twenty-four thousandths

    200.024

53. Seven hundred three ten-thousandths

54. Eight hundred and six hundredths

55. Seventy-five and thirty thousandths

56. Sixty and fifty hundredths

57. Anne read the number 4302 as "four thousand three hundred and two." Explain what is wrong with the way Anne read the number.

58. Jerry read the number 9.0106 as "nine and one hundred and six ten-thousandths." Explain the error he made.

The friends on the first page of this chapter need to select the correct fishing line for their reels. Fishing line is sold according to how many pounds of "pull" the line can withstand before breaking. Use the table to answer Exercises 59–62. Write all fractions in lowest terms. (Note: The diameter of the fishing line is its thickness.)

FISHING LINE

Test Strength (pounds)	Average Diameter (inches)
4	0.008
8	0.010
12	0.013
14	0.014
17	0.015
20	0.016

Source: Berkley Outdoor Technologies Group.

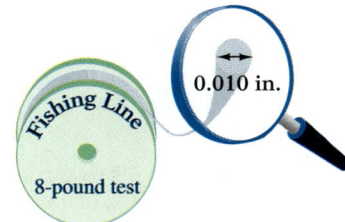

The diameter is the distance across the end of the line (or its thickness).

59. Write the diameter of 8-pound test line in words and as a fraction.

60. Write the diameter of 17-pound test line in words and as a fraction.

61. What is the test strength of the line with a diameter of $\frac{13}{1000}$ inch?

62. What is the test strength of the line with a diameter of sixteen thousandths inch?

Suppose your job is to take phone orders for precision parts. Use the table below. In Exercises 63–68, write the correct part number that matches what you hear the customer say over the phone. In Exercises 67–68, write the words you would say to the customer.

Part Number	Size in Centimeters
3-A	0.06
3-B	0.26
3-C	0.6
3-D	0.86
4-A	1.006
4-B	1.026
4-C	1.06
4-D	1.6
4-E	1.602

63. "Please send the six tenths centimeter bolt."

    Part number _____

64. "The part missing from our order was the one and six hundredths size."

    Part number _____

65. "The size we need is one and six thousandths centimeters."

    Part number _____

66. "Do you still stock the twenty-six hundredths centimeter bolt?"

    Part number _____

67. "What size is part number 4-E?" Write your answer in words.

68. "What size is part number 4-B?" Write your answer in words.

RELATING CONCEPTS (EXERCISES 69–76) For Individual or Group Work

*Use your knowledge of place value to **work Exercises 69–76 in order.***

69. Look back at the decimal place value chart on page 301 of this section. What do you think would be the names of the next four places to the *right* of hundred-thousandths? What information did you use to come up with these names?

70. A common mistake is to think that the first place to the right of the decimal point is "oneths" and the second place is "tenths." Why might someone make that mistake? How would you explain why there is no "oneths" place?

71. Use your answer from Exercise 69 to write 0.72436955 in words.

72. Use your answer from Exercise 69 to write 0.000678554 in words.

73. Write 8006.500001 in words.

74. Write 20,060.000505 in words.

75. Write this decimal using digits.
    three hundred two thousand forty ten-millionths

76. Write this decimal using digits.
    nine billion, eight hundred seventy-six million, five hundred forty-three thousand, two hundred ten and one hundred million two hundred thousand three hundred billionths

5.2 Rounding Decimal Numbers

Section 1.5 showed how to round integers. For example, 89 rounded to the nearest ten is 90, and 8512 rounded to the nearest hundred is 8500.

OBJECTIVE 1 Learn the rules for rounding decimals. It is also important to be able to **round** decimals. For example, a store is selling 2 candy mints for $0.75 but you want only one mint. The price of each mint is $0.75 ÷ 2, which is $0.375, but you cannot pay part of a cent. Is $0.375 closer to $0.37 or to $0.38? Actually, it's exactly halfway between. When this happens in everyday situations, the rule is to round *up*. The store will charge you $0.38 for the mint.

OBJECTIVES

1. Learn the rules for rounding decimals.
2. Round decimals to any given place.
3. Round money amounts to the nearest cent or nearest dollar.

Rounding a Decimal Number

Step 1 Find the place to which the rounding is being done. Draw a "cut-off" line *after* that place to show that you are cutting off and dropping the rest of the digits.

Step 2 Look *only* at the *first* digit you are cutting off.

Step 3A If this digit is **4 or less,** the part of the number you are keeping *stays the same.*

Step 3B If this digit is **5 or more,** you must **round up** the part of the number you are keeping.

Step 4 You can use the ≈ symbol or the ≐ symbol to indicate that the rounded number is now an approximation (close, but *not exact*). Both symbols mean "is approximately equal to." (In this book we will use the ≈ symbol.)

CAUTION
Do *not* move the decimal point when rounding.

OBJECTIVE 2 Round decimals to any given place. These examples show you how to round decimals.

EXAMPLE 1 Rounding a Decimal Number

Round 14.39652 to the nearest thousandth. (Is it closer to 14.396 or to 14.397?)

Step 1 Draw a "cut-off" line after the thousandths place.

$$14.396 \mid 52$$

Thousandths

You are cutting off the 5 and 2. They will be dropped.

Step 2 Look *only* at the *first* digit you are cutting off. Ignore the other digits you are cutting off.

$$14.396 \mid 52$$

Look *only* at the 5
Ignore the 2

Continued on Next Page

1 Round to the nearest thousandth.

(a) 0.33492

(b) 8.00851

(c) 265.42068

(d) 10.70180

Step 3 If the first digit you are cutting off is *5 or more*, round up the part of the number you are keeping.

$$\begin{array}{r} 14.396\,\,5\,2 \\ +\,\,\,\,0.001 \\ \hline 14.397 \end{array}$$

— First digit cut is *5 or more*, so round up by adding 1 thousandth to the part you are keeping.

So, 14.39652 rounded to the nearest thousandth is 14.397.
You can write 14.39652 ≈ 14.397.

CAUTION
When rounding integers in **Section 1.5**, you kept all the digits but changed some to zeros. With decimals, you cut off and *drop the extra digits*. In the example above, 14.39652 rounds to 14.397 *not* 14.39700.

◀◀◀ **Work Problem 1 at the Side.**

In Example 1 above, the rounded number 14.397 had *three decimal places*. **Decimal places** are the number of digits to the *right* of the decimal point. The first decimal place is tenths, the second is hundredths, the third is thousandths, and so on.

EXAMPLE 2 **Rounding Decimals to Different Places**

Round to the place indicated.

(a) Round 5.3496 to the nearest tenth. (Is it closer to 5.3 or to 5.4?)

Step 1 Draw a cut-off line after the tenths place.

5 . 3 | 4 9 6   You are cutting off the 4, 9, and 6
↑
Tenths

Step 2   5 . 3 | 4 9 6   — Look *only* at the 4
                     ‾‾‾
                     Ignore these digits.

Step 3   5 . 3 | 4 9 6   — First digit cut is *4 or less*, so the part you are keeping stays the same
              ‾‾‾
              5 . 3 ← Stays the same

So, 5.3496 rounded to the nearest tenth is 5.3 (one decimal place for tenths). You can write 5.3496 ≈ 5.3. Notice that it does *not* round to 5.3000 (which would be ten-thousandths instead of tenths).

(b) Round 0.69738 to the nearest hundredth. (Is it closer to 0.69 or to 0.70?)

Step 1   0 . 6 9 | 7 3 8   Draw a cut-off line after the hundredths place.
                ↑
                Hundredths

                                  — Look *only* at the 7.
Step 2   0 . 6 9 | 7 3 8

Continued on Next Page

ANSWERS
1. (a) 0.335 (b) 8.009
   (c) 265.421 (d) 10.702

Step 3   0.69|738   ← First digit cut is *5 or more,* so round up by adding 1 hundredth to the part you are keeping.

```
  1
  0 . 6 9    ← Keep this part.
+ 0 . 0 1    ← To round up, add 1 hundredth.
  0 . 7 0    ← 9 + 1 is 10; write 0 and carry 1 to the tenths place.
```

So, 0.69738 rounded to the nearest hundredth is 0.70. Hundredths is *two* decimal places so you *must* write the 0 in the hundredths place.

You can write 0.69738 ≈ 0.70.

CAUTION
If a *rounded* number has a 0 in the rightmost place, you *must* keep the 0. As shown above, 0.69738 rounded to the nearest hundredth is 0.70. Do *not* write 0.7, which is rounded to tenths instead of hundredths.

(c) Round 0.01806 to the nearest thousandth. (Is it closer to 0.018 or to 0.019?)

0.018|06   ← First digit cut is *4 or less,* so the part you are keeping stays the same.

0.018 ← Stays the same

So, 0.01806 rounded to the nearest thousandth is 0.018 (three decimal places for thousandths).

You can write 0.01806 ≈ 0.018.

(d) Round 57.976 to the nearest tenth. (Is it closer to 57.9 or to 58.0?)

57.9|76   ← First digit cut is *5 or more,* so round up by adding 1 tenth to the part you are keeping.

```
   1
   57.9
 + 0.1
   58.0    ← 9 + 1 is 10; write the 0 and carry 1 to the ones place.
```

So, 57.976 rounded to the nearest tenth is 58.0. You can write 57.976 ≈ 58.0. You *must* write the 0 in the tenths place to show that the number was rounded to the nearest tenth.

CAUTION
Check that your rounded answer shows *exactly* the number of decimal places asked for in the problem. Be sure your answer shows *one decimal place* if you rounded to *tenths,* two decimal places for *hundredths,* three decimal places for *thousandths,* and so on.

Work Problem 2 at the Side. ▶

OBJECTIVE 3 Round money amounts to the nearest cent or nearest dollar. When you are shopping in a store, money amounts are usually rounded to the nearest cent. There are 100 cents in a dollar.

Each cent is $\frac{1}{100}$ of a dollar.

Another way to write $\frac{1}{100}$ is 0.01. So rounding to the *nearest cent* is the same as rounding to the *nearest hundredth of a dollar.*

---

2 Round to the place indicated.

(a) 0.8988 to the nearest hundredth

(b) 5.8903 to the nearest hundredth

(c) 11.0299 to the nearest thousandth

(d) 0.545 to the nearest tenth

ANSWERS
2. (a) 0.90 (b) 5.89
   (c) 11.030 (d) 0.5

3 Round each money amount to the nearest cent.

(a) $14.595

You pay _____.

(b) $578.0663

You pay _____.

(c) $0.849

You pay _____.

(d) $0.0548

You pay _____.

EXAMPLE 3 Rounding to the Nearest Cent

Round each money amount to the nearest cent.

(a) $2.4238 (Is it closer to $2.42 or to $2.43?)

$2.42|38 ← First digit cut is *4 or less,* so the part you are keeping stays the same.
$2.42 ← You pay $2.42

$2.4238 rounded to the nearest cent is $2.42.

(b) $0.695 (Is it closer to $0.69 or to $0.70?)

$0.69|5 ← *5 or more;* round up.
$0.69
+ $0.01 ← To round up, add 1 hundredth (1 cent).
$0.70 ← You pay $0.70

$0.695 rounded to the nearest cent is $0.70.

◀◀◀ **Work Problem 3 at the Side.**

> **NOTE**
> Some stores round *all* money amounts up to the next higher cent, even if the next digit is *4 or less*. In Example 3(a) above, some stores would round $2.4238 up to $2.43, even though it is closer to $2.42.

It is also common to round money amounts to the nearest dollar. For example, you can do that on your federal and state income tax returns to make the calculations easier.

EXAMPLE 4 Rounding to the Nearest Dollar

Round to the nearest dollar.

(a) $48.69 (Is it closer to $48 or to $49?)

$48.|69 ← First digit cut is *5 or more,* so round up by adding $1
$48
+ 1
$49

$48.69 rounded to the nearest dollar is $49.

> **CAUTION**
> $48.69 rounded to the nearest dollar is $49. Be careful to write the answer as $49 to show that the rounding is to the *nearest dollar*. Writing $49.00 would show rounding to the *nearest cent*.

Continued on Next Page

ANSWERS
3. (a) $14.60 (b) $578.07
   (c) $0.85   (d) $0.05

(b) $594.36 (Is it closer to $594 or to $595?)

$594.|36 — First digit cut is *4 or less*, so the part you are keeping stays the same.

$594

$594.36 rounded to the nearest dollar is $594.

(c) $349.88 (Is it closer to $349 or to $350?)

$349.|88 — *5 or more* so round up by adding $1

$349
+   1
─────
$350

$349.88 rounded to the nearest dollar is $350.

(d) $2689.50 (Is it closer to $2689 or to $2690?)

$2689.|50 — *5 or more*, so round up by adding $1

$2689
+   1
──────
$2690

$2689.50 rounded to the nearest dollar is $2690.

> **NOTE**
> When rounding $2689.50 to the nearest dollar, above, notice that it is exactly halfway between $2689 and $2690. When this happens in everyday situations, the rule is to round *up*. (Scientists working with technical data may use a more complicated rule when rounding numbers that are exactly in the middle.)

(e) $0.61 (Is it closer to $0 or to $1?)

$0.|61 — *5 or more*, so round up.

$0.61 rounded to the nearest dollar is $1.

🖩 **Calculator Tip** Accountants and other people who work with money amounts often set their calculators to automatically round to two decimal places (nearest cent) or to round to zero decimal places (nearest dollar). Your calculator may have this feature.

Work Problem 4 at the Side.

---

4 Round to the nearest dollar.

(a) $29.10

(b) $136.49

(c) $990.91

(d) $5949.88

(e) $49.60

(f) $0.55

(g) $1.08

ANSWERS
4. (a) $29 (b) $136 (c) $991
   (d) $5950 (e) $50 (f) $1 (g) $1

Focus on Real-Data Applications

Lawn Fertilizer

Gotta Be Green

A lot's being said about personal responsibility these days, and the idea seems to be ending up on the front lawn—literally! Each spring, homeowners across the country gear up to green up their lawns, and the increased use of fertilizer has a lot of environmentalists concerned about the potential effects of chemical runoff into nearby rivers and streams.

Each year, according to a study conducted by the University of Minnesota's Department of Agriculture, each household in the Minneapolis/St. Paul metro area uses an average of 36 pounds of lawn fertilizer. That adds up to 25,529,295 pounds, or 12,765 tons. Add to that another 193,000 pounds of weed killer and you're looking at the total picture for keeping it green in the Twin Cities.

Source: Minneapolis Star Tribune.

1. According to the article,
   (a) How many pounds of lawn fertilizer are used each year in the *entire metro area?*
   (b) Do a division on your calculator to find the number of *households* in the metro area.
   (c) Would it make sense to round your answer to part (b)? If so, how would you round it?
   (d) How many pounds of *weed killer* are used each year in the entire metro area?
   (e) Do a division on your calculator to find the number of pounds of *weed killer* used by each household in the metro area. Round your answer to the nearest hundredth.

2. There are 2000 pounds in one ton.
   (a) Find the number of tons equivalent to 25,529,295 pounds of fertilizer.
   (b) Does your answer match the figure given in the article? If not, what did the author of the article do to get 12,765 tons?
   (c) Is the author's figure accurate? Why or why not?
   (d) Find the number of tons equivalent to 193,000 pounds of weed killer.
   (e) How can you write the division problem 193,000 ÷ 2,000 in a simpler way? Explain what you did.

3. According to the article,
   (a) "each household in the Minneapolis/St. Paul metro area uses an average of 36 pounds of lawn fertilizer" each year. What mathematical operation do you do to find an average?

   (b) When the calculations were done to find the average, the answer was probably not *exactly* 36 pounds. List six different values that are *less than* 36 that would round to 36. List two values with one decimal place, two values with two decimal places, and two values with three decimal places.

   (c) List six different values that are *greater than* 36 that would round to 36. List two values each with one, two, and three decimal places.

   (d) What is the *smallest* number that can be rounded to 36? What is the *largest* number?

5.2 Exercises

Round each number to the place indicated. See Examples 1 and 2.

1. 16.8974 to the nearest tenth
2. 193.845 to the nearest hundredth
3. 0.95647 to the nearest thousandth
4. 96.81584 to the nearest ten-thousandth
5. 0.799 to the nearest hundredth
6. 0.952 to the nearest tenth
7. 3.66062 to the nearest thousandth
8. 1.5074 to the nearest hundredth
9. 793.988 to the nearest tenth
10. 476.1196 to the nearest thousandth
11. 0.09804 to the nearest ten-thousandth
12. 176.004 to the nearest tenth
13. 48.512 to the nearest one
14. 3.385 to the nearest one
15. 9.0906 to the nearest hundredth
16. 30.1290 to the nearest thousandth
17. 82.000151 to the nearest ten-thousandth
18. 0.400594 to the nearest ten-thousandth

Nardos is grocery shopping. The store will round the amount she pays for each item to the nearest cent. Write the rounded amounts. See Example 3.

19. Soup is three cans for $2.45, so one can is $0.81666. Nardos pays _____.
20. Orange juice is two cartons for $3.89, so one carton is $1.945. Nardos pays _____.
21. Facial tissue is four boxes for $4.89, so one box is $1.2225. Nardos pays _____.
22. Muffin mix is three packages for $1.75, so one package is $0.58333. Nardos pays _____.
23. Candy bars are six for $2.99, so one bar is $0.4983. Nardos pays _____.
24. Boxes of spaghetti are four for $4.39, so one box is $1.0975. Nardos pays _____.

As she gets ready to do her income tax return, Ms. Chen rounds each amount to the nearest dollar. Write the rounded amounts. See Example 4.

25. Income from job, $48,649.60
26. Income from interest on bank account, $69.58
27. Union dues, $310.08
28. Federal withholding, $6064.49
29. Donations to charity, $848.91
30. Medical expenses, $609.38

Round each money amount as indicated.

31. $499.98 to the nearest dollar

32. $9899.59 to the nearest dollar

33. $0.996 to the nearest cent

34. $0.09929 to the nearest cent

35. $999.73 to the nearest dollar

36. $9999.80 to the nearest dollar

The table lists speed records for various types of transportation. Use the table to answer Exercises 37–40.

Record	Speed (miles per hour)
Land speed record (specially built car)	763.04
Motorcycle speed record (specially adapted motorcycle)	322.15
Fastest rollercoaster	106.9
Fastest military jet	2193.167
Boeing 737-300 airplane (regular passenger service)	495
Indianapolis 500 auto race (fastest average winning speed)	185.981
Daytona 500 auto race (fastest average winning speed)	177.602

Source: *The Top 10 of Everything* and *The World Almanac.*

37. Round these speed records to the nearest whole number.
(a) Motorcycle
(b) Rollercoaster

38. Round these speed records to the nearest hundredth.
(a) Daytona 500 average winning speed
(b) Indianapolis 500 average winning speed

39. Round these speed records to the nearest tenth.
(a) Indianapolis 500 average winning speed
(b) Land speed record

40. Round these speed records to the nearest hundred.
(a) military jet
(b) Boeing 737-300 airplane

RELATING CONCEPTS (EXERCISES 41–44) For Individual or Group Work

Use your knowledge about rounding money amounts to **work Exercises 41–44 in order.**

41. Explain what happens when you round $0.499 to the nearest dollar. Why does this happen?

42. Look again at Exercise 41. How else could you round $0.499 that would be more helpful? What kind of guideline does this suggest about rounding to the nearest dollar?

43. Explain what happens when you round $0.0015 to the nearest cent. Why does this happen?

44. Suppose you want to know which of these amounts is less, so you round them both to the nearest cent.

$0.5968 $0.6014

Explain what happens. Describe what you could do instead of rounding to the nearest cent.

5.3 Adding and Subtracting Signed Decimal Numbers

OBJECTIVE 1 Add and subtract positive decimals. When adding or subtracting *whole* numbers, you line up the numbers in columns so that you are adding ones to ones, tens to tens, and so on. A similar idea applies to adding or subtracting *decimal* numbers. With decimals, you line up the decimal points to be sure that you are adding tenths to tenths, hundredths to hundredths, and so on.

OBJECTIVES

1. Add and subtract positive decimals.
2. Add and subtract negative decimals.
3. Estimate the answer when adding or subtracting decimals.

Adding and Subtracting Decimal Numbers

Step 1 Write the numbers in columns with the decimal points lined up.

Step 2 If necessary, write in zeros so both numbers have the same number of decimal places. Then add or subtract as if they were whole numbers.

Step 3 Line up the decimal point in the answer directly below the decimal points in the problem.

EXAMPLE 1 Adding Decimal Numbers

Add.

(a) 16.92 and 48.34

Step 1 Write the numbers in columns with the decimal points lined up.

```
    tens
    ones
    tenths
    hundredths
   1 6 . 9 2
 + 4 8 . 3 4
```
Decimal points are lined up.

Step 2 Add as if these were whole numbers.

```
   1 1
   1 6 . 9 2
 + 4 8 . 3 4
   6 5 . 2 6
```
Step 3 *Decimal point in answer is lined up under decimal points in problem.*

(b) 5.897 + 4.632 + 12.174

Write the numbers in columns with the decimal points lined up. Then add.

```
    1 1   2 1
     5 . 8 9 7
     4 . 6 3 2
 + 1 2 . 1 7 4
   2 2 . 7 0 3
```
Decimal points are lined up.

Work Problem 1 at the Side.

In Example 1(a) above, both numbers had *two* decimal places (two digits to the right of the decimal point). In Example 1(b), all the numbers had *three* decimal places (three digits to the right of the decimal point). That made it easy to add tenths to tenths, hundredths to hundredths, and so on.

---

1 Find each sum.

(a) 2.86 + 7.09

(b) 13.761 + 8.325

(c) 0.319 + 56.007 + 8.252

(d) 39.4 + 0.4 + 177.2

ANSWERS

1. (a) 9.95 (b) 22.086 (c) 64.578
   (d) 217.0

2 Find each sum.

(a) 6.54 + 9.8

(b) 0.831 + 222.2 + 10

(c) 8.64 + 39.115 + 3.0076

(d) 5 + 429.823 + 0.76

If the numbers of decimal places do *not* match, you can write in zeros as placeholders to make them match. This is shown in Example 2 below.

EXAMPLE 2 Writing Zeros as Placeholders before Adding

Add.

(a) 7.3 + 0.85

There are two decimal places in 0.85 (tenths and hundredths), so write a 0 in the hundredths place in 7.3 so that it has two decimal places also.

$$7.3\mathbf{0} \leftarrow \text{One 0 is written in.}$$
$$+\, 0.85$$
$$\overline{8.15}$$

7.30 is equivalent to 7.3 because

$7\dfrac{30}{100}$ in lowest terms is $7\dfrac{3}{10}$

(b) 6.42 + 9 + 2.576

Write in zeros so that all the addends have three decimal places. Notice how the whole number 9 is written with the decimal point on the *right* side. (If you put the decimal point on the *left* side of the 9, you would turn it into the decimal fraction 0.9.)

$$6.42\mathbf{0} \leftarrow \text{One 0 is written in.}$$
$$9.\mathbf{000} \leftarrow \text{9 is a whole number; decimal point and three zeros are written in.}$$
$$+\, 2.576 \leftarrow \text{No zeros are needed.}$$
$$\overline{17.996}$$

> **NOTE**
> Writing zeros to the right of a *decimal* number does *not* change the value of the number, as shown in Example 2(a) above.

◀◀◀ **Work Problem 2 at the Side.**

EXAMPLE 3 Subtracting Decimal Numbers

Subtract.

(a) 15.82 from 28.93

Step 1   28.93
           − 15.82

Line up decimal points. Then you will be subtracting hundredths from hundredths and tenths from tenths.

Step 2   28.93
           − 15.82
           ─────
             13 11

Both numbers have two decimal places; no need to write in zeros.

← Subtract as if they were whole numbers.

Continued on Next Page

ANSWERS
2. (a) 16.34   (b) 233.031   (c) 50.7626
   (d) 435.583

Step 3
$$\begin{array}{r} 28.93 \\ -\ 15.82 \\ \hline 13.11 \end{array}$$
↑ Decimal point in answer is lined up.

(b) 146.35 minus 58.98
Borrowing is needed here.

Line up decimal points.
$$\begin{array}{r} 0\ 13\ 15\ \ \ 12\ 15 \\ \cancel{1}\ \cancel{4}\ \cancel{6}\ .\ \cancel{3}\ \cancel{5} \\ -\ \ \ 5\ 8\ .\ 9\ 8 \\ \hline 8\ 7\ .\ 3\ 7 \end{array}$$

Work Problem 3 at the Side.

EXAMPLE 4 Writing Zeros as Placeholders before Subtracting

Subtract.

(a) 16.5 from 28.362
Use the same steps as in Example 3 above. Remember to write in zeros so both numbers have three decimal places.

Line up decimal points.
$$\begin{array}{r} 28.362 \\ -\ 16.5\mathbf{00} \\ \hline 11.862 \end{array}$$ ← Write two zeros.
← Subtract as usual.

(b) $59.7 - 38.914$
$$\begin{array}{r} 59.7\mathbf{00} \\ -\ 38.914 \\ \hline 20.786 \end{array}$$ ← Write two zeros.

← Subtract as usual.

(c) 12 less 5.83
$$\begin{array}{r} 12.\mathbf{00} \\ -\ \ 5.83 \\ \hline 6.17 \end{array}$$ ← Write a decimal point and two zeros.
← Subtract as usual.

Work Problem 4 at the Side.

OBJECTIVE 2 Add and subtract negative decimals. The rules that you used to add integers in **Section 1.3** will also work for positive and negative decimal numbers.

Adding Signed Numbers

To add two numbers with the *same* sign, add the absolute values of the numbers. Use the common sign as the sign of the sum.

To add two numbers with *unlike* signs, subtract the smaller absolute value from the larger absolute value. Use the sign of the number with the larger absolute value as the sign of the sum.

---

3 Subtract.

(a) 22.7 from 72.9

(b) 6.425 from 11.813

(c) $20.15 − $19.67

4 Subtract.

(a) 18.651 from 25.3

(b) $5.816 − 4.98$

(c) 40 less 3.66

(d) $1 − 0.325$

ANSWERS
3. **(a)** 50.2 **(b)** 5.388 **(c)** $0.48
4. **(a)** 6.649 **(b)** 0.836 **(c)** 36.34
   (d) 0.675

5 Add.

(a) $13.245 + (-18)$

(b) $-0.7 + (-0.33)$

(c) $-6.02 + 100.5$

EXAMPLE 5 Adding Positive and Negative Decimal Numbers

Add.

(a) $-3.7 + (-16)$

Both addends are negative, so the sum will be negative. To begin, $|-3.7|$ is 3.7 and $|-16|$ is 16. Then add the absolute values.

$$\begin{array}{r} 3.7 \\ +\ 16.0 \\ \hline 19.7 \end{array} \leftarrow \text{Write a decimal point and one 0}$$

$$-3.7 + (-16) = -19.7$$
↑   ↑   ↑
Both negative   Negative sum

> **NOTE**
> In **Chapters 1–4**, the negative sign was **red** to help you distinguish it from the subtraction symbol. From now on it will be **black.** We will continue to write parentheses around negative numbers when the negative sign might be confused with other symbols. Thus in part (a) above,
>
> $-3.7 + (-16)$ means **negative** 3.7 **plus negative** 16

(b) $-5.23 + 0.792$

The addends have different signs. To begin, $|-5.23|$ is 5.23 and $|0.792|$ is 0.792. Then subtract the smaller absolute value from the larger.

$$\begin{array}{r} 5.230 \\ -\ 0.792 \\ \hline 4.438 \end{array} \leftarrow \text{Write one 0}$$

$$-5.23 + 0.792 = -4.438$$
↑   ↑
Number with larger   Answer is
absolute value   negative.
is negative.

Work Problem 5 at the Side.

In **Section 1.4** you rewrote subtraction of integers as addition of the first number to the opposite of the second number. This same strategy works with positive and negative decimal numbers.

EXAMPLE 6 Subtracting Positive and Negative Decimal Numbers

Subtract.

(a) $4.3 - 12.73$

Rewrite subtraction as adding the opposite.

$$4.3 - \mathbf{12.73} \quad \text{The opposite of 12.73 is}$$
$$ \quad -12.73$$
$$4.3 + (\mathbf{-12.73})$$

-12.73 has the larger absolute value and is negative, so the answer will be negative.

$$4.3 + (-12.73) = -8.43 \quad \text{Subtract the absolute values:}$$
↑ 12.73
Answer is $-\ 4.30$
negative. $\overline{8.43}$

Continued on Next Page

ANSWERS
5. (a) -4.755 (b) -1.03 (c) 94.48

(b) $-3.65 - (-4.8)$

Rewrite subtraction as adding the opposite.

$-3.65 - (\mathbf{-4.8})$ The opposite of -4.8 is 4.8.
$-3.65 + \mathbf{4.8}$

4.8 has the larger absolute value and is positive, so the answer will be positive.

$-3.65 + 4.8 = 1.15$ Subtract the absolute values:
$\quad\quad\quad\quad\uparrow$
$\quad\text{Answer is}\quad\quad 4.80$
$\quad\text{positive.}\quad\quad\; -3.65$
$\quad\quad\quad\quad\quad\quad\quad\; 1.15$

(c) $14.2 - (1.69 + 0.48)$ Work inside parentheses first.

$14.2 - \;\;(2.17)$ Change subtraction to adding the opposite.

$14.2 + (-2.17)$ 14.2 has the larger absolute value and is positive, so the answer will be positive.

12.03

▶▶▶ **Work Problem 6 at the Side.**

OBJECTIVE 3 Estimate the answer when adding or subtracting decimals. A common error when working decimal problems by hand is to misplace the decimal point in the answer. Or, when using a calculator, you may accidentally press the wrong key. Using *front end rounding* to estimate the answer will help you avoid these mistakes. Start by rounding each number to the highest possible place (as you did in **Section 1.5**). Here are several examples. Notice that in the rounded numbers, only the leftmost digit is something other than 0.

3.25	rounds to	3		6.812	rounds to	7
532.6	rounds to	500		26.397	rounds to	30
7094.2	rounds to	7000		351.24	rounds to	400

EXAMPLE 7 Estimating Decimal Answers

First, use front end rounding to round each number. Then add or subtract the rounded numbers to get an estimated answer. Finally, find the exact answer.

(a) Add 194.2 and 6.825

Estimate: *Exact:*

$\quad\;\; 200 \;\;\xleftarrow{\text{Rounds to}}\quad\quad 194.200$
$+\quad 7 \;\;\xleftarrow{\text{Rounds to}}\quad +\;\;\; 6.825$
$\quad\;\; 207 \quad\quad\quad\quad\quad\quad\quad\; 201.025$

The estimate goes out to the hundreds place (three places to the *left* of the decimal point), and so does the exact answer. Therefore, the decimal point is probably in the correct place in the exact answer.

(b) $\$69.42 - \$13.78$

Estimate: *Exact:*

$\;\;\$70 \xleftarrow{\text{Rounds to}} \quad \$69.42$
$-\;\; 10 \xleftarrow{\text{Rounds to}} \;-\; 13.78$
$\;\;\$60 \quad\quad\quad\quad\quad\; \$55.64$ ← Exact answer is close to estimate, so it is probably correct.

Continued on Next Page

---

6 Subtract.

(a) $-0.37 - (-6)$

(b) $5.8 - 10.03$

(c) $-312.72 - 65.7$

(d) $0.8 - (6 - 7.2)$

ANSWERS
6. (a) 5.63 (b) -4.23 (c) -378.42
(d) 2

7 Use front end rounding to estimate each answer. Then find the exact answer.

(a) $2.83 + 5.009 + 76.1$

Estimate:

Exact:

(b) $19.28 less $1.53

Estimate:

Exact:

(c) $11.365 - 38$

Estimate:

Exact:

(d) $-214.6 + 300.72$

Estimate:

Exact:

(c) $-1.861 - 7.3$

Rewrite subtraction as adding the opposite. Then use front end rounding to get an estimated answer.

$$
\begin{array}{ccc}
-1.861 & - & 7.3 \\
\downarrow & & \downarrow \\
-1.861 & + & (-7.3) \\
\downarrow & & \downarrow \\
\text{Rounded} \quad -2 & + & (-7) \quad = -9 \leftarrow \text{Estimate}
\end{array}
$$

To find the exact answer, add the absolute values.

$$
\begin{array}{r}
1.861 \leftarrow \text{Absolute value of } -1.861 \\
+ 7.300 \leftarrow \text{Absolute value of } -7.3 \text{ with two zeros written in} \\
\hline
9.161
\end{array}
$$

The answer will be negative because both numbers are negative.

$$-1.861 + (-7.3) = -9.161 \leftarrow \text{Exact}$$

The exact answer of -9.161 is reasonable because it is close to the estimated answer of -9.

◀◀◀ Work Problem 7 at the Side.

🖩 Calculator Tip If you are *adding* decimal numbers, you can enter them in any order on your calculator. Try these; jot down the answers.

9.82 (+) 1.86 (=) _____   1.86 (+) 9.82 (=) _____

The answers are the same because addition is *commutative* (see **Section 1.3**). But subtraction is *not* commutative. It *does* matter which number you enter first. Try these:

9.82 (−) 1.86 (=) _____   1.86 (−) 9.82 (=) _____

The answers are 7.96 and −7.96. As you know, positive numbers are *greater* than 0, but negative numbers are *less* than 0. So it is important to do subtraction in the correct order, particularly if it is in your checkbook!

ANSWERS

7. (a) *Estimate:* $3 + 5 + 80 = 88$;
   Exact: 83.939
   (b) *Estimate:* $20 - $2 = $18;
   Exact: $17.75
   (c) *Estimate:* $10 + (-40) = -30$;
   Exact: -26.635
   (d) *Estimate:* $-200 + 300 = 100$;
   Exact: 86.12

5.3 Exercises

Find each sum. See Examples 1 and 2.

1.   5.69
       0.24
    + 11.79

2.   372.1
       33.7
    + 42.3

3.   0.38
      7
    + 4.6

4.   3.7
      0.812
    + 55

5. $14.23 + 8 + 74.63 + 18.715 + 0.286$

6. $197.4 + 0.72 + 17.43 + 25 + 1.4$

7. $27.65 + 18.714 + 9.749 + 3.21$

8. $58.546 + 19.2 + 8.735 + 14.58$

9. Explain and correct the error that a student made when he added $0.72 + 6 + 39.5$ this way:
      0.72
     6
   + 39.50
     40.28

10. Explain and correct the error that a student made when she added $7.21 + 65 + 13.15$ this way:
       7.21
        .65
    + 13.15
      21.01

11. Show why 0.3 is equivalent to 0.3000.

12. Explain why 7 may be written as 7.0 but not as 0.7.

Find each difference. See Examples 3 and 4.

13. $90.5 - 0.8$

14. $303.72 - 0.68$

15. $0.4 - 0.291$

16. $0.35 - 0.088$

17. $6 - 5.09$

18. $80 - 16.3$

19. $15 - 8.339$

20. $44 - 0.08$

21. Explain and correct the error that a student made when he subtracted 7.45 from 15.32 this way:
       7.45
    − 15.32
      12.13

22. Explain the difference between saying "subtract 2.9 from 8" and saying "2.9 minus 8."

This drawing of a human skeleton shows the average length of the longest bones, in inches.
Use the drawing to answer Exercises 23–26.

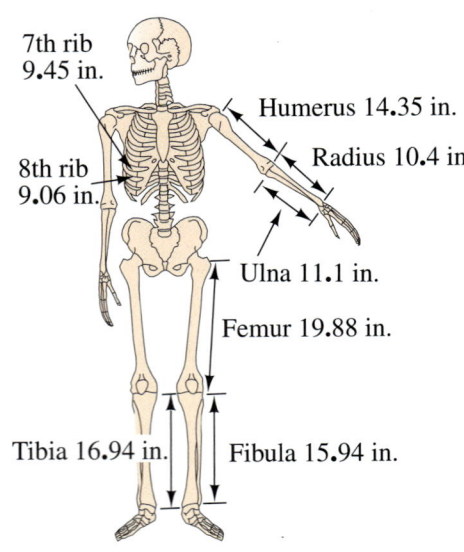

7th rib 9.45 in.
8th rib 9.06 in.
Humerus 14.35 in.
Radius 10.4 in.
Ulna 11.1 in.
Femur 19.88 in.
Tibia 16.94 in.
Fibula 15.94 in.

Source: *The Top 10 of Everything.*

23. (a) What is the combined length of the humerus and radius bones?

(b) What is the difference in the lengths of these two bones?

24. (a) What is the total length of the femur and tibia bones?

(b) How much longer is the femur than the tibia?

25. (a) Find the sum of the lengths of the humerus, ulna, femur, and tibia.

(b) How much shorter is the 8th rib than the 7th rib?

26. (a) What is the difference in the lengths of the two bones in the lower arm?

(b) What is the difference in the lengths of the two bones in the lower leg?

Find each sum or difference. See Examples 5 and 6.

27. $24.008 + (-0.995)$

28. $0.77 - 3.06$

29. $-6.05 + (-39.7)$

30. $-6.409 + 8.224$

31. $0.9 - 7.59$

32. $-489.7 - 38$

33. $-2 - 4.99$

34. $2.068 - (-32.7)$

35. $-5.009 + 0.73$

36. $-0.33 - 65$

37. $-1.7035 - (5 - 6.7)$

38. $60 + (-0.9345 + 1.4)$

39. $8000 - (8002.63 - 8)$

40. $-210 - (-0.7306 + 0.5)$

*Use your estimation skills to pick the most reasonable answer for each example. Do **not** solve the problems. Circle your choice. See Example 7.*

41. 12 − 11.725
2.75   (0.275)   27.5

42. 20 − 1.37
0.1863   1.863   18.63

43. 6.5 + 0.007
6.507   0.6507   65.07

44. 9.67 + 0.09
0.976   9.76   0.00976

45. 456.71 − 454.9
18.1   181   1.81

46. 803.25 − 0.6
802.65   0.80265   8.0265

47. 6004.003 + 52.7172
605.67202   60,567.202   6056.7202

48. 128.35 + 97.0093
2253.593   225.3593   22.53593

Use front end rounding to estimate each sum or difference. Then find the exact answer to each application problem. Use the information in the table for Exercises 49–52.

INTERNET USERS IN SELECTED COUNTRIES	
Country	Number of Users
United States	134.6 million
Japan	33.9 million
China	22.5 million
South Korea	19 million
Canada	15.4 million
France	9 million
World total	413.7 million

Source: Computer Industry Almanac.

49. How many fewer Internet users are there in Canada compared to South Korea?
Estimate:
Exact:

50. How many more users are there in Japan compared to France?
Estimate:
Exact:

51. How many Internet users are there in all the countries listed in the table?
Estimate:

Exact:

52. Using the answer from Exercise 51, calculate the number of worldwide Internet users in countries other than the ones in the table.
Estimate:
Exact:

53. The tallest known land mammal is a prehistoric ancestor of the rhino, measuring 6.4 m. Compare the rhino's height to the combined heights of these NBA basketball stars: Kevin Garnett at 2.1 m, Allen Iverson at 1.83 m, and Shaquille O'Neal at 2.16 m. Is their combined height greater or less than the prehistoric rhino's? By how much? (Source: www.NBA.com/players)

6.4 m

Estimate:

Exact:

54. Sammy works in a veterinarian's office. He weighed two kittens. One was 3.9 ounces and the other was 4.05 ounces. What was the difference in the weight of the two kittens?
Estimate:
Exact:

55. Steven One Feather gave the cashier a $20 bill to pay for $9.12 worth of groceries. How much change did he get?
Estimate:
Exact:

56. The cost of Julie's tennis racket, with tax, was $41.09. She gave the clerk two $20 bills and a $10 bill. What amount of change did Julie receive?
Estimate:
Exact:

The friends buying fishing equipment on the first page of this chapter brought along the store's sale insert from the Sunday paper. Use the information below on sale prices to answer Exercises 57–60. When estimating, round to the nearest whole number.

Source: Wal-Mart.

57. *sub* What is the difference in price between the fluorescent and regular fishing line?

    Estimate:

    Exact:

58. How much more does the least expensive spinning rod cost than the least expensive reel?

    Estimate:

    Exact:

59. Find the total *add* cost of the second-highest-priced spinning reel, two packages of tin split shot, and a three-tray tackle box. Sales tax for all the items is $2.31.

    Estimate:

    Exact:

60. One friend bought a package of bobbers on sale. He also bought some SPF15 sunscreen for $7.53 and a flotation vest for $44.96. Sales tax was $3.74. How much did he spend in all?

    Estimate:

    Exact:

Olivia Sanchez kept track of her expenses for one month. Use her list to answer Exercises 61–64.

61. What were Olivia's total *add* expenses for the month?

62. How much did Olivia pay for telephone, cable TV, *add* and Internet access?

Monthly Expenses	
Rent	$994
Car payment	$190.78
Car repairs, gas	$205
Cable TV	$39.95
Internet access	$19.95
Electricity	$40.80
Telephone	$57.32
Groceries	$186.81
Entertainment	$97.75
Clothing, laundry	$107

63. What was the difference *sub* in the amounts spent for groceries and for the car payment?

64. How much more did Olivia spend on rent than on all her car expenses? *sub*

Find the length of the dashed line in each rectangle or circle.

65.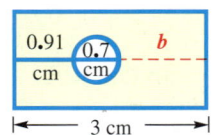

66.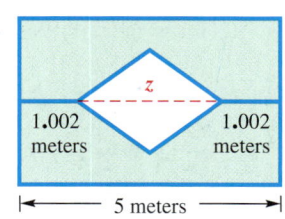

67.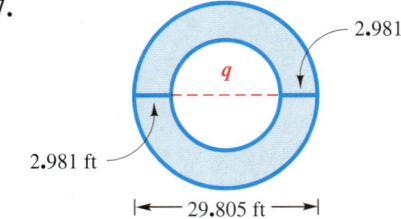

5.4 Multiplying Signed Decimal Numbers

OBJECTIVE 1 **Multiply positive and negative decimals.** The decimals 0.3 and 0.07 can be multiplied by writing them as fractions.

$$0.3 \times 0.07 = \frac{3}{10} \times \frac{7}{100} = \frac{3 \times 7}{10 \times 100} = \frac{21}{1000} = 0.021$$

1 decimal place + 2 decimal places → 3 decimal places

Can you see a way to multiply decimals without writing them as fractions? Try these steps. Remember that each number in a multiplication problem is called a *factor*, and the answer is called the *product*.

OBJECTIVES

1. Multiply positive and negative decimals.
2. Estimate the answer when multiplying decimals.

Multiplying Two Decimal Numbers

Step 1 Multiply the factors (the numbers being multiplied) as if they were whole numbers.

Step 2 Find the *total* number of decimal places in *both* factors.

Step 3 Write the decimal point in the product (the answer) so it has the same number of decimal places as the total from Step 2. You may need to write in extra zeros on the left side of the product in order to get the correct number of decimal places.

Step 4 If two factors have the *same sign*, the product is *positive*. If two factors have *different signs*, the product is *negative*.

NOTE
When multiplying decimals, you do *not* need to line up decimal points. (You *do* need to line up decimal points when adding or subtracting.)

EXAMPLE 1 Multiplying Decimal Numbers

Multiply 8.34 times (−4.2).

Step 1 Multiply the numbers as if they were whole numbers.

$$\begin{array}{r} 8.34 \\ \times\ \ 4.2 \\ \hline 1668 \\ 3336\ \ \\ \hline 35028 \end{array}$$

Step 2 Count the total number of decimal places in both factors.

$$\begin{array}{r} 8.34 \leftarrow \text{2 decimal places} \\ \times\ \ 4.2 \leftarrow \text{1 decimal place} \\ \hline 1668\ \ \ \ \text{3 total decimal places} \\ 3336\ \ \\ \hline 35028 \end{array}$$

Continued on Next Page

1 Multiply.

(a) $-2.6(0.4)$

(b) $(45.2)(0.25)$

(c) 0.104 ← 3 decimal places
$\underline{\times7}$ ← 0 decimal places
← 3 decimal places in the product

(d) $(-3.18)^2$
Hint: Recall that squaring a number means multiplying the number by itself, so this is $(-3.18)(-3.18)$.

2 Multiply.

(a) $0.04(-0.09)$

(b) $(0.2)(0.008)$

(c) $(-0.063)(-0.04)$

(d) $(0.003)^2$

ANSWERS
1. (a) -1.04 (b) 11.300 or 11.3 (c) 0.728
   (d) 10.1124
2. (a) -0.0036 (b) 0.0016 (c) 0.00252
   (d) 0.000009

Step 3 Count over 3 places in the product and write the decimal point. Count from *right to left.*

$$\begin{array}{r} 8.34 \\ \times4.2 \\ \hline 1668 \\ 3336 \\ \hline 35.028 \end{array}$$
← 2 decimal places
← 1 decimal place
3 total decimal places

← 3 decimal places in product

Count over 3 places from right to left to position the decimal point.

Step 4 The factors have *different* signs, so the product is *negative:*
8.34 times $(-4.2) = -35.028$.

▶◀ **Work Problem 1 at the Side.**

EXAMPLE 2 Writing Zeros as Placeholders in the Product

Multiply $(-0.042)(-0.03)$.

Start by multiplying, then count decimal places.

$$\begin{array}{r} 0.042 \\ \times0.03 \\ \hline 126 \end{array}$$
← 3 decimal places
← 2 decimal places
← 5 decimal places needed in product

After multiplying, the answer has only three decimal places, but five are needed. So write two zeros on the *left* side of the answer.

$$\begin{array}{r} 0.042 \\ \times0.03 \\ \hline 00126 \end{array}$$
↑↑
Write two zeros on *left* side of answer.

$$\begin{array}{r} 0.042 \\ \times0.03 \\ \hline .00126 \end{array}$$
← 3 decimal places
← 2 decimal places
← 5 decimal places

Now count over 5 places and write in the decimal point.

The final product is 0.00126, which has five decimal places. The product is *positive* because the factors have the *same* sign (both factors are negative).

▶◀ **Work Problem 2 at the Side.**

🖩 **Calculator Tip** When working with money amounts, you may need to write a 0 in your answer. For example, try multiplying $3.54 × 5 on your calculator. Write down the result.

3.54 ⊗ 5 ⊜ _____

Notice that the result is 17.7, which is *not* the way to write a money amount. You have to write the 0 in the hundredths place: $17.**70** is correct. The calculator does not show the "extra" 0 because:

$$17.70 \text{ or } 17\frac{70}{100} \quad \text{reduces to} \quad 17\frac{7}{10} \text{ or } 17.7$$

So keep an eye on your calculator—it doesn't know when you're working with money amounts.

OBJECTIVE 2 **Estimate the answer when multiplying decimals.** If you are doing multiplication problems by hand, estimating the answer helps you check that the decimal point is in the right place. When you are using a calculator, estimating helps you catch an error like pressing the ÷ key instead of the × key.

EXAMPLE 3 **Estimating before Multiplying**

First, use front end rounding to estimate $(76.34)(12.5)$. Then find the exact answer.

Estimate:

```
    80   ← Rounds to
  × 10   ← Rounds to
   800
```

Exact:

```
    76.34   ← 2 decimal places
  ×  12.5   ← 1 decimal place
    38170    3 decimal places are
   15268     in the product.
    7634
   954.250
```

Both the estimate and the exact answer go out to the hundreds place, so the decimal point in 954.250 is probably in the correct place.

Work Problem 3 at the Side.

3 First, use front end rounding to estimate the answer. Then find the exact answer.

(a) $(11.62)(4.01)$

Estimate:

Exact:

(b) $(-5.986)(-33)$

Estimate:

Exact:

(c) $8(\$4.35)$

Estimate:

Exact:

(d) $58.6(-17.4)$

Estimate:

Exact:

ANSWERS

3. (a) *Estimate:* $(10)(4) = 40$;
      Exact: 46.5962
   (b) *Estimate:* $(-6)(-30) = 180$;
      Exact: 197.538
   (c) *Estimate:* $8(\$4) = \$32$;
      Exact: $34.80
   (d) *Estimate:* $60(-20) = -1200$;
      Exact: -1019.64

Focus on Real-Data Applications

Life Insurance Benefits

Employees often must choose benefit options for programs such as medical, dental, and life insurance. Payment is made through payroll deductions. For example, BP Amoco offers group universal life insurance coverage for an employee or spouse, based on the employee's eligible pay. The rates charged to the employee depend on the age of the employee or spouse at the time enrolled in the plan, the level of coverage, and the use of tobacco. Each employee may choose coverage that is a whole number multiple of his or her eligible pay, rounded up to the next thousand dollars. For example, if an employee chooses life insurance that is triple the eligible pay, then the increment is 3.

The formula for computing the monthly cost of insurance for an employee or spouse is

$$\$\underset{\text{(eligible pay)}}{\underline{}} \times \underset{\substack{\text{(increment of}\\\text{eligible pay)}}}{\underline{}} = \$\underset{\substack{\text{(coverage}\\\text{amount)}}}{\underline{}}{}^{*} \div \$1000 \times \$\underset{\substack{\text{(monthly}\\\text{rate)}}}{\underline{}} = \$\underset{\substack{\text{(monthly cost}\\\text{of insurance)}}}{\underline{}}$$

*Round the coverage amount up to the next $1000 if it is not a whole multiple of $1000. (For example, $34,200 would be rounded up to $35,000.)

The table at the right shows the monthly rates available. Use the formula and the table to answer the problems.

If your age is ...	The monthly rate per $1000 of coverage is ...	
	Non-Tobacco User	Tobacco User
Under 25	$0.040	$0.050
25–39	$0.047	$0.056
40–44	$0.092	$0.112
45–49	$0.147	$0.178
50–54	$0.239	$0.290
55–59	$0.378	$0.458
60–64	$0.645	$0.781
65–69	$0.992	$1.201
70–74	$1.726	$2.090

Source: *Employee Benefits Handbook*, BP Amoco.

1. Suppose you work in BP Amoco's human resources department and must advise new employees of their benefit options. Calculate the monthly cost of insurance, to the nearest cent, for each of these employees.

Employee	Age	Smoker?	Eligible Pay	Increment	Monthly Cost of Insurance
Rebecca C.	25	Yes	$23,600	2	
Roger J.	58	Yes	$85,750	3	
Stan S.	49	No	$45,850	3	
Diana H.	42	No	$53,270	2	
Hulon M.	63	No	$58,100	2	

2. Eric W., aged 55, is a nonsmoker who earns $42,700 in eligible pay. He wants to limit his monthly cost for life insurance to less than $50.00 per month. How much total insurance can he purchase and what will be his monthly costs?

3. Sarah F. and Ellen S. each earn $64,250 and purchase insurance equivalent to twice her eligible salary. Sarah F. is a smoker aged 72, whereas Ellen S. is a nonsmoker aged 30. Calculate the annual premium cost for each person. Explain why you believe the difference in costs between the two policies is or is not justified.

5.4 Exercises

Multiply. See Example 1.

1. 0.042×3.2
2. 0.571×2.9
3. $-21.5(7.4)$
4. $-85.4(-3.5)$

5. $(-23.4)(-0.666)$
6. $0.896(-0.799)$
7. $\$51.88 \times 665$
8. $\$736.75 \times 118$

Use the fact that $(72)(6) = 432$ to solve Exercises 9–16 by simply counting decimal places and writing the decimal point in the correct location. Be sure to indicate the sign of the product.

9. $72(-0.6) =$ -43.2
10. $7.2(-6) =$ 432
11. $(7.2)(0.06) =$ 0.432

12. $(0.72)(0.6) =$ 432
13. $-0.72(-0.06) =$ 0.0432
14. $-72(-0.0006) =$ 432

15. $(0.0072)(0.6) =$ 0.00432
16. $(0.072)(0.006) =$ 432

Multiply. See Example 2.

17. $(0.006)(0.0052)$ 0.0000312
18. $(0.0052)(0.009)$ $52 \times 9 / 3\ 12$
19. $(-0.003)^2$ -0.003×-0.003
20. $(0.0004)^2$ $0.0004 \times 0.0004 = 0.00000016$ -0.000009

RELATING CONCEPTS (EXERCISES 21–22) For Individual or Group Work

Look for patterns as you **work Exercises 21 and 22 in order**.

21. Do these multiplications:
    $(5.96)(10) = 59.6$
    $(0.476)(10) = 4.76$
    $(722.6)(10) = 7226$
    $(3.2)(10) = 32$
    $(80.35)(10) = 803.5$
    $(0.9)(10) = 9$

    What pattern do you see? Write a "rule" for multiplying by 10. What do you think the rule is for multiplying by 100? by 1000? Write the rules and try them out on the numbers above.

22. Do these multiplications:
    $(59.6)(0.1) =$ _____
    $(0.476)(0.1) =$ _____
    $(65)(0.1) =$ _____
    $(3.2)(0.1) =$ _____
    $(80.35)(0.1) =$ _____
    $(523)(0.1) =$ _____

    What pattern do you see? Write a "rule" for multiplying by 0.1. What do you think the rule is for multiplying by 0.01? by 0.001? Write the rules and try them out on the numbers above.

First, use front end rounding to estimate the answer. Then multiply to find the exact answer.
See Example 3.

23. Estimate: Exact:
    40 ←Rounds to   39.6
    × 5 ←Rounds to   × 4.8

24. Estimate: Exact:
                 18.7
    ×   × 2.3

25. Estimate: Exact:
                 37.1
    × 4   × 42

26. Estimate: Exact:
                 5.08
    ×   × 71

27. Estimate: Exact:
                 6.53
    × 5   × 4.6

28. Estimate: Exact:
                 7.51
    ×   × 8.2

29. Estimate: Exact:
    3   2.809
    ×   × 6.85

30. Estimate: Exact:
                 73.52
    ×   × 22.34

Even with most of the problem missing, you can tell whether or not these answers are reasonable. Circle reasonable *or* unreasonable. *If the answer is unreasonable, move the decimal point or insert a decimal point to make the answer reasonable.*

31. How much was his car payment? $18.90
    reasonable
    unreasonable, should be _____

32. How many hours did she work today? 25 hours
    reasonable
    unreasonable, should be _____

33. How tall is her son? 60.5 inches
    reasonable
    unreasonable, should be _____

34. How much does he pay for rent now? $6.92
    reasonable
    unreasonable, should be _____

35. What is the price of one gallon of milk? $419
    reasonable
    unreasonable, should be _____

36. How long is the living room? 16.8 feet
    reasonable
    unreasonable, should be _____

37. How much did the baby weigh? 0.095 pounds
    reasonable
    unreasonable, should be _____

38. What was the sale price of the jacket? $1.49
    reasonable
    unreasonable, should be _____

Solve each application problem. Round money answers to the nearest cent when necessary.

39. LaTasha worked 50.5 hours over the last two weeks. She earns $18.73 per hour. How much did she make?

40. Michael's time card shows 42.2 hours at $10.03 per hour. What are his earnings?

41. Sid needs 0.6 meter of canvas material to make a carry-all bag that fits on his wheelchair. If canvas is $4.09 per meter, how much will Sid spend? (*Note:* $4.09 *per* meter means $4.09 for *one* meter.)

42. How much will Mrs. Nguyen pay for 3.5 yards of lace trim that costs $0.87 per yard?

43. Michelle filled the tank of her SUV with regular unleaded gas. Use the information shown on the pump to find how much she paid for gas.

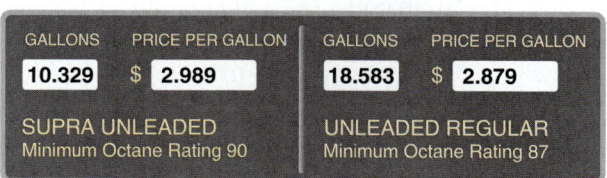

Source: Holiday.

44. Ground beef and chicken legs are on sale. Juma bought 1.7 pounds of legs. Use the information in the ad to find the amount she paid.

45. Ms. Rolack is a real estate broker who helps people sell their homes. Her fee is 0.07 times the price of the home. What was her fee for selling a $289,500 home?

46. Manny Ramirez of the Boston Red Sox had a batting average of 0.325 in the 2003 season. He went to bat 569 times. How many hits did he make? (*Hint:* Multiply the number of times at bat by his batting average.) Round to the nearest whole number. (*Source:* World Almanac.)

Paper money in the United States has not always been the same size. Shown below are the measurements of bills printed before 1929 and the measurements from 1929 on. Use this information to answer Exercises 47–50. Recall that perimeter is the total distance around the edges of a figure (see **Section 3.1**). *The area of a rectangle is found by multiplying the length by the width (see* **Section 3.2**). *(Source: www.moneyfactory.com)*

47. (a) Find the area of each $20 bill, rounded to the nearest tenth.

(b) What is the difference in the rounded areas?

48. (a) Find the perimeter of each $20 bill, to the nearest hundredth.

(b) How much less is the perimeter of today's bills than the bills printed before 1929?

49. The thickness of one piece of today's paper money is 0.0043 inch.

(a) If you had a pile of 100 bills, how high would it be?

(b) How high would a pile of 1000 bills be?

50. (a) Use your answers from Exercise 49 to find the number of bills in a pile that is 43 inches high.

(b) How much money would you have if the pile is all $20 bills?

51. Judy Lewis pays $38.96 per month for basic cable TV. The one-time installation fee was $49. How much will she pay for cable over two years? How much would she pay in two years for the deluxe cable package that costs $89.95 per month?

52. Chuck's SUV payment is $420.27 per month for four years. He also made a down payment of $5000 at the time he bought the SUV. How much will he pay altogether?

53. Paper for the copy machine at the library costs $0.015 per sheet. How much will the library pay for 5100 sheets?

54. A student group collected 2200 pounds of plastic as a fund-raiser. How much will they make if the recycling center pays $0.142 per pound?

55. Barry bought 16.5 meters of rope at $0.47 per meter and three meters of wire at $1.05 per meter. How much change did he get from three $5 bills?

56. Susan bought a 27 inch flat-screen TV that cost $549.99. She paid $55 down and $98.83 per month for six months. How much could she have saved by paying cash?

Use the information below from the Look Smart mail order catalog to answer Exercises 57–60.

Knit Shirt Ordering Information		
43–2A	short sleeved, solid colors	$14.75 each
43–2B	short sleeved, stripes	$16.75 each
43–3A	long sleeved, solid colors	$18.95 each
43–3B	long sleeved, stripes	$21.95 each
XXL size, add $2 per shirt.		
Monogram, $4.95 each. Gift box, $5 each.		

Total Price of Items (excluding monograms and gift boxes)	Shipping, Packing, and Handling
$0–25.00	$3.50
$25.01–75.00	$5.95
$75.01–125.00	$7.95
$125.01+	$9.95
Shipping to each additional address, add $4.25.	

57. Find the total cost of ordering four long-sleeved, solid-color shirts and two short-sleeved, striped shirts, all in the XXL size, and all shipped to your home.

58. What is the total cost of eight long-sleeved shirts, five in solid colors and three striped? Include the cost of shipping the solid shirts to your home and the striped shirts to your brother's home.

59. (a) What is the total cost, including shipping, of sending three short-sleeved, solid-color shirts, with monograms, in a gift box to your aunt for her birthday?

    (b) How much did the monograms, gift box, and shipping add to the cost of your gift?

60. (a) Suppose you order one of each type of shirt for yourself, adding a monogram on each of the solid-color shirts. At the same time, you order three long-sleeved striped shirts, in the XXL size, shipped to your dad in a gift box. Find the total cost of your order.

    (b) What is the difference in total cost (excluding shipping) between the shirts for yourself and the gift for your dad?

5.5 Dividing Signed Decimal Numbers

There are two kinds of decimal division problems: those in which a decimal is divided by an integer, and those in which a number is divided by a decimal. First recall the parts of a division problem.

Divisor → 2)16 ← Dividend, with 8 ← Quotient

$16 \div 2 = 8$ (Dividend ÷ Divisor = Quotient)

$\dfrac{16}{2} = 8$ (Dividend / Divisor = Quotient)

OBJECTIVES

1. Divide a decimal by an integer.
2. Divide a number by a decimal.
3. Estimate the answer when dividing decimals.
4. Use the order of operations with decimals.

OBJECTIVE 1 Divide a decimal by an integer. When the divisor is an integer, use these steps.

Dividing a Decimal Number by an Integer

Step 1 Write the decimal point in the quotient (answer) directly above the decimal point in the dividend.

Step 2 Divide as if both numbers were whole numbers.

Step 3 If both numbers have the *same sign,* the quotient is *positive.* If they have *different signs,* the quotient is *negative.*

EXAMPLE 1 Dividing Decimals by Integers

Divide.

(a) $21.93 \div (-3)$ (Dividend ÷ Divisor)

First consider $21.93 \div 3$.   3)21.93

Step 1 Write the decimal point in the quotient directly above the decimal point in the dividend.

3)21.93 — Decimal points lined up

Step 2 Divide as if the numbers were whole numbers.

7.31
3)21.93

Check by multiplying the quotient times the divisor.

Check
7.31
$\times 3$
$\overline{21.93}$

Matches, so 7.31 is correct.

Step 3 The quotient is -7.31 because the numbers had *different* signs.

$21.93 \div (-3) = -7.31$

Different signs → Negative quotient

Continued on Next Page

336 Chapter 5 Rational Numbers: Positive and Negative Decimals

1 Divide. Check your answers by multiplying.

(a) $4\overline{)93.6}$

(b) $6\overline{)6.804}$

(c) $\dfrac{278.3}{11}$

(d) $-0.51835 \div 5$

(e) $-213.45 \div (-15)$

ANSWERS
1. (a) 23.4;
   Check $(23.4)(4) = 93.6$
   (b) 1.134;
   Check $(1.134)(6) = 6.804$
   (c) 25.3;
   Check $(25.3)(11) = 278.3$
   (d) -0.10367;
   Check $(-0.10367)(5) = -0.51835$
   (e) 14.23;
   Check $(14.23)(-15) = -213.45$

(b)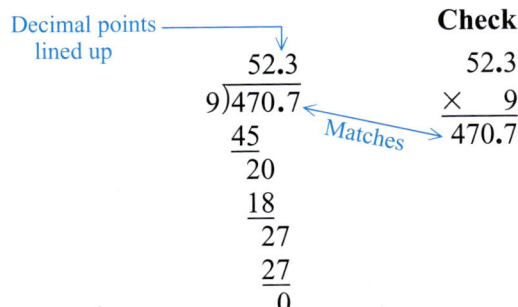

Write the decimal point in the quotient directly above the decimal point in the dividend. Then divide as if they were whole numbers.

```
Decimal points             Check
lined up
         52.3                52.3
     9)470.7              ×     9
        45         Matches  470.7
        20
        18
         27
         27
          0
```

The quotient is **52.3** and is *positive* because both numbers had the *same* sign.

◀◀◀ **Work Problem 1 at the Side.**

EXAMPLE 2 Writing Extra Zeros to Complete a Division

Divide 1.5 by 8.

Keep dividing until the remainder is 0, or until the digits in the quotient begin to repeat in a pattern. In Example 1(b) above, you ended up with a remainder of 0. But sometimes you run out of digits in the dividend before that happens. If so, write extra zeros on the right side of the dividend so you can continue dividing.

```
      0.1
   8)1.5    ← All digits have been used.
     8
     7      ← Remainder is not yet 0.
```

Write a 0 after the 5 in the dividend so you can continue dividing. Keep writing more zeros in the dividend if needed. Recall that writing zeros to the right of a decimal number does **not** change its value.

```
     0.1 8 7 5                    Check
  8)1.5 0 0 0  ← Three zeros needed    0.1875
    8 ↓          to complete the    ×      8
    7 0          division            1.5000
    6 4 ↓                         ↑
      6 0                         Matches dividend,
      5 6 ↓                       so 0.1875 is correct
        4 0
        4 0
          0  ← Stop dividing when the remainder is 0.
```

CAUTION
Notice that in decimals the dividend may *not* be the larger number. In Example 2 above, the dividend is 1.5, which is *smaller* than 8.

─── **Continued on Next Page**

> 🖩 **Calculator Tip** When *multiplying* numbers, you can enter them in any order because multiplication is commutative (see **Section 1.6**). But division is *not* commutative. It *does* matter which number you enter first. Try Example 2 on the previous page both ways; jot down your answers.
>
> 1.5 ÷ 8 = _____   8 ÷ 1.5 = _____
>
> Notice that the first answer, 0.1875, matches the result from Example 2. But the second answer is much different: 5.333333333. Be careful to enter the dividend first.

Work Problem 2 at the Side. ▶▶▶

In the next example the remainder is never 0, even if we keep dividing.

EXAMPLE 3 Rounding a Decimal Quotient

Divide 4.7 by 3. Round the quotient to the nearest thousandth.

Write extra zeros in the dividend so that you can continue dividing.

```
      1.5 6 6 6
    ┌─────────
  3 ) 4.7 0 0 0   ← Three zeros added so far
      3
      ───
      1 7
      1 5
      ───
        2 0
        1 8
        ───
          2 0
          1 8
          ───
            2 0
            1 8
            ───
              2   ← Remainder is still not 0.
```

Notice that the digit 6 in the quotient is repeating. It will continue to do so. The remainder will *never be 0.* There are two ways to show that an answer is a **repeating decimal** that goes on forever. You can write three dots after the answer, or you can write a bar above the digits that repeat (in this case, the 6).

1.5**666** . . .   or   1.5$\overline{6}$ ← Bar above repeating digit
   ↑
 Three dots

> **CAUTION**
> Do not use *both* the dots *and* the bar at the same time. Use three dots *or* the bar.

When repeating decimals occur, round the quotient according to the directions in the problem. In this example, to round to thousandths, divide out one *more* place, to ten-thousandths.

$$4.7 \div 3 = 1.5666\ldots \text{ rounds to } 1.567$$

Check the answer by multiplying 1.567 by 3. Because 1.567 is a rounded answer, the check will *not* give exactly 4.7, but it should be very close.

$$(1.567)(3) = 4.701 \quad \leftarrow \text{Does not equal exactly 4.7 because 1.567 was rounded}$$

Continued on Next Page

---

❷ Divide. Check your answers by multiplying.

(a) $\dfrac{6.4}{5}$

(b) $30.87 \div (-14)$

(c) $\dfrac{-259.5}{-30}$

(d) $0.3 \div 8$

ANSWERS
2. (a) 1.28;
   Check (1.28)(5) = 6.40 or 6.4
   (b) −2.205;
   Check (−2.205)(−14) = 30.870 or 30.87
   (c) 8.65;
   Check (8.65)(−30) = −259.50 or −259.5
   (d) 0.0375;
   Check (0.0375)(8) = 0.3000 or 0.3

3 Divide. Round quotients to the nearest thousandth. If it is a repeating decimal, also write the answer using a bar. Check your answers by multiplying.

(a) $13 \overline{)267.01}$

(b) $6 \overline{)20.5}$

(c) $\dfrac{10.22}{9}$

(d) $16.15 \div 3$

(e) $116.3 \div 11$

ANSWERS

3. (a) 20.539; no repeating digits visible on calculator;
   Check (20.539)(13) = 267.007
   (b) 3.417; $3.41\overline{6}$;
   Check (3.417)(6) = 20.502
   (c) 1.136; $1.13\overline{5}$;
   Check (1.136)(9) = 10.224
   (d) 5.383; $5.38\overline{3}$;
   Check (5.383)(3) = 16.149
   (e) 10.573; $10.5\overline{72}$;
   Check (10.573)(11) = 116.303

> **CAUTION**
> When checking quotients that you've rounded, the check will *not* match the dividend exactly, but it should be very close.

◀◀ **Work Problem 3 at the Side.**

OBJECTIVE 2 Divide a number by a decimal. To divide by a *decimal* divisor, first change the divisor to a whole number. Then divide as before. To see how this is done, write the problem in fraction form. Here is an example.

$$1.2\overline{)6.36} \quad \text{can be written} \quad \dfrac{6.36}{1.2}$$

In **Section 4.1** you learned that multiplying the numerator and denominator by the same number gives an equivalent fraction. We want the divisor (1.2) to be a whole number. Multiplying by 10 will accomplish that.

Decimal divisor → $\dfrac{6.36}{1.2} = \dfrac{(6.36)(10)}{(1.2)(10)} = \dfrac{63.6}{12}$ ← Whole number divisor

The short way to multiply by 10 is to move the decimal point *one place* to the *right* in both the divisor and the dividend.

$$1.2\overline{)6.36} \quad \text{is equivalent to} \quad 12\overline{)63.6}$$

> **NOTE**
> Moving the decimal points the *same* number of places in *both* the divisor and dividend will *not* change the answer.

Dividing by a Decimal Number

Step 1 Count the number of decimal places in the divisor and move the decimal point that many places to the *right*. (This changes the divisor to a whole number.)

Step 2 Move the decimal point in the dividend the *same* number of places to the *right*. (Write in extra zeros if needed.)

Step 3 Write the decimal point in the quotient directly above the decimal point in the dividend. Then divide as usual.

Step 4 If both numbers have the *same sign*, the quotient is *positive*. If they have *different signs*, the quotient is *negative*.

EXAMPLE 4 Dividing by Decimal Numbers

(a) $\dfrac{27.69}{0.003}$

Move the decimal point in the divisor *three* places to the *right* so that 0.003 becomes the whole number 3. In order to move the decimal point in the dividend the same number of places, write in an extra 0.

Continued on Next Page

Section 5.5 Dividing Signed Decimal Numbers

$0.003\overline{)27.690}$ Move decimal points in divisor and dividend. Then line up decimal point in the quotient.

Moving decimal point three places is the same as multiplying by 1000 ⟶ $3\overline{)27690.}$ = 9230. Divide as usual.

(b) Divide -5 by -4.2 and round the quotient to the nearest hundredth.

First consider $5 \div 4.2$. Move the decimal point in the divisor one place to the right so that 4.2 becomes the whole number 42. The decimal point in the dividend starts on the right side of 5 and is also moved one place to the right.

$$4.2\overline{)5.0000} = 1.190 \leftarrow \text{In order to round to hundredths, divide out one } more \text{ place, to thousandths.}$$

```
       1.1 9 0
4.2)5.0 0 0 0
    4 2
      8 0
      4 2
      3 8 0
      3 7 8
          2 0
```

Rounding the quotient to the nearest hundredth gives 1.19. The quotient is *positive* because both the divisor and dividend had the *same* sign (both were negative).

$$-5 \div (-4.2) \approx 1.19$$
Same sign → Positive quotient

▶▶▶ **Work Problem 4 at the Side.**

OBJECTIVE 3 Estimate the answer when dividing decimals. Estimating the answer to a division problem helps you catch errors. Compare the estimate to your exact answer. If they are very different, do the division again.

EXAMPLE 5 Estimating before Dividing

First, use front end rounding to estimate the answer. Then divide to find the exact answer.

$$580.44 \div 2.8$$

Here is how one student solved this problem. She rounded 580.44 to 600 and 2.8 to 3 to estimate the answer.

Estimate: *Exact:*

$3\overline{)600} = 200$ $2.8\overline{)580.44} = 27.3$

```
           2 7.3
2.8)5 8 0.4 4
    5 6
      2 0 4
      1 9 6
          8 4
          8 4
           0
```

Very different; need to rework the problem

Notice that the estimate, which is in the hundreds, is very different from the exact answer, which is only in the tens. This tells the student that she needs to rework the problem. Can you find the error? (The exact answer should be 207.3, which fits with the estimate of 200.)

Continued on Next Page

---

4 Divide. If the quotient does not come out even, round to the nearest hundredth.

(a) $0.2\overline{)1.04}$

(b) $0.06\overline{)1.8072}$

(c) $0.005\overline{)32}$

(d) $-8.1 \div 0.025$

(e) $\dfrac{7}{1.3}$

(f) $-5.3091 \div (-6.2)$

ANSWERS
4. (a) 5.2   (b) 30.12   (c) 6400   (d) −324
(e) 5.38 (rounded)   (f) 0.86 (rounded)

5 Decide whether each answer is reasonable by using front end rounding to estimate the answer. If the exact answer is *not* reasonable, find and correct the error.

(a) $42.75 \div 3.8 = 1.125$

Estimate:

(b) $807.1 \div 1.76 = 458.580$ to nearest thousandth

Estimate:

(c) $48.63 \div 52 = 93.519$ to nearest thousandth

Estimate:

(d) $9.0584 \div 2.68 = 0.338$

Estimate:

> **Work Problem 5 at the Side.**

OBJECTIVE 4 Use the order of operations with decimals. Use the order of operations from **Section 1.8** when a decimal problem involves more than one operation.

Order of Operations

Step 1 Work inside *parentheses* or *other grouping symbols.*

Step 2 Simplify expressions with *exponents.*

Step 3 Do the remaining *multiplications and divisions* as they occur from left to right.

Step 4 Do the remaining *additions and subtractions* as they occur from left to right.

EXAMPLE 6 Using the Order of Operations

Simplify by using the order of operations.

(a) $2.5 + (-6.3)^2 + 9.62$ Apply the exponent: $(-6.3)(-6.3)$ is 39.69

$\ \ 2.5 + 39.69 + 9.62$ Add from left to right.

$\ \ \ \ \ \ \ 42.19 + 9.62$

$\ \ \ \ \ \ \ \ \ \ \ \ \ \ \ \ 51.81$

(b) $1.82 + (5.2 - 6.7)(5.8)$ Work inside parentheses.

$\ \ 1.82 + (-1.5)(5.8)$ Multiply next.

$\ \ 1.82 + (-8.7)$ Add last.

$\ \ \ \ \ \ -6.88$

(c) $3.7^2 - 1.8 \div 5(1.5)$ Apply the exponent.

$\ \ 13.69 - 1.8 \div 5(1.5)$ Multiply and divide from *left to right*, so first divide 1.8 by 5

$\ \ 13.69 - 0.36(1.5)$ Multiply 0.36 by 1.5

$\ \ 13.69 - 0.54$ Subtract last.

$\ \ \ \ \ \ 13.15$

Continued on Next Page

ANSWERS

5. (a) Estimate is $40 \div 4 = 10$; exact answer is not reasonable, should be 11.25
   (b) Estimate is $800 \div 2 = 400$; exact answer is reasonable.
   (c) Estimate is $50 \div 50 = 1$; exact answer is not reasonable, should be 0.935
   (d) Estimate is $9 \div 3 = 3$; exact answer is not reasonable, should be 3.38

> **Work Problem 6 at the Side.**

🖩 **Calculator Tip** Most calculators that have parentheses keys ⓘ ⓘ can handle calculations like those in Example 6 on the previous page just by entering the numbers in the order given. For example, the keystrokes for Example 6(b) are:

             Parentheses

1.82 ⊕ ⓘ 5.2 ⊖ 6.7 ⓘ ⊗ 5.8 ⊜   Answer is −6.88

Standard, four-function calculators generally do *not* give the correct answer if you enter the numbers in the order given. Check the instruction manual that came with your calculator for information on "order of calculations" to see if your model has the rules for order of operations built into it. For a quick check, try entering this problem:

2 ⊕ 2 ⊗ 2 ⊜

If the result is 6, the calculator follows the order of operations. If the result is 8, it does *not* have the rules built into it. Use the space below to explain how this test works.

Answer: The test works because a calculator that follows the order of operations will automatically do the multiplication first. If the calculator does *not* have the rules built into it, it will work from left to right.

Following Order of Operations

2 + 2 × 2   Multiply
⎵⎵⎵ before adding.
2 + 4
⎵⎵⎵
6 ← Correct

Working from Left to Right

2 + 2 × 2
⎵⎵⎵
4   × 2
⎵⎵⎵
✗ 8 ← Incorrect

6 Simplify by using the order of operations.

(a) $-4.6 - 0.79 + 1.5^2$

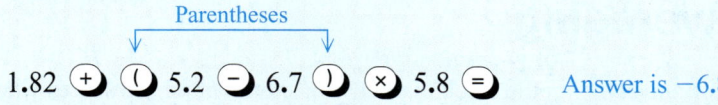

(b) $3.64 \div 1.3(3.6)$

(c) $0.08 + 0.6(2.99 - 3)$

(d) $10.85 - 2.3(5.2) \div 3.2$

ANSWERS
6. (a) −3.14 (b) 10.08 (c) 0.074
   (d) 7.1125

Focus on Real-Data Applications

Dollar-Cost Averaging

To gain the most benefit from investing in the stock market, you should purchase shares when stock prices are low and sell stocks when prices are high. Unfortunately, it is very difficult to predict whether prices will rise or fall from one day to the next. Financial advisors recommend using *dollar-cost averaging* instead of trying to guess how the market will change. By investing the same amount of money at regular intervals, such as the beginning of each month, you will be buying more shares when the stock prices are low and fewer shares when stock prices are high. Of course, *dollar-cost averaging* does not guarantee that you will make a profit. However, over time, your shares should cost less than the market average.

Many financial Web sites report current and historical data on stock performance. Suppose you invested $100 at the first of each month in 2003 and purchased shares of Best Buy at its closing price. The table shows data on Best Buy closing prices during 2003.

Best Buy Co, Inc. (BBY)	Amount Invested	Price per Share	Number of Shares Bought
January	$100	26.09	($100 ÷ 26.09) ≈ 3.8329 shares
February	$100	29.07	
March	$100	26.97	
April	$100	34.58	
May	$100	38.70	
June	$100	43.92	
July	$100	43.65	
August	$100	52.01	
September	$100	47.52	
October	$100	58.31	
November	$100	62.00	
December	$100	60.87	
Total Investment			

Source: http://finance.yahoo.com/

Use a calculator to help answer these questions.

1. Calculate the total amount invested and enter the value in the table.
2. Calculate the number of shares bought each month. Round the number of shares to the nearest ten-thousandth. The calculation for January is shown as an example.
3. Calculate the average market price per share. (*Hint:* Add the monthly prices per share and divide by 12.)
4. Calculate the average price based on *dollar-cost averaging*. (*Hint:* Divide the total amount invested by the total number of shares.)
5. Rank the following scenarios in order of which was the best investment (most profit or least loss). Show the value of each investment in December 2003 as a basis for your answer.
    (a) $1200 invested in Best Buy in January 2003.
    (b) $1200 invested in Best Buy using *dollar-cost averaging*.
    (c) $1200 invested in Best Buy in November 2003.

5.5 Exercises

Divide. See Examples 1, 2, and 4.

1. $27.3 \div (-7)$
2. $-50.4 \div 8$
3. $\dfrac{4.23}{9}$
4. $\dfrac{1.62}{6}$

5. $-20.01 \div (-0.05)$
6. $-16.04 \div (-0.08)$
7. $1.5\overline{)54}$
8. $2.4\overline{)132}$

Use the fact that $108 \div 18 = 6$ to work Exercises 9–16 simply by moving decimal points. See Examples 1, 2, and 4.

9. $1.8\overline{)0.108}$
10. $18\overline{)10.8}$
11. $0.018\overline{)108}$
12. $0.18\overline{)1.08}$

13. $0.18\overline{)10.8}$
14. $0.18\overline{)108}$
15. $18\overline{)0.0108}$
16. $1.8\overline{)0.0108}$

Divide. Round quotients to the nearest hundredth when necessary. See Examples 3 and 4.

17. $4.6\overline{)116.38}$
18. $2.6\overline{)4.992}$
19. $\dfrac{-3.1}{-0.006}$
20. $\dfrac{-1.7}{0.09}$

Divide. Round quotients to the nearest thousandth. See Examples 3 and 4.

21. $-240 \div 9.88$
22. $-7643 \div (-5.36)$
23. $0.034\overline{)342.81}$
24. $0.043\overline{)1748.4}$

RELATING CONCEPTS (EXERCISES 25–26) For Individual or Group Work

First, look back at your work in **Section 5.4**, Exercises 21 and 22. Then look for patterns as you **work Exercises 25 and 26 in order.**

25. Do these division problems.

    $3.77 \div 10 =$ _____ $9.1 \div 10 =$ _____
    $0.886 \div 10 =$ _____ $30.19 \div 10 =$ _____
    $406.5 \div 10 =$ _____ $6625.7 \div 10 =$ _____

    (a) What pattern do you see? Write a "rule" for dividing by 10. What do you think the rule is for dividing by 100? by 1000? Write the rules and try them out on the numbers above.

    (b) Compare your rules to the ones you wrote in **Section 5.4**, Exercise 21.

26. Do these division problems.

    $40.2 \div 0.1 =$ _____ $7.1 \div 0.1 =$ _____
    $0.339 \div 0.1 =$ _____ $15.77 \div 0.1 =$ _____
    $46 \div 0.1 =$ _____ $873 \div 0.1 =$ _____

    (a) What pattern do you see? Write a "rule" for dividing by 0.1. What do you think the rule is for dividing by 0.01? by 0.001? Write the rules and try them out on the numbers above.

    (b) Compare your rules to the ones you wrote in **Section 5.4**, Exercise 22.

Decide whether each answer is reasonable *or* unreasonable *by using front end rounding to estimate the answer. If the exact answer is not reasonable, find and correct the error.*
See Example 5.

27. $37.8 \div 8 = 47.25$
    Estimate: 40/8 = 5

28. $345.6 \div 3 = 11.52$
    Estimate:

29. $54.6 \div 48.1 \approx 1.135$
    Estimate: 50/50 = 1.0

30. $2428.8 \div 4.8 = 50.6$
    Estimate:

31. $307.02 \div 5.1 = 6.2$
    Estimate: 300/5

32. $395.415 \div 5.05 = 78.3$
    Estimate:

33. $9.3 \div 1.25 = 0.744$
    Estimate: 9 1

34. $78 \div 14.2 = 0.182$
    Estimate:

Solve each application problem. Round money answers to the nearest cent when necessary.

35. Rob has discovered that his daughter's favorite brand of tights are on sale. He decided to buy one pair as a surprise for her. How much did he pay?

36. The bookstore has a special price on notepads. How much did Randall pay for one notepad?

37. It will take 21 months for Aimee to pay off her credit card balance of $1408.66. How much is she paying each month?

38. Marcella Anderson bought 2.6 meters of suede fabric for $18.19. How much did she pay per meter?

39. Adrian Webb bought 619 bricks to build a barbecue pit, paying $185.70. Find the cost per brick. (*Hint:* Cost *per* brick means the cost for *one* brick.)

40. Lupe Wilson is a newspaper distributor. Last week she paid the newspaper $130.51 for 842 copies. Find the cost per copy.

41. Darren Jackson earned $476.80 for 40 hours of work. Find his earnings per hour.

42. At a CD manufacturing company, 400 CDs cost $289. Find the cost per CD.

43. It took 16.35 gallons of gas to fill the gas tank of Kim's car. She had driven 346.2 miles since her last fill-up. How many miles per gallon did she get? Round to the nearest tenth.

44. Mr. Rodriquez pays $53.19 each month to Household Finance. How many months will it take him to pay off $1436.13?

Use the table of women's longest long jumps (through June 2005) to answer Exercises 45–50. To find an average, add up the values you are interested in and then divide the sum by the number of values. Round your answer to the nearest hundredth.

Athlete	Country	Year	Length (meters)
Galina Christyakova	USSR	1988	7.52
Jackie Joyner-Kersee	U.S.	1994	7.49
Heike Drechsler	Germany	1992	7.48
Jackie Joyner-Kersee	U.S.	1987	7.45
Jackie Joyner-Kersee	U.S.	1988	7.40
Jackie Joyner-Kersee	U.S.	1991	7.32
Jackie Joyner-Kersee	U.S.	1996	7.20
Chioma Ajunwa	Nigeria	1996	7.12
Fiona May	Italy	2000	7.09
Tatyana Lebedeva	Russia	2004	7.07

Source: CNNSI.com and IAFF.org

45. Find the average length of the long jumps made by Jackie Joyner-Kersee.

46. Find the average length of all the long jumps listed in the table.

47. How much longer was the fifth-longest jump than the sixth-longest jump?

48. If the first-place athlete made five jumps of the same length, what would be the total distance jumped?

49. What was the total length jumped by the top three athletes in the table?

50. How much less was the last-place jump than the next-to-last-place jump?

Simplify by using the order of operations. See Example 6.

51. $7.2 - 5.2 + 3.5^2$

52. $6.2 + 4.3^2 - 9.72$

53. $38.6 + 11.6(10.4 - 13.4)$

54. $2.25 - 1.06(0.85 - 3.95)$

55. $-8.68 - 4.6(10.4) \div 6.4$

56. $25.1 + 11.4 \div 7.5(-3.75)$

57. $33 - 3.2(0.68 + 9) + (-1.3)^2$

58. $0.6 + (-1.89 + 0.11) \div 0.004(0.5)$

Solve each application problem.

59. Soup is on sale at six cans for $3.25, or you can purchase individual cans for $0.57. How much will you save per can if you buy six cans? Round to the nearest cent.

60. Nadia's diet says she can eat 3.5 ounces of chicken nuggets. The package weighs 10.5 ounces and contains 15 nuggets. How many nuggets can Nadia eat?

61. The U.S. Treasury prints 37,000,000 pieces of paper money each day. The printing presses run 24 hours a day. How many pieces of money are printed, to the nearest whole number:

(a) each hour

(b) each minute

(c) each second

(*Source:* www.moneyfactory.com.)

37,000,000 pieces of paper money are printed each day

62. Mach 1 is the speed of sound. Dividing a vehicle's speed by the speed of sound gives its speed on the Mach scale. In 1997, a specially built car with two 110,000-horsepower engines broke the world land speed record by traveling 763.035 miles per hour. The speed of sound changes slightly with the weather. That day it was 748.11 miles per hour. What was the car's Mach speed, to the nearest hundredth? (*Source:* Associated Press.)

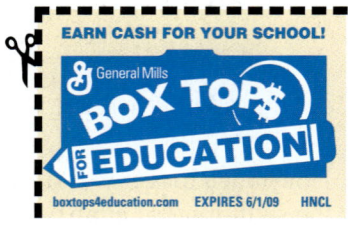

General Mills will give a school 10¢ for each box top logo from its cereals and other products. A school can earn up to $10,000 per year. Use this information to answer Exercises 63–66. Round answers to the nearest whole number. (Source: General Mills.)

63. How many box tops would a school need to collect in one year to earn the maximum amount?

64. (Complete Exercise 63 first.) If a school has 550 children, how many box tops would each child need to collect to reach the maximum?

65. How many box tops would need to be collected during each of the 38 weeks in the school year to reach the maximum amount?

66. How many box tops would each of the 550 children need to collect during each of the 38 weeks of school to reach the maximum amount?

Summary Exercises on Decimals

Write each decimal as a fraction or mixed number in lowest terms.

1. 0.8   **2.** 6.004   **3.** 0.35

Write each decimal in words.

4. 94.5   **5.** 2.0003   **6.** 0.706

Write each decimal in numbers.

7. five hundredths   **8.** three hundred nine ten-thousandths   **9.** ten and seven tenths

Round to the place indicated.

10. 6.1873 to the nearest hundredth   **11.** 0.95 to the nearest tenth   **12.** 0.42025 to the nearest thousandth

13. $0.893 to the nearest cent   **14.** $3.0017 to the nearest cent   **15.** $99.64 to the nearest dollar

Simplify.

16. $-0.27(3.5)$ **17.** $50 - 0.3801$ **18.** $0.35 \div (-0.007)$

19. $\dfrac{-90.18}{-6}$ **20.** $(0.004)(1.22)$ **21.** $1.55 - 3.7$

22. $-0.95 + 10.005$ **23.** $3.6 + 0.718 + 9 + 5.0829$

24. $32.305 - 40 + 0.7$ **25.** $-8.9 + 4^2 \div (-0.02)$

26. $(-0.18 + 2.5) + 4(-0.05)$ **27.** $0.64 \div 16.3$ Round your answer to the nearest hundredth.

Find the perimeter of each figure.

28.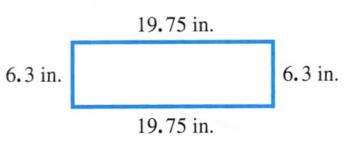

29.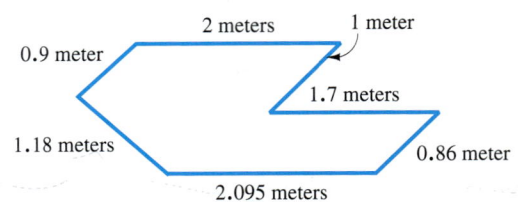

Solve each application problem.

30. At a bakery, Sue Chee bought $7.42 worth of muffins and $10.09 worth of croissants for a staff party and a $0.69 cookie for herself. How much change did she receive from two $10 bills?

31. The Bell family rented a motor home for $375 per week plus $0.35 per mile. What was the rental cost for their three-week vacation trip of 2650 miles?

The table below shows information on two tents for camping. Use the table to answer Exercises 32 and 33.

Tents	Coleman family dome	Eddie Bauer dome tent
Dimensions	13 ft × 13 ft	12 ft × 12 ft
Sleeps	8 campers	6 campers
Sale price	$127	$99

Source: target.com.

32. For the Coleman tent, find:
    (a) the area of the tent floor
    (b) the cost per square foot of floor space, to the nearest cent.

33. Find the same information for the Eddie Bauer tent as you did for the Coleman tent.

Use the information in the table to answer Exercises 34–36.

Animal	Average Weight of Animal (ounces)	Average Weight of Food Eaten Each Day (ounces)
Hamster	3.5	0.4
Queen bee	0.004	?
Hummingbird	?	0.07

Source: NCTM News Bulletin.

34. (a) In how many days will a hamster eat enough food to equal its body weight? Round to the nearest whole number of days.

    (b) If a 140-pound woman ate her body weight of food in the same number of days as the hamster, how much would she eat each day? Round to the nearest tenth.

35. While laying eggs, a queen bee eats eighty times her weight each day. Use this information to fill in one of the missing values in the table.

36. A hummingbird's body weight is about 1.6 times the weight of its daily food intake. Find its body weight.

5.6 Fractions and Decimals

Writing fractions as equivalent decimals can help you do calculations or compare the size of two numbers more easily.

OBJECTIVE 1 **Write fractions as equivalent decimals.** Recall that a fraction is one way to show division (see **Section 1.7**). For example, $\frac{3}{4}$ means $3 \div 4$. If you are doing the division by hand, write it as $4\overline{)3}$. When you do the division, the result is 0.75, the decimal equivalent of $\frac{3}{4}$.

> **Writing a Fraction as an Equivalent Decimal**
>
> *Step 1* Divide the numerator of the fraction by the denominator.
>
> *Step 2* If necessary, round the answer to the place indicated.

> **Work Problem 1 at the Side.**

EXAMPLE 1 Writing Fractions or Mixed Numbers as Decimals

(a) Write $\frac{1}{8}$ as a decimal.

$\frac{1}{8}$ means $1 \div 8$. Write it as $8\overline{)1}$. The decimal point in the dividend is on the *right* side of the 1. Write extra zeros in the dividend so you can continue dividing until the remainder is 0.

$$\frac{1}{8} \rightarrow 1 \div 8 \rightarrow 8\overline{)1} \rightarrow 8\overline{)1.000}$$

— Decimal points lined up
— Three extra zeros needed
— Remainder is 0

Therefore, $\frac{1}{8} = 0.125$

To check, write 0.125 as a fraction, then write it in lowest terms.

$$0.125 = \frac{125}{1000} \quad \text{In lowest terms:} \quad \frac{125 \div 125}{1000 \div 125} = \frac{1}{8} \quad \begin{cases} \text{Original} \\ \text{fraction} \end{cases}$$

🖩 **Calculator Tip** When using your calculator to write fractions as decimals, enter the numbers from the top down. Remember that the *order* in which you enter the numbers *does* matter in division. Example 1(a) above works like this:

$\frac{1}{8}$ ↓ Top down   Enter 1 ÷ 8 =   Answer is 0.125

What happens if you enter 8 ÷ 1 =? Do you see why that cannot possibly be correct? (Answer: $8 \div 1 = 8$. A proper fraction like $\frac{1}{8}$ cannot be equivalent to a whole number.)

Continued on Next Page

OBJECTIVES

1. Write fractions as equivalent decimals.
2. Compare the size of fractions and decimals.

❶ Rewrite each fraction so you could do the division by hand. Do *not* complete the division.

(a) $\frac{1}{9}$ is written $9\overline{)}$

(b) $\frac{2}{3}$ is written $\overline{)}$

(c) $\frac{5}{4}$ is written $\overline{)}$

(d) $\frac{3}{10}$ is written $\overline{)}$

(e) $\frac{21}{16}$ is written $\overline{)}$

(f) $\frac{1}{50}$ is written $\overline{)}$

ANSWERS
1. (a) $9\overline{)1}$ (b) $3\overline{)2}$ (c) $4\overline{)5}$
   (d) $10\overline{)3}$ (e) $16\overline{)21}$ (f) $50\overline{)1}$

2 Write each fraction or mixed number as a decimal.

(a) $\dfrac{1}{4}$

(b) $2\dfrac{1}{2}$

(c) $\dfrac{5}{8}$

(d) $4\dfrac{3}{5}$

(e) $\dfrac{7}{8}$

ANSWERS
2. (a) 0.25 (b) 2.5 (c) 0.625
   (d) 4.6 (e) 0.875

(b) Write $2\dfrac{3}{4}$ as a decimal.

One method is to divide 3 by 4 to get 0.75 for the fraction part. Then add the whole number part to 0.75.

$$\dfrac{3}{4} \rightarrow \begin{array}{r} 0.75 \\ 4\overline{)3.00} \\ \underline{2\,8} \\ 20 \\ \underline{20} \\ 0 \end{array} \quad \text{—Fraction part} \longrightarrow \begin{array}{r} 2.00 \leftarrow \text{Whole number part} \\ +\,0.75 \\ \hline 2.75 \end{array}$$

So, $2\dfrac{3}{4} = 2.75$ **Check** $2.75 = 2\dfrac{75}{100} = 2\dfrac{3}{4}$ ← Lowest terms

↑ ↑
Whole number parts match.

A second method is to write $2\dfrac{3}{4}$ as an improper fraction before dividing numerator by denominator.

$$2\dfrac{3}{4} = \dfrac{11}{4}$$

$$\dfrac{11}{4} \rightarrow 11 \div 4 \rightarrow 4\overline{)11} \rightarrow \begin{array}{r} 2.75 \\ 4\overline{)11.00} \\ \underline{8} \\ 3\,0 \\ \underline{2\,8} \\ 20 \\ \underline{20} \\ 0 \end{array} \leftarrow \text{Two extra zeros needed}$$

Whole number parts match.
So, $2\dfrac{3}{4} = 2.75$

$\dfrac{3}{4}$ is equivalent to $\dfrac{75}{100}$ or 0.75

◀◀◀ **Work Problem 2 at the Side.**

EXAMPLE 2   **Writing a Fraction as a Decimal with Rounding**

Write $\dfrac{2}{3}$ as a decimal and round to the nearest thousandth.

$\dfrac{2}{3}$ means $2 \div 3$. To round to thousandths, divide out one *more* place, to ten-thousandths.

$$\dfrac{2}{3} \rightarrow 2 \div 3 \rightarrow 3\overline{)2} \rightarrow \begin{array}{r} 0.6666 \\ 3\overline{)2.0000} \\ \underline{1\,8} \\ 20 \\ \underline{18} \\ 20 \\ \underline{18} \\ 20 \\ \underline{18} \\ 2 \end{array} \leftarrow \begin{array}{l} \text{Four zeros needed} \\ \text{for ten-thousandths} \end{array}$$

Written as a repeating decimal, $\dfrac{2}{3} = 0.\overline{6}$ ← Bar above repeating digit

Rounded to the nearest thousandth, $\dfrac{2}{3} \approx 0.667$

Continued on Next Page

🖩 Calculator Tip
Try Example 2 on the previous page on your calculator. Enter 2 ÷ 3. Which answer do you get?

| 0.6666666667 | or | 0.666666666 |

Most *scientific* and *graphing* calculators will show a 7 as the last digit. Because the 6s keep on repeating forever, the calculator automatically rounds in the last decimal place it has room to show. If you have a 10-digit display space, the calculator is rounding as shown below.

0.6666666666 (11 digits)   rounds to   0.666666667

Next digit is *5 or more,* so 6 rounds to 7

Other calculators, especially standard, four-function ones, may *not* round. They just cut off, or *truncate,* the extra digits. Such a calculator would show 0.6666666 in the display.

Would this difference in calculators show up when changing $\frac{1}{3}$ to a decimal? Why not? (Answer: The repeating digit is a 3, which is *4 or less,* so it stays as a 3 whether it's rounded or not.)

Work Problem 3 at the Side. ▶▶▶

OBJECTIVE 2 Compare the size of fractions and decimals. You can use a number line to compare fractions and decimals. For example, the number line below shows the space between 0 and 1. The locations of some commonly used fractions are marked, along with their decimal equivalents.

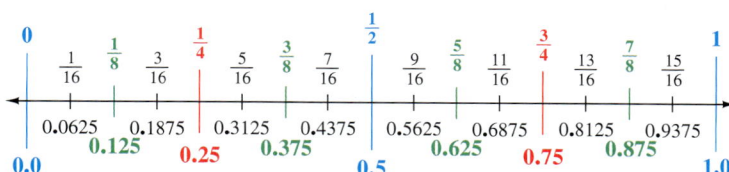

The next number line shows the locations of some commonly used fractions between 0 and 1 that are equivalent to repeating decimals. The decimal equivalents use a bar above repeating digits.

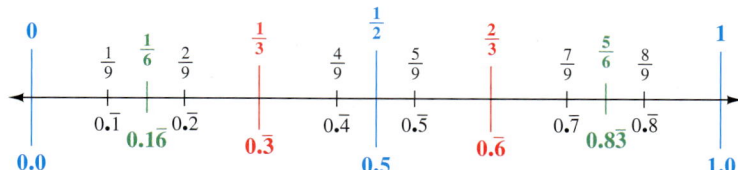

EXAMPLE 3 Using a Number Line to Compare Numbers

Use the number lines above to decide whether to write >, <, or = in the blank between each pair of numbers.

(a) 0.6875 _____ 0.625

You learned in **Section 1.2** that the number farther to the right on the number line is the greater number. On the first number line, 0.6875 is to the *right* of 0.625, so use the > symbol.

0.6875   *is greater than*   0.625   0.6875 > 0.625

Continued on Next Page

---

3 🖩 Write as decimals. Round to the nearest thousandth.

(a) $\frac{1}{3}$

(b) $2\frac{7}{9}$

(c) $\frac{10}{11}$

(d) $\frac{3}{7}$

(e) $3\frac{5}{6}$

ANSWERS

3. (a) $\frac{1}{3} \approx 0.333$ (b) $2\frac{7}{9} \approx 2.778$

(c) $\frac{10}{11} \approx 0.909$ (d) $\frac{3}{7} \approx 0.429$

(e) $3\frac{5}{6} \approx 3.833$

Chapter 5 Rational Numbers: Positive and Negative Decimals

4 Use the number lines on the previous page to help you decide whether to write <, >, or = in each blank.

(a) 0.4375 _____ 0.5

(b) 0.75 _____ 0.6875

(c) 0.625 _____ 0.0625

(d) $\frac{2}{8}$ _____ 0.375

(e) $0.8\overline{3}$ _____ $\frac{5}{6}$

(f) $\frac{1}{2}$ _____ $0.\overline{5}$

(g) $0.\overline{1}$ _____ $0.1\overline{6}$

(h) $\frac{8}{9}$ _____ $0.\overline{8}$

(i) $0.\overline{7}$ _____ $\frac{4}{6}$

(j) $\frac{1}{4}$ _____ 0.25

5 Arrange each group in order from smallest to largest.

(a) 0.7, 0.703, 0.7029

(b) 6.39, 6.309, 6.4, 6.401

(c) 1.085, $1\frac{3}{4}$, 0.9

(d) $\frac{1}{4}, \frac{2}{5}, \frac{3}{7}$, 0.428

ANSWERS
4. (a) < (b) > (c) > (d) < (e) =
   (f) < (g) < (h) = (i) > (j) =
5. (a) 0.7, 0.7029, 0.703
   (b) 6.309, 6.39, 6.4, 6.401
   (c) 0.9, 1.085, $1\frac{3}{4}$
   (d) $\frac{1}{4}, \frac{2}{5}$, 0.428, $\frac{3}{7}$

(b) $\frac{3}{4}$ _____ 0.75

On the first number line, $\frac{3}{4}$ and 0.75 are at the same point on the number line. They are equivalent, so use the = symbol.

$$\frac{3}{4} = 0.75$$

(c) 0.5 _____ $0.\overline{5}$

On the second number line, 0.5 is to the *left* of $0.\overline{5}$ (which is actually 0.555 . . .), so use the < symbol.

0.5 is less than $0.\overline{5}$ $0.5 < 0.\overline{5}$

(d) $\frac{2}{6}$ _____ $0.\overline{3}$

Write $\frac{2}{6}$ in lowest terms as $\frac{1}{3}$.
On the second number line you can see that $\frac{1}{3}$ and $0.\overline{3}$ are equivalent.

$$\frac{1}{3} = 0.\overline{3}$$

Work Problem 4 at the Side.

You can also compare fractions by first writing each one as a decimal. You can then compare the decimals by writing each one with the same number of decimal places.

EXAMPLE 4 Arranging Numbers in Order

Write each group of numbers in order, from smallest to largest.

(a) 0.49   0.487   0.4903

It is easier to compare decimals if they are all tenths, or all hundredths, and so on. Because 0.4903 has four decimal places (ten-thousandths), write zeros to the right of 0.49 and 0.487 so they also have four decimal places. Writing zeros to the right of a decimal number does *not* change its value (see **Section 5.3**). Then find the smallest and largest number of ten-thousandths.

0.49 = 0.4900 = **4900** ten-thousandths ← 4900 is in the middle.
0.487 = 0.4870 = **4870** ten-thousandths ← 4870 is the smallest.
0.4903 = **4903** ten-thousandths ← 4903 is the largest.

From smallest to largest, the correct order is shown below.

0.487   0.49   0.4903

(b) $2\frac{5}{8}$ 2.63   2.6

Write $2\frac{5}{8}$ as $\frac{21}{8}$ and divide $8\overline{)21}$ to get the decimal form, 2.625. Then, because 2.625 has three decimal places, write zeros so all the numbers have three decimal places.

$2\frac{5}{8}$ = 2.625 = 2 and **625** thousandths ← 625 is in the middle.
2.63 = 2.630 = 2 and **630** thousandths ← 630 is the largest.
2.6 = 2.600 = 2 and **600** thousandths ← 600 is the smallest.

From smallest to largest, the correct order is shown below.

2.6 $2\frac{5}{8}$ 2.63

Work Problem 5 at the Side.

5.6 Exercises

FOR EXTRA HELP: Addison-Wesley Math Tutor Center | MathXL | Digital Video Tutor CD 3 Videotape 5 | Student's Solutions Manual | MyMathLab | MathXL Tutorials on CD

Write each fraction or mixed number as a decimal. Round to the nearest thousandth when necessary. See Examples 1 and 2.

1. $\dfrac{1}{2}$
2. $\dfrac{1}{4}$
3. $\dfrac{3}{4}$
4. $\dfrac{1}{10}$
5. $\dfrac{3}{10}$

6. $\dfrac{7}{10}$
7. $\dfrac{9}{10}$
8. $\dfrac{4}{5}$
9. $\dfrac{3}{5}$
10. $\dfrac{2}{5}$

11. $\dfrac{7}{8}$
12. $\dfrac{3}{8}$
13. $2\dfrac{1}{4}$
14. $1\dfrac{1}{2}$
15. $14\dfrac{7}{10}$

16. $23\dfrac{3}{5}$
17. $3\dfrac{5}{8}$
18. $2\dfrac{7}{8}$
19. $\dfrac{1}{3}$
20. $\dfrac{2}{3}$

21. $\dfrac{5}{6}$
22. $\dfrac{1}{6}$
23. $1\dfrac{8}{9}$
24. $5\dfrac{4}{7}$

RELATING CONCEPTS (EXERCISES 25–28) For Individual or Group Work

Use your knowledge of fractions and decimals as you **work Exercises 25–28 in order.**

25. **(a)** Explain how you can tell that Keith made an error *just by looking at his final answer*. Here is his work.

$$\dfrac{5}{9} = 5\overline{)9.0}^{\,1.8} \quad \text{so} \quad \dfrac{5}{9} = 1.8$$

(b) Show the correct way to change $\dfrac{5}{9}$ to a decimal. Explain why your answer makes sense.

26. **(a)** How can you prove to Sandra that $2\dfrac{7}{20}$ is *not* equivalent to 2.035? Here is her work.

$$2\dfrac{7}{20} = 20\overline{)7.00}^{\,0.35} \quad \text{so} \quad 2\dfrac{7}{20} = 2.035$$

(b) What is the correct answer? Show how to prove that it is correct.

27. Ving knows that $\dfrac{3}{8} = 0.375$. How can he write $1\dfrac{3}{8}$ as a decimal *without* having to do a division? How can he write $3\dfrac{3}{8}$ as a decimal? $295\dfrac{3}{8}$? Explain your answer.

28. Iris has found a shortcut for writing mixed numbers as decimals.

$$2\dfrac{7}{10} = 2.7 \qquad 1\dfrac{13}{100} = 1.13$$

Does her shortcut work for all mixed numbers? Explain when it works and why it works.

Find the decimal or fraction equivalent for each number. Write fractions in lowest terms.

	Fraction	Decimal		Fraction	Decimal
29.	_____	0.4	30.	_____	0.75
31.	_____	0.625	32.	_____	0.111
33.	_____	0.35	34.	_____	0.9
35.	$\frac{7}{20}$	_____	36.	$\frac{1}{40}$	_____
37.	_____	0.04	38.	_____	0.52
39.	_____	0.15	40.	_____	0.85
41.	$\frac{1}{5}$	_____	42.	$\frac{1}{8}$	_____
43.	_____	0.09	44.	_____	0.02

Solve each application problem.

45. The average length of a newborn baby is 20.8 inches. Charlene's baby is 20.08 inches long. Is her baby longer or shorter than the average? By how much?

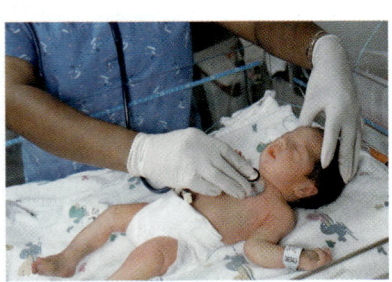

46. The patient in room 830 is supposed to get 8.3 milligrams of medicine. She was actually given 8.03 milligrams. Did she get too much or too little medicine? What was the difference?

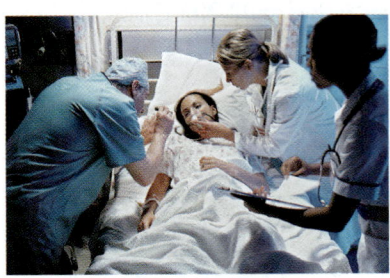

Section 5.6 Fractions and Decimals 355

47. The label on the bottle of vitamins says that each capsule contains 0.5 gram of calcium. When checked, each capsule had 0.505 gram of calcium. Was there too much or too little calcium? What was the difference?

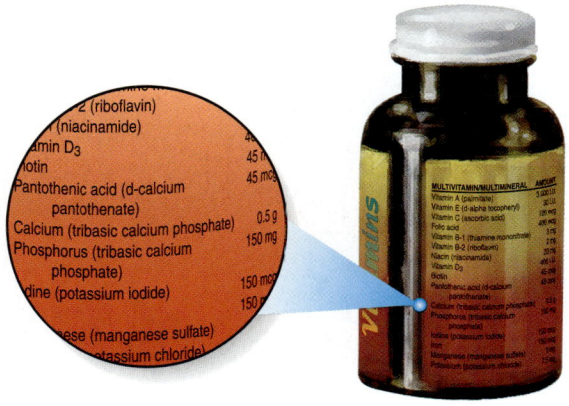

48. The glass mirror of the Hubble telescope had to be repaired in space because it would not focus properly. The problem was that the mirror's outer edge had a thickness of 0.6248 cm when it was supposed to be 0.625 cm. Was the edge too thick or too thin? By how much? (*Source:* NASA.)

49. Precision Medical Parts makes an artificial heart valve that must measure between 0.998 centimeter and 1.002 centimeters. Circle the lengths that are acceptable.

1.01 cm   0.9991 cm   1.0007 cm   0.99 cm

50. The white rats in a medical experiment must start out weighing between 2.95 ounces and 3.05 ounces. Circle the weights that can be used.

3.0 ounces   2.995 ounces   3.055 ounces
3.005 ounces

51. Ginny Brown had hoped her crops would get $3\frac{3}{4}$ inches of rain this month. The newspaper said the area received 3.8 inches of rain. Was that more or less than Ginny had hoped for? By how much?

52. The rats in the experiment in Exercise 50 gained $\frac{3}{8}$ ounce. They were expected to gain 0.3 ounce. Was their actual gain more or less than expected? By how much?

Arrange each group of numbers in order from smallest to largest. See Example 4.

53. 0.54, 0.5455, 0.5399

54. 0.76, 0.7, 0.7006

55. 5.8, 5.79, 5.0079, 5.804

56. 12.99, 12.5, 13.0001, 12.77

356 Chapter 5 Rational Numbers: Positive and Negative Decimals

57. 0.628, 0.62812, 0.609, 0.6009

0.6009 / 0.609 / 0.628 / 0.62812

58. 0.27, 0.281, 0.296, 0.3

59. 5.8751, 4.876, 2.8902, 3.88

60. 0.98, 0.89, 0.904, 0.9

61. 0.043, 0.051, 0.006, $\frac{1}{20}$

62. 0.629, $\frac{5}{8}$, 0.65, $\frac{7}{10}$

0.006 / 0.043 / 0.05 / 0.051

63. $\frac{3}{8}, \frac{2}{5}$, 0.37, 0.4001

64. 0.1501, 0.25, $\frac{1}{10}, \frac{1}{5}$

The friends on the first page of this chapter found four boxes of fishing line in a sale bin. They know that the thicker the line, the stronger it is. The diameter of the fishing line is its thickness. Use the information on the boxes to answer Exercises 65–68.

65. Which color box has the strongest line?

66. Which color box has the line with the least strength?

67. What is the difference in line diameter between the weakest and strongest lines?

68. What is the difference in line diameter between the blue and purple boxes?

Some rulers for technical drawings show each inch divided into tenths. Use this scale drawing for Exercises 69–74. Change the measurements on the drawing to decimals and round them to the nearest tenth of an inch.

69. Length **(a)** is _____

70. Length **(b)** is _____

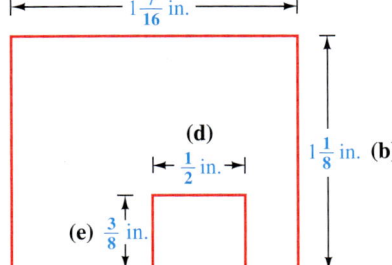

71. Length **(c)** is _____

72. Length **(d)** is _____

73. Length **(e)** is _____

74. Length **(f)** is _____

5.7 Problem Solving with Statistics: Mean, Median, Mode, and Variability

The word *statistics* originally came from words that mean *state numbers*. State numbers refer to numerical information, or *data,* gathered by the government such as the number of births, deaths, or marriages in a population. Today the word *statistics* has a much broader meaning; data from the fields of economics, social science, science, and business can all be organized and studied under the branch of mathematics called *statistics*.

OBJECTIVES

1. Find the mean of a list of numbers.
2. Find a weighted mean.
3. Find the median.
4. Find the mode.
5. Evaluate the variability of a set of data by finding the range of values.

OBJECTIVE 1 Find the mean of a list of numbers. Making sense of a long list of numbers can be difficult. So when you analyze data, one of the first things to look for is a *measure of central tendency*—a single number that you can use to represent the entire list of numbers. One such measure is the *average* or **mean**. The mean can be found with the following formula.

Finding the Mean (Average)

$$\text{mean} = \frac{\text{sum of all values}}{\text{number of values}}$$

EXAMPLE 1 Finding the Mean (Average)

David had test scores of 84, 90, 95, 98, and 88. Find his mean (average) score.

Use the formula for finding the mean. Add up all the test scores and then divide the sum by the number of tests.

$$\text{mean} = \frac{84 + 90 + 95 + 98 + 88}{5} \quad \leftarrow \text{Sum of test scores} \\ \leftarrow \text{Number of tests}$$

$$\text{mean} = \frac{455}{5} \quad \text{Divide.}$$

$$\text{mean} = 91$$

David has a mean (average) score of 91.

Work Problem 1 at the Side.

EXAMPLE 2 Applying the Mean (Average)

The sales of photo albums at Sarah's Card Shop for each day last week were $86, $103, $118, $117, $126, $158, and $149. Find the mean daily sales of photo albums.

To find the mean, add all the daily sales amounts and then divide the sum by the number of days (7).

$$\text{mean} = \frac{\$86 + \$103 + \$118 + \$117 + \$126 + \$158 + \$149}{7} \quad \leftarrow \text{Sum of sales} \\ \leftarrow \text{Number of days}$$

$$\text{mean} = \frac{\$857}{7}$$

$$\text{mean} \approx \$122.43 \quad \text{Rounded to nearest cent}$$

Work Problem 2 at the Side.

---

1 Tanya had test scores of 96, 98, 84, 88, 82, and 92. Find her mean (average) score.

2 Find the mean for each list of numbers.

(a) Monthly utility bills of $25.12, $42.58, $76.19, $32, $81.11, $26.41, $19.76, $59.32, $71.18, and $21.03

(b) The sales for one year at eight different office supply stores: $749,820; $765,480; $643,744; $824,222; $485,886; $668,178; $702,294; $525,800

ANSWERS

1. 90

2. (a) $\dfrac{\$454.70}{10} = \$45.47$

   (b) $\dfrac{\$5,365,424}{8} = \$670,678$

3 Alison Nakano works downtown. Some days she can park in cheap lots that charge $6 or $7. Other days, she has to park in lots that charge $9 or $10. Last month she kept track of the amount she spent each day for parking and the number of days she spent that amount. Find her average daily parking cost.

Parking Fee	Frequency
$ 6	2
$ 7	6
$ 8	3
$ 9	4
$10	6

OBJECTIVE 2 Find a weighted mean. Some items in a list of data might appear more than once. In this case, we find a **weighted mean**, in which each value is "weighted" by multiplying it by the number of times it occurs.

EXAMPLE 3 Finding a Weighted Mean

The table below shows the amount of contribution and the number of times the amount was given (frequency) to a food pantry. Find the weighted mean.

Contribution Value	Frequency	
$ 3	4	← 4 people each contributed $3
$ 5	2	
$ 7	1	
$ 8	5	
$ 9	3	
$10	2	
$12	1	
$13	2	

The same amount was given by more than one person: for example, $3 was given by four people, and $8 was given by five people. Other amounts, such as $12, were given by only one person.

To find the mean, multiply each contribution value by its frequency. Then add the products. Next, add the numbers in the *frequency* column to find the total number of values, that is, the total number of people who contributed money.

Value	Frequency	Product
$ 3	4	($3 • 4) = $12
$ 5	2	($5 • 2) = $10
$ 7	1	($7 • 1) = $ 7
$ 8	5	($8 • 5) = $40
$ 9	3	($9 • 3) = $27
$10	2	($10 • 2) = $20
$12	1	($12 • 1) = $12
$13	2	($13 • 2) = $26
Totals	**20**	**$154**

Finally, divide the totals.

$$\text{mean} = \frac{\$154}{20} = \$7.70$$

The mean contribution to the food pantry was $7.70.

◀◀◀ **Work Problem 3 at the Side.**

ANSWERS
3. average ≈ $8.29 (to nearest cent)

A common use of the weighted mean is to find a student's *grade point average (GPA)*, as shown in the next example.

EXAMPLE 4 Applying the Weighted Mean

Find the GPA (grade point average) for a student who earned the following grades last semester. Assume A = 4, B = 3, C = 2, D = 1, and F = 0. The number of credits determines how many times the grade is counted (the frequency).

Course	Credits	Grade	Credits • Grade
Mathematics	4	A (= 4)	4 • 4 = 16
Speech	3	C (= 2)	3 • 2 = 6
English	3	B (= 3)	3 • 3 = 9
Computer science	2	A (= 4)	2 • 4 = 8
Theater	2	D (= 1)	2 • 1 = 2
Totals	14		41

It is common to round grade point averages to the nearest hundredth. So the grade point average for this student is rounded to **2.93**.

$$\text{GPA} = \frac{41}{14} \approx 2.93$$

Work Problem 4 at the Side.

④ Find the GPA (grade point average) for a student who earned the following grades. Round to the nearest hundredth.

Course	Credits	Grade
Mathematics	5	A (= 4)
English	3	C (= 2)
Biology	4	B (= 3)
History	3	B (= 3)

OBJECTIVE 3 Find the median. Because it can be affected by extremely high or low numbers, the mean is often a poor indicator of central tendency for a list of numbers. In cases like this, another measure of central tendency, called the *median,* can be used. The **median** divides a group of numbers in half; half the numbers lie above the median, and half lie below the median.

Find the median by listing the numbers *in order* from *smallest* to *largest*. If the list contains an *odd* number of items, the median is the *middle number*.

EXAMPLE 5 Finding the Median (Odd Number of Items)

Find the median for this list of prices.

$7, $23, $15, $6, $18, $12, $24

First arrange the numbers in numerical order from smallest to largest.

Smallest → 6, 7, 12, 15, 18, 23, 24 ← Largest

Next, find the middle number in the list.

6, 7, 12, **15**, 18, 23, 24
Three are below. ↓ Three are above.
Middle number

The median price is $15.

Work Problem 5 at the Side.

⑤ Find the median for the following number of customers helped each hour at the order desk.

35, 33, 27, 31, 39, 50, 59, 25, 30

If a list contains an *even* number of items, there is no single middle number. In this case, the median is defined as the mean (average) of the *middle two* numbers.

ANSWERS

4. GPA = $\frac{47}{15}$ ≈ 3.13

5. 33 customers (the middle number when the numbers are arranged from smallest to largest)

6 Find the median for this list of measurements.

178 ft, 261 ft, 126 ft, 189 ft, 121 ft, 195 ft

7 Find the mode for each list of numbers.

(a) Ages of part-time employees (in years): 28, 16, 22, 28, 34, 22, 28

(b) Total points on a screening exam: 312, 219, 782, 312, 219, 426, 507, 600

(c) Monthly commissions of salespeople: $1706, $1289, $1653, $1892, $1301, $1782

ANSWERS

6. $\dfrac{178 + 189}{2} = 183.5$ ft

7. (a) 28 years
   (b) bimodal, 219 points and 312 points (this list has two modes)
   (c) no mode (no number occurs more than once)

EXAMPLE 6 Finding the Median (Even Number of Items)

Find the median for this list of ages, in years.

74, 7, 15, 13, 25, 28, 47, 59, 32, 68

First arrange the numbers in numerical order from smallest to largest. Then, because the list has an even number of ages, find the middle *two* numbers.

Smallest ⟶ 7, 13, 15, 25, **28, 32**, 47, 59, 68, 74 ⟵ Largest

Middle two numbers

The median age is the mean of the two middle numbers.

$$\text{median} = \dfrac{28 + 32}{2} = \dfrac{60}{2} = 30 \text{ years}$$

Work Problem 6 at the Side.

OBJECTIVE 4 Find the mode. Another statistical measure is the **mode**, which is the number that occurs *most often* in a list of numbers. For example, the test scores for ten students are shown below.

74, 81, 39, **74**, 82, 80, 100, 92, **74**, 85

The mode is 74. Three students earned a score of 74, so 74 appears more times on the list than any other score. It is *not* necessary to place the numbers in numerical order when looking for the mode, although that may help you find it more easily.

A list can have two modes; such a list is sometimes called *bimodal*. If no number occurs more frequently than any other number in a list, the list has *no mode*.

EXAMPLE 7 Finding the Mode

Find the mode for each list of numbers.

(a) 51, 32, 49, 51, 49, 90, 49, 60, 17, 60
   The number 49 occurs three times, which is more often than any other number. Therefore, **49** is the mode.

(b) 482, 485, 483, 485, 487, 487, 489, 486
   Because both **485** and **487** occur twice, each is a mode. This list is *bimodal*.

(c) $10,708; $11,519; $10,972; $12,546; $13,905; $12,182
   No price occurs more than once. This list has *no mode*.

Measures of Central Tendency

The **mean** is the sum of all the values divided by the number of values. It is the mathematical *average*.

The **median** is the *middle number* in a group of values that are listed from smallest to largest. It divides a group of numbers in half.

The **mode** is the value that occurs *most often* in a group of values.

Work Problem 7 at the Side.

Section 5.7 Problem Solving with Statistics: Mean, Median, Mode, and Variability

OBJECTIVE 5 **Evaluate the variability of a set of data by finding the range of values.** If two students each have a mean (average) score of 60 on their math tests, you might think they have done work of equal quality. However, a closer look at the test scores below gives a different impression.

	Test 1	Test 2	Test 3	Test 4	Test 5	Mean	Median
Student Y	55	60	60	60	65	60	60
Student Z	20	40	60	80	100	60	60

The means and medians are all 60. But student Y's scores are all clustered around 60. Student Z's scores, however, have greater *variability* because they are spread over a wider range, from 20 to 100. The **variability** of a set of data is the spread of the data around the mean. A quick way to evaluate the variability is to look at the *range of values*. To find the range, subtract the lowest value from the highest value.

EXAMPLE 8 Finding the Range

(a) Find the range of test scores for student Y, shown above.

Subtract student Y's lowest test score of 55 from the highest test score of 65.

$65 - 55 = 10$ ← Range of scores for student Y

(b) Find the range of test scores for student Z, shown above.

Highest score ↓
$100 - 20 = 80$ ← Range of scores for student Z
↑ Lowest score

Work Problem 8 at the Side.

Generally, the greater the *range* in a set of values, the greater the *variability*. When analyzing data, look at the mean and/or median and also look at the variability. For example, if you were the teacher of students Y and Z in Example 8 above, you would analyze their test data quite differently. Student Y has consistently gotten fairly low scores and may need some extra help. Student Z's scores, on the other hand, have improved dramatically, and he or she will probably do well on future tests.

EXAMPLE 9 Evaluating the Variability of Data

(a) Which set of data shows greater variability in the points scored in a basketball game? Show your work.

	Game 1	Game 2	Game 3	Game 4	Game 5	Mean
Player A	18	9	27	3	25	16.4
Player B	18	16	17	17	15	16.6

Range for player A = $27 - 3 = 24$

Range for player B = $18 - 15 = 3$

Player A's data has a greater range and greater variability.

Continued on Next Page

---

8 Find the mean, median, and range of values for each set of data.

(a) Ages of students in Classroom B: 21, 23, 22, 21, 22, 23

Ages of students in Classroom C: 31, 18, 25, 17, 23, 21

(b) Expenses in December: $225, $350, $100, $325, $700

Expenses in January: $300, $325, $350, $325, $350

ANSWERS

8. (a) Classroom B: mean = 22, median = 22, range = 2
   Classroom C: mean = 22.5, median = 22, range = 14
   (b) December: mean = $340, median = $325, range = $600
   January: mean = $330, median = $325, range = $50

9 (a) Which set of temperature data, in degrees, has less variability? Both cities have a mean high of 72 degrees. Show your work.

City A had daily highs of 69, 73, 70, 75, 71, 74.

City B had daily highs of 76, 55, 63, 97, 92, 49.

(b) Which city would you rather visit on vacation? Explain why.

(c) Which set of weekly sales figures has greater variability? Show your work.

Salesperson Y: $8000; $7500; $7800; $8200; $8300

Salesperson Z: $8000; $12,000; $3000; $14,500; $2500

(d) Which salesperson would you like to hire? Explain why.

(b) If you are a coach, which player would you like on your team, player A or player B? Explain why.

Answers may vary. Some things to consider are: Player A had some very good games, and the means for the two players are nearly the same. But player B is more consistent and perhaps more reliable. You would want to know more details about why player A scored so few points in games 2 and 4.

◀◀◀ **Work Problem 9 at the Side.**

ANSWERS

9. **(a)** City A: $75 - 69 = 6$;
City B: $97 - 49 = 48$
City A's temperatures have less variability.
(b) Answers will vary; be sure you explained why you chose a particular city.
(c) Salesperson Y: $8300 - $7500 = $800;
Salesperson Z: $14,500 - $2500 = $12,000
Salesperson Z's figures have greater variability.
(d) Answers will vary; be sure you explained why you chose a particular salesperson.

5.7 Exercises

FOR EXTRA HELP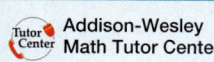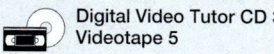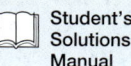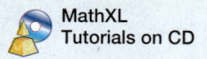

Find the mean for each list of numbers. Round answers to the nearest tenth when necessary. See Example 1.

1. Ages of infants at the child care center (in months): 4, 9, 6, 4, 7, 10, 9

2. Monthly electric bills: $53, $77, $38, $29, $49, $48

3. Final exam scores: 92, 51, 59, 86, 68, 73, 49, 80

4. Quiz scores: 18, 25, 21, 8, 16, 13, 23, 19

5. Annual salaries: $31,900; $32,850; $34,930; $39,712; $38,340, $60,000

6. Numbers of people attending baseball games: 27,500; 18,250; 17,357; 14,298; 33,110

Solve each application problem. See Examples 2 and 3.

7. The Athletic Shoe Store sold shoes at the following prices: $75.52, $36.15, $58.24, $21.86, $47.68, $106.57, $82.72, $52.14, $28.60, $72.92. Find the mean shoe sales amount.

8. In one evening, a waitress collected the following checks from her dinner customers: $30.10, $42.80, $91.60, $51.20, $88.30, $21.90, $43.70, $51.20. Find the mean dinner check amount.

9. The table below shows the face value (policy amount) of life insurance policies and the number of policies sold for each amount by the New World Life Company during one week. Find the weighted mean amount for the policies sold.

Policy Amount	Number of Policies Sold
$ 10,000	6
$ 20,000	24
$ 25,000	12
$ 30,000	8
$ 50,000	5
$100,000	3
$250,000	2

10. Detroit Metro-Sales Company prepared the table below showing the gasoline mileage obtained by each of the cars in their automobile fleet. Find the weighted mean to determine the miles per gallon for the fleet of cars.

Miles per Gallon	Number of Autos
15	5
20	6
24	10
30	14
32	5
35	6
40	4

Find the weighted mean. Round answers to the nearest tenth when necessary. See Example 3.

11.
Quiz Score	Frequency
3	4
5	2
6	5
8	5
9	2

12.
Credits per Student	Frequency
9	3
12	5
13	2
15	6
18	1

13.
Hours Worked	Frequency
12	4
13	2
15	5
19	3
22	1
23	5

14.
Students per Class	Frequency
25	1
26	2
29	5
30	4
32	3
33	5

Find the GPA (grade point average) for students earning the following grades. Assume A = 4, B = 3, C = 2, D = 1, and F = 0. Round answers to the nearest hundredth. See Example 4.

15.
Course	Credits	Grade
Biology	4	B
Biology lab	2	A
Mathematics	5	C
Health	1	F
Psychology	3	B

16.
Course	Credits	Grade
Chemistry	3	A
English	3	B
Mathematics	4	B
Theater	2	C
Astronomy	3	C

17. Look again at the grades in Exercise 15. Find the student's GPA in each of these situations.
    (a) The student earned a B instead of an F in the 1-credit class.
    (b) The student earned a B instead of a C in the 5-credit class.
    (c) Both (a) and (b) happened.

18. List the credits for the courses you're taking at this time. List the lowest grades you think you will earn in each class and find your GPA. Then list the highest grades you think you will earn and find your GPA.

Find the median for each list of numbers. See Examples 5 and 6.

19. Number of e-mail messages received:
    9, 12, 14, 15, 23, 24, 28

20. Deliveries by a newspaper distributor:
    99, 108, 109, 123, 126, 129, 146, 168, 170

21. Students enrolled in algebra each semester:
    328, 549, 420, 592, 715, 483

22. Number of cars in the parking lot each day:
    520, 523, 513, 1283, 338, 509, 290, 420

23. Number of computer service calls taken each day:
    51, 48, 96, 40, 47, 23, 95, 56, 34, 48

24. Number of gallons of paint sold per week:
    1072, 1068, 1093, 1042, 1056, 205, 1009, 1081

The table lists the cruising speed and distance flown without refueling for several types of larger airplanes used to carry passengers. Use the table to answer Exercises 25–28.

Type of Airplane	Cruising Speed (miles per hour)	Distance without Refueling (miles)
747-400	565	7650
747-200	558	6450
DC-9	505	1100
DC-10	550	5225
727	530	1550
757	530	2875

Source: Northwest Airlines *World Traveler*.

25. What is the average distance flown without refueling, to the nearest mile?

26. Find the average cruising speed.

27. (a) Find the median distance flown.

(b) Explain how you found the median.

28. (a) Find the median cruising speed.

(b) Is the median similar to the average speed from Exercise 26? Explain why or why not.

Find the mode or modes for each list of numbers. See Example 7.

29. Number of samples taken each hour:
3, 8, 5, 1, 7, 6, 8, 4, 5, 8

30. Monthly water bills:
$21, $32, $46, $32, $49, $32, $49, $25, $32

31. Ages of senior residents (in years):
74, 68, 68, 68, 75, 75, 74, 74, 70, 77

32. Patients admitted to the hospital each week:
30, 19, 25, 78, 36, 20, 45, 85, 38

33. The number of boxes of candy sold by each child:
5, 9, 17, 3, 2, 8, 19, 1, 4, 20, 10, 6

34. The weights of soccer players (in pounds):
158, 161, 165, 162, 165, 157, 163, 162

The table lists monthly normal temperatures from November through April for some of the coldest and warmest U.S. cities. Use the table to answer Exercises 35–40. Round answers to the nearest whole degree. For help in determining variability, see Examples 8 and 9.

Normal Monthly Temperatures (in Degrees Fahrenheit)						
City	Nov.	Dec.	Jan.	Feb.	Mar.	Apr.
Barrow, Alaska	−2	−11	−13	−18	−15	−2
Fairbanks, Alaska	3	−7	−10	−4	11	31
Honolulu, Hawaii	77	74	73	73	74	76
Miami, Florida	74	69	67	69	72	75

Source: National Climate Data Center.

35. (a) Find Barrow's mean temperature and Fairbanks' mean temperature for the six-month period.

(b) How much warmer is Fairbanks' mean than Barrow's mean?

36. (a) Find Barrow's median temperature and Fairbanks' median temperature for the six-month period.

(b) How much cooler is Barrow's median than Fairbanks' median?

37. Use the table on the previous page. Which set of temperatures has greater variability, Barrow or Fairbanks? Show how to calculate the range for each city.

38. (a) Use the table on the previous page. Find the mean and median for Honolulu's temperatures during the six-month period.

(b) Why are the mean and median so similar?

39. (a) Find the mean and median for Miami's temperatures during the six-month period.

(b) What do you notice about the mean and median? Explain how this is possible.

40. Which set of temperatures has less variability, Honolulu or Miami? Show how to calculate the range for each city.

RELATING CONCEPTS (EXERCISES 41–44) For Individual or Group Work

Use your knowledge of statistics as you **work Exercises 41–44 in order.**

41. (a) Find the mean, median, and range for each student's test data. Round to the nearest tenth when necessary.
Student P: 92, 80, 61, 49, 82, 53
Student Q: 70, 76, 77, 60, 67, 72

(b) Which student's data has greater variability?

(c) If you were the teacher, what advice would you give to each student?

42. (a) Find the mean, median, and range for each golfer's scores. Round to the nearest tenth when necessary.
Golfer G: 84, 87, 83, 89, 88
Golfer H: 94, 88, 76, 89, 85

(b) Which golfer had scores with less variability?

(c) If you were the coach, what advice would you give to each golfer?

43. (a) Find the set of data on ages of tenants with the least variability. Show your work.
Building A: 63, 70, 66, 62, 74, 65, 74
Building B: 23, 20, 26, 18, 21, 23, 19
Building C: 45, 32, 69, 25, 50, 37, 41

(b) Based on your work in part (a), which building would you like to live in? Explain why.

44. (a) Find the set of data on monthly salaries of computer technicians that has the greatest variability. Show your work.
Company P: $4900, $5500, $5000, $7200, $6100
Company Q: $5800, $5900, $5500, $6000, $5900
Company R: $6200, $6400, $5800, $5200, $5600

(b) Based on your work in part (a), which company would you like to work for? Explain why.

5.8 Geometry Applications: Pythagorean Theorem and Square Roots

In **Section 3.2,** you used this formula for area of a square, $A = s^2$. The blue square below has an area of 25 cm² because (5 cm)(5 cm) = 25 cm².

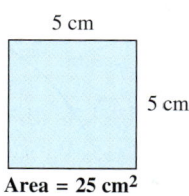
Area = 25 cm²
Area = (5 cm)(5 cm)

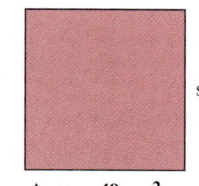
side = ? cm
Area = 49 cm²
Area = (? cm)(? cm)

OBJECTIVES

1. Find square roots using the square root key on a calculator.
2. Find the unknown length in a right triangle.
3. Solve application problems involving right triangles.

The red square above has an area of 49 cm². To find the length of a side, ask yourself, "What number can be multiplied by itself to give 49?" Because (7)(7) = 49, the length of each side is 7 cm. Also, because (7)(7) = 49 we say that 7 is the *square root* of 49, or $\sqrt{49} = 7$.

1 Find each square root.

(a) $\sqrt{36}$

Square Root
The positive **square root** of a positive number is one of two identical positive factors of that number.
For example, $\sqrt{36} = 6$ because (6)(6) = 36.

(b) $\sqrt{25}$

NOTE
There is another square root for 36. We know that
$$(6)(6) = 36 \quad \text{and} \quad (-6)(-6) = 36$$
so the *positive* square root of 36 is 6 and the *negative* square root of 36 is −6. In this section we will work only with positive square roots. You will learn more about negative square roots in Chapter 16.

(c) $\sqrt{9}$

Work Problem 1 at the Side.

A number that has a whole number as its square root is called a *perfect square*. For example, 9 is a perfect square because $\sqrt{9} = 3$, and 3 is a whole number. The first few perfect squares are listed below.

(d) $\sqrt{100}$

The First Twelve Perfect Squares

$\sqrt{1} = 1$	$\sqrt{16} = 4$	$\sqrt{49} = 7$	$\sqrt{100} = 10$
$\sqrt{4} = 2$	$\sqrt{25} = 5$	$\sqrt{64} = 8$	$\sqrt{121} = 11$
$\sqrt{9} = 3$	$\sqrt{36} = 6$	$\sqrt{81} = 9$	$\sqrt{144} = 12$

(e) $\sqrt{121}$

OBJECTIVE 1 Find square roots using the square root key on a calculator. If a number is *not* a perfect square, then you can find its *approximate* square root by using a calculator with a square root key.

ANSWERS
1. (a) 6 (b) 5 (c) 3 (d) 10 (e) 11

368 Chapter 5 Rational Numbers: Positive and Negative Decimals

2 Use a calculator with a square root key to find each square root. Round to the nearest thousandth when necessary.

(a) $\sqrt{11}$

(b) $\sqrt{40}$

(c) $\sqrt{56}$

(d) $\sqrt{196}$

(e) $\sqrt{147}$

ANSWERS

2. (a) $\sqrt{11} \approx 3.317$ (b) $\sqrt{40} \approx 6.325$
   (c) $\sqrt{56} \approx 7.483$ (d) $\sqrt{196} = 14$
   (e) $\sqrt{147} \approx 12.124$

Calculator Tip To find a square root on a *scientific* calculator, use the $\sqrt{}$ or the $\sqrt{x}$ key. (On some models you may have to press the 2nd key to access the square root function.) You do *not* need to use the $=$ key. Try these.

To find $\sqrt{16}$ press: 16 $\sqrt{x}$ Answer is 4

To find $\sqrt{7}$ press: 7 $\sqrt{x}$ Answer is 2.645751311

For $\sqrt{7}$, your calculator shows 2.645751311, which is an *approximate* answer. We will round to the nearest thousandth, so $\sqrt{7} \approx 2.646$. To check, multiply 2.646 times 2.646. Do you get 7 as the result? No, you get 7.001316, which is very close to 7. The difference is due to rounding.

EXAMPLE 1 Finding the Square Root of Numbers

Use a calculator to find each square root. Round to the nearest thousandth.

(a) $\sqrt{35}$ Calculator shows 5.916079783; round to 5.916.

(b) $\sqrt{124}$ Calculator shows 11.13552873; round to 11.136.

Work Problem 2 at the Side.

OBJECTIVE 2 Find the unknown length in a right triangle. One place you will use square roots is when working with the *Pythagorean Theorem*. This theorem applies only to *right* triangles (triangles with a 90° angle). The longest side of a right triangle is called the **hypotenuse**. It is opposite the right angle. The other two sides are called *legs*. The legs form the right angle. Here are some right triangles.

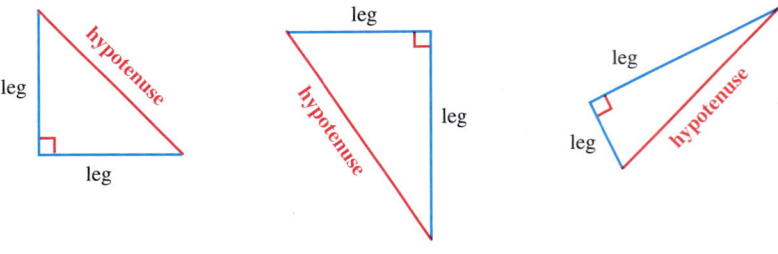

Examples of right triangles

Pythagorean Theorem

$$(\text{hypotenuse})^2 = (\text{leg})^2 + (\text{leg})^2$$

In other words, square the length of each side. After you have squared all the sides, the sum of the squares of the two legs will equal the square of the hypotenuse. An example is shown below.

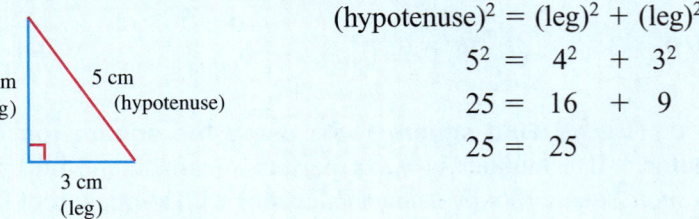

$(\text{hypotenuse})^2 = (\text{leg})^2 + (\text{leg})^2$
$5^2 = 4^2 + 3^2$
$25 = 16 + 9$
$25 = 25$

The theorem is named after Pythagoras, a Greek mathematician who lived about 2500 years ago. He and his followers may have used floor tiles to prove the theorem, as shown on the next page.

Section 5.8 Geometry Applications: Pythagorean Theorem and Square Roots   369

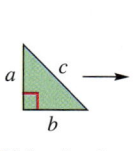
Right triangle

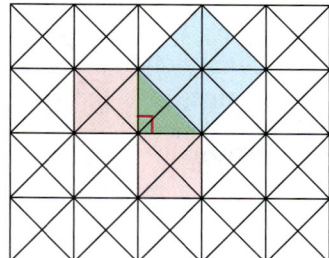

The green right triangle in the center of the floor tiles has sides a, b, and c. The pink square drawn on side a contains four triangular tiles. The pink square on side b contains four tiles. The blue square on side c contains eight tiles. The number of tiles in the square on side c equals the sum of the number of tiles in the squares on sides a and b, that is, 8 tiles = 4 tiles + 4 tiles. As a result, you often see the Pythagorean Theorem written as $c^2 = a^2 + b^2$.

If you know the lengths of any two sides in a right triangle, you can use the Pythagorean Theorem to find the length of the third side.

> **Formulas Based on the Pythagorean Theorem**
>
> To find the hypotenuse: $\text{hypotenuse} = \sqrt{(\text{leg})^2 + (\text{leg})^2}$
>
> To find a leg: $\text{leg} = \sqrt{(\text{hypotenuse})^2 - (\text{leg})^2}$

EXAMPLE 2 Finding the Unknown Length in Right Triangles

Find the unknown length in each right triangle. Round your answers to the nearest tenth when necessary.

(a)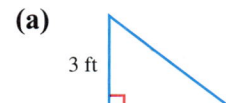

The unknown length is the side opposite the right angle, which is the hypotenuse. Use the formula for finding the hypotenuse.

$\text{hypotenuse} = \sqrt{(\text{leg})^2 + (\text{leg})^2}$ Find the hypotenuse.
$\text{hypotenuse} = \sqrt{(3)^2 + (4)^2}$ Legs are 3 and 4
$= \sqrt{9 + 16}$ (3)(3) is 9 and (4)(4) is 16
$= \sqrt{25}$
$= 5$

The hypotenuse is 5 ft long.

(b)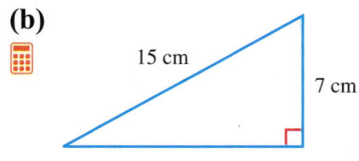

You *do* know the length of the hypotenuse (15 cm), so it is the length of one of the legs that is unknown. Use the formula for finding a leg.

$\text{leg} = \sqrt{(\text{hypotenuse})^2 - (\text{leg})^2}$ Find a leg.
$\text{leg} = \sqrt{(15)^2 - (7)^2}$ Hypotenuse is 15; one leg is 7
$= \sqrt{225 - 49}$ (15)(15) is 225 and (7)(7) is 49
$= \sqrt{176}$ Use a calculator to find $\sqrt{176}$
≈ 13.3 Round 13.26649916 to 13.3

The length of the leg is approximately 13.3 cm.

Work Problem 3 at the Side.

❸ Find the unknown length in each right triangle. Round your answers to the nearest tenth when necessary.

(a)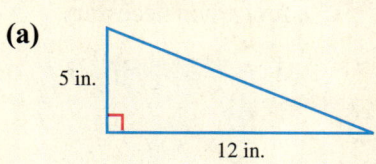

(b)

(c)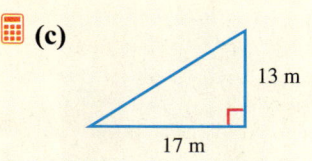

(d)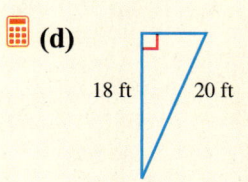

(e)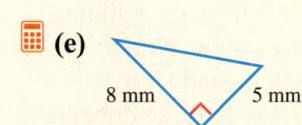

ANSWERS
3. (a) $\sqrt{169} = 13$ in. (b) $\sqrt{576} = 24$ cm
(c) $\sqrt{458} \approx 21.4$ m (d) $\sqrt{76} \approx 8.7$ ft
(e) $\sqrt{89} \approx 9.4$ mm

4 These problems show ladders leaning against buildings. Find the unknown lengths. Round answers to the nearest tenth of a foot when necessary.

(a)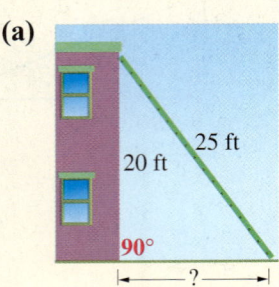

How far away from the building is the bottom of the ladder?

(b)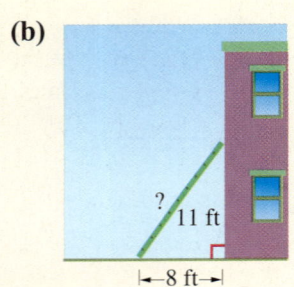

How long is the ladder?

(c) A 17 ft ladder is leaning against a building. The bottom of the ladder is 10 ft from the building. How high up on the building will the ladder reach? (*Hint:* Start by drawing a sketch of the building and the ladder.)

ANSWERS

4. (a) $\sqrt{225} = 15$ ft (b) $\sqrt{185} \approx 13.6$ ft
   (c) $\sqrt{189} \approx 13.7$ ft

CAUTION
Remember: A small square drawn in one angle of a triangle indicates a right angle (90°). You can use the Pythagorean Theorem *only* on triangles that have a right angle.

OBJECTIVE 3 Solve application problems involving right triangles.
The next example shows an application of the Pythagorean Theorem.

EXAMPLE 3 Using the Pythagorean Theorem

A television antenna is on the roof of a house, as shown. Find the length of the support wire. Round your answer to the nearest tenth of a meter if necessary.

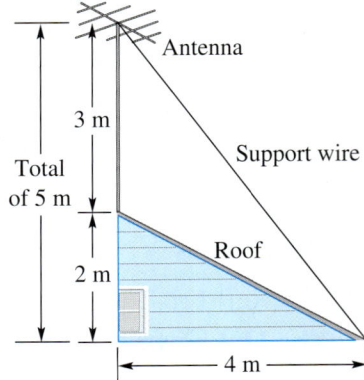

A right triangle is formed. The total length of the leg on the left is 3 m + 2 m = 5 m.

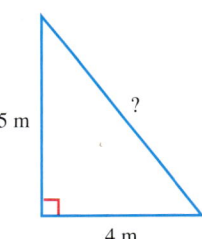

Notice that the support wire is opposite the right angle, so it is the *hypotenuse* of the right triangle.

hypotenuse = $\sqrt{(\text{leg})^2 + (\text{leg})^2}$		Find the hypotenuse.
hypotenuse = $\sqrt{(5)^2 + (4)^2}$		Legs are 5 and 4
$= \sqrt{25 + 16}$		5^2 is 25 and 4^2 is 16
$= \sqrt{41}$		Use a calculator to find $\sqrt{41}$
≈ 6.4		Round 6.403124237 to 6.4

The length of the support wire is approximately 6.4 m.

CAUTION
You use the Pythagorean Theorem to find the *length* of one side, *not* the area of the triangle. Your answer will be in linear units, such as ft, yd, cm, m, and so on (*not* ft², yd², cm², m²).

◀◀◀ **Work Problem 4 at the Side.**

5.8 Exercises

Find each square root. Starting with Exercise 5, find the square root using a calculator. Round your answers to the nearest thousandth when necessary. See Example 1.

1. $\sqrt{16}$ = 4
2. $\sqrt{4}$
3. $\sqrt{64}$ = 8
4. $\sqrt{81}$

5. $\sqrt{11}$
6. $\sqrt{23}$
7. $\sqrt{5}$
8. $\sqrt{2}$

9. $\sqrt{73}$
10. $\sqrt{80}$
11. $\sqrt{101}$
12. $\sqrt{125}$

13. $\sqrt{361}$
14. $\sqrt{729}$
15. $\sqrt{1000}$
16. $\sqrt{2000}$

17. You know that $\sqrt{25} = 5$ and $\sqrt{36} = 6$. Using just that information (no calculator), describe how you could *estimate* $\sqrt{30}$. How would you estimate $\sqrt{26}$ or $\sqrt{35}$? Now check your estimates using a calculator.

18. Explain the relationship between *squaring* a number and finding the *square root* of a number. Include two examples to illustrate your explanation.

Find the unknown length in each right triangle. Use a calculator to find square roots. Round your answers to the nearest tenth when necessary. See Example 2.

19.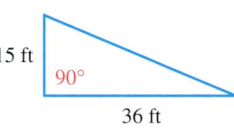
    15 ft, 90°, 36 ft

20.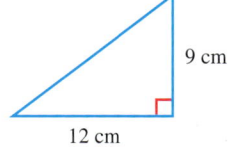
    9 cm, 12 cm

21.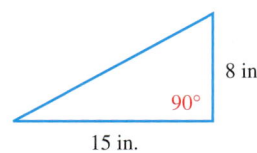
    8 in., 90°, 15 in.

22.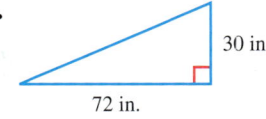
    30 in., 72 in.

23.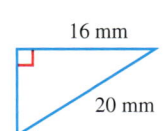
    16 mm, 20 mm

24.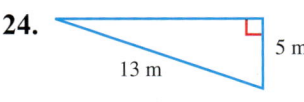
    13 m, 5 m

25.

3 in.
8 in.

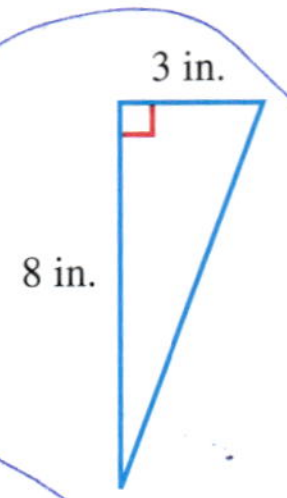

26.

5 cm
11 cm

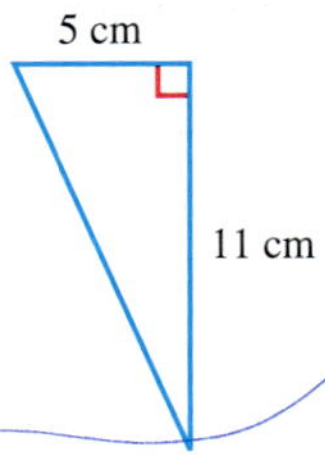

27.

7 yd
90°
4 yd

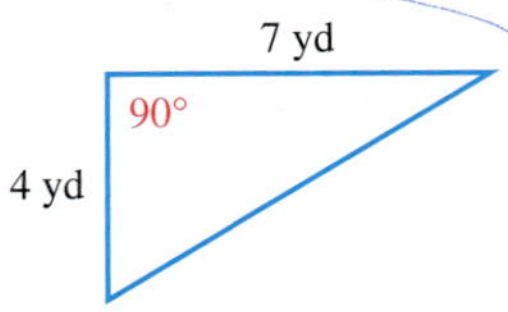

28.

7 km
10 km

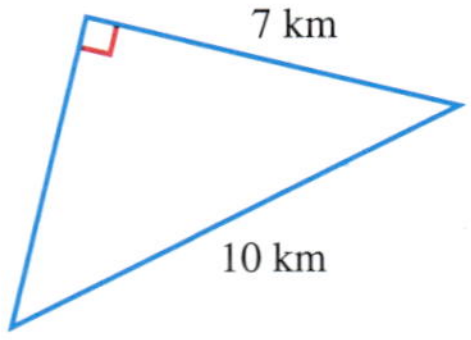

29.

22 cm
17 cm

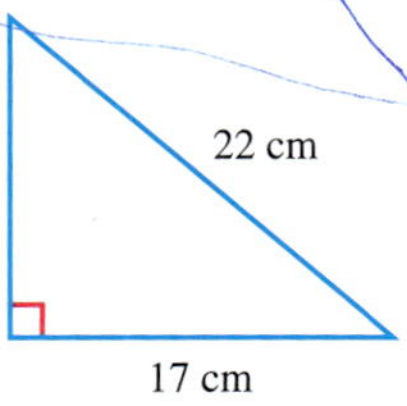

30.

16 cm
9 cm
90°

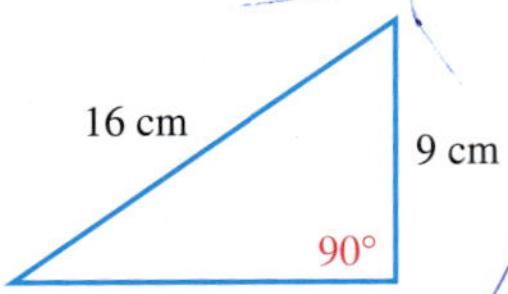

31.

1.3 m
90°
2.5 m

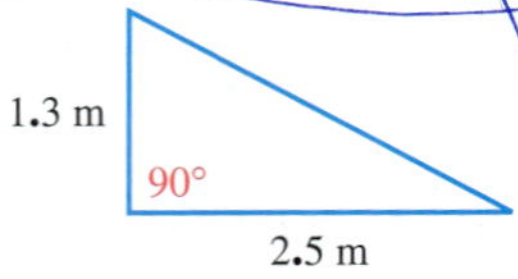

32.

4.2 mi
4.2 mi

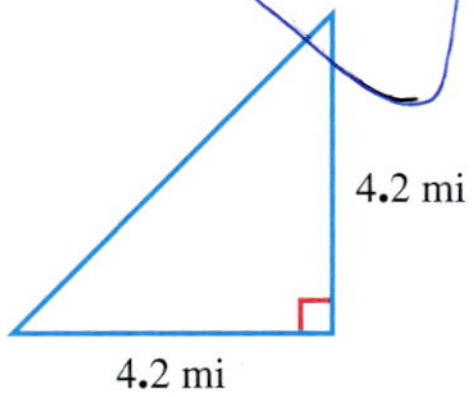

33.

11.5 cm
8.2 cm

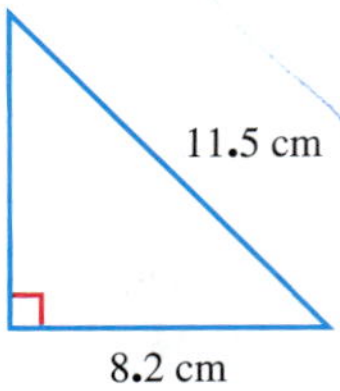

34.

9.1 mm
10.8 mm

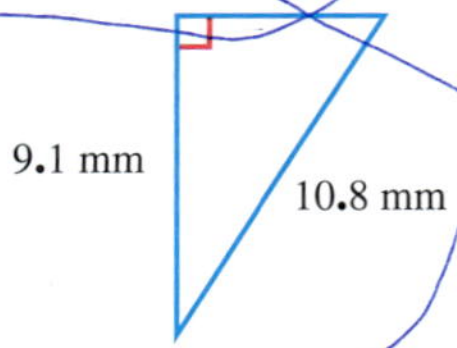

35.

13.2 km
90°
21.6 km

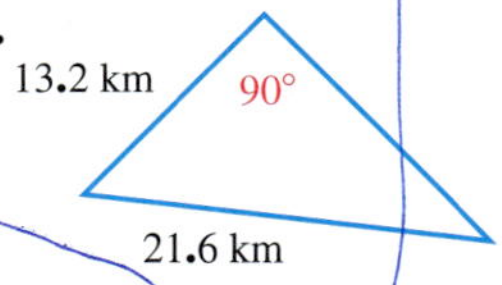

36.

26.5 ft
37.4 ft

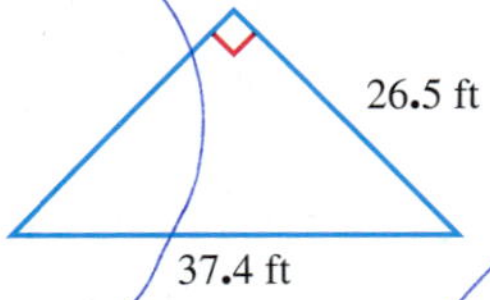

Solve each application problem. Round your answers to the nearest tenth when necessary. See Example 3.

37. Find the length of this loading ramp.

4 ft
?
7 ft

38. Find the unknown length in this roof plan.

?
3 ft
9 ft

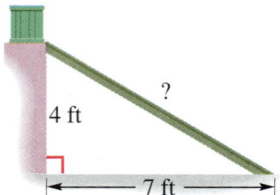

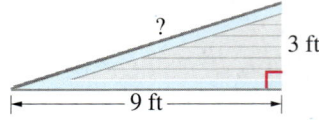

39. How high is the airplane above the ground?

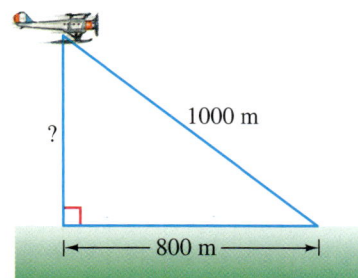

40. Find the height of this farm silo.

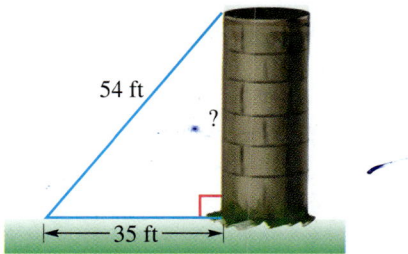

41. How long is the diagonal brace on this rectangular gate?

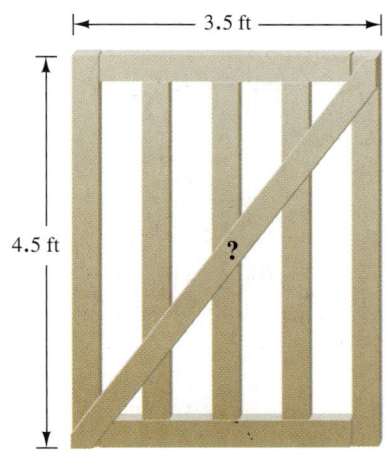

42. Find the height of this rectangular television screen. (*Source:* Sears.)

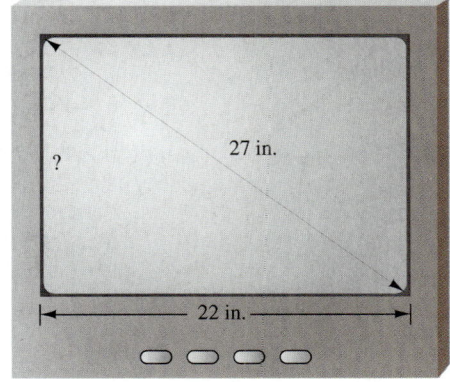

43. To reach his ladylove, a knight placed a 12 ft ladder against the castle wall. If the base of the ladder is 3 ft from the building, how high on the castle will the top of the ladder reach? Draw a sketch of the castle and ladder and solve the problem.

44. William drove his car 15 miles north, then made a right turn and drove 7 miles east. How far is he, in a straight line, from his starting point? Draw a sketch to illustrate the problem and then solve it.

45. Explain the *two* errors made by a student in solving this problem. Also find the correct answer. Round to the nearest tenth.

$? = \sqrt{(13)^2 + (20)^2}$

$= \sqrt{169 + 400}$

$= \sqrt{569} \approx 23.9 \text{ m}^2$

[Triangle with sides 13 m and 20 m, with ? as the hypotenuse]

46. Explain the *two* errors made by a student in solving this problem. Also find the correct answer. Round to the nearest tenth.

$? = \sqrt{(9)^2 + (7)^2}$

$= \sqrt{18 + 14}$

$= \sqrt{32} \approx 5.657 \text{ in.}$

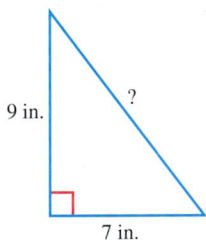

9 in.

7 in.

RELATING CONCEPTS (EXERCISES 47–50) For Individual or Group Work

Use your knowledge of the Pythagorean Theorem to **work Exercises 47–50 in order.** Round answers to the nearest tenth.

47. A major league baseball diamond is a square shape measuring 90 ft on each side. If the catcher throws a ball from home plate to second base, how far is he throwing the ball? (*Source:* American League of Professional Baseball Clubs.)

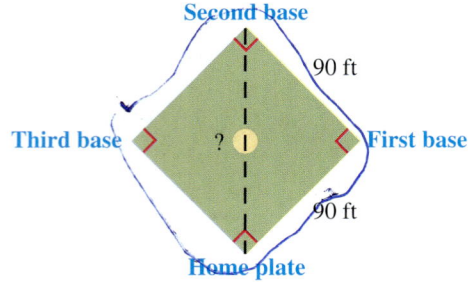

48. A softball diamond is only 60 ft on each side. (*Source:* Amateur Softball Association.)

(a) Draw a sketch of the softball diamond and label the bases and the lengths of the sides.

(b) How far is it to throw a ball from home plate to second base?

49. Look back at your answer to Exercise 47. Explain how you can tell the distance from third base to first base without doing any further calculations.

50. (a) Look back at your answer to Exercise 48. Suppose you measured the distance from home plate to second base on a softball diamond and found it was 80 ft. What would this tell you about the length of each side of the diamond? (Assume the diamond is still a square.)

(b) Bonus question: For the diamond in part (a), find the length of each side, to the nearest tenth.

5.9 Problem Solving: Equations Containing Decimals

OBJECTIVE 1 Solve equations containing decimals using the addition property of equality. In **Section 2.3** you used the addition property of equality to solve an equation like $c + 5 = 30$. The addition property says that you can add the *same* number to *both* sides of an equation and still keep it balanced. You can also use this property when an equation contains decimal numbers.

OBJECTIVES

1. Solve equations containing decimals using the addition property of equality.
2. Solve equations containing decimals using the division property of equality.
3. Solve equations containing decimals using both properties of equality.
4. Solve application problems involving equations with decimals.

EXAMPLE 1 Using the Addition Property of Equality

Solve each equation and check each solution.

(a) $w + 2.9 = -0.6$

The first step is to get the variable term (w) by itself on the left side of the equal sign. Use the addition property to "get rid of" the 2.9 on the left side by adding its opposite, -2.9, to both sides.

$$
\begin{aligned}
w + 2.9 &= -0.6 \\
\underline{-2.9 \quad -2.9} &\quad \text{Add } -2.9 \text{ to both sides.} \\
w + 0 &= -3.5 \\
w &= -3.5
\end{aligned}
$$

The solution is -3.5. To check the solution, go back to the *original* equation.

$$
\begin{aligned}
\text{Check} \quad w + 2.9 &= -0.6 \quad \text{Original equation} \\
-3.5 + 2.9 &= -0.6 \quad \text{Replace } w \text{ with } -3.5 \\
-0.6 &= -0.6 \quad \text{Balances}
\end{aligned}
$$

When w is replaced with -3.5, the equation balances, so -3.5 is the correct solution (**not** -0.6).

(b) $7 = -4.3 + x$

To get x by itself on the right side of the equal sign, add 4.3 to both sides.

$$
\begin{aligned}
7 &= -4.3 + x \\
\underline{+4.3 \quad +4.3} &\quad \text{Add 4.3 to both sides.} \\
11.3 &= 0 + x \\
11.3 &= x
\end{aligned}
$$

The solution is 11.3. To check the solution, go back to the original equation.

$$
\begin{aligned}
\text{Check} \quad 7 &= -4.3 + x \quad \text{Original equation} \\
7 &= -4.3 + 11.3 \quad \text{Replace } x \text{ with } 11.3 \\
7 &= 7 \quad \text{Balances}
\end{aligned}
$$

When x is replaced with 11.3, the equation balances, so 11.3 is the correct solution (**not** 7).

Work Problem 1 at the Side.

---

1 Solve each equation and check each solution.

(a) $8.1 = h + 9$ Check

(b) $-0.75 + y = 0$ Check

(c) $c - 6.8 = -4.8$ Check

ANSWERS

1. **(a)** $h = -0.9$ Check $8.1 = h + 9$
   $8.1 = -0.9 + 9$
   Balances $8.1 = 8.1$

   (b) $y = 0.75$ Check $-0.75 + y = 0$
   $-0.75 + 0.75 = 0$
   Balances $0 = 0$

   (c) $c = 2$ Check $c - 6.8 = -4.8$
   $2 - 6.8 = -4.8$
   Balances $-4.8 = -4.8$

2 Solve each equation and check each solution.

(a) $-3y = -0.63$ Check

(b) $2.25r = -18$ Check

(c) $1.7 = 0.5n$ Check

ANSWERS

2. (a) $y = 0.21$ Check $-3y = -0.63$
   $-3(0.21) = -0.63$
   Balances $-0.63 = -0.63$

(b) $r = -8$ Check $2.25r = -18$
   $2.25(-8) = -18$
   Balances $-18 = -18$

(c) $n = 3.4$ Check $1.7 = 0.5n$
   $1.7 = 0.5(3.4)$
   Balances $1.7 = 1.7$

OBJECTIVE 2 Solve equations containing decimals using the division property of equality. You can also use the division property of equality (from **Section 2.4**) when an equation contains decimals.

EXAMPLE 2 Using the Division Property of Equality

Solve each equation and check each solution.

(a) $5x = 12.4$

On the left side of the equation, the variable is multiplied by 5. To undo the multiplication, divide both sides by 5.

$5x$ means $5 \cdot x$.

$5x = 12.4$

$\dfrac{5 \cdot x}{5} = \dfrac{12.4}{5}$ Divide both sides by 5

On the left side, divide out the common factor of 5.

$\dfrac{\overset{1}{\cancel{5}} \cdot x}{\underset{1}{\cancel{5}}} = 2.48$ On the right side, $12.4 \div 5$ is 2.48

$x = 2.48$

The solution is 2.48. To check the solution, go back to the original equation.

Check $5x = 12.4$ Original equation
$5(2.48) = 12.4$ Replace x with 2.48
$12.4 = 12.4$ Balances

When x is replaced with 2.48, the equation balances, so 2.48 is the correct solution (*not* 12.4).

(b) $-9.3 = 1.5t$

Signs are different, so quotient is negative.

$\dfrac{-9.3}{1.5} = \dfrac{\overset{1}{\cancel{1.5}}t}{\underset{1}{\cancel{1.5}}}$ Divide both sides by the coefficient of the variable term, 1.5

$-6.2 = t$

The solution is -6.2. To check the solution, go back to the original equation.

Check $-9.3 = 1.5\,t$ Original equation
$-9.3 = 1.5\,(-6.2)$ Replace t with -6.2
$-9.3 = -9.3$ Balances

When t is replaced with -6.2, the equation balances, so -6.2 is the correct solution (*not* -9.3).

◀◀ **Work Problem 2 at the Side.**

OBJECTIVE 3 Solve equations containing decimals using both properties of equality. Sometimes you need to use both the addition and division properties to solve an equation, as shown in Example 3.

Section 5.9 Problem Solving: Equations Containing Decimals 377

> **EXAMPLE 3** Solving Equations with Several Steps
>
> **(a)** $2.5b + 0.35 = -2.65$
>
> The first step is to get the variable term, $2.5b$, by itself on the left side of the equal sign.
>
> $$
> \begin{aligned}
> 2.5b + 0.35 &= -2.65 \\
> -0.35 \quad & -0.35 \quad \text{Add } -0.35 \text{ to both sides.} \\
> \overline{2.5b + 0} &= -3.00 \\
> 2.5b &= -3
> \end{aligned}
> $$
>
> The next step is to divide both sides by the coefficient of the variable term. In $2.5b$, the coefficient is 2.5.
>
> $$\frac{2.5b}{2.5} = \frac{-3}{2.5} \quad \text{On the right side, signs do } not \text{ match, so the quotient is } negative.$$
>
> $$b = -1.2$$
>
> The solution is -1.2. To check the solution, go back to the original equation.
>
> Check $\quad 2.5b + 0.35 = -2.65 \quad$ Original equation
>
> $\quad 2.5(-1.2) + 0.35 = -2.65 \quad$ Replace b with -1.2
>
> $\quad -3 + 0.35 = -2.65$
>
> $\quad -2.65 = -2.65 \quad$ Balances, so -1.2 is the correct solution.
>
> **(b)** $5x - 0.98 = 2x + 0.4$
>
> There is a variable term on both sides of the equation. You can choose to keep the variable term on the left side, or to keep the variable term on the right side. Either way will work. Just pick the left side or the right side.
>
> Suppose that you decide to keep the variable term, $5x$, on the left side. Use the addition property to "get rid of" $2x$ on the right side by adding its opposite, $-2x$, to both sides.
>
> $$
> \begin{aligned}
> 5x - 0.98 &= 2x + 0.4 \\
> -2x & -2x \quad \text{Add } -2x \text{ to both sides.} \\
> \overline{3x - 0.98} &= 0 + 0.4
> \end{aligned}
> $$
>
> Change subtraction to adding the opposite.
>
> $$
> \begin{aligned}
> 3x + (-0.98) &= 0.4 \\
> +0.98 & +0.98 \quad \text{Add 0.98 to both sides.} \\
> \overline{3x + 0} &= 1.38
> \end{aligned}
> $$
>
> $$\frac{3x}{3} = \frac{1.38}{3} \quad \text{Divide both sides by 3}$$
>
> $$x = 0.46$$
>
> The solution is 0.46. To check the solution, go back to the original equation.
>
> Check $\quad 5x - 0.98 = 2x + 0.4 \quad$ Original equation
>
> $\quad 5(0.46) - 0.98 = 2(0.46) + 0.4 \quad$ Replace x with 0.46
>
> $\quad 2.3 - 0.98 = 0.92 + 0.4$
>
> $\quad 1.32 = 1.32 \quad$ Balances, so 0.46 is the correct solution.
>
> **Work Problem 3 at the Side.** ▶▶▶

3 Solve each equation and check each solution.

(a) $4 = 0.2c - 2.6 \quad$ Check

(b) $3.1k - 4 = 0.5k + 13.42$

Check

(c) $-2y + 3 = 3y - 6$

Check

ANSWERS

3. **(a)** $c = 33 \quad$ Check $4 = 0.2c - 2.6$

$4 = 0.2(33) - 2.6$

$4 = 6.6 - 2.6$

Balances $ 4 = 4$

(b) $k = 6.7 \quad$ Check

$3.1k - 4 = 0.5k + 13.42$

$3.1(6.7) - 4 = 0.5(6.7) + 13.42$

$20.77 - 4 = 3.35 + 13.42$

Balances $16.77 = 16.77$

(c) $y = 1.8 \quad$ Check

$-2y + 3 = 3y - 6$

$-2(1.8) + 3 = 3(1.8) - 6$

$-3.6 + 3 = 5.4 - 6$

Balances $ -0.6 = -0.6$

4 During April, a special rate was offered on air-to-ground cell phone calls. The connection fee was $1.34 and the cost per minute was $2.69. Maureen made a call that cost $39. How long did the call last? Use the six problem-solving steps.

OBJECTIVE 4 Solve application problems involving equations with decimals. Use the six problem-solving steps from **Section 3.3**.

EXAMPLE 4 Solving an Application Problem

Many larger airplanes have phones that can be used to call people on the ground. In January 2004, the cost of using the air-to-ground cell phone was $3.28 per minute plus a $2.99 connection charge. (*Source:* AT&T.) Hernando was billed $19.39 for one call. How many minutes did the call last?

Step 1   **Read** the problem. It is about the cost of a telephone call.

   Unknown: number of minutes the call lasted
   Known: Costs are $3.28 per minute plus $2.99; total cost was $19.39.

Step 2   **Assign a variable:** There is only one unknown, so let m be the number of minutes.

Step 3   **Write an equation.**

Cost per minute		Number of minutes		Connection charge		Total cost
3.28	•	m	+	2.99	=	19.39

Step 4   **Solve** the equation.

$$3.28m + 2.99 = 19.39$$
$$\underline{-2.99\ \ -2.99} \quad \text{Add } -2.99 \text{ to both sides.}$$
$$3.28m + 0 = 16.40$$

$$\frac{\overset{1}{\cancel{3.28}}m}{\underset{1}{\cancel{3.28}}} = \frac{16.40}{3.28} \quad \text{Divide both sides by 3.28}$$

$$m = 5$$

Step 5   **State the answer.** The call lasted 5 minutes.

Step 6   **Check** the solution by putting it back into the original problem.

   $3.28 per minute times 5 minutes = $16.40
   $16.40 plus $2.99 connection charge = $19.39
   Hernando was billed $19.39. ← Matches

Because 5 minutes "works" when put back into the original problem, it is the correct solution.

Work Problem 4 at the Side.

ANSWERS

4. Let m be the number of minutes.
  $1.34 + 2.69m = 39$
  The call lasted 14 minutes.

5.9 Exercises

Solve each equation and check each solution. See Examples 1 and 2.

1. $h + 0.63 = 5.1$ Check
2. $-0.2 = k - 0.7$ Check
3. $-20.6 + n = -22$ Check
4. $g - 5 = 6.03$ Check
5. $0 = b - 0.008$ Check
6. $0.18 + m = -4.5$ Check
7. $2.03 = 7a$ Check
8. $-6.2c = 0$ Check
9. $0.8p = -96$ Check
10. $-10.16 = -4r$ Check
11. $-3.3t = -2.31$ Check
12. $8.3w = -49.8$ Check

Solve each equation. Show your work. See Example 3.

13. $7.5x + 0.15 = -6$
14. $0.8 = 0.2y + 3.4$
15. $-7.38 = 2.05z - 7.38$
16. $6.2h - 0.4 = 2.7$
17. $3c + 10 = 6c + 8.65$
18. $2.1b + 5 = 1.6b + 10$
19. $8w - 6.4 = -6.4 + 5w$

20. $7r + 9.64 = -2.32 + 5r$ **21.** $-10.9 + 0.5p = 0.9p + 5.3$ **22.** $0.7x - 4.38 = x - 2.16$

Solve each application problem using the six problem-solving steps. See Example 4.

23. Most adult medication doses are for a person weighing 150 pounds. For a 45-pound child, the adult dose should be multiplied by 0.3. If the child's dose of a decongestant is 9 milligrams, what is the adult dose?

24. For a 30-pound child, an adult dose of medication should be multiplied by 0.2. If the child's dose of a cough suppressant is 3 milliliters, find the adult dose.

25. A storm blew down many trees. Several neighbors rented a chain saw for $275.80 and helped each other cut up and stack the wood. The rental company charges $65.95 per day plus a $12 sharpening fee. How many days was the saw rented? (*Source:* Central Rental.)

26. A 20-inch chain saw can be rented for $29.95 for the first two hours, and $9 for each additional hour. Steve's rental charge was $56.95. How many hours did he rent the saw? (*Source:* Central Rental.)

RELATING CONCEPTS (EXERCISES 27–30) For Individual or Group Work

When doing aerobic exercises, it is important to increase your heart rate (the number of beats per minute) so that you get the maximum benefit from the exercise. But you don't want your heart rate to be so fast that it is dangerous. Here is an expression for finding a safe maximum heart rate for a healthy person with no heart disease: $0.7(220 - a)$ *where a is the person's age.* **Work Exercises 27–30** *in order: Write an equation and solve it to find each person's age. Assume all the people are healthy.*

27. How old is a person who has a maximum safe heart rate of 140 beats per minute? *Hint:* Use the distributive property to simplify $0.7(220 - a)$.

28. How old is a person who has a maximum safe heart rate of 126 beats per minute?

29. If a person's maximum safe heart rate is 134 (rounded to the nearest whole number), how old is the person, to the nearest whole year?

30. If a person's maximum safe heart rate is 117 (rounded to the nearest whole number), how old is the person, to the nearest whole year?

5.10 Geometry Applications: Circles, Cylinders, and Surface Area

OBJECTIVE 1 Find the radius and diameter of a circle. Suppose you start with one dot on a piece of paper. Then you draw many dots that are each 2 cm away from the first dot. If you draw enough dots (points) you'll end up with a circle, as shown below on the left. Each point on the circle is exactly 2 cm away from the *center* of the circle. The 2 cm distance is called the *radius*, *r*, of the circle. The distance across the circle (passing through the center) is called the *diameter, d,* of the circle. In this circle, the diameter is 4 cm.

OBJECTIVES

1. Find the radius and diameter of a circle.
2. Find the circumference of a circle.
3. Find the area of a circle.
4. Find the volume of a cylinder.
5. Find the surface area of a rectangular solid.
6. Find the surface area of a cylinder.

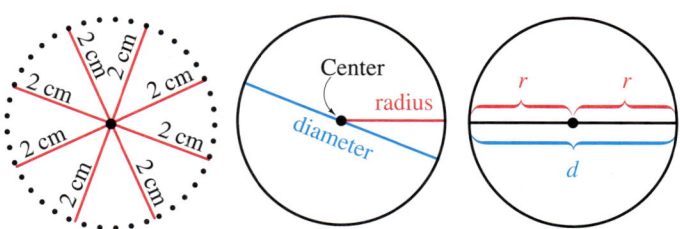

Circle, Radius, and Diameter

A **circle** is a two-dimensional (flat) figure with all points the same distance from a fixed center point.

The **radius** (*r*) is the distance from the center of the circle to any point on the circle.

The **diameter** (*d*) is the distance across the circle passing through the center.

Using the circle above on the right as a model, you can see some relationships between the radius and diameter.

Finding the Diameter and Radius of a Circle

$$\text{diameter} = 2 \cdot \text{radius}$$
$$d = 2r$$
$$\text{and} \quad r = \frac{d}{2}$$

EXAMPLE 1 Finding the Diameter and Radius of Circles

Find the unknown length of the diameter or radius in each circle.

(a)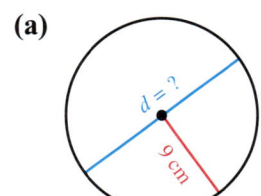

Because the radius is 9 cm, the diameter is twice as long.

$$d = 2 \cdot r$$
$$d = 2 \cdot 9 \text{ cm}$$
$$d = 18 \text{ cm}$$

Continued on Next Page

1 Find the unknown length of the diameter or radius in each circle.

(a)
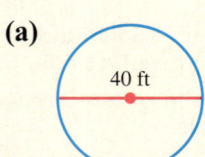
40 ft

(b)
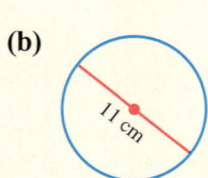
11 cm

(c)
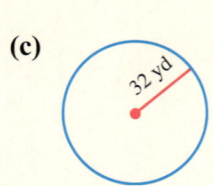
32 yd

(d)
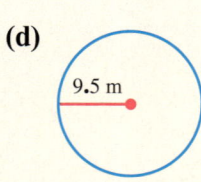
9.5 m

(b)
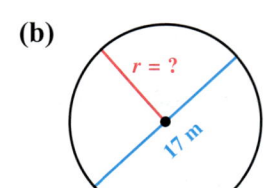
r = ?
17 m

The radius is half the diameter.

$r = \dfrac{d}{2}$ so $r = \dfrac{17 \text{ m}}{2}$

$r = 8.5$ m   or $8\dfrac{1}{2}$ m

Work Problem 1 at the Side.

OBJECTIVE 2 Find the circumference of a circle. The perimeter of a circle is called its **circumference**. Circumference is the distance around the edge of a circle.

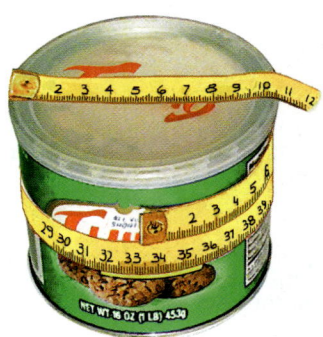

The diameter of the can in the drawing is about 10.6 cm, and the circumference of the can is about 33.3 cm. Dividing the circumference of the circle by the diameter gives an interesting result.

$$\dfrac{\text{Circumference}}{\text{diameter}} = \dfrac{33.3}{10.6} \approx 3.14 \quad \text{Rounded to the nearest hundredth}$$

Dividing the circumference of *any* circle by its diameter *always* gives an answer close to 3.14. This means that going around the edge of any circle is a little more than 3 times as far as going straight across the circle.

This ratio of circumference to diameter is called π (the Greek letter **pi**, pronounced PIE). There is no decimal that is exactly equal to π, but here is the *approximate* value.

$$\pi \approx 3.14159265359$$

Rounding the Value of Pi (π)

We usually round π to 3.14. Therefore, calculations involving π will give approximate answers and should be written using the $\approx$ symbol.

Use the following formulas to find the *circumference* of a circle.

Finding the Circumference (Distance Around a Circle)

Circumference = π · diameter

$$C = \pi d$$

or, because $d = 2r$ then $C = \pi \cdot 2r$ usually written $C = 2\pi r$

Remember to use linear units such as ft, yd, m, and cm when measuring circumference (**not** square units).

ANSWERS

1. (a) $r = 20$ ft   (b) $r = 5.5$ cm
   (c) $d = 64$ yd   (d) $d = 19$ m

Section 5.10 Geometry Applications: Circles, Cylinders, and Surface Area   383

EXAMPLE 2 Finding the Circumference of Circles

Find the circumference of each circle. Use 3.14 as the approximate value for π. Round answers to the nearest tenth.

(a)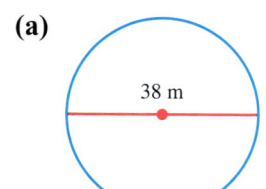

The *diameter* is 38 m, so use the formula with d in it.

$$C = \pi \cdot d$$
$$C \approx 3.14 \cdot 38 \text{ m}$$
$$C \approx 119.3 \text{ m} \quad \text{Rounded}$$

(b)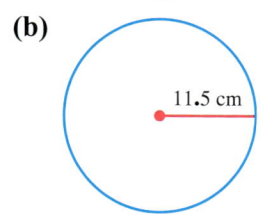

In this example, the length of the *radius* is labeled, so it is easier to use the formula with r in it.

$$C = 2 \cdot \pi \cdot r$$
$$C \approx 2 \cdot 3.14 \cdot 11.5 \text{ cm}$$
$$C \approx 72.2 \text{ cm} \quad \text{Rounded}$$

Calculator Tip Most *scientific* calculators have a π key. Try pressing it. With a 10-digit display, you'll see the value of π to the nearest billionth.

$$3.141592654$$

But this is still an approximate value, although it is more precise than rounding π to 3.14. Try finding the circumference in Example 2(a) above using the π key.

π × 38 = **119.3805208**   Rounds to 119.4

When you used 3.14 as the approximate value of π, the result rounded to 119.3, so the answers are slightly different. In this book we will use 3.14. Our measurements of radius and diameter are given as whole numbers or with tenths, so it is acceptable to round π to hundredths. Also, some students may be using a calculator without a π key.

Work Problem 2 at the Side.

OBJECTIVE 3 Find the area of a circle. To find the formula for the area of a circle, start by cutting two circles into many pie-shaped pieces.

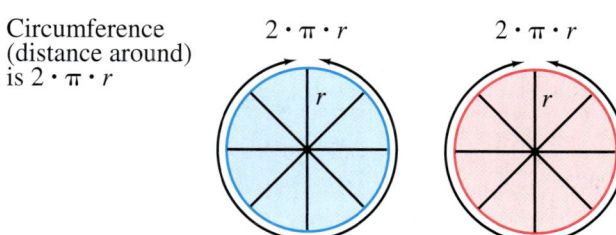

Circumference (distance around) is $2 \cdot \pi \cdot r$

Unfold the circles, much as you might "unfold" a peeled orange, and put them together as shown here.

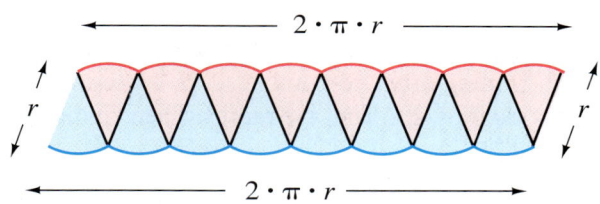

2 Find the circumference of each circle. Use 3.14 as the approximate value for π. Round answers to the nearest tenth.

(a)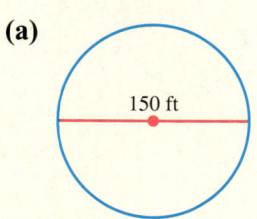

(b)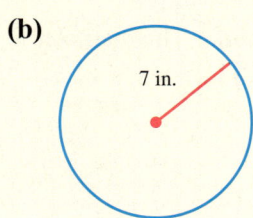

(c) diameter 0.9 km

(d) radius 4.6 m

ANSWERS

2. (a) $C \approx 471$ ft   (b) $C \approx 44.0$ in.
   (c) $C \approx 2.8$ km   (d) $C \approx 28.9$ m

384 Chapter 5 Rational Numbers: Positive and Negative Decimals

❸ Find the area of each circle. Use 3.14 for π. Round your answers to the nearest tenth.

(a)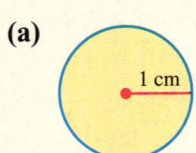

(b)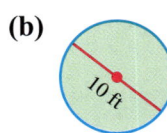

(*Hint:* The diameter is 12 m, so $r =$ _____ m)

🖩 (c)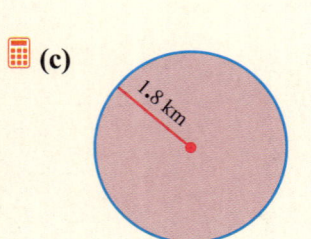

🖩 (d)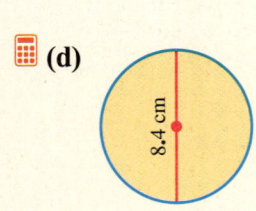

The figure is approximately a rectangle with width r (the radius of the original circle) and length $2 \cdot \pi \cdot r$ (the circumference of the original circle). The area of the "rectangle" is length times width.

$$\text{Area of "rectangle"} = \overbrace{l}^{} \cdot w$$
$$\text{Area of "rectangle"} = \overbrace{2 \cdot \pi \cdot r}^{} \cdot r$$
$$\text{Area of "rectangle"} = 2 \cdot \pi \cdot r^2 \quad \leftarrow \text{Recall that } r \cdot r \text{ is } r^2$$

Because the "rectangle" was formed from *two* circles, the area of *one* circle is half as much.

$$\frac{1}{\underset{1}{\cancel{2}}} \cdot \overset{1}{\cancel{2}} \cdot \pi \cdot r^2 = 1 \cdot \pi \cdot r^2 \quad \text{or simply} \quad \pi r^2$$

Finding the Area of a Circle

Area of a circle = π • radius • radius

$$A = \pi r^2$$

Remember to use **square units** when measuring area.

EXAMPLE 3 Finding the Area of Circles

Find the area of each circle. Use 3.14 for π. Round your answers to the nearest tenth.

(a) A circle with a radius of 8.2 cm

Use the formula $A = \pi r^2$, which means $A = \pi \cdot r \cdot r$.

$$A = \pi \cdot r \cdot r$$
$$A \approx 3.14 \cdot 8.2 \text{ cm} \cdot 8.2 \text{ cm}$$
$$A \approx 211.1 \text{ cm}^2 \quad \text{Square units for area}$$

(b)

To use the area formula $A = \pi r^2$, you need to know the radius (r). In this circle, the *diameter* is 10 ft. First find the radius.

$$r = \frac{d}{2}$$
$$r = \frac{10 \text{ ft}}{2} = 5 \text{ ft}$$

Now find the area.

$$A \approx 3.14 \cdot 5 \text{ ft} \cdot 5 \text{ ft}$$
$$A \approx 78.5 \text{ ft}^2 \quad \text{Square units for area}$$

CAUTION
When finding *circumference*, you can start with either the radius or the diameter. When finding *area*, you must use the *radius*. If you are given the diameter, divide it by 2 to find the radius. Then find the area.

Work Problem 3 at the Side.

ANSWERS
3. (a) $A \approx 3.1$ cm² (b) $A \approx 113.0$ m²
   (c) $A \approx 10.2$ km² (d) $A \approx 55.4$ cm²

Calculator Tip

You can find the area of the circle in Example 3(a) on the previous page using your calculator. The first method works on all types of calculators.

$$3.14 \; \boxed{\times} \; 8.2 \; \boxed{\times} \; 8.2 \; \boxed{=} \; 211.1336$$

You round the answer to 211.1 (nearest tenth).

On a *scientific* or *graphing* calculator you can also use the $\boxed{x^2}$ key, which automatically squares the number you enter (that is, multiplies the number times itself).

$$3.14 \; \boxed{\times} \; 8.2 \; \boxed{x^2} \; \boxed{=} \; 211.1336$$

In the next example we will find the area of a *semicircle*, which is half the area of a circle.

EXAMPLE 4 Finding the Area of a Semicircle

Find the area of the semicircle below. Use 3.14 for π. Round your answer to the nearest tenth.

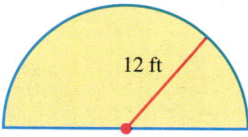

First, find the area of a whole circle with a radius of 12 ft.

$A = \pi \cdot r \cdot r$

$A \approx 3.14 \cdot 12 \text{ ft} \cdot 12 \text{ ft}$

$A \approx 452.16 \text{ ft}^2$ Do not round yet.

Divide the area of the whole circle by 2 to find the area of the semicircle.

$$\frac{452.16 \text{ ft}^2}{2} = 226.08 \text{ ft}^2$$

The *last* step is rounding 226.08 to the nearest tenth.

Area of semicircle $\approx 226.1 \text{ ft}^2$ Rounded

Work Problem 4 at the Side.

EXAMPLE 5 Applying the Concept of Circumference

A circular rug is 8 feet in diameter. The cost of fringe for the edge is $2.25 per foot. What will it cost to add fringe to the rug? Use 3.14 for π.

Circumference = $\pi \cdot d$

$C \approx 3.14 \cdot 8 \text{ ft}$

$C \approx 25.12 \text{ ft}$

cost = cost per foot • Circumference

$\text{cost} = \dfrac{\$2.25}{1 \text{ ft}} \cdot \dfrac{25.12 \text{ ft}}{1}$

cost = $56.52

The cost of adding fringe to the rug is $56.52.

Work Problem 5 at the Side.

---

4 Find the area of each semicircle. Use 3.14 for π. Round your answers to the nearest tenth.

(a)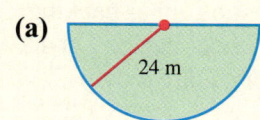

(b)

35.4 ft

(c)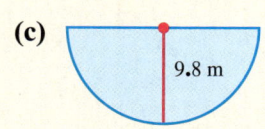

5 Find the cost of binding around the edge of a circular rug that is 3 meters in diameter. The binder charges $4.50 per meter. Use 3.14 for π.

ANSWERS

4. (a) $A \approx 904.3 \text{ m}^2$ (b) $A \approx 491.9 \text{ ft}^2$
   (c) $A \approx 150.8 \text{ m}^2$
5. $42.39

6 Find the cost of covering the underside of the rug in Margin Problem 5 on the previous page with a nonslip rubber backing. The rubber backing costs $2 per square meter.

EXAMPLE 6 Applying the Concept of Area

Find the cost of covering the underside of the rug in Example 5 (on the previous page) with a nonslip rubber backing. The rubber backing costs $1.50 per square foot. Use 3.14 for π.

First find the radius.

$$r = \frac{d}{2} = \frac{8 \text{ ft}}{2} = 4 \text{ ft}$$

Then find the area.

$$A = \pi \cdot r^2$$
$$A \approx 3.14 \cdot 4 \text{ ft} \cdot 4 \text{ ft}$$
$$A \approx 50.24 \text{ ft}^2$$

$$\text{cost} = \frac{\$1.50}{1 \text{ ft}^2} \cdot \frac{50.24 \text{ ft}^2}{1} = \$75.36$$

Work Problem 6 at the Side.

7 Find the volume of each cylinder. Use 3.14 for π. Round your answers to the nearest tenth.

(a)
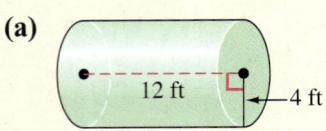
12 ft, 4 ft

OBJECTIVE 4 Find the volume of a cylinder. Several *cylinders* are shown below.

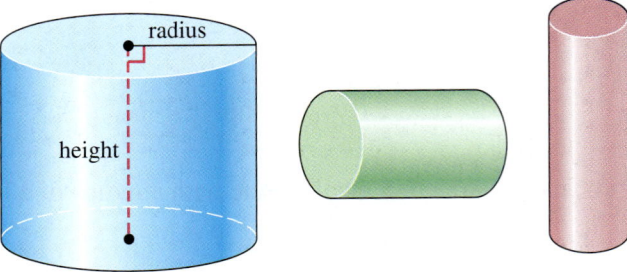

The height must be perpendicular to the circular top and bottom of the cylinder.

These are called *right circular cylinders* because the top and bottom are circles, and the side makes a right angle with the top and bottom. Examples of cylinders are a soup can, a home water heater, and a piece of pipe.

Use the following formula to find the *volume* of a *cylinder*. Notice that the first part of the formula, $\pi \cdot r \cdot r$, is the *area* of the circular base.

(b)
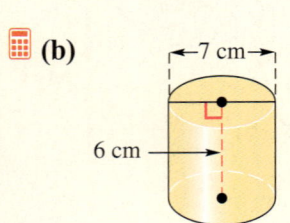
7 cm, 6 cm

> **Finding the Volume of a Cylinder**
>
> Volume of a cylinder = $\pi \cdot r \cdot r \cdot h$
>
> $$V = \pi r^2 h$$
>
> Remember to use **cubic units** when measuring volume.

EXAMPLE 7 Finding the Volume of Cylinders

Find the volume of each cylinder. Use 3.14 as the approximate value of π. Round your answers to the nearest tenth, if necessary.

(a)
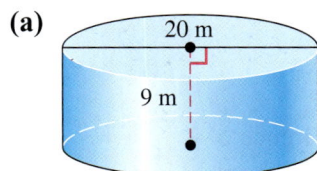
20 m, 9 m

The diameter is 20 m, so the radius is 20 m ÷ 2 = 10 m. The height is 9 m. Use the formula.

$$V = \pi \cdot r \cdot r \cdot h$$
$$V \approx 3.14 \cdot 10 \text{ m} \cdot 10 \text{ m} \cdot 9 \text{ m}$$
$$V \approx 2826 \text{ m}^3 \quad \text{Cubic units for volume}$$

(c) radius 14.5 yd, height 3.2 yd

(b)
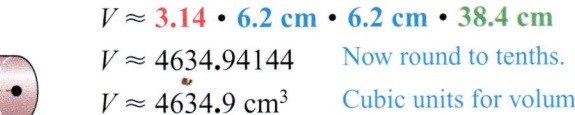
6.2 cm, 38.4 cm

$$V \approx 3.14 \cdot 6.2 \text{ cm} \cdot 6.2 \text{ cm} \cdot 38.4 \text{ cm}$$
$$V \approx 4634.94144 \quad \text{Now round to tenths.}$$
$$V \approx 4634.9 \text{ cm}^3 \quad \text{Cubic units for volume}$$

ANSWERS
6. $14.13
7. (a) $V \approx 602.9$ ft³   (b) $V \approx 230.8$ cm³
   (c) $V \approx 2112.6$ yd³

Work Problem 7 at the Side.

OBJECTIVE 5 **Find the surface area of a rectangular solid.** You have just learned how to find the *volume* of a cylinder. In **Section 4.8** you found the *volume* of a rectangular solid. For example, the volume of the cereal box shown below is $V = lwh = (7 \text{ in.})(2 \text{ in.})(10 \text{ in.}) = 140 \text{ in.}^3$ But if your company makes cereal boxes, you also need to know how much cardboard is needed for each box. You need to find the *surface area* of the box.

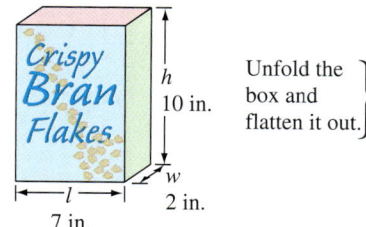
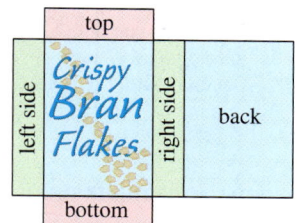

The unfolded box is made up of six rectangles: front, back, top, bottom, left side, right side.

8 Find the volume and surface area of each rectangular solid.

(a)
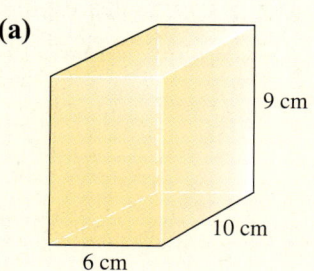

Surface area is the area on the surface of a three-dimensional object (a solid). For a rectangular solid like the cereal box, the surface area is the sum of the areas of the six rectangular sides. Notice that the top and bottom have the same area, the front and back have the same area, and the left and right sides have the same area.

Surface Area = top·w + bottom·w + Crispy Bran Flakes·h + back·h + left side·h + right side·h

$SA = l \cdot w + l \cdot w + l \cdot h + l \cdot h + w \cdot h + w \cdot h$

$SA = 2lw + 2lh + 2wh$

Finding the Surface Area of a Rectangular Solid

Surface Area = $(2 \cdot l \cdot w) + (2 \cdot l \cdot h) + (2 \cdot w \cdot h)$

$SA = 2lw + 2lh + 2wh$

Remember that area is measured in square units, so use **square units** when measuring *surface* area.

(b)
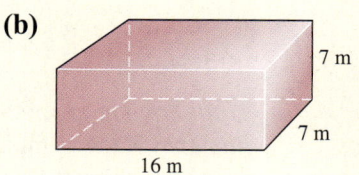

EXAMPLE 8 Finding the Volume and Surface Area of a Rectangular Solid

Find the volume and surface area of this shipping carton.

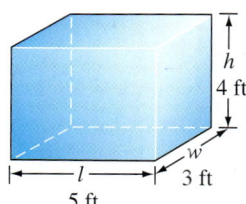

First find the volume.

$V = lwh$

$V = 5 \text{ ft} \cdot 3 \text{ ft} \cdot 4 \text{ ft}$

$V = 60 \text{ ft}^3$ ← Cubic units for volume

Next find the surface area.

$SA = 2lw + 2lh + 2wh$

$SA = (2 \cdot 5 \text{ ft} \cdot 3 \text{ ft}) + (2 \cdot 5 \text{ ft} \cdot 4 \text{ ft}) + (2 \cdot 3 \text{ ft} \cdot 4 \text{ ft})$

$SA = 30 \text{ ft}^2 + 40 \text{ ft}^2 + 24 \text{ ft}^2$

$SA = 94 \text{ ft}^2$ ← Square units for area

ANSWERS

8. (a) $V = 540 \text{ cm}^3$ (cubic cm for volume)
$SA = 408 \text{ cm}^2$ (square cm for area)
(b) $V = 784 \text{ m}^3$
$SA = 546 \text{ m}^2$

Work Problem 8 at the Side.

9 Find the volume and surface area of each cylinder. Use 3.14 for π. Round your answers to the nearest tenth.

(a)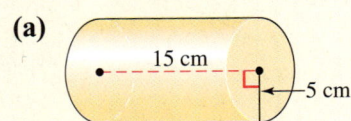

(b)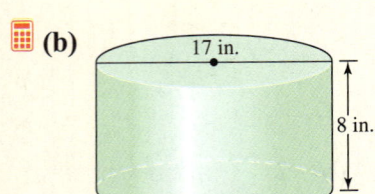

(*Hint:* You are given the *diameter* of the cylinder. Start by finding the *radius*.)

ANSWERS
9. (a) $V \approx 1177.5$ cm³ (cubic units for volume)
   $SA \approx 628$ cm² (square units for area)
   (b) $V \approx 1814.9$ in.³
   $SA \approx 880.8$ in.²

OBJECTIVE 6 **Find the surface area of a cylinder.** You can use the same idea of "unfolding" a shape to find the surface area of a cylinder, such as the soup can shown below. Finding the surface area will tell you how much aluminum you need to make the can.

The unfolded soup can is made up of a rectangular side, a circular top, and a circular bottom.

Remember that the formula for the area of a circle is πr^2.

Surface Area = [side] h + [top] r + [bottom] r

|← Circumference of can →|
($2\pi r$)

$SA = \underbrace{2\pi r \cdot h}_{2\pi rh} + \underbrace{\pi r^2 + \pi r^2}_{2\pi r^2}$

$SA = 2\pi rh + 2\pi r^2$

Finding the Surface Area of a Right Circular Cylinder

Surface Area = $(2 \cdot \pi \cdot r \cdot h) + (2 \cdot \pi \cdot r \cdot r)$

$SA = 2\pi rh + 2\pi r^2$

Remember that area is measured in square units, so use **square units** when measuring *surface* area.

EXAMPLE 9 Finding the Volume and Surface Area of a Right Circular Cylinder

Find the volume and surface area of this water tank. Use 3.14 as the approximate value for π. Round your answers to the nearest tenth when necessary.

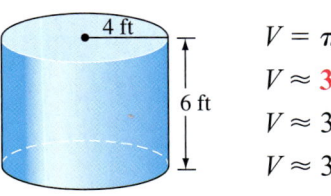

First find the volume.
$V = \pi r^2 h$
$V \approx 3.14 \cdot 4\text{ ft} \cdot 4\text{ ft} \cdot 6\text{ ft}$
$V \approx 301.44$ ft³ ← Now round to tenths.
$V \approx 301.4$ ft³ ← Cubic units for volume

Now find the surface area.
$SA = 2\pi rh + 2\pi r^2$
$SA \approx (2 \cdot 3.14 \cdot 4\text{ ft} \cdot 6\text{ ft}) + (2 \cdot 3.14 \cdot 4\text{ ft} \cdot 4\text{ ft})$
$SA \approx 150.72$ ft² + 100.48 ft²
$SA \approx 251.2$ ft² ← Square units for area

Work Problem 9 at the Side.

5.10 Exercises

FOR EXTRA HELP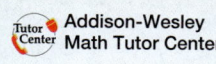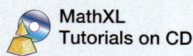

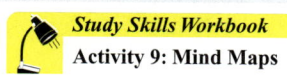

Find the unknown length in each circle. See Example 1.

1.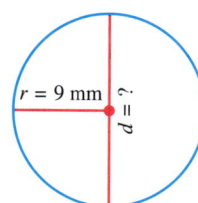
2.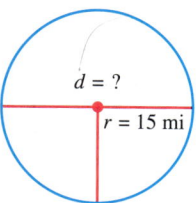
3.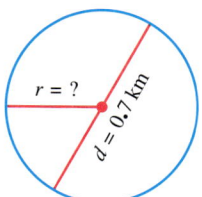
4.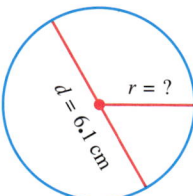

Find the circumference and area of each circle. Use 3.14 as the approximate value for π. Round your answers to the nearest tenth. See Examples 2 and 3.

5.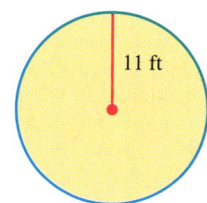
6.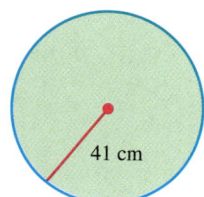
7.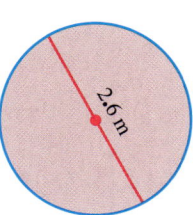
8.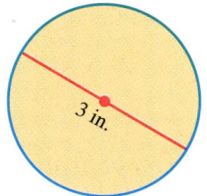

Find the circumference and area of circles having the following diameters. Use 3.14 for π. Round your answers to the nearest tenth. See Examples 2 and 3.

9. $d = 15$ cm

10. $d = 39$ ft

11. $d = 7\frac{1}{2}$ ft

12. $d = 4\frac{1}{2}$ yd

13. $d = 8.65$ km

14. $d = 19.5$ mm

Find each shaded area. Note that Exercises 15 and 18 contain semicircles. Use 3.14 as the approximate value of π. Round your answers to the nearest tenth when necessary. See Example 4.

15.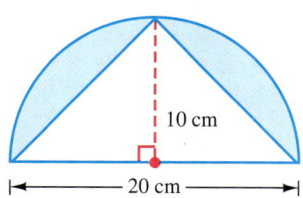

16.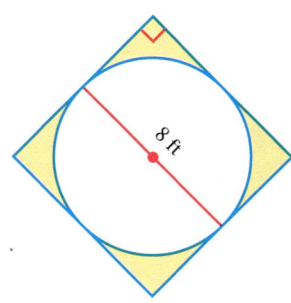

17.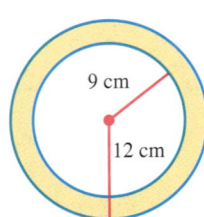

18.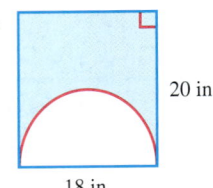

19. How would you explain π to a friend who is not in your math class? Write an explanation. Then make up a test question that requires the use of π, and show how to solve it.

20. Explain how circumference and perimeter are alike. How are they different? Make up two problems, one involving perimeter, the other circumference. Then show how to solve your problems.

Solve each application problem. Use **3.14** *as the approximate value of* π. *Round answers to the nearest tenth. See Examples 5 and 6.*

21. An irrigation system moves around a center point to water a circular area for crops. If the irrigation system is 50 yd long, how large is the watered area?

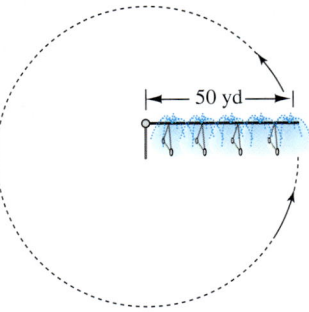

22. If you swing a ball held at the end of a string 20 cm long, how far will the ball travel on each turn?

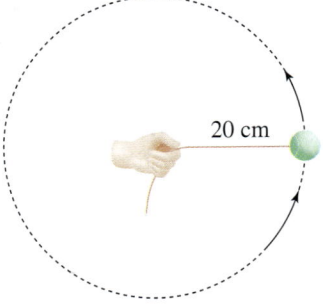

23. A Michelin Cross Terrain SUV tire has an overall diameter of 29.10 inches. How far will a point on the tire tread move in one complete turn? (*Source:* Michelin.)

Bonus question: How many revolutions does the tire make per mile? Round to the nearest whole number.

24. In September 2005, hurricane Katrina slammed into coastal Mississippi and flooded New Orleans. The diameter of the circular storm was 210 miles. In July 2005, hurricane Dennis, with a diameter of only 80 miles, hit the panhandle region of Florida. What was the area of each storm? (*Source:* Associated Press.)

For Exercises 25–30, first draw a circle and label the radius or diameter. Then solve the problem. Use 3.14 for π and round answers to the nearest tenth.

25. A radio station can be heard 150 miles in all directions during evening hours. How many square miles are in the station's broadcast area?

26. An earthquake was felt by people 900 km away in all directions from the epicenter (the source of the earthquake). How much area was affected by the quake?

27. The diameter of Diana Hestwood's wristwatch is 1 in. and the radius of the clock face on her kitchen wall is 3 in. Find the circumference and the area of each clock face.

28. The diameter of the largest known ball of twine is 12 ft 9 in. The sign posted near the ball says it has a circumference of 40 ft. Is the sign correct? *Hint:* First change 9 in. to feet and add it to 12 ft. (*Source: Guinness Book of World Records.*)

29. Blaine Fenstad wants to buy a pair of two-way radios. Some models have a range of 2 miles under ideal conditions. More expensive models have a range of 5 miles. What is the difference in the area covered by the 2-mile and 5-mile models? (*Source: Best Buy.*)

30. The National Audubon Society holds an end-of-year bird count. Volunteers count all the birds they see in a circular area during a 24-hour period. Each circle has a diameter of 15 miles. About 1700 circular areas are counted across the United States each December. What is the total area covered by the count?

31. A forester measures the circumference of a living tree at chest height, then calculates the diameter.

(a) If the circumference of one tree is 144 cm, what is the diameter?

(b) Explain how you solved part (a).

32. In Atlanta, Interstate 285 circles the city and is known as the "perimeter." If the circumference of the circle made by the highway is 62.8 miles, find:

(a) the diameter of the circle

(b) the area inside the circle.

(*Source: Greater Atlanta Newcomer's Guide.*)

33. The Mormons traveled west to Utah by covered wagon in 1847. They tied a rag to a wagon wheel to keep track of the distance they traveled. The radius of the wheel was 2.33 ft. How far did the rag travel each time the wheel made a complete revolution? (*Source: Trail of Hope.*)

34. First work Exercise 33. Then find how many wheel revolutions equaled one mile. There are 5280 ft in one mile.

35. Find the cost of sod, at $1.76 per square foot, for this playing field that has a semicircle on each end.

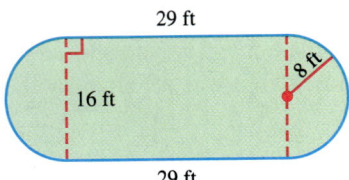

36. Find the area of this skating rink.

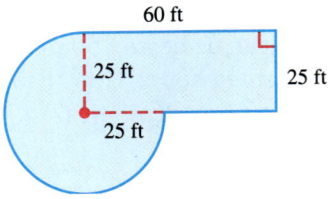

RELATING CONCEPTS (EXERCISES 37–42) For Individual or Group Work

Use the table below to **work Exercises 37–42 in order.**

Find the best buy for each type of pizza. The best buy is the lowest cost per square inch of pizza. All the pizzas are circular in shape, and the measurement given on the menu board is the diameter of the pizza in inches. Use 3.14 *as the approximate value of* π. *Round the area to the nearest tenth. Round cost per square inch to the nearest thousandth.*

Pizza Menu	Small 7½"	Medium 13"	Large 16"
Cheese only	$2.80	$ 6.50	$ 9.30
"The Works"	$3.70	$ 8.95	$14.30
Deep-dish combo	$4.35	$10.95	$15.65

37. Find the area of a small pizza.

38. Find the area of a medium pizza.

39. Find the area of a large pizza.

40. What is the cost per square inch for each size of cheese pizza? Which size is the best buy?

41. What is the cost per square inch for each size of "The Works" pizza? Which size is the best buy?

42. You have a coupon for 95¢ off any small pizza. What is the cost per square inch for each size of deep-dish combo pizza? Which size is the best buy?

Section 5.10 Geometry Applications: Circles, Cylinders, and Surface Area 393

Find the volume and surface area of each cylinder or rectangular solid. Use 3.14 as the approximate value of π. Round your answers to the nearest tenth when necessary. See Examples 7–9.

43.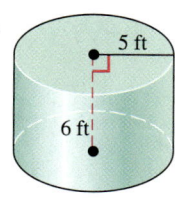

$V = \pi \cdot r^2 \cdot h$
$V = 3.14(5)^2 \cdot 6$
$V = 3.14 \cdot 25 \cdot 6$
$V = 471.0$

44.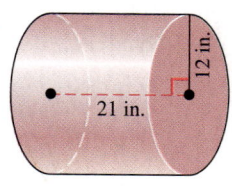

$SA = 2 \cdot \pi \cdot r^2 + 2 \cdot \pi \cdot r \cdot h$
$SA = (2 \cdot 3.14 \cdot 25) + (2 \cdot 3.14 \cdot 5 \cdot 6)$
$SA = 157 + 188.4$
$SA = 345.4$

45.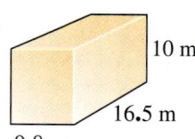

$V = 9.8 \times 16.5 \times 10$
$V = 1617 \, m$

solid area

$SA = 2(L \cdot w) + 2(L \cdot h) + 2(w \cdot h)$
$SA = (2 \cdot 9.8 \cdot 16.5) + (2 \cdot 9.8 \cdot 10) + (2 \cdot 16.5 \cdot 10)$
$SA = 323.40 + 196.0 + 330.0$
$SA = 849.4$

46.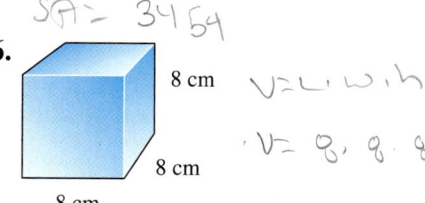

$V = L \cdot w \cdot h$
$V = 8 \cdot 8 \cdot 8$
$V = 512 \, cm$

47.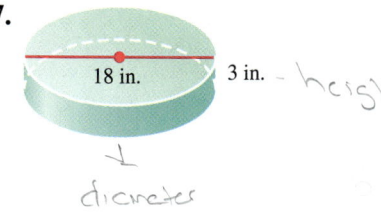

18 in. — diameter
3 in. — height

48.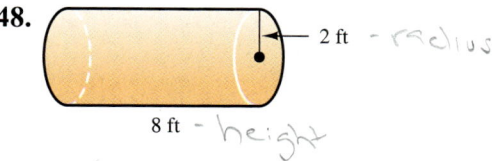

2 ft — radius
8 ft — height

49.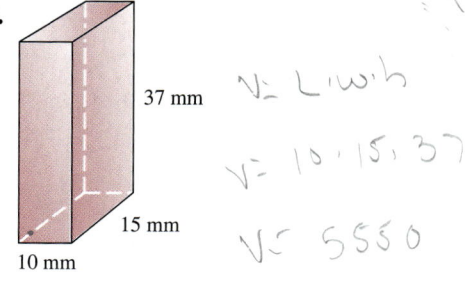

$V = L \cdot w \cdot h$
$V = 10 \cdot 15 \cdot 37$
$V = 5550$

50.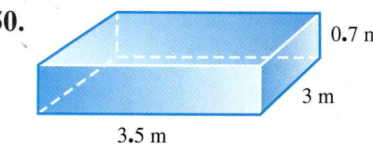

51. Explain the *two* errors made by a student in finding the volume of a cylinder with a diameter of 7 cm and a height of 5 cm. Find the correct answer.

$$V \approx 3.14 \cdot 7 \cdot 7 \cdot 5$$

$$V \approx 769.3 \text{ cm}^2$$

52. Look again at Exercise 46 on the previous page. The figure is a *cube*.

(a) What is special about the measurements of a cube?

(b) Find a shortcut you can use to calculate the surface area of a cube.

Solve each application problem. Use 3.14 *as the approximate value of* π. *Round your answers to the nearest tenth when necessary.*

53. A city sewer pipe has a diameter of 5 ft and a length of 200 ft. Find the volume of the pipe.

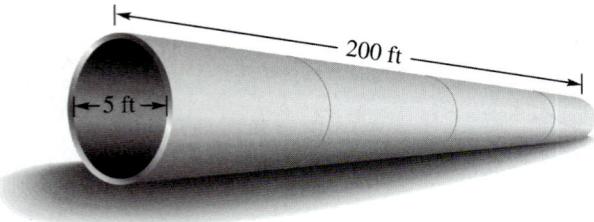

54. A cylindrical woven basket made by a Northwest Coast tribe is 8 cm high and has a diameter of 11 cm. What is the volume of the basket?

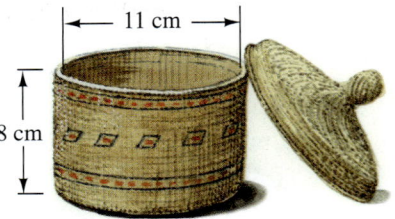

55. A box for graham crackers measures 5.5 in. by 2.8 in. by 8 in. high. Find the amount of cardboard needed to make the box.

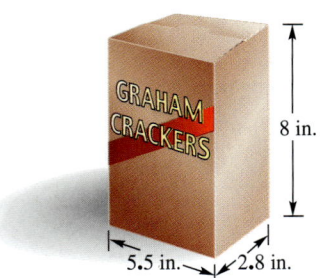

56. A soda can is 12.5 cm tall and has a diameter of 6.7 cm. How much aluminum is needed to make the can? (Assume the can has a flat top and bottom.)

Chapter 5
SUMMARY

KEY TERMS

5.1 **decimals** — Decimals, like fractions, are used to show parts of a whole.

decimal point — A decimal point is the dot that is used to separate the whole number part from the fractional part of a decimal number.

place value — Place value is the value assigned to each place to the right or left of the decimal point. Whole numbers, such as ones and tens, are to the *left* of the decimal point. Fractional parts, such as tenths and hundredths, are to the *right* of the decimal point.

5.2 **round** — To round is to "cut off" a number after a certain place, such as to round to the nearest hundredth. The rounded number is less accurate than the original number. You can use the symbol " ≈ " to mean "approximately equal to."

decimal places — Decimal places are the number of digits to the *right* of the decimal point; for example, 6.37 has two decimal places, 4.706 has three decimal places.

5.5 **repeating decimal** — A repeating decimal (like the 6 in 0.166 . . .) is a decimal number with one or more digits that repeat forever; it never ends. Use three dots to indicate that it is a repeating decimal. Or write the number with a bar above the repeating digits, as in $0.1\overline{6}$. (Use the dots or the bar, but not both.)

5.7 **mean** — The mean is the sum of all the values divided by the number of values. It is often called the *average*.

weighted mean — The weighted mean is a mean calculated so that each value is multiplied by its frequency.

median — The median is the middle number in a group of values that are listed from smallest to largest. It divides a group of values in half. If there is an even number of values, the median is the mean (average) of the two middle values.

mode — The mode is the value that occurs most often in a group of values.

variability — The variability of a set of data is the spread of the data around the mean. A quick way to evaluate variability is to find the range of values.

5.8 **square root** — A positive square root of a positive number is one of two equal positive factors of the number.

hypotenuse — The hypotenuse is the side of a right triangle opposite the 90° angle; it is the longest side.
Example: See the red side in the triangle at the right.

5.10 **circle** — A circle is a two-dimensional (flat) figure with all points the same distance from a fixed center point.
Example: See figure at the right.

radius — Radius is the distance from the center of a circle to any point on the circle.
Example: See the red radius in the circle at the right.

diameter — Diameter is the distance across a circle, passing through the center.
Example: See the blue diameter in the circle at the right.

circumference — Circumference is the distance around a circle.

π (pi) — π is the ratio of the circumference to the diameter of any circle. It is approximately equal to 3.14.

surface area — Surface area is the area on the surface of a three-dimensional object (a solid). Surface area is measured in square units.

NEW SYMBOLS

$3.8\overline{6}$ ← Bar above repeating digit(s) in a decimal number

$\sqrt{}$ square root

3.86 . . . ← Three dots indicate a repeating decimal

π Greek letter pi (pronounced PIE); ratio of the circumference of a circle to its diameter

≈ is approximately equal to

New Formulas

$$\text{mean} = \frac{\text{sum of all values}}{\text{number of values}}$$

$\text{hypotenuse} = \sqrt{(\text{leg})^2 + (\text{leg})^2}$

$\text{leg} = \sqrt{(\text{hypotenuse})^2 - (\text{leg})^2}$

diameter of a circle: $d = 2r$

radius of a circle: $r = \dfrac{d}{2}$

Circumference of a circle: $C = \pi d$ or $C = 2\pi r$

Area of a circle: $A = \pi r^2$

Volume of a cylinder: $V = \pi r^2 h$

Surface Area of a rectangular solid: $SA = 2lw + 2lh + 2wh$

Surface Area of a cylinder: $SA = 2\pi rh + 2\pi r^2$

Test Your Word Power

See how well you have learned the vocabulary in this chapter. Answers follow the Quick Review.

1. **Decimal numbers** are like fractions in that they both
   A. must be written in lowest terms
   B. need common denominators
   C. have decimal points
   D. represent parts of a whole.

2. **Decimal places** refer to
   A. the digits from 0 to 9
   B. digits to the left of the decimal point
   C. digits to the right of the decimal point
   D. the number of zeros in a decimal number.

3. The **hypotenuse** is
   A. the long base in a rectangle
   B. the height in a parallelogram
   C. the longest side in a right triangle
   D. the distance across a circle, passing through the center.

4. The **decimal point**
   A. separates the whole number part from the fractional part
   B. is always moved when finding a quotient
   C. separates tenths from hundredths
   D. is at the far left side of a whole number.

5. The number $0.\overline{3}$ is an example of
   A. an estimate
   B. a repeating decimal
   C. a rounded number
   D. a truncated number.

6. π is the ratio of
   A. the diameter to the radius of a circle
   B. the circumference to the diameter of a circle
   C. the circumference to the radius of a circle
   D. the diameter to the circumference of a circle.

7. The **median** for a set of values is
   A. the mathematical average
   B. the value that occurs most often
   C. the sum of each value times its frequency
   D. the middle value when the values are listed from smallest to largest.

8. The **circumference** of a circle is
   A. the perimeter of the circle
   B. found using the expression πr^2
   C. the distance across the circle
   D. measured in square units.

QUICK REVIEW

Concepts

5.1 *Reading and Writing Decimals*

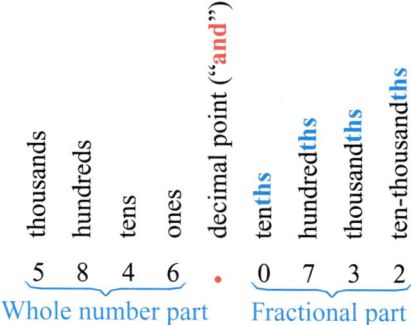

Whole number part | Fractional part

Examples

Write each decimal in words.

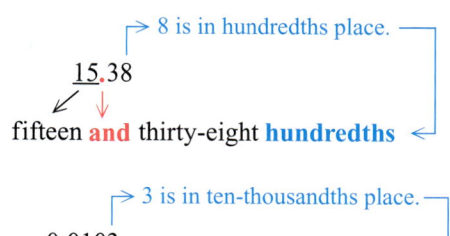

15.38

fifteen **and** thirty-eight **hundredths**

0.0103 — 3 is in ten-thousandths place.

one hundred three **ten-thousandths**

5.1 *Writing Decimals as Fractions*
The digits to the right of the decimal point are the numerator. The place value of the rightmost digit determines the denominator.

Always write the fraction in lowest terms.

Write 0.45 as a fraction in lowest terms.

The numerator is 45. The rightmost digit, 5, is in the hundredths place, so the denominator is 100. Then write the fraction in lowest terms.

$$\frac{45}{100} = \frac{45 \div 5}{100 \div 5} = \frac{9}{20} \leftarrow \text{Lowest terms}$$

5.2 *Rounding Decimals*
Find the place to which you are rounding. Draw a cut-off line to the right of that place; the rest of the digits will be dropped. Look *only* at the first digit being cut. If it is *4 or less,* the part you are keeping stays the same. If it is *5 or more,* the part you are keeping rounds up. Do not move the decimal point when rounding. Use the sign "≈" to mean "approximately equal to."

Round 0.17952 to the nearest thousandth.

```
          First digit cut is 5 or more so round up.
0.179│52
0.179    ← Keep this part.
+ 0.001  ← To round up, add 1 thousandth.
─────
0.180
```

0.17952 rounds to 0.180. Write 0.17952 ≈ 0.180.

5.3 *Adding Positive and Negative Decimals*
Estimate the answer by using front end rounding: round each number to the highest possible place.

To find the exact answer, line up the decimal points. If needed, write in zeros as placeholders. Add or subtract the absolute values as if they were whole numbers. Line up the decimal point in the answer.

If the numbers have the same sign, use the common sign as the sign of the sum. If the numbers have different signs, the sign of the sum is the sign of the number with the larger absolute value.

Add 5.68 + 785.3 + 12 + 2.007.

```
Estimate:        Exact:
      6            5.680     Use zeros as
    800          785.300     placeholders so
     10           12.000     all numbers have
  +   2         +  2.007     three decimal
  ────          ────────     places.
    818          804.987
                    ↑
                    Line up decimal points.
```

All addends are positive, so the sign of the sum is positive. The estimate and exact answer are both in hundreds, so the decimal point is probably in the correct place.

Concepts

5.3 Subtracting Positive and Negative Decimals
Rewrite subtracting as adding the opposite of the second number. Then follow the rules for adding positive and negative decimals.

5.4 Multiplying Positive and Negative Decimals
Step 1 Multiply as you would for whole numbers.
Step 2 Count the total number of decimal places in both factors.
Step 3 Write the decimal point in the answer so it has the same number of decimal places as the total from Step 2. You may need to write extra zeros on the left side of the product in order to get enough decimal places in the answer.
Step 4 If two factors have the *same sign,* the product is *positive.* If two factors have *different signs,* the product is *negative.*

5.5 Dividing by a Decimal
Step 1 Change the divisor to a whole number by moving the decimal point to the right.
Step 2 Move the decimal point in the dividend the same number of places to the right.
Step 3 Write the decimal point in the quotient directly above the decimal point in the dividend. Then divide as with whole numbers.
Step 4 If the numbers have the *same sign,* the quotient is *positive.* If they have *different signs,* the quotient is *negative.*

5.6 Writing Fractions as Decimals
Divide the numerator by the denominator. If necessary, round to the place indicated.

Examples

Subtract. $4.2 - 12.91$
$$4.2 + (-12.91)$$

$|4.2|$ is 4.2 and $|-12.91|$ is 12.91.
Subtract $12.91 - 4.2$ to get 8.71.
Because -12.91 has the larger absolute value and is negative, the answer will be negative.

$$4.2 + (-12.91) = -8.71$$

Multiply $(0.169)(-0.21)$

$$
\begin{array}{r}
0.169 \leftarrow \text{3 decimal places} \\
\times\ 0.21 \leftarrow \text{2 decimal places} \\
\hline
169 \quad\quad \text{5 total decimal places} \\
338 \\
\hline
.03549 \leftarrow \text{5 decimal places in product}
\end{array}
$$

Write in a 0 so you can count over 5 decimal places.
 The factors have different signs, so the product is negative.

$$(0.169)(-0.21) = -0.03549$$

Divide -52.8 by -0.75.
First consider $52.8 \div 0.75$.

$$
\begin{array}{r}
70.4 \\
0.75\overline{)52.800} \\
525 \\
\hline
300 \\
300 \\
\hline
0
\end{array}
$$

Move decimal point two places to the right in divisor and dividend. Write zeros in the dividend so you can move the decimal point and continue dividing until the remainder is 0.

The quotient is positive because both the divisor and dividend were negative (same signs means positive quotient).

Write $\frac{1}{8}$ as a decimal.

$\frac{1}{8}$ means $1 \div 8$. Write it as $8\overline{)1}$.
The decimal point is on the right side of 1.

$$
\begin{array}{r}
0.125 \\
8\overline{)1.000} \\
8 \\
\hline
20 \\
16 \\
\hline
40 \\
40 \\
\hline
0
\end{array}
$$

← Write the decimal point and three zeros so you can continue dividing.

Therefore, $\frac{1}{8}$ is equivalent to 0.125.

Concepts	Examples
5.6 Comparing the Size of Fractions and Decimals *Step 1* Write any fractions as decimals. *Step 2* Write zeros so that all the numbers being compared have the same number of decimal places. *Step 3* Use < to mean "is less than," > to mean "is greater than," or list the numbers from smallest to largest.	Arrange in order from smallest to largest. $0.505 \quad \frac{1}{2} \quad 0.55$ $0.505 = 505$ thousandths ← 505 is in the middle $\frac{1}{2} = 0.5 = 0.500 = 500$ thousandths ← 500 is smallest. $0.55 = 0.550 = 550$ thousandths ← 550 is largest. (smallest) $\frac{1}{2} \quad 0.505 \quad 0.55$ (largest)
5.7 Finding the Mean (Average) of a Set of Numbers *Step 1* Add all values to obtain a total. *Step 2* Divide the total by the number of values.	Here are Heather Hall's test scores in her math course. 93 76 83 93 78 82 87 85 Find Heather's mean score to the nearest tenth. $\text{mean} = \frac{93 + 76 + 83 + 93 + 78 + 82 + 87 + 85}{8}$ $= \frac{677}{8} \approx 84.6$ ← Mean test score
5.7 Finding the Median of a Set of Numbers *Step 1* Arrange the data from smallest to largest. *Step 2* If there is an odd number of values, select the middle value. If there is an even number of values, find the average of the two middle values.	Find the median for Heather Hall's scores from the previous example. List the scores from smallest to largest. 76 78 82 **83** **85** 87 93 93 Middle values The middle two values are 83 and 85. Find the average of these two values. $\frac{83 + 85}{2} = 84$ ← Median test score
5.7 Finding the Mode of a Set of Values Find the value that appears most often in the list of values. This is the mode. If no value appears more than once, there is no mode. If two different values appear the same number of times, the list is bimodal.	Find the mode for Heather's scores in the previous example. The most frequently occurring score is 93 (it occurs twice). Therefore, the mode is 93.
5.7 Evaluating the Variability of Data The variability of a set of data is the spread of the data around the mean. A quick way to evaluate the variability of data is to look at the range of values. Subtract the lowest value from the highest value. Generally, the greater the range in a set of values, the greater the variability.	Which set of test scores has greater variability? Student M: 78, 85, 82, 84, 90, 86 Student N: 87, 89, 65, 95, 58, 88 Range for student M = 90 − 78 = 12 Range for student N = 95 − 65 = 30 Student N's scores have greater range and greater variability.

Concepts	Examples
5.8 Finding the Square Root of a Number Use the square root key on a calculator, ✓ or √x. Round to the nearest thousandth when necessary.	$\sqrt{64} = 8$ $\sqrt{43} \approx 6.557$ ← 6.557438524 is rounded to the nearest thousandth.
5.8 Finding the Unknown Length in a Right Triangle To find the hypotenuse, use: $$\text{hypotenuse} = \sqrt{(\text{leg})^2 + (\text{leg})^2}$$ The hypotenuse is the side opposite the right angle. It is the longest side in a right triangle.	Find the unknown length in this right triangle. Round to the nearest tenth. 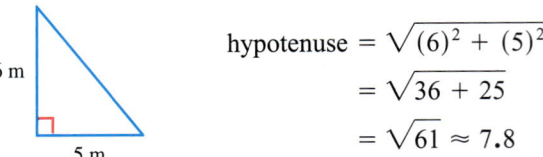 $\text{hypotenuse} = \sqrt{(6)^2 + (5)^2}$ $= \sqrt{36 + 25}$ $= \sqrt{61} \approx 7.8$ The hypotenuse is about 7.8 m long.
To find a leg, use: $$\text{leg} = \sqrt{(\text{hypotenuse})^2 - (\text{leg})^2}$$ The legs are the sides that form the right angle.	Find the unknown length in this right triangle. Round to the nearest tenth. 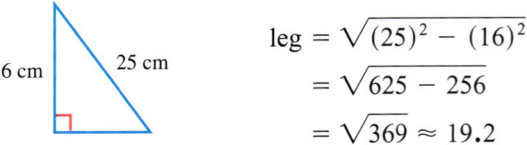 $\text{leg} = \sqrt{(25)^2 - (16)^2}$ $= \sqrt{625 - 256}$ $= \sqrt{369} \approx 19.2$ The leg is about 19.2 cm long.
5.9 Solving Equations Containing Decimals *Step 1* Use the addition property of equality to get the variable term by itself on one side of the equal sign. *Step 2* Divide both sides by the coefficient of the variable term to find the solution. *Step 3* Check the solution by going back to the original equation. Replace the variable with the solution. If the equation balances, the solution is correct.	Solve the equation and check the solution. $4.5x + 0.7 = -5.15$ $\underline{\,-0.7 -0.7}$ Add -0.7 to both sides. $4.5x + 0 = -5.85$ $\dfrac{4.5x}{4.5} = \dfrac{-5.85}{4.5}$ Divide both sides by 4.5 $x = -1.3$ The solution is -1.3. To check the solution, go back to the original equation. Check $4.5x + 0.7 = -5.15$ Original equation $4.5(-1.3) + 0.7 = -5.15$ Replace x with -1.3 $-5.85 + 0.7 = -5.15$ $-5.15 = -5.15$ Balances When x is replaced with -1.3, the equation balances, so -1.3 is the correct solution (*not* -5.15).

Concepts	Examples
5.10 Circles Use this formula to find the *diameter* of a circle when you are given the radius. $\qquad$ diameter = 2 • radius $\quad$ or $\quad d = 2r$	Find the diameter of a circle if the radius is 3 ft. $\qquad d = 2 \cdot r$ $\qquad d = 2 \cdot 3 \text{ ft} = 6 \text{ ft}$
Use this formula to find the *radius* of a circle when you are given the diameter. $\qquad$ radius = $\dfrac{\text{diameter}}{2}$ $\quad$ or $\quad r = \dfrac{d}{2}$	Find the radius of a circle if the diameter is 5 cm. 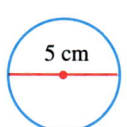 $r = \dfrac{d}{2}$ $r = \dfrac{5 \text{ cm}}{2} = 2.5 \text{ cm}$
Use these formulas to find the *circumference* of a circle. When you know the *radius,* use this formula. $\qquad C = 2 \cdot \pi \cdot \text{radius}$ $\quad$ or $\quad C = 2\pi r$ When you know the *diameter,* use this formula. $\qquad C = \pi \cdot \text{diameter}$ $\quad$ or $\quad C = \pi d$ Use 3.14 as the approximate value for π.	Find the circumference of a circle with a radius of 7 yd. Round your answer to the nearest tenth. $\qquad$ Circumference = $2 \cdot \pi \cdot r$ $\qquad C \approx 2 \cdot 3.14 \cdot 7 \text{ yd}$ $\qquad C \approx 44.0 \text{ yd}$ $\leftarrow$ Rounded
Use this formula to find the *area* of a circle. $\qquad A = \pi \cdot r \cdot r$ $\quad$ or $\quad A = \pi r^2$ Use 3.14 as the approximate value of π. Area is measured in **square units**.	Find the area of the circle. Round your answer to the nearest tenth. 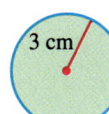 Area = $\pi \cdot r \cdot r$ $A \approx 3.14 \cdot 3 \text{ cm} \cdot 3 \text{ cm}$ $A \approx 28.3 \text{ cm}^2$ $\leftarrow$ Rounded; square units
5.10 Volume of a Cylinder Use this formula to find the volume of a cylinder. $\qquad$ Volume = $\pi \cdot r \cdot r \cdot h$ $\quad$ or $\quad V = \pi r^2 h$ where r is the radius of the circular base and h is the height of the cylinder. Volume is measured in **cubic units**.	Find the volume of this cylinder. 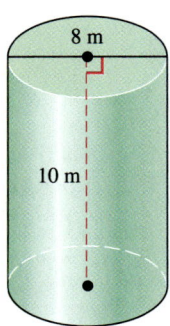 First, find the radius. $\quad r = \dfrac{8 \text{ m}}{2} = 4 \text{ m}$ $V = \pi \cdot r \cdot r \cdot h$ $V \approx 3.14 \cdot 4 \text{ m} \cdot 4 \text{ m} \cdot 10 \text{ m}$ $V \approx 502.4 \text{ m}^3$ $\leftarrow$ Cubic units for volume

Concepts

5.10 Surface Area of a Rectangular Solid
Use this formula to find the surface area of a rectangular solid.

Surface Area = (2 • l • w) + (2 • l • h) + (2 • w • h)

or $SA = 2lw + 2lh + 2wh$

where l is the length, w is the width, and h is the height of the solid.

Surface area is measured in **square units**.

Examples

Find the surface area of this packing crate.

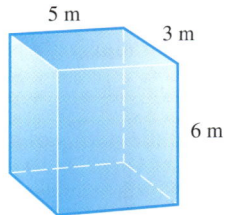

$SA = 2lw + 2lh + 2wh$

$SA = (2 \cdot \textbf{5 m} \cdot \textbf{3 m}) + (2 \cdot \textbf{5 m} \cdot \textbf{6 m}) + (2 \cdot \textbf{3 m} \cdot \textbf{6 m})$

$SA = 30 \text{ m}^2 + 60 \text{ m}^2 + 36 \text{ m}^2$

$SA = 126 \text{ m}^2$ ← Square units for surface area

5.10 Surface Area of a Cylinder
Use this formula to find the surface area of a cylinder.

Surface Area = (2 • π • r • h) + (2 • π • r • r)

or $SA = 2\pi rh + 2\pi r^2$

where r is the radius of the circular base and h is the height of the cylinder. Use 3.14 as the approximate value of π.

Surface area is measured in **square units**.

Find the surface area of a hot water tank with a height of 4.5 ft and a diameter of 1.8 ft. Round your answer to the nearest tenth.

First, find the radius. $r = \dfrac{1.8 \text{ ft}}{2} = 0.9 \text{ ft}$

$SA = 2\pi rh + 2\pi r^2$

$SA \approx (2 \cdot \textbf{3.14} \cdot \textbf{0.9 ft} \cdot \textbf{4.5 ft}) + (2 \cdot \textbf{3.14} \cdot \textbf{0.9 ft} \cdot \textbf{0.9 ft})$

$SA \approx 25.434 \text{ ft}^2 + 5.0868 \text{ ft}^2$

$SA \approx 30.5208 \text{ ft}^2$ Now round to tenths.

$SA \approx 30.5 \text{ ft}^2$ ← Square units for area

ANSWERS TO TEST YOUR WORD POWER

1. D; *Example:* For 0.7, the whole is cut into ten parts, and you are interested in 7 of the parts.
2. C; *Examples:* The number 6.87 has two decimal places; 0.309 has three decimal places.
3. C; *Example:* In triangle *ABC*, side *AC* (the red side) is the hypotenuse.

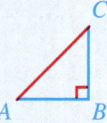

4. A; *Example:* In 5.42, the decimal point separates the whole number part, 5 ones, from the decimal part, 42 hundredths.
5. B; *Example:* The bar above the 3 in $0.\overline{3}$ indicates that the 3 repeats forever.
6. B; *Example:* The ratio of a circumference of 12.57 cm to a diameter of 4 cm is $\dfrac{12.57}{4} \approx 3.14$ (rounded).
7. D; *Example:* For this set of prices—$4, $3, $7, $2, $5, $6, $7—the median is $5.
8. A; *Example:* If the radius of a circle is 4 ft, then $C \approx 2 \cdot 3.14 \cdot 4 \approx 25.1$ ft. So the distance around the circle is about 25.1 ft.

Chapter 5
REVIEW EXERCISES

[5.1] *Name the digit that has the given place value.*

1. 243.059
   tenths
   hundredths

2. 0.6817
   ones
   tenths

3. $5824.39
   hundreds
   hundredths

4. 896.503
   tenths
   tens

5. 20.73861
   tenths
   ten-thousandths

Write each decimal as a fraction or mixed number in lowest terms.

6. 0.5

7. 0.75

8. 4.05

9. 0.875

10. 0.027

11. 27.8

Write each decimal in words.

12. 0.8

13. 400.29

14. 12.007

15. 0.0306

Write each decimal in numbers.

16. Eight and three tenths

17. Two hundred five thousandths

18. Seventy and sixty-six ten-thousandths

19. Thirty hundredths

[5.2] *Round to the place indicated.*

20. 275.635 to the nearest tenth

21. 72.789 to the nearest hundredth

22. 0.1604 to the nearest thousandth

23. 0.0905 to the nearest thousandth

24. 0.98 to the nearest tenth

Round each money amount to the nearest cent.

25. $15.8333

26. $0.698

27. $17,625.7906

Round each income or expense item to the nearest dollar.

28. The income from the pancake breakfast was $350.48.

29. Each member paid $129.50 in dues.

30. The refreshments cost $99.61.

31. The bank charges were $29.37.

[5.3] *Find each sum or difference.*

32. 0.4 − 6.07

33. −20 + 19.97

34. −1.35 + 7.229

35. 0.005 + (3 − 9.44)

First, use front end rounding to estimate each answer. Then find the exact answer.

36. Americans' favorite recreational activity is exercise walking. About 81.3 million people go walking at least once during the year. Swimming is second with 56.3 million people, and camping is third with 49.9 million people. How many more people go walking than camping? (*Source:* National Sporting Goods Association.)

    Estimate:

    Exact:

37. Today, Jasmin had $306 in her checking account. She wrote a check to the day care center for $215.53 and a check for $44.67 at the grocery store. What is the new balance in her account?

    Estimate:

    Exact:

38. Joey spent $1.59 for toothpaste, $5.33 for a gift, and $18.94 for a toaster. He gave the clerk three $10 bills. How much change did he get?

Estimate:

Exact:

39. Roseanne is training for a wheelchair race. She raced 2.3 kilometers on Monday, 4 kilometers on Wednesday, and 5.25 kilometers on Friday. How far did she race altogether?

Estimate:

Exact:

[5.4] *First, use front end rounding to estimate each answer. Then find the exact answer.*

40. Estimate: Exact:

     6.138
    × ___   × 3.7

41. Estimate: Exact:

     42.9
    × ___   × 3.3

Multiply.

42. (−5.6)(−0.002)

43. (0.071)(−0.005)

[5.5] *Decide if each answer is reasonable by rounding the numbers and estimating the answer. If the exact answer is not reasonable, find and correct the error.*

44. $706.2 \div 12 = 58.85$

　　Estimate:

45. $26.6 \div 2.8 = 0.95$

　　Estimate:

Divide. Round to the nearest thousandth when necessary.

46. $3\overline{)43.4}$

47. $\dfrac{-72}{-0.06}$

48. $-0.00048 \div 0.0012$

[5.4–5.5] *Solve each application problem.*

49. Adrienne worked 46.5 hours this week. Her hourly wage is $14.24 for the first 40 hours and 1.5 times that rate over 40 hours. Find her total earnings to the nearest dollar.

50. A book of 12 tickets costs $35.89 at the State Fair midway. What is the cost per ticket, to the nearest cent?

51. Stock in Math Tronic sells for $3.75 per share. Kenneth is thinking of investing $500. How many whole shares could he buy?

52. Ground beef is on sale at $0.99 per pound. How much will Ms. Lee pay for 3.5 pounds of hamburger, to the nearest cent?

Simplify by using the order of operations.

53. $3.5^2 + 8.7(-1.95)$

54. $11 - 3.06 \div (3.95 - 0.35)$

[5.6] *Write each fraction as a decimal. Round to the nearest thousandth when necessary.*

55. $3\dfrac{4}{5}$

56. $\dfrac{16}{25}$

57. $1\dfrac{7}{8}$

58. $\dfrac{1}{9}$

Arrange each group of numbers in order from smallest to largest.

59. 3.68, 3.806, 3.6008

60. 0.215, 0.22, 0.209, 0.2102

61. $0.17, \dfrac{3}{20}, \dfrac{1}{8}, 0.159$

[5.7] *Find the mean and the median for each set of data.*

62. Digital cameras sold:
　　18, 12, 15, 24, 9, 42, 54, 87, 21, 3

63. Number of insurance claims processed:
　　54, 28, 35, 43, 17, 37, 68, 75, 39

64. Find the weighted mean.

Dollar Value	Frequency
$42	3
$47	7
$53	2
$55	3
$59	5

65. (a) Find the mode or modes for each set of data.

Hiking boots at store J priced at $107, $69, $139, $107, $160, $84, $160

Hiking boots at store K priced at $119, $136, $99, $119, $139, $119, $95

(b) Which store's prices have greater variability? Show your work.

[5.8] *Find the unknown length in each right triangle. Use a calculator to find square roots. Round your answers to the nearest tenth when necessary.*

66.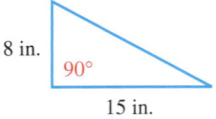

67.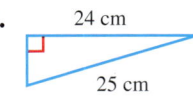

68.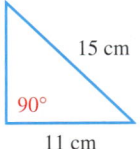

69.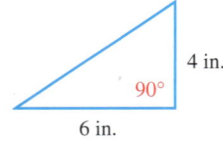

70.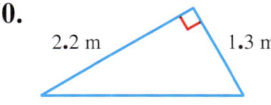

71.

[5.9] *Solve each equation. Show your work.*

72. $-0.1 = b - 0.35$

73. $-3.8x = 0$

74. $6.8 + 0.4n = 1.6$

75. $-0.375 + 1.75a = 2a$

76. $0.3y - 5.4 = 2.7 + 0.8y$

Chapter 5 Review Exercises **407**

[5.10] *Find the unknown radius or diameter.*

77. The radius of a circular irrigation field is 68.9 m. What is the diameter of the field?

78. The diameter of a juice can is 3 in. What is the radius of the can?

Find the circumference and area of each circle. Use 3.14 as the approximate value for π. Round your answers to the nearest tenth.

79.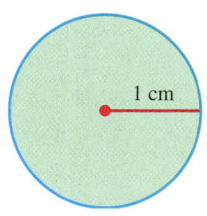

80.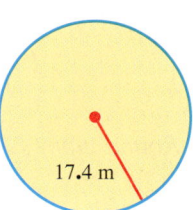

81.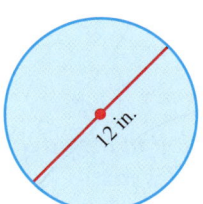

Find the volume and surface area of each solid. Use 3.14 as the approximate value for π. Round your answers to the nearest tenth when necessary.

82.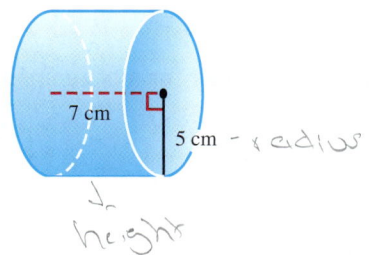

83.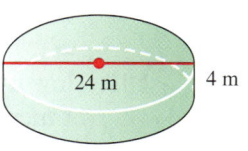

84. A rectangular cooler that measures 3.5 ft by 1.5 ft and is 1.5 ft high

---

MIXED REVIEW EXERCISES

Simplify.

85. $89.19 + 0.075 + 310.6 + 5$

86. $72.8(-3.5)$

87. $1648.3 \div 0.46$ Round to thousandths.

88. $30 - 0.9102$

89. $(4.38)(0.007)$

90. $0.005 \overline{)0.047}$

91. $72.105 + 8.2 - 95.37$

92. $\dfrac{-81.36}{9}$

93. $(0.6 - 1.22) + 4.8(-3.15)$

94. $0.455(18)$

95. $(-1.6)(-0.58)$

96. $0.218\overline{)7.63}$

97. $-21.059 - 20.8$

98. $18.3 - 3^2 \div 0.5$

Use the information in the ad to solve Exercises 99–103. Round money answers to the nearest cent. (Disregard any sales tax.)

99. How much would one pair of men's socks cost?

100. How much more would one pair of men's socks cost than one pair of children's socks?

101. How much would Fernando pay for a dozen pairs of men's socks?

102. How much would Akiko pay for five pairs of teen jeans and four pairs of women's jeans?

103. What is the difference between the cheapest sale price for athletic shoes and the highest regular price?

Solve each equation.

104. $4.62 = -6.6y$

105. $1.05x - 2.5 = 0.8x + 5$

Solve each application problem. Round answers to the nearest tenth when necessary.

106. A circular table has a diameter of 5 ft. How much rubber striping is needed to go around the edge of the table? What is the area of the tabletop?

107. Jerry missed one math test, so his test scores are 82, 0, 78, 93, 85. Find his mean and median scores.

108. LaRae drove 16 miles south, then made a 90° right turn and drove 12 miles west. How far is she, in a straight line, from her starting point?

109. A juice can that is 7 in. tall has a diameter of 4 in. What is the volume of the can?

Chapter 5
TEST

Write each decimal as a fraction or mixed number in lowest terms.

1. 18.4
2. 0.075

Write each decimal in words.

3. 60.007
4. 0.0208

First, use front end rounding to estimate each answer. Then find the exact answer.

5. $7.6 + 82.0128 + 39.59$
6. $-5.79(1.2)$

7. $-79.1 - 3.602$
8. $-20.04 \div (-4.8)$

Find the exact answer.

9. $670 - 0.996$
10. $0.15 \overline{)72}$
11. $(-0.006)(-0.007)$

12. Pat bought 3.4 meters of fabric. She paid $15.47. What was the cost per meter?

13. Davida ran a race in 3.059 minutes. Angela ran the race in 3.5 minutes. Who won? By how much?

14. Mr. Yamamoto bought 1.85 pounds of cheese at $2.89 per pound. How much did he pay for the cheese, to the nearest cent?

Solve each equation. Show your work.

15. $-5.9 = y + 0.25$
16. $-4.2x = 1.47$

17. $3a - 22.7 = 10$
18. $-0.8n + 1.88 = 2n - 6.1$

1. _____
2. _____
3. _____
4. _____
5. Estimate: _____
   Exact: _____
6. Estimate: _____
   Exact: _____
7. Estimate: _____
   Exact: _____
8. Estimate: _____
   Exact: _____
9. _____
10. _____
11. _____
12. _____
13. _____
14. _____
15. _____
16. _____
17. _____
18. _____

Arrange in order from smallest to largest.

19. $0.44, 0.451, \dfrac{9}{20}, 0.4506$

Use the order of operations to simplify.

20. $6.3^2 - 5.9 + 3.4(-0.5)$

21. Find the mean number of books loaned: 52, 61, 68, 69, 73, 75, 79, 84, 91, 98.

22. Find the mode for hot tub temperatures (Fahrenheit) of 96°, 104°, 103°, 104°, 103°, 104°, 91°, 74°, 103°.

23. Find the weighted mean.

Cost	Frequency
$6	7
$10	3
$11	4
$14	2
$19	3
$24	1

24. (a) Find the median cost of a math textbook: $54.50, $48, $39.75, $89, $56.25, $49.30, $46.90, $51.80.

 (b) Find the median cost of a biology textbook: $75, $50.45, $49.80, $30.95, $63, $97.55, $47.30.

 (c) Which set of costs has less variability? Show your work.

🖩 *Find the unknown lengths. Round your answers to the nearest tenth when necessary.*

25.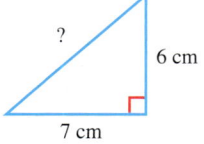

26.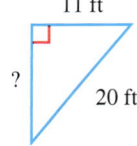

Use 3.14 as the approximate value for π and round your answers to the nearest tenth when necessary.

27. Find the radius.

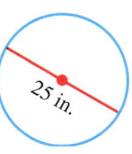

28. Find the circumference.

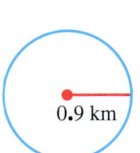

🖩 29. Find the area.

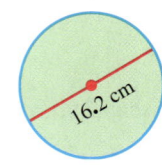

🖩 30. Find the volume.

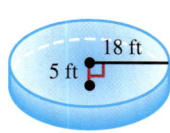

🖩 31. Find the surface area of the solid in Problem 30.

Ratio, Proportion, and Line/Angle/Triangle Relationships

6.1 Ratios
6.2 Rates
6.3 Proportions

Summary Exercises on Ratios, Rates, and Proportions

6.4 Problem Solving with Proportions
6.5 Geometry: Lines and Angles
6.6 Geometry Applications: Congruent and Similar Triangles

Seven out of every ten American households have cellular phones. Worldwide, about 1 out of 5 people use a cell phone! (*Source:* Consumer Electronics Assoc.)

You can use unit rates to find the best deal on cell phone service plans. But most plans do not cover long-distance calls to certain U.S. locations or to other countries. Then you can use unit rates to find the cheapest long-distance calling card. (See **Section 6.2,** Exercises 37–38 for calling cards and Exercises 47–50 for cell phone plans.)

6.1 Ratios

OBJECTIVES

1. Write ratios as fractions.
2. Solve ratio problems involving decimals or mixed numbers.
3. Solve ratio problems after converting units.

A **ratio** compares two quantities. You can compare two numbers, such as 8 and 4, or two measurements that have the *same* type of units, such as 3 days and 12 days. (*Rates* compare measurements with different types of units and are covered in the next section.)

Ratios can help you see important relationships. For example, if the ratio of your monthly expenses to your monthly income is 10 to 9, then you are spending $10 for every $9 you earn and going deeper into debt.

OBJECTIVE 1 Write ratios as fractions. A ratio can be written in three ways.

> **Writing a Ratio**
> The ratio of $7 to $3 can be written as follows.
>
> $$7 \text{ to } 3 \quad \text{or} \quad 7:3 \quad \text{or} \quad \frac{7}{3} \leftarrow \text{Fraction bar indicates "to."}$$
>
> ":" indicates "**to**."

Writing a ratio as a fraction is the most common method, and the one we will use here. All three ways are read, "the ratio of 7 to 3." The word **to** separates the quantities being compared.

> **Writing a Ratio as a Fraction**
> Order is important when you're writing a ratio. The quantity mentioned **first** is the **numerator**. The quantity mentioned **second** is the **denominator**. For example:
>
> $$\text{The ratio of } 5 \text{ to } 12 \quad \text{is written} \quad \frac{5}{12}$$

EXAMPLE 1 Writing Ratios

Ancestors of the Pueblo Indians built multistory apartment towns in New Mexico about 1100 years ago. A room might measure 14 feet long, 11 feet wide, and 15 feet high.

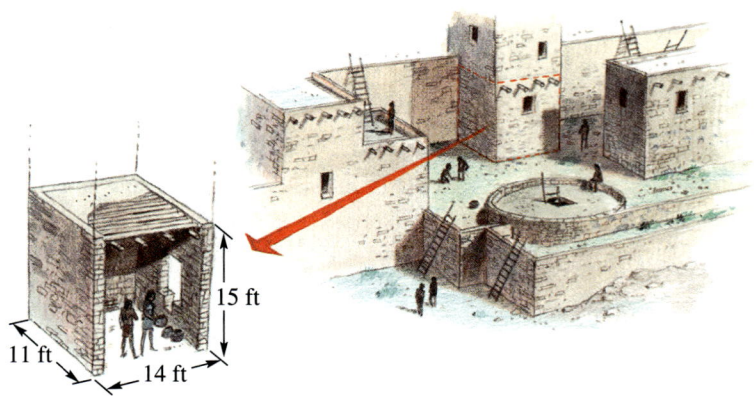

Continued on Next Page

Write each ratio as a fraction, using the room measurements.

(a) Ratio of length to width

The ratio of **length** to **width** is $\dfrac{14 \text{ feet}}{11 \text{ feet}} = \dfrac{14}{11}$

Numerator (mentioned first) — Denominator (mentioned second)

You can divide out common *units* just as you divided out common *factors* when writing fractions in lowest terms. (See **Section 4.2**.) However, do *not* rewrite the fraction as a mixed number. Keep it as the ratio of 14 to 11.

(b) Ratio of width to height

The ratio of width **to** height is $\dfrac{11 \text{ feet}}{15 \text{ feet}} = \dfrac{11}{15}$

> **CAUTION**
> Remember, the *order* of the numbers is important in a ratio. Look for the words "ratio of <u>*a*</u> to <u>*b*</u>." Write the ratio as $\dfrac{a}{b}$, **not** $\dfrac{b}{a}$. The quantity mentioned first is the numerator.

Work Problem 1 at the Side.

Any ratio can be written as a fraction. Therefore, you can write a ratio in *lowest terms*, just as you do with any fraction.

EXAMPLE 2 Writing Ratios in Lowest Terms

Write each ratio as a fraction in lowest terms.

(a) 60 days to 20 days
The ratio is $\dfrac{60}{20}$. Write this ratio in lowest terms by dividing the numerator and the denominator by 20.

$$\dfrac{60}{20} = \dfrac{60 \div 20}{20 \div 20} = \dfrac{3}{1} \quad \leftarrow \text{Ratio in lowest terms}$$

So, the ratio of 60 days to 20 days is 3 to 1, or, written as a fraction, $\dfrac{3}{1}$.

> **CAUTION**
> In the fractions chapter (**Chapter 4**), you would have rewritten $\dfrac{3}{1}$ as 3. But a *ratio* compares *two* quantities, so you need to keep both parts of the ratio and write it as $\dfrac{3}{1}$.

(b) 50 ounces of medicine to 120 ounces of medicine
The ratio is $\dfrac{50}{120}$. Divide the numerator and the denominator by 10.

$$\dfrac{50}{120} = \dfrac{50 \div 10}{120 \div 10} = \dfrac{5}{12} \quad \leftarrow \text{Ratio in lowest terms}$$

So, the ratio of 50 ounces to 120 ounces is $\dfrac{5}{12}$.

Continued on Next Page

---

1 Shane spent $14 on meat, $5 on milk, and $7 on fresh fruit. Write the following ratios as fractions in lowest terms.

(a) The ratio of amount spent on fruit to amount spent on milk

(b) The ratio of amount spent on milk to amount spent on meat

(c) The ratio of amount spent on meat to amount spent on milk

ANSWERS

1. (a) $\dfrac{7}{5}$ (b) $\dfrac{5}{14}$ (c) $\dfrac{14}{5}$

Chapter 6 Ratio, Proportion, and Line/Angle/Triangle Relationships

② Write each ratio as a fraction in lowest terms.

(a) 9 hours to 12 hours

(b) 100 meters to 50 meters

(c) The ratio of width to length for this rectangle

Length 48 ft
Width 24 ft

③ Write each ratio as a ratio of whole numbers in lowest terms.

(a) The price of Tamar's favorite brand of lipstick increased from $5.50 to $7.00. Find the ratio of the increase in price to the original price.

(b) Last week, Lance worked 4.5 hours each day. This week he cut back to 3 hours each day. Find the ratio of the decrease in hours to the original number of hours.

(c) 15 people in a large van to 6 people in a small van

The ratio is $\dfrac{15}{6} = \dfrac{15 \div 3}{6 \div 3} = \dfrac{5}{2}$ ← Ratio in lowest terms

> **NOTE**
> Although $\dfrac{5}{2} = 2\dfrac{1}{2}$, ratios are *not* written as mixed numbers. Nevertheless, in Example 2(c) above, the ratio $\dfrac{5}{2}$ does mean the large van holds $2\dfrac{1}{2}$ times as many people as the small van.

◀◀◀ **Work Problem ② at the Side.**

OBJECTIVE ② Solve ratio problems involving decimals or mixed numbers. Sometimes a ratio compares two decimal numbers or two fractions. It is easier to understand if we rewrite the ratio as a ratio of two whole numbers.

EXAMPLE 3 Using Decimal Numbers in a Ratio

The price of a Sunday newspaper increased from $1.50 to $1.75. Find the ratio of the increase in price to the original price.

The words *increase in price* are mentioned first, so the increase will be the numerator. How much did the price go up? Use subtraction.

new price − original price = increase
$1.75 − $1.50 = $0.25

The words *original price* are mentioned second, so the original price of $1.50 is the denominator.

The ratio of increase in price to original price is shown below.

$\dfrac{0.25}{1.50}$ ← increase
← original price

Now rewrite the ratio as a ratio of whole numbers. Recall that if you multiply both the numerator and denominator of a fraction by the same number, you get an equivalent fraction. The decimals in this example are hundredths, so multiply by 100 to get whole numbers. (If the decimals are tenths, multiply by 10. If thousandths, multiply by 1000.) Then write the ratio in lowest terms.

$\dfrac{0.25}{1.50} = \dfrac{(0.25)(100)}{(1.50)(100)} = \underbrace{\dfrac{25}{150}}_{\text{Ratio as two whole numbers}} = \dfrac{25 \div 25}{150 \div 25} = \dfrac{1}{6}$ ← Ratio in lowest terms

◀◀◀ **Work Problem ③ at the Side.**

EXAMPLE 4 Using Mixed Numbers in Ratios

Write each ratio as a comparison of whole numbers in lowest terms.

(a) 2 days to $2\dfrac{1}{4}$ days

Write the ratio as follows. Divide out the common units.

$\dfrac{2 \text{ days}}{2\dfrac{1}{4} \text{ days}} = \dfrac{2}{2\dfrac{1}{4}}$

Continued on Next Page

ANSWERS

2. (a) $\dfrac{3}{4}$ (b) $\dfrac{2}{1}$ (c) $\dfrac{1}{2}$

3. (a) $\dfrac{(1.50)(100)}{(5.50)(100)} = \dfrac{150 \div 50}{550 \div 50} = \dfrac{3}{11}$

 (b) $\dfrac{(1.5)(10)}{(4.5)(10)} = \dfrac{15 \div 15}{45 \div 15} = \dfrac{1}{3}$

Next, write 2 as $\frac{2}{1}$ and $2\frac{1}{4}$ as the improper fraction $\frac{9}{4}$.

$$\frac{2}{2\frac{1}{4}} = \frac{\frac{2}{1}}{\frac{9}{4}}$$

Now rewrite the problem in horizontal format, using the "÷" symbol for division. Finally, multiply by the reciprocal of the divisor, as you did in **Chapter 4**.

$$\frac{\frac{2}{1}}{\frac{9}{4}} = \frac{2}{1} \div \frac{9}{4} = \frac{2}{1} \cdot \frac{4}{9} = \frac{8}{9}$$

↑ Reciprocals ↑

The ratio, in lowest terms, is $\frac{8}{9}$.

(b) $3\frac{1}{4}$ to $1\frac{1}{2}$

Write the ratio as $\frac{3\frac{1}{4}}{1\frac{1}{2}}$. Then write $3\frac{1}{4}$ and $1\frac{1}{2}$ as improper fractions.

$$3\frac{1}{4} = \frac{13}{4} \quad \text{and} \quad 1\frac{1}{2} = \frac{3}{2}$$

The ratio is shown below.

$$\frac{3\frac{1}{4}}{1\frac{1}{2}} = \frac{\frac{13}{4}}{\frac{3}{2}}$$

Rewrite as a division problem in horizontal format, using the "÷" symbol. Then multiply by the reciprocal of the divisor.

$$\frac{13}{4} \div \frac{3}{2} = \frac{13}{4} \cdot \frac{2}{3} = \frac{13 \cdot \overset{1}{2}}{\underset{1}{2} \cdot 2 \cdot 3} = \frac{13}{6} \quad \left\{ \begin{array}{l} \text{Ratio in} \\ \text{lowest terms} \end{array} \right.$$

↑ Reciprocals ↑

> **NOTE**
> We can also work Examples 4(a) and 4(b) above by using decimals.
> **(a)** $2\frac{1}{4}$ is equivalent to 2.25, so we have the ratio shown below.
>
> $$\frac{2}{2\frac{1}{4}} = \frac{2}{2.25} = \frac{(2)(100)}{(2.25)(100)} = \frac{200}{225} = \frac{200 \div 25}{225 \div 25} = \frac{8}{9} \leftarrow \text{Same result}$$
>
> **(b)** $3\frac{1}{4}$ is equivalent to 3.25 and $1\frac{1}{2}$ is equivalent to 1.5.
>
> $$\frac{3\frac{1}{4}}{1\frac{1}{2}} = \frac{3.25}{1.5} = \frac{(3.25)(100)}{(1.5)(100)} = \frac{325}{150} = \frac{325 \div 25}{150 \div 25} = \frac{13}{6} \leftarrow \text{Same result}$$
>
> This method would *not* work for fractions that are repeating decimals, such as $\frac{1}{3}$ or $\frac{5}{6}$.

Work Problem 4 at the Side. ▶▶▶

4 Write each ratio as a ratio of whole numbers in lowest terms.

(a) $3\frac{1}{2}$ to 4

(b) $5\frac{5}{8}$ pounds to $3\frac{3}{4}$ pounds

(c) $3\frac{1}{3}$ inches to $\frac{5}{6}$ inch

ANSWERS
4. (a) $\frac{7}{8}$ (b) $\frac{3}{2}$ (c) $\frac{4}{1}$

OBJECTIVE 3 Solve ratio problems after converting units. When a ratio compares measurements, both measurements must be in the *same* units. For example, *feet* must be compared to *feet*, *hours* to *hours*, *pints* to *pints*, and *inches* to *inches*.

EXAMPLE 5 Ratio Applications Using Measurement

(a) Write the ratio of the length of the shorter board on the left to the length of the longer board on the right. Compare in inches.

First, express 2 feet in inches. Because 1 foot has 12 inches, 2 feet is

$$2 \cdot 12 \text{ inches} = 24 \text{ inches}.$$

The length of the board on the left is 24 inches, so the ratio of the lengths is shown below. The common units divide out.

$$\frac{2 \text{ ft}}{30 \text{ in.}} = \frac{24 \text{ inches}}{30 \text{ inches}} = \frac{24}{30}$$

Write the ratio in lowest terms.

$$\frac{24}{30} = \frac{24 \div 6}{30 \div 6} = \frac{4}{5} \quad \left\{ \text{Ratio in lowest terms} \right.$$

The shorter board on the left is $\frac{4}{5}$ the length of the longer board on the right.

> **NOTE**
> Notice in the example above that we wrote the ratio using the smaller unit (inches are smaller than feet). Using the smaller unit will help you avoid working with fractions. If we wrote the ratio using feet, then:
>
> $$30 \text{ inches} = 2\frac{1}{2} \text{ feet}.$$
>
> The ratio in feet is shown below.
>
> $$\frac{2 \text{ feet}}{2\frac{1}{2} \text{ feet}} = \frac{2}{1} \div \frac{5}{2} = \frac{2}{1} \cdot \frac{2}{5} = \frac{4}{5} \quad \leftarrow \text{Same result}$$
>
> The ratio is the same, but it takes more steps to get the answer. Using the smaller unit is usually easier.

(b) Write the ratio of 28 days to 3 weeks.

Since it is usually easier to write the ratio using the smaller measurement, compare in *days* because days are shorter than weeks.

First express 3 weeks in days. Because 1 week has 7 days, 3 weeks is

$$3 \cdot 7 \text{ days} = 21 \text{ days}.$$

So the ratio in days is shown below.

$$\frac{28 \text{ days}}{3 \text{ weeks}} = \frac{28 \text{ days}}{21 \text{ days}} = \frac{28}{21} = \frac{28 \div 7}{21 \div 7} = \frac{4}{3} \quad \leftarrow \left\{ \text{Ratio in lowest terms} \right.$$

The following table will help you set up ratios that compare measurements. You will work with these measurements again in **Chapter 8**.

Measurement Comparisons

Length	Capacity (Volume)
1 foot = 12 inches	1 pint = 2 cups
1 yard = 3 feet	1 quart = 2 pints
1 mile = 5280 feet	1 gallon = 4 quarts

Weight	Time
1 pound = 16 ounces	1 minute = 60 seconds
1 ton = 2000 pounds	1 hour = 60 minutes
	1 day = 24 hours
	1 week = 7 days

Work Problem 5 at the Side.

5 Write each ratio as a fraction in lowest terms. (*Hint:* Recall that it is usually easier to write the ratio using the smaller measurement unit.)

(a) 9 inches to 6 feet

(b) 2 days to 8 hours

(c) 7 yards to 14 feet

(d) 3 quarts to 3 gallons

(e) 25 minutes to 2 hours

(f) 4 pounds to 12 ounces

ANSWERS

5. (a) $\frac{1}{8}$ (b) $\frac{6}{1}$ (c) $\frac{3}{2}$ (d) $\frac{1}{4}$
   (e) $\frac{5}{24}$ (f) $\frac{16}{3}$

Focus on Real-Data Applications

Historical Ratios

The table below lists average prices and earnings from 1960, 1970, 1980, and 1990. (*Source:* Pages of Time.)

Item	1960	1970	1980	1990	Today
First-class postage stamp	$0.04	$0.06	$0.15	$0.25	
Movie ticket	$1.00	$1.50	$2.25	$4.00	
Gallon of gas	$0.26	$0.36	$1.19	$1.34	
New car (to nearest hundred)	$2600	$4000	$7200	$16,000	
Yearly income (to nearest hundred)	$5200	$9400	$19,200	$28,900	
Hourly minimum wage	$1.00	$1.60	$3.10	$3.80	

1. Fill in the amount for each item in the Today column of the table. Use local prices in your city or neighborhood for a movie ticket and a gallon of gas. Use *The World Almanac* or an on-line search engine to find the average income and new car price.

2. Use the data on the price of a movie ticket to find these ratios. Write each ratio as a ratio of whole numbers in lowest terms.

    (a) 1970 price to 1960 price
    (b) 1980 price to 1970 price
    (c) 1990 price to 1980 price
    (d) Today price to 1990 price

3. What pattern do you see in the movie ticket ratios?

4. Use the data on postage stamp prices to find each ratio involving the increase in price. Write each ratio as a ratio of whole numbers in lowest terms.

    (a) increase from 1960 to 1970 compared to 1960 price
    (b) increase from 1970 to 1980 compared to 1970 price
    (c) increase from 1980 to 1990 compared to 1980 price
    (d) increase from 1990 to Today compared to 1990 price

5. What pattern do you see in the postage stamp price increases?

6. Write a ratio in lowest terms comparing the cost of a new car to the average yearly income for each year. Then use your calculator to change each fractional ratio into a decimal. What pattern do you see?

    (a) 1960
    (b) 1970
    (c) 1980
    (d) 1990
    (e) Today

7. How many gallons of gas could you buy with one hour of minimum wage earnings in each year? Round to the nearest tenth. What pattern do you see?

    (a) 1960
    (b) 1970
    (c) 1980
    (d) 1990
    (e) Today

Section 6.1 Ratios 419

6.1 Exercises

FOR EXTRA HELP Addison-Wesley Math Tutor Center 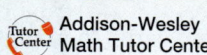 MathXL 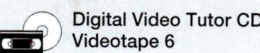 Digital Video Tutor CD 4 Videotape 6  Student's Solutions Manual 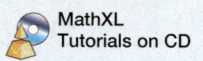 MyMathLab MathXL Tutorials on CD

Write each ratio as a fraction in lowest terms. See Examples 1 and 2.

1. 8 to 9
2. 11 to 15
3. $100 to $50
4. 35¢ to 7¢

5. 30 minutes to 90 minutes
6. 9 pounds to 36 pounds

7. 80 miles to 50 miles
8. 300 people to 450 people

9. 6 hours to 16 hours
10. 45 books to 35 books

Write each ratio as a ratio of whole numbers in lowest terms. See Examples 3 and 4.

11. $4.50 to $3.50
12. $0.08 to $0.06
13. 15 to $2\frac{1}{2}$

14. 5 to $1\frac{1}{4}$
15. $1\frac{1}{4}$ to $1\frac{1}{2}$
16. $2\frac{1}{3}$ to $2\frac{2}{3}$

Write each ratio as a fraction in lowest terms. For help, use the table of measurement relationships on page 425. See Example 5.

17. 4 feet to 30 inches
18. 8 feet to 4 yards

19. 5 minutes to 1 hour
20. 8 quarts to 5 pints

21. 15 hours to 2 days
22. 3 pounds to 6 ounces

23. 5 gallons to 5 quarts
24. 3 cups to 3 pints

The table shows the number of greeting cards that Americans buy for various occasions. Use the information to answer Exercises 25–30. Write each ratio as a fraction in lowest terms.

Holiday/Event	Cards Sold
Valentine's Day	900 million
Mother's Day	150 million
Father's Day	95 million
Graduation	60 million
Thanksgiving	30 million
Halloween	25 million

Source: Hallmark Cards.

25. Find the ratio of Thanksgiving cards to graduation cards.

26. Find the ratio of Halloween cards to Mother's Day cards.

27. Find the ratio of Valentine's Day cards to Halloween cards.

28. Find the ratio of Mother's Day cards to Father's Day cards.

29. Explain how you might use the information in the table if you owned a shop selling gifts and greeting cards.

30. Why is the ratio of Valentine's Day cards to graduation cards $\frac{15}{1}$? Give two possible reasons.

The bar graph shows worldwide sales of the most popular songs of all time. Use the graph to complete Exercises 31–34. Write each ratio as a fraction in lowest terms.

Source: The Music Information Database.

31. Write two ratios that compare the top-selling song to the second best seller and the top-selling song to the third best seller.

32. Write two ratios that compare sales of Elvis Presley's song with sales of the songs just ahead and just behind it in the graph.

33. Sales of which two songs give a ratio of $\frac{3}{1}$? There may be more than one correct answer.

34. Sales of which two songs give a ratio of $\frac{5}{6}$? There may be more than one correct answer.

Use the circle graph of one person's monthly budget to complete Exercises 35–38. Write each ratio as a fraction in lowest terms.

35. Find the ratio of taxes to transportation.

36. Find the ratio of rent to food.

37. Find the ratio of
  (a) rent to total budget.
  (b) rent and utilities to taxes and miscellaneous.

38. Find the ratio of
  (a) utilities to total budget.
  (b) food and taxes to transportation.

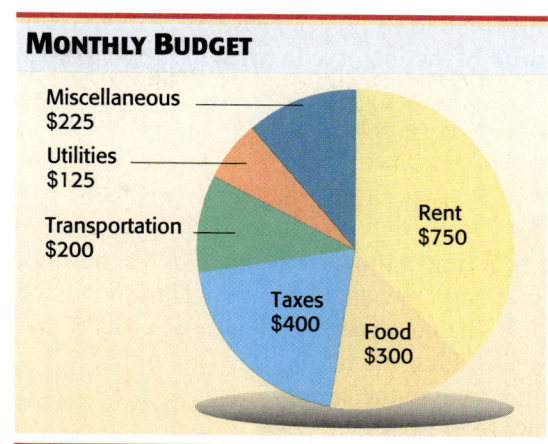

For each figure, find the ratio of the length of the longest side to the length of the shortest side. Write each ratio as a fraction in lowest terms. See Examples 2–4.

39.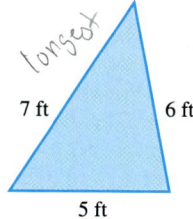

40.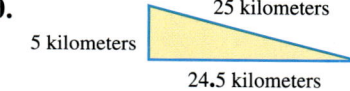

41.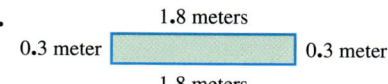

42.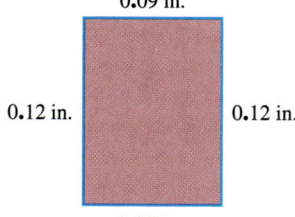

43.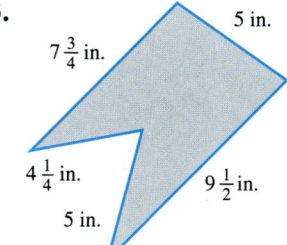

44.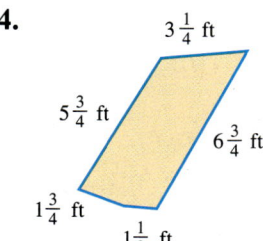

Write each ratio as a fraction in lowest terms.

45. The price of automobile engine oil recently went from $10.80 to $13.50 per case of 12 quarts. Find the ratio of the increase in price to the original price.

46. The price of an antibiotic decreased from $8.80 to $5.60 for 10 tablets. Find the ratio of the decrease in price to the original price.

47. The first time a movie was made in Minnesota, the cast and crew spent $59\frac{1}{2}$ days filming winter scenes. The next year, another movie was filmed in $8\frac{3}{4}$ weeks. Find the ratio of the first movie's filming time to the second movie's time. Compare in weeks.

48. The percheron, a large draft horse, measures about $5\frac{3}{4}$ feet at the shoulder. The prehistoric ancestor of the horse measured only $15\frac{3}{4}$ inches at the shoulder. Find the ratio of the percheron's height to its prehistoric ancestor's height. Compare in inches. (*Source: Eyewitness Books: Horse.*)

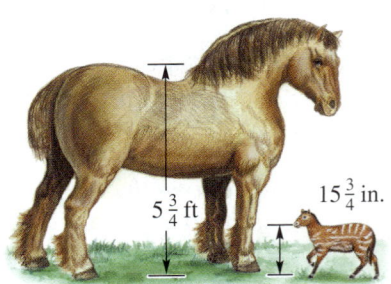

RELATING CONCEPTS (EXERCISES 49–52) For Individual or Group Work

Use your knowledge of ratios to **work Exercises 49–52 in order.**

49. In this painting, what is the ratio of the length of the longest side to the length of the shortest side? What other measurements could the painting have and still maintain the same ratio?

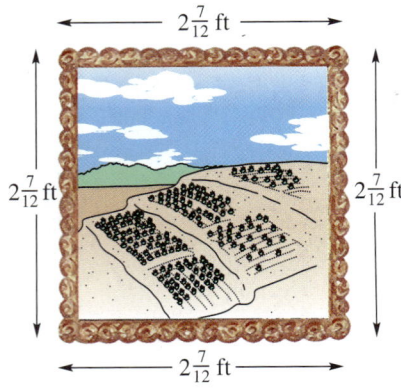

50. The ratio of my son's age to my daughter's age is 4 to 5. One possibility is that my son is 4 years old and my daughter is 5 years old. Find six other possibilities that fit the 4 to 5 ratio.

51. Amelia said that the ratio of her age to her mother's age is 5 to 3. Is this possible? Explain your answer.

52. Would you prefer that the ratio of your income to your friend's income be 1 to 3 or 3 to 1? Explain your answer.

6.2 Rates

A *ratio* compares two measurements with the same type of units, such as 9 <u>feet</u> to 12 <u>feet</u> (both length measurements). But many of the comparisons we make use measurements with different types of units, such as shown here.

$$\text{160 \underline{dollars} \textbf{for} 8 \underline{hours}} \quad \text{(money to time)}$$

$$\text{450 \underline{miles} \textbf{on} 15 \underline{gallons}} \quad \text{(distance to capacity)}$$

This type of comparison is called a **rate**.

OBJECTIVE 1 Write rates as fractions. Suppose that yesterday you hiked 18 <u>miles</u> **in** 4 <u>hours</u>. The *rate* at which you hiked can be written as a fraction in lowest terms.

$$\frac{18 \text{ miles}}{4 \text{ hours}} = \frac{18 \text{ miles} \div 2}{4 \text{ hours} \div 2} = \frac{9 \text{ miles}}{2 \text{ hours}} \quad \leftarrow \text{Rate in lowest terms}$$

In a rate, you often find one of these words separating the quantities you are comparing.

in   for   on   per   from

> **CAUTION**
> When writing a rate, always include the units, such as miles, hours, dollars, and so on. Because the units in a rate are different, the units do *not* divide out.

EXAMPLE 1 Writing Rates in Lowest Terms

Write each rate as a fraction in lowest terms.

(a) 5 gallons of chemical **for** $60

$$\frac{5 \text{ gallons} \div 5}{60 \text{ dollars} \div 5} = \frac{1 \text{ gallon}}{12 \text{ dollars}} \quad \text{Write the units: gallons and dollars.}$$

(b) $1500 wages **in** 10 weeks

$$\frac{1500 \text{ dollars} \div 10}{10 \text{ weeks} \div 10} = \frac{150 \text{ dollars}}{1 \text{ week}}$$

(c) 2225 miles **on** 75 gallons of gas

$$\frac{2225 \text{ miles} \div 25}{75 \text{ gallons} \div 25} = \frac{89 \text{ miles}}{3 \text{ gallons}}$$

Work Problem 1 at the Side.

OBJECTIVE 2 Find unit rates. When the *denominator* of a rate is 1, it is called a **unit rate**. We use unit rates frequently. For example, you earn $12.75 for *1 hour* of work. This unit rate is written:

$$\$12.75 \textbf{ per } \text{hour} \quad \text{or} \quad \$12.75 \text{ / hour}.$$

You drive 28 miles on *1 gallon* of gas. This unit rate is written

$$\text{28 miles \textbf{per} gallon} \quad \text{or} \quad \text{28 miles / gallon}.$$

Use **per** or a **/** mark when writing unit rates.

OBJECTIVES

1. Write rates as fractions.
2. Find unit rates.
3. Find the best buy based on cost per unit.

1 Write each rate as a fraction in lowest terms.

(a) $6 for 30 packages

(b) 500 miles in 10 hours

(c) 4 teachers for 90 students

(d) 1270 bushels from 30 acres

ANSWERS

1. **(a)** $\frac{\$1}{5 \text{ packages}}$ **(b)** $\frac{50 \text{ miles}}{1 \text{ hour}}$ **(c)** $\frac{2 \text{ teachers}}{45 \text{ students}}$ **(d)** $\frac{127 \text{ bushels}}{3 \text{ acres}}$

424 Chapter 6 Ratio, Proportion, and Line/Angle/Triangle Relationships

② Find each unit rate.

(a) $4.35 for 3 pounds of cheese

(b) 304 miles on 9.5 gallons of gas

(c) $850 in 5 days

(d) 24-pound turkey for 15 people

ANSWERS
2. (a) $1.45/pound (b) 32 miles/gallon
   (c) $170/day (d) 1.6 pounds/person

EXAMPLE 2 Finding Unit Rates

Find each unit rate.

(a) 337.5 miles on 13.5 gallons of gas
   Write the rate as a fraction.

   $$\frac{337.5 \text{ miles}}{13.5 \text{ gallons}} \quad \leftarrow \text{The fraction bar indicates division.}$$

Divide 337.5 by 13.5 to find the unit rate.

$$13.5\overline{)337.5} = 25.$$

$$\frac{337.5 \text{ miles} \div 13.5}{13.5 \text{ gallons} \div 13.5} = \frac{25 \text{ miles}}{1 \text{ gallon}}$$

The unit rate is 25 miles **per** gallon, or 25 miles/gallon.

(b) 549 miles in 18 hours

$$\frac{549 \text{ miles}}{18 \text{ hours}} \quad \text{Divide.} \quad 18\overline{)549.0} = 30.5$$

The unit rate is 30.5 miles per hour, or 30.5 miles/hour.

(c) $810 in 6 days

$$\frac{810 \text{ dollars}}{6 \text{ days}} \quad \text{Divide.} \quad 6\overline{)810} = 135$$

The unit rate is $135 per day, or $135/day.

◀◀◀ **Work Problem ② at the Side.**

OBJECTIVE 3 **Find the best buy based on cost per unit.** When shopping for groceries, household supplies, and health and beauty items, you will find many different brands and package sizes. You can save money by finding the lowest *cost per unit*.

Cost per Unit

Cost per unit is a rate that tells how much you pay for *one* item or *one* unit. Examples are $1.85 per gallon, $47 per shirt, and $2.98 per pound.

EXAMPLE 3 Determining the Best Buy

The local store charges the following prices for pancake syrup. Find the best buy.

Continued on Next Page

The best buy is the container with the *lowest* cost per unit. All the containers are measured in *ounces* (oz), so you first need to find the *cost per ounce* for each one. Divide the price of the container by the number of ounces in it. Round to the nearest thousandth if necessary.

Let the *order* of the *words* help you set up the rate.

cost ⟶ $1.28
per (means divide) ⟶ ―――
ounce ⟶ 12 ounces

Size	Cost per Unit (Rounded)
12 ounces	$\frac{\$1.28}{12 \text{ ounces}} \approx \$0.107$ per ounce (highest)
24 ounces	$\frac{\$1.81}{24 \text{ ounces}} \approx \$0.075$ per ounce (lowest)
36 ounces	$\frac{\$2.73}{36 \text{ ounces}} \approx \$0.076$ per ounce

The lowest cost per ounce is $0.075, so the 24-ounce container is the best buy.

NOTE
Earlier we rounded money amounts to the nearest hundredth (nearest cent). But when comparing unit costs, rounding to the nearest thousandth will help you see the difference between very similar unit costs. Notice that the 24-ounce and 36-ounce syrup containers above would both have rounded to $0.08 per ounce if we had rounded to hundredths.

Work Problem 3 at the Side.

Calculator Tip When using a calculator to find unit prices, remember that division is *not* commutative. In Example 3 above you wanted to find cost per ounce. Let the *order* of the *words* help you enter the numbers in the correct order.

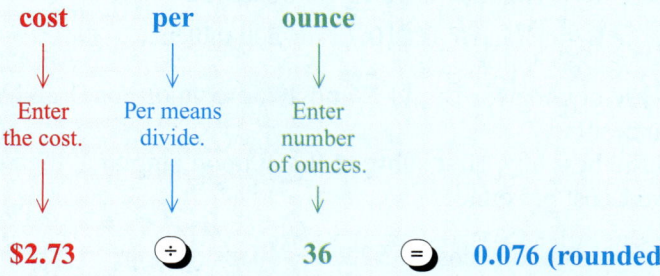

If you reversed the order and entered 36 ÷ 2.73 =, the result is the number of *ounces* per *dollar*. How could you use that information to find the best buy? (*Answer:* The best buy would be the greatest number of ounces per dollar.)

Finding the best buy is sometimes a complicated process. Things that affect the cost per unit can include "cents off" coupons and differences in how much use you'll get out of each unit.

3 Find the best buy (lowest cost per unit) for each purchase.

(a) 2 quarts for $3.25
3 quarts for $4.95
4 quarts for $6.48

(b) 6 cans of cola for $1.99
12 cans of cola for $3.49
24 cans of cola for $7

ANSWERS
3. (a) 4 quarts, at $1.62 per quart
(b) 12 cans, at about $0.291 per can

4 Solve each problem.

(a) Some batteries claim to last longer than others. If you believe these claims, which brand is the best buy?

Four-pack of AA-size batteries for $2.79

One AA-size battery for $1.19; lasts twice as long

(b) Which tube of toothpaste is the best buy? You have a coupon for 85¢ off Brand C and a coupon for 20¢ off Brand D.

Brand C is $3.89 for 6 ounces.

Brand D is $1.59 for 2.5 ounces.

ANSWERS
4. (a) One battery that lasts twice as long (like getting two) is the best buy. The cost per unit is $0.595 per battery. The four-pack is about $0.698 per battery.
(b) Brand C with the 85¢ coupon is the best buy at about $0.507 per ounce. Brand D with the 20¢ coupon is $0.556 per ounce.

EXAMPLE 4 Solving Best Buy Applications

Solve each application problem.

(a) There are many brands of liquid laundry detergent. If you feel they all do a good job of cleaning your clothes, you can base your purchase on cost per unit. But some brands are "concentrated" so you can use less detergent for each load of clothes. Which of the choices shown below is the best buy?

try **SUDZY** to clean your clothes!
50 fluid ounces for $3.99
Does same number of washloads as the old 64-ounce bottle!

WHITE-O gets out ALL the stains!
One gallon (128 ounces) for $9.89
Does twice the washloads of the old gallon bottle!

To find Sudzy's unit cost, divide $3.99 by 64 ounces, not 50 ounces. You're getting as many clothes washed as if you bought 64 ounces. Similarly, to find White-O's unit cost, divide $9.89 by 256 ounces (twice 128 ounces, or 2 • 128 ounces = 256 ounces).

$$\text{Sudzy} \quad \frac{\$3.99}{64 \text{ ounces}} \approx \$0.062 \quad \text{per ounce}$$

$$\text{White-O} \quad \frac{\$9.89}{256 \text{ ounces}} \approx \$0.039 \quad \text{per ounce}$$

White-O has the lower cost per ounce and is the best buy. (However, if you try it and it doesn't get out all the stains, Sudzy may be worth the extra cost.)

(b) "Cents-off" coupons also affect the best buy. Suppose you are looking at these choices for "extra-strength" pain reliever. Both brands have the same amount of pain reliever in each tablet.

Brand X is $2.29 for 50 tablets.

Brand Y is $10.75 for 200 tablets.

You have a 40¢ coupon for Brand X and a 75¢ coupon for Brand Y. Which choice is the best buy?

To find the best buy, first subtract the coupon amounts, then divide to find the lowest cost per ounce.

Brand X costs $2.29 − $0.40 = $1.89

$$\frac{\$1.89}{50 \text{ tablets}} \approx \$0.038 \text{ per tablet}$$

Brand Y costs $10.75 − $0.75 = $10.00

$$\frac{\$10.00}{200 \text{ tablets}} \approx \$0.05 \text{ per tablet}$$

Brand X has the lower cost per tablet and is the best buy.

Work Problem 4 at the Side.

6.2 Exercises

Write each rate as a fraction in lowest terms. See Example 1.

1. 10 cups for 6 people

2. $12 for 30 pens

3. 15 feet in 35 seconds

4. 100 miles in 30 hours

5. 14 people for 28 dresses

6. 12 wagons for 48 horses

7. 25 letters in 5 minutes

8. 68 pills for 17 people

9. $63 for 6 visits

10. 25 doctors for 310 patients

11. 72 miles on 4 gallons

12. 132 miles on 8 gallons

Find each unit rate. See Example 2.

13. $60 in 5 hours

14. $2500 in 20 days

15. 50 eggs from 10 chickens

16. 36 children from 12 families

17. 7.5 pounds for 6 people

18. 44 bushels from 8 trees

19. $413.20 for 4 days

20. $74.25 for 9 hours

Earl kept the following record of the gas he bought for his car. For each entry, find the number of miles he traveled and the unit rate. Round your answers to the nearest tenth.

	Date	Odometer at Start	Odometer at End	Miles Traveled	Gallons Purchased	Miles per Gallon
21.	2/4	27,432.3	27,758.2		15.5	
22.	2/9	27,758.2	28,058.1		13.4	
23.	2/16	28,058.1	28,396.7		16.2	
24.	2/20	28,396.7	28,704.5		13.3	

Source: Author's car records.

Find the best buy (based on the cost per unit) for each item. See Example 3. (Source: Cub Foods.)

25. Black pepper

26. Shampoo

27. Cereal
    13 ounces for $2.80
    15 ounces for $3.15
    18 ounces for $3.98

28. Soup
    2 cans for $1.90
    3 cans for $2.89
    5 cans for $4.58

29. Chunky peanut butter
    12 ounces for $1.29
    18 ounces for $1.79
    28 ounces for $3.39
    40 ounces for $4.39

30. Baked beans
    8 ounces for $0.59
    16 ounces for $0.77
    21 ounces for $0.99
    28 ounces for $1.36

31. Suppose you are choosing between two brands of chicken noodle soup. Brand A is $0.88 per can and Brand B is $0.98 per can. But Brand B has more chunks of chicken in it. Which soup is the best buy? Explain your choice.

32. A small bag of potatoes costs $0.19 per pound. A large bag costs $0.15 per pound. But there are only two people in your family, so half the large bag would probably rot before you used it up. Which bag is the best buy? Explain.

Solve each application problem. See Examples 2–4.

33. Makesha lost 10.5 pounds in six weeks. What was her rate of loss in pounds per week?

34. Enrique's taco recipe uses three pounds of meat to feed 10 people. Give the rate in pounds per person.

35. Russ works 7 hours to earn $85.82. What is his pay rate per hour?

36. Find the cost of 1 gallon of gas if 18 gallons cost $55.62.

The table lists information about three long-distance calling cards. The connection fee is charged each time you make a call, no matter how long the call lasts. Use the information in the table to answer Exercises 37–40. Round answers to the nearest thousandth when necessary.

LONG-DISTANCE CALLING CARDS (U.S.)

Card Name	Cost per Minute	Connection Fee
Radiant Penny	$0.01	$0.39
IDT Special	$0.022	$0.14
Access America	$0.047	$0.00

Source: www.1callcard.com

37. (a) Find the *actual* total cost, including the connection charge, for a five-minute call using each card.

(b) Find the cost per minute for this call using each card and select the best buy.

38. (a) Find the *actual* total cost, including the connection charge, for a 30-minute call using each card.

(b) Find the cost per minute for this call using each card and select the best buy.

39. Find the *actual* total cost per minute for a 15-minute call and a 20-minute call using each card. Then select the best buy for each call.

40. All the cards round calls up to the next full minute.

(a) Suppose you call the wrong number. How much would you pay for this 40-*second* call on each card?

(b) How much would you save on this call by using Access America instead of Radiant Penny?

41. In the 2000 Olympics, Michael Johnson ran the 400-meter event in a record time of approximately 44 seconds (actually 43.84 seconds). Give his rate in seconds per meter and in meters per second. Use 44 seconds as the time. (*Source:* www.Olympics.com)

42. Sofia can clean and adjust five hearing aids in four hours. Give her rate in hearing aids per hour and in hours per hearing aid.

43. If you believe the claims that some batteries last longer, which is the best buy?

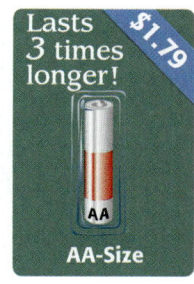

44. Which is the best buy, assuming these laundry detergents both clean equally well?

45. Three brands of cornflakes are available. Brand G is priced at $2.39 for 10 ounces. Brand K is $3.99 for 20.3 ounces and Brand P is $3.39 for 16.5 ounces. You have a coupon for 50¢ off Brand P and a coupon for 60¢ off Brand G. Which cereal is the best buy based on cost per unit?

46. Two brands of facial tissue are available. Brand K has a special price of three boxes of 175 tissues each for $5. Brand S is priced at $1.29 per box of 125 tissues. You have a coupon for 20¢ off one box of Brand S and a coupon for 45¢ off one box of Brand K. How can you get the best buy on one box of tissue?

RELATING CONCEPTS (EXERCISES 47–50) For Individual or Group Work

On the first page of this chapter, we said that unit rates can help you get the best deal on cell phone service. Use the information in the table to **work Exercises 47–50 in order.**

Company	Anytime Minutes	One-Time Activation Fee	Monthly Charge	Termination Fee
Verizon	400	$35	$59.99	$175
T-Mobile	600	$35	$39.99	$200
Nextel	500	$35	$45.99	$200
Sprint	500	$36	$55	$150

Source: Advertisements appearing in *Minneapolis Star Tribune*.

Notes:
1. All companies require that you sign a contract for one year of service and charge a one-time termination fee if you quit early.
2. Unused minutes cannot be carried over to the next month.

47. How much will each company's activation fee cost you on a monthly basis during the one-year contract?

48. All the plans allow unlimited calls on nights and weekends, so you would be using the "anytime minutes" on weekdays. Figure out the average number of weekdays per month. Then, for each plan, how many minutes could you use per weekday? Round to the nearest whole minute.

49. Find the actual average cost per "anytime minute" during the one-year contract for each company, including the activation fee. Assume you use all the minutes and no more. Decide how to round your answers so you can find the best buy.

50. Suppose that after two months you canceled your service because you found that you only used 100 "anytime minutes" per month. Under those conditions, find the actual cost per "anytime minute" for each company, to the nearest cent.

6.3 Proportions

OBJECTIVE 1 Write proportions. A **proportion** states that two ratios (or rates) are equivalent. For example,

$$\frac{\$20}{4 \text{ hours}} = \frac{\$40}{8 \text{ hours}}$$

is a proportion that says the rate $\frac{\$20}{4 \text{ hours}}$ is equivalent to the rate $\frac{\$40}{8 \text{ hours}}$. As the amount of money doubles, the number of hours also doubles. This proportion is read:

20 dollars **is to** 4 hours   **as**   40 dollars **is to** 8 hours.

EXAMPLE 1 Writing Proportions

Write each proportion.

(a) 6 feet is to 11 feet **as** 18 feet is to 33 feet.

$$\frac{6 \text{ feet}}{11 \text{ feet}} = \frac{18 \text{ feet}}{33 \text{ feet}} \quad \text{so} \quad \frac{6}{11} = \frac{18}{33} \quad \text{The common units (feet) divide out and are not written.}$$

(b) $9 is to 6 liters **as** $3 is to 2 liters.

$$\frac{\$9}{6 \text{ liters}} = \frac{\$3}{2 \text{ liters}} \quad \text{Units must be written.}$$

Work Problem 1 at the Side.

OBJECTIVE 2 Determine whether proportions are true or false. There are two ways to see whether a proportion is true. One way is to *write both of the ratios in lowest terms*.

EXAMPLE 2 Writing Both Ratios in Lowest Terms

Are the following proportions true?

(a) $\frac{5}{9} = \frac{18}{27}$

Write each ratio in lowest terms.

$$\frac{5}{9} \leftarrow \text{Already in lowest terms} \qquad \frac{18 \div 9}{27 \div 9} = \frac{2}{3} \leftarrow \text{Lowest terms}$$

Because $\frac{5}{9}$ is *not* equivalent to $\frac{2}{3}$, the proportion is *false*. The ratios are *not* proportional.

(b) $\frac{16}{12} = \frac{28}{21}$

Write each ratio in lowest terms.

$$\frac{16 \div 4}{12 \div 4} = \frac{4}{3} \quad \text{and} \quad \frac{28 \div 7}{21 \div 7} = \frac{4}{3}$$

Both ratios are equivalent to $\frac{4}{3}$, so the proportion is *true*. The ratios are proportional.

Work Problem 2 at the Side.

OBJECTIVES

1. Write proportions.
2. Determine whether proportions are true or false.
3. Find the unknown number in a proportion.

1 Write each proportion.

(a) $7 is to 3 cans as $28 is to 12 cans.

(b) 9 meters is to 16 meters as 18 meters is to 32 meters.

(c) 5 is to 7 as 35 is to 49.

(d) 10 is to 30 as 60 is to 180.

2 Determine whether each proportion is true or false by writing both ratios in lowest terms.

(a) $\frac{6}{12} = \frac{15}{30}$

(b) $\frac{20}{24} = \frac{3}{4}$

(c) $\frac{25}{40} = \frac{30}{48}$

(d) $\frac{35}{45} = \frac{12}{18}$

ANSWERS

1. (a) $\frac{\$7}{3 \text{ cans}} = \frac{\$28}{12 \text{ cans}}$ (b) $\frac{9}{16} = \frac{18}{32}$
   (c) $\frac{5}{7} = \frac{35}{49}$ (d) $\frac{10}{30} = \frac{60}{180}$
2. (a) $\frac{1}{2} = \frac{1}{2}$; true (b) $\frac{5}{6} \neq \frac{3}{4}$; false
   (c) $\frac{5}{8} = \frac{5}{8}$; true (d) $\frac{7}{9} \neq \frac{2}{3}$; false

A second way to see whether a proportion is true is to find *cross products*.

> **Using Cross Products to Determine Whether a Proportion Is True**
>
> To see whether a proportion is true, first multiply along one diagonal, then multiply along the other diagonal, as shown here.
>
> $$\frac{2}{5} = \frac{4}{10} \quad \begin{array}{l} 5 \cdot 4 = 20 \\ 2 \cdot 10 = 20 \end{array} \quad \text{Cross products are equal.}$$
>
> In this case the **cross products** are both 20. When cross products are *equal*, the proportion is *true*. If the cross products are *unequal*, the proportion is *false*.

> **NOTE**
> Why does the cross products test work? It is based on rewriting both fractions with a common denominator of 5 · 10 or 50. (We do not search for the *lowest* common denominator. We simply use the product of the two given denominators.)
>
> $$\frac{2 \cdot 10}{5 \cdot 10} = \frac{20}{50} \quad \text{and} \quad \frac{4 \cdot 5}{10 \cdot 5} = \frac{20}{50}$$
>
> We see that $\frac{2}{5}$ and $\frac{4}{10}$ are equivalent because both can be rewritten as $\frac{20}{50}$. The cross products test takes a shortcut by comparing only the two numerators (20 = 20).

EXAMPLE 3   Using Cross Products

Use cross products to see whether each proportion is true or false.

(a) $\dfrac{3}{5} = \dfrac{12}{20}$ Multiply along one diagonal and then along the other diagonal.

$$\frac{3}{5} = \frac{12}{20} \quad \begin{array}{l} 5 \cdot 12 = 60 \\ 3 \cdot 20 = 60 \end{array} \quad \text{Equal}$$

The cross products are *equal*, so the proportion is *true*.

> **CAUTION**
> Use cross products *only* when working with *proportions*. Do **not** use cross products when multiplying fractions, adding fractions, or writing fractions in lowest terms.

Continued on Next Page

(b) $\dfrac{2\frac{1}{3}}{3\frac{1}{3}} = \dfrac{9}{16}$ Find the cross products.

Changed to improper fractions

$$3\frac{1}{3} \cdot 9 = \dfrac{10}{3} \cdot \dfrac{\cancel{9}^{3}}{1} = \dfrac{30}{1} = 30$$

$$2\frac{1}{3} \cdot 16 = \dfrac{7}{3} \cdot \dfrac{16}{1} = \dfrac{112}{3} = 37\frac{1}{3}$$

Unequal

The cross products are *unequal*, so the proportion is *false*.

> **NOTE**
> The numbers in a proportion do *not* have to be whole numbers. They may be fractions, mixed numbers, decimal numbers, and so on.

Work Problem 3 at the Side.

OBJECTIVE 3 Find the unknown number in a proportion. Four numbers are used in a proportion. If any three of these numbers are known, the fourth can be found. For example, find the unknown number that will make this proportion true.

$$\dfrac{3}{5} = \dfrac{x}{40}$$

The variable x represents the unknown number. Start by finding the cross products.

$\dfrac{3}{5} = \dfrac{x}{40}$ Cross products: $5 \cdot x$ and $3 \cdot 40$

To make the proportion true, the cross products must be equal. This gives us the following equation.

$$5 \cdot x = 3 \cdot 40$$
$$5x = 120$$

Recall from **Section 2.4** that we can solve an equation of this type by dividing both sides by the coefficient of the variable term. In this case, the coefficient of $5x$ is 5.

$$\dfrac{5x}{5} = \dfrac{120}{5} \leftarrow \text{Divide both sides by 5}$$

On the left side, divide out the common factor of 5

$$\dfrac{\cancel{5} \cdot x}{\cancel{5}} = 24$$ On the right side, divide 120 by 5 to get 24

---

3 Find the cross products to see whether each proportion is true or false.

(a) $\dfrac{5}{9} = \dfrac{10}{18}$

(b) $\dfrac{32}{15} = \dfrac{16}{8}$

(c) $\dfrac{10}{17} = \dfrac{20}{34}$

(d)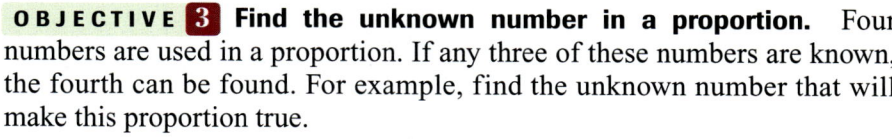
$\dfrac{2.4}{6} = \dfrac{5}{12}$
$(6)(5)=$
$(2.4)(12)=$

(e) $\dfrac{3}{4.25} = \dfrac{24}{34}$

(f) $\dfrac{1\frac{1}{6}}{2\frac{1}{3}} = \dfrac{4}{8}$

ANSWERS
3. (a) $90 = 90$; true
   (b) $240 \neq 256$; false
   (c) $340 = 340$; true
   (d) $(6)(5) = 30$; $(2.4)(12) = 28.8$; false
   (e) $102 = 102$; true
   (f) $9\frac{1}{3} = 9\frac{1}{3}$; true

Multiplying by 1 does *not* change a number, so in the numerator on the left side, $1 \cdot x$ is the same as x.

$$\frac{x}{1} = 24$$

Dividing by 1 does *not* change a number, so $\frac{x}{1}$ is the same as x.

$$x = 24$$

The unknown number in the proportion is 24. The complete proportion is shown below.

$$\frac{3}{5} = \frac{24}{40} \leftarrow x \text{ is } 24$$

Check by finding the cross products. If they are equal, you solved the problem correctly. If they are unequal, rework the problem.

$$5 \cdot 24 = \mathbf{120}$$
$$\frac{3}{5} = \frac{24}{40}$$
Equal; proportion is true.
$$3 \cdot 40 = \mathbf{120}$$

The cross products are equal, so the solution, $x = 24$, is correct.

> **CAUTION**
> The solution is 24, which is the unknown number in the proportion. 120 is *not* the solution; it is the cross product you get when *checking* the solution.

Solve a proportion for an unknown number by using the following steps.

Solving a Proportion to Find an Unknown Number
Step 1 Find the cross products.
Step 2 Show that the cross products are equivalent.
Step 3 Divide both sides of the equation by the coefficient of the variable term.
Step 4 Check by writing the solution in the *original* proportion and finding the cross products.

EXAMPLE 4 Solving Proportions

Find the unknown number in each proportion. Round answers to the nearest hundredth when necessary.

(a) $\dfrac{16}{x} = \dfrac{32}{20}$

Recall that ratios can be rewritten in lowest terms. If desired, you can do that *before* finding the cross products. In this example, write $\frac{32}{20}$ in lowest terms as $\frac{8}{5}$, which gives the proportion $\dfrac{16}{x} = \dfrac{8}{5}$.

Step 1
$$\dfrac{16}{x} = \dfrac{8}{5}$$
$x \cdot 8$
$16 \cdot 5$
Find the cross products.

Continued on Next Page

Step 2 $x \cdot 8 = 16 \cdot 5$ Show that the cross products are equivalent.
$x \cdot 8 = 80$

Step 3 $\dfrac{x \cdot \cancel{8}}{\cancel{8}} = \dfrac{80}{8}$ ← Divide both sides by 8

$x = 10$ ← Find x. (No rounding is necessary.)

Step 4   Write the solution in the *original* proportion and check by finding the cross products.

x is 10 → $\dfrac{16}{10} = \dfrac{32}{20}$

$10 \cdot 32 = \mathbf{320}$
$16 \cdot 20 = \mathbf{320}$ Equal; proportion is true.

The cross products are equal, so 10 is the correct solution (**not** 320).

NOTE
It is not necessary to write the ratios in lowest terms before solving. However, if you do, you will have smaller numbers to work with.

(b) $\dfrac{7}{12} = \dfrac{15}{x}$

Step 1 $\dfrac{7}{12} = \dfrac{15}{x}$ $\begin{array}{l}12 \cdot 15 = 180 \\ 7 \cdot x\end{array}$ Find the cross products.

Step 2 $7 \cdot x = 180$ ← Show that cross products are equivalent.

Step 3 $\dfrac{\cancel{7} \cdot x}{\cancel{7}} = \dfrac{180}{7}$ ← Divide both sides by 7

$x \approx 25.71$ ← Rounded to nearest hundredth

When the division does not come out even, check for directions on how to round your answer. Divide out one more place, then round.

$\dfrac{25.714}{7)180.000}$ ← Divide out to thousandths so you can round to hundredths.

Step 4   Write the solution in the original proportion and check by finding the cross products.

$\dfrac{7}{12} \approx \dfrac{15}{25.71}$

$(12)(15) = \mathbf{180}$
$(7)(25.71) = \mathbf{179.97}$ Very close, but not equal

The cross products are slightly different because you rounded the value of x. However, they are close enough to see that the problem was done correctly and 25.71 is the approximate solution.

Work Problem 4 at the Side.

4 Find the unknown numbers. Round to hundredths when necessary. Check your answers by finding the cross products.

(a) $\dfrac{1}{2} = \dfrac{x}{12}$

(b) $\dfrac{6}{10} = \dfrac{15}{x}$

(c) $\dfrac{28}{x} = \dfrac{21}{9}$

(d) $\dfrac{x}{8} = \dfrac{3}{5}$

(e) $\dfrac{14}{11} = \dfrac{x}{3}$

Answers
4. (a) $x = 6$ (b) $x = 25$
   (c) $x = 12$ (d) $x = 4.8$
   (e) $x \approx 3.82$ (rounded to nearest hundredth)

The next example shows how to solve for the unknown number in a proportion with fractions or decimals.

EXAMPLE 5 **Solving Proportions with Mixed Numbers and Decimals**

Find the unknown number in each proportion.

(a) $\dfrac{2\frac{1}{5}}{6} = \dfrac{x}{10}$

Step 1 $\dfrac{2\frac{1}{5}}{6} = \dfrac{x}{10}$ Find the cross products. $6 \cdot x$; $2\frac{1}{5} \cdot 10$

Find $2\frac{1}{5} \cdot 10$.

$$2\frac{1}{5} \cdot 10 = \dfrac{11}{5} \cdot \dfrac{10}{1} = \dfrac{11 \cdot 2 \cdot \cancel{5}}{\cancel{5} \cdot 1} = \dfrac{22}{1} = 22$$

Changed to improper fraction

Step 2 $\quad 6 \cdot x = 22 \quad \leftarrow$ Show that cross products are equivalent.

Step 3 $\quad \dfrac{\cancel{6} \cdot x}{\cancel{6}} = \dfrac{22}{6} \quad \leftarrow$ Divide both sides by 6.

Write the answer as a mixed number in lowest terms.

$$x = \dfrac{22 \div 2}{6 \div 2} = \dfrac{11}{3} = 3\frac{2}{3}$$

Step 4 Write the solution in the original proportion and check by finding the cross products.

$$6 \cdot 3\frac{2}{3} = \dfrac{2 \cdot \cancel{3}}{1} \cdot \dfrac{11}{\cancel{3}} = \dfrac{22}{1} = 22$$

$\dfrac{2\frac{1}{5}}{6} = \dfrac{3\frac{2}{3}}{10}$

Equal

$$2\frac{1}{5} \cdot 10 = \dfrac{11}{\cancel{5}} \cdot \dfrac{2 \cdot \cancel{5}}{1} = \dfrac{22}{1} = 22$$

The cross products are equal, so $3\frac{2}{3}$ is the correct solution (*not* 22).

Continued on Next Page

Section 6.3 Proportions   **437**

> **NOTE**
> You can use decimal numbers and your calculator to solve Example 5(a) on the previous page. $2\frac{1}{5}$ is equivalent to 2.2, so the cross products are
> $$6 \cdot x = (2.2)(10)$$
> $$\frac{\cancel{6} \cdot x}{\cancel{6}} = \frac{22}{6}$$
> When you divide 22 by 6 on your calculator, it shows 3.666666667. Write the answer using a bar to show the repeating digit: $3.\overline{6}$. Or round the answer to 3.67 (nearest hundredth).

(b) $\dfrac{1.5}{0.6} = \dfrac{2}{x}$

$(1.5)(x) = (0.6)(2)$ ← Show that cross products are equivalent.

$(1.5)(x) = 1.2$

$\dfrac{(\cancel{1.5})(x)}{\cancel{1.5}} = \dfrac{1.2}{1.5}$ ← Divide both sides by 1.5.

$x = \dfrac{1.2}{1.5}$ Complete the division.

$x = 0.8$ $\begin{array}{r} .8 \\ 1.5\overline{)1.20} \end{array}$

So the unknown number is 0.8. Write the solution in the original proportion and check by finding the cross products.

$$\dfrac{1.5}{0.6} = \dfrac{2}{0.8} \quad \begin{array}{l}(0.6)(2) = \mathbf{1.2} \\ (1.5)(0.8) = \mathbf{1.2}\end{array} \bigg] \text{Equal}$$

The cross products are equal, so 0.8 is the correct solution (*not* 1.2).

▶ **Work Problem 5 at the Side.** ▶▶

5 Find the unknown numbers. Round to hundredths on the decimal problems when necessary. Check your answers by finding the cross products.

(a) $\dfrac{3\frac{1}{4}}{2} = \dfrac{x}{8}$

(b) $\dfrac{x}{3} = \dfrac{1\frac{2}{3}}{5}$

(c) $\dfrac{0.06}{x} = \dfrac{0.3}{0.4}$

(d) $\dfrac{2.2}{5} = \dfrac{13}{x}$

(e) $\dfrac{x}{6} = \dfrac{0.5}{1.2}$

(f) $\dfrac{0}{2} = \dfrac{x}{7.092}$

ANSWERS
5. **(a)** $x = 13$ **(b)** $x = 1$ **(c)** $x = 0.08$
   (d) $x \approx 29.55$ (rounded to nearest hundredth)
   (e) $x = 2.5$ **(f)** $x = 0$

Focus on Real-Data Applications

Tour of the West

As a travel agent, you are helping some British citizens select a tour of the western United States. Highlights are visits to Mount Rushmore, Devil's Tower, Yellowstone National Park, the Grand Tetons, Bryce Canyon, Zion National Park, the Painted Desert, and the Grand Canyon. The itinerary is shown on the map to the right.

The travel distances and times between the daily stopping points are estimates, based on information from the Web site www.mapquest.com. The table below gives the daily route, the travel distances, and the travel times between the locations.

1. Calculate the average speed in miles per hour (mph) for each segment of the trip, rounded to the nearest whole number. Notice that you are working with rates that compare distance to time (miles to hours). (*Hint:* Divide the distance traveled by the time elapsed. You must first rewrite the time in decimal form. For example, on Day 4, 1 hr 40 min is $1 + \frac{40}{60}$ hr or 1.6666666 hr, which rounds to 1.67 hr. The average speed is 56 mi $\div 1.67$ hr ≈ 33.5 or 34 mph.)

Day	Location	Distance	Time	Average Speed (mph)
1	London, England, to Denver, Colorado (CO)	4693 mi	14 hr flight	
2	Denver, CO			
3	Denver, CO, to Custer, South Dakota (SD)	365 mi	8 hr	
4	Custer, SD, to Lead, SD	56 mi	1 hr, 40 min	34 mph
5	Lead, SD, to Cody, Wyoming (WY)	361 mi	7 hr, 50 min	
6	Cody, WY, to Yellowstone National Park, WY	110 mi	3 hr, 10 min	
7	Yellowstone, WY, to Jackson, WY	131 mi	3 hr, 45 min	
8	Jackson, WY, to Salt Lake City, Utah (UT)	269 mi	7 hr	
9	Salt Lake City, UT			
10	Salt Lake City, UT, to Cedar City, UT	251 mi	4 hr, 20 min	
11	Cedar City, UT, to Page, Arizona (AZ)	155 mi	4 hr, 30 min	
12	Page, AZ, to Grand Canyon, AZ	138 mi	4 hr	
13	Grand Canyon, AZ, to Las Vegas, Nevada (NV)	279 mi	6 hr, 30 min	
14	Las Vegas, NV			
15	Las Vegas, NV, to San Francisco, CA	420 mi	1 hr, 40 min flight	
	San Francisco, CA, to London, England	5376 mi	17 hr flight	
	Totals (driving portion of the tour)			

2. Calculate the total *driving* distance and total *driving* time during this tour. Find the overall average driving speed. (Do *not* include the airplane flights.)

3. Why do you think the average driving speeds for this trip are not closer to 55 mph?

4. If a friend from Scotland asked your opinion about how interesting and feasible these tourist attractions would be, what advice would you give?

6.3 Exercises

Write each proportion. See Example 1.

1. $9 is to 12 cans as $18 is to 24 cans.

2. 28 people is to 7 cars as 16 people is to 4 cars.

3. 200 adults is to 450 children as 4 adults is to 9 children.

4. 150 trees is to 1 acre as 1500 trees is to 10 acres.

5. 120 feet is to 150 feet as 8 feet is to 10 feet.

6. $6 is to $9 as $10 is to $15.

7. 2.2 hours is to 3.3 hours as 3.2 hours is to 4.8 hours.

8. 4 meters is to 4.75 meters as 6 meters is to 7.125 meters.

Determine whether each proportion is true *or* false *by writing the ratios in lowest terms. Show the simplified ratios and then write* true *or* false. *See Example 2.*

9. $\dfrac{6}{10} = \dfrac{3}{5}$

10. $\dfrac{1}{4} = \dfrac{9}{36}$

11. $\dfrac{5}{8} = \dfrac{25}{40}$

12. $\dfrac{2}{3} = \dfrac{20}{27}$

13. $\dfrac{150}{200} = \dfrac{200}{300}$

14. $\dfrac{100}{120} = \dfrac{75}{100}$

Use cross products to determine whether each proportion is true *or* false. *Show the cross products and circle* True *or* False. *See Example 3.*

15. $\dfrac{2}{9} = \dfrac{6}{27}$

    True   False

16. $\dfrac{20}{25} = \dfrac{4}{5}$

    True   False

17. $\dfrac{20}{28} = \dfrac{12}{16}$

    True   False

18. $\dfrac{16}{40} = \dfrac{22}{55}$

True   False

19. $\dfrac{110}{18} = \dfrac{160}{27}$

True   False

20. $\dfrac{600}{420} = \dfrac{20}{14}$

True   False

21. $\dfrac{3.5}{4} = \dfrac{7}{8}$

True   False

22. $\dfrac{36}{23} = \dfrac{9}{5.75}$

True   False

23. $\dfrac{18}{16} = \dfrac{2.8}{2.5}$

True   False

24. $\dfrac{0.26}{0.39} = \dfrac{1.3}{1.9}$

True   False

25. $\dfrac{6}{3\frac{2}{3}} = \dfrac{18}{11}$

True   False

26. $\dfrac{16}{13} = \dfrac{2}{1\frac{5}{8}}$

True   False

27. Suppose Ichiro Suzuki of the Seattle Mariners had 16 hits in 50 times at bat, and Sammy Sosa of the Chicago Cubs was at bat 400 times and got 128 hits. Paul is trying to convince Jamie that the two men hit equally well. Show how you could use a proportion and cross products to see whether Paul is correct.

28. Jay worked 3.5 hours and packed 91 cartons. Craig packed 126 cartons in 5.25 hours. To see if the men worked equally fast, Barry set up this proportion:

$$\dfrac{3.5}{91} = \dfrac{126}{5.25}$$

Explain what is wrong with Barry's proportion and write a correct one. Is the correct proportion true or false?

Solve each proportion to find the unknown number. Round your answers to hundredths when necessary. Check your answers by finding cross products. See Examples 4 and 5.

29. $\dfrac{1}{3} = \dfrac{x}{12}$

30. $\dfrac{x}{6} = \dfrac{15}{18}$

31. $\dfrac{15}{10} = \dfrac{3}{x}$

32. $\dfrac{5}{x} = \dfrac{20}{8}$

33. $\dfrac{x}{11} = \dfrac{32}{4}$

34. $\dfrac{12}{9} = \dfrac{8}{x}$

35. $\dfrac{42}{x} = \dfrac{18}{39}$

36. $\dfrac{49}{x} = \dfrac{14}{18}$

37. $\dfrac{x}{25} = \dfrac{4}{20}$

38. $\dfrac{6}{x} = \dfrac{4}{8}$

39. $\dfrac{8}{x} = \dfrac{24}{30}$

40. $\dfrac{32}{5} = \dfrac{x}{10}$

41. $\dfrac{99}{55} = \dfrac{44}{x}$

42. $\dfrac{x}{12} = \dfrac{101}{147}$

43. $\dfrac{0.7}{9.8} = \dfrac{3.6}{x}$

44. $\dfrac{x}{3.6} = \dfrac{4.5}{6}$

45. $\dfrac{250}{24.8} = \dfrac{x}{1.75}$

46. $\dfrac{4.75}{17} = \dfrac{43}{x}$

Find the unknown number in each proportion. Write your answers as whole or mixed numbers when possible. See Example 5.

47. $\dfrac{15}{1\frac{2}{3}} = \dfrac{9}{x}$

48. $\dfrac{x}{\frac{3}{10}} = \dfrac{2\frac{2}{9}}{1}$

49. $\dfrac{2\frac{1}{3}}{1\frac{1}{2}} = \dfrac{x}{2\frac{1}{4}}$

50. $\dfrac{1\frac{5}{6}}{x} = \dfrac{\frac{3}{14}}{\frac{6}{7}}$

Solve each proportion two different ways. First change all the numbers to decimal form and solve. Then change all the numbers to fraction form and solve; write your answers in lowest terms.

51. $\dfrac{\frac{1}{2}}{x} = \dfrac{2}{0.8}$

52. $\dfrac{\frac{3}{20}}{0.1} = \dfrac{0.03}{x}$

53. $\dfrac{x}{\frac{3}{50}} = \dfrac{0.15}{1\frac{4}{5}}$

54. $\dfrac{8\frac{4}{5}}{1\frac{1}{10}} = \dfrac{x}{0.4}$

RELATING CONCEPTS (EXERCISES 55–56) For Individual or Group Work

Work Exercises 55–56 in order. First prove that the proportions are **not** true. Then create four true proportions for each exercise by changing only one number at a time.

55. $\dfrac{10}{4} = \dfrac{5}{3}$

56. $\dfrac{6}{8} = \dfrac{24}{30}$

Summary Exercises on Ratios, Rates, and Proportions

Use the circle graph of one college's enrollment to complete Exercises 1–4. Write each ratio as a fraction in lowest terms.

1. Write the ratio of freshmen to juniors.

2. What is the ratio of freshmen to the total college enrollment?

3. Find the ratio of seniors and sophomores to juniors.

4. Write the ratio of freshmen and sophomores to juniors and seniors.

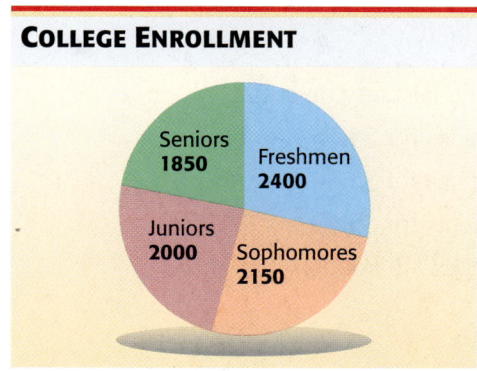

COLLEGE ENROLLMENT

Seniors 1850
Freshmen 2400
Juniors 2000
Sophomores 2150

The bar graph shows the number of Americans who play various instruments. Use the graph to complete Exercises 5 and 6. Write each ratio as a fraction in lowest terms.

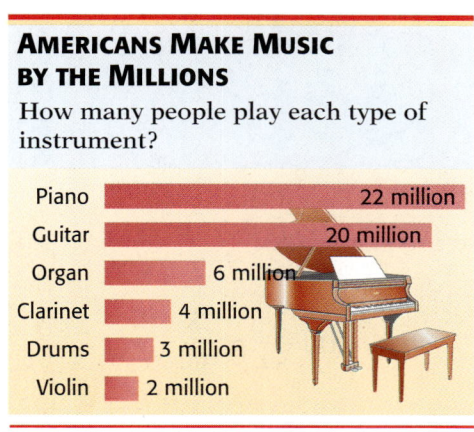

AMERICANS MAKE MUSIC BY THE MILLIONS
How many people play each type of instrument?

Piano — 22 million
Guitar — 20 million
Organ — 6 million
Clarinet — 4 million
Drums — 3 million
Violin — 2 million

Source: America by the Numbers.

5. Write six ratios that compare the least popular instrument to each of the other instruments.

6. Which two instruments give each of these ratios: **(a)** $\frac{5}{1}$; **(b)** $\frac{2}{1}$? There may be more than one correct answer.

The table lists data on the top three individual scoring NBA basketball games of all time. Use the data to answer Exercises 7 and 8. Round answers to the nearest tenth.

7. What was Wilt Chamberlain's scoring rate in points per minute and in minutes per point?

Player	Date	Points	Min
Wilt Chamberlain	3/2/62	100	48
David Thompson	4/9/78	73	43
David Robinson	4/24/94	71	44

Source: www.NBA.com

8. Find David Robinson's scoring rate in points per minute and minutes per point.

Wilt Chamberlain (#13)

9. Lucinda's paycheck showed gross pay of $652.80 for 40 hours of work and $195.84 for 8 hours of overtime work. Find her regular hourly pay rate and her overtime rate.

10. Satellite TV is being offered to new subscribers at $25 per month for 50 channels, $39 per month for 100 channels, or $48 per month for 150 channels. What is the monthly cost per channel under each plan? (*Source:* Dish1Up Satellites.)

11. Find the best buy on coffee.

    26 ounces for $4.78

    34 ounces for $5.44

    39 ounces for $7.49

    (*Source:* Cub Foods.)

12. Which brand of cat food is the best buy? You have a coupon for $2 off on Brand P, and another for $1 off on Brand N.

    Brand N is $3.75 for 3.5 pounds

    Brand P is $5.99 for 7 pounds

    Brand R is $10.79 for 18 pounds

Use either the method of writing in lowest terms or the method of finding cross products to decide whether each proportion is true *or* false. *Show your work and then write* true *or* false.

13. $\dfrac{28}{21} = \dfrac{44}{33}$

14. $\dfrac{2.3}{8.05} = \dfrac{0.25}{0.9}$

15. $\dfrac{2\frac{5}{8}}{3\frac{1}{4}} = \dfrac{21}{26}$

Solve each proportion to find the unknown number. Round your answers to hundredths when necessary.

16. $\dfrac{7}{x} = \dfrac{25}{100}$

17. $\dfrac{15}{8} = \dfrac{6}{x}$

18. $\dfrac{x}{84} = \dfrac{78}{36}$

19. $\dfrac{10}{11} = \dfrac{x}{4}$

20. $\dfrac{x}{17} = \dfrac{3}{55}$

21. $\dfrac{2.6}{x} = \dfrac{13}{7.8}$

22. $\dfrac{0.14}{1.8} = \dfrac{x}{0.63}$

23. $\dfrac{\frac{1}{3}}{8} = \dfrac{x}{24}$

24. $\dfrac{6\frac{2}{3}}{4\frac{1}{6}} = \dfrac{\frac{6}{5}}{x}$

6.4 Problem Solving with Proportions

OBJECTIVE 1 Use proportions to solve application problems. Proportions can be used to solve a wide variety of problems. Watch for problems in which you are given a ratio or rate and then asked to find part of a corresponding ratio or rate. Remember that a ratio or rate compares two quantities and often includes one of these indicator words.

in   for   on   per   from   to

Use the six problem-solving steps from **Section 3.3.** When setting up the proportion, use a variable to represent the unknown number. We have used the letter x, but you may use any letter you like.

OBJECTIVE

1 Use proportions to solve application problems.

EXAMPLE 1 Solving a Proportion Application

Mike's car can travel 163 **miles on** 6.4 **gallons** of gas. How far can it travel on a full tank of 14 **gallons** of gas? Round to the nearest mile.

Step 1 **Read** the problem. It is about how far a car can travel on a certain amount of gas.

> Unknown: miles traveled on 14 gallons of gas
> Known: 163 miles traveled on 6.4 gallons of gas

Step 2 **Assign a variable.** There is only one unknown, so let x be the number of miles traveled on 14 gallons.

Step 3 **Write an equation.** The equation is in the form of a proportion. Decide what is being compared. This problem compares **miles** to **gallons**. Write the two rates described in the problem. Be sure that *both* rates compare miles to gallons in the same order. In other words, miles is in both numerators and gallons is in both denominators.

This rate compares **miles** to **gallons**. $\dfrac{163 \text{ miles}}{6.4 \text{ gallons}} = \dfrac{x \text{ miles}}{14 \text{ gallons}}$ This rate compares **miles** to **gallons**.

(Matching units)

Step 4 **Solve** the equation. Ignore the units while finding the cross products.

$$\dfrac{163 \text{ miles}}{6.4 \text{ gallons}} = \dfrac{x \text{ miles}}{14 \text{ gallons}}$$

$(6.4)(x) = (163)(14)$ Show that cross products are equivalent.

$(6.4)(x) = 2282$

$\dfrac{(6.4)(x)}{6.4} = \dfrac{2282}{6.4}$ Divide both sides by 6.4

$x = 356.5625$ Round to 357

Step 5 **State the answer.** Mike's car can travel 357 miles, rounded to the nearest mile, on a full tank of gas.

Continued on Next Page

Chapter 6 Ratio, Proportion, and Line/Angle/Triangle Relationships

1 Set up and solve a proportion for each problem using the six problem-solving steps.

(a) If 2 pounds of fertilizer will cover 50 square feet of garden, how many pounds are needed for 225 square feet?

(b) A U.S. map has a scale of 1 inch to 75 miles. Lake Superior is 4.75 inches long on the map. What is the lake's actual length, to the nearest whole mile?

(c) Cough syrup is to be given at the rate of 30 milliliters for each 100 pounds of body weight. How much should be given to a 34-pound child? Round to the nearest whole milliliter.

ANSWERS

1. (a) $\dfrac{2 \text{ pounds}}{50 \text{ sq. feet}} = \dfrac{x \text{ pounds}}{225 \text{ sq. feet}}$
$x = 9$ pounds

(b) $\dfrac{1 \text{ inch}}{75 \text{ miles}} = \dfrac{4.75 \text{ inches}}{x \text{ miles}}$
$x \approx 356$ miles

(c) $\dfrac{30 \text{ milliliters}}{100 \text{ pounds}} = \dfrac{x \text{ milliliters}}{34 \text{ pounds}}$
$x \approx 10$ milliliters

Step 6 Check the solution by putting it back into the original problem. The car traveled 163 miles on 6.4 gallons of gas; 14 gallons is a little more than *twice as much* gas, so the car should travel a little more than *twice as far*.

$(2)(163 \text{ miles}) = 326 \text{ miles}$ ← Estimate

The solution, 357 miles, is just a little more than the estimate of 326 miles, so it is reasonable.

> **CAUTION**
> When setting up the proportion, do *not* mix up the units in the rates.
>
> compares **miles** to **gallons** $\Big\}$ $\dfrac{163 \text{ miles}}{6.4 \text{ gallons}} = \dfrac{14 \text{ gallons}}{x \text{ miles}}$ $\Big\{$ compares **gallons** to **miles**
>
> These rates do *not* compare things in the same order and *cannot* be set up as a proportion.

◀◀ **Work Problem 1 at the Side.**

EXAMPLE 2 Solving a Proportion Application

A newspaper report says that 7 out of 10 people surveyed watch the news on TV. At that rate, how many of the 3200 people in town would you expect to watch the news?

Step 1 Read the problem. It is about people watching the news on TV.

Unknown: how many people in town are expected to watch the news on TV
Known: 7 out of 10 people surveyed watched the news on TV.

Step 2 Assign a variable. There is only one unknown, so let x be the number of people in town who watch the news on TV.

Step 3 Write an equation. Set up the two rates as a proportion. You are comparing people who watch the news to people surveyed. Write the two rates described in the example. Be sure that both rates make the same comparison. "People who watch the news" is mentioned first, so it should be in the numerator of *both* rates.

People who watch the news → $\dfrac{7}{10} = \dfrac{x}{3200}$ ← People who watch the news
Total group (people surveyed) →   ← Total group (people in town)

Step 4 Solve the equation.

$$\dfrac{7}{10} = \dfrac{x}{3200}$$

$(10)(x) = (7)(3200)$ Show that cross products are equivalent.

$(10)(x) = 22{,}400$

$\dfrac{\cancel{(10)}(x)}{\cancel{10}} = \dfrac{22{,}400}{10}$ Divide both sides by 10

$x = 2240$

Continued on Next Page

Step 5 **State the answer.** You would expect 2240 people in town to watch the news on TV.

Step 6 **Check** the solution by putting it back into the original problem. Notice that 7 out of 10 people is more than half the people, but less than all the people. Half of the 3200 people in town is $3200 \div 2 = 1600$, so between 1600 and 3200 people would be expected to watch the news on TV. The solution, 2240 people, is between 1600 and 3200, so it is reasonable.

CAUTION
Always check that your answer is reasonable. If it isn't, look at the way your proportion is set up. Be sure you have matching units in the numerators and matching units in the denominators.

For example, suppose you set up the last proportion *incorrectly*, as shown below.

$$\frac{7}{10} = \frac{3200}{x} \quad \leftarrow \text{Incorrect setup}$$

$$(7)(x) = (10)(3200)$$

$$\frac{(7)(x)}{7} = \frac{32{,}000}{7}$$

$$x \approx 4571 \text{ people} \quad \leftarrow \text{Unreasonable answer}$$

This answer is *unreasonable* because there are only 3200 people in the town; it is **not** possible for 4571 people to watch the news.

Work Problem 2 at the Side.

② Solve each problem to find a reasonable answer. Then flip one side of your proportion to see what answer you get with an *incorrect* setup. Explain why the second answer is *unreasonable*.

(a) A survey showed that 2 out of 3 people would like to lose weight. At this rate, how many people in a group of 150 want to lose weight?

(b) In one state, 3 out of 5 college students receive financial aid. At this rate, how many of the 4500 students at Central Community College receive financial aid?

(c) An advertisement says that 9 out of 10 dentists recommend sugarless gum. If the ad is true, how many of the 60 dentists in our city would recommend sugarless gum?

ANSWERS
2. (a) 100 people (reasonable); incorrect setup gives 225 people (only 150 people in the group).
   (b) 2700 students (reasonable); incorrect setup gives 7500 students (only 4500 students at the college).
   (c) 54 dentists (reasonable); incorrect setup gives ≈ 67 dentists (only 60 dentists in the city).

Focus on Real-Data Applications

Feeding Hummingbirds

A recipe can be used to make as much of a mixture as you might need as long as the ingredients are kept proportional. Use the recipe for a homemade mixture of sugar water for hummingbird feeders to answer these problems.

1. What is the ratio of sugar to water in the recipe?

   What is the ratio of water to sugar in the recipe?

2. Complete each table.

Sugar	Water
1 cup	4 cups
	5 cups
	6 cups
	7 cups
2 cups	8 cups

Sugar	Water
1 cup	4 cups
	3 cups
	2 cups
	1 cup

3. How much water would you need
   (a) if you wanted to use 3 cups of sugar?
   (b) if you wanted to use 4 cups of sugar?
   (c) if you wanted to use $\frac{1}{3}$ cup of sugar?

4. One cup of sugar weighs about 200 grams and one cup of water weighs about 235 grams. If you mix 1 cup of sugar and 4 cups of water, what is the approximate weight of the resulting mixture?

5. The article says that the nectar from wildflowers visited by hummingbirds has an average sugar concentration of 21 percent. That represents a ratio of 21 to 100. If you wanted to mix a sugar solution in the same proportion as the wildflower nectar, how much water should you use for 1 cup of sugar? What proportion did you set up to solve this problem?

6. As you change the amounts of water and sugar, should you change the length of time that you boil the mixture? Explain your answer.

7. Will the length of time it takes to get the water hot enough to start boiling change? Explain your answer.

Feeding Hummingbirds

After getting a hummingbird feeder, the next step is to fill it! You have two choices at this point: you can either buy one of the commercial mixtures or you can make your own solution:

> **Recipe for Homemade Mixture:**
> 1 part sugar (not honey)
> 4 parts water
> Boil for 1 to 2 minutes. Cool.
> Store extra in refrigerator.

The concentration of the sugar is important. A 1 to 4 ratio of sugar to water is recommended because it approximates the ratio of sugar to water found in the nectar of many hummingbird flowers. A recent study of native California wildflowers visited by hummingbirds showed that their nectar had an average sugar concentration of 21 percent. This is sweet enough to attract the hummers without being too sweet. If you increase the concentration of sugar, it may be harder for the birds to digest; if you decrease the concentration, they may lose interest.

Boiling the solution helps retard fermentation. Sugar-and-water solutions are subject to rapid spoiling, especially in hot weather.

Source: The Hummingbird Book.

6.4 Exercises

Set up and solve a proportion for each application problem. See Example 1.

1. Caroline can sketch four cartoon strips in five hours. How long will it take her to sketch 18 strips?

2. The Cosmic Toads recorded eight songs on their first CD in 26 hours. How long will it take them to record 14 songs for their second CD?

3. Sixty newspapers cost $27. Find the cost of 16 newspapers.

4. Twenty-two guitar lessons cost $396. Find the cost of 12 lessons.

5. If three pounds of fescue grass seed cover about 350 square feet of ground, how many pounds are needed for 4900 square feet?

6. Anna earns $1242.08 in 14 days. How much does she earn in 260 days?

7. Tom makes $455.75 in five days. How much does he make in three days?

8. If 5 ounces of a medicine must be mixed with 8 ounces of water, how many ounces of medicine would be mixed with 20 ounces of water?

9. The bag of rice noodles shown below makes 7 servings. At that rate, how many ounces of noodles do you need for 12 servings, to the nearest ounce?

10. This can of sweet potatoes is enough for four servings. How many ounces are needed for nine servings, to the nearest ounce?

11. Three quarts of a latex enamel paint will cover about 270 square feet of wall surface. How many quarts will you need to cover 350 square feet of wall surface in your kitchen and 100 square feet of wall surface in your bathroom?

12. One gallon of clear gloss wood finish covers about 550 square feet of surface. If you need to apply three coats of finish to 400 square feet of surface, how many gallons do you need, to the nearest tenth?

Use the floor plan shown to answer Exercises 13–16. On the plan, one inch represents four feet.

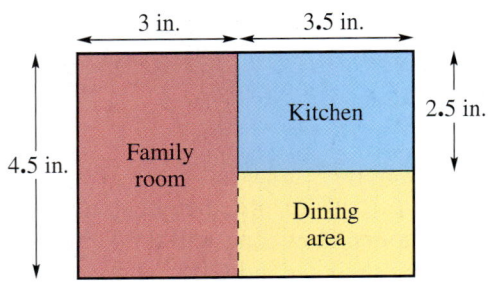

13. What is the actual length and width of the kitchen?

14. What is the actual length and width of the family room?

15. What is the actual length and width of the dining area?

16. What is the actual length and width of the entire floor plan?

The table below lists recommended amounts of food to order for 25 party guests. Use the table to answer Exercises 17 and 18. (Source: Cub Foods.)

FOOD FOR 25 GUESTS	
Item	**Amount**
Fried chicken	40 pieces
Lasagna	14 pounds
Deli meats	4.5 pounds
Sliced cheese	$2\frac{1}{3}$ pounds
Bakery buns	3 dozen
Potato salad	6 pounds

17. How much of each food item should Nathan and Amanda order for a graduation party with 60 guests?

18. Taisha is having 20 neighbors over for a Fourth of July picnic. How much food should she buy?

In Exercises 19–24, set up a proportion to solve each problem. Check to see whether your answer is reasonable. Then flip one side of your proportion to see what answer you get with an incorrect setup. Explain why the second answer is unreasonable. See Example 2.

19. About 7 out of 10 people entering our community college need to take a refresher math course. If we have 2950 entering students, how many will probably need refresher math? (*Source:* Minneapolis Community and Technical College.)

20. In a survey, only 3 out of 100 people like their eggs poached. At that rate, how many of the 60 customers who ordered eggs at Soon-Won's restaurant this morning asked to have them poached? Round to the nearest whole person.

21. About 1 out of 8 people choose vanilla as their favorite ice cream flavor. If 238 people attend an ice cream social, how many would you expect to choose vanilla? Round to the nearest whole person.

22. In a test of 200 sewing machines, only one had a defect. At that rate, how many of the 5600 machines shipped from the factory have defects?

23. About 98 out of 100 U.S. households have at least one TV set. There were 107,500,000 U.S. households in 2002. How many households had one or more TVs? (*Source:* Nielsen Media Research.)

24. In a survey, 3 out of 100 dog owners washed their pets by having the dogs go into the shower with them. If the survey is accurate, how many of the 31,200,000 dog owners in the United States use this method? (*Source:* Teledyne Water Pik; American Veterinary Medical Association.)

Set up and solve a proportion for each problem.

25. The stock market report says that five stocks went up for every six stocks that went down. If 750 stocks went down yesterday, how many went up?

26. The human body contains 90 pounds of water for every 100 pounds of body weight. How many pounds of water are in a child who weighs 80 pounds?

27. The ratio of the length of an airplane wing to its width is 8 to 1. If the length of a wing is 32.5 meters, how wide must it be? Round to the nearest hundredth.

28. The Rosebud School District wants a student-to-teacher ratio of 19 to 1. How many teachers are needed for 1850 students? Round to the nearest whole number.

29. The number of calories you burn is proportional to your weight. A 150-pound person burns 222 calories during 30 minutes of tennis. How many calories would a 210-pound person burn, to the nearest whole number? (*Source: Wellness Encyclopedia.*)

30. (Complete Exercise 29 first.) A 150-pound person burns 189 calories during 45 minutes of grocery shopping. How many calories would a 115-pound person burn, to the nearest whole number? (*Source: Wellness Encyclopedia.*)

31. At 3 P.M., Coretta's shadow is 1.05 meters long. Her height is 1.68 meters. At the same time, a tree's shadow is 6.58 meters long. How tall is the tree? Round to the nearest hundredth.

32. (Complete Exercise 31 first.) Later in the day, Coretta's shadow was 2.95 meters long. How long a shadow did the tree have at that time? Round to the nearest hundredth.

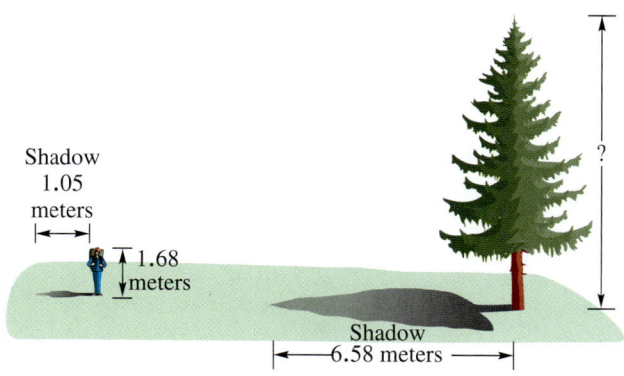

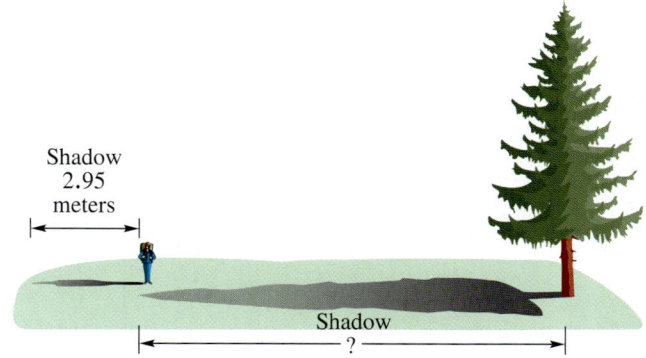

33. Can you set up a proportion to solve this problem? Explain why or why not. Jim is 25 years old and weighs 180 pounds. How much will he weigh when he is 50 years old?

34. Write your own application problem that can be solved by setting up a proportion. Also show the proportion and the steps needed to solve your problem.

35. A survey of college students shows that 4 out of 5 drink coffee. Of the students who drink coffee, 1 out of 8 adds cream to it. How many of the 48,000 students at Ohio State University would be expected to use cream in their coffee?

36. About 9 out of 10 adults think it is a good idea to exercise regularly. But of the ones who think it is a good idea, only 1 in 6 actually exercises at least three times a week. At this rate, how many of the 300 employees in our company exercise regularly?

37. The nutrition information on a bran cereal box says that a $\frac{1}{3}$ cup serving provides 80 calories and 8 grams of dietary fiber. At that rate, how many calories and grams of fiber are in a $\frac{1}{2}$ cup serving? (*Source:* Kraft Foods, Inc.)

38. A $\frac{2}{3}$ cup serving of penne pasta has 210 calories and 2 grams of dietary fiber. How many calories and grams of fiber would be in a 1 cup serving? (*Source:* Borden Foods.)

RELATING CONCEPTS (EXERCISES 39–42) For Individual or Group Work

A box of instant mashed potatoes has the list of ingredients shown in the table. Use this information to work Exercises 39–42 in order.

Ingredient	For 12 Servings
Water	$3\frac{1}{2}$ cups
Margarine	6 tablespoons
Milk	$1\frac{1}{2}$ cups
Potato flakes	4 cups

Source: General Mills.

39. Find the amount of each ingredient needed for six servings. Show *two* different methods for finding the amounts. One method should use proportions.

40. Find the amount of each ingredient needed for 18 servings. Show *two* different methods for finding the amounts, one using proportions and one using your answers from Exercise 39.

41. Find the amount of each ingredient needed for three servings, using your answers from either Exercise 39 or Exercise 40.

42. Find the amount of each ingredient needed for nine servings, using your answers from either Exercise 40 or Exercise 41.

6.5 Geometry: Lines and Angles

Geometry starts with the idea of a point. A **point** can be described as a location in space. It has no length or width. A point is represented by a dot and is named by writing a capital letter next to the dot.

Point *P*

OBJECTIVE 1 Identify lines, line segments, and rays. A **line** is a straight row of points that goes on forever in both directions. A line is drawn by using arrowheads to show that it never ends. The line is named by using the letters of any two points on the line.

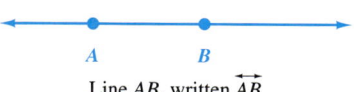
Line *AB*, written $\overleftrightarrow{AB}$

A piece of a line that has two endpoints is called a **line segment**. A line segment is named for its endpoints. The segment with endpoints *P* and *Q* is shown below. It can be named $\overline{PQ}$ or $\overline{QP}$.

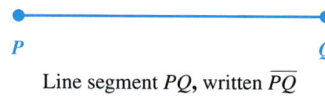
Line segment *PQ*, written $\overline{PQ}$

A **ray** is a part of a line that has only one endpoint and goes on forever in one direction. A ray is named by using the endpoint and some other point on the ray. The endpoint is always mentioned first.

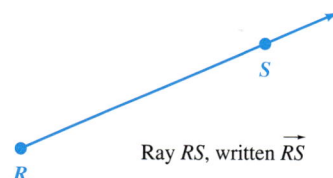
Ray *RS*, written $\overrightarrow{RS}$

EXAMPLE 1 Identifying Lines, Rays, and Line Segments

Identify each figure below as a line, line segment, or ray.

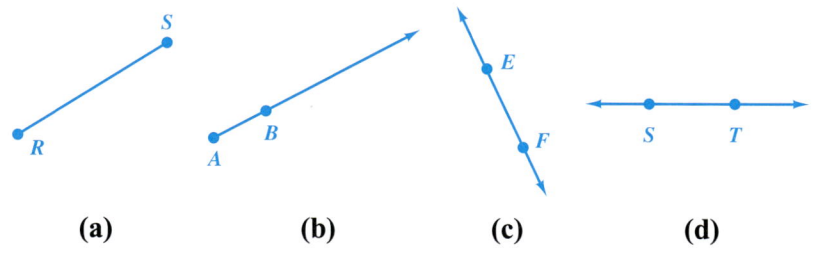
(a) (b) (c) (d)

Figure **(a)** has two endpoints, so it is a *line segment*.
Figure **(b)** starts at point *A* and goes on forever in one direction, so it is a *ray*.
Figures **(c)** and **(d)** go on forever in both directions, so they are *lines*.

Work Problem 1 at the Side.

OBJECTIVES

1. Identify lines, line segments, and rays.
2. Identify parallel and intersecting lines.
3. Identify and name angles.
4. Classify angles as right, acute, straight, or obtuse.
5. Identify perpendicular lines.
6. Identify complementary angles and supplementary angles and find the measure of a complement or supplement of a given angle.
7. Identify congruent angles and vertical angles and use this knowledge to find the measures of angles.
8. Identify corresponding angles and alternate interior angles and use this knowledge to find the measures of angles.

1 Identify each figure as a line, line segment, or ray.

(a)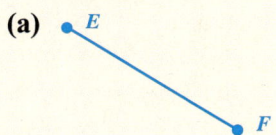

(b)

(c)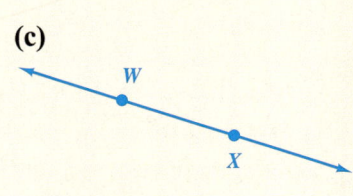

(d)

ANSWERS
1. (a) line segment (b) ray (c) line
   (d) line segment

2 Label each pair of lines as appearing to be parallel or as intersecting.

(a)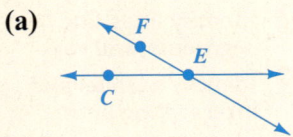

(b)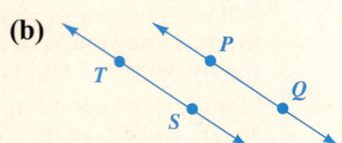

(c)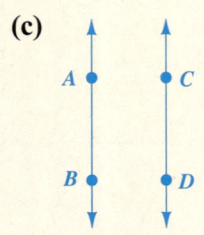

OBJECTIVE **2** **Identify parallel and intersecting lines.** A *plane* is an infinitely large, flat surface. A floor or a wall is part of a plane. Lines that are in the *same plane,* but that never intersect (never cross), are called **parallel lines,** while lines that cross are called **intersecting lines.** (Think of an intersection, where two streets cross each other.)

EXAMPLE 2 **Identifying Parallel and Intersecting Lines**

Label each pair of lines as appearing to be parallel or as intersecting.

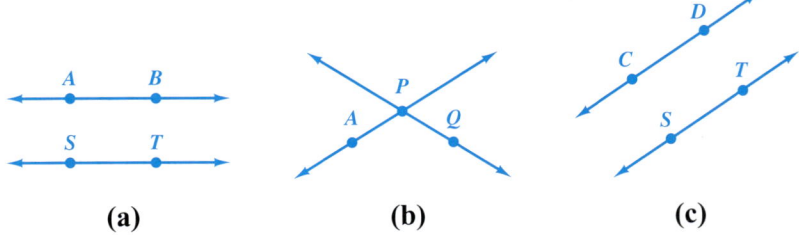

(a)   (b)   (c)

The lines in Figures **(a)** and **(c)** do not intersect; they appear to be *parallel lines.*

The lines in Figure **(b)** cross at *P*, so they are *intersecting lines.*

> **CAUTION**
> Appearances may be deceiving! Do not assume that lines are parallel unless it is stated that they are parallel.

◀◀◀ **Work Problem 2 at the Side.**

OBJECTIVE **3** **Identify and name angles.** An **angle** is made up of two rays that start at a common endpoint. This common endpoint is called the *vertex.*

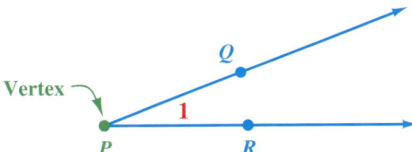

$\overrightarrow{PQ}$ and $\overrightarrow{PR}$ are called the *sides* of the angle. The angle can be named in four different ways, as shown below.

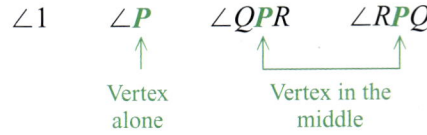

∠1   ∠P   ∠QPR   ∠RPQ
         ↑   ↑
      Vertex   Vertex in the
       alone   middle

> **Naming an Angle**
> To name an angle, write the vertex alone or write the vertex in the middle of two other points, one from each side. If two or more angles have the *same vertex,* as in Example 3 on the next page, do *not* use the vertex alone to name an angle.

ANSWERS
2. **(a)** intersecting **(b)** appear to be parallel
  (c) appear to be parallel

Section 6.5 Geometry: Lines and Angles 455

> **EXAMPLE 3** Identifying and Naming an Angle
>
> Name the highlighted angle.
>
>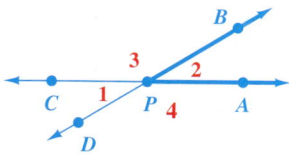
>
> The angle can be named ∠BPA, ∠APB, or ∠2. It cannot be named ∠P, using the vertex alone, because four different angles have P as their vertex.
>
> **Work Problem 3 at the Side.**

OBJECTIVE 4 Classify angles as right, acute, straight, or obtuse. Angles can be measured in **degrees**. The symbol for degrees is a small, raised circle °. Think of the minute hand on a clock as a ray of an angle. Suppose it is at 12:00. During one hour of time, the minute hand moves around in a complete circle. It moves 360 **degrees**, or 360°. In half an hour, at 12:30, the minute hand has moved halfway around the circle, or 180°. An angle of 180° is called a **straight angle**. When two rays go in opposite directions and form a straight line, then the rays form a straight angle.

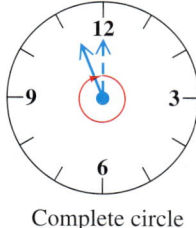

Complete circle 360°

Straight angle (half a circle) 180°

In a quarter of an hour, at 12:15, the minute hand has moved $\frac{1}{4}$ of the way around the circle, or 90°. An angle of 90° is called a **right angle**. The rays of a right angle form one corner of a square. So, to show that an angle is a **right angle**, we draw a **small square** at the vertex.

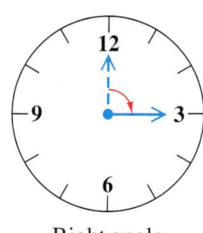

Right angle ($\frac{1}{4}$ of a circle) 90°

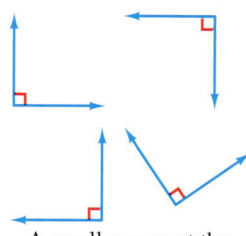

A small square at the vertex identifies right angles.

An angle that measures 1° is shown below. You can see that an angle of 1° is very small.

1° angle

3 (a) Name the highlighted angle in three different ways.

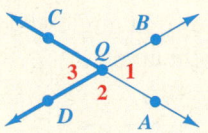

(b) Darken the rays that make up ∠ZTW.

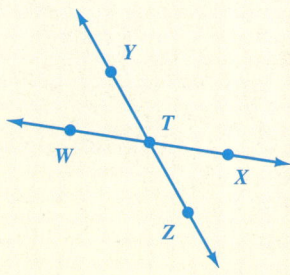

(c) Name this angle in four different ways.

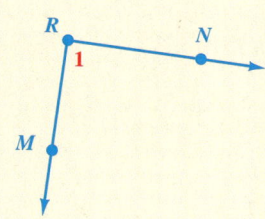

ANSWERS
3. **(a)** ∠3, ∠CQD, ∠DQC
   (b)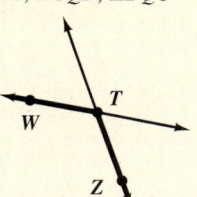
   (c) ∠1, ∠R, ∠MRN, ∠NRM

456 Chapter 6 Ratio, Proportion, and Line/Angle/Triangle Relationships

④ Label each angle as acute, right, obtuse, or straight. State the number of degrees in the right angle and in the straight angle.

(a)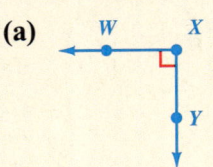

(b)

(c)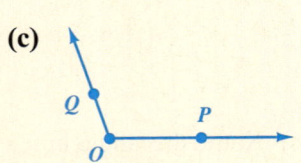

(d)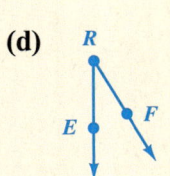

Some other terms used to describe angles are shown below.

Acute angles measure less than 90°.

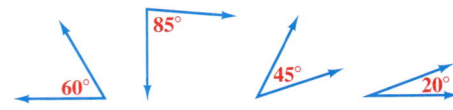

Examples of acute angles

Obtuse angles measure more than 90° but less than 180°.

Examples of obtuse angles

Section 9.2 shows you how to use a tool called a *protractor* to measure the number of degrees in an angle.

> **Classifying Angles**
> **Acute angles** measure less than 90°.
> **Right angles** measure *exactly* 90°.
> **Obtuse angles** measure more than 90° but less than 180°.
> **Straight angles** measure *exactly* 180°.

> **NOTE**
> Angles can also be measured in radians, which you will learn about in a later math course.

EXAMPLE 4 Classifying an Angle

Label each angle as acute, right, obtuse, or straight.

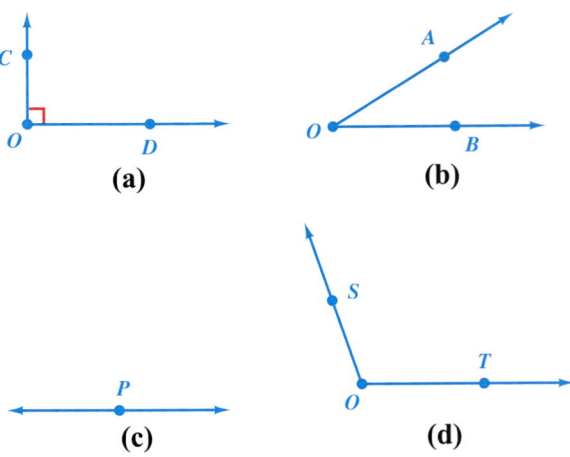

Figure **(a)** shows a *right angle* (exactly 90° and identified by a small square at the vertex).
Figure **(b)** shows an *acute angle* (less than 90°).
Figure **(c)** shows a *straight angle* (exactly 180°).
Figure **(d)** shows an *obtuse angle* (more than 90° but less than 180°).

Work Problem 4 at the Side.

ANSWERS
4. (a) right; 90° (b) straight; 180°
 (c) obtuse (d) acute

Section 6.5 Geometry: Lines and Angles 457

OBJECTIVE 5 Identify perpendicular lines. Two lines are called **perpendicular lines** if they intersect to form a right angle.

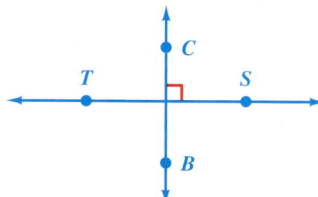

$\overleftrightarrow{CB}$ and $\overleftrightarrow{ST}$ are **perpendicular** lines because they intersect at right angles. This can be written in the following way: $\overleftrightarrow{CB} \perp \overleftrightarrow{ST}$.

EXAMPLE 5 Identifying Perpendicular Lines

Which pairs of lines are perpendicular?

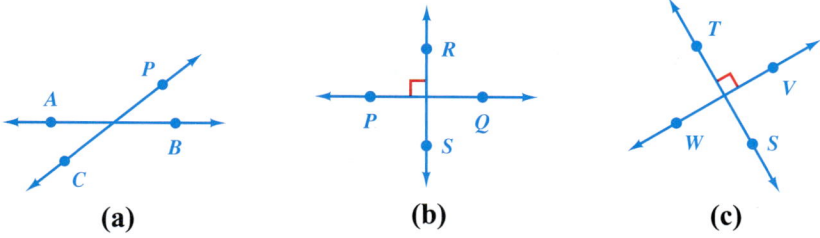

The lines in Figures **(b)** and **(c)** are *perpendicular* to each other, because they intersect at right angles.

The lines in Figure **(a)** are *intersecting lines,* but they are *not* perpendicular because they do not form a right angle.

▶ **Work Problem 5 at the Side.** ▶▶▶

OBJECTIVE 6 Identify complementary angles and supplementary angles and find the measure of a complement or supplement of a given angle. Two angles are called **complementary angles** if the sum of their measures is 90°. If two angles are complementary, each angle is the *complement* of the other.

EXAMPLE 6 Identifying Complementary Angles

Identify each pair of complementary angles.

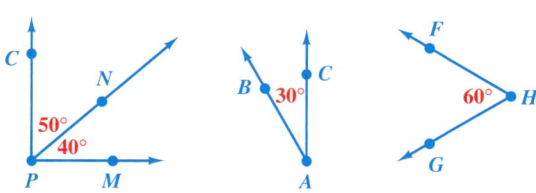

∠MPN (40°) and ∠NPC (50°) are complementary angles because

$$40° + 50° = 90°$$

∠CAB (30°) and ∠FHG (60°) are complementary angles because

$$30° + 60° = 90°$$

▶ **Work Problem 6 at the Side.** ▶▶▶

5 Which pair of lines is perpendicular? How can you describe the other pair of lines?

(a)

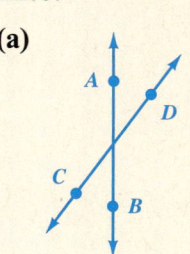

(b)

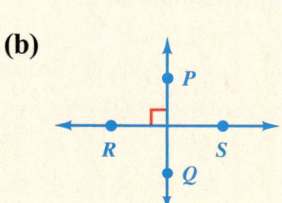

6 Identify each pair of complementary angles.

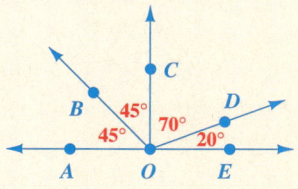

ANSWERS

5. Figure **(b)** shows perpendicular lines; Figure **(a)** shows intersecting lines.
6. ∠AOB and ∠BOC; ∠COD and ∠DOE

7 Find the complement of each angle.

(a) 35°

(b) 80°

EXAMPLE 7 Finding the Complement of Angles

Find the complement of each angle.

(a) 30°
Find the complement of 30° by subtracting. $90° - 30° = \mathbf{60°}$ ← Complement

(b) 40°
Find the complement of 40° by subtracting. $90° - 40° = \mathbf{50°}$ ← Complement

Work Problem 7 at the Side.

Two angles are called **supplementary angles** if the sum of their measures is 180°. If two angles are supplementary, each angle is the *supplement* of the other.

EXAMPLE 8 Identifying Supplementary Angles

Identify each pair of supplementary angles.

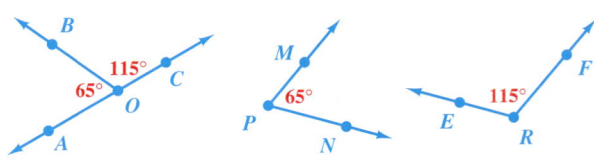

∠BOA and ∠BOC, because 65° + 115° = **180°**
∠BOA and ∠ERF, because 65° + 115° = **180°**
∠BOC and ∠MPN, because 115° + 65° = **180°**
∠MPN and ∠ERF, because 65° + 115° = **180°**

Work Problem 8 at the Side.

8 Identify each pair of supplementary angles.

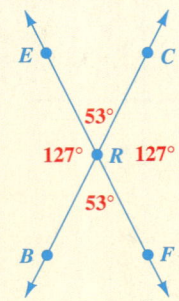

9 Find the supplement of each angle.

(a) 175°

(b) 30°

EXAMPLE 9 Finding the Supplement of Angles

Find the supplement of each angle.

(a) 70°
Find the supplement of 70° by subtracting. $180° - 70° = \mathbf{110°}$ ← Supplement

(b) 140°
Find the supplement of 140° by subtracting. $180° - 140° = \mathbf{40°}$ ← Supplement

Work Problem 9 at the Side.

OBJECTIVE 7 Identify congruent angles and vertical angles and use this knowledge to find the measures of angles. Two angles are called **congruent angles** if they measure the same number of degrees. If two angles are congruent, this is written as ∠A ≅ ∠B and read as, "angle A **is congruent to** angle B." Here is an example.

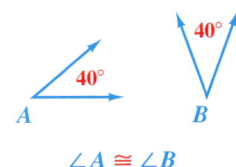

∠A ≅ ∠B

Example of congruent angles

ANSWERS
7. (a) 55° (b) 10°
8. ∠CRF and ∠BRF; ∠CRE and ∠ERB;
   ∠BRF and ∠BRE; ∠CRE and ∠CRF
9. (a) 5° (b) 150°

Section 6.5 Geometry: Lines and Angles   459

EXAMPLE 10   Identifying Congruent Angles

Identify the angles that are congruent.

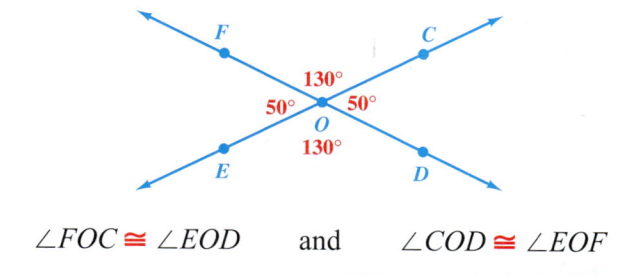

∠FOC ≅ ∠EOD   and   ∠COD ≅ ∠EOF

▶▶▶ **Work Problem 10 at the Side.**

⑩ Identify the angles that are congruent.

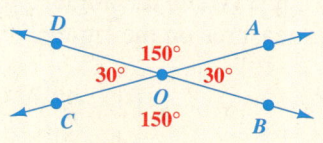

Angles that share a common side and a common vertex are called *adjacent* angles, such as ∠FOC and ∠COD in Example 10 above. Angles that do *not* share a common side are called *nonadjacent* angles. Two nonadjacent angles formed by intersecting lines are called **vertical angles.**

EXAMPLE 11   Identifying Vertical Angles

Identify the vertical angles in this figure.

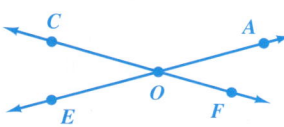

∠AOF and ∠COE are vertical angles because they do *not* share a common side and they are formed by two intersecting lines ($\overleftrightarrow{CF}$ and $\overleftrightarrow{EA}$).

∠COA and ∠EOF are also vertical angles.

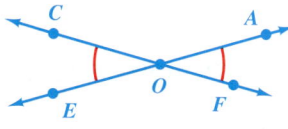   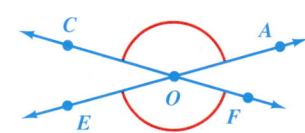

▶▶▶ **Work Problem 11 at the Side.**

⑪ Identify the vertical angles.

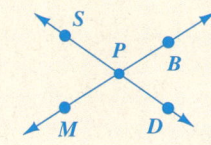

Look back at Example 10 at the top of the page. Notice that the two *congruent* angles that measure 130° are also *vertical* angles. Also, the two congruent angles that measure 50° are vertical angles. This illustrates the following property.

Vertical Angles Are Congruent

If two angles are *vertical* angles, they are *congruent;* that is, they measure the same number of degrees.

ANSWERS
10. ∠BOC ≅ ∠AOD; ∠AOB ≅ ∠DOC
11. ∠SPB and ∠MPD; ∠BPD and ∠SPM

12 In the figure below, find the measure of each unlabeled angle. Write the angle measures on the figure.

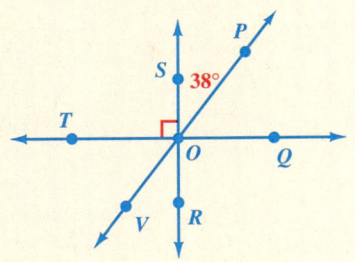

(a) ∠TOS

(b) ∠QOR

(c) ∠VOR

(d) ∠POQ

(e) ∠TOV

ANSWERS

12. (a) 90° (b) 90° (c) 38° (d) 52°
    (e) 52°

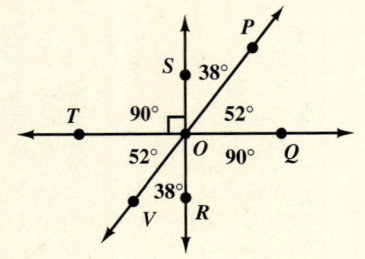

EXAMPLE 12 Finding the Measures of Vertical Angles

In the figure below, find the measure of each unlabeled angle.

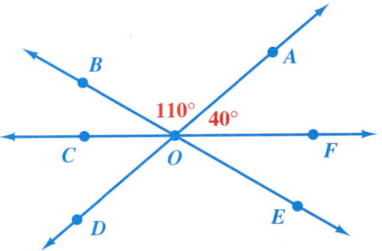

(a) ∠COD

∠COD and ∠AOF are vertical angles, so they are congruent. This means they measure the same number of degrees.

The measure of ∠AOF is 40° so the measure of ∠COD is **40°** also.

(b) ∠DOE

∠DOE and ∠BOA are vertical angles, so they are congruent.

The measure of ∠BOA is 110° so the measure of ∠DOE is **110°** also.

(c) ∠COB

Look at ∠COB, ∠BOA, and ∠AOF. Notice that $\overrightarrow{OC}$ and $\overrightarrow{OF}$ go in opposite directions. Therefore, ∠COF is a straight angle and measures 180°. To find the measure of ∠COB, subtract the sum of the other two angles from 180°.

$$180° - (110° + 40°) = 180° - (150°) = 30°$$

The measure of ∠COB is **30°**.

(d) ∠EOF

∠EOF and ∠COB are vertical angles, so they are congruent. We know from part (c) above that the measure of ∠COB is 30° so the measure of ∠EOF is **30°** also.

<<< **Work Problem 12 at the Side.**

OBJECTIVE 8 **Identify corresponding angles and alternate interior angles and use this knowledge to find the measures of angles.** We can also find congruent angles (angles with the same measure) when two *parallel lines* are crossed by a third line, called a *transversal*. When a transversal crosses two *parallel* lines, eight angles are formed, as shown below. There are special names for certain pairs of angles.

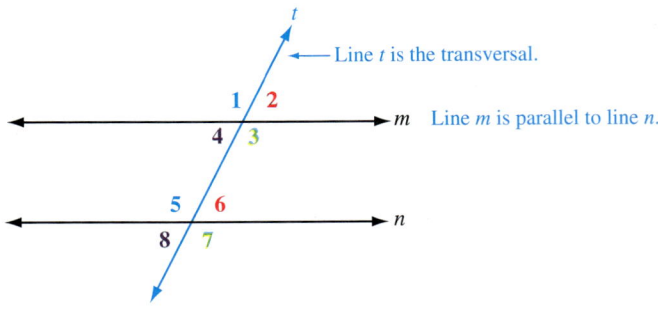

∠1 and ∠5 are called **corresponding angles**. Notice that they are both on the same side of the transversal (line *t*) and in the same relative position. *Corresponding angles are congruent,* so ∠1 and ∠5 measure the same number of degrees. There are four pairs of corresponding angles.

∠1 and ∠5 are corresponding angles, so ∠1 ≅ ∠5.
∠2 and ∠6 are corresponding angles, so ∠2 ≅ ∠6.
∠3 and ∠7 are corresponding angles, so ∠3 ≅ ∠7.
∠4 and ∠8 are corresponding angles, so ∠4 ≅ ∠8.

When a transversal crosses two parallel lines, angles 3, 4, 5, and 6 are called *interior angles*. You can see that they are "inside" the *parallel* lines.

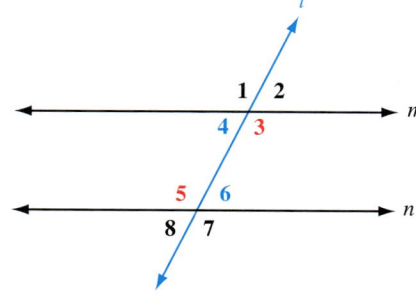

∠3 and ∠5 are alternate interior angles.
∠4 and ∠6 are alternate interior angles.

When two lines are *parallel*, then **alternate interior angles** *are congruent* (they have the same measure). Notice that alternate interior angles are on opposite sides of the transversal.

$$\angle 3 \cong \angle 5 \quad \text{and} \quad \angle 4 \cong \angle 6$$

Angles Formed by Parallel Lines and a Transversal

When two parallel lines are crossed by a transversal:
1. Corresponding angles are congruent, and
2. Alternate interior angles are congruent.

EXAMPLE 13 Identifying Corresponding Angles and Alternate Interior Angles

In each figure, line *m* is parallel to line *n*. Identify all pairs of corresponding angles and all pairs of alternate interior angles.

(a)

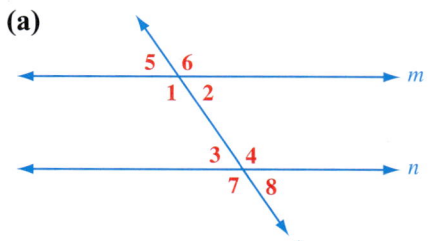

There are four pairs of corresponding angles:
∠5 and ∠3   ∠6 and ∠4
∠1 and ∠7   ∠2 and ∠8

Alternate interior angles:
∠1 and ∠4   ∠2 and ∠3

(b)

Corresponding angles:
∠1 and ∠5   ∠3 and ∠7
∠2 and ∠6   ∠4 and ∠8

Alternate interior angles:
∠3 and ∠6   ∠4 and ∠5

Work Problem 13 at the Side.

⓭ In each figure below, line *m* is parallel to line *n*. Identify all pairs of corresponding angles and all pairs of alternate interior angles.

(a)

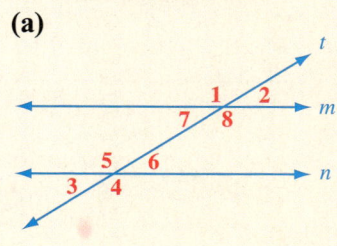

(b)

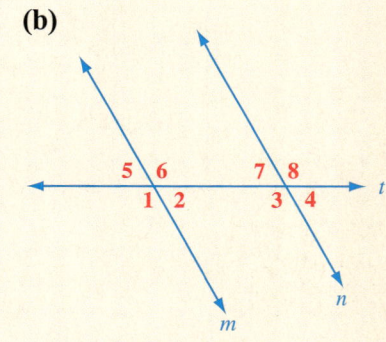

ANSWERS
13. (a) corresponding angles: ∠1 and ∠5; ∠2 and ∠6; ∠7 and ∠3; ∠8 and ∠4
alternate interior angles: ∠7 and ∠6; ∠8 and ∠5
(b) corresponding angles: ∠5 and ∠7; ∠6 and ∠8; ∠1 and ∠3; ∠2 and ∠4
alternate interior angles: ∠6 and ∠3; ∠2 and ∠7

462 Chapter 6 Ratio, Proportion, and Line/Angle/Triangle Relationships

14 In each figure below, line *m* is parallel to line *n*.

(a) The measure of ∠6 is 150°. Find the measures of the other angles.

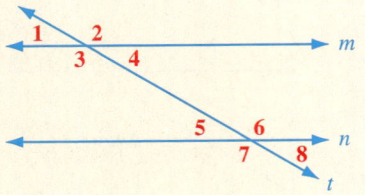

(b) The measure of ∠1 is 45°. Find the measures of the other angles.

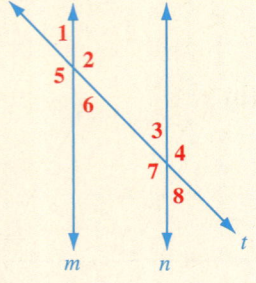

Recall that two angles are supplementary angles if the sum of their measures is 180°. Also remember that the two sides of a 180° angle form a straight line. Now you can combine your knowledge about supplementary angles with the information on parallel lines.

EXAMPLE 14 Working with Parallel Lines

In the figure at the right, line *m* is parallel to line *n* and the measure of ∠4 is 70°. Find the measures of the other angles.

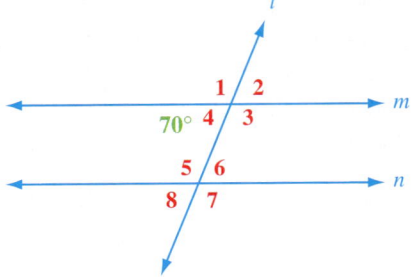

As you find the measure of each angle, write it on the figure.

∠4 ≅ ∠8 (corresponding angles), so the measure of ∠8 is also 70°.
∠4 ≅ ∠6 (alternate interior angles), so the measure of ∠6 is also 70°.
∠6 ≅ ∠2 (corresponding angles), so the measure of ∠2 is also 70°.

Notice that the exterior sides of ∠4 and ∠3 form a straight line, that is, a straight angle of 180°. Therefore, ∠4 and ∠3 are supplementary angles and the sum of their measures is 180°. If ∠4 is 70° then ∠3 must be 110° because 180° − 70° = 110°. So the measure of ∠3 is 110°.

∠3 ≅ ∠7 (corresponding angles), so the measure of ∠7 is also 110°.
∠3 ≅ ∠5 (alternate interior angles), so the measure of ∠5 is also 110°.
∠5 ≅ ∠1 (corresponding angles), so the measure of ∠1 is also 110°.

With the measures of all the angles labeled, you can double-check that each pair of angles that forms a straight angle also adds up to 180°.

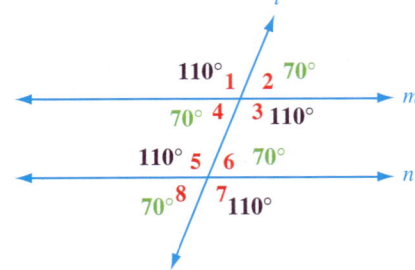

◀◀◀ **Work Problem 14 at the Side.**

ANSWERS
14. (a)

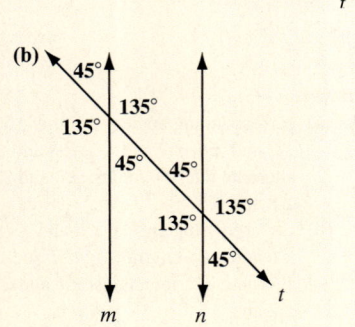

Section 6.5 Geometry: Lines and Angles   463

6.5 Exercises

FOR EXTRA HELP 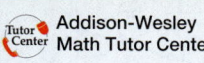  Digital Video Tutor CD 4 Videotape 6   Student's Solutions Manual    MyMathLab   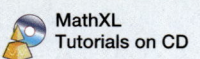 MathXL Tutorials on CD

Identify each figure as a line, line segment, *or* ray *and name it using the appropriate symbol. See Example 1.*

1.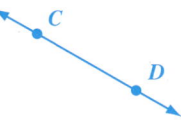

2.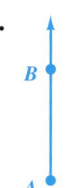

3.

4.

5.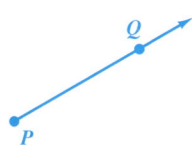

6.

Label each pair of lines as appearing to be parallel, *as* perpendicular, *or as* intersecting. *See Examples 2 and 5.*

7.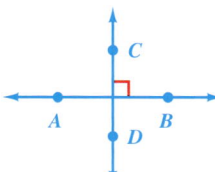

8.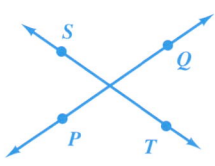

9.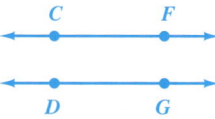

10.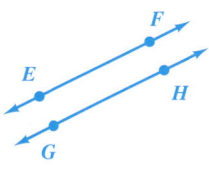

11.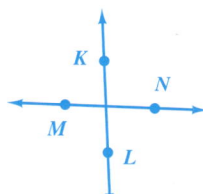

12.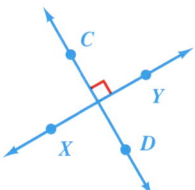

Name each highlighted angle by using the three-letter form of identification. See Example 3.

13.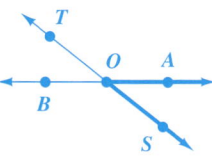

14.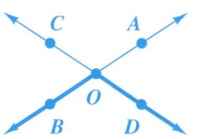

15.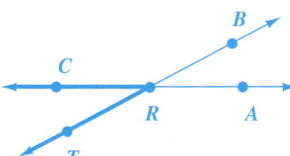

16.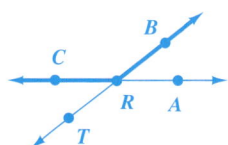

17.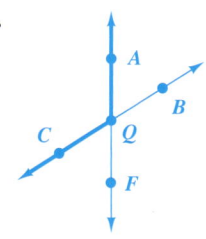

18.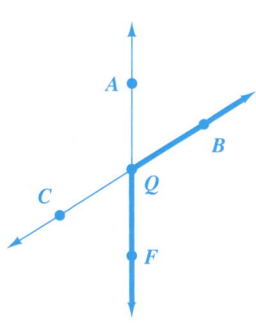

Label each angle as acute, right, obtuse, *or* straight. *For right angles and straight angles, indicate the number of degrees in the angle. See Example 4.*

19.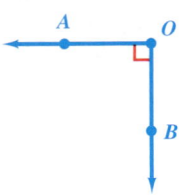

20.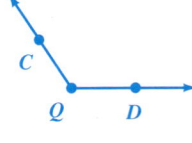

21.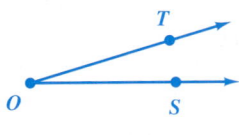

22.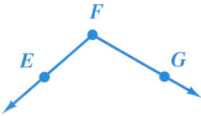

23.

24.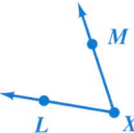

Identify each pair of complementary angles. See Example 6.

25.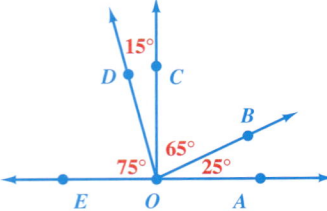

26.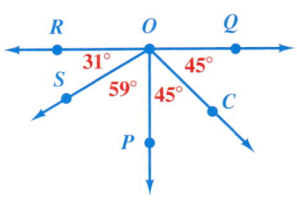

Identify each pair of supplementary angles. See Example 8.

27.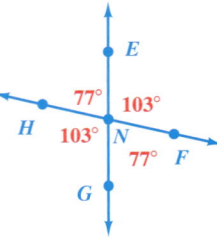

28.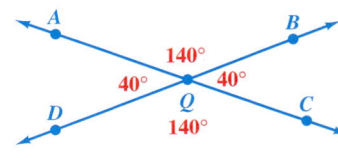

Find the complement of each angle. See Example 7.

29. 40° **30.** 35° **31.** 86° **32.** 59°

Find the supplement of each angle. See Example 9.

33. 130° **34.** 75° **35.** 90° **36.** 5°

In Exercises 37 and 38, identify the angles that are congruent. See Examples 10–12.

37.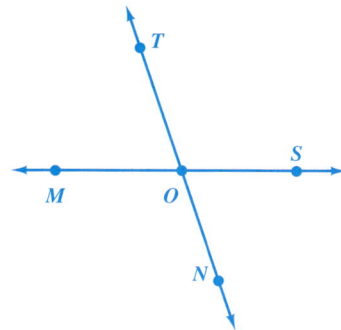

38.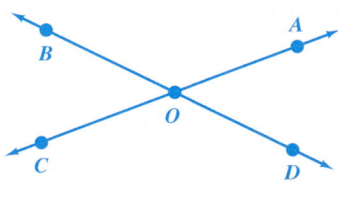

39. In the figure below, ∠AOH measures 37° and ∠COE measures 63°. Find the measure of each of the other angles.

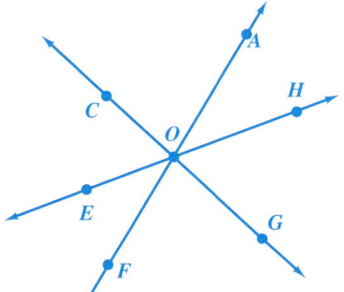

40. In the figure below, ∠POU measures 105° and ∠UOT measures 40°. Find the measure of each of the other angles.

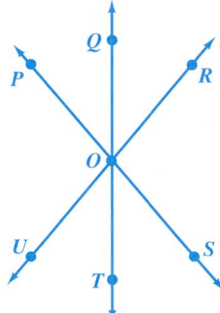

RELATING CONCEPTS (EXERCISES 41–46) For Individual or Group Work

*Use the figure to **work Exercises 41–46 in order.** Decide whether each statement is true or false. If it is true, explain why. If it is false, rewrite it to make a true statement.*

41. ∠UST is 90°.

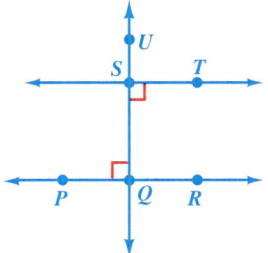

42. $\overleftrightarrow{SQ}$ and $\overleftrightarrow{PQ}$ are perpendicular.

43. The measure of ∠USQ is less than the measure of ∠PQR.

44. $\overleftrightarrow{ST}$ and $\overleftrightarrow{PR}$ are intersecting.

45. $\overleftrightarrow{QU}$ and $\overleftrightarrow{TS}$ are parallel.

46. ∠UST and ∠UQR measure the same number of degrees.

In each figure, line m is parallel to line n. Identify all pairs of corresponding angles and all pairs of alternate interior angles. See Example 13.

47.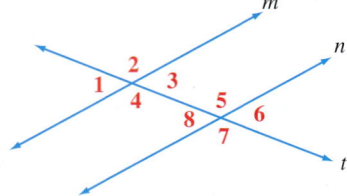

48.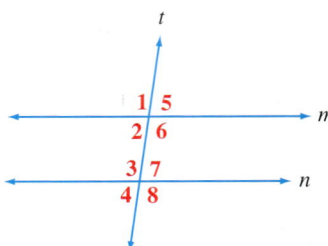

In each figure, line m is parallel to line n. Find the measure of each angle. See Example 14.

49. ∠8 measures 130°.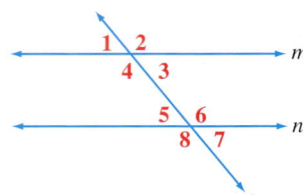

50. ∠2 measures 80°.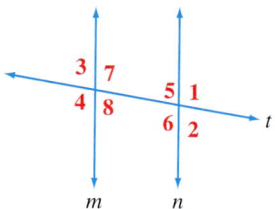

51. ∠6 measures 47°.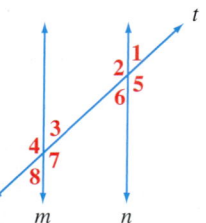

52. ∠2 measures 108°.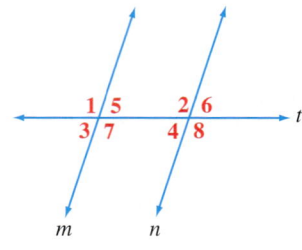

53. ∠6 measures 114°.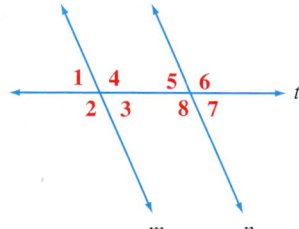

54. ∠3 measures 59°.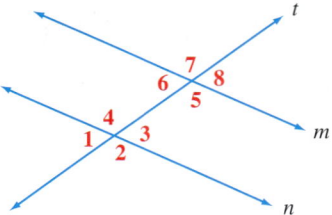

In each figure, $\overrightarrow{BA}$ is parallel to $\overrightarrow{CD}$. Find the measure of each numbered angle. See Example 14.

55.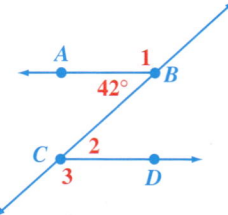

56.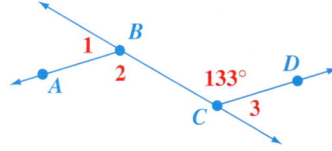

6.6 Geometry Applications: Congruent and Similar Triangles

Two useful concepts in geometry are *congruence* and *similarity*. If two figures are *identical,* both in *shape* and in *size,* we say the figures are **congruent.** In other words, the figures are perfect duplicates of each other, like getting two prints from the same negative on a roll of film. If two figures have the *same shape* but are *different sizes,* we say the figures are **similar,** like getting a print from a roll of film and then an enlargement of the same print. We'll explore the ideas of congruence and similarity using triangles.

OBJECTIVES

1. Identify corresponding parts of congruent triangles.
2. Prove that triangles are congruent using SAS, SSS, and ASA.
3. Identify corresponding parts of similar triangles.
4. Find the unknown lengths of sides in similar triangles.
5. Solve application problems involving similar triangles.

OBJECTIVE 1 Identify corresponding parts of congruent triangles.
The two triangles shown below are *congruent* because they are the *same shape* and the *same size.*

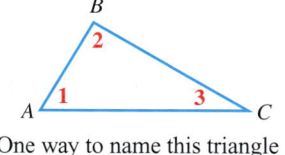
One way to name this triangle is △ABC.

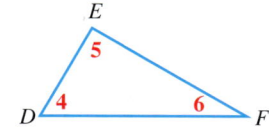
One way to name this triangle is △DEF.

Suppose you picked up △ABC and slid it over on top of △DEF. You would see that the two triangles are a perfect match. ∠1 would be on top of ∠4, so they are called *corresponding angles.* Similarly, ∠2 and ∠5 are corresponding angles, and ∠3 and ∠6 are corresponding angles. You would see that corresponding angles have the same measure, as indicated below.

$$m\angle 1 = m\angle 4 \qquad m\angle 2 = m\angle 5 \qquad m\angle 3 = m\angle 6$$

The abbreviation for measure is m, so $m\angle 1$ is read, "the measure of angle 1."

When you put △ABC on top of △DEF, you would also see that side AB is on top of side DE. We say that $\overline{AB}$ and $\overline{DE}$ are *corresponding sides.* Similarly, $\overline{BC}$ and $\overline{EF}$ are corresponding sides, and $\overline{AC}$ and $\overline{DF}$ are corresponding sides. You would see that corresponding sides have the same length.

$$AB = DE \qquad BC = EF \qquad AC = DF$$

Because corresponding angles have the same measure, and corresponding sides have the same length, we know that △ABC **is congruent to** △DEF. We can write this as △ABC ≅ △DEF.

> **Congruent Triangles**
>
> If two triangles are congruent, then
> 1. Corresponding angles have the same measure, and
> 2. Corresponding sides have the same length.

EXAMPLE 1 Identifying Corresponding Parts in Congruent Triangles

Each pair of triangles is congruent. List the corresponding angles and sides.

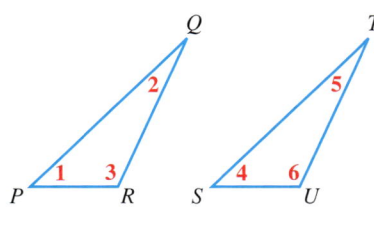

(a) If you picked up △PQR and slid it over on top of △STU, the two triangles would match.
The corresponding parts are congruent, so:

$m\angle 1 = m\angle 4 \qquad PQ = ST$
$m\angle 2 = m\angle 5 \qquad PR = SU$
$m\angle 3 = m\angle 6 \qquad QR = TU$

Continued on Next Page

1 Each pair of triangles is congruent. List the corresponding angles and the corresponding sides.

(a)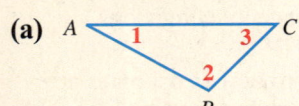

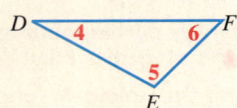

(b)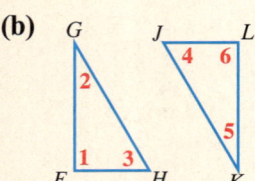

(*Hint:* Rotate △FGH, then slide it on top of △JLK.)

(c)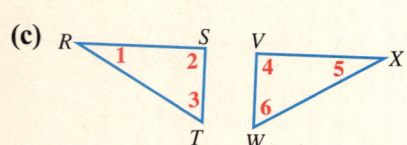

(*Hint:* Flip △RST over, then slide it on top of △VWX.)

ANSWERS

1. (a) $m\angle 1 = m\angle 4, m\angle 2 = m\angle 5,$
   $m\angle 3 = m\angle 6;$
   $AC = DF, AB = DE, BC = EF,$
   (b) $m\angle 1 = m\angle 6, m\angle 2 = m\angle 5,$
   $m\angle 3 = m\angle 4;$
   $GF = KL, FH = LJ, GH = KJ,$
   (c) $m\angle 1 = m\angle 5, m\angle 2 = m\angle 4,$
   $m\angle 3 = m\angle 6;$
   $RS = XV, RT = XW, ST = VW$

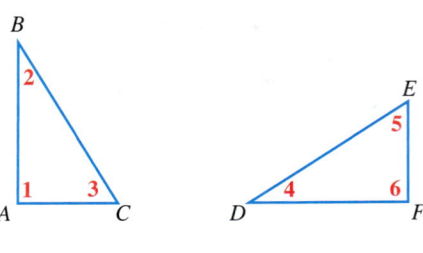

(b) If you picked up △ABC and slid it over on top of △DEF, it wouldn't match. But if you *rotate* △ABC before sliding it on top of △DEF, it *will* match.

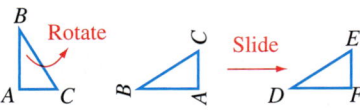

The corresponding parts are congruent, so:

$m\angle 1 = m\angle 6$ $BC = DE$
$m\angle 2 = m\angle 4$ $BA = DF$
$m\angle 3 = m\angle 5$ $CA = EF$

◀◀◀ **Work Problem 1 at the Side.**

OBJECTIVE 2 Prove that triangles are congruent using SAS, SSS, and ASA. One way to prove that two triangles are congruent would be to measure all the angles and all the sides. If the measures of the corresponding angles and sides are equal, then the triangles are congruent. But here are three quicker methods to prove that two triangles are congruent.

Proving That Two Triangles Are Congruent

1. Angle–Side–Angle (ASA) Method
If two angles and the side between them on one triangle measure the same as the corresponding parts on another triangle, the triangles are congruent.

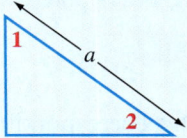 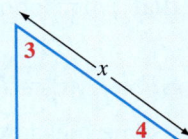

If $m\angle 1 = m\angle 3$ and $m\angle 2 = m\angle 4$ and $a = x$ then the two triangles are congruent.

2. Side–Side–Side (SSS) Method
If three sides of one triangle measure the same as the corresponding sides of another triangle, the triangles are congruent.

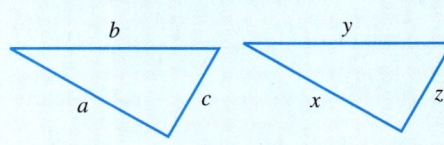

If $a = x$ and $b = y$ and $c = z$ then the two triangles are congruent.

3. Side–Angle–Side (SAS) Method
If two sides and the angle between them on one triangle measure the same as the corresponding parts on another triangle, the triangles are congruent.

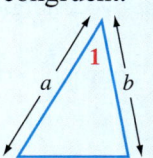 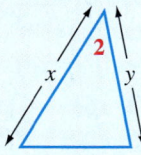

If $a = x$ and $b = y$ and $m\angle 1 = m\angle 2$ then the two triangles are congruent.

Section 6.6 Geometry Applications: Congruent and Similar Triangles   469

EXAMPLE 2 Proving That Two Triangles Are Congruent

Explain which method can be used to prove that each pair of triangles is congruent. Choose from ASA, SSS, and SAS.

(a) (b)

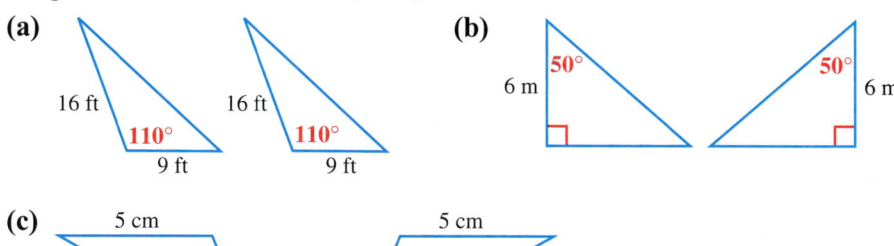

(c)

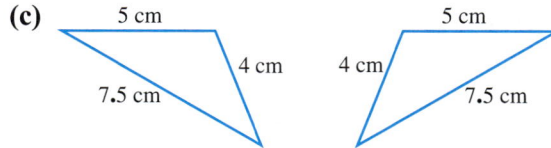

(a) On both triangles, two corresponding sides and the angle between them measure the same, so the Side–Angle–Side (SAS) method can be used to prove that the triangles are congruent.

(b) On both triangles, two corresponding angles and the side between them measure the same, so the Angle–Side–Angle (ASA) method can be used to prove that the triangles are congruent.

(c) Each pair of corresponding sides has the same length, so the Side–Side–Side (SSS) method can be used to prove that the triangles are congruent.

Work Problem 2 at the Side.

OBJECTIVE 3 Identify corresponding parts of similar triangles. Now that you've worked with *congruent* triangles, let's look at *similar* triangles. Remember that congruent triangles match exactly, both in shape and in size. Similar triangles, on the other hand, have the same shape but are *different sizes*. Three pairs of similar triangles are shown here.

Each pair of triangles has the same shape because the corresponding angles have the same measure. But the corresponding sides are *not* the same length, so the triangles are of *different sizes*.

Two similar triangles are shown to the right. Notice that corresponding angles have the same measure, but corresponding sides have different lengths.

$\overline{CB}$ corresponds to $\overline{RQ}$. Similarly, $\overline{CA}$ corresponds to $\overline{RP}$, and $\overline{BA}$ corresponds to $\overline{QP}$. Notice that each side in the larger triangle is *twice* the length of the corresponding side in the smaller triangle.

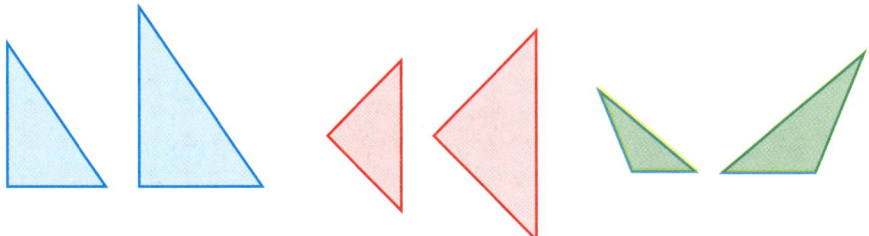

Work Problem 3 at the Side.

---

2 Determine which method can be used to prove that each pair of triangles is congruent.

(a)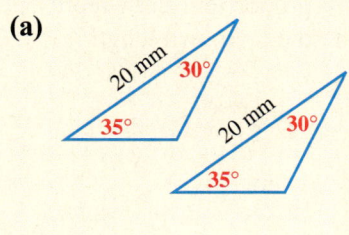

(b)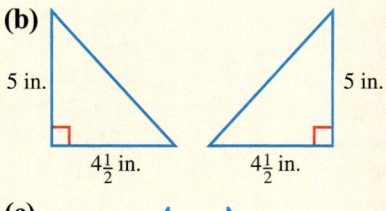

(c)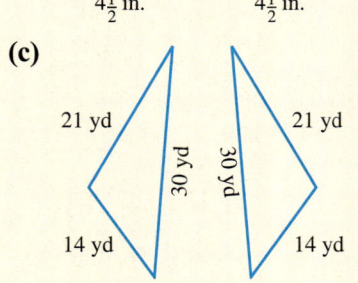

3 Identify corresponding angles and sides in these similar triangles.

(a)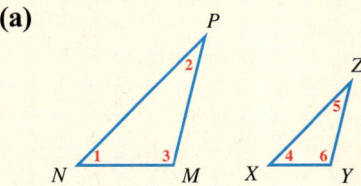

Angles: Sides:
1 and _____ $\overline{PN}$ and _____
2 and _____ $\overline{PM}$ and _____
3 and _____ $\overline{NM}$ and _____

(b)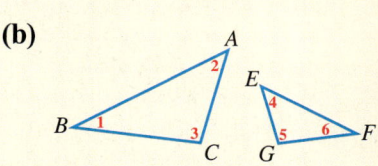

Angles: Sides:
1 and _____ $\overline{AB}$ and _____
2 and _____ $\overline{BC}$ and _____
3 and _____ $\overline{AC}$ and _____

ANSWERS
2. (a) ASA   (b) SAS   (c) SSS
3. (a) 4; 5; 6; $\overline{ZX}$; $\overline{ZY}$; $\overline{XY}$
   (b) 6; 4; 5; $\overline{EF}$; $\overline{FG}$; $\overline{EG}$

470 Chapter 6 Ratio, Proportion, and Line/Angle/Triangle Relationships

④ Find the length of $\overline{EF}$ in Example 3 at the right by setting up and solving a proportion. Let x represent the unknown length.

OBJECTIVE ④ **Find the unknown lengths of sides in similar triangles.**
Similar triangles are useful because of the following definition.

> **Similar Triangles**
>
> If two triangles are similar, then
> 1. Corresponding angles have the same measure, and
> 2. The ratios of the lengths of corresponding sides are equal.

EXAMPLE 3 Finding the Unknown Lengths of Sides in Similar Triangles

Find the length of $\overline{DF}$ in the smaller triangle. Assume the triangles are similar.

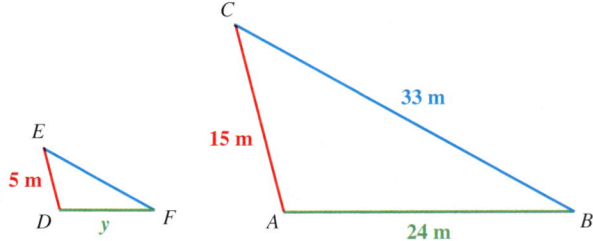

$\overline{DF}$, the length you want to find in the smaller triangle, corresponds to $\overline{AB}$ in the larger triangle. Then, notice that $\overline{ED}$ in the smaller triangle corresponds to $\overline{CA}$ in the larger triangle, and you know both of their lengths. Since the *ratios* of the lengths of corresponding sides are equal, you can set up a proportion. (Recall that a proportion states that two ratios are equal.)

Corresponding sides $\begin{Bmatrix} DF \rightarrow \\ AB \rightarrow \end{Bmatrix} \dfrac{y}{24} = \dfrac{5}{15} \begin{matrix} \leftarrow ED \\ \leftarrow CA \end{matrix}$ Corresponding sides

$$\dfrac{y}{24} = \dfrac{1}{3} \quad \text{Write } \tfrac{5}{15} \text{ in lowest terms as } \tfrac{1}{3}$$

Find the cross products.

$$\dfrac{y}{24} = \dfrac{1}{3} \quad \begin{matrix} 24 \cdot 1 = 24 \\ y \cdot 3 \end{matrix}$$

Show that the cross products are equivalent.

$$y \cdot 3 = 24$$

Divide both sides by 3.

$$\dfrac{y \cdot \cancel{3}}{\cancel{3}} = \dfrac{24}{3}$$

$$y = 8$$

$\overline{DF}$ has a length of 8 m.

◀◀ Work Problem 4 at the Side.

ANSWERS

4. $\dfrac{x}{33} = \dfrac{5}{15}$; $x = 11$ m

EXAMPLE 4 Finding an Unknown Length and the Perimeter

Find the perimeter of the smaller triangle. Assume the triangles are similar.

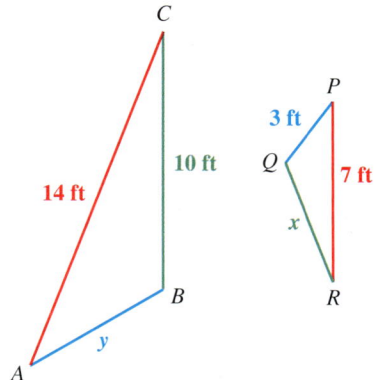

First find x, the length of $\overline{RQ}$ in the smaller triangle, then add the lengths of all three sides to find the perimeter.

The smaller triangle is turned "upside down" compared to the larger triangle, so be careful when identifying corresponding sides. $\overline{PR}$ is the longest side in the smaller triangle, and $\overline{AC}$ is the longest side in the larger triangle. So $\overline{PR}$ and $\overline{AC}$ are corresponding sides and you know both of their lengths. $\overline{QR}$, the length you want to find in the smaller triangle, corresponds to $\overline{BC}$ in the larger triangle. The ratios of the lengths of corresponding sides are equal, so you can set up a proportion.

$$\begin{array}{c} QR \rightarrow \\ BC \rightarrow \end{array} \frac{x}{10} = \frac{7}{14} \begin{array}{c} \leftarrow PR \\ \leftarrow AC \end{array}$$

$$\frac{x}{10} = \frac{1}{2} \quad \text{Write } \tfrac{7}{14} \text{ in lowest terms as } \tfrac{1}{2}$$

Find the cross products.

$$\frac{x}{10} \bowtie \frac{1}{2} \quad \begin{array}{c} 10 \cdot 1 = 10 \\ x \cdot 2 \end{array}$$

Show that the cross products are equivalent.

$$x \cdot 2 = 10$$

Divide both sides by 2.

$$\frac{x \cdot \cancel{2}}{\cancel{2}} = \frac{10}{2}$$

$$x = 5$$

$\overline{RQ}$ has a length of 5 ft.

Now add the lengths of all three sides to find the perimeter of the smaller triangle.

$$\text{Perimeter} = 5 \text{ ft} + 3 \text{ ft} + 7 \text{ ft} = 15 \text{ ft}$$

Work Problem 5 at the Side.

5 (a) Find the perimeter of triangle ABC in Example 4 at the left.

(b) Find the perimeter of each triangle. Assume the triangles are similar.

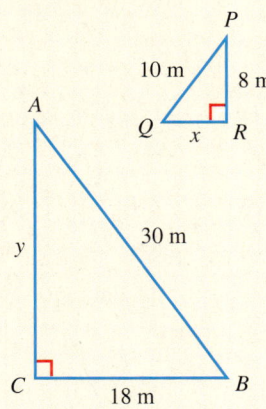

ANSWERS
5. (a) $\overline{AB}$ is 6 ft; Perimeter = 14 ft + 10 ft + 6 ft = 30 ft
(b) $x = 6$ m, Perimeter = 24 m; $y = 24$ m, Perimeter = 72 m

6 Find the height of each flagpole.

(a)
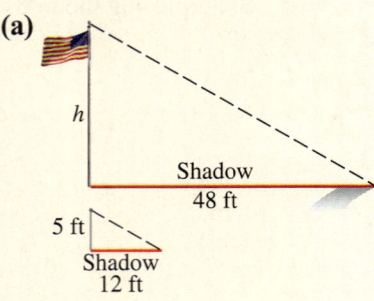

OBJECTIVE 5 Solve application problems involving similar triangles.

The next example shows an application of similar triangles.

EXAMPLE 5 Using Similar Triangles in an Application

A flagpole casts a shadow 99 m long at the same time that a pole 10 m tall casts a shadow 18 m long. Find the height of the flagpole.

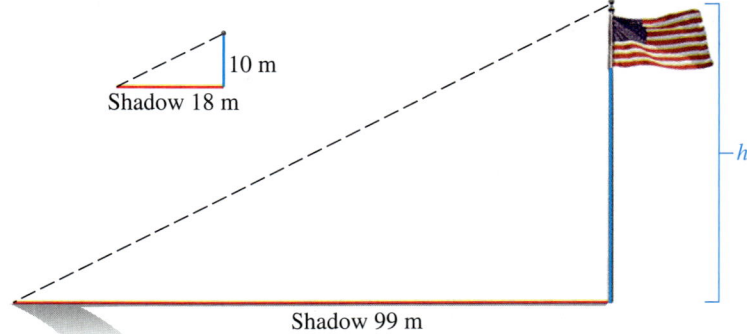

The triangles shown are similar, so write a proportion to find h.

Height in larger triangle → $\dfrac{h}{10} = \dfrac{99}{18}$ ← Shadow in larger triangle
Height in smaller triangle →   ← Shadow in smaller triangle

Find the cross products and show that they are equivalent.

$$h \cdot 18 = 10 \cdot 99$$
$$h \cdot 18 = 990$$

Divide both sides by 18.

$$\dfrac{h \cdot \cancel{18}}{\cancel{18}} = \dfrac{990}{18}$$
$$h = 55$$

The flagpole is 55 m high.

(b)
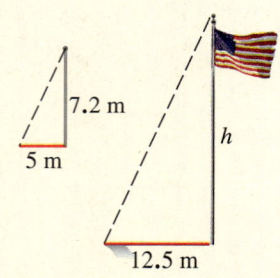

NOTE
There are several other correct ways to set up the proportion in Example 5 above. One way is to simply flip the ratios on *both* sides of the equal sign.

$$\dfrac{10}{h} = \dfrac{18}{99}$$

But there is another option, shown below.

Height in larger triangle → $\dfrac{h}{99} = \dfrac{10}{18}$ ← Height in smaller triangle
Shadow in larger triangle →   ← Shadow in smaller triangle

Notice that both ratios compare *height* to *shadow* in the same order. The ratio on the left describes the larger triangle, and the ratio on the right describes the smaller triangle.

ANSWERS
6. (a) $h = 20$ ft (b) $h = 18$ m

◀◀ Work Problem 6 at the Side.

6.6 Exercises

Which pairs of triangles appear to be congruent or similar? Label each pair as congruent, similar, *or* neither.

1.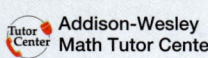
2.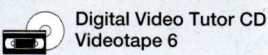
3.
4.
5.
6.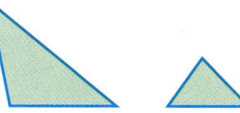

Determine which of these methods can be used to prove that each pair of triangles is congruent: Angle–Side–Angle (ASA), Side–Side–Side (SSS), or Side–Angle–Side (SAS). See Examples 1 and 2.

7.
8.
9.
10.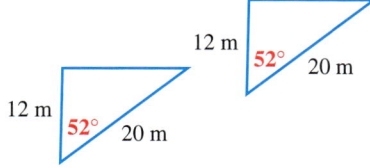
11.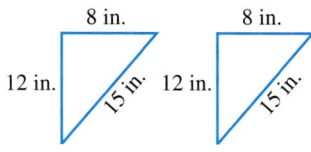
12.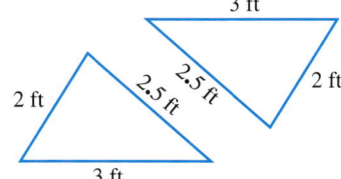

RELATING CONCEPTS (EXERCISES 13–16) For Individual or Group Work

Work Exercises 13–16 in order. *Given the information in each exercise, explain how you can prove that the indicated triangles are congruent. Note: A midpoint divides a segment into two congruent parts.*

13.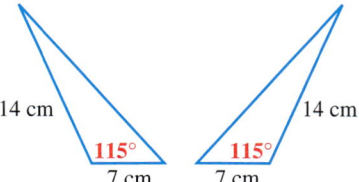

    C is the midpoint of $\overline{BE}$ and $CD = BA$.
    Prove that $\triangle ABC \cong \triangle DCE$.

14.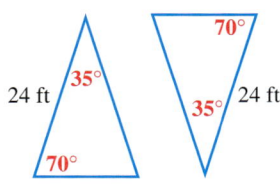

    P is the midpoint of both $\overline{WY}$ and $\overline{XZ}$; $WZ = XY$.
    Prove that $\triangle WPZ \cong \triangle YPX$.

15.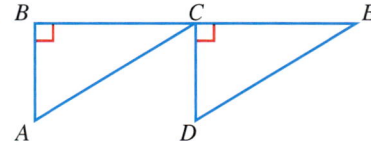

    $\overline{QS} \perp \overline{PR}$ and S is the midpoint of $\overline{PR}$. Prove that $\triangle PQS \cong \triangle RQS$.

16.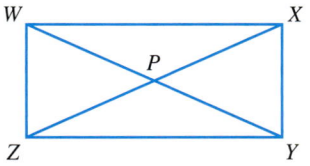

    M is the midpoint of both $\overline{LO}$ and $\overline{PN}$.
    Prove that $\triangle PLM \cong \triangle NOM$.

Write the ratio for each pair of corresponding sides in the similar triangles shown below. Write the ratios as fractions in lowest terms.

17. $\dfrac{AB}{PQ}; \dfrac{AC}{PR}; \dfrac{BC}{QR}$

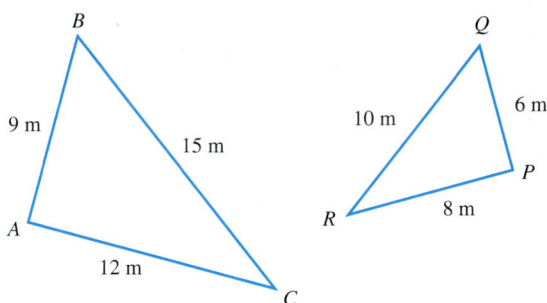

18. $\dfrac{AB}{PQ}; \dfrac{AC}{PR}; \dfrac{BC}{QR}$

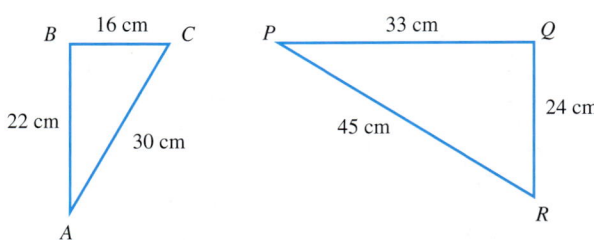

Find the unknown lengths in each pair of similar triangles. See Example 3.

19.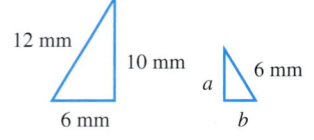

20.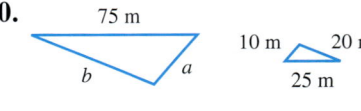

21.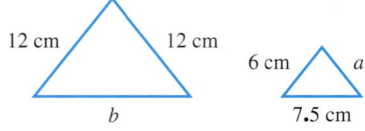

22.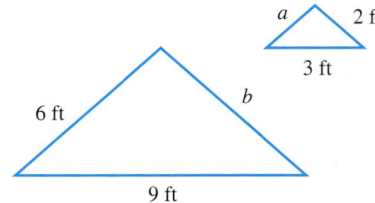

In Exercises 23 and 24, find the perimeter of each triangle. Assume the triangles are similar. See Example 4.

23.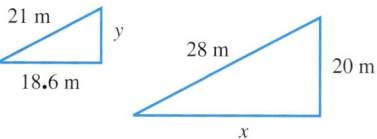

24.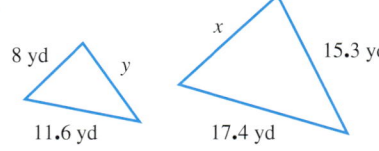

25. Triangles *CDE* and *FGH* are similar. Find the perimeter and area of triangle *FGH*. *Note:* The heights of similar triangles have the same ratio as corresponding sides. Round to the nearest tenth when necessary.

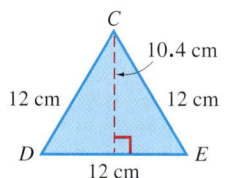 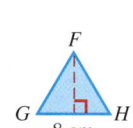

26. Triangles *JKL* and *MNO* are similar. Find the perimeter and area of triangle *MNO*.

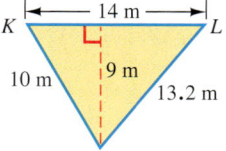 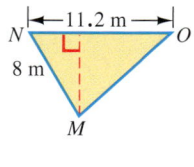

Solve each application problem. See Example 5.

27. The height of the house shown here can be found by comparing its shadow to the shadow cast by a 3-foot stick. Find the height of the house by writing a proportion and solving it.

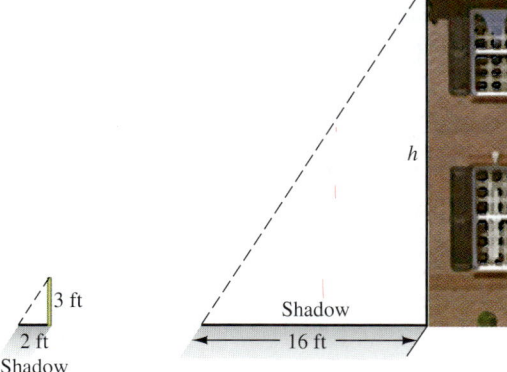

28. A fire lookout tower provides an excellent view of the surrounding countryside. The height of the tower can be found by lining up the top of the tower with the top of a 2-meter stick. Use similar triangles to find the height of the tower.

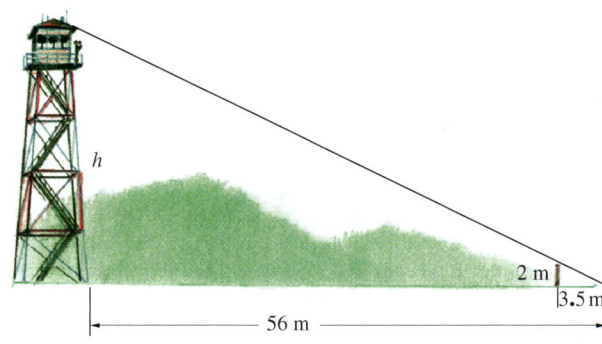

29. Look up the word *similar* in a dictionary. What is the nonmathematical definition of this word? Describe two examples of similar objects at home, school, or work.

30. Look up the word *congruent* in a dictionary. What is the nonmathematical definition of this word? Describe two examples of congruent objects at home, school, or work.

Find the unknown length in Exercises 31–34. Round your answers to the nearest tenth. Note: When a line is drawn parallel to one side of a triangle, the smaller triangle that is formed will be similar to the original triangle. In Exercises 31–32, the red segments are parallel.

31.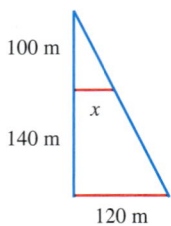

32.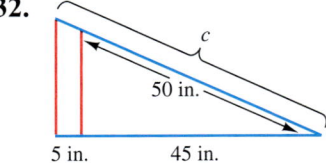

Hint: Redraw the two triangles and label the sides.

Hint: Redraw the two triangles and label the sides.

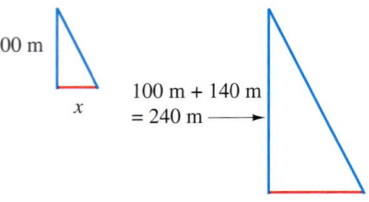

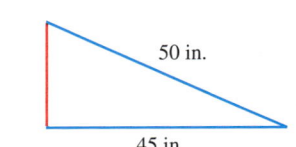

33. Use similar triangles and a proportion to find the length of the lake shown here. (*Hint:* The side 100 m long in the smaller triangle corresponds to the side of 100 m + 120 m = 220 m in the larger triangle.)

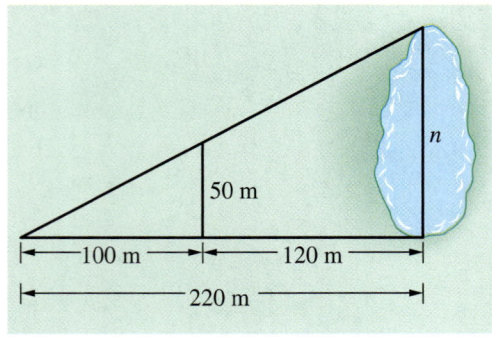

34. To find the height of the tree, find y and then add $5\frac{1}{2}$ ft for the distance from the ground to the eye level of the person.

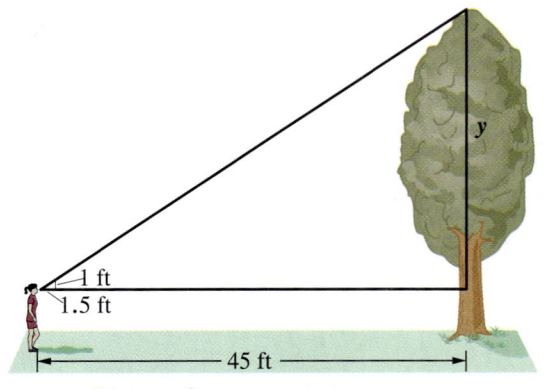

Distance from person to tree.

Chapter 6
SUMMARY

KEY TERMS

6.1 **ratio** — A ratio compares two quantities that have the same type of units. For example, the ratio of 6 apples to 11 apples is written in fraction form as $\frac{6}{11}$. The common units (apples) divide out.

6.2 **rate** — A rate compares two measurements with different types of units. Examples are 96 dollars for 8 hours, or 450 miles on 18 gallons.

unit rate — A unit rate has 1 in the denominator.

cost per unit — Cost per unit is a rate that tells how much you pay for one item or one unit. The lowest cost per unit is the best buy.

6.3 **proportion** — A proportion states that two ratios or rates are equivalent.

cross products — Multiply along one diagonal and then along the other diagonal to find the cross products of a proportion. If the cross products are equal, the proportion is true.

6.5 **point** — A point is a location in space. *Example:* Point *P* at the right.

line — A line is a straight row of points that goes on forever in both directions. *Example:* Line *AB*, written $\overleftrightarrow{AB}$, at the right.

line segment — A line segment is a piece of a line with two endpoints. *Example:* Line segment *PQ*, written $\overline{PQ}$, at the right.

ray — A ray is a part of a line that has one endpoint and extends forever in one direction. *Example:* Ray *RS*, written $\overrightarrow{RS}$, at the right.

parallel lines — Parallel lines are two lines in the same plane that never intersect (never cross). *Example:* $\overleftrightarrow{AB}$ is parallel to $\overleftrightarrow{ST}$ at the right.

intersecting lines — Intersecting lines cross. *Example:* $\overleftrightarrow{RQ}$ intersects $\overleftrightarrow{AB}$ at point *P* at the right.

angle — An angle is made up of two rays that have a common endpoint called the vertex. *Example:* Angle 1 at the right.

degrees — Degrees are used to measure angles; a complete circle is 360 degrees, written 360°.

straight angle — A straight angle is an angle that measures exactly 180°; its sides form a straight line. *Example:* Angle *G* at the right.

right angle — A right angle is an angle that measures exactly 90°. *Example:* Angle *AOB* at the right.

acute angle — An acute angle is an angle that measures less than 90°. *Example:* Angle *E* at the right.

obtuse angle — An obtuse angle is an angle that measures more than 90° but less than 180°. *Example:* Angle *F* at the right.

perpendicular lines — Perpendicular lines are two lines that intersect to form a right angle. *Example:* $\overleftrightarrow{PQ}$ is perpendicular to $\overleftrightarrow{RS}$ at the right.

(continued)

Chapter 6 Ratio, Proportion, and Line/Angle/Triangle Relationships

Key Terms

complementary angles Complementary angles are two angles whose measures add up to 90°.

supplementary angles Supplementary angles are two angles whose measures add up to 180°.

congruent angles Congruent angles are angles that measure the same number of degrees.

vertical angles Vertical angles are two nonadjacent congruent angles formed by intersecting lines. *Example:* ∠COA and ∠EOF are vertical angles at the right.

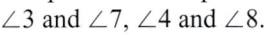

corresponding angles Corresponding angles are formed when two parallel lines are crossed by a transversal; corresponding angles are congruent and are on the same side of the transversal and in the same relative position. *Example:* In the figure at the right, line *m* is parallel to line *n*. The pairs of corresponding angles are ∠1 and ∠5, ∠2 and ∠6, ∠3 and ∠7, ∠4 and ∠8.

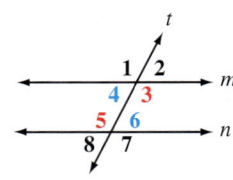

alternate interior angles When two parallel lines are crossed by a transversal, there are two pairs of alternate interior angles and each pair is congruent. They are on opposite sides of the transversal. *Example:* In the figure at the right, line *m* is parallel to line *n*. The pairs of alternate interior angles are ∠3 and ∠5, ∠4 and ∠6.

6.6 congruent figures Congruent figures are identical both in shape and in size.

similar figures Similar figures have the same shape but are different sizes.

congruent triangles Congruent triangles are triangles with the same shape and the same size; corresponding angles measure the same number of degrees and corresponding sides have the same length.

similar triangles Similar triangles are triangles with the same shape but not necessarily the same size; corresponding angles measure the same number of degrees and the *ratios* of the lengths of corresponding sides are equal.

Test Your Word Power

See how well you have learned the vocabulary in this chapter. Answers follow the Quick Review.

1. A **ratio**
   A. can be written only as a fraction
   B. compares two quantities that have the same type of units
   C. compares two quantities that have different types of units
   D. is the reciprocal of a rate.

2. A **rate**
   A. can be written only as a decimal
   B. compares two quantities that have the same type of units
   C. compares two quantities that have different types of units
   D. is the reciprocal of a ratio.

3. **Cost per unit** is
   A. the best buy
   B. a ratio written in lowest terms
   C. found by comparing cross products
   D. the price of one item or one unit.

4. A **proportion**
   A. shows that two ratios or rates are equivalent
   B. contains only whole numbers or decimals
   C. always has one unknown number
   D. states that two improper fractions are equivalent.

5. An **obtuse angle**
   A. is formed by perpendicular lines
   B. is congruent to a right angle
   C. measures more than 90° but less than 180°
   D. measures less than 90°.

6. Two angles that are **complementary**
   A. have measures that add up to 180°
   B. are always congruent
   C. form a straight angle
   D. have measures that add up to 90°.

7. **Perpendicular lines**
   A. intersect to form a right angle
   B. intersect to form an acute angle
   C. never intersect
   D. have a common endpoint called the vertex.

8. In a pair of **similar triangles**
   A. corresponding sides have the same length
   B. all the angles have the same measure
   C. the perimeters are equal
   D. the ratios of the lengths of corresponding sides are equal.

Chapter 6 Summary

QUICK REVIEW

Concepts	Examples

6.1 Writing a Ratio
A ratio compares two quantities that have the same type of units. A ratio is usually written as a fraction with the number that is mentioned first in the numerator. The common units divide out and are not written in the answer. Check that the fraction is in lowest terms.

Write this ratio as a fraction in lowest terms.

60 ounces of medicine **to** 160 ounces of medicine

$$\frac{60 \text{ ounces}}{160 \text{ ounces}} = \frac{60 \div 20}{160 \div 20} = \frac{3}{8} \quad \leftarrow \text{Ratio in lowest terms}$$

Divide out common units.

6.1 Using Mixed Numbers in a Ratio
If a ratio has mixed numbers, change the mixed numbers to improper fractions. Rewrite the problem in horizontal form, using the "÷" symbol for division. Finally, multiply by the reciprocal of the divisor.

Write as a ratio of whole numbers in lowest terms.

$$2\frac{1}{2} \text{ to } 3\frac{3}{4}$$

$$\frac{2\frac{1}{2}}{3\frac{3}{4}} \quad \text{Ratio with mixed numbers}$$

$$= \frac{\frac{5}{2}}{\frac{15}{4}} \quad \text{Ratio with improper fractions}$$

$$= \frac{5}{2} \div \frac{15}{4} = \frac{5}{2} \cdot \frac{4}{15} = \frac{\cancel{5} \cdot \cancel{2} \cdot 2}{\cancel{2} \cdot 3 \cdot \cancel{5}} = \frac{2}{3} \quad \begin{cases} \text{Ratio in} \\ \text{lowest} \\ \text{terms} \end{cases}$$

Reciprocals

6.1 Using Measurements in Ratios
When a ratio compares measurements, both measurements must be in the *same* units. It is usually easier to compare the measurements using the smaller unit, for example, inches instead of feet.

Write 8 inches to 6 feet as a ratio in lowest terms.

Compare using the smaller unit, inches. Because 1 foot has 12 inches, 6 feet is

$$6 \cdot 12 \text{ inches} = 72 \text{ inches.}$$

The ratio is shown below.

$$\frac{8 \text{ inches}}{6 \text{ feet}} = \frac{8 \text{ inches}}{72 \text{ inches}} = \frac{8 \div 8}{72 \div 8} = \frac{1}{9} \quad \begin{cases} \text{Ratio in} \\ \text{lowest terms} \end{cases}$$

6.2 Writing Rates
A rate compares two measurements with different types of units. The units do *not* divide out, so you must write them as part of the rate.

Write the rate as a fraction in lowest terms.

475 miles in 10 hours

$$\frac{475 \text{ miles} \div 5}{10 \text{ hours} \div 5} = \frac{95 \text{ miles}}{2 \text{ hours}} \quad \leftarrow \text{Must write units: miles and hours}$$

Concepts	Examples
6.2 Finding a Unit Rate A unit rate has 1 in the denominator. To find the unit rate, divide the numerator by the denominator. Write unit rates using the word **per** or a **/** mark.	Write as a unit rate: $1278 in 9 days. $\dfrac{\$1278}{9 \text{ days}}$ ← Fraction bar indicates division. $9\overline{)1278}$ gives 142, so $\dfrac{\$1278 \div 9}{9 \text{ days} \div 9} = \dfrac{\$142}{1 \text{ day}}$ Write the answer as $142 **per** day or $142/day.
6.2 Finding the Best Buy The best buy is the item with the lowest cost per unit. Divide the price by the number of units. Round to thousandths when necessary. Then compare to find the lowest cost per unit.	Find the best buy on cheese. 2 pounds for $2.25 3 pounds for $3.40 Find cost per unit (cost per pound). $\dfrac{\$2.25}{2} = \$1.125$ per pound $\dfrac{\$3.40}{3} \approx \$1.133$ per pound The lower cost per pound is $1.125, so 2 pounds of cheese is the best buy.
6.3 Writing Proportions A proportion states that two ratios or rates are equivalent. The proportion "5 is to 6 as 25 is to 30" is written as shown below. $\dfrac{5}{6} = \dfrac{25}{30}$ To see whether a proportion is true or false, multiply along one diagonal, then multiply along the other diagonal. If the two cross products are equal, the proportion is true. If the two cross products are unequal, the proportion is false.	Write as a proportion: 8 is to 40 as 32 is to 160 $\dfrac{8}{40} = \dfrac{32}{160}$ Is this proportion true or false? $\dfrac{6}{8\frac{1}{2}} = \dfrac{24}{34}$ Find the cross products. $8\frac{1}{2} \cdot 24 = \dfrac{17}{2} \cdot \dfrac{12}{1} = 204$ $6 \cdot 34 = 204$ Equal The cross products are equal, so the proportion is true.

Concepts

6.3 Solving Proportions

Solve for an unknown number in a proportion by using these steps.

Step 1 Find the cross products. (If desired, you can rewrite the ratios in lowest terms before finding the cross products.)

Step 2 Show that the cross products are equivalent.

Step 3 Divide both sides of the equation by the coefficient of the variable term.

Step 4 Check by writing the solution in the *original* proportion and finding the cross products.

Examples

Find the unknown number.

$$\frac{12}{x} = \frac{6}{8}$$
$$\frac{12}{x} = \frac{3}{4} \quad \text{Write } \tfrac{6}{8} \text{ in lowest terms.}$$

Step 1 $\dfrac{12}{x} = \dfrac{3}{4}$ Find the cross products: $x \cdot 3$ and $12 \cdot 4$

Step 2 $x \cdot 3 = 12 \cdot 4$ Show that cross products are equivalent.
$x \cdot 3 = 48$

Step 3 $\dfrac{x \cdot \cancel{3}}{\cancel{3}} = \dfrac{48}{3}$ Divide both sides by 3

Step 4 $x = 16$

x is 16 → $\dfrac{12}{16} = \dfrac{6}{8}$ $\begin{array}{l} 16 \cdot 6 = \mathbf{96} \\ 12 \cdot 8 = \mathbf{96} \end{array}$ Equal

The cross products are equal, so 16 is the correct solution.

6.4 Applications of Proportions

Use the six problem-solving steps from **Section 3.3**.

Step 1 **Read** the problem.

Step 2 **Assign a variable.**

Step 3 **Write an equation.**

If 3 pounds of grass seed cover 450 square feet of lawn, how much seed is needed for 1500 square feet of lawn?

Step 1 The problem is about the amount of grass seed needed for a lawn.

Unknown: pounds of seed needed for 1500 square feet of lawn
Known: 3 pounds cover 450 square feet of lawn

Step 2 There is only one unknown, so let x be the pounds of seed needed for 1500 square feet of lawn.

Step 3 The equation is in the form of a proportion. Be sure that both rates in the proportion compare pounds to square feet in the same order.

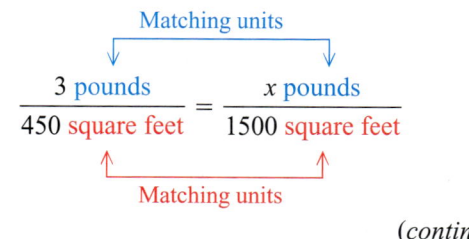

(continued)

Concepts	Examples

6.4 Applications of Proportions (continued)

Step 4 **Solve** the equation.

Step 4 Ignore the units while finding the cross products and solving for x.

$$450 \cdot x = 3 \cdot 1500 \quad \text{Show that cross products are equivalent.}$$
$$450 \cdot x = 4500$$
$$\frac{\overset{1}{\cancel{450}} \cdot x}{\cancel{450}} = \frac{4500}{450} \quad \text{Divide both sides by 450}$$
$$x = 10$$

Step 5 **State the answer.**

Step 5 10 pounds of grass seed are needed for 1500 square feet of lawn.

Step 6 **Check** the solution by putting it back into the original problem.

Step 6 450 square feet of lawn needs 3 pounds of seed; 1500 square feet is about *three times as much* lawn, so about *three times as much* seed is needed.

$$(3)(3 \text{ pounds}) = 9 \text{ pounds} \leftarrow \text{Estimate}$$

The solution, 10 pounds, is close to the estimate of 9 pounds, so it is reasonable.

6.5 Lines

A *line* is straight row of points that goes on forever in both directions. If part of a line has one endpoint, it is a *ray*. If it has two endpoints, it is a *line segment*.

Identify each figure as a line, line segment, or ray.

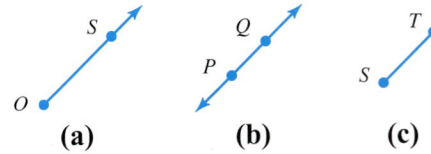

Figure **(a)** shows a ray, **(b)** shows a line, and **(c)** shows a line segment.

If two lines intersect at right angles, they are *perpendicular*. If two lines in the same plane never intersect, they are *parallel*.

Label each pair of lines as appearing to be parallel or as perpendicular.

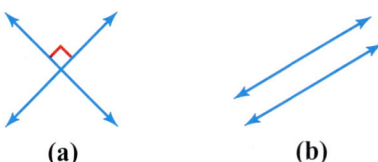

Figure **(a)** shows perpendicular lines (they intersect at 90°), and figure **(b)** shows lines that appear to be parallel (they never intersect).

Concepts

6.5 Angles
If the sum of the measures of two angles is 90°, they are *complementary*.
If the sum of the measures of two angles is 180°, they are *supplementary*.

If two angles measure the same number of degrees, the angles are *congruent*. The symbol for congruent is ≅.

Two nonadjacent angles formed by intersecting lines are called *vertical angles*. Vertical angles are congruent.

6.5 Parallel Lines
When two parallel lines are crossed by a transversal, corresponding angles are congruent, and alternate interior angles are congruent. Use this information to find the measures of the other angles.

Read $m\angle 1$ as "the measure of angle 1."

6.6 Proving That Two Triangles Are Congruent
Congruent triangles are identical both in shape and in size. This means that corresponding angles have the same measure and corresponding sides have the same length.

Here are three ways to prove that two triangles are congruent.

1. Angle–Side–Angle (ASA) method: If two angles and the side between them on one triangle measure the same as the corresponding parts on another triangle, the triangles are congruent.
2. Side–Side–Side (SSS) method: If three sides of one triangle measure the same as the corresponding sides of another triangle, the triangles are congruent.
3. Side–Angle–Side (SAS) method: If two sides and the angle between them on one triangle measure the same as the corresponding parts on another triangle, the triangles are congruent.

Examples

6.5 Angles
Find the complement and supplement of a 35° angle.

$90° - 35° = 55°$ (the complement)

$180° - 35° = 145°$ (the supplement)

Identify the vertical angles in this figure. Which angles are congruent?

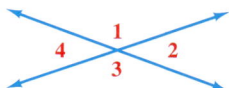

∠1 and ∠3 are vertical angles.

∠2 and ∠4 are vertical angles.

Vertical angles are congruent, so ∠1 ≅ ∠3 and ∠2 ≅ ∠4.

6.5 Parallel Lines
Line *m* is parallel to line *n* and the measure of ∠4 is 125°. Find the measures of the other angles.

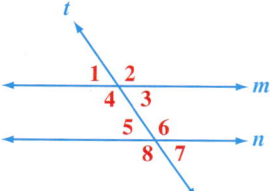

∠4 ≅ ∠8 (corresponding angles), so $m\angle 8 = 125°$.
∠4 ≅ ∠6 (alternate interior angles), so $m\angle 6 = 125°$.
∠6 ≅ ∠2 (corresponding angles), so $m\angle 2 = 125°$.
∠4 and ∠3 are supplements, so $m\angle 3 = 180° - 125° = 55°$.
∠3 ≅ ∠7 (corresponding angles), so $m\angle 7 = 55°$.
∠3 ≅ ∠5 (alternate interior angles), so $m\angle 5 = 55°$.
∠5 ≅ ∠1 (corresponding angles), so $m\angle 1 = 55°$.

6.6 Proving That Two Triangles Are Congruent
Determine which method can be used to prove that each pair of triangles is congruent.

(a) 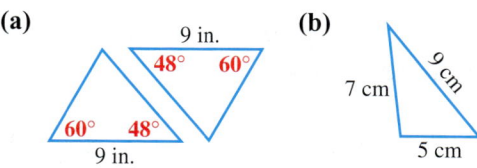 (b)

(c)

(a) On both triangles, two corresponding angles and the side between them measure the same, so use ASA.
(b) Each pair of corresponding sides has the same length, so use SSS.
(c) On both triangles, two corresponding sides and the angle between them measure the same, so use SAS.

Concepts

6.6 Finding the Unknown Lengths in Similar Triangles

Use the fact that in similar triangles, the ratios of the lengths of corresponding sides are equal. Write a proportion. Then find the cross products and show that they are equivalent. Finish solving for the unknown length.

Examples

Find x and y if the triangles are similar.

$$\frac{x}{8} = \frac{5}{10} \qquad \frac{y}{12} = \frac{5}{10}$$

$$x \cdot 10 = 8 \cdot 5 \qquad y \cdot 10 = 12 \cdot 5$$

$$\frac{x \cdot \cancel{10}}{\cancel{10}} = \frac{40}{10} \qquad \frac{y \cdot \cancel{10}}{\cancel{10}} = \frac{60}{10}$$

$$x = 4 \text{ m} \qquad y = 6 \text{ m}$$

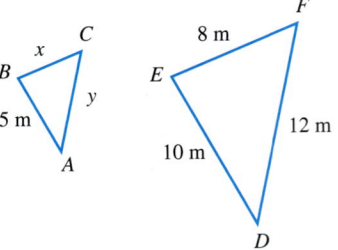

ANSWERS TO TEST YOUR WORD POWER

1. B; *Example:* The ratio of 3 miles to 4 miles is $\frac{3}{4}$; the common units (miles) divide out.
2. C; *Example:* $4.50 for 3 pounds is a rate comparing dollars to pounds.
3. D; *Example:* $3.95 per gallon tells the price of one gallon (one unit).
4. A; *Example:* The proportion $\frac{5}{6} = \frac{25}{30}$ says that $\frac{5}{6}$ is equivalent to $\frac{25}{30}$.
5. C; *Examples:* Angles that measure 91°, 120°, and 175° are all obtuse angles.
6. D; *Example:* If $\angle 1$ measures 35° and $\angle 2$ measures 55°, the angles are complementary because 35° + 55° = 90°.
7. A; *Example:* $\overleftrightarrow{EF}$ is perpendicular to $\overleftrightarrow{GH}$.

8. D; *Example:* Triangle *ABC* is similar to triangle *DEF*, so the ratios of corresponding sides are equal.

$$\frac{AB}{DE} = \frac{3 \text{ m}}{6 \text{ m}} = \frac{1}{2} \qquad \frac{BC}{EF} = \frac{2 \text{ m}}{4 \text{ m}} = \frac{1}{2} \qquad \frac{AC}{DF} = \frac{3.5 \text{ m}}{7 \text{ m}} = \frac{1}{2}$$

Chapter 6
REVIEW EXERCISES

[6.1] *Write each ratio as a fraction in lowest terms. Change to the same units when necessary, using the table of measurement comparisons in* **Section 6.1.** *Use the information in the graph to answer Exercises 1–3.*

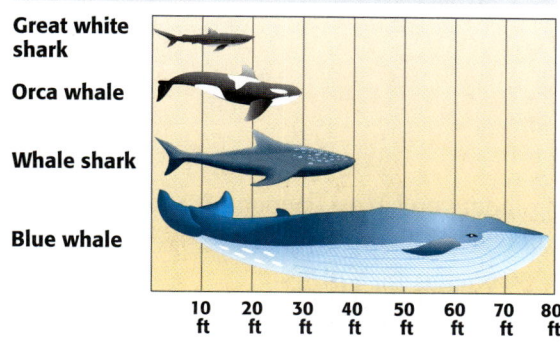

Source: Grolier Multimedia Encyclopedia.

1. Orca whale's length to whale shark's length

2. Blue whale's length to great white shark's length

3. Which two animals' lengths give a ratio of $\frac{1}{2}$? There are several answers.

4. $2.50 to $1.25

5. $0.30 to $0.45

6. $1\frac{2}{3}$ cups to $\frac{2}{3}$ cup

7. $2\frac{3}{4}$ miles to $16\frac{1}{2}$ miles

8. 5 hours to 100 minutes

9. 9 in. to 2 ft

10. 1 ton to 1500 pounds

11. 8 hours to 3 days

12. Jake sold $350 worth of his kachina figures. Ramona sold $500 worth of her pottery. What is the ratio of Ramona's sales to Jake's sales?

13. Ms. Wei's new car gets 35 miles per gallon. Her old car got 25 miles per gallon. Find the ratio of the new car's mileage to the old car's mileage.

14. This fall, 6000 students are taking math courses and 7200 students are taking English courses. Find the ratio of math students to English students.

[6.2] *Write each rate as a fraction in lowest terms.*

15. $88 for 8 dozen

16. 96 children in 40 families

17. In his keyboarding class, Patrick can type four pages in 20 minutes. Give his rate in pages per minute and minutes per page.

18. Elena made $24 in three hours. Give her earnings in dollars per hour and hours per dollar.

Find the best buy.

19. Minced onion

    13 ounces for $2.29

    8 ounces for $1.45

    3 ounces for $0.95

20. Dog food; you have a coupon for $1 off on 25 pounds or more.

    50 pounds for $19.95

    25 pounds for $10.40

    8 pounds for $3.40

[6.3] *Use either the method of writing in lowest terms or of finding cross products to decide whether each proportion is* true *or* false. *Show your work and then write* true *or* false.

21. $\dfrac{6}{10} = \dfrac{9}{15}$

22. $\dfrac{6}{48} = \dfrac{9}{36}$

23. $\dfrac{47}{10} = \dfrac{98}{20}$

24. $\dfrac{1.5}{2.4} = \dfrac{2}{3.2}$

25. $\dfrac{3\tfrac{1}{2}}{2\tfrac{1}{3}} = \dfrac{6}{4}$

Find the unknown number in each proportion. Round answers to the nearest hundredth when necessary.

26. $\dfrac{4}{42} = \dfrac{150}{x}$

27. $\dfrac{16}{x} = \dfrac{12}{15}$

28. $\dfrac{100}{14} = \dfrac{x}{56}$

29. $\dfrac{5}{8} = \dfrac{x}{20}$

30. $\dfrac{x}{24} = \dfrac{11}{18}$

31. $\dfrac{7}{x} = \dfrac{18}{21}$

32. $\dfrac{x}{3.6} = \dfrac{9.8}{0.7}$

33. $\dfrac{13.5}{1.7} = \dfrac{4.5}{x}$

34. $\dfrac{0.82}{1.89} = \dfrac{x}{5.7}$

[6.4] *Set up and solve a proportion for each application problem.*

35. The ratio of cats to dogs at the animal shelter is 3 to 5. If there are 45 dogs, how many cats are there?

36. Danielle had 8 hits in 28 times at bat during last week's games. If she continues to hit at the same rate, how many hits will she get in 161 times at bat?

37. If 3.5 pounds of ground beef cost $9.77, what will 5.6 pounds cost? Round to the nearest cent.

38. About 4 out of 10 students are expected to vote in campus elections. There are 8247 students. How many are expected to vote? Round to the nearest whole number.

39. The scale on Brian's model railroad is 1 inch to 16 feet. One of the scale model boxcars is 4.25 inches long. What is the length of a real boxcar in feet?

40. Marvette makes necklaces to sell at a local gift shop. She made 2 dozen necklaces in $16\frac{1}{2}$ hours. How long will it take her to make 40 necklaces?

41. A 180-pound person burns 284 calories playing basketball for 25 minutes. At this rate, how many calories would the person burn in 45 minutes, to the nearest whole number? (*Source: Wellness Encyclopedia.*)

42. In the hospital pharmacy, Michiko sees that a medicine is to be given at the rate of 3.5 milligrams for every 50 pounds of body weight. How much medicine should be given to a patient who weighs 210 pounds?

[6.5] *Identify each figure as a line, line segment, or ray and name it using the appropriate symbol.*

43.

44.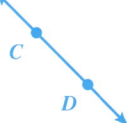

45.

Label each pair of lines as appearing to be parallel, *as* perpendicular, *or as* intersecting.

46.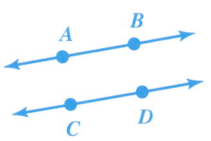

47.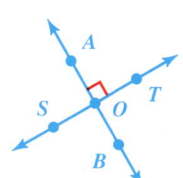

48.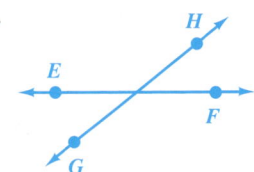

Label each angle as an acute, right, obtuse, *or* straight angle. *For right and straight angles, indicate the number of degrees in the angle.*

49.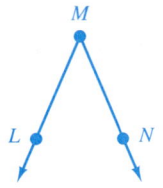

50.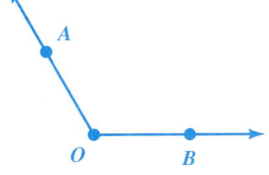

51.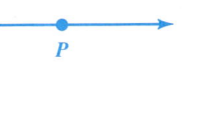

52.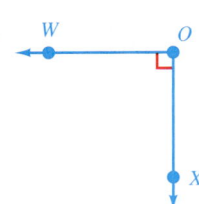

Find the complement or supplement of each angle.

53. Find each complement.
    (a) 80°
    (b) 45°
    (c) 7°

54. Find each supplement.
    (a) 155°
    (b) 90°
    (c) 33°

55. In the figure below, ∠2 measures 60°. Find the measure of each of the other angles.

    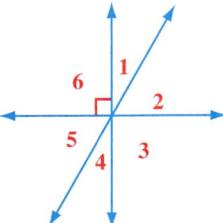

56. Line *m* is parallel to line *n* and ∠8 measures 160°. Find the measures of the other angles.

    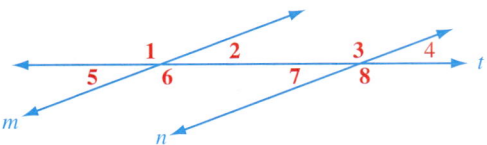

[6.6] *Determine which method can be used to prove that each pair of triangles is congruent.*

57.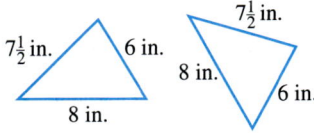

58.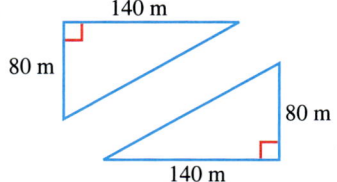

59.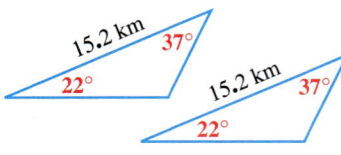

Find the unknown lengths in each pair of similar triangles. Then find the perimeter of the larger triangle in each pair.

60.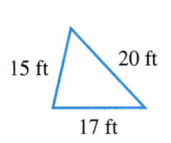

61.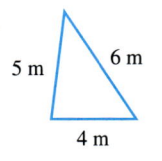

62.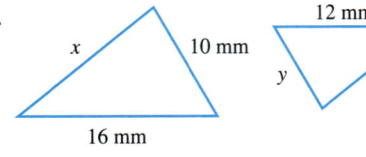

MIXED REVIEW EXERCISES

Find the unknown number in each proportion. Round to hundredths when necessary.

63. $\dfrac{x}{45} = \dfrac{70}{30}$

64. $\dfrac{x}{52} = \dfrac{0}{20}$

65. $\dfrac{64}{10} = \dfrac{x}{20}$

66. $\dfrac{15}{x} = \dfrac{65}{100}$

67. $\dfrac{7.8}{3.9} = \dfrac{13}{x}$

68. $\dfrac{34.1}{x} = \dfrac{0.77}{2.65}$

Write each ratio as a fraction in lowest terms. Change to the same units when necessary.

69. 4 dollars to 10 quarters

70. $4\dfrac{1}{8}$ inches to 10 inches

71. 10 yards to 8 feet

72. $3.60 to $0.90

73. 12 eggs to 15 eggs

74. 37 meters to 7 meters

75. 3 pints to 4 quarts

76. 15 minutes to 3 hours

77. $4\dfrac{1}{2}$ miles to $1\dfrac{3}{10}$ miles

Set up and solve a proportion for each application problem.

78. Nearly 7 out of 8 fans buy something to drink at rock concerts. How many of the 28,500 fans at today's concert would be expected to buy a beverage? Round to the nearest hundred fans.

79. Emily spent $150 on car repairs and $400 on car insurance. What is the ratio of the amount spent on insurance to the amount spent on repairs?

80. Antonio is choosing among three packages of plastic wrap. Is the best buy 25 feet for $0.78; 75 feet for $1.99; or 100 feet for $2.59? He has a coupon for 50¢ off that is good for either of the larger two packages.

81. On this scale drawing of a backyard patio, 0.5 inch represents 6 feet. If the patio measures 1.75 inches long and 1.25 inches wide on the drawing, what will be the actual length and width of the patio when it is built?

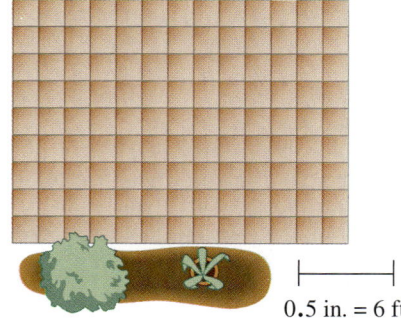

0.5 in. = 6 ft

82. A vitamin supplement for cats is to be given at the rate of 1000 milligrams for a 5-pound cat. (*Source:* St. Jon Pet Care Products.)

  (a) How much should be given to a 7-pound cat?

  (b) How much should be given to an 8-ounce kitten?

83. Charles made 251 points during 169 minutes of playing time last year. If he plays 14 minutes in tonight's game, how many points would you expect him to make? Round to the nearest whole number.

84. An antibiotic is to be given at the rate of $1\dfrac{1}{2}$ teaspoons for every 24 pounds of body weight. How much should be given to an infant who weighs 8 pounds?

490 Chapter 6 Ratio, Proportion, and Line/Angle/Triangle Relationships

Label each figure. Choose from these labels: line segment, ray, parallel lines, perpendicular lines, intersecting lines, acute angle, right angle, straight angle, obtuse angle. *Indicate the number of degrees in the right angle and the straight angle.*

85.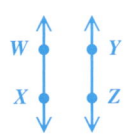

86.

87.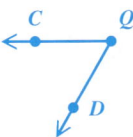

88.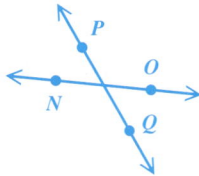

89.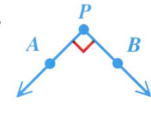

90.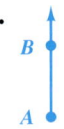

91.

92.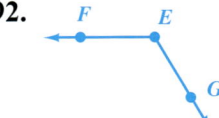

93.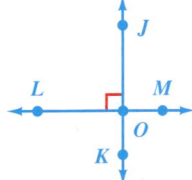

94. Explain what is happening in each sentence.
  (a) The road was so slippery that my car did a 360.

  (b) After the election, the governor's view on new taxes took a 180° turn.

95. (a) Can two obtuse angles be supplementary? Explain why or why not.

  (b) Can two acute angles be complementary? Explain why or why not.

96. In the figure below, ∠2 measures 45° and ∠7 measures 55°. Find the measure of each of the other angles.

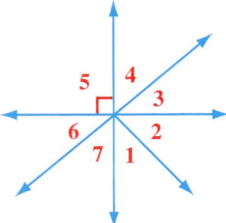

97. In the figure below, line *m* is parallel to line *n* and ∠5 measures 75°. Find the measures of the other angles.

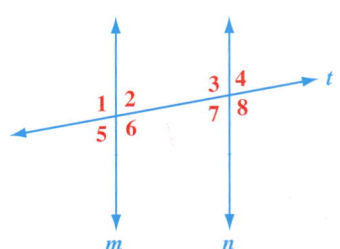

Chapter 6
TEST

Write each rate or ratio as a fraction in lowest terms. Change to the same units when necessary.

1. $15 for 75 minutes
2. 3 hours to 40 minutes

3. The little theater has 320 seats. The auditorium has 1200 seats. Find the ratio of auditorium seats to theater seats.

4. Find the best buy on spaghetti sauce. You have a coupon for 75¢ off Brand X and a coupon for 25¢ off Brand Y.
   - 28 ounces of Brand X for $3.89
   - 18 ounces of Brand Y for $1.89
   - 13 ounces of Brand Z for $1.29

Find the unknown number in each proportion. Round answers to the nearest hundredth when necessary.

5. $\dfrac{5}{9} = \dfrac{x}{45}$

6. $\dfrac{3}{1} = \dfrac{8}{x}$

7. $\dfrac{x}{20} = \dfrac{6.5}{0.4}$

8. $\dfrac{2\frac{1}{3}}{x} = \dfrac{\frac{8}{9}}{4}$

Set up and solve a proportion for each application problem.

9. Pedro typed 240 words in five minutes in his keyboarding class. At that rate, how many words could he type in 12 minutes?

10. About 2 out of every 15 people are left-handed. How many of the 650 students in our school would you expect to be left-handed? Round to the nearest whole number.

11. A medication is given at the rate of 8.2 grams for every 50 pounds of body weight. How much should be given to a 145-pound person? Round to the nearest tenth.

12. On a scale model, 1 inch represents 8 feet. If a building in the model is 7.5 inches tall, what is the actual height of the building in feet?

1. _____
2. _____
3. _____
4. _____
5. _____
6. _____
7. _____
8. _____
9. _____
10. _____
11. _____
12. _____

492 Chapter 6 Ratio, Proportion, and Line/Angle/Triangle Relationships

Choose the figure that matches each label. For right and straight angles, indicate the number of degrees in the angle.

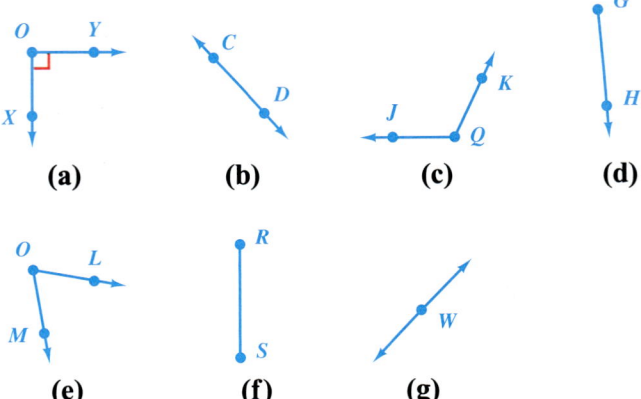

13. Acute angle is figure _____.

14. Right angle is figure _____ and its measure is _____.

15. Ray is figure _____.

16. Straight angle is figure _____ and its measure is _____.

17. Write a definition of parallel lines and a definition of perpendicular lines. Make a sketch to illustrate each definition.

18. Find the complement of an 81° angle.

19. Find the supplement of a 20° angle.

20. In the figure below, ∠4 measures 50° and ∠6 measures 95°. Find the measures of the other angles.

21. In the figure below, line m is parallel to line n and ∠3 measures 65°. Find the measures of the other angles.

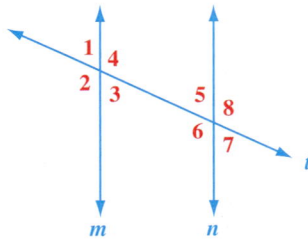

Determine which method can be used to prove that each pair of triangles is congruent.

22.

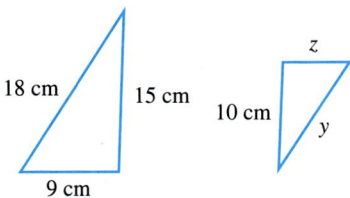

23.

24. Find the unknown lengths in these similar triangles.

25. Find the perimeter of each of these similar triangles.

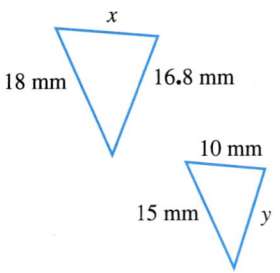

Percent

7.1 **The Basics of Percent**

7.2 **The Percent Proportion**

7.3 **The Percent Equation**

Summary Exercises on Percent

7.4 **Problem Solving with Percent**

7.5 **Consumer Applications: Sales Tax, Tips, Discounts, and Simple Interest**

Everyone likes to buy things that are "on sale." It helps if you can estimate the sale price *before* deciding to buy. Most states add on a sales tax, so that is part of your final cost as well. Find out how to calculate discounts and sales tax in **Section 7.5,** Examples 1 and 4. Then get some practice in Exercises 1–10 and 17–24.

In **Section 7.5** you'll also learn how to use percents to estimate restaurant tips and calculate the interest on loans.

7.1 The Basics of Percent

OBJECTIVES

1. Learn the meaning of percent.
2. Write percents as decimals.
3. Write decimals as percents.
4. Write percents as fractions.
5. Write fractions as percents.
6. Use 100% and 50%.

1 Write a percent to describe each situation.

(a) You leave a $20 tip for a restaurant bill of $100. What percent tip did you leave?

(b) The tax on a $100 graphing calculator is $5. What is the tax rate?

(c) You earn 94 points on a 100 point test. What percent of the points did you earn?

ANSWERS
1. (a) 20% (b) 5% (c) 94%

OBJECTIVE 1 Learn the meaning of percent. You have probably seen percents frequently in daily life. The symbol for percent is **%**. For example, during one day you may leave a 15% tip for the waitress at dinner, pay 7% sales tax on a CD player, and buy shoes at 25% off the regular price. The next day your score on a math test may be 89% correct.

The Meaning of Percent

A **percent** is a ratio with a denominator of 100. So percent means "per 100" or "how many out of 100." The symbol for percent is **%**. Read 15% as "fifteen percent."

EXAMPLE 1 Understanding Percent

Write a percent to describe each situation.

(a) If you left a $15 tip when the restaurant bill was $100, then you left $15 per $100 or $\frac{15}{100}$ or 15%.

(b) If you pay $7 in tax on a $100 CD player, then the tax rate is $7 per $100 or $\frac{7}{100}$ or 7%.

(c) If you earn 89 points on a 100-point math test, then your score is 89 out of 100 or $\frac{89}{100}$ or 89%.

Work Problem 1 at the Side.

OBJECTIVE 2 Write percents as decimals. In order to work with percents, you will need to write them as decimals or as fractions. We'll start by writing percents as equivalent decimal numbers. Twenty-five percent, or 25%, means 25 parts out of 100 parts, or $\frac{25}{100}$. Remember that the fraction bar indicates division. So we can write $\frac{25}{100}$ as $25 \div 100$. When you do the division, $25 \div 100$ is 0.25.

Indicates division $\longrightarrow \frac{25}{100}$ can be written as $25 \div 100 = 0.25$
$\uparrow$
Indicates division

Another way to remember how to write a percent as a decimal is to use the meaning of the word *percent*. The first part of the word, *per*, is an indicator word for *division*. The last part of the word, *cent*, comes from the Latin word for *hundred*. (Recall that there are 100 *cent*s in a dollar, and 100 years in a *cent*ury.)

25% is 25 **percent**

$25 \div 100 = 0.25$

Writing a Percent as an Equivalent Decimal

To write a percent as a decimal, drop the % symbol and divide by 100.

EXAMPLE 2 Writing Percents as Decimals

Write each percent as a decimal.

(a) 47%   47% = 47 ÷ 100 = 0.47   Decimal form
(b) 3%   3% = 3 ÷ 100 = 0.03   Decimal form
(c) 28.2% 28.2% = 28.2 ÷ 100 = 0.282   Decimal form
(d) 100%   100% = 100 ÷ 100 = 1.00   Decimal form
(e) 135%   135% = 135 ÷ 100 = 1.35   Decimal form

CAUTION
In Example 2(d) above, notice that 100% is 1.00, or 1, which is a whole number. Whenever you have a percent that is *100% or greater,* the equivalent decimal number will be *1 or greater.* Notice in Example 2(e) above that 135% is 1.35 (greater than 1).

Work Problem 2 at the Side.

In the exercise set for **Section 5.5,** you discovered a shortcut for *dividing* by 100: Move the decimal point *two* places to the *left*. You can use this shortcut when writing percents as decimals.

EXAMPLE 3 Changing Percents to Decimals by Moving the Decimal Point

Write each percent as a decimal by dropping the percent symbol and moving the decimal point two places to the left.

(a) 17%

17% = 17.%   ← Decimal point starts at far right side.

.17   ← Percent symbol is dropped.

Decimal point is moved two places to the left. This is a quick way to divide by 100

17% = 0.17

(b) 160%

160% = 160.% = 1.60 or 1.6   Decimal point starts at far right side. 1.60 is equivalent to 1.6

(c) 4.9%

04.9%   0 is attached so the decimal point can be moved two places to the left.

4.9% = 0.049

(d) 0.6%

00.6% = 0.006   0 is attached so the decimal point can be moved two places to the left.

Continued on Next Page

---

2 Write each percent as a decimal.

(a) 68%

(b) 5%

(c) 40.6%

(d) 200%

(e) 350%

ANSWERS
2. (a) 0.68 (b) 0.05 (c) 0.406
   (d) 2.00 or 2 (e) 3.50 or 3.5

3 Write each percent as a decimal.

(a) 90%

(b) 9%

(c) 900%

(d) 9.9%

(e) 0.9%

ANSWERS
3. (a) 0.90 or 0.9 (b) 0.09 (c) 9.00 or 9
   (d) 0.099 (e) 0.009

> **NOTE**
> In Example 3(d) on the previous page, notice that 0.6% is less than 1%. Because 1% is equivalent to 0.01 or $\frac{1}{100}$, any fraction of a percent smaller than 1% is less than 0.01. The decimal equivalent of 0.6% is 0.006, which is less than 0.01.

◀◀◀ **Work Problem 3 at the Side.**

OBJECTIVE 3 Write decimals as percents. You can write a decimal as a percent. For example, the decimal 0.25 is the same as the fraction $\frac{25}{100}$.

This fraction means 25 out of 100 parts, or 25%. Notice that multiplying 0.25 by 100 gives the same result.

$$(0.25)(100) = 25\%$$

This result makes sense because we are doing the opposite of what we did to change a percent to a decimal.

To change a percent to a decimal, we *drop* the % symbol and *divide* by 100. So, to *reverse* the process and change a decimal to a percent, we *multiply* by 100 and *attach* a % symbol.

> **Writing a Decimal as a Percent**
> To write a decimal as a percent, multiply by 100 and attach a % symbol.

> **NOTE**
> A quick way to *multiply* a number by 100 is to move the decimal point *two* places to the *right*. Notice that this is the opposite of the shortcut for *dividing* by 100.
>
> Drop % symbol; *divide* by 100.
> (Move decimal point two places *left*.)
>
> **Decimal** ⟷ **Percent**
>
> *Multiply* by 100; attach % symbol.
> (Move decimal point two places *right*.)

EXAMPLE 4 Changing Decimals to Percents by Moving the Decimal Point

Write each decimal as a percent.

(a) 0.21

   0.21
     ↑
     └── Decimal point is moved two places to the right.

   0.21 = 21% ⟵ Percent symbol is attached after decimal point is moved.
     ↑
     └── Decimal point is not written with whole number percents.

(b) 0.529 = 52.9% ⟵ Percent symbol is attached after decimal point is moved.

(c) 1.92 = 192% 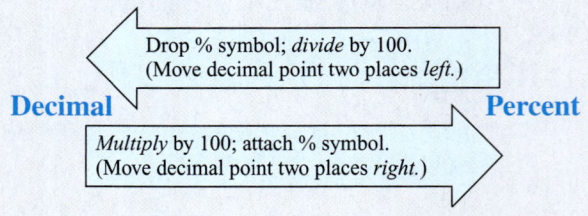 ⟵ Percent symbol is attached after decimal point is moved.

Continued on Next Page

(d) 2.5

$$2.5\underset{\curvearrowright}{0}$$ 0 is attached so the decimal point can be moved two places to the right.

$$2.5 = 250\%$$

(e) 3

$$3. = 3.\underset{\curvearrowright}{00} \quad \text{so} \quad 3 = 300\%$$

CAUTION
In Examples 4(c), 4(d), and 4(e) above, notice that 1.92, 2.5, and 3 are greater than 1. Because the number 1 is equivalent to 100%, all *numbers greater than 1* will be equivalent to *percents greater than 100%*.

> Work Problem 4 at the Side.

OBJECTIVE 4 Write percents as fractions. Percents can also be written as fractions. Recall that a percent is a ratio with a denominator of 100. For example, 89% is $\frac{89}{100}$. Because the fraction bar indicates division, we are dividing the percent by 100, just as we did when writing a percent as a decimal.

Writing a Percent as a Fraction
To write a percent as a fraction, drop the % symbol and write the number over 100. Then write the fraction in lowest terms.

EXAMPLE 5 Writing Percents as Fractions

Write each percent as a fraction or mixed number in lowest terms.

(a) 25% Drop the % symbol and write 25 over 100.

$$25\% = \frac{25}{100} \leftarrow \text{25 per 100}$$

$$= \frac{25 \div 25}{100 \div 25} = \frac{1}{4} \leftarrow \text{Lowest terms}$$

As a check, write 25% as a decimal.

$$25\% = 25 \div 100 = 0.25 \leftarrow \text{Percent sign dropped}$$

Recall that 0.25 means 25 hundredths.

$$0.25 = \frac{25}{100} = \frac{25 \div 25}{100 \div 25} = \frac{1}{4} \leftarrow \text{Same result as above}$$

(b) 76% Drop the % symbol and write 76 over 100.

$$76\% = \frac{76}{100} \begin{array}{l}\leftarrow \text{The percent becomes the numerator.} \\ \leftarrow \text{The } denominator \text{ is always 100 because percent means parts per 100}\end{array}$$

Write $\frac{76}{100}$ in lowest terms. $\frac{76 \div 4}{100 \div 4} = \frac{19}{25} \leftarrow \text{Lowest terms}$

Continued on Next Page

---

4 Write each number as a percent.

(a) 0.95

(b) 0.18

(c) 0.09

(d) 0.617

(e) 0.4

(f) 5.34

(g) 2.8

(h) 4

ANSWERS
4. (a) 95% (b) 18% (c) 9% (d) 61.7%
   (e) 40% (f) 534% (g) 280%
   (h) 400%

5 Write each percent as a fraction or mixed number in lowest terms.

(a) 50%

(b) 19%

(c) 80%

(d) 6%

(e) 125%

(f) 300%

(c) 150%

$$150\% = \frac{150}{100} = \frac{150 \div 50}{100 \div 50} = \frac{3}{2} = 1\frac{1}{2} \leftarrow \text{Mixed number}$$

(d) $100\% = \frac{100}{100} = 1 \leftarrow$ Whole number

NOTE
Remember that percent means *per 100.*

◀◀ **Work Problem 5 at the Side.**

Example 6 below shows how to write decimal percents and fraction percents as fractions.

EXAMPLE 6 Writing Decimal Percents or Fraction Percents as Fractions

Write each percent as a fraction in lowest terms.

(a) 15.5%

Drop the % symbol and write 15.5 over 100.

$$15.5\% = \frac{15.5}{100}$$

To get a whole number in the numerator, multiply the numerator and denominator by 10. (Multiplying by $\frac{10}{10}$ is the same as multiplying by 1.)

$$\frac{15.5}{100} = \frac{(15.5)(10)}{(100)(10)} = \frac{155}{1000}$$

Write the fraction in lowest terms.

$$\frac{155 \div 5}{1000 \div 5} = \frac{31}{200} \leftarrow \text{Lowest terms}$$

(b) $33\frac{1}{3}\%$

Drop the % symbol and write $33\frac{1}{3}$ over 100.

$$33\frac{1}{3}\% = \frac{33\frac{1}{3}}{100}$$

When there is a mixed number in the numerator, write it as an improper fraction. So $33\frac{1}{3}$ is $\frac{100}{3}$.

$$\frac{33\frac{1}{3}}{100} = \frac{\frac{100}{3}}{100}$$

Continued on Next Page

ANSWERS

5. (a) $\frac{1}{2}$ (b) $\frac{19}{100}$ (c) $\frac{4}{5}$ (d) $\frac{3}{50}$

(e) $1\frac{1}{4}$ (f) 3

Now you have a complex fraction (see **Section 4.6**). Rewrite the complex fraction using the ÷ symbol for division. Then follow the steps for dividing fractions.

$$\frac{\frac{100}{3}}{100} = \frac{100}{3} \div 100 = \frac{100}{3} \div \frac{100}{1} = \frac{100}{3} \cdot \frac{1}{100} = \frac{\overset{1}{\cancel{100}} \cdot 1}{3 \cdot \cancel{100}} = \frac{1}{3}$$

↑ Reciprocals ↑

> **NOTE**
> In Example 6(a) on the previous page, we could have changed 15.5% to $15\frac{1}{2}$% and then written it as the improper fraction $\frac{31}{2}$ over 100. But it is usually easier to work with decimal percents as they are.

▶ **Work Problem 6 at the Side.** ▶▶▶

OBJECTIVE 5 Write fractions as percents. Recall that to write a percent as a fraction, you *drop* the percent symbol and *divide* by 100. So, to reverse the process and change a fraction to a percent, you *multiply* by 100 and *attach* a percent symbol.

> **Writing a Fraction as a Percent**
> To write a fraction as a percent, multiply by 100 and attach a % symbol. This is the same as multiplying by 100%.

> **NOTE**
> Look back at Example 5(d) near the top of the previous page to see that 100% = 1. Recall that multiplying a number by 1 does *not* change the value of the number. So multiplying by 100% does not change the value of a number; it just gives us an *equivalent percent*.

EXAMPLE 7 Writing Fractions as Percents

Write each fraction as a percent. Round to the nearest tenth of a percent, if necessary.

(a) $\frac{2}{5}$ Multiply $\frac{2}{5}$ by 100%.

$$\frac{2}{5} = \left(\frac{2}{5}\right)(100\%) = \left(\frac{2}{5}\right)\left(\frac{100}{1}\%\right) = \left(\frac{2}{5}\right)\left(\frac{5 \cdot 20}{1}\%\right) = \frac{2 \cdot \cancel{5} \cdot 20}{\cancel{5} \cdot 1}\%$$

$$= \frac{40}{1}\% = 40\%$$

To check the result, write 40% as $\frac{40}{100}$ and reduce to lowest terms.

$$40\% = \frac{40}{100} = \frac{40 \div 20}{100 \div 20} = \frac{2}{5} \quad \leftarrow \text{Original fraction}$$

Continued on Next Page

Section 7.1 The Basics of Percent **499**

6 Write each percent as a fraction in lowest terms.

(a) 18.5%

(b) 87.5%

(c) 6.5%

(d) $66\frac{2}{3}$%

(e) $12\frac{1}{3}$%

(f) $62\frac{1}{2}$%

ANSWERS

6. (a) $\frac{37}{200}$ (b) $\frac{7}{8}$ (c) $\frac{13}{200}$ (d) $\frac{2}{3}$
   (e) $\frac{37}{300}$ (f) $\frac{5}{8}$

7 Write each fraction as a percent. If you're using a calculator, first work each one by hand. Then use your calculator and round to the nearest tenth of a percent if necessary.

(a) $\dfrac{1}{2}$

(b) $\dfrac{3}{4}$

(c) $\dfrac{1}{10}$

(d) $\dfrac{7}{8}$

(e) $\dfrac{5}{6}$

(f) $\dfrac{2}{3}$

ANSWERS

7. (a) 50% (b) 75% (c) 10%
   (d) $87\dfrac{1}{2}$% or 87.5% (Both are exact answers.)
   (e) exactly $83\dfrac{1}{3}$%, or 83.3% (rounded)
   (f) exactly $66\dfrac{2}{3}$%, or 66.7% (rounded)

(b) $\dfrac{5}{8}$ Multiply $\dfrac{5}{8}$ by 100%.

$$\dfrac{5}{8} = \left(\dfrac{5}{8}\right)(100\%) = \left(\dfrac{5}{8}\right)\left(\dfrac{100}{1}\%\right) = \left(\dfrac{5}{2 \cdot 4}\right)\left(\dfrac{4 \cdot 25}{1}\%\right) = \dfrac{5 \cdot \overset{1}{\cancel{4}} \cdot 25}{2 \cdot \underset{1}{\cancel{4}} \cdot 1}\%$$

$$= \dfrac{125}{2}\% = 62\dfrac{1}{2}\%$$

You can also do the last step of simplifying $\dfrac{125}{2}$ on your calculator. Enter $\dfrac{125}{2}$ as 125 ÷ 2 =. The result is 62.5, so $\dfrac{5}{8} = 62\dfrac{1}{2}\%$ or $\dfrac{5}{8} = 62.5\%$.

(c) $\dfrac{1}{6}$ Multiply $\dfrac{1}{6}$ by 100%.

$$\dfrac{1}{6} = \left(\dfrac{1}{6}\right)(100\%) = \left(\dfrac{1}{6}\right)\left(\dfrac{100}{1}\%\right) = \left(\dfrac{1}{2 \cdot 3}\right)\left(\dfrac{2 \cdot 50}{1}\%\right) = \dfrac{1 \cdot \overset{1}{\cancel{2}} \cdot 50}{\underset{1}{\cancel{2}} \cdot 3 \cdot 1}\%$$

$$= \dfrac{50}{3}\% = 16\dfrac{2}{3}\%$$

To simplify $\dfrac{50}{3}$ on your calculator, enter 50 ÷ 3 =. The result is 16.66666666, with the 6 continuing to repeat. The directions say to round to the nearest *tenth* of a percent.

Next digit is *5 or more*.

16.6**6**666666%   rounds to   16.**7**%
↑
Tenths place

So $\dfrac{1}{6} = 16\dfrac{2}{3}\%$ or $\dfrac{1}{6} \approx 16.7\%$.

> **Calculator Tip**   In Example 7(a) on the previous page, you can use your calculator to write $\dfrac{2}{5}$ as a percent.
>
> *Step 1*   Enter $\dfrac{2}{5}$ as 2 ÷ 5 =. Your calculator shows **0.4**
> ↑
> Decimal equivalent of $\dfrac{2}{5}$
>
> *Step 2*   Change the decimal number 0.4 to a percent by moving the decimal point two places to the right (multiplying by 100).
>
> 0.40 = 40%   ← Attach % symbol.
>
> Try this technique on Examples 7(b) and 7(c) on this page.
>
> For $\dfrac{5}{8}$, enter 5 ÷ 8 =. Your calculator shows **0.625**
>
> Move the decimal point in 0.625 two places to the right.
>
> 0.625 = 62.5%   ← Attach % symbol.

Continued on Next Page

For $\frac{1}{6}$, enter 1 ⊕ 6 ⊜. Your calculator shows 0.166666666

Some calculators show 7 in the last place.

Move the decimal point two places to the right. Then round to the nearest tenth.

$$0.166666666 = 16.66666666\% \approx 16.7\% \leftarrow \text{Attach \% symbol.}$$

With a fraction such as $\frac{1}{6}$, your calculator gives only an *approximate* answer. You can't get the exact answer of $16\frac{2}{3}\%$ by using your calculator.

> **Work Problem 7 on the Previous Page at the Side.**

OBJECTIVE 6 Use 100% and 50%. When working with percents, it is helpful to have several reference points. 100% and 50% are two helpful reference points.

100% means 100 parts out of 100 parts. That's *all* of the parts. If you pay 100% of a $245 dentist bill, you pay $245 (*all* of it).

EXAMPLE 8 Finding 100% of a Number

Fill in the blanks.

(a) 100% of $34 is _____.

100% is *all* of the money.

So 100% of $34 is $34.

(b) 100% of 4 miles is _____.

100% is *all* of the miles.

So 100% of 4 miles is 4 miles.

> **Work Problem 8 at the Side.**

50% means 50 parts out of 100 parts, which is *half* of the parts because $\frac{50}{100} = \frac{1}{2}$. So 50% of $12 is $6 (*half* of the money).

EXAMPLE 9 Finding 50% of a Number

Fill in the blanks.

(a) 50% of $20 is _____.

50% is *half* of the money.

So 50% of $20 is $10.

(b) 50% of 280 miles is _____.

50% is *half* of the miles.

So 50% of 280 miles is 140 miles.

> **Work Problem 9 at the Side.**

---

8 Fill in the blanks.

(a) 100% of $3.95 is _____.

(b) 100% of 3000 students is _____.

(c) 100% of 7 pages is _____.

(d) 100% of 305 miles is _____.

(e) 100% of $10\frac{1}{2}$ hours is _____.

9 Fill in the blanks.

(a) 50% of $10 is _____.

(b) 50% of 36 cookies is _____.

(c) 50% of 6000 women is _____.

(d) 50% of 8 hours is _____.

(e) 50% of $2.50 is _____.

ANSWERS

8. (a) $3.95 (b) 3000 students (c) 7 pages
   (d) 305 miles (e) $10\frac{1}{2}$ hours

9. (a) $5 (b) 18 cookies (c) 3000 women
   (d) 4 hours (e) $1.25

Focus on Real-Data Applications

Decimals, Percents, and Quilt Patterns

Please complete the Quilt Pattern activity in Section 4.8 before starting this activity.

 Four of the quilt patterns you worked with in Section 4.8 are repeated here. For each pattern, complete the table by copying the fractions you found and then changing the fractions to decimals and to percents. Use your calculator and round decimals to the nearest thousandth and percents to the nearest tenth when necessary. In the last row of the table, add the fractions. Then add the decimals, and finally, add the percents.

1.

Barbara Fritchie Star

Color	Fraction	Decimal	Percent
Blue			
White			
Total			

2.

Handy Andy

Color	Fraction	Decimal	Percent
Blue			
White			
Total			

3.

Peace and Plenty

Color	Fraction	Decimal	Percent
Blue			
Yellow			
Whitish			
Total			

4.

Cobwebs

Color	Fraction	Decimal	Percent
Mauve			
Purple			
Yellow			
Total			

5. What pattern do you notice in the Total row in each table?

6. Explain why some of the totals in the Cobwebs' table don't quite fit the pattern.

7.1 Exercises

Write each percent as a decimal. See Examples 2 and 3.

1. 25%
2. 35%
3. 30%
4. 20%

5. 6%
6. 3%
7. 140%
8. 250%

9. 7.8%
10. 6.7%
11. 100%
12. 600%

13. 0.5%
14. 0.2%
15. 0.35%
16. 0.076%

Write each decimal as a percent. See Example 4.

17. 0.5
18. 0.6
19. 0.62
20. 0.18

21. 0.03
22. 0.09
23. 0.125
24. 0.875

25. 0.629
26. 0.494
27. 2
28. 5

29. 2.6
30. 1.8
31. 0.0312
32. 0.0625

Write each percent as a fraction or mixed number in lowest terms. See Examples 5 and 6.

33. 20%
34. 40%
35. 50%
36. 75%

37. 55%
38. 35%
39. 37.5%
40. 87.5%

41. 6.25%
42. 43.75%
43. $16\frac{2}{3}$%
44. $83\frac{1}{3}$%

45. 130%
46. 175%
47. 250%
48. 325%

Write each fraction as a percent. If you're using a calculator, first work each one by hand. Then use your calculator and round to the nearest tenth of a percent, if necessary. See Example 7.

49. $\dfrac{1}{4}$ 50. $\dfrac{1}{5}$ 51. $\dfrac{3}{10}$ 52. $\dfrac{9}{10}$

53. $\dfrac{3}{5}$ 54. $\dfrac{3}{4}$ 55. $\dfrac{37}{100}$ 56. $\dfrac{63}{100}$

57. $\dfrac{3}{8}$ 58. $\dfrac{1}{8}$ 59. $\dfrac{1}{20}$ 60. $\dfrac{1}{50}$

61. $\dfrac{5}{9}$ 62. $\dfrac{7}{9}$ 63. $\dfrac{1}{7}$ 64. $\dfrac{5}{7}$

In each statement, write percents as decimals and decimals as percents. See Examples 2–4.

65. In 1900, only 8% of U.S. homes had a telephone. (*Source: Harper's Index.*)

66. In 1900, only 14% of homes in the United States had a bathtub. (*Source: Harper's Index.*)

67. Tornadoes can occur on any day of the year, but 42% of them appear in May and June. (*Source:* National Severe Storms Laboratory.)

68. Only 2.1% of tornadoes occur in December, making it the least likely month for twisters. (*Source:* National Severe Storms Laboratory.)

69. The property tax rate in Alpine County is 0.035.

70. A church building fund has 0.49 of the money needed.

71. The number of people taking CPR training this session is 2 times that of the last session.

72. Attendance at this year's company picnic is 3 times last year's attendance.

Write a fraction and a percent for the shaded part of each figure. Then write a fraction and a percent for the unshaded part of each figure.

73.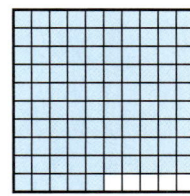

74.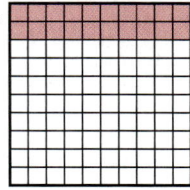

75.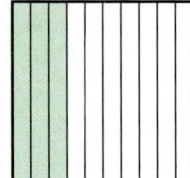

76.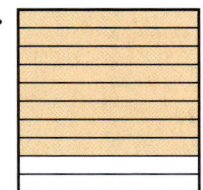

77.

78.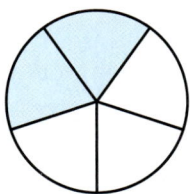

Complete this table. Write fractions and mixed numbers in lowest terms. See Examples 2–7.

	Fraction	Decimal	Percent
79.	$\frac{1}{100}$	_____	_____
80.	$\frac{1}{10}$	_____	_____
81.	_____	0.2	_____
82.	_____	0.25	_____
83.	_____	_____	30%
84.	_____	_____	40%
85.	$\frac{1}{2}$	_____	_____
86.	$\frac{3}{4}$	_____	_____
87.	_____	_____	90%
88.	_____	_____	100%
89.	_____	1.5	_____
90.	_____	2.25	_____

This bar graph shows recent trends in employee benefits. It includes the most frequently offered "work perks" and the percent of the responding companies offering them. The survey was conducted on-line and included 4800 companies ranging in size from 2 to 5000 employees. Use this graph to answer Exercises 91–94. If the answer is a percent, also write it as a fraction in lowest terms. (Source: Work Perks Survey, Ceridian Employer Services.)

WORK PERKS

- 82% Casual dress
- 60.5% Flexible hours
- 49% Personal development training
- 40% Employee entertainment/Company product discounts
- 36% Free food/Beverages
- 27% Telecommuting
- 15% Fitness centers
- 1% On-site child care

Percent of Companies / Type of Perk

91. (a) What was the fourth-most-common perk?

(b) What portion of the companies offer that perk?

92. (a) What was the top work perk?

(b) What portion of the companies offer that perk?

93. (a) Which perk was offered by *about* $\frac{1}{2}$ of the companies?

(b) Which perk was offered by *about* $\frac{1}{4}$ of the companies?

94. (a) If free food/beverages had been offered by $33\frac{1}{3}$% instead of 36% of the companies, what fraction would that be?

(b) What percent of the companies would have to offer flexible hours for the fraction to be $\frac{2}{3}$?

The diagram shows the different types of teeth in an adult's mouth. Use the diagram to answer Exercises 95–100. Write each answer in Exercises 95–98 as a fraction in lowest terms, as a decimal, and as a percent.

95. What portion of an adult's teeth are incisors, designed to bite and cut?

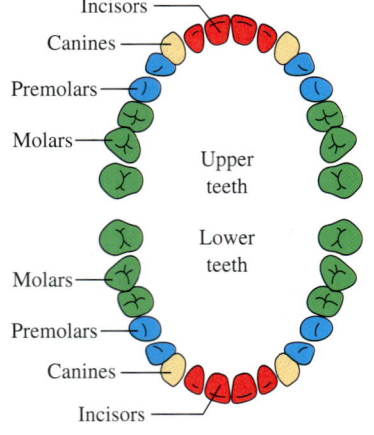

Source: Time-Life *Human Body.*

96. The pointy canine teeth tear and rip food. They are what portion of an adult's teeth?

97. The molars, which grind up food, are what portion of an adult's teeth?

98. What portion of an adult's teeth are premolars?

99. Some people have four fewer molars than shown in the tooth diagram on the previous page. For these people, the canines are what portion of their teeth? Write your answer as a fraction in lowest terms and as a percent.

100. For the adults who have four fewer molars than shown, the incisors are what portion of their teeth? Write your answer as a fraction in lowest terms and as a percent.

101. Explain and correct the errors made by students when they used their calculators on these problems.
  (a) Write $\frac{7}{20}$ as a percent.
  Student entered 7 ÷ 20 =, and the result was 0.35. So $\frac{7}{20} = 0.35\%$
  (b) Write $\frac{16}{25}$ as a percent.
  Student entered 25 ÷ 16 =, and the result was 1.5625. So $\frac{16}{25} = 156.25\%$

102. Explain and correct the errors made by students who moved decimal points to solve these problems.
  (a) Write 3.2 as a percent.
  $03.2 = 0.032$ so $3.2 = 0.032\%$
  (b) Write 60% as a decimal.
  $00.60 = 0.0060$ so $60\% = 0.0060$

Fill in the blanks in Exercises 103–116. Remember that 100% *is* all *of something and* 50% *is* half *of it. See Examples 8 and 9.*

103. (a) 100% of $78 is _____.

(b) 50% of $78 is _____.

104. (a) 100% of 5 hours is _____.

(b) 50% of 5 hours is _____.

105. (a) 100% of 15 inches is _____.

(b) 50% of 15 inches is _____.

106. (a) 100% of $6000 is _____.

(b) 50% of $6000 is _____.

107. There are 20 children in the preschool class. 100% of the children are served breakfast and lunch. How many children are served both meals?
_____.

108. The Speedy Delivery company owns 345 vans. 100% of the vans are painted white with blue lettering. How many company vans are painted white with blue lettering?
_____.

109. (a) 100% of 2.8 miles is _____.

(b) 50% of 2.8 miles is _____.

110. (a) 100% of $2.50 is _____.

(b) 50% of $2.50 is _____.

111. **(a)** John owes $285 for tuition. Financial aid will pay 50% of the cost. Financial aid will pay _____.

(b) What percent of the tuition will John have to pay?
_____

112. **(a)** The Animal Humane Society took in 20,000 animals last year. About 50% of them were dogs. The number of dogs taken in was about how many?
_____

(b) What percent of the animals were *not* dogs?
_____

113. **(a)** About 50% of the 8200 students at our college work more than 20 hours per week. How many students is this?
_____

(b) What percent of the students work 20 hours or less per week?
_____

114. **(a)** Shalayna's cell phone plan includes 1600 minutes of "off-peak" calling time per month. Last month she used 50% of her "off-peak" minutes. How many minutes did she use?
_____

(b) What percent of her "off-peak" minutes were not used?
_____

115. **(a)** Dylan's test score was 100%. How many of the 35 problems did he work correctly?
_____

(b) How many problems did Dylan miss?
_____

116. **(a)** Latrell made 100% of his 12 free throws during a practice game today. How many free throws did he make?
_____

(b) How many free throws did Latrell miss?
_____

117. Describe a shortcut way to find 50% of a number. Include two examples to illustrate the shortcut.

118. Describe a shortcut way to find 100% of a number. Include two examples to illustrate the shortcut.

7.2 The Percent Proportion

We will show you two ways to solve percent problems. One is the proportion method, which we discuss in this section. The other is the percent equation method, which we explain in **Section 7.3**.

OBJECTIVES

1. Identify the percent, whole, and part.
2. Solve percent problems using the percent proportion.

OBJECTIVE 1 Identify the percent, whole, and part. You have learned that a statement of two equivalent ratios is called a proportion (see **Section 6.3**). For example, the fraction $\frac{3}{5}$ is the same as the ratio 3 to 5, and 60% is the ratio 60 to 100. As the figure below shows, these two ratios are equivalent and make a proportion.

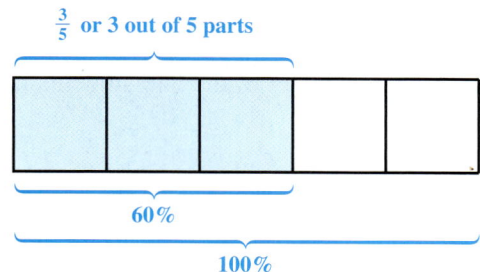

The **percent proportion** can be used to solve percent problems.

The Percent Proportion
Percent is to *100* **as** *part* is to *whole*.

$$\frac{\text{percent}}{100} = \frac{\text{part}}{\text{whole}}$$

Always 100 because percent means "per 100"

In some textbooks the percent proportion is written using the words *amount* and *base*.

$$\frac{\text{percent}}{100} = \frac{\text{amount}}{\text{base}}$$

Here is the proportion for the figure at the top of the page.

60% means 60 parts out of 100 parts. $\dfrac{60}{100} = \dfrac{3}{5}$ ← Shaded (3 parts)
← Whole (5 parts)

If we write $\frac{60}{100}$ in lowest terms, it is equal to $\frac{3}{5}$, so the proportion is true.

$$\frac{60}{100} = \frac{60 \div 20}{100 \div 20} = \frac{3}{5} \quad \leftarrow \text{Matches ratio on right side of proportion}$$

In order to use the percent proportion to solve problems, you must be able to pick out the *percent*, the *whole*, and the *part*. Look for the percent first, because it is the easiest to identify.

Identifying the Percent
The **percent** is a ratio of a part to a whole, with 100 as the denominator. In a problem, the percent appears with the word *percent* or with the symbol **%** after it.

1 Identify the percent.

(a) Of the $2000, 15% will be spent on a washing machine.

(b) 60 employees is what percent of 750 employees?

(c) The state sales tax is $6\frac{1}{2}$ percent of the $590 price.

(d) $30 is 150% of what amount of money?

(e) 75 of the 110 rental cars were rented today. What percent were rented?

2 Identify the whole.

(a) Of the $2000, 15% will be spent on a washing machine.

(b) 60 employees is what percent of 750 employees?

(c) The state sales tax is $6\frac{1}{2}$ percent of the $590 price.

(d) $30 is 150% of what amount of money?

(e) 75 of the 110 rental cars were rented today. What percent were rented?

ANSWERS

1. (a) 15 (b) unknown (c) $6\frac{1}{2}$
   (d) 150 (e) unknown
2. (a) $2000 (b) 750 (c) $590
   (d) unknown (e) 110

EXAMPLE 1   Identifying the Percent in Percent Problems

Identify the percent in each problem.

(a) 32% of the 900 women were retired. How many were retired?
   ↓
   Percent

The percent is 32. The number 32 appears with the symbol %.

(b) $150 is 25 percent of what number?
   ↓
   Percent

The percent is 25 because 25 appears with the word *percent*.

(c) If 7 students failed, what percent of the 350 students failed?
   ↓
   Percent (unknown)

The word *percent* has no number with it, so the percent is the unknown part of the problem.

Work Problem 1 at the Side.

The second thing to look for in a percent problem is the *whole* (sometimes called the *base*).

Identifying the Whole

The **whole** is the entire quantity. In a problem, the *whole* often appears after the word **of**.

EXAMPLE 2   Identifying the Whole in Percent Problems

These problems are the same as those in Example 1 above. Now identify the *whole*.

(a) 32% **of** the 900 women were retired. How many were retired?
    ↓   ↓
   Percent   Whole

The whole is 900. The number 900 appears after the word *of*.

(b) $150 is 25 percent **of** what number?
    ↓   ↓
   Percent   Whole (unknown; follows *of*)

(c) If 7 students failed, what percent **of** the 350 students failed?
                          ↓   ↓
                   Percent (unknown)   Whole (follows *of*)

Work Problem 2 at the Side.

The third and final thing to identify in a percent problem is the *part* (sometimes called the *amount*).

Identifying the Part

The **part** is the number being compared to the whole.

Section 7.2 The Percent Proportion   **511**

> **NOTE**
> If you have trouble identifying the *part*, find the *whole* and the *percent* first. The remaining number is the *part*.

EXAMPLE 3 Identifying the Part in Percent Problems

These problems are the same as those in Examples 1 and 2 on the previous page. Identify the *part*.

(a) 32% **of** the 900 women were retired. How many were retired?

   Percent — Whole — Part (unknown)

The part of the women who were retired is unknown. In other words, some part of 900 women were retired.

(b) $150 is 25 percent **of** what number?

   Part — Percent — Whole (unknown)

150 is the remaining number, so 150 is the part.

(c) If 7 students failed, what percent **of** 350 students failed?

   Part — Percent (unknown) — Whole

The part of the students who failed is 7 students.

Work Problem 3 at the Side.

OBJECTIVE 2 Solve percent problems using the percent proportion.

EXAMPLE 4 Using the Percent Proportion to Find the Part

Use the percent proportion to answer this question.

   15% of $165 is how much money?

Recall that the percent proportion is $\frac{\text{percent}}{100} = \frac{\text{part}}{\text{whole}}$. First identify the percent by looking for the % symbol or the word *percent*. Then look for the *whole* (usually follows the word *of*). Finally, identify the *part*.

   15% **of** $165 is how much money?

   Percent — Whole (follows *of*) — Part (unknown)

Set up the percent proportion. Here we use n as the variable representing the unknown part. You may use any letter you like.

$$\text{Percent} \rightarrow \frac{15}{100} = \frac{n}{165} \begin{matrix}\leftarrow \text{Part (unknown)} \\ \leftarrow \text{Whole}\end{matrix}$$
$$\text{Always 100} \rightarrow$$

Recall from **Section 6.3** that the first step in solving a proportion is to find the cross products.

Step 1 $\frac{15}{100} = \frac{n}{165}$ $\begin{matrix}100 \cdot n \\ 15 \cdot 165\end{matrix}$ Find the cross products.

Continued on Next Page

❸ Identify the part.

(a) Of the $2000, 15% will be spent on a washing machine.

(b) 60 employees is what percent of 750 employees?

(c) The state sales tax is $6\frac{1}{2}$ percent of the $590 price.

(d) $30 is 150% of what amount of money?

(e) 75 of the 110 rental cars were rented today. What percent were rented?

ANSWERS
3. (a) unknown (b) 60 (c) unknown
   (d) $30 (e) 75

512 Chapter 7 Percent

4 Use the percent proportion to answer these questions.

(a) 9% of 3250 miles is how many miles?

(b) What is 20% of 180 calories?

(c) 78% of $5.50 is how much?

(d) What is $12\frac{1}{2}$% of 400 homes? (*Hint:* Write $12\frac{1}{2}$% as 12.5%.)

5 Use the percent proportion to answer these questions.

(a) 1200 books is what percent of 5000 books?

(b) What percent of $6.50 is $0.52?

(c) 20 athletes is what percent of 32 athletes?

ANSWERS

4. (a) $\frac{9}{100} = \frac{n}{3250}$; The part is 292.5 miles.
   (b) $\frac{20}{100} = \frac{n}{180}$; The part is 36 calories.
   (c) $\frac{78}{100} = \frac{n}{5.50}$; The part is $4.29.
   (d) $\frac{12.5}{100} = \frac{n}{400}$; The part is 50 homes.

5. (a) $\frac{p}{100} = \frac{1200}{5000}$; 24%
   (b) $\frac{p}{100} = \frac{0.52}{6.50}$; 8%
   (c) $\frac{p}{100} = \frac{20}{32}$; 62.5% or $62\frac{1}{2}$%

Step 2 $100 \cdot n = 15 \cdot 165$ Show that the cross products are equivalent.
$100 \cdot n = 2475$

Step 3 $\dfrac{\overset{1}{\cancel{100}} \cdot n}{\underset{1}{\cancel{100}}} = \dfrac{2475}{100}$ Divide both sides by 100, the coefficient of the variable term. On the left side, divide out the common factor of 100.

$n = 24.75$ On the right side, $2475 \div 100$ is 24.75.

The part is **$24.75**, so 15% of $165 is $24.75.

> **CAUTION**
> When you use the percent proportion, do *not* move the decimal point in the percent or in the answer.

◀ **Work Problem 4 at the Side.**

EXAMPLE 5 Using the Percent Proportion to Find the Percent

Use the percent proportion to answer this question.

8 pounds is what percent of 160 pounds?
- Part
- Percent (unknown)
- Whole (follows *of*)

The percent proportion is $\dfrac{\text{percent}}{100} = \dfrac{\text{part}}{\text{whole}}$. Set up the proportion using p as the variable representing the unknown percent. Then find the cross products.

Percent (unknown) → $\dfrac{p}{100} = \dfrac{8}{160}$ ← Part
Always 100 → ← Whole

$\dfrac{p}{100} = \dfrac{8}{160}$ $100 \cdot 8 = 800$ Cross products
   $p \cdot 160$

$p \cdot 160 = 800$ Show that the cross products are equivalent.

$\dfrac{p \cdot \cancel{160}}{\cancel{160}} = \dfrac{800}{160}$ Divide both sides by 160

$p = 5$

The percent is **5%**. So 8 pounds is 5% of 160 pounds.

> **CAUTION**
> When you're finding an unknown percent, as in Example 5 above, be careful to label your answer with the % symbol. Do *not* add a decimal point or move the decimal point in your answer.

◀ **Work Problem 5 at the Side.**

EXAMPLE 6 Using the Percent Proportion to Find the Whole

Use the percent proportion to answer this question.

162 credits is 90% of how many credits?
- 162 → Part
- 90% → Percent
- how many credits → Whole (unknown); (follows *of*)

$$\text{Percent} \rightarrow \frac{90}{100} = \frac{162}{n} \leftarrow \text{Part}$$
$$\text{Always 100} \rightarrow \qquad\qquad \leftarrow \text{Whole (unknown)}$$

$$\frac{90}{100} = \frac{162}{n} \qquad \begin{array}{l} 100 \cdot 162 = 16{,}200 \\ 90 \cdot n \end{array} \; \text{Cross products}$$

$90 \cdot n = 16{,}200$ Show that the cross products are equivalent.

$$\frac{\cancel{90} \cdot n}{\cancel{90}} = \frac{16{,}200}{90}$$ Divide both sides by 90.

$n = 180$

The whole is **180 credits**. So 162 credits is 90% of 180 credits.

▶▶▶ **Work Problem 6 at the Side.**

So far in all the examples, the part has been *smaller* than the whole. This is because all the percents have been less than 100%. Recall that 100% of something is *all* of it. When the percent is *less* than 100%, you have *less* than all of it.

Now let's look at percents *greater* than 100%. For example,

100% of $20 is all of the money, or $20.

150% of $20 is *more* than $20.

100%	+	50%	=	150%
of the money is		of the money is		of the money is
$20	+	$10	=	$30

So 150% of $20 is $30.

When the percent is *greater* than 100%, the part is *larger* than the whole, as you'll see in Example 7 on the next page.

---

6 Use the percent proportion to answer these questions.

(a) 37 cars is 74% of how many cars?

(b) 45% of how much money is $139.59?

(c) 1.2 tons is $2\frac{1}{2}$% of how many tons?

ANSWERS

6. (a) $\frac{74}{100} = \frac{37}{n}$; 50 cars

(b) $\frac{45}{100} = \frac{139.59}{n}$; $310.20

(c) $\frac{2.5}{100} = \frac{1.2}{n}$; 48 tons

Chapter 7 Percent

7 Use the percent proportion to answer each question.

(a) 350% of $6 is how much?

(b) 23 hours is what percent of 20 hours?

(c) What percent of $47.32 is $106.47?

ANSWERS

7. (a) $\dfrac{350}{100} = \dfrac{n}{6}$; $21

(b) $\dfrac{p}{100} = \dfrac{23}{20}$; 115%

(c) $\dfrac{p}{100} = \dfrac{106.47}{47.32}$; 225%

EXAMPLE 7 Working with Percents Greater Than 100%

Use the percent proportion to answer each question.

(a) How many students is 210% of 40 students?

- Part (unknown)
- Percent
- Whole (follows *of*)

Percent → $\dfrac{210}{100} = \dfrac{n}{40}$ ← Part (unknown)
Always 100 → ← Whole

$$\dfrac{210}{100} = \dfrac{n}{40}$$

Cross products:
$100 \cdot n$
$210 \cdot 40 = 8400$

$100 \cdot n = 8400$ Show that the cross products are equivalent.

$\dfrac{\overset{1}{\cancel{100}} \cdot n}{\underset{1}{\cancel{100}}} = \dfrac{8400}{100}$ Divide both sides by 100

$n = 84$

The part is **84 students**, which is *more* than the whole of 40 students. This result makes sense because the percent is 210%. If it was exactly 200%, we would have *2 times the whole,* and 2 times 40 students is 80 students. So 210% should be even a little more than 80 students. Our answer of 84 students is reasonable.

(b) $68 is what percent *of* $50?

- Part
- Percent (unknown)
- Whole (follows *of*)

Percent (unknown) → $\dfrac{p}{100} = \dfrac{68}{50}$ ← Part
Always 100 → ← Whole

$$\dfrac{p}{100} = \dfrac{68}{50}$$

Cross products:
$100 \cdot 68 = 6800$
$p \cdot 50$

$p \cdot 50 = 6800$ Show that the cross products are equivalent.

$\dfrac{p \cdot \overset{1}{\cancel{50}}}{\underset{1}{\cancel{50}}} = \dfrac{6800}{50}$ Divide both sides by 50

$p = 136$

The percent is **136%**. This result makes sense because $68 is *more* than $50, so $68 has to be *more than 100%* of $50.

Work Problem 7 at the Side.

7.2 Exercises

Write a percent proportion and solve it to answer Exercises 1–24. If necessary, round money answers to the nearest cent and percent answers to the nearest tenth of a percent. See Examples 1–7.

1. What is 10% of 3000 runners?

2. What is 35% of 2340 volunteers?

3. 4% of 120 feet is how many feet?

4. 9% of $150 is how much money?

5. 16 pepperoni pizzas is what percent of 32 pizzas?

6. 35 hours is what percent of 140 hours?

7. What percent of 200 calories is 16 calories?

8. What percent of 350 parking spaces is 7 handicapped parking spaces?

9. 495 successful students is 90% of what number of students?

10. 84 letters is 28% of what number of letters?

11. $12\frac{1}{2}$% of what amount is $3.50?

12. $5\frac{1}{2}$% of what amount is $17.60?

13. 250% of 7 hours is how long?

14. What is 130% of 60 trees?

15. What percent of $172 is $32?

16. $14 is what percent of $398?

17. 748 books is 110% of what number of books?

18. 145% of what number of inches is 11.6 inches?

19. What is 14.7% of $274?

20. 8.3% of $43 is how much?

21. 105 employees is what percent of 54 employees?

22. What percent of 46 animals is 100 animals?

23. $0.33 is 4% of what amount?

24. 6% of what amount is $0.03?

25. A student turned in the following answers on a test. You can see that *two* of the answers are incorrect *without working the problems*. Find the incorrect answers and explain how you identified them (without actually solving the problems).

    50% of $84 is ___$42___

    150% of $30 is ___$20___

    25% of $16 is ___$32___

    100% of $217 is ___$217___

26. Name the three parts in a percent problem. For each of these three parts, write a sentence telling how you identify it.

27. Explain and correct the *two* errors that a student made when solving this problem:
    $14 is what percent of $8?

    $$\frac{p}{100} = \frac{8}{14} \qquad p \cdot 14 = 100 \cdot 8$$

    $$\frac{p \cdot \cancel{14}}{\cancel{14}} = \frac{800}{14}$$

    $$p \approx 57.1$$

    The answer is 57.1 (rounded).

28. Explain and correct the *two* errors that a student made when solving this problem: 9 children is 30% of what number of children?

    $$\frac{30}{100} = \frac{n}{9} \qquad 100 \cdot n = 30 \cdot 9$$

    $$\frac{\cancel{100} \cdot n}{\cancel{100}} = \frac{270}{100}$$

    $$n = 2.7$$

    The answer is 2.7%.

7.3 The Percent Equation

OBJECTIVES
1. Estimate answers to percent problems involving 25%.
2. Find 10% and 1% of a number by moving the decimal point.
3. Solve basic percent problems using the percent equation.

OBJECTIVE 1 Estimate answers to percent problems involving 25%. Before showing you the percent equation, we need to do some more estimation. As you have learned when working with integers, fractions, and decimals, it is always a good idea to estimate the answer. Doing so helps you catch mistakes. Also, when you're out shopping or eating in a restaurant, you will be able to estimate the sales tax, discount, or tip.

In **Section 7.1** we used shortcuts for 100% of a number (all of the number) and 50% of a number (divide the number by 2). Now let's look at a quick way to work with 25%.

25% means 25 parts out of 100 parts, or $\frac{25}{100}$, which is the same as $\frac{1}{4}$

25% of $40 would be $\frac{1}{4}$ of $40, or $10

A quick way to find $\frac{1}{4}$ of a number is to *divide it by 4*. Recall that the denominator, 4, tells you that the whole is divided into 4 equal parts.

EXAMPLE 1 Estimating 25% of a Number

Estimate the answer to each question.

(a) What is 25% of $817?
Use front end rounding to round $817 to $800. Then divide $800 by 4. The estimate is **$200**.

(b) Find 25% of 19.7 miles.
Use front end rounding to round 19.7 miles to 20 miles. Then divide 20 miles by 4. The estimate is **5 miles**.

(c) 25% of 49 days is how long?
You could round 49 days to 50 days, using front end rounding. Then divide 50 by 4 to get an estimate of **12.5 days**.

However, the division step is simpler if you notice that 48 is a multiple of 4. You can round 49 days to 48 days and divide by 4 to get an estimate of **12 days**. Either way gives you a fairly good idea of the correct answer.

Work Problem 1 at the Side.

1 Estimate the answer to each question.

(a) What is 25% of $110.38?

(b) Find 25% of 7.6 hours.

(c) 25% of 34 pounds is how many pounds?

OBJECTIVE 2 Find 10% and 1% of a number by moving the decimal point. There are also helpful shortcuts for finding 10% or 1% of a number.

Ten percent, or 10%, means 10 parts out of 100 parts or $\frac{10}{100}$, which is the same as $\frac{1}{10}$. A quick way to find $\frac{1}{10}$ of a number is to *divide it by 10*. The denominator, 10, tells you that the whole is divided into 10 equal parts. The shortcut for dividing by 10 is to move the decimal point *one* place to the *left*.

EXAMPLE 2 Finding 10% of a Number by Moving the Decimal Point

Find the *exact* answer to each question by moving the decimal point.

(a) What is 10% of $817?
To find 10% of $817, divide $817 by 10. Do the division by moving the decimal point *one* place to the *left*. The decimal point starts at the far right side of $817.

10% of $817. = $81.70 ← Exact answer

Write this 0 because it's money.

So 10% of $817 is **$81.70**.

Continued on Next Page

ANSWERS
1. (a) $100 ÷ 4 gives an estimate of $25.
   (b) 8 hours ÷ 4 gives an estimate of 2 hours.
   (c) 30 pounds ÷ 4 gives an estimate of 7.5 pounds, or 32 pounds ÷ 4 gives an estimate of 8 pounds.

2 Find the *exact* answer to each question by moving the decimal point.

(a) What is 10% of $110.38?

(b) Find 10% of 7.6 hours.

(c) 10% of 34 pounds is how many pounds?

3 Find the *exact* answer to each question by moving the decimal point.

(a) What is 1% of $110.38?

(b) Find 1% of 7.6 hours.

(c) 1% of 34 pounds is how many pounds?

(b) Find 10% of 19.7 miles.
   To find 10% of 19.7 miles, divide 19.7 by 10. Move the decimal point *one* place to the *left*.

   10% of 19.7 miles = 1.97 miles ← Exact answer

So 10% of 19.7 miles is **1.97 miles**.

◀◀ **Work Problem 2 at the Side.**

One percent, or 1%, is 1 part out of 100 parts or $\frac{1}{100}$. This time the denominator of 100 tells you that the whole is divided into 100 parts. Recall that a quick way to divide by 100 is to move the decimal point *two* places to the *left*.

EXAMPLE 3 Finding 1% of a Number by Moving the Decimal Point

Find the *exact* answer to each question by moving the decimal point.

(a) What is 1% of $817?
   To find 1% of $817, divide $817 by 100. Do the division by moving the decimal point *two* places to the *left*.

   1% of $817. = $8.17 ← Exact answer

So 1% of $817 is **$8.17**.

(b) Find 1% of 19.7 miles.
   To find 1% of 19.7 miles, divide 19.7 by 100. Move the decimal point *two* places to the *left*.

   1% of 19.7 miles = 0.197 mile ← Exact answer

So 1% of 19.7 miles is **0.197 mile**.

◀◀ **Work Problem 3 at the Side.**

Here is a summary of the shortcuts you can use with percents.

Percent Shortcuts

200% of a number is 2 times the number; **300% of a number** is 3 times the number; and so on.

100% of a number is the entire number.

To find **50% of a number,** divide the number by 2.

To find **25% of a number,** divide the number by 4.

To find **10% of a number,** divide the number by 10. To do the division, move the decimal point in the number *one* place to the *left*.

To find **1% of a number,** divide the number by 100. To do the division, move the decimal point in the number *two* places to the *left*.

ANSWERS
2. (a) $110.38 = $11.038
   (b) 7.6 hours = 0.76 hour
   (c) 34. pounds = 3.4 pounds
3. (a) $110.38 = $1.1038
   (b) 07.6 hours = 0.076 hour
   (c) 34. pounds = 0.34 pound

OBJECTIVE 3 **Solve basic percent problems using the percent equation.** In **Section 7.2** you used a proportion to solve percent problems. Now you will learn how to solve these problems using the percent equation.

> **Percent Equation**
>
> $$\text{percent } \textbf{of} \text{ whole} = \text{part}$$
>
> The word **of** indicates multiplication, so the **percent equation** becomes
>
> $$\text{percent} \cdot \text{whole} = \text{part}$$
>
> Be sure to write the percent as a decimal or fraction before using the equation.

The percent equation is just a rearrangement of the percent proportion. Recall that in the proportion you wrote the percent over 100. Because there is no 100 in the equation, you have to change the percent to a decimal or fraction by dividing by 100 *before* using the equation.

> **NOTE**
> Once you have set up a percent equation, we encourage you to use your calculator to do the multiplying or dividing needed to solve the equation. For this reason, we will always write the percent as a decimal. If you're doing the problems by hand, changing the percent to a fraction may be easier at times. Either method will work.

Examples 4, 5, and 6 below are the same percent questions that were in the examples in **Section 7.2**. There we used a proportion to answer each question. Now we will use an equation to answer them. You can then compare the equation method with the proportion method.

EXAMPLE 4 Using the Percent Equation to Find the Part

Write and solve a percent equation to answer each question.

(a) 15% of $165 is how much money?

Translate the sentence into an equation. Recall that *of* indicates multiplication and *is* translates to the equal sign. The percent must be written in decimal form. Use any letter you like to represent the unknown quantity. (We will use n for an unknown number and p for an unknown percent.)

15.% of $165 is how much money?

Write 15% as the decimal 0.15 $0.15 \cdot 165 = n$

To solve the equation, simplify the left side, multiplying 0.15 by 165.

$$(0.15)(165) = n$$
$$24.75 = n$$

So 15% of $165 is **$24.75**, which matches the answer obtained using a proportion (see Example 4 in **Section 7.2**).

Continued on Next Page

520 Chapter 7 Percent

4 Write and solve an equation to answer each question.

(a) 9% of 3250 miles is how many miles?

(b) 78% of $5.50 is how much?

(c) What is $12\frac{1}{2}$% of 400 homes? (*Hint:* Write $12\frac{1}{2}$% as 12.5%. Then move the decimal point two places to the left.)

(d) How much is 350% of $6?

ANSWERS
4. (a) $(0.09)(3250) = n$; 292.5 miles
   (b) $(0.78)(5.50) = n$; $4.29
   (c) $n = (0.125)(400)$; 50 homes
   (d) $n = (3.5)(6)$; $21

Check Use estimation to check that the solution is reasonable. First find 10% of $165 by moving the decimal point.

   10% of 165. is $16.50   and

5% of $165 would be half as much, that is, half of $16.50 or about $8.

So the *estimate* for 15% of $165 is $16.50 + $8 = $24.50. The exact answer of $24.75 is very close to this estimate, so it is reasonable.

(b) How many students is 210% of 40 students?
Translate the sentence into an equation. Write the percent in decimal form.

How many students is 210.% of 40 students?
$$n = 2.10 \cdot 40$$
Write 210% as the decimal 2.10

This time the two sides of the percent equation are reversed, so

   part = percent • whole.

Recall that the variable may be on either side of the equal sign. To solve the equation, simplify the right side, multiplying 2.10 by 40.

$$n = (2.10)(40)$$
$$n = 84$$

So **84 students** is 210% of 40 students. This matches the answer obtained by using a proportion (see Example 7(a) in **Section 7.2**).

Check Use estimation to check that the solution is reasonable. 210% is close to 200%.

   200% of 40 students is 2 times 40 students = 80 students ← Estimate

The exact answer of 84 students is close to the estimate, so it is reasonable.

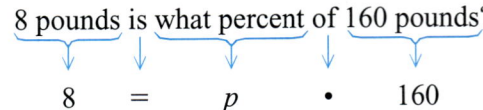 **Work Problem 4 at the Side.**

EXAMPLE 5 Using the Percent Equation to Find the Percent

Write and solve a percent equation to answer each question.

(a) 8 pounds is what percent of 160 pounds?
Translate the sentence into an equation. This time the percent is unknown. Do *not* move the decimal point in the other numbers.

   8 pounds is what percent of 160 pounds?
$$8 = p \cdot 160$$

To solve the equation, divide both sides by 160.

   On the left side, $\dfrac{8}{160} = \dfrac{p \cdot \cancel{160}}{\cancel{160}}$ On the right side, divide out the common factor of 160
   divide 8 by 160

   Solution in *decimal* form $0.05 = p$

Now **multiply the solution by 100** to change it from a *decimal* to a *percent*.

$$0.05 = 5\%$$

So 8 pounds is **5%** of 160 pounds. This matches the answer obtained by using a proportion (see Example 5 in **Section 7.2**).

Continued on Next Page

Check The solution makes sense because 10% of 160 pounds would be 16 pounds.

$$10\% \text{ of } 160 \text{ pounds is } 16 \text{ pounds} \quad \text{so}$$

5% of 160 pounds is half as much, that is, half of 16 pounds, or 8 pounds.

8 pounds matches the number given in the original problem, so 5% is the correct solution.

(b) What percent of $50 is $68?
Translate the sentence into an equation and solve it.

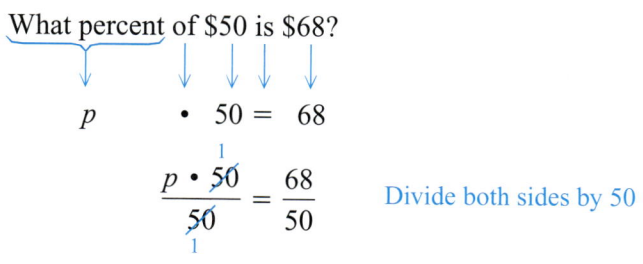

$$p \cdot 50 = 68$$

$$\frac{p \cdot 50}{50} = \frac{68}{50} \quad \text{Divide both sides by 50}$$

$$p = 1.36 \quad \leftarrow \text{Solution in } decimal \text{ form}$$

Now **multiply the solution by 100** to change it from a *decimal* to a *percent*.

$$1.36 = 136\%$$

So **136%** of $50 is $68. This matches the answer obtained by using a proportion (see Example 7(b) in **Section 7.2**).

Check The solution makes sense because 100% of $50 would be $50 (all of it), and 200% of $50 would be 2 times $50, or $100. So $68 has to be between 100% and 200%.

136% is between 100% and 200%. } → 100% of $50 = $50   { $68 is between
                                  200% of $50 = $100 $50 and $100

The solution of 136% fits the conditions.

CAUTION
When you use an equation to solve for an unknown percent, *the solution will be in decimal form.* Remember to *multiply the solution by 100* to change it from decimal form to a percent. The shortcut is to move the decimal point in the solution *two* places to the *right* and attach the % symbol.

Work Problem 5 at the Side. ▶▶▶

---

5 Write and solve an equation to answer each question.

(a) 1200 books is what percent of 5000 books?

(b) 23 hours is what percent of 20 hours?

(c) What percent of $6.50 is $0.52?

ANSWERS
5. **(a)** $1200 = p \cdot 5000$; $0.24 = 24\%$
   (b) $23 = p \cdot 20$; $1.15 = 115\%$
   (c) $p \cdot 6.50 = 0.52$; $0.08 = 8\%$

522 Chapter 7 Percent

6 Write and solve an equation to answer each question.

(a) 74% of how many cars is 37 cars?

(b) 1.2 tons is $2\frac{1}{2}$% of how many tons?

(c) 216 calculators is 160% of how many calculators?

EXAMPLE 6 Using the Percent Equation to Find the Whole

Write and solve a percent equation to answer each question.

(a) 162 credits is 90% of how many credits?
Translate the sentence into an equation. Write the percent in decimal form.

162 credits is 90.% of how many credits?

$$162 = 0.90 \cdot n \qquad \text{Write 90\% as the decimal 0.90}$$

Recall that 0.90 is equivalent to 0.9, so use 0.9 in the equation.

$$\frac{162}{0.9} = \frac{(0.9)(n)}{0.9} \qquad \text{Divide both sides by 0.9}$$

$$180 = n$$

So 162 credits is 90% of **180 credits**.

Check The solution makes sense because 90% of 180 credits should be 10% less than 100% of the credits, and 10% of 180. credits is 18 credits.

$$100\% \quad - \quad 10\% \quad = \quad 90\%$$
of 180 credits   of 180. credits   of 180 credits

180 credits − 18 credits = 162 credits ← Matches the number given in the original problem

(b) 250% of what amount is $75?
Translate the sentence into an equation. Write the percent in decimal form.

250.% of what amount is $75?

Write 250% as the decimal 2.5

$$2.5 \cdot n = 75$$

$$\frac{(2.5)(n)}{2.5} = \frac{75}{2.5} \qquad \text{Divide both sides by 2.5}$$

$$n = 30$$

So 250% of **$30** is $75.

Check The solution makes sense because 200% of $30 is 2 times $30 = $60, and 50% of $30 is $30 ÷ 2 = $15.

$$200\% \quad + \quad 50\% \quad = \quad 250\%$$
of $30   of $30   of $30

(2)($30) $30 ÷ 2

$60 + $15 = $75 ← Matches the number given in the original problem

◀◀◀ Work Problem 6 at the Side.

ANSWERS
6. (a) 0.74 • n = 37; 50 cars
  (b) 1.2 = 0.025 • n; 48 tons
  (c) 216 = 1.6 • n; 135 calculators

7.3 Exercises

FOR EXTRA HELP 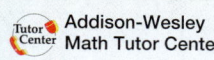 Addison-Wesley Math Tutor Center 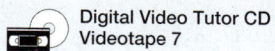 MathXL Digital Video Tutor CD 4 Videotape 7  Student's Solutions Manual MyMathLab  MathXL Tutorials on CD

*Use your estimation skills and the percent shortcuts to select the most reasonable answers. Circle your choices. Do **not** write an equation or proportion and solve it. See Examples 1–3.*

1. Find 50% of 3000 patients.

   150 patients   **1500 patients**   300 patients

2. What is 50% of 192 pages?

   48 pages   384 pages   96 pages

3. 25% of $60 is how much?

   $15 $6 $30

4. Find 25% of $2840.

   $28.40   **$710** $284

5. What is 10% of 45 pounds?

   0.45 pound   22.5 pounds   **4.5 pounds**

6. 10% of 7 feet is how many feet?

   0.7 foot   3.5 feet   14 feet

7. Find 200% of $3.50.

   $0.35 $1.75   **$7.00**

8. What is 300% of $12?

   $4   **$36** $1.20

9. 1% of 5200 students is how many students?

   520 students   **52 students**   2600 students

10. Find 1% of 460 miles.

    0.46 mile   46 miles   **4.6 miles**

11. Find 10% of 8700 cell phones.

    8700 phones   4350 phones   **870 phones**

12. 25% of 128 CDs is how many CDs?

    64 CDs   **32 CDs**   1280 CDs

13. What is 25% of 19 hours?

    4.75 hours   1.9 hours   2.5 hours

14. What is 1% of $37?

    $370 $3.70   **$0.37**

15. **(a)** Describe a shortcut for finding 10% of a number and explain *why* your shortcut works.

16. **(a)** Describe a shortcut for finding 1% of a number and explain *why* it works.

15. **(b)** Once you know 10% of a certain number, explain how you could use that information to find 20% and 30% of the same number.

16. **(b)** Once you know 1% of a certain number, explain how you could use that information to find 2% and 3% of the same number.

Write and solve an equation to answer each question in Exercises 17–48. See Examples 4–6.

17. 35% of 660 programs is how many programs?

18. 55% of 740 canisters is how many canisters?

19. 70 truckloads is what percent of 140 truckloads?

20. 30 crew members is what percent of 75 crew members?

21. 476 circuits is 70% of what number of circuits?

22. 621 tons is 45% of what number of tons?

23. $12\frac{1}{2}$% of what number of people is 135 people?

24. $6\frac{1}{2}$% of what number of bottles is 130 bottles?

25. What is 65% of 1300 species?

26. What is 75% of 360 dosages?

27. 4% of $520 is how much?

28. 7% of $480 is how much?

29. 38 styles is what percent of 50 styles?

30. 75 offices is what percent of 125 offices?

31. What percent of $264 is $330?

32. What percent of $480 is $696?

33. 141 employees is 3% of what number of employees?

34. 16 books is 8% of what number of books?

35. 32% of 260 quarts is how many quarts?

36. 44% of 430 liters is how many liters?

37. $1.48 is what percent of $74?

38. $0.51 is what percent of $8.50?

39. How many tablets is 140% of 500 tablets?

40. How many patients is 175% of 540 patients?

41. 40% of what number of salads is 130 salads?

42. 75% of what number of wrenches is 675 wrenches?

43. What percent of 160 liters is 2.4 liters?

44. What percent of 600 miles is 7.5 miles?

45. 225% of what number of gallons is 11.25 gallons?

46. 180% of what number of ounces is 6.3 ounces?

47. What is 12.4% of 8300 meters?

48. What is 13.2% of 9400 acres?

49. Explain and correct the error in each of these solutions.

  (a) 3 hours is what percent of 15 hours?

  $$3 = p \cdot 15$$

  $$\frac{3}{15} = \frac{p \cdot \cancel{15}}{\cancel{15}}$$

  $$0.2 = p$$

  The answer is 0.2%.

  (b) $50 is what percent of $20?

  $$50 \cdot p = 20$$

  $$\frac{\cancel{50} \cdot p}{\cancel{50}} = \frac{20}{50}$$

  $$p = 0.40 = 40\%$$

  The answer is 40%.

50. Explain and correct the error in each of these solutions.

  (a) 12 inches is 5% of what number of inches?

  $$(12)(0.05) = n$$

  $$0.6 = n$$

  The answer is 0.6 inch.

  (b) What is 4% of 30 pounds?

  $$n = (4)(30)$$

  $$n = 120$$

  The answer is 120 pounds.

RELATING CONCEPTS (EXERCISES 51–52) For Individual or Group Work

Use your knowledge of fractions to work Exercises 51 and 52 in order.

51. Suppose that you have this problem:
$33\frac{1}{3}\%$ of $162 is how much?

  (a) First, change $33\frac{1}{3}\%$ to a fraction. (See **Section 7.1** for help.) Then, write an equation and solve it.

  (b) Now solve the problem by changing $33\frac{1}{3}\%$ to a decimal. (*Hint:* Look at part (a) to see what fraction is equivalent to $33\frac{1}{3}\%$. Change the fraction to a decimal, using your calculator. Keep *all* the decimal places shown in the calculator's display window. Now write the equation and solve it, using your calculator.)

  (c) Compare your answers from part (a) and part (b). How different are they?

52. Now suppose that you have this problem:
22 cans is $66\frac{2}{3}\%$ of what number of cans?

  (a) First, change $66\frac{2}{3}\%$ to a fraction. (See **Section 7.1** for help.) Then, write an equation and solve it. (See **Section 4.7** for help.)

  (b) Now solve the problem by changing $66\frac{2}{3}\%$ to a decimal. Use your calculator and keep all the decimal places shown. Now write the equation and solve it.

  (c) Compare your answers from part (a) and part (b). How different are they?

Summary Exercises on Percent

1. Complete this table. Write fractions in lowest terms and as whole or mixed numbers when possible.

	Fraction	Decimal	Percent
(a)	$\frac{3}{100}$		
(b)			30%
(c)		0.375	
(d)			160%
(e)	$\frac{1}{16}$		
(f)			5%
(g)		2.0	
(h)	$\frac{4}{5}$		
(i)		0.072	

2. Use percent shortcuts to answer these questions.
   (a) 10% of 35 ft is _____.
   (b) 100% of 19 miles is _____.
   (c) 50% of 210 cows is _____.
   (d) 1% of $8 is _____.
   (e) 25% of 2000 women is _____.
   (f) 300% of $15 is _____.
   (g) 10% of $875 is _____.
   (h) 25% of 48 pounds is _____.
   (i) 1% of 9500 students is _____.

Use the percent proportion or percent equation to answer each question. If necessary, round money amounts to the nearest cent and percent answers to the nearest tenth of a percent.

3. 9 Web sites is what percent of 72 Web sites?

4. 30 DVDs is 40% of what number of DVDs?

5. 6% of $8.79 is how much?

6. 945 students is what percent of 540 students?

7. $3\frac{1}{2}$% of 168 pounds is how much, to the nearest tenth of a pound?

8. 1.25% of what number of hours is 7.5 hours?

9. What percent of 80,000 deer is 40,000 deer?

10. 465 camp sites is 93% of what number of camp sites?

11. What number of golf balls is 280% of 35 golf balls?

12. What percent of $66 is $1.80?

13. 9% of what number of apartments is 207 apartments?

14. What weight is 84% of 0.75 ounce?

15. $1160 is what percent of $800?

16. Find 3.75% of 6500 voters, to the nearest whole number.

17. What is 300% of 0.007 inch?

18. 24 minutes is what percent of 6 minutes?

19. What percent of 60 yards is 4.8 yards?

20. $0.17 is 25% of what amount?

The circle graph shows the average costs for various wedding expenses. Use the graph to answer Exercises 21–24. Round money answers to the nearest cent, if necessary.

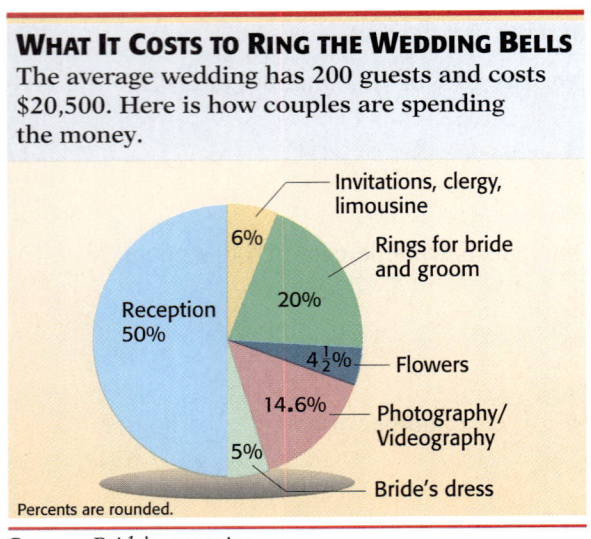

WHAT IT COSTS TO RING THE WEDDING BELLS
The average wedding has 200 guests and costs $20,500. Here is how couples are spending the money.

- Reception 50%
- Invitations, clergy, limousine 6%
- Rings for bride and groom 20%
- Flowers 4½%
- Photography/Videography 14.6%
- Bride's dress 5%

Percents are rounded.

Source: Bride's magazine.

21. Which item is least expensive, and how much is spent on it?

22. What amount is paid for invitations, clergy, and limousine?

23. Find the cost of photography and videography.

24. How much, on average, is spent for each guest at the reception?

7.4 Problem Solving with Percent

OBJECTIVES
1. Solve percent application problems.
2. Solve problems involving percent of increase or decrease.

OBJECTIVE 1 Solve percent application problems. Solving percent problems involves identifying three items: the *percent*, the *whole*, and the *part*. Then you can write a percent equation or percent proportion and solve it to answer the question in the problem. Use the six problem-solving steps from **Section 3.3**.

EXAMPLE 1 Finding the Part

A new low-income housing project charges 30% of a family's income as rent. The Smiths' family income is $1260 per month. How much will the Smiths pay for rent?

Step 1 **Read the problem.** It is about a family paying part of its income for rent.

  Unknown: amount of rent
  Known: 30% of income paid for rent; $1260 monthly income

Step 2 **Assign a variable.** There is only one unknown, so let n be the amount paid for rent.

Step 3 **Write an equation.** Use the percent equation, percent • whole = part. Recall that the *whole* often follows the word *of*.

The percent is given in the problem: 30%. The key word *of* appears *right after* 30%, which means that you can use the phrase "30% *of* a family's income" to help you write one side of the equation. Write 30% as the decimal 0.30.

30.% *of* a family's income is the rent.
$$0.30 \cdot \$1260 = n$$

Step 4 **Solve the equation.** Simplify the left side, multiplying 0.30 by 1260.

$$(0.30)(1260) = n$$
$$378 = n$$

Step 5 **State the answer.** The Smith family will pay **$378** for rent.

Step 6 **Check** the solution. The solution of $378 makes sense because 10% of $1260 is $126, so 30% would be 3 times $126, or $378.

> **NOTE**
> You could also use the percent proportion in *Step 3* above.
>
> Percent → $\dfrac{30}{100} = \dfrac{n}{1260}$ ← Part (unknown)
> Always 100 →   ← Whole
>
> The answer will be the same, $378.
>
> Throughout the rest of this chapter, we will use the percent equation. You may, if you wish, use the percent proportion instead. The final answers will be the same.

1 About 65% of the students at City Center College receive some form of financial aid. How many of the 9280 students enrolled this year are receiving aid? Use the six problem-solving steps.

► **Work Problem 1 at the Side.**

ANSWERS
1. Let n be number of students receiving aid.
   $(0.65)(9280) = n$
   6032 students receive aid.
   Check: 10% of 9280 students is 928, which rounds to 900 students. So 60% would be 6 times 900 students = 5400 students, and 70% would be 7 • 900 = 6300. The solution falls between 5400 and 6300 students, so it is reasonable.

2 There were 50 points on the first math test. Hue's score was 83% correct. How many points did Hue earn? Use the six problem-solving steps.

EXAMPLE 2 Finding the Part

When Britta received her first $180 paycheck as a math tutor, $12\frac{1}{2}\%$ was withheld for federal income tax. How much was withheld?

Step 1 **Read** the problem. It is about part of Britta's pay being withheld for taxes.

> **Unknown:** amount withheld for taxes
> **Known:** $12\frac{1}{2}\%$ of earnings withheld; $180 in pay

Step 2 **Assign a variable.** There is only one unknown, so let n be the amount withheld for taxes.

Step 3 **Write an equation.** Use the percent equation. The *percent* is given: $12\frac{1}{2}\%$. Write $12\frac{1}{2}\%$ as 12.5% and then move the decimal point two places to the left. So 12.5% becomes the decimal 0.125.

The key word *of* doesn't appear after $12\frac{1}{2}\%$. Instead, think about whether you know the *whole* or the *part*. You know Britta's *whole* paycheck is $180, but you do *not* know what *part* of it was withheld.

$$\text{percent} \cdot \text{whole} = \text{part}$$
$$12.5\% \cdot \$180 = n$$

Step 4 **Solve.**
$$(0.125)(180) = n$$
$$22.5 = n$$

Step 5 **State the answer.** **$22.50** was withheld from Britta's paycheck.

Step 6 **Check the solution.** The solution of $22.50 makes sense because 10% of $180 is $18, so a little more than $18 should be withheld.

Work Problem 2 at the Side.

EXAMPLE 3 Finding the Percent

On a 15-point quiz, Zenitia earned 13 points. What percent correct is this, to the nearest whole percent?

Step 1 **Read** the problem. It is about points earned on a quiz.

> **Unknown:** percent correct
> **Known:** earned 13 out of 15 points

Step 2 **Assign a variable.** Let p be the unknown percent.

Step 3 **Write an equation.** Use the percent equation. There is no number with a % symbol in the problem. The question, "What percent is this?" tells you that the *percent* is unknown. The *whole* is all the points on the quiz (15 points), and 13 points is the *part* of the quiz that Zenitia did correctly.

$$\text{percent} \cdot \text{whole} = \text{part}$$
$$p \cdot 15 \text{ points} = 13 \text{ points}$$

Continued on Next Page

ANSWERS

2. Let n be the points earned.
$(0.83)(50) = n$
Hue earned 41.5 points.
Check: 10% of 50 points is 5 points, so 80% would be 8 times 5 points = 40 points. Hue earned a little more than 80%, so 41.5 points is reasonable.

Step 4 **Solve** the equation.

$$\frac{p \cdot \cancel{15}}{\cancel{15}} = \frac{13}{15} \quad \text{Divide both sides by 15}$$

$$p = 0.8\overline{6} \quad \leftarrow \text{The solution is a repeating decimal.}$$

Multiply the solution by 100 to change it from a *decimal* to a *percent*.

$$0.866666667 \approx 86.6666667\% \approx 87\% \quad \leftarrow \text{Rounded}$$

Step 5 **State the answer.** Zenitia had **87%** correct, rounded to the nearest whole percent.

Step 6 **Check the solution.** The solution of 87% makes sense because she earned most of the possible points, so the percent should be fairly close to 100%.

▶ **Work Problem 3 at the Side.**

EXAMPLE 4 **Finding the Percent**

The rainfall in the Red River Valley was 33 inches this year. The average rainfall is 30 inches. This year's rainfall is what percent of the average rainfall?

Step 1 **Read the problem.** It is about comparing this year's rainfall to the average rainfall.

   Unknown: This year's rain is what percent of the average?

   Known: 33 inches this year; 30 inches is average

Step 2 **Assign a variable.** Let p be the unknown percent.

Step 3 **Write an equation.** The percent is unknown. The key word *of* appears *right after* the word *percent*, so you can use that sentence to help you write the equation.

This year's rainfall is what percent of the average rainfall?
33 inches = p • 30 inches

Step 4 **Solve the equation.**

$$33 = p \cdot 30$$

$$\frac{33}{30} = \frac{p \cdot \cancel{30}}{\cancel{30}} \quad \text{Divide both sides by 30}$$

Solution in *decimal* form → $1.1 = p$

Multiply the solution by 100 to change it from a *decimal* to a *percent*.

$$1.10 = 110\%$$

Step 5 **State the answer.** This year's rainfall is **110%** of the average rainfall.

Step 6 **Check the solution.** The solution of 110% makes sense because 33 inches is *more* than 30 inches (more than 100% of the average rainfall), so 33 inches must be *more* than 100% of 30 inches.

▶ **Work Problem 4 at the Side.**

---

❸ The Los Angeles Lakers made 47 of 80 field goal attempts in one game. What percent is this, to the nearest whole percent? Use the six problem-solving steps.

❹ Valley College predicted that 1200 new students would enroll in the fall. It actually had 1620 new students enroll. The actual enrollment is what percent of the predicted number? Use the six problem-solving steps.

ANSWERS

3. Let p be the unknown percent.
   $p \cdot 80 = 47$
   $p = 0.5875$
   $0.5875 = 58.75\% \approx 59\%$
   The Lakers made 59% of their field goals, to the nearest whole percent.

   Check: The Lakers made a little more than half of their field goals. $\frac{1}{2} = 50\%$, so the solution of 59% is reasonable.

4. Let p be the unknown percent.
   $p \cdot 1200 = 1620$
   $p = 1.35$
   $1.35 = 135\%$
   Enrollment is 135% of the predicted number.
   Check: More than 1200 students enrolled (more than 100%), so 1620 students must be more than 100% of the predicted number. The solution of 135% is reasonable.

5 Use the six problem-solving steps to answer each question.

(a) Ezra did 15 problems correctly on a test, giving him a score of $62\frac{1}{2}\%$. How many problems were on the test?

(b) A frozen dinner advertises that only 18% of its calories are from fat. If the dinner contains 55 calories from fat, what is the total number of calories in the dinner? Round to the nearest whole number.

EXAMPLE 5 Finding the Whole

A newspaper article stated that 648 pints of blood were donated at the blood bank last month, which was only 72% of the number of pints needed. How many pints of blood were needed?

Step 1 **Read the problem.** It is about blood donations.

    Unknown: number of pints needed
    Known: 648 pints were donated;
              648 pints is 72% of the number needed.

Step 2 **Assign a variable.** Let n be the number of pints needed.

Step 3 **Write an equation.** The percent is given in the problem: 72%. The key word *of* appears *right after* 72%, so you can use the phrase "72% *of* the number of pints needed" to help you write one side of the equation.

$$72\% \text{ of the number of pints needed is } 648 \text{ pints.}$$
$$0.72 \cdot n = 648$$

Step 4 **Solve.**

$$\frac{(0.72)(n)}{0.72} = \frac{648}{0.72} \quad \text{Divide both sides by 0.72}$$

$$n = 900$$

Step 5 **State the answer.** **900 pints** of blood were needed.

Step 6 **Check the solution.** The solution of 900 pints makes sense because 10% of 900 pints is 90 pints, so 70% would be 7 times 90 pints, or 630 pints, which is close to the number given in the problem (648 pints).

◀◀◀ **Work Problem 5 at the Side.**

OBJECTIVE 2 Solve problems involving percent of increase or decrease. We are often interested in looking at increases or decreases in prices, earnings, population, and many other numbers. This type of problem involves finding the percent of change. Use the following steps to find the **percent of increase**.

Finding the Percent of Increase

Step 1 Use subtraction to find the *amount* of increase.

Step 2 Use a form of the percent equation to find the *percent* of increase.

$$\text{percent } of \quad \text{whole} \quad = \quad \text{part}$$

$$\text{percent } of \text{ original value} = \text{amount of increase}$$

ANSWERS

5. (a) Let n be number of problems on the test.
$0.625 \cdot n = 15$
There were 24 problems on the test.
Check: 50% of 24 problems is $24 \div 2 = 12$ problems correct, so it is reasonable that $62\frac{1}{2}\%$ would be 15 problems correct.

(b) Let n be total number of calories.
$0.18 \cdot n = 55$
There were 306 calories (rounded) in the dinner. *Check:* 10% of 306 calories is about 30 calories, so 20% would be $2 \cdot 30 = 60$ calories, which is close to the number given (55 calories).

Section 7.4 Problem Solving with Percent   533

> **EXAMPLE 6** Finding the Percent of Increase

Brad's hourly wage as assistant manager of a fast-food restaurant was raised from $9.40 to $9.87. What was the percent of increase?

Step 1 **Read the problem.** It is about an increase in wages.

> **Unknown:** percent of increase
> **Known:** Original hourly wage was $9.40; new hourly wage is $9.87.

Step 2 **Assign a variable.** Let p be the percent of increase.

Step 3 **Write an equation.** First subtract $9.87 - $9.40 to find how much Brad's wage went up. That is the *amount* of increase. Then write an equation to find the unknown *percent* of increase. Be sure to use his *original* wage ($9.40) in the equation because we are looking for the change from his *original* wage. Do *not* use the new wage of $9.87 in the equation.

$$\$9.87 - 9.40 = \$0.47 \leftarrow \text{Amount of increase}$$

percent **of** original wage = amount of increase

$$p \cdot \$9.40 = \$0.47$$

Step 4 **Solve.**

$$\frac{(p)(9.40)}{9.40} = \frac{0.47}{9.40} \quad \text{Divide both sides by 9.40}$$

$$p = 0.05 \leftarrow \text{Solution in } decimal \text{ form}$$

Multiply the solution by 100 to change it from a *decimal* to a *percent*.

$$0.05 = 5\%$$

Step 5 **State the answer.** Brad's hourly wage increased **5%**.

Step 6 **Check the solution.** 10% of $9.40 would be a $0.94 raise. A raise of $0.47 is half as much, and half of 10% is 5%, so the solution checks.

> **CAUTION**
> When writing the percent equation in *Step 3* above, be sure to multiply the percent (p) by the *original* wage. Do *not* use the *new* wage in the percent equation.

▶▶▶ **Work Problem 6 at the Side.**

Use a similar procedure to find the **percent of decrease**.

Finding the Percent of Decrease

Step 1 Use subtraction to find the *amount* of decrease.

Step 2 Use a form of the percent equation to find the *percent* of decrease.

percent **of** whole = part

percent **of** original value = amount of decrease

---

6 Use the six problem-solving steps to answer each question.

(a) Over the last two years, Duyen's rent has increased from $650 per month to $767. What is the percent increase?

(b) A shopping mall increased the number of handicapped parking spaces from 8 to 20. What is the percent increase?

ANSWERS

6. (a) $767 - $650 = $117 increase
   Let p be percent increase.
   $$p \cdot 650 = 117$$
   $$p = 0.18 = 18\%$$
   Duyen's rent increased 18%.
   Check: 10% increase would be $650 = $65; 20% increase would be 2 · $65 = $130; so an 18% increase is reasonable.

   (b) $20 - 8 = 12$ space increase
   Let p be percent increase.
   $$p \cdot 8 = 12$$
   $$p = 1.50 = 150\%$$
   The number of parking spaces increased 150%.
   Check: 150% is $1\frac{1}{2}$, and $1\frac{1}{2} \cdot 8 =$ 12 space increase, so it checks.

Chapter 7 Percent

7 Use the six problem-solving steps to answer each question. Round answers to the nearest whole percent.

(a) During a severe winter storm, average daily attendance at an elementary school fell from 425 students to 200 students. What was the percent decrease?

(b) The makers of a brand of spaghetti sauce claim that the number of calories from fat in each serving has been reduced by 20%. Is the claim correct if the number of calories from fat dropped from 70 calories to 60 calories per serving? Explain your answer.

EXAMPLE 7 Finding the Percent of Decrease

Rozenia trained for six months to run in a marathon. Her weight dropped from 137 pounds to 122 pounds. Find the percent of decrease. Round to the nearest whole percent.

Step 1 **Read the problem.** It is about a decrease in weight.

Unknown: percent of decrease
Known: Original weight was 137 pounds; new weight is 122 pounds.

Step 2 **Assign a variable.** Let p be the percent of decrease.

Step 3 **Write an equation.** First subtract 137 pounds − 122 pounds to find how much Rozenia's weight went down. That is the *amount* of decrease. Then write an equation to find the unknown *percent* of decrease. Be sure to use her *original* weight (137 pounds) in the equation because we are looking for the change from her *original* weight. Do *not* use the new weight of 122 pounds in the equation.

137 pounds − 122 pounds = 15 pounds ← *Amount* of decrease

percent *of* original weight = amount of decrease

$$p \cdot 137 \text{ pounds} = 15 \text{ pounds}$$

Step 4 **Solve.**

$$\frac{p \cdot 137}{137} = \frac{15}{137} \quad \text{Divide both sides by 137}$$

$$p = 0.109489051 \leftarrow \text{Solution in } decimal \text{ form}$$

Multiply the solution by 100 to change it from a *decimal* to a *percent*.

$$0.109489051 = 10.9489051\% \approx 11\%$$

Rounded to nearest whole percent

Step 5 **State the answer.** Rozenia's weight decreased approximately **11%**.

Step 6 **Check the solution.** A 10% decrease would be 13.7. = 13.7 pounds, so an 11% decrease is a reasonable solution.

CAUTION
When writing the percent equation in *Step 3* above, be sure to multiply the percent (p) by the **original** weight. Do **not** use the **new** weight in the percent equation.

Work Problem 7 at the Side.

ANSWERS

7. (a) 425 − 200 = 225 student decrease
Let p be percent of decrease.
$p \cdot 425 = 225$
$p \approx 0.529 \approx 53\%$ (rounded)
Daily attendance decreased 53%.
Check: A 50% decrease would be 425 ÷ 2 ≈ 212, so a 53% decrease is reasonable.

(b) 70 − 60 = 10 calorie decrease
Let p be percent of decrease.
$p \cdot 70 = 10$
$p \approx 0.143 \approx 14\%$ (rounded)
The claim of a 20% decrease is not true; the decrease in calories is about 14%.
Check: A 10% decrease would be 70. = 7 calories; a 20% decrease would be 2 · 7 = 14 calories, so a 14% decrease is reasonable.

7.4 Exercises

Use the six problem-solving steps in Exercises 1–22. Round percent answers to the nearest tenth of a percent. See Examples 1–5.

1. Robert Garrett, who works part-time, earns $210 per week and has 18% of this amount withheld for taxes, Social Security, and Medicare. Find the amount withheld.

2. Most shampoos contain 75% to 90% water. If a 16-ounce bottle of shampoo contains 78% water, find the number of ounces of water in the bottle, to the nearest tenth. (*Source: Consumer Reports.*)

3. An ATM machine charges $2 for any size cash withdrawal. The $2 fee is what percent of a
   (a) $20 withdrawal
   (b) $40 withdrawal
   (c) $100 withdrawal
   (d) $200 withdrawal.

4. A mail-order company charges $8 for shipping and handling on any order of $50 or less. The $8 shipping charge is what percent of a
   (a) $10 order
   (b) $16 order
   (c) $25 order
   (d) $50 order.

The figure below shows, on average, what portion of the human body is made up of water, protein, fat, and so on. Use the figure to answer Exercises 5 and 6. Round answers to the nearest tenth.

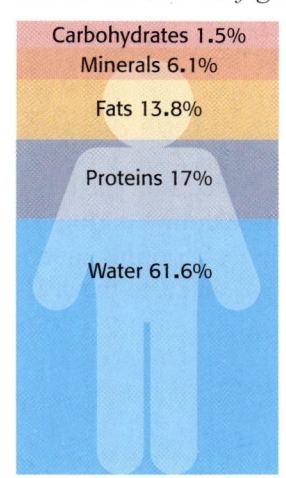

Source: Beakman & Jax, Universal Press Syndicate.

5. If an adult weighs 165 pounds, how much of that weight is
   (a) water
   (b) minerals.

6. If a teenager weighs 92 pounds, how much of that weight is
   (a) fat
   (b) carbohydrates.

7. The guided-missile destroyer USS *Sullivans* has a 335-person crew of which 44 are female. What percent of the crew is female? What percent of the crew is male? (*Source:* U.S. Navy.)

8. In a test by *Consumer Reports,* 6 of the 123 cans of tuna analyzed contained more than the 30-microgram intake limit of mercury. What percent of the cans contained an excessive level of mercury? What percent of the cans contained less than or equal to 30 micrograms of mercury?

9. The U.S. Bureau of the Census reported that Americans who are 65 years of age or older made up 12.7% of the total population in the 2000 census. It said that there were 35.7 million Americans in this group. Find the total U.S. population. (Round to the nearest million.)

10. Julie Ward has 8.5% of her earnings deposited into the credit union. If this amounts to $263.50 per month, find her monthly and annual earnings.

11. The campus honor society hoped to raise $50,000 in donations from businesses for scholarships. It actually raised $69,000. This amount was what percent of the goal?

12. Doug had budgeted $220 for textbooks but ended up spending $316.80. The amount he spent was what percent of his budget?

13. Alfonso earned a score of 95% on his test. He did 38 problems correctly. How many problems were on the test?

14. In a telephone survey, 467 U.S. women said they had done a home improvement project in the past two years. This was 45% of the women in the survey. How many women were surveyed, to the nearest whole number? (*Source:* Opinion Research Corp.)

15. During the 2004–2005 NBA season, Kevin Garnett made 683 field goals, which was 50.22% of the shots he tried. How many shots did he try, to the nearest whole number?

16. Chauncey Billups, who plays basketball for the NBA Detroit Pistons, attempted 382 free throws during the 2004–2005 season. He made 89.8% of his shots. How many free throws did Billups make, to the nearest whole number?

17. An ad for steel-belted radial tires promises 15% better mileage. If Sheera's SUV has gotten 20.6 miles per gallon in the past, what mileage can she expect after the new tires are installed? (Round to the nearest tenth of a mile.)

18. Spam and Spam Lite have yearly sales of $92 million, which is about 62% of all canned lunch meat sales. Find the total yearly sales of canned lunch meat, to the nearest million. (*Source:* Hormel Foods.)

The graph (pictograph) shows the percent of chicken noodle soup sold during the cold and flu season. Use this information to answer Exercises 19–22.

SOUP'S ON

350 million cans of chicken noodle soup are sold each year. More than half are bought during cold-and-flu season, with January being the number one month. The percent sold during each flu-season month is shown.

10% 9% 8% 15% 11% 7%
October November December January February March

Source: USA Today.

19. Which of the flu season months had the lowest sales of chicken noodle soup? How many cans were bought that month?

20. What percent of the chicken noodle soup sales take place during the flu months of October through March? What percent of sales take place in the *non-flu* season months?

21. Find the number of cans of soup sold in the highest sales month and in the second-highest sales month.

22. How many more cans of soup were sold in October than in November? How many more were sold in November than December?

Use the six problem-solving steps to find the percent increase or decrease. Round your answers to the nearest tenth of a percent. See Examples 6 and 7.

23. Henry Ford started the Ford Motor Company. The Model T first appeared in 1908 and cost $825. In 1913, Ford started using an assembly line and the price of the Model T dropped to $290. What was the percent of decrease? (*Source:* Kenneth C. Davis.)

24. In 1914, Henry Ford became a hero to his workers because he doubled the daily minimum wage. He also cut the workday from nine hours to eight hours. Find the percent of decrease in the number of hours in the workday. (*Source:* Kenneth C. Davis.)

25. Students at Lane College were charged $1449 as tuition this semester. If the tuition was $1328 last semester, find the percent of increase.

26. Americans are eating more fish. This year the average American will eat 15.5 pounds compared to only 12.5 pounds per year a decade ago. Find the percent of increase. (*Source: Consumer Reports.*)

27. Jordan's part-time work schedule has been reduced to 18 hours per week. He had been working 30 hours per week. What is the percent decrease?

28. Janis works as a hair stylist. During January, she cut her price on haircuts from $28 to $25.50 to try to get more customers. By what percent did she decrease the price?

29. In 1967 there were 78 animal species listed as threatened with extinction in the United States. In 2003 there were 517 animals on the list. What was the percent increase, to the nearest whole percent? (*Source:* U.S. Fish and Wildlife Service.)

30. The world population was estimated at 6,301,463,000 people in 2003. It is projected to reach 7,851,261,000 people by 2025. By what percent will the world's population increase in those 22 years? (*Source:* United Nations.)

31. You can have an *increase* of 150% in the price of something. Could there be a 150% *decrease* in its price? Explain why or why not.

32. Show how to use a shortcut to find 25% of $80. Then explain how to use the result to find 75% of $80 and 125% of $80 *without* solving a proportion or equation.

RELATING CONCEPTS (EXERCISES 33–36) For Individual or Group Work

As you work Exercises 33–36 in order, explain why each solution does *not* make sense. Then find and correct the error.

33. The recommended maximum daily amount of dietary fat is 65 grams. George ate 78 grams of fat today. He ate what percent of the recommended amount?

$$p \cdot 78 = 65$$
$$\frac{p \cdot 78}{78} = \frac{65}{78}$$
$$p = 0.833 = 83.3\%$$

34. The Goblers soccer team won 18 of its 25 games this season. What percent did the team win?

$$p \cdot 25 = 18$$
$$\frac{p \cdot 25}{25} = \frac{18}{25}$$
$$p = 0.72\%$$

35. The human brain is $2\frac{1}{2}\%$ of total body weight. How much would the brain of a 150-pound person weigh?

$$2\frac{1}{2}\% \text{ of } 150 = n$$
$$(2.5)(150) = n$$
$$375 = n$$

36. Yesterday, because of an ice storm, 80% of the students were absent. How many of the 800 students made it to class?

$$80\% \text{ of } 800 = n$$
$$(0.80)(800) = n$$
$$640 = n$$

The answer is 640 students.

7.5 Consumer Applications: Sales Tax, Tips, Discounts, and Simple Interest

Four of the more common uses of percent in daily life are sales taxes, tips, discounts, and simple interest on loans.

OBJECTIVES

1. Find sales tax and total cost.
2. Estimate and calculate restaurant tips.
3. Find the discount and sale price.
4. Calculate simple interest and the total amount due on a loan.

OBJECTIVE 1 Find sales tax and total cost. Most states collect **sales taxes** on the purchases you make in stores. Your county or city may also add on a small amount of sales tax. For example, your state may charge $6\frac{1}{2}\%$ on purchases and your city may add on another $\frac{1}{2}\%$ for a total of 7%. The exact percent varies from place to place but is usually from 4% to 8%. The stores collect the tax and send it to the city or state government where it is used to pay for things like road repair, public schools, parks, police and fire protection, and so on.

You can use a form of the percent equation to calculate sales tax. The **tax rate** is the *percent*. The cost of the item(s) you are buying is the *whole*. The amount of tax you pay is the *part*.

Finding Sales Tax and Total Cost

Use a form of the percent equation to find sales tax.

$$\underbrace{\text{percent}}_{\text{tax rate}} \cdot \underbrace{\text{whole}}_{\text{cost of item}} = \underbrace{\text{part}}_{\text{amount you pay in sales tax}}$$

Then add to find how much you will pay in all.

$$\text{cost of item} + \text{sales tax} = \text{total cost paid by you}$$

EXAMPLE 1 Finding Sales Tax and Total Cost

Suppose that you buy a DVD player for $289 from A-1 Electronics. The sales tax rate in your state is $6\frac{1}{2}\%$. How much is the tax? What is the total cost of the DVD player?

Step 1 **Read the problem.** It asks for the sales tax on a DVD player and the total cost.

Step 2 **Assign a variable.** Let n be the amount of tax.

Step 3 **Write an equation.** Use the sales tax equation. Write $6\frac{1}{2}\%$ as 6.5% and then move the decimal point two places to the left.

$$\underbrace{\text{tax rate}}_{06.5\%} \cdot \underbrace{\text{cost of item}}_{\$289} = \underbrace{\text{sales tax}}_{n}$$

Step 4 **Solve.**
$$(0.065)(289) = n$$
$$18.785 = n$$

The store will round the tax to the nearest cent, so $18.785 rounds to **$18.79**.

Now add the sales tax to the cost of the DVD player to find your total cost.

$$\underbrace{\text{cost of item}}_{\$289} + \underbrace{\text{sales tax}}_{\$18.79} = \underbrace{\text{total cost}}_{\mathbf{\$307.79}}$$

Continued on Next Page

① **Find the sales tax and the total cost. Round the sales tax to the nearest cent, if necessary. Check your answer by estimating the sales tax.**

(a) $495 camcorder; $5\frac{1}{2}$% sales tax

(b) $29.98 watch; 7% sales tax

(c) $1.19 candy bar; 4% sales tax

② **Find the sales tax rate on each purchase. Then use estimation to check your solution.**

(a) The tax on a $57 textbook is $3.42.

(b) The tax on a $4 notebook is $0.18.

(c) The tax on a $998 sofa is $49.90.

ANSWERS

1. (a) sales tax = $27.23 (rounded)
   total cost = $522.23
   Check: 1% of $500. is $5;
   6% is 6 times $5 = $30
   (b) sales tax = $2.10 (rounded)
   total cost = $32.08
   Check: 1% of $30. is $0.30;
   7% is 7 times $0.30 = $2.10
   (c) Sales tax = $0.05 (rounded)
   total cost = $1.24
   Check: 1% of $1 is $0.01;
   4% is 4 times $0.01 = $0.04
2. (a) 6% Check: 1% of $60. is $0.60;
   6% would be 6 times $0.60, or $3.60
   (which is close to $3.42 given in the problem).
   (b) 4.5% Check: 1% of $4 is $0.04; round 4.5% to 5%; 5% of $4 is 5 times $0.04, or $0.20 (which is close to $0.18 given in the problem).
   (c) 5% Check: 10% of $1000. is $100;
   5% is half of that, or $50 (which is close to $49.90 given in the problem).

Step 5 **State the answer.** The tax is **$18.79** and the total cost of the DVD player, including tax, is **$307.79**.

Step 6 **Check:** Use estimation to check that the amount of sales tax is reasonable. Round $289 to $300. Then 1% of $300. is $3. Round $6\frac{1}{2}$% to 7%. Then 7% would be 7 times $3 or $21 for sales tax. Our solution of $18.79 is close to $21, so it is reasonable.

▶ **Work Problem ① at the Side.**

EXAMPLE 2 Finding the Sales Tax Rate

Ms. Ortiz bought a $21,950 pickup truck. She paid an additional $1646.25 in sales tax. What was the sales tax rate?

Step 1 **Read** the problem. It asks for the sales tax rate on a truck purchase.

Step 2 **Assign a variable.** Let p be the tax rate (the percent).

Step 3 **Write an equation.** Use the sales tax equation.

tax rate • cost of item = sales tax
$$p \cdot \$21{,}950 = \$1646.25$$

Step 4 **Solve.**
$$\frac{(p)(21{,}950)}{21{,}950} = \frac{1646.25}{21{,}950} \quad \text{Divide both sides by 21,950}$$

$$p = 0.075 \quad \longleftarrow \text{Solution in } decimal \text{ form}$$

Multiply the solution by 100 to change it from a decimal to a percent: $0.075 = 7.5\%$.

Step 5 **State the answer.** The sales tax rate is **7.5%** (or $7\frac{1}{2}$%).

Step 6 **Check.** Use estimation to check that the solution is reasonable. If the tax rate was 1%, then 1% of $21950. = $219.50, or about $200.

Round 7.5% to 8%. Then 8% would be 8 times $200, or $1600.

The tax amount given in the original problem, $1646.25, is close to the estimate, so our solution of 7.5% is reasonable.

▶ **Work Problem ② at the Side.**

OBJECTIVE ② Estimate and calculate restaurant tips. Waiters and waitresses rely on tips as a major part of their income. The general rule of thumb is to leave 15% of your bill for food and beverages as a tip for the server. If you receive exceptional service or are eating in an upscale restaurant, consider leaving a 20% tip.

EXAMPLE 3 Estimating 15% and 20% Tips

First estimate each tip. Then calculate the exact amount.

(a) Kirby took his wife to dinner at a nice restaurant to celebrate her promotion at work. The bill came to $77.85. How much should he leave for a 20% tip?

Estimate: Round $77.85 to $80. Then 10% of $80. is $8.

20% would be 2 times $8, or **$16**. ← Estimate

Continued on Next Page

Section 7.5 Consumer Applications: Sales Tax, Tips, Discounts, and Simple Interest   **541**

Exact: Use the percent equation. Write 20% as a decimal. The bill for food and beverages is the *whole* and the tip is the *part*.

$$\text{percent} \cdot \text{whole} = \text{part}$$
$$20.\% \cdot \$77.85 = n$$
$$(0.20)(77.85) = n$$
$$15.57 = n$$

20% of $77.85 is **$15.57**, which is close to the estimate of $16.

A tip is usually rounded off to a convenient amount, such as the nearest quarter or nearest dollar, so Kirby left $16.

(b) Linda, Peggy, and Mary ordered similarly priced lunches and agreed to split the bill plus a 15% tip. How much should each woman pay if the bill is $21.63?

Estimate: Round $21.63 to $20. Then 10% of $20 is $2.

5% of $20 would be half as much, that is, half of $2, or $1.

So an estimate of the 15% tip is $2 + $1 = **$3**. ← Estimates

An estimate of the amount each woman should pay is ($20 + $3) ÷ 3 ≈ **$8**.

Exact: Use the percent equation to calculate the 15% tip. Add the tip to the bill. Then divide the total by 3 to find the amount each woman should pay.

$$\text{percent} \cdot \text{whole} = \text{part}$$
$$15.\% \cdot \$21.63 = n$$
$$(0.15)(21.63) = n$$
$$3.2445 = n$$

Round $3.2445 to **$3.24** (nearest cent), which is close to the estimate of $3 for the tip.

Add: $21.63 + $3.24 = $24.87 ← Total cost of lunch and tip

Divide: $24.87 ÷ 3 = **$8.29** ← Amount paid by each woman

Work Problem 3 at the Side. ▶▶▶

OBJECTIVE 3 Find the discount and sale price. Most people prefer buying things when they are on sale. A store will reduce prices, or **discount**, to attract additional customers. You can use a form of the percent equation to calculate the discount. The *rate* of discount is the *percent*. The original price is the *whole*. The amount that will be discounted (subtracted from the original price) is the *part*.

Finding the Discount and Sale Price

Use a form of the percent equation to find the discount.

$$\text{percent} \cdot \text{whole} = \text{part}$$
$$\text{rate of discount} \cdot \text{original price} = \text{amount of discount}$$

Then subtract to find the sale price.

$$\text{original price} - \text{amount of discount} = \text{sale price}$$

3 First estimate each tip. Then calculate the exact tip.

(a) 20% tip on a bill of $58.37

(b) 15% tip on a bill of $11.93

(c) A bill of $89.02 plus a 15% tip shared equally by four friends. How much will each friend pay?

ANSWERS

3. **(a)** *Estimate:* 10% of $60 is $6; 20% is 2 times $6, or $12.
   Exact: $11.67 (rounded to nearest cent)
   (b) *Estimate:* 10% of $12 is $1.20; 5% is half of $1.20, or $0.60; $1.20 + $0.60 = $1.80.
   Exact: $1.79 (rounded to nearest cent)
   (c) *Estimate* of tip: 10% of $90 is $9; 5% is half of $9, or $4.50; $9 + $4.50 = $13.50.
   Exact tip: $13.35 (rounded to nearest cent). Each friend pays: ($89.02 + $13.35) ÷ 4 = $25.59 (rounded).

4 Find the amount of the discount and the sale price for each item.

(a) An Easy-Boy leather recliner originally priced at $950 is offered at a 35% discount. What is the sale price?

(b) Eastside Department Store has women's swimsuits on sale at 40% off. One swimsuit was originally priced at $34. Another suit was originally priced at $72. What is the sale price of each suit?

EXAMPLE 4 Finding the Discount and Sale Price

The Oak Mill Furniture Store has an oak entertainment center with an original price of $840 on sale at 15% off. Find the sale price of the entertainment center.

Step 1 **Read** the problem. It asks for the sale price on an entertainment center.

Step 2 **Assign a variable.** Let n be the amount of discount.

Step 3 **Write an equation.** Use a form of the percent equation to find the discount. Write 15% as the decimal 0.15.

rate of discount • original price = amount of discount

Step 4 **Solve.**

$$(0.15)(840) = n$$
$$126 = n$$

The amount of discount is **$126**. Find the sale price of the entertainment center by subtracting the amount of the discount ($126) from the original price.

original price − amount of discount = sale price

$$\$840 - \$126 = \$714$$

Step 5 **State the answer.** During the sale, you can buy the entertainment center for **$714**.

Step 6 **Check.** Round $840 to $800. Then 10% of $800 is $80 and 5% is half as much, or $40. So 15% of $800 is $80 + $40 = $120. An estimate of the sale price is $800 − $120 = $680, so the exact answer of $714 is reasonable.

◀◀◀ Work Problem 4 at the Side.

🖩 **Calculator Tip** In Example 4 above, you can use a *scientific* calculator to find the amount of discount and subtract the discount from the original price.

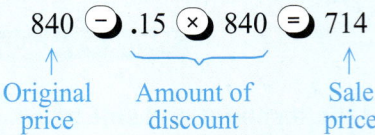

↑ Original price   Amount of discount   ↑ Sale price

Your *scientific* calculator observes the order of operations, so it will automatically do the multiplication before the subtraction. (Recall that simple, four-function calculators *may not* follow the order of operations; they might give an incorrect result.)

OBJECTIVE 4 **Calculate simple interest and the total amount due on a loan.** **Interest** is a fee paid, or a charge made, for lending or borrowing money. The amount of money borrowed is called the **principal**. The charge for interest is usually given as a percent, called the **interest rate**. The interest rate is assumed to be *per year* (for *one* year) unless stated otherwise.

In most cases, interest is calculated on the original principal and is called **simple interest**. Use the following **interest formula** to find simple interest.

ANSWERS

4. (a) Discount is $332.50; sale price is $617.50.
   (b) Discount is $13.60, sale price is $20.40; discount is $28.80, sale price is $43.20.

Section 7.5 Consumer Applications: Sales Tax, Tips, Discounts, and Simple Interest 543

Formula for Simple Interest

$$\text{Interest} = \text{principal} \cdot \text{rate} \cdot \text{time}$$

The formula is usually written using variables.

$$I = p \cdot r \cdot t \quad \text{or} \quad I = prt$$

NOTE
Simple interest calculations are used for most short-term business loans, most real estate loans, and many automobile and consumer loans.

EXAMPLE 5 Finding Simple Interest for 1 Year

Find the simple interest on a $2000 loan at 6% for 1 year.
 The amount borrowed, or principal (p), is $2000. The interest rate (r) is 6%, which is 0.06 as a decimal, and the time of the loan (t) is 1 year. Use the interest formula.

$$I = p \cdot r \cdot t$$
$$I = (2000)(0.06)(1)$$
$$I = 120$$

The interest is **$120**.

Work Problem 5 at the Side.

EXAMPLE 6 Finding Simple Interest for More Than 1 Year

Find the simple interest on a $4200 loan at $8\frac{1}{2}$% for $3\frac{1}{2}$ years.
 The principal (p) is $4200. The rate ($r$) is $8\frac{1}{2}$%, which is the same as 8.5%. Move the decimal point two places to the left to change 8.5% to a decimal.

$$8\tfrac{1}{2}\% = 8.5\% = 08.5 = 0.085$$

The time (t) is $3\frac{1}{2}$ or 3.5 years. Use the formula.

$$I = p \cdot r \cdot t$$
$$I = (4200)(0.085)(3.5)$$
$$I = 1249.50$$

The interest is **$1249.50**.

CAUTION
Be careful when changing a mixed number percent, like $8\frac{1}{2}$%, to a decimal. Writing $8\frac{1}{2}$% as 8.5% is only the first step. There is a decimal point in 8.5% but there is still a % sign. You must divide by 100 before dropping the % sign. *Remember to move the decimal point two places to the left*, as shown in Example 6 above. So $8\frac{1}{2}$% = 8.5% = 0.085.

Work Problem 6 at the Side.

5 Find the simple interest.

(a) $500 at 4% for 1 year

(b) $1850 at $9\frac{1}{2}$% for 1 year
(*Hint:* Write $9\frac{1}{2}$% as 9.5%. Then *move the decimal point* two places to the left to change 9.5% to a decimal.)

6 Find the simple interest for each loan.

(a) $340 at 5% for $3\frac{1}{2}$ years

(b) $2450 at 8% for $3\frac{1}{4}$ years
(*Hint:* Write $3\frac{1}{4}$ years as 3.25 years.)

(c) $14,200 at $7\frac{1}{2}$% for $2\frac{3}{4}$ years

ANSWERS
5. (a) $20   (b) $175.75
6. (a) $59.50   (b) $637   (c) $2928.75

7 Find the simple interest for each loan.

(a) $1600 at 7% for 4 months

(b) $25,000 at $10\frac{1}{2}$% for 3 months

Interest rates are given *per year*. For loan periods of less than one year, be careful to express the time as a fraction of a year.

If the time is given in months, use a denominator of 12, because there are 12 months in a year. A loan of 9 months would be for $\frac{9}{12}$ of a year, a loan of 7 months would be for $\frac{7}{12}$ of a year, and so on.

EXAMPLE 7 Finding Simple Interest for Less Than 1 Year

Find the simple interest on $840 at $9\frac{3}{4}$% for 7 months.
The principal is $840. The rate is $9\frac{3}{4}$% or 0.0975.

$$9\frac{3}{4}\% = 9.75\% = 09.75 = 0.0975$$

The time is $\frac{7}{12}$ of a year. Use the formula $I = prt$.

$$I = (840)(0.0975)\left(\frac{7}{12}\right) \quad \text{7 months is } \tfrac{7}{12} \text{ of a year.}$$

$$= (81.9)\left(\frac{7}{12}\right)$$

$$= \left(\frac{81.9}{1}\right)\left(\frac{7}{12}\right) \quad \begin{array}{l}\text{Multiply numerators.}\\ \text{Multiply denominators.}\end{array}$$

$$= \frac{573.3}{12} = 47.775 \quad \text{Divide 573.3 by 12}$$

The interest is **$47.78**, rounded to the nearest cent.

> **Calculator Tip** The calculator solution to Example 7 above uses chain calculations.
>
> 840 ⊗ .0975 ⊗ 7 ÷ 12 = 47.775 ← Round to $47.78

8 Find the total amount due on each loan.

(a) $2500 at $7\frac{1}{2}$% for 6 months

(b) $10,800 at 6% for 4 years

(c) $4350 at $10\frac{1}{4}$% for $2\frac{1}{2}$ years

> Work Problem 7 at the Side.

When you repay a loan, the interest is added to the original principal to find the total amount due.

Finding the Total Amount Due

amount due = principal + interest

EXAMPLE 8 Calculating the Total Amount Due

Charlesetta borrowed $1080 at 8% for three months to pay for tuition and books. Find the total amount due on her loan.

First find the interest. Use $I = prt$. Write 8% as the decimal 0.08.

$$I = (1080)(0.08)\left(\frac{3}{12}\right) \quad \text{3 months is } \tfrac{3}{12} \text{ of a year.}$$

$$I = 21.60$$

Now add the principal and the interest to find the total amount due.

amount due = principal + interest
= $1080 + $21.60
= $1101.60

The total amount due is **$1101.60**.

> Work Problem 8 at the Side.

ANSWERS
7. (a) $37.33 (rounded) (b) $656.25
8. (a) $2593.75 (b) $13,392
(c) $5464.69 (rounded)

7.5 Exercises

Find the amount of the sales tax or the tax rate and the total cost. Round money answers to the nearest cent. See Examples 1 and 2.

Cost of Item	Tax Rate	Amount of Tax	Total Cost
1. $100	6%		
2. $200	4%		
3. $68		$2.04	
4. $185		$9.25	
5. $365.98	8%		
6. $28.49	7%		
7. $2.10	$5\frac{1}{2}\%$		
8. $7.00	$7\frac{1}{2}\%$		
9. $12,600		$567	
10. $21,800		$1417	

For each restaurant bill, estimate a 15% tip and a 20% tip. Then find the exact amounts for a 15% tip and a 20% tip. Round exact amounts to the nearest cent, if necessary. See Example 3.

Bill	Estimate of 15% Tip	Exact 15% Tip	Estimate of 20% Tip	Exact 20% Tip
11. $32.17				
12. $21.94				
13. $78.33				
14. $67.85				
15. $9.55				
16. $52.61				

Find the amount or rate of discount and the sale price. Round money answers to the nearest cent if necessary. See Example 4.

Original Price	Rate of Discount	Amount of Discount	Sale Price
17. $100	15%		
18. $200	20%		
19. $180		$54	
20. $38		$9.50	
21. $17.50	25%		
22. $76	60%		
23. $37.88	10%		
24. $59.99	40%		

Section 7.5 Consumer Applications: Sales Tax, Tips, Discounts, and Simple Interest   **547**

Find the simple interest and total amount due on each loan. See Examples 5–8.

	Principal	Rate	Time	Interest	Total Amount Due
25.	$300	14%	1 year	_____	_____
26.	$600	11%	6 months	_____	_____
27.	$740	6%	9 months	_____	_____
28.	$1180	9%	2 years	_____	_____
29.	$1500	10%	18 months	_____	_____
30.	$3000	15%	5 months	_____	_____
31.	$17,800	$7\frac{1}{2}\%$	9 months	_____	_____
32.	$20,500	$5\frac{1}{2}\%$	6 months	_____	_____

Solve each application problem. Round money answers to the nearest cent, if necessary.

33. Diamonds at Discounts sells diamond engagement rings at 40% off the regular price. Find the sale price of a $\frac{1}{2}$-carat diamond ring normally priced at $1950.

34. A Toshiba pocket PC originally priced at $396 is marked down 25%. Find the price of the pocket PC after the markdown.

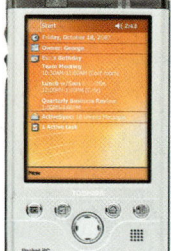

35. Evelina Jones lends $7500 to her son Rick, the owner of Rick's Limousine Service. He will repay the loan at the end of 9 months at $8\frac{1}{2}$% simple interest. What is the total amount that Rick will owe his mother?

36. The owners of Delta Trucking purchased four diesel-powered tractors for cross-country hauling at a cost of $87,500 per tractor. If they borrowed the purchase price for $1\frac{1}{2}$ years at 11% simple interest, find the total amount due.

37. A Sanyo vision picture phone with built-in camera is priced at $99.99. The sales tax rate is $6\frac{1}{2}$%. Find the total cost of the phone. (*Source:* www.BestBuy.com)

38. A weekday "golf/breakfast special" includes breakfast, 18 holes of golf, and use of a cart for $37.95 plus tax per person. If the sales tax rate is $7\frac{1}{2}$%, find the total cost per person. How much will three friends pay to play golf?

39. An Anderson wood frame French door is priced at $1980 with a sales tax of $99. Find the sales tax rate.

40. Textbooks for two classes cost $135 plus sales tax of $8.10. Find the sales tax rate.

41. A "super 45% off sale" begins today. What is the sale price of a ski parka normally priced at $135?

42. A discontinued Whirlpool model side-by-side refrigerator with in-door icemaker originally sold for $1197. What is the sale price with a 35% discount?

43. Ricia and Seitu split a $43.70 dinner bill plus 15% tip. How much did each person pay?

44. Marvette took her brother out to dinner for his birthday. The bill for food was $58.36 and for wine was $15.44. How much was her 20% tip, rounded to the nearest dollar?

45. A 32" flat-screen TV normally priced at $590 is on sale for 18% off. Find the discount and the sale price.

46. This week SUVs are offered at 15% off manufacturers' suggested price. Find the discount and the sale price of an SUV originally priced at $23,500.

47. Ms. Henderson owes $1900 in taxes. She is charged a penalty of $12\frac{1}{4}\%$ annual interest and pays the taxes and penalty after 6 months. Find the total amount she must pay.

48. Norell Di Loreto, owner of Sunset Realtors, borrows $27,000 to update her office computer system. If the loan is for 24 months at $7\frac{3}{4}\%$, find the total amount due on the loan.

49. Vincente and Samuel ordered a large deep-dish pizza for $17.98. How much did they give the delivery person to pay for the pizza and a 15% tip, rounded to the nearest dollar?

50. Cher, Maya, and Adara shared a $25.50 bill for a buffet lunch. Because the server only brought their beverages, they left a 10% tip instead of the usual 15%. How much did each person pay?

Use the information in the store ad to answer Exercises 51–54. Round sale prices and sales tax to the nearest cent if necessary.

STORE CLOSE-OUT!
All clothing is now 45% off!
All jewelry is now 30% off!
All electronics are now 65% off!
6% sales tax added to jewelry and electronics purchases.
CASH ONLY! ALL SALES ARE FINAL!

51. Danika bought a computer modem originally priced at $129 and a $60 pair of earrings. What was her bill for the two items?

52. Find David's total bill for a $189 jacket and a $75 graphing calculator.

53. Sergei purchased a television originally priced at $287.95, two pairs of $48 jeans, and a $95 ring. Find his total bill.

54. Richard picked out three pairs of $15 running shorts, two $28 shirts, and a "point and shoot" camera originally priced at $99.99. How much did he pay in all?

RELATING CONCEPTS (EXERCISES 55–56) For Individual or Group Work

Use your knowledge of percent to work Exercises 55 and 56 in order.

55. (a) College students are offered a 6% discount on a dictionary that sells for $18.50. If the sales tax is 6%, find the cost of the dictionary, including the sales tax, to the nearest cent.

(b) In part (a) the rate of discount and the sales tax rate are the same percent. Explain why the answer did *not* end up back at $18.50.

56. (a) A copier/FAX machine priced at $398 is marked down 7% to promote the new model. If the sales tax is also 7%, find the cost of the machine, including sales tax, to the nearest cent.

(b) What rate of sales tax would have made the answer in part (a) end up back at $398? Round your answer to the nearest hundredth of a percent.

Focus on Real-Data Applications

Make Your Investments Grow—Compound Interest

Simple interest is paid only on the original principal. But savings accounts and most investments earn *compound interest*. In that case, interest is paid on the principal *and* the interest earned. Calculating compound interest can be quite tedious. For this reason compound interest tables have been developed.

Suppose you deposit $1 in a savings account today that earns 4% compounded annually and you allow the deposit to remain for three years. The diagram below shows the compound amount in your account at the end of each of the 3 years.

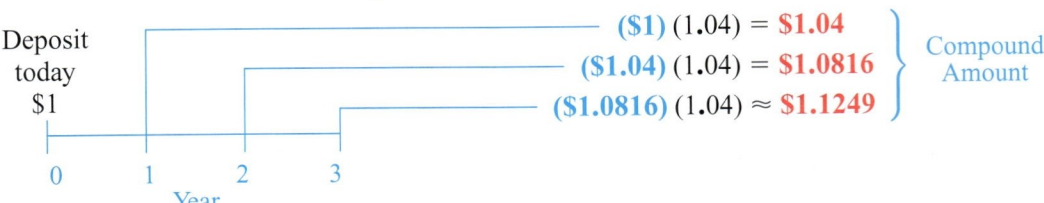

Using the compound amounts for $1, a table can be formed. Look at the table below and find the column headed 4%. The first three numbers (for years 1, 2, and 3) are the same as those we have calculated for $1 at 4% for 3 years.

COMPOUND INTEREST TABLE

Years	3.00%	3.50%	4.00%	4.50%	5.00%	5.50%	6.00%	8.00%
1	1.0300	1.0350	1.0400	1.0450	1.0500	1.0550	1.0600	1.0800
2	1.0609	1.0712	1.0816	1.0920	1.1025	1.1130	1.1236	1.1664
3	1.0927	1.1087	**1.1249**	1.1412	1.1576	1.1742	1.1910	1.2597
4	1.1255	1.1475	1.1699	1.1925	1.2155	1.2388	1.2625	1.3605
5	1.1593	1.1877	1.2167	1.2462	1.2763	1.3070	1.3382	1.4693
6	1.1941	1.2293	1.2653	1.3023	1.3401	1.3788	1.4185	1.5869
7	1.2299	1.2723	1.3159	1.3609	1.4071	1.4547	1.5036	1.7138
8	1.2668	1.3168	1.3686	1.4221	1.4775	1.5347	1.5938	1.8509
9	1.3048	1.3629	1.4233	1.4861	1.5513	1.6191	1.6895	1.9990
10	1.3439	1.4106	1.4802	1.5530	1.6289	1.7081	1.7908	2.1589

To find the compound amount for $1000 deposited at 4% interest for 3 years, multiply the number from the table (1.1249) times the principal ($1000).

($1000) (**1.1249**) = $1124.90 ← Amount in account after 3 years

1. **(a)** Use the table to find the compound amount for $1000 deposited at 6% for 3 years, and the compound amount at 8% for 3 years.
   (b) How much more do you earn in 3 years on $1000 if the interest rate is 6% instead of 4%? How much more if the interest rate is 8% instead of 4%?

2. **(a)** Use the table to find the compound amount on $15,000 invested for 10 years at 4%, at 6%, and at 8%.
   (b) How much more do you earn at 6% than 4%? How much more at 8% than 4%?

3. Compound interest makes a tremendous difference when you invest money for retirement. Suppose a person plans to retire at age 65. Use the part of the compound interest table shown at the right. Find the compound amount on $5000 invested at 8% when the person is 45 years old, and left in the account until age 65. Then find the compound amount if the investment had been made at age 35, if the investment had been made at age 25, and if the investment had been made at age 15.

Years	8.00%
20	4.6610
30	10.0627
40	21.7245
50	46.9016

Chapter 7
SUMMARY

KEY TERMS

7.1 **percent** — Percent means "per one hundred." A percent is a ratio with a denominator of 100.

7.2 **percent proportion** — The proportion used to solve percent problems is $\dfrac{\text{percent}}{100} = \dfrac{\text{part}}{\text{whole}}$.

whole — The *whole* in a percent problem is the entire quantity or the total. It is sometimes called the *base*.

part — The *part* in a percent problem is the number being compared to the *whole*.

7.3 **percent equation** — The percent equation is percent • whole = part. It can be used instead of the percent proportion to solve percent problems.

7.4 **percent of increase or decrease** — Percent of increase or decrease is the amount of change (increase or decrease) expressed as a percent of the original value.

7.5 **sales tax** — Sales tax is a percent of the total sales charged as a tax.

tax rate — The tax rate is the percent used when calculating the amount of tax.

discount — Discount is often expressed as a percent of the original price; it is then deducted from the original price, resulting in the sale price.

interest — Interest is a fee paid or a charge made for lending or borrowing money.

principal — Principal is the amount of money on which interest is charged.

interest rate — Often referred to as *rate*, it is the charge for interest and is given as a percent.

simple interest — When interest is calculated on the original principal, it is called simple interest.

interest formula — The interest formula is used to calculate simple interest. It is Interest = principal • rate • time or $I = prt$.

NEW FORMULAS

Finding simple interest: $I = prt$

NEW SYMBOLS

% percent (meaning "per 100")

TEST YOUR WORD POWER

See how well you have learned the vocabulary in this chapter. Answers follow the Quick Review.

1. **Percent** means
   A. per one thousand
   B. part divided by whole
   C. per one hundred
   D. part times whole.

2. The **whole** in a percent problem is
   A. the entire quantity or total
   B. a ratio with a denominator of 100
   C. the amount of change
   D. always 100.

3. When calculating sales tax, the **rate** is
   A. the part
   B. the whole
   C. the total cost
   D. the percent.

4. The **interest formula** is
   A. tax rate • cost of item = sales tax
   B. $I = prt$
   C. percent • whole = part
   D. $\dfrac{\text{percent}}{100} = \dfrac{\text{part}}{\text{whole}}$.

5. When calculating interest on a loan, the **principal** is the
   A. amount of money borrowed
   B. total amount due
   C. rate charged for borrowing money
   D. amount of simple interest.

6. The **percent of increase or decrease** compares the amount of change to
   A. 100
   B. the percent
   C. the original value before the change
   D. the new value after the change.

7. The **percent equation** is
   A. tax rate • cost of item = sales tax
   B. $I = prt$
   C. percent • whole = part
   D. $\dfrac{\text{percent}}{100} = \dfrac{\text{part}}{\text{whole}}$.

8. A **discount** is
   A. added to the original price
   B. divided by 100
   C. multiplied by the percent
   D. subtracted from the original price.

551

Quick Review

Concepts | **Examples**

7.1 Basics of Percent

Writing a Percent as a Decimal
To write a percent as a decimal, move the decimal point *two* places to the *left* and drop the % sign.

$50\% = 50.\% = 0.50$ or 0.5

$3\% = 03.\% = 0.03$

Writing a Decimal as a Percent
To write a decimal as a percent, move the decimal point *two* places to the *right* and attach a % sign.

$0.75 = 0.75 = 75\%$

$3.6 = 3.60 = 360\%$

Writing a Percent as a Fraction
To write a percent as a fraction, drop the % symbol and write the number over 100. Then write the fraction in lowest terms.

$35\% = \dfrac{35}{100} = \dfrac{35 \div 5}{100 \div 5} = \dfrac{7}{20}$ ← Lowest terms

$125\% = \dfrac{125}{100} = \dfrac{125 \div 25}{100 \div 25} = \dfrac{5}{4} = 1\dfrac{1}{4}$

Writing a Fraction as a Percent
To write a fraction as a percent, multiply by 100 and attach a % symbol. This is the same as multiplying by 100%.

$\dfrac{3}{5} = \dfrac{3}{5} \cdot \dfrac{100}{1}\% = \dfrac{3 \cdot 5 \cdot 20}{5 \cdot 1}\% = \dfrac{60}{1}\% = 60\%$

$\dfrac{7}{8} = \dfrac{7}{8} \cdot \dfrac{100}{1}\% = \dfrac{7 \cdot 4 \cdot 25}{2 \cdot 4 \cdot 1}\% = \dfrac{175}{2}\% = 87\dfrac{1}{2}$ or 87.5%

7.2 Using the Percent Proportion

The percent proportion is shown below.

$$\text{Always 100} \rightarrow \dfrac{\text{percent}}{100} = \dfrac{\text{part}}{\text{whole}}$$

Identify the percent first. It appears with the word *percent* or the % symbol.

The *whole* is the entire quantity or total. It often appears after the word *of*.

The *part* is the number being compared to the whole.

30 children is what percent *of* 75 children?

Percent (unknown) → $\dfrac{p}{100} = \dfrac{30}{75}$ ← Part
Always 100 → ← Whole (follows *of*)

To solve the proportion, find the cross products.

$\dfrac{p}{100} = \dfrac{30}{75}$ $100 \cdot 30 = 3000$ Cross products
$p \cdot 75$

$p \cdot 75 = 3000$ Show that the cross products are equivalent.

$\dfrac{p \cdot 75}{75} = \dfrac{3000}{75}$ Divide both sides by 75

$p = 40$

30 children is **40%** of 75 children.

7.3 Percent Shortcuts

200% of a number is 2 times the number.

200% of $35 is 2 times $35, or $70.

100% of a number is the entire number.

100% of 600 women is *all* the women (600 women).

To find 50% of a number, divide the number by 2.

50% of $8000 is $8000 ÷ 2 = $4000.

To find 25% of a number, divide the number by 4.

25% of 40 pens is 40 ÷ 4 = 10 pens.

To find 10% of a number, move the decimal point *one* place to the *left*.

10% of $92.40 is $9.24.

To find 1% of a number, move the decimal point *two* places to the *left*.

1% of 62. miles is 0.62 mile.

Concepts	Examples
7.3–7.4 *Using the Percent Equation* Use the six problem-solving steps from **Chapter 3**.	Todd's regular pay is $540 per week but $43.20 is taken out of each paycheck for medical insurance. What percent is that?
Step 1 **Read** the problem.	*Step 1* The problem asks for the percent of Todd's pay that is taken out for insurance. **Unknown:** percent taken out **Known:** Whole paycheck is $540; $43.20 taken out.
Step 2 **Assign** a variable.	*Step 2* Let p be the unknown percent.
Step 3 **Write** an equation.	*Step 3* Use the percent equation. The *whole* is Todd's entire pay of $540 and the *part* is $43.20 (the part taken out of his paycheck). percent • whole = part $p \cdot 540 = 43.20$
Step 4 **Solve** the equation.	*Step 4* $\dfrac{p \cdot 540}{540} = \dfrac{43.20}{540}$ Divide both sides by 540 $p = 0.08$ ← Decimal form Multiply 0.08 by 100 so that $0.08 = 8\%$.
Step 5 **State** the answer.	*Step 5* 8% of Todd's pay is taken out for medical insurance.
Step 6 **Check** the solution.	*Step 6* Use estimation. If 10% of Todd's pay were withheld, then 10% of $540 = $54, which is a little more than the $43.20 actually withheld. So 8% is a reasonable solution.
7.4 *Finding Percent of Increase or Decrease* Use subtraction to find the *amount* of increase or decrease. When writing the equation, be careful to use the *original* value as the whole.	Enrollment rose from 3820 students to 5157 students. Find the percent of increase. 5157 students − 3820 students = 1337 students Amount of increase percent **of** original value = amount of increase $p \cdot 3820 = 1337$ $p \cdot 3820 = 1337$ $\dfrac{p \cdot 3820}{3820} = \dfrac{1337}{3820}$ Divide both sides by 3820 $p = 0.35$ ← Decimal form $0.35 = 35\%$ ← Percent increase The enrollment increased 35%.

Concepts	Examples
7.5 Consumer Applications **Finding Sales Tax** To find sales tax, use this equation. $$\text{tax rate} \cdot \text{cost of item} = \text{sales tax}$$ Write the percent as a decimal.	Find the sales tax on an $89 pair of binoculars if the sales tax rate is 6%. $$\text{tax rate} \cdot \text{cost of item} = \text{sales tax}$$ $$06.\% \cdot \$89 = n$$ $$(0.06)(89) = n$$ $$5.34 = n$$ The sales tax is **$5.34**.

Estimating and Calculating Restaurant Tips
To estimate a 15% tip, first find 10% of the food bill by moving the decimal point *one* place to the *left*. Then add half of that amount for the other 5%.
 	
To estimate a 20% tip, first find 10% by moving the decimal point *one* place to the *left*. Then double the amount.
 	
To find the exact tip, write the percent as a decimal. The bill for food and beverages is the *whole* and the tip is the *part*.
	Estimate a 15% tip on a restaurant bill of $38.72. Then find the exact tip.
	
	Estimate: Round $38.72 to $40.
	  10% of $40. is $4.
	  Half of $4 is $2.
	So an estimate of the 15% tip is $4 + $2 = $6.
	
	Exact:
	$$\text{percent} \cdot \text{whole} = \text{part}$$
	$$15.\% \cdot 38.72 = n$$
	$$(0.15)(38.72) = n$$
	$$5.808 = n$$
	
	The exact tip is **$5.81** (rounded to nearest cent).

Finding a Discount To find a discount, use this formula. $$\text{rate of discount} \cdot \text{original price} = \text{amount of discount}$$ Then subtract to find the sale price. $$\text{original price} - \text{amount of discount} = \text{sale price}$$	All calculators are on sale at 20% off. Find the sale price of a calculator originally marked $35. $$20\% \cdot \$35 = n$$ $$(0.20)(35) = n$$ $$7 = n$$ The amount of discount is $7. Then $35 − $7 = **$28** ← Sale price
Finding Simple Interest To find the simple interest on a loan, use the formula $I = prt$. $$\text{Interest} = \text{principal} \cdot \text{rate} \cdot \text{time}$$ Time (*t*) is in years. When the time is given in months, use a fraction with 12 in the denominator because there are 12 months in a year. Write the rate (the percent) as a decimal.	$2800 is borrowed at 8% for 5 months. Find the amount of interest. $$I = p \cdot r \cdot t$$ $$= (2800)(0.08)\left(\frac{5}{12}\right)$$ $$= (224)\left(\frac{5}{12}\right) = \frac{(224)(5)}{12} \approx \$93.33$$

ANSWERS TO TEST YOUR WORD POWER

1. C; *Example:* 7% means 7 per 100, or, 7 out of 100.
2. A; *Example:* In the problem "15 computers is what percent of 75 computers," the whole is 75 computers, which is the total group of computers.
3. D; *Example:* Houston has a sales tax rate of 8.25%; Minneapolis has a sales tax rate of 7%.
4. B; *Example:* The interest formula is Interest = principal • rate • time. If you borrow $4000 for 2 years at a rate of 9%, then $I = (4000)(0.09)(2) = \$720$.
5. A; *Example:* If you borrow $4000 for 2 years at a rate of 9%, the principal is $4000.
6. C; *Example:* If your rent increased from $800 to $850, the *percent of increase* compares the amount of change ($50) to the *original* rent ($800).
7. C; *Example:* To answer the question, "What percent of 40 Web pages is 12 Web pages," let p be the unknown percent and write the equation as: $p \cdot 40 = 12$.
8. D; *Example:* If sunglasses are on sale at 10% off, then a pair of sunglasses regularly priced at $25 will have a discount of $2.50 subtracted from the price; you will pay $22.50. To find the amount of discount, multiply $(0.10)(\$25)$ to get $2.50.

Focus on Real-Data Applications

Educational Tax Incentives

The government sponsors tax incentive programs to make education more affordable. To qualify for the programs, you have to have an adjusted gross income below a certain level (typically $40,000). You can find specific information at the Internal Revenue Service Web site: www.irs.ustreas.gov.

- The Hope Scholarship offers 100% of the first $1000 spent for certain expenses, such as tuition and books, during the first year of college, plus 50% of the next $1000 incurred during the second year of college. The scholarship money is payable as a tax refund. The student cannot have completed the first two years of post-secondary education and must meet certain educational goals and workload criteria.
- Lifetime Learning Credits are based on qualified expenses, including tuition and books, and equal 20% of the first $5000 in expenses. It is not based on a student's workload and is not limited to only two years.
- Only one of the credits can be claimed for each student.

Suppose you are paying your own educational costs, and your adjusted gross income meets the guidelines to qualify for the Hope Scholarship or Lifetime Learning Credits. Your goals are to earn an Associate of Arts degree from a community college and then transfer to a state university to complete a Bachelor's degree. Tuition costs for resident students at North Harris Montgomery Community College District (NHMCCD) in Texas are used as an example of educational expenses. (Expenses vary among schools, and you can easily find that information in the college's catalog or Internet site.)

Residents of NHMCCD pay $12 registration fee for each semester enrolled plus $32 per semester hour tuition and fees. Assume that you must study a total of 15 semester hours in developmental work in mathematics, reading, and writing, and to complete an Associate of Arts degree you must study 60 additional semester hours. You decide to limit your course load to 15 credit hours each semester. Assume that one course is 3 semester hours, and you will have to purchase books at an approximate cost of $75 per course.

1. How many semesters and how many courses will it take you to finish the requirements for an Associate of Arts degree?
2. What is the total cost to complete the Associate of Arts degree for (a) books and (b) tuition and fees?
3. Calculate the total costs for tuition, fees, and books during the first two years (four semesters). What is the maximum tax incentive payable under the Hope Scholarship during (a) the first year and (b) the second year?
4. Assume that you are returning to school and do not qualify for the Hope Scholarship. What is the maximum tax incentive payable under the Lifetime Learning Credits during (a) the first year and (b) the second year?
5. How much additional tax incentive would be payable under the Lifetime Learning Credits, after the first two years, for the remaining coursework to complete the Associate of Arts degree? How much additional tax incentive would be payable under the Lifetime Learning Credits to complete a Bachelor's degree?

Chapter 7
REVIEW EXERCISES

[7.1] *Write each percent as a decimal and each decimal as a percent.*

1. 25%
2. 180%
3. 12.5%
4. 7%
5. 2.65
6. 0.02
7. 0.3
8. 0.002

Write each percent as a fraction or mixed number in lowest terms. Write each fraction as a percent.

9. 12%
10. 37.5%
11. 250%
12. 5%
13. $\frac{3}{4}$
14. $\frac{5}{8}$
15. $3\frac{1}{4}$
16. $\frac{3}{50}$

Complete this table.

Fraction	Decimal	Percent
$\frac{1}{8}$	17. _____	18. _____
19. _____	0.15	20. _____
21. _____	22. _____	180%

Use percent shortcuts to fill in the blanks.

23. 100% of $46 is _____.

24. 50% of $46 is _____.

25. 100% of 9 hours is _____.

26. 50% of 9 hours is _____.

[7.2] *Use a percent proportion to answer each question. If necessary, round percent answers to the nearest tenth of a percent.*

27. 338.8 meters is 140% of what number of meters?

28. 2.5% of what number of cases is 425 cases?

29. What is 6% of 450 cellular phones?

30. 60% of 1450 reference books is how many books?

31. What percent of 380 pairs is 36 pairs?

32. 1440 cans is what percent of 640 cans?

[7.3] *Use the percent equation to answer each question. Round money answers to the nearest cent, if necessary.*

33. 11% of $23.60 is how much?

34. What is 125% of 64 days?

35. 1.28 ounces is what percent of 32 ounces?

36. $46 is 8% of what number of dollars?

37. 8 people is 40% of what number of people?

38. What percent of 174 ft is 304.5 ft?

[7.4] *Use the six problem-solving steps to answer each question. If necessary, round percent answers to the nearest tenth of a percent.*

39. (a) A medical clinic found that 16.8% of the patients were late for their appointments in January. The number of patients who were late was 504. Find the total number of patients in January.

(b) In February, only 345 patients were late. Find the percent of decrease in late patients from January to February.

40. (a) Coreen budgeted $280 for food on her vacation. She actually spent 130% of that amount. How much did she spend on food?

(b) Coreen's vacation budget included $50 for gifts and souvenirs, but she spent $112. What percent of the budgeted amount did she spend?

41. (a) In the first part of a tree-planting project, 640 of the 800 trees planted were still living one year later. What percent of the trees planted were still living?

42. Scientists tell us that worldwide there are 9600 species of birds and that 1000 of these species are in danger of extinction. What percent of the bird species are in danger of extinction, to the nearest tenth of a percent?

The ivory-billed woodpecker, thought to be extinct, was recently found again in Arkansas.

(b) In the second part of the project, 850 trees were planted. What was the percent of increase in trees planted?

[7.5] *Find the amount of sales tax or the tax rate and the total cost. Round to the nearest cent, if necessary.*

Amount of Sale	Tax Rate	Amount of Tax	Total Cost
43. $2.79	4%	_____	_____
44. $780	_____	$58.50	_____

For each restaurant bill, estimate a 15% tip and a 20% tip. Then find the exact amount for a 15% tip and a 20% tip. Round to the nearest cent, if necessary.

Bill	Estimated 15%	Exact 15%	Estimated 20%	Exact 20%
45. $42.73	_____	_____	_____	_____
46. $8.05	_____	_____	_____	_____

Find the amount or rate of discount and the sale price.

Original Price	Rate of Discount	Amount of Discount	Sale Price
47. $37.50	10%	_____	_____
48. $252	_____	$63	_____

Find the simple interest and total amount due on each loan.

Principal	Rate	Time	Interest	Total Amount Due
49. $350	$6\frac{1}{2}\%$	3 years	_____	_____
50. $1530	16%	9 months	_____	_____

MIXED REVIEW EXERCISES

The bar graph shows the types of electronic/computer games that adults like to play. Use the information in the graph to answer Exercises 51–54.

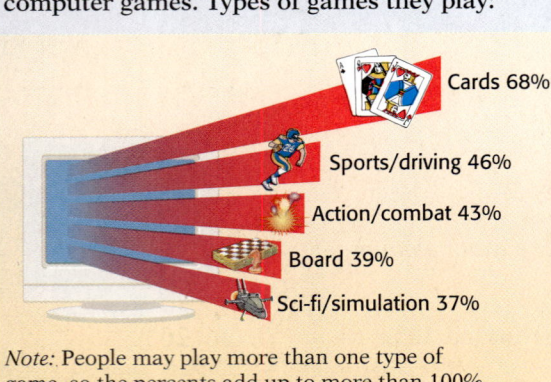

CARE FOR A GAME OF CARDS?
One in three adults say they play electronic or computer games. Types of games they play:

Cards 68%
Sports/driving 46%
Action/combat 43%
Board 39%
Sci-fi/simulation 37%

Note: People may play more than one type of game, so the percents add up to more than 100%.

Source: Cable & Telecommunications Association for Marketing.

51. Write the portion of adults who like to play electronic/computer games as a fraction and as a percent.

52. What type of game is most popular? Write the portion of players who picked this type as a percent, a decimal, and a fraction in lowest terms.

53. If 830 adults were surveyed, how many of them play electronic/computer games, to the nearest whole number?

54. Using your answer from Exercise 53, how many of the playing adults chose the most popular and least popular types of game?

The table shows the number of animals received during the first nine months of the year by the local Animal Humane Society and the number placed in new homes. Use the table to answer Exercises 55–60. Round percent answers to the nearest tenth of a percent.

Animals Received	Placed in New Homes
5371 dogs	2599 dogs
6447 cats	2346 cats
2223 other*	406 other*

* "Other" includes rabbits, birds, gerbils, mice, etc.

55. What percent of the dogs were placed in new homes?

56. The number of dogs received so far this year is 75% of the total number expected. How many dogs are expected, to the nearest whole number?

57. What percent of the cats were placed in new homes?

58. During the first nine months of last year, 2300 cats were received. What is the percent increase in cats received from last year to this year?

59. What percent of all the animals received were placed in new homes?

60. The Society's goal was to place 40% of all animals received. So far they have missed their goal by how many animals? Round to the nearest whole number.

Chapter 7
TEST

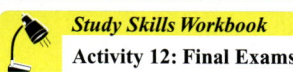
Study Skills Workbook
Activity 12: Final Exams

Write each percent as a decimal and each decimal as a percent.

1. 75%
2. 0.6
3. 1.8

4. 0.075
5. 300%
6. 2%

Write each percent as a fraction or mixed number in lowest terms.

7. 62.5%
8. 240%

Write each fraction or mixed number as a percent.

9. $\dfrac{1}{20}$
10. $\dfrac{7}{8}$
11. $1\dfrac{3}{4}$

Write and solve a proportion or an equation to answer each question. Show your work.

12. 16 laptops is 5% of what number of laptops?

13. $192 is what percent of $48?

14. Erica Green has saved 75% of the amount needed for a down payment on a condominium. If she has saved $14,625, find the total down payment needed.

15. The price of a used car is $7950 plus sales tax of $6\dfrac{1}{2}$%. Find the total cost of the car including sales tax.

16. Enrollment in mathematics courses increased from 1440 students last semester to 1925 students this semester. Find the percent of increase, to the nearest whole percent.

1. _____
2. _____
3. _____
4. _____
5. _____
6. _____
7. _____
8. _____
9. _____
10. _____
11. _____
12. _____
13. _____
14. _____
15. _____
16. _____

17. Explain a shortcut for finding 50% of a number and a shortcut for finding 25% of a number. Show an example of how to use each shortcut.

18. Explain how you would *estimate* a 15% tip on a restaurant bill of $31.94. Then explain how you would *estimate* a 20% tip on the same bill.

19. Find the exact 15% tip, to the nearest cent, for the restaurant bill in Problem 18. If you and two friends are sharing the bill and the exact tip, how much will each person pay?

Find the amount of discount and the sale price of each item in Problems 20 and 21. Round answers to the nearest cent, if necessary. Show your work.

20. Jeremy plans to use his 8% employee discount to buy a $48 clock radio so he can get to work on time.

21. A Samsung progressive scan DVD/VCR combination player regularly priced at $229.95 is on sale at 18% off.

22. Jamal found a $1089 computer on sale at 30% off because it was a "discontinued" model. The store will let him pay for it over 6 months with no interest charge. How much will each monthly payment need to be to cover the discounted price plus 7% sales tax?

23. What is the simple interest on a four-year loan of $5000 at $8\frac{1}{4}\%$?

24. Kendra borrowed $860 to pay medical expenses. The loan is for 6 months at 12% simple interest. Find the total amount due on the loan.

Measurement

8.1 Problem Solving with English Measurement

8.2 The Metric System—Length

8.3 The Metric System—Capacity and Weight (Mass)

8.4 Problem Solving with Metric Measurement

8.5 Metric–English Conversions and Temperature

A greatly enlarged photo of a computer microchip.

We are constantly measuring things, from very large to very small.

For example, on August 27, 2003, the planet Mars was closer to Earth than at any time in the last 60,000 years. Mars was "only" 34,646,437 miles away! (See **Section 8.1**, Exercises 63–66.)

On the other hand, computer microchips often measure a tiny 5 mm long by 1 mm wide. (Learn about millimeters in **Section 8.2** and then try Exercise 39.)

8.1 Problem Solving with English Measurement

OBJECTIVES

1. Learn the basic measurement units in the English system.
2. Convert among measurement units using multiplication or division.
3. Convert among measurement units using unit fractions.
4. Solve application problems using English measurement.

We measure things all the time: the distance traveled on vacation, the floor area we want to cover with carpet, the amount of milk in a recipe, the weight of the bananas we buy at the store, the number of hours we work, and many more.

In the United States we still use the **English system** of measurement for many everyday activities. Examples of English units are inches, feet, quarts, ounces, and pounds. However, the fields of science, medicine, sports, and manufacturing use the **metric system** (meters, liters, and grams). And, because the rest of the world uses only the metric system, U.S. businesses have been changing to the metric system in order to compete internationally.

OBJECTIVE 1 Learn the basic measurement units in the English system. Until the switch to the metric system is complete, we still need to know how to use the English system of measurement. The table below lists the relationships you should memorize. The time relationships are used in both the English and metric systems.

1 After memorizing the measurement conversions, answer these questions.

(a) 1 c = _____ fl oz

(b) _____ qt = 1 gal

(c) 1 wk = _____ days

(d) _____ ft = 1 yd

(e) 1 ft = _____ in.

(f) _____ oz = 1 lb

(g) 1 T = _____ lb

(h) _____ min = 1 hr

(i) 1 pt = _____ c

(j) _____ hr = 1 day

(k) 1 min = _____ sec

(l) 1 qt = _____ pt

(m) _____ ft = 1 mi

English Measurement Relationships

Length
- 1 foot (ft) = 12 inches (in.)
- 1 yard (yd) = 3 feet (ft)
- 1 mile (mi) = 5280 feet (ft)

Capacity
- 1 cup (c) = 8 fluid ounces (fl oz)
- 1 pint (pt) = 2 cups (c)
- 1 quart (qt) = 2 pints (pt)
- 1 gallon (gal) = 4 quarts (qt)

Weight
- 1 pound (lb) = 16 ounces (oz)
- 1 ton (T) = 2000 pounds (lb)

Time
- 1 minute (min) = 60 seconds (sec)
- 1 hour (hr) = 60 minutes (min)
- 1 day = 24 hours (hr)
- 1 week (wk) = 7 days

As you can see, there is no simple way to convert among these various measures. The units evolved over hundreds of years and were based on a variety of "standards." For example, one yard was the distance from the tip of a king's nose to his thumb when his arm was outstretched. An inch was three dried barleycorns laid end to end.

EXAMPLE 1 Knowing English Measurement Units

Memorize the English measurement conversions shown above. Then answer these questions.

(a) _____ day = 24 hr   Answer: 1 day

(b) 1 yd = _____ ft   Answer: 3 ft

◀◀◀ **Work Problem 1 at the Side.**

OBJECTIVE 2 Convert among measurement units using multiplication or division. You often need to convert from one unit of measure to another. Two methods of converting measurements are shown here. Study each way and use the method you prefer. The first method involves deciding whether to multiply or divide.

ANSWERS

1. (a) 8 (b) 4 (c) 7 (d) 3 (e) 12
   (f) 16 (g) 2000 (h) 60 (i) 2 (j) 24
   (k) 60 (l) 2 (m) 5280

Converting among Measurement Units

1. *Multiply* when converting from a larger unit to a smaller unit.
2. *Divide* when converting from a smaller unit to a larger unit.

Section 8.1 Problem Solving with English Measurement 565

EXAMPLE 2 Converting from One Unit of Measure to Another

Convert each measurement.

(a) 7 ft to inches

You are converting from a *larger* unit to a *smaller* unit (a *foot* is longer than an *inch*), so multiply.

Because *1 ft = 12 in.*, multiply by 12.

$$7 \text{ ft} = 7 \cdot 12 = 84 \text{ in.}$$

(b) $3\frac{1}{2}$ lb to ounces

You are converting from a *larger* unit to a *smaller* unit (a *pound* is heavier than an *ounce*), so multiply.

Because *1 lb = 16 oz*, multiply by 16.

$$3\frac{1}{2} \text{ lb} = 3\frac{1}{2} \cdot 16 = \frac{7}{2} \cdot \frac{\overset{8}{\cancel{16}}}{1} = \frac{56}{1} = 56 \text{ oz}$$

(c) 20 qt to gallons

You are converting from a *smaller* unit to a *larger* unit (a *quart* is smaller than a *gallon*), so divide.

Because *4 qt = 1 gal*, divide by 4.

$$20 \text{ qt} = \frac{20}{4} = 5 \text{ gal}$$

↑ Divide by 4

(d) 45 min to hours

You are converting from a *smaller* unit to a *larger* unit (a *minute* is less than an *hour*), so divide.

Because *60 min = 1 hr*, divide by 60 and write the fraction in lowest terms.

$$45 \text{ min} = \frac{45}{60} = \frac{45 \div 15}{60 \div 15} = \frac{3}{4} \text{ hr} \leftarrow \text{Lowest terms}$$

↑ Divide by 60

Work Problem 2 at the Side. ▶▶▶

OBJECTIVE 3 Convert among measurement units using unit fractions. If you have trouble deciding whether to multiply or divide when converting measurements, use *unit fractions* to solve the problem. You'll also find this method useful in science classes. A **unit fraction** is equivalent to 1. Here is an example.

$$\frac{12 \text{ in.}}{12 \text{ in.}} = \frac{\overset{1}{\cancel{12} \text{ in.}}}{\underset{1}{\cancel{12} \text{ in.}}} = 1$$

Use the table of measurement relationships on the previous page to find that 12 in. is the same as 1 ft. So in the numerator you can substitute 1 ft for 12 in., or in the denominator you can substitute 1 ft for 12 in. This makes two useful unit fractions.

$$\frac{\mathbf{1 \text{ ft}}}{\mathbf{12 \text{ in.}}} = 1 \quad \text{or} \quad \frac{\mathbf{12 \text{ in.}}}{\mathbf{1 \text{ ft}}} = 1$$

To convert from one measurement unit to another, just multiply by the appropriate unit fraction. Remember, a unit fraction is equivalent to 1. Multiplying something by 1 does *not* change its value.

2 Convert each measurement using multiplication or division.

(a) $5\frac{1}{2}$ ft to inches

(b) 64 oz to pounds

(c) 6 yd to feet

(d) 2 T to pounds

(e) 35 pt to quarts

(f) 20 min to hours

(g) 4 wk to days

ANSWERS
2. **(a)** 66 in. **(b)** 4 lb **(c)** 18 ft
(d) 4000 lb **(e)** $17\frac{1}{2}$ qt **(f)** $\frac{1}{3}$ hr
(g) 28 days

3 First write the unit fraction needed to make each conversion. Then complete the conversion.

(a) 36 in. to feet

unit fraction } $\dfrac{1 \text{ ft}}{12 \text{ in.}}$

(b) 14 ft to inches

unit fraction } $\dfrac{\text{in.}}{\text{ft}}$

(c) 60 in. to feet

unit fraction } _____

(d) 4 yd to feet

unit fraction } _____

(e) 39 ft to yards

unit fraction } _____

(f) 2 mi to feet

unit fraction } _____

ANSWERS

3. (a) 3 ft (b) $\dfrac{12 \text{ in.}}{1 \text{ ft}}$; 168 in.
   (c) $\dfrac{1 \text{ ft}}{12 \text{ in.}}$; 5 ft (d) $\dfrac{3 \text{ ft}}{1 \text{ yd}}$; 12 ft
   (e) $\dfrac{1 \text{ yd}}{3 \text{ ft}}$; 13 yd (f) $\dfrac{5280 \text{ ft}}{1 \text{ mi}}$; 10,560 ft

---

Use these guidelines to choose the correct unit fraction.

> **Choosing a Unit Fraction**
> The *numerator* should use the measurement unit you want in the *answer*.
> The *denominator* should use the measurement unit you want to *change*.

EXAMPLE 3 Using Unit Fractions with Length Measurements

(a) Convert 60 in. to feet.

Use a unit fraction with feet (the unit for your answer) in the numerator, and inches (the unit being changed) in the denominator. Because *1 ft = 12 in.*, the necessary unit fraction is

$\dfrac{1 \text{ ft}}{12 \text{ in.}}$ ← Unit for your answer is feet.
            ← Unit being changed is inches.

Next, multiply 60 in. times this unit fraction. Write 60 in. as the fraction $\dfrac{60 \text{ in.}}{1}$. Then divide out common units and factors wherever possible.

$$60 \text{ in.} \cdot \dfrac{1 \text{ ft}}{12 \text{ in.}} = \dfrac{\overset{5}{\cancel{60 \text{ in.}}}}{1} \cdot \dfrac{1 \text{ ft}}{\underset{1}{\cancel{12 \text{ in.}}}} = \dfrac{5 \cdot 1 \text{ ft}}{1} = 5 \text{ ft}$$

These units should match.
Divide out inches.
Divide 60 and 12 by 12

(b) Convert 9 ft to inches.

Select the correct unit fraction to change 9 ft to inches.

$\dfrac{12 \text{ in.}}{1 \text{ ft}}$ ← Unit for your answer is inches.
            ← Unit being changed is feet.

Multiply 9 ft times the unit fraction.

$$9 \text{ ft} \cdot \dfrac{12 \text{ in.}}{1 \text{ ft}} = \dfrac{9 \cancel{\text{ ft}}}{1} \cdot \dfrac{12 \text{ in.}}{1 \cancel{\text{ ft}}} = \dfrac{9 \cdot 12 \text{ in.}}{1} = 108 \text{ in.}$$

These units should match.
Divide out feet.

> **CAUTION**
> If no units will divide out, you made a mistake in choosing the unit fraction.

◀◀◀ **Work Problem 3 at the Side.**

EXAMPLE 4 Using Unit Fractions with Capacity and Weight Measurements

(a) Convert 9 pt to quarts.

First select the correct unit fraction.

$\dfrac{1 \text{ qt}}{2 \text{ pt}}$ ← Unit for your answer is quarts.
          ← Unit being changed is pints.

Continued on Next Page

Now multiply.

$$9 \text{ pt} \cdot \frac{1 \text{ qt}}{2 \text{ pt}} = \frac{9 \text{ pt}}{1} \cdot \frac{1 \text{ qt}}{2 \text{ pt}} = \frac{9}{2} \text{ qt} = 4\frac{1}{2} \text{ qt}$$

These units should match. ↑ Divide out pints.

Write as mixed number.

(b) Convert $7\frac{1}{2}$ gal to quarts.

Write as an improper fraction.

$$\frac{7\frac{1}{2} \text{ gal}}{1} \cdot \frac{4 \text{ qt}}{1 \text{ gal}} = \frac{15}{2} \cdot \frac{4}{1} \text{ qt}$$

Divide out gallons.

$$= \frac{15}{\cancel{2}} \cdot \frac{\cancel{4}^2}{1} \text{ qt}$$

$$= 30 \text{ qt}$$

(c) Convert 36 oz to pounds.

$$\frac{\cancel{36}^9 \text{ oz}}{1} \cdot \frac{1 \text{ lb}}{\cancel{16}_4 \text{ oz}} = \frac{9}{4} \text{ lb} = 2\frac{1}{4} \text{ lb}$$

NOTE
In Example 4(c) above, you get $\frac{9}{4}$ lb. Recall that $\frac{9}{4}$ means 9 ÷ 4. If you do 9 ÷ 4 on your calculator, you get 2.25 lb. English measurements usually use fractions or mixed numbers, like $2\frac{1}{4}$ lb. However, 2.25 lb is also correct and is the way grocery stores often show weights of produce, meat, and cheese.

Work Problem 4 at the Side.

EXAMPLE 5 Using Several Unit Fractions

Sometimes you may need to use two or three unit fractions to complete a conversion.

(a) Convert 63 in. to yards.

Use the unit fraction $\frac{1 \text{ ft}}{12 \text{ in.}}$ to change inches to feet and the unit fraction $\frac{1 \text{ yd}}{3 \text{ ft}}$ to change feet to yards. Notice how all the units divide out except yards, which is the unit you want in the answer.

$$\frac{63 \text{ in.}}{1} \cdot \frac{1 \text{ ft}}{12 \text{ in.}} \cdot \frac{1 \text{ yd}}{3 \text{ ft}} = \frac{63}{36} \text{ yd} = \frac{63 \div 9}{36 \div 9} \text{ yd} = \frac{7}{4} \text{ yd} = 1\frac{3}{4} \text{ yd}$$

Continued on Next Page

---

4 Convert using unit fractions.

(a) 16 qt to gallons

(b) 3 c to pints

(c) $3\frac{1}{2}$ T to pounds

(d) $1\frac{3}{4}$ lb to ounces

(e) 4 oz to pounds

ANSWERS
4. **(a)** 4 gal **(b)** $1\frac{1}{2}$ pt or 1.5 pt **(c)** 7000 lb
   (d) 28 oz **(e)** $\frac{1}{4}$ lb or 0.25 lb

5 Convert using two or three unit fractions.

(a) 4 T to ounces

(b) 3 mi to inches

(c) 36 pt to gallons

(d) 2 wk to minutes

You can also divide out common factors in the numbers.

$$\frac{\overset{21}{\cancel{63}}}{1} \cdot \frac{1}{\underset{4}{\cancel{12}}} \cdot \frac{1}{\underset{1}{\cancel{3}}} = \frac{7}{4} = 1\frac{3}{4} \text{ yd}$$

Instead of changing $\frac{7}{4}$ to $1\frac{3}{4}$, you can enter $7 \div 4$ on your calculator to get 1.75 yd. Both answers are correct because 1.75 is equivalent to $1\frac{3}{4}$.

(b) Convert 2 days to seconds.

Use three unit fractions. The first one changes days to hours, the next one changes hours to minutes, and the last one changes minutes to seconds. All the units divide out except seconds, which is what you want in your answer.

$$\frac{2 \cancel{\text{days}}}{1} \cdot \frac{24 \cancel{\text{hr}}}{1 \cancel{\text{day}}} \cdot \frac{60 \cancel{\text{min}}}{1 \cancel{\text{hr}}} \cdot \frac{60 \text{ seconds}}{1 \cancel{\text{min}}} = 172{,}800 \text{ seconds}$$

Divide out **days**.
Divide out **hr**.
Divide out **min**.

◀◀ **Work Problem 5 at the Side.**

OBJECTIVE 4 Solve application problems using English measurement. To solve measurement application problems, we will use the steps summarized here.

Step 1 **Read** the problem.
Step 2 **Work out a plan.**
Step 3 **Estimate** a reasonable answer.
Step 4 **Solve** the problem.
Step 5 **State the answer.**
Step 6 **Check** your work.

Because measurement applications often involve conversions, writing an equation may not be the most helpful way to solve the problem. Therefore, Steps 2 and 3 are different from the ones you learned in **Chapter 3**; the other steps are the same.

EXAMPLE 6 Solving English Measurement Applications

(a) A 36 oz can of coffee is on sale at Jerry's Foods for $7.89. What is the cost per pound, to the nearest cent? (*Source:* Jerry's Foods.)

Step 1 **Read** the problem. The problem asks for the cost per *pound* of coffee.

Step 2 **Work out a plan.** The weight of the coffee is given in *ounces* but the answer must be cost per *pound*. Convert ounces to pounds. The word *per* indicates division. You need to divide the cost by the number of pounds.

Step 3 **Estimate** a reasonable answer. To estimate, round $7.89 to $8. Then, there are 16 oz in a pound, so 36 oz is a little more than 2 pounds. So, $8 ÷ 2 = $4 per pound as our estimate.

Continued on Next Page

ANSWERS
5. (a) 128,000 oz   (b) 190,080 in.
   (c) $4\frac{1}{2}$ gal or 4.5 gal   (d) 20,160 min

Step 4 **Solve** the problem. Use a unit fraction to convert 36 oz to pounds.

$$\frac{\overset{9}{\cancel{36} \text{ oz}}}{1} \cdot \frac{1 \text{ lb}}{\underset{4}{\cancel{16} \text{ oz}}} = \frac{9}{4} \text{ lb} = 2.25 \text{ lb}$$

Then divide to find the *cost* per *pound*.

$$\begin{matrix}\text{cost} \rightarrow \\ \text{per} \rightarrow \\ \text{pound} \rightarrow\end{matrix} \frac{\$7.89}{2.25 \text{ lb}} = 3.50\overline{6} \approx 3.51 \quad \text{Rounded}$$

Step 5 **State the answer.** The coffee costs $3.51 per pound (to the nearest cent).

Step 6 **Check** your work. The exact answer of $3.51 is close to our estimate of $4.

(b) Bilal's favorite cake recipe uses $1\frac{2}{3}$ cups of milk. If he makes six cakes for a bake sale at his son's school, how many quarts of milk will he need?

Step 1 **Read** the problem. The problem asks for the number of *quarts* of milk needed for six cakes.

Step 2 **Work out a plan.** Multiply to find the number of *cups* of milk for six cakes. Then convert *cups* to *quarts* (the unit required in the answer).

Step 3 **Estimate** a reasonable answer. To estimate, round $1\frac{2}{3}$ cups to 2 cups. Then, 2 cups times 6 = 12 cups. There are 4 cups in a quart, so 12 cups ÷ 4 = 3 quarts as our estimate.

Step 4 **Solve** the problem. First multiply. Then use unit fractions to convert.

$$1\frac{2}{3} \cdot 6 = \frac{5}{\underset{1}{\cancel{3}}} \cdot \frac{\overset{2}{\cancel{6}}}{1} = \frac{10}{1} = 10 \text{ cups} \quad \left\{\begin{matrix}\text{Milk needed} \\ \text{for six cakes}\end{matrix}\right.$$

$$\frac{\overset{5}{\cancel{10} \text{ cups}}}{1} \cdot \frac{1 \text{ pt}}{\underset{1}{\cancel{2} \text{ cups}}} \cdot \frac{1 \text{ qt}}{2 \text{ pt}} = \frac{5}{2} \text{ qt} = 2\frac{1}{2} \text{ qt}$$

Step 5 **State the answer.** Bilal needs $2\frac{1}{2}$ qt (or 2.5 qt) of milk.

Step 6 **Check** your work. The exact answer of $2\frac{1}{2}$ qt is close to our estimate of 3 qt.

> **NOTE**
> In *Step 2* above, we *first multiplied* $1\frac{2}{3}$ cups times 6 to find the number of cups needed, then *converted* 10 cups to $2\frac{1}{2}$ quarts. It would also work to *first convert* $1\frac{2}{3}$ cups to $\frac{5}{12}$ qt, then *multiply* $\frac{5}{12}$ qt times 6 to get $2\frac{1}{2}$ qt.

Work Problem 6 at the Side. ▶▶▶

6 Solve each application problem using the six problem-solving steps.

(a) Kristin paid $3.29 for 12 oz of extra sharp cheddar cheese. What is the price per pound, to the nearest cent?

(b) A moving company estimates 11,000 lb of furnishings for an average 3-bedroom house. If the company made five such moves last week, how many tons of furnishings did they move? (*Source:* North American Van Lines.)

ANSWERS
6. **(a)** $4.39 per pound (rounded)
 (b) 27.5 T or $27\frac{1}{2}$ T

Focus on Real-Data Applications

Growing Sunflowers

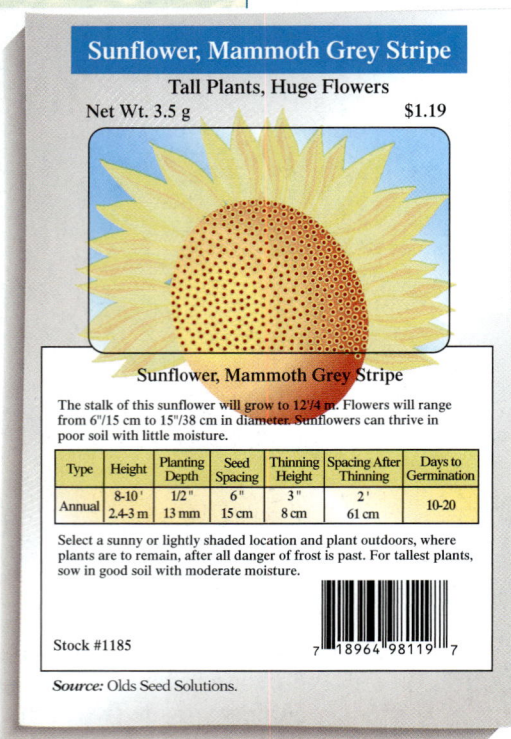

The front and back of a seed packet for sunflowers are shown at the left. Look at the top of the packet first.

1. There were 42 seeds in the packet. If 40 of the seeds sprouted, what was the cost per sprout, to the nearest cent?

2. If vegetable and flower seeds were on sale at 30% off, what was the cost per sprout, to the nearest cent?

3. What percent of the seeds sprouted, to the nearest whole percent?

4. How many seeds would weigh 1 gram?

5. The table at the bottom of the packet uses the symbol (') for feet, and the symbol (") for inches.

   (a) How tall will the plants grow, in feet?

   (b) How tall will they grow, in inches?

   (c) How tall will they grow, in yards?

6. If you plant all 42 seeds in one long row, using the spacing given on the package, how long will your row be in feet?

7. How many inches tall should the plants be when you thin them (remove less vigorous plants to give others room to grow)? How tall is that in feet?

8. What is the range in the diameter of the flowers, in inches, and in feet? Diameter is the distance across the circular flower.

9. (a) Using the information in the article on the right, how many gum wrappers are needed to make 1 foot of chain?

   (b) To make 1 inch of chain?

   (c) How many inches of chain, to the nearest hundredth, are made from one wrapper?

10. Is the article correct in saying that 125 miles is 34 million gum wrappers?

Wrapped Up in Work?

Sometimes a person's life's work makes it into a museum. So it figures that Michael Knutson's 128-foot chain of gum wrappers now resides in the Yellow Medicine County Museum in Granite Falls, Minnesota.

According to the *Redwood Gazette*, the wrapper chain started in 1974 when Knutson, now a 43-year-old woodworker, became bored during study hall. "I've never been much of a gum chewer, but I found most of the wrappers on the streets of the city," he said. (FYI: It takes 6602 wrappers to make a 128-foot chain.)

Granite Falls is about 125 miles— or 34 million gum wrappers—west of Minneapolis/ St. Paul.

Source: Minneapolis Star Tribune.

8.1 Exercises

Fill in the blanks with the measurement relationships you have memorized. See Example 1.

1. 1 yd = _____ ft

2. 1 ft = _____ in.

3. _____ fl oz = 1 c

4. _____ qt = 1 gal

5. 1 mi = _____ ft

6. 1 wk = _____ days

7. _____ lb = 1 T

8. _____ oz = 1 lb

9. 1 min = _____ sec

10. 1 day = _____ hr

Convert each measurement in Exercises 11–38 using unit fractions. See Examples 3 and 4.

11. 120 sec = _____ min

12. 180 min = _____ hr

13. 8 qt = _____ gal

14. 6 gal = _____ qt

15. An adult African elephant could weigh 7 to 8 tons. How many pounds could it weigh? (*Source: The Top 10 of Everything.*)

16. A reticulated python is the world's longest snake. It grows to a length of 18 to 33 feet. How many yards long can the snake be?

17. 9 yd = _____ ft

18. 20,000 lb = _____ T

19. 7 lb = _____ oz

20. 96 oz = _____ lb

21. 5 qt = _____ pt

22. 26 pt = _____ qt

23. 90 min = _____ hr

24. 45 sec = _____ min

25. 3 in. = _____ ft

26. 30 in. = _____ ft

572 Chapter 8 Measurement

27. 24 oz = _____ lb

28. 36 oz = _____ lb

29. 5 c = _____ pt

30. 15 qt = _____ gal

Use the information in the bar graph below to answer Exercises 31–32.

Thickness of Lake Ice Needed for Safe Walking/Driving

- 15 in. Pickup truck
- 12 in. Car
- 5 in. Snowmobile or ATV
- 4 in. Person walking

Source: Wisconsin DNR.

31. If the ice on a lake is $\frac{1}{2}$ ft thick, what will it safely support?

32. How many feet of ice are needed to safely drive a pickup truck on a lake?

33. $2\frac{1}{2}$ T = _____ lb

34. $4\frac{1}{2}$ pt = _____ c

35. $4\frac{1}{4}$ gal = _____ qt

36. $2\frac{1}{4}$ hr = _____ min

37. After 15 years a saguaro cactus is still only one-third to two-thirds of a foot tall, depending upon rainfall. How tall could the cactus be in inches? (*Source: Ecology of the Saguaro III.*)

38. Yao Ming, an NBA basketball player originally from China, is $7\frac{1}{2}$ ft tall. What is his height in inches? (*Source:* www.NBA.com)

Use two or three unit fractions to make each conversion. See Example 5.

39. 6 yd = _____ in.

40. 2 T = _____ oz

41. 112 c = _____ qt

42. 336 hr = _____ wk

43. 6 days = _____ sec

44. 5 gal = _____ c

45. $1\frac{1}{2}$ T = _____ oz

46. $3\frac{1}{3}$ yd = _____ in.

47. The statement 8 = 2 is *not* true. But with appropriate measurement units, it *is* true.

$$8 \text{ quarts} = 2 \text{ gallons}$$

Attach measurement units to these numbers to make the statements true.

(a) 1 _____ = 16 _____

(b) 10 _____ = 20 _____

(c) 120 _____ = 2 _____

(d) 2 _____ = 24 _____

(e) 6000 _____ = 3 _____

(f) 35 _____ = 5 _____

48. Explain in your own words why you can add 2 feet + 12 inches to get 3 feet, but you cannot add 2 feet + 12 pounds.

Convert each measurement. See Example 5.

49. $2\frac{3}{4}$ mi = _____ in.

50. $5\frac{3}{4}$ tons = _____ oz

51. $6\frac{1}{4}$ gal = _____ fl oz

52. $3\frac{1}{2}$ days = _____ sec

53. 24,000 oz = _____ T

54. 57,024 in. = _____ mi

Solve each application problem in Exercises 55–62. Show your work. See Example 6.

55. Geralyn bought 20 oz of strawberries for $2.29. What was the price per pound for the strawberries, to the nearest cent?

56. Zach paid $0.79 for a 1.6 oz candy bar. What was the cost per pound?

57. Dan orders supplies for the science labs. Each of the 24 stations in the chemistry lab needs 2 ft of rubber tubing. If rubber tubing sells for $8.75 per yard, how much will it cost to equip all the stations?

58. In 2001, Marquette, Michigan, had 219 in. of snowfall, while Detroit, Michigan, had 18 in. What was the difference in snowfall between the two cities, in feet? Round to the nearest tenth. (*Source: World Almanac.*)

59. Tropical cockroaches are the fastest land insects. They can run about 5 feet per second. At this rate, how long would it take the cockroach to travel one mile? (*Source: Guinness Book of World Records.*) Give your answer

(a) in seconds

(b) In minutes, to the nearest tenth.

60. A snail moves at an average speed of 2 feet every 3 minutes. At that rate, how long would it take the snail to travel one mile? (*Source: Beakman and Jax.*) Give your answer

(a) in hours

(b) in days.

61. At the day care center, each of the 15 toddlers drinks about $\frac{2}{3}$ cup of milk with lunch. The center is open 5 days a week.

(a) How many quarts of milk will the center need for one week of lunches?

(b) If the center buys milk in gallon jugs, how many jugs should be ordered for one week?

62. Bob's Candies in Albany, Georgia, makes 135,000 pounds of candy canes each day. (*Source:* Bob's Candies, Inc.)

(a) How many tons of candy canes are produced during a 5-day workweek?

(b) The plant operates 24 hours per day. How many tons of candy canes are produced each hour, to the nearest tenth?

RELATING CONCEPTS (EXERCISES 63–66) For Individual or Group Work

On the first page of this chapter, we said that the planet Mars was only 34,646,437 miles from Earth in August 2003. Use this information and learn more about Mars as you **work** *Exercises 63–66 in order.* (*Sources:* www.NASA.com *and* Associated Press.)

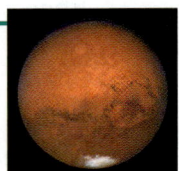

63. In order to better understand how far away Mars was in August 2003, we need to compare the distance to something more familiar.

(a) The distance around Earth is about 24,907 miles. How many times would you travel around Earth before going 34,646,437 miles? Round your answer to the nearest whole number.

(b) The driving distance from Los Angeles to New York City is 2786 miles. How many times would you have to make that drive to match the distance from Earth to Mars? Round to the nearest whole number.

64. The divisions you did in Exercise 63 were long and messy, so a calculator was very helpful. Now let's see how we can do it *without* a calculator.

(a) Round the distance from Mars to Earth to the nearest million. Round the distance around Earth to the nearest thousand. Now do the division without a calculator, using the shortcut for division by multiples of 10 in **Section 1.7**.

(b) Round the driving distance from Los Angeles to New York City to the nearest hundred. Now do the division by hand using the shortcut.

(c) Compare your answers above to those from Exercise 63. Are they just as useful as the answers in Exercise 63?

65. The tallest volcano on any planet in our solar system is on Mars. It is called Olympic Mons and is 16.7 miles high.

(a) How many feet tall is Olympic Mons?

(b) Mount Everest, the tallest place on Earth, is 29,035 ft above sea level. How many miles tall is it, to the nearest tenth?

(c) About how many *times* taller is Olympic Mons than Mount Everest?

66. The pull of gravity on Mars is less than it is on Earth. Because of this, a person who weighs 100 pounds on Earth would weigh only 36 pounds on Mars.

(a) Set up a proportion to find out how much a 180-pound person would weigh on Mars, to the nearest pound.

(b) Set up a proportion to figure out how much a 9-pound baby would weigh on Mars, to the nearest pound.

(c) How much would *you* weigh on Mars?

8.2 The Metric System—Length

Around 1790, a group of French scientists developed the metric system of measurement. It is an organized system based on multiples of 10, like our number system and our money. After you are familiar with metric units, you will see that they are easier to use than the hodgepodge of English measurement relationships you used in **Section 8.1**.

> **NOTE**
> The metric system information in this text is consistent with usage guidelines from the National Institute of Standards and Technology, www.nist.gov/metric.

OBJECTIVES

1. Learn the basic metric units of length.
2. Use unit fractions to convert among units.
3. Move the decimal point to convert among units.

OBJECTIVE 1 Learn the basic metric units of length. The basic unit of length in the metric system is the **meter** (also spelled *metre*). Use the symbol **m** for meter; do not put a period after it. If you put five of the pages from this textbook side by side, they would measure about 1 meter. Or, look at a yardstick—a meter is just a little longer. A yard is 36 inches long; a meter is about 39 inches long.

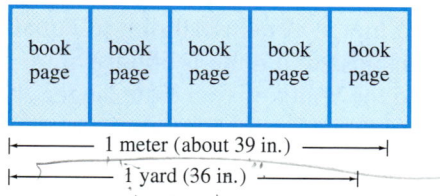

In the metric system you use meters for things like buying fabric for sewing projects, measuring the length of your living room, talking about heights of buildings, or describing track and field athletic events.

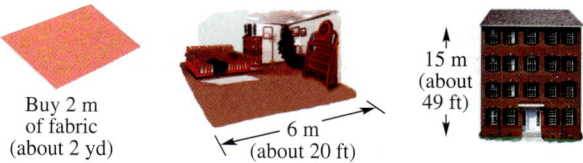

Buy 2 m of fabric (about 2 yd)

6 m (about 20 ft)

15 m (about 49 ft)

1 Circle the items that measure about 1 meter.

Length of a pencil

Length of a baseball bat

Height of doorknob from the floor

Height of a house

Basketball player's arm length

Length of a paper clip

> **Work Problem 1 at the Side.**

To make longer or shorter length units in the metric system, **prefixes** are written in front of the word *meter*. For example, the prefix *kilo* means **1000**, so a *kilo*meter is **1000** meters. The table below shows how to use the prefixes for length measurements. It is helpful to memorize the prefixes because they are also used with weight and capacity measurements. The blue boxes are the units you will use most often in daily life.

Prefix	kilo- meter	hecto- meter	deka- meter	meter	deci- meter	centi- meter	milli- meter
Meaning	1000 meters	100 meters	10 meters	1 meter	$\frac{1}{10}$ of a meter	$\frac{1}{100}$ of a meter	$\frac{1}{1000}$ of a meter
Symbol	**k**m	**h**m	**da**m	m	**d**m	**c**m	**mm**

Length units that are used most often

ANSWERS
1. baseball bat, height of doorknob, basketball player's arm length

2 Write the most reasonable metric unit in each blank. Choose from km, m, cm, and mm.

(a) The woman's height is 168 _____.

(b) The man's waist is 90 _____ around.

(c) Louise ran the 100 _____ dash in the track meet.

(d) A postage stamp is 22 _____ wide.

(e) Michael paddled his canoe 2 _____ down the river.

(f) The pencil lead is 1 _____ thick.

(g) A stick of gum is 7 _____ long.

(h) The highway speed limit is 90 _____ per hour.

(i) The classroom was 12 _____ long.

(j) A penny is about 18 _____ across.

ANSWERS
2. (a) cm (b) cm (c) m (d) mm (e) km
   (f) mm (g) cm (h) km (i) m (j) mm

Here are some comparisons to help you get acquainted with the commonly used length units: km, m, cm, mm.

Kilometers are used instead of miles. A kilometer is 1000 meters. It is about 0.6 mile (a little more than half a mile) or about 5 to 6 city blocks. If you participate in a 10 km run, you'll run about 6 miles.

A meter is divided into 100 smaller pieces called *centi*meters. Each centimeter is $\frac{1}{100}$ of a meter. Centimeters are used instead of inches. A centimeter is a little shorter than $\frac{1}{2}$ inch. The cover of this textbook is about 21 cm wide. A nickel is about 2 cm across. Measure the width and length of your little finger on this centimeter ruler. The width of your little finger is probably about 1 cm.

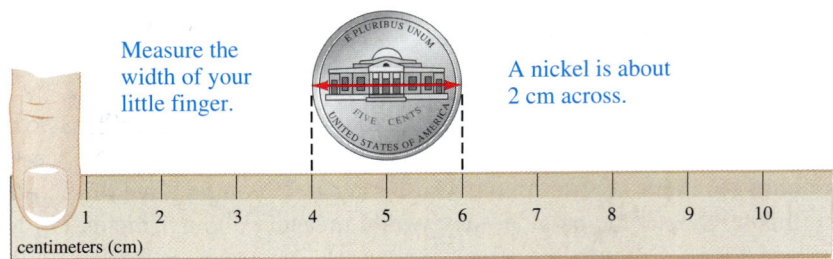

A meter is divided into 1000 smaller pieces called *milli*meters. Each millimeter is $\frac{1}{1000}$ of a meter. It takes 10 mm to equal 1 cm, so it is a very small length. The thickness of a dime is about 1 mm. Measure the width of your pen or pencil and the width of your little finger on this millimeter ruler.

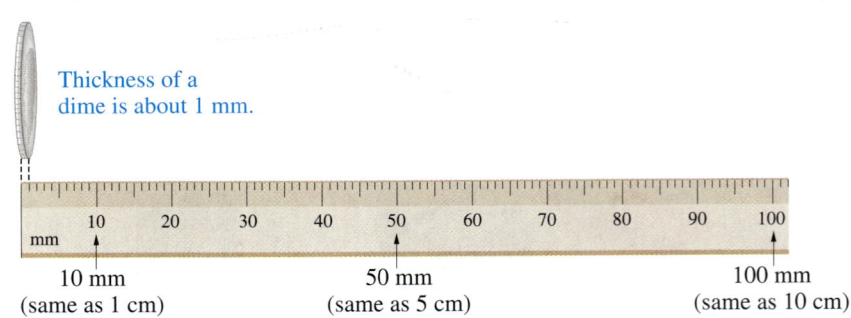

EXAMPLE 1 Using Metric Length Units

Write the most reasonable metric unit in each blank. Choose from km, m, cm, and mm.

(a) The distance from home to work is 20 _____.

   20 **km** because kilometers are used instead of miles.

   20 km is about 12 miles.

(b) My wedding ring is 4 _____ wide.

   4 **mm** because the width of a ring is very small.

(c) The newborn baby is 50 _____ long.

   50 **cm**, which is half of a meter; a meter is about 39 inches so half a meter is around 20 inches.

◀◀◀ **Work Problem 2 at the Side.**

OBJECTIVE 2 **Use unit fractions to convert among units.** You can convert among metric length units using unit fractions. Keep these relationships in mind when setting up the unit fractions.

Metric Length Relationships

1 km = 1000 m so the unit fractions are:	1 m = 1000 mm so the unit fractions are:
$\dfrac{1 \text{ km}}{1000 \text{ m}}$ or $\dfrac{1000 \text{ m}}{1 \text{ km}}$	$\dfrac{1 \text{ m}}{1000 \text{ mm}}$ or $\dfrac{1000 \text{ mm}}{1 \text{ m}}$
1 m = 100 cm so the unit fractions are:	1 cm = 10 mm so the unit fractions are:
$\dfrac{1 \text{ m}}{100 \text{ cm}}$ or $\dfrac{100 \text{ cm}}{1 \text{ m}}$	$\dfrac{1 \text{ cm}}{10 \text{ mm}}$ or $\dfrac{10 \text{ mm}}{1 \text{ cm}}$

EXAMPLE 2 **Using Unit Fractions to Convert Length Measurements**

Convert each measurement using unit fractions.

(a) 5 km to m

Put the unit for the answer (meters) in the numerator of the unit fraction; put the unit you want to change (km) in the denominator.

Unit fraction equivalent to 1 $\left\{ \dfrac{1000 \text{ m}}{1 \text{ km}} \right.$ ← Unit for answer
← Unit being changed

Multiply. Divide out common units where possible.

$$5 \text{ km} \cdot \dfrac{1000 \text{ m}}{1 \text{ km}} = \dfrac{5 \text{ km}}{1} \cdot \dfrac{1000 \text{ m}}{1 \text{ km}} = \dfrac{5 \cdot 1000 \text{ m}}{1} = 5000 \text{ m}$$

These units should match.

Thus, 5 km = 5000 m.

The answer makes sense because a kilometer is much longer than a meter, so 5 km will contain many meters.

(b) 18.6 cm to m

Multiply by a unit fraction that allows you to divide out centimeters.

Unit fraction
$$\dfrac{18.6 \text{ cm}}{1} \cdot \dfrac{1 \text{ m}}{100 \text{ cm}} = \dfrac{18.6}{100} \text{ m} = 0.186 \text{ m}$$

Thus, 18.6 cm = 0.186 m.

There are 100 cm in a meter, so 18.6 cm will be a small part of a meter. The answer makes sense.

Work Problem 3 at the Side.

3 First write the unit fraction needed to make each conversion. Then complete the conversion.

(a) 3.67 m to cm

unit fraction $\left\{ \dfrac{100 \text{ cm}}{1 \text{ m}} \right.$

(b) 92 cm to m

unit fraction $\left\{ \dfrac{\text{m}}{\text{cm}} \right.$

(c) 432.7 cm to m

unit fraction $\left\{ \rule{2cm}{0.4pt} \right.$

(d) 65 mm to cm

unit fraction $\left\{ \rule{2cm}{0.4pt} \right.$

(e) 0.9 m to mm

unit fraction $\left\{ \rule{2cm}{0.4pt} \right.$

(f) 2.5 cm to mm

unit fraction $\left\{ \rule{2cm}{0.4pt} \right.$

ANSWERS

3. **(a)** 367 cm **(b)** $\dfrac{1 \text{ m}}{100 \text{ cm}}$; 0.92 m

   (c) $\dfrac{1 \text{ m}}{100 \text{ cm}}$; 4.327 m

   (d) $\dfrac{1 \text{ cm}}{10 \text{ mm}}$; 6.5 cm

   (e) $\dfrac{1000 \text{ mm}}{1 \text{ m}}$; 900 mm

   (f) $\dfrac{10 \text{ mm}}{1 \text{ cm}}$; 25 mm

578 Chapter 8 Measurement

④ Do each multiplication or division by hand or on a calculator. Compare your answer to the one you get by moving the decimal point.

(a) $(43.5)(10) =$ _____

43.5 gives 435.

(b) $43.5 \div 10 =$ _____

43.5 gives _____.

(c) $(28)(100) =$ _____

28.00 gives _____.

(d) $28 \div 100 =$ _____

28. gives _____.

(e) $(0.7)(1000) =$ _____

0.700 gives _____.

(f) $0.7 \div 1000 =$ _____

000.7 gives _____.

ANSWERS
4. (a) 435   (b) 4.35; 4.35   (c) 2800; 2800
   (d) 0.28; 0.28   (e) 700; 700
   (f) 0.0007; 0.0007

OBJECTIVE 3 Move the decimal point to convert among units. By now you have probably noticed that conversions among metric units are made by multiplying or dividing by 10, by 100, or by 1000. A quick way to *multiply* by 10 is to move the decimal point one place to the *right*. Move it two places to the right to multiply by 100, three places to multiply by 1000. *Dividing* is done by moving the decimal point to the *left* in the same manner.

◀◀◀ **Work Problem 4 at the Side.**

An alternate conversion method to unit fractions is moving the decimal point using this **metric conversion line**.

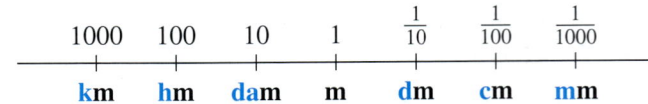

Here are the steps for using the conversion line.

> **Using the Metric Conversion Line**
> *Step 1* Find the unit you are given on the metric conversion line.
> *Step 2* Count the number of places to get from the unit you are given to the unit you want in the answer.
> *Step 3* Move the decimal point the **same number of places** and in the **same direction** as you did on the conversion line.

EXAMPLE 3 Using the Metric Conversion Line

Use the metric conversion line to make the following conversions.

(a) 5.702 km to m

Find **km** on the metric conversion line. To get to **m**, you move *three places* to the *right*. So move the decimal point in 5.702 *three places* to the *right*.

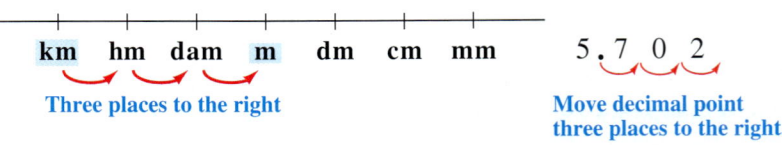

Thus, 5.702 km = 5702 m.

(b) 69.5 cm to m

Find **cm** on the conversion line. To get to **m**, move *two places* to the *left*.

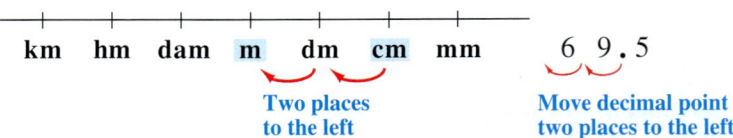

Thus, 69.5 cm = 0.695 m.

Continued on Next Page

(c) 8.1 cm to mm

From cm to mm is *one place* to the *right*.

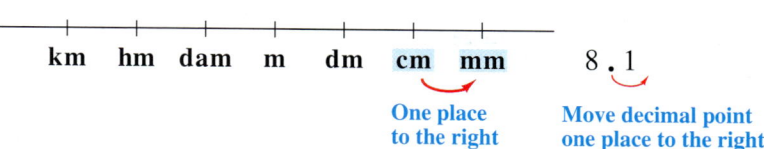

One place to the right

Move decimal point one place to the right.

Thus, 8.1 cm = 81 mm.

▶ **Work Problem 5 at the Side.** ▶▶▶

EXAMPLE 4 Practicing Length Conversions

Convert using the metric conversion line.

(a) 1.28 m to mm

Moving from m to mm is going *three places to the right*. In order to move the decimal point in 1.28 three places to the right, you must write a 0 as a placeholder.

1.28**0**   Zero is written in as a placeholder.

Move decimal point three places to the right.

Thus, 1.28 m = 1280 mm.

(b) 60 cm to m

From cm to m is two places to the left. The decimal point in 60 starts at the *far right side* because 60 is a whole number. Then move it two places to the left.

60.   60.
↑
Decimal point starts here.   Move decimal point two places to the left.

Thus, 60 cm = 0.60 m, which is equivalent to 0.6 m.

(c) 8 m to km

From m to km is three places to the left. The decimal point in 8 starts at the far right side. In order to move it three places to the left, you must write two zeros as placeholders.

Two zeros are written in as placeholders.

8.   **00**8.
↑
Decimal point starts here.   Move decimal point three places to the left.

Thus, 8 m = 0.008 km.

▶ **Work Problem 6 at the Side.** ▶▶▶

5 Convert using the metric conversion line.

(a) 12.008 km to m

(b) 561.4 m to km

(c) 20.7 cm to m

(d) 20.7 cm to mm

(e) 4.66 m to cm

(f) 85.6 mm to cm

6 Convert using the metric conversion line.

(a) 9 m to mm

(b) 3 cm to m

(c) 14.6 km to m

(d) 5 mm to cm

(e) 70 m to km

(f) 0.8 m to cm

ANSWERS
5. **(a)** 12,008 m **(b)** 0.5614 km
   (c) 0.207 m **(d)** 207 mm
   (e) 466 cm **(f)** 8.56 cm
6. **(a)** 9000 mm **(b)** 0.03 m
   (c) 14,600 m **(d)** 0.5 cm
   (e) 0.07 km **(f)** 80 cm

Focus on Real-Data Applications

Measuring Up

1. How much do nails grow in one week? one month? one year?

2. How much does scalp hair grow in one week? one month? one year? (Use metric units.)

3. When you have finished **Section 8.5,** come back to this article. Is the statement about hair growing 6 inches a year accurate? Explain your answer.

Hair and Nail Growth

Q How fast do hair and nails grow? Do they grow faster in the summer?

A Fingernails grow, on average, about one-tenth of a millimeter per day, although there is considerable variation among individuals. Fingernails grow faster than toenails, and nails on the longest fingers appear to grow the fastest.

Fingernails, as well as hair and skin, grow faster in the summer, presumably under the influence of sunlight, which expands blood vessels, bringing more oxygen and nutrients to the area and allowing for faster growth.

The rate the scalp hair grows is 0.3 to 0.4 millimeter per day, or about 6 inches a year.

Source: Minneapolis Star Tribune.

New Device Measures Distances within Billionths of an Inch

U.S. officials have unveiled "the ultimate ruler," a measuring device that can gauge distances to within billionths of an inch—the length of five individual atoms—and may help revolutionize high-tech manufacturing.

Developed at a cost of $8 million by the National Institute of Standards and Technology with other agencies, the Molecular Measuring Machine can measure distances to within 40 nanometers, or billionths of a meter. After further refinement it is expected to measure within 1 nanometer.

By way of comparison, the period at the end of this sentence is about 300,000 nanometers wide.

The machine is expected to prove a boon to U.S. manufacturers of computer chips and other tiny high-tech items that must meet exacting specifications. It can also help manufacturers better calibrate their own super-accurate equipment so that, for example, even more devices could be put onto silicon chips to increase their computing power.

"This is a key project where the United States is a long way in front of any other country," said Trevor Howe, a University of Connecticut professor of metallurgy and director of its Precision Manufacturing Center.

Source: Boston Globe.

4. The article on the left states, "the Molecular Measuring Machine can measure distances to within 40 nanometers, or billionths of a meter." This suggests that the prefix *nano* means what?

5. Write the two unit fractions you would use to convert between meters and nanometers. Use your unit fractions to change 40 nanometers to meters, and to change 300,000 nanometers to meters.

6. Why do you suppose the headline and first paragraph talk about "billionths of an inch" when the rest of the article specifies nanometers, which are billionths of a meter?

7. Use the information in the article to find the length of one atom in inches. Write your answer as a decimal number and in words.

8.2 Exercises

Use your knowledge of the meaning of metric prefixes to fill in the blanks.

1. *kilo* means _____ so

    1 km = _____ m

2. *deka* means _____ so

    1 dam = _____ m

3. *milli* means _____ so

    1 mm = _____ m

4. *deci* means _____ so

    1 dm = _____ m

5. *centi* means _____ so

    1 cm = _____ m

6. *hecto* means _____ so

    1 hm = _____ m

Use this ruler to measure the width of your hand and thumb for Exercises 7–10.

7. The width of your hand in centimeters

8. The width of your hand in millimeters

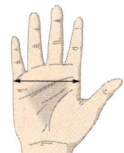

9. The width of your thumb in millimeters

10. The width of your thumb in centimeters

Write the most reasonable metric length unit in each blank. Choose from km, m, cm, and mm. See Example 1.

11. The child was 91 _____ tall.

12. The cardboard was 3 _____ thick.

13. Ming-Na swam in the 200 _____ backstroke race.

14. The bookcase is 75 _____ wide.

15. Adriana drove 400 _____ on her vacation.

16. The door is 2 _____ high.

17. An aspirin tablet is 10 _____ across.

18. Lamard jogs 4 _____ every morning.

19. A paper clip is about 3 _____ long.

20. My pen is 145 _____ long.

21. Dave's truck is 5 _____ long.

22. Wheelchairs need doorways that are at least 80 _____ wide.

23. Describe at least three examples of metric length units that you have come across in your daily life.

24. Explain one reason the metric system would be easier for a child to learn than the English system.

Convert each measurement. Use unit fractions or the metric conversion line. See Examples 2–4.

25. 7 m to cm

26. 18 m to cm

27. 40 mm to m

28. 6 mm to m

29. 9.4 km to m

30. 0.7 km to m

31. 509 cm to m

32. 30 cm to m

33. 400 mm to cm

34. 25 mm to cm

35. 0.91 m to mm

36. 4 m to mm

37. Is 82 cm greater than or less than 1 m? What is the difference in the lengths?

38. Is 1022 m greater than or less than 1 km? What is the difference in the lengths?

39. On the first page of this chapter we said that computer microchips may be only 5 mm long and 1 mm wide. Using the ruler on the previous page, draw a rectangle that measures 5 mm by 1 mm. Then convert each measurement to centimeters.

A greatly enlarged photo of a computer microchip.

40. Many cameras use film that is 35 mm wide. Film for movie theaters may be 70 mm wide. Using the ruler on the previous page, draw a line that is 35 mm long and a line 70 mm long. Then convert each measurement to centimeters.

41. The Roe River near Great Falls, Montana, is the shortest river in the world, with a north fork that is just under 18 m long. How many kilometers long is the north fork of the river? (*Source: Guinness Book of Amazing Nature.*)

42. There are 60,000 km of blood vessels in the human body. How many meters of blood vessels are in the body? (*Source: Big Book of Knowledge.*)

43. The median height for U.S. females who are 20 to 29 years old is about 1.64 m. Convert this height to centimeters and to millimeters. (*Source: U.S. National Center for Health Statistics.*)

44. The median height for 20- to 29-year-old males in the United States is about 177 cm. Convert this height to meters and to millimeters. (*Source: U.S. National Center for Health Statistics.*)

45. Use two unit fractions to convert 5.6 mm to km.

46. Use two unit fractions to convert 16.5 km to mm.

8.3 The Metric System—Capacity and Weight (Mass)

We use capacity units to measure liquids, such as the amount of milk in a recipe, the gasoline in our car tank, and the water in an aquarium. (The English capacity units used in the United States are cups, pints, quarts, and gallons.) The basic metric unit for capacity is the **liter** (also spelled *litre*). The capital letter **L** is the symbol for liter, to avoid confusion with the numeral 1.

OBJECTIVES

1. Learn the basic metric units of capacity.
2. Convert among metric capacity units.
3. Learn the basic metric units of weight (mass).
4. Convert among metric weight (mass) units.
5. Distinguish among basic metric units of length, capacity, and weight (mass).

OBJECTIVE 1 Learn the basic metric units of capacity. The liter is related to metric length in this way: A box that measures 10 cm on every side holds exactly one liter. (The volume of the box is 10 cm • 10 cm • 10 cm = 1000 cubic centimeters. Volume is discussed in **Section 4.8.**) A liter is just a little more than 1 quart.

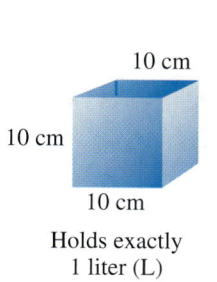
10 cm, 10 cm, 10 cm
Holds exactly 1 liter (L)

About 1 liter of milk

About 1 liter of oil for your car

A liter is a little more than one quart (just $\frac{1}{4}$ cup more).

In the metric system you use liters for things like buying milk and soda at the store, filling a pail with water, and describing the size of your home aquarium.

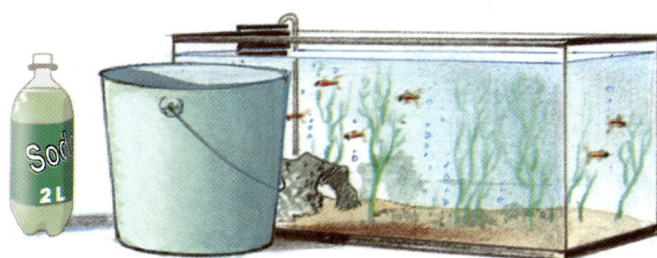

Buy a 2 L bottle of soda. Use a 12 L pail to wash floors. Watch the fish in your 40 L aquarium.

Work Problem 1 at the Side.

To make larger or smaller capacity units, we use the same **prefixes** as we did with length units. For example, *kilo* means **1000** so a *kilo*meter is **1000** meters. In the same way, a *kilo*liter is **1000** liters.

Prefix	kilo- liter	hecto- liter	deka- liter	liter	deci- liter	centi- liter	milli- liter
Meaning	1000 liters	100 liters	10 liters	1 liter	$\frac{1}{10}$ of a liter	$\frac{1}{100}$ of a liter	$\frac{1}{1000}$ of a liter
Symbol	kL	hL	daL	L	dL	cL	mL

Capacity units used most often

1 Which things can be measured in liters?

Amount of water in the bathtub

Length of the bathtub

Width of your car

Amount of gasoline you buy for your car

Weight of your car

Height of a pail

Amount of water in a pail

ANSWERS
1. water in bathtub, gasoline, water in a pail

584 Chapter 8 Measurement

2 Write the most reasonable metric unit in each blank. Choose from L and mL.

(a) I bought 8 _____ of milk at the store.

(b) The nurse gave me 10 _____ of cough syrup.

(c) This is a 100 _____ garbage can.

(d) It took 10 _____ of paint to cover the bedroom walls.

(e) My car's gas tank holds 50 _____.

(f) I added 15 _____ of oil to the pancake mix.

(g) The can of orange soda holds 350 _____.

(h) My friend gave me a 30 _____ bottle of expensive perfume.

ANSWERS
2. (a) L (b) mL (c) L
   (d) L (e) L (f) mL
   (g) mL (h) mL

The capacity units you will use most often in daily life are liters (L) and *milli*liters (mL). A tiny box that measures 1 cm on every side holds exactly one milliliter. (In medicine, this small amount is also called 1 cubic centimeter, or 1 cc for short.) It takes 1000 mL to make 1 L. Here are some useful comparisons.

Holds exactly 1 milliliter (mL)   Teaspoon holds 5 mL   One cup holds about 250 mL

EXAMPLE 1 Using Metric Capacity Units

Write the most reasonable metric unit in each blank. Choose from L and mL.

(a) The bottle of shampoo held 500 _____.
500 **mL** because 500 L would be about 500 quarts, which is too much.

(b) I bought a 2 _____ carton of orange juice.
2 **L** because 2 mL would be less than a teaspoon.

◀◀ **Work Problem 2 at the Side.**

OBJECTIVE 2 Convert among metric capacity units. Just as with length units, you can convert between milliliters and liters using unit fractions.

Metric Capacity Relationships
1 L = 1000 mL, so the unit fractions are:

$$\frac{1\ L}{1000\ mL} \quad \text{or} \quad \frac{1000\ mL}{1\ L}$$

Or you can use a metric conversion line to decide how to move the decimal point.

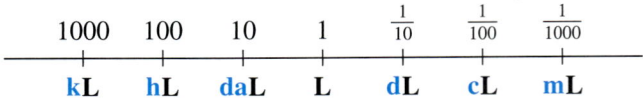

EXAMPLE 2 Converting among Metric Capacity Units

Convert using the metric conversion line or unit fractions.

(a) 2.5 L to mL

Using the metric conversion line:
From **L** to **mL** is *three places* to the *right*.

2.500 Write two zeros as placeholders.

Thus, 2.5 L = 2500 mL.

Using unit fractions:

Multiply by a unit fraction that allows you to divide out liters.

$$\frac{2.5\ \cancel{L}}{1} \cdot \frac{1000\ mL}{1\ \cancel{L}} = 2500\ mL$$

Thus, 2.5 L = 2500 mL.

Continued on Next Page

(b) 80 mL to L

Using the metric conversion line:

From **mL** to **L** is *three places* to the *left*.

80.   080.
↑
Decimal point starts here.   Move decimal point three places to the left.

Thus, 80 mL = 0.080 L or 0.08 L.

Using unit fractions:

Multiply by a unit fraction that allows you to divide out mL.

$$\frac{80 \text{ mL}}{1} \cdot \frac{1 \text{ L}}{1000 \text{ mL}}$$

$$= \frac{80}{1000} \text{ L} = 0.08 \text{ L}$$

Thus, 80 mL = 0.08 L.

> **Work Problem 3 at the Side.**

OBJECTIVE 3 Learn the basic metric units of weight (mass). The **gram** is the basic metric unit for *mass*. Although we often call it "weight," there is a difference. Weight is a measure of the pull of gravity; the farther you are from the center of the earth, the less you weigh. In outer space you become weightless, but your mass, the amount of matter in your body, stays the same regardless of where you are. In science courses, it will be important to distinguish between the weight of an object and its mass. But for everyday purposes, we will use the word *weight*.

The gram is related to metric length in this way: The weight of the water in a box measuring 1 cm on every side is 1 gram. This is a very tiny amount of water (1 mL) and a very small weight. One gram is also the weight of a dollar bill or a single raisin. A nickel weighs 5 grams. A regular-sized hamburger and bun weighs from 175 to 200 grams.

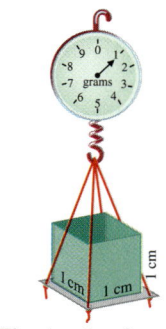

The 1 mL of water in this tiny box weighs 1 gram.

A nickel weighs 5 grams.

A dollar bill weighs 1 gram.

A hamburger weighs 175 to 200 grams.

> **Work Problem 4 at the Side.**

3 Convert.

(a) 9 L to mL

(b) 0.75 L to mL

(c) 500 mL to L

(d) 5 mL to L

(e) 2.07 L to mL

(f) 3275 mL to L

4 Which things would weigh about 1 gram?

A small paper clip

A pair of scissors

One playing card from a deck of cards

A calculator

An average-sized apple

The check you wrote at the grocery store

ANSWERS

3. (a) 9000 mL   (b) 750 mL   (c) 0.5 L
(d) 0.005 L   (e) 2070 mL   (f) 3.275 L

4. paper clip, playing card, check

5 Write the most reasonable metric unit in each blank. Choose from kg, g, and mg.

(a) A thumbtack weighs 800 _____.

(b) A teenager weighs 50 _____.

(c) This large cast-iron frying pan weighs 1 _____.

(d) Jerry's basketball weighed 600 _____.

(e) Tamlyn takes a 500 _____ calcium tablet every morning.

(f) On his diet, Greg can eat 90 _____ of meat for lunch.

(g) One strand of hair weighs 2 _____.

(h) One banana might weigh 150 _____.

ANSWERS
5. (a) mg (b) kg (c) kg
   (d) g   (e) mg (f) g
   (g) mg (h) g

To make larger or smaller weight units, we use the same **prefixes** as we did with length and capacity units. For example, *kilo* means **1000** so a *kilo*meter is **1000** meters, a *kilo*liter is **1000** liters, and a *kilo*gram is **1000** grams.

Prefix	kilo-gram	hecto-gram	deka-gram	gram	deci-gram	centi-gram	milli-gram
Meaning	1000 grams	100 grams	10 grams	1 gram	$\frac{1}{10}$ of a gram	$\frac{1}{100}$ of a gram	$\frac{1}{1000}$ of a gram
Symbol	**kg**	**hg**	**dag**	**g**	**dg**	**cg**	**mg**

Weight (mass) units that are used most often

The units you will use most often in daily life are kilograms (kg), grams (g), and milligrams (mg). *Kilo*grams are used instead of pounds. A kilogram is 1000 grams. It is about 2.2 pounds. Two packages of butter plus one stick of butter weigh about 1 kg. An average newborn baby weighs 3 to 4 kg; a college football player might weigh 100 to 130 kg.

1 kilogram is about 2.2 pounds.   100 to 130 kg   3 to 4 kg

Extremely small weights are measured in *milli*grams. It takes 1000 mg to make 1 g. Recall that a dollar bill weighs about 1 g. Imagine cutting it into 1000 pieces; the weight of one tiny piece would be 1 mg. Dosages of medicine and vitamins are given in milligrams. You will also use milligrams in science classes.

Cut a dollar bill into 1000 pieces. One tiny piece weighs 1 milligram.

EXAMPLE 3 Using Metric Weight Units

Write the most reasonable metric unit in each blank. Choose from kg, g, and mg.

(a) Ramon's suitcase weighed 20 _____.
    20 **kg** because kilograms are used instead of pounds.
    20 kg is about 44 pounds.

(b) LeTia took a 350 _____ aspirin tablet.
    350 **mg** because 350 g would be more than the weight of a hamburger, which is too much.

(c) Jenny mailed a letter that weighed 30 _____.
    30 **g** because 30 kg would be much too heavy and 30 mg is less than the weight of a dollar bill.

Work Problem 5 at the Side.

OBJECTIVE 4 Convert among metric weight (mass) units. As with length and capacity, you can convert among metric weight units by using unit fractions. The unit fractions you need are shown below.

> **Metric Weight (Mass) Relationships**
>
> 1 kg = 1000 g so the unit fractions are:
>
> $$\frac{1 \text{ kg}}{1000 \text{ g}} \quad \text{or} \quad \frac{1000 \text{ g}}{1 \text{ kg}}$$
>
> 1 g = 1000 mg so the unit fractions are:
>
> $$\frac{1 \text{ g}}{1000 \text{ mg}} \quad \text{or} \quad \frac{1000 \text{ mg}}{1 \text{ g}}$$

Or you can use a metric conversion line to decide how to move the decimal point.

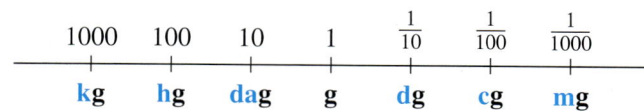

EXAMPLE 4 Converting among Metric Weight Units

Convert using the metric conversion line or unit fractions.

(a) 7 mg to g

Using the metric conversion line:
From **mg** to **g** is *three places* to the *left*.

7.
↑
Decimal point starts here.

007.
Move decimal point three places to the left.

Using unit fractions:
Multiply by a unit fraction that allows you to divide out mg.

$$\frac{7 \text{ mg}}{1} \cdot \frac{1 \text{ g}}{1000 \text{ mg}} = \frac{7}{1000} \text{ g}$$

$$= 0.007 \text{ g}$$

Thus, 7 mg = 0.007 g.

Thus, 7 mg = 0.007 g.

(b) 13.72 kg to g

Using the metric conversion line:
From **kg** to **g** is *three places* to the *right*.

13.720
Decimal point moves three places to the right.

Thus, 13.72 kg = 13,720 g.
↑
A comma (not a decimal point)

Using unit fractions:
Multiply by a unit fraction that allows you to divide out kg.

$$\frac{13.72 \text{ kg}}{1} \cdot \frac{1000 \text{ g}}{1 \text{ kg}} = 13,720 \text{ g}$$
↑
A comma (not a decimal point)

Thus, 13.72 kg = 13,720 g.

> **Work Problem 6 at the Side.**

6 Convert.

(a) 10 kg to g

(b) 45 mg to g

(c) 6.3 kg to g

(d) 0.077 g to mg

(e) 5630 g to kg

(f) 90 g to kg

ANSWERS
6. (a) 10,000 g (b) 0.045 g (c) 6300 g
   (d) 77 mg (e) 5.63 kg (f) 0.09 kg

588 Chapter 8 Measurement

7 First decide which type of units are needed: length, capacity, or weight. Then write the most appropriate unit in the blank. Choose from km, m, cm, mm, L, mL, kg, g, and mg.

(a) Gail bought a 4 _____ can of paint.

Use _____ units.

(b) The bag of chips weighed 450 _____.

Use _____ units.

(c) Give the child 5 _____ of cough syrup.

Use _____ units.

(d) The width of the window is 55 _____.

Use _____ units.

(e) Akbar drives 18 _____ to work.

Use _____ units.

(f) Each computer weighs 5 _____.

Use _____ units.

(g) A credit card is 55 _____ wide.

Use _____ units.

ANSWERS
7. (a) L; capacity (b) g; weight
   (c) mL; capacity (d) cm; length
   (e) km; length (f) kg; weight
   (g) mm; length

OBJECTIVE 5 Distinguish among basic metric units of length, capacity, and weight (mass). As you encounter things to be measured at home, on the job, or in your classes at school, be careful to use the correct type of measurement unit.

Use *length units* (kilometers, meters, centimeters, millimeters) to measure:

how long	how high	how far away
how wide	how tall	how far around (perimeter)
how deep	distance	

Use *capacity units* (liters, milliliters) to measure liquids (things that can be poured) such as:

water	shampoo	gasoline
milk	perfume	oil
soft drinks	cough syrup	paint

Also use liters and milliliters to describe how much liquid something can hold, such as an eyedropper, measuring cup, pail, or bathtub.

Use *weight units* (kilograms, grams, milligrams) to measure:

the weight of something   how heavy something is

In **Chapters 3 and 4** you used square units (such as cm^2 and m^2) to measure area, and cubic units (such as cm^3 and m^3) to measure volume.

EXAMPLE 5 Using a Variety of Metric Units

First decide which type of units are needed: length, capacity, or weight. Then write the most appropriate metric unit in the blank. Choose from km, m, cm, mm, L, mL, kg, g, and mg.

(a) The letter needs another stamp because it weighs 40 _____.

Use _____ units.

The letter weighs 40 **g** because 40 mg is less than the weight of a dollar bill and 40 kg would be about 88 pounds.

Use **weight** units because of the word "weighs."

(b) The swimming pool is 3 _____ deep at the deep end.

Use _____ units.

The pool is 3 **m** deep because 3 cm is only about an inch and 3 km is more than a mile.

Use **length** units because of the word "deep."

(c) This is a 340 _____ can of juice.

Use _____ units.

It is a 340 **ml** can because 340 liters would be more than 340 quarts.

Use **capacity** units because juice is a liquid.

Work Problem 7 at the Side.

8.3 Exercises

Write the most reasonable metric unit in each blank. Choose from L, mL, kg, g, and mg. See Examples 1 and 3.

1. The glass held 250 _____ of water.

2. Hiromi used 12 _____ of water to wash the kitchen floor.

3. Dolores can make 10 _____ of soup in that pot.

4. Jay gave 2 _____ of vitamin drops to the baby.

5. Our yellow Labrador dog grew up to weigh 40 _____.

6. A small safety pin weighs 750 _____.

7. Lori caught a small sunfish weighing 150 _____.

8. One dime weighs 2 _____.

9. Andre donated 500 _____ of blood today.

10. Barbara bought the 2 _____ bottle of cola.

11. The patient received a 250 _____ tablet of medication each hour.

12. The 8 people on the elevator weighed a total of 500 _____.

13. The gas can for the lawn mower holds 4 _____.

14. Kevin poured 10 _____ of vanilla into the mixing bowl.

15. Pam's backpack weighs 5 _____ when it is full of books.

16. One grain of salt weighs 2 _____.

Today, medical measurements are usually given in the metric system. Since we convert among metric units of measure by moving the decimal point, it is possible that mistakes can be made. Examine the following dosages and indicate whether they are reasonable or unreasonable. If a dose is unreasonable, indicate whether it is too much *or* too little.

17. Drink 4.1 L of Kaopectate after each meal.

18. Drop 1 mL of solution into the eye twice a day.

19. Soak your feet in 5 kg of Epsom salts per liter of water.

20. Inject 0.5 L of insulin each morning.

21. Take 15 mL of cough syrup every four hours.

22. Take 200 mg of vitamin C each day.

23. Take 350 mg of aspirin three times a day.

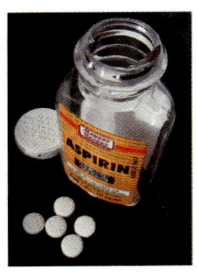

24. Buy a tube of ointment weighing 0.002 g.

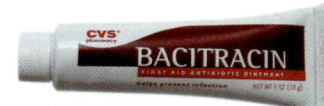

25. Describe at least two examples of metric capacity units and two examples of metric weight units that you have come across in your daily life.

26. Explain in your own words how the meter, liter, and gram are related.

27. Describe how you decide which unit fraction to use when converting 6.5 kg to grams.

28. Write an explanation of each step you would use to convert 20 mg to grams using the metric conversion line.

Convert each measurement. Use unit fractions or the metric conversion line. See Examples 2 and 4.

29. 15 L to mL

30. 6 L to mL

31. 3000 mL to L

32. 18,000 mL to L

33. 925 mL to L

34. 200 mL to L

35. 8 mL to L

36. 25 mL to L

37. 4.15 L to mL

38. 11.7 L to mL

39. 8000 g to kg

40. 25,000 g to kg

41. 5.2 kg to g

42. 12.42 kg to g

43. 0.85 g to mg

44. 0.2 g to mg

45. 30,000 mg to g

46. 7500 mg to g

47. 598 mg to g

48. 900 mg to g

49. 60 mL to L

50. 6.007 kg to g

51. 3 g to kg

52. 12 mg to g

53. 0.99 L to mL

54. 13,700 mL to L

Write the most appropriate metric unit in each blank. Choose from km, m, cm, mm, L, mL, kg, g, and mg. See Example 5.

55. The masking tape is 19 _____ wide.

56. The roll has 55 _____ of tape on it.

57. Buy a 60 _____ jar of acrylic paint for art class.

58. One onion weighs 200 _____.

59. My waist measurement is 65 _____.

60. Add 2 _____ of windshield washer fluid to your car.

61. A single postage stamp weighs 90 _____.

62. The hallway is 10 _____ long.

Solve each application problem. Show your work. (*Source for Exercises 63–68:* Top 10 of Everything.)

63. Human skin has about 3 million sweat glands, which release an average of 300 mL of sweat per day. How many liters of sweat are released each day?

64. In hot climates, the sweat glands in a person's skin may release up to 3.5 L of sweat in one day. How many milliliters is that?

65. The average weight of a human brain is 1.34 kg. How many grams is that?

66. A healthy human heart pumps about 70 mL of blood per beat. How many liters of blood does it pump per beat?

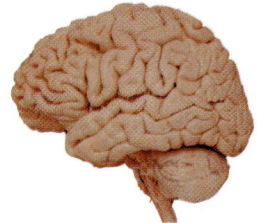

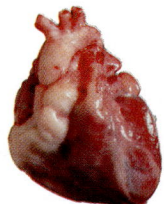

67. On average, we breathe in and out roughly 900 mL of air every 10 seconds. How many liters of air is that?

68. In the Victorian era, people believed that heavier brains meant greater intelligence. They were impressed that Otto von Bismarck's brain weighed 1907 g, which is how many kilograms?

69. A small adult cat weighs from 3000 g to 4000 g. How many kilograms is that? (*Source: Lyndale Animal Hospital.*)

70. If the letter you are mailing weighs 29 g, you must put additional postage on it. How many kilograms does the letter weigh? (*Source: U.S. Postal Service.*)

71. Is 1005 mg greater than or less than 1 g? What is the difference in the weights?

72. Is 990 mL greater than or less than 1 L? What is the difference in the amounts?

73. One nickel weighs 5 g. How many nickels are in 1 kg of nickels?

74. The ratio of the total length of all the fish to the amount of water in an aquarium can be 3 cm of fish for every 4 L of water. What is the total length of all fish you can put in a 40 L aquarium? (*Source: Tropical Aquarium Fish.*)

RELATING CONCEPTS (EXERCISES 75–78) For Individual or Group Work

*Recall that the prefix **kilo** means 1000, so a **kilo**meter is 1000 meters. You'll learn about other prefixes for numbers greater than 1000 as you **work Exercises 75–78 in order.***

75. (a) The prefix *mega* (abbreviated with a capital M) means one million. So a *mega*meter (Mm) is how many meters?

1 Mm = _____ m

(b) Figure out a unit fraction that you can use to convert megameters to meters. Then use it to convert 3.5 Mm to meters.

76. (a) The prefix *giga* (abbreviated with a capital G) means one billion. So a *giga*meter (Gm) is how many meters?

1 Gm = _____ m

(b) Figure out a unit fraction you can use to convert meters to gigameters. Then use it to convert 2500 m to gigameters.

77. (a) The prefix *tera* (abbreviated with a capital T) means one trillion. So a *tera*meter (Tm) is how many meters?

1 Tm = _____ m

(b) Think carefully before you fill in the blanks.

1 Tm = _____ Gm

1 Tm = _____ Mm

78. A computer's memory is measured in *bytes*. A byte can represent a single letter, a digit, or a punctuation mark. The memory for a desktop computer may be measured in megabytes (abbreviated MB) or gigabytes (abbreviated GB). Using the meanings of *mega* and *giga,* it would seem that

1 MB = _____ bytes and

1 GB = _____ bytes.

However, because computers use a base 2 or binary system, 1 MB is actually 2^{20} and 1 GB is 2^{30}. Use your calculator to find the actual values.

2^{20} = _____ 2^{30} = _____

8.4 Problem Solving with Metric Measurement

OBJECTIVE 1 Solve application problems involving metric measurements. One advantage of the metric system is the ease of comparing measurements in application situations. Just be sure that you are comparing similar units: mg to mg, km to km, and so on.

We will use the same problem-solving steps you used for English measurement applications in **Section 8.1**.

OBJECTIVE

1. Solve application problems involving metric measurements.

EXAMPLE 1 Solving a Metric Application

Cheddar cheese is on sale at $8.99 per kilogram. Jake bought 350 g of the cheese. How much did he pay, to the nearest cent?

Step 1 **Read** the problem. The problem asks for the cost of 350 g of cheese.

Step 2 **Work out a plan.** The price is $8.99 per *kilogram*, but the amount Jake bought is given in *grams*. Convert grams to kilograms (the unit in the price). Then multiply the weight times the cost per kilogram.

Step 3 **Estimate** a reasonable answer. Round the cost of 1 kg from $8.99 to $9. There are 1000 g in a kilogram, so 350 g is about $\frac{1}{3}$ of a kilogram. Jake is buying about $\frac{1}{3}$ of a kilogram, so $\frac{1}{3}$ of $9 = $3 as our estimate.

Step 4 **Solve** the problem. Use a unit fraction to convert 350 g to kilograms.

$$\frac{350 \text{ g}}{1} \cdot \frac{1 \text{ kg}}{1000 \text{ g}} = \frac{350}{1000} \text{ kg} = 0.35 \text{ kg}$$

Now multiply 0.35 kg times the cost per kilogram.

$$\frac{\$8.99}{1 \text{ kg}} \cdot \frac{0.35 \text{ kg}}{1} = \$3.1465 \approx \$3.15 \quad \text{Rounded}$$

Step 5 **State the answer.** Jake paid $3.15, rounded to the nearest cent.

Step 6 **Check** your work. The exact answer of $3.15 is close to our estimate of $3.

> **Work Problem 1 at the Side.**

1. Satin ribbon is on sale at $0.89 per meter. How much will 75 cm cost, to the nearest cent?

EXAMPLE 2 Solving a Metric Application

Olivia has 2.6 m of lace. How many centimeters of lace can she use to trim each of six hair ornaments? Round to the nearest tenth of a centimeter.

Step 1 **Read** the problem. The problem asks for the number of centimeters of lace for each of six hair ornaments.

Step 2 **Work out a plan.** The given amount of lace is in *meters,* but the answer must be in *centimeters*. Convert meters to centimeters, then divide by 6 (the number of hair ornaments).

Continued on Next Page

ANSWER
1. $0.67 (rounded)

2 Lucinda's doctor wants her to take 1.2 g of medication each day in three equal doses. How many milligrams should be in each dose?

Step 3 **Estimate** a reasonable answer. To estimate, round 2.6 m of lace to 3 m. Then, 3 m = 300 cm, and 300 cm ÷ 6 = 50 cm as our estimate.

Step 4 **Solve** the problem. On the metric conversion line, moving from **m** to **cm** is two places to the right, so move the decimal point in 2.6 m two places to the right. Then divide by 6.

$$2.60 \text{ m} = 260 \text{ cm} \qquad \frac{260 \text{ cm}}{6 \text{ ornaments}} \approx 43.3 \text{ cm per ornament}$$

Step 5 **State the answer.** Olivia can use about 43.3 cm of lace on each ornament.

Step 6 **Check** your work. The exact answer of 43.3 cm is close to our estimate of 50 cm.

▸ **Work Problem 2 at the Side.**

> **NOTE**
> In Example 1 on the previous page, we used a unit fraction to convert the measurement. In Example 2 above, we moved the decimal point. Use whichever method you prefer. Also, there is more than one way to solve an application problem. Another way to solve Example 2 is to divide 2.6 m by 6 to get 0.4333 m of lace for each ornament. Then convert 0.4333 m to 43.3 cm (rounded to the nearest tenth).

EXAMPLE 3 Solving a Metric Application

Rubin measured a board and found that the length was 3 m plus an additional 5 cm. He cut off a piece measuring 1 m 40 cm for a shelf. Find the length in meters of the remaining piece of board.

3 Andrea has two pieces of fabric. One measures 2 m 35 cm and the other measures 1 m 85 cm. How many meters of fabric does she have in all?

Step 1 **Read** the problem. Part of a board is cut off. The problem asks what length of board, in meters, is left over.

Step 2 **Work out a plan.** The lengths involve two units, m and cm. Rewrite both lengths in meters (the unit called for in the answer), and then subtract.

Step 3 **Estimate** a reasonable answer. To estimate, 3 m 5 cm can be rounded to 3 m, because 5 cm is less than half of a meter (less than 50 cm). Round 1 m 40 cm down to 1 m. Then, 3 m − 1 m = 2 m as our estimate.

Step 4 **Solve** the problem. Rewrite the lengths in meters. Then subtract.

$$\begin{array}{rl} 3\text{m} \rightarrow & 3.00 \text{ m} \\ \text{plus 5 cm} \rightarrow & +\,0.05 \text{ m} \\ \hline & 3.05 \text{ m} \end{array} \qquad \begin{array}{rl} 1\text{ m} \rightarrow & 1.0 \text{ m} \\ \text{plus 40 cm} \rightarrow & +\,0.4 \text{ m} \\ \hline & 1.4 \text{ m} \end{array}$$

Subtract to find leftover length.

$$\begin{array}{rl} 3.05 \text{ m} & \leftarrow \text{Board} \\ -\,1.40 \text{ m} & \leftarrow \text{Shelf} \\ \hline 1.65 \text{ m} & \leftarrow \text{Leftover piece} \end{array}$$

Step 5 **State the answer.** The length of the remaining piece of board is 1.65 m.

Step 6 **Check** your work. The exact answer of 1.65 m is close to our estimate of 2 m.

ANSWERS
2. 400 mg per dose
3. 4.2 m

▸ **Work Problem 3 at the Side.**

8.4 Exercises

Solve each application problem. Show your work. Round money answers to the nearest cent. See Examples 1–3.

1. Bulk rice at the food co-op is $0.98 per kilogram. Pam scooped some rice into a bag and put it on the scale. How much will she pay for 850 g of rice?

2. Lanh is buying a piece of plastic tubing measuring 315 cm for the science lab. The price is $4.75 per meter. How much will Lanh pay?

3. A miniature Yorkshire terrier, one of the smallest dogs, may weigh only 500 g. But a St. Bernard, the heaviest dog, could easily weigh 90 kg. What is the difference in the weights of the two dogs, in kilograms? (*Source: Big Book of Knowledge.*)

4. The world's longest insect is the giant stick insect of Indonesia, measuring 33 cm. The fairy fly, the smallest insect, is just 0.2 mm long. How much longer is the giant stick insect, in millimeters? (*Source: Big Book of Knowledge.*)

5. An adult human body contains about 5 L of blood. If each beat of the heart pumps 70 mL of blood, how many times must the heart beat to pass all the blood through the heart? Round to the nearest whole number of beats. (*Source: Harper's Index.*)

6. A floor tile measures 30 cm by 30 cm and weighs 185 g. How many kilograms would a stack of 24 tiles weigh? How much would five stacks of tile weigh? (*Source: The Tile Shop.*)

7. Each piece of lead for a mechanical pencil has a thickness of 0.5 mm and is 60 mm long. Find the total length in centimeters of the lead in a package with 30 pieces. If the price of the package is $3.29, find the cost per centimeter for the lead. (*Source: Pentel.*)

8. The apartment building caretaker puts 750 mL of chlorine into the swimming pool every day. How many liters should he order to have a one-month (30-day) supply on hand? If chlorine is sold in containers that hold 4 L, how many containers should be ordered for one month? How much chlorine will be left at the end of the month?

9. Rosa is building a bookcase. She has one board that is 2 m 8 cm long and another that is 2 m 95 cm long. How long are the two boards together, in meters?

10. Janet has 10 m 30 cm of fabric. She wants to make curtains for three windows that are all the same size. How much fabric is available for each window, to the nearest tenth of a meter?

11. In a chemistry lab, each of the 45 students needs 85 mL of acid. How many one-liter bottles of acid need to be ordered? How much acid will be left over?

12. James needs two 1.3 m pieces and two 85 cm pieces of wood molding to frame a picture. The price is $5.89 per meter plus 7% sales tax. How much will James pay?

Use the bar graph below to answer Exercises 13 and 14.

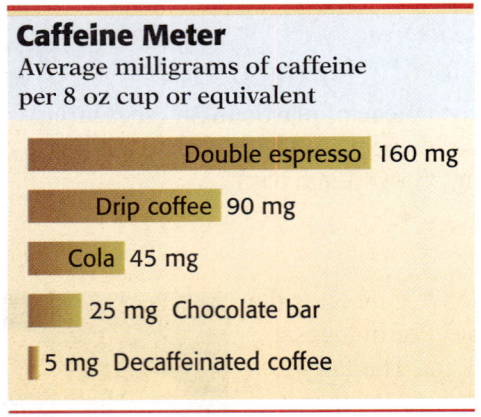

Caffeine Meter
Average milligrams of caffeine per 8 oz cup or equivalent

Double espresso 160 mg
Drip coffee 90 mg
Cola 45 mg
25 mg Chocolate bar
5 mg Decaffeinated coffee

Source: Celestial Seasonings.

13. If Agnete usually drinks three 8 oz cups of drip coffee each day, how many grams of caffeine will she consume in one week?

14. Lorenzo's doctor has suggested that he cut down on caffeine. So Lorenzo switched from drinking four 8 oz cups of cola every day to drinking two 8 oz cups of decaffeinated coffee. How many fewer grams of caffeine is he consuming each week?

15. During August 2003, Mars moved closer to Earth at a rate of about 10,000 meters per second. How much closer, in kilometers, did Mars get to Earth

   (a) in one second,

   (b) in one minute,

   (c) in one hour.

   (*Source:* www.NASA.com)

16. Some of the newest football stadiums have FieldTurf instead of grass. Use the drawing below to find the total thickness in centimeters of the top two layers of FieldTurf.

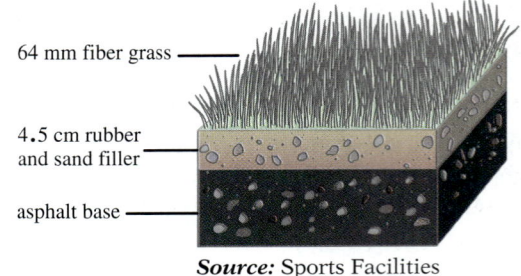

64 mm fiber grass
4.5 cm rubber and sand filler
asphalt base

Source: Sports Facilities Commission.

RELATING CONCEPTS (EXERCISES 17–21) For Individual or Group Work

It is difficult to weigh very light objects, such as a single sheet of paper or a single staple (unless you have a very expensive scientific scale). One way around this problem is to weigh a large number of the items and then divide to find the weight of one item. Of course, before dividing, you must subtract the weight of the box or wrapper that the items are packaged in to find the net weight. To complete the table, **work Exercises 17–20 in order.** *Then answer Exercise 21.*

	Item	Total Weight	Weight of Packaging	Net Weight	Weight of One Item in Grams	Weight of One Item in Milligrams
17.	Box of 50 envelopes	255 g	40 g	_____	_____	_____
18.	Box of 1000 staples	350 g	20 g	_____	_____	_____
19.	Ream of paper (500 sheets)	_____	50 g	_____	_____	3000 mg
20.	Box of 100 small paper clips	_____	5 g	_____	_____	500 mg

21. Bonus question: One million seeds from an aspen tree weigh about 0.45 kg. How much does one seed weigh in milligrams? (*Source:* Scenic State Park.)

8.5 Metric–English Conversions and Temperature

OBJECTIVE 1 Use unit fractions to convert between metric and English units. Until the United States has switched completely from the English system to the metric system, it will be necessary to make conversions from one system to the other. *Approximate* conversions can be made with the help of the table below, in which the values have been rounded to the nearest hundredth or thousandth. (The only value that is exact, not rounded, is 1 inch = 2.54 cm.)

Metric to English	English to Metric
1 kilometer ≈ 0.62 mile	1 mile ≈ 1.61 kilometers
1 meter ≈ 1.09 yards	1 yard ≈ 0.91 meter
1 meter ≈ 3.28 feet	1 foot ≈ 0.30 meter
1 centimeter ≈ 0.39 inch	1 inch = 2.54 centimeters
1 liter ≈ 0.26 gallon	1 gallon ≈ 3.78 liters
1 liter ≈ 1.06 quarts	1 quart ≈ 0.95 liter
1 kilogram ≈ 2.20 pounds	1 pound ≈ 0.45 kilogram
1 gram ≈ 0.035 ounce	1 ounce ≈ 28.35 grams

EXAMPLE 1 Converting between Metric and English Length Units

Convert 10 m to yards using unit fractions. Round your answer to the nearest tenth, if necessary.

We're changing from a *metric* unit to an *English* unit. In the "Metric to English" side of the table above, you see that 1 meter ≈ 1.09 yards. Two unit fractions can be written using that information.

$$\frac{1 \text{ m}}{1.09 \text{ yd}} \quad \text{or} \quad \frac{1.09 \text{ yd}}{1 \text{ m}}$$

Multiply by the unit fraction that allows you to divide out meters (that is, meters is in the denominator).

$$10 \text{ m} \cdot \frac{1.09 \text{ yd}}{1 \text{ m}} = \frac{10 \text{ m}}{1} \cdot \frac{1.09 \text{ yd}}{1 \text{ m}} = \frac{(10)(1.09 \text{ yd})}{1} = 10.9 \text{ yd}$$

These units should match.

Thus, 10 m ≈ 10.9 yd.

NOTE
In Example 1 above, you could also use the other numbers from the table involving meters and yards: 1 yard ≈ 0.91 meter.

$$\frac{10 \text{ m}}{1} \cdot \frac{1 \text{ yd}}{0.91 \text{ m}} = \frac{10}{0.91} \text{ yd} \approx 10.99 \text{ yd}$$

The answer is slightly different because the values in the table are rounded. Also, you have to divide instead of multiply, which is usually more difficult to do without a calculator. We will use the first method in this chapter.

Work Problem 1 at the Side.

OBJECTIVES

1. Use unit fractions to convert between metric and English units.
2. Learn common temperatures on the Celsius scale.
3. Convert temperatures using formulas and following the order of operations.

1 Convert using unit fractions. Round your answers to the nearest tenth.

(a) 23 m to yards

(b) 40 cm to inches

(c) 5 mi to kilometers (Look at the "English to Metric" side of the table.)

(d) 12 in. to centimeters

ANSWERS
1. (a) 23 m ≈ 25.1 yd (b) 40 cm ≈ 15.6 in.
   (c) 5 mi ≈ 8.1 km (d) 12 in. ≈ 30.5 cm

2 Convert. Use the values from the table on the previous page to make unit fractions. Round answers to the nearest tenth.

(a) 17 kg to pounds

(b) 5 L to quarts

(c) 90 g to ounces

(d) 3.5 gal to liters

(e) 145 lb to kilograms

(f) 8 oz to grams

ANSWERS
2. (a) 17 kg ≈ 37.4 lb (b) 5 L ≈ 5.3 qt
   (c) 90 g ≈ 3.2 oz (d) 3.5 gal ≈ 13.2 L
   (e) 145 lb ≈ 65.3 kg (f) 8 oz ≈ 226.8 g

EXAMPLE 2 Converting between Metric and English Weight and Capacity Units

Convert using unit fractions. Round your answers to the nearest tenth.

(a) 3.5 kg to pounds
Look in the "Metric to English" side of the table on the previous page to see that 1 kilogram ≈ 2.20 pounds. Use this information to write a unit fraction that allows you to divide out kilograms.

$$\frac{3.5 \text{ kg}}{1} \cdot \frac{2.20 \text{ lb}}{1 \text{ kg}} = \frac{(3.5)(2.20 \text{ lb})}{1} = 7.7 \text{ lb}$$

Thus, 3.5 kg ≈ 7.7 lb.

(b) 18 gal to liters
Look in the "English to Metric" side of the table to see that 1 gallon ≈ 3.78 liters. Write a unit fraction that will allow you to divide out gallons.

$$\frac{18 \text{ gal}}{1} \cdot \frac{3.78 \text{ L}}{1 \text{ gal}} = \frac{(18)(3.78 \text{ L})}{1} = 68.04 \text{ L}$$

68.04 rounded to the nearest tenth is 68.0.
Thus, 18 gal ≈ 68.0 L.

(c) 300 g to ounces
In the "Metric to English" side of the table, 1 gram ≈ 0.035 ounce.

$$\frac{300 \text{ g}}{1} \cdot \frac{0.035 \text{ oz}}{1 \text{ g}} = \frac{(300)(0.035 \text{ oz})}{1} = 10.5 \text{ oz}$$

Thus, 300 g ≈ 10.5 oz.

> **CAUTION**
> Because the metric and English systems were developed independently, almost all comparisons are approximate. Your answers should be written with the "≈" symbol to show they are approximate.

▶◀◀◀ **Work Problem 2 at the Side.**

OBJECTIVE 2 Learn common temperatures on the Celsius scale.
In the metric system, temperature is measured on the **Celsius** scale. On the Celsius scale, water freezes at 0 °C and boils at 100 °C. The small raised circle stands for "degrees" and the capital **C** is for Celsius. Read the temperatures like this:

Water freezes at 0 degrees Celsius (0 °C).

Water boils at 100 degrees Celsius (100 °C).

The English temperature system, used only in the United States, is measured on the **Fahrenheit** scale. On this scale:

Water freezes at 32 degrees Fahrenheit (32 °F).

Water boils at 212 degrees Fahrenheit (212 °F).

The thermometer below shows some typical temperatures in both Celsius and Fahrenheit. For example, comfortable room temperature is about 20 °C or 68 °F, and normal body temperature is about 37 °C or 98.6 °F.

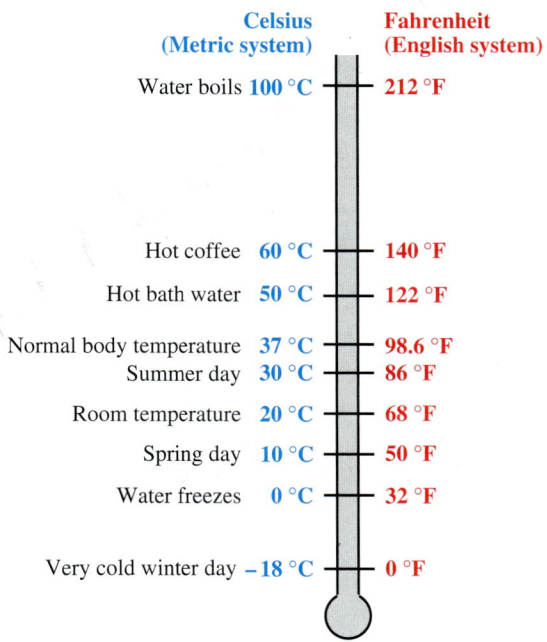

> **NOTE**
> The freezing and boiling temperatures are exact. The other temperatures are approximate. Even normal body temperature varies slightly from person to person.

EXAMPLE 3 Using Celsius Temperatures

Circle the Celsius temperature that is most reasonable for each situation.

(a) Warm summer day   29 °C   64 °C   90 °C

29 °C is reasonable. 64 °C and 90 °C are too hot; they're both above the temperature of hot bath water (above 122 °F).

(b) Inside a freezer   −10 °C   3 °C   25 °C

−10 °C is the reasonable temperature because it is the only one below the freezing point of water (0 °C). Your frozen foods would start thawing at 3 °C or 25 °C.

Work Problem 3 at the Side.

OBJECTIVE 3 Convert temperatures using formulas and following the order of operations. You can use these formulas to convert between Celsius and Fahrenheit temperatures.

Celsius–Fahrenheit Conversion Formulas

Converting from Fahrenheit (F) to Celsius (C)

$$C = \frac{5(F - 32)}{9}$$

Converting from Celsius (C) to Fahrenheit (F)

$$F = \frac{9C}{5} + 32$$

---

3 Circle the Celsius temperature that is most reasonable for each situation.

(a) Set the living room thermostat at:
    11 °C   21 °C   71 °C

(b) The baby has a fever of:
    29 °C   39 °C   49 °C

(c) Wear a sweater outside because it's:
    15 °C   25 °C   50 °C

(d) My iced tea is:
    −5 °C   5 °C   30 °C

(e) Time to go swimming! It's:
    95 °C   65 °C   35 °C

(f) Inside a refrigerator (not the freezer) it's:
    −15 °C   0 °C   3 °C

(g) There's a blizzard outside. It's:
    10 °C   0 °C   −20 °C

(h) I need hot water to get these clothes clean. It should be:
    55 °C   105 °C   200 °C

ANSWERS
3. (a) 21 °C (b) 39 °C (c) 15 °C
   (d) 5 °C (e) 35 °C (f) 3 °C
   (g) −20 °C (h) 55 °C

4 Convert to Celsius. Round your answers to the nearest degree if necessary.

(a) 72 °F

(b) 20 °F

(c) 212 °F

(d) 98.6 °F

5 Convert to Fahrenheit. Round your answers to the nearest degree if necessary.

(a) 100 °C

(b) −25 °C

(c) 32 °C

(d) 5 °C

ANSWERS
4. (a) 22 °C (rounded) (b) −7 °C (rounded)
   (c) 100 °C (d) 37 °C
5. (a) 212 °F (b) −13 °F
   (c) 90 °F (rounded) (d) 41 °F

As you use these formulas, be sure to follow the order of operations from **Section 1.8**.

Order of Operations
Step 1   Work inside **parentheses** or **other grouping symbols**.

Step 2   Simplify any expressions with **exponents**.

Step 3   Do the remaining **multiplications and divisions** as they occur from left to right.

Step 4   Do the remaining **additions and subtractions** as they occur from left to right.

EXAMPLE 4   Converting Fahrenheit to Celsius

Convert 10 °F to Celsius. Round your answer to the nearest degree.

Use the formula and the order of operations.

$$C = \frac{5(\mathbf{F} - 32)}{9} \quad \text{Replace F with 10}$$

$$= \frac{5(\mathbf{10} - 32)}{9} \quad \text{Work inside parentheses first. } 10 - 32 \text{ becomes } 10 + (-32).$$

$$= \frac{5(-22)}{9} \quad \text{Multiply in the numerator; positive times negative gives a negative product.}$$

$$= \frac{-110}{9} \quad \text{Divide; negative divided by positive gives a negative quotient.}$$

$$= -12.\overline{2} \quad \text{Round to } -12 \text{ (nearest degree).}$$

Thus, 10 °C ≈ −12 °F.

Work Problem 4 at the Side.

EXAMPLE 5   Converting Celsius to Fahrenheit

Convert 15 °C to Fahrenheit.

Use the formula and follow the order of operations.

$$F = \frac{9\mathbf{C}}{5} + 32 \quad \text{Replace C with 15}$$

$$= \frac{9 \cdot \mathbf{15}}{5} + 32$$

$$= \frac{9 \cdot 3 \cdot \cancel{5}}{\cancel{5}} + 32 \quad \text{Divide out the common factor. Multiply in the numerator.}$$

$$= 27 + 32 \quad \text{Add.}$$

$$= 59$$

Thus, 15 °C = 59 °F.

Work Problem 5 at the Side.

8.5 Exercises

Use the table on the first page of this section and unit fractions to make approximate conversions from metric to English or English to metric. Round your answers to the nearest tenth. See Examples 1 and 2.

1. 20 m to yards
2. 8 km to miles
3. 80 m to feet

4. 85 cm to inches
5. 16 ft to meters
6. 3.2 yd to meters

7. 150 g to ounces
8. 2.5 oz to grams
9. 248 lb to kilograms

10. 7.68 kg to pounds
11. 28.6 L to quarts
12. 15.75 L to gallons

13. For the 2000 Olympics, the 3M Company used 5 g of pure gold to coat Michael Johnson's track shoes. (*Source:* 3M Company.)

    (a) How many ounces of gold were used, to the nearest tenth?

    (b) Was this enough extra weight to slow him down?

14. The label on a Van Ness auto feeder for cats and dogs says it holds 1.4 kg of dry food. How many pounds of food does it hold, to the nearest tenth?

Source: Van Ness Plastics.

15. The heavy-duty wash cycle in a dishwasher uses 8.4 gal of water. How many liters does it use, to the nearest tenth? (*Source:* Frigidaire.)

16. The rinse-and-hold cycle in a dishwasher uses only 4.5 L of water. How many gallons does it use, to the nearest tenth? (*Source:* Frigidaire.)

17. The smallest pet fish are dwarf gobies, which are half an inch long. How many centimeters long is a dwarf gobie, to the nearest tenth? (*Hint:* Write half an inch in decimal form.)

18. The fastest nerve signals in the human body travel 120 m per second. How many feet per second do the signals travel? (*Source: Big Book of Knowledge.*)

19. On Northwest Airlines flights, a piece of carry-on luggage cannot exceed 40 lb in weight or measure more than 22 in. by 14 in. by 9 in. What measurements would be given to Northwest passengers from countries outside the United States? Round to the nearest tenth of a kilogram and nearest centimeter, if necessary. (*Source:* www.nwa.com)

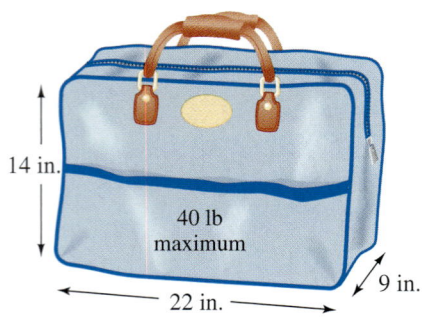

20. Northwest Airlines has a limit of 50 lb for each checked piece of luggage. The maximum size is 62 in., which is length + width + height. Convert these measurements to metric for Northwest flights from countries outside the United States. Round to the nearest tenth of a kilogram and nearest centimeter, if necessary. (*Source:* www.nwa.com)

21. Part of a tea bag packet imported from Scotland is shown below. Are the two weights on the packet equivalent? Explain your answer.

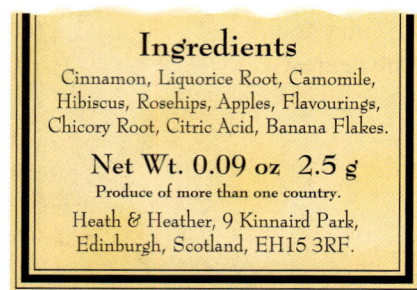

Source: Heath and Heather.

22. Part of the wrapper from a ball of twine is shown below. The manufacturer has designed the label so the twine can be sold worldwide. Are the two lengths listed on the wrapper equivalent? Explain your answer.

Source: Lehigh Group.

The BabyBjörn is a popular baby carrier imported from Sweden. Use the information from the instruction sheet that comes with the carrier to answer Exercises 23–24. (Source: BabyBjörn, Sweden.)

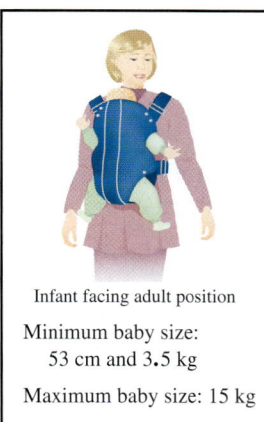

Infant facing adult position

Minimum baby size: 53 cm and 3.5 kg

Maximum baby size: 15 kg

23. Can the carrier be safely used for a newborn infant who weighs 8 lb and is 19.5 in. long? Explain your answer.

24. Can the carrier be safely used for a baby who weighs 30 lb?

Circle the more reasonable temperature for each situation. See Example 3.

25. A snowy day
28 °C   28 °F

26. Brewing coffee
80 °C   80 °F

27. A high fever
40 °C   40 °F

28. Swimming pool water
78 °C   78 °F

29. Oven temperature
150 °C   150 °F

30. Light jacket weather
10 °C   10 °F

Use the conversion formulas from this section and the order of operations to convert Fahrenheit temperatures to Celsius and Celsius temperatures to Fahrenheit. Round your answers to the nearest degree if necessary. See Examples 4 and 5.

31. 60 °F

32. 80 °F

33. −4 °F

34. 15 °F

35. 8 °C

36. 18 °C

37. −5 °C

38. 0 °C

Solve each application problem. Round your answers to the nearest degree, if necessary.

39. The highest temperature ever recorded on Earth was 136 °F at El Aziza, Libya, in 1922. The lowest was −129 °F in Antarctica. Convert these temperatures to Celsius. (*Source: The World Almanac.*)

40. Hummingbirds have a normal body temperature of 107 °F. But on cold nights they go into a state of torpor where their body temperature drops to 39 °F. What are these temperatures in Celsius? (*Source: Wildbird.*)

41. The directions for a self-stick clothes hook with adhesive on the back are as follows: "Apply to surfaces above 50 °F. Adhesive could soften and lose adhesion above 105 °F." What are these temperatures in the metric system? (*Source: 3M Company.*)

42. A box of imported Belgian chocolates carries a warning to keep the box dry and at <18 °C. Translate this warning into the English system. (*Source: Chocolaterie Guylian N.V.*)

43. Would a drop in temperature of 20 Celsius degrees be more or less than a drop of 20 Fahrenheit degrees? Explain your answer.

44. Describe one advantage of switching from the Fahrenheit temperature scale to the Celsius scale. Describe one disadvantage.

45. (a) Here is the tag on a pair of hiking boots. In what kind of weather would you be most comfortable wearing these boots?

Source: Sorel.

(b) For what Fahrenheit temperatures are the boots designed?

(c) What range of metric temperatures do you have in January where you live? Would you be comfortable in these boots?

46. Sleeping bags made by Eddie Bauer are sold around the world. Each type of sleeping bag is designed for outdoor camping in certain temperatures.

Junior bag	5 °C or warmer
Removable liner bag	0 °C to 15 °C
Conversion bag	−7 °C to 0 °C

Source: Eddie Bauer.

(a) At what Fahrenheit temperatures should you use the Junior bag?

(b) What Fahrenheit temperatures is the removable liner bag designed for?

RELATING CONCEPTS (EXERCISES 47–54) For Individual or Group Work

The article below appeared in American newspapers. However, both Newfoundland (part of Canada) and Ireland use the metric system. Their newspapers would have reported all the measurements in metric units. Complete the conversions to metric, rounding answers to the nearest tenth.

> **Q:** A recent news brief reported on some men who flew a model airplane from Newfoundland to Ireland. Can you provide some details of the flight?
>
> **A:** The model plane is 6 feet long and weighs 11 pounds. Made of balsa wood and mylar, it crossed the Atlantic—the flight path took it 1,888.3 miles—in 38 hours, 23 minutes. It soared at a cruising altitude of 1,000 feet. The plane used a souped-up piston engine and carried less than a gallon of fuel, as mandated by rules of the Federation Aeronautique Internationale, the governing body of model airplane building. When it landed in County Galway, Ireland, it had less than 2 fluid ounces of fuel left. The plane was built by Maynard Hill of Silver Spring, Maryland.
>
> (*Sources: New York Times* and *Cox News Service.*)

47. Length of model plane

48. Weight of plane

49. Length of flight path

50. Time of flight

51. Cruising altitude

52. Fuel at the start in milliliters

53. Fuel left after landing, in milliliters (*Hint:* First convert 2 fl oz to quarts.)

54. What *percent* of the fuel was left at the end of the flight?

Chapter 8
SUMMARY

KEY TERMS

8.1 **English system** — The English system of measurement (U.S. system of units) is used for many daily activities only in the United States. Commonly used units in this system include quarts, pounds, feet, miles, and degrees Fahrenheit.

metric system — The metric system of measurement is an international system used in manufacturing, science, medicine, sports, and other fields. Commonly used units in this system include meters, liters, grams, and degrees Celsius.

unit fraction — A unit fraction involves measurement units and is equivalent to 1. Unit fractions are used to convert among different measurements.

8.2 **meter** — The meter is the basic unit of length in the metric system. The symbol **m** is used for meter. One meter is a little longer than a yard.

prefixes — Attaching a prefix to meter, liter, or gram produces larger or smaller units. For example, the prefix *kilo* means 1000, so a *kilo*meter is 1000 meters.

metric conversion line — The metric conversion line is a line showing the various metric measurement prefixes and their size relationship to each other.

8.3 **liter** — The liter is the basic unit of capacity in the metric system. The symbol **L** is used for liter. One liter is a little more than one quart.

gram — The gram is the basic unit of weight (mass) in the metric system. The symbol **g** is used for gram. One gram is the weight of 1 milliliter of water or one dollar bill.

8.5 **Celsius** — The Celsius scale is used to measure temperature in the metric system. Water freezes at 0 °C and boils at 100 °C.

Fahrenheit — The Fahrenheit scale is used to measure temperature in the English system. Water freezes at 32 °F and boils at 212 °F.

NEW FORMULAS

Converting from Celsius to Fahrenheit: $F = \dfrac{9C}{5} + 32$

Converting from Fahrenheit to Celsius: $C = \dfrac{5(F - 32)}{9}$

TEST YOUR WORD POWER

See how well you have learned the vocabulary in this chapter. Answers follow the Quick Review.

1. The **metric system**
   A. uses meters, liters, and degrees Fahrenheit
   B. is based on multiples of 10
   C. is used only in the United States
   D. has evolved over centuries.

2. The **English system** of measurement
   A. is used throughout the world
   B. is based on multiples of 12
   C. uses feet, inches, quarts, and pounds
   D. was developed by a group of scientists in 1790.

3. A **unit fraction**
   A. has the unit you want to change in the numerator
   B. has a denominator of 1
   C. must be written in lowest terms
   D. is equivalent to 1.

4. A **gram** is
   A. the weight of 1 mL of water
   B. abbreviated gm
   C. equivalent to 1000 kg
   D. approximately equal to 2.2 pounds.

5. A **meter** is
   A. equivalent to 1000 cm
   B. approximately equal to $\frac{1}{2}$ inch
   C. abbreviated m with no period after it
   D. the basic unit of capacity in the metric system.

6. The **Celsius** temperature scale
   A. shows water freezing at 32°
   B. is used in the English system of measurement
   C. shows water boiling at 100°
   D. cannot be converted to the Fahrenheit temperature scale.

Quick Review

Concepts

Examples

8.1 The English System of Measurement
Memorize the basic measurement relationships. Then, to convert units, multiply when changing from a larger unit to a smaller unit; divide when changing from a smaller unit to a larger unit.

Convert each measurement.
(a) 5 ft to inches

$$5 \text{ ft} = 5 \cdot 12 = 60 \text{ in.}$$

(b) 3 lb to ounces

$$3 \text{ lb} = 3 \cdot 16 = 48 \text{ oz}$$

(c) 15 qt to gallons

$$15 \text{ qt} = \frac{15}{4} = 3\frac{3}{4} \text{ gal}$$

8.1 Using Unit Fractions
Another, more useful, conversion method is multiplying by a unit fraction. The unit you want in the answer should be in the numerator. The unit you want to change should be in the denominator.

Convert 32 oz to pounds.

$$32 \text{ oz} \cdot \frac{1 \text{ lb}}{16 \text{ oz}} \quad \}\text{Unit fraction}$$

These units should match.

$$= \frac{\overset{2}{\cancel{32} \text{ oz}}}{1} \cdot \frac{1 \text{ lb}}{\underset{1}{\cancel{16} \text{ oz}}} \quad \begin{array}{l}\text{Divide out } \textbf{ounces}.\\ \text{Divide out } \textbf{common factors}.\end{array}$$

$$= 2 \text{ lb}$$

8.1 Solving English Measurement Application Problems
To solve measurement application problems, use these problem-solving steps.

Use the six steps to solve this problem.

Mr. Green has 10 yd of rope. He is cutting it into eight pieces so his sailing class can practice knot tying. How many feet of rope will each of his eight students get?

Step 1 **Read** the problem.

Step 1 The problem asks how many feet of rope can be given to each of eight students.

Step 2 **Work out a plan.**

Step 2 Convert 10 yd to feet (the unit required in the answer). Then divide by eight students.

Step 3 **Estimate** a reasonable answer.

Step 3 There are 3 ft in one yard, so there are 30 ft in 10 yd. Then 30 ft ÷ 8 ≈ 4 ft as our estimate.

Step 4 **Solve** the problem.

Step 4 Use a unit fraction to convert 10 yd to feet, then divide.

$$\frac{10 \text{ yd}}{1} \cdot \frac{3 \text{ ft}}{1 \text{ yd}} = 30 \text{ ft}$$

$$\frac{30 \text{ ft}}{8 \text{ students}} = 3\frac{3}{4} \text{ ft or 3.75 ft per student}$$

Step 5 **State the answer.**

Step 5 Each student gets $3\frac{3}{4}$ ft or 3.75 ft of rope.

Step 6 **Check** your work.

Step 6 The exact answer of 3.75 ft is close to our estimate of 4 ft.

Chapter 8 Summary

Concepts

8.2 Basic Metric Length Units
Use approximate comparisons to judge which length units are appropriate:

    1 mm is the thickness of a dime.

    1 cm is about $\frac{1}{2}$ inch.

    1 m is a little more than 1 yard.

    1 km is about 0.6 mile.

8.2 and 8.3 Converting within the Metric System
Using Unit Fractions
One conversion method is to multiply by a unit fraction. Use a fraction with the unit you want in the answer in the numerator and the unit you want to change in the denominator.

Using the Metric Conversion Line
Another conversion method is to find the unit you are given on the metric conversion line. Count the number of places to get from the unit you are given to the unit you want. Move the decimal point the same number of places and in the same direction.

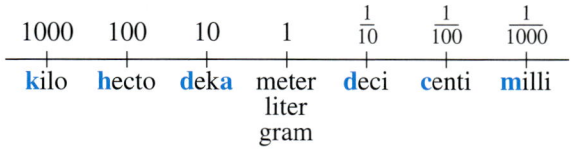

Examples

Write the most reasonable metric unit in each blank. Choose from km, m, cm, and mm.

The room is 6 __m__ long.

A paper clip is 30 __mm__ long.

He drove 20 __km__ to work.

Convert.

(a) 9 g to kg

$$\frac{9 \, \cancel{g}}{1} \cdot \frac{1 \, kg}{1000 \, \cancel{g}} = \frac{9}{1000} \, kg = 0.009 \, kg$$

Thus, 9 g = 0.009 kg.

(b) 3.6 m to cm

$$\frac{3.6 \, \cancel{m}}{1} \cdot \frac{100 \, cm}{1 \, \cancel{m}} = 360 \, cm$$

Thus, 3.6 m = 360 cm.

Convert.

(a) 68.2 kg to g

From **kg** to **g** is three places to the right.

6 8.2 0 0   Decimal point is moved three places to the right.

Thus, 68.2 kg = 68,200 g.

(b) 300 mL to L

From **mL** to **L** is three places to the left.

3 0 0.   Decimal point is moved three places to the left.

Thus, 300 mL = 0.3 L.

(c) 825 cm to m

From **cm** to **m** is two places to the left.

8 2 5.   Decimal point is moved two places to the left.

Thus, 825 cm = 8.25 m.

Concepts

8.3 Basic Metric Capacity Units
Use approximate comparisons to judge which capacity units are appropriate:

    1 L is a little more than 1 quart.

    1 mL is the amount of water in
      a cube 1 cm on each side.

    5 mL is about one teaspoon.

    250 mL is about one cup.

8.3 Basic Metric Weight (Mass) Units
Use approximate comparisons to judge which weight units are appropriate:

    1 kg is about 2.2 pounds.

    1 g is the weight of 1 mL of water
      or one dollar bill.

    1 mg is $\frac{1}{1000}$ of a gram; very tiny!

8.4 Solving Metric Application Problems
Convert units so you are comparing kg to kg, cm to cm, and so on. When a measurement involves two units, such as 6 m 20 cm, write it in terms of the unit called for in the answer (6.2 m or 620 cm).

Use these problem-solving steps.

Step 1   **Read** the problem.

Step 2   **Work out a plan.**

Step 3   **Estimate** a reasonable answer.

Step 4   **Solve** the problem.

Step 5   **State the answer.**

Step 6   **Check** your work.

Examples

Write the most appropriate metric unit in each blank. Choose from L and mL.

The pail holds 12 __L__ .

The milk carton from the vending machine holds 250 __mL__ .

Write the most appropriate metric unit in each blank. Choose from kg, g, and mg.

The wrestler weighed 95 __kg__ .

She took a 500 __mg__ aspirin tablet.

One banana weighs 150 __g__ .

Use the six steps to solve this problem.

George cut 1 m 35 cm off of a 3 m board. How long was the leftover piece, in meters?

Step 1   The problem asks for the length of the leftover piece in meters.

Step 2   Convert the cut-off measurement to meters (the unit required in the answer), and then subtract to find the "leftover."

Step 3   To estimate, round 1 m 35 cm to 1 m, because 35 cm is less than half a meter. Then, 3 m − 1 m = 2 m as our estimate.

Step 4   Convert the cut-off measurement to meters, then subtract.

```
     1 m      →    1.00 m            3.00 m  ← Board
plus 35 cm    →  + 0.35 m          − 1.35 m  ← Cut off
                   1.35 m            1.65 m  ← Left
```

Step 5   The leftover piece is 1.65 m long.

Step 6   The exact answer of 1.65 m is close to our estimate of 2 m.

Concepts	*Examples*
8.5 *Converting between Metric and English Units* Write a unit fraction using the values in the table of conversion factors on the first page of **Section 8.5.** Because the values in the table are rounded, your answers will be approximate.	Convert. Round answers to the nearest tenth. **(a)** 23 m to yards From the table, 1 meter ≈ 1.09 yards. $$\frac{23 \text{ m}}{1} \cdot \frac{1.09 \text{ yd}}{1 \text{ m}} = 25.07 \text{ yd}$$ 25.07 rounds to 25.1, so 23 m ≈ 25.1 yd. **(b)** 4 oz to grams From the table, 1 ounce ≈ 28.35 grams. $$\frac{4 \text{ oz}}{1} \cdot \frac{28.35 \text{ g}}{1 \text{ oz}} = 113.4 \text{ g}$$ Thus, 4 oz ≈ 113.4 g.
8.5 *Common Celsius Temperatures* Use approximate and exact comparisons to judge which temperatures are appropriate. **Exact comparisons:** 0 °C is the freezing point of water (32 °F). 100 °C is the boiling point of water (212 °F). **Approximate comparisons:** 10 °C for a spring day (50 °F) 20 °C for room temperature (68 °F) 30 °C for summer day (86 °F) 37 °C for normal body temperature (98.6 °F)	Circle the Celsius temperature that is most reasonable. **(a)** Hot summer day: (35 °C) 90 °C 110 °C **(b)** The first snowy day in winter: −20 °C (0 °C) 15 °C
8.5 *Converting between Fahrenheit and Celsius Temperatures* Use this formula to convert from Fahrenheit (F) to Celsius (C). $$C = \frac{5(F - 32)}{9}$$	Convert 100 °F to Celsius. Round your answer to the nearest degree if necessary. $C = \dfrac{5(\mathbf{100} - 32)}{9}$ Replace F with 100 $= \dfrac{5(68)}{9}$ $= \dfrac{340}{9}$ $= 37.\overline{7}$ Round to 38 Thus, 100 °F ≈ 38 °C.

continued

Concepts	Examples
8.5 *Converting between Fahrenheit and Celsius Temperatures (continued)* Use this formula to convert from Celsius (C) to Fahrenheit (F). $$F = \frac{9C}{5} + 32$$	Convert $-8\ °C$ to Fahrenheit. Round your answer to the nearest degree, if necessary. $F = \dfrac{9(-8)}{5} + 32$ Replace C with -8 $ = \dfrac{-72}{5} + 32$ $ = -14.4 + 32$ $ = 17.6$ Round to 18 Thus, $-8\ °C \approx 18\ °F$.

ANSWERS TO TEST YOUR WORD POWER

1. B. *Examples:* 10 meters = 1 dekameter; 100 meters = 1 hectometer; 1000 meters = 1 kilometer.
2. C. *Examples:* Feet and inches are used to measure length, quarts to measure capacity, pounds to measure weight.
3. D. *Example:* Because 12 in. = 1 ft, the unit fraction $\dfrac{12\ \text{in.}}{1\ \text{ft}}$ is equivalent to $\dfrac{12\ \text{in.}}{12\ \text{in.}} = 1$.
4. A. *Example:* A small box measuring 1 cm on every edge holds exactly 1 mL of water, and the water weighs 1 g.
5. C. *Example:* A measurement of 16 meters is written 16 m (without a period).
6. C. *Example:* In the metric system, water freezes at 0 °C and boils at 100 °C. The English system uses the Fahrenheit temperature scale where water freezes at 32 °F and boils at 212 °F.

Chapter 8
REVIEW EXERCISES

[8.1] *Fill in the blanks with the measurement relationships you have memorized.*

1. 1 lb = _____ oz
2. _____ ft = 1 yd
3. 1 T = _____ lb
4. _____ qt = 1 gal
5. 1 hr = _____ min
6. 1 c = _____ fl oz
7. _____ sec = 1 min
8. _____ ft = 1 mi
9. _____ in. = 1 ft

Convert using unit fractions.

10. 4 ft = _____ in.
11. 6000 lb = _____ T
12. 64 oz = _____ lb
13. 18 hr = _____ day
14. 150 min = _____ hr
15. $1\frac{3}{4}$ lb = _____ oz
16. $6\frac{1}{2}$ ft = _____ in.
17. 7 gal = _____ c
18. 4 days = _____ sec

19. The average depth of the world's oceans is 12,460 ft. (*Source: Handy Ocean Answer Book.*)
    (a) What is the average depth in yards?
    (b) What is the average depth in miles, to the nearest tenth?

20. During the first year of a program to recycle office paper, a company recycled 123,260 pounds of paper. The company received $40 per ton for the paper. How much money did the company make? (*Source: I. C. System.*)

[8.2] *Write the most reasonable metric length unit in each blank. Choose from km, m, cm, and mm.*

21. My thumb is 20 _____ wide.
22. Her waist measurement is 66 _____.
23. The two towns are 40 _____ apart.
24. A basketball court is 30 _____ long.
25. The height of the picnic bench is 45 _____.
26. The eraser on the end of my pencil is 5 _____ long.

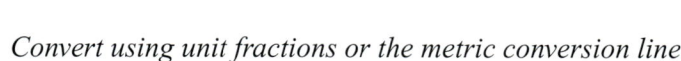

Convert using unit fractions or the metric conversion line.

27. 5 m to cm
28. 8.5 km to m
29. 85 mm to cm

30. 370 cm to m   **31.** 70 m to km   **32.** 0.93 m to mm

[8.3] *Write the most reasonable metric unit in each blank. Choose from L, mL, kg, g, and mg.*

33. The eyedropper holds 1 _____.

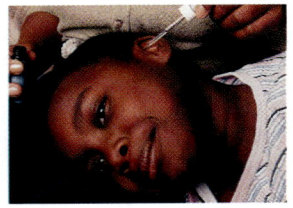

34. I can heat 3 _____ of water in this saucepan.

35. Loretta's hammer weighed 650 _____.

36. Yongshu's suitcase weighed 20 _____ when it was packed.

37. My fish tank holds 80 _____ of water.

38. I'll buy the 500 _____ bottle of mouthwash.

39. Mara took a 200 _____ antibiotic pill.

40. This piece of chicken weighs 100 _____.

Convert using unit fractions or the metric conversion line.

41. 5000 mL to L   **42.** 8 L to mL   **43.** 4.58 g to mg

44. 0.7 kg to g   **45.** 6 mg to g   **46.** 35 mL to L

[8.4] *Solve each application problem. Show your work.*

47. Each serving of punch at the wedding reception will be 180 mL. How many liters of punch are needed for 175 servings?

48. Jason is serving a 10 kg turkey to 28 people. How many grams of meat is he allowing for each person? Round to the nearest whole gram.

49. Yerald weighed 92 kg. Then he lost 4 kg 750 g. What is his weight now, in kilograms?

50. Young-Mi bought 750 g of onions. The price was $1.49 per kilogram. How much did she pay, to the nearest cent?

[8.5] *Use the table on the first page of* **Section 8.5** *and unit fractions to make approximate conversions. Round your answers to the nearest tenth, if necessary.*

51. 6 m to yards   **52.** 30 cm to inches

53. 108 km to miles

54. 800 mi to kilometers

55. 23 qt to liters

56. 41.5 L to quarts

*Write the appropriate **metric** temperature in each blank.*

57. Water freezes at _____.

58. Water boils at _____.

59. Normal body temperature is about _____.

60. Comfortable room temperature is about _____.

*Use the conversion formulas in **Section 8.5** to convert each temperature to Fahrenheit or to Celsius. Round to the nearest degree, if necessary.*

61. 77 °F

62. 5 °F

63. −2 °C

64. Water coming into a dishwasher should be at least 49 °C to clean the dishes properly. What Fahrenheit temperature is that? (*Source:* Frigidaire.)

MIXED REVIEW EXERCISES

Write the most reasonable metric unit in each blank. Choose from km, m, cm, mm, L, mL, kg, g, and mg.

65. I added 1 _____ of oil to my car.

66. The box of books weighed 15 _____.

67. Larry's shoe is 30 _____ long.

68. Jan used 15 _____ of shampoo on her hair.

69. My fingernail is 10 _____ wide.

70. I walked 2 _____ to school.

71. The tiny bird weighed 15 _____.

72. The new library building is 18 _____ wide.

73. The cookie recipe uses 250 _____ of milk.

74. Renee's pet mouse weighs 30 _____.

75. One postage stamp weighs 90 _____.

76. I bought 30 _____ of gas for my car.

Convert the following using unit fractions, the metric conversion line, or the temperature conversion formulas.

77. 10.5 cm to millimeters

78. 45 min to hours

79. 90 in. to feet

80. 1.3 m to centimeters

81. 25 °C to Fahrenheit

82. $3\frac{1}{2}$ gal to quarts

83. 700 mg to grams

84. 0.81 L to milliliters

85. 5 lb to ounces

86. 60 kg to grams

87. 1.8 L to milliliters

88. 30 °F to Celsius

89. 0.36 m to centimeters

90. 55 mL to liters

Solve each application problem. Show your work.

91. Peggy had a board measuring 2 m 4 cm. She cut off 78 cm. How long is the board now, in meters?

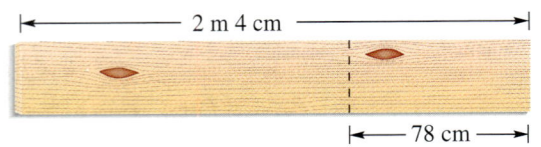

92. During the 12-day Minnesota State Fair, one of the biggest in the United States, Sweet Martha's booth sells an average of 3000 pounds of cookies per day. How many tons of cookies are sold in all? (*Source: Minneapolis Star Tribune.*)

93. Olivia is sending a recipe to her mother in Mexico. Among other things, the recipe calls for 4 oz of rice and a baking temperature of 350 °F. Convert these measurements to metric, rounding to the nearest gram and nearest degree.

94. While on vacation in Canada, Jalo became ill and went to a health clinic. They said he weighed 80.9 kg and was 1.83 m tall. Find his weight in pounds and height in feet. Round to the nearest tenth.

The largest two-axle trucks in the world are mining trucks used to haul huge loads of rock and iron ore. Some information about these $1.5 million trucks is given below. Fill in the blank spaces in the table. Then use the information to answer Exercises 101 and 102. Round answers to the nearest tenth.

WORLD'S LARGEST TWO-AXLE TRUCK

	Mining Truck	Measurements	
		Metric Units	English Units
95.	Length of truck	____ m	44 ft
96.	Height of truck	6.7 m	____ ft
97.	Height of tire	____ m	4 yd
98.	Width of tire	102 cm	____ in.
99.	Weight of load truck can carry	220,000 kg	____ lb
100.	Fuel needed to travel 1 mile	____ to ____ L	5 to 6 gal

Source: Hull Rust Mahoning Mine.

101. Using English measurements:
  (a) The truck can carry a load weighing how many tons?
  (b) How many inches high is each tire?

102. Using metric measurements:
  (a) What is the width of each tire in meters?
  (b) How tall is the truck in centimeters?

Chapter 8
TEST

Convert each measurement.

1. 9 gal = _____ qt

2. 45 ft = _____ yd

3. 135 min = _____ hr

4. 9 in. = _____ ft

5. $3\frac{1}{2}$ lb = _____ oz

6. 5 days = _____ min

Write the most reasonable metric unit in each blank. Choose from km, m, cm, mm, L, mL, kg, g, and mg.

7. My husband weighs 75 _____.

8. I hiked 5 _____ this morning.

9. She bought 125 _____ of cough syrup.

10. This apple weighs 180 _____.

11. This page is about 21 _____ wide.

12. My watch band is 10 _____ wide.

13. I bought 10 _____ of soda for the picnic.

14. The bracelet is 16 _____ long.

Convert the following measurements. Show your work.

15. 250 cm to meters

16. 4.6 km to meters

17. 5 mm to centimeters

18. 325 mg to grams

19. 16 L to milliliters

20. 0.4 kg to grams

21. 10.55 m to centimeters

22. 95 mL to liters

1. _____
2. _____
3. _____
4. _____
5. _____
6. _____
7. _____
8. _____
9. _____
10. _____
11. _____
12. _____
13. _____
14. _____
15. _____
16. _____
17. _____
18. _____
19. _____
20. _____
21. _____
22. _____

23. The rainiest place in the world is Mount Waialeale in Hawaii, which receives 460 inches of rain each year. What is the average rainfall per month, in feet, to the nearest tenth? (*Source:* National Geographic Society.)

24. A 6-inch Subway Veggie Delite sandwich has 590 mg of sodium. A 6-inch Super Subway Melt sandwich has 2.9 g of sodium. (*Source:* Subway.)
   (a) How much more sodium is in the Super Melt sandwich, in milligrams, than in the Vegie Delite?
   (b) The recommended amount of sodium is less than 2400 mg daily. How much more or less sodium does the Super Melt have than the recommended daily amount?

Pick the metric temperature that is most appropriate in each situation.

25. The water is almost boiling.
    210 °C   155 °C   95 °C

26. The tomato plants may freeze tonight.
    30 °C   20 °C   0 °C

Use the table from **Section 8.5** *and unit fractions to convert each measurement. Round your answers to the nearest tenth, if necessary.*

27. 6 ft to meters

28. 125 lb to kilograms

29. 50 L to gallons

30. 8.1 km to miles

Use the conversion formulas to convert each temperature. Round your answers to the nearest degree, if necessary.

31. 74 °F to Celsius

32. −12 °C to Fahrenheit

Solve this application problem. Show your work.

33. Denise is making five matching pillows. She needs 1 m 20 cm of braid to trim each pillow. If the braid costs $3.98 per yard, how much will she spend to trim the pillows, to the nearest cent? (First find the number of meters of braid Denise needs.)

34. Describe two benefits the United States would achieve by switching entirely to the metric system.

Graphs 9

9.1 **Problem Solving with Tables and Pictographs**

9.2 **Reading and Constructing Circle Graphs**

9.3 **Bar Graphs and Line Graphs**

9.4 **The Rectangular Coordinate System**

9.5 **Introduction to Graphing Linear Equations**

The Internet can be a great place to do research. Whether you're writing term papers and speeches for college courses, or planning on a career as a writer or reporter, you'll need to find information at various Web sites, interpret the data, and communicate it to other people. The data will often be presented in the form of a table (see **Section 9.1,** Examples 1 and 2) or a graph (see **Sections 9.2 and 9.3**). Interpreting the data accurately is critical to the success of your report, article, or speech. (See **Section 9.3,** Exercises 37–44.)

9.1 Problem Solving with Tables and Pictographs

OBJECTIVES

1. Read and interpret data presented in a table.
2. Read and interpret data from a pictograph.

Throughout this book you have used numbers, expressions, formulas, and equations to communicate rules or information. In this chapter you'll see how *tables, pictographs, circle graphs, bar graphs,* and *line graphs* are also used to communicate information.

OBJECTIVE 1 Read and interpret data presented in a table. A table presents data organized into rows and columns. The advantage of a table is that you can find very specific, exact values. The disadvantages are that you may have to spend some time searching through the table to find what you want, and it may not be easy to see trends or patterns.

The table below shows information about the performance of eight U.S. airlines during the year 2003 (January–December).

PERFORMANCE DATA FOR SELECTED U.S. AIRLINES
JANUARY–DECEMBER 2003

Airline	On-Time Performance	Luggage Handling*
Airtran	78%	2.8
American	82%	4.5
Continental	82%	3.1
Delta	82%	3.8
Northwest	83%	3.4
Southwest	88%	3.6
United	83%	3.9
US Airways	80%	3.6

*Luggage problems per 1000 passengers.
Source: Department of Transportation Air Travel Consumer Report.

For example, by reading from left to right along the row marked Delta, you first see that 82% of Delta's flights were on time during 2003. The next number is 3.8 and the heading at the top of that column is *Luggage Handling**. The little star (asterisk) tells you to look below the table for more information. Next to the asterisk below the table it says "Luggage problems per 1000 passengers." So, almost 4 passengers out of every 1000 passengers on Delta flights had some sort of problem with their luggage.

EXAMPLE 1 Reading and Interpreting Data from a Table

Use the table above on airline performance to answer these questions.

(a) What percent of US Airways' flights were on time?
Look across the row labeled US Airways to see that 80% of its flights were on time.

(b) Which airline had the worst luggage handling record?
Look down the column headed Luggage Handling. To find the *worst* record, look for the *highest* number of luggage problems. The highest number is 4.5. Then look to the left to find the airline, which is American.

(c) What was the average percent of on-time flights for the three airlines with the best performance, to the nearest whole percent?
Look down the column headed On-Time Performance to find the three highest numbers: 88%, 83%, and 83%. To find the average, add the values and divide by 3.

Continued on Next Page

$$\frac{88 + 83 + 83}{3} = \frac{254}{3} \approx 85 \text{ (rounded to nearest whole percent)}$$

The average on-time performance for the three best airlines was about 85%.

Work Problem 1 at the Side. ▶▶▶

EXAMPLE 2 **Interpreting Data from a Table**

The table below shows the maximum cab fares in five different cities in February 2004. The "flag drop" charge is made when the driver starts the meter. "Wait time" is the charge for having to wait in the middle of a ride.

MAXIMUM TAXICAB FARES ALLOWED IN SELECTED CITIES IN FEBRUARY 2004

City	Flag Drop	Price per Mile	Wait Time (per Hour)
Chicago	$1.90	$1.60	$20
Denver	$1.60	$1.80	$18
Miami	$1.70	$2.20	$21
New York	$2.50	$2	$12
Portland	$2.50	$1.50	$20

Source: Web sites for the various cities or their airports.

Use the table to answer these questions.

(a) What is the maximum fare for a 9-mile ride in New York that includes having the cab wait 15 minutes while you pick up a package at a store?

The price per mile in New York is $2, so the cost for 9 miles is 9($2) = $18. Then, add the flag drop charge of $2.50. Finally, figure out the cost of the wait time. One way to do that is to set up a proportion. Recall that 1 hour is 60 minutes, so each side of the proportion compares the cost to the number of minutes of wait time.

$$\text{Cost} \rightarrow \frac{\$12}{60 \text{ min}} = \frac{\$x}{15 \text{ min}} \leftarrow \text{Cost}$$
Wait time ↑   ↑ Wait time

$60 \cdot x = 12 \cdot 15$ Find the cross products.

$$\frac{60x}{60} = \frac{180}{60}$$ Divide both sides by 60

$x = \$3$ ← Charge for waiting 15 minutes

Total fare = $18 + $2.50 + $3 = $23.50

(b) It is customary to give the cab driver a tip. Find the total cost of the cab ride in part (a) above if the passenger added a 15% tip, rounded to the nearest quarter (nearest $0.25).

Use the percent equation to find the exact tip.

percent • whole = part   Percent equation
(15.%)($23.50) = n   Write 15% as a decimal.
(0.15)($23.50) = n
$3.525 = n   Round to $3.50 (nearest $0.25).

The total cost of the cab ride is $23.50 + $3.50 tip = $27.00.

Work Problem 2 at the Side. ▶▶▶

---

1 Use the table of airline performance on the previous page to answer these questions.

(a) What percent of American flights were on time?

(b) Which airline had the best on-time performance?

(c) Which airline had the best record for luggage handling?

(d) Which airline(s) had fewer than 3.5 luggage handling problems per 1000 passengers?

(e) What was the average number of luggage problems for all eight airlines, to the nearest tenth?

2 Use the table of cab fares at the left to answer these questions.

(a) What is the difference in the maximum fare for a cab ride of 6.5 miles in Chicago compared to New York? Assume there is no wait time.

(b) What is the maximum fare for a 12-mile cab ride in Miami that includes 10 minutes of wait time?

(c) Find the total cost for a cab ride of 4.5 miles in Denver, including 30 minutes of wait time and a 15% tip. Round the tip to the nearest $0.25.

ANSWERS
1. **(a)** 82% **(b)** Southwest **(c)** Airtran **(d)** Airtran, Continental, Northwest **(e)** $28.7 \div 8 \approx 3.6$ (rounded)
2. **(a)** Chicago $12.30; New York $15.50; New York fare is $3.20 higher. **(b)** $31.60 **(c)** $18.70 + $2.75 tip = $21.45

OBJECTIVE 2 **Read and interpret data from a pictograph.** Tables show numbers in rows and columns. Graphs, on the other hand, are a *visual* way to communicate data; that is, they show a *picture* of the information rather than a list of numbers. In this section you'll work with one type of graph, *pictographs,* and in the next two sections you'll learn about circle graphs, bar graphs, and line graphs.

The advantage of a graph is that you can easily make comparisons or see trends just by looking. The disadvantage is that the graph may not give you the specific, more exact numbers you need in some situations.

EXAMPLE 3 Reading and Interpreting a Pictograph

A **pictograph** uses symbols or pictures to represent various amounts. The pictograph below shows the population of five U.S. metropolitan areas (cities with their surrounding suburbs) in 2000. The *key* at the bottom of the graph tells you that each symbol of a person represents 2 million people. A fractional part of a symbol represents a fractional part of 2 million people.

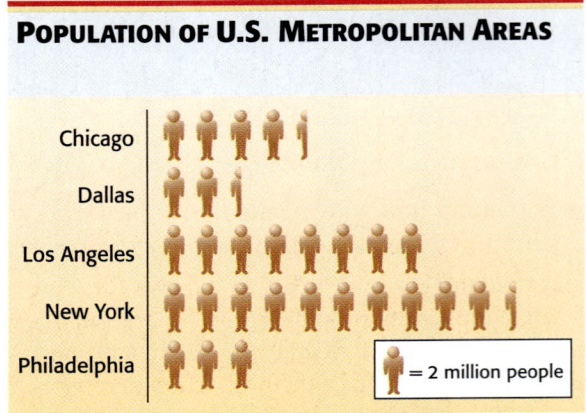

Source: U.S. Bureau of the Census, U.S. Department of Commerce.

Use the pictograph to answer these questions.

(a) What is the approximate population of Chicago?

The population of Chicago is represented by 4 whole symbols (4 • 2 million = 8 million) plus half of a symbol ($\frac{1}{2}$ of 2 million is 1 million) for a total of 9 million.

Recall that 2 million can be written as 2,000,000, so another way to get the answer is to multiply **4.5** (the number of symbols) times 2,000,000 for a total of 9,000,000. So the approximate population of Chicago is 9 million people (or 9,000,000).

(b) How much greater is the population of Los Angeles than Philadelphia?

Los Angeles shows 8 symbols and Philadelphia shows 3 symbols. So Los Angeles has 5 more symbols than Philadelphia, and 5 • 2 million = 10 million. Thus, Los Angeles has about 10 million (or 10,000,000) more people than Philadelphia.

Continued on Next Page

Section 9.1 Problem Solving with Tables and Pictographs 621

NOTE
One disadvantage of a pictograph is that it is difficult to draw and interpret fractional parts of a symbol. The data on metropolitan area populations were rounded a great deal in order to use only whole symbols and half symbols.

For example, the actual metropolitan area population of Los Angeles is 16,402,079 people. We rounded the population to the *nearest million*. The actual population of Los Angeles is closer to 16 million than to 17 million, so there are 8 whole symbols in the pictograph. That means our pictograph is off by about 400,000 people. Therefore, if you need fairly accurate numbers for a particular situation, it's better to use the information from a table than a pictograph.

Work Problem 3 at the Side.

3 Use the pictograph in Example 3 to answer these questions.

(a) What is the approximate population of Philadelphia?

(b) What is the approximate population of New York?

(c) Approximately how much greater is the population of New York than Chicago?

(d) The population of Dallas is approximately how much less than Philadelphia?

ANSWERS
3. (a) 6 million people (or 6,000,000)
   (b) 21 million people (or 21,000,000)
   (c) 12 million people (or 12,000,000)
   (d) 1 million people (or 1,000,000)

Focus on Real-Data Applications

Currency Exchange

When you travel between countries, you will exchange U.S. dollars for the local currency. The exchange rate between currencies changes daily, and you can easily find the updated rates using the Internet or any major newspaper. The table below has been extracted from the Oanda Web page, www.oanda.com. It shows how much of each country's currency was equivalent to 1 U.S. dollar on February 14, 2004.

NORTH AMERICA/CARIBBEAN CURRENCY RATES (FEBRUARY 14, 2004)

Currency	Symbol	Value
Canadian Dollar	CAD	1.3194
Cayman Islands Dollar	KYD	0.82
Jamaican Dollar	JMD	65.1
Mexican Peso	MXN	10.99
United States Dollar	USD	1.00

From the table, we can see the currency exchange rate from U.S. dollars to Mexican pesos.

$1.00 U.S. is equivalent to 10.99 Mexican pesos

You can set up a proportion to convert dollars to pesos. For example, suppose you want to determine the number of pesos that is equivalent to $50.00.

$$\frac{\$1}{10.99 \text{ pesos}} = \frac{\$50}{x \text{ pesos}} \quad \text{or} \quad \frac{1}{10.99} = \frac{50}{x}$$

$$(1)(x) = (10.99)(50)$$

$$x = 549.50 \text{ pesos}$$

So $50 buys 549.5 pesos.

1. Based on the currency exchange rates for February 14, 2004, find the amount of each local currency that is equivalent to $50 U.S. and find the number of U.S. dollars that is equivalent to 200 units of each local currency. Round your answers to the nearest cent.

   (a) $50 = _____ Canadian dollars, and 200 Canadian dollars = _____ U.S. dollars.

   (b) $50 = _____ Cayman Island dollars, and
       200 Cayman Island dollars = _____ U.S. dollars.

   (c) $50 = _____ Jamaican dollars, and 200 Jamaican dollars = _____ U.S. dollars.

2. Set up a proportion to find the number of U.S. dollars that is equivalent to 1 Mexican Peso. Round your answer to the nearest cent. 1 Mexican peso is equivalent to $ _____ (U.S.).

3. From Problem 2, you should recognize the conversion rate based on 1 Mexican peso as the expression $\frac{1}{10.99}$. What is the mathematical word that describes the relationship between the conversion rates 10.99 and $\frac{1}{10.99}$?

9.1 Exercises

FOR EXTRA HELP: 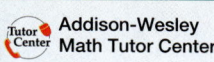 Addison-Wesley Math Tutor Center 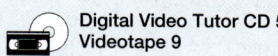 MathXL  Digital Video Tutor CD 5 Videotape 9 Student's Solutions Manual  MyMathLab 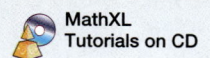 MathXL Tutorials on CD

This table lists the basketball players in the NBA with the highest scoring average at the midpoint of the 2003–2004 season. (Players must have a minimum of 10,000 points or 400 games.) Use the table to answer Exercises 1–10. See Examples 1 and 2.

ALL-TIME NBA STATISTICAL LEADERS—SCORING AVERAGE
(AT THE MIDPOINT OF THE 2003–2004 SEASON)

Player	Games	Points	Average Points Per Game
Michael Jordan	1072	32,292	30.1
Wilt Chamberlain	1045	31,419	30.1
Elgin Baylor	846	23,419	27.4
Shaquille O'Neal*	778	21,225	27.3
Allen Iverson*	527	14,267	
Jerry West	932	25,192	
Bob Pettit	792	20,880	

*Player active in 2003–2004 season.
Source: National Basketball Association.

1. (a) How many points did Wilt Chamberlain score during his NBA career?
   (b) Which player(s) scored more points than Chamberlain?

2. (a) How many games did Elgin Baylor play in during his career in the NBA?
   (b) Which player is closest to Baylor in number of games?

3. (a) Which player has been in the greatest number of games?
   (b) Which player has been in the fewest number of games?

4. Which players have scored more than 25,000 points? List them in order, starting with the player with the greatest number of points.

5. What is the difference in points scored between the player with the greatest number of points and the player with the least number of points?

6. How many fewer games has Shaquille O'Neal played in than Michael Jordan?

7. Complete the table by finding the average number of points scored per game by Allen Iverson, by Jerry West, and by Bob Pettit. Look at the other averages in the table to decide how to round your answers.

8. Find the overall scoring average (points per game) for all seven players listed in the table. Use the numbers in the *Games* column and the *Points* column to calculate your answer.

9. Which player(s) will have a different number of games and points by the *end* of the 2003–2004 season? Explain why. (Hint: Look at the note below the table.)

10. Michael Jordan and Wilt Chamberlain are tied in first place for average points per game. Explain what happens if you round their average points per game to the nearest hundredth instead of the nearest tenth.

This table shows the number of calories burned during 30 minutes of various types of exercise. The table also shows how the number of calories burned varies according to the weight of the person doing the exercise. Use the table to answer Exercises 11–20. See Examples 1 and 2.

CALORIES BURNED DURING 30 MINUTES OF EXERCISE BY PEOPLE OF DIFFERENT WEIGHTS

Activity	Calories Burned in 30 Minutes		
	110 Pounds	140 Pounds	170 Pounds
Moderate jogging	322	410	495
Moderate walking	110	140	170
Moderate bicycling	140	180	220
Aerobic dance	200	255	310
Racquetball	210	268	325
Tennis	160	205	250

Source: Fairview Health Services.

11. A person weighing 140 pounds is looking at the table.
   (a) How many calories will be burned during 30 minutes of aerobic dance?
   (b) Which activity burns the most calories?

12. A person weighing 170 pounds is looking at the table.
   (a) How many calories are burned during 30 minutes of tennis?
   (b) Which activity burns the fewest calories?

13. (a) Which activities can a 110-pound person do to burn at least 200 calories in 30 minutes?

   (b) Which activities can a 170-pound person do to burn at least 200 calories in 30 minutes?

14. (a) Which activities would burn fewer than 200 calories in 30 minutes for a 140-pound person?

   (b) Which activities would burn fewer than 300 calories in 30 minutes for a 170-pound person?

15. How many total calories will a 140-pound person burn during 15 minutes of bicycling and 60 minutes of walking?

16. How many total calories will a 110-pound person burn during 90 minutes of tennis and 15 minutes of aerobic dance?

Set up and solve proportions to answer Exercises 17–20.

17. How many calories would you expect a 125-pound person to burn
   (a) during 30 minutes of moderate jogging?
   (b) during 30 minutes of racquetball? Round to the nearest whole number.

18. How many calories would you expect a 185-pound person to burn
   (a) during 30 minutes of walking?
   (b) during 30 minutes of bicycling? Round to the nearest ten.

19. How many more calories would a 158-pound person burn during 15 minutes of aerobic dance than during 20 minutes of walking? Round to the nearest whole number.

20. How many fewer calories would a 196-pound person burn during 25 minutes of walking than during 20 minutes of tennis? Round to the nearest whole number.

This pictograph shows the approximate number of passenger arrivals and departures at selected U.S. airports in 2002. Use the pictograph to answer Exercises 21–28. See Example 3.

21. Approximately how many passenger arrivals and departures took place at the

(a) San Francisco airport?

(b) Atlanta airport?

22. Approximately how many passenger arrivals and departures took place at the

(a) Dallas airport?

(b) St. Louis airport?

23. What is the approximate total number of arrivals and departures at the two busiest airports?

24. What is the difference in the number of arrivals and departures at the busiest airport and the least-busy airport?

25. How many fewer arrivals and departures did St. Louis' airport have compared to San Francisco's airport?

26. Find the approximate total number of arrivals and departures for the three least-busy airports.

27. What is the approximate total number of arrivals and departures for all five airports?

28. Find the average number of arrivals and departures for the five airports.

RELATING CONCEPTS (EXERCISES 29–34) For Individual or Group Work

*Look back at the first table in this section, Performance Data for Selected U.S. Airlines. Use the table as you **work Exercises 29–34 in order.***

29. Suppose you are planning a business trip where you will fly to a new city each day on a tight schedule. You'll travel light, carrying one small bag on the plane rather than checking it. If you can choose any one of the airlines in the table, which one would you pick? Explain why.

30. Suppose you are planning the business trip described in Exercise 29 and the only airline that goes to the cities you want is US Airways. What could you do to minimize the problems caused by the possibility of late flights?

31. Now you are planning a two-week vacation trip to a beachfront resort. You'll be checking several bags and your expensive golf clubs. If you can choose any one of the airlines in the table, which one would you pick? Explain why.

32. Suppose you are planning the vacation trip described in Exercise 31 and American is the only airline that goes to the city you want. What could you do to minimize possible luggage handling problems?

33. Think of three possible reasons an airline might have a lower percentage of on-time flights during a particular year than they usually do. What, if anything, could the airline do to resolve each of the problems you listed?

34. Describe three other factors you might consider when selecting an airline, other than on-time performance and luggage handling problems.

9.2 Reading and Constructing Circle Graphs

A *circle graph* is another way to show a *picture* of a set of data. This picture can often be understood faster and more easily than a formula or a list of numbers.

OBJECTIVE 1 Read a circle graph. A **circle graph** is used to show how a total amount is divided into parts. The circle graph below shows you how 24 hours in the life of a college student are divided among different activities.

OBJECTIVES
1. Read a circle graph.
2. Use a circle graph.
3. Use a protractor to draw a circle graph.

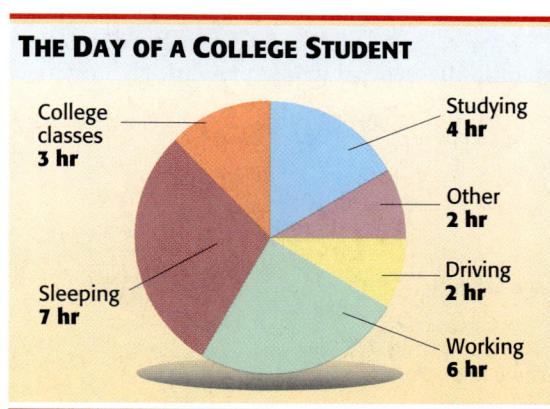

THE DAY OF A COLLEGE STUDENT
- College classes: 3 hr
- Studying: 4 hr
- Other: 2 hr
- Driving: 2 hr
- Working: 6 hr
- Sleeping: 7 hr

Work Problem 1 at the Side.

OBJECTIVE 2 Use a circle graph. The circle graph above uses pie-shaped pieces called *sectors* to show the amount of time spent on each activity (the total must be 24 hours). The circle graph can therefore be used to compare the time spent on one activity to the total number of hours in the day.

EXAMPLE 1 Using a Circle Graph

Find the ratio of time spent in college classes to the total number of hours in a day. Write the ratio as a fraction in lowest terms. (See **Section 6.1**.)
The circle graph shows that 3 of the 24 hours in a day are spent in class. The ratio of class time to the hours in a day is shown below.

$$\frac{3 \text{ hours (college classes)}}{24 \text{ hours (whole day)}} = \frac{3 \text{ hours}}{24 \text{ hours}} = \frac{\cancel{3}}{\cancel{3} \cdot 8} = \frac{1}{8} \quad \left\{ \text{Ratio in lowest terms} \right.$$

Work Problem 2 at the Side.

The circle graph above can also be used to find the ratio of the time spent on one activity to the time spent on any other activity. See the next example.

EXAMPLE 2 Finding a Ratio from a Circle Graph

Find the ratio of working time to class time.
The circle graph shows 6 hours spent working and 3 hours spent in class. The ratio of working time to class time is shown below.

$$\frac{6 \text{ hours (working)}}{3 \text{ hours (class)}} = \frac{6 \text{ hours}}{3 \text{ hours}} = \frac{\cancel{3} \cdot 2}{\cancel{3}} = \frac{2}{1} \quad \leftarrow \text{Ratio in lowest terms}$$

Work Problem 3 at the Side.

---

1 Use the circle graph at the left to answer each question.

(a) The greatest number of hours is spent in which activity?

(b) How many more hours are spent working than studying?

(c) Find the total number of hours spent studying, working, and attending classes.

2 Use the circle graph to find each ratio. Write the ratios as fractions in lowest terms.

(a) Hours spent driving to whole day

(b) Hours spent studying to whole day

(c) Hours spent sleeping and doing other to whole day

3 Use the circle graph to find each ratio. Write the ratios as fractions in lowest terms.

(a) Hours spent studying to hours spent working

(b) Hours spent working to hours spent sleeping

(c) Hours spent studying to hours spent driving

ANSWERS
1. (a) sleeping (b) 2 hours (c) 13 hours
2. (a) $\frac{1}{12}$ (b) $\frac{1}{6}$ (c) $\frac{3}{8}$
3. (a) $\frac{2}{3}$ (b) $\frac{6}{7}$ (c) $\frac{2}{1}$

4 Use the circle graph on frozen pizza sales to find the amount of sales for each company.

(a) Kraft

(b) Van De Kamps

(c) Tony's Pizza Service

(d) Pillsbury Corp.

ANSWERS
4. (a) $740,000,000 (b) $80,000,000
   (c) $600,000,000 (d) $180,000,000

A circle graph often shows data as percents. For example, total U.S. sales of frozen pizza are $2 billion each year. The circle graph below shows how sales are divided among various companies that make frozen pizza. The entire circle represents $2 billion in sales. Each sector represents the sales of one company as a percent of the total sales. The total in a circle graph must be 100%, although it may be slightly more or less due to rounding the percent for each sector.

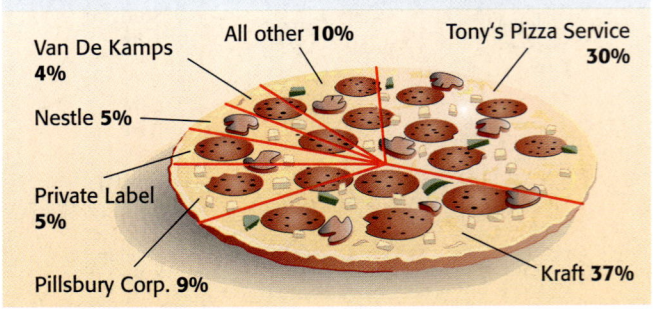

HOT SALES OF FROZEN PIZZA
Americans eat $2 billion worth of frozen pizzas each year. The percent of total sales for each company is rounded to the nearest whole percent.

Van De Kamps 4%
Nestle 5%
Private Label 5%
Pillsbury Corp. 9%
All other 10%
Tony's Pizza Service 30%
Kraft 37%

Source: Information Resources, Inc.

EXAMPLE 3 Calculating an Amount Using a Circle Graph

Use the circle graph above on frozen pizza sales to find the amount of sales for Nestle.

Recall the percent equation.

$$\text{percent} \cdot \text{whole} = \text{part}$$

The percent for Nestle is 5%. Rewrite 5% as the decimal 0.05. The *whole* is the total sales of $2 billion (the entire circle).

$$\text{percent} \cdot \text{whole} = \text{part}$$
$$05.\% \cdot \$2 \text{ billion} = n \quad \text{Write 5\% as a decimal.}$$
$$(0.05)(\$2{,}000{,}000{,}000) = n \quad \text{Write \$2 billion as \$2,000,000,000}$$
$$\$100{,}000{,}000 = n$$

The sales for Nestle are $100,000,000.

Work Problem 4 at the Side.

OBJECTIVE 3 Use a protractor to draw a circle graph. The coordinator of the Fair Oaks Youth Soccer League organizes teams in five age groups. She counts the number of registered players in each age group as shown in the table on the next page. Then she calculates what percent of the total each group represents. For example, there are 59 players in the "Under 8" group, out of 298 total players.

$$\text{percent} \cdot \text{whole} = \text{part}$$
$$p \cdot 298 = 59$$
$$\frac{p \cdot 298}{298} = \frac{59}{298}$$
$$p \approx 0.197 \approx 20\%$$

Age Group	Number of Players	Percent of Total (rounded to nearest whole percent)
Under 8 years	59	20% ← 59 players ≈ 20% of 298
Ages 8–9	46	15%
Ages 10–11	75	25%
Ages 12–13	74	25%
Ages 14–15	44	15%
Total	298	100%

You can show these percents by using a circle graph. Recall that a circle has 360 degrees (written 360°). The 360° represents the entire league, or 100% of the soccer players.

EXAMPLE 4 Drawing a Circle Graph

Using the data on *age groups,* find the number of degrees in the sector that would represent the "Under 8" group, and begin constructing a circle graph.

A complete circle has 360°. Because the "Under 8" group makes up 20% of the total number of players, the number of degrees needed for the "Under 8" sector of the circle graph is 20% of 360°.

$$20.\% \text{ of } 360° = n$$
$$(0.20)(360°) = n$$
$$72° = n$$

Use the circle drawn below and a tool called a **protractor** to make a circle graph. First, using a ruler or straightedge, draw a line from the center of the circle to the left edge. Place the hole in the protractor over the center of the circle, making sure that the 0 mark and the black line on the protractor are right over the line you drew. Find 72° and make a mark as shown in the illustration. Then remove the protractor and use the straightedge to draw a line from the 72° mark to the center of the circle. This sector is 72° and represents the "Under 8" group.

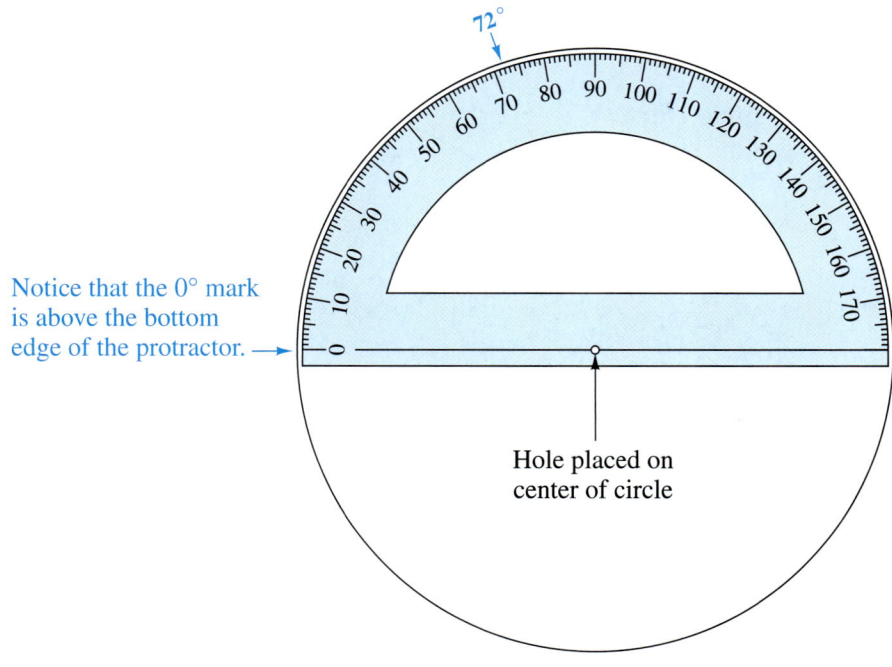

Notice that the 0° mark is above the bottom edge of the protractor.

Hole placed on center of circle

Continued on Next Page

5 Using the information on the soccer age groups in the table on the previous page, find the number of degrees needed for each sector. Then complete the circle graph at the bottom right on this page.

(a) "Ages 10–11" sector

(b) "Ages 12–13" sector

(c) "Ages 14–15" sector

To draw the "Ages 8–9" sector, begin by finding the number of degrees in the sector, which is 15% of the total circle, or 15% of 360°.

$$15\% \text{ of } 360° = n$$
$$(0.15)(360°) = n$$
$$54° = n$$

Again, place the hole of the protractor at the center of the circle, but this time align 0 with the previous 72° mark. Make a mark at 54° and draw a line from the mark to the center of the circle. This sector is 54° and represents the "Ages 8–9" group.

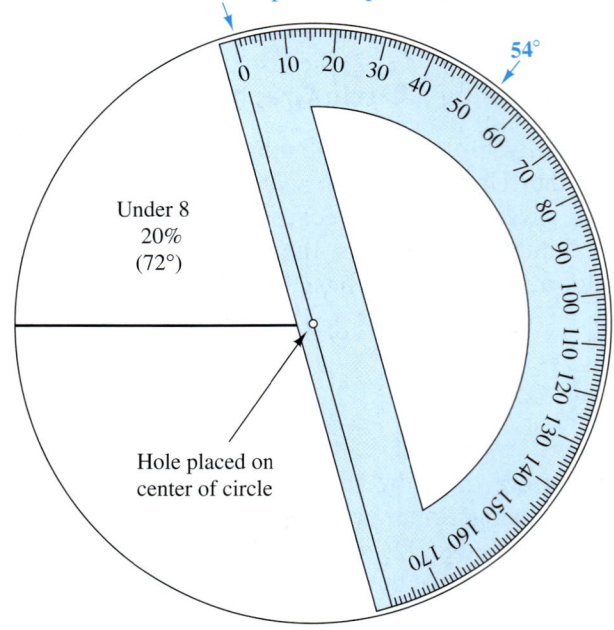

CAUTION
You must be certain that the hole in the protractor is placed on the exact center of the circle each time you measure the size of a sector.

◀◀◀ **Work Problem 5 at the Side.**

Use this circle for Problem 5 at the side.

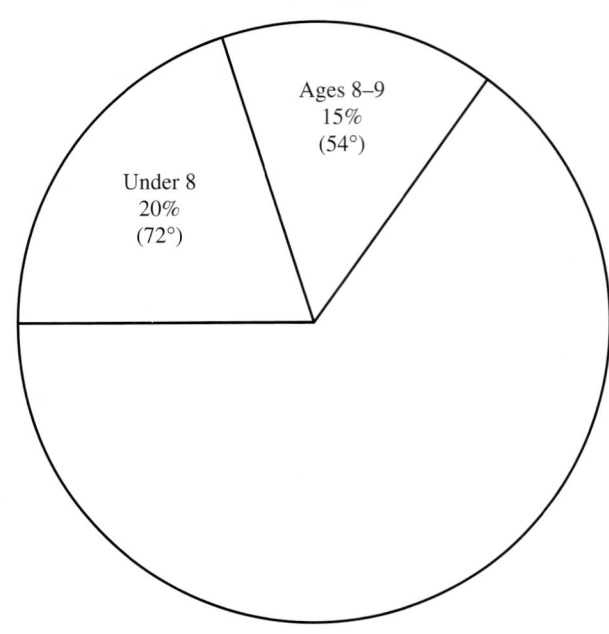

ANSWERS
5. (a) 90° (b) 90° (c) 54°

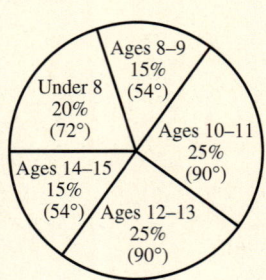

9.2 Exercises

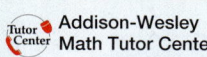 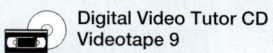   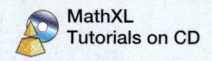

This circle graph shows the cost of adding a family room to an existing home. Use this circle graph to answer Exercises 1–6. Write ratios as fractions in lowest terms. See Examples 1 and 2.

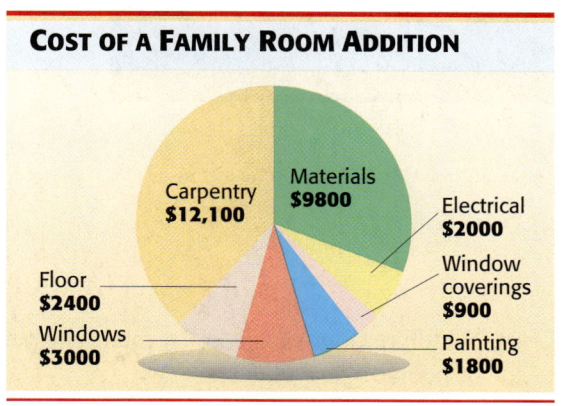

1. **(a)** Find the total cost of adding the family room.

   (b) What is the largest single expense?

2. **(a)** What is the second-largest expense in adding the family room?

   (b) What is the smallest expense?

3. **(a)** Find the ratio of the cost of materials to the total remodeling cost.

   (b) Find the ratio of the cost of windows to the cost of electrical.

4. **(a)** Find the ratio of the cost of painting to the total remodeling cost.

   (b) Find the ratio of the cost of windows to the cost of window coverings.

5. Find the ratio of the cost of carpentry, windows, and window coverings to the total remodeling cost.

6. Find the ratio of the cost of windows and electrical to the cost of the floor and painting.

This circle graph, adapted from USA Today, *shows the number of people in a survey who gave various reasons for eating dinner at restaurants. Use this circle graph to answer Exercises 7–14. See Examples 1 and 2.*

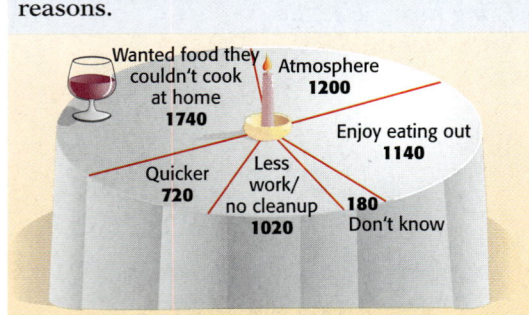

Source: Market Facts for Tyson Foods.

7. **(a)** Which reason was given by the least number of people?

   (b) Which reason was given by the second-fewest number of people?

8. **(a)** Which reason was given by the highest number of people?

   (b) Which reason was given by the second-highest number of people?

Find each ratio in Exercises 9–14. Write the ratios as fractions in lowest terms.

9. Those who said dining out is "Quicker" to total people in the survey

10. Those who said "Enjoy eating out" to the total people in the survey

11. Those who said "Less work/no cleanup" to those who said "Atmosphere"

12. Those who said "Don't know" to those who said "Quicker"

13. Those who said "Wanted food they couldn't cook at home" to those who said "Don't know"

14. Those who said "Atmosphere" to those who said "Enjoy eating out"

This circle graph shows the costs necessary to comply with the Americans with Disabilities Act (ADA) at the Dos Pueblos College. The cost of each item is expressed as a percent of the total cost of $1,740,000. Use the graph to find the dollar amount spent for each item in Exercises 15–20. See Example 3.

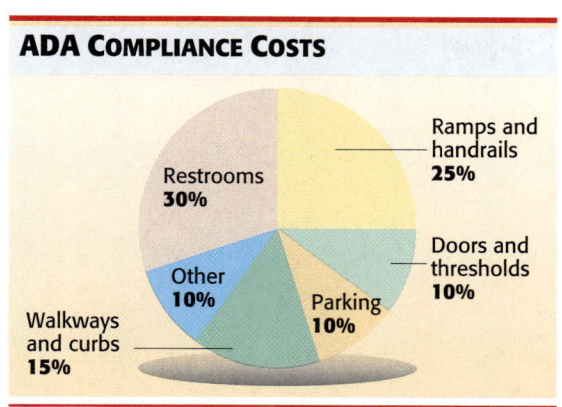

15. Restrooms

16. Ramps and handrails

17. Doors and thresholds

18. Parking

19. Walkways and curbs

20. Other

This circle graph shows the results of a survey on favorite hot dog toppings in the United States. Each topping is expressed as a percent of the 3200 people in the survey. Use the graph to find the number of people favoring each of the toppings in Exercises 21–26.

Source: National Hot Dog and Sausage Council

21. Onions

22. Sauerkraut

23. The most popular topping

24. The second-most-popular topping

25. How many more people chose chili than chose relish?

26. How many fewer people chose onions than chose mustard?

27. Describe the procedure for determining how large each sector must be to represent each of the items in a circle graph.

28. A protractor is the tool used to draw a circle graph. Give a brief explanation of what the protractor does and how you would use it to measure and draw each sector in the circle graph.

29. During one semester Kara Diano spent $5460 for school expenses as shown in this table. Find all numbers missing from the table.

Item	Dollar Amount	Percent of Total	Degrees of a Circle
(a) Rent	$1365	25%	_____
(b) Food	$1092	_____	72°
(c) Clothing	$ 546	_____	_____
(d) Books and supplies	$ 546	_____	_____
(e) Tuition and fees	$ 819	_____	_____
(f) Savings	$ 273	_____	_____
(g) Entertainment	$ 819	_____	_____

(h) Draw a circle graph using the budget information from the table. Label each sector with the budget item and the percent of total for that item. See Example 4.

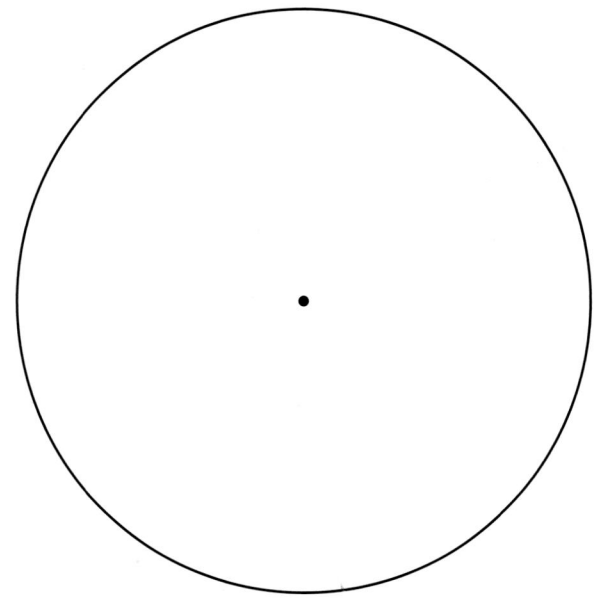

30. The Pathfinder Research Group asked 4488 Americans how they fall asleep. The results are shown in the visual on the right.

Use this information to complete the table. Round to the nearest whole percent and to the nearest degree.

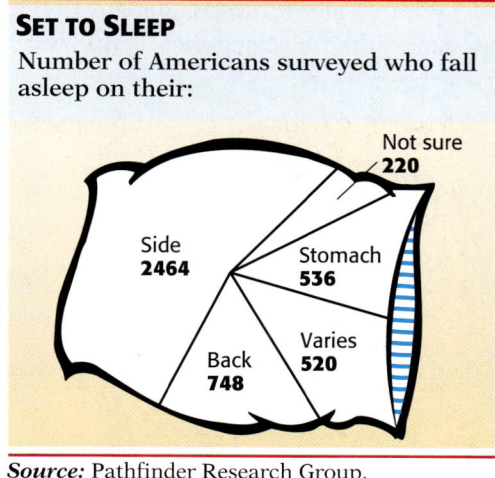

SET TO SLEEP

Number of Americans surveyed who fall asleep on their:

Source: Pathfinder Research Group.

	Sleeping Position	Number of Americans	Percent of Total	Number of Degrees
(a)	Side			
(b)	Back			
(c)	Stomach			
(d)	Varies			
(e)	Not sure			

(f) Add up the percents. Is the total 100%? Explain why or why not.

(g) Add up the degrees. Is the total 360°? Explain why or why not.

(h) Draw a circle graph using the information from the table above. Label each sector with the sleeping position and the percent of total for that position.

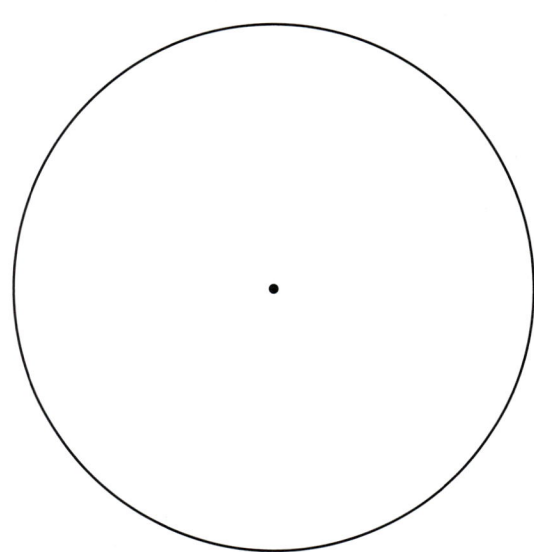

31. White Water Rafting Company divides its annual sales into five categories as follows.

Category	Annual Sales	Percent of Total
Adventure classes	$12,500	_____
Grocery/provision sales	$40,000	_____
Equipment rentals	$60,000	_____
Rafting tours	$50,000	_____
Equipment sales	$37,500	_____

(a) Find the total sales for the year.

(b) Find the percent of the total for each category and write it in the table.

(c) Find the number of degrees in a circle graph for each item.

(d) Make a circle graph showing the sales information. Label each sector with the category and the percent of total for that category.

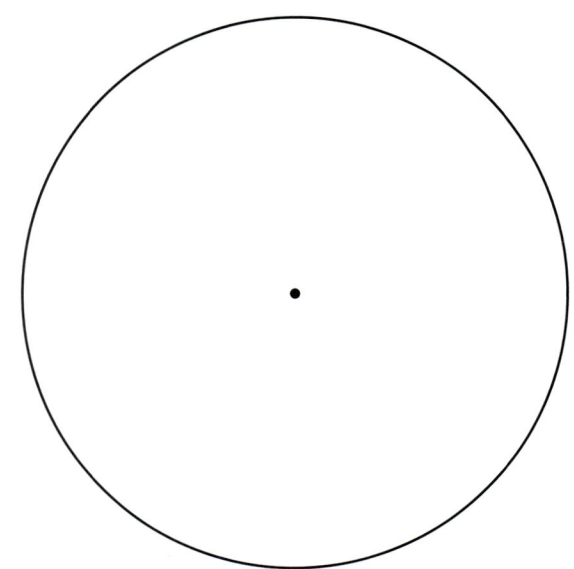

32. A book publisher had 25% of total sales in mysteries, 10% in biographies, 15% in cookbooks, 15% in romance novels, 20% in science, and the rest in travel books.

(a) Find the number of degrees in a circle graph for each type of book.

(b) Draw a circle graph, using the information given. Label each sector with the type of book and the percent of total for that type.

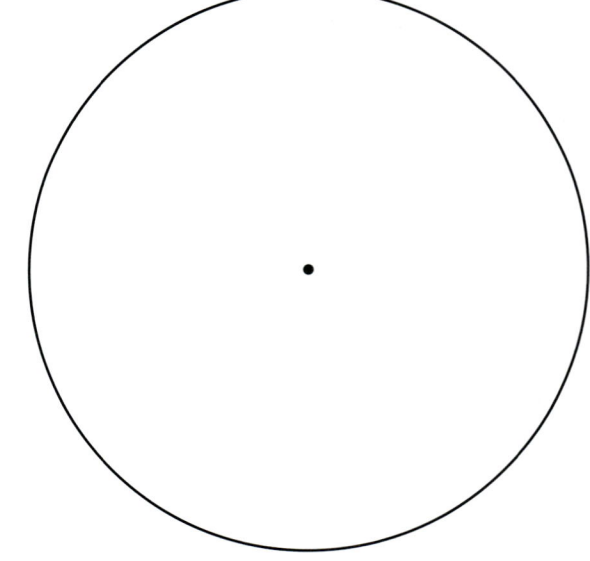

9.3 Bar Graphs and Line Graphs

OBJECTIVE 1 Read and understand a bar graph. A bar graph is useful for showing comparisons. For example, the bar graph below compares the number of college graduates who continued taking advanced courses in their major field during each of five years.

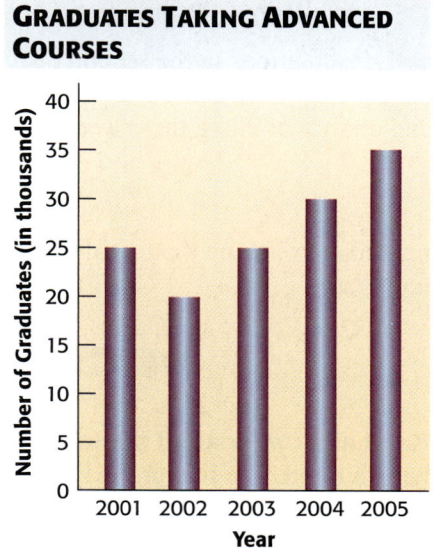

GRADUATES TAKING ADVANCED COURSES

EXAMPLE 1 Using a Bar Graph

How many college graduates took advanced classes in their major field in 2003?

The bar for 2003 rises to 25. Notice the label along the left side of the graph that says "Number of Graduates (in thousands)." The phrase *in thousands* means you have to multiply 25 by 1000 to get 25,000. So, 25,000 (not 25) graduates took advanced classes in their major field in 2003.

Work Problem 1 at the Side.

OBJECTIVE 2 Read and understand a double-bar graph.
A **double-bar graph** can be used to compare two sets of data. The graph below shows the number of DSL (digital subscriber line) installations each quarter for two different years.

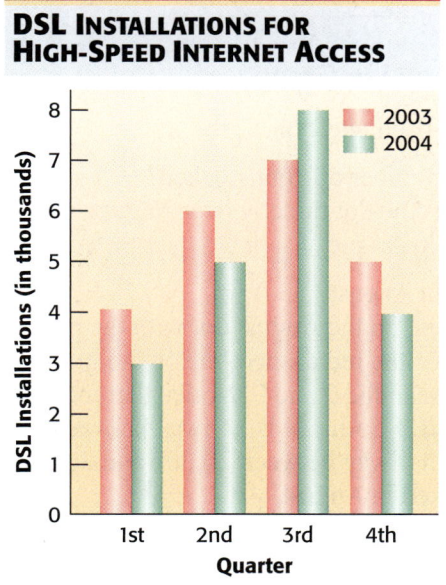

DSL INSTALLATIONS FOR HIGH-SPEED INTERNET ACCESS

Note:
1st quarter = Jan–Mar
2nd quarter = Apr–June
3rd quarter = July–Sept
4th quarter = Oct–Dec

OBJECTIVES

Read and understand
1. a bar graph;
2. a double-bar graph;
3. a line graph;
4. a comparison line graph.

1 Use the bar graph at the left to find the number of college graduates who took advanced classes in their major field in each of these years.

(a) 2001

(b) 2002

(c) 2004

(d) 2005

ANSWERS
1. (a) 25,000 graduates (b) 20,000 graduates
   (c) 30,000 graduates (d) 35,000 graduates

2 Use the double-bar graph on the previous page to find the number of DSL installations in 2003 and 2004 for each quarter.

(a) 1st quarter

(b) 3rd quarter

(c) 4th quarter

(d) Identify the quarter and year with the greatest number of installations. How many installations were made?

3 Use the line graph at the right to answer these questions.

(a) How many trout were stocked in June?

(b) How many fewer trout were stocked in May than in April?

(c) Find the total number of trout stocked during the five months.

(d) In which month were the most trout stocked? How many were stocked?

ANSWERS
2. (a) 4000; 3000 (b) 7000; 8000
   (c) 5000; 4000
   (d) 3rd quarter of 2004; 8000
3. (a) 55,000 trout (b) 10,000 fewer trout
   (c) 210,000 trout (d) July; 60,000 trout

EXAMPLE 2 Reading a Double-Bar Graph

Use the double-bar graph on the previous page to find the following.

(a) The number of DSL installations in the second quarter of 2003.
There are two bars for the second quarter. The color code in the upper right-hand corner of the graph tells you that the **red bars** represent 2003. So the **red bar** on the *left* is for the 2nd quarter of 2003. It rises to 6. Multiply 6 by 1000 because the label on the left side of the graph says *in thousands*. So there were 6000 DSL installations for the second quarter in 2003.

(b) The number of DSL installations in the second quarter of 2004.
The **green bar** for the second quarter rises to 5 and 5 times 1000 is 5000. So, in the second quarter of 2004, there were 5000 DSL installations.

> **CAUTION**
> Use a ruler or straightedge to line up the top of the bar with the number on the left side of the graph.

◀◀ **Work Problem 2 at the Side.**

OBJECTIVE 3 Read and understand a line graph. A **line graph** is often useful for showing a trend. The line graph below shows the number of trout stocked along the Feather River during five months. Each dot indicates the number of trout stocked during the month directly below that dot.

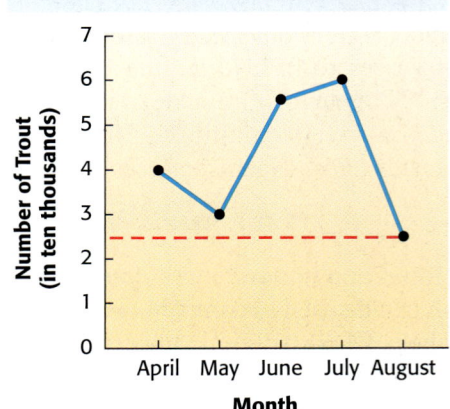

TROUT STOCKED IN FEATHER RIVER

EXAMPLE 3 Understanding a Line Graph

Use the line graph above to answer each question.

(a) In which month were the least number of trout stocked?
The lowest point on the graph is the dot directly over August, so the least number of trout were stocked in August.

(b) How many trout were stocked in August?
Use a ruler or straightedge to line up the August dot with the numbers along the left edge of the graph. (See the red dashed line on the graph.)
The August dot is halfway between the 2 and 3. Notice that the label on the left side says *in ten thousands*. So August is halfway between (2 • 10,000) and (3 • 10,000). It is halfway between 20,000 and 30,000. That means 25,000 trout were stocked in August.

◀◀ **Work Problem 3 at the Side.**

OBJECTIVE 4 **Read and understand a comparison line graph.** Two sets of data can also be compared by drawing two line graphs together as a **comparison line graph.** For example, the line graph below compares the number of minivans sold and the number of Sport Utility Vehicles (SUVs) sold during each of five years.

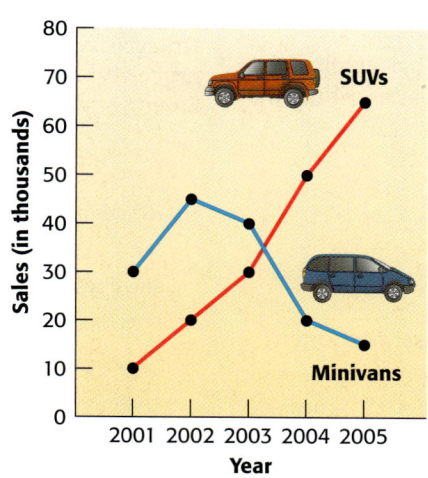

SALES OF MINIVANS AND SUVS

EXAMPLE 4 **Interpreting a Comparison Line Graph**

Use the comparison line graph above to find the following.

(a) The number of minivans sold in 2002
Find the dot on the **blue line** above 2002. Use a ruler or straightedge to line up the dot with the numbers along the left edge. The dot is halfway between 40 and 50, which is 45. Then, 45 times 1000 is 45,000 minivans sold in 2002.

(b) The number of SUVs sold in 2005
The **red line** on the graph shows that 65,000 SUVs were sold in 2005.

(c) The amount of decrease and the percent of decrease in minivan sales from 2002 to 2003. Round to the nearest whole percent.

Use subtraction to find the *amount* of decrease.

$$\text{minivan sales in 2002} - \text{minivan sales in 2003} = \text{amount of decrease}$$
$$45{,}000 - 40{,}000 = 5000$$

Now use the percent equation to find the *percent* of decrease.

$$\text{percent of original sales} = \text{amount of decrease}$$
$$p \cdot 45{,}000 = 5000$$

$$\frac{p \cdot 45{,}000}{45{,}000} = \frac{5000}{45{,}000} \quad \text{Divide both sides by 45,000}$$

$$p = 0.11\overline{1} \leftarrow \text{Solution in } decimal \text{ form}$$

Multiply by 100 to change the decimal to a percent, then round.

$$p = 0.11\overline{1} = 11.\overline{1}\% \approx 11\% \text{ to nearest whole percent}$$

The amount of decrease is 5000 minivans. The percent of decrease is 11% (rounded).

Work Problem 4 at the Side.

4 Use the comparison line graph at the left to find the following.

(a) The number of minivans sold in 2001, 2003, 2004, and 2005

(b) The number of SUVs sold in 2001, 2002, 2003, and 2004

(c) The first full year in which the number of SUVs sold was greater than the number of minivans sold

(d) Find the amount of increase and percent of increase in SUV sales from 2003 to 2004. Round to the nearest whole percent.

ANSWERS
4. **(a)** 30,000; 40,000; 20,000; 15,000
   (b) 10,000; 20,000; 30,000; 50,000
   (c) 2004
   (d) 20,000 SUVs; 67% increase (rounded)

Focus on Real-Data Applications

Grocery Shopping

1. The graph at the right is a double-bar graph. What is it about? Write a brief paragraph describing the general purpose of the graph.

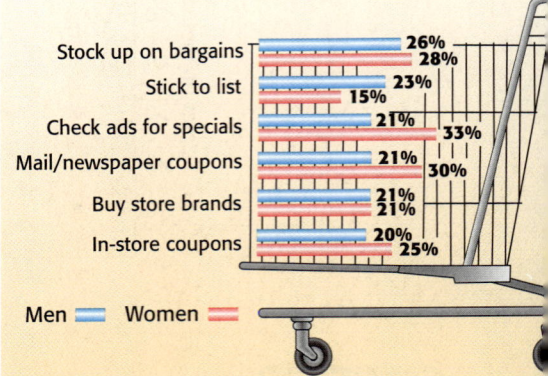

GENDER BUYS
On average, each person spends $32 weekly on groceries. Men average $35 and women $32. Saving strategies include:

Source: Food Marketing Institute.

2. Which two strategies do women use most often?

3. Which two strategies do men use most often?

4. Which two strategies show the greatest difference in use between the men and women surveyed?

5. Which two strategies show the least difference?

6. The sum of the percents for the women's responses and the sum for the men's responses do not equal 100%. Why?

7. Is it possible to decide how many of the people in the survey use *none* of the strategies listed? Why?

8. Sometimes people "jump to conclusions" without enough evidence. Which of these conclusions are reasonable, *based on the information in the graph*?
   (a) Men and women use a variety of savings strategies when grocery shopping.
   (b) Men spend more for groceries than women.
   (c) Men eat more groceries than women.
   (d) Women are better grocery shoppers than men.
   (e) There are some differences in the grocery shopping strategies used by men and women.

9. Conduct a survey of your class members. Find out how many of them regularly use each of the saving strategies shown in the graph. Complete the table below.

10. Make a double-bar graph showing your survey data. How is your data similar to the graph shown above? How is it different?

Strategy	Number of Women Using Strategy	Percent of Women Using Strategy	Number of Men Using Strategy	Percent of Men Using Strategy
Stock up on bargains				
Stick to list				
Check ads for specials				
Mail/newspaper coupons				
Buy store brands				
In-store coupons				
Total				

9.3 Exercises

This bar graph shows the top seven reasons people say they shop on-line. Use the graph to answer Exercises 1–6. See Example 1.

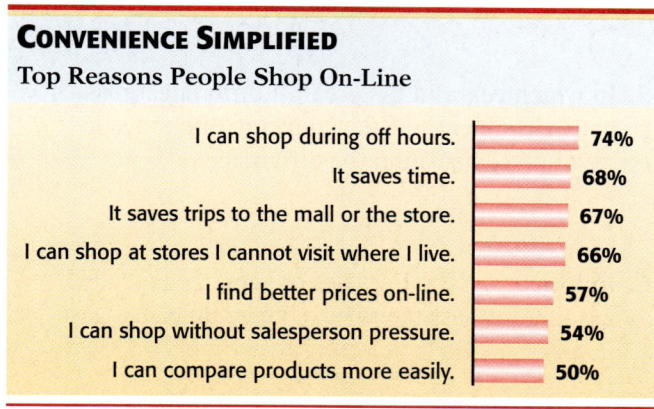

Source: EMARKETER.

1. What is the top reason people shop on-line? What percent gave this reason?

2. What is the second most popular reason for shopping on-line? What percent gave this reason?

3. What percent of the people say they find better prices on-line? If 600 people were surveyed, how many gave this answer?

4. What percent say it saves trips to the mall or store? If 600 people were surveyed, how many gave this answer?

5. Which reason(s) were given by $\frac{1}{2}$ of the people? Which reason(s) were given by nearly $\frac{3}{4}$ of the people?

6. Which reason(s) were given by about $\frac{2}{3}$ of the people?

This double-bar graph shows the number of workers who were unemployed in a city during the first six months of 2004 and 2005. Use this graph to answer Exercises 7–12. Round percents to the nearest whole number. See Example 2.

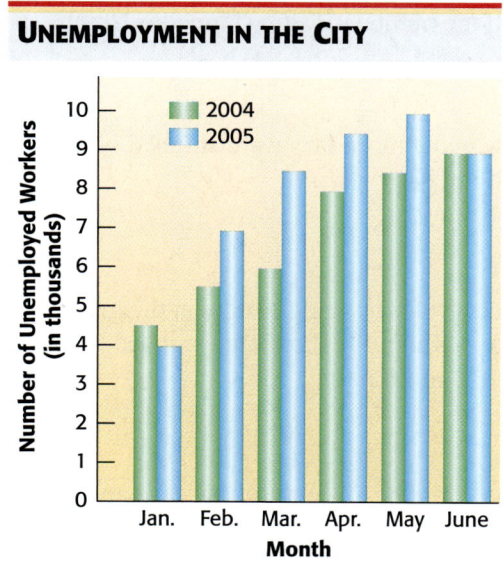

7. In the first half of 2005, which month had the greatest number of unemployed workers? What was the total number unemployed in that month?

8. How many workers were unemployed in January of 2004?

9. How many more workers were unemployed in February of 2005 than in February of 2004?

10. How many fewer workers were unemployed in March of 2004 than in March of 2005?

11. Find the amount of increase and the percent of increase in the number of unemployed workers from February 2004 to April 2004.

12. Find the amount of increase and the percent of increase in the number of unemployed workers from January 2005 to June 2005.

This double-bar graph shows sales of super unleaded and supreme unleaded gasoline at a service station for each of five years. Use this graph to answer Exercises 13–18. See Example 2.

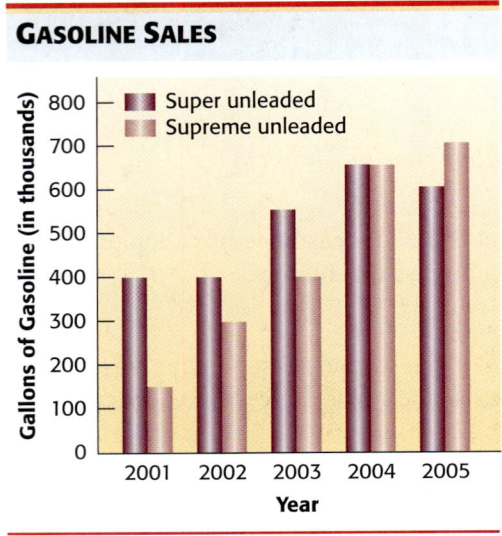

13. How many gallons of supreme unleaded gasoline were sold in 2001?

14. How many gallons of super unleaded gasoline were sold in 2004?

15. In which year did the greatest difference in sales between super unleaded and supreme unleaded gasoline occur? Find the difference.

16. In which year did the sales of supreme unleaded gasoline surpass the sales of super unleaded gasoline?

17. Find the amount of increase and percent of increase in supreme unleaded gasoline sales from 2001 to 2005. Round to the nearest whole percent.

18. Find the amount of increase and percent of increase in super unleaded gasoline sales from 2001 to 2005.

This line graph shows how sales of personal computers (PCs) have increased since they first became widely available in 1985. Use the line graph to answer Exercises 19–24. Round percents to the nearest whole percent. See Example 3.

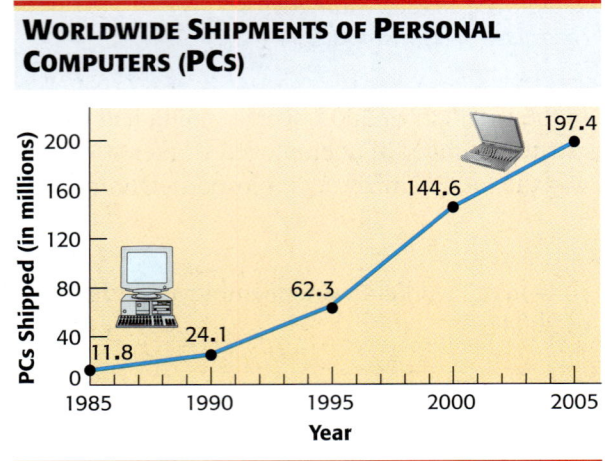

19. How many PCs were shipped in 1990?

20. Find the number of PCs shipped in 2000.

21. How many more PCs were shipped in 2005 than in 1985?

22. Place a ruler or straightedge on the graph to find the year in which PC shipments reached (a) 40 million; (b) 120 million.

23. Find the amount of increase in shipments from 1995 to 2000. What was the percent of increase, to the nearest whole percent?

24. What was the amount of increase and the percent of increase in PC shipments from 1985 to 1990?

This comparison line graph shows the number of compact discs (CDs) sold by two different chain stores during each of five years. Use this graph to find the annual number of CDs sold in each year listed in Exercises 25–28. See Example 4.

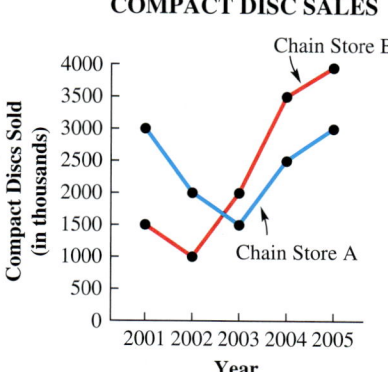

25. (a) Chain Store A in 2001
    (b) Chain Store B in 2001

26. (a) Chain Store A in 2002
    (b) Chain Store B in 2002

27. (a) Chain Store A in 2004
    (b) Chain Store A in 2005

28. (a) Chain Store B in 2003
    (b) Chain Store B in 2004

29. Describe the pattern(s) or trend(s) you see in the graph.

30. Store B used to have lower sales than Store A. What might have happened to cause this change? Give four possible explanations.

This comparison line graph shows the sales and profits of Tacos-To-Go for each of four years. Use the graph to answer Exercises 31–36.

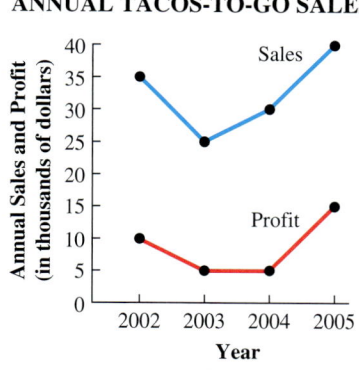

31. Find the year of lowest sales and the amount of sales.

32. Find the year of greatest profit and the amount of profit.

33. For each of the four years, find what percent the profit is of sales. Round to the nearest whole percent.

34. Find the total sales for all four years and the total profit for all four years. Total profit is what percent of total sales? Round to the nearest whole percent.

35. Give two possible explanations for the decrease in sales from 2002 to 2003 and two possible explanations for the increase in sales from 2003 to 2005.

36. *Based on the graph,* what conclusion can you make about the relationship between sales and profits for Tacos-To-Go?

RELATING CONCEPTS (EXERCISES 37–44) For Individual or Group Work

*Use the line graph on Worldwide Shipments of PCs on page 642 and the double-line graph on Compact Disc Sales on page 643 as you **work Exercises 37–44 in order.** Round percent answers to the nearest whole percent.*

37. What overall trend do you see in the graph on worldwide shipments of Personal Computers (PCs) on page 642?

38. Give at least two possible explanations for the increase in PC shipments.

39. Find the percent of increase in PC shipments from
   (a) 1985 to 1990
   (b) 1990 to 1995
   (c) 1995 to 2000
   (d) 2000 to 2005.

40. Look at your answers in Exercise 39. What trend do you see?

41. What future conditions could result in a decrease in PC shipments? Give at least two possibilities.

42. Look at the graph of Compact Disc Sales on page 643. For each store, compare the sales of compact discs in 2005 to sales in 2001. Find the percent of increase or the percent of decrease for:
   (a) Chain Store A
   (b) Chain Store B.

43. *Based on the graph,* what amount of CD sales would you predict for
   (a) Store A in 2006?

   (b) Store B in 2006?

   Explain how you arrived at your predictions.

44. (a) *Based on the graph,* which store would you like to own? Explain why.
   (b) Name at least three other things you would want to know before deciding which store to buy.

9.4 The Rectangular Coordinate System

OBJECTIVE 1 **Plot a point, given the coordinates, and find the coordinates, given a point.** A bar graph or line graph shows the relationship between two things. The line graph below is from Example 3 in **Section 9.3.** It shows the relationship between the month of the year and the number of trout stocked in the Feather River.

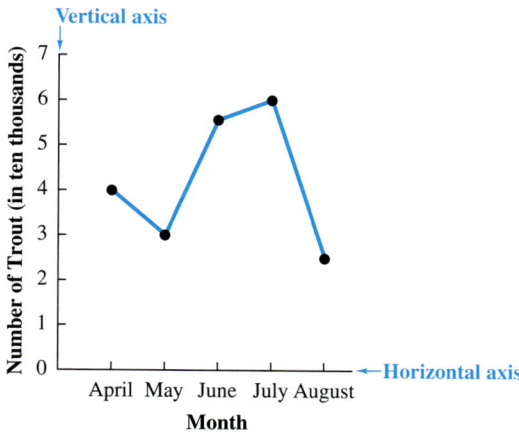

Each black dot on the graph represents a particular month paired with a particular number of trout. This is an example of **paired data.** We write each pair inside parentheses, with a comma separating the two items. To be consistent, we will always list the item on the *horizontal axis* first. In this case, the months are shown on the **horizontal axis** (the line that goes "left and right"), and the number of trout is shown along the **vertical axis** (the line that goes "up and down").

Paired Data from Line Graph on Trout Stocked in Feather River

(Apr, 40,000) (May, 30,000) (June, 55,000) (July, 60,000) (Aug, 25,000)

Each data pair gives you the location of a particular spot on the graph, and that spot is marked with a dot. This idea of paired data can be used to locate particular places on any flat surface.

Think of a small town laid out in a grid of square blocks, as shown below. To tell a taxi driver where to go, you could say, "the corner of 4th Avenue and 2nd Street" or just "4th and 2nd." As an *ordered pair*, it would be (4, 2). Of course, both you and the taxi driver need to know that the avenue is mentioned first (the number on the horizontal axis) and that the street is mentioned second (the number on the vertical axis). If the driver goes to (2, 4) instead, you'll be at the wrong corner.

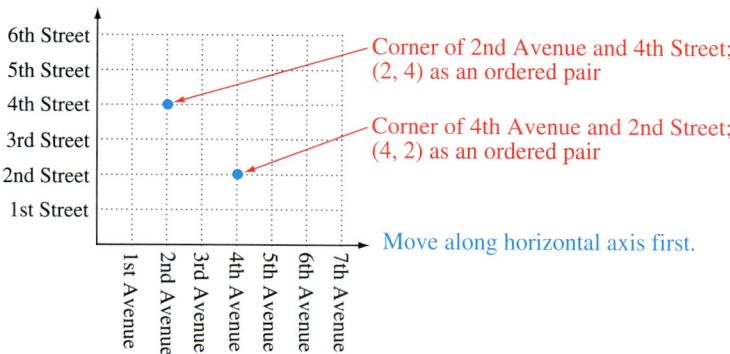

OBJECTIVES

1. Plot a point, given the coordinates, and find the coordinates, given a point.
2. Identify the four quadrants and determine which points lie within each one.

646 Chapter 9 Graphs

1 Plot each point on the grid. Write the ordered pair next to each point.

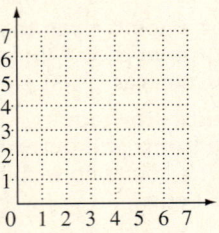

(a) (1, 4)

(b) (5, 2)

(c) (4, 1)

(d) (3, 3)

EXAMPLE 1 Plotting Points on a Grid

Use the grid at the right to plot each point.

(a) (3, 5)

Start at 0. Move *to the right* along the horizontal axis until you reach 3. Then move *up* 5 units so that you are aligned with 5 on the vertical axis. Make a dot. This is the plot, or graph, of the point (3, 5).

(b) (5, 3)

Start at 0. Move *to the right* along the horizontal axis until you reach 5. Then move *up* 3 units so that you are aligned with 3 on the vertical axis. Make a dot. This is the plot, or graph, of the point (5, 3).

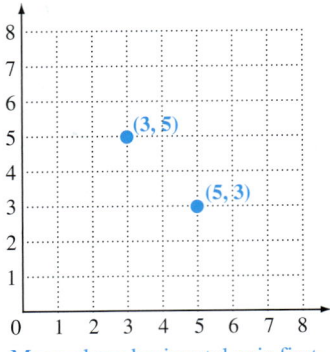

Move along horizontal axis first.

CAUTION
In Example 1, the points (3, 5) and (5, 3) are *not* the same. The "address" of a point is called an **ordered pair** because the *order* within the pair is important. Always move along the horizontal axis first.

◀◀◀ Work Problem 1 at the Side.

You have been using both positive and negative numbers throughout this book. We can extend our grid system to include negative numbers, as shown below. The horizontal axis is now a number line with 0 at the center, positive numbers extending to the right and negative numbers to the left. This horizontal number line is called the *x*-axis.

The vertical axis is also a number line, with positive numbers extending upward from 0 and negative numbers extending downward from 0. The vertical axis is called the *y*-axis. Together, the *x*-axis and the *y*-axis form a rectangular **coordinate system**. The center point (0, 0) is where the *x*-axis crosses the *y*-axis; it is called the **origin**.

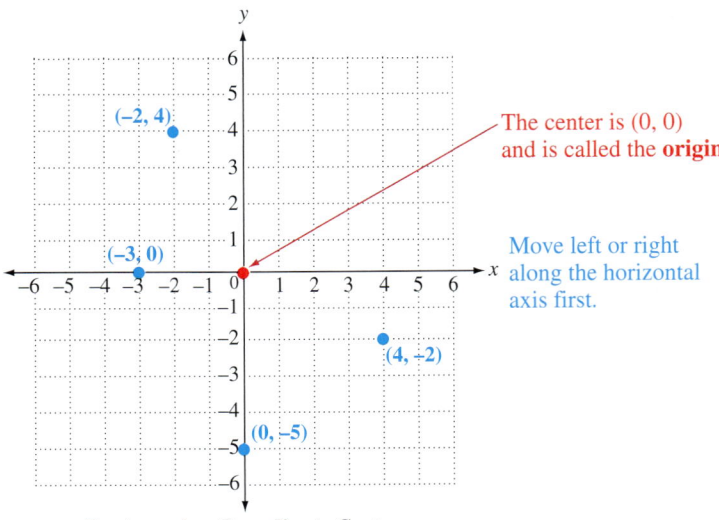

Rectangular Coordinate System

The center is (0, 0) and is called the **origin.**

Move left or right along the horizontal axis first.

ANSWERS
1.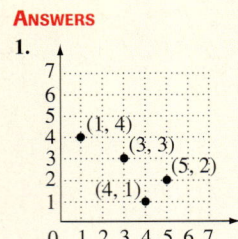

EXAMPLE 2 Plotting Points on a Rectangular Coordinate System

Plot each point on the rectangular coordinate system shown at the bottom of the previous page.

(a) $(4, -2)$

Start at 0. Then move left or right along the horizontal *x*-axis first.

Because 4 is *positive,* move *to the right* until you reach 4. Now, because the 2 is *negative,* move *down* 2 units so that you are aligned with -2 on the *y*-axis. Make a dot and label it $(4, -2)$.

(b) $(-2, 4)$

Starting at 0, move left or right along the horizontal *x*-axis first. In this case, move *to the left* until you reach -2. Then move *up* 4 units. Make a dot and label it $(-2, 4)$. Notice that $(-2, 4)$ is **not** the same as $(4, -2)$.

(c) $(0, -5)$

Move left or right along the horizontal *x*-axis first. However, because the first number is 0, stay right at the center of the coordinate system. Then move *down* 5 units. Make a dot and label it $(0, -5)$.

(d) $(-3, 0)$

Starting at 0, move *to the left* along the horizontal *x*-axis to -3. Then, because the second number is 0, stay on -3. Do *not* move up or down. Make a dot and label it $(-3, 0)$.

> **NOTE**
> When the *first* number in an ordered pair is 0, the point is on the *y*-axis, as in Example 2(c) above. When the *second* number in an ordered pair is 0, the point is on the *x*-axis, as in Example 2(d) above.

Work Problem 2 at the Side.

We can use a coordinate system and an ordered pair to show the location of any point. The numbers in the ordered pair are called the **coordinates** of the point.

EXAMPLE 3 Finding the Coordinates of Points

Find the coordinates of points *A*, *B*, *C*, and *D*.

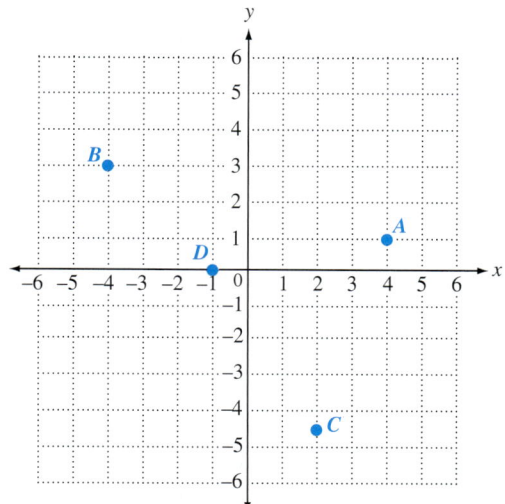

To reach point *A* from the origin, move 4 units *to the right;* then move *up* 1 unit. The coordinates are $(4, 1)$.

Continued on Next Page

---

2 Plot each point on the coordinate system shown. Write the ordered pair next to each point.

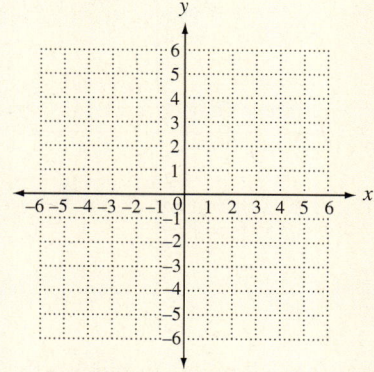

(a) $(5, -3)$

(b) $(-5, 3)$

(c) $(0, 3)$

(d) $(-4, -4)$

(e) $(-2, 0)$

ANSWERS

2.
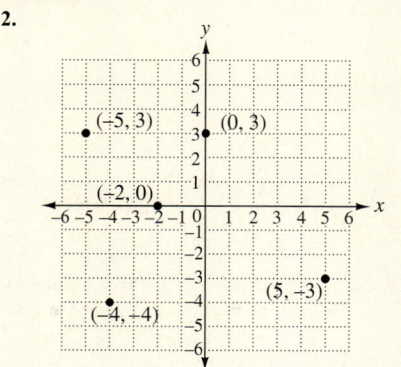

3 Find the coordinates of points A, B, C, D, and E.

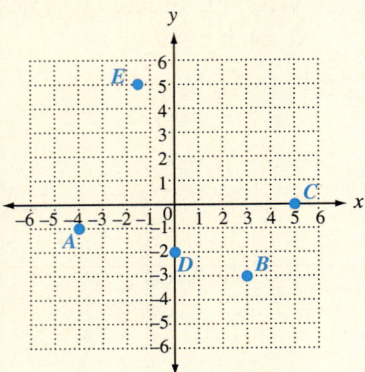

4 (a) All points in the fourth quadrant are similar in what way? Give two examples of points in the fourth quadrant.

(b) In which quadrant is each point located: $(-2, -6)$; $(0, 5)$; $(-3, 1)$; $(4, -1)$?

ANSWERS

3. A is $(-4, -1)$; B is $(3, -3)$; C is $(5, 0)$; D is $(0, -2)$; E is approximately $\left(-1\frac{1}{2}, 5\right)$.

4. (a) The pattern for all points in quadrant IV is $(+, -)$. Examples will vary; just be sure that they fit the $(+, -)$ pattern.
   (b) III; no quadrant; II; IV

To reach point B from the origin, move 4 units *to the left;* then move *up* 3 units. The coordinates are $(-4, 3)$.

To reach point C from the origin, move 2 units *to the right;* then move *down* approximately $4\frac{1}{2}$ units. The approximate coordinates are $(2, -4\frac{1}{2})$.

To reach point D from the origin, move 1 unit *to the left;* then do *not* move either up or down. The coordinates are $(-1, 0)$.

> **NOTE**
> If a point is between the lines on the coordinate system, you can use fractions to give the approximate coordinates. For example, the approximate coordinates of point C above are $(2, -4\frac{1}{2})$.

Work Problem 3 at the Side.

OBJECTIVE 2 Identify the four quadrants and determine which points lie within each one. The x-axis and y-axis divide the coordinate system into four regions, called **quadrants**. These quadrants are numbered with Roman numerals, as shown below. Points on the axes themselves are not in any quadrant.

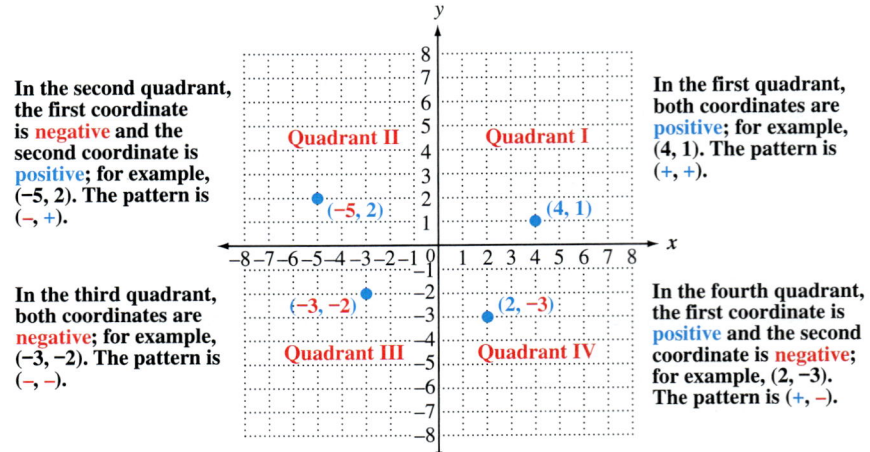

In the second quadrant, the first coordinate is **negative** and the second coordinate is **positive**; for example, $(-5, 2)$. The pattern is $(-, +)$.

In the third quadrant, both coordinates are **negative**; for example, $(-3, -2)$. The pattern is $(-, -)$.

In the first quadrant, both coordinates are **positive**; for example, $(4, 1)$. The pattern is $(+, +)$.

In the fourth quadrant, the first coordinate is **positive** and the second coordinate is **negative**; for example, $(2, -3)$. The pattern is $(+, -)$.

EXAMPLE 4 Working with Quadrants

(a) All points in the third quadrant are similar in what way? Give two examples of points in the third quadrant.

For all points in quadrant III, both coordinates are negative. The pattern is $(-, -)$. There are many possible examples, such as $(-2, -5)$ and $(-4, -4)$. Just be sure that both numbers are negative.

(b) In which quadrant is each point located: $(3, 5)$; $(1, -6)$; $(-4, 0)$?

For $(3, 5)$ the pattern is $(+, +)$, so the point is in **quadrant I**.

For $(1, -6)$ the pattern is $(+, -)$, so the point is in **quadrant IV**.

The point corresponding to $(-4, 0)$ is on the x-axis, so it isn't in any quadrant.

Work Problem 4 at the Side.

9.4 Exercises

FOR EXTRA HELP 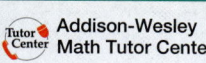 Addison-Wesley Math Tutor Center   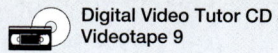 MathXL    Digital Video Tutor CD 5 Videotape 9   Student's Solutions Manual   MyMathLab   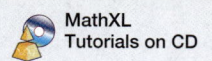 MathXL Tutorials on CD

Plot each point on the rectangular coordinate system. Label each point with its coordinates. See Examples 1 and 2.

1. $(3, 7)$ $(-2, 2)$ $(-3, -7)$ $(2, -2)$ $(0, 6)$
$(6, 0)$ $(0, -4)$ $(-4, 0)$

2. $(5, 2)$ $(-3, -3)$ $(4, -1)$ $(-4, 1)$ $(-1, 0)$
$(0, 3)$ $(2, 0)$ $(0, -5)$

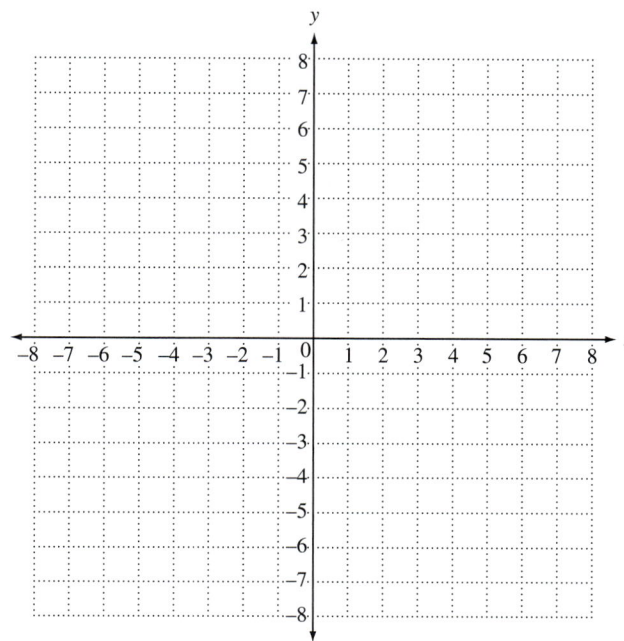
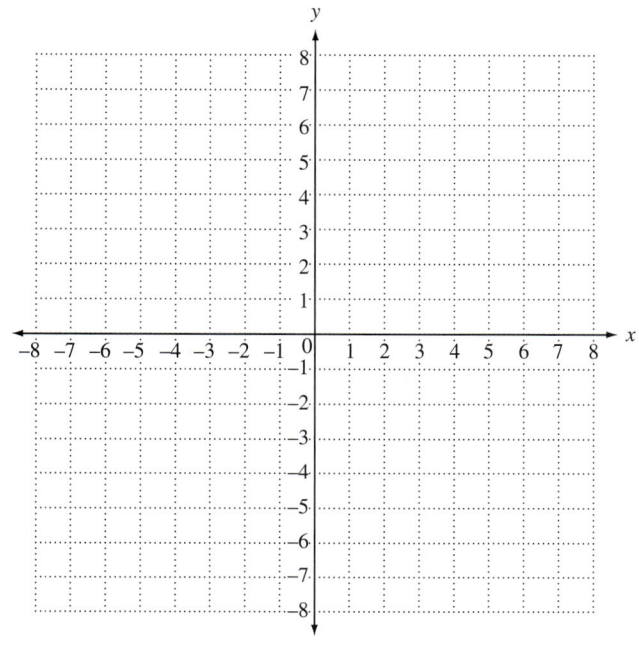

3. $(-5, 3)$ $(4, 4)$ $(-2\frac{1}{2}, 0)$ $(3, -5)$ $(0, 0)$
$(2, \frac{1}{2})$ $(-7, -5)$ $(-1, -6)$

4. $(1, 7)$ $(0, 3\frac{1}{2})$ $(-5, -1)$ $(6, -2)$ $(-2, 6)$
$(0, 0)$ $(-3, 3)$ $(-\frac{1}{2}, -2)$

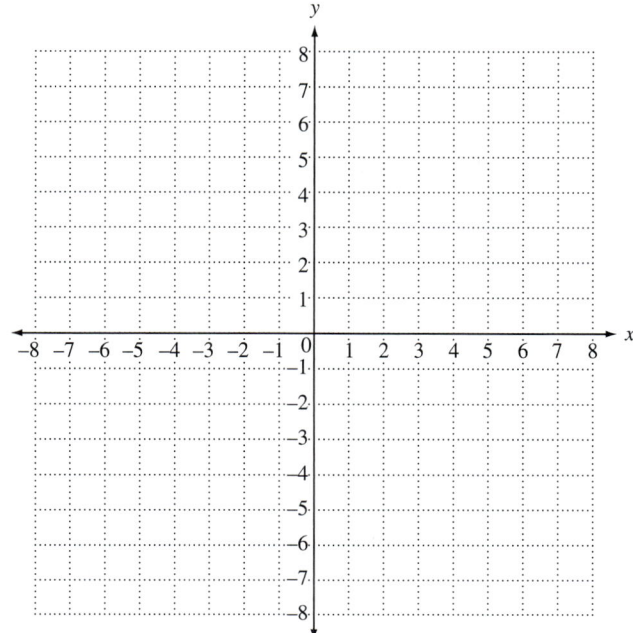
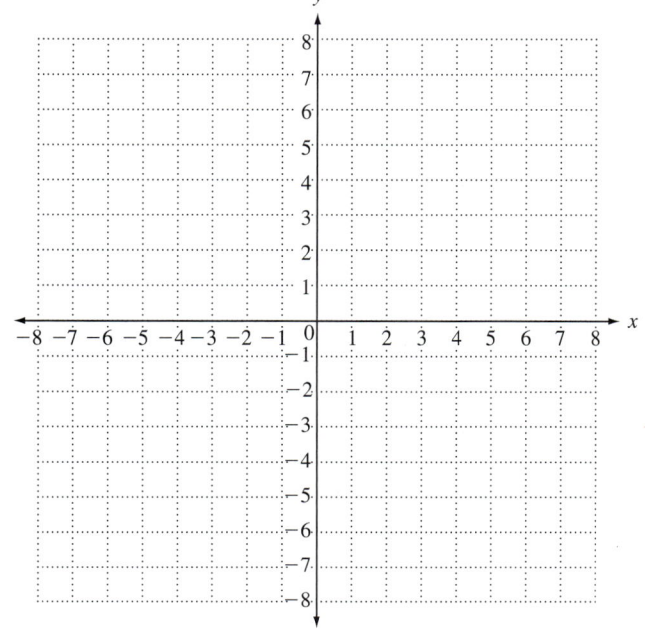

Give the coordinates of each point. See Example 3.

5.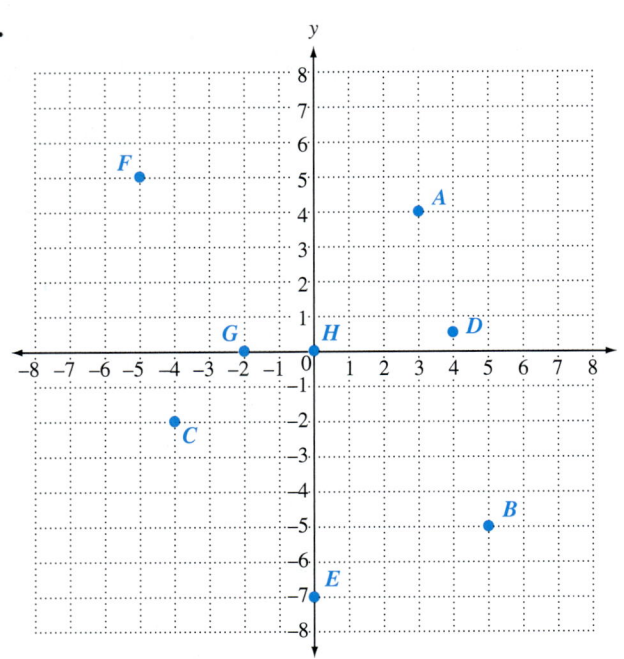

6.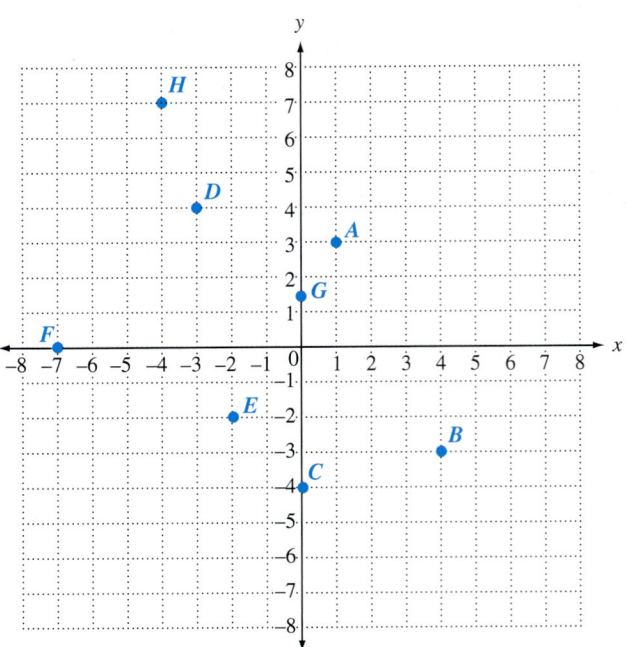

Identify the quadrant in which each point is located. See Example 4.

7. In which quadrant is each point located?
   $(-3, -7)$ $(0, 4)$ $(10, -16)$ $(-9, 5)$

8. In which quadrant is each point located?
   $(1, 12)$ $(20, -8)$ $(-5, 0)$ $(-14, 14)$

Complete each ordered pair with a number that will make the point fall in the specified quadrant.

9. (a) Quadrant II $(-4, \underline{\quad})$

   (b) Quadrant IV $(7, \underline{\quad})$

   (c) No quadrant $(\underline{\quad}, -2)$

   (d) Quadrant III $(\underline{\quad}, -1\frac{1}{2})$

   (e) Quadrant I $(3\frac{1}{4}, \underline{\quad})$

10. (a) Quadrant III $(-5, \underline{\quad})$

    (b) Quadrant I $(\underline{\quad}, 3)$

    (c) Quadrant IV $(\underline{\quad}, -\frac{1}{2})$

    (d) No quadrant $(6, \underline{\quad})$

    (e) Quadrant II $(\underline{\quad}, 1\frac{3}{4})$

11. Explain how to graph the ordered pair (a, b), where a and b are positive or negative integers.

12. Explain how to graph the ordered pair (a, b) where a is 0 and b is an integer. Explain how to graph (a, b) where a is an integer and b is 0.

9.5 Introduction to Graphing Linear Equations

In **Chapters 2–7** you solved equations that had only one variable, such as $2n - 3 = 7$ or $\frac{1}{3}x = 10$. Each of these equations had exactly one solution; n is 5 in the first equation, and x is 30 in the second equation. In other words, there was only *one* number that could replace the variable and make the equation balance. As you continue in this textbook, you will work with equations that have two variables and many different numbers that will make the equation balance. This section will get you started.

OBJECTIVES

1. Graph linear equations in two variables.
2. Identify the slope of a line as positive or negative.

OBJECTIVE 1 Graph linear equations in two variables. Suppose that you have 6 hours of study time available during a weekend. You plan to study math and psychology. For example, you could spend 4 hours on math and then 2 hours on psychology, for a total of 6 hours. Or you could spend $1\frac{1}{2}$ hours on math and then $4\frac{1}{2}$ hours on psychology, for a total of 6 hours. Here is a list of *some* of the possible combinations.

Hours on Math	+	Hours on Psychology	=	Total Hours Studying
0	+	6	=	6
1	+	5	=	6
$1\frac{1}{2}$	+	$4\frac{1}{2}$	=	6
3	+	3	=	6
4	+	2	=	6
$5\frac{1}{2}$	+	$\frac{1}{2}$	=	6
6	+	0	=	6

We can write an equation to represent this situation.

$$\text{hours studying math} + \text{hours studying psychology} = \text{total of 6 hours}$$

$$m + p = 6$$

This equation, $m + p = 6$, has *two* variables. The hours spent on math (m) can vary, and the hours spent on psychology (p) can vary.

As you can see, there is more than one solution for this equation. We can list possible solutions as *ordered pairs*. The first number in the pair is the value of m, and the second number in the pair is the corresponding value of p.

(m, p) (m, p) (m, p) (m, p) (m, p) (m, p) (m, p)

$(0, 6)$ $(1, 5)$ $\left(1\frac{1}{2}, 4\frac{1}{2}\right)$ $(3, 3)$ $(4, 2)$ $\left(5\frac{1}{2}, \frac{1}{2}\right)$ $(6, 0)$

Another way to show the solutions is to plot the ordered pairs, as you learned to do in **Section 9.4**. This method will give us a "picture" of the solutions that we listed on the previous page.

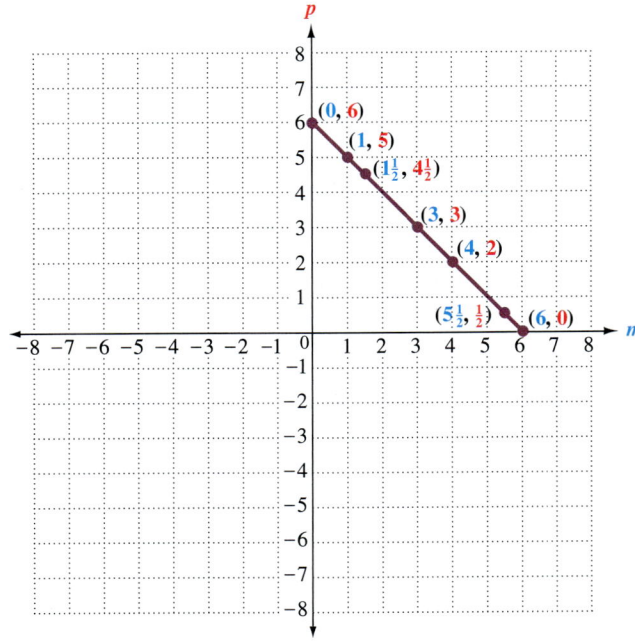

$m + p = 6$

The variables m and p represent hours, and hours can be 0 or positive numbers, but *not* negative numbers.

Notice that all the solutions (all the ordered pairs) lie on a straight line. When you draw a line connecting the ordered pairs, you have graphed the solutions. **Every point on the line is a solution.** You can use the line to find additional solutions besides the ones that we listed. For example, the point (5, 1) is on the line. This point tells you that another solution is 5 hours on math and 1 hour on psychology. The fact that the line is a *straight* line tells you that $m + p = 6$ is a *linear equation*. (The word *line* is part of the word *line*ar.) Later on in algebra you will work with equations whose solutions form a curve rather than a straight line when you graph them.

To draw the line for $m + p = 6$, we really needed only two solutions (two ordered pairs). But it's a good idea to use a third ordered pair as a check. If the three ordered pairs are *not* in a straight line, there is an error in your work.

Graphing a Linear Equation

To **graph a linear equation,** find at least three ordered pairs that satisfy the equation. Then plot the ordered pairs on a coordinate system and connect them with a straight line. *Every* point on the line is a solution of the equation.

EXAMPLE 1 Graphing a Linear Equation

Graph $x + y = 3$ by finding three solutions and plotting the ordered pairs. Then use the graph to find a fourth solution of the equation.

There are many possible solutions. Start by picking three different values for x. You can choose any numbers you like, but 0 and small numbers usually are easy to use. Then find the value of y that will make the sum equal to 3. Set up a table to organize the information.

— **Continued on Next Page**

Section 9.5 Introduction to Graphing Linear Equations

x	y	Check that $x + y = 3$	Ordered Pair (x, y)
0	3	$0 + 3 = 3$	(0, 3)
1	2	$1 + 2 = 3$	(1, 2)
2	1	$2 + 1 = 3$	(2, 1)

Plot the ordered pairs and draw a line through the points, extending it in both directions as shown below.

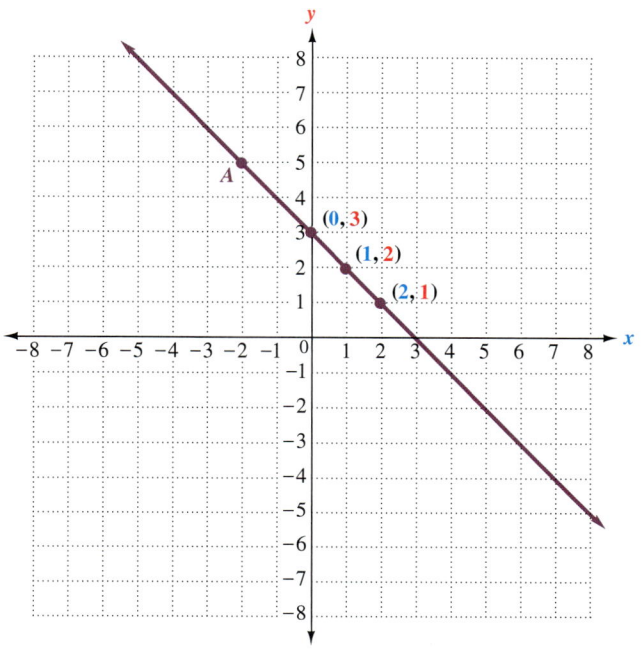

Now you can use the graph to find more solutions of $x + y = 3$. *Every* point on the line is a solution. Suppose that you pick **point A**. The coordinates are $(-2, 5)$.

To check that $(-2, 5)$ is a solution, substitute -2 for x and 5 for y in the original equation.

$$x + y = 3 \quad \text{Original equation}$$
$$-2 + 5 = 3$$
$$3 = 3 \quad \text{Balances}$$

The equation balances, so $(-2, 5)$ is another solution of $x + y = 3$.

NOTE
The line in Example 1 above was extended in both directions because *every* point on the line is a solution of $x + y = 3$. However, when we graphed the line for the hours spent studying, $m + p = 6$, we did *not* extend the line. That is because the variables m and p represented hours, and hours can only be 0 or positive numbers; all the solutions had to be in the first quadrant.

Work Problem 1 at the Side. ▷▷▷

① Graph $x + y = 5$ by finding three solutions and plotting the ordered pairs. Then use the graph to find *two* other solutions of the equation.

x	y	Check that $x + y = 5$	Ordered Pair (x, y)
0			
1			
2			

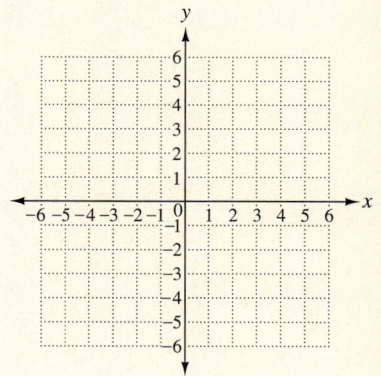

Two other solutions are (___, ___) and (___, ___).

ANSWERS
1. Plot (0, 5), (1, 4), and (2, 3).

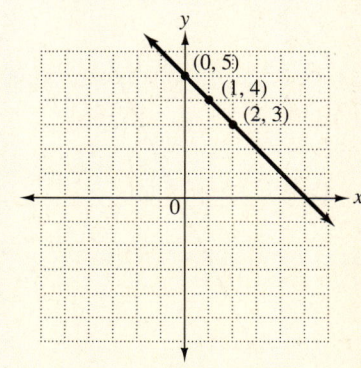

There are many other solutions. Some possibilities are $(-1, 6)$; $(3, 2)$; $(4, 1)$; $(5, 0)$; $(6, -1)$.

aphs

... by finding
... ns and plotting
... pairs. Then use
... to find *two* other
solutions of the equation.

x	y = 2 · x	Ordered Pair (x, y)
0		
1		
2		

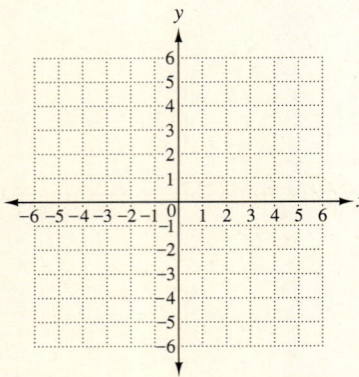

Two other solutions are (__, __)
and (__, __).

ANSWERS

2. Plot (0, 0), (1, 2), and (2, 4).

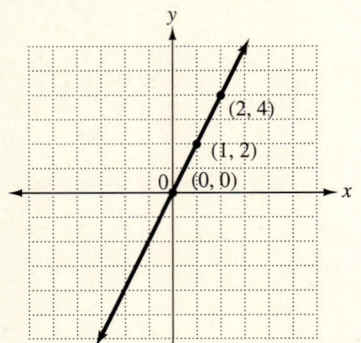

There are many other solutions. Some possibilities are (3, 6); (−1, −2); (−2, −4); (−3, −6).

EXAMPLE 2 Graphing a Linear Equation

Graph $y = -3x$ by finding three solutions and plotting the ordered pairs. Then use the graph to find a fourth solution of the equation.

You can choose any three values for x, but small numbers such as 0, 1, and 2 are easy to use. Then $y = -3x$ tells you that y is -3 times the value of x.

$$y = -3x$$
y is -3 times x

First set up a table.

x	y = −3 · x	Ordered Pair (x, y)
0	−3 · **0** is **0**	(**0, 0**)
1	−3 · **1** is **−3**	(**1, −3**)
2	−3 · **2** is **−6**	(**2, −6**)

Plot the ordered pairs and draw a line through the points. Be sure to draw arrows on both ends of the line to show that it continues in both directions.

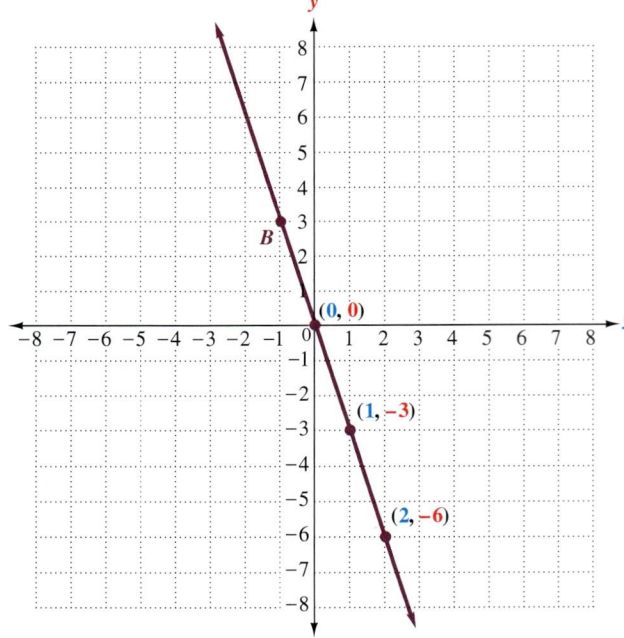

Now use the graph to find more solutions. *Every* point on the line is a solution.
Suppose that you pick **point B**. The coordinates are (−1, 3). To check that (−1, 3) is a solution, substitute −1 for x and 3 for y in the original equation.

$$y = -3x \quad \text{Original equation}$$
$$3 = -3(-1)$$
$$3 = 3 \quad \text{Balances}$$

The equation balances, so (−1, 3) is another solution of $y = -3x$.

Work Problem 2 at the Side.

Section 9.5 Introduction to Graphing Linear Equations

EXAMPLE 3 Graphing a Linear Equation

Graph $y = \dfrac{1}{2}x$ by finding three solutions and plotting the ordered pairs. Then use the graph to find a fourth solution of the equation.

Complete the table. The coefficient of x is $\tfrac{1}{2}$, so choose even numbers like 2, 4, and 6 as values for x because they are easy to divide in half. The equation $y = \tfrac{1}{2}x$ tells you that y is $\tfrac{1}{2}$ *times* the value of x.

x	$y = \tfrac{1}{2} \cdot x$	Ordered Pair (x, y)
2	$\tfrac{1}{2} \cdot 2$ is 1	(2, 1)
4	$\tfrac{1}{2} \cdot 4$ is 2	(4, 2)
6	$\tfrac{1}{2} \cdot 6$ is 3	(6, 3)

Plot the ordered pairs and draw a line through the points.

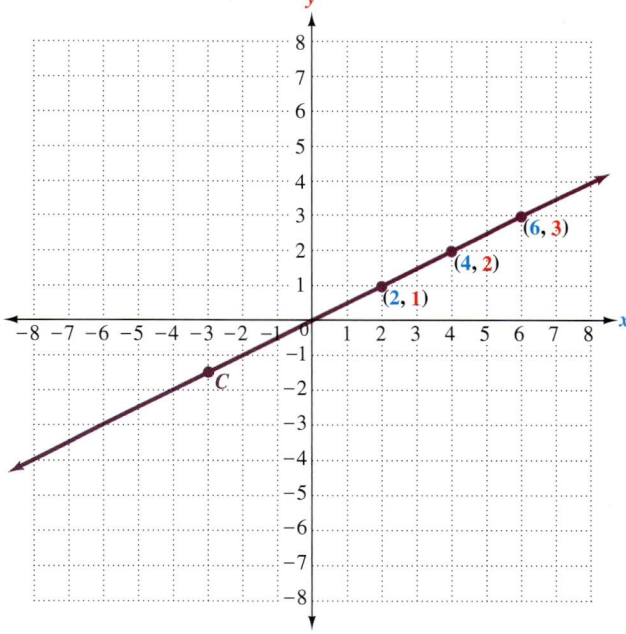

Now use the graph to find more solutions. *Every* point on the line is a solution.

Suppose that you pick **point C**. The coordinates are $(-3, -1\tfrac{1}{2})$. Check that $(-3, -1\tfrac{1}{2})$ is a solution by substituting -3 for x and $-1\tfrac{1}{2}$ for y.

$$y = \dfrac{1}{2}x \quad \text{Original equation}$$

$$-1\dfrac{1}{2} = \dfrac{1}{2}(-3)$$

$$-1\dfrac{1}{2} = -\dfrac{3}{2} \quad \text{Balances; } -1\tfrac{1}{2} \text{ is equivalent to } -\tfrac{3}{2}.$$

The equation balances, so $(-3, -1\tfrac{1}{2})$ is another solution of $y = \tfrac{1}{2}x$.

Work Problem 3 at the Side.

---

3 Graph $y = -\tfrac{1}{2}x$ by finding three solutions and plotting the ordered pairs. Then use the graph to find *two* more solutions.

x	$y = -\tfrac{1}{2} \cdot x$	Ordered Pair (x, y)
2		
4		
6		

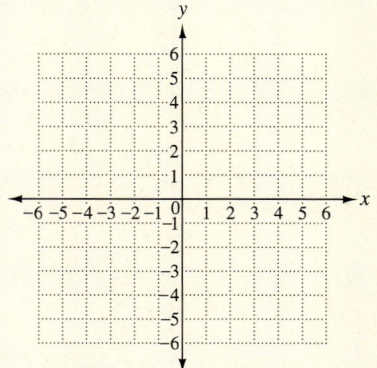

Two other solutions are (__, __) and (__, __).

ANSWERS

3. Plot $(2, -1)$, $(4, -2)$, and $(6, -3)$.

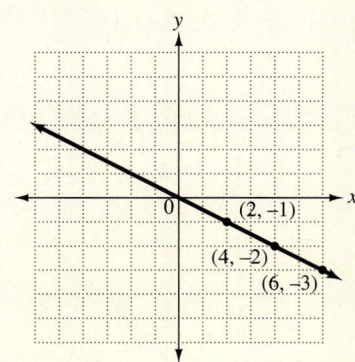

There are many other solutions. Some possibilities are $(0, 0)$; $(-2, 1)$; $(-4, 2)$; $(-6, 3)$; $\left(1, -\tfrac{1}{2}\right)$; $\left(3, -1\tfrac{1}{2}\right)$; $\left(5, -2\tfrac{1}{2}\right)$.

Graph the equation $y = x - 5$ by finding three solutions and plotting the ordered pairs. Then use the graph to find two more solutions.

x	y = x − 5	Ordered Pair (x, y)
1		
2		
3		

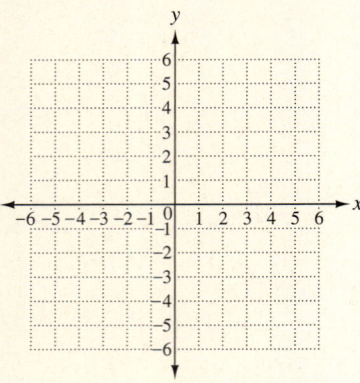

Two other solutions are (___, ___) and (___, ___).

ANSWERS

4. Plot (1, −4), (2, −3), and (3, −2).

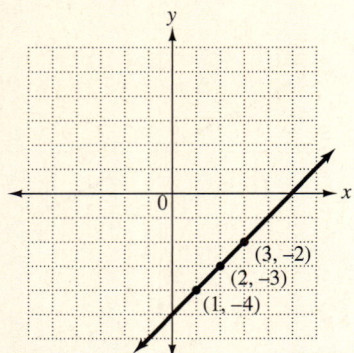

There are many other solutions. Some possibilities are (−1, −6); (0, −5); (4, −1); (5, 0); (6, 1).

EXAMPLE 4 Graphing a Linear Equation

Graph the equation $y = x + 4$ by finding three solutions and plotting the ordered pairs. Then use the graph to find two more solutions of the equation.

First set up a table. The equation $y = x + 4$ tells you that y must be 4 more than the value of x.

x	y = x + 4	Ordered Pair (x, y)
0	0 + 4 is 4	(0, 4)
1	1 + 4 is 5	(1, 5)
2	2 + 4 is 6	(2, 6)

Plot the ordered pairs and draw a line through the points.

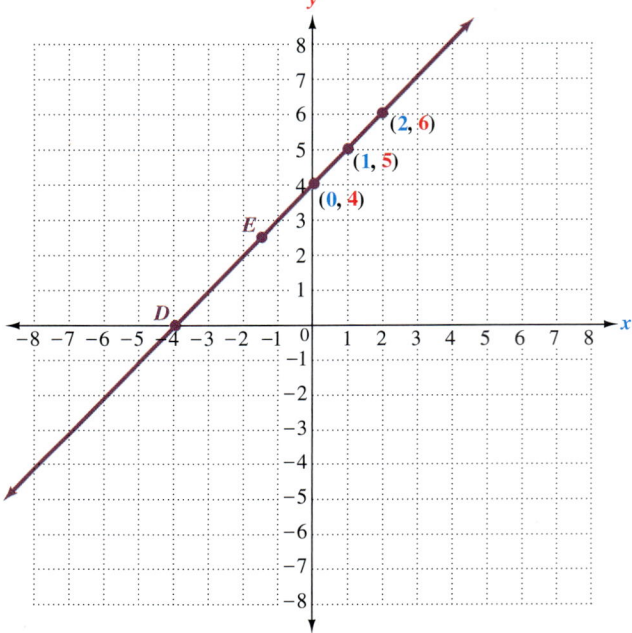

Now use the graph to find two more solutions. *Every* point on the line is a solution. Suppose that you pick **point D** at $(-4, 0)$ and **point E** at $(-1\frac{1}{2}, 2\frac{1}{2})$. Check that both ordered pairs are solutions.

Check (−4, 0)

$y = x + 4$
$0 = -4 + 4$
$0 = 0$ Balances

Check $\left(-1\frac{1}{2}, 2\frac{1}{2}\right)$

$y = x + 4$
$2\frac{1}{2} = -1\frac{1}{2} + 4$
$\frac{5}{2} = -\frac{3}{2} + \frac{8}{2}$
$\frac{5}{2} = \frac{5}{2}$ Balances

Both equations balance, so $(-4, 0)$ and $(-1\frac{1}{2}, 2\frac{1}{2})$ are also solutions of $y = x + 4$.

◀◀◀ Work Problem 4 at the Side.

OBJECTIVE 2 Identify the slope of a line as positive or negative.
Let's look again at some of the lines that we graphed for various equations. All are straight lines, but some are almost flat and some tilt steeply upward or downward.

Graph from Example 1: $x + y = 3$

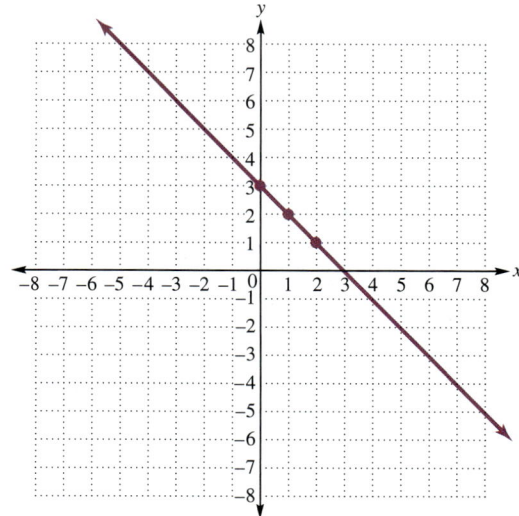

As you move from *left to right,* the line slopes downward, as if you were walking down a hill. When a line tilts downward, we say that it has a *negative slope.*

Now look at the table of solutions we used to draw the line.

The value of *x* is *increasing* from 0 to 1 to 2

x	y
0	3
1	2
2	1

The value of *y* is *decreasing* from 3 to 2 to 1

As the value of *x increases,* the value of *y* does the *opposite*—it *decreases.* Whenever one variable increases while the other variable decreases, the line will have a negative slope.

Graph from Example 3: $y = \frac{1}{2}x$

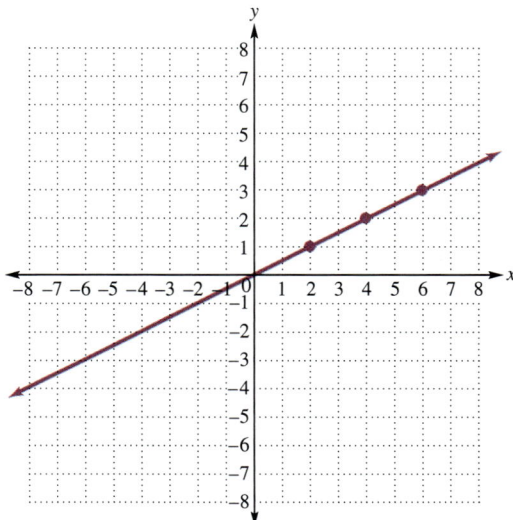

As you move from *left to right,* this line slopes upward, as if you were walking up a hill. When a line tilts upward, we say that it has a *positive slope.*

Chapter 9 Graphs

⑤ Look back at the graphs in Margin Problems 2 and 3. Then complete these sentences.

(a) The graph of $y = 2x$ has a _____ slope. As the value of x increases, the value of y _____.

(b) The graph of $y = -\frac{1}{2}x$ has a _____ slope. As the value of x increases, the value of y _____.

Now look at the table of solutions we used to draw the line.

The value of x is *increasing* from 2 to 4 to 6

x	y
2	1
4	2
6	3

The value of y is *increasing* from 1 to 2 to 3

As the value of x *increases*, the value of y does the *same* thing—it also *increases*. Whenever both variables do the same thing (both increase or both decrease), the line will have a positive slope.

Positive and Negative Slopes

As you move from left to right, a line with a *positive* slope tilts *upward* or rises. As the value of one variable increases, the value of the other variable also increases (does the same).

As you move from left to right, a line with a *negative* slope tilts *downward* or falls. As the value of one variable increases, the value of the other variable decreases (does the opposite).

EXAMPLE 5 Identifying Positive or Negative Slope in a Line

Look back at the graph of $y = -3x$ in Example 2. Then complete these sentences.

The graph of $y = -3x$ has a _____ slope.

As the value of x increases, the value of y _____.

The graph of $y = -3x$ has a <u>negative</u> slope (because it tilts downward).

As the value of x increases, the value of y <u>decreases</u> (does the opposite).

◀◀◀ **Work Problem 5 at the Side.**

ANSWERS

5. (a) positive; increases
   (b) negative; decreases

Section 9.5 Introduction to Graphing Linear Equations 659

9.5 Exercises

FOR EXTRA HELP 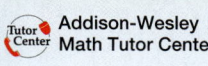 Addison-Wesley Math Tutor Center 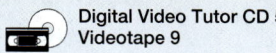 MathXL  Digital Video Tutor CD 5 Videotape 9  Student's Solutions Manual 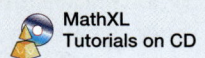 MyMathLab MathXL Tutorials on CD

Graph each equation by completing the table to find three solutions and plotting the ordered pairs. Then use the graph to find two *other solutions. See Example 1.*

1. $x + y = 4$

x	y	Ordered Pair (x, y)
0		
1		
2		

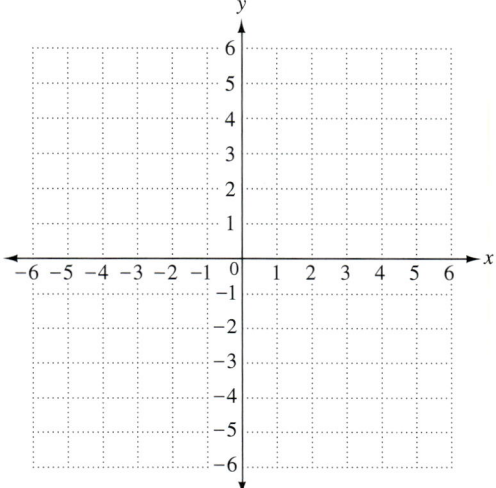

Two other solutions are (__, __) and (__, __).

2. $x + y = -4$

x	y	Ordered Pair (x, y)
0		
1		
2		

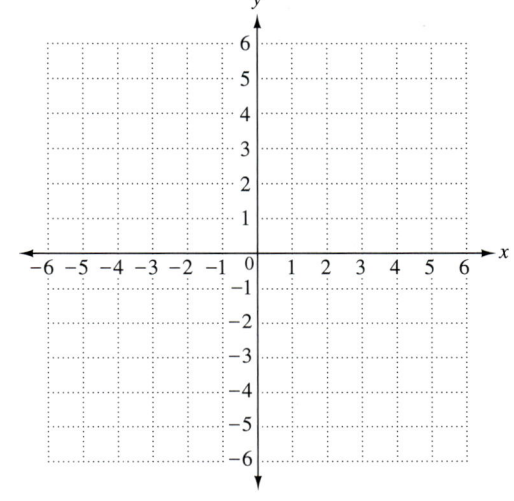

Two other solutions are (__, __) and (__, __).

3. $x + y = -1$

x	y	Ordered Pair (x, y)
0		
1		
2		

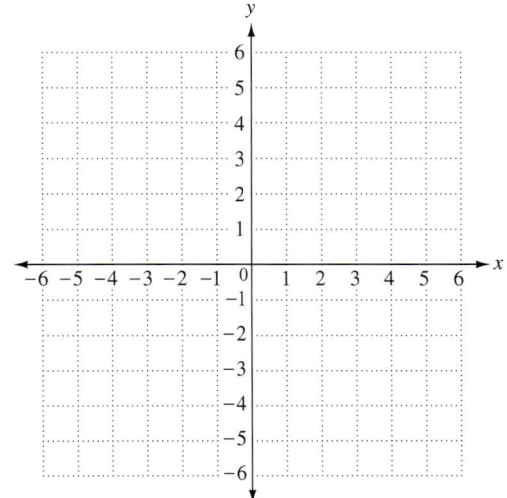

Two other solutions are (__, __) and (__, __).

4. $x + y = 1$

x	y	Ordered Pair (x, y)
0		
1		
2		

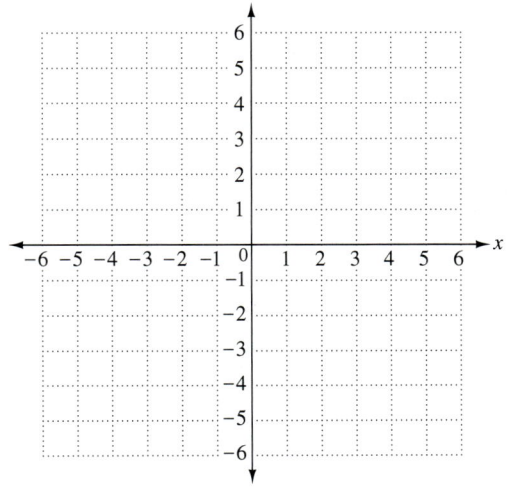

Two other solutions are (__, __) and (__, __).

5. The line in Exercise 1 crosses the y-axis at what point? _____. The line in Exercise 3 crosses the y-axis at what point? _____. Based on these examples, where would the graph of $x + y = -6$ cross the y-axis? Where would the graph of $x + y = 99$ cross the y-axis?

6. Look at where the line crosses the x-axis and where it crosses the y-axis in Exercises 2 and 4. What pattern do you see?

Graph each equation. Make your own table using the listed values of x. See Examples 2 and 4.

7. $y = x - 2$

Use 1, 2, and 3 as the values of x.

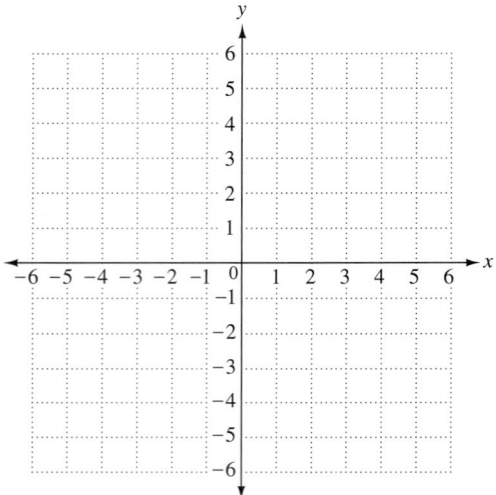

8. $y = x + 1$

Use 1, 2, and 3 as the values of x.

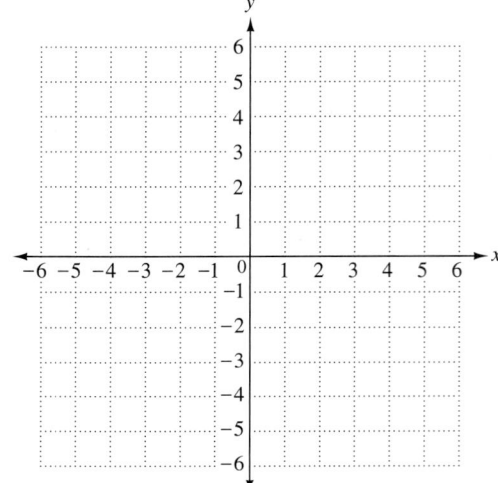

9. $y = x + 2$

Use 0, −1, and −2 as the values of x.

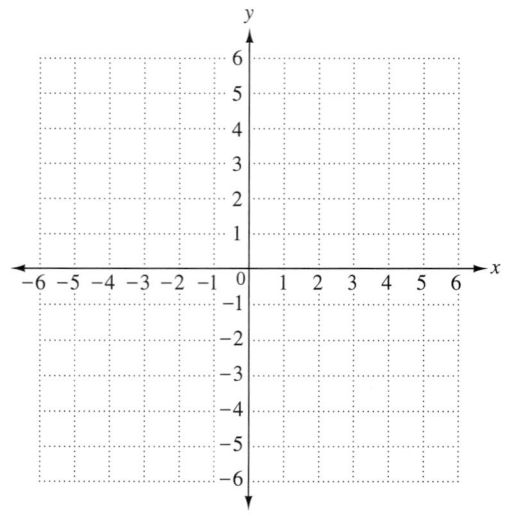

10. $y = x - 1$

Use 0, −1, and −2 as the values of x.

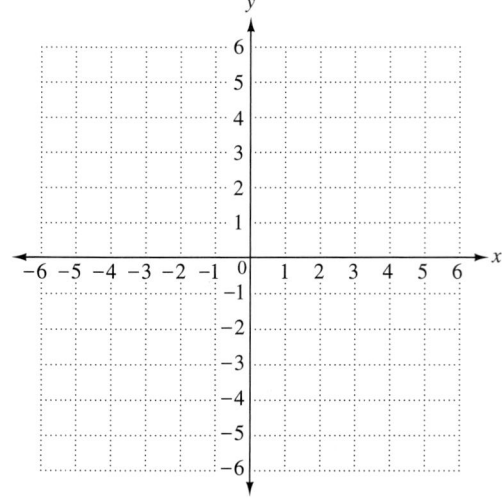

11. $y = -3x$

Use 0, 1, and 2 as the values of x.

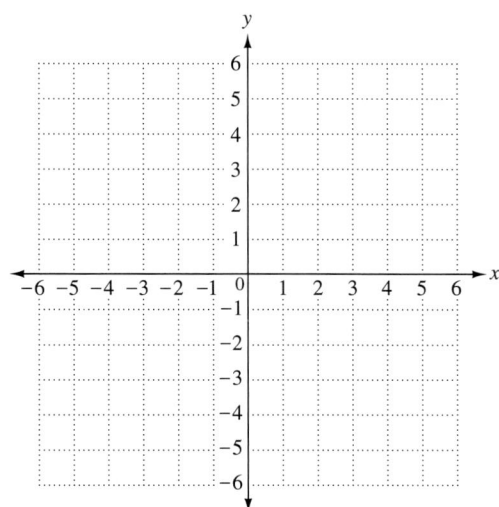

12. $y = -2x$

Use 0, 1, and 2 as the values of x.

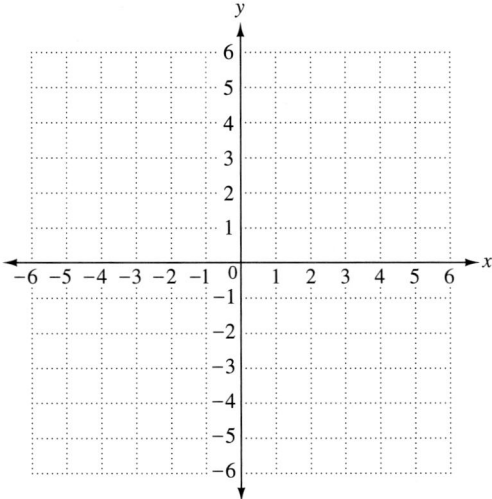

13. Look back at the graphs in Exercises 1, 3, 7, 9, and 11. Which lines have a positive slope? Which lines have a negative slope?

14. Look back at the graphs in Exercises 2, 4, 8, 10, and 12. Which lines have a positive slope? Which lines have a negative slope?

Graph each equation. Make your own table using the listed values of x. See Examples 1–4.

15. $y = \dfrac{1}{3}x$

Use 0, 3, and 6 as the values of x.

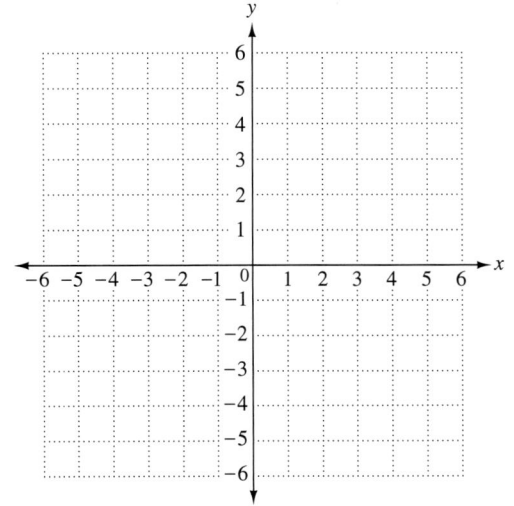

16. $y = \dfrac{1}{2}x$

Use 0, 2, and 4 as the values of x.

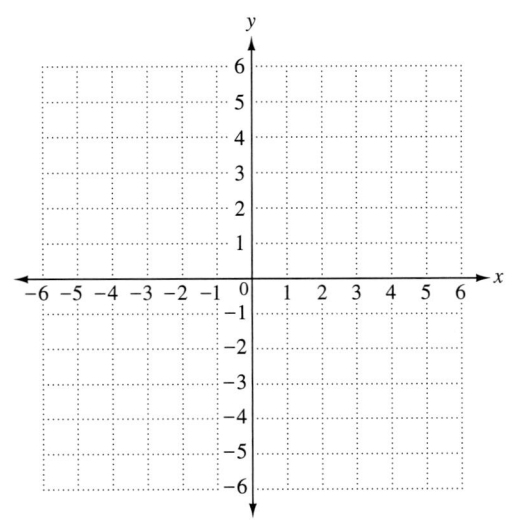

17. $y = x$

Use $-1, -2,$ and -3 as the values of x.

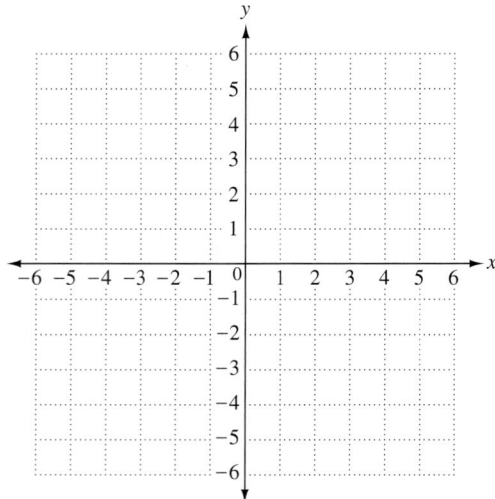

18. $x + y = 0$

Use $1, 2,$ and 3 as the values of x.

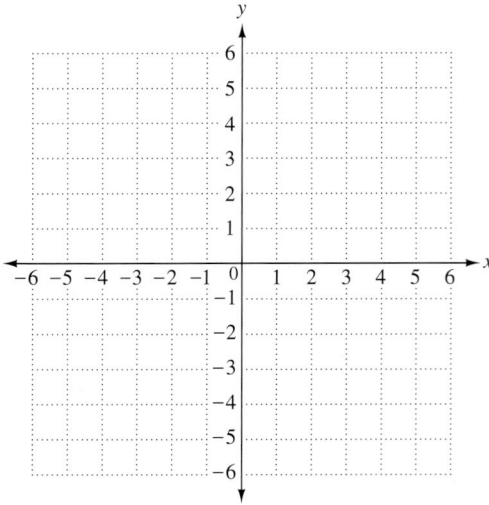

19. $y = -2x + 3$

Use $0, 1,$ and 2 as the values of x.

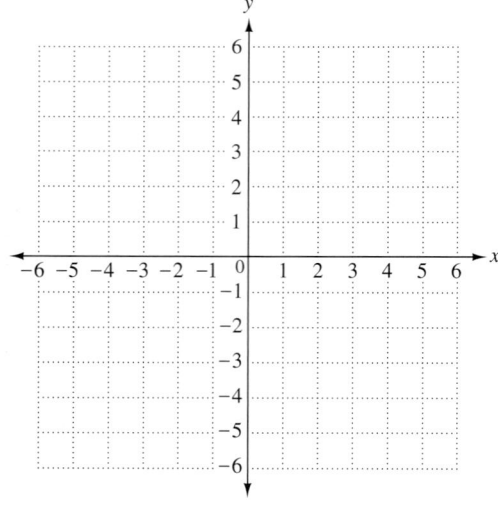

20. $y = 3x - 4$

Use $0, 1,$ and 2 as the values of x.

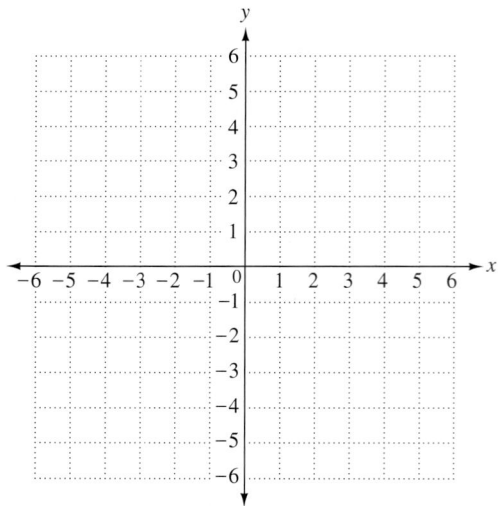

Graph each equation. Choose three values for x. Make a table showing your x values and the corresponding y values. After graphing the equation, state whether the line has a positive or negative slope. See Examples 1–5.

21. $x + y = -3$

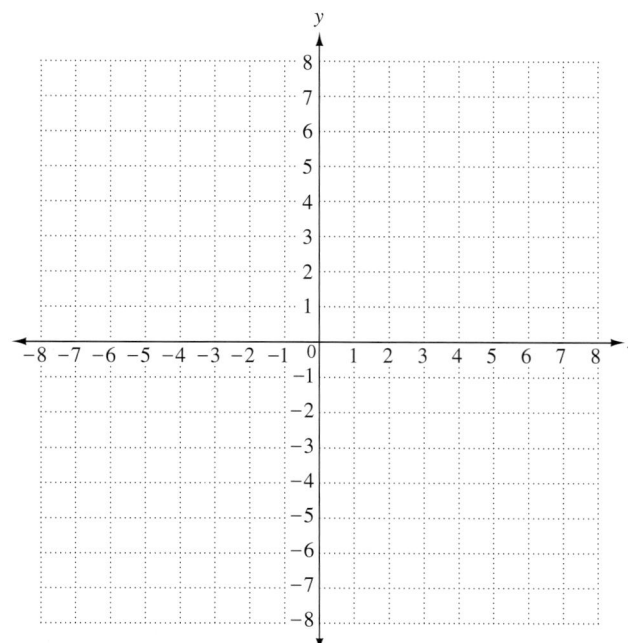

22. $x + y = 2$

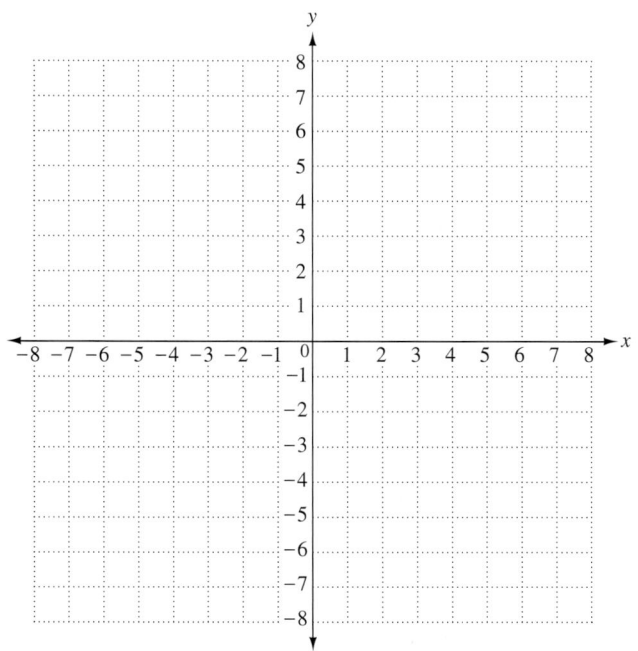

23. $y = \dfrac{1}{4}x$ (*Hint:* Try using 0 and multiples of 4 as the values of *x*.)

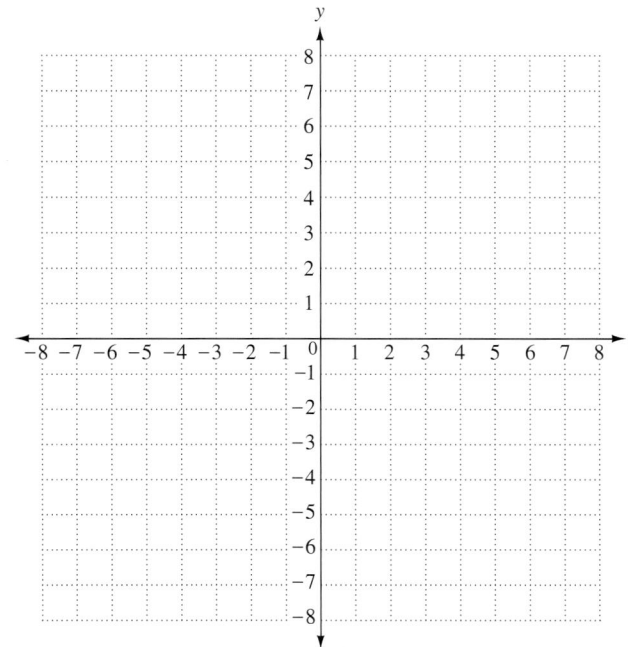

24. $y = -\dfrac{1}{3}x$ (*Hint:* Try using 0 and multiples of 3 as the values of *x*.)

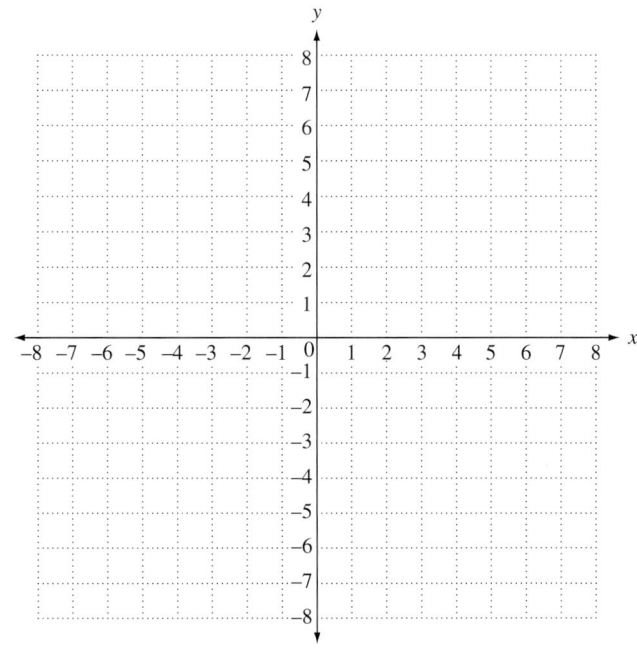

25. $y = x - 5$

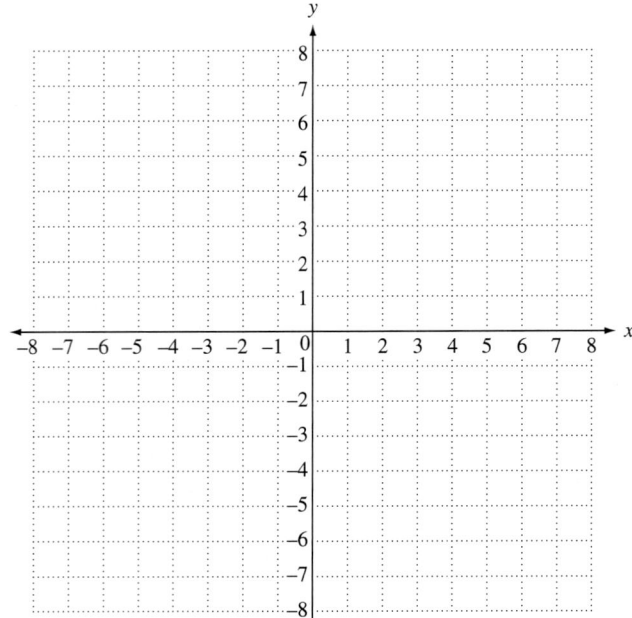

26. $y = x + 4$

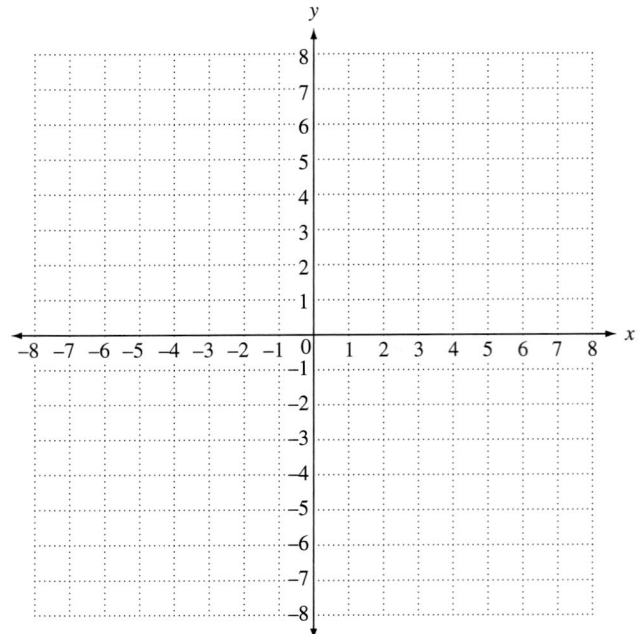

27. $y = -3x + 1$

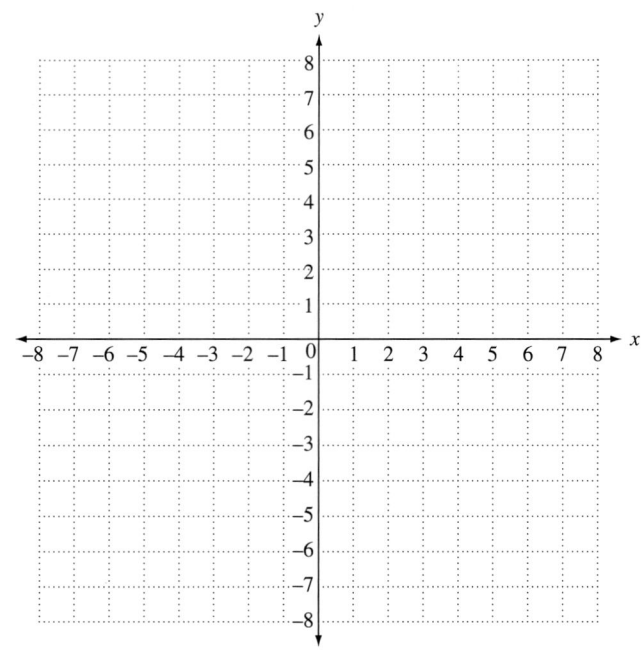

28. $y = 2x - 2$

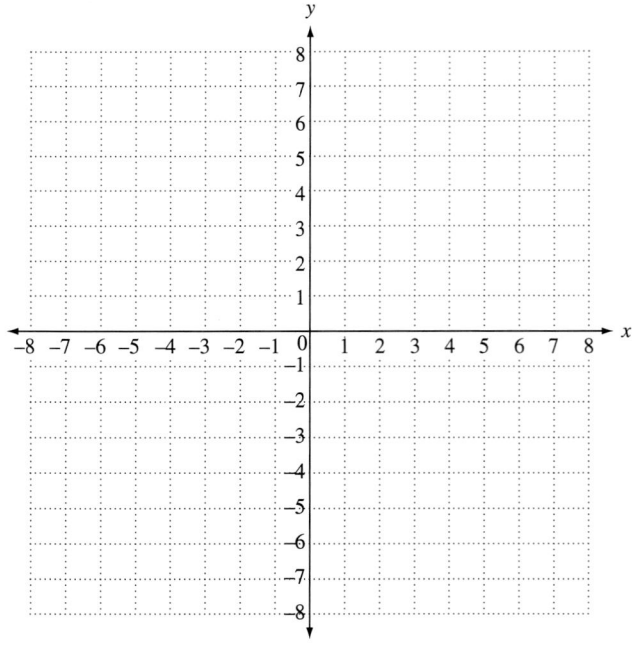

Chapter 9
SUMMARY

KEY TERMS

9.1 **table** — A table presents data organized into rows and columns.
pictograph — A pictograph uses symbols or pictures to represent various amounts.

9.2 **circle graph** — A circle graph shows how a total amount is divided into parts or sectors. It is based on percents of 360°.
protractor — A protractor is a tool (usually in the shape of a half-circle) used to measure the number of degrees in an angle or part of a circle.

9.3 **bar graph** — A bar graph uses bars of various heights to show quantity or frequency.
double-bar graph — A double-bar graph compares two sets of data by showing two sets of bars.
line graph — A line graph uses dots connected by lines to show trends.
comparison line graph — A comparison line graph shows how two or more sets of data relate to each other by showing a line graph for each set of data.

9.4 **paired data** — When each number in a set of data is matched with another number by some rule of association, we call it paired data.
horizontal axis — The horizontal axis is the number line in a coordinate system that goes "left and right."
vertical axis — The vertical axis is the number line in a coordinate system that goes "up and down."
ordered pair — An ordered pair is the "address" of a point in a coordinate system. The first number tells how far to move left or right from 0 along the horizontal axis. The second number tells how far to move up or down from 0 along the vertical axis.
x-axis — The horizontal axis is called the x-axis.
y-axis — The vertical axis is called the y-axis.
coordinate system — Together, the x-axis and the y-axis form a rectangular coordinate system. *Example:* See figure at the right.
origin — The center point of a rectangular coordinate system is (0, 0) and is called the origin. *Example:* See red dot marking (0, 0) in figure at the right.
coordinates — Coordinates are the numbers in the ordered pair that specify the location of a point on a rectangular coordinate system. *Example:* See the point $(-4, 2)$ at the right.
quadrants — The x-axis and the y-axis divide the coordinate system into four regions called quadrants; they are designated with Roman numerals. *Example:* The point $(-4, 2)$ at the right is in quadrant II.

9.5 **graph a linear equation** — All the solutions of a linear equation (all the ordered pairs that satisfy the equation) lie along a straight line. When you draw the line, you have graphed the equation.

NEW SYMBOLS

(x, y) ordered pair

Test Your Word Power

See how well you have learned the vocabulary in this chapter. Answers follow the Quick Review.

1. A **circle graph**
   A. uses symbols to represent various amounts
   B. is useful for showing trends
   C. shows how a total amount is divided into parts
   D. compares two sets of data.

2. A **line graph**
   A. uses symbols to represent various amounts
   B. is useful for showing trends
   C. shows how a total amount is divided into parts
   D. compares two sets of data.

3. A **protractor** is used to
   A. measure degrees in an angle
   B. draw circles of various sizes
   C. measure lengths
   D. calculate circumference.

4. Two sets of data can be compared using a
   A. circle graph
   B. pictograph
   C. line graph
   D. double-bar graph.

5. The **x-axis** in a rectangular coordinate system is the
   A. number line that goes "up and down"
   B. number line that goes "left and right"
   C. center point of the grid
   D. vertical axis.

6. **Quadrants** are the
   A. numbers in an ordered pair
   B. solutions of a linear equation
   C. four regions in a coordinate system
   D. paired data shown on a graph.

7. **Coordinates** are
   A. points in a straight line on a coordinate system
   B. designated with Roman numerals
   C. the solutions of a linear equation
   D. numbers used to locate a point in a coordinate system.

8. A **rectangular coordinate system**
   A. is formed by the *x*-axis and *y*-axis
   B. is divided into eight quadrants
   C. is the solution of a linear equation
   D. has only positive numbers.

Quick Review

Concepts

9.1 Reading a Table
The data in a table is organized into rows and columns. As you read from left to right along each row, check the heading at the top of each column.

Examples

Use the table below to answer the question.

PER CAPITA CONSUMPTION OF SELECTED BEVERAGES IN GALLONS

	1985	1990	1995	2000
Milk	26.7	25.7	24.3	22.6
Coffee	27.4	26.9	20.5	26.3
Bottled water	4.5	8.0	11.6	17.7
Soft drinks	35.7	46.2	47.5	49.3

Source: U.S. Dept. of Agriculture Economic Research Service.

In which years was the consumption of milk greater than the consumption of coffee?

Read across the rows labeled "milk" and "coffee" from left to right, comparing the numbers in each column. In 1995 the figure for milk (24.3 gallons) is greater than for coffee (20.5 gallons).

Concepts

9.1 *Reading a Pictograph*
A pictograph uses symbols or pictures to represent various amounts. The *key* tells you how much each symbol represents. A fractional part of a symbol represents a fractional part of the symbol's value.

Examples

Use the pictograph to answer these questions.

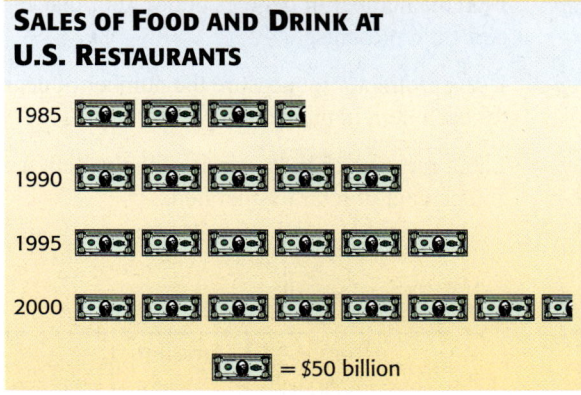

Source: National Restaurant Association.

(a) How much was spent at U.S. restaurants in 1985?

Sales for 1985 are represented by 3 whole symbols (3 • $50 billion = $150 billion) plus half of a symbol ($\frac{1}{2}$ of $50 billion = $25 billion) for a total of $175 billion.

(b) How much did restaurant expenditures increase from 1985 to 2000?

The year 2000 shows four more symbols than 1985, so the increase is 4 • $50 billion = $200 billion.

Concepts

9.2 Constructing a Circle Graph

Step 1 Determine the percent of the total for each item.

Step 2 Find the number of degrees out of 360° that each percent represents.

Step 3 Use a protractor to measure the number of degrees for each item in the circle.

Step 4 Label each sector in the circle with the item name and percent of total for that item.

Examples

Construct a circle graph for the following table, which lists expenses for a business trip.

Item	Amount
Transportation	$200
Lodging	$300
Food	$250
Entertainment	$150
Other	$100
Total	**$1000**

Item	Amount	Percent of Total			Sector Size
Transportation	$200	$\dfrac{\$200}{\$1000} = \dfrac{1}{5} = 20\%$	so	20% of 360° = (0.20)(360)	= 72°
Lodging	$300	$\dfrac{\$300}{\$1000} = \dfrac{3}{10} = 30\%$	so	30% of 360° = (0.30)(360)	= 108°
Food	$250	$\dfrac{\$250}{\$1000} = \dfrac{1}{4} = 25\%$	so	25% of 360° = (0.25)(360)	= 90°
Entertainment	$150	$\dfrac{\$150}{\$1000} = \dfrac{3}{20} = 15\%$	so	15% of 360° = (0.15)(360)	= 54°
Other	$100	$\dfrac{\$100}{\$1000} = \dfrac{1}{10} = 10\%$	so	10% of 360° = (0.10)(360)	= 36°

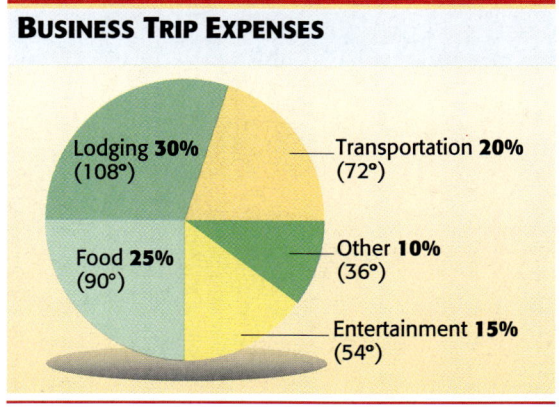

BUSINESS TRIP EXPENSES

- Lodging 30% (108°)
- Transportation 20% (72°)
- Food 25% (90°)
- Other 10% (36°)
- Entertainment 15% (54°)

Concepts

9.3 Reading a Bar Graph
The height of the bar is used to show the quantity or frequency (number) in a specific category. Use a ruler or straightedge to line up the top of each bar with the numbers on the left side of the graph.

Examples

Use the bar graph below to determine the number of students who earned each letter grade.

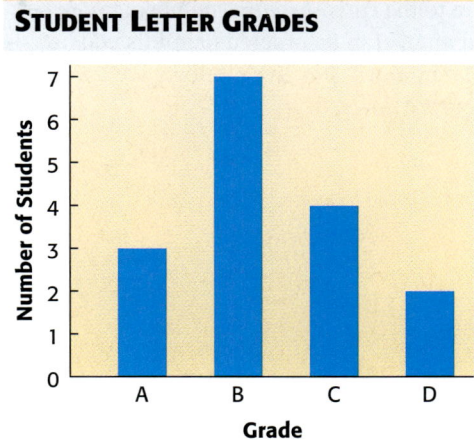

Grade	Number of Students
A	3
B	7
C	4
D	2

9.3 Reading a Line Graph
A dot is used to show the number or quantity in a specific class. The dots are connected with lines. This kind of graph is used to show a trend.

The line graph below shows the annual sales for the Fabric Supply Center for each of four years.

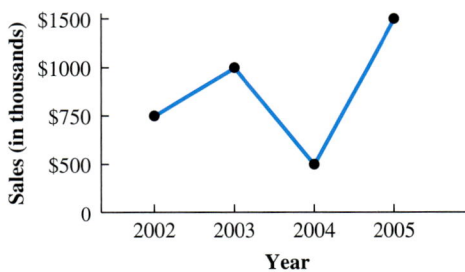

Find the sales in each year.

Year	Total Sales
2002	$750 • 1000 = $ 750,000
2003	$1000 • 1000 = $1,000,000
2004	$500 • 1000 = $ 500,000
2005	$1500 • 1000 = $1,500,000

Concepts	Examples
9.4 Plotting Points Start at the center of the coordinate system (the origin). The first number in an ordered pair tells you how far to move *left* or *right* along the horizontal axis; *positive* numbers are to the *right*, *negative* numbers to the *left*. The second number in an ordered pair tells you how far to move *up* or *down*; *positive* numbers are *up*, *negative* numbers are *down*.	To plot **(3, −2)**, start at 0, move to the *right* 3 units, and then move *down* 2 units. To plot **(−2, 3)**, start at 0, move to the *left* 2 units, and then move *up* 3 units. 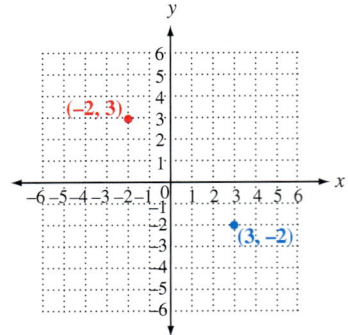
9.4 Identifying Quadrants The *x*-axis and *y*-axis divide the coordinate system into four regions called quadrants. The quadrants are designated with Roman numerals. Points in the first quadrant fit the pattern (**+, +**). Points in the second quadrant fit the pattern (**−, +**). Points in the third quadrant fit the pattern (**−, −**). Points in the fourth quadrant fit the pattern (**+, −**).	

(Quadrant diagram showing Quadrant I with (3, 2), Quadrant II with (−3, 2), Quadrant III with (−3, −2), Quadrant IV with (3, −2).)

9.5 Graphing Linear Equations
Choose any three values for *x*. Then find the corresponding values of *y*. Plot the three ordered pairs on a coordinate system. Draw a line through the points, extending it in both directions. If the three points do *not* lie on a straight line, there is an error in your work. *Every point on the line* is a solution of the given equation.

Graph $y = -2x$.

x	y = −2 • x	Ordered Pair (x, y)
1	−2 • 1 is −2	(1, −2)
2	−2 • 2 is −4	(2, −4)
3	−2 • 3 is −6	(3, −6)

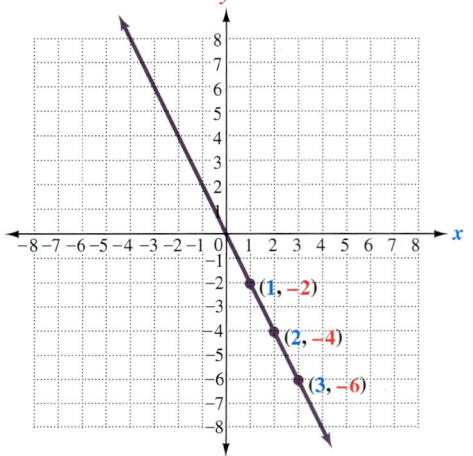

ANSWERS TO TEST YOUR WORD POWER

1. C; *Example:* A circle graph can show how a 24-hour day is divided among various activities.

2. B; *Example:* A line graph can show changes in the amount of sales over several months or several years.

3. A; *Example:* A protractor is a tool, usually in the shape of a half-circle, used to measure or draw angles of a certain number of degrees. A tool called a compass is used to draw circles of various sizes.

4. D; *Example:* Two sets of bars in different colors can compare monthly unemployment figures for two different years.

5. B; *Example:* The *x*-axis is the horizontal number line with 0 at the center, negative numbers extending to the left, and positive numbers extending to the right.

6. C; *Example:* The *x*-axis and the *y*-axis divide the coordinate system into four regions; each region is designated by a Roman numeral.

7. D; *Example:* To locate the point (2, −3), start at the origin and move 2 units to the right along the *x*-axis, then move down 3 units.

8. A; *Example:* Together, a horizontal number line (*x*-axis) and vertical number line (*y*-axis) form a rectangular coordinate system.

Focus on Real-Data Applications

Surfing the Net

1. Look at the "Source" information at the bottom of the graph. How were the numbers in the graph obtained?

2. A researcher seeks information about members of a *population*. The individuals who are polled must be representative of the *population*. Describe the population that was targeted by this survey.

3. How many people in the poll said they cut back on television viewing to find time to use the Internet?

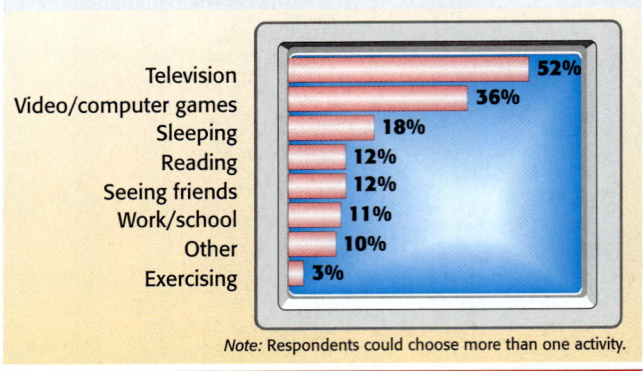

CAUGHT IN THE NET
Web users have cut back on the following activities to get more online time:

- Television 52%
- Video/computer games 36%
- Sleeping 18%
- Reading 12%
- Seeing friends 12%
- Work/school 11%
- Other 10%
- Exercising 3%

Note: Respondents could choose more than one activity.

Source: NUKE InterNETWORK poll of 500 regular users.

4. Find the number of people in the poll who cut back on each of the other activities listed in the graph.

5. Add up all the responses to the poll from Problems 3 and 4. Why is the total more than the 500 people that were in the poll?

6. Suppose you took a similar poll of 100 students at your school. Would you expect the results to be similar to those shown in the graph? Why or why not?

7. Suppose you first asked students if they regularly used the Internet, and then took a similar poll of 100 of those students. Would you expect the results to be similar to those shown in the graph? Why or why not?

8. Conduct a survey of your class members. First find out if they regularly use the Internet. Ask those who regularly use the Internet which of the activities they cut back on to have more surfing time. Each person polled can select more than one activity.
   (a) How many students are in your class poll?
   (b) Complete the table using the responses from those who regularly use the Internet.

Activity	Number Who Cut Back on the Activity	Percent Who Cut Back on the Activity
Television		
Video/computer games		
Sleeping		
Reading		
Seeing friends		
Work/school		
Other		
Exercising		

9. Make a bar graph showing your survey data. How is your data similar to the graph shown above? How is it different?

Chapter 9
REVIEW EXERCISES

[9.1] *Use the table at the right to answer Exercises 1–4.*

1. Which sport had the
   (a) fewest men's teams?
   (b) second-greatest number of women's teams?

2. Which sport had
   (a) about 10,000 female athletes?
   (b) about 7000 male athletes?

PARTICIPATION IN SELECTED NCAA SPORTS

Sport	Males			Females		
	Teams	Athletes	Average Squad	Teams	Athletes	Average Squad
Basketball	950	15,141	15.9	966	13,392	13.9
Cross country	792	10,271	13.0	838	10,141	12.1
Golf	678	7197	10.6	282	2323	8.2
Gymnastics	28	413	14.8	91	1311	14.4
Volleyball	74	1052		923	12,284	

Source: The National Collegiate Athletic Association (NCAA).

3. (a) How many more men participated in basketball than cross country?
   (b) How many fewer women participated in gymnastics than basketball?

4. Find the average squad size for men's and women's volleyball teams and complete the table. Round to the nearest tenth.

Use the pictograph to answer Exercises 5–8.

5. What is the average yearly snowfall in
   (a) Juneau?
   (b) Washington, D.C.?

6. What is the average yearly snowfall in
   (a) Minneapolis?
   (b) Cleveland?

7. Find the difference in the average yearly snowfall between
   (a) Buffalo and Cleveland.
   (b) Memphis and Minneapolis.

8. Find the difference in the average yearly snowfall between the city with the greatest amount and the city with the least amount.

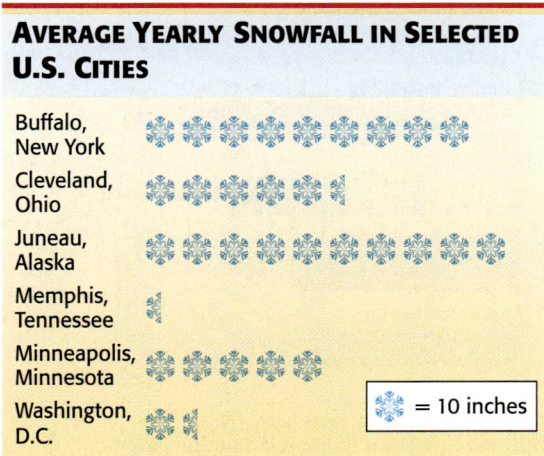

AVERAGE YEARLY SNOWFALL IN SELECTED U.S. CITIES

❄ = 10 inches

Source: National Climatic Data Center.

[9.2] *Use this circle graph for Exercises 9–13.*

9. **(a)** What was the largest single expense of the vacation? How much was that item?

   (b) What was the second-most-expensive item? How much was that item?

   (c) What was the total cost of the vacation?

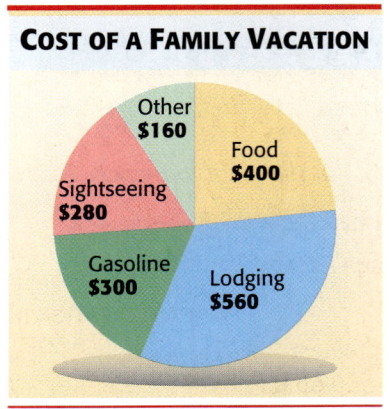

Find each ratio. Write the ratio as a fraction in lowest terms.

10. Cost of lodging to the cost of food

11. Cost of gasoline to the total cost of the vacation

12. Cost of sightseeing to the total cost of the vacation

13. Cost of gasoline to the cost of the Other category

[9.3] *This bar graph shows the top six home improvement projects that people do themselves. Use the graph to answer Exercises 14–19.*

14. What is the most popular project? What percent of the homeowners in the survey gave that answer?

15. Which project was selected the least? What percent of the homeowners in the survey gave that answer?

16. What percent of the homeowners in the survey selected carpentry projects? There were 341 people surveyed. How many of them selected carpentry, to the nearest whole number?

17. Of the 341 homeowners surveyed, how many selected landscaping or gardening projects? Round to the nearest whole number.

18. Name the project selected by about **(a)** $\frac{1}{2}$ of the homeowners **(b)** $\frac{1}{3}$ of the homeowners.

19. Give two reasons painting and wallpapering might be selected so much more than construction work.

This double-bar graph shows the number of acre-feet of water in the Lake Natoma reservoir for each of the first six months of 2004 and 2005. Use this graph to answer Exercises 20–25.

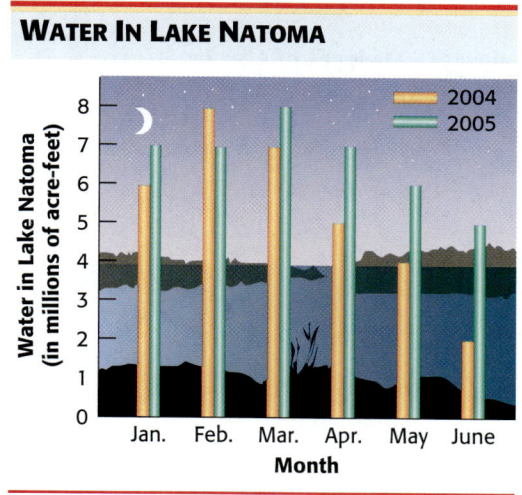

20. From January through June 2005, which month had the greatest amount of water in the lake? How much was there?

21. During which month in the first half of 2004 was the least amount of water in the lake? How much was there?

22. How many acre-feet of water were in the lake in June of 2005?

23. How many acre-feet of water were in the lake in January of 2004?

24. Find the amount of decrease and the percent of decrease in the acre-feet of water in the lake from March 2005 to June 2005.

25. Find the amount of decrease and the percent of decrease in the acre-feet of water in the lake from April 2004 to June 2004.

This comparison line graph shows the annual floor-covering sales of two different home improvement centers during each of five years. Use this graph to find the amount of sales in each year listed in Exercises 26–29 and to answer Exercises 30 and 31.

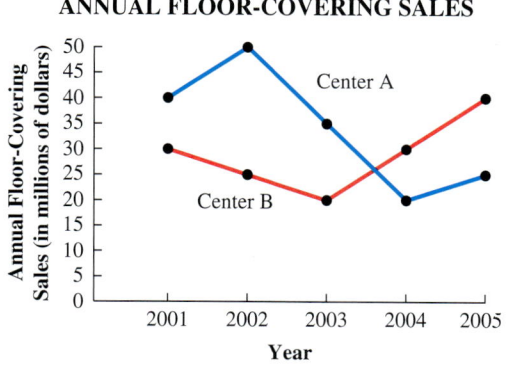

26. Center A in 2002

27. Center A in 2004

28. Center B in 2003

29. Center B in 2005

30. What trend do you see in Center A's sales from 2002 to 2005? Why might this have happened?

31. What trend do you see in Center B's sales starting in 2003? Why might this have happened?

[9.2] *The Broadway Hair Salon spent a total of* $22,400 *to open a new shop. The breakdown of expenditures for various items is shown. Find all the missing numbers in Exercises 32–36.*

Item	Dollar Amount	Percent of Total	Degrees of Circle
32. Plumbing and electrical changes	$2240	10%	_____
33. Work stations	$7840	_____	_____
34. Small appliances	$4480	_____	_____
35. Interior decoration	$5600	_____	_____
36. Supplies	_____	_____	36°

37. Draw a circle graph using the information in Exercises 32–36. Label each sector with the item and the percent of total for that item.

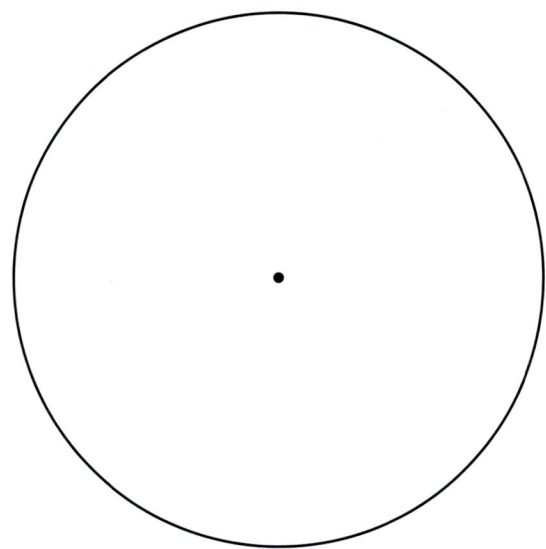

[9.4] *In Exercise 38, plot each point on the rectangular coordinate system and label it with its coordinates. In Exercise 39, give the coordinates of each point.*

38. $(1, 7)$ $(-1\frac{1}{2}, 0)$ $(-4, -2)$ $(0, 3)$ $(2, -\frac{1}{2})$

$(-7, 1)$ $(5, -5)$ $(0, -4)$

39.

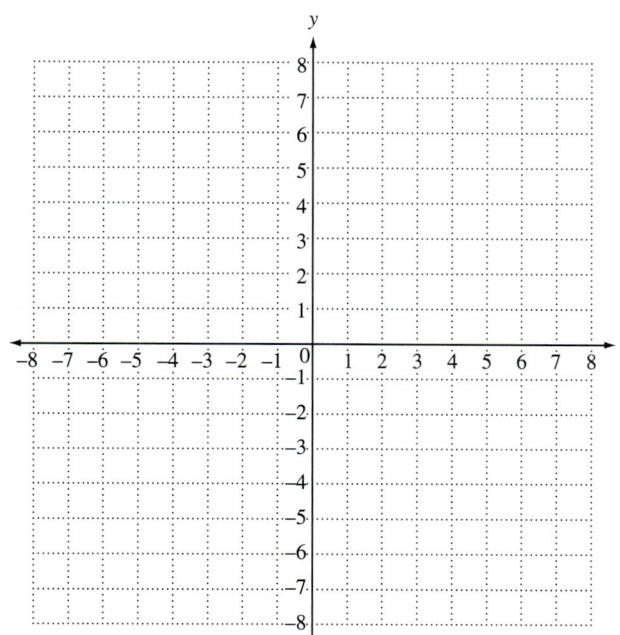
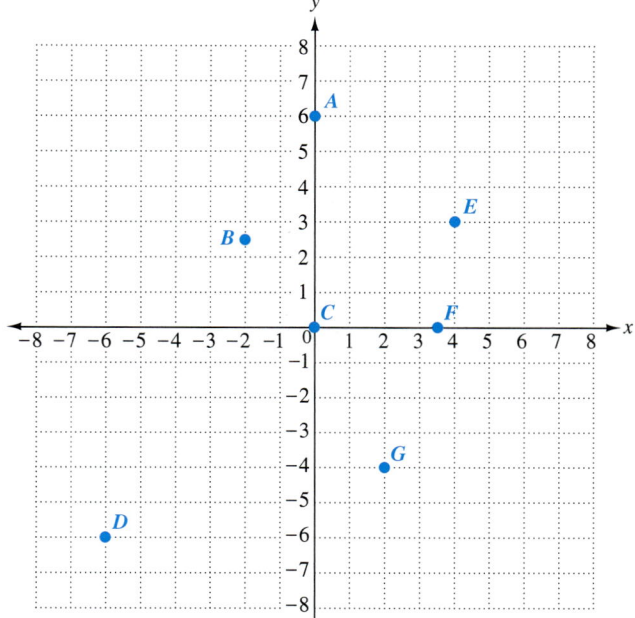

[9.5] *Graph each equation. Complete the table using the listed values of x. State whether each line has a positive or negative slope. Use the graph to find two other solutions of the equation.*

40. $x + y = -2$

Use 0, 1, and 2 as the values of x.

x	y	(x, y)

The graph of $x + y = -2$ has a _____ slope.

Two other solutions of $x + y = -2$ are

(___, ___) and (___, ___).

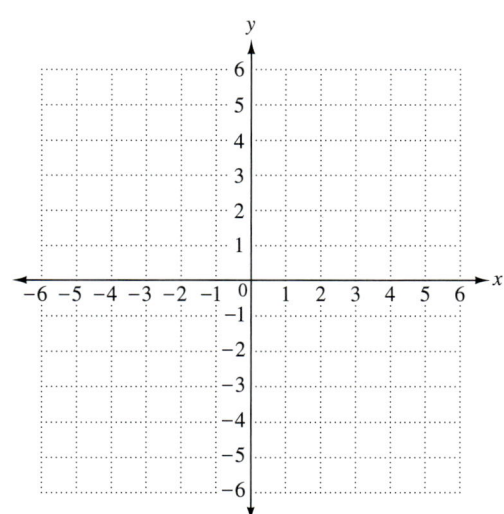

41. $y = x + 3$

Use 0, 1, and 2 as the values of x.

x	y	(x, y)

The graph of $y = x + 3$ has a _____ slope.

Two other solutions for $y = x + 3$ are
(____, ____) and (____, ____).

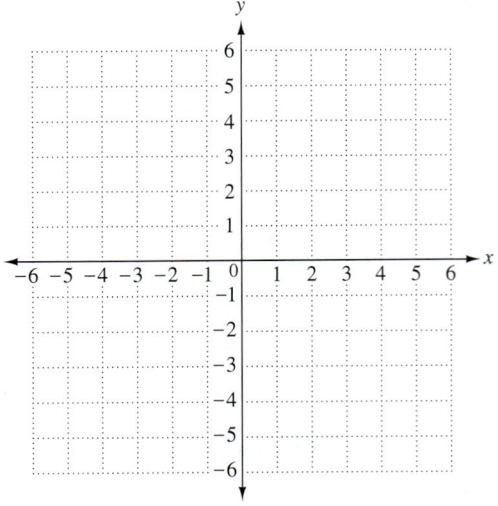

42. $y = -4x$

Use -1, 0, and 1 as the values of x.

x	y	(x, y)

The graph of $y = -4x$ has a _____ slope.

Two other solutions of $y = -4x$ are
(____, ____) and (____, ____).

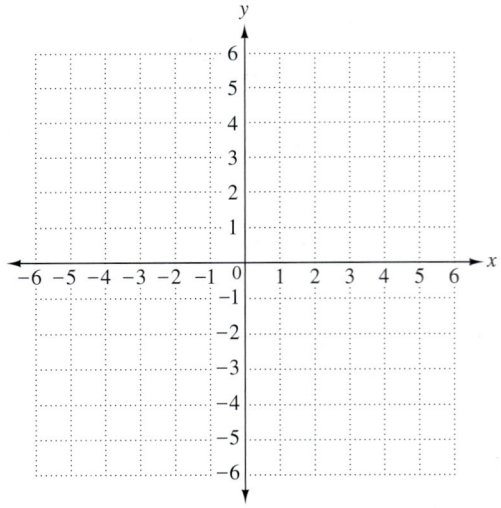

Chapter 9
TEST

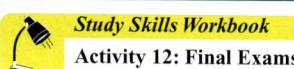
Study Skills Workbook
Activity 12: Final Exams

Use this table to answer Problems 1–3.

AMOUNT OF CALCIUM IN SELECTED FOODS

Food	Measure	Calories	Calcium (mg)
Swiss cheese	1 oz	95	219
Cream cheese	1 oz	100	23
Yogurt, fruit flavor	8 oz	230	345
Skim milk	1 cup	85	302
Sardines	3 oz	175	371

Source: U.S. Department of Agriculture.

1. Which food has **(a)** the greatest amount of calcium? **(b)** the least amount of calcium?

    1. (a) _____
       (b) _____

2. A packet of Swiss cheese contains 1.75 oz. How many calories are in that amount of Swiss cheese? Round to the nearest whole number.

    2. _____

3. Find the amount of calcium in a 6 oz container of fruit-flavored yogurt. Round to the nearest whole number.

    3. _____

Use this pictograph to answer Problems 4–6.

NUMBER OF U.S. ENDANGERED WILDLIFE SPECIES

Mammals: 🐺🐺🐺🐺🐺🐺🐺
Birds: 🐺🐺🐺🐺🐺🐺🐺🐺
Reptiles: 🐺🐺
Fish: 🐺🐺🐺🐺🐺🐺🐺🐺
Insects: 🐺🐺🐺🐺

🐺 = 10 Species

Source: U.S. Fish and Wildlife Service.

4. **(a)** Which group has the greatest number of endangered species?
   (b) How many endangered species are in that group?

    4. (a) _____
       (b) _____

5. How many more species of fish are endangered than reptiles?

    5. _____

6. What is the total number of endangered species shown in the pictograph?

    6. _____

This circle graph shows the advertising budget for Lakeland Amusement Park. Find the dollar amount budgeted for each category listed in Exercises 7–10. The total advertising budget is $2,800,000.

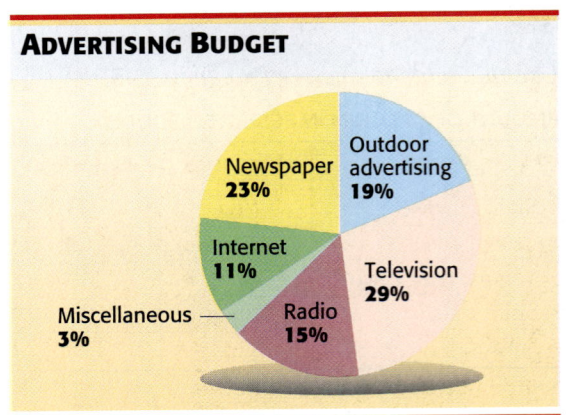

7. Which form of advertising has the largest budget? What amount is this?

8. Which form of advertising has the smallest budget? How much is this?

9. How much is budgeted for Internet advertising?

10. How much is budgeted for newspaper ads?

This graph shows one student's income and expenses for four years.

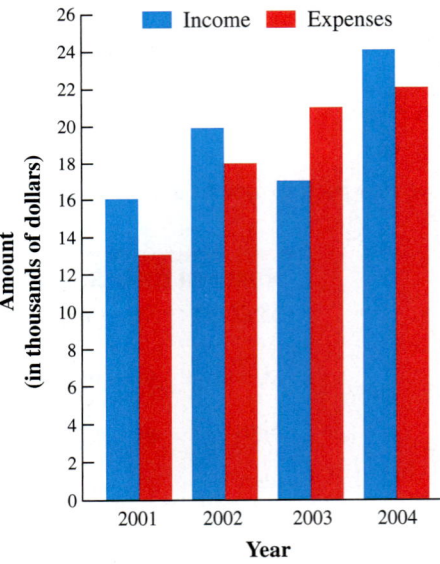

11. In what year did the student's expenses exceed income? By how much?

12. Find the amount of increase and the percent of increase in the student's expenses from 2001 to 2002. Round to the nearest whole percent.

13. In what year did the student's income decline? Give two possible explanations for the decline.

This graph shows enrollment at two community colleges.

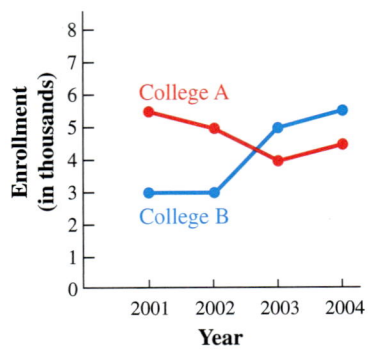

14. What was College A's enrollment in 2001? What was College B's enrollment in 2002?

15. Which college had higher enrollment in 2004? How much higher?

16. Give two possible explanations for the fact that College B's enrollment passed College A's.

During a one-year period, Oak Mill Furniture Sales had a total of $480,000 in expenses. Find all the numbers missing from the table.

Item	Dollar Amount	Percent of Total	Degrees of a Circle
17. Salaries	$168,000	_____	_____
18. Delivery expense	$24,000	_____	_____
19. Advertising	$96,000	_____	_____
20. Rent	$144,000	_____	_____
21. Other	_____	_____	36°

17. _____

18. _____

19. _____

20. _____

21. _____

22. Draw a circle graph using a protractor and the information in Problems 17–21. Label each sector with the item and the percent of total for that item.

22.

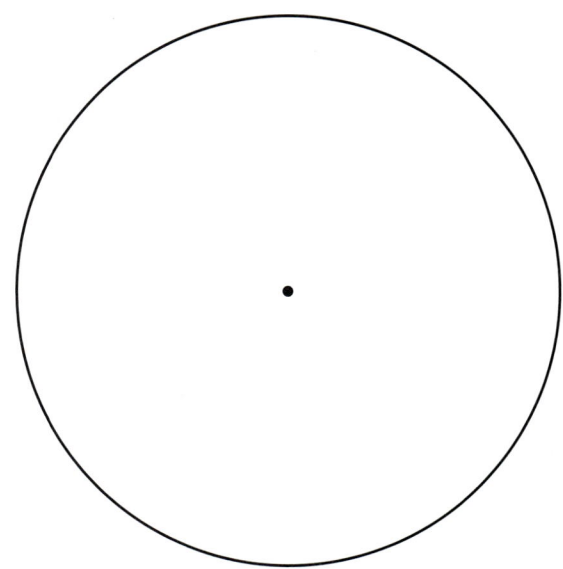

Plot each point on the coordinate system below. Label each point with its coordinates.

23. $(-5, 3)$

24. $(1, -4)$

25. $(0, 6)$

26. $(2, 0)$

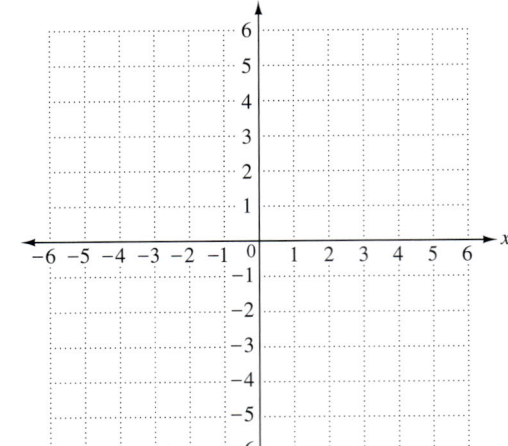

Give the coordinates of each lettered point shown below, and state which quadrant the point is in.

27. _____

28. _____

29. _____

30. _____

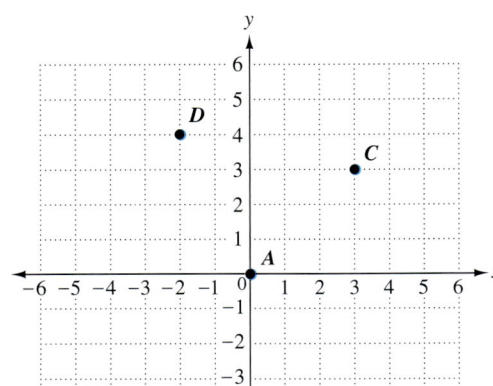

27. Point A

28. Point B

29. Point C

30. Point D

31. Graph $y = x - 4$ on the coordinate system below. Make your own table using 0, 1, and 2 as the values of x.

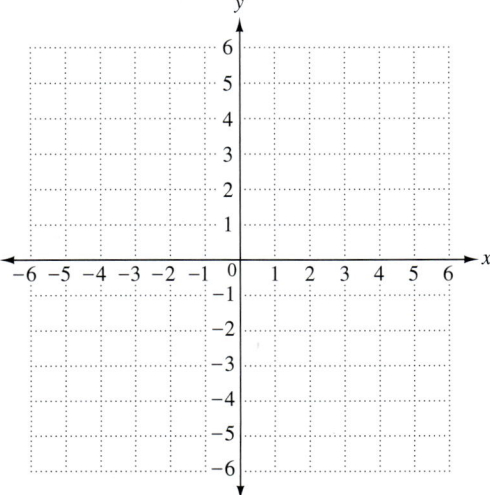

32. _____

32. Use the graph in Problem 31 to find *two* other solutions of $y = x - 4$.

Real Numbers, Equations, and Inequalities

10.1 **Real Numbers and Expressions**

10.2 **More on Solving Linear Equations**

10.3 **Formulas and Solving for a Specified Variable**

10.4 **Solving Linear Inequalities**

Americans have enjoyed "going to the movies" since the first movie theater opened in 1905. The highest number of theater tickets ever sold in one year was more than 4.49 billion tickets back in 1929. The U.S. population then was about 123 million, so, on average, each American went to a movie theater at least 36 times that year. (*Source: World Almanac, Guinness World Records.*)

We can use inequality symbols to show these comparisons.

    1929 U.S. ticket sales $>$ 4.49 billion

    1929 average theater visits per person $\geq$ 36

In Section 10.1, Exercises 35–38, you'll use these symbols to show comparisons between the all-time top money-making films.

10.1 Real Numbers and Expressions

OBJECTIVES

1. Identify rational numbers, irrational numbers, and real numbers.
2. Use the symbols $\neq$, $<$, $\leq$, $>$, and $\geq$ to compare real numbers.
3. Reverse the direction of inequality statements.
4. Use the order of operations to simplify expressions with brackets.
5. Remove parentheses and simplify expressions using the distributive property.

Throughout Chapters 1–9, you have been developing basic concepts and skills that will help you understand and use algebra. In this chapter we will strengthen your knowledge of the basics by introducing some more advanced terminology and techniques.

OBJECTIVE 1 Identify rational numbers, irrational numbers, and real numbers. You have worked with different kinds of numbers: integers in Chapter 1, fractions in Chapter 4, decimals in Chapter 5, and so on. Now let's be more specific about these different groups, or sets, of numbers.

The first numbers a young child learns are the ones used for counting, called the set of *natural numbers*.

$$1, 2, 3, 4, 5, \ldots \leftarrow \text{The three dots indicate that the list of } natural\ numbers \text{ goes on forever.}$$

Once zero is included with the natural numbers, we have the set of *whole numbers*.

$$0, 1, 2, 3, 4, 5, \ldots \leftarrow \text{The list of } whole\ numbers \text{ goes on forever.}$$

So far we have only zero and positive numbers. The next set of numbers, the *integers*, includes the natural numbers, their *opposites*, and zero. You worked with integers in Chapter 1.

$$\ldots -5, -4, -3, -2, -1, 0, 1, 2, 3, 4, 5, \ldots \leftarrow \text{The list of } integers \text{ goes on forever in both directions.}$$

Recall how we used a number line to show the position of each integer.

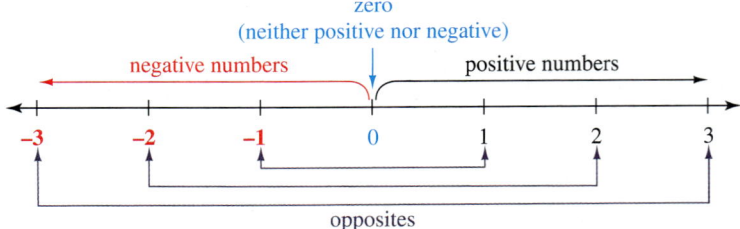

Then, in Chapters 4 and 5, you used positive and negative fractions and decimals. These numbers are also located on the number line. All the integers and fractions and some of the decimal numbers are examples of *rational numbers*.

Rational Numbers

Rational numbers are numbers that can be written as quotients of integers, with the denominator not equal to zero. In other words, rational numbers can be written in the form $\frac{a}{b}$ where a and b are integers and $b \neq 0$.

The name "*ratio*nal numbers" comes from the word *ratio*, which indicates a quotient. Some examples of rational numbers are shown here, including how to write them as a quotient of integers.

$$5 \quad \text{can be written as} \quad \frac{5}{1}$$

$$-18 \quad \text{can be written as} \quad \frac{-18}{1}$$

All integers are rational numbers.

$3\frac{1}{2}$ can be written as $\frac{7}{2}$ ⎫

$-\frac{5}{8}$ can be written as $\frac{-5}{8}$ or $\frac{5}{-8}$ ⎬ *All fractions* are rational numbers.

0.19 can be written as $\frac{19}{100}$ ⎭

2.7 can be written as $2\frac{7}{10} = \frac{27}{10}$ ⎫ These decimals are rational numbers, but *not all decimals* are rational numbers.

$0.333\ldots$ can be written as $\frac{1}{3}$ ⎬

$0.\overline{72}$ can be written as $\frac{8}{11}$ ⎭

Since any integer can be written as the quotient of itself and 1, all integers are rational numbers. A decimal number that comes to an end (terminates), such as 0.19 or 2.7 is a rational number. Decimal numbers that repeat in a *fixed block* of digits, such as 0.3333 ... and $0.\overline{72}$, are also rational numbers. (Recall that the bar above the digits 7 and 2 means that they repeat indefinitely, so 0.727272 ... can be written as $0.\overline{72}$.)

Recall that to *graph* a number, we place a dot on the number line at the point corresponding to the number.

EXAMPLE 1 Graphing Rational Numbers

Graph these rational numbers on the number line.

$$-\frac{3}{2}, \; -\frac{2}{3}, \; \frac{1}{2}, \; 1\frac{1}{3}, \; \frac{23}{8}, \; 3\frac{1}{4}$$

To locate the improper fractions on the number line, write them in the form of mixed numbers or decimals.

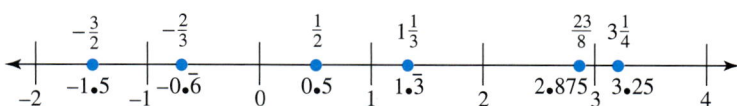

Work Problem 1 at the Side.

There are also numbers that are *not* rational. For example, in **Section 5.10** you worked with π. Recall that we found π by dividing the circumference of a circle by its diameter. In other words, π is a quotient, but it never ends and the digits never repeat in a fixed block, no matter how far you carry out the division.

$$\frac{\text{circumference of circle}}{\text{diameter of circle}} = 3.14159265\ldots$$

Powerful computers have calculated π out to millions of decimal places, but the pattern of digits never repeats and never ends. So π is an example of an *irrational number*.

Irrational Numbers

Irrational numbers are nonrational numbers represented by points on the number line. The decimal form of an irrational number does not terminate and does not repeat in a fixed block of digits.

1 Graph these rational numbers on the number line.

$$-3, \; -2.75, \; -\frac{3}{4}, \; 1\frac{1}{2}, \; \frac{17}{8}$$

```
  +---+---+---+---+---+---+---+
 -4  -3  -2  -1   0   1   2   3
```

ANSWER

1.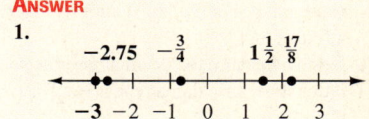

2 Identify each number as rational or irrational, and explain why.

(a) $\sqrt{36}$

(b) $0.6666\ldots$

(c) $\sqrt{13}$

(d) $0.454545\ldots$

(e) $0.131131113\ldots$

(f) 9.4375

(g) $0.\overline{27}$

ANSWERS
2. (a) rational because $\sqrt{36} = 6$
   (b) rational because the digit 6 repeats
   (c) irrational because decimal value of $\sqrt{13}$ never ends or repeats in a fixed block
   (d) rational because digits repeat in a fixed block
   (e) irrational because digits do not repeat in a fixed block
   (f) rational because the decimal terminates
   (g) rational because digits repeat in a fixed block

The decimal form of an irrational number never terminates (never ends) and never repeats in a fixed block of digits. An example is the number $0.10110111011110\ldots$ Another example is $\sqrt{7}$. In **Section 5.8** we said that the *approximate* value for $\sqrt{7}$ is 2.646. It's approximate because $(2.646)^2$ is 7.001316, not 7. The actual value of $\sqrt{7}$ in decimal form never terminates and never repeats, so it is an irrational number. These numbers lie between rational numbers on the number line. You will work with irrational numbers in Chapter 16.

> **CAUTION**
> Some square roots are irrational. Examples are $\sqrt{2}, \sqrt{3},$ and $\sqrt{7}$. However, *not all* square roots are irrational. For example, $\sqrt{9}$ is *rational*, because $\sqrt{9} = 3$.

EXAMPLE 2 Identifying Rational and Irrational Numbers

Identify each number as *rational* or *irrational,* and explain why. Use your calculator to find square roots.

(a) $0.181818\ldots$ (b) 3.125 (c) $0.20220222022220\ldots$
(d) $\sqrt{11}$ (e) $\sqrt{16}$ (f) $0.\overline{36}$

(a) Rational, because the digits repeat in a fixed block.

(b) Rational, because the decimal terminates (comes to an end).

(c) Irrational, because the digits do *not* repeat in a fixed block.

(d) Irrational, because the decimal value of $\sqrt{11}$ never terminates or repeats.

(e) Rational, because $\sqrt{16} = 4$.

(f) Rational, because the digits repeat in a fixed block.

Work Problem 2 at the Side.

Finally, *all* numbers that can be represented by points on the number line are called *real numbers.*

> **Real Numbers**
> The set of **real numbers** includes all the rational numbers *and* all the irrational numbers. All the real numbers can be represented by points on the number line.

All the numbers mentioned so far in this section are *real numbers*. The relationships between the various types of numbers are shown in two different ways on the next page. Notice that any real number is either a rational number or an irrational number.

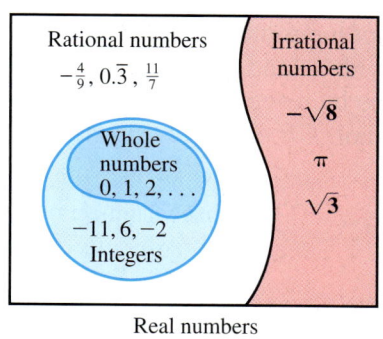

Real numbers

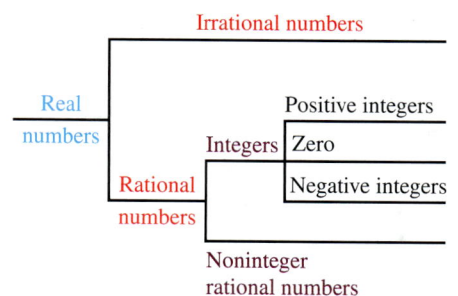

OBJECTIVE ▶ **2** **Use the symbols ≠, <, ≤, >, and ≥ to compare real numbers.** The symbols you used in earlier chapters when comparing numbers are shown below.

= is equal to	< is less than	> is greater than
$3(7) = 7(3)$	$-2 < 0$	$0.65 > 0.6$

Keep the meanings of < and > clear by remembering that the symbol always points to the lesser number. (See **Section 1.2** for more review.) Here are three other useful symbols.

≠ is not equal to	≤ is less than or equal to	≥ is greater than or equal to
$3 - 7 \neq 7 - 3$	$0 \leq 4$	$1.25 \geq 1.2$
$\frac{1}{2} \neq 0.3$	$\frac{3}{4} \leq \frac{3}{4}$	$-8 \geq -8$

Use ≠ when two numbers or quantities are *not* equal.

We read the inequality $0 \leq 4$ as "zero is less than or equal to 4." If either the < part *or* the = part is true, then the inequality is true. So $0 \leq 4$ is true because $0 < 4$. Then $\frac{3}{4} \leq \frac{3}{4}$ is true because $\frac{3}{4} = \frac{3}{4}$.

In a similar manner, $1.25 \geq 1.2$ is true because $1.25 > 1.2$. Also, $-8 \geq -8$ is true because $-8 = -8$.

EXAMPLE 3 Using the Symbols ≠, ≤, ≥

Label each statement as true or false and explain why.

(a) $-4 \leq -3$ (b) $\frac{1}{10} \geq \frac{1}{2}$ (c) $-6.2 \geq -6.2$ (d) $2 \div 1 \neq 1 \div 2$

(a) True, because $-4 < -3$. Recall that -4 is to the left of -3 on the number line.

(b) False, because $\frac{1}{10} < \frac{1}{2}$ and $\frac{1}{10} \neq \frac{1}{2}$.

(c) True, because $-6.2 = -6.2$.

(d) True, because they are not equal. $2 \div 1$ is 2, but $1 \div 2$ is $\frac{1}{2}$ or 0.5.

▶▶▶ **Work Problem 3 at the Side.** ▶▶▶

3 Label each statement as true or false and explain why.

(a) $0 \geq -\frac{3}{4}$

(b) $\sqrt{25} \leq \sqrt{25}$

(c) $3.06 \geq 3.6$

(d) $\frac{2}{3}(10) \neq \frac{3}{2}(10)$

(e) $-15 \leq -16$

(f) $\frac{1}{4} \neq \frac{12}{48}$

(g) $0.5 \geq \frac{1}{2}$

ANSWERS

3. (a) True because $0 > -\frac{3}{4}$
   (b) True because $\sqrt{25} = \sqrt{25}$
   (c) False because $3.06 < 3.6$ and $3.06 \neq 3.6$
   (d) True because $6\frac{2}{3} \neq 15$
   (e) False because $-15 > -16$ and $-15 \neq -16$
   (f) False because $\frac{12}{48}$ in lowest terms is $\frac{1}{4}$
   (g) True because $0.5 = \frac{1}{2}$

Chapter 10 Real Numbers, Equations, and Inequalities

④ Rewrite each statement with the inequality symbol reversed.

(a) $0.3 \leq 0.33$

(b) $\dfrac{2}{3} > \dfrac{2}{9}$

(c) $-5 < -1$

(d) $\dfrac{9}{10} \geq 0.7$

⑤ Simplify each expression.

(a) $4[7 + 3(6 + 1)]$

(b) $3[(-20 \div 5) - 7]$

ANSWERS

4. (a) $0.33 \geq 0.3$ (b) $\dfrac{2}{9} < \dfrac{2}{3}$
   (c) $-1 > -5$ (d) $0.7 \leq \dfrac{9}{10}$

5. (a) 112 (b) -33

OBJECTIVE 3 Reverse the direction of inequality statements. Any statement with $<$ can be converted to one with $>$, and any statement with $>$ can be converted to one with $<$. We do this by reversing both the order of the numbers and the direction of the symbol. For example, the statement $6 < 10$ can be written as $10 > 6$.

$6 < 10$ becomes $10 > 6$ (Exchange numbers. Reverse the symbol)

You can verify that both statements are true: 6 is less than 10, and 10 is greater than 6.

EXAMPLE 4 Converting Between $<$ and $>$

The list below shows each statement written in two equally correct ways.

(a) $9 < 16$ becomes $16 > 9$

(b) $0 > -2$ becomes $-2 < 0$

(c) $\dfrac{1}{2} \leq \dfrac{3}{4}$ becomes $\dfrac{3}{4} \geq \dfrac{1}{2}$

(d) $-7 \geq -8$ becomes $-8 \leq -7$

▶ **Work Problem 4 at the Side.**

OBJECTIVE 4 Use the order of operations to simplify expressions with brackets. We have been using parentheses to show several different things.

Parentheses used to indicate multiplication:	Parentheses used to indicate a negative number:	Parentheses used to indicate order of operations:
$3(5)$	$6 - (-4)$	$6 + (3 + 4)$
means multiply 3 times 5.	means 6 minus negative 4.	means add $3 + 4$ first.

An expression with double parentheses, such as the expression $2(8 + 3(6 + 5))$, can be confusing. We can avoid confusion by using square brackets, [ ], in place of one pair of parentheses.

EXAMPLE 5 Using Brackets

Simplify $2[8 + 3(6 + 5)]$.

Begin inside the parentheses. Then follow the order of operations as you complete the work inside the brackets.

$2[8 + 3(\mathbf{6 + 5})]$	Work inside parentheses: add $6 + 5$.
$2[8 + 3(\mathbf{11})]$	Multiply $3(11)$.
$2[8 + \mathbf{33}]$	Add $8 + 33$.
$2[\mathbf{41}]$	Multiply 2 times 41.
82	

▶ **Work Problem 5 at the Side.**

OBJECTIVE 5 **Remove parentheses and simplify expressions using the distributive property.** In **Section 2.2** you used the *distributive property* to simplify expressions. Here are two examples.

$$6(y + 4) \qquad 5(3x - 2)$$
$$6 \cdot y + 6 \cdot 4 \qquad 5 \cdot 3x - 5 \cdot 2$$
$$6y + 24 \qquad 15x - 10$$

We can also use the distributive property to remove parentheses when there is a *negative* number in front of them.

EXAMPLE 6 Using the Distributive Property to Remove Parentheses

Write without parentheses.

(a) $-2(2y + 3)$

$$-2(2y + 3)$$
$$(-2 \cdot 2y) + (-2 \cdot 3) \quad \text{Multiply every term inside the parentheses by } -2.$$
$$-4y + -6 \quad \text{Use the definition of subtraction "in reverse" as in Section 2.2.}$$
$$-4y - 6$$

The simplified expression is $-4y - 6$.

(b) $-5(3a - 4)$

$$-5(3a - 4)$$
$$(-5 \cdot 3a) - (-5 \cdot 4) \quad \text{Multiply every term inside the parentheses by } -5.$$
$$-15a - (-20) \quad \text{Use the definition of subtraction.}$$
$$-15a + 20$$

The simplified expression is $-15a + 20$.

CAUTION
Watch the signs carefully when there is a *negative* number in front of the parentheses, as in Example 6. Notice that every term in the simplified expression has the *opposite sign* from the original expression.

▶ **Work Problem 6 at the Side.** ▶▶▶

Sometimes an expression may have just a negative sign in front of parentheses, such as $-(3x + 5)$. We can rewrite this as $-1(3x + 5)$ and then use the distributive property to multiply every term inside the parentheses by -1.

6 Write without parentheses.

(a) $-3(4b + 1)$

(b) $-7(2x - 3)$

(c) $-4(h - 5)$

(d) $-6(-2y + 4)$

ANSWERS
6. (a) $-12b - 3$ **(b)** $-14x + 21$
     (c) $-4h + 20$ **(d)** $12y - 24$

Chapter 10 Real Numbers, Equations, and Inequalities

7 Write without parentheses.

(a) $-(6k - 5)$

(b) $-(-2 - r)$

(c) $-(-5y + 8)$

(d) $-(z + 4)$

8 Simplify.

(a) $10p + 3(5 + 2p)$

(b) $7x - 2 - (1 + x)$

(c) $-(3k^2 + 5k) + 7(k^2 - 4k)$

(d) $-2(4b - 3) - 5(2b^2 + 1)$

ANSWERS
7. (a) $-6k + 5$ (b) $2 + r$ (c) $5y - 8$
   (d) $-z - 4$
8. (a) $16p + 15$ (b) $6x - 3$ (c) $4k^2 - 33k$
   (d) $-10b^2 - 8b + 1$

EXAMPLE 7 Using the Distributive Property to Remove Parentheses

Write without parentheses.

(a) $-(3x + 5)$ can be written $-1(3x + 5)$ Multiply every term inside the parentheses by -1.

$$(-1 \cdot 3x) + (-1 \cdot 5)$$
$$-3x \quad + \quad (-5)$$ Use the definition of subtraction "in reverse."
$$-3x \quad - \quad 5$$

The simplified expression is $-3x - 5$.

(b) $-(-7r - 8)$ can be written as $-1(-7r - 8)$ Multiply every term inside the parentheses by -1.

$$(-1 \cdot -7r) - (-1 \cdot 8)$$
$$7r \quad - \quad (-8)$$ Use the definition of subtraction.
$$7r \quad + \quad 8$$

The simplified expression is $7r + 8$.

Work Problem 7 at the Side.

EXAMPLE 8 Simplifying Expressions Involving Like Terms

Simplify each expression.

(a) $14y + 2(6 + 3y)$

$14y + 2(6 + 3y)$ Multiply every term inside the parentheses by 2.
$14y + 12 + 6y$ Combine like terms.
$20y + 12$

(b) $-(2 - r) + 10r$ is written $-1(2 - r) + 10r$ Rewrite $-(2 - r)$ as $-1(2 - r)$. Then use the distributive property.
$-2 + r + 10r$
$-2 + r + 10r$ Combine like terms.
$-2 + 11r$

(c) $5(2a^2 - 6a) - 3(4a^2 - 9)$

Multiply every term inside the parentheses by 5. $5(2a^2 - 6a) -3(4a^2 - 9)$ Multiply every term inside the parentheses by -3.

$10a^2 - 30a - 12a^2 + 27$ Combine like terms.
$-2a^2 - 30a + 27$

Work Problem 8 at the Side.

10.1 Exercises

*Identify each number as **rational** or **irrational** and explain why. See Example 2.*

1. (a) -0.0625
   (b) π

2. (a) $\sqrt{5}$
   (b) 0

3. (a) $\dfrac{3}{4}$
   (b) $0.636363\ldots$

4. (a) $-5\dfrac{2}{3}$
   (b) $0.3233233323333\ldots$

5. (a) $\sqrt{2}$
   (b) $\sqrt{100}$

6. (a) $0.416666\ldots$
   (b) 12.75

*Tell whether each statement is true **always, sometimes,** or **never**.*

7. A real number is a rational number.
8. A rational number is a real number.
9. An irrational number is a real number.
10. A fraction is an irrational number.
11. An irrational number is an integer.
12. A rational number is an integer.

Write each statement in words. Then label the statement as true or false, and explain why. See Example 3.

13. $0.75 \neq \dfrac{3}{4}$
14. $0 \geq -1$
15. $-4 \leq -5$
16. $\dfrac{1}{3} \neq 0.3$
17. $0 \geq 0$
18. $4 \leq 5$

Write each statement with the inequality symbol reversed. See Example 4.

19. $12 < 19$
20. $0.55 > 0.5$
21. $\dfrac{4}{5} \geq \dfrac{1}{2}$
22. $-40 \leq -30$

First simplify each statement wherever possible. Then label the statement as true or false. See Examples 3 and 5.

23. $-17 \leq 1 - 18$
24. $-12 \geq -10 - 2$
25. $-6(8) + 9(5) \geq 0$
26. $-4(-20) - 15(5) \geq 0$
27. $6[5 - 3(4 - 2)] \neq 6$
28. $-3[(0 - 5) + 2] \neq 9$

29. $3^2[(-10 \div 5) + 7] \le 40$

30. $-(5^2)[-3 + 2(-6 \div 2)] \le -225$

31. $[6 - 4(4)] \div (-10) \ge 1$

32. $-1 \ne 4 \div [8(2) - 20]$

33. $0 \ne 12 \div [3(2) - 6]$

34. $[12 - 2(36 \div 6)] \div 9 \le 0$

RELATING CONCEPTS (EXERCISES 35–38) For Individual or Group Work

Use the table on movie gross receipts (all the income, before paying expenses) to work Exercises 35–38 in order.

ALL-TIME TOP AMERICAN MOVIES THROUGH OCTOBER 2005

Rank	Title (Year)	Gross Receipts (millions)
1	Titanic (1997)	$600.8
2	Star Wars (1977)	461.0
3	Shrek 2 (2004)	436.5
4	E.T.: The Extra-Terrestrial (1982)	435.0
5	Stars Wars: Episode I (1999)	431.1
6	Spider-Man (2002)	405.9

Source: www.boxofficereport.com

35. (a) Which films had gross receipts greater than 461 million dollars?

(b) Which films had gross receipts greater than or equal to 461 million dollars?

36. (a) Which films had gross receipts less than 435 million dollars?

(b) Which films had gross receipts less than or equal to 435 million dollars?

37. Write a statement using the $\le$ symbol that describes the gross receipts for the films ranked 5 and 6.

38. Write a statement using the $\ge$ symbol that describes the gross receipts for the top three films.

Use the distributive property to simplify each expression. See Example 7.

39. $-(4t + 5m)$

40. $-(9x + 12y)$

41. $-(-5c - 4d)$

42. $-(-13x - 15y)$

43. $-(6h - n)$

44. $-(a - 7b)$

45. $-(-3q + 5r - 8s)$

46. $-(-4z + 5w - 9y)$

Simplify each expression. See Examples 6–8.

47. $13p + 4(4 - 8p)$

48. $5x + 3(7 - 2x)$

49. $-4(y - 7) - 6$

50. $-5(t - 13) - 4$

51. $-(6 - y) + y^2 - 6$

52. $-(w + 5) - w^2 + 5$

53. $2(3b^2 - b) - 4(b - 2)$

54. $7(x^2 - 3) - 5(x + 6)$

55. $-3(-a + 1) - (2a - 4)$

56. $-8(-5 + c^2) - (10 - 7c^2)$

57. $-10(-3k - 2) + 6(-4 - k)$

58. $-7(-4 + 3x^2) + 5(x^2 - 6)$

10.2 More on Solving Linear Equations

The equations you solved in Chapters 2–7 were all *linear equations in one variable*. Such equations contain only one variable, and that variable is always raised to the first power and never appears in a denominator. Some examples of linear equations are shown below.

$$5m + 1 = 16 \qquad 2k - 2 = 5k - 11 \qquad 100 + \frac{a}{2} = 116$$

OBJECTIVES

1. Solve more difficult linear equations.
2. Solve equations that have no solution or infinitely many solutions.
3. Solve equations by first clearing fractions and decimals.

OBJECTIVE 1 Solve more difficult linear equations. We developed a 5-step process in **Sections 2.5** and **4.7** to solve linear equations. As a review, here are the steps.

Solving Linear Equations

Step 1 If possible, use the **distributive property** to remove parentheses.

Step 2 **Combine** any like terms on the left side of the equation. Combine any like terms on the right side of the equation.

Step 3 **Add** or **subtract** the same amount on both sides of the equation so that the variable term ends up by itself on one side of the equal sign and a number is by itself on the other side. You may have to do this step more than once.

Step 4 **Multiply** or **divide** both sides by the same number to find the solution.

Step 5 **Check** your solution by going back to the *original* equation. Replace the variable with your solution. Follow the order of operations to complete the calculations. If the two sides of the equation balance, your solution is correct.

EXAMPLE 1 Review of Solving Linear Equations

Solve this equation and check the solution: $-2 + 3(2x + 7) = 7 + 3x$

Step 1 Use the distributive property. $\quad -2 + 3(2x + 7) = 7 + 3x$

Step 2 Combine like terms. $\quad -2 + 6x + 21 = 7 + 3x$

$$6x + 19 = 7 + 3x$$

Step 3 Subtract 19 from both sides.

$$\underline{-19 -19}$$
$$6x + 0 = -12 + 3x$$
$$6x = -12 + 3x$$

Subtract $3x$ from both sides. $\quad \underline{-3x -3x}$

$$3x = -12 + 0$$

Step 4 Divide both sides by 3. $\quad \dfrac{3x}{3} = \dfrac{-12}{3}$

The solution is -4. $\qquad\qquad\qquad x = -4 \leftarrow$ Solution

— Continued on Next Page

Chapter 10 Real Numbers, Equations, and Inequalities

1 Solve each equation.

(a) $2p + 4 = 7(p - 2) + p$

Step 5 Check by replacing each x with -4 in the *original* equation.

$$-2 + 3(2x + 7) = 7 + 3x$$
$$-2 + 3(2 \cdot -4 + 7) = 7 + 3(-4)$$
$$-2 + 3(-8 + 7) = 7 + -12$$
$$-2 + 3(-1) = -5$$
$$-2 + -3 = -5$$
$$-5 = -5 \quad \text{Balances}$$

When x is replaced with -4, the equation balances, so **-4 is the correct solution** (*not* -5).

> **NOTE**
> Recall from **Section 2.5** that more than one sequence of steps will work. For example, in *Step 3* above, we could first subtract $3x$ from both sides, then subtract 19 from both sides. The solution would be the same.

Work Problem 1 at the Side.

(b) $-5y + 7y - 6y - 9 = 3 + 2y$

EXAMPLE 2 Removing Parentheses Before Solving an Equation

Solve this equation: $8a - (3 + 2a) = 3a + 1$

Start by removing the parentheses on the left side of the equation. Remember that the $-$ sign in this situation acts like a factor of -1, changing the sign of *every* term in the parentheses.

Step 1 Use distributive property to multiply every term inside the parentheses by -1.

$$8a - (3 + 2a) = 3a + 1$$
$$8a - 1(3 + 2a) = 3a + 1$$

Step 2 Combine like terms.

$$8a - 3 - 2a = 3a + 1$$
$$6a - 3 = 3a + 1$$

Step 3 Add 3 to both sides.

$$\underline{ + 3} \quad \underline{ + 3}$$
$$6a = 3a + 4$$

Subtract $3a$ from both sides.

$$\underline{-3a} \quad \underline{-3a}$$
$$\frac{3a}{3} = \frac{4}{3}$$

Step 4 Divide both sides by 3.

$$a = \frac{4}{3}$$

The solution is $\frac{4}{3}$. You can *check* the solution by going back to the *original* equation and replacing each a with $\frac{4}{3}$.

ANSWERS
1. (a) $p = 3$ (b) $y = -2$

> **CAUTION**
> When there is a subtraction sign in front of parentheses, be very careful to *change* the sign of *every* term inside the parentheses. In Example 2 above, the signs on the left side of the equation changed as shown below.
>
> $8a - (3 + 2a)$ Another example $-6x - (2x - 5)$
> $8a - 3 - 2a$ $-6x - 2x + 5$

② Solve each equation.

(a) $2m - (7m + 6) = 39$

Work Problem 2 at the Side.

EXAMPLE 3 Solving Linear Equations

Solve: $4(8 - 3t) = 32 - 8(t + 2)$

$$4(8 - 3t) = 32 - 8(t + 2)$$ Distributive property; notice change in signs on right side.
$$32 - 12t = 32 - 8t - 16$$ Combine like terms.
$$32 - 12t = 16 - 8t$$
$$ -32 -32$$ Subtract 32 from both sides.
$$-12t = -16 - 8t$$
$$+ 8t + 8t$$ Add $8t$ to both sides.
$$\frac{-4t}{-4} = \frac{-16}{-4}$$ Divide both sides by -4.
$$t = 4$$

The solution is 4.

(b) $6 = 4x - (3 - 2x)$

Check:

$4(8 - 3t) = 32 - 8(t + 2)$ Original equation.
$4(8 - 3 \cdot 4) = 32 - 8(4 + 2)$ Replace t with 4.
$4(8 - 12) = 32 - 8(6)$
$4(-4) = 32 - 48$
$-16 = -16$ Balances

When t is replaced with 4, the equation balances, so **4 is the correct solution** (*not* -16).

③ Solve.

(a) $2(4 - 3r) = 3(r + 1) + 14$

(b) $2 - 3(2 + 6z) = 4(z + 1)$

Work Problem 3 at the Side.

OBJECTIVE ② Solve equations that have no solution or infinitely many solutions. The equations solved so far have had exactly *one* solution. However, as you'll see in the next examples, some equations have *many* solutions or *no* solution.

ANSWERS
2. (a) $m = -9$ (b) $x = \frac{3}{2}$
3. (a) $r = -1$ (b) $z = -\frac{4}{11}$

Chapter 10 Real Numbers, Equations, and Inequalities

4 Solve each equation.

(a) $2(x - 6) = 3 + 2x - 15$

(b) $4b + 2(3 - 2b) = 6$

EXAMPLE 4 Solving an Equation That Has Infinitely Many Solutions

Solve: $5x - 15 = 5(x - 3)$

$$5x - 15 = 5(x - 3) \quad \text{Use the distributive property.}$$

$$5x - 15 = 5x - 15$$
$$\underline{+15 \qquad \quad +15} \quad \text{Add 15 to both sides.}$$
$$5x \quad = \quad 5x$$
$$\underline{-5x \qquad \quad -5x} \quad \text{Subtract } 5x \text{ from both sides.}$$
$$0 \quad = \quad 0 \quad \text{True statement; no variable.}$$

The variable has "disappeared." It is true that $0 = 0$, but what is the solution? Actually there are an *infinite* number of solutions because *any real number* is a solution. Let's try a few different possibilities.

Try 7 as a solution. Replace each x with 7.	Try -2 as a solution. Replace each x with -2.	Try $\frac{1}{2}$ as a solution. Replace each x with $\frac{1}{2}$.
$5x - 15 = 5(x - 3)$	$5x - 15 = 5(x - 3)$	$5x - 15 = 5(x - 3)$
$5 \cdot 7 - 15 = 5(7 - 3)$	$5(-2) - 15 = 5(-2 - 3)$	$5(\frac{1}{2}) - 15 = 5(\frac{1}{2} - 3)$
$35 - 15 = 5(4)$	$-10 - 15 = 5(-5)$	$\frac{5}{2} - 15 = 5(-\frac{5}{2})$
$20 = 20$	$-25 = -25$	$-\frac{25}{2} = -\frac{25}{2}$
Balances	Balances	Balances

Any real number you tried would balance. Let's see why that is so. Look again at the first step in solving the equation. After using the distributive property to remove the parentheses, both sides are exactly the same.

$$5x - 15 = 5(x - 3) \quad \text{Use the distributive property.}$$
$$5x - 15 = 5x - 15 \quad \text{Both sides are exactly the same.}$$

When both sides of an equation are exactly the same, the equation is called an **identity**. An identity is true for all replacements of the variables.

The solution for $5x - 15 = 5(x - 3)$ is **all real numbers**.

> **CAUTION**
> When solving an equation like the one in Example 4, do **not** write "0" as the solution. While 0 is *one* of the solutions, there are *infinitely many other solutions*. So, you need to write "the solution is all real numbers."

▸▸▸ **Work Problem 4 at the Side.**

ANSWERS
4. (a) all real numbers
   (b) all real numbers

Section 10.2 More on Solving Linear Equations   **697**

EXAMPLE 5 Solving an Equation That Has No Solution

Solve: $2w - 7(w + 1) = -5w + 4$

$$2w - 7(w + 1) = -5w + 4 \quad \text{Use the distributive property.}$$

$$2w - 7w - 7 = -5w + 4 \quad \text{Combine like terms.}$$

$$-5w - 7 = -5w + 4$$

$$\underline{+5w \qquad\quad +5w} \quad \text{Add } 5w \text{ to both sides.}$$

$$-7 = 4 \quad \text{False statement; no variable.}$$

As in Example 4, the variable has "disappeared." This time, however, we are left with a *false* statement: $-7 \neq 4$. Whenever this happens in solving an equation, it is a signal that the equation has *no solution*. So, you write "no solution."

Work Problem 5 at the Side.

OBJECTIVE 3 Solve equations by first clearing fractions and decimals. In **Section 4.7** we solved equations containing fractions such as $\frac{3}{4}x = 12$. We multiplied both sides by $\frac{4}{3}$ (the reciprocal of $\frac{3}{4}$).

$$\frac{\cancel{4}}{\cancel{3}}\left(\frac{\cancel{3}}{\cancel{4}}x\right) = 12\left(\frac{4}{3}\right)$$

$$x = 16$$

Another method is to first clear the equation of fractions. You can do that by multiplying both sides by the LCD (lowest common denominator) of all the fractions in the equation. (See **Section 4.4** for ways to find the LCD.)

EXAMPLE 6 Clearing an Equation of Fractions

Solve: $\frac{2}{3}x - \frac{1}{2}x = -\frac{1}{6}x + 2$

Remember that 2 can be written as $\frac{2}{1}$. The least common denominator for $\frac{2}{3}, \frac{1}{2}, -\frac{1}{6},$ and $\frac{2}{1}$ is 6. Start by multiplying both sides of the equation by 6.

$$\frac{2}{3}x - \frac{1}{2}x = -\frac{1}{6}x + 2$$

$$6\left(\frac{2}{3}x - \frac{1}{2}x\right) = 6\left(-\frac{1}{6}x + 2\right) \quad \text{Multiply both sides by 6 (the LCD).}$$

$$\cancel{6}\left(\frac{2}{\cancel{3}}x\right) - \cancel{6}\left(\frac{1}{\cancel{2}}x\right) = \cancel{6}\left(-\frac{1}{\cancel{6}}x\right) + 6(2) \quad \text{Use the distributive property to multiply each term by 6.}$$

$$4x - 3x = -1x + 12 \quad \text{Combine like terms.}$$

$$1x = -1x + 12$$

$$\underline{+1x \qquad +1x} \quad \text{Add } 1x \text{ to both sides.}$$

$$\frac{2x}{2} = \frac{12}{2} \quad \text{Divide both sides by 2.}$$

$$x = 6$$

The solution is 6. Check the solution by going back to the *original* equation and replacing each x with 6.

5 Solve each equation.

(a) $9y - 4 = 3y + 6(y + 1)$

(b) $2m - 5(m + 2) = -3(m - 5)$

ANSWERS
5. (a) $-4 \neq 6$, so no solution
   (b) $-10 \neq 15$, so no solution

6 Solve: $\frac{1}{4}x - 4 = \frac{3}{2}x + \frac{3}{4}x$

> **CAUTION**
> When clearing an equation of fractions, be careful to multiply *every* term on both sides by the LCD.

▸◀◀ **Work Problem 6 at the Side.**

You can also clear an equation of decimals by multiplying both sides by a power of 10.

EXAMPLE 7 Clearing an Equation of Decimals

Solve: $0.1t + 0.05(20 - t) = 0.09(20)$

Our goal is to multiply by some number that will change the coefficients from decimals to whole numbers. Since decimal numbers involve tenths, hundredths, thousandths, and so on, we will multiply by 10, or by 100, or by 1000, and so on. In this example there are at most two decimal places (hundredths) so we multiply by 100. A quick way to multiply a number by 100 is to move the decimal point two places to the right.

$0.10t + 0.05(20 - t) = 0.09(20)$ To multiply by 100, move the decimal point two places to the right.

$10t + 5(20 - t) = 9(20)$ Use the distributive property.

$10t + 100 - 5t = 180$ Combine like terms.

$5t + 100 = 180$

$\underline{-100 -100}$ Subtract 100 from both sides.

$\dfrac{5t}{5} = \dfrac{80}{5}$ Divide both sides by 5.

$t = 16$

The solution is 16. Check the solution by going back to the *original* equation and replacing each t with 16.

▸◀◀ **Work Problem 7 at the Side.**

7 Solve:

$0.06(100 - y) + 0.4y = 0.05(86)$

ANSWERS
6. $x = -2$
7. $y = -5$

10.2 Exercises

Solve each equation. See Examples 1–3.

1. $5m + 8 = 7 + 4m$
2. $2(2r + 1) = 3r - 6$
3. $10p + 6 = 4(3p - 1)$

4. $-5x + 8 = -3x + 10$
5. $7r - 5r + 2 = 5r - r$
6. $9p - 4p + 6 = 7p - 3p$

7. $x + 3 = -(2x + 2)$
8. $2x + 1 = -(x + 3)$

9. $4(2x - 1) = -6(x + 3)$
10. $6(3w + 5) = -2(-10w + 10)$

11. $(5y + 6) - (3 + 4y) = 10$
12. $(8r - 3) - (7r + 1) = -6$

13. $2(p + 5) - (9 + p) = -3$
14. $4(k - 6) - (3k + 2) = -5$

15. $-6(2b + 1) + (13b - 7) = 0$
16. $-5(3w - 3) + (1 + 16w) = 0$

17. $-2(8p + 2) - 3(2 - 7p) = 2(4 + 2p)$
18. $-5(1 - 2z) + 4(3 - z) = 7(3 + z)$

19. $4(7x - 1) + 3(2 - 5x) = 4(3x + 5) - 6$
20. $9(2m - 3) - 4(5 + 3m) = 5(4 + m) - 3$

Solve each equation. See Examples 4 and 5.

21. $3x + 9 = 3x + 8$

22. $-2x + 5 = -2x$

23. $8x + 1 = 1 + 8x$

24. $4w - 5 = -5 + 4w$

25. $6x + 5 - 7x + 3 = 5x - 6x - 4$

26. $4x - 3 - 8x + 1 = 5x - 9x + 7$

27. $6(4x - 1) = 12(2x + 3)$

28. $6(2x + 8) = 4(3x - 6)$

29. $3(2x - 4) = 6(x - 2)$

30. $3(6 - 4x) = 2(-6x + 9)$

31. $10(-2x + 1) = -14(x + 2) + 38 - 6x$

32. $2(2 - 3r) = 5(1 - r) - r - 1$

33. After working correctly through several steps of the solution of a linear equation, a student obtains the equation $7x = 3x$. Then the student divides both sides by x to get $7 = 3$ and gives "no solution" as the answer. Is this correct? If not, explain why.

34. If the final step in solving a linear equation leads to the statement $0 = 0$, explain why it is incorrect to say that 0 is the solution of the equation. What are the solutions of the equation?

Solve each equation by first clearing it of fractions or decimals. See Examples 6 and 7.

35. $\dfrac{3}{5}t - \dfrac{1}{10}t = t - \dfrac{5}{2}$

36. $-\dfrac{2}{7}r + 2r = \dfrac{1}{2}r + \dfrac{17}{2}$

37. $-\dfrac{1}{4}(x - 12) + \dfrac{1}{2}(x + 2) = x + 4$

38. $\dfrac{1}{9}(y + 18) + \dfrac{1}{3}(2y + 3) = y + 3$

39. $\dfrac{2}{3}k - \left(k + \dfrac{1}{4}\right) = \dfrac{1}{12}(k + 4)$

40. $-\dfrac{5}{6}q - \left(q - \dfrac{1}{2}\right) = \dfrac{1}{4}(q + 1)$

41. $5.6x + 2 = 4.6x$

42. $9.1x - 5 = 8.1x$

43. $5.2q - 4.6 - 7.1q = -2.1 - 1.9q - 2.5$

44. $-4.0x + 2.7 - 1.6x = 1.3 - 5.6x + 1.4$

45. $0.2(60) + 0.05x = 0.1(60 + x)$

46. $0.3(30) + 0.15x = 0.2(30 + x)$

47. $x + 0.05(12 - x) = 0.1(63)$

48. $0.92x + 0.98(12 - x) = 0.96(12)$

49. $0.06(10{,}000) + 0.08x = 0.072(10{,}000 + x)$

50. $0.02(5000) + 0.03x = 0.025(5000 + x)$

Solve each equation, and check your solution. See Examples 1–7.

51. $9(v + 1) - 3v = 2(3v + 1) - 8$

52. $-3(5z + 24) + 2 = 2(3 - 2z) - 4$

53. $\dfrac{1}{2}(x + 2) + \dfrac{3}{4}(x + 4) = x + 5$

54. $-(6k - 5) - (-5k + 8) = -3$

55. $-(4y + 2) - (-3y - 5) = 3$

56. $\dfrac{1}{3}(x + 3) + \dfrac{1}{6}(x - 6) = x + 3$

57. $0.10(x + 80) + 0.20x = 14$

58. $0.30(x + 15) + 0.40(x + 25) = 25$

59. $4(x + 8) = 2(2x + 6) + 20$

60. $4(x + 3) = 2(2x + 8) - 4$

61. $-2(2s - 4) - 8 = -3(4s + 4) - 1$

62. $8(t - 3) + 4t = 6(2t + 1) - 10$

10.3 Formulas and Solving for a Specified Variable

Many application problems can be solved with a formula. For example, in Chapters 3–8 you used formulas for perimeter and area of geometric figures, for money earned on bank savings, and for converting between Celsius and Fahrenheit temperatures. The formulas used in this book are shown on the inside back cover.

OBJECTIVES

1. Solve a formula for one variable, given the values of the other variables.
2. Solve a formula for a specified variable.

OBJECTIVE 1 Solve a formula for one variable given the values of the other variables. Given the values of all but one of the variables in a formula, we can find the value of the remaining variable by using the methods introduced in Chapter 3.

EXAMPLE 1 Using a Formula to Evaluate a Variable

Find the value of the remaining variable.

(a) $A = lw$; $A = 64$, $l = 10$

Recall that this formula gives the area of a rectangle with length l and width w. Substitute the given values into the formula and then solve for w.

$$A = lw$$
$$64 = 10w \quad \text{Replace } A \text{ with 64 and } l \text{ with 10.}$$
$$\frac{64}{10} = \frac{10w}{10} \quad \text{Divide both sides by 10.}$$
$$6.4 = w$$

Check that the width of the rectangle is 6.4.

Check:

$$A = l \cdot w$$
$$64 = 10\,(6.4)$$
$$64 = 64 \quad \text{Balances}$$

(b) $A = \frac{1}{2}h(b + B)$; $A = 210$, $B = 27$, $h = 10$

This formula gives the area of a *trapezoid* with parallel sides of lengths b and B and height h between the parallel sides.

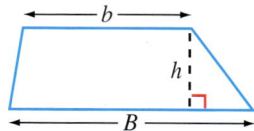

Again, begin by substituting the given values into the formula.

$$A = \frac{1}{2}h(b + B)$$

$$210 = \frac{1}{2}(10)(b + 27) \quad \text{Replace } A \text{ with 210, } h \text{ with 10, and } B \text{ with 27.}$$

Continued on Next Page

① Find the value of the remaining variable in each formula.

(a) $I = prt$; $I = \$246$, $r = 0.06$, $t = 2$

Now solve for b.

$$210 = \frac{1}{2}(10)(b + 27) \quad \text{Multiply } \frac{1}{2}(10) \text{ to get 5.}$$
$$210 = 5(b + 27) \quad \text{Use the distributive property.}$$
$$210 = 5b + 135$$
$$\underline{-135 \quad\quad -135} \quad \text{Subtract 135 from both sides.}$$
$$75 = 5b$$
$$\frac{75}{5} = \frac{5b}{5} \quad \text{Divide both sides by 5.}$$
$$15 = b$$

Check that the length of the shorter parallel side, b, is 15.

◀◀ **Work Problem 1 at the Side.**

(b) $P = 2l + 2w$; $P = 126$, $w = 25$

Formulas are often used to solve application problems. *It is a good idea to draw a sketch when a problem involves a geometric figure.*

EXAMPLE 2 Finding the Width of a Rectangular Lot

A rectangular garden plot has a perimeter of 8 meters and a length of 2.5 meters. Find the width of the lot.

Begin by drawing a rectangle and labeling the sides.

We chose w to represent the width of the garden plot in meters.

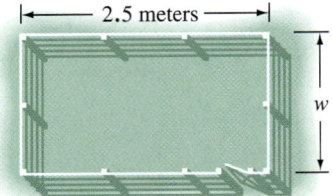

The formula for the perimeter of a rectangle is $P = 2l + 2w$.

Find the width by substituting 8 for P and 2.5 for l in the formula.

$$8 = 2(2.5) + 2w \quad \text{Replace } P \text{ with 8 and } l \text{ with 2.5}$$

② A farmer has 800 meters of fencing material to enclose a rectangular field. The width of the field is 175 meters. Find the length of the field.

Now solve the equation.

$$8 = \underbrace{2(2.5)}_{} + 2w \quad \text{Simplify.}$$
$$8 = \quad 5 \quad + 2w$$
$$\underline{-5 \quad\quad -5} \quad \text{Subtract 5 from both sides.}$$
$$\frac{3}{2} = \frac{2w}{2} \quad \text{Divide both sides by 2.}$$
$$1.5 = w$$

The width is 1.5 meters.

Check: If the length is 2.5 meters and the width is 1.5 meters, then $P = 2(2.5) + 2(1.5) = 5 + 3 = 8$ meters. This matches the perimeter given in the problem.

◀◀ **Work Problem 2 at the Side.**

ANSWERS
1. (a) $p = \$2050$ (b) $l = 38$
2. $l = 225$ meters

OBJECTIVE 2 **Solve a formula for a specified variable.** Sometimes it is necessary to solve many problems that use the same formula. For example, a surveying class might need to solve several problems that involve the formula for the area of a rectangle, $A = lw$. Suppose that in each problem the area (A) and the length (l) of a rectangle are given, and the width (w) must be found. Rather than solving for w each time the formula is used, it would be simpler to rewrite the *formula* so that it is solved for w. This process is called *solving for a specified variable*.

In solving a formula for a specified variable, we treat the specified variable as if it were the *only* variable in the equation, and *treat the other variables as if they were numbers*. Then we use the same steps to solve the equation for the specified variable that we used earlier to solve equations with just one variable.

③ (a) Solve $I = prt$ for t.

EXAMPLE 3 Solving for a Specified Variable

Solve $A = lw$ for w.

Think of "undoing" what has been done to w. Since w is *multiplied* by l, undo the multiplication by *dividing* both sides by l. (Recall that division is the opposite of multiplication.)

$$A = lw$$

$$\frac{A}{l} = \frac{lw}{l} \qquad \text{Divide both sides by } l. \text{ On the right side, divide out the common factor of } l.$$

$$\frac{A}{l} = w$$

The formula is now solved for w.

(b) Solve $P = a + b + c$ for a.

> **NOTE**
> Look back at Example 1(a) in this section. It is a specific example where we knew that A was 64 and l was 10. Notice that we divided 64 by 10 to find w. In other words, we divided the area by the length to find the width. Or, $\frac{A}{l} = w$.

Work Problem 3 at the Side.

EXAMPLE 4 Solving for a Specified Variable

(a) Solve $P = 2l + 2w$ for l.

We want to get l by itself on one side of the equation. We begin by subtracting $2w$ from both sides.

$$P = 2l + 2w$$
$$\underline{-2w \qquad -2w} \qquad \text{Subtract } 2w \text{ from both sides.}$$
$$P - 2w = 2l$$

$$\frac{P - 2w}{2} = \frac{2l}{2} \qquad \text{Divide both sides by 2.}$$

$$\frac{P - 2w}{2} = l$$

The last step gives the formula solved for l. It is in simplest form, so you need not do anything else to it.

Continued on Next Page

ANSWERS

(a) $t = \dfrac{I}{pr}$ (b) $a = P - b - c$

4 **(a)** Solve $A = p + prt$ for t.

(b) Solve $F = \frac{9}{5}C + 32$ for C.

We need to get C by itself on one side of the equation.

First "undo" the *addition* of 32 to $\frac{9}{5}C$ by *subtracting* 32 from both sides.

$$F = \frac{9}{5}C + 32$$
$$\underline{-32 \qquad\qquad -32} \quad \text{Subtract 32 from both sides.}$$
$$F - 32 = \frac{9}{5}C$$

Now multiply both sides by $\frac{5}{9}$ (the reciprocal of $\frac{9}{5}$).

$$\frac{5}{9}(F - 32) = \frac{\cancel{5}}{\cancel{9}} \cdot \frac{\cancel{9}}{\cancel{5}}C \quad \text{Multiply both sides by } \tfrac{5}{9}.$$

$$\frac{5}{9}(F - 32) = C$$

The result is the formula for converting temperatures from Fahrenheit to Celsius that you used in **Section 8.5**.

◀◀ **Work Problem 4 at the Side.**

(b) Solve $Ax + By = C$ for y.

ANSWERS

(a) $t = \dfrac{A - p}{pr}$ **(b)** $y = \dfrac{C - Ax}{B}$

10.3 Exercises

Find the value of the remaining variable in each formula. See Example 1.

1. $P = 2l + 2w$ (perimeter of a rectangle)
   $P = 20, w = 4$

2. $P = 2l + 2w$
   $P = 26, l = 8$

3. $A = \frac{1}{2}bh$ (area of triangle); $A = 70, b = 10$

4. $A = \frac{1}{2}bh$; $A = 64, h = 16$

5. $P = a + b + c$ (perimeter of a triangle)
   $P = 15, a = 3, b = 7$

6. $P = a + b + c$
   $P = 12, a = 3, c = 5$

7. $d = rt$ (distance formula); $d = 100, t = 2.5$

8. $d = rt$; $d = 252, r = 45$

9. $I = prt$ (simple interest)
   $I = 875, p = 5000, r = 0.025$

10. $I = prt$
    $I = 1575, r = 0.035, t = 6$

11. $C = 2\pi r$ (circumference of a circle)
    $C = 8.164, \pi \approx 3.14$

12. $C = 2\pi r$; $C = 16.328; \pi \approx 3.14$

13. $V = lwh$ (volume of a rectangular solid)
    $V = 384, l = 12, h = 4$

14. $V = lwh$
    $V = 150, w = 5, h = 3$

Use a formula to write an equation for each problem, then solve it. See Example 2.

15. The newspaper *The Constellation,* printed in 1859 in New York City, had a page length of 51 inches and a perimeter of 172 inches. **(a)** What was the width of the page? **(b)** What was the area of the page?

16. The *Daily Banner,* published in Roseberg, Oregon, in the nineteenth century, had a page width of 3 inches and a perimeter of 13 inches. **(a)** What was the page length? **(b)** What was the area of the page?

17. The Skydome in Toronto, Canada, is the first stadium with a hard-shell, retractable roof. The steel dome has a circumference of 1978 feet. To the nearest foot, what is the diameter of this dome? Use 3.14 as the approximate value of π.

18. The largest drum ever constructed was built in Japan in 2001. The circular face of the drum had a circumference of 49.42 ft. What was the diameter of the drum face, to the nearest hundredth of a foot? Use 3.14 as the approximate value of π.

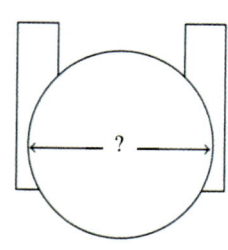

Solve the formula for the specified variable. See Examples 3–5.

19. $d = rt$ for r **20.** $d = rt$ for t **21.** $A = lw$ for l

22. $V = lwh$ for h **23.** $P = a + b + c$ for a **24.** $P = a + b + c$ for b

25. $I = prt$ for p **26.** $I = prt$ for r **27.** $A = \frac{1}{2}bh$ for b

28. $A = \frac{1}{2}bh$ for h **29.** $A = p + prt$ for r **30.** $P = 2l + 2w$ for w

31. $V = \pi r^2 h$ for h **32.** $V = \frac{1}{3}\pi r^2 h$ for h **33.** $F = \frac{9}{5}C + 32$ for C

RELATING CONCEPTS (EXERCISES 34–36) For Individual or Group Work

In many cases there are equally acceptable equivalent answers for a problem. **Work Exercises 34–36 in order** *to see how two seemingly different answers to a problem can both be correct.*

34. Solve the formula $P = 2l + 2w$ for w using the following steps.
  (a) Subtract $2l$ from both sides.
  (b) Divide both sides by 2.

35. Referring to the formula in Exercise 34, solve for w again using the following steps.
  (a) Divide each term on both sides by 2.
  (b) Subtract l from both sides.

36. Compare the results in Exercises 34(b) and 35(b). Both are acceptable answers. To show that they are equivalent, give a justification for each step below.

  (a) $\dfrac{P}{2} - l = \dfrac{P}{2} - \dfrac{2}{2} \cdot l$

  (b) $\phantom{\dfrac{P}{2} - l\ } = \dfrac{P}{2} - \dfrac{2}{2} \cdot \dfrac{l}{1}$

  (c) $\phantom{\dfrac{P}{2} - l\ } = \dfrac{P}{2} - \dfrac{2l}{2}$

  (d) $\phantom{\dfrac{P}{2} - l\ } = \dfrac{P - 2l}{2}$

10.4 Solving Linear Inequalities

The addition and multiplication properties can be extended to inequalities. **Inequalities** are statements with algebraic expressions related in the following ways:

$<$ "is less than"
$\leq$ "is less than or equal to"
$>$ "is greater than"
$\geq$ "is greater than or equal to."

We solve an inequality by finding all real number solutions for it. For example, the solutions of $x \leq 2$ include all *real numbers* that are less than or equal to 2, and not just the *integers* less than or equal to 2.

OBJECTIVES

1. Graph the solutions of inequalities on a number line.
2. Use the addition property of inequality.
3. Use the multiplication property of inequality.
4. Solve inequalities using both properties of inequality.
5. Use inequalities to solve application problems.

OBJECTIVE 1 Graph the solutions of inequalities on a number line. Graphing is a good way to show the solutions of an inequality. To graph all real numbers satisfying $x \leq 2$, we place a solid circle at 2 on a number line and draw an arrow extending from the circle to the left (to represent the fact that all numbers less than 2 are also part of the graph). The graph is shown below.

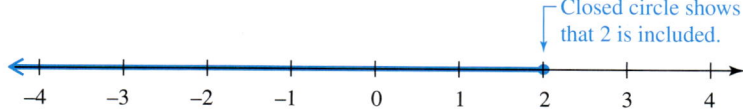

EXAMPLE 1 Graphing the Solutions of an Inequality

(a) Graph $x > -5$.

The statement $x > -5$ says that x can take any value greater than -5, but x cannot equal -5 itself. We show this on a graph by placing an *open* circle at -5 and drawing an arrow to the right. The open circle at -5 shows that -5 is *not* part of the graph.

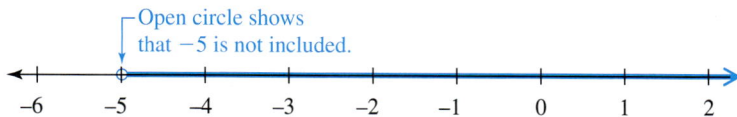

(b) Graph $3 > x$.

Recall from **Section 10.1** that $3 > x$ can be rewritten as $x < 3$. We do this by reversing the direction of the symbol from $>$ to $<$ and reversing the order of 3 and x. The graph of $x < 3$ is shown below.

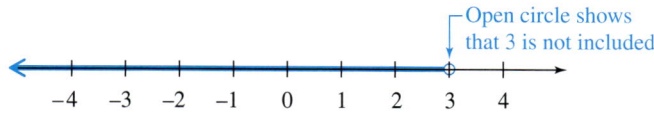

> **NOTE**
> To graph an inequality like the one in part (b) above, first rewrite it with the variable on the left. Fewer errors occur this way.

Work Problem 1 at the Side.

1 Graph each inequality.

(a) $x \leq 3$
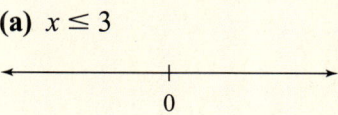

(b) $x > -4$
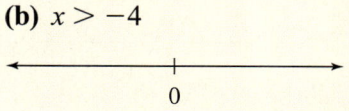

(c) $-4 \geq x$
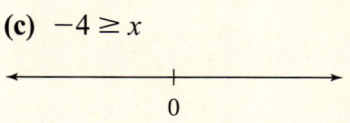

(d) $0 < x$
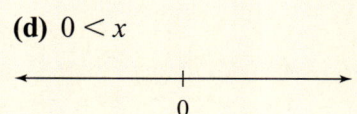

ANSWERS
1.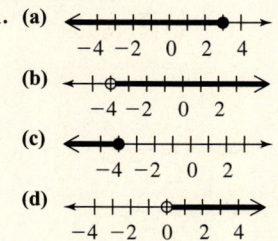

2 Graph each inequality.

(a) $-7 < x < -2$

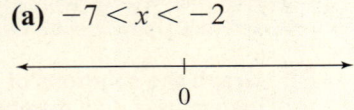

EXAMPLE 2 Graphing the Solutions of an Inequality

Graph $-3 \leq x < 2$.

The statement $-3 \leq x < 2$ is read "-3 is less than or equal to x **and** x is less than 2." We graph the solutions of this inequality by placing a *closed* circle at -3 (because -3 is part of the graph) and an *open* circle at 2 (because 2 is *not* part of the graph), then drawing a line segment between the two circles. Notice that the graph includes all points *between* -3 and 2, and includes -3 as well.

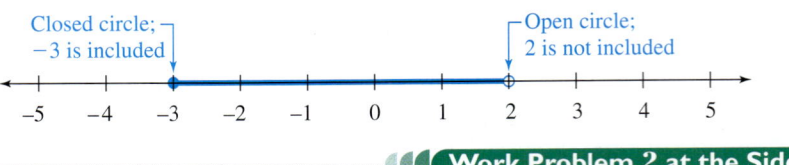

Work Problem 2 at the Side.

OBJECTIVE 2 Use the addition property of inequality. We can solve inequalities such as $x + 4 \leq 9$ in much the same way as equations.

> **Linear Inequality in One Variable**
>
> A **linear inequality in one variable** can be written in the form
>
> $$Ax + B < C,$$
>
> where A, B, and C are real numbers, with $A \neq 0$.

(b) $-6 < x \leq 4$

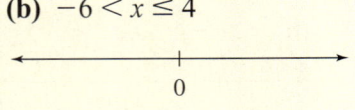

All definitions and rules are also valid for $>$, $\leq$, and $\geq$.

Examples of linear inequalities in one variable are shown below.

$$x + 5 < 2 \qquad t - 3 \geq 5 \qquad 2k + 5 \leq 10$$

Consider the inequality $2 < 5$. If 4 is added to both sides of this inequality, the result is still a true statement, as shown below.

$$2 + 4 < 5 + 4$$
$$6 < 9 \qquad \text{True}$$

Now try subtracting 8 from both sides.

$$2 - 8 < 5 - 8$$
$$-6 < -3 \qquad \text{True}$$

The result is again a true statement. These examples suggest the **addition property of inequality.**

> **Addition Property of Inequality**
>
> For any real numbers A, B, and C, the inequalities
>
> $$A < B \quad \text{and} \quad A + C < B + C$$
>
> have exactly the same solutions. In other words, the same number may be added to each side of an inequality without changing the solutions.

We can replace $<$ in the addition property of inequality with $>$, $\leq$, or $\geq$. Also, as with the addition property of equality, the same number may be *subtracted* on each side of an inequality.

Section 10.4 Solving Linear Inequalities **711**

EXAMPLE 3 Using the Addition Property of Inequality

Solve $7 + 3k > 2k - 5$.

$$7 + 3k > 2k - 5$$
$$\underline{-2k > -2k} \quad \text{Subtract } 2k \text{ from both sides.}$$
$$7 + k > -5$$
$$\underline{-7 > -7} \quad \text{Subtract 7 from both sides.}$$
$$k > -12$$

A graph of the solutions, $k > -12$, is shown below.

Work Problem 3 at the Side.

OBJECTIVE 3 **Use the multiplication property of inequality.** The addition property of inequality cannot be used to solve inequalities such as $4y \geq 28$. These inequalities require the *multiplication property of inequality*. To see how this property works, it is helpful to look at some examples.

First, write the inequality $3 < 7$ and then multiply both sides by the positive number 2.

$$3 < 7$$
$$\mathbf{2}(3) < \mathbf{2}(7) \quad \text{Multiply both sides by 2.}$$
$$6 < 14 \quad \text{True}$$

Now multiply both sides of $3 < 7$ by the negative number -5.

$$3 < 7$$
$$-\mathbf{5}(3) < -\mathbf{5}(7) \quad \text{Multiply both sides by } -5.$$
$$-15 < -35 \quad \text{False}$$

To get a true statement when multiplying both sides by -5, we must *reverse the direction of the inequality symbol*.

$$3 < 7$$
$$-\mathbf{5}(3) > -\mathbf{5}(7) \quad \text{Multiply by } -5 \text{ and reverse the symbol.}$$
$$-15 > -35 \quad \text{True}$$

Take the inequality $-6 < 2$ as another example. Multiply both sides by the positive number 4.

$$-6 < 2$$
$$\mathbf{4}(-6) < \mathbf{4}(2) \quad \text{Multiply both sides by 4.}$$
$$-24 < 8 \quad \text{True}$$

Multiply both sides of $-6 < 2$ by -5 *and at the same time reverse the direction of the inequality symbol.*

$$-6 < 2$$
$$-\mathbf{5}(-6) > -\mathbf{5}(2) \quad \text{Multiply by } -5 \text{ and reverse the symbol.}$$
$$30 > -10 \quad \text{True}$$

Work Problem 4 at the Side.

3 Solve each inequality, and graph the solutions.

(a) $-1 + 8r < 7r + 2$

(b) $4m \geq 3m - 1$

4 (a) Multiply both sides of $-2 < 8$ by 6 and then by -5. Reverse the direction of the inequality symbol when necessary to make a true statement.

(b) Multiply both sides of $-4 > -9$ by 2 and then by -8. Reverse the direction of the inequality symbol when necessary to make a true statement.

ANSWERS
3. (a) $r < 3$

(b) $m \geq -1$

4. (a) $-12 < 48$; $10 > -40$
(b) $-8 > -18$; $32 < 72$

In summary, the multiplication property of inequality has two parts.

> **Multiplication Property of Inequality**
> For any real numbers A, B, and C (where $C \neq 0$),
> 1. if C is *positive*, then the inequalities
>
> $$A < B \quad \text{and} \quad AC < BC$$
>
> have exactly the same solutions;
>
> 2. if C is *negative*, then the inequalities
>
> $$A < B \quad \text{and} \quad AC > BC$$
>
> have exactly the same solutions.
>
> In other words, both sides of an inequality may be multiplied by the same *positive* number without changing the solutions. *If the multiplier is negative, we must reverse the direction of the inequality symbol.*

We can replace $<$ in the multiplication property of inequality with $>$, $\leq$, or $\geq$. As with the multiplication property of equality, the same nonzero number may be *divided* into both sides.

> **CAUTION**
> It is important to remember the differences in the multiplication property of inequality for positive and negative numbers.
> 1. When both sides of an inequality are multiplied or divided by a positive number, the direction of the inequality symbol *does not change*. Adding or subtracting terms on both sides also *does not change* the symbol.
> 2. When both sides of an inequality are multiplied or divided by a negative number, the direction of the symbol *does change*. **Reverse the direction of the symbol of inequality only when multiplying or dividing by a negative number.**

EXAMPLE 4   Using the Multiplication Property of Inequality

Solve $3r < -18$.

Using the multiplication property of inequality, we divide both sides by 3. Since 3 is a *positive* number, the direction of the inequality symbol *does not* change. **It does not matter that the number on the right side of the inequality is negative.**

$$3r < -18$$

$$\frac{3r}{3} < \frac{-18}{3} \quad \text{Divide by 3, a }\textit{positive}\text{ number.}$$

$$r < -6 \quad \text{The direction of the inequality symbol }\textit{does not}\text{ change.}$$

The graph of the solutions is shown below.

EXAMPLE 5 Using the Multiplication Property of Inequality

Solve $-4t \geq 8$.

Here both sides of the inequality must be divided by -4, a *negative* number, which *does* change the direction of the inequality symbol.

$$-4t \geq 8$$

$$\frac{-4t}{-4} \leq \frac{8}{-4} \quad \text{Divide by } -4, \text{ a } \textit{negative} \text{ number, so } \textit{reverse} \text{ the direction of the symbol.}$$

$$t \leq -2$$

The solutions are graphed below.

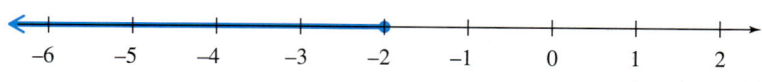

Work Problem 5 at the Side.

OBJECTIVE 4 Solve inequalities using both properties of inequality.
The steps in solving an inequality are summarized below. (Remember that $<$ can be replaced with $>$, $\leq$, or $\geq$ in this summary.)

Solving Inequalities

Step 1 Simplify each side separately. If possible, use the distributive property to remove any parentheses. Then combine like terms on each side.

Step 2 Use the addition property. Add or subtract the same amount on both sides of the inequality so that the variable term ends up by itself on one side of the inequality sign and a number is by itself on the other side. You may have to do this step more than once.

Step 3 Use the multiplication property to write the inequality in the form $x < c$ or $x > c$.

EXAMPLE 6 Solving an Inequality

Solve $5(k-3) - 7k \geq 4(k-3) + 9$.

Step 1 Use the distributive property to remove the parentheses; then combine like terms.

$$5(k-3) - 7k \geq 4(k-3) + 9 \quad \text{Use distributive property.}$$
$$5k - 15 - 7k \geq 4k - 12 + 9 \quad \text{Combine like terms.}$$
$$-2k - 15 \geq 4k - 3$$

Step 2 Use the addition property.

$$\begin{array}{rl} -2k - 15 \geq & 4k - 3 \\ \underline{-4k} & \underline{-4k} \quad \text{Subtract } 4k \text{ from both sides.} \\ -6k - 15 \geq & -3 \\ \underline{+15} & \underline{+15} \quad \text{Add 15 to both sides.} \\ -6k \quad\;\; \geq & 12 \end{array}$$

Continued on Next Page

---

5 Solve each inequality. Graph the solutions.

(a) $9y < -18$

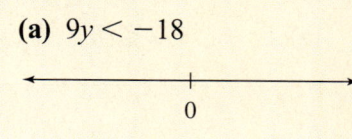

(b) $-2r > -12$

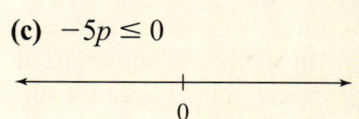

(c) $-5p \leq 0$

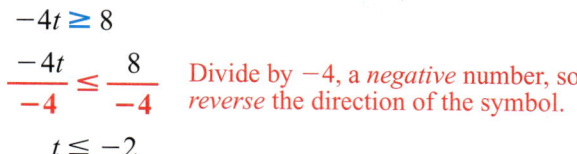

ANSWERS

5. (a) $y < -2$

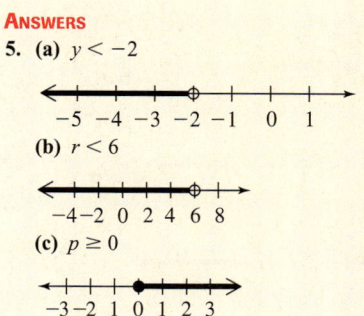

(b) $r < 6$

(c) $p \geq 0$

6 Solve each inequality and graph the solutions.

(a) $5r - r + 2 < 7r - 5$

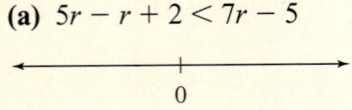

Step 3 Divide both sides by -6, a *negative* number, and *reverse the direction* of the inequality symbol.

$$\frac{-6k}{-6} \le \frac{12}{-6} \quad \text{Divide by } -6 \text{ and reverse the symbol.}$$

$$k \le -2$$

A graph of the solutions is shown below.

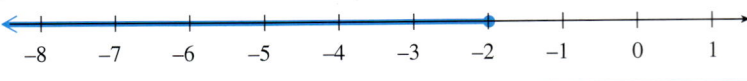

>>> **Work Problem 6 at the Side.**

OBJECTIVE 5 Use inequalities to solve application problems. Inequalities can be used to solve application problems involving phrases that suggest inequality. The chart below gives some of the more common phrases, along with examples and translations.

(b) $4(y - 1) - 3y > -15 - (2y + 1)$

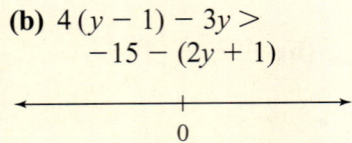

Phrase	Example	Inequality
Is greater than	A number *is greater than* 4	$x > 4$
Is less than	A number *is less than* -12	$x < -12$
Is at least	A number *is at least* 6	$x \ge 6$
Is at most	A number *is at most* 8	$x \le 8$

CAUTION
Do not confuse phrases like "5 less than a number" and statements like "5 *is* less than a number." The first of these is expressed as "$x - 5$" while the second is expressed at "$5 < x$."

7 Maggie has scores of 98, 86, and 88 on her first three tests in algebra. If she wants an average of at least 90 after her fourth test, what score must she make on her fourth test?

EXAMPLE 7 Finding an Average Test Score

Brent has scores of 86, 88, and 78 on his first three tests in geometry. If he wants an average of at least 80 after his fourth test, what score must he make on his fourth test?

Let x represent Brent's score on his fourth test. To find the average of the four scores, add them and divide the sum by 4.

$$\underbrace{\frac{86 + 88 + 78 + x}{4}}_{\text{Average}} \overset{\overset{\text{is at}}{\text{least 80.}}}{\ge} 80$$

$$4\left(\frac{252 + x}{4}\right) \ge 4(80) \quad \text{Multiply both sides by 4.}$$

$$252 + x \ge 320$$
$$-252 \qquad -252 \quad \text{Subtract 252 from both sides.}$$
$$x \ge 68$$

He must score 68 or more on the fourth test to have an average of *at least* 80.

>>> **Work Problem 7 at the Side.**

ANSWERS

6. (a) $r > \frac{7}{3}$

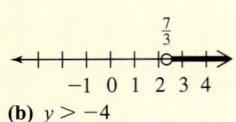

(b) $y > -4$

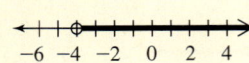

7. 88 or more

10.4 Exercises

1. Explain how you can determine whether to use an open circle or a closed circle at an endpoint when graphing an inequality on a number line.

2. Explain how the graph of $t \geq -7$ differs from the graph of $t > -7$.

3. Explain why we can't list the solutions of an inequality such as $x < 2$.

4. What inequality symbol is used to express the statement "y is at least 7"?

Graph each inequality on the given number line. See Examples 1 and 2.

5. $k \leq 4$

6. $r \leq -11$

7. $x < -3$

8. $y < 3$

9. $8 \leq x \leq 10$

10. $3 \leq x \leq 5$

11. $0 < y \leq 10$

12. $-3 \leq x < 5$

13. Why is it *wrong* to write $3 < x < -2$ to indicate that x is between -2 and 3?

14. If $p < q$ and $r < 0$, which one of the following statements is false?

   (a) $pr < qr$
   (b) $pr > qr$
   (c) $p + r < q + r$
   (d) $p - r < q - r$

Solve each inequality and graph the solutions. See Example 3.

15. $z - 8 \geq -7$

16. $p - 3 \geq -11$

17. $2k + 3 \geq k + 8$

18. $3x + 7 \geq 2x + 11$

19. $3n + 5 < 2n - 6$

20. $5x - 2 < 4x - 5$

21. Under what conditions must the inequality symbol be reversed when using the multiplication property of inequality?

22. Explain the steps you would use to solve the inequality $-5x > 20$.

23. Your friend tells you that when solving the inequality $6x < -42$, he reversed the direction of the inequality because of the presence of -42. How would you respond?

24. By what number must you *multiply* both sides of $0.2x > 6$ to get just x on the left side?

Solve each inequality and graph the solutions. See Examples 4 and 5.

25. $3x < 18$

26. $5x < 35$

27. $2y \geq -20$

28. $6m \geq -24$

29. $-8t > 24$

30. $-7x > 49$

31. $-x \geq 0$

32. $-k < 0$

33. $-\dfrac{3}{4}r < -15$

34. $-\dfrac{7}{8}t < -14$

35. $-0.02x \leq 0.06$

36. $-0.03v \geq -0.12$

Solve each inequality and graph the solutions. See Example 6.

37. $5r + 1 \geq 3r - 9$

38. $6t + 3 < 3t + 12$

39. $6x + 3 + x < 2 + 4x + 4$

40. $-4w + 12 + 9w \geq w + 9 + w$

41. $-x + 4 + 7x \leq -2 + 3x + 6$

42. $14y - 6 + 7y > 4 + 10y - 10$

43. $5(x + 3) - 6x \leq 3(2x + 1) - 4x$

44. $2(x - 5) + 3x < 4(x - 6) + 1$

45. $\frac{2}{3}(p + 3) > \frac{5}{6}(p - 4)$

46. $\frac{7}{9}(y - 5) \leq \frac{4}{3}(y + 5)$

47. $4x - (6x + 1) \leq 8x + 2(x - 3)$

48. $2y - (4y + 3) < 6y + 3(y + 4)$

49. $5(2k + 3) - 2(k - 8) > 3(2k + 4) + k - 2$

50. $2(3z - 5) + 4(z + 6) \geq 2(3z + 2) + 3z - 15$

Solve each application of inequalities. See Example 7.

51. When 8 is subtracted from the sum of three times a number and 6, the result is less than 4 more than the number. Find all such numbers.

52. When 2 is added to the difference between six times a number and 5, the result is greater than 13 added to 5 times the number. Find all such numbers.

53. Twylene Johnson has scores of 89, 78, 73, and 81 on her first four algebra tests. If she wants an average of at least 80 after her fifth test, what score must she make on her fifth test?

54. Mabimi Pampo has scores of 96 and 86 on his first two geometry tests. What must he score on his third test so that his average is at least 90?

55. The formula for converting Fahrenheit temperature to Celsius is shown below.

$$C = \frac{5}{9}(F - 32)$$

If the Celsius temperature on a certain summer day in Houston is never more than 30 degrees, how would you describe the corresponding Fahrenheit temperatures?

56. The formula for converting Celsius temperature to Fahrenheit is shown below.

$$F = \frac{9}{5}C + 32$$

The Fahrenheit temperature of Key West, Florida, has never exceeded 95 degrees. How would you describe this using Celsius temperatures?

57. For what values of x would the rectangle have perimeter of at least 400?

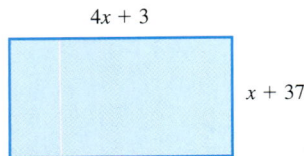

58. For what values of x would the triangle have perimeter of at least 72?

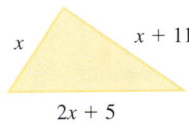

59. Audrey earned $200 at odd jobs during July, $300 during August, and $225 during September. If her average salary for the four months from July through October is to be at least $250, how much must she earn during October?

60. In order to qualify for a company pension plan, an employee must average at least $1000 per month in earnings. During the first four months of the year, an employee made $900, $1200, $1040, and $760. What amount of earnings during the fifth month will qualify the employee for the pension plan?

61. A long-distance phone call costs $2.00 for the first three minutes plus $0.30 per minute for each minute or fractional part of a minute after the first three minutes. If x represents the number of minutes of the length of the call after the first three minutes, then $2 + 0.30x$ represents the cost of the call. If Jorge has $5.60 to spend on a call, what is the maximum total time he can use the phone?

62. At the Speedy Gas 'n Go, a car wash costs $5.00, and gasoline is selling for $2.50 per gal. Terri Hoelker has $28.75 to spend, and her car is so dirty that she must have it washed. What is the maximum number of gallons of gasoline that she can purchase?

RELATING CONCEPTS (EXERCISES 63–67) For Individual or Group Work

Work Exercises 63–67 in order, to see the connection between the solution of an equation and the solutions of the corresponding inequalities. Graph the solutions in Exercises 63–65.

63. $3x + 2 = 14$

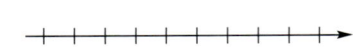

64. $3x + 2 < 14$

65. $3x + 2 > 14$

66. Now graph all the solutions together on the following number line.

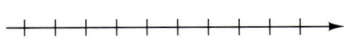

How would you describe the graph?

67. Based on your results from Exercises 63–66, if you were to graph the solutions of

$$-4x + 3 = -1, \quad -4x + 3 > -1,$$
$$\text{and} \quad -4x + 3 < -1$$

on the same number line, what do you think the graph would be?

Chapter 10
SUMMARY

KEY TERMS

10.1 **rational numbers** — Rational numbers can be written as quotients of two integers, with denominator not 0.

irrational numbers — Irrational numbers are nonrational numbers represented by points on the number line. The decimal form of an irrational number does not terminate (end) and does not repeat in a fixed block of digits.

real numbers — Real numbers include all numbers that can be represented by points on the number line, that is, all rational and irrational numbers.

10.2 **identity** — An identity is an equation that is true for all replacements of the variable. The solution of an identity is "all real numbers."

10.4 **inequality** — An inequality is a statement with algebraic expressions related by $<$, $\leq$, $>$, or $\geq$.

linear inequality in one variable — A linear inequality in one variable can be written in the form $Ax + B < C$, $Ax + B \leq C$, $Ax + B > C$, or $Ax + B \geq C$, where A, B, and C are real numbers, with $A \neq 0$.

NEW SYMBOLS

$\neq$ "is not equal to" $\leq$ "is less than or equal to" $\geq$ "is greater than or equal to"

TEST YOUR WORD POWER

See how well you have learned the vocabulary in this chapter. Answers, with examples, follow the Quick Review.

1. Which number is the decimal form of an **irrational number**?
   A. 0.63636363…
   B. −0.524
   C. 0.02022022202222…
   D. $\sqrt{0.16}$

2. All **real numbers** are
   A. quotients of two integers, with the denominator not equal to zero.
   B. rational
   C. decimals that repeat in a fixed block
   D. represented by points on the number line.

3. A **rational number** can be written as
   A. the quotient of two integers, with the denominator not equal to zero
   B. a decimal that does not repeat in a fixed block
   C. a decimal that does not terminate
   D. the product of two integers, with neither factor equal to zero.

4. An **inequality** is
   A. a statement that two algebraic expressions are equal
   B. a point on a number line
   C. an equation with no solutions
   D. a statement with algebraic expressions related by $<$, $\leq$, $>$, or $\geq$.

5. An **identity** is an equation that
   A. has no solution
   B. has all real numbers as solutions
   C. cannot be solved
   D. has 0 as the only solution.

6. Which statement is a **linear inequality**?
   A. $2x^2 + 6 > 9$
   B. $0x + 25 \geq -1$
   C. $3x + 10 \leq -5$
   D. $7x + 1 = 15$

QUICK REVIEW

Concepts

Examples

10.1 Using the Symbols $\neq$, $\leq$, $\geq$

Read the inequality $-2 \leq 0$ as "negative 2 is less than or equal to 0." If either the $<$ part or the $=$ part is true, then the inequality is true. Use the symbol $\geq$ in a similar way to mean "is greater than or equal to."

Tell whether each statement is true or false.

(a) $-3 \leq -1$ (b) $6.7 \geq 6.7$

(a) True because $-3 < -1$, that is, -3 is to the left of -1 on the number line.

(b) True because $6.7 = 6.7$.

To convert between $<$ and $>$ (or between $\leq$ and $\geq$) reverse both the order of the numbers and the direction of the symbol.

Rewrite each statement with the inequality symbol reversed.

(a) $0.1 > 0.01$

$0.01 < 0.1$
Exchange numbers

(b) $\frac{1}{2} \leq 0.7$

$0.7 \geq \frac{1}{2}$
Exchange numbers

10.1 Using Brackets

Brackets are used instead of double parentheses. Use the order of operations to complete all work inside the brackets.

Simplify.

$-3[-4 + 6(-3 + 5)]$ Work inside parentheses.
$-3[-4 + 6(2)]$ Multiply 6(2).
$-3[-4 + 12]$ Add $-4 + 12$.
$-3[8]$ Multiply $-3 \cdot 8$.
-24

10.1 Using the Distributive Property to Remove Parentheses

Multiply *every term* inside the parentheses by the number in front of the parentheses.

Watch the signs carefully if there is a *negative* number in front of the parentheses; the sign of every term inside the parentheses will change to its opposite.

Simplify each expression.

(a) $-(4a - 6)$ is written $-1(4a - 6)$

$-4a + 6$ ⟵ Simplified expression.

(b) $-6(3 + x) - 2(4x + 5)$ Use distributive property

$-18 - 6x - 8x - 10$ Combine like terms.

$-14x - 28$ ⟵ Simplified expression.

Concepts	Examples

10.2 *Solve Equations That Have No Solution or Infinitely Many Solutions*
 If, when solving an equation, the result is an *identity* (both sides of the equation are exactly the same), then there are an infinite number of solutions. Write "all real numbers" as the solution because *any* real number will work.
	Solve each equation.
	 (a) $3(a - 4) = -12 + 3a$ Use distributive property.
	     $3a - 12 = -12 + 3a$
	     $\underline{+12 \quad\quad\quad\quad +12}$ Add 12 to both sides.
	     $3a \quad = \quad 3a$
	     $\underline{-3a \quad\quad\quad -3a}$ Subtract $3a$ from both sides.
	     $0 \quad = \quad 0$ Identity
	 The solution is *all real numbers*.

If, when solving an equation, the result is a *false* statement, such as $-7 = 4$, it is a signal that the equation has *no* solution.	(b) $x - 6(x + 1) = 4 - 5x$ Use distributive property.
	     $x - 6x - 6 = 4 - 5x$ Combine like terms.
	     $-5x - 6 = 4 - 5x$
	     $\underline{+5x \quad\quad\quad\quad +5x}$ Add $5x$ to both sides.
	     $-6 = 4$ False statement
	 There is *no solution*.

10.2 *Clearing an Equation of Fractions or Decimals* To clear an equation of fractions, multiply *every term* on both sides of the equation by the lowest common denominator (LCD) of all the fractions. To clear an equation of decimals, multiply *every term* on both sides of the equation by a power of 10 that will change the coefficients from decimals to whole numbers.	Solve. $-\dfrac{5}{6}b = \dfrac{2}{3}b - \dfrac{1}{2}$ $6\left(-\dfrac{5}{6}b\right) = 6\left(\dfrac{2}{3}b - \dfrac{1}{2}\right)$ Multiply both sides by 6 (the LCD). $\overset{1}{\cancel{6}}\left(-\dfrac{5}{\cancel{6}}b\right) = \overset{2}{\cancel{6}}\left(\dfrac{2}{\cancel{3}}b\right) - \overset{3}{\cancel{6}}\left(\dfrac{1}{\cancel{2}}\right)$ Use the distributive property to multiply every term by 6. $-5b = 4b - 3$ $\underline{-4b \quad\quad -4b}$ Subtract $4b$ from both sides. $\dfrac{-9b}{-9} = \dfrac{-3}{-9}$ Divide both sides by -9. $b = \dfrac{1}{3}$

10.3 *Solving Formulas*
 To find the value of one of the variables in a formula, given values for the others, substitute the known values into the formula.
	Find l if $A = lw$, given that $A = 24$ and $w = 3$.
	     $24 = l \cdot 3$ Replace A with 24 and w with 3
	     $\dfrac{24}{3} = \dfrac{l \cdot 3}{3}$ Divide both sides by 3.
	     $8 = l$

Concepts	Examples
10.3 Solving for a Specified Variable To solve a formula for one of the variables, isolate that variable by treating the other variables as numbers, and then using the steps for solving equations.	Solve $P = 2l + 2w$ for w. $P = 2l + 2w$ $\underline{-2l \quad -2l}$ Subtract $2l$ from both sides. $P - 2l = \quad\quad 2w$ $\dfrac{P - 2l}{2} = \dfrac{2w}{2}$ Divide both sides by 2. $\dfrac{P - 2l}{2} = w$ ← Formula solved for w.
10.4 Graphing the Solutions of an Inequality On the number line, use a closed circle to indicate that a number is part of the graph. Use an open circle to indicate that a number is *not* part of the graph.	(a) Graph $x \geq -2$. 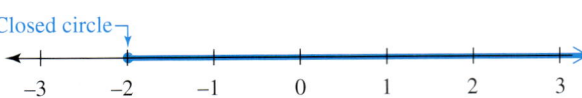 (b) Graph $0 \leq x < 4$ 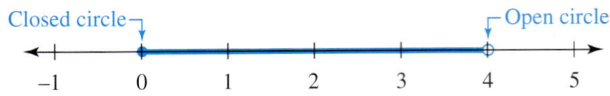
10.4 Solving Linear Inequalities To solve an inequality: 1. Simplify each side separately: clear parentheses and combine like terms. 2. Add or subtract the same number on both sides to get the variable term by itself on one side and a number on the other side. 3. Multiply or divide by the same number on both sides to get the form $x > a$ or $x < a$. *When multiplying or dividing by a negative number, reverse the direction of the inequality symbol.*	Solve $3(1 - x) + 5 - 2x > 9 - 6$ and graph the solutions. $3(1 - x) + 5 - 2x > 9 - 6$ Use distributive property. $3 - 3x + 5 - 2x > 9 - 6$ Combine like terms. $8 - 5x > 3$ $\underline{-8 \quad\quad\quad -8}$ Subtract 8 from both sides. $-5x > -5$ $\dfrac{-5x}{-5} < \dfrac{-5}{-5}$ Divide both sides by -5 and change $>$ to $<$. $x < 1$ 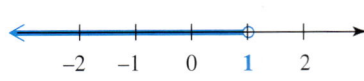

ANSWERS TO TEST YOUR WORD POWER

1. C; *Examples:* $0.1213141516\ldots, \sqrt{13}$
2. D; *Examples:* Real numbers include all the rational numbers and all the irrational numbers.
3. A; *Examples:* $3, -12, 4.1, \dfrac{5}{6}$
4. D; *Examples:* $x < 5$ and $7 + 2y \leq 11$
5. B; *Example:* $4x + 12 = 4(x + 3)$ because when the distributive property is used on the right side, the equation becomes $4x + 12 = 4x + 12$ (both sides are exactly the same)
6. C; *Example:* $2x + 5 \geq 7$

Chapter 10
REVIEW EXERCISES

[10.1] *Identify each number as rational or irrational and explain why.*

1. 24.625

2. 0.363636...

3. $\sqrt{144}$

4. $\sqrt{23}$

Label each statement as true or false, and explain why.

5. $\dfrac{2}{3} \neq 0.6$

6. $-\dfrac{5}{8} \leq -\dfrac{5}{8}$

7. $2 - 10 \geq 10 - 2$

8. $-75 \geq -50$

Simplify.

9. $6[2 + 8(3^3)]$

10. $3^2[(11 + 3) - 4]$

11. $-8 + [(-4 + 17) - (-3 - 3)]$

12. $[-9 - (2 \cdot 1) - (-3)] + [8 + (-13 + 13)]$

13. $-5(2x - 4)$

14. $-(-5 + 3p)$

15. $-(-17c - 6)$

16. $-2 - (8 - 10y)$

17. $-10 - (7 + 14r)$

18. $-8 - (-3r - 6)$

19. $-8(5k - 6) + 3(7k + 2)$

20. $-7(2t - 4) - 4(3t + 8) + 27t$

[10.2] *Solve each equation.*

21. $5x + 8 = -2(-2x - 1)$

22. $8t = 7t + \dfrac{3}{2}$

23. $(4r - 8) - (3r + 12) = 0$

24. $7(2x + 1) = 6(2x - 9)$

724 Chapter 10 Real Numbers, Equations, and Inequalities

25. $-\dfrac{6}{5}y = -18$

26. $\dfrac{1}{2}r - \dfrac{1}{6}r + 3 = 2 + \dfrac{1}{6}r + 1$

27. $3x - (-2x + 6) = 4(x - 4) + x$

28. $0.10(x + 80) + 0.20x = 14$

[10.3] *Find the value of the remaining variable in each formula.*

29. $A = \dfrac{1}{2}bh;\ A = 44,\ b = 8$

30. $C = 2\pi r;\ C = 29.83,\ \pi \approx 3.14$

Solve the formula for the specified variable.

31. $V = lwh$ for w

32. $A = \dfrac{1}{2}h(b + B)$ for h

Solve each application problem.

33. A cinema screen in Indonesia has a length of 92.75 feet and a perimeter of 326.5 feet. What is the screen's width?

34. The largest box of popcorn was filled by students in Jacksonville, Florida. The box was approximately 40 feet long, 20.7 feet wide, and had a volume of 6624 cubic feet. Find the height of the box.

[10.4] *Graph each inequality on the number line provided.*

35. $p \geq -4$

36. $x < 7$

37. $-5 \leq y < 6$

38. $r \geq \dfrac{1}{2}$

Solve each inequality. Graph the solutions.

39. $y + 6 \geq 3$

40. $5t < 4t + 2$

41. $-6x \leq -18$

42. $8(k - 5) - (2 + 7k) \geq 4$

43. $4x - 3x > 10 - 4x + 7x$

44. $3(2w + 5) + 4(8 + 3w) < 5(3w + 2) + 2w$

45. Carlotta Valdez has scores of 81, 77, and 88 on her first three calculus tests. What scores on a fourth test will give her an average of at least 85?

46. If nine times a number is added to 6, the result is at most 3. Find all such numbers.

MIXED REVIEW EXERCISES

Solve.

47. $\dfrac{y}{7} = \dfrac{y - 5}{2}$

48. $I = prt$ for r

49. $-2x > -4$

50. $2k - 5 = 4k + 13$

51. $0.05x + 0.02x = 4.9$

52. $2 - 3(y - 5) = 4 + y$

53. $9x - (7x + 2) = 3x + (2 - x)$

54. $\dfrac{1}{3}s + \dfrac{1}{2}s + 7 = \dfrac{5}{6}s + 5 + 2$

55. $4 - 5(a + 2) = 3(a + 1) - 1$

56. $P = a + b + c$ for a

57. $\dfrac{2}{3}y + \dfrac{3}{4}y = -17$

58. $2 - 6(z + 1) = 4(z - 2) + 10$

59. The area of a triangle is 182 square inches. The height is 14 inches. Find the length of the base.

60. The perimeter of a rectangle is 75 inches. The width is 17 inches. What is the length?

61. On the first two days of their vacation to Florida, Wally and Nikolas drove 430 miles and 470 miles. If they must average at least 450 miles per day, what distance must they drive on the third day?

62. Nalima has grades of 82, 91, 97, and 96 on her first four English tests. What must she make on her fifth test so that her average will be at least 90?

63. Gina used 4.7 meters of fringe around the edge of a circular tablecloth that she is making. What is the diameter of the cloth, to the nearest tenth of a meter? Use 3.14 as the approximate value of π.

64. The area of a rectangular oriental rug is 93.5 square feet. If the width is $8\tfrac{1}{2}$ feet, find the length.

Chapter 10
TEST

Identify each number as rational or irrational and explain why.

1. π
2. $3.909090\ldots$

Label each statement as true or false and explain why.

3. $-6 \leq -8$
4. $\dfrac{3}{4} \neq 0.75$
5. $1\dfrac{3}{10} \geq 0.95$

Simplify.

6. $4[-20 + 7(-2)]$
7. $-6 - [-7 + (2 - 3)]$

8. $-(-5a + 14)$
9. $-2(3x^2 + 4) - 3(x^2 + 2x)$

Solve.

10. $5x + 9 = 7x + 21$

11. $2 - 3(y - 5) = 3 + (y + 1)$

12. $2.3x + 13.7 = 1.3x + 2.9$

13. $7 - (m - 4) = -3m + 2(m + 1)$

14. $-\dfrac{4}{7}x = -1 - \dfrac{2}{3}x$

15. $0.06(x + 20) + 0.08(x - 10) = 4.6$

16. $-8(2x + 4) = -4(4x + 8)$

1. _____
2. _____
3. _____
4. _____
5. _____
6. _____
7. _____
8. _____
9. _____
10. _____
11. _____
12. _____
13. _____
14. _____
15. _____
16. _____

17. _____

17. The formula for simple interest is $I = prt$. If $I = 600$, $r = 0.08$, and $t = 3$, find p.

18. _____

18. The Ziegfield Room in Reno, Nevada has a circular turntable for performers. The circumference of the turntable is 62.5 feet. What is the diameter of the turntable, to the nearest tenth of a foot? Use 3.14 as the approximate value of π.

19. _____

19. Solve $d = 2r$ for r.

20. _____

20. Solve $A = \dfrac{1}{2} bh$ for h.

21. _____

21. Solve $A = p + prt$ for r.

Solve each inequality and graph the solutions.

22. _____

22. $-3x > -33$

23. _____

23. $-4x + 2(x - 3) \geq 4x - (3 + 5x) - 7$

24. _____

24. Paula Story has scores of 95, 84, and 100 on her first three college algebra tests. What must she score on her fourth test so that her average is at least 90?

Graphs of Linear Equations and Inequalities in Two Variables

11.1 Reading Graphs; Linear Equations in Two Variables

11.2 Graphing Linear Equations in Two Variables

11.3 Slope of a Line

11.4 Equations of Lines

Summary Exercises on Graphing Linear Equations

11.5 Graphing Linear Inequalities in Two Variables

The use and abuse of credit cards is putting college students deeper in debt than ever before. College campuses have become fertile territory as credit card companies pitch their plastic to students at bookstores, student unions, and sporting events. According to federal loan provider Nellie Mae, an estimated 83% of college students have credit cards with an average balance of more than $2300. (*Source:* ISU Extension.)

In Example 6 of Section 11.2 we examine a linear equation that models credit card debt in the United States.

11.1 Reading Graphs; Linear Equations in Two Variables

OBJECTIVES

1. Interpret graphs.
2. Write a solution as an ordered pair.
3. Decide whether a given ordered pair is a solution of a given equation.
4. Complete ordered pairs for a given equation.
5. Complete a table of values.
6. Plot ordered pairs.

① Refer to the circle graph in Figure 1.

(a) Which provider had the smallest market share in December 2003?

(b) Estimate the number of home subscribers to DirecTV.

(c) How many actual home subscribers to DirecTV were there?

ANSWERS
1. (a) C-Band
   (b) 60% of 22 million = 13.2 million
   (c) 12.1 million

We live in an age of information. As you saw in Chapter 9, graphs provide a quick way to organize and communicate much of this information. They can also be used to analyze data, make predictions, or simply entertain us. Let us review briefly.

OBJECTIVE 1 Interpret graphs. There are many ways to represent the relationship between two quantities. *Circle graphs, bar graphs,* and *line graphs* are often used for this purpose.

In a *circle graph* or *pie chart,* a circle is used to indicate the total of all the categories represented. The circle is divided into *sectors*, or wedges (like pieces of a pie), whose sizes show the relative magnitudes of the categories. The sum of all the fractional parts of the graph must be 1 (for 1 whole circle). Or, if the data is given as percents, the sum of all the parts must be 100% (because 100% = 1).

EXAMPLE 1 Interpreting a Circle Graph

The 2003 market share for satellite-TV home subscribers is shown in the circle graph in Figure 1 below.

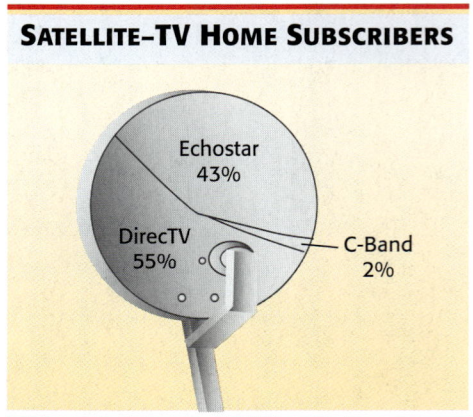

Source: Skyreport.com; *USA Today.*
Figure 1

The number of subscribers reached 22 million in December 2003.

(a) Which provider had the largest share of the home subscriber market in December 2003? What was that share?

In the circle graph, the sector for DirecTV is the largest, so DirecTV had the largest market share, 55%.

(b) Estimate the number of home subscribers to Echostar in December 2003.

A market share of 43% can be rounded to 40%, or 0.4. We multiply 0.4 by the total number of subscribers, 22 million. A good estimate for the number of Echostar subscribers would be

$$0.4(22) = 8.8 \approx 9 \text{ million.}$$

(c) How many actual home subscribers to Echostar were there?

To find the answer, we multiply the actual percent from the graph for Echostar, 43% or 0.43, by the number of subscribers, 22 million:

$$0.43(22) = 9.46$$

Thus, 9.46 million homes subscribed to Echostar. This is reasonable given our estimate in part (b).

Work Problem 1 at the Side.

A *bar graph* is used to show comparisons. It consists of a series of bars (or simulations of bars) arranged either vertically or horizontally. In a bar graph, values from two categories are paired with each other.

EXAMPLE 2 Interpreting a Bar Graph

The bar graph in Figure 2 illustrates the amount of personal savings, in billions of dollars, accumulated during the years 1997 through 2001.

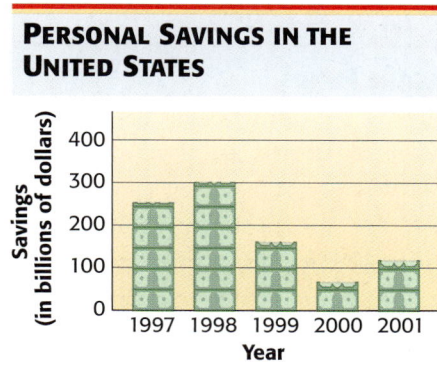

Source: U.S. Bureau of Economic Analysis.

Figure 2

(a) Which year had the greatest amount of savings? Which had the least?

The tallest bar corresponds to 1998, so the largest amount was in 1998. The year 2000, with the shortest bar, had the least.

(b) Which years had amounts greater than $200 billion?

Locate 200 on the vertical scale and follow the line across to the right. Two years, 1997 and 1998, have bars that extend above the line for 200, so they had amounts greater than $200 billion.

(c) Estimate the amounts for 1997 and 1998.

Locate the top of the bar for 1997 and move horizontally across to the vertical scale to see that it lies about halfway between 200 and 300. The amount for 1997 was about $250 billion.

Follow the top of the bar for 1998 across to the vertical scale to see that it is about 300. The amount for 1998 was about $300 billion.

(d) Estimate the difference between the amounts for the years 1997 and 1998. Interpret this result.

Based on the results of part (c), the difference is approximately

$300 billion − $250 billion = $50 billion.

The amount of personal savings increased from 1997 to 1998 by about $50 billion. (However, during the next two years, it decreased each year.)

(e) How did personal savings in 1998 compare to personal savings in 1999?

Comparing the heights of the bars for 1998 and 1999, we see that the bar for 1998 is about twice as tall as the bar for 1999. Therefore, personal savings in 1998 was about twice that for 1999.

Also, if we estimate the amounts of personal savings for 1998 and 1999 as about $300 billion and about $150 billion respectively, we see that personal savings in 1998 was about twice that for 1999.

Work Problem 2 at the Side.

2 Refer to the bar graph in Figure 2.

(a) Which years had amounts less than $200 billion?

(b) Estimate the amounts of personal savings for 2000 and 2001.

ANSWERS
2. (a) 1999, 2000, 2001
   (b) $60 billion; $120 billion

3 Refer to the line graph in Figure 3.

(a) Which year has the greatest amount of Medicare funds?

(b) Estimate Medicare funds for 2008. Is there a surplus or a deficit in 2008?

(c) About how much will funds decrease from 2006 to 2011?

A *line graph* is used to show changes or trends in data over time. To form a line graph, we connect a series of points representing data with line segments.

EXAMPLE 3 Interpreting a Line Graph

Current projections indicate that funding for Medicare will not cover its costs unless the program changes. The line graph in Figure 3 shows projections for the years 2004 through 2013.

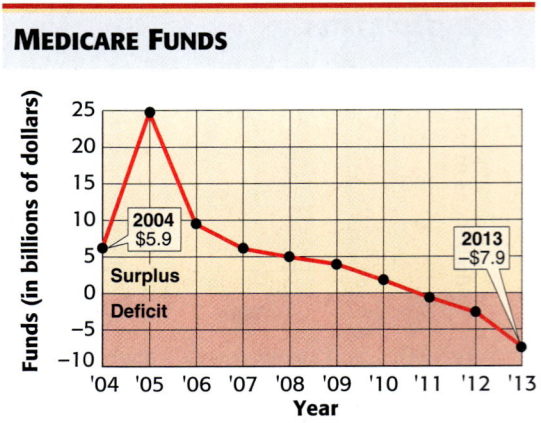

Figure 3

Source: Centers for Medicare and Medicaid Services.

(a) Which is the only period in which Medicare funds are predicted to increase?

Because the graph *rises* from 2004 to 2005 and falls in every other case, funds are predicted to increase between these two years.

(b) What will be the general trend from 2005 to 2013?

Funds will decrease, since the graph *falls* during this period.

(c) In which year will funds first show a deficit?

From 2004 to 2010, the graph is always above 0, but in 2011, it falls below 0 for the first time, indicating a deficit.

(d) Based on the figures shown in the graph, what is the difference in Medicare funds from 2004 to 2013?

$$-\$7.9 \text{ billion} - \$5.9 \text{ billion} = -\$13.8 \text{ billion}$$
$$\text{2013 amount} \quad \text{2004 amount} \quad \text{Difference}$$

The fund amount will have *decreased* $13.8 billion (as indicated by the negative sign in $-\$13.8$).

Work Problem 3 at the Side.

Many everyday situations, such as those illustrated in Examples 2 and 3, involve two quantities that are related. The equations and applications we discussed in **Chapters 2–7** had only one variable. In **Section 9.5,** we introduced *linear equations in two variables,* and this chapter extends that topic.

ANSWERS
3. (a) 2005
   (b) $5 billion; surplus
   (c) $10 billion

Linear Equation in Two Variables

A **linear equation in two variables** is an equation that can be written in the form

$$Ax + By = C,$$

where A, B, and C are real numbers and A and B are not both 0.

Some examples of linear equations in two variables in this form, called *standard form*, are

$3x + 4y = 9$, $\quad x - y = 0$, $\quad$ and $\quad x + 2y = -8$. $\quad$ Linear equations in two variables

> **NOTE**
> Other linear equations in two variables, such as
>
> $$y = 4x + 5 \quad \text{and} \quad 3x = 7 - 2y,$$
>
> are not written in standard form but could be. We will discuss the forms of linear equations in more detail in **Section 11.4**.

OBJECTIVE 2 Write a solution as an ordered pair. Recall from **Section 2.3** that a *solution* of an equation is a number that makes the equation true when it replaces the variable. For example, the linear equation in one variable $x - 2 = 5$ has solution 7, since replacing x with 7 gives a true statement.

A solution of a linear equation in *two* variables requires *two* numbers, one for each variable. For example, a true statement results when we replace x with 2 and y with 13 in the equation $y = 4x + 5$ since

$$13 = 4(2) + 5. \quad \text{Let } x = 2, y = 13.$$

The pair of numbers $x = 2$ and $y = 13$ gives one solution of the equation $y = 4x + 5$. The phrase "$x = 2$ and $y = 13$" is abbreviated

x-value $\longrightarrow$ $\quad$ $\longleftarrow$ y-value

$(2, 13)$

Ordered pair

with the x-value, 2, and the y-value, 13, given as a pair of numbers written inside parentheses. ***The x-value is always given first.*** A pair of numbers such as $(2, 13)$ is called an **ordered pair**. As the name indicates, the order in which the numbers are written is important. The ordered pairs $(2, 13)$ and $(13, 2)$ are not the same. The second pair indicates that $x = 13$ and $y = 2$. For two ordered pairs to be equal, their x-coordinates must be equal *and* their y-coordinates must be equal.

> Work Problem 4 at the Side.

OBJECTIVE 3 Decide whether a given ordered pair is a solution of a given equation. We substitute the x- and y-values of an ordered pair into a linear equation in two variables to see whether the ordered pair is a solution.

---

4 Write each solution as an ordered pair.

(a) $x = 5$ and $y = 7$

(b) $y = 6$ and $x = -1$

(c) $y = 4$ and $x = -3$

(d) $x = 3$ and $y = -12$

ANSWERS
4. (a) $(5, 7)$ (b) $(-1, 6)$ (c) $(-3, 4)$
(d) $(3, -12)$

5 Decide whether each ordered pair is a solution of the equation $5x + 2y = 20$.

(a) $(0, 10)$

$$5x + 2y = 20$$
$$5() + 2() = 20$$
$$\underline{} + 20 = 20$$
$$\underline{} = 20$$

Is $(0, 10)$ a solution?

(b) $(2, -5)$

(c) $(3, 2)$

(d) $(-4, 20)$

6 Complete each ordered pair for the equation $y = 2x - 9$.

(a) $(5,)$

$$y = 2() - 9$$
$$y = \underline{} - 9$$
$$y = \underline{}$$

The ordered pair is _____.

(b) $(2,)$

(c) $(, 7)$

(d) $(, -13)$

ANSWERS
5. (a) 0; 10; 0; 20; yes   (b) no   (c) no
   (d) yes
6. (a) 5; 10; 1; (5, 1)   (b) (2, -5)   (c) (8, 7)
   (d) (-2, -13)

EXAMPLE 4 Deciding Whether Ordered Pairs Are Solutions of an Equation

Decide whether each ordered pair is a solution of the equation $2x + 3y = 12$.

(a) $(3, 2)$

To see whether $(3, 2)$ is a solution of the given equation $2x + 3y = 12$, substitute 3 for x and 2 for y in the equation.

$$2x + 3y = 12$$
$$2(\mathbf{3}) + 3(\mathbf{2}) = 12 \quad ? \quad \text{Let } x = 3; \text{ let } y = 2.$$
$$6 + 6 = 12 \quad ?$$
$$12 = 12 \quad \text{True}$$

This result is true, so $(3, 2)$ is a solution of $2x + 3y = 12$.

(b) $(-2, -7)$

$$2x + 3y = 12$$
$$2(\mathbf{-2}) + 3(\mathbf{-7}) = 12 \quad ? \quad \text{Let } x = -2; \text{ let } y = -7.$$
$$-4 + (-21) = 12 \quad ?$$
$$-25 \neq 12 \quad \text{False}$$

This result is false, so $(-2, -7)$ is *not* a solution of $2x + 3y = 12$.

◀◀◀ **Work Problem 5 at the Side.**

OBJECTIVE 4 Complete ordered pairs for a given equation. Choosing a number for one variable in a linear equation makes it possible to find the value of the other variable.

EXAMPLE 5 Completing Ordered Pairs

Complete each ordered pair for the equation $y = 4x + 5$.

(a) $(7,)$

In this ordered pair, $x = 7$. (Remember that x always comes first.) To find the corresponding value of y, replace x with 7 in the equation.

$$y = 4x + 5$$
$$y = 4(\mathbf{7}) + 5 \quad \text{Let } x = 7.$$
$$y = 28 + 5$$
$$y = 33$$

The ordered pair is $(\mathbf{7}, \mathbf{33})$.

(b) $(, -3)$

In this ordered pair, $y = -3$. Find the value of x by replacing y with -3 in the equation; then solve for x.

$$y = 4x + 5$$
$$\mathbf{-3} = 4x + 5 \quad \text{Let } y = -3.$$
$$-8 = 4x \quad \text{Subtract 5 from both sides.}$$
$$-2 = x \quad \text{Divide both sides by 4.}$$

The ordered pair is $(-2, -3)$.

◀◀◀ **Work Problem 6 at the Side.**

OBJECTIVE 5 Complete a table of values. Ordered pairs are often displayed in a **table of values.** The table may be written either vertically or horizontally.

EXAMPLE 6 Completing Tables of Values

Complete the table of values for each equation.

(a) $x - 2y = 8$

x	y
2	
10	
	0
	-2

To complete the first two ordered pairs of the table, let $x = 2$ and $x = 10$, respectively.

If $x = 2$,
then $x - 2y = 8$
becomes $2 - 2y = 8$
$-2y = 6$
$y = -3$

If $x = 10$,
then $x - 2y = 8$
becomes $10 - 2y = 8$
$-2y = -2$
$y = 1$

Now complete the last two ordered pairs by letting $y = 0$ and $y = -2$, respectively.

If $y = 0$,
then $x - 2y = 8$
becomes $x - 2(0) = 8$
$x - 0 = 8$
$x = 8$

If $y = -2$,
then $x - 2y = 8$
becomes $x - 2(-2) = 8$
$x + 4 = 8$
$x = 4$

The completed table of values follows.

x	y
2	-3
10	1
8	0
4	-2

The corresponding ordered pairs are $(2, -3)$, $(10, 1)$, $(8, 0)$, and $(4, -2)$. Notice that each ordered pair is a solution of the given equation.

(b) $x = 5$

x	y
	-2
	6
	3

The given equation is $x = 5$. No matter which value of y is chosen, the value of x is always 5.

x	y
5	-2
5	6
5	3

The corresponding ordered pairs are $(5, -2)$, $(5, 6)$, and $(5, 3)$.

▶ **Work Problem 7 at the Side.**

7 Complete the table of values for each equation.

(a) $2x - 3y = 12$

x	y
0	
	0
3	
	-3

(b) $y = 4$

x	y
-3	
2	
5	

ANSWERS

7. (a)

x	y
0	-4
6	0
3	-2
$\frac{3}{2}$	-3

(b)

x	y
-3	4
2	4
5	4

8 Name the quadrant in which each point in the figure is located.

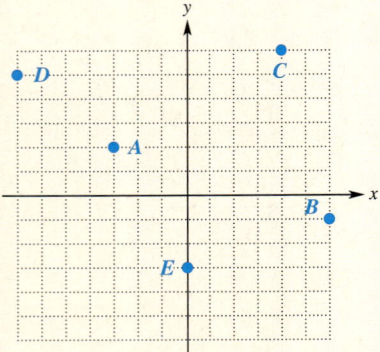

> **NOTE**
> We can think of $x = 5$ in Example 6(b) as an equation in two variables by rewriting $x = 5$ as $x + 0y = 5$. This form of the equation shows that for any value of y, the value of x is 5. Similarly, $y = 4$ in Problem 7(b) in the margin on the preceding page is the same as $0x + y = 4$.

OBJECTIVE 6 Plot ordered pairs. Recall from **Section 9.5** that every linear equation in *two* variables has an infinite number of ordered pairs as solutions. Each choice of a number for one variable leads to a particular real number for the other variable.

To graph these solutions, represented as the ordered pairs (x, y), we need *two* number lines, one for each variable. These two number lines are drawn as shown in Figure 4. To review, the horizontal number line is called the **x-axis,** and the vertical line is called the **y-axis.** Together, the x-axis and y-axis form a **rectangular coordinate system,** also called the **Cartesian coordinate system,** in honor of René Descartes (1596–1650), the French mathematician who is credited with its invention.

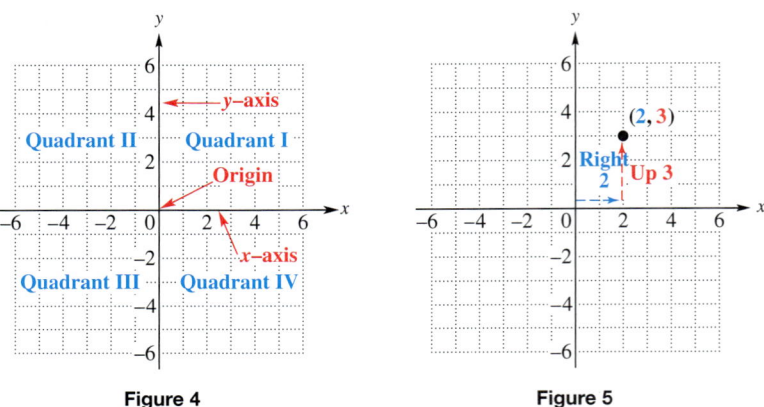

Figure 4   Figure 5

The coordinate system is divided into four regions, called **quadrants.** These quadrants are numbered counterclockwise, as shown in Figure 4. *Points on the axes themselves are not in any quadrant*. The point at which the x-axis and y-axis meet is called the **origin.** The origin, which is labeled 0 in Figure 4, is the point corresponding to (0, 0).

◀◀◀ Work Problem 8 at the Side.

The x-axis and y-axis determine a **plane,** a flat surface illustrated by a sheet of paper. By referring to the two axes, every point in the plane can be associated with an ordered pair. The numbers in the ordered pair are called the **coordinates** of the point. For example, locate the point associated with the ordered pair (2, 3) by starting at the origin. Since the x-coordinate is 2, go 2 units to the right along the x-axis. Then, since the y-coordinate is 3, turn and go up 3 units on a line parallel to the y-axis. The point (2, 3) is **plotted** in Figure 5. From now on, we will refer to the point with x-coordinate 2 and y-coordinate 3 as the point (2, 3).

ANSWERS
8. *A*, II; *B*, IV; *C*, I; *D*, II; *E*, no quadrant

NOTE
When we graph on a number line, one number corresponds to each point. On a plane, however, both numbers in the ordered pair are needed to locate a point. The ordered pair is a name for the point.

EXAMPLE 7 Plotting Ordered Pairs

Plot each ordered pair on a coordinate system.

(a) $(1, 5)$ (b) $(-2, 3)$ (c) $(-1, -4)$ (d) $(7, -2)$

(e) $\left(\frac{3}{2}, 2\right)$ (f) $(5, 0)$ (g) $(0, -3)$

See Figure 6. In part (c), locate the point $(-1, -4)$ by first going 1 unit to the left along the *x*-axis. Then turn and go 4 units down, parallel to the *y*-axis. Plot the point $\left(\frac{3}{2}, 2\right)$ in part (e) by first going $\frac{3}{2}$ (or $1\frac{1}{2}$) units to the right along the *x*-axis. Then turn and go 2 units up, parallel to the *y*-axis.

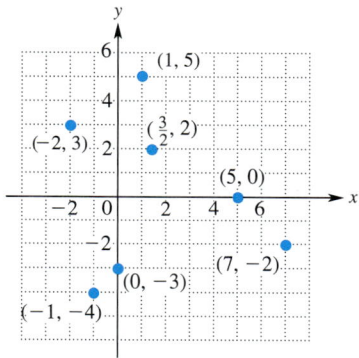

Figure 6

> Work Problem 9 at the Side.

Sometimes we can use a linear equation to mathematically describe, or *model*, a real-life situation, as shown in the next example.

EXAMPLE 8 Completing Ordered Pairs to Estimate Annual Costs of Doctors' Visits

The amount Americans pay annually for doctors' visits increased from 1990 through 2000. This amount can be approximated by the linear equation

$$y = 34.3x - 67{,}693$$

(Cost — y; Year — x)

which relates *x*, the year, and *y*, the cost in dollars. (*Source:* U.S. Health Care Financing Administration.)

(a) Complete the table of values for this linear equation.

x (Year)	y (Cost)
1990	
1996	
2000	

Continued on Next Page

9 Plot each ordered pair on a coordinate system.

(a) $(3, 5)$ (b) $(-2, 6)$

(c) $(-4, 0)$ (d) $(-5, -2)$

(e) $(5, -2)$ (f) $(0, -6)$

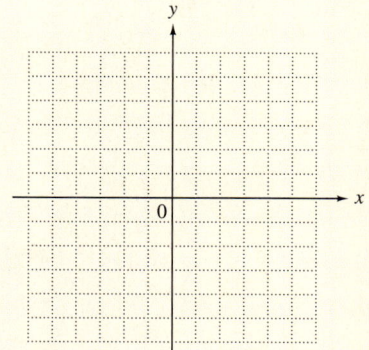

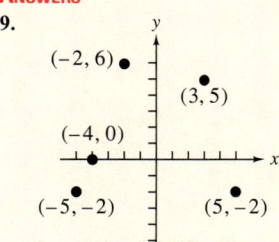

ANSWERS

9.

10 Refer to the linear equation in Example 8.

(a) Find the y-value for x = 1996. Round to the nearest whole number.

(b) Find the y-value for x = 2000. Interpret your result.

To find y when $x = 1990$, substitute into the equation.

$y = 34.3(1990) - 67,693$ Let $x = 1990$.

$y = 564$ Use a calculator.

This means that in 1990, Americans each spent about $564 on doctors' visits.

Work Problem 10 at the Side.

Including the results from Problem 10 at the side gives the completed table that follows.

x (Year)	y (Cost)
1990	564
1996	770
2000	907

We can write the results from the table of values as ordered pairs (x, y). Each year x is paired with its cost y:

(1990, 564), (1996, 770), and (2000, 907).

(b) Graph the ordered pairs found in part (a).

The ordered pairs are graphed in Figure 7. This graph of ordered pairs of data is called a **scatter diagram.** Notice how the axes are labeled: x represents the year, and y represents the cost in dollars. Different scales are used on the two axes. Here, each square represents two units in the horizontal direction and 100 units in the vertical direction. Because the numbers in the first ordered pair are so large, we show a break in the axes near the origin.

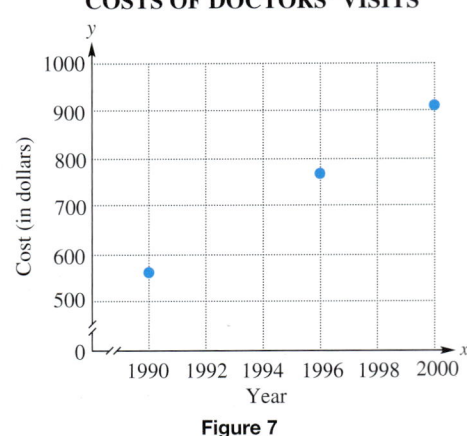

Figure 7

A scatter diagram enables us to tell whether two quantities are related to each other. In Figure 7, the plotted points could be connected to form a straight *line*, so the variables x (year) and y (cost) have a *line*ar relationship. The increase in costs is also reflected.

> **CAUTION**
> The equation in Example 8 is valid only for the years 1990 through 2000 because it was based on data for those years. Do not assume that this equation would provide reliable data for other years since the data for those years may not follow the same pattern.

ANSWERS
10. (a) $y = 770$
(b) $y = 907$; In 2000, Americans each spent about $907 on doctors' visits.

11.1 Exercises

FOR EXTRA HELP: 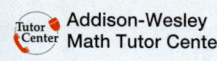 Addison-Wesley Math Tutor Center   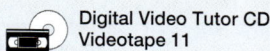 MathXL    Digital Video Tutor CD 8 / Videotape 11   Student's Solutions Manual   MyMathLab   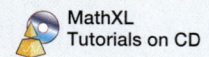 MathXL Tutorials on CD

On February 13, 2000, Peanuts *fans bid farewell to new segments of this popular comic strip (which is still being printed as* Classic Peanuts*). The circle graph shows the results of a survey of adults to determine their favorite* Peanuts *characters. Use the circle graph to work Exercises 1–4. See Example 1.*

1. Which *Peanuts* character was most popular? What percent of those surveyed named this character as their favorite?

FAVORITE PEANUTS CHARACTERS

- No opinion 15%
- Charlie Brown 26%
- Snoopy 31%
- Lucy 8%
- Linus 13%
- Pig Pen 3%
- Other 4%

Source: USA Today/CNN/Gallup Poll conducted nationwide February 4–6, 2000.

2. A random sample of 5000 adults is surveyed.
   (a) Estimate how many would be expected to name Lucy as their favorite *Peanuts* character. (*Hint:* To estimate, round 8% to 10%.)
   (b) Use the actual figure from the graph to determine how many adults would name Lucy as their favorite character.

3. Regardless of the number of adults surveyed, how many would we expect to name Charlie Brown as their favorite *Peanuts* character compared to Linus?

4. Using the group of 5000 adults, confirm your answer to Exercise 3.

The bar graph compares egg production in millions of eggs for six states in the year 2002. Use the bar graph to work Exercises 5–8. See Example 2.

5. Name the top two egg-producing states. Estimate their production.

6. Which states had egg production less than 5000 million eggs?

7. Which state had the least production? Estimate this production.

8. How did egg production in Texas (TX) compare to egg production in Iowa (IA)?

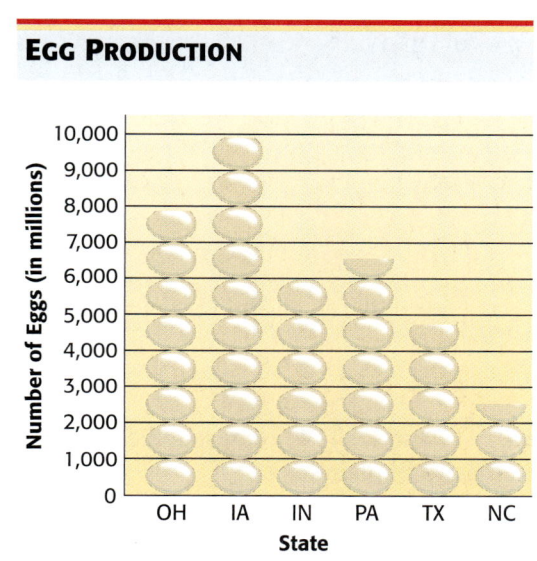

EGG PRODUCTION

Source: U.S. Department of Agriculture.

The line graph shows the average price, adjusted for inflation, that Americans have paid for a gallon of gasoline for selected years since 1970. Use the line graph to work Exercises 9–12. See Example 3.

9. Over which period of years did the greatest increase in the price of a gallon of gas occur? About how much was this increase?

10. Estimate the price of a gallon of gas during 1985, 1990, 1995, and 2000.

11. Describe the trend in gas prices from 1980 to 1995.

12. During which year(s) did a gallon of gas cost approximately $1.50?

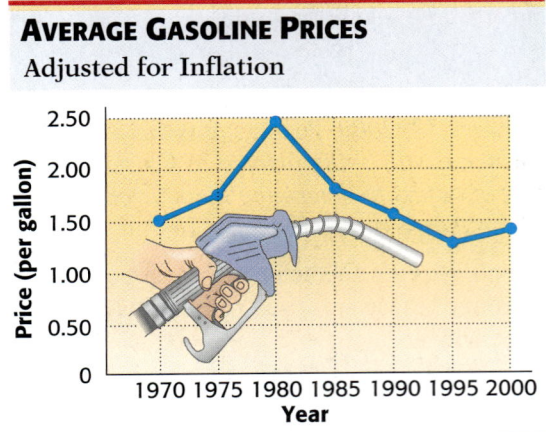

AVERAGE GASOLINE PRICES
Adjusted for Inflation

Source: American Petroleum Institute; AP research.

Use the concepts of this section to fill in each blank with the correct response.

13. The symbol (x, y) _____ represent an ordered pair, while the
    (does/does not)
    symbols $[x, y]$ and $\{x, y\}$ _____ represent ordered pairs.
    (do/do not)

14. The point whose graph has coordinates $(-4, 2)$ is in quadrant _____.

15. The point whose graph has coordinates $(0, 5)$ lies on the _____-axis.

16. The ordered pair $(4, \text{\_\_\_\_})$ is a solution of the equation $y = 3$.

17. The ordered pair $(\text{\_\_\_\_}, -2)$ is a solution of the equation $x = 6$.

18. The ordered pair $(3, 2)$ is a solution of the equation $2x - 5y = \text{\_\_\_\_}$.

Decide whether each ordered pair is a solution of the given equation. See Example 4.

19. $x + y = 9$; $(0, 9)$

20. $x + y = 8$; $(0, 8)$

21. $2x - y = 6$; $(4, 2)$

22. $2x + y = 5$; $(3, -1)$

23. $4x - 3y = 6$; $(2, 1)$

24. $5x - 3y = 15$; $(5, 2)$

25. $y = \frac{2}{3}x$; $(-6, -4)$

26. $y = -\frac{1}{4}x$; $(-8, 2)$

27. $x = -6$; $(5, -6)$

28. $y = 2$; $(2, 4)$

29. Do $(4, -1)$ and $(-1, 4)$ represent the same ordered pair? Explain.

30. Explain why it would be easier to find the corresponding y-value for $x = \frac{1}{3}$ in the equation $y = 6x + 2$ than it would be for $x = \frac{1}{7}$.

Complete each ordered pair for the equation $y = 2x + 7$. See Example 5.

31. (2,   )   **32.** (0,   )   **33.** (, 0)   **34.** (, −3)

Complete each ordered pair for the equation $y = -4x - 4$. See Example 5.

35. (0,   )   **36.** (, 0)   **37.** (, 16)   **38.** (, 24)

Complete each table of values. In Exercises 39–42, write the results as ordered pairs. See Example 6.

39. $2x + 3y = 12$

x	y
0	
	0
	8

40. $4x + 3y = 24$

x	y
0	
	0
	4

41. $3x - 5y = -15$

x	y
0	
	0
	−6

42. $4x - 9y = -36$

x	y
	0
0	
	8

43. $x = -9$

x	y
	6
	2
	−3

44. $x = 12$

x	y
	3
	8
	0

45. $y = -6$

x	y
8	
4	
−2	

46. $y = -10$

x	y
4	
0	
−4	

47. $x - 8 = 0$

x	y
	8
	3
	0

48. $y + 2 = 0$

x	y
9	
2	
0	

Give the ordered pairs for the points labeled A–F in the figure.

49. A   **50.** B   **51.** C

52. D   **53.** E   **54.** F

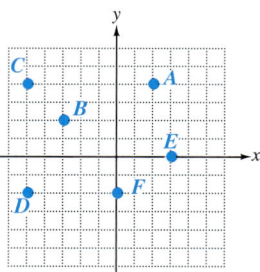

Fill in each blank with the word positive *or the word* negative.

The point with coordinates (x, y) is in

55. quadrant III if x is _____ and y is _____ .

56. quadrant II if x is _____ and y is _____ .

57. quadrant IV if x is _____ and y is _____ .

58. quadrant I if x is _____ and y is _____ .

742 Chapter 11 Graphs of Linear Equations and Inequalities in Two Variables

59. A point (x, y) has the property that $xy < 0$. In which quadrant(s) must the point lie? Explain.

60. A point (x, y) has the property that $xy > 0$. In which quadrant(s) must the point lie? Explain.

Plot each ordered pair on the rectangular coordinate system provided. See Example 7.

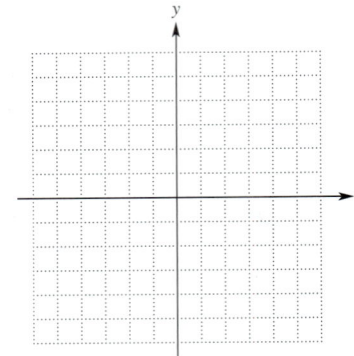

61. $(6, 2)$ **62.** $(5, 3)$ **63.** $(-4, 2)$

64. $(-3, 5)$ **65.** $\left(-\dfrac{4}{5}, -1\right)$ **66.** $\left(-\dfrac{3}{2}, -4\right)$

67. $(0, 4)$ **68.** $(0, -3)$ **69.** $(4, 0)$ **70.** $(-3, 0)$

Complete each table of values, and then plot the ordered pairs. See Examples 6 and 7.

71. $x - 2y = 6$

x	y
0	
	0
2	
	-1

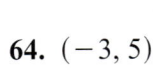

72. $2x - y = 4$

x	y
0	
	0
1	
	-6

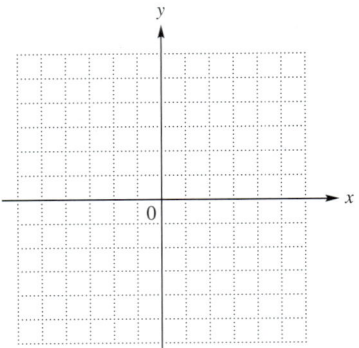

73. $3x - 4y = 12$

x	y
0	
	0
-4	
	-4

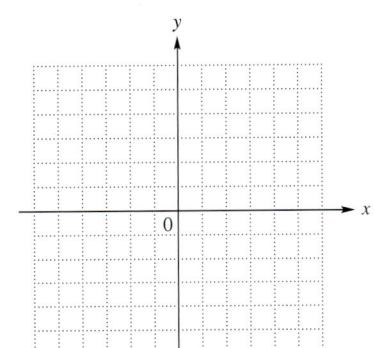

74. $2x - 5y = 10$

x	y
0	
	0
-5	
	-3

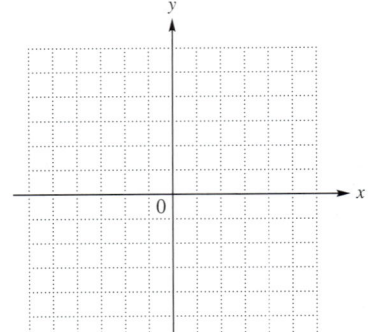

75. $y + 4 = 0$

x	y
0	
5	
-2	
-3	

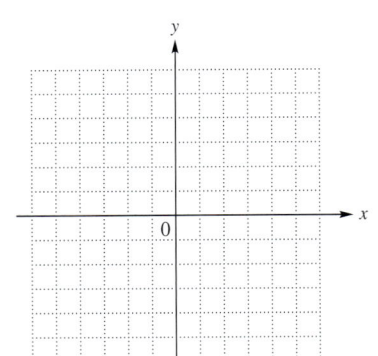

76. $x - 5 = 0$

x	y
	1
	0
	6
	-4

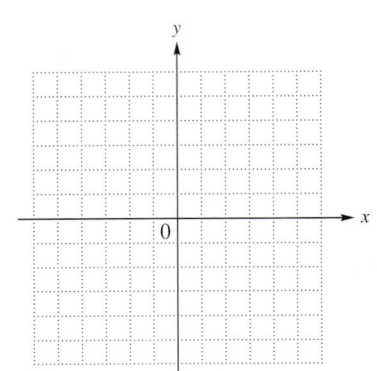

77. Look at the graphs of the ordered pairs in Exercises 71–76. Describe the pattern indicated by the plotted points.

Work each problem. See Example 8.

78. The table shows the amount of e-commerce in billions of dollars.

Year	Amount (in billions)
1996	0.7
1997	2.6
1998	7.8
1999	18.9
2000	38.2
2001	45.7

Source: Jupiter Research.

(a) Write the data from the table as ordered pairs (x, y), where x represents the year and y represents the amount of e-commerce in billions of dollars.

(b) What does the ordered pair (2002, 62.0) mean in the context of this discussion?

(c) Make a scatter diagram of the data using the ordered pairs from part (a).

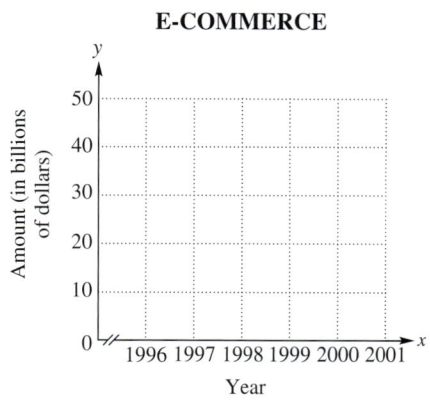
E-COMMERCE

(d) Describe what is happening to the amount of e-commerce during these years.

79. The table shows the rate (in percent) at which 4-year college students (both private and public) graduate within 5 years.

Year	Percent
1996	53.3
1997	52.8
1998	52.1
1999	51.6
2000	51.2
2001	50.9

Source: ACT.

(a) Write the data from the table as ordered pairs (x, y), where x represents the year and y represents graduation percent.

(b) What does the ordered pair (2002, 51.0) mean in the context of this discussion?

(c) Make a scatter diagram of the data using the ordered pairs from part (a).

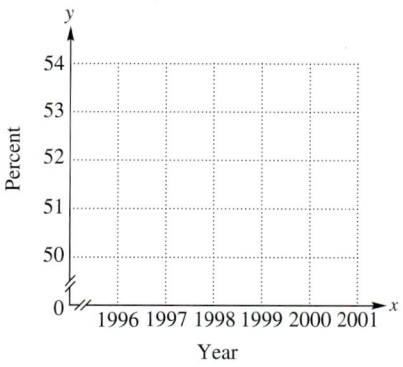
4-YEAR COLLEGE STUDENTS GRADUATING WITHIN 5 YEARS

(d) Describe the pattern indicated by the points on the scatter diagram. What is happening to graduation rates for 4-year college students within 5 years?

80. The maximum benefit for the heart from exercising occurs if the heart rate is in the target heart rate zone. The lower limit of this target zone can be approximated by the linear equation

$$y = -0.5x + 108,$$

where x represents age and y represents heartbeats per minute. (*Source:* www.fitresource.com)

(a) Complete the table of values for this linear equation.

Age	Heartbeats (per minute)
20	
40	
60	
80	

(b) Write the data from the table of values as ordered pairs.

(c) Make a scatter diagram of the data. Do the points lie in an approximately linear pattern?

TARGET HEART RATE ZONE (Lower Limit)

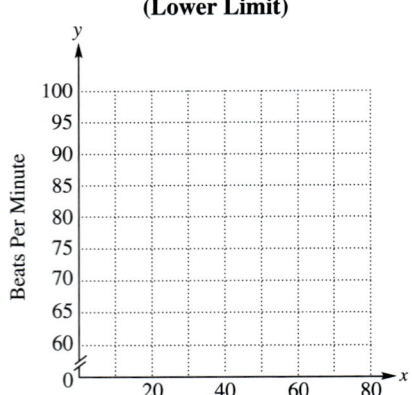

81. (See Exercise 80.) The upper limit of the target heart rate zone can be approximated by the linear equation

$$y = -0.8x + 173,$$

where x represents age and y represents heartbeats per minute. (*Source:* www.fitresource.com)

(a) Complete the table of values for this linear equation.

Age	Heartbeats (per minute)
20	
40	
60	
80	

(b) Write the data from the table of values as ordered pairs.

(c) Make a scatter diagram of the data. Describe the pattern indicated by the data.

TARGET HEART RATE ZONE (Upper Limit)

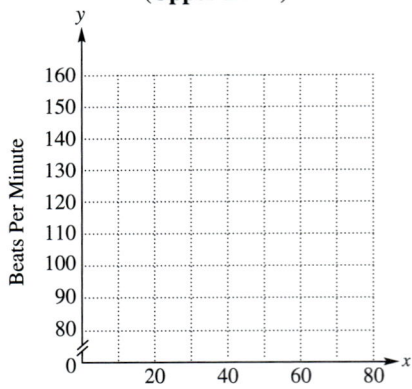

82. Refer to Exercises 80 and 81. What is the target heart rate zone for age 20? age 40?

11.2 Graphing Linear Equations in Two Variables

OBJECTIVES
1. Graph linear equations by plotting ordered pairs.
2. Find intercepts.
3. Graph linear equations where the intercepts coincide.
4. Graph linear equations of the form $y = k$ or $x = k$.
5. Use a linear equation to model data.

OBJECTIVE 1 Graph linear equations by plotting ordered pairs.
There are infinitely many ordered pairs that satisfy an equation in two variables. We find ordered pairs that are solutions of the equation $x + 2y = 7$ by choosing as many values of x (or y) as we wish and then completing each ordered pair.

For example, if we choose $x = 1$, then $y = 3$, so the ordered pair $(1, 3)$ is a solution of the equation $x + 2y = 7$.

$$1 + 2(3) = 1 + 6 = 7$$

Work Problem 1 at the Side.

Figure 8 shows a graph of all the ordered pairs found for $x + 2y = 7$ above and in Problem 1 at the side.

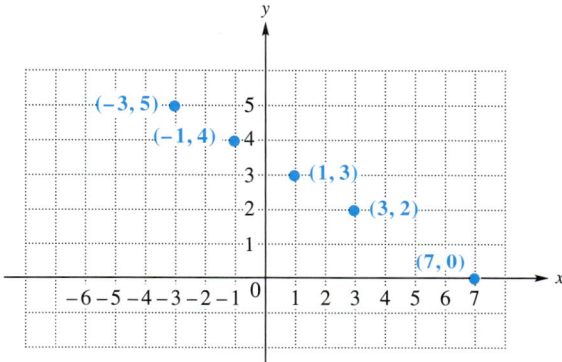

Figure 8

Notice that the points plotted in this figure all appear to lie on a straight line. The line that goes through these points is shown in Figure 9. In fact, all ordered pairs satisfying the equation $x + 2y = 7$ correspond to points that lie on this same straight line. This line gives a "picture" of all the solutions of the equation $x + 2y = 7$. Only a portion of the line is shown here, but it extends indefinitely in both directions, as suggested by the arrowhead on each end of the line.

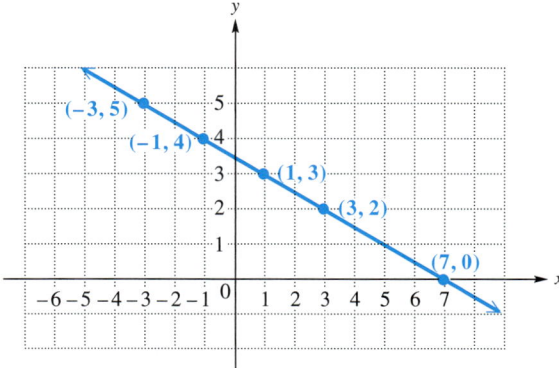

Figure 9

The line in Figure 9 is called the **graph** of the equation $x + 2y = 7$, and the process of plotting the ordered pairs and drawing the line through the corresponding points is called **graphing**. The preceding discussion can be generalized.

① Complete each ordered pair for the equation $x + 2y = 7$.

(a) $(-3,)$

(b) $(3,)$

(c) $(-1,)$

(d) $(7,)$

ANSWERS
1. (a) $(-3, 5)$ (b) $(3, 2)$
   (c) $(-1, 4)$ (d) $(7, 0)$

Chapter 11 Graphs of Linear Equations and Inequalities in Two Variables

> The graph of any linear equation in two variables is a straight line.

(Notice that the word *line* appears in the term "*line*ar equation.")

Because two distinct points determine a line, a straight line can be graphed by finding any two different points on the line. However, it is a good idea to plot a third point as a check.

EXAMPLE 1 Graphing a Linear Equation

Graph the linear equation $y = -\frac{3}{2}x + 3$.

Although this equation is not in the form $Ax + By = C$, it *could* be written in that form, so it is a linear equation. Two different points on the graph can be found by first letting $x = 0$ and then letting $y = 0$.

If $x = 0$, then

$y = -\frac{3}{2}x + 3$

$y = -\frac{3}{2}(0) + 3$ Let $x = 0$.

$y = 0 + 3$

$y = 3$.

If $y = 0$, then

$y = -\frac{3}{2}x + 3$

$0 = -\frac{3}{2}x + 3$ Let $y = 0$.

$\frac{3}{2}x = 3$ Add $\frac{3}{2}x$ to both sides.

$x = 2$.   Multiply both sides by $\frac{2}{3}$.

This gives the ordered pairs $(0, 3)$ and $(2, 0)$. Find a third point (as a check) by letting x or y equal some other number. For example, let $x = -2$. (Any number could be used, but a multiple of 2 makes multiplying by $-\frac{3}{2}$ easier.)

$y = -\frac{3}{2}x + 3$

$y = -\frac{3}{2}(-2) + 3$ Let $x = -2$.

$y = 3 + 3$

$y = 6$

These three ordered pairs are shown in the table with Figure 10. Plot the corresponding points, then draw a line through them. This line, shown in Figure 10, is the graph of $y = -\frac{3}{2}x + 3$.

x	y
0	3
2	0
−2	6

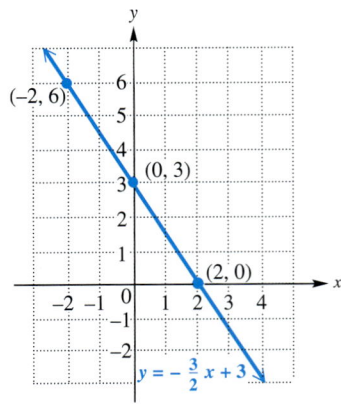

Figure 10

Continued on Next Page

CAUTION
When graphing a linear equation as in Example 1, all three points should lie on the same straight line. If they don't, double-check the ordered pairs you found.

Work Problem 2 at the Side.

② Complete the table of values, and graph the linear equation.

$x + y = 6$

x	y
0	
	0
2	

EXAMPLE 2   Graphing a Linear Equation

Graph the linear equation $4x = 5y + 20$.

As before, at least two different points are needed to draw the graph. First let $x = 0$ and then let $y = 0$ to complete two ordered pairs.

$4x = 5y + 20$ $4x = 5y + 20$
$4(0) = 5y + 20$ Let $x = 0$. $4x = 5(0) + 20$ Let $y = 0$.
$0 = 5y + 20$ $4x = 0 + 20$
$-5y = 20$ $4x = 20$
$y = -4$ $x = 5$

The ordered pairs are $(0, -4)$ and $(5, 0)$. Get a third ordered pair (as a check) by choosing some number other than 0 for x or y. We choose $y = 2$. Replacing y with 2 in the equation $4x = 5y + 20$ leads to the ordered pair $\left(\frac{15}{2}, 2\right)$, or $\left(7\frac{1}{2}, 2\right)$.

Plot the three ordered pairs $(0, -4)$, $(5, 0)$, and $\left(7\frac{1}{2}, 2\right)$, and draw a line through them. This line, shown in Figure 11, is the graph of $4x = 5y + 20$.

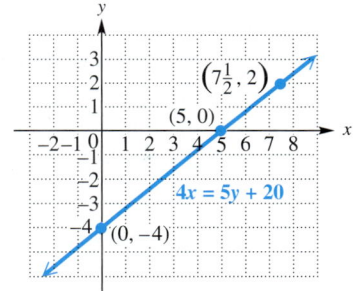

Figure 11

③ Make a table of values, and graph the linear equation.

$2x = 3y + 6$

x	y

Work Problem 3 at the Side.

OBJECTIVE 2 Find intercepts. In Figure 11, the graph crosses or intersects the y-axis at $(0, -4)$ and the x-axis at $(5, 0)$. For this reason, $(0, -4)$ is called the **y-intercept,** and $(5, 0)$ is called the **x-intercept** of the graph. The intercepts are particularly useful for graphing linear equations. (In general, any point on the y-axis has x-coordinate 0; any point on the x-axis has y-coordinate 0.) The intercepts are found by replacing, in turn, each variable with 0 in the equation and solving for the value of the other variable.

Finding Intercepts

To find the **x-intercept,** let $y = 0$ in the given equation and solve for x. Then $(x, 0)$ is the x-intercept.

To find the **y-intercept,** let $x = 0$ in the given equation and solve for y. Then $(0, y)$ is the y-intercept.

ANSWERS

2.

x	y
0	6
6	0
2	4

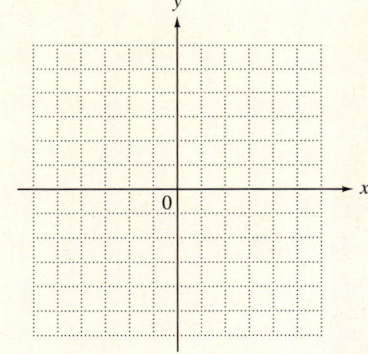

3.

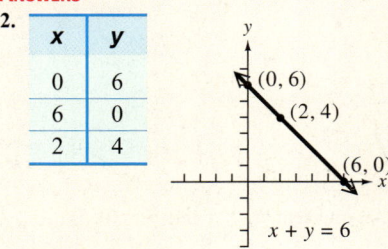

Graph $(3, 0)$ and $(0, -2)$ and one other point.

4 Find the intercepts for the graph of $5x + 2y = 10$. Then draw the graph. (Be sure to get a third point as a check.)

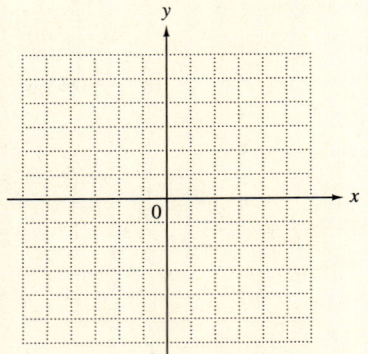

EXAMPLE 3 Finding Intercepts

Find the intercepts for the graph of $2x + y = 4$. Then draw the graph.
To find the y-intercept, let $x = 0$; to find the x-intercept, let $y = 0$.

$2x + y = 4$	$2x + y = 4$
$2(0) + y = 4$ Let $x = 0$.	$2x + 0 = 4$ Let $y = 0$.
$0 + y = 4$	$2x = 4$
$y = 4$	$x = 2$

The y-intercept is $(0, 4)$. The x-intercept is $(2, 0)$. The graph with the two intercepts shown in color is given in Figure 12. Find a third point as a check. For example, choosing $x = 1$ gives $y = 2$. Plot $(0, 4)$, $(2, 0)$, and $(1, 2)$ and draw a line through them. This line, shown in Figure 12, is the graph.

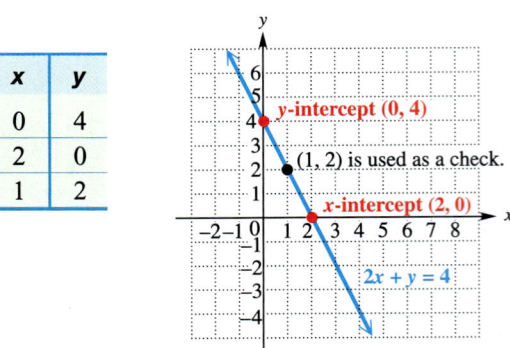

x	y
0	4
2	0
1	2

Figure 12

Work Problem 4 at the Side.

OBJECTIVE 3 Graph linear equations where the intercepts coincide. In the preceding examples, the x- and y-intercepts were used to help draw the graphs. This is not always possible. Example 4 shows what to do when the x- and y-intercepts are the same point (that is, coincide).

EXAMPLE 4 Graphing an Equation of the Form $Ax + By = 0$

Graph the linear equation $x - 3y = 0$.

If we let $x = 0$, then $y = 0$, giving the ordered pair $(0, 0)$. Letting $y = 0$ also gives $(0, 0)$. This is the same ordered pair, so choose two *other* values for x or y. Choosing 2 for y gives $x - 3 \cdot 2 = 0$, or $x = 6$, so another ordered pair is $(6, 2)$. Choosing -6 for x gives -2 for y. The ordered pairs $(-6, -2)$, $(0, 0)$, and $(6, 2)$ are used to sketch the graph in Figure 13.

x	y
0	0
6	2
-6	-2

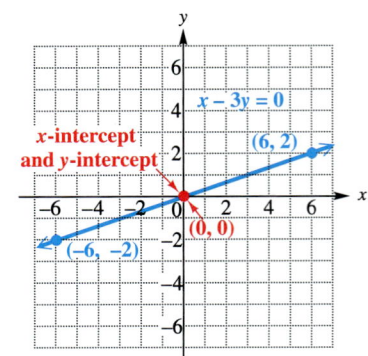

Figure 13

ANSWERS
4. x-intercept $(2, 0)$; y-intercept $(0, 5)$

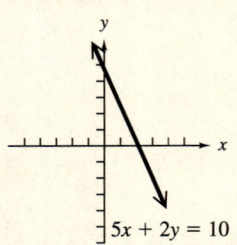

Example 4 can be generalized as follows.

> If A and B are nonzero real numbers, the graph of a linear equation of the form
>
> $$Ax + By = 0$$
>
> passes through the origin $(0, 0)$.

Work Problem 5 at the Side.

OBJECTIVE 4 Graph linear equations of the form $y = k$ or $x = k$.
The equation $y = -4$ is a linear equation in which the coefficient of x is 0. (To see this, write $y = -4$ as $0x + y = -4$.) Also, $x = 3$ is a linear equation in which the coefficient of y is 0. These equations lead to horizontal or vertical straight lines, as the next example shows.

EXAMPLE 5 Graphing Equations of the Form $y = k$ and $x = k$

(a) Graph the linear equation $y = -4$.

As the equation states, for any value of x, y is always equal to -4. Three ordered pairs that satisfy the equation are shown in the table of values. Drawing a line through these points gives the horizontal line in Figure 14. The y-intercept is $(0, -4)$; there is no x-intercept.

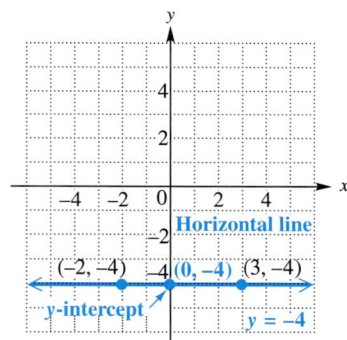

Figure 14

(b) Graph $x - 3 = 0$.

First add 3 to each side of $x - 3 = 0$ to get $x = 3$. All the ordered pairs that satisfy this equation have x-coordinate 3. Any number can be used for y. See Figure 15 for the graph of this vertical line, along with a table of values. The x-intercept is $(3, 0)$; there is no y-intercept.

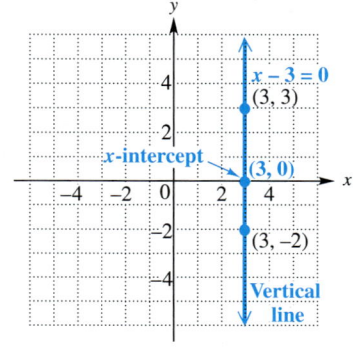

Figure 15

5 Graph each equation.

(a) $2x - y = 0$

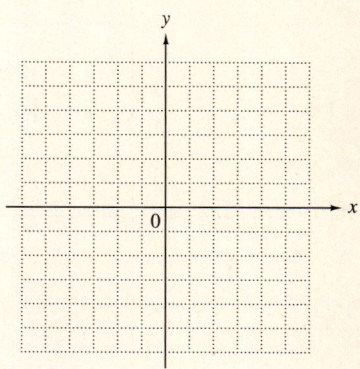

(b) $x = -4y$

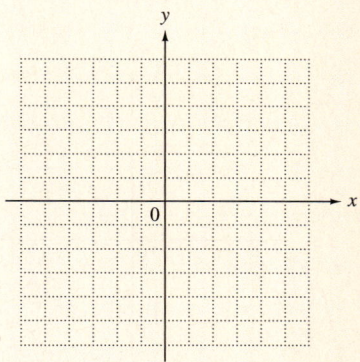

ANSWERS

5. (a)

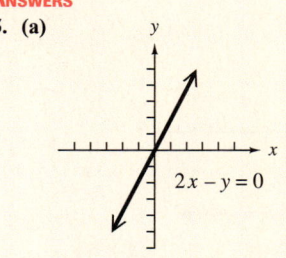

(b)

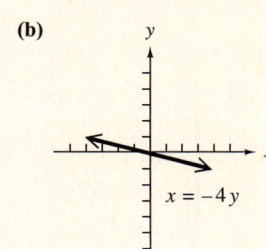

6 Graph each equation.

(a) $y = -5$

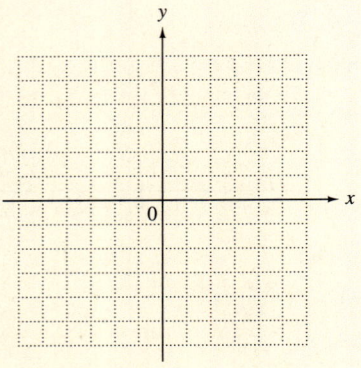

(b) $x + 4 = 6$

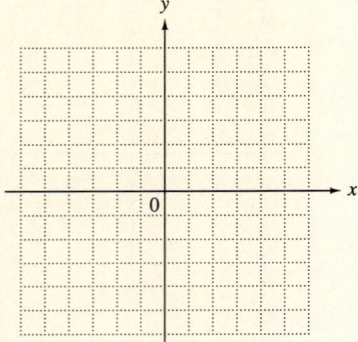

From the results in Example 5, we make the following observations.

Horizontal and Vertical Lines

The graph of the linear equation $y = k$, where k is a real number, is the horizontal line with y-intercept $(0, k)$ and no x-intercept.

The graph of the linear equation $x = k$, where k is a real number, is the vertical line with x-intercept $(k, 0)$ and no y-intercept.

Work Problem 6 at the Side.

The different forms of linear equations from this section and the methods of graphing them are summarized below.

Graphing Linear Equations

Equation	Graphing Method	Example
$y = k$	Draw a horizontal line through $(0, k)$.	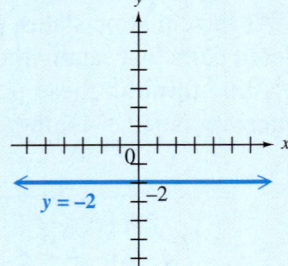
$x = k$	Draw a vertical line through $(k, 0)$.	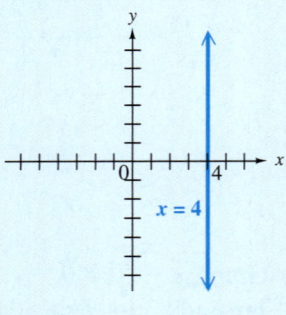
$Ax + By = 0$	Graph passes through $(0, 0)$. To get additional points that lie on the graph, choose any values for x or y, except 0.	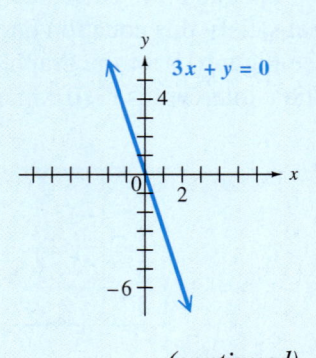

(continued)

ANSWERS

6. (a)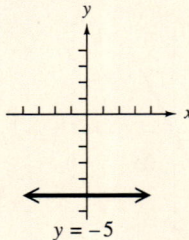

(b)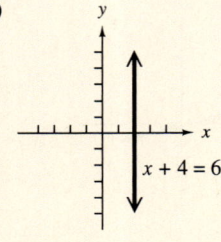

$Ax + By = C$
$A, B,$ and $C \neq 0$

Find any two points on the line. A good choice is to find the intercepts. Let $x = 0$, and find the corresponding value of y; then let $y = 0$, and find x. As a check, get a third point by choosing a value of x or y that has not yet been used.

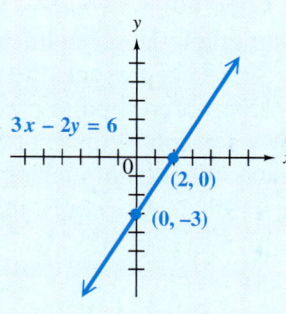

Work Problem 7 at the Side.

NOTE
Another method of graphing linear equations, using the concepts of slope and y-intercept, will be covered in Objective 2 of **Section 11.4.**

OBJECTIVE 5 Use a linear equation to model data.

EXAMPLE 6 Using a Linear Equation to Model Credit Card Debt

Credit card debt in households in the United States increased steadily during the 1990s. The amount of debt y in billions of dollars can be modeled by the linear equation

$$y = 47.3x + 281,$$

where $x = 0$ represents the year 1992, $x = 1$ represents 1993, and so on. (*Source:* Board of Governors of the Federal Reserve System.)

(a) Use the equation to approximate credit card debt in the years 1992, 1993, and 1999.

Substitute the appropriate value for each year x to find credit card debt in that year.

For 1992: $y = 47.3(\mathbf{0}) + 281$ Replace x with 0.
$y = 281$ billion dollars

For 1993: $y = 47.3(\mathbf{1}) + 281$ Replace x with 1.
$y = 328.3$ billion dollars

For 1999: $y = 47.3(\mathbf{7}) + 281$ Replace x with 7.
$y = 612.1$ billion dollars

Continued on Next Page

7 Match the information about the graphs with the linear equations in A–D.

A. $x = 5$
B. $2x - 5y = 8$
C. $y - 2 = 3$
D. $x + 4y = 0$

(a) The graph of the equation is a horizontal line.

(b) The graph of the equation passes through the origin.

(c) The graph of the equation is a vertical line.

(d) The graph of the equation passes through $(9, 2)$.

ANSWERS
7. **(a)** C   **(b)** D   **(c)** A   **(d)** B

8 Use the graph and then the equation in Example 6 to approximate credit card debt in 1997.

(b) Write the information from part (a) as three ordered pairs, and use them to graph the given linear equation.

Since x represents the year and y represents the debt in billions of dollars, the ordered pairs are (0, 281), (1, 328.3), and (7, 612.1). Figure 16 shows a graph of these ordered pairs and the line through them. (Note that arrowheads are not included with the graphed line since the data are for the years 1992 to 1999 only, that is, from $x = 0$ to $x = 7$.)

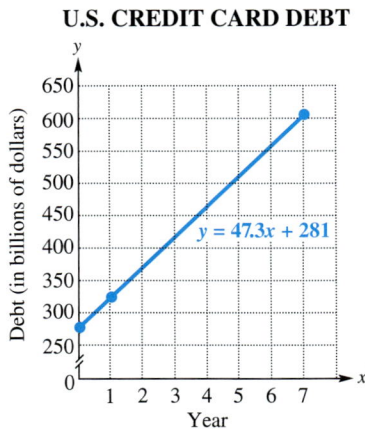

Figure 16

(c) Use the graph and then the equation to approximate credit card debt in 1996.

For 1996, $x = 4$. On the graph, find 4 on the horizontal axis and move up to the graphed line, then across to the vertical axis. It appears that credit card debt in 1996 was about 465 billion dollars.

To use the equation, substitute 4 for x.

$$y = 47.3x + 281$$
$$y = 47.3(4) + 281 \qquad \text{Let } x = 4.$$
$$y = 470.2 \text{ billion dollars}$$

This result for 1996 is close to our estimate using the graph.

◀◀ **Work Problem 8 at the Side.**

ANSWERS
8. about 510 billion dollars; 517.5 billion dollars

11.2 Exercises

Complete the given ordered pairs for each equation. Then graph each equation by plotting the points and drawing a line through them. See Examples 1 and 2.

1. $y = -x + 5$

    $(0, \ ), (\ , 0), (2, \ )$

    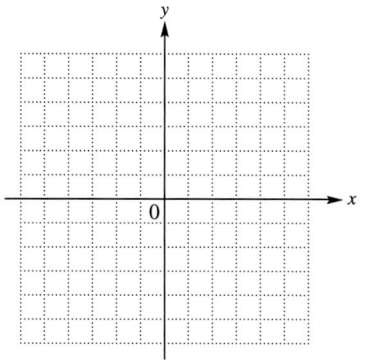

2. $y = x - 2$

    $(0, \ ), (\ , 0), (5, \ )$

    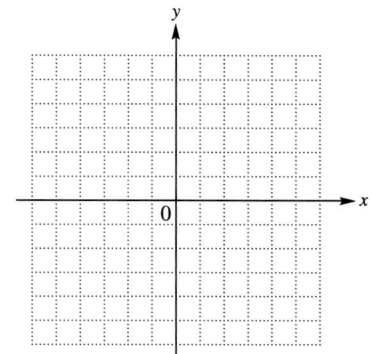

3. $y = \dfrac{2}{3}x + 1$

    $(0, \ ), (3, \ ), (-3, \ )$

    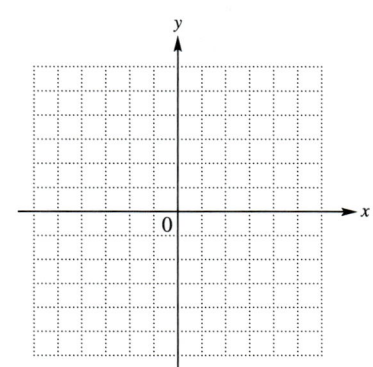

4. $y = -\dfrac{3}{4}x + 2$

    $(0, \ ), (4, \ ), (-4, \ )$

    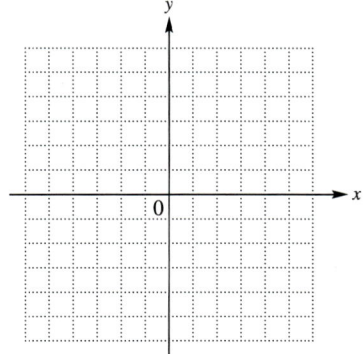

5. $3x = -y - 6$

    $(0, \ ), (\ , 0), \left(-\dfrac{1}{3}, \ \right)$

    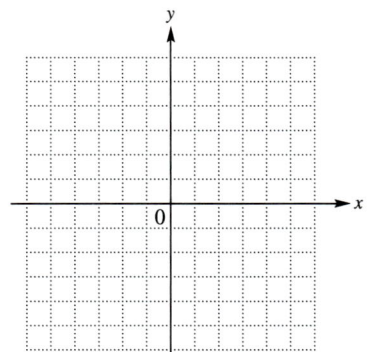

6. $x = 2y + 3$

    $(\ , 0), (0, \ ), \left(\ , \dfrac{1}{2}\right)$

    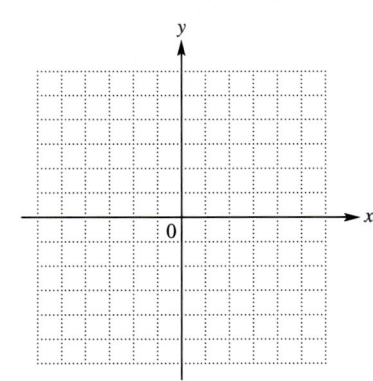

Match the information about each graph in Column I with the correct linear equation in Column II.

I	II
7. The graph of the equation has y-intercept $(0, -4)$.	A. $3x + y = -4$
8. The graph of the equation has $(0, 0)$ as x-intercept and y-intercept.	B. $x - 4 = 0$
9. The graph of the equation does not have an x-intercept.	C. $y = 4x$
10. The graph of the equation has x-intercept $(4, 0)$.	D. $y = 4$

754 Chapter 11 Graphs of Linear Equations and Inequalities in Two Variables

Find the intercepts for the graph of each equation. See Example 3.

11. $2x - 3y = 24$
    x-intercept:
    y-intercept:

12. $-3x + 8y = 48$
    x-intercept:
    y-intercept:

13. $x + 6y = 0$
    x-intercept:
    y-intercept:

14. $3x - y = 0$
    x-intercept:
    y-intercept:

15. A student attempted to graph $4x + 5y = 0$ by finding intercepts. She first let $x = 0$ and found y; then she let $y = 0$ and found x. In both cases, the resulting point was $(0, 0)$. She knew that she needed at least two points to graph the line, but was unsure what to do next because finding intercepts gave her only one point. How would you explain to her what to do next?

16. What is the equation of the *x*-axis? What is the equation of the *y*-axis?

Graph each linear equation. See Examples 1–5.

17. $x = y + 2$

18. $x = -y + 6$

19. $x - y = 4$

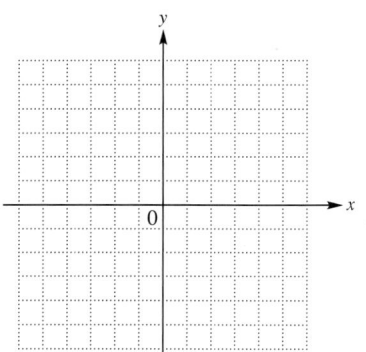
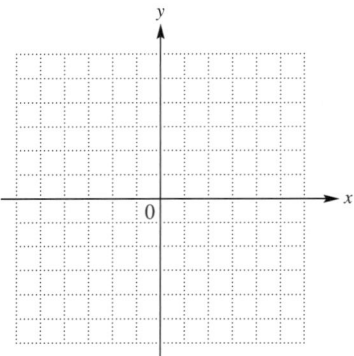

20. $x - y = 5$

21. $2x + y = 6$

22. $-3x + y = -6$

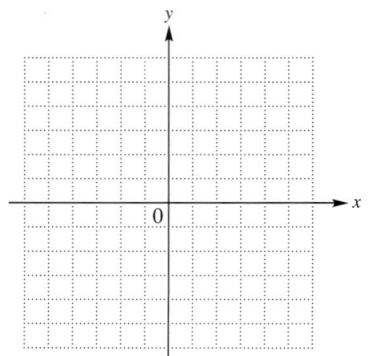
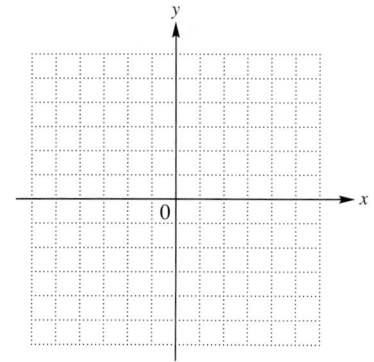
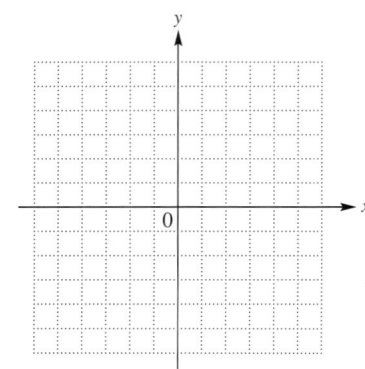

23. $3x + 7y = 14$

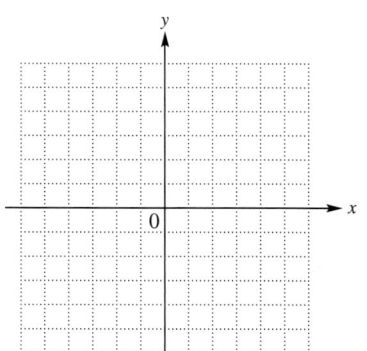

24. $6x - 5y = 18$

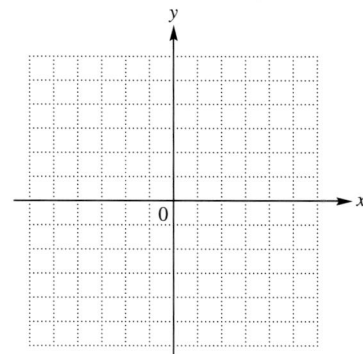

25. $y - 2x = 0$

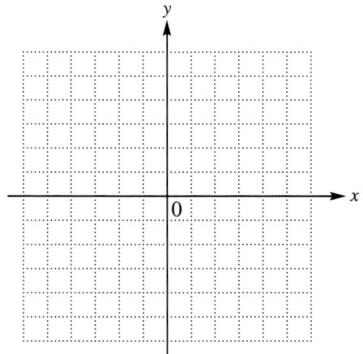

26. $y + 3x = 0$

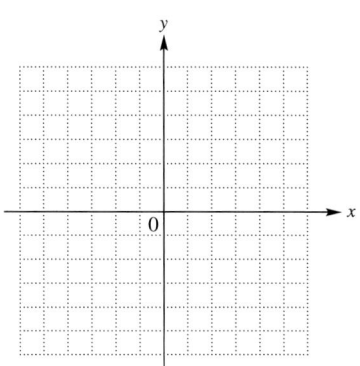

27. $y = -6x$

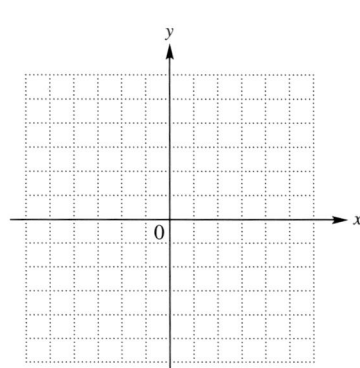

28. $x = 4$

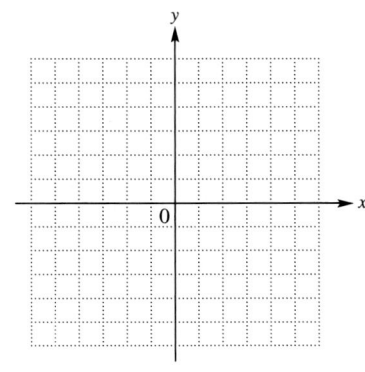

29. $x = -2$

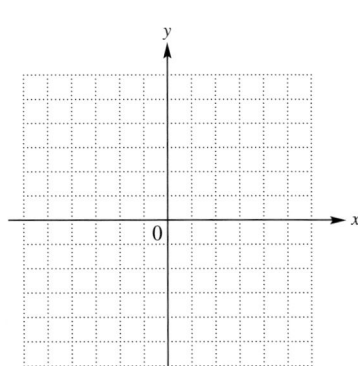

30. $y + 1 = 0$

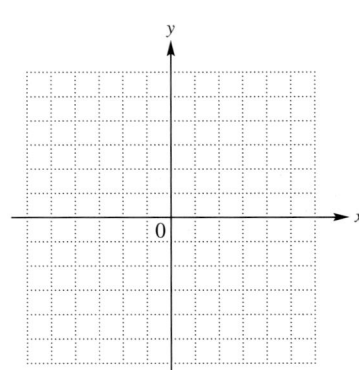

31. $y - 3 = 0$

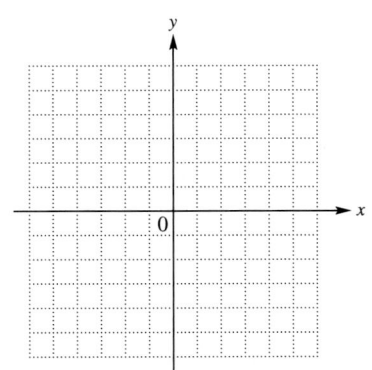

32. Write a few sentences summarizing how to graph a linear equation in two variables.

Solve each problem. See Example 6.

33. The height y (in centimeters) of a woman is related to the length of her radius bone x (from the wrist to the elbow) and is approximated by this linear equation.

$$y = 3.9x + 73.5$$

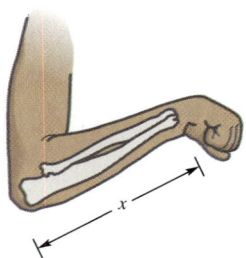

(a) Use the equation to find the approximate heights of women with radius bones of lengths 20 cm, 26 cm, and 22 cm.

(b) Graph the equation using the data from part (a).

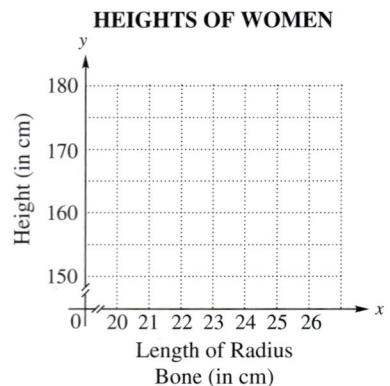

(c) Use the graph to estimate the length of the radius bone in a woman who is 167 cm tall. Then use the equation to find the length of this radius bone to the nearest centimeter. (*Hint:* Substitute for y in the equation.)

34. The weight y (in pounds) of a man taller than 60 in. can be roughly approximated by the linear equation

$$y = 5.5x - 220,$$

where x is the height of the man in inches.

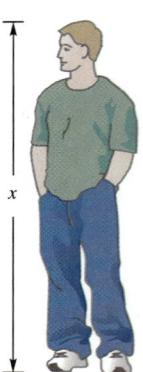

(a) Use the equation to approximate the weights of men whose heights are 62 in., 66 in., and 72 in.

(b) Graph the equation using the data from part (a).

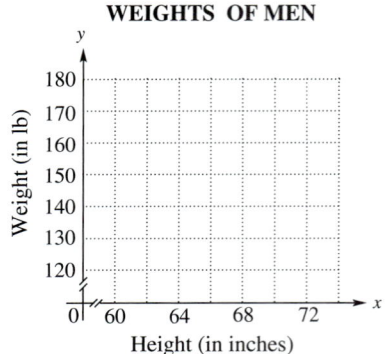

(c) Use the graph to estimate the height of a man who weighs 155 lb. Then use the equation to find the height of this man to the nearest inch. (*Hint:* Substitute for y in the equation.)

35. Refer to **Section 11.1** Exercise 80. Draw a line through the points you plotted in the scatter diagram there.

(a) Use the graph to estimate the lower limit of the target heart rate zone for age 30.

(b) Use the linear equation given there to approximate the lower limit for age 30.

(c) How does the approximation using the equation compare to the estimate from the graph?

36. Refer to **Section 11.1** Exercise 81. Draw a line through the points you plotted in the scatter diagram there.

(a) Use the graph to estimate the upper limit of the target heart rate zone for age 30.

(b) Use the linear equation given there to approximate the upper limit for age 30.

(c) How does the approximation using the equation compare to the estimate from the graph?

37. Use the results of Exercises 35(b) and 36(b) to determine the target heart rate zone for age 30.

38. Should the graphs of the target heart rate zone in the **Section 11.1** exercises be used to estimate the target heart rate zone for ages below 20 or above 80? Why or why not?

39. Per capita consumption of coffee increased for the years 1995 through 2000 as shown in the graph. If $x = 0$ represents 1995, $x = 1$ represents 1996, and so on, per capita consumption y in gallons can be modeled by this linear equation.

$$y = 1.13x + 20.67$$

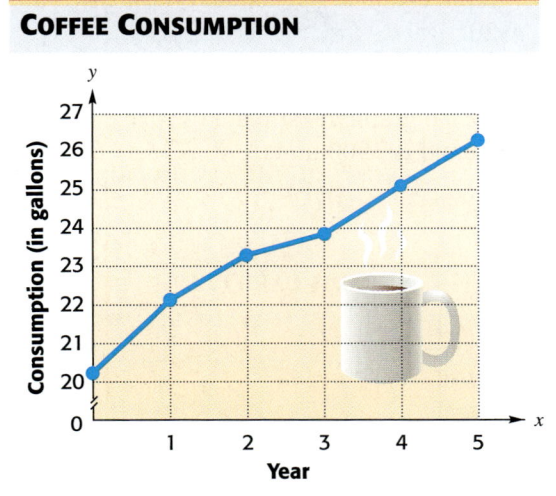

Source: U.S. Department of Agriculture.

(a) Use the equation to approximate consumption in 1995, 1996, and 1998 to the nearest tenth.

(b) Use the graph to estimate consumption for the same years.

(c) How do the approximations using the equation compare to the estimates from the graph?

40. Sporting goods sales y (in billions of dollars) from 1995 through 2000 are modeled by the linear equation

$$y = 3.606x + 41.86$$

where $x = 5$ corresponds to 1995, $x = 6$ corresponds to 1996, and so on.

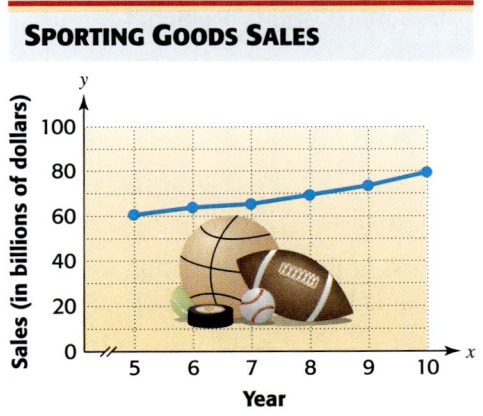

Source: U.S. Bureau of the Census.

(a) Use the equation to approximate sporting goods sales in 1995, 1997, and 2000. Round your answers to the nearest billion dollars.

(b) Use the graph to estimate sales for the same years.

(c) How do the approximations using the equation compare to the estimates using the graph?

41. The graph shows the value of a certain sport-utility vehicle (SUV) over the first 5 years of ownership.

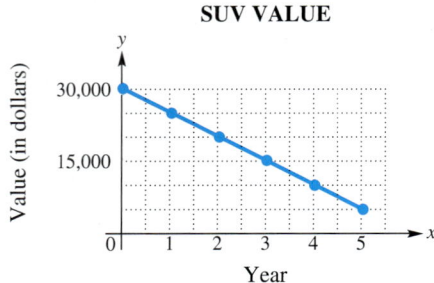

SUV VALUE

Use the graph to do the following.

(a) Determine the initial value of the SUV.

(b) Find the *depreciation* (loss in value) from the original value after the first 3 years.

(c) What is the annual or yearly depreciation in each of the first 5 years?

(d) What does the ordered pair (5, 5000) mean in the context of this problem?

42. Demand for an item is often closely related to its price. As price increases, demand decreases, and as price decreases, demand increases. Suppose demand for a video game is 2000 units when the price is $40, and demand is 2500 units when the price is $30.

(a) Let x be the price and y be the demand for the game. Graph the two given pairs of prices and demands.

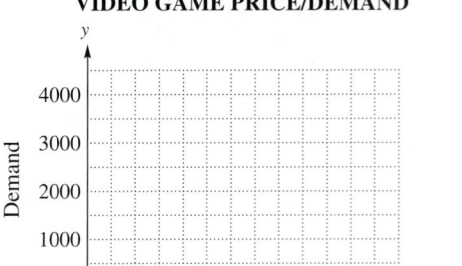

VIDEO GAME PRICE/DEMAND

(b) Assume the relationship is linear. Draw a line through the two points from part (a). From your graph, estimate the demand if the price drops to $20.

(c) Use the graph to estimate the price if the demand is 3500 units.

43. The graph of the linear equation for credit card debt from Example 6,

$$y = 47.3x + 281,$$

where $x = 0$ represents 1992, and so on, and y is in billions of dollars, is shown by the blue line in the figure. The actual data for 1992 through 1999 is also plotted as the red dots.

(a) In general, how well does the linear equation model the actual data?

(b) Use the plotted points to estimate the actual credit card debt for 1996. How does it compare to the answer in Example 6(c)?

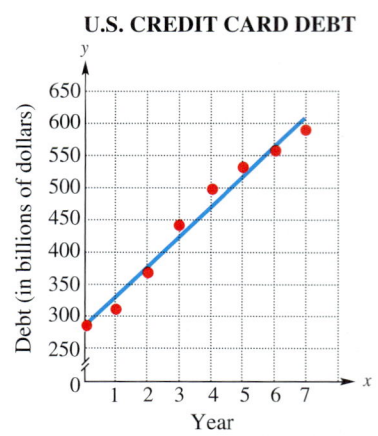

U.S. CREDIT CARD DEBT

Source: Board of Governors of the Federal Reserve System.

(c) Should this equation be used to predict credit card debt for the year 2010? Why or why not?

11.3 Slope of a Line

An important characteristic of the lines we graphed in the previous section is their slant or "steepness", as viewed from left to right. See Figure 17.

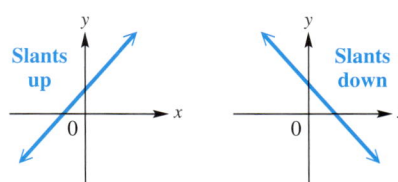

Figure 17

OBJECTIVES

1. Find the slope of a line given two points.
2. Find the slope from the equation of a line.
3. Use slope to determine whether two lines are parallel, perpendicular, or neither.

One way to measure the steepness of a line is to compare the vertical change in the line to the horizontal change while moving along the line from one fixed point to another. This measure of steepness is called the *slope* of the line.

OBJECTIVE 1 Find the slope of a line given two points. Figure 18 shows a line through two nonspecific points (x_1, y_1) and (x_2, y_2). (This notation is called **subscript notation**. Read x_1 as "x-sub-one" and x_2 as "x-sub-two.")

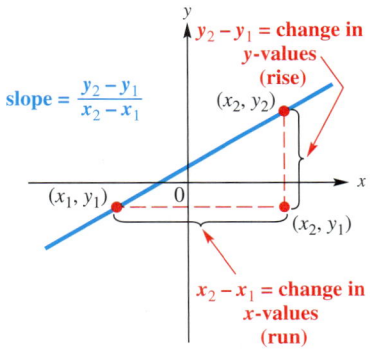

Figure 18

Moving along the line from the point (x_1, y_1) to the point (x_2, y_2) causes y to change by $y_2 - y_1$ units. This is the vertical change or **rise**. Similarly, x changes by $x_2 - x_1$ units, which is the horizontal change or **run**. (In both cases, the change is expressed as a *difference*.) Remember from **Section 6.1** that one way to compare two numbers is by using a ratio. **Slope** is the ratio of the vertical change in y to the horizontal change in x.

EXAMPLE 1 Comparing Rise to Run

Find the slope ratio of the rise to the run for the line shown in Figure 19(a) at the side and the line shown in Figure 19(b) on the next page.

We use the two points shown on each line. For the line in Figure 19(a), we find the rise from point Q to point P by determining the vertical change from -4 to 4, the difference $4 - (-4) = 8$. Similarly, we find the run by determining the horizontal change from Q to P, $5 - (-5) = 10$. Slope is the ratio

$$\frac{\text{rise}}{\text{run}} = \frac{8}{10} \text{ or } \frac{4}{5}.$$

Continued on Next Page

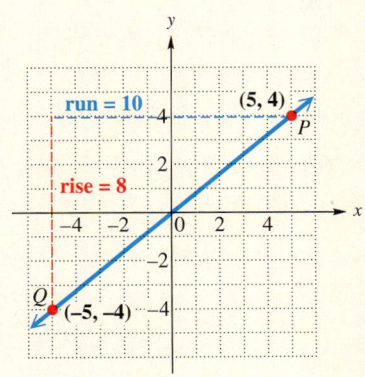

Figure 19(a)

760 Chapter 11 Graphs of Linear Equations and Inequalities in Two Variables

  Find the slope ratio of the rise to the run for each line.

(a)
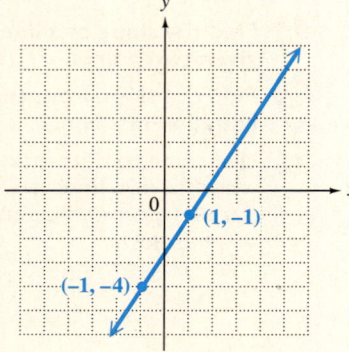

Moving from Q to P, the line in Figure 19(b) has rise $3 - (-1) = 4$ and run $2 - 6 = -4$, so the slope is the ratio

$$\frac{\text{rise}}{\text{run}} = \frac{4}{-4} = \frac{1}{-1} \text{ or } -1.$$

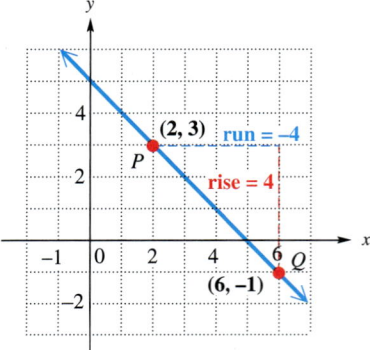

Figure 19(b)

To confirm our slope ratios, count grid squares from one point on the line to another. For example, starting at the point $(5, 0)$ in Figure 19(b), count up 1 square (the rise in the slope ratio $\frac{1}{-1}$) and then 1 square to the *left* (the run) to arrive at the point $(4, 1)$ on the line.

◄◄ Work Problem 1 at the Side.

(b)
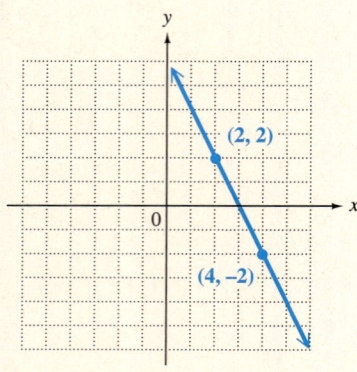

The idea of slope is used in many everyday situations. See Figure 20. For example, a highway with a 10% or $\frac{1}{10}$ grade (or slope) rises 1 m for every 10 m horizontally. Architects specify the pitch of a roof using slope; a $\frac{5}{12}$ roof means that the roof rises 5 ft for every 12 ft in the horizontal direction. The slope of a stairwell also indicates the ratio of the vertical rise to the horizontal run. In the figure, the slope of the stairwell is $\frac{8}{10}$ or $\frac{4}{5}$.

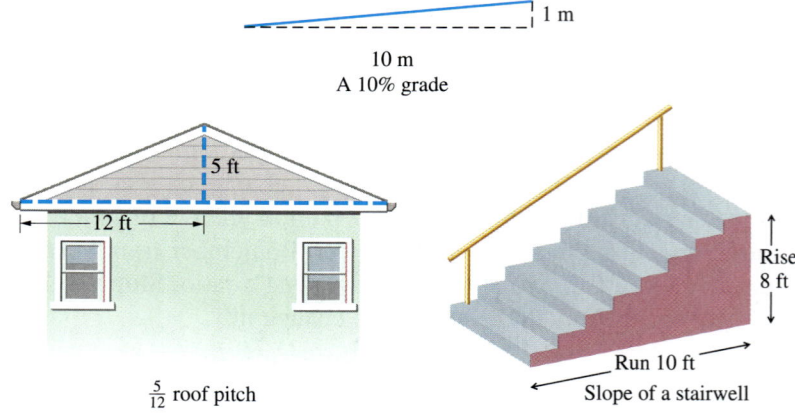

Figure 20

 Find $\dfrac{y_2 - y_1}{x_2 - x_1}$ for the following values.

(a) $y_2 = 4, y_1 = -1,$
$x_2 = 3, x_1 = 4$

If we know the coordinates of two points on a line, we can find its slope, traditionally designated m, using the slope formula.

(b) $x_1 = 3, x_2 = -5,$
$y_1 = 7, y_2 = -9$

Slope Formula

The **slope** of the line through the points (x_1, y_1) and (x_2, y_2) is

$$m = \frac{\text{change in } y}{\text{change in } x} = \frac{y_2 - y_1}{x_2 - x_1}, \quad \text{if } x_1 \neq x_2.$$

(c) $x_1 = 2, x_2 = 7,$
$y_1 = 4, y_2 = 9$

ANSWERS
1. (a) $\frac{3}{2}$ (b) -2
2. (a) -5 (b) 2 (c) 1

◄◄ Work Problem 2 at the Side.

The slope of a line tells how fast y changes for each unit of change in x; that is, the slope gives the rate of change in y for each unit of change in x.

EXAMPLE 2 Using the Slope Formula

Find the slope of each line.

(a) The line through $(1, -2)$ and $(-4, 7)$

Use the slope formula. Let $(-4, 7) = (x_2, y_2)$ and $(1, -2) = (x_1, y_1)$. Then

$$\text{slope } m = \frac{\text{change in } y}{\text{change in } x} = \frac{y_2 - y_1}{x_2 - x_1} = \frac{7 - (-2)}{-4 - 1} = \frac{9}{-5} = -\frac{9}{5}.$$

See Figure 21(a).

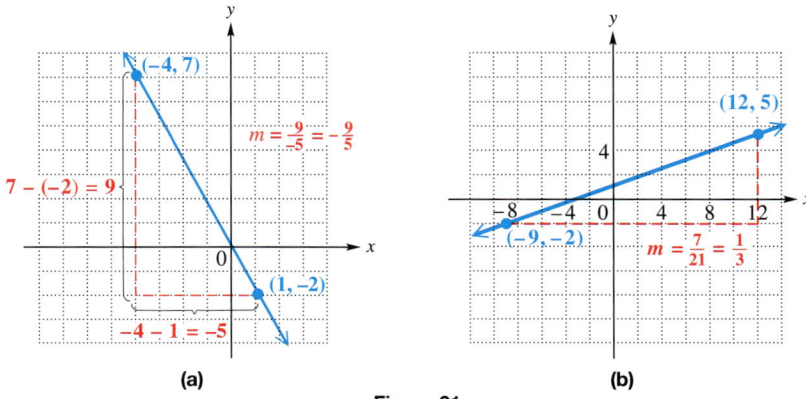

(a)   (b)

Figure 21

(b) The line through $(-9, -2)$ and $(12, 5)$

$$m = \frac{y_2 - y_1}{x_2 - x_1} = \frac{5 - (-2)}{12 - (-9)} = \frac{7}{21} = \frac{1}{3}$$

See Figure 21(b). The same slope is obtained by subtracting in reverse order.

$$m = \frac{-2 - 5}{-9 - 12} = \frac{-7}{-21} = \frac{1}{3}$$

> **CAUTION**
> *It makes no difference which point is (x_1, y_1) or (x_2, y_2); however, it is important to be consistent.* Start with the x- and y-values of one point (either one) and subtract the corresponding values of the other point. Also, the slope of a line is the same for *any* two points on the line.

Work Problem 3 at the Side.

In Example 2(a) the slope is negative and the corresponding line in Figure 21(a) falls from left to right. The slope in Example 2(b) is positive and the corresponding line in Figure 21(b) rises from left to right. These facts can be generalized.

---

3 Find the slope of each line.

(a) Through $(6, -2)$ and $(5, 4)$

(b) Through $(-3, 5)$ and $(-4, -7)$

(c) Through $(6, -8)$ and $(-2, 4)$
Find this slope in two different ways as in Example 2(b).

ANSWERS

3. **(a)** -6 **(b)** 12 **(c)** $-\dfrac{3}{2}; -\dfrac{3}{2}$

> **Positive and Negative Slopes**
>
> A line with positive slope rises from left to right.
>
> A line with negative slope falls from left to right.

EXAMPLE 3 Showing that the Slope of a Horizontal Line Is Zero

Find the slope of the line through $(-8, 4)$ and $(2, 4)$.

$$m = \frac{y_2 - y_1}{x_2 - x_1} = \frac{4 - 4}{-8 - 2} = \frac{0}{-10} = 0 \qquad \text{Zero slope}$$

As shown in Figure 22, the line through the given points is horizontal. *All horizontal lines have slope 0* since the difference in y-values is always 0.

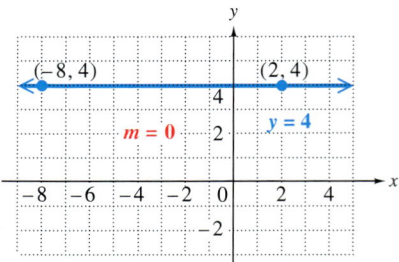

Figure 22

EXAMPLE 4 Showing that a Vertical Line Has Undefined Slope

Find the slope of the line through $(6, 2)$ and $(6, -4)$.

$$m = \frac{y_2 - y_1}{x_2 - x_1} = \frac{2 - (-4)}{6 - 6} = \frac{6}{0} \qquad \text{Undefined slope}$$

Because division by 0 is undefined, this line has undefined slope. (This is why the slope formula at the beginning of this section had the restriction $x_1 \neq x_2$.) The graph in Figure 23 shows that this line is vertical. All points on a vertical line have the same x-value, so *all vertical lines have undefined slope.*

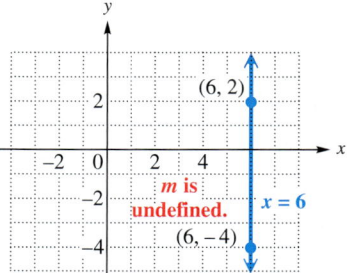

Figure 23

Slopes of Horizontal and Vertical Lines

Horizontal lines, which have equations of the form $y = k$, have **slope 0.**

Vertical lines, which have equations of the form $x = k$, have **undefined slope.**

Work Problem 4 at the Side.

OBJECTIVE 2 Find the slope from the equation of a line. The slope of a line can be found directly from its equation. For example, the slope of the line

$$y = -3x + 5$$

can be found using any two points on the line. We get these two points by first choosing two different values of x and then finding the corresponding values of y. Choose $x = -2$ and $x = 4$.

$y = -3x + 5$	$y = -3x + 5$
$y = -3(-2) + 5$ Let $x = -2$.	$y = -3(4) + 5$ Let $x = 4$.
$y = 6 + 5$	$y = -12 + 5$
$y = 11$	$y = -7$

The ordered pairs are $(-2, 11)$ and $(4, -7)$. Now use the slope formula.

$$m = \frac{11 - (-7)}{-2 - 4} = \frac{18}{-6} = -3$$

The slope, -3, is the same number as the coefficient of x in the equation $y = -3x + 5$. It can be shown that this always happens, *as long as the equation is solved for y.* This fact is used to find the slope of a line from its equation.

Finding the Slope of a Line from Its Equation

Step 1 Solve the equation for y.

Step 2 The slope is given by the coefficient of x.

NOTE
We will see in the next section that the equation $y = -3x + 5$ is written using a special form of the equation of a line,

$$y = mx + b,$$

called *slope-intercept form.*

---

④ Find the slope of each line.

(a) Through $(2, 5)$ and $(-1, 5)$

(b) Through $(3, 1)$ and $(3, -4)$

(c) With equation $y = -1$

(d) With equation $x - 4 = 0$

ANSWERS
4. (a) 0 (b) undefined
   (c) 0 (d) undefined

5 Find the slope of each line.

(a) $y = -\dfrac{7}{2}x + 1$

(b) $3x + 2y = 9$

(c) $y + 4 = 0$

(d) $x + 3 = 7$

EXAMPLE 5 Finding Slopes from Equations

Find the slope of each line.

(a) $2x - 5y = 4$

Step 1 Solve the equation for y.

$$2x - 5y = 4$$
$$-5y = -2x + 4 \qquad \text{Subtract } 2x \text{ from both sides.}$$
$$y = \dfrac{2}{5}x - \dfrac{4}{5} \qquad \text{Divide both sides by } -5.$$

Step 2 The slope is given by the coefficient of x, so the slope is $\dfrac{2}{5}$.

(b) $8x + 4y = 1$

Solve for y.

$$8x + 4y = 1$$
$$4y = -8x + 1 \qquad \text{Subtract } 8x \text{ from both sides.}$$
$$y = -2x + \dfrac{1}{4} \qquad \text{Divide both sides by 4.}$$

The slope of this line is given by the coefficient of x, which is -2.

Work Problem 5 at the Side.

OBJECTIVE 3 **Use slope to determine whether two lines are parallel, perpendicular, or neither.** Two lines in a plane that never intersect are **parallel.** We use slopes to tell whether two lines are parallel. For example, Figure 24 shows the graphs of $x + 2y = 4$ and $x + 2y = -6$. These lines appear to be parallel. Solving for y, we find that both $x + 2y = 4$ and $x + 2y = -6$ have slope $-\dfrac{1}{2}$. *Nonvertical parallel lines always have equal slopes.*

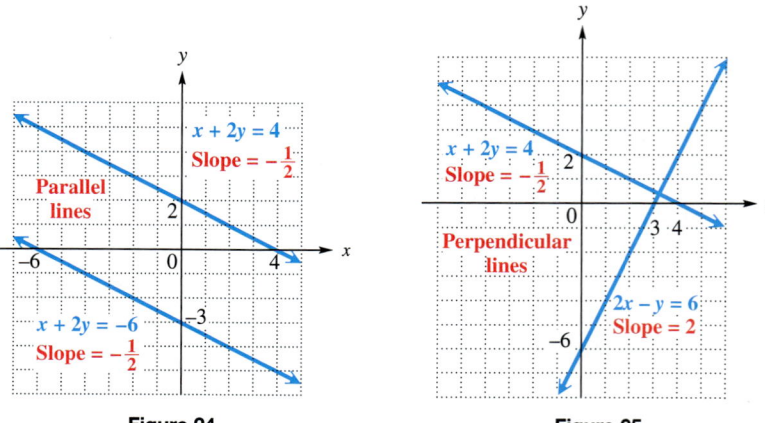

Figure 24   Figure 25

Figure 25 shows the graphs of $x + 2y = 4$ and $2x - y = 6$. These lines appear to be **perpendicular** (that is, they intersect at a 90° angle). Solving for y shows that the slope of $x + 2y = 4$ is $-\dfrac{1}{2}$, while the slope of $2x - y = 6$ is 2. The product of $-\dfrac{1}{2}$ and 2 is

$$-\dfrac{1}{2}(2) = -1.$$

This is true in general; *the product of the slopes of two perpendicular lines (neither of which is vertical) is always -1.*

ANSWERS

5. (a) $-\dfrac{7}{2}$ (b) $-\dfrac{3}{2}$ (c) 0   (d) undefined

Slopes of Parallel and Perpendicular Lines

Two lines with the same slope are parallel.

Two lines whose slopes have a product of -1 are perpendicular.

EXAMPLE 6 Deciding Whether Lines Are Parallel, Perpendicular, or Neither

Decide whether each pair of lines is *parallel, perpendicular,* or *neither*.

(a) $x + 2y = 7$
$-2x + y = 3$

Find the slope of each line by first solving each equation for y.

$x + 2y = 7$	$-2x + y = 3$
$2y = -x + 7$	$y = 2x + 3$
$y = -\dfrac{1}{2}x + \dfrac{7}{2}$	
Slope is $-\dfrac{1}{2}$.	Slope is 2.

Because the slopes are not equal, the lines are not parallel. Check the product of the slopes: $-\dfrac{1}{2}(2) = -1$. The two lines are perpendicular because the product of their slopes is -1. See Figure 26.

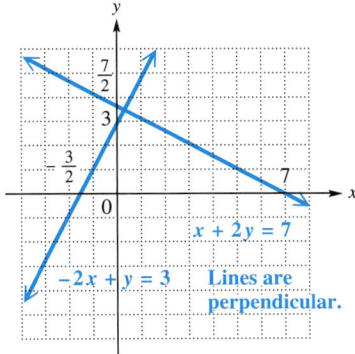

Figure 26

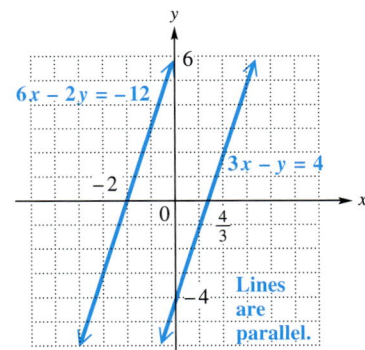

Figure 27

(b) $3x - y = 4$ $\xrightarrow{\text{Solve for } y.}$ $y = 3x - 4$
$6x - 2y = -12$ $\xrightarrow{\phantom{\text{Solve for } y.}}$ $y = 3x + 6$

Both lines have slope 3, so the lines are parallel. See Figure 27.

(c) $4x + 3y = 6$ $\xrightarrow{\text{Solve for } y.}$ $y = -\dfrac{4}{3}x + 2$
$2x - y = 5$ $\xrightarrow{\phantom{\text{Solve for } y.}}$ $y = 2x - 5$

Here the slopes are $-\dfrac{4}{3}$ and 2. Because $-\dfrac{4}{3} \neq 2$ and $-\dfrac{4}{3}(2) \neq -1$, these lines are neither parallel nor perpendicular.

(d) $5x - y = 1$ $\xrightarrow{\text{Solve for } y.}$ $y = 5x - 1$
$x - 5y = -10$ $\xrightarrow{\phantom{\text{Solve for } y.}}$ $y = \dfrac{1}{5}x + 2$

The slopes are 5 and $\dfrac{1}{5}$. The lines are not parallel, nor are they perpendicular. *(Be careful!* $5\left(\dfrac{1}{5}\right) = 1$, *not* -1.*)*

▶ **Work Problem 6 at the Side.**

---

6 Decide whether each pair of lines is *parallel, perpendicular,* or *neither*.

(a) $x + y = 6$
$x + y = 1$

(b) $3x - y = 4$
$x + 3y = 9$

(c) $2x - y = 5$
$2x + y = 3$

(d) $3x - 7y = 35$
$7x - 3y = -6$

ANSWERS
6. **(a)** parallel **(b)** perpendicular
 (c) neither **(d)** neither

Focus on Real-Data Applications

Linear or Nonlinear? That Is the Question about Windchill

The **windchill factor** measures the cooling effect that the wind has on one's skin. The table gives the windchill factor for various wind speeds and temperatures.

WINDCHILL FACTOR

Wind Speed (mph)	Air Temperature (°Fahrenheit)													
	35	30	25	20	15	10	5	0	−5	−10	−15	−20	−25	−30	−35
4	35	30	25	20	15	10	5	0	−5	−10	−15	−20	−25	−30	−35
5	32	27	22	16	11	6	0	−5	−10	−15	−21	−26	−31	−36	−42
10	22	16	10	3	−3	−9	−15	−22	−27	−34	−40	−46	−52	−58	−64
15	16	9	2	−5	−11	−18	−25	−31	−38	−45	−51	−58	−65	−72	−78
20	12	4	−3	−10	−17	−24	−31	−39	−46	−53	−60	−67	−74	−81	−88
25	8	1	−7	−15	−22	−29	−36	−44	−51	−59	−66	−74	−81	−88	−96
30	6	−2	−10	−18	−25	−33	−41	−49	−56	−64	−71	−79	−86	−93	−101
35	4	−4	−12	−20	−27	−35	−43	−52	−58	−67	−74	−82	−89	−97	−105
40	3	−5	−13	−21	−29	−37	−45	−53	−60	−69	−76	−84	−92	−100	−107
45	2	−6	−14	−22	−30	−38	−46	−54	−62	−70	−78	−85	−93	−102	−109

Source: USA Today.

The data in the table represents the relationships between two different sets of variable quantities: Windchill versus Air Temperature and Windchill versus Wind Speed. The question is whether either of these relationships is linear, that is, whether the data points, when graphed, could be connected to form a straight line.

Example 1 Windchill versus Air Temperature Choose one measure of wind speed to keep constant, such as 15 mph. Complete the table with Air Temperature (AT) as the *input* and Windchill (WC) as the *output*. Both variables are measured in degrees Fahrenheit.

AT	35	30	25	20	15	10	5	0	−5	−10	−15			
WC	16	9	2	−5	−11	−18	−25	−31	−38					

Example 2 Windchill versus Wind Speed Choose one measure of air temperature to keep constant, such as 10°F. Complete the table with Wind Speed (WS) as the *input* and Windchill (WC) as the *output*. Wind speed is measured in mph.

WS	4	5	10	15	20	25	30		
WC	10	6	−9	−18	−24				

For Group Discussion

1. Refer to the Windchill versus Air Temperature data (Example 1).
   (a) Write the data as ordered pairs.
   (b) On a sheet of graph paper, draw and label a rectangular coordinate system. Use a scale of 5 on the *x*-axis and the *y*-axis. Make a scatter diagram of the data. Does the graph represent a linear relationship?

2. Repeat Problem 1 using the Windchill versus Wind Speed data (Example 2).

11.3 Exercises

FOR EXTRA HELP 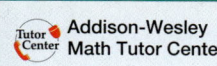 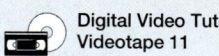 Digital Video Tutor CD 8 Videotape 11 Student's Solutions Manual MyMathLab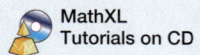

Use the coordinates of the indicated points to find the slope of each line. See Example 1.

1.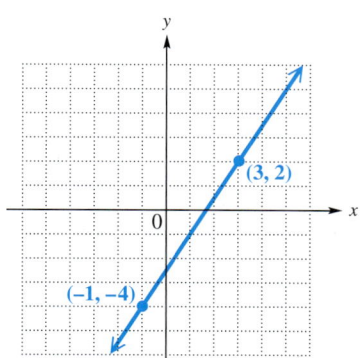

2.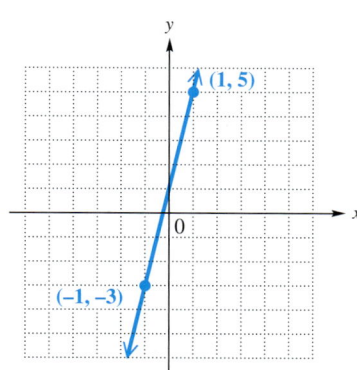

3.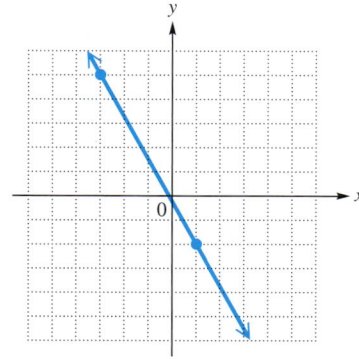

4.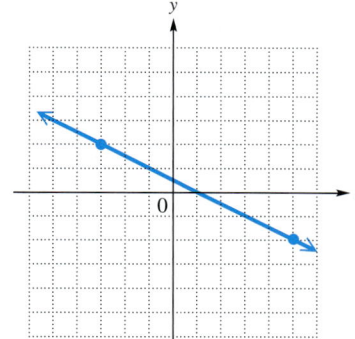

5.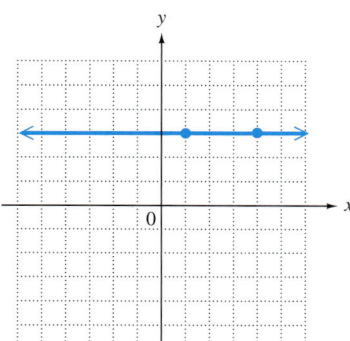

6.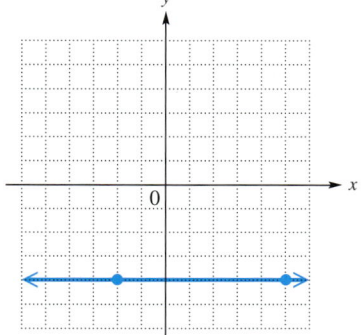

7. In the context of the graph of a straight line, what is meant by "rise"? What is meant by "run"?

8. Look at the graph in Exercise 1, and answer the following.
   (a) Start at the point $(-1, -4)$ and count vertically up to the horizontal line that goes through the other plotted point. What is this vertical change? (Remember: "up" means positive, "down" means negative.) _____
   (b) From this new position, count horizontally to the other plotted point. What is this horizontal change? (Remember: "right" means positive, "left" means negative.) _____
   (c) What is the quotient of the numbers found in parts (a) and (b)? _____
   What do we call this number? _____

9. Refer to Exercise 8. If we were to *start* at the point $(3, 2)$ and *end* at the point $(-1, -4)$, would the answer to part (c) be the same? Explain why or why not.

On the given coordinate system, sketch the graph of a straight line with the indicated slope.

10. Negative

11. Positive

12. Undefined

13. Zero

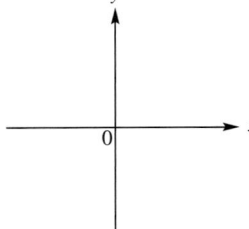

14. Explain in your own words what is meant by the *slope* of a line.

15. A student found the slope of the line through the points $(2, 5)$ and $(-1, 3)$ and got $-\frac{2}{3}$ as his answer. He showed his work as

$$\frac{3 - 5}{2 - (-1)} = \frac{-2}{3} = -\frac{2}{3}.$$

Is he correct? If not, find his error and give the correct slope.

Find the slope of the line through each pair of points. See Examples 1–4.

16. $(4, -1)$ and $(-2, -8)$ **17.** $(1, -2)$ and $(-3, -7)$ **18.** $(-8, 0)$ and $(0, -5)$

19. $(0, 3)$ and $(-2, 0)$ **20.** $(-4, -5)$ and $(-5, -8)$ **21.** $(-2, 4)$ and $(-3, 7)$

22. $(6, -5)$ and $(-12, -5)$ **23.** $(4, 3)$ and $(-6, 3)$ **24.** $(-8, 6)$ and $(-8, -1)$

25. $(-12, 3)$ and $(-12, -7)$ **26.** $(3.1, 2.6)$ and $(1.6, 2.1)$ **27.** $\left(-\frac{7}{5}, \frac{3}{10}\right)$ and $\left(\frac{1}{5}, -\frac{1}{2}\right)$

Find the slope of each line. See Example 5.

28. $y = 2x - 3$ **29.** $y = 5x + 12$ **30.** $2y = -x + 4$ **31.** $4y = x + 1$

32. $-6x + 4y = 4$ **33.** $3x - 2y = 3$ **34.** $y = 4$ **35.** $y = 6$

36. $x = 5$ **37.** $x = -2$ **38.** $x + y = 0$ **39.** $x - y = 0$

The figure at the right shows a line that has a positive slope (because it rises from left to right) and a positive y-value for the y-intercept (because it intersects the y-axis above the origin).

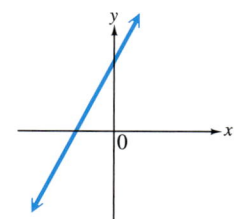

*For each figure in Exercises 40–45, decide whether **(a)** the slope is positive, negative, or 0 and whether **(b)** the y-value of the y-intercept is positive, negative, or 0.*

40. (a) _____ **41.** (a) _____ **42.** (a) _____
    (b) _____   (b) _____   (b) _____

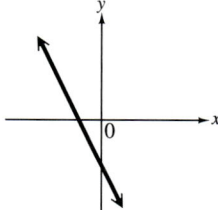 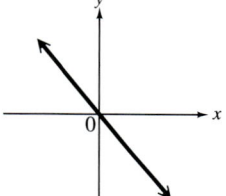 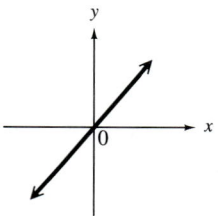

43. (a) _____ **44.** (a) _____ **45.** (a) _____
    (b) _____   (b) _____   (b) _____

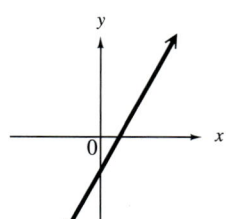 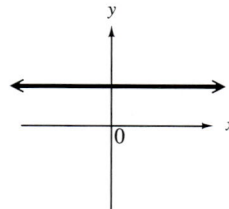 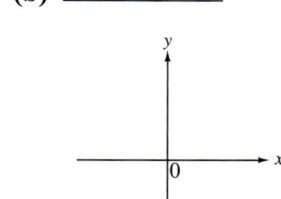

In each pair of equations, give the slope of each line, and then determine whether the two lines are parallel, perpendicular, or neither parallel nor perpendicular. See Example 6.

46. $2x + 5y = 4$ **47.** $-4x + 3y = 4$ **48.** $8x - 9y = 6$
    $4x + 10y = 1$ $-8x + 6y = 0$ $8x + 6y = -5$

49. $5x - 3y = -2$ **50.** $3x - 2y = 6$ **51.** $3x - 5y = -1$
    $3x - 5y = -8$ $2x + 3y = 3$ $5x + 3y = 2$

52. What is the slope (or pitch) of this roof?

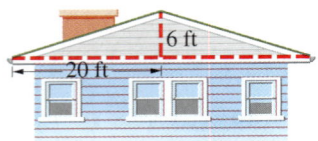

53. What is the slope (or grade) of this hill?

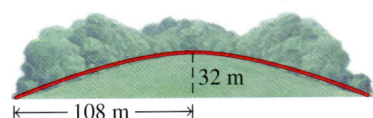

RELATING CONCEPTS (EXERCISES 54–59) For Individual or Group Work

Figure A gives public school enrollment (in thousands) in grades 9–12 in the United States. Figure B gives the (average) number of public school students per computer.

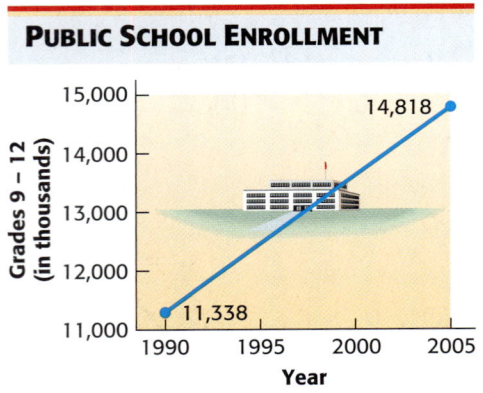

Figure A

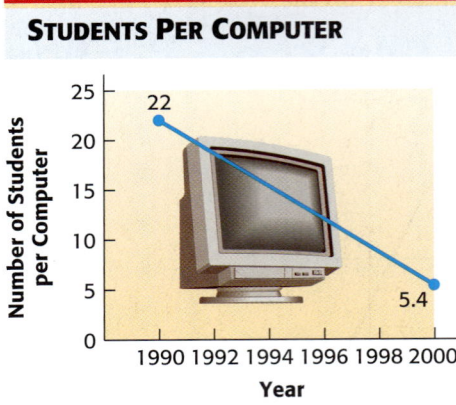

Figure B

Work Exercises 54–59 in order.

54. Use the ordered pairs (1990, 11,338) and (2005, 14,818) to find the slope of the line in Figure A.

55. The slope of the line in Figure A is _____. This means that during
(positive/negative)

the period represented, enrollment _____.
(increased/decreased)

56. The slope of a line represents its *rate of change*. Based on Figure A, what was the increase in students *per year* during the period shown?

57. Use the given information to find the slope of the line in Figure B.

58. The slope of the line in Figure B is _____. This means that during
(positive/negative)

the period represented, the number of students per computer _____.
(increased/decreased)

59. Based on Figure B, what was the decrease in students per computer *per year* during the period shown?

11.4 Equations of Lines

In **Section 11.3**, we found the slope (steepness) of a line from the equation of the line by solving the equation for *y*. In that form, the slope is the coefficient of *x*. For example, the slope of the line with equation $y = 2x + 3$ is 2, the coefficient of *x*. What does the number 3 represent? If $x = 0$, the equation becomes

$$y = 2(0) + 3 = 0 + 3 = 3.$$

Since $y = 3$ corresponds to $x = 0$, $(0, 3)$ is the *y*-intercept of the graph of $y = 2x + 3$. An equation like $y = 2x + 3$ that is solved for *y* is said to be in **slope-intercept form** because both the slope and the *y*-intercept of the line can be read directly from the equation.

Slope-Intercept Form

The slope-intercept form of the equation of a line with slope *m* and *y*-intercept $(0, b)$ is

$$y = mx + b.$$

Slope ↑   ↑ $(0, b)$ is the *y*-intercept.

Remember that the intercept in the slope-intercept form is the y-intercept.

NOTE
The slope-intercept form is the most useful form for a linear equation because of the information we can determine from it. It is also the form used by graphing calculators and the one that describes a *linear function*, an important concept in mathematics.

OBJECTIVE 1 Write an equation of a line given its slope and *y*-intercept. Given the slope and *y*-intercept of a line, we can use the slope-intercept form to find an equation of the line.

EXAMPLE 1 Finding an Equation of a Line

Find an equation of the line with slope $\frac{2}{3}$ and *y*-intercept $(0, -1)$.
Here $m = \frac{2}{3}$ and $b = -1$, so an equation is

Slope ┐   ┌ *y*-intercept
$$y = mx + b$$
$$y = \frac{2}{3}x - 1.$$

Work Problem 1 at the Side.

OBJECTIVE 2 Graph a line given its slope and a point on the line. We can use the slope and *y*-intercept to graph any line. If a linear equation is given in standard form $Ax + By = C$, it can be graphed using the following procedure.

OBJECTIVES

1. Write an equation of a line given its slope and *y*-intercept.
2. Graph a line given its slope and a point on the line.
3. Write an equation of a line given its slope and any point on the line.
4. Write an equation of a line given two points on the line.
5. Find an equation of a line that fits a data set.

1 Find an equation of the line with the given slope and *y*-intercept.

(a) slope $\frac{1}{2}$; *y*-intercept $(0, -4)$

(b) slope -1; *y*-intercept $(0, 8)$

(c) slope 3; *y*-intercept $(0, 0)$

(d) slope 0; *y*-intercept $(0, 2)$

ANSWERS
1. (a) $y = \frac{1}{2}x - 4$ (b) $y = -x + 8$
   (c) $y = 3x$ (d) $y = 2$

2 Graph $3x - 4y = 8$ using the slope and y-intercept.

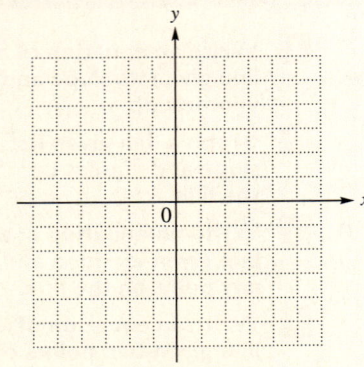

Graphing Using Slope and y-Intercept

Step 1 Solve for *y* to obtain the $y = mx + b$ form of the equation.

Step 2 Identify the y-intercept from the form just obtained. Graph the point $(0, b)$.

Step 3 The slope of the line is *m*. Use the geometric interpretation of slope ("rise over run") to find another point on the graph by counting from the y-intercept.

Step 4 Join the two points with a line to obtain the graph of the equation.

EXAMPLE 2 Graphing a Line Using Slope and y-Intercept

Graph $2x - 3y = 3$.

Step 1 Begin by solving for *y*.

$$2x - 3y = 3 \quad \text{Given equation}$$
$$-3y = -2x + 3 \quad \text{Subtract } 2x \text{ from both sides.}$$
$$y = \frac{2}{3}x - 1 \quad \text{Divide both sides by } -3.$$

Step 2 The y-intercept is $(0, -1)$. Graph this point. See Figure 28.

Step 3 The slope is $\frac{2}{3}$. By the definition of slope,

$$m = \frac{\text{change in } y}{\text{change in } x} = \frac{2}{3}.$$

Counting from the y-intercept 2 units up and 3 units to the right, we obtain another point on the graph, $(3, 1)$.

Step 4 Draw the line through the points $(0, -1)$ and $(3, 1)$ to obtain the graph of $2x - 3y = 3$. See Figure 28.

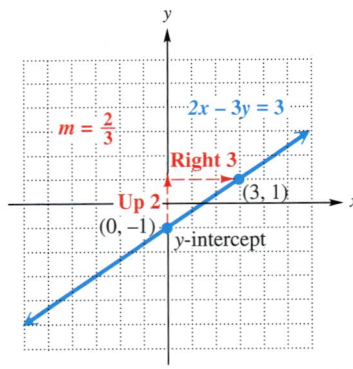

Figure 28

◀◀ Work Problem **2** at the Side.

The method of Example 2 can be extended to graph a line given its slope and *any* point on the line.

EXAMPLE 3 Graphing a Line Given a Point and the Slope

Graph the line through $(-2, 3)$ with slope -4.

First, locate the point $(-2, 3)$. See Figure 29. Write the slope -4 as

$$m = \frac{\text{change in } y}{\text{change in } x} = \frac{-4}{1}.$$

Continued on Next Page

ANSWERS
2.

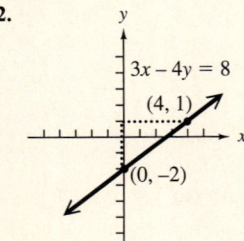

Locate another point on the line by counting 4 units down (because of the negative sign) from $(-2, 3)$ and then 1 unit to the right. Finally, draw the line through this new point P and the given point $(-2, 3)$. See Figure 29.

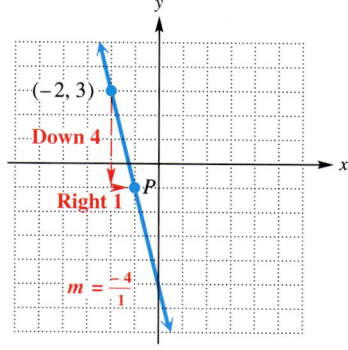

Figure 29

3 Graph the line passing through $(2, -3)$, with slope $-\frac{1}{3}$.

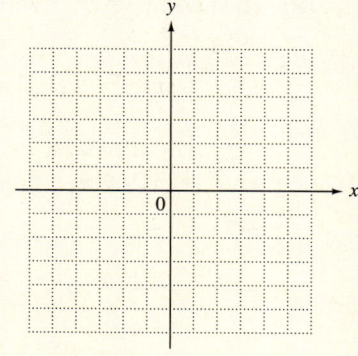

> **NOTE**
> In Example 3, we could have written the slope as $\frac{4}{-1}$ instead. In this case, we would move 4 units up from $(-2, 3)$ and then 1 unit to the left (because of the negative sign). Verify that this produces the same line.

Work Problem 3 at the Side.

OBJECTIVE 3 Write an equation of a line given its slope and any point on the line. Let m represent the slope of a line and (x_1, y_1) represent a given point on the line. Let (x, y) represent any other point on the line. Then by the slope formula,

$$\frac{y - y_1}{x - x_1} = m$$

$$y - y_1 = m(x - x_1). \quad \text{Multiply both sides by } x - x_1.$$

This result is the **point-slope form** of the equation of a line.

> **Point-Slope Form**
> The point-slope form of the equation of a line with slope m going through (x_1, y_1) is
> $$y - y_1 = m(x - x_1).$$

EXAMPLE 4 Using the Point-Slope Form to Write Equations

Find an equation of each line. Write the equation in slope-intercept form.

(a) Through $(-2, 4)$, with slope -3

The given point is $(-2, 4)$ so $x_1 = -2$ and $y_1 = 4$. Also, $m = -3$. Substitute these values into the point-slope form.

$y - y_1 = m(x - x_1)$	Point-slope form
$y - 4 = -3[x - (-2)]$	Let $x_1 = -2$, $y_1 = 4$, $m = -3$.
$y - 4 = -3(x + 2)$	Use the distributive property.
$y - 4 = -3x - 6$	Add 4 to both sides.
$y = -3x - 2$	Slope-intercept form.

Continued on Next Page

ANSWERS
3.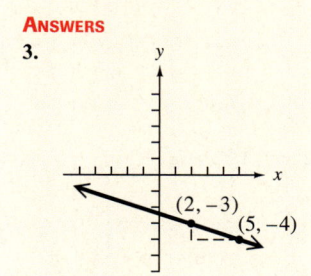

4 Find an equation for each line. Write answers in slope-intercept form.

(a) Through $(-1, 3)$, with slope -2

$y - y_1 = m(x - x_1)$

$y - \underline{\quad} = \underline{\quad}[x - (\quad)]$

$y - 3 = -2(x + \underline{\quad})$

$y - 3 = -2x - \underline{\quad}$

$y = \underline{\quad}$

(b) Through $(5, 2)$, with slope $-\frac{1}{3}$

5 Write an equation in slope-intercept form for the line through each pair of points.

(a) $(-3, 1)$ and $(2, 4)$

(b) $(2, 5)$ and $(-1, 6)$

ANSWERS
4. (a) $3; -2; -1; 1; 2; -2x + 1$
   (b) $y = -\frac{1}{3}x + \frac{11}{3}$
5. (a) $y = \frac{3}{5}x + \frac{14}{5}$ (b) $y = -\frac{1}{3}x + \frac{17}{3}$

(b) Through $(4, 2)$, with slope $\frac{3}{5}$

$y - y_1 = m(x - x_1)$ Point-slope form

$y - 2 = \frac{3}{5}(x - 4)$ Let $x_1 = 4, y_1 = 2, m = \frac{3}{5}$. Use the distributive property.

$y - 2 = \frac{3}{5}x - \frac{12}{5}$ Add 2 (which is $\frac{10}{5}$) to both sides.

$y = \frac{3}{5}x - \frac{12}{5} + \frac{10}{5}$ Combine terms.

$y = \frac{3}{5}x - \frac{2}{5}$ Slope-intercept form

We did not clear fractions after the substitution step because we want the equation in slope-intercept form—that is, solved for y.

◀◀ **Work Problem 4 at the Side.**

OBJECTIVE 4 Write an equation of a line given two points on the line. We can also use the point-slope form to find an equation of a line when two points on the line are known.

EXAMPLE 5 Finding the Equation of a Line Given Two Points

Find an equation of the line through the points $(-2, 5)$ and $(3, 4)$. Write the equation in slope-intercept form.

First, find the slope of the line, using the slope formula.

$$\text{slope } m = \frac{y_2 - y_1}{x_2 - x_1} = \frac{5 - 4}{-2 - 3} = \frac{1}{-5} = -\frac{1}{5}$$

Now use either $(-2, 5)$ or $(3, 4)$ and the point-slope form. Using $(3, 4)$ gives

$y - y_1 = m(x - x_1)$ Let $x_1 = 3, y_1 = 4, m = -\frac{1}{5}$.

$y - 4 = -\frac{1}{5}(x - 3)$ Use the distributive property.

$y - 4 = -\frac{1}{5}x + \frac{3}{5}$ Add 4 (which is $\frac{20}{5}$) to both sides.

$y = -\frac{1}{5}x + \frac{3}{5} + \frac{20}{5}$ Combine terms.

$y = -\frac{1}{5}x + \frac{23}{5}.$ Slope-intercept form

The same result would be found using $(-2, 5)$ for (x_1, y_1).

◀◀ **Work Problem 5 at the Side.**

Many of the linear equations in **Sections 11.1–11.3** were given in the form $Ax + By = C$, called **standard form**, which we define as follows.

Standard Form

A linear equation is in standard form if it is written as

$$Ax + By = C,$$

where A, B, and C are integers and $A > 0, B \neq 0$.

NOTE
The preceding definition of standard form is not the same in all texts. A linear equation can be written in this form in many different, equally correct, ways. For example,

$$3x + 4y = 12, \quad 6x + 8y = 24, \quad \text{and} \quad 9x + 12y = 36$$

all represent the same set of ordered pairs. Let us agree that $3x + 4y = 12$ is preferable to the other forms because the greatest common factor of 3, 4, and 12 is 1.

A summary of the types of linear equations follows.

Linear Equations

$x = k$	**Vertical line** Slope is undefined; x-intercept is $(k, 0)$.
$y = k$	**Horizontal line** Slope is 0; y-intercept is $(0, k)$.
$y = mx + b$	**Slope-intercept form** Slope is m; y-intercept is $(0, b)$.
$y - y_1 = m(x - x_1)$	**Point-slope form** Slope is m; line passes through (x_1, y_1).
$Ax + By = C$	**Standard form** Slope is $-\frac{A}{B}$; x-intercept is $\left(\frac{C}{A}, 0\right)$; y-intercept is $\left(0, \frac{C}{B}\right)$.

OBJECTIVE 5 Find an equation of a line that fits a data set. Earlier in this chapter, we gave linear equations that modeled real data, such as annual costs of doctors' visits and amounts of credit card debt, and then used these equations to estimate or predict values. We now develop a procedure to find such an equation if the given set of data fits a linear pattern—that is, its graph consists of points lying close to a straight line.

EXAMPLE 6 Finding an Equation of a Line That Describes Data

The table lists the average annual cost (in dollars) of tuition and fees at public 4-year colleges and universities for selected years. Year 1 represents 1994, year 3 represents 1996, and so on. Plot the data and find an equation that approximates it.

Year	Cost (in dollars)
1	2537
3	2849
5	3110
7	3349
9	3746

Source: U.S. Department of Education.

Letting y represent the cost in year x, we plot the data as shown in Figure 30 on the next page.

Continued on Next Page

6 Use the points (1, 2537) and (9, 3746) to find an equation in slope-intercept form that approximates the data of Example 6. How well does this equation approximate the cost in 1998?

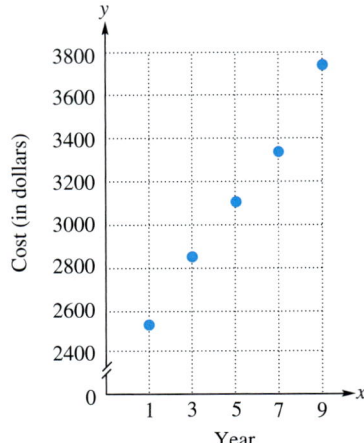

AVERAGE ANNUAL COSTS AT PUBLIC 4-YEAR COLLEGES AND UNIVERSITIES

Figure 30

The points appear to lie approximately in a straight line. We can use two of the data pairs and the point-slope form of the equation of a line to get an equation that describes the relationship between the year and the cost. We choose the ordered pairs (3, 2849) and (7, 3349) from the table on the preceding page and find the slope of the line through these points.

$$m = \frac{y_2 - y_1}{x_2 - x_1} = \frac{3349 - 2849}{7 - 3} = 125 \qquad \text{Let } (7, 3349) = (x_2, y_2) \text{ and } (3, 2849) = (x_1, y_1).$$

As we might expect, the slope, 125, is positive, indicating that tuition and fees increased $125 each year. Now use this slope and the point (3, 2849) in the point-slope form to find an equation of the line.

$y - y_1 = m(x - x_1)$ Point-slope form

$y - \mathbf{2849} = \mathbf{125}(x - \mathbf{3})$ Substitute for $x_1, y_1,$ and m.

$y - 2849 = 125x - 375$ Use the distributive property.

$y = 125x + 2474$ Add 2849 to both sides.

To see how well this equation approximates the ordered pairs in the data table, let $x = 9$ (for 2002) and find y.

$y = 125x + 2474$ Equation of the line

$y = 125(\mathbf{9}) + 2474$ Substitute 9 for x.

$y = 3599$

The corresponding value in the table for $x = 9$ is 3746, so the equation gives a value that is a bit lower than the actual value. With caution, the equation could be used to predict values for years between 1 and 9 that are not included in the table.

NOTE
In Example 6, if we had chosen two different data points, we would have gotten a slightly different equation.

ANSWERS
6. $y = 151.125x + 2385.875$; The equation gives $y \approx 3142$ when $x = 5$, which is a very good approximation.

Work Problem 6 at the Side.

11.4 Exercises

Match the correct equation in Column II with the description given in Column I.

I

1. Slope $= -2$, through the point $(4, 1)$
2. Slope $= -2$, y-intercept $(0, 1)$
3. Through the points $(0, 0)$ and $(4, 1)$
4. Through the points $(0, 0)$ and $(1, 4)$

II

A. $y = 4x$

B. $y = \dfrac{1}{4}x$

C. $y = -2x + 1$

D. $y - 1 = -2(x - 4)$

*Use the geometric interpretation of slope (rise divided by run, from **Section 11.3**) to find the slope of each line. Then, by identifying the y-intercept from the graph, write the slope-intercept form of the equation of the line.*

5.
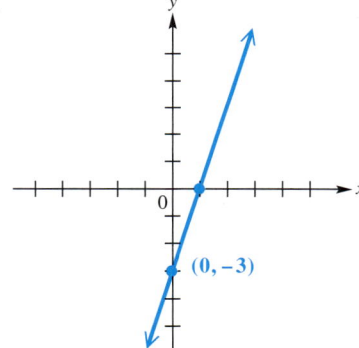

6.
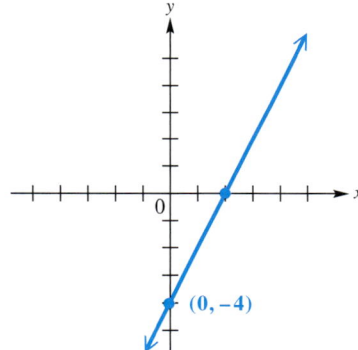

7.
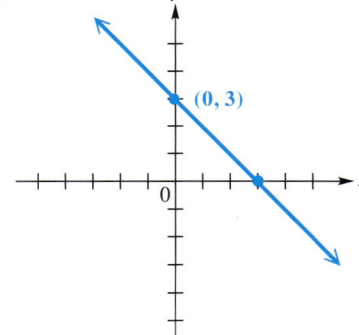

8.
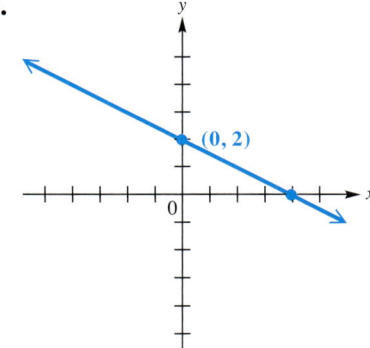

Write the equation of the line with the given slope and y-intercept. See Example 1.

9. slope 4;
   y-intercept $(0, -3)$

10. slope -5;
    y-intercept $(0, 6)$

11. slope 0;
    y-intercept $(0, 3)$

12. slope 3;
    y-intercept $(0, 0)$

778 Chapter 11 Graphs of Linear Equations and Inequalities in Two Variables

13. Explain why the equation of a vertical line cannot be written in the form $y = mx + b$.

14. Match each equation with the graph that would most closely resemble its graph.
(a) $y = x + 3$

A.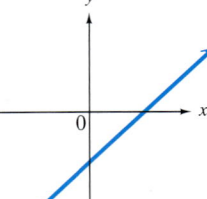

B.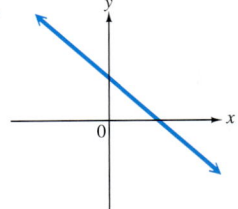

(b) $y = -x + 3$

(c) $y = x - 3$

C.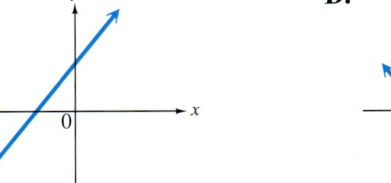

D.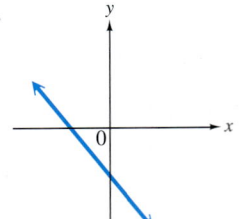

(d) $y = -x - 3$

Graph each equation by finding the slope and y-intercept, and using their definitions to find two points on the line. See Example 2.

15. $y = 3x + 2$

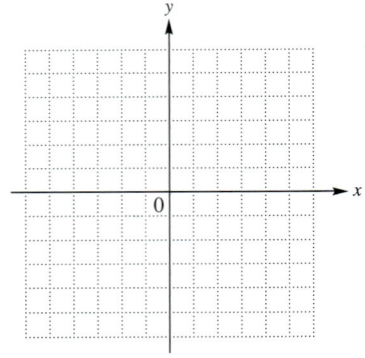

16. $y = 4x - 4$

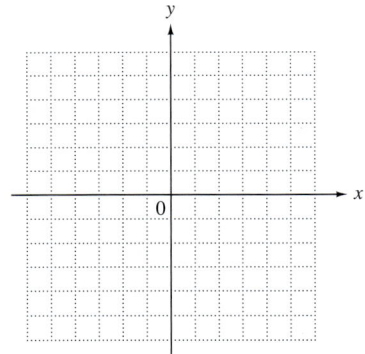

17. $2x + y = -5$

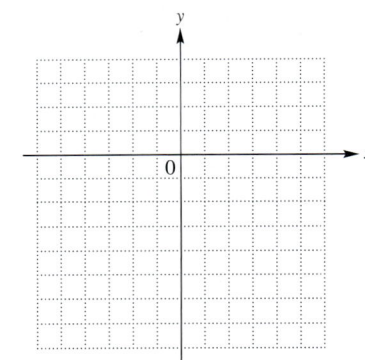

18. $3x + y = -2$

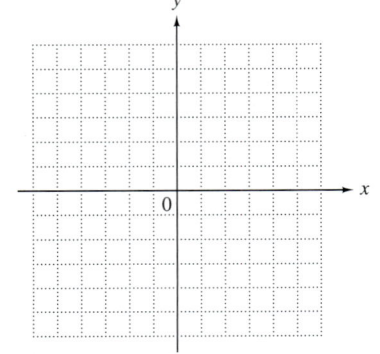

19. $x + 2y = 4$

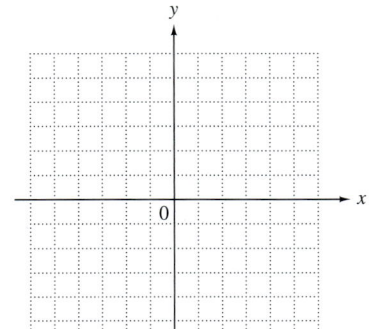

20. $x + 3y = 12$

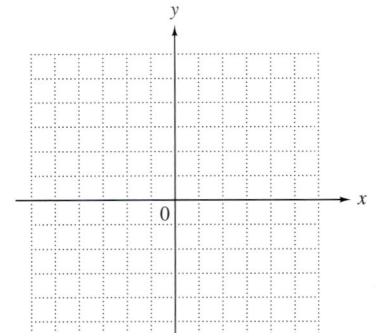

Section 11.4 Equations of Lines   779

Graph the line through the given point with the given slope. (In Exercises 25–28, recall the types of lines having slope 0 and undefined slope.) Give the slope-intercept form of the equation of the line if possible. See Example 3.

21. $(-2, 3), m = \dfrac{1}{2}$

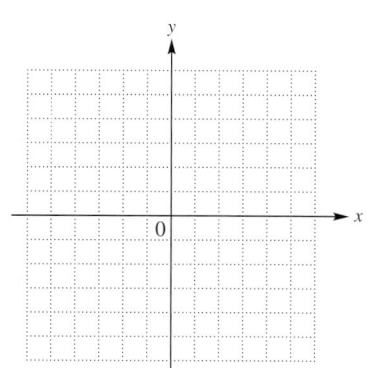

22. $(-4, -1), m = \dfrac{3}{4}$

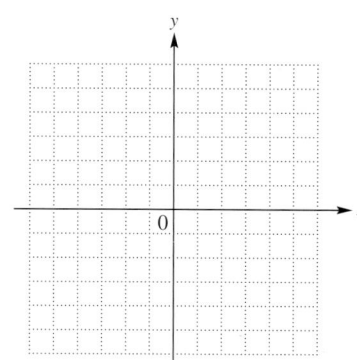

23. $(1, -5), m = -\dfrac{2}{5}$

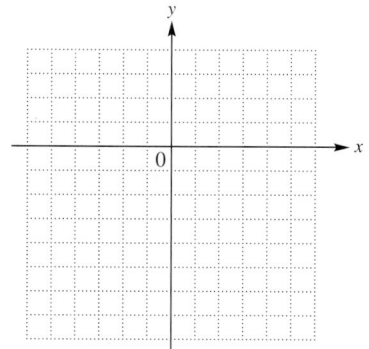

24. $(2, -1), m = -\dfrac{1}{3}$

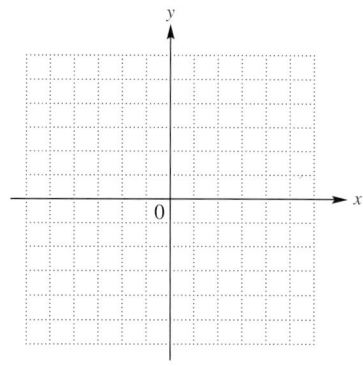

25. $(3, 2), m = 0$

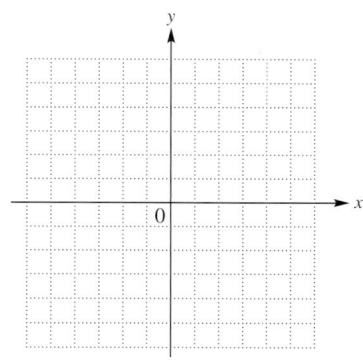

26. $(-2, 3), m = 0$

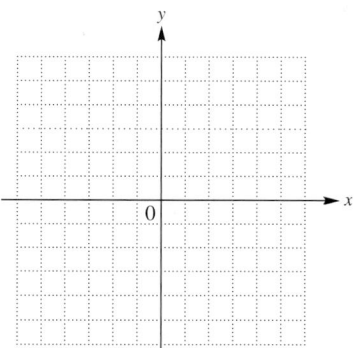

27. $(3, -2)$, undefined slope

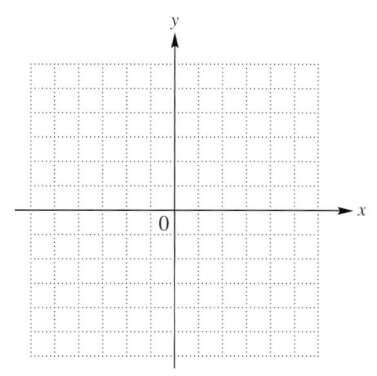

28. $(2, 4)$, undefined slope

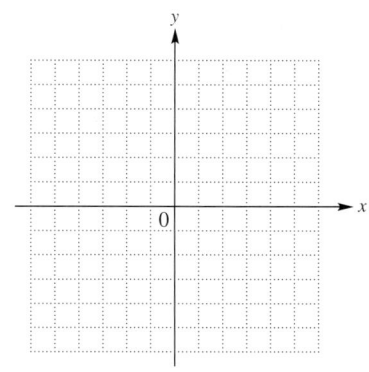

29. $(0, 0), m = \dfrac{2}{3}$

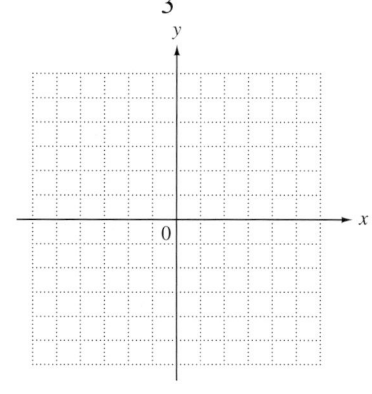

30. (a) What is the common name given to the vertical line whose x-intercept is the origin?

(b) What is the common name given to the line with slope 0 whose y-intercept is the origin?

Write an equation of the line through the given point with the given slope. Write the equation in slope-intercept form. See Example 4.

31. $(4, 1)$, $m = 2$

32. $(2, 7)$, $m = 3$

33. $(3, -10)$, $m = -2$

34. $(2, -5)$, $m = -4$

35. $(-2, 5)$, $m = \dfrac{2}{3}$

36. $(-4, 1)$, $m = \dfrac{3}{4}$

Write an equation, in slope-intercept form if possible, of the line through each pair of points. See Example 5.

37. $(8, 5)$ and $(9, 6)$

38. $(4, 10)$ and $(6, 12)$

39. $(-1, -7)$ and $(-8, -2)$

40. $(-2, -1)$ and $(3, -4)$

41. $(0, -2)$ and $(-3, 0)$

42. $(-4, 0)$ and $(0, 2)$

43. $(3, 5)$ and $(3, -2)$

44. $(3, -5)$ and $(-1, -5)$

45. $\left(\dfrac{1}{2}, \dfrac{3}{2}\right)$ and $\left(-\dfrac{1}{4}, \dfrac{5}{4}\right)$

46. $\left(-\dfrac{2}{3}, \dfrac{8}{3}\right)$ and $\left(\dfrac{1}{3}, \dfrac{7}{3}\right)$

Write an equation in slope-intercept form of the line satisfying the given conditions.

47. Through $(2, -3)$, parallel to $3x = 4y + 5$

48. Through $(-1, 4)$, perpendicular to $2x + 3y = 8$

49. Perpendicular to $x - 2y = 7$, y-intercept $(0, -3)$

50. Parallel to $5x = 2y + 10$, y-intercept $(0, 4)$

RELATING CONCEPTS (EXERCISES 51–58) For Individual or Group Work

If we think of ordered pairs of the form (C, F), then the two most common methods of measuring temperature, Celsius and Fahrenheit, can be related as follows: When C = 0, F = 32, and when C = 100, F = 212. **Work Exercises 51–58 in order.**

51. Write two ordered pairs relating these two temperature scales.

52. Find the slope of the line through the two points.

53. Use the point-slope form to find an equation of the line. (Your variables should be C and F rather than x and y.)

54. Write an equation for F in terms of C.

55. Use the equation from Exercise 54 to write an equation for C in terms of F.

56. Use the equation from Exercise 54 to find the Fahrenheit temperature when $C = 30$.

57. Use the equation from Exercise 55 to find the Celsius temperature when $F = 50$.

58. For what temperature is $F = C$?

---

*The cost to produce x items is, in some cases, expressed as $y = mx + b$. The number b gives the **fixed cost** (the cost that is the same no matter how many items are produced), and the number m is the **variable cost** (the cost to produce an additional item). Use this information to work Exercises 59 and 60.*

59. It costs $400 to start up a business of selling snow cones. Each snow cone costs $0.25 to produce.

(a) What is the fixed cost?

(b) What is the variable cost?

(c) Write the cost equation.

(d) What will be the cost to produce 100 snow cones, based on the cost equation?

(e) How many snow cones will be produced if total cost is $775?

60. It costs $2000 to purchase a copier, and each copy costs $0.02 to make.

(a) What is the fixed cost?

(b) What is the variable cost?

(c) Write the cost equation.

(d) What will be the cost to produce 10,000 copies, based on the cost equation?

(e) How many copies will be produced if total cost is $2600?

Solve each problem. See Example 6.

61. The table lists the average annual cost (in dollars) of tuition and fees at 2-year colleges for selected years, where year 1 represents 1994, year 3 represents 1996, and so on.

Year	Cost (in dollars)
1	1125
3	1239
5	1314
7	1338
9	1379

Source: U.S. Department of Education.

(a) Write five ordered pairs for the data.

(b) Plot the ordered pairs. Do the points lie approximately in a straight line?

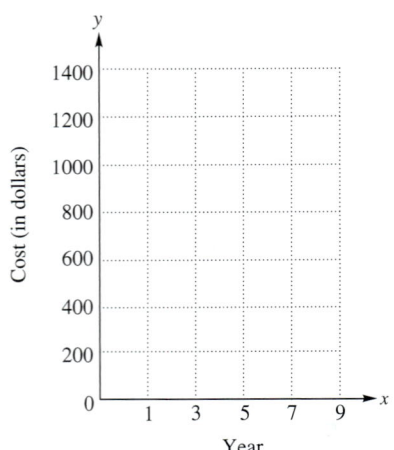

AVERAGE ANNUAL COSTS AT 2-YEAR COLLEGES

(c) Use the ordered pairs (3, 1239) and (9, 1379) to find the equation of a line that approximates the data. Write the equation in slope-intercept form. (Round the slope to the nearest hundredth and the y-intercept to the nearest whole number.)

(d) Use the equation from part (c) to predict the average annual cost at 2-year colleges in 2004 to the nearest dollar.

62. The table gives heavy-metal nuclear waste (in thousands of metric tons) from spent reactor fuel now stored temporarily at reactor sites, awaiting permanent storage. (*Source: Scientific American.*)

Year x	Waste y
1995	32
2000	42
2010*	61
2020*	76

*Estimates by the U.S. Department of Energy.

Let $x = 0$ represent 1995, $x = 5$ represent 2000 (since $2000 - 1995 = 5$), and so on.

(a) For 1995, the ordered pair is (0, 32). Write ordered pairs for the data for the other years given in the table.

(b) Plot the ordered pairs (x, y). Do the points lie approximately in a straight line?

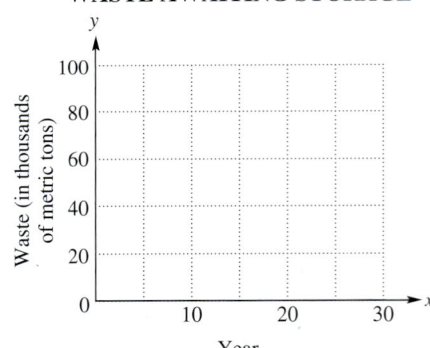

HEAVY-METAL NUCLEAR WASTE AWAITING STORAGE

(c) Use the ordered pairs (0, 32) and (25, 76) to find the equation of a line that approximates the other ordered pairs. Write the equation in slope-intercept form.

(d) Use the equation from part (c) to estimate the amount of nuclear waste in 2005. (*Hint:* What is the value of x for 2005?)

Summary Exercises on Graphing Linear Equations

Identify the slope and the y-intercept of the graph of each equation.

1. $3x + y = -6$

2. $2x + y = -4$

3. $-4x - y = 3$

4. $-5x - y = 8$

5. $-3x + 2y = 12$

6. $-5x + 3y = 15$

Graph each equation using its slope and y-intercept.

7. $m = 1, b = -2$

8. $m = 1, b = -4$

9. $m = -2, b = 6$

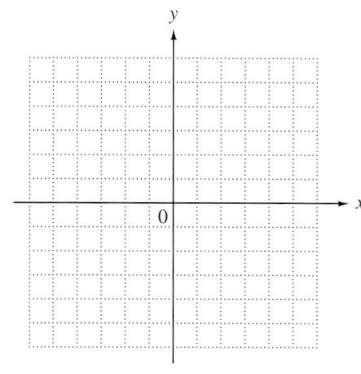 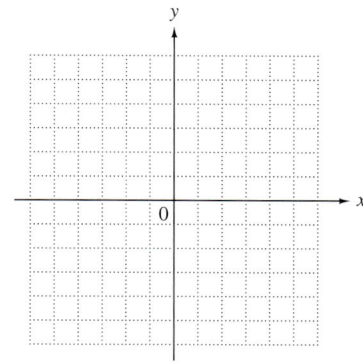 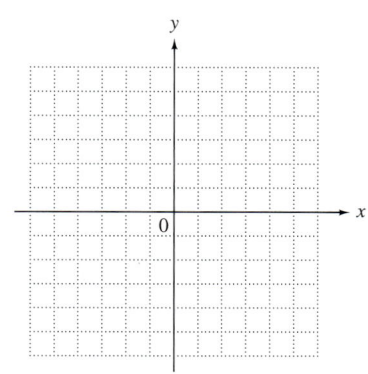

10. $m = -1, b = 6$

11. $m = -\dfrac{2}{3}, b = -2$

12. $m = -\dfrac{3}{4}, b = -1$

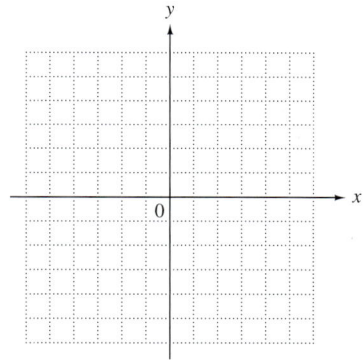 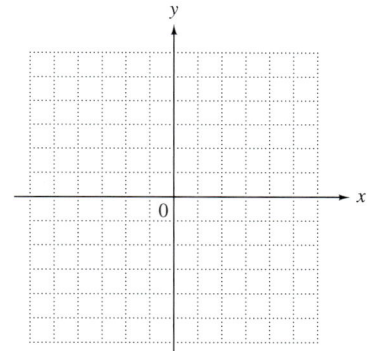 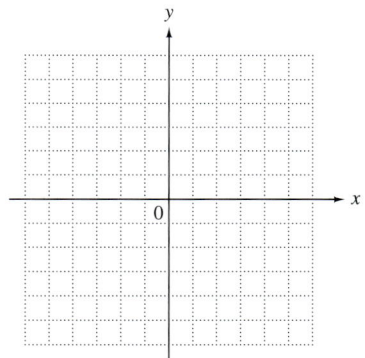

*For each equation **(a)** find the slope, **(b)** find the y-intercept, and **(c)** sketch the graph. Label two points on the graph.*

13. $-x + y = -3$

14. $-x + y = -5$

15. $x + 2y = 4$

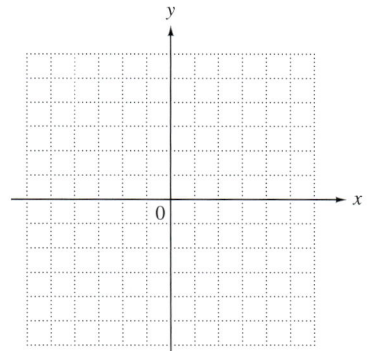 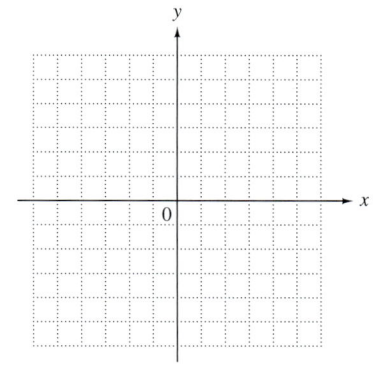 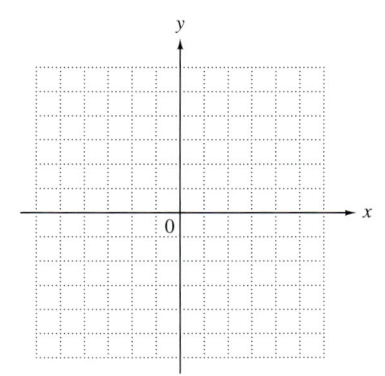

16. $x + 3y = -6$

17. $4x - 5y = 20$

18. $6x - 5y = 30$

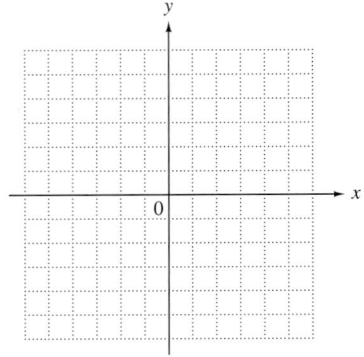 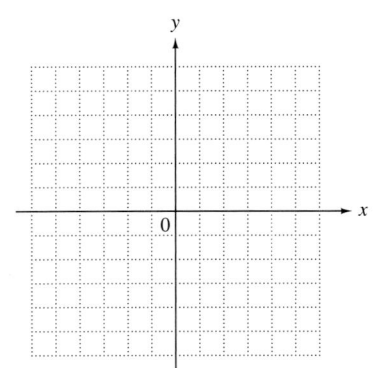 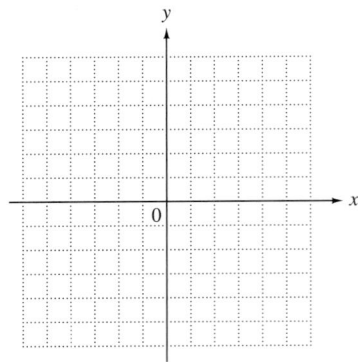

19. $2x + 3y = 12$

20. $5x + 2y = 10$

21. $x - 3y = 6$

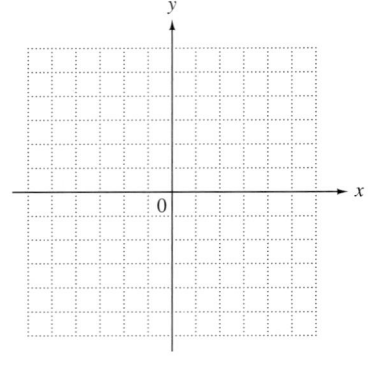 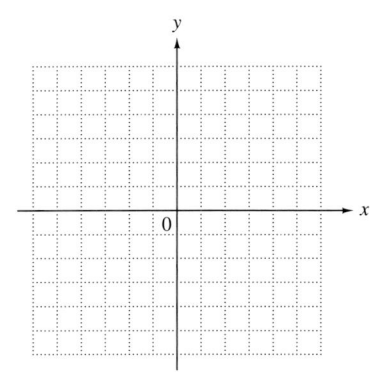 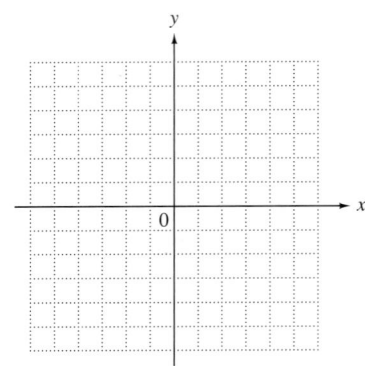

22. $x - 2y = -4$

23. $x - 4y = 0$

24. $x + 5y = 0$

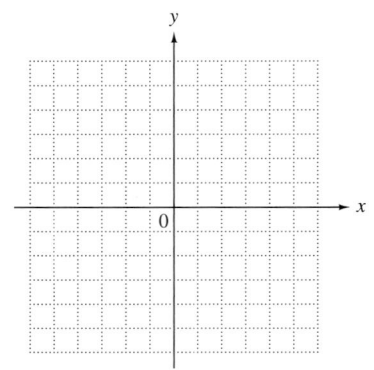 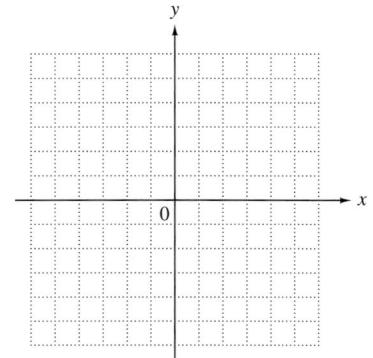 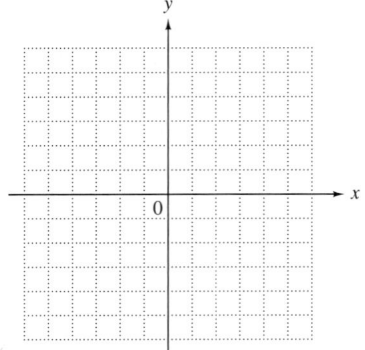

11.5 Graphing Linear Inequalities in Two Variables

In **Section 11.2** we graphed linear equations, such as $2x + 3y = 6$. Now this work is extended to **linear inequalities in two variables,** such as

$$2x + 3y \leq 6.$$

(Recall that $\leq$ is read "is less than or equal to.")

OBJECTIVE 1 Graph linear inequalities involving $\leq$ or $\geq$. The inequality $2x + 3y \leq 6$ means that

$$2x + 3y < 6 \quad \text{or} \quad 2x + 3y = 6.$$

As we found earlier, the graph of $2x + 3y = 6$ is a line. This **boundary line** divides the plane into two regions. The graph of the solutions of the inequality $2x + 3y < 6$ will include only *one* of these regions. We find the required region by solving the given inequality for y.

$$2x + 3y \leq 6 \quad \text{Subtract } 2x \text{ from both sides.}$$
$$3y \leq -2x + 6 \quad \text{Divide both sides by 3.}$$
$$y \leq -\frac{2}{3}x + 2$$

By this last statement, ordered pairs in which y is *less than or equal to* $-\frac{2}{3}x + 2$ will be solutions of the inequality. The ordered pairs in which y is equal to $-\frac{2}{3}x + 2$ are on the boundary line, so the ordered pairs in which y is less than $-\frac{2}{3}x + 2$ will be *below* that line. (This is because as we move *down* vertically, the y-values *decrease*.) To indicate the solutions, we shade the region below the line, as in Figure 31. The shaded region, along with the line, is the desired graph.

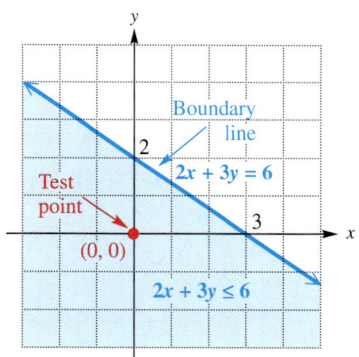

Figure 31

Work Problem 1 at the Side.

Alternatively, a test point gives a quick way to find the correct region to shade. We choose any point *not* on the boundary line. Because $(0, 0)$ is easy to substitute into an inequality, it is often a good choice, and we will use it here. We substitute 0 for x and 0 for y in the given inequality to see whether the resulting statement is true or false. In the example above,

$$2x + 3y \leq 6$$
$$2(0) + 3(0) \leq 6 \quad ? \quad \text{Let } x = 0 \text{ and } y = 0.$$
$$0 + 0 \leq 6 \quad ?$$
$$0 \leq 6. \quad \text{True}$$

Since the last statement is true, we shade the region that includes the test point $(0, 0)$. This agrees with the result shown in Figure 31.

OBJECTIVES

1. Graph linear inequalities involving $\leq$ or $\geq$.
2. Graph other linear inequalities.
3. Graph an inequality with boundary through the origin.

1 Shade the appropriate region for each linear inequality.

(a) $x + 2y \geq 6$

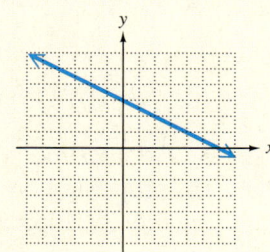

(b) $3x + 4y \leq 12$

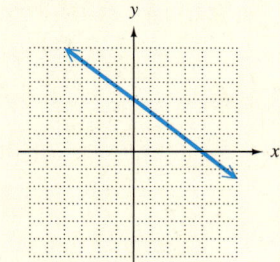

ANSWERS
1. (a)

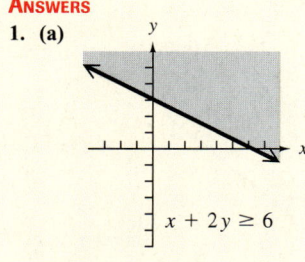

$x + 2y \geq 6$

(b)

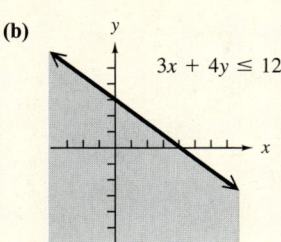

$3x + 4y \leq 12$

② Use (0, 0) as a test point to shade the proper region for the inequality $4x - 5y \leq 20$.

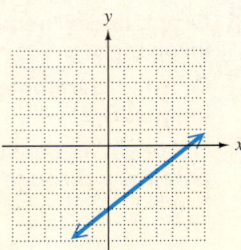

③ Use (1, 1) as a test point to shade the proper region for the inequality $3x + 5y > 15$.

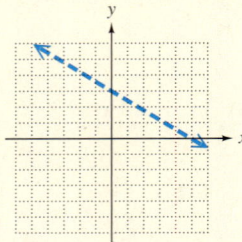

ANSWERS

2.

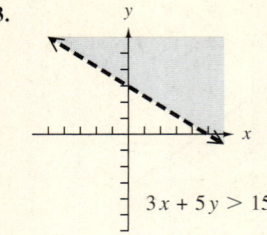

3.

> **Work Problem 2 at the Side.**

OBJECTIVE 2 Graph other linear inequalities. An inequality that does not include the equal sign is graphed in a similar way.

EXAMPLE 1 Graphing a Linear Inequality

Graph the inequality $x - y > 5$.

This inequality does not include the equal sign. Therefore, the points on the line

$$x - y = 5$$

do *not* belong to the graph. However, the line still serves as a boundary for two regions, one of which satisfies the inequality. To graph the inequality, first graph the equation $x - y = 5$. Use a *dashed line* to show that the points on the line are *not* solutions of the inequality $x - y > 5$. Then choose a test point not on the line to see which side of the line satisfies the inequality. We choose $(1, -2)$ this time.

$$x - y > 5$$
$$1 - (-2) > 5 \quad ? \quad \text{Let } x = 1 \text{ and } y = -2.$$
$$3 > 5 \quad \text{False}$$

Because $3 > 5$ is false, the graph of the inequality is the region that does *not* contain $(1, -2)$. We shade the region that does not include the test point $(1, -2)$, as in Figure 32. This shaded region is the desired graph. To check that the proper region is shaded, we select a point in the shaded region and substitute for x and y in the inequality $x - y > 5$. For example, we use $(4, -3)$ from the shaded region as follows.

$$x - y > 5$$
$$4 - (-3) > 5 \quad ? \quad \text{Let } x = 4 \text{ and } y = -3.$$
$$7 > 5 \quad \text{True}$$

This verifies that the correct region is shaded in Figure 32.

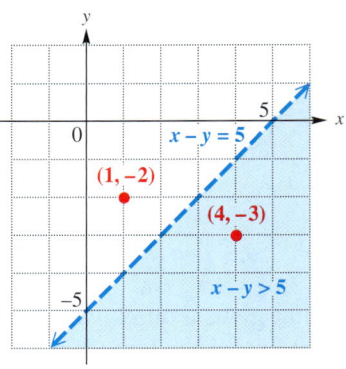

Figure 32

> **Work Problem 3 at the Side.**

A summary of the steps used to graph a linear inequality in two variables follows.

Graphing a Linear Inequality

Step 1 **Graph the boundary.** Graph the line that is the boundary of the region. Use the methods of **Section 11.2.** Draw a solid line if the inequality has ≤ or ≥; draw a dashed line if the inequality has < or >.

Step 2 **Shade the appropriate side.** Use any point not on the line as a test point. Substitute for x and y in the *inequality*. If a true statement results, shade the side containing the test point. If a false statement results, shade the other side.

4 Graph $2x - y \geq -4$.

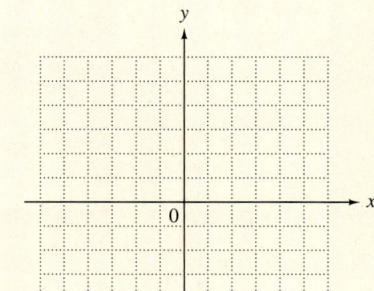

EXAMPLE 2 Graphing a Linear Inequality

Graph the inequality $2x - 5y \geq 10$.

Start by graphing the equation

$$2x - 5y = 10.$$

Use a solid line to show that the points on the line are solutions of the inequality $2x - 5y \geq 10$. Choose any test point not on the line. Again, we choose $(0, 0)$.

$$2x - 5y \geq 10$$

$2(\mathbf{0}) - 5(\mathbf{0}) \geq 10$? Let $x = 0$ and $y = 0$.

$0 - 0 \geq 10$?

$0 \geq 10$ False

Because $0 \geq 10$ is false, shade the region *not* containing $(0, 0)$. See Figure 33. Verify that a point in the shaded region satisfies the inequality.

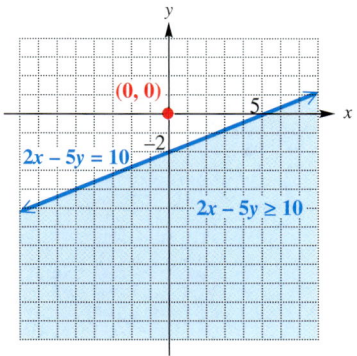

Figure 33

Work Problem 4 at the Side.

EXAMPLE 3 Graphing a Linear Inequality with a Vertical Boundary Line

Graph the inequality $x \leq 3$.

First graph $x = 3$, a vertical line through the point $(3, 0)$. Use a solid line. (Why?) Choose $(0, 0)$ as a test point.

$x \leq 3$

$0 \leq 3$? Let $x = 0$.

$0 \leq 3$ True

Continued on Next Page

ANSWERS

4.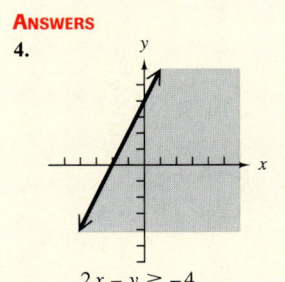

$2x - y \geq -4$

5 Graph $y < 4$.

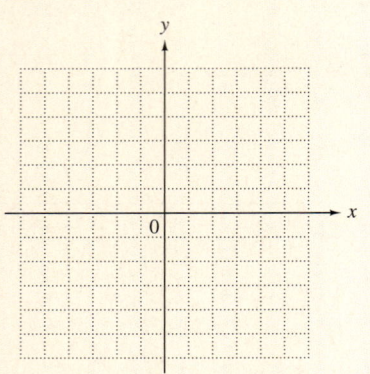

Because $0 \leq 3$ is true, shade the region containing $(0, 0)$, as in Figure 34.

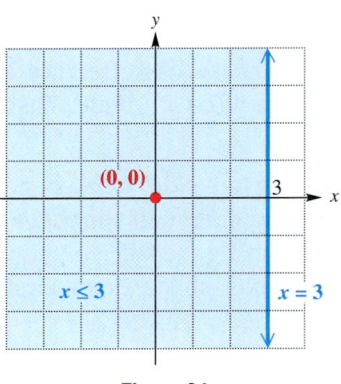

Figure 34

◀◀◀ **Work Problem 5 at the Side.**

OBJECTIVE 3 Graph an inequality with boundary through the origin. If the graph of an inequality has a boundary line through the origin, $(0, 0)$ cannot be used as a test point.

EXAMPLE 4 Graphing a Linear Inequality

Graph the inequality $x \leq 2y$.

We begin by graphing $x = 2y$, using a solid line. Some ordered pairs that can be used to graph this line are $(0, 0)$, $(6, 3)$, and $(4, 2)$. We cannot use $(0, 0)$ as a test point because $(0, 0)$ is on the line $x = 2y$. Instead, we choose a test point off the line, $(1, 3)$.

$$x \leq 2y$$
$$1 \leq 2(3) \quad ? \quad \text{Let } x = 1 \text{ and } y = 3.$$
$$1 \leq 6 \quad \quad \text{True}$$

Because $1 \leq 6$ is true, we shade the side of the graph containing the test point $(1, 3)$. See Figure 35.

6 Graph $x \geq -3y$.

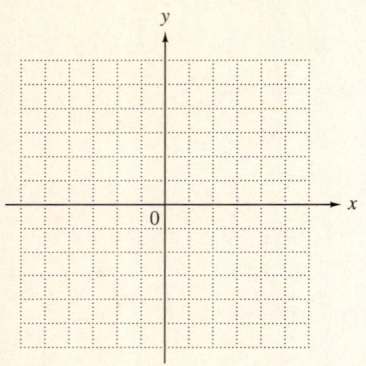

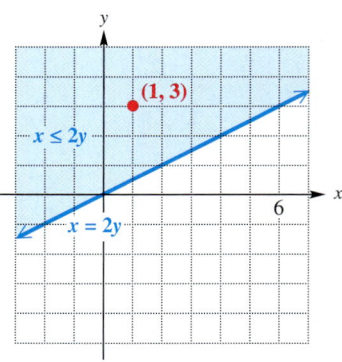

Figure 35

◀◀◀ **Work Problem 6 at the Side.**

ANSWERS

5.

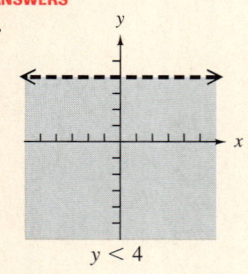

$y < 4$

6.

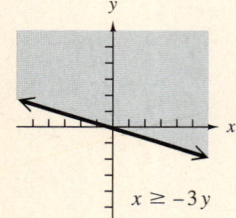

$x \geq -3y$

11.5 Exercises

Decide whether each statement is **true** *or* **false.**

1. The point (4, 0) lies on the graph of $3x - 4y < 12$.

2. The point (4, 0) lies on the graph of $3x - 4y \leq 12$.

3. Both points (4, 1) and (0, 0) lie on the graph of $3x - 2y \geq 0$.

4. The graph of $y > x$ does not contain points in quadrant IV.

The following statements were taken from articles in newspapers or magazines. Each includes a phrase that can be symbolized with one of the inequality symbols $<, \leq, >,$ *or* $\geq$. *In Exercises 5–8, give the inequality symbol for the bold-faced words.*

5. According to the Kaiser Family Foundation, in January 2001, (about) one-quarter of Medicare beneficiaries spent **more than** $2000 a year on drugs.

6. By 1937, a population of as many as a million Attwater's prairie-chickens had been cut to **less than** 9000. (*Source: National Geographic*, March 2002.)

7. In one study, 9 percent of pregnant women had active hepatitis, meaning that **at most** 9 percent of children could get it at birth. (*Source: 2000 Dow Jones & Company, Inc.*)

8. Forty percent of Americans keep **at least** one gun at home. (*Source:* Gallup Organization, quoted in *Reader's Digest*, March 2002.)

In Exercises 9–16, the straight-line boundary has been drawn. Complete each graph by shading the correct region. See Examples 1–4.

9. $x + y \geq 4$

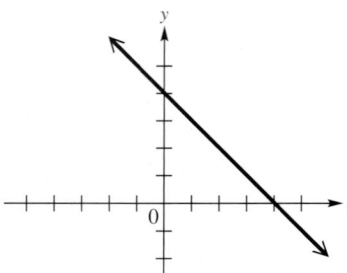

10. $x + y \leq 2$

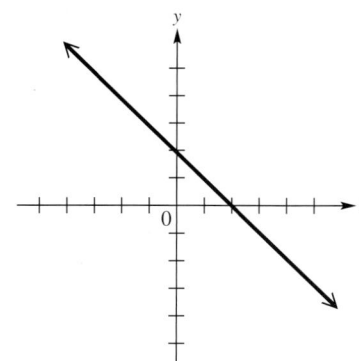

11. $x + 2y \geq 7$

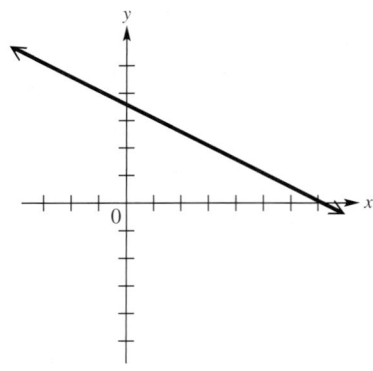

12. $2x + y \geq 5$

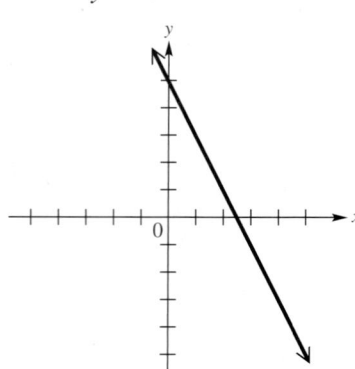

13. $-3x + 4y > 12$

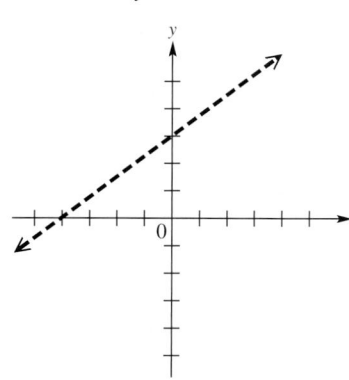

14. $4x - 5y < 20$

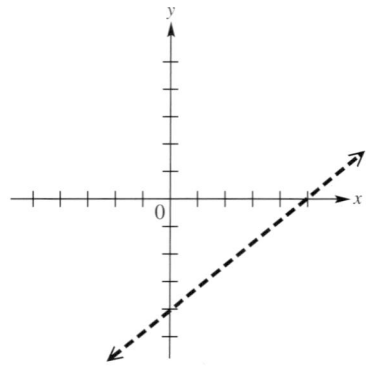

15. $x > 4$

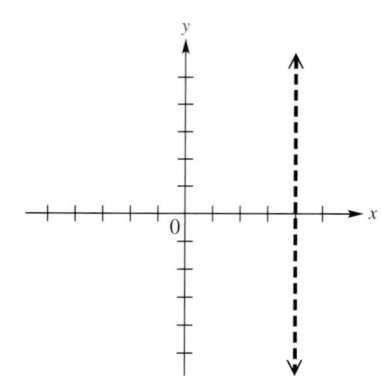

16. $y < -1$

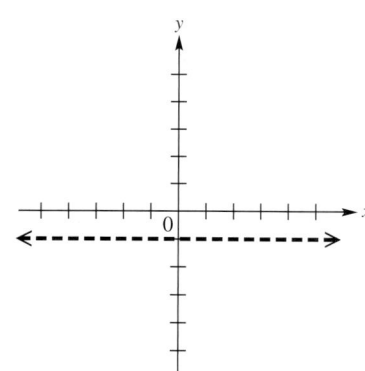

17. Explain how to determine whether to use a dashed line or a solid line when graphing a linear inequality in two variables.

18. Explain why the point (0, 0) is not an appropriate choice for a test point when graphing an inequality whose boundary goes through the origin.

Graph each linear inequality. See Examples 1–4.

19. $x + y \leq 5$

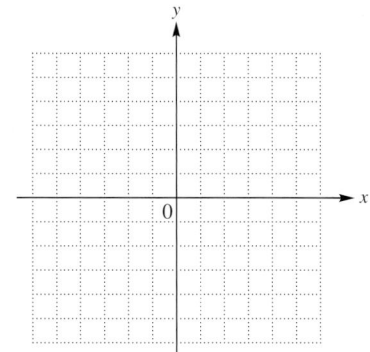

20. $x + y \geq 3$

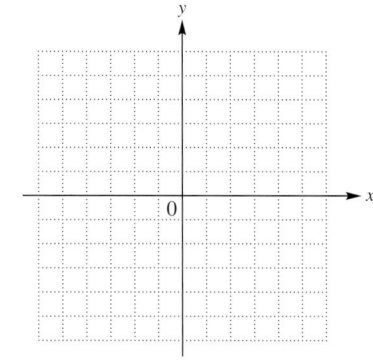

21. $x + 2y < 4$

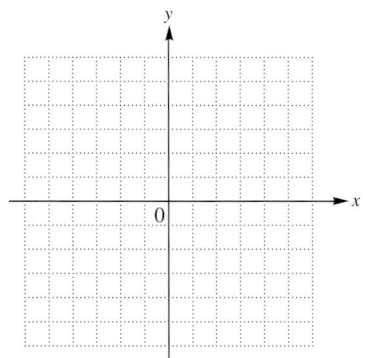

22. $x + 3y > 6$

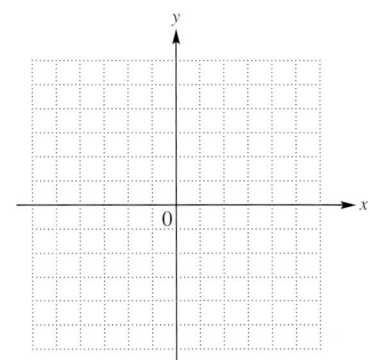

23. $2x + 6 > -3y$

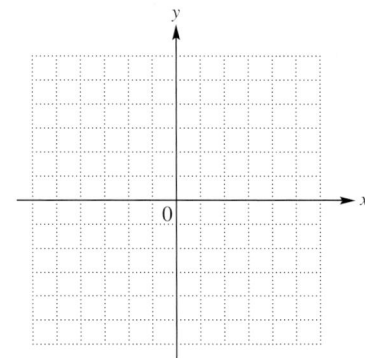

24. $-4y > 3x - 12$

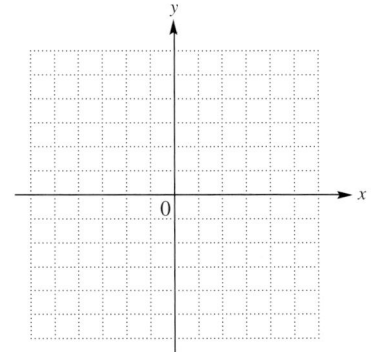

25. $y \geq 2x + 1$

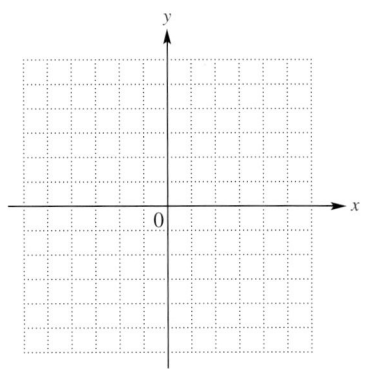

26. $y < -3x + 1$

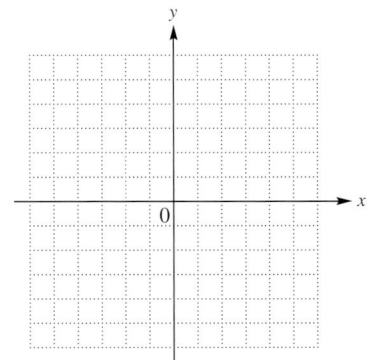

27. $x \leq -2$

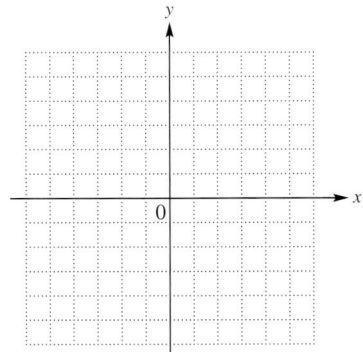

28. $x \geq 1$

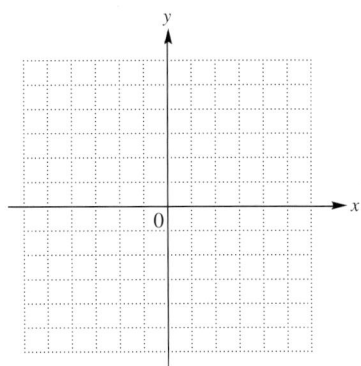

29. $y < 5$

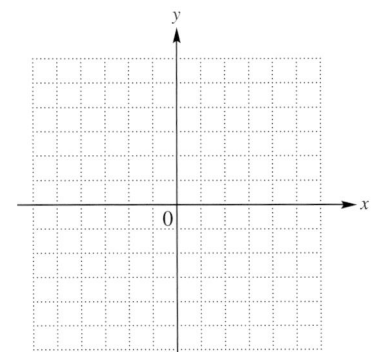

30. $y < -3$

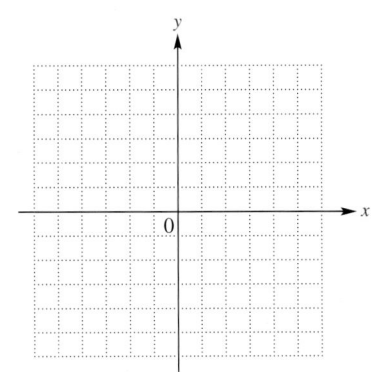

31. $y \geq 4x$

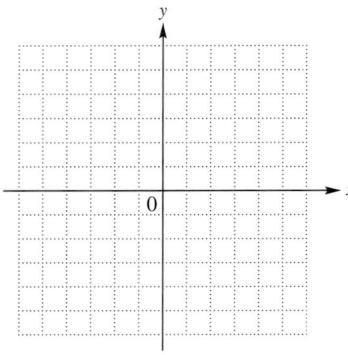

32. $y \leq 2x$

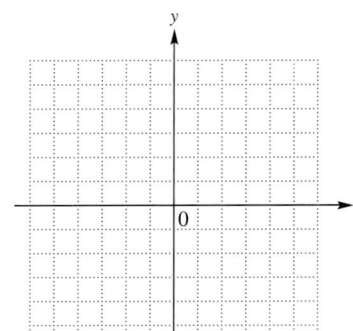

33. Explain why the graph of $y > x$ cannot lie in quadrant IV.

34. Explain why the graph of $y < x$ cannot lie in quadrant II.

Solve each problem. In part (a), $x \geq 0$ and $y \geq 0$, so graph only the part of the inequality in quadrant I.

35. A company will ship x units of merchandise to outlet I and y units of merchandise to outlet II. The company must ship a total of at least 500 units to these two outlets. This can be expressed by writing

$$x + y \geq 500.$$

(a) Graph the inequality.

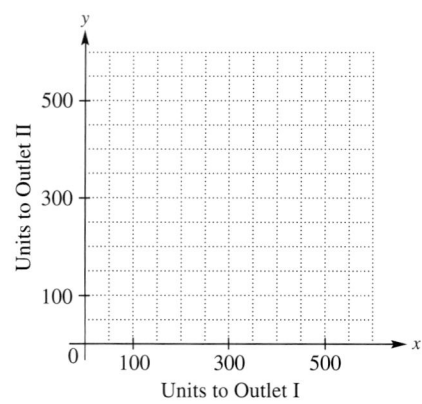

(b) Give two ordered pairs that satisfy the inequality.

36. A toy manufacturer makes stuffed bears and geese. It takes 20 min to sew a bear and 30 min to sew a goose. There is a total of 480 min of sewing time available to make x bears and y geese. These restrictions lead to the inequality

$$20x + 30y \leq 480.$$

(a) Graph the inequality.

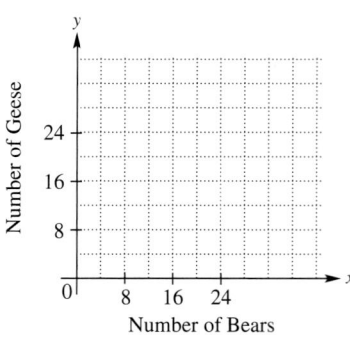

(b) Give two ordered pairs that satisfy the inequality.

Chapter 11
SUMMARY

KEY TERMS

11.1 **linear equation in two variables** — An equation that can be written in the form $Ax + By = C$ is a linear equation in two variables. (A and B are real numbers that cannot both be 0.)

ordered pair — A pair of numbers written inside parentheses, in which order is important, is called an ordered pair.

table of values — A table showing selected ordered pairs of numbers that satisfy an equation is called a table of values.

x-axis — The horizontal axis in a coordinate system is called the x-axis.

y-axis — The vertical axis in a coordinate system is called the y-axis.

rectangular (Cartesian) coordinate system — An x-axis and y-axis at right angles form a coordinate system.

quadrants — A coordinate system divides the plane into four regions called quadrants.

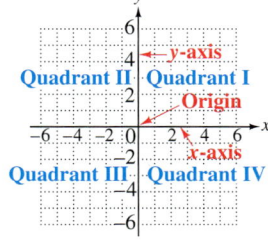

origin — The point at which the x-axis and y-axis intersect is called the origin.

plane — A flat surface determined by two intersecting lines is a plane.

coordinates — The numbers in an ordered pair are called the coordinates of the corresponding point.

plot — To plot an ordered pair is to find the corresponding point on a coordinate system.

scatter diagram — A graph of ordered pairs of data is a scatter diagram.

11.2 **graph** — The graph of an equation is the set of all points that correspond to the ordered pairs that satisfy the equation.

graphing — The process of plotting the ordered pairs that satisfy a linear equation and drawing a line through them is called graphing.

y-intercept — If a graph intersects the y-axis at k, then the y-intercept is $(0, k)$.

x-intercept — If a graph intersects the x-axis at k, then the x-intercept is $(k, 0)$.

11.3 **rise** — Rise is the vertical change between two different points on a line.

run — Run is the horizontal change between two different points on a line.

slope — The slope of a line is the ratio of the change in y compared to the change in x when moving along the line from one point to another.

parallel lines — Two lines in a plane that never intersect are parallel; the slopes of nonvertical parallel lines are equal.

perpendicular lines — Perpendicular lines intersect at a 90° angle; the product of the slopes of two perpendicular lines (neither of which is vertical) is always -1.

11.5 **linear inequality in two variables** — An inequality that can be written in the form $Ax + By < C$, $Ax + By > C$, $Ax + By \leq C$, or $Ax + By \geq C$ is a linear inequality in two variables.

boundary line — In the graph of a linear inequality, the boundary line separates the region that satisfies the inequality from the region that does not satisfy the inequality.

Chapter 11 Graphs of Linear Equations and Inequalities in Two Variables

NEW SYMBOLS

(x, y) ordered pair

(x_1, y_1) subscript notation; read as "x-sub-one, y-sub-one"

m slope of a line

TEST YOUR WORD POWER

See how well you have learned the vocabulary in this chapter. Answers, with examples, follow the Quick Review.

1. An **ordered pair** is a pair of numbers written
   A. in numerical order between brackets
   B. between parentheses or brackets
   C. between parentheses in which order is important
   D. between parentheses in which order does not matter.

2. The **coordinates** of a point are
   A. the numbers in the corresponding ordered pair
   B. the solution of an equation
   C. the values of the x- and y-intercepts
   D. the graph of the point.

3. An **intercept** is
   A. the point where the x-axis and y-axis intersect
   B. a pair of numbers written in parentheses in which order is important
   C. one of the four regions determined by a rectangular coordinate system
   D. a point where a graph intersects the x-axis or the y-axis.

4. The **slope** of a line is
   A. the measure of the run over the rise of the line
   B. the distance between two points on the line
   C. the ratio of the change in y to the change in x along the line
   D. the horizontal change compared to the vertical change of two points on the line.

5. Two nonvertical lines in a plane are **parallel** if
   A. they represent the same line
   B. their slopes are equal
   C. they intersect at a 90° angle
   D. one has a positive slope and one has a negative slope.

6. Two nonvertical lines in a plane are **perpendicular** if
   A. their slopes are equal
   B. they never intersect
   C. the product of their slopes is -1.
   D. one has a positive slope and one has a negative slope.

QUICK REVIEW

Concepts

11.1 Reading Graphs; Linear Equations in Two Variables

Circle graphs, bar graphs, and line graphs are several ways to represent the relationship between two variables.

Examples

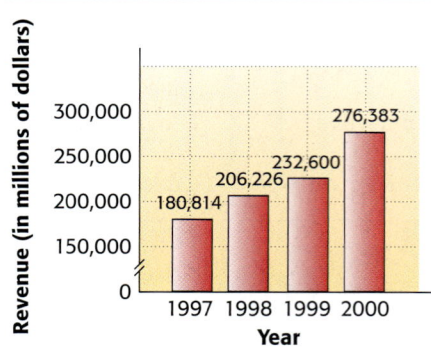

Source: Veronis Suhler Stevenson, *Communications Industry Report,* annual.

The bar graph illustrates communications industry revenue for the years 1997–2000 in millions of dollars.

Concepts

11.1 Reading Graphs; Linear Equations in Two Variables (continued)

An ordered pair is a solution of an equation if it makes the equation a true statement.

If a value of either variable in an equation is given, the value of the other variable can be found by substitution.

To plot the ordered pair $(-3, 4)$, start at the origin, go 3 units to the left, and from there go 4 units up.

11.2 Graphing Linear Equations in Two Variables

To graph a linear equation:

Step 1 Find at least two ordered pairs that are solutions of the equation.

Step 2 Plot the corresponding points.

Step 3 Draw a straight line through the points.

11.3 Slope of a Line

The slope of the line through (x_1, y_1) and (x_2, y_2) is

$$m = \frac{\text{change in } y}{\text{change in } x} = \frac{y_2 - y_1}{x_2 - x_1} \quad (x_1 \neq x_2).$$

Horizontal lines have slope 0.

Vertical lines have undefined slope.

To find the slope of a line from its equation, solve for y. The slope is the coefficient of x.

Examples

Is $(2, -5)$ or $(0, -6)$ a solution of $4x - 3y = 18$?

$$4(2) - 3(-5) = 23 \neq 18 \quad | \quad 4(0) - 3(-6) = 18$$

$(2, -5)$ is not a solution. | $(0, -6)$ is a solution.

Complete the ordered pair $(0, \quad)$ for $3x = y + 4$.

$$3(0) = y + 4 \quad \text{Let } x = 0.$$
$$0 = y + 4$$
$$-4 = y$$

The ordered pair is $(0, -4)$.

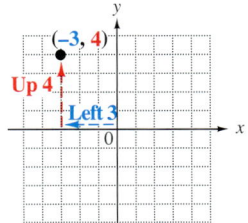

Graph $x - 2y = 4$.

x	y
0	-2
4	0

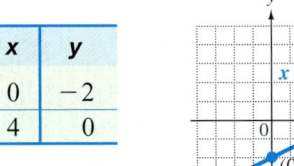

The line through $(-2, 3)$ and $(4, -5)$ has slope

$$m = \frac{-5 - 3}{4 - (-2)} = \frac{-8}{6} = -\frac{4}{3}.$$

The line $y = -2$ has slope 0.

The line $x = 4$ has undefined slope.

Find the slope of the graph of

$$3x - 4y = 12.$$
$$-4y = -3x + 12$$
$$y = \frac{3}{4}x - 3$$

The slope is $\frac{3}{4}$.

Concepts	Examples
11.4 Equations of Lines **Slope-Intercept Form** $y = mx + b$ m is the slope. $(0, b)$ is the y-intercept.	Find an equation of the line with slope **2** and y-intercept $(0, -5)$. $$y = 2x - 5$$
Point-Slope Form $y - y_1 = m(x - x_1)$ m is the slope. (x_1, y_1) is a point on the line.	Find an equation of the line with slope $-\frac{1}{2}$ through $(-4, 5)$. $$y - 5 = -\frac{1}{2}[x - (-4)]$$ $$y - 5 = -\frac{1}{2}(x + 4)$$ $$y - 5 = -\frac{1}{2}x - 2$$ $$y = -\frac{1}{2}x + 3$$
Standard Form $Ax + By = C$ A, B, and C are integers and $A > 0$, $B \neq 0$.	This equation is written in standard form as $$x + 2y = 6,$$ with $A = 1$, $B = 2$, and $C = 6$.
11.5 Graphing Linear Inequalities in Two Variables **Step 1** Graph the line that is the boundary of the region. Make it solid if the inequality is $\leq$ or $\geq$; make it dashed if the inequality is $<$ or $>$. **Step 2** Use any point not on the line as a test point. Substitute for x and y in the inequality. If the result is true, shade the side of the line containing the test point; if the result is false, shade the other side.	Graph $2x + y \leq 5$. Graph the line $2x + y = 5$. Make it solid because the symbol $\leq$ includes equality. Use $(1, 0)$ as a test point. $$2(1) + 0 \leq 5 \quad ?$$ $$2 \leq 5 \quad \text{True}$$ Shade the side of the line containing $(1, 0)$. 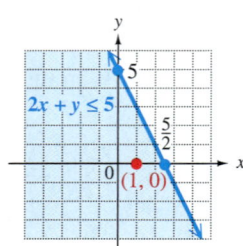

> **ANSWERS TO TEST YOUR WORD POWER**

1. C; *Examples:* (0, 3), (3, 8), (4, 0)
2. A; *Example:* The point associated with the ordered pair (1, 2) has x-coordinate 1 and y-coordinate 2.
3. D; *Example:* The graph of the equation $4x - 3y = 12$ has x-intercept at (3, 0) and y-intercept at (0, -4).
4. C; *Example:* The line through (3, 6) and (5, 4) has slope $\frac{4 - 6}{5 - 3} = \frac{-2}{2} = -1$.
5. B; *Example:* See Figure 24 in **Section 11.3**.
6. C; *Example:* See Figure 25 in **Section 11.3**.

Chapter 11
REVIEW EXERCISES

[11.1] *The percents of four-year college students in private institutions who earned a degree within five years of entry between 1996 and 2001 are shown in the graph.*

1. What was the difference in the percent of graduates between 1996 and 2001?

2. Write ordered pairs to reflect the data shown in the graph.

3. **(a)** In what year did the percent show the greatest decrease from the previous year?

   (b) In what year did the percent show the smallest decrease from the previous year?

PERCENTS OF STUDENTS GRADUATING WITHIN 5 YEARS (PRIVATE INSTITUTIONS)

Percent: 57.1 (1996), 56.6 (1997), 56.2 (1998), 55.8 (1999), 55.5 (2000), 55.1 (2001)

Source: ACT.

4. Describe the general trend seen in the graph.

Complete the given ordered pairs for each equation.

5. $y = 3x + 2$ $(-1, \ ), (0, \ ), (\ , 5)$

6. $4x + 3y = 6$ $(0, \ ), (\ , 0), (-2, \ )$

7. $x = 3y$ $(0, \ ), (8, \ ), (\ , -3)$

8. $x - 7 = 0$ $(\ , -3) (\ , 0), (\ , 5)$

Decide whether each ordered pair is a solution of the given equation.

9. $x + y = 7; (2, 5)$

10. $2x + y = 5; (-1, 3)$

11. $3x - y = 4; \left(\dfrac{1}{3}, -3\right)$

Plot each ordered pair on the given coordinate system.

12. (2, 3)

13. (−4, 2)

14. (3, 0)

15. (0, −6)

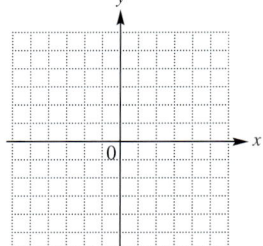

16. If $x > 0$ and $y < 0$, in what quadrant(s) must (x, y) lie? Explain.

17. On what axis does the point $(k, 0)$ lie for any real value of k? the point $(0, k)$? Explain.

Without plotting the given point, name the quadrant in which each point lies.

18. (−2, 3)

19. (−1, −4)

20. $\left(0, -5\frac{1}{2}\right)$

[11.2] *Find the intercepts for each equation.*

21. $y = 2x + 5$
x-intercept:
y-intercept:

22. $2x + y = -7$
x-intercept:
y-intercept:

23. $3x + 2y = 8$
x-intercept:
y-intercept:

Graph each linear equation.

24. $2x - y = 3$

25. $x + 2y = -4$

26. $x + y = 0$

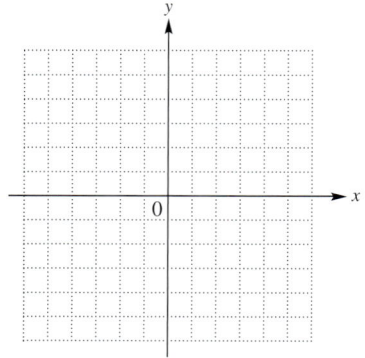
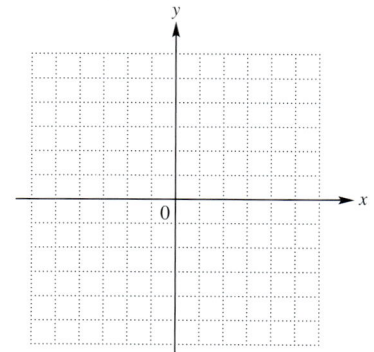
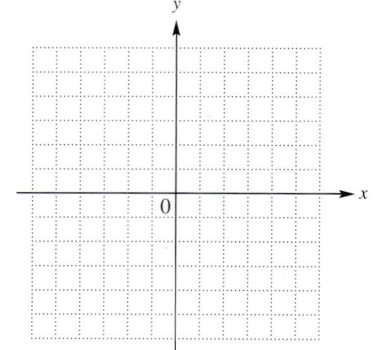

[11.3] *Find the slope of each line.*

27. Through $(2, 3)$ and $(-4, 6)$ **28.** Through $(0, 0)$ and $(-3, 2)$ **29.** Through $(0, 6)$ and $(1, 6)$

30. Through $(2, 5)$ and $(2, 8)$ **31.** $y = 3x - 4$ **32.** $y = \dfrac{2}{3}x + 1$

33. 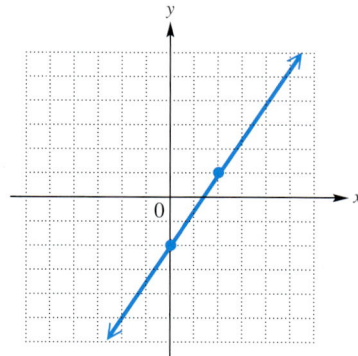   **34.**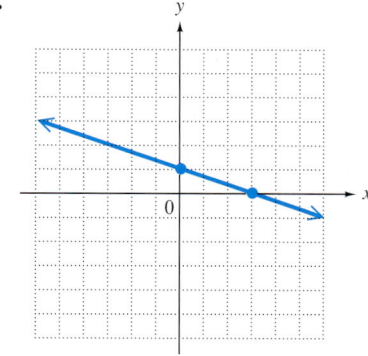

35. $x = 0$ **36.** $y = 4$

37. The line having these points

x	y
0	1
2	4
6	10

38. (a) A line parallel to the graph of $y = 2x + 3$
(b) A line perpendicular to the graph of $y = -3x + 3$

Decide whether each pair of lines is parallel, perpendicular, *or neither.*

39. $3x + 2y = 6$
$6x + 4y = 8$

40. $x - 3y = 1$
$3x + y = 4$

41. $x - 2y = 8$
$x + 2y = 8$

42. What is the slope of a line perpendicular to a line with undefined slope?

[11.4] *Write an equation in slope-intercept form (if possible) for each line.*

43. $m = -1; b = \dfrac{2}{3}$

44. The line in Exercise 34

45. Through $(4, -3); m = 1$

46. Through $(-1, 4); m = \dfrac{2}{3}$

47. Through $(1, -1); m = -\dfrac{3}{4}$

48. Through $(2, 1)$ and $(-2, 2)$

49. Through $(-4, 1)$ with slope 0

50. Through $\left(\dfrac{1}{3}, -\dfrac{3}{4}\right)$ with undefined slope

51. Consider the equation $x + 3y = 15$.

  (a) Write it in the form $y = mx + b$.

  (b) What is the slope? What is the y-intercept?

  (c) Use the slope and the y-intercept to graph the line. Indicate two points on the graph.

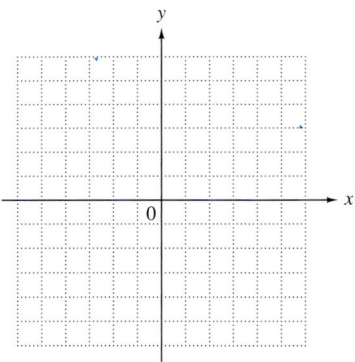

[11.5] *Graph each linear inequality.*

52. $3x + 5y > 9$

53. $2x - 3y > -6$

54. $x \geq -4$

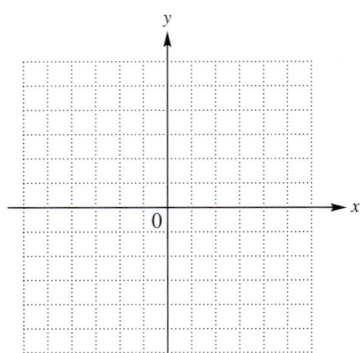 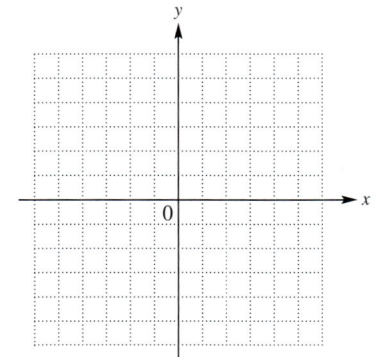 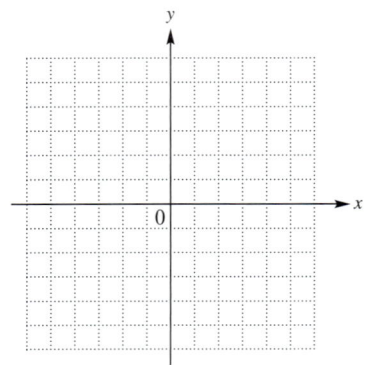

MIXED REVIEW EXERCISES

In Exercises 55–60, match each statement to the appropriate graph or graphs in A–D. Graphs may be used more than once.

A. 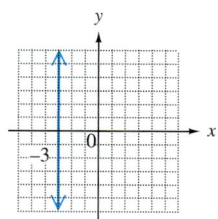 **B.** 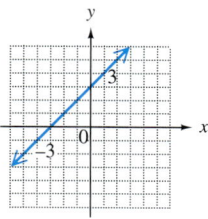 **C.** 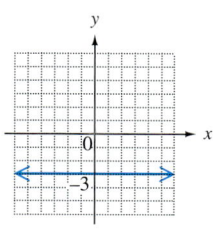 **D.**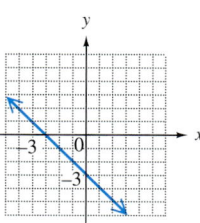

55. The line shown in the graph has undefined slope.

56. The graph of the equation has *y*-intercept $(0, -3)$.

57. The graph of the equation has *x*-intercept $(-3, 0)$.

58. The line shown in the graph has negative slope.

59. The graph is that of the equation $y = -3$.

60. The line shown in the graph has slope 1.

Find the intercepts and the slope of each line. Then graph the line.

61. $y = -2x - 5$

x-intercept:

y-intercept:

slope:

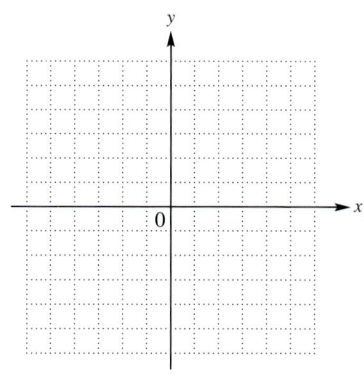

62. $x + 3y = 0$

x-intercept:

y-intercept:

slope:

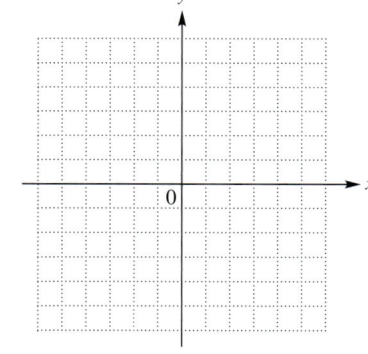

63. $y - 5 = 0$

x-intercept:

y-intercept:

slope:

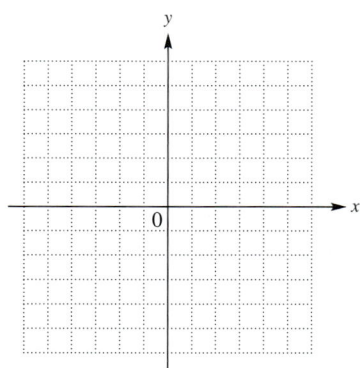

Write an equation in slope-intercept form for each line.

64. $m = -\dfrac{1}{4}; b = -\dfrac{5}{4}$

65. Through $(8, 6); m = -3$

66. Through $(3, -5)$ and $(-4, -1)$

802 Chapter 11 Graphs of Linear Equations and Inequalities in Two Variables

Graph each inequality.

67. $y < -4x$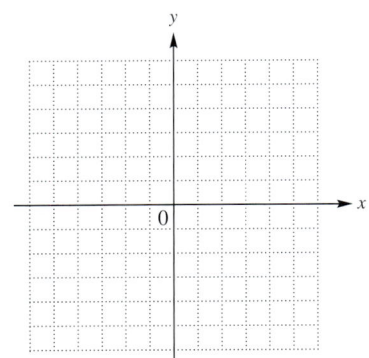

68. $x - 2y \leq 6$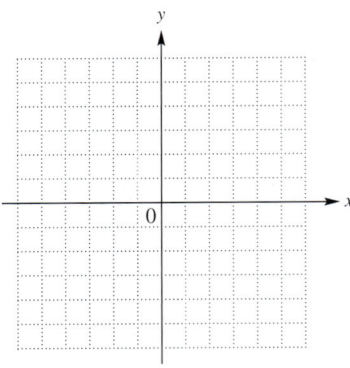

RELATING CONCEPTS (EXERCISES 69–75) For Individual or Group Work

The percents of four-year college students in public schools who earned a degree within five years of entry between 1997 and 2002 are shown in the graph. Use the graph to **work Exercises 69–75 in order.**

69. What was the difference in the percent of graduates between 1997 and 2002?

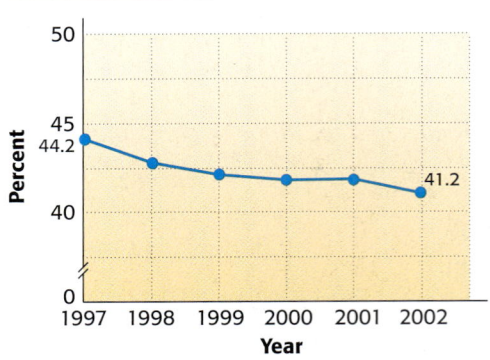

PERCENTS OF STUDENTS GRADUATING WITHIN 5 YEARS (PUBLIC INSTITUTIONS)

Source: ACT.

70. Since the points of the graph lie approximately in a linear pattern, a straight line can be used to model the data. Will this line have positive or negative slope? Explain.

71. Write two ordered pairs for the data for 1997 and 2002.

72. Use the ordered pairs from Exercise 71 to find the equation of a line that models the data. Write the equation in slope-intercept form.

73. Based on the equation you found in Exercise 72, what is the slope of the line? Does it agree with your answer in Exercise 70?

74. Use the equation from Exercise 72 to approximate the percents for 1998 through 2001, and complete the table. Round your answers to the nearest tenth.

Year	Percent
1998	
1999	
2000	
2001	

75. Use the equation from Exercise 72 to predict the percent for 2003. Can we be sure that this prediction is accurate?

Chapter 11
TEST

The line graph shows the average prices for a gallon of gasoline in the United States for the years 1998 to 2002. Use the graph to work Exercises 1–3.

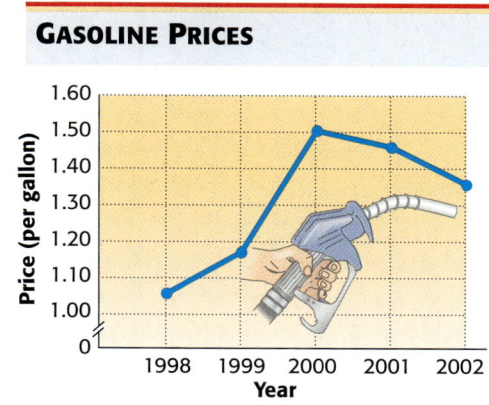

GASOLINE PRICES

Source: U.S. Department of Energy.

1. About how much did a gallon of gasoline cost in 1998?

2. About how much did the price of a gallon of gasoline increase from 1998 to 2000?

3. During which years did the price decrease?

Graph each linear equation. Give the x- and y-intercepts.

4. $3x + y = 6$

5. $y - 2x = 0$

1. _____

2. _____

3. _____

4. x-intercept: _____
   y-intercept: _____

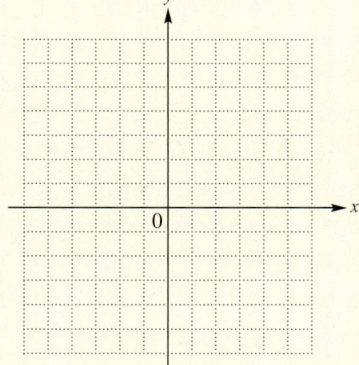

5. x-intercept: _____
   y-intercept: _____

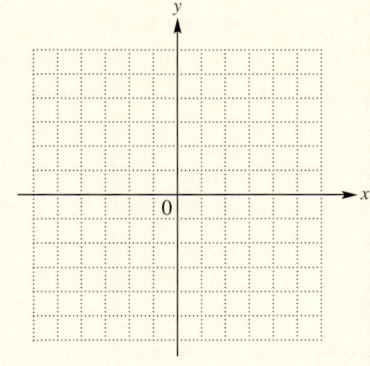

6. *x*-intercept: _____

   y-intercept: _____

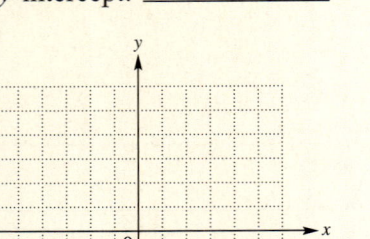

6. $x + 3 = 0$

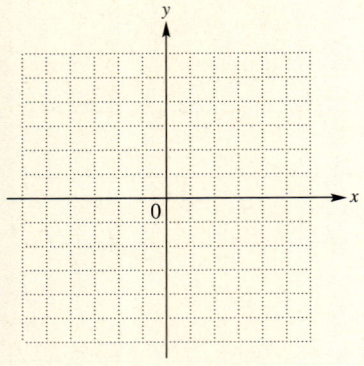

7. *x*-intercept: _____

   y-intercept: _____

7. $y = 1$

8. *x*-intercept: _____

   y-intercept: _____

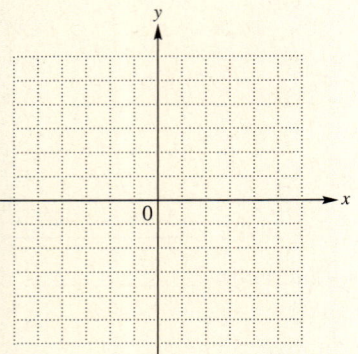

8. $x - y = 4$

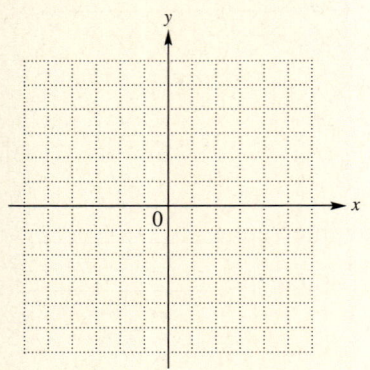

Find the slope of each line.

9. _____

9. Through $(-4, 6)$ and $(-1, -2)$

10. _____

10. $2x + y = 10$

11. _____

11. $x + 12 = 0$

12.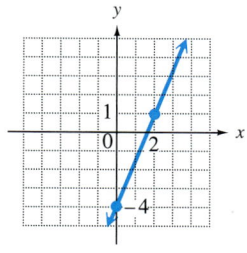

12. _____

13. A line parallel to the graph of $y - 4 = 6$

13. _____

Write an equation in slope-intercept form for each line.

14. Through $(-1, 4)$; $m = 2$

14. _____

15. The line in Exercise 12

15. _____

16. Through $(2, -6)$ and $(1, 3)$

16. _____

17. x-intercept: $(3, 0)$; y-intercept: $\left(0, \dfrac{9}{2}\right)$

17. _____

Graph each linear inequality.

18. $x + y \leq 3$

18.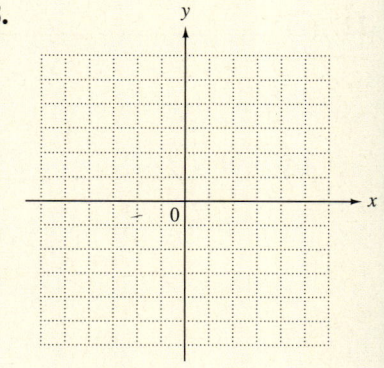

19.

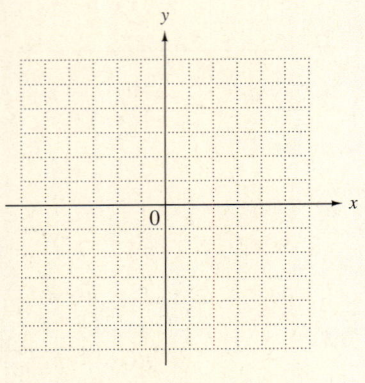

19. $3x - y > 0$

The graph shows total food and drink sales at U.S. restaurants from 1970 through 2000, where 1970 corresponds to $x = 0$. Use the graph to work Exercises 20–22.

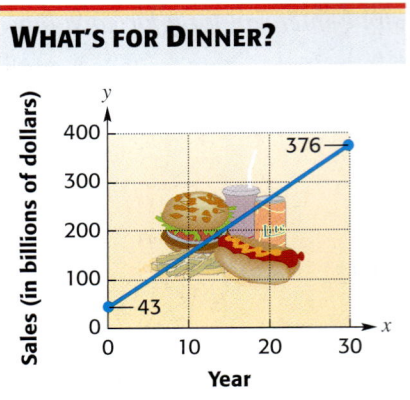

WHAT'S FOR DINNER?

Source: National Restaurant Association.

20. _____

20. Is the slope of the line in the graph positive or negative? Explain.

21. _____

21. Write two ordered pairs for the data points shown in the graph. Use them to find the slope of the line.

22. The linear equation

$$y = 11.1x + 43$$

approximates food and drink sales y in billions of dollars, where $x = 0$ again represents 1970.

22. (a) _____

(a) Use the equation to approximate food and drink sales for 1990 and 1995.

(b) _____

(b) What does the ordered pair (30, 376) mean in the context of this problem?

Systems of Linear Equations and Inequalities

12.1 Solving Systems of Linear Equations by Graphing

12.2 Solving Systems of Linear Equations by Substitution

12.3 Solving Systems of Linear Equations by Elimination

Summary Exercises on Solving Systems of Linear Equations

12.4 Applications of Linear Systems

12.5 Solving Systems of Linear Inequalities

Many people decide to start their own small business rather than work for someone else. For example, if you like bicycle riding, you might open a bicycle shop. It is encouraging to know that bicycle riding is Americans' 6th favorite sport, and that sales of bicycles and supplies more than doubled from $2.4 billion in 1990 to $5.1 billion in 2001. (*Source:* National Sporting Goods Assoc.)

But, in order to make a profit and stay in business, you need to know the *break even point* where sales revenue covers all your expenses. In Section 12.2, Exercises 30–33, we use a *system of linear equations* to find this point for a bicycle business.

12.1 Solving Systems of Linear Equations by Graphing

OBJECTIVES

1. Decide whether a given ordered pair is a solution of a system.
2. Solve linear systems by graphing.
3. Solve special systems by graphing.

A **system of linear equations**, often called a **linear system**, consists of two or more linear equations with the same variables. Examples of systems of two linear equations are shown below.

| $2x + 3y = 4$ | $x + 3y = 1$ | $x - y = 1$ | Linear |
| $3x - y = -5$ | $-y = 4 - 2x$ | $y = 3$ | systems |

In the system on the right, think of $y = 3$ as an equation in two variables by writing it as $0x + y = 3$.

OBJECTIVE 1 Decide whether a given ordered pair is a solution of a system. A **solution of a system** of linear equations is an ordered pair that makes both equations true at the same time. A solution of an equation is said to *satisfy* the equation.

1. Fill in the blanks, and decide whether the given ordered pair is a solution of the system.

    (a) (2, 5)

    $3x - 2y = -4$
    $5x + y = 15$

    $3x - 2y = -4$
    $3(\_\_) - 2(\_\_) = -4$
    $\_\_ = -4$

    $5x + y = 15$
    $5(2) + \_\_ = \_\_$
    $\_\_ = \_\_$

    (2, 5) _____ a solution.
    (is/is not)

EXAMPLE 1 Determining Whether an Ordered Pair Is a Solution

Is $(4, -3)$ a solution of each system?

(a) $x + 4y = -8$
$3x + 2y = 6$

To decide whether or not $(4, -3)$ is a solution of the system, substitute 4 for x and -3 for y in each equation.

$x + 4y = -8$		$3x + 2y = 6$	
$4 + 4(-3) = -8$	? Multiply.	$3(4) + 2(-3) = 6$	? Multiply.
$4 + (-12) = -8$	?	$12 + (-6) = 6$	?
$-8 = -8$	True	$6 = 6$	True

Because $(4, -3)$ satisfies *both* equations, it is a solution of the system.

(b) $2x - 5y = -7$
$3x + 4y = 2$

Again, substitute 4 for x and -3 for y in both equations.

$2x + 5y = -7$		$3x + 4y = 2$	
$2(4) + 5(-3) = -7$	? Multiply.	$3(4) + 4(-3) = 2$	? Multiply.
$8 + (-15) = -7$	?	$12 + (-12) = 2$	?
$-7 = -7$	True	$0 = 2$	False

The ordered pair $(4, -3)$ is not a solution of this system because it does not satisfy the second equation.

(b) $(1, -2)$

$x - 3y = 7$
$4x + y = 5$

$(1, -2)$ _____ a solution.
(is/is not)

▶ **Work Problem 1 at the Side.**

OBJECTIVE 2 Solve linear systems by graphing. One way to find the solution of a system of two linear equations is to graph both equations on the same axes. The graph of each line shows points whose coordinates satisfy the equation of that line. Any intersection point would be on both lines and would therefore be a solution of *both* equations. Thus, *the coordinates of any point where the lines intersect give a solution of the system.*

ANSWERS

1. (a) 2; 5; −4; 5; 15; 15; 15; is   (b) is not

The graph in Figure 1 shows that the solution of the system in Example 1(a) is the intersection point $(4, -3)$. Because *two different* straight lines can intersect at no more than one point, there can never be more than one solution for such a system.

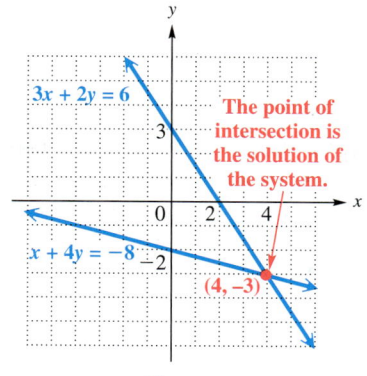

The point of intersection is the solution of the system.

Figure 1

EXAMPLE 2 Solving a System by Graphing

Solve the system of equations by graphing both equations on the same axes.

$$2x + 3y = 4$$
$$3x - y = -5$$

We graph these two equations by plotting several points for each line. Recall from **Section 11.2** that the intercepts are often convenient choices. It is a good idea to find a third ordered pair as a check.

$2x + 3y = 4$

x	y
0	$\frac{4}{3}$
2	0
-2	$\frac{8}{3}$

$3x - y = -5$

x	y
0	5
$-\frac{5}{3}$	0
-2	-1

The lines in Figure 2 suggest that the graphs intersect at the point $(-1, 2)$. We check this by substituting -1 for x and 2 for y in both equations. Because $(-1, 2)$ satisfies both equations, the solution of this system is $(-1, 2)$.

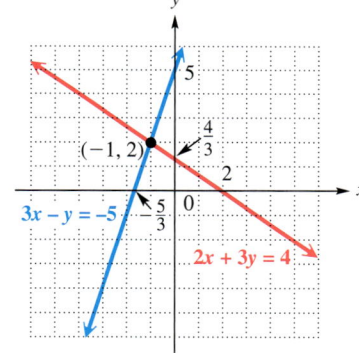

Figure 2

Work Problem 2 at the Side.

> **NOTE**
> We can also graph a linear system by writing each equation in the system in slope-intercept form and using the slope and y-intercept to graph each line. For Example 2, we have the following.
>
> $2x + 3y = 4$ becomes $y = -\frac{2}{3}x + \frac{4}{3}$ y-intercept $(0, \frac{4}{3})$; slope $-\frac{2}{3}$
>
> $3x - y = -5$ becomes $y = 3x + 5$ y-intercept $(0, 5)$; slope 3 or $\frac{3}{1}$
>
> Confirm that graphing these equations results in the same lines and the same solution shown in Figure 2.

② Solve each system of equations by graphing both equations on the same axes. Check your solutions.

(a) $5x - 3y = 9$
$x + 2y = 7$
(One of the lines is already graphed.)

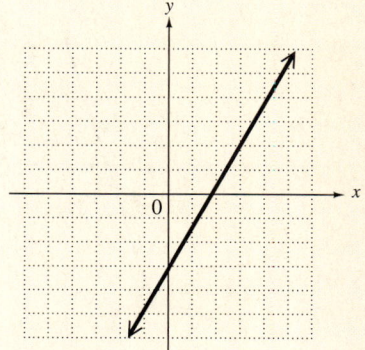

(b) $x + y = 4$
$2x - y = -1$

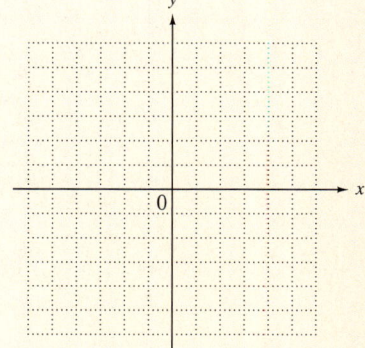

ANSWERS
2. (a) $(3, 2)$ (b) $(1, 3)$

3 Solve each system of equations by graphing both equations on the same axes.

(a) $3x - y = 4$
$6x - 2y = 12$
(One of the lines is already graphed.)

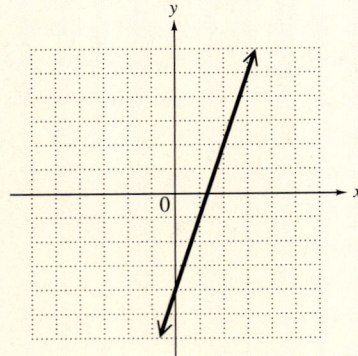

(b) $-x + 3y = 2$
$2x - 6y = -4$

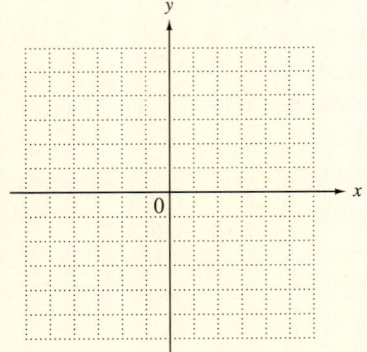

CAUTION
A difficulty with the graphing method of solution is that it may not be possible to determine from the graph the exact coordinates of the point that represents the solution, particularly if these coordinates are not integers. For this reason, algebraic methods of solution are explained later in this chapter. The graphing method does, however, show geometrically how solutions are found and is useful when approximate answers will do.

OBJECTIVE 3 Solve special systems by graphing. Sometimes the graphs of the two equations in a system either do not intersect at all or are the same line, as in the systems in Example 3.

EXAMPLE 3 Solving Special Systems

Solve each system by graphing.

(a) $2x + y = 2$
$2x + y = 8$

The graphs of these lines are shown in Figure 3. The two lines are parallel and have no points in common. For such a system, we will write "*no solution.*"

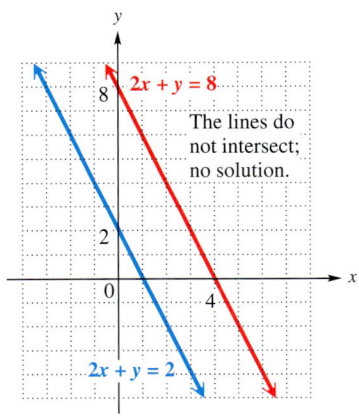

Figure 3

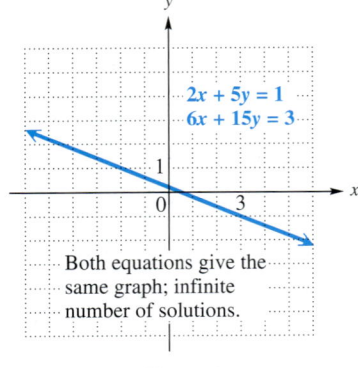

Figure 4

(b) $2x + 5y = 1$
$6x + 15y = 3$

The graphs of these two equations are the same line. See Figure 4. The second equation can be obtained by multiplying each side of the first equation by 3. In this case, every point on the line is a solution of the system, and the solutions are the infinite number of ordered pairs that satisfy the equations. We will write "*infinite number of solutions*" to indicate this case.

◀◀◀ **Work Problem 3 at the Side.**

The system in Example 2 has exactly one solution. A system with at least one solution is called a **consistent system.** A system of equations with no solutions, such as the one in Example 3(a), is called an **inconsistent system.** The equations in Example 2 are **independent equations** with different graphs. The equations of the system in Example 3(b) have the same graph and are equivalent. Because they are different forms of the same equation, these equations are called **dependent equations.**

ANSWERS
3. (a) parallel lines; no solution
   (b) same line; infinite number of solutions

Examples 2 and 3 show the three cases that may occur when solving a system of two equations with two variables.

> **Possible Types of Solutions**
>
> 1. The graphs intersect at exactly one point, which gives the (single) ordered pair solution of the system. The **system is consistent,** and the **equations are independent.**
>
>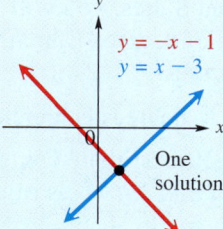
>
> 2. The graphs are parallel lines, so there is no solution. The **system is inconsistent.**
>
>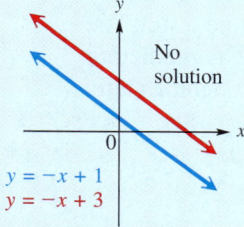
>
> 3. The graphs are the same line. The solution is an infinite number of ordered pairs. The **equations are dependent.**
>
>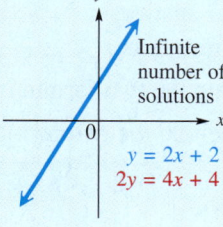

Focus on Real-Data Applications

Estimating Fahrenheit Temperature

When traveling in countries other than the United States, you will hear the daily high and low temperatures reported in degrees Celsius, instead of degrees Fahrenheit. The following information may help you interpret these Celsius temperatures.

- The linear equation to convert degrees Celsius to degrees Fahrenheit is $F = \frac{9}{5}C + 32$.
- Travel books advise you to use a *rule of thumb* to estimate Fahrenheit temperature that says "*Double the temperature (degrees Celsius) and add 30.*" This rule of thumb is written mathematically as $F = 2C + 30$.

For Group Discussion

Suppose you are interested in knowing for what temperature the rule of thumb and the actual formulas give the same result. You also want to know if the rule of thumb formula is predicting temperatures that are lower or higher than the actual temperature.

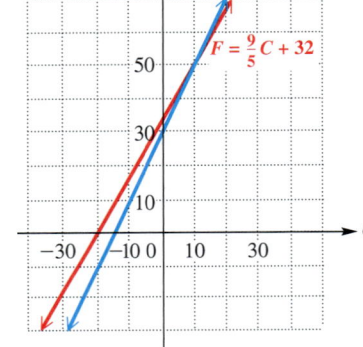

1. The two formulas can be written as the system of equations

$$F = \frac{9}{5}C + 32$$
$$F = 2C + 30.$$

   (a) Use the graph of the system of equations to find the point of intersection. (*Hint:* To check your answer, use substitution to see if it satisfies both formulas.)

   (b) For what temperature in degrees Celsius do the two formulas agree?

   (c) For what temperature in degrees Fahrenheit do the two formulas agree?

2. (a) Complete the table of values to compare the *actual* and the *rule of thumb* formulas for temperature conversion.

°C	°F (Actual)	°F (Rule of Thumb)
0		
5		
10		
15		
20		
30		

   (b) If the daily low is predicted to be 5 °C, then is the rule of thumb estimate too high or too low?

   (c) If the daily high is predicted to be 20 °C, then is the rule of thumb estimate too high or too low?

   (d) If the daily high is predicted to be 30 °C, then is the rule of thumb estimate too high or too low?

   (e) How many degrees "off" is the rule of thumb estimate for the boiling point of water?

   (f) Comment on the accuracy of using the rule of thumb as an estimate of the actual Fahrenheit temperature.

12.1 Exercises

1. Which ordered pair could be a solution of the system graphed? Why is it the only valid choice?
   A. $(2, 2)$
   B. $(-2, 2)$
   C. $(-2, -2)$
   D. $(2, -2)$

   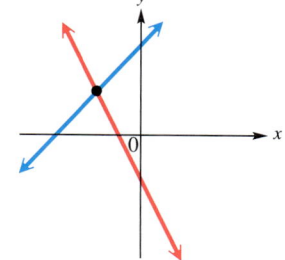

2. Which ordered pair could be a solution of the system graphed? Why is it the only valid choice?
   A. $(2, 0)$
   B. $(0, 2)$
   C. $(-2, 0)$
   D. $(0, -2)$

   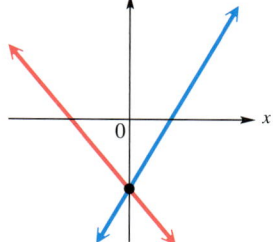

3. How can you tell without graphing that this system has no solution?
   $$x + y = 2$$
   $$x + y = 4$$

4. Explain why a system of two linear equations cannot have exactly two solutions.

Decide whether the given ordered pair is a solution of the given system. See Example 1.

5. $(2, -3)$
   $x + y = -1$
   $2x + 5y = 19$

6. $(4, 3)$
   $x + 2y = 10$
   $3x + 5y = 3$

7. $(-1, -3)$
   $3x + 5y = -18$
   $4x + 2y = -10$

8. $(-9, -2)$
   $2x - 5y = -8$
   $3x + 6y = -39$

9. $(7, -2)$
   $4x = 26 - y$
   $3x = 29 + 4y$

10. $(9, 1)$
    $2x = 23 - 5y$
    $3x = 24 + 3y$

11. $(6, -8)$
    $-2y = x + 10$
    $3y = 2x + 30$

12. $(-5, 2)$
    $5y = 3x + 20$
    $3y = -2x - 4$

814 Chapter 12 Systems of Linear Equations and Inequalities

Solve each system of equations by graphing. If the two equations produce parallel lines, write no solution. *If the two equations produce the same line, write* infinite number of solutions. *See Examples 2 and 3.*

13. $x - y = 2$
    $x + y = 6$

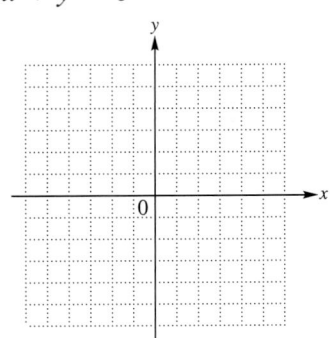

14. $x - y = 3$
    $x + y = -1$

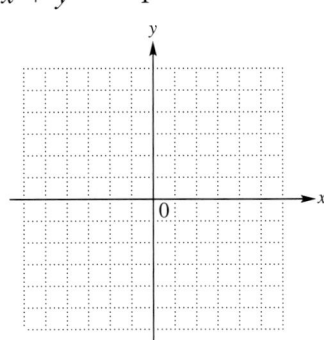

15. $x + y = 4$
    $y - x = 4$

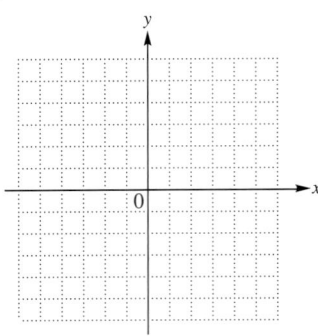

16. $x + y = -5$
    $x - y = 5$

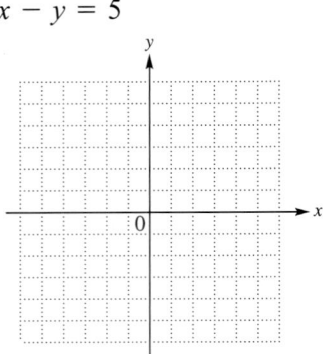

17. $x - 2y = 6$
    $x + 2y = 2$

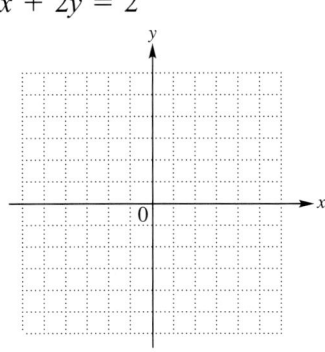

18. $2x - y = 4$
    $4x + y = 2$

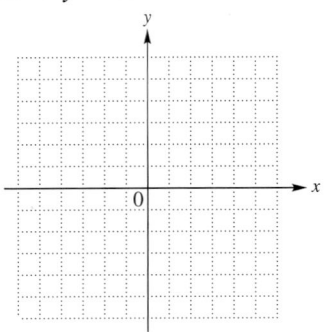

19. $3x - 2y = -3$
    $-3x - y = -6$

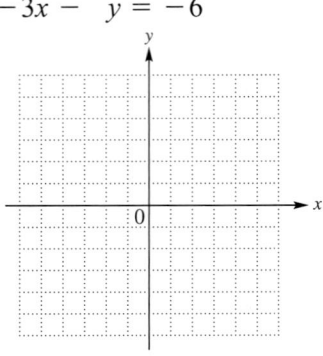

20. $2x - y = 4$
    $2x + 3y = 12$

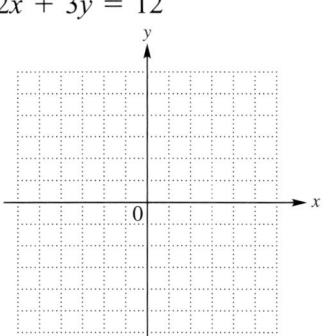

21. $2x - 3y = -6$
    $y = -3x + 2$

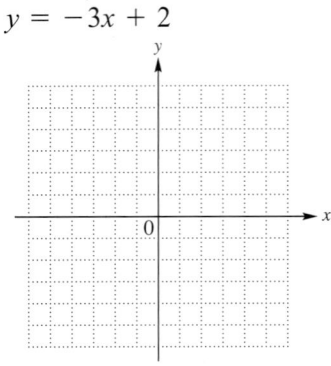

22. $-3x + y = -3$
    $y = x - 3$

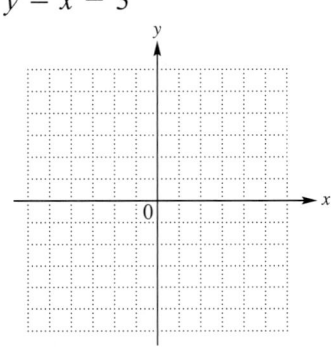

23. $x + 2y = 6$
    $2x + 4y = 8$

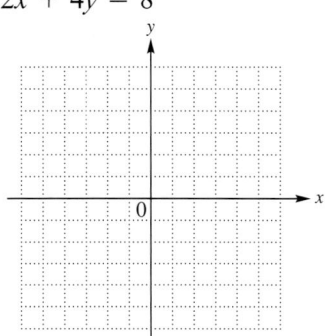

24. $2x - y = 6$
    $6x - 3y = 12$

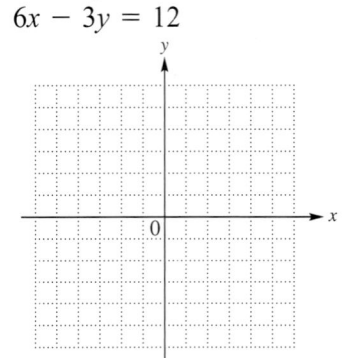

25. $2x - y = 4$
$4x = 2y + 8$

26. $3x = 5 - y$
$6x + 2y = 10$

27. $3x - 4y = 24$
$y = -\dfrac{3}{2}x + 3$

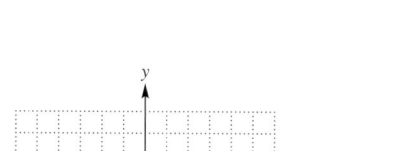
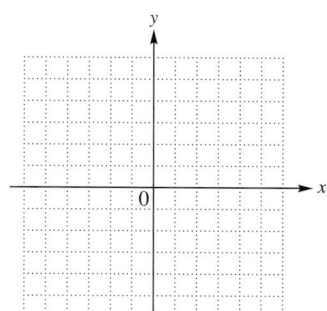
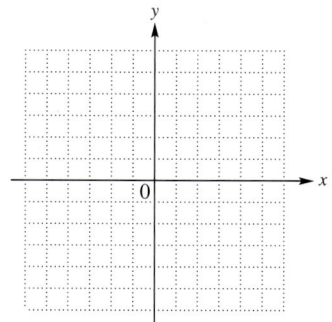

28. $3x - 2y = 12$
$y = -4x + 5$

29. $3x = y + 5$
$6x - 5 = 2y$

30. $2x = y - 4$
$4x - 2y = -4$

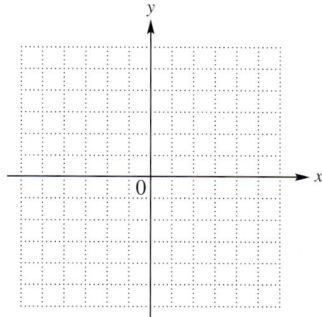
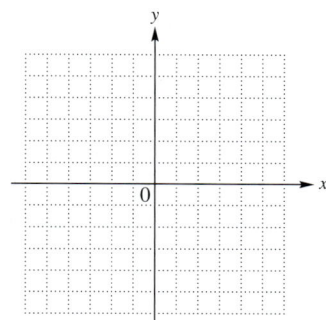
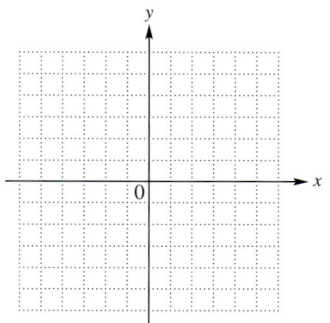

RELATING CONCEPTS (EXERCISES 31–34) For Individual or Group Work

*In Exercises 31–33, first write each equation in slope-intercept form. (See the Note following Example 2.) Then use what you learned in **Chapter 11** about slope and the y-intercept to describe the graphs of each system of equations.* **Work Exercises 31–34 in order.**

31. $3x + 2y = 6$
$-2y = 3x - 5$

32. $2x - y = 4$
$x = 0.5y + 2$

33. $x - 3y = 5$
$2x + y = 8$

34. Use the results of Exercises 31–33 to determine the number of solutions of each system.

The graph shows network share (the percentage of TV sets in use) for the early evening news programs for the three major broadcast networks from 1986 through 2000.

35. Between what years did the ABC early evening news dominate?

36. During what year did ABC's dominance end? Which network equaled ABC's share that year? What was that share?

37. During what years did ABC and CBS have equal network share? What was the share for each of these years?

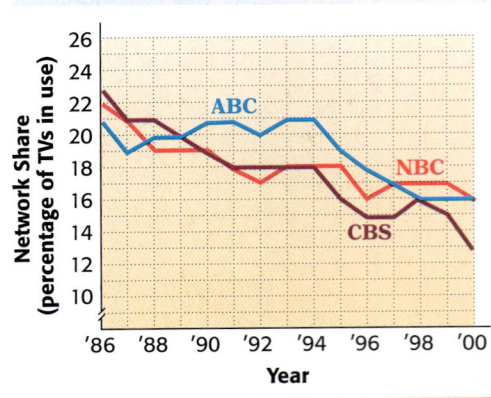

Source: Nielsen Media Research.

38. Which networks most recently had equal share? Write their share as an ordered pair of the form (year, share).

39. Explain one of the drawbacks of solving a system of equations graphically.

40. If the two lines that are the graphs of the equations in a system are parallel, how many solutions does the system have? If the two lines coincide, how many solutions does the system have?

41. Find a system of equations with the solution $(-2, 3)$, and show the graph.

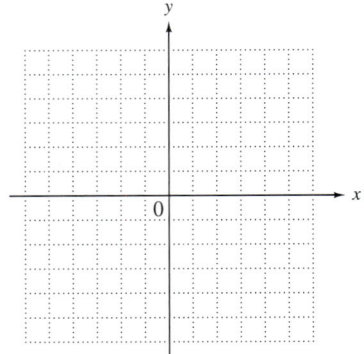

42. Solve the system

$$2x + 3y = 6$$
$$x - 3y = 5$$

by graphing. Can you check your answer? Why or why not?

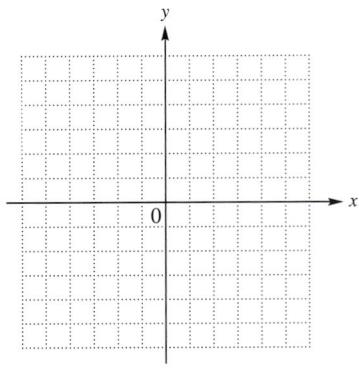

12.2 Solving Systems of Linear Equations by Substitution

OBJECTIVES
1. Solve linear systems by substitution.
2. Solve special systems.
3. Solve linear systems with fractions.

OBJECTIVE 1 Solve linear systems by substitution. Graphing to solve a system of equations has a serious drawback: It is difficult to accurately find a solution such as $\left(\frac{1}{3}, -\frac{5}{6}\right)$ from a graph. One algebraic method for solving a system of equations is the **substitution method**. This method is particularly useful for solving systems where one equation is already solved, or can be solved quickly, for one of the variables.

EXAMPLE 1 Using the Substitution Method

Solve this system.
$$3x + 5y = 26$$
$$y = 2x$$

The second equation is already solved for y. This equation says that $y = 2x$. Substituting $2x$ for y in the first equation gives the following.

$$3x + 5y = 26$$
$$3x + 5(2x) = 26 \quad \text{Let } y = 2x.$$
$$3x + 10x = 26 \quad \text{Multiply } 5(2x).$$
$$13x = 26 \quad \text{Combine like terms.}$$
$$x = 2 \quad \text{Divide both sides by 13.}$$

Because $x = 2$, we find y from the equation $y = 2x$ by substituting 2 for x.

$$y = 2(2) = 4 \quad \text{Let } x = 2.$$

Check that the solution of the given system is $(2, 4)$ by substituting 2 for x and 4 for y in *both* equations.

Work Problem 1 at the Side.

1 Fill in the blanks to solve by the substitution method. Check your solution.
$$3x + 5y = 69$$
$$y = 4x$$
$$3x + 5(\underline{\quad}) = 69$$
$$\underline{\quad} = 69$$
$$x = \underline{\quad}$$
$$y = 4(\underline{\quad}) = \underline{\quad}$$
The solution is _____.

EXAMPLE 2 Using the Substitution Method

Solve this system.
$$2x + 5y = 7$$
$$x = -1 - y$$

The second equation gives x in terms of y. Substitute $-1 - y$ for x in the first equation.

$$2x + 5y = 7$$
$$2(-1 - y) + 5y = 7 \quad \text{Let } x = -1 - y.$$
$$-2 - 2y + 5y = 7 \quad \text{Use the distributive property.}$$
$$-2 + 3y = 7 \quad \text{Combine like terms.}$$
$$3y = 9 \quad \text{Add 2 to both sides.}$$
$$y = 3 \quad \text{Divide both sides by 3.}$$

To find x, substitute 3 for y in the equation $x = -1 - y$ to get

$$x = -1 - 3 = -4.$$

Check that the solution of the given system is $(-4, 3)$.

Work Problem 2 at the Side.

2 Solve by the substitution method. Check your solution.
$$2x + 7y = -12$$
$$x = 3 - 2y$$

ANSWERS
1. $4x$; $23x$; 3; 3; 12; $(3, 12)$
2. $(15, -6)$

3 Solve each system by substitution. Check each solution.

(a) Fill in the blanks to solve
$$x + 4y = -1$$
$$2x - 5y = 11.$$
Solve the first equation for x.
$$x = -1 - \underline{}$$
Substitute into the second equation to find y.
$$2(\underline{}) - 5y = 11$$
$$-2 - 8y - 5y = 11$$
$$-2 - \underline{} y = 11$$
$$\underline{} y = 13$$
$$y = \underline{}$$
Find x.
$$x = -1 - \underline{}$$
$$x = \underline{}$$
The solution is $\underline{}$.

(b) $2x + 5y = 4$
$x + y = -1$

ANSWERS
3. (a) $4y$; $-1 - 4y$; 13; -13; -1; -4; 3; $(3, -1)$
   (b) $(-3, 2)$

> **CAUTION**
> Even though we found y first in Example 2, *the x-coordinate is always written first in the ordered pair solution of a system.*

To solve a system by substitution, follow these steps.

Solving a Linear System by Substitution
Step 1 **Solve one equation for either variable.** If one of the variables has coefficient 1 or -1, choose it.

Step 2 **Substitute** for that variable in the other equation. The result should be an equation with just one variable.

Step 3 **Solve** the equation from Step 2.

Step 4 **Substitute** the result from Step 3 into the equation from Step 1 to find the value of the other variable.

Step 5 **Check** the solution in both of the original equations. Then write the solution as an ordered pair.

EXAMPLE 3 Using the Substitution Method

Use substitution to solve this system.
$$2x = 4 - y \qquad \text{Equation (1)}$$
$$5x + 3y = 10 \qquad \text{Equation (2)}$$

Step 1 For the substitution method, we must solve one of the equations for either x or y. Because the coefficient of y in equation (1) is -1, we choose equation (1) and solve for y.
$$2x = 4 - y \qquad (1)$$
$$2x - 4 = -y \qquad \text{Subtract 4.}$$
$$-2x + 4 = y \qquad \text{Multiply by } -1.$$

Step 2 Now substitute $-2x + 4$ for y in equation (2).
$$5x + 3y = 10 \qquad (2)$$
$$5x + 3(-2x + 4) = 10 \qquad \text{Let } y = -2x + 4.$$

Step 3 Now solve the equation from Step 2.
$$5x - 6x + 12 = 10 \qquad \text{Distributive property}$$
$$-x + 12 = 10 \qquad \text{Combine terms.}$$
$$-x = -2 \qquad \text{Subtract 12.}$$
$$x = 2 \qquad \text{Multiply by } -1.$$

Step 4 Since $y = -2x + 4$ and $x = 2$, $y = -2(2) + 4 = 0$.

Step 5 Check that $(2, 0)$ is the solution.
$$2x = 4 - y \quad (1) \qquad\qquad 5x + 3y = 10 \quad (2)$$
$$2(2) = 4 - 0 \;? \qquad\qquad 5(2) + 3(0) = 10 \;?$$
$$4 = 4 \quad \text{True} \qquad\qquad 10 = 10 \quad \text{True}$$

The solution of the system is $(2, 0)$.

◀ **Work Problem 3 at the Side.**

EXAMPLE 4 Using the Substitution Method

Use substitution to solve this system.

$$2x + 3y = 10 \quad (1)$$
$$-3x - 2y = 0 \quad (2)$$

Step 1 To use the substitution method, we must solve one of the equations for one of the variables. We choose equation (1) and solve for x.

$$2x + 3y = 10 \quad (1)$$
$$2x = 10 - 3y \quad \text{Subtract } 3y.$$
$$x = 5 - \frac{3}{2}y \quad \text{Divide by 2.}$$

Step 2 Substitute this expression for x in equation (2).

$$-3x - 2y = 0 \quad (2)$$
$$-3\left(5 - \frac{3}{2}y\right) - 2y = 0 \quad \text{Let } x = 5 - \tfrac{3}{2}y.$$

Step 3
$$-15 + \frac{9}{2}y - 2y = 0 \quad \text{Distributive property}$$
$$-15 + \frac{5}{2}y = 0 \quad \text{Combine terms.}$$
$$\frac{5}{2}y = 15 \quad \text{Add 15.}$$
$$y = \frac{30}{5} = 6 \quad \text{Multiply by } \tfrac{2}{5}.$$

Step 4 Find x by substituting **6** for y in $x = 5 - \tfrac{3}{2}y$.

$$x = 5 - \frac{3}{2}(6) = -4$$

Step 5 Check that $(-4, 6)$ is the solution.

$2x + 3y = 10$ (1)	$-3x - 2y = 0$ (2)
$2(-4) + 3(6) = 10$?	$-3(-4) - 2(6) = 0$?
$-8 + 18 = 10$?	$12 - 12 = 0$?
$10 = 10$ True	$0 = 0$ True

The solution of the system is $(-4, 6)$.

NOTE

In Example 4, we could have started the solution by solving the second equation for either x or y and then substituting the result into the first equation. The solution would be the same.

> Work Problem 4 at the Side.

---

4 Solve the system by substitution. Check your solution.

$$3x + 2y = 1$$
$$3x - 4y = -11$$

ANSWERS

4. $(-1, 2)$

OBJECTIVE 2 Solve special systems. We can solve inconsistent systems with graphs that are parallel lines and systems of dependent equations with graphs that are the same line using the substitution method.

EXAMPLE 5 Solving an Inconsistent System by Substitution

Use substitution to solve this system.

$$x = 5 - 2y \quad (1)$$
$$2x + 4y = 6 \quad (2)$$

Substitute $5 - 2y$ for x in equation (2).

$$2x + 4y = 6 \quad (2)$$
$$2(5 - 2y) + 4y = 6 \quad \text{Let } x = 5 - 2y.$$
$$10 - 4y + 4y = 6 \quad \text{Distributive property}$$
$$10 = 6 \quad \text{False}$$

This false result means that the equations in the system have graphs that are parallel lines. The system is inconsistent and has no solution. See Figure 5.

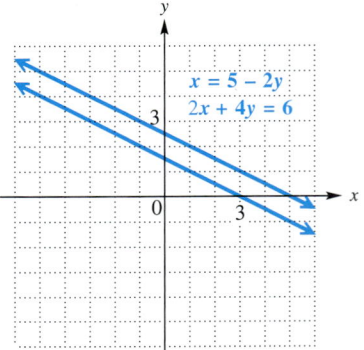

Figure 5

CAUTION
It is a common error to give "false" as the answer to an inconsistent system. The correct response is "no solution."

EXAMPLE 6 Solving a System with Dependent Equations by Substitution

Solve the system by the substitution method.

$$3x - y = 4 \quad (1)$$
$$-9x + 3y = -12 \quad (2)$$

Begin by solving equation (1) for y to get $y = 3x - 4$. Substitute $3x - 4$ for y in equation (2) and solve the resulting equation.

$$-9x + 3y = -12 \quad (2)$$
$$-9x + 3(3x - 4) = -12 \quad \text{Let } y = 3x - 4.$$
$$-9x + 9x - 12 = -12 \quad \text{Distributive property}$$
$$0 = 0 \quad \text{Add 12; combine terms.}$$

Continued on Next Page

This true result means that every solution of one equation is also a solution of the other, so the system has an infinite number of solutions: all the ordered pairs corresponding to points that lie on the common graph. A graph of the equations of this system is shown in Figure 6.

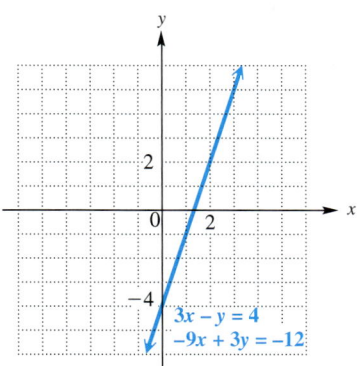

Figure 6

CAUTION
It is a common error to give "true" as the answer to a system of dependent equations. The correct response is "infinite number of solutions."

Work Problem 5 at the Side.

OBJECTIVE 3 Solve linear systems with fractions. When a system includes an equation with fractions as coefficients, eliminate the fractions by multiplying both sides of the equation by a common denominator. Then solve the resulting system.

EXAMPLE 7 Using the Substitution Method with Fractions as Coefficients

Solve the system by the substitution method.

$$3x + \frac{1}{4}y = 2 \quad (1)$$

$$\frac{1}{2}x + \frac{3}{4}y = -\frac{5}{2} \quad (2)$$

Clear equation (1) of fractions by multiplying both sides by 4.

$$4\left(3x + \frac{1}{4}y\right) = 4(2) \quad \text{Multiply by 4.}$$

$$4(3x) + 4\left(\frac{1}{4}y\right) = 4(2) \quad \text{Distributive property}$$

$$12x + y = 8 \quad (3)$$

Now clear equation (2) of fractions by multiplying both sides by the common denominator 4.

Continued on Next Page

---

5 Solve each system by substitution.

(a) $8x - y = 4$
$y = 8x + 4$

(b) $7x - 6y = 10$
$-14x + 20 = -12y$

ANSWERS
5. (a) no solution
   (b) infinite number of solutions

6 Solve the system by substitution. First clear all fractions. Check your solution.

$$\frac{2}{3}x + \frac{1}{2}y = 6$$

$$\frac{1}{2}x - \frac{3}{4}y = 0$$

$$\frac{1}{2}x + \frac{3}{4}y = -\frac{5}{2} \quad (2)$$

$$4\left(\frac{1}{2}x + \frac{3}{4}y\right) = 4\left(-\frac{5}{2}\right) \quad \text{Multiply by 4.}$$

$$4\left(\frac{1}{2}x\right) + 4\left(\frac{3}{4}y\right) = 4\left(-\frac{5}{2}\right) \quad \text{Distributive property}$$

$$2x + 3y = -10 \quad (4)$$

The given system of equations has been simplified to this equivalent system.

$$12x + y = 8 \quad (3)$$
$$2x + 3y = -10 \quad (4)$$

Solve this system by the substitution method. Equation (3) can be solved for y by subtracting $12x$ from both sides.

$$12x + y = 8 \quad (3)$$
$$y = -12x + 8 \quad \text{Subtract } 12x.$$

Now substitute this result for y in equation (4).

$$2x + 3y = -10 \quad (4)$$
$$2x + 3(-12x + 8) = -10 \quad \text{Let } y = -12x + 8.$$
$$2x - 36x + 24 = -10 \quad \text{Distributive property}$$
$$-34x = -34 \quad \text{Combine terms; subtract 24.}$$
$$x = 1 \quad \text{Divide by } -34.$$

Substitute 1 for x in $y = -12x + 8$ to get

$$y = -12(1) + 8 = -4.$$

The solution is $(1, -4)$. Check by substituting 1 for x and -4 for y in both of the original equations.

◀◀◀ **Work Problem 6 at the Side.**

ANSWERS
6. $(6, 4)$

12.2 Exercises

1. A student solves the system
$$5x - y = 15$$
$$7x + y = 21$$
and finds that $x = 3$, which is the correct value for x. The student gives the solution as "$x = 3$." Is this correct? Explain.

2. When you use the substitution method, how can you tell that a system has
   (a) no solution?
   (b) an infinite number of solutions?

Solve each system by the substitution method. Check each solution. See Examples 1–7.

3. $x + y = 12$
   $y = 3x$

4. $x + 3y = -28$
   $y = -5x$

5. $3x + 2y = 27$
   $x = y + 4$

6. $4x + 3y = -5$
   $x = y - 3$

7. $3x + 5y = 14$
   $x - 2y = -10$

8. $5x + 2y = -1$
   $2x - y = -13$

9. $3x + 4 = -y$
   $2x + y = 0$

10. $2x - 5 = -y$
    $x + 3y = 0$

11. $7x + 4y = 13$
    $x + y = 1$

12. $3x - 2y = 19$
    $x + y = 8$

13. $3x - y = 5$
    $y = 3x - 5$

14. $4x - y = -3$
    $y = 4x + 3$

15. $6x - 8y = 6$
    $2y = -2 + 3x$

16. $3x + 2y = 6$
    $6x = 8 + 4y$

17. $2x + 8y = 3$
    $x = 8 - 4y$

18. $2x + 10y = 3$
    $x = 1 - 5y$

19. $12x - 16y = 8$
    $3x = 4y + 2$

20. $6x + 9y = 6$
    $2x = 2 - 3y$

21. $5x + 4y = 40$
    $x + y = 1$

22. $x - 5y = 7$
    $2x - y = 14$

23. $3x = 6 - 4y$
    $9x + 12y = 10$

24. $\frac{1}{3}x - \frac{1}{2}y = \frac{1}{6}$
    $3x - 2y = 9$

25. $\frac{1}{5}x + \frac{2}{3}y = -\frac{8}{5}$
    $3x - y = 9$

26. $\frac{1}{6}x + \frac{1}{6}y = 2$
    $-\frac{1}{2}x - \frac{1}{3}y = -8$

27. $\frac{x}{2} - \frac{y}{3} = 9$
    $\frac{x}{5} - \frac{y}{4} = 5$

28. $\frac{x}{3} - \frac{3y}{4} = -\frac{1}{2}$
    $\frac{x}{6} + \frac{y}{8} = \frac{3}{4}$

29. $\frac{x}{5} + 2y = \frac{16}{5}$
    $\frac{3x}{5} + \frac{y}{2} = -\frac{7}{5}$

RELATING CONCEPTS (EXERCISES 30–33) For Individual or Group Work

A system of linear equations can be used to model the cost and the revenue of a business. Work Exercises 30–33 in order.

30. Suppose that you start a business manufacturing and selling bicycles, and it costs you $5000 to get started. You determine that each bicycle will cost $400 to manufacture. Explain why the linear equation $y_1 = 400x + 5000$ gives your *total* cost to manufacture x bicycles (y_1 in dollars).

31. You decide to sell each bike for $600. What expression in x represents the revenue you will take in if you sell x bikes? Write an equation using y_2 to express your revenue when you sell x bikes (y_2 in dollars).

32. Form a system from the two equations in Exercises 30 and 31, and then solve the system, assuming $y_1 = y_2$, that is, cost = revenue.

33. The value of x from Exercise 32 is the number of bikes it takes to *break even*. Fill in the blanks: When _____ bikes are sold, the break-even point is reached. At that point, you have spent _____ dollars and taken in _____ dollars.

12.3 Solving Systems of Linear Equations by Elimination

OBJECTIVE 1 Solve linear systems by elimination. An algebraic method that depends on the addition property of equality can be used to solve systems. As mentioned earlier, adding the same quantity to both sides of an equation results in equal sums.

$$\text{If } A = B, \text{ then } A + C = B + C.$$

This addition can be taken a step further. Adding *equal* quantities, rather than the *same* quantity, to both sides of an equation also results in equal sums.

$$\text{If } A = B \text{ and } C = D, \text{ then } A + C = B + D.$$

Using the addition property to solve systems is called the **elimination method**. When using this method, the idea is to *eliminate* one of the variables. To do this, one of the variables in the two equations must have coefficients that are opposites.

OBJECTIVES

1. Solve linear systems by elimination.
2. Multiply when using the elimination method.
3. Use an alternative method to find the second value in a solution.
4. Use the elimination method to solve special systems.

1 (a) Substitute 4 for x in the equation $x + y = 5$ to find the value of y.

EXAMPLE 1 Using the Elimination Method

Use the elimination method to solve this system.

$$x + y = 5$$
$$x - y = 3$$

Each equation in this system is a statement of equality, so the sum of the left sides equals the sum of the right sides. The addition is shown below.

$$(x + y) + (x - y) = 5 + 3$$

Combine terms and simplify to get the following.

$$2x = 8$$
$$x = 4 \quad \text{Divide by 2.}$$

Notice that y has been eliminated. The result, $x = 4$, gives the x-value of the solution of the given system. To find the y-value of the solution, substitute 4 for x in either of the two equations of the system.

(b) Give the solution of the system.

> **Work Problem 1 at the Side.**

The solution found at the side, (4, 1), can be checked by substituting 4 for x and 1 for y in both equations of the given system.

Check: $x + y = 5$ $x - y = 3$
          $4 + 1 = 5$? $4 - 1 = 3$?
          $5 = 5$ True $3 = 3$ True

Since both results are true, the solution of the system is (4, 1).

> **CAUTION**
> *A system is not completely solved until values for both x and y are found.* Do not stop after finding the value of only one variable. Remember to write the solution as an ordered pair.

Continued on Next Page

ANSWERS
1. (a) $y = 1$ (b) (4, 1)

❷ Solve each system by the elimination method. Check each solution.

(a) Fill in the blanks to find the solution.

$$x + y = 8$$
$$x - y = 2$$

Add.

$(x + y) + (x - y) = 8 + \underline{}$

$2\underline{} = \underline{}$

$x = \underline{}$

Find y.

$$x - y = 2$$
$$\underline{} - y = 2$$
$$-y = \underline{}$$
$$y = \underline{}$$

The solution is $\underline{}$.

(b) $3x - y = 7$
$2x + y = 3$

ANSWERS
2. (a) 2; x; 10; 5; 5; -3; 3; (5, 3)
(b) (2, -1)

Work Problem 2 at the Side.

In general, to solve a system by elimination, follow these steps.

Solving a Linear System by Elimination

Step 1 Write both equations in standard form $Ax + By = C$.

Step 2 **Transform so that the coefficients of one pair of variable terms are opposites.** Multiply one or both equations by appropriate numbers so that the sum of the coefficients of either the x- or y-terms is 0.

Step 3 **Add** the new equations to eliminate a variable. The sum should be an equation with just one variable.

Step 4 **Solve** the equation from Step 3 for the remaining variable.

Step 5 **Substitute** the result from Step 4 into either of the original equations and solve for the other variable.

Step 6 **Check** the solution in both of the original equations. Then write the solution as an ordered pair.

It does not matter which variable is eliminated first. Usually we choose the one that is more convenient to work with.

EXAMPLE 2 Using the Elimination Method

Solve this system.

$$y + 11 = 2x$$
$$5x = y + 26$$

Step 1 Rewrite both equations in the form $Ax + By = C$ to get this system.

$-2x + y = -11$ Subtract $2x$ and 11.
$5x - y = 26$ Subtract y.

Step 2 Because the coefficients of y are 1 and -1, adding will eliminate y. It is not necessary to multiply either equation by a number.

Step 3 Add the two equations. This time we use vertical addition.

$$-2x + y = -11$$
$$\underline{5x - y = 26}$$
$$3x = 15 \quad \text{Add in columns.}$$

Step 4 Solve the equation.

$$3x = 15$$
$$x = 5 \quad \text{Divide by 3.}$$

Step 5 Find the value of y by substituting 5 for x in either of the original equations. Choosing the first gives the following.

$$y + 11 = 2x$$
$$y + 11 = 2(5) \quad \text{Let } x = 5.$$
$$y + 11 = 10$$
$$y = -1 \quad \text{Subtract 11.}$$

Continued on Next Page

Step 6 Check the solution by substituting $x = 5$ and $y = -1$ into both of the original equations.

$y + 11 = 2x$	$5x = y + 26$
$(-1) + 11 = 2(5)$?	$5(5) = -1 + 26$?
$10 = 10$ True	$25 = 25$ True

The solution $(5, -1)$ is correct.

> **Work Problem 3 at the Side.**

OBJECTIVE 2 Multiply when using the elimination method. Sometimes we need to multiply one or both equations in a system by some number so that adding the equations will eliminate a variable.

EXAMPLE 3 Multiplying Both Equations When Using the Elimination Method

Solve this system.

$$2x + 3y = -15 \quad (1)$$
$$5x + 2y = 1 \quad (2)$$

Adding the two equations gives $7x + 5y = -14$, which does not eliminate either variable. However, we can multiply each equation by a suitable number so that the coefficients of one of the variables are opposites. For example, to eliminate x, multiply both sides of equation (1) by 5, and both sides of equation (2) by -2.

$$10x + 15y = -75 \quad \text{Multiply equation (1) by 5.}$$
$$\underline{-10x - 4y = -2} \quad \text{Multiply equation (2) by } -2.$$
$$11y = -77 \quad \text{Add.}$$
$$y = -7 \quad \text{Divide by 11.}$$

Substituting -7 for y in either equation (1) or (2) gives $x = 3$. Check that the solution of the system is $(3, -7)$.

> **Work Problem 4 at the Side.**

OBJECTIVE 3 Use an alternative method to find the second value in a solution. Sometimes it is easier to find the value of the second variable in a solution by using the elimination method twice.

EXAMPLE 4 Finding the Second Value Using an Alternative Method

Solve this system.

$$4x = 9 - 3y \quad (1)$$
$$5x - 2y = 8 \quad (2)$$

Rearrange the terms in equation (1) so that like terms are aligned in columns. To do this, add $3y$ to both sides to get the following system.

$$4x + 3y = 9 \quad (3)$$
$$5x - 2y = 8 \quad (2)$$

One way to proceed is to eliminate y by multiplying both sides of equation (3) by 2 and both sides of equation (2) by 3, and then adding.

Continued on Next Page

---

3 Solve each system by the elimination method. Check each solution.

(a) $2x - y = 2$
 $4x + y = 10$

(b) $8x - 5y = 32$
 $4x + 5y = 4$

4 (a) Solve the system in Example 3 by first eliminating the variable y. Check your solution.

(b) Solve

$6x + 7y = 4$
$5x + 8y = -1,$

and check your solution.

ANSWERS

3. (a) $(2, 2)$ (b) $\left(3, -\dfrac{8}{5}\right)$
4. (a) $(3, -7)$ (b) $(3, -2)$

5 Solve each system of equations.

(a) $5x = 7 + 2y$
$5y = 5 - 3x$

(b) $3y = 8 + 4x$
$6x = 9 - 2y$

6 Solve each system by the elimination method.

(a) $4x + 3y = 10$
$2x + \dfrac{3}{2}y = 12$

(b) $4x - 6y = 10$
$-10x + 15y = -25$

ANSWERS

5. (a) $\left(\dfrac{45}{31}, \dfrac{4}{31}\right)$ (b) $\left(\dfrac{11}{26}, \dfrac{42}{13}\right)$
6. (a) no solution
   (b) infinite number of solutions

$$8x + 6y = 18 \quad \text{Multiply equation (3) by 2.}$$
$$15x - 6y = 24 \quad \text{Multiply equation (2) by 3.}$$
$$\overline{23x \qquad\quad = 42} \quad \text{Add.}$$
$$x = \dfrac{42}{23} \quad \text{Divide by 23.}$$

Substituting $\dfrac{42}{23}$ for x in one of the given equations would give y, but the arithmetic involved would be messy. Instead, solve for y by starting again with the original equations and eliminating x. Multiply both sides of equation (3) by 5 and both sides of equation (2) by -4, and then add.

$$20x + 15y = 45 \quad \text{Multiply equation (3) by 5.}$$
$$-20x + 8y = -32 \quad \text{Multiply equation (2) by } -4.$$
$$\overline{23y = 13} \quad \text{Add.}$$
$$y = \dfrac{13}{23} \quad \text{Divide by 23.}$$

Check that the solution is $\left(\dfrac{42}{23}, \dfrac{13}{23}\right)$.

◀◀ **Work Problem 5 at the Side.**

When the value of the first variable is a fraction, the method used in Example 4 helps avoid arithmetic errors. Of course, this method could be used to solve any system of equations.

OBJECTIVE 4 Use the elimination method to solve special systems.

EXAMPLE 5 Using the Elimination Method for an Inconsistent System or Dependent Equations

Solve each system by the elimination method.

(a) $2x + 4y = 5$
$4x + 8y = -9$

Multiply both sides of $2x + 4y = 5$ by -2; then add to $4x + 8y = -9$.

$$-4x - 8y = -10$$
$$\underline{4x + 8y = -9}$$
$$0 = -19 \quad \text{False}$$

The false statement $0 = -19$ shows that the given system has no solution.

(b) $3x - y = 4$
$-9x + 3y = -12$

Multiply both sides of the first equation by 3; then add the two equations.

$$9x - 3y = 12$$
$$\underline{-9x + 3y = -12}$$
$$0 = 0 \quad \text{True}$$

A true statement occurs when the equations are equivalent. As before, this indicates that every solution of one equation is also a solution of the other; there are an infinite number of solutions. (See **Section 12.2**, Example 6, where the same system was solved using substitution.)

◀◀ **Work Problem 6 at the Side.**

12.3 Exercises

In Exercises 1–3, answer true *or* false *for each statement. If false, tell why.*

1. The ordered pair $(0, 0)$ *must* be a solution of a system of the form
$$Ax + By = 0$$
$$Cx + Dy = 0.$$

2. To eliminate the *y*-terms in the system
$$2x + 12y = 7$$
$$3x + 4y = 1,$$
we should multiply the bottom equation by 3 and then add.

3. The system
$$x + y = 1$$
$$x + y = 2$$
has no solution.

4. Which one of the following systems would be easier to solve using the elimination method? Why?

$$5x - 3y = 7 \qquad 7x + 2y = 4$$
$$2x + 8y = 3 \qquad -7x + 3y = 1$$

Solve each system by the elimination method. Check each solution. See Examples 1 and 2.

5. $x + y = 2$
 $2x - y = -5$

6. $3x - y = -12$
 $x + y = 4$

7. $2x + y = -5$
 $x - y = 2$

8. $2x + y = -15$
 $-x - y = 10$

9. $3x + 2y = 0$
 $-3x - y = 3$

10. $5x - y = 5$
 $-5x + 2y = 0$

11. $6x - y = -1$
 $5y = 17 + 6x$

12. $y = 9 - 6x$
 $-6x + 3y = 15$

Solve each system by the elimination method. Check each solution. See Examples 3–5.

13. $2x - y = 12$
    $3x + 2y = -3$

14. $x + y = 3$
    $-3x + 2y = -19$

15. $x + 3y = 19$
    $2x - y = 10$

16. $4x - 3y = -19$
    $2x + y = 13$

17. $x + 4y = 16$
    $3x + 5y = 20$

18. $2x + y = 8$
    $5x - 2y = -16$

19. $5x - 3y = -20$
    $-3x + 6y = 12$

20. $4x + 3y = -28$
    $5x - 6y = -35$

21. $2x - 8y = 0$
    $4x + 5y = 0$

22. $3x - 15y = 0$
    $6x + 10y = 0$

23. $x + y = 7$
    $x + y = -3$

24. $x - y = 4$
    $x - y = -3$

25. $-x + 3y = 4$
    $-2x + 6y = 8$

26. $6x - 2y = 24$
    $-3x + y = -12$

27. $4x - 3y = -19$
    $3x + 2y = 24$

28. $5x + 4y = 12$
    $3x + 5y = 15$

29. $3x - 7 = -5y$
    $5x + 4y = -10$

30. $2x + 3y = 13$
    $6 + 2y = -5x$

31. $2x + 3y = 0$
    $4x + 12 = 9y$

32. $-4x + 3y = 2$
    $5x + 3 = -2y$

33. $24x + 12y = -7$
    $16x - 17 = 18y$

34. $9x + 4y = -3$
    $6x + 7 = -6y$

35. $3x = 3 + 2y$
    $-\dfrac{4}{3}x + y = \dfrac{1}{3}$

36. $3x = 27 + 2y$
    $x - \dfrac{7}{2}y = -25$

37. $5x - 2y = 3$
    $10x - 4y = 5$

38. $3x - 5y = 1$
    $6x - 10y = 4$

39. $6x + 3y = 0$
    $-18x - 9y = 0$

40. $3x - 5y = 0$
    $9x - 15y = 0$

RELATING CONCEPTS (EXERCISES 41–46) For Individual or Group Work

Attending the movies is one of America's favorite forms of entertainment. The graph shows movie attendance from 1991 to 1999. In 1991, attendance was 1141 million, as represented by the point P(1991, 1141). In 1999, attendance was 1465 million, as represented by the point Q(1999, 1465). We can find an equation of line segment PQ using a system of equations, and then we can use the equation to approximate the attendance in any of the years between 1991 and 1999. **Work Exercises 41–46 in order.**

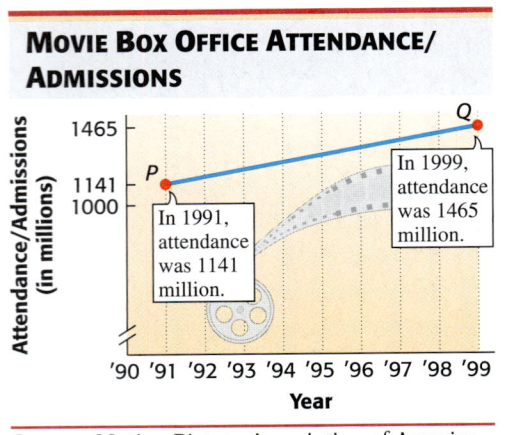

Source: Motion Picture Association of America.

41. The line segment has an equation that can be written in the form $y = ax + b$. Using the coordinates of point P with $x = 1991$ and $y = 1141$, write an equation in the variables a and b.

42. Using the coordinates of point Q with $x = 1999$ and $y = 1465$, write a second equation in the variables a and b.

43. Write the system of equations formed from the two equations in Exercises 41 and 42, and solve the system using the elimination method.

44. What is the equation of the segment PQ?

45. Let $x = 1998$ in the equation of Exercise 44, and solve for y. How does the result compare with the actual figure of 1481 million?

46. The data points for the years 1991 through 1999 do not lie in a perfectly straight line. Explain the pitfalls of relying too heavily on using the equation in Exercise 44 to predict attendance.

Summary Exercises on Solving Systems of Linear Equations

The exercises in this summary include a variety of problems on solving systems of linear equations. Since we do not usually specify the method of solution, use the following guidelines to help you decide whether to use substitution or elimination.

Choosing a Method When Solving a System of Linear Equations

1. If one of the equations of the system is already solved for one of the variables, as in the systems

$$3x + 4y = 9 \quad \text{or} \quad -5x + 3y = 9$$
$$y = 2x - 6 \qquad\qquad x = 3y - 7,$$

   the substitution method is the better choice.

2. If both equations are in standard $Ax + By = C$ form, as in

$$4x - 11y = 3$$
$$-2x + 3y = 4,$$

   and none of the variables has coefficient -1 or 1, the elimination method is the better choice.

3. If one or both of the equations are in standard form and the coefficient of one of the variables is -1 or 1, as in the systems

$$3x + y = -2 \quad \text{or} \quad -x + 3y = -4$$
$$-5x + 2y = 4 \qquad\qquad 3x - 2y = 8,$$

   either method is appropriate.

Use the information in the preceding box to solve each problem.

1. Assuming you want to minimize the amount of work required, tell whether you would use the substitution or elimination method to solve each system. Explain your answers. *Do not actually solve.*

    (a) $3x + 2y = 18$
    $y = 3x$

    (b) $3x + y = -7$
    $x - y = -5$

    (c) $3x - 2y = 0$
    $9x + 8y = 7$

2. Which one of the following systems would be easier to solve using the substitution method? Why?

$$5x - 3y = 7 \qquad 7x + 2y = 4$$
$$2x + 8y = 3 \qquad y = -3x + 1$$

*In Exercises 3 and 4, **(a)** solve the system by the elimination method, **(b)** solve the system by the substitution method, and **(c)** tell which method you prefer for that particular system and why.*

3. $4x - 3y = -8$
   $x + 3y = 13$

4. $2x + 5y = 0$
   $x = -3y + 1$

Solve each system by the method of your choice. (For Exercises 5–7, see your answers for Exercise 1.)

5. $3x + 2y = 18$
   $y = 3x$

6. $3x + y = -7$
   $x - y = -5$

7. $3x - 2y = 0$
   $9x + 8y = 7$

8. $x + y = 7$
   $x = -3 - y$

9. $5x - 4y = 15$
   $-3x + 6y = -9$

10. $4x + 2y = 3$
    $y = -x$

11. $3x = 7 - y$
    $2y = 14 - 6x$

12. $3x - 5y = 7$
    $2x + 3y = 30$

13. $3y = 4x + 2$
    $5x - 2y = -3$

14. $4x + 3y = 1$
    $3x + 2y = 2$

15. $2x - 3y = 7$
    $-4x + 6y = 14$

16. $2x + 3y = 10$
    $-3x + y = 18$

17. $6x + 5y = 13$
    $3x + 3y = 4$

18. $x - 3y = 7$
    $4x + y = 5$

Solve each system by any method. First clear all fractions.

19. $\dfrac{1}{4}x - \dfrac{1}{5}y = 9$
    $y = 5x$

20. $\dfrac{1}{2}x + \dfrac{1}{3}y = -\dfrac{1}{3}$
    $\dfrac{1}{2}x + 2y = -7$

21. $\dfrac{1}{6}x + \dfrac{1}{6}y = 1$
    $-\dfrac{1}{2}x - \dfrac{1}{3}y = -5$

22. $\dfrac{x}{5} + 2y = \dfrac{8}{5}$
    $\dfrac{3x}{5} + \dfrac{y}{2} = -\dfrac{7}{10}$

23. $\dfrac{x}{5} + y = \dfrac{6}{5}$
    $\dfrac{x}{10} + \dfrac{y}{3} = \dfrac{5}{6}$

24. $\dfrac{1}{6}x + \dfrac{1}{3}y = 8$
    $\dfrac{1}{4}x + \dfrac{1}{2}y = 12$

12.4 Applications of Linear Systems

Recall from **Section 3.4** the six-step method for solving applied problems. We modify those steps slightly to allow for two variables and two equations.

OBJECTIVES

1. Solve problems about unknown numbers.
2. Solve problems about quantities and their costs.
3. Solve problems about mixtures.
4. Solve problems about distance, rate (or speed), and time.

Solving Applied Problems with Two Variables

Step 1 **Read** the problem, several times if necessary, until you understand what is given and what is to be found.

Step 2 **Assign variables** to represent the unknown values, using diagrams or tables as needed. Write down what each variable represents.

Step 3 **Write two equations** using both variables.

Step 4 **Solve** the system of two equations.

Step 5 **State the answer** to the problem. Is the answer reasonable?

Step 6 **Check** the answer in the words of the original problem.

1 Solve this system.
$$x = 1954 + y$$
$$x + y = 24{,}098$$

OBJECTIVE 1 Solve problems about unknown numbers.

EXAMPLE 1 Solving a Problem about Two Unknown Numbers

In 2000, sales of athletic/sport footwear were $1954 million more than sales of athletic/sport clothing. Together, total sales for these items were $24,098 million. (*Source:* National Sporting Goods Association.) What were the sales for each?

Step 1 **Read** the problem carefully. We must find the 2000 sales (in millions of dollars) for athletic/sport clothing and footwear. We know how much more footwear sales were than clothing sales. Also, we know the total sales.

Step 2 **Assign variables.**

Let x = sales of footwear in millions of dollars,

and y = sales of clothing in millions of dollars.

Step 3 **Write two equations.**

$x = 1954 + y$ Sales of footwear were $1954 million more than sales of clothing.

$x + y = 24{,}098$ Total sales were $24,098 million.

Step 4 **Solve** the system from Step 3. The substitution method works well here since the first equation is already solved for x.

Work Problem 1 at the Side.

Step 5 **State the answer.** Footwear sales were $13,026 million, and clothing sales were $11,072 million.

Step 6 **Check** the answer in the original problem. Since

$$13{,}026 - 11{,}072 = 1954 \quad \text{and} \quad 13{,}026 + 11{,}072 = 24{,}098,$$

the answer satisfies the information in the problem.

Continued on Next Page

ANSWERS

1. $x = 13{,}026, y = 11{,}072$

2 Set up a system of equations for the following problem. Do not solve the system.

Two top-grossing Disney movies in 2002 were *Lilo and Stitch* and *The Santa Clause 2*. Together they grossed $284.2 million. *The Santa Clause 2* grossed $7.4 million less than *Lilo and Stitch*. How much did each movie gross? (Source: *Variety*.)

Let x = the amount (in millions) that *Lilo and Stitch* grossed,

and y = the amount (in millions) that _____ grossed.

ANSWERS

2. *The Santa Clause 2*; $x + y = 284.2$
   $y = x - 7.4$

> **CAUTION**
> If an applied problem asks for *two* values as in Example 1, be sure to give both of them in your answer.

▶◀ **Work Problem 2 at the Side.**

OBJECTIVE 2 **Solve problems about quantities and their costs.** We can also use a linear system to solve an applied problem involving two quantities and their costs.

EXAMPLE 2 Solving a Problem about Quantities and Costs

The all-time top-grossing movie *Titanic** earned more in Europe than in the United States. This may be because average movie prices in Europe exceed those of the United States.

For example, while the average movie ticket (to the nearest dollar) in 1997–1998 cost $5 in the United States, it cost an equivalent of $11 in London. Suppose that a group of 41 Americans and Londoners who paid these average prices spent a total of $307 for tickets. How many from each country were in the group?

Step 1 **Read** the problem several times.

Step 2 **Assign variables.**

Let x = the number of Americans in the group,

and y = the number of Londoners in the group.

Summarize the information given in the problem in a table. The entries in the first two rows of the Total Value column were found by multiplying the number of tickets sold by the price per ticket.

	Number of Tickets	Price per Ticket in dollars	Total Value
Americans	x	5	$5x$
Londoners	y	11	$11y$
Total	41		307

Step 3 **Write two equations.** The total number of tickets was 41, so

$$x + y = 41.$$ Total number of tickets

Since the total value was $307, the final column leads to

$$5x + 11y = 307.$$ Total value of tickets

These two equations form a system.

$$x + y = 41 \quad (1)$$
$$5x + 11y = 307 \quad (2)$$

Step 4 **Solve** the system using the elimination method. To eliminate the x-terms, multiply each side of equation (1) by -5 to get

$$-5x - 5y = -205.$$

Continued on Next Page

* Through October 2005, *Titanic* was still number one. (Source: *Variety*.)

Then add this result to equation (2).

$$-5x - 5y = -205$$
$$\underline{5x + 11y = 307} \quad (2)$$
$$6y = 102 \quad \text{Add.}$$
$$y = 17 \quad \text{Divide by 6.}$$

Substitute 17 for y in equation (1) to get

$$x + y = 41$$
$$x + 17 = 41 \quad \text{Let } y = 17.$$
$$x = 24.$$

Step 5 **State the answer.** There were 24 Americans and 17 Londoners in the group.

Step 6 **Check.** The sum of 24 and 17 is 41, so the number of moviegoers is correct. Since 24 Americans paid $5 each and 17 Londoners paid $11 each, the total of the admission prices is

$$\$5(24) + \$11(17) = \$307,$$

which agrees with the total amount stated in the problem.

Work Problem 3 at the Side.

OBJECTIVE 3 Solve problems about mixtures. In **Section 7.4** we solved percent problems using one variable. Many problems about mixtures that involve percent can be solved using a system of two equations in two variables.

EXAMPLE 3 Solving a Mixture Problem Involving Percent

A pharmacist needs 100 L of 50% alcohol solution. She has on hand 30% alcohol solution and 80% alcohol solution, which she can mix. How many liters of each will be required to make the 100 L of 50% alcohol solution?

Step 1 **Read** the problem. Note the percent of each solution and of the mixture.

Step 2 **Assign variables.**

Let x = the number of liters of 30% alcohol needed,

and y = the number of liters of 80% alcohol needed.

Summarize the information in a table. Percents are written in decimal form.

Percent	Liters of Mixture	Liters of Pure Alcohol
0.30	x	$0.30x$
0.80	y	$0.80y$
0.50	100	$0.50(100)$

Continued on Next Page

---

3 The average movie ticket (to the nearest U.S. dollar) costs $10 in Geneva and $8 in Paris. If a group of 36 people from these two cities paid $298 for tickets to see *Cheaper by the Dozen*, how many people from each city were there?

(a) Complete the table.

	Number of Tickets Sold	Price (in dollars)	Total Value
Genevans	x		
Parisians	y		
Total			

(b) Write a system of equations.

(c) Solve the system and check your answer in the words of the original problem.

ANSWERS

3. (a)

	Number of Tickets Sold	Price (in dollars)	Total Value
Genevans	x	10	$10x$
Parisians	y	8	$8y$
Total	36		298

(b) $x + y = 36$
$10x + 8y = 298$

(c) Genevans: 5; Parisians: 31

4 Solve this system.

$$x + y = 100$$
$$0.30x + 0.80y = 50$$

5 How many liters of 25% alcohol solution must be mixed with 12% solution to get 13 L of 15% solution?

(a) Complete the table.

Percent	Liters	Liters of Pure Alcohol
0.25	x	$0.25x$
0.12	y	
0.15	13	

(b) Write a system of equations, and solve it.

6 Solve the problem.

Joe needs 60 mL of 20% acid solution for a chemistry experiment. The lab has on hand only 10% and 25% solutions. How much of each should he mix to get the desired amount of 20% solution?

7 Solve using the distance formula.

A small plane traveled from Warsaw to Rome, averaging 164 mph. The trip took 2.5 hr. What is the distance from Warsaw to Rome?

ANSWERS

4. (60, 40)
5. (a)

Percent	Liters	Liters of Pure Alcohol
0.25	x	$0.25x$
0.12	y	$0.12y$
0.15	13	$0.15(13)$

(b) $x + y = 13$
$0.25x + 0.12y = 0.15(13)$

3 L of 25%, 10 L of 12%

6. 20 mL of 10%, 40 mL of 25%

7. 410 mi

Figure 7 gives an idea of what is actually happening in this problem.

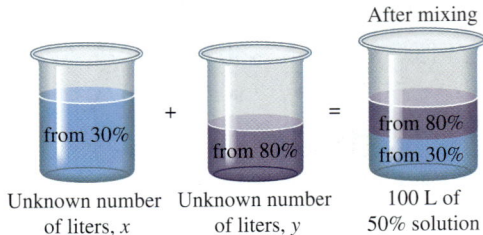

Unknown number of liters, x Unknown number of liters, y 100 L of 50% solution

Figure 7

Step 3 **Write two equations.** Since the total number of liters in the final mixture will be 100, the first equation is

$$x + y = 100.$$

To find the amount of pure alcohol in each mixture, multiply the number of liters by the concentration. The amount of pure alcohol in the 30% solution added to the amount of pure alcohol in the 80% solution will equal the amount of pure alcohol in the final 50% solution. This gives the second equation.

$$0.30x + 0.80y = 0.50(100)$$

These two equations form this system.

$$x + y = 100$$
$$0.30x + 0.80y = 50 \qquad 0.50(100) = 50$$

Step 4 **Solve** this system.

Work Problem 4 at the Side.

Step 5 **State the answer.** From Problem 4 at the side, the pharmacist should use 60 L of the 30% solution and 40 L of the 80% solution.

Step 6 Since $60 + 40 = 100$ and $0.30(60) + 0.80(40) = 50$, this mixture will give the 100 L of 50% solution, as required in the original problem.

Work Problems 5 and 6 at the Side.

OBJECTIVE 4 Solve problems about distance, rate (or speed), and time. If an automobile travels at an average rate of 50 mph for 2 hr, then it travels $50 \times 2 = 100$ mi. This is an example of the basic relationship between distance, rate, and time:

$$\text{distance} = \text{rate} \times \text{time}.$$

This relationship is given by the formula $d = rt$.

Work Problem 7 at the Side.

We can solve problems involving the distance formula with systems of two linear equations. Keep in mind that setting up a table and drawing a sketch will help you solve such problems.

EXAMPLE 4 Solving a Problem about Distance, Rate, and Time

Two executives in cities 400 mi apart drive to a business meeting at a location on the line between their cities. They meet after 4 hr. Find the speed of each car if one car travels 20 mph faster than the other.

Step 1 **Read** the problem carefully.

Step 2 **Assign variables.** Let x = the speed of the faster car,
and y = the speed of the slower car.

We use the formula $d = rt$. Since each car travels for 4 hr, the time, t, for each car is 4. See the table. The distance is found by using the formula $d = rt$ and the expressions already entered in the table.

	r	t	d
Faster Car	x	4	$4x$
Slower Car	y	4	$4y$

Find d from $d = rt$.

Figure 8 shows what is happening in the problem.

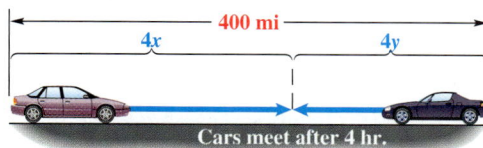
Cars meet after 4 hr.

Figure 8

Step 3 **Write two equations.** As shown in the figure, since the total distance traveled by both cars is 400 mi, one equation is

$$4x + 4y = 400.$$

Because the faster car goes 20 mph faster than the slower car, the second equation is

$$x = 20 + y.$$

Step 4 **Solve** the system of equations,

$$4x + 4y = 400 \quad (1)$$
$$x = 20 + y, \quad (2)$$

by substitution. Replace x with $20 + y$ in equation (1) and then solve for y.

$4(20 + y) + 4y = 400$ Let $x = 20 + y$.
$80 + 4y + 4y = 400$ Distributive property
$80 + 8y = 400$ Combine like terms.
$8y = 320$ Subtract 80.
$y = 40$ Divide by 8.

Since $x = 20 + y$, and $y = 40$,

$$x = 20 + 40 = 60.$$

Step 5 **State the answer.** The speeds of the two cars are 40 mph and 60 mph.

Step 6 **Check** the answer. Since each car travels for 4 hr, total distance is

$$4(60) + 4(40) = 240 + 160 = 400 \text{ mi}, \quad \text{as required.}$$

▶ **Work Problem 8 at the Side.**

---

8 Two cars that were 450 mi apart traveled toward each other. They met after 5 hr. If one car traveled twice as fast as the other, what were their speeds?

(a) Complete this table.

	r	t	d
Faster Car	x	5	
Slower Car	y	5	

(b) Write a system, and solve it.

ANSWERS
8. (a)

	r	t	d
Faster Car	x	5	$5x$
Slower Car	y	5	$5y$

(b) $5x + 5y = 450$
$x = 2y$

faster car: $x = 60$ mph
slower car: $y = 30$ mph

9 Solve this system.

$$x + y = 320$$
$$x - y = 280$$

10 Solve the problem.

In 1 hr, Ann can row 2 mi against the current or 10 mi with the current. Find the speed of the current and Ann's speed in still water. (*Hint:* Let x = the speed of the current and y = Ann's speed in still water. Then her rate against the current is $y - x$, and her rate with the current is $y + x$.)

CAUTION
Be careful! *When you use two variables to solve a problem, you must write two equations.*

EXAMPLE 5 Solving a Problem about Distance, Rate, and Time

A plane flies 560 mi in 1.75 hr traveling with the wind. The return trip against the same wind takes the plane 2 hr. Find the speed of the plane and the speed of the wind.

Step 1 **Read** the problem several times.

Step 2 **Assign variables.**

Let x = the speed of the plane,

and y = the speed of the wind.

The speed (rate) of the plane *with* the wind is $x + y$ mph, and the speed (rate) of the plane *against* the wind is $x - y$ mph. See Figure 9.

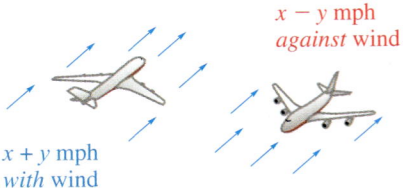

$x - y$ mph
against wind

$x + y$ mph
with wind

Figure 9

We use this information and the formula $d = rt$ (or $rt = d$) to complete a table.

	r	t	d
With Wind	$x + y$	1.75	560
Against Wind	$x - y$	2	560

Step 3 **Write two equations.** Use the table.

$1.75(x + y) = 560$ $\xrightarrow{\text{Divide by 1.75}}$ $x + y = 320$ (1)

$2(x - y) = 560$ $\xrightarrow{\text{Divide by 2}}$ $x - y = 280$ (2)

Step 4 **Solve** the system of equations (1) and (2).

◀◀◀ **Work Problem 9 at the Side.**

Step 5 **State the answer.** From Problem 9 at the side, the speed of the plane is 300 mph and the speed of the wind is 20 mph.

Step 6 **Check.** The answer seems reasonable, and true statements result when the values are substituted into the equations of the system.

◀◀◀ **Work Problem 10 at the Side.**

ANSWERS
9. (300, 20)
10. current: 4 mph;
Ann's speed in still water: 6 mph

12.4 Exercises

Choose the correct response in Exercises 1–7.

1. Which expression represents the monetary value of x 20-dollar bills?

   A. $\dfrac{x}{20}$ dollars B. $\dfrac{20}{x}$ dollars C. $20 + x$ dollars D. $20x$ dollars

2. Which expression represents the cost of t pounds of candy that sells for $1.95 per lb?

   A. $1.95t$ B. $\dfrac{\$1.95}{t}$ C. $\dfrac{t}{\$1.95}$ D. $\$1.95 + t$

3. Which expression represents the amount of interest earned on d dollars at an interest rate of 2%?

   A. $2d$ dollars B. $0.02d$ dollars C. $0.2d$ dollars D. $200d$ dollars

4. Suppose that x liters of a 40% acid solution are mixed with y liters of a 35% solution to obtain 100 L of a 38% solution. One equation in a system for solving this problem is $x + y = 100$. Which one of the following is the other equation?

   A. $0.35x + 0.40y = 0.38(100)$
   B. $0.40x + 0.35y = 0.38(100)$
   C. $35x + 40y = 38$
   D. $40x + 35y = 0.38(100)$

5. According to *Natural History* magazine, the speed of a cheetah is 70 mph. If a cheetah runs for x hours, how many miles does the cheetah cover?

   A. $70 + x$ miles B. $70 - x$ miles C. $\dfrac{70}{x}$ miles D. $70x$ miles

6. What is the speed of a plane that travels at a rate of 560 mph *against* a wind of r mph?

   A. $560 + r$ mph B. $\dfrac{560}{r}$ mph C. $560 - r$ mph D. $r - 560$ mph

7. What is the speed of a plane that travels at a rate of 560 mph *with* a wind of r mph?

   A. $\dfrac{r}{560}$ mph B. $560 - r$ mph C. $560 + r$ mph D. $r - 560$ mph

8. Using the list of steps for solving an applied problem with two variables, write a short paragraph describing the general procedure you will use to solve the problems that follow in this exercise set.

Exercises 9 and 10 are good warm-up problems. In each case, refer to the six-step problem-solving method, fill in the blanks for Steps 2 and 3, and then complete the solution by applying Steps 4–6.

9. The sum of two numbers is 98 and the difference between them is 48. Find the two numbers.

   Step 1 **Read** the problem carefully.

   Step 2 **Assign variables.**

   Let $x =$ the first number and let

   $y =$ _____ .

   Step 3 **Write two equations.**

   First equation: $x + y = 98$

   Second equation: _____

10. The sum of two numbers is 201 and the difference between them is 11. Find the two numbers.

    Step 1 **Read** the problem carefully.

    Step 2 **Assign variables.**

    Let $x =$ the first number and let

    $y =$ _____ .

    Step 3 **Write two equations.**

    First equation: $x + y = 201$

    Second equation: _____

Write a system of equations for each problem, and then solve the problem. See Example 1.

11. During 2002, two of the top-grossing concert tours were by the Rolling Stones and Cher. Together the two tours visited 117 cities. Cher visited 51 more cities than the Rolling Stones. How many cities did each tour visit? (*Source:* Pollstar.)

12. In 2003, the two top formats of U.S. commercial radio stations were country and news/talk. There were 288 fewer news/talk stations than country stations, and together they comprised a total of 3912 stations. How many stations of each format were there? (*Source:* M Street Corporation.)

13. The two top-grossing movies of 2003 were *The Lord of the Rings: The Return of the King* and *Finding Nemo*. *Finding Nemo* grossed $21.4 million less than *The Lord of the Rings: The Return of the King*, and together the two films took in $700.8 million. How much did each of these movies earn? (*Source:* Nielsen EDI.)

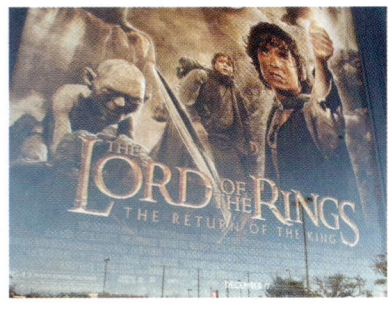

14. The life span of a $100 bill is 80 months longer than the life span of a $1 bill. If the combined life span of a $1 bill and a $100 bill is 124 months, what is the life span of each bill? (*Source:* Bureau of Engraving and Printing.)

15. The Terminal Tower in Cleveland, Ohio, is 242 ft shorter than the Key Tower, also in Cleveland. The total of the heights of the two buildings is 1658 ft. Find the heights of the buildings. (*Source: World Almanac and Book of Facts.*)

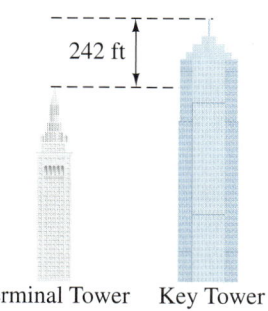

Terminal Tower   Key Tower

16. In 2003, a total of 943.8 thousand people lived in the cities of Las Vegas, Nevada, and Sacramento, California. Sacramento had 73.4 thousand fewer residents than Las Vegas. What was the population of each city? (*Source:* Bureau of the Census.)

If x units of a product cost C dollars to manufacture and earn revenue of R dollars, the value of x where the expressions for C and R are equal is called the **break-even quantity,** *the number of units that produce 0 profit. In Exercises 17 and 18,* **(a)** *find the break-even quantity, and* **(b)** *decide whether the product should be produced based on whether it will earn a profit. (Profit equals revenue minus cost.)*

17. $C = 85x + 900$; $R = 105x$; no more than 38 units can be sold.

18. $C = 105x + 6000$; $R = 255x$; no more than 400 units can be sold.

Write a system of equations for each problem, and then solve the system. See Example 2.

19. A motel clerk counts his $1 and $10 bills at the end of a day. He finds that he has a total of 74 bills having a combined monetary value of $326. Find the number of bills of each denomination that he has.

Denomination of Bill	Number of Bills	Total Value
$1	x	
$10	y	
Totals	74	$326

20. Letarsha is a bank teller. At the end of a day, she has a total of 69 $5 and $10 bills. The total value of the money is $590. How many of each denomination does she have?

Denomination of Bill	Number of Bills	Total Value
$5	x	$5x$
$10	y	
Totals		

21. A newspaper advertised DVDs and CDs. Tracy Sudak went shopping and bought each of her seven nephews a gift, either a DVD of the movie *Miracle* or the latest Linkin Park CD. The DVD cost $14.95 and the CD cost $16.88, and she spent a total of $114.30. How many DVDs and how many CDs did she buy?

22. Terry Wong saw the ad (see Exercise 21) and he, too, went shopping. He bought each of his five nieces a gift, either a DVD of *Home on the Range* or the CD soundtrack to *Scooby Doo 2: Monsters Unleashed*. The DVD cost $14.99 and the soundtrack cost $13.88, and he spent a total of $70.51. How many DVDs and CDs did he buy?

23. Maria Lopez has twice as much money invested at 5% simple annual interest as she does at 4%. If her yearly income from these two investments is $350, how much does she have invested at each rate?

24. Charles Miller invested his textbook royalty income in two accounts, one paying 3% annual simple interest and the other paying 2% interest. He earned a total of $11 interest. If he invested three times as much in the 3% account as he did in the 2% account, how much did he invest at each rate?

25. Average movie ticket prices in the United States are, in general, lower than in other countries. It would cost $77.87 to buy three tickets in Japan plus two tickets in Switzerland. Three tickets in Switzerland plus two tickets in Japan would cost $73.83. How much does an average movie ticket cost in each of these countries? (*Source:* Business Traveler International.)

26. (See Exercise 25.) Four movie tickets in Germany plus three movie tickets in France would cost $62.27. Three tickets in Germany plus four tickets in France would cost $62.19. How much does an average movie ticket cost in each of these countries? (*Source:* Business Traveler International.)

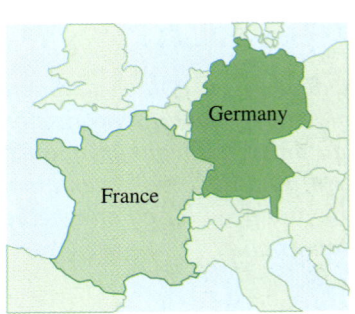

Write a system of equations for each problem, and then solve the system. See Example 3.

27. A 40% dye solution is to be mixed with a 70% dye solution to get 120 L of a 50% solution. How many liters of the 40% and 70% solutions will be needed?

Percent (as a Decimal)	Liters of Solution	Liters of Pure Dye
0.40	x	
0.70	y	
0.50	120	

28. A 90% antifreeze solution is to be mixed with a 75% solution to make 120 L of a 78% solution. How many liters of the 90% and 75% solutions will be used?

Percent (as a Decimal)	Liters of Solution	Liters of Pure Antifreeze
0.90	x	
0.75	y	
0.78	120	

29. A merchant wishes to mix coffee worth $6 per lb with coffee worth $3 per lb to get 90 lb of a mixture worth $4 per lb. How many pounds of the $6 and the $3 coffees will be needed?

Dollars per Pound	Pounds	Cost
6	x	
	y	
	90	

30. A grocer wishes to blend candy selling for $1.20 per lb with candy selling for $1.80 per lb to get a mixture that will be sold for $1.40 per lb. How many pounds of the $1.20 and the $1.80 candies should be used to get 45 lb of the mixture?

Dollars per Pound	Pounds	Cost
	x	
1.80	y	
	45	

31. How many barrels of pickles worth $40 per barrel and pickles worth $60 per barrel must be mixed to obtain 50 barrels of a mixture worth $48 per barrel?

32. The owner of a nursery wants to mix some fertilizer worth $70 per bag with some worth $90 per bag to obtain 40 bags of mixture worth $77.50 per bag. How many bags of each type should she use?

Write a system of equations for each problem, and then solve the system. See Examples 4 and 5.

33. RAGBRAI®, the Des Moines **R**egister's **A**nnual **G**reat **B**icycle **R**ide **A**cross **I**owa, is the longest and oldest touring bicycle ride in the world. Suppose a cyclist began the 490 mi ride on July 25, 2004 in western Iowa at the same time that a car traveling toward it left eastern Iowa. If the bicycle and the car met after 7 hr and the car traveled 40 mph faster than the bicycle, find the average speed of each. (*Source:* www.ragbrai.org)

RAGBRAI XXXII ROUTE
July 25 – 31, 2004

34. In 2002, Atlanta's Hartsfield Airport was the nation's busiest. Suppose two planes leave the airport at the same time, one traveling east and the other traveling west. If the planes are 2100 mi apart after 2 hr and one plane travels 50 mph faster than the other, find the speed of each plane. (*Source:* Airports Council International—North America.)

35. Toledo and Cincinnati are 200 mi apart. A car leaves Toledo traveling toward Cincinnati, and another car leaves Cincinnati at the same time, traveling toward Toledo. The car leaving Toledo averages 15 mph faster than the other, and they meet after 1 hr and 36 min. What are the rates of the cars?

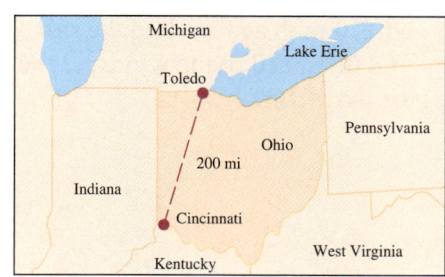

36. Kansas City and Denver are 600 mi apart. Two cars start from these cities, traveling toward each other. They meet after 6 hr. Find the rate of each car if one travels 30 mph slower than the other.

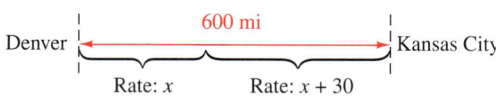

37. At the beginning of a bicycle ride for charity, Roberto and Juana are 30 mi apart. If they leave at the same time and ride in the same direction, Roberto overtakes Juana in 6 hr. If they ride toward each other, they meet in 1 hr. What are their speeds?

38. Mr. Abbot left Farmersville in a plane at noon to travel to Exeter. Mr. Baker left Exeter in his automobile at 2 P.M. to travel to Farmersville. It is 400 mi from Exeter to Farmersville. If the sum of their speeds was 120 mph, and if they crossed paths at 4 P.M., find the speed of each.

39. A boat takes 3 hr to go 24 mi upstream. It can go 36 mi downstream in the same time. Find the speed of the current and the speed of the boat in still water if x = the speed of the boat in still water and y = the speed of the current.

	d	r	t
Downstream	36	$x + y$	
Upstream	24	$x - y$	

40. It takes a boat $1\frac{1}{2}$ hr to go 12 mi downstream, and 6 hr to return. Find the speed of the boat in still water and the speed of the current. Let x = the speed of the boat in still water and y = the speed of the current.

	d	r	t
Downstream	12	$x + y$	$\frac{3}{2}$
Upstream			6

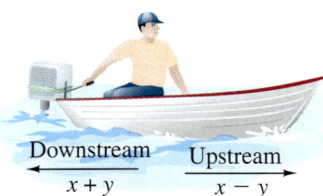

41. If a plane can travel 440 mph against the wind and 500 mph with the wind, in the same amount of time (1 hour), find the speed of the wind and the speed of the plane in still air.

42. A small plane travels 200 mph with the wind in one hour and 120 mph against it in the same amount of time. Find the speed of the wind and the speed of the plane in still air.

12.5 Solving Systems of Linear Inequalities

OBJECTIVE

1 Solve systems of linear inequalities by graphing.

We graphed the solutions of a linear inequality in **Section 11.5.** Recall that to graph the solutions of $x + 3y > 12$, for example, we first graph $x + 3y = 12$ by finding and plotting a few ordered pairs that satisfy the equation. Because the points on the line do *not* satisfy the inequality, we use a dashed line. To decide which side of the line includes the points that are solutions, we choose a test point not on the line, such as $(0, 0)$. Substituting these values for x and y in the inequality gives

$$x + 3y > 12$$
$$0 + 3(0) > 12$$
$$0 > 12,$$

a false result. This indicates that the solutions are those points on the side of the line that does not include $(0, 0)$, as shown in Figure 10.

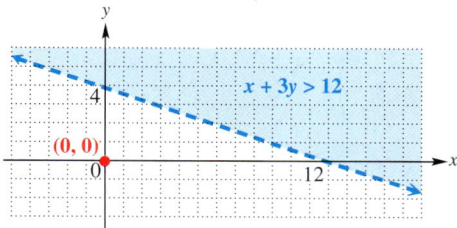

Figure 10

Now we use the same techniques to solve systems of linear inequalities.

OBJECTIVE 1 Solve systems of linear inequalities by graphing.
A **system of linear inequalities** consists of two or more linear inequalities. The **solution of a system of linear inequalities** includes all points that make all inequalities of the system true at the same time. To solve a system of linear inequalities, use the following steps.

Solving a System of Linear Inequalities

Step 1 **Graph the inequalities.** Graph each inequality using the method of **Section 11.5.**

Step 2 **Choose the intersection.** Indicate the solution of the system by shading the intersection of the graphs (the region where the graphs overlap).

EXAMPLE 1 Solving a System of Two Linear Inequalities

Graph the solution of this system.

$$3x + 2y \leq 6$$
$$2x - 5y \geq 10$$

To graph $3x + 2y \leq 6$, graph the solid boundary line $3x + 2y = 6$ and shade the region containing $(0, 0)$, as shown in Figure 11(a) on the next page. Then graph $2x - 5y \geq 10$ with the solid boundary line $2x - 5y = 10$. The test point $(0, 0)$ makes this inequality false, so shade the region on the other side of the boundary line. See Figure 11(b).

— **Continued on Next Page**

848 Chapter 12 Systems of Linear Equations and Inequalities

1 Graph the solution of this system.

$$x - 2y \leq 8$$
$$3x + y \geq 6$$

To get you started, the graphs of $x - 2y = 8$ and $3x + y = 6$ are shown.

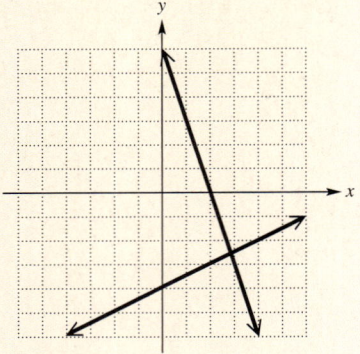

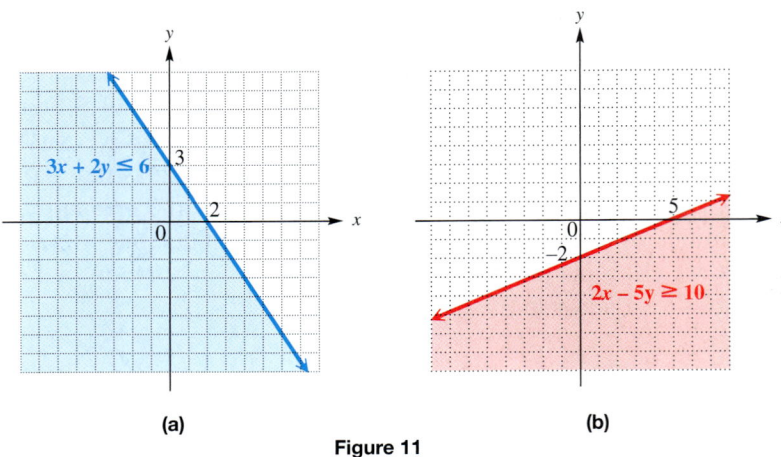

Figure 11

The solution of this system includes all points in the intersection (overlap) of the graphs of the two inequalities. It includes the gray shaded region and portions of the two boundary lines shown in Figure 12.

Figure 12

◀◀◀ **Work Problem 1 at the Side.**

NOTE
We usually do all the work on one set of axes. In the following examples, only one graph is shown. Be sure that the region of the final solution is clearly indicated.

EXAMPLE 2 Solving a System of Two Linear Inequalities

Graph the solution of this system.

$$x - y > 5$$
$$2x + y < 2$$

Figure 13 shows the graphs of both $x - y > 5$ and $2x + y < 2$. Dashed lines show that the graphs of the inequalities do not include their boundary lines. The solution of the system is the region with the gray shading. The solution does not include either boundary line.

Continued on Next Page

ANSWERS
1.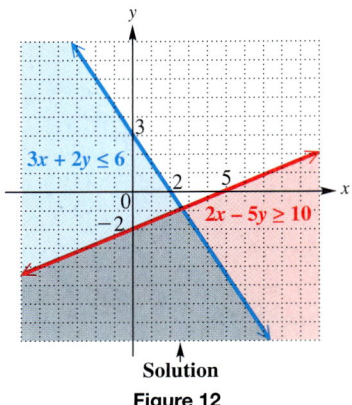

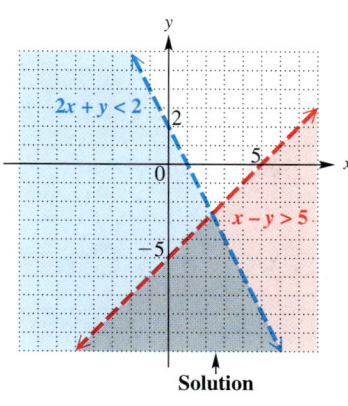

Figure 13

2 Graph the solution of each system.

(a) $x + 2y < 0$
$3x - 4y < 12$

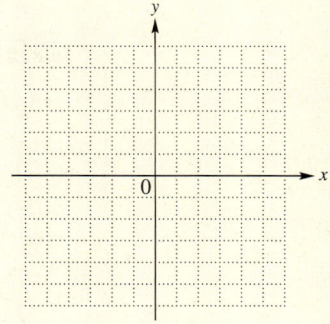

EXAMPLE 3 Solving a System of Three Linear Inequalities

Graph the solution of this system.

$$4x - 3y \leq 8$$
$$x \geq 2$$
$$y \leq 4$$

Recall that $x = 2$ is a vertical line through the point $(2, 0)$, and $y = 4$ is a horizontal line through $(0, 4)$. The graph of the solution is the shaded region in Figure 14, including all boundary lines.

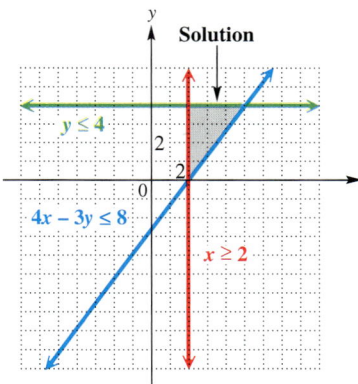

Figure 14

Work Problem 2 at the Side.

(b) $3x + 2y \leq 12$
$x \leq 2$
$y \leq 4$

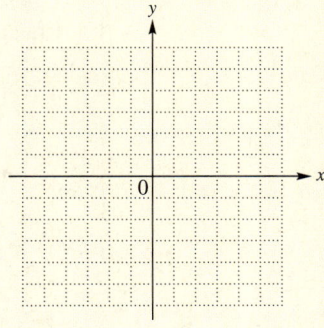

ANSWERS

2. (a)

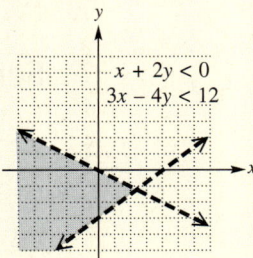

(b)

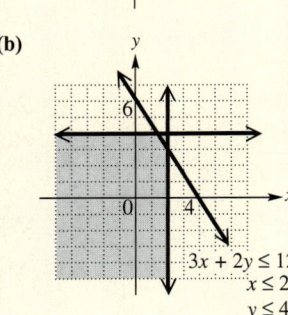

Focus on Real-Data Applications

Sales of Compact Discs versus Cassettes

Sales of compact discs (CDs) and cassettes for the years 1990 through 2000 are given in the table. The number of years since 1990 is represented by x. Sales of compact discs and cassettes can be modeled by linear equations.

- A linear model for the sales of compact discs:

  $y = 69.57x + 303.805$

- A linear model for the sales of cassettes:

  $y = -35.042x + 437.227$

y is sales in millions. The actual data and the linear models are graphed below.

Year	x	CD Sales (in millions)	Cassette Sales (in millions)
1990	0	286.5	442.2
1991	1	333.4	360.1
1992	2	407.5	366.4
1993	3	495.4	339.5
1994	4	662.1	345.4
1995	5	722.9	272.6
1996	6	778.9	225.3
1997	7	753.1	172.6
1998	8	847.0	158.5
1999	9	938.9	123.6
2000	10	942.5	76.0

Source: *World Almanac and Book of Facts*, 2004.

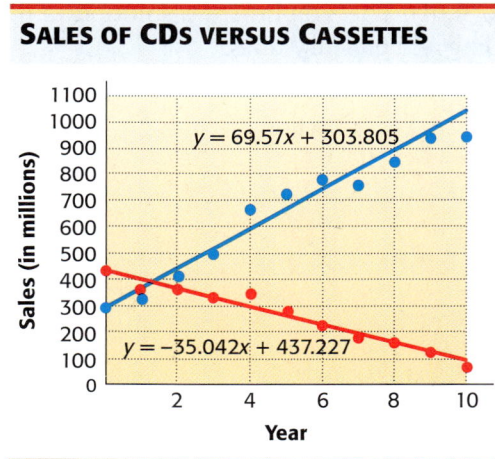

For Group Discussion

1. Shade the region on the graph that corresponds to the solution of the following system.

   $y \leq 69.57x + 303.805$
   $y \geq -35.042x + 437.227$

   Interpret the solution in the context of sales of compact discs and cassettes.

2. Solve the linear inequality $69.57x + 303.805 > -35.042x + 437.227$. Round your answer to the nearest hundredth.

3. What does the solution to the inequality in Problem 2 represent in the context of sales of CDs and cassettes?

12.5 Exercises

Match each system of inequalities with the correct graph from choices A–D.

1. $x \geq 5$
   $y \leq -3$

2. $x \leq 5$
   $y \geq -3$

3. $x > 5$
   $y < -3$

4. $x < 5$
   $y > -3$

A.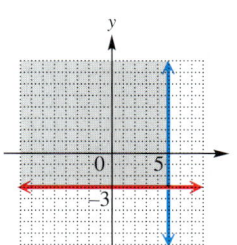

B.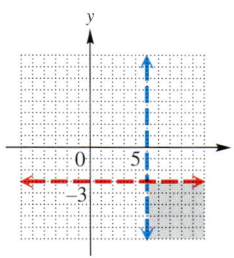

C.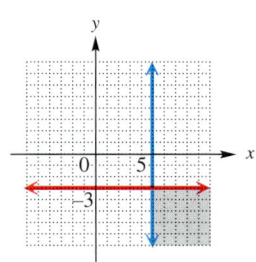

D.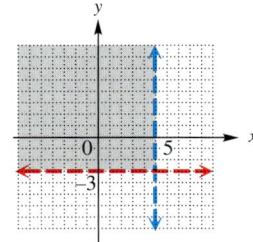

Graph the solution of each system of linear inequalities. See Examples 1–3.

5. $x + y \leq 6$
   $x - y \geq 1$

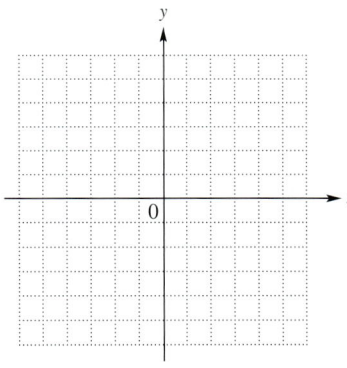

6. $x + y \leq 2$
   $x - y \geq 3$

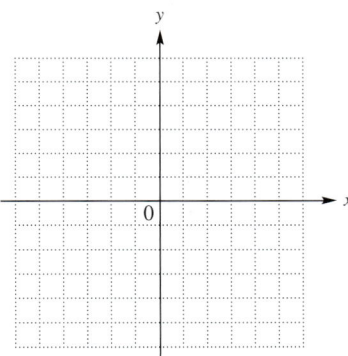

7. $4x + 5y \geq 20$
   $x - 2y \leq 5$

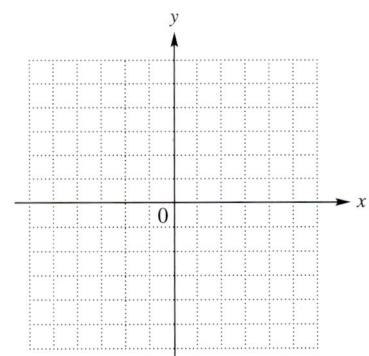

8. $x + 4y \leq 8$
   $2x - y \geq 4$

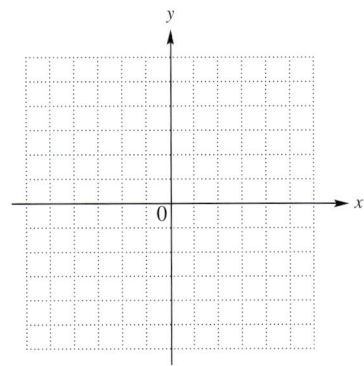

9. $2x + 3y < 6$
   $x - y < 5$

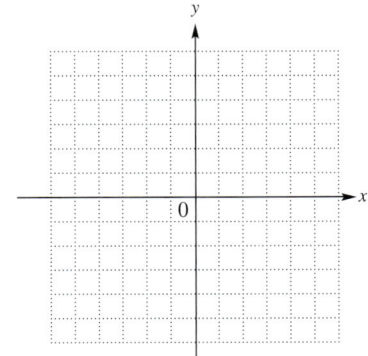

10. $x + 2y < 4$
    $x - y < -1$

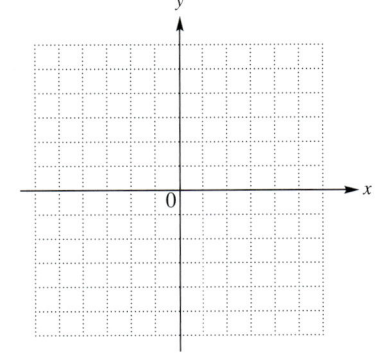

852 Chapter 12 Systems of Linear Equations and Inequalities

11. $y \leq 2x - 5$
$x < 3y + 2$

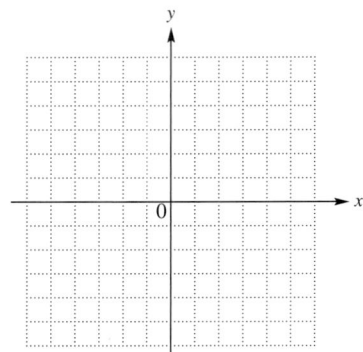

12. $x \geq 2y + 6$
$y > -2x + 4$

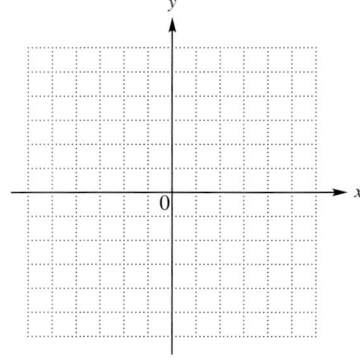

13. $4x + 3y < 6$
$x - 2y > 4$

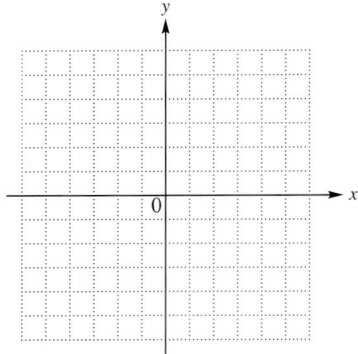

14. $3x + y > 4$
$x + 2y < 2$

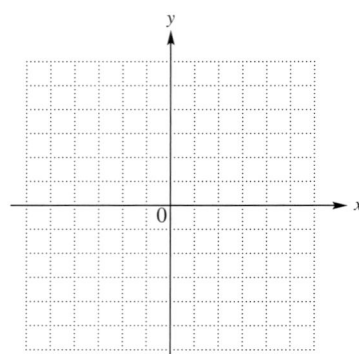

15. $x \leq 2y + 3$
$x + y < 0$

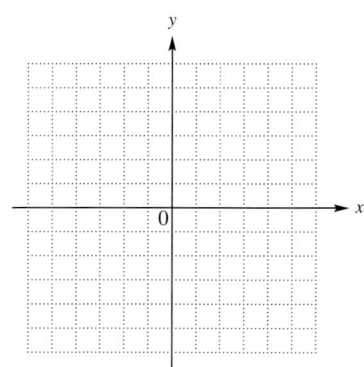

16. $x \leq 4y + 3$
$x + y > 0$

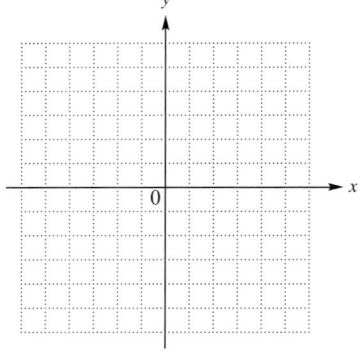

17. $4x + 5y < 8$
$y > -2$
$x > -4$

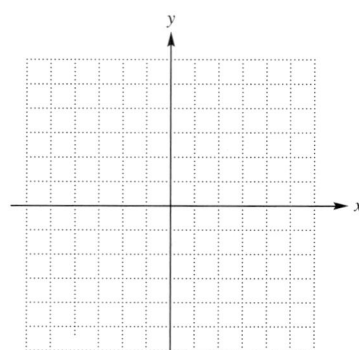

18. $x + y \geq -3$
$x - y \leq 3$
$y \leq 3$

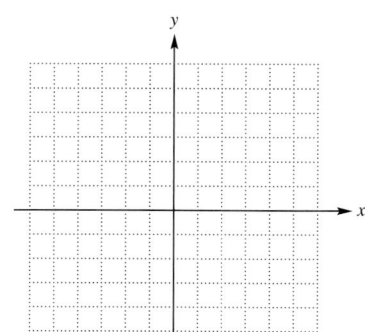

19. $3x - 2y \geq 6$
$x + y \leq 4$
$x \geq 0$
$y \geq -4$

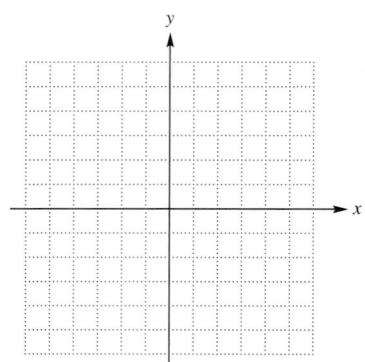

20. Every system of inequalities illustrated in the examples of this section has infinitely many solutions. Explain why this is so. Does this mean that *any* ordered pair is a solution?

Chapter 12
SUMMARY

KEY TERMS

12.1 **system of linear equations**   A system of linear equations (or **linear system**) consists of two or more linear equations with the same variables.

solution of a system   The solution of a system of linear equations includes all the ordered pairs that make all the equations of the system true at the same time.

consistent system   A system of equations with at least one solution is a consistent system.

inconsistent system   An inconsistent system of equations is a system with no solution.

independent equations   Equations of a system that have different graphs are called independent equations.

dependent equations   Equations of a system that have the same graph (because they are different forms of the same equation) are called dependent equations.

12.5 **system of linear inequalities**   A system of linear inequalities contains two or more linear inequalities (and no other kinds of inequalities).

solution of a system of linear inequalities   The solution of a system of linear inequalities includes all points that make all inequalities of the system true at the same time.

TEST YOUR WORD POWER

See how well you have learned the vocabulary in this chapter. Answers, with examples, follow the Quick Review.

1. A **system of linear equations** consists of
   A. at least two linear equations with different variables
   B. two or more linear equations that have an infinite number of solutions
   C. two or more linear equations with the same variables
   D. two or more linear inequalities.

2. A **solution of a system** of linear equations is
   A. an ordered pair that makes one equation of the system true
   B. an ordered pair that makes all the equations of the system true at the same time
   C. any ordered pair that makes one or the other or both equations of the system true
   D. the set of values that make all the equations of the system false.

3. A **consistent system** is a system of equations
   A. with one solution
   B. with no solution
   C. with two solutions
   D. that has parallel lines at its graph.

4. An **inconsistent system** is a system of equations
   A. with one solution
   B. with no solution
   C. with an infinite number of solutions
   D. that have the same graph.

5. **Dependent equations**
   A. have different graphs
   B. have no solution
   C. have one solution
   D. are different forms of the same equation.

853

Quick Review

Concepts

Examples

12.1 Solving Systems of Linear Equations by Graphing

An ordered pair is a solution of a system if it makes all equations of the system true at the same time.

Is $(4, -1)$ a solution of the system $\begin{array}{l} x + y = 3 \\ 2x - y = 9 \end{array}$?

Yes, because $4 + (-1) = 3$ and $2(4) - (-1) = 9$ are both true.

If the graphs of the equations of a system are both sketched on the same axes, then the points of intersection, if any, are solutions of the system.

If the graphs of the equations do not intersect (that is, the lines are parallel), then the system has no solution.

If the graphs of the equations are the same line, then the system has an infinite number of solutions.

Solve by graphing.

$$x + y = 5$$
$$2x - y = 4$$

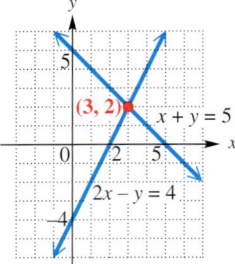

The ordered pair $(3, 2)$ satisfies both equations, so $(3, 2)$ is the solution of the system.

12.2 Solving Systems of Linear Equations by Substitution

Step 1 Solve one equation for either variable.

Step 2 Substitute for that variable in the other equation to get an equation in one variable.

Step 3 Solve the equation from Step 2.

Step 4 Substitute the result into the equation from Step 1 to get the value of the other variable.

Step 5 Check. Write the solution as an ordered pair.

Solve by substitution.

$$x + 2y = -5 \quad (1)$$
$$y = -2x - 1 \quad (2)$$

Equation (2) is already solved for y.
Substitute $-2x - 1$ for y in equation (1).

$$x + 2(-2x - 1) = -5$$
$$x - 4x - 2 = -5$$
$$-3x - 2 = -5$$
$$-3x = -3$$
$$x = 1$$

To find y, let $x = 1$ in equation (2):

$$y = -2(1) - 1 = -3.$$

The solution $(1, -3)$ checks.

12.3 Solving Systems of Linear Equations by Elimination

Step 1 Write both equations in standard form $Ax + By = C$.

Step 2 If necessary, multiply one or both equations by appropriate numbers so that the sum of the coefficients of either the x- or y-terms is 0.

Step 3 Add the equations to get an equation with only one variable (or no variable).

Step 4 Solve the equation from Step 3.

Solve by elimination.

$$x + 3y = 7 \quad (1)$$
$$3x - y = 1 \quad (2)$$

Multiply equation (1) by -3 to eliminate the x-terms.

$$\begin{array}{r} -3x - 9y = -21 \\ 3x - y = 1 \\ \hline -10y = -20 \end{array} \quad \text{Add.}$$

$$y = 2 \qquad \text{Divide by } -10.$$

Concepts

12.3 Solving Systems of Linear Equations by Elimination (continued)

Step 5 Substitute the solution from Step 4 into either of the original equations to find the value of the remaining variable.

Step 6 Check. Write the solution as an ordered pair.

12.4 Applications of Linear Systems
Use the modified six-step method.

Step 1 **Read** the problem carefully.

Step 2 **Assign variables** for each unknown value. Use diagrams or tables as needed.

Step 3 **Write two equations** using both variables.

Step 4 **Solve** the system.

Step 5 **State the answer.**

Step 6 **Check** the answer in the words of the original problem.

12.5 Solving Systems of Linear Inequalities
To solve a system of two or more linear inequalities, graph the inequalities on the same axes. (This was explained in **Section 11.5.**) The solution of the system is the intersection (overlap) of the regions of the graphs. The portions of the boundary lines that bound the region of solutions are included for a $\leq$ or $\geq$ inequality and excluded for a $<$ or $>$ inequality.

Examples

Substitute to get the value of x.

$$x + 3(\mathbf{2}) = 7 \quad (1)$$
$$x + 6 = 7$$
$$x = \mathbf{1}$$

Since $\mathbf{1} + 3(\mathbf{2}) = 7$ and $3(\mathbf{1}) - \mathbf{2} = 1$, the solution $(1, 2)$ checks.

The sum of two numbers is 30. Their difference is 6. Find the numbers.

Let x represent one number.
Let y represent the other number.

$$\begin{aligned} x + y &= 30 \\ x - y &= 6 \\ \hline 2x &= 36 \quad \text{Add.} \\ x &= 18 \quad \text{Divide by 2.} \end{aligned}$$

Let $x = 18$ in the first equation: $18 + y = 30$. Solve to get $y = 12$. The numbers are 18 and 12.

The sum of 18 and 12 is 30, and the difference between 18 and 12 is 6, so the answer checks.

The shaded region is the solution of this system.

$$2x + 4y \geq 5$$
$$x \geq 1$$

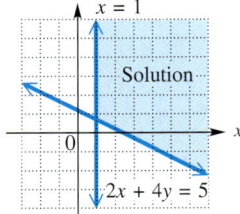

ANSWERS TO TEST YOUR WORD POWER

1. C; *Example:* $2x + y = 7, \quad 3x - y = 3$
2. B; *Example:* The ordered pair $(2, 3)$ satisfies both equations of the system in the Answer 1 example, so it is a solution of the system.
3. A; *Example:* The system in the Answer 1 example is consistent. The graphs of the equations intersect at exactly one point, in this case the solution $(2, 3)$.
4. B; *Example:* The equations of two parallel lines make up an inconsistent system; their graphs never intersect, so there is no solution to the system.
5. D; *Example:* The equations $4x - y = 8$ and $8x - 2y = 16$ are dependent because their graphs are the same line.

Focus on Real-Data Applications

Systems of Linear Equations and Modeling

A system of linear equations is an efficient tool for finding a linear *model*, or the equation for data that is known to be linear. Recall that the slope-intercept form of the equation of a line is $y = mx + b$. Once we know the values of the slope, m, and the y-intercept, b, we can then write the exact model. If we know two ordered pairs, then we can write a system of linear equations in which the unknown quantities are m and b.

For example, to find the formula to convert from Kelvin (K), the most commonly used thermodynamic temperature scale, as the *input* to degrees Fahrenheit (°F) as the *output*, we only need to know the data presented in the table. The linear equation has the form $F = mK + b$, and ordered pairs have the format (K, F).

	K	°F
Water Freezes	273	32
Water Boils	373	212

The system of linear equations to be solved is shown below.

$$32 = m(273) + b$$
$$212 = m(373) + b$$

For Group Discussion

Use systems of linear equations to find the model for each problem.

1. Solve the system of linear equations given above to find the conversion formula from K to °F.

2. Water freezes at 0 °C and boils at 100 °C. Write the system of linear equations to find the model for converting from degrees Celsius as the input to degrees Fahrenheit as the output.

3. Suppose you begin a carefully managed weight-loss program in which you expect your weight to decline steadily. (A constant weight loss means that it is reasonable to assume that the relationship between number of weeks on the program and weight is linear.) After two weeks you weigh 179 lb, and after six weeks you weigh 169 lb. Use x to represent the number of weeks on the program and y to represent your weight in pounds at the end of x weeks. Write the linear model for your weight-loss program.

Extension: Suppose a line contains the distinct points (x_1, y_1) and (x_2, y_2), where $x_1 \neq x_2$. Use the method of this activity to derive the slope formula.

Chapter 12
REVIEW EXERCISES

[12.1] *Decide whether the given ordered pair is a solution of the given system.*

1. $(3, 4)$
$4x - 2y = 4$
$5x + y = 19$

2. $(-5, 2)$
$x - 4y = -13$
$2x + 3y = 4$

Solve each system by graphing.

3. $x + y = 4$
$2x - y = 5$

4. $x - 2y = 4$
$2x + y = -2$

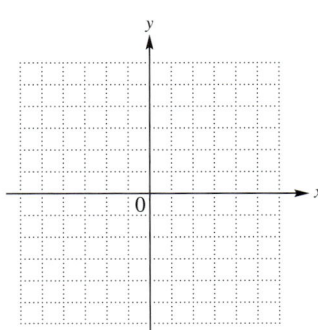

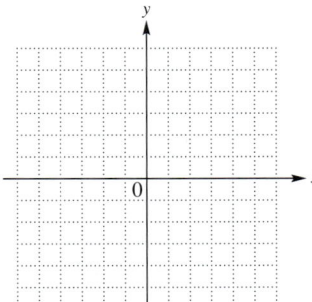

5. $x - 2 = 2y$
$2x - 4y = 4$

6. $2x + 4 = 2y$
$y - x = -3$

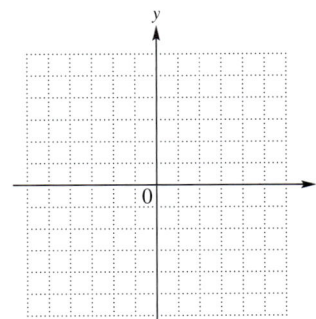

7. When a student was asked to determine whether the ordered pair $(1, -2)$ is a solution of the system

$$x + y = -1$$
$$2x + y = 4,$$

he answered "yes." His reasoning was that the ordered pair satisfies the equation $x + y = -1$; that is, $1 + (-2) = -1$ is true. Why is the student's answer wrong?

[12.2] *Solve each system by the substitution method.*

8. $3x + y = 7$
   $x = 2y$

9. $2x - 5y = -19$
   $y = x + 2$

10. $4x + 5y = 44$
    $x + 2 = 2y$

11. $5x + 15y = 3$
    $x + 3y = 2$

[12.3] *Solve each system by the elimination method.*

12. $2x - y = 13$
    $x + y = 8$

13. $3x - y = -13$
    $x - 2y = -1$

14. $-4x + 3y = 25$
    $6x - 5y = -39$

15. $3x - 4y = 9$
    $6x - 8y = 18$

16. For the system
$$2x + 12y = 7$$
$$3x + 4y = 1,$$
if we were to multiply the first (top) equation by -3, by what number would we have to multiply the second (bottom) equation in order to

(a) eliminate the x-terms when solving by the elimination method?

(b) eliminate the y-terms when solving by the elimination method?

Solve each system by any method.

17. $x - 2y = 5$
    $y = x - 7$

18. $5x - 3y = 11$
    $2y = x - 4$

19. $\dfrac{x}{2} + \dfrac{y}{3} = 7$
    $\dfrac{x}{4} + \dfrac{2y}{3} = 8$

20. $\dfrac{3x}{4} - \dfrac{y}{3} = \dfrac{7}{6}$
    $\dfrac{x}{2} + \dfrac{2y}{3} = \dfrac{5}{3}$

[12.4] *Solve each problem by using a system of equations.*

21. At the end of 2001, Subway topped McDonald's as the largest restaurant chain in the United States. Subway operated 148 more restaurants than McDonald's, and together the two chains had 26,346 restaurants. How many restaurants did each company operate? (*Source:* USA Today.)

22. Two of the most popular magazines in the United States are *Modern Maturity* and *Reader's Digest*. Together, the average total circulation for these two magazines during a recent 6-month period was 35.6 million. *Reader's Digest* circulation was 5.4 million less than that of *Modern Maturity*. What were the circulation figures for each magazine? (*Source:* Audit Bureau of Circulations and Magazine Publishers of America.)

23. The perimeter of a rectangle is 90 m. Its length is $1\frac{1}{2}$ times its width. Find the length and width of the rectangle.

24. A cashier has 20 bills, all of which are $10 or $20 bills. The total value of the money is $330. How many of each type does the cashier have?

Denomination of Bill	Number of Bills	Total Value
$10	x	$10x$
$20		
Totals		$330

25. Candy that sells for $1.30 per lb is to be mixed with candy selling for $0.90 per lb to get 100 lb of a mix that will sell for $1 per lb. How much of each type should be used?

26. A certain plane flying with the wind travels 540 mi in 2 hr. Later, flying against the same wind, the plane travels 690 mi in 3 hr. Find the speed of the plane in still air and the speed of the wind.

27. After taxes, Ms. Cesar's game show winnings were $18,000. She invested part of it at 3% annual simple interest and the rest at 4%. Her interest income for the first year was $650. How much did she invest at each rate?

Percent	Amount of Principal	Interest
0.03	x	
0.04	y	
Totals	$18,000	

28. A 40% antifreeze solution is to be mixed with a 70% solution to get 90 L of a 50% solution. How many liters of the 40% and 70% solutions will be needed?

Percent	Number of Liters	Amount of Pure Antifreeze
0.40	x	
0.70	y	
0.50	90	

[12.5] *Graph the solution for each system of linear inequalities.*

29. $x + y \geq 2$
$x - y \leq 4$

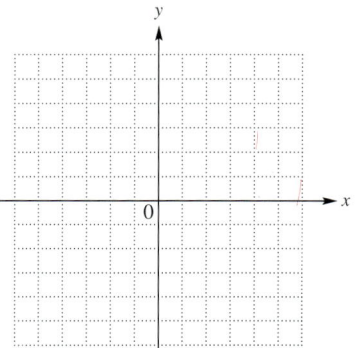

30. $y \geq 2x$
$2x + 3y \leq 6$

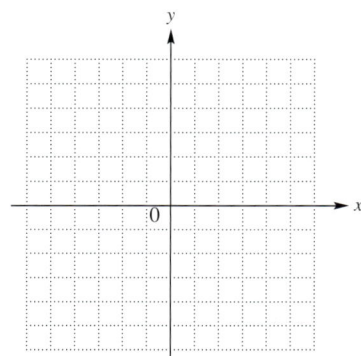

31. $x + y < 3$
$2x > y$

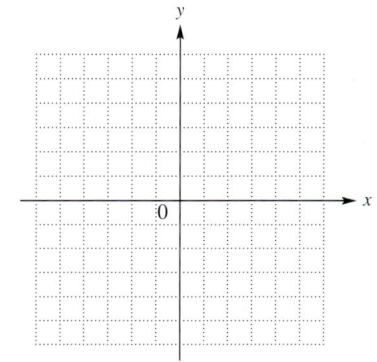

860 Chapter 12 Systems of Linear Equations and Inequalities

32. Which system of linear inequalities is graphed in the figure?

A. $x \leq 3$
   $y \leq 1$

B. $x \leq 3$
   $y \geq 1$

C. $x \geq 3$
   $y \leq 1$

D. $x \geq 3$
   $y \geq 1$

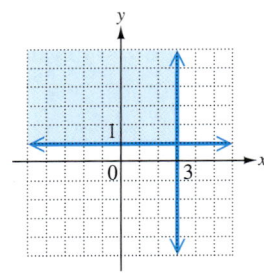

33. Without actually graphing, determine which system of inequalities has no solution.

A. $x \geq 4$
   $y \leq 3$

B. $x + y > 4$
   $x + y < 3$

C. $x > 2$
   $y < 1$

D. $x + y > 4$
   $x - y < 3$

MIXED REVIEW EXERCISES

Solve each system.

34. $3x + 4y = 6$
    $4x - 5y = 8$

35. $\dfrac{3x}{2} + \dfrac{y}{5} = -3$
    $4x + \dfrac{y}{3} = -11$

36. $x + 6y = 3$
    $2x + 12y = 2$

37. $x + y < 5$
    $x - y \geq 2$

38. $y \leq 2x$
    $x + 2y > 4$

39. $y < -4x$
    $y < -2$

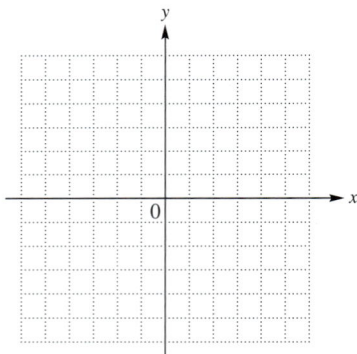 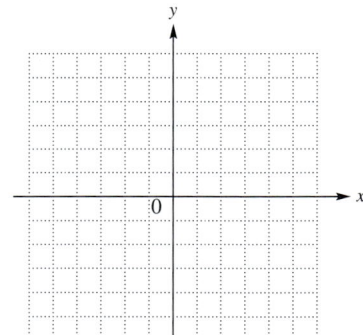 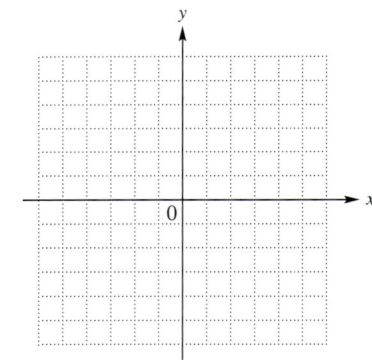

40. The perimeter of an isosceles triangle is 29 in. One side of the triangle is 5 in. longer than each of the two equal sides. Find the lengths of the sides of the triangle.

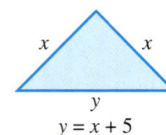

41. In the 2004 National Football League AFC finals, the New England Patriots beat the Indianapolis Colts by 10 points, and the winning score was 4 less than twice the losing score. What was the final score of the game? (*Source:* NFL.)

42. Eboni Perkins compared the monthly payments she would incur for two types of mortgages; fixed-rate and variable-rate. Her observations led to the following graph.

(a) For which years would the monthly payment be more for the fixed-rate mortgage than for the variable-rate mortgage?

(b) In what year would the payments be the same, and what would those payments be?

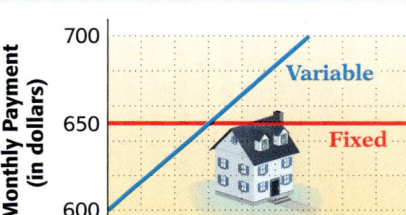

Chapter 12
TEST

1. Solve the system by graphing.

    $2x + y = 1$
    $3x - y = 9$

    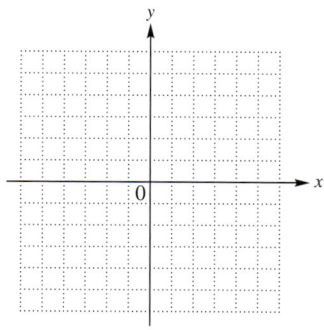

2. Suppose that the graph of a system of two linear equations consists of lines that have the same slope but different y-intercepts. How many solutions does the system have?

Solve each system by the substitution method.

3. $2x + y = -4$
   $x = y + 7$

4. $4x + 3y = -35$
   $x + y = 0$

Solve each system by the elimination method.

5. $2x - y = 4$
   $3x + y = 21$

6. $4x + 2y = 2$
   $5x + 4y = 7$

7. $6x - 5y = 0$
   $-2x + 3y = 0$

8. $4x + 5y = 2$
   $-8x - 10y = 6$

Solve each system by any method.

9. $3x = 6 + y$
   $6x - 2y = 12$

10. $\dfrac{x}{2} - \dfrac{y}{4} = 7$
    $\dfrac{2x}{3} + \dfrac{5y}{4} = 3$

1. _____

2. _____

3. _____

4. _____

5. _____

6. _____

7. _____

8. _____

9. _____

10. _____

Solve each problem.

11. The distance between Memphis and Atlanta is 782 mi less than the distance between Minneapolis and Houston. Together, the two distances total 1570 mi. How far is it between Memphis and Atlanta? How far is it between Minneapolis and Houston? (*Source: Rand McNally Road Atlas.*)

12. In 2002, the two most popular amusement parks in the United States were Disneyland and the Magic Kingdom at Walt Disney World. Disneyland had 1.3 million fewer visitors than the Magic Kingdom, and together they had 26.7 million visitors. How many visitors did each park have? (*Source: Amusement Business.*)

13. A 15% solution of alcohol is to be mixed with a 40% solution to get 50 L of a final mixture that is 30% alcohol. How much of each of the original solutions should be used?

14. Two cars leave from Perham, Minnesota, and travel in the same direction. One car travels $1\frac{1}{3}$ times as fast as the other. After 3 hr they are 45 mi apart. What are the speeds of the cars?

Graph the solution of each system of inequalities.

15. $2x + 7y \leq 14$
$x - y \geq 1$

16. $2x - y > 6$
$4y + 12 \geq -3x$

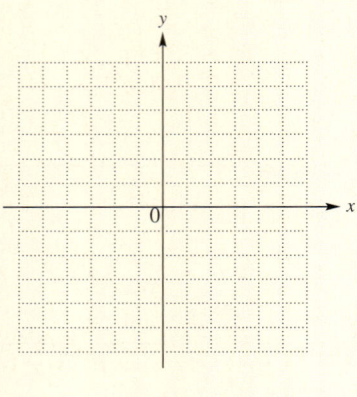

Exponents and Polynomials

13.1 Adding and Subtracting Polynomials

13.2 The Product Rule and Power Rules for Exponents

13.3 Multiplying Polynomials

13.4 Special Products

13.5 Integer Exponents and the Quotient Rule

Summary Exercises on the Rules for Exponents

13.6 Dividing a Polynomial by a Monomial

13.7 Dividing a Polynomial by a Polynomial

13.8 An Application of Exponents: Scientific Notation

Just how much is a trillion? A trillion, which is written 1,000,000,000,000, is an incredibly huge number. A trillion is a million million or a thousand billion. A trillion seconds would last more than 31,000 years, that is, 310 centuries. To be a trillionaire, a person would need a stack of $1000 bills over 67 miles high.

The U.S. budget first exceeded $1 trillion in 1987 and topped $2 trillion for the 2003 fiscal year. (*Source: The Gazette.*) In Exercises 1–3 and 45–50 of Section 13.8, we use exponents and scientific notation to write large numbers related to budgets, planets, pumpkins, and people.

13.1 Adding and Subtracting Polynomials

OBJECTIVES

1. Review combining like terms.
2. Know the vocabulary for polynomials.
3. Evaluate polynomials.
4. Add polynomials.
5. Subtract polynomials.
6. Add and subtract polynomials with more than one variable.

Recall from **Section 2.2** that in an expression such as

$$4x^3 + 6x^2 + 5x + 8,$$

the quantities that are added, $4x^3$, $6x^2$, $5x$, and 8 are called *terms*. In the term $4x^3$, the number 4 is called the *numerical coefficient*, or simply the *coefficient*, of x^3. In the same way, 6 is the coefficient of x^2 in the term $6x^2$, 5 is the coefficient of x in the term $5x$, and 8 is the *constant* term.

OBJECTIVE 1 Review combining like terms. In Section 2.2, we saw that *like terms* are terms with exactly the same variables, with the same exponents on the variables, such as $2x^2$ and $-5x^2$. Only the coefficients may differ. Like terms are combined, or *added*, by adding their coefficients using the distributive property.

EXAMPLE 1 Adding Like Terms

Simplify each expression by adding like terms.

(a) $-4x^3 + 6x^3 = (-4 + 6)x^3$ Distributive property
$= 2x^3$

(b) $9x^6 - 14x^6 + x^6 = (9 - 14 + 1)x^6$ $x^6 = 1x^6$
$= -4x^6$

(c) $12m^2 + 5m + 4m^2 = (12 + 4)m^2 + 5m$
$= 16m^2 + 5m$

(d) $3x^2y + 4x^2y - x^2y = (3 + 4 - 1)x^2y$
$= 6x^2y$

In Example 1(c), we cannot combine $16m^2$ and $5m$. These two terms are unlike because the exponents on the variables are different. *Unlike terms* have different variables or different exponents on the same variables.

◀◀◀ **Work Problem 1 at the Side.**

OBJECTIVE 2 Know the vocabulary for polynomials. A **polynomial in x** is a term or the sum of a finite number of terms of the form ax^n, for any real number a and any whole number n. For example,

$$16x^8 - 7x^6 + 5x^4 - 3x^2 + 4$$ Polynomial

is a polynomial in x. This polynomial is written in **descending powers**, because the exponents on x decrease from left to right. On the other hand,

$$2x^3 - x^2 + \frac{4}{x}$$ Not a polynomial

is not a polynomial, since a variable appears in a denominator. Of course, we could define a *polynomial* using any variable, not just x, as in Example 1(c). In fact, polynomials may have terms with more than one variable, as in Example 1(d).

◀◀◀ **Work Problem 2 at the Side.**

---

1 Add like terms.

(a) $5x^4 + 7x^4$

(b) $9pq + 3pq - 2pq$

(c) $r^2 + 3r + 5r^2$

(d) $8t + 6w$

2 Choose all descriptions that apply for each of the expressions in parts (a)–(d).

A. Polynomial
B. Polynomial written in descending powers
C. Not a polynomial

(a) $3m^5 + 5m^2 - 2m + 1$

(b) $2p^4 + p^6$

(c) $\dfrac{1}{x} + 2x^2 + 3$

(d) $x - 3$

ANSWERS

1. (a) $12x^4$ (b) $10pq$ (c) $6r^2 + 3r$
   (d) cannot be added—unlike terms
2. (a) A and B (b) A (c) C (d) A and B

The **degree of a term** is the sum of the exponents on the variables. A constant term has degree 0. For example, $3x^4$ has degree 4, while $6x^{17}$ has degree 17. The term $5x$ (or $5x^1$) has degree 1, -7 has degree 0, and $2x^2y$ has degree $2 + 1 = 3$ (y has an exponent of 1). The **degree of a polynomial** is the greatest degree of any nonzero term of the polynomial. For example, $3x^4 - 5x^2 + 6$ is of degree 4, the polynomial $5x + 7$ is of degree 1, 3 is of degree 0, and $x^2y + xy - 5xy^2$ is of degree 3.

Three types of polynomials are very common and are given special names. A polynomial with only one term is called a **monomial**. (*Mono-* means "one," as in *mono*rail.) Examples are

$$9m, \quad -6y^5, \quad a^2, \quad \text{and} \quad 6. \quad \text{Monomials}$$

A polynomial with exactly two terms is called a **binomial**. (*Bi-* means "two," as in *bi*cycle.) Examples are

$$-9x^4 + 9x^3, \quad 8m^2 + 6m, \quad \text{and} \quad 3m^5 - 9m^2. \quad \text{Binomials}$$

A polynomial with exactly three terms is called a **trinomial**. (*Tri-* means "three," as in *tri*angle.) Examples are

$$9m^3 - 4m^2 + 6, \quad \frac{19}{3}y^2 + \frac{8}{3}y + 5, \quad \text{and} \quad -3m^5 - 9m^2 + 2. \quad \text{Trinomials}$$

EXAMPLE 2 Classifying Polynomials

For each polynomial, first simplify if possible by combining like terms. Then give the degree and tell whether the simplified polynomial is a *monomial*, a *binomial*, a *trinomial*, or *none of these*.

(a) $2x^3 + 5$

The polynomial cannot be simplified. The degree is 3. The polynomial is a binomial.

(b) $4x - 5x + 2x$

Add like terms to simplify: $4x - 5x + 2x = x$. The degree is 1 (since $x = x^1$). The simplified polynomial is a monomial.

> **Work Problem 3 at the Side.**

OBJECTIVE 3 Evaluate polynomials. A polynomial usually represents different numbers for different values of the variable.

EXAMPLE 3 Evaluating a Polynomial

Find the value of $3x^4 + 5x^3 - 4x - 4$ when $x = -2$ and when $x = 3$.

First, substitute -2 for x.

$$3x^4 + 5x^3 - 4x - 4 = 3(-2)^4 + 5(-2)^3 - 4(-2) - 4$$
$$= 3 \cdot 16 + 5(-8) - 4(-2) - 4 \quad \text{Apply exponents.}$$
$$= 48 - 40 + 8 - 4 \quad \text{Multiply.}$$
$$= 12 \quad \text{Add and subtract.}$$

Next, replace x with 3.

$$3x^4 + 5x^3 - 4x - 4 = 3(3)^4 + 5(3)^3 - 4(3) - 4$$
$$= 3 \cdot 81 + 5 \cdot 27 - 4(3) - 4$$
$$= 243 + 135 - 12 - 4$$
$$= 362$$

---

3 For each polynomial, first simplify if possible. Then give the degree and tell whether the simplified polynomial is a *monomial*, *binomial*, *trinomial*, or *none of these*.

(a) $3x^2 + 2x - 4$

(b) $x^3 + 4x^3$

(c) $x^8 - x^7 + 2x^8$

ANSWERS
3. **(a)** degree 2; trinomial
   (b) simplify to $5x^3$; degree 3; monomial
   (c) simplify to $3x^8 - x^7$; degree 8; binomial

866 Chapter 13 Exponents and Polynomials

④ Find the value of $2y^3 + 8y - 6$ in each case.

(a) when $y = -1$

(b) when $y = 4$

⑤ Add each pair of polynomials vertically.

(a) $4x^3 - 3x^2 + 2x$
$6x^3 + 2x^2 - 3x$

(b) $x^2 - 2x + 5$
and $4x^2 - 2$

ANSWERS
4. (a) -16 (b) 154
5. (a) $10x^3 - x^2 - x$ (b) $5x^2 - 2x + 3$

> **CAUTION**
> Notice the use of parentheses around the numbers that are substituted for the variable in Example 3. This is particularly important when substituting a negative number for a variable that is raised to a power, so the sign of the product is correct.

◀◀ Work Problem 4 at the Side.

OBJECTIVE 4 Add polynomials. Polynomials may be added, subtracted, multiplied, and divided.

> **Adding Polynomials**
> To add two polynomials, add like terms.

EXAMPLE 4 Adding Polynomials Vertically

(a) Add $6x^3 - 4x^2 + 3$ and $-2x^3 + 7x^2 - 5$ vertically.
Write like terms in columns.

$$6x^3 - 4x^2 + 3$$
$$-2x^3 + 7x^2 - 5$$

Now add, column by column.

$$\begin{array}{ccc} 6x^3 & -4x^2 & 3 \\ -2x^3 & 7x^2 & -5 \\ \hline 4x^3 & 3x^2 & -2 \end{array}$$

Add the three sums together.

$$4x^3 + 3x^2 + (-2) = 4x^3 + 3x^2 - 2$$

(b) Add $2x^2 - 4x + 3$ and $x^3 + 5x$ vertically.
Write like terms in columns and add column by column.

$$\begin{array}{r} 2x^2 - 4x + 3 \\ x^3 + 5x \\ \hline x^3 + 2x^2 + x + 3 \end{array}$$ Leave spaces for missing terms.

◀◀ Work Problem 5 at the Side.

The polynomials in Example 4 also could be added horizontally.

EXAMPLE 5 Adding Polynomials Horizontally

(a) Add $6x^3 - 4x^2 + 3$ and $-2x^3 + 7x^2 - 5$ horizontally.
Combine like terms.

$$(6x^3 - 4x^2 + 3) + (-2x^3 + 7x^2 - 5) = 4x^3 + 3x^2 - 2$$

The sum is the same as in Example 4(a) above.

Continued on Next Page

(b) Add $2x^2 - 4x + 3$ and $x^3 + 5x$ horizontally.

$$(2x^2 - 4x + 3) + (x^3 + 5x) = 2x^2 - 4x + 3 + x^3 + 5x$$
$$= x^3 + 2x^2 + x + 3 \quad \text{Combine like terms.}$$

Work Problem 6 at the Side. ▶▶▶

6 Find each sum horizontally.

(a) $(2x^4 - 6x^2 + 7)$
    $+ (-3x^4 + 5x^2 + 2)$

OBJECTIVE 5 **Subtract polynomials.** In **Section 1.4,** the difference $x - y$ was defined as $x + (-y)$. That is, we find the difference $x - y$ by adding x and the opposite of y. For example,

$$7 - 2 = 7 + (-2) = 5 \quad \text{and} \quad -8 - (-2) = -8 + 2 = -6.$$

A similar method is used to subtract polynomials.

Subtracting Polynomials

To subtract two polynomials, change all the signs of the second polynomial and add the result to the first polynomial.

(b) $(3x^2 + 4x + 2)$
    $+ (6x^3 - 5x - 7)$

EXAMPLE 6 Subtracting Polynomials

(a) Perform the subtraction $(5x - 2) - (3x - 8)$.
Change the signs in the second polynomial and add like terms.

$$(5x - 2) - (3x - 8) = (5x - 2) + (-3x + 8)$$
$$= 2x + 6$$

(b) Subtract $6x^3 - 4x^2 + 2$ from $11x^3 + 2x^2 - 8$.
Rewrite the problem with the polynomials in the correct order.

$$(11x^3 + 2x^2 - 8) - (6x^3 - 4x^2 + 2)$$

Change all the signs in the second polynomial and add the two polynomials.

$$(11x^3 + 2x^2 - 8) + (-6x^3 + 4x^2 - 2) = 5x^3 + 6x^2 - 10$$

To check a subtraction problem, use the fact that if

$$a - b = c, \quad \text{then} \quad a = b + c.$$

For example, $6 - 2 = 4$, so we check by writing $6 = 2 + 4$, which is correct. Check the polynomial subtraction above by adding $6x^3 - 4x^2 + 2$ and $5x^3 + 6x^2 - 10$.

$$(6x^3 - 4x^2 + 2) + (5x^3 + 6x^2 - 10) = 11x^3 + 2x^2 - 8$$

Since the sum is $11x^3 + 2x^2 - 8$, the subtraction was performed correctly.

7 Subtract, and check your answers by addition.

(a) $(14y^3 - 6y^2 + 2y - 5)$
    $- (2y^3 - 7y^2 - 4y + 6)$

(b) Subtract
$$\left(-\frac{3}{2}y^2 + \frac{4}{3}y + 6\right)$$
from $\left(\frac{7}{2}y^2 - \frac{11}{3}y + 8\right)$.

Work Problem 7 at the Side. ▶▶▶

Subtraction also can be done in columns. We will use vertical subtraction in **Section 13.7** when we study polynomial division.

ANSWERS
6. (a) $-x^4 - x^2 + 9$
   (b) $6x^3 + 3x^2 - x - 5$
7. (a) $12y^3 + y^2 + 6y - 11$
   (b) $5y^2 - 5y + 2$

Chapter 13 Exponents and Polynomials

8 Subtract, using the method of subtracting by columns.

$(4y^3 - 16y^2 + 2y)$
$- (12y^3 - 9y^2 + 16)$

EXAMPLE 7 Subtracting Polynomials Vertically

Use the method of subtracting by columns to find

$$(14y^3 - 6y^2 + 2y - 5) - (2y^3 - 7y^2 - 4y + 6).$$

Arrange like terms in columns.

$$14y^3 - 6y^2 + 2y - 5$$
$$2y^3 - 7y^2 - 4y + 6$$

Change all signs in the second row, and then add.

$$14y^3 - 6y^2 + 2y - 5$$
$$\underline{-2y^3 + 7y^2 + 4y - 6} \quad \text{Change signs.}$$
$$12y^3 + y^2 + 6y - 11 \quad \text{Add.}$$

◀◀ **Work Problem 8 at the Side.**

9 Perform the indicated operations.

$(6p^4 - 8p^3 + 2p - 1)$
$- (-7p^4 + 6p^2 - 12)$
$+ (p^4 - 3p + 8)$

Either the horizontal or the vertical method may be used for adding or subtracting polynomials.

EXAMPLE 8 Adding and Subtracting More Than Two Polynomials

Perform the indicated operations to simplify the expression

$$(4 - x + 3x^2) - (2 - 3x + 5x^2) + (8 + 2x - 4x^2).$$

Rewrite, changing the subtraction to adding the opposite.

$(4 - x + 3x^2) - (2 - 3x + 5x^2) + (8 + 2x - 4x^2)$
$= (4 - x + 3x^2) + (-2 + 3x - 5x^2) + (8 + 2x - 4x^2)$
$= (2 + 2x - 2x^2) + (8 + 2x - 4x^2)$ Combine like terms.
$= 10 + 4x - 6x^2$ Combine like terms.

◀◀ **Work Problem 9 at the Side.**

10 Add or subtract.

(a) $(3mn + 2m - 4n)$
$+ (-mn + 4m + n)$

OBJECTIVE 6 Add and subtract polynomials with more than one variable. Polynomials in more than one variable are added and subtracted by combining like terms, just as with single-variable polynomials.

(b) $(5p^2q^2 - 4p^2 + 2q)$
$- (2p^2q^2 - p^2 - 3q)$

EXAMPLE 9 Adding and Subtracting Multivariable Polynomials

Add or subtract as indicated.

(a) $(4a + 2ab - b) + (3a - ab + b)$
$= 4a + 2ab - b + 3a - ab + b$
$= 7a + ab$ Combine like terms.

(b) $(2x^2y + 3xy + y^2) - (3x^2y - xy - 2y^2)$
$= 2x^2y + 3xy + y^2 - 3x^2y + xy + 2y^2$
$= -x^2y + 4xy + 3y^2$

◀◀ **Work Problem 10 at the Side.**

ANSWERS
8. $-8y^3 - 7y^2 + 2y - 16$
9. $14p^4 - 8p^3 - 6p^2 - p + 19$
10. (a) $2mn + 6m - 3n$
    (b) $3p^2q^2 - 3p^2 + 5q$

13.1 Exercises

Fill in each blank with the correct response.

1. In the term $7x^5$, the coefficient is _____ and the exponent is _____.
2. The expression $5x^3 - 4x^2$ has _____ term(s).
   (how many?)
3. The degree of the term $-4x^8$ is _____.
4. The polynomial $4x^2 - y^2$ _____ an example of a trinomial.
   (is/is not)
5. When $x^2 + 10$ is evaluated for $x = 4$, the result is _____.
6. _____ is an example of a monomial with coefficient 5, in the variable x, having degree 9.

For each polynomial, state the number of terms, and name the coefficient of each term.

7. $6x^4$
8. $-9y^5$
9. t^4
10. s^7
11. $-19r^2 - r$
12. $2y^3 - y$
13. $x + 8x^2$
14. $v - 2v^3$

In each polynomial, combine like terms whenever possible. Write the result with descending powers.

15. $-3m^5 + 5m^5$
16. $-4y^3 + 3y^3$
17. $2r^5 + (-3r^5)$
18. $-19y^2 + 9y^2$
19. $0.2m^5 - 0.5m^2$
20. $-0.9y + 0.9y^2$
21. $-3x^5 + 2x^5 - 4x^5$
22. $6x^3 - 8x^3 + 9x^3$
23. $-4p^7 + 8p^7 + 5p^9$
24. $-3a^8 + 4a^8 - 3a^2$
25. $-4y^2 + 3y^2 - 2y^2 + y^2$
26. $3r^5 - 8r^5 + r^5 + 2r^5$

For each polynomial, first simplify, if possible, and write it with descending powers. Then give the degree of the simplified polynomial, and tell whether it is a monomial, binomial, trinomial, *or* none of these. *See Example 2.*

27. $6x^4 - 9x$
28. $7t^3 - 3t$
29. $5m^4 - 3m^2 + 6m^5 - 7m^3$

30. $6p^5 + 4p^3 - 8p^4 + 10p^2$

31. $\dfrac{5}{3}x^4 - \dfrac{2}{3}x^4 + \dfrac{1}{3}x^2 - 4$

32. $\dfrac{4}{5}r^6 + \dfrac{1}{5}r^6 - r^4 + \dfrac{2}{5}r$

33. $0.8x^4 - 0.3x^4 - 0.5x^4 + 7$

34. $1.2t^3 - 0.9t^3 - 0.3t^3 + 9$

*Find the value of each polynomial **(a)** when $x = 2$ and **(b)** when $x = -1$. See Example 3.*

35. $-2x + 3$

36. $5x - 4$

37. $2x^2 + 5x + 1$

38. $-3x^2 + 14x - 2$

39. $2x^5 - 4x^4 + 5x^3 - x^2$

40. $x^4 - 6x^3 + x^2 + 1$

41. $-4x^5 + x^2$

42. $2x^6 - 4x$

RELATING CONCEPTS (EXERCISES 43–46) For Individual or Group Work

A polynomial can model the distance in feet that a car going approximately 68 mph will skid in t seconds. If we let D represent this distance, then

$$D = 100t - 13t^2.$$

*Each time we evaluate this polynomial for a value of t, we get one and only one output value D. This idea is basic to the concept of a **function**, an important concept in mathematics. Exercises 43–46 illustrate this idea with this polynomial and two others. **Work them in order.***

43. Use the given polynomial to approximate the skidding distance in feet if $t = 5$ sec.

44. Use the polynomial equation $D = 100t - 13t^2$ to find the distance the car will skid in 1 sec. Write an ordered pair of the form (t, D).

45. If gasoline costs $2.45 per gal, then the monomial $2.45x$ gives the cost, in dollars, of x gallons. How much would 4 gal cost?

46. If it costs $15 plus $2 per day to rent a chain saw, the binomial $2x + 15$ gives the cost in dollars to rent the chain saw for x days. How much would it cost to rent the saw for 6 days?

Add or subtract as indicated. See Examples 4 and 7.

47. Add.
$$3m^2 + 5m$$
$$2m^2 - 2m$$

48. Add.
$$4a^3 - 4a^2$$
$$6a^3 + 5a^2$$

49. Subtract.
$$12x^4 - x^2$$
$$8x^4 + 3x^2$$

50. Subtract.
$$13y^5 - y^3$$
$$7y^5 + 5y^3$$

51. Add.
$$\frac{2}{3}x^2 + \frac{1}{5}x + \frac{1}{6}$$
$$\frac{1}{2}x^2 - \frac{1}{3}x + \frac{2}{3}$$

52. Add.
$$\frac{4}{7}y^2 - \frac{1}{5}y + \frac{7}{9}$$
$$\frac{1}{3}y^2 - \frac{1}{3}y + \frac{2}{5}$$

53. Subtract.
$$12m^3 - 8m^2 + 6m + 7$$
$$5m^2 - 4$$

54. Subtract.
$$5a^4 - 3a^3 + 2a^2 - a + 6$$
$$-6a^4 - a^2 + a - 1$$

Perform the indicated operations. See Examples 5, 6, and 8.

55. $(2r^2 + 3r - 12) + (6r^2 + 2r)$

56. $(3r^2 + 5r - 6) + (2r - 5r^2)$

57. $(8m^2 - 7m) - (3m^2 + 7m - 6)$

58. $(x^2 + x) - (3x^2 + 2x - 1)$

59. $(16x^3 - x^2 + 3x) + (-12x^3 + 3x^2 + 2x)$

60. $(-2b^6 + 3b^4 - b^2) + (b^6 + 2b^4 + 2b^2)$

61. $(7y^4 + 3y^2 + 2y) - (18y^5 - 5y^3 + y)$

62. $(8t^5 + 3t^3 + 5t) - (19t^4 - 6t^2 + t)$

63. $[(8m^2 + 4m - 7) - (2m^3 - 5m + 2)] - (m^2 + m)$

64. $[(9b^3 - 4b^2 + 3b + 2) - (-2b^3 + b)] - (8b^3 + 6b + 4)$

Find the perimeter of each geometric figure.

65.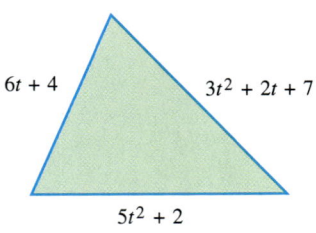

rectangle: top $4x^2 + 3x + 1$, side $x + 2$

66.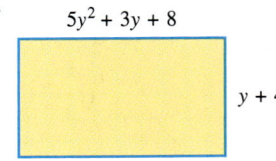

rectangle: top $5y^2 + 3y + 8$, side $y + 4$

67.

triangle: left side $6t + 4$, right side $3t^2 + 2t + 7$, bottom $5t^2 + 2$

68.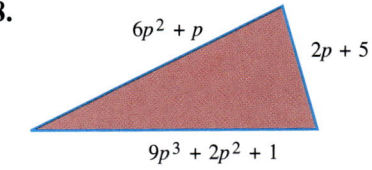

triangle: top-left $6p^2 + p$, top-right $2p + 5$, bottom $9p^3 + 2p^2 + 1$

69. Subtract $9x^2 - 3x + 7$ from $-2x^2 - 6x + 4$.

70. Subtract $-5w^3 + 5w^2 - 7$ from $6w^3 + 8w + 5$.

71. Explain why the degree of the term 3^4 is not 4. What is its degree?

72. Can the sum of two polynomials in x, both of degree 3, be of degree 2? If so, give an example.

Add or subtract as indicated. See Example 9.

73. $(9a^2b - 3a^2 + 2b) + (4a^2b - 4a^2 - 3b)$

74. $(4xy^3 - 3x + y) + (5xy^3 + 13x - 4y)$

75. $(2c^4d + 3c^2d^2 - 4d^2) - (c^4d + 8c^2d^2 - 5d^2)$

76. $(3k^2h^3 + 5kh + 6k^3h^2) - (2k^2h^3 - 9kh + k^3h^2)$

77. Subtract.
$$\begin{array}{r} 9m^3n - 5m^2n^2 + 4mn^2 \\ -3m^3n + 6m^2n^2 + 8mn^2 \\ \hline \end{array}$$

78. Subtract.
$$\begin{array}{r} 12r^5t + 11r^4t^2 - 7r^3t^3 \\ -8r^5t + 10r^4t^2 + 3r^3t^3 \\ \hline \end{array}$$

13.2 The Product Rule and Power Rules for Exponents

OBJECTIVES
1. Review the use of exponents.
2. Use the product rule for exponents.
3. Use the rule $(a^m)^n = a^{mn}$.
4. Use the rule $(ab)^m = a^m b^m$.
5. Use the rule $\left(\dfrac{a}{b}\right)^m = \dfrac{a^m}{b^m}$.
6. Use combinations of the rules for exponents.
7. Use the rules for exponents in an application of geometry.

OBJECTIVE 1 **Review the use of exponents.** In Section 1.8 we used exponents to write repeated products. Recall that in the expression 5^2, the number 5 is called the *base* and 2 is called the *exponent* or *power*. The expression 5^2 is called an *exponential expression*. Although we do not usually write a quantity with an exponent of 1, in general, for any quantity a, we can write it as a^1.

EXAMPLE 1 Review of Using Exponents

Write $3 \cdot 3 \cdot 3 \cdot 3 \cdot 3$ in exponential form and evaluate.

Since 3 occurs as a factor five times, the base is **3** and the exponent is **5**. The exponential expression is 3^5, read "3 to the fifth power" or simply "3 to the fifth." To evaluate, actually do the multiplication.

$$3^5 = 3 \cdot 3 \cdot 3 \cdot 3 \cdot 3 = 243$$

▶ **Work Problem 1 at the Side.**

① Write $2 \cdot 2 \cdot 2 \cdot 2$ in exponential form and evaluate.

EXAMPLE 2 Evaluating Exponential Expressions

Evaluate each exponential expression. Name the base and the exponent.

	Base	Exponent
(a) $5^4 = 5 \cdot 5 \cdot 5 \cdot 5 = 625$	5	4
(b) $-5^4 = -1 \cdot 5^4 = -1 \cdot (5 \cdot 5 \cdot 5 \cdot 5) = -625$	5	4
(c) $(-5)^4 = (-5)(-5)(-5)(-5) = 625$	-5	4

CAUTION
Notice the difference between Examples 2(b) and (c). In -5^4 the lack of parentheses shows that the exponent 4 applies only to the base 5, and not -5. In $(-5)^4$ the parentheses show that the exponent 4 applies to the base -5. In summary, $-a^n$ and $(-a)^n$ are not always the same.

Expression	Base	Exponent	Example
$-a^n$	a	n	$-3^2 = -(3 \cdot 3) = -9$
$(-a)^n$	$-a$	n	$(-3)^2 = (-3)(-3) = 9$

② Evaluate each exponential expression. Name the base and the exponent.

(a) $(-2)^5$ (b) -2^5

(c) -4^2 (d) $(-4)^2$

▶ **Work Problem 2 at the Side.**

OBJECTIVE 2 **Use the product rule for exponents.** To develop the product rule, we use the definition of an exponential expression.

$$2^4 \cdot 2^3 = \underbrace{(2 \cdot 2 \cdot 2 \cdot 2)}_{\text{4 factors}} \underbrace{(2 \cdot 2 \cdot 2)}_{\text{3 factors}}$$

$$= \underbrace{2 \cdot 2 \cdot 2 \cdot 2 \cdot 2 \cdot 2 \cdot 2}_{\text{4 factors + 3 factors = 7 factors}}$$

$$= 2^7$$

ANSWERS
1. $2^4 = 16$
2. (a) -32; -2; 5 (b) -32; 2; 5
   (c) -16; 4; 2 (d) 16; -4; 2

3 Find each product by the product rule, if possible.

(a) $8^2 \cdot 8^5$

(b) $(-7)^5 \cdot (-7)^3$

(c) $y^3 \cdot y$

(d) $4^2 \cdot 3^5$

(e) $6^4 + 6^2$

Also, $6^2 \cdot 6^3 = (6 \cdot 6)(6 \cdot 6 \cdot 6)$
$= 6 \cdot 6 \cdot 6 \cdot 6 \cdot 6$
$= 6^5.$

Generalizing from these examples, $2^4 \cdot 2^3 = 2^{4+3} = 2^7$ and $6^2 \cdot 6^3 = 6^{2+3} = 6^5$. In each case, adding the exponents gives the exponent of the product, suggesting the **product rule for exponents.**

> **Product Rule for Exponents**
>
> For any positive integers m and n,
>
> $$a^m \cdot a^n = a^{m+n}$$
>
> (Keep the same base and add the exponents.)
>
> *Example:* $6^2 \cdot 6^5 = 6^{2+5} = 6^7$

> **CAUTION**
>
> Avoid the common error of multiplying the bases when using the product rule.
>
> $6^2 \cdot 6^5 \neq 36^7$ ← **Error** $6^2 \cdot 6^5 = 6^7$ ← **Correct**
>
> *Keep the same base and add the exponents.*

EXAMPLE 3 Using the Product Rule

Use the product rule for exponents to find each product, if possible.

(a) $6^3 \cdot 6^5 = 6^{3+5} = 6^8$ by the product rule.

(b) $(-4)^7(-4)^2 = (-4)^{7+2} = (-4)^9$ by the product rule.

(c) $x^2 \cdot x = x^2 \cdot x^1 = x^{2+1} = x^3$

(d) $m^4 \cdot m^3 = m^{4+3} = m^7$

(e) $2^3 \cdot 3^2$

The product rule does *not* apply to the product $2^3 \cdot 3^2$ because the *bases are different.*

$$2^3 \cdot 3^2 = 8 \cdot 9 = 72$$

> **CAUTION**
>
> *The bases of the factors must be the same* before we can apply the product rule for exponents.

(f) $2^3 + 2^4$

The product rule does *not* apply to $2^3 + 2^4$ because it is a *sum,* not a *product.*

$$2^3 + 2^4 = 8 + 16 = 24$$

◀ **Work Problem 3 at the Side.**

ANSWERS
3. (a) 8^7 (b) $(-7)^8$ (c) y^4
   (d) different bases; cannot use the product rule (product: 3888)
   (e) it's a sum; cannot use the product rule

Section 13.2 The Product Rule and Power Rules for Exponents **875**

EXAMPLE 4 Using the Product Rule

Multiply $2x^3$ and $3x^7$.

We use the associative and commutative properties and the product rule.

$2x^3 \cdot 3x^7 = 2 \cdot 3 \cdot x^3 \cdot x^7 = 6x^{10}$ $2x^3 = 2 \cdot x^3; 3x^7 = 3 \cdot x^7$

CAUTION
Be sure you understand the difference between *adding* and *multiplying* exponential expressions. Here is a comparison.

$8x^3 + 5x^3 = (8 + 5)x^3 = 13x^3$ **Adding expressions**

$(8x^3)(5x^3) = (8 \cdot 5)x^{3+3} = 40x^6$ **Multiplying expressions**

Work Problem 4 at the Side.

OBJECTIVE 3 Use the rule $(a^m)^n = a^{mn}$. We can simplify an expression such as $(8^3)^2$ with the product rule for exponents, as follows.

$(8^3)^2 = (8^3)(8^3) = 8^{3+3} = 8^6$

The product of the exponents in $(8^3)^2$, $3 \cdot 2$, gives the exponent in 8^6. Also,

$(5^2)^4 = 5^2 \cdot 5^2 \cdot 5^2 \cdot 5^2$ Definition of exponent
$= 5^{2+2+2+2}$ Product rule
$= 5^8,$

and $2 \cdot 4 = 8$. These examples suggest **power rule (a) for exponents.**

Power Rule (a) for Exponents
For any positive integers m and n,
$$(a^m)^n = a^{mn}$$
(Raise a power to a power by multiplying exponents.)
Example: $(3^2)^4 = 3^{2 \cdot 4} = 3^8$

EXAMPLE 5 Using Power Rule (a)

Use power rule (a) for exponents to simplify each expression.

(a) $(2^5)^3 = 2^{5 \cdot 3} = 2^{15}$ **(b)** $(5^7)^2 = 5^{7 \cdot 2} = 5^{14}$
(c) $(x^2)^5 = x^{2 \cdot 5} = x^{10}$ **(d)** $(n^3)^2 = n^{3 \cdot 2} = n^6$

Work Problem 5 at the Side.

OBJECTIVE 4 Use the rule $(ab)^m = a^m b^m$. We can rewrite the expression $(4x)^3$ as shown below.

$(4x)^3 = (4x)(4x)(4x)$ Definition of exponent
$= 4 \cdot 4 \cdot 4 \cdot x \cdot x \cdot x$ Commutative and associative properties
$= 4^3 x^3$ Definition of exponent

This example suggests **power rule (b) for exponents.**

4 Multiply.

(a) $5m^2 \cdot 2m^6$

(b) $3p^5 \cdot 9p^4$

(c) $-7p^5 \cdot (3p^8)$

5 Simplify each expression.

(a) $(5^3)^4$

(b) $(6^2)^5$

(c) $(3^2)^4$

(d) $(a^6)^5$

ANSWERS
4. (a) $10m^8$ (b) $27p^9$ (c) $-21p^{13}$
5. (a) 5^{12} (b) 6^{10} (c) 3^8 (d) a^{30}

6 Simplify.

(a) $5(mn)^3$

(b) $(3a^2b^4)^5$

(c) $(-5m^2)^3$

Power Rule (b) for Exponents

For any positive integer m,
$$(ab)^m = a^m b^m.$$
(Raise a product to a power by raising each factor to the power.)

Example: $(2p)^5 = 2^5 p^5$

EXAMPLE 6 Using Power Rule (b)

Use power rule (b) to simplify each expression.

(a) $(3xy)^2 = 3^2 x^2 y^2$ Power rule (b)
$ = 9x^2 y^2$

(b) $9(pq)^2 = 9(p^2 q^2)$ Power rule (b)
$ = 9p^2 q^2$

(c) $5(2m^2 p^3)^4 = 5[2^4 (m^2)^4 (p^3)^4]$ Power rule (b)
$ = 5(2^4 m^8 p^{12})$ Power rule (a)
$ = 5 \cdot 2^4 m^8 p^{12}$
$ = 80 m^8 p^{12}$ $5 \cdot 2^4 = 5 \cdot 16 = 80$

(d) $(-5^6)^3 = (-1 \cdot 5^6)^3$ $-a = -1 \cdot a$
$ = (-1)^3 (5^6)^3$ Power rule (b)
$ = -1 \cdot 5^{18}$ Power rule (a)
$ = -5^{18}$

CAUTION
Power rule (b) does not apply to a sum.
$$(x + 4)^2 \neq x^2 + 4^2 \leftarrow \text{Error}$$
You will learn how to work with $(x + 4)^2$ in **Section 13.4**.

◀◀◀ **Work Problem 6 at the Side.**

OBJECTIVE 5 Use the rule $\left(\frac{a}{b}\right)^m = \frac{a^m}{b^m}$. Since the quotient $\frac{a}{b}$ can be written as $a \cdot \frac{1}{b}$, we can use power rule (b), together with some of the properties of real numbers, to get **power rule (c) for exponents**.

Power Rule (c) for Exponents

For any positive integer m,
$$\left(\frac{a}{b}\right)^m = \frac{a^m}{b^m} \quad \text{as long as } b \neq 0.$$

(Raise a quotient to a power by raising both the numerator and the denominator to the power.)

Example: $\left(\frac{5}{3}\right)^2 = \frac{5^2}{3^2}$

ANSWERS
6. (a) $5m^3 n^3$ (b) $3^5 a^{10} b^{20}$ or $243 a^{10} b^{20}$
    (c) $-5^3 m^6$ or $-125 m^6$

EXAMPLE 7 Using Power Rule (c)

Simplify each expression.

(a) $\left(\dfrac{2}{3}\right)^5 = \dfrac{2^5}{3^5}$ or $\dfrac{32}{243}$

(b) $\left(\dfrac{m}{n}\right)^4 = \dfrac{m^4}{n^4}$ when $n \neq 0$

Work Problem 7 at the Side.

7 Simplify. Assume all variables represent nonzero real numbers.

(a) $\left(\dfrac{5}{2}\right)^4$

Next we list the rules for exponents discussed in this section. These rules are basic to the study of algebra and should be *memorized*.

Rules for Exponents

For positive integers m and n:

		Examples
Product rule	$a^m \cdot a^n = a^{m+n}$	$6^2 \cdot 6^5 = 6^{2+5} = 6^7$
Power rule (a)	$(a^m)^n = a^{mn}$	$(3^2)^4 = 3^{2 \cdot 4} = 3^8$
Power rule (b)	$(ab)^m = a^m b^m$	$(2p)^5 = 2^5 p^5$
Power rule (c)	$\left(\dfrac{a}{b}\right)^m = \dfrac{a^m}{b^m}$ when $b \neq 0$	$\left(\dfrac{5}{3}\right)^2 = \dfrac{5^2}{3^2}$

(b) $\left(\dfrac{p}{q}\right)^2$

OBJECTIVE 6 Use combinations of the rules for exponents. In the next example, more than one rule is needed to simplify an expression.

EXAMPLE 8 Using Combinations of Rules

Simplify each expression.

(a) $\left(\dfrac{2}{3}\right)^2 \cdot 2^3 = \dfrac{2^2}{3^2} \cdot \dfrac{2^3}{1}$ Power rule (c)

$= \dfrac{2^2 \cdot 2^3}{3^2 \cdot 1}$ Multiply fractions.

$= \dfrac{2^5}{3^2}$ Product rule

(c) $\left(\dfrac{r}{t}\right)^3$

(b) $(5x)^3 (5x)^4 = (5x)^7$ Product rule

$= 5^7 x^7$ Power rule (b)

(c) $(2x^2 y^3)^4 (3xy^2)^3 = 2^4 (x^2)^4 (y^3)^4 \cdot 3^3 x^3 (y^2)^3$ Power rule (b)

$= 2^4 \cdot x^8 \cdot y^{12} \cdot 3^3 \cdot x^3 \cdot y^6$ Power rule (a)

$= 2^4 \cdot 3^3 x^8 x^3 y^{12} y^6$ Commutative and associative properties

$= 16 \cdot 27 x^{11} y^{18}$ Product rule

$= 432 x^{11} y^{18}$

Be careful in the first step when using power rule b.

$(2x^2 y^3)^4 = 2^4 x^{2 \cdot 4} y^{3 \cdot 4}$ ***not*** $(2 \cdot 4) x^{2 \cdot 4} y^{3 \cdot 4}$

Do not multiply the coefficient 2 and the exponent 4.

Continued on Next Page

ANSWERS

7. (a) $\dfrac{5^4}{2^4}$ or $\dfrac{625}{16}$ (b) $\dfrac{p^2}{q^2}$ (c) $\dfrac{r^3}{t^3}$

8 Simplify.

(a) $(2m)^3(2m)^4$

(b) $\left(\dfrac{5k^3}{3}\right)^2$

(c) $\left(\dfrac{1}{5}\right)^4 (2x)^2$

(d) $(-3xy^2)^3(x^2y)^4$

9 Find the area.

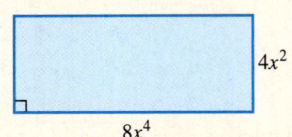

(d) $(-x^3y)^2(-x^5y^4)^3$

Think of the negative sign in each factor as -1.

$$\begin{aligned}(-1x^3y)^2(-1x^5y^4)^3 &= (-1)^2(x^3)^2y^2 \cdot (-1)^3(x^5)^3(y^4)^3 && \text{Power rule (b)}\\ &= (-1)^2(x^6)(y^2)(-1)^3(x^{15})(y^{12}) && \text{Power rule (a)}\\ &= (-1)^5(x^{21})(y^{14}) && \text{Product rule}\\ &= -1x^{21}y^{14}\\ &= -x^{21}y^{14}\end{aligned}$$

▶◀ **Work Problem 8 at the Side.**

OBJECTIVE 7 Use the rules for exponents in an application of geometry.

EXAMPLE 9 Using an Area Formula

Find an expression that represents the area of each geometric figure in Figure 1.

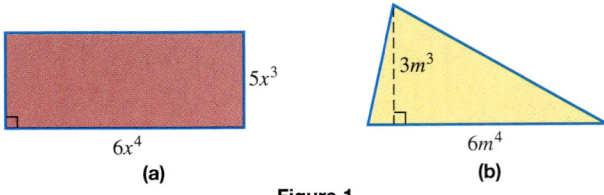

Figure 1

(a) Use the formula for the area of a rectangle, $A = lw$.

$$A = (6x^4)(5x^3) \quad \text{Area formula}$$
$$A = 30x^7 \quad \text{Product rule}$$

(b) This is a triangle with base $6m^4$ and height $3m^3$.

$$\begin{aligned}A &= \dfrac{1}{2}bh && \text{Area formula}\\ &= \dfrac{1}{2}(6m^4)(3m^3) && \text{Substitute.}\\ &= \dfrac{1}{2}(18m^7) && \text{Product rule}\\ &= 9m^7\end{aligned}$$

▶◀ **Work Problem 9 at the Side.**

ANSWERS

8. (a) 2^7m^7 or $128m^7$ (b) $\dfrac{5^2k^6}{3^2}$ or $\dfrac{25k^6}{9}$
(c) $\dfrac{2^2x^2}{5^4}$ or $\dfrac{4x^2}{625}$
(d) $-3^3x^{11}y^{10}$ or $-27x^{11}y^{10}$

9. $32x^6$

13.2 Exercises

1. What exponent is understood on the base x in the expression xy^2?

2. How are the expressions 3^2, 5^3, and 7^4 read?

Decide whether each statement is true or false.

3. $3^3 = 9$
4. $(-2)^4 = 2^4$
5. $(a^2)^3 = a^5$
6. $\left(\dfrac{1}{4}\right)^2 = \dfrac{1}{4^2}$

Write each expression using exponents. See Example 1.

7. $(-2)(-2)(-2)(-2)(-2)$
8. $w \cdot w \cdot w \cdot w \cdot w \cdot w$
9. $\left(\dfrac{1}{2}\right)\left(\dfrac{1}{2}\right)\left(\dfrac{1}{2}\right)\left(\dfrac{1}{2}\right)\left(\dfrac{1}{2}\right)\left(\dfrac{1}{2}\right)$
10. $\left(-\dfrac{1}{4}\right)\left(-\dfrac{1}{4}\right)\left(-\dfrac{1}{4}\right)\left(-\dfrac{1}{4}\right)\left(-\dfrac{1}{4}\right)$
11. $(-8p)(-8p)$
12. $(-7x)(-7x)(-7x)(-7x)$

13. Explain how the expressions $(-3)^4$ and -3^4 are different.

14. Explain how the expressions $(5x)^3$ and $5x^3$ are different.

Identify the base and the exponent for each exponential expression. In Exercises 15–18, also evaluate the expression. See Example 2.

15. 3^5
16. 2^7
17. $(-3)^5$
18. $(-2)^7$
19. $(-6x)^4$
20. $(-8x)^4$
21. $-6x^4$
22. $-8x^4$

23. Explain why the product rule does not apply to the expression $5^2 + 5^3$. Then evaluate the expression by finding the individual powers and adding the results.

24. Explain why the product rule does not apply to the expression $3^2 \cdot 4^3$. Then evaluate the expression by finding the individual powers and multiplying the results.

Use the product rule to simplify each expression. Write each answer in exponential form. See Examples 3 and 4.

25. $5^2 \cdot 5^6$
26. $3^6 \cdot 3^7$
27. $4^2 \cdot 4^7 \cdot 4^3$
28. $5^3 \cdot 5^8 \cdot 5^2$
29. $(-7)^3(-7)^6$
30. $(-9)^8(-9)^5$
31. $t^3 \cdot t^8 \cdot t^{13}$
32. $n^5 \cdot n^6 \cdot n^9$
33. $(-8r^4)(7r^3)$
34. $(10a^7)(-4a^3)$
35. $(-6p^5)(-7p^5)$
36. $(-5w^8)(-9w^8)$

880 Chapter 13 Exponents and Polynomials

Use the power rules for exponents to simplify each expression. Write each answer in exponential form. See Examples 5–7.

37. $(4^3)^2$
38. $(8^3)^6$
39. $(t^4)^5$
40. $(y^6)^5$

41. $(7r)^3$
42. $(11x)^4$
43. $(5xy)^5$
44. $(9pq)^6$

45. $8(qr)^3$
46. $4(vw)^5$
47. $\left(\dfrac{1}{2}\right)^3$
48. $\left(\dfrac{1}{3}\right)^5$

49. $\left(\dfrac{a}{b}\right)^3$ $(b \neq 0)$
50. $\left(\dfrac{r}{t}\right)^4$ $(t \neq 0)$
51. $\left(\dfrac{9}{5}\right)^8$
52. $\left(\dfrac{12}{7}\right)^3$

53. $(-2x^2y)^3$
54. $(-5m^4p^2)^3$
55. $(3a^3b^2)^2$
56. $(4x^3y^5)^4$

Use a combination of the rules for exponents introduced in this section to simplify each expression. See Example 8.

57. $\left(\dfrac{5}{2}\right)^3 \cdot \left(\dfrac{5}{2}\right)^2$
58. $\left(\dfrac{3}{4}\right)^5 \cdot \left(\dfrac{3}{4}\right)^6$
59. $\left(\dfrac{9}{8}\right)^3 \cdot 9^2$
60. $\left(\dfrac{8}{5}\right)^4 \cdot 8^3$

61. $(2x)^9(2x)^3$
62. $(6y)^5(6y)^8$
63. $(-6p)^4(-6p)$
64. $(-13q)^3(-13q)$

65. $(6x^2y^3)^5$
66. $(5r^5t^6)^7$
67. $(x^2)^3(x^3)^5$
68. $(y^4)^5(y^3)^5$

69. $(2w^2x^3y)^2(x^4y)^5$
70. $(3x^4y^2z)^3(yz^4)^5$
71. $(-r^4s)^2(-r^2s^3)^5$

72. $(-ts^6)^4(-t^3s^5)^3$
73. $\left(\dfrac{5a^2b^5}{c^6}\right)^3$ when $c \neq 0$
74. $\left(\dfrac{6x^3y^9}{z^5}\right)^4$ when $z \neq 0$

75. $(-5m^3p^4q)^2(p^2q)^3$
76. $(-a^4b^5)(-6a^3b^3)^2$
77. $(2x^2y^3z)^4(xy^2z^3)^2$

Find the area of each figure. See Example 9.

78.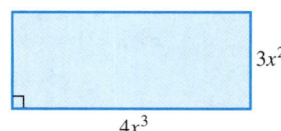

79.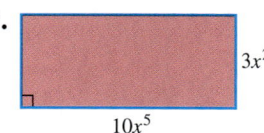

80.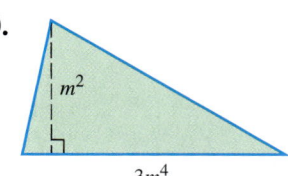

13.3 Multiplying Polynomials

OBJECTIVE 1 Multiply a monomial and a polynomial. As shown in Section 13.2, we find the product of two monomials by using the rules for exponents and the commutative and associative properties. For example,

$$(-8m^6)(-9n^6) = (-8)(-9)(m^6)(n^6) = 72m^6n^6.$$

OBJECTIVES

1. Multiply a monomial and a polynomial.
2. Multiply two polynomials.
3. Multiply binomials by the FOIL method.

CAUTION
Do not confuse addition of terms with multiplication of terms.
$7q^5 + 2q^5 = 9q^5$, but $(7q^5)(2q^5) = 7 \cdot 2q^{5+5} = 14q^{10}$.

To find the product of a monomial and a polynomial with more than one term, we use the distributive property and multiplication of monomials.

① Find each product.

(a) $5m^3(2m + 7)$

EXAMPLE 1 Multiplying a Monomial and a Polynomial

Use the distributive property to find each product.

(a) $4x^2(3x + 5)$

$4x^2(3x + 5) = 4x^2(3x) + 4x^2(5)$ Distributive property
$= 12x^3 + 20x^2$ Multiply monomials.

(b) $-8m^3(4m^3 + 3m^2 + 2m - 1)$
$= -8m^3(4m^3) + (-8m^3)(3m^2)$
$\quad + (-8m^3)(2m) + (-8m^3)(-1)$ Distributive property
$= -32m^6 - 24m^5 - 16m^4 + 8m^3$ Multiply monomials.

(b) $2x^4(3x^2 + 2x - 5)$

Work Problem 1 at the Side.

OBJECTIVE 2 Multiply two polynomials. We can use the distributive property repeatedly to find the product of any two polynomials. For example, to find the product of the polynomials $x^2 + 3x + 5$ and $x - 4$, think of $x - 4$ as a single quantity and use the distributive property as follows.

$$(x^2 + 3x + 5)(x - 4) = x^2(x - 4) + 3x(x - 4) + 5(x - 4)$$

Now use the distributive property three more times to find the products $x^2(x - 4)$, $3x(x - 4)$, and $5(x - 4)$.

$x^2(x - 4) + 3x(x - 4) + 5(x - 4)$
$= x^2(x) + x^2(-4) + 3x(x) + 3x(-4) + 5(x) + 5(-4)$
$= x^3 - 4x^2 + 3x^2 - 12x + 5x - 20$ Multiply monomials.
$= x^3 - x^2 - 7x - 20$ Combine terms.

This example suggests the following rule.

(c) $-4y^2(3y^3 + 2y^2 - 4y + 8)$

Multiplying Polynomials
To multiply two polynomials, multiply each term of the second polynomial by each term of the first polynomial and add the products.

ANSWERS
1. **(a)** $10m^4 + 35m^3$
   (b) $6x^6 + 4x^5 - 10x^4$
   (c) $-12y^5 - 8y^4 + 16y^3 - 32y^2$

Chapter 13 Exponents and Polynomials

② Multiply.

(a) $(m^3 - 2m + 1)$
 $\cdot (2m^2 + 4m + 3)$

(b) $(6p^2 + 2p - 4)(3p^2 - 5)$

③ Find the product.

$3x^2 + 4x - 5$
$x + 4$

EXAMPLE 2 Multiplying Two Polynomials

Multiply $(m^2 + 5)(4m^3 - 2m^2 + 4m)$.

Multiply each term of the second polynomial by each term of the first.

$(\boldsymbol{m^2} + \boldsymbol{5})(4m^3 - 2m^2 + 4m)$
$= \boldsymbol{m^2}(4m^3) + \boldsymbol{m^2}(-2m^2) + \boldsymbol{m^2}(4m) + \boldsymbol{5}(4m^3) + \boldsymbol{5}(-2m^2) + \boldsymbol{5}(4m)$
$= 4m^5 - 2m^4 + 4m^3 + 20m^3 - 10m^2 + 20m$
$= 4m^5 - 2m^4 + 24m^3 - 10m^2 + 20m$ Combine like terms.

◀◀ Work Problem 2 at the Side.

When at least one of the factors in a product of polynomials has three or more terms, it is often easier to write one polynomial above the other vertically.

EXAMPLE 3 Multiplying Polynomials Vertically

Multiply $(x^3 + 2x^2 + 4x + 1)(3x + 5)$ using the vertical method.

Write the polynomials as follows.

$$\begin{array}{r} x^3 + 2x^2 + 4x + 1 \\ 3x + 5 \\ \hline \end{array}$$

It is not necessary to line up terms in columns, because any terms may be multiplied (not just like terms). Begin by multiplying each of the terms in the top row by 5.

$$\begin{array}{r} x^3 + 2x^2 + 4x + 1 \\ 3x + \boldsymbol{5} \\ \hline 5x^3 + 10x^2 + 20x + 5 \end{array}$$ $5(x^3 + 2x^2 + 4x + 1)$

Notice how this process is similar to multiplication of whole numbers. Now multiply each term in the top row by $3x$. Be careful to place like terms in columns, since the final step will involve addition (as in multiplying two whole numbers).

$$\begin{array}{r} \color{red}{x^3 + 2x^2 + 4x + 1} \\ \boldsymbol{3x} + 5 \\ \hline 5x^3 + 10x^2 + 20x + 5 \\ \color{red}{3x^4 + 6x^3 + 12x^2 + 3x} \end{array}$$ $\color{red}{3x(x^3 + 2x^2 + 4x + 1)}$

Finally, add like terms.

$$\begin{array}{r} x^3 + 2x^2 + 4x + 1 \\ 3x + 5 \\ \hline 5x^3 + 10x^2 + 20x + 5 \\ 3x^4 + 6x^3 + 12x^2 + 3x \\ \hline 3x^4 + 11x^3 + 22x^2 + 23x + 5 \end{array}$$

The product is $3x^4 + 11x^3 + 22x^2 + 23x + 5$.

◀◀ Work Problem 3 at the Side.

ANSWERS

2. (a) $2m^5 + 4m^4 - m^3 - 6m^2 - 2m + 3$
 (b) $18p^4 + 6p^3 - 42p^2 - 10p + 20$
3. $3x^3 + 16x^2 + 11x - 20$

EXAMPLE 4 Multiplying Polynomials Vertically

Find the product of $4m^3 - 2m^2 + 4m$ and $\frac{1}{2}m^2 + \frac{5}{2}$.

$$\begin{array}{r} 4m^3 - 2m^2 + 4m \\ \frac{1}{2}m^2 + \frac{5}{2} \\ \hline 10m^3 - 5m^2 + 10m \\ 2m^5 - m^4 + 2m^3 \\ \hline 2m^5 - m^4 + 12m^3 - 5m^2 + 10m \end{array}$$

Terms of top row multiplied by $\frac{5}{2}$
Terms of top row multiplied by $\frac{1}{2}m^2$
Add.

Work Problem 4 at the Side. ▶▶▶

We can use a rectangle to model polynomial multiplication. For example, to find the product

$$(2x + 1)(3x + 2),$$

label a rectangle with each term as shown below on the left. Then put the product of each pair of monomials in the appropriate box as shown on the right.

	3x	2
2x		
1		

	3x	2
2x	$6x^2$	$4x$
1	$3x$	2

The product of the binomials is the sum of these four monomial products.

$$(2x + 1)(3x + 2) = 6x^2 + 4x + 3x + 2$$
$$= 6x^2 + 7x + 2$$

Work Problem 5 at the Side. ▶▶▶

OBJECTIVE 3 Multiply binomials by the FOIL method. In algebra, many of the polynomials to be multiplied are both binomials (with just two terms). For these products, the **FOIL method** reduces the rectangle method to a systematic approach without the rectangle. To develop the FOIL method, we use the distributive property to find $(x + 3)(x + 5)$.

$$(x + 3)(x + 5) = (x + 3)x + (x + 3)5$$
$$= x(x) + 3(x) + x(5) + 3(5)$$
$$= x^2 + 3x + 5x + 15$$
$$= x^2 + 8x + 15$$

Here is where the letters of the word FOIL originate.

$(x + 3)(x + 5)$ Multiply the **First terms**: $x(x)$.   **F**

$(x + 3)(x + 5)$ Multiply the **Outer terms**: $x(5)$.   **O**
   This is the **outer product**.

$(x + 3)(x + 5)$ Multiply the **Inner terms**: $3(x)$.   **I**
   This is the **inner product**.

$(x + 3)(x + 5)$ Multiply the **Last terms**: $3(5)$.   **L**

The inner product and the outer product should be added mentally so that the three terms of the answer can be written without extra steps as

$$(x + 3)(x + 5) = x^2 + 8x + 15.$$

---

4 Find each product.

(a) $k^3 - k^2 + k + 1$
  $\frac{2}{3}k - \frac{1}{3}$

(b) $a^3 + 3a - 4$
  $2a^2 + 6a + 5$

5 Use the rectangle method to find each product.

(a) $(4x + 3)(x + 2)$

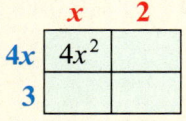

(b) $(x + 5)(x^2 + 3x + 1)$

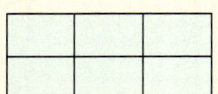

ANSWERS
4. (a) $\frac{2}{3}k^4 - k^3 + k^2 + \frac{1}{3}k - \frac{1}{3}$
   (b) $2a^5 + 6a^4 + 11a^3 + 10a^2 - 9a - 20$
5. (a) $4x^2 + 11x + 6$
   (b) $x^3 + 8x^2 + 16x + 5$

884 Chapter 13 Exponents and Polynomials

❻ For the product $(2p - 5)(3p + 7)$, find the following.

(a) Product of first terms

(b) Outer product

(c) Inner product

(d) Product of last terms

(e) Complete product in simplified form

❼ Use the FOIL method to find each product.

(a) $(m + 4)(m - 3)$

(b) $(y + 7)(y + 2)$

(c) $(r - 8)(r - 5)$

ANSWERS
6. (a) $2p(3p) = 6p^2$ (b) $2p(7) = 14p$
   (c) $-5(3p) = -15p$ (d) $-5(7) = -35$
   (e) $6p^2 - p - 35$
7. (a) $m^2 + m - 12$ (b) $y^2 + 9y + 14$
   (c) $r^2 - 13r + 40$

A summary of the steps in the FOIL method follows.

> **Multiplying Binomials by the FOIL Method**
>
> *Step 1* Multiply the two **F**irst terms of the binomials to get the first term of the answer.
>
> *Step 2* Find the **O**uter product and the **I**nner product and combine them (when possible) to get the middle term of the answer.
>
> *Step 3* Multiply the two **L**ast terms of the binomials to get the last term of the answer.
>
>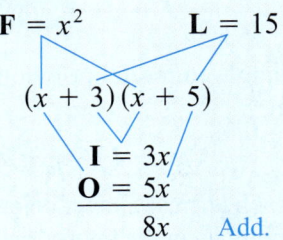
>
> Product is $x^2 + 8x + 15$.

◀◀◀ **Work Problem 6 at the Side.**

EXAMPLE 5 Using the FOIL Method

Use the FOIL method to find the product $(x + 8)(x - 6)$.

Step 1 **F** Multiply the **first** terms.
$$x(x) = x^2$$

Step 2 **O** Find the **outer** product.
$$x(-6) = -6x$$

I Find the **inner** product.
$$8(x) = 8x$$

Add the outer and inner products mentally.
$$-6x + 8x = 2x$$

Step 3 **L** Multiply the **last** terms.
$$8(-6) = -48$$

The product of $x + 8$ and $x - 6$ is the sum of the terms found in the three steps above, so
$$(x + 8)(x - 6) = x^2 + 2x - 48.$$

As a shortcut, this product can be found in the following manner.

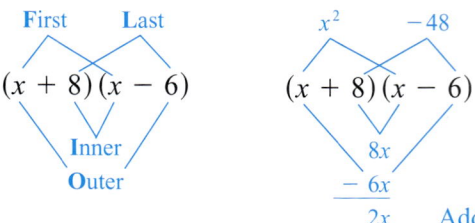

The product is again $x^2 + 2x - 48$.

◀◀◀ **Work Problem 7 at the Side.**

It is not possible to add the inner and outer products of the FOIL method if unlike terms result, as shown in the next example.

EXAMPLE 6 Using the FOIL Method

Multiply $(9x - 2)(3y + 1)$.

First	$(\mathbf{9x} - 2)(\mathbf{3y} + 1)$	$27xy$
Outer	$(\mathbf{9x} - 2)(3y + \mathbf{1})$	$9x$
Inner	$(9x - \mathbf{2})(\mathbf{3y} + 1)$	$-6y$
Last	$(9x - \mathbf{2})(3y + \mathbf{1})$	-2

⎤ Unlike terms (for Outer and Inner)

$$\text{F}\text{O}\text{I}\text{L}$$
$$(9x - 2)(3y + 1) = 27xy + 9x - 6y - 2$$

Work Problem 8 at the Side.

EXAMPLE 7 Using the FOIL Method

Find each product.

$$\text{F}\text{O}\text{I}\text{L}$$

(a) $(2k + 5y)(k + 3y) = 2k(k) + 2k(3y) + 5y(k) + 5y(3y)$
$$= 2k^2 + 6ky + 5ky + 15y^2$$
$$= 2k^2 + 11ky + 15y^2$$

(b) $(7p + 2q)(3p - q) = 21p^2 - pq - 2q^2$ FOIL

(c) $2x^2(x - 3)(3x + 4) = 2x^2(3x^2 - 5x - 12)$ FOIL
$$= 6x^4 - 10x^3 - 24x^2$$ Distributive property

Work Problem 9 at the Side.

NOTE

Example 7(c) showed one way to multiply three polynomials. We could have multiplied $2x^2$ and $x - 3$ first, then multiplied that product and $3x + 4$ as follows.

$$2x^2(x - 3)(3x + 4) = (2x^3 - 6x^2)(3x + 4)$$
$$= 6x^4 - 10x^3 - 24x^2$$

---

8 Find the product.

$(4x - 3)(2y + 5)$

9 Find each product.

(a) $(6m + 5)(m - 4)$

(b) $(3r + 2t)(3r + 4t)$

(c) $y^2(8y + 3)(2y + 1)$

ANSWERS
8. $8xy + 20x - 6y - 15$
9. **(a)** $6m^2 - 19m - 20$
   (b) $9r^2 + 18rt + 8t^2$
   (c) $16y^4 + 14y^3 + 3y^2$

Focus on Real-Data Applications

Algebra as Generalized Arithmetic

The rules of algebra are consistent with those for arithmetic. To learn a method for multiplying binomial factors, such as $(3x + 2)(4x + 1)$, we can observe the method for multiplying two-digit numbers, such as $32 \cdot 41$. The number 32 is shorthand for the expanded number $3 \cdot 10 + 2$, and the number 41 is shorthand for $4 \cdot 10 + 1$. So, multiplying $32 \cdot 41$ is the same as multiplying $(3 \cdot 10 + 2)(4 \cdot 10 + 1)$.

1. An expanded version of the usual algorithm is shown in the first column below. Each partial product, such as 1×2 and 40×30, is shown to clarify how it contributes to the process.
2. In the standard algorithm, it is clear that the 32 represents the sum of 1×2 and 1×30, and 1280 represents the sum of 40×2 and 40×30.
3. When FOIL is used to multiply the numbers, the term $11 \cdot 10$ represents the sum of the partial products 1×30 and 40×2. When we simplify $12 \cdot 10^2 + 11 \cdot 10 + 2$, the result is 1312.
4. When FOIL is used to multiply the binomials, each term exactly matches the corresponding term from the multiplication of the numbers.

Partial Products Multiplication Algorithm	Standard Algorithm	FOIL Method for Numbers	FOIL Method for Variables
32	32	$(3 \cdot 10 + 2)(4 \cdot 10 + 1)$	$(3x + 2)(4x + 1)$
$\times\ 41$	$\times\ 41$	$12 \cdot 10^2 + 3 \cdot 10 + 8 \cdot 10 + 2$	$12x^2 + 3x + 8x + 2$
$2 = 1 \times 2$	32	$12 \cdot 10^2 + 11 \cdot 10 + 2$	$12x^2 + 11x + 2$
$30 = 1 \times 30$	1280	$1200 + 110 + 2$	
$80 = 40 \times 2$	1312	1312	
$1200 = 40 \times 30$			
1312			

For Group Discussion

Use FOIL to compute each binomial product and corresponding arithmetic product. Verify that the results of the arithmetic product are valid. For the numerical problems, write the correct *signed* product for each term.

1. $(2x + 1)(2x - 3)$ and $(2 \cdot 10 + 1)(2 \cdot 10 - 3)$, which is $21 \cdot 17$
   Does the arithmetic FOIL result simplify to the correct answer?

2. $(4x - 2)(2x + 5)$ and $(4 \cdot 10 - 2)(2 \cdot 10 + 5)$, which is $38 \cdot 25$
   Does the arithmetic FOIL result simplify to the correct answer?

3. $(7x - 3)(5x - 4)$ and $(7 \cdot 10 - 3)(5 \cdot 10 - 4)$, which is $67 \cdot 46$
   Does the arithmetic FOIL result simplify to the correct answer?

4. A mental trick for multiplying two-digit numbers, such as $27 \cdot 18$, follows: "Multiply the ones $(7 \times 8 = 56)$. Write the 6 in the ones place, and carry the 5. Add the inner product (7×1), the outer product (2×8), and the carried 5 $(7 + 16 + 5 = 28)$. Write the 8 in the tens place, and carry the 2. Multiply the tens (2×1), and add to the carried 2. Write 4 in the hundreds place. The answer is 486." Why does the trick work?

13.3 Exercises

Find each product using the rectangle method shown in the text.

1. $(x + 3)(x + 4)$
2. $(x + 5)(x + 2)$
3. $(2x + 1)(x^2 + 3x + 2)$
4. $(x + 4)(3x^2 + 2x + 1)$

5. In multiplying a monomial by a polynomial, such as in $4x(3x^2 + 7x^3) = 4x(3x^2) + 4x(7x^3)$, the first property that is used is the _____ property.

6. Match each product in parts (a)–(d) with the correct polynomial in choices A–D.
   (a) $(x - 5)(x + 3)$ (b) $(x + 5)(x + 3)$ (c) $(x - 5)(x - 3)$ (d) $(x + 5)(x - 3)$

   A. $x^2 + 8x + 15$ B. $x^2 - 8x + 15$ C. $x^2 - 2x - 15$ D. $x^2 + 2x - 15$

Find each product. See Example 1.

7. $-2m(3m + 2)$
8. $-5p(6 + 3p)$
9. $\frac{3}{4}p(8 - 6p + 12p^3)$
10. $\frac{4}{3}x(3 + 2x + 5x^3)$
11. $2y^5(3 + 2y + 5y^4)$
12. $2m^4(3m^2 + 5m + 6)$

Find each product. See Examples 2–4.

13. $(6x + 1)(2x^2 + 4x + 1)$
14. $(9y - 2)(8y^2 - 6y + 1)$
15. $(4m + 3)(5m^3 - 4m^2 + m - 5)$
16. $(y + 4)(3y^3 - 2y^2 + y + 3)$
17. $(2x - 1)(3x^5 - 2x^3 + x^2 - 2x + 3)$
18. $(2a + 3)(a^4 - a^3 + a^2 - a + 1)$
19. $(5x^2 + 2x + 1)(x^2 - 3x + 5)$
20. $(2m^2 + m - 3)(m^2 - 4m + 5)$

Find each binomial product using the FOIL method. See Examples 5–7.

21. $(n - 2)(n + 3)$
22. $(r - 6)(r + 8)$
23. $(4r + 1)(2r - 3)$
24. $(5x + 2)(2x - 7)$
25. $(3x + 2)(3x - 2)$
26. $(7x + 3)(7x - 3)$

27. $(3q + 1)(3q + 1)$

28. $(4w + 7)(4w + 7)$

29. $(3t + 4s)(2t + 5s)$

30. $(8v + 5w)(2v + 3w)$

31. $(-0.3t + 0.4)(t + 0.6)$

32. $(-0.5x + 0.9)(x - 0.2)$

33. $\left(x - \dfrac{2}{3}\right)\left(x + \dfrac{1}{4}\right)$

34. $\left(-\dfrac{8}{3} + 3k\right)\left(-\dfrac{2}{3} - k\right)$

35. $\left(-\dfrac{5}{4} + 2r\right)\left(-\dfrac{3}{4} - r\right)$

36. $2m^3(4m - 1)(2m + 3)$

37. $3y^3(2y + 3)(y - 5)$

38. $5t^4(t + 3)(3t - 1)$

RELATING CONCEPTS (EXERCISES 39–44) For Individual or Group Work

Work Exercises 39–44 in order. (All units are in yards.) *Refer to the figure as necessary.*

A rectangle with width 10 and length $3x + 6$.

39. Find a polynomial that represents the area of the rectangle.

40. Suppose you know that the area of the rectangle is 600 yd². Use this information and the polynomial from Exercise 39 to write an equation in x, and solve it.

41. What are the dimensions of the rectangle?

42. Suppose the rectangle represents a lawn and it costs $3.50 per square yard to lay sod on the lawn. How much will it cost to sod the entire lawn?

43. Use the result of Exercise 41 to find the perimeter of the lawn.

44. Again, suppose the rectangle represents a lawn and it costs $9.00 per yard to fence the lawn. How much will it cost to fence the lawn?

45. Perform the following multiplications: $(x + 4)(x - 4)$; $(y + 2)(y - 2)$; $(r + 7)(r - 7)$. Observe your answers, and explain the pattern that can be found in the answers.

46. Repeat Exercise 45 for the following: $(x + 4)(x + 4)$; $(y - 2)(y - 2)$; $(r + 7)(r + 7)$.

13.4 Special Products

In this section, we develop shortcuts to find certain binomial products that occur frequently.

OBJECTIVES

1. Square binomials.
2. Find the product of the sum and difference of two terms.
3. Find greater powers of binomials.

OBJECTIVE 1 Square binomials. The square of a binomial can be found quickly by using the method shown in Example 1.

EXAMPLE 1 Squaring a Binomial

Find $(m + 3)^2$.

Squaring $m + 3$ by the FOIL method gives

$$(m + 3)(m + 3) = m^2 + 3m + 3m + 9$$
$$= m^2 + 6m + 9.$$

This result has the squares of the first and the last terms of the binomial:

$$m^2 = m^2 \quad \text{and} \quad 3^2 = 9.$$

The middle term, $6m$, is twice the product of the two terms of the binomial, since the outer and inner products are $m(3)$ and $3(m)$, and

$$m(3) + 3(m) = 2(m)(3) = 6m.$$

Work Problem 1 at the Side.

Example 1 suggests the following rules.

Square of a Binomial

The square of a binomial is a trinomial consisting of the square of the first term, plus twice the product of the two terms, plus the square of the last term of the binomial. For a and b,

$$(a + b)^2 = a^2 + 2ab + b^2.$$

Also,

$$(a - b)^2 = a^2 - 2ab + b^2.$$

EXAMPLE 2 Squaring Binomials

Use the rules to square each binomial.

$$(a - b)^2 = a^2 - 2 \cdot a \cdot b + b^2$$

(a) $(5z - 1)^2 = (5z)^2 - 2(5z)(1) + (1)^2$
$= 25z^2 - 10z + 1$ $(5z)^2 = 5^2 z^2 = 25z^2$

(b) $(3b + 5r)^2 = (3b)^2 + 2(3b)(5r) + (5r)^2$
$= 9b^2 + 30br + 25r^2$

(c) $(2a - 9x)^2 = 4a^2 - 36ax + 81x^2$

(d) $\left(4m + \dfrac{1}{2}\right)^2 = (4m)^2 + 2(4m)\left(\dfrac{1}{2}\right) + \left(\dfrac{1}{2}\right)^2$

$= 16m^2 + 4m + \dfrac{1}{4}$

---

1 Consider the binomial $x + 4$.

(a) What is the first term of the binomial? Square it.

(b) What is the last term of the binomial? Square it.

(c) Find twice the product of the two terms of the binomial.

(d) Find $(x + 4)^2$.

ANSWERS

1. (a) x; x^2 (b) 4; 16   (c) $8x$
   (d) $x^2 + 8x + 16$

2 Find each square by using the rules for the square of a binomial.

(a) $(t + u)^2$

(b) $(2m - p)^2$

(c) $(4p + 3q)^2$

(d) $(5r - 6s)^2$

(e) $\left(3k - \dfrac{1}{2}\right)^2$

ANSWERS
2. (a) $t^2 + 2tu + u^2$
 (b) $4m^2 - 4mp + p^2$
 (c) $16p^2 + 24pq + 9q^2$
 (d) $25r^2 - 60rs + 36s^2$
 (e) $9k^2 - 3k + \dfrac{1}{4}$

Notice that in the square of a sum, all of the terms are positive, as in Examples 2(b) and (d). In the square of a difference, the middle term is negative, as in Examples 2(a) and (c).

> **CAUTION**
> A common error when squaring a binomial is to forget the middle term of the product. In general,
> $$(a + b)^2 \ne a^2 + b^2.$$

◀◀◀ **Work Problem 2 at the Side.**

OBJECTIVE 2 Find the product of the sum and difference of two terms. Binomial products of the form $(a + b)(a - b)$ also occur frequently. In these products, one binomial is the sum of two terms, and the other is the difference of the same two terms. For example, the product of $x + 2$ and $x - 2$ is

$$(x + 2)(x - 2) = x^2 - 2x + 2x - 4$$
$$= x^2 - 4.$$

As this example suggests, the product of $a + b$ and $a - b$ is the difference between two squares.

> **Product of the Sum and Difference of Two Terms**
> $$(a + b)(a - b) = a^2 - b^2$$

EXAMPLE 3 Finding the Product of the Sum and Difference of Two Terms

Find each product.

(a) $(x + 4)(x - 4)$

Use the rule for the product of the sum and difference of two terms.
$$(x + 4)(x - 4) = x^2 - 4^2$$
$$= x^2 - 16$$

(b) $\left(\dfrac{2}{3} - w\right)\left(\dfrac{2}{3} + w\right)$

By the commutative property, this product is the same as $\left(\dfrac{2}{3} + w\right)\left(\dfrac{2}{3} - w\right)$.

$$\left(\dfrac{2}{3} - w\right)\left(\dfrac{2}{3} + w\right) = \left(\dfrac{2}{3} + w\right)\left(\dfrac{2}{3} - w\right)$$
$$= \left(\dfrac{2}{3}\right)^2 - w^2$$
$$= \dfrac{4}{9} - w^2$$

(c) $x(x + 2)(x - 2) = x(x^2 - 4)$
$$= x^3 - 4x$$

Section 13.4 Special Products

EXAMPLE 4 Finding the Product of the Sum and Difference of Two Terms

Find each product.

$$(a + b)(a - b)$$

(a) $(5m + 3)(5m - 3)$

Use the rule for the product of the sum and difference of two terms.

$$(5m + 3)(5m - 3) = (5m)^2 - 3^2$$
$$= 25m^2 - 9$$

(b) $(4x + y)(4x - y) = (4x)^2 - y^2$
$$= 16x^2 - y^2$$

(c) $\left(z - \dfrac{1}{4}\right)\left(z + \dfrac{1}{4}\right) = z^2 - \dfrac{1}{16}$

(d) $2p(p^2 + 3)(p^2 - 3) = 2p(p^4 - 9)$
$$= 2p^5 - 18p$$

> Work Problem 3 at the Side.

The product rules of this section will be important later, particularly in **Chapters 14** and **15**. Therefore, it is important to learn these rules and practice using them.

OBJECTIVE 3 Find greater powers of binomials. The methods used in the previous section and this section can be combined to find greater powers of binomials.

EXAMPLE 5 Finding Greater Powers of Binomials

Find each product.

(a) $(x + 5)^3 = (x + 5)^2(x + 5)$ $a^3 = a^2 \cdot a$
$$= (x^2 + 10x + 25)(x + 5) \quad \text{Square the binomial.}$$
$$= x^3 + 10x^2 + 25x + 5x^2 + 50x + 125 \quad \text{Multiply polynomials.}$$
$$= x^3 + 15x^2 + 75x + 125 \quad \text{Combine like terms.}$$

(b) $(2y - 3)^4 = (2y - 3)^2(2y - 3)^2$ $a^4 = a^2 \cdot a^2$
$$= (4y^2 - 12y + 9)(4y^2 - 12y + 9) \quad \text{Square each binomial.}$$
$$= 16y^4 - 48y^3 + 36y^2 - 48y^3 + 144y^2 \quad \text{Multiply polynomials.}$$
$$ - 108y + 36y^2 - 108y + 81$$
$$= 16y^4 - 96y^3 + 216y^2 - 216y + 81 \quad \text{Combine like terms.}$$

(c) $-2r(r + 2)^3 = -2r(r + 2)(r + 2)^2$
$$= -2r(r + 2)(r^2 + 4r + 4)$$
$$= -2r(r^3 + 4r^2 + 4r + 2r^2 + 8r + 8)$$
$$= -2r(r^3 + 6r^2 + 12r + 8)$$
$$= -2r^4 - 12r^3 - 24r^2 - 16r$$

> Work Problem 4 at the Side.

---

3 Find each product by using the rule for the sum and difference of two terms.

(a) $(6a + 3)(6a - 3)$

(b) $(10m + 7)(10m - 7)$

(c) $(7p + 2q)(7p - 2q)$

(d) $\left(3r - \dfrac{1}{2}\right)\left(3r + \dfrac{1}{2}\right)$

(e) $3x(x^3 - 4)(x^3 + 4)$

4 Find each product.

(a) $(m + 1)^3$

(b) $(3k - 2)^4$

(c) $-3x(x - 4)^3$

ANSWERS
3. (a) $36a^2 - 9$ (b) $100m^2 - 49$
    (c) $49p^2 - 4q^2$ (d) $9r^2 - \dfrac{1}{4}$
    (e) $3x^7 - 48x$
4. (a) $m^3 + 3m^2 + 3m + 1$
    (b) $81k^4 - 216k^3 + 216k^2 - 96k + 16$
    (c) $-3x^4 + 36x^3 - 144x^2 + 192x$

Focus on Real-Data Applications

Using a Rule of Thumb

A Dynamic Homes neighborhood features variations of three models of houses. Each buyer has an option to purchase a concrete patio extension, constructed to his or her choice of size. Each of the three models is designed so that the patio will be in the shape of a large square that is missing a corner square, similar to the diagram shown on the left. The size of the removed corner and the length of the patio vary from model to model, as does the size of the entire patio. To determine the dimensions of the patio, the rectangle of size $b \times (a - b)$ in the middle patio diagram can be rotated and repositioned to form the rectangle shown on the right, with length $a + b$ and width $a - b$.

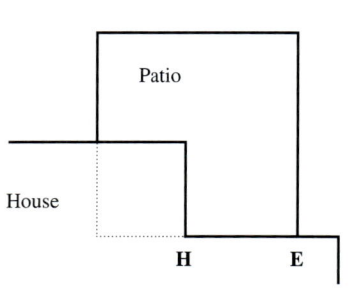
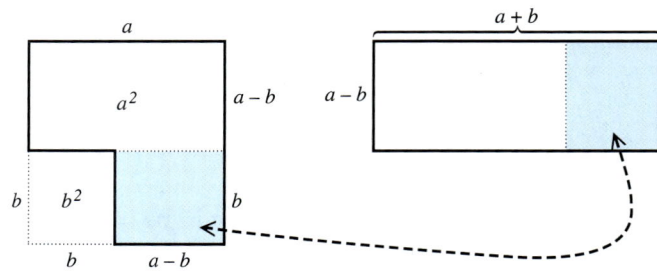

Since every time a house is sold the salesperson must ask the construction supervisor to calculate the cost of the patio extension, the supervisor devises a *rule of thumb* based on the quantity $(a + b)(a - b)$. There are only three possible choices for the value of b, one for each model house. This means that the salesperson needs to know only one additional measurement—the distance from the corner H to the edge of the patio E, which is $a - b$. The thickness of the patio is $\frac{1}{3}$ ft. The cost of concrete is $58.00 per yd^3. (*Source:* Dunham Price, Inc., Westlake, LA.) There is a 100% markup in price to cover profit as well as costs for ground preparation, building forms, and finishing the concrete, giving a cost factor of $116.00 per yd^3.

For Group Discussion

1. Why is knowing the values for b and $a - b$ sufficient for finding the quantity $a + b$?

2. Explain why the area of the patio can be written both as $(a + b)(a - b)$ and $a^2 - b^2$.

3. Apply the construction supervisor's rule of thumb to the three examples given in the table. To calculate the total cost for a patio with long side of length a feet, use the table and work from left to right. The second-to-last column is the volume of concrete needed; round up to the nearest tenth of a cubic yard. The last column is the total cost of the patio.

House Model	Corner Length b ft	Corner Length Doubled $2b$	Distance from Corner H to Patio Edge E $a - b$	Add Previous Two Columns $2b + (a - b)$	Multiply Previous Two Columns $(a + b)(a - b)$	VOLUME Multiply Previous Column by $\frac{1}{3}$ and Divide by 27	COST Multiply Previous Column by Cost Factor of $116.00
A	5		10				
B	6		12				
C	8		14				

13.4 Exercises

1. Consider the square $(2x + 3)^2$.
   (a) What is the square of the first term, $(2x)^2$? _____
   (b) What is twice the product of the two terms, $2(2x)(3)$? _____
   (c) What is the square of the last term, 3^2? _____
   (d) Write the final product, which is a trinomial, using your results from parts (a)–(c). _____

2. Repeat Exercise 1 for the square $(3x - 2)^2$.

Find each square. See Examples 1 and 2.

3. $(a - c)^2$
4. $(p - y)^2$
5. $(p + 2)^2$
6. $(r + 5)^2$

7. $(4x - 3)^2$
8. $(5y + 2)^2$
9. $(0.8t + 0.7s)^2$
10. $(0.7z - 0.3w)^2$

11. $\left(5x + \frac{2}{5}y\right)^2$
12. $\left(6m - \frac{4}{5}n\right)^2$
13. $\left(4a - \frac{3}{2}b\right)^2$
14. $x(2x + 5)^2$

15. $-(4r - 2)^2$
16. $-(3y - 8)^2$

17. Consider the product $(7x + 3y)(7x - 3y)$.
   (a) What is the product of the first terms, $7x(7x)$? _____
   (b) Multiply the outer terms, $7x(-3y)$. Then multiply the inner terms, $3y(7x)$. Add the results. What is this sum? _____
   (c) What is the product of the last terms, $3y(-3y)$? _____
   (d) Write the complete product using your answers in parts (a) and (c). _____
   Why is the sum found in part (b) omitted here?

18. Repeat Exercise 17 for the product $(5x + 7y)(5x - 7y)$.

Find each product. See Examples 3 and 4.

19. $(q + 2)(q - 2)$
20. $(x + 8)(x - 8)$
21. $(2w + 5)(2w - 5)$
22. $(3z + 8)(3z - 8)$

23. $(10x + 3y)(10x - 3y)$
24. $(13r + 2z)(13r - 2z)$
25. $(2x^2 - 5)(2x^2 + 5)$
26. $(9y^2 - 2)(9y^2 + 2)$

27. $\left(7x + \frac{3}{7}\right)\left(7x - \frac{3}{7}\right)$
28. $\left(9y + \frac{2}{3}\right)\left(9y - \frac{2}{3}\right)$
29. $p(3p + 7)(3p - 7)$
30. $q(5q - 1)(5q + 1)$

RELATING CONCEPTS (EXERCISES 31–40) For Individual or Group Work

Special products can be illustrated by using areas of rectangles. Use the figure and **work Exercises 31–36 in order,** *to justify the special product* $(a + b)^2 = a^2 + 2ab + b^2$.

31. Express the area of the large square as the square of a binomial.

32. Give the monomial that represents the area of the red square.

33. Give the monomial that represents the sum of the areas of the blue rectangles.

34. Give the monomial that represents the area of the yellow square.

35. What is the sum of the monomials you obtained in Exercises 32–34?

36. Explain why the binomial square you found in Exercise 31 must equal the polynomial you found in Exercise 35.

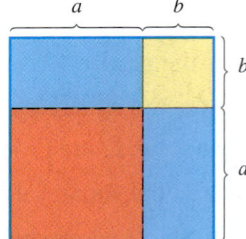

To understand how the special product $(a + b)^2 = a^2 + 2ab + b^2$ *can be applied to a purely numerical problem,* **work Exercises 37–40 in order.**

37. Evaluate 35^2 using either traditional paper-and-pencil methods or a calculator.

38. The number 35 can be written as $30 + 5$. Therefore, $35^2 = (30 + 5)^2$. Use the special product for squaring a binomial with $a = 30$ and $b = 5$ to write an expression for $(30 + 5)^2$. Do not simplify at this time.

39. Use the order of operations to simplify the expression you found in Exercise 38.

40. How do the answers in Exercises 37 and 39 compare?

Find each product. See Example 5.

41. $(m - 5)^3$ **42.** $(p + 3)^3$ **43.** $(2a + 1)^3$

44. $(3m - 1)^3$ **45.** $(3r - 2t)^4$ **46.** $(2z + 5y)^4$

47. $3x^2(x - 3)^3$ **48.** $4p^3(p + 4)^3$ **49.** $-8x^2y(x + y)^4$

In Exercises 50 and 51, refer to the figure shown here.

50. Find a polynomial that represents the volume of the cube.

51. If the value of x is 6, what is the volume of the cube?

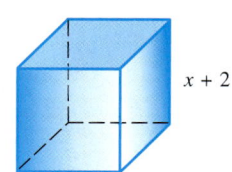

$x + 2$

13.5 Integer Exponents and the Quotient Rule

In all our earlier work, exponents were positive integers. Now we want to develop meaning for exponents that are not positive integers.

Consider the following list of exponential expressions.

$$2^4 = 16$$
$$2^3 = 8$$
$$2^2 = 4$$

Do you see the pattern in the values? Each time we reduce the exponent by 1, the value is divided by 2 (the base). Using this pattern, we can continue the list to smaller and smaller integer exponents.

$$2^1 = 2$$
$$2^0 = 1$$
$$2^{-1} = \frac{1}{2}$$

Work Problem 1 at the Side.

From the preceding list and the answers to Problem 1, it appears that we should define 2^0 as 1 and negative exponents as reciprocals.

OBJECTIVE 1 Use 0 as an exponent. We want the definitions of 0 and negative exponents to satisfy the rules for exponents from **Section 13.2**. For example, if $6^0 = 1$,

$$6^0 \cdot 6^2 = 1 \cdot 6^2 = 6^2 \quad \text{and} \quad 6^0 \cdot 6^2 = 6^{0+2} = 6^2,$$

so the product rule is satisfied. Check that the power rules are also valid for a 0 exponent. Thus, we define a 0 exponent as follows.

Zero Exponent

For any nonzero real number a, $a^0 = 1$.

Example: $17^0 = 1$.

EXAMPLE 1 Using Zero Exponents

Evaluate each exponential expression.

(a) $60^0 = 1$ (b) $(-60)^0 = 1$
(c) $-60^0 = -(1) = -1$ (d) $y^0 = 1, \quad y \neq 0$
(e) $6y^0 = 6(1) = 6, \quad y \neq 0$ (f) $(6y)^0 = 1, \quad y \neq 0$

CAUTION
Look again at Examples 1(b) and (c). In $(-60)^0$, the base is -60 and the exponent is 0. Any nonzero base raised to the exponent 0 is 1. In -60^0, the base is 60. Then $60^0 = 1$, and $-60^0 = -1$.

Work Problem 2 at the Side.

OBJECTIVES

1. Use 0 as an exponent.
2. Use negative numbers as exponents.
3. Use the quotient rule for exponents.
4. Use combinations of rules.

---

1 Continue the list of exponentials using $-2, -3,$ and -4 as exponents.

$2^{-2} =$ _____

$2^{-3} =$ _____

$2^{-4} =$ _____

2 Evaluate.

(a) 28^0

(b) $(-16)^0$

(c) -7^0

(d) $m^0, \quad m \neq 0$

(e) $-p^0, \quad p \neq 0$

ANSWERS

1. $2^{-2} = \frac{1}{4}; 2^{-3} = \frac{1}{8}; 2^{-4} = \frac{1}{16}$
2. (a) 1 (b) 1 (c) -1 (d) 1 (e) -1

OBJECTIVE 2 **Use negative numbers as exponents.** From the lists at the beginning of this section and margin problem 1, since $2^{-2} = \frac{1}{4}$ and $2^{-3} = \frac{1}{8}$, we can deduce that 2^{-n} should equal $\frac{1}{2^n}$. Is the product rule valid in such cases? For example, if we multiply 6^{-2} by 6^2, we get

$$6^{-2} \cdot 6^2 = 6^{-2+2} = 6^0 = 1.$$

The expression 6^{-2} behaves as if it were the reciprocal of 6^2, because their product is 1. The reciprocal of 6^2 may be written $\frac{1}{6^2}$, leading us to define 6^{-2} as $\frac{1}{6^2}$. This is a particular case of the definition of negative exponents.

> **Negative Exponents**
>
> For any nonzero real number a and any integer n,
>
> $$a^{-n} = \frac{1}{a^n}$$
>
> *Example:* $3^{-2} = \frac{1}{3^2}$

By definition, a^{-n} and a^n are reciprocals, since

$$a^n \cdot a^{-n} = a^n \cdot \frac{1}{a^n} = 1.$$

Since $1^n = 1$, the definition of a^{-n} can also be written

$$a^{-n} = \frac{1}{a^n} = \frac{1^n}{a^n} = \left(\frac{1}{a}\right)^n.$$

For example, $6^{-3} = \left(\frac{1}{6}\right)^3$ and $\left(\frac{1}{3}\right)^{-2} = 3^2$.

EXAMPLE 2 Using Negative Exponents

Simplify by writing each expression with positive exponents.

(a) $3^{-2} = \frac{1}{3^2} = \frac{1}{9}$

(b) $5^{-3} = \frac{1}{5^3} = \frac{1}{125}$

(c) $\left(\frac{1}{2}\right)^{-3} = 2^3 = 8$ $\frac{1}{2}$ and 2 are reciprocals.

Notice that we can change the base to its reciprocal if we also change the sign of the exponent.

(d) $\left(\frac{2}{5}\right)^{-4} = \left(\frac{5}{2}\right)^4$ $\frac{2}{5}$ and $\frac{5}{2}$ are reciprocals.

(e) $\left(\frac{4}{3}\right)^{-5} = \left(\frac{3}{4}\right)^5$

(f) $4^{-1} - 2^{-1} = \frac{1}{4} - \frac{1}{2} = \frac{1}{4} - \frac{2}{4} = -\frac{1}{4}$

Apply the exponents first, then subtract.

Continued on Next Page

(g) $p^{-2} = \dfrac{1}{p^2}, \quad p \neq 0$

(h) $\dfrac{1}{x^{-4}}, \quad x \neq 0$

$$\dfrac{1}{x^{-4}} = \dfrac{1^{-4}}{x^{-4}} \qquad 1^{-4} = 1$$

$$= \left(\dfrac{1}{x}\right)^{-4} \qquad \text{Power rule (c)}$$

$$= x^4 \qquad \tfrac{1}{x} \text{ and } x \text{ are reciprocals.}$$

CAUTION
A negative exponent does not indicate a negative number; negative exponents lead to reciprocals.

Expression	Example	
a^{-n}	$3^{-2} = \dfrac{1}{3^2} = \dfrac{1}{9}$	Not negative
$-a^{-n}$	$-3^{-2} = -\dfrac{1}{3^2} = -\dfrac{1}{9}$	Negative

Work Problem 3 at the Side.

The definition of negative exponents allows us to move factors in a fraction if we also change the signs of the exponents. Here is an example.

$$\dfrac{2^{-3}}{3^{-4}} = \dfrac{\tfrac{1}{2^3}}{\tfrac{1}{3^4}} = \dfrac{1}{2^3} \div \dfrac{1}{3^4} = \dfrac{1}{2^3} \cdot \dfrac{3^4}{1} = \dfrac{3^4}{2^3}$$

The final result is shown below.

$$\dfrac{2^{-3}}{3^{-4}} = \dfrac{3^4}{2^3}$$

Changing from Negative to Positive Exponents
For any nonzero numbers a and b, and any integers m and n,

$$\dfrac{a^{-m}}{b^{-n}} = \dfrac{b^n}{a^m} \quad \text{and} \quad \left(\dfrac{a}{b}\right)^{-m} = \left(\dfrac{b}{a}\right)^m$$

Examples: $\dfrac{3^{-5}}{2^{-4}} = \dfrac{2^4}{3^5}$ and $\left(\dfrac{4}{5}\right)^{-3} = \left(\dfrac{5}{4}\right)^3$

---

3 Write with positive exponents.

(a) 4^{-3}

(b) 6^{-2}

(c) $\left(\dfrac{2}{3}\right)^{-2}$

(d) $2^{-1} + 5^{-1}$

(e) $m^{-5}, \quad m \neq 0$

(f) $\dfrac{1}{z^{-4}}, \quad z \neq 0$

ANSWERS
3. (a) $\dfrac{1}{4^3}$ (b) $\dfrac{1}{6^2}$ (c) $\left(\dfrac{3}{2}\right)^2$
   (d) $\dfrac{1}{2} + \dfrac{1}{5} = \dfrac{7}{10}$ (e) $\dfrac{1}{m^5}$ (f) z^4

Chapter 13 Exponents and Polynomials

4 Write with only positive exponents. Assume all variables represent nonzero real numbers.

(a) $\dfrac{7^{-1}}{5^{-4}}$

(b) $\dfrac{x^{-3}}{y^{-2}}$

(c) $\dfrac{4h^{-5}}{m^{-2}k}$

(d) $p^2 q^{-5}$

(e) $\left(\dfrac{3m}{p}\right)^{-2}$

ANSWERS

4. (a) $\dfrac{5^4}{7}$ (b) $\dfrac{y^2}{x^3}$ (c) $\dfrac{4m^2}{h^5 k}$
   (d) $\dfrac{p^2}{q^5}$ (e) $\dfrac{p^2}{3^2 m^2}$

EXAMPLE 3 Changing from Negative to Positive Exponents

Write with only positive exponents. Assume all variables represent nonzero real numbers.

(a) $\dfrac{4^{-2}}{5^{-3}} = \dfrac{5^3}{4^2}$

(b) $\dfrac{m^{-5}}{p^{-1}} = \dfrac{p^1}{m^5} = \dfrac{p}{m^5}$

(c) $\dfrac{a^{-2}b}{3d^{-3}} = \dfrac{bd^3}{3a^2}$

Notice that b in the numerator and the coefficient 3 in the denominator were not affected.

(d) $x^3 y^{-4} = \dfrac{x^3 y^{-4}}{1} = \dfrac{x^3}{y^4}$ (e) $\left(\dfrac{x}{2y}\right)^{-4} = \left(\dfrac{2y}{x}\right)^4 = \dfrac{2^4 y^4}{x^4}$

▶ **Work Problem 4 at the Side.**

CAUTION

Be careful. We cannot change negative exponents to positive exponents using this rule if the exponents occur in a *sum* of terms. For example,

$$\dfrac{5^{-2} + 3^{-1}}{7 - 2^{-3}}$$

cannot be written with positive exponents using the rule given here. We would have to use the definition of a negative exponent to rewrite this expression with positive exponents, as shown below.

$$\dfrac{\dfrac{1}{5^2} + \dfrac{1}{3}}{7 - \dfrac{1}{2^3}}$$

OBJECTIVE 3 Use the quotient rule for exponents. We now consider the quotient of exponential expressions with the same base. We know that

$$\dfrac{6^5}{6^3} = \dfrac{6 \cdot 6 \cdot 6 \cdot 6 \cdot 6}{6 \cdot 6 \cdot 6} = 6^2.$$

Notice that the difference between the exponents, $5 - 3 = 2$, is the exponent in the quotient. Also,

$$\dfrac{6^2}{6^4} = \dfrac{6 \cdot 6}{6 \cdot 6 \cdot 6 \cdot 6} = \dfrac{1}{6^2} = 6^{-2}.$$

Here, $2 - 4 = -2$. These examples suggest the **quotient rule for exponents.**

Quotient Rule for Exponents

For any nonzero real number a and any integers m and n,

$$\frac{a^m}{a^n} = a^{m-n}$$

(Keep the same base and subtract the exponents.)

Example: $\dfrac{5^8}{5^4} = 5^{8-4} = 5^4$

CAUTION

A common **error** is to write $\dfrac{5^8}{5^4} = 1^{8-4} = 1^4$. Notice that by the quotient rule, the quotient should have the **same base**. The base here is 5.

$$\frac{5^8}{5^4} = 5^{8-4} = 5^4$$

If you are not sure, use the definition of an exponent to write out the factors:

$$5^8 = 5 \cdot 5 \cdot 5 \cdot 5 \cdot 5 \cdot 5 \cdot 5 \cdot 5 \quad \text{and} \quad 5^4 = 5 \cdot 5 \cdot 5 \cdot 5$$

Then it is clear that the quotient is 5^4.

EXAMPLE 4 Using the Quotient Rule for Exponents

Simplify, using the quotient rule for exponents. Write answers with positive exponents. Assume all variables represent nonzero real numbers.

(a) $\dfrac{5^8}{5^6} = 5^{8-6} = 5^2$

(b) $\dfrac{4^2}{4^9} = 4^{2-9} = 4^{-7} = \dfrac{1}{4^7}$

(c) $\dfrac{5^{-3}}{5^{-7}} = 5^{-3-(-7)} = 5^4$

(d) $\dfrac{q^5}{q^{-3}} = q^{5-(-3)} = q^8$

(e) $\dfrac{3^2 x^5}{3^4 x^3} = \dfrac{3^2}{3^4} \cdot \dfrac{x^5}{x^3} = 3^{2-4} \cdot x^{5-3} = 3^{-2} x^2 = \dfrac{x^2}{3^2}$

(f) $\dfrac{(m+n)^{-2}}{(m+n)^{-4}} = (m+n)^{-2-(-4)} = (m+n)^{-2+4} = (m+n)^2$, $m \neq -n$

(g) $\dfrac{7x^{-3}y^2}{2^{-1}x^2y^{-5}} = \dfrac{7 \cdot 2^1 y^2 y^5}{x^2 x^3} = \dfrac{14y^7}{x^5}$

▶▶▶ **Work Problem 5 at the Side.**

5 Simplify. Write answers with positive exponents. Assume all variables represent nonzero real numbers.

(a) $\dfrac{5^{11}}{5^8}$

(b) $\dfrac{4^7}{4^{10}}$

(c) $\dfrac{6^{-5}}{6^{-2}}$

(d) $\dfrac{8^4 m^9}{8^5 m^{10}}$

(e) $\dfrac{3^{-1}(x+y)^{-3}}{2^{-2}(x+y)^{-4}}$, $x \neq -y$

ANSWERS

5. (a) 5^3 (b) $\dfrac{1}{4^3}$ (c) $\dfrac{1}{6^3}$ (d) $\dfrac{1}{8m}$

   (e) $\dfrac{4}{3}(x+y)$

The definitions and rules for exponents given in this section and **Section 13.2** are summarized below.

Definitions and Rules for Exponents

For any integers m and n: Examples

Product rule	$a^m \cdot a^n = a^{m+n}$		$7^4 \cdot 7^5 = 7^9$
Zero exponent	$a^0 = 1$ when $a \neq 0$		$(-3)^0 = 1$
Negative exponent	$a^{-n} = \dfrac{1}{a^n}$ when $a \neq 0$		$5^{-3} = \dfrac{1}{5^3}$
Quotient rule	$\dfrac{a^m}{a^n} = a^{m-n}$ when $a \neq 0$		$\dfrac{2^2}{2^5} = 2^{2-5} = 2^{-3} = \dfrac{1}{2^3}$
Power rule (a)	$(a^m)^n = a^{mn}$		$(4^2)^3 = 4^6$
Power rule (b)	$(ab)^m = a^m b^m$		$(3k)^4 = 3^4 k^4$
Power rule (c)	$\left(\dfrac{a}{b}\right)^m = \dfrac{a^m}{b^m}$ when $b \neq 0$		$\left(\dfrac{2}{3}\right)^2 = \dfrac{2^2}{3^2}$
Negative-to-positive rules	$\dfrac{a^{-m}}{b^{-n}} = \dfrac{b^n}{a^m}$ when $a, b \neq 0$		$\dfrac{2^{-4}}{5^{-3}} = \dfrac{5^3}{2^4}$
	$\left(\dfrac{a}{b}\right)^{-m} = \left(\dfrac{b}{a}\right)^m$		$\left(\dfrac{4}{7}\right)^{-2} = \left(\dfrac{7}{4}\right)^2$

OBJECTIVE 4 Use combinations of rules. We sometimes need to use more than one rule to simplify an expression.

EXAMPLE 5 Using a Combination of Rules

Use a combination of the rules for exponents to simplify each expression. Assume all variables represent nonzero real numbers.

(a) $\dfrac{(4^2)^3}{4^5} = \dfrac{4^6}{4^5}$ Power rule (a)

$= 4^{6-5}$ Quotient rule

$= 4^1 = 4$

(b) $(2x)^3 (2x)^2 = (2x)^5$ Product rule

$= 2^5 x^5$ or $32 x^5$ Power rule (b)

(c) $\left(\dfrac{2x^3}{5}\right)^{-4} = \left(\dfrac{5}{2x^3}\right)^4$ Negative-to-positive rule

$= \dfrac{5^4}{2^4 x^{12}}$ or $\dfrac{625}{16 x^{12}}$ Power rules (a)–(c)

Continued on Next Page

(d) $\left(\dfrac{3x^{-2}}{4^{-1}y^3}\right)^{-3} = \dfrac{3^{-3}x^6}{4^3 y^{-9}}$ Power rules (a)–(c)

$\phantom{\left(\dfrac{3x^{-2}}{4^{-1}y^3}\right)^{-3}} = \dfrac{x^6 y^9}{4^3 \cdot 3^3}$ or $\dfrac{x^6 y^9}{1728}$ Negative-to-positive rule

(e) $\dfrac{(4m)^{-3}}{(3m)^{-4}} = \dfrac{4^{-3}m^{-3}}{3^{-4}m^{-4}}$ Power rule (b)

$\phantom{\dfrac{(4m)^{-3}}{(3m)^{-4}}} = \dfrac{3^4 m^4}{4^3 m^3}$ Negative-to-positive rule

$\phantom{\dfrac{(4m)^{-3}}{(3m)^{-4}}} = \dfrac{3^4 m^{4-3}}{4^3}$ Quotient rule

$\phantom{\dfrac{(4m)^{-3}}{(3m)^{-4}}} = \dfrac{3^4 m}{4^3}$ or $\dfrac{81m}{64}$

NOTE
Since the steps can be done in several different orders, there are many equally correct ways to simplify expressions like those in Examples 5(d) and 5(e).

Work Problem 6 at the Side.

6 Simplify. Assume all variables represent nonzero real numbers.

(a) $12^5 \cdot 12^{-7} \cdot 12^6$

(b) $y^{-2} \cdot y^5 \cdot y^{-8}$

(c) $\dfrac{(6x)^{-1}}{(3x^2)^{-2}}$

(d) $\dfrac{3^9 \cdot (x^2 y)^{-2}}{3^3 \cdot x^{-4} y}$

ANSWERS

6. (a) 12^4 or $20{,}736$ **(b)** $\dfrac{1}{y^5}$ **(c)** $\dfrac{3x^3}{2}$

    (d) $\dfrac{3^6}{y^3}$ or $\dfrac{729}{y^3}$

Focus on Real-Data Applications

Numbers BIG and SMALL

The ancient Egyptians used the *astonished man* symbol to represent a million. We now use exponents to write very big and very small numbers efficiently. Columbia University professor Edward Kasner asked his nine-year-old nephew to think of a name for the exceedingly large number 10^{100}, which is a 1 followed by 100 zeros. His nephew proclaimed it to be a **googol**. Kasner then called the unimaginably large number 10^{googol} a **googolplex**.

> A googol is 10^{100} or
> 10,000,000,000,000,000,000,000,000,000,000,
> 000,000,000,000,000,000,000,000,000,000,000,
> 000,000,000,000,000,000,000,000,000,000.

Real-world examples of very big and very small numbers are described in Howard Eves's book *Mathematical Circles Revisited*. A few such examples are listed here.

- The total number of electrons in the universe is, according to an estimate by Sir Arthur Eddington, about 10^{79}.
- The number of grains of sand on the beach at Coney Island, New York, is about 10^{20}.
- The total number of printed words since the Gutenberg Bible appeared is approximately 10^{16}.
- The temperature at the center of an atomic bomb explosion is 2×10^8 degrees Fahrenheit.
- The diameter of a human hair is about 10^{-4} millimeters.
- The diameter of a nucleus of a cell is about 10^{-6} millimeters (1 micron).
- The probability of winning a lottery by choosing 6 numbers from among 50 is about 6.3×10^{-8}.
- The size of a quark is approximately 10^{-18} millimeters.

Interestingly enough, the Web search engine Google™ is named after a googol. Sergey Brin, president and co-founder of Google, Inc., was a math major. He chose the name Google to describe the vast reach of this search engine.

For Group Discussion

Suppose that you have been offered a new job with "salary negotiable." You present to your new employer the following offer: You will work for 30 days and will be paid 1¢ on day one, 2¢ on day two, 4¢ on day three, 8¢ on day four, and so on, doubling your pay each day for the month. Your employer accepts your proposal, and you begin work on the first day of the next month.

1. On which day will you have received half of your month's wages?

2. When will you have received one-fourth of your total month's wages?

3. What do you think your approximate monthly salary will be?

4. How many days would it take you to become a "googolaire"?

13.5 Exercises

Decide whether each expression is positive, negative, or 0.

1. $(-2)^{-3}$
2. $(-3)^{-2}$
3. -2^4
4. -3^6
5. $\left(\dfrac{1}{4}\right)^{-2}$
6. $\left(\dfrac{1}{5}\right)^{-2}$
7. $1 - 5^0$
8. $1 - 7^0$

Each expression is equal to either 0, 1, or -1. Decide which is correct. See Example 1.

9. $(-4)^0$
10. $(-10)^0$
11. -9^0
12. -5^0
13. $(-2)^0 - 2^0$
14. $(-8)^0 - 8^0$
15. $\dfrac{0^{10}}{10^0}$
16. $\dfrac{0^5}{5^0}$

Evaluate each expression. See Examples 1 and 2.

17. $7^0 + 9^0$
18. $8^0 + 6^0$
19. 4^{-3}
20. 5^{-4}
21. $\left(\dfrac{1}{2}\right)^{-4}$
22. $\left(\dfrac{1}{3}\right)^{-3}$
23. $\left(\dfrac{6}{7}\right)^{-2}$
24. $\left(\dfrac{2}{3}\right)^{-3}$
25. $5^{-1} + 3^{-1}$
26. $6^{-1} + 2^{-1}$
27. $-2^{-1} + 3^{-2}$
28. $(-3)^{-2} + (-4)^{-1}$

RELATING CONCEPTS (EXERCISES 29–32) For Individual or Group Work

*In Objective 1, we used the product rule to develop a definition of an exponent of 0. We can also use the quotient rule. To see this, **work Exercises 29–32 in order.***

29. Consider the expression $\dfrac{25}{25}$. What is its simplest form?

30. Write the quotient in Exercise 29 using the fact that $25 = 5^2$.

31. Apply the quotient rule for exponents to your answer for Exercise 30. Give the answer as a power of 5.

32. Because your answers for Exercises 29 and 31 both represent $\dfrac{25}{25}$, they must be equal. Write this equality. What definition does it support?

Use the quotient rule to simplify each expression. Write each expression with positive exponents. Assume that all variables represent nonzero real numbers. See Examples 2–4.

33. $\dfrac{9^4}{9^5}$

34. $\dfrac{7^3}{7^4}$

35. $\dfrac{6^{-3}}{6^2}$

36. $\dfrac{4^{-2}}{4^3}$

37. $\dfrac{1}{6^{-3}}$

38. $\dfrac{1}{5^{-2}}$

39. $\dfrac{2}{r^{-4}}$

40. $\dfrac{3}{s^{-8}}$

41. $\dfrac{4^{-3}}{5^{-2}}$

42. $\dfrac{6^{-2}}{5^{-4}}$

43. $p^5 q^{-8}$

44. $x^{-8} y^4$

45. $\dfrac{r^5}{r^{-4}}$

46. $\dfrac{a^6}{a^{-4}}$

47. $\dfrac{6^4 x^8}{6^5 x^3}$

48. $\dfrac{3^8 y^5}{3^{10} y^2}$

49. $\dfrac{6y^3}{2y}$

50. $\dfrac{5m^2}{m}$

51. $\dfrac{3x^5}{3x^2}$

52. $\dfrac{10p^8}{2p^4}$

Use a combination of the rules for exponents to simplify each expression. Write answers with only positive exponents. Assume that all variables represent nonzero real numbers. See Example 5.

53. $\dfrac{(7^4)^3}{7^9}$

54. $\dfrac{(5^3)^2}{5^2}$

55. $x^{-3} \cdot x^5 \cdot x^{-4}$

56. $y^{-8} \cdot y^5 \cdot y^{-2}$

57. $\dfrac{(3x)^{-2}}{(4x)^{-3}}$

58. $\dfrac{(2y)^{-3}}{(5y)^{-4}}$

59. $\left(\dfrac{x^{-1} y}{z^2}\right)^{-2}$

60. $\left(\dfrac{p^{-4} q}{r^{-3}}\right)^{-3}$

61. $(6x)^4 (6x)^{-3}$

62. $(10y)^9 (10y)^{-8}$

63. $\dfrac{(m^7 n)^{-2}}{m^{-4} n^3}$

64. $\dfrac{(m^8 n^{-4})^2}{m^{-2} n^5}$

65. $\dfrac{5x^{-3}}{(4x)^2}$

66. $\dfrac{-3k^5}{(2k)^2}$

67. $\left(\dfrac{2p^{-1} q}{3^{-1} m^2}\right)^2$

68. $\left(\dfrac{4xy^2}{x^{-1} y}\right)^{-2}$

Summary Exercises on the Rules for Exponents

Use the rules for exponents to simplify each expression. Use only positive exponents in your answers. Assume that all variables represent nonzero real numbers.

1. $\left(\dfrac{6x^2}{5}\right)^{12}$

2. $\left(\dfrac{rs^2t^3}{3t^4}\right)^6$

3. $(10x^2y^4)^2(10xy^2)^3$

4. $(-2ab^3c)^4(-2a^2b)^3$

5. $\left(\dfrac{9wx^3}{y^4}\right)^3$

6. $(4x^{-2}y^{-3})^{-2}$

7. $\dfrac{c^{11}(c^2)^4}{(c^3)^3(c^2)^{-6}}$

8. $\left(\dfrac{k^4t^2}{k^2t^{-4}}\right)^{-2}$

9. $5^{-1} + 6^{-1}$

10. $\dfrac{(3y^{-1}z^3)^{-1}(3y^2)}{(y^3z^2)^{-3}}$

11. $\dfrac{(2xy^{-1})^3}{2^3x^{-3}y^2}$

12. $-8^0 + (-8)^0$

13. $(z^4)^{-3}(z^{-2})^{-5}$

14. $\left(\dfrac{r^2st^5}{3r}\right)^{-2}$

15. $\dfrac{(3^{-1}x^{-3}y)^{-1}(2x^2y^{-3})^2}{(5x^{-2}y^2)^{-2}}$

16. $\left(\dfrac{5x^2}{3x^{-4}}\right)^{-1}$

17. $\left(\dfrac{-2x^{-2}}{2x^2}\right)^{-2}$

18. $\dfrac{(x^{-4}y^2)^3(x^2y)^{-1}}{(xy^2)^{-3}}$

19. $\dfrac{(a^{-2}b^3)^{-4}}{(a^{-3}b^2)^{-2}(ab)^{-4}}$

20. $(2a^{-30}b^{-29})(3a^{31}b^{30})$

21. $5^{-2} + 6^{-2}$

22. $\left(\dfrac{(x^{47}y^{23})^2}{x^{-26}y^{-42}}\right)^0$

23. $\left(\dfrac{7a^2b^3}{2}\right)^3$

24. $-(-12^0)$

25. $-(-12)^0$

26. $\dfrac{0^{12}}{12^0}$

27. $\dfrac{(2xy^{-3})^{-2}}{(3x^{-2}y^4)^{-3}}$

28. $\left(\dfrac{a^2b^3c^4}{a^{-2}b^{-3}c^{-4}}\right)^{-2}$

29. $(6x^{-5}z^3)^{-3}$

30. $(2p^{-2}qr^{-3})(2p)^{-4}$

31. $\dfrac{(xy)^{-3}(xy)^5}{(xy)^{-4}}$

32. $42^0 - (-12)^0$

33. $\dfrac{(7^{-1}x^{-3})^{-2}(x^4)^{-6}}{7^{-1}x^{-3}}$

34. $\left(\dfrac{3^{-4}x^{-3}}{3^{-3}x^{-6}}\right)^{-2}$

35. $(5p^{-2}q)^{-3}(5pq^3)^4$

36. $8^{-1} + 6^{-1}$

37. $\left(\dfrac{4r^{-6}s^{-2}t}{2r^8s^{-4}t^2}\right)^{-1}$

38. $(13x^{-6}y)(13x^{-6}y)^{-1}$

39. $\dfrac{(8pq^{-2})^4}{(8p^{-2}q^{-3})^3}$

40. $\left(\dfrac{mn^{-2}p}{m^2np^4}\right)^{-2}\left(\dfrac{mn^{-2}p}{m^2np^4}\right)^3$

41. $-(-3^0)^0$

42. $5^{-1} - 8^{-1}$

13.6 Dividing a Polynomial by a Monomial

OBJECTIVE 1 Divide a polynomial by a monomial. We add two fractions with a common denominator as follows.

$$\frac{a}{c} + \frac{b}{c} = \frac{a+b}{c}$$

Looking at this statement in reverse gives us a rule for dividing a polynomial by a monomial.

Dividing a Polynomial by a Monomial
To divide a polynomial by a monomial, divide each term of the polynomial by the monomial:

$$\frac{a+b}{c} = \frac{a}{c} + \frac{b}{c} \quad \text{when } c \neq 0$$

Examples: $\frac{2+5}{3} = \frac{2}{3} + \frac{5}{3}$ and $\frac{x+3z}{2y} = \frac{x}{2y} + \frac{3z}{2y}$

The parts of a division problem are named here.

Dividend → $\frac{12x^2 + 6x}{6x} = 2x + 1$ ← Quotient
Divisor →

EXAMPLE 1 Dividing a Polynomial by a Monomial

Divide $5m^5 - 10m^3$ by $5m^2$.
Use the preceding rule, with + replaced by −. Then use the quotient rule.

$$\frac{5m^5 - 10m^3}{5m^2} = \frac{5m^5}{5m^2} - \frac{10m^3}{5m^2} = m^3 - 2m$$

Check by multiplying: $5m^2(m^3 - 2m) = 5m^5 - 10m^3$.
Because division by 0 is undefined, the quotient

$$\frac{5m^5 - 10m^3}{5m^2}$$

is undefined if $m = 0$. From now on, we assume that no denominators are 0.

Work Problem 1 at the Side.

EXAMPLE 2 Dividing a Polynomial by a Monomial

Divide: $\frac{16a^5 - 12a^4 + 8a^2}{4a^3}$.

Divide each term of $16a^5 - 12a^4 + 8a^2$ by $4a^3$.

$$\frac{16a^5 - 12a^4 + 8a^2}{4a^3} = \frac{16a^5}{4a^3} - \frac{12a^4}{4a^3} + \frac{8a^2}{4a^3}$$

$$= 4a^2 - 3a + \frac{2}{a} \quad \text{Quotient rule}$$

Continued on Next Page

OBJECTIVE

1. Divide a polynomial by a monomial.

1 Divide.

(a) $\frac{6p^4 + 18p^7}{3p^2}$

(b) $\frac{12m^6 + 18m^5 + 30m^4}{6m^2}$

(c) $(18r^7 - 9r^2) \div (3r)$

ANSWERS
1. (a) $2p^2 + 6p^5$ (b) $2m^4 + 3m^3 + 5m^2$
(c) $6r^6 - 3r$

Chapter 13 Exponents and Polynomials

② Divide.

(a) $\dfrac{20x^4 - 25x^3 + 5x}{5x^2}$

(b) $\dfrac{50m^4 - 30m^3 + 20m}{10m^3}$

③ Divide.

(a) $\dfrac{8y^7 - 9y^6 - 11y - 4}{y^2}$

(b) $\dfrac{12p^5 + 8p^4 + 3p^3 - 5p^2}{3p^3}$

④ Divide.

$\dfrac{45x^4y^3 + 30x^3y^2 - 60x^2y}{-15x^2y}$

ANSWERS

2. (a) $4x^2 - 5x + \dfrac{1}{x}$ (b) $5m - 3 + \dfrac{2}{m^2}$

3. (a) $8y^5 - 9y^4 - \dfrac{11}{y} - \dfrac{4}{y^2}$

   (b) $4p^2 + \dfrac{8p}{3} + 1 - \dfrac{5}{3p}$

4. $-3x^2y^2 - 2xy + 4$

The quotient $4a^2 - 3a + \dfrac{2}{a}$ is not a polynomial because of the expression $\dfrac{2}{a}$, which has a variable in the denominator. While the sum, difference, and product of two polynomials are always polynomials, the quotient of two polynomials may not be.

Again, *check* by multiplying.

$$4a^3\left(4a^2 - 3a + \dfrac{2}{a}\right) = 4a^3(4a^2) - 4a^3(3a) + 4a^3\left(\dfrac{2}{a}\right)$$

$$= 16a^5 - 12a^4 + 8a^2$$

◀◀ **Work Problem 2 at the Side.**

EXAMPLE 3 Dividing a Polynomial by a Monomial

Divide $12x^4 - 7x^3 + 4x$ by $4x$.

$$\dfrac{12x^4 - 7x^3 + 4x}{4x} = \dfrac{12x^4}{4x} - \dfrac{7x^3}{4x} + \dfrac{4x}{4x}$$

$$= 3x^3 - \dfrac{7x^2}{4} + 1 \qquad \text{Quotient rule}$$

Check by multiplying.

CAUTION

In Example 3, notice that the quotient $\dfrac{4x}{4x} = 1$. It is a common error to leave the 1 out of the answer. Multiplying to check will show that the answer $3x^3 - \dfrac{7x^2}{4}$ is not correct.

◀◀ **Work Problem 3 at the Side.**

EXAMPLE 4 Dividing a Polynomial by a Monomial

Divide the polynomial $180x^4y^{10} - 150x^3y^8 + 120x^2y^6 - 90xy^4 + 100y$ by the monomial $-30xy^2$.

$$\dfrac{180x^4y^{10} - 150x^3y^8 + 120x^2y^6 - 90xy^4 + 100y}{-30xy^2}$$

$$= \dfrac{180x^4y^{10}}{-30xy^2} - \dfrac{150x^3y^8}{-30xy^2} + \dfrac{120x^2y^6}{-30xy^2} - \dfrac{90xy^4}{-30xy^2} + \dfrac{100y}{-30xy^2}$$

$$= -6x^3y^8 + 5x^2y^6 - 4xy^4 + 3y^2 - \dfrac{10}{3xy}$$

◀◀ **Work Problem 4 at the Side.**

13.6 Exercises

Fill in each blank with the correct response.

1. In the statement $\dfrac{6x^2 + 8}{2} = 3x^2 + 4$, _____ is the dividend, _____ is the divisor, and _____ is the quotient.

2. The expression $\dfrac{3x + 12}{x}$ is undefined if $x =$ _____.

3. To check the division shown in Exercise 1, multiply _____ by _____ and show that the product is _____.

4. The expression $5x^2 - 3x + 6 + \frac{2}{x}$ _____ a polynomial.
   (is/is not)

5. Explain why the division problem $\dfrac{16m^3 - 12m^2}{4m}$ can be performed using the method of this section, while the division problem $\dfrac{4m}{16m^3 - 12m^2}$ cannot.

6. Evaluate $\dfrac{5y + 6}{2}$ when $y = 2$. Evaluate $5y + 3$ when $y = 2$. Does $\dfrac{5y + 6}{2}$ equal $5y + 3$?

Perform each division. See Examples 1–4.

7. $\dfrac{60x^4 - 20x^2 + 10x}{2x}$

8. $\dfrac{120x^6 - 60x^3 + 80x^2}{2x}$

9. $\dfrac{20m^5 - 10m^4 + 5m^2}{-5m^2}$

10. $\dfrac{12t^5 - 6t^3 + 6t^2}{-6t^2}$

11. $\dfrac{8t^5 - 4t^3 + 4t^2}{2t}$

12. $\dfrac{8r^4 - 4r^3 + 6r^2}{2r}$

13. $\dfrac{4a^5 - 4a^2 + 8}{4a}$

14. $\dfrac{5t^8 + 5t^7 + 15}{5t}$

15. $\dfrac{12x^5 - 4x^4 + 6x^3}{-6x^2}$

16. $\dfrac{24x^6 - 12x^5 + 30x^4}{-6x^2}$

17. $\dfrac{4x^2 + 20x^3 - 36x^4}{4x^2}$

18. $\dfrac{5x^2 - 30x^4 + 30x^5}{5x^2}$

19. $\dfrac{4x^4 + 3x^3 + 2x}{3x^2}$

20. $\dfrac{5x^4 - 6x^3 + 8x}{3x^2}$

21. $\dfrac{27r^4 - 36r^3 - 6r^2 + 3r - 2}{3r}$

22. $\dfrac{8k^4 - 12k^3 - 2k^2 - 2k - 3}{2k}$

23. $\dfrac{2m^5 - 6m^4 + 8m^2}{-2m^3}$

24. $\dfrac{6r^5 - 8r^4 + 10r^2}{-2r^4}$

25. $(20a^4b^3 - 15a^5b^2 + 25a^3b) \div (-5a^4b)$

26. $(16y^5z - 8y^2z^2 + 12yz^3) \div (-4y^2z^2)$

27. $(120x^{11} - 60x^{10} + 140x^9 - 100x^8) \div (10x^{12})$

28. $(120x^{12} - 84x^9 + 60x^8 - 36x^7) \div (12x^9)$

29. The quotient in Exercise 19 is $\dfrac{4x^2}{3} + x + \dfrac{2}{3x}$. Notice how the third term is written with x in the denominator. Would $\tfrac{2}{3}x$ be an acceptable form for this term? Explain why or why not. Is $\tfrac{4}{3}x^2$ an acceptable form for the first term? Why or why not?

30. What expression represents the length of the rectangle?

    $2x$ | Area $= 12x^2 - 4x + 2$

31. What polynomial, when divided by $5x^3$, yields $3x^2 - 7x + 7$ as a quotient?

32. The quotient of a certain polynomial and $-12y^3$ is $6y^3 - 5y^2 + 2y - 3 + \tfrac{7}{y}$. Find the polynomial.

RELATING CONCEPTS (EXERCISES 33–36) For Individual or Group Work

Our system of numeration is called a decimal system. It is based on powers of ten. In a whole number such as 2846, each digit is understood to represent the number of powers of ten for its place value. The 2 represents two thousands (2×10^3), the 8 represents eight hundreds (8×10^2), the 4 represents four tens (4×10^1), and the 6 represents six ones (or units) (6×10^0). In expanded form we write

$$2846 = (2 \times 10^3) + (8 \times 10^2) + (4 \times 10^1) + (6 \times 10^0).$$

Keeping this information in mind, **work Exercises 33–36 in order.**

33. Divide 2846 by 2, using paper-and-pencil methods: $2\overline{)2846}$.

34. Write your answer in Exercise 33 in expanded form.

35. Use the methods of this section to divide the polynomial $2x^3 + 8x^2 + 4x + 6$ by 2.

36. Compare your answers in Exercises 34 and 35. How are they similar? How are they different? For what value of x does the answer in Exercise 35 equal the answer in Exercise 34?

13.7 Dividing a Polynomial by a Polynomial

OBJECTIVE 1 Divide a polynomial by a polynomial. We use a method of "long division" to divide a polynomial by a polynomial (other than a monomial), similar to that used for dividing two whole numbers. *Both polynomials must be written in descending powers.*

OBJECTIVES

1. Divide a polynomial by a polynomial.
2. Apply division to a geometry problem.

Dividing Whole Numbers

Step 1
Divide 6696 by 27.
$$27 \overline{)6696}$$

Step 2
66 divided by 27 = **2**;
$2 \cdot 27 = \mathbf{54}$.

$$\begin{array}{r} 2 \\ 27 \overline{)6696} \\ 54 \end{array}$$

Step 3
Subtract; then bring down the next digit.

$$\begin{array}{r} 2 \\ 27 \overline{)6696} \\ 54 \\ \hline 129 \end{array}$$

Step 4
129 divided by 27 = **4**;
$4 \cdot 27 = \mathbf{108}$.

$$\begin{array}{r} 24 \\ 27 \overline{)6696} \\ 54 \\ \hline 129 \\ 108 \end{array}$$

Step 5
Subtract; then bring down the next digit.

$$\begin{array}{r} 24 \\ 27 \overline{)6696} \\ 54 \\ \hline 129 \\ 108 \\ \hline 216 \end{array}$$

Dividing Polynomials

Divide $8x^3 - 4x^2 - 14x + 15$ by $2x + 3$.
$$2x + 3 \overline{)8x^3 - 4x^2 - 14x + 15}$$

$8x^3$ divided by $2x = \mathbf{4x^2}$;
$4x^2(2x + 3) = \mathbf{8x^3 + 12x^2}$.

$$\begin{array}{r} 4x^2 \\ 2x + 3 \overline{)8x^3 - 4x^2 - 14x + 15} \\ 8x^3 + 12x^2 \end{array}$$

Subtract; then bring down the next term.

$$\begin{array}{r} 4x^2 \\ 2x + 3 \overline{)8x^3 - 4x^2 - 14x + 15} \\ 8x^3 + 12x^2 \\ \hline -16x^2 - 14x \end{array}$$

(To subtract two polynomials, change the signs of the second and then add.)

$-16x^2$ divided by $2x = \mathbf{-8x}$;
$-8x(2x + 3) = \mathbf{-16x^2 - 24x}$.

$$\begin{array}{r} 4x^2 - 8x \\ 2x + 3 \overline{)8x^3 - 4x^2 - 14x + 15} \\ 8x^3 + 12x^2 \\ \hline -16x^2 - 14x \\ -16x^2 - 24x \end{array}$$

Subtract; then bring down the next term.

$$\begin{array}{r} 4x^2 - 8x \\ 2x + 3 \overline{)8x^3 - 4x^2 - 14x + 15} \\ 8x^3 + 12x^2 \\ \hline -16x^2 - 14x \\ -16x^2 - 24x \\ \hline 10x + 15 \end{array}$$

(continued)

Step 6

| 216 divided by 27 = **8**; | 10x divided by 2x = **5**; |
| 8 · 27 = **216**. | 5(2x + 3) = **10x + 15**. |

$$\begin{array}{r} 24\mathbf{8} \\ 27{\overline{\smash{\big)}\,6696}} \\ \underline{54} \\ 129 \\ \underline{108} \\ 216 \\ \underline{\mathbf{216}} \\ 0 \end{array} \qquad \begin{array}{r} 4x^2 - 8x + \mathbf{5} \\ 2x + 3{\overline{\smash{\big)}\,8x^3 - 4x^2 - 14x + 15}} \\ \underline{8x^3 + 12x^2} \\ -16x^2 - 14x \\ \underline{-16x^2 - 24x} \\ 10x + 15 \\ \underline{\mathbf{10x + 15}} \\ 0 \end{array}$$

6696 divided by 27 is 248. The remainder is zero.

$8x^3 - 4x^2 - 14x + 15$ divided by $2x + 3$ is $4x^2 - 8x + 5$. The remainder is zero.

Step 7

Check by multiplying.

$27 \cdot 248 = 6696$

Check by multiplying.

$(2x + 3)(4x^2 - 8x + 5)$
$= 8x^3 - 4x^2 - 14x + 15$

EXAMPLE 1 Dividing a Polynomial by a Polynomial

Divide $5x + 4x^3 - 8 - 4x^2$ by $2x - 1$.

Both polynomials must be written with the exponents in descending order. Rewrite the first polynomial as $4x^3 - 4x^2 + 5x - 8$. Then begin the division process.

Divide $4x^3 - 4x^2 + 5x - 8$ by $2x - 1$.

$$\begin{array}{r} 2x^2 - x + 2 \\ 2x - 1{\overline{\smash{\big)}\,4x^3 - 4x^2 + 5x - 8}} \\ \underline{4x^3 - 2x^2} \\ -2x^2 + 5x \\ \underline{-2x^2 + x} \\ 4x - 8 \\ \underline{4x - 2} \\ -6 \leftarrow \text{Remainder} \end{array}$$

Step 1 $4x^3$ divided by $2x$ = $\mathbf{2x^2}$; $2x^2(2x - 1) = 4x^3 - 2x^2$.

Step 2   Subtract; bring down the next term.

Step 3 $-2x^2$ divided by $2x$ = $\mathbf{-x}$; $-x(2x - 1) = -2x^2 + x$.

Step 4   Subtract; bring down the next term.

Step 5 $4x$ divided by $2x$ = $\mathbf{2}$; $2(2x - 1) = 4x - 2$.

Step 6   Subtract. The remainder is -6. Write the remainder as the numerator of a fraction that has $2x - 1$ as its denominator. The answer is not a polynomial because of the nonzero remainder.

$$\frac{4x^3 - 4x^2 + 5x - 8}{2x - 1} = 2x^2 - x + 2 + \frac{-6}{2x - 1}$$

Continued on Next Page

Step 7 Check by multiplying.

$$(2x - 1)\left(2x^2 - x + 2 + \frac{-6}{2x - 1}\right)$$
$$= (2x - 1)(2x^2) + (2x - 1)(-x) + (2x - 1)(2)$$
$$+ (2x - 1)\left(\frac{-6}{2x - 1}\right)$$
$$= 4x^3 - 2x^2 - 2x^2 + x + 4x - 2 - 6$$
$$= 4x^3 - 4x^2 + 5x - 8$$

Work Problem 1 at the Side.

EXAMPLE 2 Dividing into a Polynomial with Missing Terms

Divide $x^3 - 1$ by $x - 1$.

Here the polynomial $x^3 - 1$ is missing the x^2-term and the x-term. When terms are missing, use 0 as the coefficient for each missing term. (Zero acts as a placeholder here, just as it does in our number system.)

$x^3 - 1$ can be written as $x^3 + 0x^2 + 0x - 1$

Now divide.

$$\begin{array}{r} x^2 + x + 1 \\ x - 1 \overline{\smash{)}x^3 + 0x^2 + 0x - 1} \\ \underline{x^3 - x^2 } \\ x^2 + 0x \\ \underline{x^2 - x } \\ x - 1 \\ \underline{x - 1} \\ 0 \end{array}$$

The remainder is 0. The quotient is $x^2 + x + 1$. *Check* by multiplying.

$$(x - 1)(x^2 + x + 1) = x^3 - 1$$

Work Problem 2 at the Side.

EXAMPLE 3 Dividing by a Polynomial with Missing Terms

Divide $x^4 + 2x^3 + 2x^2 - x - 1$ by $x^2 + 1$.

Since $x^2 + 1$ has a missing x term, write it as $x^2 + 0x + 1$. Then go through the division process as follows.

$$\begin{array}{r} x^2 + 2x + 1 \\ x^2 + 0x + 1 \overline{\smash{)}x^4 + 2x^3 + 2x^2 - x - 1} \\ \underline{x^4 + 0x^3 + x^2 } \\ 2x^3 + x^2 - x \\ \underline{2x^3 + 0x^2 + 2x } \\ x^2 - 3x - 1 \\ \underline{x^2 + 0x + 1} \\ -3x - 2 \leftarrow \text{Remainder} \end{array}$$

Continued on Next Page

---

1 Divide.

(a) $(x^3 + x^2 + 4x - 6) \div (x - 1)$

(b) $\dfrac{p^3 - 2p^2 - 5p + 9}{p + 2}$

2 Divide.

(a) $\dfrac{r^2 - 5}{r + 4}$

(b) $(x^3 - 8) \div (x - 2)$

ANSWERS
1. (a) $x^2 + 2x + 6$
   (b) $p^2 - 4p + 3 + \dfrac{3}{p + 2}$
2. (a) $r - 4 + \dfrac{11}{r + 4}$
   (b) $x^2 + 2x + 4$

3 Divide.

(a)

$(2x^4 + 3x^3 - x^2 + 6x + 5)$
$\div (x^2 - 1)$

When the result of subtracting ($-3x - 2$, in this case) is a polynomial of lesser degree than the divisor ($x^2 + 0x + 1$), that polynomial is the remainder. Write the answer as

$$x^2 + 2x + 1 + \frac{-3x - 2}{x^2 + 1}.$$

Multiply to check that this is the correct quotient.

▶◀ **Work Problem 3 at the Side.**

EXAMPLE 4 Dividing a Polynomial with a Quotient That Has Fractional Coefficients

Divide $4x^3 + 2x^2 + 3x + 1$ by $4x - 4$.

$$\begin{array}{r} x^2 + \frac{3}{2}x + \frac{9}{4} \\ 4x - 4 \overline{\smash{)}4x^3 + 2x^2 + 3x + 1} \\ \underline{4x^3 - 4x^2} \\ 6x^2 + 3x \\ \underline{6x^2 - 6x} \\ 9x + 1 \\ \underline{9x - 9} \\ 10 \end{array}$$

The answer is $x^2 + \frac{3}{2}x + \frac{9}{4} + \frac{10}{4x - 4}$.

(b)

$$\frac{2m^5 + m^4 + 6m^3 - 3m^2 - 18}{m^2 + 3}$$

4 Divide $3x^3 + 7x^2 + 7x + 10$ by $3x + 6$.

▶◀ **Work Problem 4 at the Side.**

OBJECTIVE 2 Apply division to a geometry problem.

EXAMPLE 5 Using an Area Formula

The area of the rectangle in Figure 2 is $x^3 + 4x^2 + 8x + 8$ square units and the width is $x + 2$ units. What is its length?

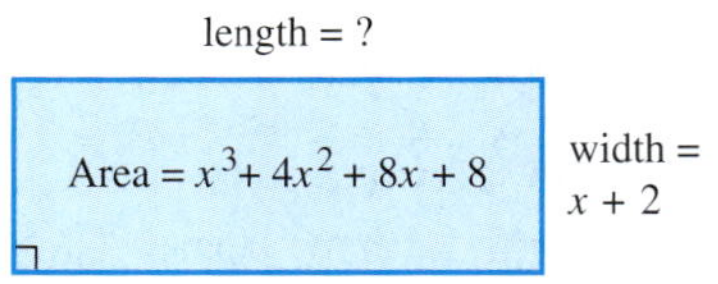

length = ?

Area = $x^3 + 4x^2 + 8x + 8$ width = $x + 2$

Figure 2

5 Divide $x^3 + 4x^2 + 8x + 8$ by $x + 2$.

Since $A = lw$, solving for l gives $l = \frac{A}{w}$. Divide $x^3 + 4x^2 + 8x + 8$ by the width, $x + 2$.

▶◀ **Work Problem 5 at the Side.**

The quotient from Problem 5 at the side, $x^2 + 2x + 4$, represents the length of the rectangle in units.

ANSWERS

3. **(a)** $2x^2 + 3x + 1 + \frac{9x + 6}{x^2 - 1}$
   (b) $2m^3 + m^2 - 6$
4. $x^2 + \frac{1}{3}x + \frac{5}{3}$
5. $x^2 + 2x + 4$

13.7 Exercises

1. In the division problem $(4x^4 + 2x^3 - 14x^2 + 19x + 10) \div (2x + 5) = 2x^3 - 4x^2 + 3x + 2$, which polynomial is the divisor? Which is the quotient?

2. When dividing one polynomial by another, how do you know when to stop dividing?

3. In dividing $12m^2 - 20m + 3$ by $2m - 3$, what is the first step?

4. In the division in Exercise 3, what is the second step?

Perform each division. See Example 1.

5. $\dfrac{x^2 - x - 6}{x - 3}$

6. $\dfrac{m^2 - 2m - 24}{m - 6}$

7. $\dfrac{2y^2 + 9y - 35}{y + 7}$

8. $\dfrac{2y^2 + 9y + 7}{y + 1}$

9. $\dfrac{p^2 + 2p + 20}{p + 6}$

10. $\dfrac{x^2 + 11x + 16}{x + 8}$

11. $(r^2 - 8r + 15) \div (r - 3)$

12. $(t^2 + 2t - 35) \div (t - 5)$

13. $\dfrac{4a^2 - 22a + 32}{2a + 3}$

14. $\dfrac{9w^2 + 6w + 10}{3w - 2}$

15. $\dfrac{8x^3 - 10x^2 - x + 3}{2x + 1}$

16. $\dfrac{12t^3 - 11t^2 + 9t + 18}{4t + 3}$

Perform each division. See Examples 2–4.

17. $\dfrac{3y^3 + y^2 + 2}{y + 1}$

18. $\dfrac{2r^3 - 6r - 36}{r - 3}$

19. $\dfrac{3k^3 - 4k^2 - 6k + 10}{k^2 - 2}$

20. $\dfrac{5z^3 - z^2 + 10z + 2}{z^2 + 2}$

21. $(x^4 - x^2 - 2) \div (x^2 - 2)$

22. $(r^4 + 2r^2 - 3) \div (r^2 - 1)$

23. $\dfrac{6p^4 - 15p^3 + 14p^2 - 5p + 10}{3p^2 + 1}$

24. $\dfrac{6r^4 - 10r^3 - r^2 + 15r - 8}{2r^2 - 3}$

25. $\dfrac{2x^5 + x^4 + 11x^3 - 8x^2 - 13x + 7}{2x^2 + x - 1}$

26. $\dfrac{4t^5 - 11t^4 - 6t^3 + 5t^2 - t + 3}{4t^2 + t - 3}$

27. $\dfrac{x^4 - 1}{x^2 - 1}$

28. $\dfrac{y^3 + 1}{y + 1}$

29. $(10x^3 + 13x^2 + 4x + 1) \div (5x + 5)$

30. $(6x^3 - 19x^2 - 19x - 4) \div (2x - 8)$

Work each problem. See Example 5.

31. What is the length of the rectangle if the area is $5x^3 + 7x^2 - 13x - 6$ square units?

32. Find the measure of the base of the parallelogram if the area is $2x^3 + 2x^2 - 3x - 1$ square units.

RELATING CONCEPTS (EXERCISES 33–36) For Individual or Group Work

We can find the value of a polynomial in x for a given value of x by substituting that number for x. Surprisingly, we can accomplish the same thing by division. For example, to find the value of $2x^2 - 4x + 3$ for $x = -3$, we would divide $2x^2 - 4x + 3$ by $x - (-3)$. The remainder will give the value of the polynomial for $x = -3$. **Work Exercises 33–36 in order.**

33. Find the value of $2x^2 - 4x + 3$ for $x = -3$ by substitution.

34. Divide $2x^2 - 4x + 3$ by $x + 3$. Give the remainder.

35. Compare your answers to Exercises 33 and 34. What do you notice?

36. Choose another polynomial and evaluate it both ways for some value of the variable. Do the answers agree?

13.8 An Application of Exponents: Scientific Notation

OBJECTIVE 1 Express numbers in scientific notation. One example of the use of exponents comes from science. The numbers occurring in science are often extremely large (such as the distance from Earth to the sun, 93,000,000 mi) or extremely small (the wavelength of yellow-green light, approximately 0.0000006 m). Because of the difficulty of working with many zeros, scientists often express such numbers with exponents. Each number is written as

$$a \times 10^n, \text{ where } 1 \leq |a| < 10 \text{ and } n \text{ is an integer.}$$

This form is called **scientific notation.** There is always one nonzero digit before the decimal point. This is shown in the following examples.

OBJECTIVES
1. Express numbers in scientific notation.
2. Convert numbers in scientific notation to numbers without exponents.
3. Use scientific notation in calculations.
4. Solve application problems using scientific notation.

> **NOTE**
> In work with scientific notation, the times symbol, ×, is commonly used.

$3.19 \times 10^1 = 3.19 \times 10 = 31.9$ Decimal point moves 1 place to the right.
$3.19 \times 10^2 = 3.19 \times 100 = 319.$ Decimal point moves 2 places to the right.
$3.19 \times 10^3 = 3.19 \times 1000 = 3190.$ Decimal point moves 3 places to the right.
$3.19 \times 10^{-1} = 3.19 \times 0.1 = 0.319$ Decimal point moves 1 place to the left.
$3.19 \times 10^{-2} = 3.19 \times 0.01 = 0.0319$ Decimal point moves 2 places to the left.
$3.19 \times 10^{-3} = 3.19 \times 0.001 = 0.00319$ Decimal point moves 3 places to the left.

A number in scientific notation is always written with the decimal point after the first nonzero digit and then multiplied by the appropriate power of 10. For example, 35 is written 3.5×10^1, or 3.5×10; 56,200 is written 5.62×10^4, since

$$56,200 = 5.62 \times \mathbf{10,000} = 5.62 \times \mathbf{10^4}.$$

To write a number in scientific notation, follow these steps.

> **Writing a Number in Scientific Notation**
> *Step 1* Move the decimal point to the right of the first nonzero digit.
> *Step 2* Count the number of places you moved the decimal point.
> *Step 3* The number of places in Step 2 is the absolute value of the exponent on 10.
> *Step 4* The exponent on 10 is positive if the original number is greater than the number in Step 1; the exponent is negative if the original number is less than the number in Step 1. If the decimal point is not moved, the exponent is 0.

For negative numbers, follow these steps using the absolute value of the number; then make the result negative.

1 Write each number in scientific notation.

(a) 63,000

(b) 5,870,000

(c) 0.0571

(d) −0.000062

EXAMPLE 1 Using Scientific Notation

Write each number in scientific notation.

(a) 93,000,000

Move the decimal point to follow the first nonzero digit. Count the number of places the decimal point was moved.

$$9.3\,000\,000 \quad \text{7 places}$$

The number will be written in scientific notation as 9.3×10^n. To find the value of n, first compare 93,000,000 with 9.3. Since 93,000,000 is *greater* than 9.3, we must multiply by a *positive* power of 10 so the product 9.3×10^n will equal the larger number.

Since the decimal point was moved 7 places, and since n is positive,

$$93{,}000{,}000 = 9.3 \times 10^7.$$

(b) $463{,}000{,}000{,}000{,}000 = 4.63\,000\,000\,000\,000 \quad \text{14 places}$

$$= 4.63 \times 10^{14}$$

(c) $3.021 = 3.021 \times 10^0$

(d) 0.00462

Move the decimal point to the right of the first nonzero digit and count the number of places the decimal point was moved.

$$0\,004.62 \quad \text{3 places}$$

Because 0.00462 is *less* than 4.62, the exponent must be *negative*.

$$0.00462 = 4.62 \times 10^{-3}$$

(e) $-0.0000762 = -7.62 \times 10^{-5}$

Work Problem 1 at the Side.

OBJECTIVE 2 Convert numbers in scientific notation to numbers without exponents. To convert a number written in scientific notation to a number without exponents, work in reverse. *Multiplying a number by a positive power of 10 will make the number greater; multiplying by a negative power of 10 will make the number less.*

EXAMPLE 2 Writing Numbers without Exponents

Write each number without exponents.

(a) 6.2×10^3

Since the exponent is positive, make 6.2 greater by moving the decimal point 3 places to the right. It is necessary to attach two zeros.

$$6.2 \times 10^3 = 6.200 = 6200$$

(b) $4.283 \times 10^5 = 4.28300 = 428{,}300$ Move 5 places to the right; attach zeros as necessary.

(c) $-9.73 \times 10^{-2} = -09.73 = -0.0973$ Move 2 places to the left.

As these examples show, *the exponent tells the number of places and the direction that the decimal point is moved.*

2 Write without exponents.

(a) 4.2×10^3

(b) 8.7×10^5

(c) 6.42×10^{-3}

ANSWERS

1. (a) 6.3×10^4 (b) 5.87×10^6
   (c) 5.71×10^{-2} (d) -6.2×10^{-5}
2. (a) 4200 (b) 870,000 (c) 0.00642

Work Problem 2 at the Side.

OBJECTIVE 3 **Use scientific notation in calculations.** The next example shows how scientific notation can be used with products and quotients.

EXAMPLE 3 Multiplying and Dividing with Scientific Notation

Write each product or quotient without exponents.

(a) $(6 \times 10^3)(5 \times 10^{-4})$
$= (6 \times 5)(10^3 \times 10^{-4})$ Commutative and associative properties
$= 30 \times 10^{-1}$ Product rule for exponents
$= 30. = 3$ Write without exponents.

(b) $\dfrac{6 \times 10^{-5}}{2 \times 10^3} = \dfrac{6}{2} \times \dfrac{10^{-5}}{10^3} = 3 \times 10^{-8} = 0.00000003$

Work Problem 3 at the Side.

OBJECTIVE 4 **Solve application problems using scientific notation.** Sometimes application problems may involve one or more numbers written in scientific notation and other numbers written without exponents. By writing all the numbers in scientific notation, we can use the product and quotient rules for exponents to help solve the problem.

EXAMPLE 4 Solving an Application Problem

There are approximately 3.03×10^{23} molecules in one gram of hydrogen gas. How many molecules are in 2000 g of hydrogen gas? Write your answer both in scientific notation and as a number without exponents.

First write 2000 in scientific notation.

$2000 = 2.000$ Move 3 places; 2000 is *greater* than 2.0 so the exponent is *positive*.
$= 2.0 \times 10^3$

Now multiply the number of molecules in one gram times the number of grams.

$(3.03 \times 10^{23})(2.0 \times 10^3)$
$= (3.03 \times 2.0)(10^{23} \times 10^3)$ Use the product rule for exponents.
$= 6.06 \times 10^{26}$

Written in scientific notation, there are 6.06×10^{26} molecules in 2000 g of hydrogen gas. Written without an exponent, the number is 606,000,000,000,000,000,000,000,000 molecules.

Work Problem 4 at the Side.

3 Simplify, and write without exponents.

(a) $(2.6 \times 10^4)(2 \times 10^{-6})$

(b) $\dfrac{4.8 \times 10^2}{2.4 \times 10^{-3}}$

4 There are 4.356×10^4 square feet in one acre. How many square feet are in 1500 acres? Write your answer both in scientific notation and as a number without exponents. (*Source: World Almanac.*)

ANSWERS
3. **(a)** 0.052 **(b)** 200,000
4. 6.534×10^7 square feet, or 65,340,000 square feet

5 In a laboratory experiment, a test group of 950 common house spiders weighed 2.09×10^{-1} pound. Find the weight of each spider. Write your answer both in scientific notation and as a number without exponents.

Calculator Tip Most scientific calculators have a key that allows you to enter numbers in scientific notation. Usually this key is labeled [EE] or [EXP]. Then you could solve Example 4, above, by pressing:

$$3.03 \; \boxed{EXP} \; 23 \; \boxed{\times} \; 2000 \; \boxed{=} \qquad \boxed{6.06^{26}}$$

The exponent in the answer is 26. **You have to write the × 10 part of the answer** to get 6.06×10^{26}.

If you need to enter a number in scientific notation that has a negative exponent, use the *change of sign* key. For example, to enter 8.45×10^{-6}, press:

$$8.45 \; \boxed{EXP} \; 6 \; \boxed{+/-} \qquad \boxed{8.45^{-06}}$$

EXAMPLE 5 Solving an Application Problem

At a printing shop, a stack of 5×10^3 sheets of paper measures 0.38 meter high. What is the thickness of one sheet of paper? Write your answer both in scientific notation and as a number without exponents.

First write 0.38 in scientific notation.

$$0.38 = 0\_3.8 \qquad \text{Move 1 place; 0.38 is \textit{less} than 3.8 so the exponent is negative.}$$
$$= 3.8 \times 10^{-1}$$

Now divide the total height of the stack by the number of sheets in the stack.

$$\frac{3.8 \times 10^{-1}}{5 \times 10^3} = \frac{3.8}{5} \times \frac{10^{-1}}{10^3} = 0.76 \times 10^{-4} \qquad \text{Use the quotient rule for exponents.}$$

The answer 0.76×10^{-4} is *not* in proper scientific notation yet, because 0.76 is not between 1 and 10. The decimal point in 0.76 must be moved *1 place* to get 7.6. Then, because 0.76 is *less* than 7.6, add *negative* 1 to the exponent.

$$0.76 \times 10^{-4} = 7.6 \times 10^{-5} \qquad (-4) + (-1)$$

Move decimal point 1 place.

The thickness of one sheet of paper is 7.6×10^{-5} meter. Written without an exponent, the thickness is 0.000076 meter.

Work Problem 5 at the Side.

ANSWERS
5. 2.2×10^{-4} pound, or 0.00022 pound

13.8 Exercises

FOR EXTRA HELP Addison-Wesley Math Tutor Center   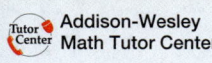 MathXL   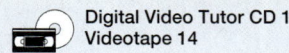 Digital Video Tutor CD 10 Videotape 14    Student's Solutions Manual   MyMathLab   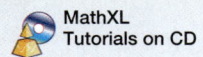 MathXL Tutorials on CD

Write the numbers (other than dates) mentioned in the following statements in scientific notation.

1. The mass of Pluto, the smallest planet, is 0.0021 times that of Earth; the mass of Jupiter, the largest planet, is 317.83 times that of Earth. (*Source: World Almanac*.)

2. NASA had a budget of $15,190,000,000 in 2004 for the international space station and other space projects. (*Source:* NASA.)

3. There are 831,200,000 pounds of pumpkins grown every year in the United States. Illinois leads the nation with 319,000,000 pounds.

4. The human body contains trace amounts of several minerals. For example, a 150-pound body has about 0.00023 pound of copper in it.

State whether or not the given number is written in scientific notation. If it is not, write it as such.

5. 4.56×10^3 6. 7.34×10^5 7. 5,600,000   8. 34,000

9. 0.004   10. 0.0007   11. 0.8×10^2 12. 0.9×10^3

13. Explain in your own words how to write a very large number in scientific notation, and how to write a very small number in scientific notation.

14. Explain how to multiply a number by a positive power of ten. Then explain how to multiply a number by a negative power of ten.

Write each number in scientific notation. See Example 1.

15. 5,876,000,000   16. 9,994,000,000   17. 82,350   18. 78,330

19. 0.000007   20. 0.0000004   21. 0.00203   22. 0.0000578

Write each number without exponents. See Example 2.

23. 7.5×10^5 24. 8.8×10^6 25. 5.677×10^{12} 26. 8.766×10^9

27. 6.21×10^0

28. 8.56×10^0

29. 7.8×10^{-4}

30. 8.9×10^{-5}

31. 5.134×10^{-9}

32. 7.123×10^{-10}

Perform the indicated operations. Write the answers both in scientific notation and as numbers without exponents. See Example 3.

33. $(2 \times 10^8)(3 \times 10^3)$

34. $(4 \times 10^7)(3 \times 10^3)$

35. $(5 \times 10^4)(3 \times 10^2)$

36. $(8 \times 10^5)(2 \times 10^3)$

37. $(3.15 \times 10^{-4})(2.04 \times 10^8)$

38. $(4.92 \times 10^{-3})(2.25 \times 10^7)$

39. $\dfrac{9 \times 10^{-5}}{3 \times 10^{-1}}$

40. $\dfrac{12 \times 10^{-4}}{4 \times 10^{-3}}$

41. $\dfrac{8 \times 10^3}{2 \times 10^2}$

42. $\dfrac{5 \times 10^4}{1 \times 10^3}$

43. $\dfrac{(2.6 \times 10^{-3})(7.0 \times 10^{-1})}{(2 \times 10^2)(3.5 \times 10^{-3})}$

44. $\dfrac{(9.5 \times 10^{-1})(2.4 \times 10^4)}{(5 \times 10^3)(1.2 \times 10^{-2})}$

Work each problem. Write answers both in scientific notation and without exponents. See Examples 4 and 5.

45. The population of China in 2003 was about 1.304×10^9 people. The country's land area is 3,700,000 square miles. What is the average number of people per square mile in China? (*Source:* United Nations World Population Prospects.)

46. The mass of Earth is approximately 5.98×10^{24} kg, and the mass of the moon is approximately 7.36×10^{22} kg. Earth's mass is how many times greater than the moon's mass? (*Source:* Time Life *The Universe.*)

47. The average diameter of a hydrogen atom is 1×10^{-10} meter. (*Source: FactFinder.*)

 (a) If one million hydrogen atoms are lined up in a row, how long would the row be?

 (b) Find the length of a row of one billion hydrogen atoms.

48. The speed of light is about 1.86×10^5 miles per second. (*Source:* Time Life *The Universe.*)

 (a) How far does light travel in one minute?

 (b) How far does light travel in one hour?

49. Americans mailed about 2×10^{10} cards, letters, and packages during December 2002. The U.S. population is about 2.88×10^8 people. On average, how many items did each person mail during that month? (*Source:* United States Postal Service and Bureau of the Census.)

50. American golfers buy more than 5×10^8 golf balls each year. On average, how many golf balls are sold each month? (*Source:* National Sporting Goods Association.)

Chapter 13
SUMMARY

KEY TERMS

13.1 **polynomial** — A polynomial is a term or the sum of a finite number of terms with whole number exponents.
descending powers — A polynomial in x is written in descending powers if the exponents on x in its terms are in decreasing order.
degree of a term — The degree of a term is the sum of the exponents on the variables.
degree of a polynomial — The degree of a polynomial is the greatest degree of any term of the polynomial.
monomial — A monomial is a polynomial with one term.
binomial — A binomial is a polynomial with two terms.
trinomial — A trinomial is a polynomial with three terms.

13.3 **FOIL** — FOIL is a shortcut method for finding the product of two binomials.
outer product — The outer product of $(2x + 3)(x - 5)$ is $2x(-5)$.
inner product — The inner product of $(2x + 3)(x - 5)$ is $3x$.

13.8 **scientific notation** — Scientific notation is used to write very large or very small numbers using exponents. The numbers are written as $a \times 10^n$, where $1 \leq |a| < 10$ and n is an integer.

NEW SYMBOLS

x^{-n} x to the negative n power x^0 x to the zero power

TEST YOUR WORD POWER

See how well you have learned the vocabulary in this chapter. Answers, with examples, follow the Quick Review.

1. A **polynomial** is an algebraic expression made up of
   A. a term or a finite product of terms with positive coefficients and exponents
   B. a term or a finite sum of terms with real coefficients and whole number exponents
   C. the product of two or more terms with positive exponents
   D. the sum of two or more terms with whole number coefficients and exponents.

2. The **degree of a term** is
   A. the number of variables in the term
   B. the product of the exponents on the variables
   C. the least exponent on the variables
   D. the sum of the exponents on the variables.

3. A **trinomial** is a polynomial with
   A. only one term
   B. exactly two terms
   C. exactly three terms
   D. more than three terms.

4. A **binomial** is a polynomial with
   A. only one term
   B. exactly two terms
   C. exactly three terms
   D. more than three terms.

5. A **monomial** is a polynomial with
   A. only one term
   B. exactly two terms
   C. exactly three terms
   D. more than three terms.

6. **FOIL** is a method for
   A. adding two binomials
   B. adding two trinomials
   C. multiplying two binomials
   D. multiplying two trinomials.

923

Quick Review

Concepts

Examples

13.1 Adding and Subtracting Polynomials

Addition
Add like terms.

Add.
$$2x^2 + 5x - 3$$
$$\underline{5x^2 - 2x + 7}$$
$$7x^2 + 3x + 4$$

Subtraction
Change the signs of the terms in the second polynomial and add to the first polynomial.

Subtract.
$(2x^2 + 5x - 3) - (5x^2 - 2x + 7)$
$= (2x^2 + 5x - 3) + (-5x^2 + 2x - 7)$
$= -3x^2 + 7x - 10$

13.2 The Product Rule and Power Rules for Exponents

For any integers m and n:

Product rule $a^m \cdot a^n = a^{m+n}$

Power rule (a) $(a^m)^n = a^{mn}$

Power rule (b) $(ab)^m = a^m b^m$

Power rule (c) $\left(\dfrac{a}{b}\right)^m = \dfrac{a^m}{b^m}$ when $b \neq 0$

$2^4 \cdot 2^5 = 2^{4+5} = 2^9$

$(3^4)^2 = 3^{4 \cdot 2} = 3^8$

$(6a)^5 = 6^5 a^5$

$\left(\dfrac{2}{3}\right)^4 = \dfrac{2^4}{3^4}$

13.3 Multiplying Polynomials

Multiply each term of the first polynomial by each term of the second polynomial. Then add like terms.

Multiply.
$$3x^3 - 4x^2 + 2x - 7$$
$$\underline{ 4x + 3}$$
$$9x^3 - 12x^2 + 6x - 21$$
$$\underline{12x^4 - 16x^3 + 8x^2 - 28x }$$
$$12x^4 - 7x^3 - 4x^2 - 22x - 21$$

FOIL Method

Step 1 Multiply the two first terms to get the first term of the answer.

Step 2 Find the outer product and the inner product and mentally add them, when possible, to get the middle term of the answer.

Step 3 Multiply the two last terms to get the last term of the answer.

Add the terms found in Steps 1–3.

Multiply $(2x + 3)(5x - 4)$.

$2x(5x) = \mathbf{10x^2}$

$2x(-4) + 3(5x) = \mathbf{7x}$

$3(-4) = \mathbf{-12}$

$(2x + 3)(5x - 4) = \mathbf{10x^2 + 7x - 12}$

13.4 Special Products

Square of a Binomial

$$(a + b)^2 = a^2 + 2ab + b^2$$
$$(a - b)^2 = a^2 - 2ab + b^2$$

$(3x + 1)^2 = 9x^2 + 6x + 1$

$(2m - 5n)^2 = 4m^2 - 20mn + 25n^2$

Product of the Sum and Difference of Two Terms

$$(a + b)(a - b) = a^2 - b^2$$

$(4a + 3)(4a - 3) = 16a^2 - 9$

Concepts	Examples
13.5 Integer Exponents and the Quotient Rule If $a, b \neq 0$, for integers m and n: **Zero exponent** $\quad a^0 = 1$ **Negative exponent** $\quad a^{-n} = \dfrac{1}{a^n}$ **Quotient rule** $\quad \dfrac{a^m}{a^n} = a^{m-n}$ **Negative-to-positive rules** $\quad \dfrac{a^{-m}}{b^{-n}} = \dfrac{b^n}{a^m} \quad$ and $\quad \left(\dfrac{a}{b}\right)^{-m} = \left(\dfrac{b}{a}\right)^m$	$15^0 = 1$ $5^{-2} = \dfrac{1}{5^2} = \dfrac{1}{25}$ $\dfrac{4^8}{4^3} = 4^{8-3} = 4^5$ $\dfrac{6^{-2}}{7^{-3}} = \dfrac{7^3}{6^2} \qquad \left(\dfrac{5}{3}\right)^{-4} = \left(\dfrac{3}{5}\right)^4$
13.6 Dividing a Polynomial by a Monomial Divide each term of the polynomial by the monomial. $$\dfrac{a+b}{c} = \dfrac{a}{c} + \dfrac{b}{c}$$	Divide. $\dfrac{4x^3 - 2x^2 + 6x - 8}{2x} = \dfrac{4x^3}{2x} - \dfrac{2x^2}{2x} + \dfrac{6x}{2x} - \dfrac{8}{2x}$ $\qquad\qquad\qquad\qquad\quad = 2x^2 - x + 3 - \dfrac{4}{x}$
13.7 Dividing a Polynomial by a Polynomial Use "long division."	Divide. $3x + 4 \overline{\smash{\big)}\, 6x^2 - 7x - 21}$ gives $2x - 5 + \dfrac{-1}{3x+4}$ $\qquad\qquad\qquad\;\; \underline{6x^2 + 8x}$ $\qquad\qquad\qquad\qquad -15x - 21$ $\qquad\qquad\qquad\qquad \underline{-15x - 20}$ $\qquad\qquad\qquad\qquad\qquad\;\; -1 \;\leftarrow$ Remainder
13.8 An Application of Exponents: Scientific Notation To write a number in scientific notation (as $a \times 10^n$), move the decimal point to the right of the first nonzero digit. If the decimal point is moved n places, and this makes the number smaller, n is positive; otherwise, n is negative. If the decimal point is not moved, n is 0.	$247 = 2.47 \times 10^2$ $0.0051 = 5.1 \times 10^{-3}$ $4.8 = 4.8 \times 10^0$ $3.25 \times 10^5 = 325{,}000$ $8.44 \times 10^{-6} = 0.00000844$

Answers to Test Your Word Power

1. B; *Example:* $5x^3 + 2x^2 - 7$
2. D; *Examples:* The term 6 has degree 0, the term $3x$ has degree 1, the term $-2x^8$ has degree 8, and the term $5x^2y^4$ has degree 6.
3. C; *Example:* $2a^2 - 3ab + b^2$
4. B; *Example:* $3t^3 + 5t$
5. A; *Examples:* -5 and $4xy^5$ are both monomials.
6. C; *Example:* $(m+4)(m-3) = \overset{F}{m(m)} - \overset{O}{3m} + \overset{I}{4m} + \overset{L}{4(-3)} = m^2 + m - 12$

Focus on Real-Data Applications

Earthquake Intensities Measured by the Richter Scale

Charles F. Richter devised a scale in 1935 to compare the intensities, or relative power, of earthquakes. The **intensity** of an earthquake is measured relative to the intensity of a standard **zero-level** earthquake of intensity I_0. The relationship is equivalent to $I = I_0 \times 10^R$, where R is the **Richter scale** measure. For example, if an earthquake has magnitude 5.0 on the Richter scale, then its intensity is calculated as $I = I_0 \times 10^{5.0} = I_0 \times 100{,}000$, which is 100,000 times as intense as a zero-level earthquake. The following diagram illustrates the intensities of earthquakes and their Richter scale magnitudes.

Intensity	I_0	$I_0 \times 10^1$	$I_0 \times 10^2$	$I_0 \times 10^3$	$I_0 \times 10^4$	$I_0 \times 10^5$	$I_0 \times 10^6$	$I_0 \times 10^7$	$I_0 \times 10^8$
Richter Scale	0	1	2	3	4	5	6	7	8

To compare two earthquakes to each other, a ratio of the intensities is calculated. For example, to compare an earthquake that measures 8.0 on the Richter scale to one that measures 5.0, simply find the ratio of the intensities.

$$\frac{\text{intensity } 8.0}{\text{intensity } 5.0} = \frac{I_0 \times 10^{8.0}}{I_0 \times 10^{5.0}} = \frac{10^8}{10^5} = 10^{8-5} = 10^3 = 1000$$

Therefore an earthquake that measures 8.0 on the Richter scale is 1000 times as intense as one that measures 5.0.

For Group Discussion

The table gives Richter scale measurements for several earthquakes.

	Earthquake	Richter Scale Measurement
1960	Concepción, Chile	9.5
1906	San Francisco, California	8.3
1939	Erzincan, Turkey	8.0
1998	Sumatra, Indonesia	7.0
1998	Adana, Turkey	6.3

Source: World Almanac and Book of Facts.

1. Compare the intensity of the 1939 Erzincan earthquake to the 1998 Sumatra earthquake.

2. Compare the intensity of the 1998 Adana earthquake to the 1906 San Francisco earthquake.

3. Compare the intensity of the 1939 Erzincan earthquake to the 1998 Adana earthquake.

4. Suppose an earthquake measures 7.2 on the Richter scale. How would the intensity of a second earthquake compare if its Richter scale measure differed by +3.0? By −1.0?

Chapter 13
REVIEW EXERCISES

[13.1] *Simplify each polynomial if possible, and write it with descending powers of the variable. Give the degree of the simplified polynomial and tell whether it is a monomial, binomial, trinomial, or none of these.*

1. $9m^2 + 11m^2 + 2m^2$

2. $-4p + p^3 - p^2 + 8p + 2$

3. $12a^5 - 9a^4 + 8a^3 + 2a^2 - a + 3$

4. $-7y^5 - 8y^4 - y^5 + y^4 + 9y$

Add or subtract as indicated.

5. Add.

$$-2a^3 + 5a^2$$
$$\underline{-3a^3 - a^2}$$

6. Add.

$$4r^3 - 8r^2 + 6r$$
$$\underline{-2r^3 + 5r^2 + 3r}$$

7. Subtract.

$$6y^2 - 8y + 2$$
$$\underline{-5y^2 + 2y - 7}$$

8. Subtract.

$$-12k^4 - 8k^2 + 7k - 5$$
$$\underline{k^4 + 7k^2 + 11k + 1}$$

9. $(2m^3 - 8m^2 + 4) + (8m^3 + 2m^2 - 7)$

10. $(-5y^2 + 3y + 11) + (4y^2 - 7y + 15)$

11. $(6p^2 - p - 8) - (-4p^2 + 2p + 3)$

12. $(12r^4 - 7r^3 + 2r^2) - (5r^4 - 3r^3 + 2r^2 + 1)$

[13.2] *Use the product rule or power rules to simplify each expression. Write the answer in exponential form.*

13. $4^3 \cdot 4^8$

14. $(-5)^6(-5)^5$

15. $(-8x^4)(9x^3)$

16. $(2x^2)(5x^3)(x^9)$

17. $(19x)^5$

18. $(-4y)^7$

19. $5(pt)^4$

20. $\left(\dfrac{7}{5}\right)^6$

21. $(3x^2y^3)^3$

22. $(t^4)^8(t^2)^5$

23. $(6x^2z^4)^2(x^3yz^2)^4$

24. Explain why the product rule for exponents does not apply to the expression $7^2 + 7^4$.

[13.3] *Find each product.*

25. $5x(2x + 14)$

26. $-3p^3(2p^2 - 5p)$

27. $(3r - 2)(2r^2 + 4r - 3)$

28. $(2y + 3)(4y^2 - 6y + 9)$

29. $(5p^2 + 3p)(p^3 - p^2 + 5)$

30. $(3k - 6)(2k + 1)$

31. $(6p - 3q)(2p - 7q)$

32. $(m^2 + m - 9)(2m^2 + 3m - 1)$

[13.4] *Find each product.*

33. $(a + 4)^2$

34. $(3p - 2)^2$

35. $(2r + 5s)^2$

36. $(r + 2)^3$

37. $(2x - 1)^3$

38. $(6m - 5)(6m + 5)$

39. $(2z + 7)(2z - 7)$

40. $(5a + 6b)(5a - 6b)$

41. $(2x^2 + 5)(2x^2 - 5)$

42. Explain why $(a + b)^2$ is not equal to $a^2 + b^2$.

[13.5] *Evaluate each expression.*

43. $5^0 + 8^0$

44. 2^{-5}

45. $\left(\dfrac{6}{5}\right)^{-2}$

46. $4^{-2} - 4^{-1}$

Simplify. Write each answer in exponential form, using only positive exponents. Assume all variables represent nonzero numbers.

47. $\dfrac{6^{-3}}{6^{-5}}$

48. $\dfrac{x^{-7}}{x^{-9}}$

49. $\dfrac{p^{-8}}{p^4}$

50. $\dfrac{r^{-2}}{r^{-6}}$

51. $(2^4)^2$

52. $(9^3)^{-2}$

53. $(5^{-2})^{-4}$

54. $(8^{-3})^4$

55. $\dfrac{(m^2)^3}{(m^4)^2}$

56. $\dfrac{y^4 \cdot y^{-2}}{y^{-5}}$

57. $\dfrac{r^9 \cdot r^{-5}}{r^{-2} \cdot r^{-7}}$

58. $(-5m^3)^2$

59. $(2y^{-4})^{-3}$

60. $\dfrac{ab^{-3}}{a^4 b^2}$

61. $\dfrac{(6r^{-1})^2 \cdot (2r^{-4})}{r^{-5}(r^2)^{-3}}$

62. $\dfrac{(2m^{-5}n^2)^3(3m^2)^{-1}}{m^{-2}n^{-4}(m^{-1})^2}$

[13.6] *Perform each division.*

63. $\dfrac{-15y^4}{-9y^2}$

64. $\dfrac{-12x^3y^2}{6xy}$

65. $\dfrac{6y^4 - 12y^2 + 18y}{-6y}$

66. $\dfrac{2p^3 - 6p^2 + 5p}{2p^2}$

67. $(5x^{13} - 10x^{12} + 20x^7 - 35x^5) \div (-5x^4)$

68. $(-10m^4n^2 + 5m^3n^3 + 6m^2n^4) \div (5m^2n)$

[13.7] *Perform each division.*

69. $(2r^2 + 3r - 14) \div (r - 2)$

70. $\dfrac{12m^2 - 11m - 10}{3m - 5}$

71. $\dfrac{10a^3 + 5a^2 - 14a + 9}{5a^2 - 3}$

72. $\dfrac{2k^4 + 4k^3 + 9k^2 - 8}{2k^2 + 1}$

[13.8] *Write each number in scientific notation.*

73. 48,000,000

74. 28,988,000,000

75. 0.000065

76. 0.0000000824

Write each number without exponents.

77. 2.4×10^4

78. 7.83×10^7

79. 8.97×10^{-7}

80. 9.95×10^{-12}

Simplify. Write the answers both in scientific notation and as numbers without exponents.

81. $(2 \times 10^{-3})(4 \times 10^5)$

82. $\dfrac{8 \times 10^4}{2 \times 10^{-2}}$

83. $\dfrac{12 \times 10^{-8}}{4 \times 10^{-3}}$

84. $\dfrac{(2.5 \times 10^5)(4.8 \times 10^{-4})}{(7.5 \times 10^8)(1.6 \times 10^{-5})}$

Solve each application problem. Write your answers both in scientific notation and as numbers without exponents.

85. The population of the United States in 2002 was about 2.88×10^8 people spread over 3.54×10^6 square miles of land. What is the average number of people per square mile? (*Source:* U.S. Bureau of the Census.)

86. Light travels at a speed of 3.0×10^5 kilometers per second. It takes about 36 seconds for light from the sun to reach Venus. How far is Venus from the sun? (*Source:* World Almanac.)

MIXED REVIEW EXERCISES

Perform the indicated operations. Write answers with positive exponents. Assume that no denominators are equal to 0.

87. $19^0 - 3^0$

88. $(3p)^4(3p^{-7})$

89. 7^{-2}

90. $(-7 + 2k)^2$

91. $\dfrac{2y^3 + 17y^2 + 37y + 7}{2y + 7}$

92. $\left(\dfrac{6r^2s}{5}\right)^4$

93. $-m^5(8m^2 + 10m + 6)$

94. $\left(\dfrac{1}{2}\right)^{-5}$

95. $(25x^2y^3 - 8xy^2 + 15x^3y) \div (5x)$

96. $(6r^{-2})^{-1}$

97. $(2x + y)^3$

98. $2^{-1} + 4^{-1}$

99. $(a + 2)(a^2 - 4a + 1)$

100. $(5y^3 - 8y^2 + 7) - (-3y^3 + y^2 + 2)$

101. $(2r + 5)(5r - 2)$

102. $(12a + 1)(12a - 1)$

103. Find a polynomial that represents the area of the rectangle shown.

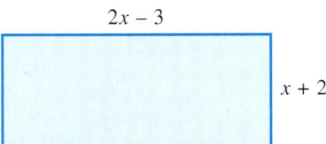

104. If the side of a square has a measure represented by $5x^4 + 2x^2$, what polynomial represents its area?

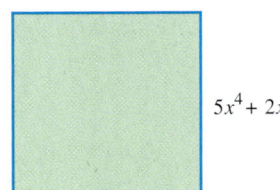

Chapter 13
TEST

Perform the indicated operations.

1. $(5t^4 - 3t^2 + 7t + 3) - (t^4 - t^3 + 3t^2 + 8t + 3)$

    1. _____

2. $(2y^2 - 8y + 8) + (-3y^2 + 2y + 3) - (y^2 + 3y - 6)$

    2. _____

3. Subtract.

    $9t^3 - 4t^2 + 2t + 2$
    $\underline{9t^3 + 8t^2 - 3t - 6}$

    3. _____

Simplify, and write each answer with only positive exponents.

4. $(-2)^3(-2)^2$

    4. _____

5. $\left(\dfrac{6}{m^2}\right)^3, \quad m \neq 0$

    5. _____

6. $3x^2(-9x^3 + 6x^2 - 2x + 1)$

    6. _____

7. $(2r - 3)(r^2 + 2r - 5)$

    7. _____

8. $(t - 8)(t + 3)$

    8. _____

9. $(4x + 3y)(2x - y)$

    9. _____

10. $(5x - 2y)^2$

    10. _____

11. $(10v + 3w)(10v - 3w)$

    11. _____

12. $(x + 1)^3$

    12. _____

Evaluate each expression.

13. 5^{-4}

    13. _____

14. $(-3)^0 + 4^0$

    14. _____

15. $4^{-1} + 3^{-1}$

    15. _____

Perform the indicated operations. In Exercises 16 and 17, write each answer using only positive exponents. Assume that variables represent nonzero numbers.

16. $\dfrac{8^{-1} \cdot 8^4}{8^{-2}}$

17. $\dfrac{(x^{-3})^{-2}(x^{-1}y)^2}{(xy^{-2})^2}$

18. $\dfrac{8y^3 - 6y^2 + 4y + 10}{2y}$

19. $(-9x^2y^3 + 6x^4y^3 + 12xy^3) \div (3xy)$

20. $\dfrac{2x^2 + x - 36}{x - 4}$

21. $(3x^3 - x + 4) \div (x - 2)$

Write each number in scientific notation.

22. (a) 344,000,000,000

(b) 0.00000557

Write each number without exponents.

23. (a) 2.96×10^7

(b) 6.07×10^{-8}

24. The mass of the sun is about 330,000 times the mass of Earth. If Earth's mass is about 5.98×10^{24} kg, find the approximate mass of the sun. Write your answer in scientific notation.

25. What polynomial expression represents the area of this square?

$3x + 9$

Factoring and Applications

14.1 Factors; The Greatest Common Factor

14.2 Factoring Trinomials

14.3 Factoring Trinomials by Grouping

14.4 Factoring Trinomials Using FOIL

14.5 Special Factoring Techniques

Summary Exercises on Factoring

14.6 Solving Quadratic Equations by Factoring

14.7 Applications of Quadratic Equations

Wireless communication uses radio waves to carry signals and messages across distances. Cellular phones, one of the most popular forms of wireless communication, have become an invaluable tool for people to stay connected to family, friends, and work while on the go. In 2002 alone, U.S. sales of cellular phones totaled some $8835 million and 66% of all U.S. households used cell phones. (*Source: Microsoft Encarta Encyclopedia;* Consumer Electronics Association.)

In Exercise 31 of Section 14.7, we use a *quadratic equation* to model the number of cell phones owned by Americans in various years.

14.1 Factors; The Greatest Common Factor

OBJECTIVES

1. Find the greatest common factor of a list of numbers.
2. Find the greatest common factor of a list of variable terms.
3. Factor out the greatest common factor.
4. Factor by grouping.

Recall from **Section 4.2** that to **factor** a number means to write it as the product of two or more numbers. The product is called the **factored form** of the number. Here is an example.

$$12 = \underbrace{6 \cdot 2}_{\text{Factored form}}$$
(Factors)

Factoring is a process that "undoes" multiplying. We multiply $6 \cdot 2$ to get 12, but we factor 12 by writing it as $6 \cdot 2$.

OBJECTIVE 1 Find the greatest common factor of a list of numbers. An integer that is a factor of two or more integers is a **common factor** of those integers. For example, 6 is a common factor of 18 and 24 because 6 is a factor of both 18 and 24. Other common factors of 18 and 24 are 1, 2, and 3. The **greatest common factor (GCF)** of a list of integers is the largest common factor of those integers. This means 6 is the greatest common factor of 18 and 24, since it is the largest of their common factors.

> **NOTE**
> Factors of a number are also divisors of the number. The greatest common factor is the same as the greatest common divisor.

EXAMPLE 1 Finding the Greatest Common Factor for Numbers

Find the greatest common factor (GCF) for each list of numbers.

(a) 30, 45

First write each number in prime factored form.

$$30 = 2 \cdot 3 \cdot 5$$
$$45 = 3 \cdot 3 \cdot 5$$

Use each prime the *least* number of times it appears in *all* the factored forms. There is no 2 in the prime factored form of 45, so there will be no 2 in the greatest common factor. The least number of times 3 appears in all the factored forms is 1; the least number of times 5 appears is also 1. From this, the

$$\text{GCF} = 3^1 \cdot 5^1 = 3 \cdot 5 = 15.$$

(b) 72, 120, 432

Find the prime factored form of each number.

$$72 = 2 \cdot 2 \cdot 2 \cdot 3 \cdot 3$$
$$120 = 2 \cdot 2 \cdot 2 \cdot 3 \cdot 5$$
$$432 = 2 \cdot 2 \cdot 2 \cdot 2 \cdot 3 \cdot 3 \cdot 3$$

The least number of times 2 appears in all the factored forms is 3, and the least number of times 3 appears is 1. There is no 5 in the prime factored form of either 72 or 432, so the

$$\text{GCF} = 2^3 \cdot 3^1 = 24.$$

Continued on Next Page

(c) 10, 11, 14

Write the prime factored form of each number.

$$10 = 2 \cdot 5$$
$$11 = 11$$
$$14 = 2 \cdot 7$$

There are no primes common to all three numbers, so the GCF is 1.

▶▶▶ **Work Problem 1 at the Side.**

OBJECTIVE 2 Find the greatest common factor of a list of variable terms. The terms $x^4, x^5, x^6,$ and x^7 have x^4 as the greatest common factor because the least exponent on the variable x is 4.

$$x^4 = 1 \cdot \mathbf{x^4}, \quad x^5 = x \cdot \mathbf{x^4}, \quad x^6 = x^2 \cdot \mathbf{x^4}, \quad x^7 = x^3 \cdot \mathbf{x^4}$$

> **NOTE**
> The exponent on a variable in the GCF is the *least* exponent that appears on that variable in *all* the terms.

EXAMPLE 2 Finding the Greatest Common Factor for Variable Terms

Find the greatest common factor (GCF) for each list of terms.

(a) $21m^7, -18m^6, 45m^8$

$$21m^7 = \mathbf{3} \cdot 7 \cdot m^7$$
$$-18m^6 = -1 \cdot 2 \cdot \mathbf{3} \cdot 3 \cdot m^6$$
$$45m^8 = \mathbf{3} \cdot 3 \cdot 5 \cdot m^8$$

First, 3 is the greatest common factor of the coefficients 21, −18, and 45. The least exponent on m is 6, so the

$$\text{GCF} = 3m^6.$$

(b) $x^4y^2, x^7y^5, x^3y^7, y^{15}$

$$x^4\mathbf{y^2}, \quad x^7\mathbf{y^5}, \quad x^3\mathbf{y^7}, \quad \mathbf{y}^{15}$$

There is no x in the last term, y^{15}, so x will not appear in the greatest common factor. There is a y in each term, however, and 2 is the least exponent on y. The GCF is y^2.

(c) $-a^2b, -ab^2$

$$-a^2b = -1a^2b = -1 \cdot 1 \cdot a^2b$$
$$-ab^2 = -1ab^2 = -1 \cdot 1 \cdot ab^2$$

The factors of −1 are −1 and 1. Since $1 > -1$, the GCF is $1ab$ or ab.

> **NOTE**
> In a list of negative terms, sometimes a negative common factor is preferable (even though it is not the greatest common factor). In Example 2(c), for instance, we might prefer $-ab$ as the common factor. In factoring exercises, either answer will be acceptable.

---

1 Find the greatest common factor (GCF) for each list of numbers.

(a) 30, 20, 15

$$30 = 2 \cdot 3 \cdot 5$$
$$20 = 2 \cdot \underline{} \cdot \underline{}$$
$$15 = 3 \cdot \underline{}$$
$$\text{GCF} = \underline{}$$

(b) 42, 28, 35

(c) 12, 18, 26, 32

(d) 10, 15, 21

ANSWERS
1. **(a)** 2; 5; 5; GCF-5 **(b)** 7 **(c)** 2 **(d)** 1

Chapter 14 Factoring and Applications

② Find the greatest common factor for each list of terms.

(a) $6m^4, 9m^2, 12m^5$

$6m^4 = 2 \cdot \underline{} \cdot m^4$

$9m^2 = 3 \cdot \underline{} \cdot \underline{}$

$12m^5 = 2 \cdot 2 \cdot \underline{} \cdot \underline{}$

GCF = $\underline{}$

(b) $-12p^5, -18q^4$

(c) y^4z^2, y^6z^8, z^9

(d) $12p^{11}, 17q^5$

ANSWERS

2. (a) 3; 3; m^2; 3; m^5; GCF-$3m^2$ (b) 6
   (c) z^2 (d) 1

We find the greatest common factor of a list of terms as follows.

Finding the Greatest Common Factor (GCF)

Step 1 **Factor.** Write each number in prime factored form.

Step 2 **List common factors.** List each prime number or each variable that is a factor of every term in the list. (If a prime does not appear in one of the prime factored forms, it cannot appear in the greatest common factor.)

Step 3 **Choose least exponents.** Use as exponents on the common prime factors the *least* exponents from the prime factored forms.

Step 4 **Multiply.** Multiply the primes from Step 3. If there are no primes left after Step 3, the greatest common factor is 1.

◀◀◀ Work Problem 2 at the Side.

OBJECTIVE 3 Factor out the greatest common factor. The polynomial

$$3m + 12$$

has two terms, $3m$ and 12. The greatest common factor of these two terms is 3. We can write $3m + 12$ so that each term is a product with 3 as one factor.

$$3m + 12 = 3 \cdot m + 3 \cdot 4$$
$$= 3(m + 4) \qquad \text{Distributive property}$$

The factored form of $3m + 12$ is $3(m + 4)$. This process is called **factoring out the greatest common factor.**

> **CAUTION**
> The polynomial $3m + 12$ is *not* in factored form when written as a *sum*.
>
> $3 \cdot m + 3 \cdot 4$ Not in factored form
>
> The *terms* are factored, but the polynomial is not. The factored form of $3m + 12$ is the *product* shown below.
>
> $3(m + 4)$ In factored form

Writing a polynomial as a product, that is, in factored form, is called **factoring** the polynomial.

EXAMPLE 3 Factoring Out the Greatest Common Factor

Factor out the greatest common factor.

(a) $5y^2 + 10y = 5y(y) + 5y(2)$ GCF = $5y$

$ = 5y(y + 2)$ Distributive property

Check by multiplying. $5y(y + 2) = 5y(y) + 5y(2)$

$ = 5y^2 + 10y$ Original polynomial

Continued on Next Page

(b) $20m^5 + 10m^4 - 15m^3$
$= 5m^3(4m^2) + 5m^3(2m) - 5m^3(3)$ GCF = $5m^3$
$= 5m^3(4m^2 + 2m - 3)$ Factor out $5m^3$.

Check: $5m^3(4m^2 + 2m - 3) = 20m^5 + 10m^4 - 15m^3$ Original polynomial

(c) $x^5 + x^3 = x^3(x^2) + x^3(1) = x^3(x^2 + 1)$ Don't forget the 1.

(d) $20m^7p^2 - 36m^3p^4 = 4m^3p^2(5m^4) - 4m^3p^2(9p^2)$ GCF = $4m^3p^2$
$= 4m^3p^2(5m^4 - 9p^2)$

(e) $\dfrac{1}{6}n^2 + \dfrac{5}{6}n = \dfrac{1}{6}n(n) + \dfrac{1}{6}n(5)$ GCF = $\tfrac{1}{6}n$
$= \dfrac{1}{6}n(n + 5)$

> **CAUTION**
> Be sure to include the **1** in a problem like Example 3(c). *Always check that the factored form can be multiplied out to give the original polynomial.*

Work Problem 3 at the Side. ▶▶▶

EXAMPLE 4   Factoring Out the Greatest Common Factor

Factor out the greatest common factor.

(a) $a(a + 3) + 4(a + 3)$
The binomial $a + 3$ is the greatest common factor here.

Same
$a(a + 3) + 4(a + 3) = (a + 3)(a + 4)$

(b) $x^2(x + 1) - 5(x + 1) = (x + 1)(x^2 - 5)$ Factor out $x + 1$.

Work Problem 4 at the Side. ▶▶▶

OBJECTIVE 4   Factor by grouping. When a polynomial has four terms, common factors can sometimes be used to **factor by grouping**.

EXAMPLE 5   Factoring by Grouping

Factor by grouping.

(a) $2x + 6 + ax + 3a$
Group the first two terms and the last two terms, since the first two terms have a common factor of 2 and the last two terms have a common factor of a.

$2x + 6 + ax + 3a = (2x + 6) + (ax + 3a)$
$= 2(x + 3) + a(x + 3)$

The expression is still *not* in factored form because it is the **sum** of two terms. Now, however, $x + 3$ is a common factor and can be factored out.

Continued on Next Page

---

3 Factor out the greatest common factor.

(a) $4x^2 + 6x$

(b) $10y^5 - 8y^4 + 6y^2$

(c) $m^7 + m^9$

(d) $8p^5q^2 + 16p^6q^3 - 12p^4q^7$

(e) $\dfrac{1}{3}b^2 - \dfrac{2}{3}b$

(f) $13x^2 - 27$

4 Factor out the greatest common factor.

(a) $r(t - 4) + 5(t - 4)$

(b) $y^2(y + 2) - 3(y + 2)$

(c) $x(x - 1) - 5(x - 1)$

ANSWERS
3. **(a)** $2x(2x + 3)$
   (b) $2y^2(5y^3 - 4y^2 + 3)$
   (c) $m^7(1 + m^2)$
   (d) $4p^4q^2(2p + 4p^2q - 3q^5)$
   (e) $\dfrac{1}{3}b(b - 2)$
   (f) no common factor (except 1)
4. **(a)** $(t - 4)(r + 5)$
   (b) $(y + 2)(y^2 - 3)$
   (c) $(x - 1)(x - 5)$

5 Factor by grouping.

(a) $pq + 5q + 2p + 10$

(b) $2xy + 3y + 2x + 3$

(c) $2a^2 - 4a + 3ab - 6b$

(d) $x^3 + 3x^2 - 5x - 15$

$$2x + 6 + ax + 3a = (2x + 6) + (ax + 3a) \quad \text{Group terms.}$$
$$= 2(x + 3) + a(x + 3) \quad \text{Factor each group.}$$
$$= (x + 3)(2 + a) \quad \text{Factor out } x + 3.$$

The final result is in factored form because it is a *product*. Note that the goal in factoring by grouping is to get a common factor, $x + 3$ here, so that the last step is possible. Check by multiplying the binomials using the FOIL method from **Section 13.3**.

Check: $(x + 3)(2 + a) = 2x + ax + 6 + 3a \quad$ FOIL
$$= 2x + 6 + ax + 3a, \quad \text{Rearrange terms.}$$

which is the original polynomial.

(b) $6ax + 24x + a + 4 = (6ax + 24x) + (a + 4) \quad$ Group terms.
$$= 6x(a + 4) + 1(a + 4) \quad \text{Factor each group; remember the 1.}$$
$$= (a + 4)(6x + 1) \quad \text{Factor out } a + 4.$$

Check: $(a + 4)(6x + 1) = 6ax + a + 24x + 4 \quad$ FOIL
$$= 6ax + 24x + a + 4, \quad \text{Rearrange terms.}$$

which is the original polynomial.

(c) $2x^2 - 10x + 3xy - 15y = (2x^2 - 10x) + (3xy - 15y) \quad$ Group terms.
$$= 2x(x - 5) + 3y(x - 5) \quad \text{Factor each group.}$$
$$= (x - 5)(2x + 3y) \quad \text{Factor out the common factor, } x - 5.$$

Check: $(x - 5)(2x + 3y) = 2x^2 + 3xy - 10x - 15y \quad$ FOIL
$$= 2x^2 - 10x + 3xy - 15y \quad \text{Original polynomial}$$

(d) $t^3 + 2t^2 - 3t - 6 = (t^3 + 2t^2) + (-3t - 6) \quad$ Group terms.
$$= t^2(t + 2) - 3(t + 2) \quad \text{Factor out } -3 \text{ so there is a common factor, } t + 2; -3(t + 2) = -3t - 6.$$
$$= (t + 2)(t^2 - 3) \quad \text{Factor out } t + 2.$$

Check: $(t + 2)(t^2 - 3) = t^3 - 3t + 2t^2 - 6 \quad$ FOIL
$$= t^3 + 2t^2 - 3t - 6 \quad \text{Original polynomial}$$

> **CAUTION**
> *Be careful with signs when grouping* in a problem like Example 5(d). It is wise to check the factoring in the second step, as shown in the example side comment, before continuing.

Work Problem 5 at the Side.

ANSWERS
5. (a) $(p + 5)(q + 2)$
 (b) $(2x + 3)(y + 1)$
 (c) $(a - 2)(2a + 3b)$
 (d) $(x + 3)(x^2 - 5)$

Use these steps to factor a polynomial with four terms by grouping.

> **Factoring by Grouping**
>
> *Step 1* **Group terms.** Collect the terms into two groups so that each group has a common factor.
>
> *Step 2* **Factor within groups.** Factor out the greatest common factor from each group.
>
> *Step 3* **Factor the entire polynomial.** Factor a common binomial factor from the results of Step 2.
>
> *Step 4* **If necessary, rearrange terms.** If Step 2 does not result in a common binomial factor, try a different grouping.

EXAMPLE 6 Rearranging Terms Before Factoring by Grouping

Factor by grouping.

(a) $10x^2 - 12y + 15x - 8xy$

Factoring out the common factor of 2 from the first two terms and the common factor of x from the last two terms gives

$$10x^2 - 12y + 15x - 8xy = 2(5x^2 - 6y) + x(15 - 8y).$$

This does not lead to a common factor, so we try rearranging the terms. There is usually more than one way to do this. We try

$$10x^2 - 8xy - 12y + 15x,$$

and group the first two terms and the last two terms as follows.

$$10x^2 - 8xy - 12y + 15x = 2x(5x - 4y) + 3(-4y + 5x)$$
$$= 2x(5x - 4y) + 3(5x - 4y)$$
$$= (5x - 4y)(2x + 3)$$

Check: $(5x - 4y)(2x + 3) = 10x^2 + 15x - 8xy - 12y$ FOIL
$$= 10x^2 - 12y + 15x - 8xy$$ Original polynomial

(b) $2xy + 12 - 3y - 8x$

We need to rearrange these terms to get two groups that each have a common factor. Trial and error suggests the following grouping.

$2xy + 12 - 3y - 8x = (2xy - 3y) + (-8x + 12)$ Group terms.
$ = y(2x - 3) - 4(2x - 3)$ Factor each group; be careful with signs.
$ = (2x - 3)(y - 4)$ Factor out $2x - 3$.

Since the quantities in parentheses in the second step must be the same, we factored out -4 rather than 4. *Check* by multiplying.

> **CAUTION**
> Use negative signs carefully when grouping, as in Example 6(b), or a sign error will occur. *Always check by multiplying.*

Work Problem 6 at the Side. ▶

6 Factor by grouping.

(a) $6y^2 - 20w + 15y - 8yw$

(b) $9mn - 4 + 12m - 3n$

ANSWERS
6. (a) $(2y + 5)(3y - 4w)$
   (b) $(3m - 1)(3n + 4)$

Real-Data Applications

Idle Prime Time

A positive integer greater than 1 is a **prime number** if its only factors are 1 and itself. Every positive integer can be written as a product of prime numbers in a unique way, except for the order of the factors. Finding new primes has intrigued people from ancient Greece to modern times. The *Great Internet Mersenne Prime Search* is a consortium headed by George Woltman and Scott Kurowski that has discovered seven world record primes. On May 15, 2004, Josh Findley discovered the prime number, $2^{24,036,583} - 1$. His calculation took over two weeks on his 2.4 GHz Pentium 4 computer. This prime number is nearly a million digits larger than the last prime found and has 7,235,733 decimal digits when written out. (*Source:* http://www.mersenne.org/prime.htm)

Prime numbers are essential in the development of unbreakable codes that, in an era of Internet commerce, ensure security in transmitting and storing computer data.

The oldest known method for finding prime numbers is the Sieve of Eratosthenes, similar to the version shown below. Numbers that are not prime (composite numbers) are eliminated and only the prime numbers are left. Begin with 2. Two is prime but multiples of 2 are not, so delete the remaining numbers in Column 2 and all of Columns 4 and 6. Three is prime, but multiples of 3 are not, so delete the remaining numbers in Column 3. Examine the remaining numbers and eliminate any that are composite (such as 25 or 91). The prime numbers are highlighted.

For Group Discussion

1. **Twin primes** occur in pairs that differ by 2. List all the twin primes from the table.

2. Observe that all prime numbers larger than 3 are in Columns 1 and 5. Each number in Column 5 is 1 less than a multiple of 6, and therefore has the form $6n - 1$. Each number in Column 1 has a similar structure, $6n + 1$. Show that the larger of each of these twin primes, found in the year 2000, has the form $6n + 1$:

   $$1{,}693{,}965 \times 2^{66{,}443} \pm 1$$

   and

   $$4{,}648{,}619{,}711{,}505 \times 2^{60{,}000} \pm 1.$$

   (*Hint:* Show that the leading term is divisible by both 2 and 3.)

SIEVE OF ERATOSTHENES

Col 1	Col 2	Col 3	Col 4	Col 5	Col 6
1	2	3	4	5	6
7	8	9	10	11	12
13	14	15	16	17	18
19	20	21	22	23	24
25	26	27	28	29	30
31	32	33	34	35	36
37	38	39	40	41	42
43	44	45	46	47	48
49	50	51	52	53	54
55	56	57	58	59	60
61	62	63	64	65	66
67	68	69	70	71	72
73	74	75	76	77	78
79	80	81	82	83	84
85	86	87	88	89	90
91	92	93	94	95	96
97	98	99	100	101	102

3. **Mersenne primes,** named for the 17th-century French monk Marin Mersenne, have the form $2^p - 1$, where p is a prime number. Not all such numbers are prime. Show that $2^{11} - 1$ is composite and $2^5 - 1$ is prime.

4. A **Sophie Germain prime,** named for an 18th-century French mathematician, is an odd prime p for which $2p + 1$ is also prime. For example, 5 is a Sophie Germain prime since 11 ($2 \cdot 5 + 1$) is prime, but 13 is not since 27 ($2 \cdot 13 + 1$) is composite. List the Sophie Germain primes from the table.

14.1 Exercises

Find the greatest common factor for each list of numbers. See Example 1.

1. 12, 16
2. 18, 24
3. 40, 20, 4
4. 50, 30, 5
5. 18, 24, 36, 48
6. 15, 30, 45, 75
7. 4, 9, 12
8. 9, 16, 24

Find the greatest common factor for each list of terms. See Example 2.

9. $16y, 24$
10. $18w, 27$
11. $30x^3, 40x^6, 50x^7$
12. $60z^4, 70z^8, 90z^9$
13. $-x^4y^3, -xy^2$
14. $-a^4b^5, -a^3b$
15. $42ab^3, -36a, 90b, -48ab$
16. $45c^3d, 75c, 90d, -105cd$

Complete each factoring.

17. $9m^4 = 3m^2(\quad)$
18. $12p^5 = 6p^3(\quad)$
19. $-8z^9 = -4z^5(\quad)$
20. $-15k^{11} = -5k^8(\quad)$
21. $6m^4n^5 = 3m^3n(\quad)$
22. $27a^3b^2 = 9a^2b(\quad)$
23. $12y + 24 = 12(\quad)$
24. $18p + 36 = 18(\quad)$
25. $10a^2 - 20a = 10a(\quad)$
26. $15x^2 - 30x = 15x(\quad)$
27. $8x^2y + 12x^3y^2 = 4x^2y(\quad)$
28. $18s^3t^2 + 10st = 2st(\quad)$

Factor out the greatest common factor. See Examples 3 and 4.

29. $x^2 - 4x$
30. $m^2 - 7m$
31. $6t^2 + 15t$
32. $8x^2 + 6x$
33. $\frac{1}{4}d^2 - \frac{3}{4}d$
34. $\frac{1}{5}z^2 + \frac{3}{5}z$
35. $12x^3 + 6x^2$
36. $21b^3 - 7b^2$
37. $65y^{10} + 35y^6$
38. $100a^5 + 16a^3$
39. $11w^3 - 100$
40. $13z^5 - 80$
41. $8m^2n^3 + 24m^2n^2$
42. $19p^2y - 38p^2y^3$
43. $4x^3 - 10x^2 + 6x$
44. $9z^3 - 6z^2 + 12z$
45. $13y^8 + 26y^4 - 39y^2$

46. $5x^5 + 25x^4 - 20x^3$

47. $45q^4p^5 + 36qp^6 + 81q^2p^3$

48. $125a^3z^5 + 60a^4z^4 - 85a^5z^2$

49. $c(x + 2) + d(x + 2)$

50. $r(5 - x) + t(5 - x)$

51. $a^2(2a + b) - b(2a + b)$

52. $3x(x^2 + 5) - y(x^2 + 5)$

53. $q(p + 4) - 1(p + 4)$

54. $y^2(x - 4) + 1(x - 4)$

Factor by grouping. See Examples 5 and 6.

55. $5m + mn + 20 + 4n$

56. $ts + 5t + 2s + 10$

57. $6xy - 21x + 8y - 28$

58. $2mn - 8n + 3m - 12$

59. $3xy + 9x + y + 3$

60. $6n + 4mn + 3 + 2m$

61. $7z^2 + 14z - az - 2a$

62. $2b^2 + 3b - 8ab - 12a$

63. $18r^2 + 12ry - 3xr - 2xy$

64. $5m^2 + 15mp - 2mr - 6pr$

65. $w^3 + w^2 + 9w + 9$

66. $y^3 + y^2 + 6y + 6$

67. $3a^3 + 6a^2 - 2a - 4$

68. $10x^3 + 15x^2 - 8x - 12$

69. $16m^3 - 4m^2p^2 - 4mp + p^3$

70. $10t^3 - 2t^2s^2 - 5ts + s^3$

71. $y^2 + 3x + 3y + xy$

72. $m^2 + 14p + 7m + 2mp$

73. $2z^2 + 6w - 4z - 3wz$

74. $2a^2 + 20b - 8a - 5ab$

RELATING CONCEPTS (EXERCISES 75–78) For Individual or Group Work

In many cases, the choice of which pairs of terms to group when factoring by grouping can be made in different ways. To see this for Example 6(b), work Exercises 75–78 in order.

75. Start with the polynomial from Example 6(b), $2xy + 12 - 3y - 8x$, and rearrange the terms as follows: $2xy - 8x - 3y + 12$. What property from **Section 1.3** allows this?

76. Group the first two terms and the last two terms of the rearranged polynomial in Exercise 75. Then factor each group.

77. Is your result from Exercise 76 in factored form? Explain your answer.

78. If your answer to Exercise 77 is *no*, factor the polynomial. Is the result the same as the one shown for Example 6(b)?

14.2 Factoring Trinomials

Using FOIL, the product of the binomials $k - 3$ and $k + 1$ is

$$(k - 3)(k + 1) = k^2 - 2k - 3. \quad \text{Multiplying}$$

Suppose instead that we are given the polynomial $k^2 - 2k - 3$ and want to rewrite it as the product $(k - 3)(k + 1)$. That is,

$$k^2 - 2k - 3 = (k - 3)(k + 1). \quad \text{Factoring}$$

Recall from **Section 14.1** that this process is called factoring the polynomial. Factoring reverses or "undoes" multiplying.

OBJECTIVE 1 Factor trinomials with a coefficient of 1 for the squared term. When factoring polynomials with integer coefficients, we use only integers in the factors. For example, we can factor $x^2 + 5x + 6$ by finding integers m and n such that

$$x^2 + 5x + 6 = (x + m)(x + n).$$

To find these integers m and n, we first use FOIL to multiply the two binomials on the right side of the equation:

$$(x + m)(x + n) = x^2 + nx + mx + mn.$$

By the distributive property,

$$x^2 + nx + mx + mn = x^2 + (n + m)x + mn.$$

Comparing this result with $x^2 + 5x + 6$ shows that we must find integers m and n having a sum of 5 and a product of 6.

$$x^2 + 5x + 6 = x^2 + (n + m)x + mn$$

Product of m and n is 6.
Sum of m and n is 5.

Because many pairs of integers have a sum of 5, it is best to begin by listing those pairs of integers whose product is 6. Both 5 and 6 are positive, so we consider only pairs in which both integers are positive.

▶ **Work Problem 1 at the Side.** ▶▶▶

From Problem 1 at the side, we see that the numbers 1 and 6 and the numbers 2 and 3 both have a product of 6, but only the pair 2 and 3 has a sum of 5. So 2 and 3 are the required integers, and

$$x^2 + 5x + 6 = (x + 2)(x + 3).$$

Check by multiplying the binomials using FOIL. *Make sure that the sum of the outer and inner products produces the correct middle term.*

Check: $(x + 2)(x + 3) = x^2 + 5x + 6$

$\quad\quad\quad\quad\quad 2x$
$\quad\quad\quad\quad\quad \underline{3x}$
$\quad\quad\quad\quad\quad 5x \quad$ Add.

This method of factoring can be used only for trinomials that have 1 as the coefficient of the squared term. Methods for factoring other trinomials will be given in the next two sections.

OBJECTIVES

1. Factor trinomials with a coefficient of 1 for the squared term.
2. Factor trinomials after factoring out the greatest common factor.

1 (a) List all pairs of positive integers whose product is 6.

(b) Find the pair from part (a) whose sum is 5.

ANSWERS
1. (a) 1, 6; 2, 3   (b) 2, 3

Chapter 14 Factoring and Applications

② Factor each trinomial.

(a) $y^2 + 12y + 20$

First complete the given list of numbers.

Factors of 20	Sums of Factors
20, 1	20 + 1 = 21
10, __	10 + __ = __
5, __	5 + __ = __

(b) $x^2 + 9x + 18$

③ Factor each trinomial.

(a) $t^2 - 12t + 32$

First complete the given list of numbers.

Factors of 32	Sums of Factors
−32, −1	−32 + (−1) = −33
−16, __	−16 + (__) = __
−8, __	−8 + (__) = __

(b) $y^2 - 10y + 24$

ANSWERS
2. (a) 2; 2; 12; 4; 4; 9; $(y + 10)(y + 2)$
   (b) $(x + 3)(x + 6)$
3. (a) −2; −2; −18; −4; −4; −12; $(t - 8)(t - 4)$
   (b) $(y - 6)(y - 4)$

EXAMPLE 1 Factoring a Trinomial with All Positive Terms

Factor $m^2 + 9m + 14$.

Look for two integers whose product is **14** and whose sum is **9**. List the pairs of integers whose products are 14. Then examine the sums. Only positive integers are needed since all signs in $m^2 + 9m + 14$ are positive.

Factors of 14	Sums of Factors	
14, 1	14 + 1 = 15	
7, 2	7 + 2 = 9	Sum is 9.

From the list, 7 and 2 are the required integers, since $7 \cdot 2 = 14$ and $7 + 2 = 9$. Thus,

$$m^2 + 9m + 14 = (m + 2)(m + 7).$$

Check: $(m + 2)(m + 7) = m^2 + 7m + 2m + 14$ FOIL
$= m^2 + 9m + 14$ Original polynomial

NOTE
In Example 1, the answer $(m + 2)(m + 7)$ also could have been written
$$(m + 7)(m + 2).$$
Because of the commutative property of multiplication, the order of the factors does not matter. *Always check by multiplying.*

◀◀ **Work Problem 2 at the Side.**

EXAMPLE 2 Factoring a Trinomial with a Negative Middle Term

Factor $x^2 - 9x + 20$.

Find two integers whose product is 20 and whose sum is −9. Since the numbers we are looking for have a *positive product* and a *negative sum*, we consider only pairs of negative integers.

Factors of 20	Sums of Factors	
−20, −1	−20 + (−1) = −21	
−10, −2	−10 + (−2) = −12	
−5, −4	−5 + (−4) = −9	Sum is −9.

The required integers are −5 and −4, so
$$x^2 - 9x + 20 = (x - 5)(x - 4).$$

Check: $(x - 5)(x - 4) = x^2 - 4x - 5x + 20$
$= x^2 - 9x + 20$

◀◀ **Work Problem 3 at the Side.**

EXAMPLE 3 Factoring a Trinomial with Two Negative Terms

Factor $p^2 - 2p - 15$.

Find two integers whose product is -15 and whose sum is -2. If these numbers do not come to mind right away, find them (if they exist) by listing all the pairs of integers whose product is -15. Because the last term, -15, is negative, we need pairs of integers with different signs.

Factors of -15	Sums of Factors
$15, -1$	$15 + (-1) = 14$
$-15, 1$	$-15 + 1 = -14$
$5, -3$	$5 + (-3) = 2$
$-5, 3$	$-5 + 3 = -2$ Sum is -2.

The required integers are -5 and 3, so

$$p^2 - 2p - 15 = (p - 5)(p + 3).$$

Check: Multiply $(p - 5)(p + 3)$.

NOTE
In Examples 1–3, notice that we listed factors in descending order (disregarding sign) when we were looking for the required pair of integers. This helps avoid skipping the correct combination.

▶ Work Problem 4 at the Side.

As shown in the next example, some trinomials cannot be factored using only integers. We call such trinomials **prime polynomials.**

EXAMPLE 4 Deciding whether Polynomials Are Prime

Factor each trinomial.

(a) $x^2 - 5x + 12$

As in Example 2, both factors must be negative to give a positive product and a negative sum. First, list all pairs of negative integers whose product is 12. Then examine the sums.

Factors of 12	Sums of Factors
$-12, -1$	$-12 + (-1) = -13$
$-6, -2$	$-6 + (-2) = -8$
$-4, -3$	$-4 + (-3) = -7$

None of the pairs of integers has a sum of -5. Therefore, the trinomial $x^2 - 5x + 12$ *cannot be factored using only integers;* it is a *prime polynomial.*

(b) $k^2 - 8k + 11$

There is no pair of integers whose product is 11 and whose sum is -8, so $k^2 - 8k + 11$ is a prime polynomial.

▶ Work Problem 5 at the Side.

4 Factor each trinomial.

(a) $a^2 - 9a - 22$

(b) $r^2 - 6r - 16$

5 Factor each trinomial, if possible.

(a) $r^2 - 3r - 4$

(b) $m^2 - 2m + 5$

ANSWERS
4. (a) $(a - 11)(a + 2)$ **(b)** $(r - 8)(r + 2)$
5. (a) $(r - 4)(r + 1)$ **(b)** prime

6 Factor each trinomial.

(a) $b^2 - 3ab - 4a^2$

(b) $r^2 - 6rs + 8s^2$

The procedure for factoring a trinomial of the form $x^2 + bx + c$ follows.

> **Factoring $x^2 + bx + c$**
>
> Find two integers whose product is c and whose sum is b.
>
> 1. Both integers must be positive if b and c are positive.
> 2. Both integers must be negative if c is positive and b is negative.
> 3. One integer must be positive and one must be negative if c is negative.

EXAMPLE 5 Factoring a Trinomial with Two Variables

Factor $z^2 - 2bz - 3b^2$.

Here, the coefficient of z in the middle term is $-2b$, so we need to find two expressions whose product is $-3b^2$ and whose sum is $-2b$. The expressions are $-3b$ and b, so

$$z^2 - 2bz - 3b^2 = (z - 3b)(z + b).$$

Check: $(z - 3b)(z + b) = z^2 + zb - 3bz - 3b^2$
$= z^2 + 1bz - 3bz - 3b^2$
$= z^2 - 2bz - 3b^2$

◀◀ Work Problem 6 at the Side.

7 Factor each trinomial completely.

(a) $2p^3 + 6p^2 - 8p$

OBJECTIVE 2 Factor trinomials after factoring out the greatest common factor. The trinomial in the next example does not have a coefficient of 1 for the squared term. (In fact, there is no squared term.) However, there may be a common factor.

EXAMPLE 6 Factoring a Trinomial with a Common Factor

Factor $4x^5 - 28x^4 + 40x^3$.

First, factor out the greatest common factor, $4x^3$.

$$4x^5 - 28x^4 + 40x^3 = \mathbf{4x^3}(x^2 - 7x + 10)$$

Now factor $x^2 - 7x + 10$. The integers -5 and -2 have a product of 10 and a sum of -7. The complete factored form is

(b) $3x^4 - 15x^3 + 18x^2$

$$4x^5 - 28x^4 + 40x^3 = \mathbf{4x^3}(x - 5)(x - 2). \quad \text{Include } 4x^3.$$

Check: $4x^3(x - 5)(x - 2) = 4x^3(x^2 - 7x + 10)$
$= 4x^5 - 28x^4 + 40x^3$

> **CAUTION**
> When factoring, *always look for a common factor first*. Remember to include the common factor as part of the answer. As a check, multiplying out the complete factored form should give the original polynomial.

ANSWERS
6. (a) $(b - 4a)(b + a)$
   (b) $(r - 4s)(r - 2s)$
7. (a) $2p(p + 4)(p - 1)$
   (b) $3x^2(x - 3)(x - 2)$

◀◀ Work Problem 7 at the Side.

14.2 Exercises

1. When factoring a trinomial in x as $(x + a)(x + b)$, what must be true of a and b, if the last term of the trinomial is negative?

2. In Exercise 1, what must be true of a and b if the last term is positive?

3. What is meant by a *prime polynomial*?

4. How can you check your work when factoring a trinomial? Does the check ensure that the trinomial is *completely* factored?

In Exercises 5–8, list all pairs of integers with the given product. Then find the pair whose sum is given. See the tables in Examples 1–4.

5. Product: 12; Sum: 7

6. Product: 18; Sum: 9

7. Product: -24; Sum: -5

8. Product: -36; Sum: -16

9. Which one of the following is the correct factored form of $x^2 - 12x + 32$?
   A. $(x - 8)(x + 4)$
   B. $(x + 8)(x - 4)$
   C. $(x - 8)(x - 4)$
   D. $(x + 8)(x + 4)$

10. What would be the first step in factoring $2x^3 + 8x^2 - 10x$?

Complete each factoring.

11. $x^2 + 15x + 44 = (x + 4)()$

12. $r^2 + 15r + 56 = (r + 7)()$

13. $x^2 - 9x + 8 = (x - 1)()$

14. $t^2 - 14t + 24 = (t - 2)()$

15. $y^2 - 2y - 15 = (y + 3)()$

16. $t^2 - t - 42 = (t + 6)()$

17. $x^2 + 9x - 22 = (x - 2)()$

18. $x^2 + 6x - 27 = (x - 3)()$

19. $y^2 - 7y - 18 = (y + 2)()$

20. $y^2 - 2y - 24 = (y + 4)()$

Factor completely. If a polynomial cannot be factored, write prime. *See Examples 1–4.*

21. $y^2 + 9y + 8$ **22.** $a^2 + 9a + 20$ **23.** $b^2 + 8b + 15$

24. $x^2 + 6x + 8$ **25.** $m^2 + m - 20$ **26.** $p^2 + 4p - 5$

27. $x^2 + 3x - 40$ **28.** $d^2 + 4d - 45$ **29.** $y^2 - 8y + 15$

30. $y^2 - 6y + 8$ **31.** $z^2 - 15z + 56$ **32.** $x^2 - 13x + 36$

33. $r^2 - r - 30$ **34.** $q^2 - q - 42$ **35.** $a^2 - 8a - 48$

36. $m^2 - 10m - 24$ **37.** $x^2 + 4x + 5$ **38.** $t^2 + 11t + 12$

Factor completely. See Examples 5 and 6.

39. $r^2 + 3ra + 2a^2$ **40.** $x^2 + 5xa + 4a^2$ **41.** $x^2 + 4xy + 3y^2$

42. $p^2 + 9pq + 8q^2$ **43.** $t^2 - tz - 6z^2$ **44.** $a^2 - ab - 12b^2$

45. $v^2 - 11vw + 30w^2$ **46.** $v^2 - 11vx + 24x^2$ **47.** $4x^2 + 12x - 40$

48. $5y^2 - 5y - 30$ **49.** $2t^3 + 8t^2 + 6t$ **50.** $3t^3 + 27t^2 + 24t$

51. $2x^6 + 8x^5 - 42x^4$ **52.** $4y^5 + 12y^4 - 40y^3$ **53.** $a^5 + 3a^4b - 4a^3b^2$

54. $z^{10} - 4z^9y - 21z^8y^2$ **55.** $m^3n - 10m^2n^2 + 24mn^3$ **56.** $y^3z + 3y^2z^2 - 54yz^3$

57. Use the FOIL method from **Section 13.3** to show that $(2x + 4)(x - 3) = 2x^2 - 2x - 12$. Why, then, is it incorrect to completely factor $2x^2 - 2x - 12$ as $(2x + 4)(x - 3)$?

58. Why is it incorrect to completely factor $3x^2 + 9x - 12$ as the product $(x - 1)(3x + 12)$?

14.3 Factoring Trinomials by Grouping

Trinomials like $2x^2 + 7x + 6$, in which the coefficient of the squared term is *not* 1, are factored with extensions of the methods from the previous sections. One such method uses factoring by grouping from **Section 14.1**.

OBJECTIVE 1 Factor trinomials by grouping when the coefficient of the squared term is not 1. Recall that a trinomial such as $m^2 + 3m + 2$ is factored by finding two integers whose product is 2 and whose sum is 3. To factor $2x^2 + 7x + 6$, we look for two integers whose product is $2 \cdot 6 = 12$ and whose sum is 7.

$$2x^2 + 7x + 6$$
(Sum is 7. Product is $2 \cdot 6 = 12$.)

By considering pairs of positive integers whose product is 12, the necessary integers are found to be 3 and 4. We use these integers to write the middle term, $7x$, as $7x = 3x + 4x$. The trinomial $2x^2 + 7x + 6$ becomes

$$2x^2 + 7x + 6 = 2x^2 + \underbrace{3x + 4x}_{7x} + 6.$$

$$= (2x^2 + 3x) + (4x + 6) \quad \text{Group terms.}$$
$$= x(2x + 3) + 2(2x + 3) \quad \text{Factor each group.}$$

Must be the same

$$2x^2 + 7x + 6 = (2x + 3)(x + 2) \quad \text{Factor out } 2x + 3.$$

Check: $(2x + 3)(x + 2) = 2x^2 + 7x + 6$ ✓

In the preceding example, we could have written $7x$ as $4x + 3x$. Factoring by grouping this way would give the same answer.

Work Problem 1 at the Side.

EXAMPLE 1 Factoring Trinomials by Grouping

Factor each trinomial.

(a) $6r^2 + r - 1$

We must find two integers with a product of $6(-1) = -6$ and a sum of 1.

$$6r^2 + r - 1 = 6r^2 + 1r - 1$$
(Sum is 1. Product is $6(-1) = -6$.)

The integers are -2 and 3. We write the middle term, r, as $-2r + 3r$.

$$6r^2 + r - 1 = 6r^2 - 2r + 3r - 1 \quad r = -2r + 3r$$
$$= (6r^2 - 2r) + (3r - 1) \quad \text{Group terms.}$$
$$= 2r(3r - 1) + 1(3r - 1) \quad \text{The binomials must be the same.}$$
$$= (3r - 1)(2r + 1) \quad \text{Factor out } 3r - 1.$$

Check: $(3r - 1)(2r + 1) = 6r^2 + r - 1$ ✓

Continued on Next Page

OBJECTIVE

1 Factor trinomials by grouping when the coefficient of the squared term is not 1.

① **(a)** Factor $2x^2 + 7x + 6$ by writing $7x$ as $4x + 3x$. Complete the following.

$2x^2 + 7x + 6$
$= 2x^2 + 4x + 3x + 6$
$= (2x^2 + \underline{}) + (3x + \underline{})$
$= 2x(x + \underline{}) + 3(x + \underline{})$
$= (\underline{})(2x + 3)$

(b) Is the answer in part (a) the same as in the example? (Remember that the order of the factors does not matter.)

ANSWERS
1. (a) $4x$; 6; 2; 2; $x + 2$ **(b)** yes

2 Factor each trinomial by grouping.

(a) $2m^2 + 7m + 3$

(b) $5p^2 - 2p - 3$

(c) $15k^2 - km - 2m^2$

3 Factor each trinomial completely.

(a) $4x^2 - 2x - 30$

(b) $18p^4 + 63p^3 + 27p^2$

(c) $6a^2 + 3ab - 18b^2$

(b) $12z^2 - 5z - 2$

Look for two integers whose product is $12(-2) = -24$ and whose sum is -5. The required integers are 3 and -8. Rewrite the middle term, $-5z$, as $3z - 8z$.

$$12z^2 - 5z - 2 = 12z^2 + 3z - 8z - 2 \qquad -5z = 3z - 8z$$
$$= (12z^2 + 3z) + (-8z - 2) \qquad \text{Group terms.}$$
$$= 3z(4z + 1) - 2(4z + 1) \qquad \text{Factor each group; be careful with signs.}$$
$$= (4z + 1)(3z - 2) \qquad \text{Factor out } 4z + 1.$$

Check: $(4z + 1)(3z - 2) = 12z^2 - 5z - 2$

(c) $10m^2 + mn - 3n^2$

Two integers whose product is $10(-3) = -30$ and whose sum is 1 are -5 and 6. Rewrite the trinomial with four terms.

$$10m^2 + mn - 3n^2 = 10m^2 - 5mn + 6mn - 3n^2 \qquad mn = -5mn + 6mn$$
$$= 5m(2m - n) + 3n(2m - n) \qquad \text{Group terms; factor each group.}$$
$$= (2m - n)(5m + 3n) \qquad \text{Factor out } 2m - n.$$

Check by multiplying.

◀◀ **Work Problem 2 at the Side.**

EXAMPLE 2 Factoring a Trinomial with a Common Factor by Grouping

Factor $28x^5 - 58x^4 - 30x^3$.

First factor out the greatest common factor, $2x^3$.

$$28x^5 - 58x^4 - 30x^3 = \mathbf{2x^3}(14x^2 - 29x - 15)$$

To factor $14x^2 - 29x - 15$, find two integers whose product is $14(-15) = -210$ and whose sum is -29. Begin by factoring 210 into prime factors.

$$210 = 2 \cdot 3 \cdot 5 \cdot 7$$

Combine these prime factors in pairs in different ways, using one positive and one negative to get -210. The factors 6 and -35 have the correct sum, -29. Now rewrite the given trinomial and factor it.

$$28x^5 - 58x^4 - 30x^3 = 2x^3(14x^2 + 6x - 35x - 15)$$
$$= 2x^3[(14x^2 + 6x) + (-35x - 15)]$$
$$= 2x^3[2x(7x + 3) - 5(7x + 3)]$$
$$= 2x^3[(7x + 3)(2x - 5)]$$
$$= 2x^3(7x + 3)(2x - 5)$$

Check by multiplying.

CAUTION
Remember to include the common factor in the final result.

◀◀ **Work Problem 3 at the Side.**

ANSWERS
2. (a) $(2m + 1)(m + 3)$
   (b) $(5p + 3)(p - 1)$
   (c) $(5k - 2m)(3k + m)$
3. (a) $2(2x + 5)(x - 3)$
   (b) $9p^2(2p + 1)(p + 3)$
   (c) $3(2a - 3b)(a + 2b)$

14.3 Exercises

The middle term of each trinomial has been rewritten. Now factor by grouping.
See Example 1.

1. $m^2 + 8m + 12$
   $= m^2 + 6m + 2m + 12$

2. $x^2 + 9x + 14$
   $= x^2 + 7x + 2x + 14$

3. $a^2 + 3a - 10$
   $= a^2 + 5a - 2a - 10$

4. $y^2 - 2y - 24$
   $= y^2 + 4y - 6y - 24$

5. $10t^2 + 9t + 2$
   $= 10t^2 + 5t + 4t + 2$

6. $6x^2 + 13x + 6$
   $= 6x^2 + 9x + 4x + 6$

7. $15z^2 - 19z + 6$
   $= 15z^2 - 10z - 9z + 6$

8. $12p^2 - 17p + 6$
   $= 12p^2 - 9p - 8p + 6$

9. $8s^2 + 2st - 3t^2$
   $= 8s^2 - 4st + 6st - 3t^2$

10. $3x^2 - xy - 14y^2$
    $= 3x^2 - 7xy + 6xy - 14y^2$

11. $15a^2 + 22ab + 8b^2$
    $= 15a^2 + 10ab + 12ab + 8b^2$

12. $25m^2 + 25mn + 6n^2$
    $= 25m^2 + 15mn + 10mn + 6n^2$

13. Which pair of integers would be used to rewrite the middle term when factoring $12y^2 + 5y - 2$ by grouping?
    A. $-8, 3$ **B.** $8, -3$ **C.** $-6, 4$ **D.** $6, -4$

14. Which pair of integers would be used to rewrite the middle term when factoring $20b^2 - 13b + 2$ by grouping?
    A. $10, 3$ **B.** $-10, -3$ **C.** $8, 5$ **D.** $-8, -5$

Complete the steps to factor each trinomial by grouping.

15. $2m^2 + 11m + 12$
    (a) Find two integers whose product is
    _____ · _____ = _____ and whose sum is _____ .
    (b) The required integers are _____ and _____ .
    (c) Write the middle term $11m$ as _____ + _____ .
    (d) Rewrite the given trinomial using four terms.
    (e) Factor the polynomial in part (d) by grouping.
    (f) Check by multiplying.

16. $6y^2 - 19y + 10$
    (a) Find two integers whose product is
    _____ · _____ = _____ and whose sum is _____ .
    (b) The required integers are _____ and _____ .
    (c) Write the middle term $-19y$ as _____ + _____ .
    (d) Rewrite the given trinomial using four terms.
    (e) Factor the polynomial in part (d) by grouping.
    (f) Check by multiplying.

Chapter 14 Factoring and Applications

Factor each trinomial by grouping. See Examples 1 and 2.

17. $2x^2 + 7x + 3$ **18.** $3y^2 + 13y + 4$ **19.** $4r^2 + r - 3$

20. $4r^2 + 3r - 10$ **21.** $8m^2 - 10m - 3$ **22.** $20x^2 - 28x - 3$

23. $21m^2 + 13m + 2$ **24.** $38x^2 + 23x + 2$ **25.** $6b^2 + 7b + 2$

26. $6w^2 + 19w + 10$ **27.** $12y^2 - 13y + 3$ **28.** $15a^2 - 16a + 4$

29. $24x^2 - 42x + 9$ **30.** $48b^2 - 74b - 10$ **31.** $2m^3 + 2m^2 - 40m$

32. $3x^3 + 12x^2 - 36x$ **33.** $32z^5 - 20z^4 - 12z^3$ **34.** $18x^5 + 15x^4 - 75x^3$

35. $12p^2 + 7pq - 12q^2$ **36.** $6m^2 - 5mn - 6n^2$ **37.** $6a^2 - 7ab - 5b^2$

38. $25g^2 - 5gh - 2h^2$ **39.** $5 - 6x + x^2$ **40.** $7 + 8x + x^2$

41. On a quiz, a student factored $16x^2 - 24x + 5$ by grouping as follows.

$16x^2 - 24x + 5$
$= 16x^2 - 4x - 20x + 5$
$= 4x(4x - 1) - 5(4x - 1)$ His answer

He thought his answer was correct since it checked by multiplying. Why was the answer marked wrong? What is the correct factored form?

42. On the same quiz, another student factored $3k^3 - 12k^2 - 15k$ by first factoring out the common factor $3k$ to get $3k(k^2 - 4k - 5)$. Then she wrote

$k^2 - 4k - 5 = k^2 - 5k + k - 5$
$= k(k - 5) + 1(k - 5)$
$= (k - 5)(k + 1).$ Her answer

Why was the answer marked wrong? What is the correct factored form?

14.4 Factoring Trinomials Using FOIL

OBJECTIVE 1 Factor trinomials using FOIL. This section shows an alternative method of factoring trinomials in which the coefficient of the squared term is not 1. This method uses trial and error.

To factor $2x^2 + 7x + 6$ (the same trinomial factored at the beginning of **Section 14.3**) by trial and error, we use FOIL backwards. We want to write $2x^2 + 7x + 6$ as the product of two binomials.

$$2x^2 + 7x + 6 = (\quad)(\quad)$$

The product of the two first terms of the binomials is $2x^2$. The possible factors of $2x^2$ are $2x$ and x or $-2x$ and $-x$. Since all terms of the trinomial are positive, we consider only positive factors. Thus, we have

$$2x^2 + 7x + 6 = (2x\quad)(x\quad).$$

The product of the two last terms, 6, can be factored as $1 \cdot 6, 6 \cdot 1, 2 \cdot 3,$ or $3 \cdot 2$. Try each pair to find the pair that gives the correct middle term, $7x$.

> **Work Problem 1 at the Side.**

In part (b) at the side, since $2x + 6 = 2(x + 3)$, the binomial $2x + 6$ has a common factor of 2, while $2x^2 + 7x + 6$ has no common factor other than 1. The product $(2x + 6)(x + 1)$ cannot be correct. (Part (c) also has one binomial factor with a common factor.)

> **NOTE**
> If the original polynomial has no common factor, then none of its binomial factors will either.

Now try the remaining numbers 3 and 2 as factors of 6.

$$(2x + 3)(x + 2) = 2x^2 + 7x + 6 \quad \text{Correct}$$

$3x$
$4x$
$7x$ Add.

Finally, we see that $2x^2 + 7x + 6$ factors as

$$2x^2 + 7x + 6 = (2x + 3)(x + 2).$$

Check by multiplying: $(2x + 3)(x + 2) = 2x^2 + 7x + 6.$

EXAMPLE 1 Factoring a Trinomial with All Positive Terms Using FOIL

Factor $8p^2 + 14p + 5$.

The number 8 has several possible pairs of factors, but 5 has only 1 and 5 or -1 and -5. For this reason, it is easier to begin by considering the factors of 5. Ignore the negative factors since all coefficients in the trinomial are positive. If $8p^2 + 14p + 5$ can be factored, the factors will have the form

$$(\quad + 5)(\quad + 1).$$

Continued on Next Page

OBJECTIVE

1 Factor trinomials using FOIL.

1 Multiply to decide whether each factored form is correct or incorrect for

$$2x^2 + 7x + 6.$$

(a) $(2x + 1)(x + 6)$

(b) $(2x + 6)(x + 1)$

(c) $(2x + 2)(x + 3)$

ANSWERS
1. (a) incorrect (b) incorrect (c) incorrect

② Factor each trinomial.

(a) $2p^2 + 9p + 9$

(b) $6p^2 + 19p + 10$

(c) $8x^2 + 14x + 3$

③ Factor each trinomial.

(a) $4y^2 - 11y + 6$

(b) $9x^2 - 21x + 10$

ANSWERS
2. (a) $(2p + 3)(p + 3)$
   (b) $(3p + 2)(2p + 5)$
   (c) $(4x + 1)(2x + 3)$
3. (a) $(4y - 3)(y - 2)$
   (b) $(3x - 5)(3x - 2)$

When factoring $8p^2 + 14p + 5$, the possible pairs of factors of $8p^2$ are $8p$ and p, or $4p$ and $2p$. Try various combinations, checking to see if the middle term is $14p$ in each case.

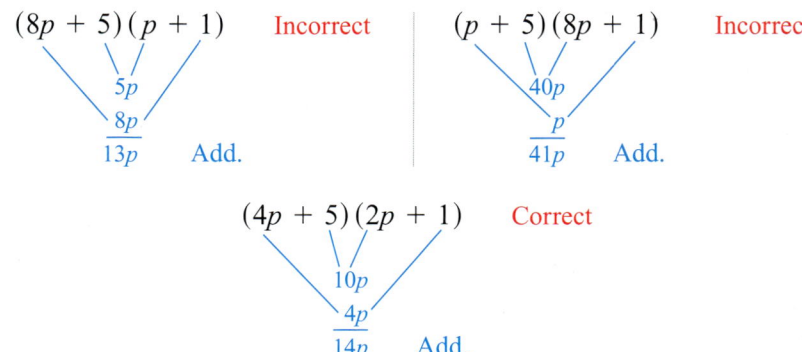

Since $14p$ is the correct middle term,
$$8p^2 + 14p + 5 = (4p + 5)(2p + 1).$$
Check: $(4p + 5)(2p + 1) = 8p^2 + 14p + 5$

◀◀◀ **Work Problem 2 at the Side.**

EXAMPLE 2 Factoring a Trinomial with a Negative Middle Term Using FOIL

Factor $6x^2 - 11x + 3$.

Since 3 has only 1 and 3 or -1 and -3 as factors, it is better here to begin by factoring 3. The last term of the trinomial $6x^2 - 11x + 3$ is positive and the middle term has a negative coefficient, so we consider only negative factors. We need two negative factors because the *product* of two negative factors is positive and their *sum* is negative, as required.

Try -3 and -1 as factors of 3:
$$(\quad - 3)(\quad - 1).$$

The factors of $6x^2$ may be either $6x$ and x, or $2x$ and $3x$.

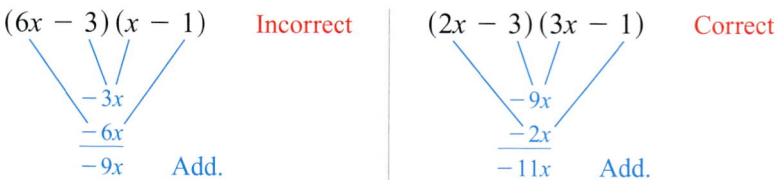

The factors $2x$ and $3x$ produce $-11x$, the correct middle term, so
$$6x^2 - 11x + 3 = (2x - 3)(3x - 1).$$

Check by multiplying.

> **NOTE**
> In Example 2, we might also realize that our initial attempt to factor $6x^2 - 11x + 3$ as $(6x - 3)(x - 1)$ *cannot* be correct since $6x - 3$ has a common factor of 3 and the original polynomial does not.

◀◀◀ **Work Problem 3 at the Side.**

EXAMPLE 3 Factoring a Trinomial with a Negative Last Term Using FOIL

Factor $8x^2 + 6x - 9$.

The integer 8 has several possible pairs of factors, as does -9. Since the last term is negative, one positive factor and one negative factor of -9 are needed. Since the coefficient of the middle term is small, it is wise to avoid large factors such as 8 or 9. We try 4 and 2 as factors of 8, and 3 and -3 as factors of -9, and check the middle term.

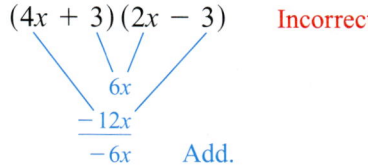

Now we try interchanging 3 and -3, since only the sign of the middle term is incorrect.

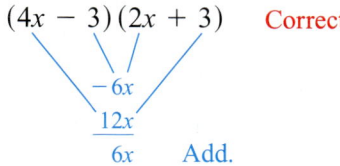

This combination produces $6x$, the correct middle term, so

$$8x^2 + 6x - 9 = (4x - 3)(2x + 3).$$

Work Problem 4 at the Side.

EXAMPLE 4 Factoring a Trinomial with Two Variables

Factor $12a^2 - ab - 20b^2$.

There are several pairs of factors of $12a^2$, including

$$12a \text{ and } a, \quad 6a \text{ and } 2a, \quad \text{and} \quad 4a \text{ and } 3a,$$

just as there are many possible pairs of factors of $-20b^2$, including

$$20b \text{ and } -b, \quad -20b \text{ and } b, \quad 10b \text{ and } -2b,$$
$$-10b \text{ and } 2b, \quad 4b \text{ and } -5b, \quad \text{and} \quad -4b \text{ and } 5b.$$

Once again, since the coefficient of the middle term is small, avoid the larger factors. Try the factors $6a$ and $2a$ and $4b$ and $-5b$.

$$(6a + 4b)(2a - 5b)$$

This cannot be correct, as mentioned before, since $6a + 4b$ has 2 as a common factor, while the given trinomial does not. Try $3a$ and $4a$ with $4b$ and $-5b$.

$$(3a + 4b)(4a - 5b) = 12a^2 + ab - 20b^2 \quad \text{Incorrect}$$

Here the middle term is ab, rather than $-ab$. Interchange the signs of the last two terms in the factors.

$$(3a - 4b)(4a + 5b) = 12a^2 - ab - 20b^2 \quad \text{Correct}$$

Work Problem 5 at the Side.

---

4 Factor each trinomial, if possible.

(a) $6x^2 + 5x - 4$

(b) $6m^2 - 11m - 10$

(c) $4x^2 - 3x - 7$

(d) $3y^2 + 8y - 6$

5 Factor each trinomial.

(a) $2x^2 - 5xy - 3y^2$

(b) $8a^2 + 2ab - 3b^2$

ANSWERS
4. (a) $(3x + 4)(2x - 1)$
   (b) $(2m - 5)(3m + 2)$
   (c) $(4x - 7)(x + 1)$
   (d) prime
5. (a) $(2x + y)(x - 3y)$
   (b) $(4a + 3b)(2a - b)$

6 Factor each trinomial.

(a) $36z^3 - 6z^2 - 72z$

(b) $-24x^3 + 32x^2y + 6xy^2$

> **EXAMPLE 5** Factoring Trinomials with Common Factors
>
> Factor each trinomial.
>
> (a) $15y^3 + 55y^2 + 30y$
>
> First factor out the greatest common factor, $5y$.
>
> $$15y^3 + 55y^2 + 30y = 5y(3y^2 + 11y + 6)$$
>
> Now factor $3y^2 + 11y + 6$. Try $3y$ and y as factors of $3y^2$ and 2 and 3 as factors of 6.
>
> $$(3y + 2)(y + 3) = 3y^2 + 11y + 6 \quad \text{Correct}$$
>
> The complete factored form of $15y^3 + 55y^2 + 30y$ is
>
> $$15y^3 + 55y^2 + 30y = 5y(3y + 2)(y + 3).$$
>
> *Check* by multiplying.
>
> (b) $-24a^3 - 42a^2 + 45a$
>
> The common factor could be $3a$ or $-3a$. If we factor out $-3a$, the first term of the trinomial will be positive, which makes it easier to factor.
>
> $$-24a^3 - 42a^2 + 45a = -3a(8a^2 + 14a - 15) \quad \text{Factor out } -3a.$$
> $$= -3a(4a - 3)(2a + 5) \quad \text{Use trial and error.}$$
>
> *Check* by multiplying.

CAUTION
This caution bears repeating: *Remember to include the common factor in the final factored form.*

◀◀◀ **Work Problem 6 at the Side.**

ANSWERS
6. (a) $6z(2z - 3)(3z + 4)$
   (b) $-2x(6x + y)(2x - 3y)$

14.4 Exercises

Decide which is the correct factored form of the given polynomial.

1. $2x^2 - x - 1$
   A. $(2x - 1)(x + 1)$ B. $(2x + 1)(x - 1)$

2. $3a^2 - 5a - 2$
   A. $(3a + 1)(a - 2)$ B. $(3a - 1)(a + 2)$

3. $4y^2 + 17y - 15$
   A. $(y + 5)(4y - 3)$ B. $(2y - 5)(2y + 3)$

4. $12c^2 - 7c - 12$
   A. $(6c - 2)(2c + 6)$ B. $(4c + 3)(3c - 4)$

5. $4k^2 + 13mk + 3m^2$
   A. $(4k + m)(k + 3m)$ B. $(4k + 3m)(k + m)$

6. $2x^2 + 11x + 12$
   A. $(2x + 3)(x + 4)$ B. $(2x + 4)(x + 3)$

Complete each factoring.

7. $6a^2 + 7ab - 20b^2 = (3a - 4b)()$

8. $9m^2 - 3mn - 2n^2 = (3m + n)()$

9. $2x^2 + 6x - 8 = 2()$
   $= 2()()$

10. $3x^2 - 9x - 30 = 3()$
    $= 3()()$

11. $4z^3 - 10z^2 - 6z = 2z()$
    $= 2z()()$

12. $15r^3 - 39r^2 - 18r = 3r()$
    $= 3r()()$

13. For the polynomial $12x^2 + 7x - 12$, 2 is not a common factor. Explain why the binomial $2x - 6$, then, cannot be a factor of the polynomial.

14. Explain how the signs of the last terms of the two binomial factors of a trinomial are determined.

Factor each trinomial completely. See Examples 1–5.

15. $3a^2 + 10a + 7$

16. $7r^2 + 8r + 1$

17. $2y^2 + 7y + 6$

18. $5z^2 + 12z + 4$

19. $15m^2 + m - 2$

20. $6x^2 + x - 1$

21. $12s^2 + 11s - 5$

22. $20x^2 + 11x - 3$

23. $10m^2 - 23m + 12$

24. $6x^2 - 17x + 12$

25. $8w^2 - 14w + 3$

26. $9p^2 - 18p + 8$

27. $20y^2 - 39y - 11$

28. $10x^2 - 11x - 6$

29. $3x^2 - 15x + 16$

30. $2t^2 + 13t - 18$ **31.** $20x^2 + 22x + 6$ **32.** $36y^2 + 81y + 45$

33. $40m^2q + mq - 6q$ **34.** $15a^2b + 22ab + 8b$ **35.** $15n^4 - 39n^3 + 18n^2$

36. $24a^4 + 10a^3 - 4a^2$ **37.** $15x^2y^2 - 7xy^2 - 4y^2$ **38.** $14a^2b^3 + 15ab^3 - 9b^3$

39. $5a^2 - 7ab - 6b^2$ **40.** $6x^2 - 5xy - y^2$ **41.** $12s^2 + 11st - 5t^2$

42. $25a^2 + 25ab + 6b^2$ **43.** $6m^6n + 7m^5n^2 + 2m^4n^3$ **44.** $12k^3q^4 - 4k^2q^5 - kq^6$

If a trinomial has a negative coefficient for the squared term, such as $-2x^2 + 11x - 12$, *it may be easier to factor by first factoring out the common factor* -1:

$$-2x^2 + 11x - 12 = -1(2x^2 - 11x + 12)$$
$$= -1(2x - 3)(x - 4).$$

Use this method to factor the trinomials in Exercises 45–50.

45. $-x^2 - 4x + 21$ **46.** $-x^2 + x + 72$ **47.** $-3x^2 - x + 4$

48. $-5x^2 + 2x + 16$ **49.** $-2a^2 - 5ab - 2b^2$ **50.** $-3p^2 + 13pq - 4q^2$

RELATING CONCEPTS (EXERCISES 51–56) For Individual or Group Work

One of the most common problems that beginning algebra students face is this: If an answer obtained doesn't look exactly like the one given in the back of the book, is it necessarily incorrect? Often there are several different equivalent forms of an answer that are all correct. **Work Exercises 51–56 in order,** *to see how and why this is possible for factoring problems.*

51. Factor the integer 35 as the product of two prime numbers.

52. Factor the integer 35 as the product of the negatives of two prime numbers.

53. Verify the following factored form: $6x^2 - 11x + 4 = (3x - 4)(2x - 1)$.

54. Verify the following factored form: $6x^2 - 11x + 4 = (4 - 3x)(1 - 2x)$.

55. Compare the two valid factored forms in Exercises 53 and 54. How do the factors in each case compare?

56. Suppose you know that the correct factored form of a particular trinomial is $(7t - 3)(2t - 5)$. Based on your observations in Exercises 51–55, what is another valid factored form?

14.5 Special Factoring Techniques

By reversing the rules for multiplication of binomials from **Section 13.4**, we obtain rules for factoring polynomials in certain forms.

OBJECTIVES

1. Factor a difference of squares.
2. Factor a perfect square trinomial.

OBJECTIVE 1 Factor a difference of squares. The formula for the product of the sum and difference of the same two terms is

$$(a + b)(a - b) = a^2 - b^2.$$

Reversing this rule leads to the following special factoring rule.

> **Factoring a Difference of Squares**
> $$a^2 - b^2 = (a + b)(a - b)$$

For example,

$$m^2 - 16 = m^2 - 4^2 = (m + 4)(m - 4).$$

As the next examples show, the following conditions must be true for a binomial to be a difference of squares.

1. Both terms of the binomial must be squares, such as

$$x^2, \quad 9y^2, \quad 25, \quad 1, \quad m^4.$$

2. The terms of the binomial must have different signs (one positive and one negative).

EXAMPLE 1 Factoring Differences of Squares

Factor each binomial, if possible.

$$a^2 - b^2 = (a + b)(a - b)$$

(a) $x^2 - 49 = x^2 - 7^2 = (x + 7)(x - 7)$

(b) $y^2 - m^2 = (y + m)(y - m)$

(c) $z^2 - \dfrac{9}{16} = z^2 - \left(\dfrac{3}{4}\right)^2 = \left(z + \dfrac{3}{4}\right)\left(z - \dfrac{3}{4}\right)$

(d) $x^2 - 8$

Because 8 is not the square of an integer, this binomial is not a difference of squares. It is a prime polynomial.

(e) $p^2 + 16$

Since $p^2 + 16$ is a *sum* of squares, it is not equal to $(p + 4)(p - 4)$. Also, using FOIL,

$$(p - 4)(p - 4) = p^2 - 8p + 16 \neq p^2 + 16$$

and

$$(p + 4)(p + 4) = p^2 + 8p + 16 \neq p^2 + 16,$$

so $p^2 + 16$ is a prime polynomial.

Chapter 14 Factoring and Applications

1 Factor, if possible.

(a) $p^2 - 100$

(b) $x^2 - \dfrac{25}{36}$

(c) $x^2 + y^2$

(d) $9m^2 - 49$

(e) $64a^2 - 25$

2 Factor completely.

(a) $50r^2 - 32$

(b) $27y^2 - 75$

(c) $25a^2 - 64b^2$

(d) $k^4 - 49$

(e) $81r^4 - 16$

ANSWERS
1. (a) $(p + 10)(p - 10)$
   (b) $\left(x + \dfrac{5}{6}\right)\left(x - \dfrac{5}{6}\right)$
   (c) prime
   (d) $(3m + 7)(3m - 7)$
   (e) $(8a + 5)(8a - 5)$
2. (a) $2(5r + 4)(5r - 4)$
   (b) $3(3y + 5)(3y - 5)$
   (c) $(5a + 8b)(5a - 8b)$
   (d) $(k^2 + 7)(k^2 - 7)$
   (e) $(9r^2 + 4)(3r + 2)(3r - 2)$

> **CAUTION**
> As Example 1(e) suggests, *after any common factor is removed, a sum of squares cannot be factored.*

EXAMPLE 2 Factoring Differences of Squares

Factor each difference of squares.

$$a^2 - b^2 = (a + b)(a - b)$$

(a) $25m^2 - 16 = (5m)^2 - 4^2 = (5m + 4)(5m - 4)$

(b) $49z^2 - 64 = (7z)^2 - 8^2 = (7z + 8)(7z - 8)$

> **NOTE**
> As in previous sections, you should *always check a factored form by multiplying.*

Work Problem 1 at the Side.

EXAMPLE 3 Factoring More Complex Differences of Squares

Factor completely.

(a) $81y^2 - 36$

First factor out the common factor, 9.

$81y^2 - 36 = 9(9y^2 - 4)$ Factor out 9.

$ = 9[(3y)^2 - 2^2]$

$ = 9(3y + 2)(3y - 2)$ Difference of squares

(b) $9x^2 - 4z^2 = (3x)^2 - (2z)^2 = (3x + 2z)(3x - 2z)$

(c) $p^4 - 36 = (p^2)^2 - 6^2 = (p^2 + 6)(p^2 - 6)$

Neither $p^2 + 6$ nor $p^2 - 6$ can be factored further.

(d) $m^4 - 16 = (m^2)^2 - 4^2$

$ = (m^2 + 4)(m^2 - 4)$ Difference of squares

$ = (m^2 + 4)(m + 2)(m - 2)$ Difference of squares again

> **CAUTION**
> Remember to factor again when any of the factors is a difference of squares, as in Example 3(d). Check by multiplying.

Work Problem 2 at the Side.

OBJECTIVE 2 **Factor a perfect square trinomial.** The expressions 144, $4x^2$, and $81m^6$ are called *perfect squares* because

$$144 = 12^2, \quad 4x^2 = (2x)^2, \quad \text{and} \quad 81m^6 = (9m^3)^2.$$

A **perfect square trinomial** is a trinomial that is the square of a binomial. For example, $x^2 + 8x + 16$ is a perfect square trinomial because it is the square of the binomial $x + 4$, as shown below.

$$x^2 + 8x + 16 = (x + 4)(x + 4)$$
$$= (x + 4)^2$$

For a trinomial to be a perfect square, *two of its terms must be perfect squares.* For this reason, $16x^2 + 4x + 15$ is not a perfect square trinomial because only the term $16x^2$ is a perfect square.

On the other hand, even if two of the terms are perfect squares, the trinomial may not be a perfect square trinomial. For example, $x^2 + 6x + 36$ has two perfect square terms, but it is not a perfect square trinomial. (Try to find a binomial that can be squared to give $x^2 + 6x + 36$.)

We can multiply to see that the square of a binomial gives one of the following perfect square trinomials.

Factoring Perfect Square Trinomials

$$a^2 + 2ab + b^2 = (a + b)^2$$
$$a^2 - 2ab + b^2 = (a - b)^2$$

The middle term of a perfect square trinomial is always twice the product of the two terms in the squared binomial. (This was shown in **Section 13.4.**) Use this to check any attempt to factor a trinomial that appears to be a perfect square.

EXAMPLE 4 Factoring a Perfect Square Trinomial

Factor $x^2 + 10x + 25$.

The term x^2 is a perfect square, and so is 25. Try to factor the trinomial as

$$x^2 + 10x + 25 = (x + 5)^2.$$

To check, take twice the product of the two terms in the squared binomial.

$$2 \cdot x \cdot 5 = 10x$$

Twice First term Last term
of binomial of binomial

Since $10x$ is the middle term of the trinomial, the trinomial is a perfect square and can be factored as $(x + 5)^2$. Thus,

$$x^2 + 10x + 25 = (x + 5)^2.$$

Work Problem 3 at the Side.

EXAMPLE 5 Factoring Perfect Square Trinomials

Factor each trinomial.

(a) $x^2 - 22x + 121$

The first and last terms are perfect squares ($121 = 11^2$ or $(-11)^2$). Check to see whether the middle term of $x^2 - 22x + 121$ is twice the product of the first and last terms of the binomial $x - 11$.

Continued on Next Page

---

3 Factor each trinomial.

(a) $p^2 + 14p + 49$

(b) $m^2 + 8m + 16$

(c) $x^2 + 2x + 1$

ANSWERS
3. **(a)** $(p + 7)^2$ **(b)** $(m + 4)^2$ **(c)** $(x + 1)^2$

4 Factor each trinomial.

(a) $p^2 - 18p + 81$

(b) $16a^2 + 56a + 49$

(c) $121p^2 + 110p + 100$

(d) $64x^2 - 48x + 9$

(e) $27y^3 + 72y^2 + 48y$

$$2 \cdot x \cdot (-11) = -22x$$
Twice — First term — Last term

Since twice the product of the first and last terms of the binomial is the middle term, $x^2 - 22x + 121$ is a perfect square trinomial and

$$x^2 - 22x + 121 = (x - 11)^2.$$
Same sign

Notice that the sign of the second term in the squared binomial is the same as the sign of the middle term in the trinomial.

(b) $9m^2 - 24m + 16 = (3m)^2 + 2(3m)(-4) + (-4)^2 = (3m - 4)^2$
Twice — First term — Last Term

(c) $25y^2 + 20y + 16$

The first and last terms are perfect squares.

$$25y^2 = (5y)^2 \quad \text{and} \quad 16 = 4^2$$

Twice the product of the first and last terms of the binomial $5y + 4$ is

$$2 \cdot 5y \cdot 4 = 40y,$$

which is not the middle term of $25y^2 + 20y + 16$. This trinomial is not a perfect square. In fact, the trinomial cannot be factored even with the methods of the previous sections; it is a prime polynomial.

(d) $12z^3 + 60z^2 + 75z = 3z(4z^2 + 20z + 25)$ Factor out $3z$.
$= 3z[(2z)^2 + 2(2z)(5) + 5^2]$
$= 3z(2z + 5)^2$

NOTE
1. The sign of the second term in the squared binomial is always the same as the sign of the middle term in the trinomial.
2. The first and last terms of a perfect square trinomial must be *positive,* because they are squares. For example, the polynomial $x^2 - 2x - 1$ cannot be a perfect square because the last term is negative.
3. Perfect square trinomials can also be factored using grouping or FOIL, although using the method of this section is often easier.

◀◀◀ **Work Problem 4 at the Side.**

The methods of factoring discussed in this section are summarized here.

Special Factoring Rules

Difference of squares	$a^2 - b^2 = (a + b)(a - b)$
Perfect square trinomials	$a^2 + 2ab + b^2 = (a + b)^2$
	$a^2 - 2ab + b^2 = (a - b)^2$

ANSWERS
4. (a) $(p - 9)^2$ (b) $(4a + 7)^2$ (c) prime
(d) $(8x - 3)^2$ (e) $3y(3y + 4)^2$

14.5 Exercises

1. To help you factor a difference of squares, complete the following list of squares.

   $1^2 =$ _____ $2^2 =$ _____ $3^2 =$ _____ $4^2 =$ _____ $5^2 =$ _____
   $6^2 =$ _____ $7^2 =$ _____ $8^2 =$ _____ $9^2 =$ _____ $10^2 =$ _____
   $11^2 =$ _____ $12^2 =$ _____ $13^2 =$ _____ $14^2 =$ _____ $15^2 =$ _____
   $16^2 =$ _____ $17^2 =$ _____ $18^2 =$ _____ $19^2 =$ _____ $20^2 =$ _____

2. To use the factoring techniques described in this section, you will sometimes need to recognize fourth powers of integers. Complete the following list of fourth powers.

   $1^4 =$ _____ $2^4 =$ _____ $3^4 =$ _____ $4^4 =$ _____ $5^4 =$ _____

3. The following powers of x are all perfect squares: $x^2, x^4, x^6, x^8, x^{10}$. Based on this observation, we may make a conjecture (an educated guess) that if the power of a variable is divisible by _____ (with 0 remainder), then it is a perfect square.

4. Which of the following are differences of squares?
   A. $x^2 - 4$ **B.** $y^2 + 9$ **C.** $2a^2 - 25$ **D.** $9m^2 - 1$

Factor each binomial completely. Use your answers in Exercises 1 and 2 as necessary. See Examples 1–3.

5. $y^2 - 25$

6. $t^2 - 16$

7. $p^2 - \dfrac{1}{9}$

8. $q^2 - \dfrac{1}{4}$

9. $m^2 - 12$

10. $k^2 - 18$

11. $9r^2 - 4$

12. $4x^2 - 9$

13. $36m^2 - \dfrac{16}{25}$

14. $100b^2 - \dfrac{4}{49}$

15. $36x^2 - 16$

16. $32a^2 - 8$

17. $196p^2 - 225$

18. $361q^2 - 400$

19. $16r^2 - 25a^2$

20. $49m^2 - 100p^2$

21. $100x^2 + 49$

22. $81w^2 + 16$

23. $p^4 - 49$

24. $r^4 - 25$

25. $x^4 - 1$

26. $y^4 - 16$

27. $p^4 - 256$

28. $16k^4 - 1$

29. When a student was directed to factor $x^4 - 81$ completely, his teacher did not give him full credit when he answered $(x^2 + 9)(x^2 - 9)$. The student argued that because his answer does indeed give $x^4 - 81$ when multiplied out, he should be given full credit. Was the teacher justified in her grading of this item? Why or why not?

30. The binomial $4x^2 + 16$ is a sum of squares that *can* be factored. How is this binomial factored? When can a sum of squares be factored?

31. In the polynomial $9y^2 + 14y + 25$, the first and last terms are perfect squares. Can the polynomial be factored? If it can, factor it. If it cannot, explain why it is not a perfect square trinomial.

32. Which of the following are perfect square trinomials?

   A. $y^2 - 13y + 36$ **B.** $x^2 + 6x + 9$ **C.** $4z^2 - 4z + 1$ **D.** $16m^2 + 10m + 1$

Factor each trinomial completely. It may be necessary to factor out the greatest common factor first. See Examples 4 and 5.

33. $w^2 + 2w + 1$

34. $p^2 + 4p + 4$

35. $x^2 - 8x + 16$

36. $x^2 - 10x + 25$

37. $t^2 + t + \dfrac{1}{4}$

38. $m^2 + \dfrac{2}{3}m + \dfrac{1}{9}$

39. $x^2 - 1.0x + 0.25$

40. $y^2 - 1.4y + 0.49$

41. $2x^2 + 24x + 72$

42. $3y^2 - 48y + 192$

43. $16x^2 - 40x + 25$

44. $36y^2 - 60y + 25$

45. $49x^2 - 28xy + 4y^2$

46. $4z^2 - 12zw + 9w^2$

47. $64x^2 + 48xy + 9y^2$

48. $9t^2 + 24tr + 16r^2$

49. $50h^3 - 40h^2y + 8hy^2$

50. $18x^3 + 48x^2y + 32xy^2$

RELATING CONCEPTS (EXERCISES 51–54) For Individual or Group Work

We have seen that multiplication and factoring are reverse processes. We know that multiplication and division are also related. To check a division problem, we multiply the quotient by the divisor to get the dividend. To see how factoring and division are related, **work Exercises 51–54 in order.**

51. Factor $10x^2 + 11x - 6$.

52. Use long division from **Section 13.7** to divide $10x^2 + 11x - 6$ by $2x + 3$.

53. Could we have predicted the result in Exercise 52 from the result in Exercise 51? Explain.

54. Divide $x^3 - 1$ by $x - 1$. Use your answer to factor $x^3 - 1$.

Summary Exercises on Factoring

As you factor a polynomial, ask yourself these questions to decide on a suitable factoring technique.

> **Factoring a Polynomial**
> 1. **Is there a common factor?** If so, factor it out.
> 2. **How many terms are in the polynomial?**
>   *Two terms:* Check to see whether it is a difference of squares.
>   *Three terms:* Is it a perfect square trinomial? If the trinomial is not a perfect square, check to see whether the coefficient of the second-degree term is 1. If so, use the method of **Section 14.2**. If the coefficient of the squared term of the trinomial is not 1, use the general factoring methods of **Sections 14.3** and **14.4**.
>   *Four terms:* Try to factor the polynomial by grouping.
> 3. **Can any factors be factored further?** If so, factor them.

Factor each polynomial completely. Remember to check by multiplying.

1. $32m^9 + 16m^5 + 24m^3$

2. $2m^2 - 10m - 48$

3. $14k^3 + 7k^2 - 70k$

4. $9z^2 + 64$

5. $6z^2 + 31z + 5$

6. $m^2 - 3mn - 4n^2$

7. $49z^2 - 16y^2$

8. $100n^2r^2 + 30nr^3 - 50n^2r$

9. $16x^2 + 20x$

10. $20 + 5m + 12n + 3mn$

11. $10y^2 - 7yz - 6z^2$

12. $y^4 - 81$

13. $m^2 + 2m - 15$

14. $6y^2 - 5y - 4$

15. $32z^3 + 56z^2 - 16z$

16. $15y^2 + 5y$

17. $z^2 - 12z + 36$

18. $9m^2 - 64$

19. $y^2 - 4yk - 12k^2$

20. $16z^2 - 8z + 1$

21. $6y^2 - 6y - 12$

22. $x^2 + \dfrac{1}{2}x + \dfrac{1}{16}$

23. $p^2 - 17p + 66$

24. $a^2 + 17a + 72$

25. $k^2 + 9$

26. $108m^2 - 36m + 3$

27. $z^2 - 3za - 10a^2$

28. $2a^3 + a^2 - 14a - 7$

29. $4k^2 - 12k + 9$

30. $a^2 - 3ab - 28b^2$

31. $16r^2 + 24rm + 9m^2$

32. $3k^2 + 4k - 4$

33. $n^2 - 12n - 35$

34. $a^4 - 625$

35. $16k^2 - 48k + 36$

36. $8k^2 - 10k - 3$

37. $36y^6 - 42y^5 - 120y^4$

38. $5z^3 - 45z^2 + 70z$

39. $8p^2 + 23p - 3$

40. $8k^2 - 2kh - 3h^2$

41. $54m^2 - 24z^2$

42. $4k^2 - 20kz + 25z^2$

43. $6a^2 + 10a - 4$

44. $15h^2 + 11hg - 14g^2$

45. $m^2 - 81$

46. $10z^2 - 7z - 6$

47. $125m^4 - 400m^3n + 195m^2n^2$

48. $9y^2 + 12y - 5$

49. $m^2 - 4m + 4$

50. $36x^2 + 32x + 9$

51. $27p^{10} - 45p^9 - 252p^8$

52. $10m^2 + 25m - 60$

53. $4 - 2q - 6p + 3pq$

54. $k^2 - \dfrac{64}{121}$

55. $64p^2 - 100m^2$

56. $m^3 + 4m^2 - 6m - 24$

57. $100a^2 - 81y^2$

58. $8a^2 + 23ab - 3b^2$

59. $a^2 + 8a + 16$

60. $4y^2 - 25$

14.6 Solving Quadratic Equations by Factoring

Galileo Galilei (1564–1642) developed theories to explain physical phenomena and set up experiments to test his ideas. According to legend, Galileo dropped objects of different weights from the Leaning Tower of Pisa to disprove the belief that heavier objects fall faster than lighter objects. He developed a formula for freely falling objects described by $d = 16t^2$, where d is the distance in feet that an object falls (disregarding air resistance) in t seconds, regardless of weight.

The equation $d = 16t^2$ is a *quadratic equation*. A quadratic equation contains a squared term and no terms of higher degree.

OBJECTIVES

1. Solve quadratic equations by factoring.
2. Solve other equations by factoring.

Quadratic Equation

A **quadratic equation** is an equation that can be written in the form

$$ax^2 + bx + c = 0,$$

where a, b, and c are real numbers, and $a \neq 0$. This form is called **standard form**.

$x^2 + 5x + 6 = 0 \qquad 2t^2 - 5t = 3 \qquad y^2 = 4$ Examples of quadratic equations

In these examples, only $x^2 + 5x + 6 = 0$ is in *standard form*.

> **Work Problems 1 and 2 at the Side.**

Up to now, we have factored *expressions*, including many quadratic expressions of the form $ax^2 + bx + c$. In this section, we use factored quadratic expressions to solve quadratic *equations*.

OBJECTIVE 1 **Solve quadratic equations by factoring.** We use the **zero-factor property** to solve a quadratic equation by factoring.

Zero-Factor Property

If a and b are real numbers and $ab = 0$, then $a = 0$ or $b = 0$.

In words, if the product of two numbers is 0, then at least one of the numbers must be 0. One number *must* be 0, but both *may* be 0.

1 Which of the following equations are quadratic equations?

A. $y^2 - 4y - 5 = 0$

B. $x^3 - x^2 + 16 = 0$

C. $2z^2 + 7z = -3$

D. $x + 2y = -4$

2 Write each quadratic equation in standard form.

(a) $x^2 - 3x = 4$

(b) $y^2 = 9y - 8$

EXAMPLE 1   Using the Zero-Factor Property

Solve each equation.

(a) $(x + 3)(2x - 1) = 0$

The product $(x + 3)(2x - 1)$ is equal to 0. By the zero-factor property, the only way that the product of these two factors can be 0 is if at least one of the factors equals 0. Therefore, either $x + 3 = 0$ or $2x - 1 = 0$.

$x + 3 = 0$ or $2x - 1 = 0$ Zero-factor property
$x = -3$ $2x = 1$ Solve each equation.
                          $x = \dfrac{1}{2}$

Continued on Next Page

ANSWERS
1. A, C
2. (a) $x^2 - 3x - 4 = 0$
   (b) $y^2 - 9y + 8 = 0$

3 Solve each equation. Check your solutions.

(a) $(x - 5)(x + 2) = 0$

(b) $(3x - 2)(x + 6) = 0$

(c) $z(2z + 5) = 0$

The given equation, $(x + 3)(2x - 1) = 0$, has two solutions, -3 and $\frac{1}{2}$. *Check* these solutions by substituting -3 for x in the original equation, $(x + 3)(2x - 1) = 0$. Then start over and substitute $\frac{1}{2}$ for x.

If $x = -3$, then

$$(x + 3)(2x - 1) = 0$$
$$(-3 + 3)[2(-3) - 1] = 0 \quad ?$$
$$0(-7) = 0. \quad \text{True}$$

If $x = \frac{1}{2}$, then

$$(x + 3)(2x - 1) = 0$$
$$\left(\frac{1}{2} + 3\right)\left(2 \cdot \frac{1}{2} - 1\right) = 0 \quad ?$$
$$\frac{7}{2}(1 - 1) = 0 \quad ?$$
$$\frac{7}{2} \cdot 0 = 0. \quad \text{True}$$

Both -3 and $\frac{1}{2}$ result in true equations, so they are solutions to the original equation.

(b) $\qquad y(3y - 4) = 0$

$y = 0 \quad$ or $\quad 3y - 4 = 0 \qquad$ Zero-factor property

$\qquad\qquad\qquad 3y = 4 \qquad$ Finish solving right-hand equation.

$$y = \frac{4}{3}$$

Check these solutions by substituting each one in the original equation. The solutions are 0 and $\frac{4}{3}$.

◀◀◀ Work Problem 3 at the Side.

> **NOTE**
> The word *or* as used in Example 1 means "one or the other or both."

In Example 1, each equation to be solved was given with the polynomial in factored form. If the polynomial in an equation is not already factored, first make sure that the equation is in standard form. Then factor.

EXAMPLE 2 Solving Quadratic Equations

Solve each equation.

(a) $x^2 - 5x = -6$

First, write the equation in standard form by adding 6 to both sides.

$$x^2 - 5x = -6$$
$$x^2 - 5x + 6 = 0 \qquad \text{Add 6 to both sides.}$$

Now factor $x^2 - 5x + 6$. Find two numbers whose product is 6 and whose sum is -5. These two numbers are -2 and -3, so the equation becomes:

$$(x - 2)(x - 3) = 0 \qquad \text{Factor.}$$

$x - 2 = 0 \quad$ or $\quad x - 3 = 0 \qquad$ Zero-factor property

$x = 2 \qquad\qquad\quad x = 3 \qquad$ Solve each equation.

Continued on Next Page

ANSWERS

3. (a) $-2, 5$ (b) $-6, \frac{2}{3}$ (c) $-\frac{5}{2}, 0$

Check: If $x = 2$, then

$2^2 - 5(2) = -6$?
$4 - 10 = -6$?
$-6 = -6.$ True

If $x = 3$, then

$3^2 - 5(3) = -6$?
$9 - 15 = -6$?
$-6 = -6.$ True

Both solutions check, so the solutions are 2 and 3.

(b)
$$y^2 = y + 20$$
$$y^2 - y - 20 = 0 \quad \text{Write in standard form.}$$
$$(y - 5)(y + 4) = 0 \quad \text{Factor.}$$
$$y - 5 = 0 \quad \text{or} \quad y + 4 = 0 \quad \text{Zero-factor property}$$
$$y = 5 \qquad\qquad y = -4 \quad \text{Solve each equation.}$$

Check the solutions 5 and -4 by substituting in the original equation.

Work Problem 4 at the Side.

In summary, follow these steps to solve quadratic equations by factoring.

Solving a Quadratic Equation by Factoring

Step 1 **Write the equation in standard form,** that is, with all terms on one side of the equal sign in descending powers of the variable and 0 on the other side.

Step 2 **Factor** completely.

Step 3 **Use the zero-factor property** to set each factor with a variable equal to 0, and solve the resulting equations.

Step 4 **Check** each solution in the original equation.

EXAMPLE 3 Solving a Quadratic Equation with a Common Factor

Solve $4p^2 + 40 = 26p$.

$$4p^2 - 26p + 40 = 0 \quad \text{Standard form}$$
$$2(2p^2 - 13p + 20) = 0 \quad \text{Factor out 2.}$$
$$2p^2 - 13p + 20 = 0 \quad \text{Divide both sides by 2.}$$
$$(2p - 5)(p - 4) = 0 \quad \text{Factor.}$$
$$2p - 5 = 0 \quad \text{or} \quad p - 4 = 0 \quad \text{Zero-factor property}$$
$$2p = 5 \qquad\qquad p = 4 \quad \text{Solve each equation.}$$
$$p = \frac{5}{2}$$

Check the solutions $\frac{5}{2}$ and 4 by substituting in the original equation.

CAUTION
A common error is to include the common factor 2 as a solution in Example 3. **Only factors containing variables lead to solutions.**

Work Problem 5 at the Side.

4 Solve each equation. Check your solutions.

(a) $m^2 - 3m - 10 = 0$

(b) $r^2 + 2r = 8$

5 Solve each equation. Check your solutions.

(a) $10a^2 - 5a - 15 = 0$

(b) $4x^2 - 2x = 42$

ANSWERS
4. (a) $-2, 5$ (b) $-4, 2$
5. (a) $-1, \frac{3}{2}$ (b) $-3, \frac{7}{2}$

Chapter 14 Factoring and Applications

6 Solve each equation. Check your solutions.

(a) $49m^2 - 9 = 0$

(b) $p(4p + 7) = 2$

(c) $m^2 = 3m$

EXAMPLE 4 Solving Quadratic Equations

Solve each equation.

(a) $16m^2 - 25 = 0$

We can factor the left side of the equation as the difference of squares (**Section 14.5**).

$$16m^2 - 25 = 0$$
$$(4m + 5)(4m - 5) = 0 \qquad \text{Factor.}$$
$$4m + 5 = 0 \quad \text{or} \quad 4m - 5 = 0 \qquad \text{Zero-factor property}$$
$$4m = -5 \qquad\qquad 4m = 5 \qquad \text{Solve each equation.}$$
$$m = -\frac{5}{4} \qquad\qquad m = \frac{5}{4}$$

Check the solutions $-\frac{5}{4}$ and $\frac{5}{4}$ in the original equation.

(b) $k(2k + 5) = 3$

We first need to write this equation in standard form.

$$k(2k + 5) = 3 \qquad \text{Distributive property.}$$
$$2k^2 + 5k = 3 \qquad \text{Subtract 3 from both sides.}$$
$$2k^2 + 5k - 3 = 0 \qquad \text{Standard form}$$
$$(2k - 1)(k + 3) = 0 \qquad \text{Factor.}$$
$$2k - 1 = 0 \quad \text{or} \quad k + 3 = 0 \qquad \text{Zero-factor property}$$
$$2k = 1 \qquad\qquad k = -3 \qquad \text{Solve each equation.}$$
$$k = \frac{1}{2}$$

Check that the solutions are $\frac{1}{2}$ and -3.

(c) $y^2 = 2y$

$$y^2 - 2y = 0 \qquad \text{Standard form}$$
$$y(y - 2) = 0 \qquad \text{Factor.}$$
$$y = 0 \quad \text{or} \quad y - 2 = 0 \qquad \text{Zero-factor property}$$
$$y = 2 \qquad \text{Finish solving right-hand equation.}$$

Check that the solutions are 0 and 2.

CAUTION

In Example 4(b), the zero-factor property could not be used to solve the equation $k(2k + 5) = 3$ in its given form because of the 3 on the right. *The zero-factor property applies only to a product that equals 0.*

In Example 4(c), it is tempting to begin by dividing both sides of the equation $y^2 = 2y$ by y to get $y = 2$. Note that we do not get the other solution, 0, if we divide by a variable. (We *may* divide both sides of an equation by a *nonzero* real number, however. For instance, in Example 3 we divided both sides by 2.)

ANSWERS

6. (a) $-\frac{3}{7}, \frac{3}{7}$ (b) $-2, \frac{1}{4}$ (c) $0, 3$

Work Problem 6 at the Side.

> **NOTE**
> Not all quadratic equations can be solved by factoring. A more general method for solving such equations is given in **Chapter 17**.

OBJECTIVE 2 Solve other equations by factoring. We can also extend the zero-factor property to solve equations that involve more than two factors with variables, as shown in Examples 5 and 6. (These equations are *not* quadratic equations. Why not?)

EXAMPLE 5 Solving an Equation with More Than Two Variable Factors

Solve $6z^3 - 6z = 0$.

$$6z^3 - 6z = 0$$
$$6z(z^2 - 1) = 0 \quad \text{Factor out } 6z.$$
$$6z(z + 1)(z - 1) = 0 \quad \text{Factor } z^2 - 1.$$

By an extension of the zero-factor property, this product can equal 0 only if at least one of the factors equals 0. Write and solve three equations, one for each factor with a variable.

$$6z = 0 \quad \text{or} \quad z + 1 = 0 \quad \text{or} \quad z - 1 = 0$$
$$z = 0 \quad\quad\quad z = -1 \quad\quad\quad z = 1$$

Check by substituting, in turn, 0, -1, and 1 in the original equation. The solutions are 0, -1, and 1.

> **Work Problem 7 at the Side.**

EXAMPLE 6 Solving an Equation with a Quadratic Factor

Solve $(2x - 1)(x^2 - 9x + 20) = 0$.

$$(2x - 1)(x^2 - 9x + 20) = 0$$
$$(2x - 1)(x - 5)(x - 4) = 0 \quad \text{Factor } x^2 - 9x + 20.$$
$$2x - 1 = 0 \quad \text{or} \quad x - 5 = 0 \quad \text{or} \quad x - 4 = 0 \quad \text{Zero-factor property}$$
$$x = \frac{1}{2} \quad\quad\quad x = 5 \quad\quad\quad x = 4$$

Check. The solutions are $\frac{1}{2}$, 5, and 4.

> **Work Problem 8 at the Side.**

> **CAUTION**
> In Example 6, it would be unproductive to begin by multiplying the two factors together. Keep in mind that the zero-factor property and its extension requires the product of two or more factors to equal 0. Always consider first whether an equation is given in the appropriate form to apply the zero-factor property.

---

7 Solve each equation. Check your solutions.

(a) $r^3 - 16r = 0$

(b) $x^3 - 3x^2 - 18x = 0$

8 Solve each equation. Check your solutions.

(a) $(m + 3)(m^2 - 11m + 10) = 0$

(b) $(2x + 5)(4x^2 - 9) = 0$

ANSWERS

7. (a) $-4, 0, 4$ (b) $-3, 0, 6$
8. (a) $-3, 1, 10$ (b) $-\frac{5}{2}, -\frac{3}{2}, \frac{3}{2}$

Focus on Real-Data Applications

Factoring Trinomials Made Easy

To factor a trinomial using FOIL, we must find the *Outer* and *Inner* coefficients that sum to give the coefficient of the middle term. Our approach begins with a **key number,** found by multiplying the coefficients of the first and last terms of the trinomial. For the trinomial $6x^2 - x - 2$, for instance, the key number is -12 since $6(-2) = -12$.

Step 1 Display the factors of -12 by entering $Y_1 = -12/X$ in a graphing calculator (Screen 1), and using an automatic table (Screen 2). Factors of -12 are automatically displayed in pairs as 1, -12; 2, -6; 3, -4; 4, -3; and 6, -2 (Screen 3). You could scroll up or down to find other factors. Note that 5, -2.4 and 7, -1.714 are not acceptable factor pairs since -2.4 and -1.714 are not integers.

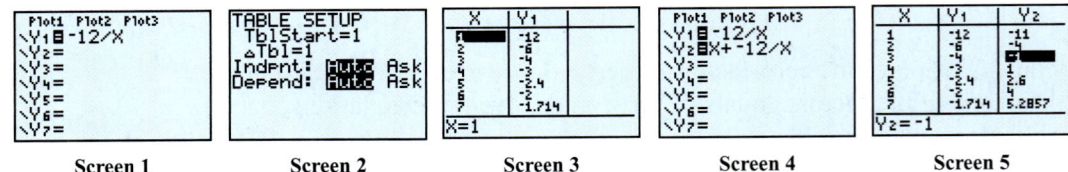

Screen 1　　Screen 2　　Screen 3　　Screen 4　　Screen 5

Step 2 Find the pair of factors that sum to the *middle* term coefficient, -1. We can let the calculator do this, too. Enter $Y_2 = X + -12/X$ (Screen 4). In this case, X is one of the factors, and $-12/X$ is the other, so Y_2 will give the sum. Look for -1 in the Y_2 column in Screen 5. (You may have to scroll up or down to find it.)

Step 3 Screen 5 shows that the coefficients of the outer and inner products are 3 and -4. Write $6x^2 - x - 2$ as $6x^2 + 3x - 4x - 2$. Using factoring by grouping,

$$(6x^2 + 3x) + (-4x - 2) = 3x(2x + 1) - 2(2x + 1)$$
$$= (2x + 1)(3x - 2).$$

For Group Discussion

Factor each trinomial given in the column heads of the table. First find the key number, and then use a calculator to find the coefficients of the outer and inner products of FOIL. Finally, factor by grouping.

Trinomial	$3x^2 - 2x - 8$	$2x^2 - 11x + 15$	$10x^2 + 11x - 6$	$4x^2 + 5x + 3$
Key Number	-24 (Why?)			
Outer, Inner Coefficients				
Factor by Grouping				(*Hint:* What does it mean if the middle term coefficient is *not* listed in the Y_2 column?)

14.6 Exercises

Solve each equation, and check your solutions. See Example 1.

1. $(x + 5)(x - 2) = 0$
2. $(x - 1)(x + 8) = 0$
3. $(2m - 7)(m - 3) = 0$
4. $(6k + 5)(k + 4) = 0$
5. $t(6t + 5) = 0$
6. $w(4w + 1) = 0$
7. $2x(3x - 4) = 0$
8. $6x(4x + 9) = 0$
9. $\left(x + \frac{1}{2}\right)\left(2x - \frac{1}{3}\right) = 0$
10. $\left(a + \frac{2}{3}\right)\left(5a - \frac{1}{2}\right) = 0$
11. $(0.5z - 1)(2.5z + 2) = 0$
12. $(0.25x + 1)(x - 0.5) = 0$
13. $(x - 9)(x - 9) = 0$
14. $(2x + 1)(2x + 1) = 0$

15. What is wrong with this "solution"?
$$2x(3x - 4) = 0$$
$$x = 2 \quad \text{or} \quad x = 0 \quad \text{or} \quad 3x - 4 = 0$$
$$x = \frac{4}{3}$$
The solutions are 2, 0, and $\frac{4}{3}$.

16. What is wrong with this "solution"?
$$x(7x - 1) = 0$$
$$7x - 1 = 0 \quad \text{Zero-factor property}$$
$$x = \frac{1}{7}$$
The solution is $\frac{1}{7}$.

Solve each equation, and check your solutions. See Examples 2–6.

17. $y^2 + 3y + 2 = 0$
18. $p^2 + 8p + 7 = 0$
19. $y^2 - 3y + 2 = 0$
20. $r^2 - 4r + 3 = 0$
21. $x^2 = 24 - 5x$
22. $t^2 = 2t + 15$
23. $x^2 = 3 + 2x$
24. $m^2 = 4 + 3m$
25. $z^2 + 3z = -2$
26. $p^2 - 2p = 3$
27. $m^2 + 8m + 16 = 0$
28. $b^2 - 6b + 9 = 0$
29. $3x^2 + 5x - 2 = 0$
30. $6r^2 - r - 2 = 0$
31. $6p^2 = 4 - 5p$

32. $6x^2 = 4 + 5x$

33. $9s^2 + 12s = -4$

34. $36x^2 + 60x = -25$

35. $y^2 - 9 = 0$

36. $m^2 - 100 = 0$

37. $16k^2 - 49 = 0$

38. $4w^2 - 9 = 0$

39. $n^2 = 121$

40. $x^2 = 400$

41. $x^2 = 7x$

42. $t^2 = 9t$

43. $6r^2 = 3r$

44. $10y^2 = -5y$

45. $g(g - 7) = -10$

46. $r(r - 5) = -6$

47. $z(2z + 7) = 4$

48. $b(2b + 3) = 9$

49. $2(y^2 - 66) = -13y$

50. $3(t^2 + 4) = 20t$

51. $5x^3 - 20x = 0$

52. $3x^3 - 48x = 0$

53. $9y^3 - 49y = 0$

54. $16r^3 - 9r = 0$

55. $(2r + 5)(3r^2 - 16r + 5) = 0$

56. $(3m + 4)(6m^2 + m - 2) = 0$

57. $(2x + 7)(x^2 + 2x - 3) = 0$

58. $(x + 1)(6x^2 + x - 12) = 0$

59. Galileo's formula for freely falling objects, $d = 16t^2$, was given at the beginning of this section. The distance d in feet an object falls depends on the time elapsed t in seconds. (This is an example of an important mathematical concept, a *function*.)

(a) Use Galileo's formula and complete the following table. (*Hint:* Substitute each given value into the formula and solve for the unknown value.)

t in seconds	0	1	2	3		
d in feet	0	16			256	576

(b) When $t = 0$, $d = 0$. Explain this in the context of the problem.

(c) When you substituted 256 for d and solved for t, you should have found two solutions, 4 and -4. Why doesn't -4 make sense as an answer?

14.7 Applications of Quadratic Equations

We can use factoring to solve quadratic equations that arise in applications. We follow the same six problem-solving steps given in **Section 3.3**.

OBJECTIVES
1. Solve problems about geometric figures.
2. Solve problems about consecutive integers.
3. Solve problems using the Pythagorean formula.
4. Solve problems using given quadratic models.

Solving an Applied Problem

Step 1 **Read** the problem, several times if necessary, until you *understand* what is given and what is to be found.

Step 2 **Assign a variable** to represent the unknown value, using diagrams or tables as needed. Write down what the variable represents. Express any other unknown values in terms of the variable.

Step 3 **Write an equation** using the variable expression(s).

Step 4 **Solve** the equation.

Step 5 **State the answer.** Does it seem reasonable?

Step 6 **Check** the answer in the words of the original problem.

OBJECTIVE 1 Solve problems about geometric figures. Some of the applied problems in this section require one of the formulas given on the inside back cover of the text.

EXAMPLE 1 Solving an Area Problem

The Monroes want to plant a rectangular garden in their yard. The width of the garden will be 4 ft less than its length, and they want it to have an area of 96 ft^2. (ft^2 means square feet.) Find the length and width of the garden.

Step 1 **Read** the problem carefully. We need to find the dimensions of a garden with area 96 ft^2.

Step 2 **Assign a variable.** You know the least about the length, so

Let x = the length of the garden.

Then $x - 4$ = the width. (The width is 4 ft less than the length.)

See Figure 1.

Figure 1

Step 3 **Write an equation.** The area of a rectangle is given by

$$\text{Area} = lw = \text{length} \times \text{width}. \quad \text{Area formula}$$

Substitute 96 for the area, x for the length, and $x - 4$ for the width.

$$96 = x(x - 4) \quad \text{Let } A = 96, l = x, w = x - 4.$$

Continued on Next Page

① Solve each problem.

(a) The length of a rectangular room is 2 m more than the width. The area of the floor is 48 m². Find the length and width of the room.

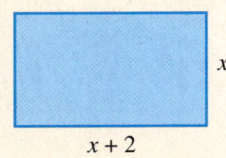

(b) The length of each side of a square is increased by 4 in. The sum of the areas of the original square and the larger square is 106 in². What is the length of a side of the original square?

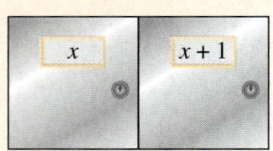

Figure 2

ANSWERS

1. (a) length: 8 m; width: 6 m   (b) 5 in.

Step 4 **Solve.**

$$96 = x^2 - 4x \qquad \text{Distributive property}$$
$$0 = x^2 - 4x - 96 \qquad \text{Standard form}$$
$$0 = (x - 12)(x + 8) \qquad \text{Factor.}$$
$$x - 12 = 0 \quad \text{or} \quad x + 8 = 0 \qquad \text{Zero-factor property}$$
$$x = 12 \qquad x = -8 \qquad \text{Solve each equation.}$$

Step 5 **State the answer.** The solutions are 12 and -8. A rectangle cannot have a side of negative length, so discard -8. The length of the garden will be 12 ft. The width will be $12 - 4 = 8$ ft.

Step 6 **Check.** The width is 4 ft less than the length; the area is $12 \cdot 8 = 96$ ft², as required.

> **CAUTION**
> When solving applied problems, *always check solutions* against physical facts and discard any answers that are not appropriate.

◀◀◀ **Work Problem 1 at the Side.**

OBJECTIVE 2 Solve problems about consecutive integers. Two **consecutive integers** are integers that are next to each other on a number line, such as 5 and 6, or -11 and -10. **Consecutive odd integers** are *odd* integers that are next to each other, such as 5 and 7, or -13 and -11. **Consecutive even integers** are defined similarly; for example, 4 and 6 are consecutive even integers, as are -10 and -8.

> **PROBLEM-SOLVING HINT**
> In consecutive integer problems, if x represents the first integer, then for
> two consecutive integers, use $\qquad x, \; x + 1;$
> three consecutive integers, use $\qquad x, \; x + 1, \; x + 2;$
> two consecutive *even* or *odd* integers, use $\qquad x, \; x + 2;$
> three consecutive *even* or *odd* integers, use $\qquad x, \; x + 2, \; x + 4.$

EXAMPLE 2 Solving a Consecutive Integer Problem

The product of the numbers on two consecutive post-office boxes is 210. Find the box numbers.

Step 1 **Read** the problem. Note that the boxes are numbered consecutively.

Step 2 **Assign a variable.**

Let $x =$ the first box number.
Then $x + 1 =$ the next consecutive box number.

See Figure 2 at the left.

Step 3 **Write an equation.** The product of the box numbers is 210, so

$$x(x + 1) = 210.$$

────── **Continued on Next Page**

Step 4 **Solve.**

$$x(x + 1) = 210$$
$$x^2 + x = 210 \quad \text{Distributive property}$$
$$x^2 + x - 210 = 0 \quad \text{Standard form}$$
$$(x + 15)(x - 14) = 0 \quad \text{Factor.}$$
$$x + 15 = 0 \quad \text{or} \quad x - 14 = 0 \quad \text{Zero-factor property}$$
$$x = -15 \quad \text{or} \quad x = 14$$

Step 5 **State the answer.** The solutions are -15 and 14. Discard the solution -15 since a box number cannot be negative. When $x = 14$, then $x + 1 = 15$, so the post-office boxes have the numbers 14 and 15.

Step 6 **Check.** The numbers 14 and 15 are consecutive and their product is $14 \cdot 15 = 210$, as required.

> **Work Problem 2 at the Side.**

② Solve the problem.
The product of the numbers on two consecutive lockers at a health club is 132. Find the locker numbers.

EXAMPLE 3 Solving a Consecutive Integer Problem

The product of two consecutive odd integers is 1 less than five times their sum. Find the integers.

Step 1 **Read** carefully. This problem is a little more complicated.

Step 2 **Assign a variable.** We must find two consecutive *odd* integers.

Let $x =$ the smaller integer.
Then $x + 2 =$ the next larger odd integer.

Step 3 **Write an equation.** According to the problem, the product is 1 less than five times the sum.

The product is five times the sum less 1.
$$x(x + 2) = 5(x + x + 2) - 1$$

Step 4 **Solve.**

$$x^2 + 2x = 5x + 5x + 10 - 1 \quad \text{Distributive property}$$
$$x^2 + 2x = 10x + 9 \quad \text{Combine like terms.}$$
$$x^2 - 8x - 9 = 0 \quad \text{Standard form}$$
$$(x - 9)(x + 1) = 0 \quad \text{Factor.}$$
$$x - 9 = 0 \quad \text{or} \quad x + 1 = 0 \quad \text{Zero-factor property}$$
$$x = 9 \quad \text{or} \quad x = -1$$

Step 5 **State the answer.** We need to find two consecutive odd integers.

If $x = 9$ is the smaller, then $x + 2 = 9 + 2 = 11$ is the larger.
If $x = -1$ is the smaller, then $x + 2 = -1 + 2 = 1$ is the larger.

There are two sets of answers here since integers can be positive or negative.

Step 6 **Check.** The product of the first pair of integers is $9 \cdot 11 = 99$. One less than five times their sum is $5(9 + 11) - 1 = 99$. Thus 9 and 11 satisfy the problem. Repeat the check with -1 and 1.

> **Work Problem 3 at the Side.**

③ Solve each problem.

(a) The product of two consecutive even integers is 4 more than two times their sum. Find the integers.

(b) Find three consecutive odd integers such that the product of the smallest and largest is 16 more than the middle integer.

ANSWERS
2. 11 and 12
3. (a) 4 and 6 or -2 and 0   **(b)** 3, 5, 7

> **CAUTION**
> Do *not* use $x, x+1, x+3$, and so on to represent consecutive odd integers. To see why, let $x = 3$. Then $x + 1 = 3 + 1 = 4$ and $x + 3 = 3 + 3 = 6$, and 3, 4, and 6 are not consecutive odd integers.

OBJECTIVE 3 Solve problems using the Pythagorean formula. The next example uses the Pythagorean formula from **Section 5.8**.

> **Pythagorean Formula**
> If a right triangle (a triangle with a 90° angle) has longest side of length c and two other sides of lengths a and b, then
> $$a^2 + b^2 = c^2.$$
>
>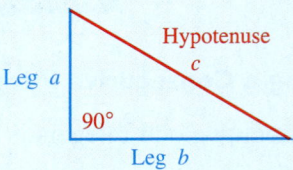
>
> Recall that the longest side, the *hypotenuse,* is opposite the right angle. The two shorter sides are the *legs* of the triangle.

EXAMPLE 4 Using the Pythagorean Formula

Ed and Mark leave their office, with Ed traveling north and Mark traveling east. When Mark is 1 mi farther than Ed from the office, the distance between them is 2 mi more than Ed's distance from the office. Find their distances from the office and the distance between them.

Step 1 **Read** the problem again. We must find three distances.

Step 2 **Assign a variable.** You know the least about Ed's distance, so let x represent Ed's distance from the office. Then $x + 1$ represents Mark's distance from the office, and $x + 2$ represents the distance between them. Place these on a right triangle, as in Figure 3.

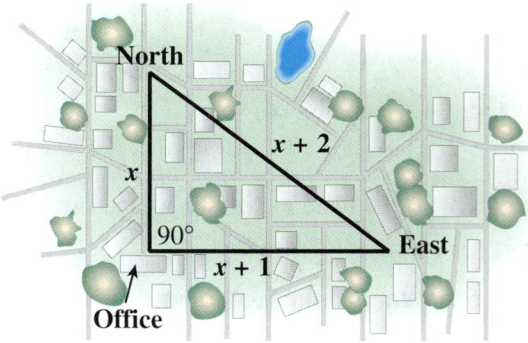

Figure 3

Step 3 **Write an equation.** Substitute into the Pythagorean formula.
$$a^2 + b^2 = c^2$$
$$x^2 + (x + 1)^2 = (x + 2)^2$$

Continued on Next Page

Step 4 Solve.
$$x^2 + x^2 + 2x + 1 = x^2 + 4x + 4$$
$$x^2 - 2x - 3 = 0 \quad \text{Standard form}$$
$$(x - 3)(x + 1) = 0 \quad \text{Factor.}$$
$$x - 3 = 0 \quad \text{or} \quad x + 1 = 0 \quad \text{Zero-factor property}$$
$$x = 3 \quad \text{or} \quad x = -1$$

Step 5 **State the answer.** Since -1 cannot represent a distance, 3 is the only possible answer. Ed's distance is 3 mi, Mark's distance is $3 + 1 = 4$ mi, and the distance between them is $3 + 2 = 5$ mi.

Step 6 **Check.** Since $3^2 + 4^2 = 5^2$, the answer is correct.

> **CAUTION**
> When solving a problem involving the Pythagorean formula, be sure that the expressions for the sides are properly placed.
> $$\text{leg}^2 + \text{leg}^2 = \text{hypotenuse}^2$$

Work Problem 4 at the Side.

OBJECTIVE 4 **Solve problems using given quadratic models.** In Examples 1–4, we wrote quadratic equations to model, or mathematically describe, various situations and then solved the equations. Now we are given the quadratic models and must use them to determine data.

EXAMPLE 5 Finding the Height of a Ball

A tennis player can hit a ball 180 feet per second. If she hits a ball directly upward, the height h of the ball in feet at time t in seconds is modeled by the quadratic equation
$$h = -16t^2 + 180t + 6.$$
How long will it take for the ball to reach a height of 206 feet?

A height of 206 feet means $h = 206$, so we substitute 206 for h in the equation and then solve for t.

$$\mathbf{206} = -16t^2 + 180t + 6 \quad \text{Let } h = 206.$$
$$-16t^2 + 180t + 6 = 206 \quad \text{Interchange sides.}$$
$$-16t^2 + 180t - 200 = 0 \quad \text{Standard form}$$
$$4t^2 - 45t + 50 = 0 \quad \text{Divide by } -4.$$
$$(4t - 5)(t - 10) = 0 \quad \text{Factor.}$$
$$4t - 5 = 0 \quad \text{or} \quad t - 10 = 0 \quad \text{Zero-factor property}$$
$$t = \frac{5}{4} \quad \text{or} \quad t = 10$$

Since we found two acceptable answers, the ball will be 206 feet above the ground *twice*: once on its way up at $\frac{5}{4}$ seconds, and once on its way down at 10 seconds. See Figure 4.

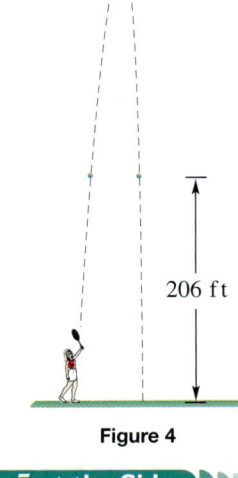

Figure 4

Work Problem 5 at the Side.

---

4 Solve the problem.

The hypotenuse of a right triangle is 3 in. longer than the longer leg. The shorter leg is 3 in. shorter than the longer leg. Find the lengths of the sides of the triangle.

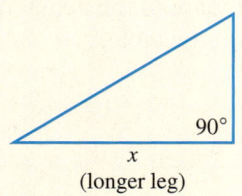

x (longer leg)

5 Solve the problem.

The number of impulses fired after a nerve has been stimulated is modeled by
$$I = -x^2 + 2x + 60,$$
where x is in milliseconds (ms) after the stimulation. When will 45 impulses occur? Do you get two solutions? Why is only one answer given?

ANSWERS
4. 9 in., 12 in., 15 in.
5. After 5 ms; There are two solutions, -3 and 5; Only one answer makes sense here because time cannot be a negative number.

6 Solve the problem.

Use the model in Example 6 to find the annual percent increase in spending on hospital services in 1995. Round your answer to the nearest tenth. How does it compare to the actual data from the table?

EXAMPLE 6 Modeling Increases in Hospital Costs

The annual percent increase y in spending on hospital services in the years 1994–2001 can be modeled by the quadratic equation

$$y = 0.37x^2 - 4.1x + 12,$$

where $x = 4$ represents 1994, $x = 5$ represents 1995, and so on. (*Source: Center for Studying Health System Change.*)

(a) Use the model to find the annual percent increase to the nearest tenth in 1997.

Since $x = 4$ represents 1994, $x = 7$ represents 1997. Substitute 7 for x in the equation.

$y = 0.37(7)^2 - 4.1(7) + 12$ Let $x = 7$.

$y \approx 1.4$ Round to the nearest tenth.

Spending on hospital services increased about 1.4% in 1997.

(b) Repeat part (a) for 2001.

$y = 0.37(11)^2 - 4.1(11) + 12$ For 2001, let $x = 11$.

$y \approx 11.7$

In 2001, spending on hospital services increased about 11.7%.

(c) The model used in parts (a) and (b) was developed using the data in the table below. How do the results in parts (a) and (b) compare to the actual data from the table?

Year	Percent Increase
1994	1.8
1995	0.8
1996	0.5
1997	1.3
1998	3.4
1999	5.8
2000	7.1
2001	12.0

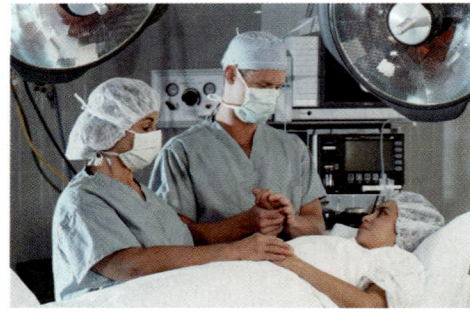

From the table, the actual data for 1997 is 1.3%. Our answer, 1.4%, is slightly high. For 2001, the actual data is 12.0%, so our answer of 11.7% in part (b) is a little low.

◀◀ **Work Problem 6 at the Side.**

NOTE
A graph of the quadratic equation from Example 6 is shown in Figure 5. Notice the basic shape of this graph, which follows the general pattern of the data in the table—it decreases from 1994 to 1996 and then increases from 1997 to 2001. We consider such graphs of quadratic equations, called *parabolas*, in **Chapter 17**.

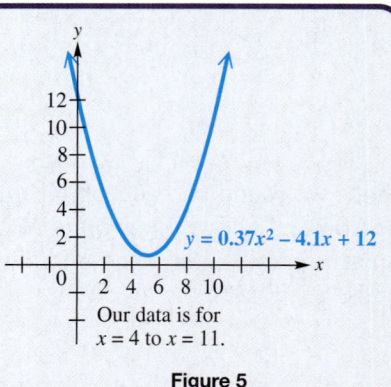

Our data is for $x = 4$ to $x = 11$.

Figure 5

ANSWER
6. 0.8%; The actual data for 1995 is 0.8%, so our answer using the model is the same.

Section 14.7 Applications of Quadratic Equations   981

14.7 Exercises

FOR EXTRA HELP 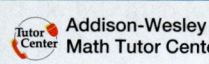 Addison-Wesley Math Tutor Center   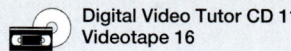 MathXL    Digital Video Tutor CD 11 Videotape 16   Student's Solutions Manual   MyMathLab   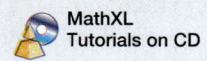 MathXL Tutorials on CD

1. To review the six problem-solving steps first introduced in **Section 3.3**, complete each statement.

   Step 1: _____ the problem, several times if necessary, until you understand what is given and what must be found.

   Step 2: Assign a _____ to represent the unknown value.

   Step 3: Write a(n) _____ using the variable expression(s).

   Step 4: _____ the equation.

   Step 5: State the _____.

   Step 6: _____ the answer in the words of the _____ problem.

2. A student solves an applied problem and gets 6 or -3 for the length of the side of a square. Which of these answers is reasonable? Explain.

In Exercises 3–6, a figure and a corresponding geometric formula are given. Using x as the variable, complete Steps 3–6 for each problem. (Refer to the steps in Exercise 1 as needed.)

3.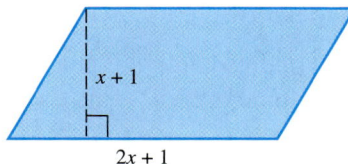

   Area of a parallelogram: $A = bh$

   The area of this parallelogram is 45 sq. units. Find its base and height.

4.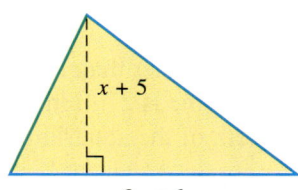

   Area of a triangle: $A = \frac{1}{2}bh$

   The area of this triangle is 60 sq. units. Find its base and height.

5.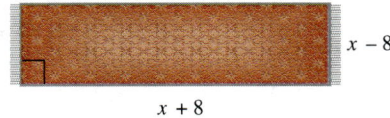

   Area of a rectangular rug: $A = lw$

   The area of this rug is 80 sq. units. Find its length and width.

6.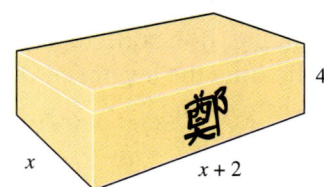

   Volume of a rectangular Chinese box: $V = lwh$

   The volume of this box is 192 cu. units. Find its length and width.

Solve each problem. Check your answers to be sure they are reasonable. Refer to the formulas on the inside back cover. See Example 1.

7. The length of a VHS videocassette case is 3 in. more than its width. The area of the rectangular top side of the case is 28 in.². Find the length and width of the case.

8. A plastic box that holds a standard audiocassette has length 4 cm longer than its width. The area of the rectangular top of the box is 77 cm². Find the length and width of the box.

9. The dimensions of a Gateway EV700 computer monitor screen are such that its length is 3 in. more than its width. If the length is increased by 1 in. while the width remains the same, the area is increased by 10 in.². What are the dimensions of the screen? (*Source:* Author's computer.)

10. The keyboard of the computer in Exercise 9 is 11 in. longer than it is wide. If both its length and width are increased by 2 in., the area of the top of the keyboard is increased by 54 in.². Find the length and width of the keyboard. (*Source:* Author's computer.)

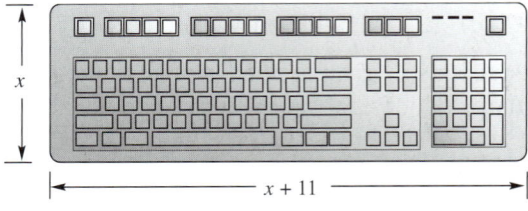

11. A ten-gallon aquarium is 3 in. higher than it is wide. Its length is 21 in., and its volume is 2730 in.³. What are the height and width of the aquarium?

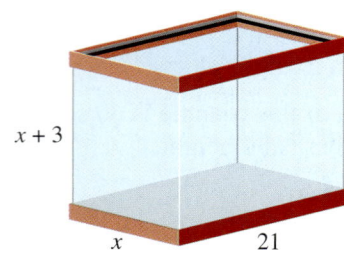

12. A toolbox is 2 ft high, and its width is 3 ft less than its length. If its volume is 80 ft³, find the length and width of the box.

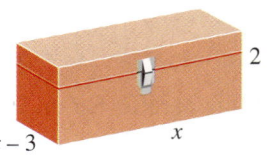

13. A square mirror has sides measuring 2 ft less than the sides of a square painting. If the difference between their areas is 32 ft², find the lengths of the sides of the mirror and the painting.

 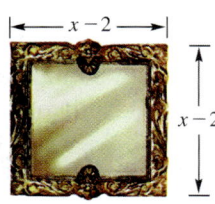

14. The sides of one square have length 3 m more than the sides of a second square. If the area of the larger square is subtracted from 4 times the area of the smaller square, the result is 36 m². What are the lengths of the sides of each square?

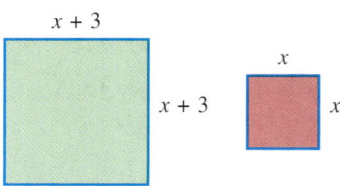

Solve each problem about consecutive integers. See Examples 2 and 3.

15. The product of the numbers on two consecutive volumes of research data is 420. Find the volume numbers.

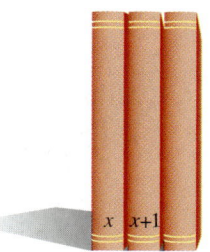

16. The product of the page numbers on two facing pages of a book is 600. Find the page numbers.

17. The product of two consecutive integers is 11 more than their sum. Find the integers.

18. The product of two consecutive integers is 4 less than four times their sum. Find the integers.

19. Find two consecutive odd integers such that their product is 15 more than three times their sum.

20. Find two consecutive odd integers such that five times their sum is 23 less than their product.

21. Find three consecutive even integers such that the sum of the squares of the smaller two is equal to the square of the largest.

22. Find three consecutive even integers such that the square of the sum of the smaller two is equal to twice the largest.

Use the Pythagorean formula to solve each problem. See Example 4.

23. The hypotenuse of a right triangle is 1 cm longer than the longer leg. The shorter leg is 7 cm shorter than the longer leg. Find the length of the longer leg of the triangle.

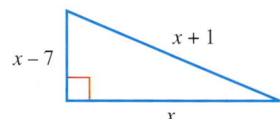

24. The longer leg of a right triangle is 1 m longer than the shorter leg. The hypotenuse is 1 m shorter than twice the shorter leg. Find the length of the shorter leg of the triangle.

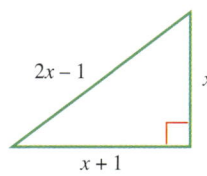

25. Terri works due north of home. Her husband Denny works due east. They leave for work at the same time. By the time Terri is 5 mi from home, the distance between them is 1 mi more than Denny's distance from home. How far from home is Denny?

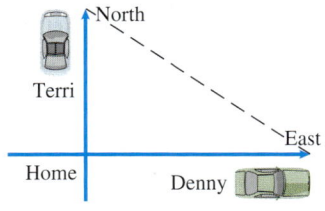

26. Two cars left an intersection at the same time. One traveled north. The other traveled 14 mi farther, but to the east. How far apart were they then, if the distance between them was 4 mi more than the distance traveled east?

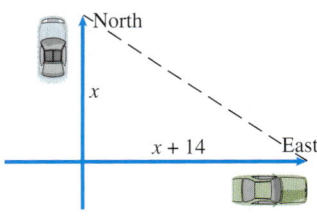

27. A ladder is leaning against a building. The distance from the bottom of the ladder to the building is 4 ft less than the length of the ladder. How high up the side of the building is the top of the ladder if that distance is 2 ft less than the length of the ladder?

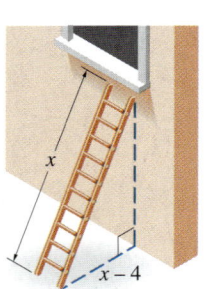

28. A lot has the shape of a right triangle with one leg 2 m longer than the other. The hypotenuse is 2 m less than twice the length of the shorter leg. Find the length of the shorter leg.

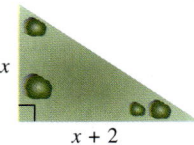

Solve each problem. See Examples 5 and 6.

29. An object propelled from a height of 48 ft with an initial velocity of 32 feet per second after t seconds has height

$$h = -16t^2 + 32t + 48.$$

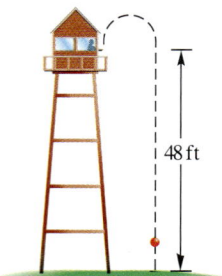

(a) After how many seconds is the height 64 ft? (*Hint:* Let $h = 64$ and solve.)

(b) After how many seconds is the height 60 ft?

(c) After how many seconds does the object hit the ground? (*Hint:* When the object hits the ground, $h = 0$.)

(d) The quadratic equation from part (c) has two solutions, yet only one of them is appropriate for answering the question. Why is this so?

30. If an object is propelled upward from ground level with an initial velocity of 64 feet per second, its height h in feet t seconds later is

$$h = -16t^2 + 64t.$$

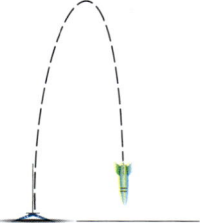

(a) After how many seconds is the height 48 ft?

(b) The object reaches its maximum height 2 sec after it is propelled. What is this maximum height?

(c) After how many seconds does the object hit the ground?

(d) The quadratic equation from part (c) has two solutions, yet only one of them is appropriate for answering the question. Why is this so?

31. The table shows the number of cellular phones (in millions) owned by Americans.

Year	Cellular Phones (in millions)
1990	5
1992	11
1994	24
1996	44
1998	62
2000	109
2002	140

Source: Cellular Telecommunications Industry Association.

We used the data to develop the quadratic equation
$$y = 0.866x^2 + 1.02x + 5.29,$$
which models the number of cellular phones y (in millions) in the year x, where $x = 0$ represents 1990, $x = 2$ represents 1992, and so on.

(a) Use the model to find the number of cellular phones in 1996 to the nearest tenth. How does the result compare to the actual data in the table?

(b) What value of x corresponds to 2002?

(c) Use the model to find the number of cellular phones in 2002 to the nearest tenth. How does the result compare to the actual data in the table?

(d) Assuming that the trend in the data continues, use the quadratic equation to predict the number of cellular phones in 2005 to the nearest tenth.

RELATING CONCEPTS (EXERCISES 32–40) For Individual or Group Work

The U.S. trade deficit represents the amount by which exports are less than imports. It provides not only a sign of economic prosperity but also a warning of potential decline. The data in the table shows the U.S. trade deficit for 1995 through 2000.

Year	Deficit (in billions of dollars)
1995	97.5
1996	104.3
1997	104.7
1998	164.3
1999	271.3
2000	378.7

Source: U.S. Department of Commerce.

Use the data to **work Exercises 32–40 in order.**

32. How much did the trade deficit increase from 1999 to 2000? What percent increase is this (to the nearest percent)?

33. The U.S. trade deficit might be approximated by the linear equation

$$y = 40.8x + 66.9,$$

where y is the deficit in billions of dollars. Here $x = 0$ represents 1995, $x = 1$ represents 1996, and so on. Use this equation to approximate the trade deficits in 1997, 1999, and 2000.

34. How do your answers from Exercise 33 compare to the actual data in the table?

35. The trade deficit y (in billions of dollars) might also be approximated by the quadratic equation

$$y = 18.5x^2 - 33.4x + 104,$$

where $x = 0$ again represents 1995, $x = 1$ represents 1996, and so on. Use this equation to approximate the trade deficits in 1997, 1999, and 2000.

36. Compare your answers from Exercise 35 to the actual data in the table. Which equation, the linear one in Exercise 33 or the quadratic one in Exercise 35, models the data better?

37. We can also see graphically why the linear equation is not a very good model for the data. To do so, write the data from the table as a set of ordered pairs (x, y), where x represents the years since 1995 and y represents the trade deficit in billions of dollars.

38. Plot the ordered pairs from Exercise 37 on the graph.

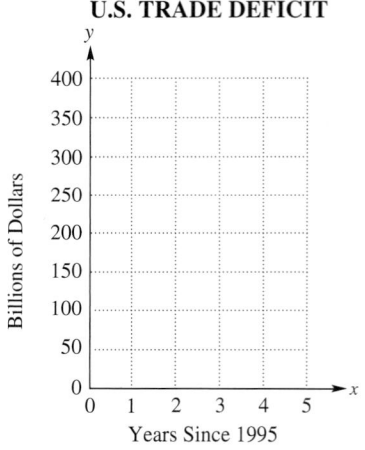

U.S. TRADE DEFICIT

Recall from **Chapter 11** that a linear equation has a straight line for its graph. Do the ordered pairs you plotted lie in a linear pattern?

39. Assuming that the trend in the data continues and since the quadratic equation modeled that data fairly well, use the quadratic equation to predict the trade deficit for the year 2002.

40. The actual trade deficit for 2002 was 417.9 billion dollars.

(a) How does the actual deficit for 2002 compare to your prediction from Exercise 39?

(b) Should the quadratic equation be used to predict the U.S. trade deficit for years after 2000? Explain.

Chapter 14
SUMMARY

KEY TERMS

14.1	factor	An expression *A* is a factor of an expression *B* if *B* can be divided by *A* with 0 remainder.
	factored form	An expression is in factored form when it is written as a product.
	greatest common factor (GCF)	The greatest common factor is the largest quantity that is a factor of each of a group of quantities.
	factoring	The process of writing a polynomial as a product is called factoring.
14.2	prime polynomial	A prime polynomial is a polynomial that cannot be factored using only integers.
14.5	perfect square trinomial	A perfect square trinomial is a trinomial that can be factored as the square of a binomial.
14.6	quadratic equation	A quadratic equation is an equation that can be written in the form $ax^2 + bx + c = 0$, with $a \neq 0$.
	standard form	The form $ax^2 + bx + c = 0$ is the standard form of a quadratic equation.
14.7	consecutive integers	Consecutive integers are next to each other on a number line; they differ by 1.

TEST YOUR WORD POWER

See how well you have learned the vocabulary in this chapter. Answers, with examples, follow the Quick Review.

1. **Factoring** is
   A. a method of multiplying polynomials
   B. the process of writing a polynomial as a product
   C. the answer in a multiplication problem
   D. a way to add the terms of a polynomial.

2. A polynomial is in **factored form** when
   A. it is prime
   B. it is written as a sum
   C. the squared term has a coefficient of 1
   D. it is written as a product.

3. The **greatest common factor** of a polynomial is
   A. the least integer that divides evenly into all the terms of the polynomial
   B. the least term that is a factor of all the terms in the polynomial
   C. the greatest term that is a factor of all the terms in the polynomial
   D. the variable that is common to all the terms in the polynomial.

4. A **perfect square trinomial** is a trinomial
   A. that can be factored as the square of a binomial
   B. that cannot be factored
   C. that is multiplied by a binomial
   D. where all terms are perfect squares.

5. A **quadratic equation** is an equation that can be written in the form
   A. $y = mx + b$
   B. $ax^2 + bx + c = 0$ $(a \neq 0)$
   C. $Ax + By = C$
   D. $x = k$.

6. **Consecutive integers** are two integers
   A. whose product is 1
   B. whose sum is 0
   C. that differ by 1
   D. that have a quotient of -1.

Quick Review

Concepts

Examples

14.1 Factors; The Greatest Common Factor
Finding the Greatest Common Factor (GCF)

Step 1 Write each number in prime factored form.

Step 2 List each prime number or each variable that is a factor of every term in the list.

Step 3 Use as exponents on the common prime factors the least exponents from the prime factored forms.

Step 4 Multiply the primes from Step 3.

Factoring by Grouping

Step 1 Group the terms.

Step 2 Factor out the greatest common factor from each group.

Step 3 Factor a common binomial factor from the results of Step 2.

Step 4 If necessary, rearrange terms and try a different grouping.

Find the greatest common factor of $4x^2y$, $-6x^2y^3$, and $2xy^2$.

$$4x^2y = \mathbf{2} \cdot 2 \cdot \mathbf{x^2} \cdot \mathbf{y}$$
$$-6x^2y^3 = -1 \cdot \mathbf{2} \cdot 3 \cdot \mathbf{x^2} \cdot \mathbf{y^3}$$
$$2xy^2 = \mathbf{2} \cdot \mathbf{x} \cdot \mathbf{y^2}$$

The greatest common factor is $2xy$.

Factor by grouping.

$$2a^2 + 2ab + a + b = (2a^2 + 2ab) + (a + b)$$
$$= 2a(a + b) + 1(a + b)$$
$$= (a + b)(2a + 1)$$

14.2 Factoring Trinomials
To factor $x^2 + bx + c$, find m and n such that $mn = c$ and $m + n = b$.

$$x^2 + bx + c$$
$$mn = c$$
$$m + n = b$$

Then $x^2 + bx + c = (x + m)(x + n)$.

Check by multiplying.

Factor $x^2 + 6x + 8$.

$$x^2 + 6x + 8$$
$$mn = 8$$
$$m + n = 6$$

$m = 2$ and $n = 4$

$x^2 + 6x + 8 = (x + 2)(x + 4)$

Check: $(x + 2)(x + 4) = x^2 + 4x + 2x + 8$
$$= x^2 + 6x + 8$$

14.3 Factoring Trinomials by Grouping
To factor $ax^2 + bx + c$ by grouping, first find m and n.

$$ax^2 + bx + c$$
$$m + n = b$$
$$mn = ac$$

Then factor $ax^2 + mx + nx + b$ by grouping.

Factor $3x^2 + 14x - 5$.

-15

Find two integers with a product of $3(-5) = -15$ and a sum of 14. The integers are -1 and 15.

$$3x^2 + 14x - 5 = 3x^2 - x + 15x - 5$$
$$= (3x^2 - x) + (15x - 5)$$
$$= x(3x - 1) + 5(3x - 1)$$
$$= (3x - 1)(x + 5)$$

Concepts

14.4 Factoring Trinomials Using FOIL
To factor $ax^2 + bx + c$ by trial and error, use FOIL backwards.

14.5 Special Factoring Techniques
Difference of Squares
$$a^2 - b^2 = (a + b)(a - b)$$

Perfect Square Trinomials
$$a^2 + 2ab + b^2 = (a + b)^2$$
$$a^2 - 2ab + b^2 = (a - b)^2$$

14.6 Solving Quadratic Equations by Factoring
Zero-Factor Property
If a and b are real numbers and $ab = 0$, then $a = 0$ or $b = 0$.

Solving a Quadratic Equation by Factoring

Step 1 Write the equation in standard form.

Step 2 Factor.

Step 3 Use the zero-factor property.

Step 4 Check.

14.7 Applications of Quadratic Equations
Pythagorean Formula
In a right triangle, the square of the hypotenuse equals the sum of the squares of the legs.
$$a^2 + b^2 = c^2$$

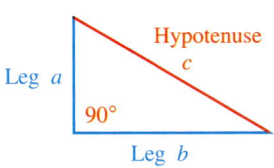

Examples

By trial and error,
$$3x^2 + 14x - 5 = (3x - 1)(x + 5).$$

Factor.
$$4x^2 - 9 = (2x + 3)(2x - 3)$$

$$9x^2 + 6x + 1 = (3x + 1)^2$$
$$4x^2 - 20x + 25 = (2x - 5)^2$$

If $(x - 2)(x + 3) = 0$, then $x - 2 = 0$ or $x + 3 = 0$.

Solve $2x^2 = 7x + 15$.
$$2x^2 - 7x - 15 = 0$$
$$(2x + 3)(x - 5) = 0$$
$$2x + 3 = 0 \quad \text{or} \quad x - 5 = 0$$
$$2x = -3 \qquad\qquad x = 5$$
$$x = -\frac{3}{2}$$

The solutions $-\frac{3}{2}$ and 5 satisfy the original equation.

In a right triangle, one leg measures 2 ft longer than the other. The hypotenuse measures 4 ft longer than the shorter leg. Find the lengths of the three sides of the triangle.

Let $x = $ the length of the shorter leg. Then
$$x^2 + (x + 2)^2 = (x + 4)^2.$$

Solve this equation to get $x = 6$ or $x = -2$. Discard -2 as a solution. Check that the sides measure 6 ft, $6 + 2 = 8$ ft, and $6 + 4 = 10$ ft.

ANSWERS TO TEST YOUR WORD POWER

1. B; *Example:* $x^2 - 5x - 14 = (x - 7)(x + 2)$
2. D; *Example:* The factored form of $x^2 - 5x - 14$ is $(x - 7)(x + 2)$.
3. C; *Example:* The greatest common factor of $8x^2$, $22xy$, and $16x^3y^2$ is $2x$.
4. A; *Example:* $a^2 + 2a + 1$ is a perfect square trinomial; its factored form is $(a + 1)^2$.
5. B; *Examples:* $y^2 - 3y + 2 = 0$, $x^2 - 9 = 0$, $2m^2 = 6m + 8$
6. C; *Examples:* 4 and 5 are consecutive integers; so are -19 and -18.

Focus on Real-Data Applications

Stopping Distance

The overall *stopping distance* is the sum of the *thinking distance* (how far the car travels once you realize you have to brake) and the *braking distance* (how far the car travels after you apply the brakes).

The data in the table represents three distinct relationships. The *input* is speed in miles per hour for all three relationships. The *output* is thinking distance in feet for the first relationship, braking distance in feet for the second relationship, and overall stopping distance in feet for the third relationship.

Speed	Thinking Distance	Braking Distance	Overall Stopping Distance
20 mph	20 ft	20 ft	40 ft
30 mph	30 ft	45 ft	75 ft
40 mph	40 ft	80 ft	120 ft
50 mph	50 ft	125 ft	175 ft
60 mph	60 ft	180 ft	240 ft
70 mph	70 ft	245 ft	315 ft

Source: *Pass Your Driving Theory Test,* British School of Motoring.

For Group Discussion

1. In the relationship between thinking distance and speed, the *output* (y) is numerically the same as the *input* (x).
   (a) Write the equation that expresses this relationship.
   (b) When the speed is doubled from 20 mph to 40 mph, how does the thinking distance change?
   (c) Does the same pattern hold true if a 30 mph speed is doubled?
   (d) Is the equation linear or quadratic? Explain.

2. The relationship between braking distance and speed is given by the equation $y = \frac{1}{20}x^2$.
   (a) Show that this equation corresponds to the table values for speeds of 20 mph and 40 mph.
   (b) When the speed is doubled from 20 mph to 40 mph, how does the braking distance change?
   (c) Does the same pattern hold true if a 30 mph speed is doubled?
   (d) Is the equation linear or quadratic? Explain.

3. The relationship between overall stopping distance and speed is based on the equations in Problems 1 and 2.
   (a) Use those results to write the equation that expresses the relationship between overall stopping distance and speed.
   (b) Is the equation linear or quadratic? Explain.
   (c) A *rule of thumb* for calculating overall stopping distance is to take speed in *tens* of miles per hour, divide by 2, add 1, and multiply the result by the speed in miles per hour. For example, at 40 mph, the rule says: $4 \div 2 + 1 = 3$; $3 \times 40 = 120$ ft. Show why this rule of thumb works. [*Hint:* Factor the right side of the equation in part (a).]

Chapter 14
REVIEW EXERCISES

[14.1] *Factor out the greatest common factor or factor by grouping.*

1. $7t + 14$
2. $60z^3 + 30z$
3. $35x^3 + 70x^2$
4. $100m^2n^3 - 50m^3n^4 + 150m^2n^2$
5. $2xy - 8y + 3x - 12$
6. $6y^2 + 9y + 4xy + 6x$

[14.2] *Factor completely.*

7. $x^2 + 5x + 6$
8. $y^2 - 13y + 40$
9. $q^2 + 6q - 27$
10. $r^2 - r - 56$
11. $r^2 - 4rs - 96s^2$
12. $p^2 + 2pq - 120q^2$
13. $8p^3 - 24p^2 - 80p$
14. $3x^4 + 30x^3 + 48x^2$
15. $m^2 - 3mn - 18n^2$
16. $y^2 - 8yz + 15z^2$
17. $p^7 - p^6q - 2p^5q^2$
18. $3r^5 - 6r^4s - 45r^3s^2$
19. $x^2 + x + 1$
20. $3x^2 + 6x + 6$

[14.3–14.4]

21. To begin factoring $6r^2 - 5r - 6$, what are the possible first terms of the two binomial factors, if we consider only positive integer coefficients?

22. What is the first step you would use to factor $2z^3 + 9z^2 - 5z$?

Factor completely.

23. $2k^2 - 5k + 2$

24. $3r^2 + 11r - 4$

25. $6r^2 - 5r - 6$

26. $10z^2 - 3z - 1$

27. $5t^2 - 11t + 12$

28. $24x^5 - 20x^4 + 4x^3$

29. $-6x^2 + 3x + 30$

30. $10r^3s + 17r^2s^2 + 6rs^3$

[14.5]

31. Which one of the following is a difference of squares?
    A. $32x^2 - 1$
    B. $4x^2y^2 - 25z^2$
    C. $x^2 + 36$
    D. $25y^3 - 1$

32. Which one of the following is a perfect square trinomial?
    A. $x^2 + x + 1$
    B. $y^2 - 4y + 9$
    C. $4x^2 + 10x + 25$
    D. $x^2 - 20x + 100$

Factor completely.

33. $n^2 - 64$

34. $25b^2 - 121$

35. $49y^2 - 25w^2$

36. $144p^2 - 36q^2$

37. $x^2 + 100$

38. $x^2 - \dfrac{49}{100}$

39. $z^2 + 10z + 25$

40. $r^2 - 12r + 36$

41. $9t^2 - 42t + 49$

42. $16m^2 + 40mn + 25n^2$

43. $54x^3 - 72x^2 + 24x$

44. $x^2 + \dfrac{2}{3}x + \dfrac{1}{9}$

[14.6] *Solve each equation, and check the solutions.*

45. $(4t + 3)(t - 1) = 0$

46. $(x + 7)(x - 4)(x + 3) = 0$

47. $x(2x - 5) = 0$

48. $z^2 + 4z + 3 = 0$

49. $m^2 - 5m + 4 = 0$

50. $x^2 = -15 + 8x$

51. $3z^2 - 11z - 20 = 0$

52. $81t^2 - 64 = 0$

53. $y^2 = 8y$

54. $n(n - 5) = 6$

55. $t^2 - 14t + 49 = 0$

56. $t^2 = 12(t - 3)$

57. $(5z + 2)(z^2 + 3z + 2) = 0$

58. $x^2 = 9$

[14.7] *Solve each problem.*

59. The length of a rug is 6 ft more than the width. The area is 40 ft². Find the length and width of the rug.

60. In Section 5.10, you used this formula for the surface area (*SA*) of a rectangular solid.

$$SA = 2lw + 2lh + 2wh$$

A treasure chest from a sunken galleon has dimensions as shown in the figure. Its surface area is 650 ft². Find its width.

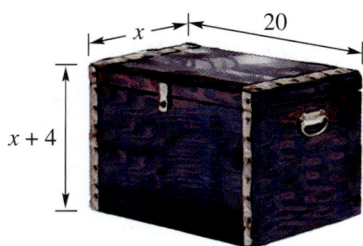

61. The length of a rectangle is three times the width. If the width were increased by 3 m while the length remained the same, the new rectangle would have an area of 30 m². Find the length and width of the original rectangle.

62. The volume of a rectangular box is 120 m³. The width of the box is 4 m, and the height is 1 m less than the length. Find the length and height of the box.

63. The product of two consecutive integers is 29 more than their sum. What are the integers?

64. Two cars left an intersection at the same time. One traveled west, and the other traveled 14 miles less, but to the south. How far apart were they then, if the distance between them was 16 miles more than the distance traveled south?

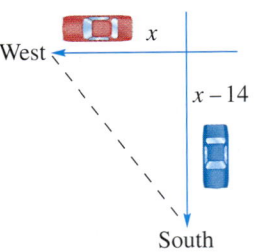

If an object is propelled upward with an initial velocity of 128 feet per second, its height h in feet after t seconds is given by this formula: $h = 128t - 16t^2$. *Find the height of the object after each period of time.*

65. 1 sec **66.** 2 sec **67.** 4 sec

68. For the object described above, when does it return to the ground?

69. Annual revenue in millions of dollars for eBay is shown in the table.

Year	Annual Revenue (in millions of dollars)
1997	5.1
1998	47.4
1999	224.7
2000	431.4
2001	748.8
2002	1214.1

Source: eBay.

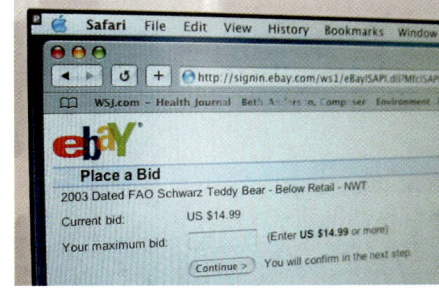

Using the data, we developed this quadratic equation: $y = 47.8x^2 - 0.135x + 7.65$ to model eBay revenues y in year x, where $x = 0$ represents 1997, $x = 1$ represents 1998, and so on.

(a) Use the model to find eBay revenue (to the nearest tenth) in 2001. How does your answer compare to the actual data from the table?

(b) Use the model to predict annual revenue (to the nearest tenth) for eBay in 2003.

(c) Revenue for eBay through the third quarter of 2003 was $1516.7 million. Given this information, do you think your prediction in part (b) is reliable? Explain.

MIXED REVIEW EXERCISES

70. Which of the following is *not* factored completely?

A. $3(7t)$ B. $3x(7t + 4)$ C. $(3 + x)(7t + 4)$ D. $3(7t + 4) + x(7t + 4)$

71. Although $(2x + 8)(3x - 4) = 6x^2 + 16x - 32$ is a true statement, the polynomial is not factored completely. Explain why and give the complete factored form.

Factor completely.

72. $z^2 - 11zx + 10x^2$

73. $3k^2 + 11k + 10$

74. $15m^2 + 20mp - 12m - 16p$

75. $y^4 - 625$

76. $6m^3 - 21m^2 - 45m$

77. $24ab^3c^2 - 56a^2bc^3 + 72a^2b^2c$

78. $25a^2 + 15ab + 9b^2$

79. $12x^2yz^3 + 12xy^2z - 30x^3y^2z^4$

80. $2a^5 - 8a^4 - 24a^3$

81. $12r^2 + 8rq - 15q^2$

82. $100a^2 - 9$

83. $49t^2 + 56t + 16$

Solve.

84. $t(t - 7) = 0$

85. $x^2 + 3x = 10$

86. $25x^2 + 20x + 4 = 0$

Solve each problem. Use the geometry formulas on the inside back cover as needed.

87. A lot is shaped like a right triangle. The hypotenuse is 3 m longer than the longer leg. The longer leg is 6 m longer than twice the length of the shorter leg. Find the lengths of the sides of the lot.

88. A pyramid has a rectangular base with a length that is 2 m more than the width. The height of the pyramid is 6 m, and its volume is 48 m³. Find the length and width of the base.

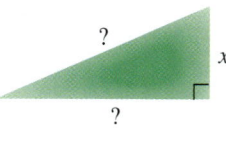

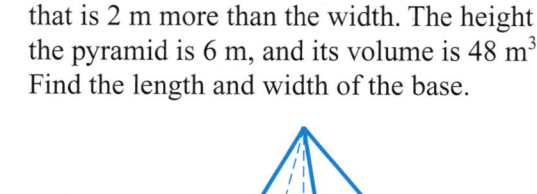

89. The product of the smaller two of three consecutive integers is equal to 23 plus the largest. Find the integers.

90. If an object is dropped, the distance d in feet it falls in t seconds (disregarding air resistance) is given by the quadratic equation
$$d = 16t^2.$$
Find the distance an object would fall in the following times.

(a) 4 sec **(b)** 8 sec

91. The floor plan for a house is a rectangle with length 7 m more than its width. The area is 170 m². Find the width and length of the house.

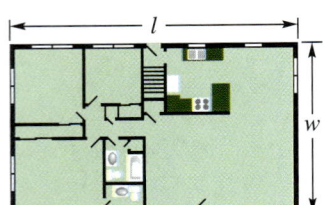

92. The triangular sail of a schooner has an area of 30 m². The height of the sail is 4 m more than the base. Find the base of the sail.

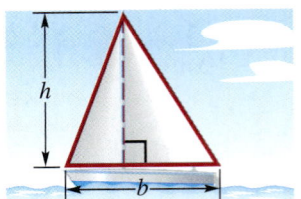

93. The numbers of alternative-fueled vehicles, in thousands, in use for the years 1998–2001 are given in the table.

Year	Number (in thousands)
1998	384
1999	407
2000	432
2001	456

Source: Energy Information Administration, Alternatives to Traditional Fuels, 2001.

Using the data, we developed this quadratic equation: $y = 0.25x^2 - 25.65x + 496.6$ to model the number of vehicles y in year x. Here we used $x = 98$ for 1998, $x = 99$ for 1999, and so on.

(a) What prediction for 2002 is given by the equation?

(b) Why might the prediction for 2002 be unreliable?

Chapter 14
TEST

1. Which one of the following is the correct, completely factored form of $2x^2 - 2x - 24$?

    A. $(2x + 6)(x - 4)$ **B.** $(x + 3)(2x - 8)$
    C. $2(x + 4)(x - 3)$ **D.** $2(x + 3)(x - 4)$

Factor each polynomial completely.

2. $12x^2 - 30x$

3. $2m^3n^2 + 3m^3n - 5m^2n^2$

4. $2ax - 2bx + ay - by$

5. $x^2 - 9x + 14$

6. $2x^2 + x - 3$

7. $6x^2 - 19x - 7$

8. $3x^2 - 12x - 15$

9. $10z^2 - 17z + 3$

10. $t^2 + 2t + 3$

11. $x^2 + \dfrac{1}{36}$

12. $y^2 - 49$

13. $9y^2 - 64$

14. $x^2 + 16x + 64$

15. $4x^2 - 28xy + 49y^2$

16. $-2x^2 - 4x - 2$

17. $6t^4 + 3t^3 - 108t^2$

18. $4r^2 + 10rt + 25t^2$

1. _____
2. _____
3. _____
4. _____
5. _____
6. _____
7. _____
8. _____
9. _____
10. _____
11. _____
12. _____
13. _____
14. _____
15. _____
16. _____
17. _____
18. _____

19. $4t^3 + 32t^2 + 64t$

20. $x^4 - 81$

21. Why is $(p + 3)(p + 3)$ *not* the correct factored form of $p^2 + 9$?

Solve each equation.

22. $(x + 3)(x - 9) = 0$

23. $2r^2 - 13r + 6 = 0$

24. $25x^2 - 4 = 0$

25. $x(x - 20) = -100$

26. $t^2 = 3t$

Solve each problem.

27. The length of a rectangular flower bed is 3 ft less than twice its width. The area of the bed is 54 ft². Find the dimensions of the flower bed.

28. Find two consecutive integers such that the square of the sum of the two integers is 11 more than the smaller integer.

29. A carpenter needs to cut a brace to support a wall stud, as shown in the figure. The brace should be 7 ft less than three times the length of the stud. If the brace will be anchored on the floor 15 ft away from the stud, how long should the brace be?

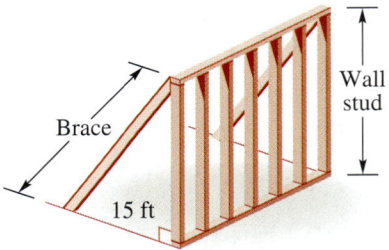

30. The number of U.S. cable TV networks y from 1984 through 2003 can be approximated by the quadratic equation

$$y = 1.06x^2 - 4.77x + 47.9$$

where $x = 0$ represents 1984, $x = 1$ represents 1985, and so on. (*Source:* National Cable Television Association.) Use the model to estimate the number of cable TV channels in 2000. Round your answer to the nearest whole number.

Rational Expressions and Applications

15

15.1 **The Fundamental Property of Rational Expressions**

15.2 **Multiplying and Dividing Rational Expressions**

15.3 **Least Common Denominators**

15.4 **Adding and Subtracting Rational Expressions**

15.5 **Complex Fractions**

15.6 **Solving Equations with Rational Expressions**

Summary Exercises on Rational Expressions and Equations

15.7 **Applications of Rational Expressions**

15.8 **Variation**

In the 1994 movie *Little Big League,* the young Billy Heywood inherits the Minnesota Twins baseball team and becomes manager. He leads the team to the Division Championship and then to the playoffs. But before the final playoff game, the biggest game of the year, he can't keep his mind on his job because a homework problem is giving him trouble.

If Joe can paint a house in 3 hours, and Sam can paint the same house in 5 hours, how long does it take for them to do it together?

With the help of one of his players, he is able to solve the problem, and the team goes on to victory. Its solution requires the use of algebraic fractions, or *rational expressions*. In Section 15.7 we show how to solve such problems, and in the Chapter Review Exercises, we ask you to solve Billy Heywood's problem.

15.1 The Fundamental Property of Rational Expressions

OBJECTIVES

1. Find the values of the variable for which a rational expression is undefined.
2. Find the numerical value of a rational expression.
3. Write rational expressions in lowest terms.
4. Recognize equivalent forms of rational expressions.

The quotient of two integers (with denominator not 0), such as $\frac{2}{3}$ or $-\frac{3}{4}$, is called a rational number. In the same way, the quotient of two polynomials with denominator not equal to 0 is called a *rational expression*.

Rational Expression

A **rational expression** is an expression of the form

$$\frac{P}{Q}$$

where P and Q are polynomials, with $Q \neq 0$.

Examples of rational expressions include

$$\frac{-6x}{x^3 + 8}, \quad \frac{9x}{y + 3}, \quad \text{and} \quad \frac{2m^3}{8}. \qquad \text{Rational expressions}$$

Our work with rational expressions will require much of what we learned in **Chapters 13 and 14** on polynomials and factoring, as well as the rules for fractions from **Chapter 4**.

OBJECTIVE 1 Find the values of the variable for which a rational expression is undefined. A fraction with denominator 0 is *not* a rational expression because division by 0 is undefined. Be careful when substituting a number in the denominator of a rational expression. For example, in

$$\frac{8x^2}{x - 3} \quad \leftarrow \text{Denominator cannot equal 0.}$$

the variable x can take on any value except 3. When $x = 3$, the denominator becomes $3 - 3 = \mathbf{0}$, making the expression undefined.

To determine the values for which a rational expression is undefined, use the following procedure.

Determining When a Rational Expression Is Undefined

Step 1 Set the denominator of the rational expression equal to 0.

Step 2 Solve this equation.

Step 3 The solutions of the equation are the values that make the rational expression undefined.

EXAMPLE 1 Finding Values That Make Rational Expressions Undefined

Find any values of the variable for which each rational expression is undefined.

(a) $\dfrac{p + 5}{3p + 2}$

Remember that the *numerator* may be any number; we must find any value of p that makes the *denominator* equal to 0 since division by 0 is undefined.

Continued on Next Page

Step 1 Set the denominator equal to 0.
$$3p + 2 = 0$$
Step 2 Solve this equation.
$$3p = -2$$
$$p = -\frac{2}{3}$$
Step 3 Since $p = -\frac{2}{3}$ will make the denominator 0, the given expression is undefined for $-\frac{2}{3}$.

(b) $\dfrac{9m^2}{m^2 - 5m + 6}$

Set the denominator equal to 0, and then find the solutions of the equation.

$$m^2 - 5m + 6 = 0$$
$$(m - 2)(m - 3) = 0 \quad \text{Factor.}$$
$$m - 2 = 0 \quad \text{or} \quad m - 3 = 0 \quad \text{Zero-factor property}$$
$$m = 2 \quad \text{or} \quad m = 3$$

The original expression is undefined for $m = 2$ and for $m = 3$.

(c) $\dfrac{2r}{r^2 + 1}$

This denominator cannot equal 0 for any value of r because r^2 is always greater than or equal to 0, and adding 1 makes the sum greater than 0. Thus, there are no values for which this rational expression is undefined.

> **Work Problem 1 at the Side.**

① Find all values for which each rational expression is undefined.

(a) $\dfrac{x + 2}{x - 5}$

(b) $\dfrac{3r}{r^2 + 6r + 8}$

(c) $\dfrac{-5m}{m^2 + 4}$

OBJECTIVE 2 Find the numerical value of a rational expression.
We use substitution to evaluate a rational expression for a given value of the variable.

EXAMPLE 2 Evaluating Rational Expressions

Find the numerical value of $\dfrac{3x + 6}{2x - 4}$ for each value of x.

(a) $x = 1$

$$\frac{3x + 6}{2x - 4} = \frac{3(1) + 6}{2(1) - 4} \quad \text{Let } x = 1.$$
$$= \frac{9}{-2} = -\frac{9}{2}$$

(b) $x = 2$

$$\frac{3x + 6}{2x - 4} = \frac{3(2) + 6}{2(2) - 4} = \frac{12}{0} \quad \text{Let } x = 2.$$

Substituting 2 for x makes the denominator 0, so the expression is undefined when $x = 2$.

> **Work Problem 2 at the Side.**

② Find the value of each rational expression when $x = 3$.

(a) $\dfrac{x}{2x + 1}$

(b) $\dfrac{2x + 6}{x - 3}$

ANSWERS
1. (a) 5 (b) $-4, -2$ (c) never undefined
2. (a) $\dfrac{3}{7}$ (b) undefined

③ Write each rational expression in lowest terms.

(a) $\dfrac{5x^4}{15x^2}$

OBJECTIVE 3 Write rational expressions in lowest terms. A fraction such as $\frac{2}{3}$ is said to be in *lowest terms*. How can "lowest terms" be defined? We use the idea of greatest common factor for this definition, which applies to all rational expressions.

> **Lowest Terms**
> A rational expression $\frac{P}{Q}$ (where $Q \neq 0$) is in **lowest terms** if the greatest common factor of its numerator and denominator is 1.

The properties of rational numbers also apply to rational expressions. We use the **fundamental property of rational expressions** to write a rational expression in lowest terms.

> **Fundamental Property of Rational Expressions**
> If $\frac{P}{Q}$ (where $Q \neq 0$) is a rational expression and if K represents any polynomial, where $K \neq 0$, then we have the following.
> $$\frac{PK}{QK} = \frac{P}{Q}$$

This property is based on the identity property of multiplication which says that multiplication by 1 leaves any number unchanged.

$$\frac{PK}{QK} = \frac{P}{Q} \cdot \frac{K}{K} = \frac{P}{Q} \cdot 1 = \frac{P}{Q}$$

(b) $\dfrac{6p^3}{2p^2}$

The next example shows how to write both a rational number and a rational expression in lowest terms. Notice the similarity in the procedures. In both cases, we factor and then divide out the greatest common factor.

EXAMPLE 3 Writing in Lowest Terms

Write each expression in lowest terms.

(a) $\dfrac{30}{72}$

Begin by factoring.

$$\frac{30}{72} = \frac{2 \cdot 3 \cdot 5}{2 \cdot 2 \cdot 2 \cdot 3 \cdot 3}$$

(b) $\dfrac{14k^2}{2k^3}$

Write k^2 as $k \cdot k$ and k^3 as $k \cdot k \cdot k$.

$$\frac{14k^2}{2k^3} = \frac{2 \cdot 7 \cdot k \cdot k}{2 \cdot k \cdot k \cdot k}$$

Group any factors common to the numerator and denominator.

$$\frac{30}{72} = \frac{5 \cdot (2 \cdot 3)}{2 \cdot 2 \cdot 3 \cdot (2 \cdot 3)} \qquad \frac{14k^2}{2k^3} = \frac{7(2 \cdot k \cdot k)}{k(2 \cdot k \cdot k)}$$

Use the fundamental property to divide out the common factors.

$$\frac{30}{72} = \frac{5}{2 \cdot 2 \cdot 3} = \frac{5}{12} \qquad \frac{14k^2}{2k^3} = \frac{7}{k}$$

◀◀◀ **Work Problem ③ at the Side.**

ANSWERS

3. (a) $\dfrac{x^2}{3}$ (b) $3p$

Writing a Rational Expression in Lowest Terms

Step 1 **Factor** the numerator and denominator completely.

Step 2 **Use the fundamental property** to divide out any common factors.

EXAMPLE 4 Writing in Lowest Terms

Write each rational expression in lowest terms.

(a) $\dfrac{3x - 12}{5x - 20}$

Begin by factoring both numerator and denominator. Then use the fundamental property of rational expressions to divide out any common factors.

$$\dfrac{3x - 12}{5x - 20} = \dfrac{3(x - 4)}{5(x - 4)} = \dfrac{3}{5}$$

Step 1   Step 2

The given expression is equal to $\frac{3}{5}$ for all values of x, where $x \neq 4$ (since the denominator of the original rational expression is 0 when x is 4).

(b) $\dfrac{m^2 + 2m - 8}{2m^2 - m - 6}$

$$\dfrac{m^2 + 2m - 8}{2m^2 - m - 6} = \dfrac{(m + 4)(m - 2)}{(2m + 3)(m - 2)} \quad \text{Factor. (Step 1)}$$

$$= \dfrac{m + 4}{2m + 3} \quad \text{Fundamental property (Step 2)}$$

Here $m \neq -\frac{3}{2}$ and $m \neq 2$, since the denominator of the original expression is 0 for these values of m.

From now on, we will write statements of equality of rational expressions with the understanding that they apply only to those real numbers that make neither denominator equal to 0.

CAUTION

Rational expressions cannot be written in lowest terms until after the numerator and denominator have been factored. Only common factors can be divided out, not common terms. For example,

$$\dfrac{6x + 9}{4x + 6} = \dfrac{3(2x + 3)}{2(2x + 3)} = \dfrac{3}{2} \quad \Big| \quad \dfrac{6 + x}{4x} \leftarrow \text{Numerator cannot be factored.}$$

↑ Divide out the common factor.   Already in lowest terms

Work Problem 4 at the Side.

---

4 Write each rational expression in lowest terms.

(a) $\dfrac{4y + 2}{6y + 3}$

(b) $\dfrac{8p + 8q}{5p + 5q}$

(c) $\dfrac{x^2 + 4x + 4}{4x + 8}$

(d) $\dfrac{a^2 - b^2}{a^2 + 2ab + b^2}$

ANSWERS

4. (a) $\dfrac{2}{3}$ (b) $\dfrac{8}{5}$ (c) $\dfrac{x + 2}{4}$ (d) $\dfrac{a - b}{a + b}$

EXAMPLE 5 Writing in Lowest Terms (Factors Are Opposites)

Write $\dfrac{x-y}{y-x}$ in lowest terms.

At first glance, there does not seem to be any way in which $x - y$ and $y - x$ can be factored to get a common factor. However, $y - x$ can be factored as

$$y - x = -1(-y + x) = -1(x - y).$$

Now, use the fundamental property to simplify.

$$\frac{x-y}{y-x} = \frac{1(x-y)}{-1(x-y)} = \frac{1}{-1} = -1$$

In Example 5, notice that $y - x$ is the opposite of $x - y$. A general rule for this situation follows.

> If the numerator and the denominator of a rational expression are opposites, such as in $\dfrac{x-y}{y-x}$, then the rational expression is equal to -1.

CAUTION
Although x and y appear in both the numerator and denominator in Example 5, we cannot use the fundamental property right away because they are *terms*, not *factors*. **Terms are added, while factors are multiplied.**

EXAMPLE 6 Writing in Lowest Terms (Factors Are Opposites)

Write each rational expression in lowest terms.

(a) $\dfrac{2-m}{m-2}$

Since $2 - m$ and $m - 2$ (or $-2 + m$) are opposites,

$$\frac{2-m}{m-2} = -1.$$

(b) $\dfrac{4x^2 - 9}{6 - 4x}$

Factor the numerator and denominator.

$$\frac{4x^2 - 9}{6 - 4x} = \frac{(2x+3)(2x-3)}{2(3-2x)}$$

$$= \frac{(2x+3)(2x-3)}{2(-1)(2x-3)} \quad \text{Write } 3 - 2x \text{ as } -1(2x-3).$$

$$= \frac{2x+3}{2(-1)} \quad \text{Fundamental property}$$

$$= \frac{2x+3}{-2} \quad \text{or} \quad -\frac{2x+3}{2} \quad \tfrac{a}{-b} = -\tfrac{a}{b}$$

Continued on Next Page

(c) $\dfrac{3+r}{3-r}$

The quantity $3-r$ *is not* the opposite of $3+r$. This rational expression is already in lowest terms.

Work Problem 5 at the Side.

OBJECTIVE 4 Recognize equivalent forms of rational expressions.
When working with rational expressions, it is important to be able to recognize equivalent forms of an expression. For example, the common fraction $-\dfrac{5}{6}$ can also be written as $\dfrac{-5}{6}$ and as $\dfrac{5}{-6}$. Consider also the final rational expression from Example 6(b),

$$-\dfrac{2x+3}{2}.$$

The $-$ sign representing the -1 factor is in front of the expression, on the same line as the fraction bar. The -1 factor may be placed in front of the expression, in the numerator, or in the denominator. Some other acceptable forms of this rational expression are shown below.

$\dfrac{-(2x+3)}{2}$ Use parentheses in the numerator.

$\dfrac{-2x-3}{2}$ Distribute.

$\dfrac{2x+3}{-2}$ Multiply $-\dfrac{2x+3}{2}$ by $\dfrac{-1}{-1}$.

CAUTION
$\dfrac{-2x+3}{2}$ is *not* an equivalent form of $\dfrac{-(2x+3)}{2}$. The sign preceding 3 in the numerator of $\dfrac{-2x+3}{2}$ should be $-$ rather than $+$. *Be careful to apply the distributive property correctly.*

EXAMPLE 7 Writing Equivalent Forms of a Rational Expression

Write four equivalent forms of this rational expression.

$$-\dfrac{3x+2}{x-6}$$

If we apply the negative sign to the numerator, we have this equivalent form.

$$\dfrac{-(3x+2)}{x-6}$$

Continued on Next Page

---

5 Write each rational expression in lowest terms.

(a) $\dfrac{5-y}{y-5}$

(b) $\dfrac{m-n}{n-m}$

(c) $\dfrac{25x^2-16}{12-15x}$

(d) $\dfrac{9-k}{9+k}$

ANSWERS

5. (a) -1 (b) -1 (c) $\dfrac{5x+4}{-3}$ or $-\dfrac{5x+4}{3}$
   (d) already in lowest terms

6 Decide whether each rational expression is equivalent to
$$-\frac{2x-6}{x+3}.$$

(a) $\dfrac{-(2x-6)}{x+3}$

(b) $\dfrac{-2x+6}{x+3}$

(c) $\dfrac{-2x-6}{x+3}$

(d) $\dfrac{2x-6}{-(x+3)}$

(e) $\dfrac{2x-6}{-x-3}$

(f) $\dfrac{2x-6}{x-3}$

By distributing the negative sign in $\dfrac{-(3x+2)}{x-6}$, we have another equivalent form.

$$\frac{-3x-2}{x-6}$$

If we apply the negative sign to the denominator of the given fraction $-\dfrac{3x+2}{x-6}$, we get

$$\frac{3x+2}{-(x-6)}$$

or, distributing once again,

$$\frac{3x+2}{-x+6}.$$

CAUTION
Recall that $-\dfrac{5}{6} \neq \dfrac{-5}{-6}$. Thus, in Example 7, it would be incorrect to distribute the negative sign in $-\dfrac{3x+2}{x-6}$ to *both* the numerator *and* the denominator. This would lead to the *opposite* of the original expression.

◀◀◀ **Work Problem 6 at the Side.**

ANSWERS
6. (a) equivalent (b) equivalent
(c) not equivalent (d) equivalent
(e) equivalent (f) not equivalent

15.1 Exercises

1. Fill in each blank with the correct response.
   (a) The rational expression $\dfrac{x+5}{x-3}$ is undefined when $x =$ _____, and is equal to 0 when $x =$ _____.
   (b) The rational expression $\dfrac{p-q}{q-p}$ is undefined when $p =$ _____, and in all other cases when written in lowest terms is equal to _____.

2. Make the correct choice for each blank.
   (a) $\dfrac{4-r^2}{4+r^2}$ _____ (is/is not) equal to -1.
   (b) $\dfrac{5+2x}{3-x}$ and $\dfrac{-5-2x}{x-3}$ _____ (are/are not) equivalent rational expressions.

3. Define *rational expression* in your own words, and give an example.

4. Give an example of a rational expression that is not in lowest terms, and then show the steps required to write it in lowest terms.

Find any value(s) for which each rational expression is undefined. See Example 1.

5. $\dfrac{2}{5y}$
6. $\dfrac{7}{3z}$
7. $\dfrac{4x^2}{3x-5}$
8. $\dfrac{2x^3}{3x-4}$
9. $\dfrac{m+2}{m^2+m-6}$
10. $\dfrac{r-5}{r^2-5r+4}$
11. $\dfrac{3x}{x^2+2}$
12. $\dfrac{4q}{q^2+9}$

Find the numerical value of each rational expression when (a) $x = 2$ and (b) $x = -3$. See Example 2.

13. $\dfrac{5x-2}{4x}$
14. $\dfrac{3x+1}{5x}$
15. $\dfrac{2x^2-4x}{3x}$
16. $\dfrac{4x^2-1}{5x}$

17. $\dfrac{(-3x)^2}{4x+12}$
18. $\dfrac{(-2x)^3}{3x+9}$
19. $\dfrac{5x+2}{2x^2+11x+12}$
20. $\dfrac{7-3x}{3x^2-7x+2}$

21. If 2 is substituted for x in the rational expression $\dfrac{x-2}{x^2-4}$, the result is $\dfrac{0}{0}$. We often hear the statement "Any number divided by itself is 1." Does this mean that this expression is equal to 1 for $x = 2$? If not, explain.

22. For $x \neq 2$, the rational expression $\dfrac{2(x-2)}{x-2}$ is equal to 2. Can $\dfrac{2x-2}{x-2}$ also be simplified to 2? Explain.

Write each rational expression in lowest terms. See Examples 3 and 4.

23. $\dfrac{18r^3}{6r}$
24. $\dfrac{27p^2}{3p}$
25. $\dfrac{4(y-2)}{10(y-2)}$

26. $\dfrac{15(m-1)}{9(m-1)}$
27. $\dfrac{(x+1)(x-1)}{(x+1)^2}$
28. $\dfrac{(t+5)(t-3)}{(t-1)(t+5)}$

29. $\dfrac{7m+14}{5m+10}$
30. $\dfrac{8z-24}{4z-12}$
31. $\dfrac{m^2-n^2}{m+n}$

32. $\dfrac{a^2-b^2}{a-b}$
33. $\dfrac{12m^2-3}{8m-4}$
34. $\dfrac{20p^2-45}{6p-9}$

35. $\dfrac{3m^2-3m}{5m-5}$
36. $\dfrac{6t^2-6t}{2t-2}$
37. $\dfrac{9r^2-4s^2}{9r+6s}$

38. $\dfrac{16x^2-9y^2}{12x-9y}$
39. $\dfrac{2x^2-3x-5}{2x^2-7x+5}$
40. $\dfrac{3x^2+8x+4}{3x^2-4x-4}$

41. $\dfrac{zw+4z-3w-12}{zw+4z+5w+20}$
42. $\dfrac{km+4k+4m+16}{km+4k+5m+20}$

Write each rational expression in lowest terms. See Examples 5 and 6.

43. $\dfrac{6-t}{t-6}$
44. $\dfrac{2-k}{k-2}$
45. $\dfrac{m^2-1}{1-m}$

46. $\dfrac{a^2-b^2}{b-a}$
47. $\dfrac{q^2-4q}{4q-q^2}$
48. $\dfrac{z^2-5z}{5z-z^2}$

Write four equivalent expressions for each expression. See Example 7.

49. $-\dfrac{x+4}{x-3}$
50. $-\dfrac{x+6}{x-1}$

51. $-\dfrac{2x-3}{x+3}$
52. $-\dfrac{5x-6}{x+4}$

53. $\dfrac{-3x+1}{5x-6}$
54. $\dfrac{-2x-9}{3x+1}$

55. The area of the rectangle is represented by $x^4 + 10x^2 + 21$. What is the width? (*Hint*: Use $w = \frac{A}{l}$.)

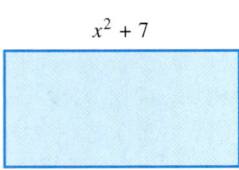

$x^2 + 7$

56. The volume of the box is represented by $(x^2 + 8x + 15)(x + 4)$. Find the polynomial that represents the area of the bottom of the box.

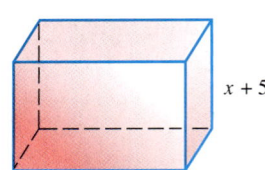

$x + 5$

15.2 Multiplying and Dividing Rational Expressions

OBJECTIVES
1. Multiply rational expressions.
2. Find reciprocals.
3. Divide rational expressions.

OBJECTIVE 1 **Multiply rational expressions.** The product of two fractions is found by multiplying the numerators and multiplying the denominators. Rational expressions are multiplied in the same way.

Multiplying Rational Expressions
The product of the rational expressions $\frac{P}{Q}$ and $\frac{R}{S}$ is shown below.

$$\frac{P}{Q} \cdot \frac{R}{S} = \frac{PR}{QS}$$

In words: To multiply rational expressions, multiply the numerators and multiply the denominators.

In the following example, the parallel discussion with rational numbers and rational expressions lets you compare the steps.

EXAMPLE 1 Multiplying Rational Expressions
Multiply. Write each answer in lowest terms.

(a) $\dfrac{3}{10} \cdot \dfrac{5}{9}$

(b) $\dfrac{6}{x} \cdot \dfrac{x^2}{12}$

Indicate the product of the numerators and the product of the denominators.

$$\frac{3}{10} \cdot \frac{5}{9} = \frac{3 \cdot 5}{10 \cdot 9} \qquad \frac{6}{x} \cdot \frac{x^2}{12} = \frac{6 \cdot x^2}{x \cdot 12}$$

Leave the products in factored form because common factors are needed to write the product in lowest terms. Factor the numerator and denominator to further identify any common factors. Then use the fundamental property to write each product in lowest terms.

$$\frac{3}{10} \cdot \frac{5}{9} = \frac{3 \cdot 5}{2 \cdot 5 \cdot 3 \cdot 3} = \frac{1}{6} \qquad \frac{6}{x} \cdot \frac{x^2}{12} = \frac{6 \cdot x \cdot x}{2 \cdot 6 \cdot x} = \frac{x}{2}$$

Work Problem 1 at the Side.

EXAMPLE 2 Multiplying Rational Expressions
Multiply $\dfrac{x+y}{2x} \cdot \dfrac{x^2}{(x+y)^2}$. Write the answer in lowest terms.

Use the definition of multiplication. Indicate the products in the first step so that common factors are easily identified.

$$\frac{x+y}{2x} \cdot \frac{x^2}{(x+y)^2} = \frac{(x+y)x^2}{2x(x+y)^2} \qquad \text{Multiply numerators.}$$
$$\text{Multiply denominators.}$$

$$= \frac{(x+y)x \cdot x}{2x(x+y)(x+y)} \qquad \text{Factor; identify common factors.}$$

$$= \frac{x}{2(x+y)} \qquad \frac{(x+y)x}{x(x+y)} = 1; \text{ lowest terms}$$

Work Problem 2 at the Side.

---

1 Multiply. Write each answer in lowest terms.

(a) $\dfrac{2}{7} \cdot \dfrac{5}{10}$

(b) $\dfrac{3m^2}{2} \cdot \dfrac{10}{m}$

(c) $\dfrac{8p^2q}{3} \cdot \dfrac{9}{q^2p}$

2 Multiply. Write each answer in lowest terms.

(a) $\dfrac{a+b}{5} \cdot \dfrac{30}{2(a+b)}$

(b) $\dfrac{3(p-q)}{q^2} \cdot \dfrac{q}{2(p-q)^2}$

ANSWERS

1. (a) $\dfrac{1}{7}$ (b) $15m$ (c) $\dfrac{24p}{q}$

2. (a) 3 (b) $\dfrac{3}{2q(p-q)}$

Chapter 15 Rational Expressions and Applications

3 Multiply. Write each answer in lowest terms.

(a) $\dfrac{x^2 + 7x + 10}{3x + 6} \cdot \dfrac{6x - 6}{x^2 + 2x - 15}$

(b) $\dfrac{m^2 + 4m - 5}{m + 5} \cdot \dfrac{m^2 + 8m + 15}{m - 1}$

4 Find each reciprocal.

(a) $\dfrac{6b^5}{3r^2 b}$

(b) $\dfrac{t^2 - 4t}{t^2 + 2t - 3}$

ANSWERS

3. (a) $\dfrac{2(x - 1)}{x - 3}$ (b) $(m + 5)(m + 3)$

4. (a) $\dfrac{3r^2 b}{6b^5}$ (b) $\dfrac{t^2 + 2t - 3}{t^2 - 4t}$

EXAMPLE 3 Multiplying Rational Expressions

Multiply. Write the answer in lowest terms.

$$\dfrac{x^2 + 3x}{x^2 - 3x - 4} \cdot \dfrac{x^2 - 5x + 4}{x^2 + 2x - 3} = \dfrac{(x^2 + 3x)(x^2 - 5x + 4)}{(x^2 - 3x - 4)(x^2 + 2x - 3)}$$ Definition of multiplication

$$= \dfrac{x(x + 3)(x - 4)(x - 1)}{(x - 4)(x + 1)(x + 3)(x - 1)}$$ Factor.

$$= \dfrac{x}{x + 1}$$ Lowest terms

The quotients

$$\dfrac{x + 3}{x + 3}, \quad \dfrac{x - 4}{x - 4}, \quad \text{and} \quad \dfrac{x - 1}{x - 1}$$

are all equal to 1, justifying the final product $\dfrac{x}{x + 1}$.

Work Problem 3 at the Side.

OBJECTIVE 2 Find reciprocals. If the product of two rational expressions is 1, the rational expressions are called **reciprocals** (or **multiplicative inverses**) of each other. The reciprocal of a rational expression is found by interchanging the numerator and the denominator. For example,

$$\dfrac{2x - 1}{x - 5} \quad \text{has reciprocal} \quad \dfrac{x - 5}{2x - 1}.$$

EXAMPLE 4 Finding Reciprocals of Rational Expressions

Find the reciprocal of each rational expression.

(a) $\dfrac{4p^3}{9q}$

Interchange the numerator and denominator. The reciprocal is $\dfrac{9q}{4p^3}$.

(b) $\dfrac{k^2 - 9}{k^2 - k - 20}$ has reciprocal $\dfrac{k^2 - k - 20}{k^2 - 9}$.

Work Problem 4 at the Side.

OBJECTIVE 3 Divide rational expressions. To develop a method for dividing rational numbers and rational expressions, consider the following problem. Suppose that you have $\tfrac{7}{8}$ gal of milk and you wish to find how many quarts you have. Since 1 qt is $\tfrac{1}{4}$ gal, you must ask yourself, "How many $\tfrac{1}{4}$s are there in $\tfrac{7}{8}$?" This would be interpreted as

$$\dfrac{7}{8} \div \dfrac{1}{4} \quad \text{or} \quad \dfrac{\tfrac{7}{8}}{\tfrac{1}{4}}$$

since the fraction bar means division.

The fundamental property of rational expressions discussed earlier can be applied to rational number values of P, Q, and K. When $P = \frac{7}{8}$, $Q = \frac{1}{4}$, and $K = 4$ (where K is the reciprocal of $Q = \frac{1}{4}$), we have the following.

$$\frac{P}{Q} = \frac{P \cdot K}{Q \cdot K} = \frac{\frac{7}{8} \cdot 4}{\frac{1}{4} \cdot 4} = \frac{\frac{7}{8} \cdot 4}{1} = \frac{7}{8} \cdot \frac{4}{1}$$

So, to divide $\frac{7}{8}$ by $\frac{1}{4}$, we must multiply $\frac{7}{8}$ by the reciprocal of $\frac{1}{4}$, namely $\frac{4}{1}$ or 4. Since $\frac{7}{8}(4) = \frac{7}{2}$, there are $\frac{7}{2}$ or $3\frac{1}{2}$ qt in $\frac{7}{8}$ gal.

The preceding discussion illustrates the rule for dividing common fractions. To divide $\frac{a}{b}$ by $\frac{c}{d}$, multiply $\frac{a}{b}$ by the reciprocal of $\frac{c}{d}$. Division of rational expressions is defined in the same way.

> **Dividing Rational Expressions**
>
> If $\frac{P}{Q}$ and $\frac{R}{S}$ are any two rational expressions, with $\frac{R}{S} \neq 0$, then
>
> $$\frac{P}{Q} \div \frac{R}{S} = \frac{P}{Q} \cdot \frac{S}{R} = \frac{PS}{QR}$$
>
> ↑ Reciprocals
>
> In words: To divide one rational expression by another rational expression, multiply the first rational expression (the dividend) by the reciprocal of the second rational expression (the divisor).

The next example compares the division of two rational numbers and the division of two rational expressions. Notice the similarity in the steps.

EXAMPLE 5 Dividing Rational Expressions

Divide. Write each answer in lowest terms.

(a) $\dfrac{5}{8} \div \dfrac{7}{16}$ (b) $\dfrac{y}{y + 3} \div \dfrac{4y}{y + 5}$

Multiply the first expression by the reciprocal of the second.

$$\frac{5}{8} \div \frac{7}{16} = \frac{5}{8} \cdot \frac{16}{7} \quad \text{Reciprocal of } \tfrac{7}{16} \text{ is } \tfrac{16}{7}$$

$$= \frac{5 \cdot 16}{8 \cdot 7}$$

$$= \frac{5 \cdot 8 \cdot 2}{8 \cdot 7}$$

$$= \frac{10}{7}$$

$$\frac{y}{y + 3} \div \frac{4y}{y + 5}$$

$$= \frac{y}{y + 3} \cdot \frac{y + 5}{4y} \quad \text{Reciprocal of } \tfrac{4y}{y+5} \text{ is } \tfrac{y+5}{4y}$$

$$= \frac{y(y + 5)}{(y + 3)(4y)}$$

$$= \frac{y + 5}{4(y + 3)}$$

Work Problem 5 at the Side.

5 Divide. Write each answer in lowest terms.

(a) $\dfrac{3}{4} \div \dfrac{5}{16}$

(b) $\dfrac{r}{r - 1} \div \dfrac{3r}{r + 4}$

(c) $\dfrac{6x - 4}{3} \div \dfrac{15x - 10}{9}$

ANSWERS

5. (a) $\dfrac{12}{5}$ (b) $\dfrac{r + 4}{3(r - 1)}$ (c) $\dfrac{6}{5}$

Chapter 15 Rational Expressions and Applications

6 Divide. Write each answer in lowest terms.

(a) $\dfrac{5a^2b}{2} \div \dfrac{10ab^2}{8}$

(b) $\dfrac{(3t)^2}{w} \div \dfrac{3t^2}{5w^4}$

EXAMPLE 6 Dividing Rational Expressions

Divide. Write the answer in lowest terms.

$\dfrac{(3m)^2}{(2p)^3} \div \dfrac{6m^3}{16p^2} = \dfrac{(3m)^2}{(2p)^3} \cdot \dfrac{16p^2}{6m^3}$ Multiply by the reciprocal.

$= \dfrac{9m^2}{8p^3} \cdot \dfrac{16p^2}{6m^3}$ Power rule for exponents

$= \dfrac{9 \cdot 16m^2p^2}{8 \cdot 6p^3m^3}$ Multiply numerators. Multiply denominators.

$= \dfrac{3}{mp}$ Lowest terms

Work Problem 6 at the Side.

7 Divide. Write each answer in lowest terms.

(a) $\dfrac{y^2 + 4y + 3}{y + 3} \div \dfrac{y^2 - 4y - 5}{y - 3}$

(b) $\dfrac{4x(x + 3)}{2x + 1} \div \dfrac{-x^2(x + 3)}{4x^2 - 1}$

EXAMPLE 7 Dividing Rational Expressions

Divide. Write the answer in lowest terms.

$\dfrac{x^2 - 4}{(x + 3)(x - 2)} \div \dfrac{(x + 2)(x + 3)}{-2x}$

$= \dfrac{x^2 - 4}{(x + 3)(x - 2)} \cdot \dfrac{-2x}{(x + 2)(x + 3)}$ Multiply by the reciprocal.

$= \dfrac{(x + 2)(x - 2)}{(x + 3)(x - 2)} \cdot \dfrac{-2x}{(x + 2)(x + 3)}$ Factor.

$= \dfrac{-2x(x + 2)(x - 2)}{(x + 3)(x - 2)(x + 2)(x + 3)}$ Multiply numerators. Multiply denominators.

$= \dfrac{-2x}{(x + 3)^2}$ or $-\dfrac{2x}{(x + 3)^2}$ Lowest terms

Work Problem 7 at the Side.

8 Divide. Write each answer in lowest terms.

(a) $\dfrac{ab - a^2}{a^2 - 1} \div \dfrac{a - b}{a - 1}$

(b) $\dfrac{x^2 - 9}{2x + 6} \div \dfrac{9 - x^2}{4x - 12}$

EXAMPLE 8 Dividing Rational Expressions (Factors Are Opposites)

Divide. Write the answer in lowest terms.

$\dfrac{m^2 - 4}{m^2 - 1} \div \dfrac{2m^2 + 4m}{1 - m}$

$= \dfrac{m^2 - 4}{m^2 - 1} \cdot \dfrac{1 - m}{2m^2 + 4m}$ Multiply by the reciprocal.

$= \dfrac{(m + 2)(m - 2)}{(m + 1)(m - 1)} \cdot \dfrac{1 - m}{2m(m + 2)}$ Factor; $1 - m$ and $m - 1$ differ only in sign.

$= \dfrac{-1(m - 2)}{2m(m + 1)}$ From **Section 15.1,** $\dfrac{1 - m}{m - 1} = -1$.

$= \dfrac{2 - m}{2m(m + 1)}$ Distribute -1 in the numerator.

Work Problem 8 at the Side.

ANSWERS

6. (a) $\dfrac{2a}{b}$ (b) $15w^3$

7. (a) $\dfrac{y - 3}{y - 5}$ (b) $-\dfrac{4(2x - 1)}{x}$

8. (a) $\dfrac{-a}{a + 1}$ (b) $\dfrac{-2x + 6}{3 + x}$

In summary, follow these steps to multiply or divide rational expressions.

Multiplying or Dividing Rational Expressions

Step 1 **Note the operation.** If the operation is division, use the definition of division to rewrite as multiplication.

Note: Steps 2 and 3 may be interchanged. It is a matter of personal preference.

Step 2 **Factor** all numerators and denominators completely.

Step 3 **Multiply** numerators and multiply denominators.

Step 4 **Write in lowest terms** using the fundamental property.

Focus on Real-Data Applications

Is 5 or 10 Minutes Really Worth the Risk?

A poignant e-mail entitled *The Drive Home* recounted the story of the habitual speeder named Jack, who was stopped by a policeman. The policeman happened to be his friend. Rather than writing a speeding ticket, the policeman handed Jack a note that described the loss of his daughter to a speeding driver. Jack was deeply affected and changed his attitude toward life on the road. The e-mail then gave some facts about speed.

- For a trip of 20 miles in a 40 mile-per-hour (mph) zone, speeding by 10 mph will save 6 minutes.
- For a trip of 50 miles in a 55 mph zone, speeding by 10 mph will save 8.3 minutes and speeding by 15 mph will save 11.6 minutes.

The question is, "Is 5 or 10 minutes really worth the risk?"

The relationship between time, distance, and rate is given by $t = \dfrac{d}{r}$. In the example, the time it takes to travel 20 miles in a 40 mph zone is $\dfrac{20 \text{ mi}}{40 \text{ mph}} = \dfrac{1}{2}$ hr, or 30 minutes. At a speed of 50 mph, the time is $\dfrac{20 \text{ mi}}{50 \text{ mph}} = \dfrac{2}{5}$ hr, or 24 minutes, since $\dfrac{2}{5}(60 \text{ min}) = 24$ min. The author's claim that only 6 minutes were saved is valid.

For Group Discussion

1. Verify the claims that traveling 50 miles at 65 mph in a 55 mph zone saves only 8.3 minutes and traveling at 70 mph saves only 11.6 minutes.

2. To save 10 minutes in a trip of 20 miles in a 40 mph zone, by how many miles per hour would Jack have to exceed the speed limit? (*Hint:* To save 10 minutes, the time for the trip would have to be only 20 minutes, which is $\dfrac{20}{60} = \dfrac{1}{3}$ hr. Determine the fraction of the form $\dfrac{20}{40 + x}$ that is equivalent to $\dfrac{1}{3}$ and then determine the value for x.)

3. Suppose you have to drive 10 miles to school in a 30 mph zone and you are running late. To save 5 minutes on your trip, at what speed would you have to travel? If stopped by a police officer, would you likely be given a ticket?

15.2 Exercises

1. Match each multiplication problem in Column I with the correct product in Column II.

   I

   (a) $\dfrac{5x^3}{10x^4} \cdot \dfrac{10x^7}{2x}$

   (b) $\dfrac{10x^4}{5x^3} \cdot \dfrac{10x^7}{2x}$

   (c) $\dfrac{5x^3}{10x^4} \cdot \dfrac{2x}{10x^7}$

   (d) $\dfrac{10x^4}{5x^3} \cdot \dfrac{2x}{10x^7}$

   II

   A. $\dfrac{2}{5x^5}$

   B. $\dfrac{5x^5}{2}$

   C. $\dfrac{1}{10x^7}$

   D. $10x^7$

2. Match each division problem in Column I with the correct quotient in Column II.

   I

   (a) $\dfrac{5x^3}{10x^4} \div \dfrac{10x^7}{2x}$

   (b) $\dfrac{10x^4}{5x^3} \div \dfrac{10x^7}{2x}$

   (c) $\dfrac{5x^3}{10x^4} \div \dfrac{2x}{10x^7}$

   (d) $\dfrac{10x^4}{5x^3} \div \dfrac{2x}{10x^7}$

   II

   A. $\dfrac{5x^5}{2}$

   B. $10x^7$

   C. $\dfrac{2}{5x^5}$

   D. $\dfrac{1}{10x^7}$

Multiply. Write each answer in lowest terms. See Examples 1 and 2.

3. $\dfrac{10m^2}{7} \cdot \dfrac{14}{15m}$

4. $\dfrac{36z^3}{6z} \cdot \dfrac{28}{z^2}$

5. $\dfrac{16y^4}{18y^5} \cdot \dfrac{15y^5}{y^2}$

6. $\dfrac{20x^5}{-2x^2} \cdot \dfrac{8x^4}{35x^3}$

7. $\dfrac{2(c+d)}{3} \cdot \dfrac{18}{6(c+d)^2}$

8. $\dfrac{4(y-2)}{x} \cdot \dfrac{3x}{6(y-2)^2}$

Find the reciprocal of each rational expression. See Example 4.

9. $\dfrac{3p^3}{16q}$

10. $\dfrac{6x^4}{9y^2}$

11. $\dfrac{r^2 + rp}{7}$

12. $\dfrac{16}{9a^2 + 36a}$

13. $\dfrac{z^2 + 7z + 12}{z^2 - 9}$

14. $\dfrac{p^2 - 4p + 3}{p^2 - 3p}$

Divide. Write each answer in lowest terms. See Examples 5, 6, and 7.

15. $\dfrac{9z^4}{3z^5} \div \dfrac{3z^2}{5z^3}$

16. $\dfrac{35q^8}{9q^5} \div \dfrac{25q^6}{10q^5}$

17. $\dfrac{4t^4}{2t^5} \div \dfrac{(2t)^3}{-6}$

18. $\dfrac{-12a^6}{3a^2} \div \dfrac{(2a)^3}{27a}$

19. $\dfrac{3}{2y - 6} \div \dfrac{6}{y - 3}$

20. $\dfrac{4m + 16}{10} \div \dfrac{3m + 12}{18}$

21. Explain in your own words how to multiply rational expressions.

22. Explain in your own words how to divide rational expressions.

Multiply or divide. Write each answer in lowest terms. See Examples 3, 7, and 8.

23. $\dfrac{5x - 15}{3x + 9} \cdot \dfrac{4x + 12}{6x - 18}$

24. $\dfrac{8r + 16}{24r - 24} \cdot \dfrac{6r - 6}{3r + 6}$

25. $\dfrac{2 - t}{8} \div \dfrac{t - 2}{6}$

26. $\dfrac{4}{m - 2} \div \dfrac{16}{2 - m}$

27. $\dfrac{5 - 4x}{5 + 4x} \cdot \dfrac{4x + 5}{4x - 5}$

28. $\dfrac{5 - x}{5 + x} \cdot \dfrac{x + 5}{x - 5}$

29. $\dfrac{6(m - 2)^2}{5(m + 4)^2} \cdot \dfrac{15(m + 4)}{2(2 - m)}$

30. $\dfrac{7(q - 1)}{3(q + 1)^2} \cdot \dfrac{6(q + 1)}{3(1 - q)^2}$

31. $\dfrac{p^2 + 4p - 5}{p^2 + 7p + 10} \div \dfrac{p - 1}{p + 4}$

32. $\dfrac{z^2 - 3z + 2}{z^2 + 4z + 3} \div \dfrac{z - 1}{z + 1}$

33. $\dfrac{2k^2 - k - 1}{2k^2 + 5k + 3} \div \dfrac{4k^2 - 1}{2k^2 + k - 3}$

34. $\dfrac{2m^2 - 5m - 12}{m^2 + m - 20} \div \dfrac{4m^2 - 9}{m^2 + 4m - 5}$

35. $\dfrac{2k^2 + 3k - 2}{6k^2 - 7k + 2} \cdot \dfrac{4k^2 - 5k + 1}{k^2 + k - 2}$

36. $\dfrac{2m^2 - 5m - 12}{m^2 - 10m + 24} \div \dfrac{4m^2 - 9}{m^2 - 9m + 18}$

37. $\dfrac{m^2 + 2mp - 3p^2}{m^2 - 3mp + 2p^2} \div \dfrac{m^2 + 4mp + 3p^2}{m^2 + 2mp - 8p^2}$

38. $\dfrac{r^2 + rs - 12s^2}{r^2 - rs - 20s^2} \div \dfrac{r^2 - 2rs - 3s^2}{r^2 + rs - 30s^2}$

39. $\left(\dfrac{x^2 + 10x + 25}{x^2 + 10x} \cdot \dfrac{10x}{x^2 + 15x + 50}\right) \div \dfrac{x + 5}{x + 10}$

40. $\left(\dfrac{m^2 - 12m + 32}{8m} \cdot \dfrac{m^2 - 8m}{m^2 - 8m + 16}\right) \div \dfrac{m - 8}{m - 4}$

41. Consider the division problem $\dfrac{x - 6}{x + 4} \div \dfrac{x + 7}{x + 5}$.
We know that division by 0 is undefined, so the restrictions on x are $x \neq -4$, $x \neq -5$, and $x \neq -7$. Why is the last restriction needed?

42. If the rational expression $\dfrac{5x^2y^3}{2pq}$ represents the area of a rectangle and $\dfrac{2xy}{p}$ represents the length, what rational expression represents the width?

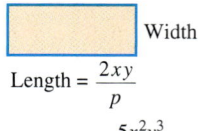 Width

Length = $\dfrac{2xy}{p}$

The area is $\dfrac{5x^2y^3}{2pq}$.

15.3 Least Common Denominators

OBJECTIVE 1 **Find the least common denominator for a group of fractions.** Just as with common fractions, adding or subtracting rational expressions (to be discussed in the next section) often requires a **least common denominator (LCD)**, the simplest expression that is divisible by all denominators. For example, the least common denominator for $\frac{2}{9}$ and $\frac{5}{12}$ is 36 because 36 is the smallest positive number divisible by both 9 and 12.

Least common denominators can often be found by inspection. For example, the LCD for $\frac{1}{6}$ and $\frac{2}{3m}$ is $6m$. In other cases, the LCD can be found by a procedure similar to that used in **Section 14.1** for finding the greatest common factor.

OBJECTIVES

1. Find the least common denominator for a group of fractions.
2. Rewrite rational expressions with given denominators.

1 Find the LCD for each pair of fractions.

(a) $\dfrac{7}{10}, \dfrac{1}{25}$

(b) $\dfrac{7}{20p}, \dfrac{11}{30p}$

(c) $\dfrac{4}{5x}, \dfrac{13}{10x}$

Finding the Least Common Denominator (LCD)

Step 1 **Factor** each denominator into prime factors.

Step 2 **List each different denominator factor** the *greatest* number of times it appears in any of the denominators.

Step 3 **Multiply** the denominator factors from Step 2 to get the LCD.

When each denominator is factored into prime factors, every prime factor must be a factor of the least common denominator.

In Example 1, we find the LCD for both numerical and algebraic denominators.

EXAMPLE 1 Finding the LCD

Find the LCD for each pair of fractions.

(a) $\dfrac{1}{24}, \dfrac{7}{15}$ (b) $\dfrac{1}{8x}, \dfrac{3}{10x}$

Step 1 Write each denominator in factored form with numerical coefficients in prime factored form.

$24 = 2 \cdot 2 \cdot 2 \cdot 3 = 2^3 \cdot 3$ $8x = 2 \cdot 2 \cdot 2 \cdot x = 2^3 \cdot x$

$15 = 3 \cdot 5$ $10x = 2 \cdot 5 \cdot x$

Step 2 We find the LCD by taking each different factor the *greatest* number of times it appears as a factor in any of the denominators.

The factor 2 appears three times in one product and not at all in the other, so the greatest number of times 2 appears is three. The greatest number of times both 3 and 5 appear is one.

Here 2 appears three times in one product and once in the other, so the greatest number of times 2 appears is three. The greatest number of times 5 appears is one, and the greatest number of times x appears in either product is one.

Step 3 LCD $= 2 \cdot 2 \cdot 2 \cdot 3 \cdot 5$ LCD $= 2 \cdot 2 \cdot 2 \cdot 5 \cdot x$
$= 2^3 \cdot 3 \cdot 5$ $= 2^3 \cdot 5 \cdot x$
$= 120$ $= 40x$

Work Problem 1 at the Side.

ANSWERS
1. (a) 50 (b) $60p$ (c) $10x$

2 Find the LCD.

(a) $\dfrac{15}{16m^3n}, \dfrac{5}{9m^5}$

(b) $\dfrac{3}{25a^2}, \dfrac{9}{10a^3b}$

3 Find the LCD.

(a) $\dfrac{7}{3a}, \dfrac{11}{a^2 - 4a}$

(b) $\dfrac{2m}{m^2 - 3m + 2}, \dfrac{5m - 3}{m^2 + 3m - 10}, \dfrac{4m + 7}{m^2 + 4m - 5}$

(c) $\dfrac{6}{x - 4}, \dfrac{3x - 1}{4 - x}$

EXAMPLE 2 Finding the LCD

Find the LCD for $\dfrac{5}{6r^2}$ and $\dfrac{3}{4r^3}$.

Step 1 Factor each denominator.

$$6r^2 = 2 \cdot 3 \cdot r^2$$
$$4r^3 = 2^2 \cdot r^3$$

Step 2 The greatest number of times 2 appears is two, the greatest number of times 3 appears is one, and the greatest number of times r appears is three; therefore,

Step 3 $\quad\quad\quad\quad \text{LCD} = 2^2 \cdot 3 \cdot r^3 = 12r^3$.

⮜⮜ **Work Problem 2 at the Side.**

EXAMPLE 3 Finding the LCD

Find the LCD.

(a) $\dfrac{6}{5m}, \dfrac{4}{m^2 - 3m}$

Factor each denominator.

$$5m = 5 \cdot m$$
$$m^2 - 3m = m(m - 3)$$

Use each different factor the greatest number of times it appears.

$$\text{LCD} = 5 \cdot m \cdot (m - 3) = 5m(m - 3)$$

Because m is not a *factor* of $m - 3$, both factors, m and $m - 3$, must appear in the LCD.

(b) $\dfrac{1}{r^2 - 4r - 5}, \dfrac{3}{r^2 - r - 20}, \dfrac{1}{r^2 - 10r + 25}$

$$\left. \begin{array}{l} r^2 - 4r - 5 = (r - 5)(r + 1) \\ r^2 - r - 20 = (r - 5)(r + 4) \\ r^2 - 10r + 25 = (r - 5)^2 \end{array} \right\} \text{Factor each denominator.}$$

Use each different factor the greatest number of times it appears as a factor. The LCD is

$$(r - 5)^2(r + 1)(r + 4).$$

(c) $\dfrac{1}{q - 5}, \dfrac{3}{5 - q}$

The expressions $q - 5$ and $5 - q$ are opposites of each other because

$$-(q - 5) = -q + 5 = 5 - q.$$

Therefore, either $q - 5$ or $5 - q$ can be used as the LCD.

⮜⮜ **Work Problem 3 at the Side.**

OBJECTIVE 2 Rewrite rational expressions with given denominators. Once the LCD has been found, the next step in preparing to add or subtract two rational expressions is to use the fundamental property to write equivalent rational expressions. The next example shows how to do this with both numerical and algebraic fractions.

ANSWERS
2. (a) $144m^5n$ (b) $50a^3b$
3. (a) $3a(a - 4)$
   (b) $(m - 1)(m - 2)(m + 5)$
   (c) either $x - 4$ or $4 - x$

EXAMPLE 4 Writing Rational Expressions with Given Denominators

Rewrite each rational expression with the indicated denominator.

(a) $\dfrac{3}{8} = \dfrac{}{40}$

(b) $\dfrac{9k}{25} = \dfrac{}{50k}$

For each example, first factor the denominator on the right. Then compare the denominator on the left with the one on the right to decide what factors are missing. (It may be necessary to factor both denominators.)

$$\dfrac{3}{8} = \dfrac{}{5 \cdot 8} \qquad\qquad \dfrac{9k}{25} = \dfrac{}{25 \cdot 2k}$$

A factor of 5 is missing. Using the fundamental property, multiply $\dfrac{3}{8}$ by $\dfrac{5}{5}$.

Factors of 2 and k are missing. Multiply by $\dfrac{2k}{2k}$.

$$\dfrac{3}{8} = \dfrac{3}{8} \cdot \dfrac{5}{5} = \dfrac{15}{40}$$

$\dfrac{5}{5} = 1$

$$\dfrac{9k}{25} = \dfrac{9k}{25} \cdot \dfrac{2k}{2k} = \dfrac{18k^2}{50k}$$

$\dfrac{2k}{2k} = 1$

Work Problem 4 at the Side.

EXAMPLE 5 Writing Rational Expressions with Given Denominators

Rewrite each rational expression with the indicated denominator.

(a) $\dfrac{8}{3x+1} = \dfrac{}{12x+4}$

Factor the denominator on the right.

$\dfrac{8}{3x+1} = \dfrac{}{4(3x+1)}$ Factor.

The missing factor is 4, so multiply the fraction on the left by $\dfrac{4}{4}$.

$$\dfrac{8}{3x+1} \cdot \dfrac{4}{4} = \dfrac{32}{12x+4} \quad \text{Fundamental property}$$

(b) $\dfrac{12p}{p^2+8p} = \dfrac{}{p^3+4p^2-32p}$

Factor $p^2 + 8p$ as $p(p+8)$. Compare with the denominator on the right, which factors as $p(p+8)(p-4)$. The factor $p-4$ is missing, so multiply $\dfrac{12p}{p(p+8)}$ by $\dfrac{p-4}{p-4}$.

$$\dfrac{12p}{p^2+8p} = \dfrac{12p}{p(p+8)} \cdot \dfrac{p-4}{p-4} \quad \text{Fundamental property}$$

$$= \dfrac{12p(p-4)}{p(p+8)(p-4)} \quad \text{Multiplication of rational expressions}$$

$$= \dfrac{12p^2-48p}{p^3+4p^2-32p} \quad \text{Multiply the factors.}$$

---

4 Rewrite each rational expression with the indicated denominator.

(a) $\dfrac{3}{4} = \dfrac{}{36}$

(b) $\dfrac{7k}{5} = \dfrac{}{30p}$

ANSWERS

4. (a) $\dfrac{27}{36}$ (b) $\dfrac{42kp}{30p}$

5 Rewrite each rational expression with the indicated denominator.

(a) $\dfrac{9}{2a+5} = \dfrac{}{6a+15}$

(b) $\dfrac{5k+1}{k^2+2k} = \dfrac{}{k^3+k^2-2k}$

> **NOTE**
> In the next section we add and subtract rational expressions, which sometimes requires the steps illustrated in Examples 4 and 5. While it is beneficial to leave the denominator in factored form, we multiplied the factors in the denominator in Example 5(b),
>
> $$\dfrac{12p(p-4)}{p(p+8)(p-4)},$$
>
> to give the answer,
>
> $$\dfrac{12p^2 - 48p}{p^3 + 4p^2 - 32p},$$
>
> in the same form as the original problem.

◀◀◀ **Work Problem 5 at the Side.**

ANSWERS

5. (a) $\dfrac{27}{6a+15}$ (b) $\dfrac{(5k+1)(k-1)}{k^3+k^2-2k}$

15.3 Exercises

Choose the correct response in Exercises 1–4.

1. Suppose that the greatest common factor of a and b is 1. Then the least common denominator for $\frac{1}{a}$ and $\frac{1}{b}$ is

    A. a **B.** b **C.** ab **D.** 1.

2. If a is a factor of b, then the least common denominator for $\frac{1}{a}$ and $\frac{1}{b}$ is

    A. a **B.** b **C.** ab **D.** 1.

3. The least common denominator for $\frac{11}{20}$ and $\frac{1}{2}$ is

    A. 40 **B.** 2 **C.** 20 **D.** none of these.

4. Suppose that we wish to write the fraction $\dfrac{1}{(x-4)^2(y-3)}$ with denominator $(x-4)^3(y-3)^2$. We must multiply both the numerator and the denominator by

    A. $(x-4)(y-3)$ **B.** $(x-4)^2$
    C. $x-4$ **D.** $(x-4)^2(y-3)$.

Find the least common denominator for the fractions in each list. See Examples 1–3.

5. $\dfrac{2}{15}, \dfrac{3}{10}, \dfrac{7}{30}$

6. $\dfrac{5}{24}, \dfrac{7}{12}, \dfrac{9}{28}$

7. $\dfrac{3}{x^4}, \dfrac{5}{x^7}$

8. $\dfrac{2}{y^5}, \dfrac{3}{y^6}$

9. $\dfrac{5}{36q}, \dfrac{17}{24q}$

10. $\dfrac{4}{30p}, \dfrac{9}{50p}$

11. $\dfrac{6}{21r^3}, \dfrac{13}{12r^5}$

12. $\dfrac{9}{35t^2}, \dfrac{5}{49t^6}$

13. If the denominators of two fractions in prime factored form are $2^3 \cdot 3$ and $2^2 \cdot 5$, what is the factored form of their LCD?

14. Suppose two rational expressions have denominators $(t+4)^3(t-3)$ and $(t+4)^2(t+8)$. Find the factored form of their LCD. What is the similarity between the answers for this problem and for Exercise 13?

15. If two denominators have greatest common factor equal to 1, how can you easily find their least common denominator?

16. Suppose two fractions have denominators a^k and a^r, where k and r are natural numbers, with $k > r$. What is their least common denominator?

Find the least common denominator for each group of fractions. See Examples 1–3.

17. $\dfrac{9}{28m^2}, \dfrac{3}{12m-20}$

18. $\dfrac{16}{27a^3}, \dfrac{8}{9a-45}$

19. $\dfrac{7}{5b-10}, \dfrac{11}{6b-12}$

20. $\dfrac{3}{7x^2+21x}, \dfrac{1}{5x^2+15x}$

21. $\dfrac{5}{c-d}, \dfrac{8}{d-c}$

22. $\dfrac{4}{y-x}, \dfrac{7}{x-y}$

23. $\dfrac{3}{k^2 + 5k}, \dfrac{2}{k^2 + 3k - 10}$

24. $\dfrac{1}{z^2 - 4z}, \dfrac{4}{z^2 - 3z - 4}$

25. $\dfrac{5}{p^2 + 8p + 15}, \dfrac{3}{p^2 - 3p - 18}, \dfrac{2}{p^2 - p - 30}$

26. $\dfrac{10}{y^2 - 10y + 21}, \dfrac{2}{y^2 - 2y - 3}, \dfrac{5}{y^2 - 6y - 7}$

Rewrite each rational expression with the given denominator. See Examples 4 and 5.

27. $\dfrac{4}{11} = \dfrac{}{55}$

28. $\dfrac{6}{7} = \dfrac{}{42}$

29. $\dfrac{-5}{k} = \dfrac{}{9k}$

30. $\dfrac{-3}{q} = \dfrac{}{6q}$

31. $\dfrac{13}{40y} = \dfrac{}{80y^3}$

32. $\dfrac{5}{27p} = \dfrac{}{108p^4}$

33. $\dfrac{5t^2}{6r} = \dfrac{}{42r^4}$

34. $\dfrac{8y^2}{3x} = \dfrac{}{30x^3}$

35. $\dfrac{5}{2(m + 3)} = \dfrac{}{8(m + 3)}$

36. $\dfrac{7}{4(y - 1)} = \dfrac{}{16(y - 1)}$

37. $\dfrac{-4t}{3t - 6} = \dfrac{}{12 - 6t}$

38. $\dfrac{-7k}{5k - 20} = \dfrac{}{40 - 10k}$

39. $\dfrac{14}{z^2 - 3z} = \dfrac{}{z(z - 3)(z - 2)}$

40. $\dfrac{12}{x(x + 4)} = \dfrac{}{x(x + 4)(x - 9)}$

41. $\dfrac{2(b - 1)}{b^2 + b} = \dfrac{}{b^3 + 3b^2 + 2b}$

42. $\dfrac{3(c + 2)}{c(c - 1)} = \dfrac{}{c^3 - 5c^2 + 4c}$

15.4 Adding and Subtracting Rational Expressions

To add and subtract rational expressions, we find least common denominators and write equivalent fractions with the LCD.

OBJECTIVE 1 Add rational expressions having the same denominator. We find the sum of two rational expressions with the same procedure that we used for adding two fractions in **Section 4.4**.

Adding Rational Expressions
If $\frac{P}{Q}$ and $\frac{R}{Q}$ (where $Q \neq 0$) are rational expressions, then
$$\frac{P}{Q} + \frac{R}{Q} = \frac{P+R}{Q}.$$

In words: To add rational expressions with the same denominator, add the numerators and keep the same denominator.

The first example shows how the addition of rational expressions compares with that of rational numbers.

EXAMPLE 1 Adding Rational Expressions with the Same Denominator

Add. Write each answer in lowest terms.

(a) $\dfrac{4}{9} + \dfrac{2}{9}$ (b) $\dfrac{3x}{x+1} + \dfrac{3}{x+1}$

The denominators are the same, so the sum is found by adding the two numerators and keeping the same (common) denominator.

$$\frac{4}{9} + \frac{2}{9} = \frac{4+2}{9}$$
$$= \frac{6}{9}$$
$$= \frac{2}{3}$$

$$\frac{3x}{x+1} + \frac{3}{x+1} = \frac{3x+3}{x+1}$$
$$= \frac{3(x+1)}{x+1}$$
$$= 3$$

Work Problem 1 at the Side.

OBJECTIVE 2 Add rational expressions having different denominators. We use the following steps to add two rational expressions with different denominators.

Adding with Different Denominators
Step 1 Find the least common denominator (LCD).

Step 2 Rewrite each rational expression as an equivalent rational expression with the LCD as the denominator.

Step 3 Add the numerators to get the numerator of the sum. The LCD is the denominator of the sum.

Step 4 Write in lowest terms using the fundamental property.

OBJECTIVES

1. Add rational expressions having the same denominator.
2. Add rational expressions having different denominators.
3. Subtract rational expressions.

1 Add. Write each answer in lowest terms.

(a) $\dfrac{7}{15} + \dfrac{3}{15}$

(b) $\dfrac{3}{y+4} + \dfrac{2}{y+4}$

(c) $\dfrac{x}{x+y} + \dfrac{1}{x+y}$

(d) $\dfrac{a}{a+b} + \dfrac{b}{a+b}$

(e) $\dfrac{x^2}{x+1} + \dfrac{x}{x+1}$

ANSWERS
1. (a) $\dfrac{2}{3}$ (b) $\dfrac{5}{y+4}$ (c) $\dfrac{x+1}{x+y}$
   (d) 1 (e) x

Chapter 15 Rational Expressions and Applications

2 Add. Write each answer in lowest terms.

(a) $\dfrac{1}{10} + \dfrac{1}{15}$

(b) $\dfrac{6}{5x} + \dfrac{9}{2x}$

(c) $\dfrac{m}{3n} + \dfrac{2}{7n}$

ANSWERS

2. (a) $\dfrac{1}{6}$ (b) $\dfrac{57}{10x}$ (c) $\dfrac{7m+6}{21n}$

EXAMPLE 2 Adding Rational Expressions with Different Denominators

Add. Write each answer in lowest terms.

(a) $\dfrac{1}{12} + \dfrac{7}{15}$

(b) $\dfrac{2}{3y} + \dfrac{1}{4y}$

Step 1 First find the LCD using the methods of the previous section.

$$\text{LCD} = 2^2 \cdot 3 \cdot 5 = 60 \qquad \text{LCD} = 2^2 \cdot 3 \cdot y = 12y$$

Step 2 Now rewrite each rational expression as a fraction with the LCD (either 60 or $12y$) as the denominator.

$$\dfrac{1}{12} + \dfrac{7}{15} = \dfrac{1(5)}{12(5)} + \dfrac{7(4)}{15(4)} \qquad \dfrac{2}{3y} + \dfrac{1}{4y} = \dfrac{2(4)}{3y(4)} + \dfrac{1(3)}{4y(3)}$$

$$= \dfrac{5}{60} + \dfrac{28}{60} \qquad\qquad\qquad = \dfrac{8}{12y} + \dfrac{3}{12y}$$

Step 3 Since the fractions now have common denominators, add the numerators.

Step 4 Write in lowest terms if necessary.

$$\dfrac{5}{60} + \dfrac{28}{60} = \dfrac{5+28}{60} \qquad \dfrac{8}{12y} + \dfrac{3}{12y} = \dfrac{8+3}{12y}$$

$$= \dfrac{33}{60} = \dfrac{11}{20} \qquad\qquad\qquad = \dfrac{11}{12y}$$

Work Problem 2 at the Side.

EXAMPLE 3 Adding Rational Expressions

Add. Write the answer in lowest terms.

$$\dfrac{2x}{x^2 - 1} + \dfrac{-1}{x+1}$$

Step 1 Since the denominators are different, find the LCD.

$$x^2 - 1 = (x+1)(x-1)$$

$$x + 1 \text{ is prime.}$$

The LCD is $(x+1)(x-1)$.

Step 2 Rewrite each rational expression as a fraction with common denominator $(x+1)(x-1)$.

$$\dfrac{2x}{x^2-1} + \dfrac{-1}{x+1} = \dfrac{2x}{(x+1)(x-1)} + \dfrac{-1(x-1)}{(x+1)(x-1)} \quad \text{Multiply the second fraction by } \tfrac{x-1}{x-1}.$$

$$= \dfrac{2x}{(x+1)(x-1)} + \dfrac{-x+1}{(x+1)(x-1)} \quad \text{Distributive property}$$

Step 3
$$= \dfrac{2x - x + 1}{(x+1)(x-1)} \quad \text{Add numerators; keep the same denominator.}$$

$$= \dfrac{x+1}{(x+1)(x-1)} \quad \text{Combine like terms in the numerator.}$$

Continued on Next Page

Step 4 $= \dfrac{1(x+1)}{(x+1)(x-1)}$ Use the identity property for multiplication: multiplying by 1 does not change the result.

$= \dfrac{1}{x-1}$ Fundamental property

Work Problem 3 at the Side.

EXAMPLE 4 Adding Rational Expressions

Add. Write the answer in lowest terms.

$$\dfrac{2x}{x^2+5x+6} + \dfrac{x+1}{x^2+2x-3}$$

$$= \dfrac{2x}{(x+2)(x+3)} + \dfrac{x+1}{(x+3)(x-1)}$$ Factor the denominators.

The LCD is $(x+2)(x+3)(x-1)$. Use the fundamental property.

$$= \dfrac{2x(x-1)}{(x+2)(x+3)(x-1)} + \dfrac{(x+1)(x+2)}{(x+2)(x+3)(x-1)}$$

$$= \dfrac{2x(x-1) + (x+1)(x+2)}{(x+2)(x+3)(x-1)}$$ Add numerators; keep the same denominator.

$$= \dfrac{2x^2 - 2x + x^2 + 3x + 2}{(x+2)(x+3)(x-1)}$$ Multiply.

$$= \dfrac{3x^2 + x + 2}{(x+2)(x+3)(x-1)}$$ Combine like terms.

It is usually more convenient to leave the denominator in factored form. The numerator cannot be factored here, so the expression is in lowest terms.

Work Problem 4 at the Side.

EXAMPLE 5 Adding Rational Expressions with Denominators That Are Opposites

Add. Write the answer in lowest terms.

$$\dfrac{y}{y-2} + \dfrac{8}{2-y}$$

One way to get a common denominator is to multiply the second expression by -1 in both the numerator and the denominator, giving $y-2$ as a common denominator.

$$\dfrac{y}{y-2} + \dfrac{8}{2-y} = \dfrac{y}{y-2} + \dfrac{8(-1)}{(2-y)(-1)}$$ Fundamental property

$$= \dfrac{y}{y-2} + \dfrac{-8}{y-2}$$ Distributive property

$$= \dfrac{y-8}{y-2}$$ Add numerators; keep the same denominator.

If we had chosen to use $2-y$ as the common denominator, the final answer would be in the form $\dfrac{8-y}{2-y}$, which is equivalent to $\dfrac{y-8}{y-2}$.

Work Problem 5 at the Side.

---

❸ Add. Write each answer in lowest terms.

(a) $\dfrac{2p}{3p+3} + \dfrac{5p}{2p+2}$

(b) $\dfrac{4}{y^2-1} + \dfrac{6}{y+1}$

(c) $\dfrac{-2}{p+1} + \dfrac{4p}{p^2-1}$

❹ Add. Write each answer in lowest terms.

(a) $\dfrac{2k}{k^2-5k+4} + \dfrac{3}{k^2-1}$

(b) $\dfrac{4m}{m^2+3m+2} + \dfrac{2m-1}{m^2+6m+5}$

❺ Add. Write the answer in lowest terms.

$$\dfrac{m}{2m-3n} + \dfrac{n}{3n-2m}$$

ANSWERS

3. (a) $\dfrac{19p}{6(p+1)}$ (b) $\dfrac{2(3y-1)}{(y+1)(y-1)}$

(c) $\dfrac{2}{p-1}$

4. (a) $\dfrac{(2k-3)(k+4)}{(k-4)(k-1)(k+1)}$

(b) $\dfrac{6m^2+23m-2}{(m+2)(m+1)(m+5)}$

5. $\dfrac{m-n}{2m-3n}$ or $\dfrac{n-m}{3n-2m}$

Chapter 15 Rational Expressions and Applications

6 Subtract. Write each answer in lowest terms.

(a) $\dfrac{3}{m^2} - \dfrac{2}{m^2}$

(b) $\dfrac{x}{2x+3} - \dfrac{3x+4}{2x+3}$

ANSWERS

6. (a) $\dfrac{1}{m^2}$ (b) $\dfrac{-2(x+2)}{2x+3}$

OBJECTIVE 3 Subtract rational expressions. To subtract rational expressions, use the following rule.

Subtracting Rational Expressions

If $\dfrac{P}{Q}$ and $\dfrac{R}{Q}$ (where $Q \neq 0$) are rational expressions, then

$$\dfrac{P}{Q} - \dfrac{R}{Q} = \dfrac{P - R}{Q}.$$

In words: To subtract rational expressions with the same denominator, subtract the numerators and keep the same denominator.

EXAMPLE 6 Subtracting Rational Expressions with the Same Denominator

Subtract. Write the answer in lowest terms.

$$\dfrac{2m}{m-1} - \dfrac{m+3}{m-1} = \dfrac{2m - (m+3)}{m-1}$$ Use parentheses around the quantity being subtracted. Subtract numerators; keep the same denominator.

$$= \dfrac{2m - m - 3}{m-1}$$ Distributive property

$$= \dfrac{m-3}{m-1}$$ Combine like terms.

CAUTION
Sign errors often occur in subtraction problems like the one in Example 6. Remember that the numerator of the fraction being subtracted must be treated as a single quantity. *Be sure to use parentheses after the subtraction sign* to avoid this common error.

Work Problem 6 at the Side.

EXAMPLE 7 Subtracting Rational Expressions with Different Denominators

Subtract. Write the answer in lowest terms.

$$\dfrac{9}{x-2} - \dfrac{3}{x}$$

The LCD is $x(x-2)$.

$$\dfrac{9}{x-2} - \dfrac{3}{x} = \dfrac{9x}{x(x-2)} - \dfrac{3(x-2)}{x(x-2)}$$ Rewrite each expression with the LCD.

$$= \dfrac{9x - 3(x-2)}{x(x-2)}$$ Subtract numerators; keep the same denominator.

Continued on Next Page

$$= \frac{9x - 3x + 6}{x(x - 2)}$$ Distributive property; be careful with signs.

$$= \frac{6x + 6}{x(x - 2)}$$ Combine like terms.

$$= \frac{6(x + 1)}{x(x - 2)}$$ Factor.

> **NOTE**
> We factor the final numerator in Example 7 to get an answer in the form $\frac{6(x + 1)}{x(x - 2)}$. The fundamental property does not apply, since there are no common factors. The answer is in lowest terms.

Work Problem 7 at the Side.

7 Subtract. Write each answer in lowest terms.

(a) $\dfrac{1}{k + 4} - \dfrac{2}{k}$

(b) $\dfrac{6}{a + 2} - \dfrac{1}{a - 3}$

EXAMPLE 8 Subtracting Rational Expressions with Denominators That Are Opposites

Subtract. Write the answer in lowest terms.

$$\frac{3x}{x - 5} - \frac{2x - 25}{5 - x}$$

The denominators are opposites, so either may be used as the common denominator. We will choose $x - 5$.

$$\frac{3x}{x - 5} - \frac{2x - 25}{5 - x} = \frac{3x}{x - 5} - \frac{2x - 25}{5 - x} \cdot \frac{-1}{-1}$$ Fundamental property

$$= \frac{3x}{x - 5} - \frac{-2x + 25}{x - 5}$$ Multiply.

$$= \frac{3x - (-2x + 25)}{x - 5}$$ Subtract numerators; use parentheses.

$$= \frac{3x + 2x - 25}{x - 5}$$ Distributive property

$$= \frac{5x - 25}{x - 5}$$ Combine like terms.

$$= \frac{5(x - 5)}{x - 5}$$ Factor.

$$= 5$$ Lowest terms

Work Problem 8 at the Side.

8 Subtract. Write each answer in lowest terms.

(a) $\dfrac{5}{x - 1} - \dfrac{3x}{1 - x}$

(b) $\dfrac{2y}{y - 2} - \dfrac{1 + y}{2 - y}$

EXAMPLE 9 Subtracting Rational Expressions

Subtract. Write the answer in lowest terms.

$$\frac{6x}{x^2 - 2x + 1} - \frac{1}{x^2 - 1}$$

Begin by factoring the denominators.

$$x^2 - 2x + 1 = (x - 1)^2 \quad \text{and} \quad x^2 - 1 = (x - 1)(x + 1)$$

Continued on Next Page

ANSWERS

7. (a) $\dfrac{-k - 8}{k(k + 4)}$ (b) $\dfrac{5(a - 4)}{(a + 2)(a - 3)}$

8. (a) $\dfrac{5 + 3x}{x - 1}$ (b) $\dfrac{3y + 1}{y - 2}$

Chapter 15 Rational Expressions and Applications

9 Subtract. Write each answer in lowest terms.

(a) $\dfrac{4y}{y^2 - 1} - \dfrac{5}{y^2 + 2y + 1}$

(b) $\dfrac{3r}{r^2 - 5r} - \dfrac{4}{r^2 - 10r + 25}$

10 Subtract. Write each answer in lowest terms.

(a) $\dfrac{2}{p^2 - 5p + 4} - \dfrac{3}{p^2 - 1}$

(b) $\dfrac{q}{2q^2 + 5q - 3} - \dfrac{3q + 4}{3q^2 + 10q + 3}$

From the factored denominators, identify the LCD,

$$(x - 1)^2(x + 1).$$

Use the factor $x - 1$ twice because it appears twice in the first denominator.

$$\dfrac{6x}{(x - 1)^2} - \dfrac{1}{(x - 1)(x + 1)}$$

$$= \dfrac{6x(x + 1)}{(x - 1)^2(x + 1)} - \dfrac{1(x - 1)}{(x - 1)(x - 1)(x + 1)} \quad \text{Fundamental property}$$

$$= \dfrac{6x(x + 1) - 1(x - 1)}{(x - 1)^2(x + 1)} \quad \text{Subtract numerators.}$$

$$= \dfrac{6x^2 + 6x - x + 1}{(x - 1)^2(x + 1)} \quad \text{Distributive property}$$

$$= \dfrac{6x^2 + 5x + 1}{(x - 1)^2(x + 1)} \text{ or } \dfrac{(2x + 1)(3x + 1)}{(x - 1)^2(x + 1)} \quad \text{Combine like terms.}$$

Verify that the final expression is in lowest terms.

◀◀◀ Work Problem 9 at the Side.

EXAMPLE 10 Subtracting Rational Expressions

Subtract. Write the answer in lowest terms.

$$\dfrac{q}{q^2 - 4q - 5} - \dfrac{3}{2q^2 - 13q + 15}$$

To find the LCD, factor each denominator.

$$q^2 - 4q - 5 = (q + 1)(q - 5)$$
$$2q^2 - 13q + 15 = (q - 5)(2q - 3)$$

The LCD is $(q + 1)(q - 5)(2q - 3)$. Rewrite each rational expression with the LCD, using the fundamental property.

$$\dfrac{q}{(q + 1)(q - 5)} - \dfrac{3}{(q - 5)(2q - 3)}$$

$$= \dfrac{q(2q - 3)}{(q + 1)(q - 5)(2q - 3)} - \dfrac{3(q + 1)}{(q + 1)(q - 5)(2q - 3)}$$

$$= \dfrac{q(2q - 3) - 3(q + 1)}{(q + 1)(q - 5)(2q - 3)} \quad \text{Subtract numerators.}$$

$$= \dfrac{2q^2 - 3q - 3q - 3}{(q + 1)(q - 5)(2q - 3)} \quad \text{Distributive property}$$

$$= \dfrac{2q^2 - 6q - 3}{(q + 1)(q - 5)(2q - 3)} \quad \text{Combine like terms.}$$

Verify that the final expression is in lowest terms.

◀◀◀ Work Problem 10 at the Side.

ANSWERS

9. (a) $\dfrac{4y^2 - y + 5}{(y + 1)^2(y - 1)}$

(b) $\dfrac{3r - 19}{(r - 5)^2}$

10. (a) $\dfrac{14 - p}{(p - 4)(p - 1)(p + 1)}$

(b) $\dfrac{-3q^2 - 4q + 4}{(2q - 1)(q + 3)(3q + 1)}$

15.4 Exercises

Match the expression in Column I with the correct sum or difference in Column II.

I

1. $\dfrac{x}{x+6} + \dfrac{6}{x+6}$

2. $\dfrac{2x}{x-6} - \dfrac{12}{x-6}$

3. $\dfrac{6}{x-6} - \dfrac{x}{x-6}$

4. $\dfrac{6}{x+6} - \dfrac{x}{x+6}$

5. $\dfrac{x}{x+6} - \dfrac{6}{x+6}$

6. $\dfrac{1}{x} + \dfrac{1}{6}$

7. $\dfrac{1}{6} - \dfrac{1}{x}$

8. $\dfrac{1}{6x} - \dfrac{1}{6x}$

II

A. 2

B. $\dfrac{x-6}{x+6}$

C. -1

D. $\dfrac{6+x}{6x}$

E. 1

F. 0

G. $\dfrac{x-6}{6x}$

H. $\dfrac{6-x}{x+6}$

Note: When adding and subtracting rational expressions, several different equivalent forms of the answer often exist. If your answer does not look exactly like the one given in the back of the book, check to see whether you have written an equivalent form.

Add or subtract. Write each answer in lowest terms. See Examples 1 and 6.

9. $\dfrac{4}{m} + \dfrac{7}{m}$

10. $\dfrac{5}{p} + \dfrac{11}{p}$

11. $\dfrac{a+b}{2} - \dfrac{a-b}{2}$

12. $\dfrac{x-y}{2} - \dfrac{x+y}{2}$

13. $\dfrac{x^2}{x+5} + \dfrac{5x}{x+5}$

14. $\dfrac{t^2}{t-3} + \dfrac{-3t}{t-3}$

15. $\dfrac{y^2 - 3y}{y+3} + \dfrac{-18}{y+3}$

16. $\dfrac{r^2 - 8r}{r-5} + \dfrac{15}{r-5}$

17. Explain with an example how to add or subtract rational expressions with the same denominator.

18. Explain with an example how to add or subtract rational expressions with different denominators.

Add or subtract. Write each answer in lowest terms. See Examples 2, 3, 4, and 7.

19. $\dfrac{z}{5} + \dfrac{1}{3}$

20. $\dfrac{p}{8} + \dfrac{3}{5}$

21. $\dfrac{5}{7} - \dfrac{r}{2}$

22. $\dfrac{10}{9} - \dfrac{z}{3}$

23. $-\dfrac{3}{4} - \dfrac{1}{2x}$

24. $-\dfrac{5}{8} - \dfrac{3}{2a}$

25. $\dfrac{x+1}{6} + \dfrac{3x+3}{9}$

26. $\dfrac{2x-6}{4} + \dfrac{x+5}{6}$

27. $\dfrac{x+3}{3x} + \dfrac{2x+2}{4x}$

28. $\dfrac{x+2}{5x} + \dfrac{6x+3}{3x}$

29. $\dfrac{2}{x+3} + \dfrac{1}{x}$

30. $\dfrac{3}{x-4} + \dfrac{2}{x}$

31. $\dfrac{x}{x-2} + \dfrac{4}{x+2}$

32. $\dfrac{2x}{x-1} + \dfrac{3}{x+1}$

33. $\dfrac{t}{t+2} + \dfrac{5-t}{t} - \dfrac{4}{t^2+2t}$

34. $\dfrac{2p}{p-3} + \dfrac{2+p}{p} - \dfrac{-6}{p^2-3p}$

35. What are the two possible LCDs that could be used for the sum

$$\frac{10}{m-2} + \frac{5}{2-m}?$$

36. If one form of the correct answer to a sum or difference of rational expressions is $\frac{4}{k-3}$, what would be an alternative form of the answer if the denominator is $3 - k$?

Add or subtract. Write each answer in lowest terms. See Examples 5 and 8.

37. $\dfrac{4}{x-5} + \dfrac{6}{5-x}$

38. $\dfrac{10}{m-2} + \dfrac{5}{2-m}$

39. $\dfrac{-1}{1-y} + \dfrac{3-4y}{y-1}$

40. $\dfrac{-4}{p-3} - \dfrac{p+1}{3-p}$

41. $\dfrac{2}{x-y^2} + \dfrac{7}{y^2-x}$

42. $\dfrac{-8}{p-q^2} + \dfrac{3}{q^2-p}$

43. $\dfrac{x}{5x-3y} - \dfrac{y}{3y-5x}$

44. $\dfrac{t}{8t-9s} - \dfrac{s}{9s-8t}$

45. $\dfrac{3}{4p-5} + \dfrac{9}{5-4p}$

46. $\dfrac{8}{3-7y} - \dfrac{2}{7y-3}$

In each subtraction problem, the rational expression that follows the subtraction sign has a numerator with more than one term. Be very careful with signs and find each difference. See Examples 6–10.

47. $\dfrac{2m}{m-n} - \dfrac{5m+n}{2m-2n}$

48. $\dfrac{5p}{p-q} - \dfrac{3p+1}{4p-4q}$

49. $\dfrac{5}{x^2 - 9} - \dfrac{x + 2}{x^2 + 4x + 3}$

50. $\dfrac{1}{a^2 - 1} - \dfrac{a - 1}{a^2 + 3a - 4}$

51. $\dfrac{2q + 1}{3q^2 + 10q - 8} - \dfrac{3q + 5}{2q^2 + 5q - 12}$

52. $\dfrac{4y - 1}{2y^2 + 5y - 3} - \dfrac{y + 3}{6y^2 + y - 2}$

Perform the indicated operations. See Examples 1–10.

53. $\dfrac{4}{r^2 - r} + \dfrac{6}{r^2 + 2r} - \dfrac{1}{r^2 + r - 2}$

54. $\dfrac{6}{k^2 + 3k} - \dfrac{1}{k^2 - k} + \dfrac{2}{k^2 + 2k - 3}$

55. $\dfrac{x + 3y}{x^2 + 2xy + y^2} + \dfrac{x - y}{x^2 + 4xy + 3y^2}$

56. $\dfrac{m}{m^2 - 1} + \dfrac{m - 1}{m^2 + 2m + 1}$

57. $\dfrac{r + y}{18r^2 + 9ry - 2y^2} + \dfrac{3r - y}{36r^2 - y^2}$

58. $\dfrac{2x - z}{2x^2 + xz - 10z^2} - \dfrac{x + z}{x^2 - 4z^2}$

59. Refer to the rectangle in the figure.
   (a) Find an expression that represents its perimeter. Give the simplified form.
   (b) Find an expression that represents its area. Give the simplified form.

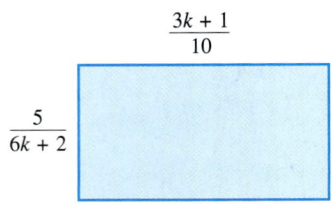

60. Refer to the triangle in the figure. Find an expression that represents its perimeter.

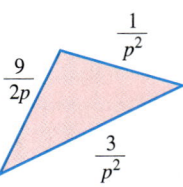

15.5 Complex Fractions

The quotient of two mixed numbers in arithmetic, such as $2\frac{1}{2} \div 3\frac{1}{4}$, can be written as a fraction:

$$2\frac{1}{2} \div 3\frac{1}{4} = \frac{2\frac{1}{2}}{3\frac{1}{4}} = \frac{2 + \frac{1}{2}}{3 + \frac{1}{4}}.$$

The last expression is the quotient of expressions that involve fractions. In algebra, some rational expressions also have fractions in the numerator, or denominator, or both.

OBJECTIVES

1. Simplify a complex fraction by writing it as a division problem (Method 1).
2. Simplify a complex fraction by multiplying numerator and denominator by the least common denominator (Method 2).

Complex Fraction

A rational expression with one or more fractions in the numerator, denominator, or both, is called a **complex fraction.**

Examples of complex fractions include

$$\frac{2 + \frac{1}{2}}{3 + \frac{1}{4}}, \quad \frac{\frac{3x^2 - 5x}{6x^2}}{2x - \frac{1}{x}}, \quad \text{and} \quad \frac{3 + x}{5 - \frac{2}{x}}. \quad \text{Complex fractions}$$

The parts of a complex fraction are named as follows.

$$\frac{\dfrac{2}{p} - \dfrac{1}{q}}{\dfrac{3}{p} + \dfrac{5}{q}}$$

$\leftarrow$ Numerator of complex fraction
$\leftarrow$ Main fraction bar
$\leftarrow$ Denominator of complex fraction

OBJECTIVE 1 **Simplify a complex fraction by writing it as a division problem (Method 1).** Since the main fraction bar represents division in a complex fraction, one method of simplifying a complex fraction involves division. This is the method we introduced in **Section 4.6.**

Method 1

To simplify a complex fraction:

Step 1 Write both the numerator and denominator as single fractions.

Step 2 Change the complex fraction to a division problem.

Step 3 Perform the indicated division.

Once again, the first example shows complex fractions from both arithmetic and algebra.

1 Simplify each complex fraction using Method 1.

(a) $\dfrac{\dfrac{2}{5}+\dfrac{1}{4}}{\dfrac{1}{2}+\dfrac{1}{3}}$

(b) $\dfrac{6+\dfrac{1}{x}}{5-\dfrac{2}{x}}$

(c) $\dfrac{9-\dfrac{4}{p}}{\dfrac{2}{p}+1}$

2 Simplify each complex fraction using Method 1.

(a) $\dfrac{\dfrac{rs^2}{t}}{\dfrac{r^2s}{t^2}}$

(b) $\dfrac{\dfrac{m^2n^3}{p}}{\dfrac{m^4n}{p^2}}$

ANSWERS

1. (a) $\dfrac{39}{50}$ (b) $\dfrac{6x+1}{5x-2}$ (c) $\dfrac{9p-4}{2+p}$

2. (a) $\dfrac{st}{r}$ (b) $\dfrac{n^2p}{m^2}$

EXAMPLE 1 Simplifying Complex Fractions (Method 1)

Simplify each complex fraction.

(a) $\dfrac{\dfrac{2}{3}+\dfrac{5}{9}}{\dfrac{1}{4}+\dfrac{1}{12}}$

(b) $\dfrac{6+\dfrac{3}{x}}{\dfrac{x}{4}+\dfrac{1}{8}}$

Step 1 First, write each numerator as a single fraction.

$$\dfrac{2}{3}+\dfrac{5}{9}=\dfrac{2(3)}{3(3)}+\dfrac{5}{9} \qquad 6+\dfrac{3}{x}=\dfrac{6}{1}+\dfrac{3}{x}$$

$$=\dfrac{6}{9}+\dfrac{5}{9}=\dfrac{11}{9} \qquad =\dfrac{6x}{x}+\dfrac{3}{x}=\dfrac{6x+3}{x}$$

Do the same thing with each denominator.

$$\dfrac{1}{4}+\dfrac{1}{12}=\dfrac{1(3)}{4(3)}+\dfrac{1}{12} \qquad \dfrac{x}{4}+\dfrac{1}{8}=\dfrac{x(2)}{4(2)}+\dfrac{1}{8}$$

$$=\dfrac{3}{12}+\dfrac{1}{12}=\dfrac{4}{12} \qquad =\dfrac{2x}{8}+\dfrac{1}{8}=\dfrac{2x+1}{8}$$

Step 2 The original complex fraction can now be written as follows.

$$\dfrac{\dfrac{11}{9}}{\dfrac{4}{12}} \qquad \dfrac{\dfrac{6x+3}{x}}{\dfrac{2x+1}{8}}$$

Step 3 Write the division in horizontal format. Use the definition of division and multiply by the reciprocal. Then write in lowest terms.

$$\dfrac{11}{9}\div\dfrac{4}{12}=\dfrac{11}{9}\cdot\dfrac{12}{4} \qquad \dfrac{6x+3}{x}\div\dfrac{2x+1}{8}=\dfrac{6x+3}{x}\cdot\dfrac{8}{2x+1}$$

$$=\dfrac{11\cdot 3\cdot 4}{3\cdot 3\cdot 4} \qquad =\dfrac{3(2x+1)}{x}\cdot\dfrac{8}{2x+1}$$

$$=\dfrac{11}{3} \qquad =\dfrac{24}{x}$$

Work Problem 1 at the Side.

EXAMPLE 2 Simplifying a Complex Fraction (Method 1)

Simplify the complex fraction.

$$\dfrac{\dfrac{xp}{q^3}}{\dfrac{p^2}{qx^2}}$$

The numerator and denominator are already single fractions, so write in horizontal format, use the definition of division, and write in lowest terms.

$$\dfrac{xp}{q^3}\div\dfrac{p^2}{qx^2}=\dfrac{xp}{q^3}\cdot\dfrac{qx^2}{p^2}=\dfrac{x^3}{q^2p}$$

Work Problem 2 at the Side.

EXAMPLE 3 Simplifying a Complex Fraction (Method 1)

Simplify the complex fraction.

$$\frac{\frac{3}{x+2} - 4}{\frac{2}{x+2} + 1} = \frac{\frac{3}{x+2} - \frac{4(x+2)}{x+2}}{\frac{2}{x+2} + \frac{1(x+2)}{x+2}}$$ Write both second terms with a denominator of $x + 2$.

$$= \frac{\frac{3 - 4(x+2)}{x+2}}{\frac{2 + 1(x+2)}{x+2}}$$ Subtract in the numerator.

Add in the denominator.

$$= \frac{\frac{3 - 4x - 8}{x+2}}{\frac{2 + x + 2}{x+2}}$$ Distributive property

$$= \frac{\frac{-5 - 4x}{x+2}}{\frac{4 + x}{x+2}}$$ Combine like terms.

$$= \frac{-5 - 4x}{x+2} \cdot \frac{x+2}{4+x}$$ Write in horizontal format; multiply by the reciprocal.

$$= \frac{-5 - 4x}{4 + x}$$ Lowest terms

CAUTION

Watch out for this common error.

$$\frac{\frac{a}{b} + \frac{c}{d}}{\frac{e}{f} + \frac{g}{h}} \neq \left(\frac{a}{b} + \frac{c}{d}\right) \cdot \left(\frac{f}{e} + \frac{h}{g}\right)$$

Work Problem 3 at the Side.

OBJECTIVE 2 **Simplify a complex fraction by multiplying numerator and denominator by the least common denominator (Method 2).** Since any expression can be multiplied by a form of 1 to get an equivalent expression, we may multiply both the numerator and the denominator of a complex fraction by the same nonzero expression to get an equivalent complex fraction. If we choose the expression to be the LCD of all the fractions within the complex fraction, the complex fraction will be simplified. This is Method 2.

---

3 Simplify by Method 1.

$$\frac{\frac{2}{x-1} + \frac{1}{x+1}}{\frac{3}{x-1} - \frac{4}{x+1}}$$

ANSWERS

3. $\dfrac{3x + 1}{-x + 7}$

Chapter 15 Rational Expressions and Applications

4 Simplify by Method 2.

(a) $\dfrac{\dfrac{2}{3} - \dfrac{1}{4}}{\dfrac{4}{9} + \dfrac{1}{2}}$

(b) $\dfrac{2 - \dfrac{6}{a}}{3 + \dfrac{4}{a}}$

(c) $\dfrac{\dfrac{p}{5-p}}{\dfrac{4p}{2p+1}}$

Method 2

To simplify a complex fraction:

Step 1 Find the LCD of all fractions within the complex fraction.

Step 2 Multiply both the numerator and the denominator of the complex fraction by this LCD using the distributive property as necessary. Write in lowest terms.

In the next example, Method 2 is used to simplify the complex fractions from Example 1.

EXAMPLE 4 Simplifying Complex Fractions (Method 2)

Simplify each complex fraction.

(a) $\dfrac{\dfrac{2}{3} + \dfrac{5}{9}}{\dfrac{1}{4} + \dfrac{1}{12}}$

(b) $\dfrac{6 + \dfrac{3}{x}}{\dfrac{x}{4} + \dfrac{1}{8}}$

Step 1 Find the LCD for all denominators in the complex fraction.

The LCD for 3, 9, 4, and 12 is 36. | The LCD for x, 4, and 8 is $8x$.

Step 2 Multiply the numerator and denominator of the complex fraction by the LCD.

$\dfrac{\dfrac{2}{3} + \dfrac{5}{9}}{\dfrac{1}{4} + \dfrac{1}{12}} = \dfrac{36\left(\dfrac{2}{3} + \dfrac{5}{9}\right)}{36\left(\dfrac{1}{4} + \dfrac{1}{12}\right)}$ $\qquad$ $\dfrac{6 + \dfrac{3}{x}}{\dfrac{x}{4} + \dfrac{1}{8}} = \dfrac{8x\left(6 + \dfrac{3}{x}\right)}{8x\left(\dfrac{x}{4} + \dfrac{1}{8}\right)}$

$= \dfrac{36\left(\dfrac{2}{3}\right) + 36\left(\dfrac{5}{9}\right)}{36\left(\dfrac{1}{4}\right) + 36\left(\dfrac{1}{12}\right)}$ $\qquad$ $= \dfrac{8x(6) + 8x\left(\dfrac{3}{x}\right)}{8x\left(\dfrac{x}{4}\right) + 8x\left(\dfrac{1}{8}\right)}$

$= \dfrac{24 + 20}{9 + 3}$ $\qquad$ $= \dfrac{48x + 24}{2x^2 + x}$

$= \dfrac{44}{12} = \dfrac{4 \cdot 11}{4 \cdot 3}$ $\qquad$ $= \dfrac{24(2x + 1)}{x(2x + 1)}$

$= \dfrac{11}{3}$ $\qquad$ $= \dfrac{24}{x}$

◀◀◀ **Work Problem 4 at the Side.**

EXAMPLE 5 Simplifying a Complex Fraction (Method 2)

Simplify the complex fraction.

$$\dfrac{\dfrac{3}{5m} - \dfrac{2}{m^2}}{\dfrac{9}{2m} + \dfrac{3}{4m^2}}$$

Continued on Next Page

ANSWERS

4. (a) $\dfrac{15}{34}$ (b) $\dfrac{2a - 6}{3a + 4}$ (c) $\dfrac{2p + 1}{4(5 - p)}$

Section 15.5 Complex Fractions

The LCD for $5m$, m^2, $2m$, and $4m^2$ is $20m^2$. Multiply the numerator and denominator by $20m^2$.

$$\frac{\dfrac{3}{5m} - \dfrac{2}{m^2}}{\dfrac{9}{2m} + \dfrac{3}{4m^2}} = \frac{20m^2\left(\dfrac{3}{5m} - \dfrac{2}{m^2}\right)}{20m^2\left(\dfrac{9}{2m} + \dfrac{3}{4m^2}\right)}$$

$$= \frac{20m^2\left(\dfrac{3}{5m}\right) - 20m^2\left(\dfrac{2}{m^2}\right)}{20m^2\left(\dfrac{9}{2m}\right) + 20m^2\left(\dfrac{3}{4m^2}\right)} \quad \text{Distributive property}$$

$$= \frac{12m - 40}{90m + 15}$$

Work Problem 5 at the Side.

⑤ Simplify by Method 2.

$$\frac{\dfrac{2}{5x} - \dfrac{3}{x^2}}{\dfrac{7}{4x} + \dfrac{1}{2x^2}}$$

Either of the two methods shown in this section can be used to simplify a complex fraction. You may want to choose one method and stick with it to eliminate confusion. However, some students prefer to use Method 1 for problems like Example 2, which is the quotient of two fractions. They prefer Method 2 for problems like Examples 1, 3, 4, and 5, which have sums or differences in the numerators or denominators or both.

EXAMPLE 6 Deciding on a Method and Simplifying Complex Fractions

Simplify.

(a) $\dfrac{\dfrac{1}{y} + \dfrac{2}{y + 2}}{\dfrac{4}{y} - \dfrac{3}{y + 2}}$

Although either method will work, we will use Method 2 since there are sums and differences in the numerator and denominator. The LCD is $y(y + 2)$. Multiply the numerator and denominator by the LCD.

$$\frac{\dfrac{1}{y} + \dfrac{2}{y + 2}}{\dfrac{4}{y} - \dfrac{3}{y + 2}} = \frac{\left(\dfrac{1}{y} + \dfrac{2}{y + 2}\right)y(y + 2)}{\left(\dfrac{4}{y} - \dfrac{3}{y + 2}\right)y(y + 2)}$$

$$= \frac{1(y + 2) + 2y}{4(y + 2) - 3y} \quad \text{Distributive property; fundamental property}$$

$$= \frac{y + 2 + 2y}{4y + 8 - 3y} \quad \text{Distributive property}$$

$$= \frac{3y + 2}{y + 8} \quad \text{Combine like terms.}$$

Continued on Next Page

ANSWERS

5. $\dfrac{8x - 60}{35x + 10}$

6 Simplify. Use either method.

(a) $\dfrac{\dfrac{1}{x} + \dfrac{2}{x-1}}{\dfrac{2}{x} - \dfrac{4}{x-1}}$

(b) $\dfrac{1 - \dfrac{2}{x} - \dfrac{15}{x^2}}{1 + \dfrac{5}{x} + \dfrac{6}{x^2}}$

(c) $\dfrac{\dfrac{2x+3}{x-4}}{\dfrac{4x^2-9}{x^2-16}}$

(b) $\dfrac{1 - \dfrac{2}{x} - \dfrac{3}{x^2}}{1 - \dfrac{5}{x} + \dfrac{6}{x^2}}$

Use Method 2.

$\dfrac{1 - \dfrac{2}{x} - \dfrac{3}{x^2}}{1 - \dfrac{5}{x} + \dfrac{6}{x^2}} = \dfrac{\left(1 - \dfrac{2}{x} - \dfrac{3}{x^2}\right)x^2}{\left(1 - \dfrac{5}{x} + \dfrac{6}{x^2}\right)x^2}$ Multiply numerator and denominator by the LCD, x^2.

$= \dfrac{x^2 - 2x - 3}{x^2 - 5x + 6}$ Distributive property

$= \dfrac{(x-3)(x+1)}{(x-3)(x-2)}$ Factor.

$= \dfrac{x+1}{x-2}$ Lowest terms

(c) $\dfrac{\dfrac{x+2}{x-3}}{\dfrac{x^2-4}{x^2-9}}$

Since this is simply a quotient of two rational expressions, we will use Method 1.

$\dfrac{\dfrac{x+2}{x-3}}{\dfrac{x^2-4}{x^2-9}} = \dfrac{x+2}{x-3} \div \dfrac{x^2-4}{x^2-9}$

$= \dfrac{x+2}{x-3} \cdot \dfrac{x^2-9}{x^2-4}$ Definition of division

$= \dfrac{x+2}{x-3} \cdot \dfrac{(x+3)(x-3)}{(x+2)(x-2)}$ Factor.

$= \dfrac{x+3}{x-2}$ Multiply.

◀◀ **Work Problem 6 at the Side.**

ANSWERS

6. (a) $\dfrac{3x+1}{-2x-2}$ (b) $\dfrac{x-5}{x+2}$ (c) $\dfrac{x+4}{2x-3}$

15.5 Exercises

FOR EXTRA HELP: Addison-Wesley Math Tutor Center, MathXL, Digital Video Tutor CD 12 / Videotape 18, Student's Solutions Manual, MyMathLab, MathXL Tutorials on CD

Note: In many problems involving complex fractions, several different equivalent forms of the answer often exist. If your answer does not look exactly like the one given in the back of the book, check to see whether you have written an equivalent form.

1. Consider the complex fraction $\dfrac{\frac{1}{2} - \frac{1}{3}}{\frac{5}{6} - \frac{1}{12}}$. Answer each part, outlining Method 1 for simplifying this complex fraction.

   (a) To combine the terms in the numerator, we must find the LCD of $\frac{1}{2}$ and $\frac{1}{3}$. What is this LCD? Determine the simplified form of the numerator of the complex fraction.

   (b) To combine the terms in the denominator, we must find the LCD of $\frac{5}{6}$ and $\frac{1}{12}$. What is this LCD? Determine the simplified form of the denominator of the complex fraction.

   (c) Now use the results from parts (a) and (b) to write the complex fraction as a division problem using the symbol $\div$.

   (d) Perform the operation from part (c) to obtain the final simplification.

2. Consider the same complex fraction given in Exercise 1, $\dfrac{\frac{1}{2} - \frac{1}{3}}{\frac{5}{6} - \frac{1}{12}}$. Answer each part, outlining Method 2 for simplifying this complex fraction.

   (a) We must determine the LCD of all the fractions within the complex fraction. What is this LCD?

   (b) Multiply every term in the complex fraction by the LCD found in part (a), but at this time do not combine the terms in the numerator and the denominator.

   (c) Now combine the terms from part (b) to obtain the simplified form of the complex fraction.

Simplify each complex fraction. Use either method. See Examples 1–6.

3. $\dfrac{-\frac{4}{3}}{\frac{2}{9}}$

4. $\dfrac{-\frac{5}{6}}{\frac{5}{4}}$

5. $\dfrac{\frac{p}{q^2}}{\frac{p^2}{q}}$

6. $\dfrac{\frac{a}{x}}{\frac{a^2}{2x}}$

7. $\dfrac{\frac{x}{y^2}}{\frac{x^2}{y}}$

8. $\dfrac{\frac{p^4}{r}}{\frac{p^2}{r^2}}$

9. $\dfrac{\frac{4a^4b^3}{3a}}{\frac{2ab^4}{b^2}}$

10. $\dfrac{\frac{2r^4t^2}{3t}}{\frac{5r^2t^5}{3r}}$

11. $\dfrac{\dfrac{m+2}{3}}{\dfrac{m-4}{m}}$

12. $\dfrac{\dfrac{q-5}{q}}{\dfrac{q+5}{3}}$

13. $\dfrac{\dfrac{2}{x}-3}{\dfrac{2-3x}{2}}$

14. $\dfrac{6+\dfrac{2}{r}}{\dfrac{3r+1}{4}}$

15. $\dfrac{\dfrac{1}{x}+x}{\dfrac{x^2+1}{8}}$

16. $\dfrac{\dfrac{3}{m}-m}{\dfrac{3-m^2}{4}}$

17. $\dfrac{a-\dfrac{5}{a}}{a+\dfrac{1}{a}}$

18. $\dfrac{q+\dfrac{1}{q}}{q+\dfrac{4}{q}}$

19. $\dfrac{\dfrac{1}{2}+\dfrac{1}{p}}{\dfrac{2}{3}+\dfrac{1}{p}}$

20. $\dfrac{\dfrac{3}{4}-\dfrac{1}{r}}{\dfrac{1}{5}+\dfrac{1}{r}}$

21. $\dfrac{\dfrac{t}{t+2}}{\dfrac{4}{t^2-4}}$

22. $\dfrac{\dfrac{m}{m+1}}{\dfrac{3}{m^2-1}}$

23. $\dfrac{\dfrac{1}{k+1}-1}{\dfrac{1}{k+1}+1}$

24. $\dfrac{\dfrac{2}{p-1}+2}{\dfrac{3}{p-1}-2}$

25. $\dfrac{2+\dfrac{1}{x}-\dfrac{28}{x^2}}{3+\dfrac{13}{x}+\dfrac{4}{x^2}}$

26. $\dfrac{4-\dfrac{11}{x}-\dfrac{3}{x^2}}{2-\dfrac{1}{x}-\dfrac{15}{x^2}}$

27. $\dfrac{\dfrac{1}{m-1}+\dfrac{2}{m+2}}{\dfrac{2}{m+2}-\dfrac{1}{m-3}}$

28. $\dfrac{\dfrac{5}{r+3}-\dfrac{1}{r-1}}{\dfrac{2}{r+2}+\dfrac{3}{r+3}}$

29. $2-\dfrac{2}{2+\dfrac{2}{2+2}}$

30. $3-\dfrac{2}{4+\dfrac{2}{4-2}}$

15.6 Solving Equations with Rational Expressions

In **Section 10.2** we solved equations with fractions as coefficients. By using the multiplication property of equality, we cleared the fractions by multiplying by the LCD. We continue this work here.

OBJECTIVES
1. Distinguish between operations with rational expressions and equations with terms that are rational expressions.
2. Solve equations with rational expressions.
3. Solve a formula for a specified variable.

OBJECTIVE 1 Distinguish between operations with rational expressions and equations with terms that are rational expressions. Before solving equations with rational expressions, you must understand the difference between sums and differences of terms with rational coefficients, or rational *expressions,* and *equations* with terms that are rational expressions. Sums and differences lead to *expressions,* while *equations* are solved.

EXAMPLE 1 Distinguishing between Expressions and Equations

Identify each of the following as an *expression* or an *equation.* Then simplify the expression or solve the equation.

(a) $\dfrac{3}{4}x - \dfrac{2}{3}x$

This is a difference of two terms. It represents an *expression* since there is no equal sign. Find the LCD, write each coefficient with this LCD, and combine like terms.

$$\dfrac{3}{4}x - \dfrac{2}{3}x = \dfrac{9}{12}x - \dfrac{8}{12}x \qquad \text{Find a common denominator.}$$

$$= \dfrac{1}{12}x \qquad \text{Combine like terms.}$$

(b) $\dfrac{3}{4}x - \dfrac{2}{3}x = \dfrac{1}{2}$

Because of the equal sign, this is an *equation* to be solved. We proceed as in **Section 10.2,** using the multiplication property of equality to clear fractions. The LCD is 12.

$$\dfrac{3}{4}x - \dfrac{2}{3}x = \dfrac{1}{2}$$

$$12\left(\dfrac{3}{4}x - \dfrac{2}{3}x\right) = 12\left(\dfrac{1}{2}\right) \qquad \text{Multiply both sides by 12, the LCD.}$$

$$12\left(\dfrac{3}{4}x\right) - 12\left(\dfrac{2}{3}x\right) = 12\left(\dfrac{1}{2}\right) \qquad \text{Distributive property}$$

$$9x - 8x = 6 \qquad \text{Multiply.}$$

$$x = 6 \qquad \text{Combine like terms.}$$

Continued on Next Page

1 Identify each as an *expression* or an *equation*. Then perform the operation to simplify the expression or solve the equation.

(a) $\dfrac{x}{3} + \dfrac{x}{5} = 7 + x$

(b) $\dfrac{2x}{3} - \dfrac{4x}{9}$

2 Solve each equation, and check your solutions.

(a) $\dfrac{x}{5} + 3 = \dfrac{3}{5}$

(b) $\dfrac{x}{2} - \dfrac{x}{3} = \dfrac{5}{6}$

ANSWERS

1. (a) equation; -15 (b) expression; $\dfrac{2x}{9}$

2. (a) -12 (b) 5

Check: $\dfrac{3}{4}x - \dfrac{2}{3}x = \dfrac{1}{2}$ Original equation

$\dfrac{3}{4}(6) - \dfrac{2}{3}(6) = \dfrac{1}{2}$?   Let $x = 6$.

$\dfrac{9}{2} - 4 = \dfrac{1}{2}$?   Multiply.

$\dfrac{1}{2} = \dfrac{1}{2}$ True

The check shows that 6 is the solution of the equation.

The ideas of Example 1 can be summarized as follows.

> When adding or subtracting rational expressions, the LCD must be kept throughout the simplification. When solving an equation, the LCD is used to multiply both sides so that denominators are eliminated.

◀◀ Work Problem 1 at the Side.

OBJECTIVE 2 Solve equations with rational expressions. When an equation involves fractions as in Example 1(b), we use the multiplication property of equality to clear it of fractions. Choose as multiplier the LCD of all denominators in the fractions of the equation.

EXAMPLE 2 Solving an Equation with Rational Expressions

Solve $\dfrac{x}{3} + \dfrac{x}{4} = 10 + x$. Check the solution.

We begin by multiplying both sides of the equation by 12, the LCD.

$12\left(\dfrac{x}{3} + \dfrac{x}{4}\right) = 12(10 + x)$

$12\left(\dfrac{x}{3}\right) + 12\left(\dfrac{x}{4}\right) = 12(10) + 12x$ Distributive property

$4x + 3x = 120 + 12x$

$7x = 120 + 12x$ Combine like terms.

$-5x = 120$ Subtract $12x$.

$x = -24$ Divide by -5.

Check: $\dfrac{x}{3} + \dfrac{x}{4} = 10 + x$ Original equation

$\dfrac{-24}{3} + \dfrac{-24}{4} = 10 - 24$?   Let $x = -24$.

$-8 - 6 = -14$?

$-14 = -14$ True

The solution is -24.

◀◀ Work Problem 2 at the Side.

CAUTION
Note that the use of the LCD here is different from its use in the previous section. Here, we use the multiplication property of equality to multiply each side of an *equation* by the LCD. Earlier, we used the fundamental property to multiply a *fraction* by another fraction that had the LCD as both its numerator and denominator. Be careful not to confuse these two methods.

3 Solve each equation, and check your solutions.

(a) $\dfrac{k}{6} - \dfrac{k+1}{4} = -\dfrac{1}{2}$

EXAMPLE 3 Solving an Equation with Rational Expressions

Solve $\dfrac{p}{2} - \dfrac{p-1}{3} = 1$.

$$6\left(\dfrac{p}{2} - \dfrac{p-1}{3}\right) = 6 \cdot 1 \quad \text{Multiply by the LCD, 6.}$$

$$6\left(\dfrac{p}{2}\right) - 6\left(\dfrac{p-1}{3}\right) = 6 \quad \text{Distributive property}$$

$$3p - 2(p-1) = 6$$

Be very careful to put parentheses around $p - 1$; otherwise, you may find an incorrect solution. Continue simplifying and solve.

$$3p - 2p + 2 = 6 \quad \text{Distributive property}$$
$$p + 2 = 6 \quad \text{Combine like terms.}$$
$$p = 4 \quad \text{Subtract 2.}$$

Check to see that 4 is correct by replacing p with 4 in the original equation.

Work Problem 3 at the Side.

(b) $\dfrac{2m-3}{5} - \dfrac{m}{3} = -\dfrac{6}{5}$

When solving an equation that has a variable in the denominator, remember that the number 0 cannot be used as a denominator. Therefore, *the solution cannot be a number that will make the denominator equal 0.*

EXAMPLE 4 Solving an Equation with Rational Expressions

Solve $\dfrac{x}{x-2} = \dfrac{2}{x-2} + 2$. Check the proposed solution.

The common denominator is $x - 2$. (*Note:* Because $x = 2$ makes a denominator in the given equation equal 0, x cannot equal 2.) Solve the equation by multiplying each side of the equation by $x - 2$.

$$(x-2)\left(\dfrac{x}{x-2}\right) = (x-2)\left(\dfrac{2}{x-2} + 2\right)$$

$$(x-2)\left(\dfrac{x}{x-2}\right) = (x-2)\left(\dfrac{2}{x-2}\right) + (x-2)(2)$$

$$x = 2 + 2x - 4$$
$$x = -2 + 2x \quad \text{Combine like terms.}$$
$$-x = -2 \quad \text{Subtract } 2x.$$
$$x = 2 \quad \text{Divide by } -1.$$

Continued on Next Page

ANSWERS
3. (a) 3 (b) -9

4 Solve the equation, and check your solution.

$$1 - \frac{2}{x+1} = \frac{2x}{x+1}$$

Check: The proposed solution is 2. To check this solution, we substitute 2 in the original equation.

$$\frac{x}{x-2} = \frac{2}{x-2} + 2 \quad \text{Original equation}$$

$$\frac{2}{2-2} = \frac{2}{2-2} + 2 \quad ?$$

$$\frac{2}{0} = \frac{2}{0} + 2 \quad ?$$

Notice that 2 makes both denominators equal 0. Because 0 cannot be the denominator, there is no solution.

While it is always a good idea to check solutions to guard against arithmetic and algebraic errors, *it is essential to check proposed solutions when variables appear in denominators in the original equation.* Some students like to determine which numbers cannot be solutions *before* solving the equation.

◀◀ **Work Problem 4 at the Side.**

The steps used to solve an equation with rational expressions follow.

Solving An Equation with Rational Expressions

Step 1 **Multiply** both sides of the equation by the LCD. (This clears the equation of fractions.)

Step 2 **Solve** the resulting equation.

Step 3 **Check** each proposed solution by substituting it in the original equation. Reject any that cause a denominator to equal 0.

EXAMPLE 5 Solving an Equation with Rational Expressions

Solve $\dfrac{2}{x^2 - x} = \dfrac{1}{x^2 - 1}$. Check the proposed solution.

Step 1 Begin by finding the LCD.

$$\frac{2}{x(x-1)} = \frac{1}{(x+1)(x-1)} \quad \text{Factor the denominators to find the LCD.}$$

Since $x^2 - x$ can be factored as $x(x-1)$, and $x^2 - 1$ can be factored as $(x+1)(x-1)$, the LCD is $x(x+1)(x-1)$.

Step 2 Notice that 0, -1, and 1 cannot be solutions of this equation. Multiply each side of the equation by $x(x+1)(x-1)$.

$$x(x+1)(x-1)\frac{2}{x(x-1)} = x(x+1)(x-1)\frac{1}{(x+1)(x-1)}$$

$$2(x+1) = x$$

$$2x + 2 = x \quad \text{Distributive property}$$

$$2 = -x \quad \text{Subtract } 2x.$$

$$x = -2 \quad \text{Multiply by } -1; \text{rewrite.}$$

Continued on Next Page

ANSWERS
4. no solution (When the equation is solved, -1 is found. However, because $x = -1$ leads to a 0 denominator in the original equation, there is no solution.)

Step 3 The proposed solution is -2, which does not make any denominator equal 0.

Check:
$$\frac{2}{x^2 - x} = \frac{1}{x^2 - 1} \qquad \text{Original equation}$$

$$\frac{2}{(-2)^2 - (-2)} = \frac{1}{(-2)^2 - 1} \quad ? \qquad \text{Let } x = -2.$$

$$\frac{2}{4 + 2} = \frac{1}{4 - 1} \quad ?$$

$$\frac{1}{3} = \frac{1}{3} \qquad \text{True}$$

The solution is indeed -2.

Work Problem 5 at the Side.

⑤ Solve each equation, and check your solutions.

(a) $\dfrac{4}{x^2 - 3x} = \dfrac{1}{x^2 - 9}$

(b) $\dfrac{2}{p^2 - 2p} = \dfrac{3}{p^2 - p}$

EXAMPLE 6 Solving an Equation with Rational Expressions

Solve $\dfrac{2m}{m^2 - 4} + \dfrac{1}{m - 2} = \dfrac{2}{m + 2}$.

Factor the first denominator on the left.

$$\frac{2m}{(m + 2)(m - 2)} + \frac{1}{m - 2} = \frac{2}{m + 2}$$

Multiply by the LCD, $(m + 2)(m - 2)$. (Notice that -2 and 2 cannot be solutions.)

$$(m + 2)(m - 2)\left(\frac{2m}{(m + 2)(m - 2)} + \frac{1}{m - 2}\right)$$
$$= (m + 2)(m - 2)\frac{2}{m + 2}$$

$$(m + 2)(m - 2)\frac{2m}{(m + 2)(m - 2)} + (m + 2)(m - 2)\frac{1}{m - 2}$$
$$= (m + 2)(m - 2)\frac{2}{m + 2}$$

$$2m + m + 2 = 2(m - 2)$$
$$3m + 2 = 2m - 4 \qquad \text{Combine like terms; distributive property}$$
$$m + 2 = -4 \qquad \text{Subtract } 2m.$$
$$m = -6 \qquad \text{Subtract 2.}$$

Check to see that -6 is a solution of the given equation.

Work Problem 6 at the Side.

⑥ Solve each equation, and check your solutions.

(a) $\dfrac{2p}{p^2 - 1} = \dfrac{2}{p + 1} - \dfrac{1}{p - 1}$

(b) $\dfrac{8r}{4r^2 - 1} = \dfrac{3}{2r + 1} + \dfrac{3}{2r - 1}$

EXAMPLE 7 Solving an Equation with Rational Expressions

Solve $\dfrac{1}{x - 1} + \dfrac{1}{2} = \dfrac{2}{x^2 - 1}$.

The denominator $x^2 - 1$ factors as $(x + 1)(x - 1)$. Multiply each side of the equation by the LCD, $2(x + 1)(x - 1)$. (Notice that -1 and 1 cannot be solutions.)

Continued on Next Page

ANSWERS
5. (a) -4 (b) 4
6. (a) -3 (b) 0

7 Solve the equation, and check your solution.

$$\frac{2}{3x+1} - \frac{1}{x} = \frac{-6x}{3x+1}$$

8 Solve each equation, and check your solutions.

(a) $\dfrac{1}{x-2} + \dfrac{1}{5} = \dfrac{2}{5(x^2-4)}$

(b) $\dfrac{6}{5a+10} - \dfrac{1}{a-5} = \dfrac{4}{a^2-3a-10}$

$$2(x+1)(x-1)\left(\frac{1}{x-1} + \frac{1}{2}\right) = 2(x+1)(x-1)\frac{2}{(x+1)(x-1)}$$

$$2(x+1)(x-1)\frac{1}{x-1} + 2(x+1)(x-1)\frac{1}{2}$$

$$= 2(x+1)(x-1)\frac{2}{(x+1)(x-1)}$$

$$2(x+1) + (x+1)(x-1) = 4$$

$2x + 2 + x^2 - 1 = 4$ Distributive property
$x^2 + 2x + 1 = 4$ Combine like terms.
$x^2 + 2x - 3 = 0$ Subtract 4.
$(x+3)(x-1) = 0$ Factor.

Solving this equation suggests that $x = -3$ or $x = 1$. But 1 makes a denominator of the original equation equal 0, so 1 is not a solution. However, -3 is a solution, as shown by substituting -3 for x in the original equation.

Check: $\dfrac{1}{x-1} + \dfrac{1}{2} = \dfrac{2}{x^2-1}$ Original equation

$\dfrac{1}{-3-1} + \dfrac{1}{2} = \dfrac{2}{(-3)^2-1}$?   Let $x = -3$.

$\dfrac{1}{-4} + \dfrac{1}{2} = \dfrac{2}{9-1}$?   Simplify.

$\dfrac{1}{4} = \dfrac{1}{4}$ True

The check shows that -3 is a solution.

▶▶ **Work Problem 7 at the Side.**

EXAMPLE 8 Solving an Equation with Rational Expressions

Solve $\dfrac{1}{k^2+4k+3} + \dfrac{1}{2k+2} = \dfrac{3}{4k+12}$.

Factor the three denominators to get the common denominator, $4(k+1)(k+3)$. (Notice that -1 and -3 cannot be solutions.)

$$4(k+1)(k+3)\left(\frac{1}{(k+1)(k+3)} + \frac{1}{2(k+1)}\right)$$

$$= 4(k+1)(k+3)\frac{3}{4(k+3)}$$ Multiply by the LCD.

$$4(k+1)(k+3)\frac{1}{(k+1)(k+3)} + 2 \cdot 2(k+1)(k+3)\frac{1}{2(k+1)}$$

$$= 4(k+1)(k+3)\frac{3}{4(k+3)}$$

$4 + 2(k+3) = 3(k+1)$ Simplify.
$4 + 2k + 6 = 3k + 3$ Distributive property
$2k + 10 = 3k + 3$ Combine like terms.
$7 = k$ Subtract $2k$ and 3.

Check to see that 7 is a solution of the given equation.

▶▶ **Work Problem 8 at the Side.**

ANSWERS

7. $\dfrac{1}{2}$

8. (a) $-4, -1$ (b) 60

OBJECTIVE 3 **Solve a formula for a specified variable.** Solving a formula for a specified variable was first discussed in **Section 10.3**. *Remember to treat the variable for which you are solving as if it were the only variable, and all others as if they were constants.*

EXAMPLE 9 Solving for a Specified Variable

Solve $a = \dfrac{v - w}{t}$ for v.

Our goal is to get v alone on one side of the equation.

$$a = \frac{v - w}{t} \qquad \text{Given equation}$$
$$at = v - w \qquad \text{Multiply both sides by } t.$$
$$at + w = v \qquad \text{Add } w \text{ to both sides.}$$
$$v = at + w \qquad \text{Rewrite equation.}$$

To check this, substitute $at + w$ for v in the original equation. The final result will be the identity $a = a$, indicating that the result obtained is correct.

Work Problem 9 at the Side.

EXAMPLE 10 Solving for a Specified Variable

Solve the formula $\dfrac{1}{a} = \dfrac{1}{b} + \dfrac{1}{c}$ for c.

The LCD of all the fractions in the equation is abc, so multiply both sides by abc.

$$\frac{1}{a} = \frac{1}{b} + \frac{1}{c}$$
$$abc\left(\frac{1}{a}\right) = abc\left(\frac{1}{b} + \frac{1}{c}\right) \qquad \text{Multiply by the LCD, } abc.$$
$$abc\left(\frac{1}{a}\right) = abc\left(\frac{1}{b}\right) + abc\left(\frac{1}{c}\right) \qquad \text{Distributive property}$$
$$bc = ac + ab$$

Since we are solving for c, transform so that all terms with c are on one side of the equation. Do this by subtracting ac from both sides.

$$bc - ac = ab \qquad \text{Subtract } ac.$$

Factor out the common factor c on the left side.

$$c(b - a) = ab \qquad \text{Factor out } c.$$

Finally, divide both sides by the coefficient of c, which is $b - a$.

$$c = \frac{ab}{b - a}$$

---

⑨ Solve $z = \dfrac{x}{x + y}$ for y.

ANSWERS

9. $y = \dfrac{x - zx}{z}$

10 Solve $\dfrac{2}{x} = \dfrac{1}{y} + \dfrac{1}{z}$ for z.

> **CAUTION**
> Students often have trouble in the step that involves factoring out the variable for which they are solving. In Example 10, we had to factor out c on the left side so that we could divide both sides by $b - a$.
> $$bc - ac = ab$$
> $$c(b - a) = ab$$
> $$c = \dfrac{ab}{b - a}$$
>
> When solving an equation for a specified variable, *be sure that the specified variable appears alone on only one side of the equal sign in the final equation.*

◀◀◀ Work Problem 10 at the Side.

ANSWERS

10. $z = \dfrac{xy}{2y - x}$ or $z = \dfrac{-xy}{x - 2y}$

15.6 Exercises

Identify each as an expression *or an* equation. *Then simplify the expression or solve the equation. See Example 1.*

1. $\dfrac{7}{8}x + \dfrac{1}{5}x$

2. $\dfrac{4}{7}x + \dfrac{3}{5}x$

3. $\dfrac{7}{8}x + \dfrac{1}{5}x = 1$

4. $\dfrac{4}{7}x + \dfrac{3}{5}x = 1$

5. $\dfrac{3}{5}y - \dfrac{7}{10}y$

6. $\dfrac{3}{5}y - \dfrac{7}{10}y = 1$

7. Explain how the LCD is used in a different way when adding and subtracting rational expressions compared to solving equations with rational expressions.

8. If we multiply both sides of the equation $\dfrac{6}{x+5} = \dfrac{6}{x+5}$ by $x + 5$, we get $6 = 6$. Are all real numbers solutions of this equation? Explain.

Solve each equation, and check your solutions. See Examples 2 and 3.

9. $\dfrac{2}{3}x + \dfrac{1}{2}x = -7$

10. $\dfrac{1}{4}x - \dfrac{1}{3}x = 1$

11. $\dfrac{p}{3} - \dfrac{p}{6} = 4$

12. $\dfrac{x}{15} + \dfrac{x}{5} = 4$

13. $\dfrac{3x}{5} - 6 = x$

14. $\dfrac{5t}{4} + t = 9$

15. $\dfrac{4m}{7} + m = 11$

16. $a - \dfrac{3a}{2} = 1$

17. $\dfrac{z-1}{4} = \dfrac{z+3}{3}$

18. $\dfrac{r-5}{2} = \dfrac{r+2}{3}$

19. $\dfrac{3p+6}{8} = \dfrac{3p-3}{16}$

20. $\dfrac{2z+1}{5} = \dfrac{7z+5}{15}$

21. $\dfrac{2x+3}{-6} = \dfrac{3}{2}$

22. $\dfrac{4y+3}{6} = \dfrac{5}{2}$

23. $\dfrac{q+2}{3} + \dfrac{q-5}{5} = \dfrac{7}{3}$

24. $\dfrac{b+7}{8} - \dfrac{b-2}{3} = \dfrac{4}{3}$

25. $\dfrac{t}{6} + \dfrac{4}{3} = \dfrac{t-2}{3}$

26. $\dfrac{x}{2} = \dfrac{5}{4} + \dfrac{x-1}{4}$

27. $\dfrac{3m}{5} - \dfrac{3m-2}{4} = \dfrac{1}{5}$

28. $\dfrac{8p}{5} = \dfrac{3p-4}{2} + \dfrac{5}{2}$

29. What values of x would have to be rejected as possible solutions of the equation $\dfrac{1}{x-4} = \dfrac{3}{2x}$?

30. What is wrong with this problem: Solve $\dfrac{2}{3x} + \dfrac{1}{5x}$.

Solve each equation, and check your solutions. See Examples 4–8.

31. $\dfrac{2x+3}{x} = \dfrac{3}{2}$

32. $\dfrac{5-2y}{y} = \dfrac{1}{4}$

33. $\dfrac{k}{k-4} - 5 = \dfrac{4}{k-4}$

34. $\dfrac{-5}{a+5} = \dfrac{a}{a+5} + 2$

35. $\dfrac{3}{x-1} + \dfrac{2}{4x-4} = \dfrac{7}{4}$

36. $\dfrac{2}{p+3} + \dfrac{3}{8} = \dfrac{5}{4p+12}$

37. $\dfrac{y}{3y+3} = \dfrac{2y-3}{y+1} - \dfrac{2y}{3y+3}$

38. $\dfrac{2k+3}{k+1} - \dfrac{3k}{2k+2} = \dfrac{-2k}{2k+2}$

39. $\dfrac{2}{m} = \dfrac{m}{5m+12}$

40. $\dfrac{x}{4-x} = \dfrac{2}{x}$

41. $\dfrac{-2}{z+5} + \dfrac{3}{z-5} = \dfrac{20}{z^2-25}$

42. $\dfrac{3}{r+3} - \dfrac{2}{r-3} = \dfrac{-12}{r^2-9}$

43. $\dfrac{3y}{y^2+5y+6} = \dfrac{5y}{y^2+2y-3} - \dfrac{2}{y^2+y-2}$

44. $\dfrac{x+4}{x^2-3x+2} - \dfrac{5}{x^2-4x+3} = \dfrac{x-4}{x^2-5x+6}$

45. $\dfrac{5x}{14x+3} = \dfrac{1}{x}$

46. $\dfrac{m}{8m+3} = \dfrac{1}{3m}$

47. $\dfrac{2}{z-1} - \dfrac{5}{4} = \dfrac{-1}{z+1}$

48. $\dfrac{5}{p-2} = 7 - \dfrac{10}{p+2}$

49. If you are solving a formula for the letter k, and your steps lead to the equation $kr - mr = km$, what would be your next step?

50. If you are solving a formula for the letter k, and your steps lead to the equation $kr - km = mr$, what would be your next step?

Solve each formula for the specified variable. See Example 9.

51. $m = \dfrac{kF}{a}$ for F

52. $I = \dfrac{kE}{R}$ for E

53. $m = \dfrac{kF}{a}$ for a

54. $I = \dfrac{kE}{R}$ for R

55. $I = \dfrac{E}{R + r}$ for R

56. $I = \dfrac{E}{R + r}$ for r

57. $h = \dfrac{2A}{B + b}$ for A

58. $d = \dfrac{2S}{n(a + L)}$ for S

59. $d = \dfrac{2S}{n(a + L)}$ for a

60. $h = \dfrac{2A}{B + b}$ for B

Solve each equation for the specified variable. See Example 10.

61. $\dfrac{2}{r} + \dfrac{3}{s} + \dfrac{1}{t} = 1$ for t

62. $\dfrac{5}{p} + \dfrac{2}{q} + \dfrac{3}{r} = 1$ for r

63. $\dfrac{1}{a} - \dfrac{1}{b} - \dfrac{1}{c} = 2$ for c

64. $\dfrac{-1}{x} + \dfrac{1}{y} + \dfrac{1}{z} = 4$ for y

65. $9x + \dfrac{3}{z} = \dfrac{5}{y}$ for z

66. $-3t - \dfrac{4}{p} = \dfrac{6}{s}$ for p

Summary Exercises on Rational Expressions and Equations

We have performed the four operations of arithmetic with rational expressions and solved equations with rational expressions. The exercises in this summary include a mixed variety of problems of these types.

Students often confuse *simplifying* rational expressions with the *solution of equations* with rational expressions. For example, the four possible operations to simplify the rational expressions $\frac{1}{x}$ and $\frac{1}{x-2}$ are performed as follows.

Simplify by adding expressions:

$$\frac{1}{x} + \frac{1}{x-2} = \frac{1(x-2)}{x(x-2)} + \frac{x(1)}{x(x-2)} \quad \text{Write with a common denominator.}$$

$$= \frac{x-2+x}{x(x-2)} \quad \text{Add numerators; keep the same denominator.}$$

$$= \frac{2x-2}{x(x-2)} \quad \text{Combine like terms.}$$

Simplify by subtracting expressions:

$$\frac{1}{x} - \frac{1}{x-2} = \frac{1(x-2)}{x(x-2)} - \frac{x(1)}{x(x-2)} \quad \text{Write with a common denominator.}$$

$$= \frac{x-2-x}{x(x-2)} \quad \text{Subtract numerators; keep the same denominator.}$$

$$= \frac{-2}{x(x-2)} \quad \text{Combine like terms.}$$

Simplify by multiplying expressions:

$$\frac{1}{x} \cdot \frac{1}{x-2} = \frac{1}{x(x-2)} \quad \text{Multiply numerators and multiply denominators.}$$

Simplify by dividing expressions:

$$\frac{1}{x} \div \frac{1}{x-2} = \frac{1}{x} \cdot \frac{x-2}{1} = \frac{x-2}{x} \quad \text{Multiply by the reciprocal of the divisor.}$$

On the other hand, *solve* this *equation*. An equation has an $=$ sign.

$$\frac{1}{x} + \frac{1}{x-2} = \frac{3}{4}$$

Neither 0 nor 2 can be a solution of this equation, since each will cause a denominator to equal 0. We use the multiplication property of equality to multiply both sides by the LCD, $4x(x-2)$, leading to an equation with no denominators.

$$4x(x-2)\frac{1}{x} + 4x(x-2)\frac{1}{x-2} = 4x(x-2)\frac{3}{4} \quad \text{Multiply both sides by the LCD, } 4x(x-2).$$

$$4(x-2) + 4x = 3x(x-2)$$

$$4x - 8 + 4x = 3x^2 - 6x \quad \text{Distributive property}$$

$$0 = 3x^2 - 14x + 8 \quad \text{Standard form}$$

$$0 = (3x-2)(x-4) \quad \text{Factor.}$$

$$x - 4 = 0 \quad \text{or} \quad 3x - 2 = 0 \quad \text{Zero-factor property}$$

$$x = 4 \quad \text{or} \quad x = \frac{2}{3}$$

Both $\frac{2}{3}$ and 4 are solutions since neither makes a denominator equal 0.

In conclusion, remember the following points when working exercises involving rational expressions.

Points to Remember When Working with Rational Expressions

1. The fundamental property is applied only after numerators and denominators have been *factored*.
2. When adding and subtracting rational expressions, the common denominator must be kept throughout the problem and in the final result.
3. Always look to see if the answer is in lowest terms; if it is not, use the fundamental property.
4. When solving equations, the LCD is used to clear the equation of fractions. Multiply each side by the LCD. (Notice how this differs from the use of the LCD in Point 2.)
5. When solving equations with rational expressions, reject any proposed solution that causes an original denominator to equal 0.

For each exercise, indicate "expression" if an expression is to be simplified or "equation" if an equation is to be solved. Then simplify the expression or solve the equation.

1. $\dfrac{4}{p} + \dfrac{6}{p}$

2. $\dfrac{x^3 y^2}{x^2 y^4} \cdot \dfrac{y^5}{x^4}$

3. $\dfrac{1}{x^2 + x - 2} \div \dfrac{4x^2}{2x - 2}$

4. $\dfrac{8}{m - 5} = 2$

5. $\dfrac{2y^2 + y - 6}{2y^2 - 9y + 9} \cdot \dfrac{y^2 - 2y - 3}{y^2 - 1}$

6. $\dfrac{2}{k^2 - 4k} + \dfrac{3}{k^2 - 16}$

7. $\dfrac{x - 4}{5} = \dfrac{x + 3}{6}$

8. $\dfrac{3t^2 - t}{6t^2 + 15t} \div \dfrac{6t^2 + t - 1}{2t^2 - 5t - 25}$

9. $\dfrac{4}{p + 2} + \dfrac{1}{3p + 6}$

10. $\dfrac{1}{y} + \dfrac{1}{y - 3} = -\dfrac{5}{4}$

11. $\dfrac{3}{t - 1} + \dfrac{1}{t} = \dfrac{7}{2}$

12. $\dfrac{6}{y} - \dfrac{2}{3y}$

13. $\dfrac{5}{4z} - \dfrac{2}{3z}$

14. $\dfrac{k + 2}{3} = \dfrac{2k - 1}{5}$

15. $\dfrac{1}{m^2 + 5m + 6} + \dfrac{2}{m^2 + 4m + 3}$

16. $\dfrac{2k^2 - 3k}{20k^2 - 5k} \div \dfrac{2k^2 - 5k + 3}{4k^2 + 11k - 3}$

17. $\dfrac{2}{x + 1} + \dfrac{5}{x - 1} = \dfrac{10}{x^2 - 1}$

18. $\dfrac{x}{x - 2} + \dfrac{3}{x + 2} = \dfrac{8}{x^2 - 4}$

15.7 Applications of Rational Expressions

In **Section 15.6** we solved equations with rational expressions; now we can solve applications that involve this type of equation. The six-step problem-solving method of **Section 3.3** still applies.

OBJECTIVES

1. Solve problems about numbers.
2. Solve problems about distance, rate, and time.
3. Solve problems about work.

OBJECTIVE 1 Solve problems about numbers. We begin with an example about an unknown number.

EXAMPLE 1 Solving a Problem about an Unknown Number

If the same number is added to both the numerator and the denominator of the fraction $\frac{2}{5}$, the result is equivalent to $\frac{2}{3}$. Find the number.

Step 1 **Read** the problem carefully. We are trying to find a number.

Step 2 **Assign a variable.** Here, let x = the number added to the numerator and the denominator.

Step 3 **Write an equation.** The fraction

$$\frac{2+x}{5+x}$$

represents the result of adding the same number to both the numerator and the denominator. Since this result is equivalent to $\frac{2}{3}$, the equation is

$$\frac{2+x}{5+x} = \frac{2}{3}.$$

Step 4 **Solve** this equation by multiplying each side by the LCD, $3(5+x)$.

$$3(5+x)\frac{2+x}{5+x} = 3(5+x)\frac{2}{3}$$

$$3(2+x) = 2(5+x)$$

$$6 + 3x = 10 + 2x \quad \text{Distributive property}$$

$$x = 4 \quad \text{Subtract } 2x; \text{ subtract } 6.$$

Step 5 **State the answer.** The number is 4.

Step 6 **Check** the solution in the words of the original problem. If 4 is added to both the numerator and the denominator of $\frac{2}{5}$, the result is $\frac{6}{9}$ which is equivalent to $\frac{2}{3}$, as required.

Work Problem 1 at the Side.

1 Solve each problem.

(a) A certain number is added to the numerator and subtracted from the denominator of $\frac{5}{8}$. The new fraction equals the reciprocal of $\frac{5}{8}$. Find the number.

(b) The denominator of a fraction is 1 more than the numerator. If 6 is added to the numerator and subtracted from the denominator, the result is $\frac{15}{4}$. Find the original fraction.

OBJECTIVE 2 Solve problems about distance, rate, and time. If an automobile travels at an average rate of 65 mph for 2 hr, then it travels $65 \times 2 = 130$ mi. Recall from **Section 12.4** that this is an example of the basic relationship between distance, rate, and time given by the formula $d = rt$. By solving, in turn, for r and t in the formula, we obtain two other equivalent forms of the formula. The three forms are given below.

Distance, Rate, and Time Relationship

$$d = rt \qquad r = \frac{d}{t} \qquad t = \frac{d}{r}$$

ANSWERS

1. (a) 3   (b) $\dfrac{9}{10}$

2 Solve each problem.

(a) The world record in the men's 100-meter dash was set in 1999 by Maurice Green, who ran it in 9.79 seconds. What was his speed in meters per second?

(b) The world record for the women's 3000-meter run was set by Junxia Wang in 1993. Her speed was 6.173 meters per second. What was her time in seconds?

(c) A small plane flew from Chicago to St. Louis averaging 145 mph. The trip took 2 hr. What is the distance between Chicago and St. Louis?

The next example illustrates the uses of these formulas.

EXAMPLE 2 Finding Distance, Rate, or Time

(a) The speed of sound is 1088 feet per second at sea level at 32°F. In 5 seconds under these conditions, how far will sound travel?

$$1088 \times 5 = 5440 \text{ ft}$$
$$\text{rate} \times \text{time} = \text{distance}$$

Here, we found distance given rate and time. We used the formula $d = rt$ rewritten as $rt = d$.

(b) The winner of the first Indianapolis 500 race (in 1911) was Ray Harroun, driving a Marmon Wasp at an average speed of 74.602 mph. (*Source: World Almanac and Book of Facts.*) How long did it take him to complete the 500 miles?

$$\frac{\text{distance} \rightarrow 500}{\text{rate} \rightarrow 74.602} = 6.70 \text{ hr (rounded)} \leftarrow \text{time}$$

Here, we found time given rate and distance using $t = \frac{d}{r}$. To convert 0.70 hr to minutes, multiply by 60 to get $0.70(60) = 42$. It took Harroun about 6 hr, 42 min to complete the race.

(c) At the 2004 Olympic Games in Athens, Greece, Dutch swimmer Inge de Bruijn won the women's 50-meter freestyle swimming event in 24.58 seconds. (*Source:* www.olympics.com) What was her swimming speed (rate)?

$$\text{rate} = \frac{\text{distance} \rightarrow 50}{\text{time} \rightarrow 24.58} = 2.03 \text{ meters per second (rounded)}$$

Work Problem 2 at the Side.

PROBLEM-SOLVING HINT
Many applied problems use the formulas just discussed. The next two examples show how to solve typical applications of the formula $d = rt$. A helpful strategy for solving such problems is to **first make a sketch** showing what is happening in the problem. **Then make a table** using the information given, along with the unknown quantities. The table will help you organize the information, and the sketch will help you set up the equation.

EXAMPLE 3 Solving a Motion Problem about Distance, Rate, and Time

Two cars leave Baton Rouge, Louisiana, at the same time and travel east on Interstate 10. One travels at a constant speed of 55 mph and the other travels at a constant speed of 63 mph. In how many hours will the distance between them be 24 miles?

Continued on Next Page

ANSWERS
2. (a) 10.21 meters per second
(b) 486 seconds (c) 290 miles

Step 1 **Read** the problem. We are trying to find the time when the distance between the cars will be 24 miles.

Step 2 **Assign a variable.** Since we are looking for time, let t = the number of hours until the distance between them is 24 miles. The sketch in Figure 1 shows what is happening in the problem.

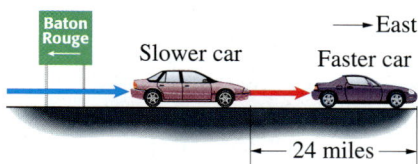

Figure 1

Now, construct a table like the one that follows. Fill in the information given in the problem, and use t for the time traveled by each car. Multiply rate by time to get the expressions for distances traveled.

	Rate	× Time	= Distance
Faster Car	63	t	$63t$
Slower Car	55	t	$55t$

⬅ Difference is 24 miles

The quantities $63t$ and $55t$ represent the two distances. Refer to Figure 1, and notice that the *difference* between the larger distance and the smaller distance is 24 miles.

Step 3 **Write an equation.**

$$63t - 55t = 24$$

Step 4 **Solve.**

$$63t - 55t = 24$$
$$8t = 24 \quad \text{Combine like terms.}$$
$$t = 3 \quad \text{Divide by 8.}$$

Step 5 **State the answer.** It will take the cars 3 hr to be 24 miles apart.

Step 6 **Check.** After 3 hr the faster car will have traveled $63 \times 3 = 189$ miles, and the slower car will have traveled $55 \times 3 = 165$ miles. Since $189 - 165 = 24$, the conditions of the problem are satisfied.

PROBLEM-SOLVING HINT
In motion problems like the one in Example 3, once you have filled in two pieces of information in each row of the table, you should automatically fill in the third piece of information, using the appropriate form of the formula relating distance, rate, and time. Set up the equation based on your sketch and the information in the table.

❸ Solve each problem.

(a) From a point on a straight road, Lupe and Maria ride bicycles in opposite directions. Lupe rides 10 mph and Maria rides 12 mph. In how many hours will they be 55 miles apart?

(b) At a given hour, two steamboats leave a city in the same direction on a straight canal. One travels at 18 mph, and the other travels at 25 mph. In how many hours will the boats be 35 miles apart?

ANSWERS

3. (a) $2\frac{1}{2}$ hr (b) 5 hr

EXAMPLE 4 Solving a Problem about Distance, Rate, and Time

The Tickfaw River has a current of 3 mph. A motorboat takes as long to go 12 miles downstream as to go 8 miles upstream. What is the speed of the boat in still water?

Step 1 **Read** the problem again. We are looking for the speed of the boat in still water.

Step 2 **Assign a variable.** Let x = the speed of the boat in still water. Because the current pushes the boat when the boat is going downstream, the speed of the boat downstream will be the sum of the speed of the boat and the speed of the current, $x + 3$ mph. Because the current slows down the boat when the boat is going upstream, the boat's speed upstream is given by the difference between the speed of the boat in still water and the speed of the current, $x - 3$ mph. See Figure 2.

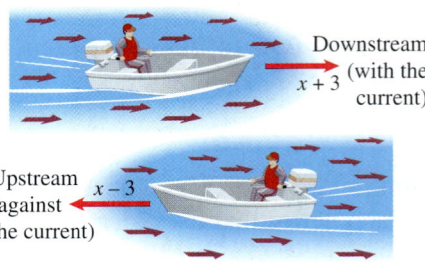

Figure 2

This information is summarized in the following table.

	d	r	t
Downstream	12	$x + 3$	
Upstream	8	$x - 3$	

Fill in the column representing time by using the formula $t = \frac{d}{r}$. Then the time upstream is the distance divided by the rate, or

$$t = \frac{d}{r} = \frac{8}{x - 3}$$

and the time downstream is also the distance divided by the rate, or

$$t = \frac{d}{r} = \frac{12}{x + 3}.$$

Now complete the table.

	d	r	t	
Downstream	12	$x + 3$	$\frac{12}{x + 3}$	⎫ Times are equal.
Upstream	8	$x - 3$	$\frac{8}{x - 3}$	⎭

Step 3 **Write an equation.** According to the original problem, the time upstream equals the time downstream. The two times from the table must therefore be equal, giving the equation

$$\frac{12}{x + 3} = \frac{8}{x - 3}.$$

Continued on Next Page

Step 4 **Solve.** Begin by multiplying each side by $(x + 3)(x - 3)$.

$$(x + 3)(x - 3) \frac{12}{x + 3} = (x + 3)(x - 3) \frac{8}{x - 3}$$

$$12(x - 3) = 8(x + 3)$$

$12x - 36 = 8x + 24$ Distributive property

$4x = 60$ Subtract $8x$; add 36.

$x = 15$ Divide by 4.

Step 5 **State the answer.** The speed of the boat in still water is 15 mph.

Step 6 **Check.** First find the speed of the boat downstream, which is $15 + 3 = 18$ mph. Traveling 12 miles would take

$$t = \frac{d}{r} = \frac{12}{18} = \frac{2}{3} \text{ hr.}$$

The speed of the boat upstream is $15 - 3 = 12$ mph, and traveling 8 miles would take

$$t = \frac{d}{r} = \frac{8}{12} = \frac{2}{3} \text{ hr.}$$

The time upstream equals the time downstream, as required.

Work Problem 4 at the Side.

OBJECTIVE 3 Solve problems about work. Suppose that you can mow your lawn in 4 hr. Then after 1 hr, you will have mowed $\frac{1}{4}$ of the lawn. After 2 hr, you will have mowed $\frac{2}{4}$ or $\frac{1}{2}$ of the lawn, and so on. This idea is generalized as follows.

> **Rate of Work**
> If a job can be completed in t units of time, then the rate of work is
> $$\frac{1}{t} \text{ job per unit of time.}$$

> **PROBLEM-SOLVING HINT**
> The relationship between problems involving work and problems involving distance is a very close one. Recall that the formula $d = rt$ says that distance traveled is equal to rate of travel multiplied by time traveled. Similarly, the fractional part of a job accomplished is equal to the rate of work multiplied by the time worked. In the lawn mowing example, after 3 hr, the fractional part of the job done is shown below.
>
>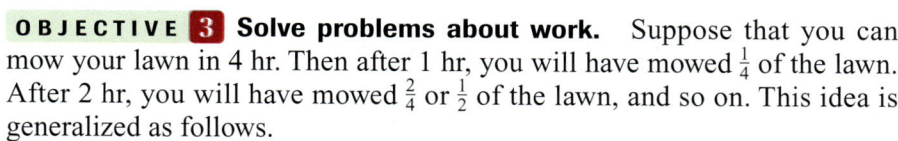
>
> After 4 hr, $\frac{1}{4}(4) = 1$ whole job has been done.

4 Solve each problem.

(a) A boat can go 20 miles against the current in the same time it can go 60 miles with the current. The current is flowing at 4 mph. Find the speed of the boat with no current.

(b) An airplane, maintaining a constant airspeed, takes as long to go 450 miles with the wind as it does to go 375 miles against the wind. If the wind is blowing at 15 mph, what is the speed of the plane?

ANSWERS
4. (a) 8 mph   (b) 165 mph

EXAMPLE 5 Solving a Problem about Work Rates

With spraying equipment, Mateo can paint the woodwork in a small house in 8 hr. His assistant, Chet, needs 14 hr to complete the same job painting by hand. If both Mateo and Chet work together, how long will it take them to paint the woodwork?

Step 1 **Read** the problem again. We are looking for time working together.

Step 2 **Assign a variable.** Let $x =$ the number of hours it will take for Mateo and Chet to paint the woodwork, working together.

Certainly, x will be less than 8, since Mateo alone can complete the job in 8 hr. Begin by making a table as shown. Remember that based on the previous discussion, Mateo's rate alone is $\frac{1}{8}$ job per hour, and Chet's rate is $\frac{1}{14}$ job per hour.

	Rate	Time Working Together	Fractional Part of the Job Done When Working Together
Mateo	$\frac{1}{8}$	x	$\frac{1}{8}x$
Chet	$\frac{1}{14}$	x	$\frac{1}{14}x$

Sum is 1 whole job.

Step 3 **Write an equation.** Since together Mateo and Chet complete 1 whole job, we must add their individual fractional parts and set the sum equal to 1.

$$\underbrace{\frac{1}{8}x}_{\text{Fractional part done by Mateo}} + \underbrace{\frac{1}{14}x}_{\text{Fractional part done by Chet}} = \underbrace{1}_{\text{1 whole job}}$$

Step 4 **Solve.**

$$56\left(\frac{1}{8}x + \frac{1}{14}x\right) = 56(1) \quad \text{Multiply by the LCD, 56.}$$

$$56\left(\frac{1}{8}x\right) + 56\left(\frac{1}{14}x\right) = 56(1) \quad \text{Distributive property}$$

$$7x + 4x = 56$$

$$11x = 56 \quad \text{Combine like terms.}$$

$$x = \frac{56}{11} \quad \text{Divide by 11.}$$

Step 5 **State the answer.** Working together, Mateo and Chet can paint the woodwork in $\frac{56}{11}$ hr, or $5\frac{1}{11}$ hr.

Step 6 **Check** to be sure the answer is correct.

NOTE
An alternative approach in work problems is to consider the part of the job that can be done in 1 hr. For instance, in Example 5 Mateo can do the entire job in 8 hr, and Chet can do it in 14 hr. Thus, their work rates, as we saw in Example 5, are $\frac{1}{8}$ and $\frac{1}{14}$, respectively. Since it takes them x hr to complete the job when working together, in 1 hr they can paint $\frac{1}{x}$ of the woodwork. The amount painted by Mateo in 1 hr plus the amount painted by Chet in 1 hr must equal the amount they can do together. This leads to the equation shown here.

$$\underset{\text{Amount by Mateo in one hour}}{\rightarrow} \frac{1}{8} + \underset{\text{Amount by Chet in one hour}}{\frac{1}{14}} = \frac{1}{x} \underset{\text{Amount together in one hour}}{\leftarrow}$$

Compare this with the equation in Example 5. Multiplying each side by $56x$ leads to

$$7x + 4x = 56$$

the same equation found in the third line of Step 4 in the example. The final solution is the same.

Work Problem 5 at the Side.

5 Solve each problem.

(a) Michael can paint a room, working alone, in 8 hr. Lindsay can paint the same room, working alone, in 6 hr. How long will it take them if they work together?

(b) Roberto can detail his Camaro in 2 hr working alone. His brother Marco can do the job in 3 hr working alone. How long would it take them if they worked together?

ANSWERS

5. (a) $3\frac{3}{7}$ hr **(b)** $1\frac{1}{5}$ hr

Focus on Real-Data Applications

Upward Mobility*

As a struggling college student you have been driving the "Wimp," a 1989 Honda CRX, which has one saving grace—it gets 30 miles per gallon in the city. Now that you are graduating and are being recruited for your dream job, your first major purchase will be a new truck or sport-utility vehicle. In addition to car payments, you also must consider increased gasoline costs.

Based on past experience, you anticipate driving 15,000 miles a year. The new vehicle requires premium unleaded gasoline at $1.40 per gallon, instead of the $1.35 cost for regular unleaded that you use now. Currently, you use 500 gallons of gasoline per year, and you spend $675 per year on gasoline.

Gasoline usage $\quad \dfrac{15{,}000 \text{ mi}}{1 \text{ yr}} \div \dfrac{30 \text{ mi}}{1 \text{ gal}} = \dfrac{15{,}000 \text{ mi}}{1 \text{ yr}} \cdot \dfrac{1 \text{ gal}}{30 \text{ mi}} = 500 \text{ gallons per year}$

Current gasoline costs $\quad \dfrac{500 \text{ gal}}{1 \text{ yr}} \cdot \dfrac{\$1.35 \text{ gal}}{1 \text{ gal}} = \$675 \text{ per year}$

You test drove a 4.8 Liter, V8, Chevy Tahoe that is rated to get 15 miles per gallon (mpg). To calculate the additional costs for gasoline, you must first compute the cost for gasoline usage in the Tahoe and then subtract your current gasoline costs. Observe the *process* so that you can copy it to devise a **cost equation** that you can use to evaluate costs for all the other vehicles that you want to test drive. The *variable* quantity (the number of miles per gallon) is shown in blue.

Chevy Tahoe gasoline usage $\quad \dfrac{15{,}000 \text{ mi}}{1 \text{ yr}} \div \dfrac{\textcolor{blue}{15 \text{ mi}}}{1 \text{ gal}} = \dfrac{15{,}000 \text{ mi}}{1 \text{ yr}} \cdot \dfrac{1 \text{ gal}}{\textcolor{blue}{15 \text{ mi}}}$

Projected gasoline cost increase $\quad \dfrac{15{,}000 \text{ gal}}{\textcolor{blue}{15} \text{ yr}} \cdot \dfrac{\$1.40}{1 \text{ gal}} - \$675 = \dfrac{(15{,}000)(1.40)}{\textcolor{blue}{15}} - 675$

For Group Discussion

1. Write a cost equation that computes the additional costs, y, for gasoline for a truck or SUV that gets x mpg. Assume that the new vehicle requires premium gasoline. (*Hint:* Replace the variable quantity in the process above with x.)

2. Evaluate the cost equation for the following vehicles to predict increased gasoline costs.
   (a) Honda CR-V, 4 L, 23 mpg
   (b) Ford Expedition, 5.4 L, V8, 14 mpg
   (c) Chevy Yukon, 5.3 L, V8, 12 mpg

3. A calculator graph of the cost equation is shown here.
   (a) What can you conclude about the effect of *decreasing* mileage rating on the additional gasoline costs?

   (b) The graph crosses the x-axis at approximately $x = 31$. Explain why.

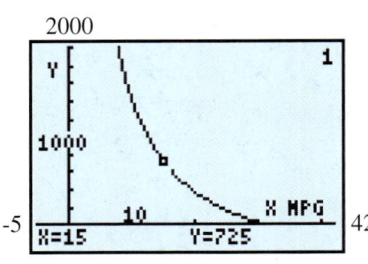

---

* Based on *Driving Rationally* by Patricia Stone, Tomball College. Original vehicle models and gasoline prices are used.

15.7 Exercises

Use Steps 2 and 3 of the six-step problem solving method to set up the equation you would use to solve each problem. (Remember that Step 1 is to read the problem carefully.) Do not actually solve the equation. See Example 1.

1. The numerator of the fraction $\frac{5}{6}$ is increased by an amount so that the value of the resulting fraction is equivalent to $\frac{13}{3}$. By what amount was the numerator increased?
   (a) Let $x = $ _____. *(Step 2)*
   (b) Write an expression for "the numerator of the fraction $\frac{5}{6}$ is increased by an amount."
   (c) Set up an equation to solve the problem.
   (Step 3)

2. If the same number is added to the numerator and subtracted from the denominator of $\frac{23}{12}$, the resulting fraction is equivalent to $\frac{3}{2}$. What is the number?
   (a) Let $x = $ _____. *(Step 2)*
   (b) Write an expression for "a number is added to the numerator of $\frac{23}{12}$." Then write an expression for "the same number is subtracted from the denominator of $\frac{23}{12}$."
   (c) Set up an equation to solve the problem.
   (Step 3)

Use the six-step method to solve each problem. See Example 1.

3. In a certain fraction, the denominator is 4 less than the numerator. If 3 is added to both the numerator and the denominator, the resulting fraction is equivalent to $\frac{3}{2}$. What was the original fraction?

4. In a certain fraction, the denominator is 6 more than the numerator. If 3 is added to both the numerator and the denominator, the resulting fraction is equivalent to $\frac{5}{7}$. What was the original fraction?

5. The denominator of a certain fraction is three times the numerator. If 2 is added to the numerator and subtracted from the denominator, the resulting fraction is equivalent to 1. What was the original fraction?

6. The numerator of a certain fraction is four times the denominator. If 6 is added to both the numerator and the denominator, the resulting fraction is equivalent to 2. What was the original fraction?

7. One-sixth of a number is 5 more than the same number. What is the number?

8. One-third of a number is 2 more than one-sixth of the same number. What is the number?

9. A quantity, its $\frac{3}{4}$, its $\frac{1}{2}$, and its $\frac{1}{3}$, added together, become 93. What is the quantity? (*Source: Rhind Mathematical Papyrus.*)

10. A quantity, its $\frac{2}{3}$, its $\frac{1}{2}$, and its $\frac{1}{7}$, added together, become 33. What is the quantity? (*Source: Rhind Mathematical Papyrus.*)

Solve each problem. See Example 2.

11. At the Tyson Foods Invitational, Gail Devers of the United States won the indoor 60-meter hurdle event for women in 7.10 seconds. What was her rate? (*Source:* www.usatoday/sports.com)

12. In the 2002 Winter Games, Catriona LeMay Doan of Canada won the 500-meter speed skating event for women. Her rate was 13.4048 meters per second. What was her time (to the nearest hundredth of a second)? (*Source: Sports Illustrated Sports Almanac.*)

13. The winner of the 2003 Daytona 500 (mile) race was Michael Waltrip, who drove his Chevrolet to victory with a rate of 133.870 mph. What was his time? (*Source: World Almanac and Book of Facts.*)

14. In 2003, Gil de Ferran drove his Dallara-Toyota to victory in the Indianapolis 500 (mile) race. His rate was 156.291 mph. What was his time? (*Source: World Almanac and Book of Facts.*)

15. Meseret Defar of Ethiopia won the women's 5000-meter race in the 2004 Olympics with a time of 14.761 minutes. What was her rate? (*Source:* www.olympics.com)

16. The winner of the women's 1500-meter race in the 2004 Olympics was Kelly Holmes of Great Britain with a time of 3.965 minutes. What was her rate? (*Source:* www.olympics.com)

Set up the equation you would use to solve each problem. Do not actually solve the equation. See Examples 3 and 4.

17. Luvenia can row 4 mph in still water. It takes as long to row 8 miles upstream as 24 miles downstream. How fast is the current? (Let x = speed of the current.)

	d	r	t
Upstream	8	$4 - x$	
Downstream	24	$4 + x$	

18. Julio flew his airplane 500 miles against the wind in the same time it took him to fly it 600 miles with the wind. If the speed of the wind was 10 mph, what was the average speed of his plane? (Let x = speed of the plane in still air.)

	d	r	t
Against the Wind	500	$x - 10$	
With the Wind	600	$x + 10$	

Solve each problem. See Examples 3 and 4.

19. If a migrating hawk travels m mph in still air, what is its rate when it flies into a steady headwind of 5 mph? What is its rate with a tailwind of 5 mph?

20. Suppose Stephanie walks D miles at R mph in the same time that Wally walks d miles at r mph. Give an equation relating D, R, d, and r.

21. A plane flies 350 miles with the wind in the same time that it can fly 310 miles against the wind. The plane has a still-air speed of 165 mph. Find the speed of the wind.

22. A boat can go 20 miles against a current in the same time that it can go 60 miles with the current. The current is 4 mph. Find the speed of the boat in still water.

23. The distance from Seattle, Washington, to Victoria, British Columbia, is about 148 miles by ferry. It takes about 4 hours less time to travel by the same ferry to Vancouver, British Columbia, a distance of about 74 miles. What is the average speed of the ferry?

24. Alexis Glaser flew from Dallas to Indianapolis at 180 mph and then flew back at 150 mph. The trip at the slower speed took 1 hour longer than the trip at the higher speed. Find the distance between the two cities.

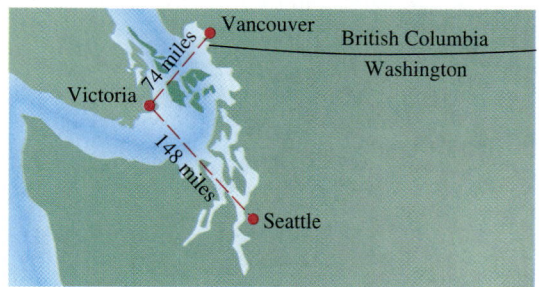

In Exercises 25 and 26, set up the equation you would use to solve each problem. Do not actually solve the equation. See Example 5.

25. Edwin Bedford can tune up his Chevy in 2 hr working alone. His son, Beau, can do the job in 3 hr working alone. How long would it take them if they worked together? (Let t represent the time working together.)

	r	t	w
Edwin		t	
Beau		t	

26. Working alone, Jorge can paint a room in 8 hr. Caterina can paint the same room working alone in 6 hr. How long will it take them if they work together? (Let x represent the time working together.)

	r	t	w
Jorge		x	
Caterina		x	

Solve each problem. See Example 5.

27. Lea can groom the horses in her boarding stable in 5 hr, while Tran needs 4 hr. How long will it take them to groom the horses if they work together?

28. Geraldo and Luisa Hernandez operate a small laundry. Luisa, working alone, can clean a day's laundry in 9 hr. Geraldo can clean a day's laundry in 8 hr. How long would it take them if they work together?

29. Todd's copier can do a printing job in 7 hr. Scott's copier can do the same job in 12 hr. How long would it take to do the job using both copiers?

30. A pump can pump the water out of a flooded basement in 10 hr. A smaller pump takes 12 hr. How long would it take to pump the water from the basement using both pumps?

31. Hilda can paint a room in 6 hr. Working together with Brenda, they can paint the room in $3\frac{3}{4}$ hr. How long would it take Brenda to paint the room by herself?

32. Grant can completely mess up his room in 15 min. If his cousin Wade helps him, they can completely mess up the room in $8\frac{4}{7}$ min. How long would it take Wade to mess up the room by himself?

33. An inlet pipe can fill a swimming pool in 9 hr, and an outlet pipe can empty the pool in 12 hr. Through an error, both pipes are left open. How long will it take to fill the pool?

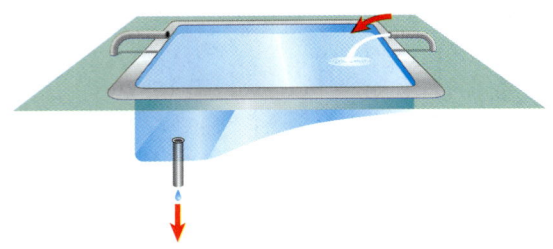

34. One pipe can fill a swimming pool in 6 hr, and another pipe can do it in 9 hr. How long will it take the two pipes working together to fill the pool $\frac{3}{4}$ full?

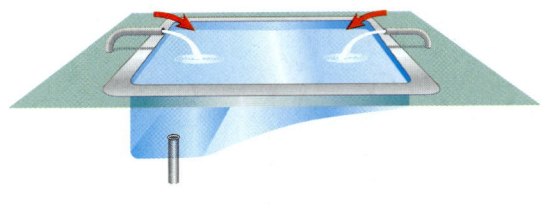

35. Refer to Exercise 33. Assume the error was discovered after both pipes had been running for 3 hr, and the outlet pipe was then closed. How much more time would then be required to fill the pool? (*Hint:* Consider how much of the job had been done when the error was discovered.)

36. A cold water faucet can fill a sink in 12 min, and a hot water faucet can fill it in 15 min. The drain can empty the sink in 25 min. If both faucets are on and the drain is open, how long will it take to fill the sink?

37. Refer to the table in Exercise 17. Suppose that a student made the error of interchanging the positions of the expressions $4 - x$ and $4 + x$, but used the correct method of setting up the equation, applying the formula $t = \frac{d}{r}$. Solve the equation the student used, and explain how the student should immediately know that there is something wrong in the setup.

15.8 Variation

OBJECTIVE 1 Solve problems about direct variation. Suppose that gasoline costs $2.80 per gallon. Then 1 gallon costs $2.80, 2 gallons cost 2($2.80) = $5.60, 3 gallons cost 3($2.80) = $8.40, and so on. Each time, the total cost is obtained by multiplying the number of gallons by the price per gallon. In general, if k equals the price per gallon and x equals the number of gallons, then the total cost y is equal to kx. Notice that as the number of gallons increases, the total cost increases.

The preceding discussion is an example of variation. Equations with fractions often result when discussing variation. As in the gasoline example, two variables *vary directly* if one is a constant multiple of the other.

OBJECTIVES

1. Solve problems about direct variation.
2. Solve problems about inverse variation.

1 Solve each problem.

(a) If z varies directly as t, and $z = 11$ when $t = 4$, find z when $t = 32$.

Direct Variation

y varies directly as x if there exists a constant k such that

$$y = kx.$$

The constant k in the equation for direct variation is a numerical value, such as 2.80 in the gasoline price discussion.

EXAMPLE 1 Using Direct Variation

Suppose y varies directly as x, and $y = 20$ when $x = 4$. Find y when $x = 9$.

Since y varies directly as x, there is a constant k such that $y = kx$. We know that $y = 20$ when $x = 4$. Substitute these values into $y = kx$ and solve for k.

$$y = kx$$
$$20 = k \cdot 4$$
$$k = 5$$

Since $y = kx$ and $k = 5$, then

$$y = 5x. \qquad \text{Let } k = 5.$$

When $x = 9$,

$$y = 5x = 5 \cdot 9 = 45. \qquad \text{Let } x = 9.$$

Thus, $y = 45$ when $x = 9$.

(b) The circumference of a circle varies directly as the radius. A circle with a radius of 7 cm has a circumference of 43.96 cm. Find the circumference if the radius is 11 cm.

▶▶▶ **Work Problem 1 at the Side.** ▶▶▶

OBJECTIVE 2 Solve problems about inverse variation. Another common type of variation, in which the value of one variable increases while the value of another decreases, is called *inverse variation*. For example, an increase in the supply of an item causes a decrease in the price of the item.

ANSWERS
1. (a) 88   (b) 69.08 cm

2 Solve the problem.

Suppose z varies inversely as t, and $z = 8$ when $t = 2$. Find z when $t = 32$.

> **Inverse Variation**
>
> y **varies inversely as** x if there exists a constant k such that
>
> $$y = \frac{k}{x}$$

EXAMPLE 2 Using Inverse Variation

Suppose y varies inversely as x, and $y = 3$ when $x = 8$. Find y when $x = 6$.

Since y varies inversely as x, there is a constant k such that $y = \frac{k}{x}$. We know that $y = 3$ when $x = 8$, so we can find k.

$$y = \frac{k}{x}$$

$$3 = \frac{k}{8}$$

$$k = 24$$

Since $y = \frac{24}{x}$, we let $x = 6$ and solve for y.

$$y = \frac{24}{x} = \frac{24}{6} = 4$$

Therefore, when $x = 6$, $y = 4$.

◀◀ **Work Problem 2 at the Side.**

3 Solve the problem.

The current in a simple electrical circuit varies inversely as the resistance. If the current is 80 amps when the resistance is 10 ohms, find the current if the resistance is 16 ohms.

EXAMPLE 3 Using Inverse Variation

In the manufacturing of a certain medical syringe, the cost of producing the syringe varies inversely as the number produced. If 10,000 syringes are produced, the cost is $2 per unit. Find the cost per unit to produce 25,000 syringes.

Let x = the number of syringes produced
and c = the cost per unit.

Since c varies inversely as x, there is a constant k such that

$$c = \frac{k}{x}$$

Find k by replacing c with 2 and x with 10,000.

$$2 = \frac{k}{10,000}$$

$$20,000 = k \qquad \text{Multiply by 10,000.}$$

Since $c = \frac{k}{x}$,

$$c = \frac{20,000}{25,000} = 0.80 \qquad \text{Let } k = 20,000 \text{ and } x = 25,000.$$

The cost per unit to make 25,000 syringes is $0.80.

◀◀ **Work Problem 3 at the Side.**

ANSWERS

2. $\frac{1}{2}$
3. 50 amps

15.8 Exercises

1. **(a)** If the constant of variation is positive and y varies directly as x, then as x increases, y _____.
   (increases/decreases)

   (b) If the constant of variation is positive and y varies inversely as x, then as x increases, y _____.
   (increases/decreases)

2. Bill Veeck was the owner of several major league baseball teams in the 1950s and 1960s. He was known to often sit in the stands and enjoy games with his paying customers. Here is a quote attributed to him:

   "I have discovered in 20 years of moving around a ballpark, that the knowledge of the game is usually in inverse proportion to the price of the seats."

   Explain in your own words the meaning of this statement. (To prove his point, Veeck once allowed the fans to vote on managerial decisions.)

Solve each problem involving direct variation. See Example 1.

3. If z varies directly as x, and $z = 30$ when $x = 8$, find z when $x = 4$.

4. If y varies directly as x, and $x = 27$ when $y = 6$, find x when $y = 2$.

5. If d varies directly as r, and $d = 200$ when $r = 40$, find d when $r = 60$.

6. If d varies directly as t, and $d = 150$ when $t = 3$, find d when $t = 5$.

Solve each problem involving inverse variation. See Example 2.

7. If z varies inversely as x, and $z = 50$ when $x = 2$, find z when $x = 25$.

8. If x varies inversely as y, and $x = 3$ when $y = 8$, find y when $x = 4$.

9. If m varies inversely as r, and $m = 12$ when $r = 8$, find m when $r = 16$.

10. If p varies inversely as q, and $p = 7$ when $q = 6$, find p when $q = 2$.

Solve each variation problem. See Examples 1–3.

11. For a given base, the area of a triangle varies directly as its height. Find the area of a triangle with a height of 6 in., if the area is 10 in.² when the height is 4 in.

12. The interest on an investment varies directly as the rate of interest. If the interest is $48 when the interest rate is 5%, find the interest when the rate is 4.2%.

13. Hooke's law for an elastic spring states that the distance a spring stretches varies directly with the force applied. If a force of 75 lb stretches a certain spring 16 in., how much will a force of 200 lb stretch the spring?

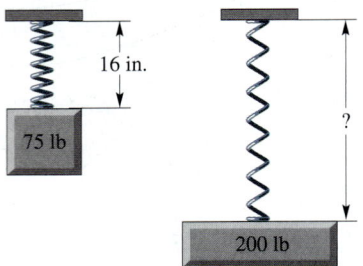

14. The pressure exerted by water at a given point varies directly with the depth of the point beneath the surface of the water. Water exerts 4.34 pounds per square inch for every 10 feet traveled below the water's surface. What is the pressure exerted on a scuba diver at 20 feet?

15. For a constant area, the length of a rectangle varies inversely as the width. The length of a rectangle is 27 ft when the width is 10 ft. Find the width of a rectangle with the same area if the length is 18 ft.

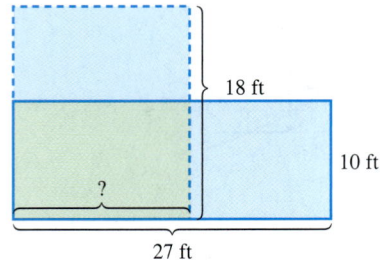

16. Over a specified distance, speed varies inversely with time. If a Dodge Viper on a test track goes a certain distance in one-half minute at 160 mph, what speed is needed to go the same distance in three-fourths minute?

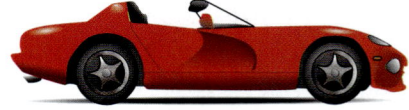

17. If the temperature is constant, the pressure of a gas in a container varies inversely as the volume of the container. If the pressure is 10 pounds per square foot in a container with a volume of 3 cubic feet, what is the pressure in a container with a volume of 1.5 cubic feet?

18. The current in a simple electrical circuit varies inversely as the resistance. If the current is 20 amps when the resistance is 5 ohms, find the current when the resistance is 8 ohms.

19. In the inversion of raw sugar, the rate of change of the amount of raw sugar varies directly as the amount of raw sugar remaining. The rate is 200 kilograms per hour when there are 800 kilograms left. What is the rate of change per hour when only 100 kilograms are left?

20. The force required to compress a spring varies directly as the change in the length of the spring. If a force of 12 lb is required to compress a certain spring 3 in., how much force is required to compress the spring 5 in.?

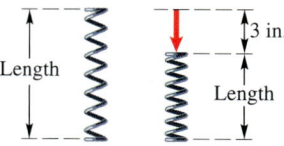

Use personal experience or intuition to determine whether the situation suggests direct or inverse variation.*

21. The rate and the distance traveled by a pickup truck in 3 hours

22. The number of different lottery tickets you buy and your probability of winning that lottery

23. The number of days from now until December 25 and the magnitude of the frenzy of Christmas shopping

24. The amount of pressure put on the accelerator of a car and the speed of the car

25. The amount of gasoline that you pump and the amount of empty space left in your tank

26. The surface area of a balloon and its diameter

27. The amount of gasoline you pump and the amount you will pay

28. The number of days until the end of the baseball season and the number of home runs that Alex Rodriguez has

---

* The authors thank Linda Kodama of Kapi'olani Community College for suggesting the inclusion of exercises of this type.

*Recall from **Section 6.6** that two triangles are **similar** if they have the same shape (but not necessarily the same size). Similar triangles have side lengths that vary directly. The figure shows two similar triangles. Notice that the ratios of the corresponding sides are all equal to $\frac{3}{2}$.*

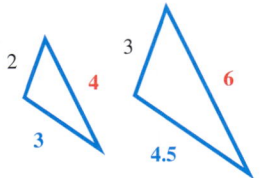

$$\frac{3}{2} = \frac{3}{2} \qquad \frac{4.5}{3} = \frac{3}{2} \qquad \frac{6}{4} = \frac{3}{2}$$

If we know that two triangles are similar, we can set up a direct variation equation to solve for the length of an unknown side.

Find the length x, given that each pair of triangles is similar.

29.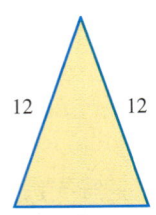

30.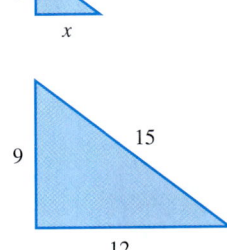

31.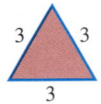

32.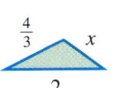

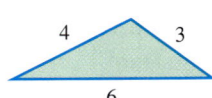

Use similar triangles and direct variation to solve each problem. (Source: The Guinness Book of World Records.)

33. One of the tallest candles ever constructed was exhibited at the 1897 Stockholm Exhibition. If it cast a shadow 5 ft long at the same time a vertical pole 32 ft high cast a shadow 2 ft long, how tall was the candle?

34. An enlarged version of the chair used by George Washington at the Constitutional Convention casts a shadow 18 ft long at the same time a vertical pole 12 ft high casts a shadow 4 ft long. How tall is the chair?

Chapter 15
SUMMARY

KEY TERMS

15.1 rational expression — The quotient of two polynomials with denominator not 0 is called a rational expression.

lowest terms — A rational expression is written in lowest terms if the greatest common factor of its numerator and denominator is 1.

15.3 least common denominator (LCD) — The simplest expression that is divisible by all denominators is called the least common denominator.

15.5 complex fraction — A rational expression with one or more fractions in the numerator, denominator, or both, is called a complex fraction.

15.8 direct variation — y varies directly as x if there is a constant k such that $y = kx$.

inverse variation — y varies inversely as x if there is a constant k such that $y = \frac{k}{x}$.

TEST YOUR WORD POWER

See how well you have learned the vocabulary in this chapter. Answers, with examples, follow the Quick Review.

1. A **rational expression** is
   A. an algebraic expression made up of a term or the sum of a finite number of terms with real coefficients and whole number exponents
   B. a polynomial equation of degree 2
   C. an expression with one or more fractions in the numerator, denominator, or both
   D. the quotient of two polynomials with denominator not 0.

2. A **complex fraction** is
   A. an algebraic expression made up of a term or the sum of a finite number of terms with real coefficients and whole number exponents
   B. a polynomial equation of degree 2
   C. a rational expression with one or more fractions in the numerator, denominator, or both
   D. the quotient of two polynomials with denominator not 0.

3. In a given set of fractions, the **least common denominator** is
   A. the smallest denominator of all the denominators
   B. the smallest expression that is divisible by all the denominators
   C. the largest integer that evenly divides the numerator and denominator of all the fractions
   D. the largest denominator of all the denominators.

4. If two positive quantities x and y are in **direct variation,** and the constant of variation is positive, then
   A. as x increases, y decreases
   B. as x increases, y increases
   C. as x increases, y remains constant
   D. as x decreases, y remains constant.

Chapter 15 Rational Expressions and Applications

QUICK REVIEW

Concepts | Examples

15.1 The Fundamental Property of Rational Expressions
To find the value(s) for which a rational expression is undefined, set the denominator equal to 0 and solve the equation.

Find the values for which the expression is undefined.
$$\frac{x-4}{x^2-16}$$
$$x^2 - 16 = 0$$
$$(x-4)(x+4) = 0$$
$$x - 4 = 0 \quad \text{or} \quad x + 4 = 0$$
$$x = 4 \quad \text{or} \quad x = -4$$

The rational expression is undefined for 4 and -4.

Writing a Rational Expression in Lowest Terms

Write $\dfrac{x^2-1}{(x-1)^2}$ in lowest terms.

Step 1 Factor the numerator and denominator.

$$\frac{x^2-1}{(x-1)^2} = \frac{(x-1)(x+1)}{(x-1)(x-1)} = \frac{x+1}{x-1}$$

Step 2 Use the fundamental property to divide out common factors from the numerator and denominator.

There are often several different equivalent forms of a rational expression.

Give two equivalent forms of $-\dfrac{x-1}{x+2}$.

Distribute the $-$ sign in the numerator to get $\dfrac{-(x-1)}{x+2}$

or $\dfrac{-x+1}{x+2}$; do so in the denominator to get $\dfrac{x-1}{-x-2}$.

(There are other forms as well.)

15.2 Multiplying and Dividing Rational Expressions
Multiplying Rational Expressions

Multiply. $\dfrac{3x+9}{x-5} \cdot \dfrac{x^2-3x-10}{x^2-9}$

Step 1 Factor.

$$= \frac{3(x+3)}{x-5} \cdot \frac{(x-5)(x+2)}{(x+3)(x-3)}$$

Step 2 Multiply numerators and multiply denominators.

$$= \frac{3(x+3)(x-5)(x+2)}{(x-5)(x+3)(x-3)}$$

Step 3 Write in lowest terms.

$$= \frac{3(x+2)}{x-3}$$

Dividing Rational Expressions

Divide. $\dfrac{2x+1}{x+5} \div \dfrac{6x^2-x-2}{x^2-25}$

Step 1 Multiply the first rational expression by the reciprocal of the second.

$$= \frac{2x+1}{x+5} \cdot \frac{x^2-25}{6x^2-x-2}$$

Step 2 Factor.

$$= \frac{2x+1}{x+5} \cdot \frac{(x+5)(x-5)}{(2x+1)(3x-2)}$$

Step 3 Multiply the numerators and the denominators,
Step 4 and write in lowest terms.

$$= \frac{x-5}{3x-2}$$

Concepts	Examples
15.3 Least Common Denominators **Finding the LCD** *Step 1* Factor each denominator into prime factors. *Step 2* List each different factor the greatest number of times it appears. *Step 3* Multiply the factors from Step 2 to get the LCD.	Find the LCD for $\dfrac{3}{k^2 - 8k + 16}$ and $\dfrac{1}{4k^2 - 16k}$. $k^2 - 8k + 16 = (k-4)^2$ $4k^2 - 16k = 4k(k-4)$ LCD $= (k-4)^2 \cdot 4 \cdot k$ $= 4k(k-4)^2$
Writing a Rational Expression with a Specified Denominator *Step 1* Factor both denominators. *Step 2* Decide what factors the denominator must be multiplied by to equal the specified denominator. *Step 3* Multiply the rational expression by that factor divided by itself (multiply by 1).	Find the numerator: $\dfrac{5}{2z^2 - 6z} = \dfrac{}{4z^3 - 12z^2}$ $\dfrac{5}{2z(z-3)} = \dfrac{}{4z^2(z-3)}$ $2z(z-3)$ must be multiplied by $2z$. $\dfrac{5}{2z(z-3)} \cdot \dfrac{2z}{2z} = \dfrac{10z}{4z^2(z-3)} = \dfrac{10z}{4z^3 - 12z^2}$
15.4 Adding and Subtracting Rational Expressions **Adding Rational Expressions** *Step 1* Find the LCD. *Step 2* Rewrite each rational expression with the LCD as denominator. *Step 3* Add the numerators to get the numerator of the sum. The LCD is the denominator of the sum. *Step 4* Write in lowest terms.	Add. $\dfrac{2}{3m+6} + \dfrac{m}{m^2 - 4}$ $3m + 6 = 3(m+2)$ $m^2 - 4 = (m+2)(m-2)$ The LCD is $3(m+2)(m-2)$. $= \dfrac{2(m-2)}{3(m+2)(m-2)} + \dfrac{3m}{3(m+2)(m-2)}$ $= \dfrac{2m - 4 + 3m}{3(m+2)(m-2)}$ $= \dfrac{5m - 4}{3(m+2)(m-2)}$
Subtracting Rational Expressions Follow the same steps as for addition, but subtract in Step 3.	Subtract. $\dfrac{6}{k+4} - \dfrac{2}{k}$ The LCD is $k(k+4)$. $\dfrac{6k}{(k+4)k} - \dfrac{2(k+4)}{k(k+4)} = \dfrac{6k - 2(k+4)}{k(k+4)}$ $= \dfrac{6k - 2k - 8}{k(k+4)}$ Be careful with signs when subtracting the numerators. $= \dfrac{4k - 8}{k(k+4)}$ or $\dfrac{4(k-2)}{k(k+4)}$

Concepts	Examples
15.5 Complex Fractions **Simplifying Complex Fractions**	Simplify.
Method 1 Simplify the numerator and denominator separately. Then divide the simplified numerator by the simplified denominator.	*Method 1* $\dfrac{\dfrac{1}{a} - a}{1 - a} = \dfrac{\dfrac{1}{a} - \dfrac{a^2}{a}}{1 - a} = \dfrac{\dfrac{1 - a^2}{a}}{1 - a}$ $= \dfrac{1 - a^2}{a} \cdot \dfrac{1}{1 - a}$ $= \dfrac{(1-a)(1+a)}{a(1-a)} = \dfrac{1+a}{a}$
Method 2 Multiply the numerator and denominator of the complex fraction by the LCD of all the denominators in the complex fraction. Write in lowest terms.	*Method 2* $\dfrac{\dfrac{1}{a} - a}{1 - a} = \dfrac{\left(\dfrac{1}{a} - a\right)a}{(1-a)a} = \dfrac{\dfrac{a}{a} - a^2}{(1-a)a}$ $= \dfrac{1 - a^2}{(1-a)a} = \dfrac{(1+a)(1-a)}{(1-a)a}$ $= \dfrac{1+a}{a}$
15.6 Solving Equations with Rational Expressions **Solving Equations with Rational Expressions**	Solve $\dfrac{x}{x-3} + \dfrac{4}{x+3} = \dfrac{18}{x^2-9}$.
Step 1 Find the LCD of all denominators in the equation.	The LCD is $(x-3)(x+3)$. Note that 3 and -3 cannot be solutions, as they cause a denominator to equal 0. $\dfrac{x}{x-3} + \dfrac{4}{x+3} = \dfrac{18}{(x-3)(x+3)}$ Factor.
Step 2 Multiply each side of the equation by this LCD.	$x(x+3) + 4(x-3) = 18$ Multiply by $(x-3)(x+3)$.
Step 3 Solve the resulting equation, which should have no fractions.	$x^2 + 3x + 4x - 12 = 18$ Distributive property $x^2 + 7x - 30 = 0$ Subtract 18; standard form $(x-3)(x+10) = 0$ Factor. $x - 3 = 0$ or $x + 10 = 0$ Zero-factor property
Step 4 Check each proposed solution, and reject any value that causes an original denominator to equal 0.	Reject $\rightarrow x = 3$ or $x = -10$ The only solution is -10.

Concepts

15.7 Applications of Rational Expressions

Solving Problems about Distance

Use the six-step method.

Step 1 **Read** the problem carefully.

Step 2 **Assign a variable.** Use a table to identify distance, rate, and time. Solve $d = rt$ for the unknown quantity in the table.

Step 3 **Write an equation.** From the wording in the problem, decide the relationship between the quantities. Use those expressions to write an equation.

Step 4 **Solve** the equation.

Step 5 **State the answer.**

Step 6 **Check** the solution.

Solving Problems about Work

Step 1 **Read** the problem carefully.

Step 2 **Assign a variable.** State what the variable represents. Put the information from the problem in a table. If a job is done in t units of time, then the rate is $\frac{1}{t}$.

Step 3 **Write an equation.** The sum of the fractional parts should equal 1 (whole job).

Step 4 **Solve** the equation.

Steps 5 and 6 **State the answer** and **check** the solution.

Examples

On a trip from Sacramento to Monterey, Marge traveled at an average speed of 60 mph. The return trip, at an average speed of 64 mph, took $\frac{1}{4}$ hr less. How far did she travel between the two cities?

Let x = the unknown distance.

	d	r	$t = \dfrac{d}{r}$
Going	x	60	$\dfrac{x}{60}$
Returning	x	64	$\dfrac{x}{64}$

Since the time for the return trip was $\frac{1}{4}$ hr less, the time going equals the time returning plus $\frac{1}{4}$.

$$\frac{x}{60} = \frac{x}{64} + \frac{1}{4}$$

$16x = 15x + 240$ Multiply by 960.

$x = 240$ Subtract $15x$.

She traveled 240 mi.

The trip there took $\frac{240}{60} = 4$ hr, while the return trip took $\frac{240}{64} = 3\frac{3}{4}$ hr, which is $\frac{1}{4}$ hr less time. The solution checks.

It takes the regular mail carrier 6 hr to cover her route. A substitute takes 8 hr to cover the same route. How long would it take them to cover the route together?

Let x = the number of hours to cover the route together.

The rate of the regular carrier is $\frac{1}{6}$ job per hour; the rate of the substitute is $\frac{1}{8}$ job per hour. Multiply rate by time to get the fractional part of the job done.

	Rate	Time	Part of the Job Done
Regular	$\dfrac{1}{6}$	x	$\dfrac{1}{6}x$
Substitute	$\dfrac{1}{8}$	x	$\dfrac{1}{8}x$

The equation is $\dfrac{1}{6}x + \dfrac{1}{8}x = 1$.

The solution of the equation is $\frac{24}{7}$. The solution checks, because $\frac{1}{6}\left(\frac{24}{7}\right) + \frac{1}{8}\left(\frac{24}{7}\right) = 1$ is true.

It would take them $\frac{24}{7}$ or $3\frac{3}{7}$ hr to cover the route together.

Concepts	Examples
15.8 Variation **Solving Variation Problems** *Step 1* Write the variation equation. Use $$y = kx \quad \text{for direct variation,}$$ $$y = \frac{k}{x} \quad \text{for inverse variation.}$$ *Step 2* Find k by substituting the given values of x and y into the equation. *Step 3* Write the equation with the value of k from Step 2 and the given value of x or y. Solve for the remaining variable.	If y varies inversely as x, and $y = 4$ when $x = 9$, find y when $x = 6$. The equation for inverse variation is $$y = \frac{k}{x}.$$ $$4 = \frac{k}{9}$$ $$k = \mathbf{36}$$ $$y = \frac{\mathbf{36}}{x} \qquad k = 36$$ $$y = \frac{36}{6} \qquad \text{Let } x = 6.$$ $$y = 6$$

ANSWERS TO TEST YOUR WORD POWER

1. D; *Examples:* $-\dfrac{3}{4y}$, $\dfrac{5x^3}{x+2}$, $\dfrac{a+3}{a^2-4a-5}$

2. C; *Examples:* $\dfrac{\frac{2}{3}}{\frac{4}{7}}$, $\dfrac{x - \frac{1}{y}}{x + \frac{1}{y}}$, $\dfrac{\frac{2}{a+1}}{a^2 - 1}$

3. B; *Examples:* The least common denominator of $\dfrac{1}{2}, \dfrac{1}{3},$ and $\dfrac{1}{4}$ is 12. The least common denominator of $\dfrac{1}{x}$ and $\dfrac{1}{x+1}$ is $x(x+1)$.

4. B; *Example:* The equation $y = 3x$ represents direct variation. When $x = 2, y = 6$. If x increases to 3, then y increases to $3(3) = 9$.

Chapter 15
REVIEW EXERCISES

[15.1] *Find the value(s) of the variable for which each rational expression is undefined.*

1. $\dfrac{4}{x-3}$

2. $\dfrac{y+3}{2y}$

3. $\dfrac{m-2}{m^2-2m-3}$

4. $\dfrac{2k+1}{3k^2+17k+10}$

Find the numerical value of each rational expression when **(a)** $x = -2$ *and* **(b)** $x = 4$.

5. $\dfrac{x^2}{x-5}$

6. $\dfrac{4x-3}{5x+2}$

7. $\dfrac{3x}{x^2-4}$

8. $\dfrac{x-1}{x+2}$

Write each rational expression in lowest terms.

9. $\dfrac{5a^3b^3}{15a^4b^2}$

10. $\dfrac{m-4}{4-m}$

11. $\dfrac{4x^2-9}{6-4x}$

12. $\dfrac{4p^2+8pq-5q^2}{10p^2-3pq-q^2}$

Write four equivalent expressions for each fraction.

13. $-\dfrac{4x-9}{2x+3}$

14. $\dfrac{8-3x}{3+6x}$

[15.2] *Find each product or quotient. Write each answer in lowest terms.*

15. $\dfrac{8x^2}{12x^5} \cdot \dfrac{6x^4}{2x}$

16. $\dfrac{9m^2}{(3m)^4} \div \dfrac{6m^5}{36m}$

17. $\dfrac{x-3}{4} \cdot \dfrac{5}{2x-6}$

18. $\dfrac{2r+3}{r-4} \cdot \dfrac{r^2-16}{6r+9}$

19. $\dfrac{3q+3}{5-6q} \div \dfrac{4q+4}{2(5-6q)}$

20. $\dfrac{y^2-6y+8}{y^2+3y-18} \div \dfrac{y-4}{y+6}$

21. $\dfrac{2p^2+13p+20}{p^2+p-12} \cdot \dfrac{p^2+2p-15}{2p^2+7p+5}$

22. $\dfrac{3z^2+5z-2}{9z^2-1} \cdot \dfrac{9z^2+6z+1}{z^2+5z+6}$

[15.3] *Find the least common denominator for each list of fractions.*

23. $\dfrac{1}{8}, \dfrac{5}{12}, \dfrac{7}{32}$

24. $\dfrac{4}{9y}, \dfrac{7}{12y^2}, \dfrac{5}{27y^4}$

25. $\dfrac{1}{m^2 + 2m}, \dfrac{4}{m^2 + 7m + 10}$

26. $\dfrac{3}{x^2 + 4x + 3}, \dfrac{5}{x^2 + 5x + 4}, \dfrac{2}{x^2 + 7x + 12}$

Rewrite each rational expression with the given denominator.

27. $\dfrac{5}{8} = \dfrac{}{56}$

28. $\dfrac{10}{k} = \dfrac{}{4k}$

29. $\dfrac{3}{2a^3} = \dfrac{}{10a^4}$

30. $\dfrac{9}{x - 3} = \dfrac{}{18 - 6x}$

31. $\dfrac{-3y}{2y - 10} = \dfrac{}{50 - 10y}$

32. $\dfrac{4b}{b^2 + 2b - 3} = \dfrac{}{(b + 3)(b - 1)(b + 2)}$

[15.4] *Add or subtract as indicated. Write each answer in lowest terms.*

33. $\dfrac{10}{x} + \dfrac{5}{x}$

34. $\dfrac{6}{3p} - \dfrac{12}{3p}$

35. $\dfrac{9}{k} - \dfrac{5}{k - 5}$

36. $\dfrac{4}{y} + \dfrac{7}{7 + y}$

37. $\dfrac{m}{3} - \dfrac{2 + 5m}{6}$

38. $\dfrac{12}{x^2} - \dfrac{3}{4x}$

39. $\dfrac{5}{a - 2b} + \dfrac{2}{a + 2b}$

40. $\dfrac{4}{k^2 - 9} - \dfrac{k + 3}{3k - 9}$

41. $\dfrac{8}{z^2 + 6z} - \dfrac{3}{z^2 + 4z - 12}$

42. $\dfrac{11}{2p - p^2} - \dfrac{2}{p^2 - 5p + 6}$

[15.5] *Simplify each complex fraction.*

43. $\dfrac{\dfrac{a^4}{b^2}}{\dfrac{a^3}{b}}$

44. $\dfrac{\dfrac{y - 3}{y}}{\dfrac{y + 3}{4y}}$

45. $\dfrac{\dfrac{3m + 2}{m}}{\dfrac{2m - 5}{6m}}$

46. $\dfrac{\dfrac{1}{p} - \dfrac{1}{q}}{\dfrac{1}{q - p}}$

47. $\dfrac{x + \dfrac{1}{w}}{x - \dfrac{1}{w}}$

48. $\dfrac{\dfrac{1}{r + t} - 1}{\dfrac{1}{r + t} + 1}$

[15.6] *Solve each equation. Check your solutions.*

49. $\dfrac{k}{5} - \dfrac{2}{3} = \dfrac{1}{2}$

50. $\dfrac{4-z}{z} + \dfrac{3}{2} = \dfrac{-4}{z}$

51. $\dfrac{x}{2} - \dfrac{x-3}{7} = -1$

52. $\dfrac{3y-1}{y-2} = \dfrac{5}{y-2} + 1$

53. $\dfrac{3}{m-2} + \dfrac{1}{m-1} = \dfrac{7}{m^2 - 3m + 2}$

Solve for the specified variable.

54. $m = \dfrac{Ry}{t}$ for t

55. $x = \dfrac{3y-5}{4}$ for y

56. $\dfrac{1}{r} - \dfrac{1}{s} = \dfrac{1}{t}$ for t

[15.7] *Solve each problem. Use the six-step method.*

57. One-fourth of a number is 9 less than the same number. What is the number?

58. In a certain fraction, the denominator is 5 less than the numerator. If 5 is added to both the numerator and the denominator, the resulting fraction is equivalent to $\tfrac{5}{4}$. Find the original fraction.

59. The denominator of a certain fraction is six times the numerator. If 3 is added to the numerator and subtracted from the denominator, the resulting fraction is equivalent to $\tfrac{2}{5}$. Find the original fraction.

60. On August 18, 1996, Scott Sharp won the True Value 200-mile Indy race driving a Ford with an average speed of 130.934 mph. What was his time? (*Source: Sports Illustrated Sports Almanac.*)

61. A man can plant his garden in 5 hr, working alone. His daughter can do the same job in 8 hr. How long would it take them if they worked together?

62. At a given hour, two steamboats leave a city in the same direction on a straight canal. One travels at 18 mph, and the other travels at 25 mph. In how many hours will the boats be 70 miles apart?

[15.8] *Solve each problem.*

63. If a parallelogram has a fixed area, the height varies inversely as the base. A parallelogram has a height of 8 cm and a base of 12 cm. Find the height if the base is changed to 24 cm.

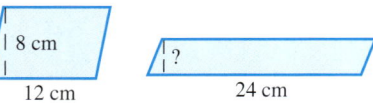

64. If y varies directly as x, and $x = 12$ when $y = 5$, find x when $y = 3$.

MIXED REVIEW EXERCISES

Perform the indicated operations.

65. $\dfrac{4}{m-1} - \dfrac{3}{m+1}$

66. $\dfrac{8p^5}{5} \div \dfrac{2p^3}{10}$

67. $\dfrac{r-3}{8} \div \dfrac{3r-9}{4}$

68. $\dfrac{\dfrac{5}{x} - 1}{\dfrac{5-x}{3x}}$

69. $\dfrac{4}{z^2 - 2z + 1} - \dfrac{3}{z^2 - 1}$

Solve.

70. $F = \dfrac{k}{d - D}$ for d

71. $\dfrac{2}{z} - \dfrac{z}{z+3} = \dfrac{1}{z+3}$

72. Anne Kelly flew her plane 400 km with the wind in the same time it took her to go 200 km against the wind. The speed of the wind is 50 kilometers per hour. Find the speed of the plane in still air.

73. "If Joe can paint a house in 3 hours, and Sam can paint the same house in 5 hours, how long does it take for them to do it together?" (From the movie *Little Big League*; see chapter opening.)

74. In rectangles of constant area, length and width vary inversely. When the length is 24, the width is 2. What is the width when the length is 12?

75. If w varies inversely as z, and $w = 16$ when $z = 3$, find w when $z = 2$.

Chapter 15
TEST

1. Find any values for which $\dfrac{3x-1}{x^2-2x-8}$ is undefined.

2. Find the numerical value of $\dfrac{6r+1}{2r^2-3r-20}$ when
   (a) $r = -2$ and (b) $r = 4$.

3. Write four rational expressions equivalent to $-\dfrac{6x-5}{2x+3}$.

Write each rational expression in lowest terms.

4. $\dfrac{-15x^6y^4}{5x^4y}$

5. $\dfrac{6a^2+a-2}{2a^2-3a+1}$

Multiply or divide. Write each answer in lowest terms.

6. $\dfrac{5(d-2)}{9} \div \dfrac{3(d-2)}{5}$

7. $\dfrac{6k^2-k-2}{8k^2+10k+3} \cdot \dfrac{4k^2+7k+3}{3k^2+5k+2}$

8. $\dfrac{4a^2+9a+2}{3a^2+11a+10} \div \dfrac{4a^2+17a+4}{3a^2+2a-5}$

Find the least common denominator for each list of fractions.

9. $\dfrac{-3}{10p^2}, \dfrac{21}{25p^3}, \dfrac{-7}{30p^5}$

10. $\dfrac{r+1}{2r^2+7r+6}, \dfrac{-2r+1}{2r^2-7r-15}$

Rewrite each rational expression with the given denominator.

11. $\dfrac{15}{4p} = \dfrac{}{64p^3}$

12. $\dfrac{3}{6m-12} = \dfrac{}{42m-84}$

Add or subtract. Write each answer in lowest terms.

13. $\dfrac{4x+2}{x+5} + \dfrac{-2x+8}{x+5}$

14. $\dfrac{-4}{y+2} + \dfrac{6}{5y+10}$

15. $\dfrac{x+1}{3-x} - \dfrac{x^2}{x-3}$

16. $\dfrac{3}{2m^2-9m-5} - \dfrac{m+1}{2m^2-m-1}$

1. _____
2. (a) _____
   (b) _____
3. _____
4. _____
5. _____
6. _____
7. _____
8. _____
9. _____
10. _____
11. _____
12. _____
13. _____
14. _____
15. _____
16. _____

Simplify each complex fraction.

17. $\dfrac{\frac{2p}{k^2}}{\frac{3p^2}{k^3}}$

18. $\dfrac{\frac{1}{x+3} - 1}{1 + \frac{1}{x+3}}$

19. Solve the equation $\dfrac{2x}{x-3} + \dfrac{1}{x+3} = \dfrac{-6}{x^2-9}$. Be sure to check your answer(s).

20. Solve the formula $F = \dfrac{k}{d-D}$ for D.

Solve each problem.

21. If the same number is added to the numerator and subtracted from the denominator of $\frac{5}{6}$, the resulting fraction is equivalent to $\frac{1}{10}$. What is the number?

22. A boat goes 7 mph in still water. It takes as long to go 20 miles upstream as 50 miles downstream. Find the speed of the current.

23. A man can paint a room in his house, working alone, in 5 hr. His wife can do the job in 4 hr. How long will it take them to paint the room if they work together?

24. If x varies directly as y, and $x = 12$ when $y = 4$, find x when $y = 9$.

25. Under certain conditions, the length of time that it takes for fruit to ripen during the growing season varies inversely as the average maximum temperature during the season. If it takes 25 days for fruit to ripen with an average maximum temperature of 80°F, find the number of days it would take at 75°F. Round your answer to the nearest whole number.

Roots and Radicals

16.1 **Evaluating Roots**

16.2 **Multiplying, Dividing, and Simplifying Radicals**

16.3 **Adding and Subtracting Radicals**

16.4 **Rationalizing the Denominator**

16.5 **More Simplifying and Operations with Radicals**

Summary Exercises on Operations with Radicals

16.6 **Solving Equations with Radicals**

The London Eye opened on New Year's Eve in 1999. This unique Ferris wheel features 32 observation capsules and has a diameter of 135 meters. Located on the bank of the Thames River, it faces the Houses of Parliament and is the sixth tallest structure in London. (*Source:* www.londoneye.com)

In the Focus on Real-Data Applications at the end of Section 16.6, we use a formula involving *radicals,* the subject of this chapter, to determine the truth of the claim that passengers on the London Eye can see Windsor Castle, 25 miles away.

16.1 Evaluating Roots

OBJECTIVES

1. Find square roots.
2. Decide whether a given root is rational, irrational, or not a real number.
3. Find decimal approximations for irrational square roots.
4. Use the Pythagorean formula.
5. Find higher roots.

In **Section 1.8**, we discussed the idea of the *square* of a number. Recall that squaring a number means multiplying the number by itself.

If $a = 8$, then $a^2 = 8 \cdot 8 = 64$

If $a = -4$, then $a^2 = (-4)(-4) = 16$

If $a = -\dfrac{1}{2}$, then $a^2 = \left(-\dfrac{1}{2}\right)\left(-\dfrac{1}{2}\right) = \dfrac{1}{4}$

In this chapter, the opposite process is considered.

If $a^2 = 64$, then $a = ?$

If $a^2 = 16$, then $a = ?$

If $a^2 = \dfrac{1}{4}$, then $a = ?$

❶ Find all square roots.

(a) 100

(b) 25

(c) 36

(d) $\dfrac{25}{36}$

OBJECTIVE 1 Find square roots. To find a in the three preceding statements, we must find a number that when multiplied by itself results in the given number. The number a is called a **square root** of the number a^2.

EXAMPLE 1 Finding All Square Roots of a Number

Find all square roots of 49.

To find a square root of 49, think of a number that when multiplied by itself gives 49. As discussed in **Section 5.8**, one square root is 7 because $7 \cdot 7 = 49$. But there is another square root of 49, which is -7, because $(-7)(-7) = 49$. The number 49 has *two* square roots, 7 and -7; one is positive, and one is negative.

◀◀◀ **Work Problem 1 at the Side.**

The **positive** or **principal square root** of a number is written with the symbol $\sqrt{}$. For example, the positive square root of 121 is 11, written

$$\sqrt{121} = 11$$

The symbol $-\sqrt{}$ is used for the **negative square root** of a number. For example, the negative square root of 121 is -11, written

$$-\sqrt{121} = -11$$

The symbol $\sqrt{}$, called a **radical sign**, always represents the positive square root (except that $\sqrt{0} = 0$). The number inside the radical sign is called the **radicand**, and the entire expression, radical sign and radicand, is called a **radical**.

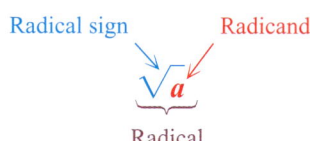

Radical sign   Radicand

Radical

An algebraic expression containing a radical is called a **radical expression**.

Radicals have a long mathematical history. The radical sign $\sqrt{}$ has been used since sixteenth-century Germany and was probably derived from the letter R. The radical symbol in the margin comes from the Latin word for root, *radix*. It was first used by Leonardo da Pisa (Fibonacci) in 1220.

Early radical symbol

ANSWERS

1. (a) $10, -10$ (b) $5, -5$ (c) $6, -6$ (d) $\dfrac{5}{6}, -\dfrac{5}{6}$

Our discussion of square roots is summarized as follows.

Square Roots

If a is a positive real number, then

$\sqrt{a}$ is the positive or principal square root of a,

and $\quad -\sqrt{a}$ is the negative square root of a

For nonnegative a,

$$\sqrt{a} \cdot \sqrt{a} = (\sqrt{a})^2 = a \quad \text{and} \quad -\sqrt{a} \cdot (-\sqrt{a}) = (-\sqrt{a})^2 = a$$

Also, $\sqrt{0} = 0$.

▦ **Calculator Tip** Recall from **Section 5.8** that most calculators have a square root key, usually labeled √x̄, that allows us to find the square root of a number. On some models, you may have to press the INV or 2nd key to access the square root function.

EXAMPLE 2 Finding Square Roots

Find each square root.

(a) $\sqrt{144}$

The radical $\sqrt{144}$ represents the positive or principal square root of 144. Think of a positive number whose square is 144.

$$12^2 = 144, \quad \text{so} \quad \sqrt{144} = 12$$

(b) $-\sqrt{1024}$

This symbol represents the negative square root of 1024. A calculator with a square root key can be used to find $\sqrt{1024} = 32$. Then, $-\sqrt{1024} = -32$.

(c) $\sqrt{\dfrac{4}{9}} = \dfrac{2}{3}$ **(d)** $-\sqrt{\dfrac{16}{49}} = -\dfrac{4}{7}$ **(e)** $\sqrt{0.81} = 0.9$

▶ **Work Problem 2 at the Side.**

As noted above, when the square root of a positive real number is squared, the result is that positive real number. Also, $(\sqrt{0})^2 = 0$.

EXAMPLE 3 Squaring Radical Expressions

Find the *square* of each radical expression.

(a) $\sqrt{13}$

$(\sqrt{13})^2 = 13$ Definition of square root

(b) $-\sqrt{29}$

$(-\sqrt{29})^2 = 29$ The square of a *negative* number is positive.

(c) $\sqrt{p^2 + 1}$

$(\sqrt{p^2 + 1})^2 = p^2 + 1$

▶ **Work Problem 3 at the Side.**

Section 16.1 Evaluating Roots **1087**

❷ Find each square root.

(a) $\sqrt{16}$

(b) $-\sqrt{169}$

(c) $-\sqrt{225}$

(d) $\sqrt{729}$

(e) $\sqrt{\dfrac{36}{25}}$

(f) $\sqrt{0.49}$

❸ Find the *square* of each radical expression.

(a) $\sqrt{41}$

(b) $-\sqrt{39}$

(c) $\sqrt{2x^2 + 3}$

ANSWERS

2. (a) 4 (b) -13 (c) -15
   (d) 27 (e) $\dfrac{6}{5}$ (f) 0.7

3. (a) 41 (b) 39 (c) $2x^2 + 3$

4 Tell whether each square root is *rational*, *irrational*, or *not a real number*.

(a) $\sqrt{9}$

(b) $\sqrt{7}$

(c) $\sqrt{\dfrac{4}{9}}$

(d) $\sqrt{72}$

(e) $\sqrt{-43}$

ANSWERS
4. (a) rational (b) irrational (c) rational
   (d) irrational (e) not a real number

OBJECTIVE 2 Decide whether a given root is rational, irrational, or not a real number. All numbers with square roots that are rational are called **perfect squares**.

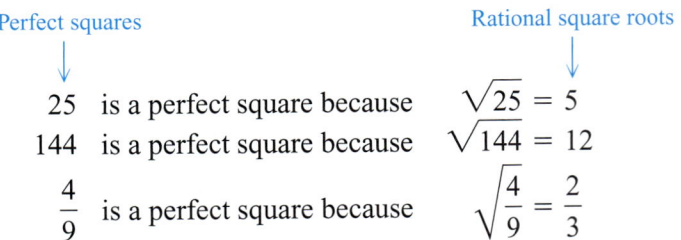

25 is a perfect square because $\sqrt{25} = 5$
144 is a perfect square because $\sqrt{144} = 12$
$\dfrac{4}{9}$ is a perfect square because $\sqrt{\dfrac{4}{9}} = \dfrac{2}{3}$

A number that is not a perfect square has a square root that is not a rational number. For example, $\sqrt{5}$ is not a rational number because it cannot be written as the ratio of two integers. Its decimal equivalent (or approximation) neither terminates nor repeats. However, $\sqrt{5}$ is a real number and corresponds to a point on the number line. As mentioned in **Section 10.1**, a real number that is not rational is called an *irrational number*. The number $\sqrt{5}$ is irrational. Many square roots of integers are irrational.

> If a is a *positive* real number that is *not* a perfect square, then
> $\sqrt{a}$ is irrational.

Not every number has a real number square root. For example, there is no real number that can be squared to obtain -36. (The square of a real number can never be negative.) Because of this, $\sqrt{-36}$ *is not a real number.*

> If a is a *negative* real number, then $\sqrt{a}$ is *not* a real number.

CAUTION
Be careful not to confuse $\sqrt{-36}$ and $-\sqrt{36}$. $\sqrt{-36}$ is not a real number since there is no real number that can be squared to obtain -36. However, $-\sqrt{36}$ is the negative square root of 36, which is -6.

EXAMPLE 4 Identifying Types of Square Roots

Tell whether each square root is *rational*, *irrational*, or *not a real number*.

(a) $\sqrt{17}$
Because 17 is not a perfect square, $\sqrt{17}$ is irrational.

(b) $\sqrt{64}$
The number 64 is a perfect square, 8^2, so $\sqrt{64} = 8$, a rational number.

(c) $\sqrt{-25}$
There is no real number whose square is -25. Therefore, $\sqrt{-25}$ is not a real number.

Work Problem 4 at the Side.

> **NOTE**
> Not all irrational numbers are square roots of integers. For example, π (approximately 3.14159) is an irrational number that is not a square root of any integer.

OBJECTIVE 3 Find decimal approximations for irrational square roots. Even if a number is irrational, a decimal that approximates the number can be found using a calculator, as we did in **Section 5.8.** For example, if we use a calculator to find $\sqrt{10}$, the display might show 3.16227766, which is only an *approximation* of $\sqrt{10}$, not an exact rational value.

EXAMPLE 5 Approximating Irrational Square Roots

Use a calculator to find a decimal approximation for each square root. Round answers to the nearest thousandth.

(a) $\sqrt{11}$

Using the square root key of a calculator gives 3.31662479, which rounds to 3.317, so $\sqrt{11} \approx 3.317$, where $\approx$ means "is approximately equal to."

(b) $\sqrt{39} \approx 6.245$ Use a calculator.   (c) $-\sqrt{740} \approx -27.203$

Work Problem 5 at the Side.

OBJECTIVE 4 Use the Pythagorean formula. Many applications of square roots use the Pythagorean formula. Recall from **Section 14.7** that by this formula if c is the length of the hypotenuse of a right triangle, and a and b are the lengths of the two legs, as shown in Figure 1, then

$$a^2 + b^2 = c^2$$

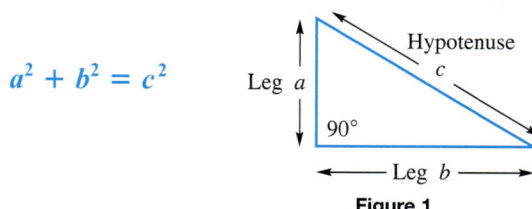

Figure 1

EXAMPLE 6 Using the Pythagorean Formula

Find the length of the unknown side of each right triangle with sides a, b, and c, where c is the hypotenuse. Round answers to the nearest thousandth.

(a) $a = 3, b = 4$

Use the Pythagorean formula to find c^2 first.

$c^2 = a^2 + b^2$
$c^2 = 3^2 + 4^2$ Let $a = 3$ and $b = 4$.
$c^2 = 9 + 16$ Square.
$c^2 = 25$ Add.

Since the length of a side of a triangle must be a positive number, find the positive square root of 25 to get c.

$$c = \sqrt{25} = 5$$

Continued on Next Page

5 Find a decimal approximation for each square root. Round answers to the nearest thousandth.

(a) $\sqrt{28}$

(b) $\sqrt{63}$

(c) $-\sqrt{190}$

(d) $\sqrt{1000}$

ANSWERS
5. (a) 5.292 (b) 7.937 (c) −13.784
   (d) 31.623

6 Find the length of the unknown side in each right triangle. Round any decimal approximations to the nearest thousandth.

(a) $a = 7, b = 24$

(b) $c = 9, b = 5$

Substitute the given values in the Pythagorean formula. Then solve for a^2.

$$c^2 = a^2 + b^2$$
$$9^2 = a^2 + 5^2 \quad \text{Let } c = 9 \text{ and } b = 5.$$
$$81 = a^2 + 25 \quad \text{Square.}$$
$$56 = a^2 \quad \text{Subtract 25.}$$

Use a calculator to find $a = \sqrt{56} \approx 7.483$.

CAUTION
Be careful not to make the common mistake of thinking that $\sqrt{a^2 + b^2}$ equals $a + b$. As Example 6(a) shows, $\sqrt{9 + 16} = \sqrt{25} = 5$. However, $\sqrt{9} + \sqrt{16} = 3 + 4 = 7$. Since $5 \neq 7$, in general,
$$\sqrt{a^2 + b^2} \neq a + b.$$

(b) $c = 20, b = 13$

◀◀◀ **Work Problem 6 at the Side.**

The Pythagorean formula can be used to solve applied problems that involve right triangles. Use the same six problem-solving steps that we have been using throughout the text.

EXAMPLE 7 Using the Pythagorean Formula to Solve an Application

A ladder 10 ft long leans against a wall. The foot of the ladder is 6 ft from the base of the wall. How high up the wall does the top of the ladder rest?

Step 1 **Read** the problem again.

Step 2 **Assign a variable.** As shown in Figure 2, a right triangle is formed with the ladder as the hypotenuse. Let a represent the height of the top of the ladder when measured straight down to the ground.

(c)

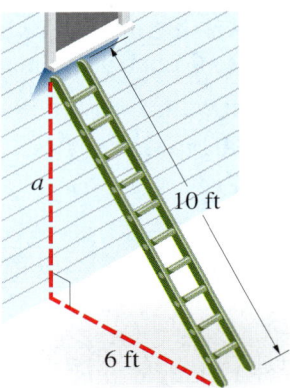

Triangle with sides 8, 11, and ?

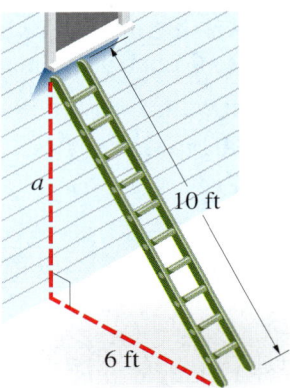

Figure 2

Continued on Next Page

ANSWERS
6. (a) $\sqrt{625} = 25$ (b) $\sqrt{231} \approx 15.199$
(c) $\sqrt{57} \approx 7.550$

Step 3 **Write an equation** using the Pythagorean formula.

$$c^2 = a^2 + b^2$$
$$10^2 = a^2 + 6^2 \quad \text{Let } c = 10 \text{ and } b = 6.$$

Step 4 **Solve.**
$$100 = a^2 + 36 \quad \text{Square.}$$
$$64 = a^2 \quad \text{Subtract 36.}$$
$$\sqrt{64} = a$$
$$a = 8 \quad \sqrt{64} = 8$$

Choose the positive square root of 64 since *a* represents a length.

Step 5 **State the answer.** The top of the ladder rests 8 ft up the wall.

Step 6 **Check.** From Figure 2, we see that we must have
$$8^2 + 6^2 = 10^2 \quad ?$$
$$64 + 36 = 100. \quad \text{True}$$

The check confirms that the top of the ladder rests 8 ft up the wall.

▶ **Work Problem 7 at the Side.** ▶▶▶

7 A rectangle has dimensions 5 ft by 12 ft. Find the length of its diagonal.

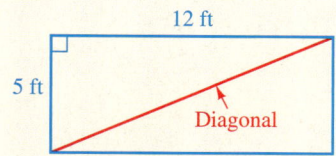

(Note that the diagonal divides the rectangle into two right triangles with itself as the hypotenuse.)

OBJECTIVE 5 **Find higher roots.** Finding the square root of a number is the inverse (reverse) of squaring a number. In a similar way, there are inverses to finding the cube of a number, or finding the fourth or higher power of a number. These inverses are the **cube root**, written $\sqrt[3]{a}$, and the **fourth root**, written $\sqrt[4]{a}$. Similar symbols are used for higher roots. In general, we have the following.

$$\sqrt[n]{a}$$

The *n*th root of *a* is written $\sqrt[n]{a}$

In $\sqrt[n]{a}$, the number *n* is the **index** or **order** of the radical.

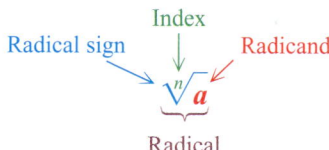

We could write $\sqrt[2]{a}$ instead of $\sqrt{a}$, but the simpler symbol $\sqrt{a}$ is customary since the square root is the most commonly used root.

8 Complete the following list of perfect cubes and perfect fourth powers.

Perfect Cubes	Perfect Fourth Powers
$1^3 = 1$	$1^4 = 1$
$2^3 = 8$	$2^4 = 16$
$3^3 = 27$	$3^4 = 81$
$4^3 = \underline{}$	$4^4 = \underline{}$
$5^3 = \underline{}$	$5^4 = \underline{}$
$6^3 = \underline{}$	$6^4 = \underline{}$
$7^3 = \underline{}$	$7^4 = \underline{}$
$8^3 = \underline{}$	$8^4 = \underline{}$
$9^3 = \underline{}$	$9^4 = \underline{}$
$10^3 = \underline{}$	$10^4 = \underline{}$

🖩 **Calculator Tip** A calculator that has a key marked $\sqrt[x]{y}$, x^y, or y^x can be used to find higher roots. (On some models you must first press the INV or 2nd key.)

When working with cube roots or fourth roots, it is helpful to memorize the first few *perfect cubes* ($1^3 = 1$, $2^3 = 8$, $3^3 = 27$, and so on) and the first few *perfect fourth powers* ($1^4 = 1$, $2^4 = 16$, $3^4 = 81$, and so on).

▶ **Work Problem 8 at the Side.** ▶▶▶

ANSWERS
7. 13 ft
8. Perfect cubes: 64; 125; 216; 343; 512; 729; 1000
   Perfect fourth powers: 256; 625; 1296; 2401; 4096; 6561; 10,000

Chapter 16 Roots and Radicals

9 Find each cube root.

(a) $\sqrt[3]{27}$

(b) $\sqrt[3]{64}$

(c) $\sqrt[3]{-125}$

10 Find each root.

(a) $\sqrt[4]{81}$

(b) $\sqrt[4]{-81}$

(c) $-\sqrt[4]{625}$

(d) $\sqrt[5]{243}$

(e) $\sqrt[5]{-243}$

EXAMPLE 8 Finding Cube Roots

Find each cube root.

(a) $\sqrt[3]{8}$

Look for a number that can be cubed to give 8. Because $2^3 = 8$, $\sqrt[3]{8} = 2$.

(b) $\sqrt[3]{-8} = -2$ because $(-2)^3 = -8$.

(c) $\sqrt[3]{216} = 6$ because $6^3 = 216$.

Notice in Example 8(b) that we can find the cube root of a negative number. (Contrast this with the square root of a negative number, which is not real.) In fact, the cube root of a positive number is positive, and the cube root of a negative number is negative. ***There is only one real number cube root for each real number.***

Work Problem 9 at the Side.

When a radical has an *even index* (square root, fourth root, and so on), *the radicand must be nonnegative* to yield a real number root. Also,

$$\sqrt{a}, \sqrt[4]{a}, \sqrt[6]{a}, \text{ and so on are positive (principal) roots;}$$

$$-\sqrt{a}, -\sqrt[4]{a}, -\sqrt[6]{a}, \text{ and so on are negative roots.}$$

EXAMPLE 9 Finding Higher Roots

Find each root.

(a) $\sqrt[4]{16}$

$\sqrt[4]{16} = 2$ because 2 is positive and $2^4 = 16$.

(b) $-\sqrt[4]{16}$

From part (a), $\sqrt[4]{16} = 2$, so the negative root $-\sqrt[4]{16} = -2$.

(c) $\sqrt[4]{-16}$

For a real number fourth root, the radicand must be nonnegative. There is no real number that equals $\sqrt[4]{-16}$.

(d) $-\sqrt[5]{32}$

First find $\sqrt[5]{32}$. Because 2 is the number whose fifth power is 32, $\sqrt[5]{32} = 2$. Since $\sqrt[5]{32} = 2$, it follows that

$$-\sqrt[5]{32} = -2.$$

(e) $\sqrt[5]{-32}$

Because $(-2)^5 = -32$, $\sqrt[5]{-32} = -2$.

Work Problem 10 at the Side.

ANSWERS

9. (a) 3   (b) 4   (c) −5
10. (a) 3   (b) not a real number
    (c) −5   (d) 3   (e) −3

16.1 Exercises

Decide whether each statement is true or false. If false, tell why.

1. Every positive number has two real square roots.
2. A negative number has negative square roots.
3. Every nonnegative number has two real square roots.
4. The positive square root of a positive number is its principal square root.
5. The cube root of every real number has the same sign as the number itself.
6. Every positive number has three real cube roots.

Find all square roots of each number. See Example 1.

7. 9
8. 16
9. 64
10. 100
11. 169
12. 225
13. $\dfrac{25}{196}$
14. $\dfrac{81}{400}$
15. 900
16. 1600

Find each square root. See Examples 2 and 4(c).

17. $\sqrt{1}$
18. $\sqrt{4}$
19. $\sqrt{49}$
20. $\sqrt{81}$
21. $-\sqrt{256}$
22. $-\sqrt{196}$
23. $-\sqrt{\dfrac{144}{121}}$
24. $-\sqrt{\dfrac{49}{36}}$
25. $\sqrt{0.64}$
26. $\sqrt{0.16}$
27. $\sqrt{-121}$
28. $\sqrt{-64}$
29. $-\sqrt{-49}$
30. $-\sqrt{-100}$

Find the square of each radical expression. See Example 3.

31. $\sqrt{100}$
32. $\sqrt{36}$
33. $-\sqrt{19}$
34. $-\sqrt{99}$
35. $\sqrt{\dfrac{2}{3}}$
36. $\sqrt{\dfrac{5}{7}}$
37. $\sqrt{3x^2 + 4}$
38. $\sqrt{9y^2 + 3}$

What must be true about the value of x for each statement in Exercises 39–42 to be true?

39. $\sqrt{x}$ represents a positive number.

40. $-\sqrt{x}$ represents a negative number.

41. $\sqrt{x}$ is not a real number.

42. $-\sqrt{x}$ is not a real number.

Write rational, irrational, *or* not a real number *for each number. If a number is rational, give its exact value. If a number is irrational, give a decimal approximation to the nearest thousandth. Use a calculator as necessary. See Examples 4 and 5.*

43. $\sqrt{25}$

44. $\sqrt{169}$

45. $\sqrt{29}$

46. $\sqrt{33}$

47. $-\sqrt{64}$

48. $-\sqrt{81}$

49. $-\sqrt{300}$

50. $-\sqrt{500}$

51. $\sqrt{-29}$

52. $\sqrt{-47}$

53. $\sqrt{1200}$

54. $\sqrt{1500}$

Work Exercises 55 and 56 without using a calculator.

55. Choose the best estimate for the length and width (in meters) of this rectangle.

A. 11 by 6   **B.** 11 by 7   **C.** 10 by 7   **D.** 10 by 6

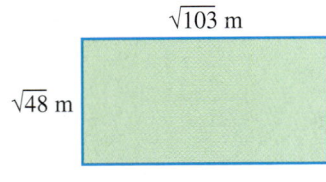

56. Choose the best estimate for the base and height (in feet) of this triangle.

A. $b = 8, h = 5$ **B.** $b = 8, h = 4$
C. $b = 9, h = 5$ **D.** $b = 9, h = 4$

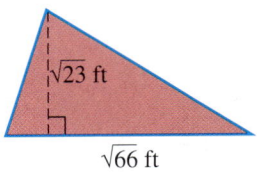

Find the length of the unknown side of each right triangle with sides a, b, and c, where c is the hypotenuse. See Figure 1 and Example 6. Give any decimal approximations to the nearest thousandth.

57. $a = 8, b = 15$

58. $a = 24, b = 10$

59. $a = 6, c = 10$

60. $b = 12, c = 13$

61. $a = 11, b = 4$

62. $a = 13, b = 9$

Solve each problem. See Example 7.

63. The diagonal of a rectangle measures 25 cm. The width of the rectangle is 7 cm. Find the length of the rectangle.

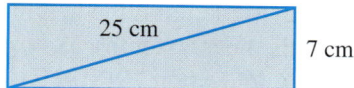

64. The length of a rectangle is 40 m, and the width is 9 m. Find the measure of the diagonal of the rectangle.

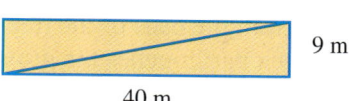

65. Tyler is flying a kite on 100 ft of string. How high is it above his hand (vertically) if the horizontal distance between Tyler and the kite is 60 ft?

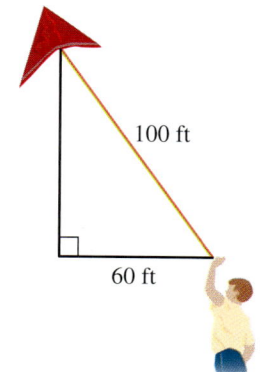

66. A guy wire is attached to the mast of a short-wave transmitting antenna. It is attached 96 ft above ground level. If the wire is staked to the ground 72 ft from the base of the mast, how long is the wire?

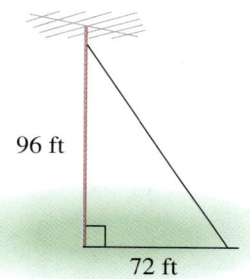

67. A surveyor measured the distances shown in the figure. Find the distance across the lake between points R and S.

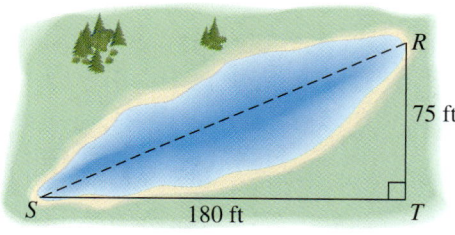

68. A boat is being pulled toward a dock with a rope attached at water level. When the boat is 24 ft from the dock, 30 ft of rope is extended. What is the height of the dock above the water?

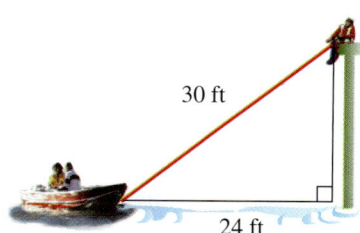

69. Find the length of the unknown side (to the nearest thousandth) in the figure.

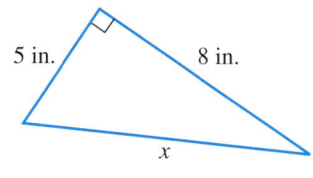

70. What is the length of the unknown side (to the nearest thousandth) in the figure?

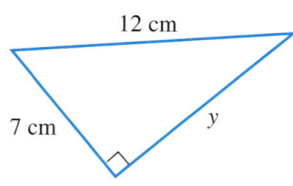

71. One of the authors of this text took this photo of a broken tree in a field near his home. The vertical distance from the base of the broken tree to the point of the break is 6.5 ft. The length of the broken part is 15 ft. How far along the ground (to the nearest tenth) is it from the base of the tree to the point where the broken part touches the ground?

72. A television set is "sized" according to the diagonal measurement of the viewing screen. One of the authors of this text purchased a 19-in. TV, so the TV measures 19 in. from one corner of the viewing screen diagonally to the other corner. The viewing screen is 15.5 in. wide. Find the height of the viewing screen (to the nearest tenth). (*Source:* Phillips Magnavox color television 19PR21C1.)

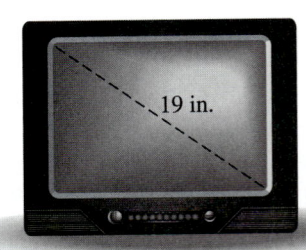

73. Use specific values for a and b different from those given in the "Caution" following Example 6 to show that $\sqrt{a^2 + b^2} \neq a + b$.

74. Why would the values $a = 0$ and $b = 1$ *not* be satisfactory in Exercise 73?

Find each root. See Examples 8 and 9.

75. $\sqrt[3]{1}$

76. $\sqrt[3]{8}$

77. $\sqrt[3]{125}$

78. $\sqrt[3]{1000}$

79. $\sqrt[3]{-27}$

80. $\sqrt[3]{-64}$

81. $\sqrt[3]{-216}$

82. $\sqrt[3]{-343}$

83. $-\sqrt[3]{8}$

84. $-\sqrt[3]{216}$

85. $\sqrt[4]{256}$

86. $\sqrt[4]{625}$

87. $\sqrt[4]{1296}$

88. $\sqrt[4]{10{,}000}$

89. $\sqrt[4]{-1}$

90. $\sqrt[4]{-625}$

91. $-\sqrt[4]{81}$

92. $-\sqrt[4]{256}$

93. $\sqrt[5]{-1024}$

94. $\sqrt[5]{-100{,}000}$

16.2 Multiplying, Dividing, and Simplifying Radicals

OBJECTIVE 1 **Multiply radicals.** We develop several rules for finding products and quotients of radicals in this section. To illustrate the rule for products, notice that

$$\sqrt{4} \cdot \sqrt{9} = 2 \cdot 3 = 6 \quad \text{and} \quad \sqrt{4 \cdot 9} = \sqrt{36} = 6$$

showing that

$$\sqrt{4} \cdot \sqrt{9} = \sqrt{4 \cdot 9}$$

This result is a particular case of the **product rule for radicals**.

Product Rule for Radicals
For nonnegative real numbers a and b,

$$\sqrt{a} \cdot \sqrt{b} = \sqrt{a \cdot b} \quad \text{and} \quad \sqrt{a \cdot b} = \sqrt{a} \cdot \sqrt{b}$$

In words, the product of two radicals is the radical of the product, and the radical of a product is the product of the radicals.

EXAMPLE 1 Using the Product Rule to Multiply Radicals

Use the product rule for radicals to find each product.

(a) $\sqrt{2} \cdot \sqrt{3} = \sqrt{2 \cdot 3} = \sqrt{6}$ Product rule

(b) $\sqrt{7} \cdot \sqrt{5} = \sqrt{35}$ Product rule

(c) $\sqrt{11} \cdot \sqrt{a} = \sqrt{11a}$ Assume $a \geq 0$.

Work Problem 1 at the Side.

OBJECTIVE 2 **Simplify radicals using the product rule.** A square root radical is *simplified* when no perfect square factor remains under the radical sign. This is accomplished by using the product rule.

EXAMPLE 2 Using the Product Rule to Simplify Radicals

Simplify each radical.

(a) $\sqrt{20}$

Because 20 has a perfect square factor of 4, we can write

$$\sqrt{20} = \sqrt{4 \cdot 5} \quad \text{4 is a perfect square.}$$
$$= \sqrt{4} \cdot \sqrt{5} \quad \text{Product rule}$$
$$= 2\sqrt{5} \quad \sqrt{4} = 2$$

Thus, $\sqrt{20} = 2\sqrt{5}$. Because 5 has no perfect square factor (other than 1), $2\sqrt{5}$ is called the *simplified form* of $\sqrt{20}$. Note that $2\sqrt{5}$ represents a product, where the factors are 2 and $\sqrt{5}$.

We could also factor 20 into prime factors and look for pairs of like factors. Each pair of like factors produces one factor outside the radical.

$$\sqrt{20} = \sqrt{2 \cdot 2 \cdot 5} = 2\sqrt{5}$$

Continued on Next Page

OBJECTIVES

1. Multiply radicals.
2. Simplify radicals using the product rule.
3. Simplify radicals using the quotient rule.
4. Simplify radicals involving variables.
5. Simplify higher roots.

1 Use the product rule for radicals to find each product.

(a) $\sqrt{6} \cdot \sqrt{11}$

(b) $\sqrt{2} \cdot \sqrt{5}$

(c) $\sqrt{10} \cdot \sqrt{r}, \quad r \geq 0$

ANSWERS
1. (a) $\sqrt{66}$ (b) $\sqrt{10}$ (c) $\sqrt{10r}$

Chapter 16 Roots and Radicals

2 Simplify each radical.

(a) $\sqrt{8}$

(b) $\sqrt{27}$

(c) $\sqrt{50}$

(d) $\sqrt{60}$

(e) $\sqrt{30}$

(b) $\sqrt{72}$

Begin by looking for the *largest* perfect square factor of 72. This number is 36, so

$\sqrt{72} = \sqrt{36 \cdot 2}$ 36 is a perfect square.

$= \sqrt{36} \cdot \sqrt{2}$ Product rule

$= 6\sqrt{2}$ $\sqrt{36} = 6$

We could also factor 72 into its prime factors and look for pairs of like factors.

$\sqrt{72} = \sqrt{2 \cdot 2 \cdot 2 \cdot 3 \cdot 3} = 2 \cdot 3 \cdot \sqrt{2} = 6\sqrt{2}$

In either case, we obtain $6\sqrt{2}$ as the simplified form of $\sqrt{72}$. However, our work is simpler if we begin with the largest perfect square factor.

(c) $\sqrt{300} = \sqrt{100 \cdot 3}$ 100 is a perfect square.

$= \sqrt{100} \cdot \sqrt{3}$ Product rule

$= 10\sqrt{3}$ $\sqrt{100} = 10$

(d) $\sqrt{15}$

The number 15 has no perfect square factors (except 1), so $\sqrt{15}$ cannot be simplified further.

Work Problem 2 at the Side.

Sometimes the product rule can be used to simplify a product, as Example 3 shows.

EXAMPLE 3 Multiplying and Simplifying Radicals

Find each product and simplify.

(a) $\sqrt{9} \cdot \sqrt{75} = 3\sqrt{75}$ $\sqrt{9} = 3$

$= 3\sqrt{25 \cdot 3}$ 25 is a perfect square.

$= 3\sqrt{25} \cdot \sqrt{3}$ Product rule

$= 3 \cdot 5 \cdot \sqrt{3}$ $\sqrt{25} = 5$

$= 15\sqrt{3}$ Multiply.

Notice that we could have used the product rule to get $\sqrt{9} \cdot \sqrt{75} = \sqrt{675}$, and then simplified. However, the product rule as used here allows us to obtain the final answer without using a large number like 675.

(b) $\sqrt{8} \cdot \sqrt{12} = \sqrt{8 \cdot 12}$ Product rule

$= \sqrt{4 \cdot 2 \cdot 4 \cdot 3}$ Factor; 4 is a perfect square.

$= \sqrt{4} \cdot \sqrt{4} \cdot \sqrt{2 \cdot 3}$ Product rule

$= 2 \cdot 2 \cdot \sqrt{6}$ $\sqrt{4} = 2$

$= 4\sqrt{6}$ Multiply.

Continued on Next Page

ANSWERS

2. (a) $2\sqrt{2}$ (b) $3\sqrt{3}$ (c) $5\sqrt{2}$
    (d) $2\sqrt{15}$ (e) cannot be simplified further

(c) $2\sqrt{3} \cdot 3\sqrt{6} = 2 \cdot 3 \cdot \sqrt{3 \cdot 6}$ Product rule

$\phantom{2\sqrt{3} \cdot 3\sqrt{6}} = 6\sqrt{18}$ Multiply.

$\phantom{2\sqrt{3} \cdot 3\sqrt{6}} = 6\sqrt{9 \cdot 2}$ Factor; 9 is a perfect square.

$\phantom{2\sqrt{3} \cdot 3\sqrt{6}} = 6\sqrt{9} \cdot \sqrt{2}$ Product rule

$\phantom{2\sqrt{3} \cdot 3\sqrt{6}} = 6 \cdot 3 \cdot \sqrt{2}$ $\sqrt{9} = 3$

$\phantom{2\sqrt{3} \cdot 3\sqrt{6}} = 18\sqrt{2}$ Multiply.

> **NOTE**
> We could also simplify Example 3(b) as follows.
> $$\sqrt{8} \cdot \sqrt{12} = \sqrt{4 \cdot 2} \cdot \sqrt{4 \cdot 3}$$
> $$= 2\sqrt{2} \cdot 2\sqrt{3}$$
> $$= 2 \cdot 2 \cdot \sqrt{2} \cdot \sqrt{3}$$
> $$= 4\sqrt{6} \quad \text{Same result}$$
> There is often more than one way to find such a product.

Work Problem 3 at the Side.

OBJECTIVE 3   **Simplify radicals using the quotient rule.** The **quotient rule for radicals** is very similar to the product rule. It, too, can be used either way.

> **Quotient Rule for Radicals**
> If a and b are nonnegative real numbers and $b \neq 0$, then
> $$\sqrt{\frac{a}{b}} = \frac{\sqrt{a}}{\sqrt{b}} \quad \text{and} \quad \frac{\sqrt{a}}{\sqrt{b}} = \sqrt{\frac{a}{b}}.$$
> In words, the radical of a quotient is the quotient of the radicals, and the quotient of two radicals is the radical of the quotient.

EXAMPLE 4   **Using the Quotient Rule to Simplify Radicals**

Use the quotient rule to simplify each radical.

(a) $\sqrt{\frac{25}{9}} = \frac{\sqrt{25}}{\sqrt{9}} = \frac{5}{3}$ Quotient rule

(b) $\frac{\sqrt{288}}{\sqrt{2}} = \sqrt{\frac{288}{2}} = \sqrt{144} = 12$ Quotient rule

(c) $\sqrt{\frac{3}{4}} = \frac{\sqrt{3}}{\sqrt{4}} = \frac{\sqrt{3}}{2}$ Quotient rule

Work Problem 4 at the Side.

3 Find each product and simplify.

(a) $\sqrt{3} \cdot \sqrt{15}$

(b) $\sqrt{10} \cdot \sqrt{50}$

(c) $\sqrt{12} \cdot \sqrt{2}$

(d) $\sqrt{7} \cdot \sqrt{14}$

(e) $3\sqrt{5} \cdot 4\sqrt{10}$

4 Use the quotient rule to simplify each radical.

(a) $\sqrt{\frac{81}{16}}$

(b) $\frac{\sqrt{192}}{\sqrt{3}}$

(c) $\sqrt{\frac{10}{49}}$

ANSWERS

3. (a) $3\sqrt{5}$ (b) $10\sqrt{5}$ (c) $2\sqrt{6}$
    (d) $7\sqrt{2}$ (e) $60\sqrt{2}$

4. (a) $\frac{9}{4}$ (b) 8   (c) $\frac{\sqrt{10}}{7}$

Chapter 16 Roots and Radicals

5 Simplify $\dfrac{8\sqrt{50}}{4\sqrt{5}}$.

EXAMPLE 5 Using the Quotient Rule to Divide Radicals

Simplify $\dfrac{27\sqrt{15}}{9\sqrt{3}}$.

Use multiplication of fractions and the quotient rule as follows.

$$\frac{27\sqrt{15}}{9\sqrt{3}} = \frac{27}{9} \cdot \frac{\sqrt{15}}{\sqrt{3}} = \frac{27}{9} \cdot \sqrt{\frac{15}{3}} = 3\sqrt{5}$$

Work Problem 5 at the Side.

Some problems require both the product and quotient rules.

EXAMPLE 6 Using Both the Product and Quotient Rules

Simplify $\sqrt{\dfrac{3}{5}} \cdot \sqrt{\dfrac{1}{5}}$.

$$\sqrt{\frac{3}{5}} \cdot \sqrt{\frac{1}{5}} = \sqrt{\frac{3}{5} \cdot \frac{1}{5}} \quad \text{Product rule}$$

$$= \sqrt{\frac{3}{25}} \quad \text{Multiply fractions.}$$

$$= \frac{\sqrt{3}}{\sqrt{25}} \quad \text{Quotient rule}$$

$$= \frac{\sqrt{3}}{5} \quad \sqrt{25} = 5$$

6 Simplify.

(a) $\sqrt{\dfrac{5}{6}} \cdot \sqrt{120}$

(b) $\sqrt{\dfrac{3}{8}} \cdot \sqrt{\dfrac{7}{2}}$

Work Problem 6 at the Side.

OBJECTIVE 4 Simplify radicals involving variables. Radicals can also involve variables, such as $\sqrt{x^2}$. Simplifying such radicals can get a little tricky. If x represents a nonnegative number, then $\sqrt{x^2} = x$. If x represents a negative number, then $\sqrt{x^2} = -x$, the opposite of x (which is positive). For example,

$$\sqrt{5^2} = 5, \quad \text{but} \quad \sqrt{(-5)^2} = \sqrt{25} = 5, \quad \text{the } \textit{opposite} \text{ of } -5.$$

This means that the square root of a squared number is always nonnegative. We can use absolute value to express this.

$\sqrt{a^2}$

For any real number a,

$$\sqrt{a^2} = |a|$$

The product and quotient rules apply when variables appear under the radical sign, as long as the variables represent only *nonnegative* real numbers. **To avoid negative radicands, variables under radical signs are assumed to be nonnegative in this text.** Therefore, absolute value bars are not necessary, since for $x \geq 0$, $|x| = x$.

ANSWERS

5. $2\sqrt{10}$

6. (a) 10   (b) $\dfrac{\sqrt{21}}{4}$

EXAMPLE 7 Simplifying Radicals Involving Variables

Simplify each radical. Assume that all variables represent nonnegative real numbers.

(a) $\sqrt{x^4} = x^2$ since $(x^2)^2 = x^4$

(b) $\sqrt{25m^6} = \sqrt{25} \cdot \sqrt{m^6}$ Product rule
$= 5m^3$ $(m^3)^2 = m^6$

(c) $\sqrt{8p^{10}} = \sqrt{4 \cdot 2 \cdot p^{10}}$ 4 is a perfect square
$= \sqrt{4} \cdot \sqrt{2} \cdot \sqrt{p^{10}}$ Product rule
$= 2 \cdot \sqrt{2} \cdot p^5$ $(p^5)^2 = p^{10}$
$= 2p^5\sqrt{2}$

(d) $\sqrt{r^9} = \sqrt{r^8 \cdot r}$
$= \sqrt{r^8} \cdot \sqrt{r}$ Product rule
$= r^4\sqrt{r}$ $(r^4)^2 = r^8$

(e) $\sqrt{\dfrac{5}{x^2}} = \dfrac{\sqrt{5}}{\sqrt{x^2}}$ Quotient rule
$= \dfrac{\sqrt{5}}{x}$ $x \neq 0$

NOTE
A quick way to find the square root of a variable raised to an even power is to divide the exponent by the index, 2. For example:

$$\sqrt{x^6} = x^3 \quad \text{and} \quad \sqrt{x^{10}} = x^5$$
$\quad\quad\uparrow \quad\quad\quad\quad\quad\quad\quad\quad\uparrow$
$\quad 6 \div 2 = 3 \quad\quad\quad 10 \div 2 = 5$

Work Problem 7 at the Side.

---

7 Simplify each radical. Assume that all variables represent nonnegative real numbers.

(a) $\sqrt{x^8}$

(b) $\sqrt{36y^6}$

(c) $\sqrt{100p^{12}}$

(d) $\sqrt{12z^2}$

(e) $\sqrt{a^5}$

(f) $\sqrt{\dfrac{10}{n^4}}, \; n \neq 0$

OBJECTIVE 5 Simplify higher roots. The product and quotient rules for radicals also work for other roots. To simplify cube roots, look for factors that are *perfect cubes*. A **perfect cube** is a number with a rational cube root. For example, $\sqrt[3]{64} = 4$, and because 4 is a rational number, 64 is a perfect cube. Higher roots are handled in a similar manner.

Properties of Radicals
For all real numbers where the indicated roots exist,

$$\sqrt[n]{a} \cdot \sqrt[n]{b} = \sqrt[n]{ab} \quad \text{and} \quad \dfrac{\sqrt[n]{a}}{\sqrt[n]{b}} = \sqrt[n]{\dfrac{a}{b}} \quad (\text{where } b \neq 0).$$

ANSWERS
7. (a) x^4 (b) $6y^3$ (c) $10p^6$ (d) $2z\sqrt{3}$
(e) $a^2\sqrt{a}$ (f) $\dfrac{\sqrt{10}}{n^2}$

Chapter 16 Roots and Radicals

8 Simplify each radical.

(a) $\sqrt[3]{108}$

(b) $\sqrt[4]{160}$

(c) $\sqrt[4]{\dfrac{16}{625}}$

EXAMPLE 8 Simplifying Higher Roots

Simplify each radical.

(a) $\sqrt[3]{32} = \sqrt[3]{8 \cdot 4}$ 8 is a perfect cube.
$= \sqrt[3]{8} \cdot \sqrt[3]{4}$ Product rule
$= 2\sqrt[3]{4}$

(b) $\sqrt[4]{32} = \sqrt[4]{16 \cdot 2}$ 16 is a perfect fourth power.
$= \sqrt[4]{16} \cdot \sqrt[4]{2}$ Product rule
$= 2\sqrt[4]{2}$

(c) $\sqrt[3]{\dfrac{27}{125}} = \dfrac{\sqrt[3]{27}}{\sqrt[3]{125}} = \dfrac{3}{5}$ Quotient rule

Work Problem 8 at the Side.

Higher roots of radicals involving variables can also be simplified. To simplify cube roots with variables, use the fact that for any real number a,

$$\sqrt[3]{a^3} = a.$$

This is true whether a is positive or negative. (Why?)

9 Simplify each radical.

(a) $\sqrt[3]{z^9}$

(b) $\sqrt[3]{8x^6}$

(c) $\sqrt[3]{54t^5}$

(d) $\sqrt[3]{\dfrac{a^{15}}{64}}$

EXAMPLE 9 Simplifying Cube Roots Involving Variables

Simplify each radical.

(a) $\sqrt[3]{m^6} = m^2$ $(m^2)^3 = m^6$

(b) $\sqrt[3]{27x^{12}} = \sqrt[3]{27} \cdot \sqrt[3]{x^{12}}$ Product rule
$= 3x^4$ $3^3 = 27$; $(x^4)^3 = x^{12}$

(c) $\sqrt[3]{32a^4} = \sqrt[3]{8a^3 \cdot 4a}$ 8 is a perfect cube.
$= \sqrt[3]{8a^3} \cdot \sqrt[3]{4a}$ Product rule
$= 2a\sqrt[3]{4a}$ $(2a)^3 = 8a^3$

(d) $\sqrt[3]{\dfrac{y^3}{125}} = \dfrac{\sqrt[3]{y^3}}{\sqrt[3]{125}}$ Quotient rule
$= \dfrac{y}{5}$

Work Problem 9 at the Side.

ANSWERS

8. (a) $3\sqrt[3]{4}$ (b) $2\sqrt[4]{10}$ (c) $\dfrac{2}{5}$

9. (a) z^3 (b) $2x^2$ (c) $3t\sqrt[3]{2t^2}$ (d) $\dfrac{a^5}{4}$

16.2 Exercises

Decide whether each statement is true or false. If false, show why.

1. $\sqrt{(-6)^2} = -6$
2. $\sqrt[3]{(-6)^3} = -6$

Use the product rule for radicals to find each product. See Example 1.

3. $\sqrt{3} \cdot \sqrt{5}$
4. $\sqrt{3} \cdot \sqrt{7}$
5. $\sqrt{2} \cdot \sqrt{11}$
6. $\sqrt{2} \cdot \sqrt{15}$

7. $\sqrt{6} \cdot \sqrt{7}$
8. $\sqrt{5} \cdot \sqrt{6}$
9. $\sqrt{13} \cdot \sqrt{r}, r \geq 0$
10. $\sqrt{19} \cdot \sqrt{k}, k \geq 0$

11. Which one of the following radicals is simplified? See Example 2.

    A. $\sqrt{47}$ B. $\sqrt{45}$ C. $\sqrt{48}$ D. $\sqrt{44}$

12. If p is a prime number, is $\sqrt{p}$ in simplified form? Explain your answer.

Simplify each radical. See Example 2.

13. $\sqrt{45}$
14. $\sqrt{27}$
15. $\sqrt{24}$
16. $\sqrt{44}$

17. $\sqrt{90}$
18. $\sqrt{56}$
19. $\sqrt{75}$
20. $\sqrt{18}$

21. $\sqrt{125}$
22. $\sqrt{80}$
23. $\sqrt{145}$
24. $\sqrt{110}$

25. $\sqrt{160}$
26. $\sqrt{128}$
27. $-\sqrt{700}$
28. $-\sqrt{600}$

Find each product and simplify. See Example 3.

29. $\sqrt{3} \cdot \sqrt{18}$
30. $\sqrt{3} \cdot \sqrt{21}$
31. $\sqrt{12} \cdot \sqrt{48}$
32. $\sqrt{50} \cdot \sqrt{72}$

33. $\sqrt{12} \cdot \sqrt{30}$
34. $\sqrt{30} \cdot \sqrt{24}$
35. $2\sqrt{10} \cdot 3\sqrt{2}$

36. $5\sqrt{6} \cdot 2\sqrt{10}$
37. $5\sqrt{3} \cdot 2\sqrt{15}$
38. $4\sqrt{6} \cdot 3\sqrt{2}$

39. Simplify the product $\sqrt{8} \cdot \sqrt{32}$ in two ways. First, multiply 8 by 32 and simplify the square root of this product. Second, simplify $\sqrt{8}$, simplify $\sqrt{32}$, and then multiply. How do the answers compare? Make a conjecture (an educated guess) about whether the correct answer can always be obtained using either method when simplifying a product such as this.

40. Simplify the radical $\sqrt{288}$ in two ways. First, factor 288 as $144 \cdot 2$ and then simplify. Second, factor 288 as $48 \cdot 6$ and then simplify. How do the answers compare? Make a conjecture concerning the quickest way to simplify such a radical.

Simplify each radical expression. See Examples 4–6.

41. $\sqrt{\dfrac{16}{225}}$
42. $\sqrt{\dfrac{9}{100}}$
43. $\sqrt{\dfrac{7}{16}}$
44. $\sqrt{\dfrac{13}{25}}$

45. $\dfrac{\sqrt{75}}{\sqrt{3}}$
46. $\dfrac{\sqrt{200}}{\sqrt{2}}$
47. $\sqrt{\dfrac{5}{2}} \cdot \sqrt{\dfrac{125}{8}}$

48. $\sqrt{\dfrac{8}{3}} \cdot \sqrt{\dfrac{512}{27}}$
49. $\dfrac{30\sqrt{10}}{5\sqrt{2}}$
50. $\dfrac{50\sqrt{20}}{2\sqrt{10}}$

Simplify each radical. Assume that all variables represent nonnegative real numbers. See Example 7.

51. $\sqrt{m^2}$ **52.** $\sqrt{k^2}$ **53.** $\sqrt{y^4}$ **54.** $\sqrt{s^4}$

55. $\sqrt{36z^2}$ **56.** $\sqrt{49n^2}$ **57.** $\sqrt{400x^6}$ **58.** $\sqrt{900y^8}$

59. $\sqrt{18x^8}$ **60.** $\sqrt{20r^{10}}$ **61.** $\sqrt{45c^{14}}$ **62.** $\sqrt{50d^{20}}$

63. $\sqrt{z^5}$ **64.** $\sqrt{y^3}$ **65.** $\sqrt{a^{13}}$ **66.** $\sqrt{p^{17}}$

67. $\sqrt{64x^7}$ **68.** $\sqrt{25t^{11}}$ **69.** $\sqrt{x^6 y^{12}}$ **70.** $\sqrt{a^8 b^{10}}$

71. $\sqrt{81m^4 n^2}$ **72.** $\sqrt{100c^4 d^6}$ **73.** $\sqrt{\dfrac{7}{x^{10}}},\ x \neq 0$ **74.** $\sqrt{\dfrac{14}{z^{12}}},\ z \neq 0$

75. $\sqrt{\dfrac{y^4}{100}}$ **76.** $\sqrt{\dfrac{w^8}{144}}$ **77.** $\sqrt{\dfrac{x^6}{y^8}},\ y \neq 0$ **78.** $\sqrt{\dfrac{a^4}{b^6}},\ b \neq 0$

Simplify each radical. See Example 8.

79. $\sqrt[3]{40}$ **80.** $\sqrt[3]{48}$ **81.** $\sqrt[3]{54}$ **82.** $\sqrt[3]{135}$

83. $\sqrt[3]{128}$ **84.** $\sqrt[3]{192}$ **85.** $\sqrt[4]{80}$ **86.** $\sqrt[4]{243}$

87. $\sqrt[3]{\dfrac{8}{27}}$ **88.** $\sqrt[3]{\dfrac{64}{125}}$ **89.** $\sqrt[3]{-\dfrac{216}{125}}$ **90.** $\sqrt[3]{-\dfrac{1}{64}}$

Simplify each radical. See Example 9.

91. $\sqrt[3]{p^3}$

92. $\sqrt[3]{w^3}$

93. $\sqrt[3]{x^9}$

94. $\sqrt[3]{y^{18}}$

95. $\sqrt[3]{64z^6}$

96. $\sqrt[3]{125a^{15}}$

97. $\sqrt[3]{343a^9b^3}$

98. $\sqrt[3]{216m^3n^6}$

99. $\sqrt[3]{16t^5}$

100. $\sqrt[3]{24x^4}$

101. $\sqrt[3]{\dfrac{m^{12}}{8}}$

102. $\sqrt[3]{\dfrac{n^9}{27}}$

The volume of a cube is found with the formula $V = s^3$, where s is the length of an edge of the cube. Use this information in Exercises 103 and 104.

103. A container in the shape of a cube has a volume of 216 cm³. What is the depth of the container?

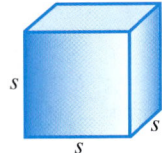

104. A cube-shaped box must be constructed to contain 128 ft³. What should the dimensions (height, width, and length) of the box be?

The volume of a sphere is found with the formula $V = \tfrac{4}{3}\pi r^3$, where r is the length of the radius of the sphere. Use this information in Exercises 105 and 106.

105. A ball in the shape of a sphere has a volume of 288π in.³. What is the radius of the ball?

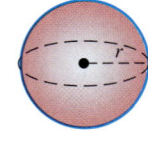

106. Suppose that the volume of the ball described in Exercise 105 is multiplied by 8. How is the radius affected?

Work Exercises 107 and 108 without using a calculator.

107. Choose the best estimate for the area (in square inches) of this rectangle.

    A. 45 B. 72 C. 80 D. 90

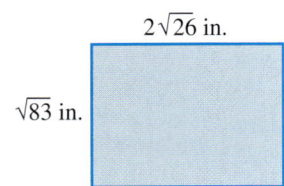

108. Choose the best estimate for the area (in square feet) of the triangle.

    A. 20 B. 40 C. 60 D. 80

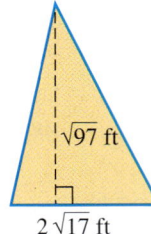

16.3 Adding and Subtracting Radicals

OBJECTIVE 1 Add and subtract radicals. We add or subtract radicals by using the distributive property. Here are two examples.

$8\sqrt{3} + 6\sqrt{3} = (8+6)\sqrt{3} = 14\sqrt{3}$ Adding radicals

$2\sqrt{11} - 7\sqrt{11} = -5\sqrt{11}$ Subtracting radicals

Only **like radicals**, those that are *multiples of the same root of the same number,* can be combined in this way. In the examples above, $8\sqrt{3}$ and $6\sqrt{3}$ are like radicals, as are $2\sqrt{11}$ and $-7\sqrt{11}$. On the other hand, examples of **unlike radicals** are shown below.

$2\sqrt{5}$ and $2\sqrt{3}$ Unlike because radicands are different.

$2\sqrt{3}$ and $2\sqrt[3]{3}$ Unlike because indexes are different.

Work Problem 1 at the Side.

EXAMPLE 1 Adding and Subtracting Like Radicals

Add or subtract, as indicated.

(a) $3\sqrt{6} + 5\sqrt{6} = (3+5)\sqrt{6} = 8\sqrt{6}$ Distributive property

(b) $5\sqrt{10} - 7\sqrt{10} = (5-7)\sqrt{10} = -2\sqrt{10}$

(c) $\sqrt{7} + 2\sqrt{7} = 1\sqrt{7} + 2\sqrt{7} = (1+2)\sqrt{7} = 3\sqrt{7}$

(d) $\sqrt{5} + \sqrt{5} = 1\sqrt{5} + 1\sqrt{5} = 2\sqrt{5}$

(e) $\sqrt{3} + \sqrt{7}$ cannot be added using the distributive property.

Work Problem 2 at the Side.

OBJECTIVE 2 Simplify radical sums and differences. Sometimes one or more radical expressions in a sum or difference must first be simplified. Any like radicals that result can then be added or subtracted.

EXAMPLE 2 Adding and Subtracting Radicals That Must Be Simplified

Add or subtract, as indicated.

(a) $3\sqrt{2} + \sqrt{8} = 3\sqrt{2} + \sqrt{4 \cdot 2}$ Factor.

$\qquad = 3\sqrt{2} + \sqrt{4} \cdot \sqrt{2}$ Product rule

$\qquad = 3\sqrt{2} + 2\sqrt{2}$ $\sqrt{4} = 2$

$\qquad = 5\sqrt{2}$ Add like radicals.

(b) $\sqrt{18} - \sqrt{27} = \sqrt{9 \cdot 2} - \sqrt{9 \cdot 3}$ Factor.

$\qquad = \sqrt{9} \cdot \sqrt{2} - \sqrt{9} \cdot \sqrt{3}$ Product rule

$\qquad = 3\sqrt{2} - 3\sqrt{3}$ $\sqrt{9} = 3$

Since $\sqrt{2}$ and $\sqrt{3}$ are unlike radicals, the answer cannot be simplified.

Continued on Next Page

OBJECTIVES

1. Add and subtract radicals.
2. Simplify radical sums and differences.
3. Simplify more complicated radical expressions.

1 Indicate whether the radicals in each pair are *like* or *unlike*.

(a) $5\sqrt{6}$ and $4\sqrt{6}$

(b) $2\sqrt{3}$ and $3\sqrt{2}$

(c) $\sqrt{10}$ and $\sqrt[3]{10}$

(d) $7\sqrt{2x}$ and $8\sqrt{2x}$

(e) $\sqrt{3y}$ and $\sqrt{6y}$

2 Add or subtract, as indicated.

(a) $8\sqrt{5} + 2\sqrt{5}$

(b) $-4\sqrt{3} + 9\sqrt{3}$

(c) $12\sqrt{11} - 3\sqrt{11}$

(d) $\sqrt{15} + \sqrt{15}$

(e) $2\sqrt{7} + 2\sqrt{10}$

ANSWERS
1. (a) like (b) unlike (c) unlike
   (d) like (e) unlike
2. (a) $10\sqrt{5}$ (b) $5\sqrt{3}$ (c) $9\sqrt{11}$
   (d) $2\sqrt{15}$ (e) cannot be added

1108 Chapter 16 Roots and Radicals

3 Add or subtract, as indicated.

(a) $\sqrt{8} + 4\sqrt{2}$

(b) $\sqrt{27} + \sqrt{12}$

(c) $5\sqrt{200} - 6\sqrt{18}$

4 Simplify each radical expression. Assume that all variables represent non-negative real numbers.

(a) $\sqrt{7} \cdot \sqrt{21} + 2\sqrt{27}$

(b) $\sqrt{3r} \cdot \sqrt{6} + \sqrt{8r}$

(c) $y\sqrt{72} - \sqrt{18y^2}$

(d) $\sqrt[3]{81x^4} + 5\sqrt[3]{24x^4}$

ANSWERS
3. (a) $6\sqrt{2}$ (b) $5\sqrt{3}$ (c) $32\sqrt{2}$
4. (a) $13\sqrt{3}$ (b) $5\sqrt{2r}$ (c) $3y\sqrt{2}$
   (d) $13x\sqrt[3]{3x}$

(c) $2\sqrt{12} + 3\sqrt{75} = 2(\sqrt{4} \cdot \sqrt{3}) + 3(\sqrt{25} \cdot \sqrt{3})$ Product rule
$= 2(2\sqrt{3}) + 3(5\sqrt{3})$ $\sqrt{4} = 2; \sqrt{25} = 5$
$= 4\sqrt{3} + 15\sqrt{3}$ Multiply.
$= 19\sqrt{3}$ Add like radicals.

◀◀◀ **Work Problem 3 at the Side**

OBJECTIVE 3 Simplify more complicated radical expressions. When simplifying, the order of operations from **Section 1.8** still applies.

EXAMPLE 3 Simplifying Radical Expressions

Simplify each radical expression. Assume that all variables represent non-negative real numbers.

(a) $\sqrt{5} \cdot \sqrt{15} + 4\sqrt{3} = \sqrt{5 \cdot 15} + 4\sqrt{3}$ Product rule
$= \sqrt{75} + 4\sqrt{3}$ Multiply.
$= \sqrt{25 \cdot 3} + 4\sqrt{3}$ 25 is a perfect square.
$= \sqrt{25} \cdot \sqrt{3} + 4\sqrt{3}$ Product rule
$= 5\sqrt{3} + 4\sqrt{3}$ $\sqrt{25} = 5$
$= 9\sqrt{3}$ Add like radicals.

(b) $\sqrt{2} \cdot \sqrt{6k} + \sqrt{27k} = \sqrt{12k} + \sqrt{27k}$ Product rule
$= \sqrt{4 \cdot 3k} + \sqrt{9 \cdot 3k}$ Factor.
$= \sqrt{4} \cdot \sqrt{3k} + \sqrt{9} \cdot \sqrt{3k}$ Product rule
$= 2\sqrt{3k} + 3\sqrt{3k}$ $\sqrt{4} = 2; \sqrt{9} = 3$
$= 5\sqrt{3k}$ Add like radicals.

(c) $3x\sqrt{50} + \sqrt{2x^2} = 3x\sqrt{25 \cdot 2} + \sqrt{x^2 \cdot 2}$ Factor.
$= 3x\sqrt{25} \cdot \sqrt{2} + \sqrt{x^2} \cdot \sqrt{2}$ Product rule
$= 3x \cdot 5\sqrt{2} + x\sqrt{2}$ $\sqrt{25} = 5; \sqrt{x^2} = x$
$= 15x\sqrt{2} + x\sqrt{2}$ Multiply.
$= 16x\sqrt{2}$ Add like radicals.

(d) $2\sqrt[3]{32m^3} - \sqrt[3]{108m^3} = 2\sqrt[3]{(8m^3)4} - \sqrt[3]{(27m^3)4}$ Factor.
$= 2(2m)\sqrt[3]{4} - 3m\sqrt[3]{4}$ $\sqrt[3]{8m^3} = 2m;$ $\sqrt[3]{27m^3} = 3m$
$= 4m\sqrt[3]{4} - 3m\sqrt[3]{4}$ Multiply.
$= m\sqrt[3]{4}$ Subtract like radicals.

> **CAUTION**
> A sum or difference of radicals can be simplified only if the radicals are **like radicals**. Thus, $\sqrt{5} + 3\sqrt{5} = 4\sqrt{5}$, but $\sqrt{5} + 5\sqrt{3}$ cannot be simplified further. Also, $2\sqrt{3} + 5\sqrt[3]{3}$ cannot be simplified further.

◀◀◀ **Work Problem 4 at the Side.**

16.3 Exercises

Fill in each blank with the correct response.

1. $5\sqrt{2} + 6\sqrt{2} = (5+6)\sqrt{2} = 11\sqrt{2}$ is an example of the _____ property.

2. Like radicals have the same _____ of the same _____.

3. $\sqrt{5} + 5\sqrt{3}$ cannot be simplified because the _____ are different.

4. $4\sqrt[3]{2} + 3\sqrt{2}$ cannot be simplified because the _____ are different.

Simplify and add or subtract wherever possible. See Examples 1 and 2.

5. $14\sqrt{7} - 19\sqrt{7}$

6. $16\sqrt{2} - 18\sqrt{2}$

7. $\sqrt{17} + 4\sqrt{17}$

8. $5\sqrt{19} + \sqrt{19}$

9. $6\sqrt{7} - \sqrt{7}$

10. $11\sqrt{14} - \sqrt{14}$

11. $\sqrt{45} + 4\sqrt{20}$

12. $\sqrt{24} + 6\sqrt{54}$

13. $5\sqrt{72} - 3\sqrt{50}$

14. $6\sqrt{18} - 5\sqrt{32}$

15. $-5\sqrt{32} + 2\sqrt{98}$

16. $-4\sqrt{75} + 3\sqrt{12}$

17. $5\sqrt{7} - 3\sqrt{28} + 6\sqrt{63}$

18. $3\sqrt{11} + 5\sqrt{44} - 8\sqrt{99}$

19. $2\sqrt{8} - 5\sqrt{32} - 2\sqrt{48}$

20. $5\sqrt{72} - 3\sqrt{48} + 4\sqrt{128}$

21. $4\sqrt{50} + 3\sqrt{12} - 5\sqrt{45}$

22. $6\sqrt{18} + 2\sqrt{48} + 6\sqrt{28}$

23. $\frac{1}{4}\sqrt{288} + \frac{1}{6}\sqrt{72}$

24. $\frac{2}{3}\sqrt{27} + \frac{3}{4}\sqrt{48}$

Find the perimeter of each figure.

25.

26.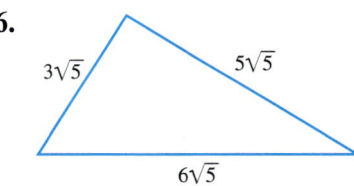

Perform the indicated operations. Assume that all variables represent nonnegative real numbers. See Example 3.

27. $\sqrt{6} \cdot \sqrt{2} + 9\sqrt{3}$

28. $4\sqrt{15} \cdot \sqrt{3} + 4\sqrt{5}$

29. $\sqrt{9x} + \sqrt{49x} - \sqrt{25x}$

30. $\sqrt{4a} - \sqrt{16a} + \sqrt{100a}$

31. $\sqrt{6x^2} + x\sqrt{24}$

32. $\sqrt{75x^2} + x\sqrt{108}$

33. $3\sqrt{8x^2} - 4x\sqrt{2} - x\sqrt{8}$

34. $\sqrt{2b^2} + 3b\sqrt{18} - b\sqrt{200}$

35. $-8\sqrt{32k} + 6\sqrt{8k}$

36. $4\sqrt{12x} + 2\sqrt{27x}$

37. $2\sqrt{125x^2z} + 8x\sqrt{80z}$

38. $\sqrt{48x^2y} + 5x\sqrt{27y}$

39. $4\sqrt[3]{16} - 3\sqrt[3]{54}$

40. $5\sqrt[3]{128} + 3\sqrt[3]{250}$

41. $6\sqrt[3]{8p^2} - 2\sqrt[3]{27p^2}$

42. $8k\sqrt[3]{54k} + 6\sqrt[3]{16k^4}$

43. $5\sqrt[4]{m^3} + 8\sqrt[4]{16m^3}$

44. $5\sqrt[4]{m^5} + 3\sqrt[4]{81m^5}$

RELATING CONCEPTS (EXERCISES 45–48) For Individual or Group Work

Adding and subtracting like radicals is no different than adding and subtracting like terms.
Work Exercises 45–48 in order.

45. Combine like terms: $5x^2y + 3x^2y - 14x^2y$.

46. Combine like terms: $5(p - 2q)^2(a + b) + 3(p - 2q)^2(a + b) - 14(p - 2q)^2(a + b)$.

47. Combine like radicals: $5a^2\sqrt{xy} + 3a^2\sqrt{xy} - 14a^2\sqrt{xy}$.

48. Compare your answers in Exercises 45–47. How are they alike? How are they different?

16.4 Rationalizing the Denominator

OBJECTIVES
1. Rationalize denominators with square roots.
2. Write radicals in simplified form.
3. Rationalize denominators with cube roots.

OBJECTIVE 1 Rationalize denominators with square roots. Although calculators now make it fairly easy to divide by a radical in an expression such as $\frac{1}{\sqrt{2}}$, it is sometimes easier to work with radical expressions if the denominators do not contain any radicals. For example, the radical in the denominator of $\frac{1}{\sqrt{2}}$ can be eliminated by multiplying the numerator and denominator by $\sqrt{2}$, since $\sqrt{2} \cdot \sqrt{2} = \sqrt{4} = 2$.

$$\frac{1}{\sqrt{2}} = \frac{1 \cdot \sqrt{2}}{\sqrt{2} \cdot \sqrt{2}} = \frac{\sqrt{2}}{2} \qquad \text{Multiply by } \frac{\sqrt{2}}{\sqrt{2}} = 1.$$

This process of changing the denominator from a radical (irrational number) to a rational number is called **rationalizing the denominator.** The value of the radical expression is not changed; only the form is changed, because the expression has been multiplied by 1 in the form of $\frac{\sqrt{2}}{\sqrt{2}}$.

EXAMPLE 1 Rationalizing Denominators

Rationalize each denominator.

(a) $\dfrac{9}{\sqrt{6}}$

We must eliminate the radical in the denominator.

$$\frac{9}{\sqrt{6}} = \frac{9 \cdot \sqrt{6}}{\sqrt{6} \cdot \sqrt{6}} \qquad \text{Multiply by } \frac{\sqrt{6}}{\sqrt{6}} = 1.$$

$$= \frac{9\sqrt{6}}{6} \qquad \text{In the denominator, } \sqrt{6} \cdot \sqrt{6} = \sqrt{36} = 6$$

$$= \frac{3\sqrt{6}}{2} \qquad \text{Lowest terms}$$

(b) $\dfrac{12}{\sqrt{8}}$

The denominator could be rationalized by multiplying by $\sqrt{8}$. However, first simplifying the denominator is more direct.

$$\sqrt{8} = \sqrt{4} \cdot \sqrt{2} = 2\sqrt{2}$$

Then multiply the numerator and denominator by $\sqrt{2}$.

$$\frac{12}{\sqrt{8}} = \frac{12}{2\sqrt{2}} \qquad \sqrt{8} = 2\sqrt{2}$$

$$= \frac{12 \cdot \sqrt{2}}{2\sqrt{2} \cdot \sqrt{2}} \qquad \text{Multiply by } \frac{\sqrt{2}}{\sqrt{2}} = 1.$$

$$= \frac{12 \cdot \sqrt{2}}{2 \cdot 2} \qquad \text{In the denominator, } \sqrt{2} \cdot \sqrt{2} = \sqrt{4} = 2$$

$$= \frac{12\sqrt{2}}{4} \qquad \text{Multiply.}$$

$$= 3\sqrt{2} \qquad \text{Lowest terms}$$

Chapter 16 Roots and Radicals

1 Rationalize each denominator.

(a) $\dfrac{3}{\sqrt{5}}$

(b) $\dfrac{-6}{\sqrt{11}}$

(c) $-\dfrac{\sqrt{7}}{\sqrt{2}}$

(d) $\dfrac{20}{\sqrt{18}}$

2 Simplify.

(a) $\sqrt{\dfrac{16}{11}}$

(b) $\sqrt{\dfrac{5}{18}}$

(c) $\sqrt{\dfrac{8}{32}}$

ANSWERS

1. (a) $\dfrac{3\sqrt{5}}{5}$ (b) $\dfrac{-6\sqrt{11}}{11}$
   (c) $-\dfrac{\sqrt{14}}{2}$ (d) $\dfrac{10\sqrt{2}}{3}$

2. (a) $\dfrac{4\sqrt{11}}{11}$ (b) $\dfrac{\sqrt{10}}{6}$ (c) $\dfrac{1}{2}$

> **NOTE**
> In Example 1(b), we could also have rationalized the original denominator $\sqrt{8}$ by multiplying by $\sqrt{2}$, since $\sqrt{8}\cdot\sqrt{2} = \sqrt{16} = 4$.
>
> $$\dfrac{12}{\sqrt{8}} = \dfrac{12\cdot\sqrt{2}}{\sqrt{8}\cdot\sqrt{2}} = \dfrac{12\sqrt{2}}{\sqrt{16}} = \dfrac{12\sqrt{2}}{4} = 3\sqrt{2}$$
>
> Both approaches are correct.

◀◀ **Work Problem 1 at the Side.**

OBJECTIVE 2 Write radicals in simplified form. A radical is considered to be in simplified form if the following three conditions are met.

Simplified Form of a Radical
1. The radicand contains no factor (except 1) that is a perfect square (when dealing with square roots), a perfect cube (when dealing with cube roots), and so on.
2. The radicand has no fractions.
3. No denominator contains a radical.

EXAMPLE 2 Simplifying a Radical

Simplify $\sqrt{\dfrac{27}{5}}$.

This violates condition 2. To begin, use the quotient rule for radicals.

$\sqrt{\dfrac{27}{5}} = \dfrac{\sqrt{27}}{\sqrt{5}}$ Quotient rule

$= \dfrac{\sqrt{27}\cdot\sqrt{5}}{\sqrt{5}\cdot\sqrt{5}}$ Rationalize the denominator.

$= \dfrac{\sqrt{27}\cdot\sqrt{5}}{5}$ In the denominator, $\sqrt{5}\cdot\sqrt{5} = \sqrt{25} = 5$

$= \dfrac{\sqrt{9\cdot 3}\cdot\sqrt{5}}{5}$ Factor.

$= \dfrac{\sqrt{9}\cdot\sqrt{3}\cdot\sqrt{5}}{5}$ Product rule

$= \dfrac{3\cdot\sqrt{3}\cdot\sqrt{5}}{5}$ $\sqrt{9} = 3$

$= \dfrac{3\sqrt{15}}{5}$ Product rule

◀◀ **Work Problem 2 at the Side.**

Section 16.4 Rationalizing the Denominator 1113

EXAMPLE 3 Simplifying a Product of Radicals

Simplify $\sqrt{\dfrac{5}{8}} \cdot \sqrt{\dfrac{1}{6}}$

Use both the product and quotient rules.

$\sqrt{\dfrac{5}{8}} \cdot \sqrt{\dfrac{1}{6}} = \sqrt{\dfrac{5}{8} \cdot \dfrac{1}{6}}$ Product rule

$= \sqrt{\dfrac{5}{48}}$ Multiply fractions.

$= \dfrac{\sqrt{5}}{\sqrt{48}}$ Quotient rule

First simplify the denominator and then rationalize it.

$= \dfrac{\sqrt{5}}{\sqrt{16} \cdot \sqrt{3}}$ Product rule

$= \dfrac{\sqrt{5}}{4\sqrt{3}}$ $\sqrt{16} = 4$

$= \dfrac{\sqrt{5} \cdot \sqrt{3}}{4\sqrt{3} \cdot \sqrt{3}}$ Rationalize the denominator.

$= \dfrac{\sqrt{15}}{4 \cdot 3}$ Product rule; $\sqrt{3} \cdot \sqrt{3} = 3$

$= \dfrac{\sqrt{15}}{12}$ Multiply.

Work Problem 3 at the Side.

EXAMPLE 4 Simplifying a Quotient of Radicals

Simplify $\dfrac{\sqrt{4x}}{\sqrt{y}}$. Assume that x and y represent positive real numbers.

Multiply the numerator and denominator by $\sqrt{y}$.

$\dfrac{\sqrt{4x}}{\sqrt{y}} = \dfrac{\sqrt{4x} \cdot \sqrt{y}}{\sqrt{y} \cdot \sqrt{y}}$ Rationalize the denominator.

$= \dfrac{\sqrt{4xy}}{y}$ Product rule; $\sqrt{y} \cdot \sqrt{y} = y$

$= \dfrac{2\sqrt{xy}}{y}$ $\sqrt{4} = 2$

Work Problem 4 at the Side.

3 Simplify.

(a) $\sqrt{\dfrac{1}{2}} \cdot \sqrt{\dfrac{5}{6}}$

(b) $\sqrt{\dfrac{1}{10}} \cdot \sqrt{20}$

(c) $\sqrt{\dfrac{5}{8}} \cdot \sqrt{\dfrac{24}{10}}$

4 Simplify $\dfrac{\sqrt{5p}}{\sqrt{q}}$. Assume that p and q represent positive real numbers.

ANSWERS

3. (a) $\dfrac{\sqrt{15}}{6}$ (b) $\sqrt{2}$ (c) $\dfrac{\sqrt{6}}{2}$

4. $\dfrac{\sqrt{5pq}}{q}$

5 Simplify $\sqrt{\dfrac{5r^2t^2}{7}}$. Assume that r and t represent nonnegative real numbers.

EXAMPLE 5 Simplifying a Radical Quotient

Simplify $\sqrt{\dfrac{2x^2y}{3}}$. Assume that x and y represent nonnegative real numbers.

$$\sqrt{\dfrac{2x^2y}{3}} = \dfrac{\sqrt{2x^2y}}{\sqrt{3}} \qquad \text{Quotient rule}$$

$$= \dfrac{\sqrt{2x^2y} \cdot \sqrt{3}}{\sqrt{3} \cdot \sqrt{3}} \qquad \text{Rationalize the denominator.}$$

$$= \dfrac{\sqrt{6x^2y}}{3} \qquad \text{Product rule; } \sqrt{3} \cdot \sqrt{3} = 3$$

$$= \dfrac{\sqrt{x^2}\sqrt{6y}}{3} \qquad \text{Product rule}$$

$$= \dfrac{x\sqrt{6y}}{3} \qquad \sqrt{x^2} = x, \text{ since } x \geq 0.$$

◀◀ Work Problem 5 at the Side.

OBJECTIVE 3 Rationalize denominators with cube roots. A denominator with a cube root is rationalized by changing the radicand in the denominator to a perfect cube, as shown in the next example.

EXAMPLE 6 Rationalizing Denominators with Cube Roots

Rationalize each denominator.

(a) $\sqrt[3]{\dfrac{3}{2}}$

First write the expression as a quotient of radicals. Then multiply the numerator and denominator by the appropriate number of factors of 2 to make the denominator a perfect cube. This will eliminate the radical in the denominator. Here, multiply by $\sqrt[3]{2^2}$.

$$\sqrt[3]{\dfrac{3}{2}} = \dfrac{\sqrt[3]{3}}{\sqrt[3]{2}} = \dfrac{\sqrt[3]{3} \cdot \sqrt[3]{2^2}}{\sqrt[3]{2} \cdot \sqrt[3]{2^2}} = \dfrac{\sqrt[3]{3 \cdot 2^2}}{\sqrt[3]{2^3}} = \dfrac{\sqrt[3]{12}}{2} \qquad \sqrt[3]{2^3} = \sqrt[3]{8} = 2$$

↑ Denominator is a perfect cube.

(b) $\dfrac{\sqrt[3]{3}}{\sqrt[3]{4}}$

Since $\sqrt[3]{4} \cdot \sqrt[3]{2} = \sqrt[3]{2^2} \cdot \sqrt[3]{2} = \sqrt[3]{2^3} = 2$, multiply the numerator and denominator by $\sqrt[3]{2}$.

$$\dfrac{\sqrt[3]{3}}{\sqrt[3]{4}} = \dfrac{\sqrt[3]{3} \cdot \sqrt[3]{2}}{\sqrt[3]{2^2} \cdot \sqrt[3]{2}} = \dfrac{\sqrt[3]{6}}{\sqrt[3]{2^3}} = \dfrac{\sqrt[3]{6}}{2}$$

Continued on Next Page

ANSWERS

5. $\dfrac{rt\sqrt{35}}{7}$

(c) $\dfrac{\sqrt[3]{2}}{\sqrt[3]{3x^2}}$ where $x \neq 0$

Multiply the numerator and denominator by the appropriate number of factors of 3 and of x to get a perfect cube in the denominator. Here, multiply by $\sqrt[3]{3^2 x}$ (that is, $\sqrt[3]{9x}$) since $\sqrt[3]{3x^2} \cdot \sqrt[3]{3^2 x} = \sqrt[3]{(3x)^3} = 3x$.

$$\dfrac{\sqrt[3]{2}}{\sqrt[3]{3x^2}} = \dfrac{\sqrt[3]{2} \cdot \sqrt[3]{3^2 x}}{\sqrt[3]{3x^2} \cdot \sqrt[3]{3^2 x}} = \dfrac{\sqrt[3]{18x}}{\sqrt[3]{(3x)^3}} = \dfrac{\sqrt[3]{18x}}{3x}$$

↳ Denominator is a perfect cube.

CAUTION

A common error in a problem like the one in Example 6(a) is to multiply by $\sqrt[3]{2}$ instead of $\sqrt[3]{2^2}$. Doing this would give a denominator of $\sqrt[3]{2} \cdot \sqrt[3]{2} = \sqrt[3]{4}$. Because 4 is not a perfect cube, the denominator is still not rationalized.

Work Problem 6 at the Side.

6 Rationalize each denominator.

(a) $\sqrt[3]{\dfrac{5}{7}}$

(b) $\dfrac{\sqrt[3]{5}}{\sqrt[3]{9}}$

(c) $\dfrac{\sqrt[3]{4}}{\sqrt[3]{25y}}$ where $y \neq 0$

ANSWERS

6. (a) $\dfrac{\sqrt[3]{245}}{7}$ (b) $\dfrac{\sqrt[3]{15}}{3}$ (c) $\dfrac{\sqrt[3]{20y^2}}{5y}$

Focus on Real-Data Applications

The Golden Ratio—A Star Number

The **Golden Ratio**, the number $\dfrac{1+\sqrt{5}}{2}$, is called phi, ϕ. The number has been known since ancient times and is called the *sacred ratio* in the Rhind Papyrus from 1600 B.C. The Egyptians used ϕ to build the Great Pyramids, and the ancient Greeks used ϕ in art and architecture, striving for the proportion that was most pleasing to the eye.

The Golden Ratio is widespread in the star formed by connecting the vertices (corners) of a regular pentagon inscribed in a circle. In the figure shown, ABCDE forms a regular pentagon. Each of the following ratios forms the Golden Ratio, ϕ.

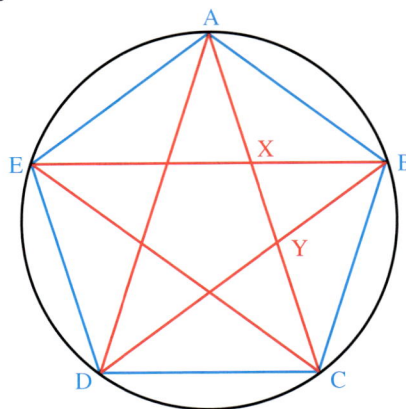

$$\dfrac{AC}{AY} \qquad \dfrac{AY}{AX}$$

Using symmetry, you should be able to identify other similar relationships.

For Group Discussion

1. List two more ratios in the star that form the Golden Ratio. You may label additional points.

2. The Golden Ratio has some curious properties. Use $\phi = \dfrac{1+\sqrt{5}}{2}$ to find the following quantities. Write your answers in exact form (using radicals). Rationalize denominators when appropriate.

   (a) $\dfrac{1}{\phi}$

   (b) $\phi - 1$

   (c) ϕ^2

   (d) $\phi + 1$

3. Based on your results from Problem 2, which forms in (a)–(d) are equivalent?

16.4 Exercises

Rationalize each denominator. See Examples 1 and 2.

1. $\dfrac{8}{\sqrt{2}}$
2. $\dfrac{12}{\sqrt{3}}$
3. $\dfrac{-\sqrt{11}}{\sqrt{3}}$
4. $\dfrac{-\sqrt{13}}{\sqrt{5}}$

5. $\dfrac{7\sqrt{3}}{\sqrt{5}}$
6. $\dfrac{4\sqrt{6}}{\sqrt{5}}$
7. $\dfrac{24\sqrt{10}}{16\sqrt{3}}$
8. $\dfrac{18\sqrt{15}}{12\sqrt{2}}$

9. $\dfrac{16}{\sqrt{27}}$
10. $\dfrac{24}{\sqrt{18}}$
11. $\dfrac{-3}{\sqrt{50}}$
12. $\dfrac{-5}{\sqrt{75}}$

13. $\dfrac{63}{\sqrt{45}}$
14. $\dfrac{27}{\sqrt{32}}$
15. $\dfrac{\sqrt{24}}{\sqrt{8}}$
16. $\dfrac{\sqrt{36}}{\sqrt{18}}$

17. $\sqrt{\dfrac{1}{2}}$
18. $\sqrt{\dfrac{1}{3}}$
19. $\sqrt{\dfrac{13}{5}}$
20. $\sqrt{\dfrac{17}{11}}$

21. When we rationalize the denominator in an expression such as $\dfrac{4}{\sqrt{3}}$, we multiply both the numerator and denominator by $\sqrt{3}$. By what number are we actually multiplying the given expression, and what property of real numbers justifies the fact that our result is equal to the given expression?

22. In Example 1(a), we show algebraically that $\dfrac{9}{\sqrt{6}}$ is equal to $\dfrac{3\sqrt{6}}{2}$. Support this result numerically by finding the decimal approximation of $\dfrac{9}{\sqrt{6}}$ on your calculator, and then finding the decimal approximation of $\dfrac{3\sqrt{6}}{2}$. What do you notice?

Simplify each product of radicals. See Example 3.

23. $\sqrt{\dfrac{7}{13}} \cdot \sqrt{\dfrac{13}{3}}$
24. $\sqrt{\dfrac{19}{20}} \cdot \sqrt{\dfrac{20}{3}}$
25. $\sqrt{\dfrac{21}{7}} \cdot \sqrt{\dfrac{21}{8}}$
26. $\sqrt{\dfrac{5}{8}} \cdot \sqrt{\dfrac{5}{6}}$

27. $\sqrt{\dfrac{1}{12}} \cdot \sqrt{\dfrac{1}{3}}$
28. $\sqrt{\dfrac{1}{8}} \cdot \sqrt{\dfrac{1}{2}}$
29. $\sqrt{\dfrac{2}{9}} \cdot \sqrt{\dfrac{9}{2}}$
30. $\sqrt{\dfrac{4}{3}} \cdot \sqrt{\dfrac{3}{4}}$

Simplify each radical. Assume that all variables represent positive real numbers. See Examples 4 and 5.

31. $\dfrac{\sqrt{7}}{\sqrt{x}}$

32. $\dfrac{\sqrt{19}}{\sqrt{y}}$

33. $\dfrac{\sqrt{4x^3}}{\sqrt{y}}$

34. $\dfrac{\sqrt{9t^3}}{\sqrt{s}}$

35. $\sqrt{\dfrac{5x^3z}{6}}$

36. $\sqrt{\dfrac{3st^3}{5}}$

37. $\sqrt{\dfrac{9a^2r^5}{7t}}$

38. $\sqrt{\dfrac{16x^3y^2}{13z}}$

39. Which one of the following would be an appropriate choice for multiplying the numerator and denominator of $\dfrac{\sqrt[3]{2}}{\sqrt[3]{5}}$ by in order to rationalize the denominator?

    A. $\sqrt[3]{5}$ B. $\sqrt[3]{25}$ C. $\sqrt[3]{2}$ D. $\sqrt[3]{4}$

40. In Example 6(b), we multiplied the numerator and denominator of $\dfrac{\sqrt[3]{3}}{\sqrt[3]{4}}$ by $\sqrt[3]{2}$ to rationalize the denominator. Suppose we had chosen to multiply by $\sqrt[3]{16}$ instead. Would we have obtained the correct answer after all simplifications were done?

Rationalize each denominator. Assume that variables in the denominator represent nonzero real numbers. See Example 6.

41. $\sqrt[3]{\dfrac{3}{2}}$

42. $\sqrt[3]{\dfrac{2}{5}}$

43. $\dfrac{\sqrt[3]{4}}{\sqrt[3]{7}}$

44. $\dfrac{\sqrt[3]{5}}{\sqrt[3]{10}}$

45. $\sqrt[3]{\dfrac{3}{4y^2}}$

46. $\sqrt[3]{\dfrac{3}{25x^2}}$

47. $\dfrac{\sqrt[3]{7m}}{\sqrt[3]{36n}}$

48. $\dfrac{\sqrt[3]{11p}}{\sqrt[3]{49q}}$

*In Exercises 49 and 50, **(a)** give the answer as a simplified radical and **(b)** use a calculator to give the answer correct to the nearest thousandth.*

49. The period p of a pendulum is the time it takes for it to swing from one extreme to the other and back again. The value of p in seconds is given by

    $$p = k \cdot \sqrt{\dfrac{L}{g}}$$

    where L is the length of the pendulum, g is the acceleration due to gravity, and k is a constant. Find the period when $k = 6$, $L = 9$ feet, and $g = 32$ feet per second squared.

    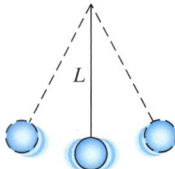

50. The velocity v of a meteorite approaching Earth is given in kilometers per second by

    $$v = \dfrac{k}{\sqrt{d}}$$

    where d is its distance from the center of Earth and k is a constant. What is the velocity of a meteorite that is 6000 kilometers away from the center of Earth, if $k = 450$?

16.5 More Simplifying and Operations with Radicals

The conditions for which a radical is in simplest form were listed in Section 16.4. Below is a set of guidelines to follow when you are simplifying radical expressions.

OBJECTIVES

1. Simplify products of radical expressions.
2. Use conjugates to rationalize denominators of radical expressions.
3. Write radical expressions with quotients in lowest terms.

Simplifying Radical Expressions

1. If a radical represents a rational number, use that rational number in place of the radical.

    Examples: $\sqrt{49} = 7 \qquad \sqrt{\dfrac{169}{9}} = \dfrac{13}{3}$

2. If a radical expression contains products of radicals, use the product rule for radicals, $\sqrt[n]{x} \cdot \sqrt[n]{y} = \sqrt[n]{xy}$, to get a single radical.

    Examples: $\sqrt{3} \cdot \sqrt{2} = \sqrt{6} \qquad \sqrt[3]{5} \cdot \sqrt[3]{x} = \sqrt[3]{5x}$

3. If a radicand has a factor that is a perfect square, express the radical as the product of the positive square root of the perfect square and the remaining radical factor. A similar statement applies to higher roots.

    Examples: $\sqrt{20} = \sqrt{4 \cdot 5} = \sqrt{4} \cdot \sqrt{5} = 2\sqrt{5}$

    $\sqrt[3]{16} = \sqrt[3]{8 \cdot 2} = \sqrt[3]{8} \cdot \sqrt[3]{2} = 2\sqrt[3]{2}$

4. If a radical expression contains sums or differences of radicals, use the distributive property to combine like radicals.

    Examples: $3\sqrt{2} + 4\sqrt{2} = 7\sqrt{2}$

    $3\sqrt{2} + 4\sqrt{3}$ cannot be simplified further.

5. Rationalize any denominator containing a radical.

    Examples: $\dfrac{5}{\sqrt{3}} = \dfrac{5 \cdot \sqrt{3}}{\sqrt{3} \cdot \sqrt{3}} = \dfrac{5\sqrt{3}}{3}$

    $\sqrt{\dfrac{3}{2}} = \dfrac{\sqrt{3}}{\sqrt{2}} = \dfrac{\sqrt{3} \cdot \sqrt{2}}{\sqrt{2} \cdot \sqrt{2}} = \dfrac{\sqrt{6}}{2}$

    $\sqrt[3]{\dfrac{1}{4}} = \dfrac{\sqrt[3]{1}}{\sqrt[3]{4}} = \dfrac{\sqrt[3]{1} \cdot \sqrt[3]{2}}{\sqrt[3]{4} \cdot \sqrt[3]{2}} = \dfrac{\sqrt[3]{2}}{\sqrt[3]{8}} = \dfrac{\sqrt[3]{2}}{2}$

OBJECTIVE 1 Simplify products of radical expressions. Use the above guidelines.

EXAMPLE 1 Multiplying Radical Expressions

Find each product, and simplify.

(a) $\sqrt{5}(\sqrt{8} - \sqrt{32})$

Start by simplifying $\sqrt{8}$ and $\sqrt{32}$.

$\sqrt{8} = 2\sqrt{2}$ and $\sqrt{32} = 4\sqrt{2}$

Continued on Next Page

1 Find each product, and simplify.

(a) $\sqrt{7}(\sqrt{2} + \sqrt{5})$

(b) $\sqrt{2}(\sqrt{8} + \sqrt{20})$

(c) $(\sqrt{2} + 5\sqrt{3})(\sqrt{3} - 2\sqrt{2})$

(d) $(\sqrt{2} - \sqrt{5})(\sqrt{10} + \sqrt{2})$

$$\sqrt{5}(\sqrt{8} - \sqrt{32}) = \sqrt{5}(2\sqrt{2} - 4\sqrt{2}) \quad \text{$\sqrt{8} = 2\sqrt{2}$; $\sqrt{32} = 4\sqrt{2}$}$$
$$= \sqrt{5}(-2\sqrt{2}) \quad \text{Subtract like radicals.}$$
$$= -2\sqrt{5 \cdot 2} \quad \text{Product rule}$$
$$= -2\sqrt{10} \quad \text{Multiply.}$$

(b) $(\sqrt{3} + 2\sqrt{5})(\sqrt{3} - 4\sqrt{5})$

We can find the products of sums of radicals in the same way that we found the product of binomials in **Section 13.3,** using the FOIL method.

$$(\sqrt{3} + 2\sqrt{5})(\sqrt{3} - 4\sqrt{5})$$
$$= \underbrace{\sqrt{3}(\sqrt{3})}_{\text{First}} + \underbrace{\sqrt{3}(-4\sqrt{5})}_{\text{Outer}} + \underbrace{2\sqrt{5}(\sqrt{3})}_{\text{Inner}} + \underbrace{2\sqrt{5}(-4\sqrt{5})}_{\text{Last}}$$
$$= 3 - 4\sqrt{15} + 2\sqrt{15} - 8 \cdot 5 \quad \text{Product rule}$$
$$= 3 - 2\sqrt{15} - 40 \quad \text{Add like radicals; multiply.}$$
$$= -37 - 2\sqrt{15} \quad \text{Combine terms.}$$

(c) $(\sqrt{3} + \sqrt{21})(\sqrt{3} - \sqrt{7})$
$$= \sqrt{3}(\sqrt{3}) + \sqrt{3}(-\sqrt{7}) + \sqrt{21}(\sqrt{3})$$
$$\quad + \sqrt{21}(-\sqrt{7}) \quad \text{FOIL}$$
$$= 3 - \sqrt{21} + \sqrt{63} - \sqrt{147} \quad \text{Product rule}$$
$$= 3 - \sqrt{21} + \sqrt{9} \cdot \sqrt{7} - \sqrt{49} \cdot \sqrt{3} \quad \text{9 and 49 are perfect squares.}$$
$$= 3 - \sqrt{21} + 3\sqrt{7} - 7\sqrt{3} \quad \text{$\sqrt{9} = 3$; $\sqrt{49} = 7$}$$

Since there are no like radicals, no terms can be combined.

Work Problem 1 at the Side.

The special products of binomials discussed in **Section 13.4** can be applied to radicals. Example 2 uses the rules for the square of a binomial,
$$(a + b)^2 = a^2 + 2ab + b^2 \quad \text{and} \quad (a - b)^2 = a^2 - 2ab + b^2.$$

EXAMPLE 2 Using Special Products with Radicals

Find each product.

(a) $(\sqrt{10} - 7)^2$

Follow the second pattern given above. Let $a = \sqrt{10}$ and $b = 7$.

$$(\sqrt{10} - 7)^2 = (\sqrt{10})^2 - 2(\sqrt{10})(7) + 7^2$$
$$= 10 - 14\sqrt{10} + 49 \quad \text{$(\sqrt{10})^2 = 10$; $7^2 = 49$}$$
$$= 59 - 14\sqrt{10} \quad \text{Combine like terms.}$$

(b) $(2\sqrt{3} + 4)^2 = (2\sqrt{3})^2 + 2(2\sqrt{3})(4) + 4^2 \quad \text{$a = 2\sqrt{3}$; $b = 4$}$
$$= 12 + 16\sqrt{3} + 16 \quad \text{$(2\sqrt{3})^2 = 4 \cdot 3 = 12$}$$
$$= 28 + 16\sqrt{3} \quad \text{Combine like terms.}$$

Continued on Next Page

ANSWERS
1. (a) $\sqrt{14} + \sqrt{35}$
   (b) $4 + 2\sqrt{10}$
   (c) $11 - 9\sqrt{6}$
   (d) $2\sqrt{5} + 2 - 5\sqrt{2} - \sqrt{10}$

(c) $(5 - \sqrt{x})^2 = 5^2 - 2(5)(\sqrt{x}) + (\sqrt{x})^2$
$= 25 - 10\sqrt{x} + x, \quad x \geq 0$

CAUTION
Be careful! In Examples 2(a) and (b),
$59 - 14\sqrt{10} \neq 45\sqrt{10}$ and $28 + 16\sqrt{3} \neq 44\sqrt{3}$.
Only like radicals can be combined.

Work Problem 2 at the Side.

Example 3 uses the rule for the product of the sum and difference of two terms,

$$(a + b)(a - b) = a^2 - b^2.$$

EXAMPLE 3 Using a Special Product with Radicals

Find each product.

(a) $(4 + \sqrt{3})(4 - \sqrt{3})$
Follow the pattern given above. Let $a = 4$ and $b = \sqrt{3}$.
$(4 + \sqrt{3})(4 - \sqrt{3}) = 4^2 - (\sqrt{3})^2$
$= 16 - 3 \qquad 4^2 = 16; (\sqrt{3})^2 = 3$
$= 13$

(b) $(\sqrt{x} - \sqrt{6})(\sqrt{x} + \sqrt{6}) = (\sqrt{x})^2 - (\sqrt{6})^2$
$= x - 6, \quad x \geq 0 \qquad (\sqrt{x})^2 = x; (\sqrt{6})^2 = 6$

Work Problem 3 at the Side.

Notice that the results in Example 3 do not contain radicals. The pairs of expressions being multiplied, $4 + \sqrt{3}$ and $4 - \sqrt{3}$, and $\sqrt{x} - \sqrt{6}$ and $\sqrt{x} + \sqrt{6}$, are called **conjugates** of each other.

OBJECTIVE 2 Use conjugates to rationalize denominators of radical expressions. Conjugates similar to those in Example 3 can be used to rationalize the denominators in more complicated quotients, such as

$$\frac{2}{4 - \sqrt{3}}.$$

By Example 3(a), if this denominator, $4 - \sqrt{3}$, is multiplied by $4 + \sqrt{3}$, then the product $(4 - \sqrt{3})(4 + \sqrt{3})$ is the rational number 13. Multiplying the numerator and denominator of the quotient by $4 + \sqrt{3}$ gives

$$\frac{2}{4 - \sqrt{3}} = \frac{2(4 + \sqrt{3})}{(4 - \sqrt{3})(4 + \sqrt{3})} = \frac{2(4 + \sqrt{3})}{13}.$$

The denominator has now been rationalized and contains no radicals.

---

2 Find each product. Simplify the answers.

(a) $(\sqrt{5} - 3)^2$

(b) $(4\sqrt{2} + 5)^2$

(c) $(6 + \sqrt{m})^2, \quad m \geq 0$

3 Find each product. Simplify the answers.

(a) $(3 + \sqrt{5})(3 - \sqrt{5})$

(b) $(\sqrt{3} - 2)(\sqrt{3} + 2)$

(c)
$(\sqrt{5} + \sqrt{3})(\sqrt{5} - \sqrt{3})$

(d)
$(\sqrt{10} - \sqrt{y})(\sqrt{10} + \sqrt{y}),$
$y \geq 0$

ANSWERS
2. (a) $14 - 6\sqrt{5}$ (b) $57 + 40\sqrt{2}$
   (c) $36 + 12\sqrt{m} + m$
3. (a) 4 (b) -1 (c) 2 (d) $10 - y$

Chapter 16 Roots and Radicals

4 Rationalize each denominator.

(a) $\dfrac{5}{4 + \sqrt{2}}$

(b) $\dfrac{\sqrt{5} + 3}{2 - \sqrt{5}}$

(c) $\dfrac{1}{\sqrt{6} + \sqrt{3}}$

(d) $\dfrac{7}{5 - \sqrt{x}}$

Answers

4. (a) $\dfrac{5(4 - \sqrt{2})}{14}$ (b) $-11 - 5\sqrt{5}$

(c) $\dfrac{\sqrt{6} - \sqrt{3}}{3}$ (d) $\dfrac{7(5 + \sqrt{x})}{25 - x}$

Using Conjugates to Simplify Radical Expressions

To simplify a radical expression with two terms in the denominator, where at least one of those terms is a radical, multiply both the numerator and the denominator by the conjugate of the denominator.

EXAMPLE 4 Using Conjugates to Rationalize Denominators

Simplify by rationalizing each denominator.

(a) $\dfrac{5}{3 + \sqrt{5}}$

We can eliminate the radical in the denominator by multiplying both the numerator and denominator by $3 - \sqrt{5}$, the conjugate of the denominator.

$$\dfrac{5}{3 + \sqrt{5}} = \dfrac{5(3 - \sqrt{5})}{(3 + \sqrt{5})(3 - \sqrt{5})} \quad \text{Multiply by } \dfrac{3 - \sqrt{5}}{3 - \sqrt{5}} = 1.$$

$$= \dfrac{5(3 - \sqrt{5})}{3^2 - (\sqrt{5})^2} \quad (a + b)(a - b) = a^2 - b^2$$

$$= \dfrac{5(3 - \sqrt{5})}{9 - 5} \quad 3^2 = 9; (\sqrt{5})^2 = 5$$

$$= \dfrac{5(3 - \sqrt{5})}{4} \quad \text{Subtract.}$$

(b) $\dfrac{6 + \sqrt{2}}{\sqrt{2} - 5}$

Multiply the numerator and denominator by $\sqrt{2} + 5$, the conjugate of the denominator.

$$\dfrac{6 + \sqrt{2}}{\sqrt{2} - 5} = \dfrac{(6 + \sqrt{2})(\sqrt{2} + 5)}{(\sqrt{2} - 5)(\sqrt{2} + 5)} \quad \text{Multiply by } \dfrac{\sqrt{2} + 5}{\sqrt{2} + 5} = 1.$$

$$= \dfrac{6\sqrt{2} + 30 + 2 + 5\sqrt{2}}{2 - 25} \quad \text{FOIL}; (a + b)(a - b) = a^2 - b^2$$

$$= \dfrac{11\sqrt{2} + 32}{-23} \quad \text{Combine like terms.}$$

$$= \dfrac{-11\sqrt{2} - 32}{23} \quad \dfrac{a}{-b} = \dfrac{-a}{b}$$

(c) $\dfrac{4}{3 - \sqrt{x}} = \dfrac{4(3 + \sqrt{x})}{(3 - \sqrt{x})(3 + \sqrt{x})} \quad \text{Multiply by } \dfrac{3 + \sqrt{x}}{3 + \sqrt{x}} = 1.$

$$= \dfrac{4(3 + \sqrt{x})}{9 - x} \quad 3^2 = 9; (\sqrt{x})^2 = x$$

(We assume here that $x > 0$ and $x \neq 9$.)

◀◀◀ **Work Problem 4 at the Side.**

OBJECTIVE 3 Write radical expressions with quotients in lowest terms.

EXAMPLE 5 Writing a Radical Quotient in Lowest Terms

Write $\dfrac{3\sqrt{3} + 9}{12}$ in lowest terms.

Factor the numerator and denominator, and then use the fundamental property from **Section 15.1** to divide out common factors.

$$\frac{3\sqrt{3} + 9}{12} = \frac{3(\sqrt{3} + 3)}{3(4)} = 1 \cdot \frac{\sqrt{3} + 3}{4} = \frac{\sqrt{3} + 3}{4}$$

CAUTION

An expression like the one in Example 5 can only be simplified by factoring a common factor from the denominator and *each* term of the numerator. Here is an example.

$$\frac{4 + 8\sqrt{5}}{4} \neq 1 + 8\sqrt{5} \quad \text{Incorrect}$$

Factor the numerator first, then simplify.

$$\frac{4 + 8\sqrt{5}}{4} = \frac{4(1 + 2\sqrt{5})}{4} = 1 + 2\sqrt{5} \quad \text{Correct}$$

▶ **Work Problem 5 at the Side.**

5 Write each quotient in lowest terms.

(a) $\dfrac{5\sqrt{3} - 15}{10}$

(b) $\dfrac{12 + 8\sqrt{5}}{16}$

ANSWERS

5. (a) $\dfrac{\sqrt{3} - 3}{2}$ (b) $\dfrac{3 + 2\sqrt{5}}{4}$

Focus on Real-Data Applications

Spaceship Earth—A Geodesic Sphere

Geodesic domes became a popular design base for houses in the 1960s. The original patent was awarded to R. Buckminster Fuller in 1951. For the same square footage of interior space, a geodesic dome has less surface area and, thus, both reduces energy costs and is less prone to storm damage. One of the most famous geodesic domes is the full geodesic sphere Spaceship Earth, which is a major attraction at Epcot Center at Walt Disney World, Florida.

The sphere is made up of pyramids, shaped from three equilateral **facets**, or faces. A set of four pyramids forms a **panel** that is also an equilateral triangle. There are 954 panels, each supporting 12 facets. Some of the facets are removed for support beams, so there are actually only 11,324 facets on the sphere. The structure of the panels and the facets are easily seen in a close-up view of Spaceship Earth.

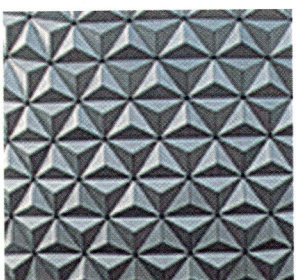

The outside **surface area** of Spaceship Earth is approximately 150,000 ft². To envision the idea of surface area, think about the task of painting the geodesic sphere. The painted surface is the surface area.

Each of the facets is an equilateral triangle. The area A of an equilateral triangle that has a side of length s is given by this formula.

$$A = \frac{\sqrt{3}}{4}s^2$$

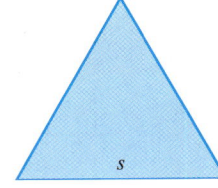

For Group Discussion

1. Use the given data for the outside surface area and the number of facets to find the surface area of one of Spaceship Earth's triangular facets. Round your answer to the nearest hundredth of a square foot. (Record the actual result for use in calculations for subsequent problems.)

2. Use the formula for the area of an equilateral triangle to find the length of the side of one of the triangular facets, to the nearest tenth of a foot.

3. How long is the side of one of the equilateral triangular panels?

4. Derive the formula for the area of an equilateral triangle. (*Hint:* Envision the right triangle that makes up half of the equilateral triangle, and use the Pythagorean formula to write an equation that relates the height h and the side s. Then use the formula for the area of a triangle.)

16.5 Exercises

In Exercises 1–4, perform the operations mentally, and write the answers without doing intermediate steps.

1. $\sqrt{49} + \sqrt{36}$
2. $\sqrt{100} - \sqrt{81}$
3. $\sqrt{2} \cdot \sqrt{8}$
4. $\sqrt{8} \cdot \sqrt{8}$

Simplify each expression. Use the five guidelines given in this section. Assume that all variables represent nonnegative real numbers. See Examples 1–3.

5. $\sqrt{5}(\sqrt{3} - \sqrt{7})$
6. $\sqrt{7}(\sqrt{10} + \sqrt{3})$
7. $2\sqrt{5}(\sqrt{2} + 3\sqrt{5})$

8. $3\sqrt{7}(2\sqrt{7} + 4\sqrt{5})$
9. $3\sqrt{14} \cdot \sqrt{2} - \sqrt{28}$
10. $7\sqrt{6} \cdot \sqrt{3} - 2\sqrt{18}$

11. $(2\sqrt{6} + 3)(3\sqrt{6} + 7)$
12. $(4\sqrt{5} - 2)(2\sqrt{5} - 4)$
13. $(5\sqrt{7} - 2\sqrt{3})(3\sqrt{7} + 4\sqrt{3})$

14. $(2\sqrt{10} + 5\sqrt{2})(3\sqrt{10} - 3\sqrt{2})$
15. $(8 - \sqrt{7})^2$
16. $(6 - \sqrt{11})^2$

17. $(2\sqrt{7} + 3)^2$
18. $(4\sqrt{5} + 5)^2$
19. $(\sqrt{a} + 1)^2$

20. $(\sqrt{y} + 4)^2$
21. $(5 - \sqrt{2})(5 + \sqrt{2})$
22. $(3 - \sqrt{5})(3 + \sqrt{5})$

23. $(\sqrt{8} - \sqrt{7})(\sqrt{8} + \sqrt{7})$
24. $(\sqrt{12} - \sqrt{11})(\sqrt{12} + \sqrt{11})$
25. $(\sqrt{y} - \sqrt{10})(\sqrt{y} + \sqrt{10})$

26. $(\sqrt{t} - \sqrt{13})(\sqrt{t} + \sqrt{13})$
27. $(\sqrt{2} + \sqrt{3})(\sqrt{6} - \sqrt{2})$
28. $(\sqrt{3} + \sqrt{5})(\sqrt{15} - \sqrt{5})$

29. $(\sqrt{10} - \sqrt{5})(\sqrt{5} + \sqrt{20})$
30. $(\sqrt{6} - \sqrt{3})(\sqrt{3} + \sqrt{18})$
31. $(\sqrt{5} + \sqrt{30})(\sqrt{6} + \sqrt{3})$

32. $(\sqrt{10} - \sqrt{20})(\sqrt{2} - \sqrt{5})$
33. $(\sqrt{5} - \sqrt{10})(\sqrt{x} - \sqrt{2})$
34. $(\sqrt{x} + \sqrt{6})(\sqrt{10} + \sqrt{3})$

35. In Example 1(b), the original expression simplifies to $-37 - 2\sqrt{15}$. Students often try to simplify such expressions by combining -37 and -2 to get $-39\sqrt{15}$, which is incorrect. Explain why.

36. If you try to rationalize the denominator of $\dfrac{2}{4 + \sqrt{3}}$ by multiplying the numerator and denominator by $4 + \sqrt{3}$, what problem arises? What should you multiply by?

Rationalize each denominator. Write quotients in lowest terms. Assume that all variables represent nonnegative real numbers. See Examples 4 and 5.

37. $\dfrac{1}{3 + \sqrt{2}}$
38. $\dfrac{1}{4 - \sqrt{3}}$
39. $\dfrac{14}{2 - \sqrt{11}}$
40. $\dfrac{19}{5 - \sqrt{6}}$

41. $\dfrac{\sqrt{2}}{2-\sqrt{2}}$

42. $\dfrac{\sqrt{7}}{7-\sqrt{7}}$

43. $\dfrac{\sqrt{5}}{\sqrt{2}+\sqrt{3}}$

44. $\dfrac{\sqrt{3}}{\sqrt{2}+\sqrt{3}}$

45. $\dfrac{\sqrt{5}+2}{2-\sqrt{3}}$

46. $\dfrac{\sqrt{7}+3}{4-\sqrt{5}}$

47. $\dfrac{12}{\sqrt{x}+1}$

48. $\dfrac{10}{\sqrt{x}-4}$

49. $\dfrac{3}{7-\sqrt{x}}$

50. $\dfrac{1}{6+\sqrt{z}}$

Write each quotient in lowest terms. See Example 5.

51. $\dfrac{6\sqrt{11}-12}{6}$

52. $\dfrac{12\sqrt{5}-24}{12}$

53. $\dfrac{2\sqrt{3}+10}{16}$

54. $\dfrac{4\sqrt{6}+24}{20}$

55. $\dfrac{12-\sqrt{40}}{4}$

56. $\dfrac{9-\sqrt{72}}{12}$

RELATING CONCEPTS (EXERCISES 57–62) For Individual or Group Work

Work Exercises 57–62 in order, *to see why a common student error is indeed an error.*

57. Use the distributive property to write $6(5 + 3x)$ as a sum.

58. Your answer in Exercise 57 should be $30 + 18x$. Why can't we combine these two terms to get $48x$?

59. Repeat Exercise 14 from earlier in this exercise set.

60. Your answer in Exercise 59 should be $30 + 18\sqrt{5}$. Many students will, in error, try to combine these terms to get $48\sqrt{5}$. Why is this wrong?

61. Write the expression similar to $30 + 18x$ that simplifies to $48x$. Then write the expression similar to $30 + 18\sqrt{5}$ that simplifies to $48\sqrt{5}$.

62. Write a short paragraph explaining the similarities between combining like terms and combining like radicals.

Solve each problem.

63. Here is the formula for the radius of the circular top or bottom of a tin can with a surface area S and a height h.

$$r = \frac{-h + \sqrt{h^2 + 0.64S}}{2}$$

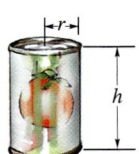

What radius should be used to make a can with a height of 12 in. and a surface area of 400 in.²?

64. If an investment of P dollars grows to A dollars in 2 years, the annual rate of return on the investment is given by this formula.

$$r = \frac{\sqrt{A} - \sqrt{P}}{\sqrt{P}}$$

Rationalize the denominator. Then find the annual rate of return r (as a percent) if $50,000 increases to $58,320.

Summary Exercises on Operations with Radicals

Perform all indicated operations and express each answer in simplest form. Assume that all variables represent positive real numbers.

1. $5\sqrt{10} - 8\sqrt{10}$

2. $\sqrt{5}(\sqrt{5} - \sqrt{3})$

3. $(1 + \sqrt{3})(2 - \sqrt{6})$

4. $\sqrt{98} - \sqrt{72} + \sqrt{50}$

5. $(3\sqrt{5} - 2\sqrt{7})^2$

6. $\dfrac{3}{\sqrt{6}}$

7. $\sqrt[3]{16t^2} - \sqrt[3]{54t^2} + \sqrt[3]{128t^2}$

8. $\dfrac{8}{\sqrt{7} - \sqrt{5}}$

9. $\dfrac{1 + \sqrt{2}}{1 - \sqrt{2}}$

10. $(1 + \sqrt[3]{3})(1 - \sqrt[3]{3} + \sqrt[3]{9})$

11. $(\sqrt{3} + 6)(\sqrt{3} - 6)$

12. $\dfrac{1}{\sqrt{t} + \sqrt{3}}$

13. $\sqrt[3]{8x^3y^5z^6}$

14. $\dfrac{12}{\sqrt[3]{9}}$

15. $\dfrac{5}{\sqrt{6} - 1}$

16. $\sqrt{\dfrac{2}{3x}}$

17. $\dfrac{6\sqrt{3}}{5\sqrt{12}}$

18. $\dfrac{8\sqrt{50}}{2\sqrt{25}}$

19. $\dfrac{-4}{\sqrt[3]{4}}$

20. $\dfrac{\sqrt{6} - \sqrt{5}}{\sqrt{6} + \sqrt{5}}$

21. $\sqrt{75x} - \sqrt{12x}$

22. $(5 + 3\sqrt{3})^2$

23. $(\sqrt{7} - \sqrt{6})(\sqrt{7} + \sqrt{6})$

24. $\sqrt[3]{\dfrac{16}{81}}$

25. $x\sqrt[4]{x^5} - 3\sqrt[4]{x^9} + x^2\sqrt[4]{x}$

26. $\sqrt{6} + \sqrt{6}$

27. $\sqrt{14} + \sqrt{17}$

28. $9\sqrt{24} - 2\sqrt{54} + 3\sqrt{20}$

29. $\sqrt{\dfrac{3}{4}} \cdot \sqrt{\dfrac{1}{5}}$

30. $\dfrac{5}{\sqrt{5}}$

31. $\sqrt[3]{24} + 6\sqrt[3]{81}$

32. $\dfrac{8}{4 - \sqrt{x}}$

33. $\sqrt[3]{4}\left(\sqrt[3]{2} - 3\right)$

34. $\sqrt{32x} - \sqrt{18x}$

35. $\sqrt{\dfrac{5}{8}}$

36. $(7 + \sqrt{x})^2$

37. A biologist has shown that the number of different plant species S on a Galápagos Island is related to the area of the island, A (in square miles), by this formula.

$$S = 28.6\sqrt[3]{A}$$

How many plant species (to the nearest whole number) would exist on such an island with the following areas?

(a) 8 mi²   **(b)** 27,000 mi²

16.6 Solving Equations with Radicals

A **radical equation** is an equation with a variable in the radicand, such as

$$\sqrt{x+1} = 3 \quad \text{or} \quad 3\sqrt{x} = \sqrt{8x+9}. \quad \text{\textcolor{blue}{Radical equations}}$$

OBJECTIVE 1 **Solve radical equations.** The addition and multiplication properties of equality are not enough to solve radical equations. We need a new property, called the *squaring property*.

> **Squaring Property of Equality**
>
> If each side of a given equation is squared, all solutions of the original equation are *among* the solutions of the squared equation.

> **CAUTION**
>
> Be very careful with the squaring property: Using this property can give a new equation with *more* solutions than the original equation. For example, starting with the equation $x = 4$ and squaring each side gives
>
> $$x^2 = 4^2 \quad \text{or} \quad x^2 = 16.$$
>
> This last equation, $x^2 = 16$, has *two* solutions, 4 or -4, while the original equation, $x = 4$, has only *one* solution, 4. Because of this possibility, checking is more than just a guard against algebraic errors when solving an equation with radicals. It is an essential part of the solution process. *All potential solutions from the squared equation must be checked in the original equation.*

EXAMPLE 1 Using the Squaring Property of Equality

Solve $\sqrt{p+1} = 3$.

Use the squaring property of equality to square each side of the equation.

$$(\sqrt{p+1})^2 = 3^2$$
$$p + 1 = 9 \qquad \textcolor{blue}{(\sqrt{p+1})^2 = p+1}$$
$$p = 8 \qquad \textcolor{blue}{\text{Subtract 1.}}$$

Now check this potential solution in the original equation.

Check:
$$\sqrt{p+1} = 3$$
$$\sqrt{8+1} = 3 \quad \textcolor{blue}{?} \quad \textcolor{blue}{\text{Let } p = 8.}$$
$$\sqrt{9} = 3 \quad \textcolor{blue}{?}$$
$$3 = 3 \quad \textcolor{blue}{\text{True}}$$

Because this statement is true, 8 is the solution of $\sqrt{p+1} = 3$. In this case the equation obtained by squaring had just one solution, which also satisfied the original equation.

Work Problem 1 at the Side.

OBJECTIVES

1. Solve radical equations.
2. Identify equations with no solutions.
3. Solve equations by squaring a binomial.

1 Solve each equation. Be sure to check your solutions.

(a) $\sqrt{k} = 3$

(b) $\sqrt{x-2} = 4$

(c) $\sqrt{9-t} = 4$

ANSWERS
1. (a) 9 (b) 18 (c) -7

2 Solve each equation.

(a) $\sqrt{3x+9} = 2\sqrt{x}$

(b) $5\sqrt{x} = \sqrt{20x+5}$

EXAMPLE 2 Using the Squaring Property with a Radical on Each Side

Solve $3\sqrt{x} = \sqrt{x+8}$.

Square each side of the equation.

$$(3\sqrt{x})^2 = (\sqrt{x+8})^2$$
$$3^2(\sqrt{x})^2 = (\sqrt{x+8})^2 \qquad (ab)^2 = a^2b^2$$
$$9x = x + 8 \qquad (\sqrt{x})^2 = x; (\sqrt{x+8})^2 = x+8$$
$$8x = 8 \qquad \text{Subtract } x.$$
$$x = 1 \qquad \text{Divide by 8.}$$

Check:
$$3\sqrt{x} = \sqrt{x+8} \qquad \text{Original equation}$$
$$3\sqrt{1} = \sqrt{1+8} \quad ? \qquad \text{Let } x = 1.$$
$$3(1) = \sqrt{9} \qquad ?$$
$$3 = 3 \qquad \text{True}$$

The solution of $3\sqrt{x} = \sqrt{x+8}$ is 1.

CAUTION
Do not write the final result obtained in the check as the solution. In Example 2, the solution is 1, *not* 3.

Work Problem 2 at the Side.

OBJECTIVE 2 Identify equations with no solutions. Not all radical equations have solutions, as shown in Examples 3 and 4.

EXAMPLE 3 Using the Squaring Property When One Side Is Negative

Solve $\sqrt{x} = -3$.

Square each side of the equation.

$$(\sqrt{x})^2 = (-3)^2$$
$$x = 9 \quad \longleftarrow \text{Potential solution}$$

Check:
$$\sqrt{x} = -3$$
$$\sqrt{9} = -3 \quad ? \qquad \text{Let } x = 9.$$
$$3 = -3 \qquad \text{False}$$

Because the statement $3 = -3$ is false, the number 9 is *not* a solution of the given equation and is said to be an **extraneous solution**; it must be discarded. In fact, $\sqrt{x} = -3$ has no solution.

NOTE
Because $\sqrt{x}$ represents the *principal* or *nonnegative* square root of x in Example 3, we might have seen immediately that there is no solution.

ANSWERS
2. (a) 9 (b) 1

EXAMPLE 4 Using the Squaring Property with a Quadratic Equation

Solve $p = \sqrt{p^2 + 5p + 10}$.

Square each side.

$$p^2 = (\sqrt{p^2 + 5p + 10})^2$$
$$p^2 = p^2 + 5p + 10 \qquad (\sqrt{p^2 + 5p + 10})^2 = p^2 + 5p + 10$$
$$0 = 5p + 10 \qquad \text{Subtract } p^2.$$
$$-10 = 5p \qquad \text{Subtract 10.}$$
$$p = -2 \qquad \text{Divide by 5.}$$

Check this potential solution in the original equation.

Check: $\quad p = \sqrt{p^2 + 5p + 10}$

$-2 = \sqrt{(-2)^2 + 5(-2) + 10}$? Let $p = -2$.

$-2 = \sqrt{4 - 10 + 10}$?

$-2 = 2$ False

Because $p = -2$ leads to a false result, the equation has no solution.

> **Work Problem 3 at the Side.**

OBJECTIVE 3 Solve equations by squaring a binomial. The next examples use the following rules from **Section 13.4**.

$$(a + b)^2 = a^2 + 2ab + b^2$$

and

$$(a - b)^2 = a^2 - 2ab + b^2.$$

By these patterns, for example,

$$(x - 3)^2 = x^2 - 2x(3) + 3^2$$
$$= x^2 - 6x + 9.$$

> **Work Problem 4 at the Side.**

EXAMPLE 5 Using the Squaring Property When One Side Has Two Terms

Solve $\sqrt{2x - 3} = x - 3$.

Square each side, using the preceding result to square the binomial on the right side of the equation.

$$(\sqrt{2x - 3})^2 = (x - 3)^2$$
$$2x - 3 = x^2 - 6x + 9$$

This equation is quadratic because of the x^2-term. As shown in **Section 14.6**, to solve this equation requires that one side be equal to 0. Subtract $2x$ and add 3 on each side, giving

$$0 = x^2 - 8x + 12. \qquad \text{Standard form}$$
$$0 = (x - 6)(x - 2) \qquad \text{Factor.}$$
$$x - 6 = 0 \quad \text{or} \quad x - 2 = 0 \qquad \text{Zero-factor property}$$
$$x = 6 \quad \text{or} \quad x = 2 \qquad \text{Solve.}$$

Continued on Next Page

---

3 Solve each equation. (*Hint:* In part (a), subtract 4 from each side.)

(a) $\sqrt{x} + 4 = 0$

(b) $x = \sqrt{x^2 - 4x - 16}$

4 Square each expression.

(a) $w - 5$

(b) $2k - 5$

(c) $3m - 2p$

ANSWERS
3. (a) no solution (b) no solution
4. (a) $w^2 - 10w + 25$ (b) $4k^2 - 20k + 25$
   (c) $9m^2 - 12mp + 4p^2$

Chapter 16 Roots and Radicals

5 Solve each equation.

(a) $\sqrt{6w + 6} = w + 1$

Check *both* of these potential solutions in the original equation.

Check:

If $x = 6$, then

$\sqrt{2x - 3} = x - 3$
$\sqrt{2(6) - 3} = 6 - 3$?
$\sqrt{12 - 3} = 3$?
$\sqrt{9} = 3$?
$3 = 3.$ **True**

If $x = 2$, then

$\sqrt{2x - 3} = x - 3$
$\sqrt{2(2) - 3} = 2 - 3$?
$\sqrt{4 - 3} = -1$?
$\sqrt{1} = -1$?
$1 = -1.$ **False**

Only 6 is a valid solution of the equation; 2 is extraneous.

◀◀◀ Work Problem 5 at the Side.

Sometimes we must write an equation in a different form before squaring each side. For example, suppose we want to solve $3\sqrt{x} - 1 = 2x$. Squaring each side gives the following.

$$(3\sqrt{x} - 1)^2 = (2x)^2$$
$$9x - 6\sqrt{x} + 1 = 4x^2$$

Squaring results in a more complicated equation that still contains a radical. In a case like this it is better to rewrite the original equation so that the radical is alone on one side of the equal sign, as shown in Example 6.

(b) $2u - 1 = \sqrt{10u + 9}$

EXAMPLE 6 Rewriting an Equation before Using the Squaring Property

Solve $3\sqrt{x} - 1 = 2x$.

Isolate the radical by adding 1 to each side.

$3\sqrt{x} = 2x + 1$
$(3\sqrt{x})^2 = (2x + 1)^2$ Square each side.
$9x = 4x^2 + 4x + 1$
$0 = 4x^2 - 5x + 1$ Subtract $9x$.
$0 = (4x - 1)(x - 1)$ Factor.
$4x - 1 = 0$ or $x - 1 = 0$ Zero-factor property
$x = \dfrac{1}{4}$ or $x = 1$ Solve.

Check:

If $x = \dfrac{1}{4}$, then

$3\sqrt{x} - 1 = 2x$
$3\sqrt{\dfrac{1}{4}} - 1 = 2\left(\dfrac{1}{4}\right)$?
$\dfrac{1}{2} = \dfrac{1}{2}$ **True**

If $x = 1$, then

$3\sqrt{x} - 1 = 2x$
$3\sqrt{1} - 1 = 2(1)$?
$2 = 2$ **True**

Both solutions check, so the solutions to the original equation are $\dfrac{1}{4}$ and 1.

ANSWERS
5. (a) 5, −1   (b) 4

CAUTION
Errors often occur when each side of an equation is squared. For instance, in Example 6 when each side of

$$3\sqrt{x} = 2x + 1$$

is squared, *the entire binomial on the right must be squared.* Here, $(2x + 1)^2 = 4x^2 + 4x + 1$. It would be incorrect to square the $2x$ and the 1 separately to get $4x^2 + 1$.

Work Problem 6 at the Side.

Some radical equations require squaring twice, as in the next example.

EXAMPLE 7 Using the Squaring Property Twice

Solve $\sqrt{21 + x} = 3 + \sqrt{x}$.

$$(\sqrt{21 + x})^2 = (3 + \sqrt{x})^2 \quad \text{Square each side.}$$
$$21 + x = 9 + 6\sqrt{x} + x$$
$$12 = 6\sqrt{x} \quad \text{Subtract 9; subtract } x.$$
$$2 = \sqrt{x} \quad \text{Divide by 6.}$$
$$2^2 = (\sqrt{x})^2 \quad \text{Square each side again.}$$
$$4 = x$$

Check: If $x = 4$, then

$$\sqrt{21 + x} = 3 + \sqrt{x} \quad \text{Original equation}$$
$$\sqrt{21 + 4} = 3 + \sqrt{4} \quad ?$$
$$5 = 5. \quad \text{True}$$

The solution is 4.

Work Problem 7 at the Side.

In summary, use the following steps to solve a radical equation.

Solving a Radical Equation

Step 1 **Isolate a radical.** Arrange the terms so that a radical is alone on one side of the equation.

Step 2 **Square each side.**

Step 3 **Combine like terms.**

Step 4 **Repeat Steps 1–3, if necessary.** If there is still a term with a radical, repeat Steps 1–3.

Step 5 **Solve the equation.** Find all potential solutions.

Step 6 **Check.** All potential solutions *must* be checked in the original equation.

---

6 Solve each equation.

(a) $\sqrt{x - 3} = x - 15$

(b) $\sqrt{z + 5} + 2 = z + 5$

7 Solve each equation.

(a) $\sqrt{p + 1} - \sqrt{p - 4} = 1$

(b) $\sqrt{2x + 1} + \sqrt{x + 4} = 3$

ANSWERS
6. (a) 16 (b) −1
7. (a) 8 (b) 0

Focus on Real-Data Applications

On a Clear Day

The Empire State Building, the Eiffel Tower, and the world's tallest buildings evoke images of sitting on top of the world. The prospect of viewing sights from such lofty heights stirs our imaginations. But how far can we really see? On a clear day, the maximum distance in kilometers that you can see from a tall building is given by this formula.

$$\text{sight distance} = 111.7 \sqrt{\text{height of building in kilometers}}$$

(*Source: A Sourcebook of Applications of School Mathematics*, NCTM.)

For Group Discussion

On a clear day, how far could you see from the top of each of these famous buildings and structures? Round answers to the nearest mile. Use these relationships: **1 ft ≈ 0.3048 m** and **1 km ≈ 0.621371 mi.**

1. The London Eye, which opened on New Year's Eve 1999, is a unique form of a Ferris wheel that features 32 observation capsules. It is located on the bank of the Thames River facing the Houses of Parliament and is the sixth tallest structure in London, with a diameter of 135 m. (*Source:* www.londoneye.com) Does the formula justify the claim that on a clear day, passengers on the London Eye can see Windsor Castle, 25 mi away?

2. The Empire State Building opened in 1931 on 5th Avenue in New York City. The building is 1250 ft high. (The antenna reaches to 1454 ft.) The observation deck, located on the 102nd floor, is at a height of 1050 ft. (*Source:* www.esbnyc.com) How far could you see on a clear day from the observation deck?

3. The twin Petronas Towers in Kuala Lumpur, Malaysia, are 1483 ft high (including the spires). (*Source: World Almanac and Book of Facts*, 2004.) How far would one of the builders have been able to see on a clear day from the top of a spire?

4. The Khufu Pyramid in Giza (also known as Cheops Pyramid) was built in about 2566 B.C. to a height, at that time, of 482 ft. It is now only about 450 ft high. (*Source:* www.touregypt.net/cheop.htm) How far would one of the original builders of the pyramid have been able to see from the top of the pyramid?

16.6 Exercises

Solve each equation. See Examples 1–4.

1. $\sqrt{x} = 7$
2. $\sqrt{k} = 10$
3. $\sqrt{t+2} = 3$
4. $\sqrt{x+7} = 5$

5. $\sqrt{r-4} = 9$
6. $\sqrt{k-12} = 3$
7. $\sqrt{4-t} = 7$
8. $\sqrt{9-s} = 5$

9. $\sqrt{2t+3} = 0$
10. $\sqrt{5x-4} = 0$
11. $\sqrt{3x-8} = -2$
12. $\sqrt{6x+4} = -3$

13. $\sqrt{w} - 4 = 7$
14. $\sqrt{t} + 3 = 10$
15. $\sqrt{10x-8} = 3\sqrt{x}$

16. $\sqrt{17t-4} = 4\sqrt{t}$
17. $5\sqrt{x} = \sqrt{10x+15}$
18. $4\sqrt{z} = \sqrt{20z-16}$

19. $\sqrt{3x-5} = \sqrt{2x+1}$
20. $\sqrt{5x+2} = \sqrt{3x+8}$
21. $k = \sqrt{k^2 - 5k - 15}$

22. $s = \sqrt{s^2 - 2s - 6}$
23. $7x = \sqrt{49x^2 + 2x - 10}$
24. $6x = \sqrt{36x^2 + 5x - 5}$

25. The first step in solving the equation $\sqrt{2x + 1} = x - 7$ is to square each side of the equation. Errors often occur in solving equations such as this one when the right side of the equation is squared incorrectly. What is the square of the right side?

26. What is wrong with the following "solution"?

$$-\sqrt{x - 1} = -4$$
$$-(x - 1) = 16 \quad \text{Square each side.}$$
$$-x + 1 = 16 \quad \text{Distributive property}$$
$$-x = 15 \quad \text{Subtract 1.}$$
$$x = -15 \quad \text{Multiply by } -1.$$

Solve each equation. See Examples 5 and 6.

27. $\sqrt{2x + 1} = x - 7$

28. $\sqrt{3x + 3} = x - 5$

29. $\sqrt{3k + 10} + 5 = 2k$

30. $\sqrt{4t + 13} + 1 = 2t$

31. $\sqrt{5x + 1} - 1 = x$

32. $\sqrt{x + 1} - x = 1$

33. $\sqrt{6t + 7} + 3 = t + 5$

34. $\sqrt{10x + 24} = x + 4$

35. $x - 4 - \sqrt{2x} = 0$

36. $x - 3 - \sqrt{4x} = 0$

37. $\sqrt{x + 6} = 2x$

38. $\sqrt{k + 12} = k$

Solve each equation. See Example 7.

39. $\sqrt{x + 1} - \sqrt{x - 4} = 1$

40. $\sqrt{2x + 3} + \sqrt{x + 1} = 1$

41. $\sqrt{x} = \sqrt{x - 5} + 1$

42. $\sqrt{2x} = \sqrt{x + 7} - 1$

43. $\sqrt{3x + 4} - \sqrt{2x - 4} = 2$

44. $\sqrt{1 - x} + \sqrt{x + 9} = 4$

45. $\sqrt{2x + 11} + \sqrt{x + 6} = 2$

46. $\sqrt{x + 9} + \sqrt{x + 16} = 7$

Solve each problem.

47. The square root of the sum of a number and 4 is 5. Find the number.

48. A certain number is the same as the square root of the product of 8 and the number. Find the number.

49. Three times the square root of 2 equals the square root of the sum of some number and 10. Find the number.

50. The negative square root of a number equals that number decreased by 2. Find the number.

Solve each problem. Give answers to the nearest tenth.

51. To estimate the speed at which a car was traveling at the time of an accident, a police officer drives the car involved in the accident under conditions similar to those during which the accident took place and then skids to a stop. If the car is driven at 30 mph, then the speed at the time of the accident is given by

$$s = 30\sqrt{\frac{a}{p}}$$

where a is the length of the skid marks left at the time of the accident and p is the length of the skid marks in the police test. Find s for the following values of a and p.

(a) $a = 862$ ft; $p = 156$ ft
(b) $a = 382$ ft; $p = 96$ ft
(c) $a = 84$ ft; $p = 26$ ft

52. A formula for calculating the distance, d, one can see from an airplane to the horizon on a clear day is

$$d = 1.22\sqrt{x}$$

where x is the altitude of the plane in feet and d is given in miles.

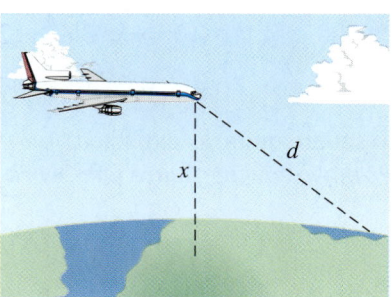

How far can one see to the horizon in a plane flying at the following altitudes?

(a) 15,000 ft
(b) 18,000 ft
(c) 24,000 ft

53. A surveyor wants to find the height of a building. At a point 110.0 ft from the base of the building he sights to the top of the building and finds the distance to be 193.0 ft. How high is the building?

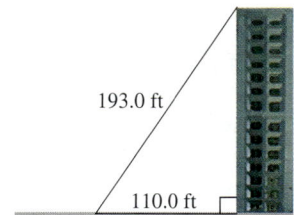

54. Two towns are separated by dense woods. To go from Town B to Town A, it is necessary to travel due west for 19.0 mi, then turn due north and travel for 14.0 mi. How far apart are the towns?

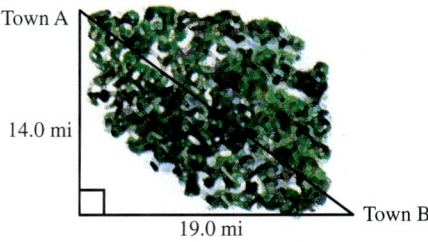

RELATING CONCEPTS (EXERCISES 55–60) For Individual or Group Work

The most common formula for the area of a triangle is $A = \frac{1}{2}bh$, where b is the length of the base and h is the height. What if the height is not known? What if we know only the lengths of the sides? Another formula, known as **Heron's formula,** allows us to calculate the area of a triangle if we know the lengths of the sides a, b, and c. First let s equal the **semiperimeter,** which is one-half the perimeter.

$$s = \frac{1}{2}(a + b + c)$$

The area A is given by the formula

$$A = \sqrt{s(s-a)(s-b)(s-c)}.$$

For example, the familiar 3–4–5 right triangle has area

$$A = \frac{1}{2}(3)(4) = 6 \text{ square units,}$$

using the familiar formula. Using Heron's formula, $s = \frac{1}{2}(3 + 4 + 5) = 6$, and

$$\begin{aligned} A &= \sqrt{6(6-3)(6-4)(6-5)} \\ &= \sqrt{6 \cdot 3 \cdot 2 \cdot 1} \\ &= \sqrt{36} = 6 \end{aligned}$$

The area is 6 square units, as expected.

Consider the following figure, and **work Exercises 55–60 in order.**

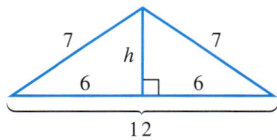

55. The lengths of the sides of the entire triangle are 7, 7, and 12. Find the semiperimeter s.

56. Now use Heron's formula to find the area of the entire triangle. Write it as a simplified radical.

57. Find the value of h by using the Pythagorean formula.

58. Find the area of each of the congruent right triangles forming the entire triangle by using the formula $A = \frac{1}{2}bh$.

59. Double your result from Exercise 58 to determine the area of the entire triangle.

60. How do your answers in Exercises 56 and 59 compare? (*Note:* They should be equal, since the area of the entire triangle is unique.)

Chapter 16
SUMMARY

KEY TERMS

16.1 square root — The number b is a square root of a if $b^2 = a$.
principal square root — The positive square root of a number is its principal square root.
radicand — The number or expression inside a radical sign is called the radicand.
radical — A radical sign with a radicand is called a radical.
radical expression — An algebraic expression containing a radical is called a radical expression.
perfect square — A number with a rational square root is called a perfect square.
cube root — The number b is a cube root of a if $b^3 = a$.
index (order) — In a radical of the form $\sqrt[n]{a}$, the number n is the index or order.

16.2 perfect cube — A number with a rational cube root is called a perfect cube.

16.3 like radicals — Like radicals are multiples of the same root of the same number.

16.4 rationalizing the denominator — The process of changing the denominator of a fraction from a radical (irrational number) to an expression not involving a radical is called rationalizing the denominator.

16.5 conjugate — The conjugate of $a + b$ is $a - b$.

16.6 radical equation — An equation with a variable in the radicand is a radical equation.
extraneous solution — A potential solution that does not satisfy the given equation is an extraneous solution.

Radical sign, Index → $\sqrt[n]{a}$ ← Radicand
Radical

NEW SYMBOLS

 radical sign $\approx$ is approximately equal to   cube root of a nth root of a

TEST YOUR WORD POWER

See how well you have learned the vocabulary in this chapter. Answers, with examples, follow the Quick Review.

1. A **square root** of a number is
   A. the number raised to the second power
   B. the number under a radical sign
   C. a number that when multiplied by itself gives the original number
   D. the inverse of the number.

2. A **radicand** is
   A. the index of a radical
   B. the number or expression inside the radical sign
   C. the positive root of a number
   D. the radical sign.

3. A **radical** is
   A. a symbol that indicates the nth root
   B. an algebraic expression containing a square root
   C. the positive nth root of a number
   D. a radical sign and the number or expression inside it.

4. The **principal root** of a positive number with even index n is
   A. the positive nth root of the number
   B. the negative nth root of the number

   C. the square root of the number
   D. the cube root of the number.

5. **Like radicals** are
   A. radicals in simplest form
   B. algebraic expressions containing radicals
   C. multiples of the same root of the same number
   D. radicals with the same index.

(continued)

1141

TEST YOUR WORD POWER (CONTINUED)

6. **Rationalizing the denominator** is the process of
   A. eliminating fractions from a radical expression
   B. changing the denominator of a fraction from a radical to an expression not involving a radical.
   C. clearing a radical expression of radicals
   D. multiplying radical expressions.

7. The **conjugate** of $a + b$ is
   A. $a - b$
   B. $a \cdot b$
   C. $a \div b$
   D. $(a + b)^2$.

8. An **extraneous solution** is a value
   A. that makes an equation false and must be discarded
   B. that makes an equation true
   C. that makes an expression equal 0
   D. that checks in the original equation.

QUICK REVIEW

Concepts	Examples

16.1 Evaluating Roots

If a is a positive real number, then

$\sqrt{a}$ is the positive or principal square root of a;

$-\sqrt{a}$ is the negative square root of a; $\sqrt{0} = 0$.

If a is a negative real number, then $\sqrt{a}$ is not a real number.

$\sqrt{49} = 7$

$-\sqrt{81} = -9$

$\sqrt{-25}$ is not a real number.

If a is a positive rational number, then $\sqrt{a}$ is rational if a is a perfect square. $\sqrt{a}$ is irrational if a is not a perfect square.

$\sqrt{\dfrac{4}{9}}$ and $\sqrt{16}$ are rational. $\sqrt{\dfrac{2}{3}}$ and $\sqrt{21}$ are irrational.

Every real number has exactly one real cube root.

$\sqrt[3]{27} = 3 \qquad \sqrt[3]{-8} = -2$

Pythagorean Formula

If c is the length of the longest side (hypotenuse) of a right triangle and a and b are the lengths of the shorter sides (legs), then

$$a^2 + b^2 = c^2.$$

Find b for the triangle in the figure.

$10^2 + b^2 = (2\sqrt{61})^2$

$100 + b^2 = 4(61)$

$100 + b^2 = 244$

$b^2 = 144$

$b = 12$

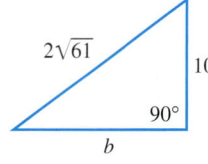

Concepts

16.2 Multiplying, Dividing, and Simplifying Radicals

Product Rule for Radicals
For nonnegative real numbers a and b,
$$\sqrt{a} \cdot \sqrt{b} = \sqrt{ab} \quad \text{and} \quad \sqrt{ab} = \sqrt{a} \cdot \sqrt{b}.$$

Quotient Rule for Radicals
If a and b are nonnegative real numbers and $b \neq 0$, then
$$\sqrt{\frac{a}{b}} = \frac{\sqrt{a}}{\sqrt{b}} \quad \text{and} \quad \frac{\sqrt{a}}{\sqrt{b}} = \sqrt{\frac{a}{b}}$$

If all indicated roots are real, then
$$\sqrt[n]{a} \cdot \sqrt[n]{b} = \sqrt[n]{ab} \quad \text{and} \quad \frac{\sqrt[n]{a}}{\sqrt[n]{b}} = \sqrt[n]{\frac{a}{b}} \quad (\text{where } b \neq 0)$$

16.3 Adding and Subtracting Radicals
Add and subtract like radicals by using the distributive property. *Only like radicals can be combined in this way.*

16.4 Rationalizing the Denominator
The denominator of a radical can be rationalized by multiplying both the numerator and denominator by a number that will eliminate the radical from the denominator.

16.5 More Simplifying and Operations with Radicals
When appropriate, use the rules for adding and multiplying polynomials to simplify radical expressions.

These formulas are useful when simplifying radical expressions.
$$(a+b)^2 = a^2 + 2ab + b^2$$
$$(a-b)^2 = a^2 - 2ab + b^2$$
$$(a+b)(a-b) = a^2 - b^2$$

Examples

$$\sqrt{5} \cdot \sqrt{7} = \sqrt{5 \cdot 7} = \sqrt{35}$$
$$\sqrt{8} \cdot \sqrt{2} = \sqrt{16} = 4$$
$$\sqrt{48} = \sqrt{16 \cdot 3} = \sqrt{16} \cdot \sqrt{3} = 4\sqrt{3}$$

$$\sqrt{\frac{25}{64}} = \frac{\sqrt{25}}{\sqrt{64}} = \frac{5}{8} \qquad \frac{\sqrt{8}}{\sqrt{2}} = \sqrt{\frac{8}{2}} = \sqrt{4} = 2$$

$$\sqrt[3]{5} \cdot \sqrt[3]{3} = \sqrt[3]{15} \qquad \frac{\sqrt[4]{12}}{\sqrt[4]{4}} = \sqrt[4]{\frac{12}{4}} = \sqrt[4]{3}$$

$$2\sqrt{5} + 4\sqrt{5} = (2+4)\sqrt{5}$$
$$= 6\sqrt{5}$$
$$\sqrt{8} + \sqrt{32} = 2\sqrt{2} + 4\sqrt{2}$$
$$= 6\sqrt{2}$$

$$\frac{2}{\sqrt{3}} = \frac{2 \cdot \sqrt{3}}{\sqrt{3} \cdot \sqrt{3}} = \frac{2\sqrt{3}}{3}$$

$$\sqrt[3]{\frac{5}{121}} = \frac{\sqrt[3]{5} \cdot \sqrt[3]{11}}{\sqrt[3]{11^2} \cdot \sqrt[3]{11}} = \frac{\sqrt[3]{55}}{11}$$

$$\sqrt{6}(\sqrt{5} - \sqrt{7}) = \sqrt{30} - \sqrt{42}$$

$$(\sqrt{3}+1)(\sqrt{3}-2) = 3 - 2\sqrt{3} + \sqrt{3} - 2 \quad \text{FOIL}$$
$$= 1 - \sqrt{3} \quad \text{Combine terms.}$$

$$(\sqrt{13} - \sqrt{2})^2 = (\sqrt{13})^2 - 2(\sqrt{13})(\sqrt{2}) + (\sqrt{2})^2$$
$$= 13 - 2\sqrt{26} + 2$$
$$= 15 - 2\sqrt{26}$$

$$(\sqrt{5} + \sqrt{3})(\sqrt{5} - \sqrt{3}) = 5 - 3 = 2$$

(continued)

Concepts	Examples
16.5 More Simplifying and Operations with Radicals *(continued)* Any denominators with radicals should be rationalized.	$\dfrac{3}{\sqrt{6}} = \dfrac{3 \cdot \sqrt{6}}{\sqrt{6} \cdot \sqrt{6}} = \dfrac{3\sqrt{6}}{6} = \dfrac{\sqrt{6}}{2}$
If a radical expression contains two terms in the denominator and at least one of those terms is a square root radical, multiply both the numerator and denominator by the conjugate of the denominator.	$\dfrac{6}{\sqrt{7} - \sqrt{2}} = \dfrac{6(\sqrt{7} + \sqrt{2})}{(\sqrt{7} - \sqrt{2})(\sqrt{7} + \sqrt{2})}$ $\phantom{\dfrac{6}{\sqrt{7} - \sqrt{2}}} = \dfrac{6(\sqrt{7} + \sqrt{2})}{7 - 2}$ Multiply. $\phantom{\dfrac{6}{\sqrt{7} - \sqrt{2}}} = \dfrac{6(\sqrt{7} + \sqrt{2})}{5}$ Subtract.

16.6 Solving Equations with Radicals

Solving a Radical Equation

Step 1 Isolate a radical.

Step 2 Square each side. (By the squaring property of equality, all solutions of the original equation are *among* the solutions of the squared equation.)

Step 3 Combine like terms.

Step 4 If there is still a term with a radical, repeat Steps 1–3.

Step 5 Solve the equation for potential solutions.

Step 6 Check all potential solutions from Step 5 in the original equation.

Solve $\sqrt{2x - 3} + x = 3$

$\sqrt{2x - 3} = 3 - x$ Isolate the radical.

$(\sqrt{2x - 3})^2 = (3 - x)^2$ Square each side.

$2x - 3 = 9 - 6x + x^2$

$0 = x^2 - 8x + 12$ Standard form

$0 = (x - 2)(x - 6)$ Factor.

$x - 2 = 0$ or $x - 6 = 0$ Zero-factor property.

$x = 2$ or $x = 6$ Solve.

A check is essential here. Verify that 2 is the only solution. (6 is extraneous.)

> **ANSWERS TO TEST YOUR WORD POWER**

1. C; *Examples:* 6 is a square root of 36 since $6^2 = 6 \cdot 6 = 36$; -6 is also a square root of 36.
2. B; *Example:* In $\sqrt{3xy}$, $3xy$ is the radicand.
3. D; *Examples:* $\sqrt{144}$, $\sqrt{4xy^2}$, and $\sqrt{4 + t^2}$
4. A; *Examples:* $\sqrt{36} = 6$, $\sqrt[4]{81} = 3$, and $\sqrt[6]{64} = 2$
5. C; *Examples:* $\sqrt{7}$ and $3\sqrt{7}$ are like radicals; so are $2\sqrt[3]{6k}$ and $5\sqrt[3]{6k}$.
6. B; *Example:* To rationalize the denominator of $\dfrac{5}{\sqrt{3} + 1}$, multiply the numerator and denominator by $\sqrt{3} - 1$ to get $\dfrac{5(\sqrt{3} - 1)}{2}$.
7. A; *Example:* The conjugate of $\sqrt{3} + 1$ is $\sqrt{3} - 1$.
8. A; *Example:* The potential solution 2 is extraneous when $\sqrt{5q - 1} + 3 = 0$ is solved using the method of **Section 16.6**.

Chapter 16
REVIEW EXERCISES

[16.1] *Find all square roots of each number.*

1. 49 **2.** 81 **3.** 196 **4.** 121 **5.** 225 **6.** 729

Find each root.

7. $\sqrt{16}$ **8.** $-\sqrt{0.36}$ **9.** $\sqrt[3]{1000}$ **10.** $\sqrt[4]{81}$

11. $\sqrt{-8100}$ **12.** $-\sqrt{4225}$ **13.** $\sqrt{\dfrac{49}{36}}$ **14.** $\sqrt{\dfrac{100}{81}}$

Match each radical in Column I with the equivalent choice in Column II. Choices may be used once, more than once, or not at all.

I	II
15. $\sqrt{64}$	**A.** 4
16. $-\sqrt{64}$	**B.** 8
17. $\sqrt{-64}$	**C.** -4
18. $\sqrt[3]{64}$	**D.** Not a real number
19. $\sqrt[3]{-64}$	**E.** 16
20. $-\sqrt[3]{-64}$	**F.** -8

21. Find the length of side x.

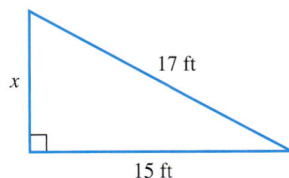

22. A Gateway EV700 computer monitor has viewing screen dimensions as shown in the figure. Find the diagonal measure of the viewing screen to the nearest tenth. (*Source:* Author's computer.)

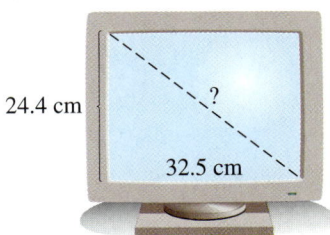

Chapter 16 Roots and Radicals

Write rational, irrational, or not a real number for each number. If a number is rational, give its exact value. If a number is irrational, give a decimal approximation for the number. Round approximations to the nearest thousandth.

23. $\sqrt{23}$

24. $\sqrt{169}$

25. $-\sqrt{25}$

26. $\sqrt{-4}$

[16.2] Simplify each expression.

27. $\sqrt{2} \cdot \sqrt{7}$

28. $\sqrt{5} \cdot \sqrt{15}$

29. $-\sqrt{27}$

30. $\sqrt{48}$

31. $\sqrt{160}$

32. $\sqrt{12} \cdot \sqrt{27}$

33. $\sqrt{32} \cdot \sqrt{48}$

34. $\sqrt{50} \cdot \sqrt{125}$

35. $\sqrt{\dfrac{9}{4}}$

36. $-\sqrt{\dfrac{121}{400}}$

37. $\sqrt{\dfrac{7}{169}}$

38. $\sqrt{\dfrac{1}{6}} \cdot \sqrt{\dfrac{5}{6}}$

39. $\sqrt{\dfrac{2}{5}} \cdot \sqrt{\dfrac{2}{45}}$

40. $\dfrac{3\sqrt{10}}{\sqrt{5}}$

41. $\dfrac{24\sqrt{12}}{6\sqrt{3}}$

42. $\dfrac{8\sqrt{150}}{4\sqrt{75}}$

Simplify each expression. Assume that all variables represent nonnegative real numbers.

43. $\sqrt{p} \cdot \sqrt{p}$

44. $\sqrt{k} \cdot \sqrt{m}$

45. $\sqrt{r^{18}}$

46. $\sqrt{x^{10}y^{16}}$

47. $\sqrt{x^9}$

48. $\sqrt{\dfrac{36}{p^2}}, \quad p \neq 0$

49. $\sqrt{a^{15}b^{21}}$

50. $\sqrt{121x^6y^{10}}$

51. $\sqrt[3]{y^6}$

52. $\sqrt[3]{216x^{15}}$

53. Use a calculator to find approximations for $\sqrt{0.5}$ and $\dfrac{\sqrt{2}}{2}$. Based on your results, do you think that these two expressions represent the same number? If so, verify it *algebraically*.

[16.3] *Simplify and combine terms where possible.*

54. $\sqrt{11} + \sqrt{11}$

55. $3\sqrt{2} + 6\sqrt{2}$

56. $3\sqrt{75} + 2\sqrt{27}$

57. $4\sqrt{12} + \sqrt{48}$

58. $4\sqrt{24} - 3\sqrt{54} + \sqrt{6}$

59. $2\sqrt{7} - 4\sqrt{28} + 3\sqrt{63}$

60. $\dfrac{2}{5}\sqrt{75} + \dfrac{3}{4}\sqrt{160}$

61. $\dfrac{1}{3}\sqrt{18} + \dfrac{1}{4}\sqrt{32}$

62. $\sqrt{15} \cdot \sqrt{2} + 5\sqrt{30}$

Simplify each expression. Assume that all variables represent nonnegative real numbers.

63. $\sqrt{4x} + \sqrt{36x} - \sqrt{9x}$

64. $\sqrt{16p} + 3\sqrt{p} - \sqrt{49p}$

65. $\sqrt{20m^2} - m\sqrt{45}$

66. $3k\sqrt{8k^2n} + 5k^2\sqrt{2n}$

[16.4] *Perform the indicated operations, and write all answers in simplest form. Rationalize all denominators. Assume that all variables represent nonnegative real numbers.*

67. $\dfrac{10}{\sqrt{3}}$

68. $\dfrac{8\sqrt{2}}{\sqrt{5}}$

69. $\dfrac{12}{\sqrt{24}}$

70. $\sqrt{\dfrac{2}{5}}$

71. $\sqrt{\dfrac{5}{14}} \cdot \sqrt{28}$

72. $\sqrt{\dfrac{2}{7}} \cdot \sqrt{\dfrac{1}{3}}$

73. $\sqrt{\dfrac{r^2}{16x}},\ x \neq 0$

74. $\sqrt[3]{\dfrac{1}{3}}$

Solve each problem.

75. The radius r of a cone in terms of its volume V is given by this formula.

$$r = \sqrt{\frac{3V}{\pi h}}$$

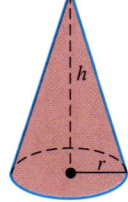

Rationalize the denominator of the radical expression.

76. The radius r of a sphere in terms of its surface area S is given by this formula.

$$r = \sqrt{\frac{S}{4\pi}}$$

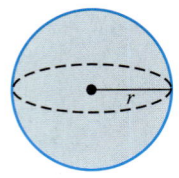

Rationalize the denominator of the radical expression.

[16.5] *Simplify each expression.*

77. $-\sqrt{3}(\sqrt{5} + \sqrt{27})$

78. $3\sqrt{2}(\sqrt{3} + 2\sqrt{2})$

79. $(2\sqrt{3} - 4)(5\sqrt{3} + 2)$

80. $(\sqrt{7} + 2\sqrt{6})(\sqrt{12} - \sqrt{2})$

81. $(2\sqrt{3} + 5)(2\sqrt{3} - 5)$

82. $(\sqrt{x} + 2)^2$, $x \geq 0$

Rationalize each denominator.

83. $\dfrac{1}{2 + \sqrt{5}}$

84. $\dfrac{2}{\sqrt{2} - 3}$

85. $\dfrac{3}{1 + \sqrt{x}}$, $x \geq 0$

86. $\dfrac{\sqrt{8}}{\sqrt{2} + 6}$

87. $\dfrac{\sqrt{5} - 1}{\sqrt{2} + 3}$

88. $\dfrac{2 + \sqrt{6}}{\sqrt{3} - 1}$

Write each quotient in lowest terms.

89. $\dfrac{15 + 10\sqrt{6}}{15}$

90. $\dfrac{3 + 9\sqrt{7}}{12}$

91. $\dfrac{6 + \sqrt{192}}{2}$

[16.6] *Solve each equation.*

92. $\sqrt{x} + 5 = 0$

93. $\sqrt{k+1} = 7$

94. $\sqrt{5t+4} = 3\sqrt{t}$

95. $\sqrt{2p+3} = \sqrt{5p-3}$

96. $\sqrt{4x+1} = x - 1$

97. $\sqrt{13+4t} = t + 4$

98. $\sqrt{2-x} + 3 = x + 7$

99. $\sqrt{x} - x + 2 = 0$

100. $\sqrt{x+2} - \sqrt{x-3} = 1$

MIXED REVIEW EXERCISES

Simplify each expression if possible. Assume that all variables represent nonnegative real numbers.

101. $\sqrt{3} \cdot \sqrt{27}$

102. $2\sqrt{27} + 3\sqrt{75} - \sqrt{300}$

103. $\sqrt{\dfrac{121}{t^2}}, \quad t \neq 0$

104. $\dfrac{1}{5 + \sqrt{2}}$

105. $\sqrt{\dfrac{1}{3}} \cdot \sqrt{\dfrac{24}{5}}$

106. $\sqrt{50y^2}$

107. $\sqrt[3]{-125}$

108. $-\sqrt{5}(\sqrt{2} + \sqrt{75})$

109. $\sqrt{\dfrac{16r^3}{3s}}, \quad s \neq 0$

110. $\dfrac{12 + 6\sqrt{13}}{12}$

111. $-\sqrt{162} + \sqrt{8}$

112. $(\sqrt{5} - \sqrt{2})^2$

113. $(6\sqrt{7} + 2)(4\sqrt{7} - 1)$

114. $-\sqrt{121}$

115. $\sqrt{98}$

Solve.

116. $\sqrt{x+2} = x - 4$ **117.** $\sqrt{k} + 3 = 0$ **118.** $\sqrt{1 + 3t} - t = -3$

119. The *fall speed,* in miles per hour, of a vehicle running off the road into a ditch is given by the formula

$$S = \frac{2.74D}{\sqrt{h}}$$

where D is the horizontal distance traveled from the level surface to the bottom of the ditch and h is the height (or depth) of the ditch. What is the fall speed (to the nearest tenth) of a vehicle that traveled 32 ft horizontally into a ditch 5 ft deep?

RELATING CONCEPTS (EXERCISES 120–124) For Individual or Group Work

*In **Chapter 11** we plotted points in the rectangular coordinate plane. In all cases our points had coordinates that were rational numbers. However, ordered pairs may have irrational coordinates as well. Consider the points $A(2\sqrt{14}, 5\sqrt{7})$ and $B(-3\sqrt{14}, 10\sqrt{7})$. Work Exercises 120–124 in order.*

120. Write an expression that represents the slope of the line containing points A and B. Do not simplify yet.

121. Simplify the numerator and the denominator in the expression from Exercise 120 by combining like radicals.

122. Write the fraction from Exercise 121 as the square root of a fraction in lowest terms.

123. Rationalize the denominator of the expression found in Exercise 122.

124. Based on your answer in Exercise 123, does line AB rise or fall from left to right?

Chapter 16
TEST

On this test, assume that all variables represent nonnegative real numbers.

1. Find all square roots of 196.

2. Consider $\sqrt{142}$.
   - (a) Determine whether it is rational or irrational.
   - (b) Find a decimal approximation to the nearest thousandth.

3. If $\sqrt{a}$ is not a real number, then what kind of number must a be?

Simplify where possible.

4. $\sqrt[3]{216}$

5. $-\sqrt{27}$

6. $\sqrt{\dfrac{128}{25}}$

7. $\sqrt[3]{32}$

8. $\dfrac{20\sqrt{18}}{5\sqrt{3}}$

9. $3\sqrt{28} + \sqrt{63}$

10. $3\sqrt{27x} - 4\sqrt{48x} + 2\sqrt{3x}$

11. $\sqrt{32x^2y^3}$

12. $(6 - \sqrt{5})(6 + \sqrt{5})$

13. $(2 - \sqrt{7})(3\sqrt{2} + 1)$

14. $(\sqrt{5} + \sqrt{6})^2$

Solve each problem.

15. Find the length of the unknown side in this right triangle.

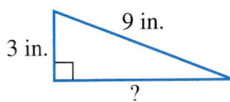

(a) Give its length in simplified radical form.

(b) Round the answer to the nearest thousandth.

16. In electronics, the impedance Z of an alternating current circuit is given by the formula

$$Z = \sqrt{R^2 + X^2}$$

where R is the resistance and X is the reactance, both in ohms. Find the value of the impedance Z if $R = 40$ ohms and $X = 30$ ohms.

Rationalize each denominator.

17. $\dfrac{5\sqrt{2}}{\sqrt{7}}$

18. $\sqrt{\dfrac{2}{3x}}$, $x \neq 0$

19. $\dfrac{-2}{\sqrt[3]{4}}$

20. $\dfrac{-3}{4 - \sqrt{3}}$

21. Write in lowest terms: $\dfrac{\sqrt{12} + 3\sqrt{128}}{6}$

Solve each equation.

22. $\sqrt{p} + 4 = 0$

23. $\sqrt{x + 1} = 5 - x$

24. $3\sqrt{x} - 2 = x$

25. What is wrong with the following "solution"?

$$\sqrt{2x + 1} + 5 = 0$$
$$\sqrt{2x + 1} = -5 \quad \text{Subtract 5.}$$
$$2x + 1 = 25 \quad \text{Square both sides.}$$
$$2x = 24 \quad \text{Subtract 1.}$$
$$x = 12 \quad \text{Divide by 2.}$$

The solution is 12.

Quadratic Equations

17.1 Solving Quadratic Equations by the Square Root Property

17.2 Solving Quadratic Equations by Completing the Square

17.3 Solving Quadratic Equations by the Quadratic Formula

Summary Exercises on Quadratic Equations

17.4 Graphing Quadratic Equations

17.5 Introduction to Functions

In this chapter we develop methods of solving quadratic equations. The graphs of such equations in two variables, called *parabolas*, have many applications. For example, the Parkes radio telescope, pictured here, has a *parabolic* dish shape with a diameter of 210 ft and a depth of 32 ft. (*Source: Structure Technology for Large Radio and Radar Telescope Systems,* The MIT Press.)

In Example 5 of Section 17.4, we use a graph to model this parabolic shape and find its quadratic equation.

17.1 Solving Quadratic Equations by the Square Root Property

OBJECTIVES

1. Solve equations of the form $x^2 = k$, where $k > 0$.
2. Solve equations of the form $(ax + b)^2 = k$, where $k > 0$.
3. Use formulas involving squared variables.

In **Section 14.6** we solved quadratic equations by factoring. However, since not all quadratic equations can be solved by factoring, we need to develop other methods. In this chapter we do just that. Recall that a *quadratic equation* is an equation that can be written in the form

$$ax^2 + bx + c = 0 \quad \text{Standard form}$$

for real numbers a, b, and c, with $a \neq 0$. As we saw in **Section 14.6,** to solve $x^2 + 4x + 3 = 0$ by the zero-factor property, we begin by factoring the left side and then setting each factor equal to 0.

$$x^2 + 4x + 3 = 0$$
$$(x + 3)(x + 1) = 0 \quad \text{Factor.}$$
$$x + 3 = 0 \quad \text{or} \quad x + 1 = 0 \quad \text{Zero-factor property}$$
$$x = -3 \quad \text{or} \quad x = -1 \quad \text{Solve each equation.}$$

The solutions are -3 and -1.

OBJECTIVE 1 Solve equations of the form $x^2 = k$, where $k > 0$. We can solve equations such as $x^2 = 9$ by factoring as follows.

$$x^2 = 9$$
$$x^2 - 9 = 0 \quad \text{Subtract 9.}$$
$$(x + 3)(x - 3) = 0 \quad \text{Factor.}$$
$$x + 3 = 0 \quad \text{or} \quad x - 3 = 0 \quad \text{Zero-factor property}$$
$$x = -3 \quad \text{or} \quad x = 3 \quad \text{Solve each equation.}$$

We might also solve $x^2 = 9$ by noticing that x must be a number whose square is 9. Thus, $x = \sqrt{9} = 3$ or $x = -\sqrt{9} = -3$. This approach is generalized as the **square root property of equations.**

> **Square Root Property of Equations**
> If k is a positive number and if $a^2 = k$, then
> $$a = \sqrt{k} \quad \text{or} \quad a = -\sqrt{k}$$

EXAMPLE 1 Solving Quadratic Equations of the Form $x^2 = k$

Solve each equation. Write radicals in simplified form.

(a) $x^2 = 16$

By the square root property, if $x^2 = 16$, then

$$x = \sqrt{16} = 4 \quad \text{or} \quad x = -\sqrt{16} = -4.$$

An abbreviation for $x = 4$ or $x = -4$ is $x = \pm 4$ (read "positive or negative 4"). Check each solution by substituting it for x in the original equation.

(b) $z^2 = 5$

The solutions $\sqrt{5}$ and $-\sqrt{5}$ may be written as $\pm\sqrt{5}$.

Continued on Next Page

(c)
$$5m^2 - 32 = 8$$
$$5m^2 = 40 \qquad \text{Add 32.}$$
$$m^2 = 8 \qquad \text{Divide by 5.}$$
$$m = \sqrt{8} \quad \text{or} \quad m = -\sqrt{8} \qquad \text{Square root property}$$
$$m = 2\sqrt{2} \quad \text{or} \quad m = -2\sqrt{2} \qquad \sqrt{8} = \sqrt{4} \cdot \sqrt{2} = 2\sqrt{2}$$

The solutions are $\pm 2\sqrt{2}$.

(d) $x^2 = -4$

Because -4 is a negative number and because the square of a real number cannot be negative, there is no real number solution for this equation. (The square root property cannot be used because of the requirement that k must be positive.)

Work Problem 1 at the Side.

OBJECTIVE 2 **Solve equations of the form $(ax + b)^2 = k$, where $k > 0$.** In each equation in Example 1, the exponent 2 had a single variable as its base. The square root property can be extended to solve equations where the base is a binomial, as shown in the next example.

EXAMPLE 2 Solving Quadratic Equations of the Form $(x + b)^2 = k$

Solve each equation.

(a) $(x - 3)^2 = 16$

Apply the square root property, using $x - 3$ as the base.
$$(x - 3)^2 = 16$$
$$x - 3 = \sqrt{16} \quad \text{or} \quad x - 3 = -\sqrt{16}$$
$$x - 3 = 4 \qquad \qquad x - 3 = -4 \qquad \sqrt{16} = 4$$
$$x = 7 \qquad \qquad x = -1 \qquad \text{Add 3.}$$

Check both answers in the original equation.

$(x - 3)^2 = 16$	$(x - 3)^2 = 16$
$(7 - 3)^2 = 16$? Let $x = 7$.	$(-1 - 3)^2 = 16$? Let $x = -1$.
$4^2 = 16$?	$(-4)^2 = 16$?
$16 = 16$ True	$16 = 16$ True

The solutions are 7 and -1.

(b)
$$(x + 1)^2 = 6$$
$$x + 1 = \sqrt{6} \quad \text{or} \quad x + 1 = -\sqrt{6} \qquad \text{Square root property}$$
$$x = -1 + \sqrt{6} \quad \text{or} \quad x = -1 - \sqrt{6} \qquad \text{Add } -1.$$

Check:
$$(-1 + \sqrt{6} + 1)^2 = (\sqrt{6})^2 = 6$$
$$(-1 - \sqrt{6} + 1)^2 = (-\sqrt{6})^2 = 6$$

The solutions are $-1 + \sqrt{6}$ and $-1 - \sqrt{6}$.

Work Problem 2 at the Side.

---

1 Solve each equation. Write radicals in simplified form.

(a) $x^2 = 49$

(b) $x^2 = 11$

(c) $2x^2 + 8 = 32$

(d) $x^2 = -9$

2 Solve each equation.

(a) $(x + 2)^2 = 36$

(b) $(x - 4)^2 = 3$

ANSWERS
1. (a) $-7, 7$ (b) $-\sqrt{11}, \sqrt{11}$
 (c) $-2\sqrt{3}, 2\sqrt{3}$
 (d) no real number solution
2. (a) $-8, 4$ (b) $4 + \sqrt{3}, 4 - \sqrt{3}$

3 Solve $(2x - 5)^2 = 18$.

4 Solve each equation.

(a) $(5x + 1)^2 = 7$

(b) $(7x - 1)^2 = -1$

5 Use the formula in Example 5 to approximate the length of a bass weighing 2.80 lb and having a girth of 11 in.

ANSWERS

3. $\dfrac{5 + 3\sqrt{2}}{2}, \dfrac{5 - 3\sqrt{2}}{2}$

4. (a) $\dfrac{-1 + \sqrt{7}}{5}, \dfrac{-1 - \sqrt{7}}{5}$

(b) no real number solution

5. approximately 17.48 in.

EXAMPLE 3 Solving a Quadratic Equation of the Form $(ax + b)^2 = k$

Solve $(3r - 2)^2 = 27$.

$3r - 2 = \sqrt{27}$	or	$3r - 2 = -\sqrt{27}$	Square root property
$3r - 2 = 3\sqrt{3}$		$3r - 2 = -3\sqrt{3}$	$\sqrt{27} = \sqrt{9} \cdot \sqrt{3} = 3\sqrt{3}$
$3r = 2 + 3\sqrt{3}$		$3r = 2 - 3\sqrt{3}$	Add 2.
$r = \dfrac{2 + 3\sqrt{3}}{3}$		$r = \dfrac{2 - 3\sqrt{3}}{3}$	Divide by 3.

The solutions are $\dfrac{2 + 3\sqrt{3}}{3}$ and $\dfrac{2 - 3\sqrt{3}}{3}$.

Work Problem 3 at the Side.

CAUTION
The solutions in Example 3 are fractions that cannot be simplified, since 3 is *not* a common factor in the numerator.

EXAMPLE 4 Recognizing a Quadratic Equation with No Real Solution

Solve $(x + 3)^2 = -9$.

Because the square root of -9 is not a real number, there is no real number solution for this equation.

Work Problem 4 at the Side.

OBJECTIVE 3 Use formulas involving squared variables.

EXAMPLE 5 Finding the Length of a Bass

The formula $w = \dfrac{L^2 g}{1200}$ is used to approximate the weight of a bass, in pounds, given its length L and its girth g, where both are measured in inches. Approximate the length of a bass weighing 2.20 lb and having a girth of 10 in. (Source: *Sacramento Bee.*)

$w = \dfrac{L^2 g}{1200}$ Given formula

$2.20 = \dfrac{L^2 \cdot 10}{1200}$ $w = 2.20, g = 10$

$2640 = 10L^2$ Multiply by 1200.

$L^2 = 264$ Divide by 10; interchange the sides.

$L \approx 16.25$ Use a calculator; $L > 0$.

The length of the bass is a little more than 16 in. (We discard the negative solution -16.25 since L represents length.)

Work Problem 5 at the Side.

17.1 Exercises

Decide whether each statement is true *or* false. *If false, tell why.*

1. If k is a prime number, then $x^2 = k$ has two irrational solutions.

2. If k is a positive perfect square, then $x^2 = k$ has two rational solutions.

3. If k is a positive integer, then $x^2 = k$ must have two rational solutions.

4. If $-10 < k < 0$, then $x^2 = k$ has no real solution.

5. If $-10 < k < 10$, then $x^2 = k$ has no real solution.

6. If k is an integer greater than 24 and less than 26, then $x^2 = k$ has two solutions, -5 and 5.

Solve each equation by using the square root property. Write all radicals in simplest form. See Example 1.

7. $x^2 = 81$

8. $x^2 = 121$

9. $k^2 = 14$

10. $m^2 = 22$

11. $t^2 = 48$

12. $x^2 = 54$

13. $x^2 = \dfrac{25}{4}$

14. $m^2 = \dfrac{36}{121}$

15. $x^2 = -100$

16. $x^2 = -64$

17. $z^2 = 2.25$

18. $w^2 = 56.25$

19. $r^2 - 3 = 0$

20. $x^2 - 13 = 0$

21. $7x^2 = 4$

22. $3x^2 = 10$

23. $5x^2 + 4 = 8$

24. $4x^2 - 3 = 7$

25. $3x^2 - 8 = 64$

26. $2x^2 + 7 = 61$

Solve each equation by using the square root property. Express all radicals in simplest form. See Examples 2–4.

27. $(x - 3)^2 = 25$

28. $(x - 7)^2 = 16$

29. $(x + 5)^2 = -13$

30. $(x + 2)^2 = -17$

31. $(x - 8)^2 = 27$

32. $(x - 5)^2 = 40$

33. $(3x + 2)^2 = 49$

34. $(5x + 3)^2 = 36$

35. $(4x - 3)^2 = 9$

36. $(7x - 5)^2 = 25$

37. $(5 - 2x)^2 = 30$

38. $(3 - 2x)^2 = 70$

39. $(3x + 1)^2 = 18$

40. $(5x + 6)^2 = 75$

41. $\left(\dfrac{1}{2}x + 5\right)^2 = 12$

42. $\left(\dfrac{1}{3}x + 4\right)^2 = 27$

43. $(4x - 1)^2 - 48 = 0$

44. $(2x - 5)^2 - 180 = 0$

45. Michael solved the equation in Exercise 37 and wrote his solutions as $\dfrac{5 + \sqrt{30}}{2}, \dfrac{5 - \sqrt{30}}{2}$.

Lindsay solved the same equation and wrote her solutions as $\dfrac{-5 + \sqrt{30}}{-2}, \dfrac{-5 - \sqrt{30}}{-2}$.

The teacher gave them both full credit. Explain why both students were correct, although their answers seem to differ.

46. In the solutions found in Example 3 of this section, why is it not valid to simplify by dividing out the 3s in the numerators and denominators?

Solve each problem. See Example 5.

47. One expert at marksmanship can hold a silver dollar at forehead level, drop it, draw his gun, and shoot the coin as it passes waist level. The distance traveled by a falling object is given by

$$d = 16t^2$$

where d is the distance (in feet) the object falls in t seconds. If the coin falls about 4 ft, use the formula to estimate the time that elapses between the dropping of the coin and the shot.

48. The illumination produced by a light source depends on the distance from the source. For a particular light source, this relationship can be expressed as

$$d^2 = \frac{4050}{I}$$

where d is the distance from the source (in feet) and I is the amount of illumination in foot-candles. How far from the source is the illumination equal to 50 foot-candles?

49. The area A of a circle with radius r is given by this formula.

$$A = \pi r^2$$

If a circle has an area of 81π in.², what is its radius?

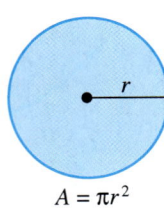

$A = \pi r^2$

50. The surface area S of a sphere with radius r is given by this formula.

$$S = 4\pi r^2$$

If a sphere has a surface area of 36π ft², what is its radius?

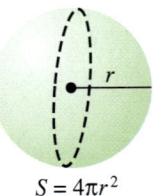

$S = 4\pi r^2$

The amount A that P dollars invested at an annual rate of interest r will grow to in 2 years is $A = P(1 + r)^2$.

51. At what interest rate will $100 grow to $110.25 in 2 years?

52. At what interest rate will $500 grow to $572.45 in 2 years?

17.2 Solving Quadratic Equations by Completing the Square

OBJECTIVES

1. Solve quadratic equations by completing the square when the coefficient of the squared term is 1.
2. Solve quadratic equations by completing the square when the coefficient of the squared term is not 1.
3. Simplify an equation before solving.
4. Solve applied problems that require quadratic equations.

OBJECTIVE 1 Solve quadratic equations by completing the square when the coefficient of the squared term is 1. The methods we have studied so far are not enough to solve this equation.

$$x^2 + 6x + 7 = 0$$

If we could write the equation in the form $(x + 3)^2$ equals a constant, we could solve it with the square root property discussed in **Section 17.1**. To do that, we need to have a perfect square trinomial on one side of the equation.

Recall from **Section 14.5** that a perfect square trinomial has the form

$$x^2 + 2kx + k^2 \quad \text{or} \quad x^2 - 2kx + k^2$$

where k represents a positive number.

EXAMPLE 1 Creating Perfect Square Trinomials

Complete each trinomial so it is a perfect square.

(a) $x^2 +$ _____ $+ 16$

Here $k^2 = 16$, so $k = 4$ and $2kx = 2(4)(x) = 8x$. The perfect square trinomial is $x^2 + 8x + 16$.

(b) $4m^2 -$ _____ $+ 9$

Replace x^2 by $4m^2 = (2m)^2$ and k^2 by $9 = 3^2$, to get the middle term.

$$2(3)(2m) = 12m$$

The perfect square trinomial is $4m^2 - 12m + 9$.

(c) $x^2 + 18x +$ _____

Here the middle term $18x$ must equal $2kx$.

$$2kx = 18x$$
$$k = 9 \qquad \text{Divide by } 2x.$$

Thus, $k = 9$ and $k^2 = 81$. The required trinomial is $x^2 + 18x + 81$.

① Complete each trinomial so it is a perfect square.

(a) $x^2 +$ _____ $+ 36$

(b) $25x^2 -$ _____ $+ 4$

(c) $x^2 + 14x +$ _____

▶ Work Problem 1 at the Side.

EXAMPLE 2 Rewriting an Equation to Use the Square Root Property

Solve $x^2 + 6x + 7 = 0$.

Start by subtracting 7 from each side of the equation.

$$x^2 + 6x = -7$$

The quantity on the left side of $x^2 + 6x = -7$ must be made into a perfect square trinomial. Here $2kx = 6x$, so $k = 3$ and $k^2 = 9$. The expression $x^2 + 6x + 9$ is a perfect square, since

$$x^2 + 6x + 9 = (x + 3)^2.$$

Therefore, if 9 is added to each side, the equation will have a perfect square trinomial on the left side, as needed.

$$x^2 + 6x + 9 = -7 + 9 \qquad \text{Add 9.}$$
$$(x + 3)^2 = 2 \qquad \text{Factor.}$$

Continued on Next Page

ANSWERS
1. (a) $12x$ (b) $20x$ (c) 49

2 Solve $x^2 - 4x - 1 = 0$.

Now use the square root property to complete the solution.

$$(x + 3)^2 = 2$$
$$x + 3 = \sqrt{2} \quad \text{or} \quad x + 3 = -\sqrt{2}$$
$$x = -3 + \sqrt{2} \quad \text{or} \quad x = -3 - \sqrt{2}$$

The solutions of the original equation are $-3 + \sqrt{2}$ and $-3 - \sqrt{2}$. Check by substituting $-3 + \sqrt{2}$ and $-3 - \sqrt{2}$ for x in the original equation.

◀◀◀ **Work Problem 2 at the Side.**

The process of changing the form of the equation in Example 2 from

$$x^2 + 6x + 7 = 0 \quad \text{to} \quad (x + 3)^2 = 2$$

is called **completing the square**. Completing the square changes only the form of the equation. To see this, multiply out the left side of $(x + 3)^2 = 2$ and combine terms. Then subtract 2 from each side to see that the result is $x^2 + 6x + 7 = 0$.

Look again at the original equation,

$$x^2 + 6x + 7 = 0.$$

3 Solve by completing the square.

(a) $x^2 + 4x = 1$

Notice that the coefficient of x is 6; half of 6 is 3, and 3^2 is 9.

$$\frac{1}{2}(6) = 3 \quad \text{and} \quad 3^2 = 9$$
↑
Coefficient of x

To complete the square in Example 2, we added 9 to each side.

EXAMPLE 3 Completing the Square to Solve a Quadratic Equation

Complete the square to solve $x^2 - 8x = 5$.
To complete the square on $x^2 - 8x$, take half the coefficient of x and square it.

(b) $z^2 + 6z - 3 = 0$

$$\frac{1}{2}(-8) = -4 \quad \text{and} \quad (-4)^2 = 16$$
↑
Coefficient of x

Add the result, 16, to each side of the equation.

$$x^2 - 8x = 5 \quad \text{Given equation}$$
$$x^2 - 8x + 16 = 5 + 16 \quad \text{Add 16.}$$
$$(x - 4)^2 = 21 \quad \text{Factor the left side as the square of a binomial.}$$

Now apply the square root property.

$$x - 4 = \sqrt{21} \quad \text{or} \quad x - 4 = -\sqrt{21} \quad \text{Square root property}$$
$$x = 4 + \sqrt{21} \quad \text{or} \quad x = 4 - \sqrt{21} \quad \text{Add 4.}$$

A check indicates that the solutions are $4 + \sqrt{21}$ and $4 - \sqrt{21}$.

◀◀◀ **Work Problem 3 at the Side.**

ANSWERS
2. $2 + \sqrt{5}, 2 - \sqrt{5}$
3. (a) $-2 + \sqrt{5}, -2 - \sqrt{5}$
   (b) $-3 + 2\sqrt{3}, -3 - 2\sqrt{3}$

OBJECTIVE 2 Solve quadratic equations by completing the square when the coefficient of the squared term is not 1. If an equation has the form

$$ax^2 + bx + c = 0, \quad \text{where } a \neq 1,$$

then to obtain 1 as the coefficient of x^2, we first divide each side of the equation by a.

EXAMPLE 4 Solving a Quadratic Equation by Completing the Square

Solve $4x^2 + 16x = 9$.

Before completing the square, the coefficient of x^2 must be 1, not 4. We make the coefficient 1 by dividing each side of the equation by 4.

$$4x^2 + 16x = 9$$

$$x^2 + 4x = \frac{9}{4} \quad \text{Divide by 4.}$$

Next, we complete the square by taking half the coefficient of x, or $\frac{1}{2}(4) = 2$, and squaring the result: $2^2 = 4$. We add 4 to each side of the equation, combine terms on the right side, and factor on the left.

$$x^2 + 4x + 4 = \frac{9}{4} + 4 \quad \text{Add 4.}$$

$$x^2 + 4x + 4 = \frac{25}{4} \quad \text{Combine terms.}$$

$$(x + 2)^2 = \frac{25}{4} \quad \text{Factor.}$$

Use the square root property and solve for x.

$$x + 2 = \sqrt{\frac{25}{4}} \quad \text{or} \quad x + 2 = -\sqrt{\frac{25}{4}} \quad \text{Square root property}$$

$$x + 2 = \frac{5}{2} \quad\quad\quad x + 2 = -\frac{5}{2} \quad \text{Take square roots.}$$

$$x = -2 + \frac{5}{2} \quad\quad x = -2 - \frac{5}{2} \quad \text{Add } -2.$$

$$x = \frac{1}{2} \quad\quad\quad\quad x = -\frac{9}{2} \quad \text{Combine terms.}$$

Check:

$$4x^2 + 16x = 9 \quad\quad\quad\quad\quad 4x^2 + 16x = 9$$

$$4\left(\frac{1}{2}\right)^2 + 16\left(\frac{1}{2}\right) = 9 \ ? \quad\quad 4\left(-\frac{9}{2}\right)^2 + 16\left(-\frac{9}{2}\right) = 9 \ ?$$

$$4\left(\frac{1}{4}\right) + 8 = 9 \ ? \quad\quad\quad 4\left(\frac{81}{4}\right) - 72 = 9 \ ?$$

$$1 + 8 = 9 \ ? \quad\quad\quad\quad\quad 81 - 72 = 9 \ ?$$

$$9 = 9 \quad \text{True} \quad\quad\quad\quad 9 = 9 \quad \text{True}$$

The two solutions are $\frac{1}{2}$ and $-\frac{9}{2}$.

Chapter 17 Quadratic Equations

④ Solve by completing the square.

(a) $9x^2 + 18x = -5$

(b) $4t^2 - 24t + 11 = 0$

The steps in solving a quadratic equation $ax^2 + bx + c = 0$ by completing the square are summarized here.

> **Solving a Quadratic Equation by Completing the Square**
>
> *Step 1* **Be sure the squared term has coefficient 1.** If the coefficient of the squared term is 1, proceed to Step 2. If the coefficient of the squared term is not 1 but some other nonzero number a, divide each side of the equation by a.
>
> *Step 2* **Write in correct form.** Make sure that all terms with variables are on one side of the equal sign and that all constants are on the other side.
>
> *Step 3* **Complete the square.** Take half the coefficient of the first-degree term, and square the result. Add the square to each side of the equation. Factor the variable side, and combine terms on the other side.
>
> *Step 4* **Solve.** Apply the square root property to solve the equation.

◀◀◀ **Work Problem 4 at the Side.**

EXAMPLE 5 Solving a Quadratic Equation by Completing the Square

⑤ Solve by completing the square.

(a) $3x^2 + 5x = 2$

Solve $2x^2 - 7x = 9$.

Step 1 Divide each side of the equation by 2 to get a coefficient of 1 for the x^2-term.

$$x^2 - \frac{7}{2}x = \frac{9}{2} \quad \text{Divide by 2.}$$

(Step 2 is not needed for this equation.)

Step 3 Now take half the coefficient of x and square it. Half of $-\frac{7}{2}$ is $-\frac{7}{4}$, and $\left(-\frac{7}{4}\right)^2 = \frac{49}{16}$. Add $\frac{49}{16}$ to each side of the equation, and write the left side as a perfect square.

$$x^2 - \frac{7}{2}x + \frac{49}{16} = \frac{9}{2} + \frac{49}{16} \quad \text{Add } \frac{49}{16}.$$

(b) $2x^2 - 4x - 1 = 0$

$$\left(x - \frac{7}{4}\right)^2 = \frac{121}{16} \quad \text{Factor on the left; add on the right.}$$

Step 4 Use the square root property.

$$x - \frac{7}{4} = \sqrt{\frac{121}{16}} \quad \text{or} \quad x - \frac{7}{4} = -\sqrt{\frac{121}{16}}$$

$$x = \frac{7}{4} + \frac{11}{4} \quad\quad x = \frac{7}{4} - \frac{11}{4} \quad \text{Add } \frac{7}{4}; \sqrt{\frac{121}{16}} = \frac{11}{4}.$$

$$x = \frac{18}{4} = \frac{9}{2} \quad\quad x = -\frac{4}{4} = -1$$

ANSWERS

4. (a) $-\frac{1}{3}, -\frac{5}{3}$ (b) $\frac{11}{2}, \frac{1}{2}$

5. (a) $-2, \frac{1}{3}$ (b) $\frac{2+\sqrt{6}}{2}, \frac{2-\sqrt{6}}{2}$

Check that the solutions are $\frac{9}{2}$ and -1.

◀◀◀ **Work Problem 5 at the Side.**

Section 17.2 Solving Quadratic Equations by Completing the Square 1165

EXAMPLE 6 Solving a Quadratic Equation by Completing the Square

Solve $4p^2 + 8p + 5 = 0$.

$$4p^2 + 8p + 5 = 0$$

$$p^2 + 2p + \frac{5}{4} = 0 \qquad \text{Divide by 4.}$$

$$p^2 + 2p = -\frac{5}{4} \qquad \text{Subtract } \tfrac{5}{4}.$$

The coefficient of p is 2. Take half of 2, square the result, and add this square to each side: $\left[\tfrac{1}{2}(2)\right]^2 = 1$. The left side can then be written as a perfect square.

$$p^2 + 2p + 1 = -\frac{5}{4} + 1 \qquad \text{Add 1.}$$

$$(p + 1)^2 = -\frac{1}{4} \qquad \text{Factor; add.}$$

The square root of $-\tfrac{1}{4}$ is not a real number, so the square root property does not apply. This equation has no real number solution.*

Work Problem 6 at the Side.

OBJECTIVE 3 Simplify an equation before solving. The next example shows how to simplify a quadratic equation before solving it.

EXAMPLE 7 Simplifying an Equation before Completing the Square

Solve $(x + 3)(x - 1) = 2$.

$$(x + 3)(x - 1) = 2 \qquad \text{Given equation}$$
$$x^2 + 2x - 3 = 2 \qquad \text{Use FOIL.}$$
$$x^2 + 2x = 5 \qquad \text{Add 3.}$$
$$x^2 + 2x + 1 = 5 + 1 \qquad \text{Add } \left[\tfrac{1}{2}(2)\right]^2 = 1.$$
$$(x + 1)^2 = 6 \qquad \text{Factor on the left; add on the right.}$$
$$x + 1 = \sqrt{6} \quad \text{or} \quad x + 1 = -\sqrt{6} \qquad \text{Square root property}$$
$$x = -1 + \sqrt{6} \quad \text{or} \quad x = -1 - \sqrt{6} \qquad \text{Add } -1.$$

The solutions are $-1 + \sqrt{6}$ and $-1 - \sqrt{6}$.

Work Problem 7 at the Side.

> **NOTE**
> The solutions given in Example 7 are *exact*. In applications, decimal solutions are more appropriate. Using the square root key of a calculator, $\sqrt{6} \approx 2.449$. Evaluating the two solutions gives these results.
>
> $$x \approx 1.449 \quad \text{and} \quad x \approx -3.449$$

---

*The equation in Example 6 has no *real number* solution. In the context of another number system, called the *complex numbers,* however, this equation does have solutions. The complex numbers include numbers whose squares are negative. These numbers are discussed in intermediate and college algebra courses.

6 Solve $5x^2 + 3x + 1 = 0$ by completing the square.

7 Solve each equation.

(a) $r(r - 3) = -1$

(b) $(x + 2)(x + 1) = 5$

ANSWERS

6. no real number solution

7. (a) $\dfrac{3 + \sqrt{5}}{2}, \dfrac{3 - \sqrt{5}}{2}$

   (b) $\dfrac{-3 + \sqrt{21}}{2}, \dfrac{-3 - \sqrt{21}}{2}$

Chapter 17 Quadratic Equations

8 Suppose a ball is propelled upward with an initial velocity of 128 feet per second. Its height at time t (in seconds) is given by

$$s = -16t^2 + 128t$$

where s is in feet. At what times will it be 48 ft above the ground? Give answers to the nearest tenth.

OBJECTIVE 4 Solve applied problems that require quadratic equations. There are many practical applications of quadratic equations. The next example illustrates an application from physics.

EXAMPLE 8 Solving a Velocity Problem

If a ball is propelled into the air from ground level with an initial velocity of 64 feet per second, its height s (in feet) in t seconds is given by this formula.

$$s = -16t^2 + 64t$$

How long will it take the ball to reach a height of 48 feet?

Since s represents the height, substitute **48** for s in the formula.

$$48 = -16t^2 + 64t$$

We solve this equation for time, t, by completing the square. First we divide each side by -16. We also interchange the sides of the equation.

$-3 = t^2 - 4t$	Divide by -16.
$t^2 - 4t = -3$	Interchange the sides.
$t^2 - 4t + 4 = -3 + 4$	Add $[\frac{1}{2}(-4)]^2 = 4$.
$(t-2)^2 = 1$	Factor.
$t - 2 = 1$ or $t - 2 = -1$	Square root property
$t = 3$ or $t = 1$	Add 2.

You may wonder how we can get two correct answers for the time required for the ball to reach a height of 48 feet. The ball reaches that height twice, once on the way up and again on the way down. So it takes 1 second to reach 48 feet on the way up, and then after 3 seconds, the ball reaches 48 feet again on the way down.

Work Problem 8 at the Side.

ANSWERS
8. 0.4 and 7.6 seconds

17.2 Exercises

Complete each trinomial so that it is a perfect square. See Example 1.

1. $x^2 + \underline{} + 25$

2. $9x^2 + \underline{} + 4$

3. $z^2 - 14z + \underline{}$

4. $4a^2 - 32a + \underline{}$

5. Which step is an appropriate way to begin solving the quadratic equation
$$2x^2 - 4x = 9$$
by completing the square?
   A. Add 4 to each side of the equation.
   B. Factor the left side as $2x(x - 2)$.
   C. Factor the left side as $x(2x - 4)$.
   D. Divide each side by 2.

6. In Example 3 of **Section 14.6**, we solved the quadratic equation
$$4p^2 - 26p + 40 = 0$$
by factoring. If we were to solve by completing the square, would we get the same solutions, $\frac{5}{2}$ and 4?

Find the number that should be added to each expression to make it a perfect square. See Examples 1–3.

7. $x^2 + 20x$

8. $z^2 + 18z$

9. $x^2 - 5x$

10. $m^2 - 9m$

11. $r^2 + \frac{1}{2}r$

12. $s^2 - \frac{1}{3}s$

Solve each equation by completing the square. See Examples 2 and 3.

13. $x^2 - 4x = -3$

14. $x^2 - 2x = 8$

15. $x^2 + 5x + 6 = 0$

16. $x^2 + 6x + 5 = 0$

17. $x^2 + 2x - 5 = 0$

18. $x^2 + 4x + 1 = 0$

19. $t^2 + 6t + 9 = 0$

20. $k^2 - 8k + 16 = 0$

21. $x^2 + x - 1 = 0$

22. $x^2 + x - 3 = 0$

Solve each equation by completing the square. See Examples 4–7.

23. $4x^2 + 4x - 3 = 0$

24. $9x^2 + 3x - 2 = 0$

25. $2x^2 - 4x = 5$

26. $2x^2 - 6x = 3$

27. $2p^2 - 2p + 3 = 0$

28. $3q^2 - 3q + 4 = 0$

29. $3k^2 + 7k = 4$

30. $2k^2 + 5k = 1$

31. $(x + 3)(x - 1) = 5$

32. $(y - 8)(y + 2) = 24$

33. $-x^2 + 2x = -5$

34. $-r^2 + 3r = -2$

Solve each problem. See Example 8.

35. If an object is propelled upward from ground level with an initial velocity of 96 feet per second, its height s (in feet) in t seconds is given by the formula $s = -16t^2 + 96t$. In how many seconds will the object be at a height of 80 feet?

36. How much time will it take the object described in Exercise 35 to be at a height of 100 feet? Round your answers to the nearest tenth.

37. If an object is propelled upward on the surface of Mars from ground level with an initial velocity of 104 feet per second, its height s (in feet) in t seconds is given by the formula $s = -13t^2 + 104t$. How long will it take for the object to be at a height of 195 feet?

38. How long will it take the object in Exercise 37 to return to the surface? (*Hint:* When it returns to the surface, $s = 0$.)

39. A farmer has a rectangular cattle pen with perimeter 350 ft and area 7500 ft². What are the dimensions of the pen? (*Hint:* Use the figure to set up the equation.)

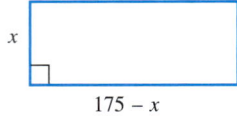

40. The base of a triangle measures 1 m more than three times the height of the triangle. Its area is 15 m². Find the lengths of the base and the height.

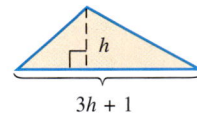

17.3 Solving Quadratic Equations by the Quadratic Formula

We can solve any quadratic equation by completing the square, but the method is tedious. In this section we complete the square on the general quadratic equation

$$ax^2 + bx + c = 0, \quad a \neq 0,$$

to obtain the *quadratic formula*, a formula that gives the solution(s) for any quadratic equation.

> **NOTE**
> In $ax^2 + bx + c = 0$, there is a restriction that a is not zero. If it were, the equation would be linear, not quadratic.

OBJECTIVES

1. Identify the values of a, b, and c in a quadratic equation.
2. Use the quadratic formula to solve quadratic equations.
3. Solve quadratic equations with only one solution.
4. Solve quadratic equations with fractions as coefficients.

OBJECTIVE 1 Identify the values of a, b, and c in a quadratic equation. The first step in solving a quadratic equation by this new method is to identify the values of a, b, and c in the standard form of the quadratic equation.

EXAMPLE 1 Identifying Values of a, b, and c in Quadratic Equations

Match the coefficients of each quadratic equation with the letters a, b, and c of the standard quadratic equation $ax^2 + bx + c = 0$.

(a) $\overset{a}{\underset{\downarrow}{2x^2}} + \overset{b}{\underset{\downarrow}{3x}} - \overset{c}{\underset{\downarrow}{5}} = 0$

In this example $a = 2$, $b = 3$, and $c = -5$.

(b) $-x^2 + 2 = 6x$

First rewrite the equation with 0 on one side to match the standard form $ax^2 + bx + c = 0$.

$$-x^2 + 2 = 6x$$
$$-x^2 - 6x + 2 = 0 \quad \text{Subtract } 6x.$$

Here, $a = -1$, $b = -6$, and $c = 2$. (Notice that the coefficient of x^2 is understood to be -1.)

(c) $5x^2 - 12 = 0$

The x-term is missing, so write the equation as

$$5x^2 + 0x - 12 = 0.$$

Then $a = 5$, $b = 0$, and $c = -12$.

(d) $(2x - 7)(x + 4) = -23$

Write the equation in standard form.

$$(2x - 7)(x + 4) = -23$$
$$2x^2 + x - 28 = -23 \quad \text{Use FOIL on the left.}$$
$$2x^2 + x - 5 = 0 \quad \text{Add 23.}$$

Now, identify the values: $a = 2$, $b = 1$, and $c = -5$.

Work Problem 1 at the Side.

---

1 Match the coefficients of each quadratic equation with the letters a, b, and c of the standard quadratic equation $ax^2 + bx + c = 0$.

(a) $5x^2 + 2x - 1 = 0$

(b) $3x^2 = x - 2$

(c) $9x^2 - 13 = 0$

(d) $(3x + 2)(x - 1) = 8$

ANSWERS
1. (a) $a = 5, b = 2, c = -1$
   (b) $a = 3, b = -1, c = 2$
   (c) $a = 9, b = 0, c = -13$
   (d) $a = 3, b = -1, c = -10$

Chapter 17 Quadratic Equations

OBJECTIVE 2 Use the quadratic formula to solve quadratic equations. To develop the quadratic formula, we follow the steps for completing the square on $ax^2 + bx + c = 0$ ($a > 0$) given in **Section 17.2**. For comparison, we also show the corresponding steps for solving $2x^2 + x - 5 = 0$ from Example 1(d).

Step 1 Transform so that the coefficient of the squared term equals 1.

$2x^2 + x - 5 = 0$		$ax^2 + bx + c = 0$	Standard form
$x^2 + \dfrac{1}{2}x - \dfrac{5}{2} = 0$	Divide by 2.	$x^2 + \dfrac{b}{a}x + \dfrac{c}{a} = 0$	Divide by a.

Step 2 Write so that the variable terms are alone on the left side.

$$x^2 + \frac{1}{2}x = \frac{5}{2} \quad \text{Add } \tfrac{5}{2}. \qquad x^2 + \frac{b}{a}x = -\frac{c}{a} \quad \text{Subtract } \tfrac{c}{a}.$$

Step 3 Add the square of half the coefficient of x to each side, factor the left side, and combine terms on the right.

$$x^2 + \frac{1}{2}x + \frac{1}{16} = \frac{5}{2} + \frac{1}{16} \quad \text{Add } \tfrac{1}{16}. \qquad x^2 + \frac{b}{a}x + \frac{b^2}{4a^2} = -\frac{c}{a} + \frac{b^2}{4a^2} \quad \text{Add } \tfrac{b^2}{4a^2}.$$

$$\left(x + \frac{1}{4}\right)^2 = \frac{41}{16} \quad \text{Factor; add on right.} \qquad \left(x + \frac{b}{2a}\right)^2 = \frac{b^2 - 4ac}{4a^2} \quad \text{Factor; add on right.}$$

Step 4 Use the square root property to complete the solution.

$$x + \frac{1}{4} = \pm\sqrt{\frac{41}{16}} \qquad x + \frac{b}{2a} = \pm\sqrt{\frac{b^2 - 4ac}{4a^2}}$$

$$x + \frac{1}{4} = \pm\frac{\sqrt{41}}{4} \qquad x + \frac{b}{2a} = \pm\frac{\sqrt{b^2 - 4ac}}{2a}$$

$$x = -\frac{1}{4} \pm \frac{\sqrt{41}}{4} \qquad x = -\frac{b}{2a} \pm \frac{\sqrt{b^2 - 4ac}}{2a}$$

$$x = \frac{-1 \pm \sqrt{41}}{4} \qquad x = \frac{-b \pm \sqrt{b^2 - 4ac}}{2a}$$

The final result in the column on the right is called the **quadratic formula**. (It is also valid for $a < 0$.) *It is a key result that should be memorized.* Notice that there are two values, one for the $+$ sign and one for the $-$ sign.

Quadratic Formula

The solutions of the quadratic equation $ax^2 + bx + c = 0$, $a \neq 0$, are

$$x = \frac{-b + \sqrt{b^2 - 4ac}}{2a} \quad \text{and} \quad x = \frac{-b - \sqrt{b^2 - 4ac}}{2a}$$

or in compact form:

$$x = \frac{-b \pm \sqrt{b^2 - 4ac}}{2a}$$

> **CAUTION**
> Notice in the quadratic formula that the fraction bar is under $-b$ as well as the radical. When using this formula, *be sure to find the values of $-b \pm \sqrt{b^2 - 4ac}$ first, then divide those results by the value of $2a$.*

EXAMPLE 2 Solving a Quadratic Equation by the Quadratic Formula

Use the quadratic formula to solve $2x^2 - 7x - 9 = 0$. (This equation was solved by completing the square in **Section 17.2.** See Example 5.)

Match the coefficients of the variables with those of the standard quadratic equation

$$ax^2 + bx + c = 0.$$

Here, $a = 2$, $b = -7$, and $c = -9$. Substitute these numbers into the quadratic formula, and simplify the result.

$$x = \frac{-b \pm \sqrt{b^2 - 4ac}}{2a}$$

$$x = \frac{-(-7) \pm \sqrt{(-7)^2 - 4(2)(-9)}}{2(2)} \quad \text{Let } a = 2, b = -7, c = -9.$$

$$x = \frac{7 \pm \sqrt{49 + 72}}{4}$$

$$x = \frac{7 \pm \sqrt{121}}{4}$$

$$x = \frac{7 \pm 11}{4} \quad \sqrt{121} = 11$$

Find the two individual solutions by first using the plus sign, and then using the minus sign:

$$x = \frac{7 + 11}{4} = \frac{18}{4} = \frac{9}{2} \quad \text{or} \quad x = \frac{7 - 11}{4} = \frac{-4}{4} = -1.$$

Check:

$$2x^2 - 7x - 9 = 0 \qquad\qquad 2x^2 - 7x - 9 = 0$$

$$2\left(\frac{9}{2}\right)^2 - 7\left(\frac{9}{2}\right) - 9 = 0 \ ? \qquad 2(-1)^2 - 7(-1) - 9 = 0 \ ?$$

$$\frac{81}{2} - \frac{63}{2} - 9 = 0 \ ? \qquad\qquad 2 + 7 - 9 = 0 \ ?$$

$$0 = 0 \quad \text{True} \qquad\qquad\qquad 0 = 0 \quad \text{True}$$

The solutions are $\frac{9}{2}$ and -1, the same solutions obtained by completing the square in **Section 17.2.**

▶ **Work Problem 2 at the Side.**

---

2 Solve by using the quadratic formula.

(a) $2x^2 + 3x - 5 = 0$

(b) $6x^2 + x - 1 = 0$

ANSWERS

2. (a) $1, -\frac{5}{2}$ (b) $-\frac{1}{2}, \frac{1}{3}$

Chapter 17 Quadratic Equations

③ Solve $x^2 + 1 = -8x$ by the quadratic formula.

EXAMPLE 3 Rewriting a Quadratic Equation before Using the Quadratic Formula

Solve $x^2 = 2x + 1$.

Find a, b, and c by rewriting the equation in standard form (with 0 on one side). Add $-2x - 1$ to each side of the equation to get

$$x^2 - 2x - 1 = 0.$$

Then $a = 1$, $b = -2$, and $c = -1$. Substitute these values into the quadratic formula.

$$x = \frac{-b \pm \sqrt{b^2 - 4ac}}{2a}$$

$$x = \frac{-(-2) \pm \sqrt{(-2)^2 - 4(1)(-1)}}{2(1)} \quad \text{Let } a = 1, b = -2, c = -1.$$

$$x = \frac{2 \pm \sqrt{4 + 4}}{2}$$

$$x = \frac{2 \pm \sqrt{8}}{2}$$

$$x = \frac{2 \pm 2\sqrt{2}}{2} \quad \sqrt{8} = \sqrt{4} \cdot \sqrt{2} = 2\sqrt{2}$$

Write these solutions in lowest terms by factoring $2 \pm 2\sqrt{2}$ as $2(1 \pm \sqrt{2})$.

$$x = \frac{2 \pm 2\sqrt{2}}{2} = \frac{2(1 \pm \sqrt{2})}{2} = 1 \pm \sqrt{2}$$

The two solutions of the original equation are $1 + \sqrt{2}$ and $1 - \sqrt{2}$.

Work Problem 3 at the Side.

④ Solve $9x^2 - 12x + 4 = 0$.

OBJECTIVE 3 Solve quadratic equations with only one solution. In the quadratic formula, when the quantity under the radical, $b^2 - 4ac$, equals 0, the equation has just one rational number solution. In this case, the trinomial $ax^2 + bx + c$ is a perfect square.

EXAMPLE 4 Solving a Quadratic Equation with Only One Solution

Solve $4x^2 + 25 = 20x$.

Rewrite the equation in standard form by subtracting $20x$ from each side.

$$4x^2 - 20x + 25 = 0$$

Here, $a = 4$, $b = -20$, and $c = 25$. Use the quadratic formula.

$$x = \frac{-(-20) \pm \sqrt{(-20)^2 - 400}}{2(4)} = \frac{20 \pm 0}{8} = \frac{5}{2}$$

In this case, $b^2 - 4ac = 0$, and the trinomial $4x^2 - 20x + 25$ is a perfect square. There is just one solution, $\frac{5}{2}$.

Work Problem 4 at the Side.

ANSWERS
3. $-4 + \sqrt{15}, -4 - \sqrt{15}$
4. $\frac{2}{3}$

NOTE
The single solution of the equation in Example 4 is a rational number. If all solutions of a quadratic equation are rational, the equation can be solved by factoring as well.

OBJECTIVE 4 Solve quadratic equations with fractions as coefficients. It is usually easier to clear quadratic equations of fractions before solving them. This is accomplished by multiplying all terms in the equation by the LCD of the fractions.

EXAMPLE 5 Solving a Quadratic Equation with Fractions as Coefficients

Solve $\frac{1}{10}t^2 = \frac{2}{5}t - \frac{1}{2}$.

Eliminate the denominators by multiplying each side of the equation by the least common denominator, 10.

$$10\left(\frac{1}{10}t^2\right) = 10\left(\frac{2}{5}t - \frac{1}{2}\right)$$

$$10\left(\frac{1}{10}t^2\right) = 10\left(\frac{2}{5}t\right) - 10\left(\frac{1}{2}\right) \quad \text{Distributive property}$$

$$t^2 = 4t - 5 \quad \text{Multiply.}$$

$$t^2 - 4t + 5 = 0 \quad \text{Standard form.}$$

From this form, we see that $a = 1$, $b = -4$, and $c = 5$. Use the quadratic formula to complete the solution.

$$t = \frac{-(-4) \pm \sqrt{(-4)^2 - 4(1)(5)}}{2(1)}$$

$$t = \frac{4 \pm \sqrt{16 - 20}}{2}$$

$$t = \frac{4 \pm \sqrt{-4}}{2}$$

The radical $\sqrt{-4}$ is not a real number, so the equation has no real number solution.

Work Problem 5 at the Side.

5 Solve.

(a) $x^2 - \frac{4}{3}x + \frac{2}{3} = 0$

(b) $x^2 - \frac{9}{5}x = \frac{2}{5}$

ANSWERS
5. (a) no real number solution
   (b) $-\frac{1}{5}, 2$

Focus on Real-Data Applications

Estimating Interest Rates

Banks offer savings accounts in the form of money market funds that pay variable interest rates, depending on the status of the prime interest rate in effect at the beginning of each month. The prime interest rate is the rate that the Federal Reserve pays its largest customers.

The compound interest formula, $A = P(1 + r)^n$, computes the amount A that P dollars invested at a periodic rate of interest r will grow to in n periods. The period for compound interest can be a year (annual), six months (semi-annual), three months (quarterly), one month (monthly), or even daily. The periodic interest rate is the annual rate divided by the number of periods per year. For example, to find the monthly rate, you divide the annual rate by 12. Similarly, if you know the monthly rate, you can multiply it by 12 to find the annual rate. If you borrow money to purchase a car or a house, the quantities P, r, and n are typically determined for the entire purchase.

Suppose you are saving money on a month-to-month basis at your bank. On March 1 you open a money market fund account with a deposit of $250; on April 1 you deposit $175; and on May 1 you deposit $300. Your deposit slip on May 1 shows a balance of $726.20. You are curious about the average interest rate earned from the date you opened the account. The investment can be thought of as three separate transactions, added together.

March 1:	The investment earns interest for 2 months (March, April).	$A = 250(1 + r)^2$
April 1:	The investment earns interest for 1 month (April).	$A = 175(1 + r)$
May 1:	The investment has not yet earned interest.	$A = 300$

The total investment value can be written as a quadratic equation.

$$250(1 + r)^2 + 175(1 + r) + 300 = 726.20$$

To solve this equation, you could expand and collect like terms, but the algebra is tedious. A better option is to replace $1 + r$ with x and solve the quadratic equation.

$$250x^2 + 175x + 300 = 726.20$$

Once you have found x, you simply subtract 1 to get r, the monthly interest rate.

For Group Discussion

1. Consider the quadratic equation $250x^2 + 175x + 300 = 726.20$.
   (a) Use the quadratic formula to solve the equation for x. Remember to write it in standard form. (*Hint:* x must have a value between 1 and 2 since x is $1 + r$ and r is an interest rate.)

   (b) Calculate the periodic (monthly) rate, r, and the annual rate, rounded to ten-thousandths. Write the annual interest rate as a percent.

2. After talking to an investment advisor, you decide to invest in another fund. On the first of June, July, and August you again invest $250, $175, and $300, respectively. After making the August 1 deposit, your account value is $732.96. To find the average interest rate, you must now solve this quadratic equation.

   $$250x^2 + 175x + 300 = 732.96$$

   (a) Use the quadratic formula to solve the equation for x.

   (b) Calculate the periodic (monthly) rate, r, and the annual rate, rounded to ten-thousandths. Write the annual interest rate as a percent.

17.3 Exercises

Write each equation in the form $ax^2 + bx + c = 0$, if necessary. Then identify the values of a, b, and c. Do not actually solve the equation. See Example 1.

1. $4x^2 + 5x - 9 = 0$
   $a = \underline{}$ $b = \underline{}$ $c = \underline{}$

2. $8x^2 + 3x - 4 = 0$
   $a = \underline{}$ $b = \underline{}$ $c = \underline{}$

3. $3x^2 = 4x + 2$
   $a = \underline{}$ $b = \underline{}$ $c = \underline{}$

4. $5x^2 = 3x - 6$
   $a = \underline{}$ $b = \underline{}$ $c = \underline{}$

5. $3x^2 = -7x$
   $a = \underline{}$ $b = \underline{}$ $c = \underline{}$

6. $9x^2 = 8x$
   $a = \underline{}$ $b = \underline{}$ $c = \underline{}$

Use the quadratic formula to solve each equation. Write all radicals in simplified form, and write all answers in lowest terms. See Examples 2–4.

7. $k^2 = -12k + 13$

8. $r^2 = 8r + 9$

9. $p^2 - 4p + 4 = 0$

10. $9x^2 + 6x + 1 = 0$

11. $2x^2 + 12x = -5$

12. $5m^2 + m = 1$

13. $2x^2 = 5 + 3x$

14. $2z^2 = 30 + 7z$

15. $6x^2 + 6x = 0$

16. $4n^2 - 12n = 0$

17. $-2x^2 = -3x + 2$

18. $-x^2 = -5x + 20$

19. $3x^2 + 5x + 1 = 0$

20. $6x^2 - 6x + 1 = 0$

21. $7x^2 = 12x$

22. $9r^2 = 11r$

23. $x^2 - 24 = 0$

24. $z^2 - 96 = 0$

25. $25x^2 - 4 = 0$

26. $16x^2 - 9 = 0$

27. $3x^2 - 2x + 5 = 10x + 1$

28. $4x^2 - x + 4 = x + 7$

29. $2x^2 + x + 5 = 0$

30. $3x^2 + 2x + 8 = 0$

31. If we apply the quadratic formula and find that the value of $b^2 - 4ac$ is negative, what can we conclude?

32. If we were to solve the quadratic equation $-2x^2 - 4x + 3 = 0$, we might choose to use $a = -2$, $b = -4$, and $c = 3$. On the other hand, we might decide to multiply each side by -1 to begin, obtaining the equation $2x^2 + 4x - 3 = 0$, and then use $a = 2$, $b = 4$, and $c = -3$. Show that in either case, we obtain the same solutions.

Use the quadratic formula to solve each equation. See Example 5.

33. $\dfrac{3}{2}k^2 - k - \dfrac{4}{3} = 0$

34. $\dfrac{2}{5}x^2 - \dfrac{3}{5}x - 1 = 0$

35. $\dfrac{1}{2}x^2 + \dfrac{1}{6}x = 1$

36. $\dfrac{2}{3}t^2 - \dfrac{4}{9}t = \dfrac{1}{3}$

37. $0.5x^2 = x + 0.5$

38. $0.25x^2 = -1.5x - 1$

39. $\dfrac{3}{8}x^2 - x + \dfrac{17}{24} = 0$

40. $\dfrac{1}{3}x^2 + \dfrac{8}{9}x + \dfrac{7}{9} = 0$

Solve each problem.

41. A frog is sitting on a stump 3 ft above the ground. He hops off the stump and lands on the ground 4 ft away. During his leap, his height h is given by the equation
$$h = -0.5x^2 + 1.25x + 3$$
where x is the distance in feet from the base of the stump, and h is in feet. How far was the frog from the base of the stump when he was 1.25 ft above the ground?

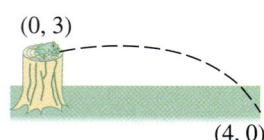

42. An astronaut on the moon throws a baseball upward. The height h of the ball, in feet, x seconds after he throws it, is given by this equation.
$$h = -2.7x^2 + 30x + 6.5$$
After how many seconds is the ball 12 ft above the moon's surface? Give answer(s) to the nearest tenth.

43. A rule for estimating the number of board feet of lumber that can be cut from a log depends on the diameter of the log. To find the diameter d required to get 9 board feet of lumber, we use this equation.
$$\left(\dfrac{d-4}{4}\right)^2 = 9$$
Solve this equation for d. Are both answers reasonable?

44. An old Babylonian problem asks for the length of the side of a square, given that the area of the square minus the length of a side is 870. Find the length of the side. (*Source: An Introduction to the History of Mathematics.*)

Summary Exercises on Quadratic Equations

Four algebraic methods have now been introduced for solving quadratic equations written in the form $ax^2 + bx + c = 0$. The following chart shows some advantages and some disadvantages of each method.

Method	Advantages	Disadvantages
1. Factoring	It is usually the fastest method.	Not all equations can be solved by factoring. Some factorable polynomials are difficult to factor.
2. Square root property	It is the simplest method for solving equations of the form $(ax + b)^2 =$ a number.	Few equations are given in this form.
3. Completing the square	It can always be used. (Also, the procedure is useful in other areas of mathematics.)	It requires more steps than other methods.
4. Quadratic formula	It can always be used.	It is more difficult than factoring because of the $\sqrt{b^2 - 4ac}$ expression.

Solve each quadratic equation by the method of your choice.

1. $x^2 = 36$

2. $x^2 + 3x = -1$

3. $x^2 - \dfrac{100}{81} = 0$

4. $81t^2 = 49$

5. $z^2 - 4z + 3 = 0$

6. $w^2 + 3w + 2 = 0$

7. $z(z - 9) = -20$

8. $x^2 + 3x - 2 = 0$

9. $(3k - 2)^2 = 9$

10. $(2s - 1)^2 = 10$

11. $(x + 6)^2 = 121$

12. $(5k + 1)^2 = 36$

13. $(3r - 7)^2 = 24$

14. $(7p - 1)^2 = 32$

15. $(5x - 8)^2 = -6$

16. $2t^2 + 1 = t$

17. $-2x^2 = -3x - 2$

18. $-2x^2 + x = -1$

19. $8z^2 = 15 + 2z$

20. $3k^2 = 3 - 8k$

21. $0 = -x^2 + 2x + 1$

22. $3x^2 + 5x = -1$

23. $5x^2 - 22x = -8$

24. $x(x + 6) + 4 = 0$

25. $(x + 2)(x + 1) = 10$

26. $16x^2 + 40x + 25 = 0$

27. $4x^2 = -1 + 5x$

28. $2p^2 = 2p + 1$

29. $3x(3x + 4) = 7$

30. $5x - 1 + 4x^2 = 0$

31. $\dfrac{x^2}{2} + \dfrac{7x}{4} + \dfrac{11}{8} = 0$

32. $t(15t + 58) = -48$

33. $9k^2 = 16(3k + 4)$

34. $\dfrac{1}{5}x^2 + x + 1 = 0$

35. $x^2 - x + 3 = 0$

36. $4x^2 - 11x + 8 = -2$

37. $-3x^2 + 4x = -4$

38. $z^2 - \dfrac{5}{12}z = \dfrac{1}{6}$

39. $5k^2 + 19k = 2k + 12$

40. $\dfrac{1}{2}x^2 - x = \dfrac{15}{2}$

41. $x^2 - \dfrac{4}{15} = -\dfrac{4}{15}x$

42. $(x + 2)(x - 4) = 16$

17.4 Graphing Quadratic Equations

OBJECTIVES
1. Graph quadratic equations.
2. Find the vertex of a parabola.
3. Solve an application involving a parabola.

OBJECTIVE 1 Graph quadratic equations. In **Chapter 11** we saw that the graph of a linear equation in two variables is a straight line that represents all the solutions of the equation. Quadratic equations in two variables, of the form $y = ax^2 + bx + c$, are graphed in this section.

The simplest quadratic equation is $y = x^2$ (or $y = 1x^2 + 0x + 0$). The graph of this equation is not a straight line, as seen in Example 1 that follows. The equation can be graphed in much the same way that straight lines were graphed, by finding ordered pairs that satisfy the equation $y = x^2$.

EXAMPLE 1 Graphing a Quadratic Equation

Graph $y = x^2$.

Select several values for x; then find the corresponding y-values. For example, selecting $x = 2$ gives

$$y = 2^2 = 4,$$

and so the point (2, 4) is on the graph of $y = x^2$. (Recall that in an ordered pair such as (2, 4), the x-value comes first and the y-value second.)

Work Problem 1 at the Side.

If the points from Problem 1 at the side are plotted on a coordinate system and a smooth curve is drawn through them, the graph is as shown in Figure 1. The table of values completed in Problem 1 is shown with the graph.

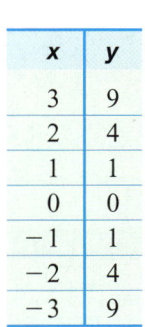 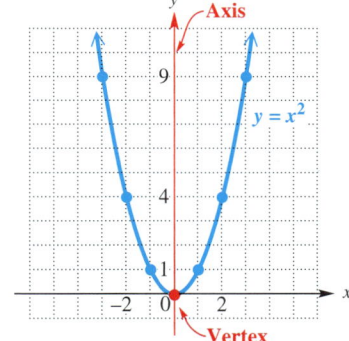

Figure 1

Work Problem 2 at the Side.

The curve in Figure 1 is called a **parabola**. The point (0, 0), the *lowest* point on this graph, is called the **vertex** of the parabola. The vertical line through the vertex (the y-axis here) is called the **axis** of the parabola. The axis of a parabola is a **line of symmetry** for the graph, because if the graph is folded on this line, the two halves will coincide.

Every equation of the form

$$y = ax^2 + bx + c,$$

with $a \neq 0$, has a graph that is a parabola. Because of its many useful properties, the parabola occurs frequently in real-life applications. For example, if an object is thrown into the air, the path that the object follows is a parabola (ignoring wind resistance). The cross sections of radar, spotlight, and telescope reflectors also form parabolas.

① Complete the table of values for $y = x^2$.

x	y
3	
2	4
1	
0	
−1	
−2	
−3	

② Graph $y = \frac{1}{2}x^2$ by first completing the table of values.

x	y
−2	
−1	
0	
1	
2	

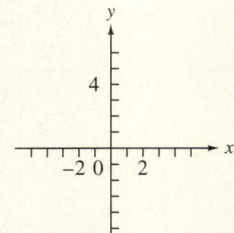

ANSWERS

1.
x	y
3	9
2	4
1	1
0	0
−1	1
−2	4
−3	9

2.
x	y
−2	2
−1	$\frac{1}{2}$
0	0
1	$\frac{1}{2}$
2	2

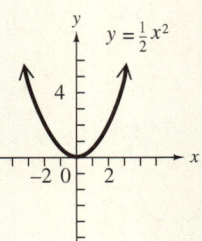

3 Complete each ordered pair for $y = -x^2 - 1$.

$(-2,), \quad (-1,),$
$(1,), \quad (2,)$

4 Graph each equation, and identify each vertex.

(a) $y = -x^2 - 3$

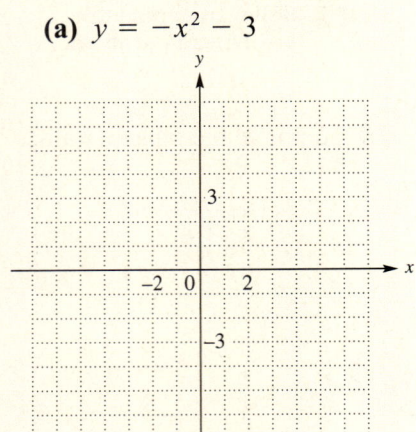

(b) $y = x^2 + 3$

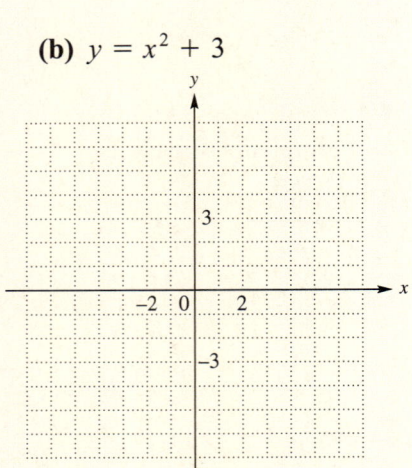

ANSWERS
3. $(-2, -5), (-1, -2), (1, -2), (2, -5)$
4. (a)

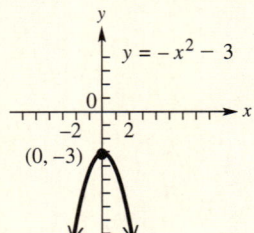

(b)

EXAMPLE 2 Graphing a Parabola

Graph $y = -x^2 - 1$.

Find several ordered pairs. To begin, check for intercepts. Let $x = 0$ to find the y-intercept.

$$y = -x^2 - 1 = -0^2 - 1 = -1,$$

giving the ordered pair $(0, -1)$. Let $y = 0$ to find the x-intercepts.

$$y = -x^2 - 1$$
$$0 = -x^2 - 1$$
$$x^2 = -1$$

This equation has no real number solution, so there are no x-intercepts. Choose additional x-values near $x = 0$ to find other ordered pairs.

◀◀◀ **Work Problem 3 at the Side.**

The ordered pair $(0, -1)$ and the ordered pairs from Problem 3 at the side are listed in the table shown with Figure 2. Plot these points and connect them with a smooth curve as shown in Figure 2. The vertex of this parabola is $(0, -1)$. The graph opens downward because x^2 has a negative coefficient, so the vertex is the *highest* point of the graph.

x	y
-2	-5
-1	-2
0	-1
1	-2
2	-5

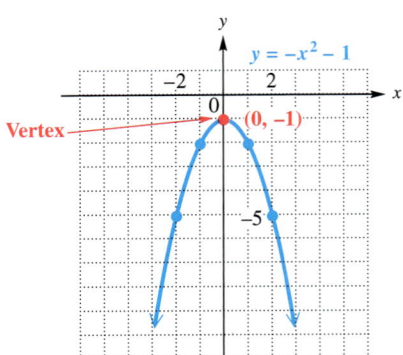

Figure 2

◀◀◀ **Work Problem 4 at the Side.**

OBJECTIVE 2 Find the vertex of a parabola. The vertex is the most important point to locate when you are graphing a quadratic equation. The next example shows how to find the vertex in a more general case.

EXAMPLE 3 Finding the Vertex to Graph a Parabola

Graph $y = x^2 - 2x - 3$.

We want to find the vertex of the graph. *If a parabola has two x-intercepts, the vertex is exactly halfway between them.* Therefore, we begin by finding the x-intercepts. We let $y = 0$ in the equation, and solve for x.

$$0 = x^2 - 2x - 3$$
$$0 = (x + 1)(x - 3) \qquad \text{Factor.}$$
$$x + 1 = 0 \quad \text{or} \quad x - 3 = 0 \qquad \text{Zero-factor property}$$
$$x = -1 \quad \text{or} \quad x = 3$$

There are two x-intercepts, $(-1, 0)$ and $(3, 0)$.

─── **Continued on Next Page**

As previously mentioned, the x-value of the vertex is halfway between the x-values of the two x-intercepts. Thus, it is $\frac{1}{2}$ their sum.

$$x = \frac{1}{2}(-1 + 3) = \frac{1}{2}(2) = 1$$

We find the corresponding y-value by substituting 1 for x in the equation.

$$y = 1^2 - 2(1) - 3 = -4$$

The vertex is $(1, -4)$. Here, the axis is the line $x = 1$.

Now we find the y-intercept.

$$y = 0^2 - 2(0) - 3 \quad \text{Let } x = 0.$$
$$y = -3$$

The y-intercept is $(0, -3)$.

We plot the three intercepts and the vertex, and find additional ordered pairs as needed. For example, if $x = 2$, then

$$y = 2^2 - 2(2) - 3 = -3,$$

leading to the ordered pair $(2, -3)$. A table with the ordered pairs we have found is shown with the graph in Figure 3.

x	y	
-2	5	
-1	0	← x-intercept
0	-3	← y-intercept
1	-4	← Vertex
2	-3	
3	0	← x-intercept
4	5	

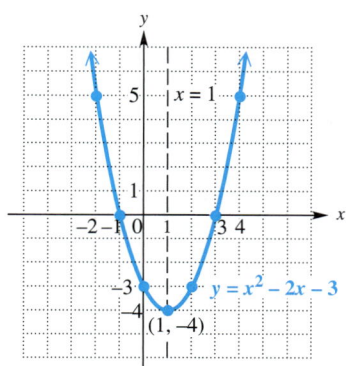

Figure 3

Work Problem 5 at the Side.

⑤ Graph $y = x^2 + 2x - 8$

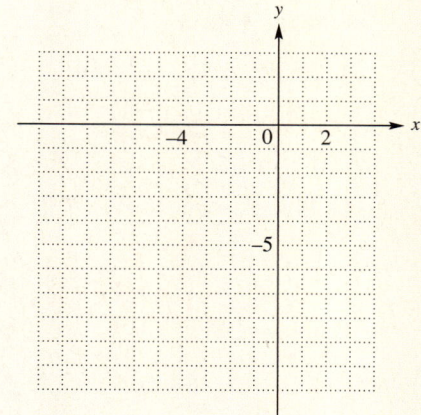

We can generalize from Example 3. The x-coordinates of the x-intercepts for $y = ax^2 + bx + c$, by the quadratic formula, are

$$x = \frac{-b + \sqrt{b^2 - 4ac}}{2a} \quad \text{and} \quad x = \frac{-b - \sqrt{b^2 - 4ac}}{2a}.$$

Thus, we can find the x-value of the vertex as shown below.

$$x = \frac{1}{2}\left(\frac{-b + \sqrt{b^2 - 4ac}}{2a} + \frac{-b - \sqrt{b^2 - 4ac}}{2a}\right)$$

$$x = \frac{1}{2}\left(\frac{-b + \sqrt{b^2 - 4ac} - b - \sqrt{b^2 - 4ac}}{2a}\right)$$

$$x = \frac{1}{2}\left(\frac{-2b}{2a}\right)$$

$$x = -\frac{b}{2a}$$

ANSWERS

5.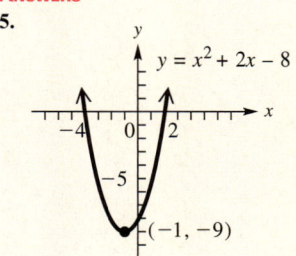

6 Complete the following ordered pairs for
$y = x^2 - 4x + 1$.

(5,), (1,), (4,),
(3,), (−1,)

For the equation in Example 3, $y = x^2 - 2x - 3$, $a = 1$, and $b = -2$. Thus, the x-value of the vertex is

$$x = -\frac{b}{2a} = -\frac{-2}{2(1)} = 1,$$

which is the same x-value for the vertex we found in Example 3. (The x-value of the vertex is $x = -\frac{b}{2a}$, even if the graph has no x-intercepts.) A procedure for graphing quadratic equations follows.

Graphing the Parabola $y = ax^2 + bx + c$

Step 1 **Find the vertex.** Let $x = -\frac{b}{2a}$, and find the corresponding y-value by substituting for x in the equation.

Step 2 **Find the y-intercept.**

Step 3 **Find the x-intercepts** (if they exist).

Step 4 **Plot** the intercepts and the vertex.

Step 5 **Find and plot additional ordered pairs** near the vertex and intercepts as needed, using the symmetry about the axis of the parabola.

EXAMPLE 4 Graphing a Parabola

Graph $y = x^2 - 4x + 1$.
  The x-value of the vertex is

$$x = -\frac{b}{2a} = -\frac{-4}{2(1)} = 2.$$

The y-value of the vertex is

$$y = 2^2 - 4(2) + 1 = -3$$

so the vertex is $(2, -3)$. The axis is the line $x = 2$. Now find the intercepts. Let $x = 0$ in $y = x^2 - 4x + 1$ to get the y-intercept $(0, 1)$. Let $y = 0$ to get the x-intercepts. If $y = 0$, the equation is $0 = x^2 - 4x + 1$, which cannot be solved by factoring. Use the quadratic formula to solve for x.

$$x = \frac{4 \pm \sqrt{16 - 4}}{2} \qquad \text{Let } a = 1, b = -4, c = 1.$$

$$x = \frac{4 \pm \sqrt{12}}{2}$$

$$x = \frac{4 \pm 2\sqrt{3}}{2} \qquad \sqrt{12} = \sqrt{4} \cdot \sqrt{3} = 2\sqrt{3}$$

$$x = \frac{2(2 \pm \sqrt{3})}{2} = 2 \pm \sqrt{3} \qquad \text{Factor; lowest terms}$$

Use a calculator to find that the x-intercepts are approximately $(3.7, 0)$ and $(0.3, 0)$, with x-values rounded to the nearest tenth.

Work Problem 6 at the Side.

Continued on Next Page

ANSWERS
6. $(5, 6), (1, -2), (4, 1), (3, -2), (-1, 6)$

Plot the intercepts, vertex, and the points found in Problem 6. Connect these points with a smooth curve. The graph is shown in Figure 4.

x	y
−1	6
0	1
$2 - \sqrt{3} \approx 0.3$	0
1	−2
2	−3
3	−2
$2 + \sqrt{3} \approx 3.7$	0
4	1
5	6

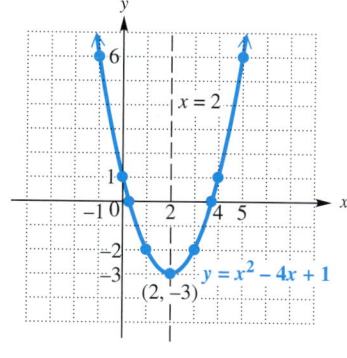

Figure 4

Work Problem 7 at the Side.

OBJECTIVE 3 Solve an application involving a parabola. Parabolic shapes are found all around us. Satellite dishes that receive television signals are becoming more popular each year. Radio telescopes use parabolic reflectors to track incoming signals. The final example discusses how to describe a cross section of a parabolic dish using an equation.

EXAMPLE 5 Finding the Equation of a Parabolic Satellite Dish

The Parkes radio telescope on the opening page of this chapter has a parabolic dish shape with a diameter of 210 ft and a depth of 32 ft. Figure 5(a) shows a diagram of such a dish. Figure 5(b) shows how a cross section of the dish can be modeled by a graph, with the vertex of the parabola at the origin of a coordinate system. Find the equation of this graph.

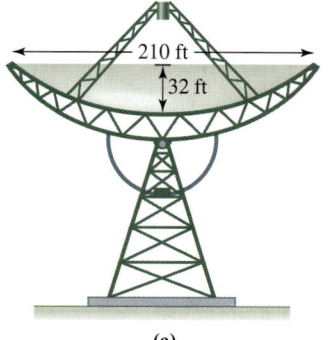

(a)

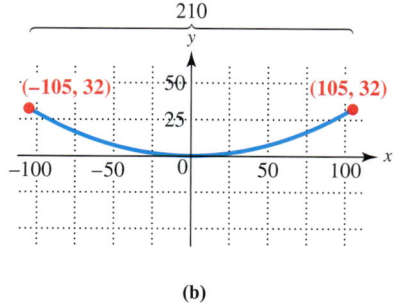

(b)

Figure 5

Continued on Next Page

7 Graph each parabola. Identify the vertex and y-intercept, and give the coordinates of the x-intercepts to the nearest tenth.

(a) $y = x^2 - 3x - 3$

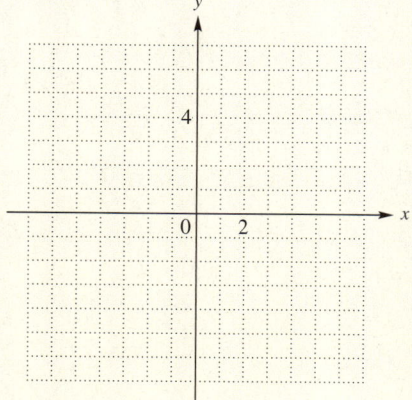

(b) $y = -x^2 + 2x + 4$

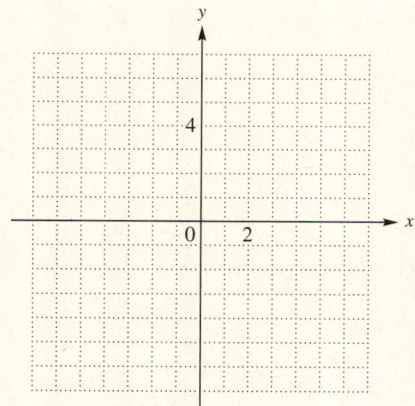

ANSWERS

7. (a)

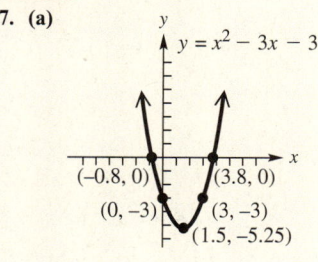

(b)

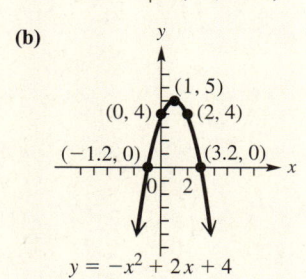

⑧ Suppose that a radio telescope has a parabolic dish shape with a diameter of 350 ft and a depth of 48 ft. Find the equation of the graph of a cross section.

Because the vertex is at the origin, the equation will be of the form $y = ax^2$. (See Example 1 and Problem 1, for example.) As shown in Figure 5(b), one point on the graph has coordinates (105, 32). Letting $x = 105$ and $y = 32$, we can solve for a.

$$y = ax^2 \quad \text{General equation}$$
$$32 = a(105)^2 \quad \text{Substitute for } x \text{ and } y.$$
$$32 = 11{,}025a \quad 105^2 = 11{,}025$$
$$a = \frac{32}{11{,}025} \quad \text{Divide by 11,025.}$$

Thus the equation is $y = \frac{32}{11{,}025}x^2$.

◀◀◀ **Work Problem 8 at the Side.**

ANSWERS

8. $y = \dfrac{48}{30{,}625}x^2$

17.4 Exercises

1. In your own words, explain what is meant by the vertex of a parabola.

2. In your own words, explain what is meant by the line of symmetry of a parabola that opens upward or downward.

Graph each equation. Give the coordinates of the vertex in each case. See Examples 1–4.

3. $y = 2x^2$

4. $y = 3x^2$

5. $y = x^2 - 4$

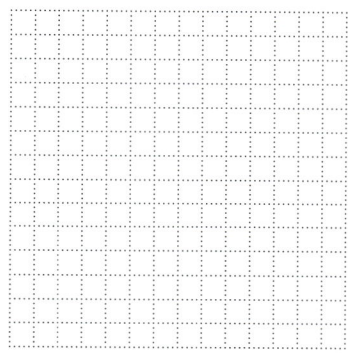

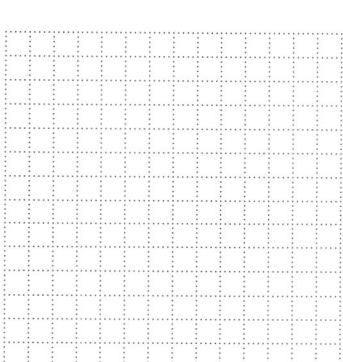

6. $y = x^2 - 6$

7. $y = -x^2 + 2$

8. $y = -x^2 + 4$

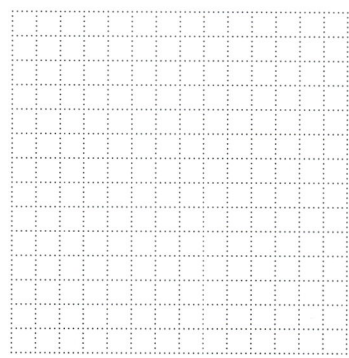

9. $y = (x + 3)^2$

10. $y = (x - 4)^2$

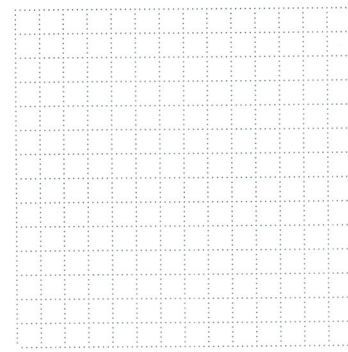

11. $y = x^2 + 2x + 3$

12. $y = x^2 - 4x + 3$

13. $y = -x^2 + 6x - 5$

14. $y = -x^2 - 4x - 3$

15. Based on your work in Exercises 3–14, what seems to be the direction in which the parabola $y = ax^2 + bx + c$ opens if $a > 0$? if $a < 0$?

16. See Examples 1–3. How many real solutions does a quadratic equation have if its corresponding graph has

  (a) no x-intercepts

  (b) one x-intercept

  (c) two x-intercepts?

Solve each problem. See Example 5.

17. The U.S. Naval Research Laboratory designed a giant radio telescope that had a diameter of 300 ft and maximum depth of 44 ft. The graph on the right below describes a cross section of this telescope. Find the equation of this parabola. (*Source: Structure Technology for Large Radio and Radar Telescope Systems*, The MIT Press.)

18. Suppose the telescope in Exercise 17 had a diameter of 400 ft and maximum depth of 50 ft. Find the equation of this parabola.

17.5 Introduction to Functions

If gasoline costs $2.50 per gal and you buy **1** gal, then you must pay $2.50(**1**) = $2.50. If you buy **2** gal, your cost is $2.50(**2**) = $5.00; for **3** gal, your cost is $2.50(**3**) = $7.50, and so on. Generalizing, if x represents the number of gallons, then the cost is $2.50x$. If we let y represent the cost, then the equation $y = 2.50x$ *relates* the number of gallons, x, to the cost in dollars, y. The ordered pairs (x, y) that satisfy this equation form a *relation*.

OBJECTIVES

1. Understand the definition of a relation.
2. Understand the definition of a function.
3. Decide whether an equation defines a function.
4. Use function notation.
5. Apply the function concept in an application.

OBJECTIVE 1 Understand the definition of a relation. In an ordered pair (x, y), x and y are called the **components** of the ordered pair. Any set of ordered pairs is called a **relation**. The set of all first components in the ordered pairs of a relation is the **domain** of the relation, and the set of all second components in the ordered pairs is the **range** of the relation.

EXAMPLE 1 Using Ordered Pairs to Define Relations*

(a) The relation $\{(0, 1), (2, 5), (3, 8), (4, 2)\}$ has domain $\{0, 2, 3, 4\}$ and range $\{1, 2, 5, 8\}$. The correspondence between the elements of the domain and the elements of the range is shown in Figure 6.

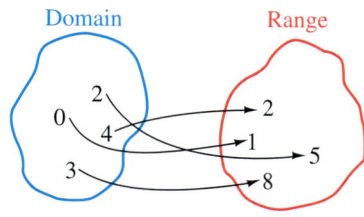

Figure 6

(b) The relation $\{(3, 5), (3, 6), (3, 7), (3, 8)\}$ has domain $\{3\}$ and range $\{5, 6, 7, 8\}$.

▶▶▶ **Work Problem 1 at the Side.**

OBJECTIVE 2 Understand the definition of a function. We now investigate an important type of relation, called a *function*.

Function

A **function** is a set of ordered pairs in which each first component corresponds to exactly one second component.

By this definition, the relation in Example 1(a) is a function, but the relation in Example 1(b) is not a function. Notice, however, that if the components of the ordered pairs in Example 1(b) were interchanged, giving the relation

$$\{(5, 3), (6, 3), (7, 3), (8, 3)\},$$

then the relation *would* be a function; in this case, each domain element (first component) corresponds to *exactly one* range element (second component).

1 Give the domain and the range of each relation.

(a) $\{(5, 10), (15, 20), (25, 30), (35, 40)\}$

(b) $\{(1, 4), (2, 4), (3, 4)\}$

ANSWERS

1. (a) domain: $\{5, 15, 25, 35\}$;
   range: $\{10, 20, 30, 40\}$
   (b) domain: $\{1, 2, 3\}$;
   range: $\{4\}$

---

* It is standard notation to use braces around the ordered pairs that form a relation.

2 Decide whether each relation is a function.

(a) $\{(-2, 8), (-1, 1), (0, 0), (1, 1), (2, 8)\}$

(b) $\{(5, 2), (5, 1), (5, 0)\}$

EXAMPLE 2 Determining Whether Relations Are Functions

Determine whether each relation is a function.

(a) $\{(-2, 4), (-1, 1), (0, 0), (1, 1), (2, 4)\}$

Notice that each first component appears once and only once. Because of this, the relation is a function.

(b) $\{(9, 3), (9, -3), (4, 2)\}$

The first component 9 appears in two ordered pairs, and corresponds to two different second components. Therefore, this relation is not a function.

Work Problem 2 at the Side.

The simple relations given in Examples 1 and 2 were defined by listing the ordered pairs or by showing the correspondence with a figure. Most useful functions have an infinite number of ordered pairs and are usually defined with equations that tell how to get the second components, given the first. We have been using equations with x and y as the variables, where x represents the first component (input) and y the second component (output) in the ordered pairs.

Here are some everyday examples of functions.

1. The **cost y** in dollars charged by an express mail company is a function of the **weight in pounds x** determined by the equation $y = 1.5(x - 1) + 9$.

2. In one state, the sales tax is 6% of the price of an item. The **tax y** on a particular item is a function of the **price x**, because $y = 0.06x$.

3. The **distance d** traveled by a car moving at a constant speed of 45 mph is a function of the **time t**. Thus, $d = 45t$.

The function concept can be illustrated by an input-output "machine," as seen in Figure 7. It shows how the express mail company equation $y = 1.5(x - 1) + 9$ provides an output (the cost, represented by y) for a given input (the weight in pounds, given by x).

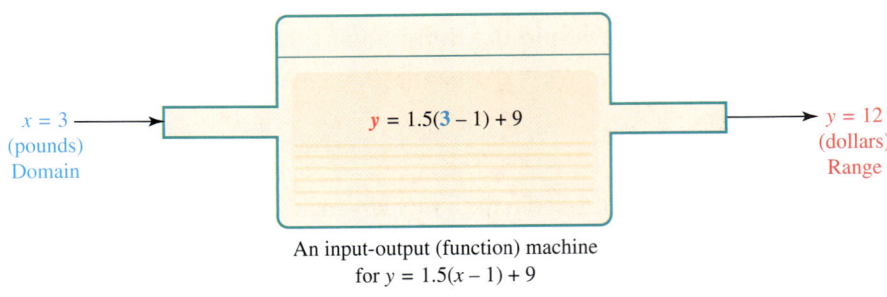

An input-output (function) machine for $y = 1.5(x - 1) + 9$

Figure 7

OBJECTIVE 3 Decide whether an equation defines a function. Given the graph of an equation, the definition of a function can be used to decide whether or not the graph represents a function. By the definition of a function, each x-value must lead to exactly one y-value. In Figure 8(a) on the next page, the indicated x-value leads to two y-values, so this graph is not the graph of a function. A vertical line can be drawn that intersects this graph in more than one point.

ANSWERS

2. (a) function (b) not a function

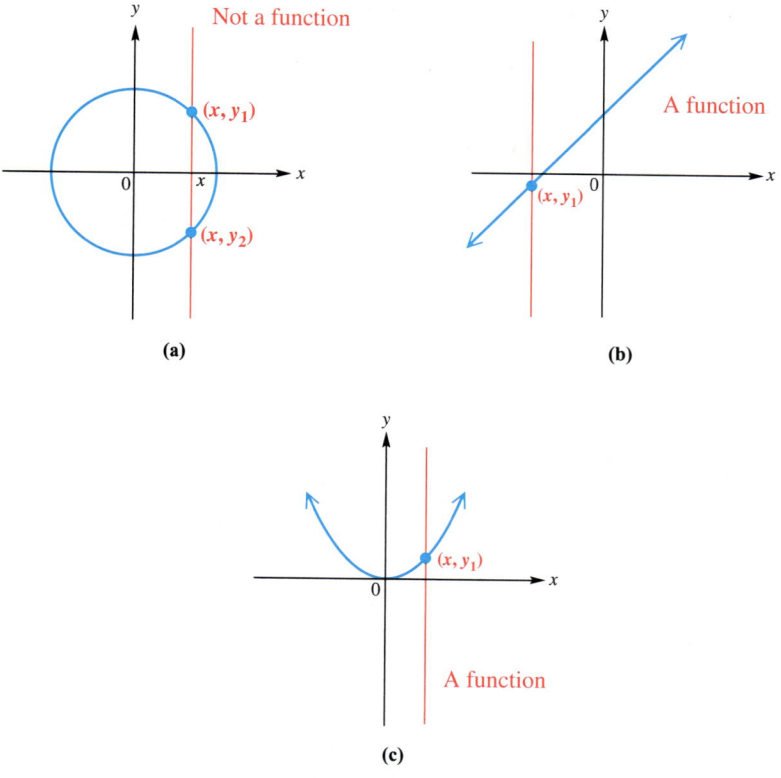

Figure 8

On the other hand, in Figures 8(b) and 8(c) any vertical line will intersect each graph in no more than one point. Because of this, the graphs in Figures 8(b) and 8(c) are graphs of functions. This idea leads to the **vertical line test** for a function.

> **Vertical Line Test**
>
> If a vertical line intersects a graph in more than one point, the graph is not the graph of a function.

As Figure 8(b) suggests, any nonvertical line is the graph of a function. For this reason, any linear equation of the form $y = mx + b$ defines a function. (Recall that a vertical line has undefined slope.) Also, any vertical parabola, as in Figure 8(c), is the graph of a function, so any quadratic equation of the form $y = ax^2 + bx + c$, where $a \neq 0$, defines a function.

EXAMPLE 3 **Deciding Whether Relations Define Functions**

Decide whether each relation graphed or defined is a function.

(a)

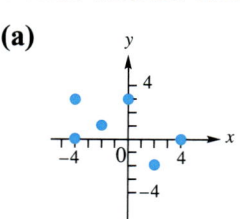

Because there are two ordered pairs with first component -4, this is not the graph of a function.

Continued on Next Page

3 Decide whether each relation graphed or defined is a function.

(a)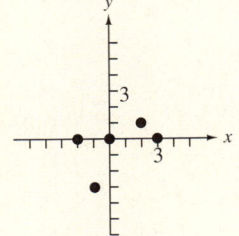

(b)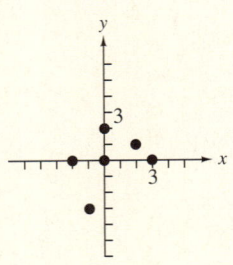

(c)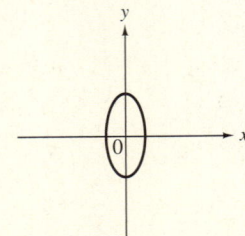

(d)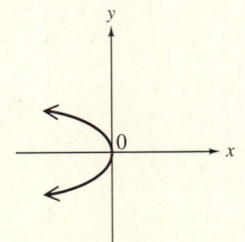

(e) $y = 3$

ANSWERS
3. (a) function (b) not a function
   (c) not a function (d) not a function
   (e) function

(b)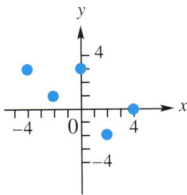

Every first component is paired with one and only one second component, and so no vertical line intersects the graph in more than one point. Therefore, this is the graph of a function.

(c) $y = 2x - 9$

This linear equation is in the form $y = mx + b$. Since the graph of this equation is a line that is not vertical, the equation defines a function.

(d)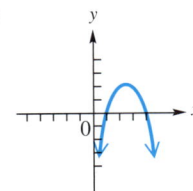

(e)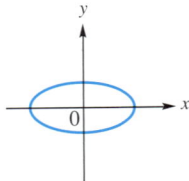

Use the vertical line test. Any vertical line will cross the graph of a vertical parabola just once, so this is the graph of a function.

The vertical line test shows that this graph is not the graph of a function; a vertical line could cross the graph twice.

(f) $x = 4$

The graph of $x = 4$ is a vertical line, so the equation does not define a function.

▶◀ **Work Problem 3 at the Side.**

OBJECTIVE 4 Use function notation. The letters f, g, and h are commonly used to name functions. For example, the function $y = 3x + 5$ may be written

$$f(x) = 3x + 5,$$

where $f(x)$ is read "f of x." The notation $f(x)$ is another way of writing y in a function. For the function defined by $f(x) = 3x + 5$, if $x = 7$ then

$$f(7) = 3 \cdot 7 + 5 \qquad \text{Let } x = 7.$$
$$= 21 + 5 = 26$$

Read this result, $f(7) = 26$, as "f of 7 equals 26." The notation $f(7)$ means the value of y when x is 7. The statement $f(7) = 26$ says that the value of y is 26 when x is 7. It also indicates that the point $(7, 26)$ lies on the graph of f.
  To find $f(-3)$, substitute -3 for x.

$$f(-3) = 3(-3) + 5 \qquad \text{Let } x = -3.$$
$$= -9 + 5 = -4$$

CAUTION
The notation $f(x)$ does *not* mean f times x. **The symbol $f(x)$ represents the function value of x. It also represents the y-value that corresponds to x.**

Function Notation

In the notation $f(x)$,

- f is the name of the function,
- x is the domain value,

and $f(x)$ is the range value y for the domain value x.

EXAMPLE 4 Using Function Notation

For the function defined by $f(x) = x^2 - 3$, find the following.

(a) $f(2)$

Substitute 2 for x.

$$f(x) = x^2 - 3$$
$$f(2) = 2^2 - 3 \qquad \text{Let } x = 2.$$
$$= 4 - 3 = 1$$

(b) $f(0) = 0^2 - 3 = 0 - 3 = -3$

(c) $f(-3) = (-3)^2 - 3 = 9 - 3 = 6$

▶▶▶ **Work Problem 4 at the Side.**

OBJECTIVE 5 Apply the function concept in an application. Because a function assigns to each element in its domain exactly one element in its range, the function concept is used in real-data applications where two quantities are related. Our final example discusses such an application, using a table of values.

EXAMPLE 5 Applying the Function Concept to Population

Asian-American populations (in millions) are shown in the table.

ASIAN-AMERICAN POPULATION

Year	Population (in millions)
1996	9.7
1998	10.5
2000	11.2
2002	12

Source: U.S. Bureau of the Census.

(a) Use the table to write a set of ordered pairs that defines a function f.

If we choose the years as the domain elements and the populations as the range elements, the information in the table can be written as a set of four ordered pairs. In set notation, the function f is shown below.

$$f = \{(1996, 9.7), (1998, 10.5), (2000, 11.2), (2002, 12)\}$$

(b) What is the domain of f? What is the range?

The domain is the set of years, or x-values:

$$\{1996, 1998, 2000, 2002\}$$

The range is the set of populations, in millions, or y-values:

$$\{9.7, 10.5, 11.2, 12\}$$

Continued on Next Page

---

4 For $f(x) = 6x - 2$, find each function value.

(a) $f(-1)$

(b) $f(0)$

(c) $f(1)$

ANSWERS

4. **(a)** -8 **(b)** -2 **(c)** 4

Chapter 17 Quadratic Equations

5 The numbers of U.S. children (in millions) educated at home for selected years are given in the table.

School Year	Number of Children
1997	1.1
1998	1.2
1999	1.3
2000	1.5
2001	1.7

Source: National Home Education Research Institute, Salem, OR.

(a) Write a set of ordered pairs that defines a function f for the data.

(b) Give the domain and range of f.

(c) Find $f(1999)$.

(d) In what year did the number of children equal 1.5 million?

(c) Find $f(1996)$ and $f(2000)$.

We repeat the table and the set of ordered pairs from part (a).

ASIAN-AMERICAN POPULATION

Year	Population (in millions)
1996	9.7
1998	10.5
2000	11.2
2002	12

Source: U.S. Bureau of the Census.

$$f = \{(1996, 9.7), (1998, 10.5), (2000, 11.2), (2002, 12)\}$$

Thus, $f(1996) = 9.7$ million and $f(2000) = 11.2$ million.

(d) For what x-value does $f(x)$ equal 12 million? 10.5 million?

We use the table or the ordered pairs found in part (a) to determine $f(2002) = 12$ million and $f(1998) = 10.5$ million.

Work Problem 5 at the Side.

ANSWERS

5. (a) $f = \{(1997, 1.1), (1998, 1.2), (1999, 1.3), (2000, 1.5), (2001, 1.7)\}$
   (b) domain: $\{1997, 1998, 1999, 2000, 2001\}$; range: $\{1.1, 1.2, 1.3, 1.5, 1.7\}$
   (c) 1.3 million
   (d) 2000

17.5 Exercises

FOR EXTRA HELP: 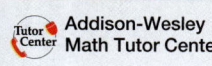 Addison-Wesley Math Tutor Center   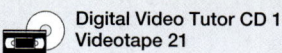 MathXL   Digital Video Tutor CD 14 Videotape 21   Student's Solutions Manual   MyMathLab   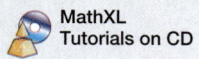 MathXL Tutorials on CD

Complete the following table for the function defined by $f(x) = x + 2$.

x	x + 2	f(x)	(x, y)
0	2	2	(0, 2)
1. 1			
2. 2			
3. 3			
4. 4			

5. Describe the graph of function f in Exercises 1–4 if the domain is {0, 1, 2, 3, 4}.

6. Describe the graph of function f in Exercises 1–4 if the domain is the set of all real numbers.

Determine whether each relation is a function. Give the domain and the range in Exercises 7–12. See Examples 1–3.

7. $\{(-4, 3), (-2, 1), (0, 5), (-2, -8)\}$

8. $\{(3, 7), (1, 4), (0, -2), (-1, -1), (-2, 5)\}$

9.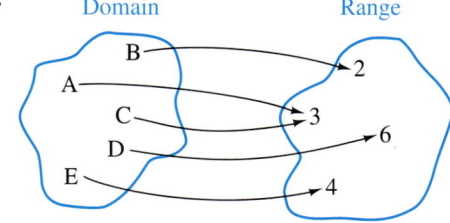

10.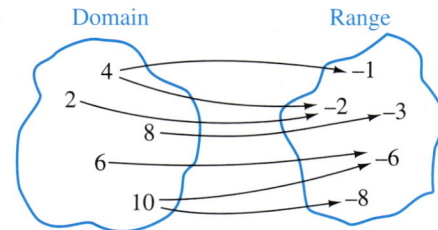

11.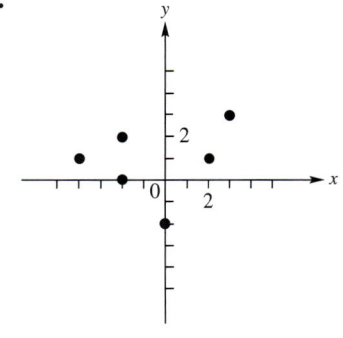

12.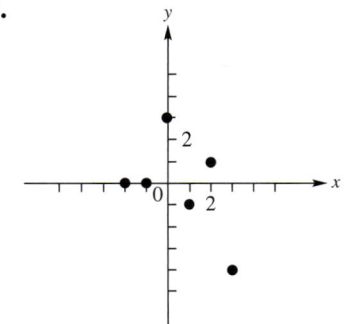

13.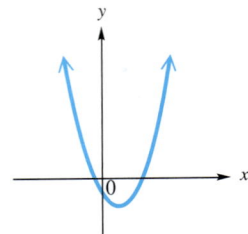

14.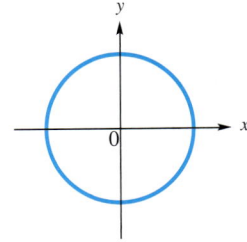

15.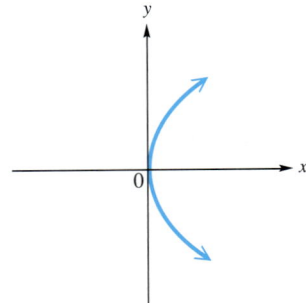

16.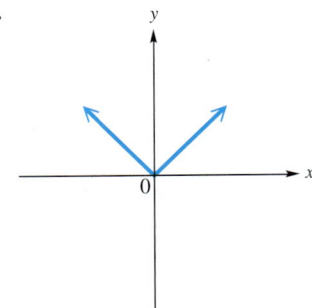

Decide whether each equation defines y as a function of x. (Remember that to be a function, every value of x must give one and only one value of y.) See Example 3.

17. $y = 5x + 3$ **18.** $y = -7x + 12$ **19.** $x = |y|$ **20.** $x = y^2$

RELATING CONCEPTS (EXERCISES 21–24) For Individual or Group Work

*A function defined by $f(x) = 3x - 4$, called a **linear function** because its graph is a straight line, can be graphed by replacing $f(x)$ with y and then using the methods described earlier. Let us assume that some function is written in the form $f(x) = mx + b$, for particular values of m and b.* **Work Exercises 21–24 in order.**

21. If $f(2) = 4$, name the coordinates of one point on the line.

22. If $f(-1) = -4$, name the coordinates of another point on the line.

23. Use the results of Exercises 21 and 22 to find the slope of the line.

24. Use the slope-intercept form of the equation of a line to write the function in the form $f(x) = mx + b$.

For each function f, find **(a)** *f*(2), **(b)** *f*(0), *and* **(c)** *f*(−3). *See Example 4.*

25. $f(x) = 4x + 3$

26. $f(x) = -3x + 5$

27. $f(x) = x^2 - x + 2$

28. $f(x) = x^3 + x$

29. $f(x) = |x|$

30. $f(x) = |x + 7|$

The number of U.S. foreign-born residents has grown by more than 43% since 1990. The graph shows the number of such residents (in millions) for selected years. Use the information in the graph for Exercises 31–35. See Example 5.

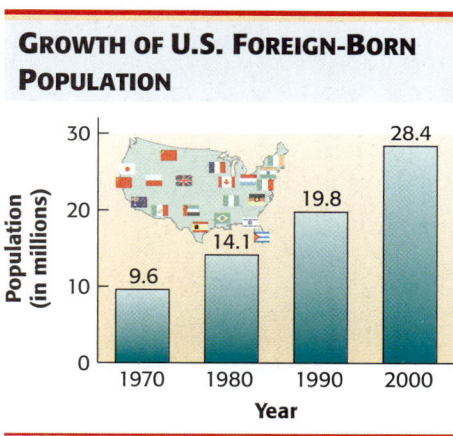

GROWTH OF U.S. FOREIGN-BORN POPULATION

Source: U.S. Bureau of the Census.

31. Write the information in the graph as a set of ordered pairs. Does this set define a function?

32. Suppose that *g* is the name given to this relation. Give the domain and range of *g*.

33. Find $g(1980)$ and $g(1990)$.

34. For what value of *x* does $g(x) = 28.4$ (million)?

35. Suppose $g(2002) = 30.3$ (million). What does this tell you in the context of the application?

RELATING CONCEPTS (EXERCISES 36–40) For Individual or Group Work

The data give the percent of U.S. active-duty female military personnel during selected years from 1987 to 2003. Use the information in the table to **work Exercises 36–40 in order.**

ACTIVE-DUTY FEMALE MILITARY PERSONNEL

Year	Percent
1987	10.2
1993	11.6
2000	14.4
2003	15.1

Source: U.S. Department of Defense.

36. Plot the ordered pairs (year, percent) from the table. Do the points suggest that a linear function would give a reasonable approximation of the data?

37. Use the first and last data pairs in the table to write an equation relating the data.

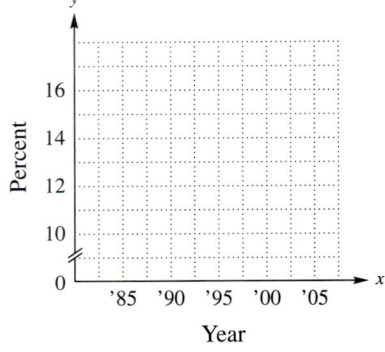

38. Use your equation from Exercise 37 to approximate the percent of active-duty female military personnel in 1993 and 2000. Round to the nearest tenth of a percent.

39. Use the second and third data pairs to write an equation relating the data.

40. Use your equation from Exercise 39 to approximate the number of active-duty female military personnel in 1987 and 2003. Which of the results in Exercise 38 and this exercise give better approximations?

Chapter 17
SUMMARY

KEY TERMS

17.4 parabola — The graph of the quadratic equation $y = ax^2 + bx + c$ is called a parabola.

vertex — The vertex of a parabola that opens upward or downward is the lowest or highest point on the graph.

axis — The axis of a parabola that opens upward or downward is a vertical line through the vertex.

line of symmetry — If a graph is folded on its line of symmetry, the two sides coincide.

17.5 components — In an ordered pair (x, y), x and y are the components.

relation — Any set of ordered pairs is called a relation.

domain — The set of all first components in the ordered pairs of a relation is the domain of the relation.

range — The set of all second components in the ordered pairs of a relation is the range of the relation.

function — A function is a set of ordered pairs in which each first component corresponds to exactly one second component.

NEW SYMBOLS

$\pm$ positive or negative

$f(x)$ function f of x

TEST YOUR WORD POWER

See how well you have learned the vocabulary in this chapter. Answers, with examples, follow the Quick Review.

1. A **relation** is
   A. any set of ordered pairs
   B. a set of ordered pairs in which each first component corresponds to exactly one second component
   C. two sets of ordered pairs that are related
   D. a graph of ordered pairs.

2. The **domain** of a relation is
   A. the set of all x- and y-values in the ordered pairs of the relation
   B. the difference between the components in an ordered pair of the relation
   C. the set of all first components in the ordered pairs of the relation
   D. the set of all second components in the ordered pairs of the relation.

3. The **range** of a relation is
   A. the set of all x- and y-values in the ordered pairs of the relation
   B. the difference between the components in an ordered pair of the relation
   C. the set of all first components in the ordered pairs of the relation
   D. the set of all second components in the ordered pairs of the relation.

4. A **function** is
   A. any set of ordered pairs
   B. a set of ordered pairs in which each first component corresponds to exactly one second component
   C. two sets of ordered pairs that are related
   D. a graph of ordered pairs.

Quick Review

Concepts

17.1 Solving Quadratic Equations by the Square Root Property

Square Root Property of Equations
If k is positive, and if $a^2 = k$, then
$$a = \sqrt{k} \quad \text{or} \quad a = -\sqrt{k}.$$

Examples

Solve $(2x + 1)^2 = 5$.

$$2x + 1 = \sqrt{5} \quad \text{or} \quad 2x + 1 = -\sqrt{5}$$
$$2x = -1 + \sqrt{5} \quad \quad 2x = -1 - \sqrt{5}$$
$$x = \frac{-1 + \sqrt{5}}{2} \quad \quad x = \frac{-1 - \sqrt{5}}{2}$$

The solutions are $\dfrac{-1 + \sqrt{5}}{2}$ and $\dfrac{-1 - \sqrt{5}}{2}$.

17.2 Solving Quadratic Equations by Completing the Square

Solving a Quadratic Equation by Completing the Square

Step 1 If the coefficient of the squared term is 1, go to Step 2. If it is not 1, divide each side of the equation by this coefficient.

Step 2 Make sure that all variable terms are on one side of the equation, and all constant terms are on the other.

Step 3 Take half the coefficient of x, square it, and add the square to each side of the equation. Factor the variable side, and combine terms on the other side.

Step 4 Use the square root property to solve the equation.

Solve $2x^2 + 4x - 1 = 0$.

$$x^2 + 2x - \frac{1}{2} = 0 \quad \text{Divide by 2.}$$

$$x^2 + 2x = \frac{1}{2} \quad \text{Add } \tfrac{1}{2}.$$

$$x^2 + 2x + 1 = \frac{1}{2} + 1 \quad \left[\tfrac{1}{2}(2)\right]^2 = 1$$

$$(x + 1)^2 = \frac{3}{2} \quad \text{Factor; combine like terms.}$$

$$x + 1 = \sqrt{\frac{3}{2}} \quad \text{or} \quad x + 1 = -\sqrt{\frac{3}{2}}$$

$$x + 1 = \frac{\sqrt{3} \cdot \sqrt{2}}{\sqrt{2} \cdot \sqrt{2}} \quad \quad x + 1 = -\frac{\sqrt{3} \cdot \sqrt{2}}{\sqrt{2} \cdot \sqrt{2}}$$

$$x + 1 = \frac{\sqrt{6}}{2} \quad \quad x + 1 = -\frac{\sqrt{6}}{2}$$

$$x = -1 + \frac{\sqrt{6}}{2} \quad \quad x = -1 - \frac{\sqrt{6}}{2}$$

$$x = \frac{-2 + \sqrt{6}}{2} \quad \quad x = \frac{-2 - \sqrt{6}}{2}$$

The solutions are $\dfrac{-2 + \sqrt{6}}{2}$ and $\dfrac{-2 - \sqrt{6}}{2}$.

Concepts	Examples

17.3 Solving Quadratic Equations by the Quadratic Formula

Quadratic Formula

The solutions of $ax^2 + bx + c = 0$ $(a \neq 0)$ are

$$x = \frac{-b \pm \sqrt{b^2 - 4ac}}{2a}.$$

Solve $3x^2 - 4x - 2 = 0$.

$$x = \frac{-(-4) \pm \sqrt{(-4)^2 - 4(3)(-2)}}{2(3)}$$

$$x = \frac{4 \pm \sqrt{16 + 24}}{6}$$

$$x = \frac{4 \pm \sqrt{40}}{6} = \frac{4 \pm 2\sqrt{10}}{6}$$

$$x = \frac{2(2 \pm \sqrt{10})}{2(3)} = \frac{2 \pm \sqrt{10}}{3}$$

The solutions are $\frac{2 + \sqrt{10}}{3}$ and $\frac{2 - \sqrt{10}}{3}$.

17.4 Graphing Quadratic Equations

To graph $y = ax^2 + bx + c$:

Step 1 Find the vertex: $x = -\frac{b}{2a}$; find y by substituting this value for x in the equation.

Graph $y = 2x^2 - 5x - 3$.

$$x = -\frac{b}{2a} = -\frac{-5}{2(2)} = \frac{5}{4}$$

$$y = 2\left(\frac{5}{4}\right)^2 - 5\left(\frac{5}{4}\right) - 3 = 2\left(\frac{25}{16}\right) - \frac{25}{4} - 3$$

$$= \frac{25}{8} - \frac{50}{8} - \frac{24}{8} = -\frac{49}{8}$$

The vertex is $\left(\frac{5}{4}, -\frac{49}{8}\right)$.

Step 2 Find the y-intercept.

$$y = 2(0)^2 - 5(0) - 3 = -3$$

The y-intercept is $(0, -3)$.

Step 3 Find the x-intercepts (if they exist).

$$0 = 2x^2 - 5x - 3$$
$$0 = (2x + 1)(x - 3)$$
$$2x + 1 = 0 \quad \text{or} \quad x - 3 = 0$$
$$2x = -1 \quad\quad\quad x = 3$$
$$x = -\frac{1}{2}$$

The x-intercepts are $\left(-\frac{1}{2}, 0\right)$ and $(3, 0)$.

Step 4 Plot the intercepts and the vertex.

Step 5 Find and plot additional ordered pairs near the vertex and intercepts as needed.

x	y
$-\frac{1}{2}$	0
0	-3
1	-6
$\frac{5}{4}$	$-\frac{49}{8}$
2	-5
3	0

Concepts	Examples
17.5 Introduction to Functions **Vertical Line Test** If a vertical line intersects a graph in more than one point, the graph is not the graph of a function.	By the vertical line test, the graph shown is not the graph of a function. 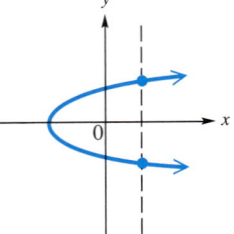
Domain and Range The set of all first components in the ordered pairs of a relation is the *domain* of the relation. The set of all second components is the *range* of the relation.	The function $$\{(10, 5), (20, 15), (30, 25)\}$$ has domain $\{10, 20, 30\}$ and range $\{5, 15, 25\}$.
To find a particular function value, substitute the given *x*-value in the function.	If $f(x) = 2x + 7$, find $f(3)$. $$f(3) = 2(3) + 7$$ $$= 13$$

> **ANSWERS TO TEST YOUR WORD POWER**

1. A; *Example:* $\{(0, 2), (2, 4), (3, 6), (-1, 3)\}$
2. C; *Example:* The domain in the relation given in Problem 1 is the set of *x*-values, that is, $\{0, 2, 3, -1\}$.
3. D; *Example:* The range of the relation given in Problem 1 is the set of *y*-values, that is, $\{2, 4, 6, 3\}$.
4. B; *Example:* The relation given in Problem 1 is a function since each *x*-value corresponds to exactly one *y*-value.

Chapter 17
REVIEW EXERCISES

[17.1] *In Exercises 1–8, solve each equation by using the square root property. Express all radicals in simplest form.*

1. $y^2 = 144$
2. $x^2 = 37$
3. $m^2 = 128$
4. $(k + 2)^2 = 25$
5. $(r - 3)^2 = 10$
6. $(2p + 1)^2 = 14$
7. $(3k + 2)^2 = -3$
8. $(3x + 5)^2 = 0$

[17.2] *Solve each equation by completing the square.*

9. $m^2 + 6m + 5 = 0$
10. $p^2 + 4p = 7$
11. $-x^2 + 5 = 2x$
12. $2x^2 - 3 = -8x$
13. $4(x^2 + 7x) + 29 = -20$
14. $(4x + 1)(x - 1) = -7$

Solve each problem.

15. If an object is propelled upward on Earth from a height of 50 feet, with an initial velocity of 32 feet per second, then its height after t seconds is given by $h = -16t^2 + 32t + 50$, where h is in feet. After how many seconds will it reach a height of 30 ft?

16. Find the lengths of the three sides of the right triangle shown.

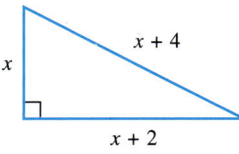

17. What must be added to $x^2 + kx$ to make it a perfect square?

[17.3]

18. Consider the equation $x^2 - 9 = 0$.
    (a) Solve the equation by factoring.
    (b) Solve the equation by the square root property.
    (c) Solve the equation by the quadratic formula.
    (d) Compare your answers. If a quadratic equation can be solved by both factoring and the quadratic formula, should we always get the same results? Explain.

Solve each equation by using the quadratic formula.

19. $-4x^2 - 2x + 7 = 0$

20. $2x^2 + 8 = 4x + 11$

21. $x(5x - 1) = 1$

22. $\dfrac{1}{4}x^2 = 2 - \dfrac{3}{4}x$

23. $\dfrac{1}{2}x^2 + 3x = 5$

24. Why is this not the statement of the quadratic formula for $ax^2 + bx + c = 0$?

$$x = -b \pm \dfrac{\sqrt{b^2 - 4ac}}{2a}$$

[17.4] Sketch the graph of each equation. Identify each vertex.

25. $y = -3x^2$

26. $y = -x^2 + 5$

27. $y = x^2 - 2x + 1$

28. $y = -x^2 + 2x + 3$ **29.** $y = x^2 + 4x + 2$ **30.** $y = (x + 4)^2$

31. Refer to Example 5 and Exercise 17 in **Section 17.4**. Suppose that a telescope has a diameter of 200 ft and a maximum depth of 30 ft. Find the equation for a cross section of the parabolic dish.

[17.5] *Decide whether each relation is or is not a function. In Exercises 32 and 33, give the domain and the range.*

32. $\{(-2, 4), (0, 8), (2, 5), (2, 3)\}$ **33.** $\{(8, 3), (7, 4), (6, 5), (5, 6), (4, 7)\}$

34. 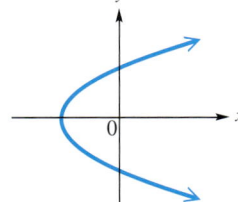   **35.**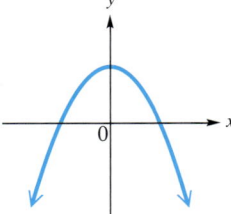

36. $2x + 3y = 12$ **37.** $y = x^2$ **38.** $x = 2|y|$

Find (a) $f(2)$ and (b) $f(-1)$.

39. $f(x) = 3x + 2$ **40.** $f(x) = 2x^2 - 1$ **41.** $f(x) = |x + 3|$

Becky and Brad are the owners of Cole's Baseball Cards. They have found that the price y, in dollars, of a particular Brad Radke baseball card depends on the demand x, in hundreds, for the card, according to the function defined by

$$y = -x^2 + 12x - 26.$$

42. What demand produces a price of $6 for the card?

43. Find the vertex of the parabola $y = -x^2 + 12x - 26$.

44. Give the demand and price that correspond to the vertex.

MIXED REVIEW EXERCISES

Solve by any method.

45. $(2t - 1)(t + 1) = 54$

46. $(2p + 1)^2 = 100$

47. $(k + 2)(k - 1) = 3$

48. $6t^2 + 7t - 3 = 0$

49. $2x^2 + 3x + 2 = x^2 - 2x$

50. $x^2 + 2x + 5 = 7$

51. $m^2 - 4m + 10 = 0$

52. $k^2 - 9k + 10 = 0$

53. $(5x + 6)^2 = 0$

54. $\frac{1}{2}r^2 = \frac{7}{2} - r$

55. $x^2 + 4x = 1$

56. $7x^2 - 8 = 5x^2 + 8$

Chapter 17
TEST

Solve by using the square root property.

1. $x^2 = 39$

2. $(x + 3)^2 = 64$

3. $(4x + 3)^2 = 24$

Solve by completing the square.

4. $x^2 - 4x = 6$

5. $2x^2 + 12x - 3 = 0$

Solve by the quadratic formula.

6. $2x^2 + 5x - 3 = 0$

7. $3w^2 + 2 = 6w$

8. $4x^2 + 8x + 11 = 0$

9. $t^2 - \frac{5}{3}t + \frac{1}{3} = 0$

Solve by the method of your choice.

10. $p^2 - 2p - 1 = 0$

11. $(2x + 1)^2 = 18$

12. $(x - 5)(2x - 1) = 1$

13. $t^2 + 25 = 10t$

1. _____

2. _____

3. _____

4. _____

5. _____

6. _____

7. _____

8. _____

9. _____

10. _____

11. _____

12. _____

13. _____

Solve each problem.

14. If an object is propelled into the air from ground level with an initial velocity of 64 feet per second, its height s (in feet) after t seconds is given by the formula $s = -16t^2 + 64t$. After how many seconds will the object reach a height of 64 feet?

15. Find the lengths of the three sides of the right triangle.

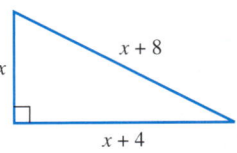

Sketch the graph of each equation. Identify each vertex.

16. $y = (x - 3)^2$

17. $y = -x^2 - 2x - 4$

18. Decide whether each relation represents a function. If it does, give the domain and the range.

(a) $\{(2, 3), (2, 4), (2, 5)\}$

(b) $\{(0, 2), (1, 2), (2, 2)\}$

19. Use the vertical line test to determine whether the graph is that of a function.

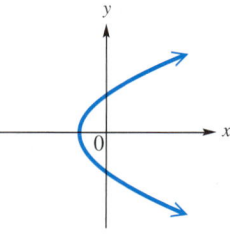

20. If $f(x) = 3x + 7$, find $f(-2)$.

Whole Numbers Computation: Pretest

This test will check your skills in doing whole numbers computation, using paper and pencil. Each part of the test is keyed to a section in the Review Chapter, which follows this test. Based on your test results, work the appropriate section(s) in the Review Chapter *before you start Chapter 1.*

Adding Whole Numbers   *(Do not use a calculator.)*

1. 368
    + 22

2. 7093
    + 6073

3. 85
    + 2968

4. 57,208
    915
    + 59,387

5. 714 + 3728 + 9 + 683,775

1. _____
2. _____
3. _____
4. _____
5. _____

Subtracting Whole Numbers   *(Do not use a calculator.)*

1. 426
    − 76

2. 3358
    − 2729

3. 30,602
    − 5708

4. 4006 − 97

5. 679,420 − 88,033

1. _____
2. _____
3. _____
4. _____
5. _____

Multiplying Whole Numbers   *(Do not use a calculator.)*

1. 3 × 3 × 0 × 6

2. 3841
    ×   7

3. (520) (3000)

1. _____
2. _____
3. _____

(continued)

Do not use a calculator; show your work.

4. 71
    × 26

5. Multiply 359 and 48.

6. 853 × 609

4. _____

5. _____

6. _____

Dividing Whole Numbers (Do not use a calculator; show your work.)

1. $3\overline{)69}$

2. $12 \div 0$

3. $\dfrac{25{,}036}{4}$

4. $7\overline{)5655}$

5. $52\overline{)1768}$

6. $45{,}000 \div 900$

7. $38\overline{)2300}$

8. $83\overline{)44{,}799}$

1. _____

2. _____

3. _____

4. _____

5. _____

6. _____

7. _____

8. _____

Now check your answers on page A–40 in the Answers section at the back of the book. Record the number of problems you worked correctly in each part of the test.

Adding Whole Numbers: _____ correct out of 5.

  If you got 0, 1, or 2 correct, work **Section R.1** in the Review Chapter.

Subtracting Whole Numbers: _____ correct out of 5.

  If you got 0, 1, or 2 correct, work **Section R.2** in the Review Chapter.

Multiplying Whole Numbers: _____ correct out of 6.

  If you got 0, 1, 2, or 3 correct, work **Section R.3** in the Review Chapter.

Dividing Whole Numbers: _____ correct out of 8.

  If you got 0, 1, or 2 correct, work **Sections R.4** and **R.5** in the Review Chapter.

  If you got 3 or 4 correct, work **Section R.5** in the Review Chapter.

Whole Numbers Review

> **NOTE**
> Use the **Whole Numbers Computation: Pretest** on pages 1207–1208 to determine which sections you need to work in this chapter.

R.1 Adding Whole Numbers
R.2 Subtracting Whole Numbers
R.3 Multiplying Whole Numbers
R.4 Dividing Whole Numbers
R.5 Long Division

R.1 Adding Whole Numbers

There are 4 triangles at the left and 2 at the right. In all, there are 6 triangles.

The process of finding the total is called *addition*. Here 4 and 2 were added to get 6. Addition is written with a + sign, as shown below.

$$4 + 2 = 6$$

OBJECTIVES

1. Add two single-digit numbers.
2. Add more than two numbers.
3. Add when carrying is not required.
4. Add with carrying.
5. Use addition to solve application problems.
6. Check the sum in addition.

OBJECTIVE 1 Add two single-digit numbers. In addition, the numbers being added are called **addends,** and the resulting answer is called the **sum** or **total.**

$$\begin{array}{r} 4 \leftarrow \text{Addend} \\ +\,2 \leftarrow \text{Addend} \\ \hline 6 \leftarrow \text{Sum (answer)} \end{array}$$

Addition problems can also be written horizontally as follows.

$$\underset{\text{Addend}}{4} \;+\; \underset{\text{Addend}}{2} \;=\; \underset{\text{Sum}}{6}$$

> **Commutative Property of Addition**
> The **commutative property of addition** states that changing the *order* of the addends in an addition problem does not change the sum.

For example, the sum of 4 + 2 is the same as the sum of 2 + 4. Both sums are 6. This allows the addition of the same numbers in a different order.

Chapter R Whole Numbers Review

1 Add. Then use the commutative property to write another addition problem and find the sum.

(a) $3 + 4$

(b) $9 + 9$

(c) $7 + 8$

(d) $6 + 9$

2 Add each column of numbers.

(a) 5
     4
     6
     9
    + 2

(b) 7
     5
     1
     2
    + 6

(c) 9
     2
     1
     3
    + 4

(d) 3
     8
     6
     4
    + 8

ANSWERS

1. (a) 7; $4 + 3 = 7$ (b) 18; no change
   (c) 15; $8 + 7 = 15$ (d) 15; $9 + 6 = 15$
2. (a) 26 (b) 21 (c) 19 (d) 29

EXAMPLE 1 Adding Two Single-Digit Numbers

Add. Then use the commutative property to write another addition problem and find the sum.

(a) $6 + 2 = 8$ and $2 + 6 = 8$

(b) $5 + 9 = 14$ and $9 + 5 = 14$

(c) $8 + 3 = 11$ and $3 + 8 = 11$

(d) $8 + 8 = 16$ (No change occurs when commutative property is used.)

◀◀ **Work Problem 1 at the Side.**

Associative Property of Addition

By the **associative property of addition,** changing the *grouping* of addends does not change the sum.

For example, the sum of $3 + 5 + 6$ may be found in several ways.

$(3 + 5) + 6 = 8 + 6 = 14$ Parentheses tell you to add $3 + 5$ first.

$3 + (5 + 6) = 3 + 11 = 14$ Parentheses tell you to add $5 + 6$ first.

Either method gives a sum of 14.

OBJECTIVE 2 Add more than two numbers. To add several numbers, first write them in a column. Add the first number to the second. Add this sum to the third number; continue until all the numbers are used.

EXAMPLE 2 Adding More Than Two Numbers

Add 2, 5, 6, 1, and 4.

```
  2  ┐ 2 + 5 = 7
  5  ┘
  6      7 + 6 = 13
  1         13 + 1 = 14
+ 4            14 + 4 = 18
 18
```

◀◀ **Work Problem 2 at the Side.**

NOTE
By the commutative and associative properties of addition, you may also add numbers by starting at the bottom of a column. Adding down from the top or adding up from the bottom will give the same sum.

OBJECTIVE 3 Add when carrying is not required. If numbers have two or more digits, first you must arrange the numbers in columns so that the ones digits are in the same column, tens are in the same column, hundreds are in the same column, and so on. Next, you add column by column, starting at the right.

Section R.1 Adding Whole Numbers **1211**

EXAMPLE 3 Adding without Carrying

Add 511 + 23 + 154 + 10.

First line up the numbers in columns, with the ones column at the right.

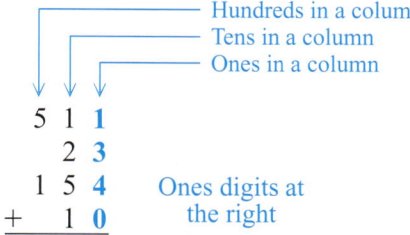

```
  5 1 1
    2 3
  1 5 4      Ones digits at
+   1 0      the right
```

Now start at the right and add the ones digits. Add the tens digits next, and finally, the hundreds digits.

```
    5 1 1
      2 3
    1 5 4
  +   1 0
    6 9 8
    ↑ ↑ ↑
    │ │ └── Sum of ones
    │ └──── Sum of tens
    └────── Sum of hundreds
```

The sum of the four numbers is 698.

▶ **Work Problem 3 at the Side.**

OBJECTIVE 4 Add with carrying. If the sum of the digits in a column is more than 9, use **carrying** (also called *regrouping*).

EXAMPLE 4 Adding with Carrying

Add 47 and 29.

Add the digits in the ones column.

$$\begin{array}{r} 47 \\ + 2\mathbf{9} \\ \hline \end{array}$$

↑── Sum of ones is 16

Because 16 is 1 ten and 6 ones, write 6 ones in the ones column and *carry* 1 ten to the tens column.

```
        Carry 1 ten.
   1 ←──────┐
  47        │
+ 29   7 + 9 = 16
 ───        │
   6 ←──────┘
        Write 6 ones in ones column.
```

Add the digits in the tens column, including the carried 1.

```
   1
  47
+ 29
 ───
  76
  ↑
  └── Sum of digits in tens column
```

▶ **Work Problem 4 at the Side.**

3 Add.

(a) 25
    + 73

(b) 364
    + 532

(c) 42,305
    + 11,563

4 Add by using carrying.

(a)   69
    + 26

(b)   76
    + 18

(c)   56
    + 37

(d)   34
    + 49

ANSWERS
3. (a) 98 (b) 896 (c) 53,868
4. (a) 95 (b) 94 (c) 93 (d) 83

5 Add, carrying when necessary.

(a) 481
      79
      38
    +395

(b) 4271
     372
    8976
    + 162

(c)   57
       4
     392
     804
      51
    + 27

(d) 7821
     435
      72
     305
    +1693

6 Add with mental carrying.

(a) 278
     825
      14
       3
       7
    +9275

(b) 3305
     650
     708
      29
      40
       6
    +   3

(c) 15,829
       765
        78
        15
         9
         7
    +13,179

ANSWERS
5. (a) 993 (b) 13,781 (c) 1335
   (d) 10,326
6. (a) 10,402 (b) 4741 (c) 29,882

EXAMPLE 5 Adding with Carrying

Add 324 + 7855 + 23 + 7 + 86.

Step 1   Add the digits in the ones column.

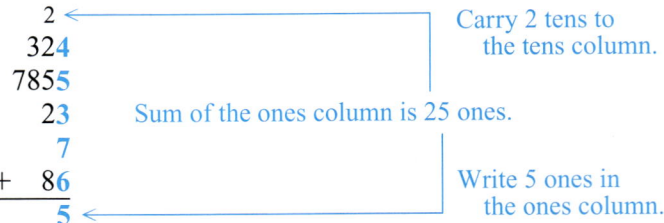

Sum of the ones column is 25 ones.
Carry 2 tens to the tens column.
Write 5 ones in the ones column.

In 25, the 5 represents 5 ones and is written in the ones column, while the 2 represents 2 tens and is carried to the tens column.

Step 2   Now add the digits in the tens column, including the carried 2.

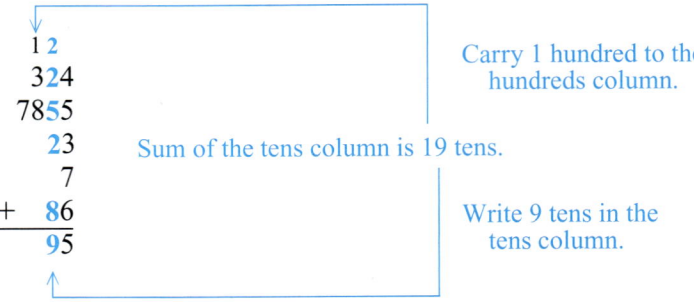

Sum of the tens column is 19 tens.
Carry 1 hundred to the hundreds column.
Write 9 tens in the tens column.

Step 3   Add the hundreds column, including the carried 1.

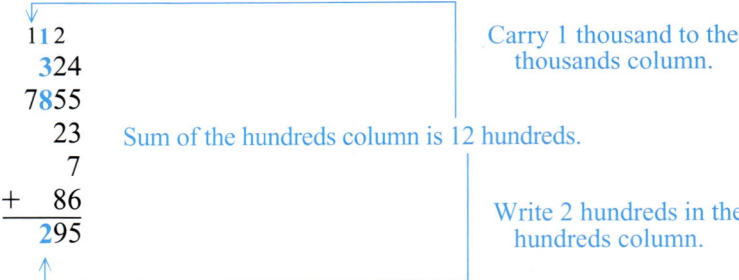

Sum of the hundreds column is 12 hundreds.
Carry 1 thousand to the thousands column.
Write 2 hundreds in the hundreds column.

Step 4   Add the thousands column, including the carried 1.

```
 112
 324
7855
  23     Sum of the thousands
   7       column is 8 thousands.
+ 86
8295
```

Thus, 324 + 7855 + 23 + 7 + 86 = 8295.

Work Problem 5 at the Side.

NOTE
For additional speed, try to carry mentally. Do not write the carried numbers, but just remember them. Try this method in Problem 6 at the side. If it works for you, use it.

Work Problem 6 at the Side.

Section R.1 Adding Whole Numbers **1213**

OBJECTIVE 5 Use addition to solve application problems.

EXAMPLE 6 Applying Addition Skills

On this map, the distance in miles from one location to another is written alongside the road. Find the shortest route from Altamonte Springs to Clear Lake.

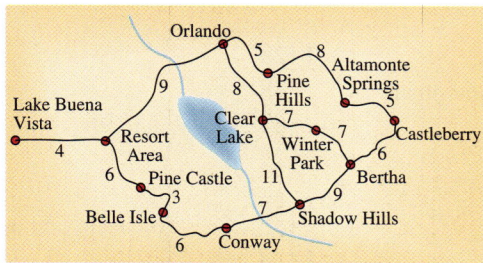

Approach Add the mileage along various routes from Altamonte Springs to Clear Lake. Then select the shortest route.

Solution One way from Altamonte Springs to Clear Lake is through Orlando. Add the mileage numbers along this route.

```
   8    Altamonte Springs to Pine Hills
   5    Pine Hills to Orlando
 + 8    Orlando to Clear Lake
  21  → miles from Altamonte Springs to Clear Lake,
            going through Orlando
```

Another way is through Bertha and Winter Park. Add the mileage numbers along this route.

```
   5    Altamonte Springs to Castleberry
   6    Castleberry to Bertha
   7    Bertha to Winter Park
 + 7    Winter Park to Clear Lake
  25  → miles from Altamonte Springs to Clear Lake
            through Bertha and Winter Park
```

The shortest route from Altamonte Springs to Clear Lake is 21 miles through Orlando.

Work Problem 7 at the Side.

EXAMPLE 7 Applying Addition Skills

Using the map in Example 6 above, find the total mileage from Shadow Hills to Castleberry to Orlando and back to Shadow Hills through Clear Lake.

Approach Add the mileage from Shadow Hills to Castleberry to Orlando and back to Shadow Hills to find the total.

Solution Use the numbers from the map.

```
    9    Shadow Hills to Bertha
    6    Bertha to Castleberry
    5    Castleberry to Altamonte Springs
    8    Altamonte Springs to Pine Hills
    5    Pine Hills to Orlando
    8    Orlando to Clear Lake
 + 11    Clear Lake to Shadow Hills
   52  → miles from Shadow Hills to Castleberry to
            Orlando and back to Shadow Hills
```

Work Problem 8 at the Side.

7 Use the map to find the shortest route from Conway to Pine Hills.

8 The road is closed between Orlando and Clear Lake, so this route cannot be used. Find the next shortest route from Orlando to Clear Lake.

ANSWERS
7. 29 miles through Belle Isle
8. Orlando to Pine Hills   5
   Pine Hills to Altamonte Springs   8
   Altamonte Springs to Castleberry   5
   Castleberry to Bertha   6
   Bertha to Winter Park   7
   Winter Park to Clear Lake   + 7
                                         38 miles

1214 Chapter R Whole Numbers Review

9 Check each addition. If the sum is incorrect, find the correct sum.

(a)
32
8
5
$+14$
59

(b)
872
539
46
$+152$
1609

(c)
79
218
7
$+639$
953

(d)
$21{,}892$
$11{,}746$
$+43{,}925$
$79{,}563$

OBJECTIVE 6 Check the sum in addition. Checking the answer is an important part of problem solving. A common method for checking addition is to re-add from the bottom to top. This is an application of the commutative and associative properties of addition.

EXAMPLE 8 Checking Addition

Check this sum.

Add down.
1428
738
63
125
17
$+485$
1428 — To check, add up.

Adding down and adding up should give the same sum. In this case, the answers agree, so the sum is probably correct.

EXAMPLE 9 Checking Addition

Check each sum. If the sum is incorrect, find the correct sum.

(a)
$785 1033$ — Correct, because both answers are the same
$63 785$
$+185 63$
$1033 +185$ — To check, add up.
1033

(b)
$635 2454$ — Error, because answers are different
$73 635$
$831 73$
$+915 831$
$2444 +915$ — To check, add up.
2444

Re-add to find that the correct sum is 2454.

◀◀◀ Work Problem 9 at the Side.

ANSWERS

9. (a) correct (b) correct
(c) incorrect; should be 943
(d) incorrect; should be 77,563

R.1 Exercises

FOR EXTRA HELP Addison-Wesley Math Tutor Center 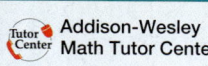 MathXL Student's Solutions Manual  MyMathLab 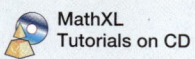 MathXL Tutorials on CD

Add. Then use the commutative property to write another addition problem and find the sum. See Example 1.

1. 3213 + 5715

2. 6344 + 1655

3. 38,204 + 21,020

4. 63,251 + 36,305

Add, carrying as necessary. See Examples 2–5.

5. 67
    + 83

6. 78
    + 36

7. 746
    + 905

8. 621
    + 359

9. 798
    + 206

10. 172
     + 156

11. 7968
     + 1285

12. 1768
     + 8275

13. 7896
     + 3728

14. 9382
     + 7586

15. 3705
       3916
     + 9037

16. 6629
       6076
     + 8218

17.   32
     + 4977

18. 402
     + 9938

19. 3077
          8
     + 421

20. 56
       7721
     + 172

21. 9056
         78
       6089
     + 731

22. 4022
        709
       8621
     + 37

23. 18
        708
       9286
     + 636

24. 1708
        321
         61
     + 8926

Check each sum by adding from bottom to top. If an answer is incorrect, find the correct sum. See Examples 8 and 9.

25. ____
       179
       214
     + 376
       759

26. ____
        17
       296
       713
     + 94
      1220

27. ____
      4713
        28
       615
     + 64
      5420

28. ____
      6 215
        744
         36
     + 4 284
      11,279

Using the map below, find the shortest route between each pair of cities. See Examples 6 and 7.

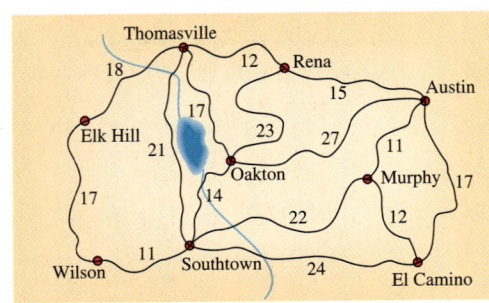

29. Southtown and Rena

30. Elk Hill and Oakton

31. Thomasville and Murphy

32. Austin and Wilson

Solve each application problem.

33. The sale price for a basic auto tune-up is $79, a tire rotation is $24, and an oil change is $19. Find the total cost for all the services.

34. Jane Lim bought an 11-piece set of golf clubs for $120, a dozen golf balls for $9, golf shoes for $45, and a golf glove for $12. How much did she spend in all? (*Source:* Sportmart.)

35. There are 413 women and 286 men on the sales staff. How many people are on the sales staff?

36. One department in an office building has 283 employees while another department has 218 employees. How many employees are in the two departments?

37. According to the latest census, the two states with the highest populations are California with 33,871,648 people and Texas with 20,851,820 people. How many people live in those two states? (*Source:* U.S. Bureau of the Census.)

38. The two states with the smallest populations are Wyoming with 493,782 people and Vermont with 608,827 people. What is the total population of the two states? (*Source:* U.S. Bureau of the Census.)

Find the perimeter of (total distance around) each figure.

39.

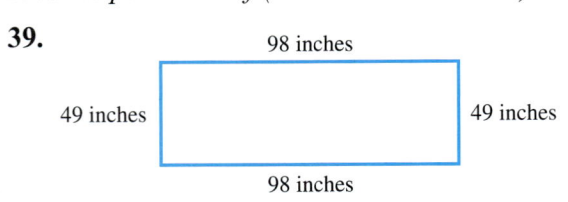

40.

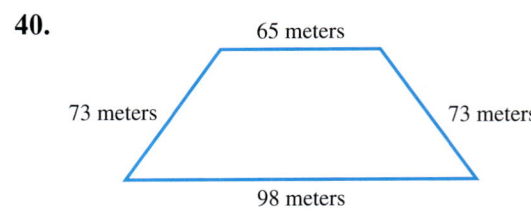

41.

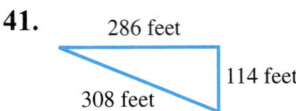

42.

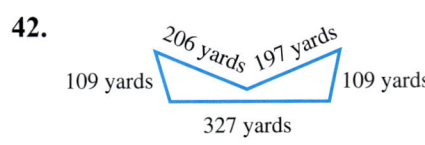

R.2 Subtracting Whole Numbers

Suppose you have $18, and you spend $15 for gasoline. You then have $3 left. There are two different ways of looking at these numbers.

As an addition problem:

$$\underset{\substack{\uparrow \\ \text{Amount} \\ \text{spent}}}{\$15} + \underset{\substack{\uparrow \\ \text{Amount} \\ \text{left}}}{\$3} = \underset{\substack{\uparrow \\ \text{Original} \\ \text{amount}}}{\$18}$$

As a subtraction problem:

$$\underset{\substack{\uparrow \\ \text{Original} \\ \text{amount}}}{\$18} - \underset{\substack{\uparrow \\ \text{Subtraction} \\ \text{symbol}}}{} \underset{\substack{\uparrow \\ \text{Amount} \\ \text{spent}}}{\$15} = \underset{\substack{\uparrow \\ \text{Amount} \\ \text{left}}}{\$3}$$

OBJECTIVES

1. Change addition problems to subtraction problems or the reverse.
2. Identify the minuend, subtrahend, and difference.
3. Subtract when no borrowing is needed.
4. Use addition to check subtraction answers.
5. Subtract with borrowing.
6. Use subtraction to solve application problems.

OBJECTIVE 1 Change addition problems to subtraction problems or the reverse. As this example shows, an addition problem can be changed to a subtraction problem and a subtraction problem can be changed to an addition problem.

EXAMPLE 1 Changing Addition Problems to Subtraction

Change each addition problem to a subtraction problem.

(a) $4 + 1 = 5$

Two subtraction problems are possible, as shown below.

$$5 - 1 = 4 \quad \text{or} \quad 5 - 4 = 1$$

These figures show each subtraction problem.

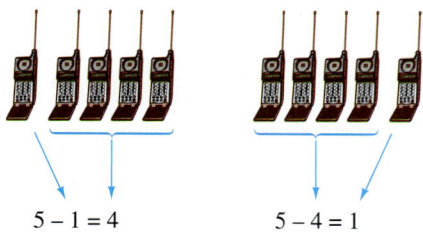

$5 - 1 = 4 \qquad 5 - 4 = 1$

(b) $10 + 17 = 27$

$$27 - 17 = 10 \quad \text{or} \quad 27 - 10 = 17$$

Work Problem 1 at the Side.

EXAMPLE 2 Changing Subtraction Problems to Addition

Change each subtraction problem to an addition problem.

(a) $8 - 3 = 5$

$8 = 3 + 5$

It is also correct to write $8 = 5 + 3$ (using the commutative property).

Continued on Next Page

---

1 Write two subtraction problems for each addition problem.

(a) $4 + 3 = 7$

(b) $6 + 5 = 11$

(c) $150 + 220 = 370$

(d) $623 + 55 = 678$

ANSWERS
1. (a) $7 - 3 = 4$ or $7 - 4 = 3$
   (b) $11 - 5 = 6$ or $11 - 6 = 5$
   (c) $370 - 220 = 150$ or $370 - 150 = 220$
   (d) $678 - 55 = 623$ or $678 - 623 = 55$

2 Write an addition problem for each subtraction problem.

(a) $5 - 3 = 2$

(b) $8 - 3 = 5$

(c) $21 - 15 = 6$

(d) $58 - 42 = 16$

3 Subtract.

(a) $\begin{array}{r} 56 \\ -31 \\ \hline \end{array}$

(b) $\begin{array}{r} 38 \\ -14 \\ \hline \end{array}$

(c) $\begin{array}{r} 378 \\ -235 \\ \hline \end{array}$

(d) $\begin{array}{r} 3927 \\ -2614 \\ \hline \end{array}$

(e) $\begin{array}{r} 5464 \\ -324 \\ \hline \end{array}$

ANSWERS
2. (a) $5 = 3 + 2$ (b) $8 = 3 + 5$
   (c) $21 = 15 + 6$ (d) $58 = 42 + 16$
3. (a) 25   (b) 24   (c) 143   (d) 1313
   (e) 5140

(b) $19 - 14 = 5$
$19 = 14 + 5$

(c) $290 - 130 = 160$
$290 = 130 + 160$

Work Problem 2 at the Side.

OBJECTIVE 2 Identify the minuend, subtrahend, and difference. In subtraction, as in addition, the numbers in a problem have names. For example, in the problem $8 - 5 = 3$, the number 8 is the **minuend**, 5 is the **subtrahend**, and 3 is the **difference** or answer.

$$8 - 5 = 3 \leftarrow \text{Difference (answer)}$$
$$\uparrow \qquad \uparrow$$
$$\text{Minuend} \quad \text{Subtrahend}$$

$\begin{array}{r} 8 \\ -5 \\ \hline 3 \end{array}$ $\leftarrow$ Minuend
$\leftarrow$ Subtrahend
$\leftarrow$ Difference

OBJECTIVE 3 Subtract when no borrowing is needed. Subtract two numbers by lining up the numbers in columns so that the digits in the ones place are in the same column. Next, subtract by columns, starting at the right with the ones column.

EXAMPLE 3 Subtracting Two Numbers without Borrowing

Subtract.

(a) $\begin{array}{r} 53 \\ -21 \\ \hline 32 \end{array}$ — Ones digits are lined up in the same column.

— $3 - 1 = 2$
— 5 tens $-$ 2 tens $=$ 3 tens

(b) $\begin{array}{r} 385 \\ -161 \\ \hline 224 \end{array}$ — Ones digits are lined up.

$\leftarrow 5 - 1 = 4$
— 8 tens $-$ 6 tens $=$ 2 tens
— 3 hundreds $-$ 1 hundred $=$ 2 hundreds

(c) $\begin{array}{r} 9431 \\ -210 \\ \hline 9221 \end{array}$ $\leftarrow 1 - 0 = 1$

— 3 tens $-$ 1 ten $=$ 2 tens
— 4 hundreds $-$ 2 hundreds $=$ 2 hundreds
— 9 thousands $-$ 0 thousands $=$ 9 thousands

Work Problem 3 at the Side.

OBJECTIVE 4 Use addition to check subtraction answers. You can check $8 - 3 = 5$ by *adding* 3 and 5.

$3 + 5 = 8$, so $8 - 3 = 5$ is correct.

EXAMPLE 4 Checking Subtraction

Check each answer.

(a) $\begin{array}{r} 89 \\ -47 \\ \hline 42 \end{array}$

Rewrite as an addition problem, as shown earlier in Example 2.

Subtraction problem $\left\{ \begin{array}{r} 89 \\ -47 \\ \hline 42 \end{array} \right\}$ Addition problem $\begin{array}{r} 47 \\ +42 \\ \hline 89 \end{array}$

Because $47 + 42 = 89$, the subtraction was done correctly.

(b) $72 - 41 = 21$

Rewrite as an addition problem.

$$72 = 41 + 21$$

But, $41 + 21 = 62$, *not* 72, so the subtraction was done *incorrectly*. Rework the original subtraction to get the correct answer of 31. Then, $41 + 31 = 72$.

(c) $\begin{array}{r} 374 \\ -141 \\ \hline 233 \end{array}$ ← Match → $141 + 233 = 374$

The answer checks.

Work Problem 4 at the Side.

OBJECTIVE 5 Subtract with borrowing. If a digit in the minuend is *less* than the one directly below it, **borrowing** is necessary (also called *regrouping*).

EXAMPLE 5 Subtracting with Borrowing

Subtract 19 from 57.

Write the problem in vertical format.

$\begin{array}{r} 5\;7 \\ -1\;9 \end{array}$

In the ones column, 7 is *less* than 9, so, in order to subtract, you must borrow 1 ten from the 5 tens.

5 tens − 1 ten = 4 tens ⟶ $\begin{array}{r} 4\;17 \\ \cancel{5}\;\cancel{7} \\ -1\;9 \end{array}$ ← 1 ten = 10 ones, and 10 + 7 = 17

Now subtract $17 - 9$ in the ones column and then subtract 4 tens − 1 ten in the tens column.

$\begin{array}{r} 4\;17 \\ \cancel{5}\;\cancel{7} \\ -1\;9 \\ \hline 3\;8 \end{array}$ ← Difference

Thus, $57 - 19 = 38$. Check by adding 19 and 38. You should get 57.

Work Problem 5 at the Side.

---

4 Do an addition to decide whether each answer is correct. If incorrect, find the correct answer.

(a) $\begin{array}{r} 65 \\ -23 \\ \hline 42 \end{array}$

(b) $\begin{array}{r} 46 \\ -32 \\ \hline 24 \end{array}$

(c) $\begin{array}{r} 374 \\ -251 \\ \hline 113 \end{array}$

(d) $\begin{array}{r} 7531 \\ -4301 \\ \hline 3230 \end{array}$

5 Subtract.

(a) $\begin{array}{r} 67 \\ -38 \end{array}$

(b) $\begin{array}{r} 97 \\ -29 \end{array}$

(c) $\begin{array}{r} 31 \\ -17 \end{array}$

(d) $\begin{array}{r} 863 \\ -47 \end{array}$

(e) $\begin{array}{r} 762 \\ -157 \end{array}$

ANSWERS
4. (a) correct (b) incorrect; should be 14
   (c) incorrect; should be 123 (d) correct
5. (a) 29 (b) 68 (c) 14 (d) 816 (e) 605

Chapter R Whole Numbers Review

6 Subtract.

(a) 354
   − 82

(b) 457
   − 68

(c) 874
   − 486

(d) 1437
   − 988

(e) 8739
   − 3892

ANSWERS
6. (a) 272 (b) 389 (c) 388 (d) 449
   (e) 4847

EXAMPLE 6 Subtracting with Borrowing

Subtract, borrowing when necessary.

(a) 7856
   − 137

Borrow 1 ten. ⟶ 1 ten = 10 ones; 10 + 6 = 16

```
       4 16
   7 8 5̸ 6̸
  −  1 3 7
  ─────────
   7 7 1 9   ← Difference
```

(b) 635
   − 546

Borrow 1 ten. ⟶ 1 ten = 10 ones; 10 + 5 = 15

```
     2 15
   6 3̸ 5̸           Need to borrow
  − 5 4 6           further because
  ───────           2 is less than 4
         9         in tens column.
```

Borrow 1 hundred. ⟶ 1 hundred = 10 tens; 10 tens + 2 tens = 12 tens

```
   5 12 15
   6̸ 3̸ 5̸
  − 5 4 6
  ───────
       8 9   ← Difference
```

(c) 412
   − 225

```
     0 12
   4 1̸ 2̸           Need to borrow
  − 2 2 5           further because
  ───────           0 is less than 2
         7         in tens column.
```

```
   3 10 12
   4̸ 1̸ 2̸
  − 2 2 5
  ───────
     1 8 7   ← Difference
```

Work Problem 6 at the Side.

Sometimes a minuend has zeros in some of the positions. In such cases, borrowing may be a little more complicated than what we have shown so far.

EXAMPLE 7 Borrowing with Zeros

Subtract.

$$\begin{array}{r} 4607 \\ -\ 3168 \end{array}$$

It is not possible to borrow from the tens position. Instead, you must first borrow from the hundreds position.

Borrow 1 hundred. ⟶ 1 hundred is 10 tens.

```
     5 10
   4 6̸ 0̸ 7
  − 3 1 6 8
```

Now borrow from the tens position.

```
           9          ← 10 tens − 1 ten = 9 tens
       5 10 17        ← 1 ten = 10 ones; 10 + 7 = 17
   4 6̸ 0̸ 7̸
  − 3 1 6 8
  ─────────
           9
```

Continued on Next Page

Complete the problem.

$$\begin{array}{r} \overset{9}{\overset{51017}{4\cancel{6}\cancel{0}\cancel{7}}} \\ -\;3168 \\ \hline 1439 \end{array}$$ ← Difference

Check by adding 1439 and 3168; you should get 4607.

▶ **Work Problem 7 at the Side.** ▶▶▶

EXAMPLE 8 Borrowing with Zeros

Subtract.

(a)   708
       − 149

1 hundred is 10 tens. → Borrow 1 ten.
Borrow 1 hundred. → 1 ten is 10 ones; 10 + 8 = 18

$$\begin{array}{r} 61018 \\ \cancel{7}\cancel{0}\cancel{8} \\ -\;149 \\ \hline 559 \end{array}$$

(b)   380
       − 276

Borrow 1 ten. — 1 ten is 10 ones.

$$\begin{array}{r} 710 \\ 3\cancel{8}\cancel{0} \\ -\;276 \\ \hline 104 \end{array}$$

(c)   9000
       − 6999

$$\begin{array}{r} 99 \\ 8101010 \\ \cancel{9}\cancel{0}\cancel{0}\cancel{0} \\ -\;6999 \\ \hline 2001 \end{array}$$

▶ **Work Problem 8 at the Side.** ▶▶▶

Recall that an answer to a subtraction problem can be checked by adding.

EXAMPLE 9 Checking Subtraction

Use addition to check each answer.

    Check

(a)   613
       − 275
       ─────
         338   Matches

        275
       + 338
       ─────
         613   Correct

— **Continued on Next Page**

---

7 Subtract.

(a)   308
       − 285

(b)   206
       − 148

(c)   5073
       − 1632

8 Subtract.

(a)   405
       − 267

(b)   370
       − 163

(c)   1570
       − 983

(d)   7001
       − 5193

(e)   4000
       − 1782

ANSWERS
7. **(a)** 23 **(b)** 58 **(c)** 3441
8. **(a)** 138 **(b)** 207 **(c)** 587 **(d)** 1808
   (e) 2218

1222 Chapter R Whole Numbers Review

9 Use addition to check each answer. If an answer is incorrect, find the correct answer.

(a) 425
   − 368
   ─────
     57

(b) 670
   − 439
   ─────
    241

(c) 14,726
   − 8 839
   ───────
      5 887

(b) 1915   Check
   − 1635   1635
   ──────   Matches + 280
     280   ──────
                   1915   Correct

(c) 15,803   Check
   − 7 325   7 325
   ─────── Does not match + 8 578
     8 578   ──────
                            15,903   Error

Rework the original problem to get the correct answer, 8478.

◀◀◀ **Work Problem 9 at the Side.**

OBJECTIVE 6 Use subtraction to solve application problems.

EXAMPLE 10 Applying Subtraction Skills

Diana Lopez drives a United Parcel Service delivery truck. Using the table below, decide how many more deliveries were made by Lopez on Monday than on Thursday.

PACKAGE DELIVERY (LOPEZ)

Day	Number of Deliveries
Monday	137
Tuesday	126
Wednesday	119
Thursday	89
Friday	147
Saturday	0

10 Use the table from Example 10 at the right. How many more deliveries did Lopez make

(a) on Friday than on Tuesday?

(b) on Tuesday than on Wednesday?

(c) on Thursday than on Saturday?

Lopez made 137 deliveries on Monday, but had only 89 deliveries on Thursday. Find how many more deliveries were made on Monday than on Thursday by subtracting 89 from 137.

  137 ← Deliveries on Monday
− 89 ← Deliveries on Thursday
─────
   48 ← More deliveries on Monday

Lopez made 48 more deliveries on Monday than she made on Thursday.

◀◀◀ **Work Problem 10 at the Side.**

ANSWERS
9. (a) correct   (b) incorrect; should be 231
   (c) correct
10. (a) 21 more deliveries
    (b) 7 more deliveries
    (c) 89 more deliveries

R.2 Exercises

FOR EXTRA HELP Addison-Wesley Math Tutor Center  MathXL Student's Solutions Manual 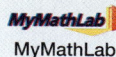 MyMathLab 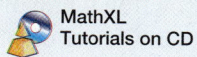 MathXL Tutorials on CD

Use addition to check each subtraction. If an answer is incorrect, find the correct answer. See Examples 3 and 4.

1. 89
    − 27

    63

2. 47
    − 35

    13

3. 382
    − 261

    131

4. 838
    − 516

    322

Subtract, borrowing when necessary. See Examples 5–8.

5. 36
    − 28

6. 97
    − 39

7. 83
    − 58

8. 65
    − 28

9. 45
    − 29

10. 93
    − 37

11. 719
    − 658

12. 916
    − 618

13. 771
    − 252

14. 973
    − 788

15. 9861
    − 684

16. 6171
    − 1182

17. 9988
    − 2399

18. 3576
    − 1658

19. 38,335
    − 29,476

20. 61,278
    − 3 559

21. 40
    − 37

22. 80
    − 73

23. 60
    − 37

24. 70
    − 27

25. 6020
    − 4078

26. 7050
    − 6045

27. 8503
    − 2816

28. 16,004
    − 5 087

29. 80,705
    − 61,667

30. 72,000
    − 44,234

31. 66,000
    − 444

32. 77,000
    − 308

33. 20,080
    − 96

34. 80,056
    − 69

Use addition to check each subtraction. If an answer is incorrect, find the correct answer. See Example 9.

35. 3070
    − 576

    2596

36. 1439
    − 1169

    270

37. 27,600
    − 807

    26,793

38. 34,021
    − 33,708

    727

Solve each application problem. See Example 10.

39. A man burns 103 calories during 30 minutes of bowling while a woman burns 88 calories. How many fewer calories does a woman burn than a man?

40. Lynn Couch had $553 in her checking account. She wrote a check for $308 for school fees. How much is left in her account?

41. An airplane was carrying 254 passengers. When it landed in Atlanta, 133 passengers got off. How many passengers were left on the plane?

42. On Tuesday, 5822 people went to a soccer game, and on Friday, 7994 people went to a soccer game. How many more people went to the game on Friday?

This table shows the median yearly earnings for various occupations. Use the table to answer Exercises 43 and 44.

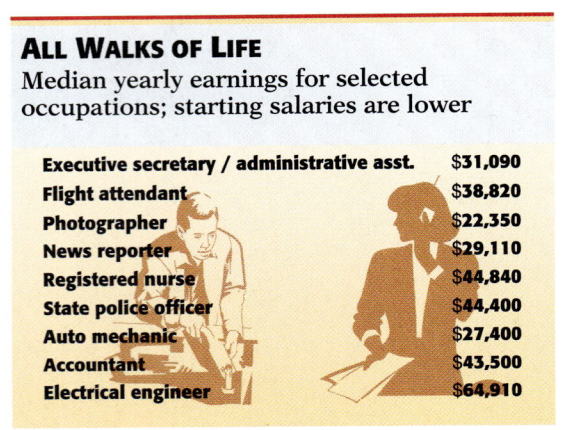

ALL WALKS OF LIFE
Median yearly earnings for selected occupations; starting salaries are lower

Occupation	Earnings
Executive secretary / administrative asst.	$31,090
Flight attendant	$38,820
Photographer	$22,350
News reporter	$29,110
Registered nurse	$44,840
State police officer	$44,400
Auto mechanic	$27,400
Accountant	$43,500
Electrical engineer	$64,910

Source: U.S. Department of Labor. *Occupational Outlook Handbook.*

43. (a) Identify the occupations with the highest and lowest earnings.

(b) What is the difference in the earnings for these two occupations?

44. (a) How much less are the earnings of an auto mechanic than an executive secretary?

(b) How much more are the earnings of an accountant than a flight attendant?

45. Downtown Toronto's skyline is dominated by the CN Tower, which rises 1815 feet. The Sears Tower in Chicago is 1450 feet high. Find the difference in height between the two structures. (*Source:* SkyscraperPage.com)

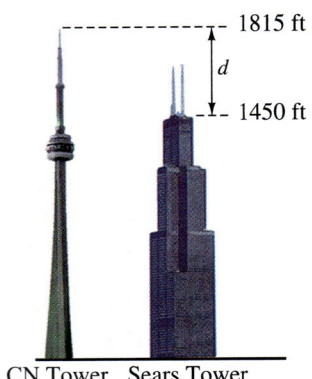

CN Tower   Sears Tower

46. The fastest animal in the world, the peregrine falcon, dives at 185 miles per hour. A Boeing 747 cruises at 580 miles per hour. How much faster does the plane fly than the falcon? (*Source: Top 10 of Everything.*)

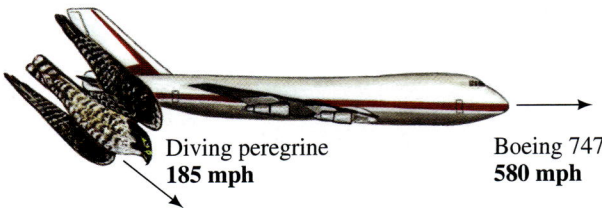

Diving peregrine
185 mph

Boeing 747
580 mph

R.3 Multiplying Whole Numbers

OBJECTIVES
1. Identify the parts of a multiplication problem.
2. Do chain multiplications.
3. Multiply by single-digit numbers.
4. Multiply quickly by numbers ending in zeros.
5. Multiply by numbers having more than one digit.
6. Use multiplication to solve application problems.

Suppose we want to know the total number of computers in a computer lab. The computers are arranged in four rows with three computers in each row. Adding the number 3 a total of 4 times gives 12.

$$3 + 3 + 3 + 3 = 12$$

This result can be illustrated by the figure below.

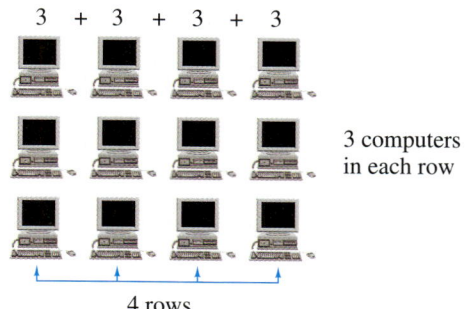

3 computers in each row

4 rows

OBJECTIVE 1 Identify the parts of a multiplication problem. Multiplication is a shortcut for repeated addition. The numbers being multiplied are called **factors**. The answer is called the **product**. For example, the product of 3 and 4 can be written with the symbol ×, a raised dot, parentheses, or, in computer work, an asterisk.

$$\begin{array}{r} 3 \\ \times\ 4 \\ \hline 12 \end{array}$$ ← Factor (also called *multiplicand*)
← Factor (also called *multiplier*)
← Product (answer)

$3 \times 4 = 12$ $3 \cdot 4 = 12$ $(3)(4) = 12$ $3 * 4 = 12$

Raised dot   In computer work

Work Problem 1 at the Side.

Commutative Property of Multiplication

By the **commutative property of multiplication**, changing the *order* of two factors does not change the product.

For example: $3 \times 5 = 15$ and $5 \times 3 = 15$

Both products are 15.

CAUTION
Remember, addition also has a commutative property. For example, $4 + 2$ has the same sum as $2 + 4$. Subtraction, however, is *not* commutative.

EXAMPLE 1 Multiplying Two Numbers

Multiply. (Remember that a raised dot means to multiply.) Do the work mentally.

(a) $3 \times 4 = \mathbf{12}$ By the commutative property, $4 \times 3 = 12$ also.

(b) $6 \cdot 0 = \mathbf{0}$ The product of any number and 0 is 0; if you give no money to each of 6 relatives, you give no money.

(c) $(4)(8) = \mathbf{32}$ By the commutative property, $(8)(4) = 32$ also.

Work Problem 2 at the Side.

---

1 Identify the factors and the product in each multiplication problem.

(a) $3 \times 6 = 18$

(b) $32 = 8 \times 4$

(c) $5 \cdot 7 = 35$

(d) $(3)(9) = 27$

2 Multiply. Do the work mentally. Then use the commutative property to write another multiplication problem and find the product.

(a) 4×7

(b) 0×9

(c) $8 \cdot 6$

(d) $5 \cdot 5$

(e) $(3)(8)$

ANSWERS
1. (a) factors: 3, 6; product: 18
   (b) factors: 8, 4; product: 32
   (c) factors: 5, 7; product: 35
   (d) factors: 3, 9; product: 27
2. (a) 28; $7 \times 4 = 28$ (b) 0; $9 \times 0 = 0$
   (c) 48; $6 \cdot 8 = 48$ (d) 25; no change
   (e) 24; $(8)(3) = 24$

3 Multiply.

(a) $2 \times 3 \times 4$

(b) $6 \cdot 1 \cdot 5$

(c) $(8)(3)(0)$

(d) $3 \times 3 \times 7$

(e) $4 \cdot 2 \cdot 8$

(f) $(2)(2)(9)$

ANSWERS
3. (a) 24   (b) 30   (c) 0   (d) 63
   (e) 64   (f) 36

OBJECTIVE 2 Do chain multiplications. Some multiplications involve more than two factors.

> **Associative Property of Multiplication**
> By the **associative property of multiplication,** changing the *grouping* of factors does not change the product.

EXAMPLE 2 Multiplying Three Numbers

Multiply $2 \times 3 \times 5$.

$$(2 \times 3) \times 5 \quad \text{Parentheses tell what to do first.}$$
$$6 \times 5 = 30$$

Also,

$$2 \times (3 \times 5)$$
$$2 \times 15 = 30$$

By the associative property, either grouping results in the same product.

> **Calculator Tip** The calculator approach to Example 2 above uses chain calculations. Notice that you can enter *all* the factors before pressing the $=$ key.
>
> $2 \; \boxed{\times} \; 3 \; \boxed{\times} \; 5 \; \boxed{=} \; 30$

◀◀◀ **Work Problem 3 at the Side.**

OBJECTIVE 3 Multiply by single-digit numbers. Carrying may be needed in multiplication problems with larger factors.

EXAMPLE 3 Carrying with Multiplication

Multiply.

(a)   53
    × 4

Start by multiplying in the ones column. Multiply 4 times 3 ones.

```
   1
  53
 × 4       4 × 3 = 12
   2
```
Carry 1 ten to the tens column.
Write 2 ones in the ones column.

Next, multiply 4 times 5 tens.

```
   1
  53
 × 4       4 × 5 tens = 20 tens
   2
```

Add the 1 that was carried to the tens column.

```
   1
  53
 × 4
 212       20 tens + 1 ten = 21 tens
```

Continued on Next Page

(b) 724
   × 5

Work as shown below.

```
    12
   724
 ×   5
  3620  ← 5 × 4 = 20 ones; write 0 ones and carry 2 tens.
         ── 5 × 2 tens = 10 tens; add the 2 carried tens to get 12 tens;
            write 2 tens and carry 1 hundred.
         ── 5 × 7 hundreds = 35 hundreds; add the
            1 carried hundred to get 36 hundreds.
```

Work Problem 4 at the Side.

OBJECTIVE 4 Multiply quickly by numbers ending in zeros. The product of two whole number factors is also called a **multiple** of either factor. For example, since 4 • 2 = 8, the number 8 is a multiple of 4 and 8 is also a multiple of 2. Multiples of 10 are very useful when multiplying. A *multiple of 10* is a whole number that ends in 0, such as 10, 20, or 30; 100, 200, or 300; 1000, 2000, or 3000; and so on. There is a short way to multiply by multiples of 10. Look at the following examples.

$$26 \times 1 = 26$$
$$26 \times 10 = 260$$
$$26 \times 100 = 2600$$
$$26 \times 1000 = 26{,}000$$

Do you see a pattern in the multiplications? These examples suggest the following rule.

> **Multiplying by Multiples of 10**
>
> To multiply a whole number by 10, by 100, or by 1000, attach one, two, or three zeros to the right of the whole number.

EXAMPLE 4 Multiplying by Multiples of 10

Multiply.

(a) 59 × 10 = 59**0**
              ↑
              └── Attach 0

(b) 74 × 100 = 74**00**
                ↑↑
                └── Attach 00

(c) 803 × 1000 = 803**,000**
                  ↑↑↑
                  └── Attach 000

Work Problem 5 at the Side.

You can also find the product of other multiples of ten by attaching zeros.

---

4 Multiply.

(a) 52
    × 5

(b) 79
    × 0

(c) 862
    × 9

(d) 2831
    ×   7

(e) 4714
    ×   8

5 Multiply by attaching zeros.

(a) 45 × 10

(b) 102 × 100

(c) 571 × 1000

ANSWERS
4. (a) 260 (b) 0 (c) 7758 (d) 19,817
   (e) 37,712
5. (a) 450 (b) 10,200 (c) 571,000

Chapter R Whole Numbers Review

6 Multiply by attaching zeros.

(a) 14 × 50

(b) 68 × 400

(c) 180
    × 30

(d) 6100
    × 90

(e)   800
    × 200

7 Complete each multiplication.

(a)   35
    × 54
    ———
    140
    175
    ———

(b)   76
    × 49
    ————
     684
     304
    ————

ANSWERS
6. (a) 700 (b) 27,200 (c) 5400
   (d) 549,000 (e) 160,000
7. (a) 1890 (b) 3724

EXAMPLE 5 Multiplying with Other Multiples of 10

Multiply.

(a) 75 × 3000

Multiply 75 by 3, and then attach three zeros.

$$\begin{array}{r} 75 \\ \times\ 3 \\ \hline 225 \end{array}$$

75 × 3000 = 225,000

(b) 150 × 70

Multiply 15 by 7, and then attach two zeros.

$$\begin{array}{r} 15 \\ \times\ 7 \\ \hline 105 \end{array}$$

150 × 70 = 10,500

▶◀◀◀ **Work Problem 6 at the Side.**

OBJECTIVE 5 Multiply by numbers having more than one digit.

The next example shows multiplication when both factors have more than one digit.

EXAMPLE 6 Multiplying with More Than One Digit

Multiply 46 and 23.
First multiply 46 by 3.

$$\begin{array}{r} 1 \\ 46 \\ \times\ \ 3 \\ \hline 138 \end{array}$$ ← 46 × 3 = 138

Now multiply 46 by 20.

$$\begin{array}{r} 1 \\ 46 \\ \times\ 20 \\ \hline 138 \\ 920 \end{array}$$ ← 46 × 20 = 920

Add the results.

$$\begin{array}{r} 46 \\ \times\ 23 \\ \hline 138 \\ +\ 920 \\ \hline 1058 \end{array}$$

← 46 × 3
← 46 × 20
← Add to find the product.

Both 138 and 920 are called *partial products*. To save time, the 0 in 920 is usually not written.

$$\begin{array}{r} 46 \\ \times\ 23 \\ \hline 138 \\ 92 \\ \hline 1058 \end{array}$$

← 0 not written. Be very careful to place the 2 in the tens column.

▶◀◀◀ **Work Problem 7 at the Side.**

EXAMPLE 7 Using Partial Products

Multiply.

(a)
```
      2 3 3
  ×   1 3 2
  ─────────
      4 6 6
    6 9 9        Tens lined up
  2 3 3          Hundreds lined up
  ─────────
  3 0,7 5 6  ← Product
```

(b)
```
    5 3 8
  ×   4 6
```

First multiply by 6.
```
        2 4
      5 3 8
  ×     4 6      Carrying is
  ─────────     needed here.
      3 2 2 8
```

Now multiply by 4, being careful to line up the tens.
```
      1 3
      2 4
      5 3 8
  ×     4 6
  ─────────
      3 2 2 8  ┐
      2 1 5 2  ┘ — Add the partial products.
  ─────────
      2 4,7 4 8
```

Work Problem 8 at the Side.

When 0 appears in the multiplier, be sure to move the partial product to the left to account for the position held by the 0.

EXAMPLE 8 Multiplication with Zeros

Multiply.

(a)
```
      1 3 7
  ×   3 0 6
  ─────────
      8 2 2
    0 0 0        Tens lined up
  4 1 1          Hundreds lined up
  ─────────
  4 1,9 2 2
```

(b)
```
      1 4 0 6                  1 4 0 6
  ×   2 0 0 1              ×   2 0 0 1
  ─────────                ─────────
      1 4 0 6                  1 4 0 6
    0 0 0 0  ← 0 to line up tens
  0 0 0 0    ← 0 to line up hundreds   2 8 1 2 0 0  ← Zeros are written
  2 8 1 2                                             so that the
  ─────────                ─────────                  partial product
  2,8 1 3,4 0 6            2,8 1 3,4 0 6              2812 starts in
                                                      the thousands
                                                      column.
```

> **CAUTION**
> In Example 8(b) above, in the solution on the right, be careful to insert zeros so that thousands are lined up in the thousands column.

Continued on Next Page

---

8 Multiply.

(a)
```
      3 8
  ×   1 5
```

(b)
```
      3 1
  ×   4 3
```

(c)
```
      6 7
  ×   5 9
```

(d)
```
    2 3 4
  ×   7 3
```

(e)
```
    8 3 5
  × 1 8 9
```

ANSWERS
8. (a) 570 (b) 1333 (c) 3953
   (d) 17,082 (e) 157,815

Chapter R Whole Numbers Review

9 Multiply.

(a) $\begin{array}{r} 28 \\ \times\ 60 \\ \hline \end{array}$

(b) $\begin{array}{r} 817 \\ \times\ 30 \\ \hline \end{array}$

(c) $\begin{array}{r} 481 \\ \times\ 206 \\ \hline \end{array}$

(d) $\begin{array}{r} 3526 \\ \times\ 6002 \\ \hline \end{array}$

> **Work Problem 9 at the Side.**

OBJECTIVE 6 Use multiplication to solve application problems.

EXAMPLE 9 Applying Multiplication Skills

Find the total cost of 24 cell phones priced at $59 each.

Approach To find the cost of all the phones, multiply the cost of one phone ($59) by the number of phones (24).

Solution Multiply $59 by 24.

$$\begin{array}{r} 59 \\ \times\ 24 \\ \hline 236 \\ 118 \\ \hline 1416 \end{array}$$

The total cost of 24 cell phones is $1416.

> **Calculator Tip** If you are using a calculator for Example 9 above, press the following keys.
>
> 59 ⊗ 24 ⊜ 1416

> **Work Problem 10 at the Side.**

10 Find the total cost of these items.

(a) 36 months of cable TV costing $48 each month

(b) 28 laptop computers priced at $1090 each

(c) 60 months of car payments at $289 per month

ANSWERS

9. (a) 1680 (b) 24,510 (c) 99,086
   (d) 21,163,052
10. (a) $1728 (b) $30,520 (c) $17,340

R.3 Exercises

Find each product. Try to do the work mentally. See Examples 1 and 2.

1. $3 \times 1 \times 3$
2. $2 \times 8 \times 2$
3. $9 \times 1 \times 7$
4. $2 \times 4 \times 5$
5. $9 \cdot 5 \cdot 0$

6. $6 \cdot 0 \cdot 8$
7. $4 \cdot 1 \cdot 6$
8. $1 \cdot 5 \cdot 7$
9. $(2)(3)(6)$
10. $(4)(1)(9)$

Multiply. See Examples 3–8.

11. 35×7
12. 76×9
13. 28×6
14. 83×5
15. 3182×6

16. 7326×5
17. $36{,}921 \times 7$
18. $28{,}116 \times 4$
19. 125×30
20. 246×50

21. 1485×30
22. 8522×50
23. 900×300
24. 400×700
25. $43{,}000 \times 2000$

26. $11{,}000 \times 9000$
27. 68×22
28. 82×32
29. 83×45
30. $(43)(27)$

31. $(32)(475)$
32. $(67)(218)$
33. $(729)(45)$
34. $(681)(47)$
35. 538×342

36. 3228×751
37. 8162×407
38. 528×106
39. 6310×3008
40. 3533×5001

Solve each application problem. See Example 9.

41. Giant kelp plants in the ocean can grow 18 inches each day. How much could kelp grow in two weeks? How much could it grow in a 30-day month? (*Source: Natural Bridges State Park, CA.*)

42. A hospital has 20 bottles of thyroid medication, with each bottle containing 2500 tablets. How many of these tablets does the hospital have in all?

43. There are 12 tomato plants to a flat. If a garden center has 48 flats, find the total number of tomato plants.

44. A hummingbird's wings beat about 65 times per second, a chickadee's wings about 27 times per second. How many times does each bird's wings beat in one minute? (*Source: Birder's Handbook.*)

45. A new Saturn automobile gets 38 miles per gallon on the highway. How many miles can it travel on 11 gallons of gas?

46. Find the total cost of 16 gallons of paint at $18 per gallon.

Ruby-throated hummingbird
65 wing beats per second

Use addition, subtraction, or multiplication to solve each problem.

47. The distance from Reno, Nevada, to the Atlantic Ocean is 2695 miles, while the distance from Reno to the Pacific Ocean is 255 miles. How much farther is it to the Atlantic Ocean than it is to the Pacific Ocean? If you make three round trips from Reno to the Atlantic Ocean, how many frequent flier miles will you earn?

48. The largest living land mammal is the African elephant, and the largest mammal of all time is the blue whale. An African bull elephant may weigh 15,225 pounds and a blue whale may weigh 12 to 25 times that amount. Find the range of weights for the blue whale and the difference between the lightest and heaviest. (*Source: Big Book of Knowledge.*)

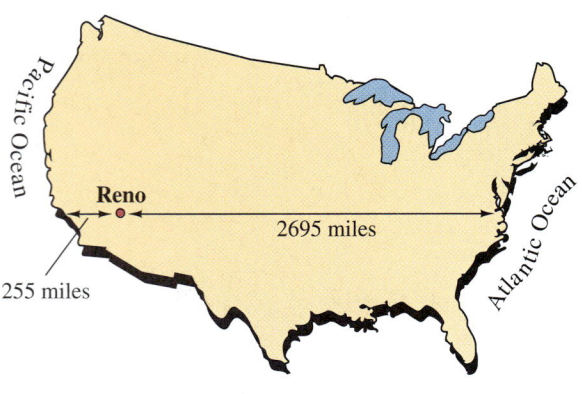

49. A high-fat meal contains 1406 calories, while a low-fat meal contains 348 calories. How many more calories are in seven high-fat meals than in seven low-fat meals?

50. Dannie Sanchez bought four tires at $110 each, two seat covers at $49 each, and six socket wrenches at $3 each. Find the total amount that he spent.

R.4 Dividing Whole Numbers

Suppose $12 is to be divided into 3 equal parts. Each part would be $4, as shown here.

3 equal parts

OBJECTIVES

1. Write division problems in three ways.
2. Identify the parts of a division problem.
3. Divide 0 by a number.
4. Recognize that division by 0 is undefined.
5. Divide a number by itself.
6. Use short division.
7. Use multiplication to check quotients.
8. Use tests for divisibility.

OBJECTIVE 1 Write division problems in three ways. Just as $3 \cdot 4$, 3×4, and $(3)(4)$ are different ways of writing the multiplication of 3 and 4, there are several ways to write 12 divided by 3.

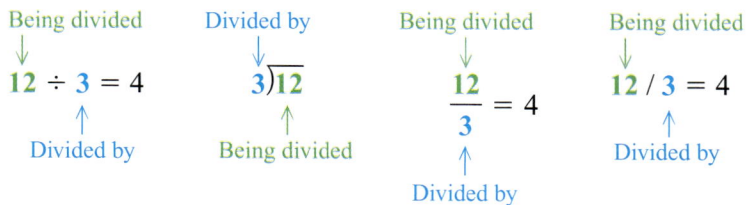

We will use three division symbols, $\div$, $\overline{)}$, and —. In algebra the bar, —, is frequently used. In computer science the slash, /, is used.

EXAMPLE 1 Using Division Symbols

Rewrite each division using two other symbols.

(a) $20 \div 4 = 5$

This division can also be written $4\overline{)20}^{\,5}$ or $\dfrac{20}{4} = 5$

(b) $\dfrac{18}{6} = 3$ can also be written $18 \div 6 = 3$ or $6\overline{)18}^{\,3}$

(c) $5\overline{)40}^{\,8}$ can also be written $40 \div 5 = 8$ or $\dfrac{40}{5} = 8$

▶▶▶ **Work Problem 1 at the Side.**

OBJECTIVE 2 Identify the parts of a division problem. In division, the number being divided is the **dividend**, the number divided by is the **divisor**, and the answer is the **quotient**.

$$\text{dividend} \div \text{divisor} = \text{quotient}$$

$$\text{divisor}\overline{)\text{dividend}}^{\,\text{quotient}} \qquad \dfrac{\text{dividend}}{\text{divisor}} = \text{quotient}$$

EXAMPLE 2 Identifying the Parts in a Division Problem

Identify the dividend, divisor, and quotient.

(a) $35 \div 7 = 5$

$35 \div 7 = 5 \leftarrow \text{Quotient}$

↗ ↖

Dividend Divisor

Continued on Next Page

---

1 Rewrite each division using two other symbols.

(a) $48 \div 6 = 8$

(b) $24 \div 6 = 4$

(c) $9\overline{)36}^{\,4}$

(d) $\dfrac{42}{6} = 7$

ANSWERS

1. (a) $6\overline{)48}^{\,8}$ and $\dfrac{48}{6} = 8$

   (b) $6\overline{)24}^{\,4}$ and $\dfrac{24}{6} = 4$

   (c) $36 \div 9 = 4$ and $\dfrac{36}{9} = 4$

   (d) $6\overline{)42}^{\,7}$ and $42 \div 6 = 7$

2 Identify the dividend, divisor, and quotient.

(a) $10 \div 2 = 5$

(b) $6 = 30 \div 5$

(c) $\dfrac{28}{7} = 4$

(d) $2\overline{)36}^{\,18}$

3 Divide.

(a) $0 \div 9$

(b) $\dfrac{0}{36}$

(c) $57\overline{)0}$

4 Write each division problem as a multiplication problem.

(a) $6\overline{)18}^{\,3}$

(b) $\dfrac{28}{4} = 7$

(c) $48 \div 8 = 6$

ANSWERS
2. (a) dividend: 10; divisor: 2; quotient: 5
   (b) dividend: 30; divisor: 5; quotient: 6
   (c) dividend: 28; divisor: 7; quotient: 4
   (d) dividend: 36; divisor: 2; quotient: 18
3. all 0
4. (a) $6 \cdot 3 = 18$ (b) $4 \cdot 7 = 28$
   (c) $8 \cdot 6 = 48$

(b) $\dfrac{100}{20} = 5$ Dividend

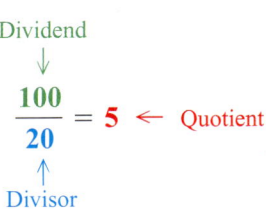

(c) $12\overline{)72}^{\,6}$

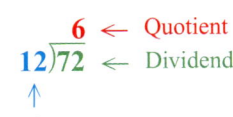

Work Problem 2 at the Side.

OBJECTIVE 3 Divide 0 by a number. If no money, or $0, is divided equally among five people, each person gets $0. There is a general rule for dividing 0.

Dividing 0
When 0 is divided by any other number (except 0), the quotient is 0.

EXAMPLE 3 Dividing 0 by a Number

Divide.

(a) $0 \div 12 = \mathbf{0}$

(b) $0 \div 1728 = \mathbf{0}$

(c) $\dfrac{0}{375} = \mathbf{0}$

(d) $129\overline{)0}^{\,\mathbf{0}}$

Work Problem 3 at the Side.

Recall that a subtraction such as $8 - 3 = 5$ can be written as the addition $8 = 3 + 5$. In a similar way, any division can be written as a multiplication. For example, $12 \div 3 = 4$ can be written as

$3 \times 4 = 12$ or, by the commutative property, $4 \times 3 = 12$

EXAMPLE 4 Changing Division to Multiplication

Change each division to multiplication.

(a) $\dfrac{20}{4} = 5$ becomes $4 \cdot 5 = 20$

(b) $8\overline{)48}^{\,6}$ becomes $8 \cdot 6 = 48$

(c) $72 \div 9 = 8$ becomes $9 \cdot 8 = 72$

Work Problem 4 at the Side.

Section R.4 Dividing Whole Numbers **1235**

OBJECTIVE 4 Recognize that division by 0 is undefined. Division by 0 cannot be done. To see why, try to find the answer to this division.

$$9 \div 0 = ?$$

As we have just seen, any division problem can be changed to a multiplication problem so that

$$\text{divisor} \cdot \text{quotient} = \text{dividend}.$$

If you convert the problem $9 \div 0 = ?$ to its multiplication counterpart, it reads as follows.

$$0 \cdot ? = 9$$

You already know that 0 times any number must always equal 0. Try any number you like to replace the **?** and you'll always get 0 instead of 9. Therefore, the division problem $9 \div 0$ *cannot* be done. Mathematicians say it is *undefined* and have agreed never to divide by 0. However, $0 \div 9$ *can* be done. Check by rewriting it as a multiplication problem.

$$0 \div 9 = 0 \quad \text{because} \quad 0 \cdot 9 = 0 \quad \text{is true.}$$

> **Dividing by 0**
>
> Dividing by 0 cannot be done. We say that division by 0 is *undefined*. It is impossible to compute an answer.

EXAMPLE 5 Dividing by 0 Is Undefined

All of these divisions are *undefined*.

(a) $\dfrac{6}{0}$ is undefined.

(b) $0\overline{)8}$ is undefined.

(c) $18 \div 0$ is undefined.

(d) $\dfrac{0}{0}$ is undefined.

> **Division Involving 0**
>
> $$\dfrac{0}{\text{nonzero number}} = 0 \quad \textbf{but} \quad \dfrac{\text{number}}{0} \text{ is } undefined.$$

> **CAUTION**
>
> When 0 is the divisor in a problem, write *undefined*. You can never divide by 0.

Work Problem 5 at the Side.

Calculator Tip Try these two problems on your calculator. Jot down your answers.

$$9 \div 0 = \underline{\qquad} \qquad 0 \div 9 = \underline{\qquad}$$

When you try to divide by 0, the calculator cannot do it, so it shows the word "Error" in the display, or the letters "ERR" or "E" (for "error").

---

5 Find the quotient whenever possible.

(a) $\dfrac{8}{0}$

(b) $\dfrac{0}{8}$

(c) $0\overline{)32}$

(d) $32\overline{)0}$

(e) $100 \div 0$

(f) $0 \div 100$

ANSWERS

5. (a) undefined (b) 0 (c) undefined
   (d) 0 (e) undefined (f) 0

Chapter R Whole Numbers Review

6 Divide.

(a) $5 \div 5$

(b) $14\overline{)14}$

(c) $\dfrac{37}{37}$

7 Divide.

(a) $2\overline{)18}$

(b) $3\overline{)39}$

(c) $4\overline{)88}$

(d) $2\overline{)462}$

OBJECTIVE 5 Divide a number by itself. What happens when a number is divided by itself? For example, $4 \div 4$ or $97 \div 97$?

Dividing a Number by Itself
When a nonzero number is divided by itself, the quotient is 1.

EXAMPLE 6 Dividing a Nonzero Number by Itself

Divide.

(a) $16 \div 16 = \mathbf{1}$

(b) $32\overline{)32}$ with quotient **1**

(c) $\dfrac{57}{57} = \mathbf{1}$

Work Problem 6 at the Side.

OBJECTIVE 6 Use short division. Short division is a quick method of dividing a number by a one-digit divisor.

EXAMPLE 7 Using Short Division

Divide $2\overline{)86}$.

First, divide 8 by 2.

$2\overline{)86}$ with 4 above the 8; $\dfrac{8}{2} = 4$

Next, divide 6 by 2.

$2\overline{)86}$ with 43 above; $\dfrac{6}{2} = 3$

The quotient is 43.

Work Problem 7 at the Side.

When two numbers do *not* divide evenly, the leftover portion is called the **remainder.** The remainder is always *less* than the divisor.

EXAMPLE 8 Using Short Division with a Remainder

Divide 147 by 4.

Rewrite the problem. $4\overline{)147}$

Because 1 cannot be divided by 4, divide 14 by 4.

$4\overline{)14^27}$ with 3 above; $\dfrac{14}{4} = 3$ with 2 left over

Continued on Next Page

ANSWERS
6. all 1
7. (a) 9   (b) 13   (c) 22   (d) 231

Next, divide 27 by 4. The final number left over is the remainder. Write the remainder to the side. "R" stands for remainder.

$$4\overline{)14^27} \quad \frac{3\ 6\ \mathbf{R}3}{} \qquad \frac{27}{4} = 6 \text{ with 3 left over}$$

The quotient is 36 **R**3.

▶ **Work Problem 8 at the Side.** ▶▶▶

EXAMPLE 9 Dividing with a Remainder

Divide 1809 by 7.
   Divide 7 into 18.

$$7\overline{)18^409} \quad \frac{2}{} \qquad \frac{18}{7} = 2 \text{ with 4 left over}$$

Divide 7 into 40.

$$7\overline{)18^40^59} \quad \frac{2\ 5}{} \qquad \frac{40}{7} = 5 \text{ with 5 left over}$$

Divide 7 into 59.

$$7\overline{)18^40^59} \quad \frac{2\ 5\ 8\ \mathbf{R}3}{} \qquad \frac{59}{7} = 8 \text{ with 3 left over}$$

▶ **Work Problem 9 at the Side.** ▶▶▶

OBJECTIVE 7 Use multiplication to check quotients. Check the answer to a division problem as follows.

> **Checking Division**
>
> (divisor × quotient) + remainder = dividend
>
> Parentheses tell you what to do first. In this case, multiply the divisor by the quotient first and then add the remainder.

EXAMPLE 10 Using Multiplication to Check Division

Check each quotient.

(a) $5\overline{)458}$ with quotient 91 **R**3

(divisor × quotient) + remainder = dividend

$$(5 \times 91) + 3$$

$$455 + 3 = 458$$

↑ Matches original dividend
so the division was done correctly

Continued on Next Page

---

8 Divide.

(a) $2\overline{)225}$

(b) $3\overline{)275}$

(c) $4\overline{)538}$

(d) $\dfrac{819}{5}$

9 Divide.

(a) $5\overline{)937}$

(b) $\dfrac{675}{7}$

(c) $3\overline{)1885}$

(d) $8\overline{)1135}$

ANSWERS
8. (a) 112 **R**1 (b) 91 **R**2 (c) 134 **R**2
   (d) 163 **R**4
9. (a) 187 **R**2 (b) 96 **R**3 (c) 628 **R**1
   (d) 141 **R**7

10 Check each division. If a quotient is incorrect, find the correct quotient.

(a) $3\overline{)115}$ quotient 38 **R**1

(b) $8\overline{)743}$ quotient 92 **R**2

(c) $4\overline{)1312}$ quotient 328

(d) $5\overline{)2033}$ quotient 46 **R**3

ANSWERS

10. (a) correct
    (b) incorrect; should be 92 **R**7
    (c) correct
    (d) incorrect; should be 406 **R**3

(b) $6\overline{)1258}$ quotient 29 **R**4

(divisor × quotient) + remainder = dividend

$(6 \times 29) + 4$

$174 + 4 = 178$

Does **not** match original dividend of 1258

The quotient does *not* check. Rework the original problem to get the correct quotient, 209 **R**4. Then, to check, $(6 \times 209) + 4 = 1254 + 4 = 1258$, the original dividend.

> **CAUTION**
> A common error is forgetting to add the remainder. Be sure to add any remainder when checking a division problem.

◀◀◀ Work Problem 10 at the Side.

OBJECTIVE 8 Use tests for divisibility. It is often important to know whether a number is divisible by another number. You will find this useful in **Chapter 4** when working with fractions.

Divisibility
A whole number is *divisible* by another whole number if the remainder is 0.

There are some quick tests you can use to decide whether one number is divisible by another.

Tests for Divisibility
A number is divisible by

2	if it ends in 0, 2, 4, 6, or 8.
3	if the sum of its digits is divisible by 3.
4	if the last two digits make a number that is divisible by 4.
5	if it ends in 0 or 5.
6	if it is divisible by both 2 and 3.
8	if the last three digits make a number that is divisible by 8.
9	if the sum of its digits is divisible by 9.
10	if it ends in 0.

The most commonly used divisibility tests are those for 2, 3, 5, and 10.

Divisibility by 2
A number is divisible by **2** if the number ends in 0, 2, 4, 6, or 8.

EXAMPLE 11 Testing for Divisibility by 2

Are the following numbers divisible by 2?

(a) 986
↑
└──── Ends in 6

Because the number ends in 6, which is in the list (0, 2, 4, 6, or 8), the number 986 is divisible by 2.

(b) 3255 is *not* divisible by 2.
↑
└──── Ends in 5, and *not* in 0, 2, 4, 6, or 8

Work Problem 11 at the Side.

Divisibility by 3
A number is divisible by **3** if the sum of its digits is divisible by **3**.

EXAMPLE 12 Testing for Divisibility by 3

Are the following numbers divisible by 3?

(a) 4251

Add the digits.

$$4 + 2 + 5 + 1 = 12$$

Because 12 is divisible by 3, the number 4251 is divisible by 3.

(b) 29,806

Add the digits.

$$2 + 9 + 8 + 0 + 6 = 25$$

Because 25 is *not* divisible by 3, the number 29,806 is *not* divisible by 3.

> **CAUTION**
> Be careful when testing for divisibility by *adding the digits*. This method works only when testing for divisibility by 3 or by 9.

Work Problem 12 at the Side.

---

11 Which numbers are divisible by 2?

(a) 612

(b) 315

(c) 2714

(d) 36,000

12 Which numbers are divisible by 3?

(a) 836

(b) 7545

(c) 242,913

(d) 102,484

ANSWERS
11. (a), (c), and (d)
12. (b) and (c)

13 Which numbers are divisible by 5?

(a) 160

(b) 635

(c) 3381

(d) 108,605

14 Which numbers are divisible by 10?

(a) 290

(b) 218

(c) 2020

(d) 11,670

Divisibility by 5 and by 10

A number is divisible by **5** if it ends in **0** or **5**.

A number is divisible by **10** if it ends in 0.

EXAMPLE 13 Determining Divisibility by 5

Are the following numbers divisible by 5?

(a) 12,900   ends in 0, so it is divisible by 5.

(b) 4325   ends in 5, so it is divisible by 5.

(c) 392   ends in 2, so it is *not* divisible by 5.

◀◀ **Work Problem 13 at the Side.**

EXAMPLE 14 Determining Divisibility by 10

Are the following numbers divisible by 10?

(a) 700 and 9140   end in 0, so both numbers are divisible by 10.

(b) 355 and 18,743   do *not* end in 0, so these numbers are *not* divisible by 10.

◀◀ **Work Problem 14 at the Side.**

ANSWERS
13. (a), (b), (d)
14. (a), (c), (d)

R.4 Exercises

Divide. Then rewrite each division using two other division symbols. See Examples 1, 3, 5, and 6.

1. $\dfrac{12}{12}$
2. $\dfrac{9}{0}$
3. $24 \div 0$
4. $4 \div 4$
5. $\dfrac{0}{4}$

6. $0 \div 8$
7. $0 \div 12$
8. $\dfrac{0}{7}$
9. $0\overline{)21}$
10. $2 \div 0$

Divide by using short division. See Examples 7–9. Also, in Exercises 11–14, identify the dividend, the divisor, and the quotient. See Example 2.

11. $4\overline{)108}$
12. $5\overline{)135}$
13. $9\overline{)324}$
14. $8\overline{)176}$

15. $6\overline{)9137}$
16. $9\overline{)8371}$
17. $6\overline{)1854}$
18. $8\overline{)856}$

19. $4024 \div 4$
20. $16{,}024 \div 8$
21. $15{,}018 \div 3$
22. $32{,}008 \div 8$

23. $\dfrac{26{,}684}{4}$
24. $\dfrac{16{,}398}{9}$
25. $\dfrac{74{,}751}{6}$
26. $\dfrac{72{,}543}{5}$

27. $\dfrac{71{,}776}{7}$
28. $\dfrac{77{,}621}{3}$
29. $\dfrac{128{,}645}{7}$
30. $\dfrac{172{,}255}{4}$

Check each quotient. If a quotient is incorrect, find the correct quotient. See Example 10.

31. $7\overline{)4692}$ quotient 67 R2
32. $9\overline{)5974}$ quotient 663 R5
33. $6\overline{)21{,}409}$ quotient $3\,568$ R2
34. $4\overline{)103{,}516}$ quotient $25{,}879$

35. $6\overline{)18{,}023}$ quotient $3\,003$ R5
36. $8\overline{)33{,}664}$ quotient $4\,208$
37. $6\overline{)69{,}140}$ quotient $11{,}523$ R2
38. $3\overline{)82{,}598}$ quotient $27{,}532$ R1

Solve each application problem.

39. It is important to drink enough water. A rule of thumb is to divide your weight by 2 and drink that number of ounces of water each day. How many ounces of water should a 148-pound person drink daily? A 206-pound person?
(*Source:* Fairview Health Services.)

40. Ken Griffey, Jr., has a nine-year baseball contract for $117,000,000. Find his pay for each year.
(*Source:* USA Today.)

41. Six identical delivery vans for Rosita's Bakery cost a total of $99,600. Find the cost of each van.

42. Ted Slauson, coordinator of Toys for Tots, has collected 2628 toys. If his group gives four toys to each child, how many children will receive the gifts?

A college theater group is looking at two different budgets to cover the cost of props, costumes, and programs for the spring play. They are not sure whether to charge $5, $7, or $9 per ticket. The play's director organized the information in the table below. Use the table for Exercises 43–46.

Budget Amount	Number of $5 Tickets	Number of $7 Tickets	Number of $9 Tickets
$1890			
$2205			

43. For the $1890 budget, find the number of tickets that would need to be sold at $5 each; at $7 each; at $9 each. Write your answers in the table at the left.

44. For the $2205 budget, find the number of tickets that would need to be sold at $5 each; at $7 each; at $9 each. Write your answers in the table at the left.

45. If the director thinks that 300 tickets can be sold, which ticket price will cover the $1890 budget? How much extra money will there be?

46. If the director thinks that 300 tickets can be sold, which ticket price will cover the $2205 budget? How much money will be left over?

47. Kaci Salmon, a supervisor at Albany Electric, earns $184 for an 8-hour shift. The workers she supervises each earn $112 or $152 for an 8-hour shift. What are the hourly wages for Kaci and for her workers?

48. In a hospital weight loss program, Patient A lost 72 pounds in six months, Patient B lost 81 pounds in nine months, and Patient C lost 91 pounds in seven months. On average, how much did each patient lose each month?

Put a ✓ mark in the blank if the number at the left is divisible by the number at the top of each column. Use the divisibility tests from Examples 11–14.

	2	3	5	10			2	3	5	10
49. 30	___	___	___	___		**50.** 25	___	___	___	___
51. 184	___	___	___	___		**52.** 192	___	___	___	___
53. 445	___	___	___	___		**54.** 897	___	___	___	___
55. 903	___	___	___	___		**56.** 500	___	___	___	___
57. 5166	___	___	___	___		**58.** 8302	___	___	___	___
59. 21,763	___	___	___	___		**60.** 32,472	___	___	___	___

R.5 Long Division

Long division is used to divide by a number with more than one digit.

OBJECTIVE 1 Do long division. In long division, estimate the various numbers by using a *trial divisor* to get a *trial quotient*.

EXAMPLE 1   Using a Trial Divisor and a Trial Quotient

Divide. $42\overline{)3066}$

Because 42 is closer to 40 than to 50, use 4 as a trial divisor.

$$42$$
$$\uparrow$$
$$\text{Trial divisor is 4}$$

Try to divide the first digit of the dividend by 4. Because 3 cannot be divided by 4, use the first *two* digits, 30.

$$\frac{30}{4} = 7 \text{ with remainder 2}$$

$$\begin{array}{r} 7 \leftarrow \text{Trial quotient} \\ 42\overline{)3066} \\ \uparrow \end{array}$$

7 goes over the 6, because $\frac{306}{42}$ is about 7

Multiply 7 and 42 to get 294; next, subtract 294 from 306.

$$\begin{array}{r} 7 \\ 42\overline{)3066} \\ \underline{294} \leftarrow 7 \times 42 = 294 \\ 12 \leftarrow 306 - 294 = 12 \end{array}$$

Bring down the 6 at the right.

$$\begin{array}{r} 7 \\ 42\overline{)3066} \\ \underline{294\downarrow} \\ 126 \leftarrow 6 \text{ brought down} \end{array}$$

Use the trial divisor, 4.

$$\begin{array}{r} 73 \\ 42\overline{)3066} \\ \underline{294} \\ 126 \leftarrow \text{First two digits of 126} \rightarrow \frac{12}{4} = 3 \\ \underline{126} \leftarrow 3 \times 42 = 126 \\ 0 \end{array}$$

Check the quotient by multiplying 42 and 73. The product should be 3066, which matches the original dividend.

> **CAUTION**
> The first digit in the answer in long division must be placed in the proper position over the dividend.

Work Problem 1 at the Side.

OBJECTIVES

1. Do long division.
2. Divide multiples of 10.
3. Use multiplication to check quotients.

1 Divide.

(a) $64\overline{)4608}$ *Hint:* 64 is closer to 60 than 70, so use 6 as the trial divisor.

(b) $32\overline{)1792}$ Use _____ as the trial divisor.

(c) $51\overline{)2295}$ Use _____ as the trial divisor.

(d) $\dfrac{6391}{83}$ Use _____ as the trial divisor.

ANSWERS
1. (a) 72 (b) Use 3; 56 (c) Use 5; 45
   (d) Use 8; 77

2 Divide.

(a) $56\overline{)2352}$

(b) $38\overline{)1599}$

(c) $65\overline{)5416}$

(d) $89\overline{)6649}$

EXAMPLE 2 Dividing to Find a Trial Quotient

Divide. $58\overline{)2730}$

Use 6 as a trial divisor, because 58 is closer to 60 than to 50.

First two digits of dividend → $\dfrac{27}{6}$ = **4** with remainder 3

$$\begin{array}{r} 4 \leftarrow \text{Trial quotient} \\ 58\overline{)2730} \\ \underline{232} \leftarrow 4 \times 58 = 232 \\ 41 \leftarrow 273 - 232 = 41 \text{ (smaller than 58, the divisor)} \end{array}$$

Bring down the 0.

$$\begin{array}{r} 4 \\ 58\overline{)2730} \\ \underline{232}\downarrow \\ 410 \leftarrow \text{0 brought down} \end{array}$$

$$\begin{array}{r} 46 \\ 58\overline{)2730} \\ \underline{232} \\ 410 \\ \underline{348} \leftarrow 6 \times 58 = 348 \\ 62 \leftarrow \text{Greater than 58} \end{array}$$

First two digits of 410 → $\dfrac{41}{6}$ = **6** with remainder 5

The remainder, 62, is *greater than the divisor*, 58, so **7** should be used in the quotient instead of **6**.

$$\begin{array}{r} 47 \text{ R4} \\ 58\overline{)2730} \\ \underline{232} \\ 410 \\ \underline{406} \leftarrow 7 \times 58 = 406 \\ 4 \leftarrow 410 - 406 = 4 \end{array}$$

Work Problem 2 at the Side.

Sometimes it is necessary to write a 0 in the quotient.

EXAMPLE 3 Writing Zeros in the Quotient

Divide. $42\overline{)8734}$

Start as above in Example 2.

$$\begin{array}{r} 2 \\ 42\overline{)8734} \\ \underline{84} \leftarrow 2 \times 42 = 84 \\ 3 \leftarrow 87 - 84 = 3 \end{array}$$

Bring down the 3.

$$\begin{array}{r} 2 \\ 42\overline{)8734} \\ \underline{84}\downarrow \\ 33 \leftarrow \text{3 brought down} \end{array}$$

Continued on Next Page

ANSWERS

2. (a) 42 (b) 42 R3 (c) 83 R21
   (d) 74 R63

Since 33 cannot be divided by 42, write a 0 in the quotient as a placeholder.

$$\begin{array}{r} 2\,\mathbf{0} \\ 42\overline{)8734} \\ \underline{84} \\ 33 \end{array}$$ ← 0 in quotient

Bring down the final digit, the 4.

$$\begin{array}{r} 20 \\ 42\overline{)873\mathbf{4}} \\ \underline{84\downarrow} \\ 33\mathbf{4} \end{array}$$ ← 4 brought down

Complete the problem.

$$\begin{array}{r} 207 \;\mathbf{R}40 \\ 42\overline{)8734} \\ \underline{84} \\ 334 \\ \underline{294} \\ 40 \end{array}$$
← 7 × 42 = 294
← 334 − 294 = 40

The quotient is 207 **R**40.

> **CAUTION**
> There must be a digit in the quotient (answer) above *every* digit in the dividend *once the answer has begun*. Notice that in Example 3 above, a 0 was used to assure an answer digit above every digit in the dividend.

Work Problem 3 at the Side.

OBJECTIVE 2 Divide multiples of 10. When the divisor and dividend both contain zeros at the far right, recall that these numbers are multiples of 10. There is a short way to divide multiples of 10. Look at the following examples.

$$26{,}000 \div 1 = 26{,}000$$
$$26{,}000 \div 10 = 2600$$
$$26{,}000 \div 100 = 260$$
$$26{,}000 \div 1000 = 26$$

Do you see a pattern in these divisions using multiples of 10? These examples suggest the following rule.

Dividing by Multiples of 10
Divide a whole number by 10, by 100, or by 1000 by dropping one, two, or three zeros from the whole number.

EXAMPLE 4 Dividing by Multiples of 10

Divide.

(a) $6\underline{0} \div 1\mathbf{0} = 6$ — 0 in divisor; Drop 0 from dividend.

Continued on Next Page

3 Divide.

(a) $24\overline{)3127}$

(b) $52\overline{)10{,}660}$

(c) $39\overline{)15{,}933}$

(d) $78\overline{)23{,}462}$

ANSWERS
3. (a) 130 **R**7 (b) 205 (c) 408 **R**21
   (d) 300 **R**62

4 Divide by dropping zeros.

(a) 50 ÷ 10

(b) 1800 ÷ 100

(c) 305,000 ÷ 1000

5 Drop zeros and then divide.

(a) 60)7200

(b) 130)131,040

(c) 2600)195,000

6 Check each division. If the quotient is incorrect, find the correct quotient.

(a)
```
     43
18)774
   72
   ---
    54
    54
    ---
     0
```

(b)
```
        42 R178
426)19,170
    1 704
    -----
    1 130
      952
      ---
      178
```

ANSWERS
4. (a) 5   (b) 18   (c) 305
5. (a) 120   (b) 1008   (c) 75
6. (a) correct   (b) incorrect; should be 45

(b) 3500 ÷ 100 = 35   — 00 in divisor
                         — Drop 00 from dividend.

(c) 915,000 ÷ 1000 = 915   — 000 in divisor
                           — Drop 000 from dividend.

▶ **Work Problem 4 at the Side.**

You can also find the quotient for other multiples of 10 by dropping zeros.

EXAMPLE 5 Dividing by Multiples of 10

Divide.

(a) 40)11,000   Drop 0 from the divisor and the dividend.

```
      275
   4)1100
      8
      ---
      30
      28
      ---
       20
       20
       ---
        0
```

(b) 3500)31,500   Drop two zeros from the divisor and the dividend.

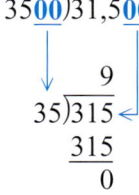
```
     9
  35)315
     315
     ---
       0
```

▶ **Work Problem 5 at the Side.**

OBJECTIVE 3 Use multiplication to check quotients. Quotients in long division can be checked just as quotients in short division were checked. Multiply the quotient and divisor, then add any remainder. The result should match the original dividend.

EXAMPLE 6 Checking Division

Check the quotient.

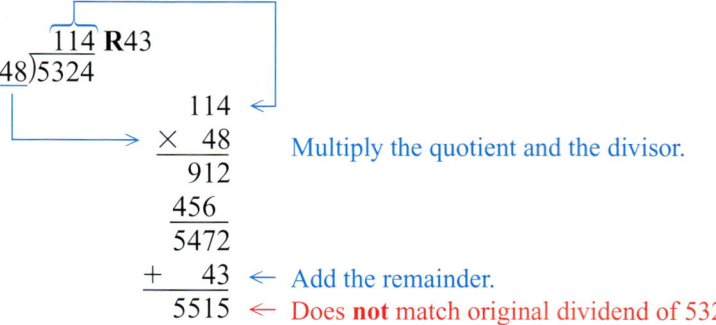

The quotient does *not* check. Rework the original problem to get 110 **R**44.

▶ **Work Problem 6 at the Side.**

R.5 Exercises

*First, indicate what number to use as the trial divisor. Then use the trial divisor to find the first digit in the quotient and write it in the correct position. Do **not** finish the division.*

1. $24\overline{)768}$

   Use _____ as the trial divisor.

2. $35\overline{)805}$

   Use _____ as the trial divisor.

3. $18\overline{)4500}$

   Use _____ as the trial divisor.

4. $28\overline{)3500}$

   Use _____ as the trial divisor.

5. $86\overline{)10,327}$

   Use _____ as the trial divisor.

6. $51\overline{)24,026}$

   Use _____ as the trial divisor.

7. $52\overline{)38,025}$

   Use _____ as the trial divisor.

8. $63\overline{)34,400}$

   Use _____ as the trial divisor.

9. $77\overline{)249,826}$

   Use _____ as the trial divisor.

10. $92\overline{)247,892}$

    Use _____ as the trial divisor.

11. $420\overline{)470,800}$

    Use _____ as the trial divisor.

12. $190\overline{)901,050}$

    Use _____ as the trial divisor.

Divide by using long division. See Examples 1–3. In Exercises 21–24, drop zeros before dividing. See Examples 4 and 5.

13. $42\overline{)8699}$

14. $58\overline{)2204}$

15. $47\overline{)11,121}$

16. $83\overline{)39,692}$

17. $26\overline{)62,583}$

18. $28\overline{)84,249}$

19. $63\overline{)78,072}$

20. $55\overline{)43,223}$

21. $150\overline{)499,760}$

22. $720\overline{)52,560}$

23. $400\overline{)340,000}$

24. $900\overline{)153,000}$

Check each division. If a quotient is incorrect, find the correct quotient. See Example 6.

25. 56)5943 106 R17

26. 87)3254 37 R37

27. 600)394,800 658 R9

28. 300)139,100 463 R200

29. 410)25,420 62 R3

30. 760)132,600 174 R360

31. 72)32,465 450 R65

32. 47)9570 23 R29

Solve each application problem by using addition, subtraction, multiplication, or division.

33. In 1900, the average workweek was 59 hours. Today it is 38 hours. How many more hours were worked each year in 1900 than today? Assume 50 weeks of work per year. (*Source:* Reiman Publications.)

34. The U.S. Government Printing Office uses 255,000 pounds of ink each year. If it does an equal amount of printing on each of 200 workdays in a year, find the weight of the ink used each day.

35. The most expensive standard hotel room in a recent study was the Ritz-Carlton at $375 per night, while the least expensive was Motel 6 at $32 per night. Find the amount saved by staying four nights at Motel 6 instead of the Ritz-Carlton. (*Source: USA Today.*)

36. Two divorced parents share their child's education costs, which amount to $3718 per year. If one parent pays $1880 each year, find the amount paid by the other parent over five years.

37. Judy Martinez owes $11,088 on a loan. Find her monthly payment if the loan is to be paid off in 36 months.

38. A consultant charged $19,360 for studying a school's compliance with the Americans with Disabilities Act. If the consultant worked 220 hours, find the rate charged per hour.

39. Clarence Hanks can assemble 42 circuits in one hour. How many circuits can he assemble in a 5-day workweek of 8 hours per day?

40. There are two conveyer lines in a factory, each of which packages 240 sacks of salt per hour. If the lines operate for 8 hours, find the total number of sacks of salt packaged by the two lines.

41. A youth soccer association brought in $7588 from fund-raising projects. There were expenses of $838 that had to be paid first, with the balance of the money divided evenly among the 18 teams. How much did each team receive?

42. Feather Farms Egg Ranch collected 3545 eggs in the morning and 2575 eggs in the afternoon. If the eggs are packed in flats containing 30 eggs each, find the number of flats needed for packing.

Chapter R
SUMMARY

KEY TERMS

R.1 **addends** — Addends are the numbers being added in an addition problem.
sum (total) — The answer in an addition problem is the sum (total).
commutative property of addition — The commutative property of addition states that changing the *order* of two addends in an addition problem does not change the sum.
associative property of addition — The associative property of addition states that changing the *grouping* of addends does not change the sum.
carrying — The process of carrying is used in an addition problem when the sum of the digits in a column is greater than 9; also called *regrouping*.

R.2 **minuend** — In the subtraction problem $8 - 5 = 3$, the 8 is the minuend. It is the number from which another number is subtracted.
subtrahend — In the subtraction problem $8 - 5 = 3$, the 5 is the subtrahend. It is the number being subtracted.
difference — The answer in a subtraction problem is the difference.
borrowing — Borrowing is used in subtraction if a digit is less than the one directly below it; also called *regrouping*.

R.3 **factors** — The numbers being multiplied are factors. For example, in $3 \times 4 = 12$, both 3 and 4 are factors.
product — The answer in a multiplication problem is the product.
commutative property of multiplication — The commutative property of multiplication states that changing the *order* of two factors in a multiplication problem does not change the product.
associative property of multiplication — The associative property of multiplication states that changing the *grouping* of factors does not change the product.
multiple — The product of two whole number factors is a multiple of both those numbers.

R.4 **dividend** — In division, the number being divided is the dividend.
divisor — In division, the number being used to divide another number is the divisor.
quotient — The answer in a division problem is the quotient.
short division — Short division is a quick method of dividing a number by a one-digit divisor.
remainder — The remainder is the number left over when two numbers do not divide evenly. The remainder is always less than the divisor.

R.5 **long division** — The process of long division is used to divide by a number with more than one digit.

QUICK REVIEW

Concepts	Examples
R.1 *Addition of Whole Numbers* Add from top to bottom, starting with the ones column and working left. To check, add from bottom to top.	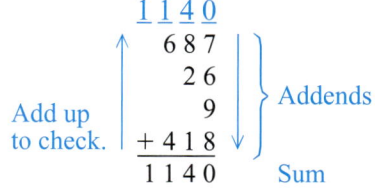

Concepts	Examples
R.1 Commutative Property of Addition Changing the *order* of two addends in an addition problem does not change the sum.	$2 + 4 = 6$ $4 + 2 = 6$ By the commutative property, the sum is the same.
R.1 Associative Property of Addition Changing the *grouping* of addends does not change the sum.	$(2 + 3) + 4 = 5 + 4 = 9$ $2 + (3 + 4) = 2 + 7 = 9$ By the associative property, the sum is the same.
R.2 Subtraction of Whole Numbers Subtract the subtrahend from the minuend to get the difference; use borrowing when necessary. To check, add the difference to the subtrahend to get the minuend.	Problem: 4738 ← Minuend, -649 ← Subtrahend, 4089 ← Difference. Check: $649 + 4089 = 4738$ (Matches).
R.3 Multiplication of Whole Numbers The numbers being multiplied are called *factors*. The multiplicand is being multiplied by the multiplier, giving the product. When the multiplier has more than one digit, partial products must be found and then added to find the product.	78 ← Multiplicand (Factor) $\times\ 24$ ← Multiplier (Factor) 312 ← Partial product 156 ← Partial product (move one position left) 1872 ← Product
R.3 Commutative Property of Multiplication Changing the *order* of two factors in a multiplication problem does not change the product.	$3 \times 4 = 12$ $4 \times 3 = 12$ By the commutative property, the product is the same.
R.3 Associative Property of Multiplication Changing the *grouping* of factors does not change the product.	$(2 \times 3) \times 4 = 6 \times 4 = 24$ $2 \times (3 \times 4) = 2 \times 12 = 24$ By the associative property, the product is the same.
R.4 Division of Whole Numbers $\div$ and $\overline{)}\ $ mean divide. Also a —, as in $\frac{25}{5}$, means to divide the top number (dividend) by the bottom number (divisor).	Divisor → $4\overline{)88}$ ← Dividend, Quotient = 22; $\underline{88}$, 0. $88 \div 4 = 22$ Dividend → $\dfrac{88}{4} = 22$ ← Quotient, Divisor

Chapter R
REVIEW EXERCISES

If you need help with any of these review exercises, look in the section indicated in the red brackets.

[R.1] *Add.*

1. 74
   $+18$

2. 35
   $+78$

3. 807
   4606
   $+51$

4. 8215
   9
   $+7433$

[R.2] *Subtract.*

5. 238
   -199

6. 573
   -389

7. 2210
   -1986

8. $99{,}704$
   $-73{,}838$

[R.3] *Multiply. Do the work mentally.*

9. $2 \times 4 \times 6$

10. $9 \times 1 \times 5$

11. $(6)(1)(8)$

12. $7 \cdot 7 \cdot 0$

Multiply.

13. 43
    $\times4$

14. 781
    $\times7$

15. 5440
    $\times6$

16. $93{,}105$
    $\times5$

Multiply by using the shortcuts for multiples of 10.

17. 320
    $\times60$

18. 280
    $\times90$

19. 517
    $\times400$

20. $16{,}000$
    $\times8\,000$

Multiply.

21. 34
    $\times18$

22. 52
    $\times36$

23. 655
    $\times21$

24. 392
    $\times77$

[R.4] *Divide.*

25. $42 \div 7$

26. $18 \div 18$

27. $\dfrac{125}{0}$

28. $\dfrac{0}{35}$

[R.4–R.5] *Divide.*

29. $4\overline{)432}$

30. $9\overline{)216}$

31. $76\overline{)26{,}752}$

32. $2704 \div 18$

MIXED REVIEW EXERCISES

Solve each application problem.

33. There are 52 playing cards in a deck. How many cards are there in two decks? In six decks?

34. The college bookstore received mathematics textbooks packed 12 books per carton. How many textbooks were in a delivery of 238 cartons?

35. Raoul rented a car for four days at $47 per day. There was also a $24 charge to fill the gas tank when he returned the car. Raoul had budgeted $200 for the rental. Was he over or under his budget? By how much?

36. The Village Middle School wants to raise $315,280 for a new library. If $87,340 has already been raised, how much more is needed to reach the goal?

37. It takes 2000 hours of work to build one home. How many hours of work are needed to build five homes? To build 12 homes?

38. A Japanese bullet train travels 80 miles in one hour. Find the number of miles traveled in three hours, and the number of miles traveled in seven hours.

39. If an acre needs 250 pounds of fertilizer, how many acres can be fertilized with 5750 pounds of fertilizer?

40. Each home in a subdivision requires 180 feet of fencing. Find the number of homes that can be fenced with 5760 feet of fencing material.

41. Susan Hessney had $382 in her checking account. She wrote a check for $135 and deposited her $563 tax refund. What is the new balance in her account?

42. Find the total cost of four T-shirts at $14 each and three sweatshirts at $29 each.

43. The Houston Space Center charges $18 for each adult admission, $14 for each child, and $17 for each senior. Find the total cost to admit 18 adults, 6 children, and 15 seniors.

44. A newspaper carrier has 56 customers who take the paper daily and 23 customers who take the paper on weekends only. A daily customer pays $18 per month and a weekend-only customer pays $11 per month. Find the total monthly collections.

Chapter R
TEST

Fill in the blanks to complete each sentence.

1. In an addition problem, the numbers being added are called _____ and the answer is called the _____.

2. In a multiplication problem, the numbers being multiplied are called _____ and the answer is called the _____.

3. In a subtraction problem the answer is called the _____ and in a division problem the answer is called the _____.

*Add, subtract, multiply, or divide, as indicated. Do **not** use a calculator.*

4. $984 + 65 + 7561$

5. $\quad 17{,}063$
   $\quad\quad\quad 7$
   $\quad\quad\; 12$
   $\quad\; 1\;505$
   $\; 93{,}710$
   $+\quad 333$

6. $17{,}002 - 54$

7. $\quad 5062$
   $-\; 1978$

8. $5 \times 7 \times 4$

9. $57 \cdot 3000$

10. $(85)(21)$

11. $\quad 7381$
    $\times\quad 603$

12. $6\overline{)1236}$

13. $\dfrac{791}{0}$

14. $38{,}472 \div 84$

15. $280\overline{)44{,}800}$

1. _____
2. _____
3. _____
4. _____
5. _____
6. _____
7. _____
8. _____
9. _____
10. _____
11. _____
12. _____
13. _____
14. _____
15. _____

Solve each application problem.

16. Find the cost of 48 glass beakers for a chemistry lab at $11 per beaker.

17. In a consumer survey of cell phones, the most expensive model cost $350 and the least expensive model was $59. Find the difference in price between the most expensive and the least expensive phone.

18. A stamping machine produces 900 license plates each hour. How long will it take to produce 31,500 license plates?

19. Kenée Shadbourne paid $690 for tuition, $185 for books, and $68 for supplies. These amounts were withdrawn from her checking account, which had a balance of $1108. Find the new balance in her account.

20. An appliance manufacturer assembles 118 self-cleaning ovens each hour during the first four hours and 139 standard ovens each hour during the next four hours. Find the total number of ovens assembled in the 8-hour period.

21. The monthly rents collected from the four units in an apartment building are $785, $800, $815, and $725. After expenses of $1085 are paid, find the amount that remains.

22. Describe the divisibility tests for 2, 5, and 10. Also give two examples for each test: one number that is divisible and one number that is not divisible.

Appendix A: Inductive and Deductive Reasoning

Appendix A Inductive and Deductive Reasoning

OBJECTIVE 1 Use inductive reasoning to analyze patterns. In many scientific experiments, conclusions are drawn from specific outcomes. After many repetitions and similar outcomes, the findings are generalized into statements that appear to be true. When general conclusions are drawn from specific observations, we are using a type of reasoning called **inductive reasoning.** The next several examples illustrate this type of reasoning.

OBJECTIVES

1. Use inductive reasoning to analyze patterns.
2. Use deductive reasoning to analyze arguments.
3. Use deductive reasoning to solve problems.

EXAMPLE 1 Using Inductive Reasoning

Find the next number in the sequence 3, 7, 11, 15, ...

To discover a pattern, calculate the difference between each pair of successive numbers.

$$7 - 3 = 4$$
$$11 - 7 = 4$$
$$15 - 11 = 4$$

Notice that the difference is always 4. Each number is 4 greater than the previous one. Thus, the next number in the pattern is $15 + 4$, or 19.

1 Find the next number in the sequence 2, 8, 14, 20, . . . Describe the pattern.

▶▶▶ **Work Problem 1 at the Side.** ▶▶▶

EXAMPLE 2 Using Inductive Reasoning

Find the next number in this sequence.

$$7, 11, 8, 12, 9, 13, \ldots$$

The pattern in this example involves addition and subtraction.

$$7 + 4 = 11$$
$$11 - 3 = 8$$
$$8 + 4 = 12$$
$$12 - 3 = 9$$
$$9 + 4 = 13$$

To get the second number, we add 4 to the first number. To get the third number, we subtract 3 from the second number. To obtain subsequent numbers, we continue the pattern. The next number is $13 - 3 = 10$.

2 Find the next number in the sequence 6, 11, 7, 12, 8, 13, . . . Describe the pattern.

▶▶▶ **Work Problem 2 at the Side.** ▶▶▶

ANSWERS
1. 26; add 6 each time.
2. 9; add 5, subtract 4.

A–1

A-2 Appendix A: Inductive and Deductive Reasoning

❸ Find the next number in the sequence 2, 6, 18, 54, ... Describe the pattern.

EXAMPLE 3 Using Inductive Reasoning

Find the next number in the sequence 1, 5, 25, 125, ...

Each number after the first is obtained by multiplying the previous number by 5. So the next number would be 125 · 5 = 625.

◀◀◀ Work Problem 3 at the Side.

EXAMPLE 4 Using Inductive Reasoning

(a) Find the next geometric shape in this sequence.

The figures alternate between a blue circle and a red triangle. Also, the number of dots increases by 1 in each subsequent figure. Thus, the next figure should be a blue circle with five dots inside it.

(b) Find the next geometric shape in this sequence.

The first two shapes consist of vertical lines with horizontal lines at the bottom extending first *left* and then *right*. The third shape is a vertical line with a horizontal line at the top extending to the *left*. Therefore, the next shape should be a vertical line with a horizontal line at the top extending to the *right*.

❹ Find the next shape in this sequence.

◀◀◀ Work Problem 4 at the Side.

OBJECTIVE ❷ **Use deductive reasoning to analyze arguments.** In the previous discussion, specific cases were used to find patterns and predict the next event. There is another type of reasoning called **deductive reasoning**, which moves from general cases to specific conclusions.

EXAMPLE 5 Using Deductive Reasoning

Does the conclusion follow from the premises in this argument?

All Buicks are automobiles. ← Premise
All automobiles have horns. ← Premise
∴ All Buicks have horns. ← Conclusion

In this example, the first two statements are called *premises* and the third statement (below the line) is called a *conclusion*. The symbol ∴ is a mathematical symbol meaning "**therefore**." The entire set of statements is called an *argument*.

Continued on Next Page

ANSWERS
3. 162; multiply by 3.
4.

The focus of deductive reasoning is to determine whether the conclusion follows (is valid) from the premises. A set of circles called **Euler circles** is used to analyze the argument.

In Example 5, the statement "All Buicks are automobiles" can be represented by two circles, one for Buicks and one for automobiles. Note that the circle representing Buicks is totally inside the circle representing automobiles because the first premise states that *all* Buicks are automobiles.

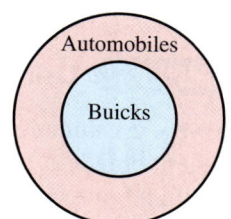

Now, a circle is added to represent the second statement, vehicles with horns. This circle must completely surround the circle representing automobiles because the second premise states that *all* automobiles have horns.

To analyze the conclusion, notice that the circle representing Buicks is *completely* inside the circle representing vehicles with horns. Therefore, it must follow that all Buicks have horns. ***The conclusion is valid.***

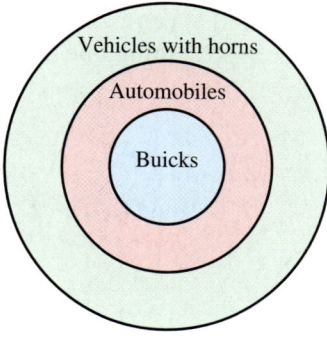

Work Problem 5 at the Side.

5 Does the conclusion follow from the premises? Draw a set of Euler circles to analyze the argument.

All cars have four wheels.
All Fords are cars.
∴ All Fords have four wheels.

6 Does each conclusion follow from the premises? Draw a set of Euler circles to analyze each argument.

(a) All animals are wild.
All cats are animals.
∴ All cats are wild.

(b) All students use math.
All adults use math.
∴ All adults are students.

EXAMPLE 6 Using Deductive Reasoning

Does the conclusion follow from the premises? Draw a set of Euler circles to analyze the argument.

All tables are round.
All glasses are round.
∴ All glasses are tables.

Use Euler circles. Draw a circle representing tables *inside* a circle representing round objects, because the first premise states that *all* tables are round.

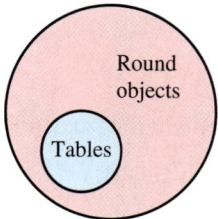

The second statement requires that a circle representing glasses must now be drawn inside the circle representing round objects, but not necessarily inside the circle representing tables. Therefore, the conclusion does *not* follow from the premises. This means that ***the conclusion is invalid.***

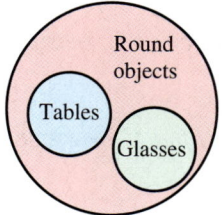

Work Problem 6 at the Side.

ANSWERS

5. The conclusion is valid.

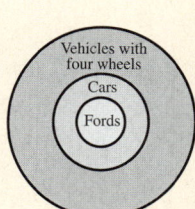

6. (a) The conclusion is valid.

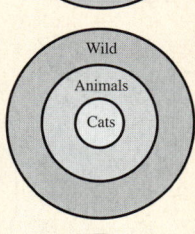

(b) The conclusion is valid.

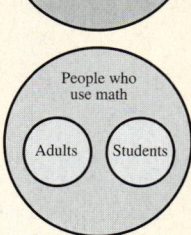

7 In a college class of 100 students, 35 take both math and history, 50 take history, and 40 take math. How many take neither math nor history? Draw a Venn diagram.

8 A Chevy, BMW, Honda, and Ford are parked side by side. The known facts are:

(a) The Ford is on the right end.

(b) The BMW is next to the Honda.

(c) The Chevy is between the Ford and the Honda.

Which car is parked on the left end?

OBJECTIVE 3 Use deductive reasoning to solve problems. Another type of deductive reasoning problem occurs when a set of facts is given in a problem and a conclusion must be drawn using these facts.

EXAMPLE 7 Using Deductive Reasoning

There were 25 students enrolled in a ceramics class. During the class, 10 of the students made a bowl and 8 students made a birdbath. Three students made both a bowl and a birdbath. How many students did not make either a bowl or a birdbath?

This type of problem is best solved by organizing the data using a drawing called a **Venn diagram**. Two overlapping circles are drawn, with each circle representing one item made by the students, as shown below.

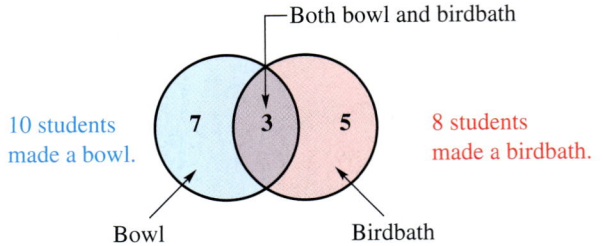

In the region where the circles overlap, write the number of students who made *both* items, namely, 3. In the remaining portion of the birdbath circle, write the number 5, which when added to 3 will give the total number of students who made a birdbath, namely, 8. In a similar manner, write 7 in the remaining portion of the bowl circle, since $7 + 3 = 10$, the total number of students who made a bowl. The total of all three numbers written in the circles is 15. Since there were 25 students in the class, this means $25 - 15$ or 10 students did not make either a birdbath or a bowl.

Work Problem 7 at the Side.

EXAMPLE 8 Using Deductive Reasoning

Four cars in a race finish first, second, third, and fourth. The following facts are known.

(a) Car A beat Car C.

(b) Car D finished between Cars C and B.

(c) Car C beat Car B.

In which order did the cars finish?

To solve this type of problem, it is helpful to use a line diagram.

1. *Write A before C,* because Car A beat Car C (fact **a**).

   A   C

2. *Write B after C,* because Car C beat Car B (fact **c**).

   A   C   B

3. *Write D between C and B,* because Car D finished between Car C and Car B (fact **b**).

The correct order of finish is shown below.

A   C   D   B

Work Problem 8 at the Side.

ANSWERS

7.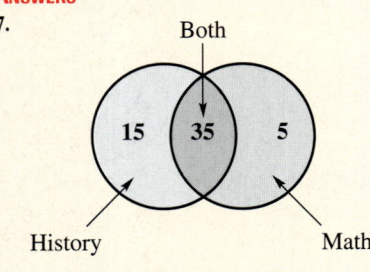

   $15 + 35 + 5 = 55$
   $100 - 55 = 45$
   45 students take neither math nor history.

8. BMW

Appendix A Exercises

Find the next number in each sequence. Describe the pattern in each sequence. See Examples 1–3.

1. 2, 9, 16, 23, 30, . . .

2. 5, 8, 11, 14, 17, . . .

3. 0, 10, 8, 18, 16, . . .

4. 3, 9, 7, 13, 11, . . .

5. 1, 2, 4, 8, . . .

6. 1, 4, 16, 64, . . .

7. 1, 3, 9, 27, 81, . . .

8. 3, 6, 12, 24, 48, . . .

9. 1, 4, 9, 16, 25, . . .

10. 6, 7, 9, 12, 16, . . .

Find the next shape in each sequence. See Example 4.

11.

12.

13.

14.

For each argument, draw Euler circles and then state whether or not the conclusion follows from the premises. See Examples 5 and 6.

15. All animals are wild.
All lions are animals.
∴ All lions are wild.

16. All students are hard workers.
All business majors are students.
∴ All business majors are hard workers.

17. All teachers are serious.
All mathematicians are serious.
∴ All mathematicians are teachers.

18. All boys ride bikes.
All Americans ride bikes.
∴ All Americans are boys.

Solve each application problem. See Examples 7 and 8.

19. In a given 30-day period, a husband watched television 20 days and his wife watched television 25 days. If they watched television together 18 days, how many days did neither watch television? Draw a Venn diagram.

20. In a class of 40 students, 21 students take both calculus and physics. If 30 students take calculus and 25 students take physics, how many do not take either calculus or physics? Draw a Venn diagram.

21. Tom, Dick, Mary, and Joan all work for the same company. One is a secretary, one is a computer operator, one is a receptionist, and one is a mail clerk.

(a) Tom and Joan eat dinner with the computer operator.

(b) Dick and Mary carpool with the secretary.

(c) Mary works on the same floor as the computer operator and the mail clerk.

Who is the computer operator?

22. Four cars—a Ford, a Buick, a Mercedes, and an Audi—are parked in a garage in four spaces.

(a) The Ford is in the last space.

(b) The Buick and Mercedes are next to each other.

(c) The Audi is next to the Ford but not next to the Buick.

Which car is in the first space?

Appendix B: Graphing Direct and Inverse Variation

Appendix B Graphing Direct and Inverse Variation

In **Section 15.8** we introduced the concept of *variation,* the idea that a change in one variable causes a change in another variable. Now we will look at the graphs of equations that model direct variation and inverse variation.

OBJECTIVES

1. Draw the graph of an equation that models direct variation and interpret information from it.
2. Draw the graph of an equation that models inverse variation and interpret information from it.

OBJECTIVE 1 **Draw the graph of an equation that models direct variation and interpret information from it.** Recall that *direct variation* between two variables means that as the value of one variable *increases,* the value of the other variable *also increases* (does the same thing). For example, as the number of gallons of gas you pump into your car increases, the total cost also increases. Or, as the number of hours that you drive your car increases, the total distance you travel also increases.

EXAMPLE 1 The Graph for Direct Variation

Suppose you are driving your car at an average rate of 50 miles per hour. How far you travel depends upon how many hours you drive. Write the equation of variation and draw a graph showing the relationship between the distance you travel (in miles) and the time (in hours).

Recall from **Section 15.7** that the basic distance formula is $d = rt$ where d is distance, r is rate, and t is time.

$$d = rt \quad \text{Replace } r \text{ with 50 miles per hour.}$$
$$d = 50t$$

The equation of direct variation is $d = 50t$.

In **1** hour you travel 50 miles. $d = 50 \cdot 1 = 50$
In **2** hours you travel 100 miles. $d = 50 \cdot 2 = 100$
In **3** hours you travel 150 miles. $d = 50 \cdot 3 = 150$

Continued on Next Page

Appendix B: Graphing Direct and Inverse Variation

1 Jeff drives a test car at an average speed of 100 miles per hour on a racetrack. Write the equation of direct variation for the relationship between the distance Jeff travels (in miles) and the time he drives (in hours). Then draw a graph by plotting at least three ordered pairs.

t (hours)	d (miles)
1	
2	
3	

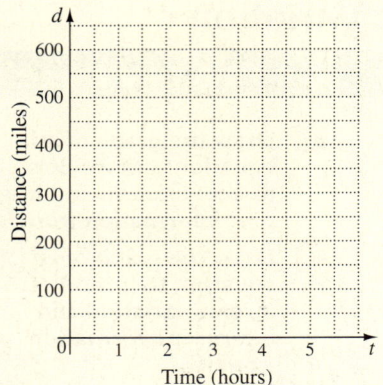

A table of possible values for t and d is shown below. The ordered pairs are plotted on the graph, and then connected with a straight line.

t (hours)	d = 50t (miles)
1	50
2	100
3	150
4	200

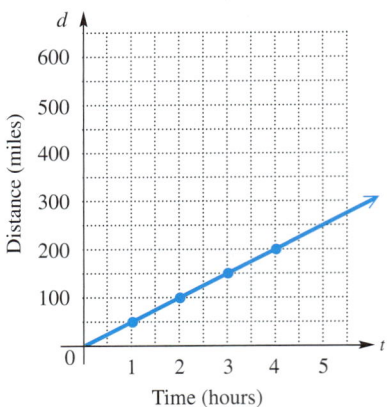

> **NOTE**
> We did *not* use 1, 2, 3, and so on, for the vertical scale (distance). We used 50, 100, 150, and so on, because those numbers better fit the situation. Also note that time and distance can only be positive numbers, so all the points on the graph are in the first quadrant.

◀◀◀ **Work Problem 1 at the Side.**

In **Section 15.8** we saw that the general equation for direct variation is $y = kx$, and we know that this is the equation of a straight line. Notice that the graph of $d = 50t$ in Example 1 and the graph of $d = 100t$ in margin problem 1 are both straight lines. These graphs illustrate the following.

2 Use the graph in margin problem 1 above to answer these questions.

(a) How many hours will it take Jeff to travel 400 miles?

(b) How far will Jeff travel in $1\frac{1}{2}$ hours?

Graphs of Equations Describing Direct Variation
The graph of an equation describing *direct variation* is a *straight line*.

EXAMPLE 2 Interpreting Information from a Graph of Direct Variation

Use the graph in Example 1 to answer these questions.

(a) How far will you travel in 5 hours?

(b) How many hours will it take you to travel 125 miles?

(a) The point on the graph corresponding to 5 hours has coordinates **(5, 250)** so you will travel 250 miles.

(b) The point on the graph corresponding to 125 miles has coordinates **(2.5, 125)** so it will take 2.5 or $2\frac{1}{2}$ hours.

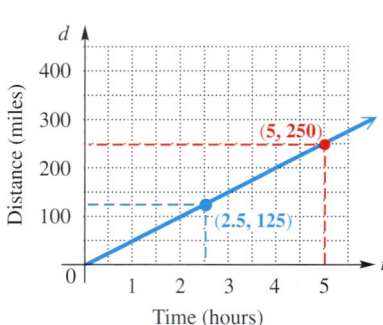

◀◀◀ **Work Problem 2 at the Side.**

ANSWER
1. $d = 100t$

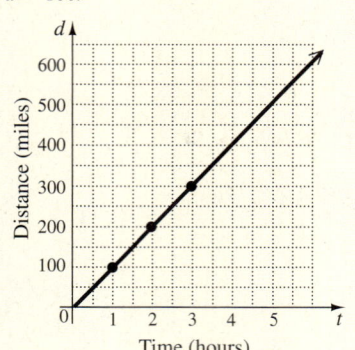

2. (a) 4 hours (b) 150 miles

OBJECTIVE 2 Draw the graph of an equation that models inverse variation and interpret information from it. Recall that *inverse variation* between two variables means that as the value of one variable *increases,* the value of the other variable *decreases* (does the opposite).

EXAMPLE 3 The Graph for Inverse Variation

Suppose you drive 200 miles from your home to visit your parents. How long it takes you to get there depends upon the average rate (speed) at which you drive. You may have to drive more slowly due to bad weather, time of day, construction delays, and so on. Write the equation of variation and draw a graph showing the relationship between the time it takes (in hours) and your rate (in miles per hour).

First solve the basic distance formula $d = rt$ for t.

$$\frac{d}{r} = \frac{rt}{r} \quad \text{Divide both sides by } r.$$

$$\frac{d}{r} = t \quad \text{or} \quad t = \frac{d}{r} \quad \text{Replace } d \text{ with 200 miles.}$$

$$t = \frac{200}{r}$$

The equation of variation is $t = \dfrac{200}{r}$

At **20** miles per hour, it will take 10 hours. $\quad t = \dfrac{200}{20} = 10$

At **40** miles per hour, it will take 5 hours. $\quad t = \dfrac{200}{40} = 5$

At **50** miles per hour, it will take 4 hours. $\quad t = \dfrac{200}{50} = 4$

A table of possible values for r and t is shown below. The ordered pairs are plotted on the graph and then connected with a *curve*.

r (miles per hour)	$t = \dfrac{200}{r}$ (hours)
10	20
20	10
40	5
50	4
100	2

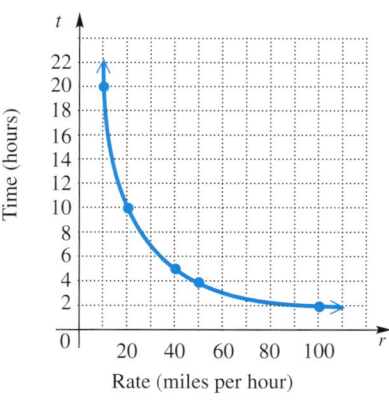

As your rate of speed *increases,* the time it takes to travel to your parents' home *decreases.* But notice that the points on the graph cannot be connected by a *straight* line. They are connected by a *curve* that comes close to the axes but never touches them.

Work Problem 3 at the Side.

3 Lisa drives 60 miles to attend a Saturday class at a state university. Write the equation of variation for the relationship between the time it takes her to get to the university and the rate (speed) at which she drives. Then draw a graph by plotting at least five ordered pairs.

r (miles per hour)	t (hours)
5	
10	
20	
30	
60	

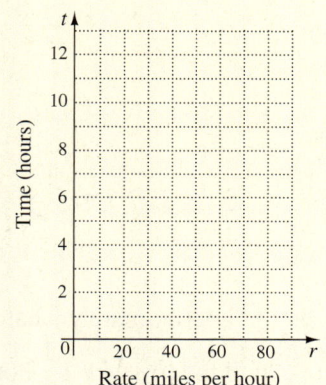

ANSWER

3. $t = \dfrac{60}{r}$

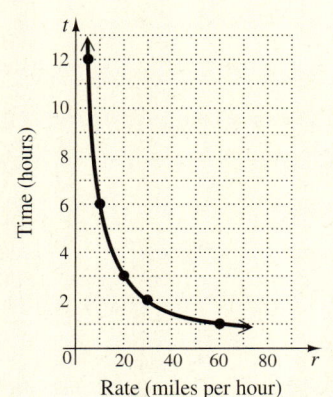

A-10 Appendix B: Graphing Direct and Inverse Variation

4 Use the graph in margin problem 3 on the previous page to answer these questions.

(a) Approximately how long will the trip take at a rate of 50 miles per hour?

(b) At approximately what rate will the trip take $1\frac{1}{2}$ hours?

> **NOTE**
> The curve in Example 3 is part of a special figure called a *hyperbola*. You will learn more about hyperbolas in later algebra courses.

The graph for $t = \dfrac{200}{r}$ in Example 3 and the graph for $t = \dfrac{60}{r}$ in margin problem 3 illustrate the following.

> **Graphs of Equations Describing Inverse Variation**
> The graph of an equation describing *inverse variation* is a *curve* that approaches the axes but does not touch them.

EXAMPLE 4 Interpreting Information from a Graph of Inverse Variation

Use the graph in Example 3 to answer these questions.

(a) At approximately what rate will the trip take 6 hours?

(b) Approximately how long will the trip take at a rate of 65 miles per hour?

(a) You must estimate the rate at the point on the graph corresponding to 6 hours. It is between 30 and 40, but closer to 30, so a reasonable estimate is **33 miles per hour**.

(b) You must estimate the time at the point on the graph that corresponds to 65 miles per hour. It is about half way between 2 hours and 4 hours, so a reasonable estimate is **3 hours**.

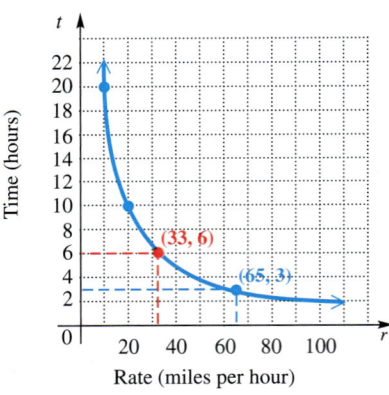

▶◀◀ **Work Problem 4 at the Side.**

ANSWER

4. (a) About $1\frac{1}{4}$ hours. (b) 40 miles per hour

Appendix B Exercises

FOR EXTRA HELP — Addison-Wesley Math Tutor Center · MathXL · Student's Solutions Manual · MyMathLab · MathXL Tutorials on CD

In Exercises 1–6, write the equation of variation. Then draw a graph by plotting at least three points. See Example 1.

1. Michael drives his UPS delivery truck at an average rate (r) of 20 miles per hour. Show the relationship between the distance he travels (d) and the time (t).

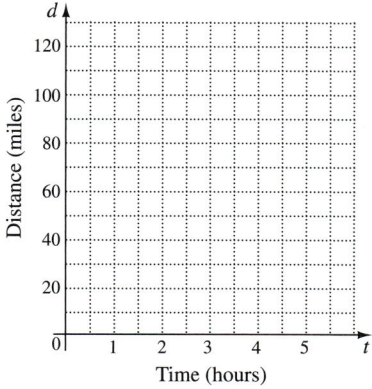

2. In winter Michael's average rate (r) is only 10 miles per hour. Show the relationship between distance (d) and time (t).

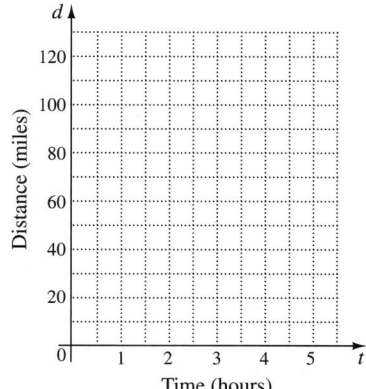

3. Tara has a part-time job with an hourly wage (w) of $12. Show the relationship between her total pay (P) and the time (t) she works in hours.

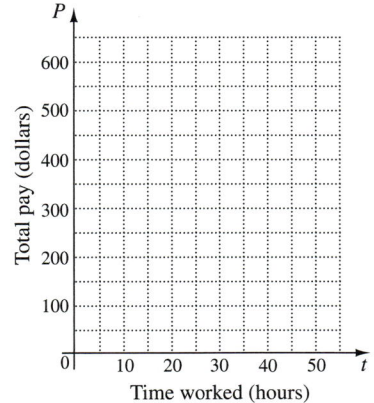

4. If Tara gets a job with an hourly wage (w) of $15, show the relationship between her total pay (P) and the time (t) she works.

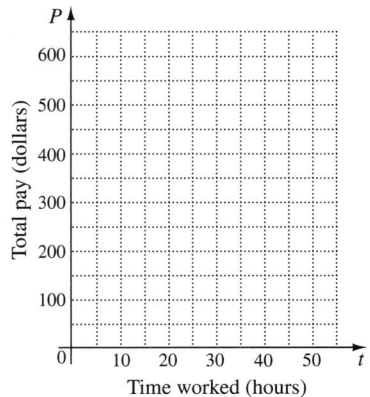

A–12 Appendix B: Graphing Direct and Inverse Variation

5. If gasoline costs $2.50 per gallon, show the relationship between the total cost (C) and the number of gallons (g) you buy.

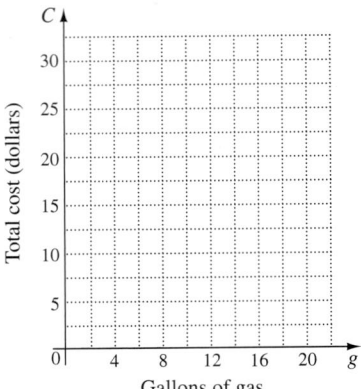

6. If gasoline drops to $2.25 per gallon, show the relationship between the total cost (C) and the number of gallons (g) you buy.

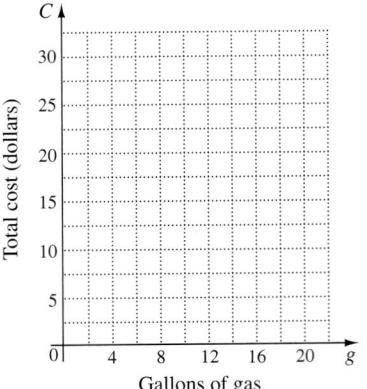

In Exercises 7–10, write the equation of variation. Then draw a curve by plotting at least five points. See Example 3.

7. Marcella covers a 100-mile delivery route. Show the relationship between the time it takes (t) and her average rate (r).

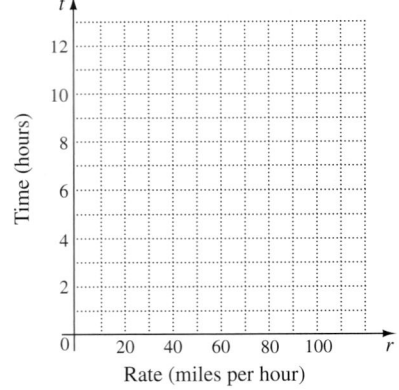

8. Malik has a 50-mile delivery route. Show the relationship between the time it takes (t) and his average rate (r).

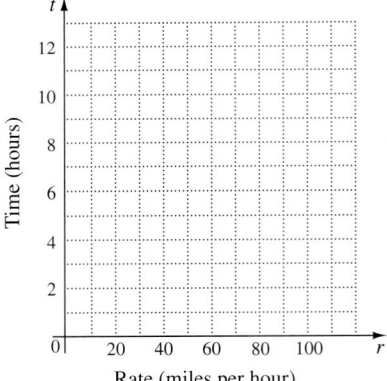

9. Peggy wants to plant 60 square feet of strawberries in a rectangular plot. There are many different combinations of length (*l*) and width (*w*) for the garden plot that result in an area of 60 square feet. Show the relationship between length and width of the plot.

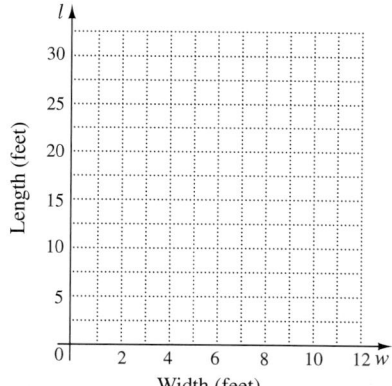

10. See Exercise 9. If Peggy also wants 48 square feet of lettuce in a rectangular plot, show the relationship between length (*l*) and width (*w*) of the plot.

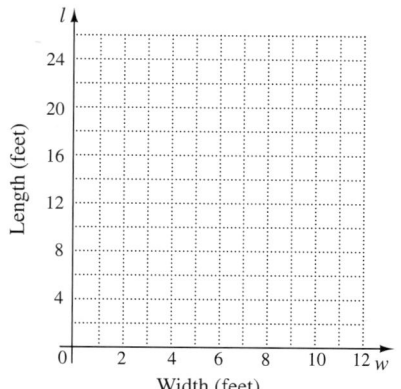

Use the graphs in Exercises 3–10 to answer these questions. Your answers may vary slightly when the graphed line falls between the grid lines. See Examples 2 and 4.

11. Use the graph in Exercise 3. What number of hours will give Tara total pay of $300?

12. Use the graph in Exercise 4. What is Tara's total pay for 15 hours of work?

13. Use the graph in Exercise 5. What is the approximate cost of 11 gallons of gas?

14. Use the graph in Exercise 6. Approximately how many gallons of gas can be purchased with $15?

15. Use the graph in Exercise 7. About how long will it take to do the route at a rate of 30 miles per hour?

16. Use the graph in Exercise 8. If Malik completes the route in 3.5 hours, what was his approximate rate?

17. Use the graph in Exercise 9. If the length is $7\frac{1}{2}$ feet, what is the width?

18. Use the graph in Exercise 10. If the length is 9 feet, what is the width?

19. (a) Describe the difference between graphs of direct variation and graphs of inverse variation.
    (b) Identify the exercises that illustrate direct variation and the exercises that illustrate inverse variation.

20. Think of an example of direct variation. Describe the relationship and draw a graph to illustrate the relationship.

Appendix C: Graphing Quadratic Inequalities

Appendix C Graphing Quadratic Inequalities

OBJECTIVE 1 Graph quadratic inequalities. In **Section 11.5**, we graphed *linear* inequalities by graphing the straight line and then using a test point to decide which side of the line to shade. We can use the same steps to graph a *quadratic* inequality. The parabola is the boundary between two regions. The graph will include the region enclosed by the parabola, *or* the region outside the parabola.

OBJECTIVE

1 Graph quadratic inequalities.

Graphing a Quadratic Inequality

Step 1 Graph the parabola using the methods in **Section 17.4**. Make the parabola a *solid* curve if the inequality has $\leq$ or $\geq$. Make the parabola a *dashed* curve if the inequality has $<$ or $>$.

Step 2 Choose a test point (x, y) not on the parabola; either *inside* the parabola, or *outside* the parabola. Substitute the values of x and y into the inequality. If a *true* statement results, shade the region *containing* the test point. If a *false* statement results, shade the *other* region.

EXAMPLE 1 Graphing a Quadratic Inequality

Graph $y \geq x^2 + 2x - 3$

Step 1 Graph the parabola $y = x^2 + 2x - 3$. First find the vertex. The x-value of the vertex is shown below.

$$x = -\frac{b}{2a} = -\frac{2}{2(1)} = -1$$

The y-value of the vertex is found by letting $x = -1$ in the equation.

$$y = (-1)^2 + 2(-1) - 3 = -4$$

The vertex is $(-1, -4)$.

Continued on Next Page

A-16 Appendix C: Graphing Quadratic Inequalities

1 Graph $y < x^2 - x - 2$.

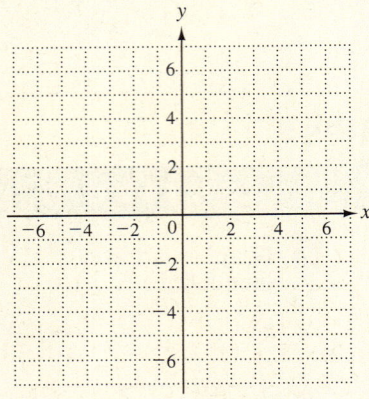

ANSWER

1. Parabola is a dashed curve; shade the region outside the parabola.

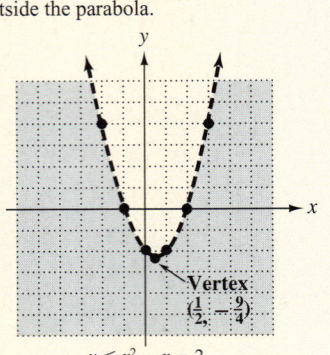

Next, find the *y*-intercept by letting $x = 0$ in the equation.

$$y = 0^2 + 2(0) - 3 = -3$$

The *y*-intercept is $(0, -3)$.

Finally, find the *x*-intercepts by letting $y = 0$ in the equation.

$$0 = x^2 + 2x - 3$$
$$0 = (x - 1)(x + 3)$$

$x - 1 = 0$ or $x + 3 = 0$
$x = 1$ $x = -3$

The *x*-intercepts are $(1, 0)$ and $(-3, 0)$.

List the vertex and intercepts in a table, as shown below. Plot the points. Then find several more ordered pairs near the vertex to complete the parabola. Draw a *solid* parabola connecting the points because the inequality uses $\geq$.

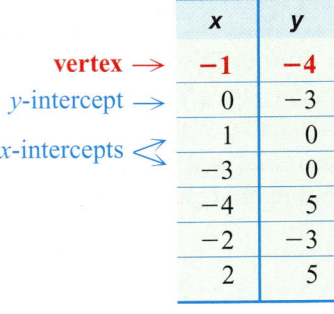

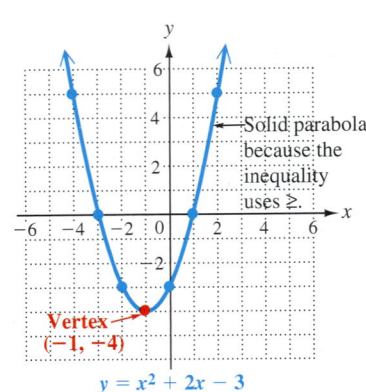

Step 2 Choose a test point not on the curve. We choose $(0, 0)$ which is *inside* the parabola. Substitute $(0, 0)$ in the inequality.

$y \geq x^2 + 2x - 3$ Let $x = 0$ and $y = 0$.
$0 \geq 0^2 + 2(0) - 3$?
$0 \geq -3$ True

The result is a *true* statement, so we shade the region *containing* the test point $(0, 0)$. That is, we shade the region inside the parabola. *All* the points in the shaded region, *and all* the points on the parabola, are solutions of $y \geq x^2 + 2x - 3$, as shown below.

All points *on* the parabola and *all* points *inside* the parabola are solutions for $y \geq x^2 + 2x - 3$.

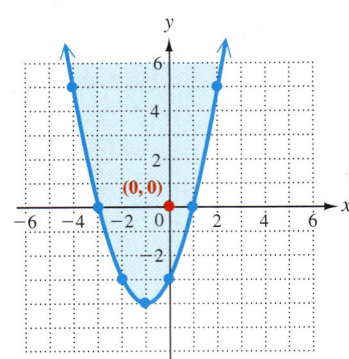

Work Problem 1 at the Side.

Appendix C Exercises

Graph each quadratic inequality by finding the vertex, the intercepts, and several other points and then shading the appropriate region. Identify the vertex. See Example 1.

1. $y < -x^2 - 4x + 5$

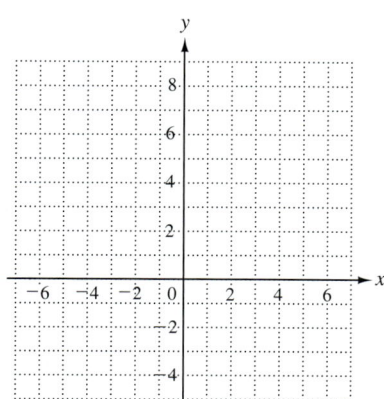

2. $y \geq x^2 + 6x + 8$

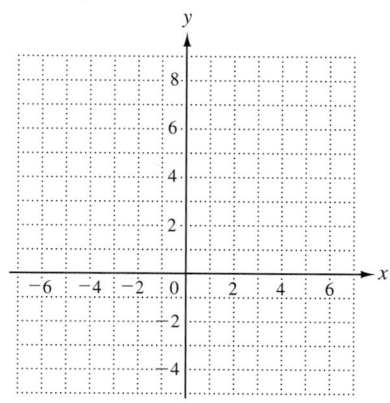

3. $y \leq 2x^2 - x - 1$

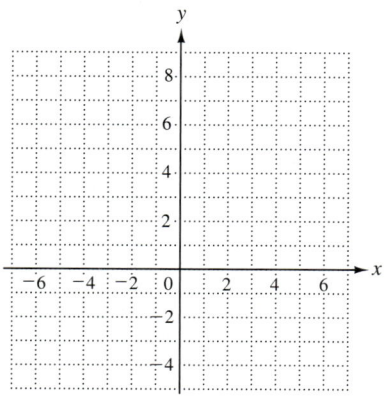

4. $y \leq -x^2 - 2x + 1$

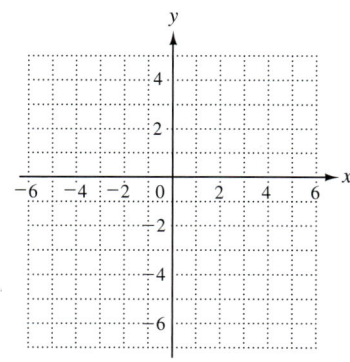

5. $y > -7 + 6x - x^2$

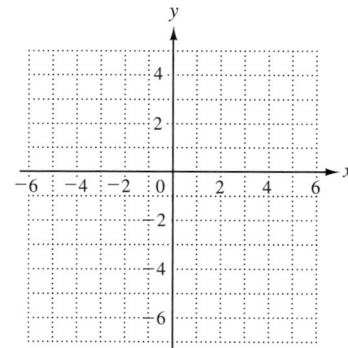

6. $y < x^2 + 6x + 7$

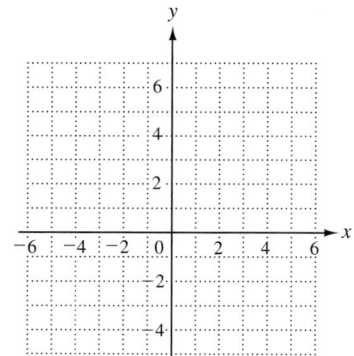

7. In your own words explain how you decide whether to shade the region inside the parabola or outside the parabola in a quadratic inequality.

8. Suppose someone had drawn just the left half of the parabola in Exercise 4. Describe how you could draw the rest of the parabola *without* doing any calculations.

A–18 Appendix C: Graphing Quadratic Inequalities

9. $y \geq x^2 - 9$

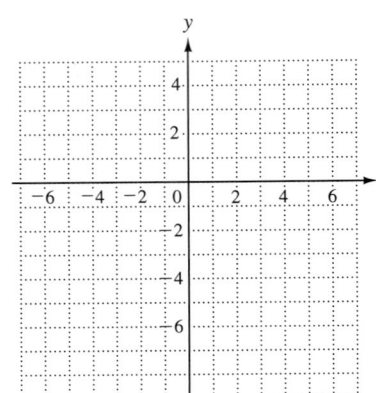

10. $y > 1 - x^2$

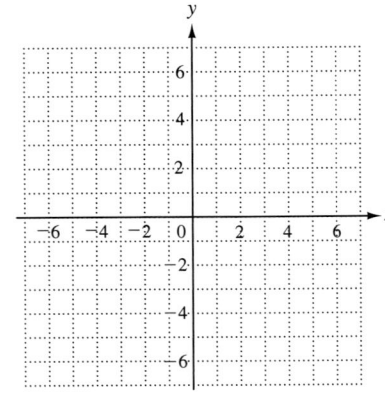

11. $y \leq -x^2 + x + 6$

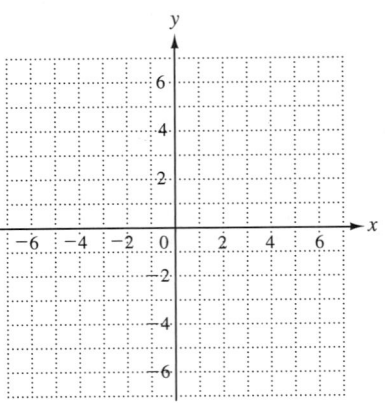

12. $y \geq x^2 - x - 6$

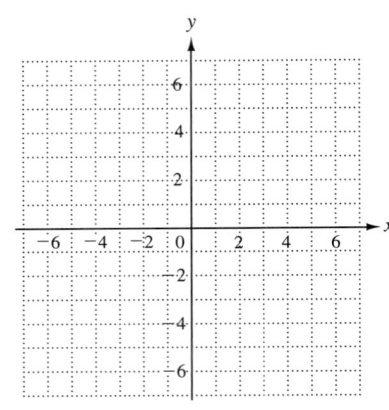

13. $y > -x^2 - 2x + 1$

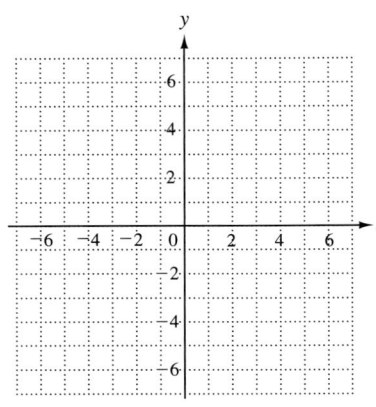

14. $y > x^2 - 2x - 4$

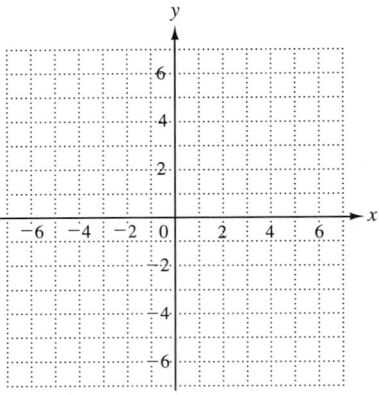

15. $y \geq x^2 + 4x + 2$

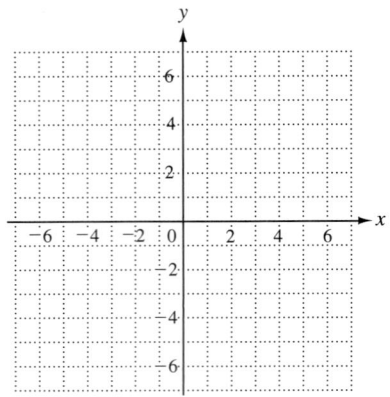

16. $y < 2x^2 - x - 1$

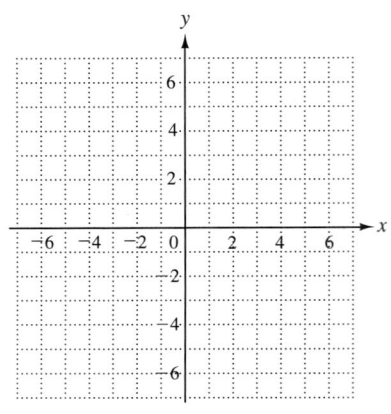

Answers to Selected Exercises

In this section we provide the answers that we think most students will obtain when they work the exercises using the methods explained in the text. If your answer does not look exactly like the one given here, it is not necessarily wrong. In many cases there are equivalent forms of the answer that are correct. For example, if the answer section shows $\frac{3}{4}$ and your answer is 0.75, you have obtained the correct answer but written it in a different (yet equivalent) form. Unless the directions specify otherwise, 0.75 is just as valid an answer as $\frac{3}{4}$.

In general, if your answer does not agree with the one given in the text, see whether it can be transformed into the other form. If it can, then it is the correct answer. If you still have doubts, talk with your instructor.

DIAGNOSTIC PRETEST

(page xix)
1. 25,003,701 **2.** 6 **3.** 4,400,000 **4.** $-\$18$ **5.** -40
6. $4r - 33$ **7.** $t = 5$ **8.** $n = 1$ **9.** 58 cm **10.** 66 in.²
11. $4n + 13 = n - 8$; $n = -7$ **12.** $x + x + 12 = 238$; 113 ft, 125 ft
13. $\frac{3}{4}$ **14.** $-\frac{1}{4}$ **15.** $x = -20$ **16.** 30 cm² **17.** $18.17 (rounded)
18. $r = 2.3$ **19.** $10.20 **20.** 16.3 cm (rounded) **21.** $x = 108$
22. 121.6 miles **23.** 117° **24.** $x = 8$ ft; $y = 7.5$ ft **25.** (a) 2.5
(b) 30% **26.** (a) $\frac{19}{50}$ (b) 7.5% **27.** $249.10 **28.** $13.80
29. (a) 4.32 m (b) 80 g **30.** 5 ft **31.** 10.4 L **32.** (a) m
(b) mg **33.** $252 **34.** (a) 2001, 2002 (b) 2004; 30 sections
35. (a) quadrant II (b) no quadrant (c) quadrant III
(d) quadrant I
36.

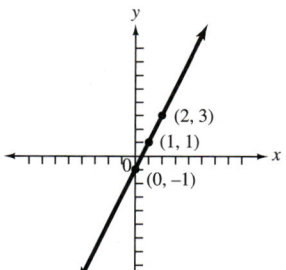

x	y	(x, y)
0	−1	(0, −1)
1	1	(1, 1)
2	3	(2, 3)

37. $81 \geq 81$; true **38.** no solution **39.** $t = \dfrac{A - p}{pr}$
40. $x \leq 5$

41. x-intercept: (5, 0); y-intercept: (0, −2)

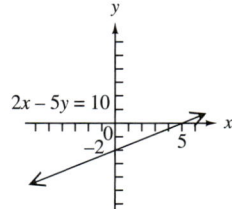

42. -4 **43.** $y = -x + 1$

44.

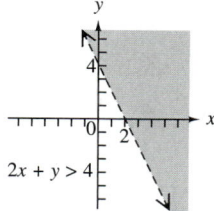

45. (2, 3) **46.** no solution **47.** Marla: 61 mph; Rick: 53 mph
48.

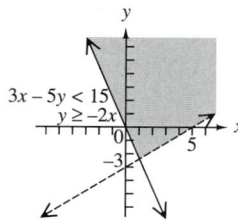

49. $-2m^3 - 9m - 5$ **50.** $49z^2 + 42zw + 9w^2$ **51.** $-\dfrac{3}{10}$
52. (a) 4.45×10^8 (b) 0.000234 **53.** $(3x - 4)(x + 2)$
54. $(4n + 7)(4n - 7)$ **55.** $-3, 5$ **56.** 15 in. by 11 in.
57. $\dfrac{x + 5}{x + 4}$ **58.** $-\dfrac{1}{3r}$ **59.** $\dfrac{z^3 + 2z^2 + 3z}{(z - 3)(z + 3)}$ **60.** $y + 1$ **61.** 0
62. $12 - 2\sqrt{35}$ **63.** $2\sqrt{10}$ **64.** 27 **65.** $3 + 2\sqrt{5}, 3 - 2\sqrt{5}$
66. $\dfrac{-3 + \sqrt{17}}{4}, \dfrac{-3 - \sqrt{17}}{4}$
67. vertex: (0, 5)

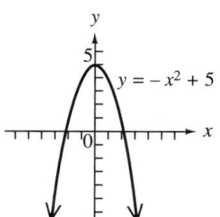

68. function; domain: $\{-2, -1, 0, 1, 2\}$; range: $\{0, 1, 4\}$

CHAPTER 1

Section 1.1 (page 5)
1. 15; 0; 83,001 **3.** 7; 362,049 **5.** hundreds
7. hundred-thousands **9.** ten-millions **11.** hundred-billions
13. ten-trillions, hundred-billions, millions, hundred-thousands, ones
15. eight thousand, four hundred twenty-one **17.** forty-six thousand, two hundred five **19.** three million, sixty-four thousand, eight hundred one **21.** eight hundred forty million, one hundred eleven thousand, three **23.** fifty-one billion, six million, eight hundred eighty-eight thousand, three hundred twenty-one
25. three trillion, seven hundred twelve million **27.** 46,805
29. 5,600,082 **31.** 271,900,000 **33.** 12,417,625,310
35. 600,000,071,000,400 **37.** six thousand, three hundred seventy-five **39.** 101,280,000 **41.** ninety-three million, nine hundred seventy-two thousand, six hundred two dollars

A-19

43. 55,000,800 **45.** six million, four hundred thousand every day; two billion, three hundred thirty-six million in one year **47.** largest: 97,651,100; ninety-seven million, six hundred fifty-one thousand, one hundred; smallest: 10,015,679; ten million, fifteen thousand, six hundred seventy-nine **48.** Answers will vary. **49.** sixty-fours; thirty-twos; sixteens; eights **(a)** 101 **(b)** 1010 **(c)** 1111 **50. (a)** Answers will vary but should mention that the location or place in which a digit is written gives it a different value. **(b)** 8 = VIII; 38 = XXXVIII; 275 = CCLXXV; 3322 = MMMCCCXXII **(c)** The Roman system is *not* a place value system because no matter what place it's in, M = 1000, C = 100, etc. One disadvantage is that it takes much more space to write many large numbers; another is that there is no symbol for zero.

Section 1.2 (page 13)

1. ⁺29,035 feet or 29,035 feet **3.** ⁻128.6 degrees **5.** ⁻18 yards **7.** ⁺$100 or $100 **9.** ⁻6½ pounds

11. [number line: point at -2]
13. [number line: points at -3, 1, 4]
15. [number line: points at -6, -2, 0, 1]

17. > **19.** < **21.** < **23.** > **25.** < **27.** > **29.** > **31.** < **33.** 15 **35.** 3 **37.** 0 **39.** 200 **41.** 75 **43.** 8042 **45.** A C D B on number line at -2, -1, 0, 1 **46.** ⁻1.5, ⁻1, 0, 0.5 **47.** A: may be at risk; B: above normal; C: normal; D: normal **48. (a)** The patient would think the interpretation was "above normal" and wouldn't get treatment. **(b)** Patient D's score of 0; zero is neither positive nor negative.

Section 1.3 (page 21)

1. 3

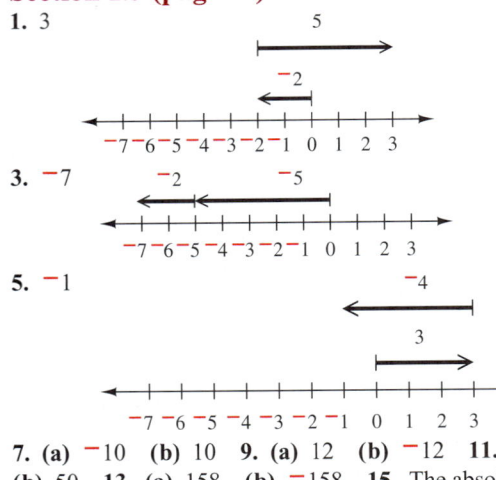

3. ⁻7

5. ⁻1

7. (a) ⁻10 **(b)** 10 **9. (a)** 12 **(b)** ⁻12 **11. (a)** ⁻50 **(b)** 50 **13. (a)** 158 **(b)** ⁻158 **15.** The absolute values are the same in each pair of answers so the only difference in the sums is the common sign. **17. (a)** 2 **(b)** ⁻2 **19. (a)** ⁻7 **(b)** 7 **21. (a)** ⁻5 **(b)** 5 **23. (a)** 150 **(b)** ⁻150 **25.** Each pair of answers differs only in the sign of the answer. This occurs because the signs of the addends are reversed. **27.** ⁻3 **29.** 7 **31.** ⁻7 **33.** 1 **35.** ⁻8 **37.** ⁻20 **39.** ⁻17 **41.** ⁻22 **43.** ⁻19 **45.** 6 **47.** ⁻5 **49.** 0 **51.** 5 **53.** ⁻32 **55.** 13 + ⁻17 = ⁻4 yards **57.** ⁻62 + 50 = ⁻$12 **59.** ⁻88 + 35 = ⁻$53 **61.** Jeff: ⁻20 + 75 + ⁻55 = 0 points; Terry: 42 + ⁻15 + 20 = 47 points **63.** ⁻1 **65.** ⁻1 **67.** ⁻5 + ⁻18; ⁻23 **69.** 15 + ⁻4; 11 **71.** 6 + (⁻14 + 14); 6 + 0 = 6 **73.** (⁻14 + ⁻6) + ⁻7;
⁻20 + ⁻7 = ⁻27 **75.** Some possibilities are: ⁻6 + 0 = ⁻6; 10 + 0 = 10; 0 + 3 = 3 **77.** ⁻4116 **79.** 8686 **81.** ⁻96,077

Section 1.4 (page 27)

1. ⁻6; 6 + ⁻6 = 0 **3.** 13; ⁻13 + 13 = 0 **5.** 0; 0 + 0 = 0 **7.** 14 **9.** ⁻2 **11.** ⁻12 **13.** ⁻25 **15.** ⁻23 **17.** 5 **19.** 20 **21.** 11 **23.** ⁻60 **25.** 0 **27.** 0 **29.** ⁻6 **31. (a)** 8 **(b)** ⁻2 **(c)** 2 **(d)** ⁻8 **33. (a)** ⁻3 **(b)** 11 **(c)** ⁻11 **(d)** 3 **35.** ⁻6 **37.** ⁻5 **39.** 3 **41.** ⁻10 **43.** 12 **45.** ⁻5 **47. (a)** 21 °F; 30 − 21 = 9 degrees difference; **(b)** 0 °F; 15 − 0 = 15 degrees difference **(c)** ⁻17 °F; 5 − ⁻17 = 22 degrees difference; **(d)** ⁻41 °F; ⁻10 − ⁻41 = 31 degrees difference **49.** The student forgot to change 6 to its opposite, ⁻6. It should be ⁻6 + ⁻6 = ⁻12. **51.** ⁻11 **53.** ⁻5 **55.** ⁻10 **57.** Answers on left: ⁻8; 8. On right: ⁻1; 1 Subtraction is *not* commutative; the absolute value of the answer is the same, but the sign changes. **58.** Subtracting 0 from a number does *not* change the number. For example ⁻5 − 0 = ⁻5. But subtracting a number from 0 *does* change the number to its opposite. For example, 0 − ⁻5 = 5.

Section 1.5 (page 35)

1. 630 **3.** ⁻1080 **5.** 7900 **7.** ⁻86,800 **9.** 42,500 **11.** ⁻6000 **13.** 15,800 **15.** ⁻78,000 **17.** 6000 **19.** 53,000 **21.** 600,000 **23.** ⁻9,000,000 **25.** 140,000,000 **27.** 30,000 miles **29.** ⁻60 degrees **31.** $10,000 **33.** 60,000,000 Americans **35.** ⁻300 feet **37.** 600,000 people in Alaska; 30,000,000 people in California **39.** Answers will vary but should mention looking only at the second digit, rounding first digit up when second digit is 5 or more, leaving first digit unchanged when second digit is 4 or less. Examples will vary. **41.** Estimate: ⁻40 + 90 = 50; Exact: ⁻42 + 89 = 47 **43.** Estimate: 20 + ⁻100 = ⁻80; Exact: 16 + ⁻97 = ⁻81 **45.** Estimate: ⁻300 + ⁻400 = ⁻700; Exact: ⁻273 + ⁻399 = ⁻672 **47.** Estimate: 3000 + 7000 = 10,000; Exact: 3081 + 6826 = 9907 **49.** Estimate: 20 + ⁻80 = ⁻60; Exact: 23 − 81 = ⁻58 **51.** Estimate: ⁻40 + ⁻40 = ⁻80; Exact: ⁻39 − 39 = ⁻78 **53.** Estimate: ⁻100 + 30 + ⁺70 = 0; Exact: ⁻106 + 34 − ⁻72 = 0 **55.** Estimate: 80,000 − 50,000 = $30,000; Exact: 78,650 − 52,882 = $25,768 **57.** Estimate: 2000 − 700 − 300 − 300 − 200 − 200 = $300; Exact: 1920 − 685 − 325 − 320 − 182 − 150 = $258 **59.** Estimate: ⁻100 + 40 + 50 = ⁻10 degrees; Exact: ⁻102 + 37 + 52 = ⁻13 degrees **61.** Estimate: 400 + 100 = 500 doors and windows; Exact: 412 + 147 = 559 doors and windows

Section 1.6 (page 45)

1. (a) 63 **(b)** 63 **(c)** ⁻63 **(d)** ⁻63 **3. (a)** ⁻56 **(b)** ⁻56 **(c)** 56 **(d)** 56 **5.** ⁻35 **7.** ⁻45 **9.** ⁻18 **11.** ⁻50 **13.** ⁻40 **15.** ⁻56 **17.** 32 **19.** 77 **21.** 0 **23.** 133 **25.** 13 **27.** 0 **29.** 48 **31.** ⁻56 **33.** ⁻160 **35.** 5 **37.** ⁻3 **39.** ⁻1 **41.** 0 **43.** 5 **45.** ⁻4 **47.** Commutative property: changing the *order* of the factors does not change the product. Associative property: changing the *grouping* of the factors does not change the product. Examples will vary. **49.** Examples will vary. Some possibilities are: **(a)** 6 • ⁻1 = ⁻6; 2 • ⁻1 = ⁻2; 15 • ⁻1 = ⁻15 **(b)** ⁻6 • ⁻1 = 6; ⁻2 • ⁻1 = 2; ⁻15 • ⁻1 = 15 The result of multiplying any nonzero number times ⁻1 is the number with the opposite sign. **50.** The products are 4, ⁻8, 16, ⁻32. The absolute value doubles each time and the sign changes. The next three products are 64, ⁻128, 256. **51.** = 9 • ⁻3 + 9 • 5 Both results are 18. **53.** = 8 • 25 Both products are 200. **55.** = (⁻3 • 2) • 5 Both products are ⁻30. **57.** Estimate: 300 • 50 = $15,000; Exact: 324 • 52 = $16,848 **59.** Estimate: ⁻10,000 • 10 = ⁻$100,000; Exact: ⁻9950 • 12 = ⁻$119,400 **61.** Estimate: 200 • 10 = $2000; Exact: 182 • 13 = $2366 **63.** Estimate: 20 • 400 = 8000 hours; Exact: 24 • 365 = 8760 hours **65.** ⁻512 **67.** 0 **69.** ⁻355,299 **71.** $247 **73.** ⁻22 degrees **75.** 772 points

Section 1.7 (page 55)

1. (a) 7 (b) 7 (c) −7 (d) −7   3. (a) −7 (b) 7 (c) −7 (d) 7   5. (a) 1 (b) 35 (c) −13 (d) 1   7. (a) 0 (b) undefined (c) undefined (d) 0   9. −4   11. −3   13. 6   15. −11   17. undefined   19. −14   21. 10   23. 4   25. −1   27. 0   29. 191   31. −499   33. 2   35. −4   37. 40   39. −48   41. 5   43. 0   45. $2 \div 1 = 2$ but $1 \div 2 = 0.5$, so division is not commutative.   46. $(12 \div 6) \div 2 = 2 \div 2 = 1; 12 \div (6 \div 2) = 12 \div 3 = 4$; different quotients. Division is not associative.   47. Similar: If the signs match, the result is positive. If the signs are different, the result is negative. Different: Multiplication is commutative, division is not. You can multiply by 0, but dividing by 0 is undefined.   48. Examples will vary. The properties are: Any nonzero number divided by itself is 1. Any number divided by 1 is the number. Division by 0 is undefined. Zero divided by any other number (except 0) is 0.

49. (a) $\frac{-6}{-1} = 6; \frac{-2}{-1} = 2; \frac{-15}{-1} = 15$

(b) $\frac{6}{-1} = -6; \frac{2}{-1} = -2; \frac{15}{-1} = -15$ When dividing by −1, change the sign of the dividend to its opposite to get the quotient.

50. Division is not commutative. $\frac{0}{-3} = 0$ because $0 \cdot {-3} = 0$.
But $\frac{-3}{0}$ is undefined because when $\frac{-3}{0} = ?$ is rewritten as $? \cdot 0 = -3$, no number can replace ? and make a true statement.   51. *Estimate:* $-40{,}000 \div 20 = -2000$ feet; *Exact:* $-35{,}836 \div 17 = -2108$ feet   53. *Estimate:* $-200 + 500 = \$300$; *Exact:* $-238 + 450 = \$212$ 55. *Estimate:* $400 - 100 = 300$ days; *Exact:* $365 - 106 = 259$ days   57. *Estimate:* $-700 \cdot 40 = -28{,}000$ feet; *Exact:* $-730 \cdot 37 = -27{,}010$ feet   59. *Estimate:* $300 \div 5 = 60$ miles; *Exact:* $315 \div 5 = 63$ miles   61. Average score of 168   63. The back shows 520 grams, which is 10 grams more than the front.   65. −$15   67. 16 hours, with 40 minutes left over   69. 33 rooms, with space for 2 people unused.   71. −10   73. undefined   75. 31.70979198 rounds to 32 years.

Section 1.8 (page 65)

1. $4 \cdot 4 \cdot 4$; 4 cubed or 4 to the third power   3. 2^7; 128; 2 to the seventh power   5. 5^4; 625; 5 to the fourth power   7. 7^2; $7 \cdot 7$; 49   9. 10^1; 10; 10   11. (a) 10 (b) 100 (c) 1000 (d) 10,000   13. (a) 4 (b) 16 (c) 64 (d) 256   15. 9,765,625   17. 4096   19. 4   21. 25   23. −64   25. 81   27. −1000   29. 1   31. 108   33. 200   35. −750   37. −32   39. (a) The answers are 4, −8, 16, −32, 64, −128, 256, −512. When a negative number is raised to an even power, the answer is positive; when raised to an odd power, the answer is negative. (b) negative; positive   41. −6   43. 0   45. −39   47. 16   49. 23   51. −43   53. 7   55. −3   57. 0   59. −38   61. 41   63. −2   65. 13   67. 126   69. 8   71. $\frac{27}{-3} = -9$ 73. $\frac{-48}{-4} = 12$ 75. $\frac{-60}{-1} = 60$ 77. −4050   79. 7   81. $\frac{27}{0}$ is undefined.

Summary Exercises on Operations with Integers (page 69)

1. −6   2. 0   3. −7   4. −7   5. 63   6. −1   7. −56   8. −22   9. 12   10. 8   11. −13   12. 0   13. 0   14. −17   15. −48   16. −10   17. −50   18. undefined   19. −14   20. −6   21. 0   22. 16   23. −30   24. undefined   25. 48   26. −19   27. 2   28. −20   29. 0   30. 16   31. −3   32. −36   33. 6   34. −7   35. −5   36. −31   37. −2   38. −32   39. −5   40. −4

41. −9732   42. 100   43. 4   44. −343   45. 5   46. −6   47. −10   48. −5   49. 8   50. 1   51. $\frac{-22}{-11} = 2$ 52. $\frac{8}{-2} = -4$ 53. $\frac{27}{0}$ is undefined.   54. $\frac{-8}{8} = -1$

Chapter 1 Review Exercises (page 77)

1. 86, 0, 35,600   2. eight hundred six   3. three hundred nineteen thousand, twelve   4. sixty million, three thousand, two hundred   5. fifteen trillion, seven hundred forty-nine billion, six   6. 504,100   7. 620,080,000   8. 99,007,000,356
9. number line from −5 to 5 with points at −4, −1, 2, 4
10. >   11. <   12. >   13. <   14. 5   15. 9   16. 0   17. 125   18. −1   19. −13   20. −3   21. 0   22. 1   23. −24   24. −7   25. 3   26. 0   27. −17   28. 5; $-5 + 5 = 0$ 29. −18; $18 + {-18} = 0$ 30. −7   31. 17   32. −16   33. 13   34. 18   35. −22   36. 0   37. −20   38. −1   39. −3   40. 14   41. 15   42. −16   43. 3   44. −8   45. 0   46. 210   47. 59,000   48. 85,000,000   49. −3000   50. −7,060,000   51. 400,000   52. −200 pounds   53. −1000 feet   54. 400,000,000 directories   55. 9,000,000,000 people   56. −54   57. 56   58. −100   59. 0   60. 24   61. 17   62. −48   63. 125   64. −36   65. 50   66. −72   67. 9   68. −7   69. undefined   70. 5   71. −18   72. 0   73. 15   74. −1   75. −5   76. 18   77. 0   78. 156 days and 2 extra hours   79. 10,000   80. 32   81. 27   82. 16   83. −125   84. 8   85. 324   86. −200   87. −25   88. −2   89. 10   90. −28   91. $\frac{8}{-8} = -1$ 92. $\frac{11}{0}$ is undefined.   93. associative property of addition   94. commutative property of multiplication   95. addition property of 0   96. multiplication property of 0   97. distributive property   98. associative property of multiplication   99. *Estimate:* $\$10{,}000 \cdot 200 = \$2{,}000{,}000$; *Exact:* $\$11{,}900 \cdot 192 = \$2{,}284{,}800$ 100. *Estimate:* $\$200 + \$400 - \$700 = -\$100$; *Exact:* $\$185 + \$428 - \$706 = -\$93$ 101. *Estimate:* $800 \div 20 = 40$ miles; *Exact:* $840 \div 24 = 35$ miles   102. *Estimate:* $(\$40 \cdot 20) + (\$90 \cdot 10) = \$1700$; *Exact:* $(\$39 \cdot 19) + (\$85 \cdot 12) = \$1761$ 103. −$700, −$100, $700, $900, $0, $700   104. January; April   105. $2100   106. −$1850

Chapter 1 Test (page 81)

1. twenty million, eight thousand, three hundred seven
2. 30,000,700,005
3. number line from −3 to 3 with points at −2, 0, 1, 3
4. >; <   5. 10; 14   6. −6   7. −5   8. 7   9. −40   10. 10   11. 64   12. −50   13. undefined   14. −60   15. −5   16. −45   17. 6   18. 25   19. 0   20. 9   21. 128   22. −2   23. 8   24. −16   25. An exponent shows how many times to use a factor in repeated multiplication. Examples will vary. Some possibilities are $(2)^4 = 2 \cdot 2 \cdot 2 \cdot 2 = 16$ and $(-3)^2 = (-3)(-3) = 9$.   26. Commutative property: changing the *order* of addends does not change the sum. Associative property: changing the *grouping* of addends does not change the sum. Examples will vary.   27. 900   28. 36,420,000,000   29. 350,000   30. *Estimate:* $\$200 + \$300 + -\$500 = \$0$; *Exact:* $\$184 + \$293 + -\$506 = -\$29$ 31. *Estimate:* $-1000 \div 10 = -100$ yards; *Exact:* $-1104 \div 12 = -95$ yards   32. *Estimate:* $30(200 - 100) = 3000$ calories; *Exact:* $31(220 - 110) = 3410$ calories   33. $-10 - {-100} = -10 + {^+100} = 90$ degrees difference   34. 27 cartons because 26 cartons would leave 28 pounds of books unpacked

CHAPTER 2

Section 2.1 (page 91)

1. c is variable; 4 is constant. 3. h is variable; 5 is coefficient.
5. m is variable; $^-3$ is constant. 7. c is variable; 2 is coefficient; 10 is constant. 9. x and y are variables. 11. g is variable; $^-6$ is coefficient; 9 is constant 13. (a) $654 + 10$ is 664 robes.
(b) $208 + 10$ is 218 robes. (c) $95 + 10$ is 105 robes.
15. (a) $3 \cdot 11$ inches is 33 inches. (b) $3 \cdot 3$ feet is 9 feet.
17. (a) $3 \cdot 12 - 5$ is 31 brushes. (b) $3 \cdot 16 - 5$ is 43 brushes.
19. (a) $\frac{332}{4}$ is 83 points. (b) $\frac{637}{7}$ is 91 points.
21. $12 + 12 + 12 + 12$ is 48, $4 \cdot 12$ is 48; $0 + 0 + 0 + 0$ is 0, $4 \cdot 0$ is 0; $^-5 + ^-5 + ^-5 + ^-5 = ^-20$, $4 \cdot ^-5$ is $^-20$
23. $^-2(^-4) + 5$ is $8 + 5$ is 13; $^-2(^-6) + ^-2$ is $12 + ^-2$ is 10; $^-2(0) + ^-8$ is $0 + ^-8$ is $^-8$ 25. A variable is a letter that represents the part of a rule that varies or changes depending on the situation. An expression expresses, or tells, the rule for doing something. For example, $c + 5$ is an expression, and c is the variable.
27. $b \cdot 1 = b$ or $1 \cdot b = b$ 29. $\frac{b}{0}$ is undefined or $b \div 0$ is undefined. 31. $c \cdot c \cdot c \cdot c \cdot c \cdot c$ 33. $x \cdot x \cdot x \cdot x \cdot y \cdot y \cdot y$
35. $^-3 \cdot a \cdot a \cdot a \cdot b$ 37. $9 \cdot x \cdot y \cdot y$
39. $^-2 \cdot c \cdot c \cdot c \cdot c \cdot c \cdot d$ 41. $a \cdot a \cdot a \cdot b \cdot c \cdot c$
43. 16 45. $^-24$ 47. $^-18$ 49. $^-128$ 51. $^-18,432$
53. 311,040 55. 56 57. $\frac{36}{0}$ is undefined. 59. (a) 3 miles
(b) 2 miles (c) 1 mile 60. (a) $\frac{1}{2}$ mile; take half of the distance for 5 seconds (b) $7\frac{1}{2}$ seconds; find the number halfway between 5 seconds and 10 seconds (c) $12\frac{1}{2}$ seconds; find the number halfway between 10 seconds and 15 seconds

Section 2.2 (page 103)

1. $2b^2$ and b^2; The coefficients are 2 and 1. 3. ^-xy and $2xy$; The coefficients are $^-1$ and 2. 5. 7, 3, and $^-4$; The like terms are constants. 7. $12r$ 9. $6x^2$ 11. ^-4p 13. $^-3a^3$ 15. 0
17. xy 19. $6t^4$ 21. $4y^2$ 23. ^-8x 25. $12a + 4b$ 27. $7rs + 14$
29. $a + 2ab^2$ 31. $^-2x + 2y$ 33. $7b^2$ 35. cannot be simplified
37. $^-15r + 5s + t$ 39. $30a$ 41. $^-8x^2$ 43. $^-20y^3$ 45. $18cd$
47. $21a^2bc$ 49. $12w$ 51. $6b + 36$ 53. $7x - 7$ 55. $21t + 3$
57. $^-10r - 6$ 59. $^-9k - 36$ 61. $50m - 300$ 63. $8y + 16$
65. $6a^2 + 3$ 67. $9m - 34$ 69. $^-25$ 71. $24x$ 73. $5n + 13$
75. $11p - 1$ 77. A simplified expression still has variables, but is written in a simpler way. When evaluating an expression, the variables are all replaced by specific numbers. 79. Like terms have matching variable parts, that is, matching letters and exponents. The coefficients do not have to match. Examples will vary.
81. Keep the variable part unchanged when combining like terms. The correct answer is $5x + 8$. 83. $^-2y + 9$ 85. 0 87. ^-9x

Section 2.3 (page 113)

1. 58 is the solution. 3. $^-16$ is the solution.
5. $^-12$ is the solution.
7. $p = 4$ Check $\underbrace{4 + 5}_{9} = 9$
$\phantom{p = 4 \text{ Check }}9 = 9$
9. $r = 10$ Check $8 = \underbrace{10 - 2}_{8}$
$\phantom{r = 10 \text{ Check }}8 = 8$

11. $n = ^-8$ Check $^-5 = n + 3$
$\phantom{n = ^-8 \text{ Check }}^-5 = \underbrace{^-8 + 3}_{^-5}$
$\phantom{n = ^-8 \text{ Check }}^-5 = ^-5$
13. $k = 18$ Check $^-4 + k = 14$
$\phantom{k = 18 \text{ Check }}\underbrace{^-4 + 18}_{14} = 14$
$\phantom{k = 18 \text{ Check }}14 = 14$
15. $y = 6$ Check $y - 6 = 0$
$\phantom{y = 6 \text{ Check }}\underbrace{6 - 6}_{0} = 0$
$\phantom{y = 6 \text{ Check }}0 = 0$
17. $r = ^-6$ Check $7 = r + 13$
$\phantom{r = ^-6 \text{ Check }}7 = \underbrace{^-6 + 13}_{7}$
$\phantom{r = ^-6 \text{ Check }}7 = 7$
19. $x = 11$ Check $x - 12 = ^-1$
$\phantom{x = 11 \text{ Check }}\underbrace{11 + ^-12}_{^-1} = ^-1$
$\phantom{x = 11 \text{ Check }}^-1 = ^-1$
21. $t = ^-3$ Check $^-5 = ^-2 + t$
$\phantom{t = ^-3 \text{ Check }}^-5 = \underbrace{^-2 + ^-3}_{^-5}$
$\phantom{t = ^-3 \text{ Check }}^-5 = ^-5$
23. Does not balance $\underbrace{^-2 + ^-5}_{^-7} = 3$
 The correct solution is 8. $^-7 \neq 3$
 $\underbrace{8 - 5}_{3} = 3$
 Balances $3 = 3$
25. $7 + x = ^-11$
 $\underbrace{7 + ^-18}_{^-11} = ^-11$
 Balances $^-11 = ^-11$
 $^-18$ is the correct solution.
27. $^-10 = ^-10 + b$
 $^-10 = \underbrace{^-10 + 10}_{0}$
 Does not balance $^-10 \neq 0$
 The correct solution is 0. $^-10 = \underbrace{^-10 + 0}_{^-10}$
 Balances $^-10 = ^-10$
29. $c = 6$; see *Student's Solutions Manual* for **Checks** of odd-numbered Exercises 29–39. 31. $y = 5$ 33. $b = ^-30$ 35. $t = 0$
37. $z = ^-7$ 39. $w = 3$ 41. $x = ^-10$ 43. $a = 0$
45. $y = ^-25$ 47. $x = 15$ 49. $k = 113$ 51. $b = 18$
53. $r = ^-5$ 55. $n = ^-105$ 57. $h = ^-5$ 59. No, the solution is $^-14$, the number used to replace x in the original equation.
61. $g = 295$ graduates 63. $c = 55$ chirps 65. $p = \$110$ per month in winter 67. $m = ^-19$ 69. $x = 2$ 71. Equations will vary. Some possibilities are (a) $n - 1 = ^-3$ and $8 = x + 10$.
(b) $y + 6 = 6$ and $^-5 = ^-5 + b$. 72. (a) $x = \frac{1}{2}$
(b) $y = \frac{5}{4}$ (c) $n = \$0.85$ (d) Equations will vary.

Section 2.4 (page 123)

1. $z = 2$ Check $\underbrace{6 \cdot 2}_{12} = 12$
$\phantom{z = 2 \text{ Check }}12 = 12$

3. $r = 4$ Check $48 = 12r$
$48 = 12 \cdot 4$
$48 = 48$

5. $y = 0$ Check $3y = 0$
$3 \cdot 0 = 0$
$0 = 0$

7. $k = {}^-10$ Check ${}^-7k = 70$
${}^-7 \cdot {}^-10 = 70$
$70 = 70$

9. $r = 6$ Check ${}^-54 = {}^-9r$
${}^-54 = {}^-9 \cdot 6$
${}^-54 = {}^-54$

11. $b = {}^-5$ Check ${}^-25 = 5b$
${}^-25 = 5 \cdot {}^-5$
${}^-25 = {}^-25$

13. $r = 3$ Check $2 \cdot 3 = 6$
$6 = 6$

15. $p = {}^-3$ Check ${}^-12 = 5p - p$
${}^-12 = 5 \cdot {}^-3 - {}^-3$
${}^-12 = {}^-15 + {}^+3$
${}^-12 = {}^-12$

17. $a = {}^-5$ **19.** $x = {}^-10$ **21.** $w = 0$ **23.** $t = 3$ **25.** $t = 0$
27. $m = 9$ **29.** $y = {}^-1$ **31.** $z = {}^-5$ **33.** $p = {}^-2$ **35.** $k = 7$
37. $b = {}^-3$ **39.** $x = {}^-32$ **41.** $w = 2$ **43.** $n = 50$
45. $p = {}^-10$ **47.** Each solution is the opposite of the number in the equation. So the rule is: When you change the variable from negative to positive, then change the number in the equation to its opposite. In ${}^-x = 5$, the opposite of 5 is ${}^-5$, so $x = {}^-5$.
49. Divide by the coefficient of x, which is 3, *not* by the opposite of 3. The correct solution is 5. **51.** $s = 15$ ft **53.** $s = 24$ meters
55. $y = 27$ **57.** $x = 1$

Section 2.5 (page 131)

1. $p = 1$ Check $7(1) + 5 = 12$
$7 + 5 = 12$
$12 = 12$

2. $y = 1$ Check $2 = 8y - 6$
$2 = 8(1) - 6$
$2 = 8 - 6$
$2 = 2$

5. $m = 0$ Check ${}^-3m + 1 = 1$
${}^-3(0) + 1 = 1$
$0 + 1 = 1$
$1 = 1$

7. $a = {}^-2$ Check $28 = {}^-9a + 10$
$28 = {}^-9({}^-2) + 10$
$28 = 18 + 10$
$28 = 28$

9. $x = {}^-4$ Check ${}^-5x - 4 = 16$
${}^-5({}^-4) - 4 = 16$
$20 - 4 = 16$
$16 = 16$

11. $p = 4; 4 = p$ Check $6(4) - 2 = 4(4) + 6$
$24 - 2 = 16 + 6$
$22 = 22$

13. $k = {}^-2; {}^-2 = k$ Check ${}^-2k - 6 = 6k + 10$
${}^-2({}^-2) - 6 = 6({}^-2) + 10$
$4 + {}^-6 = {}^-12 + 10$
${}^-2 = {}^-2$

15. $a = 5; 5 = a$ Check ${}^-18 + 7a = 2a + 7$
${}^-18 + 7(5) = 2(5) + 7$
${}^-18 + 35 = 10 + 7$
$17 = 17$

17. $w = 6$ **19.** $y = {}^-9$ **21.** $t = {}^-5$ **23.** $x = 0$ **25.** $h = 1$
27. $y = {}^-2$ **29.** $m = {}^-3$ **31.** $w = 2$ **33.** $x = 5$ **35.** $a = 3$
37. $b = {}^-3$ **39.** $k = 4$ **41.** $c = 0$ **43.** $y = {}^-5$ **45.** $n = 21$
47. $c = 30$ **49.** $p = {}^-2$ **51.** $b = {}^-2$
53. The series of steps may vary. One possibility is:

${}^-2t - 10 = 3t + 5$ Change subtraction to adding the opposite.
${}^-2t + {}^-10 = 3t + 5$ Add $2t$ to both sides
$\underline{2t2t}$ (addition property).
$0 + {}^-10 = 5t + 5$ Add ${}^-5$ to both sides
$\underline{{}^-5{}^-5}$ (addition property).
$\dfrac{{}^-15}{5} = \dfrac{5t}{5}$ Divide both sides by 5 (division property).
${}^-3 = t$

55. Check ${}^-8 + 4(3) = 2(3) + 2$
${}^-8 + 12 = 6 + 2$
$4 \neq 8$

The check does not balance, so 3 is not the correct solution. The student added ${}^-2a$ to ${}^-8$ on the left side, instead of adding ${}^-2a$ to $4a$. The correct solution is 5. **57. (a)** It must be negative. **(b)** The sum of x and a positive number is negative, so x must be negative. **58. (a)** It must be positive. **(b)** The sum of d and a negative number is positive, so d must be positive. **59. (a)** It must be positive; when the signs are different, the product is negative. **(b)** The product of n and a negative number is negative, so n must be positive. **60. (a)** It must be negative also; when the signs match, the product is positive. **(b)** The product of y and a negative number is positive, so y must be negative.

Chapter 2 Review Exercises (page 143)

1. (a) Variable is k; coefficient is 4; constant is ${}^-3$.
(b) ${}^-9y + 20$ **2. (a)** 70 test tubes **(b)** 106 test tubes
3. (a) $x \cdot x \cdot y \cdot y \cdot y \cdot y$ **(b)** $5 \cdot a \cdot b \cdot b \cdot b$ **4. (a)** 9
(b) ${}^-27$ **(c)** ${}^-128$ **(d)** 720 **5.** $ab^2 + 3ab$ **6.** ${}^-4x + 2y - 7$
7. $16g^3$ **8.** $12r^2t$ **9.** $5k + 10$ **10.** ${}^-6b - 8$ **11.** $6y$
12. $20x + 2$ **13.** Expressions will vary. One possibility is $6a^3 + a^2 + 3a - 6$.

14. $n = -11$ Check $\underbrace{16 + -11}_{5} = 5$
$\phantom{14. n = -11 \text{ Check } 16 + -11} = 5$

15. $a = 4$ Check $\underbrace{-4 + 2}_{-2} = \underbrace{2(4) - 6 - 4}_{8 + -6 + -4}$
$\phantom{15. a = 4 \text{ Check } -4 + 2 = }-2 = -2$

16. $m = -8$ 17. $k = 10$ 18. $t = 0$ 19. $p = -6$ 20. $r = 2$
21. $h = -12$ 22. $w = 4$ 23. $c = -2$ 24. $n = 15$ employees
25. $a = -5$ 26. $p = 10$ 27. $y = 5$ 28. $m = 3$ 29. $x = 9$
30. $b = 7$ 31. $z = -3$ 32. $n = 4$ 33. $t = 0$ 34. $d = 5$
35. $b = -2$

Chapter 2 Test (page 145)
1. -7 is coefficient; w is variable; 6 is constant.
2. Buy 177 hot dogs.   3. $x \cdot x \cdot x \cdot x \cdot x \cdot y \cdot y \cdot y$
4. $4 \cdot a \cdot b \cdot b \cdot b \cdot b$ 5. -200 6. $-4w^3$ 7. 0   8. c
9. cannot be simplified   10. $-40b^2$ 11. $15k$ 12. $21t + 28$
13. $-4a - 24$ 14. $6x - 15$ 15. $-9b + c + 6$
16. $x = 5$ Check $-4 = \underbrace{5 - 9}_{-4}$
$\phantom{16. x = 5 \text{ Check }} -4 = -4$

17. $w = -11$ Check $\underbrace{-7(-11)}_{77} = 77$
$\phantom{17. w = -11 \text{ Check }} 77 = 77$

18. $p = -14$ Check $\underbrace{-1(-14)}_{14} = 14$
$\phantom{18. p = -14 \text{ Check }} 14 = 14$

19. $a = 3$ Check $-15 = -3(3 + 2)$
$\phantom{19. a = 3 \text{ Check }} -15 = -3(5)$
$\phantom{19. a = 3 \text{ Check }} -15 = -15$

20. $n = -8$ 21. $m = 15$ 22. $x = -1$ 23. $m = 2$
24. $b = 54$ 25. $c = 0$ 26. Equations will vary. Two possibilities are $x - 5 = -9$ and $-24 = 6y$.

CHAPTER 3

Section 3.1 (page 153)
1. $P = 36$ cm   3. $P = 100$ in.   5. $P = 4$ miles   7. $P = 88$ mm
9. $s = 30$ ft   11. $s = 1$ mm   13. $s = 23$ yards   15. $s = 2$ ft
17. $P = 28$ yd   19. $P = 70$ cm   21. $P = 72$ ft   23. $P = 26$ in.
25. $l = 9$ cm   Check $9 \text{ cm} + 9 \text{ cm} + 6 \text{ cm} + 6 \text{ cm} = 30 \text{ cm}$
27. $w = 1$ mile   Check $4 \text{ mi} + 4 \text{ mi} + 1 \text{ mi} + 1 \text{ mi} = 10 \text{ mi}$
29. $w = 2$ ft   Check $6 \text{ ft} + 6 \text{ ft} + 2 \text{ ft} + 2 \text{ ft} = 16 \text{ ft}$
31. $l = 2$ m   Check $2 \text{ m} + 2 \text{ m} + 1 \text{ m} + 1 \text{ m} = 6 \text{ m}$
33. $P = 208$ m   35. $P = 320$ ft   37. $P = 54$ mm   39. $P = 48$ ft
41. $P = 78$ in.   43. $P = 125$ m   45. ? $= 40$ cm   47. ? $= 12$ in.
49. (a) Sketches will vary.   (b) Formula for perimeter of an equilateral triangle is $P = 3s$, where s is the length of one side.   (c) The formula will *not* work for other kinds of triangles because the sides will have different lengths.   51. (a) 140 miles
(b) 350 miles   (c) 560 miles   52. (a) 70 miles
(b) 175 miles   (c) 280 miles   (d) The rate is half of 70 miles per hour, so the distance will be half as far; divide each result in Exercise 51 by 2.   53. (a) 50 hours   (b) 60 hours   (c) 150 hours
54. (a) 61 miles per hour   (b) 57 miles per hour
(c) 65 miles per hour

Section 3.2 (page 163)
1. $A = 77$ ft^2 3. $A = 100$ m^2 5. $A = 775$ mm^2 7. $A = 36$ in.^{2}
9. $A = 105$ cm^2 11. $A = 72$ ft^2 13. $A = 625$ mi^2
15. $A = 1$ m^2 17. $l = 6$ ft   Check $A = 6 \text{ ft} \cdot 3 \text{ ft}; A = 18 \text{ ft}^2$
19. $w = 80$ yd   Check $A = 90 \text{ yd} \cdot 80 \text{ yd}; A = 7200 \text{ yd}^2$
21. $l = 14$ in.   Check $A = 14 \text{ in.} \cdot 11 \text{ in.}; A = 154 \text{ in.}^2$

23. $s = 6$ m   25. $s = 2$ ft   27. $h = 20$ cm
Check $A = 25 \text{ cm} \cdot 20 \text{ cm}; A = 500 \text{ cm}^2$ 29. $b = 17$ in.
Check $A = 17 \text{ in.} \cdot 13 \text{ in.} A = 221 \text{ in.}^2$ 31. $h = 1$ m
Check $A = 9 \text{ m} \cdot 1 \text{ m}; A = 9 \text{ m}^2$ 33. Height is not part of perimeter; square units are used for area, not perimeter
$P = 25 \text{ cm} + 25 \text{ cm} + 25 \text{ cm} + 25 \text{ cm}; P = 100 \text{ cm}$.
35. Rectangle; $P = 32$ m; $A = 39$ m^2 37. Parallelogram; $P = 34$ yd; $A = 56$ yd^2 39. Square; $P = 36$ in.; $A = 81$ in.^{2}
41. $P = 48$ m; $A = 144$ m^2 43. $108   45. $725   47. 53 yd
49. Panoramic: $P = 36$ in., $A = 56$ in.^{2}; 4 in. $\times$ 6 in.: $P = 20$ in., $A = 24$ in.^{2}; 4 in. $\times$ 7 in.: $P = 22$ in., $A = 28$ in.^{2}

51. [5 ft × 1 ft]   [4 ft × 2 ft]   [3 ft × 3 ft]

52. (a) 5 ft by 1 ft has area of 5 ft^2; 4 ft by 2 ft has area of 8 ft^2; 3 ft by 3 ft has area of 9 ft^2 (b) The square plot 3 ft by 3 ft has the greatest area.

53. [7 ft × 1 ft]   [6 ft × 2 ft]   [5 ft × 3 ft]   [4 ft × 4 ft]

54. (a) 7 ft^2, 12 ft^2, 15 ft^2, 16 ft^2 (b) Square plots have the greatest area.

Section 3.3 (page 173)
1. $14 + x$ or $x + 14$ 3. $-5 + x$ or $x + -5$ 5. $20 - x$
7. $x - 9$ 9. $x - 4$ 11. $-6x$ 13. $2x$ 15. $\dfrac{x}{2}$ 17. $2x + 8$ or $8 + 2x$ 19. $7x - 10$ 21. $2x + x$ or $x + 2x$
23. $4n - 2 = 26; n = 7$ Check $4 \cdot 7 - 2$ does equal 26.
$\phantom{23. 4n - 2 = 26; n = 7 \text{ Check }} \underbrace{28 - 2}_{26}$

25. $2n + n = -15$ or $n + 2n = -15; n = -5$
Check $2 \cdot -5 + -5$ does equal -15.
$ \underbrace{-10 + -5}_{-15}$

27. $5n + 12 = 7n; n = 6$ Check $\underbrace{5 \cdot 6 + 12}_{30 + 12 = 42} = \underbrace{7 \cdot 6}_{42}$
$\phantom{27. 5n + 12 = 7n; n = 6 \text{ Check }} 42 = 42$

29. $30 - 3n = 2 + n; n = 7$ Check $\underbrace{30 - 3 \cdot 7}_{30 - 21 = 9} = \underbrace{2 + 7}_{9}$
$\phantom{29. 30 - 3n = 2 + n; n = 7 \text{ Check }} 9 = 9$

31. Let w be Ricardo's original weight. $w + 15 - 28 + 5 = 177$. He weighed 185 pounds originally.   33. Let c be the number of cookies the children ate. $18 - c + 36 = 49$.   Her children ate 5 cookies.   35. Let p be the number of pens in each box. $6p - 32 - 35 = 5$. There were 12 pens in each box.   37. Let d be each member's dues. $14d + 340 - 575 = -25$.   Each member paid $15.   39. Let a be Tamu's age. $4a - 75 = a$. Tamu is 25 years old.   41. Let m be the amount Brenda spent. $2m - 3 = 81$. Brenda spent $42.   43. Let b be the number of pieces in each bag. $5b - 3 \cdot 48 = b$.   There were 36 pieces of candy in each bag.
45. Let d be the daily amount for an infant. $2d + 3 = 15$.   An infant should receive 6 mg of iron.

Section 3.4 (page 181)
1. a is my age; $a + 9$ is my sister's age. $a + a + 9 = 51$. I am 21; my sister is 30.   3. m is husband's earnings; $m + 1500$ is Lien's.

$m + m + 1500 = 37,500$. Husband earned \$18,000; Lien earned \$19,500. **5.** m is printer's cost; $5m$ is computer's cost; $5m + m = \$1320$. Printer cost \$220; computer cost \$1100.
7. Shorter piece is x; longer piece is $x + 10$. $x + x + 10 = 78$. Shorter piece is 34 cm; longer piece is 44 cm. **9.** Longer piece is x; shorter piece is $x - 7$. $x + x - 7 = 31$. Longer piece is 19 ft; shorter piece is 12 ft. **11.** s is number of Senators; $5s - 65$ is number of Representatives. $s + 5s - 65 = 535$. 100 Senators; 435 Representatives **13.** First part is x; second part is x; third part is $x + 25$. $x + x + x + 25 = 706$. First part is 227 m; second part is 227 m; third part is 252 m. **15.** Length is 19 yd.
17. Length is 12 ft; width is 6 ft. **19.** Length is 13 in.; width is 5 in.
21. $A = 88$ in.^{2}; $P = 52$ in.

Chapter 3 Review Exercises (page 189)

1. Square; $P = 112$ cm **2.** Rectangle; $P = 22$ mi
3. Parallelogram; $P = 42$ yd **4.** $P = 141$ m **5.** 12 ft = $4s$; $s = 3$ ft **6.** 128 yd = $2l + 2(31$ yd); $l = 33$ yd
7. 72 in. = $2(21$ in.) + $2w$; $w = 15$ in. **8.** $A = 40$ ft^2
9. $A = 625$ m^2 **10.** $A = 208$ yd^2 **11.** 126 ft^2 = 14 ft • w; $w = 9$ ft **12.** 88 cm^2 = 11 cm • h; $h = 8$ cm **13.** 100 mi^2 = $s \cdot s$; $s = 10$ mi **14.** $57 - x$ **15.** $15 + 2x$ or $2x + 15$ **16.** $-9x$
17. $4n + 6 = {-}30$; $n = {-}9$ **18.** $10 - 2n = 4 + n$; $n = 2$
19. m is money originally in account. $m - 600 + 750 + 75 = 309$. \$84 was originally in Grace's account. **20.** c is number of candles in each box. $4c - 25 = 23$. There were 12 candles in each box.
21. p is Reggie's prize money; $p + 300$ is Donald's prize money. $p + p + 300 = 1000$. Reggie gets \$350; Donald gets \$650.
22. w is the width; $2w$ is the length. $84 = 2(2w) + 2(w)$. The width is 14 cm; the length is 28 cm. **23. (a)** 36 ft = $4s$; $s = 9$ ft
(b) $A = 9$ ft • 9 ft; $A = 81$ ft^2 **24.** Rectangles will vary. Two possibilities are: $A = 7$ ft • 3 ft, $A = 21$ ft^2; $A = 6$ ft • 4 ft, $A = 24$ ft^2
25. Let f be the fencing for the garden. $36 - f + 20 = 41$. 15 ft of fencing was used on the garden. **26.** w is the width; $w + 2$ is the length. $36 = 2(w + 2) + 2 \cdot w$. Width is 8 ft; length is 10 ft.

Chapter 3 Test (page 191)

1. $P = 262$ m **2.** $P = 40$ in. **3.** $P = 12$ miles **4.** $P = 12$ ft
5. $P = 110$ cm **6.** $A = 486$ mm^2 **7.** $A = 140$ cm^2
8. $A = 3740$ mi^2 **9.** $A = 36$ m^2 **10.** 12 ft = $4s$; $s = 3$ ft
11. 34 ft = $2l + 2(6$ ft); $l = 11$ ft **12.** 65 in.^{2} = 13 in. • h; $h = 5$ in.
13. 12 cm^2 = 4 cm • w; $w = 3$ cm **14.** 16 ft^2 = s^2; $s = 4$ ft
15. Linear units like ft are used to measure length, width, height, and perimeter. Area is measured in square units like ft^2 (squares that measure 1 ft on each side). **16.** $4n + 40 = 0$; $n = {-}10$
17. $7n - 23 = n + 7$; $n = 5$ **18.** Son spent \$15.
19. Daughter is 7 years old. **20.** One piece is 57 cm; one piece is 61 cm. **21.** Sketches may vary; length is 168 ft; width is 42 ft.
22. Marcella worked 11 hours; Tim worked 8 hours.

Chapter 4

Section 4.1 (page 201)

1. $\dfrac{5}{8}, \dfrac{3}{8}$ **3.** $\dfrac{2}{3}, \dfrac{1}{3}$ **5.** $\dfrac{3}{2}, \dfrac{1}{2}$ **7.** $\dfrac{11}{6}, \dfrac{1}{6}$ **9.** $\dfrac{2}{11}, \dfrac{3}{11}, \dfrac{4}{11}$
11. $\dfrac{8}{25}, \dfrac{17}{25}$ **13.** $\dfrac{13}{71}, \dfrac{58}{71}$ **15.** $\dfrac{6}{20}$ **17.** $\dfrac{19}{20}$ **19.** N: 3; D: 4
21. N: 12; D: 7 **23.** Proper: $\dfrac{1}{3}, \dfrac{5}{8}, \dfrac{7}{16}$; Improper: $\dfrac{8}{5}, \dfrac{6}{6}, \dfrac{12}{2}$
25. Proper: $\dfrac{3}{4}, \dfrac{9}{11}, \dfrac{7}{15}$; Improper: $\dfrac{3}{2}, \dfrac{5}{5}, \dfrac{19}{18}$

27. Fractions will vary. The denominator shows the number of equal parts in the whole and the numerator shows how many of the parts are being considered. The fraction bar separates the numerator from the denominator. Show your drawing to your instructor.

29.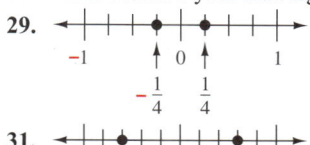

31. number line from −1 to 1 with points at $-\dfrac{3}{5}$ and $\dfrac{3}{5}$

33. number line from −1 to 1 with points at $-\dfrac{7}{8}$ and $\dfrac{7}{8}$

35. $-\dfrac{3}{4}$ pound **37.** $-\dfrac{1}{2}$ quart **39.** $\dfrac{3}{10}$ mile
41. $\dfrac{2}{5}$ **43.** $\dfrac{9}{10}$ **45.** $\dfrac{13}{6}$ **47. (a)** $\dfrac{12}{24}$ **(b)** $\dfrac{8}{24}$ **(c)** $\dfrac{16}{24}$ **(d)** $\dfrac{6}{24}$
(e) $\dfrac{18}{24}$ **(f)** $\dfrac{4}{24}$ **(g)** $\dfrac{20}{24}$ **(h)** $\dfrac{3}{24}$ **(i)** $\dfrac{9}{24}$ **(j)** $\dfrac{15}{24}$ **49. (a)** $-\dfrac{1}{3}$
(b) $-\dfrac{2}{3}$ **(c)** $-\dfrac{2}{3}$ **(d)** $-\dfrac{1}{3}$ **(e)** $-\dfrac{2}{3}$ **(f)** Some possibilities are: $-\dfrac{4}{12} = -\dfrac{1}{3}; -\dfrac{8}{24} = -\dfrac{1}{3}; -\dfrac{20}{30} = -\dfrac{2}{3}; -\dfrac{24}{36} = -\dfrac{2}{3}$.
51. (a) $\dfrac{1467}{3912}$ **(b)** Divide 3912 by 8 to get 489; multiply 3 by 489 to get 1467. **52. (a)** $\dfrac{4256}{5472}$ **(b)** Divide 5472 by 9 to get 608; multiply 7 by 608 to get 4256. **53. (a)** $-\dfrac{1}{5}$ **(b)** Divide 3485 by 2, by 3, and by 5 to see that dividing by 5 gives 697. Or divide 3485 by 697 to get 5. **54. (a)** $-\dfrac{1}{6}$ **(b)** Divide 4902 by 4, by 6, and by 8 to see that dividing by 6 gives 817. Or divide 4902 by 817 to get 6. **55.** 7 **56.** 5 **57.** 15 **58.** 16 **59.** Multiply or divide the numerator and denominator by the same nonzero number. Some possibilities are: $\dfrac{2}{3} = \dfrac{2 \cdot 4}{3 \cdot 4} = \dfrac{8}{12}$ and $\dfrac{10}{16} = \dfrac{10 \div 2}{16 \div 2} = \dfrac{5}{8}$
61. You cannot do it if you want the numerator to be a whole number, because 5 does not divide into 18 evenly. You could use multiples of 5 as the denominator, such as 10, 15, 20, etc. **63.** 10
65. -1 **67.** -6 **69.** 3 **71.** 1 **73.** 2 **75.** -5 **77.** -5
79. 1 **81.** -8 **83.** $\dfrac{2}{5}$ is unshaded. **85.** $\dfrac{5}{8}$ is unshaded.
87. $\dfrac{1}{4}$ is unshaded. **89.** $\dfrac{2}{3}$ is unshaded. **91.** $\dfrac{0}{6}$ is unshaded.
93. Divide each rectangle into 5 equal parts, then shade both rectangles completely. **95.** Drawings will vary.
97. One possibility is shown. ○■■■■■■■△△△
99. One possibility is shown. ⊙!)!!,...???

Section 4.2 (page 215)

1. comp. prime comp. neither prime prime comp. comp.
3. $2 \cdot 3$ **5.** $2 \cdot 2 \cdot 5$ **7.** $5 \cdot 5$ **9.** $2 \cdot 2 \cdot 3 \cdot 3$
11. $2 \cdot 2 \cdot 11$ **13.** $2 \cdot 2 \cdot 2 \cdot 11$ **15.** $3 \cdot 5 \cdot 5$
17. A composite number has a factor(s) other than itself or 1. A prime number is a whole number that has exactly two *different* factors, itself and 1. The whole numbers 0 and 1 are neither prime nor composite.

19. $\dfrac{\cancel{2}\cdot\cancel{2}\cdot\cancel{2}}{\cancel{2}\cdot\cancel{2}\cdot\cancel{2}\cdot 2}=\dfrac{1}{2}$ **21.** $\dfrac{\cancel{2}\cdot\cancel{2}\cdot\cancel{2}\cdot\cancel{2}\cdot 2}{\cancel{2}\cdot\cancel{2}\cdot\cancel{2}\cdot\cancel{2}\cdot 3}=\dfrac{2}{3}$ **23.** $\dfrac{2\cdot\cancel{7}}{3\cdot\cancel{7}}=\dfrac{2}{3}$ **25.** $\dfrac{\cancel{2}\cdot 2\cdot\cancel{3}\cdot 3}{\cancel{2}\cdot\cancel{3}\cdot 7}=\dfrac{6}{7}$ **27.** $\dfrac{3\cdot 3\cdot\cancel{7}}{2\cdot 5\cdot\cancel{7}}=\dfrac{9}{10}$ **29.** $\dfrac{\cancel{3}\cdot\cancel{3}\cdot 3}{\cancel{3}\cdot\cancel{3}\cdot 5}=\dfrac{3}{5}$ **31.** $\dfrac{\cancel{2}\cdot 2\cdot\cancel{3}}{\cancel{2}\cdot\cancel{3}\cdot 3}=\dfrac{2}{3}$ **33.** $\dfrac{\cancel{5}\cdot 7}{2\cdot 2\cdot 2\cdot\cancel{5}}=\dfrac{7}{8}$ **35.** $\dfrac{\cancel{2}\cdot\cancel{3}\cdot\cancel{3}\cdot\cancel{5}}{\cancel{2}\cdot 2\cdot\cancel{3}\cdot\cancel{3}\cdot\cancel{5}}=\dfrac{1}{2}$ **37.** $\dfrac{2\cdot\cancel{3}\cdot\cancel{5}\cdot\cancel{7}}{3\cdot\cancel{3}\cdot\cancel{5}\cdot\cancel{7}}=\dfrac{2}{3}$ **39.** $\dfrac{\cancel{3}\cdot\cancel{11}\cdot 13}{\cancel{3}\cdot 3\cdot 5\cdot\cancel{11}}=\dfrac{13}{15}$ **41. (a)** $\dfrac{1}{4}$ **(b)** $\dfrac{1}{2}$ **(c)** $\dfrac{1}{10}$ **(d)** $\dfrac{60}{60}=1$ **43. (a)** $\dfrac{1}{3}$ **(b)** $\dfrac{1}{5}$ **(c)** $\dfrac{7}{15}$ **45. (a)** $\dfrac{5}{24}$ **(b)** $\dfrac{2}{3}$ **(c)** $\dfrac{1}{8}$ **47. (a)** The result of dividing 3 by 3 is 1, so 1 should be written above and below all the slashes. The numerator is $1\cdot 1$, so the correct answer is $\tfrac{1}{4}$. **(b)** You must divide numerator and denominator by the *same* number. The fraction is already in lowest terms because 9 and 16 have no common factor besides 1. **49.** $\dfrac{2c}{5}$ **51.** $\dfrac{4}{7}$ **53.** $\dfrac{6r}{5s}$ **55.** $\dfrac{1}{7n^2}$ **57.** already in lowest terms **59.** $\dfrac{7}{9y}$ **61.** $\dfrac{7k}{2}$ **63.** $\dfrac{1}{3}$ **65.** $\dfrac{c}{d}$ **67.** $\dfrac{6ab}{c}$ **69.** already in lowest terms **71.** $3eg^2$

Section 4.3 (page 227)

1. $-\dfrac{3}{16}$ **3.** $\dfrac{9}{10}$ **5.** $\dfrac{1}{2}$ **7.** -6 **9.** 36 **11.** $\dfrac{15}{4y}$ **13.** $\dfrac{1}{2}$ **15.** $\dfrac{6}{5}$ **17.** -9 **19.** $-\dfrac{1}{6}$ **21.** $\dfrac{11}{15d}$ **23.** b **25. (a)** Forgot to write 1s in numerator when dividing out common factors. Answer is $\tfrac{1}{6}$. **(b)** Used reciprocal of $\tfrac{2}{3}$ in multiplication, but the reciprocal is used only in division. Correct answer is $\tfrac{16}{3}$. **27. (a)** Forgot to use reciprocal of $\tfrac{4}{1}$; correct answer is $\tfrac{1}{6}$. **(b)** Used reciprocal of $\tfrac{5}{6}$ instead of reciprocal of $\tfrac{10}{9}$; correct answer is $\tfrac{3}{4}$. **29.** Rewrite division as multiplication. Leave the first number (dividend) the same. Change the second number (divisor) to its reciprocal by "flipping" it. Then multiply. **31.** $\dfrac{4}{15}$ **33.** $-\dfrac{9}{32}$ **35.** 21 **37.** 15 **39.** undefined **41.** $-\dfrac{55}{12}$ **43.** $8b$ **45.** $\dfrac{3}{5d}$ **47.** $\dfrac{2x^2}{w}$ **49.** $\dfrac{3}{10}$ yd^2 **51.** 80 dispensers **53.** earn $9300; borrow $3100 **55.** 9 trips **57.** 70 infield players **59.** 3200 times **61. (a)** $58,000 **(b)** $11,600 **63.** $25,375 **65.** 15 million birds **67.** 129 million dogs and cats

Section 4.4 (page 239)

1. $\dfrac{7}{8}$ **3.** $-\dfrac{1}{2}$ **5.** $\dfrac{1}{2}$ **7.** $-\dfrac{9}{40}$ **9.** $\dfrac{13}{24}$ **11.** $-\dfrac{3}{5}$ **13.** $-\dfrac{7}{18}$ **15.** $\dfrac{8}{7}$ **17.** $-\dfrac{3}{8}$ **19.** $\dfrac{3+5c}{15}$ **21.** $\dfrac{10-m}{2m}$ **23.** $\dfrac{8}{b^2}$ **25.** $\dfrac{bc+21}{7b}$ **27.** $\dfrac{-4-cd}{c^2}$ **29.** $-\dfrac{44}{105}$ **31.** You cannot add or subtract until all the fractional pieces are the same size. **33. (a)** $\dfrac{15}{20}+\dfrac{8}{20}=\dfrac{23}{20}$ Cannot add fractions with unlike denominators; use 20 as the LCD. **(b)** When rewriting fractions with 18 as a denominator, you must multiply denominator and numerator by the same number. The correct answer is $\dfrac{15}{18}-\dfrac{8}{18}=\dfrac{7}{18}$. **35. (a)** $\dfrac{1}{12};\dfrac{1}{12}$ Addition is commutative. **(b)** $\dfrac{1}{3};-\dfrac{1}{3}$ Subtraction is *not* commutative. **(c)** $-\dfrac{3}{5};-\dfrac{3}{5}$ Multiplication is commutative. **(d)** $6;\dfrac{1}{6}$ Division is *not* commutative. **36. (a)** 0; 0; The sum of a number and its opposite is 0. **(b)** 1; 1; When a nonzero number is divided by itself, the quotient is 1. **(c)** $\dfrac{5}{6};-\dfrac{17}{20}$; Multiplying by 1 leaves a number unchanged. **(d)** 1; 1; A number times its reciprocal is 1. **37.** $\dfrac{47}{60}$ in. **39.** $\dfrac{23}{24}$ cubic yard **41.** $\dfrac{3}{4}$ acre **43.** $\dfrac{23}{50}$ of workers **45.** $\dfrac{4}{25}$ of workers **47.** $\dfrac{7}{24}$ of the day; 7 hours **49.** $\dfrac{1}{8}$ of the day **51.** $\dfrac{5}{16}$ inch **53.** $\dfrac{1}{12}$ mile

Section 4.5 (page 253)

1. [number line with points at $-4,-3,-2,-1,0,1,2,3,4$; dots at -3 and 2] **3.** [number line with points at $-4,-3,-2,-1,0,1,2,3,4$; dots at -2 and 1] **5.** $\dfrac{9}{2}$ **7.** $-\dfrac{8}{5}$ **9.** $\dfrac{19}{8}$ **11.** $-\dfrac{57}{10}$ **13.** $\dfrac{161}{15}$ **15.** $4\dfrac{1}{3}$ **17.** $-2\dfrac{1}{2}$ **19.** $3\dfrac{2}{3}$ **21.** $-5\dfrac{2}{3}$ **23.** $11\dfrac{3}{4}$ **25.** $7\dfrac{7}{8}$; $2\cdot 4=8$ **27.** $1\dfrac{5}{21}$; $3\div 3=1$ **29.** $5\dfrac{1}{2}$; $4+2=6$ **31.** $3\dfrac{2}{3}$; $4-1=3$ **33.** $\dfrac{17}{18}$; $6\div 6=1$ **35.** $6\dfrac{1}{5}$; $8-2=6$ **37.** $P=7$ in.; $A=3\dfrac{1}{16}$ in.^{2} **39.** $P=19\dfrac{1}{2}$ yd; $A=21\dfrac{1}{8}$ yd^2 **41.** $13+9=22$ ft; $21\dfrac{1}{6}$ ft **43.** $2\cdot 6=12$ ounces; $9\dfrac{5}{8}$ ounces **45.** $4-2=2$ miles; $2\dfrac{3}{10}$ miles **47.** $4\cdot 5=20$ yd; $18\dfrac{3}{4}$ yd **49.** $10-2-3=5$ cubic yards; $5\dfrac{1}{8}$ cubic yards **51.** $19\div 5$ is about 4 hours; $3\dfrac{3}{4}$ hours **53.** $30-6-2=22$ in.; $21\dfrac{3}{8}$ in. **55.** $24+35+24+35=118$ in.; $116\dfrac{1}{2}$ in. **57.** $25{,}730\div 10=2573$ anchors; 2480 anchors **59.** $(4\cdot 23)+(5\cdot 24)+(3\cdot 25)=287$ in.; $280\dfrac{1}{8}$ in.

Summary Exercises on Fractions (page 257)

1. (a) $\dfrac{3}{8};\dfrac{5}{8}$ **(b)** $\dfrac{4}{5};\dfrac{1}{5}$ **2.** [number line from -1 to 1 with marks at $-\tfrac{2}{3}$ and $\tfrac{2}{3}$] **3. (a)** 24 **(b)** 4 **4. (a)** 1 **(b)** -4 **(c)** 9

Answers to Selected Exercises A-27

5. (a) $2 \cdot 2 \cdot 2 \cdot 3 \cdot 3$ (b) $3 \cdot 5 \cdot 7$ **6.** (a) $\frac{4}{5}$ (b) $\frac{7}{8}$ **7.** $\frac{1}{2}$
8. $-\frac{5}{24}$ **9.** $\frac{17}{16}$ **10.** $\frac{5}{6}$ **11.** $-\frac{2}{15}$ **12.** $-\frac{3}{8}$ **13.** 56 **14.** $\frac{11}{24}$
15. $-\frac{7}{6}$ **16.** $-\frac{19}{12}$ **17.** $\frac{25}{12}$ **18.** 35 **19.** $7\frac{7}{12}$; $5 + 3 = 8$
20. $11\frac{3}{7}$; $2 \cdot 5 = 10$ **21.** $3\frac{3}{10}$; $6 - 3 = 3$
22. $\frac{16}{35}$; $2 \div 4 = \frac{2}{4}$ or $\frac{1}{2}$ **23.** 4; $5 \div 1 = 5$ **24.** $2\frac{2}{3}$; $3 - 1 = 2$
25. (a) $2\frac{1}{16}$ in. (b) $\frac{5}{16}$ in. **26.** $P = 3\frac{1}{2}$ in.; $A = \frac{49}{64}$ in.^{2}
27. 12 batches **28.** 6¢ **29.** Not sure, 225 adults; Real, 675 adults;
Imaginary, 600 adults **30.** diameter $= \frac{1}{4}$ in.; $P = 4\frac{1}{2}$ in.
31. 23 bottles **32.** 11 lots

Section 4.6 (page 263)
1. $\frac{9}{16}$ **3.** $\frac{8}{125}$ **5.** $-\frac{1}{27}$ **7.** $\frac{1}{32}$ **9.** $\frac{49}{100}$ **11.** $\frac{36}{25}$ or $1\frac{11}{25}$
13. $\frac{12}{25}$ **15.** $\frac{1}{100}$ **17.** $-\frac{3}{2}$ or $-1\frac{1}{2}$ **19.** (a) The answers are
$\frac{1}{4}, -\frac{1}{8}, \frac{1}{16}, -\frac{1}{32}, \frac{1}{64}, -\frac{1}{128}, \frac{1}{256}, -\frac{1}{512}$. (b) When a negative
number is raised to an even power, the answer is positive.
When a negative number is raised to an odd power, the answer
is negative. **20.** (a) Ask yourself, "What number, times itself,
is 4?" This is the numerator. Then ask, "What number, times itself,
is 9?" This is the denominator. The number under the ketchup is
either $\frac{2}{3}$ or $-\frac{2}{3}$. (b) The number under the ketchup is $-\frac{1}{3}$ because
$\left(-\frac{1}{3}\right)\left(-\frac{1}{3}\right)\left(-\frac{1}{3}\right) = -\frac{1}{27}$. (c) Either $\frac{1}{2}$ or $-\frac{1}{2}$.
(d) No real number works, because both $\left(\frac{3}{4}\right)^2$ and $\left(-\frac{3}{4}\right)^2$ give a
positive result. (e) Either $\frac{1}{3}$ or $-\frac{1}{3}$ inside one set of parentheses
and $\frac{1}{2}$ or $-\frac{1}{2}$ inside the other. **21.** -4 **23.** $\frac{5}{16}$ **25.** $\frac{1}{3}$ **27.** $\frac{1}{6}$
29. $-\frac{17}{24}$ **31.** $-\frac{4}{27}$ **33.** $\frac{1}{36}$ **35.** $\frac{9}{64}$ in.^{2} **37.** $\frac{19}{10}$ or $1\frac{9}{10}$ miles
39. 4 **41.** $-\frac{25}{2}$ or $-12\frac{1}{2}$ **43.** $\frac{1}{14}$ **45.** $\frac{5}{18}$ **47.** -6 **49.** $\frac{9}{100}$

Section 4.7 (page 273)
1. $a = 30$ Check $\frac{1}{3}(30) = 10$
$10 = 10$
3. $b = -24$ Check $-20 = \frac{5}{6}(-24)$
$-20 = -20$
5. $c = 6$ Check $-\frac{7}{2}(6) = -21$
$-21 = -21$
7. $m = \frac{3}{4}$ Check $\frac{9}{16} = \frac{3}{4}\left(\frac{3}{4}\right)$
$\frac{9}{16} = \frac{9}{16}$
9. $d = -\frac{6}{5}$ Check $\frac{3}{10} = -\frac{1}{4}\left(-\frac{6}{5}\right)$
$\frac{3}{10} = \frac{3}{10}$

11. $n = 12$ Check $\frac{1}{6}(12) + 7 = 9$
$2 + 7 = 9$
$9 = 9$
13. $r = -9$ Check $-10 = \frac{5}{3}(-9) + 5$
$-10 = -15 + 5$
$-10 = -10$
15. $x = 24$ Check $\frac{3}{8}(24) - 9 = 0$
$9 - 9 = 0$
$0 = 0$
17. $y = 45$ **19.** $n = -18$ **21.** $x = \frac{1}{12}$ **23.** $b = -\frac{1}{8}$
25. (a) $\frac{1}{6}(18) + 1 = -2$
$3 + 1 = -2$
$4 \ne -2$
Does *not* balance; correct solution is -18.
(b) $-\frac{3}{2} = \frac{9}{4}\left(-\frac{2}{3}\right)$
$-\frac{3}{2} = -\frac{3}{2}$
Balances, so $-\frac{2}{3}$ is correct solution.
27. Some possibilities are: $\frac{1}{2}x = 4$; $-\frac{1}{4}a = -2$; $\frac{3}{4}b = 6$.
29. Let a be the man's age.
$109 = 100 + \frac{a}{2}$
The man is 18 years old.
31. Let a be the woman's age.
$122 = 100 + \frac{a}{2}$
The woman is 44 years old.
33. Let p be the penny size. $\frac{p}{4} + \frac{1}{2} = 3$. The penny size is 10.
35. Let p be the penny size. $\frac{p}{4} + \frac{1}{2} = \frac{5}{2}$. The penny size is 8.
37. Let h be the man's height. $\frac{11}{2}h - 220 = 209$. The man is
78 in. tall. **39.** Let h be the woman's height. $\frac{11}{2}h - 220 = 132$.
The woman is 64 in. tall.

Section 4.8 (page 283)
1. $P = 202$ m; $A = 1914$ m^2 **3.** $P = 5$ ft; $A = \frac{27}{32}$ ft^2
5. $P = 26\frac{1}{4}$ yd; $A = 30\frac{3}{4}$ yd^2 **7.** $P = \frac{454}{15}$ yd or $30\frac{4}{15}$ yd;
$A = \frac{115}{3}$ yd^2 or $38\frac{1}{3}$ yd^2 **9.** $A = 198$ m^2 **11.** $A = 1716$ m^2
13. Perimeter is the distance around the outside edges of a flat shape
and is measured in linear units. Area is the space inside a flat shape
and is measured in square units. Volume is the space inside a solid
shape and is measured in cubic units. **15.** $A = \frac{63}{8}$ ft^2 or $7\frac{7}{8}$ ft^2
17. 132 m of curbing; 726 m^2 of sod **19.** (a) 175 yd of frontage
(b) $A = 8750$ yd^2 **21.** Rectangular solid; $V = 528$ cm^3

23. Rectangular solid or cube; $V = 15\frac{5}{8}$ in.^{3}
25. Pyramid; $V = 800$ cm^3 27. $V = 18$ in.^{3} 29. $V = 651{,}775$ m^3
31. $V = 513$ cm^3 33. The correct answers are $A = 135$ ft^2 (square feet, not squaring 135) and $5\frac{1}{2}$ in. (perimeter is in linear units, not square units).

Chapter 4 Review Exercises (page 295)

1. $\frac{2}{5}; \frac{1}{5}$ 2. $\frac{3}{10}; \frac{7}{10}$

3.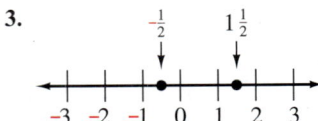

4. (a) -4 (b) 8 (c) -1 5. $\frac{7}{8}$ 6. $\frac{3}{5}$ 7. already in lowest terms   8. $\frac{3x}{8}$ 9. $\frac{1}{5b}$ 10. $\frac{4n}{7m^2}$ 11. $\frac{1}{16}$ 12. -12 13. $\frac{8}{27}$
14. $\frac{1}{6x}$ 15. $2a^2$ 16. $\frac{6}{7k}$ 17. $\frac{5}{24}$ 18. $-\frac{2}{15}$ 19. $\frac{19}{6}$ or $3\frac{1}{6}$
20. $\frac{3}{2}$ or $1\frac{1}{2}$ 21. $\frac{4n+15}{20}$ 22. $\frac{3y-70}{10y}$ 23. $1\frac{5}{13}$; $2 \div 2 = 1$
24. $2\frac{1}{2}$; $7 - 5 = 2$ 25. $4\frac{1}{20}$; $2 + 2 = 4$ 26. $-\frac{27}{64}$ 27. $\frac{1}{36}$
28. $-\frac{4}{5}$ 29. $\frac{7}{6}$ or $1\frac{1}{6}$ 30. 10 31. $-\frac{4}{27}$ 32. $w = 20$
33. $r = -15$ 34. $x = \frac{1}{2}$ 35. $A = 14$ ft^2 36. Rectangular solid; $V = 32\frac{1}{2}$ in.^{3} 37. Pyramid; $V = 93\frac{1}{3}$ m^3 38. $\frac{1}{4}$ pound; $7\frac{1}{2}$ pounds   39. $\frac{5}{6}$ hour; $10\frac{11}{12}$ hours   40. 12 preschoolers, 40 toddlers, 8 infants   41. $P = 2\frac{1}{10}$ miles; $A = \frac{9}{40}$ mi^2

Chapter 4 Test (page 297)

1. $\frac{5}{6}; \frac{1}{6}$ 2. 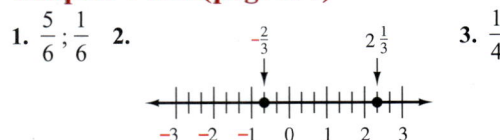   3. $\frac{1}{4}$

4. already in lowest terms   5. $\frac{2a^2}{3b}$ 6. $\frac{13}{15}$ 7. -2 8. $-\frac{7}{40}$
9. 14 10. $-\frac{2}{27}$ 11. $\frac{25}{8}$ or $3\frac{1}{8}$ 12. $\frac{4}{9}$ 13. $\frac{9}{16}$ 14. $\frac{4}{7y}$
15. $\frac{24-n}{4n}$ 16. $\frac{10+3a}{15}$ 17. $\frac{1}{18b}$ 18. $-\frac{1}{18}$ 19. $-\frac{31}{30}$ or $-1\frac{1}{30}$ 20. $5 \div 1 = 5$; $\frac{64}{15}$ or $4\frac{4}{15}$ 21. $3 - 2 = 1$; $\frac{3}{2}$ or $1\frac{1}{2}$
22. $d = 35$ 23. $t = -\frac{15}{7}$ or $-2\frac{1}{7}$ 24. $b = 8$ 25. $x = -15$
26. $A = 52$ m^2 27. $A = \frac{117}{2}$ yd^2 or $58\frac{1}{2}$ yd^2 28. Rectangular solid; $V = 6480$ m^3 29. Pyramid; $V = 16$ yd^3 30. $14\frac{3}{4}$ hours; $1\frac{5}{6}$ hours   31. $\frac{7}{2}$ days or $3\frac{1}{2}$ days   32. 7392 students work

CHAPTER 5

Section 5.1 (page 305)

1. 7; 0; 4   3. 5; 1; 8   5. 4; 7; 0   7. 1; 6; 3   9. 1; 8; 9
11. 6; 2; 1   13. 410.25   15. 6.5432   17. 5406.045   19. $\frac{7}{10}$
21. $13\frac{2}{5}$ 23. $\frac{7}{20}$ 25. $\frac{33}{50}$ 27. $10\frac{17}{100}$ 29. $\frac{3}{50}$ 31. $\frac{41}{200}$
33. $5\frac{1}{500}$ 35. $\frac{343}{500}$ 37. five tenths   39. seventy-eight hundredths
41. one hundred five thousandths   43. twelve and four hundredths
45. one and seventy-five thousandths   47. 6.7   49. 0.32
51. 420.008   53. 0.0703   55. 75.030   57. Anne should not say "and" because that denotes a decimal point.   59. ten thousandths inch; $\frac{10}{1000} = \frac{1}{100}$ inch   61. 12 pounds   63. 3-C   65. 4-A
67. one and six hundred two thousandths centimeters
69. millionths, ten-millionths, hundred-millionths, billionths; these match the words on the left side of the chart with "ths" attached.
70. First place to the left of the decimal point is ones, so the first place to the right could be one*ths*, like tens and ten*ths*. But anything that is 1 or more is to the left of the decimal point.   71. seventy-two million four hundred thirty-six thousand nine hundred fifty-five hundred-millionths   72. six hundred seventy-eight thousand five hundred fifty-four billionths   73. eight thousand six and five hundred thousand one millionths   74. twenty thousand, sixty and five hundred five millionths   75. 0.0302040   76. 9,876,543,210.100200300

Section 5.2 (page 315)

1. 16.9   3. 0.956   5. 0.80   7. 3.661   9. 794.0   11. 0.0980
13. 49   15. 9.09   17. 82.0002   19. $0.82   21. $1.22
23. $0.50   25. $48,650   27. $310   29. $849   31. $500
33. $1.00   35. $1000   37. (a) 322 miles per hour
(b) 107 miles per hour   39. (a) 186.0 miles per hour
(b) 763.0 miles per hour   41. Rounds to $0 (zero dollars) because $0.499 is closer to $0 than to $1.   42. Round $0.499 to the nearest cent to get $0.50. Guideline: Round amounts less than $1.00 to nearest cent instead of nearest dollar.   43. Rounds to $0.00 (zero cents) because $0.0015 is closer to $0.00 than to $0.01.
44. Both round to $0.60. Rounding to nearest thousandth (tenth of a cent) would allow you to identify $0.597 as less than $0.601.

Section 5.3 (page 323)

1. 17.72   3. 11.98   5. 115.861   7. 59.323
9. 6 should be written 6.00; sum is 46.22.
11. $0.3000 = \frac{3000 \div 1000}{10{,}000 \div 1000} = \frac{3}{10} = 0.3$ 13. 89.7
15. 0.109   17. 0.91   19. 6.661   21. The student subtracted in the wrong order; 15.32 should be on top; correct answer is 7.87
23. (a) 24.75 in.   (b) 3.95 in.   25. (a) 62.27 in.   (b) 0.39 in.
27. 23.013   29. -45.75 31. -6.69 33. -6.99 35. -4.279
37. -0.0035 39. 5.37   41. 0.275   43. 6.507   45. 1.81
47. 6056.7202   49. Estimate: $19 - 15 = 4$ million people; Exact: 3.6 million people
51. Estimate: $100 + 30 + 20 + 20 + 20 + 9 = 199$ million people; Exact: 234.4 million people   53. Estimate: $2 + 2 + 2 = 6$ m; Exact: 6.09 m, which is 0.31 m less than the rhino's height.
55. Estimate: $\$20 - \$9 = \$11$; Exact: $10.88
57. Estimate: $\$5 - \$5 = \$0$; Exact: $0.30
59. Estimate: $\$19 + 2 + 2 + 10 + 2 = \$35$; Exact: $35.25
61. $1939.36   63. $3.97   65. $b = 1.39$ cm   67. $q = 23.843$ ft

Section 5.4 (page 331)
1. 0.1344   3. −159.10   5. 15.5844   7. $34,500.20   9. −43.2
11. 0.432   13. 0.0432   15. 0.00432   17. 0.0000312
19. 0.000009   21. 59.6; 4.76; 7226; 32; 803.5; 9. Multiplying by 10, decimal point moves one place to the right; by 100, two places to the right; by 1000, three places to the right.   22. 5.96; 0.0476; 6.5; 0.32; 8.035; 52.3. Multiplying by 0.1, decimal point moves one place to the left; by 0.01, two places to the left; by 0.001, three places to the left.   23. *Estimate:* 40 × 5 = 200; *Exact:* 190.08
25. *Estimate:* 40 × 40 = 1600; *Exact:* 1558.2
27. *Estimate:* 7 × 5 = 35; *Exact:* 30.038   29. *Estimate:* 3 × 7 = 21; *Exact:* 19.24165   31. unreasonable; $189.00   33. reasonable
35. unreasonable; $4.19   37. unreasonable; 9.5 pounds
39. $945.87 (rounded)   41. $2.45 (rounded)   43. $53.50 (rounded)   45. $20,265   47. (a) Area before 1929 ≈ 23.2 in.²; Area today ≈ 16.0 in.²   (b) 7.2 in.²   49. (a) 0.43 inch
(b) 4.3 inches   51. $984.04; $2207.80   53. $76.50
55. $4.09 (rounded)   57. $129.25   59. (a) $70.05   (b) $25.80

Section 5.5 (page 343)
1. −3.9   3. 0.47   5. 400.2   7. 36   9. 0.06   11. 6000
13. 60   15. 0.0006   17. 25.3   19. 516.67 (rounded)
21. −24.291 (rounded)   23. 10,082.647 (rounded)   25. 0.377; 0.0886; 40.65; 0.91; 3.019; 662.57   (a) Dividing by 10, decimal point moves one place to the left; by 100, two places to the left; by 1000, three places to the left.   (b) The decimal point moved to the *right* when multiplying by 10, by 100, or by 1000; here it moves to the *left* when dividing by 10, by 100, or by 1000.   26. 402; 3.39; 460; 71; 157.7; 8730   (a) Dividing by 0.1, decimal point moves one place to the right; by 0.01, two places to the right; by 0.001, three places to the right.   (b) The decimal point moved to the *left* when multiplying by 0.1, 0.01, or 0.001; here it moves to the *right* when dividing by 0.1, 0.01, or 0.001.   27. unreasonable; *Estimate:* 40 ÷ 8 = 5; 8)37.8 → 4.725   29. reasonable; *Estimate:* 50 ÷ 50 = 1   31. unreasonable; *Estimate:* 300 ÷ 5 = 60; 5.1)307.02 → 60.2   33. unreasonable; *Estimate:* 9 ÷ 1 = 9; 1.25)9.3 → 7.44
35. $4.00 (rounded)   37. $67.08 (rounded)   39. $0.30
41. $11.92 per hour   43. 21.2 miles per gallon (rounded)
45. 7.37 meters (rounded)   47. 0.08 meter   49. 22.49 meters
51. 14.25   53. 3.8   55. −16.155   57. 3.714   59. $0.03 (rounded)
61. (a) 1,541,667 pieces (rounded)   (b) 25,694 pieces (rounded)
(c) 428 pieces (rounded)   63. 100,000 box tops   65. 2632 box tops (rounded)

Summary Exercises on Decimals (page 347)
1. $\frac{4}{5}$ 2. $6\frac{1}{250}$ 3. $\frac{7}{20}$ 4. ninety-four and five tenths   5. two and three ten-thousandths   6. seven hundred six thousandths
7. 0.05   8. 0.0309   9. 10.7   10. 6.19   11. 1.0   12. 0.420
13. $0.89   14. $3.00   15. $100   16. −0.945   17. 49.6199
18. −50   19. 15.03   20. 0.00488   21. −2.15   22. 9.055
23. 18.4009   24. −6.995   25. −808.9   26. 2.12   27. 0.04
28. $P = 52.1$ in.   29. $P = 9.735$ meters   30. $1.80
31. $2052.50   32. (a) 169 ft²   (b) $0.75 (rounded)   33. 144 ft²; $0.69 (rounded)   34. (a) 9 days (rounded)   (b) 15.6 pounds (rounded)   35. Average weight of food eaten by bee is 0.32 ounce.
36. 0.112 ounce

Section 5.6 (page 353)
1. 0.5   3. 0.75   5. 0.3   7. 0.9   9. 0.6   11. 0.875   13. 2.25
15. 14.7   17. 3.625   19. 0.333 (rounded)   21. 0.833 (rounded)

23. 1.889 (rounded)   25. (a) A proper fraction is less than 1, so $\frac{5}{9}$ cannot be equivalent to a mixed number.   (b) $\frac{5}{9}$ means 5 ÷ 9 or 9)5 so correct answer is 0.556 (rounded). This makes sense because both the fraction and decimal are less than 1.
26. (a) $2.035 = 2\frac{35}{1000} = 2\frac{7}{200}$, not $2\frac{7}{20}$ (b) Adding the whole number part gives 2 + 0.35, which is 2.35 not 2.035. To check, $2.35 = 2\frac{35}{100} = 2\frac{7}{20}$ but $2.035 = 2\frac{35}{1000} = 2\frac{7}{200}$.
27. Just add the whole number part to 0.375. So $1\frac{3}{8} = 1.375$; $3\frac{3}{8} = 3.375$; $295\frac{3}{8} = 295.375$.   28. It works only when the fraction part has a one-digit numerator and a denominator of 10, or a two-digit numerator and a denominator of 100, and so on.
29. $\frac{2}{5}$ 31. $\frac{5}{8}$ 33. $\frac{7}{20}$ 35. 0.35   37. $\frac{1}{25}$ 39. $\frac{3}{20}$ 41. 0.2
43. $\frac{9}{100}$ 45. shorter; 0.72 inch   47. too much; 0.005 gram
49. 0.9991 cm, 1.0007 cm   51. more; 0.05 inch   53. 0.5399, 0.54, 0.5455   55. 5.0079, 5.79, 5.8, 5.804   57. 0.6009, 0.609, 0.628, 0.62812   59. 2.8902, 3.88, 4.876, 5.8751   61. 0.006, 0.043, $\frac{1}{20}$, 0.051   63. 0.37, $\frac{3}{8}, \frac{2}{5}$, 0.4001   65. red box   67. 0.01 inch
69. 1.4 in. (rounded)   71. 0.3 in. (rounded)   73. 0.4 in. (rounded)

Section 5.7 (page 363)
1. 7 months   3. 69.8 (rounded)   5. $39,622   7. $58.24
9. $35,500   11. 6.1 (rounded)   13. 17.2 hours (rounded)
15. 2.60   17. (a) 2.80   (b) 2.93 (rounded)   (c) 3.13 (rounded)
19. 15 messages   21. 516 students   23. 48 calls
25. 4142 miles (rounded)   27. (a) 4050 miles   (b) List the values from smallest to largest. There are an even number of values, so find the average of the middle two values: (2875 + 5525) ÷ 2 = 4050.   29. 8 samples
31. 68 and 74 years (bimodal)   33. no mode
35. (a) −10 degrees; 4 degrees   (b) 14 degrees warmer
37. Barrow's range is −2 − (−18) = 16 degrees; Fairbanks' range is 31 − (−10) = 41 degrees; Fairbanks' temperatures have greater variability.   39. (a) mean = 71 degrees; median ≈ 71 degrees
(b) They are nearly identical because there is so little variation in the temperatures.   41. (a) Student P: mean = 69.5; median = 70.5; range = 43. Student Q: mean ≈ 70.3; median = 71; range = 17
(b) Student P   (c) Answers will vary.   42. (a) Golfer G: mean = 86.2; median = 87; range = 6; Golfer H: mean = 86.4; median = 88; range = 18.   (b) Golfer G   (c) Answers will vary.
43. (a) A's range = 74 − 62 = 12; B's range = 26 − 18 = 8; C's range = 69 − 25 = 44; Building B's ages have the least variability.
(b) Answers will vary.   44. (a) P's range = $7200 − 4900 = $2300; Q's range = $6000 − 5500 = $500; R's range = $6400 − 5200 = $1200; Company P's salaries have the greatest variability.   (b) Answers will vary.

Section 5.8 (page 371)
1. 4   3. 8   5. 3.317 (rounded)   7. 2.236 (rounded)   9. 8.544 (rounded)   11. 10.050 (rounded)   13. 19   15. 31.623 (rounded)
17. 30 is about halfway between 25 and 36, so $\sqrt{30}$ should be about halfway between 5 and 6, or about 5.5. Using a calculator, $\sqrt{30} \approx 5.477$. Similarly, $\sqrt{26}$ should be a little more than $\sqrt{25}$; by calculator $\sqrt{26} \approx 5.099$. And $\sqrt{35}$ should be a little less than $\sqrt{36}$; by calculator $\sqrt{35} \approx 5.916$.   19. $\sqrt{1521} = 39$ ft

21. $\sqrt{289} = 17$ in.   **23.** $\sqrt{144} = 12$ mm   **25.** $\sqrt{73} \approx 8.5$ in.
27. $\sqrt{65} \approx 8.1$ yd   **29.** $\sqrt{195} \approx 14.0$ cm
31. $\sqrt{7.94} \approx 2.8$ m   **33.** $\sqrt{65.01} \approx 8.1$ cm
35. $\sqrt{292.32} \approx 17.1$ km   **37.** $\sqrt{65} \approx 8.1$ ft
39. $\sqrt{360{,}000} = 600$ m   **41.** $\sqrt{32.5} \approx 5.7$ ft
43. $\sqrt{135} \approx 11.6$ ft.   **45.** The student used the formula for finding the hypotenuse but the unknown side is a leg, so use $? = \sqrt{(20)^2 - (13)^2}$. Also, the final answer should be m, not m². Correct answer is $\sqrt{231} \approx 15.2$ m.   **47.** $\sqrt{16{,}200} \approx 127.3$ ft
48. (a)   **(b)** $\sqrt{7200} \approx 84.9$ ft

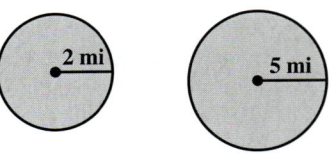

49. The distance from third to first is the same as the distance from home to second because the baseball diamond is a square.
50. (a) The side length is less than 60 ft.   **(b)** $80^2 = 6400$; $6400 \div 2 = 3200$; $\sqrt{3200} \approx 56.6$ ft

Section 5.9 (page 379)

Please see *Student's Solutions Manual* for a sample *check* for Exercises 1–12.   **1.** $h = 4.47$ **3.** $n = -1.4$ **5.** $b = 0.008$
7. $a = 0.29$ **9.** $p = -120$ **11.** $t = 0.7$ **13.** $x = -0.82$
15. $z = 0$ **17.** $c = 0.45$ **19.** $w = 0$ **21.** $p = -40.5$
23. Let d be the adult dose. $0.3\, d = 9$; The adult dose is 30 milligrams.
25. Let d be the number of days. $65.95 d + 12 = 275.80$; The saw was rented for 4 days.   **27.** $0.7(220 - a) = 140$; The person is 20 years old.   **28.** $0.7(220 - a) = 126$; The person is 40 years old.
29. $0.7(220 - a) = 134$; $a \approx 28.57$, which rounds to 29. The person is about 29 years old.   **30.** $0.7(220 - a) = 117$; $a \approx 52.86$, which rounds to 53. The person is about 53 years old.

Section 5.10 (page 389)

1. $d = 18$ mm   **3.** $r = 0.35$ km   **5.** $C \approx 69.1$ ft; $A \approx 379.9$ ft²
7. $C \approx 8.2$ m; $A \approx 5.3$ m²   **9.** $C \approx 47.1$ cm, $A \approx 176.6$ cm²
11. $C \approx 23.6$ ft; $A \approx 44.2$ ft²   **13.** $C \approx 27.2$ km; $A \approx 58.7$ km²
15. $A \approx 57$ cm²   **17.** $A \approx 197.8$ cm²   **19.** π is the ratio of the circumference of a circle to its diameter. If you divide the circumference of any circle by its diameter, the answer is always a little more than 3. The approximate value is 3.14, which we call π (pi). Your test question could involve finding the circumference or the area of a circle.   **21.** $A \approx 7850$ yd²
23. $C \approx 91.4$ in.; Bonus: 693 revolutions/mile (rounded)
25. $A \approx 70{,}650$ mi²

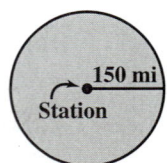

27. watch: $C \approx 3.1$ in.; $A \approx 0.8$ in.²; wall clock: $C \approx 18.8$ in.; $A \approx 28.3$ in.²

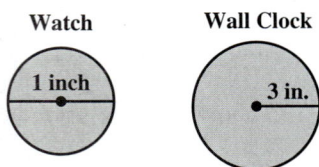

29. $A \approx 78.5$ mi² $- 12.6$ mi²; difference ≈ 65.9 mi²

31. (a) $d \approx 45.9$ cm   **(b)** Divide the circumference by π (144 cm $\div$ 3.14).   **33.** $C \approx 14.6$ ft   **35.** $\$1170.33$ (rounded)
37. $A \approx 44.2$ in.²   **38.** $A \approx 132.7$ in.²   **39.** $A \approx 201.0$ in.²
40. small: $\$0.063$ (rounded); medium: $\$0.049$ (rounded); large: $\$0.046$ (rounded) Best Buy   **41.** small: $\$0.084$ (rounded); medium: $\$0.067$ (rounded) Best Buy; large: $\$0.071$ (rounded)
42. small: $\$0.077$ (rounded) Best Buy; medium: $\$0.083$ (rounded); large: $\$0.078$ (rounded)   **43.** $V \approx 471$ ft³; $SA \approx 345.4$ ft²
45. $V \approx 1617$ m³; $SA \approx 849.4$ m²   **47.** $V \approx 763.0$ in.³; $SA \approx 678.2$ in.²   **49.** $V \approx 5550$ mm³; $SA \approx 2150$ mm²
51. Student should use radius of 3.5 cm instead of diameter of 7 cm in the formula; units for volume are cm³ not cm². Correct answer is $V \approx 192.3$ cm³   **53.** $V \approx 3925$ ft³   **55.** $SA \approx 163.6$ in.²

Chapter 5 Review Exercises (page 403)

1. 0; 5   **2.** 0; 6   **3.** 8; 9   **4.** 5; 9   **5.** 7; 6   **6.** $\dfrac{1}{2}$ **7.** $\dfrac{3}{4}$
8. $4\dfrac{1}{20}$ **9.** $\dfrac{7}{8}$ **10.** $\dfrac{27}{1000}$ **11.** $27\dfrac{4}{5}$ **12.** eight tenths   **13.** four hundred and twenty-nine hundredths   **14.** twelve and seven thousandths   **15.** three hundred six ten-thousandths   **16.** 8.3
17. 0.205   **18.** 70.0066   **19.** 0.30   **20.** 275.6   **21.** 72.79
22. 0.160   **23.** 0.091   **24.** 1.0   **25.** $\$15.83$ **26.** $\$0.70$
27. $\$17{,}625.79$ **28.** $\$350$ **29.** $\$130$ **30.** $\$100$ **31.** $\$29$
32. -5.67 **33.** -0.03 **34.** 5.879   **35.** -6.435
36. Estimate: 80 million $-$ 50 million $=$ 30 million; Exact: 31.4 million people   **37.** Estimate: $\$300 - \$200 - \$40 = \$60$; Exact: $\$45.80$ **38.** Estimate: $\$2 + \$5 + \$20 = \$27$; $\$30 - \$27 = \$3$; Exact: $\$4.14$ **39.** Estimate: $2 + 4 + 5 = 11$ kilometers; Exact: 11.55 kilometers   **40.** $6 \times 4 = 24$; 22.7106
41. $40 \times 3 = 120$; 141.57   **42.** 0.0112   **43.** -0.000355
44. reasonable; $700 \div 10 = 70$ **45.** unreasonable; $30 \div 3 = 10$; $2.8\overline{)26.6}\; 9.5$ **46.** 14.467 (rounded)   **47.** 1200   **48.** -0.4
49. $\$708$ (rounded)   **50.** $\$2.99$ (rounded)   **51.** 133 shares (rounded)
52. $\$3.47$ (rounded)   **53.** -4.715 **54.** 10.15   **55.** 3.8
56. 0.64   **57.** 1.875   **58.** 0.111 (rounded)   **59.** 3.6008, 3.68, 3.806   **60.** 0.209, 0.2102, 0.215, 0.22   **61.** $\dfrac{1}{8}, \dfrac{3}{20}, 0.159, 0.17$
62. mean: 28.5 digital cameras; median: 19.5 digital cameras
63. mean: 44 claims; median: 39 claims   **64.** $\$51.05$
65. (a) Store J: $\$107$ and $\$160$ (bimodal); Store K: $\$119$
(b) Range for Store J $= \$160 - 69 = \$91$; range for Store K $= \$139 - 95 = \$44$; Store J's prices have greater variability.
66. $\sqrt{289} = 17$ in.   **67.** $\sqrt{49} = 7$ cm   **68.** $\sqrt{104} \approx 10.2$ cm
69. $\sqrt{52} \approx 7.2$ in.   **70.** $\sqrt{6.53} \approx 2.6$ m
71. $\sqrt{71.75} \approx 8.5$ km   **72.** $b = 0.25$ **73.** $x = 0$ **74.** $n = -13$
75. $a = -1.5$ **76.** $y = -16.2$ **77.** $d = 137.8$ m
78. $r = 1\dfrac{1}{2}$ in. or 1.5 in.   **79.** $C \approx 6.3$ cm; $A \approx 3.1$ cm²
80. $C \approx 109.3$ m; $A \approx 950.7$ m²   **81.** $C \approx 37.7$ in.; $A \approx 113.0$ in.²
82. $V \approx 549.5$ cm³; $SA \approx 376.8$ cm²
83. $V \approx 1808.6$ m³; $SA \approx 1205.8$ m²   **84.** $V \approx 7.9$ ft³; $SA \approx 25.5$ ft²

85. 404.865 **86.** −254.8 **87.** 3583.261 (rounded) **88.** 29.0898
89. 0.03066 **90.** 9.4 **91.** −15.065 **92.** −9.04 **93.** −15.74
94. 8.19 **95.** 0.928 **96.** 35 **97.** −41.859 **98.** 0.3
99. $3.00 (rounded) **100.** $2.17 (rounded) **101.** $35.96
102. $199.71 **103.** $78.50 **104.** $y = -0.7$ **105.** $x = 30$
106. 15.7 ft of rubber striping; $A \approx 19.6$ ft^2 **107.** mean is 67.6; median is 82 **108.** 20 miles **109.** $V \approx 87.9$ in.^{3}

Chapter 5 Test (page 409)

1. $18\frac{2}{5}$ **2.** $\frac{3}{40}$ **3.** sixty and seven thousandths **4.** two hundred eight ten-thousandths **5.** *Estimate:* $8 + 80 + 40 = 128$; *Exact:* 129.2028 **6.** *Estimate:* $-6(1) = -6$; *Exact:* -6.948
7. *Estimate:* $-80 - 4 = -84$; *Exact:* -82.702 **8.** *Estimate:* $-20 \div (-5) = 4$; *Exact:* 4.175 **9.** 669.004 **10.** 480
11. 0.000042 **12.** $4.55 per meter **13.** Davida, by 0.441 minute
14. $5.35 (rounded) **15.** $y = -6.15$ **16.** $x = -0.35$
17. $a = 10.9$ **18.** $n = 2.85$ **19.** $0.44, \frac{9}{20}, 0.4506, 0.451$
20. 32.09 **21.** 75 books **22.** 103° and 104° (bimodal)
23. $11.25 **24. (a)** $50.55 **(b)** $50.45 **(c)** math books: $89 − $39.75 = $49.25; biology books: $97.55 − $30.95 = $66.60; math books have less variability **25.** $\sqrt{85} \approx 9.2$ cm
26. $\sqrt{279} \approx 16.7$ ft **27.** $r = 12.5$ in. **28.** $C \approx 5.7$ km
29. $A \approx 206.0$ cm^2 **30.** $V \approx 5086.8$ ft^3 **31.** $SA \approx 2599.9$ ft^2

CHAPTER 6

Section 6.1 (page 419)

1. $\frac{8}{9}$ **3.** $\frac{2}{1}$ **5.** $\frac{1}{3}$ **7.** $\frac{8}{5}$ **9.** $\frac{3}{8}$ **11.** $\frac{9}{7}$ **13.** $\frac{6}{1}$ **15.** $\frac{5}{6}$
17. $\frac{8}{5}$ **19.** $\frac{1}{12}$ **21.** $\frac{5}{16}$ **23.** $\frac{4}{1}$ **25.** $\frac{1}{2}$ **27.** $\frac{36}{1}$
29. Answers will vary. One possibility is stocking cards of various types in the same ratios as those in the table. **31.** $\frac{6}{5}; \frac{36}{17}$
33. *White Christmas* to *It's Now or Never*; *White Christmas* to *I Will Always Love You*; *Candle in the Wind* to *I Want to Hold Your Hand*
35. $\frac{2}{1}$ **37. (a)** $\frac{3}{8}$ **(b)** $\frac{7}{5}$ **39.** $\frac{7}{5}$ **41.** $\frac{6}{1}$ **43.** $\frac{38}{17}$ **45.** $\frac{1}{4}$
47. $\frac{34}{35}$ **49.** $\frac{1}{1}$; as long as the sides all have the same length, any measurement you choose will maintain the ratio.
50. Answers will vary. Some possibilities are:
$\frac{4}{5} = \frac{8}{10} = \frac{12}{15} = \frac{16}{20} = \frac{20}{25} = \frac{24}{30} = \frac{28}{35}$ **51.** It is not possible. Amelia would have to be older than her mother to have a ratio of 5 to 3. **52.** Answers will vary, but a ratio of 3 to 1 means your income is 3 times your friend's income.

Section 6.2 (page 427)

1. $\frac{5 \text{ cups}}{3 \text{ people}}$ **3.** $\frac{3 \text{ feet}}{7 \text{ seconds}}$ **5.** $\frac{1 \text{ person}}{2 \text{ dresses}}$ **7.** $\frac{5 \text{ letters}}{1 \text{ minute}}$
9. $\frac{\$21}{2 \text{ visits}}$ **11.** $\frac{18 \text{ miles}}{1 \text{ gallon}}$ **13.** $12 per hour or $12/hour
15. 5 eggs per chicken or 5 eggs/chicken **17.** 1.25 pounds/person
19. $103.30/day **21.** 325.9; 21.0 (rounded) **23.** 338.6; 20.9 (rounded) **25.** 4 ounces for $3.09 **27.** 15 ounces for $3.15
29. 18 ounces for $1.79 **31.** Answers will vary. For example, you might choose Brand B because you like more chicken, so the cost per chicken chunk may actually be the same or less than Brand A.

33. 1.75 pounds/week **35.** $12.26/hour **37. (a)** Radiant $0.44; IDT $0.25; Access $0.235 **(b)** Radiant $0.088/min, IDT $0.05/min, Access $0.047/min; Access America is the best buy.
39. For a 15-minute call: Radiant $0.036/min; IDT $0.031/min; Access $0.047/min; IDT is the best buy. For a 20-minute call: Radiant $0.030/min; IDT $0.029/min; Access $0.047/min; IDT is the best buy. **41.** 0.11 second/meter; $9.\overline{09}$ or 9.1 meters/second (rounded) **43.** One battery for $1.79; like getting 3 batteries so $1.79 ÷ 3 ≈ $0.597 per battery **45.** Brand P with the 50¢ coupon is the best buy. ($3.39 − $0.50 = $2.89 ÷ 16.5 ounces ≈ $0.175 per ounce) **47.** $2.92 (rounded) per month for Verizon, T-Mobile, and Nextel; $3 per month for Sprint. **48.** Average weekdays per month is about 21.7; Verizon ≈ 18 min/weekday; T-Mobile ≈ 28 min/weekday; Nextel and Sprint ≈ 23 min/weekday.
49. Round to hundredths (nearest cent) to see that T-mobile is the best buy at $0.07 per "anytime minute"; Verizon ≈ $0.16; Sprint ≈ $0.12; Nextel ≈ $0.10. **50.** Verizon ≈ $1.65 per "anytime minute"; T-mobile ≈ $1.57; Nextel ≈ $1.63; Sprint = $1.48

Section 6.3 (page 439)

1. $\frac{\$9}{12 \text{ cans}} = \frac{\$18}{24 \text{ cans}}$ **3.** $\frac{200 \text{ adults}}{450 \text{ children}} = \frac{4 \text{ adults}}{9 \text{ children}}$
5. $\frac{120}{150} = \frac{8}{10}$ **7.** $\frac{2.2}{3.3} = \frac{3.2}{4.8}$ **9.** $\frac{3}{5} = \frac{3}{5}$; true
11. $\frac{5}{8} = \frac{5}{8}$; true **13.** $\frac{3}{4} \neq \frac{2}{3}$; false **15.** $54 = 54$; true
17. $336 \neq 320$; false **19.** $2880 \neq 2970$; false
21. $28 = 28$; true **23.** $44.8 \neq 45$; false **25.** $66 = 66$; true
27. $\frac{16 \text{ hits}}{50 \text{ at bats}} = \frac{128 \text{ hits}}{400 \text{ at bats}}$ $\begin{matrix}50 \cdot 128 = 6400 \\ 16 \cdot 400 = 6400\end{matrix}$ Cross products are equal so the proportion is *true*; they hit equally well. **29.** $x = 4$
31. $x = 2$ **33.** $x = 88$ **35.** $x = 91$ **37.** $x = 5$ **39.** $x = 10$
41. $x \approx 24.44$ (rounded) **43.** $x = 50.4$ **45.** $x \approx 17.64$ (rounded)
47. $x = 1$ **49.** $x = 3\frac{1}{2}$ **51.** $x = 0.2$ or $x = \frac{1}{5}$
53. $x = 0.005$ or $x = \frac{1}{200}$ **55.** Find the cross products: $20 \neq 30$, so the proportion is false.
$\frac{6\frac{2}{3}}{4} = \frac{5}{3}$ or $\frac{10}{6} = \frac{5}{3}$ or $\frac{10}{4} = \frac{7.5}{3}$ or $\frac{10}{4} = \frac{5}{2}$
56. Find the cross products: $192 \neq 180$, so the proportion is false.
$\frac{6.4}{8} = \frac{24}{30}$ or $\frac{6}{7.5} = \frac{24}{30}$ or $\frac{6}{8} = \frac{22.5}{30}$ or $\frac{6}{8} = \frac{24}{32}$

Summary Exercises on Ratios, Rates, and Proportions (page 443)

1. $\frac{6}{5}$ **2.** $\frac{2}{7}$ **3.** $\frac{2}{1}$ **4.** $\frac{13}{11}$ **5.** Comparing the violin to piano, guitar, organ, clarinet and drums gives ratios of $\frac{1}{11}, \frac{1}{10}, \frac{1}{3}, \frac{1}{2}$, and $\frac{2}{3}$, respectively. **6. (a)** guitar to clarinet **(b)** organ to drums, or clarinet to violin **7.** 2.1 points/min; 0.5 min/point
8. 1.6 points/min; 0.6 min/point **9.** $16.32/hour; $24.48/hour of overtime **10.** $0.50/channel; $0.39/channel; $0.32/channel
11. 34 ounces, at $0.16 per ounce **12.** Brand P with the $2 coupon is the best buy at $0.57 per pound.
13. $\frac{4}{3} = \frac{4}{3}$ or $924 = 924$; true **14.** $2.0125 \neq 2.07$; false
15. $68\frac{1}{4} = 68\frac{1}{4}$; true **16.** $x = 28$ **17.** $x = 3.2$ **18.** $x = 182$

19. $x \approx 3.64$ (rounded)   **20.** $x \approx 0.93$ (rounded)   **21.** $x = 1.56$ **22.** $x \approx 0.05$ (rounded)   **23.** $x = 1$ **24.** $x = \dfrac{3}{4}$

Section 6.4 (page 449)

1. 22.5 hours   **3.** $7.20   **5.** 42 pounds   **7.** $273.45
9. 10 ounces (rounded)   **11.** 5 quarts   **13.** 14 ft, 10 ft
15. 14 ft, 8 ft   **17.** 96 pieces of chicken; 33.6 pounds lasagna; 10.8 pounds deli meats; $5\dfrac{3}{5}$ pounds cheese; 7.2 dozen (about 86) buns; 14.4 pounds potato salad.   **19.** 2065 students (reasonable); about 4214 students with incorrect setup (only 2950 students in the group)   **21.** about 30 people (reasonable); about 1904 people with incorrect setup (only 238 people attended)   **23.** 105,350,000 households (reasonable); about 109,693,878 households with incorrect setup (only 107,500,000 U.S. households)   **25.** 625 stocks
27. 4.06 meters (rounded)   **29.** 311 calories (rounded)
31. 10.53 meters (rounded)   **33.** You cannot solve this problem using a proportion because the ratio of age to weight is not constant. As Jim's age increases, his weight may decrease, stay the same, or increase.   **35.** 4800 students use cream.   **37.** 120 calories and 12 grams of fiber   **39.** $1\dfrac{3}{4}$ cups water, 3 tablespoons margarine, $\dfrac{3}{4}$ cup milk, 2 cups flakes   **40.** $5\dfrac{1}{4}$ cups water, 9 tablespoons margarine, $2\dfrac{1}{4}$ cups milk, 6 cups flakes   **41.** $\dfrac{7}{8}$ cup water, $1\dfrac{1}{2}$ tablespoons margarine, $\dfrac{3}{8}$ cup milk, 1 cup flakes
42. $2\dfrac{5}{8}$ cups water, $4\dfrac{1}{2}$ tablespoons margarine, $1\dfrac{1}{8}$ cups milk, 3 cups flakes

Section 6.5 (page 463)

1. line named $\overleftrightarrow{CD}$ or $\overleftrightarrow{DC}$ **3.** line segment named $\overline{GF}$ or $\overline{FG}$ **5.** ray named $\overrightarrow{PQ}$ **7.** perpendicular   **9.** parallel
11. intersecting   **13.** $\angle AOS$ or $\angle SOA$ **15.** $\angle CRT$ or $\angle TRC$
17. $\angle AQC$ or $\angle CQA$ **19.** right (90°)   **21.** acute
23. straight (180°)   **25.** $\angle EOD$ and $\angle COD$; $\angle AOB$ and $\angle BOC$
27. $\angle HNE$ and $\angle ENF$; $\angle HNG$ and $\angle GNF$; $\angle HNE$ and $\angle HNG$; $\angle ENF$ and $\angle GNF$ **29.** 50°   **31.** 4°   **33.** 50°   **35.** 90°
37. $\angle SON \cong \angle TOM$; $\angle TOS \cong \angle MON$ **39.** $\angle GOH$ measures 63°; $\angle EOF$ measures 37°; $\angle AOC$ and $\angle GOF$ both measure 80°.
41. True, because $\overleftrightarrow{UQ}$ is perpendicular to $\overleftrightarrow{ST}$.   **42.** True, because they form a 90° angle, as indicated by the small red square.
43. False; the angles have the same measure (both are 180°).
44. False; $\overleftrightarrow{ST}$ and $\overleftrightarrow{PR}$ are parallel.   **45.** False; $\overleftrightarrow{QU}$ and $\overleftrightarrow{TS}$ are perpendicular.   **46.** True, because both angles are formed by perpendicular lines, so they both measure 90°.   **47.** corresponding angles; $\angle 1$ and $\angle 8$, $\angle 2$ and $\angle 5$, $\angle 3$ and $\angle 6$, $\angle 4$ and $\angle 7$; alternate interior angles; $\angle 4$ and $\angle 5$, $\angle 3$ and $\angle 8$.   **49.** $\angle 2$, $\angle 4$, $\angle 6$, $\angle 8$ all measure 130°; $\angle 1$, $\angle 3$, $\angle 5$, $\angle 7$ all measure 50°.   **51.** $\angle 6$, $\angle 1$, $\angle 3$, $\angle 8$ all measure 47°; $\angle 5$, $\angle 2$, $\angle 7$, $\angle 4$ all measure 133°.
53. $\angle 6$, $\angle 8$, $\angle 4$, $\angle 2$ all measure 114°; $\angle 7$, $\angle 5$, $\angle 3$, $\angle 1$ all measure 66°.   **55.** $\angle 1 \cong \angle 3$, both are 138°; $\angle 2 \cong \angle ABC$, both are 42°.

Section 6.6 (page 473)

1. congruent   **3.** neither   **5.** similar   **7.** SAS   **9.** SSS
11. ASA   **13.** use SAS: $BC = CE$, $\angle ABC \cong \angle DCE$, $BA = CD$
14. use SSS: $WP = YP$, $ZP = XP$, $WZ = YX$
15. use SAS: $PS = RS$, $m\angle QSP = m\angle QSR = 90°$, $QS = QS$ (common side)   **16.** use SAS: $LM = OM$, $PM = NM$,

$\angle LMP \cong \angle OMN$ (vertical angles)   **17.** $\dfrac{3}{2}; \dfrac{3}{2}; \dfrac{3}{2}$ **19.** $a = 5$ mm; $b = 3$ mm   **21.** $a = 6$ cm; $b = 15$ cm   **23.** $x = 24.8$ m; Perimeter $= 72.8$ m; $y = 15$ m; Perimeter $= 54.6$ m
25. Perimeter $= 8$ cm $+ 8$ cm $+ 8$ cm $= 24$ cm; Area $= (0.5)(8$ cm$)(6.9$ cm$) = 27.6$ cm^2 **27.** $h = 24$ ft
29. One dictionary definition is "resembling, but not identical." Examples of similar objects are sets of different size pots or measuring cups; small and large size cans of beans; child's tennis shoe and adult's tennis shoe.   **31.** $x = 50$ m   **33.** $n = 110$ m

Chapter 6 Review Exercises (page 485)

1. $\dfrac{3}{4}$ **2.** $\dfrac{4}{1}$ **3.** great white shark to whale shark; whale shark to blue whale   **4.** $\dfrac{2}{1}$ **5.** $\dfrac{2}{3}$ **6.** $\dfrac{5}{2}$ **7.** $\dfrac{1}{6}$ **8.** $\dfrac{3}{1}$ **9.** $\dfrac{3}{8}$ **10.** $\dfrac{4}{3}$
11. $\dfrac{1}{9}$ **12.** $\dfrac{10}{7}$ **13.** $\dfrac{7}{5}$ **14.** $\dfrac{5}{6}$ **15.** $\dfrac{\$11}{1 \text{ dozen}}$ **16.** $\dfrac{12 \text{ children}}{5 \text{ families}}$
17. 0.2 page/minute or $\dfrac{1}{5}$ page/minute; 5 minutes/page   **18.** $8/hour; 0.125 hour/dollar or $\dfrac{1}{8}$ hour/dollar   **19.** 13 ounces for $2.29
20. 25 pounds for $10.40 − $1 coupon   **21.** $\dfrac{3}{5} = \dfrac{3}{5}$ or $90 = 90$; true
22. $\dfrac{1}{8} \neq \dfrac{1}{4}$ or $432 \neq 216$; false   **23.** $\dfrac{47}{10} \neq \dfrac{49}{10}$ or $980 \neq 940$; false
24. $4.8 = 4.8$; true   **25.** $14 = 14$; true   **26.** $x = 1575$ **27.** $x = 20$
28. $x = 400$ **29.** $x = 12.5$ **30.** $x \approx 14.67$ (rounded)
31. $x \approx 8.17$ (rounded)   **32.** $x = 50.4$ **33.** $x \approx 0.57$ (rounded)
34. $x \approx 2.47$ (rounded)   **35.** 27 cats   **36.** 46 hits
37. $15.63 (rounded)   **38.** 3299 students (rounded)
39. 68 ft   **40.** $27\dfrac{1}{2}$ hours or 27.5 hours   **41.** 511 calories (rounded)
42. 14.7 milligrams   **43.** line segment named $\overline{AB}$ or $\overline{BA}$
44. line named $\overleftrightarrow{CD}$ or $\overleftrightarrow{DC}$ **45.** ray named $\overrightarrow{OP}$
46. parallel   **47.** perpendicular   **48.** intersecting   **49.** acute
50. obtuse   **51.** straight; 180°   **52.** right; 90°   **53.** (a) 10°
(b) 45°   (c) 83°   **54.** (a) 25°   (b) 90°   (c) 147°   **55.** $\angle 1$ and $\angle 4$ measure 30°; $\angle 3$ and $\angle 6$ measure 90°; $\angle 5$ measures 60°
56. $\angle 8$, $\angle 3$, $\angle 6$, $\angle 1$ all measure 160°; $\angle 4$, $\angle 7$, $\angle 2$, $\angle 5$ all measure 20°   **57.** SSS   **58.** SAS   **59.** ASA   **60.** $y = 30$ ft; $x = 34$ ft; $P = 104$ ft   **61.** $y = 7.5$ m; $x = 9$ m; $P = 22.5$ m
62. $x = 12$ mm; $y = 7.5$ mm; $P = 38$ mm   **63.** $x = 105$ **64.** $x = 0$
65. $x = 128$ **66.** $x \approx 23.08$ (rounded)   **67.** $x = 6.5$
68. $x \approx 117.36$ (rounded)   **69.** $\dfrac{8}{5}$ **70.** $\dfrac{33}{80}$ **71.** $\dfrac{15}{4}$ **72.** $\dfrac{4}{1}$
73. $\dfrac{4}{5}$ **74.** $\dfrac{37}{7}$ **75.** $\dfrac{3}{8}$ **76.** $\dfrac{1}{12}$ **77.** $\dfrac{45}{13}$
78. 24,900 fans (rounded)   **79.** $\dfrac{8}{3}$ **80.** 75 ft for $1.99 − $0.50 coupon   **81.** 21 ft long; 15 ft wide
82. (a) 1400 milligrams   (b) 100 milligrams
83. 21 points (rounded)   **84.** $\dfrac{1}{2}$ or 0.5 teaspoon   **85.** parallel lines
86. line segment   **87.** acute angle   **88.** intersecting lines
89. right angle; 90°   **90.** ray   **91.** straight angle; 180°
92. obtuse angle   **93.** perpendicular lines   **94.** (a) The car turned around in a complete circle.   (b) The governor took the opposite view, for example, having once opposed taxes but now supporting them.   **95.** (a) No; because obtuse angles are >90°,

their sum would be >180°. (b) Yes; because acute angles are <90°, their sum could equal 90°. **96.** ∠1 measures 45°; ∠3 and ∠6 measure 35°; ∠4 measures 55°; ∠5 measures 90° **97.** ∠5, ∠2, ∠7, ∠4 all measure 75°; ∠6, ∠1, ∠3, ∠8 all measure 105°.

Chapter 6 Test (page 491)
1. $\frac{\$1}{5 \text{ minutes}}$ **2.** $\frac{9}{2}$ **3.** $\frac{15}{4}$ **4.** 18 ounces for $\$1.89 - \$0.25$ coupon **5.** $x = 25$ **6.** $x \approx 2.67$ (rounded)
7. $x = 325$ **8.** $x = 10\frac{1}{2}$ **9.** 576 words
10. 87 students (rounded) **11.** 23.8 grams (rounded)
12. 60 feet **13.** (e) **14.** (a); 90° **15.** (d) **16.** (g); 180°
17. Parallel lines are lines in the same plane that never intersect. Perpendicular lines intersect to form a right angle. Sketches will vary. **18.** 9° **19.** 160° **20.** $m\angle 1 = 50°, m\angle 2 = 35°, m\angle 3 = 95°, m\angle 5 = 35°$ **21.** measures of ∠1, ∠3, ∠5, and ∠7 are all 65°; measures of ∠2, ∠4, ∠6, and ∠8 are all 115°. **22.** ASA (Angle–Side–Angle) **23.** SAS (Side–Angle–Side)
24. $y = 12$ cm; $z = 6$ cm **25.** In the larger triangle, $x = 12$ mm, $P = 46.8$ mm; in the smaller triangle, $y = 14$ mm, $P = 39$ mm

CHAPTER 7

Section 7.1 (page 503)
1. 0.25 **3.** 0.30 or 0.3 **5.** 0.06 **7.** 1.40 or 1.4 **9.** 0.078
11. 1.00 or 1 **13.** 0.005 **15.** 0.0035 **17.** 50% **19.** 62%
21. 3% **23.** 12.5% **25.** 62.9% **27.** 200% **29.** 260%
31. 3.12% **33.** $\frac{1}{5}$ **35.** $\frac{1}{2}$ **37.** $\frac{11}{20}$ **39.** $\frac{3}{8}$ **41.** $\frac{1}{16}$ **43.** $\frac{1}{6}$
45. $1\frac{3}{10}$ **47.** $2\frac{1}{2}$ **49.** 25% **51.** 30% **53.** 60% **55.** 37%
57. $37\frac{1}{2}$% or 37.5% **59.** 5% **61.** exactly $55\frac{5}{9}$%, or 55.6% (rounded) **63.** exactly $14\frac{2}{7}$%, or 14.3% (rounded) **65.** 0.08
67. 0.42 **69.** 3.5% **71.** 200% **73.** $\frac{95}{100}$ or 95% shaded; $\frac{5}{100}$ or 5% unshaded **75.** $\frac{3}{10}$ or 30% shaded; $\frac{7}{10}$ or 70% unshaded
77. $\frac{3}{4}$ or 75% shaded; $\frac{1}{4}$ or 25% unshaded **79.** 0.01; 1%
81. $\frac{1}{5}$; 20% **83.** $\frac{3}{10}$; 0.3 **85.** 0.5; 50% **87.** $\frac{9}{10}$; 0.9
89. $1\frac{1}{2}$; 150% **91.** (a) employee entertainment/company product discounts (b) 40%; $\frac{2}{5}$ **93.** (a) personal development training
(b) telecommuting **95.** $\frac{1}{4}$; 0.25; 25% **97.** $\frac{3}{8}$; 0.375; 37.5%
99. $\frac{1}{7}$; exactly $14\frac{2}{7}$%, or 14.3% (rounded) **101.** (a) The student forgot to move the decimal point in 0.35 two places to the right. So $\frac{7}{20} = 35\%$. (b) The student did the division in the wrong order. Enter $16 \div 25$ to get 0.64 and then move the decimal point two places to the right. So $\frac{16}{25} = 0.64 = 64\%$. **103.** (a) $78
(b) $39 **105.** (a) 15 inches
(b) $7\frac{1}{2}$ inches **107.** 20 children **109.** (a) 2.8 miles

(b) 1.4 miles **111.** (a) $142.50 (b) 50%
113. (a) 4100 students (b) 50% **115.** (a) 35 problems
(b) 0 problems **117.** 50% means 50 out of 100 parts. That's half of the number. A shortcut for finding 50% of a number is to divide the number by 2.

Section 7.2 (page 515)
1. $\frac{10}{100} = \frac{n}{3000}$; 300 runners **3.** $\frac{4}{100} = \frac{n}{120}$; 4.8 feet
5. $\frac{p}{100} = \frac{16}{32}$; 50% **7.** $\frac{p}{100} = \frac{16}{200}$; 8% **9.** $\frac{90}{100} = \frac{495}{n}$; 550 students **11.** $\frac{12.5}{100} = \frac{3.50}{n}$; $28 **13.** $\frac{250}{100} = \frac{n}{7}$; 17.5 hours
15. $\frac{p}{100} = \frac{32}{172}$; 18.6% (rounded) **17.** $\frac{110}{100} = \frac{748}{n}$; 680 books
19. $\frac{14.7}{100} = \frac{n}{274}$; $40.28 (rounded) **21.** $\frac{p}{100} = \frac{105}{54}$; 194.4% (rounded) **23.** $\frac{4}{100} = \frac{0.33}{n}$; $8.25
25. 150% of $30 cannot be *less* than $30; 25% of $16 cannot be *greater than* $16. **27.** The correct proportion is $\frac{p}{100} = \frac{14}{8}$. The answer should be labeled with the % symbol. Correct answer is 175%.

Section 7.3 (page 523)
1. 1500 patients **3.** $15 **5.** 4.5 pounds **7.** $7.00
9. 52 students **11.** 870 phones **13.** 4.75 hours **15.** (a) 10% means $\frac{10}{100}$ or $\frac{1}{10}$. The denominator tells you to divide the whole by 10. The shortcut for dividing by 10 is to move the decimal point one place to the left. (b) Once you find 10% of a number, multiply the result by 2 for 20% and by 3 for 30%.
17. 231 programs **19.** 50% **21.** 680 circuits **23.** 1080 people
25. 845 species **27.** $20.80 **29.** 76% **31.** 125%
33. 4700 employees **35.** 83.2 quarts **37.** 2% **39.** 700 tablets
41. 325 salads **43.** 1.5% **45.** 5 gallons **47.** 1029.2 meters
49. (a) Multiply 0.2 by 100 to change it from a decimal to a percent. So, 0.20 = 20%. (b) The correct equation is $50 = p \cdot 20$, so the solution is 250%. **51.** (a) $\frac{1}{3} \cdot 162 = n$; the solution is $54.
(b) $(0.333333333)(162) = n$; depending upon how your calculator rounds numbers, the solution is either $54 or $53.99999995.
(c) There is no difference or the difference is insignificant.
52. (a) $22 = \frac{2}{3} \cdot n$; the solution is 33 cans.
(b) $22 = (0.666666667)(n)$; depending upon how your calculator rounds numbers, the solution is either 33 cans or 32.99999998 cans.
(c) There is no difference or the difference is insignificant.

Summary Exercises on Percent (page 527)
1. (a) 0.03; 3% (b) $\frac{3}{10}$; 0.3 (c) $\frac{3}{8}$; 37.5% (d) $1\frac{3}{5}$; 1.6
(e) 0.0625; 6.25% (f) $\frac{1}{20}$; 0.05 (g) 2; 200% (h) 0.8; 80%
(i) $\frac{9}{125}$; 7.2% **2.** (a) 3.5 ft (b) 19 miles (c) 105 cows
(d) $0.08 (e) 500 women (f) $45 (g) $87.50 (h) 12 pounds
(i) 95 students **3.** $12\frac{1}{2}$% or 12.5% **4.** 75 DVDs **5.** $0.53 (rounded) **6.** 175% **7.** 5.9 pounds (rounded) **8.** 600 hours
9. 50% **10.** 500 camp sites **11.** 98 golf balls **12.** 2.7% (rounded) **13.** 2300 apartments **14.** 0.63 ounce **15.** 145%

A-34 Answers to Selected Exercises

16. 244 voters (rounded) **17.** 0.021 inch **18.** 400% **19.** 8%
20. $0.68 **21.** flowers; $922.50 **22.** $1230 **23.** $2993
24. $51.25

Section 7.4 (page 535)
1. $37.80 **3. (a)** 10% **(b)** 5% **(c)** 2% **(d)** 1%
5. (a) 101.6 pounds (rounded) **(b)** 10.1 pounds (rounded)
7. 13.1% female; 86.9% male (both rounded) **9.** 281 million people (rounded) **11.** 138% **13.** 40 problems
15. 1360 shots (rounded) **17.** 23.7 miles per gallon (rounded)
19. March; 24.5 million cans, or 24,500,000 cans
21. 52.5 million cans (January); 38.5 million cans (February)
23. 64.8% (rounded) **25.** 9.1% (rounded) **27.** 40%
29. 563% (rounded) **31.** No. 100% is the entire price, so a decrease of 100% would take the price down to 0. Therefore, 100% is the maximum possible decrease in the price of something.
33. George ate more than 65 grams, so the percent must be >100%. Use $p \cdot 65 = 78$ to get 120%. **34.** The team won more than half the games, so the percent must be >50%. Correct solution is $0.72 = 72\%$.
35. The brain could not weigh 375 pounds, which is more than the person weighs. $2\frac{1}{2}\% = 2.5\% = 0.025$, so $(0.025)(150) = n$ and $n = 3.75$ pounds. **36.** If 80% were absent, then only 20% made it to class. $800 - 640 = 160$ students, or use $(0.20)(800) = n$.

Section 7.5 (page 545)
1. $6; $106 **3.** 3%; $70.04 **5.** $29.28 (rounded); $395.26
7. $0.12 (rounded); $2.22 **9.** $4\frac{1}{2}\%$; $13,167
11. $3 + $1.50 = $4.50; $4.83 (rounded); 2($3) = $6; $6.43 (rounded) **13.** $8 + $4 = $12; $11.75 (rounded); 2($8) = $16; $15.67 (rounded) **15.** $1 + $0.50 = $1.50; $1.43 (rounded); 2($1) = $2; $1.91 **17.** $15; $85 **19.** 30%; $126 **21.** $4.38 (rounded); $13.12 **23.** $3.79 (rounded); $34.09
25. $42; $342 **27.** $33.30; $773.30 **29.** $225; $1725
31. $1001.25; $18,801.25 **33.** $1170 **35.** $7978.13 (rounded)
37. $106.49 (rounded) **39.** 5% **41.** $74.25 **43.** $25.13
45. $106.20; $483.80 **47.** $2016.38 (rounded) **49.** $21 (rounded)
51. $92.38 **53.** $230.12 **55. (a)** $18.43 (rounded to nearest cent)
(b) When calculating the discount, the *whole* is $18.50. But when calculating the sales tax, the *whole* is only $17.39 (the discounted price). **56. (a)** $396.05 (rounded to nearest cent).
(b) 7.53% sales tax (rounded to the nearest hundredth) would give a final cost of $398.01.

Chapter 7 Review Exercises (page 557)
1. 0.25 **2.** 1.8 **3.** 0.125 **4.** 0.07 **5.** 265% **6.** 2%
7. 30% **8.** 0.2% **9.** $\frac{3}{25}$ **10.** $\frac{3}{8}$ **11.** $2\frac{1}{2}$ **12.** $\frac{1}{20}$ **13.** 75%
14. 62.5% or $62\frac{1}{2}\%$ **15.** 325% **16.** 6% **17.** 0.125
18. 12.5% **19.** $\frac{3}{20}$ **20.** 15% **21.** $1\frac{4}{5}$ **22.** 1.8 **23.** $46
24. $23 **25.** 9 hours **26.** $4\frac{1}{2}$ hours **27.** 242 meters
28. 17,000 cases **29.** 27 cellular phones **30.** 870 reference books **31.** 9.5% (rounded) **32.** 225% **33.** $2.60 (rounded)
34. 80 days **35.** 4% **36.** $575 **37.** 20 people **38.** 175%
39. (a) 3000 patients **(b)** 31.5% decrease (rounded)
40. (a) $364; **(b)** 224% **41. (a)** 80%
(b) 6.3% increase (rounded) **42.** 10.4% (rounded)
43. $0.11 (rounded); $2.90 **44.** $7\frac{1}{2}\%$; $838.50
45. $4 + $2 = $6; $6.41 (rounded); 2($4) = $8; $8.55 (rounded)
46. $0.80 + $0.40 = $1.20; $1.21 (rounded); 2($0.80) = $1.60; $1.61 **47.** $3.75; $33.75 **48.** 25%; $189 **49.** $68.25; $418.25
50. $183.60; $1713.60 **51.** $\frac{1}{3}$; $33\frac{1}{3}\%$ (exact) or 33.3% (rounded)
52. cards; 68%; 0.68; $\frac{17}{25}$ **53.** 277 adults (rounded) **54.** cards, 188 adults (rounded); sci-fi/simulation, 102 adults (rounded)
55. 48.4% (rounded) **56.** 7161 dogs (rounded)
57. 36.4% (rounded) **58.** 180.3% increase (rounded)
59. 38.1% (rounded) **60.** 265 animals (rounded)

Chapter 7 Test (page 561)
1. 0.75 **2.** 60% **3.** 180% **4.** 7.5% or $7\frac{1}{2}\%$ **5.** 3.00 or 3
6. 0.02 **7.** $\frac{5}{8}$ **8.** $2\frac{2}{5}$ **9.** 5% **10.** 87.5% or $87\frac{1}{2}\%$
11. 175 **12.** 320 laptops **13.** 400% **14.** $19,500
15. $8466.75 **16.** 34% increase (rounded) **17.** To find 50% of a number, divide the number by 2. To find 25% of a number, divide the number by 4. Examples will vary. **18.** Round $31.94 to $30. Then 10% of $30 is $3 and 5% of $30 is half of $3 or $1.50, so a 15% tip estimate is $3 + $1.50 = $4.50. A 20% tip estimate is 2($3) = $6. **19.** Exact tip is $4.79 (rounded). Each person pays $12.24 (rounded). **20.** $3.84; $44.16 **21.** $41.39 (rounded); $188.56 **22.** $815.66 ÷ 6 ≈ $135.94 **23.** $1650 **24.** $911.60

CHAPTER 8

Section 8.1 (page 571)
1. 3 **3.** 8 **5.** 5280 **7.** 2000 **9.** 60 **11.** 2 **13.** 2
15. 14,000 to 16,000 lb **17.** 27 **19.** 112 **21.** 10 **23.** $1\frac{1}{2}$ or 1.5
25. $\frac{1}{4}$ or 0.25 **27.** $1\frac{1}{2}$ or 1.5 **29.** $2\frac{1}{2}$ or 2.5 **31.** snowmobile/ATV; person walking **33.** 5000 **35.** 17 **37.** 4 to 8 in.
39. 216 **41.** 28 **43.** 518,400 **45.** 48,000 **47. (a)** pound/ounces **(b)** quarts/pints or pints/cups **(c)** minutes/hours or seconds/minutes **(d)** feet/inches **(e)** pounds/tons **(f)** days/weeks **49.** 174,240
51. 800 **53.** 0.75 or $\frac{3}{4}$ **55.** $1.83 (rounded) **57.** $140
59. (a) 1056 sec **(b)** 17.6 min **61. (a)** $12\frac{1}{2}$ qt **(b)** 4 jugs, because you can't buy part of a jug **63. (a)** Travel 1391 times around Earth (rounded). **(b)** Drive 12,436 times from Los Angeles to New York (rounded). **64. (a)** 35,000,000 ÷ 25,000 becomes 35,000 ÷ 25 = 1400 times around Earth **(b)** 12,500 times from Los Angeles to New York **(c)** They are very close and just as useful for comparing such large numbers. **65. (a)** 88,176 ft tall **(b)** 5.5 mi (rounded) **(c)** about 3 times taller
66. (a) 65 lb (rounded) **(b)** 3 lb (rounded) **(c)** Answers will vary but should be about $\frac{1}{3}$ of your weight on Earth.

Section 8.2 (page 581)
1. 1000; 1000 **3.** $\frac{1}{1000}$ or 0.001; $\frac{1}{1000}$ or 0.001
5. $\frac{1}{100}$ or 0.01; $\frac{1}{100}$ or 0.01 **7.** Answers will vary; about 8 to 10 cm.
9. Answers will vary; about 15 to 25 mm. **11.** cm **13.** m
15. km **17.** mm **19.** cm **21.** m **23.** Some possible answers are: 35 mm film for cameras, track and field events, metric auto parts, and lead refills for mechanical pencils. **25.** 700 cm
27. 0.040 m or 0.04 m **29.** 9400 m **31.** 5.09 m **33.** 40 cm

35. 910 mm   **37.** less; 18 cm or 0.18 m
39. 5 mm = 0.5 cm
     1 mm = 0.1 cm
41. 0.018 km   **43.** 164 cm; 1640 mm   **45.** 0.0000056 km

Section 8.3 (page 589)
1. mL **3.** L **5.** kg **7.** g **9.** mL **11.** mg **13.** L **15.** kg
17. unreasonable; too much   **19.** unreasonable; too much
21. reasonable   **23.** reasonable   **25.** Some capacity examples are 2 L bottles of soda and shampoo bottles marked in mL; weight examples are grams of fat listed on cereal boxes and vitamin doses in milligrams.   **27.** Unit for your answer (g) is in numerator; unit being changed (kg) is in denominator so it will divide out. The unit fraction is $\frac{1000 \text{ g}}{1 \text{ kg}}$.   **29.** 15,000 mL   **31.** 3 L   **33.** 0.925 L
35. 0.008 L   **37.** 4150 mL   **39.** 8 kg   **41.** 5200 g   **43.** 850 mg
45. 30 g   **47.** 0.598 g   **49.** 0.06 L   **51.** 0.003 kg   **53.** 990 mL
55. mm   **57.** mL   **59.** cm   **61.** mg   **63.** 0.3 L   **65.** 1340 g
67. 0.9 L   **69.** 3 kg to 4 kg   **71.** greater; 5 mg or 0.005 g
73. 200 nickels   **75. (a)** 1,000,000
(b) $\frac{3.5 \text{ Mm}}{1} \cdot \frac{1,000,000 \text{ m}}{1 \text{ Mm}} = 3,500,000 \text{ m}$
76. (a) 1,000,000,000
(b) $\frac{2500 \text{ m}}{1} \cdot \frac{1 \text{ Gm}}{1,000,000,000 \text{ m}} = 0.0000025 \text{ Gm}$
77. (a) 1,000,000,000,000   **(b)** 1000; 1,000,000
78. 1,000,000; 1,000,000,000; $2^{20} = 1,048,576$; $2^{30} = 1,073,741,824$

Section 8.4 (page 595)
1. $0.83 (rounded)   **3.** 89.5 kg   **5.** 71 beats (rounded)
7. 180 cm; $0.02/cm (rounded)   **9.** 5.03 m   **11.** 4 bottles; 0.175 L or 175 mL   **13.** 1.89 g   **15. (a)** 10 km   **(b)** 600 km
(c) 36,000 km   **17.** 215 g; 4.3 g; 4300 mg   **18.** 330 g; 0.33 g; 330 mg   **19.** 1550 g; 1500 g; 3 g   **20.** 55 g; 50 g; 0.5 g
21. 0.45 mg

Section 8.5 (page 601)
1. 21.8 yd   **3.** 262.4 ft   **5.** 4.8 m   **7.** 5.3 oz   **9.** 111.6 kg
11. 30.3 qt   **13. (a)** about 0.2 oz   **(b)** probably not   **15.** about 31.8 L   **17.** about 1.3 cm   **19.** 18 kg; 56 cm by 36 cm by 23 cm (all rounded)   **21.** Converting ounces to grams, 0.09 oz ≈ 2.55 g, which rounds to 2.6 g. However, converting grams to ounces, 2.5 g ≈ 0.0875 oz which does round to 0.09 oz.   **23.** 3.5 kg ≈ 7.7 lb so the baby is heavy enough. But 53 cm ≈ 20.7 in. so the baby is not long enough to be in the carrier.   **25.** 28 °F   **27.** 40 °C
29. 150 °C   **31.** 16 °C (rounded)   **33.** −20 °C   **35.** 46 °F (rounded)   **37.** 23 °F   **39.** 58 °C (rounded); −89 °C (rounded)
41. 10 °C and 41 °C (rounded)   **43.** More. There are 180 degrees between freezing and boiling on the Fahrenheit scale, but only 100 degrees on the Celsius scale, so each Celsius degree is a greater change in temperature.   **45. (a)** pleasant weather, above freezing but not hot   **(b)** 75 °F to 39 °F (rounded)   **(c)** Answers will vary. In Minnesota, it's 0 °C to −40 °C; in California, 24 °C to 0 °C.
47. about 1.8 m   **48.** about 5.0 kg   **49.** about 3040.2 km
50. 38 hr 23 min   **51.** about 300 m   **52.** less than 3780 mL
53. about 59.4 mL   **54.** about 1.6%

Chapter 8 Review Exercises (page 611)
1. 16   **2.** 3   **3.** 2000   **4.** 4   **5.** 60   **6.** 8   **7.** 60
8. 5280   **9.** 12   **10.** 48   **11.** 3   **12.** 4   **13.** $\frac{3}{4}$ or 0.75
14. $2\frac{1}{2}$ or 2.5   **15.** 28   **16.** 78   **17.** 112   **18.** 345,600
19. (a) $4153\frac{1}{3}$ yd (exact) or $4153.\overline{3}$ (rounded)
(b) 2.4 mi (rounded)   **20.** $2465.20   **21.** mm   **22.** cm   **23.** km
24. m   **25.** cm   **26.** mm   **27.** 500 cm   **28.** 8500 m   **29.** 8.5 cm
30. 3.7 m   **31.** 0.07 km   **32.** 930 mm   **33.** mL   **34.** L
35. g   **36.** kg   **37.** L   **38.** mL   **39.** mg   **40.** g   **41.** 5 L
42. 8000 mL   **43.** 4580 mg   **44.** 700 g   **45.** 0.006 g   **46.** 0.035 L
47. 31.5 L   **48.** 357 g (rounded)   **49.** 87.25 kg
50. $1.12 (rounded)   **51.** 6.5 yd (rounded)
52. 11.7 in. (rounded)   **53.** 67.0 mi (rounded)   **54.** 1288 km
55. 21.9 L (rounded)   **56.** 44.0 qt (rounded)   **57.** 0 °C
58. 100 °C   **59.** 37 °C   **60.** 20 °C   **61.** 25 °C   **62.** −15 °C
63. 28 °F (rounded)   **64.** 120 °F (rounded)   **65.** L   **66.** kg
67. cm   **68.** mL   **69.** mm   **70.** km   **71.** g   **72.** m   **73.** mL
74. g   **75.** mg   **76.** L   **77.** 105 mm   **78.** $\frac{3}{4}$ hr or 0.75 hr
79. $7\frac{1}{2}$ ft or 7.5 ft   **80.** 130 cm   **81.** 77 °F   **82.** 14 qt
83. 0.7 g   **84.** 810 mL   **85.** 80 oz   **86.** 60,000 g
87. 1800 mL   **88.** −1 °C   **89.** 36 cm   **90.** 0.055 L
91. 1.26 m   **92.** 18 tons   **93.** 113 g; 177 °C (both rounded)
94. 178.0 lb; 6.0 ft (both rounded)   **95.** 13.2 m   **96.** 22.0 ft
97. 3.6 m   **98.** 39.8 in.   **99.** 484,000 lb   **100.** 18.9 to 22.7 L
101. (a) 242 tons   **(b)** 144 in.   **102. (a)** 1.02 m   **(b)** 670 cm

Chapter 8 Test (page 615)
1. 36 qt   **2.** 15 yd   **3.** 2.25 hr or $2\frac{1}{4}$ hr   **4.** 0.75 ft or $\frac{3}{4}$ ft
5. 56 oz   **6.** 7200 min   **7.** kg   **8.** km   **9.** mL   **10.** g   **11.** cm
12. mm   **13.** L   **14.** cm   **15.** 2.5 m   **16.** 4600 m   **17.** 0.5 cm
18. 0.325 g   **19.** 16,000 mL   **20.** 400 g   **21.** 1055 cm
22. 0.095 L   **23.** 3.2 ft (rounded)   **24. (a)** 2310 mg
(b) 500 mg or 0.5 g more   **25.** 95 °C   **26.** 0 °C
27. 1.8 m (rounded)   **28.** 56.3 kg (rounded)   **29.** 13 gal
30. 5.0 mi (rounded)   **31.** 23 °C (rounded)   **32.** 10 °F (rounded)
33. 6 m ≈ 6.54 yd; $26.03 (rounded)   **34.** Possible answers: Use same system as rest of the world; easier system for children to learn; less use of fractional numbers; compete internationally.

CHAPTER 9

Section 9.1 (page 623)
1. (a) 31,419 points   **(b)** Michael Jordan   **3. (a)** Michael Jordan
(b) Allen Iverson   **5.** 18,025 points   **7.** Iverson, 27.1; West, 27.0; Pettit, 26.4; round to nearest tenth.   **9.** The asterisks next to O'Neal's and Iverson's names mean that they played in the entire 2003–2004 season. During the rest of the season, they added games and points to the midseason numbers shown in the table.
11. (a) 255 calories   **(b)** moderate jogging   **13. (a)** moderate jogging, aerobic dance, racquetball   **(b)** moderate jogging, moderate bicycling, aerobic dance, racquetball, tennis
15. 370 calories   **17. (a)** 366 calories (rounded)
(b) 239 calories   **19.** about 39 calories   **21. (a)** 30 million or 30,000,000   **(b)** 75 million or 75,000,000   **23.** 140 million or 140,000,000   **25.** 5 million or 5,000,000   **27.** 245 million or 245,000,000   **29.** Answers will vary. One possibility: choose Southwest because it has the best on-time performance.
30. Answers will vary. Possibilities include planning more time between each flight, or doing some or all of your business via conference calls or e-mail.   **31.** Answers will vary. One possibility: choose Airtran because it has the fewest luggage problems.   **32.** Answers will vary. Possibilities include buying heavy-duty luggage or shipping the golf clubs via a delivery service.   **33.** Answers will vary. Possibilities include a lot of

bad weather, maintenance problems, new computer system.
34. Answers will vary. Possibilities include availability of nonstop flights, convenience of departure times, type and size of aircraft, availability of low-cost fares.

Section 9.2 (page 631)

1. (a) $32,000 **(b)** carpentry, $12,100 **3. (a)** $\frac{\$9800}{\$32,000} = \frac{49}{160}$
(b) $\frac{\$3000}{\$2000} = \frac{3}{2}$ **5.** $\frac{\$16,000}{\$32,000} = \frac{1}{2}$ **7. (a)** don't know
(b) quicker **9.** $\frac{720}{6000} = \frac{3}{25}$ **11.** $\frac{1020}{1200} = \frac{17}{20}$ **13.** $\frac{1740}{180} = \frac{29}{3}$
15. $522,000 **17.** $174,000 **19.** $261,000 **21.** 160 people
23. mustard; 960 people **25.** 64 people **27.** First find the percent of the total that is to be represented by each item. Next, multiply the percent by 360° to find the size of each sector. Finally, use a protractor to draw each sector. **29. (a)** 90° **(b)** 20°
(c) 10%; 36° **(d)** 10%; 36° **(e)** 15%; 54° **(f)** 5%; 18°
(g) 15%; 54° **(h)** see circle graph below.

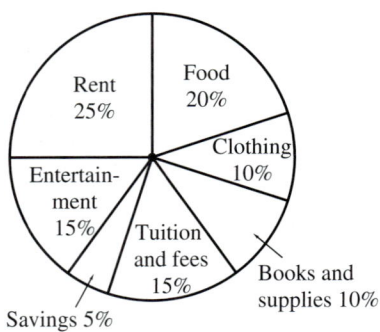

31. (a) $200,000 **(b)** 6.25%; 20%; 30%; 25%; 18.75%
(c) 22.5°; 72°; 108°; 90°; 67.5°
(d) see circle graph below.

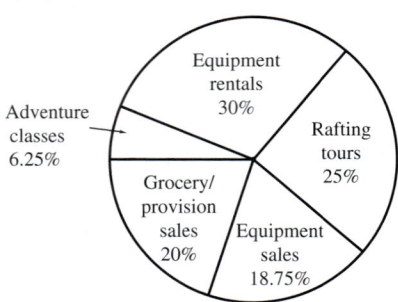

Section 9.3 (page 641)

1. can shop during off hours; 74% **3.** 57%; 342 people **5.** $\frac{1}{2}$, compare products more easily; nearly $\frac{3}{4}$, shop during off hours
7. May; 10,000 unemployed **9.** 1500 workers **11.** 2500 workers; 45% increase (rounded) **13.** 150,000 gallons **15.** 2001; 250,000 gallons **17.** 550,000 gallons; 367% increase (rounded)
19. 24.1 million or 24,100,000 PCs **21.** 185.6 million or 185,600,000 more PCs **23.** 82.3 million or 82,300,000 PCs; 132% increase (rounded) **25. (a)** 3,000,000 CDs **(b)** 1,500,000 CDs **27. (a)** 2,500,000 CDs **(b)** 3,000,000 CDs **29.** Answers will vary. Possibilities include: Both stores had decreased sales from 2001 to 2002 and increased sales from 2003 to 2005; Store B had lower sales than Store A in 2001–02 but higher sales than Store A in 2003–2005. **31.** 2003; $25,000 **33.** 2002, 29% (rounded); 2003, 20%; 2004, 17% (rounded); 2005, 38% (rounded) **35.** Answers will vary. Possibilities include: The decrease in sales may have resulted from poor service or greater competition; the increase in sales may have been a result of more advertising or better service.
37. Shipments have increased at a rapid rate since 1985.
38. Answers will vary. Some possibilities are lower prices; more uses and applications for students, home use, and businesses; improved technology. **39. (a)** 104% (rounded) **(b)** 159% (rounded)
(c) 132% (rounded) **(d)** 37% (rounded) **40.** Since 1990, the percent of increase for each 5-year period has been dropping.
41. Answers will vary. Some possibilities are: More people will already own a computer and not want to buy another; some new invention will replace computers. **42. (a)** no change; 0% **(b)** 167% increase (rounded) **43. (a)** Answers will vary; perhaps 3,500,000 CDs in 2006. **(b)** Answers will vary; perhaps 4,500,000 CDs in 2006. **44. (a)** most people will probably pick Store B because of its greater sales and more consistent upward trend. **(b)** Answers will vary. Some possibilities include: age and physical condition of the store, annual expenses, sales of other products, annual profit.

Section 9.4 (page 649)

1.

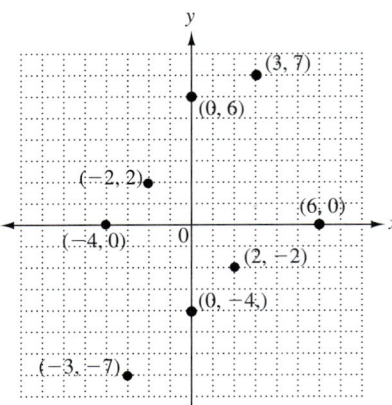

3.

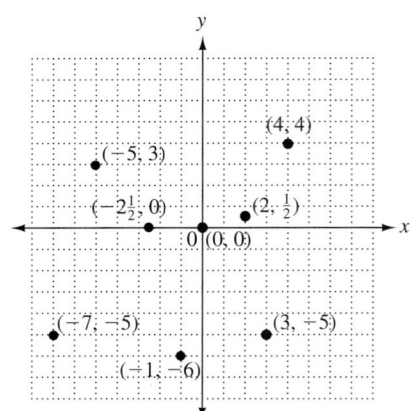

5. A is $(3, 4)$; B is $(5, -5)$; C is $(-4, -2)$; D is approximately $\left(4, \frac{1}{2}\right)$; E is $(0, -7)$; F is $(-5, 5)$; G is $(-2, 0)$; H is $(0,0)$.
7. III, none, IV, II **9. (a)** any positive number **(b)** any negative number **(c)** 0 **(d)** any negative number **(e)** any positive number
11. Starting at the origin, move left or right along the x-axis to the number a; then move up if b is positive or move down if b is negative.

Section 9.5 (page 659)

1. 4; (0, 4); 3; (1, 3); 2; (2, 2); All points on the line are solutions.

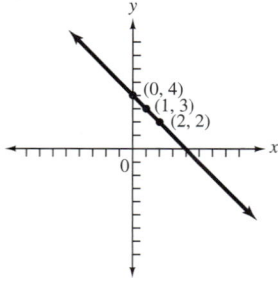

3. −1; (0, −1); −2; (1, −2); −3; (2, −3); All points on the line are solutions.

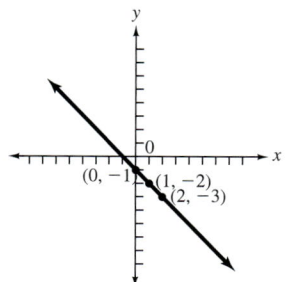

5. 4; −1; −6; 99

7.

x	y	(x, y)
1	−1	(1, −1)
2	0	(2, 0)
3	1	(3, 1)

9.

x	y	(x, y)
0	2	(0, 2)
−1	1	(−1, 1)
−2	0	(−2, 0)

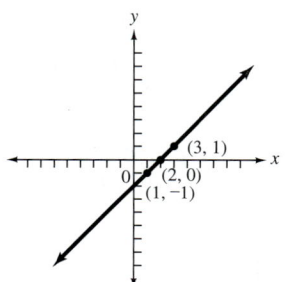
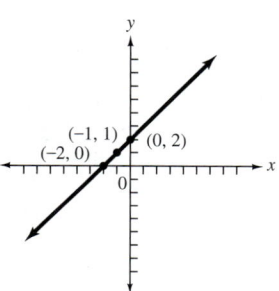

11.

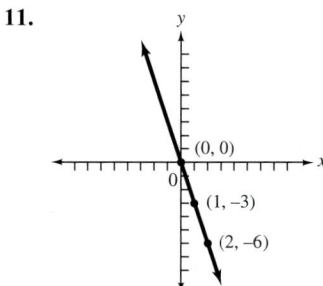

x	y	(x, y)
0	0	(0, 0)
1	−3	(1, −3)
2	−6	(2, −6)

13. The lines in Exercises 7 and 9 have a positive slope. The lines in Exercises 1, 3, and 11 have a negative slope.

15.

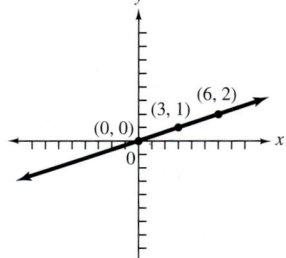

x	y	(x, y)
0	0	(0, 0)
3	1	(3, 1)
6	2	(6, 2)

17.

x	y	(x, y)
−1	−1	(−1, −1)
−2	−2	(−2, −2)
−3	−3	(−3, −3)

19.

x	y	(x, y)
0	3	(0, 3)
1	1	(1, 1)
2	−1	(2, −1)

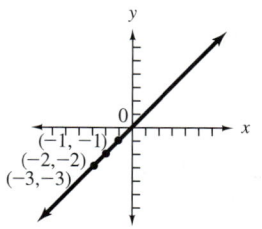
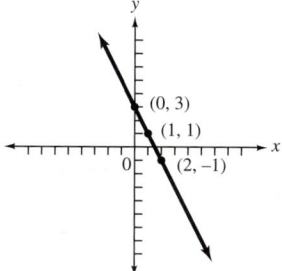

21. **23.**

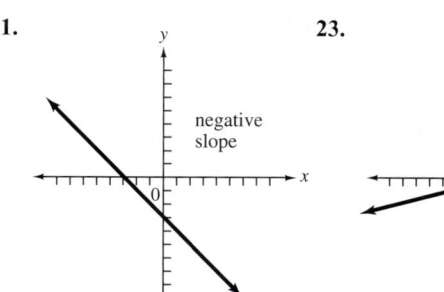

25. **27.**

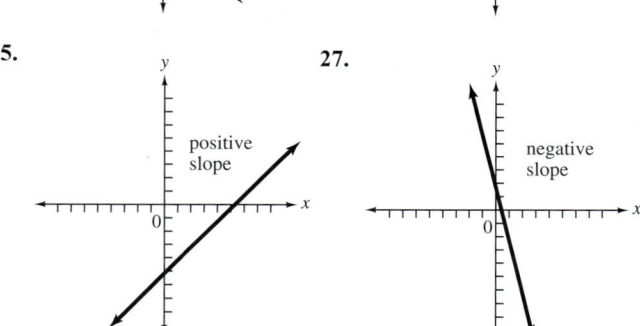

Chapter 9 Review Exercises (page 673)

1. (a) gymnastics **(b)** volleyball **2. (a)** cross country **(b)** golf
3. (a) 4870 men **(b)** 12,081 women **4.** men's, 14.2 (rounded); women's, 13.3 (rounded) **5. (a)** 100 inches **(b)** 15 inches
6. (a) 50 inches **(b)** 55 inches **7. (a)** 35 inches **(b)** 45 inches
8. 95 inches difference between Juneau and Memphis
9. (a) lodging; $560 **(b)** food; $400 **(c)** $1700
10. $\frac{560}{400} = \frac{7}{5}$ **11.** $\frac{300}{1700} = \frac{3}{17}$ **12.** $\frac{280}{1700} = \frac{14}{85}$
13. $\frac{300}{160} = \frac{15}{8}$ **14.** painting and wall papering; 63%
15. construction work; 33% **16.** 43%; 147 homeowners (rounded)
17. 184 homeowners (rounded) **18. (a)** interior decorating
(b) construction work **19.** Answers will vary. Possibilities include: painting and wall papering are easier to do, take less time, or cost less than construction work. **20.** March; 8,000,000 acre-feet
21. June; 2,000,000 acre-feet **22.** 5,000,000 acre-feet
23. 6,000,000 acre-feet **24.** 3,000,000 acre-feet; 37.5% decrease
25. 3,000,000 acre-feet; 60% decrease **26.** $50,000,000
27. $20,000,000 **28.** $20,000,000 **29.** $40,000,000
30. Sales decreased for two years and then moved up slightly. Answers will vary. Perhaps there is less new construction, remodeling, and home

improvement in the area near Center A. Or, better product selection and service may have reversed the decline in sales.
31. Sales are increasing. Answers will vary. New construction may have increased in the area near Center B, or greater advertising may attract more attention. **32.** 36° **33.** 35%; 126° **34.** 20% 72°
35. 25% 90° **36.** $2240; 10% **37.** See graph below.

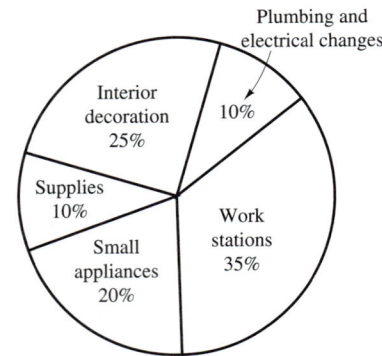

38.

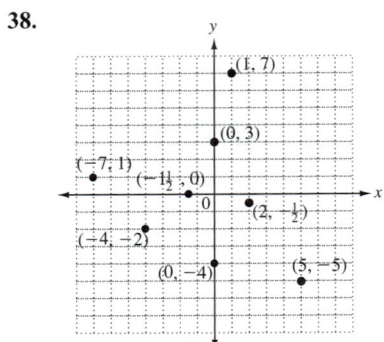

39. A is (0, 6); B is approximately $\left(-2, 2\frac{1}{2}\right)$; C is (0, 0); D is $(-6, -6)$; E is (4, 3); F is approximately $\left(3\frac{1}{2}, 0\right)$; G is (2, −4).

40.

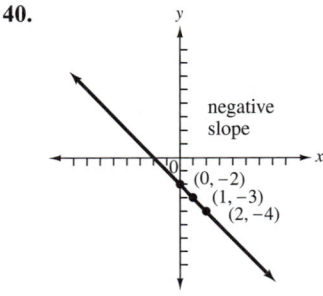

x	y	(x, y)
0	−2	(0, −2)
1	−3	(1, −3)
2	−4	(2, −4)

There are many answers because all points on the line are solutions. Some possibilities are (−1, −1) and (−2, 0).

41.

x	y	(x, y)
0	3	(0, 3)
1	4	(1, 4)
2	5	(2, 5)

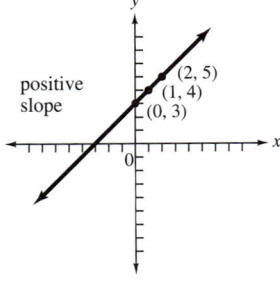

All points on the line are solutions. Some possibilities are (−1, 2) and (−2, 1).

42.

x	y	(x, y)
−1	4	(−1, 4)
0	0	(0, 0)
1	−4	(1, −4)

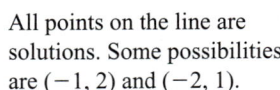

All points on the line are solutions. Some possibilities are $\left(-\frac{1}{2}, 2\right)$ and $\left(\frac{1}{2}, -2\right)$.

Chapter 9 Test (page 679)
1. (a) sardines (b) cream cheese **2.** 166 calories (rounded)
3. 259 mg (rounded) **4.** (a) birds (b) 80 bird species
5. 55 species **6.** 265 species **7.** television; $812,000
8. miscellaneous; $84,000 **9.** $308,000 **10.** $644,000
11. 2003; $4000 **12.** $5000; 38% increase (rounded)
13. 2003; explanations will vary. Some possibilities are: laid off from work, changed jobs, was ill, cut down on hours worked.
14. 5500 students; 3000 students **15.** College B; 1000 students
16. Explanations will vary. For example, College B may have added new courses or lowered tuition or added child care.
17. 35%; 126° **18.** 5%; 18° **19.** 20%; 72° **20.** 30%; 108°
21. $48,000; 10%
22.

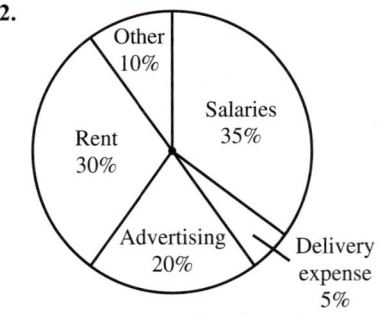

23–26.

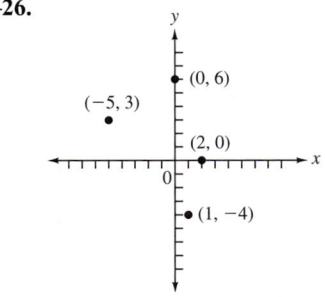

27. (0, 0); no quadrant **28.** (−5, −4); quadrant III
29. (3, 3); quadrant I **30.** (−2, 4); quadrant II
31.

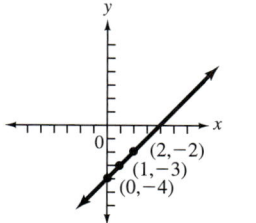

x	y	(x, y)
0	−4	(0, −4)
1	−3	(1, −3)
2	−2	(2, −2)

32. Answers will vary; all points on the line are solutions. Some possibilities are (3, −1) and (4, 0).

Chapter 10

Section 10.1 (page 691)

1. (a) rational because it terminates; (b) irrational because decimal form never ends or repeats in a fixed block
3. (a) rational because it's the quotient of integers; (b) rational because digits repeat in a fixed block
5. (a) irrational because decimal form never ends or repeats in a fixed block; (b) rational because $\sqrt{100} = 10$
7. sometimes **9.** always **11.** never
13. "75 hundredths does not equal three-fourths." False; $0.75 = \frac{3}{4}$.
15. "Negative 4 is less than or equal to negative 5." False; $-4 > -5$ and $-4 \neq -5$.
17. "Zero is greater than or equal to zero." True, because $0 = 0$.
19. $19 > 12$ **21.** $\frac{1}{2} \leq \frac{4}{5}$ **23.** $-17 \leq -17$; True
25. $-3 \geq 0$; False **27.** $-6 \neq 6$; True **29.** $45 \leq 40$; False
31. $1 \geq 1$; True **33.** $0 \neq$ undefined; True **35.** (a) *Titanic*; (b) *Titanic* and *Star Wars* **36.** (a) *Star Wars: Episode I* and *Spider-Man*; (b) *E.T.*, *Star Wars: Episode I*, and *Spider-Man*
37. Answers will vary. One possibility is: gross receipts ≤ 431.1 million dollars. **38.** Answers will vary. One possibility is: gross receipts ≥ 436.5 million dollars. **39.** $-4t - 5m$ **41.** $5c + 4d$
43. $-6h + n$ **45.** $3q - 5r + 8s$ **47.** $-19p + 16$
49. $-4y + 22$ **51.** $y^2 + y - 12$ **53.** $6b^2 - 6b + 8$
55. $a + 1$ **57.** $24k - 4$

Section 10.2 (page 699)

1. $m = -1$ **3.** $p = 5$ **5.** $r = 1$ **7.** $x = -\frac{5}{3}$ **9.** $x = -1$
11. $y = 7$ **13.** $p = -4$ **15.** $b = 13$ **17.** $p = 18$ **19.** $x = 12$
21. no solution **23.** all real numbers **25.** no solution
27. no solution **29.** all real numbers **31.** all real numbers
33. No, it is incorrect to divide both sides by a variable. If $-3x$ is added to both sides, the equation becomes $4x = 0$, so $x = 0$ is the correct solution. **35.** $t = 5$ **37.** $x = 0$ **39.** $k = -\frac{7}{5}$
41. $x = -2$ **43.** all real numbers **45.** $x = 120$ **47.** $x = 6$
49. $x = 15{,}000$ **51.** no solution **53.** $x = 4$ **55.** $y = 0$
57. $x = 20$ **59.** all real numbers **61.** $s = -\frac{13}{8}$

Section 10.3 (page 707)

1. $l = 6$ **3.** $h = 14$ **5.** $c = 5$ **7.** $r = 40$ **9.** $t = 7$
11. $r \approx 1.3$ **13.** $w = 8$ **15.** (a) width = 35 inches
(b) area = 1785 square inches **17.** $d \approx 630$ ft **19.** $r = \frac{d}{t}$
21. $l = \frac{A}{w}$ **23.** $a = P - b - c$ **25.** $p = \frac{I}{rt}$ **27.** $b = \frac{2A}{h}$
29. $r = \frac{A - p}{pt}$ **31.** $h = \frac{V}{\pi r^2}$ **33.** $C = \frac{5}{9}(F - 32)$ or $C = \frac{5F - 160}{9}$ **34.** (a) $P - 2l = 2w$ (b) $\frac{P - 2l}{2} = w$
35. (a) $\frac{P}{2} = l + w$ (b) $\frac{P}{2} - l = w$
36. (a) Multiplicative identity property (b) A number divided by 1 is equal to itself. (c) Multiplication of fractions (d) Subtraction of fractions

Section 10.4 (page 715)

1. Use an open circle if the symbol is > or <. Use a closed circle if the symbol is ≥ or ≤. **3.** Every real number less than 2 is a solution. There are an infinite number of solutions, so it is not possible to list them all. **5.** ⟵————•——⟶ at 4
7. ⟵———◦————⟶ at −3
9. ⟵——◦———•⟶ at 8, 10
11. ⟵——◦———•⟶ at 0, 10
13. It would imply that $3 < -2$, which is false.
15. $z \geq 1$ at 1
17. $k \geq 5$ at 5
19. $n < -11$ at −11
21. It must be reversed when multiplying or dividing by a negative number. **23.** His method is incorrect. He divided *by* the positive number 6. The sign of the number divided *into* does not matter.
25. $x < 6$ at 6
27. $y \geq -10$ at −10
29. $t < -3$ at −3
31. $x \leq 0$ at 0
33. $r > 20$ at 20
35. $x \geq -3$ at −3
37. $r \geq -5$ at −5
39. $x < 1$ at 1
41. $x \leq 0$ at 0
43. $x \geq 4$ at 4
45. $p < 32$ at 32
47. $x \geq \frac{5}{12}$ at $\frac{5}{12}$, 0
49. $k > -21$ at −21
51. all numbers less than 3 **53.** 79 or more **55.** It is never more than 86 degrees Fahrenheit. **57.** 32 or greater **59.** at least $275
61. 15 minutes **63.** ⟵————•———⟶ at 0, 4

64.

65.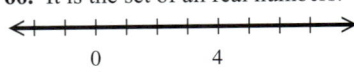

66. It is the set of all real numbers.

67. The graph would be the set of all real numbers.

Chapter 10 Review Exercises (page 723)
1. rational because it terminates **2.** rational because it repeats in a fixed block **3.** rational because $\sqrt{144} = 12$ **4.** irrational because decimal form does not terminate or repeat in a fixed block.
5. True because $\frac{2}{3}$ does not equal $\frac{6}{10}$. **6.** True because $-\frac{5}{8} = -\frac{5}{8}$.
7. False because $-8 < 8$ and $-8 \neq 8$ **8.** False because $-75 < -50$ and $-75 \neq -50$ **9.** 1308 **10.** 90 **11.** 11 **12.** 0
13. $-10x + 20$ **14.** $5 - 3p$ **15.** $17c + 6$ **16.** $-10 + 10y$
17. $-17 - 14r$ **18.** $-2 + 3r$ **19.** $-19k + 54$ **20.** $t - 4$
21. $x = -6$ **22.** $t = \frac{3}{2}$ **23.** $r = 20$ **24.** $x = -\frac{61}{2}$ **25.** $y = 15$
26. $r = 0$ **27.** no solution **28.** $x = 20$ **29.** $h = 11$
30. $r = 4.75$ **31.** $w = \frac{V}{lh}$ **32.** $h = \frac{2A}{b+B}$ **33.** 70.5 feet
34. 8 feet **35.** [number line with point at -4]
36. [number line with point at 7]
37. [number line from -5 to 6]
38. [number line with point at $\frac{1}{2}$, from 0]
39. $y \geq -3$ [number line at -3]
40. $t < 2$ [number line at 2]
41. $x \geq 3$ [number line at 3]
42. $k \geq 46$ [number line at 46]
43. $x < -5$ [number line at -5]
44. $w < -37$ [number line at -37]
45. 94 or more **46.** all numbers less than or equal to $-\frac{1}{3}$
47. $y = 7$ **48.** $r = \frac{I}{pt}$ **49.** $x < 2$ **50.** $k = -9$ **51.** $x = 70$
52. $y = \frac{13}{4}$ **53.** no solution **54.** all real numbers **55.** $a = -1$
56. $a = P - b - c$ **57.** $y = -12$ **58.** $z = -\frac{3}{5}$ **59.** 26 inches
60. $20\frac{1}{2}$ inches or 20.5 inches **61.** 450 miles or more
62. 84 or more **63.** 1.5 meters (rounded) **64.** 11 feet

Chapter 10 Test (page 727)
1. irrational because the decimal form never ends or repeats in a fixed block **2.** rational because it repeats in a fixed block
3. False because $-6 > -8$ and $-6 \neq -8$ **4.** False because $\frac{3}{4}$ in

decimal form is 0.75 and $0.75 = 0.75$ **5.** True because $1.3 > 0.95$ or $1\frac{3}{10} > \frac{95}{100}$ **6.** -136 **7.** 2 **8.** $5a - 14$ **9.** $-9x^2 - 6x - 8$
10. $x = -6$ **11.** $y = \frac{13}{4}$ **12.** $x = -10.8$ **13.** no solution
14. $x = -\frac{21}{2}$ **15.** $x = 30$ **16.** all real numbers
17. $p = \$2500$ **18.** 19.9 ft (rounded) **19.** $r = \frac{d}{2}$ **20.** $h = \frac{2A}{b}$
21. $r = \frac{A - p}{pt}$ **22.** $x < 11$ [number line at 11]
23. $x \leq 4$ [number line at 4] **24.** 81 or more

CHAPTER 11

Section 11.1 (page 739)
1. Snoopy; 31% **3.** Since 26% is twice as much as 13%, we can expect twice as many adults to favor Charlie Brown.
5. Ohio (OH): about 8000 million eggs; Iowa (IA): about 10,000 million eggs **7.** North Carolina (NC); about 2500 million eggs
9. from 1975 to 1980; about $0.75 **11.** The price of a gallon of gas was decreasing. **13.** does; do not **15.** y **17.** 6 **19.** yes
21. yes **23.** no **25.** yes **27.** no **29.** No. For two ordered pairs (x, y) to be equal, the x-values must be equal and the y-values must be equal. Here we have $4 \neq -1$ and $-1 \neq 4$. **31.** 11
33. $-\frac{7}{2}$ **35.** -4 **37.** -5 **39.** 4; 6; -6; $(0, 4)$; $(6, 0)$; $(-6, 8)$
41. 3; -5; -15; $(0, 3)$; $(-5, 0)$; $(-15, -6)$ **43.** -9; -9; -9
45. -6; -6; -6 **47.** 8; 8; 8 **49.** $(2, 4)$ **51.** $(-5, 4)$
53. $(3, 0)$ **55.** negative; negative **57.** positive; negative
59. If $xy < 0$, then either $x < 0$ and $y > 0$ or $x > 0$ and $y < 0$. If $x < 0$ and $y > 0$, then the point lies in quadrant II. If $x > 0$ and $y < 0$, then the point lies in quadrant IV.

61.–70.

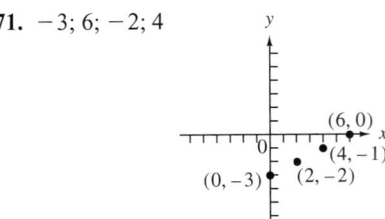

71. $-3; 6; -2; 4$

73. $-3; 4; -6; -\frac{4}{3}$

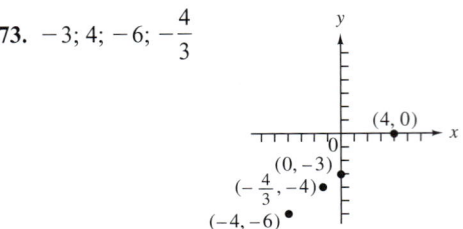

75. $-4; -4; -4; -4$

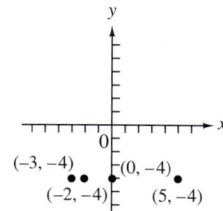

77. The points in each graph appear to lie on a straight line.
79. (a) (1996, 53.3), (1997, 52.8), (1998, 52.1), (1999, 51.6), (2000, 51.2), (2001, 50.9) **(b)** (2002, 51.0) indicates that 51.0 percent of college students in 2002 graduated within 5 years.

(c) 4-YEAR COLLEGE STUDENTS GRADUATING WITHIN 5 YEARS

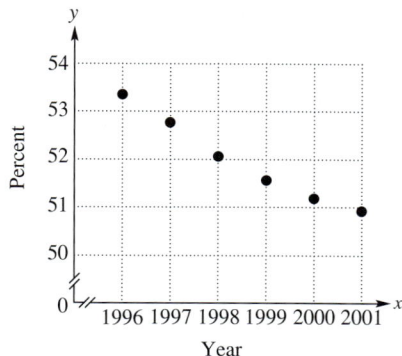

(d) The points appear to be approximated by a straight line. Graduation rates for 4-year college students within 5 years are decreasing.
81. (a) 157, 141, 125, 109 **(b)** (20, 157), (40, 141), (60, 125), (80, 109)

(c) TARGET HEART RATE ZONE (Upper Limit)

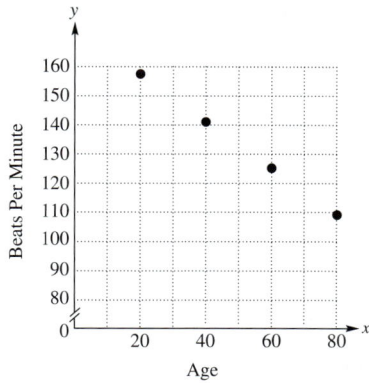

The points lie in a linear pattern.

Section 11.2 (page 753)

1. 5; 5; 3   **3.** 1; 3; -1

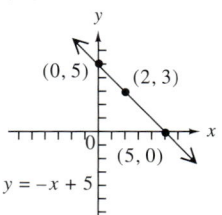 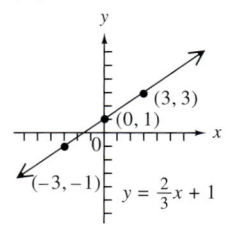

5. $-6; -2; -5$

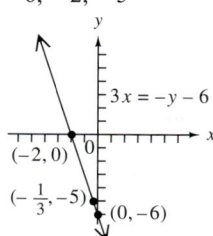

7. A   **9.** D   **11.** (12, 0); (0, -8)   **13.** (0, 0); (0, 0)
15. Choose a value *other than* 0 for either x or y. For example, if $x = -5, y = 4$.

17. 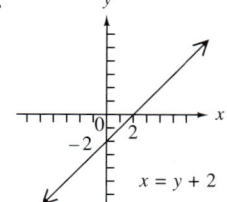   **19.**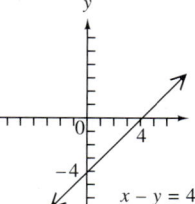

21. 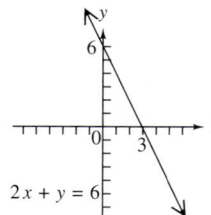   **23.**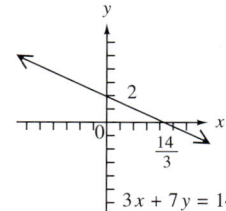

25. 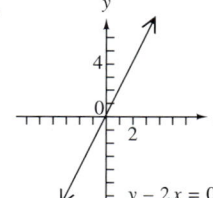   **27.**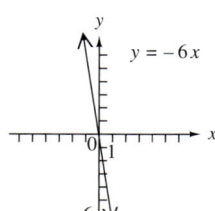

29. 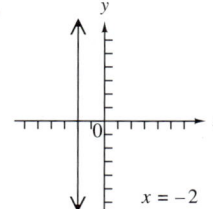   **31.**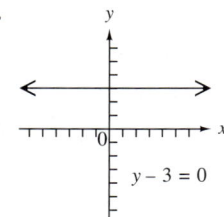

33. (a) 151.5 cm, 174.9 cm, 159.3 cm
(b) HEIGHTS OF WOMEN

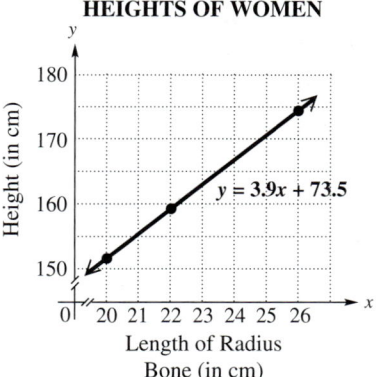

(c) 24 cm; 24 cm   **35. (a)** 93   **(b)** 93   **(c)** They are the same.

37. between 93 and 149 **39. (a)** 1995: 20.7; 1996: 21.8; 1998: 24.1 (all in gallons) **(b)** 1995: 20.3; 1996: 22.1; 1998: 23.9 (all in gallons) **(c)** They differ by 0.4, 0.3, and 0.2 gallon, respectively. **41. (a)** $30,000 **(b)** $15,000 **(c)** $5000 **(d)** After 5 years, the SUV has a value of $5000. **43. (a)** The equation is a fairly good model. **(b)** The actual debt for 1996 is about $500 billion dollars; this is about $30 billion more than the amount given by the equation. **(c)** No. Data for future years might not follow the same pattern, so the linear equation would not be a reliable model.

Section 11.3 (page 767)

1. $\frac{3}{2}$ **3.** $-\frac{7}{4}$ **5.** 0 **7.** Rise is the vertical change between two different points on a line. Run is the horizontal change between two different points on a line. **9.** Yes, the answer would be the same. It doesn't matter which point you start with. The slope would be expressed as the quotient of -6 and -4, which simplifies to $\frac{3}{2}$. **10.–13.** Answers will vary.

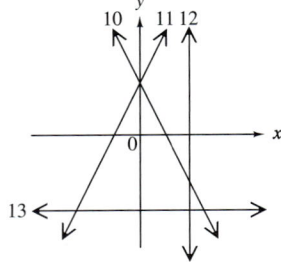

15. His answer is incorrect. Because he found the difference $3 - 5 = -2$ in the numerator, he should have subtracted in the same order in the denominator to get $-1 - 2 = -3$. The correct slope is $\frac{-2}{-3} = \frac{2}{3}$.

17. $\frac{5}{4}$ **19.** $\frac{3}{2}$ **21.** -3 **23.** 0 **25.** undefined **27.** $-\frac{1}{2}$
29. 5 **31.** $\frac{1}{4}$ **33.** $\frac{3}{2}$ **35.** 0 **37.** undefined **39.** 1
41. (a) negative **(b)** 0 **43. (a)** positive **(b)** negative
45. (a) 0 **(b)** negative **47.** $\frac{4}{3}; \frac{4}{3}$; parallel **49.** $\frac{5}{3}; \frac{3}{5}$; neither
51. $\frac{3}{5}; -\frac{5}{3}$; perpendicular **53.** $\frac{8}{27}$ **54.** 232 thousand or 232,000
55. positive; increased **56.** 232,000 students **57.** -1.66
58. negative; decreased **59.** 1.66 students per computer

Section 11.4 (page 777)

1. D **3.** B **5.** $m = 3; y = 3x - 3$ **7.** $m = -1; y = -x + 3$
9. $y = 4x - 3$ **11.** $y = 3$ **13.** A vertical line has undefined slope, so there is no value for m. Also, there is no y-intercept, so there can be no value for b.

15. 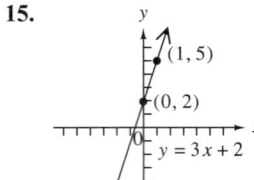   **17.**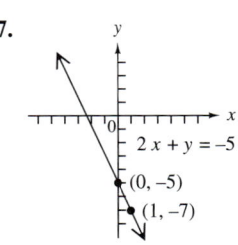

19. 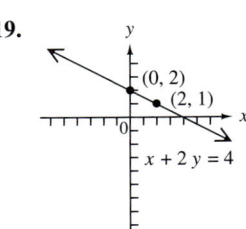   **21.** $y = \frac{1}{2}x + 4$

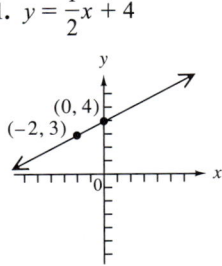

23. $y = -\frac{2}{5}x - \frac{23}{5}$ **25.** $y = 2$

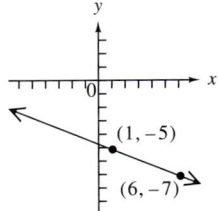   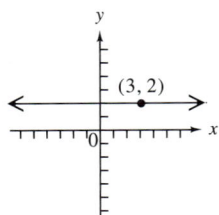

27. $x = 3$ (no slope-intercept form)   **29.** $y = \frac{2}{3}x$

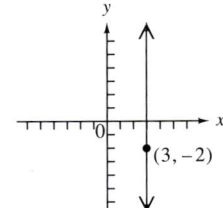   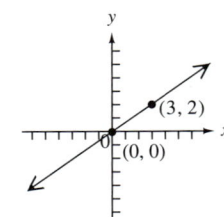

31. $y = 2x - 7$ **33.** $y = -2x - 4$ **35.** $y = \frac{2}{3}x + \frac{19}{3}$
37. $y = x - 3$ **39.** $y = -\frac{5}{7}x - \frac{54}{7}$ **41.** $y = -\frac{2}{3}x - 2$
43. $x = 3$ (no slope-intercept form) **45.** $y = \frac{1}{3}x + \frac{4}{3}$
47. $y = \frac{3}{4}x - \frac{9}{2}$ **49.** $y = -2x - 3$ **51.** $(0, 32); (100, 212)$
52. $\frac{9}{5}$ **53.** $F - 32 = \frac{9}{5}(C - 0)$ **54.** $F = \frac{9}{5}C + 32$
55. $C = \frac{5}{9}(F - 32)$ **56.** 86° **57.** 10° **58.** $-40°$
59. (a) $400 **(b)** $0.25 **(c)** $y = 0.25x + 400$ **(d)** $425
(e) 1500 snow cones **61. (a)** $(1, 1125), (3, 1239), (5, 1314),$
$(7, 1338), (9, 1379)$

(b) yes

AVERAGE ANNUAL COSTS AT 2-YEAR COLLEGES

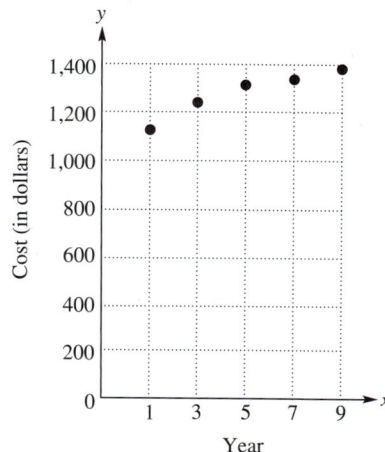

(c) $y = 23.33x + 1169$ (d) $1426

Summary Exercises on Graphing Linear Equations (page 783)

1. $-3; (0, -6)$ **2.** $-2; (0, -4)$ **3.** $-4; (0, -3)$
4. $-5; (0, -8)$ **5.** $\frac{3}{2}; (0, 6)$ **6.** $\frac{5}{3}, (0, 5)$

7. 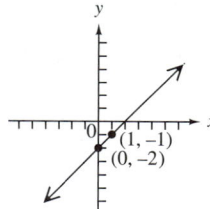 **8.**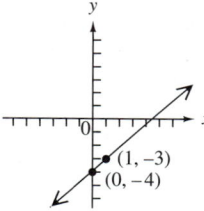

9. 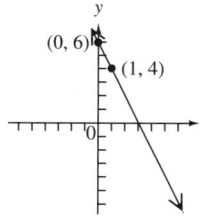 **10.**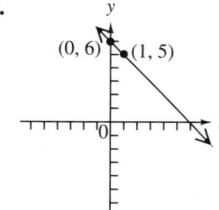

11. 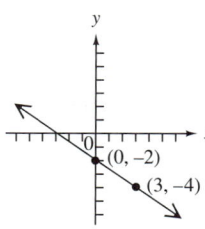 **12.**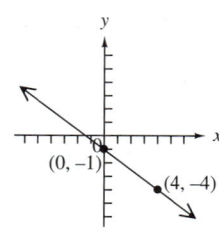

13. (a) 1 (b) $(0, -3)$ (c)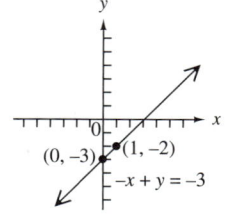

14. (a) 1 (b) $(0, -5)$ (c)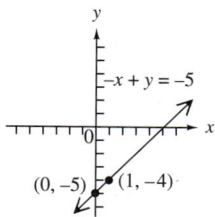

15. (a) $-\frac{1}{2}$ (b) $(0, 2)$ (c)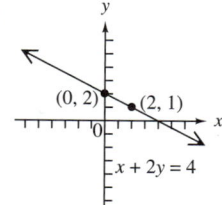

16. (a) $-\frac{1}{3}$ (b) $(0, -2)$ (c)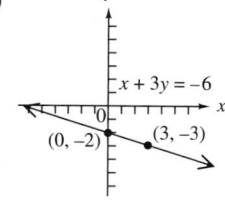

17. (a) $\frac{4}{5}$ (b) $(0, -4)$ (c)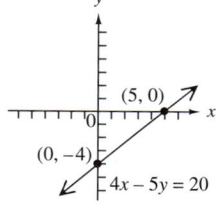

18. (a) $\frac{6}{5}$ (b) $(0, -6)$ (c)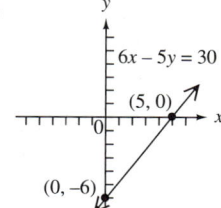

19. (a) $-\frac{2}{3}$ (b) $(0, 4)$ (c)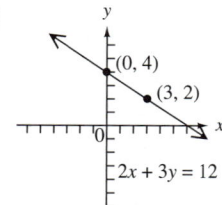

20. (a) $-\frac{5}{2}$ (b) $(0, 5)$ (c)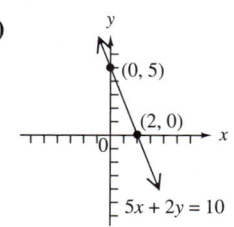

21. (a) $\dfrac{1}{3}$ (b) $(0, -2)$ (c)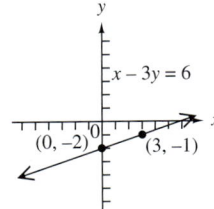

22. (a) $\dfrac{1}{2}$ (b) $(0, 2)$ (c)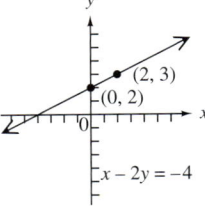

23. (a) $\dfrac{1}{4}$ (b) $(0, 0)$ (c)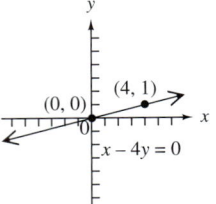

24. (a) $-\dfrac{1}{5}$ (b) $(0, 0)$ (c)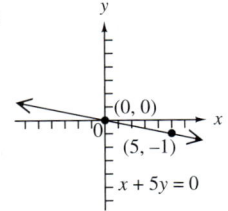

Section 11.5 (page 789)
1. false **3.** true **5.** $>$ **7.** $\leq$

9. 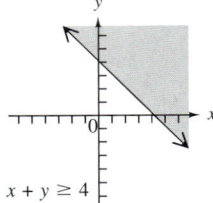 **11.**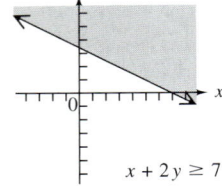

13. 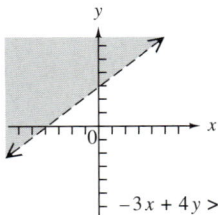 **15.**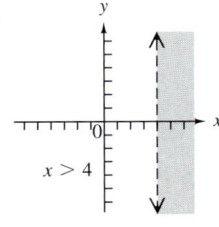

17. Use a dashed line if the symbol is $<$ or $>$. Use a solid line if the symbol is $\leq$ or $\geq$.

19. 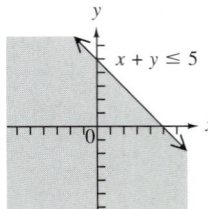 **21.**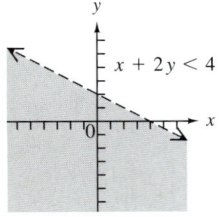

23. 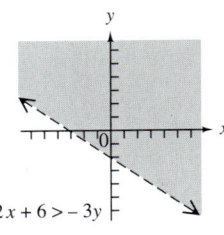 **25.**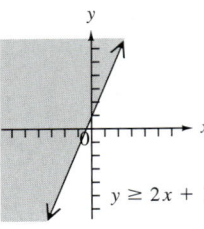

27. 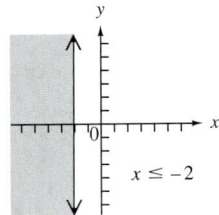 **29.**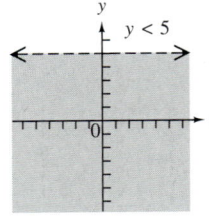

31.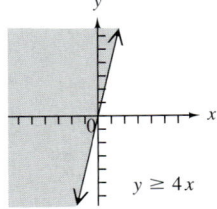

33. Every point in quadrant IV has a positive x-value and a negative y-value. Substituting into $y > x$ would imply that a negative number is greater than a positive number, which is always false. Thus, the graph of $y > x$ cannot lie in quadrant IV.

35. (a)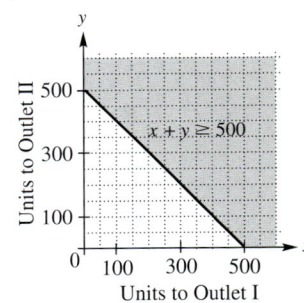

(b) $(500, 0)$ and $(200, 400)$; Other answers are possible.

Chapter 11 Review Exercises (page 797)
1. 2.0% **2.** $(1996, 57.1), (1997, 56.6), (1998, 56.2), (1999, 55.8), (2000, 55.5), (2001, 55.1)$ **3.** (a) 1997; (b) 2000 **4.** There is a general trend of decreasing percents of private school students earning a degree within five years. **5.** $-1; 2; 1$ **6.** $2; \dfrac{3}{2}; \dfrac{14}{3}$

7. $0; \dfrac{8}{3}; -9$ **8.** 7; 7; 7 **9.** yes **10.** no **11.** yes

12.–15.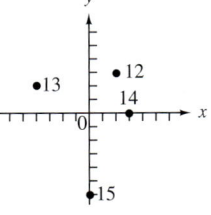

16. x is positive in quadrants I and IV; y is negative in quadrants III and IV. Thus, if x is positive and y is negative, (x, y) must lie in quadrant IV. **17.** In the ordered pair $(k, 0)$, the y-value is 0, so the point lies on the x-axis. In the ordered pair $(0, k)$, the x-value is 0, so the point lies on the y-axis. **18.** quadrant II **19.** quadrant III **20.** no quadrant

21. $\left(-\frac{5}{2}, 0\right); (0, 5)$ **22.** $\left(-\frac{7}{2}, 0\right); (0, -7)$ **23.** $\left(\frac{8}{3}, 0\right); (0, 4)$ **61.** $\left(-\frac{5}{2}, 0\right); (0, -5); -2$

24. 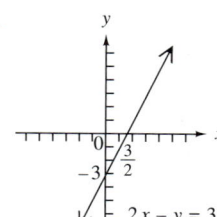 **25.**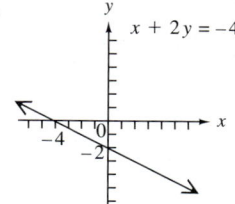

26.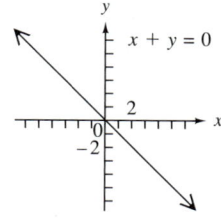

62. $(0, 0); (0, 0); -\frac{1}{3}$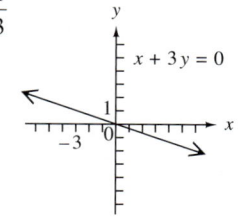

27. $-\frac{1}{2}$ **28.** $-\frac{2}{3}$ **29.** 0 **30.** undefined **31.** 3 **32.** $\frac{2}{3}$
33. $\frac{3}{2}$ **34.** $-\frac{1}{3}$ **35.** undefined **36.** 0 **37.** $\frac{3}{2}$ **38.** (a) 2
(b) $\frac{1}{3}$ **39.** parallel **40.** perpendicular **41.** neither **42.** 0
43. $y = -x + \frac{2}{3}$ **44.** $y = -\frac{1}{3}x + 1$ **45.** $y = x - 7$
46. $y = \frac{2}{3}x + \frac{14}{3}$ **47.** $y = -\frac{3}{4}x - \frac{1}{4}$ **48.** $y = -\frac{1}{4}x + \frac{3}{2}$
49. $y = 1$ **50.** $x = \frac{1}{3}$ **51.** (a) $y = -\frac{1}{3}x + 5$ (b) slope: $-\frac{1}{3}$;
y-intercept: (0, 5) (c)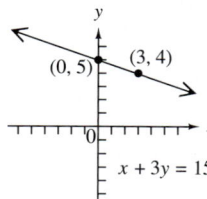

63. no x-intercept; (0, 5); 0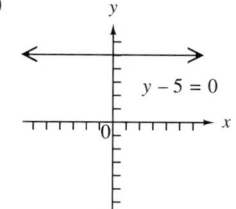

64. $y = -\frac{1}{4}x - \frac{5}{4}$ **65.** $y = -3x + 30$ **66.** $y = -\frac{4}{7}x - \frac{23}{7}$
67. 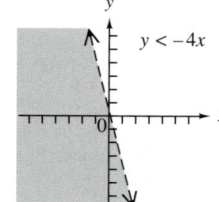 **68.**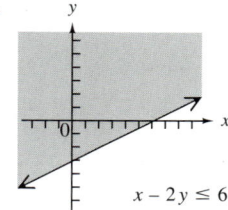

69. 3.0% **70.** Since the graph falls from left to right, the slope is negative. **71.** (1997, 44.2), (2002, 41.2)
72. $y = -0.6x + 1242.4$ **73.** -0.6; yes **74.** 43.6, 43.0, 42.4, 41.8 **75.** 40.6%; No. The equation is based on data only for 1997 through 2002.

Chapter 11 Test (page 803)
1. $1.06 **2.** $0.45 **3.** from 2000 to 2002
4. x-intercept: (2, 0); y-intercept: (0, 6)
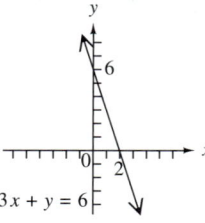

52. 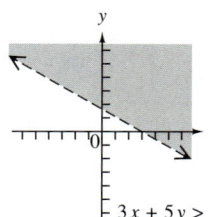 **53.**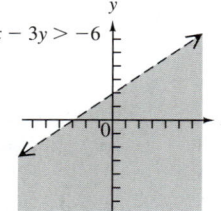

54.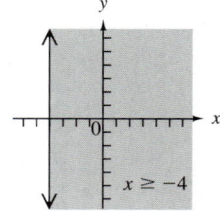

5. x-intercept: (0, 0); y-intercept: (0, 0)
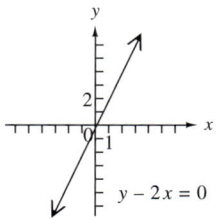

55. A **56.** C, D **57.** A, B, D **58.** D **59.** C **60.** B

6. x-intercept: $(-3, 0)$; y-intercept: none

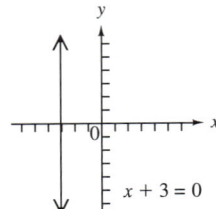

7. x-intercept: none; y-intercept: $(0, 1)$

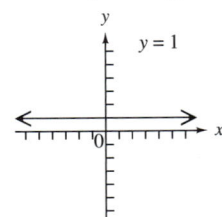

8. x-intercept: $(4, 0)$; y-intercept: $(0, -4)$

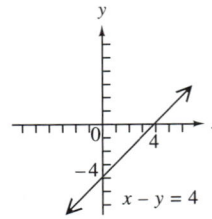

9. $-\dfrac{8}{3}$ **10.** -2 **11.** undefined **12.** $\dfrac{5}{2}$ **13.** 0 **14.** $y = 2x + 6$
15. $y = \dfrac{5}{2}x - 4$ **16.** $y = -9x + 12$ **17.** $y = -\dfrac{3}{2}x + \dfrac{9}{2}$

18. 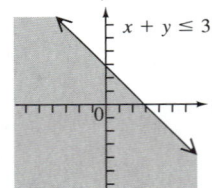   **19.**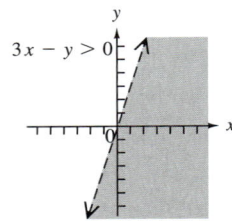

20. The slope is positive since food and drink sales are increasing.
21. (0, 43), (30, 376); 11.1 **22. (a)** 1990: $265 billion; 1995: $320.5 billion **(b)** In 2000, food and drink sales were $376 billion.

Chapter 12

Section 12.1 (page 813)

1. B, because the ordered pair must be in quadrant II. **3.** There is no way that the sum of two numbers can be both 2 and 4 at the same time. **5.** no **7.** yes **9.** yes **11.** no
We show the graphs here only for Exercises 13–17.
13. $(4, 2)$ 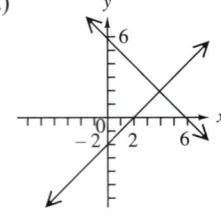   **15.** $(0, 4)$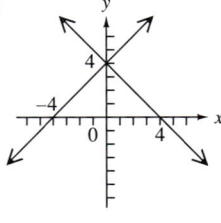

17. $(4, -1)$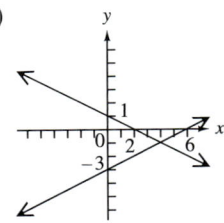

19. $(1, 3)$ **21.** $(0, 2)$ **23.** no solution **25.** infinite number of solutions **27.** $(4, -3)$ **29.** no solution
31. $y = -\dfrac{3}{2}x + 3$; $y = -\dfrac{3}{2}x + \dfrac{5}{2}$; The graphs are parallel lines.
32. $y = 2x - 4$; $y = 2x - 4$; The graphs are the same line.
33. $y = \dfrac{1}{3}x - \dfrac{5}{3}$; $y = -2x + 8$; The graphs are intersecting lines.
34. Exercise 31: no solution; Exercise 32: infinite number of solutions; Exercise 33: one solution **35.** 1989–1997 **37.** 1989: share 20%; 1998: share 16% **39.** If the coordinates of the point of intersection are not integers, the solution will be difficult to determine from a graph. **41.** Answers will vary, but the lines must intersect at $(-2, 3)$.

Section 12.2 (page 823)

1. No, it is not correct, because the solution is the ordered pair $(3, 0)$. The y-value must also be determined. **3.** $(3, 9)$ **5.** $(7, 3)$
7. $(-2, 4)$ **9.** $(-4, 8)$ **11.** $(3, -2)$ **13.** infinite number of solutions **15.** $\left(\dfrac{1}{3}, -\dfrac{1}{2}\right)$ **17.** no solution
19. infinite number of solutions **21.** $(36, -35)$ **23.** no solution
25. $(2, -3)$ **27.** $(10, -12)$ **29.** $(-4, 2)$ **30.** To find the total cost, multiply the number of bicycles (x) by the cost per bicycle (400 dollars) and add the fixed cost (5000 dollars). Thus, $y_1 = 400x + 5000$ gives this total cost (in dollars).
31. $y_2 = 600x$ **32.** $y_1 = 400x + 5000, y_2 = 600x$, solution: $(25, 15{,}000)$ **33.** 25; 15,000; 15,000

Section 12.3 (page 829)

1. true **3.** true **5.** $(-1, 3)$ **7.** $(-1, -3)$ **9.** $(-2, 3)$
11. $\left(\dfrac{1}{2}, 4\right)$ **13.** $(3, -6)$ **15.** $(7, 4)$ **17.** $(0, 4)$ **19.** $(-4, 0)$
21. $(0, 0)$ **23.** no solution **25.** infinite number of solutions
27. $(2, 9)$ **29.** $(-6, 5)$ **31.** $\left(-\dfrac{6}{5}, \dfrac{4}{5}\right)$ **33.** $\left(\dfrac{1}{8}, -\dfrac{5}{6}\right)$
35. $(11, 15)$ **37.** no solution **39.** infinite number of solutions
41. $1141 = 1991a + b$ **42.** $1465 = 1999a + b$
43. $1991a + b = 1141, 1999a + b = 1465$; solution: $(40.5, -79{,}494.5)$ **44.** $y = 40.5x - 79{,}494.5$ **45.** 1424.5 (million); This is slightly less than the actual figure. **46.** Since the data do not lie in a perfectly straight line, the quantity obtained from an equation determined in this way will probably be "off" a bit. We cannot put too much faith in models such as this one, because not all sets of data points are linear in nature.

Summary Exercises on Solving Systems of Linear Equations (page 833)

1. (a) Use substitution since the second equation is solved for y. **(b)** Use elimination since the coefficients of the y-terms are opposites. **(c)** Use elimination since the equations are in standard form with no coefficients of 1 or -1. Solving by substitution would involve fractions. **2.** The system on the right is easier to solve by substitution because the second equation is already solved for y.
3. (a) $(1, 4)$ **(b)** $(1, 4)$ **(c)** Answers will vary. **4. (a)** $(-5, 2)$
(b) $(-5, 2)$ **(c)** Answers will vary. **5.** $(2, 6)$ **6.** $(-3, 2)$

7. $\left(\dfrac{1}{3}, \dfrac{1}{2}\right)$ **8.** no solution **9.** (3, 0) **10.** $\left(\dfrac{3}{2}, -\dfrac{3}{2}\right)$
11. infinite number of solutions **12.** (9, 4) **13.** $\left(-\dfrac{5}{7}, -\dfrac{2}{7}\right)$
14. (4, −5) **15.** no solution **16.** (−4, 6) **17.** $\left(\dfrac{19}{3}, -5\right)$
18. $\left(\dfrac{22}{13}, -\dfrac{23}{13}\right)$ **19.** (−12, −60) **20.** (2, −4) **21.** (18, −12)
22. (−2, 1) **23.** $\left(13, -\dfrac{7}{5}\right)$ **24.** infinite number of solutions

Section 12.4 (page 841)
1. D **3.** B **5.** D **7.** C **9.** the second number; $x - y = 48$; The two numbers are 73 and 25. **11.** Cher: 84; Rolling Stones: 33 **13.** *The Lord of the Rings: The Return of the King:* $361.1 million; *Finding Nemo:* $339.7 million **15.** Terminal Tower: 708 ft; Key Tower: 950 ft **17. (a)** 45 units **(b)** Do not produce; the product will lead to a loss. **19.** 46 ones; 28 tens **21.** 2 DVDs of *Miracle*; 5 Linkin Park CDs **23.** $2500 at 4%; $5000 at 5% **25.** Japan: $17.19; Switzerland: $13.15 **27.** 80 L of 40% solution; 40 L of 70% solution **29.** 30 lb at $6 per lb; 60 lb at $3 per lb **31.** 30 barrels at $40 per barrel; 20 barrels at $60 per barrel **33.** bicycle: 15 mph; car: 55 mph **35.** car leaving Cincinnati: 55 mph; car leaving Toledo: 70 mph **37.** Roberto: 17.5 mph; Juana: 12.5 mph **39.** boat: 10 mph; current: 2 mph **41.** plane: 470 mph; wind: 30 mph

Section 12.5 (page 851)
1. C **3.** B

5. 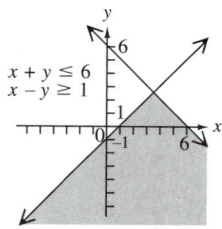 **7.**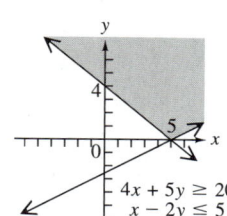

9. 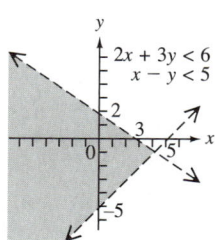 **11.**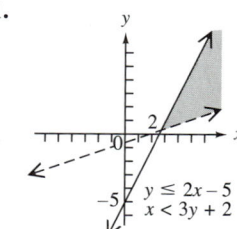

13. 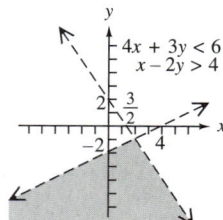 **15.**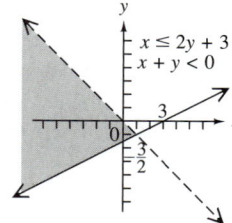

17. 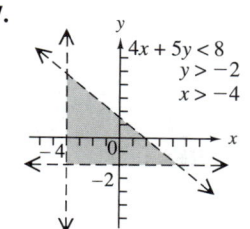 **19.**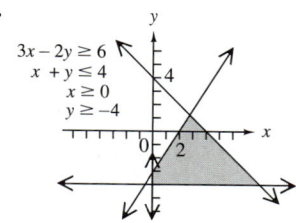

Chapter 12 Review Exercises (page 857)
1. yes **2.** no **3.** (3, 1) **4.** (0, −2) **5.** infinite number of solutions **6.** no solution **7.** It is not a solution of the system because it is not a solution of the second equation, $2x + y = 4$. **8.** (2, 1) **9.** (3, 5) **10.** (6, 4) **11.** no solution **12.** (7, 1) **13.** (−5, −2) **14.** (−4, 3) **15.** infinite number of solutions **16. (a)** 2 **(b)** 9 **17.** (9, 2) **18.** $\left(\dfrac{10}{7}, -\dfrac{9}{7}\right)$ **19.** (8, 9) **20.** (2, 1) **21.** Subway: 13,247 restaurants; McDonald's: 13,099 restaurants **22.** *Modern Maturity:* 20.5 million; *Reader's Digest:* 15.1 million **23.** length: 27 m; width: 18 m **24.** 13 twenties; 7 tens **25.** 25 lb of $1.30 candy; 75 lb of $0.90 candy **26.** plane: 250 mph; wind: 20 mph **27.** $7000 at 3%; $11,000 at 4% **28.** 60 L of 40% solution; 30 L of 70% solution

29. 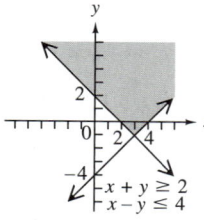 **30.**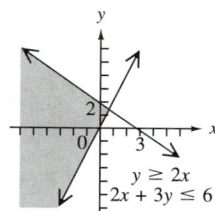

31.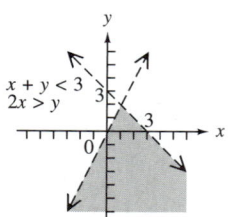

32. B **33.** B **34.** (2, 0) **35.** (−4, 15) **36.** no solution

37. 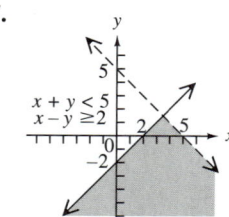 **38.**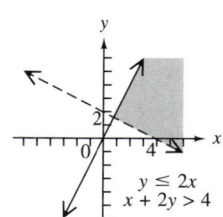

39.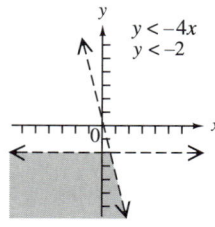

40. 8 in., 8 in., and 13 in. **41.** New England 24, Indianapolis 14 **42. (a)** years 0–6 **(b)** year 6; about $650

Chapter 12 Test (page 861)

1. $(2, -3)$ **2.** It has no solution.   **3.** $(1, -6)$ **4.** $(-35, 35)$ **5.** $(5, 6)$ **6.** $(-1, 3)$ **7.** $(0, 0)$ **8.** no solution   **9.** infinite number of solutions   **10.** $(12, -4)$ **11.** Memphis and Atlanta: 394 mi; Minneapolis and Houston: 1176 mi   **12.** Disneyland: 12.7 million; Magic Kingdom: 14.0 million   **13.** 20 L of 15% solution; 30 L of 40% solution   **14.** slower car: 45 mph; faster car: 60 mph

15. 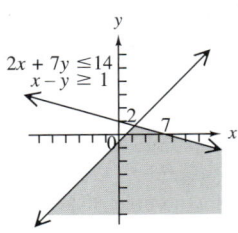   **16.**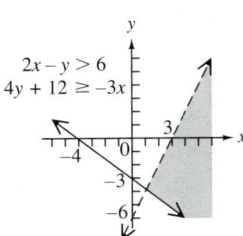

CHAPTER 13

Section 13.1 (page 869)

1. 7; 5   **3.** 8   **5.** 26   **7.** 1; 6   **9.** 1; 1   **11.** 2; -19, -1 **13.** 2; 1, 8   **15.** $2m^5$ **17.** $-r^5$ **19.** cannot be simplified; $0.2m^5 - 0.5m^2$ **21.** $-5x^5$ **23.** $5p^9 + 4p^7$ **25.** $-2y^2$ **27.** already simplified; degree 4; binomial   **29.** already simplified; $6m^5 + 5m^4 - 7m^3 - 3m^2$; degree 5; none of these   **31.** $x^4 + \frac{1}{3}x^2 - 4$; degree 4; trinomial   **33.** 7; degree 0; monomial   **35.** (a) -1 (b) 5   **37.** (a) 19 (b) -2 **39.** (a) 36 (b) -12 **41.** (a) -124 (b) 5   **43.** 175 ft   **44.** 87 ft; (1, 87)   **45.** $9.80   **46.** $27   **47.** $5m^2 + 3m$ **49.** $4x^4 - 4x^2$ **51.** $\frac{7}{6}x^2 - \frac{2}{15}x + \frac{5}{6}$ **53.** $12m^3 - 13m^2 + 6m + 11$ **55.** $8r^2 + 5r - 12$ **57.** $5m^2 - 14m + 6$ **59.** $4x^3 + 2x^2 + 5x$ **61.** $-18y^5 + 7y^4 + 5y^3 + 3y^2 + y$ **63.** $-2m^3 + 7m^2 + 8m - 9$ **65.** $8x^2 + 8x + 6$ **67.** $8t^2 + 8t + 13$ **69.** $-11x^2 - 3x - 3$ **71.** The degree of a term is determined by the exponents on the variables, but 3 is not a variable. The degree of $3^4 = 3^4 x^0$ is 0.   **73.** $13a^2b - 7a^2 - b$ **75.** $c^4d - 5c^2d^2 + d^2$ **77.** $12m^3n - 11m^2n^2 - 4mn^2$

Section 13.2 (page 879)

1. 1   **3.** false   **5.** false   **7.** $(-2)^5$ **9.** $\left(\frac{1}{2}\right)^6$ **11.** $(-8p)^2$ **13.** The expression $(-3)^4$ means $(-3)(-3)(-3)(-3) = 81$, while -3^4 means $-(3 \cdot 3 \cdot 3 \cdot 3) = -81$.   **15.** base: 3; exponent: 5; 243   **17.** base: -3; exponent: 5; -243 **19.** base: $-6x$; exponent: 4   **21.** base: x; exponent: 4   **23.** The product rule does not apply to $5^2 + 5^3$ because it is a *sum*, not a product. $5^2 + 5^3 = 25 + 125 = 150$ **25.** 5^8 **27.** 4^{12} **29.** $(-7)^9$ **31.** t^{24} **33.** $-56r^7$ **35.** $42p^{10}$ **37.** 4^6 **39.** t^{20} **41.** $7^3 r^3$ **43.** $5^5 x^5 y^5$ **45.** $8q^3 r^3$ **47.** $\frac{1}{2^3}$ **49.** $\frac{a^3}{b^3}$ **51.** $\frac{9^8}{5^8}$ **53.** $(-2)^3 x^6 y^3$ **55.** $3^2 a^6 b^4$ **57.** $\frac{5^5}{2^5}$ **59.** $\frac{9^5}{8^3}$ **61.** $2^{12} x^{12}$ **63.** $(-6)^5 p^5$ **65.** $6^5 x^{10} y^{15}$ **67.** x^{21} **69.** $2^2 w^4 x^{26} y^7$ or $4w^4 x^{26} y^7$ **71.** $-r^{18} s^{17}$ **73.** $\frac{5^3 a^6 b^{15}}{c^{18}}$ or $\frac{125 a^6 b^{15}}{c^{18}}$ **75.** $25 m^6 p^{14} q^5$ **77.** $16 x^{10} y^{16} z^{10}$ **79.** $30x^7$

Section 13.3 (page 887)

1. $x^2 + 7x + 12$ **3.** $2x^3 + 7x^2 + 7x + 2$ **5.** distributive   **7.** $-6m^2 - 4m$ **9.** $6p - \frac{9}{2}p^2 + 9p^4$ **11.** $6y^5 + 4y^6 + 10y^9$ **13.** $12x^3 + 26x^2 + 10x + 1$ **15.** $20m^4 - m^3 - 8m^2 - 17m - 15$ **17.** $6x^6 - 3x^5 - 4x^4 + 4x^3 - 5x^2 + 8x - 3$ **19.** $5x^4 - 13x^3 + 20x^2 + 7x + 5$ **21.** $n^2 + n - 6$ **23.** $8r^2 - 10r - 3$ **25.** $9x^2 - 4$ **27.** $9q^2 + 6q + 1$ **29.** $6t^2 + 23st + 20s^2$ **31.** $-0.3t^2 + 0.22t + 0.24$ **33.** $x^2 - \frac{5}{12}x - \frac{1}{6}$ **35.** $\frac{15}{16} - \frac{1}{4}r - 2r^2$ **37.** $6y^5 - 21y^4 - 45y^3$ **39.** $30x + 60$ yd^2 **40.** $30x + 60 = 600$; 18   **41.** 10 yd by 60 yd   **42.** $2100   **43.** 140 yd   **44.** $1260   **45.** The answers are $x^2 - 16$, $y^2 - 4$, and $r^2 - 49$. Each product is the difference of the square of the first term and the square of the last term of the binomials.

Section 13.4 (page 893)

1. (a) $4x^2$ (b) $12x$ (c) 9 (d) $4x^2 + 12x + 9$ **3.** $a^2 - 2ac + c^2$ **5.** $p^2 + 4p + 4$ **7.** $16x^2 - 24x + 9$ **9.** $0.64t^2 + 1.12ts + 0.49s^2$ **11.** $25x^2 + 4xy + \frac{4}{25}y^2$ **13.** $16a^2 - 12ab + \frac{9}{4}b^2$ **15.** $-16r^2 + 16r - 4$ **17.** (a) $49x^2$ (b) 0 (c) $-9y^2$ (d) $49x^2 - 9y^2$; Because 0 is the identity element for addition, it is not necessary to write "+ 0."   **19.** $q^2 - 4$ **21.** $4w^2 - 25$ **23.** $100x^2 - 9y^2$ **25.** $4x^4 - 25$ **27.** $49x^2 - \frac{9}{49}$ **29.** $9p^3 - 49p$ **31.** $(a+b)^2$ **32.** a^2 **33.** $2ab$ **34.** b^2 **35.** $a^2 + 2ab + b^2$ **36.** They both represent the area of the entire large square.   **37.** 1225   **38.** $30^2 + 2(30)(5) + 5^2$ **39.** 1225   **40.** They are equal.   **41.** $m^3 - 15m^2 + 75m - 125$ **43.** $8a^3 + 12a^2 + 6a + 1$ **45.** $81r^4 - 216r^3 t + 216r^2 t^2 - 96rt^3 + 16t^4$ **47.** $3x^5 - 27x^4 + 81x^3 - 81x^2$ **49.** $-8x^6 y - 32x^5 y^2 - 48x^4 y^3 - 32x^3 y^4 - 8x^2 y^5$ **51.** 512 cubic units

Section 13.5 (page 903)

1. negative   **3.** negative   **5.** positive   **7.** 0   **9.** 1   **11.** -1 **13.** 0   **15.** 0   **17.** 2   **19.** $\frac{1}{64}$ **21.** 16   **23.** $\frac{49}{36}$ **25.** $\frac{8}{15}$ **27.** $-\frac{7}{18}$ **29.** 1   **30.** $\frac{5^2}{5^2}$ **31.** 5^0 **32.** $5^0 = 1$; This supports the definition of a 0 exponent.   **33.** $\frac{1}{9}$ **35.** $\frac{1}{6^5}$ **37.** 6^3 **39.** $2r^4$ **41.** $\frac{5^2}{4^3}$ **43.** $\frac{p^5}{q^8}$ **45.** r^9 **47.** $\frac{x^5}{6}$ **49.** $3y^2$ **51.** x^3 **53.** 7^3 **55.** $\frac{1}{x^2}$ **57.** $\frac{4^3 x}{3^2}$ or $\frac{64x}{9}$ **59.** $\frac{x^2 z^4}{y^2}$ **61.** $6x$ **63.** $\frac{1}{m^{10} n^5}$ **65.** $\frac{5}{16x^5}$ **67.** $\frac{36 q^2}{m^4 p^2}$

Summary Exercises on the Rules for Exponents (page 905)

1. $\frac{6^{12} x^{24}}{5^{12}}$ **2.** $\frac{r^6 s^{12}}{729 t^6}$ **3.** $10^5 x^7 y^{14}$ **4.** $-128 a^{10} b^{15} c^4$ **5.** $\frac{729 w^3 x^9}{y^{12}}$ **6.** $\frac{x^4 y^6}{16}$ **7.** c^{22} **8.** $\frac{1}{k^4 t^{12}}$ **9.** $\frac{11}{30}$ **10.** $y^{12} z^3$ **11.** $\frac{x^6}{y^5}$ **12.** 0   **13.** $\frac{1}{z^2}$ **14.** $\frac{9}{r^2 s^2 t^{10}}$ **15.** $\frac{300 x^3}{y^3}$ **16.** $\frac{3}{5x^6}$ **17.** x^8 **18.** $\frac{y^{11}}{x^{11}}$ **19.** $\frac{a^6}{b^4}$ **20.** $6ab$ **21.** $\frac{61}{900}$ **22.** 1   **23.** $\frac{343 a^6 b^9}{8}$ **24.** 1   **25.** -1 **26.** 0   **27.** $\frac{27 y^{18}}{4x^8}$ **28.** $\frac{1}{a^8 b^{12} c^{16}}$ **29.** $\frac{x^{15}}{216 z^9}$ **30.** $\frac{q}{8 p^6 r^3}$

31. x^6y^6 **32.** 0 **33.** $\dfrac{343}{x^{15}}$ **34.** $\dfrac{9}{x^6}$ **35.** $5p^{10}q^9$ **36.** $\dfrac{7}{24}$
37. $\dfrac{r^{14}t}{2s^2}$ **38.** 1 **39.** $8p^{10}q$ **40.** $\dfrac{1}{mn^3p^3}$ **41.** -1 **42.** $\dfrac{3}{40}$

Section 13.6 (page 909)
1. $6x^2 + 8$; 2; $3x^2 + 4$ **3.** $3x^2 + 4$; 2 (These may be reversed.); $6x^2 + 8$ **5.** To use the method of this section, the divisor must be just one term. This is true of the first problem, but not the second.
7. $30x^3 - 10x + 5$ **9.** $-4m^3 + 2m^2 - 1$ **11.** $4t^4 - 2t^2 + 2t$
13. $a^4 - a + \dfrac{2}{a}$ **15.** $-2x^3 + \dfrac{2x^2}{3} - x$ **17.** $1 + 5x - 9x^2$
19. $\dfrac{4x^2}{3} + x + \dfrac{2}{3x}$ **21.** $9r^3 - 12r^2 - 2r + 1 - \dfrac{2}{3r}$
23. $-m^2 + 3m - \dfrac{4}{m}$ **25.** $-4b^2 + 3ab - \dfrac{5}{a}$
27. $\dfrac{12}{x} - \dfrac{6}{x^2} + \dfrac{14}{x^3} - \dfrac{10}{x^4}$ **29.** No, $\dfrac{2}{3}x$ means $\dfrac{2x}{3}$, which is not the same as $\dfrac{2}{3x}$. In the first case we multiply by x; in the second case we divide by x. Yes, $\dfrac{4}{3}x^2 = \dfrac{4x^2}{3}$. In both cases we are multiplying by x^2.
31. $15x^5 - 35x^4 + 35x^3$ **33.** 1423
34. $(1 \times 10^3) + (4 \times 10^2) + (2 \times 10^1) + (3 \times 10^0)$
35. $x^3 + 4x^2 + 2x + 3$ **36.** They are similar in that the coefficients of the powers of ten are equal to the coefficients of the powers of x. They are different in that one is a number while the other is a polynomial. They are equal if $x = 10$.

Section 13.7 (page 915)
1. The divisor is $2x + 5$; the quotient is $2x^3 - 4x^2 + 3x + 2$.
3. Divide $12m^2$ by $2m$ to get $6m$. **5.** $x + 2$ **7.** $2y - 5$
9. $p - 4 + \dfrac{44}{p + 6}$ **11.** $r - 5$ **13.** $2a - 14 + \dfrac{74}{2a + 3}$
15. $4x^2 - 7x + 3$ **17.** $3y^2 - 2y + 2$ **19.** $3k - 4 + \dfrac{2}{k^2 - 2}$
21. $x^2 + 1$ **23.** $2p^2 - 5p + 4 + \dfrac{6}{3p^2 + 1}$ **25.** $x^3 + 6x - 7$
27. $x^2 + 1$ **29.** $2x^2 + \dfrac{3}{5}x + \dfrac{1}{5}$ **31.** $x^2 + x - 3$ units **33.** 33
34. 33 **35.** They are the same. **36.** The answers should agree.

Section 13.8 (page 921)
1. 2.1×10^{-3}; 3.1783×10^2 **3.** 8.312×10^8; 3.19×10^8
5. in scientific notation **7.** not in scientific notation; 5.6×10^6
9. not in scientific notation; 4×10^{-3} **11.** not in scientific notation; 8×10^1 **13.** To write a number in scientific notation $(a \times 10^n)$ move the decimal point to the right of the first nonzero digit. The exponent, n, is the number of places moved. If the original number is *greater* than a, then n is positive. If the original number is *less* than a, then n is negative. **15.** 5.876×10^9
17. 8.235×10^4 **19.** 7×10^{-6} **21.** 2.03×10^{-3}
23. 750,000 **25.** 5,677,000,000,000 **27.** 6.21 **29.** 0.00078
31. 0.000000005134 **33.** 6×10^{11}; 600,000,000,000
35. 1.5×10^7; 15,000,000 **37.** 6.426×10^4; 64,260
39. 3×10^{-4}; 0.0003 **41.** 4×10^1; 40 **43.** 2.6×10^{-3}; 0.0026
45. 3.52×10^2 (rounded) or 352 people per square mile
47. (a) 1×10^{-4} or 0.0001 meter **(b)** 1×10^{-1} or 0.1 meter
49. 6.94×10^1 (rounded) or about 69 items

Chapter 13 Review Exercises (page 927)
1. $22m^2$; degree 2; monomial **2.** $p^3 - p^2 + 4p + 2$; degree 3; none of these **3.** already in descending powers; degree 5; none of these **4.** $-8y^5 - 7y^4 + 9y$; degree 5; trinomial
5. $-5a^3 + 4a^2$ **6.** $2r^3 - 3r^2 + 9r$ **7.** $11y^2 - 10y + 9$
8. $-13k^4 - 15k^2 - 4k - 6$ **9.** $10m^3 - 6m^2 - 3$
10. $-y^2 - 4y + 26$ **11.** $10p^2 - 3p - 11$ **12.** $7r^4 - 4r^3 - 1$
13. 4^{11} **14.** $(-5)^{11}$ **15.** $-72x^7$ **16.** $10x^{14}$ **17.** 19^5x^5
18. $(-4)^7y^7$ **19.** $5p^4t^4$ **20.** $\dfrac{7^6}{5^6}$ **21.** $3^3x^6y^9$ **22.** t^{42}
23. $6^2x^{16}y^4z^{16}$ **24.** The product rule for exponents does not apply here because we want the sum of 7^2 and 7^4, not their product.
25. $10x^2 + 70x$ **26.** $-6p^5 + 15p^4$ **27.** $6r^3 + 8r^2 - 17r + 6$
28. $8y^3 + 27$ **29.** $5p^5 - 2p^4 - 3p^3 + 25p^2 + 15p$
30. $6k^2 - 9k - 6$ **31.** $12p^2 - 48pq + 21q^2$
32. $2m^4 + 5m^3 - 16m^2 - 28m + 9$ **33.** $a^2 + 8a + 16$
34. $9p^2 - 12p + 4$ **35.** $4r^2 + 20rs + 25s^2$
36. $r^3 + 6r^2 + 12r + 8$ **37.** $8x^3 - 12x^2 + 6x - 1$
38. $36m^2 - 25$ **39.** $4z^2 - 49$ **40.** $25a^2 - 36b^2$ **41.** $4x^4 - 25$
42. $(a + b)^2 = (a + b)(a + b) = a^2 + 2ab + b^2$. The term $2ab$ is not in $a^2 + b^2$. **43.** 2 **44.** $\dfrac{1}{32}$ **45.** $\dfrac{5^2}{6^2}$ or $\dfrac{25}{36}$ **46.** $-\dfrac{3}{16}$
47. 6^2 **48.** x^2 **49.** $\dfrac{1}{p^{12}}$ **50.** r^4 **51.** 2^8 **52.** $\dfrac{1}{9^6}$ **53.** 5^8
54. $\dfrac{1}{8^{12}}$ **55.** $\dfrac{1}{m^2}$ **56.** y^7 **57.** r^{13} **58.** $(-5)^2m^6$ **59.** $\dfrac{y^{12}}{2^3}$
60. $\dfrac{1}{a^3b^5}$ **61.** $2 \cdot 6^2 \cdot r^5$ **62.** $\dfrac{2^3n^{10}}{3m^{13}}$ **63.** $\dfrac{5y^2}{3}$ **64.** $-2x^2y$
65. $-y^3 + 2y - 3$ **66.** $p - 3 + \dfrac{5}{2p}$
67. $-x^9 + 2x^8 - 4x^3 + 7x$ **68.** $-2m^2n + mn^2 + \dfrac{6n^3}{5}$
69. $2r + 7$ **70.** $4m + 3 + \dfrac{5}{3m - 5}$ **71.** $2a + 1 + \dfrac{-8a + 12}{5a^2 - 3}$
72. $k^2 + 2k + 4 + \dfrac{-2k - 12}{2k^2 + 1}$ **73.** 4.8×10^7
74. 2.8988×10^{10} **75.** 6.5×10^{-5} **76.** 8.24×10^{-8}
77. 24,000 **78.** 78,300,000 **79.** 0.000000897
80. 0.00000000995 **81.** 8×10^2; 800 **82.** 4×10^6; 4,000,000 **83.** 3×10^{-5}; 0.00003
84. 1×10^{-2}; 0.01 **85.** 8.1×10^1 (rounded) or about 81 people per square mile **86.** 1.08×10^7 or 10,800,000 kilometers
87. 0 **88.** $\dfrac{3^5}{p^3}$ **89.** $\dfrac{1}{7^2}$ **90.** $49 - 28k + 4k^2$ **91.** $y^2 + 5y + 1$
92. $\dfrac{6^4r^8s^4}{5^4}$ **93.** $-8m^7 - 10m^6 - 6m^5$ **94.** 2^5
95. $5xy^3 - \dfrac{8y^2}{5} + 3x^2y$ **96.** $\dfrac{r^2}{6}$ **97.** $8x^3 + 12x^2y + 6xy^2 + y^3$
98. $\dfrac{3}{4}$ **99.** $a^3 - 2a^2 - 7a + 2$ **100.** $8y^3 - 9y^2 + 5$
101. $10r^2 + 21r - 10$ **102.** $144a^2 - 1$ **103.** $2x^2 + x - 6$
104. $25x^8 + 20x^6 + 4x^4$

Chapter 13 Test (page 931)
1. $4t^4 + t^3 - 6t^2 - t$ **2.** $-2y^2 - 9y + 17$ **3.** $-12t^2 + 5t + 8$
4. $(-2)^5$ or -2^5 **5.** $\dfrac{6^3}{m^6}$ **6.** $-27x^5 + 18x^4 - 6x^3 + 3x^2$
7. $2r^3 + r^2 - 16r + 15$ **8.** $t^2 - 5t - 24$ **9.** $8x^2 + 2xy - 3y^2$
10. $25x^2 - 20xy + 4y^2$ **11.** $100v^2 - 9w^2$

12. $x^3 + 3x^2 + 3x + 1$ **13.** $\dfrac{1}{625}$ **14.** 2   **15.** $\dfrac{7}{12}$ **16.** 8^5
17. x^2y^6 **18.** $4y^2 - 3y + 2 + \dfrac{5}{y}$ **19.** $-3xy^2 + 2x^3y^2 + 4y^2$
20. $2x + 9$ **21.** $3x^2 + 6x + 11 + \dfrac{26}{x-2}$ **22. (a)** 3.44×10^{11}
(b) 5.57×10^{-6} **23. (a)** 29,600,000   **(b)** 0.0000000607
24. 1.9734×10^{30} kg   **25.** $9x^2 + 54x + 81$ square units

CHAPTER 14

Section 14.1 (page 941)
1. 4   **3.** 4   **5.** 6   **7.** 1   **9.** 8   **11.** $10x^3$ **13.** xy^2 **15.** 6
17. $3m^2$ **19.** $2z^4$ **21.** $2mn^4$ **23.** $y + 2$ **25.** $a - 2$
27. $2 + 3xy$ **29.** $x(x - 4)$ **31.** $3t(2t + 5)$ **33.** $\dfrac{1}{4}d(d - 3)$
35. $6x^2(2x + 1)$ **37.** $5y^6(13y^4 + 7)$
39. no common factor (except 1)   **41.** $8m^2n^2(n + 3)$
43. $2x(2x^2 - 5x + 3)$ **45.** $13y^2(y^6 + 2y^2 - 3)$
47. $9qp^3(5q^3p^2 + 4p^3 + 9q)$ **49.** $(x + 2)(c + d)$
51. $(2a + b)(a^2 - b)$ **53.** $(p + 4)(q - 1)$
55. $(5 + n)(m + 4)$ **57.** $(2y - 7)(3x + 4)$
59. $(y + 3)(3x + 1)$ **61.** $(z + 2)(7z - a)$
63. $(3r + 2y)(6r - x)$ **65.** $(w + 1)(w^2 + 9)$
67. $(a + 2)(3a^2 - 2)$ **69.** $(4m - p^2)(4m^2 - p)$
71. $(y + 3)(y + x)$ **73.** $(z - 2)(2z - 3w)$
75. commutative property   **76.** $2x(y - 4) - 3(y - 4)$
77. No, because it is not a product. It is the difference between $2x(y - 4)$ and $3(y - 4)$.   **78.** $(2x - 3)(y - 4)$; yes

Section 14.2 (page 947)
1. a and b must have different signs.   **3.** A prime polynomial is one that cannot be factored using only integers in the factors.
5. 1 and 12, -1 and -12, 2 and 6, -2 and -6, 3 and 4, -3 and -4; the pair with a sum of 7 is 3 and 4.   **7.** 1 and -24, -1 and 24, 2 and -12, -2 and 12, 3 and -8, -3 and 8, 4 and -6, -4 and 6; the pair with a sum of -5 is 3 and -8.   **9.** C   **11.** $x + 11$
13. $x - 8$ **15.** $y - 5$ **17.** $x + 11$ **19.** $y - 9$
21. $(y + 8)(y + 1)$ **23.** $(b + 3)(b + 5)$
25. $(m + 5)(m - 4)$ **27.** $(x + 8)(x - 5)$
29. $(y - 5)(y - 3)$ **31.** $(z - 8)(z - 7)$ **33.** $(r - 6)(r + 5)$
35. $(a - 12)(a + 4)$ **37.** prime   **39.** $(r + 2a)(r + a)$
41. $(x + y)(x + 3y)$ **43.** $(t + 2z)(t - 3z)$
45. $(v - 5w)(v - 6w)$ **47.** $4(x + 5)(x - 2)$
49. $2t(t + 1)(t + 3)$ **51.** $2x^4(x - 3)(x + 7)$
53. $a^3(a + 4b)(a - b)$ **55.** $mn(m - 6n)(m - 4n)$
57. The factored form $(2x + 4)(x - 3)$ is incorrect because $2x + 4$ has a common factor of 2, which must be factored out for the trinomial to be *completely* factored.

Section 14.3 (page 951)
1. $(m + 6)(m + 2)$ **3.** $(a + 5)(a - 2)$ **5.** $(2t + 1)(5t + 2)$
7. $(3z - 2)(5z - 3)$ **9.** $(2s - t)(4s + 3t)$
11. $(3a + 2b)(5a + 4b)$ **13.** B   **15. (a)** 2; 12; 24; 11
(b) 3; 8 (Order is irrelevant.)   **(c)** $3m$; $8m$
(d) $2m^2 + 3m + 8m + 12$ **(e)** $(2m + 3)(m + 4)$
(f) $(2m + 3)(m + 4) = 2m^2 + 11m + 12$
17. $(2x + 1)(x + 3)$ **19.** $(4r - 3)(r + 1)$
21. $(4m + 1)(2m - 3)$ **23.** $(3m + 1)(7m + 2)$
25. $(2b + 1)(3b + 2)$ **27.** $(4y - 3)(3y - 1)$
29. $3(4x - 1)(2x - 3)$ **31.** $2m(m - 4)(m + 5)$
33. $4z^3(8z + 3)(z - 1)$ **35.** $(3p + 4q)(4p - 3q)$
37. $(3a - 5b)(2a + b)$ **39.** $(5 - x)(1 - x)$ **41.** The student stopped too soon. He needs to factor out the common factor $4x - 1$ to get $(4x - 1)(4x - 5)$ as the correct answer.

Section 14.4 (page 957)
1. B   **3.** A   **5.** A   **7.** $2a + 5b$ **9.** $x^2 + 3x - 4$; $x + 4$, $x - 1$, or $x - 1$, $x + 4$ **11.** $2z^2 - 5z - 3$; $2z + 1$, $z - 3$, or $z - 3$, $2z + 1$ **13.** The binomial $2x - 6$ cannot be a factor because it has a common factor of 2, but the polynomial does not.
15. $(3a + 7)(a + 1)$ **17.** $(2y + 3)(y + 2)$
19. $(3m - 1)(5m + 2)$ **21.** $(3s - 1)(4s + 5)$
23. $(5m - 4)(2m - 3)$ **25.** $(4w - 1)(2w - 3)$
27. $(4y + 1)(5y - 11)$ **29.** prime   **31.** $2(5x + 3)(2x + 1)$
33. $q(5m + 2)(8m - 3)$ **35.** $3n^2(5n - 3)(n - 2)$
37. $y^2(5x - 4)(3x + 1)$ **39.** $(5a + 3b)(a - 2b)$
41. $(4s + 5t)(3s - t)$ **43.** $m^4n(3m + 2n)(2m + n)$
45. $-1(x + 7)(x - 3)$ **47.** $-1(3x + 4)(x - 1)$
49. $-1(a + 2b)(2a + b)$ **51.** $5 \cdot 7$ **52.** $(-5)(-7)$
53. The product of $3x - 4$ and $2x - 1$ is $6x^2 - 11x + 4$.
54. The product of $4 - 3x$ and $1 - 2x$ is $6x^2 - 11x + 4$.
55. The factors in Exercise 53 are the opposites of the factors in Exercise 54.   **56.** $(3 - 7t)(5 - 2t)$

Section 14.5 (page 963)
1. 1; 4; 9; 16; 25; 36; 49; 64; 81; 100; 121; 144; 169; 196; 225; 256; 289; 324; 361; 400   **3.** 2   **5.** $(y + 5)(y - 5)$
7. $\left(p + \dfrac{1}{3}\right)\left(p - \dfrac{1}{3}\right)$ **9.** prime   **11.** $(3r + 2)(3r - 2)$
13. $\left(6m + \dfrac{4}{5}\right)\left(6m - \dfrac{4}{5}\right)$ **15.** $4(3x + 2)(3x - 2)$
17. $(14p + 15)(14p - 15)$ **19.** $(4r + 5a)(4r - 5a)$
21. prime   **23.** $(p^2 + 7)(p^2 - 7)$ **25.** $(x^2 + 1)(x + 1)(x - 1)$
27. $(p^2 + 16)(p + 4)(p - 4)$ **29.** The teacher was justified, because it was not factored *completely*; $x^2 - 9$ can be factored as $(x + 3)(x - 3)$. The complete factored form is $(x^2 + 9)(x + 3)(x - 3)$.   **31.** No, it is not a perfect square since the middle term would have to be $30y$.   **33.** $(w + 1)^2$
35. $(x - 4)^2$ **37.** $\left(t + \dfrac{1}{2}\right)^2$ **39.** $(x - 0.5)^2$
41. $2(x + 6)^2$ **43.** $(4x - 5)^2$ **45.** $(7x - 2y)^2$
47. $(8x + 3y)^2$ **49.** $2h(5h - 2y)^2$ **51.** $(2x + 3)(5x - 2)$
52. $5x - 2$ **53.** Yes. We saw in Exercise 51 that $(2x + 3)(5x - 2) = 10x^2 + 11x - 6$.   **54.** The quotient is $x^2 + x + 1$, so $x^3 - 1 = (x - 1)(x^2 + x + 1)$.

Summary Exercises on Factoring (page 965)
1. $8m^3(4m^6 + 2m^2 + 3)$ **2.** $2(m + 3)(m - 8)$
3. $7k(2k + 5)(k - 2)$ **4.** prime   **5.** $(6z + 1)(z + 5)$
6. $(m + n)(m - 4n)$ **7.** $(7z + 4y)(7z - 4y)$
8. $10nr(10nr + 3r^2 - 5n)$ **9.** $4x(4x + 5)$
10. $(4 + m)(5 + 3n)$ **11.** $(5y - 6z)(2y + z)$
12. $(y^2 + 9)(y + 3)(y - 3)$ **13.** $(m - 3)(m + 5)$
14. $(2y + 1)(3y - 4)$ **15.** $8z(4z - 1)(z + 2)$
16. $5y(3y + 1)$ **17.** $(z - 6)^2$ **18.** $(3m + 8)(3m - 8)$
19. $(y - 6k)(y + 2k)$ **20.** $(4z - 1)^2$ **21.** $6(y - 2)(y + 1)$
22. $\left(x + \dfrac{1}{4}\right)^2$ **23.** $(p - 6)(p - 11)$ **24.** $(a + 8)(a + 9)$
25. prime   **26.** $3(6m - 1)^2$ **27.** $(z + 2a)(z - 5a)$
28. $(2a + 1)(a^2 - 7)$ **29.** $(2k - 3)^2$ **30.** $(a - 7b)(a + 4b)$
31. $(4r + 3m)^2$ **32.** $(3k - 2)(k + 2)$ **33.** prime
34. $(a^2 + 25)(a + 5)(a - 5)$ **35.** $4(2k - 3)^2$
36. $(4k + 1)(2k - 3)$ **37.** $6y^4(3y + 4)(2y - 5)$
38. $5z(z - 2)(z - 7)$ **39.** $(8p - 1)(p + 3)$
40. $(4k - 3h)(2k + h)$ **41.** $6(3m + 2z)(3m - 2z)$
42. $(2k - 5z)^2$ **43.** $2(3a - 1)(a + 2)$
44. $(3h - 2g)(5h + 7g)$ **45.** $(m + 9)(m - 9)$

46. $(5z - 6)(2z + 1)$ **47.** $5m^2(5m - 13n)(5m - 3n)$
48. $(3y - 1)(3y + 5)$ **49.** $(m - 2)^2$ **50.** prime
51. $9p^8(3p + 7)(p - 4)$ **52.** $5(2m - 3)(m + 4)$
53. $(2 - q)(2 - 3p)$ **54.** $\left(k + \dfrac{8}{11}\right)\left(k - \dfrac{8}{11}\right)$
55. $4(4p + 5m)(4p - 5m)$ **56.** $(m + 4)(m^2 - 6)$
57. $(10a + 9y)(10a - 9y)$ **58.** $(8a - b)(a + 3b)$
59. $(a + 4)^2$ **60.** $(2y + 5)(2y - 5)$

Section 14.6 (page 973)

1. $-5, 2$ **3.** $3, \dfrac{7}{2}$ **5.** $-\dfrac{5}{6}, 0$ **7.** $0, \dfrac{4}{3}$ **9.** $-\dfrac{1}{2}, \dfrac{1}{6}$ **11.** $-0.8, 2$
13. 9 **15.** Set each *variable* factor equal to 0, to get $2x = 0$ or $3x - 4 = 0$. The solutions are 0 and $\dfrac{4}{3}$. **17.** $-2, -1$ **19.** 1, 2
21. $-8, 3$ **23.** $-1, 3$ **25.** $-2, -1$ **27.** -4 **29.** $-2, \dfrac{1}{3}$
31. $-\dfrac{4}{3}, \dfrac{1}{2}$ **33.** $-\dfrac{2}{3}$ **35.** $-3, 3$ **37.** $-\dfrac{7}{4}, \dfrac{7}{4}$ **39.** $-11, 11$
41. 0, 7 **43.** $0, \dfrac{1}{2}$ **45.** 2, 5 **47.** $-4, \dfrac{1}{2}$ **49.** $-12, \dfrac{11}{2}$
51. $-2, 0, 2$ **53.** $-\dfrac{7}{3}, 0, \dfrac{7}{3}$ **55.** $-\dfrac{5}{2}, \dfrac{1}{3}, 5$ **57.** $-\dfrac{7}{2}, -3, 1$
59. (a) 64; 144; 4; 6 **(b)** No time has elapsed, so the object hasn't fallen (been released) yet. **(c)** Time cannot be negative.

Section 14.7 (page 981)

1. Read; variable; equation; Solve; answer; Check; original
3. *Step 3:* $45 = (2x + 1)(x + 1)$; *Step 4:* $x = 4$ or $x = -\dfrac{11}{2}$; *Step 5:* base: 9 units; height: 5 units; *Step 6:* $9 \cdot 5 = 45$
5. *Step 3:* $80 = (x + 8)(x - 8)$; *Step 4:* $x = 12$ or $x = -12$; *Step 5:* length: 20 units; width: 4 units; *Step 6:* $20 \cdot 4 = 80$
7. length: 7 in.; width: 4 in. **9.** length: 13 in.; width: 10 in.
11. height: 13 in.; width: 10 in. **13.** mirror: 7 ft; painting: 9 ft
15. 20, 21 **17.** $-3, -2$ or 4, 5 **19.** $-3, -1$ or 7, 9
21. $-2, 0, 2$ or 6, 8, 10 **23.** 12 cm **25.** 12 mi **27.** 8 ft
29. (a) 1 sec **(b)** $\dfrac{1}{2}$ sec and $1\dfrac{1}{2}$ sec **(c)** 3 sec **(d)** The negative solution, -1, does not make sense since t represents time, which cannot be negative. **31. (a)** 42.6 million; The result using the model is a little less than 44 million, the actual number for 1996. **(b)** 12 **(c)** 142.2 million; The result is a little more than 140 million, the actual number for 2002. **(d)** 215.4 million
32. 107.4 billion dollars; 40% **33.** 1997: 148.5 billion dollars; 1999: 230.1 billion dollars; 2000: 270.9 billion dollars. **34.** The answers using the linear equation are not at all close to the actual data. **35.** 1997: 111.2 billion dollars; 1999: 266.4 billion dollars; 2000: 399.5 billion dollars **36.** The answers in Exercise 35 are fairly close to the actual data. The quadratic equation models the data better.
37. (0, 97.5), (1, 104.3), (2, 104.7), (3, 164.3), (4, 271.3), (5, 378.7)

38. no

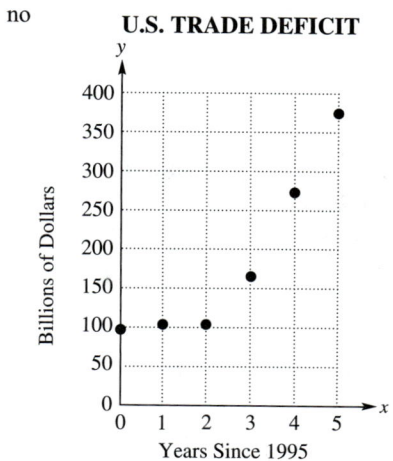

U.S. TRADE DEFICIT

39. 776.7 billion dollars **40. (a)** The actual deficit is quite a bit less than the prediction. **(b)** No, data for later years might not follow the same pattern.

Chapter 14 Review Exercises (page 991)

1. $7(t + 2)$ **2.** $30z(2z^2 + 1)$ **3.** $35x^2(x + 2)$
4. $50m^2n^2(2n - mn^2 + 3)$ **5.** $(x - 4)(2y + 3)$
6. $(2y + 3)(3y + 2x)$ **7.** $(x + 3)(x + 2)$ **8.** $(y - 5)(y - 8)$
9. $(q + 9)(q - 3)$ **10.** $(r - 8)(r + 7)$
11. $(r + 8s)(r - 12s)$ **12.** $(p + 12q)(p - 10q)$
13. $8p(p + 2)(p - 5)$ **14.** $3x^2(x + 2)(x + 8)$
15. $(m + 3n)(m - 6n)$ **16.** $(y - 3z)(y - 5z)$
17. $p^5(p - 2q)(p + q)$ **18.** $3r^3(r + 3s)(r - 5s)$ **19.** prime
20. $3(x^2 + 2x + 2)$ **21.** r and $6r$, $2r$ and $3r$ **22.** Factor out z.
23. $(2k - 1)(k - 2)$ **24.** $(3r - 1)(r + 4)$
25. $(3r + 2)(2r - 3)$ **26.** $(5z + 1)(2z - 1)$ **27.** prime
28. $4x^3(3x - 1)(2x - 1)$ **29.** $-3(x + 2)(2x - 5)$
30. $rs(5r + 6s)(2r + s)$ **31.** B **32.** D **33.** $(n + 8)(n - 8)$
34. $(5b + 11)(5b - 11)$ **35.** $(7y + 5w)(7y - 5w)$
36. $36(2p + q)(2p - q)$ **37.** prime **38.** $\left(x + \dfrac{7}{10}\right)\left(x - \dfrac{7}{10}\right)$
39. $(z + 5)^2$ **40.** $(r - 6)^2$ **41.** $(3t - 7)^2$ **42.** $(4m + 5n)^2$
43. $6x(3x - 2)^2$ **44.** $\left(x + \dfrac{1}{3}\right)^2$ **45.** $-\dfrac{3}{4}, 1$ **46.** $-7, -3, 4$
47. $0, \dfrac{5}{2}$ **48.** $-3, -1$ **49.** 1, 4 **50.** 3, 5 **51.** $-\dfrac{4}{3}, 5$
52. $-\dfrac{8}{9}, \dfrac{8}{9}$ **53.** 0, 8 **54.** $-1, 6$ **55.** 7 **56.** 6
57. $-\dfrac{2}{5}, -2, -1$ **58.** $-3, 3$ **59.** length: 10 ft; width: 4 ft
60. 5 ft **61.** length: 6 m; width: 2 m
62. length: 6 m; height: 5 m **63.** 6, 7 or $-5, -4$ **64.** 26 miles
65. 112 ft **66.** 192 ft **67.** 256 ft **68.** after 8 sec
69. (a) $771.9 million; the answer using the model is a little high. **(b)** $1727.6 million **(c)** No, the prediction seems low. If eBay revenues were $1516.7 million through three quarters, they were approximately $500 million per quarter, which would lead to annual revenue in 2003 of about $2000 million. **70.** D
71. The factor $2x + 8$ has a common factor of 2. The complete factored form is $2(x + 4)(3x - 4)$. **72.** $(z - x)(z - 10x)$
73. $(3k + 5)(k + 2)$ **74.** $(3m + 4p)(5m - 4)$
75. $(y^2 + 25)(y + 5)(y - 5)$ **76.** $3m(2m + 3)(m - 5)$
77. $8abc(3b^2c - 7ac^2 + 9ab)$ **78.** prime
79. $6xyz(2xz^2 + 2y - 5x^2yz^3)$ **80.** $2a^3(a + 2)(a - 6)$
81. $(2r + 3q)(6r - 5q)$ **82.** $(10a + 3)(10a - 3)$

83. $(7t + 4)^2$ **84.** $0, 7$ **85.** $-5, 2$ **86.** $-\dfrac{2}{5}$
87. 15 m, 36 m, 39 m **88.** length: 6 m; width: 4 m
89. $-5, -4, -3$ or $5, 6, 7$ **90.** **(a)** 256 ft **(b)** 1024 ft
91. width: 10 m; length: 17 m **92.** 6 m
93. (a) 481 thousand vehicles **(b)** The estimate may be unreliable because the conditions that prevailed in the years 1998–2001 may have changed, causing either a greater increase or a greater decrease in the numbers of alternative-fueled vehicles.

Chapter 14 Test (page 997)
1. D **2.** $6x(2x - 5)$ **3.** $m^2n(2mn + 3m - 5n)$
4. $(2x + y)(a - b)$ **5.** $(x - 7)(x - 2)$ **6.** $(2x + 3)(x - 1)$
7. $(3x + 1)(2x - 7)$ **8.** $3(x + 1)(x - 5)$
9. $(5z - 1)(2z - 3)$ **10.** prime **11.** prime
12. $(y + 7)(y - 7)$ **13.** $(3y + 8)(3y - 8)$ **14.** $(x + 8)^2$
15. $(2x - 7y)^2$ **16.** $-2(x + 1)^2$ **17.** $3t^2(2t + 9)(t - 4)$
18. prime **19.** $4t(t + 4)^2$ **20.** $(x^2 + 9)(x + 3)(x - 3)$
21. $(p + 3)(p + 3) = p^2 + 6p + 9 \ne p^2 + 9$ **22.** $-3, 9$
23. $\dfrac{1}{2}, 6$ **24.** $-\dfrac{2}{5}, \dfrac{2}{5}$ **25.** 10 **26.** $0, 3$ **27.** 6 ft by 9 ft
28. $-2, -1$ **29.** 17 ft **30.** 243 cable channels (rounded)

CHAPTER 15

Section 15.1 (page 1007)
1. (a) $3; -5$ **(b)** $q; -1$ **3.** A rational expression is a quotient of polynomials, such as $\dfrac{x + 3}{x^2 - 4}$. **5.** 0 **7.** $\dfrac{5}{3}$ **9.** $-3, 2$
11. never undefined **13. (a)** 1 **(b)** $\dfrac{17}{12}$ **15. (a)** 0 **(b)** $-\dfrac{10}{3}$
17. (a) $\dfrac{9}{5}$ **(b)** undefined **19. (a)** $\dfrac{2}{7}$ **(b)** $\dfrac{13}{3}$ **21.** No, not if the number is 0. Division by 0 is undefined. **23.** $3r^2$ **25.** $\dfrac{2}{5}$
27. $\dfrac{x - 1}{x + 1}$ **29.** $\dfrac{7}{5}$ **31.** $m - n$ **33.** $\dfrac{3(2m + 1)}{4}$ **35.** $\dfrac{3m}{5}$
37. $\dfrac{3r - 2s}{3}$ **39.** $\dfrac{x + 1}{x - 1}$ **41.** $\dfrac{z - 3}{z + 5}$ **43.** -1 **45.** $-(m + 1)$
47. -1

Answers may vary in Exercises 49–53.
49. $\dfrac{-(x + 4)}{x - 3}, \dfrac{-x - 4}{x - 3}, \dfrac{x + 4}{-(x - 3)}, \dfrac{x + 4}{-x + 3}$
51. $\dfrac{-(2x - 3)}{x + 3}, \dfrac{-2x + 3}{x + 3}, \dfrac{2x - 3}{-(x + 3)}, \dfrac{2x - 3}{-x - 3}$
53. $\dfrac{3x - 1}{5x - 6}, \dfrac{-(3x - 1)}{5x - 6}, \dfrac{-3x + 1}{-5x + 6}, \dfrac{3x - 1}{-5x + 6}$
55. $x^2 + 3$

Section 15.2 (page 1015)
1. (a) B **(b)** D **(c)** C **(d)** A **3.** $\dfrac{4m}{3}$ **5.** $\dfrac{40y^2}{3}$ **7.** $\dfrac{2}{c + d}$
9. $\dfrac{16q}{3p^3}$ **11.** $\dfrac{7}{r^2 + rp}$ **13.** $\dfrac{z^2 - 9}{z^2 + 7z + 12}$ **15.** 5 **17.** $-\dfrac{3}{2t^4}$
19. $\dfrac{1}{4}$ **21.** To multiply two rational expressions, multiply the numerators and multiply the denominators. Write the answer in lowest terms. **23.** $\dfrac{10}{9}$ **25.** $-\dfrac{3}{4}$ **27.** -1 **29.** $\dfrac{-9(m - 2)}{m + 4}$
31. $\dfrac{p + 4}{p + 2}$ **33.** $\dfrac{(k - 1)^2}{(k + 1)(2k - 1)}$ **35.** $\dfrac{4k - 1}{3k - 2}$ **37.** $\dfrac{m + 4p}{m + p}$
39. $\dfrac{10}{x + 10}$ **41.** Division requires multiplying by the reciprocal of the second rational expression. In the reciprocal, $x + 7$ is in the denominator, so $x \ne -7$.

Section 15.3 (page 1021)
1. C **3.** C **5.** 30 **7.** x^7 **9.** $72q$ **11.** $84r^5$ **13.** $2^3 \cdot 3 \cdot 5$
15. The least common denominator is their product.
17. $28m^2(3m - 5)$ **19.** $30(b - 2)$ **21.** $c - d$ or $d - c$
23. $k(k + 5)(k - 2)$ **25.** $(p + 3)(p + 5)(p - 6)$ **27.** $\dfrac{20}{55}$
29. $\dfrac{-45}{9k}$ **31.** $\dfrac{26y^2}{80y^3}$ **33.** $\dfrac{35t^2r^3}{42r^4}$ **35.** $\dfrac{20}{8(m + 3)}$ **37.** $\dfrac{8t}{12 - 6t}$
39. $\dfrac{14(z - 2)}{z(z - 3)(z - 2)}$ **41.** $\dfrac{2(b - 1)(b + 2)}{b^3 + 3b^2 + 2b}$

Section 15.4 (page 1029)
1. E **3.** C **5.** B **7.** G **9.** $\dfrac{11}{m}$ **11.** b **13.** x **15.** $y - 6$
17. To add or subtract rational expressions with the same denominator, combine the numerators and keep the same denominator. For example, $\dfrac{3x + 2}{x - 6} + \dfrac{-2x - 8}{x - 6} = \dfrac{x - 6}{x - 6}$.
Then write in lowest terms: $\dfrac{x - 6}{x - 6} = 1$. **19.** $\dfrac{3z + 5}{15}$
21. $\dfrac{10 - 7r}{14}$ **23.** $\dfrac{-3x - 2}{4x}$ **25.** $\dfrac{x + 1}{2}$ **27.** $\dfrac{5x + 9}{6x}$
29. $\dfrac{3x + 3}{x(x + 3)}$ **31.** $\dfrac{x^2 + 6x - 8}{(x - 2)(x + 2)}$ **33.** $\dfrac{3}{t}$
35. $m - 2$ or $2 - m$ **37.** $\dfrac{-2}{x - 5}$ or $\dfrac{2}{5 - x}$ **39.** -4
41. $\dfrac{-5}{x - y^2}$ or $\dfrac{5}{y^2 - x}$ **43.** $\dfrac{x + y}{5x - 3y}$ or $\dfrac{-x - y}{3y - 5x}$
45. $\dfrac{-6}{4p - 5}$ or $\dfrac{6}{5 - 4p}$ **47.** $\dfrac{-(m + n)}{2(m - n)}$
49. $\dfrac{-x^2 + 6x + 11}{(x + 3)(x - 3)(x + 1)}$ **51.** $\dfrac{-5q^2 - 13q + 7}{(3q - 2)(q + 4)(2q - 3)}$
53. $\dfrac{9r + 2}{r(r + 2)(r - 1)}$
55. $\dfrac{2x^2 + 6xy + 8y^2}{(x + y)(x + y)(x + 3y)}$ or $\dfrac{2x^2 + 6xy + 8y^2}{(x + y)^2(x + 3y)}$
57. $\dfrac{15r^2 + 10ry - y^2}{(3r + 2y)(6r - y)(6r + y)}$ **59. (a)** $\dfrac{9k^2 + 6k + 26}{5(3k + 1)}$ **(b)** $\dfrac{1}{4}$

Section 15.5 (page 1039)
1. (a) $6; \dfrac{1}{6}$ **(b)** $12; \dfrac{3}{4}$ **(c)** $\dfrac{1}{6} \div \dfrac{3}{4}$ **(d)** $\dfrac{2}{9}$ **3.** -6 **5.** $\dfrac{1}{pq}$ **7.** $\dfrac{1}{xy}$
9. $\dfrac{2a^2b}{3}$ **11.** $\dfrac{m(m + 2)}{3(m - 4)}$ **13.** $\dfrac{2}{x}$ **15.** $\dfrac{8}{x}$ **17.** $\dfrac{a^2 - 5}{a^2 + 1}$
19. $\dfrac{3(p + 2)}{2(2p + 3)}$ **21.** $\dfrac{t(t - 2)}{4}$ **23.** $\dfrac{-k}{2 + k}$ **25.** $\dfrac{2x - 7}{3x + 1}$
27. $\dfrac{3m(m - 3)}{(m - 1)(m - 8)}$ **29.** $\dfrac{6}{5}$

Section 15.6 (page 1049)
1. expression; $\dfrac{43}{40}x$ **3.** equation; $\dfrac{40}{43}$ **5.** expression; $-\dfrac{1}{10}y$
7. When solving an equation, we multiply both sides by the LCD, which eliminates all denominators. When adding or subtracting fractions, we multiply by 1 in the form $\dfrac{\text{LCD}}{\text{LCD}}$. The denominators are not eliminated. **9.** -6 **11.** 24 **13.** -15 **15.** 7

17. -15 **19.** -5 **21.** -6 **23.** 5 **25.** 12 **27.** 2 **29.** 0 and 4 **31.** -6 **33.** no solution **35.** 3 **37.** 3 **39.** $-2, 12$ **41.** no solution **43.** $-6, \dfrac{1}{2}$ **45.** $-\dfrac{1}{5}, 3$ **47.** $-\dfrac{3}{5}, 3$ **49.** Transform the equation so that the terms with k are on one side and the remaining term is on the other. **51.** $F = \dfrac{ma}{k}$ **53.** $a = \dfrac{kF}{m}$ **55.** $R = \dfrac{E - Ir}{I}$ **57.** $A = \dfrac{h(B + b)}{2}$ **59.** $a = \dfrac{2S - ndL}{nd}$ **61.** $t = \dfrac{rs}{rs - 2s - 3r}$ or $t = \dfrac{-rs}{-rs + 2s + 3r}$ **63.** $c = \dfrac{ab}{b - a - 2ab}$ or $c = \dfrac{-ab}{-b + a + 2ab}$ **65.** $z = \dfrac{3y}{5 - 9xy}$ or $z = \dfrac{-3y}{9xy - 5}$

Summary Exercises on Rational Expressions and Equations (page 1053)

1. expression; $\dfrac{10}{p}$ **2.** expression; $\dfrac{y^3}{x^3}$ **3.** expression; $\dfrac{1}{2x^2(x + 2)}$ **4.** equation; 9 **5.** expression; $\dfrac{y + 2}{y - 1}$ **6.** expression; $\dfrac{5k + 8}{k(k - 4)(k + 4)}$ **7.** equation; 39 **8.** expression; $\dfrac{t - 5}{3(2t + 1)}$ **9.** expression; $\dfrac{13}{3(p + 2)}$ **10.** equation; $-1, \dfrac{12}{5}$ **11.** equation; $\dfrac{1}{7}, 2$ **12.** expression; $\dfrac{16}{3y}$ **13.** expression; $\dfrac{7}{12z}$ **14.** equation; 13 **15.** expression; $\dfrac{3m + 5}{(m + 2)(m + 3)(m + 1)}$ **16.** expression; $\dfrac{k + 3}{5(k - 1)}$ **17.** equation; no solution (Reject 1 as a solution because it causes denominators to equal 0.) **18.** equation; -7

Section 15.7 (page 1063)

1. (a) the amount (b) $5 + x$ (c) $\dfrac{5 + x}{6} = \dfrac{13}{3}$ **3.** $\dfrac{9}{5}$ **5.** $\dfrac{2}{6}$ **7.** -6 **9.** 36 **11.** 8.45 meters per second **13.** 3.735 hr **15.** 338.730 meters per minute **17.** $\dfrac{8}{4 - x} = \dfrac{24}{4 + x}$ **19.** into a headwind: $m - 5$ mph; with a tailwind: $m + 5$ mph **21.** 10 mph **23.** 18.5 mph **25.** $\dfrac{1}{2}t + \dfrac{1}{3}t = 1$ or $\dfrac{1}{2} + \dfrac{1}{3} = \dfrac{1}{t}$ **27.** $2\dfrac{2}{9}$ hr **29.** $4\dfrac{8}{19}$ hr **31.** 10 hr **33.** 36 hr **35.** $8\dfrac{1}{4}$ hr **37.** The equation would be erroneously written $\dfrac{8}{4 + x} = \dfrac{24}{4 - x}$. Solving for x gives $x = -2$. Because x represents the speed of the current, it cannot be negative. Therefore, the student should realize that there is a problem in the setup.

Section 15.8 (page 1069)

1. (a) increases (b) decreases **3.** 15 **5.** 300 **7.** 4 **9.** 6 **11.** 15 in.^{2} **13.** $42\dfrac{2}{3}$ in. **15.** 15 ft **17.** 20 pounds per square foot **19.** 25 kilograms per hour **21.** direct **23.** inverse **25.** inverse **27.** direct **29.** $x = 8$ **31.** $x = 2$ **33.** 80 ft

Chapter 15 Review Exercises (page 1079)

1. 3 **2.** 0 **3.** $-1, 3$ **4.** $-5, -\dfrac{2}{3}$ **5.** (a) $-\dfrac{4}{7}$ (b) -16 **6.** (a) $\dfrac{11}{8}$ (b) $\dfrac{13}{22}$ **7.** (a) undefined (b) 1 **8.** (a) undefined (b) $\dfrac{1}{2}$ **9.** $\dfrac{b}{3a}$ **10.** -1 **11.** $\dfrac{-(2x + 3)}{2}$ **12.** $\dfrac{2p + 5q}{5p + q}$ Answers may vary in Exercises 13 and 14. **13.** $\dfrac{-(4x - 9)}{2x + 3}, \dfrac{-4x + 9}{2x + 3}, \dfrac{4x - 9}{-(2x + 3)}, \dfrac{4x - 9}{-2x - 3}$ **14.** $\dfrac{-8 + 3x}{-3 - 6x}, \dfrac{-(-8 + 3x)}{3 + 6x}, \dfrac{8 - 3x}{-(-3 - 6x)}, \dfrac{-8 + 3x}{3 + 6x}$ **15.** 2 **16.** $\dfrac{2}{3m^6}$ **17.** $\dfrac{5}{8}$ **18.** $\dfrac{r + 4}{3}$ **19.** $\dfrac{3}{2}$ **20.** $\dfrac{y - 2}{y - 3}$ **21.** $\dfrac{p + 5}{p + 1}$ **22.** $\dfrac{3z + 1}{z + 3}$ **23.** 96 **24.** $108y^4$ **25.** $m(m + 2)(m + 5)$ **26.** $(x + 3)(x + 1)(x + 4)$ **27.** $\dfrac{35}{56}$ **28.** $\dfrac{40}{4k}$ **29.** $\dfrac{15a}{10a^4}$ **30.** $\dfrac{-54}{18 - 6x}$ **31.** $\dfrac{15y}{50 - 10y}$ **32.** $\dfrac{4b(b + 2)}{(b + 3)(b - 1)(b + 2)}$ **33.** $\dfrac{15}{x}$ **34.** $-\dfrac{2}{p}$ **35.** $\dfrac{4k - 45}{k(k - 5)}$ **36.** $\dfrac{28 + 11y}{y(7 + y)}$ **37.** $\dfrac{-2 - 3m}{6}$ **38.** $\dfrac{3(16 - x)}{4x^2}$ **39.** $\dfrac{7a + 6b}{(a - 2b)(a + 2b)}$ **40.** $\dfrac{-k^2 - 6k + 3}{3(k + 3)(k - 3)}$ **41.** $\dfrac{5z - 16}{z(z + 6)(z - 2)}$ **42.** $\dfrac{-13p + 33}{p(p - 2)(p - 3)}$ **43.** $\dfrac{a}{b}$ **44.** $\dfrac{4(y - 3)}{y + 3}$ **45.** $\dfrac{6(3m + 2)}{2m - 5}$ **46.** $\dfrac{(q - p)^2}{pq}$ **47.** $\dfrac{xw + 1}{xw - 1}$ **48.** $\dfrac{1 - r - t}{1 + r + t}$ **49.** $\dfrac{35}{6}$ **50.** -16 **51.** -4 **52.** no solution **53.** 3 **54.** $t = \dfrac{Ry}{m}$ **55.** $y = \dfrac{4x + 5}{3}$ **56.** $t = \dfrac{rs}{s - r}$ **57.** 12 **58.** $\dfrac{20}{15}$ **59.** $\dfrac{3}{18}$ **60.** 1.527 hr **61.** $3\dfrac{1}{13}$ hr **62.** 10 hr **63.** 4 cm **64.** $\dfrac{36}{5}$ **65.** $\dfrac{m + 7}{(m - 1)(m + 1)}$ **66.** $8p^2$ **67.** $\dfrac{1}{6}$ **68.** 3 **69.** $\dfrac{z + 7}{(z + 1)(z - 1)^2}$ **70.** $d = \dfrac{k + FD}{F}$ or $d = \dfrac{k}{F} + D$ **71.** $-2, 3$ **72.** 150 kilometers per hour **73.** $1\dfrac{7}{8}$ hr **74.** 4 **75.** 24

Chapter 15 Test (page 1083)

1. $-2, 4$ **2.** (a) $\dfrac{11}{6}$ (b) undefined **3.** (Answers may vary.) $\dfrac{-(6x - 5)}{2x + 3}, \dfrac{-6x + 5}{2x + 3}, \dfrac{6x - 5}{-(2x + 3)}, \dfrac{6x - 5}{-2x - 3}$ **4.** $-3x^2y^3$ **5.** $\dfrac{3a + 2}{a - 1}$ **6.** $\dfrac{25}{27}$ **7.** $\dfrac{3k - 2}{3k + 2}$ **8.** $\dfrac{a - 1}{a + 4}$ **9.** $150p^5$ **10.** $(2r + 3)(r + 2)(r - 5)$ **11.** $\dfrac{240p^2}{64p^3}$ **12.** $\dfrac{21}{42m - 84}$ **13.** 2 **14.** $\dfrac{-14}{5(y + 2)}$ **15.** $\dfrac{x^2 + x + 1}{3 - x}$ or $\dfrac{-x^2 - x - 1}{x - 3}$ **16.** $\dfrac{-m^2 + 7m + 2}{(2m + 1)(m - 5)(m - 1)}$ **17.** $\dfrac{2k}{3p}$ **18.** $\dfrac{-2 - x}{4 + x}$ **19.** $-\dfrac{1}{2}$

20. $D = \dfrac{dF - k}{F}$ or $D = \dfrac{k - dF}{-F}$ **21.** -4 **22.** 3 mph   **23.** $2\dfrac{2}{9}$ hr   **24.** 27   **25.** 27 days

CHAPTER 16

Section 16.1 (page 1093)
1. true   **3.** false; Zero has only one square root.   **5.** true
7. $-3, 3$ **9.** $-8, 8$ **11.** $-13, 13$ **13.** $-\dfrac{5}{14}, \dfrac{5}{14}$ **15.** $-30, 30$
17. 1   **19.** 7   **21.** -16 **23.** $-\dfrac{12}{11}$ **25.** 0.8   **27.** not a real number   **29.** not a real number   **31.** 100   **33.** 19   **35.** $\dfrac{2}{3}$
37. $3x^2 + 4$ **39.** x must be positive.   **41.** x must be negative.
43. rational; 5   **45.** irrational; 5.385   **47.** rational; -8
49. irrational; -17.321 **51.** not a real number   **53.** irrational; 34.641   **55.** C   **57.** $c = 17$ **59.** $b = 8$ **61.** $c \approx 11.705$
63. 24 cm   **65.** 80 ft   **67.** 195 ft   **69.** 9.434 in.   **71.** 13.5 ft
73. Answers will vary. For example, if $a = 2$ and $b = 7$, $\sqrt{a^2 + b^2} = \sqrt{2^2 + 7^2} = \sqrt{53}$, while $a + b = 2 + 7 = 9$. Therefore, $\sqrt{a^2 + b^2} \neq a + b$ because $\sqrt{53} \neq 9$.   **75.** 1
77. 5   **79.** -3 **81.** -6 **83.** -2 **85.** 4   **87.** 6   **89.** not a real number   **91.** -3 **93.** -4

Section 16.2 (page 1103)
1. false; $\sqrt{(-6)^2} = \sqrt{36} = 6$ **3.** $\sqrt{15}$ **5.** $\sqrt{22}$ **7.** $\sqrt{42}$
9. $\sqrt{13r}$ **11.** A   **13.** $3\sqrt{5}$ **15.** $2\sqrt{6}$ **17.** $3\sqrt{10}$ **19.** $5\sqrt{3}$
21. $5\sqrt{5}$ **23.** cannot be simplified   **25.** $4\sqrt{10}$ **27.** $-10\sqrt{7}$
29. $3\sqrt{6}$ **31.** 24   **33.** $6\sqrt{10}$ **35.** $12\sqrt{5}$ **37.** $30\sqrt{5}$
39. $\sqrt{8} \cdot \sqrt{32} = \sqrt{8 \cdot 32} = \sqrt{256} = 16$. Also, $\sqrt{8} = 2\sqrt{2}$ and $\sqrt{32} = 4\sqrt{2}$, so $\sqrt{8} \cdot \sqrt{32} = 2\sqrt{2} \cdot 4\sqrt{2} = 8 \cdot 2 = 16$. Both methods give the same answer, and the correct answer can always be obtained using either method.   **41.** $\dfrac{4}{15}$ **43.** $\dfrac{\sqrt{7}}{4}$
45. 5   **47.** $\dfrac{25}{4}$ **49.** $6\sqrt{5}$ **51.** m **53.** y^2 **55.** $6z$ **57.** $20x^3$
59. $3x^4\sqrt{2}$ **61.** $3c^7\sqrt{5}$ **63.** $z^2\sqrt{z}$ **65.** $a^6\sqrt{a}$ **67.** $8x^3\sqrt{x}$
69. x^3y^6 **71.** $9m^2n$ **73.** $\dfrac{\sqrt{7}}{x^5}$ **75.** $\dfrac{y^2}{10}$ **77.** $\dfrac{x^3}{y^4}$ **79.** $2\sqrt[3]{5}$
81. $3\sqrt[3]{2}$ **83.** $4\sqrt[3]{2}$ **85.** $2\sqrt[4]{5}$ **87.** $\dfrac{2}{3}$ **89.** $-\dfrac{6}{5}$ **91.** p
93. x^3 **95.** $4z^2$ **97.** $7a^3b$ **99.** $2t\sqrt[3]{2t^2}$ **101.** $\dfrac{m^4}{2}$ **103.** 6 cm
105. 6 in.   **107.** D

Section 16.3 (page 1109)
1. distributive   **3.** radicands   **5.** $-5\sqrt{7}$ **7.** $5\sqrt{17}$ **9.** $5\sqrt{7}$
11. $11\sqrt{5}$ **13.** $15\sqrt{2}$ **15.** $-6\sqrt{2}$ **17.** $17\sqrt{7}$
19. $-16\sqrt{2} - 8\sqrt{3}$ **21.** $20\sqrt{2} + 6\sqrt{3} - 15\sqrt{5}$ **23.** $4\sqrt{2}$
25. $22\sqrt{2}$ **27.** $11\sqrt{3}$ **29.** $5\sqrt{x}$ **31.** $3x\sqrt{6}$ **33.** 0
35. $-20\sqrt{2k}$ **37.** $42x\sqrt{5z}$ **39.** $-\sqrt[3]{2}$ **41.** $6\sqrt[3]{p^2}$
43. $21\sqrt[4]{m^3}$ **45.** $-6x^2y$ **46.** $-6(p - 2q)^2(a + b)$
47. $-6a^2\sqrt{xy}$ **48.** The answers are alike because the numerical coefficient of the three answers is the same: -6. Also, the first variable factor is raised to the second power, and the second variable factor is raised to the first power. The answers are different because the variables are different: x and y, then $p - 2q$ and $a + b$, and then a and $\sqrt{xy}$.

Section 16.4 (page 1117)
1. $4\sqrt{2}$ **3.** $\dfrac{-\sqrt{33}}{3}$ **5.** $\dfrac{7\sqrt{15}}{5}$ **7.** $\dfrac{\sqrt{30}}{2}$ **9.** $\dfrac{16\sqrt{3}}{9}$
11. $\dfrac{-3\sqrt{2}}{10}$ **13.** $\dfrac{21\sqrt{5}}{5}$ **15.** $\sqrt{3}$ **17.** $\dfrac{\sqrt{2}}{2}$ **19.** $\dfrac{\sqrt{65}}{5}$
21. We are actually multiplying by 1. The identity property of multiplication justifies our result.   **23.** $\dfrac{\sqrt{21}}{3}$ **25.** $\dfrac{3\sqrt{14}}{4}$ **27.** $\dfrac{1}{6}$
29. 1   **31.** $\dfrac{\sqrt{7x}}{x}$ **33.** $\dfrac{2x\sqrt{xy}}{y}$ **35.** $\dfrac{x\sqrt{30xz}}{6}$ **37.** $\dfrac{3ar^2\sqrt{7rt}}{7t}$
39. B   **41.** $\dfrac{\sqrt[3]{12}}{2}$ **43.** $\dfrac{\sqrt[3]{196}}{7}$ **45.** $\dfrac{\sqrt[3]{6y}}{2y}$ **47.** $\dfrac{\sqrt[3]{42mn^2}}{6n}$
49. (a) $\dfrac{9\sqrt{2}}{4}$ seconds   (b) 3.182 seconds

Section 16.5 (page 1125)
1. 13   **3.** 4   **5.** $\sqrt{15} - \sqrt{35}$ **7.** $2\sqrt{10} + 30$ **9.** $4\sqrt{7}$
11. $57 + 23\sqrt{6}$ **13.** $81 + 14\sqrt{21}$ **15.** $71 - 16\sqrt{7}$
17. $37 + 12\sqrt{7}$ **19.** $a + 2\sqrt{a} + 1$ **21.** 23   **23.** 1
25. $y - 10$ **27.** $2\sqrt{3} - 2 + 3\sqrt{2} - \sqrt{6}$ **29.** $15\sqrt{2} - 15$
31. $\sqrt{30} + \sqrt{15} + 6\sqrt{5} + 3\sqrt{10}$
33. $\sqrt{5x} - \sqrt{10} - \sqrt{10x} + 2\sqrt{5}$ **35.** Because multiplication must be performed before addition, it is incorrect to add -37 and -2. Only like radicals can be combined.   **37.** $\dfrac{3 - \sqrt{2}}{7}$
39. $-4 - 2\sqrt{11}$ **41.** $1 + \sqrt{2}$ **43.** $-\sqrt{10} + \sqrt{15}$
45. $2\sqrt{5} + \sqrt{15} + 4 + 2\sqrt{3}$ **47.** $\dfrac{12(\sqrt{x} - 1)}{x - 1}$
49. $\dfrac{3(7 + \sqrt{x})}{49 - x}$ **51.** $\sqrt{11} - 2$ **53.** $\dfrac{\sqrt{3} + 5}{8}$ **55.** $\dfrac{6 - \sqrt{10}}{2}$
57. $30 + 18x$ **58.** They are not like terms.   **59.** $30 + 18\sqrt{5}$
60. They are not like radicals.   **61.** Make the first term $30x$, so that $30x + 18x = 48x$; make the first term $30\sqrt{5}$, so that $30\sqrt{5} + 18\sqrt{5} = 48\sqrt{5}$.   **62.** Both like terms and like radicals are combined by adding their numerical coefficients. The variables in like terms are replaced by radicals in like radicals.   **63.** 4 in.

Summary Exercises on Operations with Radicals (page 1129)
1. $-3\sqrt{10}$ **2.** $5 - \sqrt{15}$ **3.** $2 - \sqrt{6} + 2\sqrt{3} - 3\sqrt{2}$
4. $6\sqrt{2}$ **5.** $73 - 12\sqrt{35}$ **6.** $\dfrac{\sqrt{6}}{2}$ **7.** $3\sqrt[3]{2t^2}$
8. $4\sqrt{7} + 4\sqrt{5}$ **9.** $-3 - 2\sqrt{2}$ **10.** 4   **11.** -33
12. $\dfrac{\sqrt{t} - \sqrt{3}}{t - 3}$ **13.** $2xyz^2\sqrt[3]{y^2}$ **14.** $4\sqrt[3]{3}$ **15.** $\sqrt{6} + 1$
16. $\dfrac{\sqrt{6x}}{3x}$ **17.** $\dfrac{3}{5}$ **18.** $4\sqrt{2}$ **19.** $-2\sqrt[3]{2}$ **20.** $11 - 2\sqrt{30}$
21. $3\sqrt{3x}$ **22.** $52 + 30\sqrt{3}$ **23.** 1   **24.** $\dfrac{2\sqrt[3]{18}}{9}$ **25.** $-x^2\sqrt[4]{x}$
26. $2\sqrt{6}$ **27.** cannot be simplified further   **28.** $12\sqrt{6} + 6\sqrt{5}$
29. $\dfrac{\sqrt{15}}{10}$ **30.** $\sqrt{5}$ **31.** $20\sqrt[3]{3}$ **32.** $\dfrac{8(4 + \sqrt{x})}{16 - x}$
33. $2 - 3\sqrt[3]{4}$ **34.** $\sqrt{2x}$ **35.** $\dfrac{\sqrt{10}}{4}$ **36.** $49 + 14\sqrt{x} + x$
37. (a) 57 species   (b) 858 species

Section 16.6 (page 1137)

1. 49 **3.** 7 **5.** 85 **7.** -45 **9.** $-\dfrac{3}{2}$ **11.** no solution
13. 121 **15.** 8 **17.** 1 **19.** 6 **21.** no solution **23.** 5
25. $x^2 - 14x + 49$ **27.** 12 **29.** 5 **31.** 0, 3 **33.** $-1, 3$
35. 8 **37.** 4 **39.** 8 **41.** 9 **43.** 4, 20 **45.** -5 **47.** 21
49. 8 **51.** (a) 70.5 mph (b) 59.8 mph (c) 53.9 mph
53. 158.6 ft **55.** $s = 13$ units **56.** $6\sqrt{13}$ square units
57. $h = \sqrt{13}$ units **58.** $3\sqrt{13}$ square units
59. $6\sqrt{13}$ square units **60.** They are both $6\sqrt{13}$ square units.

Chapter 16 Review Exercises (page 1145)

1. $-7, 7$ **2.** $-9, 9$ **3.** $-14, 14$ **4.** $-11, 11$ **5.** $-15, 15$
6. $-27, 27$ **7.** 4 **8.** -0.6 **9.** 10 **10.** 3 **11.** not a real number **12.** -65 **13.** $\dfrac{7}{6}$ **14.** $\dfrac{10}{9}$ **15.** B **16.** F **17.** D
18. A **19.** C **20.** A **21.** 8 ft **22.** 40.6 cm **23.** irrational; 4.796 **24.** rational; 13 **25.** rational; -5 **26.** not a real number
27. $\sqrt{14}$ **28.** $5\sqrt{3}$ **29.** $-3\sqrt{3}$ **30.** $4\sqrt{3}$ **31.** $4\sqrt{10}$
32. 18 **33.** $16\sqrt{6}$ **34.** $25\sqrt{10}$ **35.** $\dfrac{3}{2}$ **36.** $-\dfrac{11}{20}$ **37.** $\dfrac{\sqrt{7}}{13}$
38. $\dfrac{\sqrt{5}}{6}$ **39.** $\dfrac{2}{15}$ **40.** $3\sqrt{2}$ **41.** 8 **42.** $2\sqrt{2}$ **43.** p
44. $\sqrt{km}$ **45.** r^9 **46.** $x^5 y^8$ **47.** $x^4 \sqrt{x}$ **48.** $\dfrac{6}{p}$
49. $a^7 b^{10} \sqrt{ab}$ **50.** $11 x^3 y^5$ **51.** y^2 **52.** $6x^5$ **53.** Yes, because both approximations are 0.7071067812. **54.** $2\sqrt{11}$ **55.** $9\sqrt{2}$
56. $21\sqrt{3}$ **57.** $12\sqrt{3}$ **58.** 0 **59.** $3\sqrt{7}$ **60.** $2\sqrt{3} + 3\sqrt{10}$
61. $2\sqrt{2}$ **62.** $6\sqrt{30}$ **63.** $5\sqrt{x}$ **64.** 0 **65.** $-m\sqrt{5}$
66. $11 k^2 \sqrt{2n}$ **67.** $\dfrac{10\sqrt{3}}{3}$ **68.** $\dfrac{8\sqrt{10}}{5}$ **69.** $\sqrt{6}$ **70.** $\dfrac{\sqrt{10}}{5}$
71. $\sqrt{10}$ **72.** $\dfrac{\sqrt{42}}{21}$ **73.** $\dfrac{r\sqrt{x}}{4x}$ **74.** $\dfrac{\sqrt[3]{9}}{3}$ **75.** $r = \dfrac{\sqrt{3V\pi h}}{\pi h}$
76. $r = \dfrac{\sqrt{S\pi}}{2\pi}$ **77.** $-\sqrt{15} - 9$ **78.** $3\sqrt{6} + 12$
79. $22 - 16\sqrt{3}$ **80.** $2\sqrt{21} - \sqrt{14} + 12\sqrt{2} - 4\sqrt{3}$
81. -13 **82.** $x + 4\sqrt{x} + 4$ **83.** $-2 + \sqrt{5}$ **84.** $\dfrac{-2\sqrt{2} - 6}{7}$
85. $\dfrac{3(1 - \sqrt{x})}{1 - x}$ **86.** $\dfrac{-2 + 6\sqrt{2}}{17}$
87. $\dfrac{-\sqrt{10} + 3\sqrt{5} + \sqrt{2} - 3}{7}$ **88.** $\dfrac{2\sqrt{3} + 2 + 3\sqrt{2} + \sqrt{6}}{2}$
89. $\dfrac{3 + 2\sqrt{6}}{3}$ **90.** $\dfrac{1 + 3\sqrt{7}}{4}$ **91.** $3 + 4\sqrt{3}$ **92.** no solution
93. 48 **94.** 1 **95.** 2 **96.** 6 **97.** $-3, -1$ **98.** -2 **99.** 4
100. 7 **101.** 9 **102.** $11\sqrt{3}$ **103.** $\dfrac{11}{t}$ **104.** $\dfrac{5 - \sqrt{2}}{23}$
105. $\dfrac{2\sqrt{10}}{5}$ **106.** $5y\sqrt{2}$ **107.** -5 **108.** $-\sqrt{10} - 5\sqrt{15}$
109. $\dfrac{4r\sqrt{3rs}}{3s}$ **110.** $\dfrac{2 + \sqrt{13}}{2}$ **111.** $-7\sqrt{2}$ **112.** $7 - 2\sqrt{10}$
113. $166 + 2\sqrt{7}$ **114.** -11 **115.** $7\sqrt{2}$ **116.** 7
117. no solution **118.** 8 **119.** 39.2 mph
120. $\dfrac{10\sqrt{7} - 5\sqrt{7}}{-3\sqrt{14} - 2\sqrt{14}}$ or $\dfrac{5\sqrt{7} - 10\sqrt{7}}{2\sqrt{14} + 3\sqrt{14}}$ **121.** $-\dfrac{5\sqrt{7}}{5\sqrt{14}}$
122. $-\sqrt{\dfrac{1}{2}}$ **123.** $-\dfrac{\sqrt{2}}{2}$ **124.** It falls from left to right.

Chapter 16 Test (page 1151)

1. $-14, 14$ **2.** (a) irrational (b) 11.916 **3.** a must be negative.
4. 6 **5.** $-3\sqrt{3}$ **6.** $\dfrac{8\sqrt{2}}{5}$ **7.** $2\sqrt[3]{4}$ **8.** $4\sqrt{6}$ **9.** $9\sqrt{7}$
10. $-5\sqrt{3x}$ **11.** $4xy\sqrt{2y}$ **12.** 31
13. $6\sqrt{2} + 2 - 3\sqrt{14} - \sqrt{7}$ **14.** $11 + 2\sqrt{30}$
15. (a) $6\sqrt{2}$ in. (b) 8.485 in. **16.** 50 ohms **17.** $\dfrac{5\sqrt{14}}{7}$
18. $\dfrac{\sqrt{6x}}{3x}$ **19.** $-\sqrt[3]{2}$ **20.** $\dfrac{-3(4 + \sqrt{3})}{13}$
21. $\dfrac{\sqrt{3} + 12\sqrt{2}}{3}$ **22.** no solution **23.** 3 **24.** 1, 4
25. 12 is an extraneous solution. A check shows that 12 does not satisfy the original equation. So the equation has *no solution*.

CHAPTER 17

Section 17.1 (page 1157)

1. true **3.** false; If k is a positive integer that is not a perfect square, then the solutions will be irrational. **5.** false; For values of k that satisfy $0 \leq k < 10$, there are real solutions.
7. $-9, 9$ **9.** $-\sqrt{14}, \sqrt{14}$ **11.** $-4\sqrt{3}, 4\sqrt{3}$ **13.** $-\dfrac{5}{2}, \dfrac{5}{2}$
15. no real number solution **17.** $-1.5, 1.5$ **19.** $-\sqrt{3}, \sqrt{3}$
21. $-\dfrac{2\sqrt{7}}{7}, \dfrac{2\sqrt{7}}{7}$ **23.** $-\dfrac{2\sqrt{5}}{5}, \dfrac{2\sqrt{5}}{5}$ **25.** $-2\sqrt{6}, 2\sqrt{6}$
27. $-2, 8$ **29.** no real number solution
31. $8 + 3\sqrt{3}, 8 - 3\sqrt{3}$ **33.** $-3, \dfrac{5}{3}$ **35.** $0, \dfrac{3}{2}$
37. $\dfrac{5 + \sqrt{30}}{2}, \dfrac{5 - \sqrt{30}}{2}$ **39.** $\dfrac{-1 + 3\sqrt{2}}{3}, \dfrac{-1 - 3\sqrt{2}}{3}$
41. $-10 + 4\sqrt{3}, -10 - 4\sqrt{3}$ **43.** $\dfrac{1 + 4\sqrt{3}}{4}, \dfrac{1 - 4\sqrt{3}}{4}$
45. The answers are equivalent. If the answer of either student is multiplied by $\dfrac{-1}{-1}$, it will look like the answer of the other student.
47. about $\dfrac{1}{2}$ second **49.** 9 in. **51.** 5%

Section 17.2 (page 1167)

1. $10x$ **3.** 49 **5.** D **7.** 100 **9.** $\dfrac{25}{4}$ **11.** $\dfrac{1}{16}$ **13.** 1, 3
15. $-3, -2$ **17.** $-1 + \sqrt{6}, -1 - \sqrt{6}$ **19.** -3
21. $\dfrac{-1 + \sqrt{5}}{2}, \dfrac{-1 - \sqrt{5}}{2}$ **23.** $-\dfrac{3}{2}, \dfrac{1}{2}$
25. $\dfrac{2 + \sqrt{14}}{2}, \dfrac{2 - \sqrt{14}}{2}$ **27.** no real number solution
29. $\dfrac{-7 + \sqrt{97}}{6}, \dfrac{-7 - \sqrt{97}}{6}$ **31.** $-4, 2$ **33.** $1 + \sqrt{6}, 1 - \sqrt{6}$
35. 1 second and 5 seconds **37.** 3 seconds and 5 seconds
39. 75 ft by 100 ft

Section 17.3 (page 1175)

1. $4; 5; -9$ 3. $3; -4; -2$ 5. $3; 7; 0$ 7. $-13, 1$ 9. 2
11. $\dfrac{-6 + \sqrt{26}}{2}, \dfrac{-6 - \sqrt{26}}{2}$ 13. $-1, \dfrac{5}{2}$ 15. $-1, 0$
17. no real number solution 19. $\dfrac{-5 + \sqrt{13}}{6}, \dfrac{-5 - \sqrt{13}}{6}$
21. $0, \dfrac{12}{7}$ 23. $-2\sqrt{6}, 2\sqrt{6}$ 25. $-\dfrac{2}{5}, \dfrac{2}{5}$
27. $\dfrac{6 + 2\sqrt{6}}{3}, \dfrac{6 - 2\sqrt{6}}{3}$ 29. no real number solution
31. There is no real number solution. 33. $-\dfrac{2}{3}, \dfrac{4}{3}$
35. $\dfrac{-1 + \sqrt{73}}{6}, \dfrac{-1 - \sqrt{73}}{6}$ 37. $1 + \sqrt{2}, 1 - \sqrt{2}$
39. no real number solution 41. 3.5 ft
43. $-8, 16$; Only 16 board feet is a reasonable answer.

Summary Exercises on Quadratic Equations (page 1177)

1. $-6, 6$ 2. $\dfrac{-3 + \sqrt{5}}{2}, \dfrac{-3 - \sqrt{5}}{2}$ 3. $-\dfrac{10}{9}, \dfrac{10}{9}$ 4. $-\dfrac{7}{9}, \dfrac{7}{9}$
5. $1, 3$ 6. $-2, -1$ 7. $4, 5$ 8. $\dfrac{-3 + \sqrt{17}}{2}, \dfrac{-3 - \sqrt{17}}{2}$
9. $-\dfrac{1}{3}, \dfrac{5}{3}$ 10. $\dfrac{1 + \sqrt{10}}{2}, \dfrac{1 - \sqrt{10}}{2}$ 11. $-17, 5$ 12. $-\dfrac{7}{5}, 1$
13. $\dfrac{7 + 2\sqrt{6}}{3}, \dfrac{7 - 2\sqrt{6}}{3}$ 14. $\dfrac{1 + 4\sqrt{2}}{7}, \dfrac{1 - 4\sqrt{2}}{7}$
15. no real number solution 16. no real number solution
17. $-\dfrac{1}{2}, 2$ 18. $-\dfrac{1}{2}, 1$ 19. $-\dfrac{5}{4}, \dfrac{3}{2}$ 20. $-3, \dfrac{1}{3}$
21. $1 + \sqrt{2}, 1 - \sqrt{2}$ 22. $\dfrac{-5 + \sqrt{13}}{6}, \dfrac{-5 - \sqrt{13}}{6}$ 23. $\dfrac{2}{5}, 4$
24. $-3 + \sqrt{5}, -3 - \sqrt{5}$ 25. $\dfrac{-3 + \sqrt{41}}{2}, \dfrac{-3 - \sqrt{41}}{2}$
26. $-\dfrac{5}{4}$ 27. $\dfrac{1}{4}, 1$ 28. $\dfrac{1 + \sqrt{3}}{2}, \dfrac{1 - \sqrt{3}}{2}$
29. $\dfrac{-2 + \sqrt{11}}{3}, \dfrac{-2 - \sqrt{11}}{3}$ 30. $\dfrac{-5 + \sqrt{41}}{8}, \dfrac{-5 - \sqrt{41}}{8}$
31. $\dfrac{-7 + \sqrt{5}}{4}, \dfrac{-7 - \sqrt{5}}{4}$ 32. $-\dfrac{8}{3}, -\dfrac{6}{5}$
33. $\dfrac{8 + 8\sqrt{2}}{3}, \dfrac{8 - 8\sqrt{2}}{3}$ 34. $\dfrac{-5 + \sqrt{5}}{2}, \dfrac{-5 - \sqrt{5}}{2}$
35. no real number solution 36. no real number solution
37. $-\dfrac{2}{3}, 2$ 38. $-\dfrac{1}{4}, \dfrac{2}{3}$ 39. $-4, \dfrac{3}{5}$ 40. $-3, 5$ 41. $-\dfrac{2}{3}, \dfrac{2}{5}$
42. $-4, 6$

Section 17.4 (page 1185)

1. The vertex of a parabola is the lowest or highest point on the graph.
3. vertex: $(0, 0)$

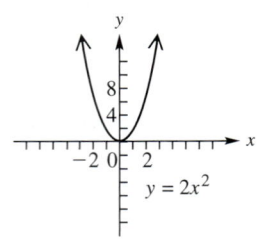

5. vertex: $(0, -4)$

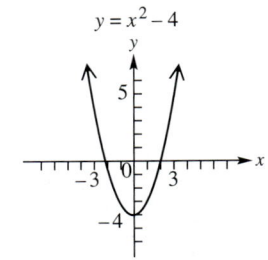

7. vertex: $(0, 2)$

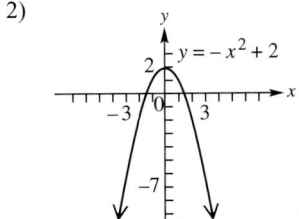

9. vertex: $(-3, 0)$

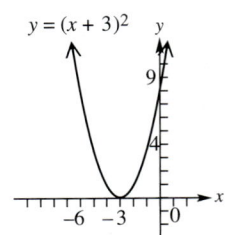

11. vertex: $(-1, 2)$

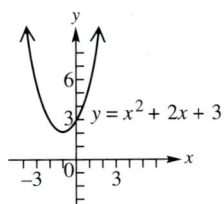

13. vertex: $(3, 4)$

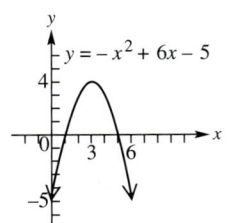

15. If $a > 0$, it opens upward, and if $a < 0$, it opens downward.
17. $y = \dfrac{11}{5625}x^2$

Section 17.5 (page 1193)

1. $3; 3; (1, 3)$ 3. $5; 5; (3, 5)$ 5. The graph consists of the five points $(0, 2), (1, 3), (2, 4), (3, 5),$ and $(4, 6)$. 7. not a function; domain: $\{-4, -2, 0\}$; range: $\{3, 1, 5, -8\}$
9. function; domain: $\{A, B, C, D, E\}$; range: $\{2, 3, 6, 4\}$ 11. not a function; domain: $\{-4, -2, 0, 2, 3\}$; range: $\{-2, 0, 1, 2, 3\}$
13. function 15. not a function 17. function
19. not a function 21. $(2, 4)$ 22. $(-1, -4)$ 23. $\dfrac{8}{3}$
24. $f(x) = \dfrac{8}{3}x - \dfrac{4}{3}$ 25. (a) 11 (b) 3 (c) -9
27. (a) 4 (b) 2 (c) 14 29. (a) 2 (b) 0 (c) 3
31. $\{(1970, 9.6), (1980, 14.1), (1990, 19.8), (2000, 28.4)\}$; yes
33. $g(1980) = 14.1$ (million); $g(1990) = 19.8$ (million)
35. For the year 2002, the function predicts 30.3 million foreign-born residents in the United States.

36. yes

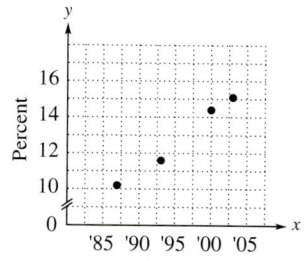

37. $y = 0.30625x - 598.32$ **38.** 1993: 12.0%; 2000: 14.2%
39. $y = 0.4x - 785.6$ **40.** 1987: 9.2%; 2003: 15.6%; The equation from Exercise 38 gives better approximations. The results in Exercise 38 vary by 0.4% and 0.2% from the data. The results here give answers that vary by 1.0% and 0.5%.

Chapter 17 Review Exercises (page 1201)

1. $-12, 12$ **2.** $-\sqrt{37}, \sqrt{37}$ **3.** $-8\sqrt{2}, 8\sqrt{2}$ **4.** $-7, 3$
5. $3 + \sqrt{10}, 3 - \sqrt{10}$ **6.** $\dfrac{-1+\sqrt{14}}{2}, \dfrac{-1-\sqrt{14}}{2}$
7. no real number solution **8.** $-\dfrac{5}{3}$ **9.** $-5, -1$
10. $-2 + \sqrt{11}, -2 - \sqrt{11}$ **11.** $-1 + \sqrt{6}, -1 - \sqrt{6}$
12. $\dfrac{-4+\sqrt{22}}{2}, \dfrac{-4-\sqrt{22}}{2}$ **13.** $-\dfrac{7}{2}$ **14.** no real number solution **15.** 2.5 seconds **16.** 6, 8, 10 **17.** $\left(\dfrac{k}{2}\right)^2$ or $\dfrac{k^2}{4}$
18. (a) $-3, 3$ (b) $-3, 3$ (c) $-3, 3$ (d) We will always get the same results, no matter which method of solution is used.
19. $\dfrac{-1+\sqrt{29}}{4}, \dfrac{-1-\sqrt{29}}{4}$ **20.** $\dfrac{2+\sqrt{10}}{2}, \dfrac{2-\sqrt{10}}{2}$
21. $\dfrac{1+\sqrt{21}}{10}, \dfrac{1-\sqrt{21}}{10}$ **22.** $\dfrac{-3+\sqrt{41}}{2}, \dfrac{-3-\sqrt{41}}{2}$
23. $-3 + \sqrt{19}, -3 - \sqrt{19}$ **24.** The $-b$ term should be above the fraction bar.
25. vertex: (0, 0)

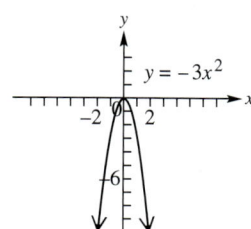

26. vertex: (0, 5)

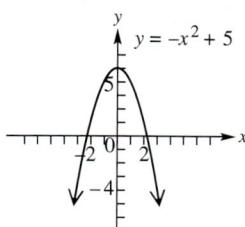

27. vertex: (1, 0)

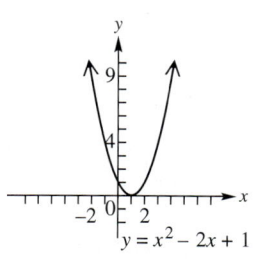

28. vertex: (1, 4)

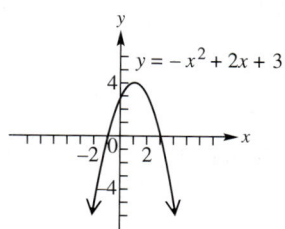

29. vertex: $(-2, -2)$

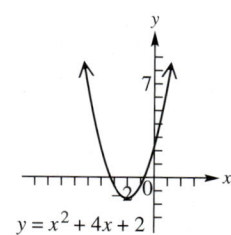

30. vertex: $(-4, 0)$

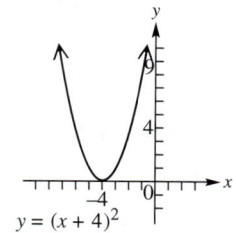

31. $y = \dfrac{3}{1000}x^2$ **32.** not a function; domain: $\{-2, 0, 2\}$; range: $\{4, 8, 5, 3\}$ **33.** function; domain: $\{8, 7, 6, 5, 4\}$; range: $\{3, 4, 5, 6, 7\}$ **34.** not a function **35.** function
36. function **37.** function **38.** not a function **39.** (a) 8 (b) -1 **40.** (a) 7 (b) 1 **41.** (a) 5 (b) 2 **42.** 400 or 800
43. (6, 10) **44.** demand: 600; price: $10 **45.** $-\dfrac{11}{2}, 5$
46. $-\dfrac{11}{2}, \dfrac{9}{2}$ **47.** $\dfrac{-1+\sqrt{21}}{2}, \dfrac{-1-\sqrt{21}}{2}$ **48.** $-\dfrac{3}{2}, \dfrac{1}{3}$
49. $\dfrac{-5+\sqrt{17}}{2}, \dfrac{-5-\sqrt{17}}{2}$ **50.** $-1 + \sqrt{3}, -1 - \sqrt{3}$
51. no real number solution **52.** $\dfrac{9+\sqrt{41}}{2}, \dfrac{9-\sqrt{41}}{2}$ **53.** $-\dfrac{6}{5}$
54. $-1 + 2\sqrt{2}, -1 - 2\sqrt{2}$ **55.** $-2 + \sqrt{5}, -2 - \sqrt{5}$
56. $-2\sqrt{2}, 2\sqrt{2}$

Chapter 17 Test (page 1205)

1. $-\sqrt{39}, \sqrt{39}$ **2.** $-11, 5$ **3.** $\dfrac{-3+2\sqrt{6}}{4}, \dfrac{-3-2\sqrt{6}}{4}$
4. $2 + \sqrt{10}, 2 - \sqrt{10}$ **5.** $\dfrac{-6+\sqrt{42}}{2}, \dfrac{-6-\sqrt{42}}{2}$ **6.** $-3, \dfrac{1}{2}$
7. $\dfrac{3+\sqrt{3}}{3}, \dfrac{3-\sqrt{3}}{3}$ **8.** no real number solution
9. $\dfrac{5+\sqrt{13}}{6}, \dfrac{5-\sqrt{13}}{6}$ **10.** $1 + \sqrt{2}, 1 - \sqrt{2}$
11. $\dfrac{-1+3\sqrt{2}}{2}, \dfrac{-1-3\sqrt{2}}{2}$ **12.** $\dfrac{11+\sqrt{89}}{4}, \dfrac{11-\sqrt{89}}{4}$
13. 5 **14.** 2 seconds **15.** 12, 16, 20

16. vertex: (3, 0)

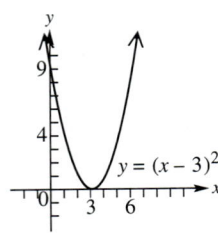

17. vertex: (−1, −3)

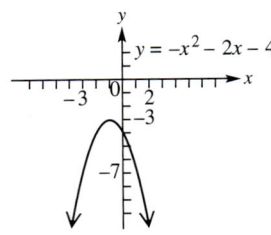

18. (a) not a function (b) function; domain: {0, 1, 2}; range: {2}
19. not a function **20.** 1

WHOLE NUMBERS COMPUTATION: PRETEST

(page 1207)

Adding Whole Numbers
1. 390 **2.** 13,166 **3.** 3053 **4.** 117,510 **5.** 688,226

Subtracting Whole Numbers
1. 350 **2.** 629 **3.** 24,894 **4.** 3909 **5.** 591,387

Multiplying Whole Numbers
1. 0 **2.** 26,887 **3.** 1,560,000 **4.** 1846 **5.** 17,232
6. 519,477

Dividing Whole Numbers
1. 23 **2.** undefined **3.** 6259 **4.** 807 **R6** **5.** 34 **6.** 50
7. 60 **R20** **8.** 539 **R62**

CHAPTER R

Section R.1 (page 1215)
1. 8928; 5715 + 3213 = 8928 **3.** 59,224; 21,020 + 38,204 = 59,224 **5.** 150 **7.** 1651 **9.** 1004 **11.** 9253 **13.** 11,624
15. 16,658 **17.** 5009 **19.** 3506 **21.** 15,954 **23.** 10,648
25. incorrect; should be 769 **27.** correct **29.** 33 miles
31. 38 miles **33.** $122 **35.** 699 people **37.** 54,723,468 people
39. 294 inches **41.** 708 feet

Section R.2 (page 1223)
1. incorrect; should be 62 **3.** incorrect; should be 121 **5.** 8
7. 25 **9.** 16 **11.** 61 **13.** 519 **15.** 9177 **17.** 7589
19. 8859 **21.** 3 **23.** 23 **25.** 1942 **27.** 5687 **29.** 19,038
31. 65,556 **33.** 19,984 **35.** incorrect; should be 2494
37. correct **39.** 15 calories **41.** 121 passengers
43. (a) Highest is electrical engineer; lowest is photographer.
(b) $42,560 **45.** 365 feet

Section R.3 (page 1231)
1. 9 **3.** 63 **5.** 0 **7.** 24 **9.** 36 **11.** 245 **13.** 168
15. 19,092 **17.** 258,447 **19.** 3750 **21.** 44,550 **23.** 270,000
25. 86,000,000 **27.** 1496 **29.** 3735 **31.** 15,200 **33.** 32,805
35. 183,996 **37.** 3,321,934 **39.** 18,980,480
41. 252 inches; 540 inches **43.** 576 plants **45.** 418 miles
47. 2440 miles; 16,170 miles **49.** 7406 calories

Section R.4 (page 1241)
1. 1; $12\overline{)12}$; 12 ÷ 12 **3.** undefined; $\frac{24}{0}$; $0\overline{)24}$ **5.** 0; $4\overline{)0}$; 0 ÷ 4
7. 0; $\frac{0}{12}$; $12\overline{)0}$ **9.** undefined; $\frac{21}{0}$; 21 ÷ 0 **11.** 27; dividend: 108;
divisor: 4; quotient: 27. **13.** 36; dividend: 324; divisor: 9; quotient: 36
15. 1522 **R5** **17.** 309 **19.** 1006 **21.** 5006 **23.** 6671
25. 12,458 **R3** **27.** 10,253 **R5** **29.** 18,377 **R6** **31.** incorrect; should be 670 **R2** **33.** incorrect; should be 3568 **R1** **35.** correct
37. correct **39.** 74 ounces; 103 ounces **41.** $16,600
43. 378 $5 tickets; 270 $7 tickets; 210 $9 tickets **45.** $7; $210
47. $23; $14; $19 **49.** 2, 3, 5, 10 **51.** 2 **53.** 5 **55.** 3
57. 2, 3 **59.** none of the above

Section R.5 (page 1247)
1. $24\overline{)768}$ $\overset{3}{}$ use 2 as the trial divisor. **3.** $18\overline{)4500}$ $\overset{2}{}$ use 2 as the trial divisor.
5. $86\overline{)10,327}$ $\overset{1}{}$ use 9 as the trial divisor. **7.** $52\overline{)38,025}$ $\overset{7}{}$ use 5 as the trial divisor.
9. $77\overline{)249,826}$ $\overset{3}{}$ use 8 as the trial divisor. **11.** $420\overline{)470,800}$ $\overset{1}{}$ use 4 as the trial divisor.
13. 207 **R5** **15.** 236 **R29** **17.** 2407 **R1** **19.** 1239 **R15**
21. 3331 **R110** **23.** 850 **25.** incorrect; should be 106 **R7**
27. incorrect; should be 658 **29.** incorrect; should be 62
31. correct **33.** 1050 hours **35.** $1372 **37.** $308
39. 1680 circuits **41.** $375

Chapter R Review Exercises (page 1251)
1. 92 **2.** 113 **3.** 5464 **4.** 15,657 **5.** 39 **6.** 184 **7.** 224
8. 25,866 **9.** 48 **10.** 45 **11.** 48 **12.** 0 **13.** 172 **14.** 5467
15. 32,640 **16.** 465,525 **17.** 19,200 **18.** 25,200 **19.** 206,800
20. 128,000,000 **21.** 612 **22.** 1872 **23.** 13,755 **24.** 30,184
25. 6 **26.** 1 **27.** undefined **28.** 0 **29.** 108 **30.** 24 **31.** 352
32. 150 **R4** **33.** 104 cards; 312 cards **34.** 2856 textbooks
35. $12 over his budget **36.** $227,940 **37.** 10,000 hours; 24,000 hours **38.** 240 miles; 560 miles **39.** 23 acres
40. 32 homes **41.** $810 **42.** $143 **43.** $663 **44.** $1261

Chapter R Test (page 1253)
1. addends; sum or total **2.** factors; product **3.** difference; quotient
4. 8610 **5.** 112,630 **6.** 16,948 **7.** 3084 **8.** 140 **9.** 171,000
10. 1785 **11.** 4,450,743 **12.** 206 **13.** undefined **14.** 458
15. 160 **16.** $528 **17.** $291 **18.** 35 hours **19.** $165
20. 1028 ovens **21.** $2040 **22.** A number is divisible by 2 if it ends in 0, 2, 4, 6, or 8. A number is divisible by 5 if it ends in 0 or 5. A number is divisible by 10 if it ends in 0. Examples will vary.

APPENDIX A

Appendix A Exercises (page A-5)
1. 37; add 7 **3.** 26; add 10, subtract 2 **5.** 16; multiply by 2
7. 243; multiply by 3 **9.** 36; add 3, add 5, add 7, add 9, etc.; or $1^2, 2^2, 3^2, 4^2$, etc.

11. 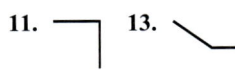 **13.**

15. Conclusion is valid.

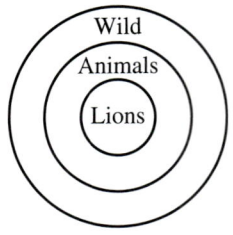

17. Conclusion is invalid.

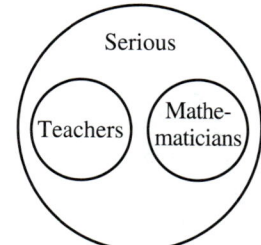

19.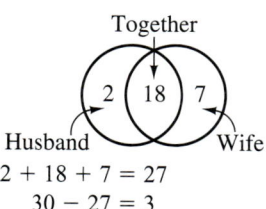
$2 + 18 + 7 = 27$
$30 - 27 = 3$
Neither watched TV on 3 days.

21. Dick

APPENDIX B

Appendix B Exercises (page A-11)
1. $d = 20t$

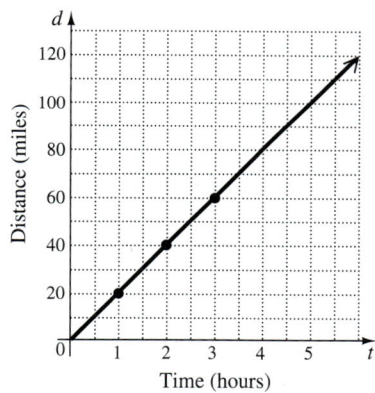

3. $P = 12t$

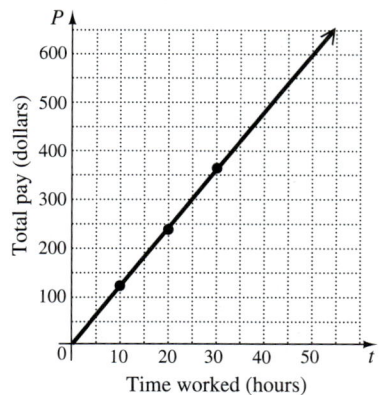

5. $C = 2.50g$

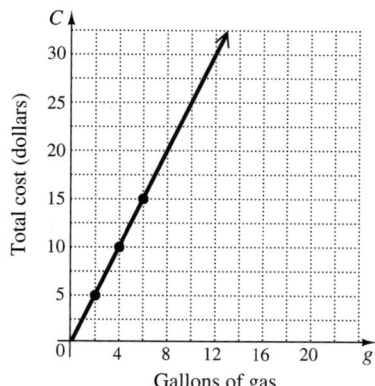

7. $t = \dfrac{100}{r}$

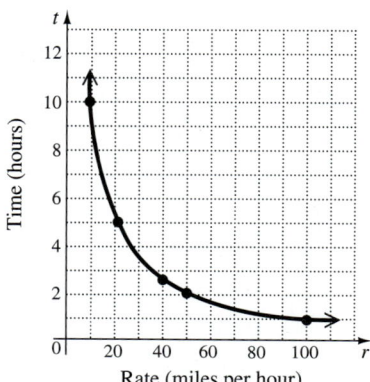

9. $l = \dfrac{60}{w}$

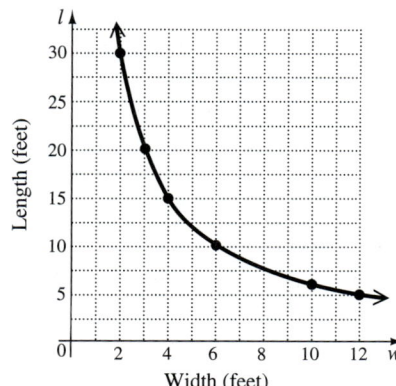

11. 25 hours   **13.** About $27
15. About 3.5 hours   **17.** 8 feet
19. (a) Graphs of direct variation are straight lines. Graphs of inverse variation are curves. (b) Exercises 1–6 illustrate direct variation: Exercises 7–10 illustrate inverse variation.

Appendix C

Appendix C Exercises (page A-17)

1. $y < -x^2 - 4x + 5$

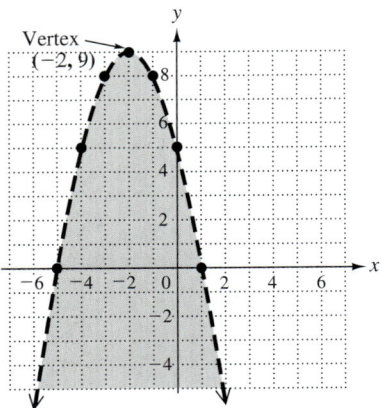

3. $y \leq 2x^2 - x - 1$

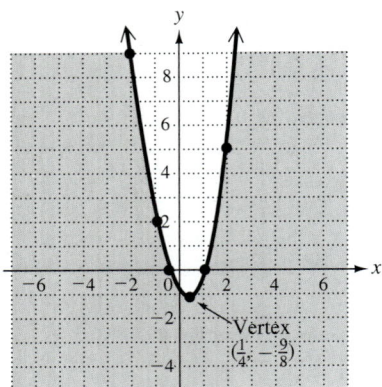

5. $y > -7 + 6x - x^2$

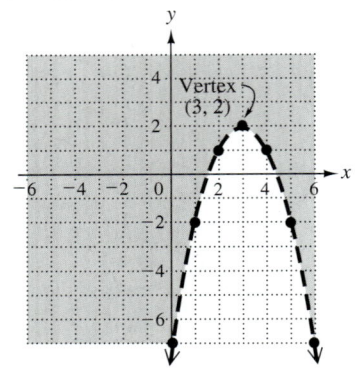

7. Select a test point not on the parabola. Substitute for x and y in the inequality. If the result is a true statement, shade the region containing the test point. If the result is false, shade the other region.

9. $y \geq x^2 - 9$

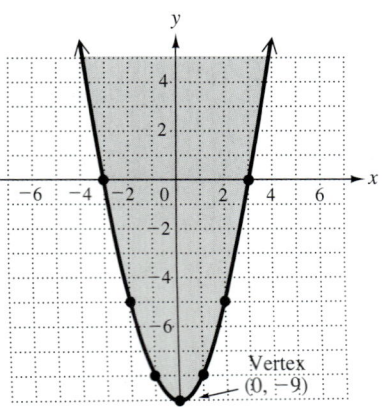

11. $y \leq -x^2 + x + 6$

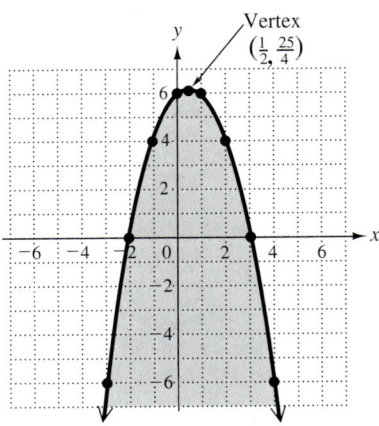

13. $y > -x^2 - 2x + 1$

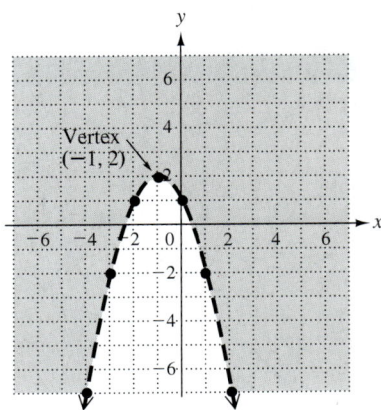

15. $y \geq x^2 + 4x + 2$

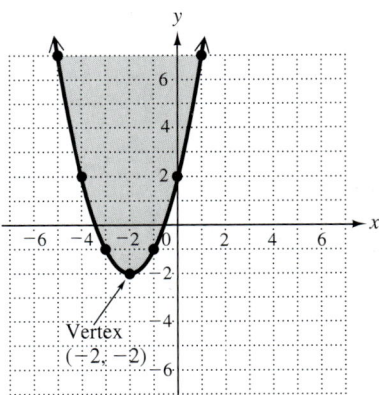

Index

Note: page numbers with A indicate appendix pages; page numbers with n indicate a note

A

Absolute value
  addition of, 15–17
  defined, 11, 71
  fractions, 197
  integers, 11–12, 15–17
  on number line, 11, 73
  symbol for, 71
Acute angles, 456, 477
Addends
  defined, 15, 71, 1209, 1249
  grouping, 18–19, 1210
Addition. *See also* Addition property of equality
  applications and application problems, 1213
  associative property, 18–19, 39, 71, 73, 1210, 1249, 1250
  on calculators, 26
  carrying, 1210–1212, 1249
  changing with subtraction, 25–26, 1217–1218
  checking answers, 1214
  checking subtraction answers, 1218–1219, 1221–1222
  commutative property, 18, 71, 73, 1209–1210, 1249, 1250
  decimals, 317–322, 397
  equations, solving with, 107–112
  estimation of, 32, 74
  exponential expressions, 875, 881
  fractions, 231–237
  integers, 15–19, 26
  like fractions, 231–232
  like terms, 864, 866
  mixed numbers, 249–250
  multivariable polynomials, 868–869
  on number line, 15–16
  polynomials, 864–868, 881, 924
  product of sum and difference of two terms, 890–891, 924
  property of inequality, 710–711, 713–714
  property of zero, 17–18, 71, 73
  radicals, 1107, 1143
  rational expressions, 1023–1025, 1075
  signed fractions, 231–237
  signed numbers, 15–19, 73
  symbol for, in algebra, 86
  two integers with different sign, 16–17, 73
  two integers with same sign, 15–16, 73
  unlike fractions, 233, 235–237
  whole numbers, 317, 1209–1214
Addition property of equality
  decimals, 375–377
  defined, 109, 137
  fractions, 269–270
  integers, 109–118, 127–129, 139
Addition property of inequality, 710–711, 713–714
Addition property of zero, 17–18, 71, 73
Additive inverse of integers, 25
Adjacent angles, 459
Aerobic fitness and heart rate, 262
Algebra
  algebraic expressions, 102, 167–168, 1086
  arithmetic connections, 180, 886
  expression of multiplication, 39, 875
Algebraic equations, 180
Alternate interior angles, 460–462, 478
Altitude. *See* Height
Amount, percent proportion, 509
Angle-Side-Angle (ASA) method of congruence, 468–469, 483
Angles
  acute, 456, 477
  adjacent, 459
  alternate interior, 460–462, 478
  applications and application problems, 453–462
  classification, 455–456
  complementary, 457–458, 478, 483
  congruent, 458–459, 478, 483
  corresponding, 460–462, 478
  defined, 454, 477
  geometric applications, 453–462
  identifying and naming, 454–455
  measurement, 455–456, 628–630, 665
  nonadjacent, 459
  obtuse, 456, 477
  right, 148, 183, 370, 455, 456, 477
  straight, 455, 456, 477
  supplementary, 458, 478, 483
  vertical, 458–460, 478, 483
Applications and application problems, 147–192. *See also* Problem solving; Real-Data Applications
  addition, 1213
  area, 157–162, 386
  arithmetic to algebra connections, 180
  circle, area of, 386
  circumference, 385
  completing the square, quadratic equations, 1166
  consumer, 539–542, 551, 554
  decimals, 378
  discounts, 493, 541–542, 551, 554
  distance, rate and time problems, 1055–1059, 1077
  division, with remainders, 52–53
  English measurement, 568–569, 606
  estimation of, 44
  exponents, 878, 919–920
  formula to estimate variables, 704
  formulas, 172, 704, 1090–1091
  fractions, 226
  front end rounding, 44
  functions, 1191–1192
  geometric application of exponents, 878
  indicator words to solve, 226
  inequalities, 714
  key terms and summary, 183–188
  linear inequalities, 714
  linear systems, 835–840, 855
  metric measurement, 593–594, 608
  mixed numbers, 251
  multiplication, 44, 1230
  one unknown quantity, 167–171, 187
  parabolas, 979, 980, 1183–1184
  with parabolas, 1183–1184
  percent, 529–534, 553
  perimeter, 148–152, 162
  population and functions, 1191–1192
  proportions, 445–447, 481–482
  Pythagorean formula, 370, 1090–1091
  quadratic equations, 975–980, 989, 1166
  rational expressions, 999, 1059–1061, 1065, 1067–1068, 1077, 1078
  review exercises, 189–190
  right triangles, 369–370
  sales price, 493, 541–542, 546, 551, 554
  sales tax, 493, 539–540, 545, 551, 554
  scientific notation, 863, 902, 919–920, 921
  similar triangles, 472
  subtraction, 1222

I-1

test, 191–192
two unknown quantities, 177–179, 188
variation problems, 1067–1068, 1078, A–7 to A–10
velocity problems, 1166
work problems, 999, 1059–1061, 1065, 1077
"Approximately equal to" sign, 32, 309, 395, 1141
Area. *See also* Surface area
applications and application problems, 157–162, 277–280
circle, 382–385, 388, 396, 401
defined, 157, 183
division of polynomial by polynomial, 914
exponent rules, 878
to find missing length, 186
formula, 183, 186
geometric applications, 277–280
parallelogram, 161–162
quadratic equations, 975–976
Real-Date Application, 892
rectangle, 157–159, 161
of semicircle, 385
square, 159–160
triangle, 277–279, 287
unit symbol, 183
vs. perimeter, 157
Argument, deductive reasoning, A–2 to A–4
Arithmetic
connections to algebra, 180, 886
division symbol, 49
multiplication symbol, 39
ASA (Angle-Side-Angle) method of congruence, 468–469, 483
Associative property
of addition, 18–19, 39, 71, 73, 1210, 1249, 1250
of multiplication, 43, 71, 74, 98–99, 875, 877, 881, 1226, 1249, 1250
Asterisk, as multiplication symbol, 1225
Average (mean), 357–360, 395, 396, 399
Average (mean), weighted, 358–359, 395
Axis. *See also* x-axis; y-axis
of coordinate system, 645–647, 670
defined, 1179, 1197
horizontal or vertical, 665
of parabola, 1179

B

Bar graphs, 637–638, 665, 669, 730, 731
Bar notation for repeating digits, 395
Base
of exponents, 59, 873, 874
of parallelogram, 160–162
of percent proportion, 509
powers of, 59
product rule for exponents, 874
of pyramid, 280–281
quotient rule for exponents, 899
of triangle, 277–278

Best buy, based on cost per unit, 424–426, 480
Billions, 2, 3, 4
Bimodal list, 360
Binomials. *See also* Polynomials
defined, 923
FOIL method, 883–885
greater powers of, 891
multiplication of, 883–885, 924
square of, 889–890, 924
square root property of equations, 1155–1156
squaring a binomial, radical equations, 1133–1135
Boiling point, 598–599
Borrowing, 1218, 1219–1222, 1249
Boundary line, graph of inequality, 785, 787–788, 793
Braces, in relations, 1187
Brackets, order of operations, 688, 720
Break even point, 807, 824
Brin, Sergey, 902
Business break even point, 807, 824

C

Calculators
addition, 26
area of circle, 385
change of sign key, 26, 42
circles, 383, 385
circumference of circle, 383
completing the square, quadratic equations, 1165
decimals, 304, 322, 328, 337, 341, 349, 351
discounts, 542
division, 42, 49, 50, 52, 53, 1235
exponents, 60
expressions, 87–88
fractions, 221–222, 349, 500–501
fractions, to percents, 500–501
front end rounding, 34
higher roots, 1091
linear function, 771
mixed numbers, 250
money amounts, 313, 328
multiplication, 42, 221–222, 1226, 1230
negative signs, 9
order of operations, 61
per unit rates, 425
percents from fractions, 500–501
prime factorization, 212
rounding, 34, 313
sales price, 542
scientific notation, 919, 920
signed numbers, 26
simple interest, 544
slope-intercept form, 771
square roots, 367–368, 1087
subtraction on, 26
unit rates, 425

Capacity, measurement of. *See* Volume
Carrying
with addition, 1210–1212, 1249
defined, 1249
with multiplication, 1226–1227
Cartesian (rectangular) coordinate system
defined, 645, 736, 793
plotting points, 646–648, 670, 736–738, 745–747, 795
Cassette vs. compact discs sales, 850
Cell phones
call plans, 411, 429, 430
numbers owned by Americans, 933, 985
Celsius, 598–600, 605, 609–610
Celsius-Fahrenheit conversion formula, 599, 605, 609
Center of circle, 381
Centimeter (cm), 575, 576, 577
Central tendency. *See* Measures of central tendency
Chain calculations, multiplication, 1226
Check the solutions
addition, 1214
defined, 137
division, 1237–1238, 1246
linear equations, 694
multiplication, to check division, 1237–1238, 1246
solving equations, 111–112, 694
subtraction, 1218–1219, 1221–1222
Circle graph, 627–630
construction using protractor, 628–630, 668
defined, 627, 665
finding ratio from, 627–628
read and interpret, 627–628, 730
Circles, 381–386
area, 382–385, 388, 395, 396, 401
center of, 381
circumference, 382–383, 395, 396, 401
defined, 381, 395, 401
diameter, 381–382, 395, 396, 401
Euler, A–3
radius, 381–382, 384, 386, 395, 396, 401
Circumference, 382–383, 395, 396, 401
Classifying polynomials, 865
Coefficients
completing the square, 1163–1165
defined, 86, 137, 864
factoring trinomials, 943–944, 949–950
fractional, division of polynomials with, 914
fractional, quadratic equations with, 1173
identifying like terms, 96–97, 138
multiplication expressions, 86
numerical, 864
quadratic formula with, 1173
substitution method, linear system equations, 821–822
College student credit card debt, 729, 752
Combining like terms, 95–98, 138

Common denominator. *See also* Least common denominator (LCD)
  addition and subtraction, 231–232
  defined, 231
  like fractions, 231–232
  unlike fractions, 233
Common factors. *See also* Greatest common factor (GCF)
  defined, 207, 934
  division of, 208, 224–225, 413
  fractions in lowest terms, 207–214
Common sign, 15
Commutative property
  of addition, 18, 71, 73, 89, 1209–1210, 1249, 1250
  expression simplification, 98
  of multiplication, 43, 71, 74, 89, 98, 210, 875, 877, 881, 1225, 1249, 1250
Compact discs vs. cassette sales, 850
Comparison line graph, 639, 665
Complementary angles, 457–458, 478, 483
Completing the square, quadratic equations, 1161–1166
  applications and application problems, 1166
  on calculator, 1165
  with coefficients, 1163–1165
  defined, 1162, 1198
  simplify before solving, 1165
  square root property, 1161–1162, 1198
  steps in solving, 1164
Complex fractions, 261, 283, 1033–1038, 1076
Complex numbers, 1165n
Components of an ordered pair, 1187, 1197
Composite numbers, 209, 287
Compound interest, 550
Conclusion, deductive reasoning, A–2 to A–4
Congruence, Angle-Side-Angle (ASA) method of, 468–469, 483
Congruent angles, 458–459, 478, 483
Congruent figures, 467, 478
Congruent triangles, 467–469, 478, 483
Conjugates, 1121–1122, 1141
Consecutive integers, 976–978, 987
Consistent system of linear equations, 810, 811, 820, 853
Constant terms, 95, 98, 864
Constants, 84–85, 137
Consumer applications, 539–542
  discounts, 493, 541–542, 551, 554
  sales price, 493, 541–542, 546, 551, 554
  sales tax, 493, 539–540, 545, 551, 554
  simple interest, 542–544, 551, 554
  tips, 540–541, 551, 554
Conversions. *See* Measurement conversions
Coordinate system
  axes of, 645–647, 670
  defined, 646, 665
  ordered pairs, 645–648, 665, 670
  plotting points, 646–648, 670, 736–738, 745–747, 795
  quadrants, 648, 670
  rectangular, 645–648, 736, 793
Coordinates, 665, 736, 793. *See also* Rectangular coordinate system
Corresponding angles, 460–462, 478
Corresponding parts, triangles, 467–469
Cost and quantity problems, linear system, 836–837
Cost comparisons, wedding, 54
Cost per unit, 424–426, 477
Counting (natural) numbers, 684
Credit card debt, college student, 729, 752
Cricket chirps and temperature, 83, 91
Cross products, 432–437, 477
Cube roots, 59, 1091, 1114–1115, 1148
Cube roots, perfect, 1091, 1101, 1141
Cubes (rectangular solid), 279–280
Cubic centimeter (cc), 584
Cubic units, 279, 287
Currency exchange, 622
Curves, graph of, A–9 to A–10. *See also* Parabolas
Cylinders, 386–388
  right circular, 386
  surface area, 388, 396, 402
  volume, 386, 396, 401

D

Data. *See also* Tables
  interpretation, 617, 644
  paired, 645, 665
  range of, 361
  variability of, 361–362
Decimal places, 309–313, 395
Decimal points
  in addition, 317
  defined, 300, 395
  integers vs. decimals, 302
  metric unit conversion, 578
  multiplication, 327
  percents, 496
  reading, 303, 397
  rounding, 309
Decimals, 299–352
  addition of, 317–322, 397
  applications and application problems, 378
  arranged in order, smallest to largest, 352
  on calculator, 304, 322, 328, 337, 341, 349, 351
  clearing, to solve linear equations, 698–699, 721
  defined, 300, 395
  division of, 335–341, 376–377, 398
  equations with, 375–378, 400
  estimation of, 321–322, 329, 339
  fractions, decimals written as, 303–304
  fractions written as decimals, 300–301, 349–352, 397, 398
  as irrational, 686
  irrational number approximation, 1089
  key terms and summary, 395–400
  linear equations with, 698–699, 721
  mixed number form, 303–304, 543
  multiplication of, 327–329, 398
  on number line, 351–352
  order of operations, 340–341
  percents, decimals written as, 496–497, 552
  percents written as decimals, 494–496, 552
  place value chart, 301–302, 397
  proportions, 436–437
  as rational numbers, 685
  ratios, 414–415
  reading and writing, 4, 300–304, 397
  Real-Data Applications, 314, 330, 342
  repeating, 337, 395, 685
  review exercises, 403–408
  rounding, 309–313, 337–338, 350, 397
  scientific notation, 917
  signed, 317–322, 327–329, 335–341, 397–398
  size, vs. fractions, 351–352, 399
  subtraction of, 317, 320–322, 398
  test, 409–410
Decrease, percent of, 532–534, 551
Deductive reasoning, A–2 to A–4
Degrees
  of an angle, 455–456
  defined, 477
  first-degree equation (*See* Linear equations)
  of a polynomial, 865, 923
  second-degree equation (*See* Quadratic equations)
  of a term, 865, 923
Denominators. *See also* Fractions; Least common denominator (LCD); Rationalizing the denominator
  common, 231–232
  defined, 195, 287
  ratios, 412–413
  unit fractions, 566, 606
Dependent equations, 810, 811, 820–821, 853
Dependent systems. *See* Linear equations
Descartes, Rene, 736
Descending powers, 864, 911–913, 923
Diameter, 381–382, 395, 396, 401
Difference. *See also* Subtraction
  defined, 1218, 1249
  product of sum and difference of two terms, 890–891, 924
  of squares, factoring, 959–960, 962, 989
Digits
  defined, 2, 71
  place value, 2–4, 301–302
  repeating, 337, 395, 685
  writing numbers in, 3–4

I-3

Direct variation, 1067, 1073, A–7 to A–8
Discounts, 493, 541–542, 546, 551, 554
Distance, rate, and time, 147, 156, 438, 838–840, 1055–1059, 1077, A–7 to A–9
Distance formula, 147, 156, 838
Distance measure, 20
Distributive property
   binomials, multiplication of, 883–884
   defined, 43–44, 71
   equation solving with, 129–130, 139
   linear equations in one variable, 694–699
   polynomials, multiplication of, 881, 883
   signed, 689–690, 696
   simplification, 96, 99–101, 689–691, 720
Dividend, 222, 1233–1234, 1249
Divisibility, tests for, 211, 1238–1240
Division. *See also* Division property of equality
   applications and application problems, 52–53
   on calculator, 42, 49, 50, 52, 53, 1235
   changed to multiplication, 1234
   checking answers, 1237–1238, 1246
   common factors, 208, 224–225, 413
   complex fractions, 261, 283, 1033–1035, 1076
   decimals, 335–341, 376–377, 398
   divisibility tests, 211, 1238–1240
   equation solving with, 119–122
   estimation of, 51–52
   exponents, quotient rule for, 898–901, 925
   fractions, 199–200, 222–226, 261, 283, 1033–1035, 1076
   indicator words for, 226
   integers, 49–53, 75, 335–338
   long, 1243–1246, 1249
   mixed numbers, 246–248
   by multiples of ten, 1240, 1245–1246
   of number by itself, 1236
   parts of a problem, 1233–1234
   polynomial by monomial, 907–908, 925
   polynomial by polynomial, 911–914, 925
   properties of, 50
   quotient rule for exponents, 898–901, 925
   radical expression quotients in lowest terms, 1123
   radicals and quotient rule, 1099–1100, 1112, 1143
   rational expressions, 1010–1013, 1074
   rational numbers, 685–685
   reciprocal of fractions, 222–223, 283
   remainders, 52–53, 75, 1236
   with scientific notation, 919
   short, 1236–1237, 1249
   signed decimals, 335–341, 398
   signed fractions, 222–226
   signed numbers, 50
   simplification of complex fractions, 1033–1035, 1076
   symbols for, 86, 1233
   trial quotients and divisors, 1243–1244
   unit conversion measurement problems, 564–565
   whole numbers, 1233–1246, 1250
   with zero, 50, 336, 1234–1235
Division method for factoring, 210
Division property of equality
   decimals, 376–377
   defined, 137, 287
   fractions, 267–270
   integers, 119–122, 127–129, 137, 140
Divisors
   defined, 222, 1233–1234, 1249
   division by zero, 1235
   factors, and greatest common divisor, 934
   trial, 1243
Dollar-cost averaging, 342
Dollars and cents, 311–313, 328
Domain of relation, 1187, 1188, 1191, 1197
Dot, as multiplication symbol, 39, 1225
Double-bar graphs, 637–638, 665
Drawing. *See* Graphs

E

Earthquake intensity, 926
Educational tax incentives, 556
Elimination method, linear system equations, 825–828
   alternative method to find second value, 827–828
   choice of, 833
   defined, 825, 854–855
   multiplication and, 827
   to solve special systems, 828
   steps in, 826
   use of, 825–827
English system of measurement
   applications and application problems, 568–569, 606
   defined, 564, 605, 606
   length, 417, 564
   temperature, 598–600, 609–610
   time, 417, 564
   unit conversion, 416–417, 564–569
   volume, 417, 564
   weight, 417, 564
Equal sign, 107, 687
Equality. *See also* Addition property of equality; Division property of equality; Inequalities
   "approximately equal to" sign, 32, 309, 395
   multiplication property of, 267–270, 287, 1041–1046, 1076
   principle for equation solving, 119
   squaring property of, 1131–1135
Equations, 107–146, 683–728. *See also* Equality; Inequalities; Linear equations; Percent equations; Quadratic equations; Radical equations
   with addition, 107–112
   checking, 111–112, 694
   checking ordered pair as solution, 733–734
   with decimals, 375–378, 400
   defined, 107, 108, 137
   dependent, 810, 811, 820–821, 853
   distributive property, 129–130
   with division, 119–122
   equal signs, 107
   equality principle for, 119
   equations of lines, 771–776, 796
   extraneous solution of, 1132–1133
   with fractions, 267–271
   goals of, 109
   identify, 108
   independent, 810, 811, 853
   key terms and summary, 137–141, 719–722
   of lines, 771–776, 796
   with percent, 517–522, 552
   point-slope form, 772–775, 796
   with radicals, 1131–1135, 1144
   with rational expressions, 1041–1046, 1076
   Real-Data Applications, 172, 180
   with real numbers and expressions, 684–690, 720
   review exercises, 143–144, 723–726
   from sentences, 168, 187
   with several steps, 127–130, 141
   simplification of, 95–101, 111–112, 121
   slope, 763–764
   slope-intercept form, 763–765, 771–776, 796
   square root property of, 1154–1156, 1161–1162, 1177, 1198
   test, 145–146, 727–728
   variables, 84–90
   variables, solving for specified, 703–706, 721–722
Equivalence
   fractions, 198–200, 283, 287
   rational expression forms, 1005–1006, 1018–1020
Estimates
   addition answers, 32, 74
   decimals, 321–322, 329, 339
   defined, 32, 71
   division, 51–52
   front end rounding, 32–34, 44, 52
   integers, 32–34
   mixed numbers, 246–251
   multiplication, 44, 75
   percents, 517
   subtraction answers, 32, 74
Euler circles, A–3
Evaluate the expression, defined, 137
Even consecutive integers, 976–978
Eves, Howard, 902
Exponential expressions, 138, 873, 898–900, 925
Exponential form, 59

Exponential notation, 59
Exponents. *See also* Negative exponents; Polynomials
   applications and application problems, 919–920
   base of, 59, 873
   on calculators, 60
   defined, 59, 71, 74, 873
   descending powers, 864, 911–913, 923
   expressions with, 59–60, 91, 138, 873, 898–900, 925
   fractional, 259
   greatest common factor (GCF), 936
   identifying like terms, 138
   integers, 59–60, 895–898, 925
   numbers, big and small, 863, 902, 921
   order of operations, 63
   power rule for, 875–878, 900, 924
   product rule for, 873–878, 896, 900, 924
   quotient rule for, 898–901, 925
   Real-Data Applications, 902, 926
   review of use, 873
   rule summary, 877, 924
   scientific notation, 863, 902, 917–920, 921, 923, 925
   signed numbers, 59
   simplification of expressions with, 59–60, 259
   with variables, 89–90, 138
   zero, 895, 900, 923, 925
Expressions. *See also* Radical expressions; Rational expressions; Simplification of expressions
   algebraic, 102, 167–168, 1086
   on calculators, 87–88
   defined, 84, 137
   evaluating, 85–88, 138
   exponential, 59–60, 91, 138, 873, 898–900, 925
   identifying, 84–85
   Real-Data Applications, 102
   real numbers, 684–690, 720
   scientific notation, 917
   substitution, 817–822, 833–834, 854
   terms of, 95–96, 864, 865
   with two variables, 88
Extraneous solution, 1132–1133, 1141

F

Facets, geodesic domes, 1124
Factor tree, 211–212
Factored form, 934, 987
Factoring, 933–998. *See also* Factoring trinomials; Factors; Prime factorization
   defined, 934, 987
   difference of squares, 959–960, 962, 989
   division method, 210
   equation with more than two variable factors, 971
   equation with quadratic factor, 971
   factor tree, 211–212
   by grouping, 937–939, 988
   key terms and summary, 987–989
   perfect square trinomials, 960–962, 989
   polynomials, 936–937, 943–971
   quadratic equation solving, 967–971, 989, 1177
   rational expressions in lowest terms, 1003
   Real-Data Applications, 940, 972, 990
   review exercises, 991–996
   special techniques, 959–962, 989
   test, 997–998
   write fractions in lowest terms, 212–214
Factoring trinomials, 943–972
   all positive terms, 944, 953–954
   coefficient of squared term equal to one, 943–944
   coefficient of squared term not equal to one, 949–950
   common factors, 946, 950, 956
   factoring out greatest common factor (GCF), 946
   FOIL, 953–956, 989
   by grouping, 949–950, 988
   negative last term with FOIL, 955
   negative middle term, 944, 954
   perfect square trinomials, 960–962, 989
   prime polynomials, 945
   steps of, 946
   two negative terms, 945
   two variables, 946, 955
Factors. *See also* Common factors; Greatest common factor (GCF)
   common, 207, 934
   defined, 39, 71, 207, 934, 987, 1249
   factor tree, 211–212
   greatest common factor (GCF), 934–937, 988
   by grouping, 98, 937–939
   of multiplication, 1225
   multiplication of several, 41–42
Fahrenheit
   defined, 598–599, 605
   estimation of, 812
   Fahrenheit-Celsius conversion formula, 599–600, 605, 609–610
Fibonnaci, 1086
Figures
   congruent, 467, 478
   similar, 467, 478
Film ticket sales, 683, 692, 832
First-degree equation. *See* Linear equations
Fishing equipment, 299, 307, 326, 356
Five, divisibility by, 1240
FOIL method
   defined, 883, 923
   factored form of polynomial, 938
   factoring trinomials, 943, 953–956, 989
   multiplication of binomials, 883–885
   steps in, 884, 924
Formulas
   area, 183, 186
   defined, 148, 183
   distance, 147, 156, 838
   exponent rule summary, 877, 924
   Fahrenheit-Celsius conversion, 599–600, 605, 609–610
   Heron's, 1140
   hypotenuse, 396
   inequalities, solving for specified variable, 703–706, 721–722
   interest, 542–543, 551, 554
   linear equations, 703–706, 721–722
   perimeter, 183
   power rule for exponents, 877, 924
   product rule for exponents, 877, 924
   Pythagorean, 978–979, 989, 1089–1091, 1142
   quadratic, 1169–1173, 1199
   rational expressions, 1047–1048
   slope, 761–762, 795
   slope-intercept form, 763–765, 771–776, 796
   variables, solving for specified, 703–706, 721–722, 1047–1048
Fourth root, 1091
Fraction bar, 64, 195
Fractions, 193–298. *See also* Ratios
   absolute value, 197
   addition of, 231–237
   applications and application problems, 226
   on calculators, 221–222, 349, 500–501
   complex, 261, 283, 1033–1038, 1076
   decimals, fractions written as, 300–301, 349–352, 397, 398
   decimals written as fractions, 303–304
   defined, 194, 287
   denominator, 195, 287
   division, 199–200, 222–226, 267–270, 283, 1033–1035, 1076
   equations with, 267–271
   equivalent, 198–200, 283, 287
   exponents, 259, 876–877
   graphing on a number line, 196–197
   improper, 195–196, 244–246, 283, 287
   key terms and summary, 287–294
   like fractions, 231–232, 283, 287
   linear equations with, 698–699, 721, 821–822
   lowest terms, 207–214
   multiplication, 219–222, 226
   as music time signatures, 238
   numerator, 195, 287
   order of operations, 259–260
   percents, fractions written as, 499–501, 552
   percents written as fractions, 497–499, 552
   power rule for exponents, 876–877
   prime factorization to multiply, 220–221

I-5

proper, 195–196, 283, 287
quadratic formula with, 1173
rates, 423, 479
as rational numbers, 684–685
Real-Data Applications, 238, 252, 262, 272, 282
reciprocals, 222–223
repeating decimals, 337, 395, 685
review exercises, 295–296
signed, 194–200
simplification, 200
size, vs. decimals, 351–352, 399
substitution method, linear system equations, 821–822
subtraction of, 231–237
test, 297–298
types of, 195–196
unit, 565–568
unlike fractions, 231, 233–237, 283, 287
Freezing point, 598–599
Front end rounding
  on calculators, 34
  defined, 33, 71, 74
  division, 52
  estimation of, 32–34, 44, 52
  multiplication, 44
Fuller, R. Buckminster, 1124
Functional notation, 1190–1191, 1197
Functions, 1187–1192
  applications and application problems, 1191–1192
  defined, 1187–1188, 1197
  domain, 1200
  input-output machine, 1188
  notation for, 1190–1191, 1197
  range, 1200
  symbol for, 1197
  vertical line test, 1188–1190, 1200
Fundamental property of rational expressions, 1000–1006, 1017–1020, 1074
$f(x)$, 1190–1191, 1197

G

g (gram), 585–586
Galilei, Galileo, 967
Gasoline cost, 1062
GCF. See Greatest common factor (GCF)
Gender and grocery shopping, 640
Geodesic domes, 1124
Geometry and geometric applications. See also Area; Perimeter
  area, 277–280
  circles, 381–386
  congruent and similar triangles, 467, 469–472
  cylinders, 386–388
  defined, 148
  division of polynomial by polynomial, 914
  exponent rules in, 878

key terms and summary, 395–396, 401–402, 477–478, 482–484, 490
lines and angles, 453–462
Pythagorean Theorem, 368–370
quadratic equations, 975–976
review exercises, 406–407, 487–488
square root, 367–368
surface area, 387–388
test, 410, 491–492
two unknown quantities, 178–179, 188
volume, 280–281
Germain, Sophie, 940
Golden Ratio (sacred ratio), 1116
Googol, 902
Googolplex, 902
Gram (g), 585–586, 605
Graphing, defined, 745, 793
Graphs, 617–682. See also Graphs of inequalities; Graphs of linear equations
  bar graph, 637–638, 665, 669, 730, 731
  circle graph, 627–630, 668
  of curves, A–9 to A–10 (See also Parabolas)
  defined, 745, 793
  for direct variation, A–7 to A–8
  double-bar graph, 637–638, 665
  of horizontal lines, 774–775
  for inverse variation, A–8 to A–9
  line graph, 638–639, 665, 669
  numbers on a number line, 10
  ordered pairs, 645–648, 665, 670
  pictograph, 620–621, 665, 667
  pie graph, 627–630, 668
  rational numbers, 685
  Real-Data Applications, 622, 640, 672
  rectangular coordinate system, 645–648, 670, 736–738, 745–747, 795
  signed numbers, 72
  for variation, A–7 to A–10
  of vertical lines, 774–775
Graphs of inequalities. See also Parabolas
  boundary line, 785, 787–788, 793
  linear inequalities, solving, 709–710, 785–788, 796, 847–849, 855
  on number line, 709–710, 722
  quadratic inequalities, A–15 to A–16
  in two variables, 785–788, 793, 796
Graphs of linear equations
  defined, 665, 670
  direct variation, A–8
  intercepts, 748–749
  key terms and summary, 665–670
  linear system equations, 808–811, 854
  rectangular coordinate system, 645–648, 670, 736–738, 745–747, 795
  review exercises, 673–678
  slope as negative/positive, 657–658
  slope-intercept form, 763–765, 771–776, 796
  solution method, 808–810, 854

tables and variables for, 618–619, 651–658, 665, 666, 670
test, 679–682
two variable equations, 651–656, 730–738, 745–752, 785–788, 794–796
variation, A–8
Greater powers of binomials, 891
Greater than. See also Inequalities
  to compare real numbers, 687
  converting to less than, 688
  defined, 72
  linear inequalities, solving, 688, 847–849, 855
  symbol for, 10–11, 71, 687, 719, 720
Greater than or equal to
  graph linear inequalities, 785–788, 796
  symbol for, 70, 687–688, 719
Greatest common factor (GCF), 934–937
  defined, 934, 987
  factoring out, 936–937
  factoring trinomials, 946
  of numbers, 934–935
  steps to find, 936, 988
  of variable terms, 935
Greeks, historic influence on mathematics, 368–370
Grocery shopping, 640
Grouping
  of addends, 18–19, 1210
  associative property of addition, 18–19, 39, 71, 73, 1210, 1249, 1250
  associative property of multiplication, 43, 71, 74, 98–99, 875, 877, 881, 1226, 1249, 1250
  factoring by, 937–939, 988
  factoring trinomials by, 949–950, 988
  order of operations, 61

H

Hair growth (scalp), 580
Heart rate and aerobic fitness, 262
Height
  of cylinder, 386
  of parallelogram, 160–162
  of pyramid, 280–281
  of triangle, 277–278
Heron's formula, 1140
Higher roots, 1091–1092, 1101–1102
Historical ratios, 418
Horizontal axis, 665
Horizontal lines. See also Lines
  graph of, 774–775
  intercepts, 750–751
  linear equations for, 775
  slope of, 762, 763, 795
Hotel expenses, 272
Hummingbird feeding, 448
Hundreds, place value, 2
Hyperbola, A–10
Hypotenuse of a right triangle, 368, 395, 396, 978–979, 989, 1089

I

Identity equation, 697, 719
Identity property of multiplication, 1002
Improper fractions, 195–196, 244–246, 283, 287
Inconsistent system of equations, 810, 820, 853
Increase of percent, 532–533, 551
Independent equations, 810, 811, 853
Index of a radical, 1091
Index (order), 1141
Indicator words, 226
Inductive reasoning, A–1 to A–2
Inequalities, 683–862. *See also* Equality; Graphs of inequalities; Linear inequalities
   addition property of inequality, 710–711, 713–714
   applications and application problems, 714
   converting to greater than or less than, 688
   defined, 72, 709, 719
   key terms and summary, 719–722, 793–796, 853–855
   linear, solving, 847–849, 855
   multiplication property of inequality, 711–714
   quadratic inequalities, A–15 to A–16
   Real-Data Applications, 766, 812, 850, 856
   for real number comparison, 687
   real numbers and expressions, 684–690, 720
   review exercises, 723–726, 797–802, 857–860
   solving linear, 709–714, 722, 847–849, 855
   statement reversal, 688
   symbols for, 10–11, 70, 71, 72, 687–688, 719, 720
   test, 727–728, 803–806, 861–862
   variables, solving for specified, 703–706, 721–722, 1047–1048
Infinite number of solutions, 810
Inner product, of binomials, 883–885, 923
Input-output machine, functions, 1188
Inspection, to find LCD, 234
Integers, 1–81
   absolute value of, 11–12, 15–17
   addition of, 15–17, 15–19, 26, 73
   additive inverse of, 25
   common factor of, 934
   comparing, 10–11, 72
   consecutive, 976–978, 987
   defined, 10, 71, 684
   divide decimal by, 335–338
   division, 49–53, 75, 335–338
   estimation of, 32–34
   exponents, 59–60
   as exponents, 59–60, 895–898, 925
   key terms and summary, 71–76
   multiplication, 39–44, 51, 74, 222
   negative of, 684, 689
   on number line, 10–11, 15–16
   order of operations, 60–64
   place value, 2–4
   as rational numbers, 684–685
   Real-Data Applications, 20, 54
   review exercises, 77–80
   rounding, 29–34, 74, 310
   signed numbers, 9–12
   subtraction of, 25–26, 73
   test, 81–82
   vs. decimals, 302
Intensity of earthquakes, 926
Intercepts. *See also* x-intercept; y-intercept
   defined, 747, 793
   finding, 747–748
   horizontal lines, 750–751
   linear equations, 748–749
   of parabola, 1180–1182
   slope-intercept form, 763–765, 771–776, 796
Interest
   compound, 550
   defined, 551
   formula for, 542–543, 551, 554
   rates of, 542, 551, 1174
   simple, 542–544, 551, 554
Interior angles, alternate, 460–462, 478
Intersecting lines, 454, 477, 847–849, 855
Inverse
   additive, of integers, 25
   multiplicative, of a rational expression, 1010
   variation of, 1067–1068, 1073, A–8 to A–9
Invert and multiply, reciprocals, 223–224
Investment growth, 550
Irrational numbers
   defined, 685, 719
   identifying, 686–687
   pi (π) as, 685
   radicals, 1088–1089, 1111
   square roots, 1089
Irregular shape, perimeter, 152

K

Key number, 972
kg (Kilogram), 575, 586
Kilogram (kg), 575, 586
Kilometer (km), 575–576, 577
km (Kilometer), 575–576, 577
Kurowski, Scott, 940

L

L (liter), 583, 584, 605
Lawn fertilizer, 314
LCD. *See* Least common denominator (LCD)
Least common denominator (LCD)
   addition of rational expressions, 1023–1025, 1075
   clearing fractions, to solve linear equations, 698–699, 721
   complex fractions, 1035–1038, 1076
   defined, 233, 234, 283, 1073
   by inspection, 234
   multiplication property of equality, 1041–1046, 1076
   prime factors, 234
   rational expressions, 1017–1020, 1023–1028, 1041–1046, 1075–1076
   simplification of complex fractions, 1035–1038, 1076
   solving equations with rational expressions, 1041–1046, 1076
   subtraction of rational expressions, 1026–1028, 1075–1076
   unlike fraction addition and subtraction with variable, 237
Legs
   formula for, 396
   of right triangle, 368–370, 978–979, 989, 1089
Length
   English system measurements, 417, 564, 588, 597
   metric system measurements, 575–579, 588, 597, 607
   parallelogram, 186
   rectangle, 150–151, 159, 185, 186
   of right triangle, 368–370
   square, 159–160, 186
   unit fractions, 565–566, 577, 597, 607
Less than. *See also* Inequalities
   converting to greater than, 688
   defined, 72
   linear inequalities, solving, 688, 847–849, 855
   for real number comparison, 687
   symbol for, 10–11, 71, 72, 687, 719, 720
Less than or equal to
   graph linear inequalities, 785–788, 796
   symbol for, 687–688, 719, 720
Life insurance benefits, 330
Lightening distance calculation, 83, 94
Like fractions, 231–232, 283, 287
Like radicals, 1107, 1141
Like terms
   addition of, 864, 866
   combining, 95–98, 138
   defined, 96, 137
   identifying, 96–97, 138
   linear equations in one variable, 694–696
   polynomials, 864
   subtraction of, 867–868, 924
Line graph, 638–639, 665, 669, 730, 732
Line segments, 453, 477, 482
Linear equations. *See also* Graphs of linear equations; Slope; System of linear equations
   check the solution, 694

clearing decimals or fractions, 698–699, 721
with decimals, 698–699, 721
distributive property, 694–699
elimination, 825–828, 833–834, 854–855
equations of lines, 771–776, 796
with fractions, 698–699, 721, 821–822
for horizontal lines, 775
inequalities, solving, 847–849, 855
key terms and summary, 853–855
many solutions, 696–697, 721
no solutions, 696, 698, 721, 1132–1133
in one variable, 694–699
ordered pair as solution, 808–811, 854
point-slope form, 772–775, 796
Real-Data Applications, 766, 812, 856
review exercises, 857–860
slope-intercept form, 763–765, 771–776, 796
slope of a line, 759–765, 771–774, 795
solving, 694–699, 721
standard form of, 733, 774–775, 796
tests, 861–862
variables, solving for specified, 703–706, 721–722, 1047–1048
for vertical line, 775
Linear equations in two variables
defined, 795
graphs of, 651–656, 730–738, 745–752, 785–788, 794–796
ordered pair as solution, 733–734, 795
ordered pair plotting, 646–648, 670, 736–738, 745–747, 795
table of values, 735–736
Linear function on calculators, 771
Linear inequalities. *See also* Graphs of inequalities
addition property of inequality, 710–711, 713–714
applications and application problems, 835–840, 855
defined, 72
greater than, 688, 847–849, 855
greater than or equal to, 785–788, 796
key terms and summary, 719–722, 793–796, 853–855
less than, 688, 847–849, 855
less than or equal to, 785–788, 796
multiplication property of inequality, 711–714
in one variable, 719
Real-Data Applications, 766, 812, 850, 856
for real number comparison, 687
review exercises, 723–726, 797–802, 857–860
slope of a line, 759–765, 771–774, 795
solving, 709–714, 722, 847–849, 855
symbol for, 10–11, 70, 71, 72, 687–688, 719, 720
systems of, 847–849, 853, 855

tests, 727–728, 803–806, 861–862
in two variables, 785–788, 793, 796
Linear model, 856
Linear units, 158
Lines. *See also* Horizontal lines; Number lines; Parallel lines; Slope; Vertical lines
defined, 453, 477, 482
equations of, 771–776, 796
geometric applications, 453–462
intersecting, 454, 477, 847–849, 855
line segments, 453, 477, 482
perpendicular, 457, 477, 482, 764–765, 793
slope, 657–658
slope-intercept form, 763–765, 771–776, 796
of symmetry, 1179, 1197
transversal, 460–461
Liter (L), 583, 584, 605
London Eye sight distance, 1085, 1136
Long-distance calling card, 411, 429, 430
Long division, 1243–1246, 1249
Lowest terms
defined, 207, 287, 1002, 1073
fractions, 207–214, 287
prime factorization, 212–214
proportions, 431–432
radical expressions, 1123
rates, 423
rational expressions, 1002–1005, 1074
ratios, 413–414

M

m (meter), 575, 577, 605
Mars, 563, 574
Mass, 585. *See also* Weight (mass)
Matches, ordered (pairs), 645–648, 665, 670
Mathematical Circles Revisited (Eves), 902
Mean, 357–360, 395, 396, 399
Mean, weighted, 358–359, 395
Measurement, 563–616. *See also* Measurement conversions
angles, 455–456, 628–630, 665
applications and application problems, 416–417, 479, 570
area units, 157–158
English system, 417, 564–569
key terms and summary, 605
length, 417, 564, 565–566, 577, 597, 607
metric system, 575–594, 605, 607–609
perimeter units, 158
ratio applications, 416–417, 479
Real-Data Applications, 570, 580
review exercises, 611–614
right angles, 455, 456, 477
temperature, 598–600, 609–610
test, 615–616
time, 417, 564
unit fractions, 565–566, 577, 597, 607
volume, 417, 564, 566–567, 583–585
weight, 417, 564, 566–567

Measurement conversions
capacity units, 584–585, 588, 608
English measurement units, 416–417, 564–569
length units, 588
metric units, 576–579, 605, 607
temperature units, 598–600, 605, 609–610
unit fractions, 565–568
weight units, 587
Measures of central tendency, 357–362
defined, 357, 360
mean (average), 357–360, 395, 396, 399
mean (average), weighted, 358–359, 395
median, 359–360, 395, 399
mode, 360, 395, 399
Median, 359–360, 395, 399
Mersenne, Marin, 940
Mersenne primes, 940
Meter (m), 575, 577, 605
Metric conversion line, 578–579, 605, 607
Metric measurement system
applications and application problems, 593–594, 608
capacity, 583–585, 608
defined, 565, 605
length, 575–579, 607
Metric-English conversion, 597–600
prefixes, 575, 583, 586
problem solving, 593–594, 608
temperature, 598–600, 609–610
unit conversion, 576–579, 607
for weight (mass), 585–588, 608
mg (milligram), 586
Microchips, 563, 582
Milligram (mg), 586
Milliliter (mL), 583, 584
Millimeter (mm), 575, 576, 577
Millions, place value, 2, 4
Minuend, 1218, 1249, 1250
Mixed numbers, 243–251
addition of, 249–250
applications and application problems, 251
on calculator, 250
from decimals, 303–304, 543
decimals written as, 303–304
defined, 243, 287
division, 246–248
estimation with, 246–251
as improper fractions, 244–246
multiplication, 246–248
on a number line, 243
problem solving, 243–251
ratios, 414–415, 479
recipes with, 252
rounding, 246
subtraction of, 249–250
Mixture problems, 837–838
mL (milliliter), 583, 584
mm (millimeter), 575, 576, 577
Mode, 360, 395, 399

Models
- linear equations, 737–738, 751–752
- linear model, 856
- quadratic model applications, 979–980

Money, 311–313, 328

Monomials. *See also* Polynomials
- defined, 865, 923
- division of polynomial by, 907–908, 925
- multiplication of, 881

Movie ticket sales, 683, 692, 832

Multiples of ten, 1227–1228, 1240, 1245–1246

Multiplicand, 1225

Multiplication. *See also* Distributive property; Exponents
- applications and application problems, 44, 1230
- associative property of, 43, 71, 74, 98–99, 875, 877, 881, 1226, 1249, 1250
- of binomials, 883–885, 924
- on calculator, 42, 221–222, 1226, 1230
- to check division, 1237–1238, 1246
- coefficients in, 86
- commutative property, 43, 71, 74, 89, 210, 875, 877, 881, 1225, 1249, 1250
- complex fraction, 1035–1038, 1076
- cross products, 432–437, 477
- decimals, 327–329, 398
- division changed to, 1234
- elimination method, linear system equations, 827
- estimation of, 44, 75
- exponential expressions, 873, 875
- expressions of, 86–89, 99, 139
- FOIL method for binomials, 883–885, 924
- fractions, 219–222, 226, 1035–1038, 1076
- indicator words for, 226
- integers, 39–44, 51, 74, 222
- inverse of rational expression, 1010
- mixed numbers, 246–248
- monomials, 881
- by multiples of ten, 1227–1228
- of numbers ending in zeros, 42–43, 1227
- by numbers with more than one digit, 1228
- polynomials, 763–765, 771–773, 774–776, 796
- power rule for exponents, 875–878, 900, 924
- properties of, 42–44
- radicals and product rule, 1097–1099, 1113–1114, 1119–1121, 1143
- rational expressions, 1009–1010, 1074
- reciprocals, 223–224
- scientific notation, 919
- of several factors, 41–42
- signed decimals, 327–329, 398
- signed fractions, 219–222, 226
- simplification of complex fractions, 1035–1038, 1076

- slope of perpendicular lines, 764–765
- of sum and difference of two terms, 1121
- symbols for, 39, 86, 1225
- of two signed numbers, 40–41
- unit conversion measurement problems, 564–565
- whole numbers, 1225–1230
- with zeros, 1229

Multiplication properties
- of equality, 267–270, 287, 1041–1046, 1076
- of inequality, 711–714
- of one, 42–43, 71, 74
- of zero, 42–43, 71, 74

Multiplicative inverse, of a rational expression, 1010

Multiplier, 1225

Multivariable polynomial, 868–869

Music time signature, 238

N

Nail growth (fingers and toes), 580

Nails, size and type, 193, 276

Natural (counting) numbers, 684

Natural numbers, 684

Negative exponents. *See also* Exponents
- defined, 896
- negative-to-positive rules, 896–898, 900, 925
- product rule with, 896
- reciprocals, 896–898
- symbol for, 923
- vs. negative numbers, 897

Negative fractions, 196–197

Negative numbers, 9–12. *See also* Negative exponents; Signed numbers
- addition of, 15–19, 73
- on calculators, 26
- defined, 9
- division of, 50
- equation solving with, 122, 140
- graphing, 10, 72
- greatest common factor (GCF), 935
- as integers, 684, 689
- mixed numbers, 243
- multiplication of, 39–44, 51, 74
- on number line, 10, 15–16
- opposite of, 25
- roots, 1091
- rounding, 29
- as signed number, 9–12
- square roots, 367, 1086
- subtraction of, 25–26
- writing, 9–10

Negative sign, 9, 122, 196

Negative slope, 657–658, 762

Negative terms, factoring trinomials, 944–945

Nonadjacent angles, 459

Not equal, 29

"Not equal to" symbol, 107, 687, 719, 720

Notation, exponential, 59

Notation, scientific. *See* Scientific notation

nth root symbol, 1141

Number lines. *See also* Graphs
- absolute value, 11, 73
- addition on, 15–19
- comparing integers, 10–11
- decimals, 351–352
- defined, 10, 71
- equivalent fractions, 198
- graph linear inequalities, 709–710, 722
- graphing fractions on, 196–197
- graphing numbers on, 10–11
- greater than, 11
- integers on, 684
- less than, 11
- mixed numbers on, 243
- opposite of signed numbers, 25
- subtraction on, 25

Numbers. *See also* Decimals; Fractions; Integers; Irrational numbers; Mixed numbers; Negative numbers; Positive numbers; Signed numbers; Square roots; Whole numbers
- big and small, 863, 902, 921
- common factors, 207–214
- complex numbers, 1165n
- composite, 209, 283
- counting, 684
- divisibility tests, 211, 1238–1240
- factored form of, 934
- greatest common factor (GCF), 934–937, 988
- natural, 684
- opposite of, 25, 71, 684, 689
- prime, 209–210, 283, 287, 936, 940
- rational, 200, 1088
- real, 684–690, 719–720, 1156
- in scientific notation, 917–918
- square of, 59, 1086
- in words, 3–4

Numerators. *See also* Fractions
- defined, 195, 287
- ratios, 412–413
- unit fractions, 566, 606

Numerical coefficients, 864

O

Obtuse angles, 456, 477

Odd consecutive integers, 976–978

Ones, place value, 2

Operations. *See* Order of operations

Opposite of a number, 25, 71, 684, 689

Order of a radical, 1091

Order of operations, 60–64
- addition, 60, 61
- addition, commutative property of, 18, 71, 73, 1209–1210, 1249, 1250

on calculators, 61
decimals, 340–341
defined, 76
descending powers, 864, 923
division, 61
exponents, 61, 63
fractions, 64, 76, 259–260
integers, 60–64
memory mnemonic, 63
multiplication, 61
multiplication, commutative property of, 43, 71, 74, 89, 210, 875, 877, 881, 1225, 1249, 1250
radicals, 1091
simplification of expressions, 688
steps in, 61, 76
subtraction, 60, 61
temperature conversion, 600
with variables, 89
whole numbers, 62
Ordered pairs
checking as solution, 733–734
components of, 1187
coordinate system, 645–648, 665, 670
defined, 651, 733, 793, 1187
graph linear inequalities, 785–788, 796
intercepts, 747–748
parabolas as quadratic equations graphs, 1179–1184, 1199
plotting, linear equations in two variables, 645–648, 670, 736–738, 745–747, 795
in relations, 1187
slope, 759–765
slope-intercept form, 763–765, 771–776, 796
as solution, linear equations in two variables, 733–734, 795
as solution of a system, 808–811, 854
symbol for, 794
system solution, 818
table of values for, 735–737
Origin, 646, 665, 736, 788, 793
Outer product, of binomials, 883–885, 923

P

Paired data, 645, 665
Pairs, ordered, 645–648, 665, 670
Panel, geodesic domes, 1124
Parabolas. *See also* Graphs; Quadratic equations
applications and application problems, 979, 980, 1183–1184
axis of, 1179
defined, 1179, 1197
intercepts, 1180–1182
parabolic dish, 1153, 1183–1184, 1186
quadratic equations, 1179–1184, 1199
quadratic inequalities, A–15 to A–16
quadratic model applications, 979, 980
vertex of, 1179

Parallel lines
angle formation, 460–462
defined, 454, 477, 482, 793
slope comparison, 764–765
Parallelograms
area, 161–162, 183
base of, 160–162
defined, 151, 183
height of, 160–162
perimeter, 151–152
Parentheses. *See also* Distributive property
associative property of addition, 18–19, 39
brackets, 688, 720
as multiplication symbol, 39, 1225
negative signs, 196
order of operations, 61
simplification of expressions, 688
Part. *See also* Whole
defined, 510, 551
in percent equations, 509–511, 517–520, 529–530, 552–553
Partial products, 1229
Patio plan rules of thumb, 892
Pattern analysis, inductive reasoning, A–1 to A–2
Per unit rates, 424–426, 477
Percent, 493–462. *See also* Percent equations; Percent proportion
applications and application problems, 529–534, 553
on calculator, fractions to percents, 500–501
decimals, percents written as, 494–496, 552
decimals written as percents, 496–497, 552
of decrease, 532–534, 551
defined, 494, 509, 551
estimation of, 517
finding 1% and 10%, 517–518
finding 50% and 100%, 501
fractions, percents written as, 497–499, 552
fractions written as percents, 499–501, 552
greater than 100%, 513–514
of increase, 532–533, 551
key terms and summary, 551–554
percent equations, 519–521, 530–531, 552–553
percent proportion, 509–514, 551
problem solving, 529–534, 553
Real-Data Applications, 502, 550
review exercises, 557–560
shortcuts, 518
test, 561–562
Percent equations, 517–522
defined, 517, 551, 552–553
to find the part, 517–520, 529–530
to find the percent, 519–521, 530–531
to find the whole, 522, 532

Percent proportion, 509–514
base, 509
defined, 509, 551
to find percent, 512
identify part, 510–511
identify whole, 510–511, 513
Perfect cubes, 1091, 1101, 1141
Perfect square trinomial, 960–962, 987, 989
Perfect squares, 367, 1088, 1141, 1161
Perimeter
applications and application problems, 148–152, 162
defined, 148, 157, 183, 184
to find missing length, 185
formula, 183
irregular shape, 152
parallelogram, 151–152
rectangle, 149–151, 183–184
semiperimeter, 1140
square, 148–149, 183–184
triangle, 152, 471
vs. area, 157
Perpendicular lines, 457, 477, 482, 764–765, 793
Pi (π), 382, 395, 685, 1089
Pictographs, 620–621, 665, 667
Pie chart, 627–630
construction using protractor, 628–630, 668
defined, 627, 665
finding ratio from, 627–628
read and interpret, 627–628, 730
Pisa, Leonardo da, 1086
Place value, 2–4
decimal chart, 301–302, 397
defined, 71, 395
of digits, 3, 301–302
identifying, through hundred-trillions, 3
system for, 2, 71
Plane, 454, 736, 737, 793
Plot, defined, 793
Plotting points on a coordinate system, 646–648, 670, 736–738, 745–747, 795
Point-slope form, 772–775, 796
Points. *See also* Decimal points
coordinates of, 665
defined, 453, 477
linear equations, graphing, 652
plotting, on coordinate system, 646–648, 670, 736–738, 745–747, 795
slope-intercept form, 763–765, 771–776, 796
standard form of linear equations, 774–775
Polynomials, 863–932. *See also* Binomials; Monomials; Trinomials
addition of, 864–868, 881, 924
binomials, 865
classifying, 865
defined, 864, 923
degrees of, 865, 923

descending powers, 864, 923
division by monomial, 907–908, 925
division by polynomial, 911–914, 925
evaluating, 865–866
factoring, 936–937, 943–971
key terms and summary, 923–925
multiplication of, 881–885, 889–891, 924
multivariable, 868–869
prime, 945–946, 987
Real-Data Applications, 886, 892
review exercises, 927–930
subtraction of, 865–868, 924
terms of, 723, 864–865
test, 931–932
trinomials, 865
trinomials, factoring, 943–971
in x, 864
Positive numbers, 9–12
 defined, 9
 fractions, 196–197
 on number line, 10
 as signed number, 9–12
 whole numbers, 684
 writing, 9–10
Positive or negative, symbol for, 1197
Positive (principal) square root, 367, 1086, 1087, 1091, 1141
Positive slope, 657–658, 762
Power rule for exponents, 875–878, 900, 924
Powers. *See also* Exponents
 of base, 59
 descending, 864, 911–913, 923
 greater powers of binomials, 891
Prefixes, metric, 575, 583, 586
Premises, deductive reasoning, A–2 to A–4
Price per unit, 424–426, 477
Prime factorization
 on calculators, 212
 defined, 209, 287
 fractions in lowest terms, 212–214
 least common denominator (LCD), 234
 to multiply fractions, 220–221
 of number, 209–210
Prime numbers
 defined, 209, 287, 940
 greatest common factor (GCF), 936
 prime factorization, 209–210
Prime polynomials, 945–946, 987
Principal, loans, 542, 551, 554
Principal (positive) square root, 367, 1086, 1087, 1091, 1141
Problem solving. *See also* Applications and application problems; Real-Data Applications
 area, 157–162
 cricket chirps and temperature, 83, 91
 deductive reasoning, A–2 to A–4
 distance, 147, 156, A–7 to A–9
 English measurement, 564–569
 equations with decimals, 375–378, 400
 estimation, 32–34, 243–251

metric measurement, 593–594, 608
mixed numbers and estimating, 243–251
one unknown quantity, 167–171
percent, 529–534, 553
with perimeters, 148–152
pictographs, 620–621, 667
proportions, 445–447, 481–482
rounding, 29–34, 246
sequences and inductive reasoning, A–1 to A–2
statistics, 357–362
tables, 618–619, 666
two unknown quantities, 177–179
variation, A–7 to A–10
windchill, 1, 28, 766
Product rule
 for exponents, 873–878, 896, 900, 924
 for radicals, 1097–1099, 1143
Products. *See also* Multiplication
 of binomials, 883–885, 923
 defined, 39, 71, 1249
 factored form of polynomial, 936, 938
 FOIL method, 883–885
 of multiplication, 1225
 partial, 1229
 of sum and difference of two terms, 890–891, 924
Proper fractions, 195–196, 283, 287
Properties. *See also* Distributive property; Division property of equality
 addition property of inequality, 710–711, 713–714
 addition property of zero, 17–18, 71, 73
 associative property of addition, 18–19, 39, 71, 73, 1210, 1249, 1250
 associative property of multiplication, 43, 71, 74, 98–99, 875, 877, 881, 1226, 1249, 1250
 commutative property of addition, 18, 71, 73, 89, 1209–1210, 1249, 1250
 commutative property of multiplication, 43, 71, 74, 89, 98, 210, 875, 877, 881, 1225, 1249, 1250
 of division, 50
 fundamental property of rational expressions, 1000–1006, 1017–1020, 1074
 identity property of multiplication, 1002
 of multiplication, 42–44
 multiplication property of equality, 267–270, 287, 1041–1046, 1076
 multiplication property of inequality, 711–714
 multiplication property of one, 42–43, 71, 74
 multiplication property of zero, 42–43, 71, 74
 square root property of equations, 1154–1156, 1161–1162, 1177, 1198
 squaring property of equality, 1131–1135
 zero-factor property, 967–968, 969, 970, 989, 1154

Proportions, 431–447
 applications and application problems, 445–447
 cross products, 432–437, 477
 decimals, 436–437
 defined, 431, 477
 finding unknown number, 433–437
 key terms and summary, 477, 480–482
 in lowest terms, 431–432
 percent, 509–514, 551
 problem solving, 445–447, 481–482
 Real-Data Applications, 438, 448
 review exercises, 486–487
 terms of, 431–432
 test, 491–492
 true or false, 431–433
 writing, 431, 480
Protractors, 456, 628–630, 665
Pyramids, 280–281, 283
Pythagoras, 368
Pythagorean formula, 978–979, 989, 1089–1091, 1142
Pythagorean theorem, 368–370

Q

Quadrants, 648, 665, 670, 736, 793
Quadratic equations, 1153–1206
 applications and application problems, 975–980, 989
 completing the square method for solving, 1161–1166, 1177, 1198
 consecutive integer applications, 976–978
 defined, 967, 987
 factoring method for solving, 967–971, 989, 1177
 formulas with squared variables, 1156
 with fractional coefficient, 1173
 graphing, 1179–1184, 1199
 with no solution, 1132–1133
 with one solution, 1173
 parabolas as graphs of, 1179–1184, 1199
 Pythagorean formula, 978–979, 989
 quadratic formula method for solving, 1169–1173, 1177, 1199
 quadratic model applications, 979–980
 Real-Data Applications, 1174
 review exercise, 1201–1204
 square root property of equations, 1154–1156, 1161–1162, 1177, 1198
 standard form of, 967, 987, 1154
 steps to solve applications, 975
 test, 1205–1206
 in two variables, 1179
 zero-factor property, 967–968, 969, 970, 989, 1154
Quadratic formula, 1169–1173
 with coefficients, 1173
 defined, 1169, 1170
 with fractions, 1173
 to solve quadratic equations, 1170–1173
 value identification, 1169

I-11

Quadratic inequalities, graph of, A–15 to A–16
Quantity and cost problems, 836–837
Quilt patterns, 282, 502
Quotient rule
 for exponents, 898–901, 925
 for radicals, 1099–1100, 1143
Quotients. *See also* Division
 checking, 1237–1238, 1246
 defined, 49, 71, 1233–1234, 1249
 trial, 1243–1244
 zeros in, 1244–1245

R

Radians, 456
Radical equations
 defined, 1131, 1141
 with no solutions, 1132–1133
 solving, 1131–1135, 1144
 squaring a binomial, 1133–1135
 squaring property method for solving, 1131–1135
 steps to solve, 1135
Radical expressions. *See also* Square roots
 defined, 1086, 1141
 quotients in lowest terms, 1123
 rationalizing the denominator, 1119, 1121–1122
 simplification of radicals, 1108
 squaring, 1087
Radical sign, 1086, 1141
Radicals, 1085–1152. *See also* Simplification of radicals
 addition, 1107, 1143
 defined, 1086, 1141
 division of, 1099–1100, 1143
 equations with, 1131–1135, 1144
 index of, 1091–1092
 key terms and summary, 1141–1144
 multiplication of, 1097–1099, 1143
 order of, 1091
 product rule for, 1097–1099, 1143
 quotient rule for, 1099–1100, 1143
 rationalizing the denominator, 1111–1115, 1143
 Real-Data Applications, 1116, 1124, 1136
 review exercises, 1145–1150
 roots, 1086–1092, 1142
 special products of, 1119–1122
 subtraction, 1107–1108, 1143
 test, 1151–1152
Radicand, 1086, 1141
Radius, 381–382, 384, 386, 395, 396, 401
Range
 function notation, 1191
 of input-output machine, 1188
 of a relation, 1187, 1197
 of values, 361
Rates
 best buy, based on cost per unit, 424–426, 480

 defined, 423, 477
 distance formula, 147, 156
 as fractions, 423, 479
 interest, 542, 551, 554
 lowest terms, 423
 tax, 539–540, 551, 554
 time and distance problems, 147, 156, 438, 838–840, 1055–1059, 1077, A–7 to A–9
 unit, 423–424, 480
 writing, 423, 479
Ratio
 key terms and summary, 477, 479–480
 review exercises, 486–487
 slope, 759
 test, 491–492
Rational equations, 1053
Rational expressions, 999–1084. *See also* Fractions
 addition of, 1023–1025, 1075
 applications and application problems, 1055–1061, 1067–1068, 1077
 complex fractions, 1076
 defined, 1073
 distance, rate and time problems, 1055–1059, 1077
 division of, 1010–1013, 1074
 equivalent forms of, 1005–1006, 1018–1020
 fundamental property of, 1000–1006, 1017–1020, 1074
 with given denominators, 1018–1020
 least common denominator (LCD), 1017–1020, 1023–1028, 1041–1046, 1075–1076
 in lowest terms, 1002–1005, 1074
 multiplication of, 1009–1010, 1074
 multiplicative inverses of, 1010
 numerical value of, 1001
 Real-Data Applications, 1014, 1062
 reciprocals of, 1010
 review exercises, 1079–1082
 simplification summary, 1053
 solving equations with, 1041–1046, 1076
 with specified denominator, 1075
 subtraction of, 1026–1028, 1075–1076
 summary of operations, 1054
 test, 1083–1084
 as undefined, 1000–1001
 unknown number problem, 1055
 variables, solving for specified, 1047–1048
 variation problems, 1067–1068, 1078
 vs. rational equations, 1053
 work problems, 999, 1059–1061, 1065, 1077
Rational numbers, 299–410. *See also* Decimals; Fractions
 applications and application problems, 367–370, 381–388

 circles, 381–386
 cylinders, 386–388
 defined, 200, 684, 719
 geometry applications, 367–370, 381–388
 graphing, 685
 identifying, 686–687
 key terms and summary, 395–408
 Pythagorean theorem, 368–370
 quadratic formula solution as, 1173
 Real-Data Applications, 238, 252, 262, 272, 282, 330, 342
 square roots, 367–370, 1088
 statistics, 357–362
 surface area, 387–388
 test, 409–410
Rational square root, 1088
Rationalizing the denominator, 1111–1115
 conjugates, 1121–1122
 cube roots, 1114–1115
 defined, 1111, 1141, 1143
 radical expressions, 1119, 1121–1122
 square roots, 1111–1114
Ratios, 412–426. *See also* Fractions
 applications and application problems, 416–417, 479
 from circle graphs, 627–628
 decimals, 414–415
 defined, 420, 423, 477
 as fractions, 412–413
 in lowest terms, 413–414
 measurement application, 416–417, 479
 mixed numbers, 414–415, 479
 order of numbers, 413
 Real-Data Applications, 418
 writing, 479
Rays, 453, 477, 482
Real-Data Applications. *See also* Applications and application problems; Problem solving
 algebra, arithmetic connections to, 180, 886
 algebraic expressions, 102
 arithmetic connections to algebra, 180, 886
 auto service, 20
 compact discs vs. cassette sales, 850
 compound interest, 550
 cost comparison, 54
 currency exchange, 622
 decimals, 314, 330, 342
 distance, 20, 990
 dollar-cost averaging, 342
 earthquake measurements, 926
 educational tax incentives, 556
 equations, 172, 180
 exponents, 886, 902, 926
 expressions, 102, 142
 factoring, 940, 972, 990
 Fahrenheit estimation, 812
 formulas, 172

I-12

fractions, 238, 252, 262, 272, 282
gasoline cost, 1062
geodesic domes, 1124
Golden Ratio, 1116
graphs, 622, 640, 672
grocery shopping, 640
growing sunflowers, 570
hair and nail growth, 580
heart rate, 262
hotel expenses, 272
hummingbird feeding, 448
inequalities, 766, 812, 850, 856
integer operations, 20, 54
integers, 20, 54
interest rates, 1174
investment growth, 550
lawn fertilizer, 314
life insurance benefits, 330
linear equations, 766, 812, 856
linear inequalities, 766, 812, 850, 856
linear model, 856
measurement, 570, 580
metric system, 580
music time signature, 238
numbers, big and small, 902
patio plan rules of thumb, 892
percent, 502, 550
polynomials, 886, 892
prime numbers, 940
proportion, 438, 448
quadratic equations, 1174
quilt patterns, 282, 502
radicals, 1116, 1124, 1136
rational expressions, 1014, 1062
rational numbers, 238, 252, 262, 272, 282, 330, 342
ratios, 418
recipes with mixed numbers, 252
Richter scale, 926
signed numbers, 20
speeding and time saved, 1014
stopping distance, 990
surfing the net, 672
system of linear equations, 856
table, 622
trinomial factoring, 972
tuition costs, 142
variables, 102
wedding expenses, 54
windchill factor, 766
Real numbers, 684–690, 719–720, 1156
Reasoning, A–1 to A–4
   deductive, A–2 to A–4
   inductive, A–1 to A–2
Recipes, 252
Reciprocals
   defined, 222–223, 283
   fractions, 222–223
   negative exponents, 896–898
   of rational expressions, 1010
   zero exponents, 896–897

Rectangles
   area, 157–159, 161, 183
   defined, 150, 183
   finding width, 704
   length, 150–151, 159, 185, 186
   perimeter, 149–151, 183–184, 184
   width, 150–151, 159
Rectangular coordinate system
   defined, 645, 736, 793
   plotting points, 646–648, 670, 736–738, 745–747, 795
Rectangular solid
   surface area, 387–388, 396, 402
   volume, 280, 283, 387
Regrouping (carrying or borrowing), 1211–1212, 1218–1222, 1226–1227, 1249
Relation, domain of, 1187, 1188, 1191, 1197
Remainders, in division, 52–53, 75, 1236, 1249
Repeating decimals, 337, 395, 685
Restaurant tips, 540–541, 551, 554
Richter scale, 926
Right angles
   defined, 148
   measurement, 455, 456, 477
   Pythagorean theorem, 370
   symbol for, 148, 183
Right circular cylinders, 386, 388
Right triangles
   applications and application problems, 370
   hypotenuse, 368–370, 978–979, 989, 1089
   Pythagorean formula, 978–979, 989, 1089–1091
   Pythagorean theorem, 368–370
Rise, 759–761, 793
Road trip, 438
Roots, 1086–1092. *See also* Square roots
   cube roots, 1091–1092, 1114–1115, 1148
   evaluation, 1086–1092, 1142
   fourth, 1091
   higher roots, 1091–1092, 1101–1102
   negative roots, 1091
   principal (positive) roots, 1091
Rounding. *See also* Front end rounding
   decimals, 309–313, 337–338, 350, 397
   defined, 29, 71, 395
   fractions as decimals, 350
   integers, 29–34, 74
   mixed numbers, 246
   money, 311–313
   money on calculators, 313
   rules for, 29–32
   truncated decimals, 351
   value of pi (π), 382
Rules
   negative-to-positive exponent rules, 898
   power rule for exponents, 875–878, 900, 924

   product rule for exponents, 873–878, 896, 900, 924
   product rule for radicals, 1097–1099, 1143
   quotient rule for exponents, 898–901, 925
   quotient rule for radicals, 1099–1100, 1143
   for rounding, 29–32
Run, 759–761, 793

S

Sacred ratio (Golden Ratio), 1116
Sales price, 493, 541–542, 546, 551, 554
Sales tax, 493, 539–540, 545, 551, 554
SAS (Side-Angle-Side) method of congruence, 468–469, 483
Satellite dish, 1153, 1183–1184, 1186
Scatter diagram, 738, 793
Scientific notation
   applications and application problems, 863, 902, 919–920, 921
   on calculators, 919, 920
   defined, 917, 923, 925
   division with, 919
   multiplication with, 919
   numbers with exponents, 917–918
   numbers without exponents, 918
Second-degree equation. *See* Quadratic equations
Sectors of circle graph, 627, 730
Segments, line, 453, 477, 482
Semicircles, 385
Semiperimeter, 1140
Sentences, writing as equations, 168, 187
Sequences and inductive reasoning, A–1 to A–2
Sets, deductive reasoning, A–2 to A–4
Short division, 1236–1237, 1249
Side-Angle-Side (SAS) method of congruence, 468–469, 483
Side-Side-Side (SSS) method of congruence, 468–469, 483
Sides
   of angle, 454
   Angle-Side-Angle (ASA) method of congruence, 468–469, 483
   Side-Angle-Side (SAS) method of congruence, 468–469, 483
   Side-Side-Side (SSS) method of congruence, 468–469, 483
   of square, 149, 159–160, 185, 186
   unknown length, of triangles, 470–471
Sieve of Eratosthenes, 940
Sight distance, London Eye, 1085, 1136
Signed decimals
   addition of, 317–320, 321–322, 397
   division, 335–341, 398
   multiplication, 327–329, 398
   subtraction of, 317, 320–322, 398
Signed fractions, 194–200
   addition of, 231–237

I-13

division, 222–226
mixed numbers, 243
multiplication, 219–222, 226
subtraction of, 231–237
Signed numbers, 9–12
addition of, 15–19, 73
on calculators, 26
defined, 9
division, 50
equation solving with, 122, 140
exponential expressions, 59, 873
graphing, 10, 72
greatest common factor (GCF), 935
as integers, 684
mixed numbers, 243
multiplication of, 39–44, 51, 74
on number line, 10, 15–16
opposite of, 25
square root, 367
subtraction of, 25–26
Similar figures, 467, 478
Similar terms. *See* Like terms
Similar triangles, 467, 469–472, 478, 484
Simple interest, 542–544, 551, 554
Simplification. *See also* Simplification of expressions; Simplification of radicals
complex fractions, 261
of complex fractions, 1033–1038, 1076
distributive property, 96, 99–101, 689–691, 720
division property of equality, 121
of exponents, 59–60, 259
fractions, 64, 200, 259
Simplification of expressions, 688–692
brackets, 688, 720
commutative property, 98
complicated radical expressions, 1108
defined, 95, 137
distributive property, 689–691, 720
equation solving with, 95–101
with exponents, 59–60
of multiplication expressions, 99, 139
order of operations, 688
parentheses, 688
rational expressions, 1053
Simplification of radicals
addition and subtraction, 1107–1108
complicated radical expressions, 1108
higher roots, 1101–1102
overview, 1112–1114, 1143–1144
product rule, 1097–1099, 1113–1114, 1119–1121, 1143
quotient rule, 1099–1100, 1112, 1143
rationalizing the denominator, 1112–1115
with variables, 1100–1101
Size, fractions vs. decimals, 351–352, 399
Slope
defined, 793
finding slopes from equations, 763–764
formula for, 761–762, 795

of horizontal lines, 762, 763, 795
of a line, 657–658, 759–765, 771–774, 795
line comparison, 764–765
negative, 657–658, 762
perpendicular lines, 764–765
positive, 762
rise and run, 759–761
slope-intercept form, 763–765, 771–776, 796
symbol for, 794
vertical lines, 762–763, 795
Slope-intercept form, 763–765, 771–776, 796
Solution, defined, 137
Solution of a system
defined, 808, 853
by graphing, 808–810, 854
ordered pair as, 808, 854
special system solving, 810–811
Solution to equation, 137, 733.
See also Equations
Solving for a specified variable, 703–706, 721–722, 1047–1048
Sophie Germain prime, 940
Speeding and time saved, 1014
Sphere, geodesic domes, 1124
Square root property of equations, 1154–1156, 1161–1162, 1177, 1198
Square roots
approximating, 1089
on calculators, 367–368, 1087
defined, 367, 395, 1086, 1141
finding, 367–368, 400, 1087
irrational, 1088–1089
as irrational, 686
negative, 1086, 1087
not a real number, 1088
of a number, 1086–1087
positive (principal), 367, 1086, 1087, 1091, 1141
principal, 1086, 1087, 1141
quadratic equations, 1154–1155, 1198
rational, 1088
rationalizing the denominator, 1111–1114
signed, 367
symbol for, 395, 1086
Square units, 157–158, 183, 277, 384
Squares. *See also* Exponents
of binomials, 889–890, 924
completing the square, quadratic equations, 1161–1166, 1198
difference of squares, factoring, 959–960, 962, 989
of a number, 59, 1086
perfect, 367, 1088
perfect square trinomial, 960–962, 987, 989
perfect squares, 367, 1088, 1141, 1161
reading exponents, 59
squaring a binomial, radical equations, 1133–1135

Squares (geometric)
area, 159–160, 183, 186
defined, 148, 183
perimeter, 148–149, 183–184
sides of, 149, 159–160, 185, 186
Squaring property of equality, 1131–1135
SSS (Side-Side-Side) method of congruence, 468–469, 483
Standard form
linear equations, 733, 774–775, 796
quadratic equations, 967, 987, 1154
Statistics, 357–362
defined, 357
linear equation to fit data, 775–776
problem solving with, 357–362
Stopping distance problem, 990
Straight angles, 455, 456, 477
Subscript notation, 794
Substitution method, linear system equations
choice of, 833–834
defined, 817, 854
dependent equations, 820–821
fractions, 821–822
inconsistent systems, 820
special systems, 820–821
steps in, 818
use of, 817–819
Subtraction
applications and application problems, 1222
borrowing, 1218, 1219–1222, 1249
on calculators, 26
changing addition to, 25–26, 1217–1218
changing to addition, 25–26, 1217–1218
checking answers, 1218–1219, 1221–1222
decimals, 317, 320–322, 398
defined, 25
estimation of, 32, 74
fractions, 231–237
integers, 25–26, 73
like fractions, 231–232
like terms, 867–868
mixed numbers, 249–250
multivariable polynomials, 868–869
on number line, 25
polynomials, 865–868, 924
product of sum and difference of two terms, 890–891, 924
radicals, 1107–1108, 1143
rational expressions, 1026–1028, 1075–1076
signed fractions, 231–237
signed numbers, 25–26
symbol for, 86
unlike fractions, 233, 235–237
whole numbers, 317, 1217–1222
Subtraction sign vs. negative number, 9
Subtrahend, 1218, 1249, 1250
Sums. *See also* Addition
defined, 15, 71, 1209, 1249

negative-to-positive exponent rules, 898
power rule for exponents, 876
product of sum and difference of two terms, 890–891, 924
total, defined, 1209, 1249
Supplementary angles, 458, 478, 483
Surface area
   cylinder, 387–388, 396, 402
   defined, 395
   of geodesic dome, 1124
   geodesic domes, 1124
   rectangular solid, 387–388, 396, 402
   of a sphere, 1124
Symbols
   absolute value, 71
   addition, 86
   area, 183
   division, 49, 86, 1233
   functions, 1197
   greater than, 10–11, 71, 687, 719, 720
   greater than or equal to, 70, 687–688, 719
   inequalities, 10–11, 70, 71, 72, 687–688, 719, 720
   less than, 10–11, 71, 72, 687, 719, 720
   less than or equal to, 687–688, 719, 720
   linear inequalities, 10–11, 70, 71, 72, 687–688, 719, 720
   multiplication, 39, 86, 1225
   negative exponents, 923
   not equal to, 107, 687, 719, 720
   nth root, 1141
   ordered pairs, 794
   positive or negative, 1197
   right angles, 148, 183
   slope, 794
   square root, 395, 1086
   subtraction, 86
   therefore, A–2
Symmetry, lines of, 1179, 1197
System of linear equations. *See also* Elimination method, linear system equations; Linear equations
   applications and application problems, 835–840, 855
   consistent, 810, 811, 820, 853
   cost and quantity problems, 836–837
   defined, 808, 853
   distance, rate and time problems, 838–840
   equations for, 808–811, 817–822, 825–828, 854
   graphs of, 808–811, 817–822, 825–828, 854
   inconsistent, 810, 820, 853
   inequalities, 847–849, 853, 855
   infinite number of solutions, 810
   mixture problems, 837–838
   with no solution, 696, 698, 721, 1132–1133
   plotting points on coordinate system, 646–648, 670, 736–738, 745–747, 795
   quantity and cost problems, 836–837
   Real-Data Applications, 856
   substitution method, 817–822, 833–834, 854
   with two variables, 835–840
   unknown number problems, 835–836
Systems for place value, 2, 71

T

Tables
   defined, 665, 793
   graphs of linear equations, 618–619, 665, 666
   linear equation graphing, 651–658, 665, 670
   ordered pairs, 735–737
   read and interpret, 618–619, 666
   of values, 735–737, 793, 795
Tax rate, 539–540, 551, 554, 556
Temperature, 598–600, 609–610
Temperature and cricket chirps, 83, 91
Tens
   divisibility by, 1240
   multiple of, 1227–1228, 1240, 1245–1246
   place value, 2
Terms. *See also* Like terms; Lowest terms; Unlike terms
   algebraic expressions, 167–168
   combining, 864
   constant, 95, 98
   defined, 95, 137, 864
   degrees of a, 865, 923
   of expressions, 95–96, 864
   factored form of polynomial, 936
   missing, division of polynomials, 913–914
   of polynomials, 723, 864–865
   of proportion, 431–432
   unlike terms, 96, 864
   variable, 127–128, 137
Theorem, Pythagorean, 368–370. *See also* Rules
Therefore symbol, A–2
Thousands, place value, 2, 4
Three, divisibility by, 1239
Thunderstorm distance calculation, 83, 94
Time, rate and distance, 147, 156, 438, 838–840, 1055–1059, 1077
Time measurement, 417, 564
Tip calculation, 539–540, 551, 554
Total, 1209, 1249. *See also* Sums
Transversals, 460–461
Trapezoids, area of, 703–704
Trial divisor, 1243
Trial quotient, 1243–1244
Triangles. *See also* Right triangles
   area, 277–279, 287
   base of, 277–278
   congruent, 467–469, 478, 483
   defined, 152, 183
   facets, geodesic domes, 1124
   height of, 277–278
   perimeter, 152, 471
   similar, 467, 469–472, 478, 484
Trillions, 2, 4, 863
Trinomials. *See also* Factoring trinomials; Polynomials
   completing the square, 1161
   defined, 923
   as square of binomial, 889–890
Truncated decimals, 351
Two variables. *See also* Linear equations in two variables
   expressions with, 88
   factoring equation with more than two variable factors, 971
   factoring trinomials, 946, 955
   graphs of inequalities, 785–788, 793, 796
   quadratic equations, 1179
Two, divisibility by, 1239

U

Undefined, division by zero as, 50, 1235
Undefined slope, 762, 763
Unit, cost per, 424–426, 477
Unit fractions
   capacity (volume), 566–567
   defined, 565, 605, 606
   length, 565–566, 577, 597, 607
   metric capacity, 584–585, 608
   using several, 567–568
   weight, 417, 566–567, 598
Unit rates
   best buy, based on cost per unit, 424–426, 480
   defined, 423–424, 477
   using a calculator, and, calculators, 425
Unit ratios, conversion, 413, 416–417
Unknown number problems, linear system, 835–836
Unknowns. *See* Variables
Unlike fractions, 233–237
   addition of, 233, 235–237
   common denominator, 233–234
   defined, 231, 287
   subtraction of, 233, 235–237
Unlike radicals, 1107
Unlike terms, 96, 864

V

Value, absolute. *See* Absolute value
Values, table of, 735–737, 793, 795
Variability of data, 361–362, 395, 399
Variable terms, 95, 127–128, 137, 935
Variables, 84–90
   defined, 84, 137
   division of fractions with, 225
   exponent use with, 89–90, 138
   expressions with two variables, 88
   formula to estimate and solve for, 703–706, 722
   identifying, 84–85

key terms and summary, 137–141
linear equation graphing, 651–658, 665, 670
multiplication of fractions with, 222
power rule for exponents, 875–878, 900, 924
properties of operations using, 89
Real-Data Applications, 102
simplification of radicals with, 1100–1101
solving for specified, 703–706, 721–722, 1047–1048
test, 145–146
unlike fraction addition and subtraction, 236–237
writing fractions in lowest terms, 213–214
Variation
   defined, 1073
   direct, 1067, 1073, A–7 to A–8
   graph for, A–7 to A–10
   inverse, 1067–1068, 1073, A–8 to A–9
   problems, 1067–1068, 1078, A–7 to A–10
Velocity problems, 1166
Venn diagrams, A–4
Vertex
   of angle, 454
   of parabola, 1179, 1180–1183
Vertical angles, 458–460, 478, 483
Vertical axis, 665
Vertical lines. *See also* Lines
   graphs of, 774–775
   intercepts, 750–751
   linear equations for, 775
   slope, 762–763, 795
   vertical line test for functions, 1188–1190, 1200
Vertical multiplication of polynomials, 882–883
Volume
   cylinder, 386, 396, 401
   defined, 279, 287
   English system, 564
   of a cube, 279–280
   of geometric solids, 280–281
   measurement conversions, 584–585, 588, 608
   metric system, 583–585, 608
   pyramid, 280–281, 283
   rectangular solid, 280, 283, 387
   unit fractions, 566–567

W

Weight (mass)
   English measurement system, 417, 564, 598
   metric measurement system, 585–588, 598, 608
   unit fractions, 417, 566–567, 598
Weighted mean, 358–359, 395

Whole. *See also* Part
   defined, 510, 551
   fraction as part of or more than, 194–195
   percent equations to find, 522, 532, 552–553
   in percent problems, 509–511, 513
   in percent proportion, 513
Whole numbers, 1209–1254
   addition, 317, 1209–1214, 1249
   carrying in addition, 1210–1212, 1249
   carrying in multiplication, 1226–1227
   computation pretest, 1207–1208
   in decimal place value, 301–302
   defined, 2, 71, 684
   divisibility tests, 211, 1238–1240
   division, 911–912, 1233–1246, 1250
   identifying, 2–3
   key terms and summary, 1249–1250
   multiplication, 1225–1230
   order of operations, 62
   place value, 2–4
   place value chart, 2
   reading and writing, 72
   review exercises, 1251–1252
   subtraction of, 317, 1217–1222, 1250
   test, 1253–1254
Width of rectangle, 150–151, 185
Windchill, 1, 28, 766
Woltman, George, 940
Words
   into algebraic expressions, 167–168
   indicator, to solve problems, 226
   for numbers, 3–4
Work problems, 999, 1059–1061, 1065, 1077
Writing
   decimals, 4, 300–304, 496–497, 552
   decimals as percents, 496–497, 552
   decimals written as fractions, 303–304
   equivalent fractions, 198–199
   fractions, 198–199, 207–214, 300–301, 303–304, 349–352, 397, 398
   fractions as decimals, 300–301, 349–352, 397, 398
   fractions in lowest terms, 207–214
   negative numbers, 9–10
   numbers, 3–4
   percents as decimals, 494–496, 552
   positive numbers, 9–10
   proportions, 431, 480
   rates, 423, 479
   signed numbers, 9–10
   whole numbers, 72

X

x-axis
   defined, 646, 665, 793
   functions, 1187–1192
   ordered pair, 733, 736–738, 795, 818
   plotting points, 646–648, 670, 736–738, 745–747, 795

rectangular coordinate system, 645, 646–648, 670
substitution method, 818
x-intercept
   defined, 747, 793
   finding, 747–748
   linear equations, 748–749
   of parabola, 1180–1182
   quadratic equation, 1180–1182, 1199
   quadratic inequalities, A–15 to A–16
   slope-intercept form, 763–765, 771–776, 796

Y

y-axis
   defined, 646, 665, 793
   functions, 1187–1192
   ordered pair, 733, 736–738, 795
   plotting points, 646–648, 670, 736–738, 745–747, 795
   rectangular coordinate system, 645, 646–648, 670
y-intercept
   defined, 747, 793
   finding, 747–748
   linear equations, 748–749
   of parabola, 1180–1182
   quadratic equation, 1180–1182, 1199
   quadratic inequalities, A–15 to A–16
   slope-intercept form, 763–765, 771–776, 796

Z

Zero
   addition property of, 17–18, 71, 73
   borrowing with, 1220–1221
   as decimal placeholders, 311
   division, 50, 336, 913, 1234–1235, 1244–1245
   division of polynomial by polynomial, 913
   as exponents, 59, 895, 900, 923, 925
   multiplication, 42–43, 1227, 1229
   multiplication property of zero, 42–43, 71, 74
   on number line, 10
   as placeholder, 318–319, 328, 913, 1228
   in quotients, 1244–1245
   slope of horizontal line, 762, 763
   whole numbers, 684
Zero-factor property, quadratic equations, 967–968, 969, 970, 989, 1154

THEA Test Correlation Index

Skill	Text Reference

Fundamental Mathematics
1. Solve word problems involving integers, fractions, decimals, and units of measurement.
   - Solve word problems involving integers. — Sections 1.2–1.7
   - Solve word problems involving fractions. — Sections 4.1–4.7
   - Solve word problems involving decimals (including percents). — Sections 5.1–5.6, 7.4–7.5
   - Solve word problems involving ratio and proportions. — Sections 6.1–6.4
   - Solve word problems involving units of measurement and conversions (including scientific notation). — Chapter 8, Section 13.8

2. Solve problems involving data interpretation and analysis.
   - Interpret information from line graphs, bar graphs, pictographs, and pie charts. — Sections 9.1–9.3, 11.1
   - Interpret data from tables. — Section 9.1, many of the "Focus on Real-Data Application" pages
   - Recognize appropriate graphic representations of various data. — Sections 9.1–9.3
   - Analyze and interpret data using measures of central tendency (mean, median, and mode). Analyze and interpret data using the concept of variability. — Section 5.7

Algebra
3. Graph numbers or number relationships.
   - Identify the graph of a given equation. — Sections 11.1–11.2
   - Identify the graph of a given inequality. — Sections 10.4, 11.5
   - Find the slope and/or intercepts of a given line. — Sections 11.2–11.3
   - Find the equation of a line. — Section 11.4
   - Recognize and interpret information from the graph of a function (including direct and inverse variation). — Section 11.2, Appendix B

4. Solve one- and two-variable equations.
   - Find the value of the unknown in a given one-variable equation. — Sections 2.3–2.5, 4.7, 5.9, 7.3, 10.2, 15.6, 16.6
   - Express one variable in terms of a second variable in two-variable equations. — Section 12.2
   - Solve systems of two equations in two variables (including graphical solutions). — Sections 12.1–12.3

5. Solve word problems involving one and two variables.
   - Identify the algebraic equivalent of a stated relationship. — Sections 3.3–3.4, 6.4, 7.4, 12.4, 15.7
   - Solve word problems involving one and two unknowns. — Sections 3.3–3.4, 6.4, 7.4–7.5, 10.3, 12.4, 15.7

6. Understand operations with algebraic expressions and functional notation.
   - Factor quadratics and polynomials. — Chapter 14
   - Perform operations on and simplify polynomial expressions. — Sections 2.2, 10.1, 13.1, 13.3–13.4, 13.6–13.7
   - Perform operations on and simplify rational expressions. — Sections 4.3–4.4, 4.6, 15.1–15.5
   - Perform operations on and simplify radical expressions. — Sections 16.1–16.5
   - Apply principles of functions and functional notation. — Section 17.5

Skill	Text Reference

7. Solve problems involving quadratic equations.
   - Graph quadratic functions.
   - Graph quadratic inequalities.
   - Solve quadratic equations using factoring, completing the square, or the quadratic formula.
   - Solve problems involving quadratic models.

Sections 17.4–17.5
Appendix C

Sections 14.6, 17.1–17.3
Sections 14.7, 17.1–17.3

Geometry
8. Solve problems involving geometric figures.
   - Solve problems involving two-dimensional geometric figures (e.g., perimeter and area problems).
   - Solve problems involving three-dimensional geometric figures (e.g., volume and surface area problems).
   - Solve problems involving Pythagorean theorem.

Sections 3.1–3.2, 4.8, 5.10, 10.3

Sections 4.8, 5.10
Section 5.8, 14.7, 16.1

9. Solve problems involving geometric concepts.
   - Solve problems using principles of similarity and congruence.
   - Solve problems using principles of parallelism and perpendicularity.

Section 6.6
Sections 6.5, 11.3

Problem Solving
10. Apply reasoning skills.
    - Draw conclusions using inductive reasoning.
    - Draw conclusions using deductive reasoning.

Appendix A
Appendix A

11. Solve applied problems involving a combination of mathematical skills.
    - Apply combinations of mathematical skills to solve problems.
    - Apply combinations of mathematical skills to solve a series of related problems.

Throughout

Throughout, especially in "Relating Concepts" exercises and "Focus on Real-Data Application" pages